Contents

ASM Handbook®

Volume 14B
Metalworking: Sheet Forming

Prepared under the direction of the
ASM International Handbook Committee

S.L. Semiatin, Volume Editor

Steven R. Lampman, Project Editor
Bonnie R. Sanders, Manager of Production
Diane Grubbs, Editorial Assistant
Madrid Tramble, Senior Production Coordinator
Pattie Pace, Production Coordinator
Diane Wilkoff, Production Coordinator
Kathryn Muldoon, Production Assistant
Scott D. Henry, Senior Manager, Product and Service Development

Editorial Assistance
Elizabeth Marquard
Heather Lampman
Cindy Karcher
Beverly Musgrove
Kathleen Dragolich
Carol Polakowski
William P. Riley III

ASM INTERNATIONAL
®
**The Materials
Information Society**

Materials Park, Ohio 44073-0002
www.asminternational.org

Library of Congress Cataloging-in-Publication Data

ASM International

ASM Handbook
Includes bibliographical references and indexes
Contents: v.1. Properties and selection—irons, steels, and high-performance alloys—v.2. Properties and selection—nonferrous alloys and special-purpose materials—[etc.]—v.21. Composites

1. Metals—Handbooks, manuals, etc. 2. Metal-work—Handbooks, manuals, etc. I. ASM International. Handbook Committee.
II. Metals Handbook.
TA459.M43 1990 620.1´6 90-115
SAN: 204-7586

ISBN-13: 978-0-87170-710-9
ISBN-10: 0-87170-710-1

ASM International®
Materials Park, OH 44073-0002
www.asminternational.org

Printed in the United States of America

Multiple copy reprints of individual articles are available from Technical Department, ASM International.

Foreword

ASM Handbook Volume 14B on sheet forming is the second of two volumes on metalworking technology, following the publication of Volume 14A on forging and bulk forming in 2005. These two volumes reflect the continuing mission of the *ASM Handbook*® series to provide in-depth and practical engineering knowledge in areas of technological significance.

Like many major manufacturing operations, the technology of sheet-metal fabrication is being transformed in response to the competitive demands of a global economy and computer-aided engineering. Product and process design are becoming more integrated, and all stages of processing are being enhanced by computer technologies that help implement process-control strategies to reduce scrap and achieve net-shape forming capability on the shop floor. These advances involve the efforts of various technical communities, and ASM International is pleased to help disseminate their knowledge for the benefit of others in the economical manufacturing of effective products.

Thanks are extended to all the contributors and especially to Lee Semiatin, who, as Volume Editor, has championed this entire effort with his tireless devotion. Dr. Semiatin is to be congratulated and lauded for all his efforts in identifying and recruiting authors, directing the editorial activities of review and revision, and responding effectively in the development of both Volume 14A and 14B. His volunteer commitment is enormous, and we are indebted to Lee Semiatin.

Reza Abbaschian
President
ASM International

Stanley C. Theobald
Managing Director
ASM International

Policy on Units of Measure

By a resolution of its Board of Trustees, ASM International has adopted the practice of publishing data in both metric and customary U.S. units of measure. In preparing this Handbook, the editors have attempted to present data in metric units based primarily on Système International d'Unités (SI), with secondary mention of the corresponding values in customary U.S. units. The decision to use SI as the primary system of units was based on the aforementioned resolution of the Board of Trustees and the widespread use of metric units throughout the world.

For the most part, numerical engineering data in the text and in tables are presented in SI-based units with the customary U.S. equivalents in parentheses (text) or adjoining columns (tables). For example, pressure, stress, and strength are shown both in SI units, which are pascals (Pa) with a suitable prefix, and in customary U.S. units, which are pounds per square inch (psi). To save space, large values of psi have been converted to kips per square inch (ksi), where 1 ksi = 1000 psi. The metric tonne (kg×10³) has sometimes been shown in megagrams (Mg). Some strictly scientific data are presented in SI units only.

To clarify some illustrations, only one set of units is presented on artwork. References in the accompanying text to data in the illustrations are presented in both SI-based and customary U.S. units. On graphs and charts, grids corresponding to SI-based units usually appear along the left and bottom edges. Where appropriate, corresponding customary U.S. units appear along the top and right edges.

Data pertaining to a specification published by a specification-writing group may be given in only the units used in that specification or in dual units, depending on the nature of the data. For example, the typical yield strength of steel sheet made to a specification written in customary U.S. units would be presented in dual units, but the sheet thickness specified in that specification might be presented only in inches.

Data obtained according to standardized test methods for which the standard recommends a particular system of units are presented in the units of that system. Wherever feasible, equivalent units are also presented. Some statistical data may also be presented in only the original units used in the analysis.

Conversions and rounding have been done in accordance with IEEE/ASTM SI-10, with attention given to the number of significant digits in the original data. For example, an annealing temperature of 1570 °F contains three significant digits. In this case, the equivalent temperature would be given as 855 °C; the exact conversion to 854.44 °C would not be appropriate. For an invariant physical phenomenon that occurs at a precise temperature (such as the melting of pure silver), it would be appropriate to report the temperature as 961.93 °C or 1763.5 °F. In some instances (especially in tables and data compilations), temperature values in °C and °F are alternatives rather than conversions.

The policy of units of measure in this Handbook contains several exceptions to strict conformance to IEEE/ASTM SI-10; in each instance, the exception has been made in an effort to improve the clarity of the Handbook. The most notable exception is the use of g/cm³ rather than kg/m³ as the unit of measure for density (mass per unit volume).

SI practice requires that only one virgule (diagonal) appear in units formed by combination of several basic units. Therefore, all of the units preceding the virgule are in the numerator and all units following the virgule are in the denominator of the expression; no parentheses are required to prevent ambiguity.

Preface

Since the 1988 publication of Volume 14, *Forming and Forging* (of the 9th Edition *Metals Handbook* series, subsequently renamed the *ASM Handbook* series in 1991), advances in the forming of sheet metals have focused on a number of new or improved processes, new materials, increasing utilization of flexible-manufacturing and rapid-prototyping techniques, and the application of sophisticated process models and process-control strategies. A number of these advances have been driven by the needs of mass production in the automotive industry, but also partly by niche markets such as aerospace. Inexpensive yet powerful, computing resources have emerged as an important element in process design and control, tooling development, and product-process integration.

Innumerable configurations can be produced from sheet by various fabrication operations such as bending, stretching, deep drawing, hole-making, and flanging. These distinct manufacturing processes are performed in various combinations to produce a finished part along with considerations for material savings and manufacturing ease. Currently, improvements in computational capability are having a significant impact on the cost-effective application, *integrated* engineering evaluation, and robust production of sheet-metal products. The increasing utilization of process-control strategies to reduce scrap and achieve net-shape forming capability during all stages of processing is also enhanced by computer technologies implemented on the shop floor.

This Volume provides a broad overview of sheet-metal fabrication technologies and applications. The intent is to cover basic concepts and methods of sheet forming and developments in forming technology. Since the late 1980s, a number of processes have been introduced and/or undergone substantial improvement. These processes include high-production superplastic forming of aluminum, the use of tailor-welded blanks in automotive manufacturing, increasing utilization of rubber-pad (hydro-) forming, and high-velocity metal forming. Recent advances in the forming of sheet metals also include increasing utilization of flexible-manufacturing and rapid-prototyping techniques. New advances are also being made in the forming of advanced high-strength steels and magnesium alloys. In addition, the evaluation and analysis of material formability is improving with new techniques, such as stress-based forming-limit criteria.

It is hoped that this publication provides a useful reference for the many practitioners in this vital industry. Many thanks go to the contributors, who volunteered their time and expertise in this endeavor. This work would not have been possible without them.

S.L. Semiatin
Volume Editor

Authors and Contributors

Sean R. Agnew
University of Virginia

Debbie Aliya
Aliya Analytical

Brian Allen
AK Steel Corporation

Brian Baker
Special Metals Corporation

Frédéric Barlat
Alcoa Technical Center

Joseph D. Beal
The Boeing Company

William T. Becker
(deceased)

B.-A. Behrens
University of Hanover

David Berardis
Fenn Technologies

B.P. Bewlay
General Electric Global Research

Rodney Boyer
The Boeing Company

Jian Cao
Northwestern University

L.C. Chan
The Hong Kong Polytechnic University

L. Chen
Technical Materials, Inc.

Raymond Cribb
Brush Wellman Inc.

Paul Crook
Haynes International Inc.

Glenn S. Daehn
The Ohio State University

Mahmoud Y. Demeri
FormSys, Inc.

E. Doege
(deceased)

Joseph A. Douthett
AK Steel Corporation, Inc.

H. Lee Flower
Haynes International Inc.

Peter Friedman
Ford Research Laboratory

D.U. Furrer
Ladish Company

A.K. Ghosh
University of Michigan

C.H. Hamilton

Ed Herman
Creative Concepts Company, Inc.

Louis E. Huber, Jr.
Cabot Supermetals Corporation

Dennis Huffman
Timken (retired)

Kent L. Johnson
Engineering Systems Inc.

Jacob A. Kallivayalil
Alcoa Technical Center

Serope Kalpakjian
Illinois Institute of Technology (retired)

Pawel Kazanowski
Hydro Aluminum Cedar Tools

Stuart Keeler
Keeler Technologies LLC

Menachem Kimchi
Edison Welding Institute

Brad Kinsey
University of New Hampshire

Gary L. Kinzel
The Ohio State University

Dwaine Klarstrom
Haynes International Inc.

R. Kopp
Aachen University

Howard Kuhn
Consultant

G. Kurz
University of Hanover

Rein Küttner
Tallinn Technical University, Estonia

Rob Larsen
The Boeing Company

T.C. Lee
The Hong Kong Polytechnic University

Joe Lemsky
Ladish Co., Inc.

Donald R. Lesuer
Lawrence Livermore National Laboratory

M. Li
The Ohio State University

Huimin Liu
Ford Motor Company

Peter P. Liu
Eastern Illinois University

Terry Lowe
U.S. Department of Energy,
Los Alamos National Laboratory

James C. Malas
Air Force Research Laboratory

Alan Male
University of Kentucky

Frank Mandigo
Olin Corporation

Steve Matthews
Haynes International Inc.

Christopher A. Michaluk
Michaluk and Associates

Michael Miles
Brigham Young University

Matt Miller
Cornell University

Wojciech Z. Misiolek
Lehigh University

Toby Padfield
ZF Sachs Automotive of America

Henry Rack
Clemson University

Chung-Yeh Sa
General Motors

Daniel Sanders
The Boeing Company

Daniel J. Schaeffler
Engineering Quality Solutions, Inc.

Berthold Scholtes
Universität Kassel

J. Schulz
Aachen University

S.L. Semiatin
Air Force Research Laboratory

Howard W. Sizek
Air Force Research Laboratory

Philip Smith
Alcoa Technical Center

Krishna Srivastava
Haynes International Inc.

Edgar A. Starke, Jr.
University of Virginia

Torgeir Svinning
SINTEF, Norway

C.Y. Tang
The Hong Kong Polytechnic University

Don Tillack
Tillack Metallurgical Consulting

Derek Tyler
Olin Corporation

Peter Ulintz
Anchor Manufacturing Group Inc.

Ravi Venugopal
Sysendes, Inc.

Lotta Lamminen Vihtonen
Helsinki University of Technology

Evan J. Vineberg
Engineering Quality Solutions, Inc.

Otmar Voehringer
Universität Karlsruhe

O. Vogt
University of Hanover

Boel Wadman
IVF, Sweden

R.H. Wagoner
The Ohio State University

J.F. Wang
The Ohio State University

Jyhwen Wang
Texas A&M University

Michael L. Wenner
General Motors R&D Center

Cedric Xia
Ford Research Laboratory

Jeong Whan Yoon
Alcoa Technical Center

Gunter Zittel
Elmag, Inc.

Contents

Introduction

Introduction to Sheet-Forming Processes

S.L. Semiatin, Air Force Research Laboratory

SHEET FORMING comprises deformation processes in which a metal blank is shaped by tools or dies, primarily under the action of tensile stresses. The design and control of such processes depend on the characteristics of the workpiece material, the conditions at the tool/workpiece interface, the mechanics of plastic deformation (metal flow), the equipment used, and the finished-product requirements. These factors influence the selection of tool geometry and material as well as processing conditions (workpiece and tooling temperatures, lubrication, etc.). Because of the complexity of many sheet-forming operations, models of various types, such as analytic, physical, or numerical models, are often relied on to design such processes.

This Volume presents the state of the art in sheet-forming processes. A companion volume (Volume 14A, *Metalworking: Bulk Forming,* 2005) describes the state of the art in bulk-forming processes. Various major sections of this Volume deal with descriptions of specific processes, selection of equipment and die materials, forming practice for specific alloys, and various aspects of process design and control. This article provides a brief historical perspective, a classification of sheet-forming processes, and a summary of some of the more recent developments in the field. Since the publication in 1988 of the last edition of *ASM Handbook* Volume 14, *Forming and Forging,* sheet-forming practice has seen a number of notable advances. Recent materials developments include the introduction or increased application of advanced high-strength steels, magnesium alloys, and superplastic sheet materials. New or improved process developments include application of flexible manufacturing and rapid prototyping techniques and the increased use of sophisticated process simulation and control tools. Some of these technological advances are briefly summarized in this article.

Historical Perspective

Bulk-forming techniques such as forging have been used in various forms for thousands of years. By comparison, sheet-forming processes, except for the bending of narrow workpieces, are relatively recent. This is because techniques to roll wide sheets of metal of uniform thickness were not developed until the 1500s. It was not until the middle part of the 19th century, however, that mass production via sheet forming became a reality with the advent of processes for forming tin-plated sheet steels for the canning industry (Ref 1). Subsequently, two major commercial sectors set the pace for advances in sheet forming, namely, the automotive industry, beginning at approximately the turn of the 20th century, and the home-appliance industry after World War I. Both of these industries required large quantities of low-carbon sheet steel. Such needs were met by the development of tandem-mill rolling, a technology pioneered in the paper industry and subsequently adopted in the steel industry by the American Rolling Mill Company (Armco) and others.

During the latter half of the 20th century, the need to reduce the weight of automobiles to improve fuel economy spurred the development of various grades of high-strength sheet steels in addition to low-carbon and ultralow-carbon steels with improved formability. These developments were aided by increased knowledge of the effect of alloying, rolling, and annealing practices on ductility, plastic anisotropy, and thus stamping performance as well as on post-formed properties. The transition from glass (primarily used prior to the 1970s) to metallic containers for beverages also led to a boom in the sheet-metal rolling and forming industries, first for steel and then for aluminum.

More recently, warm and superplastic forming techniques for aluminum and magnesium sheet alloys have been developed and implemented in the automotive and other industries. Furthermore, the conventional and superplastic forming of titanium, nickel-base, and refractory alloys has become commonplace for aerospace applications. These developments have been aided by increased knowledge related to formability, the effect of prior processing on microstructure and texture evolution in sheet materials, and the constitutive response of metals during forming. The development of powerful user-friendly finite-element codes for process simulation and design, the advent of rapid prototyping applications, the adoption of flexible binder technology, and the use of adaptive process controls have been among the most recent advances in the area of sheet forming.

Classification of Sheet-Forming Processes

In metalworking, an initially simple part—a billet or a blanked sheet, for example—is plastically deformed between tools (or dies) to obtain the desired final configuration. Metal-forming processes are usually classified according to two broad categories:

- Bulk, or massive, forming operations
- Sheet-forming operations

Sheet forming is also referred to as forming. In the broadest and most accepted sense, however, the term *forming* can be used to describe both bulk-deformation processes and sheet-forming processes. Forming processes are also typically classified further into categories, as described in Fig. 1.

In both bulk and sheet deformation, forming is done with some contact between the surfaces of the deforming metal and a tool. Depending on the nature of the contact, friction between the tool and workpiece also may have a major influence on material flow. The major distinction is that deformation occurs in bulk or in a more localized or directional fashion. Bulk-forming operations typically involve multidirectional deformation throughout the volume of a worked mass, as in the cases of forging, extrusion drawing, rolling, coining, sizing, and thread forming. In bulk forming, the input material is in billet, rod, or slab form, and the surface-to-volume ratio in the formed part increases considerably under the action of largely compressive loading. Bulk-forming processes are described in *Metalworking: Bulk Forming,* Volume 14A of *ASM Handbook,* 2005.

In contrast to bulk forming, sheet forming often involves local deformation. During sheet forming, a piece of sheet metal is plastically deformed by tensile loads into a three-dimensional shape, often without a significant change in its thickness or surface characteristics. The characteristics of sheet-metal-forming processes are (Ref 3):

- The workpiece is a sheet or a part fabricated from a sheet.
- The deformation usually causes significant changes in the shape, but not necessarily the cross-sectional area, of the sheet.

- In some cases, the magnitudes of the plastic and the elastic (recoverable) deformations are comparable; therefore, elastic recovery or springback may be significant.

Examples of sheet-forming processes include deep drawing, stretching, bending, rubber-pad forming, and other methods (Fig. 2). Sheet-forming methods can also be classified according to suitable methods in obtaining desired dimensional features, such as surface contours or deep recesses (Table 1). Some methods of local deformation also extend beyond sheet forming to the bending and forming of solid sections and tubular products. These forming methods are discussed in the Section "Forming of Bar, Tube, and Wire" in this Volume.

Process-Related Developments

During the last decade, a number of processes have been introduced and/or undergone substantial improvement. These processes include high-production superplastic forming of aluminum, the forming of tailor-welded blanks, rubber-pad (hydro) forming, and high-velocity metal forming.

Superplastic forming (SPF) of aluminum has recently been commissioned for high-production applications in the automotive industry. Unlike previous niche applications of the technology, in which forming rates were very slow (~10^{-4} to 10^{-3} s^{-1}) and workpiece temperatures were relatively high (~500 °C, or 930 °F), newer uses of aluminum SPF comprise moderate strain rates (10^{-3} to 10^{-2} s^{-1}) and somewhat lower temperatures. Perhaps the largest-volume use of the new approaches is that recently introduced by General Motors for the forming of 5083 aluminum-magnesium alloy sheet and referred to as quick plastic forming (QPF) (Ref 5, 6). As in other SPF techniques, SPF shaping usually involves the blow forming of prelubricated panels into a female die. Typical forming times are of the order of 5 min, or a fraction of the time used for more conventional SPF of aluminum (and titanium) sheet-metal parts. The QPF manufacturing cell (Fig. 3) uses sheet that has been preblanked and coated with boron nitride to enhance separation of finished parts from the tooling following forming. Using primarily robotic means for improved safety and control, blanked sheets are then passed through successive stages comprising convection oven preheating outside the press, forming, cooling, and transfer from the cell to subsequent operations. Dimensional and distortion control are enhanced by the use of tool-based extraction of parts following forming and the prevention of lubricant buildup by cleaning the hot tooling using dry ice. An additional important design enhancement comprises the use of specially designed heated tooling rather than a more conventional system in which the entire die set/die stack is heated.

Additional information on superplastic forming is contained in the article "Superplastic Sheet Forming" in this Volume.

Forming of tailor-welded blanks is a technology that was first used in the 1980s but received substantial attention only in the last 5 to 10 years. Most tailored blanks consist of steels of different thicknesses, grades/strengths, and sometimes coatings that are welded prior to forming (Ref 7). Welding processes include

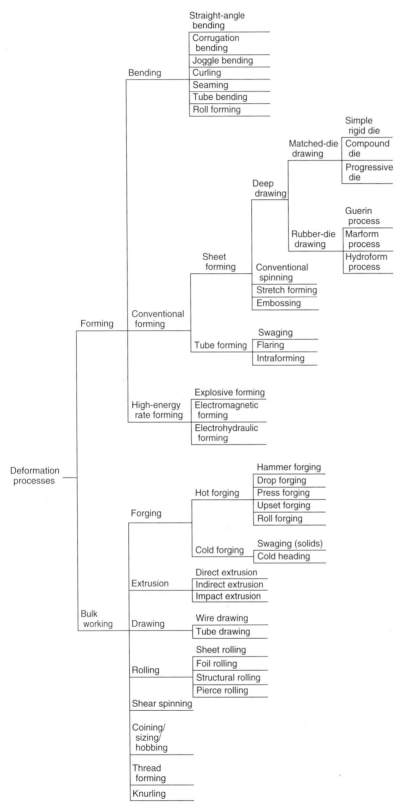

Fig. 1 Taxonomy of deformation processes. Source: Ref 2

seam, high-frequency butt, friction, and laser welding techniques. Finished parts with a desirable variation in properties are thus produced. For example, specific areas of a structure requiring higher strength or load-carrying capability may use steel of a thicker gage or different composition; other areas may require a steel with higher corrosion resistance. Employed primarily by the automotive industry, the approach can yield property blends that can lead to substantial vehicle weight savings and thus improved fuel economy. Other applications have been proposed in the construction industry; these include prefabricated building panels, steel decking, and highway crash barriers. Manufacturing issues relate to the forming characteristics of tailor-welded blanks, the ability to hold dimensional tolerances, and the aesthetics of the weld for parts in which the joint is exposed. For instance, even when blanks are joined using laser welding, the seam is still usually visible after painting. Therefore, tailor-welded-blank applications in the automotive industry tend to be restricted to nonvisible areas or locations in which body moldings can be used to cover the seam. More information in this area can be found in the article "Forming of Steel Tailor-Welded Blanks" in this Volume.

Rubber-pad (hydro) forming in its simplest form comprises the forming of sheet metal using (1) a flexible rubber pad and rigid punch around which the sheet is shaped, or (2) a die into which the sheet is formed under the action of a pressurized fluid acting on a rubber diaphragm. Thus, tooling costs are considerably reduced because a number of different part geometries can be made using the same pad/fluid medium. In addition, the process often provides better dimensional control, less springback, and the capability of producing parts with greater complexity or using fewer forming steps. In the 1980s, rubber-pad forming of sheet metals was primarily applied to make automotive and aircraft skin panels that required limited production runs for which the cost of conventional hard tooling was difficult to justify. More recently, various process modifications, such as warm hydroforming, integral hydrobulge forming, combined drawing and bulging, and viscous pressure forming, have been introduced (Ref 8). In addition, methods for enhancing flange draw-in, such as the introduction of fluid pressure against the edge of the blank, have been reported to lead to significant increases in drawability during hydroforming. More information on this technology is summarized in the article "Rubber-Pad Forming and Hydroforming" in this Volume.

High-velocity metal forming (HVMF) consists of those processes in which high-strain-rate

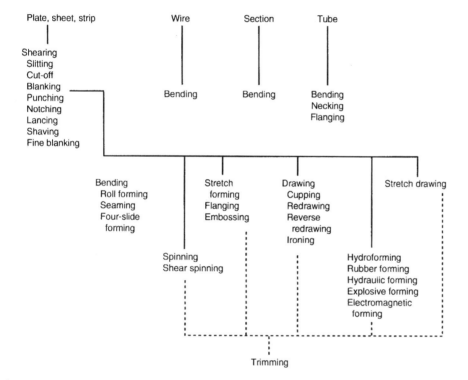

Fig. 2 General classification of sheet-forming processes. Source: Ref 4

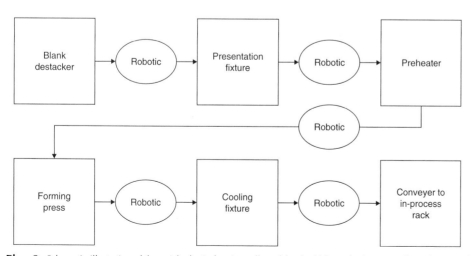

Fig. 3 Schematic illustration of the quick plastic forming cell used for the high-production superplastic forming of aluminum-magnesium sheet alloys in the automotive industry. Source: Ref 5

Table 1   Classification of sheet metal forming processes

Bending and straight flanging

  Brake bending
  Roll bending

Surface contouring of sheet

  Contour stretch forming (stretch forming)
  Androforming
  Age forming
  Creep forming
  Die-quench forming
  Bulging
  Vacuum forming

Linear contouring

  Linear stretch forming (stretch forming)
  Linear roll forming (roll forming)

Deep recessing and flanging

  Spinning (and roller flanging)
  Deep drawing
  Rubber-pad forming
  Marform process
  Rubber-diaphragm hydroforming (fluid cell forming or fluid forming)

Shallow recessing

  Dimpling
  Drop hammer forming
  Electromagnetic forming
  Explosive forming
  Joggling

Source: Ref 3

impulsive loading is used to deform sheet metal into the shape of a die into which it is accelerated. The advantages of HVMF relative to conventional sheet-forming processes include improved formability and reduced wrinkling, more uniform strain distribution in a single forming step, reduced springback, and the ability to impart fine details over large areas. Including techniques such as explosive, electromagnetic, and electrohydraulic forming, these methods have been used for some time, primarily for very specialized applications with small lot sizes. Although the main features of modern HVMF equipment were well established in the 1960s and 1970s, the increased potential of HVMF via its incorporation into hybrid systems has been realized only during the last few years. Hybrid systems include punch-and-die or hydroforming setups in which an HVMF subsystem has been embedded into the tooling. In so doing, forming operations that would typically require multiple forming steps and sets of dies can be accomplished in a single, modified set of tooling. For example, Ref 9 describes how electromagnetic forming (EMF) coils can be inserted into conventional tooling to aid in forming sharp-corner radii or other fine features that require the improved formability that HVMF provides (Fig. 4); typically, the EMF impulse is applied near the end of the forming process in such cases. As a demonstration, hybrid hard tooling/EMF forming has been applied to the manufacture of an aluminum inner door panel, a part that could not be fabricated using traditional methods (Ref 10). In this case, the electromagnetic actuator was used to put a sharp corner into the sheet to which a large radius of curvature had been imparted by conventional stamping. Another hybrid approach based on combining hydroforming and electrohydraulic or explosive forming has also been proposed. In this case, the press tonnage required to develop the high fluid pressures needed to ensure the flow of the sheet into fine details at the end of forming is reduced by using local HVMF. More information on HVMF is contained in the article "High-Velocity Metal Forming" in this Volume.

Rapid Prototyping and Flexible Manufacturing Techniques

These include cost-effective approaches to evaluate tooling designs prior to the manufacture of expensive steel dies and dieless forming techniques such as thermal forming and peen forming. Additional information in this area is contained in the articles "Rapid Prototyping for Sheet-Metal Forming," "Thermal Forming of Sheet and Plate," and "Peen Forming" in this Volume.

Rapid prototyping (RP) in the sheet-forming industry most commonly involves the fabrication of dies from inexpensive materials in order to evaluate metal flow and possible defect generation (e.g., excessive thinning, wrinkling, etc.) prior to the sinking of steel dies. Rapid prototyping dies are an economical alternative for making parts when lot sizes are relatively small. Techniques for making RP dies include stereolithography, laminated object manufacture, and fused deposition processes. These techniques are similar in that they make use of a computer-aided design file to produce an object, such as a die, in a layer-by-layer fashion. For example, in a typical stereolithography approach, an ultraviolet source selectively scans the surface of a liquid (light-sensitive) photopolymer resin in order to cure the desired pattern describing the die cross section. The substrate is then lowered a small amount (typically ~0.1 to 0.5 mm, or 0.004 to 0.02 in.). The process is then repeated to produce the next and subsequent layers in a similar fashion. Using this technique, a die can typically be made in several hours, which is very short in comparison to the time for making the corresponding steel die (~50 to 100 h). Furthermore, dies made by this means can produce a modest number (~3 to 30) of sheet-metal parts from aluminum alloys, mild steel, and stainless steel using both conventional and hydroforming presses (Ref 11).

Thermal forming is a dieless, flexible manufacturing technique for forming sheet metals. It consists of the local heating of sheet using laser, plasma, or other sources to effect controlled shapes. The mechanism underlying thermal forming is the development and control of temperature gradients both through the thickness and across the plane of the sheet. These gradients produce local stresses that primarily cause bending and, to a lesser extent, upsetting and buckling (Ref 12). Such forming approaches are best suited for making parts in small quantities or to evaluate the form and fit of a new part. Thus, thermal forming can also be classified as an emerging RP technology. Originally used in the 1980s for the forming of plate, the method has been explored for the forming of sheet since the 1990s. Initial work used lasers as the heating source. Because of the expense, low energy-conversion efficiency, and safety concerns (due to multidirectional reflections of the laser beam), among other factors associated with laser-based systems, the use of other heating methods, such as plasma-arc sources, is currently being examined. The design of one plasma-arc apparatus (Ref 13), for instance, comprises four components: a plasma-arc torch, a positioning system, a cooling system, and control software (Fig. 5a). The principal control variables include the torch power, the heating trajectory along the sheet surface, the tilt angle between the torch and the workpiece, and the torch stand-off distance. Figure 5(b) illustrates the forming of a stainless steel box using a plasma-based system.

Peen forming is another dieless, flexible manufacturing technique for forming sheet metals. It is used to make large plan-area parts that are produced in small lot sizes, primarily for the aerospace and ship building industries. Forming relies on the deformation imparted by the impact of balls, which replace conventional forming dies. Unlike shot peening, in which deformation is usually limited to very shallow depths (~100 to 200 μm, or 4 to 8 mils) in a bulk workpiece, peen forming introduces plastic deformation to depths that correspond to a large portion of or the entire cross section of a sheet. At modest shot velocities, only the surface layers are elongated, resulting in a convex curvature of the sheet (Fig. 6). At higher shot velocities, the

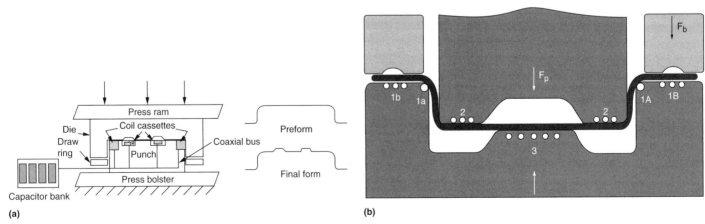

(a)

(b)

Fig. 4 Tooling designs for hybrid systems combining conventional tooling and electromagnetic forming. (a) General setup. (b) Closeup in one section. Source: Ref 9

entire cross section is deformed; the deformation is highest on the surface opposite to the impact surface, and, in these instances, a concave curvature is developed in the sheet. Although peen forming was originally developed and applied in the 1950s for simple shapes, a number of major advances and new applications have recently been made. These include double-sided, simultaneous peen forming, which enables the formation of curvatures that are either more complex or with considerably smaller radii (Ref 15). The technique has also been applied to straighten laser-welded stringer reinforcements in aircraft fuselages (Ref 16).

Dieless numerical-control (NC) forming is yet another dieless flexible manufacturing technique for making sheet-metal parts, usually in small lot sizes. The process uses simple tooling to form small, local regions of sheet material in an incremental fashion until the entire part is made. In practice, the geometry of the sheet-metal part is converted from computer-aided design data through a computer-aided manufacturing system to NC data. These data are then downloaded to a three-axis computer numerical-control machine, which drives the forming tooling (Fig. 7). The blank is clamped to a square blankholder (workholder) such that

there is no draw-in from the binder. The so-called z-tool moves in the z (vertical) and y directions; the workholder is counterbalanced to the vertical movement of the z-tool and is translated in the vertical direction according to the descent of the z-tool and along the x and y directions. The x- and y-plane movements are synchronized with the tool motion to produce the desired form on the sheet. Often, a support tool is placed under the workholder, and the sheet-metal blank is formed against it. By this means, the stretching and bending deformations are well controlled.

Materials-Related Developments

Recent materials developments include the introduction or increased application of advanced high-strength steels, magnesium alloys, and various ultrafine-grain materials for superplastic sheet forming.

Advanced high-strength steels (AHSS) are materials that are receiving much attention in the automotive industry as a means to reduce vehicle weight and improve fuel economy. Spurred by initiatives such as the Ultra-Light Steel Auto Body program, the implementation of AHSS is aimed at providing material properties that bridge the range between conventional microalloyed high-strength steel grades (with strengths of 210 to 550 MPa, or 30 to 80 ksi) and ultrahigh-strength steels (with strengths >550 MPa, or 80 ksi), while maintaining adequate sheet formability (Fig. 8). This property combination is due to the higher strain-hardening rates of AHSS, which possess multiphase microstructures of ferrite, martensite, bainite, and/or retained austenite in various

amounts. Specific types of AHSS include dual-phase, transformation-induced plasticity, complex-phase, and martensite grades. The strength of these grades is also enhanced during the bake-hardening cycle following painting. The increase in strength increases with prior forming strain, unlike conventional bake-hardenable grades for which little additional hardening occurs in regions that have undergone strains greater than ~2% during forming. The specific microstructures and mechanical behavior of each of the AHSS is summarized in Ref 17 and in the article "Forming of Advanced High-Strength Steels" in this Volume. Application guidelines are in Ref 18.

Magnesium alloys exhibit a relatively high strength-to-density ratio, making them excellent candidates for lightweight structural application, but possess poor formability at room temperature. The low ductility arises from the limited number of slip systems in the hexagonal close-packed crystal structure of magnesium that restrict the ability to accommodate uniform deformation in each grain in a polycrystalline material. At temperatures between approximately 225 and 300 °C (440 and 570 °F), however, additional slip modes become active, and formability increases substantially. Thus, the warm forming of magnesium sheet has recently been developed to make items such as computer cases and, perhaps in the future, automotive parts. The successful application of warm forming requires a careful consideration of the stress-strain behavior of the specific alloy to be used, requirements related to the type of press and press-speed controls, and tooling/heating system design for a high-production environment. These considerations are described in

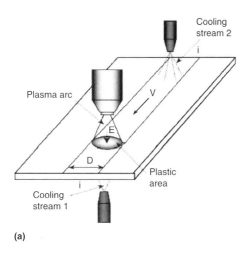

(a)

(b)

Fig. 5 Thermal forming of sheet metals using a plasma-arc heating source. (a) Schematic illustration of the equipment. (b) Stainless steel sheet-metal blank and box that was thermally formed. Source: Ref 13

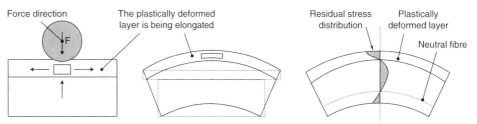

Fig. 6 Schematic illustration of the deformation during peen forming that produces a convex sheet curvature. Source: Ref 14

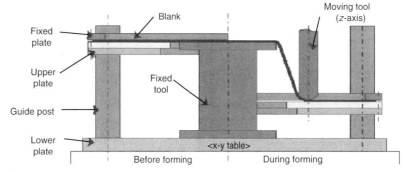

Fig. 7 Schematic illustration of dieless numerical-control forming. Courtesy of Amino Corporation

detail in the article "Forming of Magnesium Alloys" in this Volume.

Ultrafine-grain sheet materials have enabled superplastic sheet forming at lower temperatures and/or higher strain rates than materials with conventional microstructures. Such materials typically contain a submicrocrystalline microstructure produced by severe plastic-deformation techniques such as equal-channel angular extrusion, multiaxial (*abc*) forging, and/or flat rolling, usually at warm working temperatures. For example, warm working of beta-annealed and water-quenched alpha/beta-titanium alloys has been used to develop ultrafine microstructures that can be superplastically formed at a temperature of the order of 775 °C (1425 °F), which is ~100 to 150 °C (180 to 270 °F) lower than typical (Ref 19–21). The lower processing temperature reduces sheet contamination (alpha-case formation), enables the use of less-expensive tool materials, and gives rise to faster cycle times, all of which improve process economics.

Superalloy sheet materials for superplastic forming were commercialized in the 1990s. Similar to aluminum and titanium superplastic sheet alloys, these materials have a very fine grain size and moderately high values of the strain-rate sensitivity index (*m* value). One of the materials in this class is the SPF version of alloy 718. This material has been made using conventional hot working followed by a series of cold working and annealing steps to produce sheet with a thickness of 2 mm (0.08 in.) or less (Ref 22). The sheet has a grain size of the order of ASTM 12 (~5 μm, or 0.2 mil), which is stable at SPF temperatures of ~925 to 980 °C (1700 to 1800 °F) and strain rates of ~10^{-4} to 10^{-3} s^{-1}. Under these processing conditions, the material

exhibits a tensile ductility of at least 250% and can be used to make parts in 1 to 3 h using a gas pressure of ~2 MPa (0.3 ksi).

Process Simulation, Design, and Control

Substantial progress in the design and control of sheet-forming processes has been realized since the 1980s. As for bulk forming, much of this progress has been paced by the development and application of powerful computer codes for process simulation.

Process simulation tools for sheet forming have been largely finite-element-method (FEM) based, as has been the case for the simulation of bulk-forming processes. However, there are a number of special challenges associated with the application of FEM for sheet forming. These include the selection of element type, the treatment of contact conditions, and the description of the constitutive behavior of the workpiece material (Ref 23). Shell elements that provide information on through-thickness variations of stress and strain have been found to provide the most accurate simulation predictions. The complexity of the variable contact between the workpiece and dies during sheet forming as well as the friction conditions specific to different forming processes also pose a challenge in FEM simulations.

Because the majority of sheet forming is conducted at cold or warm working temperatures using rolled sheet materials with varying degrees of crystallographic texture, the description of constitutive behavior in the FEM code is usually considerably more complex than that for the simulation of hot, bulk-forming operations.

Hence, various internal-state variable models have been found to be useful to describe the evolution of dislocation density, substructure, and strength. In addition, the most accurate simulation results are obtained when texture evolution is simulated simultaneously with deformation and then used to formulate the yield function and flow rule that are important parts of the FEM formulation. The complete simulation of sheet-forming operations in which texture is tracked and updated continuously with strain, however, can be very complex and computationally intensive (Ref 24, 25). Hence, various methods have been used to simplify the problem depending on the precise output that is desired. If detailed information of the final texture is not needed, for example, the most straightforward approach is to use a yield function and flow rule for the starting material that have been derived from a crystal-plasticity analysis, but not update these functions during each increment of deformation. This approach has been applied, for example, to simulate the warm forming of aluminum-magnesium alloys (Ref 26).

Accurate sheet-forming FEM simulations are being used increasingly to establish regions in which undesirably large amounts of thinning or workpiece fracture may occur. In addition, FEM has been used to quantify the local deformation and fracture behavior and hence edge quality produced during the blanking of sheet metal (Ref 27). For instance, FEM predictions in conjunction with the McClintock ductile failure criterion (Ref 28) were used to establish the occurrence of material fracture during the blanking of low-carbon steel.

Process control technology represents an increasingly important aspect in sheet forming, especially with regard to controlling dimensional accuracy and avoiding failure. For example, the control of part springback following forming is important with regard to subsequent forming or assembly operations. Springback may be mitigated via changes in the forming process itself or by die design alterations. The former method relies on reducing stress gradients through the sheet thickness (which give rise to bending moments and thus springback) by applying tensions during or following stamping. Specifically, control of the blankholder force may be used during forming, or reverse-bending tooling may be used following forming. In the former case, care is required to avoid deforming the sheet to the point at which localized thinning and tearing may occur, as may be the case for complex or severely drawn parts that are processed close to the forming limit of the workpiece material.

Compensation for springback can also be made via changes in die design such that the desired geometry is achieved following unloading after forming. These approaches typically involve a series of FEM simulations in which tooling, lubrication, and process variables are varied. At the end of each simulation, the values of an objective function (e.g., deviation from

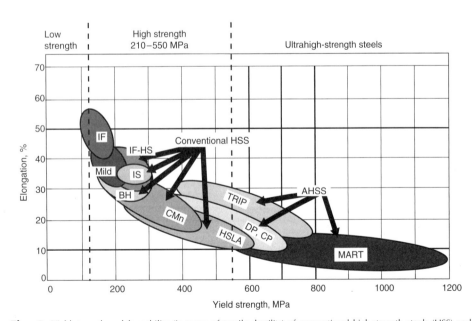

Fig. 8 Yield strength and formability (in terms of tensile ductility) of conventional high-strength steels (HSS) and advanced high-strength steels (AHSS). Types of steels: BH, bake-hardening; CMn, carbon-manganese; CP, complex phase; DP, dual-phase; HSLA, high-strength, low-alloy steel; IF-HS, interstitial-free high strength; IF, interstitial-free; IS, isotropic steels; MART, martensitic; TRIP, transformation-induced plasticity. Adapted from Ref 17 and 18

desired part shape) and their gradients (i.e., changes with respect to the change in each control parameter) are determined in order to pick a new search direction in control-parameter or die-shape space. Other methods to establish appropriate die designs to compensate for springback (e.g., the so-called springforward and displacement adjustment techniques) have also been investigated. These methods, as well as a detailed discussion of the modeling and control of springback, are discussed in the article "Springback" in this Volume.

Closed-loop feedback systems and flexible binders are currently being considered and implemented for the control of sheet-forming processes. In traditional stamping, sheet panels are drawn using rigid binders (blankholders) with constant preset binder forces under open-loop press control. The primary function of the blankholder is to control the amount of metal that flows into the die cavity so that wrinkling, localized thinning, and tearing are avoided. Binders may be flat or contain a draw-bead. In flat binders, restraining forces in the drawn sheet are provided by friction between the binder plates and the sheet metal. In fixed drawbead binders, the restraining forces result from a combination of plastic deformation and friction. Variations in material and process variables (lubrication/friction behavior, blank thickness, material behavior, blank placement, die wear) may introduce irregularities that give rise to erratic forming response.

It has been found that failure by wrinkling or tearing is highly dependent on the magnitude and trajectory of the binder force. The dynamic, closed-loop control of one or several process variables, such as the binder force, may be used to compensate for material/process variations and thus help to reduce or eliminate scrap. For example, researchers (Ref 29, 30) have demonstrated the closed-loop approach for the deep drawing of drawing-quality low-carbon steel

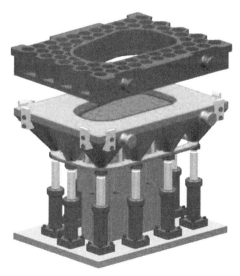

Fig. 9 Sheet-metal forming system containing a flexible binder with individually controlled hydraulic cylinders. Source: Ref 31

cups. It was found that variations in lubrication were the most likely reason for variations in forming performance. The objective of this work, therefore, was to develop a strategy to control the binder force in such a way as to produce a punch-force-versus-stroke trajectory that had been found to produce a sound part in a consistent fashion. Hence, the punch force was used as the control parameter for modulating the binder force during the drawing process.

In related work, the Institute of Metal Forming Technology of the University of Stuttgart in Germany has developed an advanced flexible binder control system with individually controlled hydraulic cylinders (Fig. 9). Cone-shaped segments localize the applied pressure in specific areas without influencing the pressure in adjacent areas. Individual control of local binder areas thus allows precise control of metal flow into the die cavity. Unlike rigid binders, a flexible binder produces the right amount of elastic deformation in the binder area and thus the optimal friction conditions for metal flow. The flexible binder technology has been successfully demonstrated on a full-sized aluminum front fender using both open-loop control and an iterative process for pin-force selection guided by computer simulation. Results from the trials showed that flexible binder technology improves the quality of fenders by allowing the manipulation of pressure settings in binder areas that control metal flow to the locations at which defect formation is most likely. Flexible binder control technology has also been applied for the drawing of double-sided stainless steel sinks (Ref 31).

Computer-aided engineering (CAE) systems developed during recent years have incorporated various elements of product design, process design, and process control. Designed primarily for the automotive industry, some of the newer commercial systems allow the design of die faces, binders, and so on for the complete die line, including blank development/blanking, die structural analysis, forming, trimming, flanging, and springback compensation. In some cases, geometrical features, friction, and so on can be entered parametrically to allow optimal overall designs to be determined rapidly via an automated iterative procedure and with less intervention by the process analyst. The analysis of the metal-forming processes themselves have benefited from improved descriptions of material properties (e.g., better yield functions for anisotropic sheet materials) and failure criteria (e.g., the development of stress-based forming-limit criteria). Results from CAE systems can also be used to evaluate the quality of the final product and its suitability for the intended service application. For instance, the variation in local thickness and yield strength (due to strain hardening) in various regions of a part is readily ascertained.

Additional information in the areas of process simulation, design, and control is contained in the articles "Computer-Aided Engineering in Sheet-Metal Forming," "Process Modeling and

Simulation for Sheet Forming of Aluminum Alloys," "CAD/CAM and Die Face Design in Sheet Metal Forming," and "Process and Feedback Control for Manufacturing" in this Volume.

Future Outlook

Recent advances in the forming of sheet metals have focused on the development of a number of new or improved processes, the increasing use of flexible manufacturing and rapid prototyping techniques, and the application of sophisticated process models and process control strategies. A number of these advances have been driven by the needs of mass production in the automotive industry, but also partly by niche markets such as aerospace. The application of inexpensive yet powerful computing resources has and will continue to be an important element in process design and control. From a design standpoint, computer-aided techniques for tooling and heating system design, tooling manufacture, and overall process design will continue to play a vital role in maintaining the economic viability of the sheet-metal-forming industry in light of the inroads made by various nonmetallic materials. The increasing use of process control strategies to reduce scrap and achieve net shape forming capability during all stages of processing will also be enhanced by computer technologies implemented on the shop floor.

REFERENCES

1. S.S. Hecker and A.K. Ghosh, The Forming of Sheet Metal, *Sci. Am.,* Vol 235 (No. 5), Nov 1976, p 100–108
2. R. Todd, D. Allen, and L. Alting, *Manufacturing Processes Reference Guide,* Industrial Press Inc., 1994
3. T. Altan, S.I. Oh, and H.L. Gegel, *Metal Forming: Fundamentals and Applications,* American Society for Metals, 1983
4. J. Schey, Manufacturing Processes and Their Selection, *Materials Selection and Design,* Vol 20, *ASM Handbook,* G. Dieter, Ed., ASM International, 1997, p 691
5. J.G. Schroth, General Motors' Quick Plastic Forming Process, *Advances in Superplasticity and Superplastic Forming,* E.M. Taleff, P.A. Friedman, P.E. Krajewski, R.S. Mishra, and J.G. Schroth, Ed., TMS, 2004, p 9–20
6. C.-M. Kim, J.G. Schroth, G.A. Kruger, and M. Konopnicki, Double-Action QPF Tools, *Advances in Superplasticity and Superplastic Forming,* E.M. Taleff, P.A. Friedman, P.E. Krajewski, R.S. Mishra, and J.G. Schroth, Ed., TMS, 2004, p 65–76
7. R.J. Pallett and R.J. Lark, The Use of Tailor-Welded Blanks in the Manufacture of Construction Components, *J. Mater. Process. Technol.,* Vol 117, 2001, p 249–254
8. S.H. Zhang, Z.R. Wang, Y. Xu, Z.T. Wang, and L.X. Zhou, Recent Developments in

Sheet Hydroforming Technology, *J. Mater. Process. Technol.,* Vol 151, 2004, p 237–241

9. G.S. Daehn, J. Shang, and V.J. Vohnout, Electromagnetically-Assisted Sheet Forming: Enabling Difficult Shapes and Materials by Controlled Energy Distribution, *Energy-Efficient Manufacturing Processes,* I.E. Anderson, T.G. Marchaux, and C. Cockrill, Ed., TMS, 2003, p 117–128

10. V.J. Vohnout, "A Hybrid Quasi-Static Dynamic Process for Forming Large Sheet Metal Parts from Aluminum Alloys," Ph.D. thesis, The Ohio State University, 1998

11. B. Fritz and R. Noorani, Form Sheet Metal with RP Tooling, *Adv. Mater. Process.,* April 1999, p 37–39

12. M. Marya and G.R. Edwards, A Study on the Laser Forming of Near-Alpha and Metastable Beta Titanium Alloy Sheets, *J. Mater. Process. Technol.,* Vol 108, 2001, p 376–383

13. A.T. Male, P.J. Li, Y.W. Chen, and Y.M. Zhang, Flexible Forming of Sheet Metal Using Plasma Arc, *J. Mater. Process. Technol.,* Vol 115, 2001, p 61–64

14. R. Kopp, Ein Analytischer Beitrag zum Kugelstrahlumformen, *Bänder Bleche Rohre,* Vol 12, 1974, p 512–522

15. R. Kopp and J. Schulz, Flexible Sheet Forming Technology by Double-Sided, Simultaneous Shot-Peen Forming, *CIRP Ann.,* Vol 51 (No. 1), 2002, p 195–198

16. A. Friese, J. Lohmar, and F. Wüstefeld, Current Applications of Advanced Peen Forming Implementation, *Proc. Eighth International Conference on Shot Peening,* Garmisch-Partenkirchen, Germany, 2002

17. Ultra-Light-Steel Auto-Body Advanced-Vehicle Concepts, Technical Transfer Dispatch 6: ULSAB-AVC Body Structure Materials, Porsche Engineering Services, Troy, MI, May 2001

18. *Advanced High-Strength Steel (AHSS) Application Guidelines,* International Iron and Steel Institute, March 2005, online at www.worldautosteel.org

19. H. Inagaki, Enhanced Superplasticity in High Strength Ti Alloys. *Z. Metallk.,* Vol 86, 1995, p 643–650

20. G.A. Salishchev, R.M. Galeyev, O.R. Valiakhmetov, R.V. Safiullin, R.Y. Lutfullin, O.N. Senkov, F.H. Froes, and O.A. Kaibyshev, Development of Ti-6A1-4V Sheet with Low-Temperature Superplastic Properties, *J. Mater. Process. Technol.,* Vol 116, 2001, p 265–268

21. P.N. Comley, Lowering the Heat—The Development of Reduced SPF Temperature Titanium Alloys for Aircraft Production, *Mater. Sci. Forum,* Vol 447–448, 2004, p 233–238

22. G.D. Smith and H.L. Flower, Superplastic Forming of Alloy 718, *Adv. Mater. Process.,* 1994, p 32–34

23. K.J. Bathe, On the State of Finite Element Procedures for Forming Processes, *NUMIFORM 2004,* S. Ghosh, J.M. Castro, and J.K. Lee, Ed., American Institute of Physics, 2004, p 34–38

24. A. Reis, A.D. Santos, J. Ferreira Duarte, A.B. Rocha, S.-Y. Li, E. Hofelin, A. Van Bael, P. Van Houtte, and C. Teodosiu, Experimental Validation of a New Plasticity Model of Texture and Strain-Induced Anisotropy, *Proc. Fourth International Esaform Conference,* A.M. Habraken, Ed., Department MSM, University of Liege, Liege, Belgium, 2002, p 433–436

25. L. Neumann, H. Aretz, R. Kopp, M. Crumbach, M. Goerdeler, and G. Gottstein, Modelling Set Up for Through-Process Simulation of Aluminum Cup Production, *NUMIFORM 2004,* S. Ghosh, J.M. Castro, and J.K. Lee, Ed., American Institute of Physics, 2004, p 388–393

26. J. Huetink, Implementation of Microstructural Material Phenomena in Macro Scale Simulations of Forming Processes, *NUMIFORM 2004,* S. Ghosh, J.M. Castro, and J.K. Lee, Ed., American Institute of Physics, 2004, p 52–55

27. E. Taupin, J. Breitling, W.T. Wu, and T. Altan, Material Fracture and Burr Formation in Blanking: Results of FEM Simulations and Comparisons with Experiments, *J. Mater. Process. Technol.,* Vol 59 (No. 1–2), 1996, p 68–78

28. F.A. McClintock, A Criterion for Ductile Fracture by the Growth of Holes, *J. Appl. Mech. (Trans. ASME),* Vol 90, 1968, p 363–371

29. M.Y. Demeri, C.-W. Hsu, and A.G. Ulsoy, Application of Real-Time Process Control in Sheet Metal Forming, *New Developments in Sheet Metal Forming,* K. Siegert, Ed., Institute for Metal Forming Technology, University of Stuttgart, Stuttgart, Germany, 2000, p 213–228

30. C.-W. Hsu, A.G. Ulsoy, and M. Demeri, Optimization of the Reference Punch Force Trajectory for Process Control in Sheet-Metal Forming, *Proc. Japan-USA Symposium on Flexible Automation,* S.Y. Liang and T. Arai, Ed., ASME, 2000

31. M.Y. Demeri, "Flexible Binder Control System for Robust Stamping," report for USAMP/DOE Project AMD-301, USCAR, Southfield, MI, 2004

Design for Sheet Forming

Howard Kuhn, Consultant

MECHANICAL PARTS are designed to meet specific functional requirements, such as transmission of loads (mechanical, thermal, or electromagnetic), geometric connectivity, and filling of space. At the same time, design should attempt to satisfy objectives for sustainability, including ease of manufacture, prolonged reliability, and recyclability. Sheet-forming processes provide considerable geometric and material flexibility in meeting these requirements, and design of parts for sheet forming must take into account these benefits as well as the limitations of the processes.

Production of sheet metal parts starts with flat stock (generally cold-rolled strip). A two-dimensional pattern is cut from the sheet, and then a sequence of unit operations is used to form the pattern into a part (generally three-dimensional). Unit sheet-forming operations include punching, bending, stretching, and drawing and are usually performed with hard tooling, but may include rubber pad tooling in combination with hard tools. These operations are used to form a wide variety of geometric features such as holes, flanges, beads and ribs, bosses, and deep or shallow recesses. Such features can be combined in various ways to meet the design requirements of the part, such as strength and attachment locations for joining with other parts. While meeting the functional design requirements, designers can also exercise options that reduce the amount of material used and/or reduce die maintenance and other manufacturing considerations.

The variety and combination of geometric features that can be produced by sheet forming is virtually endless, and comprehensive examination of the design benefits and limits of each feature is beyond the scope of this article. Reviewing each of the basic forming operations and their general geometric features, however, will illustrate the general approach to design for sheet forming and the considerations that must be made for material savings and manufacturing ease, in addition to part function. The basic operations include hole making, flanging, bead and rib forming, and stretching and drawing for shallow or deep recesses.

Hole Punching

Holes and slots are used in sheet metal parts for attachment and assembly to other parts or components. In some cases, multiple holes are used for passage of fluids, as in filters and grates. Hole making in sheet metal involves use of a punch and die to separate a slug from the sheet by shearing. The sequence of localized deformation in hole punching is shown in Fig. 1. After a small amount of plastic indentation by the punch, severe shearing deformation occurs between the punch edge and the die corner. Cracks begin to form within this shear zone at the punch and die corners and then progress toward each other. Final separation of the slug from the sheet involves tearing, resulting in a burr around the exit perimeter of the hole. This same sequence of events occurs in all shearing operations, including blanking of the initial two-dimensional pattern for subsequent forming steps, as well as in trimming excess metal from a formed part.

As a result of the shearing operation in Fig. 1, part of the surface of the hole is burnished (due to pure shear deformation) and part of it is a ragged fracture surface. Such surfaces are not suitable for load bearing or for threading without further processing, but they can be used as clearance holes for fasteners or shafts.

Hole Flanging. Holes can be flanged, or collared, to strengthen them and to provide an area for threading, if required. Collaring is accomplished by hole punching and expansion or by piercing and expansion, as shown in Fig. 2(b) and (c). In punching and expansion, a double action punch is used that first punches a small hole and then expands it by passing the tapered section of the punch through the hole. Because of the raggedness of the punched (sheared) edge, expansion of the hole circumference by the tapered die may cause cracking at the edge. The amount of expansion achievable without tearing can be increased by using a punch having a logarithmic profile, called a tractrix (Ref 2) (Fig. 2c). Piercing and expansion (Fig. 2c) accomplishes both hole making and collaring in one continuous stroke but leaves a very ragged hole edge.

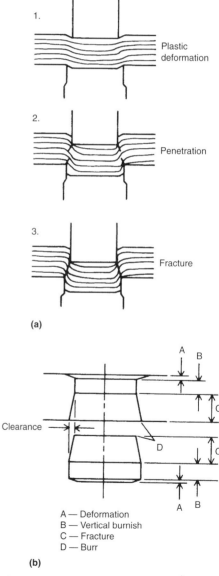

A — Deformation
B — Vertical burnish
C — Fracture
D — Burr

(b)

Fig. 1 Shearing deformation during hole punching (a); initial indentation by the punch is followed by intense shearing between the punch and die and then fracture for final separation. These stages are manifest in the appearance and surface finish of the hole (b). Source: Ref 1

Hole Size. The size of holes that can be punched is limited by the sheet material strength and thickness. As is evident in Fig. 1, the hole-making punch is subjected to a compressive load, which is equal to the force required to cause the shearing action. The resulting punch stress for a round punch is given by:

$$\sigma = 4\tau \, T/D \qquad \text{(Eq 1)}$$

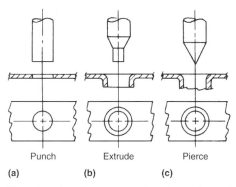

Fig. 2 Hole making by (a) punching, (b) punching and expansion, and (c) piercing. Source: Ref 1

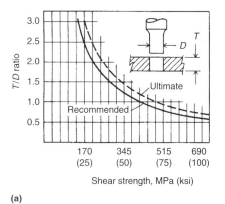

Fig. 3 Hole punching parameters. (a) Allowable sheet thickness-to-punch diameter for hole punching as a function of material shear strength. (b) Minimum punch diameter (hole size) as a function of material thickness for three materials: A, 195 MPa (28 ksi) copper; B, 345 MPa (50 ksi) steel; C, 690 MPa (100 ksi) heat treated steel. Note that, for a given sheet thickness, the minimum allowable hole diameter increases as the material strength increases. *T*, thickness; *D*, diameter. Source: Ref 1

where σ is the compressive stress in the punch, τ is the shear strength of the sheet metal, *T* is the sheet thickness, and *D* is the punch diameter. For a given punch material, the allowable *T/D* ratio is inversely proportional to the shear strength of the sheet material, so the minimum hole diameter that can be punched increases with increasing strength of the sheet material (Fig. 3).

A further constraint on punch diameter is imposed by the buckling failure mode. Taking the punch length to be at least as large as the sheet thickness, then *T/D* also represents the punch length to diameter ratio. When this ratio exceeds 2.5, buckling under compressive load is a strong possibility. Therefore, in Fig. 3(a), the allowable *T/D* ratio plateaus at 2.5, even for sheet materials having low shear strength.

Hole Location. In addition to the shear deformation shown in Fig. 1, hole punching causes deformation of metal in the vicinity of the hole, as depicted in Fig. 4. This has two consequences: distortion of the edges of sheet blanks, and cambering of long parts if a row of holes is punched off center. To avoid distortion of the blank edges, the edge of the hole should be at least 2*T* away from the edge. This same limitation applies to the distance between the edge of the hole and any other feature, such as another hole or flange. The residual stress leading to cambering, however, can be prevented only by punching holes on the centerline.

Filters and grates are often formed by punching an array of holes in the sheet (Fig. 5).

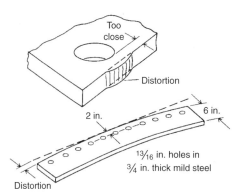

Fig. 4 Edge distortion and cambering caused by hole punching near an edge or in an off-center row. Source: Ref 1

Within the constraints of functionality of the filter or grate, the hole diameter and number of holes should be considered from the point of view of die maintenance (Ref 3). For example, for a given area of hole openings, the total circumference length of holes will increase with decreasing hole diameter. However, the shearing action during punching (Fig. 1) causes wear on the punch and die along this total circumferential length. The total shearing force also increases with increase in this circumferential length. For a given area of hole openings, therefore, increasing the hole diameter will reduce the total wear land and total punching force. Increasing hole diameter will also avoid the punch force limits given by Fig. 3.

Flanging

Part strengthening and attachment surfaces can be accomplished in sheet metal parts by forming a flange at the edge of the part. The bending process used to form flanges subjects the material to tension on the outer convex surface and compression on the inside concave surface (Fig. 6). A neutral axis, separating tensile and compressive deformation zones, is located at the centerline of thin sheet but shifts closer to the inside radius in bending of thick sheet. Plastic deformation occurs only within the curved section.

The maximum strain (*e*) at the outer surface of the sheet during bending is given by:

$$e = \frac{T}{2R} \qquad \text{(Eq 2)}$$

where *T* is the sheet thickness and *R* is the bend radius. This strain level is usually achieved after the bend angle reaches about 90°. Further bending, even to 180°, does not increase the strain beyond this value, but it does increase the width of the plastic deformation zone (Ref 4).

Defects. One consequence of the stresses in bending is the possibility of fracture on the convex surface (Fig. 7). That is, the strain expressed in Eq 2 must be less than the fracture limit strain of the material. Generally, the fracture limit in bending is given as a minimum bend radius, expressed as a multiple of the sheet

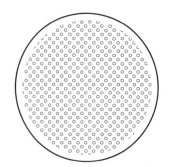
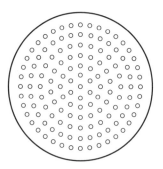

Fig. 5 An array of holes punched in sheet metal to form a filter or grate. Reducing the number of holes and increasing their size reduces tool wear and punching loads.

thickness, i.e., the inverse of Eq 2. Materials having a large strain at fracture, therefore, have small R/T ratios, permitting small bend radii for a given sheet thickness. Tables of these limits are available in a number of references (Ref 4, 5). Typically, the limiting values for R/T are 0.3 to 0.7 for soft metals, 0.6 to 1.2 for harder metals, and 2 or greater for hard-to-form materials. For example, R/T values for titanium and zirconium exceed 3.

Another consequence of the stress and deformation in Fig. 6 is elastic recovery of the plastic deformation. This results in springback of the sheet so that the final bend angle of the sheet will be slightly larger than the angle of the bending die. Die designers have various means for compensating for elastic springback; however, the part designer should be aware that the degree of springback, and the difficulty of accommodating it, increases with increasing yield strength of the sheet, increasing bend radius, and increasing sheet thickness.

When straight-line bends are made in a part, localized stresses occur at the intersection of the bend and the plane of the part. In these regions, tensile stresses during bending may cause small tears and compressive stresses may lead to localized wrinkling. Notches added to the part pattern, as illustrated in Fig. 8, relieve these localized stresses and should be used if they don't interfere with the part function. Typical notch radii are $2T$. Such radii not only avoid

localized tearing or wrinkling during bending, they also serve to reduce stress concentrations due to applied loads during subsequent use of the part, decreasing the probability of fatigue failures at those points.

Contoured Flanges. Often it is useful to form a flange on a contoured profile (Fig. 9). As in the case of straight flanges described previously, the strains due to bending must be in the formability range of the material, in accordance with the strain described in Eq 2. In contoured flanges, in addition to this bending strain, a strain also occurs at the edge of the flange. For convex (compression) flanges, the outer radius of the blank decreases, causing compression and possible wrinkling in the flange. The strain at the flange edge is:

$$e = -W/R_2 \qquad (Eq\ 3)$$

where W is the flange width and R_2 is the flange profile radius (not the bend radius). In this case, the strain is compressive and should not exceed 3% to avoid wrinkling. An example of a convex flanged part is shown as type B in Fig. 9.

For a concave (stretch) flange, the flange edge is in tension because the profile radius increases during the flange formation. Equation 3 applies for stretch flanges, as well, except that the strain is tensile. The ratio of flange width to profile radius in this case should not exceed the fracture strain of the sheet material with a sheared edge, which is always less than the fracture strain of the rolled sheet surface during bending. Typically, this limit is 10% for easy-to-form alloys.

In Fig. 9 the flange is formed perpendicular to the plane of the sheet, but along a curved line in the plane of the sheet. Other applications require strengthening flanges on a part that is

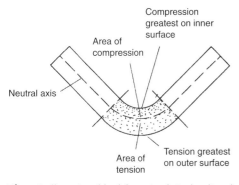

Fig. 6 Illustration of the deformation during bending of sheet. The convex surface undergoes tension and the concave surface undergoes compression, with the neutral axis (zero deformation) at or near the centerline. Source: Ref 1

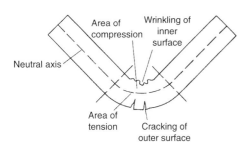

Fig. 7 Cracking defects on the convex surface and wrinkling on the concave surface during bending. Source: Ref 1

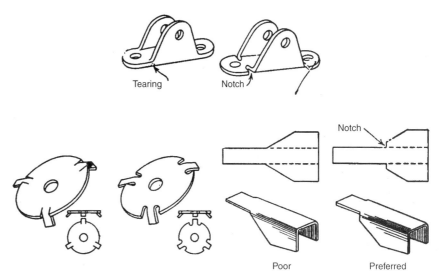

Fig. 8 Use of notches at the intersections between bend lines and sheet planes relieves stresses that can cause failure at those points. Typically the notch radii are $2T$. Reprinted with permission. Source: Ref 3

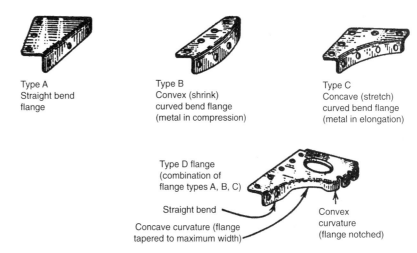

Type A
Straight bend flange

Type B
Convex (shrink) curved bend flange (metal in compression)

Type C
Concave (stretch) curved bend flange (metal in elongation)

Type D flange (combination of flange types A, B, C)

Straight bend

Concave curvature (flange tapered to maximum width)

Convex curvature (flange notched)

Fig. 9 Flange formation on a curved profile in the flat plane of the part. Convex profiles lead to compression in the flange and possibly wrinkling; concave profiles involve stretching and possibly cracking. Reprinted with permission. Source: Ref 3

also formed into a contour out of the plane of the sheet, as shown in Fig. 10. In this case, the flanging operation takes place first, followed by contouring in steel dies, or in a roll-forming operation. In either case, the flange undergoes tension or compression, depending on the orientation of the flange relative to the contoured sheet. If the flange is on the convex side of the contour curvature, the resulting tensile stresses may cause cracking at the edges of the flanges. If the flanges are on the concave side, the compressive stresses may lead to buckling and wrinkling.

For flanges on a curved profile (Fig. 9) as well as on a contoured part (Fig. 10), round notches cut into the undeformed blank will alleviate the tensile and compressive stresses that may lead to defects. These notches are illustrated in Fig. 11 for convex and concave flanges. Alternatively,

in convex compression flanges, flutes may be formed into the flange, as shown in Fig. 12, to prevent buckling.

Bend Orientation. A further influence of material on bending involves the mechanical texture in the material. Sheet metal is produced by hot rolling an ingot of the metal in numerous rolling passes to a typical thickness of 2.0 to 2.5 mm (0.08 to 0.1 in.). During hot rolling, any inclusions and impurities in the material become plastic and elongate along with the base metal. The resulting hot-rolled sheet contains elongated stringers of impurities in the rolling direction. The hot-rolled sheet is then descaled and cold rolled to finished thickness (<1 mm, or 0.04 in.) for sheet forming. The impurity stringers remain during cold rolling, giving the sheet metal a mechanical texture or fiber in the rolling direction. As a result, ductility of the sheet is lower when the stress is perpendicular to the stringers than when the stress is parallel to the stringers. When the sheet metal undergoes bending deformation, the tensile stress shown in Fig. 6 will lead to cracking more readily if it is perpendicular to the stringer direction. Therefore, bend axes in parts should be oriented perpendicular to the stringer direction so that the tensile stress is parallel to the stringers. Taking this factor into account, Fig. 13

illustrates preferred orientations of the bend axis relative to the rolling direction of the sheet.

Beads and Ribs

Indentations in the form of beads or ribs (long beads) can be formed in sheet metal to increase the localized strength and stiffness of the part. They are commonly used on flat sections of sheet metal parts for this purpose. In addition, ribs are used around holes to strengthen the area and provide a stiffer base for attachment of mating parts. For example, a circular bead that is $2T$ in height increases the stiffness of the part in the vicinity of the bead by six times and reduces the maximum stress by 66%. Examples of beads and ribs are shown in Fig. 14. Another example was shown previously in the flutes formed in the flange of Fig. 12.

Beads and ribs are formed by stretching metal to fill grooves in the die. A rubber pad or a metal punch is used for this operation. Because the metal is stretched, deep narrow ribs may exceed the formability limit of the sheet. In addition, in rubber pad forming, the pressure required to form the material increases with increasing sheet thickness and decreasing radius of the rib. For highly formable materials such as annealed low-carbon steel and aluminum, the maximum depth of ribs is approximately 5 times the sheet thickness, and the minimum bottom radius is approximately 10 times the sheet thickness. A table of data describing the forming limits for beads and ribs is given in Ref 6.

Large Recesses

Large recesses or cavities are formed in sheet metal by stretching and drawing. Such operations are the most taxing of the sheet-forming operations because the strains are great and the combinations of strains that occur in complex shapes make simple rules of thumb on formability limits difficult to formulate. In their simplest forms, stretching and drawing are illustrated in Fig. 15.

Stretching. During stretching (Fig. 15a), the sheet metal blank is clamped and a metal punch or hydraulic pressure is applied to stretch the

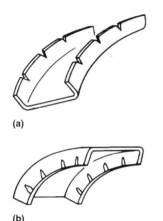

(a)

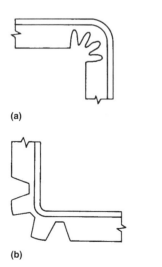

(b)

Fig. 10 Flange formation on a contoured part. Tearing can occur on the edges of flanges on the convex side of the contour (a), while wrinkling can occur in flanges on the concave side (b). Source: Ref 1

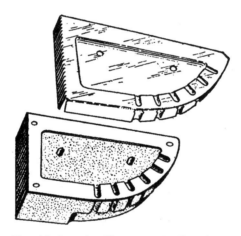

Fig. 12 Formation of flutes on a convex flanged part to control wrinkling. Reprinted with permission. Source: Ref 3

(a)

(b)

Fig. 11 Notching to avoid (a) wrinkling or (b) cracking during flange forming on contoured parts. Source: Ref 1

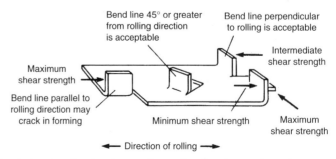

Fig. 13 Preferred orientation of bends in sheet metal to avoid cracking due to inclusion alignment in the sheet. Source: Ref 1

sheet into a partial hemispherical shape. During stretching, the metal thins as it expands and may reach the ductility limit of the material. Stretching operations are used to form large hemispherical recesses or large parts having

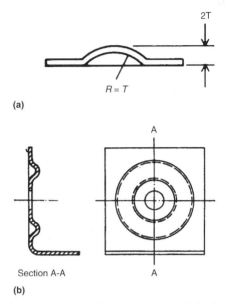

(a)

(b)

Section A-A

Fig. 14 Beads and ribs. (a) Cross section of a bead or rib formed in sheet metal for strengthening. (b) Concentric ribs formed around a hole to strengthen and stiffen the part. *R*, radius; *T*, stock thickness. Source: Ref 1

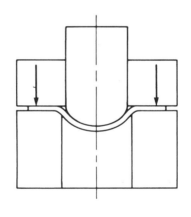

(a)

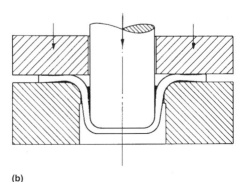

(b)

Fig. 15 Stretching and drawing. (a) Stretch forming. (b) Deep drawing. Source: Ref 1

curvature in one or two planes. Stretching, on a much smaller scale, occurs during bead and rib forming, as described in Fig. 12 and 14.

The dominant material parameters affecting stretchability are the material's work-hardening index and strain-rate sensitivity. Greater work-hardening ability and rate sensitivity spreads the tendency for thinning to material surrounding the thinning region, thus delaying the formation of tears in the material. Materials with low work-hardening and rate-sensitivity indexes, however, allow rapid thinning of material at any point of discontinuity, leading to tearing at low strains.

Deep Drawing. Drawing (Fig. 15b), is a much more complex deformation process that involves a variety of tensile and compressive stress states. In this case, the sheet blank is pressed in a hold-down plate but not prevented from sliding. As the punch contacts the sheet, the sheet begins to stretch from around the punch radius. At the same time, material in the flange region starts to draw in from beneath the hold-down plate. Because the material in the flange area is decreasing in diameter, it undergoes compression in the circumferential direction. The hold-down plate provides vertical constraint that prevents wrinkling of the flange due to this circumferential compression. Meanwhile, as metal flows around the die radius, it undergoes bending and unbending. Once around the die radius, the metal continues to stretch in the axial direction and collapse onto the punch. This deformation mode is most severe because the metal must stretch in the axial direction without contracting in the circumferential direction. This plane-strain mode of deformation may lead to tears at the bottom of the cup where the thinning is greatest.

Increasing the ratio of blank diameter to cup diameter increases the plane-strain stretching stress at the bottom of the cup. As a result, each material has a limiting draw ratio (LDR), or ratio of maximum blank diameter to cup diameter. For example, under standard drawing tool conditions (die radius, punch radius, clamping pressure), the LDR for aluminum alloys is 2.1, for copper alloys 2.2, for steels 2.3, and for titanium alloys 2.6. The reason for this wide

discrepancy is the anisotropy of the sheet material, as measured by the *R*-value (ratio of width strain to axial strain in a tension test). Materials having high *R*-values will deform more in the plane of the sheet and less through the thickness, thus allowing more deformation before thinning and tearing.

In addition to materials parameters, tooling parameters also affect the LDR. As shown in Fig. 16, the LDR increases as the die radius increases. This effect is the result of less concentration of bending stresses as the material moves around the die radius from the flange into the wall of the cup. The punch radius also affects LDR, as shown by the family of curves in Fig. 16. The optimum value is a punch radius-to-sheet thickness ratio of 10.

As a result of this limitation on the amount of diameter reduction that can be accomplished in one drawing operation (LDR), deep cups require several drawing steps, each one requiring its own tool set. Alternatively, the cans used extensively for beverages are made by one drawing operation to form a shallow cup, followed by ironing processes that thin and extend the wall of the can.

Drawing of box shapes involves further complications in metal flow than described in Fig. 15(b). The corners of drawn boxes involve combined stretching and drawing deformation that can exceed the formability limits of the material. This effect limits the box depth and corner radius that can be accomplished in boxes, as illustrated in Fig. 17. In this result (for stainless steel), the allowable depth of the box decreases and the allowable corner radius increases as the strength of the material increases (and ductility decreases). Formability is also influenced by stock thickness, as discussed in the following example.

Example: Effect of Metal Thickness on Deep Drawing of Shells (Ref 7). The stock thickness and its relation to the diameter or width of the blank affects the formability of round, square, and rectangular shells. Stock thickness also affects the depth of shell that can be drawn in one operation.

Round Shells. The number of drawing operations needed to produce a finished round shell

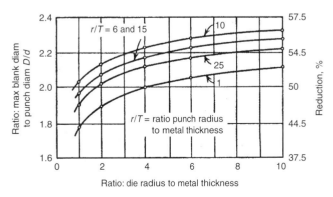

Fig. 16 Limiting draw ratio (LDR) as a function of die radius in deep drawing of brass cups. Also shown is the effect of punch radius on LDR; The optimum value of punch radius is 10*T*. Reprinted with permission. Source: Ref 3

depends on blank diameter, depth and diameter of the shell, properties and thickness of the work metal, and draw-die and punch-nose radii.

The ratio of the punch diameter to the blank diameter (d/D) is a simple and practical measure for use in determining formability. The maximum d/D ratio depends markedly on the metal thickness and the draw-die radius. The ratio of metal thickness to blank diameter (t/D) reflects the principle of volume constancy between the blank and the shell and, as shown in Tables 1 and 4, affects the formability factor d/D. The formability factors in Table 1 are multiplied by the blank diameter, or by the previous shell diameter, to obtain the shell diameter. For each redraw, the factors increase, indicating a decrease in formability. The size of draw-die radius also has an effect on the formability factors, as footnoted in Table 1. When the factors given in Table 1 are not exceeded, three or four draws can be made in 1010 steel without annealing if the punch-nose radius is at least 75% of the draw-die radius.

The ratio of draw depth to shell diameter, and the size of the draw-die radius, depend on the t/D ratio, as shown in Table 2.

When drawing flanged round shells, the ratio of the flange diameter to the shell diameter (D_f/d) and the t/D ratio both affect the maximum depth-to-diameter ratio of the shell (Table 3). The formability factors d/D shown in Table 4 should not be exceeded during the first drawing operation for a flanged round shell.

Square or Rectangular Shells. As noted, drawing of square or rectangular shells subjects the metal in the corners to compressive stresses. Metal in the sidewalls is subject to tensile stresses after the metal has been drawn over the die radius. Compressive stresses in the sidewalls are caused by the flow of metal from the corner areas.

Compression of the metal from the corner areas into the sidewalls depends on the ratio of the inside vertical corner radius of the shell to the width of the shell (r/w). Resistance of the metal to bending and wrinkling depends on the ratio of metal thickness to blank width (t/W). Formability factors ($0.71\sqrt{r/h}$) for square or rectangular shells drawn in one operation from 1010 steel are listed in Table 5; formability factors for less ductile metals are up to 10% larger.

Maximum ratios of depth to width for square or rectangular shells drawn in one operation are given in Table 6, these values are based on the ratio of shell corner radius to shell width (r/w), and on the ratio of metal thickness to blank width (t/W).

The number of drawing operations needed is influenced by the ratio of shell depth to shell width (h/w), and the ratio of the inside vertical corner radius of the shell to the width of the shell (r/w), as shown in Fig. 18. Areas A, B, and C below curve 2 in Fig. 18 represent the limits

Table 2 Maximum depth-to-diameter (h/d) ratios for drawing round shells without flanges from 1010 steel(a)

Drawing operation	Maximum h/d ratio for steel with thickness of the following percentages of blank diameter:					
	2.0–1.5%	1.5–1.0%	1.0–0.6%	0.6–0.3%	0.3–0.15%	0.15–0.08%
First	0.94–0.77	0.84–0.65	0.70–0.57	0.62–0.50	0.52–0.45	0.46–0.38
Second	1.88–1.54	1.60–1.32	1.36–1.1	1.13–0.94	0.96–0.83	0.90–0.70
Third	3.5–2.7	2.8–2.2	2.3–1.8	1.9–1.5	1.6–1.3	1.3–1.1
Fourth	5.6–4.3	4.3–3.5	3.6–2.9	2.9–2.4	2.4–2.0	2.0–1.5
Fifth	8.9–6.6	6.6–5.1	5.2–4.1	4.1–3.3	3.3–2.7	2.7–2.0

(a) The larger h/d ratios in each range are for draw-die radii of $4t$ to $8t$ (t is stock thickness); the smaller ratios are for draw-die radii of $8t$ to $15t$. Also, draw-die radii increase as ratios of stock thickness to blank diameter (t/D ratios) decrease.

Table 3 Maximum depth-to-diameter (h/d) ratios for drawing round shells with flanges in one operation from 1010 steel(a)

D_f/d ratio(b)	Maximum h/d ratio for steel with thickness of the following percentages of blank diameter:				
	2.0–1.5%	1.5–1.0%	1.0–0.6%	0.6–0.3%	0.3–0.15%
1.1	0.90–0.75	0.82–0.65	0.70–0.57	0.62–0.50	0.52–0.45
1.3	0.80–0.65	0.72–0.56	0.60–0.50	0.53–0.45	0.47–0.40
1.5	0.70–0.58	0.63–0.50	0.53–0.45	0.48–0.40	0.42–0.35
1.8	0.58–0.48	0.53–0.42	0.44–0.37	0.39–0.34	0.35–0.29
2.0	0.51–0.42	0.46–0.36	0.38–0.32	0.34–0.29	0.30–0.25
2.2	0.45–0.35	0.40–0.31	0.33–0.27	0.29–0.25	0.26–0.22
2.5	0.35–0.28	0.32–0.25	0.27–0.22	0.23–0.20	0.21–0.17
2.8	0.27–0.22	0.24–0.19	0.21–0.17	0.18–0.15	0.16–0.13
3.0	0.22–0.18	0.20–0.16	0.17–0.14	0.15–0.12	0.13–0.10

(a) The larger h/d ratios in each range are for draw-die radii of $4t$ to $8t$ (t is stock thickness); the smaller ratios are for draw-die radii of $8t$ to $15t$. Also, draw-die radii increase as t/D ratios decrease. (b) Ratio of flange diameter (D_f) to shell diameter (d).

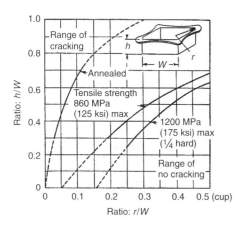

Fig. 17 Limitations in drawing box shapes in stainless steel. Limitation occurs in depth of the box as well as in the corner radius. Reprinted with permission. Source: Ref 3

Table 1 Formability factors (d/D ratios) for drawing round shells without flanges from 1010 steel(a)

Stock thickness, % of blank diam	Formability factor (d/D ratio)				
	First draw	Second draw	Third draw	Fourth draw	Fifth draw
2.0	0.48	0.73	0.76	0.78	0.80
1.5	0.50	0.75	0.78	0.80	0.82
1.0	0.53	0.76	0.79	0.81	0.84
0.6	0.55	0.78	0.80	0.82	0.85
0.3	0.58	0.79	0.81	0.83	0.86
0.15	0.60	0.80	0.82	0.85	0.87
0.08	0.63	0.82	0.84	0.86	0.88

(a) For draw-die radii of $8t$ to $15t$ (t is stock thickness); factors may increase about 2% for draw-die radii of $4t$ to $8t$.

Table 4 Formability factors (d/D ratios) for drawing round shells with flanges in one operation from 1010 steel

D_f/d ratio(a)	Formability factor (d/D ratio) for steel with thickness of the following percentages of blank diameter:				
	2.0–1.5%	1.5–1.0%	1.0–0.6%	0.6–0.3%	0.3–0.15%
1.1	0.51	0.53	0.55	0.57	0.59
1.3	0.49	0.51	0.53	0.54	0.55
1.5	0.47	0.49	0.50	0.51	0.52
1.8	0.45	0.46	0.47	0.48	0.48
2.0	0.42	0.43	0.44	0.45	0.45
2.2	0.40	0.41	0.42	0.42	0.42
2.5	0.37	0.38	0.38	0.38	0.38
2.8	0.34	0.35	0.35	0.35	0.35
3.0	0.32	0.33	0.33	0.33	0.33

(a) Ratio of flange diameter to shell diameter

Table 5 Formability factors for square or rectangular shells drawn in one operation from 1010 steel

r/w ratio(a)	Formability factor ($0.71\sqrt{r/h}$) for steel with thickness of the following percentages of blank width:			
	2.0%	1.0%	0.6%	0.3%
0.4	0.40	0.42	0.44	0.48
0.3	0.36	0.38	0.40	0.42
0.2	0.33	0.34	0.36	0.38
0.1	0.25	0.25	0.25	0.25
0.05	0.15	0.15	0.15	0.15

(a) Ratio of inside vertical corner radius of shell to width of shell

for drawing square or rectangular shells in one operation. Shells in areas D, E, and F above curve 1 need two or more operations. The space between the curves 1 and 2 represents shells that are borderline for drawing in one operation. Shells in areas A, B, and C have a stock thickness of no more than 0.6% of the blank width; in areas D, E, and F, the stock thickness is not less than 2% of blank width.

Each area in Fig. 18 represents shells with certain characteristics. Area A represents shallow shells with small inside corner radii and an $r/(w-h)$ ratio of no more than 0.17. With these shells, only a small amount of metal moves into the sidewalls from the corner areas; thus, the sidewall height is constant. Typical blank development for area-A shells is shown in Fig. 19.

Area B in Fig. 18 represents shallow shells with medium-size corner radii and an $r/(w-h)$ ratio of 0.17 to 0.4. The movement of metal into the sidewalls from the corner areas causes some earing in this type of shell. Figures 19(c) and 20 show a typical corner of a blank for an area-B shell.

Area C in Fig. 18 represents shells with medium to large corner radii and with an $r/(w-h)$ ratio of more than 0.4. The compression of metal into the sidewalls of these shells causes

considerable earing. One method of developing the corner of the blank for wide or shallow area-C shells is illustrated in Fig. 20.

Workpieces in area D in Fig. 18 have an h/w ratio no greater than 0.65, and a blank shaped as shown in Fig. 19(d). Area F represents shells

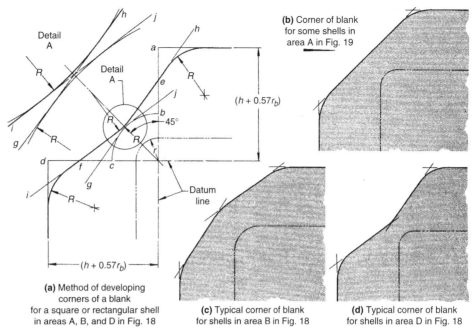

(a) Method of developing corners of a blank for a square or rectangular shell in areas A, B, and D in Fig. 18

(b) Corner of blank for some shells in area A in Fig. 19

(c) Typical corner of blank for shells in area B in Fig. 18

(d) Typical corner of blank for shells in area D in Fig. 18

Fig. 19 Corner-layout method, and typical corners, for blanks for square or rectangular shells in areas A, B, and D in Fig. 18. Corner shape varies with shell dimensions, and the shape typical of area B and that typical of area D extend into adjacent areas.

Table 6 Maximum depth-to-width (h/w) ratios for square or rectangular shells drawn in one operation from 1010 steel

r/w ratio(a)	Maximum h/w ratio for steel with thickness of the following percentages of blank width:			
	2.0–1.5%	1.5–1.0%	1.0–0.6%	0.6–0.3%
0.3	1.0	0.95	0.90	0.85
0.2	0.90	0.82	0.76	0.70
0.15	0.75	0.70	0.65	0.60
0.10	0.60	0.55	0.50	0.45
0.05	0.40	0.35	0.30	0.25

(a) Ratio of inside vertical corner radius of shell to width of shell

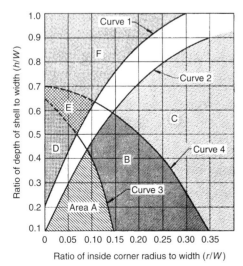

Fig. 18 Dimensional relations defining six areas of formability of square or rectangular drawn shells of 1010 steel. See text for details.

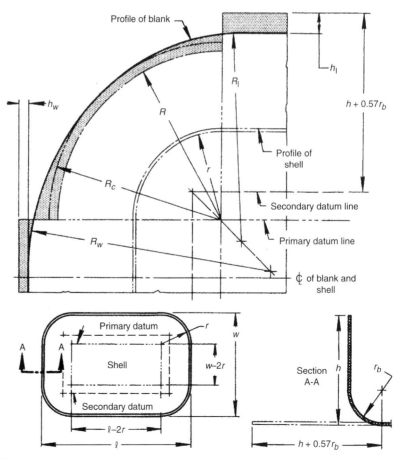

Fig. 20 Corner-layout method for wide shells with large corner radii in areas B and C in Fig. 18

with an h/w ratio greater than 0.7. The ranges of areas D and F overlap into area E.

The procedure for laying out the corners of blanks for shells in areas A, B, and D in Fig. 18 with reference to Fig. 19(a), is:

1. Lay out a square or rectangle by drawing lines through the loci of the workpiece corner radii. These lines are the primary datum lines.
2. The width of metal needed for the sidewall and bottom radius is $h + 0.57r_b$, where h is the shell depth, in inches, and r_b is the bottom inside radius, in inches. This width is added to each side of the layout of step 1, from the primary datum lines.
3. The corner radius R of the blank, which provides the metal needed from each corner of the shell, is:

$$R = \sqrt{r^2 + 2hr - 0.86r_b(r + 0.16r_b)}$$

where r is the inside corner radius of the workpiece, in inches. For parts in area D, radius R is increased by 10 to 20%.
4. Lay out corner radius R, using the loci of the workpiece corner radii as center points, so as to intersect the lines drawn in step 1 at points b and c.
5. Bisect ab and cd. Through the dividing points e and f, draw gh and ij tangent to arc of corner radius R drawn in step 4. (These tangents may coincide (Fig. 19b), they may cross each other outwardly (Fig. 19c), or they may cross each other inwardly (Fig. 19d).)
6. Blend the intersections of the tangents and the edges of the blank with an arc of radius R. When the tangents intersect inwardly, blend the intersections with an outward arc of radius R (detail A in Fig. 19a).

Shells in area B in Fig. 18 and shallow or wide shells in area C have a larger amount of metal compressed into the sidewalls than in area A. The blank is adjusted for this by increasing the metal for the corners and decreasing the metal for the sidewalls, as described subsequently, with reference to Fig. 20:

1. Start the blank layout using steps 1, 2, 3, and 4 in the preceding procedure. If the corner radius is not the same as the bottom radius, the loci of the bottom radii establish secondary datum lines from which the metal for the sidewalls found in step 2 in the six-step procedure previously listed is projected.
2. The corrected corner radius R_c is calculated by the formula:

$$R_c = R[0.074(R/2r)^2 + 0.982]$$

3. The width of metal for the sidewalls is reduced by h_w and h_l, calculated by the formulas:

$$h_w = yR^2/(w - 2r) \text{ and } h_l = yR^2/(l - 2r)$$

Values for factor y are given in Table 7.
4. Blend the intersections of the width, length, and corner radii of the blank by making R_w tangent to R_c and the blank width; and R_l tangent to R_c and the blank length.

Square shells in area F, and in area C with a large h/w ratio, can be drawn form round blanks. The blank diameter is determined according to the following formula:

$$D = 1.13$$
$$\times \sqrt{w^2 + 4w(h - 0.43r) - 1.72r(h + 0.33r)}$$

Rectangular shells in areas C and F with large h/w ratios usually are drawn from blanks with semicircular ends connected by two parallel sides; their length is the difference between the length and width of the shell.

Scrap Reduction

Reducing the amount of scrap in sheet-forming operations can be accomplished in two ways:

- By clever nesting of the blanked shapes so that the unused material is minimized
- By prevention of defects and other causes for rejection of finished parts

The first approach involves careful consideration in process preplanning and may be enhanced by part redesign. The second approach involves use of advanced techniques for analysis of deformation during sheet forming to prevent defects. Although the scrap can be recycled, handling and reprocessing of the scrap are additional steps with costs that detract from profitability. Therefore, using both approaches in early part design stages enables scrap minimization to be an important element in the product realization system.

Blank Layout. Part redesign can be used to reduce the amount of scrap, as shown by the examples in Fig. 21. In Fig. 21(a), the original design involved tab ends that were best arranged in a nesting pattern of alternating orientation of the blanks. By altering the design to decrease the angle of the tabs and orienting the edge of the tabs to be perpendicular to the main portion of the blank, the blanks can now be nested for complete utilization of

Table 7 Factor y for corner radius of blank for a square or rectangular shell in area C in Fig. 19

r/w ratio(a)	Factor y for shell with h/w ratio(b) of:			
	0.3	0.4	0.6	0.6
0.10	...	0.15	0.20	0.27
0.15	0.08	0.11	0.17	0.20
0.20	0.06	0.10	0.12	0.17
0.25	0.05	0.08	0.10	0.12
0.30	0.04	0.06	0.08	...

(a) Ratio of inside vertical corner radius of shell to width of shell. (b) Ratio of shell depth to shell width

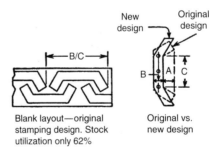
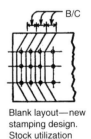

Blank layout—original stamping design. Stock utilization only 62%

Original vs. new design

Blank layout—new stamping design. Stock utilization nearly 100%

(a)

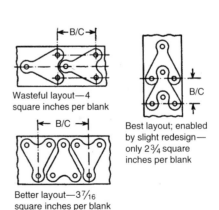

Wasteful layout—4 square inches per blank

Better layout—3⁷⁄₁₆ square inches per blank

Best layout; enabled by slight redesign—only 2¾ square inches per blank

(b)

Fig. 21 Examples of part redesign to allow blank nesting for improved material utilization. Reprinted with permission. Source: Ref 3

material with no scrap. In Fig. 21(b), alternating orientation of the blanks reduced scrap somewhat, but a slight redesign of the part (and resulting blank shape) allowed the tab of one blank to be nested in a recess of the next blank.

Process Analysis. Applying the best analytical tools during the part design stage of product development will increase the likelihood of success and reduce the probability of defects during production, with its scrap loss. One major tool is the diemaker's experience, generally the result of many years of observations and mini-experiments. Some of this experience is captured in sheet-forming and die-making handbooks. This approach is limited, however, to the geometric features and materials that make up the diemaker's experiential database.

When new materials or unusual shapes are encountered, a well-refined technique has been established that uses circular grids, stamped on the sheet metal blanks, which are measured after the forming process is complete (Ref 8). Grid measurements indicate the amount of strain that has occurred and are compared with a formability diagram of the material. Strains that are close to the materials' forming limits

may be reduced by alteration of the tooling or lubrication. Of course, this approach requires manufacture of the tooling before the process can be examined, which is a costly and time-consuming task.

Another approach involves the use of computerized simulation of the sheet forming process in conjunction with a forming limit diagram for the material (see the articles in the section "Process Design for Sheet Forming" in this Volume). Sophisticated analysis programs that solve the equations for plastic deformation of material are used to simulate the metal flow during sheet forming and calculate the strains in the material. Comparing these results with the material forming limit diagram indicates the areas of potential failure. Such programs can be used to determine alterations in the tooling and/or lubrication method that reduce the probability of failure.

Both the grid analysis and the computer simulation approaches to sheet-forming process analysis are described in more detail elsewhere in this *Handbook*. From a product design perspective, however, both methods should be considered in the early part design stages to evaluate design alternatives that will reduce the risk of failure during processing,

and the subsequent scrap loss in defective parts.

REFERENCES

1. O.D. Lascoe, *Handbook of Fabrication Processes,* ASM International, 1987
2. K. Lange, Ed., *Handbook of Metal Forming,* McGraw-Hill, 1985, p 22.3–22.5
3. D.A. Smith and R. Bakerjian, *Die Design Handbook,* Society of Manufacturing Engineers, 1990
4. G. Sachs, *Principles and Methods of Sheet-Metal Fabricating,* Reinhold Publishers, New York, 1954; see also D.A. Smith, Ed., *Die Design Handbook,* Society of Manufacturing Engineers, 1990, p 6-3
5. J. Datsko and C.T. Yang, Correlation of Bendability of Materials with Their Tensile Properties, *Trans. ASME, J. Eng. Ind.,* Vol 82 B, 1960, p 309–314
6. D.A. Smith, Ed., *Die Design Handbook,* Society of Manufacturing Engineers, 1990, p 8-8
7. H.E. Ihle, *Forming,* Vol 4, *Metals Handbook,* 8th ed., 1969, p 192
8. S.P. Keeler, Sheet Metal Forming in the 80s, *Met. Prog.,* July 1980, p 25–29

Shearing, Cutting, Blanking, and Piercing

Cutting Operations

CUTTING OPERATIONS in metal fabrication industries include various mechanical, chemical, and thermal methods. Mechanical methods involve fabrication equipment such as shears, iron workers, and nibblers. Punch presses and press brakes also are sometimes used for shearing a few pieces or are used temporarily when more efficient equipment is not available. Production shearing, however, is usually done in machines that are designed for this operation (see the article "Shearing of Sheet, Strip, and Plate" in this Volume). Mechanical cutting is also done with machining equipment such as band saws or abrasive waterjet cutting. Nonmechanical methods of cutting include material removal by chemical and/or thermal processes. These methods include gas cutting, electric arc cutting, and laser cutting. This article provides a general overview on these mechanical and nonmechanical cutting methods, with additional information in separate articles in this Volume.

Mechanical Methods of Cutting

The most prevalent equipment used for mechanical cutting includes shears, iron workers, nibblers, and band saws. Each has specific applications for which it is efficient and cost-effective.

Shears are normally used to prepare large, relatively thin materials ($\leq$25 mm, or 1 in., thick) by cutting them to final dimensions. The two principal shearing variations are straight-knife shearing and rotary shearing. In straight-knife shearing, the work metal is placed between a stationary lower knife and a movable upper knife. As the upper knife is forced down, the work metal is penetrated to a specific portion of its thickness. Straight-knife shearing is used for squaring and cutting flat stock to the required shape and size. It is usually used for producing square and rectangular shapes, although triangles and other straight-sided shapes are also sheared with straight knives. Rotary shearing is used for producing circular or other contoured shapes (see the article "Shearing of Sheet, Strip, and Plate" in this Volume).

Straight-knife shearing is the most economical method of cutting straight-sided blanks from stock no more than 50 mm (2 in.) thick. The process is also widely used for cutting sheet into blanks that will subsequently be formed or drawn. Because shear gaging can be set within ±0.13 mm (±0.005 in.), the shearing process is generally limited to ±0.4 mm (±$^{1}/_{64}$ in.) tolerances in 16-gage material. The tolerance range increases with thickness.

Straight-knife shearing is seldom used for shearing metal harder than approximately 30 HRC. When extremely soft, ductile metal (especially thin sheet) is sheared, the edges of the metal roll, and large burrs result. As the hardness of the work metal increases, knife life decreases for shearing a given thickness of metal.

In general, it is practical to shear flat stock up to 38 mm ($1^{1}/_{2}$ in.) thick in a squaring shear. Squaring shears up to 9 m (30 ft) long are available (even longer shears have been built), and some types are equipped with a gap that permits shearing of work metal longer than the shear knife.

Iron workers (heavy-duty shears) are used primarily for preparing structural shapes such as bars, angles, and T-bars (see the article "Shearing of Bars and Bar Sections" in this Volume). Like the shears used for plate and flat sheet, iron workers leave a square butt edge that may require further preparation for heavier sections before welding. The production shearing of bars and bar sections is usually done in machines with a throat opening designed for large, bulky workpieces. Guillotine and multipurpose (combination) machines are widely used for welding preparation. The multipurpose machines feature interchangeable punches and dies for shearing, punching, and coping. Squaring shears, normally used for sheet and plate, can also be used for cutting bar stock to length. Punch presses and press brakes can be provided with appropriate tooling for cutting operations.

Nibblers are power tools that "nibble" small pieces away from the material being worked. They range in size from hand-held machines to large floor-mounted equipment. Because the hand-held machines are portable, they are suitable for preparation of large parts on site, where thermal cutting may be impractical.

Nibblers can cut flat sheet of both ferrous and nonferrous metals, as well as of fiberglass and plastic. Corrugated and trapezoidal metal sheets can also be cut. Nibblers are frequently used to make edge preparations for welding of stainless steels and nonferrous metals. Unlike shears, which can curl or distort the separated pieces, nibblers remove material, creating a kerf or cutting slot; therefore, distortion of the sheet is eliminated.

Nibblers use a reciprocating punch and stationary die to accomplish the cutting process. During operation, the punch approaches and makes contact with the material being worked. The stroke of the punch compresses the material; the increased compression causes the material to deflect. As the shear point of the material is exceeded, the material is penetrated by the punch. The material then breaks (Fig. 1), and the break continues to widen until a cone-shaped chip is ejected. The punch is then withdrawn. The resulting edges are usually free of burrs. If burrs do occur, the gap between the punch and die is too wide (Fig. 2). On some equipment, this gap can be varied.

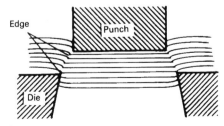

Fig. 1 Schematic of nibbler cutting action

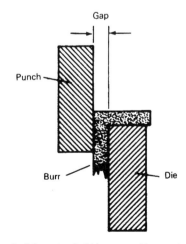

Fig. 2 Schematic of nibbler setup with excessive gap between the punch and die, resulting in the formation of burrs on the workpiece

Because of their unique cutting action, nibblers do not need to be forced into the material. This so-called "natural feed" requires only light operator pressure. Also, the weight of the nibbler does not need to be supported solely by the operator: Once the nibbler has started cutting, the weight of the tool is supported by the sheet being worked. Nibblers are considered to be very safe to operate.

Some equipment is available with guide fences (similar to those used with circular saws), circle-cutting attachments, and template guides. Punches and dies on some nibblers can be resharpened; however, in many cases these parts are disposable. For optimal punch and die life, cutting oil should be used to lubricate the cutting process.

Band Saws. Sawing of a metal can be done with power band saws, hacksaws, and circular saws (see the article "Sawing" in *Machining,* Volume 16 of *ASM Handbook,* 1989). Each of these methods is used in cutoff operations (cutting pieces to a required length), although band sawing also provides a method of cutting contours. The most satisfactory range of material hardness for sawing is approximately 180 to 250 HB, but steels up to 400 HB and some free-machining steels up to 450 HB can be sawed at reduced cutting rates.

In general, hacksawing and band sawing involve lower-cost machines with greater versatility and a larger cutting range. Circular sawing produces the smoothest finishes and the closest dimensional tolerances, especially on nonferrous materials. Table 1 compares the general characteristics of power band sawing, hacksawing, and circular sawing. The tool life and tool cost ratings in Table 1 are only general and may change completely for different applications.

Band saws have a thinner cutting tool than the other sawing machines. Band sawing produces a kerf of approximately 1.6 mm ($^1/_{16}$ in.); hacksawing and circular sawing produce kerfs of 8 and 9.5 mm ($^5/_{16}$ and $^3/_8$ in.), respectively. Because the saw band is thinner than the other sawing tools, less power is required for cutting, and binding in the kerf is less likely. Another advantage of band

Table 1 Cost and performance comparisons of hacksawing, band sawing, and circular sawing

Characteristics	Type of machine(a)		
	Lowest		Highest
Machine cost	H	→ B →	C
Power requirements	B	→ H →	C
Productivity	H	→ C →	B
Versatility	H	→ C →	B
Tool cost	C	→ H →	B
Tool life	H	→ B →	C
Accuracy and finish	H	→ B →	C
Kerf loss	B	→ H →	C

(a) H, hacksawing machine; B, band sawing machine; C, circular sawing machine

sawing is that the uniform, continuous cutting action produces even wear over the entire length of the saw band. The blades for band saws are also changed less frequently, because they are longer than the blades in other types of saws.

Band saws can be used to cut contours or straight lines, depending on the type of machine. Most band saws are designed for either vertical or horizontal movement of the saw band, although some manufacturers offer combination vertical-horizontal band saws for light- to medium-duty cutting. The band saws available include contour band saws, cutoff band saws, tilt-frame universal band saws, and plate band saws.

Contour band saws are vertical types that allow an operator to cut various contoured shapes by guiding the blade along a layout line. These machines can use very narrow bands for cutting small radii as well as wider bands for larger radii and straight cuts. Cutoff band saws can be either horizontal or vertical. They cannot cut contours but can make straight cuts very efficiently. Some machines can cut angles and compound miters in solids as well as structural shapes.

The two most widely used types of band saw blades available are carbon and bimetal. Both types have hardened tooth tips for good wear resistance. Carbon steel blades are made from very high carbon tool steel and have either flexible backs or hard backs. The hard-back type has a heat treated back for better strength and straighter cuts. Bimetal blades are made with high-speed steel that is electron-beam welded to a lower-alloy backing steel. The result is a blade with the cutting properties of high-speed steel at the tooth tips and the flexibility and strength of a heat treated alloy on the backing or carrier band.

Bimetal blades should be used when cutting difficult-to-machine materials and for high-production sawing. Bimetal blades will last longer than carbon blades, and they typically cut at ten times faster rates than those achieved with carbon blades. Carbon steel blades are most widely used for cutting of nonmetallic materials. However, they also do a satisfactory job when cutting low-carbon and low-alloy steels and perform well when cutting nonferrous alloys (for example, aluminum).

Table 2 lists the recommended speeds and cutting rates for many commonly sawed materials. The numbers are based on sawing of solid materials in the fully annealed condition. The data given should only be considered as a starting point; actual speeds and cutting rates will vary due to material variations and the type and condition of the saw that is used. Table 2 only applies when using bimetal blades. Carbon steel blades cannot be run at the band speeds shown because the heat generated at these speeds will cause them to soften and fail immediately.

When cutting pipes, tubes, or structural shapes, it is necessary to compute slower cutting rates than those given in Table 2. The cutting rates should be modified as follows:

Part size		Percentage of speed shown in Table 2
mm	in.	
<4.8	$<^3/_{16}$	40
4.8–9.5	$^3/_{16}–^3/_8$	50
9.5–16	$^3/_8–^5/_8$	60
>16	$>^5/_8$	70

Proper break-in is important to ensure optimal blade performance. When using a new blade, cutting at the full rate will cause fracturing of the very sharp tooth edges. Breaking in a new blade removes the ultrasharp edges and allows the blade to retain its cutting ability longer. A new blade can be broken in by running at the recommended band speed, but reducing the feed pressure for the initial cuts. When cutting work-hardening materials (for example, nickel alloys), it is important to use enough feed pressure to generate a chip and to prevent work hardening of the cut surface.

Knowledge of chip formation is helpful in finding the optimal speeds and cutting rates to use when band sawing. When a blade cuts, each tooth tip penetrates the workpiece and shears off a chip of material. Parameters such as band speed, feed, lubrication, and blade tip design all affect chip formation.

Visual examination of the chips can be used to determine proper speeds and cutting rates (Table 3). In general, the optimal conditions are present when the chips are curled and silver in color, indicating they were properly formed without excessive heat generation.

Gas Cutting

Gas cutting is essentially a process of controlled oxidation of metal. Because of the properties of iron and its oxidation at elevated temperature, gas cutting is most efficient in cutting carbon, low-alloy, and some high-alloy steels. Many high-alloy steels and most nonferrous metals either resist continuous oxidation or in other ways upset the delicate balance of physical and chemical reactions required for uniform cutting. As a result, cutting of such materials becomes more a matter of progressive melting than of controlled oxidation, and the cut edges generally are round and irregular. To overcome this, chemical flux or metal powder (principally iron powder) can be added to the oxygen jet to provide the reactions necessary for controlled cutting. Chemical fluxes and metal powders are used for cutting stainless steel, cast iron, and nonferrous metals in mills, foundries, and fabricating shops. Equipment for such cutting is somewhat awkward, and speeds are lower than those of some arc cutting processes.

Various gas cutting processes have been known as flame cutting, burning, oxyfuel gas cutting, and oxygen cutting. The terms *oxyfuel gas cutting* and *oxygen cutting* may be more accurate than other terms. However, all the terms mentioned previously have been used. Moreover, there can be some shift in the boundaries of cutting processes. For example, at least two arc cutting methods use oxygen to assist cutting. In another form of oxygen cutting, a laser beam is passed through the window of an oxygen-pressurized chamber and out a small orifice at the other end.

The various processes for gas cutting of ferrous metals are identified by the fuel gas that is used to produce the preheat flame. The commercially important processes are oxyacetylene cutting, oxynatural gas cutting, oxypropane cutting, and oxy-MAPP gas cutting. (MAPP is a trade name for a proprietary mixture of stabilized methylacetylene and propadiene.) Other fuel gases used include hydrogen, ethylene, methane, ethane, and butane.

For gas cutting of stainless steel and non-ferrous metals, and for faster cutting of cast iron and oxidation-resistant ferrous alloys, two other methods are available: chemical flux cutting and metal powder cutting. Other gas cutting methods may be identified by the special purposes for which they are intended, such as oxyfuel gas gouging, oxygen lance cutting, and underwater gas cutting. These various types of oxygen cutting processes using oxyfuel gas and its modifications (i.e., flux cutting and metal powder cutting to cut oxidation-resistant materials) are discussed in more detail in the article "Oxyfuel Gas Cutting" in this Volume.

Electric Arc Cutting

Arc cutting melts metal by heat generated from an electric arc. Because extremely high temperatures are developed, arc cutting can be used to cut almost any metal. Modifications of the process include the use of compressed gases to cause rapid oxidation (or to prevent oxidation) of the workpiece, thus incorporating aspects of the gas cutting process. Arc cutting methods include air carbon arc, gas metal arc, gas tungsten arc, shielded metal arc, plasma arc, and oxygen arc cutting. The two methods of industrial importance are plasma arc cutting and air carbon arc cutting. Other methods include electric arc cutting using consumable tubular electrodes (the Exo-Process) and oxygen arc cutting (see the article "Electric Arc Cutting" in this Volume).

The past decade has seen a great increase in the use of plasma arc cutting, because of its high cutting speed. Technical features of oxyfuel cutting and plasma arc cutting are compared in Table 4. The process increases the productivity of cutting machines over oxyfuel gas cutting without increasing space or machinery requirements. The basic plasma arc cutting torch is similar in design to that of a plasma arc welding torch (Ref 1). (See also the article "Plasma Arc

Table 2 Speed and cutting rates for cutoff band sawing with bimetal blades of selected ferrous and nonferrous metals

Work metal	m/min	sfm	mm²×10³/min	in.²/min	Work metal	m/min	sfm	mm²×10³/min	in.²/min
Ferrous metals					**Ferrous metals (continued)**				
Carbon and steel low-alloy steels					321, 347	37–27	120–90	2.6–1.3	4–2
1008–1013	100–84	325–275	9.0–6.4	14–10	410, 420, 420F	43–30	140–100	2.6–1.3	4–2
1015–1035	106–90	350–300	9.7–7.1	15–11	416, 430F	55–43	180–140	4.5–3.2	7–5
1036–1064	68–58	225–190	5.8–4.5	9–7	430, 446	27–18	90–60	2.6–1.9	4–3
1065–1095	52–44	170–145	5.2–3.9	8–6	440A, B, C	33–21	110–70	2.6–1.3	4–2
1108–1132	106–84	350–275	9.7–7.7	15–12	440F, 443	40–30	130–100	2.6–1.3	4–2
1137–1151	80–68	260–225	6.4–5.2	10–8	17-7 PH, 17-4 PH	27–15	90–50	2.6–1.3	4–2
1212–1213	106–90	350–300	9.7–7.7	15–12	**Nonferrous metals**				
1330–1345	65–58	210–190	5.2–3.9	8–6	Copper alloys				
4023–4047	80–70	260–230	5.2–3.9	8–6	170, beryllium copper	84–60	275–200	5.2–3.9	8–6
4130–4140	75–67	250–220	5.8–4.5	9–7		68–53	225–175	3.9–2.6	6–4
4320–4340	70–55	230–180	4.5–3.2	7–5		43–27	140–90	1.9–1.3	3–2
4815–4820	58–53	190–175	3.9–2.9	6–4.5	510 phosphor bronze 5% A	90–75	300–250	6.4–5.2	10–8
5046	75–67	250–220	5.8–4.5	9–7		53–38	175–125	3.2–1.9	5–3
5140–5160	70–60	230–200	4.2–3.2	6.5–5	614, aluminum bronze D	106–90	350–300	9.0–6.4	14–10
50100–52100	52–37	170–120	3.9–2.6	6–4		53–38	175–125	3.2–1.9	5–3
6118–6150	68–45	225–150	4.8–2.6	7.5–4	656, high-silicon bronze	100–84	325–275	9.7–7.7	15–12
8615–8645	70–53	230–175	4.5–3.2	7–5		53–38	175–125	3.9–1.9	6–3
8720–8740	68–53	225–175	4.5–3.2	7–5	675, manganese bronze A	100–84	325–275	9.7–7.7	15–12
9310	53–45	175–150	2.6–1.9	4–3		60–45	200–150	3.9–2.6	6–4
Tool steels					Nickel alloys				
W1	67–55	220–180	3.9–3.2	6–5	Inconel	30–18	100–60	1.9–1.3	3–2
S2, S5	45–33	150–110	2.6–1.9	4–3	Inconel X-750	24–18	80–60	1.0–0.3	1.5–0.5
O1, O2	65–55	210–180	3.9–2.6	6–4	Monel 400	30–18	100–60	1.9–0.6	3–1
A2	60–52	200–170	2.6–1.9	4–3	Monel R-405	45–23	150–75	2.6–1.3	4–2
D2, D3	37–27	120–90	1.9–1.3	3–2	Monel K-500	24–18	80–60	1.3–0.3	2–0.5
D7	27–18	90–60	1.3–0.6	2–1	Monel 501	30–18	100–60	1.9–0.6	3–1
H12, H13, H21	58–49	190–160	3.2–2.6	5–4	Hastelloy A	37–23	120–75	1.9–1.0	3–1.5
T1, T2	40–30	130–100	2.3–1.3	3.5–2	Hastelloy B	30–23	100–75	1.6–0.6	2.5–1
T6, T8	30–21	100–70	1.6–0.6	2.5–1	Hastelloy C	27–18	90–60	1.0–0.45	1.5–0.7
T15	23–15	75–50	1.3–0.6	2–1	Titanium alloys				
M1	45–37	150–120	3.2–1.9	5–3	Ti; Ti-1.5Fe-2.5Cr	27–18	90–60	0.6–0.2	1–0.3
M2, M3	33–24	110–80	2.6–1.3	4–2	Ti-4Al-4Mn; Ti-6Al-4V	33–21	110–70	1.3–3.9	2–6
M4, M10, M15	27–18	90–60	1.6–0.6	2.5–1	Ti-2Fe-2Cr-2Mo	27–18	90–60	1.0–0.3	1.5–0.5
L6	55–49	180–160	3.9–2.6	6–4					
Stainless steels									
201, 202, 302, 304	37–24	120–80	2.6–1.3	4–2					
303, 303F	40–27	130–90	3.2–1.3	5–2					
308, 309, 310, 330	24–18	80–60	1.3–0.6	2–1					
314, 316, 317	23–15	75–50	1.3–0.6	2–1					

(a) Based on the use of a 25 mm (1 in.) wide high-speed steel band, regular tooth form (except hook tooth form for metal thicker than approximately 250 mm, or 10 in.), raker set, to cut scale-free, solid bar stock up to 460 mm (18 in.) thick; based on the use of a cutting fluid, except for D2, D3, and D7 tool steels, which are cut dry

Welding" in *Welding, Brazing, and Soldering,* Volume 6 of *ASM Handbook,* 1993). For the cutting of metals, increased gas flows create a high-velocity plasma gas jet that is used to melt the metal and blow it away to form a kerf.

Laser Cutting

Laser cutting melts a material by focusing a coherent beam of monochromatic light on the workpiece. The effectiveness of laser cutting in a particular application depends on the pulsing and focusing of the beam and on the reflectivity, absorption coefficient, thermal conductivity, specific heat, and heat of vaporization of the workpiece. The greatest use of lasers in metal cutting is primarily two-axis profiling of sheet goods that may otherwise have been blanked out by punch press or fabricated by hand after laborious layout of the pattern. Currently, most metal cutting falls within 9.5 mm (0.375 in.) and thinner, although CO_2 lasers are now competitive with plasma arc cutting for metal thicknesses of 13 mm (0.500 in.) and greater. The principal factor in the use of laser metal cutting is manufacturing methodology. Laser cutting is ideal for batch processes, just-in-time, or low- to medium-volume production. Most laser cutting work is performed on generic or multipurpose materials handling systems, as opposed to dedicated automation-controlled systems.

Table 5 compares the technical features of laser cutting with other cutting methods. Additional information is provided in the article "Laser Cutting" in this Volume.

Abrasive Waterjet Cutting

Waterjet cutting uses a high-velocity stream of water to cut materials. Cutting of hard materials such as metals is done when the waterjet contains abrasives, in which case the method is referred to as abrasive waterjet cutting. Waterjet cutting has several advantages over other two-dimensional cutting methods and is an alternative to laser cutting, gas cutting, and plasma cutting (see Table 5 for a comparison of technical features). Advantages include:

- High cutting accuracy without leaving any frayed edges or burrs
- No distortion, because heat is not used in the cutting process
- Complex patterns can be cut without tooling
- Capable of cutting many different types of material, such as steel, aluminum, titanium, plastic, stone, rubber, glass, ceramic, and fiberglass

Figure 3 summarizes the capabilities of the laser beam, oxyfuel, plasma arc, and the abrasive waterjet in cutting a variety of metals. Current technology limits the abrasive waterjet to cutting metals having a maximum thickness of approximately 150 mm (6 in.). More information is available in the article "Abrasive Waterjet Cutting" in this Volume.

Table 3 Visual examination of chips to determine adjustments in band saw parameters that are required to optimize machinability

Form	Color	Appearance or condition	Blade speed	Blade feed	Other
	Blue or brown	Thick and hard; thick and short	Decrease	Decrease	Check cutting fluid and mix
	Blue or brown	Thick and hard; brittle	Decrease	Decrease	Check cutting fluid and mix
	Silver or light straw	Thick and hard; springy (stiff)	Suitable	Decrease slightly	Check blade for proper pitch
	Silver	Thin and hard; springy (loose)	Increase	Decrease	Check blade for proper pitch
	Silver	Thin and curled (loose)	Suitable	Suitable	. . .
	Silver	Thin and straight; springy (loose)	Suitable	Increase	. . .
	Silver	Powdered	Decrease	Increase	. . .
	Silver	Thin and curled (very tight)	Suitable	Decrease	Use blade with coarser pitch

Source: "Operator's Guide and Applications Guide," DoALL Co., 1990

Table 4 Comparison of oxyfuel gas cutting and plasma arc cutting processes

	Oxyfuel	Plasma arc
Flame temperature	3040 °C (5500 °F)	28,000 °C (50,000 °F)
Action	Oxidation, melting, expulsion	Melting, expulsion
Preheat	Yes	No
Kerf	Narrow	Wide
Cut	Both sides square	One side square
Speed	Moderate	High
Heat-affected zone	Moderate	Narrow
Cutting ability:		
Carbon steel	Yes	Yes
Stainless steel	Requires special process	Yes
Aluminum	No	Yes
Copper	No	Yes
Special alloys	Some	Yes
Nonmetallics	No	Yes

Table 5 Typical technical features of cutting processes

	Laser	Abrasive waterjet	Plasma arc	Oxyfuel
Materials	All homogeneous	All	Metallic	Metallic
Max thickness (steel), mm (in.)	30 (1.2)	100 (4)	50 (2)	300 (12)
Kerf width, mm (in.)	0.1–1.0 (0.004–0.04)	0.7–2.5 (0.03–0.10)	>1 (>0.04)	>2 (>0.08)
Heat-affected zone width, mm (in.)	0.05 (0.002)	0 (0)	>0.4 (>0.016)	>0.6 (>0.024)
Edge quality (relative)	Square, smooth	Square, smooth	Bevelled (~17°)	Square, rough
Edge roughness (R_a), μm (μin.)	1–10 (40–400)	2.0–6.5 (80–255)	. . .	. . .
Smallest hole diameter, mm (in.)	0.5 (0.02)	>1.5 (>0.06)	>1.5 (>0.06)	20 (0.8)
Energy input (relative)	Low	Low	High	Medium
Capital cost (relative)	1	1	0.1	0.01
Productivity (relative)	High	Medium-low	Medium	Low

Source: Ref 2

REFERENCES

1. W.H. Kearns, Ed., *Welding Handbook,* Vol 2, *Welding Processes—Arc and Gas*

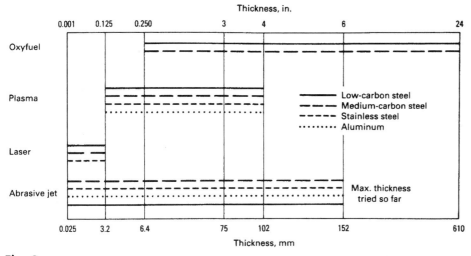

Fig. 3 Cutting thickness ranges for various cutting processes

Welding and Cutting, Brazing, and Soldering, 7th ed., American Welding Society, 1978, p 499–507

2. J.C. Ion, *Laser Processing of Engineering Materials: Principles, Procedures and Industrial Applications,* Elsevier, 2005, p 352

Principles of Shearing, Blanking, and Piercing

Many shearing, blanking, and piercing operations are based on the same underlying principles of shear mechanisms, and an understanding of shearing theory enables one to understand the following operations (Fig. 1):

shearing. Cutting material with dies or blades. *Shear* is an inclination between two cutting edges used for the purpose of reducing the required shear force. A shear is a tool for cutting metal and other materials by a closing motion of two sharp, closely adjoining edges.

punching. A general term describing the process of die cutting a hole in material such as sheet metal, plate, or some structural shape. A *punch* is the male part of a die set and usually the upper member.

perforating. A more specific term used in the stamping industry for die cutting of holes in material.

piercing. Penetration of material using a sharp-pointed punch, leaving a jagged hole similar to a bullet hole.

extruding. Forming of a flange around a hole in sheet metal.

blanking. Cutting or shearing of material to a predetermined contour from sheet or strip stock.

notching. Cutting of various shapes from the corner or edge of a strip, sheet, or part.

nibbling. Progressive notching at a high rate of speed, making either a smooth finished edge or a scalloped edge.

lancing. Cutting into a workpiece without producing a detached slug. Usually combined with forming, such as in production of louvres.

Shear Action in Metalcutting

The shear cutting or punching action results from a closing motion of two sharp, closely adjoined edges on material placed between them. The material is stressed in shear to the point of fracture while going through three phases (Fig. 2):

1. *Deformation:* As the cutting edges begin to close on the material, deformation occurs on both sides of the material next to the cut edge.
2. *Penetration:* The cutting edges cut or penetrate the material, causing fracture lines.
3. *Fracture:* The point where the upper and lower fracture lines meet. At this point the work is done, but in punching, the punch must continue to move through the material to clear the slug.

The shear cutting action produces four inherent characteristics found on both the parent material and the cut-off (or punched-out) part (Fig. 3). These characteristics are:

- Plastic deformation
- Vertical burnish-cut band
- Angular fracture
- Burr caused by the fracture starting above the cutting edge

The amount of each of these four characteristics depends on:

- Material thickness
- Material type and hardness

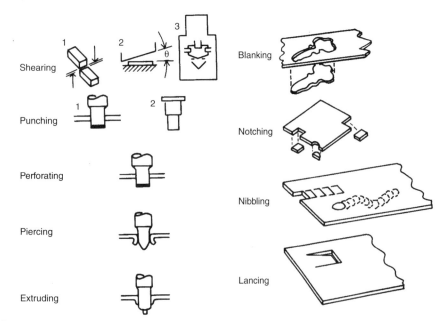

Fig. 1 Pressworking operations employing shearing principles. (See definitions in text.)

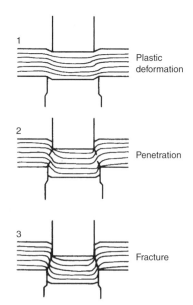

Fig. 2 Three phases of shear cutting

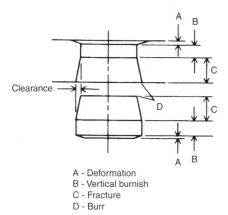

A - Deformation
B - Vertical burnish
C - Fracture
D - Burr

Fig. 3 Four characteristics of shear cutting

- Amount of clearance between cutting edges
- Condition of cutting edges
- Firmness of support of material on both sides of the cut
- Diameter of hole or blank in relation to material thickness

If all the preceding conditions are satisfied but the edge condition is still not acceptable, there are other methods that may be employed. One of the most common finishing methods is shaving, in which a small amount of material is removed to eliminate the fracture angle. Another is the fine blanking process, in which a special press or die set is used to compress the material in the shear plain during the cutting cycle and thus eliminate the fracture angle.

Deformation

The type, hardness, and thickness of the material all have effects on the amount of deformation. Softer and thicker materials deform the most. Clearance between the cutting edges and the support of the material on both sides of the cut also have strong effects. Figure 3 shows a

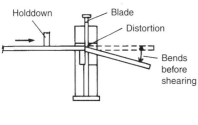

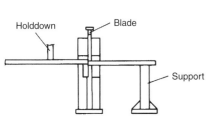

Fig. 4 Single-cut flat bar shearing with holddown and no rear support, and with rear support

hole punched with excessive clearance: note the deformation around the top surface. Supporting the material is generally no problem in hole punching, because the punch holds the slug or blank firmly against the punch face. However, in making a straight shear cut on a long bar, it is necessary to use a holddown to keep the bar from twisting around the lower blade. The combination of tensile and compressive stresses that occur in metal cutting causes this twisting. A support opposite the holddown produces the best results. Note that, without the support, the material bends before fracturing, and an angular impression from the blade is left on the end of the bar (Fig. 4).

Penetration

Penetration is the sum distance of the deformation and vertical burnish height. It is expressed as a percentage of the material thickness and is defined as the distance the punch must travel before the metal fractures. The percentage of penetration varies with the type and hardness of the material. As the material becomes harder, the percentage of penetration decreases (Fig. 5).

Penetration increases where holes are less than 1.5 times the material thickness due to the high compressive stress in the material cut zone.

Clearance

The angular fracture and the quality of the cut, the punched hole, or the blank are greatly dependent on the amount of clearance between the two opposed cutting edges. Figure 3 shows the clearance as being the distance between the mating cutting edges. Without proper clearance,

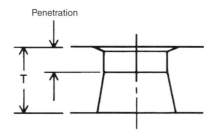

Material	Penetration, %T
Aluminum	60
Copper	55
Brass	50
Bronze	25
Steel, 0.10 C:	
Annealed	50
Cold rolled	38
Steel, 0.30 C:	
Annealed	33
Cold rolled	22
Silicon steel	30
Nickel	55

Fig. 5 Penetration as a percentage of stock thickness, showing variation with material hardness

the material will not fracture cleanly. Figure 3 shows how upper and lower fractures should meet; if they do, a clean hole is produced with a minimum power requirement. With insufficient clearance, a defect known as secondary shear is produced, as shown in Fig. 6. The two fracture lines were not permitted to meet, leaving a ring of material that must be stressed to its point of fracture with a further expenditure of energy. The amount of secondary shear decreases as the clearance increases toward the proper clearance. Dull tools result in insufficient clearances as well as burrs.

Excessive clearance between the mating edges causes extreme plastic deformation, a large burr, and a high angle of fracture, as shown in Fig. 6. Therefore, proper clearance may be defined as that clearance which causes no secondary shear and a minimum plastic deformation and burring.

Proper clearance varies with the thickness and type of material. As shown earlier in Fig. 5, the amount of penetration or fracture depends on the type and hardness of the material. Clearance is expressed as a percentage of the material thickness and must be qualified as to whether it means clearance per side or total diametral (overall) clearance. Proper clearances are best found by trial and error. No formulas or tables are available that will give exact clearances, but the values given in Table 1 have worked well in short-run punching where it is important to select standard dies for punching materials in a range of thicknesses.

Where to Apply Clearances. Where the clearance is applied must be considered in blanking or punching to close tolerances. Figure 3 is a good illustration. The die is larger than the punch by the amount of clearance required to produce a clean fracture. If the hole is to be held accurately to size, the punch must be

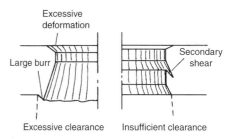

Fig. 6 Effects of excessive and insufficient clearances

Table 1 Recommended clearances for punching

Material	Total diametral clearance, % of material thickness		
	Minimum	Best	Maximum
Copper, 1/2 hard	8	12	16
Brass, 1/2 hard	6	11	16
Mild steel	10	15	20
Steel, 0.50% C	12	18	24
Aluminum, soft	5	10	15
Stainless steel	12	18	24

that size and the die must be oversize to allow proper clearance. Conversely, if the operation is blanking of a disk that has to be accurately held, the die must be that size and the punch must be undersize to allow proper clearance.

A punched hole will shrink very slightly, which must be allowed for if high precision is required. The amount varies with the type and hardness of the material but normally does not exceed 0.05 mm (0.002 in.). Punching or blanking consistently produces very accurate parts. (Additional discussion of die and punch clearance is provided in the section "Die Clearances and Stripping Forces" in this article.

Tool Life. Clearance has a great effect on tool life. Maximum life is attained only when clearance is proper. Insufficient clearance greatly reduces tool life due to springback and cold welding of the material to the punch and die. In punching harder materials, insufficient clearance creates a much higher stress on the cutting edge and causes it to chip and break down.

Excessive clearance also has an adverse effect on tool life, caused by the material being stretched over the cutting edges, and causes the sharp edges to break down prematurely.

A good grade of die-cutting lubricant greatly extends punch and die life.

Punches wear at twice the rate of dies because a punch has to pass through the material twice, in and out, whereas only the slug passes through the die.

Stripping Force

The punching process requires two actions, punching and stripping. Forcing the punch through the material is simple in comparison with stripping or extracting the punch. A stripping force due to the resiliency or springback of the punched material grips the punch. Additional friction is created by cold welding and galling that occur on the punch surface.

Stripping force is generally expressed as a percentage of the force required to punch the hole. This percentage greatly changes with the type of material being punched and with the amount of clearance between the cutting edges (Fig. 7). Figure 8 presents graphs showing the effects of these two conditions.

The surface finish of the punch changes with the number of holes being punched, and if a lubricant is not used, the effect of punch finish can be very severe, as shown in Fig. 9. The upper curve shows that dry punching produces a stripping force after seven holes of 25% of the punching force. After seven holes, the surface of the punch smoothens and the force drops. (Figure 10 shows the punch after each hole.)

A good grade of die-cutting lubricant substantially reduces the stripping force, as shown in Fig. 9 (lower curve).

Stripping (Fig. 11) can be accomplished with a spring-loaded (urethane or die rubber) plate that holds the material firmly against the die or a positive stripper plate that is set a short distance above the material. With a positive stripper, the material is lifted with the ascending punch until it contacts the stripper and is freed of the punch.

Shearing Force

The formula for the force F (area times shearing stress) required to shear, blank, or punch a given material, assuming there is no shear on the punch or die, is:

$$F = LTS \text{ (for any shape cut)}$$

$$F = DTS \text{ (for round holes)}$$

where L is sheared length, in inches; T is material thickness, in inches; S is shear strength of material, in pounds per square inch; and D is diameter, in inches.

The shear strength of the material, in pounds per square inch, is used, and values for some materials can be found in Table 2. This represents the force required to shear or cut a 25.4 mm (1 in.) square bar of metal (Fig. 12). The same rule is applied to bars of various shapes and for punching and blanking, as shown. For punching and blanking, the perimeter of cut is multiplied by the thickness of the material to find the area (Fig. 13). Note in Table 2 that the shear and tensile strengths are not the same. Also, the yield strength of a material cannot be used for shear strength, because they are not the same. If the shear strength of a material is not given or known, it can be calculated by conducting a simple test on a hydraulic press by punching a hole.

The tonnages required for punching or blanking round holes in mild steel plate of various thicknesses are presented in Table 3 for convenience. For example, punching a 12.7 mm (1/2 in.) diameter hole through 6.4 mm (1/4 in.) thick mild steel requires 9.8 tons. Table 3 can also be used for determining tonnages for other materials by multiplying the value in the table by the chart multiplier given in Table 2 for the material in question. For example, the chart multiplier for aluminum alloy 6061-T6 is 0.58, which means that its shear strength is 58% of that for mild steel. Thus, punching a 12.7 mm (1/2 in.) diameter hole through 6.4 mm (1/4 in.) thick 6061-T6 would require 9.8 × 0.58, or 5.68 tons. Tonnage requirements for punching holes greater than 25.4 mm (1 in.) in diameter can be calculated by adding the tonnages for two or more diameters, the total of which equals the desired diameter. For example, the tonnage required for punching a 38.1 mm (1½ in.) diameter hole through 6.4 mm (1/4 in.) thick

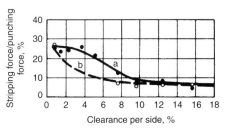

Punch diameter, 0.600 in.; no lubricant
a Thickness, 5/16 in.; tensile strength, 58,000 psi
b Thickness, 5/16 in.; tensile strength, 90,000 psi

Fig. 7 Variation of stripping force with clearance

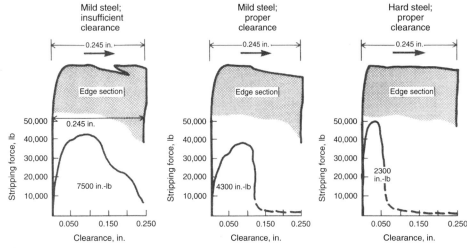

Fig. 8 Force curves showing effects of clearance and material hardness

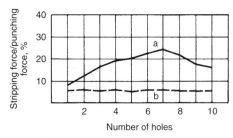

a No lubricant
b Molycote Paste G

Fig. 9 Variation of stripping force with lubrication and number of holes. Mild steel, 12.7 mm (1/2 in.) thick; punch diameter, 15.2 mm (0.600 in.); 10% overall clearance

mild steel would be the sum of the tonnages for punching a 12.7 mm (½ in.) diameter hole and a 25.4 mm (1 in.) diameter hole through 6.4 mm (¼ in.) thick plate, or 9.8 + 19.7 = 29.5 tons.

There are times when the shear strength of a material is unknown but the tensile strength is known. In such instances, the shear strength can be estimated by taking a percentage of the tensile strength, but this percentage varies with the type and thickness of the material. Figure 14 gives percentages found by sampling three types of material.

Effect of Clearance on Force and Power Requirements. Clearance has little or no effect on the amount of force required for shearing or punching, provided there is no shear on the blades, punch, or die. Many studies have been conducted to support this theory, and the results of one are given in Fig. 15. The two curves represent mild steel of equal thickness, one with insufficient clearance and the other with proper clearance. The heights of these two curves are approximately the same, the one for insufficient clearance being slightly higher.

The areas under the curves represent the power requirements. Insufficient clearance greatly increases the work requirement (850 J, or 7500 in.·lbf, compared with 485 J, or 4300 in.·lbf) as a result of the secondary shear. With insufficient clearance the shearing force is present throughout the material thickness, whereas with proper clearance, the force drops off approximately halfway through the material. It is quite obvious at this point why tool life is shortened by insufficient clearance—the tool performs 75% more work.

Reducing Shear Forces. A progressive shearing action is commonly used to reduce the required force. This is accomplished by stepping or staggering of punches where more than one punch is used so that they do not cut at the same time, or by grinding an angle on the punch and die edges. In both methods, the work is done over a greater distance with less force, but the total work performed is the same as if there were no stepping or shear on the punches.

Figure 16 illustrates three punches, each stepped equal to half the material thickness. With proper clearance, in punching of mild steel, the maximum force required would be that required for just one punch. The work curve at the top of the figure shows the maximum force plus the fact that three times as much energy must be delivered from the flywheel of a mechanical press, which can be a problem for some presses. Most hydraulic presses deliver full tonnage throughout the stroke with sufficient energy.

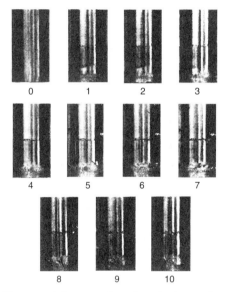

Fig. 10 Galling on surface of punch after punching each of ten holes

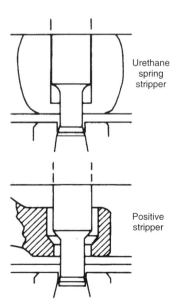

Fig. 11 Stripping with spring-loaded and positive stripper plates

Table 2 Average ultimate strengths of selected materials

Material	Chart multiplier	Ultimate strength, psi	
		Shear	Tensile
Aluminum:			
1100-O	0.19	9,500	13,000
1100-H14	0.22	11,000	18,000
3003-H14	0.28	14,000	22,000
2024-T4	0.82	41,000	68,000
5005-H18	0.32	16,000	29,000
6063-T5	0.36	18,000	30,000
6061-T4	0.48	24,000	35,000
6061-T6	0.58	29,000	41,000
7075-T6	0.98	49,000	82,000
Brass, rolled sheet:			
Soft	0.64	32,000	46,000
½ hard	0.88	44,000	65,000
Hard	1.00	50,000	78,000
Copper:			
¼ hard	0.50	25,000	38,000
Hard	0.70	35,000	50,000
Steel:			
Mild A-7 structural	1.00	50,000	65,000
Boiler plate	1.10	55,000	70,000
Structural A-36	1.20	60,000	85,000
Structural CORTEN (ASTM A242)	1.28	64,000	90,000
Cold rolled C-1018	1.20	60,000	85,000
Hot rolled C-1050	1.40	70,000	100,000
Hot rolled C-1095	2.20	110,000	150,000
Hot rolled C-1095, annealed	1.64	82,000	110,000
Stainless 302, annealed	1.40	70,000	90,000
Stainless 304, cold rolled	1.40	70,000	90,000
Stainless 316, cold rolled	1.40	70,000	90,000

Flats $F = LTS$

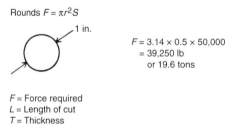

$F = 1 \times 1 \times 50,000$
= 50,000 lb
or 25 tons

$F = 4 \times \frac{1}{4} \times 50,000$
= 50,000 lb
or 25 tons

Rounds $F = \pi r^2 S$

$F = 3.14 \times 0.5 \times 50,000$
= 39,250 lb
or 19.6 tons

F = Force required
L = Length of cut
T = Thickness
r = Radius of round bar
S = Shear strength

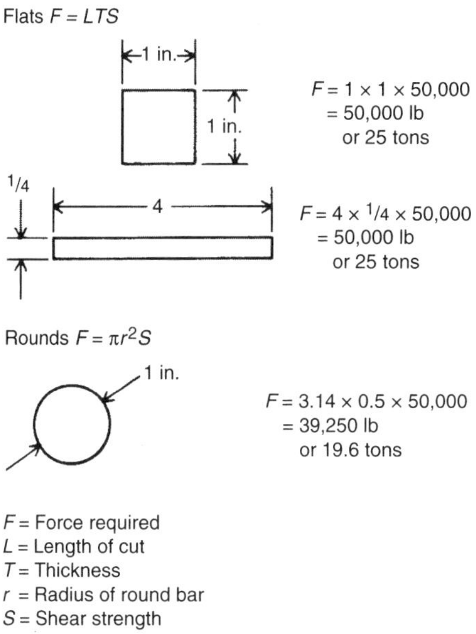

Fig. 12 Shearing-force calculations

Round holes $F = \pi DTS$

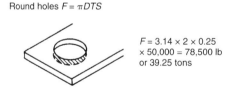

$F = 3.14 \times 2 \times 0.25 \times 50,000 = 78,500$ lb
or 39.25 tons

Squares and rectangles $F = 2(A + B)TS$

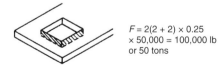

$F = 2(2 + 2) \times 0.25 \times 50,000 = 100,000$ lb
or 50 tons

F = Force required
D = Diameter of hole
T = Thickness
S = Shear strength
A = Base length
B = Width

Fig. 13 Punching-force calculations

Table 3 Required force for punching mild steel plate

Metal gage	Thickness, in.	Force, tons, required for punching hole diameters, in., of:														
		1/8	3/16	1/4	5/16	3/8	7/16	1/2	9/16	5/8	11/16	3/4	13/16	7/8	15/16	1
20	0.036	0.35	0.53	0.71	0.88	1.1	1.2	1.4	1.6	1.8	1.9	2.1	2.3	2.5	2.7	2.8
18	0.048	0.47	0.71	0.94	1.2	1.4	1.7	1.9	2.1	2.4	2.6	2.8	3.1	3.3	3.5	3.8
1/16 or 16	0.060	0.59	0.89	1.2	1.5	1.8	2.1	2.4	2.7	2.9	3.2	3.5	3.8	4.1	4.4	4.7
14	0.075	0.74	1.1	1.5	1.9	2.2	2.6	2.9	3.3	3.7	4.1	4.4	4.8	5.2	5.5	5.9
12	0.105	1.0	1.6	2.1	2.6	3.1	3.6	4.1	4.7	5.2	5.7	6.2	6.7	7.2	7.7	8.3
1/8 or 11	0.120	1.2	1.8	2.4	3.0	3.5	4.1	4.7	5.3	5.9	6.5	7.1	7.7	8.3	8.8	9.4
10	0.135	...	2.0	2.7	3.3	4.0	4.6	5.3	6.0	6.6	7.3	8.0	8.6	9.3	10.0	10.6
3/16	0.187	...	2.8	3.7	4.6	5.5	6.5	7.4	8.3	9.2	10.2	11.1	12.0	12.9	13.8	14.8
1/4	0.250	...	...	4.9	6.2	7.4	8.6	9.8	11.0	12.3	13.5	14.8	16.0	17.2	18.5	19.7
5/16	0.312	...	...	...	7.8	9.2	10.8	12.3	13.8	15.4	16.9	18.4	20.0	21.5	23.0	24.6
3/8	0.375	...	...	...	...	11.1	13.0	14.8	16.6	18.5	20.3	22.1	24.0	25.8	27.7	29.5
1/2	0.500	...	...	...	...	...	17.2	19.7	22.1	24.6	27.1	29.5	32.0	34.4	36.9	39.4
5/8	0.625	...	...	...	...	...	...	...	...	30.8	33.8	36.9	40.0	43.0	46.1	49.2
3/4	0.750	...	...	...	...	...	...	...	...	...	40.6	44.3	48.0	51.9	55.4	59.0
7/8	0.875	...	...	...	...	...	...	...	...	...	...	51.6	56.0	60.2	64.6	69.0
1	1.00	...	...	...	...	...	...	...	...	...	...	...	64.0	68.8	73.8	78.8

Chart multiplier: Values in chart are for plate with a shear strength of 50,000 psi. For punching of materials with different shear strengths, it is necessary to use a multiplier (Table 2) for calculating the proper amount of force required to punch the hole. Example: To calculate the required force for punching a 15/16 in. diam hole through ASTM A-36 steel (60,000 psi shear strength) 1 in. thick, multiply 73.8 tons by the multiplier for A-36 (1.20) to arrive at 73.8 × 1.20 = 88.6 tons. Recommended press size, 96 ton series.

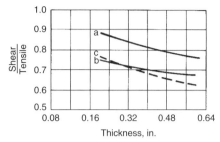

Fig. 14 Relationship of shear strength to tensile strength

a – 58,000 psi tensile
b – 90,000 psi tensile
c – 104,000 psi tensile

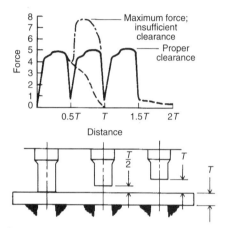

Fig. 16 Stepped punches

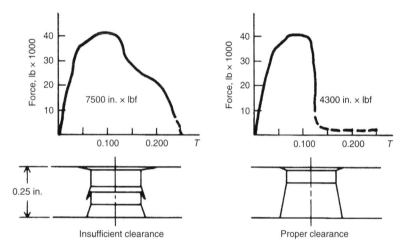

Fig. 15 Variation of work requirement (area under curve) with clearance

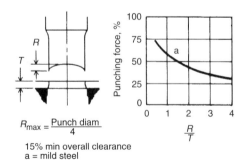

$R_{max} = \dfrac{\text{Punch diam}}{4}$

15% min overall clearance
a = mild steel

Fig. 17 Effect of punch-face shear on punching force

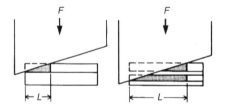

Fig. 18 Effect of clearance on force required for punching with sheared tooling. Force F varies with cutting length L; L depends on clearance and on the type and hardness of material.

If the die clearance for punching the three holes were insufficient, the maximum required force would be greatly affected, as shown by the dotted curve. The force requirement of one punch must diminish before the next punch contacts the material. (Figure 15 illustrates the problem of insufficient clearance.)

Calculating the effect of applying a shear angle to the face of a punch is difficult due to many variables. Figure 17 shows the effect of shear angle as related to the thickness of material being punched. For example, if the punch has an R (as defined in Fig. 17) of 6.4 mm (1/4 in.) and the thickness (T) of the material being punched is 6.4 mm (1/4 in.), the calculated punching force using the formula in Fig. 13 would be reduced by 40%.

The effect of shear shown in Fig. 17 holds true only if a 15% overall clearance is used. Less clearance increases the area being sheared, as shown in Fig. 18. Insufficient clearance can increase the punching force by as much as 50%.

An example of the necessity for proper clearance is where a 30 ton hydraulic press will punch a 51 mm (2 in.) diameter hole through 6.4 mm (1/4 in.) thick mild steel with a punch having a 3.2 mm (1/8 in.) shear (1/2 T). However, with insufficient clearance, the required tonnage approaches the full tonnage for that of a punch and die without shear (40 ton), and the 30 ton press will not do the job. A die 0.79 mm (1/32 in.) larger than the punch should be used in this case to provide proper clearance.

There are many ways of applying shear to a blade, punch, and die. The standard practice on a squaring shear is to have the blade inclined from one side to the other, normally at a rate of 0.3 mm/cm (3/8 in. per foot). Figure 19 illustrates various methods of applying shear to the punch and die. It is important to maintain balanced loading of the punch and die to prevent side thrusts on the punch that would cause the punch to crowd over and hit the die. This is the same problem encountered in punching part of a hole at the edge of a sheet.

Use of shear on a punch and die is an inexpensive way of stretching press capacity, provided the press has sufficient energy to accept the additional work. Many mechanical (flywheel-type) presses do not have sufficient energy to take advantage of shear. In fact, shear can decrease capacity if not properly applied.

Diameter-to-Thickness Ratios

Very small holes cannot be punched through very thick material—for example, a 6.4 mm (1/4 in.) diameter hole through 25.4 mm (1 in.) thick mild steel. In this case, the hole would have to be drilled, but where is the limiting point?

The old rule of thumb that the punch diameter must be at least equal to the thickness of the material has cost industry thousands of dollars. One steel fabricator had the architect increase the hole diameter specification from 20.6 to 23.8 mm (13/16 to 15/16 in.) so he could punch several thousand holes in 25.4 mm (1 in.) thick beams. He consented to pay the extra cost for larger fasteners, which was far less than the cost of drilling. Unfortunately, he did not know that with his portable press he could punch 20.6 mm (13/16 in.) diameter holes through 25.4 mm (1 in.) thick mild steel consistently.

When all the factors involved in the thickness-diameter ratio limitation are considered, it is possible to come up with a new, more realistic set of ratios for nonshock applications.

The diameter of the punch must be such that the punch compressive strength is greater than the force required to punch the hole. This punching force can be found by multiplying the material thickness by its shear strength (in pounds per square inch), then multiplying by the length of cut.

To determine if a punch will endure when used in a hydraulic press, the following factors must be considered:

- A = Cross-sectional area of the punch, as determined by hole size and shape (Fig. 20)
- T = Thickness of material being punched
- S_s = Shear strength of material being punched
- S_c = Compressive stress in the punch
- L = Length of cut

The compressive stress in the punch can be calculated from the formula:

$$S_c = \frac{T \times S_s \times L}{A}$$

The maximum allowable compressive stress (S_c) depends on the type and hardness of the tool steel from which the punch is made. A good grade of oil-hardened shock-resistant tool steel will withstand a compressive stress of 300,000 psi before breaking and can be used safely at 250,000 psi with good tool life.

The curves in Fig. 21 are based on these punching strength values, and for a known shear strength (of the material being punched), the curves give the recommended thickness-to-diameter ratio.

The curve shown as a dotted line represents the ultimate strength (300,000 psi); the solid curve, the recommended working stress (250,000 psi). For example, for punching of mild steel with a shear strength of 50,000 psi, the recommended thickness-to-diameter ratio is 1 1/4 to 1; the ultimate ratio is 1 1/2 to 1. Therefore, it is safe to punch a 25.4 mm (1 in.) diameter hole through 31.8 mm (1 1/4 in.) thick mild steel.

Quite often, a ratio between 1 1/4 and 1 1/2 to 1 is used for mild steel, but punch life is shortened.

Minimum Hole Size. The graph in Fig. 22 shows the minimum diameter of hole that can be punched through a given thickness of material. Three different materials are illustrated.

To use this graph, locate the thickness of material along the vertical scale and follow across horizontally to the lower edge of the indicated area for the material being punched. From this point of intersection, drop down to the horizontal scale and read the minimum recommended hole diameter.

The upper edge of the indicated area for each material represents the breaking point for the punch. Working within the indicated areas will therefore result in shortened punch life. For example, in punching 19 mm (3/4 in.) thick mild

steel, 15.1 mm (19/32 in.) is the recommended minimum hole diameter; if a 12.7 mm (1/2 in.) diameter punch is used, it will fail.

A punch will fail in one of two ways when overloaded. If its elastic limit is slightly exceeded, the punch will expand as it is pushed through the material. A very high force is then required during stripping, causing the punch to break either at the punch end or under the head.

The second type of failure occurs when the compressive stress is greatly exceeded, and the punch simply buckles before penetrating the material.

Special guided punches (Fig. 23) are available for piercing material up to 12.7 mm (1/2 in.) thick; these punches are supported to prevent buckling and offer thickness-to-diameter ratios as high as 2 to 1 in mild steel.

Limitations of Punching

Punching can cause distortion. In some parts, there is nothing that can be done to overcome it, but in many cases, steps can be taken to minimize it.

One common problem is the closeness of a hole to the edge of a part, which, if too close, causes bulging along the edge. Preferably, two times the thickness of the material should be allowed from the edge of the hole to the edge of the part.

Another frequent problem that occurs on strip and bar stock is camber due to off-center punching—for example, punching a row of 24 mm (15/16 in.) diameter holes on 75 mm (3 in.) centers in a 150 mm (6 in.) wide, 20 mm (3/4 in.) thick bar of mild steel where the common centerline is 50 mm (2 in.) in from one side. Figure 24 shows the type of distortion that will result. The holes should be on-center where possible, and the bar should not be too narrow. Insufficient clearance will cause increased distortion, due to the increased outward forces produced.

Die Clearances and Stripping Forces

Relationship of Die Clearance to Stress-Strain Curves. One of the most important

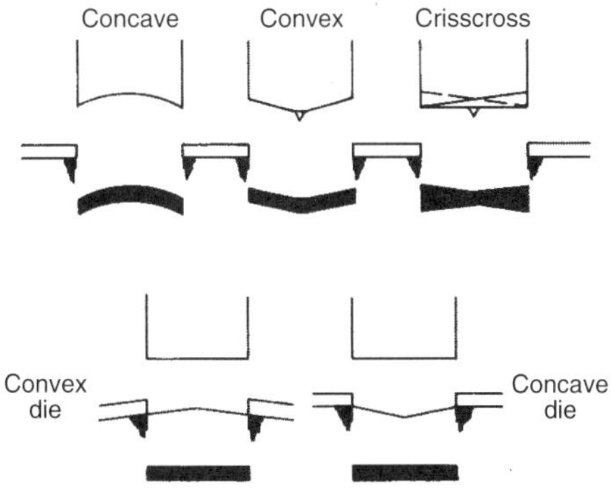

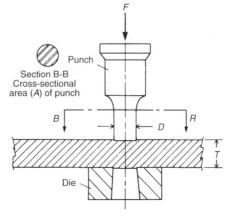

Fig. 19 Methods of applying shear to punches and dies

Fig. 20 Relationship of punch diameter to material thickness

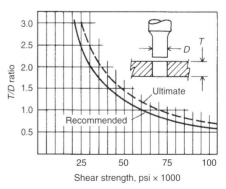

Fig. 21 Ratio of material thickness to punch diameter as a function of shear strength

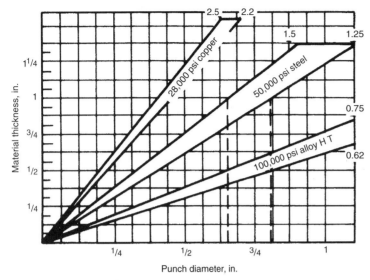

Fig. 22 Minimum punch diameter (hole size) as a function of material thickness

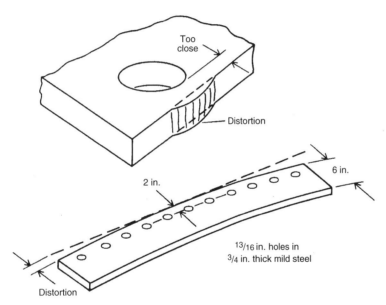

Fig. 24 Distortion resulting from off-center punching

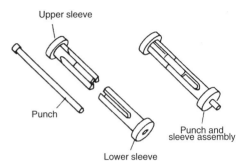

Fig. 23 Guided punch. Courtesy of Durable Punch and Die Company

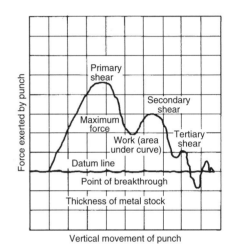

Fig. 25 Key for interpreting plot of loads from punching tests

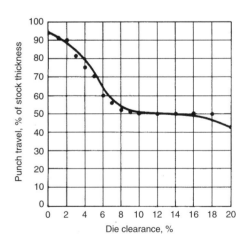

Fig. 26 Distance of punch travel through 3.2 mm (1/8 in.) thick flat naval brass stock (hardness, 62 HRB) for complete severance

factors in the design of punches and dies is the amount of clearance necessary between the punch and die. The amount of clearance, to a great extent, determines the life of the die, the interval of time between successive sharpenings of the die and punch, the quality of the work being sheared or stamped, the power requirements for the punch press producing the work, and the stresses built up in the press during a cycle. Die clearance varies with the thickness, type, hardness, and complexity of the work. No exact information giving the correct amount of die clearance for a set of known conditions is available. The determination of die clearance has been largely a matter of guesswork and rule-of-thumb procedures.

The subject of die clearance and its many aspects covers an extremely large field, and

prolonged and varied investigations have been necessary to give complete explanation to the variety of phenomena involved. One example is data acquisition directed toward investigation of instrumentation for die clearance studies conducted at Purdue University, as described here by Lascoe (Ref 1). A stress-strain indicator was used to record two electrical input signals plotting the force applied to the material and the distance through which the punch progressed. By examination of the load curve—that is, a plot of force applied versus the position of the punch—it was possible to determine the dynamic properties of the metal being worked at that particular clearance (Fig. 25). By analyzing a series of load curves resulting from the use of different die clearances, an accurate determination of the optimal die clearance for a given set of

conditions was made (Fig. 26, 27). The stress-strain indicator produces voltages that, when applied to an electronic display or digital recording, plot the load curve. Strain gages and a differential transformer were used as the input transducers.

The purpose of this study was to investigate the possibility, usefulness, and reliability of such an electronic approach to the accurate determination of optimal die clearances. The investigation was of a general nature in which various types of metals and clearances were used. The oscilloscope traces of the load curves were then analyzed for possible analytical relationships between die clearance and maximum force required to stamp out the work, work required to sever the part, amount of "dish" of the resulting work, and the distance of travel of the punch through the metal stock before complete severance occurred. From this study it was established that die clearance definitely affects the aforementioned factors and that such an electronic approach to the determination of an optimal die clearance for a given set of conditions is practical and useful.

Interpretation of Stress-Strain Curves. The functioning of a great many blanking dies may be likened to a tensile test for determining physical properties of a material; that is, the material is stressed at a concentrated point until rupture occurs. The pattern of this rupture—stress versus strain—closely approaches that of a metal slug being separated from its "mother" sheet by means of a blanking operation. The resulting stress-strain curve provides a wealth of information. The explanation of the procedure is in interpreting the stress-strain curves:

- The integral of the curve represents the work required to blank out the slug. When the vertical displacement conversion factor to stress and the horizontal conversion factor to strain are known, the areas under the curve can be determined and the work determined directly.
- The maximum height of the curve represents the maximum amount of force required to sever the blank from the sheet. The vertical-displacement calibration factor is known because the press ram can be calibrated on a materials testing machine (refer to Fig. 25).

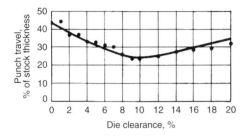

Fig. 27 Distance of punch travel through 3.2 mm (1/8 in.) thick flat aluminum stock (hardness, 73 HRB) for complete severance

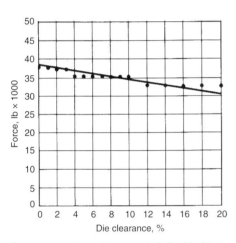

Fig. 28 Maximum force required for blanking a 25.4 mm (1 in.) diameter slug from 3.2 mm (1/8 in.) thick flat naval brass stock (hardness, 62 HRB)

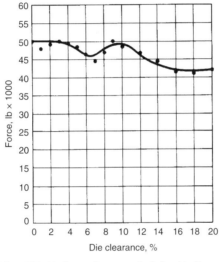

Fig. 29 Maximum force required for blanking a 25.4 mm (1 in.) diameter slug from 3.2 mm (1/8 in.) thick flat stainless steel stock (hardness, 85 HRB)

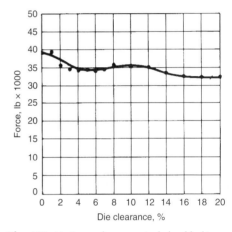

Fig. 30 Maximum force required for blanking a 25.4 mm (1 in.) diameter slug from 3.2 mm (1/8 in.) thick flat aluminum stock (hardness, 73 HRB)

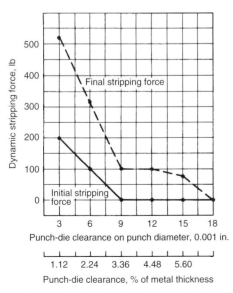

Fig. 31 Dynamic stripping force as a function of punch-die clearance for stainless steel. Hardness: 86 HRB. Thickness of metal strip: 3.5 mm (0.136 in.)

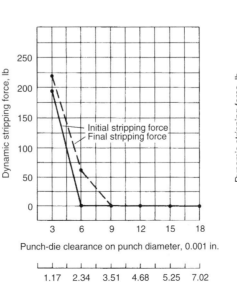

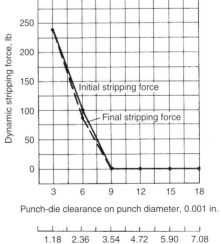

Fig. 32 Dynamic stripping force as a function of punch-die clearance. (a) Brass, 1/2 hard. Hardness: 77 HRB. Thickness of metal strip: 3.25 mm (0.128 in.). (b) Brass. Hardness: 69 HRB. Thickness of metal strip: 3.23 mm (0.127 in.)

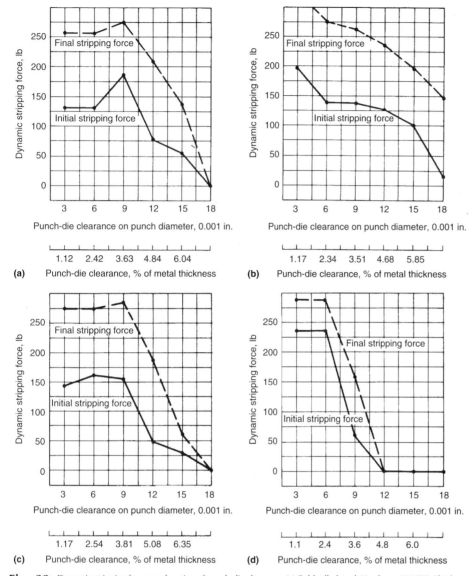

Fig. 33 Dynamic stripping force as a function of punch-die clearance. (a) Cold rolled steel. Hardness: 92 HRB. Thickness of metal strip: 3.15 mm (0.124 in.). (b) Hot rolled steel. Hardness: 65 HRB. Thickness of metal strip: 3.25 mm (0.128 in.). (c) Cold rolled steel. Hardness: 93 HRB. Thickness of metal strip: 3.0 mm (0.118 in.). (d) SAE 1020 cold rolled steel. Hardness: 99 HRB. Thickness of metal strip: 3.18 mm (0.125 in.).

- The amounts of secondary and tertiary shear can be determined from the areas under the second and third peaks, respectively.
- The differential of the curve at any point indicates the rate of change in the loading or unloading force at that time. If the curve is approaching the vertical, it means that the loading is applied almost instantaneously without any appreciable plastic deformation in the blanked slug, whereas a gradually sloping curve indicates plastic deformation or burring.
- Likewise, if the curve drops off almost vertically, it means that the severed piece has been cut "clean." If the dropoff is gradual, it means that a number of shear lines have overlapped each other and that, when the complete breakthrough does occur, the resulting cut is rough and ragged.

- The point at which the trace dips down to or below the datum line is the point of breakthrough or sudden load reversal, as was the case with aluminum. In some instances, the propagation of a crack will move faster than the punch, causing the trace to fall below the datum line even though the piece has not been completely severed.
- Any forces indicated beyond the horizontal distance representing the metal thickness are caused by the resistance of the slugs to movement between the die walls.

Results. The results of this test were highly satisfactory from the standpoint of a visual graphical record of the data secured. The data showed a definite relationship between the die clearances and the dynamic characteristics of the materials being tested. Perhaps the major

contribution of this investigation was to prove that the dynamic properties of metals are much different from the static properties in many instances. One author states that press-tonnage requirements are estimated by using dies with sharp edges and that the maximum pressure normally required equals the area adjacent to the perimeter of the cut. This is true for static conditions but is not applicable to dynamic conditions. Examination of the force-versus-die-clearance curve indicated that the dynamic force required for blanking is approximately 20 to 25% above the static force necessary to blank the same type and thickness of metal. This excessive amount of force required for blanking was apparent in all the metals used.

An increase in die clearance up to 8 to 12% gave work curves that showed a decided decrease in the amount of work required for blanking. A clearance below 6%, for 3 mm (1/8 in.) material, generally indicated excessive tool wear and excessive stresses in the press.

The nature of the blanked edge depended on the clearance. As clearance increases, the finish or quality of the blanked edge increased up to approximately 6 to 10%. Then plastic deformation occurs.

The "dish" deformation of the blanked slug tends to increase with increasing clearance. Dishing is much more prevalent in the softer materials; consequently, die clearance is more critical for softer metals. Die clearance has to be large enough to reduce work and dynamic forces, yet small enough to allow the product to fall within acceptable dish specifications. Die clearance can be excessive for the harder materials because of their ability to withstand dishing.

Metal hardness is an important factor in determining stress-strain characteristics. Because of the inability of the harder metals to absorb the inertial effects of the punch-press ram, the maximum force required for blanking is higher for harder metals than for softer metals of the same ultimate shearing stress. Figures 28, 29, and 30 show forces versus die clearances for various materials.

Dynamic Stripping Forces in Blanking. When a hole is punched, the material exerts a radial clamping force on the punch surface, and so a certain amount of force is required to withdraw the punch. This is called the stripping force.

The strength design of the stripper plate, which holds the material down during the upstroke of the punch, is based on the maximum value of the stripping force. Because experience formulas provide only a rough approximation method for the computation of this force, it is advisable to use a large safety factor. It is easy to design the stripper plate to be strong enough to withstand the largest stripping forces that occur.

There is a more important aspect of this force: its characteristics provide important information about the punching condition and about certain influential variables.

With an electronic recording method it is possible to measure the stripping force under actual operating conditions, thus eliminating the

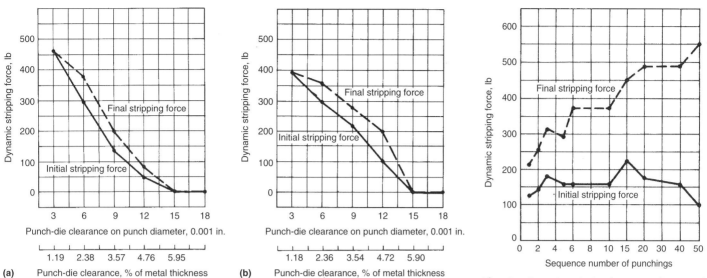

Fig. 34 Dynamic stripping force as a function of punch-die clearance. (a) Aluminum 2024-T3. Hardness: 61 HRB. Thickness of metal strip: 3.2 mm (0.126 in.). (b) Aluminum 2024-T. Hardness: 64 HRB. Thickness of metal strip: 3.23 mm (0.127 in.)

Fig. 35 Dynamic stripping force as a function of punch wear for tool steel. Hardness: 62 HRB

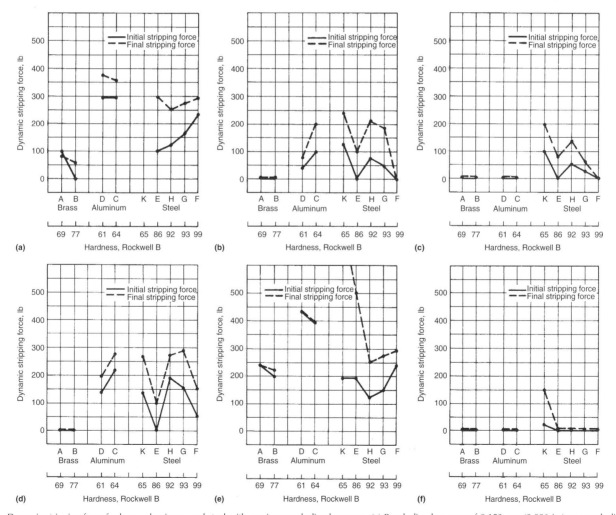

Fig. 36 Dynamic stripping force for brass, aluminum, and steel with varying punch-die clearances. (a) Punch-die clearances of 0.152 mm (0.006 in.) on punch diameter and 0.076 mm (0.003 in.) per cutting edge. (b) Punch-die clearances of 0.305 mm (0.012 in.) on punch diameter and 0.244 mm (0.0096 in.) per cutting edge. (c) Punch-die clearances of 0.381 mm (0.015 in.) on punch diameter and 0.191 mm (0.0075 in.) per cutting edge. (d) Punch-die clearances of 0.229 mm (0.009 in.) on punch diameter and 0.114 mm (0.0045 in.) per cutting edge. (e) Punch-die clearances of 0.076 mm (0.003 in.) on punch diameter and 0.038 mm (0.0015 in.) per cutting edge. (f) Punch-die clearances of 0.457 mm (0.018 in.) on punch diameter and 0.244 mm (0.0096 in.) per cutting edge

approximations of formulas based on experience. This investigation method results in a load curve, which is a plot of what is called the dynamic stripping force versus the position of the punch relative to the material being punched. Electric strain gages attached to the stripper plate record the size of the dynamic stripping force. The horizontal travel of this beam is related to the movement of the punch. This is achieved through a differential transformer with the core bar attached to the ram of the punch press.

The primary purpose of this investigation was to design and build a test setup based on the recording method outlined previously. Its secondary purpose was to perform some limited investigations in order to determine its usefulness and reliability.

Two major investigations were performed:

- For nine different materials, the relation between the dynamic stripping force and the amount of punch-die clearance was established.

- The effect of punch wear on the dynamic stripping force was recorded.

Results. The best results indicated a strong dependence of punch load on the amount of punch-die clearance (Fig. 31 to 36). The same was found to be true for punch wear, when it is caused by material buildup near the cutting edge of the punch. Several points are also worthy of consideration:

- A rectangular shape of the load curve for the dynamic stripping force indicates a favorable punching condition. The more triangular the form of the curve becomes, the more material buildup occurs near the cutting edge of the punch, and the worse the general punching condition.
- The characteristics of the insides of the holes are directly related to the size of the dynamic stripping force. It is possible to develop theoretical explanations for most of the characteristics of the dynamic stripping force.

- A microscopic analysis of the area near the cutting edge of the punch reveals interesting aspects of punch wear. If this wear is caused by material buildup near the cutting edge, then the load curve clearly shows this condition with an increase of the terminal stripping force values.
- A tentative theory can be developed that provides an explanation for the material buildup near the cutting edge of the punch.

ACKNOWLEDGMENT

This article was adapted from O.D. Lascoe, *Handbook of Fabrication Processes,* ASM International, 1988.

REFERENCE

1. O.D. Lascoe, *Handbook of Fabrication Processes,* ASM International, 1988, p 132–133

Shearing of Sheet, Strip, and Plate

Revised by L. Chen, Engineered Materials Solutions, Inc.

SHEARING is a method for cutting a material piece into smaller pieces using a shear knife to force the material past an opposition shear knife in a progression form. Shearing is widely used to divide large, flat stock such as sheet, strip, and plate. Shearing of sheet, strip, and plate is broadly classified according to the type of blade (knife or cutter) used as either straight or rotary. Straight-knife shearing is used for squaring and cutting flat stock to the required shape and size. It is usually used for producing square and rectangular shapes, although triangles and other straight-sided shapes are also sheared with straight knives. Rotary shearing (which should not be confused with slitting, as discussed in the article "Flattening, Leveling, Slitting, and Shearing of Coiled Product" in this Volume) is used for producing circular or other contoured shapes.

Straight-Knife Shearing

In straight-knife shearing, the flat workpiece is placed between a stationary lower knife and a movable upper knife. As the upper knife is forced downward, it cuts the metal into two parts. Straight-knife shearing is the most economical method of cutting straight-sided blanks from flat sheet, strip, and plate with thickness no more than 50 mm (2 in.). The process is also widely used for cutting sheet into blanks that will subsequently be formed or drawn. Because shear gaging can be set within ±0.13 mm (±0.005 in.), the shearing process is generally limited to ±0.4 mm (±1/64 in.) tolerances in 16 gage material. The tolerance range increases with thickness.

Straight-knife shearing is seldom used for shearing metal harder than approximately 40 HRC. As the hardness of the work metal increases, knife life decreases for shearing a given thickness of metal. In general, it is practical to shear flat stock up to 38 mm (1 1/2 in.) thick in a squaring shear. Squaring shears up to 9 m (30 ft) long are available (even longer shears have been built), and some types are equipped with a gap that permits shearing of metal longer than the shear knife. When extremely soft, ductile metal (especially thin sheet) is sheared, the edges of the metal roll, and large burrs result.

Principle of Shearing

The principle of shearing is simple. When the upper knife is forced to move downward, both the upper and lower knives gradually come together and contact the metal being sheared. The knives penetrate to a certain portion of the metal thickness until its shear strength is reached, at which point the unpenetrated portion of the metal fractures, and the work metal separates (Fig. 1). At the final stage of shearing, the upper knife continuously moves downward and results in freeing the sheared pieces from the original workpiece. The upper knife wall rubs against the metal edge to cause the cutting edge area to burnish, while the lower knife wall rubs against the sheared piece edge to cause a second burnish area. A burr occurs on both the sheared piece and the original workpiece. The amount of penetration depends largely on the shear strength and thickness of the work metal. The knife will penetrate 30 to 60% of the metal thickness for low-carbon steel, depending on thickness (see the section "Capacity" in this article). The penetration will be greater for a more ductile metal such as copper. Conversely, the penetration will be less for metals that are harder than low-carbon steel.

A sheared edge is characterized by the smoothness of the penetrated portion and the relative roughness of the fractured portion. Sheared edges do not have the quality of machined edges. However, when knives are kept sharp and in proper adjustment, it is possible to obtain sheared edges acceptable for a wide range of applications. The quality of sheared edges generally improves as workpiece thickness decreases.

Machines for Straight-Knife Shearing

Punch presses and press brakes can be used for shearing a few pieces or are used temporarily when more efficient equipment is not available. Production shearing, however, is usually done in machines that are designed for this operation.

Squaring shears are usually used for trimming and cutting sheet, strip, or plate to specific size (Fig. 2). These shears (also called resquaring or guillotine shears) are available in a wide range of sizes and designs. Some types of lines also permit slitting when the workpiece moves for shearing.

The sheet, strip, or plate is held firmly by hold-down devices while the upper knife moves down past the lower knife. Most sheet or plate is sheared by setting the upper knife at an angle (Fig. 1a). The position of one of the knives can be adjusted to maintain optimal clearance between the knives. Squaring shears can be actuated mechanically, hydraulically, or pneumatically.

Mechanical Shears. The power train of a mechanical shear consists of a motor, the flywheel, a worm shaft that is gear driven by a flywheel, a clutch that connects the worm gear drive to the driven shaft, and a ram actuated by the driven shaft through eccentrics and connecting links. Under most operating conditions, a mechanical shear can deliver more strokes per minute (spm) than a hydraulic shear. Some mechanical shears cycle as fast as 100 spm.

Another advantage of the mechanical shear is that, because of the energy stored in the flywheel, a smaller motor can be used for intermittent shearing. For example, a mechanical shear with a no-cutting or free-running speed of 65 spm can make approximately six full shearing strokes (i.e., shearing maximum thickness and length of cut) per minute with a standard motor. However, when the same shear is cutting at full capacity in a rapid shear mode, a much larger motor is required, because in rapid cutting, there is not enough time between cuts for the smaller motor to restore the speed of the flywheel.

An additional advantage of the mechanical shear is that its moving knife travels faster than the moving knife of a hydraulic shear. In some cases, greater knife speed can decrease work metal twist, bow, and camber.

Most mechanical shears are provided with enough horsepower to build up the flywheel speed after each cut but not enough to allow the operator to run full-capacity cuts in a high-speed mode. It should be noted that some of the mechanical shears are not of the flywheel type; rather, their motor directly drives the shear blade beam to move down to make a cut. Mechanical

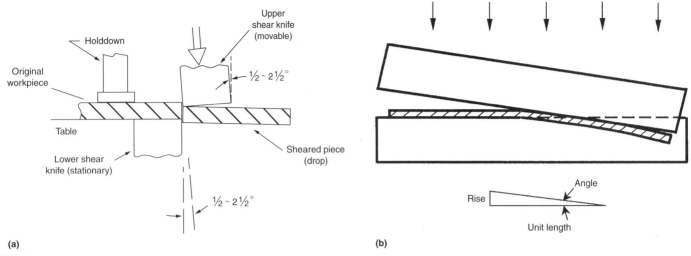

Fig. 1 Schematic illustration of straight-knife shearing. (a) Side view. (b) Rake angle

Fig. 2 Typical squaring shear. Courtesy of Cincinnati Inc.

shears are rated in strokes per minute, not cuts per minute. Most shearing applications do not require high-speed cutting.

Hydraulic shears are actuated by a motor-driven pump that forces oil into a cylinder against a piston; the movement of the piston energizes the ram holding the upper knife. A hydraulic shear can make longer strokes than a mechanical shear. In general, long shears and shears with low-carbon steel capacities above 13 mm ($\frac{1}{2}$ in.) are almost all hydraulic. Hydraulic shears are designed with a fixed load capacity. This prevents the operator from shearing material that exceeds capacity and, therefore, saves costly damage to the machine structure; this is a basic advantage of hydraulic shears.

The total load during shearing is related to the shear knife rake angle, sharpness of the knives, knife clearance and type, and the mechanical properties and thickness of the material being sheared. It is possible to stall the machine during shearing within a rated capacity if the clearance is incorrect, the knife is dull, or the back piece is excessively deep. In this way, the hydraulic shear is protected from damage caused by overloading. A mechanical shear would not be constrained by an overload prevention system and would continue to cut under nearly all conditions.

Pneumatic shears are used almost exclusively for shearing thin metal (seldom thicker than 1.50 mm, or 0.060 in.) in relatively short pieces (seldom longer than 1.5 m, or 5 ft). Activation of air cylinders makes the shear blade beam move to make a cut. Shop compressed air or a freestanding air compressor is used to provide power to air cylinders.

Alligator shears have a shearing action similar to that of a pair of scissors. The lower knife is stationary, and the upper knife, held securely in an arm, moves in an arc around a fulcrum pin. This type of machine is most widely used for shearing bars and bar sections and for preparing scrap. Alligator shears are available in various sizes, including those that can shear plate up to 32 mm ($1\frac{1}{4}$ in.) thick by 760 mm (30 in.) long and plate up to 50 mm (2 in.) thick in shorter lengths. The lighter machines can be made portable; the heavier machines, however, must be firmly anchored in concrete, especially if they will be used in conjunction with roller conveyor tables in the shearing of plate.

Accessory Equipment for Straight-Knife Shearing

Some accessories have been incorporated into most shear designs and are required for efficient and accurate straight-knife shearing.

Holddowns (Fig. 1a) are mechanical or hydraulic devices that hold the work metal firmly in position to prevent movement during shearing. The holddown pressure must be greater than the forces generated in cutting the work material. These forces depend on the knife clearance, rake angle, and depth of material back piece. The most efficient holddown system is a series of independent units that securely clamps stock of varying thickness automatically and without adjustment. The force on each holddown foot must be substantial, ranging from several hundred pounds on a machine for shearing sheet to several tons for shearing plate. Holddowns must be timed automatically with the ram stroke, so that they clamp the work metal securely before the knife makes contact and release their hold instantly after shearing is completed.

Back gages are adjustable stops that permit reproducibility of dimensions of sheared workpieces in a production run. Most gages are controlled electrically. Push-button control provides a selection of high traverse speeds and slow locating movements for accurate final positioning. The addition of a computer numerical control system permits dimensional accuracy and repeatability, increased productivity, and hands-off safe operations.

For thin sheet, magnetic overhead rollers eliminate sag and support the sheet for accurate gaging. For rapid and accurate cutting, back gages are equipped with electronic sensors that automatically trip the shear only when the sheet is accurately positioned.

Pneumatic supports are used to support thin sheet. Support arms are designed to elevate into a horizontal position that is flush with the shear table, permitting material to be supported in the correct position against the back-gage stop. Blank inaccuracies due to unsupported and poorly positioned sheets are virtually eliminated.

Back gages are also equipped with retractable stops for shearing mill plate. With the stops out of the way, mill plate of almost any length can be fed into the shear and cut to the desired length. When stops are not used, the workpiece can be notched or scribed to indicate the cutoff position.

Front Gages. When gaging from the front of the machine, the operator locates the work metal by means of stops secured in the table or in the front support arms. Power operation of the front support arms allows the blank dimensions to be entered digitally using a computer control. Front gaging is often done by means of a squaring arm.

Squaring arms (Fig. 3) are extensions attached to the entrance side of a shearing machine that are used to locate long sections of work metal in the proper position for shearing. Each arm is provided with a linear scale and with stops for accurate, consistent positioning of the work metal. Squaring arms are reversible to allow use of the shear at either end and to more evenly distribute the wear on the shear knives. Power operation can be added to the squaring arm for increasing blank accuracy and reducing setup time. This type of squaring arm prevents the arm from moving from side to side.

Straight Shear Knives

Material used for shear knives depends on specific shearing operations. Most shear knives are made in one piece from tool steel; some are made of carbon or alloy steel. The composition, hardness, thickness, and quantity of metal being sheared are the most important factors in the selection of knife material. An AISI D2 tool steel is often recommended for cold shearing metals up to 6.4 mm ($\frac{1}{4}$ in.) thick (Table 1). Knives made of modified A8, H13, or S5 tool steels are recommended for low-volume production or for occasional shearing of metals up to 6.4 mm ($\frac{1}{4}$ in.) thick (except the more highly abrasive metals such as silicon steel). Knives made of A2 tool steel have been satisfactory for high-production cold shearing of soft nonferrous metals. However, D2 knives are usually more economical because of better wear resistance, but they are not usually recommended for cold shearing materials more than 6.4 mm ($\frac{1}{4}$ in.) thick because they are likely to break under impact loads. Nevertheless, depending mainly on knife design and length of cut, knives made of D2 tool steel have been successfully used for cold shearing aluminum alloys up to 32 mm ($1\frac{1}{4}$ in.) thick.

Modified A8 or H13 tool steels are suitable for some cold-shearing applications in which the work metal is more than 6.4 mm ($\frac{1}{4}$ in.) thick, as indicated in Table 1. However, the shock-resistant grades S2 and S5 are usually recommended for shearing heavy sections of all metals.

The length and design of the knife can influence the selection of knife material. Although water-hardened tool steels such as W1 and W2 are suitable for many cold-shearing applications, the rapid cooling during heat treatment causes greater distortion than that in knives made from the oil-hardened or air-hardened steels such as D2. For example, a bar 4.17 m (164 in.) long is needed to make a shear knife 4.16 m (164 in.) long from D2 tool steel. The same knife made from W2 requires a bar 4.17 m (164 in.) long. Both steels elongate when heat treated, but W2 will bow more readily than D2. Because straightening is difficult, the W2 knife therefore must have more grinding stock. The additional grinding decreases the depth of the hardened shell and shortens the useful life of the knife.

Hardness. The rate at which a knife wears in cold shearing depends primarily on its carbon content, alloy content, and hardness. Insufficient hardness in a knife used for cold shearing will shorten its service life. In one application, a knife made of S5 tool steel with a hardness of 44 HRC wore three times as fast as one with a hardness of 54 HRC used under the same conditions. Despite the desirability of having shear knives as hard as possible to minimize wear, it is often necessary to sacrifice some hardness to prevent knife breakage as the hardness or thickness of the metal being sheared increases.

Hardness alone does not always guarantee knife performance and life, because different wear resistance, ductility, or toughness can be obtained at the same hardness. Improper quenching and tempering can result in a unsuitable microstructure and thus insufficient wear resistance, ductility, or toughness for a specific shearing application even though the steel may have the same hardness as obtained in the same steel from proper processes.

Recommendations for the hardness of knives for cold shearing cannot always be made without knowledge of the details of the shearing operations. Such details include type, mechanical properties, and thickness range of the material to be sheared as well as shearing speed and knife dimensions. For example, knives made of D2 tool steel have often been successfully used at 58 to 60 HRC for shearing low-carbon steel up to 6.3 mm ($\frac{1}{4}$ in.) thick. However, the hardness of a D2 knife must be kept below 58 HRC to prevent knife break if the material to be sheared is high-strength low-alloy steel. The shock-resistant S-grade tool steels are used in the hardness range between 50 and 58 HRC. The hardness of the knives should be decreased from the higher end of the range to the lower end when the hardness or thickness of the materials to be sheared increases. The higher end of this range is applicable to shearing of steels 6.4 to 13 mm ($\frac{1}{4}$ to $\frac{1}{2}$ in.) thick and to nonferrous metals. As shock loading increases with the shearing of harder or thicker metals, knife hardness is decreased toward the low end of the aforementioned hardness range.

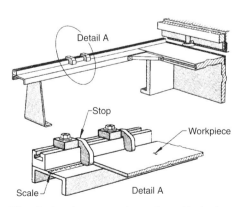

Fig. 3 Squaring arm attachment for positioning long pieces in a shearing machine

Table 1 Recommended materials for straight shear knives for cold shearing of flat metals

Metal to be sheared	Thickness, ≤6.4 mm ($\frac{1}{4}$ in.) Low production	Thickness, ≤6.4 mm ($\frac{1}{4}$ in.) High production	Thickness, 6.4–12.7 mm ($\frac{1}{4}$–$\frac{1}{2}$ in.) Low production	Thickness, 6.4–12.7 mm ($\frac{1}{4}$–$\frac{1}{2}$ in.) High production	Thickness, >12.7 mm ($\frac{1}{2}$ in.)
Carbon and low-alloy steels (up to 0.35% C)	Modified A8; H13; L6	D2	Modified A8; H13; L6	A2	S5(a)
Carbon and low-alloy steels (0.35% C)	Modified A8; H13; L6	D2	Modified A8; H13; L6	S5	S5(a)
Stainless steels and heat-resistant alloys	Modified A8; H13; L6	D2	S5	A2	S5(a)
Silicon electrical steels	D2	D2; carbide	S5	S5	(b)
Copper and alloys; aluminum and alloys	Modified A8; H13; L6	A2; D2	Modified A8; H13; L6	A2	S5(a)
Titanium and titanium alloys	D2	D2	...	...	...

(a) S5 is preferred for stock thicker than 19 mm ($\frac{3}{4}$ in.) (b) Seldom sheared thicker than 12.7 mm ($\frac{1}{2}$ in.)

Operating Parameters

Capacity. Most shearing machines are rated according to the section size of low-carbon steel they can cut. The tensile strength of low-carbon steel sheet and plate is generally below 520 MPa (75 ksi) and the yield strength below 350 MPa (51 ksi). Shears are frequently rated in terms of their ability to cut low-carbon steel with a tensile strength of 415 MPa (60 ksi) and yield strength of 275 MPa (40 ksi). An allowance for normal over-tolerance material thickness is included in the capacity rating of the machine. The use of a machine for shearing other metals is primarily based on the relationship of the tensile strength and ductility of low-carbon steel to that of the particular metal to be sheared. Metals with a tensile strength higher than that of low-carbon steel almost always reduce the capacity of the machine. For example, the machine capacity for shearing high-strength low-alloy steels is reduced to approximately two-thirds to three-quarters of the rated capacity for low-carbon steel. Conversely, for shearing aluminum alloys, machine capacity can range from $1\frac{1}{4}$ to $1\frac{1}{2}$ times the rated capacity for low-carbon steel.

Table 2 compares the shearing capacities of various metals with those of low-carbon steel. The metal thicknesses given in the table are based on the thickness of low-carbon steel that can be sheared with the same shearing capacity. For example, a specific force is required to shear 6.4 mm ($\frac{1}{4}$ in.) thick low-carbon steel. Table 2 shows that the same force can shear only a 4.8 mm ($\frac{3}{16}$ in.) thickness of type 302 stainless steel but can shear a 6.4 mm ($\frac{1}{4}$ in.) thick aluminum.

Ductility also can affect machine capacity. For example, annealed copper, because of its high elongation, requires as much shearing effort as low-carbon steel, even though copper has considerably lower tensile strength. Similarly, carbon steel with very low carbon (<0.1% C) and higher-than-normal elongation will require a higher-capacity machine.

Power Requirements. The energy consumed during shearing is a function of the average stress, the cross-sectional area to be sheared, and the depth of maximum knife penetration at the instant of final fracture of the work metal. For any metal, the amount of energy consumed is thus proportional to the area under the shearing stress-strain curve up to the point of fracture.

Figure 4 shows typical shearing stress-strain curves for hot-rolled and cold-rolled steels. The distance through which the force acts (knife penetration) is approximately 35% of the work metal thickness for hot-rolled steel and 18.5% for cold-rolled steel. For example, in the curve for hot-rolled steel, the average stress under the curve is 73.5% of the maximum shearing stress S_{max}, and the distance through which the force acts is 35% of the work metal thickness. Therefore, the energy E used in shearing hot-rolled steel is:

$$E = 0.735\,S_{max} \times Wt \times 0.35\,t = 0.257\,S_{max} \times Wt^2 \quad \text{(Eq 1)}$$

where W is work metal width, and t is its thickness. Applying Eq 1 to the curve for cold-rolled steel in Fig. 4 yields an energy consumption of $0.136\,S_{max} \times Wt^2$.

The maximum instantaneous horsepower HP_{max} required for cutting work metal in a shear is determined by:

$$HP_{max} = \frac{Wt \times S_{max} \times V}{33{,}000} \quad \text{(Eq 2)}$$

where V is the speed of the shear knife, and the other variables are as previously defined.

The average power requirement HP_{avg} for a shear making n cuts per minute in hot-rolled steel is:

$$HP_{avg} = \frac{Wt^2 \times n \times S_{max}}{1{,}540{,}000} \quad \text{(Eq 3)}$$

Equations 2 and 3 determine the net power required for actual shearing of the workpiece.

The amount of power needed to operate the hold-down system and to overcome friction must be added to net power.

Friction depends on the design of the shearing machine and the knife, type of bearings, alignment, lubrication, temperature of operation, and size of the machine in relation to the area of the section to be sheared. When shearing metal of nearly the maximum size for which a shear is designed, the loss of horsepower by friction for well-designed machines seldom exceeds 25% of the gross horsepower.

Shearing Force. It should be noted that the shearing force F required to cut a piece of metal follows a proportional relationship:

$$F \propto spt^2(1 - p/2)/R \quad \text{(Eq 4)}$$

where s is the shear strength of the metal, t is the thickness of the material, p is the percentage of penetration of the knife into the material, and R is the rake of the shear knife blade.

It is not possible, therefore, to calculate the required shearing force for different work metal thicknesses based solely on change of rake. Even for low-carbon steel, the amount of knife penetration prior to fracture can be as great as 60% of the work metal thickness for 3.4 mm (0.135 in.) thick stock and as little as 30% for 19 mm ($\frac{3}{4}$ in.) thick stock by the same set of a shear.

Rake, R, is the tangent of the angle formed between the lower (fixed) shear knife and the upper (movable) shear knife (Fig. 1b). It normally expressed as the upper knife rise per a unit length of the lower knife. For example, a rake of 21 mm/m ($\frac{1}{4}$ in./ft) means that the upper knife rises 21 mm for each meter ($\frac{1}{4}$ in. for each foot) of linear distance along the knives. Rakes below 21 mm/m ($\frac{1}{4}$ in./ft) are rarely used; a rake of 42 mm/m ($\frac{1}{2}$ in./ft) or higher is typical of many plate shears.

A certain degree of rake is used to permit progressive shearing of the work metal along the length of the knife. As the rake decreases, the amount of upper knife blade engagement increases as it travels through the material; this results in a greater required shearing force. A higher rake reduces the shearing force and

Table 2 Shearing capacities for various metals compared to those for low-carbon steel

Thickness of low-carbon steel(a)		AISI type 302 stainless steel(b)		Full-hard steel strip		Aluminum alloys	
mm	in.	mm	in.	mm	in.	mm	in.
1.52	0.060	0.91	0.036	1.22	0.048	1.90	0.075
1.90	0.075	1.22	0.048	1.52	0.060	3.05	0.120
3.05	0.120	1.52	0.060	1.90	0.075	3.40	0.134
3.40	0.134	1.90	0.075	2.67	0.105	4.8	$\frac{3}{16}$
4.8	$\frac{3}{16}$	3.40	0.134	3.9	$\frac{5}{32}$	5.6	$\frac{7}{32}$
6.4	$\frac{1}{4}$	4.8	$\frac{3}{16}$	4.8	$\frac{3}{16}$	6.4	$\frac{1}{4}$
7.9	$\frac{5}{16}$	5.6	$\frac{7}{32}$	5.6	$\frac{7}{32}$	9.5	$\frac{3}{8}$
9.5	$\frac{3}{8}$	6.4	$\frac{1}{4}$	6.4	$\frac{1}{4}$	11.1	$\frac{7}{16}$
11.1	$\frac{7}{16}$	7.9	$\frac{5}{16}$	7.9	$\frac{5}{16}$	12.7	$\frac{1}{2}$
12.7	$\frac{1}{2}$	9.5	$\frac{3}{8}$	9.5	$\frac{3}{8}$	15.9	$\frac{5}{8}$
15.9	$\frac{5}{8}$	11.1	$\frac{7}{16}$	11.1	$\frac{7}{16}$	19.0	$\frac{3}{4}$
19.0	$\frac{3}{4}$	12.7	$\frac{1}{2}$	12.7	$\frac{1}{2}$	25.4	1
22.2	$\frac{7}{8}$	15.9	$\frac{5}{8}$	15.9	$\frac{5}{8}$	31.8	$1\frac{1}{4}$
25.4	1	19.0	$\frac{3}{4}$	19.0	$\frac{3}{4}$	38.1	$1\frac{1}{2}$
31.8	$1\frac{1}{4}$	25.4	1	25.4	1	50.8	2

(a) Also applicable to soft to half-hard strip steel, alclad steel, and copper and copper alloys. (b) Also applies to most other austenitic stainless steels, normalized alloy steels such as 4130 or 8630, annealed high-carbon steels, and annealed tool steels

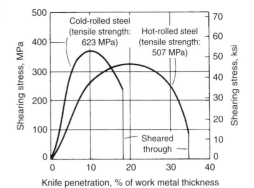

Fig. 4 Typical curves of shear stress and shear strain for hot-rolled and cold-rolled steels

allows the use of a smaller machine to shear the material than would be necessary if the cutting edges of the knives were parallel. However, a high rake increases the distortion in the sheared materials. A high rake also can cause slippage and therefore require high hold-down forces. A rake should be as low as possible to reduce the amount of distortion in the sheared materials.

Knife clearance is another important parameter; it is the gap between the upper knife and the lower knife as they pass each other. Too tight a clearance will result in secondary fracture after the first fracture, and the cut edge displays a characteristic ragged shape. On the other hand, too much clearance will result in a tear and high burr rather than a clean cut on the cut edge. A more serious consequence of excessive clearance is that it can cause the workpiece to be pulled between the knives. This, in turn, causes overloading of the machine and may result in failure of machine components or shear knives.

The major effects of knife clearance are the appearance of the sheared edge and the squareness of the cut. For example, for annealed mild steel, when sheared edge condition and appearance are critical, the knife clearance is between 4 and 10% of the material thickness. In contrast, it is 9 to 15% when the edge condition and appearance are not critical. The knife clearance also affects the degree of twist of the drop sheet and the shearing force required. Because of knife deflection during shearing in some shears, the clearance at the center of the knife is usually set less than that at the ends. Knife clearance (except on machines using a fixed clearance) is generally increased as work metal thickness increases.

When soft metals are sheared, insufficient clearance causes double (secondary) shearing, which appears as a burnished area at the top and bottom of a sheared edge with a rough area between the burnished edges. Knife clearance usually should be increased as the hardness of the metal being sheared decreases. Some mechanical shears are constructed to operate with a fixed clearance, and no adjustments are made for variations in work metal composition or thickness. The knife clearance is set for the thinnest material to be sheared. If the range of thicknesses sheared is not too large, then double shear may be avoided.

Ram speed in straight-knife shearing (and, in turn, the knife speed) has some effect on results in the shearing of flat sheet, strip, and plate. Low linear speed produces a rough sheared edge. As speed is increased, a cleaner sheared edge is obtained. In general, speeds to 21 to 24 m/min (70 to 80 ft/min) can be used without difficulty when shearing annealed metals. Regardless of the speed used, adequate hold-down force is mandatory.

Shearing Attributes and Defects

Dimensional accuracy obtained in straight-knife shearing is influenced by the capacity and condition of the machine, condition of the knives, knife clearance, and work metal thickness and condition. For example, a total tolerance of 0.25 mm (0.010 in.) usually can be achieved in sheets with a thickness no more than 3.4 mm (0.135 in.) when they are sheared to a size up to 3.7 m (12 ft) long. This tolerance applies to sheets that are essentially free from residual stress and flat within commercial limits. Sheets that are not flat or have residual stress, or both, cannot be sheared with the same accuracy.

Greater tolerances are required in shearing plate. A total tolerance of 0.5 to 1.0 mm (0.020 to 0.040 in.) can be maintained when plate is sheared in squaring shears. Dimensions can be held to a tolerance of approximately ± 1.6 mm ($\pm^{1}/_{16}$ in.) when shearing in alligator shears.

Sheared edge condition already has been discussed in this section. In summary, good sheared edge quality needs proper knife sharpness, knife clearance, and sufficiently high ram speed.

Bow, or cross bow, is one of three major shape defects introduced in short sheared pieces during shearing (Fig. 5). Bow is nearly proportional to the rake angle of the upper knife when a short piece is being sheared. A high rake causes the materials to bend during shearing. Reducing the rake angle will minimize a bow in a short sheet being cut when the incoming material has a good shape. The bow will become negligible when the sheet being sheared is more than 100 mm (4 in.) long.

Twist in a short sheared piece is generally proportional to the rake of the upper knife. Thick plates twist more than thin sheets, if both of them have the same good quality of incoming materials in shape. Soft materials tend to have more twist than hard materials. Wide incoming material has a higher tendency to twist in a short sheared piece than narrow incoming material. Short sheared pieces appear more twisted than long sheared pieces. When 6.4 mm ($^{1}/_{4}$ in.) plate is sheared, there is some twist in a 25 mm (1 in.) long sheared piece (drop). If the length of the sheared piece increases to 100 mm (4 in.), no twist occurs visually. When 25 mm (1 in.) plate is sheared, the length of the sheared piece must be greater than 125 mm (5 in.) for no measurable twist.

Camber cannot be eliminated in short sheared pieces (<100 mm, or 4 in.) but sometimes can be reduced by lowering the rake angle of the upper knife. Sufficient hold-down pressure close to the shear may help reduce camber.

Rotary Shearing

Rotary shearing, or circle shearing (not to be confused with slitting), is a process for cutting sheet and plate in a straight line or in contours by means of two revolving, tapered circular cutters. Table 3 lists recommended cutter materials.

For conventional cutting to produce a perpendicular edge, the cutters approach each other and line up vertically at one point (Fig. 6a). The point of cutting is also a pivot point for the workpiece. Because of the round shape of the knives, they offer no obstruction to movement of the workpiece to the right or left. This feature permits the cutting of circles and irregular shapes that have small radii, as well as cutting along straight lines.

Overlapping of the cutters to the position shown in Fig. 6(b) permits the shearing of smooth beveled edges in straight lines or circular shapes. With the cutters positioned as shown in Fig. 6(b), a bevel can be cut across the entire thickness of the workpiece, resulting in a sharp edge on the bottom of the workpiece, or (by varying the overlap of the cutters) only a corner of the workpiece can be sheared off, leaving a vertical edge (or land) for approximately half the workpiece thickness.

The shearing of workpieces into circular blanks requires the use of a holding fixture that permits rotation of the workpiece to generate the desired circle. For straight-line cutting in a rotary shear, a straight-edge fixture is used, mounted in the throat of the machine behind the cutter heads.

Applicability. Any metal composition or hardness that can be sheared with straight knives can be sheared with rotary cutters. In general, rotary shearing in commercially available machines is limited to work metal 25 mm (1 in.) thick or less. There is no minimum thickness. For example, wire cloth made from 0.025 mm (0.001 in.) diameter wire can be successfully sheared by the rotary method.

Circles up to 3 m (10 ft) in diameter or larger can be produced by using special clamping equipment. Minimum diameters depend on the thickness of the work metal and the size of the rotary cutters. Typically, with material up to 3.2 mm ($^{1}/_{8}$ in.) thick, the minimum circle that can normally be cut is 150 mm (6 in.) in diameter. For 6.4 mm ($^{1}/_{4}$ in.) thick stock, the minimum diameter is 230 mm (9 in.), and for 25 mm (1 in.) thick stock, the minimum diameter is 610 mm (24 in.).

Rotary shearing is limited to cutting one workpiece at a time. As in straight-knife shearing, multiple layers cannot be sheared, because each layer prevents the necessary breakthrough of the preceding workpiece.

Rotary shearing, plasma cutting, laser cutting, waterjet cutting, gas cutting, and electric arc cutting are competitive for some operations (see the article "Cutting Operations" in this Volume). Each can produce straight or beveled edges of comparable accuracy. The selection of one of these processes depends largely on the thickness

Bow   Twist   Camber

Fig. 5 Three major shape defects introduced by shearing of short sheared pieces

Table 3 Recommended knife materials for rotary shearing of flat metals

Metal to be sheared	Thickness to be sheared		
	4.8 mm (3/16 in.) or less	4.8–6.4 mm (3/16–1/4 in.)	6.4 mm (1/4 in.)
Carbon, alloy, and stainless steels	D2(a)	A2(b)	S4; S5
Silicon electrical steels	M2(c); D2(d)	D2	...
Copper and aluminum alloys	A2; D2	A2; D2	A2(e)
Titanium and titanium alloys	D2(f); A2(g)	...	...

(a) L6 is also recommended for shearing carbon and alloy sheet containing >0.35% C. (b) D2 is also recommended for low-carbon and low-alloy sheet. (c) For sheet >0.8 mm (1/32 in.) thick. (d) For sheet >0.8 mm (1/32 in.) thick. (e) S5 is recommended for sheet >12.7 mm (1/2 in.) thick. (f) For sheet <3.2 mm (1/8 in.) thick. (g) For sheet >3.2 mm (1/8 in.) thick

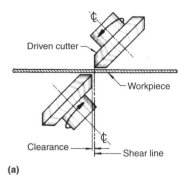

(a)

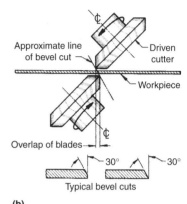

(b)

Fig. 6 Two types of rotary shearing. (a) Conventional arrangement of cutters for producing a perpendicular edge. (b) Overlap of cutters for producing a beveled edge

of the work metal. In general, rotary shearing is used for sheet and plate less than 13 mm (1/2 in.) thick, and laser cutting is used for sheet and plate less than 25 mm (1 in.) thick; gas cutting is used for thicknesses of 13 mm (1/2 in.) or more.

Gas cutting is less suitable for cutting a single thickness of sheet or thin plate because the heat causes excessive distortion, but it is often feasible to minimize this problem by stack cutting (cutting several thicknesses at a time). Gas cutting is more versatile than rotary shearing; it can produce smaller circles than rotary shearing and can produce rings in one operation. Gas cutting, however, produces a relatively large heat-affected zone on the workpiece.

Laser cutting produces an extremely narrow kerf that provides unmatched precision for cutting small holes, narrow slots, and closely spaced patterns. Complex openings, contours, and patterns, which are impossible to cut with conventional tools, are routinely cut using a laser and require little if any additional processing. A smaller heat-affected zone than traditional thermal cutting processes minimizes distortion and improves part quality.

Circle Generation. For cutting circles, the workpiece is placed in a special fixture consisting of a C-shaped, deep-throated frame having a rotating pin or clamp point at its outer extremity. The maximum circle that can be sheared is governed by the depth of the clamp throat and by the amount of clearance necessary to permit the rotating workpiece to clear the deep part of the C-frame on the machine. Thus, when using square blanks, removal of the corners allows a larger circle to be cut.

There are two methods of holding the center point of the work metal during circular shearing. In one method, the work metal is clamped by a screw-type handwheel or by an air cylinder, each of which incorporates two pivoting pressure disks—one above and one below the workpiece. The disks permit the workpiece to rotate in a horizontal plane. The other method is by center pinning, in which a hole is drilled or punched in the work metal for locating and rotating it on a pin in the center clamping attachment. The hole is at the predetermined center of the circle to be produced.

Of the two methods, center pinning provides greater rigidity because the work metal cannot slip off center during shearing. The circle generated when the work metal is held by the clamping method may not be perfectly round if

the clamping fixture has not been properly located or if it has shifted because of pressure on the cutters. The disadvantage of center pinning is that a hole must be made in the work and must be closed by plug welding if it is not wanted in the finished product.

Adjustment of Rotary Cutters. The upper cutter head and drive of a rotary shear is raised and lowered by power; a clutch mechanism limits the upward and downward travel. Power movement of the upper cutter is essential (especially when cutting plate stock) because the shearing edges of the cutters must be moved toward each other in proper alignment in order to create the initial shearing action. In setting up, the work metal is often rotated in the clamp attachment, with the cutter exerting light pressure, to determine whether a true circle is being generated. Additional pressure is then applied by the vertical screwdown of the upper cutter to cause shearing.

Only the upper cutter is rotated by the power-drive mechanism. The pinching and rotating action of the upper cutter causes the work metal to rotate between the cutters, and the metal causes the bottom cutter to rotate. The position of the upper cutter in relation to the lower cutter is important. Figure 6(a) shows the setting for shearing a straight edge. Clearance between the cutters is as important as it is with straight knives. Overlap of the cutters, as shown in Fig. 6(b), produces a bevel cut. The degree of bevel up to a maximum of 30° can be adjusted by changing the amount of overlap of the cutters.

Accuracy of the sheared circle depends on the rigidity of the center clamping device, sharpness of the cutters, maintenance of optimal clearance between the cutters, thickness of the work metal, and cutting speed. For work metal up to approximately 3.2 mm (1/8 in.) thick, dimensional accuracy within ±0.8 mm (±1/32 in.) can be obtained when generating a 760 mm (30 in.) diameter circle. With proper setup of equipment, the sheared edge will show only a slight indication of the initial penetration.

Speeds of 2.4 to 6.7 m/min (8 to 22 ft/min) are most commonly used for the rotary shearing of metal up to 6.4 mm (1/4 in.) thick. Speeds of 1.5 to 3 m/min (5 to 10 ft/min) are used for rotary shearing metal that is 6.4 to 25 mm (1/4 to 1 in.) thick.

Flanging and Joggling. With cutters replaced by forming tools, the rotary shear can be used to form flanges and joggles on flat stock. The

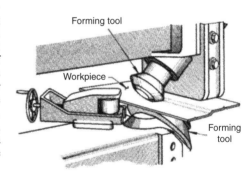

Fig. 7 Setup and tooling for forming a joggle in plate stock using a rotary shear

maximum joggle that can be produced is usually limited to the thickness of the work metal. Because the work metal is made to flow into a different shape during flanging or joggling, the amount of energy required reduces the capacity of the machine to 75% of the rated capacity for shearing. Figure 7 shows a typical setup for forming a joggle.

Safety

Shearing machines must be equipped with devices for protecting personnel from the hazards of shear knives, flywheels, gears, and other moving parts. The guards and safety devices used must be sufficiently rigid enough to withstand damage from operating personnel moving heavy material into position. The squaring shears used for sheet metal should have guards on all moving parts, including fly-wheels, hold-downs, and knives. The treadle, whether mechanical or electrical, should have a lock for supervisory control. Knife and hold-down guard openings should be large enough to provide visibility but small enough to keep the operator's fingers out of the danger area.

Proper opening dimensions are outlined in ANSI standard B11.4, "Safety Requirements for Shears."

The shears used for shearing plate are more difficult to safeguard because of the greater clearances needed under the hold-downs and upper knife to permit entry of the plate (especially when it is bowed or buckled). Guards on shears for plate should be of the type that raises only when the plate is inserted and then rests on the surface of the plate. When there is no workpiece in the machine, the guards rest within 6.4 mm ($^1/_4$ in.) of the surface of the table.

Shears should comply with the construction requirements of the Occupational Safety and Health Act and National Safety Standards. Additional safety information can be obtained from the loss prevention group of major insurance carriers for Workman's Compensation and the National Safety Council. Safety regulations also cover the maximum noise level permitted from a shearing operation to prevent permanent impairment of hearing.

Flattening, Leveling, Slitting, and Shearing of Coiled Product

L. Chen, Technical Materials, Inc.

METAL PRODUCTION MILLS produce flat metal sheet and strip products into coil form, and the coiled product will be processed further for subsequent fabrication processes: flattening, leveling, slitting, and cut-to-length, normally called final coil processes. Flattening or leveling is used to correct shape in the coiled sheet, strip, or plate. Slitting divides a wide coil longitudinally into narrow coils for desirable width, while cut-to-length is to shear coiled plate, sheet, or strip transversely into flat pieces in specific length before further metal part or component forming.

Flattening and Leveling

Flattening and leveling are processes used to correct shape defects in coiled sheet, strip, and plate to achieve a required flat-shape condition. They are widely applied in stamping, roll forming, cut-to-length, and slitting lines, in addition to their own separate coil process lines.

Shape Defects

There are six major shape defects in coiled sheet, strip, or plate (Fig. 1): coil set; cross bow; camber; wavy edges; buckles within a sheet or strip such as center buckles or quarter buckles; and, finally, twist. Coil set and cross bow are attributed to a residual strained length difference between the top and the bottom surfaces. That is a minor difference in length from surface to surface that causes a curvature at the longitudinal direction (coil set) or at transverse direction (cross bow). Camber, wavy edges, and buckles are considered to be a result of a longitudinal length differential from edge to edge across the width. For example, the edges of the strip or sheet longer than the center will cause the wavy edges; vice versa, a center buckle will be generated when the center is longer. A combination of residual strained length difference in both surface-to-surface and edge-to-edge causes a twist in a strip.

Buckles are localized pockets or regions of excessive length or surface that must rise and fall from the nominal plane of the strip in order to be accommodated by the surrounding metal. They look like a series of waves or domed areas and are sometimes called "oil-canning" in reference to the domed, snap-through bottom of metal oil cans. Buckles can occur anywhere across the rolled width and tend to repeat along the length. Buckles that occur in the central portion of the strip are referred to as *center buckles* (Fig. 2, 3) and those that are off-center are called *quarter buckles*. The term *full center* refers to a center buckle that has edge-to-edge coverage and extends for some distance along the length of the strip.

Center buckles, quarter buckles, and full centers can be created during cold rolling as a result of mechanical misalignment or of cross-sectional irregularities in supply coils.

Wavy edges consist of bucklelike distortions that exist whenever the edge is longer than other portions of the strip and usually occur at the immediate edge. Two examples of wavy edges are shown in Fig. 4 and 5.

Herringbone (Fig. 6) may occur during cold rolling and is recognized as a series of long, overlapping waves running at varying angles between the rolling direction and the transverse direction. A herringbone pattern generally covers the central portion of the metal width and results from application of insufficient tension for the cross-sectional area and strength level of the metal being rolled.

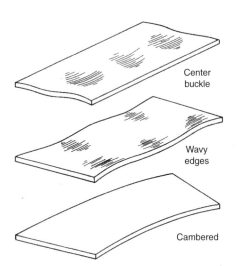

Fig. 1 Major types of shape defects in coiled flat product

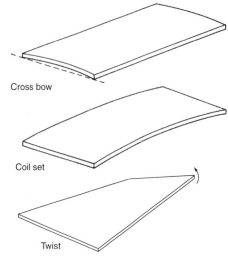

Fig. 2 Center buckles in repetitive, elongated pattern. Similar, but usually rounder displacements, are called oil-can buckles. Source: Ref 1

Transverse (cross) bow is a curvature across the full width of the metal that renders it somewhat canoe-shaped, or gutterlike, along its length.

Longitudinal corrugation (Fig. 7) is a longitudinal condition similar to transverse bow except that the sense of the curvature changes sign at least once across the width of the metal. Corrugation is usually caused by application of excessive tension, to which foil is especially susceptible during continuous annealing.

Curl appears as a longitudinal curvature or sweep along the length of coiled metal (Fig. 8). Curl may be induced during rolling, or by passing the metal over a small-diameter roll, such that the combined tensile and bending stresses exceed the yield stress of the metal, leaving a degree of permanent curvature in the metal.

Coil set, similar in appearance to curl, is also a longitudinal curvature along the coil length. Coil set can occur when metal is wound onto a small-inside diameter (ID) arbor with excess tension. The smaller the ID, the thicker the metal, and the softer the temper, the greater is the likelihood of inducing coil set. Because the smallest ID exists at the core, this is where the greatest tendency for coil set exists. However, each succeeding wrap of coil is actually wound onto the diameter represented by the prior wrap, thereby reducing the bending stresses wrap-by-wrap and reducing the tendency for coil set wrap-by-wrap. Thus, coil set, within the coil where it is first formed, is highly likely to be most severe at the coil ID and decreases to a lower value or to zero at the coil OD. In practice, curl and coil set may not be easily distinguishable although resulting from different causes. Curl can have a higher degree of curvature than coil set. The curvature of the latter is normally no less than the radius of the coil as the material wraps.

Twist is a condition wherein a transverse axis held in the plane of the strip would rotate about the longitudinal axis when moved along the strip. Such a condition is evident in a short length of material if, when the material is freely placed on a flat surface, only three of the four corners touch the surface.

Camber is a side-wise curvature along the length of a strip. Although camber may be apparent in full-width material, it is often a problem in narrower material. It results from incoming strip shape defects, rolling residual stresses, excessive slit burr at one side edge of a slit strip, or excessive tension on one side edge of a slit strip exiting the slitter. The result, however, is always a strip with one edge longer than the other. In general, the narrower the width and the thicker the metal, the greater is the tendency for camber to develop.

Camber is a measure of curvature, which can be described as part of a circle with camber as a square function of the radial distance (L) over which it is measured. Camber measured over different lengths can be expressed as:

$$C_2 = C_1 \left[\frac{L_2}{L_1} \right]^2$$

where C_1 is the camber measured over a length L_1, and C_2 is the camber of the same strip piece measured over a different length of L_2.

Flatteners and Levelers

Flatteners and levelers are widely applied in stamping, roll forming, cut-to-length and slitting lines, in addition to their own separate coil processing lines. Both flatteners and levelers are used for shape correction and control. However, their designs and thus effectiveness and scope of usage are different. It is also noted that the terminology used for different shape correction machines is not standardized. In this article, terms generally accepted in North America are used.

Flatteners

The flattener, also called a straightener, for sheet, strip, and plate, incorporates a series of parallel upper and lower work rolls in a staggered position. The entry side and exit side of the upper frame housing the upper set of work rolls can be independently adjusted up and down. In some flatteners, the upper frame may also be adjustable for tilting within a small range along its longitudinal centerline. The work rolls can either have no backup rolls, as in a two-high flattener (Fig. 9), or be backed-up by a set of backup rolls, as in a four-high flattener (Fig. 10). The two-high flatteners typically have five to nine work rolls and their work-roll diameter is normally

Fig. 3 Center buckles in a random pattern. Source: Ref 1

Fig. 6 Herringbone. Source: Ref 1

Fig. 4 Wavy edges in a regularly spaced pattern. Source: Ref 1

Fig. 7 Longitudinal corrugation. Source: Ref 1

Fig. 5 Wavy edges in a random pattern. Source: Ref 1

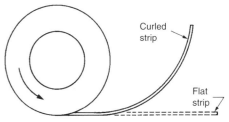

Fig. 8 Curl. Source: Ref 1

sufficiently large to avoid roll deflection. Two-high flatteners can be either a self-driven or a pull-through style. The four-high flattener allows the use of small-diameter work rolls because work-roll deflection is restrained by backup rolls. The four-high flatteners are typically self-driven and have seven to thirteen work rolls.

Levelers

A roller leveler (or simply referred to as a leveler here) is similar to a four-high flattener in that the design involves four-high small-diameter work rolls. Unlike the four-high flattener, however, each work roll in the leveler is supported by a number of narrow backup rollers, instead of straight solid backup rolls (Fig. 11). This arrangement allows small work rolls and a close work roll spacing in the leveler for more capability in shape correction. A series of backup rollers at the same transverse position for all work rolls under the same frame are called a flight, and they have a common support housing extending from the entry to the exit of the roller leveler. Each flight of backup rollers can be vertically adjusted, independently from other flights, by either a mechanical or a hydraulic mechanism. A deflection of work rolls in the leveler therefore can be deliberately adjusted in a controlled manner. Roller levelers usually have seven to nineteen work rolls.

Some levelers have the five-high or six-high backup design for high surface finish quality strip. The six-high levelers have two additional rows of straight solid intermediate rolls between work rolls and adjustable backup rollers, each at the top and bottom frames. This arrangement prevents marking on the top and bottom surfaces of the strip but limits the capability to correct poor shape because the adjustable roller flights act on the intermediate rolls. The five-high leveler has only one row of the intermediate rolls between work rolls and backup roller, normally in the lower frame, while its upper frame contains only work rolls and backup rollers as in four-high levelers. Therefore, it gives better capability for shape correction than the six-high leveler, while preventing marking on one side of the strip surfaces. A roller leveler has a certain capability range in strip thickness for a given work roll diameter and roll spacing. Normally, the upper strip thickness limit is 3 to 4 times that of the lower limit.

Roller levelers have the ability to control the deflection of work rolls so that one portion of the material across the strip width can be subjected to more bending than another portion. The bending causes a tension in the material of the outer layers in the strip and compression in the material of the inner layers near the work roll. At a certain bending radius, the stress in the material at the outermost layer will exceed the material yield strength and permanent plastic elongated deformation occurs. Similarly, the material at the innermost layer will experience compressive plastic deformation when the stress exceeds the material compressive yield strength. Higher-degree bending with a smaller bending radius causes more material to yield and makes plastic deformation shift inward to the neutral plane (Fig. 12).

Multiple interchangeable roll cassettes are used in a single roller leveler to extend its thickness capability range. Each cassette has its own specified roll diameter and roll spacing different from those in other cassettes.

Cluster levelers have a design with different diameter work rolls housed in clusters and provide a very wide range in strip thickness that the single cluster leveler is capable of leveling.

A tension leveler is a roller leveler with a mechanism for applying tension on the strip by bridle or pinch rolls. Simplified roller levelers without independent adjustments on each flight but with a tension mechanism are also called tension levelers by some manufacturers. During the tension leveling process, external tension is applied on the strip as it alternately passes over a series of very small-diameter work rolls, which are backed-up by bigger-diameter rolls. The strip is elongated under applied tension during the process. The range of the elongation in strip during tension leveling is typically from 0.1 to 5%.

Principle of Shape Correction

The principle of shape correction by a flattener and a leveler is simple. In each case, a strip or sheet bends alternatively up and down as it passes between the upper and lower sets of work rolls. This causes selective plastic elongation in a portion of the strip or sheet. A local region of shorter length takes more external stress than a longer-length region in a strip, in addition to the fact that its residual stress is tensile. Therefore,

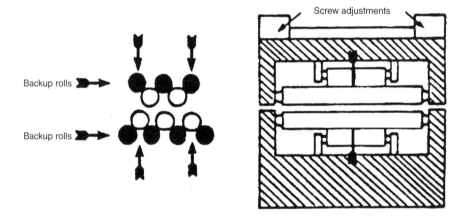

Fig. 10 Four-high flattener with backup rolls to support work rolls

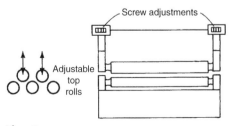

Fig. 9 Two-high flattener with five work rolls

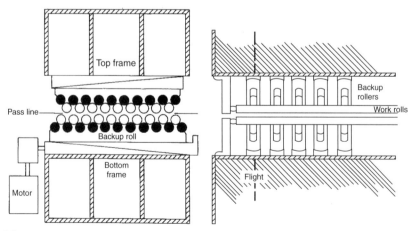

Fig. 11 Four-high roller leveler with adjustable backup rolls

the short-length region will yield first and undergo more plastic elongation than the original longer-length region, where the residual stress is compressive and offsets a portion of the external stress. Then, nominal uniform strip length in different regions, surface-to surface or edge-to-edge, is achieved as a consequence of a series of reversed bending. The residual stresses in the strip are also reduced and redistributed in a more uniform pattern after a series of bends. During tension leveling, an external tensile stress applied on the strip makes the neutral plane shift toward the inner layer (Fig. 12). Higher tension results in tensile stress across the entire thickness of the strip and, finally, plastic tensile deformation across the entire thickness.

Figure 13 illustrates the principle on stress-strain curves in a simplified case, where curve A represents a short-length region material and curve B represents a longer-length region material. Initially, the difference in length between A and B is δ_i, and the residual stresses are assumed to be zero. If a tensile stress from bending or an external tension exceeds the material yield strength, plastic deformation occurs, in addition to elastic deformation. When the stress is released after bending, both curves A and B will return along the slope of the elastic modulus. It can be seen that the difference in length between A and B at this point, represented by δ_{f1}, is smaller than δ_i (Fig. 13a). If the materials undergo a reversed bending at the next set of work rolls and there is no external tension, the stress on the materials becomes compression instead of tension; the stress-strain curves A and B will continue to the negative side—compression until the stress releases (Fig. 13b). If stress is

high enough to cause permanent (plastic) deformation, the difference in length between A and B, δ_{f2}, at the zero stress becomes even smaller than δ_{f1}. This process is repeated as multiple reversed bending continues when the material moves through a series of work rolls and, finally, the difference in length between A and B becomes near zero. It should be noticed that because of Bauschinger effect, the material yield strength in the regions A and B will not increase after plastic deformation during repeat reverse bending if no external tension is applied.

The outermost layer of a thin strip has less elongation than that in a thick strip at the same bending radius. To achieve the same amount of flattening or leveling results, a higher degree of bending (e.g., a smaller bending radius with a smaller bending wrapping angle) in a thin strip is required than the degree of bending required for a thick strip. This requires small-diameter work rolls as well as a shorter work roll spacing and a closer gap (e.g. more penetration) between upper and lower work rolls. However, small-diameter work rolls tend to bend under loading during processing. Therefore, backup rolls in four-high flatteners and backup rollers in levelers are used to control work roll deflection.

Shape Corrections

Shape Correction with Flatteners. Both two-high and four-high flatteners can control and eliminate coil set and cross bow but cannot eliminate buckle and wavy edge. A flattener can reduce twist in a narrow strip, but eliminating

high-degree twist may also introduce another type of shape defect. Normally, flatteners cannot correct camber. However, a flattener with tilting adjustment ability along the longitudinal centerlines of its upper frame may reduce and eliminate a small degree of camber in a strip. A deliberate misalignment of a pull-through flattener to the line may be able to eliminate a camber in a narrow strip (<50 mm, or 2 in.). Auxiliary side push rolls near the entry side of a flattener are used for camber correction in a narrow strip (<100 mm, or 4 in. wide with thickness > 0.6 mm, or 0.025 in.).

For a coil set strip, length difference between the top and bottom surfaces diminishes after a series of reversal bending in strip when it is passed between upper and lower rolls, and therefore, coil set is corrected to the flat shape in a properly set flattener. To correct cross bow, a much higher-degree bending is needed because the reduction in length difference between the top and the bottom surfaces in the transverse direction depends on lateral contraction resulting from longitudinal deformation. High-degree bending needs small-diameter work rolls for a small bending radius, and a closer work roll spacing for a small bending wrapping angle. This can be readily achieved in a four-high flattener.

The gap or penetration depth of work rolls at the entry side of a flattener should be tight enough for shape correction, and the initial gap setting at the exit side should be approximately equal to a strip thickness (Fig. 14). In most cases, especially with more than seven work rolls, the setting at the exit side will determine a final coil set direction (zero, down or up linear curvature)

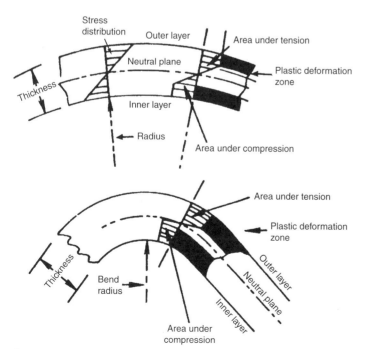

Fig. 12 Bending mechanism during flattening and leveling showing an effect on plastic deformation

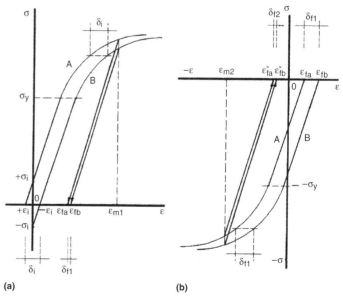

Fig. 13 Stress-strain curve illustration of reduction in a localized length difference by plastic deformation during flattening and leveling. (a) First imposed state of tension before it is released. (b) Continuing from the state of applied tension to a state of applied compression before the stress is released. Scale of strain is exaggerated for clarity.

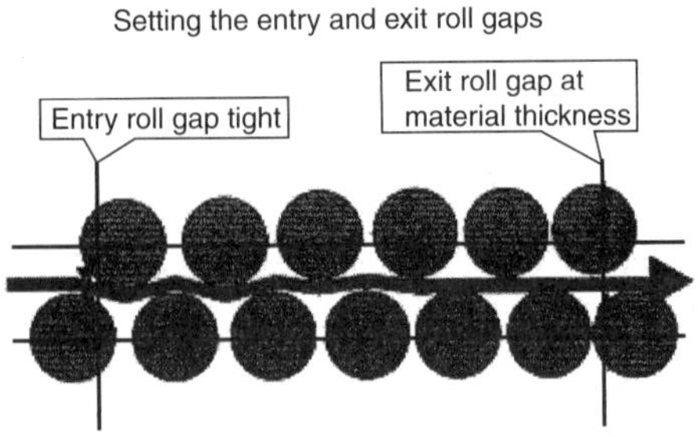

Setting the entry and exit roll gaps

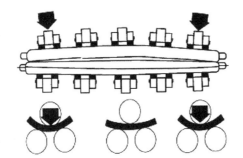

Fig. 14 Setup of work roll gaps

when the entry gap is tight enough to remove the original coil set. Excessively tight gap of work rolls in a two-high flattener may cause overloading and thus deflection (bending) of work rolls. This may cause wavy edge in a wide sheet during flattening. The upper limit of flattener capability in material thickness and strength is normally restricted by unintended roll bending, while the lower limit (in material thickness) of flattener capability is determined by a work roll diameter and a work roll spacing.

Shape Correction with Roller Levers. Roller levelers can control and correct not only coil set and cross bow, but also wavy edges and buckles. This is because they have the ability to control the deflection of their work rolls so that one portion of the strip can be subjected to more deformation than another. In the case of the center buckle or longer center than the edges, the edges should be stretched more than the center until the same length is achieved between them. The backup rollers in the leveler should be set tighter on the edges than the center so that bending on the edges has a higher degree of bending than that in the center (Fig. 15). In contrast, if the strip has wavy or loose edges, the center should be stretched more until it has the same length as the edges. The backup rollers should be set up closer in the center than on the edges (Fig. 16).

Shape Correction with Tension Levelers. Tension levelers can correct coil set, cross bow, buckle, and wavy edges, as well as twist and camber. As shown schematically in Fig. 17, a tension-leveler applies tension to the strip as it alternately passes over and under a series of very small-diameter, backed-up rolls. When used

Fig. 15 Roller setting in a leveler for correcting center buckle: tight for high degree bending at the edges

correctly, tension leveling has proved to be the most versatile tool commercially used for improving shape of coiled material (Fig. 18).

Tension leveling is the only effective method to correct camber. The set-up parameters of tension leveler include: diameter of work rolls, roll spacing in case an exchangeable roll cassette is used, adjustment at the entry side of each flight of backup rolls, adjustment at the exit side of each flight of backup roller, and tension or stretch elongation.

Slitting

Slitting is a process to cut a single, wide strip, sheet or plate, normally in coil form, lengthwise into several narrower strips. During slitting, a moving strip passes between a number of circular blades or knives mounted on two parallel rotating arbors. Slitting can be applied to materials such as paper, plastics, composite, and fiber, as well as ferrous and nonferrous metals, from thin foil less than 0.01 mm (0.0004 in.) thick to over 25 mm (1 in.) thick plate. The discussions in this article focus mainly on metal coil slitting.

Slitting Lines. In simple terms, a slitting line consists of three essential units:

- An uncoiler, also known as a payoff or unwinder, for holding the coil and feeding the strip to the slitter
- A slitter with two parallel arbors holding the knives and other tooling for slitting the strip

- A recoiler, also called takeup or rewinder, for rewinding the slit strips (Fig. 19).

Other equipment can be added to the line for coil handling, strip feeding and guiding, shape correction, gaging, shearing, end joining, edge selvage disposal, packaging, and so forth.

The slitter itself consists of supporting structure and two parallel arbors, one above the other, mounted with shear slitter knives. The distance between two parallel slitter arbors is vertically adjustable to accommodate different diameter knives and different thicknesses of materials. This is normally done during slitter tooling setup or beginning of slitting. However, some computer numerical control precision slitters can perform a continuous minor adjustment according to the conditions of an incoming strip during slitting.

Slitting lines commonly have either one or two recoilers, although some are equipped with individual recoilers for each cut produced. On lines with only one recoiler, all cuts are wound on this recoiler and metal-separating disks or overarm separators are used to keep the individual cuts separated during winding. A disadvantage of single recoilers arises in slitting many cuts because the separator thickness must be accommodated between adjacent cuts, resulting in a fan-out pattern of the cuts from the tight-line slitter (Fig. 20). This situation may induce undesired camber and rippled edges in the slit metal. To avoid this problem, two recoilers are generally used such that each takes up every other cut across the width, thus eliminating fan-out, as shown in Fig. 21.

Modern slitting lines can slit a coil at a line speed of 300 to 400 m/min (980 to 1310 ft/min). Some modern slitter lines for thin strips are able to run up to 900 m/min (2950 ft/min). These slitting lines are normally designed for high productivity with fully automatic loading and feeding of an incoming coil and unloading of finish slit coils in very short time. Either exchangeable second slitter head or programmable automatic knife-position slitter head can be found in these slitter lines. The setup of slitter tooling on the second slitter head can be verified and adjusted by offline trial slitting. The exchange between the pre-setup second slitter head and the inline slitter head can be completed automatically within two or three minutes. The

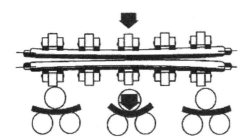

Fig. 16 Roller setting in a leveler for correcting wavy edges: tight for high-degree bending at the center

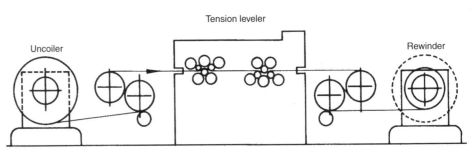

Fig. 17 Schematic of tension-leveling line

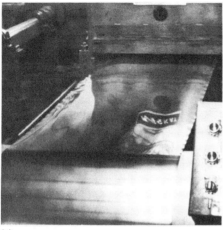

(a)

(b)

Fig. 18 Sheet (a) before and (b) after tension leveling

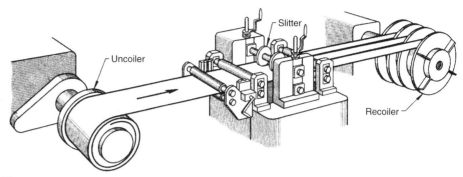

Fig. 19 Three essential units of a slitting line to slit a wide coiled strip into narrower width coiled strips

slitting time of these slitting lines, therefore, can be readily above 65% of total operation time, which includes slitter set up time, incoming coil loading time, slitting time, and unloading time of slit coils.

Slitting lines are generally classified into three broad categories:

- Pull-through
- Driven
- Help-driven

The choice between pull-through and driven lines depends largely on strip shear strength and thickness, number of slit strips, slitting speed, and slit quality requirements. In general, when the metal strip to be slit is less than 0.25 mm (0.010 in.) thick, a drive or helper-drive slitter line is preferred because thin metal strip is likely to tear if it is pulled through a slitter head.

Pull-through slitting lines (Fig. 22) use a motor-driven recoiler to pull the strip through the slitter. The slitter arbors are not driven in this type of slitter. This type of slitting line is normally a tight line; that is, the strip is under tension from the uncoiler through the slitter to the recoiler. However, slack strands (Fig. 23) may

occur due to differences in speeds of different strip sections.

Driven slitters (Fig. 24) have motors that drive the recoiler and arbors in the slitter. The payoff may also be driven. These motors are synchronized to maintain approximately constant speed of the metal as it travels through the slitting line. Driven slitters are preferred for thin strip because they allow winding of slit strips under lower unit tension. Driven slitter lines may operate with one or two slack strands between the slitter and recoiler, especially for thin metal strip, to allow for minor speed differences in different slit strips wound on a common winding arbor. The speed variation in the different slit sections is from variations in the slit-coil diameter due to poor shape and/or nonuniform thickness across the width of an incoming coil.

Looping is especially useful in slitting soft metal or thin-thickness material. Synchronization of the motors in a driven system can allow a reduction in the applied tension between two given points in the line. In driven systems, looping allows the metal to sag under its own weight, which allows it to find the path of least resistance. Three types of looping are illustrated in Fig. 25. In pre-looping, the loop is established between the coil payoff and the slitter, which reduces wrinkling and creasing of the metal as it enters the slitter. In post-looping, the loop is

established between the slitter and the rewinder, which minimizes distortion and damage to the edges of the slit widths as they scrape against the separators on the rewinder. This configuration is illustrated in Fig. 26.

Double-looping (Fig. 25), a combination of pre-looping and post-looping, offers the advantages of both to ensure minimal metal damage. A precision driven slitting line can be also a tight line or operated in the tight-line mode, especially when equipped with a tension leveler placed before or after the slitter and/or with a slip core recoiler.

Help-driven-type slitters. In the help-driven-type slitting line, the torque applied to the slitter arbors, from the slitter driven motor, reduces the tension on the pulled strip to avoid snagging at the entry slitter knives. The helper torque is insufficient to drive the slitters alone, thus avoiding the speed-match issue in a purely driven-type slitting line. The help-driven-type slitting line is in the tight-line configuration.

Interactions between Knives and Strip during Slitting. While dish knives are primarily used for slitting plastic film, paper, and metal foil, rotary slitter knives are most commonly used in metal strip slitting and are discussed here. The interactions between slitter knives and metal strip during slitting with proper slitter tooling and setup normally can be divided into the following four steps: (a) At the beginning of slitting,

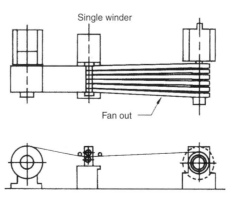

Fig. 20 Slitting line with one rewinder showing fan-out problem

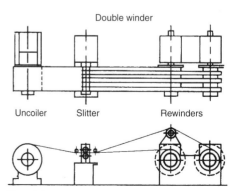

Fig. 21 Slitting line with two rewinders

when a metal strip moves forward to the slitter knives before a shearing point, plastic deformation occurs on the material by compressive force at knife contact surface. This can result in edge distortion or "rollover" along the slit strip edge, especially in soft materials. (b) As the metal strip moves further to the knife shearing point, the knives penetrate into the strip and create the flat, shiny or so-called shear or "burnished" surface on the portion of the slit edge. The depth of burnished shear surface portion depends on the material strength, ductility, and thickness of the strip as well as the slitter knife clearance. (c) As the knife penetrates further into the strip, the maximum shear strength of the material is exceeded, and fracture takes place, which results in a total separation of the material. The fracture surface on the edge appears dull, in contrast with the shiny, burnished shear area in the second slitting stage. The angle of fracture generally falls between 6 and 12° depending on the type and mechanical properties of the material. (d) Finally, a burr is formed on the strip edge on the opposite side of the rollover. Burr is attributed mainly to a compressive plastic deformation displacement around the edge corner of the knife. Figure 27 illustrates the exaggerated cross-section view of a single slit edge introduced by these four slitting stages.

Slitter Tooling and Setup. Slitters may be equipped with either standard (male/female) tooling or solid tooling setups. Solid setups may be used when tighter tolerances are required than those possible with standard setups. In the standard setup, slitter knives of a fixed or standard thickness (width) are used, and spacers are located between the knives to establish the desired slit width (Fig. 28). In this setup, the slit width is equal to the full spacer width of the female load, and the slit strip will have a burr pointing up. A standard slitter setup for three cuts is shown in Fig. 29, with alternating male and female loads on each arbor. These cuts, from left to right, will have edges with burr down, up, and down, respectively.

Figure 30 shows basic slitter tooling and one type (male/female) of slit setup in a single slit strip. The slitter tooling basically consists of knives, spacers, steel or rubber stripper rings over the spacers. It may also include metal or plastic shims of a thickness between 0.127 and 0.013 mm (0.005 and 0.0005 in.) for width match, but this practice is being replaced by shimless tooling in industry. The spacers and stripper ringers in the male side can be simply substituted by a knife or knives, or whole male set of knives, spacers and stripper ringers substituted by a single knife. Wood fingers, a type of

setup used in the past, may still be found in place of stripper rings in some slitters.

Appropriate slitter tooling setup is essential to obtain proper slit width, slit edge, and other slit qualities. The slit width is the distance between female knives as shown in Fig. 30. Knife horizontal clearance and vertical clearance (or overlap) are very important to slit quality, especially to edge condition. Too much horizontal clearance can cause excessive edge rollover and burr, while too tight horizontal clearance may cause bad edge as well as premature wearing or damage of slit knives. As a guideline, Table 1 lists the horizontal clearance and burnished percentage for different metals. When a strip thickness becomes thinner than 0.38 mm (0.015 in.), a horizontal clearance tends to be at the lower end of the listed percentage value or even lower, especially when edge quality is critical for burr-free. In most cases when metal strip thickness is below 0.127 mm (0.005 in.), a zero horizontal clearance is used with thin knives. Knife overlap is also called knife vertical clearance or positive vertical clearance. It depends not only on the thickness and strength of material strip to be slit, but also on other factors such as number of slit strips, horizontal clearance, knife sharpness, rubber ring diameter or steel ring pressure, and slitter arbor deflection.

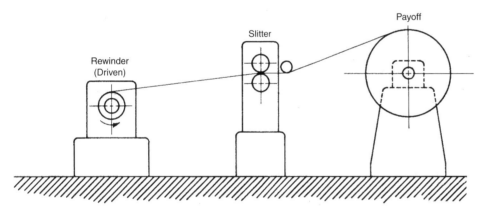

Fig. 22 Pull-through slitting line. Source: Ref 1

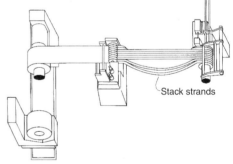

Fig. 23 Pull-through slitting line with slack strands to accommodate minor differences in speed

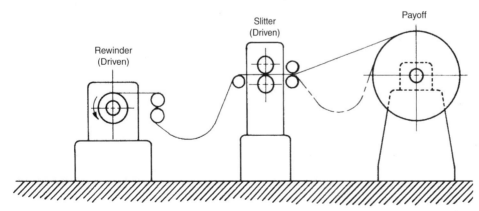

Fig. 24 Driven slitting line

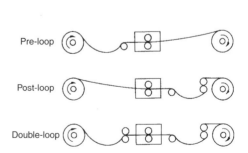

Fig. 25 Methods of looping in a slitting line. Source: Ref 1

A formula to decide knife overlap is difficult to give. Usually, for the same slitter, slitter tooling, slit material, slit width, and incoming width, the knife overlap increases with strip thickness from near zero for ultrathin foil to a maximum value for the strip at a thickness between 0.75 to 1.60 mm (0.030 to 0.063 in.), then decreases with strip thickness from a positive value to the negative side (as in Fig. 30). In practice, it is determined by operator experience using a trial and error approach. At the beginning of slitting, the arbors are adjusted to bring them together until all slit strips have clean cuts. Excessive knife overlap can cause excessive knife wear, coil set, camber, and wavy edges.

It is useful to use a stereoscope or common optical magnifier to check slit edge quality. Examination of shear/fracture ratio, appearance, and burr on slit edge and their consistency can determine whether the slitter tooling clearance setup is correct. Figure 31 illustrates a typical slit edge under stereoscope.

The selection of proper thickness knives, spacers, and their combination for required slit widths can be made using manual calculation or

readily by a commercial computer software program. For slitter tooling setup, Table 1 already gives the guideline for horizontal clearance. These are typical process parameters that can be applied universally with some math calculations, either by a computer program or manually. Other process parameters are specific for different slitter machines, different sets of slitter tooling, different materials, different materials thickness, and different width. A slitter may have hundreds of setup sheets, and different slitters with different sets of slitter tooling can have different process parameters for tooling setup even for the same material and same size.

Slit Quality Attributes and Troubleshooting. Quality attributes of slit strip

include: slit width, burr, edge quality, cross bow, and camber.

Slit width is the most important slitting attribute in slitting operations. A slit width tolerance of ±0.075 mm (±0.003 in.) can be readily achieved for a slit width less than 75 mm (3 in.) and ±0.125 mm (±0.005 in.) for a width less than 175 mm (7 in.) on a modern precision slitter. Slit width is equal to the distance between female knives if the strip is held flat at the slit knives. Verification of correct slitter tooling setup should be performed to check whether the width is within the required tolerance. A loose arbor nut can cause the slit width to become out of tolerance. Excessive arbor deflection, poor slitter arbor maintenance condition, or other slitter machine issues can be the reasons for poor width accuracy. Use of pressure boards or multiple deflection rolls at the entry side of the slitter knife can overcome an inconsistency in slit width due to poor shape of a thin strip.

Burr is normally found to some extent on slit edge. Dull slitter knife or too much horizontal clearance of the knives is a primary cause of excessive burr. Knife vertical clearance, machine rigidity, machine maintenance, and precision of knives also have effects on severity of burr. Checking the actual knife clearance by filler gages (measure/check the distance of a gap by inserting different thick metal "sheets" into the gap), inspecting knife sharpness, and readjustment of slitter tooling setting are among the first things to do for troubleshooting of excessive burr. Measure the distance of a gap by inserting different thickness of metal sheet into the gap. Plastic shim used in slitter tooling setup is not desirable because it may become soft and compressed and, thus, change the horizontal clearance when slitter tooling becomes hot during high-speed slitting. If burr-free edges are required, a subsequent device such as a deburr roll or file can be installed in the line to roll down the burr or to remove the burr.

Poor edge quality can be caused by too-tight slitter knife clearance when the edge appears rough or there are metal fines or slivers on the

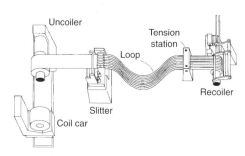

Fig. 26 Schematic of driven-type slitting line with single loop after the slitter. The uncoiler, slitter, and recoiler in the line are driven by separate motors synchronized to maintain constant speed of material being slit. Minor differences in speed may still require the use of slack loops in the pit.

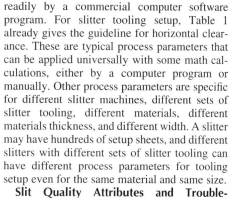

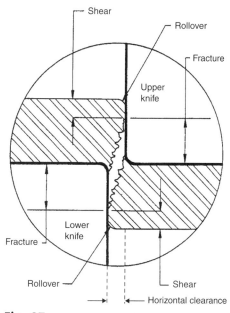

Fig. 27 Cross section view of a slit edge

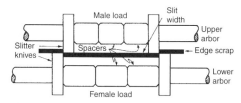

Fig. 28 Standard setup for one cut

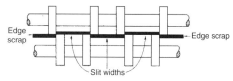

Fig. 29 Standard setup for three cuts

Table 1 Guideline of horizontal clearance for slitter knife setup

Strip thickness >0.50 mm (0.0197 in.)

Materials	Ultimate tensile strength MPa (ksi)	Horizontal clearance % of thickness	Shear/fracture ratio
Annealed aluminum	103 (15)	5–7%	50/50
Annealed low-carbon steel, soft copper, medium alloy aluminum	207–276 (30–40)	8–12%	25/75
High-strength, low-alloy steel, alloy copper, annealed stainless steel, 70XX aluminum	414–621 (60–90)	15–18%	15/85
Cold-rolled stainless steel, high-carbon steel, beryllium copper, high alloy steel	690–1241 (100–180)	20–25%	<10/90

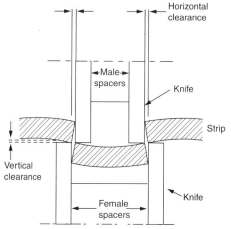

Fig. 30 Detail view of slit edge with vertical clearance (overlap) and horizontal clearance

strip edge. Too-tight horizontal clearance causes double shear and break. Speed mismatch between the slitter and the recoiler in a tight slitter line can also result in poor edge. Inspection and adjustment of separators near or at the recoiler and knife vertical overlap are also recommended when poor slit edge is present.

Turnover. Another feature that may accompany a slit edge is called turnover, or pulldown. As shown in Fig. 32, turnover is a slight curvature along the slit edge created by the shearing action of the slitting process. Turnover, usually found only in very soft metal, is almost negli-

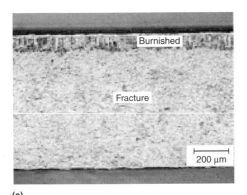

(a)

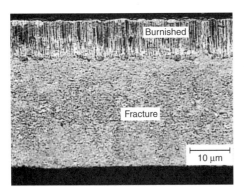

(b)

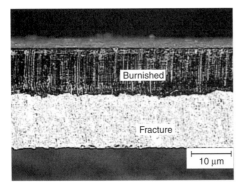

(c)

Fig. 31 Typical appearance of slit edge showing shear penetration burnished zone and facture zone. (a) Cold-rolled 304 stainless, ~10% burnished (sheared) vs. 90% fracture zone. (b) Annealed 304 stainless, 27% burnished (sheared) vs. 73% fracture zone. (c) Annealed pure silver ~50% burnished, 50% fracture

gible on the slit edges of precision sheet, strip, and foil.

Cross bow is a transverse curvature across the width in slit metal strip (Fig. 1). It can either come from an incoming material strip due to residual stress or poor shape or be introduced by poor slitter tooling setup. Normally, high cross bow has the same direction in adjacent slit strips if it comes from the incoming material. If the cross bow is introduced by poor slitter tooling setup, there is a need to check whether slitter knife horizontal clearance is too tight, slit knife overlap is too much, rubber ring diameter is too large, or steel ring pressure is too high.

Camber is width-wise curvature present longitudinally in the same plane of metal strip (Fig. 1). Camber can be introduced by the slitter or be attributed to an incoming material strip itself. To verify whether the camber comes from the incoming strip, one can simply flip over the incoming strip to observe whether the camber of the slit strip is in the same direction as the one before flipping. If the camber has the same direction from the slitter head, the camber is most likely introduced by the slitter tooling setup, and the slitter tooling then needs to be reset to correct the problem. Excessive burr on one edge of the slit strip also can cause camber when the slit strip is wound into a coil. The other slitter-introduced camber is due to the "fan-out" of slit strips from the slitter to the recoiler in a tight slitter line when there are many narrow slit strips. In this case, the slit strips near the edges may have more severe camber than the center slit strips. Troubleshooting of reverse direction camber along the same slit strip may require checking the slitter tooling and its setup, slitter arbors, housing bearings, and slitter design capability as well as an incoming material strip.

Rewinding uniformity describes how even the side edge of the wound finish slit coils is. The rewinding uniformity depends on strip shape condition, strip thickness uniformity across width, slit burr, residual stress in the material, and rewinding tension control during slitting. For an intermediate thickness between 0.40 and 1.5 mm (0.016 and 0.059 in.), separating plates between each slit cuts at the rewinder may be used to force the slit strip being wound to wobble only within a small range move during rewinding in order to achieve even rewinding uniformity for each slit coil. However, caution must be taken, otherwise a separating plate may damage the slit edge during rewinding.

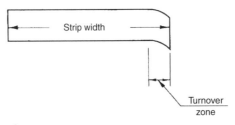

Fig. 32 Turnover in slit strip

Edge Conditioning. There are many applications that require reorientation or elimination of the edge burr on slit strip to facilitate handling, accommodate tooling clearances, or eliminate cracking during severe forming operations. In order to meet these requirements, the slit strip must be processed through specialized edge-conditioning equipment after slitting.

In cases where the intention is merely to reorient or mash down the burr to prevent personal injury or jamups in processing equipment, the edge-conditioning unit utilizes a pair or a series of rolls that flatten the burr, or turn it 90° from a vertical to a horizontal plane, as shown in Fig. 33(b). This method still leaves the strip with a burr, except that it sticks out instead of down, which is the direction of a slitting burr on an unconditioned edge (Fig. 33a).

A modification of this method is one in which the burr is turned 90° to a horizontal plane and then mashed along the edge of the strip, as shown in Fig. 33(c). Again, because the burr has not actually been removed, the strip cannot be thought of as deburred, although the resulting edge is satisfactory for some applications.

A truly deburred edge is one that has been processed through some type of cutting tool that actually removes the burr, resulting in a cross section such as that shown in Fig. 34.

There are some high-performance applications that require edges from which have been removed not only the burr, but also the brittle metal in the fracture zone of the cut, to prevent the possibility of strip breakage emanating from edge cracks. This type of edge, generally called a rounded edge (Fig. 35), is produced by processing the strip through side-mounted shaving, filing, or grinding equipment.

Cut-to-Length Lines

Cut-to-length lines are used to cut a metal plate, sheet, or strip transversely to a certain required length from coil. Various technologies, such as laser cutting, plasma cutting, and water jet cutting, can be used in cutting plate or sheet to a required length. However, cut-to-length lines are normally referred to the coil process lines using shearing or knife cutting. This section discusses only cut-to-length lines using straight knife shearing. Nevertheless, some fundamentals of line configurations in cut-to-length lines may be applicable when an alternative cutting method is used to replace knife shearing (see the section "Straight-Knife Shearing" in the article "Shearing of Sheet, Strip, and Plate" in this Volume for shearing principles, machines, operating parameters, sheared attributes, and other related topics).

All cut-to-length lines normally have at least four functions:

- The coiled product (sheet, plate, or strip) is first uncoiled and sent to a leveler or flattener.
- The metal is leveled or flattened by the leveler or flattener to ensure its flatness (see section

"Flattening and Leveling" in this article) before it is fed to a shear.

- The metal is then cut by the shear to a prescribed length.
- Finally, the sheared plates or sheets are stacked by a stacker.

The lines usually have feeding or conveyor systems to support these four functions. Some cut-to-length lines may also incorporate a slitter for edge trimming or strip slitting before the shear. In modern lines, the length-measuring or gaging system can be computer numerically controlled.

Stationary Shear Lines. There are two basic types of shear lines: stationary and flying shear types. Most cut-to-length lines employ stationary shears. The strip is stopped at the shear during the cut. These lines are generally more accurate than other types. There are three types of feeding arrangements for stationary shear lines: start/stop, hump table, and looping pit.

Stop/start cut-to-length lines (Fig. 36) are usually arranged so that the coil is fed into the shear to a prescribed length and then stopped during the cut. Upon completion of the cut, the

coil feed accelerates until it stops for the next cut. Close tolerances can be obtained, and line configuration is simple. However, the throughput (meter/min or sheet/min) is low. Therefore, these lines are generally confined to heavy-gage requirements. Stopping the coiled sheet in the flattener or leveler prior to shearing may leave a mark on lighter-gage materials.

Hump Table. Stationary shear lines with hump tables (Fig. 37) consist of an uncoiler, a flattener, and/or a leveler to correct for strip shape and to feed the strip over a hump table, a stationary shear, a gage table with retractable stop, and a stacker that stacks the sheared plates or sheets as they are delivered from the gage table.

The retractable stop with gage table, which follows the shear sequence, is used to control the length of the cut sheets. When the uncoiled strip touches the gage stop, it triggers a limit switch that actuates the shear to cut a sheet. Because the strip continues to flow from the uncoiler, it causes a hump to form above the hump table in front of the shear. When shearing is completed, the gage stop retracts, and the cut sheet is delivered to the stacker. As the cut sheet is removed, it activates a limit switch that resets the gage stop. Then, the shear opens, permitting the strip to slide forward through the shear and onto the gage table against the gage stop again, and the cycle is repeated.

Looping Pit. As shown in Fig. 38, many stationary shear lines have precision measuring feeder rolls just before the shear, instead of hump and gage tables. In these lines, there is a looping pit below the pass line just before the feed rolls and the shear. The flattener and/or leveler runs continuously. A slitter for edge trimming or

normal strip slitting can be added after the flattener or leveler and before the loop pit. The strip can be accumulated in the pit during a shearing cycle. Side guides control the feed angle of the strip for maximum cut squareness as it exits the looping pit and enters the shear. In some lines, the shear knives can be pivoted to cut trapezoidal blanks. The stacker is equipped at the end of the line to pick up and stack the sheared pieces for packing or subsequent operations. An air cushion may be used in some stackers for high surface finish quality cut-to-length products.

A stationary shear line generally provides the best squareness and length tolerance and most productivity for lighter-gage materials. Because of the nature of the loop, this feeding method is not practical for strip over about 6.4 mm ($^{1}/_{4}$ in.) thick. A hump line may be somewhat less expensive when the costs of installation are included to consider, but tolerance, squareness, and productivity may be sacrificed.

Flying Shear Lines. There are several different types of flying shear lines such as rocker (flying die), rotary drum, and oscillating. The lines are designed to shear metal strip or sheet in a way to synchronize a speed match with the moving metal. Because the shear action matches the speed of the moving strip, sheet, or plate, an accurate sheared length can be achieved without having to stop the strip and to restart each time of shearing. This continuous flow results in high line productivity. However, the flying lines normally cost more than stationary shear lines when a certain designed requirement in accuracy is needed. The lines may include an uncoiler, a flattener, measuring rolls, a flying shear, a runout conveyer, and a stacker. The rocker shear line is the least-expensive type in the flying shear lines;

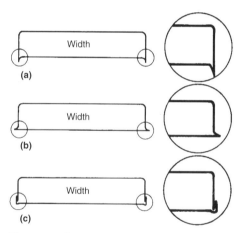

Fig. 33 Schematic of unconditioned and conditioned burrs. (a) Unconditioned edge with burr sticking down. (b) Conditioned edge with burr rolled 90° to stick out. (c) Conditioned edge with burr rolled 90° and then pressed along the edge

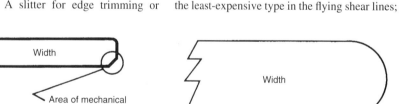

Fig. 34 Edge with burr removed

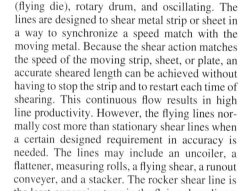

Fig. 35 Rounded edge after burr removal

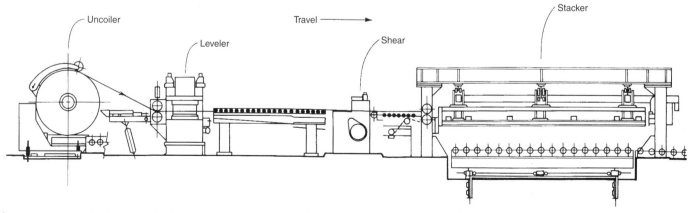

Fig. 36 Stop/start cut-to-length line. A strip travel is stopped to allow material to be cut to a prescribed length in the shear and then restarted until material is again at the prescribed length, when the stop/start cycle is repeated.

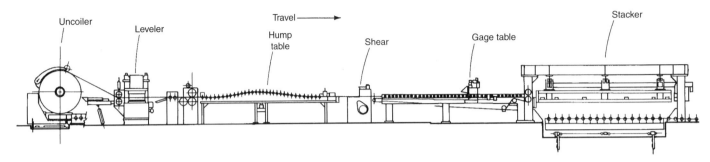

Fig. 37 Stationary shear lines with a hump table. A strip continuously moves from uncoiler even during shearing sequence, causing the coil sheet to form a loop above the pass line and over the hump table. A limit switch actuates shear when the uncoiled strip touches a retractable stop.

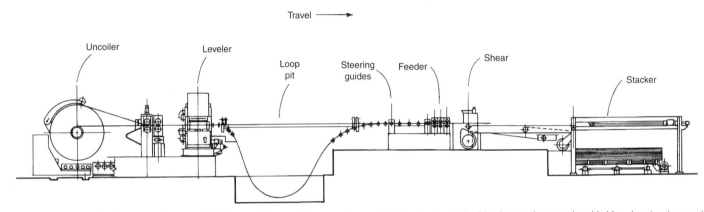

Fig. 38 Stationary shear lines with a looping pit. The loop pit is located below pass line and just before the feed rolls of the shear. A slitter may be added for edge trimming or strip slitting after a flattener or leveler.

it is best suited to light-gage sheet or strip shearing at slow speeds. The rotary drum shear is often used for light-gage narrow strip applications. It usually offers a high shearing accuracy at high speeds. The oscillating shear line is normally used for high-quality finished parts with higher speeds than rocker shear lines.

Blanking lines are a special type of shearing line. Blanking lines have the special capability of rapidly cutting and stacking relatively short and dimensionally accurate blanks as final parts.

Cut-to-Length Line Capability. The capability of cut-to-length line in maximum plate/strip thickness is limited by the shear capability but may also be affected by the thickness capability of a flattener or leveler in the line for tight requirements in flatness and dimensional accuracy. Dimensional accuracy of the cut sheet length depends on the line configuration, condition of the equipment, speed and length of sheet, and the condition of the master coil. Most modern lines generally can achieve accuracy to ±0.8 mm (±1/32 in.), and up to ±0.4 mm (±1/64 in.) when equipped with sophisticated devices.

ACKNOWLEDGMENTS

Portions of this article were adapted from content in Ref 1.

REFERENCE

1. H.B. Bowman, *Handbook of Precision Sheet, Strip, and Foil,* American Society for Metals, 1980

SELECTED REFERENCES

- P. Brook et al., A Machine for Correcting the Shape of Strip, *J. Inst. Metals,* Vol 90, 1961–1962, p 1–6
- *Coil Processing,* section 5, Slitting, p 87–143; section 6, Blanking and Cut-to-Length, p 169–177, FMA, 2001
- T. Eckhardt, "Rotary Slitting," technical paper F-5010, FAM, 1978
- M.R. Edwards, Engineering and Operating Aspects of Leveling, *J. Iron Steel Inst.,* Dec 1965, p 1237–1239
- H. Kerns, Burr-Free Slitting and Knife Set-Up, Coil Processing for Red Metals Workshop (Hartford, CT), May 1998, *FMA*
- R. Lofatrom, Light-Gauge and Narrow Slitting Lines, *Handbook of Metalforming Processes,* H. Thies, Ed., Marcel Dekker, p 89–116
- D.S. Matsunaga, J.W. Slabowski, and T. Marecki, Tension Roller, U.S. Patent 5,007,272, Nov 9, 1989
- M. Marincic, Getting Control of Your Cut-to-Length Line, *Fabricator,* July 2002
- F. Remirez, Leveling Defects—Cause and Cures, *Met. Form.,* July 2000, p 32–34
- J. Rogers and W. Millan, *Coil Slitting,* Pergamon Press, 1972
- T. Sheppard, A Mathematical Analysis of the Roller-Leveling Process, *J. Inst. Metals,* Vol 95, 1967, p 225–231
- K. Shoop, What's New with Multiblanking Lines, *Stamp. J.,* May/June 2002
- J.W. Slabowski et al., Adjustable Levelers, U.S. Patent 5,097,691, March 24, 1992
- R. Taylor, Precision Slitting Using Proper Techniques to Improve Quality, *Coil Processing for Red Metals Workshop* (Hartford, CT), May 1998, *FMA*
- *The Book of Leveling* (a company technical brochure), Herr-Voss Corp., Callery, PA, 1994
- E. Theis, Everything You Need to Know about Flatteners and Levelers for Coil Processing, Part III: How Coil Processors Can Make Metal Flat so It Stays That Way, *Fabricator,* Dec 2002, p 44–47
- B. Weber, Correcting Shape Problems in Flat-Rolled Coil, Part II: Taking a Look at Flattener and Leveler Designs, *Fabricator,* Oct 2001
- A.W. Winterman, Understanding the Importance of Slitting Knives, *Fabricator,* Jan/Feb 1979

Selection of Materials for Shearing, Blanking, and Piercing Tools

SHEARING IS A PROCESS of cutting flat product with blades, rotary cutters, or with the aid of a blanking or punching die. Cropping of bars is a related process. The shearing process creates severe wear conditions, and this article describes some of the wear and material factors for tools used to shear flat product, principally sheet. The process is described in general terms, and the methods of wear control are briefly reviewed in terms of tool materials and lubrication. This is followed by sections on cold and hot shearing with tool steels. Most shear blades are solid, one-piece blades made of tool steel. However, carbide tooling is also used for long runs, and shear blades may be composite tools that consist of tool-material inserts in heat treated medium-carbon or low-alloy steel backings. Coatings also are instrumental in the economical improvement of tool life.

Shearing Process

The process of shearing may involve cutting along a straight line with the aid of a shear blade, slitting with rotary cutters (Fig. 1), or cutting of contours with a blanking or punching die. Cutting straight lines with shear blades is discussed in more detail in the article "Shearing of Sheet, Strip, and Plate" in this Volume. General aspects are also described in the article "Principles of Shearing, Blanking, and Piercing" in this Volume.

For this article, the basic process is illustrated by an example of hole punching (Fig. 2). When the tool first contacts the workpiece, elastic deformation occurs, the magnitude of which is a function of the elastic properties of the material, the cutting zone geometry, and the restraint provided by the adjacent material and die elements such as holddowns and strippers. Then plastic deformation (a slight extrusion) begins and, after some critical penetration, which is a function of the die clearance and the ductility of the material, cracks are generated at the die and punch edges (Fig. 2a). On further penetration the cracks propagate to complete the cut, while the cutting force reaches its maximum, which is only slightly dependent on clearance and not at all dependent on friction. The part must be pushed farther to remove it from the cutting zone; in this phase the force drops steeply and is partly attributable to friction on the die land and punch side surfaces. When cutting is done with a less-than-optimal clearance, secondary shearing takes place (Fig. 2b), resulting in a second peak in the force-displacement curve, the

magnitude of which is slightly affected by friction on the punch side surface.

The resulting cut surfaces are neither perpendicular nor very smooth—at best, peak-to-valley roughness (R_t) of about 3 to 6 μm transverse to the punch movement and 0.1 to 2 μm parallel to it. Much-improved accuracy, perpendicular edges, and a smooth finish may be ensured by various special processes such as fine blanking (precision blanking), in which a specially shaped blankholder (Fig. 2c) imposes compressive stresses on the cutting zone, delays crack initiation, and ensures that the whole thickness is plastically sheared.

Wear

Shearing creates severe wear conditions summarized as follows (Schey, Ref 1):

- Die and tool edges and the die land and punch side surfaces (the flanks) are exposed to the virgin surfaces generated in the course of

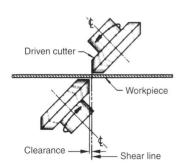

Fig. 1 Conventional arrangement of cutters in a rotary shearing machine, for production of a perpendicular edge

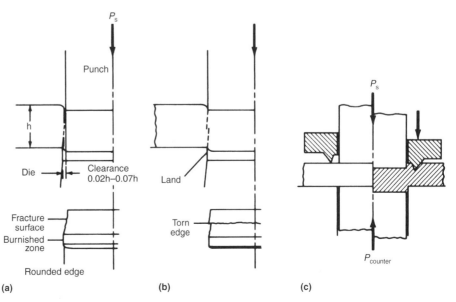

Fig. 2 Schematic of the shearing process. (a) Side view of cut edge. (b) Torn edge with less than optimum clearance. (c) Blankholder for fine blanking

shearing. Relative sliding creates ideal conditions for adhesive wear of the edges and flanks (Fig. 3).

- The workpiece material suffers severe strain hardening. Even though the depth of this heavily deformed zone is shallow (typically, 30 to 50% of the sheet thickness), it increases local tool pressures and gives increased support to any abrasive particles that may be present, thus making them more damaging.
- Shearing forces and, particularly, localized pressures can be high. Forces can be reduced by selection of an adequate clearance (typically around 8%), but only slightly. High production rates involve impact conditions imposed at high rates of repetition. Repeated loading, especially in the presence of adhesive bonds, leads to spalling (fatigue) of the edges, which terminates the run. Crater wear is also possible on the punch end.
- Elastic deformation of the workpiece results in relative movement along the punch end face, first radially outward and then, after crack initiation, inward. This leads to abrasive wear on the punch face. By limiting elastic deflections, a stripper reduces this form of wear. In combination with high normal loads, crater wear is observed.
- Elastic springback of the workpiece material increases the pressure acting on the punch during retraction and thus increases flank wear, particularly in piercing of a hole.
- High production rates contribute to a temperature increase. In punching of stainless steel, the temperature rise was 80 °C (175 °F) under dry and 55 °C (130 °F) under lubricated conditions, values sufficient to increase adhesive and oxidative wear.
- Thermoelectric currents generated in shearing promote wear. This source of wear can be neutralized with a compensating circuit.

Wear patterns include the following:

Flank wear can be characterized by length (*l* in Fig. 3) or area. Flank wear is important because it determines the length that is lost in regrinding. Its origin is in adhesion and abrasion and increases with the number of strokes—at a higher rate when the clearance is too small (Fig. 4). The wear length increases asymptotically to the maximum given by the punch penetration. A semiquantitative assessment is possible by inspecting the flank for scoring and pickup.

Edge (tip) wear, even though difficult to separate from flank wear (Fig. 3), is important in that it determines burr height. This, too, increases with the number of parts and is at a minimum at some fairly generous clearance (12% in Fig. 5). Excessive clearance (17% in Fig. 5) leads to a large burr also, but only because the part is finally separated by tensile fracture, and not because of wear.

Face wear is mostly abrasive in origin and as such increases linearly with the number of parts.

Die life is usually defined by the maximum tolerable burr height. However, wear cannot be judged from absolute burr height alone because this is a function also of materials properties. Burr height increases with increasing ductility and is thus generally less on cold-rolled than on annealed material. Similarly, alloying that reduces ductility also reduces burr height; on silicon steels burr height increased with decreasing silicon content and decreasing *r* value. Because of these effects, face wear may be considered as another indicator. Tool life generally follows a Weibull distribution—i.e., it has a normal distribution but with a finite lower limit. Excessive face wear causes plastic deformation in the form of dishing of the part due to contact prior to the initiation of fracture. The deformed blank shape can then be taken as a criterion for critical tool wear.

Wear Control

Friction does not affect the process of shearing (blanking, punching) itself, and therefore the prime purpose of lubrication is to reduce die wear. Adhesive, abrasive, fatigue, and chemical wear mechanisms contribute to a loss of punch and die profile and thus to an increase in clearance, with the consequent formation of a burr on the sheared part. Wear is reduced by appropriate choice of die materials and lubricants. Die materials must be hard and yet be of adequate ductility and also low adhesion. Die surface coatings play an important role. Lubricants are invariably compounded to minimize adhesive and abrasive wear through formation of boundary and extreme-pressure (EP) layers. However, because lubrication does not directly affect the shearing process, it is more meaningful to consider the various factors that influence wear control and die life.

General factors influencing tool life in cold-work applications are listed in Fig. 6. In general, selection of tool material and/or appropriate surface treatment should be based on the most likely failure mechanisms that may occur. Depending on the application (shearing, blanking, bending, drawing, etc.), cold working dies may experience wear, galling, indentation, chipping, and cracking damage characterized by:

- Abrasive or adhesive wear resulting in continuous or discontinuous material loss, respectively, which is related to sheet material, application, and process conditions.
- Galling due to physical and/or chemical adhesion of sheet material to the tool material. The severity of galling is dependent on surface smoothness, chemical composition of tool and sheet material and type, size, and distribution of hard phases in tool material.

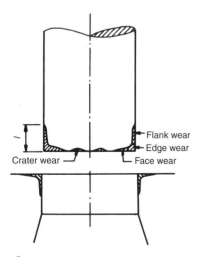

Fig. 3 Wear of shearing punch and die

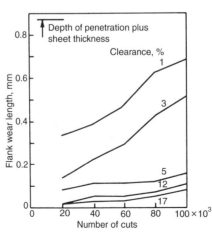

Fig. 4 Dependence of flank wear on punch-to-die clearance in cutting of steel blanks

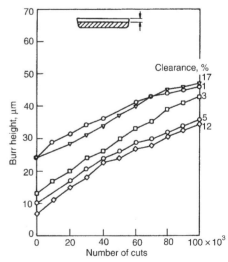

Fig. 5 Effect of clearance on burr development in blanking of steel sheet

Surface coating may reduce or eliminate galling.

- Plastic deformation if the working stresses are higher than the compressive yield strength of the tool material.
- Chipping due to the stresses from the process and the fatigue resistance of the tool material.
- Cracking due to the stresses in the process, the length, and the geometrical configuration of an existing crack result in a stress intensity higher than the fracture toughness of the material.

These factors depend on the cold-work application. In blanking, all of the failure mechanisms may appear, whereas in press forming applications, galling, adhesive wear, and plastic deformation are the normal failure mechanisms encountered. Damages caused by plastic deformation, chipping, and cracking are often not controllable and should be avoided to prevent costly production interruptions. Wear and galling are more predictable failure mechanisms and can be controlled by a scheduled maintenance of the die sets.

Tool Materials. Because of the prominence of adhesive and abrasive wear, both tool material composition and hardness are important die-life factors. However, excessive hardness increases spalling. The danger of spalling, chipping, and cracking demands some measure of ductility and toughness. Wear resistance also depends on the hardness of the carbide phase (H_c) that is dispersed over an effective area fraction of the surface (α) dispersed. Wear resistance of tool steels thus may not be simply a matter of hardness of just the tool-steel matrix (H_m) but rather an effective hardness (H_e) such that: $H_e = H_c + (1 - \alpha) H_m$.

Commonly used tool steels for sheet-metal fabrication include 5 wt%-Cr tool steel (AISI A2) and 12 wt%-Cr tool steel (AISI D2). However, these materials (cast or wrought form) may not be the most effective choice in more demanding forming or cutting applications with higher-strength sheet materials or thicker high-strength materials. Ductility and toughness of A2 and D5 tool steel may also limit die life, especially in cutting applications that put higher demands on wear resistance, strength, and ductility as compared with forming operations (see the article "Selection of Materials for Press-Forming Dies" in this Volume for general information on press-work tool materials).

Wear resistance of tools for shearing, blanking, and piercing is achieved with dispersed quantities of hard particulates, typically carbides, in various types of materials, including:

- Cemented carbides (tungsten carbides in a cobalt matrix)
- Steel-bonded carbides (titanium carbides in a steel matrix)
- Powder metallurgy tool steels

Carbide tooling is commonly for long runs. Steel-bonded carbides have an intermediate level of wear resistance between tool steels and cemented carbides based on WC-Co. They consist of 25 to 45 vol% TiC homogeneously dispersed in a steel matrix. Matrices include tool steels, maraging steels, and martensitic stainless steels. Steel-bonded TiCs respond to heat treatment and are machinable by conventional methods when the binder is in the annealed condition. Fully hardened steel-bonded carbides can be tempered at varying temperatures, thereby obtaining greater toughness than WC-Co. However, this gain in toughness is accompanied by some sacrifice in hardness. Steel-bonded carbides are not recommended for cutting tools in machining because the hardness drops off rapidly at the high temperatures developed during machining. They are, however, used for blades, knives, and stamping dies in cold-working applications (see the article "Selection of Materials for Press-Forming Dies" for more information on steel-bonded carbides).

Powder metallurgy (P/M) tool steels are used in cold-work operations where a combination of good wear resistance and compressive strength (high hardness) is required. Powder metallurgy methods allow production of tool steels with vanadium carbides for wear resistance. Vanadium P/M tool steels are used for punches and dies in cold-forming and stamping operations, and they are considered a cost-effective replacement for tungsten carbide and steel-bonded carbide tools, particularly where these materials are prone to chipping and breakage or where the cost of these materials in prohibitive. Vanadium P/M tool steels are particularly well suited for cold-working applications because of their excellent combination of wear resistance, toughness, and grinding characteristics over a wide range of hardnesses.

The first P/M tool steel with high vanadium for high-performance wear applications was Crucible CPM 10V (or PM 10V) (Crucible Materials Corp., Syracuse, NY), which was introduced commercially in 1978. Also developed at about the same time were modified grades with a lower-matrix carbon and slightly lower vanadium content for better toughness. More recent P/M tool steel development work has focused on the following:

- Compositions containing 15 to 18% V with up to 30% by volume of primary metal carbide-(MC)-type carbides for even more wear resistance
- Low-to-intermediate carbide volume fraction materials moderately alloyed with vanadium and chromium to optimize the toughness properties while still maintaining good wear resistance
- High-vanadium, high-chromium compositions for wear applications that also require good corrosion resistance

More information on P/M tool steels is contained in the article "Selection of Materials for Press-Forming Dies" in this Volume. These high-performance P/M tool steels are used for a wide range of cold-work punch-and-die applications such as long-run stamping, fineblanking, shearing, and piercing. For example, Fig. 7 shows a typical high-production-rate, progressive stamping die application in which CPM 10V punches were compared with D2 steel and an M4 P/M tool steel (Crucible CPM Rex M4). The material being stamped was a 0.381 mm (0.015 in.)-thick copper-beryllium strip. In one operation, D2 piercing punches at 60 to 62 HRC averaged 75,000 parts before losing size; CPM Rex M4 at 64 HRC showed some signs of wear after 200,000 parts. The CPM 10V punch showed no wear after 400,000 parts and ultimately produced over 1.5 million parts. Other

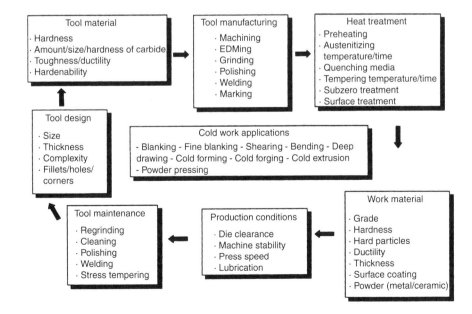

Fig. 6 Factors influencing tool life in cold work applications. Source: Ref 2

parts produced by punching and blanking with P/M tool steels are shown Fig. 8 and 9.

Coatings and surface treatments to improve tool performance include nitriding, hard facing with nickel-base alloys, boronizing, vapor deposition coating, and, hard-chromium coating. At least a doubling and as much as a quadrupling of tool life can be expected. Chemical vapor deposition (CVD) coatings with TiC is well known. The benefits increase with increasing sheet thickness, and up to a hundredfold increase in die life is found. Tools coated with TiC also greatly improve the surface roughness of fine-blanked parts. Physical vapor deposition (PVD) of hard coatings is also useful with better dimensional control, lower treatment temperature, but with less resistance to abrasive wear than CVD coatings (Fig. 10).

Since the first introduction of hard coatings, many types of single and multiple CVD and PVD coatings have been developed. The first real large-scale application was CVD coatings of TiC for cutting tools and, later, forming tools. Deposition of thick, hard coatings by PVD methods was initially more difficult, but now PVD methods permit coatings with thicknesses exceeding 10 μm. This allows the benefits of PVD for applications that used to require CVD. Physical vapor deposition techniques can eliminate the need for an additional heat treatment and risk for change in dimensions can be eliminated. In addition to wear protection, modern PVD coatings can also enable dry or nearly dry metal forming due to very advantageous low friction properties. The inherent low friction of many carbon-base coatings not only gives a low friction, but it also prevents cold welding (galling) of work material on the tool. Current industrial trends to reduce the consumption of lubricants in forming operations, are expected to further increase the application of these coatings.

CVD (Ref 2). The CVD method for tool coatings commonly include TiC, TiN, and TiCN. Process temperatures for adequate growth rate are about 1000 °C (1830 °F). This temperature is far above the tempering temperature of tool steels, and so a post heat treatment is necessary.

Fig. 8 Parts with holes punched by P/M tool steels. Parts include: punching of fine slots in brass parts for clocks; punching of 3 mm (0.118 in.) holes in high-carbon steel (0.7% C) sheet (240 HB); punching of holes in chain links made of 0.55% carbon steel with a thickness of 2 mm (0.079 in.) and a hardness of 290 HB. Courtesy of Uddeholm Corp.

This has to be performed in a reducing atmosphere, or vacuum, because the Ti-based coatings oxidize above ~550 °C (~1020 °F). The post treatment establishes some limitations on which parts can be coated with CVD. These are mainly related to demands involving very high tolerances and/or when long and slim objects are to be treated. There is no real limitation for CVD and coatings into cavities and holes (compared with PVD) because it is only a gas distribution related limitation. Normal coating thickness with CVD is in the range of 4 to 10 μm with process time typically between 5 and 8 hours.

PVD (Ref 2). Tool coatings by the PVD method include TiN, TiAlN, and CrN. Titanium

Fig. 9 Blanked parts produced with P/M tooling. Parts include blanking of watch cases for wrist watches made from 18-8 stainless steel; fine blanking of components for automobile safety belts. Courtesy of Uddeholm Corp.

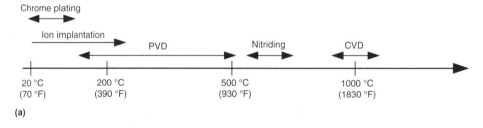

Fig. 10 Comparison of surface treatment techniques. (a) Process temperature of techniques. (b) Surface finish and dimensional tolerance. (c) Abrasive wear resistance. CVD, chemical vapor deposition; PVD, physical vapor deposition. Source: Ref 2

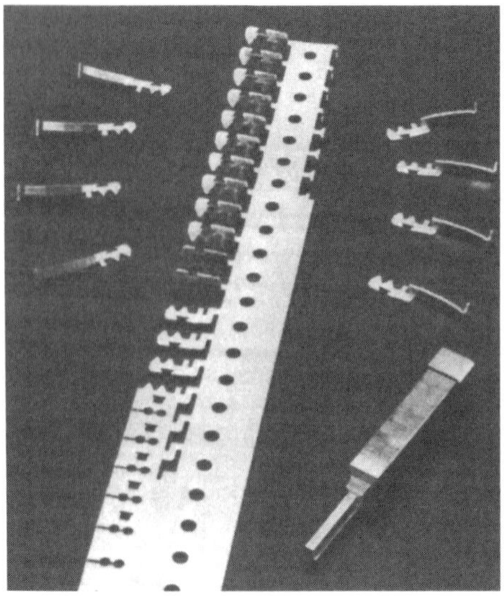

Fig. 7 CPM 10V punch and copper-beryllium blank used in a progressive stamping operation. Courtesy of Crucible Materials Corporation

nitride is the classic and most universally used coating and is relatively easy to apply as compared with other coating materials. Physical vapor deposition TiCN coatings were found to offer better wear resistance than TiN. The relatively high hardness, high compressive stress, and the fact that TiCN often has a gradient as well as a layered structure enhances the wear resistance of the coating. Titanium aluminum nitride has been found to form a thin layer of alumina (Al_2O_3) on the surface. This prevents degradation at high temperatures, which is why TiAlN outperforms TiN as well as TiCN in applications where the contact temperature is relatively high.

The most favorable properties of CrN are its very high toughness, in combination with a relatively high hardness (2000 HV) and a low adhesion to some engineering materials, e.g., Cu. The latter property has been utilized in metal forming of copper with very good results. CrN has also showed good results as a wear-resistant coating in corrosive environments. Physical vapor deposition coatings display unique properties compared with the corresponding bulk materials (if these even are possible to make). Extremely high hardness (ranging from 2000 for CrN to 3500 HV for TiAlN), high toughness, excellent wear resistance, and chemical inertness are typical examples. These are due mostly to the fact that PVD processes tend to yield fine-grained coatings, with fairly high residual (compressive) stresses.

A group of PVD coatings that frequently are used in applications where wear resistance in combination with low friction is required are diamond-like carbide coatings (DLC). These coatings typically are doped with metals, e.g., Cr or metal carbides such as WC. Typical applications are ball bearings and other wear parts where low friction is required. The beneficial low friction property of DLC coatings makes them increasingly interesting also for different tooling applications in which low friction is an important property. Dry metal forming, deep hole drilling, dry machining in soft materials such as aluminum and so forth are examples of applications in which DLC coatings already have proved to be a successful choice.

Carbon coatings containing metal (WC/C) and the combination of hard coatings (TiN, TiCN, TiAlN) with WC/C reduce the erosion tendency, and in some cases, they even permit a lubricant-free manufacturing process. Ductile materials such as aluminum, austenitic steel, or galvanized materials with carbon coatings can be correctly processed.

Workpiece Materials (Ref 1). The workpiece material is of importance from the points of view of both adhesive and abrasive wear. The presence of abrasive oxides or coatings is, obviously, damaging. Thus, a softer bainitic steel covered with a thicker oxide film suffers greater wear than a harder strip covered with a thinner blue oxide film, whereas a polished (bright) strip minimizes abrasive wear. Similarly, the SiO_2 powder found in some insulating coatings

on transformer steel contributes to increased wear.

The absence of an oxide is, however, not always desirable. When adhesion is a problem, some oxide is preferable. Adhesion becomes more important with smaller clearances, and below 5% clearance the harder strip suffered greater wear because its thin oxide could not prevent adhesion. Chipping then becomes a danger. Adhesion is governed by the tool/workpiece materials combination. With an austenitic stainless steel, TiC reduced wear substantially but suffered from spalling. In contrast, a TiN coating virtually eliminated wear (Fig. 11). However, with a ferritic stainless steel it was TiC that eliminated wear while TiN only reduced it.

Lubrication. Various lubricants are effective in reducing wear, especially wear that is due primarily to adhesion. Liquid lubricants form a squeeze film when the punch hits the sheet. Most of the film squeezes out, but in doing so it also lubricates the end face and reduces face wear. The squeezed-out oil is available to seep into the cutting zone, reducing adhesion on the punch flank. As the punch retracts, the oil lubricates the flank, reduces the retracting force, and minimizes scoring and adhesive wear, but also causes the part to stick to the punch face.

Because of the need to reduce adhesion, most lubricants are compounded. Repeated contact and elevated temperatures activate EP additives, and these are found in all formulations except those destined to be used with aluminum alloys. Heavier sheet imposes more onerous duties, and therefore both base oil viscosity and additive levels are higher. Aqueous lubricants can be used, although oil-base ones are more suitable for severe duties. Liquid lubricants are helpful also in removing wear debris, which becomes a major cause of wear especially when clearances are large, and thus adhesive wear is less severe.

Of the most frequently used sheet metals, austenitic stainless steel is prone to adhesion and requires a minimum of 7.5% Cl in the oil.

Ferritic stainless steel is next in severity. Carbon steel, especially in thicker gages and at higher hardness levels, creates severe conditions, and both chlorine and sulfur are effective. MoS_2 and/or polytetrafluoroethylene (PTFE) powder also can be effective as additives. Very low-carbon steel, with its tendency to smear, necessitates frequent regrinding of carbide tooling; the use of a lubricant with active sulfur dramatically increases die life. Oils with boundary additives are usual for aluminum alloys, while both boundary and EP additives are often employed for copper alloys. If the part is cleaned soon after forming, sulfur compounds are allowable. Various other substances, such as caprolactam-base lubricants, have been reported.

Coatings can be effective too. In blanking of austenitic stainless steel, a PVC coating has been used. Similarly, some of the insulating coatings deposited on transformer steel, such as an isophthalic-acid-base alkyl film or a fused Mg-phosphate-chromate-borate coating, reduce wear. Inorganic coatings, especially those containing silica, increase wear and spalling.

Lubricants are effective also in fine blanking, especially in reducing flank wear. Transverse roughness of the part improves with chlorine and sulfur additives and with higher oil viscosity. By appropriate shaping of the die elements, lubricant reservoirs can be formed Fig. 12 and the quality of cut improved.

Thick plate is sometimes punched hot. Hot forging lubricants are generally suitable, ranging from graphitic formulations to glass cloth.

Shear-Blade Tool Steels

Most shear blades are solid, one-piece blades made of tool steel. However, some are composite tools that consist of tool-material inserts in heat treated medium-carbon or low-alloy steel backings.

Cold Shearing and Slitting of Metals. Tool steels recommended for cold shearing of various metals are presented in Table 1, and blade materials for rotary slitting are given in Table 2. The composition and thickness of the material being sheared are the most important factors in the selection of a blade material. Other significant factors are cost, availability, heat treating characteristics, and previous experience in similar applications.

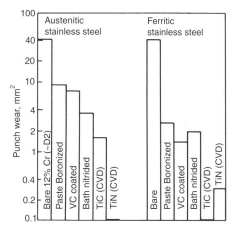

Fig. 11 Effect of punch surface treatments on wear measured in a simulation test

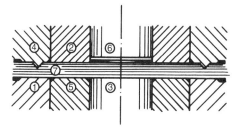

Fig. 12 Lubricant pockets developed in fine blanking by use of appropriately shaped tools

Tool steels vary in toughness and wear resistance, and the metals being sheared vary in hardness and resistance to shearing. If the material to be sheared is very thin and of relatively low hardness, the shear-blade material can be low in toughness but must have optimum wear resistance. For shearing material of greater thickness and higher hardness, it may be necessary to decrease blade hardness, or change to a less wear-resistant blade material or to a shock-resistant tool steel having a hard case over a tough core, to obtain the toughness needed to resist edge chipping. For example, D2 tool steel is recommended for shearing any metal substantially less than 6 mm (1/4 in.) thick, but for thicknesses of 6 mm and greater, tool steels of greater toughness must be used (successively, A2; A9; and the shock-resisting types S2, S5, and S6).

Effect of Hardness on Wear. The rate at which a blade wears in cold shearing depends chiefly on its carbon content and hardness. As shown in Fig. 13(a) and (b), shallow-hardening steels such as W2 may equal D2 in performance until several sharpenings have ground the hardened stock away; then hardness is lower, and life between grinds decreases accordingly. In these tests, shallow-hardening W2 was superior to deeper-hardening S2 until about 1.5 mm (0.060 in.) had been removed, because of the higher initial surface hardness of W2.

Use of a low-hardness blade for cold shearing will result in short blade life. A blade made of S4 or S5 tool steel with a hardness of 44 HRC wore three times as fast as one with a hardness of 54 HRC when other conditions were equal. A similar comparison of blade wear as a function of hardness for S4 and D2 blades is given in Fig. 13(c).

Recommendations on hardness of blades for cold shearing cannot always be made without knowledge of operational details. For example, a D2 blade performed satisfactorily at 61 HRC in one application, but blades at this hardness broke under similar operating conditions in a different plant. Blades made of D2 tool steel at 58 to 60 HRC usually perform satisfactorily in shearing mild steel up to 6 mm (1/4 in.) thick, and in many instances, D2 shear blades have been used successfully at 60 to 62 HRC. However, in shearing metals such as high-strength low-alloy steels, hardness of D2 blades must be kept below 58 HRC to prevent breakage.

The shock-resisting group S tool steels are used at hardness levels from 50 to 58 HRC. Blades with hardnesses near the higher end of this range are recommended for shearing steels 6 to 13 mm (1/4 to 1/2 in.) thick and for nonferrous metals. For shearing harder or thicker metals, blades with hardnesses nearer the low end of this range are used in order to compensate for increased shock loading. Table 3 gives some typical hardnesses of blades used in ten different plants for cold shearing of specific ferrous and nonferrous products.

Tool Steels for Hot Shearing of Metals. Shearing is done at elevated temperatures when the work material is thick and resistant to shearing or when hot shearing is otherwise desirable as part of the manufacturing process. The strong secondary hardening of group H tool steels provides sufficient resistance to softening to make them useful for shear blades operating at temperatures up to 425 °C (800 °F).

Type H11 tool steel is satisfactory for most hot shearing operations. The slightly more expensive types H12 and H13 also have been recommended. No data are available that prove any of these three steels to be superior to the other two. More costly steels such as H21 and H25 are recommended only when H11 has been tried and found to be inadequate. In many instances, however, higher-alloy tool steels are used for shear blades. For example, one large mill used H25 as the standard blade material for hot shearing aluminum at 150 to 425 °C (300 to 800 °F), as indicated in Table 3.

Hardness of blades for hot shearing varies considerably with conditions such as thickness and temperature of the metal being sheared, and type and condition of available equipment. However, hardness is usually kept within the range from 38 to 48 HRC. In one steel mill, H12 tool steel at 45 to 47 HRC is used for slab shear blades 1.45 m (57 in.) long, 125 mm (5 in.) high, 75.4 mm (2.969 in.) thick at the top and 73.0 mm (2.875 in.) thick at the bottom. In another mill, slabs and blooms are sheared with blades made from a steel containing 0.35 C, 0.90 Mn, 1.00 Cr, 1.25 Mo, and 0.25 V, hardened to 38 to 42 HRC.

The temperature of the metal being sheared influences blade life. Higher-alloy blades may be required if the metal being sheared has high strength at elevated temperatures.

Hard-faced blades are satisfactory, and in some plants are used exclusively, for hot

Table 1 Recommended blade materials for cold shearing flat metals

Material to be sheared	Blade material(s) for work-metal thickness of:		
	6 mm (1/4 in.) or less	6 to 13 mm (1/4 to 1/2 in.)	13 mm (1/2 in.) and over
Carbon and low-alloy steels up to 0.35% C	D2, A2, CPM 10V	A2, A9	S2, S5, S6, S7
Carbon and low-alloy steels, 0.35% C and over	D2, A2, CPM 10V	A9, S5, S7	S2, S5, S6, S7
Stainless steels and heat-resisting alloys	D2, A2, CPM 10V	A2, A9, S2	S2, S5, S6, S7
High-silicon electrical steels	D2, T15, CPM 10V, cemented carbide inserts(a)	S2, S5, S7	(b)
Copper and aluminum alloys	D2, A2	A2	S2, S5, S6, S7
Titanium alloys	D2	. . .	. . .

(a) Carbide inserts usually are brazed to heat treated medium-carbon or low-alloy steel backings. (b) Seldom sheared in these thicknesses

Table 2 Recommended blade materials for rotary slitting of flat metals

Material to be sheared	Blade material for work-metal thickness of:		
	4.5 mm (3/16 in.) or less	4.5 to 6.5 mm (3/16 to 1/4 in.)	6.5 mm (1/4 in.) or more
Carbon, alloy and stainless steels	D2, CPM 10V	D2, A2, A9	A9, S5, S6, S7
High-silicon electrical steels	D2, M2, CPM 10V cemented carbide inserts	D2	. . .
Copper and aluminum alloys	A2, D2, CPM 10V	A2, D2	A2, S5, S6, S7
Titanium alloys	D2, A2, CPM 10V	. . .	. . .

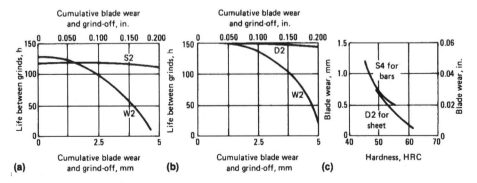

Fig. 13 Comparison of wear and life of different tool steels in cold shearing of steel. (a) Cold shearing of 19 mm-thick (3/4 in.-thick) low-carbon steel plate. (b) Cold shearing of 6 mm-thick (1/4 in.-thick) low-carbon steel plate. (a) and (b) As more and more of the edge is worn away and ground off in resharpening, the life of W2 between grinds continuously decreases until all of the hardened shell is gone and the blade must be scrapped or rehardened. (c) Effect of hardness of cutting edge on wear of shear blades

Table 3   Service data for shear blades

Type of shear	Material sheared	Thickness of material sheared mm	Thickness of material sheared in.	Blade steel	Blade hardness, HRC	Blade service before regrinding
Cold shearing of steel						
Sheet metal	Low-carbon steel	5	Up to $^{3}/_{16}$	W2	58 to 60	30,000 cuts
Sheet metal	Low-carbon steel	5	Up to $^{3}/_{16}$	A2	58 to 60	55,000 cuts
Sheet metal	Low-carbon steel	5	Up to $^{3}/_{16}$	D2	58 to 60	100,000 cuts
Bar shear	1025 & 1040 steel	25	1	W2	58 to 60	20,000 cuts
Bar shear	1025 & 1040 steel	25	1	L6	...	40,000 cuts
Bar shear	1025 & 1040 steel	25	1	S5	...	100,000 cuts
Sheet and strip	1010 steel	5	Up to $^{3}/_{16}$	D2	58 to 60	150,000 cuts
Sheet and strip	Stainless steel	12	0.478	D2	58 to 60	65,000 cuts
Sheet	2 to 5% silicon steel	0.8	0.032	D2	58 to 60	45,000 cuts
Slitter	Carbon, silicon and galvanized steels	0.16 to 4	0.005 to 0.160	D2	58 to 60	1 week(a)
Slitter	Stainless steel	0.8 to 1.5	0.030 to 0.060	D2	58 to 60	15,000 ft
Blade 1.5 m by 100 mm by 25 mm (60 by 4 by 1 in.)	Stainless and silicon steels	2 to 2.5	0.080 to 0.100	D2	60 to 62	2 weeks(b)
Hot shearing of steel						
Slab shear	Various steels	...	...	4340 mod	...	30,000 to 40,000 tons
Cold shearing of copper alloys						
Sheet	Brass	0.16 to 5	0.005 to 0.187	S1	54 to 58	5,000 cuts
Slab	Brass	25 to 60	1.0 to 2.25	S1	54 to 58	25,000 cuts
Hot shearing of copper alloys						
Slab	Brass	Up to 45	Up to 1.75	H11	42	15,000 cuts
Hot shearing of aluminum alloys						
Automatic flying cutoff shear	Aluminum alloys at 150 to 250 °C (300 to 500 °F)	7	0.280	H25	43 to 46	10,000 cuts
75 mm (3 in.) shear	Aluminum alloys at 315 to 425 °C (600 to 800 °F)	76 to 140	3 to 3.5	H25	43 to 46	17,000 to 20,000 cuts

(a) Blades were reground weekly, with about 0.025 mm (0.001 in.) of stock being removed. (b) Maximum. Blades usually were changed weekly

---

Table 4   Life of hard-faced blades for hot shearing of steel in a specific steel mill

Type of shear	Steel for blade body	Hard facing alloy(a)	Blade life, tons of steel sheared
Billet, 300 by 300 mm (12 by 12 in.)	1030 cast	1B new, 2B repair, 3A ribs	29,000
Billet, 250 by 250 mm (10 by 10 in.)	1030 forged	2B	5,800
Billet duplex	1030 cast	None; inserts of H21 or M2	26,000
Billet	H21	None	3,584
Bloomer, 1 m (40 in.)	1045 plate	1B new, 2B repair	7,680
Bloomer, 1.1 m (44 in.)	1045 plate	1B	87,000
Slab, 900 mm (36 in.)	1045 plate	1B	71,000
Plate	6150 (mod)	None	6,000
Rail, 250 by 250 mm (10 by 10 in.)	1030 cast	1B new, 2B repair, 3A ribs	12,960
Rail	1045 plate	2B	5,400
Billet	1045 plate	2B	10,800

(a) Nominal compositions. Alloy 1B: 0.5 C, 0.9 Si, 4.75 Cr, 1.2 W, 1.4 Mo, rem Fe. Alloy 2B: 0.75 C, 0. 5 Mn, 0.65 Si, 4 Cr, 1 V, 1.2 W, 8 Mo, rem Fe. Alloy 3A: 3 C, 1 Si, 28 Cr, 4 Mo, rem Fe

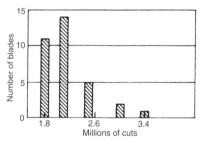

Fig. 14 Statistical distribution of life for shear blades with tungsten carbide inserts. Variation in life of 33 identical shear blades used for the same operation. Blades were made of carbon steel with cemented tungsten carbide inserts (15 to 30% Co) for the cutting edges. Material sheared was 4620 steel rod, 8.7 mm (0.343 in.) in diam, at 149 to 187 HB.

shearing. Table 4 summarizes data on blade life in one steel plant where hard-faced blades are used for most hot shearing operations.

Shear-Blade Life. Service data on life of shear blades are scarce because maintenance programs employed in most high-production mills call for removal and redressing of blades, regardless of condition, during scheduled shutdowns. Available data report blade life in terms of number of cuts, linear footage (in slitting), tonnage, or time between redressings. The service data in Table 3 encompass all of these variables.

The number of cuts per edge before regrinding is the most common basis of evaluation. For different types of shearing, the number of cuts reported has varied from 5000 to more than 2 million. Therefore, meaningful comparisons

can be made only when conditions are nearly identical. Even when identical blades are used for the same shearing operation, life may vary by as much as 100%. This is substantiated in Fig. 14, which presents data on performance of 33 blades with carbide inserts used for shearing small rods. Wide variation in blade life also is characteristic of hot and cold shearing of nonferrous metals (see Fig. 15).

Materials for Machine Knives

Unlike materials for metal-slitting applications, materials for knives used in cutting papers, films, and foils usually are selected on the basis of cost and wear resistance, without considera-

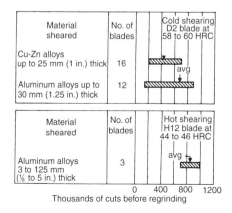

Fig. 15 Variation in life of identical blades used for shearing nonferrous metals

tion of toughness. This allows wider latitude in selection of tool materials, which often results in better performance with fewer compromises. The final choice must be influenced by specific production requirements, maintenance considerations, and the projected life of the tool being used.

Selection of tool materials for shearing and slitting papers, films, foils, and textiles is complicated by the vast array of materials to be cut and by the equally numerous types of tools and methods employed for such operations.

It is common to use the term *machine knives* to identify those tools that are most commonly used to slit, score, shear, trim, perforate, or otherwise cut very thin stock. There are two main categories of machine knives:

- Circular knives, which include score cutters, shear cutters, burst cutters, and single knife cutters (such as slicers and disc knives)
- Straight knife cutters such as trimmer knives, guillotine knives, sheeter knives, cutoff knives, and specialty knives.

Materials for these various knives are described next.

Circular Cutters

Circular-knife design includes coating or impregnation of low-alloy or carbon steels with wear-resistant materials and use of laminated steels produced by cladding low-cost cores on one or both sides, depending on bevel configuration, with high-alloy, wear-resistant alloys such as group D and group M tool steels. Knives tipped with cemented carbides have become popular for certain types of service, especially in shear slitting of coated films. The advantages of extremely long life frequently outweigh the expense and difficulties encountered in proper setup and maintenance of such tools.

Score Cutters are circular knives that exert pressure on the material to be slit by crushing it against a surface (usually a roll or sleeve on a shaft) that is harder than the cutter. Such knives usually have a double bevel at the cutting edge.

Score cutters are most commonly made of 52100 steel because this material is widely available in bar form and is well suited to mass production and easy maintenance by the user. Score cutters are hardened to 60 to 62 HRC, a hardness level that is adequate for most operations and that will not result in scoring of an opposing platen sleeve or hardened roll. Such sleeves and rolls are made of 52100 or carburized steel, hardened to not less than 60 HRC. Alternative materials for score cutters are O7 and D2 tool steels, both of which are available as bar stock. These steels are heat treated to 60 HRC or higher, the same hardness as that specified for 52100 steel cutters.

Score cutters generally should be made of bar stock, not of sheet. The more uniform directional properties of bar stock yield more consistent service life. The angle and radius of the double bevel are important determinants of performance, as is the pressure with which the knife cuts against the platen sleeve. Group M and group T high-speed steels are not recommended: score cutting cannot make full use of their high hardness and resistance to softening at elevated temperatures, and their higher cost is not justified. In addition, brittle, high-hardness tool materials should not be used for score cutters because they will give unsatisfactory service.

Finally, because most score cutters are dry ground by the user, materials must be selected from only those steels that will retain their hardness under these conditions. A properly adjusted score cutter will slice about 350 km (1 million feet) of paper between sharpenings.

Shear Cutters are circular knives that utilize the scissor-cutting principle, comprise pairs of opposing knives through which the material to be slit is passed.

Shear cutters can be produced from a much wider array of alloys, the choice being influenced by such factors as tool design, material to be cut, machine design, and maintenance limitations. The entire range of tool steels (including the popular 52100 and other low-alloy types plus the higher-alloy group D, group M, and group T steels), as well as specialty tool steels and cemented carbides, can be considered. For standard applications, 52100 and O1 are selected most often, chiefly because they are readily available as sheet, bar stock, or tubing. Most types of paper can be cut efficiently and economically with these tool steels. Cutting of coated and impregnated papers sometimes results in premature wear of low-alloy tool steel cutters, in which case D2 is a suitable alternative. Films and foils—whether plain, laminated, coated, or metallized—usually can be cut more efficiently with knives made from tool steels of higher alloy content, such as group D or group M tool steels, or specialty tool steels such as CPM 10V, hardened to 62 to 64 HRC.

In many instances, other considerations also have strong influences on selection of materials for cutting tools and on tool performance, including the design of the knife or machine, the surface finish of the tool, and other limitations imposed by maintenance or production practices. High-alloy tools are necessarily more sensitive to misuse and require greater skills in setup, operation, and maintenance.

Certain films, such as polyester coated with iron oxide or chromium dioxide (magnetic tape), are particularly destructive to knife blades. For these applications, expense becomes secondary to performance, and group M and group T high-speed steels (including the cobalt-containing types M42 and T15) are used to obtain maximum hardness and wear resistance. High-speed steels are preferred because they seem better able to preserve the high degrees of accuracy and surface finish required for such knives. CPM 10V also is well suited for these applications, offering a higher degree of wear resistance than most of the alternative materials.

Burst cutters either penetrate the material with a progressive motion (as in core cutting) or pass through one or more layers of material without contacting another tool surface. In burst cutting, the knife may be used in conjunction with a grooved roll, or it may be entirely free-cutting in a manner similar to razor-blade slitting.

Burst cutters are, by design and function, fairly thin knives. Material selection thus is restricted to alloys that are readily available in thin sheet, a group that includes 52100 steel, 1075 steel, and razor-blade stock. Except in those applications where resistance to elevated temperature is required (as in core cutting with a stationary blade), use of high-speed steels is rarely economical.

Single knives, slicers, and disc knives are used for cutting rolls of textile, foam, or film in a progressive fashion without unwinding (and rewinding) the roll.

Single-knife cutters are used to reduce wide rolls of paper, foam, or textile to narrower rolls without unwinding and rewinding. These knives have long, thin, one-sided or two-sided bevels, are kept sharp by means of one or more grinding wheels situated on the machine, and are activated either automatically (for grinding at specific intervals) or manually (for grinding as required).

Under normal circumstances, such knives are made from L2, L6, or D2 tool steel. Certain applications of single knives, including slicing of foam or impregnated fabrics, cause the thin bevel to heat up considerably as it penetrates the material being slit. In these cases, an alloy that has higher resistance to elevated temperatures, such as M2 or T1 tool steel, may be required. The choice may be restricted by availability of tool material or by production limitations. For instance, some single-blade slicer knives are 750 mm (30 in.) or more in diameter and would require heat treating facilities not generally available. Slicer knives are also particularly sensitive to flutter and straightness problems, so that selection of tool steel must be based on the ability to keep these factors within acceptable bounds.

Slicer knives. Group D, group M, and group T tool steels should be selected for slicer knives only where it is possible to use grinding wheels specifically intended for grinding materials other than standard group L tool steels. Large-diameter D2 knives, for instance, respond well to on-machine grinding with cubic boron nitride wheels, but these wheels represent a considerable investment and must be set accurately to achieve optimal performance from both wheel and knife.

Straight Knives

Types of straight knives include trimmer knives, guillotine knives, sheeter knives, and cutoff knives.

Trimmer and guillotine knives are used for cutting one or more layers of material by virtue of a bevel that exerts a downward force through the material.

Sheeter knives and cutoff knives are used in pairs for cutting roll stock into sheet or for cutting sheet to length. One or both of the knives in a pair can be mounted on a rotating drum or reciprocating holder. The lower (bed) knife frequently is stationary.

Specialty Knives. There are many other applications involving cutting of thin materials for which there is a wide variety of specially designed knives that do not fit into any of the foregoing categories. The selection of materials for specialty knives is not included in this article.

The choice of material for a straight knife is governed by several factors, including the cross section and length of the tool, production and maintenance facilities, and the cost of the knife itself.

A standard paper-trimmer knife consists of a carbon steel backing and an insert that provides the actual cutting edge. The backing, or carrier, remains soft after heat treatment. This allows the manufacturer or user of the knife to drill the necessary mounting holes without having to compensate for distortion during heat treatment and gives the knife a certain amount of shock resistance. The insert material can range from O1 to M2 to one of the group T high-speed steels. For most applications, O1 is ideal with respect to initial cost, ease of maintenance, and adequate performance between resharpenings. For cutting certain materials, such as coated papers, films, and foils, use of a more highly alloyed insert frequently can be justified in terms of quality of cut and service life, even though it may double the original cost of the knife. Trimmer knives also are available in solid steel, in which case the material is selectively hardened or induction hardened to produce a zone in which hardness diminishes from the cutting edge to the back of the blade. Care must be taken to ensure that a solid knife has uniform hardness along the entire length of the cutting edge and that the hardened zone is sufficiently deep to allow for numerous resharpenings. Such knives are considerably cheaper than insert-type knives but are also necessarily limited to steels that can be locally hardened.

Sheeter and cutoff knives are similar in function to trimmer knives and frequently are made with cutting-edge inserts of M2 high-speed steel hardened to 62 to 64 HRC. Under optimal conditions, life of these blades is very long—50 to 100 million cuts between resharpenings.

Some sheeter or cutoff blade designs are too narrow in cross section to allow for use of inserts; such knives are most often made of solid 52100, O1, or D2 tool steel.

Selection of Material for Blanking and Piercing Dies

Blanking and piercing dies include the punches, dies, and related components used to blank, pierce, and shape metallic and non-metallic sheet and plate in a stamping press. Blanking is a process of cutting by shear, as described in the articles "Principles of Shearing, Blanking, and Piercing" and "Blanking and Piercing" in this Volume. Sectional views of the blanking dies and the blanking and piercing punches used for making simple parts are shown in Fig. 16. Parts that are more complex require notching and compound dies.

Selection of tool material depends primarily on the number of acceptable parts to be produced by blanking and piercing. A common indication of tool deterioration is the production of a burr along the sheared edge of the workpiece. When tools are new, there is minimal clearance between punch and die, and the cutting edges are sharp. Under these conditions, the break in the stock begins at the underside (the side not in contact with the punch) because there the stock is subjected to the greatest tensile stress from stretching of the outer fiber. As more and more parts are produced, the cutting edges of the punch and the die become rounded by wear (Fig. 3), and the stress distribution in the stock is changed. Stress on the underside is reduced, breaking at that point is delayed, and deformation accompanied by work hardening occurs. When breaking starts, it nucleates from both sides simultaneously, and a burr develops on both the slug and the surrounding area of the sheet from which it was cut. The height of this burr increases with tool wear. Acceptable burr height varies with the application but is usually between 0.025 and 0.125 mm (0.001 and 0.005 in.).

Tool materials include tool steels, cemented carbides, steel-bonded carbides, or powder metallurgy tools (see "Tool Materials" in this article). Tool coatings are also instrumental in improving tool life, especially for high-volume blanking or piercing of thicker sections or high-strength sheet. Coating and surface treatments of tool steels are described in the section "Coatings and Surface Treatments" in this article.

Tool Materials for Punches and Dies. Table 5 lists typical materials for punches and dies used for blanking parts of different sizes and degrees of severity from several different work materials about 1.3 mm (0.050 in.) thick in various quantities. Illustrations of typical parts are presented in Fig. 17. Typical materials for the punches and dies used to shave several work materials of this same thickness in various quantities are listed in Table 6.

Tables 5 and 6 can be used to select punch and die materials for parts made of sheet that is thicker or thinner than the 1.3 mm (0.050 in.) used for the parts illustrated in Fig. 17. For thicker sheet, the punch and die material recommended for the next greater production quantity should be used instead of the material recommended for the production quantity that will actually be made (in Tables 5 and 6, the column to the right of the production quantity that will actually be made). For thinner sheet, the punch and die material recommended for the next lower production quantity should be used instead of the material recommended for the production quantity that will actually be made (in Tables 5 and 6, the column to the left of the production quantity that will actually be made).

Table 7 lists typical materials for perforator punches used on several different work materials. The usual limiting slenderness ratio (punch diameter to sheet thickness) for piercing aluminum, brass, and steel is 2.5:1 for unguided punches and 1:1 for guided punches. For piercing spring steel and stainless steel, this ratio ranges from 3:1 to 1.5:1 for unguided punches and from 1:1 to 0.5:1 for accurately guided punches. Typical hardnesses for these perforator punches are given in Fig. 18.

Table 8 lists typical materials for perforator bushings of all three types (punch holder, guide or stripper, and perforator or die). These recommendations are particularly applicable to precision bushings—for example, where the outside diameter is ground to a tolerance of -0, $+0.008$ mm (-0, $+0.0003$ in.) and is concentric with the inside diameter within 0.005 mm (0.0002 in.) total indicator reading. The hardness of W1 bushings should be 62 to 64 HRC, and that of D2 bushings, 61 to 63 HRC.

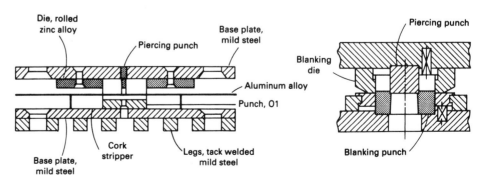

Fig. 16 Sectional views illustrating typical tools used for blanking and piercing simple shapes. Tooling at left is for short-run production of parts similar to parts 1 and 2 in Fig. 17 made from relatively thin-gage metal. Tooling at right is for longer production runs.

Tool Materials for Die plates. Die plates and die parts that hold inserts are usually made of gray iron, alloy steel, or tool steel. For stamping thick sheet or hard materials, either class 50 gray iron or 4140 steel heat treated to a hardness of 30 to 40 HRC should be used. For long-run die plates for stamping thick or hard materials, steels such as 4340 and H11 are preferred when inserts are pressed into the die plates, and 4340 is nearly always used when inserts are screwed in. Die plates for stamping thin or soft sheet can be made of class 25 or class 30 gray iron or carbon steel.

Secondary Tooling. Punch holders and die shoes for carbide dies are made of high-strength gray iron or low-carbon steel plate. Yokes for retaining carbide sections are usually made of O1 tool steel hardened to 55 to 60 HRC. Backup plates for carbide tools are preferably made of O1 hardened to 48 to 52 HRC. Strippers can usually be made of low- or medium-carbon steel (1020 or 1035) plate. Where a hardened plate is used for medium-production work, 4140 flame hardened, W1 conventionally hardened, or W1 cyanided and oil quenched are often preferred. Hardened strippers for carbide dies and high-production D2, D4, or CPM 10V dies are made of O1 or A2, hardened to 50 to 54 HRC.

Custom-made hardened guides and locator pins are usually made of W1 or W2 for most medium- or long-run dies or of alloy steels such as 4140 for low-cost short-run dies. Commercial guide pins are often made of SAE 1117 and then carburized, hardened, and finished to a surface roughness of 0.4 μm (16 μin.) rms.

Applications of Specific Materials. Rolled zinc alloy tooling plate is available in the form of 6.4 mm (1/4 in.) plate from the principal suppliers of zinc-base die-casting alloys. Dies of this material are sheared in with a flame-hardened O1 punch, and strippers of 9.5 mm (3/8 in.) sheet cork are invariably used with them.

Tools of hot-rolled low-carbon steel plate (0.10 to 0.20% C) can be used for short runs of small parts if these tools have been surface hardened, either by carburizing to a depth of 0.25 to 0.50 mm (0.010 to 0.020 in.) or by cyaniding to a depth of 0.1 to 0.2 mm (0.004 to 0.008 in.). Because of distortion in heat treatment, use of this material is limited to the blanking of small, symmetrical shapes.

For the long-run blanking of soft materials, various sizes of aircraft-quality 4140 steel plate have been used. In this application, 4140 is normally flame hardened to about 50 HRC. Flame hardening the working edge of a large die has an advantage over through hardening in that very little warping or change of size occurs. However, tools with inside or outside corners

Table 5 Typical punch and die materials for blanking 1.3 mm (0.050 in.) sheet

See Fig. 17 for illustrations of typical parts

Work material	Tool material for production quantity of:				
	1000	10,000	100,000	1,000,000	10,000,000
Part 1 and similar 75 mm (3 in.) parts					
Aluminum, copper, and magnesium alloys	Zn(a), O1, A2	O1, A2	O1, A2	D2, CPM 10V	Carbide
Carbon and alloy steel, up to 0.70% C, and ferritic stainless steel	O1, A2	O1, A2	O1, A2	D2, CPM 10V	Carbide
Stainless steel, austenitic, all tempers	O1, A2	O1, A2	A2, D2	D4, CPM 10V	Carbide
Spring steel, hardened, 52 HRC max	A2	A2, D2	D2	D4, CPM 10V	Carbide
Electrical sheet, transformer grade, 0.64 mm (0.025 in.)	A2	A2, D2	A2, D2	D4, CPM 10V	Carbide
Paper, gaskets, and similar soft materials	W1(b)	W1(b)	W1(c), A2(d)	W1(d), A2(d)	D2, CPM 10V
Plastic sheet, not reinforced	O1	O1	O1, A2	D2, CPM 10V	Carbide
Plastic sheet, reinforced	O1(e), A2	A2(f)	A2(f)	D2(f), CPM 10V	Carbide
Part 2 and similar 305 mm (12 in.) parts					
Aluminum, copper, and magnesium alloys	Zn(a), 4140(g)	4140(h), A2	A2	A2, D2, CPM 10V	Carbide
Carbon and alloy steel, up to 0.70% C, and stainless steels up to quarter hard	4140(h), A2	4140(h), A2	A2	A2, D2, CPM 10V	Carbide
Stainless steel, austenitic, more than quarter hard	A2	A2, D2	D2	D2, D4, CPM 10V	Carbide
Spring steel, hardened, 52 HRC max	A2	A2, D2	D2	D2, D4, CPM 10V	Carbide
Electrical sheet, transformer grade, 0.64 mm (0.025 in.)	A2	A2, D2	A2, D2	D2, D4, CPM 10V	Carbide
Paper, gaskets, and similar soft materials	4140(j)	4140(j)	A2	A2	D2, CPM 10V
Plastic sheet, not reinforced	4140(j)	4140(h), A2	A2	D2, CPM 10V	Carbide
Plastic sheet, reinforced	A2(e)	A2(e)	D2(e)	D2(e), CPM 10V	Carbide
Part 3 and similar 75 mm (3 in.) parts					
Aluminum, copper, and magnesium alloys	O1, A2	O1, A2	O1, A2	A2, D2, CPM 10V	Carbide
Carbon and alloy steel, up to 0.70% C, and ferritic stainless steel	O1, A2	O1, A2	O1, A2	A2, D2, CPM 10V	Carbide
Stainless steel, austenitic, all tempers	A2	A2, D2	A2, D2	D2, D4, CPM 10V	Carbide
Spring steel, hardened, 52 HRC max	A2	A2, D2	D2, D4	D2, D4, CPM 10V	Carbide
Electrical sheet, transformer grade, 0.64 mm (0.025 in.)	A2	A2, D2	D2, D4	D2, D4, CPM 10V	Carbide
Paper, gaskets, and other soft materials	W1(b)	W1(b)	W1(k), A2	W1(k), A2	D2, CPM 10V
Plastic sheet, not reinforced	O1	O1	A2	A2, D2, CPM 10V	Carbide
Plastic sheet, reinforced	O1(m)	A2(f)	A2(f)	D2(f), CPM 10V	Carbide
Part 4 and similar 305 mm (12 in.) parts					
Aluminum, copper, and magnesium alloys	A2	A2	A2, D2	A2, D2, CPM 10V	Carbide
Carbon and alloy steel, up to 0.70% C, and ferritic stainless steel	A2	A2	A2, D2	A2, D2, CPM 10V	Carbide
Stainless steel, austenitic, up to quarter hard	A2	A2	A2, D2	D2, D4, CPM 10V	Carbide
Stainless steel, austenitic, more than quarter hard	A2	D2	D2	D2, D4, CPM 10V	Carbide
Spring steel, hardened, 52 HRC max	A2	A2, D2	D2	D2, D4, CPM 10V	Carbide
Electrical sheet, transformer grade, 0.64 mm (0.025 in.)	A2	A2, D2	D2	D2, D4, CFM 10V	Carbide
Paper, gaskets, and other soft materials	W1(b)	W1(b)	W1(n)	W1, A2	D2, CPM 10V
Plastic sheet, not reinforced	A2	A2	A2	A2, D2, CPM 10V	Carbide
Plastic sheet, reinforced	A2(f)	A2(f)	D2(f)	D2(f), CPM 10V	Carbide

Note: Although carbide is recommended in this table only for 10 million pieces, it should usually be considered also for runs of 1–10 million pieces. (a) Zn refers to a die made of zinc alloy plate and a punch of hardened tool steel. (b) For punching up to 10,000 parts, the W1 punch and die would be left soft and the punch peened to compensate for wear if necessary. (c) For punching 10,000–1,000,000 pieces, the W1 punch can be soft so that it can be peened to compensate for wear, or it can be hardened and ground to size. (d) Of the two alternatives listed, A2 tool steel is preferred if compound cooling is to be used for quantities of 10,000–1,000,000. (e) This O1 punch may have to be cyanided 0.1 to 0.2 mm (0.004 to 0.008 in.) deep to make even 1000 pieces. (f) For the application indicated, the punch and die should be gas nitrided 12 h at 540–565 °C (1000–1050 °F). (g) Soft. (h) Working edges are flame hardened in this application. (j) May be soft or flame hardened. (k) For punching 10,000–1,00,000 pieces, the punch would be W1, left soft so that it can be peened to compensate for wear, and the die would be O1, hardened. (m) Cyaniding of the punch is advisable, even for 1000 pieces. (n) For punching 10,000–1,000,000 pieces, the W1 die would be hardened and the W1 punch would be soft so that it can be peened to compensate for wear.

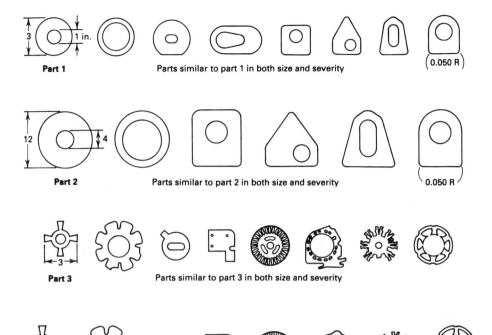

Fig. 17 Typical parts of varying severity that are commonly produced by blanking and piercing. Parts 1 and 2 are relatively simple parts that require dies similar to those illustrated in Fig. 16. Parts 3 and 4 are more complex, requiring notching and the use of compound or progressive dies. Dimensions given in inches

Table 6 Typical punch and die materials for shaving 1.3 mm (0.050 in.) sheet

Work material	Tool material for production quantity of:			
	1000	10,000	100,000	1,000,000
Aluminum, copper, and magnesium alloys	O1(a)	A2	A2	D4(b), CPM 10V
Carbon and alloy steel, up to 0.30% C, and ferritic stainless steel	A2	A2	D2	D4(b), CPM 10V
Carbon and alloy steel, 0.30–0.70% C	A2	D2	D2	D4(b), CPM 10V
Stainless steel, austenitic, up to quarter hard	A2	D2	D4(b)	D4(b), CPM 10V
Stainless steel, austenitic, more than quarter hard, and spring steel hardened to 52 HRC max	A2	D2	D4(b)	M2(b), CPM 10V

(a) Type O2 is preferred for dies that must be made by broaching. (b) On frail or intricate sections, D2 should be used in preference to D4 or M2. Carbide shaving punches may also be practical for this quantity.

Table 7 Typical materials for perforator punches

Work material	Punch material for production quantity of:		
	10,000	100,000	1,000,000
Punch diameters up to 6.4 mm (1/4 in.)			
Aluminum, brass, carbon steel, paper, and plastics	M2	M2, CPM 10V	M2, CPM 10V
Spring steel, stainless steel, electrical sheet, and reinforced plastics	M2	M2, CPM 10V	M2, CPM 10V
Punch diameters over 6.4 mm (1/4 in.)			
Aluminum, brass, carbon steel, paper, and plastics	W1	W1	D2, CPM 10V
Spring steel, stainless steel, electrical sheet, and reinforced plastics	M2	M2, CPM 10V	M2, CPM 10V

Table 8 Typical materials for perforator bushings

Work material	Bushing material for production quantity of:		
	10,000	100,000	1,000,000
Aluminum, brass, carbon steel, paper, and plastics	W1(a)	W1(a)	D2
Spring steel, stainless steel, electrical sheet, and reinforced plastics	D2	D2	D2 or carbide

(a) When bushings are of a shape that cannot be ground after hardening, an oil- or air-hardening steel is recommended to minimize distortion.

may have soft spots after flame hardening and, if so, will perform poorly.

The tool steels in Table 5 are assumed to have been hardened and tempered by conventional methods to their maximum usable hardness (58 to 61 HRC). In addition to these tool steels, type O6 has given satisfactory service in multiple-stage progressive dies, and type A10, because of its low austenitizing temperature, high dimensional accuracy, and good dimensional stability, is often used to make large dies for stamping laminations.

In some applications, M2 high-speed steel tools may produce smaller burrs than D2 tools (for equal numbers of parts). In addition, steel-bonded carbides and high-vanadium-carbide powder metallurgy tool materials such as CPM 10V should be considered for critical applications. Steel-bonded carbides (steel matrix with TiC) belong to the family of cemented carbide produced by powder metallurgy, but differ from cemented carbides in that the steel matrix can be altered by heat treatment. In the annealed conditions, steel-bonded carbides can be machined, then heat treated. As previously noted (see "Tool Materials" in this article), wear resistance is between that of tool steels and Co-WC cemented carbides.

Cemented tungsten-carbide tooling should be considered where production life must be four or more times that possible with D4 tool steel. Partial or complete carbide inserts in tool steel dies may be considered for lower quantities, especially where close tolerances and minimum burr height are desired or where tool life between resharpenings needs to be extended. However, brazed inserts are hazardous, and the cost of

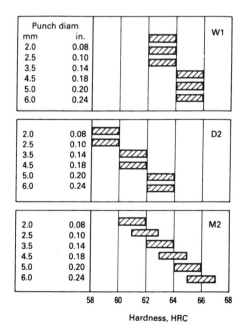

Fig. 18 Typical hardnesses for tool steel perforator punches. Regardless of material, punches should be tempered back to 56 to 60 HRC if they are to be subjected to heavy shock or used to pierce thick material.

dovetailed or mechanically held inserts approaches that of complete carbide dies.

Composition and hardness of carbides frequently used in blanking and piercing dies are as follows:

Composition, %			Hardness, HRA
W	C	Co	
75.1	4.9	20.0	86
78.9	5.1	16.0	86
81.7	5.3	13.0	88
88.3	5.7	6.0	91

The first material listed in the preceding table should be used where shock is appreciable. The second material combines toughness and wear resistance and is preferred for heavy-duty service, such as the piercing of silicon steel. Where close tolerances must be held in piercing silicon steel laminations, the third material is useful. The fourth material is best for guides and guide rolls and for applications involving very light shock. The data in Fig. 19 show that the difference in wear life between two different tool steels at the same hardness is negligible compared with the difference between the average life of conventional tool steel dies and the life of a carbide die or a CPM 10V die.

Die life in the blanking and piercing of high-carbon steel varies with different applications, depending greatly on the dimensional accuracy that must be maintained and the burr height that can be tolerated on the blanked parts. The most important difference between the blanking and piercing of high-carbon and of low-carbon steel is that greater clearance between punch and die is required for high-carbon steels (see Table 1 in the article "Blanking of Low-Carbon Steel" in this Volume).

Along with the increase in punch-to-die clearance, other tool changes are required for efficiency in the blanking and piercing of high-carbon steel. The following examples illustrate various types of tooling changes for blanking high-carbon steel. The first example describes an application in which a change in tool material increased die life by a factor of about 28. The second example illustrates the common practice to have a punch mechanism with a second, chisel-pointed punch placed as close as possible to the blanking edge. With high-carbon steel (more so than with low-carbon steel), the scrap skeleton from blanking and piercing operations is likely, in springing back, to adhere to the punch, sometimes causing spalling of the punch edges. This second punch serves to spread the scrap skeleton, minimizing its adherence to the blanking punch (see example 2).

Example 1: Change from Tool Steel to Carbide Punches and Dies. The part shown in Fig. 20 was blanked and pierced from 48 mm (1⁷⁄₈ in.) wide coils of bright-finish 1045 steel having a hardness of 70 to 75 HR15N (equivalent to 20 to 30 HRC). Thickness ranged from 0.48 to 0.986 mm (0.019 to 0.0388 in.); tolerance for all thicknesses was +0.000, −0.03 mm (+0.000, −0.001 in.).

The punch and the die were both originally made of D2 tool steel at 62 to 63 HRC. To keep burr height on the parts at or below the maximum of 0.08 mm (0.003 in.), it was necessary to grind the die after each 25,000 pieces. The steel die had a usable depth of 16 mm (⁵⁄₈ in.), and 0.20 mm

(0.008 in.) was removed with each grinding. Therefore, with 78 grindings, the total die life was 1.95 million pieces.

To improve die life, the tool material for both the punch and the die was changed to carbide. The carbide tools cost three times as much as the steel punch and die, but production between grinds increased to 350,000 pieces and only 0.10 mm (0.004 in.) of stock was removed per grind. Therefore, with the same amount of usable die, the total die life would be 54.6 million pieces.

The die was a compound-type unit. It was operated in a 445 kN (50 tonf) open-back inclinable mechanical press having a 75 mm (3 in.) stroke and mechanical feed.

Example 2: Cost Reduction When Blanking Replaced Milling. The blanked workpiece shown in Fig. 21 replaced a machined part of slightly different dimensions. For the machined part, starting blanks 75 mm (3 in.) long by 17 mm (¹¹⁄₁₆ in.) wide by 4.0 mm (⁵⁄₃₂ in.) thick were sawed from ground flat stock of A2 tool steel spheroidize-annealed to 14 to 18 HRC. The long edges of the blanks were ground to reduce the width of the blank from 17 to 15.9 mm (¹¹⁄₁₆ to 0.625 in.). Grinding was followed by four separate milling operations. This change in production method from milling to blanking reduced machining time by an order of magnitude and decreased cost per piece by a factor of about 120 for 100 pieces.

The blanking was done in a 280 kN (32 tonf) open-back inclinable mechanical press with a punch and die made from A2 tool steel at 60 to 62 HRC. To obtain acceptable edges on the workpiece, a punch-to-die clearance of 8 to 10% of stock thickness per side was used. Because of the force required for stripping, it was necessary to add a chisel-edge punch to spread the scrap skeleton and to prevent damage to the blanking punch. In addition, a minimum corner radius of 1.6 mm (0.062 in.) was necessary for efficient stripping, and the overall width of the part was increased to 17.7 mm (0.696 in.).

ACKNOWLEDGMENTS

Portions of this article were adapted from Ref 1 and 2.

REFERENCES

1. J. Schey, *Tribology in Metalworking,* American Society for Metals, 1983
2. G. Håkansson, Experiences of Surface Coated Steels, *International Conference on Recent Advances in Manufacture of Tools and Dies and Stamping of Sheet Steels,* Volvo, 2004

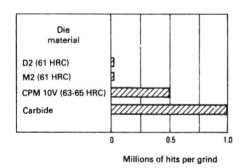

Fig. 19 Relative service lives of three steel dies and one carbide die. Die life was obtained under the same operating conditions; that is, the blanking of 3.25% Si electrical steel sheet 0.36 mm (0.014 in.) thick. Dies were reground when they had worn sufficiently to produce a burr 0.13 mm (0.005 in.) high.

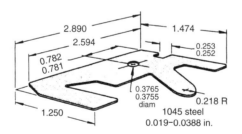

Fig. 20 Blanked and pierced textile machine part for which carbide compound dies had 28 times the total life of those made of D2 tool steel. Dimensions given in inches

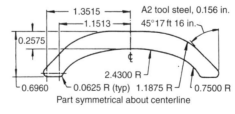

Fig. 21 Part produced by blanking for less than 1% of the per-piece cost of producing it by milling. The punch was modified to solve a stripping problem. Dimensions given in inches

Trimming Operations

FOR MANY TYPES of stamped parts, excess metal must be allowed for holding purposes during operations that shape the metal into the form of the part. This excess metal must, of course, be removed; such removal is done by trimming. Trimming is the removal of excess metal from a stamped part to allow the part to reach the finished stage or to prepare it for subsequent operations. For nearly all parts requiring drawing operations, trimming is necessary. Formed parts other than those produced from fully developed blanks also must undergo some trimming before they are brought to final size.

Analysis of Parts to be Trimmed

Trimming of a part is usually performed immediately after a drawing operation. Therefore, it is necessary to consider what the part will look like after it is removed from the drawing die.

The process engineer must provide answers to many questions, such as:

- Should the part be trimmed open side up or open side down?
- Can the part be trimmed in one operation, or should it be trimmed in two operations?
- Is it necessary to separate two or more pieces if parts are made in multiples?
- Is any notching required?
- Are both rough and finished trimming needed?
- Should two trimming dies be built, or can a cam trimming die be used?
- Are there any operations that can be combined with trimming?

A plan view of a plaster model can reveal that the shape of the trim outline that is going to require external cutting will be square, rectangular, oval, irregular but symmetrical, or irregular and nonsymmetrical (Fig. 1). A side elevation view may reveal that the binder metal is in a flat plane, has a single contour, or has multiple contours.

Internal cutting may also be required on the panel if it is necessary to separate parts (such as for the right-hand and left-hand sides of a car) or if large openings must be made somewhere in the panel for the purpose of allowing an extruded section to be made such as a window opening, a lightening hole, or an access hole.

The thickness of the metal to be cut may have some bearing on the trimming operation. In addition, consideration must be given to the possible creation of knifelike edges on the part that might injure employees who must handle the panels in later operations.

Selection of Trimming Dies

Due to the wide variety of stamped parts, a number of different types of trimming dies have been developed. The edge requirements and size of the part will generally indicate what kind of die should be used.

Conventional Trimming Dies

The flush trimming die is the type of die most widely used in automotive stamping (Fig. 2). This type of die has two opposing cutting steels, operating much in the same fashion as a pair of scissors, that cut vertically on the downstroke of the press. The part is usually supported open side down on a solid pad mounted on the lower shoe; this pad is very similar to the type used on a solid forming die. There is usually a spring-loaded pad fastened to the punch. This pad holds the part firmly in position on the downstroke of the press. The upper cutting steels surrounding the spring pad contact the part after the pad has descended, and the scissors-like action takes place when the upper trimming steels pass by the lower trimming steels. After the press has completed its cycle, the part is ejected or lifted out of the die. Flush trimming results in a small burr around the edge of the cut because of the necessary clearance allowed between the punch and die.

The location of this burr is frequently important in flanging the part in later operations. The burr may be located on the top or the bottom of the flange depending on the position of the part in the die. If the part is positioned open side up, the burr will be on the underside of the flange. If the part is inverted, the burr will be located on the top of the flange.

After the press has completed its cycle, the metal cut away from the part, which is the offal, is usually in the form of a ring. This material must be disposed of in one way or another. If allowed to remain in ring form, it would require manual or mechanical removal from the die. To avoid this, scrap cutters are designed and built as integral units of the lower trimming steels. These are knifelike blades mounted perpendicular to the trim line. Mating steels are also mounted on the punch. These scrap cutters are usually designed to part the material at distances from 305 to 455 mm (12 to 18 in.), as described

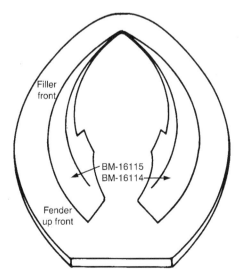

Filler front

BM-16115
BM-16114

Fender up front

Fig. 1 Trim line on plaster die model

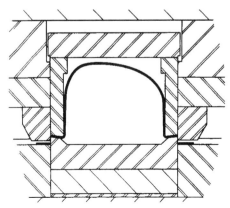

Fig. 2 Flush trimming die

later. Provision is made for disposal of this stock in the design of the die and layout of the press department.

A parting trimming die is used for separating two or more parts formed or drawn from a single sheet of metal (Fig. 3). The cutting steels in these dies must be accurately aligned, and the parts must be firmly held during cutting to ensure that separation is accomplished at exactly the right place. Usually when parts are separated, a thin strip of metal is cut from between them. Consideration must be given to the removal of this scrap metal. Parting is frequently combined with flush trimming.

A horn trimming die, which is sometimes called a "hang-up" die, is a die generally mounted on a horn-type press that has an overhanging mechanism for separating the slide. The die is mounted on the front of the press. A portion of the part hangs over the die, and then the press is cycled. Such dies are generally used when trimming of relatively short length is involved.

Cam Trimming Dies

A cam trimming die (Fig. 4) may be used where the operation cannot be performed entirely with a vertical stroke of the press. One or more cams are used to convert the vertical action of the ram into an action in a more suitable direction, such as 70°, 85°, 90°, or some other angle. The use of cams, while providing perhaps the only solution to a difficult trimming application, poses certain problems for pressworking. These problems should be thoroughly considered before a final decision to employ a cam trimming die is made. A cam die is justified if a saving in piece price can be achieved or if a limited number of presses are available for running the parts, which would necessitate the combination of cams on a flush trimming die. The design of the cam mechanism must be such that parts can be loaded and unloaded without interference. Provision must be made for easy removal of scrap generated by the cam trimming steels. Cam construction must be done with consideration of employee safety. Cams should be located where they are not in the way of the employee. Where

cams move out from the press bed, guards should be installed to prevent injury to plant personnel.

The cam slide contains the cutting steel and is guided by steel gibs with hardened steel wear plates. Provision should be made for shimming of the wear plates to adjust the clearance of the cutting tools to compensate for wear.

A cam mounted on the die is usually driven by a wedge-shape driver attached to the punch. The action is obtained by a wedge-shape block on the punch, called the driver, which moves a wedge-shape block on the cam. On the upstroke of the press, the cam is returned by springs or by an air cylinder. In some cases, the advance and return of the cam are accomplished with a positive-type driver that is designed with several offsets to accomplish the desired action. Stop

blocks are used to limit the back travel of the cam. The back travel must be great enough, however, to permit easy loading and unloading of parts.

A cam mounted on the punch is used for certain applications where the path of cam action is between perpendicular and 45% from the vertical action of the press. Such a cam will float on the punch and is actuated by a driver mounted on the die.

A shimmy trimming die, which is a form of cam trimming die, is often used in trimming cup-shape parts where exceptionally clean cut edges are required. In this type of die, when it is desired to trim the part open side up, the part is pushed into position by the ram of the press. A lower floating die, activated by positive-motion

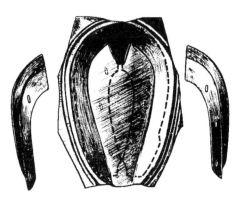

Fig. 3 Parting trimming dies used to separate parts drawn in multiples

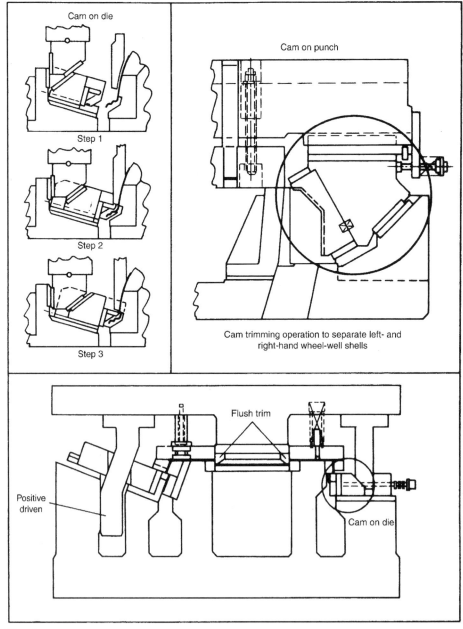

Cam on die

Step 1

Step 2

Step 3

Cam on punch

Cam trimming operation to separate left- and right-hand wheel-well shells

Flush trim

Positive driven

Cam on die

Fig. 4 Cam trimming dies

cams mounted on the punch, reciprocates from side to side and from front to back to trim the part. If it is desired to trim the part open side down, the floating section may be mounted on the upper shoe with the drivers on the lower shoe.

A pinch trimming die is frequently used for trimming cup-shape parts with only the vertical action of the press (Fig. 5). This operation is often combined with a drawing operation and may even be combined with blanking. In such a compound die, the punch first draws the part to the proper depth. After this, the punch continues its travel, and the flange is pinched off by the action of the cutting steels on the punch against the cutting steels on the die. Clearances between punch and die must be held to a minimum to ensure an even pinch-off. Although every effort is made to prevent the formation of ragged edges on the trim line, such an objective is seldom obtained with a pinch-trimming die.

Rough and Finish Trimming

Rough Trimming. There are instances in which certain areas of a part may be trimmed away prior to completion of other operations. This is called rough trimming, because the edge produced will not be the final edge of the part.

Rough trimming is often done prior to a re-striking operation. Restriking operations have as their purpose the final sizing of the curved areas of drawn parts. The rough trimming is necessary to prevent the scrap around the part from restricting metal movement or causing springback.

Rough trimming is used in other instances to permit convenient handling of a part after a first drawing operation. This is particularly true of large panels drawn from square-sheared stock, where large holding flanges represent a hazardous and awkward handling problem.

Rough trimming may also be necessary when a very large amount of scrap must be cut away. Cutting all the scrap in one finish trimming operation may cause problems in removing the

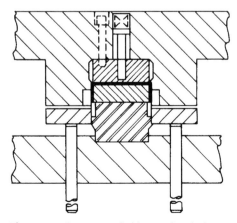

Fig. 5 Pinch-trimming die for combined drawing and trimming

scrap from the die. In such a case, a portion of the scrap may be trimmed off in one operation and the part finish trimmed in another operation. This may provide a more suitable arrangement and a solution to the scrap-disposal problem.

Finish trimming generally is done as the final operation in bringing a part to final size. The location of the rough trim line in relation to the finish trim line is of major importance in setting up a rough trimming operation. If insufficient metal is left for finish trimming, the finish cut will not be clean. An excessively narrow strip of metal will crush or draw down during trimming, leaving chips or hairlike burrs. In most cases, these chips or burrs will be deposited on the die surface, causing trouble in later operations. The chips will build up on the die, causing scoring of parts and injury to the die. Generally speaking, at least 3 mm ($^1/_8$ in.) of metal should be left for removal in a finish trimming operation to avoid burrs and chips.

Construction Details of Trimming Dies

The production of parts to specifications is the primary consideration in the construction of a trimming die. Before die design is begun, the shape and contour of the finished part (as shown by part print, prototype model, and die model) are carefully considered. The locations of holes and their proximity to the trim line are noted. If holes are too close to the trim line, distortion or breakage may occur unless provision is made for proper holding. The shape and contour of the part will also provide clues as to the amount of metal that will be provided for holding during operations preceding trimming and will indicate the amount of metal that will be left for trimming.

Die design must also be considered in relation to the specifications of the press to be used. A determination is made as to whether the press has the proper shut height and area to accommodate the die needed for trimming, and whether it will accommodate the stamping with the binder scrap attached. Consideration is also given to the possibility of designing a die that will be amenable to automation once it is installed in the press.

There are numerous details that must be worked out in the design and construction of a trimming die if proper performance is to be obtained. Although many of these details are of interest only to the die designer and diemaker, several should also be of concern to the process engineer.

One of the factors influencing construction of trimming dies is the extreme care that must be given to the guiding of cutting tools. While some side thrusts and unbalanced conditions may be tolerated in forming or drawing operations, positive guiding in trimming operations is imperative because of the necessity for proper mating between the upper and lower cutting

steels. If cutting steels are not properly aligned, they will certainly suffer damage caused by the opposing edges striking together. To guard against such damage, guide pins are used to ensure accurate travel of the moving member of the die. Trimming dies for small parts may be adequately guided by two guide pins. For larger parts, more pins (up to four) may be required. These guide pins are supplied whether or not the cutting job to be done is evenly balanced.

In cases where unbalanced forces are present in trimming, heel blocks should be used to absorb the side thrust. These heel blocks consist of built-up steel bearing surfaces usually located at all four corners of the die. Ordinarily mounted on the lower die or die shoe, the blocks restrain the punch from moving sideways as it descends. This provision prevents tool breakage and helps to ensure a satisfactory edge on the trimmed part. It is important to locate heel blocks so that they will not interfere with loading or unloading of parts.

The selection of steels for trimming dies will depend on the complexity of the operation and the production volume involved.

The shoes for trimming dies are made of the same material as the other dies in the press lines. The cutting steels on both the punch and die are usually made of composite sections rather than from solid steel blocks, because it is easier to mount the sections at various angles on the mounting blocks and to remove them for resharpening.

As in all types of cutting, the pressure requirements in trimming operations can be reduced by grinding shear on the cutting edges of the tools. In trimming, the shear is ground on the punch. The shear may be either balanced or unbalanced. Where a long cut is to be made, it is generally advantageous to provide balanced shear to reduce the tendency of the blank to shift in the die.

Another consideration in setting up trimming operations is the clearances provided between the cutting steels. If insufficient clearance is allowed, the edges will be ragged. Secondary shearing will take place, adding to the load on the press and the wear on the cutting steels. Excessive clearance will cause the edge of the cut to draw down, leaving a burred edge. Many of the same clearances for piercing operations apply generally to trimming.

When parts are trimmed, there is a tendency for them to stick to the sides of the cutting tools. To prevent the edges of the part from lifting up and perhaps buckling as the punch ascends, strippers are provided. These strippers are usually spring-powered. Springs must be strong enough to strip the part from the cutting steels.

Trimming operations call for relatively close observance of tolerances. To ensure that tolerances are met, the part being trimmed must be accurately and firmly located in the dies. Special locators are placed over spacer plates where necessary for locating the part. If the shape of the part is not conducive to locating, pin locators or gages may be used in holes in the part. Gages should be adequate for holding the part firmly. Proper gaging (locating) is vital to accurate and fast trimming operations. Stripper plates frequently assist in holding the part by contacting the work prior to the cutting action.

Selection of Presses for Trimming

Because of the requirements for exceptional control and relatively high precision, special care must be exercised in selection of a press for a trimming operation. The press as well as the die must be equipped to ensure accurate mating of the cutting steels. Older presses should be examined carefully to make certain they will provide good alignment.

Gap-frame presses are frequently used for trimming small or medium-size parts in a wide range of volume requirements. Because of their accessibility, they are particularly adaptable to the use of automation for handling of parts. Such presses are usually of the inclinable type, and thus gravity can be utilized to facilitate unloading.

There are, however, certain features of gap-frame presses that may serve to complicate trimming operations if press adjustments are not made. The feature of these presses that is most important is their gap construction. This gap tends to deflect, and spring open, under prolonged use and/or heavy loading. Where trimming is involved, this factor has important implications for the maintenance of the punch and die.

The deflection tends to cause the edges of the cutting steels to contact each other, resulting in breakage or undue wear. To counteract this disadvantage, tie rods may be installed across the gap to give the press greater rigidity. Another means of minimizing this problem is to employ presses of greater tonnage than is required by the cutting operation involved. This is called a press tonnage safety allowance and is a frequently used solution to the problem of deflection.

Guiding and stabilization of the press slide itself is another important factor. In addition to carefully aligned gibs on the press, guide pins and heel blocks should be provided in the die itself.

Straight-Side Presses. Deflection is more easily controlled in a straight-side press, where gibs at all four corners of the press are frequently used. The gibs tend to neutralize any side thrusts caused by uneven cutting pressures. Presses with two- and four-point suspension resist deflection from unequal loading. Nevertheless, heel blocks are often used in trimming dies run in four-column straight-side presses.

Scrap Handling

The handling and disposal of scrap have an important bearing on the efficiency of trimming operations. In any trimming application, some scrap or offal is generated. Provision must be made for removal of this material from the area of press.

Implications for Overall Press Operations. Removal of scrap in trimming operations has implications for the total effectiveness of all press operations. If adequate means of scrap disposal are not provided, an entire stamping operation could be easily bogged down. The removal of scrap often determines the speed of a trimming operation, and trimming speed often governs the entire press line. Thus, adequate facilities for scrap removal constitute one of the initial considerations in establishing effective stamping operations. In addition to the scrap involved in trimming, thought must also be given to the removal of scrap from blanking, piercing, and other types of cutting operations.

There are two aspects of the problem of scrap removal that stand out. One is the removal of scrap from the die itself, and the other is the removal of scrap from the area of operations. Several principles for solving these problems have been worked out. Consideration is given to some of the steps that should be taken.

Scrap Size. One of the first steps in effective scrap handling is to reduce the scrap to manageable size. Trimming involves cutting away portions of material that vary considerably with respect to size. The size of scrap will determine the ease with which it may be removed from the die and also from the press area. The maximum size of scrap that can be handled successfully is 610 mm (24 in.) in length. Smaller pieces are preferable. A length of 455 mm (18 in.) is considered to be optimum for most efficient removal from the press and from the work area.

Removal of Scrap from Dies. The methods to be used in removing trimmings and other scrap from dies should be planned concurrently with die design. This will permit the designer and the process engineer to make certain that adequate space is allowed for any die attachments necessary for removing scrap.

The die should be designed with sufficient height to allow the scrap to fall away from the die surface. This height will depend on the size of scrap and the provisions for receiving it outside the press.

In many cases, chutes should be incorporated in the die to carry the scrap to stock boxes or conveyors. The chutes should be designed to provide ample space for trimmings to fall through. If trimmings become wedged in a chute, the press will have to be shut down until they are cleared. In some cases, the design of the die will permit the scrap to fall directly into stock boxes without the use of chutes. It is necessary that the die design allow each piece of scrap to be completely removed from the die area, or at least to a space where it can be allowed to pile up for occasional removal. Operating speed is seriously reduced in cases where operators must clear away scrap after each press cycle.

Where scrap is large, or where it is trimmed from all around the part, scrap cutters should be provided to reduce it to manageable size. These cutters should be provided with shear to reduce load requirements. They should also be installed with consideration of possible changes in the binder surface of the die.

It is sometimes difficult to direct the fall of scrap to the proper area. This is particularly true of internal trimmings that fall through the die. To remove this type of scrap, small conveyors may be used. These may be of the belt variety, or, where trimmings are very small and tend to slip under a belt, vibration-type conveyors may be used.

Removal of Scrap from Work Area. The most effective means of removing scrap from the work area is to shed it through chutes to a conveyor that will move it directly to the scrap baler. This conveyor is usually situated under the press floor and away from the stamping operations. Where it is unfeasible to use this system, scrap bins or trailers can be used. These too, can often be located in a basement area where they will not interfere with work on the press floor. When these bins and containers are filled, they are taken away to the baler by industrial trucks and empty containers are substituted.

Use of Scrap and Offal. The metal cut away from parts in trimming and blanking operations is often referred in press-room terminology as either scrap or offal. While there is no particular distinction between these two terms in standard dictionaries, each has a special meaning for the process engineer and each serves an important purpose in overall company operations.

Offal is generally considered to be the type of trimmings that can be put to further use in the stamping plant. This material is often used for making small parts in low-, medium-, and in some cases high-volume production. Since offal usually is unevenly shaped, hand methods of feeding it to blanking presses are generally used. An operator, by using good judgment, can often obtain good blanks from a large percentage of a piece of offal. Offal is also used as blanks for some small parts that can be successfully blanked and formed from the material in one operation. A process engineer, by giving attention to the trimmings that must be cut away from some parts, can often find means of achieving significant material economy by using the trimmings for producing other stampings.

Many trimmings, especially those from deeply drawn parts, are wrinkled and distorted, and it is impractical to use them in further operations. Other trimmings are too small or too oddly shaped to lend themselves to further use in stamping. Such material is baled and returned to the steel plant where it is used as the very necessary scrap charge in making new steel.

Material Handling in Trimming

Automatic Loading. Trimming frequently sets the pace for a group of related press

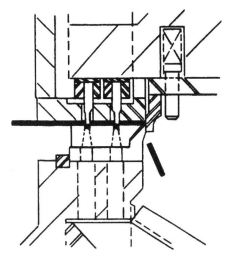

Fig. 6 Combined trimming and piercing

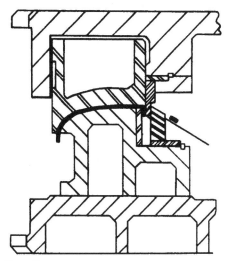

Fig. 7 Combined trimming and flanging

operations, and thus it is important that trimming dies be built to meet the production-volume requirements of the part. One method of attaining satisfactory speed in a press operation is to cut down the time required for loading the press. While loading takes place, the press is idle. Reduction in loading time can therefore yield a higher degree of utilization of press capacity. To reduce loading time, consideration should be given to the use of automatic loading devices. This possibility should be considered during construction of the die so that space can be allowed for any necessary attachments. For small

parts, an automatic loading device may consist of a simple gravity slide that will bring the part to rest properly in the die. For larger parts, a shuttle conveyor may be used.

Automatic Unloading. Another means of reducing waiting time in press operations is to provide speedy automatic means of unloading parts. For small parts, air blasts should be used. For medium parts, an effective unloading system may use lifters to free the part from the die and a positive kicker to eject the part. Provision may also be made in some cases to use the incoming part as an effective tool for forcing the completed

part out of the die. Where parts are very large, iron hands are frequently used to extract them. To expedite handling, belt conveyors or stock boxes should be used to receive the parts as they issue from the die.

Combined Operations

Trimming operations are frequently combined with other types of cutting operations that must for various reasons follow the drawing of a part. Some forming operations also may be successfully combined with trimming.

Trimming and Piercing. Drawn parts frequently must be pierced after drawing in order for the holes not to be deformed. In these cases, it is economical to combine the piercing and trimming operations (Fig. 6). Holes may be pierced in the flange of a drawn part in the same die that trims the flange.

Trimming and flanging are also frequently combined (Fig. 7). The trimming is accomplished first as the punch depresses the part on a spring pad. When trimming is completed, the punch continues down and forms the flange. Trimming is also combined with many other types of operations.

ACKNOWLEDGMENT

Adapted from "Trimming Operations" by O.D. Lascoe in *Handbook of Fabrication Processes,* ASM International, 1988, p 171–180.

Oxyfuel Gas Cutting

OXYFUEL GAS CUTTING (OFC) includes a group of cutting processes that use controlled chemical reactions to remove preheated metal by rapid oxidation in a stream of pure oxygen. A fuel gas/oxygen flame heats the workpiece to ignition temperature, and a stream of pure oxygen feeds the cutting (oxidizing) action. The OFC process, which is also referred to as burning or flame cutting, can cut carbon and low-alloy plate of virtually any thickness. Castings more than 760 mm (30 in.) thick commonly are cut by OFC processes. With oxidation-resistant materials, such as stainless steels, either a chemical flux or metal powder is added to the oxygen stream to promote the exothermic reaction.

The simplest oxyfuel gas cutting equipment consists of two cylinders (one for oxygen and one for the fuel gas), gas flow regulators and gages, check valves, flashback arrestors, gas supply hoses, and a cutting torch with a set of exchangeable cutting tips. Such manually operated equipment is portable and inexpensive. Cutting machines, employing one or several cutting torches guided by solid template pantographs, optical line tracers, numerical controls, or computers, improve production rates and provide superior cut quality. Machine cutting is important for profile cutting, that is, the cutting of regular and irregular shapes from flat stock.

Principles of Operation

Oxyfuel gas cutting begins by heating a small area on the surface of the metal to the ignition temperature of 760 to 870 °C (1400 to 1600 °F) with an oxyfuel gas flame. Upon reaching this temperature, the surface of the metal will appear bright red. A cutting-oxygen stream is then directed at the preheated spot, causing rapid oxidation of the heated metal and generating large amounts of heat. This heat supports continued oxidation of the metal as the cut progresses. Combusted gas and the pressurized oxygen jet flush the molten oxide away, exposing fresh surfaces for cutting. The metal in the path of the oxygen jet burns. The cut progresses, making a narrow slot, or kerf, through the metal.

To start a cut at the edge of a plate, the edge of the preheat flame is placed just over the edge to heat the material. When the plate heats to red, the cutting oxygen is turned on, and the torch moves over the plate to start the cut.

During cutting, oxygen and fuel gas flow through separate lines to the cutting torch at pressures controlled by pressure regulators, adjusted by the operator. The cutting torch contains gas ducts, a mixing chamber, and valves to supply an oxyfuel gas mixture of the proper ratio for preheat and a pure oxygen stream for cutting to the torch tip. By adjusting the control valves on the torch handle or at the cutting machine controller, the operator sets the precise oxyfuel gas mixture desired. Depressing the cutting-oxygen lever on the torch during manual operation initiates the cutting-oxygen flow. For machine cutting, oxygen is normally controlled by the operator at a remote station or by numerical control. Cutting tips have a single cutting-oxygen orifice centered within a ring of smaller oxyfuel gas exit ports. The operator changes the cutting capacity of the torch by changing the cutting tip size and by resetting pressure regulators and control valves. Because different fuel gases have different combustion and flow characteristics, the construction of cutting tips, and sometimes of mixing chambers, varies according to the type of gas.

Oxyfuel gas flames initiate the oxidation action and sustain the reaction by continuously heating the metal at the line of the cut. The flame also removes scale and dirt that may impede or distort the cut.

The rate of heat transfer in the workpiece influences the heat balance for cutting. As the thickness of the metal to be cut increases, more heat is needed to keep the metal at its ignition temperature. Increasing the preheat gas flow and reducing the cutting speed maintains the necessary heat balance.

Oxygen flow also must increase as the thickness of the metal to be cut increases. The jet of cutting oxygen must have sufficient volume and velocity to penetrate the depth of the cut and still maintain its shape and effective oxygen content.

Quality of Cut. Oxyfuel gas cutting operations combine more than 20 variables. Suppliers of cutting equipment provide tables that give approximate gas pressures for various sizes and styles of cutting torches and tips and recommended cutting speeds; these variables are operator controlled. Other variables include type and condition (scale, oil, dirt, flatness) of material, thickness of cut, type of fuel gas, and quality and angle of cut.

Where dimensional accuracy and squareness of the cut edge are important, the operator must adjust the process to minimize the kerf, the width of metal removed by cutting, and to increase smoothness of the cut edge. Careful balancing of all cutting variables helps attain a narrow kerf and smooth cut edge. The thicker the work material, the greater the oxygen volume required and therefore, the wider the cutting nozzle and kerf.

Process Capabilities

Oxyfuel gas cutting processes are used primarily for severing carbon and low-alloy steels. Other iron-base alloys and some nonferrous metals can be oxyfuel gas cut, although process modification may be required, and cut quality may not be as high as is obtained in cutting the more widely used grades of steel. High-alloy steels, stainless steels, cast iron, and nickel alloys do not readily oxidize and therefore do not provide enough heat for a continuous reaction. As the carbon and alloy contents of the steel to be cut increase, preheating or postheating, or both, often are necessary to overcome the effect of the heat cycle, particularly the quench effect of cooling.

Some of the high-alloy steels, such as stainless steel, and cast iron can be cut successfully by injecting metal powder (usually iron) or a chemical additive into the oxygen jet. The metal powder supplies combustion heat and breaks up oxide films. Chemical additives combine with oxides to form lower-temperature-melting products that flush away.

Applications. Large-scale applications of oxyfuel cutting are found in shipbuilding, structural fabrication, manufacture of earth-moving equipment, machinery construction, and in the fabrication of pressure vessels and storage tanks. Many machine structures, originally made from forgings and castings, can be made at less cost by redesigning them for OFC and welding with the advantages of quick delivery of plate material from steel suppliers, low cost of oxyfuel gas cutting equipment, and flexibility of design.

Structural shapes, pipe, rod, and similar materials can be cut to length for construction or cut up in scrap and salvage operations. In steel mills and foundries, projections such as caps, gates, and risers can be severed from billets and castings. Mechanical fasteners can be quickly cut for disassembly using OFC. Holes can be made in steel components by piercing and cutting. Machine OFC is used to cut steel plate to size, to cut various shapes from plate, and to prepare plate edges (bevel cutting) for welding.

Gears, sprockets, handwheels, clevises and frames, and tools such as wrenches can be cut out of wrought materials by oxyfuel gas torches. Often, these oxyfuel-cut products can be used without further finishing. However, when cutting medium- or high-carbon steel or other metal that hardens by rapid cooling, the hardening effect must be considered, especially if the workpiece is to be subsequently machined.

Thickness Limits. Gas can cut steels less than 3.2 mm ($^1/_8$ in.) thick to over 1525 mm (60 in.) thick, although some sacrifice in quality occurs near both ends of this range. With very thin material, operators may have some difficulty in keeping heat input low to avoid melting the kerf edges and to minimize distortion. Steel under 6.4 mm ($^1/_4$ in.) thick often is stacked for cutting of several parts in a single torch pass.

There are a number of advantages and disadvantages when OFC is compared to other cutting operations such as arc cutting, milling, shearing, or sawing. The advantages of OFC are:

- Metal can be cut faster by OFC. Setup is generally simpler and faster than for machining and approximatley equal to that of mechanical severing (sawing and shearing).
- Oxyfuel gas cutting patterns are not confined to straight lines as in sawing and shearing, or to fixed patterns as in die cutting processes. Cutting direction can be changed rapidly on a small radius during operation.
- Manual OFC equipment costs are low compared to machine tools. Such equipment is portable and self-contained, requiring no outside power and well suited for field use.
- When properties and dimensional accuracy of gas cut plate are acceptable, OFC can replace costly machining operations. It offers reduced labor and overhead costs, reduced material costs, reduced tooling costs, and faster delivery.
- With advanced machinery, OFC lends itself to high-volume parts production.
- Large plates can be cut quickly in place by moving the gas torch rather than the plate.
- Two or more pieces can be cut simultaneously using stack cutting methods and multiple-torch cutting machines.

The limitations of the OFC process include:

- Dimensional tolerances are poorer than for machining and shearing.
- Because OFC relies on oxidation of iron, it is limited to cutting low-alloy steels without the use of powder or flux additions to the cutting oxygen.
- Heat generated by OFC can degrade the metallurgical properties of the work material adjacent to the cut edges. Hardenable steels may require preheat and/or postheat to control

microstructural and mechanical properties and to avoid cracking.
- Preheat flames and the expelled red hot slag pose a very real fire hazard to plant and personnel.

Oxygen consumption and flow rates vary depending on whether economy, speed, or accuracy of cut is desired. For average straight-line cutting of low-carbon steel, consumption of cutting oxygen per pound of metal removed varies with thickness of the metal and is lowest at a thickness of 100 to 125 mm (4 to 5 in.).

By assuming that for every unit mass of iron oxidized an equal mass of iron melts, one can calculate the amount of heat generated by the cutting reaction—heat emitted is 6680 kJ/kg (2870 Btu/lb) of iron oxidized. Melting of 0.5 kg (1 lb) of iron takes 715 kJ (680 Btu), based on a melting point of 1540 °C (2800 °F), 0.8 kJ/kg · K (0.2 Btu/lb · °F) as the specific heat, and 272 kJ/kg (117 Btu/lb) as the heat of fusion. Only a small amount of the heat melts the iron; most of it, approximately 4900 kJ/kg (2100 Btu/lb), goes into the reaction. Some of this superheats the molten metal, some soaks into the workpiece, and some leaves by radiation and convection. Most of it leaves with the slag and hot exhaust gases.

As cutting oxygen flows down through the cut, the quantity available for reaction decreases. If the flow of oxygen is large and well collimated, the rate of cutting through the depth of the cut is approximately constant. The cutting face remains vertical if the oxygen is in excess and if cutting speed is not excessive.

If oxygen flow is insufficient, or cutting speed too high, the lower portions of the cut react more slowly, and the cutting face curves behind the torch. The horizontal distance between point of entry and exit is called drag (Fig. 1).

Drag influences edge quality. Optimal edge quality results from zero drag—the oxygen stream enters and leaves the cut in a straight line along the cutting tip axis. This is called a drop-cut, designating a clean, fully severed edge. Increasing cutting speed or reducing oxygen flow makes less oxygen available at the bottom of the cut, causing the bottom of the cut to "drag" behind the top of the cut. A drag of 20% means that the bottom of the cut edge lags the top surface by 20% of the material thickness. Drag lines appear as curved ripples on the cut edge. For fast, rough cuts, some drag is acceptable.

Drag is a rough measure of cut quality and of economy in oxygen consumption. In metal thicknesses up to 50 to 75 mm (2 or 3 in.), 10 to 15% drag indicates good quality of cut and economy. The key to quality cuts is control of heat input. Higher quality demands less drag; more drag indicates poorer quality and low oxygen consumption. Excessive drag may lead to incomplete cutting.

In very thin sections, drag has little significance. In very thick sections, the goal is to avoid excessive drag.

Preheating may consist of merely warming a cold workpiece with a torch or may require furnace heating of the work beyond 540 °C (1000 °F). For some alloy steels, preheat temperatures are 205 to 315 °C (400 to 600 °F). Carbon steel billets and other sections occasionally are cut at 870 °C (1600 °F) and higher. As is the case in welding, the carbon equivalent of the steel being cut is a major factor in determining the need for preheating.

In OFC, preheating is accomplished by means of the oxyfuel gas flame, which surrounds the cutting-oxygen stream. At cut initiation, the preheat flame, the result of oxygen and fuel gas combustion, brings a small amount of material to ignition temperature so that combustion can proceed. After cutting begins, the preheat flame merely adds heat to compensate for heat lost by convection and radiation or through gas exhausted during cutting. The flame also helps to remove or burn off scale and dirt on the plate surface; the hot combusted gases protect the stream of cutting oxygen from the atmosphere.

Preheating may also be applied over a broader area of the work. It may include soaking the entire workpiece in a furnace to bring it up to 95 to 205 °C (200 to 400 °F), or a simple overall warmup with a torch to bring cold plate to room temperature. A preheat improves cutting speed significantly, allowing faster torch travel for greater productivity and reduced consumption of fuel gas. Broader preheat smooths the temperature gradient between the base metal and the cut edge, possibly reducing thermal stress and minimizing hardening effects in some steels.

Properties of Fuel Gases

Each cutting job entails a different type or volume of work to be completed. Consequently, the best gas for all cutting in a fabricating plant is found through experimentation. Evaluating a gas for a single job requires a test run that monitors fuel gas and oxygen flow rate, labor costs, overhead, and the amount of work performed. If plant production varies from week to week, gas performance should be measured over a long enough period to achieve an accurate cost

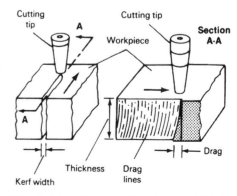

Fig. 1 Schematic cross section of work metal during oxyfuel gas cutting showing drag on cutting face

analysis. Any of the fuel gases may perform well over a range of flow rates. When comparing gases, performance should be rated at the lowest flow rate that gives acceptable results for each gas.

Gas manufacturers provide comprehensive data on flame temperature, heat of combustion, oxygen-to-gas consumption ratio, heat transfer, and heat distribution in the flame. Gas properties can be rated on a weight or volume basis. Flow rates, for example, commonly are given in terms of volume. Gas may be sold, however, by weight or by volume, and specifications may state properties in terms of one or the other to make the product seem as attractive as possible. For example, acetylene occupies more volume per pound than does propane. Consequently, cost based on weight makes propane liquid seem inexpensive. Cost based on volume makes the lighter fuels, such as acetylene, seem less costly. The user must understand these ratings and comparisons.

Heat Distribution in the Flame. Combustion of fuel gas produces the preheat flame, which initiates cutting action and helps sustain the operation. These flames heat the surface of the work to ignition temperature to (1) initiate cutting, (2) descale and clean the work surface, (3) supply heat to the work and the cutting-oxygen stream to maintain the heat needed for continuous cutting, and (4) shield the cutting-oxygen stream from surrounding air. Preheat gases consist of hydrocarbons, which produce water vapor and carbon dioxide as products of chain reactions. These reactions occur in cones within the flame, usually visible as an inner and outer flame. A gas whose inner flame has a high temperature and high heat release provides the most concentrated heat. These gases are superior for fast starts in flame cutting of high-alloy steels that are difficult to cut.

When low heat release is accompanied by high flame temperature, even though it is above the melting point of steel, heat is diffused and gives slow starts in flame cutting. Gases, such as natural gas, that release most of their heat in the outer flame are well suited to heating and heavy cutting. Heat distribution is a good indication of the potential performance of a particular gas.

Fuel Types

Acetylene (C_2H_2) is the most widely used cutting gas. It burns hotter than any of the other common fuel gases, making it indispensable for certain jobs. Despite some disadvantages, acetylene has been used for cutting for a longer time than any other gas. Its performance is well understood, equipment for its use is perfected and widely marketed, and it is readily available. It has become the standard against which all other gases are compared.

Acetylene combustion produces a hot, short flame with a bright inner cone at each cutting-tip port; the hottest point is at the tip of this inner cone. Combustion starts in the inner cone and is brought to completion in a cooler, blue, outer flame. The sharp distinction between the two flames helps the cutting operator to adjust the ratio of oxygen to acetylene.

Depending on this ratio, the flame may be carburizing (reducing), neutral, or oxidizing. A neutral flame results when just enough oxygen is supplied for primary combustion, yielding carbon monoxide (CO) and hydrogen (H_2). These products then combine with oxygen in ambient air to form the blue, outer flame, yielding carbon dioxide (CO_2) and water (H_2O). The neutral ratio of oxygen to acetylene is approximately 1 to 1, and the flame temperature at the tip of the inner cone is approximately 3040 °C (5500 °F). This flame is used for manual cutting.

When the oxygen-to-acetylene ratio is reduced to approximately 0.9 to 1, a bright streamer begins to appear, and the flame becomes carburizing, or reducing. A carburizing flame is sometimes used for rough cutting of cast iron. When the oxygen-to-acetylene ratio is increased to more than 1 to 1, the inner cones are shorter, "necked in" at the sides, and more sharply defined; this flame is oxidizing. Flame temperature increases until, at a ratio of approximately 1.7 to 1, the temperature is maximum, or somewhat over 3095 °C (5600 °F) at the tip of the cones. An oxidizing flame can be used for preheating at the start of the cut, and for cutting very thick sections.

According to the reaction:

$$2C_2H_2 + 5O_2 \rightarrow 4CO_2 + 2H_2O$$

an oxygen-to-acetylene ratio of 2.5 to 1 is required for a complete reaction. Of the 2.5 parts of oxygen needed for acetylene to burn completely, approximately 1.5 parts come from the air and 1 part from line oxygen. Total consumption of the line oxygen is relatively low, an advantage of acetylene over all other fuel gases. Operation of any oxyfuel equipment in confined spaces, such as the inside of a closed tank or vessel, requires strict safety considerations. Normally forced ventilation (both into and out of the enclosed space) to supply the additional air needed for both breathing and for flame combustion is provided.

Combustion occurs in stages. In the small inner cone at the tip of the torch, acetylene burns with feed-line oxygen. This reaction gives off a blistering amount of heat; the tip of the inner cone is the hottest part. Combustion starts in the inner cone and is brought to completion in a cooler, blue, outer flame. The sharp distinction between the two flames helps the cutting operator to adjust the ratio of oxygen to acetylene.

A neutral flame, recommended for manual cutting, consumes equal volumes of line oxygen and acetylene. In practice, operators may raise the line oxygen ratio to 1.5, a proportion that produces an oxidizing flame. The inner cone contracts and becomes sharply defined. This is the hottest flame attainable with acetylene or with any other raw fuel gas. The triple bond in acetylene makes it the hottest burning gas. In cutting of steel, operators should adjust the

acetylene-to-oxygen ratio to produce a neutral flame, even though it is below the highest possible temperature. A neutral flame minimizes oxidation and carburization of the base metal. Users sometimes use the hotter oxidizing flame for piercing at the start of a cut.

Acetylene is unstable at room temperature and moderate pressure and can explode under a blow even without the presence of air or oxygen. For safe storage and handling, it is dissolved in acetone. A typical cylinder is filled with heavy porous clay, which holds the acetone. Acetylene, forced into the cylinder, saturates the acetone. When the cylinder valve opens, acetylene vaporizes above the acetone solution. The regulator limits outlet pressure to 105 kPa gage pressure (15 psig) or less.

Acetylene must be used at pressures below 105 kPa (15 psi), which is a stable operating range. Safety codes specify equipment and handling practices for acetylene. When supplied in special cylinders, acetylene is dissolved in acetone, which is contained in a porous mass that fills the cylinder. This technique eliminates the sensitivity of acetylene at pressures over 105 kPa (15 psi). Such cylinders can be filled to pressures exceeding 105 kPa (15 psi) but not greater than 1725 kPa (250 psi). Acetylene may also be supplied from generators. With either means of supply, strict safety regulations must be observed to avoid sudden decomposition and explosion.

Cylinders for acetylene should be handled with care. Safety requires them to be placed upright to prevent acetone solvent from entering the regulator. The ratio of line oxygen consumption to acetylene consumption is lower than that for the other fuel gases. Acetylene uses less oxygen at its maximum flame temperature, so that handling of oxygen cylinders may be reduced, thus offsetting the inconvenience of the acetylene cylinder.

Cost of acetylene as a fuel gas changes with consumption. Large-volume users of acetylene often install bulk trailers that hook up to manifold piping systems. When volume justifies, users may install acetylene-generating plants, thus reducing handling costs.

Heat content of acetylene (kJ/m^3, or Btu/ft^3) is lower than all gases except natural gas. Although acetylene burns at a high temperature, it has to burn at higher flow rates than other gases to deliver the same amount of heat. Acetylene releases heat rapidly in a small, concentrated area. Acetylene is a poor choice for cutting a large block of metal, because of high fuel cost and the need for large volumes of the fuel to obtain required total heat. To concentrate heat on a limited area, as in cutting thin plate, acetylene is a good choice. It burns faster than other gases, and it burns close to the torch tip. Heat energy concentrates at the tip of the tiny cone.

Acetylene gives a narrow heat-affected zone (HAZ) and possibly less distortion (depending on the volume and shape of the work) than a wider-spreading heat. It is particularly well suited to fast cutting of plate under 13 mm

($1/2$ in.) thick. The hot flame cuts through heavy mill scale or rust and can make a bevel quickly with good edge quality. Acetylene facilitates short, stop-and-start cutting jobs, such as cutting structural members and reinforcing bars, because of its short preheat time. For stack cutting and for cutting heavy sections—operations requiring large heat input into a deep kerf—acetylene does not perform as well as other gases. This type of work calls for a fuel gas with high heat output in the secondary flame, such as propane or methylacetylene-propadiene-stabilized gas.

However, manual cutting is cooler with acetylene than with other gases; less heat rises into the operator's face. It is one of the most versatile and convenient fuels for shops that perform a wide range of fuel gas operations and for shops that cannot justify handling more than one gas. Data for both machine and manual oxyacetylene cutting of carbon steel plate of varying thickness are given in Tables 1 and 2.

Natural gas is a mixture of gases, depending on the composition at the well. One source defines the most widely used mixture as 85% methane (CH_4), 4% ethane (C_2H_6), and 11% other (N_2, H_2, O_2, H_2O). Some wells produce natural gas with large proportions of ethane and propane. However, the main component is methane (CH_4) and is therefore usually given the chemical symbol for methane (CH_4).

The chemical reaction for complete combustion is:

$$CH_4 + 2O_2 \rightarrow CO_2 + 2H_2O$$

This indicates an oxygen-to-methane ratio of 2 to 1; this ratio is used for the preheat flame. Maximum flame temperature at the tip of the inner cones is approximately 2760 °C (5000 °F). Both higher and lower temperatures have been reported; also, the optimal oxygen-to-gas ratio is approximately 2 to 1.

Although available in pressurized cylinders, natural gas usually is supplied through low-pressure lines from a local utility. Torchmakers overcome low line pressure through torch design. A siphon stream in the mixing chamber pulls the gas in at the required rate for complete combustion. Torches operate at pressures as low as 2 kPa (4 ozf/in.^{2}) gage line pressure.

A natural gas flame is more diffuse than an acetylene flame; heat intensity is lower; and adjustment for carburizing, neutral, and oxidizing flame is less clearly defined. Initial cutting speeds are slower, and oxygen consumption is greater. Also, more time is required for preheating with natural gas than with acetylene. An excess of oxygen shortens preheat time but increases consumption of oxygen. Neither acetylene nor natural gas accumulates in low pockets. When burned alone in air, the flame of natural gas does not produce soot, as is the case with acetylene.

Table 1 Recommended parameters for manual oxyacetylene cutting of carbon steels

| Plate thickness | | Diameter of cutting orifice(a) | | Oxygen pressure(a) | | Cutting speed(a)(b) | | Gas consumption | | | | | |
| | | | | | | | | Oxygen | | | Acetylene(c) | | |
mm	in.	mm	in.	kPa	psi	mm/min	in./min	m³/h	ft³/h	ft³/linear ft	m³/h	ft³/h	ft³/linear ft
3.2	1/8	0.965–1.016	0.0380–0.0400	103–159	15–23	510–760	20–30	1.3–1.6	45–55	0.37–0.45	0.20–0.25	7–9	0.06–0.07
6.4	1/4	0.965–1.511	0.0380–0.0595	76–138	11–20	405–660	16–26	1.4–2.6	50–93	0.63–0.72	0.25–0.31	9–11	0.08–0.11
9.5	3/8	0.965–1.511	0.0380–0.0595	117–173	17–25	380–610	15–24	1.7–3.26	60–115	0.80–0.96	0.28–0.34	10–12	0.10–0.13
13	1/2	1.181–1.511	0.0465–0.0595	138–207	20–30	305–560	12–22	1.9–3.54	66–125	1.10–1.14	0.28–0.37	10–13	0.12–0.17
19	3/4	1.181–1.511	0.0465–0.0595	165–241	24–35	305–510	12–20	3.31–4.04	117–143	1.43–1.95	0.34–0.42	12–15	0.15–0.20
25	1	1.181–1.511	0.0465–0.0595	193–276	28–40	230–460	9–18	3.7–4.5	130–160	1.78–2.89	0.37–0.45	13–16	0.18–0.29
38	1 1/2	1.511–2.057	0.0595–0.0810	241–331	35–48	150–355	6–14	4.05–5.04	143–178	1.96–3.18	0.42–0.51	15–18	0.21–0.33
50	2	1.702–2.057	0.0670–0.0810	152–345	22–50	150–330	6–13	5.23–6.54	185–231	3.55–6.16	0.45–0.57	16–20	0.31–0.53
75	3	1.702–2.057	0.0670–0.0810	228–379	33–55	100–255	4–10	6.8–8.2	240–290	5.80–12.00	0.54–0.65	19–23	0.46–0.95
100	4	2.057–2.184	0.0810–0.0860	290–414	42–60	100–205	4–8	8.29–10.9	293–388	9.70–14.64	0.59–0.74	21–26	0.65–1.05
125	5	2.057–2.184	0.0810–0.0860	365–483	53–70	90–165	3.5–6.4	9.82–12.4	347–437	13.66–19.83	0.68–0.82	24–29	0.91–1.37
150	6	2.489–2.527	0.0980–0.0995	310–552	45–80	76–137	3.0–5.4	11.3–16.1	400–567	21.00–26.70	0.76–0.91	27–32	1.19–1.80
205	8	2.527	0.0995	414–531	60–77	66–107	2.6–4.2	14.3–17.4	505–615	29.30–38.84	0.892–1.09	31.5–38.5	1.83–2.42
255	10	2.527	0.0995	517–662	75–96	48–81	1.9–3.2	17.3–21.2	610–750	46.90–64.20	1.04–1.28	36.9–45.1	2.57–3.84
305	12	3.048	0.1200	476–593	69–86	36–66	1.4–2.6	20.4–24.9	720–880	67.70–103.00	1.20–1.46	42.3–51.7	3.98–6.05

(a) Values do not necessarily vary in exact proportion to plate thickness, because straight-line relations do not exist among pressure, speed, and orifice sizes. (b) Lowest speeds and highest gas consumptions are for inexperienced operators, short cuts, dirty or nonuniform material. Highest speeds and lowest gas consumptions are for experienced operators, long cuts, clean and uniform material. (c) Pressure of acetylene for the preheating flames is more a function of torch design than of the thickness of the part being cut. For acetylene pressure data, see charts from manufacturers of apparatus.

Table 2 Recommended parameters for machine oxyacetylene cutting of carbon steels

| Plate thickness | | Diameter of cutting orifice(a) | | Oxygen pressure(a) | | Cutting speed(b) | | Gas consumption | | | | | |
| | | | | | | | | Oxygen | | | Acetylene(c) | | |
mm	in.	mm	in.	kPa	psi	mm/min	in./mm	m³/h	ft³/h	ft³/linear ft	m³/h	ft³/h	ft³/linear ft
3.2	1/8	0.635–1.016	0.0250–0.0400	103–159	15–23	560–815	22–32	1.13–1.56	40–55	0.34–0.36	0.20–0.25	7–9	0.05–0.06
6.4	1/4	0.787–1.511	0.0310–0.0595	76–241	11–35	510–710	20–28	1.27–2.63	45–93	0.34–0.66	0.23–0.31	8–11	0.07–0.08
9.5	3/8	0.787–1.511	0.0310–0.0595	117–276	17–40	485–660	19–26	2.32–3.26	82–115	0.86–0.89	0.25–0.34	9–12	0.08–0.09
13	1/2	0.787–1.511	0.0310–0.0595	138–379	20–55	430–610	17–24	2.97–3.54	105–125	1.04–1.24	0.28–0.37	10–13	0.11–0.12
19	3/4	0.965–1.511	0.0380–0.0595	165–345	24–50	380–560	15–22	3.31–4.50	117–159	1.45–1.56	0.34–0.42	12–15	0.14–0.16
25	1	1.181–1.511	0.0465–0.0595	193–379	28–55	355–485	14–19	3.68–4.93	130–174	1.83–1.86	0.37–0.45	13–16	0.17–0.19
38	1 1/2	1.702–2.057	0.0670–0.0810	172–379	25–55	305–380	12–15	5.24–6.79	185–240	3.20	0.40–0.51	14–18	0.23–0.24
50	2	1.702–2.057	0.0670–0.0810	152–415	22–60	255–355	10–14	5.24–7.36	185–260	3.70–3.72	0.45–0.57	16–20	0.29–0.32
75	3	2.057–2.184	0.0810–0.0860	228–345	33–50	205–280	8–11	6.79–9.40	240–332	6.00–6.04	0.51–0.65	18–23	0.42–0.45
100	4	2.057–2.184	0.0810–0.0860	290–415	42–60	165–230	6.5–9	8.29–10.9	293–384	8.53–9.02	0.59–0.74	21–26	0.58–0.65
125	5	2.057–2.184	0.0810–0.0860	365–448	53–65	140–190	5.5–7.5	9.82–11.6	347–411	10.97–12.62	0.65–0.82	23–29	0.77–0.84
150	6	2.489–2.527	0.0980–0.0995	310–448	45–65	115–165	4.5–6.5	11.3–13.9	400–490	15.10–17.78	0.74–0.91	26–32	0.98–1.16
205	8	2.489–2.527	0.0980–0.0995	415–620	60–90	94–124	3.7–4.9	14.3–17.7	505–625	25.52–27.30	0.88–1.10	31–39	1.59–1.68
255	10	2.527–2.794	0.0995–0.1100	517–620	75–90	74–102	2.9–4.0	17.3–21.2	610–750	37.50–42.10	1.05–1.27	37–45	2.25–2.55
305	12	2.794–3.048	0.1100–0.1200	476–724	69–105	61–89	2.4–3.5	20.4–24.9	720–880	49.70–60.00	1.19–1.47	42–52	2.97–3.50

(a) Values do not necessarily vary in exact proportion in plate thickness, because straight-line relations do not exist among pressure, speed, and orifice sizes. (b) Lowest speeds and highest gas consumptions are for inexperienced operators, short cuts, dirty or nonuniform material. Highest speeds and lowest gas consumptions are for experienced operators, long cuts, clean and uniform material. (c) Pressure of acetylene for the preheating flames is more a function of torch design than of the thickness of the part being cut. For acetylene pressure data, see charts from manufacturers of apparatus.

Because heat content is low and heat emission is diffuse, natural gas cannot be used to weld steel. Consequently, extra installations are needed if oxyfuel gas welding is to be performed in addition to cutting. Despite these disadvantages, use of natural gas for cutting has increased. It is the lowest-cost commercial fuel gas and, with careful torch adjustment, produces excellent cuts in light- to heavy-gage material. Data for manual oxynatural gas straight-line cutting of carbon steel plate of varying thickness are given in Table 3. Table 4 gives data for oxynatural gas shape cutting of carbon steel plate. Table 5 gives data for machine oxynatural gas drop cutting of shapes from thick carbon steel plate.

Propane (C_3H_8) is a petroleum-based fuel usually supplied as a liquid in storage tanks from which it is drawn off as a gas. The gas is dispensed from bulk storage tanks through pipelines. It has a narrow range of flammability and is relatively stable. However, propane is heavier than air and can therefore collect in low-lying areas, resulting in a fire or explosion hazard. Complete combustion requires an oxygen-to-propane ratio of 5 to 1. However, approximately 30% of the oxygen needed is taken from the ambient air. When the ratio of oxygen to propane is 4.5 to 1, the flame temperature is approximately 2760 °C (5000 °F) at the tip of the inner cones. At 4.25 to 1, the flame temperature is approximately 2650 °C (4800 °F). Flame properties are similar to those of natural gas, with respect to diffuseness, heat intensity, flame adjustment, and cutting speed. When burned alone in air, the flame is soot-free.

Propylene is a liquefied gas similar to propane. It has a higher flame temperature than propane. The flame temperature of propylene is approximately equal to that of methylacetylene-propadiene-stabilized (MPS) gas, although its heat content is slightly less. However, propylene is also heavier than air and can therefore collect in low-lying areas, resulting in a fire or explosion hazard. On a volume basis, propylene is usually less expensive than acetylene; it does, however, consume more oxygen during combustion. The combustion reaction for propylene is:

$$2C_3H_6 + 9O_2 \rightarrow 6CO_2 + 6H_2O$$

The combustion ratio for propylene is 4.5 to 1. Line oxygen for a neutral flame is approximately 3.5 to 1. Distributors sell propylene under various trade names, either pure or as improved mixtures with propane and other hydrocarbon additives.

Methylacetylene-Propadiene-Stabilized Gas. The common trade name of MPS gas is MAPP gas. Methylacetylene-propadiene-stabilized mixtures are by-products of the manufacture of chemicals such as ethylene. These mixtures combine the qualities of an acetylene flame with a more even heat distribution in a fuel that is less prone to explosion and less costly than acetylene. It is

Table 3 Recommended parameters for manual oxynatural gas straight-line cutting of carbon steels

Plate thickness		Diameter of cutting orifice(a)		Oxygen pressure		Minimum natural gas pressure		Cutting speed(b)		Approximate gas consumption			
										Oxygen		Natural gas	
mm	in.	mm	in.	kPa	psi	kPa	psi	mm/min	in./min	m³/h	ft³/h	m³/h	ft³/h
3.2	1/8	1.17	0.046	103	15	21	3	510–710	20–28	0.71	25	0.23	8
6.4	1/4	1.17	0.046	124	18	21	3	455–710	18–28	0.99	35	0.28	10
9.5	3/8	1.17	0.046	138	20	21	3	405–510	16–20	1.27	45	0.40	14
13	1/2	1.50	0.059	207	30	21	3	330–430	13–17	1.42	50	0.51	18
19	3/4	1.50	0.059	241	35	21	3	255–380	10–15	2.12	75	0.57	20
25	1	1.50	0.059	275	40	21	3	230–330	9–13	2.83	100	0.68	24
38	1 1/2	1.70	0.067	275	40	21	3	180–305	7–12	4.10	145	0.74	26
50	2	1.70	0.067	310	45	21	3	150–255	6–10	5.38	190	0.79	28
64	2 1/2	1.70	0.067	345	50	21	3	150–230	6–9	6.94	245	0.85	30
75	3	2.36	0.093	345	50	21	3	125–205	5–8	7.64	270	0.91	32
100	4	2.36	0.093	379	55	21	3	125–180	5–7	9.06	320	1.02	36
125	5	2.36	0.093	414	60	21	3	100–150	4–6	11.3	400	1.13	40
150	6	2.79	0.110	414	60	21	3	100–150	4–6	13.3	470	1.36	48
180	7	2.79	0.110	483	70	21	3	75–125	3–5	14.7	520	1.47	52
205	8	2.79	0.110	552	80	21	3	75–100	3–4	16.4	580	1.59	56
255	10	2.79	0.110	620	90	21	3	75–100	3–4	24.1	850	1.70	60
305	12	2.79	0.110	690	100	21	3	50–75	2–3	28.3	1000	1.81	64

(a) Using injector-type torch and two-piece tips. (b) Variations in cutting speeds may be caused by mill scale on plate, variation in oxygen purity, flame adjustment, condition of equipment, impurities in steel, and variation in heat content of natural gas.

Table 4 Recommended parameters for machine oxynatural gas shape cutting of carbon steels

Plate thickness		Diameter of cutting orifice(a)		High preheat				Low preheat						Natural gas		Cutting speed(b)		Width of kerf (approx)	
				Oxygen		Natural gas		Oxygen				Natural gas							
mm	in.	mm	in.	kPa	psi	kPa	psi	kPa	psi	Pa	ozf/in.²	kPa	psi	kPa	psi	mm/min	in./min	mm	in.
6.4	1/4	0.91	0.036	193	28	21	3	48–83	7–12	430	1	482–517	70–75	455–710	18–28	2.0	0.08		
9.5	3/8	0.94	0.037	193	28	21	3	48–83	7–12	430	1	482–551	70–80	455–660	18–26	2.0	0.08		
13	1/2	0.99	0.039	234	34	21	3	48–83	7–12	430	1	517–551	75–80	405–610	16–24	2.3	0.09		
16	5/8	0.99	0.039	234	34	21	3	48–83	7–12	430	1	517–551	75–80	405–585	16–23	2.3	0.09		
19	3/4	1.17	0.046	317	46	21	3	55–83	8–12	430	1	551–586	80–85	405–560	16–22	2.5	0.10		
25	1	1.17	0.046	331	48	21	3	55–83	8–12	430	1	551–586	80–85	355–510	14–20	2.5	0.10		
32	1 1/4	1.37	0.054	303	44	21	3	55–83	8–12	430	1	551–586	80–85	355–455	14–18	3.0	0.12		
38	1 1/2	1.37	0.054	303	44	21	3	55–83	8–12	430	1	620–655	90–95	330–455	13–18	3.0	0.12		
44	1 3/4	1.37	0.054	310	45	21	3	55–83	8–12	430	1	620–655	90–95	305–430	12–17	3.0	0.12		
50	2	1.37	0.054	310	45	21	3	55–83	8–12	430	1	690	100	255–380	10–15	3.0	0.12		
57	2 1/4	1.40	0.055	317	46	21	3	55–83	8–12	430	1	690	100	230–380	9–15	3.3	0.13		
64	2 1/2	1.40	0.055	317	46	21	3	55–83	8–12	430	1	690	100	205–355	8–14	3.3	0.13		
70	2 3/4	1.70	0.067	317	46	21	3	55–83	8–12	430	1	690	100	205–330	8–13	3.6	0.14		
75	3	1.70	0.067	317	46	21	3	55–83	8–12	430	1	690	100	180–330	7–13	3.6	0.14		
89	3 1/2	1.85	0.073	345	50	21	3	69–97	10–14	430	1	724	105	150–305	6–12	3.8	0.15		
100	4	1.85	0.073	345	50	21	3	69–97	10–14	430	1	758	110	150–280	6–11	3.8	0.15		
114	4 1/2	2.08	0.082	345	50	21	3	69–97	10–14	430	1	758	110	125–255	5–10	4.3	0.17		
125	5	2.08	0.082	345	50	21	3	69–97	10–14	430	1	793	115	125–255	5–10	4.3	0.17		
140	5 1/2	2.44	0.096	345	50	21	3	69–97	10–14	430	1	793	115	125–230	5–9	4.6	0.18		
150	6	2.44	0.096	345	50	21	3	69–97	10–14	430	1	827	120	125–230	5–9	4.6	0.18		
165	6 1/2	2.44	0.096	345	50	21	3	69–97	10–14	430	1	827	120	100–205	4–8	4.6	0.18		
190	7 1/2	2.44	0.096	345	50	21	3	69–97	10–14	430	1	827	120	100–205	4–8	4.6	0.18		
205	8	2.44	0.096	345	50	21	3	69–97	10–14	430	1	827	120	75–180	3–7	4.6	0.18		

(a) Two-piece tips for high-speed machine cutting. (b) Variations in cutting speed may be caused by mill scale on plate, variation in oxygen purity, flame adjustment, condition of equipment, impurities in steel, and variation in heat content of natural gas.

supplied as a liquid, in large tanks or in portable cylinders.

Methylacetylene-propadiene-stabilized gases contain a mixture of several hydrocarbons, including propadiene (allene), propane, butane, butadiene, and methylacetylene. Methylacetylene, like acetylene, is a high-energy triple-bond compound. It is unstable, but other compounds in the mixture dilute it sufficiently to enable safe handling.

Compositions of MPS mixtures are proprietary and may vary; consequently, an exact combustion equation cannot be specified. The mixture burns hotter than propane or propylene. It also affords a high release of heat energy in the primary flame cone, characteristic of acetylene. The outer flame gives relatively high heat release, similar to propane and propylene. The overall heat distribution in the flame is the most even of any of the other gases. The inner cone releases 19.3 MJ/m^3 (517 Btu/ft^3), and the outer flame releases 70.37 MJ/m^3 (1889 Btu/ft^3). The coupling distance is therefore less exacting than for acetylene. The best coupling distance for MPS fuel places the outer cone on the plate; however, a shorter coupling distance also delivers considerable heat.

The neutral MPS gas-oxygen flame generates 89.62 MJ/m^3 (2406 Btu/ft^3) with a 2925 °C (5300 °F) flame with a ratio of 3.5 to 4 parts oxygen to 1 part fuel. Values vary with the composition of the gas. The carburizing flame, 2.2 parts or less oxygen to 1 part fuel, can weld alloys, such as aluminum, that oxidize readily. A neutral flame, with a ratio of 2.3 parts line oxygen to 1 part fuel gas, can weld steel. At 2.8 line oxygen ratio, the flame becomes oxidizing, unsuitable for welding. Methylacetylene-propadiene-stabilized gas produces its hottest flame at 3.3 line oxygen-to-gas ratio.

In comparing the cost of MPS gas with the cost of acetylene, differences in cylinder yield and consumption rate must be considered. A 54 kg (120 lb) cylinder of MPS gas yields 17.5 m^3 (620 ft^3) of gas; a 115 kg (240 lb) cylinder of acetylene yields only 7.4 m^3 (260 ft^3) of gas. In addition, acetylene burns faster. Thus, storage, transportation, and time and labor for changing cylinders become important cost factors.

Methylacetylene-propadiene-stabilized gas competes with acetylene for almost every job that uses fuel gas. Its most unusual use perhaps is in deep-water cutting. Because acetylene outlet pressure is limited to 105 kPa gage pressure (15 psig), it cannot be used below 9 m (30 ft) of water. The Navy and several shipyards use MPS gas for underwater work. Table 6 provides data for oxy/MPS gas cutting of carbon steel plate.

Effect of Oxyfuel Cutting on Base Metal

During cutting of steel, the temperature of a narrow zone adjacent to the cut face is raised considerably above the transformation range. As the cut progresses, the steel cools back through this range. Cooling rate depends on the heat conductivity, on the mass of the surrounding material, on loss of heat by radiation and convection, and on speed of cutting. When steel is at room temperature, the rate of cooling at the cut is sufficient to produce a quenching effect on the cut edges, particularly in heavier cuts in large masses of cold metal. Depending on the amount of carbon and alloying elements present (i.e., the carbon equivalent), and on the rate of cooling, pearlitic steel transforms into microstructures ranging from acicular or spheroidized carbides in ferrite to the much harder bainitic or martensitic constituents. The HAZ may be 0.8 to 6.4 mm (^{1}/$_{32}$ to ^{1}/$_4$ in.) deep for steels 9.5 to 150 mm (^{3}/$_8$ to 6 in.) thick. Approximate depths of the HAZ in oxyfuel gas cut carbon steels are given in Table 7. Some increase in hardness usually occurs at the outer margin of the HAZ of nearly all steels.

Carbon and Low-Alloy Steels

Low-Carbon Steel. For steels containing 0.25% C or less cut at room temperature, the hardening effect usually is negligible, although at the upper carbon limit it may be significant if subsequent machining is required. Short of preheating or annealing the workpiece, hardening may be lessened by ensuring that (1) the cutting flame is neutral to slightly oxidizing, (2) the

flame is burning cleanly, and (3) the inner cones of the flame are at the correct height. By increasing the machining allowance slightly, the first cut usually can be made deep enough to penetrate below the hardened zone in most steels. Mechanical properties of low-carbon steels generally are not adversely affected by OFC. Typical data for OFC of low-carbon steel plate are given in Tables 1 through 6.

Medium-Carbon Steels. Steels having carbon contents of 0.25 to 0.45% are affected only slightly by hardening caused by OFC. Up to 0.30% C, steels with very low alloy content show some hardening of the cut edges but generally not enough to cause cracking. Over 0.35% C, preheating to 260 to 315 °C (500 to 600 °F) is needed to avoid cracking. All medium-carbon steels should be preheated if the gas cut edges are to be machined and may require additional postheating as well.

High-Carbon and Alloy Steels. Oxyfuel gas cutting of steels containing over 0.45% C and of hardenable alloy steels at room temperature may produce a thin layer of hard, brittle material on the cut surface that may crack from the stress of cooling. Preheating and annealing may alleviate hardening and the formation of residual stress.

Preheating to 260 to 315 °C (500 to 600 °F) is sufficient for high-carbon steels; alloy steels may require preheating as high as 540 °C (1000 °F). Preheat temperature should be maintained during cutting. Thick preheated sections should be cut as soon as possible after the piece has been withdrawn from the furnace.

Postheating and annealing also controls the effects of gas cutting in carbon and low-alloy steels. Postheating can reduce the hardness and strength level of any bainitic or martensitic microstructures formed during the cutting process as well as relieve locked-in residual stresses. Postheating immediately after oxyfuel cutting can also reduce the tendency for cracking, especially for cast irons and for alloy steels with medium-to-high carbon content. Annealing restores the original structure of the steel, whether it be predominantly pearlitic or predominantly ferritic with spheroidized carbide, and it also provides stress relief. Many steels do not require annealing if they have been properly preheated or preheated and postheated. (See *Heat Treating*, Volume 4 of *ASM Handbook*, 1991, for annealing practices for specific steels.)

Local annealing is a localized postheat treatment that can be used to prevent hardening or to soften an already hardened cut surface. Either the preheating flame of the cutting torch or a special oxyfuel heating torch may be used for local annealing, as well as electrical resistance heating elements and quartz lamps, depending on the mass of the workpiece and the area to be covered. The heat-affected portion of the workpiece should be heated uniformly, and the temperature gradient at the boundary of the heated mass should be gradual enough to avoid distortion of the workpiece.

Local annealing is not a substitute for preheating; it cannot correct damage done

Table 5 Typical parameters for machine oxynatural gas drop cutting of shapes from low-carbon steel plate 255 to 545 mm (10 to 21^1/$_2$ in.) thick

Torch constructed with two-piece recessed tips that have milled preheat flutes (heavy preheat) and straight-bore cutting-oxygen orifices

| Plate thickness | | Cutting orifice diameter | | Preheat | | | | Cutting oxygen(a) | | Cutting speed | | Cutting oxygen | |
| | | | | Oxygen | | Natural gas | | | | | | | |
mm	in.	mm	in.	kPa	psi	kPa	psi	kPa	psi	mm/min	in./min	m^3/h	ft^3/h
255	10	6.35	0.250	240–310	35–45	34	5	186	27	82.6	3.25	39.6	1400
325	12^3/$_4$	7.14	0.281	240–310	35–45	34	5	193	28	82.6	3.25	49.5	1750
395	15^1/$_2$	7.14	0.281	240–310	35–45	34	5	207	30	76.2	3.0	53.8	1900
455	18	7.92	0.312	240–310	35–45	34	5	193	28	108	4.25	63.6	2245
535	21	7.92	0.312	240–310	35–45	34	5	193	28	76.2	3.0	63.6	2245
545	21^1/$_2$	7.92	0.312	240–310	35–45	34	5	207	30	95.3	3.75	67.0	2365

(a) Pressure measured at torch inlet

Table 6 Recommended parameters for oxy/methylacetylene-propadiene-stabilized (MPS) gas cutting of carbon steels

Plate thickness mm	in.	Cutting tip no.(a)	Cutting speed mm/min	in./min	Oxygen Cutting pressure(b) kPa	psi	Cutting rate of flow m³/h	ft³/h	Preheat pressure kPa	psi	Preheat rate of flow m³/h	ft³/h	MPS gas Cutting pressure kPa	psi	Cutting rate of flow m³/h	ft³/h	Kerf width mm	in.
Cutting with standard-pressure tips																		
3.2	1/8	75	760–915	30–36	275–345	40–50	0.34–0.42	12–15	35–70	5–10	0.20–0.71	7–25	14–70	2–10	0.06–0.28	2–10	0.64	0.025
4.7	3/16	72	660–815	26–32	275–345	40–50	0.57–0.85	20–30	35–70	5–10	0.20–0.71	7–25	14–70	2–10	0.06–0.28	2–10	0.76	0.03
6.4	1/4	68	610–760	24–30	275–345	40–50	0.85–1.13	30–40	35–70	5–10	0.20–0.71	7–25	14–70	2–10	0.06–0.28	2–10	1.02	0.04
13	1/2	61	560–710	22–28	275–345	40–50	1.56–1.84	55–65	35–70	5–10	0.34–0.71	12–25	14–70	2–10	0.14–0.28	5–10	1.27	0.05
19	3/4	56	405–560	16–22	275–345	40–50	1.70–2.12	60–75	35–70	5–10	0.34–0.71	12–25	14–70	2–10	0.14–0.28	5–10	1.52	0.06
25	1	56	355–510	14–20	275–345	40–50	1.70–2.12	60–75	35–70	5–10	0.34–0.71	12–25	14–70	2–10	0.14–0.28	5–10	1.52	0.06
32	1 1/4	54	330–430	13–17	345–415	50–60	2.97–3.40	105–120	70–140	10–20	0.57–0.99	20–35	14–70	2–10	0.23–0.42	8–15	2.03	0.08
38	1 1/2	54	305–405	12–16	345–415	50–60	2.97–3.40	105–120	70–140	10–20	0.57–0.99	20–35	14–70	2–10	0.23–0.42	8–15	2.03	0.08
50	2	52	255–355	10–14	345–415	50–60	4.10–5.38	145–190	70–140	10–20	0.57–0.99	20–35	14–70	2–10	0.23–0.42	8–15	2.29	0.09
64	2 1/2	48	230–330	9–13	345–415	50–60	5.95–7.50	210–265	70–205	10–30	0.57–1.42	20–50	41–70	6–10	0.23–0.57	8–20	2.54	0.10
75	3	48	205–330	8–13	345–415	50–60	5.95–7.50	210–265	70–205	10–30	0.57–1.42	20–50	41–70	6–10	0.23–0.57	8–20	2.54	0.10
100	4	46	180–305	7–12	415–485	60–70	8.21–9.34	290–330	70–205	10–30	0.71–1.42	25–50	41–70	6–10	0.28–0.57	10–20	3.81	0.15
125	5	46	150–255	6–10	485–550	70–80	9.34–11.46	330–405	70–205	10–30	0.71–1.42	25–50	41–70	6–10	0.28–0.57	10–20	3.81	0.15
150	6	42	125–205	5–8	415–485	60–70	10.62–13.31	375–470	70–205	10–30	0.71–1.42	25–50	41–105	6–15	0.28–0.57	10–20	4.06	0.16
205	8	35	100–180	4–7	415–485	60–70	13.73–16.70	485–590	205–345	30–50	1.13–2.83	40–100	70–105	10–15	0.57–1.27	20–45	4.83	0.19
255	10	30	75–150	3–6	275–485	40–70	14.16–17.69	500–625	205–345	30–50	1.13–2.83	40–100	70–105	10–15	0.57–1.27	20–45	5.08	0.20
305	12	30	75–125	3–5	345–585	50–85	18.26–24.49	645–865	205–345	30–50	1.70–4.25	60–150	70–105	10–15	0.85–1.70	30–60	5.33	0.21
Cutting with high-speed tips																		
3.2	1/8	75	815–965	32–38	415–485	60–70	0.57–0.71	20–25	35–70	5–10	0.20–0.71	7–25	14–70	2–10	0.08–0.28	3–10	0.64	0.025
4.7	3/16	72	710–815	28–32	485–550	70–80	0.85–1.13	30–40	35–70	5–10	0.20–0.71	7–25	14–70	2–10	0.08–0.28	3–10	0.76	0.03
6.4	1/4	68	660–815	26–32	485–550	70–80	1.56–1.84	55–65	35–70	5–10	0.20–0.71	7–25	14–70	2–10	0.08–0.28	3–10	1.27	0.05
13	1/2	61	610–760	24–30	550–620	80–90	2.12–2.70	75–95	35–70	5–10	0.34–0.71	12–25	14–70	2–10	0.14–0.28	5–10	1.52	0.06
19	3/4	56	510–660	20–26	550–620	80–90	3.26–3.68	115–130	35–70	5–10	0.34–0.71	12–25	14–70	2–10	0.14–0.28	5–10	1.78	0.07
25	1	56	455–610	18–24	550–620	80–90	3.26–3.68	115–130	35–70	5–10	0.34–0.71	12–25	14–70	2–10	0.14–0.28	5–10	1.78	0.07
32	1 1/4	54	405–510	16–20	485–550	70–80	4.39–4.81	155–170	70–140	10–20	0.57–0.99	20–35	14–70	2–10	0.23–0.42	8–15	2.03	0.08
38	1 1/2	54	380–485	15–19	550–620	80–90	4.81–5.10	170–180	70–140	10–20	0.57–0.99	20–35	14–70	2–10	0.23–0.42	8–15	2.03	0.08
50	2	52	355–455	14–18	550–620	80–90	6.09–7.22	215–255	70–140	10–20	0.57–0.99	20–35	14–70	2–10	0.23–0.42	8–15	2.29	0.09
64	2 1/2	52	305–430	12–17	550–620	80–90	6.09–7.22	215–255	70–140	10–20	0.57–0.99	20–35	14–70	2–10	0.23–0.42	8–15	2.29	0.09
75	3	48	255–380	10–15	550–620	80–90	9.48–11.32	335–400	70–140	10–20	0.57–0.99	20–35	41–70	6–10	0.28–0.42	10–15	2.54	0.10
100	4	46	230–355	9–14	550–620	80–90	10.61–12.03	375–425	70–140	10–20	0.57–0.99	20–35	41–70	6–10	0.28–0.42	10–15	3.05	0.12

Note: All recommendations are for straight-line cutting with a three-hose torch perpendicular to work. (a) All tips are of design recommended by the supplier. (b) Pressure of cutting oxygen measured at the torch

during cutting, such as upsetting of the metal or cracking at the cut edges. Local annealing is limited to steel plate up to 40 mm (1.5 in.) thick. From 40 to 75 mm (1.5 to 3 in.) thick, heat should be applied to both sides of the plate. This method is not suitable for thicknesses over 75 mm (3 in.). If local annealing cannot be done simultaneously with cutting, the cut edges should be tempered after cutting with a suitable heating torch.

Local preheating heats the volume of the workpiece enclosing the HAZ of the cut. If

Table 7 Approximate depths of heat-affected zone (HAZ) in oxygen-cut carbon steels

Plate thickness mm	in.	HAZ depth mm	in.
Low-carbon steels			
<13	<1/2	<0.8	<1/32
13	1/2	0.8	1/32
150	6	3.2	1/8
High-carbon steels			
<13	<1/2	<0.8	<1/32
13	1/2	0.8–1.6	1/32–1/16
150	6	3.2–6.4	1/8–1/4

Note: The depth of the fully hardened zone is considerably less than the depth of the HAZ. For most applications of gas cutting, the affected metal does not have to be removed

the volume of material to be heated is small, the flame of a cutting torch can be used for preheating. When the workpiece is thick and broad, a special heating torch may be necessary. Workpieces must be heated uniformly through the section to be cut, without excessive temperature gradient.

Distortion, which is the result of heating by the gas flame, can cause considerable damage during (1) cutting of thin plate (less than 7.9 mm, or 5/16 in., thick), (2) cutting of long narrow widths, (3) close-tolerance profile cutting, and (4) cutting of plates that contain high residual stresses. The heat may release some of the restraint to locked-in stress or may add new stress. In either case, deformation (warpage) may occur, thus causing inaccurate finished cuts. Plates in the annealed condition have little or no residual stress.

Preheating the workpiece can reduce distortion by reducing differential expansion, thereby decreasing stress gradients. Careful planning of the cutting sequence also may help. For example, when trimming opposite sides of a plate, both sides should be cut in the same direction at the same time. When cutting rings, the inside diameter should be cut first; the remaining plate restrains the material for the outside-diameter cut. In general, the larger portion of material should be used to retain a shape for as long as possible; the cutting sequence should be

balanced to maintain even heat input and resultant residual stresses about the neutral axis of plate or part.

Deformation. In cuts made from large plates, the cutting thermal cycle changes the shape of narrow sections and leaves residual stress in the large section (Fig. 2). The

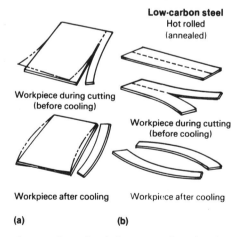

Fig. 2 Effects of oxyfuel gas cutting thermal cycle on shape of sections. (a) Plate with large restraint on one side of kerf, little restraint on the other side. Phantom lines indicate direction of residual stress that would cause deformation except for restraint. (b) Plate with little restraint on either side

temperature gradient near the cut is steep, ranging from melting point at the cut to room temperature a short distance from it. Plate does not return to its original shape unless the entire plate is uniformly heated and cooled.

As the metal heats, it expands, and its yield strength decreases; the weakened heated material is compressed by the surrounding cooler, stronger metal. The hotter metal continues to expand elastically in all directions until its compressive yield strength is reached, at which point it yields plastically in directions not under restraint. The portion of this upset metal above 870 °C (1600 °F) is virtually stress-free; the remainder is under compressive stress that is equal to its yield strength. Metal that expands but does not upset is under compressive strength below yield. The net stress on the heated side of the neutral axis causes bowing of a narrow plate during cutting, as shown in Fig. 2.

As the heated metal begins to cool, it contracts, and its strength increases. First, the contraction reduces the compressive stress in the still-expanded metal. When the compressive stress reaches zero and the plate regains its original shape, previously upset metal also has regained strength. This metal is now in tension as it cools, and its tensile yield strength increases. Tension increases until the metal reaches room temperature. Residual tensile stress in the cooling side of the neutral axis causes the bowing of narrow plates after cooling (Fig. 2). Controlled upsetting is the basis of flame straightening.

Stainless Steels

Stainless steels do not support oxyfuel combustion and therefore require metal powder cutting, flux cutting, or plasma are cutting processes. Except for stabilized types, stainless steels degrade under the heat of metal powder or flux cutting processes. Carbide precipitation occurs in the HAZ approximately 3 mm (1/8 in.) from the edge, where the metal has been heated to 425 to 870 °C (800 to 1600 °F) long enough for dissolved carbon to migrate to the grain boundaries and combine with the chromium to form chromium carbide. The chromium-poor (sensitized) regions near grain boundaries are subject to corrosion in service. This type of corrosion can be prevented by a high-temperature stabilizing anneal after cutting, which puts the carbon back into solution. However, the required quench through the sensitizing temperature range may distort the material.

Water quenching of the cut edge directly behind the cutting torch may also avert sensitization. Because it takes approximately 2 min at sensitizing temperature for carbide precipitation to occur, water quenching must be done immediately. Distortion is more likely with this method than with the stabilizing anneal. Still another procedure is to remove the sensitizing zone entirely by chipping, grinding, or machining.

Cast Irons

Because of high carbon content, cast iron resists ordinary gas cutting. Gray iron contains some carbon in the form of flakes of graphite and some in the form of iron carbide, both of which hinder oxidation of the iron. For this reason, gray iron is classed as oxidation-resistant with respect to gas cutting. High-quality production cuts typical of steel are not expected; iron castings usually are gas cut to remove gates, risers, and defects; to repair or alter castings in service; or for scrap.

Techniques. Cutting is done manually, using more preheat and cutting oxygen than is used in cutting equal thicknesses of steel. The increased gas flow is obtained by using a larger tip. The preheat flames are adjusted to be carburizing, with the excess acetylene streamer approximately equal to the thickness being cut, as shown in Fig. 3. This adjustment helps to maintain preheat in the cut, because the excess acetylene combines with cutting oxygen beyond the tip. (The same principle is used to some extent in cutting thick steel sections.)

Before the cut is started, the area of the initial cut is preheated, the point of starting is heated to melting, and then the cutting oxygen is released.

In cutting, the torch is advanced in half-circles, as shown in Fig. 4. The size of the half-circles and the speed of advance depend on the thickness of the cut. This oscillating technique helps the cutting jet to get behind and blow out the slag and molten metal at the cut. The kerf is wider and the cut edges are considerably rougher than in cutting steel. Also, oxygen and acetylene consumptions are greater.

Other methods for cutting cast iron, more effective than ordinary gas cutting, are metal powder cutting and flux cutting. These methods are frequently used in foundries for removing gates, risers, and sprues or for breaking up ladle skulls.

Equipment

Commercial gases usually are stored in high-pressure cylinders. Natural gas—primarily methane—is supplied by pipeline from gas wells. The user taps into local gas lines. Acetylene, dissolved in acetone, is available in clay-filled cylinders. High-volume users often have

acetylene generators on site. For heavy consumption or where many welding and cutting stations use fuel gas, banks of gas cylinders are maintained at a central location in the plant, and the gas is manifolded and piped to the point of use.

Manual gas cutting equipment consists of gas regulators, gas hoses, cutting torches, cutting tips, and multipurpose wrenches. Auxiliary equipment may include a hand truck, tip cleaners, torch ignitors, and protective goggles. Machine cutting equipment varies from simple rail-mounted "bug" carriages to large bridge-mounted torches that are driven by computer-directed drives.

Gas regulators reduce gas pressure and moderate gas flow rate between the source of gas and its entry into the cutting torch to deliver gas to the cutting apparatus at the required operating pressure. Gas enters the regulating device at a wide range of pressures. Gas flows through the regulator and is delivered to the hose-torch-tip system at the operating pressure, which is preset by manual adjustment at the regulator and at the torch. When pressure at the regulator drops below the preset pressure, regulator valves open to restore pressure to the required level. During cutting, the regulator maintains pressure within a narrow range of the pressure setting.

Regulators should be selected for use with specific types of gas and for specific pressure ranges. Portable oxyacetylene equipment requires an oxygen regulator on the oxygen cylinder and an acetylene regulator on the acetylene cylinder, which are not interchangeable.

High-low regulators conserve preheat oxygen when natural gas or liquefied petroleum gas (LPG) is the preheat fuel used in OFC. These gases require a longer time to start a cut than acetylene or MPS. High-low regulators reduce preheat flow to a predetermined level when the flow of cutting oxygen is initiated. When the regulator switches from high to low, preheat cutback may range from 75 to 25% as plate thicknesses increase from 9.5 to 205 mm (3/8 to 8 in.). High-low regulators are used for manual and automatic cutting with natural gas and with LPG.

Hose. Flexible hose, usually 3.2 to 13 mm (1/8 to 1/2 in.) in diameter, rated at 1.4 MPa gage pressure (200 psig) maximum, carries gas from the regulator to the cutting torch. Oxygen hoses are green; the fittings have right-hand threads. Fuel gas hoses are red; the fittings have left-hand threads and a groove cut around the fittings. For heavy cutting, two oxygen hoses may be necessary—one for preheat and one for cutting

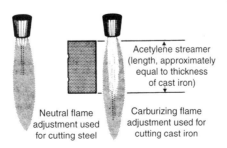

Fig. 3 Flame adjustments for cutting steel and cast iron

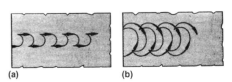

Fig. 4 Movement of torch when cutting cast iron. (a) Cutting thin cast iron. (b) Cutting heavy cast iron

oxygen. Multitorch cutting machines often have three-hose torches.

Cutting torches, such as the one shown in Fig. 5, control the mixture and flow of preheat oxygen and fuel gas and the flow of cutting oxygen. The curing torch discharges these gases through a cutting tip at the proper velocity and flow rate. Pressure of the gases at the torch inlets, as well as size and design of the cutting tip, limits these functions, which are operator controlled.

Oxygen inlet control valves and fuel gas inlet control valves permit operator adjustment of gas flow. Fuel gas flows through a duct and mixes with the preheat oxygen; the mixed gases then flow to the preheating flame orifices in the cutting tip. The oxygen flow is divided—a portion of the flow mixes with the fuel gas, and the remainder flows through the cutting-oxygen orifice in the cutting tip. A lever-actuated valve on the manual torch starts the flow of cutting oxygen; machine cutting starts the oxygen from a panel control.

Fuel gases supplied at low pressure, such as natural gas tapped from a city line, require an injector-mixer (Fig. 5b) to increase fuel gas flow above normal operating pressures. Optimal torch performance relies on proper matching of the mixer to the available fuel gas pressure. The use of reverse-flow check valves on both the fuel and oxygen gas inlets of the cutting torch is an excellent safety precaution. These check valves reduce the possibility of mixing gases in the hoses and regulators. Many of these check valves are also supplied by flashback arrestors. A flashback is the burning of the preheat flame in or behind the torch mixing chamber (i.e., a shrill hissing sound when the flame is burning inside the welding torch). Flashback is usually caused by an overheated cutting tip or by the failure to purge the gas hose lines before lighting the torch. This is a serious condition requiring immediate action. The oxygen valve should be turned off first, followed by turning off the fuel valve on the torch.

Cutting tips are precision-machined nozzles, produced in a range of sizes and types. Figure 6(a) shows a single-piece acetylene cutting tip. A two-piece tip used for natural gas (methane) or LPG is shown in Fig. 6(b). A tip nut holds the tip in the torch. For a given type of cutting tip, the diameters of the central hole, the cutting-oxygen orifice, and the preheat ports increase with the thickness of the metal to be cut. Cutting tip selection should match the fuel gas: hole diameters must be balanced to ensure an adequate preheat-to-cutting-oxygen ratio. Preheat gas flows through ports that surround the cutting-oxygen orifice. Smoothness of bore and accuracy of size and shape of the oxygen orifice are important to efficiency. Worn, dirty bores reduce cut quality by causing turbulence in the cutting-oxygen stream.

The size of the cutting tip orifice determines the rate of flow and velocity of the preheat gases and cutting oxygen. Flow to the cutting tip can be varied by adjustment at the torch inlet valve or at the regulator, or both.

Increasing cutting-oxygen flow solely by increasing the oxygen pressure results in turbulence and reduces cutting efficiency. Turbulence in the cutting oxygen causes wide kerfs, slows cutting, increases oxygen consumption, and lowers quality of cut. Consequently, larger cutting tips are required for making heavier cuts.

Standard tips, as shown in Fig. 7(a), have a straight-bore oxygen port. Oxygen pressures range from 205 to 415 kPa (30 to 60 psi) and are used for manual cutting. High-speed tips, or divergent cutting tips (Fig. 7b), use a converging, diverging orifice to achieve high gas velocities. The oxygen orifice flares outward. High-speed tips operate at cutting-oxygen pressures of approximately 690 kPa (100 psi) and provide cutting jets of supersonic velocity. These tips are precision made and are more costly than straight-drilled tips, but they produce superior results—improved edge quality and cutting speeds 20% higher than standard tips. Best suited to machine cutting, high-speed tips produce superior cuts in

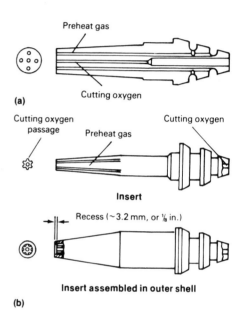

Fig. 6 Types of cutting tips. (a) Single-piece acetylene cutting tip. (b) Two-piece tip for natural gas or liquefied petroleum gas. Fuel gas and preheat oxygen mix in tip. Recessed bore helps promote laminar flow of gas.

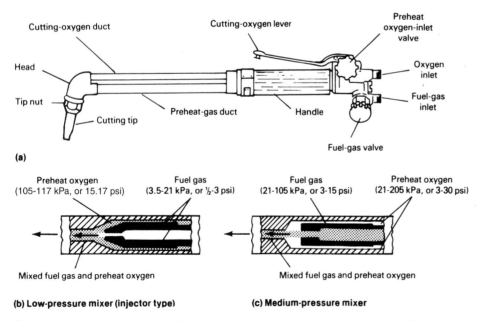

Fig. 5 Typical manual cutting torch in which preheat gases are mixed before entering torch head. (a) Schematic view of torch. (b) and (c) Sections through preheat gas duct showing two types of mixers commonly used with the torch shown. After the workpiece is sufficiently preheated, the operator depresses the lever to start the flow of cutting oxygen. Valves control the flow of oxygen and fuel gas to achieve required flow and mixture at the cutting tip.

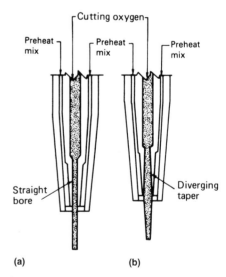

Fig. 7 Oxyfuel cutting tips. (a) Standard cutting tip with straight-bore oxygen orifice. (b) High-speed cutting tip with divergent-bore oxygen orifice

plate up to approximately 150 mm (6 in.) thick. Above this thickness, advantages of their use decrease; they are not recommended for cutting metal more than 255 mm (10 in.) thick.

Equipment Selection Factors. Natural gas and LPGs operate most efficiently with high-low gas regulators, injector-type cutting torches, and two-piece, divergent, recessed cutting tips. Acetylene cutting is most efficient with divergent single-piece tips. If acetylene is supplied by low-pressure generators, an injector-type torch is best suited to most cutting applications.

Two-piece, divergent cutting tips are best suited for use with MPS gas; the tip recess should be less than for use of natural gas or propane. Injector-type torches and high-low regulators are not required with MPS gas.

Guidance Equipment. In freehand cutting, the operator can usually follow a layout accurately at low speeds, but the cut edges may be ragged. For accurate manual cutting at speeds over 250 mm/min (10 in./min), the torch tip should be guided with a metal straightedge or template. Circles and arcs are cut smoothly with the aid of a radius bar, a light rod clamped and adjusted to the torch at one end, while the other end is held at the circle center.

Machine guidance equipment includes magnetic tracing of a metal template, manual spindle tracing, optical tracing of a line drawing, guidance by numerically controlled tape or by programmable controllers, and computer-programmed guidance equipment (Fig. 8).

Portable cutting machines are used primarily for straight-line and circular cutting. Components include a torch mounted on a portable motor-driven carriage that travels on a track or some other torch guidance device. The operator adjusts travel speed and monitors the operation.

Machine cutting torches are of heavy in-line tubular construction. The outer barrel of the torch has a rack, which fits into a gear on the torch holder, for raising and lowering the torch over the work. Gas ducts and valves are encased in the cutting torch barrel. The cutting tip is mounted in line with the outer barrel axis. A valve knob or a lever-operated poppet valve replaces the spring-loaded cutting-oxygen lever of the manual cutting torch that would require continued hand pressure.

On some portable machines, gases are supplied to connections on the carriage, rather than directly to the torch to avoid hose drag on the torch. Short hoses are used from machine connections to the torch. Some carriages can accommodate two or more torches operating simultaneously, for such operations as squaring and weld joint beveling.

The operator follows the carriage to make adjustments. When plates are wavy or distorted, the operator may need to adjust torch height to avoid losing the cut. When carefully operated, track-guided torches can produce cuts at speeds and quality approaching those obtainable with large computer-controlled stationary cutting machines.

Stationary cutting machines, as shown in Fig. 8 and 9, are used for straight-line and circular cuts, but their primary use is for cutting complex parts, that is, for cutting shapes. Plate to be cut is moved to the machine. The plate is positioned on a large table, where the plate is supported at a large number of points with very little surface area of the supports exposed to the cutting action. Some of these tables are filled with water to help reduce the fume generation from the cutting operation. Sometimes, the cutting table is equipped with a down-draft exhaust system to remove the fume from the cutting area.

On shape-cutting machines, cutting torches move left and right on a bridge mounted over the cutting table. The bridge moves back and forth on supports that ride on floor-mounted tracks. The combined movement of the torches on the bridge and the bridge on the track allows the torch to cut any shape in the x-y plane. Bridges are either of cantilever or gantry design. Suppliers classify cutting-machine capacity by the maximum width of plate that can be cut.

Machine Directions. Methods for directing the motion of shape-cutting machines have become increasingly sophisticated and include manual, magnetic, and electronic means of control. The simplest machines have one or two torches and use manual or magnetic tracing.

For manual tracing, the operator either steers an idler wheel or spindle around a template or guides a wheel or focused light beam around an outline on paper. Cutting speed is controlled by setting the speed of the tracing head (pantograph director) or by setting the speed of the torch carriage (coordinate drive). Cutting speed in manual tracing is approximately 350 mm/min (14 in./min), depending on operator skill.

Magnetic tracing is done with a knurled magnetized spindle that rotates against the edge of a steel template. The spindle is linked to a pantograph. Direct-reading tachometers, showing cutting speed in inches per minute, assist in

(a)   (b)

Fig. 8 Gantry shape-cutting system. (a) Computer numerical controlled (CNC) cutting tool incorporating oxyfuel torches, plasma arc torches, 90° indexing triple-torch oxyfuel stations for straight-line beveling, and zinc powder or punch markers. (b) Closeup of CNC control console. Courtesy of ESAB North America, Inc.

adjusting cutting speed. These control methods are relatively slow.

Faster, electronic tracers use a photoelectric cell that scans the reflection of a beam of light directed on the outline of a template. Templates are line drawings on paper, white-on-black paper cutouts, or photonegatives of a part outline. To hold tolerances closer than 1.6 mm ($\frac{1}{16}$ in.) continuously, templates of plastic film, glass cloth, or some other durable, dimensionally stable material should be used.

In scanning the edge of a white-on-black template, the circuit through the photoelectric cell balances when the cell senses an equal amount of black and white. A change in this balance sends an impulse to a motor that moves the tracing head back to balance. In line tracing, the photoelectric cell scans the line from side to side. As long as the light reflects equally from both sides of the line, the steering signals balance. When the photocell scans more light on one side of the line than on the other, the scanner rotates to balance.

Some machines adjust to permit parts to be cut approximately 0.8 mm ($\frac{1}{32}$ in.) larger or smaller than the template. This feature, called kerf compensation, is useful for cutting to close tolerances, especially when the template has insufficient kerf allowance.

Coordinate-drive machines translate motion 1 to 1 or in other ratios. Such ratio cutting permits the use of templates in any proportion, from full-scale to one-tenth of part size.

Tape Control. Cutting-machine movement may be controlled by programmed electronic signals stored on magnetic tape (numerical control). This form of control started with the use of punched tapes, similar to punched computer cards. These machines do not require templates, and the tapes may be easily stored and used many times.

Some cutting machines receive directions from a microprocessor, programmed directly or from a program stored on a tape. The most sophisticated machines take directions directly from a computer (computerized numerical control) and use computer graphics (Fig. 8).

Nesting of Shapes. Savings in material, labor, and gas consumption can be gained by nesting parts in the stock layout for single-torch or multiple-torch operation. Savings can be realized whenever one cut can be made instead of two. Sometimes, a shape can be modified for better nesting. The advent of computer graphics allows cutting-machine programmers to create layouts of part patterns on cathode ray tube screens, manipulating cutting patterns for greatest plate use. Several firms offer programs that optimize parts nesting.

Starting the Cut

To start a cut at the outer edge of the workpiece, the operator should place the cutting torch so that the ends of the inner cones in the preheating flame just clear the work metal. When a spot of metal at the top of the edge is heated to bright red, the stream of cutting oxygen is turned on. The reactions form a slot in the plate edge, and the torch can move along the desired line of cut. Piercing starts are necessary to cut holes, slots, and shapes.

In manual piercing, the spot should be heated to bright red. Then, the torch should be raised to 13 mm ($\frac{1}{2}$ in.) above the normal cutting distance, and cutting oxygen should be turned on slowly. As soon as the metal is penetrated, the torch is lowered to cutting height, and cutting action is started.

Machine torch piercing is similar to manual operations, except that torch travel begins after the cutting oxygen is slowly turned on and the flame starts to eject molten slag. Initial piercing action does not completely penetrate the plate, so that the completed pierce covers a short distance (depending on plate thickness) and forms a sloping trough with molten slag ejected from the torch. Length of the lead-in should be increased for heavy plate.

Light Cutting

Light cutting (material less than 9.5 mm, or $\frac{3}{8}$ in., thick) requires extra care in tip selection, tip cleanliness, and control of gas pressure. Quality of cut when oxyfuel gas cutting steel thinner than 3.2 mm ($\frac{1}{8}$ in.) is virtually impossible to control. Wide, uneven kerf, ragged edge, and distortion result from heat buildup.

For straight or large-radius cuts in thicknesses 3.2 mm ($\frac{1}{8}$ in.) and over, angling the cutting torch forward increases the distance through the cut and may help to deflect some of the heat input. When cutting shapes, however, the torch must be vertical.

One of the most important techniques for cutting of thin sections is stack cutting (see the section "Stack Cutting" in this article for more detailed information on this process). This technique avoids the difficulties inherent in light

Fig. 9 Stationary oxyfuel gas cutting machine

cutting and also increases production. Sections stacked for cutting must be flat and tightly clamped together.

Medium Cutting

In cutting plate 9.5 to 255 mm (³⁄₈ to 10 in.) thick, OFC achieves its highest efficiency and produces the most satisfactory cuts. Conditions for medium cutting of carbon steel plate using a variety of oxyfuel combinations are presented in Tables 1 through 6. Commercial cuts must meet these requirements:

- Drag is short; markings on the face of the cut approach vertical.
- Side of the cut is smooth, not fluted, grooved, or ragged.
- Slag should not adhere to the bottom of the cut.
- Upper and lower edges should be sharp.
- Cost is moderate.

The following variables require adjustment to obtain satisfactory commercial cuts:

- Suitable cutting tip, with cutting orifice of correct type and size, and proper degree of preheat
- Suitable oxygen and fuel gas pressures
- Correct cutting speed
- Uniform torch movement
- Clean, smooth-bore cutting orifice and pre-heat holes
- High-purity oxygen
- Proper angle of the cutting jet in relation to the upper edge of the cut

Surface finish for gas-cut production parts is compared with sample cuts. One such group of machine-cut samples is shown in Fig. 10. Sample A represents high quality, with drag varying approximately 2 to 7%. Samples B, C, D, and E show typical defects resulting from improper cutting speed or torch settings. The American Welding Society (AWS) provides standard plastic samples of cut edges.

Tip Design. General-purpose cylindrical-bore tips operate best with cutting-oxygen pressures of 205 to 415 kPa (30 to 60 psi); a pressure of 275 kPa (40 psi) is a good initial setting. Divergent-orifice tips operating with cutting-oxygen pressures of approximately 690 kPa (100 psi), with variations of ±140 kPa (±20 psi), may provide better results. Because tip size controls kerf width, a smaller tip should be used for a narrower kerf; the part may not drop free, and slower speed must be used. Larger-than-normal tips may be used when the metal is covered with thick scale.

Cutting Speed. Within recommended speed ranges, slow cutting produces the best results; higher speeds are used for maximum output. Optimal cutting speed may be determined by observing the material that is expelled from the underside of the workpiece. When speed is too slow, the cutting stream forms slag drippings. At excessive speed, a stream trails at a sharp angle

to the bottom of the workpiece. Highest-quality cutting occurs when a single, continuous stream is expelled from the underside for approximately 6.4 mm (¼ in.), then breaks into a uniform spray. This condition is accompanied by a sound similar to that of canvas ripping.

Preheat. Less preheat is needed to continue a cut than to start it. When optimal quality is desired, the amount of preheat should be reduced after starting the cut. Commercially available devices automatically reduce preheat after cutting begins.

In thicknesses up to 50 mm (2 in.), excessive preheat may cause the top corner to melt and roll over. Insufficient preheat results in loss of cut. Correct preheat provides a continuous reaction with a sharp, clean top edge of the cut.

Kerf compensation, or the allowance of half the kerf width on the outline of the template, is the amount added for outside cuts and subtracted for inside cuts. In straight-line cutting or in following a layout, the operator may compensate for this measurement; in machine cutting, kerf compensation should be incorporated in the template. Many automatic machines incorporate kerf compensation in their controllers.

Machine Accuracy. When a machine is properly maintained, the guidance system controls accuracy. Mechanical followers are very accurate when properly adjusted. Compensation for the diameter of the follower wheel or spindle must be made in the same manner as for kerf width when making a template.

Kerf angle, the small deviation of the kerf wall from a right angle, may be caused by failure to set the torch perpendicular to the work. It may also be the result of widening of the cutting-oxygen stream. In the latter case, the width of the kerf increases from the top to the bottom of the cut; this may be corrected in straight cutting by angling the torch slightly to make one wall of the kerf perpendicular. In shape cutting, where this cannot be done, kerf angle can be corrected by slightly reducing cutting-oxygen pressure.

Plate movement may be vertical, caused by warp, or lateral, caused by planar expansion and contraction of the part. Vertical movement requires either manual or automatic control of torch height. A very slight allowance can be made in the initial setting of the tip-to-work distance. Lateral movement can be controlled by fixing the part to be cut on the plate support, wedging the cut within the stock plate, or allowing only a small amount of scrap trim around the part so that the scrap will move instead of the workpiece.

Heavy Cutting

Cutting of metal 255 mm (10 in.) or more in thickness requires attention to gas flow, preheat flame setting, drag, and starting technique. Standard cutting torches may be used to cut steel up to 455 to 510 mm (18 to 20 in.) thick; heavy-duty torches are necessary for sections up to 1525 mm (60 in.) thick.

Gas Flow Requirements. Heavy cutting requires a uniform supply of high-purity oxygen and fuel gas at constant pressure, and an adequate volume of cutting oxygen. Torches should have three hoses—one for the cutting oxygen, one for the preheat oxygen, and one for the fuel gas. This

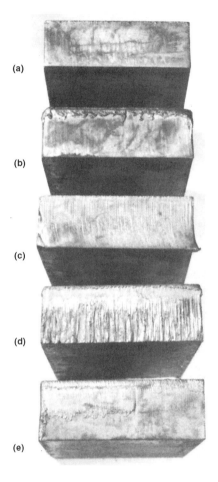

(a)
(b)
(c)
(d)
(e)

Torch settings for bar A(a)

Preheat-oxygen pressure:
High (edge starting) — 310 kPa (45 psi)
Low (cutting) — 55–83 kPa (8–12 psi)
Preheat natural gas pressure:
High (edge cutting) — 21 kPa (3 psi)
Low (cutting) — Under 430 Pa (1 oz/in.²)
Cutting-oxygen pressure — 690 kPa (100 psi)
Cutting speed — 255–380 mm/min (10–15 in./min)

(a) Settings for bars B, C, D, and E were varied as noted in the caption. All bars were cut with two-piece divergent-nozzle (high-speed) tips with 1.37 mm (0.054 in.) diam cutting-oxygen orifice. All cutting was done with low preheat settings, except for bar B, which was cut with both preheat gases set higher than for starting.

Fig. 10 Comparison of surface finish of gas-cut specimens. (a) Proper speed, preheat, and cutting-oxygen pressure. Note clean face and nearly straight drag lines. (b) Proper speed and cutting-oxygen pressure, too much preheat. Note excessive slag and rounding of top edge. (c) Proper preheat and cutting-oxygen pressure, but too much speed. Note increase in drag and uncut corner at lower left. (d) Proper speed and preheat, but too much cutting-oxygen pressure. Surface is rough, and top edge is melted. (e) Proper preheat and cutting-oxygen pressure, but speed too low. Burned slag adheres to cut surface.

allows setting of proper pressures independently for preheating and cutting. Cutting-oxygen hose should be at least 13 mm (½ in.) in diameter to ensure an adequate volume of oxygen at low pressure settings. Regulators should be used that are capable of providing the large volumes of gas required. Cutting-oxygen flow requirements (ft³/h) vary from 80 to 120 times ($80t$ to $120t$) the thickness to be cut measured in inches. Cutting-oxygen consumption for shape cutting approaches $120t$ or higher; for straight-line cutting, it is between $80t$ and $100t$. Pressures are 140 to 415 kPa (20 to 60 psi) at the torch.

Preheating. In heavy cutting, the preheat flame must extend almost to the bottom of the thickness to be cut to avoid excessive drag. If the cutting torch cannot deliver enough heat, an extra torch may be needed, or the workpiece may be furnace heated. A temperature boost of 95 to 150 °C (200 to 300 °F) assists cutting action considerably. Tip height above the work is important; if the tip is too close, excessive melting and rounding of the top edge occur. If too high, preheat will be insufficient and cutting will be slowed. Generally, 25 to 50 mm (1 to 2 in.) of separation is satisfactory.

The starting edge should be thoroughly preheated, with the preheated zone extending far down the face, as shown in Fig. 11(a). As soon as the metal begins to melt, the flow of cutting oxygen and the cutting motion should proceed simultaneously, cutting at normal speed. The cut should progress down the face at a constant rate until it breaks through the bottom. Drag will be long at first, but it will shorten as soon as the cut is confined. Too slow a speed causes a shelf part way through the cut edge (Fig. 11e). Too high a speed results in incomplete penetration or in extremely long drag, as shown in Fig. 11(f).

Drag. To successfully complete the cut (drop cutting), minimum drag conditions must exist, as shown in Fig. 12(a). Too much drag leaves an uncut corner at the end of the cut (Fig. 12b and c). Angled cutting may be used in straight-line work to counter drag (Fig. 12d and e), but for shape cutting, the torch must be perpendicular to the work.

Starting. Extra care is needed in starting a heavy cut to avoid leaving an uncut corner or pocketing the flame in the lower portion of the cut. Starting sometimes is facilitated by first undercutting the forward edge of the material with a hand torch (Fig. 13).

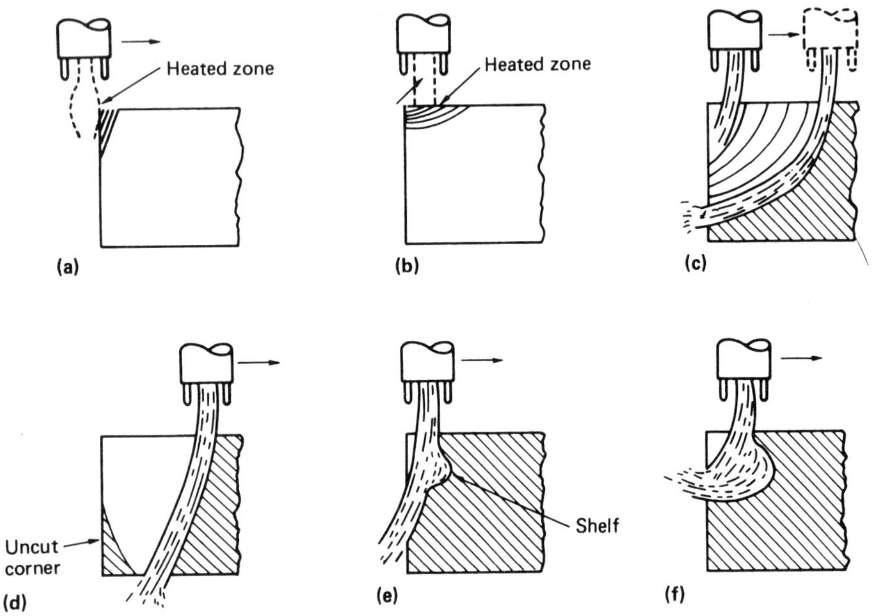

Stack Cutting

Stack cutting is a practical means of cutting sheets too thin for ordinary cutting, but it also can reduce the cost of cutting thicker material. Savings are multiplied by using multitorch machines for stack cutting. Several torches can be used, each cutting up to 30 pieces of 3.2 mm (⅛ in.) thick material simultaneously. Thicknesses most practical for this technique are 20 gage (0.91 mm, or 0.036 in.) to 13 mm (½ in.), although material 19 mm (¾ in.) thick has been stack cut. Shapes made by stack cutting usually have sharper edges than those cut singly.

Material for stack cutting should be clean and free from loose or heavy scale. The sheets or plates must be clamped tightly in a pile, with edges aligned where the cut is to start. Air gaps between pieces will interfere with the cutting flame. The most practical total thickness for the stack is 125 or 150 mm (5 or 6 in.), although 250 mm (10 in.) stacks have been cut. The risk

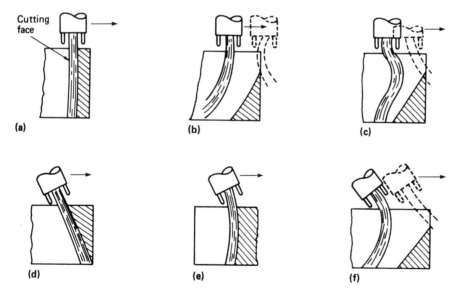

Fig. 11 Proper and improper techniques for heavy cutting. (a) Proper torch position; preheat is primarily on starting face. (b) Improper start; oxygen stream is too far onto work, which results in action of cut as shown in (c) and in uncut corner as shown in (d). (e) Excessive oxygen pressure or action of cut at too low a speed. (f) Insufficient oxygen pressure or action of cut at too high a speed

Fig. 12 Proper and improper techniques for completing a heavy cut. (a) Minimum drag (typical at balanced conditions) permits flame to break through cutting face uniformly at all points. (b) Excessive drag, typically caused by insufficient oxygen or excessive speed, results in undercut. (c) Forward drag resulting from excessive oxygen pressure or too little speed. (d) and (e) Forward angling of torch to minimize drag and thus avoid an uncut corner. (f) Excessive forward angling of torch, resulting in undercut

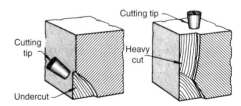

Fig. 13 Undercutting as an aid in starting a heavy cut

of a costly mishap increases with the thickness to be cut and the number of pieces.

When stack cutting pieces less than 4.8 mm ($^3/_{16}$ in.) thick, a waster plate of 6.4 mm ($^1/_4$ in.) material is clamped atop the stack. The waster plate ensures better starting and a sharper edge for the top. It also prevents buckling.

In addition to increasing productivity, stack cutting also reduces materials handling, grinding and cleaning costs, gas consumption, and floor-space requirements. However, savings must justify the time for stacking and clamping. The method is best used when the flat sheets are bundled in stacks at the steel mill.

An application of stack cutting, as practiced in a diesel engine plant, is described in the first example that follows.

When comparing stack cutting with multi-torch machine cutting, the deciding factor in making the choice is not the actual cutting time per piece. The saving includes reduced materials handling, less grinding and cleaning, and lower gas consumption. The second example that follows describes an application in which stack cutting required less time than multitorch cutting of small lots of a simple shape.

Example 1: Stack Cutting of Cooling-Fan Side Plates (Fig. 14). The cooling fan of a direct-current main generator for a diesel locomotive consisted of two side plates, 132 cm (52 in.) outside diameter by 76 cm (30 in.) inside diameter by 4.8 mm ($^3/_{16}$ in.) thick, between which 24 fan blades were riveted. The assembled fan was 15.9 cm ($6^1/_4$ in.) wide. The side plates (Fig. 14) were cut in stacks of 20 to

30 in a two-torch gas cutting machine with an electronic tracer, using only one cutting torch. An injector-type torch with a two-piece divergent cutting tip was used with the oxynatural gas process. Torch settings and other processing details are shown in the table with Fig. 14, together with alternative production methods.

Example 2: Stack Cutting versus Multitorch Machine Cutting (Fig. 15). A six-torch gas cutting machine with an electronic tracer was used to cut the 46 cm (18 in.) perimeter of the 9.5 mm ($^3/_8$ in.) thick triangle shown in Fig. 15. At a cutting speed of 48 cm/min (19 in./min), six pieces were cut in 0.95 min, or 0.158 min per piece. However, each piece required 0.5 min of light grinding to clean the edges. Total time for these two operations was 0.658 min per piece. With setup time of 2.3 min per piece added, total time per piece was 2.958 min.

When stack cutting was used, the material came from the mill in bundles of 13 plates. The plates were supplied in sizes for convenient multiples of the nested shape, the total quantity being based on estimated needs for several months. Bundles were stored as received and used as required. Each bundle of 13 plates made a stack 12.4 cm ($4^7/_8$ in.) high.

A single torch was used for stack cutting. At a cutting speed of 18 cm/min (7 in./min), 2.571 min was required for cutting the stack, or 0.198 min per piece. This was approximately 25% longer than it took with the multitorch machine. However, only the bottom plate of each stack required light grinding, making the cleaning time per piece 0.038 min. Setup time was 1.662 min per piece. Total time per piece for all operations was therefore 1.898 min—or 36% less than by the six-torch method.

Preparation of Weld Edges

Steel plates to be welded are resquared, trimmed to size, cut to shape, and beveled by gas cutting, often in one operation. Some cutting also is done on forgings and castings to prepare them for welding. Cutting accuracy can be held to 6 or 3 mm ($^1/_4$ or $^1/_8$ in.), and squareness and smoothness of the cut edges are usually satisfactory. Adherence of slag to the underside of the

cut can be closely controlled; however, a light cleaning operation may be needed. If so, a wide-blade chisel, a grinding wheel, or a wire brush may be used.

Bevel cutting differs from ordinary profile cutting in that the cut edge, instead of being square with the plate surface, is at an angle. Bevel cutting is used primarily to prepare the edges of plate for welding. Some bevels are cut to remove a sharp corner (see "Example 7: Gas Cutting versus Band Sawing" in this article) or to cut shapes.

Single and double bevel cuts usually vary from 30° to 45°, measured from the square, uncut edge. Low-quality cut surfaces may result for bevels of less than 10° unless an extra scrap allowance is made. For bevels greater than 60°, it is difficult to maintain proper preheat temperature.

Standard cutting equipment is used for cutting bevels in straight-line cutting and in cutting circles. Most shapes cannot be bevel cut by standard machines, because there is no provision for rotating the torch with a change in direction. Some tape-controlled machines have a rotating-torch mechanism for cutting beveled shapes. Circular bevel cutting can be done with portable equipment that can use a radius bar. Sometimes, the workpiece can be rotated past a stationary cutting torch.

Cutting-tip selection for beveling requires special consideration, for two reasons:

- When the cutting tip is inclined to the metal surface, heat transfer to the plate is reduced.
- Because of the angle, the actual cutting thickness is greater than the plate thickness.

Preheating for Bevel Cutting. Torch inclination does not affect the oxidizing action of the cutting-oxygen stream if preheat is adequate. Therefore, tip size (size of cutting-oxygen orifice) and cutting-oxygen flow are selected on the basis of the as-cut thickness. Preheat gas flow, however, is affected in two ways:

- As the angle of inclination increases, heat transfer becomes less efficient, and more preheat gas flow is required.
- In contrast to perpendicular cutting, more preheat is required for thinner materials.

(The latter effect may be related to the percentage of heat deflected and to the higher speeds at which thinner materials are cut; the effect disappears in metal thicknesses greater than 10 cm, or 4 in.)

Because preheat is important in bevel cutting, several techniques, in addition to increased preheat gas flow, are used to obtain proper preheat. The best position for oxyacetylene cutting is that in which the uppermost cones of flame touch the plate surface while the lower cones extend into the cut. A second torch may be used to precede the cutting torch, for additional preheat. Also, extra preheat tips, called bevel adapters, are available. These adapters have a special heating nozzle offset from the cutting tip so that, when

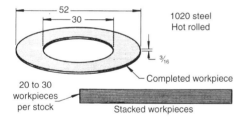

Torch settings

Cutting tip	Two-piece divergent type
Diameter of cutting-oxygen orifice	1.7 mm (0.067 in.)
Flow rates (approx):	
Natural gas preheat	0.96 m³/h (34 ft³/h)
Preheat oxygen	1.9 m³/h (68 ft³/h)
Cutting oxygen	10 m³/h (356 ft³/h)
Cutting speed (25 pieces)	150 mm/min (6 in./min)

Alternative production methods

Up to 10 pieces	Nibble, saw, machine turn(a)
10 to 25	Stack gas cut (portable machine)
25 to 1000	Stack gas cut (stationary machine)
Continuous production(b)	Die blank

(a) Methods listed in descending order of preference. (b) Minimum quantity, 1000 pieces per year, two years' production

Fig. 14 Stack-cut flat ring (example 1). Dimensions in inches

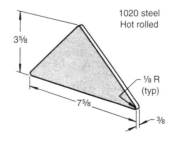

Fig. 15 Triangle that was produced in 36% less time when gas cut in stacks of 13 with a single torch than when six pieces were individually gas cut at one time with a six-torch machine (example 2). Dimensions in inches

the cutting tip is inclined at 45°, the heating tip is normal to the plate, as shown in Fig. 16. Bevel adapters are right-hand or left-hand; therefore, two may be needed. Much beveling is done, however, without special equipment.

Data for bevel cutting with adapters, using natural gas preheat, are shown in Table 8. This table also can be used when cutting bevels without the use of extra preheat tips, by selecting the next larger tip to provide the extra preheat needed, due to torch inclination. For example, a 45° bevel having a 9.5 mm ($^3/_8$ in.) A and B dimension requires a 0.93 mm (0.037 in.) orifice tip with an adapter; the same bevel without an adapter would require a 1.37 mm (0.054 in.) orifice tip and the settings listed for it. The cutting-tip height for oxynatural gas beveling should be approximately 3 to 6 mm ($^1/_8$ to $^1/_4$ in.) at the closest point above the work. For bevels larger than those for which dimensions are given in Table 8, the torch-to-work distance usually is increased.

When a plate must be trimmed to size as well as partly beveled, both operations can be done at the same time, using two torches close together, with the perpendicular torch leading. Special multitip bevel-cutting heads are also available for this purpose. Two torches may similarly be used for cutting a double bevel intersecting at an edge, with the leading torch cutting the underside bevel. Three torches can be used to cut a double bevel with a short vertical face between bevels; the torch cutting the underside bevel leads, the perpendicular torch follows, and the torch cutting the upper bevel is last.

Because preheat is so important in bevel cutting, and the acetylene flame has better heat transfer, oxyacetylene bevel cutting is more efficient than oxynatural gas or oxypropane bevel cutting in thicknesses up to 10 cm (4 in.). Methylacetylene-propadiene-stabilized gas has also proved advantageous for bevel cutting.

J-Grooving. A type of gas cutting known as J-grooving sometimes is used to prepare the edges of heavy pressure vessels for welding. The purpose is to provide a groove with less cross-sectional areas than an angular bevel would provide in a heavy plate, thus saving time, labor, and filler metal in welding. The cutting tips used in this operation are gouging tips; hence, J-grooving is discussed in the section "Oxyfuel Gouging."

Applicable Shapes

Irregular shapes may be cut by special techniques, or may be beyond the practical limits of gas cutting.

Cutting of Bars and Structural Shapes. To cut round steel bars, gas pressure should be adjusted for the maximum thickness (diameter), and the cut should be started at the outside. As the cut progresses, the torch should be raised and lowered to follow the circumference. The bar should be nicked with a chisel to make a burr at the point where the cut is to begin. Oxyfuel gas cutting of round bars is usually manual, so the cut surface is rough. The chief advantage of OFC over sawing is that the torch can be brought to the work.

The same advantage applies to gas cutting of various structural shapes, although cutting can be smoother and sharper because it can be guided on the flanges and webs. Extra care must be taken when cutting thick filleted sections. These cuts usually are satisfactory for welding but may

require light grinding if exposed to view. Columns that require end capping usually are machined to exact size.

Circular holes can be gas cut, rather than drilled, if the hole wall does not require a machined finish, and if the roundness and concentricity have wide tolerances. The following types of holes can be made by gas cutting: clearance holes for piping and bolting; concrete grouting holes; holes for piping where the pipe is welded to the piece containing the hole; inspection or access holes; and holes for plug welds.

Holes can best be gas cut with the aid of electronic tracers. Accuracy depends on template accuracy. Hole location and hole size in plate up to 7.6 cm (3 in.) thick can be held readily to the tolerances listed in Table 9. As shown in Table 10, the cut is advanced from a pierced starting hole. In piercing, the slag must be removed from any area that the flame will later touch. This is particularly important for small holes.

Recommended minimum hole diameters are shown in Table 10. The two levels of quality

Table 8 Torch settings for various bevels, using extra preheat tip (bevel adapter) in oxynatural gas cutting

Bevel dimensions							Cutting-oxygen orifice diam(a)		Pressure				Speed	
A		B		C					Oxygen		Natural gas			
mm	in.	mm	in.	mm	in.	mm	in.		kPa	psi	kPa	psi	mm/min	in./min
Bevel angle (α) of 45°														
6.4	$^1/_4$	6.4	$^1/_4$	9.5	$^3/_8$	0.94	0.037		310	45	69	10	330	13
9.5	$^3/_8$	9.5	$^3/_8$	13	$^1/_2$	0.94	0.037		310	45	69	10	330	13
13	$^1/_2$	13	$^1/_2$	17	$^{11}/_{16}$	1.37	0.054		414	60	69	10	320	12$^1/_2$
19	$^3/_4$	19	$^3/_4$	27	$^{1}/_{16}$	1.37	0.054		414	60	69	10	280	11
25	1	25	1	36	$^{17}/_{16}$	1.37	0.054		414	60	69	10	280	11
32	$1^1/_4$	32	$1^1/_4$	44	$1^3/_4$	1.40	0.055		414	60	69	10	280	11
38	$1^1/_2$	38	$1^1/_2$	53	$2^1/_8$	1.40	0.055		448	65	69	10	255	10
50	2	50	2	71	$2^{13}/_{16}$	1.85	0.073		483	70	69	10	230	9
64	$2^1/_2$	64	$2^1/_2$	89	$3^1/_2$	1.85	0.073		483	70	69	10	165	$6^1/_2$
75	3	75	3	107	$4^1/_4$	1.85	0.073		552	80	69	10	150	6
Bevel angle (α) of 30°														
4.0	$^5/_{32}$	6.4	$^1/_4$	8	$^5/_{16}$	1.37	0.054		414	60	69	10	380	15
5.6	$^7/_{32}$	9.5	$^3/_8$	11	$^7/_{16}$	1.37	0.054		414	60	69	10	355	14
6.4	$^1/_4$	11	$^7/_{16}$	13	$^1/_2$	1.37	0.054		414	60	69	10	355	14
7.1	$^9/_{32}$	13	$^1/_2$	14	$^9/_{16}$	1.37	0.054		414	60	69	10	355	14
9.5	$^3/_8$	17	$2^1/_{32}$	19	$^3/_4$	1.37	0.054		414	60	69	10	355	14
11	$^7/_{16}$	19	$^3/_4$	22	$^7/_8$	1.37	0.054		414	60	69	10	355	14
13	$^1/_2$	21	$^{27}/_{32}$	25	1	1.37	0.054		414	60	69	10	330	13
14	$^9/_{16}$	24.6	$^{31}/_{32}$	28	$1^1/_8$	1.37	0.054		414	60	69	10	330	13
19	$^3/_4$	32	$1^1/_4$	38	$1^1/_2$	1.37	0.054		414	60	69	10	305	12
22	$^7/_8$	38	$1^1/_2$	44	$1^3/_4$	1.37	0.054		414	60	69	10	280	11
25	1	44	$1^3/_4$	50	2	1.37	0.054		414	60	69	10	265	10$^1/_2$
29	$1^5/_{32}$	50	2	58	$2^5/_{16}$	1.40	0.055		448	65	69	10	255	10
32	$1^1/_4$	54	$2^5/_{32}$	64	$2^1/_2$	1.40	0.055		448	65	69	10	230	9
36	$1^7/_{16}$	64	$2^1/_2$	72	$2^7/_8$	1.40	0.055		483	70	69	10	230	9
38	$1^1/_2$	65	$2^{19}/_{32}$	75	3	1.40	0.055		483	70	69	10	180	7
44	$1^3/_4$	76	$3^1/_{32}$	89	$3^1/_2$	1.85	0.073		517	75	69	10	180	7
50	2	87	$3^{15}/_{32}$	100	4	1.85	0.073		517	75	69	10	165	$6^1/_2$

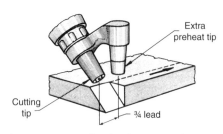

Fig. 16 Cutting a bevel with an extra preheat tip, or bevel adapter. For beveling in the opposite direction, an opposite-hand adapter is needed. Dimensions in inches

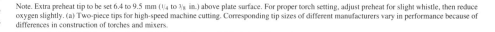

Partial bevel   Full bevel

Note. Extra preheat tip to be set 6.4 to 9.5 mm ($^1/_4$ to $^3/_8$ in.) above plate surface. For proper torch setting, adjust preheat for slight whistle, then reduce oxygen slightly. (a) Two-piece tips for high-speed machine cutting. Corresponding tip sizes of different manufacturers vary in performance because of differences in construction of torches and mixers.

shown indicate the influence of hole size on accuracy. Normal production tolerances (Table 9) are difficult to maintain in thicknesses over 7.5 cm (3 in.), because of kerf angle. Holes larger than those listed can be gas cut to normal production tolerances.

Slots. The procedure for gas cutting of slots is the same as that for gas cutting of holes. Table 10 lists minimum slot sizes that can be cut to normal tolerances, and contains a sketch that illustrates torch movement. Slots can be made at less cost by gas cutting than by profile end milling, if a machined edge is not mandatory.

Borderline Shapes. The next four examples illustrate shapes that are borderline for gas cutting. The first example concerns a shape in which two sides had to be exactly perpendicular. In the second example, three sides of a shape were required to hold a tolerance of $+0$, -0.8 mm $(-\frac{1}{32}$ in.).

In the third example, a shape with a narrow pointed end was required to maintain sharp edges and corners. When gas-cut shapes have rounded top edges (rollover), the cause usually is excessive preheat. However, this condition can also occur when the cutting path surrounds too small a mass of the shape being cut, causing excessive heat accumulation, as in the application described in the example that follows.

The fourth example describes a shape that contained a long slot, narrower than the minimum recommended for normal accuracy in Table 10. In this example, the narrow re-entrant cut is the opposite of a projection in Example 5. Here, it was questionable whether normal production tolerances could be maintained if the part was flame cut.

Example 3: Two Perpendicular Sides. The gusset shown in Fig. 17 was gas cut from 13 mm $(\frac{1}{2}$ in.) thick low-carbon steel so that sides A and B would be 90° to each other. By positioning the part a minimum of 50 mm (2 in.) from any edge of the plate and cutting in a clockwise direction, the restraint of the uncut portion of the plate was used to keep the two adjacent sides of the gusset in position during machine gas cutting. As shown in Fig. 17, the cut was started clockwise, from a pierced start, along the noncritical side. The shorter side A, then side B, were cut so that they could not move in relation to each other. The noncritical sides were cut last.

The cutting machine was equipped with an injector torch for use with low-pressure natural gas. A high-speed, two-piece cutting tip was used, together with high-low gas regulators. The operating procedure and torch settings used are given in the table that accompanies Fig. 17.

Example 4: Three Close-Tolerance Sides. Figure 18 shows the layout for gas cutting a part that had to fit a welding fixture. Two methods were used to hold the contour bounded by surfaces A, B, and C to a tolerance of $+0$, -0.8 mm $-\frac{1}{32}$ in.):

- Wedges were inserted into the kerf, locking the shape to the plate for maximum, restraint.
- The effects of distortion that remained were overcome by making allowances in the template.

The latter was accomplished by making cutting tests and measuring and adjusting the master aluminum template until the proper shape was developed. The master template was used only for preparing the template used on the job.

The equipment for cutting was the same as that described in the previous example. Operating procedure and torch settings are given in the table that accompanies Fig. 18. Although the procedure mentions the use of a paper template, a plastic template was substituted later, to extend template life to more than 4 h.

Example 5: Shape with Narrow Pointed End. The 19 cm (7.5 in.) thick shape shown in Fig. 19 included a narrow portion where the cutting path doubled back. Although the finished shape was within the tolerances given in Table 9, a slight rounding of the top edge of the 25 mm (1 in.) wide projection and of point A (Fig. 19) was observed. From top to bottom, the kerf width increased 1.6 mm $(\frac{1}{16}$ in.) per side, making the 25 mm (1 in.) projection 22 mm $(\frac{7}{8}$ in.) wide at the bottom with a kerf angle of approximately $\frac{1}{2}°$.

To start the cut for this shape, a hole was flame pierced to the left of the centerline at the top, and from it the cut progressed as shown in the sketch. Because of the deviations described previously, this shape was considered borderline insofar as process capability was concerned, in spite of the fact that tolerances were not exceeded.

The minimum width for a projection, such as on the shape that was cut in this example (Fig. 19), is a function of plate thickness. This relation cannot be stated precisely, mainly because the rounding of corners is also a function

Table 9   Dimensional tolerances normally obtainable in production gas cutting of plate

Plate thickness, cm (in.)	Tolerance, mm (in.)
Up to 1.3 ($\frac{1}{2}$) inclusive	± 1.6 ($\pm \frac{1}{16}$)
Over 1.3 to 2.8 ($\frac{1}{2}$ to $1\frac{1}{8}$)	± 2.4 ($\pm \frac{3}{32}$)
Over 2.8 to 7.6 ($1\frac{1}{8}$ to 3)	± 3.2 ($\pm \frac{1}{8}$)
Over 7.6 to 13 (3 to 5)	± 4.0 ($\pm \frac{5}{32}$)
Over 13 to 15 (5 to 6)	± 4.8 ($\pm \frac{3}{16}$)
Over 15 to 20 (6 to 8)	± 5.6 ($\pm \frac{7}{32}$)
Over 20 to 25 (8 to 10) inclusive	± 6.4 ($\pm \frac{1}{4}$)

For machine cutting with equipment in good working condition, and with adequate control for minimizing distortion in cutting shapes and long, narrow pieces.

Table 10   Recommended minimum dimensions for machine gas-cut holes and slots

Work metal thickness		Minimum hole diameter				Minimum slot size(b)	
		Rough cuts(a)		Accurate cuts(b)			
mm	in.	mm	in.	mm	in.	mm	in.
6.4–11	$\frac{1}{4}$–$\frac{7}{16}$	25	1	32	$1\frac{1}{4}$	13 by 38	$\frac{1}{2}$ by $1\frac{1}{2}$
13–24	$\frac{1}{2}$–$\frac{15}{16}$	19	$\frac{3}{4}$	32	$1\frac{1}{4}$	13 by 19	$\frac{1}{2}$ by $\frac{3}{4}$
25–47	1–$1\frac{7}{8}$	19	$\frac{3}{4}$	32	$1\frac{1}{4}$	19 by 25	$\frac{3}{4}$ by 1
50–72	2–$2\frac{7}{8}$	25	1	50	2	25 by 38	1 by $1\frac{1}{2}$
75–122	3–$4\frac{7}{8}$	50	2	...	...	38 by 50	$1\frac{1}{2}$ by 2
127–197	5–$7\frac{3}{4}$	75	3	...	...	50 by 75	$1\frac{1}{2}$ by 3

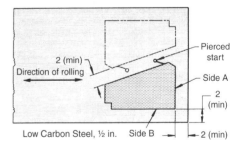

Note: Holes or slots are cut from pierced starts as illustrated above. Torch settings are as indicated in tables of cutting conditions for the fuel gas used. (a) Accuracy will be somewhat less than normal tolerances shown in Table 9. (b) Normal production tolerances shown in Table 9 will be maintained. Minimum sizes of holes are larger than those of slots because circular holes are more difficult to cut.

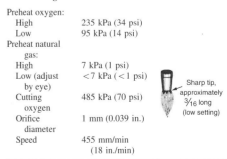

Operating procedure

1. Ink trace paper template from master template. Use magnets to hold template for tracing and cutting.
2. Locate and space items on template and plate as shown.
3. Set torch tip 13 to 19 mm ($\frac{1}{2}$ to $\frac{3}{4}$ in.) above plate.
4. Cut in clockwise direction.
5. Inspect first piece on each plate.
6. Destroy paper template 4 h after it is made.

Torch settings

Preheat oxygen:
High   235 kPa (34 psi)
Low   95 kPa (14 psi)
Preheat natural gas:
High   7 kPa (1 psi)
Low (adjust by eye)   <7 kPa (<1 psi)
Cutting oxygen   485 kPa (70 psi)
Orifice diameter   1 mm (0.039 in.)
Speed   455 mm/min (18 in./min)

Sharp tip, approximately $\frac{3}{16}$ long (low setting)

Fig. 17 Gas cutting a gusset with two perpendicular sides (example 3). Dimensions in inches

of the size of the tracing follower and the diameter of the cutting flame. The following table, however, may be used as a guide:

Plate thickness		Minimum projection width	
mm	in.	mm	in.
6.4	1/4	9.5	3/8
13	1/2	13	1/2
25	1	19	3/4
38	1 1/2	25	1
51	2	32	1 1/4
76	3	38	1 1/2
76–197	3–7 1/4	51	2

The opposite of a projection is the narrow re-entrant cut described in the next example.

Example 6: Narrow Re-entrant Cut. The 7.6 cm (3 in.) thick shape shown in Fig. 20 was cut within ±3.2 mm (±1/8 in.) (the tolerance given in Table 9), even though the 13 mm (1/2 in.) wide slot was narrower than the minimum width recommended in Table 10, and the 25 mm (1 in.) projection was less than the 38 mm (1.5 in.) nominal minimum for 7.6 cm (3 in.) stock. The shape was cut by machine, using natural gas preheat fuel.

Two problems were anticipated. One was that the heat buildup in the 13 mm (1/2 in.) portion of the slot would burn up the scrap, causing an irregular path for the cutting flame. The other

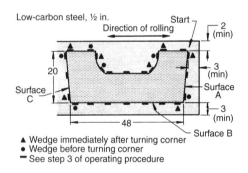

Low-carbon steel, 1/2 in.

▲ Wedge immediately after turning corner
● Wedge before turning corner
─ See step 3 of operating procedure

Operating procedure

1. Ink trace paper template from master template. Use magnets to hold template for tracing and cutting
2. Set torch tip 13 to 19 mm (1/2 to 3/4 in.) above plate.
3. Cut in clockwise direction with single torch.
4. Wedge approximately 13 mm (1/2 in.) behind cutting tip.
5. Reset previously placed wedges, where possible, after setting new wedges.
6. Inspect first piece on each plate.
7. Destroy paper template 4 h after it is made.

Torch settings

Preheat oxygen:
　High　　　　235 kPa (34 psi)
　Low　　　　95 kPa (14 psi)
Preheat natural
　gas:
　High　　　　7 kPa (1 psi)
　Low (adjust　<7 kPa (<1 psi)
　by eye)
Cutting oxygen　485 kPa (70 psi)
Orifice diameter　1 mm (0.039 in.)
Speed　　　　455 mm/min
　　　　　　　(18 in./min)

Sharp tip, approximately 3/16 long (low setting)

Fig. 18 Gas cutting a part with three sides held to close tolerances (example 4). Dimensions in inches

was that melting would occur at the edges of the 25 mm (1 in.) wide area of the projection.

However, by using a high-speed cutting tip at the high end of the cutting-speed range (18 to 33 cm/min, or 7 to 13 in./min), both problems were overcome. A narrow kerf was cut in the slot, leaving a scrap width of 6.4 mm (3/4 in.), which was sufficient to confine the flame. Heat buildup at the end of the projection was minimized by the cutting speed.

Gas Cutting versus Alternate Methods

There are few gas cutting operations that cannot be performed by some type of machining. Machining may produce more accurate results, but time and cost may be saved by gas cutting. The two examples that follow describe applications in which gas cutting was used to remove a large quantity of metal, either to eliminate machining or to minimize the time required for machining to final size.

Example 7: Gas Cutting versus Band Sawing. The cross shown in Fig. 21(a) originally was rough cut by band sawing and then rough and finish milled to size. The cross was cut from a piece of forged 4340 steel, 48.9 by 29.2 by 13.3 cm thick (19 1/4 by 11 1/2 by 5 1/4 in. thick).

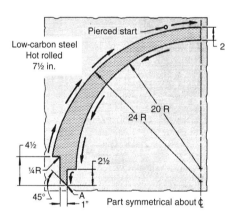

Low-carbon steel
Hot rolled
7 1/2 in.

Part symmetrical about ₵

Fig. 19 Gas-cut shape on which the narrow projection was subject to melting at the edges (example 5). Dimensions in inches

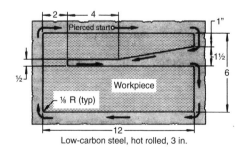

Low-carbon steel, hot rolled, 3 in.

Fig. 20 Shape that was gas cut to normal tolerance (+3.2 mm, or +1/8 in.) for 7.6 cm (3 in.) plate thickness, despite narrow re-entrant slot and thin projection, which ordinarily are subject to melting at the edges (example 6). Dimensions in inches

Because the shape could be traced by a single line set at the proper angle, gas cutting was substituted for band sawing, to save time. The cross was cut 7.6 mm (0.30 in.) oversize to allow for machining. A pantograph-type machine with a magnetic tracer was used for gas cutting. After a few trials to establish the best cutting sequence, the following operations were set up:

1. Set the torch to a compound angle consisting of a 32.5° bevel and left-to-right inclination of 13.75°. (The inclination angle caused some concern because the flame would lead the cut

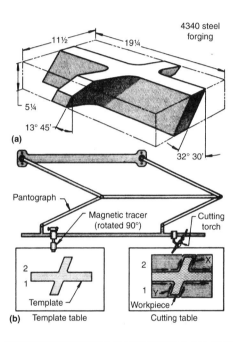

4340 steel forging

(a)

Pantograph
Magnetic tracer (rotated 90°)
Cutting torch

Template
Workpiece

(b)　Template table　　　Cutting table

Operating conditions for gas cutting

Cutting tip	Heavy-duty
Diameter of cutting-oxygen orifice	2.7 mm (0.1065 in.)
Diameter of preheat orifices (six)	1.2 mm (0.0465 in.)
Cutting-oxygen pressure	414–483 kPa (60–70 psi)
Cutting-oxygen flow rate	17–19 m³/h (600–680 ft³/h)
Preheat-oxygen flow rate (2-hose torch)	2.2 m³/h (77 ft³/h)
Acetylene pressure	41 kPa (6 psi)
Acetylene flow rate	2.0 m³/h (70 ft³/h)
Cutting speed	102–152 mm/min (4–6 in./min)

Man-hours per piece, band sawing

Drill starter holes	1.557
Band saw (rough size)	3.780
Rough mill (size, comparable to gas cut)	3.617
Total	8.954

Man-hours per piece, gas cutting

Load and unload furnace; set up block	0.113
Gas cut to size	0.495
Grind (clean up); handling	0.128
Total	0.736

Fig. 21 Cross (a) that was gas cut, by method shown in (b), in approximately 8% of the time formerly required for producing it by band sawing (example 7). Dimensions in inches

at faces X and Y, Fig. 21b, and such a condition could possibly cause gouging.)

2. Preheat the block to 425 °C (800 °F) and soak for 6 min. (Permissible range was 320 to 480 °C, or 600 to 900 °F.)
3. Place the preheated block on the cutting table and align with a fixed steel template.
4. Make cuts 1 and 2 from left to right, as shown in Fig. 21(b). (Details of cutting are listed in the table with Fig. 21.)
5. Cool in air to room temperature, allowing complete transformation of the cut faces. (Immediate postheat treatment caused increased hardness of the cut faces.)

Actual time saved by gas cutting over band sawing was determined by comparing man-hours for both methods. This comparison, shown in the table that accompanies Fig. 21, indicates that gas cutting required less than one-twelfth the time needed for band sawing.

Example 8: Gas Cutting versus Planing. Many boiler drums for steam generators in power plants are made from two semicylindrical half shells welded together longitudinally to form a cylinder. The ends of the cylinder are then closed with formed heads joined to the shell by a circumferential weld. The half shells are hot formed in a press from flat steel plate. After forming, the excess material at the longitudinal joints must be removed before the shells are welded together; either planing or gas cutting can be used for this removal.

This example illustrates a saving that was achieved by gas cutting. To establish the cost advantage in removing excess metal from the longitudinal joints of boiler-drum half shells (Fig. 22) by gas cutting, as compared with planing, data from previous operations on shells of various lengths and thicknesses were compared.

Data for the planing operation were averaged from 83 half-shell operations. The planing cut was made at a speed of 7 m/min (23 sfm), feed of 1 mm/min (0.040 in./min), and 19 mm (0.75 in.) depth of cut, using a high-speed steel cutting tool.

For gas cutting, the half shells were submerged in water to a depth 25 mm (1 in.) below the line of cut, to prevent distortion. Two cutting torches were used in a special cutting machine, so that both edges of the shell could be cut simultaneously. Torch settings and travel speed were based on standard straight-line oxyacetyl-ene cutting of the same thickness. Over 60 cuts were made in the same range of thicknesses used for planing.

Data were translated into cost per foot, and the data indicated that, gas cutting cost approximately 64% less per half shell, on the average.

Close-Tolerance Cutting

Shapes can be oxyfuel gas cut to tolerances of 0.8 mm ($1/32$ in.) in plate up to 64 mm ($2\frac{1}{2}$ in.) thick with multitorch pantograph or coordinate-drive machines. Conditions must be optimum:

- Operators must be well trained and experienced.
- Machines must be properly adjusted so that they can trace and retrace a pattern with an accuracy of ±0.13 mm (±0.005 in.) in a given direction.
- Workpiece must be flat and free of dirt, scale, grease, and oil.
- Workpiece should be supported or restrained to prevent movement of the shape during cutting.
- Cutting tip must be clean (selected for high-quality, narrow-kerf cutting).
- Cutting speed and gas pressures should be tuned to specific production conditions.

Optical tracing devices can follow an ink line 0.64 mm (0.025 in.) wide on paper. If cutting speed does not exceed 760 mm/min (30 in./min), a template with an outside corner radius of 2.4 mm ($3/32$ in.) will produce a sharp corner. A 2.4 mm ($3/32$ in.) inside radius on the template produces a somewhat larger radius in the workpiece, depending on thickness and cutting tip. Numerical control tracing, whether punched tape or computer assisted, offers the greatest accuracy. Inaccuracies introduced by distortion of templates are avoided, and electronic directions are more exacting than those generated by mechanical cutting machines, which set limits on cutting speed.

Oxyfuel Gouging

Oxyfuel gas gouging differs from ordinary gas cutting in that, instead of cutting through the material in a single pass, the process makes grooves or surface cuts. Special cutting torches are required for most applications, although some gouging is done with standard torches and a special tip.

Special cutting tips for gouging vary in design, to suit the size and shape of the desired groove or surface cut.

A gouging tip for cutting flat grooves is shown in Fig. 23(a). Tips of this type are made in various sizes and shapes for manual or mechanical scarfing of billets and slabs, or for the removal of defects, pads, and fins from castings. Torches for these applications may include attachments for dispensing iron powder to increase the speed of cutting or to permit the scarfing of stainless steel. Gas consumption, especially of oxygen, is much greater than in ordinary gas cutting.

A gouging tip for cutting round grooves is shown in Fig. 23(b). The bent tip makes it possible to hold the torch so that the cutting flame will strike at a low angle. This type of gouging tip is used to remove defects from metal, such as in welds, or to cut a weld groove. A weld groove can be cut on the back of a welded joint to expose clean, sound metal for the final welding pass. Similarly, a weld groove can be cut in a butt joint between two plates.

J-Grooves. The tip shown in Fig. 23(a) also has been used to prepare the edges of thick-wall pressure vessels for welding by cutting a J-groove. Where plate edges with opposing J-grooves are butted together, a weld groove is formed (Fig. 24) that is more efficient, in thick sections, than a conventional V-groove. J-grooves are often made by planing, but oxyfuel gas gouging can be used to cut these grooves by a single pass in material up to 10 cm (4 in.) thick, and by two passes in thicker material.

J-grooves have been oxyfuel gas gouged in flat plates supported horizontally, so that dross from

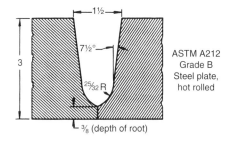

ASTM A212 Grade B Steel plate, hot rolled

Pressure of preheat oxygen	207 kPa (30 psi)
Pressure of preheat acetylene	76 kPa (11 psi)
Pressure of cutting oxygen at nozzle	152 kPa (22 psi)
Consumption of cutting oxygen	184 m³/h (6500 ft³/h)
Vertical tip angle	19°
Horizontal tip angle	11$\frac{1}{2}$°
Tip rotation angle	0°
Tip height	0
Tip offset	7.9 mm ($5/16$ in.)
Cutting speed	889 mm/min (35 in./min)

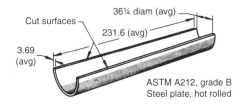

Fig. 22 Half shell from which excess metal of 7.7 cm (3.05 in.) average depth was removed by gas cutting at less cost than by planing (example 8). Dimensions in inches

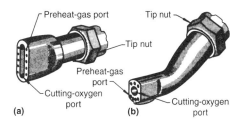

Fig. 23 Two types of gouging tips. (a) Tip for cutting flat grooves. (b) Tip for cutting round grooves

Fig. 24 Weld groove formed from two J-grooves made by oxyacetylene gouging. Dimensions in inches

the cut at the vertical edge is driven out by the combined forces of the flame and gravity. However, the following example describes an attempt to cut the grooves in formed half shells, supported in a U-position with the joint edges horizontal. This was done to reduce setup cost, because the half shells had already been sized by gas cutting.

Example 9: Cutting J-Grooves by Gouging. After excess stock had been removed from boiler-drum half shells by gas cutting (example 8, Fig. 22), preparation of weld edges was required. Edge preparation consisted of cutting a J-groove along the longitudinal edges of the half shells. This was usually done by planing. In this application, however, large cost savings were possible if oxyfuel gas gouging could be used. The cost advantage would accrue mainly from using the same equipment and setup as was used for gas cutting to remove the excess stock.

The groove was changed slightly to accommodate the gouging technique. The objective was to provide a groove large enough to permit weld deposit into the root of the joint, but to minimize the cross-sectional area of the groove to reduce welding costs. Critical variables included gas flow and the position of the cutting tip relative to the cut. The tip was of the type shown in Fig. 23(a).

The three important angles of approach lay in planes taken through the axis of the tip. The vertical angle was in a plane at right angles to the surface to be cut, and it determined the angle of incidence of the flame to this surface. The horizontal angle determined the direction in which the molten slag would be driven from one side of the plate. The tip angle was the angle of rotation of the tip about its axis, and it determined the slope of the groove.

Other important dimensions were tip height (the distance of the tip from the surface), and tip offset, which determined the size of the root or uncut "land" on the surface to be cut.

A typical weld groove formed by abutting J-grooves gouged in this operation is shown in Fig. 24; conditions used for oxyacetylene gouging are listed in the accompanying table.

Gas gouging proved to be borderline for a trade-off with planing. A very high degree of skill was required to control all the variables in order to produce repetitive high-quality cuts. Material costs and the critical nature of the joints placed a heavy penalty on errors and defects. Because planing depended less on skill, and was quite dependable, this plant continued to use it for cutting J-grooves.

Other Gas Cutting Methods

The basic oxyfuel method can be modified to allow gas cutting of metals, such as stainless steel and most nonferrous alloys, that resist continuous oxidation. Modifications of the oxyfuel method include metal powder cutting, chemical flux cutting (or flux cutting), and oxygen lance cutting.

Metal Powder Cutting

Finely divided iron-rich powder suspended in a jet of moving air or dispensed by a vibratory device is directed into the gas flame in metal powder cutting. The iron powder passes through and is heated by the preheat flame, so that it burns in the oxygen stream. Heat generated by the burning iron particles improves cutting action. Cuts can be made in stainless steel and cast iron at speeds only slightly lower than those used for equal thicknesses of carbon steel. By adding a small amount of aluminum powder, cuts can be made through copper and brass.

Thickness. In plate 25 to 100 mm (1 to 4 in.) thick, powder cuts can be produced by machine to an accuracy of 0.8 to 1.6 mm ($^1/_{32}$ to $^1/_{16}$ in.). Heavier sections are seldom cut, except for the trimming of castings; in this application, hand cutting requires greater allowances to avoid damage. Typical metal powder cutting applications include removal of risers; cutting of bars, plates, and slabs to size; and scrapping.

Quality of Cut. The kerf has a layer of scale that, on stainless steel, flakes off as the workpiece cools. The surface exposed after scale removal has the texture of sandpaper. Light grinding is normally sufficient to smooth high

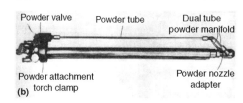

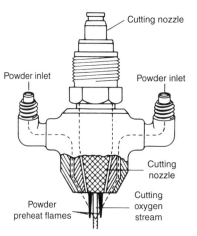

Fig. 25 Powder cutting attachments. (a) Single-tube attachment. (b) Multijet attachment. (c) Enlargement of powder nozzle adapter

spots and remove iron particles and oxide. Unstabilized austenitic stainless steel may become sensitized by the heat of cutting. Powder-cut cast iron develops a hardened case at the surface, which may require annealing or removal by grinding.

Equipment. In metal powder cutting, a gas torch with an external powder attachment (Fig. 25), or a torch with built-in powder passages, is used. A vibratory or pneumatic powder dispenser (Fig. 26), air supply, and powder hose are required, in addition to fuel and oxygen lines. The equipment may be used manually for removing metal, such as risers from castings, or mechanized for straight-line or shape cutting by machine. Powder cutting torches mounted on gas cutting machines are capable of cutting stainless steel.

Powder dispensers may be of two general types: pneumatic and vibratory. Generally, all dispensers consist of a compressed air/nitrogen filter and regulator, a powder hopper, and a means of expelling the powder from the dispenser.

During operation of a pneumatic dispenser (Fig. 26), powder flow to the ejector unit is the result of gravitational force and the pressure differential between the discharge end of the air line in the ejector and the surface of the powder when the torch powder valve is opened. Compressed air (or nitrogen) passing through the ejector draws powder from the ejector baffle plate into the gas stream, forming an air-powder (or nitrogen-powder) mixture that then passes through the ejector outlet tube to the torch or

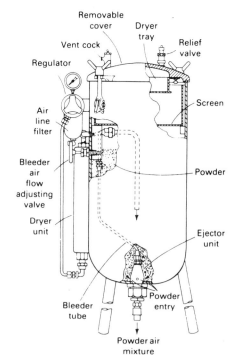

Fig. 26 Pneumatic powder dispenser for use in metal powder cutting

lance. The ejector adjusting screw controls the amount of powder falling on the baffle plate; the regulator and the air throttling value control the compressed air or nitrogen supplied to the ejector and the hopper. By manipulating the ejector adjusting screw, the regulator pressure, and the air throttling valve, the proper compressed air-powder (or nitrogen-powder) mixture can be obtained. This powder-conveying gas mixture is carried through a hose (6.4 mm, or 1/4 in., maximum diameter by 7.5 m, or 25 ft, long) connected to the cutting torch.

A removable cover that is fitted with a pressure-relief valve allows the hopper to be filled. A drying tray and a removable screen for eliminating oversized powder particles are also fitted into the top of the hopper.

Vibratory dispensers allow a quantity of powder to be dispensed from a hopper that is governed by a vibrator. Powder flow can be controlled by adjusting the amplitude of vibration.

Vibratory dispensers are generally used where uniform and accurate powder flow is required. Typical applications include precision cutting of materials, such as stainless steel, as well as production of high-quality, sharp top edges on cuts in carbon steels.

Powder Selection and Production. Many metallic and nonmetallic powders have been used for cutting operations. The ideal powdered solid fuel, whether metallic or nonmetallic:

- Exhibits nonhygroscopic action (does not readily absorb moisture)
- Liberates considerable heat during combustion
- Generates combustion products that flux refractory oxides
- Generates nonirritating and nontoxic combustion products
- Provides good process economy

Nonmetallic powders, such as sodium bicarbonate, remove refractory oxides by forming liquid slags that have a fluxing effect. Such powders, however, give off little or no heat. Consequently, cutting speeds are significantly slower than for metallic powders.

The most suitable metallic powders for cutting applications are iron and iron-aluminum blends. For metal powder cutting of stainless steel, iron powder can be used. For cutting oxidation-resistant materials other than stainless steels, blends containing iron and atomized −100 +325 mesh aluminum powder are preferable.

Iron powder used for metal-cutting applications is produced by water atomization, or by a combination of granulation of molten iron by high-pressure water jets and chemical reduction.

Aluminum powder is sometimes added to iron powder to form blends used for cutting and lancing concrete, firebrick, difficult-to-cut nonferrous metals, and refractory-laden scrap, such as slag deposits on ladle pouring spouts. Table 11 provides data on cutting-grade aluminum powders that are produced by air atomization of molten aluminum.

Powder blends consisting of iron and 10 to 40 wt% Al are used to cut a variety of materials. Although preblended compositions consisting of iron and 15 wt% Al are available, iron and aluminum powders can be custom blended for specific applications, as shown in Table 12.

To blend powders properly, components should be poured simultaneously into a dry, nitrogen-filled container and tumbled for approximately 5 min. Additional blending occurs as the iron-aluminum powder is screened in the powder dispenser. Slight nonuniformities in the mixture have little effect on cutting performance.

Powder-Cutting Apparatus. Attachments for metal powder cutting are designed for use with conventional oxyfuel cutting torches. Attachments are available in three types: single-tube, multijet, and dual-tube. In addition to powder cutting attachments, specially designed powder cutting torches are available.

Single-tube attachments, which are clamped to a conventional machine or hand-held oxyfuel cutting torch, as shown in Fig. 25(a), externally discharge a single, uniform stream of powder into the cutting-oxygen stream at an angle of approximately 25°.

Multijet attachments, as shown in Fig. 25(b), are equipped with an adapter that fits over the standard cutting nozzle. The adapter (Fig. 25c) completely encircles the nozzle and feeds powder from a ring of ports through the preheat flames and into the cutting-oxygen stream.

Dual-tube attachments are used for high-capacity machine cutting torch applications in which the size and cost of a multijet attachment are prohibitive. Two powder nozzles are used to feed powder into the high-velocity cutting-oxygen stream.

Powder cutting torches are designed for manual powder cutting. Figure 27 shows a typical powder cutting torch. Depressing the cutting lever of the torches initiates the flow of both cutting oxygen and powder. The nozzles used in these torches are made from chrome-plated pure copper inserted into an abrasion-resistant sleeve.

Applications. Metal powder cutting is frequently used in cutting, scarfing, and lancing of oxidation-resistant materials. This type of metal cutting was originally developed for cutting stainless steels. The alloying elements (chromium and nickel) that impart desirable properties to stainless steels have made these steels difficult to process by conventional oxyfuel cutting. Chromium is the most detrimental; when oxyfuel cutting is applied to stainless steels, chromium, which has a high affinity for oxygen at elevated temperatures, immediately forms the highly refractory chromium oxide on the faces of the kerf and prevents further oxidation. These refractory oxides are melted and fluxed by the combustion of the powder particles during metal powder cutting. Table 13 gives typical operating conditions for cutting of 18-8 stainless steel.

Table 11 Typical properties of atomized aluminum powder for metal powder cutting and lancing

Screen analysis

U.S. mesh	%
On 100	Trace
−100 +200	52–67
−200 +325	31–42
−325	0–4

Chemical analysis(a), %

Aluminum	99.7
Iron	0.18–0.25 max
Silicon	0.12–0.15 max
Other metallics, each	0.01–0.03 max
Other metallics, total	0.15 max

Physical properties

Apparent density	1.1 g/cm^3
Tap density	1.4 g/cm^3
Surface area	0.10–0.20 m^2/g

(a) Chemical analysis excludes 0.4% aluminum oxide, which exists on the surface of the particles.

Table 12 Iron-aluminum blends used for cutting and lancing difficult-to-cut materials

Nominal composition	Applications
Fe-10–15Al	Refractory scrap, thin nonferrous sections up to 25 mm (1 in.) thick
Fe-25Al	Aluminum, brass, bronze
Fe-30Al	Nickel, Monel, Inconel, Hastelloy, and concrete
Fe-40Al	Copper, brass, and bronze heavy sections 15 cm (6 in.) thick and up

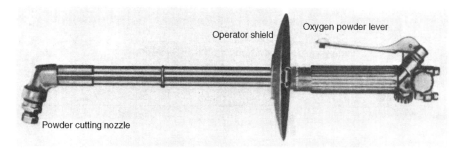

Fig. 27 Metal powder cutting torch

Nonferrous metals with the following thicknesses are suitable candidates for powder cutting:

	Thickness	
Material	cm	in.
Aluminum	25	10
Nickel	15	6
Brass and bronze	25	10
Copper	15	6
Hastelloy	16.5	6.5
Inconel and Monel	25	10

Powder Scarfing. In steel mills, powder metal cutting is used to scarf metals with alloy contents that are too high for oxyfuel scarfing. Typical applications include powder scarfing of large, bloomed ingots and small, unrolled ingots, slabs, and billets. Conditioning of these product forms by powder scarfing is less expensive and less time-consuming than mechanical methods such as grinding. Table 14 provides typical operating conditions for scarfing of 18-8 stainless steel.

Powder lancing is a piercing process in which the energy of the powder reaction is applied to oxygen lancing, thus permitting rapid, effective piercing of many materials that are difficult to pierce with a standard oxygen lance. These include iron and steel containing inclusions, reinforced concrete, firebrick, cinder block, aluminum billets, and sand and metal incrustations inside large castings.

Typical powder lancing applications include:

- Removal of blast furnace bosh plates
- Removal of large masses of iron (salamanders) that are deposited at the base of a blast furnace
- Cleaning of furnace linings
- Furnace tapping to remove slag
- Cleaning of soaking pits
- Removal of ladle skulls
- Piercing holes in reinforced concrete walls and floors

The efficiency of the powder lance in piercing reinforced concrete is preferred to the time-consuming, costly alternative of drilling with pneumatic tools.

Flux Cutting

Flux cutting processes are well suited to materials that form refractory oxides, such as stainless steels. Finely pulverized flux is injected into the cutting oxygen before it enters the cutting torch. The torch has separate ducts for preheat oxygen, fuel gas, and cutting oxygen. When the flux strikes the refractory oxides that are formed when the cutting oxygen is turned on, it reacts with them to form a slag of lower-melting-temperature compounds. This slag is driven out, enabling oxidation of the metal to proceed. The operator should have an approved respirator for protection from toxic fumes generated by the process.

Oxygen Lance Cutting

Oxygen lance cutting uses oxygen supplied at pressure through a consumable steel pipe or lance. Typically, an oxyfuel torch is used to heat the cutting end of the lance to a red heat, and then the oxygen flow is started through the lance. The lance or pipe begins to burn in a self-sustaining exothermic reaction at the burning end, and the preheat torch is removed. When the burning end of the lance is brought near the workpiece, the work is melted by the heat of the flame. Lance tubes are normally sold in lengths of 3.2 m (10½ ft) with diameters of 16 mm (⅝ in.). More recent versions of the lance tubes have many small-diameter carbon steel wires packed into the steel tube to increase the cutting life and capability of the lance.

An experienced operator can pierce or cut through sections several feet thick by oxygen lancing. Protection is needed from flying sparks and molten slag or metal. Oxygen lancing can pierce nearly every material, including metals, reinforced concrete, refractory brick, mortar, and slag. The oxygen lance has also been used in underwater applications, having first been lighted before being placed underwater. Cutting oxygen is supplied at 550 to 830 kPa gage pressure (80 to 120 psig) and is controlled by a simple on-off valve at the end of the lance. The oxygen lance, in addition to being used for piercing, is also used to cut apart large steel plate structures. After September 11, 2001, this process was used extensively in cutting large fallen portions of the World Trade Center's structural steel into smaller, more easily moved pieces.

The discharge end of the iron pipe is preheated to ignition temperature, the valve at the opposite end of the pipe is opened gradually, and the oxygen burns off the end of the pipe, which is pressed gently against the metal to be cut or pierced. As the metal starts to react with the oxygen or melt from the heat generated by the burning pipe, the pipe is advanced to cut a hole somewhat larger in diameter than the pipe.

The action is continued until the work metal is completely penetrated or the pipe is burned to the minimum length for safe operation. Then, a new length of pipe is attached, and the process is resumed.

Severing requires a series of piercing operations along the desired line of cut, or washing along one side of the initial hole to form a slot. In the latter method, care must be taken to provide continual drain of molten slag and metal from the cut; otherwise, the entering pipe will overheat and burn up too fast, or the dross will solidify. Restarting the cut on solidified dross is difficult.

An oxygen lance is often used to make starting holes for cutting thick sections of steel and in heavy scrapping operations. The addition of powder to the oxygen has greatly improved the effectiveness of the lance on oxidation-resistant materials, such as cast iron, refractories, and concrete.

Table 13 Typical oxyfuel gas powder cutting conditions for 18-8 austenitic stainless steel

Metal thickness		Cutting-oxygen orifice diameter		Cutting-oxygen pressure		Cutting speed		Gas consumption				Powder flow	
								Oxygen		Acetylene			
mm	in.	mm	in.	kPa	psi	mm/s	in./min	L/min	ft³/h	L/min	ft³/h	g/min	oz./min
13	½	1.02	0.040	344	50	5.9	14	59	125	7	15	113	4
25	1	1.52	0.060	344	50	5.1	12	106	225	11	23	113	4
51	2	1.52	0.060	344	50	4.2	10	142	300	11	23	113	4
76	3	2.03	0.080	344	50	3.8	9	260	550	15	32	142	5
102	4	2.54	0.100	344	50	3.4	8	319	675	18	38	170	6
127	5	3.05	0.120	414	60	3.0	7	378	800	21	45	198	7
152	6	3.56	0.140	414	60	2.5	6	425	900	30	63	227	8
203	8	3.56	0.140	483	70	1.7	4	472	1000	30	63	227	8
254	10	4.06	0.160	517	75	1.5	3.5	520	1100	35	75	227	8
305	12	4.06	0.160	517	75	1.3	3	566	1200	35	75	227	8

Table 14 Typical powder scarfing conditions for 18-8 stainless steel

Cutting-oxygen orifice diameter		Oxygen pressure		Acetylene pressure		Powder flow	
mm	in.	kPa	psi	kPa	psi	kg/h	lb/h
0.5	0.020	345–552	50–80	55–83	8–12	7–14	15–30
1.02	0.040	448–517	65–75	55–83	8–12	14–34	30–75

Note: Distance of nozzle from work: 64 mm (2½ in.); impingement angle: 30° (approximate); lateral angle: 50 to 10° (approximate); speed: 41 to 71 mm/s (96 to 168 in./min)

Underwater Gas Cutting

In underwater gas cutting, an oxyhydrogen flame usually is preferred to oxyacetylene for preheat. Hydrogen is stable, and the flame is easily controlled at the higher pressures required for underwater operation. Oxynatural gas cutting is also used in underwater gas cutting, for the same reason.

The flame and the oxygen jet are shielded from the water by a cup that surrounds the cutting tip and into which compressed air is fed. The cup shield also serves as a guide by establishing the proper tip-to-work cutting distance. Air and gaseous products escape through slits in the shield. The higher heat input of oxygen-arc cutting is more efficient in cutting thicknesses of 9.5 mm (3/8 in.) and less, because of the cooling effect of the water.

Safety

The hazards of combustion or possible explosion associated with the gases used in OFC, as well as the presence of toxic gases and dust, make it necessary for the user to follow established safety precautions. The three areas of most concern, dealt with in this section, are protective clothing, handling and storage of gas cylinders, and the working environment.

Protective Clothing. Size, nature, and location of the work to be cut dictate the necessary protective clothing. The operator may require some or all of the following:

- Tinted goggles or face shield for light cutting, up to 25 mm (1 in.)—shade No. 3 or 4; for medium cutting, 25 to 150 mm (1 to 6 in.)—shade No. 4 or 5; and for heavy cutting, thicker than 150 mm (6 in.)—shade No. 5 or 6
- Hard hat and sometimes a close-fitting hat beneath
- Safety glasses
- Flame-resistant jacket, coat, hood, or apron; wear woolen clothing, not cotton or synthetic fiber; keep sleeves and collars buttoned, no cuffs; button-down pockets
- Protective gloves, wristlets, leggings, and spats. When cutting plated or coated stock, respiratory protection should be used if necessary.
- Clothing should be free of grease and oil, and free of ragged edges.

Cylinders. Fuel gas cylinders should be placed in a location where they cannot be knocked over. Cylinders should be kept upright, with valve ends up to avoid hazardous liquid withdrawal; the protective caps should never be removed, except when the cylinders are connected for use. Cylinders should never be exposed to high heat or open flame, which could cause them to explode.

Cylinders should not be placed where they may complete an electric circuit, such as against a welding bench. They should not be used as an electrical ground or as a surface on which to strike an arc.

Damage to cylinders by impact should be avoided. Regulators should be removed before cylinders are moved, except on portable outfits. Slings or magnets should not be used as supports or rollers.

Damaged or defective cylinders should be removed from service immediately, tagged with a statement of the problem, and returned to the supplier. Leaking cylinders should be transported to an isolated location outdoors, the supplier notified, and his instructions for disposal followed.

No attempt should be made to fill a small cylinder from a large one. Filling requires special equipment and training.

Before storing, valves should be closed and caps replaced on empty or unused cylinders. Empty cylinders should be marked accordingly and kept separate from filled ones. Unused oxygen or fuel gas should not be drained from cylinders, and oxygen cylinders should be kept away from fuel gas cylinders.

Working Environment. Torches should not be used around chlorinated solvent vapors; heat forms phosgene and other corrosive, toxic products from them. Adequate ventilation must be provided, especially when alloys containing lead, cadmium, zinc, mercury, beryllium, or other toxic elements are being cut. A flame should not be used in a closed vessel or pipeline that has held flammable or explosive material, such as gasoline. Adequate ventilation should also be provided when work is being performed in a confined area. Under such conditions, the person doing the cutting should have an assistant, in case trouble occurs.

Fire-prevention warnings in "Safety in Welding and Cutting" (ANSI/AWS 249.1), which have legal status as Occupational Safety and Health Administration (OSHA) regulations, should be heeded. Workers should know the locations and operating procedures of the nearest fire extinguishers.

If cutting is to be done over concrete, it should be protected with metal or another suitable material; concrete spalls explosively when overheated. Hoses should be protected from sparks, hot slag, hot objects, sharp edges, and open flame.

ACKNOWLEDGMENTS

This article was adapted from "Oxyfuel Gas Cutting" in *Welding, Brazing, and Soldering,* Volume 6 of *ASM Handbook,* 1993. The original content is from the article "Gas Cutting" in *Forming,* Volume 4 of *Metals Handbook,* 8th ed., 1969, p 278–300. Special thanks to Kent Johnson of Engineering Systems Incorporated for his help in reviewing content for this volume. The section "Safety" in this article is adapted from Welding Engineering Data Sheet No. 491, "Safety Checklist for Oxyfuel Gas Cutting."

Electric Arc Cutting

ELECTRIC-ARC CUTTING is a method of melting or oxidizing metal by applying heat from an electric arc to the work-metal surface along a line of cut. Because of the extremely high temperature developed, the electric arc can be used to cut any metal that conducts electricity. Modifications of the basic process include the use of compressed gases to cause rapid oxidation (or to prevent oxidation) of the work metal, thus incorporating some aspects of gas cutting.

Electric-arc cutting includes several processes, of which the following are of commercial importance: air-carbon arc cutting, oxygen arc cutting, plasma arc cutting, and the Exo-Process. Other, seldom-used processes, which largely have been superseded by the aforementioned, are briefly described in the last section of this article.

Electric arc cutting can be used on ferrous and nonferrous metals for rough severing, such as removing risers or scrap cutting, as well as for more closely controlled operations. Each process has particular capabilities and limitations. Special applications include shape cutting, grooving, gouging, and underwater cutting. Maintaining the squareness of cut edges, particularly in sections more than 50 mm (2 in.) thick, is generally more difficult in electric arc cutting than in gas cutting.

Plasma Arc Cutting*

Plasma arc cutting (PAC) is an erosion process that utilizes a constricted arc in the form of a high-velocity jet of ionized (electrically conductive) gas to melt and sever metal in a narrow, localized area. The arc is concentrated by a nozzle onto a small area of the workpiece. The metal is continuously melted by the intense heat of the arc and then removed by the jetlike gas stream issuing from the torch nozzle. The PAC process relies on heat generated from electrical arcing between the torch electrode and the workpiece. It generates very high temperatures (28,000 °C, or 50,000 °F, compared to 3000 °C, or 5500 °F, for oxyfuel cutting) and does not depend on a chemical reaction between the gas and the work metal. Therefore, plasma arc cutting can be used on almost any material

*Adapted from C. Landry, Plasma Arc Cutting, *Welding, Brazing, and Soldering,* Vol 6, *ASM Handbook,* ASM International, 1993, p 1166–1171

that conducts electricity, including those that are resistant to oxyfuel gas cutting. This traditional mode is referred to as transferred arc cutting. Another mode is the nontransferred arc method, which can cut nonmetallic objects (such as rubber, plastic, styrofoam, fabric, and wood) with a good quality surface to within 0.50 to 0.75 mm (0.020 to 0.030 in.) tolerances.

Plasma arc cutting was originally developed for cutting nonferrous metals using inert gases. Modifications of the process with the use of oxygen or compressed air in the orifice gas also permit the cutting of carbon and alloy steel with improved cutting speeds and a cut quality similar to that obtained with oxyfuel cutting. Plasma arc cutting can reach high cutting speeds (Fig. 1) with productivity increases over that of oxyfuel gas cutting without increasing space or machinery requirements. Plasma arc cutting of carbon steel plate can be done faster than with oxyfuel gas cutting processes in thicknesses below 75 mm (3 in.) if the appropriate equipment is used. For thicknesses under 25 mm (1 in.), plasma arc cutting speed can be up to five to eight times greater than that for oxyfuel gas cutting (Fig. 2). For thicknesses greater than 38 mm (1½ in.), the choice of plasma arc or oxyfuel

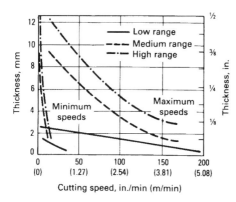

Fig. 1 Typical cutting speeds for PAC of carbon steel or stainless using 6.8 m³/h (240 ft³/h) of air at 345 kPa (50 psi) from a single source. This information represents realistic expectations using recommended practices and well maintained systems. Other factors such as parts wear, air quality, line voltage fluctuations, and operator experience may also affect system performance.

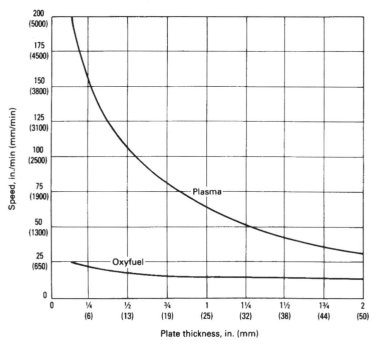

Fig. 2 Comparison of oxyfuel gas cutting and PAC of plain carbon steel

gas cutting depends on other factors such as equipment costs, load factor, and applications for cutting thinner plates and nonferrous metals. Characteristics of plasma arc cutting and oxyfuel gas cutting are compared in Table 1.

Operating Principles

The basic plasma arc cutting torch is similar in design to that of a plasma arc welding torch (Ref 1). In its simplest form, a PAC system includes a gas-cooled torch with leads that connect to a power supply (Fig. 3). The basic components include:

- A torch designed for either manual or mechanized operation
- Torch leads that carry the electric power, gas, and coolant to the torch
- A coolant system for the torch
- A power supply that provides the proper current and voltage for the plasma arc
- A manifold system for combining power, gas, and coolant for connection to the torch
- A control system that consists of switches, gages, dials, meters, and other controls for operating the PAC system and the gas supply

In the PAC process, a cool and inert gas, such as compressed air, is forced under pressure through a small orifice in the front of the cutting torch. This torch is connected by leads to a direct current (dc) power supply. In the torch, a portion of the inert gas is changed into a plasma (ionized gas) by heat created by the discharge of a high-voltage arc from the power supply. This arc is created between an electrode (dc negative) in the torch and the tip (nozzle) of the torch through which the gas flows.

The basic plasma arc cutting circuitry is shown in Fig. 4. The process operates on direct current, straight polarity (dcsp), electrode negative, with a constricted transferred arc. In the transferred arc mode, an arc is struck between the electrode in the torch and the workpiece.

The arc is initiated by a pilot arc between the electrode and the constricting nozzle. The nozzle is connected to ground (positive) through a current-limiting resistor and a pilot arc relay contact. The pilot arc is initiated by a high-frequency generator connected to the electrode and nozzle. The welding power supply then maintains this low-current arc inside the torch. Ionized orifice gas from the pilot arc is blown through the constricting nozzle orifice. This forms a low-resistance path to ignite the main arc between the electrode and the workpiece. When the main arc ignites, the pilot arc relay may be opened automatically to avoid unnecessary heating of the constricting nozzle.

When a small amount of dc from the power supply is imposed on this high-voltage arc, it achieves a pilot arc that extends approximately 13 mm (1/2 in.), as a plasma jet from the tip orifice. This pilot arc acts as a path that establishes the main cutting arc to the workpiece. When the pilot arc is brought in contact with the metal workpiece, which is connected to the positive side of the dc power supply, the transferred (main) arc occurs and the pilot arc is shut off.

The cutting torch is usually operated with a standoff, that is, the distance between torch and workpiece. Standoff distance may vary from approximately 3 to 10 mm (1/8 to 3/8 in.) to about 5 to 20 mm (1/4 to 5/8 in.) for some machine-operated plasma arc torches. In some applications, the torch tip can actually be dragged along the workpiece. The transferred plasma arc heats the metal by the combined effects of the electric arc from the power supply, the high-temperature plasma created by constriction of the arc by the orifice, and the reassociation of the gas molecules at the workpiece. The molten metal is blown out of the cut, or kerf, by the high-velocity, well-columnated plasma jet, resulting in fast, clean cuts.

Equipment

Plasma Arc Cutting Torches. All plasma arc torches constrict the arc by passing it through an orifice as it travels away from the electrode and toward the workpiece. The basic design and terminology for a PAC torch are shown in Fig. 5. The tip, or nozzle, of the torch contains the orifice that constricts the arc. The diameter of the orifice, which ranges from 1 to 6.4 mm (0.038 to 0.250 in.), determines the maximum cutting current that can be used with that tip. The larger the orifice, the more current that can be forced through it. As the orifice gas passes through the arc, it is heated rapidly to high temperature, expands, and accelerates as it passes through the constricting orifice. The intensity and velocity of the arc plasma gas are determined by such variables as the type of orifice gas and its entrance pressure, constricting orifice shape and diameter, and the plasma energy density on the work.

Because the plasma constricting nozzle is exposed to the high plasma flame temperatures (estimated at 10,000 to 14,000 °C, or 18,000 to 25,000 °F), the nozzle is sometimes made of water-cooled copper. In addition, the torch should be designed to produce a boundary layer of gas between the plasma and the nozzle. Cooling of PAC torches that operate at less than 150 A can be simply accomplished by properly channeling the gases used in torch operation. High-powered systems require a water-based coolant system, which consists of a reservoir, pump, and heat-exchanger assembly. Because this type of coolant is often in contact with both negative and positive electric potentials within the torch, it is important to use a deionized (nonconductive) coolant.

Tips are designed so that the orifice provides maximum arc constriction. If a tip orifice is overpowered, then it will become gouged or washed out, resulting in a change of arc characteristics and in the deterioration of cut quality and cutting speed. To attain optimal cutting conditions, tips are used at or very near their maximum cutting rating. Torch tips are either water cooled, through internal passages connected to a closed-loop coolant system, or indirectly cooled, by good thermal contact with other torch parts that are either water or gas cooled.

Table 1   Comparison of oxyfuel gas and PAC processes

	Oxyfuel	Plasma arc
Flame temperature	3040 °C (5500 °F)	28,000 °C (50,000 °F)
Action	Oxidation, melting, expulsion	Melting, expulsion
Preheat	Yes	No
Kerf	Narrow	Wide
Cut	Both sides square	One side square
Speed	Moderate	High
Heat-affected zone	Moderate	Narrow
Cutting ability:		
Carbon steel	Yes	Yes
Stainless steel	Requires special process	Yes
Aluminum	No	Yes
Copper	No	Yes
Special alloys	Some	Yes
Nonmetallics	No	Yes

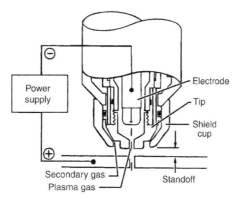

Fig. 3 Typical PAC torch

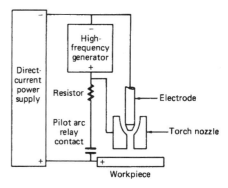

Fig. 4 Plasma arc cutting circuit. The process operates on direct current electrode negative (straight polarity). The arc is initiated by a pilot arc between the electrode and torch nozzle. Pilot arc is initiated by the high-frequency generator, which is connected to the electrode nozzle.

The electrode, or cathode, in a PAC torch is the surface from which the plasma arc is generated. This electrode is in electrical contact with the negative potential in the power supply. Both electrode material and design are quite varied. For most applications that use nitrogen or argon-hydrogen as the plasma gas, a 2% thoriated tungsten is used as the electrode. Hafnium and zirconium are commonly used as electrode materials for both air and oxygen plasma cutting.

In terms of its construction, the electrode can be solid, such as a rod of tungsten or a tungsten element held in a solid-copper holder, or it can be hollow and have two pieces (copper and the electrode element) that allow direct water or gas cooling. Because proper cooling will promote both tip and electrode life, PAC torches designed for cutting above 150 A are usually water cooled.

In the area between the electrode and the tip orifice, which is called the plenum, the initial ionization or plasma formation takes place. The plasma gas enters the plenum in either a straight (laminar) flow pattern or a tangentially swirling pattern. Although torch design depends on several factors, pointed electrodes usually utilize laminar flow patterns, whereas flat or blunt-ended electrodes utilize a swirling gas to keep the arc attachment point moving around the center of the electrode. The swirl is most commonly imparted on the gas by a gas distributor, or swirl ring, that is located near the end of the electrode.

Other parts that are common to most PAC torches include internal ceramic or high-temperature plastic insulators for electrical insulation, collets or other mechanisms for retaining the electrode, and shield cups to insulate the front end of the torch from the workpiece.

Leads. The PAC torch is connected to the other system components by means of a leads package that consists of hoses, fittings, and wires. The leads supply the electric power, gases, and cooling medium to the torch. Lead lengths vary from 3 to 68 m (10 to 225 ft), depending on the application. Simple extension kits can be used to join leads packages together for increased length.

The power supply for a PAC system must have a dc output with the negative side connected to the electrode in the torch and the positive side connected to the workpiece. Systems that include a pilot arc also have a positive connection, limited in current through a resistor, to the tip in the torch. Power supplies are usually of a constant-current, dropping volt-amp curve design, and they have higher open-circuit voltages (up to 400 V) than those found in common welding power supplies, in order to accommodate the high operating voltage (90 to 200 V) of the torches. The amperage output of power supplies can be designed to have one fixed level, or several switchable fixed settings, or an infinitely adjustable output by means of a control potentiometer.

The manifold assembly combines the dc power, gas, coolant, and high-voltage starting circuitry and provides it to the cutting torch via the leads package.

Controls for starting and stopping the cutting operation, selecting the proper amperage output, adjusting gas flows, and regulating other system functions can be located on the manifold box, on the power supply, or in a dedicated control cabinet for convenient mounting near an operator station.

All-in-one packaged PAC systems that combine the coolant system, power supply, manifold assembly, and controls in one unit are available. These systems offer improved portability, simplicity of operation, and reduced costs.

Optional equipment for manual PAC systems includes drag-type shield cups with protrusions, or feet, that enable direct contact with the workpiece to allow the operator to maintain a constant standoff. Wheels that can be attached to the torch for manual guidance, as well as circle-cutting attachments that include an adjustable radius rod to allow the cutting of various-sized holes, are also available.

Optional equipment for mechanized PAC systems includes torch-standoff controls that automatically find and maintain the proper standoff for the cutting operation. Water-shrouded systems are also available. When used in conjunction with water tables, these systems provide a blanket of water around the front of the torch during the cut, in order to minimize smoke, fumes, noise, and arc glare.

Machine Ratings. In selecting a plasma cutting unit, the thickness of plate to be cut and the required cutting speed should be considered. Table 2 shows thickness capacity and midrange cutting speeds of four plasma units. These ratings are an average of speeds quoted by two manufacturers of plasma arc cutting equipment.

The thickness capacity of a cutting unit should first be examined to determine the cutting speed it can achieve for a given application. Next the speeds quoted for the next-larger unit should be studied to see whether the greater speed justifies its higher cost. For example, a 30 A unit cuts 6 mm (¼ in.) stainless steel plate at 125 to 250 mm/min (5 to 10 in./min); the 50 A unit cuts 6 mm (¼ in.) plate at 635 to 1270 mm/min (25 to 50 in./min), a significant increase. If the average required cutting thickness exceeds 75% of the maximum thickness capacity of a unit, the next larger size should be considered.

Pierce capacity is usually half the cutting thickness capacity, an important consideration in selection of plasma equipment. To cut plate thicker than 75 mm (3 in.), connecting 400

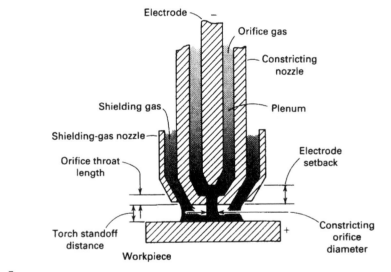

Fig. 5 Components of a PAC torch

Table 2 Cutting speed of PAC machines for stainless steel

Machine rating, A	Cutting speed, m/min (in./min)							
	Plate thickness, mm (in.)							
	1.5 (¹/₁₆)	3 (¹/₈)	6 (¹/₄)	9 (³/₈)	13 (¹/₂)	25 (1.0)	50 (2.0)	75 (3.0)
30	0.75–1.5 (30–60)	0.5–0.75 (20–30)	0.13–0.25 (5–10)	...	...	...	...	...
50	1.5–3.0 (60–120)	1.3–2.5 (50–100)	0.6–1.3 (25–50)	0.13–0.25 (5–10)	0.025–0.13 (1–5)	...	...	...
100	...	1.5–2.8 (60–110)	0.75–1.5 (30–60)	0.5–1.0 (20–40)	0.4–0.5 (15–20)	0.13–0.25 (5–10)	...	...
400	...	3–4 (120–150)	4–4.3 (150–170)	3–3.5 (120–140)	2.5–3 (100–120)	1.0–1.5 (40–60)	0.25–0.5 (10–20)	0.08–0.2 (3–8)

or 500 A units in parallel extends thickness capacity.

Gases

Any gas or gas mixture that does not degrade the properties of the tungsten electrode or the workpiece can serve as a plasma gas. The gas mixture varies according to the plasma equipment design criteria. The most commonly used gas is compressed air; all common metals, such as carbon and alloy steels, stainless steels, and aluminum, can be cut with compressed air. As the metal thickness increases (over 25 mm, or 1 in., with steels and stainless steels), benefits are derived from the use of nitrogen plasma with CO_2 shielding. Aluminum cut quality is improved using argon-hydrogen plasma and nitrogen as the secondary gas blanket. When high-duty cycles are used, a change from compressed air to nitrogen/CO_2 prolongs the consumable life.

The plasma or secondary shielding gases that are used in PAC systems are inert, and they are usually supplied from high-pressure cylinders, bulk tanks, or air compressors. Pressure regulators are adjusted to provide from 140 to 560 kPa (20 to 80 psi) of gas, depending on system requirements.

If water is used in place of a secondary gas, then it is usually supplied from a filtered, in-plant water system at a flow rate of 0.63 to 1.9 L/min (10 to 30 gal/h), again, depending on system requirements.

The purity of plasma gases is usually of a welding grade or higher. The plasma gas can be compressed air, which is inexpensive and provides fast, clean, high-quality cuts on carbon steels, but a relatively short electrode life, because of the rapid oxidation that occurs at high temperatures. If purity from an in-plant compressor is a problem, then either charcoal filters that remove oil vapors and moisture traps or dryers can be added into the line. As an alternative, clean, dry, compressed air can be obtained in cylinder form.

Another choice is nitrogen, which is relatively inexpensive and provides good-quality cuts on stainless steels and aluminum. It also provides good electrode life.

Argon-hydrogen, which is usually supplied in a 65 or 70% Ar and a 35 or 30% H mixture, operates at a higher temperature. It can reduce the quantity of smoke and fumes created during the cutting operation and provide a higher-quality surface finish on aluminum and stainless steels that are more than 13 mm (0.5 in.) thick. However, argon-hydrogen is usually from two to three times more expensive than nitrogen. In addition, it is not as readily available and requires more-critical adjustment of the operating parameters.

Oxygen plasma provides the same benefits as air plasma when cutting carbon steels. Because of the absence of nitrogen, it can minimize the amount of nitrides in the cut surface, thereby enabling better-quality welds on "as-cut" pieces that have not undergone additional finishing operations. Like air plasma, oxygen causes more rapid electrode erosion than does nitrogen or argon-hydrogen plasma.

Operating Sequence

A routine check of the PAC system before power is applied involves examining the torch to ensure proper assembly, verifying adequate coolant and gas supplies, and inspecting all components of the system to verify that they are in order.

The disconnect switch can then be turned on, which applies power to the system. Gas supplies should be turned on and adjusted according to the recommended pressure or flow. The plasma gas should be allowed to purge for an additional 2 to 3 min in order to eliminate any traces of moisture that accumulated within the plasma lines. This moisture is usually the result of condensation that forms when the system is off for prolonged periods of time.

The "start-cut" signal can be supplied by either an operator who simply pushes a start button on a mechanized control station to energize the switch on a manually operated torch or by a computer-generated signal on a fully automated system. The start-cut signal will cause the main contactor in the power supply to energize and initiate a gas prepurge. The prepurge ensures that gas is at the torch before an arc is struck and can last from 2 to 10 s if extra-long torch leads are used. At the end of the prepurge, the high-frequency arc and pilot arc circuits are energized, establishing the pilot arc at the torch. The pilot arc duration can be controlled by a timing circuit, limiting it to approximately 5 s, or it can be continuous, thereby shutting off only after an actual cut is established. Some PAC systems include circuitry for an automatic restart pilot, which enables automatic cycling between the pilot and cut modes as long as the start signal is present. This type of operation is valuable when cutting expanded metal, grating, or other material with an interrupted cut line.

The cutting arc will begin as soon as the torch and pilot arc are brought close enough to the workpiece (usually, 13 to 25 mm, or 1/2 to 1 in.) to allow main arc transfer. However, some systems do not use a pilot arc to start the torch. Rather, they only use high frequency and require that the torch be placed within 1 mm (0.038 in.) to establish a cutting arc. The cutting action will continue as long as there is metal to which the arc can transfer. If the cutting speed is too fast, then the arc will not fully penetrate the metal, and an incomplete cut or gouge will occur. If the cutting speed is too slow, then the kerf will widen as available metal is blown away and, eventually, the arc will be extinguished.

When the "start signal" is removed, or when a stop signal is provided, the cutting arc or pilot arc will shut off and the system will enter a post-purge mode, which is a continuation of plasma gas flow for 2 to 40 s to ensure that the electrode cools in a good atmosphere and to remove any residual heat within the torch. The system is then ready for the next start signal to begin the cutting sequence again. Some systems include circuitry that allows immediate starting of the pilot or cutting sequence at any time during the post-purge.

For manual operation, the torch is usually held with two hands to provide optimum comfort and speed control. An exception to this rule occurs when small, lightweight torches are used in conjunction with templates or a straight edge that requires one hand to be used for positioning while the other hand operates the torch. In manual operation, the current and rate of gas flow are set, and the arc is struck by pressing a button on the torch, which is guided manually over the work.

There are two means of starting the cut in manual PAC systems: edge starts and pierce starts. For edge starting, the torch tip is usually placed directly over the edge of the workpiece. Resting the torch at the start position ensures that the cut will begin at the proper location. The torch switch is pressed, and the torch is moved along the cut line as soon as the arc is established. At the end of the cut, the arc is automatically extinguished, and the control opens the contactor and closes the gas valves. The operator can extinguish the arc at any time by moving the torch away from the work metal.

Piercing is best accomplished by slightly angling the torch over the starting point so that the molten metal is directed away from the torch. After starting the arc, the torch is moved to the vertical position and the cut proceeds along the cut line. This technique prevents molten metal from blowing back at the torch and causing damage. Thicker metals may require that the torch be left stationary over the start point for a few seconds to ensure full arc penetration. Because a pierce will sometimes result in a blow hole that is larger than the normal kerf width, it is good practice to start the pierce off the cut line in the scrap piece to eliminate discontinuities in the cut surface of a good piece.

In mechanized applications, edge starts are accomplished by lining the torch up with the cut line. Pierce starts cannot be done in the same way as they are in manual cutting, because the torch cannot be conveniently angled and then brought back to a vertical position. Best results are achieved with a slow-moving pierce, in which the torch motion begins when the arc is struck. This allows the molten metal to flow back along the cut line until full penetration is achieved. Some allowance should be made for the start of a moving pierce to ensure that the entire piece is cut out.

A third means of starting the arc in mechanized applications is the running-edge start. In this mode, the torch is positioned off the workpiece, the pilot arc is established, and then the torch is put in motion. The cutting arc is established as soon as the pilot hits the edge of the work.

For either manual or mechanized operation, the best cut quality is usually achieved with a torch standoff of 3 to 10 mm ($\frac{1}{8}$ to $\frac{1}{2}$ in.). Thinner metals, under 3 mm ($\frac{1}{8}$ in.) thick, require a closer standoff. At low power levels (below 40 A), good-quality cuts can be obtained by dragging the torch tip in contact with the work.

The proper cutting speed for achieving the best quality is determined by the angle that the arc makes with a vertical plane as it exits the workpiece (Fig. 6). The optimal speed for air cutting is usually achieved when the arc is vertical ($0°$). Nitrogen and argon-hydrogen cuts are optimized by adjusting speed to obtain a 5 to $10°$ trailing arc. The best oxygen cuts are made using a speed that provides a slight (1 to $2°$) leading arc. The PAC torch should always be kept square with the workpiece to provide maximum arc penetration and to obtain consistent quality when contour cutting.

Process Variations

The conventional PAC torch, utilizing the basic system components described earlier, uses only a single gas. Because conventional torches do not incorporate any secondary gas, they must be liquid cooled.

Several process variations involve auxiliary shielding in the form of gas or water to improve the plasma arc cutting quality for particular applications. They are generally applicable to materials in the 3 to 38 mm ($\frac{1}{8}$ to $1\frac{1}{2}$ in.) thickness range, depending on the current rating of the plasma machine.

Dual-flow plasma cutting, also referred to as gas-shielded PAC, provides a secondary gas blanket around the arc plasma, as shown in Fig. 7. The usual orifice gas is nitrogen or compressed air. The shielding gas is selected for the material to be cut. It may be compressed air for mild steel, CO_2 for stainless steels, and an argon-hydrogen mixture for aluminum. The torch design provides special passages for this secondary gas. The functions of the secondary gas are to:

- Assist the plasma in blowing away molten metal
- Enable faster and cleaner cuts
- Provide better cooling of the torch front end
- Reduce accumulation of metal spatter on the torch
- Reduce top-edge rounding of the workpiece

- Permit easier piercing
- Minimize double arcing

In gas-cooled, rather than liquid-cooled, torches, it is the secondary gas that provides the cooling action. The choice of secondary gas can also affect cut quality, cutting speed, and the amount of smoke and fumes produced during the PAC process.

Water-shielded PAC is a variation of the gas-shielded process that is only used in mechanized applications. The water, which can be provided from standard plant water systems at a rate of 0.63 to 0.95 L/min (10 to 15 gal/h), is substituted for the secondary gas in torches designed for this type of operation. Using a water shield, rather than a gas shield, is less expensive. The water shield helps to significantly reduce the level of noise, arc radiation, and fume generated from the plasma cutting operation. It also results in reduced top-edge rounding and fumes, and a clean, shiny, cut surface.

Water-injection PAC is similar to the water-shielded process, except that the water is forced into the plasma stream, at approximately 1.26 L/min (20 gal/h), just after the arc exits the tip orifice and just before it leaves the ceramic piece on the front end of the torch (Fig. 8). The impinging water increases the constriction and density of the plasma arc, resulting in improved cut squareness and increased cutting speed. Because of the entry of water at the tip orifice in a water-injection torch, the quality of water is more critical than that used in water-shielded cutting. A plant water system that has high levels of dissolved minerals may require water softeners in conjunction with the system in order to avoid rapid deterioration of part lifetimes. Water-injection cutting is only used in mechanized setups.

Process Capabilities

Plasma arc cutting is unique in its ability to make fast, clean cuts on both ferrous and

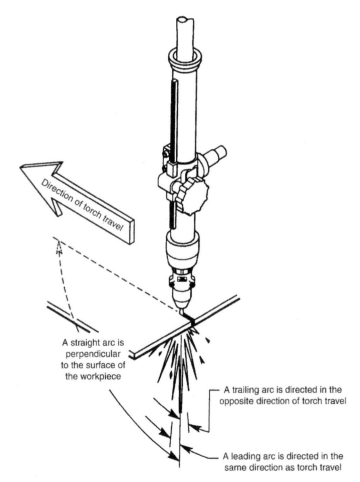

A straight arc is perpendicular to the surface of the workpiece

A trailing arc is directed in the opposite direction of torch travel

A leading arc is directed in the same direction as torch travel

Fig. 6 Arc characteristics of PAC torch

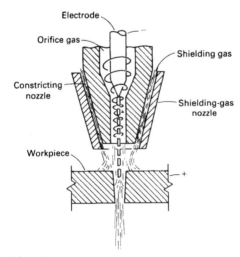

Electrode

Orifice gas

Constricting nozzle

Workpiece

Shielding gas

Shielding-gas nozzle

Fig. 7 Dual-flow PAC

nonferrous metals. When compared with the oxyfuel cutting of ferrous metals, the major advantage of this process is its speed, which can be from four to five times greater than the speed of oxyfuel cutting. For example, when cutting 6 mm (1/4 in.) carbon steel with an oxyfuel process, a speed of 0.68 m/min (27 in./min) is used, but when using a 150 A PAC system with air plasma, the speed goes up to 3.81 m/min (150 in./min) (Fig. 9). Because labor and overhead represent the primary costs of operating a mechanized cutting system, reducing the time to cut a part will, in turn, reduce the cost of the part itself. The fivefold increase in cutting speed of the PAC process, versus that of the oxyfuel process on carbon steel, will translate to a part cost of only one-fifth that achieved with the oxyfuel process.

Because the speed of the PAC process, like that of the oxyfuel process, is dependent on metal thickness, increasing thickness will diminish the speed advantage. Therefore, other factors that favor the PAC process have to be determined when comparing it with the oxyfuel process for cutting carbon steel that is more than 50 mm (2 in.) thick.

Another advantage of the PAC process, versus the oxyfuel process, is its ability to cut carbon steel with virtually no distortion, because of the fast travel speed and narrow, localized heating of the metal. This same high-speed localized heating results in a much narrower heat-affected zone (HAZ), that is, the depth of metallurgical change from the cut surface, than that of the oxyfuel process. Typical HAZs of up to 1.78 mm (0.070 in.) from oxyfuel cuts on carbon steel have been measured, whereas the PAC process produces HAZs of only 0.51 mm (0.020 in.). The HAZ will have an effect on the overall strength, corrosion resistance, and susceptibility to surface cracking of the cut part. If the application is such that the cut part must undergo machining to remove the HAZ, then the depth of machining required is much less with the PAC process.

Normally, PAC surfaces do not require any additional finishing operations before they are welded, although this usually depends on the final application and the codes to which the cut part being welded must conform.

The PAC process has the ability to make both edge starts and pierce starts without a preheat, as is required for oxyfuel cutting. This advantage becomes quite dramatic when the application calls for interrupted cuts, such as on grating or expanded metal.

These same capabilities apply to the cutting of stainless steel and aluminum. No preheat is required, and there is no distortion of the cut part. For plasma arc cuts on stainless steel and aluminum, respectively, the HAZ has been measured at 0.25 and 2.03 mm (0.010 and 0.080 in.).

Process Mechanization

Plasma arc cutting can be mechanized in several ways, the simplest of which takes the form of portable carriages or tractors. These devices can travel along a track to achieve straight-line cutting or along an arc guided either by a radius rod or by hand to cut curves. Although tractors are primarily used for simple cutoff, squaring, and trimming operations, they are also used as parts of large fixtures for cutting cast and rolled metals.

Pattern/follower wheel-shaped cutting machines or pantographs are used for the plasma arc cutting of relatively small, simple shapes. A scale model of the desired piece is held and traced by a follower that is connected, through a series of arms, to the torch, which cuts either an identical or a larger-scale part.

Optical tracing or electric-eye machines are used to trace part shapes that are in the form of templates or, more commonly, in the form of line drawings that are the same scale as the desired part. A mechanized PAC setup that uses optical tracing is shown in Fig. 10. The speed of an optical tracing machine is limited by the capability of the electric eye to follow a line. Low-speed machines that operate at speeds up to 1.52 m/min (60 in./min), which were originally designed for use with oxyfuel torches, are compatible with lower-amperage PAC systems. New high-speed machines, with speeds of up to 12.7 m/min (500 in./min), have been developed for operation with high-amperage plasma cutting systems.

Computer numerical control (CNC) shape-cutting machines with speeds of up to 25.4 m/min (1000 in./min) represent the state-of-the-art in automated cutting. CNC machines are much more reliable, have greater capabilities, and are lower priced than their numerical (tape) control machine predecessors, which are now considered obsolete. CNC machines can be connected directly into the central processor (direct numerical control) of a plant for complete integration of product design, material requirements, work in process, and other functions. Because the evolution of CNC machines paralleled that of mechanized PAC systems, their combination represents the highest level of productivity available in metal cutting today.

Automated material handling systems are sometimes used to keep up with the speed of CNC PAC systems. Typical systems utilize a

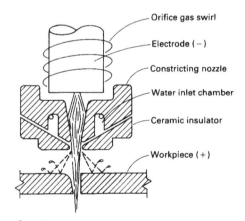

Fig. 8 Water injection PAC arrangement

Orifice gas swirl
Electrode (−)
Constricting nozzle
Water inlet chamber
Ceramic insulator
Workpiece (+)

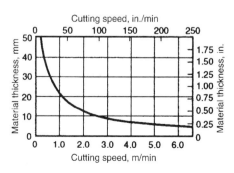

Fig. 9 Cutting speeds versus mild steel thickness (150 A, air plasma)

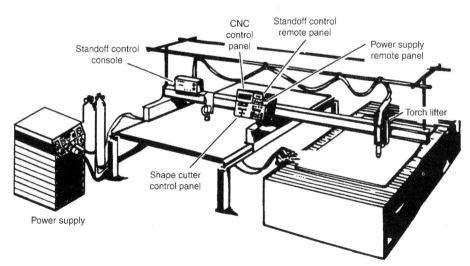

Fig. 10 Mechanized PAC system that uses optical tracing of part shapes

raw material loading station, a cutting station, and a finished parts station. While the cutting operation is taking place, the operator is loading raw material and unloading finished parts. Upon completion of the cutting sequence, a conveyor moves the finished parts off the cutting area as it moves raw material onto the cut station. The operator signals the start-cut sequence and the cycle is repeated.

Robots are being used in increasing numbers for PAC operations involving the contour cutting of pipe and vessels, the removal of sprues and risers from castings, and the cutting of shapes in various planes of cast or formed parts. Material handling is also a great concern in robotics applications. Multiple workstations are the rule, because they allow an operator or another robot to load unfinished work and remove finished parts while the cutting operation is ongoing.

Quality of Cut

Most plasma cutting torches impose a swirl on the orifice gas flow pattern by injecting gas through tangential holes or slots (Fig. 7, 8). As a result of the swirl of the plasma gas, walls of plasma arc cuts have a V-shaped included angle of 2 to 4° on one of the cut edges. When a straight edge is required on the cut part, the operator must operate the torch carefully so that the bevel is on the scrap side of the cut. When the operator is facing the direction of torch travel, if the gas swirls clockwise, the bevel will be on the left side of the cut. In many cases, a small bevel is acceptable; it may even be used as a weld preparation. The relationship of torch travel direction to the part with clockwise swirl of the orifice as is illustrated in Fig. 11.

Quality of cut includes surface smoothness, kerf width, degree of parallelism of the cut faces, dross adhesion on the bottom of the cut, and sharpness of top and bottom faces. Table 3 provides data on the causes of imperfections in plasma arc cutting of low-carbon steel, stainless steel, and aluminum. Besides the HAZ, the quality of a plasma arc cut involves the following factors (Fig. 12):

- Cut angle
- Dross
- Surface finish

- Top-edge squareness
- Kerf width

Cut Angle. Plasma arc cutting will usually result in an angle on the cut surface of approximately 1 to 3° on the "good" side and 3 to 8° on the "bad" side, when using torches that swirl the plasma. With laminar-flow torches, the angle on both sides is usually about 4 to 8°. These angles are most noticeable in mechanized applications where the torch is square to the workpiece. The good and bad sides of the cut are determined by torch travel direction and plasma gas swirl. With a clockwise gas swirl, the good side of the cut will be on the right with respect to torch motion. Therefore, in the cutting of a ring or flange in which the minimum angle is desired on both the inside and outside diameters, the outside would be cut in the clockwise direction, while the inside would be cut counterclockwise (Fig. 11). This change in angle from good side to bad side of the cut is the result of the arc attachment point on the cut surface and the release of energy on the good side before contact with the bad side occurs.

Changing an internal torch part, such as the electrode or gas distributor, can enable those plasma cutting torches that create a gas swirl to change from a clockwise to a counterclockwise swirl direction. This then enables mirror-image cutting when using two torches in a mechanized setup, such as in strip-cutting applications.

Dross is the resolidified metal that adheres to the bottom edge of the plasma cut. The concentration of dross will be heavier on the bad side of the cut. The amount of dross that forms is a result of the type of metal being cut, the cutting speed, and the arc current. Dross can be formed from either too high or too low a cutting speed, but there is a "window" between these two extremes in which dross-free cuts can usually be achieved. The dross-free range is greater on stainless steel and aluminum than it is on carbon steel and copper alloys. If dross-free cuts cannot be achieved, then a minimum amount of low-speed dross is more desirable, because it is more easily removed than the high-speed variety.

Surface Finish. The cut surface from the plasma arc process is normally rougher than that achieved by oxyfuel cutting on carbon steels, and it is definitely more rough than most machining

processes. In most metals, this roughness usually appears as a ripple along the cut surface. This is partly due to the output waveform of the dc power supply (the smoother the output, the smoother the cut), but it is also determined to an extent by the gases used and the torch design. Water-shielded and water-injection plasma arc cutting provide much smoother cuts than do gas-shielded or conventional plasma techniques. The use of argon-hydrogen as a plasma or a shielding gas will result in smoother cut surfaces on most metals than when using nitrogen plasma or any other gas shield.

Top-Edge Squareness. Most metals experience some top-edge rounding when the PAC process is applied. Top-edge rounding is more pronounced on the thinner metals. This rounding is due to a higher heat concentration at the top of the cut and can be minimized by using a gas-shielded PAC process. It can be even better controlled when water-shielded or water-injection plasma arc cutting is used.

The kerf width obtained by plasma arc cutting will be greater than that achieved by oxyfuel cutting on carbon steel, but not as great as that obtained by other processes, such as abrasive cutting or arc gouging. Width of kerf

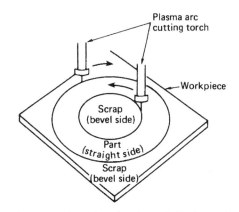

Fig. 11 Relationship of torch travel direction to the part with clockwise swirl of the orifice gas. With the clockwise swirling plasma gas, the bevel side of the cut is on the left when the operator is looking in the direction of the torch travel. To achieve straight cuts on the inner diameter and the outer diameter of the ring, torch directions must reverse to keep the right side of the cut on the part edge.

Table 3 Causes of imperfections in plasma arc cuts

Type of imperfection	Cause of imperfection		
	Low-carbon steel	Stainless steel	Aluminum
Top edge rounding	Excessive speed, excessive standoff	Excessive speed, excessive standoff	Seldom occurs
Top edge dross	Excessive standoff, dross easily removed	Excessive standoff, excessive hydrogen	Excessive standoff, dross easily removed
Top side roughness	Seldom occurs	Excessive hydrogen or standoff, insufficient speed	Insufficient hydrogen
Side bevel—positive	Excessive speed, excessive standoff	Excessive speed, excessive standoff	Excessive speed, insufficient hydrogen
Side bevel—negative	Seldom occurs	Seldom occurs	Excessive hydrogen
Top side undercut	Excessive hydrogen	Excessive hydrogen	Insufficient speed, insufficient hydrogen
Bottom side undercut	Seldom occurs	Slight effect at near-optimum conditions	Seldom occurs
Concave surface	Seldom occurs	Excessive hydrogen	Excessive hydrogen, insufficient speed
Convex surface	Excessive speed	Insufficient hydrogen, excessive speed	Seldom occurs
Bottom edge rounding	Excessive speed	Seldom occurs	Seldom occurs
Bottom dross	Excessive hydrogen or speed, insufficient standoff	Insufficient speed, excessive hydrogen	Excessive speed
Bottom side roughness	Insufficient standoff	Seldom occurs	Insufficient hydrogen

is about one to two times the kerf of conventional oxyfuel gas cutting. The range is usually 5 to 10 mm ($^3/_{16}$ to $^3/_8$ in.), although some users achieve 0.8 mm ($^1/_{32}$ in.). For thick work metal, width of kerf may exceed 9 mm ($^3/_8$ in.). The rule of thumb for estimating the kerf in plasma arc cutting is that its width will be approximately 1.5 to 2 times the tip orifice diameter.

Heat-Affected Zone. The high speeds possible with plasma arc cutting result in relatively low heat input to the workpiece. Heat-affected zones are therefore narrow. The HAZ on stainless steel plate 25 mm (1 in.) thick cut at 1270 mm/min (50 in./min) is 0.08 to 0.13 mm (0.003 to 0.005 in.). Sensitization in unstabilized stainless steels is usually avoided, and distortion from cutting is normally avoided.

Bevel cutting for weld preparation is an important application of PAC. The intense heat of the process makes it suitable for all types of beveling at a higher efficiency than oxyfuel gas cutting.

Applications

Plasma arc cutting is used in a variety of industries. Plasma arc cutting can be used to cut any metal. Most applications are for carbon steel, aluminum, and stainless steel. It can be used for stack cutting, plate beveling, shape cutting, and piercing.

In stack cutting, the plates should be clamped together as closely as possible. However, plasma arc cutting can usually tolerate wider gaps between carbon steel plates than can oxyfuel gas cutting. When high plasma arc cutting speeds are used, there is less distortion of the top plate. Several plates of 1.5 to 6 mm ($^1/_{16}$ to $^1/_4$ in.) thickness can be economically stack cut.

For shape cutting, PAC torches are used on shape cutting machines similar to those used for oxyfuel gas cutting (see Fig. 6 in the article "Oxyfuel Gas Cutting" in this Volume).

Generally, plasma arc shape cutting machines can operate at higher travel speeds than is possible with oxyfuel gas cutting machines. Because of the fumes and heat produced by the cutting action, water tables are normally used with plasma arc shape cutting machines. The water just touches the bottom of the plate, where it traps the fumes, slag, and dross as they emerge from the bottom of the kerf. It also helps reduce noise.

For smaller systems with output capabilities of 100 A, typical end users include:

- Small job shops that must fabricate specialty and subcontracted items from stainless steel, aluminum, brass, copper, carbon steel, galvanized metal, and other metals
- Manufacturers and installers of ductwork for heating, ventilating, and air conditioning, who must cut galvanized metals
- Manufacturers of food processing and institutional kitchen equipment, who must cut stainless steels
- Automobile manufacturers, auto body repair shops, and dismantlers, who must cut high-strength steels used in unibody construction
- Manufacturers and rebuilders of over-the-road liquid transport trailers used to haul petroleum, chemicals, milk, and fertilizers, who must cut stainless steels and aluminum
- Contractors and maintenance departments that erect and maintain all kinds of plants, including food processing, petroleum, chemical, pulp and paper mills, power plants, and mining operations

For systems with output capabilities greater than 100 A, end users include:

- Large shops that fabricate specialty and subcontracted parts from virtually any metal
- Steel warehouses and service centers that size and shape-cut both ferrous and nonferrous metals, as well as those that size scrap

- Pipe fabricators and piping contractors, who must size, bevel, and make intersection cutouts on carbon and stainless steel pipe
- Structural fabricators, who must perform high-speed slitting of H and I beams and the sizing and slitting of plate
- Manufacturers of ships, barges, and offshore rig platforms, who must perform plate sizing, pipe cutoff, bulkhead fabrication, and propeller trimming
- Aluminum, steel, and copper mills that size and strip rolled plate and sheet, or that size scrap, or that perform plant equipment maintenance
- Scrap yards that perform high-speed sizing of stainless steel, copper, brass, aluminum, and other metals

Safety

Operators of PAC equipment should be aware of seven areas of potential hazards that are associated with the process. The proper operation of the equipment and adherence to certain precautions will minimize the risks involved.

Gases and Fumes. Potentially hazardous gases associated with the PAC process include ozone, nitrogen dioxide, and phosgene (caused by the breakdown of chlorinated solvents in ultraviolet light). Metal fumes vary with the composition of the metal being cut. Fumes of beryllium, cadmium, cobalt, copper, lead, mercury, silver, vanadium, and several other elements may be hazardous to individuals in the PAC area. Various limits imposed on the levels of gases and fumes in the work area are listed in a handbook published by the Occupational Safety and Health Administration (OSHA) and in the American Welding Society document A6.3-69, "Recommended Safe Practices for Plasma Arc Cutting."

Gases and fumes created by the PAC process can be effectively controlled by properly installed ventilation systems in the work area. Water tables that are used either alone, for surface or underwater cutting, or in conjunction with water-shrouded systems can effectively reduce gases and fumes. Respirators can be used when ventilation systems, and water tables are impractical. Chlorinated solvents and other sources of phosgene gas should be removed from the PAC work area.

Noise. The high level of noise associated with some PAC operations (usually high power) can lead to hearing damage. Maximum exposure times to various noise levels are printed in the OSHA handbook. The operator and any other personnel within the vicinity of the PAC operation should wear proper ear protection. Underwater cutting and the use of water-shrouded systems are effective means for reducing noise levels.

Radiant Energy. The ultraviolet infrared and visible light from the plasma arc can lead to injury of the eyes and skin. Operators and anyone else in the vicinity of the PAC operation should

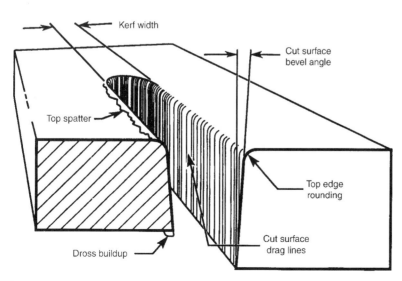

Fig. 12 Characteristics of a plasma arc cut

be protected by proper clothing that covers all skin and by welding helmets or shields that contain the proper shaded lens. For high-powered operations, the use of a number 10 (or greater) shaded lens may be required by the operator. If the intensity of the PAC process causes eye discomfort to the operator, a lens with a darker shade should be utilized.

Electric Shock. Both the input power to and the output from a PAC system is sufficient to be hazardous if an individual receives a shock from the system. Output voltages in the PAC process usually range from 100 to 200 V dc; typical welding processes run at 20 to 40 V dc. All equipment should be installed and maintained according to the U.S. Standard National Electrical Code and any local standards. All components of a PAC system should be properly earth grounded. System power should be shut off at a wall disconnect before the torch or any other system component is serviced. Any cracked insulators, power leads, or exposed live parts should be corrected immediately. Operators of PAC systems should be insulated from the workpiece and ground, and those who work in wet or damp areas should take precautions to ensure adequate protection from electric shock.

Fire. The sparks and the hot molten metal that is blown out of the cut can cause fires. Sparks that result from piercing can travel a considerable distance. All combustibles should be removed from the work area, and a fire extinguisher should be located in the vicinity. Because the pilot arc from a PAC torch is very hot, extra care should be taken to prevent it from contacting clothing and other flammable materials.

Compressed Gas Cylinders. High-pressure gas cylinders, which are often used to supply PAC systems, can be hazardous if they are not treated properly. Proper handling requires that valve covers always be in place when transporting cylinders, and that cylinders be secured to stationary objects by chains or lashing straps to prevent them from falling over.

Explosions. When hydrogen is used as a plasma or secondary gas, in conjunction with argon, extra care should be taken because it is highly explosive. In some instances, explosions

of hydrogen gas can occur even if hydrogen is not supplied as a portion of the plasma or secondary gas. This is because hydrogen can form when some metals, particularly aluminum, are being cut over a water table, by a reaction between the metal and the water, and by the dissociation of water that is due to the plasma arc. During the cutting operation, the plasma arc can ignite this hydrogen, which will explode if it is confined. Adequate ventilation of these hydrogen pockets is necessary for the safe operation of PAC systems. The water in the water table should not be any deeper than 50.8 mm (2 in.) below the bottom of the metal workpiece support bars. A blower system should be used to flush out all hydrogen from beneath the workpiece.

Air-Carbon Arc Cutting*

Air-carbon arc (ACA) cutting removes metal physically, rather than by a chemical (oxidation) process in oxyfuel gas cutting. Gouging or cutting occurs when the intense heat of the arc between the carbon electrode and the workpiece melts part of the workpiece (Fig. 13). A stream of compressed air simultaneously blows through the arc quickly enough to blow the molten material from the kerf or groove. The air must lift the molten metal clear of the arc before the metal solidifies. Because this process does not require oxidation to maintain the cut, it can gouge metals that the oxyfuel cutting process cannot. The metal removal rate depends on the melting rate and efficiency of the air jet in removing the molten metal.

Its most common uses are for weld joint preparation, removal of defective welds, removal of welds and attachments when dismantling tanks and steel structures, and removal of gates, risers, and defects from castings. The low heat input of ACA gouging makes this process ideal for joint preparation and for weld removal on high-strength steels. Base-metal temperatures rise very little, about 80 °C (150 °F) in most applications. Industry has enthusiastically adopted ACA gouging for numerous applications, such as metal fabrication and casting finishing, chemical

and petroleum technology, construction, mining, and general repair and maintenance.

The process is used throughout the world to create consistent grooves requiring little or no additional cleaning on square-butt plate seams to prepare them for welding. The process can then be used to backgouge the seam to sound metal to ensure 100% penetration of the welded joint. If a problem arises during the welding process and an area of the weld does not meet specifications, then the process can be used to remove the defective area without damaging or detrimentally affecting the base metal.

The ACA process is also used in foundries to remove fins and risers from castings. It can then be used to smooth areas that contact the surface to prepare the casting for shipment. The process is flexible, efficient, and cost effective to use on numerous metals, such as carbon steel, stainless steel, and other ferrous alloys; gray, malleable, and ductile iron; aluminum; nickel; copper alloys; and other nonferrous metals.

Equipment

Typical components of an ACA system are shown in Fig. 14. As in electric arc welding, the ACA process requires an arc of intense heat to develop a molten pool on the workpiece. Compressed air then blows away this molten metal. The process requires a welding power source, an air compressor, a carbon electrode, and a gouging torch. A holder clamps the carbon-graphite electrode in position parallel to an air stream, which exists from orifices in the electrode holder to strike the molten metal immediately behind the arc. The electrode holder contains an air flow control valve, an air hose, and a cable. The cable connects to the welding machine; the air hose connects to a source of compressed air. Cuts or gouges should be made only in the direction of the air flow. The electrode angle will vary, depending on the application. The cutter should maintain the correct arc length to enable air to remove molten metal (Fig. 14b).

Power Sources. Constant-voltage direct current with a flat to slightly rising voltage characteristic is best for most ACA cutting applications. Direct current is preferred; copper alloys, however, cut better with alternating current. Table 4 provides data on power sources for ACA cutting and gouging.

Single-phase machines with low open-circuit voltage may not work for ACA gouging. However, any three-phase welding power source of sufficient capacity can be used. The open-circuit voltage should be higher than the required arc voltage to allow for voltage drop in the circuit. Because the arc voltage typically ranges from a low of 35 to a high of 56 V, the open-circuit voltage should be at least 60 V. The actual arc voltage used in the process is governed by arc length and the type of gouging.

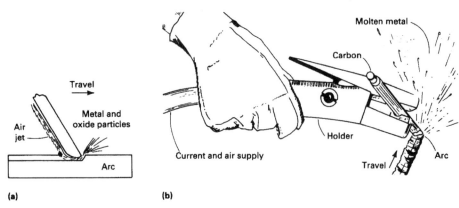

Fig. 13 (a) Air-carbon arc cutting action. (b) Manual air-carbon arc cutting

*Adapted from R.L. Strohl, Air-Carbon Arc Cutting, *Welding, Brazing, and Soldering,* Vol 6, *ASM Handbook,* ASM International, 1993, p 1172–1177

Direct current, electrode positive (reverse polarity, DCEP) is used with the process, except in special applications discussed later. The electrode should extend, at most, 180 mm (7 in.) from the gouging torch, with the air jet between the electrode and workpiece. A minimum extension of 50 mm (2 in.) should be used. Damage of the torch parts will occur if the electrode sticks out less than 50 mm (2 in.).

Absorption of Carbon. Reverse polarity (dc electrode positive) air-carbon arc cutting removes metal faster than does straight polarity (dc electrode negative). However, the dc reverse polarity current carries ionized carbon atoms from the electrode to the base metal, with the potential for increasing the carbon content of the cut surfaces. To minimize this pickup and potential for high hardness, the air stream must be adjusted to ensure removal of all molten metal and the use of proper electrode manipulation techniques is essential.

Compressed Air. Ordinary compressed air should be used for ACA gouging. Normal pressures range from 550 to 700 kPa (80 to 100 psi) at the torch. Higher pressures can be used, but they do not remove metal more efficiently. Deep grooves in thick metal require pressures up to 860 kPa (125 psi). A pressure as low as 275 to 415 kPa (40 to 60 psi) may be suitable for light work with a light-duty manual torch. This low level of pressure should not be used with general-duty torches. Air pressure is not critical in ACA cutting; the process requires a sufficient volume of air to ensure a clean, slag-free surface. The amount of air required depends on the type of work (0.08 to 0.9 m^3/min, or 3 to 33 ft^3/min, for manual operations and 0.7 to 1.4 m^3/min, or 25 to 50 ft^3/min, for mechanized operations).

Air Supply. Regardless of the pressure used with manual torches, air hoses should have a minimum inside diameter of 6 mm (1/4 in.) with

no constrictions. Mechanized torches with automatic arc length control should have an air-supply hose with minimum inside diameter of 12 mm (1/2 in.). Attention should be given to the field connections to ensure that this minimum is maintained without severely restricting the air-flow volume. Table 5 gives the consumption rates of compressed air for manual and mechanized torches, as well as compressor power rating for intermittent and continuous use. The compressor receiver tank must be large enough to accommodate the compressor rating.

Air-carbon arc cutting electrodes are made from mixtures of carbon and graphite. The three basic types of air-carbon arc cutting electrodes are described below. Cross sections vary; round electrode rods are most common. Electrodes also come in flat, half-round, and special shapes to produce specially designed groove shapes.

Direct current copper-coated electrodes are used most frequently because of long life, stable arc characteristics, and groove uniformity. These electrodes are made by mixing carbon and graphite with a binder. When this mixture is baked, it produces dense, homogeneous graphite electrodes of low electrical resistance. These electrodes are then coated with a controlled thickness of copper. The available diameters of these electrodes are 3.2 mm (1/8 in.), 4.0 mm

(5/32 in.), 4.8 mm (3/16 in.), 6.4 mm (1/4 in.), 7.9 mm (5/16 in.), 9.5 mm (3/8 in.), 12.7 mm (1/2 in.), 15.9 mm (5/8 in.), and 19.1 mm (3/4 in.).

Jointed electrodes work without stub loss. They are furnished with a female socket and a matching male tang and are available in diameters of 7.9 mm (5/16 in.), 9.5 mm (3/8 in.), 12.7 mm (1/2 in.), 15.9 mm (5/8 in.), 19.1 mm (3/4 in.), and 25.4 mm (1 in.). Jointed carbons are always used with the male tang pointing up (that is, away from the gouge). This ensures proper joint performance.

Flat, or rectangular, coated electrodes are available in dimensions of 4.0 by 9.5 mm (5/32 by 3/8 in.) and 4.8 by 15.9 mm (3/16 by 5/8 in.). These electrodes are used to make rectangular grooves and to remove weld reinforcements.

Direct current electrodes are generally restricted to diameters of less than 9.5 mm (3/8 in.) and are therefore limited in use. During cutting, these electrodes erode more than the coated electrodes. They are manufactured in the same way as the coated electrodes, but without the copper coating. Plain electrodes are available in diameters of 4.0 mm (5/32 in.), 4.8 mm (3/16 in.), 6.4 mm (1/4 in.), 7.9 mm (5/16 in.), and 9.5 mm (3/8 in.).

Alternating current (ac) coated electrodes are made from a mixture of carbon, graphite, and

Table 4 Power sources for air-carbon arc cutting and gouging

Equipment	Polarity	Use
Variable-voltage motor-generator, resistor, and resistor grid	Direct current	All electrode sizes
Constant-voltage motor generator, rectifier	Direct current	Electrodes > 1/4 in. in diameter
Transformer	Alternating current	Alternating current electrodes only
Rectifier	Alternating current, direct current	Direct current from three-phase transformer only; single-phase source not recommended. Use alternating current with alternating current electrodes only.

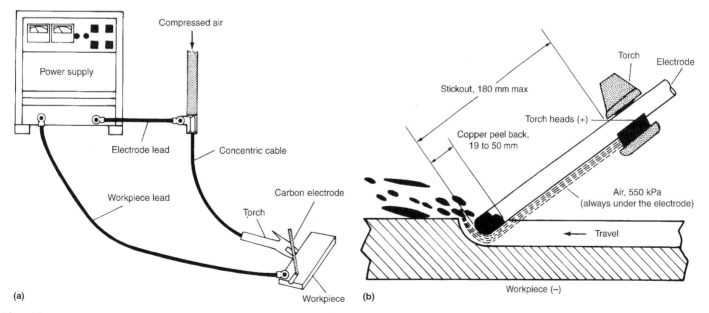

Fig. 14 Carbon-air arc system. (a) Arrangement of basic components. (b) Details of carbon-air cutting torch

a special binder. Rare-earth materials are added to ensure arc stabilization when using alternating current. These electrodes, coated with copper, are available in diameters of 4.8 mm ($^3/_{16}$ in.), 6.4 mm ($^1/_4$ in.), 9.5 mm ($^3/_8$ in.), and 13 mm ($^1/_2$ in.).

A manual gouging torch is shown in Fig. 15. The electrode is held in a swivel head that contains one or more air holes, so that the air jet stays aligned with the electrode, regardless of the angle of the electrode to the gouging torch. Torches with two heads (that is, the air jet is on two sides of the electrode) or with a fixed angle between the electrode and the holder are better for certain applications, such as removing pads and risers from large castings (pad washing).

Manual torches are usually air cooled. In high-current applications, water-cooled cable assemblies can be used with heavy-duty torches.

Control of Automatic Gouging Torches. There are two methods of controlling automatic ACA gouging torches. Either method can make grooves of consistent depth to a tolerance of ± 0.64 mm (± 0.025 in.). These automatic units are used to achieve high-quality gouges and to increase production (Fig. 16).

The amperage-controlled method maintains the arc current by amperage signals through solid-state controls. This method controls the electrode feed speed, which maintains the preset

amperage. It is run with constant-voltage power sources only.

A voltage-controlled method maintains arc length by voltage signals through solid-state electronic controls. This method controls the electrode feed speed, which maintains the preset voltage. It can run with constant-current power sources.

Operating Techniques

The angle of the electrode, speed of cut, and amount of current determine depth and contour of the cut or groove. The electrode is held at an angle, and an arc is struck between the end of the electrode and the work metal. The electrode is then pushed forward. Data on groove depth, electrode size, current, and travel speed for air-carbon arc gouging is available from various equipment manufacturers.

For through-cutting, the electrode is placed at a steeper angle, almost vertically inclined. Plate thicknesses greater than 13 mm ($^1/_2$ in.) may require multiple passes.

Grooves as deep as 25 mm (1 in.) can be made in a single pass. A steep angle, approaching that used for through-cutting, and rapid advance produce a deep, narrow groove; a flatter angle and slower advance produce a wide, shallow groove. Electrode diameter directly influences groove width. Operators should use a wash or

weave action to remove excess metal such as risers and pad stubs, or in surfacing. Smoothness of the gouged or cut surface depends on the stability of electrode positioning, as well as on the steadiness of the electrode as it advances during the cutting operation. Mechanized gouging, with the electrode and holder traveling in a carriage on a track, produces smoother surfaces five times faster than does manual work.

Gouging with Manual Torches. As shown in Fig. 14(b), the electrode should be gripped so that a maximum of 180 mm (7 in.) extends from the torch. For aluminum, this extension should be 75 mm (3 in.). Table 6 shows suggested currents for the different electrode types and sizes.

The air jet should be turned on before striking the arc, and the torch should be held as shown in Fig. 17. The electrode should slope back from the direction of travel, with the air jet behind the electrode. During gouging, the air jet sweeps beneath the electrode end and removes all molten metal. The arc can be struck by lightly touching the electrode to the workpiece. The electrode should not be drawn back once the arc is struck.

Gouging is different from arc welding in that metal is removed, rather than deposited. A short arc should be maintained in the direction of cut by working fast enough to keep up with metal removal. Steadiness of movement controls the smoothness of the resulting cut.

Vertical gouging should be conducted downhill, permitting gravity to help remove the molten metal. Although vertical gouging can

Table 5   Recommended minimum air supply requirements for air-carbon arc cutting

	Air pressure(a)		Air consumption		Recommended compressor rating				ASME receiver size	
					Intermittent use		Continuous use			
Type of torch	kPa	psi	L/min	ft³/min	kW	hp	kW	hp	L	gal
Light duty(b)	280	40	227	8	0.4	0.5	1.1	1.5	227	60
General duty(b)	550	80	708	25	3.7	5	5.6	7.5	303	80
Multipurpose(c)	550	80	934	33	5.6	7.5	7.5	10	303	80
Automatic(d)	414	60	1303	46	...	...	11.2	15	303	80

(a) Pressure while torch is in operation. (b) Accommodates flat electrodes. (c) Generally considered to be a foundry torch. (d) Requires some kind of mechanical manipulation

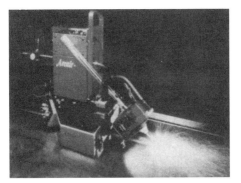

Fig. 16 Automatic gouging unit

Table 6   Suggested current ranges for various electrode sizes

Electrode size		Minimum amperage	Maximum amperage
mm	in.		
3	$^1/_8$	60	90
4	$^5/_{32}$	90	150
5	$^3/_{16}$	200	250
6	$^1/_4$	300	400
8	$^5/_{16}$	350	450
10	$^3/_8$	450	600
13	$^1/_2$	800	1000
16	$^5/_8$	1000	1250
19	$^3/_4$	1250	1600
25	1	1600	2200

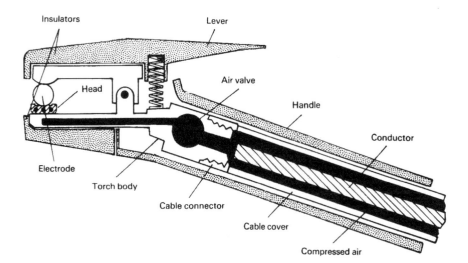

Fig. 15 Cutaway view of a manual gouging torch with air flow on bottom of electrode

be accomplished uphill, it is difficult. Horizontal gouging can be done either to the right or to the left, but always with forehand gouging.

The air jet should always be positioned under the electrode. When gouging in the overhead position, the electrode and torch should be held to prevent molten metal from dripping on the glove of the operator. The depth of the groove is controlled by the travel speed. Grooves up to 25 mm (1 in.) deep can be made. However, the deeper the groove, the more experienced the operator needs to be. Slow travel speeds will produce a deep groove. Fast speeds will produce a shallow groove. The width of the groove is determined by the size of the electrode used. The groove is usually about 3.2 mm ($^1/_8$ in.) wider than the diameter of the electrode. A wider groove can be made with a small electrode that is oscillated in either a circular or weave motion. When gouging, a push angle that is 35° from the surface of the workpiece should be used for most applications. A steady rest will ensure a smoothly gouged surface, especially in the overhead position. Proper travel speed depends on electrode size, the base metal being used, amperage, and air pressure. The proper speed, which produces a smooth hissing sound, will result in a good gouge.

The severing technique is like gouging, except that the operator holds the electrode at a steeper angle that is between 70 and 80° to the workpiece surface. When cutting thick, non-ferrous metals, the electrode should be held perpendicular to the workpiece, with the air jet favoring the desired side. With the electrode in this position, the operator can sever the metal by moving the arc up and down using a sawing motion.

Washing. When using the ACA process to remove metal from large areas (for example, surfacing metal or riser pads on castings), the operator should position the electrode as shown in Fig. 18. The electrode should weave from side to side while it is being pushed forward at the depth desired. In the pad-washing operation, an angle that is 15 to 70° to the workpiece should

be used. The 15° angle is used for light finishing passes, whereas steeper angles allow deeper rough gouging with greater ease.

Gouging torches with fixed-angle heads that hold the electrode at the correct angle are well-suited for this application. When other types of torches are used, the air jet should be kept behind the electrode. Operator steadiness determines the smoothness of the surface produced.

Beveling. In one beveling method that is used for thick plates, the electrode should be held with a travel angle of 90° and a work angle that is equal to the bevel angle. The air jet should be positioned between the electrode and the workpiece. A second method, usually used for thin plates, requires that the torch be held parallel to the edge being beveled. Its position should equal the bevel angle. The air jet should be positioned between the electrode and the workpiece surface.

Equipment Selection

Gouging torches are selected for the specific application. They range from light-duty sizes appropriate for farm and body shop applications to extra-heavy-duty torches that are best suited to foundry applications.

General-purpose torches are available in four duty levels. Light-duty torches accept round electrodes that range from 3.2 to 6.5 mm ($^1/_8$ to $^1/_4$ in.) and 9.5 mm ($^3/_8$ in.) flat electrodes. Their maximum current level is 450 A. Medium-duty torches accept round electrodes that range from 3.97 to 9.5 mm ($^5/_{32}$ to $^3/_8$ in.) and 9.5 mm ($^3/_8$ in.) flat electrodes. Their maximum current level is 600 A. Heavy-duty torches accept round electrodes that range from 3.97 to 12.7 mm ($^5/_{32}$ to $^1/_2$ in.) and flat electrodes of 9.5 or 15.9 mm ($^3/_8$ or $^5/_8$ in.). Their maximum current level is 1000 A. Extra-heavy-duty torches accept round electrodes that range from 3.97 to 15.9 mm ($^5/_{32}$

to $^5/_8$ in.) and flat electrodes of 9.5 or 15.9 mm ($^3/_8$ or $^5/_8$ in.). Their maximum current level is 1250 A.

Foundry heavy-duty torches are appropriate for general foundry work and heavy-duty fabrication. They are limited to 1600 A when used with air-cooled cables and 2000 A when used with water-cooled cables.

Automated gouging torches are used for edge preparations and backgouging, as well as high-quality and high-productivity applications. Jointed carbon electrodes ranging from 7.9 to 19.1 mm ($^5/_{16}$ to $^3/_4$ in.) are accepted.

Power Sources. Any three-phase welding power source with enough capacity can be used for the ACA gouging process if the open-circuit voltage is high enough to allow for a voltage drop in the circuit. Some constant-voltage power sources with drooping characteristics require a very high open-circuit voltage to run ACA gouging equipment. Mechanized gouging and other applications that require maximum arc time should utilize a power source with a 100% duty cycle for the required amperage.

A dc constant-current power source motor generator, rectifier, or resistor grid unit is preferred for use with all electrode sizes. A dc constant-voltage power source motor generator or rectifier is usable only for 7.9 mm ($^5/_{16}$ in.) diam and larger electrodes. It may cause carbon deposits when used with small electrodes, and it is not suitable for automatic torches with voltage control only. An ac constant-current power source (transformer) is recommended for ac electrodes only. With ac/dc transformer rectifiers, the direct current should be supplied by a three-phase source, because single-phase sources give unsatisfactory arc characteristics. Alternating current output from ac/dc units is satisfactory, if ac electrodes are used.

Automatic systems are often used in modern fabrication facilities. These systems offer a high-quality, high-productivity alternative to manual

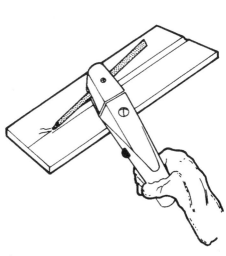

Fig. 17 Flat gouging position

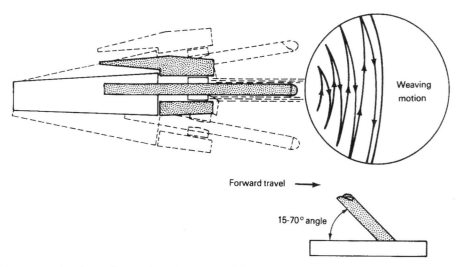

Fig. 18 Pad-washing technique. The torch remains parallel to the workpiece, and a weaving motion is used forward across the width of the area being cleaned. The electrode-to-workpiece angle should be from 15 to 70°. The more shallow the angle, the smoother the finish; the 70° angle is used mainly on cast iron.

gouging. Two types of systems can be considered. Both operate on a signal from the arc to control the gouging.

Dual-Signal System. Either constant-current or constant-voltage power supplies can be used with this type of automatic system. When the system uses constant current, the arc length is maintained through a voltage-signal system. A predetermined voltage setting is set on the system controller, which then advances or retracts the electrode through a stepping motor to maintain the arc length. With a constant-voltage power source, amperage sensing controls the feeding or retracting of the electrode to maintain the right arc current.

The single-signal system, like the dual-signal system, maintains arc length through voltage signal. However, it will not operate with an amperage signal. This type of system only operates with a constant-current power source.

Advantages. Automatic ACA systems ensure better productivity and quality. The systems can perform out-of-position gouging. They perform well when making long gouges in flat workpieces using a moving gouging apparatus and when making circular gouges in pipes and tanks using stationary gouging apparatus. They produce an even U-shaped groove and can control the depth of the groove to within ±0.64 mm (±0.025 in.). Table 7 shows typical operating information for U-shaped grooves.

Vacuum gouging represents a new variation of gouging that was developed in the late 1980s. This process replaces the air jet that is used to

evacuate the molten slag from the groove area with a high-volume vacuum that not only captures the slag and fume associated with the gouging process, but also reduces the noise level created when performing the gouging process.

The vacuum gouging process utilizes a specially designed nozzle that attaches to an automatic gouging head. By water cooling the nozzle and the attached vacuum hose, the slag and fume from the gouging process are pulled by the vacuum into a capture dram. The slag is knocked into the bottom of the catch tank, which is partially filled with water, and the fume is caught in a filter before the air is exhausted from the vacuum. Currently, this new vacuum system is limited to automated gouging operations on flat plate or pressure vessel circumferential seams.

Important Process Variables

Like any thermal-cutting process, the ACA process is sensitive to variables in operation. These variables can cause changes in the finished gouge that range from being undetectable to being unacceptable.

Electrode diameter and type are variables that determine the groove size, as well as productivity, groove quality, and metal-removal rates. The width of the groove will be about 3.2 mm ($\frac{1}{8}$ in.) wider than the diameter of the electrode. The choice of a proper electrode can be based on the size of the desired groove. Available power will dictate the outer limit. For example, a 13 mm ($\frac{1}{2}$ in.) wide, 6.4 mm ($\frac{1}{4}$ in.) deep groove that is 254 mm (10 in.) long could be made manually in two passes using a 6.4 mm ($\frac{1}{4}$ in.) diam electrode, or in one pass using a 9.5 mm ($\frac{3}{8}$ in.) diam electrode. In the first case, the best gouging rate would be 254 mm/min (10 in./min), divided by 2, or 125 mm/min (5 in./min). The travel speed of the second case (one pass) is 430 mm/min (17 in./min).

The 9.5 mm ($\frac{3}{8}$ in.) diam electrode increases the gouging rate by 200%, which could offset the additional cost of the electrode. Automatic systems increase the productivity rate even further, through finite control of the arc voltage.

The gouging amperage is the melting force of the process. It is affected by electrode size. If the amperage were set too low for the electrode size, then the melting rate of the base metal would be inadequate and free-carbon deposits would occur. Although the base metal would melt, a setting that is too high would rapidly deteriorate the electrode while reducing the metal removed per electrode. Too high a setting can also substantially reduce torch life.

Voltage is the pressure, or arc force, that enables the current to flow across the arc gap. The ACA process often requires a higher voltage than do most welding processes. To ensure proper operation, a power source with a high enough open-circuit voltage to maintain a 28 V operating minimum should be used. Inadequate voltage can create a sputtering arc or it can

prevent arc establishment, resulting in uneven grooves and, probably, free-carbon deposits that would require excessive grinding to remove.

Adequate air pressure and flow rate are required for the adequate removal of molten metal by the air jet. Air volume is as important as air pressure. Pressure is the speed of the air that moves the molten metal from the groove. If there is an inadequate volume of air to lift the molten material out of the groove, then it cannot be removed by the pressure or velocity. The result is excessive slag adhesion and unnecessary grinding to clean up the groove.

Travel speed affects the depth of the gouge and the quality of the groove. The faster the travel speed of an electrode, the shallower the gouge. A smaller electrode can be used if the travel speed is too fast for the comfort of the operator, or automatic gouging can be used. A groove that is too deep for the diameter of the electrode results in a poor-quality groove that requires much grinding.

The electrode push angle can vary somewhat. When gouging manually, a steeper angle tends to give a more V-shaped groove. When gouging automatically, a steeper angle gives a slightly deeper groove at the same travel speed.

Base Metals. The gouging procedures that are recommended for specific base metals are:

- For carbon steel and low-alloy steel, such as ASTM A 514 and A 517, use dc electrodes with DCEP. Although ac electrodes with an ac transformer can be used, their efficiency is half that of dc electrodes.
- For stainless steel, the same guidelines apply as for carbon steel.
- For cast iron, including malleable and ductile (nodular) forms, use 13 mm ($\frac{1}{2}$ in.) diam or larger electrodes at the highest rated amperage. Special techniques are needed when gouging these metals. The push angle should be at least 70° off the workpiece, and the depth of the cut should not exceed 13 mm ($\frac{1}{2}$ in.) per pass.
- For copper alloys (copper content ≤60%), use dc electrodes with direct current electrode negative (DCEN) at the maximum amperage rating of the electrode.
- For copper alloys (copper content >60% or large workpiece), use dc electrodes with DCEN at the maximum amperage rating of the electrode, or use ac electrodes with alternating current.
- For aluminum bronze and aluminum-nickel-bronze (special naval propeller alloys), use dc electrodes with DCEN.
- For nickel alloys (nickel content >80% of mass), use ac electrodes with alternating current.
- For nickel alloys (nickel content <80% of mass), use dc electrodes with DCEP.
- For magnesium alloys, use dc electrodes with DCEP and wire-brush the groove before welding.
- For aluminum alloys, use dc electrodes with DCEP and wire-brush the groove with

Table 7 Automatic air-carbon arc U-shaped groove operating data(a)

Electrode diameter		Desired depth		dc current, A	Travel speed	
mm	in.	mm	in.		mm/min	in./min
7.9	$\frac{5}{16}$	3.2	$\frac{1}{8}$	400	1651	65
		4.8	$\frac{3}{16}$	400	1143	45
		6.4	$\frac{1}{4}$	450	914	36
		7.9	$\frac{5}{16}$	450	838	33
		11.1	$\frac{7}{16}$	450	572	22.5
9.5	$\frac{3}{8}$	3.2	$\frac{1}{8}$	500	1778	70
		4.8	$\frac{3}{16}$	500	1118	44
		6.4	$\frac{1}{4}$	500	889	35
		9.5	$\frac{3}{8}$	500	508	20
		14.3	$\frac{9}{16}$	500	445	17.5
12.7	$\frac{1}{2}$	3.2	$\frac{1}{8}$	850	2438	96
		6.4	$\frac{1}{4}$	850	1448	57
		9.5	$\frac{3}{8}$	850	889	35
		12.7	$\frac{1}{2}$	850	610	24
		19.1	$\frac{3}{4}$	850	445	17.5
15.9	$\frac{5}{8}$	6.4	$\frac{1}{4}$	1250	1829	72
		9.5	$\frac{3}{8}$	1250	1219	48
		12.7	$\frac{1}{2}$	1250	940	37
		15.9	$\frac{5}{8}$	1250	762	30
		23.8	$\frac{15}{16}$	1250	495	19.5
19.1	$\frac{3}{4}$	6.4	$\frac{1}{4}$	1400	1829	72
		9.5	$\frac{3}{8}$	1400	1068	42
		12.7	$\frac{1}{2}$	1400	865	34
		15.9	$\frac{5}{8}$	1400	687	27
		19.1	$\frac{3}{4}$	1400	560	22
		28.6	$1\frac{1}{8}$	1400	330	13

Note: If a groove depth greater than 1.5 times the diameter of the electrode being used is desired, then the groove should be made in two or more passes. (a) Based on laboratory conditions; information should be used as a guide and adjusted for field variance as required

stainless brushes before welding. Electrode extension (length of electrode between torch and work) should not exceed 75 mm (3 in.). Direct current electrodes with DCEN can also be used.

- For titanium, zirconium, hafnium, and their alloys, there are no current procedures. If these metals are prepared for welding using the ACA process, then the groove should be cleaned before welding. These metals can be cut for remelting. In cases where preheat is required for welding, a similar preheat should be used for gouging.

Effects on Base Metal

Metallurgically, the base metal is not affected by the carbon particles being carried toward it from the electrode, because it remains solid and the atoms are very tightly packed. However, because the material being removed is molten, the atoms are loosely spread. This material, usually referred to as slag, easily absorbs the carbon particles.

If the ACA process is properly used, no surface carburization will occur. Due to the rapid quenching effect of the air used to remove the metal, the HAZ will undergo hardening. The HAZ for ACA is typically 2.5 mm (0.10 in.) or less in thickness. This HAZ hardening must be considered when a finish machining operation follows the gouging process, especially when working on high-carbon steels. Some materials (for example, nickel) will undergo surface cracking when gouged. These conditions should be considered when choosing the parameters for any required machining operations.

When the ACA process is used under improper conditions, this molten carburized metal may be left on the workpiece surface. Its color is a dull gray-black, in contrast to the bright blue of the properly made groove. Inadequate air flow may leave small pools of carburized metal in the bottom of the groove. Irregular electrode travel, especially in a manual operation, may cause ripples in the groove wall that trap the carburized metal. Finally, an improper electrode push angle may cause small beads of carburized metal to remain on the edge of the groove.

Copper from copper-coated electrodes does not transfer to the cut surface in base metal, unless the process is improperly used.

Carburized metal can be removed from a cut surface by grinding, but the necessity to do so can be prevented by gouging properly in the right condition.

When compared with oxyfuel cutting, the ACA process requires less heat input. Therefore, a workpiece that is gouged or cut by the ACA process is less distorted. The machining of low-carbon and nonhardenable steel is not affected by the ACA process. However, when used on cast iron and high-carbon steels, this process may cause enough hardening to make the cut surface tough to machine. Still, because the hardened zone is shallow (approximately 0.15 mm, or

0.006 in.), a cutting tool can penetrate it to remove the hardened surface.

Safety and Health

Safe practices in welding and cutting processes are described in ANSI Z49.1, "Safety in Welding and Cutting," and ANSI Z49.2, "Fire Prevention in Use of Welding and Cutting Processes" (American National Standards Institute). The operators of the ACA process and their supervisors should adhere to the practices discussed in these documents.

Electrodes. An electrode that carries electric current will arc if it touches the workpiece or any grounded metal object. Proper handling of the torch and electrode will prevent accidental arcing. Electrodes should always be used within their proper amperage range in order to effectively remove metal and to avoid damage to the torch. Carbon electrodes should be kept dry. Damp electrodes should be baked for 10 h at 150 °C (300 °F). Wet electrodes may shatter.

Torches. Oxygen should never be used in an air-carbon arc torch. Neither the torch nor the electrode should ever be immersed in water. All electrical connections should be wrench tight before the torches are used.

Electrical Power. The ACA process uses electric energy from a welding power source. The same safety precautions should be used as when welding. An operator should never stand in water while using ACA equipment, and the torches and electrodes should never be cooled by water. To prevent accidental arcing, partially used electrodes should be removed from torches when work is interrupted.

Personal Protective Equipment and Clothing. A No. 12 shade-filter lens should be used for eye protection against arc radiation. When conducting heavy metal removal operations with large electrodes, a No. 14 shade-filter lens or an equivalent combination of filter lenses provides the best protection. Flash goggles should be used, especially when two or more welders are working in the same area. Proper protective clothing must be worn to provide enough protection from the infrared and ultraviolet radiations of the arc. Leather aprons, sleeves, leggings, and so on should be used for out-of-position gouging or for heavy metal removal operations with large electrodes. Ear protection is recommended when noise exceeds permissible levels.

Fire and Burn Hazards. The ACA process may cause fire. All flammable materials should be removed from the work area. Booths, metal screens, and other deflectors should be placed to catch the hot metal and particle spray ejected by the compressed air stream. Protective clothing should be used to prevent burns caused by contact with hot metal or electrodes. Out-of-position or heavy metal removal operations require that many precautions be taken.

Ventilation Hazards. The fumes from the ACA process can be harmful. Ventilation and/or an exhaust around the arc is necessary. When

using gouging torches, the operator should keep his head away from the fumes. Protective-breathing or air-circulation gear may be needed.

Other Electric Arc Cutting Methods

Exo-Process. A relatively new electric arc cutting process, called the Exo-Process, has been developed. Similar to flux-cored processes, it uses a consumable tubular electrode and a specially designed gun that feeds high-speed compressed air to the arc (Fig. 19). The air flow functions to push molten metal from the gouge cavity, to constrict the arc for more precise control, and to cool the electrode. The system can be adapted to conventional gas metal arc welding equipment (it requires a direct current constant-voltage power source—150 A minimum—and a conventional wire feeder).

The velocity of the air flow at the arc is the key factor for straight cutting. A 1.5 mm (^{1}/16 in.) wire size can cut up to 6 mm (^{1}/4 in.) thick carbon steel. Speed and edge cut quality on most commercial metals and alloys is good, particularly for sheet metal thicknesses. Gouge quality on carbon steels is also good. The process would be well suited for automated equipment, in that high travel speeds may be attained. An obvious benefit of the process is that it can be mounted on a gas metal arc dual-wire feed system to provide the operator with a multifunctional welding and cutting unit.

Oxygen arc cutting uses a flux-covered tubular steel electrode. The covering insulates the electrode from arcing between it and the sides of the cut. The arc raises the work material to combustion temperature; the oxygen stream burns the material away. Oxidation, or combustion, liberates additional heat to support continuing combustion of sidewall material as the cut progresses. The electric arc supplies the preheat necessary to obtain and maintain ignition at the point where the oxygen jet strikes the

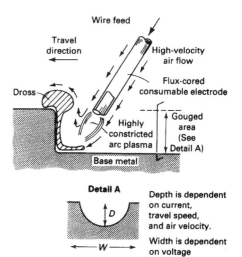

Fig. 19 The Exo-Process for gouging

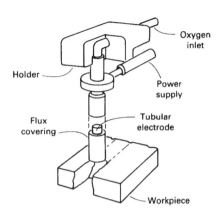

Fig. 20 Components of an oxygen arc electrode

surface of the work. The process finds greatest use in underwater cutting.

Oxygen arc cutting has been used effectively for cutting steels, stainless steels, cast iron, and nonferrous metals in any position. When cutting oxidation-resistant metals, melting action occurs. The covering on the electrode acts as a flux; it functions in a manner similar to that of powdered flux or powdered metal injected into the gas flame in the flux-injection method of oxyfuel gas cutting of stainless steel.

Oxygen arc cutting uses direct or alternating current, although DCEN, straight polarity, is preferred. The electrode and the electrode holder convey the electric current and oxygen to the arc. Electrode holders must be fully insulated; underwater cutting requires a flashback arrester, and the electrode must have a watertight plastic coating. Components of an oxygen arc electrode are shown in Fig. 20.

Seldom-Used Electric Arc Cutting Methods. The four electric arc cutting processes described in this section either are almost obsolete or have never been used extensively. However, these processes are sometimes used for highly specialized applications or when equipment for more suitable methods is not available.

Carbon arc cutting is one of the oldest electric arc cutting methods. This process uses the heat generated by the carbon arc to melt metal progressively to produce a kerf. The melted metal is removed from the cut by gravity or by the force of the arc, or by both.

The use of carbon arc cutting is limited to cutting up of scrap and dismantling of structural steel assemblies.

Shielded Metal Arc Cutting. Metal can be cut by a shielded metal arc, using flux-covered stick electrodes. The molten metal is removed from the cut by gravity or by the force of the arc, or both.

Shielded metal arc cutting is most commonly used for removing gates and risers from castings and for cutting up scrap. The process can also be used for underwater cutting with specially processed (waterproof) electrodes.

Gas Metal Arc Cutting (MIG). In this process, heat is obtained from an electric arc formed between a continuously fed wire and the work metal. The molten metal is continuously blown from the kerf by the shielding gas. The process is used mainly on stainless steel and aluminum, although other metals can be cut by this method.

Gas metal arc cutting is most often used in fabricating shops where gas metal arc welding is done, as the same equipment can be used for both cutting and welding. Gas metal arc cutting is limited in application, because the rate of wire consumption is high, and the thickness of sections that can be cut is limited to about $1\frac{1}{2}$ in. for stainless steel and about 3 in. for aluminum.

Gas tungsten arc cutting (TIG) is similar to gas tungsten arc welding, except that higher amperages and voltages are used. The same basic circuit and similar shielding gases are used for both processes.

Although TIG can be used for cutting most ferrous and nonferrous metals, it is largely confined to the cutting of stainless steel and aluminum $\frac{1}{2}$ in. or less in thickness. The quality of the cut is usually good enough to permit welding without machining the cut surfaces.

REFERENCE

1. W.H. Kearns, Ed., *Welding Handbook,* Vol 2, *Welding Processes—Arc and Gas Welding and Cutting, Brazing, and Soldering,* 7th ed., American Welding Society, 1978, p 499–507

Laser Cutting

INDUSTRIAL LASERS are being used in numerous materials processing applications. They can weld microswitches and auto transmission gears, scribe and machine ceramic substrates, and drill jet engine turbine blades and baby bottle nipples. They are also used in heat treating, cladding, ablating, and marking. However, cutting represents their largest single application, primarily two-axis profiling of sheet goods that may otherwise have been blanked out by punch press or fabricated by hand after laborious layout of the pattern.

The versatility of the laser in cutting operations is responsible for its widespread use. The same laser can be used to cut men's suits, newspapers, circuit boards, motorcycle fenders, circular saw blades, stainless steel auto exhaust tubing, and 13 mm (0.5 in.) thick alloy steel for aircraft disc brakes.

Its flexibility makes the laser an ideal tool for prototype or production work. Because laser cutting is a noncontact process, no tool wear occurs. Laser systems can cut intricate parts with accuracies of ±0.025 mm (±0.001 in.) and with surface finishes better than 1.3 μm (50 μin.) for some steels, and better than 0.50 μm (20 μin.) for some nonmetals. Because laser cutting gives a clean edge, secondary operations such as deburring can be reduced or eliminated.

Laser cutting of metal typically falls within 9.5 mm (0.375 in.) and thinner, although CO_2 lasers are now competitive with plasma arc cutting for metal thicknesses of 13 mm (0.500 in.) and greater. The principal factor in the use of laser metal cutting is manufacturing methodology. Laser cutting is ideal for batch processes, just-in-time, or low to medium-volume production. Most laser-cutting work is performed on generic or multipurpose materials handling systems, as opposed to dedicated automation-controlled systems. Technical features of laser cutting are compared with other methods of cutting in Table 1 (Ref 1).

Definition of a Laser

The word *laser* is an acronym for light amplification by stimulated emission of radiation. There are essentially three components that are necessary for lasing action. There must be an active media that can be excited, a method of exciting the media, such as an electrical discharge between an anode and cathode, and a resonator.

The molecules of the active media must be excited in order to stimulate the emission of radiant energy or light. As the molecules are charged to a higher energy state, they excite their electrons into a higher energy level. Because unstable electrons seek their lowest energy state, they release this added energy as light particles, known as photons.

The resonator consists of two parallel mirrors that reflect light particles between them, thereby amplifying the stimulated emission of light. Of the two mirrors in the resonator, the rear mirror is 100% reflective, while the front, or output, mirror is typically only 50% reflective and therefore 50% transmissive. The percentage of light that is transmitted through the front mirror is commonly known as the laser beam. It is this parallel, monochromatic, intense beam of light particles that is used for materials processing. The percentage of light remaining inside the resonator is necessary to maintain the continuous stimulated emission of photons.

Laser Types

Generally, there is a clear-cut reason for using one type of laser over another. Carbon dioxide (CO_2) lasers are capable of delivering much higher average power than neodymium: yttrium-aluminum-garnet (Nd:YAG) lasers. As such, CO_2 lasers can cut faster and produce deeper weld penetration than Nd:YAG lasers. Pulsed Nd:YAG lasers develop a high pulse energy that allows percussion drilling and the cutting of metals at angles and thicknesses not possible with CO_2 lasers. There are some applications—spot welding and hole cutting, for example—where either laser type can provide acceptable results at comparable speeds. Capabilities and operating ranges for CO_2 and Nd:YAG lasers are shown in Tables 2 and 3.

The CO_2 laser and the Nd:YAG laser are by far the most commonly used materials-processing lasers. The CO_2 laser relies on a gas mixture of CO_2, helium (He), and nitrogen (N) as its active media. It usually uses an electrical discharge between an anode and cathode as the method of

Table 1 Typical technical features of cutting processes

	Laser	Abrasive waterjet	Plasma arc	Oxyfuel
Materials	All homogeneous	All	Metallic	Metallic
Max thickness (steel), mm (in.)	30 (1.2)	100 (4)	50 (2)	300 (12)
Kerf width, mm (in.)	0.1–1.0 (0.004–0.04)	0.7–2.5 (0.03–0.10)	>1 (>0.04)	>2 (>0.08)
Heat-affected zone width, mm (in.)	0.05 (0.002)	0 (0)	>0.4 (>0.016)	>0.6 (>0.024)
Edge quality (relative)	Square, smooth	Square, smooth	Bevelled (~17°)	Square, rough
Edge roughness (R_a), μm (μin.)	1–10 (40–400)	2.0–6.5 (80–255)	...	...
Smallest hole diameter, mm (in.)	0.5 (0.02)	>1.5 (>0.06)	>1.5 (>0.06)	20 (0.8)
Energy input (relative)	Low	Low	High	Medium
Capital cost (relative)	1	1	0.1	0.01
Productivity (relative)	High	Medium-low	Medium	Low

Source: Ref 1

Table 2 Process capabilities of CO_2 lasers by power range

	200–300 W	300–500 W	500–800 W	800–1500 W	1500–3000 W	>3000 W
Seam welds, spot welds, maximum penetration, mm (in.)	0.75 (0.030)	1.3 (0.050)	2.0 (0.080)	3.2 (0.125)	6.4 (0.250)	19.0 (0.750)
Cutting, maximum thickness, mm (in.)	1.5 (0.060)	5.0 (0.20)	9.5 (0.375)	12.7 (0.500)	19.0 (0.750)	<25.0 (<1.00)
Heat treating case depth, mm (in.)	...	...	0.75 (0.030)	1.3 (0.050)	1.3 (0.050)	1.3 (0.050)
Cladding, surfacing, glazing, annealing	...	...	Yes	Yes	Yes	Yes

Source: Ref 2

media excitation, and the standard two-mirror resonator. The CO_2 laser produces a wavelength of 10.6 μm (420 μin.), which is invisible to the human eye. Visible light falls between 0.4 and 0.7 μm (15 and 28 μin.). Although practical materials-processing lasers have powers ranging from 150 W to 8 kW, CO_2 lasers have been built with powers above 25 kW. Most lasers used for cutting have powers that range from 150 W to 3 kW. Figure 1 shows a typical CO_2 gas laser design.

Cutting with lasers is accomplished with CO_2 and Nd:YAG lasers. More often, cutting is done with CO_2 lasers because of the faster cutting rates. Typical cutting rates for a 1250 W CO_2 laser are illustrated in Fig. 2. Process variables (Ref 2) in laser cutting are described as follows.

Average power for cutting with a continuous-wave CO_2 laser is typically 250 to 5000 W. In the pulsed mode, lower average powers can cut metal because of higher peak instantaneous powers. The power ranges for cutting metal in the pulsed mode range from less than 100 to 2000 W for CO_2 lasers. Cutting with Nd:YAG lasers is accomplished with a power range of less than 100 to more than 400 W.

Pulse length is selected to optimize the quality of the cut surface. Shorter pulse lengths (<0.75 ms) are used for the intricate cutting of thin metals. Shorter pulse lengths may limit the maximum energy achievable in a single pulse. Longer pulse lengths (up to 2 ms) provide greater pulse energies, allowing thicker metals to be cut. Similar pulse lengths would be used for both CO_2 and Nd:YAG lasers.

Pulse frequency is adjusted to give the maximum cutting speed for the quality required. In general, higher frequencies are used to cut thinner metals. For CO_2 lasers, higher frequencies range from 200 to 500 Hz. In Nd:YAG lasers, higher frequencies range from 30 to 100 Hz. Lower frequencies are used to cut thicker metal sections.

Pulse energy needed in cutting is related to material thickness. As metal thickness increases, greater pulse energy is required. In practice, CO_2 lasers deliver maximum pulse energies up to 2 J at longer pulse lengths and lower pulse frequencies. Nd:YAG lasers can produce much higher pulse energies up to 80 J, with pulse frequency limited by the maximum power rating of the laser.

Lens choice is based on metal thickness, composition, and quality requirements and on beam diameter. Wider kerf widths are obtained by using longer focal-length lenses, and some materials, such as aluminum, require a larger kerf width for good results. The following guidelines are for beam diameters from 13 to 25 mm (0.5 to 1 in.).

For CO_2 lasers, a general rule is to use a 65 mm (2.5 in.) focal length for cutting metals up to 6 mm (0.25 in.) thick. A 125 mm (5 in.) focal-length lens can be used for metals from 5 to 15.8 mm (0.2 to 0.625 in.) thick. A 190 or 250 mm (7.5 or 10 in.) lens can be used for materials thicker than 13 mm (0.5 in.).

Typical lenses for Nd:YAG lasers have focal lengths of 100 mm (4 in.) for metals thinner than 3 mm (0.125 in.). For cutting metals up to 25 mm (1 in.), the focal lengths range from 150 to 250 mm (6 to 10 in.). Thicker metals require even longer focal lengths.

Gas jets provide a coaxial, columnar flow of gas through the cut slot to remove molten metal. The gas type can be oxygen, inert gas, or air, depending on material type and quality requirements. Oxygen is most commonly used for cutting steels. When an oxide-free surface is desired, an inert gas, such as helium, is used.

Gas pressures for oxygen range from 100 to 350 kPa (15 to 50 psi), while pressures for inert gases range from 200 to 620 kPa (30 to 90 psi). Typical nozzle orifices range from 0.75 to 2.5 mm (0.030 to 0.100 in.). Nozzle standoff,

Table 3 Process capabilities of Nd:YAG lasers by power range

	<100 W	150–200 W	200–400 W
Microwelding, soldering, marking	Yes	...	...
Seam welds, spot welds, maximum penetration, mm (in.)	...	1.3 (0.050)	2.0 (0.080)
Cutting, maximum thickness, mm (in.)	...	5.0 (0.2)	38 (1.5)
Drilling, maximum thickness, mm (in.)	...	5.0 (0.2)	38 (1.5)

Source: Ref 2

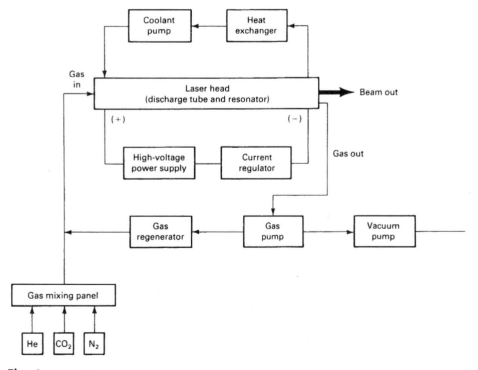

Fig. 1 General CO_2 laser design

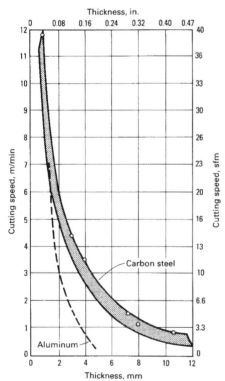

Fig. 2 Cutting rates for 1250 W CO_2 laser

depending on laser type, nozzle design, and gas flow rates, varies from 0 to 1.5 mm (0.0 to 0.06 in.) for CO_2 lasers and up to 5 mm (0.200 in.) for Nd:YAG lasers. Nozzle standoff and gas pressure to a large degree determine the quality of the cut. Using a CO_2 laser, a clean cut (oxide free and dross free) can be achieved in thinner sections of metals such as stainless steel, aluminum, or titanium when using inert coaxial gas at 620 kPa (90 psi) with minimum nozzle standoff.

CO_2 Lasers

There are three basic types of CO_2 gas lasers: slow axial flow, transverse flow, and fast axial flow.

The slow axial flow laser is an older, proven design that has excellent mode, stability, and pulsing capabilities. They are available with powers up to approximately 800 W, allowing them to cut steels up to 6.5 mm (0.25 in.) thick. Figure 3 shows the typical design theory of a slow axial flow laser. The lasing action occurs by injecting laser gas at a pressure of approximately 2.7 kPa (20 torr) into an evacuated glass tube. The tube has a rear mirror and an output coupler at either end. A high-voltage glow discharge is then passed down the tube between an anode and cathode, causing lasing to occur. This design, which can only create 70 W/m (20 W/ft) of tube length, requires a long resonator cavity to produce high power levels.

The transverse-flow laser was developed in response to the size versus power limitations of the slow axial flow laser. Transverse-flow laser design allows power to be generated to approximately 25 kW. The transverse laser accomplishes this by using a tangential blower to move a high volume of laser gas transversely across the anode-cathode electrical path, as shown in Fig. 4. The basis of this design is that power is directly related to the volume of laser gas that is excited at any one time. The draw backs are that these lasers cannot be electronically pulsed for cutting very intricate geometries, and the beam quality, or mode, is not ideal for high-quality cutting applications.

The fast axial flow laser is the newest CO_2 type. It combines the high power of the transverse-flow laser with the beam quality and some of the pulsing capabilities of the slow axial flow laser. Its design is similar to the slow axial flow laser, except that a tangential blower

forces a large amount of laser gas axially down the resonator, as shown in Fig. 5. This increases the available power to 700 W/m (210 W/ft) of active laser resonator. This laser type is currently being built with power up to 3 kW and is becoming increasingly popular as a cutting laser because of its increased power, excellent beam quality, and pulsing capabilities.

Both the slow and fast axial flow lasers can operate in either the continuous wave (CW) or electronically pulsed modes. In CW operation, the laser operates at a continuous power level. This type of operation provides the highest cutting travel speeds. Because of the high speeds required to cut thinner material above 3175 mm/min (125 in./min), a loss of accuracy can occur on intricate parts because of motion

system limitations. Overheating will also occur on any thickness if the part is very complex. The solution to these problems is to pulse the laser electronically, thereby allowing intermittent high peak powers (approximately two to eight times the peak CW power) and overall lower average powers, as shown in Fig. 6. This results in controlled heat input and higher accuracies, but at reduced feed rates, compared to CW operation. Typical pulse rates for CO_2 laser cutting range from 100 to 1000 Hz.

Nd:YAG Laser

The Nd:YAG is the second type of industrial laser. It is a solid-state laser in which the active

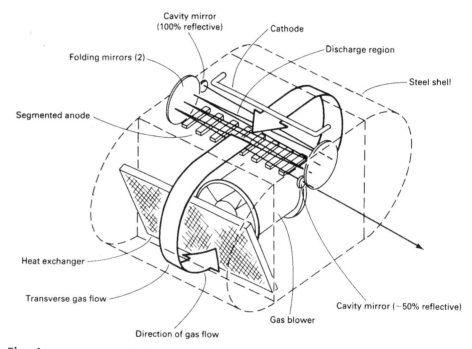

Fig. 4 Transverse or cross-flow CO_2 laser

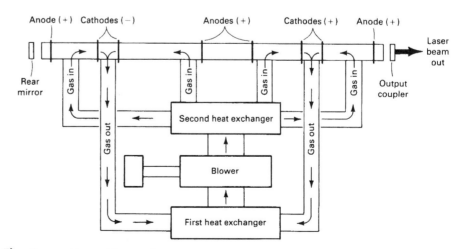

Fig. 5 Typical fast axial flow CO_2 laser

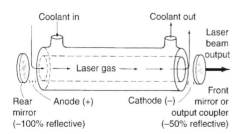

Fig. 3 Slow axial flow CO_2 laser

medium is neodymium, which is dissolved in a matrix of yttrium-aluminum-garnet. The resulting crystal is formed into a rod, which is excited by external flash lamps using xenon or krypton. As the lamps flash, the light is absorbed by the rod, exciting the medium to emit photons. This laser uses the same type of two-mirror resonator described earlier. The Nd:YAG laser is a pulse-only laser with a cutting speed limited to approximately 760 mm/min (30 in./min). This laser also emits invisible infrared light but with a wave length of 1.06 µm (41.7 µin.), compared to 10.6 µm (417 µin.) for a CO_2 laser. Because of this shorter wavelength, the Nd:YAG beam is more readily absorbed by metals and is therefore used to cut gold, silver, copper, platinum, and other metals that would be very reflective to a CO_2 laser. The Nd:YAG laser also does a superb job of drilling and trepanning small holes in metals and is typically used on aircraft jet engine parts made from high-temperature superalloys and titanium. Figure 7 shows a typical Nd:YAG laser system design.

Competing Cutting Methods

The advantages and disadvantages of several conventional metal-shaping processes that compete with the laser are shown in Table 4. While the laser does not displace any of these processes in terms of their special capabilities, laser cutting does fill a very important void. For example, the laser is the ideal method for producing short-run or prototype blanked parts of large, complex or small, intricate shapes (Fig. 8). The choice to laser cut these parts is based on cost. The expense of temporary tooling or edge finishing far exceeds the cost of laser cutting.

A second, broader example is the cutting of industrial-quality circular saw blades. Several years ago, a major manufacturer of such blades purchased a two-axis laser system to blank out the entire saw blade. These saw blades are being produced in sizes ranging from 50 to 915 mm (2 to 36 in.) in diameter, with an average order quantity of 22 pieces. Because of the small order quantity and the highly customized tooth profiles, many setups and operations were required to punch the geometries. However, by using the laser, which cut the saw geometry in just one operation, the manufacturer was able to reduce the cost on many items up to 30% over

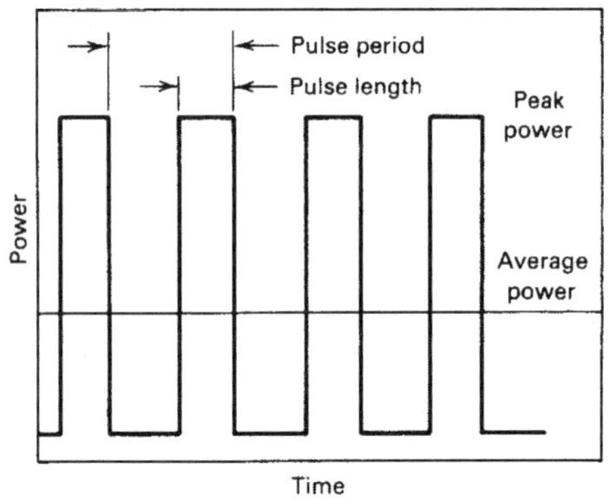

Fig. 6 Pulsed waveform

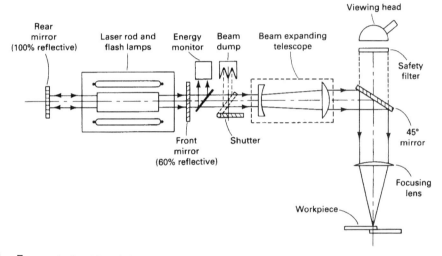

Fig. 7 Typical Nd:YAG laser design

Table 4   Advantages and disadvantages of laser cutting versus traditional metal-cutting methods

Cutting method	Advantages	Disadvantages
Laser	Good edge quality, good accuracy, small kerf, narrow HAZ, no distortion, little noise; cuts nonmetals, cuts small and complex shapes	High equipment cost, limited to under 13 mm (0.5 in.) thick, slower feed rates over 6.4 mm (0.25 in.); cuts single layer
Plasma arc	Lower equipment costs, faster feed rates over 6.4 mm (0.25 in.); cuts over 13 mm (0.5 in.) thick	Lower accuracy, decreased edge quality, larger kerf, wider HAZ, noisy, higher operating costs; only cuts metal
Laser	Good edge quality, good accuracy, lower scrap rate, no distortion, small kerf, no tooling or tool wear, increased part nesting; cuts complex shapes, cuts up to 13 mm (0.5 in.) thick, cuts tempered materials and nonmetals	Higher equipment costs, lower process rate, higher costs on larger part quantities
Nibbling (turret punch press)	Good process rate, lower equipment costs, economical on medium- to high-production runs	Lower edge quality, high tool wear, high tooling costs, low accuracy, distortion, scrap; only cuts 10 mm (0.38 in.) thick
Laser	Good edge quality, no tooling or dies, short setup times, rapid and low-cost design changes, noncontact cutting; cuts complex shapes and three-dimensional shapes, cuts tempered materials	Higher equipment costs, low rate on high volumes
Punch press	High volume rates; lower costs at high volumes; cuts over 13 mm (0.5 in.) thick	Greater tool fabrication time, higher tool costs and maintenance, more setup time, poorer tool design, part stresses, lower edge quality; only cuts annealed steel
Laser	High feed rate, economical on small and medium quantities; cuts nonmetals and nonconductive metals	Lower edge quality, higher equipment costs, thickness limitations
Wire electric discharge machining	Good edge quality, good accuracy, lower equipment costs; cuts over 13 mm (0.5 in.) thick, cuts very fine and complex shapes	Very slow on any thickness, fixturing
Laser	Narrow kerf, faster feed rates, good accuracy, good edge quality	Small HAZ, fumes; cuts limited materials
Abrasive waterjet	No HAZ; cuts up to 152 mm (6 in.) material thickness, no distortion, cuts all materials	Abrasive disposal, noisy, high-pressure plumbing, slow feed rates
Laser	Flexibility, faster feed rate, short setup time	Small HAZ, accuracy, high equipment costs
Numerically controlled milling	No HAZ, good accuracy, good edge quality, low equipment costs	Limited feed rate, high tool costs and maintenance

HAZ, heat-affected zone

conventional punching, milling, turning, and blanking methods.

Laser cutting does have limitations for many applications. In the case of the heat-affected zone and fusion layer present in some aerospace hardware, laser cutting can only be used to produce a semifinished part that requires further processing using a different method.

Laser-cutting advantages and disadvantages must be carefully considered for each particular application before deciding whether use of the laser is advantageous. Because applications are too far-ranging to generalize, in-depth information should be obtained from a laser manufacturer, systems house, laser job shop, or consultant.

General Cutting Principles

The mechanism for cutting steel with a laser is basically the same as cutting steel with an oxygen-fuel process in which the fuel gas acts to heat the material so the oxygen can oxidize and react exothermically with the steel to produce the cutting action. The oxygen also helps to sweep the molten material out of the kerf. In laser cutting, the fuel gas is replaced with a laser beam focused to approximately 0.1 mm (0.004 in.), resulting in a power density of 1 MW/cm² (6.5 MW/in.²). It is this characteristic that allows a 3.94 mm (0.155 in.) thick alloy steel chain saw bar to be cut out at 2.54 m/min (100 in./min) with a 0.15 mm (0.006 in.) kerf width and without part distortion.

Laser cutting can also be accomplished on nonferrous and nonmetallic materials using assist gases. In the case of nonferrous metals (aluminum, copper, brass, and bronze), which have high thermal conductivities but do not react with assist gases and are reflective to the laser beam, cutting occurs when the laser beam heats the material well above its melting point, and an assist gas, such as air, argon, or helium, is used to sweep the molten material out of the cut. The inert gases are used only when the cut edge of the material must be free of any impurities that would reduce its serviceability in a very harsh environment.

In the case of nonmetallic materials with low thermal conductivities and high beam absorption, such as wood, cloth, and paper, vaporization of the material upon cutting is nearly 100%. Because material vapors have a tendency to rise, an assist gas, such as compressed air, is used at a low pressure to protect the focusing lens from the damaging vapors. Organic materials such as plastics and woods are also cut with lower gas pressures. Acrylic plastic will yield a fine polished edge if cut with an inert gas or compressed air at 70 kPa (10 psi) pressure. Table 5 summarizes the general purposes of assist gases.

Process Variables

There are seven basic parameters in the laser-cutting process: beam quality (mode), power (CW or pulse), travel speed, assist gas, nozzles, focusing lens, and focal-point position. Slight

changes in any one of these parameters can yield significant changes in cut quality.

To fine-tune excellent cut quality, it is recommended that only one parameter at a time be varied, while the others remain constant. Laser processing maps (Ref 1) are also a method in determining appropriate parameters. Additional guidelines on laser cutting are available in Ref 3.

Beam quality, or mode, is a very important parameter. A laser should be used with a TEM_{00} (TEM, transmission electron microscopy) or Gaussian mode, which is ideal for cutting (Fig. 9a). A Gaussian mode has most of its energy in its center. Figures 9(b) and (c) illustrate the power density versus the radial distance from the beam centerline for both TEM_{00} and TEM_{01*} modes. The TEM_{00} mode can be focused to a smaller spot size than can TEM_{01*} and has greater energy density of power per unit area, thereby increasing cutting efficiency. This TEM_{00} mode decreases the heat input to the part, which allows faster cutting speeds and a smaller heat-affected zone (HAZ). Because the beam can be focused to a smaller spot size, kerf width is also reduced.

Power plays a significant role in respect to both feed rates and thicknesses. For example, Fig. 10 shows that increasing power increases speed and thickness for a given material, while holding all other parameters constant. Figure 10 also shows that the maximum material thickness to be cut also increases with increasing power. When pulse cutting is used, the average powers decrease, while the peak powers of each pulse increase from two to eight times the maximum CW power. This reduction of average power reduces the feed rate by approximately 60 to 80%.

An optimal travel speed exists that yields the best cut quality, while holding all other parameters constant. Because feed rate is very material dependent, experimentation is needed in order to obtain the best results. However, for most metals, the best cut is obtained at the

Table 5 Assist gas usage

Gas	Major function	Usage
Oxygen	Promotes chemical reaction	Cutting ferrous metals
Argon, helium, and nitrogen	Inhibit chemical reaction	Cutting thin metal for oxide-free edge; cutting chemically reactive metals and materials
Air and inert gases	Remove excess by-products	Lens protection, absorptive plume removal; cooling to clear hot gases away from thermally sensitive materials; cutting nonmetals

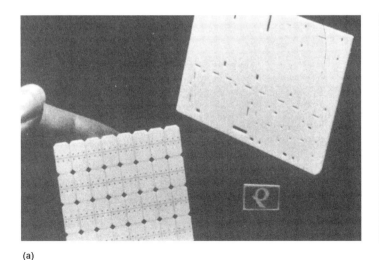

(a)

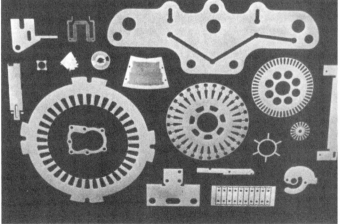

(b)

Fig. 8 Typical laser-cut parts. (a) Ceramic. (b) Metal

maximum achievable speed, which also helps to minimize the HAZ. If the feed rate is too low, burning will occur, and a large HAZ with slag formation will be present. In nonmetals such as wood or cloth, excessive charring will be present on the cut face. If the feed rate is too fast, the

(a)

(b)

(c)

Fig. 9 (a) TEM_{00} or Gaussian mode energy distribution. (b) TEM_{00} power density (sharp tool). (c) TEM_{01*} power density (blunt tool)

beam will climb out of the cut, and only partial cutting will occur.

Use of oxygen as the assist gas for cutting steel and stainless steel increases cutting speeds by 20 to 40% compared to use of air. Air is used mostly to cut nonmetals because it helps reduce oxidation and burning. When cutting stainless steel with an oxygen assist gas, slag or a fusion layer occurs, which is not tolerable in some applications. Although argon or helium assist gases eliminate oxidation, they reduce travel speeds by up to 50%.

The gas nozzle design and standoff distance, which is the distance between the workpiece and the nozzle, can significantly affect cut quality. A properly designed nozzle will produce a laminar, high-flow-rate assist gas through the cut. The laminar flow can be affected by the standoff distance. If the standoff distance is too great, the smoothly flowing gas tends to break up. This disrupts the flow through the kerf and decreases the edge quality. Typical nozzle diameters are 1 to 2 mm (0.040 to 0.080 in.) with 0.5 to 3 mm (0.020 to 0.12 in.) standoff distances.

A focusing lens is used to focus the beam on the workpiece. This increases the power density of the beam. The lens is used because the output beam of a laser is typically 11 to 21 mm ($7/16$ to $13/16$ in.) in diameter and does not possess enough energy per unit area to melt and vaporize materials. The lens can focus the beam to a spot size of 0.1 mm (0.004 in.) in diameter. The same principle is in operation as when using a magnifying glass to focus sunlight on a piece of paper, causing it to burn.

Lenses are classified by focal length, or the distance from the lens to the point at which the

spot size is smallest (Fig. 11). Typical lenses come with focal lengths that range from 38 to 254 mm (1.5 to 10 in.). Shorter focal-length lenses have higher energy densities because they have a smaller spot size, as shown in Fig. 12. These lenses, however, have a limited depth of field, or usable beam. Depth of field is the area of the focused beam that has enough energy density to process materials. This spot-size limitation can be offset by using a larger input beam on a longer focal-length lens, as shown in Fig. 13. Short lenses are typically used on reflective materials, such as aluminum, or on thin materials, for faster feed rates and an improved surface finish. Longer focal-length lenses are usually used on materials 6.4 mm (0.25 in.) or more thick because of their greater depth of field, which produces a squarer cut and sharper geometry definition at the bottom of the cut. Because of increased spot size, these lenses decrease feed rate and overall surface finish quality, while increasing the HAZ. Most lenses in use are between 64 and 127 mm (2.5 and 5.0 in.) in focal lengths.

The focal point position greatly affects the surface finish and dimensional accuracy of the part. Accuracy is affected because the beam is never parallel (Fig. 12); hence, as the focal point is moved to different positions, the kerf width can increase or decrease. This effect must be considered when cutting high-accuracy parts.

Surface finish is very dependent on the focal point. For the majority of metals, the focal point

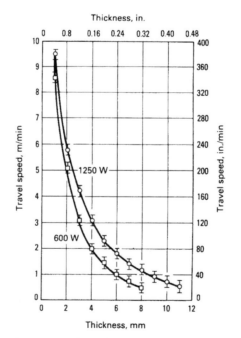

Fig. 10 Travel speed versus thickness for 600 and 1250 W CO_2 lasers. Focused power at workpiece using 65 mm ($2^1/2$ in.) focal length lens. Oxygen assist gas at 350 kPa (50 psi). Carbon and alloy steels used

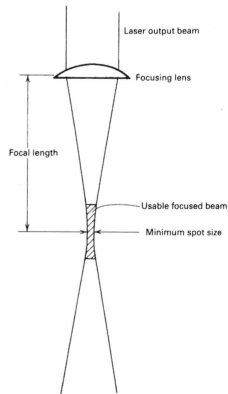

Fig. 11 Geometry of focused beam

is set slightly below the work surface. For other materials, such as stainless steels, the focus is set well below the surface of the material to yield the best cut. However, the focal position itself is very dependent on the type of material and other process parameters and therefore must be manipulated to find the optimal setting.

Material Conditions

Several material conditions can affect the quality of the laser cut. First, surface cleanliness can have a detrimental effect on edge quality. All steel alloys should be either hot rolled, pickled, and oiled; or cold rolled. Mill scale, which can interfere with the beam, greatly reduces edge quality and dimensional accuracy. Rusted steel also decreases cut quality. No other special cleaning methods are required.

Second, flatness of the material to be cut affects the focal point of the beam. The focal point must be controlled to ± 0.25 mm

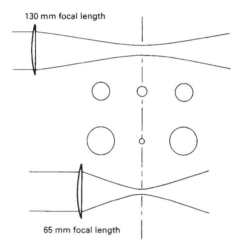

Fig. 12 Beam diameters at equal distances from focus position for lenses of different focal length

(0.010 in.) to achieve the best possible cut. Increasing surface roughness can deflect the assist gas, resulting in a nonlaminar assist gas flow, thereby decreasing edge quality. Most sheet steel is smooth enough to be laser cut.

Third, coatings on the material surface usually have no adverse effect on cutting. Thin layers of plastic on metal surfaces are cut without problem, although proper fume collection procedures must be used. Electrogalvanized steel can be laser cut, provided that the galvanized layer is thin. Again, it is recommended that fume collection be used. Steels with paint on one surface should be cut without oxygen because a paint-oxygen reaction produces a very poor cut.

Finally, the ambient temperature of materials must be taken into consideration. Because carbon steels react exothermically with oxygen at low temperatures of approximately 40 °C (104 °F), high ambient temperatures result in a cut with wide, rough, low dimensional accuracy. This effect also must be taken into account when cutting very intricate geometries in carbon steels. However, pulse cutting at reduced feed rates reduces the heat input and allows satisfactory cutting of very small and intricate shapes.

System Equipment

The basic laser-cutting system consists of a laser, a motion system, a controller that is computer numerically controlled (CNC), and a beam delivery system. Optional equipment includes water chillers, dust/fume collectors, compressed air equipment, transformers and related electrical equipment, and a computer-aided design/computer-aided manufacturing (CAD/CAM) system. Typical costs for an installed system can range from approximately $100,000 for a small, two-axis cutting system to over a million dollars for a large, five-axis cutting system.

Many styles and sizes of motion systems are available. The most popular is the x-y axis table system. The x-y motion is coordinated by a CNC controller and is used to blank contours out of flat sheet stock. These systems range in size from 300 by 300 mm (12 by 12 in.) to 1.6 by 3.0 m (63 by 120 in.) for a large-bed sheet-cutting system. These two-axis motion systems are available in three styles, each of which manipulates the workpiece and/or beam differently in the course of contouring a part, as shown in Fig. 14.

The first system, shown in Fig. 14(a), has a moving workpiece and a stationary beam, which is the simplest and least costly system. Motion systems of this type usually provide the highest accuracies for laser cutting. For cutting areas

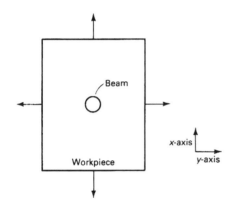

(a)

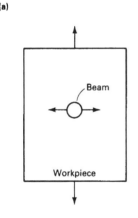

(b)

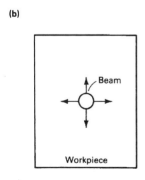

(c)

Fig. 14 Three types of two-axis motion systems. (a) Moving x-y table, stationary beam. (b) Moving x-axis table, moving y-axis beam. (c) Moving x-y beam, stationary table

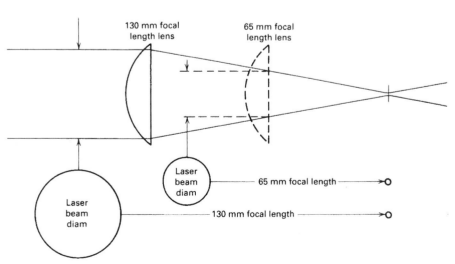

Fig. 13 Beam expansion for same spot size

larger than 1.2 by 1.2 m (48 by 48 in.), the mass of a two-axis table system becomes large and therefore reduces travel speeds and accuracies. For large-sheet cutting requirements, it is advantageous to move the workpiece in one axis, while moving the beam and cutting head assembly in the other axis, as shown in Fig. 14(b). This allows for significantly less mass movement on the motion system, while incorporating some floor space savings. For systems that are 1.2 by 2.4 m (48 by 96 in.) or larger, it is more common to move the beam and optics on a gantry-style motion system, as shown in Fig. 14(c). These systems do, however, require more attention to beam alignment. A moving optics system can also be used on smaller systems in which the parts are too heavy to move accurately, or on assembly lines.

Most laser-cutting systems usually employ some type of floating third, or z, axis to allow the focused beam to be introduced to the workpiece, because a workpiece is not usually perfectly flat. These floating cutting heads sense height changes through contact and noncontact methods. Figure 15 shows a contact method using roller balls.

Other types of motion systems have three, five, seven, and nine axes of programmable motion. The three-axis systems have either a programmable z or a rotary axis. The z-axis allows cutting on parts with different heights, whereas a rotary axis allows contour cutting of tubing. The five-axis gantry-style motion systems incorporate x, y, and z axes, with a wrist on the z-axis. A wrist is a set of orthogonal rotating mirrors providing increased beam manipulation. This system allows contour cutting on formed parts and is necessary because the beam must be normal to the surface being cut. Typical applications are

the trimming of stamped parts such as automotive and motorcycle fenders. In specialized cases, a nine-axis robot can manipulate either the beam or the workpiece. Some applications include cutting holes for sunroofs in auto assembly lines or other types of ventilation.

A CNC controller is used to coordinate the motion and the laser operation. The controller reads standard numerically controlled (NC) programs that describe the contour to be cut. The NC programs are prepared with a CAD/CAM system. Then, either the NC file is punched on paper tape and read into the controller by a tape reader or a number of other methods, including floppy disks, erasable programmable read-only memories, and a direct numerical control between the controller and the off-line computer.

A dust/fume collecting system should be installed: Laser cutting of metallics produces dust, which should be removed, while nonmetallics can produce smoke and fumes, which can be irritating and/or hazardous and likewise should be removed from the work area. Optional equipment includes a water chiller to cool the laser and external optics, transformers for proper electrical supply, and compressed dry air for the beam delivery and assist gas.

Laser-Cutting Applications

Virtually any metal can be laser cut. The conditions for cutting some metals, however, are difficult to obtain. Following is a general description of how well each metal group can be cut. Additional information on laser cutting of engineering materials is in Ref 1.

All carbon and alloy steels can be laser cut to over 13 mm (0.50 in.) thick with an oxygen assist gas. As Fig. 10 shows, the feed rates are dependent on available power at the workpiece. A general rule of thumb is that for a given thickness, the feed rate increases approximately 50% when power is doubled. The resulting kerfs can be as small as 0.1 mm (0.004 in.) thick with a 65 mm ($2\frac{1}{2}$ in.) focal-length lens. The resulting HAZ is small (approximately 0.1 to 0.3 mm, or 0.004 to 0.012 in.), depending on thickness and speed. The edges are clean, smooth, and square.

Higher-carbon steel exhibits an improved edge quality, although the HAZ is slightly larger and harder. Impurities of phosphorus and sulfur can cause some edge burning: Lower-quality steels exhibit this edge burning, whereas alloy steels generally do not. In fact, the alloy steels, such as chrome-nickel-molybdenum (for example, 4340) and chrome-molybdenum (for example, 4130), are perfect candidates for extremely high-quality edges that are very smooth and clean. These steels do, however, require approximately 50% more time to pulse a starting hole than is the case with a similar thickness of plain carbon steel. Another consideration is that the HAZ is slightly larger and harder (approximately 45 to 60 HRC), depending on the alloy and carbon content. This means that further

processing can be difficult on the laser-cut edges unless they are annealed.

Laser-cut tool steels have almost the same results. The exceptions are the alloys that have very dense alloying elements, such as tungsten. These alloys retain a large amount of heat when molten, which helps to limit thickness and produce very rough cuts having heavy slag deposits. The air- and oil-hardened alloys exhibit very good edge quality and can be cut up to 10.2 mm (0.400 in.) thick at approximately 0.8 m/min (32 in./min) at 1300 W. Alloys such as D-2 and M-2 can only be cut up to approximately 4 mm (0.160 in.) thick with the same power.

Stainless steel alloys are also readily cut using a laser. The feed rates are reduced, however, because these alloys do not react as effectively with oxygen as carbon steel alloys do. An inert assist gas may be used to obtain a weld-ready edge, free of all oxides, at the expense of approximately one-half the oxygen-assisted speed. Stainless steels maintain their corrosion resistance because the HAZ is small.

In general, the ferritic (400-series) stainless steels produce smoother cuts with less slag than austenitic (300-series) stainless steels do. These stainless alloys do not contain nickel and can be cut to approximately 6.5 mm (0.260 in.) thick.

The added presence of nickel does affect the energy coupling and heat transfer in the alloy. This means that these alloys are not effectively laser cut above 5 mm (0.197 in.) thick. The viscosity of the molten nickel is very high and has a tendency to migrate and adhere to the bottom of the cut. This increases the heat in the metal and produces a large HAZ, as well as a rough cut starting approximately one-third to one-half the way through the material. This effect can be reduced by using high-pressure assist gas jets or rapid-cooling methods (such as water) in the cut.

Nonferrous Alloys. Aluminum alloys can readily be laser cut but only to approximately 4 mm (0.160 in.) thick. The thickness is limited because aluminum has high reflectivity at infrared wavelengths and high thermal conductivity. To overcome these effects, the laser must have a TEM_{00} mode, which allows for tighter focusing and higher power outputs, that is, 500 W or more. To further improve the cutting, short focal-length lenses and high assist gas pressures help reduce and/or eliminate slag that forms when molten material is blown to the back side of the cut and solidifies; the slag is very easily removed. Feed rates are generally 25% slower than when laser cutting stainless steels.

Copper, brass, and bronze are even more reflective and heat conductive than is aluminum. Because brass and bronze are alloys of copper, they can be laser cut, but with limited thickness and speed. The cuts can be rough, and a slag is present on the bottom of the cut. Copper has been cut under ideal conditions in thicknesses up to 6.35 mm (0.25 in.) with 1200 W. The speeds are extremely slow, so practical upper cutting is limited to approximately 2.6 mm (0.100 in.).

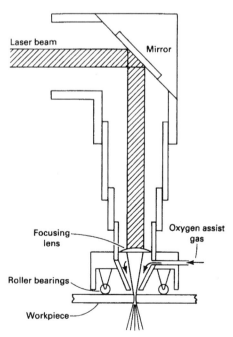

Fig. 15 Cutting head design

Other high-tech alloys can also be laser cut. Titanium cuts very well up to approximately 7 mm (0.275 in.) thick. An inert assist gas is usually employed to prevent an oxide layer from forming. Oxygen can be used, but because titanium reacts very strongly with it, high feed rates are needed to prevent burning the material.

Nonmetal Cutting. Most nonmetals rely on the same technique for laser cutting. The cutting action is dependent on the ability of the laser beam to be absorbed by and to vaporize the material being cut. Because the cut does not rely on an oxidation reaction, as in steel cutting, the laser vaporizes the material with the aid of an assist gas (either inert gas or compressed air) to help purge the cut of the vaporized material. The CO_2 laser is typically preferred for nonmetal cutting because of its good absorption by many materials and its high cutting speeds.

Acrylic is easily cut with the CO_2 laser. Thicknesses up to 50 mm (2 in.) have been cut for aircraft windshields, but most work is done at thicknesses of approximately 6.4 mm (0.25 in.), while a 150 W unit provides speeds of 1.25 m/min (50 in./min). The cut edge is usually frosted at these speeds. A fine polished edge is attainable in acrylic at approximately half speed with very low assist gas pressure.

Quartz and nontempered glass are commonly cut for a variety of applications. The edge quality of glass is rough compared to laser cuts in most other materials, at speeds up to 2.5 m/min (100 in./min) for a 3 mm (0.10 in.) thickness using a 600 W CO_2 laser, and approximately 5 m/min (200 in./min) for a 1200 W unit.

Quartz, on the other hand, cuts with a fine, clean edge with speeds of 5 m/min (200 in./min) for 1 mm (0.040 in.) thick material and 1 m/min (40 in./min) for 3 mm (0.10 in.) at the 600 W level.

Ceramic of 96% Al_2O_3 (alumina) is used for electronic substrates and is very brittle; the noncontact cutting characteristic of the laser therefore makes it the preferred method. The laser is used in the pulse mode to minimize heat input to the part. A high-quality edge is produced at 0.5 m/min (20 in./min) for alumina substrates up to 1 mm (0.04 in.) and 0.25 m/min (10 in./min) in the 1 to 2 mm (0.04 to 0.08 in.) range with a 500 W CO_2 laser in the pulsed mode.

Aramid fibers have tensile strengths that exceed those of steel. The fiber is woven into sheets for use as bullet-proof vests and is bonded in layers with epoxy for structural aircraft parts. Its high strength makes it very difficult to cut with conventional cutting methods. The CO_2 laser cuts with high travel speeds and can bond the cut edges of aramid and most other woven fabrics to prevent fraying. Aramid, bonded in layers with epoxy, can be cut in thicknesses up to 10 mm (0.40 in.). Because of the nature of the epoxy, the cut edge of the aramid-epoxy composite is darkened upon laser cutting. This is more pronounced in thicker samples with heavy epoxy layers. At the 600 W level, typical speeds are 12.75 m/min (500 in./min) for 1.25 mm (0.05 in.) thick material and approximately 7.75 m/min (300 in./min) for 3 mm (0.10 in.) aramid-epoxy composites.

Intricate shapes can be cut from wood with thicknesses well over 25 mm (1.0 in.). A solid wood product cuts very nicely and leaves a smooth but darkened edge. A popular application is cutting patterns in 15 to 19 mm (0.60 to 0.75 in.) beech plywood for steel rule die boards. Speed and focus changes create different width slots for different width blades (up to approximately 2 mm, or 0.08 in.). Therefore, speeds range from 0.5 to 1.8 m/min (20 to 70 in./min) for a 600 W CO_2 laser.

ACKNOWLEDGMENT

This article was adapted from "Laser Cutting" by Gregg P. Simpson and Thomas J. Culkin in *Forging and Forming*, Vol 14 of *ASM Handbook*, 1988.

REFERENCES

1. J.C. Ion, *Laser Processing of Engineering Materials*, Elsevier Butterworth-Heinnemann, 2005
2. D. Elza and G. White, Laser Beam Machining, *Machining*, Vol 16, *ASM Handbook*, ASM International, 1989, p 572
3. J. Powell, *LIA Guide to Laser Cutting*, Laser Institute of America, 1999

Abrasive Waterjet Cutting

ABRASIVE WATERJET CUTTING operates by the impingement of a high-velocity abrasive-laden fluid jet against the workpiece, yet it produces no heat (and therefore no heat-affected zone) to degrade metals or other materials. The finished edge obtained by the process often eliminates the need for postmachining to improve surface finish.

A coherent fluid jet is formed by forcing high-pressure abrasive-laden water through a tiny sapphire orifice. The accelerated jet exiting the nozzle travels at more than twice the speed of sound and cuts as it passes through the workpiece. Cuts can be initiated at any point on the workpiece and can be made in any direction of contour—linear or tangential. The narrow kerf produced by the stream results in neither delamination nor thermal or nonthermal stresses along the cutting path.

In addition to applications in the machining of superalloys; armor plate; titanium; and high-nickel, -chromium, and -molybdenum alloys, abrasive waterjet machining can also be used to cut concrete, rock, glass, ceramics, composites, and plastics. The ability of the abrasive waterjet to cut most metals without any thermal or mechanical distortion places this innovative process on the leading edge of material cutting technology.

Process Development

In 1968, Dr. Norman Franz filed his first patents on the use of high-pressure water streams to cut materials. The first commercial application of this process, in 1971, involved the cutting of 9.5 mm ($^3/_8$ in.) thick pressed board for manufacturing furniture forms. Since then, numerous waterjet units have been installed by various manufacturers worldwide. Waterjet cutting technology, which involves pumping a 0.08 to 0.46 mm (0.003 to 0.018 in.) diameter water stream at 207 to 414 MPa (30 to 60 ksi), was initially developed to cut or slit nonwoven materials, fiberglass building products, corrugated box materials, and plastics.

It was later found that hard or extremely dense materials such as metals and aerospace composites could be cut when particles of dry abrasives such as garnet and silica were added to the waterjet. This modification produced the abrasive waterjet and is responsible for the ability to cut advanced materials much more efficiently than with standard mechanical or thermal cutting methods. With abrasives added to the waterjet, the liquid stream itself is merely the medium that propels the abrasive instead of being the primary cutting force.

Abrasive waterjet cutting is used to cut metals and composite materials, such as boron/aluminum honeycomb, aluminum/boron carbide, and graphite composites, into intricate shapes and curves with virtually no heat input into the workpiece. It has been in use in industrial applications since 1983.

Advantages and Limitations

Advantages. The advantages of abrasive waterjet machining are summarized as follows:

- Ability to cut through most sections of dense or hard materials, such as metals and glass, leaving a clean, finished edge—3.2 to 6.3 μm (125 to 250 μin.) roughness with 60-, 80-, or 100-grit abrasive at 0.22 to 1.1 kg/min (0.5 to 2.5 lb/min)—without the need for secondary machining
- Ability to produce contours, shape-cutting, bevels of any angle, and three-dimensional profiling, because the process is omnidirectional
- Easy integration into computer-controlled systems, optical tracers, and full-scale six-axis robots. The cutting head weighs as little as 4.5 kg (10 lb) for easy mounting on robotic arms; precision robotics can accommodate cutting heads weighing 23 to 32 kg (50 to 70 lb)
- Wide availability and low cost of garnet and silica, the most common abrasive materials used
- Low water consumption (0.473 L/min, or 0.125 gal/min), which translates to 28 L/h (7.5 gal/h) despite the high pressures used
- No heat-affected zone
- Simple workpiece tooling, because the jet delivers approximately 2 to 9 N (0.5 to 2 lbf) of vertical force
- Finished part tolerance of 0.075 to 0.25 mm (0.003 to 0.010 in.) with conventional abrasive waterjet cutting of materials under 50 mm (2 in.) thick

Limitations. This device cannot replace tools that mill, turn, or drill blind holes or perform other operations that involve cutting or drilling to a partial depth. Glass and composite materials should be pierced at low pressures (70 to 83 MPa, or 10 to 12 ksi) to minimize chipping and delamination. Tempered glass is an example of one material that should not be machined with an abrasive waterjet. Other limitations are:

- Cutting through air gaps, such as cutting a pipe, causes the jet to flare out as it cuts through the first layer and produces a distorted, wider kerf cut through the second layer.
- The abrasive waterjet is typically used to cut all the way through a material, and only with special equipment and for special applications can it mill a blind hole or drill to partial depth.
- Glass and composite materials should be pierced at low pressures (70 to 85 MPa, or 10 to 12 ksi) to minimize chipping and delamination. Tempered glass is an example of one material that should not be machined with an abrasive waterjet, or any other process for that matter, because tempered glass requires the stresses to be sealed at the glass edge—relieving that seal will result in the glass breaking.
- The surface produced is a satin-smooth, sandblasted finish. Finishes better than 1.5 μm (60 μin.) RA (roughness average) are difficult or not possible.
- The smallest internal raidus that can be produced on a material is 0.25 mm (0.010 in.).

Cutting Principle

The abrasive waterjet cuts material by the action of abrasive solids (entrained by the waterjet) on the workpiece. Depending on the properties of the material, cutting occurs by erosion, shearing, failure under rapidly changing localized stress fields, or micromachining effects. A small abrasive jet nozzle is used (Fig. 1). Water is pressurized to 414 MPa (60 ksi) and expelled through a sapphire nozzle to form a coherent high-velocity (914 m/s, or 3000 ft/s) jet. A stream of abrasive particles is introduced into the nozzle to form a concentrated abrasive jet slurry (0.5 to 2.5 kg/L, or 4 to 20 lb/gal). The momentum of the waterjet as it travels toward the nozzle is transferred to the solid particles, and thus their velocities are rapidly increased.

This momentum transfer between the waterjet and the abrasive is a complex phenomenon. There is a limited dynamic stability of the high-pressure waterjet, and it breaks into droplets that accelerate the solid particles. In addition, the solid particles impose drag forces on the waterjet.

The result of this momentum transfer between the water and the abrasive particles is a focused high-velocity stream of abrasive. The cutting rate is controlled by changing the feed rate, the standoff distance, the waterjet pressure, or the abrasives.

Abrasive System Components

The primary components of an abrasive waterjet cutting system are the high-pressure water pump, the nozzle assembly, and the abrasive catcher assembly. These components are connected by a network of hoses and swivels and are controlled by a system of control valves and sensors (Fig. 2).

High-Pressure Water Pump. Two types of high-pressure water pumps are used: a triplex pump and an intensifier pump.

A triplex pump operates in the same manner as a low-pressure "pressure washer" used to pressure wash. It is a pump that operates from three plungers connected to a crank. These pumps are simpler than intensifier pumps and deliver 140 to 380 MPa (20 to 55 kpsi) water pressure and come in a variety of horsepower and gallon per minute output capacities. A popular size is a 22 kW (30 hp) triplex pump delivering 3 L/min (0.82 gal/min) or 380 MPa (55 kpsi) of water. Although the triplex pump is gaining use, the intensifier pump is the original method of pressurizing water for waterjet cutting.

Intensifier pumps create water pressures of 210 to 415 MPa (30 to 60 kpsi) and flow rates to 13 L/min (3.5 gal/min). A motor drives a hydraulic radial piston with an intensification ratio of 20 to 1; that is, the water pressure is twenty times the oil pressure.

The pressure intensifier principle is best illustrated by the force equilibrium of the double-acting piston (Fig. 3). Hydraulic oil pressure acting on the piston results in a force on the plunger pressurizing the water in the small chamber. Force equilibrium is achieved when the water pressure equals the hydraulic oil pressure multiplied by the effective area of the piston divided by the area of the plunger (assuming no friction losses). The ratio of the effective piston area to the plunger area is termed the pressure intensification ratio. Because the intensification ratio is constant by virtue of the fixed piston-to-plunger-diameter ratio, water pressure can be regulated by controlling the hydraulic oil pressure.

To dampen pressure oscillations in the high-pressure output, the high-pressure water is routed to a shock attenuator. Sensors are installed throughout the intensifier pump to monitor flow

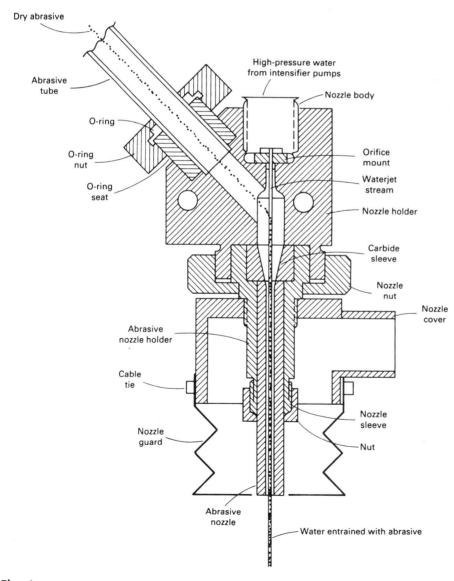

Fig. 1 Cross-sectional view of abrasive nozzle assembly showing path of water and abrasives

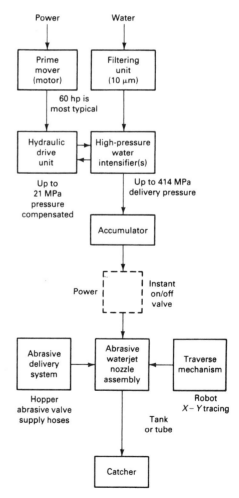

Fig. 2 Block diagram of abrasive waterjet system components

rates, temperatures, fluid levels, and operating conditions. Before the water enters the intensifier pumping system, it is filtered in three stages from 10 to 1 to 0.05 μm (400 to 40 to 2 μin.) to remove small particles of matter and minerals that could damage seals and valves of the pump. It is necessary to prevent cavitation of the water as it enters the intensifier pump systems. The inlet water pressure is increased to a minimum of 410 kPa (60 psi).

The microprocessor-based controller, through operator-actuated keypad commands, determines the oil pressure delivered to the intensifier by the electric motor driven hydraulic pump. Incoming filtered water is pressurized through the principle of water intensification and routed to the waterjet cutting equipment. Figure 4 illustrates the principal parts of the intensifier pump.

Particle Stream Erosion Nozzle. The high-pressure water is directed through a tiny sapphire, ruby, or diamond orifice to form a coherent fluid jet. Then, after this supersonic pure waterjet has been formed, a carefully metered amount of

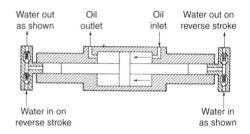

Fig. 3 Cross-sectional view of the pressurization of water to 414 MPa (60 ksi) using the fluid pressure intensifier principle

dry abrasive is pulled into the cutting stream via a venturi effect and accelerated by the waterjet. The tiny orifice is approximately 0.13 to 0.5 mm (0.005 to 0.020 in.) in diameter. After the dry abrasive particles are fed into the mixing chamber to become entrained in the water stream (by the venturi effect), the abrasive stream is then directed through a 0.5 to 2.3 mm (0.020 to 0.090 in.) diameter tungsten carbide nozzle. At this point, cutting of material takes place directly under the nozzle as both water and abrasives exit the nozzle at velocities of approximately 640 m/s (2100 ft/s) in a coherent, focused steam ranging in diameter from 1.0 to 1.5 mm (0.040 to 0.060 in.).

Operating life of a synthetic sapphire orifice, which has a 5 min replacement time, is 250 to 500 h. The operating life of the tungsten carbide abrasive nozzle is limited to 0.50 to 6 h because of the erosive effects of the accelerated water/abrasive stream.

The abrasive waterjet catcher system collects the spent fluid after it passes through the material being cut. The design of the catcher system is based on whether the cutting system uses a stationary nozzle or a moving nozzle. For a stationary nozzle, the workpiece is fed to the cutting operation, and a tank is used to collect the spent fluid (Fig. 5). A moving nozzle can be used with the same type of setup if the cutting area is contained within the tank area. The tank should be lined with ceramic pieces to suppress the cutting or piercing of the tank lining by the abrasive waterjet. Multiple pieces of concrete block, brick, thick slate, and white iron have been used to alleviate this problem. The pieces work well with a moving nozzle but must be moved or replaced at varied intervals. Abrasives

settle to the bottom, and the tank requires periodic cleaning. The accumulated water is drawn off through a value placed low in the tankwall.

A system incorporating a funnel-shaped catcher containing metallic shot to disperse the energy of the liquid has been designed for use with a movable nozzle. This device has a relatively long life expectancy as a catcher.

Abrasives

Surface finish is an important part of the performance, wear, and appearance of a product. Parts may perform better or may have a higher fatigue strength or a better appearance with a higher degree of surface finish.

Two distinct surface textures are produced when cutting with an abrasive waterjet. The top half of the thickness may have a surface roughness of 3.3 to 10 μm (130 to 400 μin.). The bottom half of the cut may have striations formed by the exit of the garnet from the workpiece (Fig. 6).

Garnet versus Silica. Tables 1 and 2 list surface finish ranges for garnet and silica abrasives. The following conclusions can be drawn from these data:

- Traverse cutting speed is the most important variable affecting surface finish.
- Flow rate is the second most significant variable affecting surface finish.
- Pressure is an important variable affecting surface finish but is dependent on the specific garnet size.

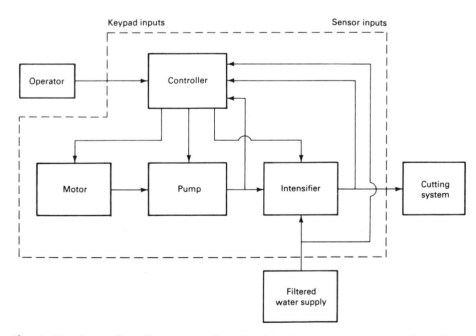

Fig. 4 Block diagram of intensifier pump assembly and its relationship to the microprocessor-based controller and cutting system

Fig. 5 Stationary waterjet nozzle cutting through a movable workpiece (René 100 alloy gating contacts). Courtesy of Department of Industrial and Manufacturing Engineering, University of Rhode Island

- Garnet will cut faster than a comparably sized silica abrasive.
- A 60-grit garnet will cut 23.5% faster than a 60- to 80-grit silica abrasive.
- Silica sands produce a higher-quality surface finish than garnet.
- The highest-quality surface finish occurs with the slowest comparable cutting speed and at the highest flow rate for both garnet and silica.

Thus, garnet is an effective abrasive. Silica sand may be adequate for cleaning operations involving the cutting of thin metals or composites. Abrasives such as aluminum oxide or silicon carbide should be used for tough materials such as ceramics. However, these harder abrasives generally decrease the life of the nozzle.

Angular grain shapes will shear the workpiece much more efficiently than round grain abrasives. The abrasive particles should be of uniform size.

Reclaiming of abrasives is not feasible, because much time and effort would be required in drying and then regarding the particle sizes. The amount of abrasive that would be reusable does not warrant the expense. New graded garnet and silicon sands are readily available and inexpensive.

Cleanup and disposal of the abrasive fluid can be a problem. Because the spent abrasive slurry has the consistency of mud, it is advisable to wait until the slurry is dry before attempting cleanup and disposal. Garnet is environmentally safe to dispose of. However, the particles mixed in with it from the material being cut may be classified as hazardous waste, and its disposal may be subject to both state and federal environmental regulations.

Abrasive Waterjet Speeds

As the water pressure, flow rate, and therefore the horsepower increase, so does the cutting speed. As cut speed increases, the cost per inch decreases and abrasive flow rate increases. Eventually, a point is reached where more abrasive will overburden the process and reduce the cut speed and increase cost per inch. Therefore, a peak performance point exists for the abrasive flow rate. Today's (2005) systems are designed to fix the system at this peak point for all but the most custom applications. Maximum separation cut speeds with a standard abrasive waterjet system delivering 3 L/min (0.8 gal/min) of 415 MPa (60 kpsi) water and 0.5 kg/min (1.2 lb/min) of 80-mesh abrasive are listed in Table 3 for various materials. To obtain a high-quality finish, multiply the maximum speeds in Table 3 by 0.3 to 0.4. Various grades of a metal

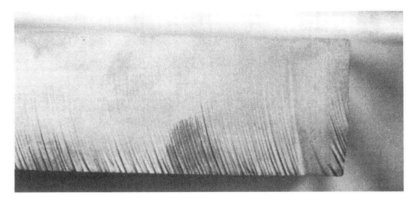

Fig. 6 Striations formed by the exit of garnet abrasives at the bottom of a workpiece. Material is 22 mm (7/8 in.) thick 4340 steel cut at 38 mm/min (1½ in./min). Top surface has a 3.6 μm (140 μin.) surface finish; bottom surface has a 4.7 μm (185 μin.) surface finish. Courtesy of Department of Industrial and Manufacturing Engineering, University of Rhode Island

Table 1 Surface finish range for various grits of garnet abrasives

Flow rate		Nozzle pressure		Cutting speed, %	Surface finish, grit									
					36		60		80		100		150	
kg/min	lb/min	MPa	ksi		μm	μin.	μm	μin.	μm	μin.	μm	μin.	μm	μin.
0.45	1.0	207	30	100	8.9–10.2	350–400	8.1–8.9	320–350	7.1–8.6	280–340	4.1–4.6	160–180	4.3–4.6	170–180
				80	8.6–9.7	340–380	8.6–9.4	340–370	6.6–7.4	260–290	4.3–4.6	170–180	4.2–4.6	165–180
				50	7.6–8.4	300–330	6.9–7.4	270–290	5.6–6.4	220–250	4.3–4.6	170–180	2.9–3.2	115–125
		241	35	100	8.1–9.4	320–370	8.1–8.9	320–350	6.9–7.6	270–300	4.8–5.6	190–220	4.1–4.3	160–170
				80	5.8–6.6	230–260	7.1–8.1	280–320	6.9–7.6	270–300	4.6–5.1	180–200	3.4–3.8	135–150
				50	5.8–6.6	230–260	7.4–8.1	290–320	5.1–5.8	200–230	4.1–4.4	160–175	3.3–3.6	130–140
		276	40	100	7.1–8.4	280–330	7.1–7.9	280–310	7.1–8.4	280–330	4.3–4.6	170–180	3.7–4.1	145–160
				80	7.1–7.4	280–290	8.1–9.1	320–360	6.4–7.4	250–290	4.7–5.1	185–200	3.3–3.7	130–145
				50	6.6–7.1	260–280	6.9–7.6	270–300	5.8–6.4	230–250	4.1–4.3	160–170	2.7–3.0	105–120
0.68	1.5	207	30	100	7.1–7.9	280–310	8.1–8.9	320–350	5.3–6.1	210–240	4.4–5.3	175–210	4.1–4.4	160–175
				80	6.9–7.9	270–310	7.6–8.4	300–330	5.1–5.8	200–230	4.8–5.5	190–215	3.9–4.4	155–175
				50	5.6–6.4	220–250	6.4–7.1	250–280	4.6–5.1	180–200	4.4–4.7	175–185	2.9–3.2	115–125
		241	35	100	7.1–7.6	280–300	7.1–8.1	280–320	6.6–7.1	260–280	4.3–4.6	170–180	3.0–3.4	120–135
				80	7.4–7.9	290–310	7.4–8.4	290–330	5.3–6.1	210–240	4.4–5.6	175–220	3.3–3.6	130–140
				50	6.1–6.6	240–260	6.4–7.1	250–280	4.8–5.6	190–220	4.6–4.8	180–190	3.0–3.4	120–135
		276	40	100	7.1–8.6	280–340	6.6–7.9	260–310	6.1–6.6	240–260	4.3–4.6	170–180	3.4–3.7	135–145
				80	7.1–7.9	280–310	8.1–8.9	320–350	5.3–6.1	210–240	4.4–4.7	175–185	3.3–3.7	130–145
				50	6.1–7.1	240–280	6.4–7.1	250–280	5.1–5.8	200–230	3.6–3.9	140–155	2.9–3.2	115–125
0.91	2.0	207	30	100	7.6–8.6	300–340	6.4–7.1	250–280	5.1–5.8	200–230	4.3–4.6	170–180	4.4–5.3	175–210
				80	6.6–7.1	260–280	6.4–7.4	250–290	4.8–5.6	190–220	4.8–6.1	190–240	3.8–4.3	150–170
				50	6.1–6.6	240–260	6.1–6.9	240–270	4.8–5.3	190–210	3.7–4.1	145–160	2.4–2.8	95–110
		241	35	100	7.4–8.1	290–320	7.4–8.4	290–330	5.8–6.9	230–270	4.1–4.4	160–175	4.1–4.3	160–170
				80	6.4–7.4	250–290	7.1–8.1	280–320	5.1–5.8	200–230	3.7–4.3	145–170	4.4–4.7	175–185
				50	5.6–6.1	220–240	6.6–7.1	260–280	4.8–5.6	190–220	3.2–3.6	125–140	3.3–3.8	130–150
		276	40	100	6.9–7.4	270–290	7.4–8.1	290–320	4.8–5.6	190–220	3.8–4.4	150–175	3.4–4.1	135–160
				80	5.8–6.9	230–270	7.1–8.1	280–320	4.6–5.3	180–210	3.7–4.2	145–165	4.2–4.7	165–185
				50	5.8–6.6	230–260	6.4–6.9	250–270	4.8–5.6	190–220	3.2–3.6	125–140	2.9–3.2	115–125

36-grit surface finish values read off 1000 unit scale; 60-, 80-, 100-, and 150-grit values read off 300 unit scale

Table 2 Surface finish range for various grits of silica abrasives

Flow rate		Nozzle pressure		Cutting speed, %	Surface finish, grit					
					35–60		60–80		80–120	
kg/min	lb/min	MPa	ksi		μm	μin.	μm	μin.	μm	μin.
0.45	1.0	207	30	100	4.1–4.8	160–190	5.2–5.7	205–225	4.8–5.7	190–225
				80	7.0–7.2	275–285	5.6–6.0	220–235	4.3–4.7	170–185
				50	5.7–6.1	225–240	5.0–5.7	195–225	3.9–4.3	155–170
		241	35	100	6.5–7.2	255–285	5.5–5.8	215–230	4.1–4.4	160–175
				80	6.2–7.0	245–275	4.6–5.5	180–215	4.1–4.4	160–175
				50	5.5–6.1	215–240	4.6–5.0	180–195	3.3–3.8	130–150
		276	40	100	6.5–7.1	255–280	5.3–5.7	210–225	4.3–4.6	170–180
				80	5.8–6.4	230–250	4.8–5.3	190–210	3.9–4.4	155–175
				50	5.7–6.5	225–255	4.6–4.8	180–190	3.6–3.9	140–155
0.68	1.5	207	30	100	5.8–6.2	230–245	5.8–6.1	230–240	4.4–4.8	175–190
				80	6.0–6.5	235–255	4.6–5.3	180–210	4.2–4.6	165–180
				50	6.0–6.7	235–265	4.3–4.7	170–185	3.0–3.4	120–135
		241	35	100	5.7–6.1	225–240	5.3–5.8	210–230	3.6–3.9	140–155
				80	5.8–6.2	230–245	4.6–5.1	180–200	3.4–3.8	135–150
				50	5.6–6.1	220–240	4.4–4.7	175–185	3.2–3.6	125–140
		276	40	100	6.0–6.9	235–270	4.8–5.3	190–210	3.7–4.2	145–165
				80	6.0–6.5	235–255	4.7–5.2	185–205	3.6–3.9	140–155
				50	5.8–6.5	230–255	4.3–4.6	170–180	3.3–3.8	130–150
0.91	2.0	207	30	100	6.0–6.5	235–255	4.4–4.7	175–185	5.1–5.5	200–215
				80	6.1–6.7	240–265	5.0–5.3	195–210	4.2–4.6	165–180
				50	5.6–6.0	220–235	4.1–4.4	160–175	3.3–3.6	130–140
		241	35	100	5.8–6.6	230–260	4.3–4.6	170–180	4.4–4.8	175–190
				80	6.1–6.6	240–260	4.2–4.6	165–180	3.6–3.9	140–155
				50	5.2–5.8	205–230	3.9–4.2	155–165	3.2–3.3	125–130
		276	40	100	6.0–6.6	235–260	4.2–4.4	165–175	3.6–4.1	140–160
				80	5.7–6.2	225–245	4.2–4.4	165–175	3.6–3.9	140–155
				50	5.7–6.2	225–245	3.3–3.7	130–145	3.0–3.2	120–125

All surface finish values read off 300 unit scale.

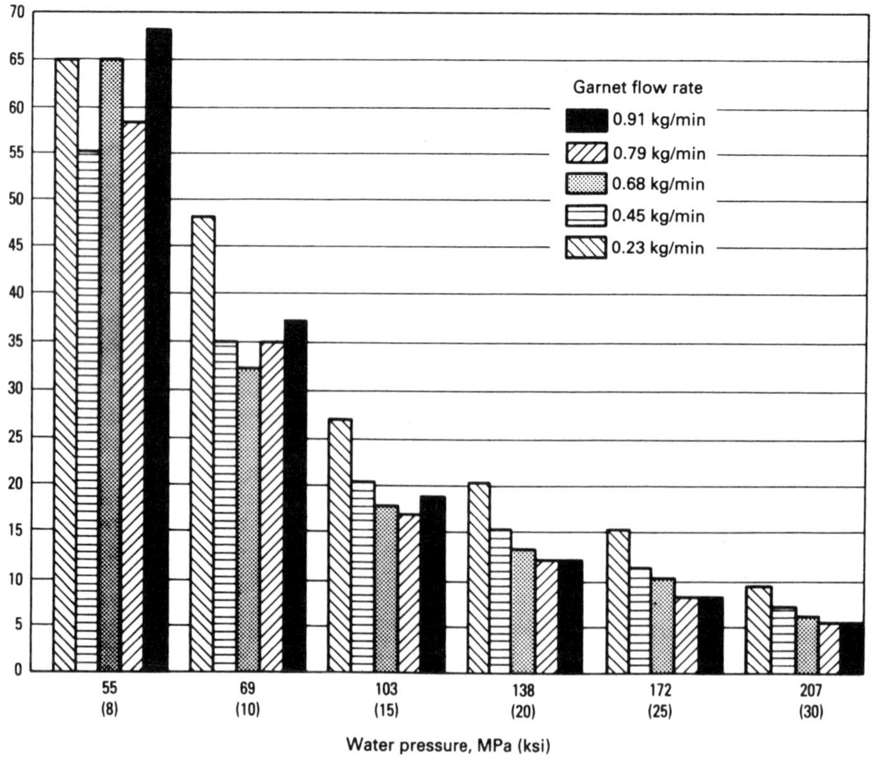

Fig. 7 Water pressures and garnet abrasive flow rates versus piercing times for 9.5 mm (³⁄₈ in.) thick 6061-T6 aluminum. Source: Department of Industrial and Manufacturing Engineering, University of Rhode Island

(e.g., type 304 stainless steel, mild steel, etc.) have little effect on cut speed.

In addition to empirical testing, the abrasive waterjet process has been successfully mathematically modeled based on the abrasive waterjet process parameters, material type, material thickness, and desired edge quality. Intelligent control systems and software products allow users to enter these parameters and obtain the resulting cut speed. Operators simply type in the current jet setup, material type, thickness, and desired surface quality, and the machine does the rest. Some sample calculations follow.

Calculation of Abrasive Waterjet Speeds. The following parameters are assumed:

- Sand flow rate of 0.68 kg/min (1.5 lb/min)
- Water flow rate of 4.5 L/min (1.2 gal/min)
- Jewel size of 0.46 mm (0.018 in.) diameter

Using the continuity of momentum equation and assuming that the water is not compressed at all, the theoretical (ideal) velocity of the water exiting the jewel orifice is 461 m/s (1513 ft/s), and the theoretical velocity of the garnet and water at the nozzle is 402 m/s (1319 ft/s).

If, instead of 4.5 L/min (1.2 gal/min), it is assumed that the water flow rate is 3.84 L/min (1 gal/min)* and that the water is compressed by 8%, the actual flow rate is 3.5 L/min (0.92 gal/min). Therefore, the actual velocity of the water exiting the jewel orifice is 354 m/s (1160 ft/s), and the actual velocity of the garnet and water at the nozzle is 297 m/s (973 ft/s). The speed of sound under normal atmospheric conditions is approximately 335 m/s (1100 ft/s).

The following examples illustrate the effect of different water pressures and garnet flow rates on a single thickness of two different metals as well as the effects of these variables on different thickness of one metal.

Example 1: Analysis of Varying Water Pressures and Garnet Flow Rates on 9.5 mm (³⁄₈ in.) Thick 6061-T6 Aluminum and 6.4 mm (¹⁄₄ in.) Thick Type 304 Stainless Steel. All piercings were made with a 3.2 mm (¹⁄₈ in.) nozzle standoff having a 1.6 mm (0.062 in.) nozzle diameter. The garnet flow rates were 0.23, 0.45, 0.68, 0.79, and 0.91 kg/min (0.5, 1.0, 1.5, 1.75, and 2.0 lb/min). The water pressures were 55, 70, 105, 140, 170, and 205 MPa (8, 10, 15, 20, 25, and 30 ksi). The water pressure was held constant at 205 MPa (30 ksi) for the cutting, and the garnet flow rates were varied at 0.45, 0.68, and 0.91 kg/min (1.0, 1.5, and 2.0 lb/min).

Figures 7 and 8 show that much more time is required for piercing stainless steel than for aluminum at low water pressure. This reflects the hardness of the stainless steel. Although the two metals are not the same thickness, similarities between Fig. 7 and 8 are evident.

*This flow rate is calculated by counting the number of times the intensifier pump cycled in 1 min. With this particular configuration, it cycled 50 times. The result obtained is (50 cycles/min) (0.076 L/cycle) = 3.8 L/min, or in English units: (50 cycles/min) (0.02 gal/cycle) = 1.0 gal/min.

The garnet flow rate affects the piercing time. High garnet flow rate versus low garnet flow rate, with pressure being constant, requires a longer time for piercing.

Figure 9 shows that piercing time acts as an exponential function relating to thickness. The same trend can be seen with garnet flow rates of 0.45, 0.68, and 0.79 kg/min (1.0, 1.5, and 1.75 lb/min).

Pieces that were not cut completely through had small ridges in the half kerf of the cut. The number of ridges in the cut corresponded to an equal number of cycles of the intensifier pump.

At cutting speeds greater than 100 mm/min (4 in./min), a noticeable taper was evident in the kerf width. At cutting speeds less than 100 mm/min (4 in./min), no significant taper was evident. Varying the garnet flow rates had no significant effect on the width of the kerf.

Similar characteristics were found in the cutting of stainless steel and aluminum. There is a rounding effect, and the kerf is larger for cuts at the same speed but increasing garnet flow. As aluminum thickness increased to 38 mm (1 1/2 in.), higher water pressure was needed with an increase in the garnet flow rate.

Example 2: Analysis of Varying Water Pressures and Garnet Flow Rates on Several Thicknesses of 6061-T6 Aluminum Bar Stock. A piece of aluminum 6061-T6 bar stock 64 mm (2 1/2 in.) wide by 38 mm (1 1/2 in.) thick and 610 mm (2 ft) long was machined down to provide step thicknesses of 6.4, 9.5, 13, 19, 25, 32, and 38 mm (1/4, 3/8, 1/2, 3/4, 1, 1 1/4, and 1 1/2 in.), as shown in Fig. 10(a).

Each section was pierced at several garnet flow rates with varying water pressures. The water pressures were 83, 103, 138, 172, and 207 MPa (12, 15, 20, 25, and 30 ksi). The garnet flow rates were 0.23, 0.45, 0.68, 0.79, and 0.91 kg/min (0.5, 1.0, 1.5, 1.75, and 2.0 lb/min).

The times were recorded using a digital timer connected to the robot controller. A new 1.6 mm (0.062 in.) nozzle was used with a 3.2 mm (1/8 in.) standoff.

Table 3 Abrasive waterjet cutting speeds of various materials

Material	Thickness		Maximum cutting speed(a)	
	mm	in.	cm/min	in./min
Aluminum	6	0.25	178	70
	13	0.5	100	40
	25	1.0	45	18
Marble	6	0.25	380	150
	13	0.5	178	70
	25	1.0	76	30
Steel (mild or stainless)	6	0.25	63	25
	13	0.5	28	11
	25	1.0	11	4.5
Titanium	6	0.25	76	30
	13	0.5	35	14
	25	1.0	15	6

(a) Cutting speeds with a standard waterjet system delivering 3 L/min (0.8 gal/min) of 415 MPa (60 kpsi) pressurized water with 0.5 kg/min (1.2 lb/min) of 80-mesh abrasive. Courtesy of Flow International Corp.

An evaluation of the piercings illustrated that the entrance and exit hole sizes reflected the water pressure, sand flow, and section thickness (Fig. 10b). With the 0.23 kg/min (0.5 lb/min) garnet flow rate, there was a slight overall decrease in hole size as the section thicknesses were reduced.

Factors Affecting Cut Quality

Bending of the kerf can be readily seen in the Plexiglas (Rohm and Haas Company) piece shown in Fig. 11. This bending motion occurs because the abrasive stream loses energy as it cuts and travels further down into the target piece and becomes progressively less efficient. Meanwhile, the nozzle is still moving and cutting the material at the top, and this forces the kerf to bend away from the direction of the motion of the nozzle. Slowing down traverse speed or increasing cutting power will mitigate this phenomenon.

At high traverse speeds—especially in thick materials—this bending of the kerf will be most noticeable at the very end of the material. This is the case with the 50 mm (2 in.) thick semihardened steel section shown in Fig. 12. Because the bend is a smooth one, the cutting action at the bottom depends on the continual cutting at the top. When the end of the material is reached, there is a sudden interruption in this bend, and the jet stream is no longer deflected onto the uncut bottom section. This leaves an uncut crescent-shaped piece at the very bottom of the end of the cut that must be cut again or broken apart by hand. This problem can be alleviated if the computer program of the robot includes provisions for either slowing the traverse rate at the very end of the cut so that the stream is barely bent and/or changing the approach angle of the nozzle at the very end of the cut.

Incomplete Initial Cuts. The effects of an incomplete first cut can be seen in the Plexiglas shown in Fig. 11 and in the cut tantalum-silicon

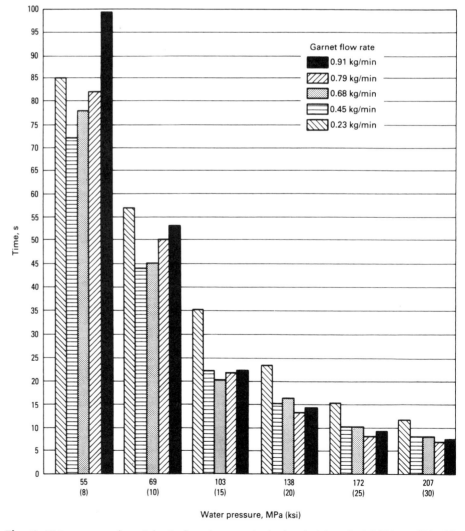

Fig. 8 Water pressure and garnet abrasive flow rates versus piercing times for 6.4 mm (1/4 in.) thick type 304 stainless steel. Source: Department of Industrial and Manufacturing Engineering, University of Rhode Island

piece shown in Fig. 13. A very poor surface finish can be expected at the bottom of the cut if the abrasive waterjet stream does not cut completely through the material on the first attempt. At the bottom of the cut, where the cutting power of the garnet is expended, there are rough ridges, and the kerf width is greatly expanded because of the bouncing around of the excess garnet. This wider kerf and expanded area are especially noticeable on the cut surfaces of the tantalum-silicon piece (Fig. 13).

Special Precautions for Thicknesses Over 6.4 mm ($1/4$ in.). Tantalum-silicon and Plexiglas samples up to 102 mm (4 in.) thick have been successfully cut with an abrasive waterjet. In cutting any material over approximately 6.4 mm ($1/4$ in.) thick, special precautions should be taken if a corner cut is to be made in the piece or if the material is to be pierced and then line cut.

When the Plexiglas piece shown in Fig. 11 was pierced, the hole itself was very smooth. However, when a traverse motion was initiated after piercing, the bending of the stream caused a section to the left of the hole to be cut away. This was not the desired effect. The same problem can occur when attempting a turn in the middle of the piece. The stream cannot simply bend around the turn. The nozzle must be stopped for a sufficient amount of time to allow the bend to

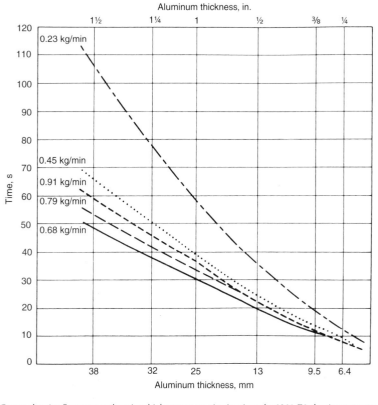

Fig. 9 Garnet abrasive flow rates and section thickness versus piercing times for 6061-T6 aluminum at a water pressure of 207 MPa (30 ksi). Source: Department of Industrial and Manufacturing Engineering, University of Rhode Island

(a)

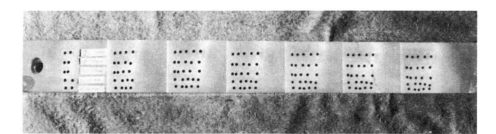

(b)

Fig. 10 Two views of various thicknesses of 6061-T6 aluminum used in piercing studies with varying water pressure and garnet flow rates used on each thickness. (a) Three-dimensional view of workpiece showing varying thickness of 640 mm (2 ft) long slab before piercing. From left to right, slab thicknesses are 38, 32, 25, 19, 13, 9.5, and 6.4 mm ($1/2$, $1/4$, 1, $3/4$, $1/2$, $3/8$, and $1/4$ in.). (b) Top view of piece after piercing was performed. Left side is 38 mm ($1/2$ in.) section. Courtesy of Department of Industrial and Manufacturing Engineering, University of Rhode Island

Fig. 11 Bending of kerf in a Plexiglas workpiece subjected to excessively high traverse cutting speeds. The portion at the bottom was pierced before cutting, and this caused a section of the hole to be eroded away as it was subsequently cut by the abrasive waterjet nozzle. Courtesy of Department of Industrial and Manufacturing Engineering, University of Rhode Island

straighten out, and then the nozzle can continue its motion.

Work Hardening. Abrasive waterjet machining alters the hardness of the cut surface of a number of metals. This slight work hardening is indicated in Table 4.

Abrasive Grit Size. Figure 14 shows how abrasive flow rates and grit number affect the depth of cut when machining cast iron. Figure 14 also indicates that neither coarse (36 and 16 grit) nor fine particles (100 and 150 grit) of garnet are the most effective abrasives. Medium abrasive of 60 and 80 grit has been found to be the most effective cutting media for a wide variety of metals.

Pressure is the most important parameter to be optimized in considering the depth of cut required in metals. Soft metals are insensitive to particle size but are very much affected by pressure. Figure 15 indicates the effect of increasing water pressure on 11 metals and a ceramic (Al_2O_3).

Applications

A hand-held abrasive waterjet unit can be used, but accuracy and quality are compromised because of human instability. The abrasive nozzle must be held firmly and accurately, and the standoff of the nozzle, as well as the rate of cut, must be closely controlled. Servo-type robot arms (Fig. 16) can be programmed for point-to-point linear movements in which arcs or circles

Table 4 Effect of abrasive waterjet cutting on surface hardening of metals

Metal	Base	Abrasive waterjet cut
	\multicolumn{2}{c}{Hardness(a)}	
Titanium	34 HRC	34.3 HRC
Aluminum 6061-T6	54.8 HRB	58.7 HRB
Magnesium	54.1 HRB	58.4 HRB
Carbon steel A-572	82.8 HRB	84.5 HRB
Tool steel	91.5 HRB	93.3 HRB

(a) Average of five measurements. Source: Ref 1

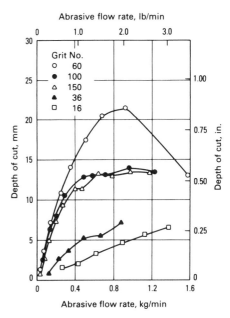

Fig. 14 Effect of abrasive flow rate and grit number on depth of cut (garnet abrasive; 220 MPa, or 32 ksi water pressure; 0.46 mm, or 0.018 in., waterjet diameter; 152 mm/min, or 6 in./min, traverse speed; cast iron). Source: Department of Industrial and Manufacturing Engineering, University of Rhode Island

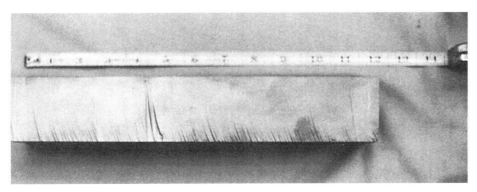

Fig. 12 Crescent-shaped striations in a 50 mm (2 in.) thick piece of semihardened steel cut at 22.3 mm/min (0.88 in./min). Direction of cut is from top of picture to bottom. Waterjet failed to cut completely through the steel in two places (for left as well as at the 108 mm, or $4\frac{1}{4}$ in., mark). Courtesy of Department of Industrial and Manufacturing Engineering, University of Rhode Island

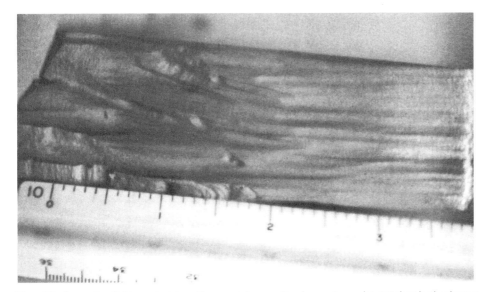

Fig. 13 Poor surface finish on a tantalum-silicon workpiece resulting from an incomplete initial cut by the abrasive waterjet stream. Left side is bottom of cut. Courtesy of Department of Industrial and Manufacturing Engineering, University of Rhode Island

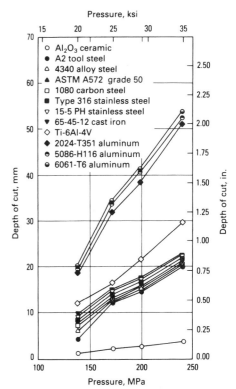

Fig. 15 Depth of cut results for different materials (60-grit garnet abrasive; 0.91 kg/min, or 2 lb/min, abrasive flow rate; 0.51 mm, or 0.020 in., waterjet diameter; 152 mm/min, or 6 in./min, traverse speed). Source: Department of Industrial and Manufacturing Engineering, University of Rhode Island

are approximated by numerous short straight-line segments.

Machining of Metals and Metal-Matrix Composites. The following list summarizes the various applications for abrasive waterjet cutting:

- Job shops of various materials
- Foundries (removal of burned-in sand, cutting gates, and risers from cast parts)
- Naval and commercial shipyards (high-strength steel, lead, and so on)
- Railroad cars (manufacture and repair)
- Metal fabrication shops
- Aircraft manufacturers (titanium, Inconel, stacked metals)
- Heavy equipment manufacturers (tractors, hoists, cranes, industrial winches, derricks)
- Industrial vehicles (trucks, tankers, construction vehicles)
- Structural fabrications (bridges, skyscrapers) and heavy aluminum works
- Specialty metal fabrication (titanium, nickel alloys, chromium alloys)
- Military vehicles (tanks, armored personnel carriers, landing craft)

- Oil and gas (oil well casings, pipeline repair, platform repair)
- Mining (metal structures)

The cutting speed of the abrasive waterjet has made it a valuable tool for high-volume applications. Table 5 indicates the effect of material thickness on cutting speed for eight widely used metals and an aluminum oxide ceramic.

Abrasive waterjet cutting is also effective in fabrication of labor-intensive aircraft components, such as wingsweep fairings, upper and lower panels, bay doors, and miscellaneous bonded panels. These are large and complex parts—mostly 6.4 to 9.5 mm (1/4 to 3/8 in.) thick and averaging 6400 to 7600 linear mm (250 to 300 in.) of periphery—that are ideally suited to a robot system capable of precision five-axis cutting over an extensive area. One example is cutting of 1.6 mm (0.063 in.) titanium using 0.68 kg/min (1.5 lb/min) of 60-grit red garnet abrasive at 300 mm/min (12 in./min). Another aerospace manufacturer has used abrasive waterjet cutting to cut through aluminum and titanium up to 64 mm (2.5 in.) thick with garnet abrasives.

Where metallurgical contamination is a concern, as in the cutting of bimetallics, abrasive waterjet machining overcomes the problems of distortion, delamination, and contamination. Abrasive waterjet cutting is also cost-effective for composites. Metal-matrix composites that are cut mechanically at 25 mm/min (1 in./min) can be cut by an abrasive waterjet at 380 to 760 mm/min (15 to 30 in./min).

Figures 17 through 19 illustrate the wide variety of metals (including difficult-to-machine materials such as titanium and alloy 100) and intricate shapes (specifically the bracket shown in Fig. 18 and the turbine rotor in Fig. 19) that can be easily cut with the abrasive waterjet. The stainless steel guide bracket shown in Fig. 18 illustrates profiling around corners that would be difficult to achieve with standard machining methods. This part would require 16 or more hours to complete using conventional methods; with the abrasive waterjet, such a part can be machined within a few hours.

Machining of Nonmetallics. Nonmetallic materials that are readily cut by abrasive waterjet include various plastics and composite materials, such as polyurethane rubber, Styrofoam (Dow

(a)

(b)

Fig. 16 Abrasive waterjet nozzle assembly (a) mounted on robot arm to interface with robot. (b) Closeup of nozzle. Black air line on top is connected to electric solenoid located on robot arm. High-pressure water line attaches at center-left of nozzle and flexes as robot arm moves and rotates. Courtesy of Flow Systems, Inc.

Table 5 Effect of material thickness on abrasive waterjet cutting speed for various metals and a ceramic material
Abrasive used is garnet.

Diameters, mm (in.)		Abrasive Feed rate kg/min (lb/min)	Mesh size	Nozzle pressure		Material thickness		Maximum cutting rate, mm/min (in./min)								Ceramic 99.6% aluminum oxide
Orifice	Nozzle			MPa	ksi	mm	in.	Aluminum and aluminum alloys	Brass	Carbon steel	Copper	Alloy 718	Stainless steel	Titanium	Tool steel, 38 HRC	
0.23 (0.009)	0.79 (0.031)	0.23 (0.5)	100	310	45	0.8	0.031	4570 (180)	1270 (50)	1520 (60)	1270 (50)	1520 (60)	1140 (45)	2030 (80)	890 (35)	127 (5)
						1.6	0.063	2030 (80)	762 (30)	1270 (50)	1020 (40)	1140 (45)	762 (30)	1520 (60)	762 (30)	61 (2.4)
						3.2	0.125	1270 (50)	457 (18)	762 (30)	559 (22)	559 (22)	610 (24)	1140 (45)	635 (25)	38 (1.5)
0.33 (0.013)	1.19 (0.047)	0.68 (1.5)	80	240	35	6.4	0.250	762 (30)	254 (10)	508 (20)	305 (12)	305 (12)	406 (16)	762 (30)	432 (17)	23 (0.9)
						12.7	0.500	457 (18)	102 (4)	305 (12)	152 (6)	152 (6)	254 (10)	457 (18)	330 (13)	15 (0.6)
0.46 (0.018)	1.19 (0.047)	0.91 (2.0)	80	240	35	19.0	0.750	305 (12)	25 (1)	203 (8)	75 (3)	75 (3)	152 (6)	305 (12)	254 (10)	8 (0.3)
						25.4	1.00	203 (8.0)	13 (0.5)	152 (6)	38 (1.5)	38 (1.5)	102 (4)	152 (6)	191 (7.5)	...
0.56 (0.022)	1.57 (0.062)	1.46 (3.2)	60	240	35	50.8	2.00	152 (6.0)	8 (0.3)	75 (3)	15 (0.6)	5 (0.2)	57 (2.25)	75 (3)	127 (5)	...
						76.2	3.00	127 (5.0)	5 (0.2)	50 (2)	8 (0.3)	3 (0.1)	38 (1.5)	50 (2)	50 (2)	...
						102	4.00	102 (4.0)	3 (0.1)	25 (1)	3 (0.1)	...	25 (1)	25 (1)	25 (1)	...

Table 6 Effect of material thickness on abrasive waterjet cutting speed for nonmetallic materials

Abrasive used is garnet.

Parameters									Maximum cutting rate, mm/min (in./min)									
Abrasive						Material thickness		Plastics and composites								Glass		
Diameters, mm (in.)		Feed rate		Nozzle pressure						Carbon/ carbon composite	Epoxy/ glass composite	Graphite/ epoxy composite	Aramid fiber composite	Polypropylene				
Orifice	Nozzle	kg/min (lb/min)	Mesh size	MPa	ksi	mm	in.	Acetal	Acrylic						Laminate	Plate	Stained
0.23 (0.009)	0.79 (0.031)	0.23 (0.5)	100	310	45	0.8	0.031	3180 (125)	3050 (120)	2540 (100)	6350 (250)	4450 (175)	2540 (100)	2540 (100)	...	7620 (300)	7620 (300)
						1.6	0.063	2290 (90)	2030 (80)	1910 (75)	5720 (225)	3810 (150)	1520 (60)	1910 (75)	...	6350 (250)	6350 (250)
						3.2	0.125	1780 (70)	1400 (55)	1400 (55)	4570 (180)	3180 (125)	1020 (40)	1220 (48)	660 (26)	5080 (200)	5080 (200)
						6.4	0.250	1270 (50)	915 (36)	1020 (40)	2540 (100)	2540 (100)	510 (20)	915 (36)	559 (22)	3810 (150)	3810 (150)
						12.7	0.500	890 (35)	457 (18)	508 (20)	1020 (40)	1270 (50)	279 (11)	610 (24)	457 (18)	2540 (100)	2540 (100)
0.33 (0.013)	1.19 (0.047)	0.68 (1.5)	80	240	35	19.0	0.750	610 (24)	305 (12)	254 (10)	711 (28)	635 (25)	152 (6)	381 (15)	305 (12)	1270 (50)	1270 (50)
						25.4	1.00	381 (15)	254 (10)	127 (5)	559 (22)	508 (20)	75 (3)	203 (8)	203 (8)	635 (25)	...
0.46 (0.018)	1.19 (0.047)	0.91 (2.0)	80	240	35	50.8	2.00	152 (6)	102 (4)	20 (0.8)	305 (12)	254 (10)	25 (1.0)	102 (4)	127 (5)	381 (15)	...
						76.2	3.00	64 (2.5)	25 (1.0)	8 (0.3)	127 (5)	127 (5)	13 (0.5)	50 (2)	25 (1)	127 (5)	...
						102	4.00	25 (1)	8 (0.3)	3 (0.1)	25 (1)	50 (2)	3 (0.1)	38 (1.5)	20 (0.8)	50 (2)	...

Chemical Company), Kevlar (E.I. du Pont de Nemours and Company), foam rubber, electrical wire, laminated paper, cardboard, presintered ceramics, and acrylonitrile-butadiene-styrene plastic. Table 6 lists the cutting speeds obtainable for typical nonmetallic materials.

Fig. 17 Profiling of titanium for aerospace applications using an abrasive waterjet. Courtesy of Flow Systems, Inc.

Fig. 18 Stainless steel aircraft guide bracket showing the profiling capabilities of the abrasive waterjet. Courtesy of Flow Systems, Inc.

Listed subsequently are some representative examples of abrasive waterjet technology used to cut nonmetallics for automotive applications:

• Trimming of thermoformed wood-fiber substrate interior door panels at speeds of 20 m/min (800 in./min)
• Finish cutting to size of 3.9 kg (8½ lb) rear door panels after initial rough cutting with more conventional methods; the panels were made of two layers of sheet molding compound coated with continuous layers of polyester resin paste and then rolled to remove trapped air.

Domestic and international electronics firms are using waterjet technology to cut the plastic laminates used in printed circuit boards. The hairlike size of the waterjet kerf and the omnidirectional, sharp cutting capabilities of the waterjet are ideally suited to the precision cutting and trimming of the boards even when they

Fig. 19 Turbine rotor machined from a solid alloy 100 blank using an abrasive waterjet that accelerated a 60-mesh garnet abrasive at 207 MPa (30 ksi). Courtesy of Flow Systems, Inc.

are loaded with their electronic components. The absence of any lateral forces or mechanical pressure, usually associated with mechanical cutters, eliminates the board flexure that could break solder joints.

Safety

Safety problems caused by such conditions as fire hazards and dust and noise pollution are minimized through the use of abrasive waterjet cutting, as follows:

• Safety is increased in an already hazardous atmosphere, particularly in comparison to flame and/or plasma cutting torches. Because there is no heat buildup with abrasive waterjet cutting, fire hazards are eliminated. There is no radiation emission or danger from flying slag particles.
• Airborne dust is virtually eliminated, making operation less hazardous to personnel working in close proximity to the machine. Containment or other methods of airborne dust control are unnecessary.
• Noise levels range from 85 to 95 dBA, which is consistent with Occupational Safety and Health Administration (OSHA) regulations.

ACKNOWLEDGMENTS

This article was adapted from "Abrasive Waterjet Cutting" by J. Gerin Sylvia in *Forging and Forming*, Volume 14, *Metals Handbook*, 9th ed., ASM International, 1988, with additional review and changes provided by Chip Burnham of Flow International Corporation.

REFERENCE

1. M. Hashish, "Application of Abrasive Waterjets to Metal Cutting," Flow Industries, Inc., 1986

Blanking and Piercing

VARIOUS CUTTING OPERATIONS are often used to provide parts that are subsequently formed to final shape by such operations as bending, drawing, coining, and spinning. Cutting operations are frequently conducted in the same press tooling used to form and shape the final part geometry. The principal objective of any cutting operation is to produce a workpiece that has the correct geometric shape, is free of distortion, and possesses sheared edges that are of sufficient quality to allow subsequent forming, finishing, and/or handling operations. The most common group of cutting operations includes blanking, piercing, parting, shearing, notching, trimming, perforating, and so on. This article focuses on blanking and piercing operations.

Fundamentals of Cutting

In blanking, piercing, and related cutting operations (such as trimming, notching, parting, and so on), the metal is stressed in shear between approaching cutting edges until a fracture occurs. It is desirable to estimate the maximum blanking or shearing pressure required for the various tools, the effect of this maximum pressure or shear on the punch or die (grinding of one or the other at an angle), and the power and flywheel energy needed for the work.

If the tools are ground flat and parallel and friction on the punch and die is neglected, the maximum pressure required to blank out a punching is normally the product of the shear strength of the metal and the cross-sectional area around the periphery of the blank. Thus, the formula for maximum pressure (P) is:

$$P = L \times t \times S, \text{ or } P = \pi D \times t \times S \qquad \text{(Eq 1)}$$

where S is the shear strength of the material, t is the sheet thickness, and L is the length of the cut or the periphery of the blank. If the blank is round, L may be replaced by D, which is the diameter of the punching (or the sum of several diameters). Figure 1 is a nomograph derived from the aforementioned formula, with its limits selected to cover the normal range of power-press (and squaring-shear) capacities. An example is shown in dashed lines. The shear-strength values indicated at the left in the chart depend, of course, on material condition and prior processing.

As a flat punch progresses through a sheet of metal, it deforms the metal plastically to a point

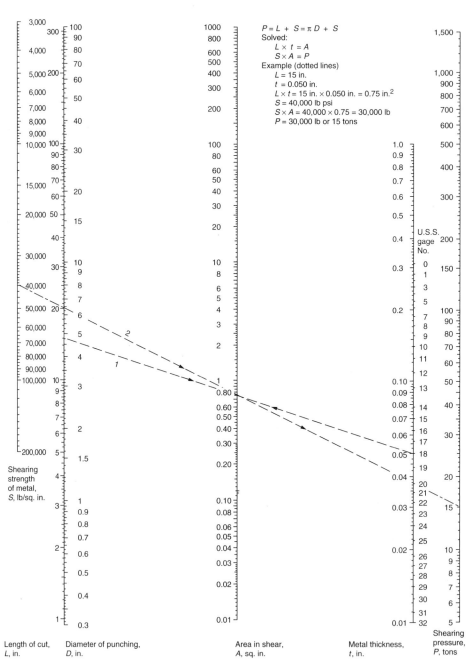

Fig. 1 Chart (derived from Eq 1) for determining blanking pressure to cover the normal range of power-press (and squaring-shear) capacities. In reading this chart, lay a straightedge or ruler on the chart so that it connects the desired values of metal thickness with the length of cut (L) or the diameter (D) for cutting a round. Mark or note the point where it crosses the centerline. Move the straightedge to connect this point on the centerline with the proper value for the shearing strength of the metal to be cut. The line through these two points crosses the shearing-pressure line at the tonnage required to punch the blank or hole. The shear-strength values indicated at the left depend on material condition and prior processing.

where the ultimate strength is exceeded, and fractures start from the opposing cutting edges of the punch and die (Fig. 2). If there is proper clearance between the tools, the fractures meet almost instantly, and the fractured portion of the sheared edge appears quite clean. If the clearance is not sufficient, the first fractures do not meet, and secondary shearing takes place, with a characteristically irregular and ragged appearance around the edge. When the cut is clean, a brightly burnished band around the edge of the blank indicates how far the punch had to penetrate before the fracture occurred. This is of interest in connection with Fig. 3 and 4, because the work is entirely or nearly completed at this point.

Figure 5 illustrates diagrammatically the effect of shear on the blanking load. The term *shear,* as used here, refers to grinding either the punch or the die at an angle, the other member remaining flat so as not to distort the product. The amount of shear is the difference between the high and low points of the angular face, even though the punch may be ground down both ways from one or more high points.

The effect of shear is clearly to reduce the peak load of blanking by shearing a little work metal at a time instead of making the whole cut at once. The total energy required to do the job is not changed, being merely that a lower pressure is continued for a longer time. A small amount of shear (less than the distance a flat punch must

travel to effect shearing) does not appreciably reduce the load but does ease the snapback, especially of C-frame presses, which occurs when the whole load is released instantly by the fracture.

Figure 3 is designed for use in approximating the extent to which the total or maximum load is reduced by shear on the tools. In the case of ample clearance and a clean fracture, the result obtained is high, because the calculation is based on the assumption that the working pressure without shear is equal to the maximum pressure recorded, throughout the working distance. Actually, however, the pressure rises gradually to the maximum and then usually falls somewhat to the point at which fracture occurs. The average pressure for the penetration distance, which would naturally be lower than the maximum pressure, would give an accurate result. An accurate pressure diagram is rarely available,

and the chart in Fig. 3 is close enough for ordinary purposes. Note that it is always desirable to have considerable excess press capacity for blanking work, for the sake of the tools.

In the case of insufficient clearance in the tools and a ragged fracture, secondary shearing occurs so that the pressure curve does not drop sharply. The amount of secondary shearing increases as the clearance becomes less and less. The square-root scale for percent penetration is an arbitrary method of compensating for the effect of secondary shearing but is ample for ordinary cases. In many cases, of course, the result will be high.

Figure 3 is read from the top down, and, as indicated by the example in dashed lines, the third and fifth lines are used only as pivot points for the straightedge. The formula is:

$$P = \left(\frac{t \times \% \text{ penetration}}{\text{Amount of shear}}\right) \times P_{max} \qquad \text{(Eq 2)}$$

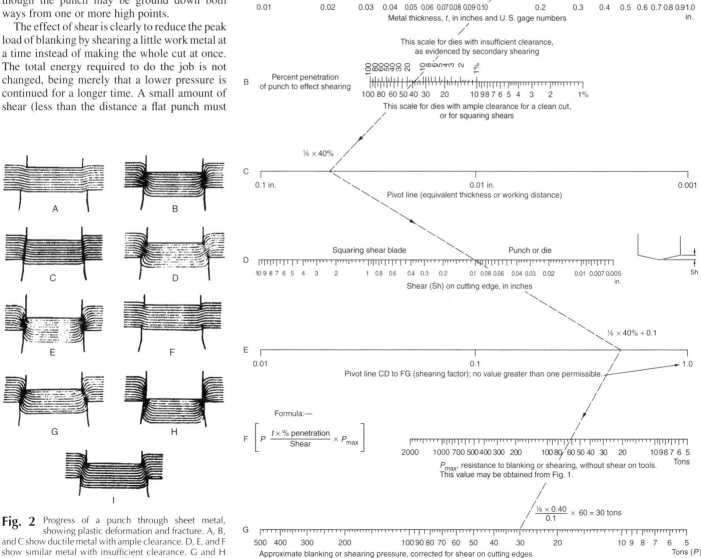

Fig. 2 Progress of a punch through sheet metal, showing plastic deformation and fracture. A, B, and C show ductile metal with ample clearance. D, E, and F show similar metal with insufficient clearance. G and H show hard metal with sufficient clearance. I shows the effect of dull cutting edges.

Fig. 3 Chart for determining effect of angular shear

The thickness multiplied by the percent penetration to effect shearing gives the actual working distance in inches. This value divided by the amount of shear on the tools, in inches, gives the proportion of the length of the cutting edge that is actually working at any instant (Fig. 5). This value multiplied by the maximum pressure (without shear) from Fig. 1 gives the approximate maximum working pressure.

In checking power requirements of flywheel capacities, it is necessary to know the actual energy absorbed in doing the work. In the case of shearing, this is properly the working distance multiplied by the average pressure, or, graphically, the area under the pressure-distance curve. Because such curves are not easily obtained and therefore average pressure cannot be measured, the approximate methods used previously are once again employed. Thus, in Fig. 4, the max-

imum pressure (P, from Fig. 1) is used in place of average pressure, and the working distance is taken as the product of the metal thickness, t, and the percent penetration, using two scales for the latter. Then, the energy or work, W, in inch-tons is:

$$W = T \times \% \text{ penetration} \times P \times 1.16 \qquad \text{(Eq 3)}$$

The 1.16 adds a 16% allowance for machine friction. This can only be a general case, of course, because the arrangement and condition of machines vary widely.

It should also be noted that an allowance should be made for heavy stripper springs, when used, and for wall friction in pushing slugs through the dies when there are long straight walls, as is occasionally the case.

In using Fig. 4, read thickness and percent penetration first, to obtain the pivot point for the

straightedge. Then, set the straightedge to connect this point and the pressure value, to obtain the energy requirement.

Blanking Operations

Blanking is the process of cutting or shearing, from sheet-metal stock, a piece of metal of predetermined contour to prepare it for subsequent operations. The piece of metal resulting from this operation is commonly called a blank. The prime objective in planning a blanking operation is to select a blank that will produce a part of prescribed quality, within reasonable cost, that can be produced in the quantity required to maintain production schedules. The blank must be designed to provide enough metal to shape the part and, at the same time, prevent excessive scrap loss due to trimming.

The design of the blank also depends on the desired part and the subsequent method of forming. For example, when a part is drawn, deformation occurs over a major portion of its surface. The metal flows and bends and sometimes stretches. To accomplish this severe deformation, the metal must be held around the edges to prevent wrinkling and tearing. In contrast, deformation during bending is usually confined to the area immediately adjacent to a bend line without the metal flowing or stretching. In this case, a blank may be laid out for bending without the necessity of trimming. In the design of the blank for a drawn part, however, metal must be provided at proper places to permit the part to be shaped to specifications. Metal also must be provided for holding.

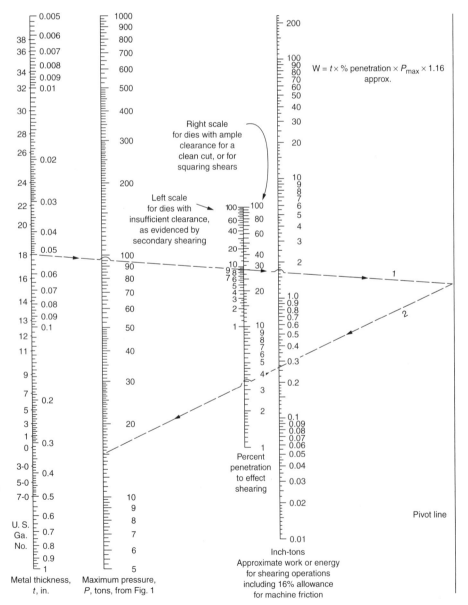

Fig. 4 Chart for determining energy required in shearing

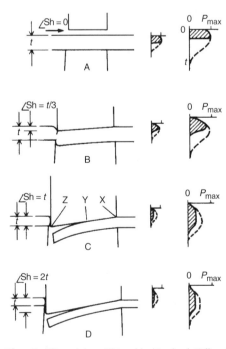

Fig. 5 Effect of shear (Sh) on blanking load. Different angles of shear on the punch (or die) reduce the amount of work it is doing at any instant.

The types of blanks used in sheet-forming operations fall into four convenient classifications. These classifications are rectangular blanks, rough blanks, partially developed blanks, and fully developed blanks (Fig. 6 and Table 1). This classification of blanks has permitted the application of some general rules to follow in setting up the blanking operation for a part. These rules deal with the type of a blank that should be used for a given type of part, the form of material from which the blanks should be cut, and the method and equipment that should be used in making the blank.

Rectangular Blanks

A rectangular blank is the least complex type of blank and the easiest to produce. As its name implies, it is rectangular in shape, having four straight sides. Its size is determined by the amount of metal required in the part and the amount of metal needed for trim allowances and holding. Rectangular blanks are usually restricted to production of drawn parts, although it would be possible to produce symmetrical, simple channeled formed parts from such blanks.

Only a portion of a rectangular blank appears in the finished part, because the outer edges of the blank provide surfaces for holding during drawing and are trimmed away. A fuel-tank half for an automobile is an example of a part made from a rectangular blank.

Why Rectangular Blanks Are Used. While the rectangular blank is somewhat limited in its application to various types of parts, certain definite advantages can be obtained through its use. These advantages result primarily from the type of equipment used in making rectangular blanks.

A squaring shear is the most prominent type of equipment used in making rectangular blanks. This machine consists of an appropriate frame and a moving cutting blade that acts on metal positioned over a lower cutting edge. A blank is produced at each stroke of the blade, which can be adjusted to operate at a very high speed. Thus, a rectangular blank can be produced at a high rate of speed. In addition, a squaring shear can be set up with automatic feed and cutoff, resulting in a very low direct labor cost per piece.

The squaring shear can be adjusted to accommodate blanks of varying length with very

little difficulty. A wide range of blank widths also can be accommodated. The ease of adjustment of the squaring shear makes it a very versatile machine. This makes rectangular blanks advantageous, because very little time is involved in setting up for their production.

One of the chief advantages of the rectangular blank is that no die is required. This results in a very considerable savings in the cost of establishing a blanking operation.

One of the unfavorable aspects of the rectangular blank is the large amount of waste material that results from its use. Except in the case of a rectangular-shaped part, such as a gas tank, a sizeable portion of a rectangular blank is lost to immediate production in the form of unusable scrap. On low-production jobs, the scrap loss is often offset by a reduction in die costs. In other jobs, however, especially high-production jobs, scrap loss is often one of the most important factors. For this reason, the choice of a rectangular blank should be weighted carefully in terms of the cost of scrap and its relation to the savings possible through use of the square-shearing method.

Materials Used in Making Rectangular Blanks. The form of material used in producing rectangular blanks is an important consideration in setting up blanking operations. Two forms of material are generally used: sheet stock and coil stock (Fig. 7 and 8).

Sheet stock may be used for producing rectangular blanks for low-, medium-, and high-volume requirements. Where this form of stock is used, manual operation of the squaring shear is usually involved. A package of sheets is delivered to the squaring shear by crane or lift truck.

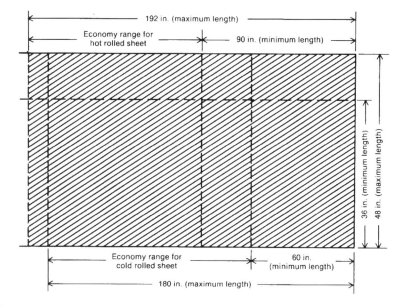

Fig. 6 Types of blanks

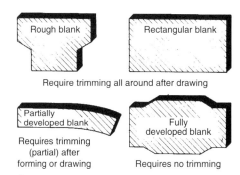

Fig. 7 Preferred sizes for ordering sheet stock

Table 1 Summary of application of blank types for different blanking methods

Type of blank to be produced	Blanking method			
	Squaring shear	Push-through blanking die	Compound blanking and piercing die	Pierce, notch, and cutoff die
Rectangular blank	Good	Poor	Poor	Poor
Rough blank	Good. For straight-side symmetrical blanks; unsuitable for blanks with curved edges	Fair. For small parts only	Fair. For small parts only	Good. For small and large parts
Partially developed blank	Poor	Fair	Good. For small and large parts and parts with holes that can be pierced before drawing or forming	Good. For small and large parts and parts with holes that can be pierced before forming
Fully developed blank	Poor	Good. For small parts that do not require holes	Good. For large parts and those requiring holes	Fair. Precision sizing usually not obtainable with notch-and-cutoff die

Each sheet is removed from the bundle and fed manually to the shear, where it is cut to a predetermined layout.

Using the proper size of sheet for making blanks is important from the standpoint of material cost. First, a sheet should be of a size that will yield the most blanks with minimum scrap loss. A no-scrap layout is most desirable.

Second, the size of the sheet should be planned with consideration of the variations in steel prices that are based on width and length of sheet. The most economical sizes of hot rolled low-carbon steel sheet range from 0.9 to 1.2 m (36 to 48 in.) in width and from 2.3 to 4.9 m (90^1/$_{16}$ to 192 in.) in length. For cold rolled low-carbon sheet, the economical range of widths is the same as for hot rolled sheet. The length range, however, is from 1.5 to 4.6 m (60^1/$_{16}$ to 180 in.). The purchase price for sheet outside these size ranges is greater because of additional operations required at the steel plant. Narrower sheet must be supplied by adjusting the mills or by slitting wider sheet. Greater widths require special adjustment of the reducing mills. Lengths outside the economy-sized range cost more because of the increased effort involved in operating the flying shear that is used to cut the sheet.

Coil stock, when used, should also be ordered with consideration of the relation of price to size. Ordinarily, coil length does not affect price. Width, however, does, and the most economical width range is from 0.9 to 1.2 m (36 to 48 in.).

Coil stock is used for producing rectangular blanks for high-volume requirements. The coil is delivered by crane and clamped between the arbors of an uncoiling machine. The coil is passed to the shear automatically through feed rolls. The stock advances to a stop on the back of the shear, and the cut is made by the blade, which is automatically activated by a limit switch. This type of operation is very economical from the standpoint of direct labor costs, because manual handling is greatly reduced and production is greater.

Rough Blanks

A rough blank is a workpiece cut roughly to the size and general contour of the finished part but having a margin of extra metal completely around the trim line. This extra metal is supplied for holding in subsequent operations and is cut off in the trim operation. A rough blank may be cut in a number of shapes. It may have straight sides and be in the shape of a triangle, diamond, trapezoid, hexagon, octagon, or even a "T." It also may have a combination of straight and contoured sides, giving it a nonsymmetrical shape conforming to no particular shape classification.

Rough blanks are used for making drawn parts. The surplus metal outside the trim line permits the binder surface of the die and the blankholder to restrain the edges of the blank for desired metal flow and/or stretching action. In rare instances, a rough blank may be used for a formed part that presents difficult metal-holding problems, such as a part with a difficult contoured flange.

Why Rough Blanks Are Used. The chief purpose of using rough blanks is to obtain material economy. As compared with rectangular blanks, rough blanks achieve this purpose very well. Because the blank conforms roughly to the shape of the part, less metal is trimmed away as scrap. A rough blank, properly nested, will also require less metal per piece than a comparable rectangular blank. ("Nesting" refers to the layout of the blank in the sheet or coil stock.)

Another advantage of rough blanks is that they generally can be handled more easily through successive operations than can rectangular blanks. The binder surface of a rough blank is generally smaller, and it is usually free from extended protruding corners. This permits easier loading of the part in operations subsequent to the initial draw and reduces the hazard of protruding corners being struck against press or handling machines.

How Rough Blanks Are Made. Depending on the needs of the situation, one of two methods may be used in making a rough blank. For blanks having symmetrical shapes and straight edges, such as triangular, diamond-shaped, and trapezoidal blanks, the squaring shear may be used. For nonsymmetrical parts, presses are used with dies of the cutoff type. These dies are sometimes called progressive blanking dies.

A plain cutoff die (Fig. 9) cuts the blank in two or more steps in a manner similar to the squaring shear. The die set includes upper and lower shearing steels. In operation, the stock is moved forward to a predetermined stop, the press is cycled, and the part is cut off. Each cut creates the final edge of one blank and the initial edge of another blank. One of the principal differences between a progressive cutoff die with a die and press and a squaring shear is that the cutoff die often produces an irregular cut, while the squaring shear always cuts on a straight line with simple shear blades.

The cutoff die is often used to produce two blanks simultaneously. This is made possible by incorporating two sets of cutting steels in the die. When this method is employed, one blank may be ejected off the end of the die and the other off the side of the die for effective disposal. Feeding must be arranged to advance the coil or strip as required.

The notch-and-cutoff die is also used for making rough blanks. This die is similar in operation to the plain cutoff die. However, two separate work stations are incorporated in the notch-and-cutoff die. In the initial work station, notching punches are mounted on the upper shoe with mating die sections on the lower shoe. These punches may be designed to notch triangles, semicircles, and many other blank-edge shapes that would be difficult or impossible to produce in a cutoff die alone. The notching tools provide only a part of the shape of a blank's edges; the remainder of the shape is supplied by the cutoff tools. The cutoff takes place in the last work station. The cutoff tool is usually lined up with the apex of the notch to ensure a clean cut.

Pierce, notch, and cutoff dies (Fig. 10) are used in some cases for producing rough blanks for formed parts. The major characteristics of this type of die are discussed in relation to developed blanks.

The presses used for the foregoing progressive-type blanking operations are usually of the gap-frame inclinable variety for easy feeding of stock and quick disposal of blanks. For some large blanks, such as for fenders and quarter panels, where pressure requirements warrant, large presses of the straight-side type are used.

Materials Used in Making Rough Blanks. The principles cited for selecting an economical form of material for making rectangular blanks apply equally to rough blanks. In fact, rough

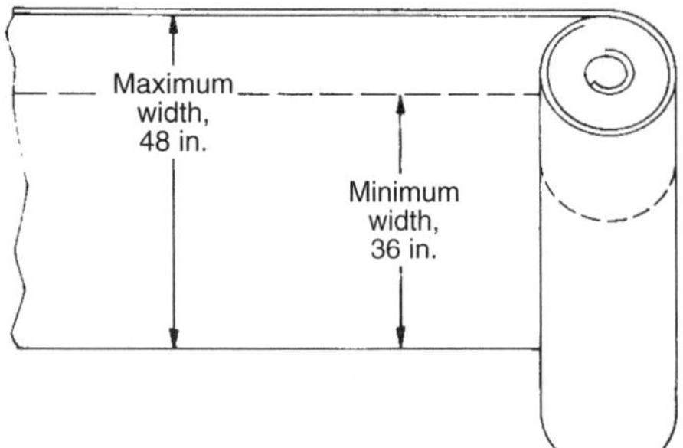

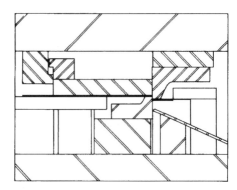

Fig. 8 Preferred sizes for ordering coil stock

Fig. 9 Plain cutoff die

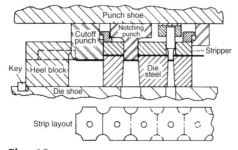

Fig. 10 Pierce, notch, and cutoff die

blanks require special consideration of layout for material economy. Because rough blanks are often nonsymmetrical, they must be planned and designed carefully to achieve material economy. The choice of strip or coil stock should be made with a view toward volume requirements and the reduction of offal (Fig. 11).

Developed Blanks

A developed blank may be described as a blank of any shape, symmetrical or non-symmetrical, that is precision sized so that no trimming is required after the part has been shaped. Developed blanks are used almost exclusively for parts that are formed. Practically all stampings used in chassis frame construction are in this category, as well as many internal parts of automobile bodies and front-end assemblies (Fig. 12).

Why Developed Blanks Are Used. Developed blanks offer greater opportunities for savings in material and operating costs than any other type of blank used in stamping operations. There is no scrap resulting from trimming, and the need for expensive trim dies is avoided, where developed blanks can be used. Developed blanks cannot be used for drawn parts because of the requirement for extra metal for holding purposes in drawing.

Some scrap does result when developed blanks are used. This scrap is generated when the blank is cut from the parent stock. However, by proper nesting, scrap can be minimized. The scrap can also be used in production of blanks for small parts.

How Developed Blanks Are Made. A developed blank having mainly contoured edges is made in a push-through blanking die mounted in a suitable press. One advantage of this type of setup is the ease with which disposal of the blank is accomplished. The blank is merely pushed through an opening in the lower die shoe and the press bolster to a container below. In some instances, the blank is allowed to fall onto a conveyor that then carries it to a subsequent operation.

A push-through blanking die (Fig. 13) may be defined as one in which a sheet of metal to be cut is laid over a sharp-edged die opening, and a punch, with cutting edges to fit the die, is forced through the sheet metal into the die opening.

The punch assembly consists of a tool steel punch and a stripper mounted on a shoe with guide bushings. The stripper is provided to remove the scrap from the punch on the upstroke of the press.

The die assembly consists of a conventional die shoe that is equipped with guide pins to ensure accurate alignment. The guide pins are important in making a developed blank because of the precision sizing necessary. Careful alignment is also necessary to prevent damage to cutting steels. Additional components of the push-through die are the composite steels that comprise the cutting edges, backing plates, and stock guides.

Push-through dies are not considered practical for punching out blanks for large parts, because a large die opening would allow the stock to sag and become distorted before the cut was made.

For low- or medium-volume work, a single push-through blanking die may be used with offal or sheet stock hand fed to the die. For high-volume work, double dies are often used to make two blanks, either separate or connected, with each stroke of the press. In some cases, dies for two different parts may be used in the same press at the same time. Coil stock and automatic feed mechanisms are usually used when production requirements are high.

A developed blank may also be made in dies other than the push-through type. These dies are the compound blanking and piercing die, and the notch, pierce, and cutoff die.

The compound blanking and piercing die (Fig. 14) performs more than one operation. Both blanking and piercing are completed in one stroke of the press and in one station in the die. The compound blanking die is situated in an upside-down position, with the sized cutting section on the upper shoe. This upside-down position provides support for large parts to prevent buckling or sagging, permitting large

developed blanks to be made. However, many small developed blanks are also produced in this type of die.

Combining piercing and blanking in one operation enhances the economy of a press operation. However, care should be taken that holes pierced will not become undesirably deformed in later forming operations.

The important parts of the punch assembly are the die section, the piercing-punch assembly, the stripper, and the knockout pad or pin.

The principal components of the die assembly are the tool steel punch and the stripper. The punch includes die buttons for receiving the piercing punches that are mounted on the upper shoe.

The actions that take place in a blanking and piercing die may occur simultaneously or in sequence. Piercing usually occurs first, because the metal can be held more firmly before blanking takes place. In some cases, however,

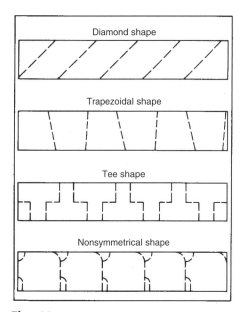

Fig. 11 Layouts for rough blanks

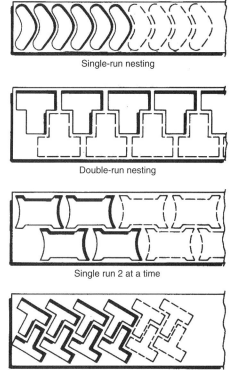

Fig. 12 Layouts for developed blanks

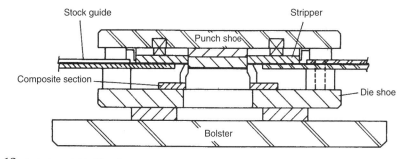

Fig. 13 Push-through blanking die

the blanking may be done first, with the face of the punch being used to hold the material while piercing takes place.

The notch, pierce, and cutoff die can also be used for producing a developed blank for a forming operation. This die is a modification of the progressive blanking die and includes the piercing function where holes produced will not be distorted in later forming operations.

Presses used in making developed blanks come in a range of sizes proportionate to the size of the blank being made. Most are of the straight-side type to ensure accurate alignment of cutting steels.

Materials Used in Making Developed Blanks. In making a developed blank, the choice of the form of material to be used is a highly important factor. The choice will be influenced by the size of the blank and the production volume of the part.

Offal stock is used for small, low-volume parts, that is, where requirements per 16 h day are less than 1400 parts. Offal is restricted to such work because of the difficulty in handling the unevenly shaped pieces. It is used only with blanking and piercing, blanking and forming, or push-through blanking dies. The progressive type of die cannot use offal for blanks because of difficulties in moving the part through such a die. The use of offal in all possible applications is desirable from the standpoint of material economy but must be analyzed very closely.

Sheet stock is used for low-volume work, and for medium-volume work where the requirements are from 1400 to 2000 parts per 16 h day, or for large parts. All sizes of blanks may be cut from this type of material. It may be used in push-through blanking dies or in compound blanking and piercing dies. It can also be used in progressive blanking dies. Heavy-gage sheet stock is used for high-volume work where requirements are over 2000 per day, because such metal is of too-heavy gage, and coils would be too short. Stock of this type is used for making generator frames, chassis frame parts, and other heavy-duty applications.

Coil stock may be used for medium-volume work in a push-through, blanking and piercing, or progressive blanking die. Coil stock is used on all high-production stampings whenever it is practical to secure and use it in such form.

The layout of material for a developed blank is another important aspect of establishing an effective operation. The form of a developed blank is usually complex, and nesting it economically in a sheet or coil is sometimes more difficult than for a rectangular or rough blank.

One method of arriving at economical nesting is to prepare several paper templates of blank and place them on the stock. By shifting them around on the stock, the nesting that results in the least scrap may be determined.

If, in using strip stock, two nestings show equal promise, it may be necessary to lay out the entire strip to determine which nesting will yield the least scrap at the beginning and end of the strip. End scrap is less important when coil stock is used, because of the greater length of material. The method selected for cutting the blanks from strip or coil may also provide means for material economy. By feeding a strip through the die twice, it may be possible to obtain blanks from the scrap left after the first pass. Blanking in multiples (cutting more than one blank at a time) may offer additional material savings. It is generally practical to blank multiples from the widest stock possible. It is easier to feed wide strips of short multiples than it is to feed narrow strips of long multiples. Use of excessively wide strips must be avoided to escape payment of premium steel prices.

One method of determining the best nesting arrangement among several choices is to calculate the total area of stock required in square inches, to produce each piece. The scrap involved is included in the calculation. To determine the area for each piece from a given layout, multiply the distance between the centers of two blanks by the width of the strip. A comparison of the figures obtained from analyzing several nesting possibilities by this method will readily reveal the best course to follow.

Certain basic rules in blank development for push-through die work should be followed in order to ensure production of blanks having clean-cut edges. These rules pertain to the amount of scrap that should be left between developed blanks to ensure a clean cut. For stock lighter than 11 gage, at least 3.2 mm ($\frac{1}{8}$ in.) of stock should be allowed between the blanks and at the edge of the stock. For stock of 11 gage or heavier, at least 4.8 mm ($\frac{3}{16}$ in.) of stock should be allowed for scrap. This scrap allowance is necessary for a clean cut and to prevent the edges of the blank from being pinched to a sharp edge or being drawn down into an excessive burr. Sometimes, where edge conditions are less important and blanks are fairly symmetrical, no scrap allowance is necessary. If such is the case, blanks may be laid out adjoining one another.

In some cases, the grain direction of the metal may be a factor in laying out blanks. This factor is prevalent in electric-furnace and bessemer steels. There is no appreciable grain direction in open-hearth steel. Where definite grain direction does exist, however, blanks for forming operations should be cut in such a way that bends will be made across the grain by at least 30°. Bends made parallel with the grain direction will cause fractures at the radius of the bend.

Balancing the cuts in blanking operations is another important factor in ensuring effective blanking operations. Where cuts are unbalanced, shifting of the cutting steels is apt to take place. This will result in malformed blanks and excessive wear on cutting tools. This problem may be overcome, in some instances, by cutting blanks two at a time. The two unbalanced cutting edges will offset one another to create a balanced condition.

Wherever possible, blanking operations should be set up so as to safeguard the ease with which subsequent operations can be accomplished. This again may call for blanking of two parts at a time and leaving them fastened together to provide balanced conditions in forming or piercing operations that follow.

Partially Developed Blanks

A partially developed blank (Fig. 15) is defined as a blank that is precision sized on two opposite sides of a part, with the remaining two sides providing excess metal for holding purposes. This type of blank is used for drawing of parts where holding is required on only two sides. Roof rails are typical parts for which partially developed blanks may be used. There are only a few automotive stampings of this type, and hence, partially developed blanks are rarely used. Where they are used, however, the complexity of trimming is slightly reduced.

A partially developed blank is usually made on a cutoff die, a notch-and-cutoff die, or a compound blanking die.

The general rules for selecting forms of material for partially developed blanks correspond to those cited for developed blanks.

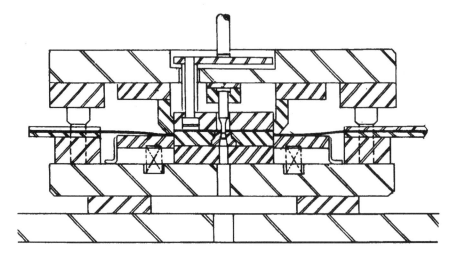

Fig. 14 Blanking and piercing die

Blanking Clearances and Shear

Certain mechanical problems that arise in cutting of metals also must be considered in establishing blanking operations. These mechanical problems include clearances between cutting steels and the pressure requirements for performing the shearing operation.

Clearances. In blanking, the die steels are built to the dimensional specifications of the blank, and any clearance required is machined on the punch. Clearances must be sufficient to avoid secondary shearing and close enough to prevent excessive burrs and deformation of blank edges. Pressure requirements for blanking are also affected by clearance. Usually, the greater the clearance provided, the less the required pressure. For best quality in blank edges, cutting tools should be kept sharp.

Shear is ground on the die section in blanking operations. The sheet metal in a stamping operation always assumes the shape of the punch. Hence, if the shear were on the punch, the blank would be deformed. By providing shear on the die, the blank is produced flat, and the scrap is deformed.

Shear is provided to reduce pressure requirements by permitting the cut to be made a little at a time. Shear provides the additional benefit of facilitating accurate blanking, because the reduction in pressure reduces the chances of distorting the part. Balanced or unbalanced shear may be provided, depending on the cut to be made and the precision required in the part.

Piercing Operations

Production of a sheet-metal stamping requires application of a variety of types of stamping operations. In the preparatory stages, blanking is used to cut out a workpiece of proper design and with sufficient metal to yield the desired part. This workpiece, or blank, is forced to assume the required shape by means of a forming or drawing operation. The specified edge dimensions of the part are obtained either in the blanking operation or in a trimming operation. However, in nearly all cases, a stamping requires more than sizing and shaping operations before it becomes a finished part. Holes, slots, tabs, or notches are incorporated in the part to bring it into conformity with specifications. The creation of holes or openings in sheet metal is done by the process of piercing.

Pierced holes serve many important purposes in the overall design and use of a part. They provide means of fastening the part to other components of an assembly and sometimes provide means of making a component lighter in weight. They serve as locating points to facilitate subsequent stamping operations and provide access to oil holes, drain plugs, and other parts through firewalls and other panels of the car.

Although piercing is regarded as one of the least complex of stamping operations, the establishment of successful piercing operations requires careful planning by the process engineer. The great extent to which piercing is employed is, in itself, enough to support this conclusion. In addition, tools must be provided that will prevent excessive cost due to breakage. A good deal of standardization in tooling is called for. Operational sequences must be established properly in order to prevent deformation of pierced holes and the production of defective parts. In short, the success of a total stamping process is equally as dependent on efficient piercing operations as it is on other types of operations.

Types of Piercing Operations

The various types of piercing operations employed in metal stamping (Fig. 16) derive their names from the kind of hole or cut that is to be made. Each of these holes or cuts is designed to accomplish a specific purpose in the design or use of the part.

Conventional piercing usually consists of driving a round flat-face punch through the part into a die section or button. The portion of metal under the punch is totally removed from the part. This piece of metal is called a slug. The resulting hole is normally used as a means of fastening parts together with rivets, bolts, or screws.

The majority of piercing operations are of this type, and the holes so produced cover a broad range of sizes. As a result, punches of various sizes are necessary to provide the tools required to meet piercing needs. This range of punch sizes has been modified by the application of punch-sized standards. Care should always be taken to provide a standard-size punch wherever possible.

Piercing with a Pointed Punch. Another operation commonly referred to as piercing is the creation of a hole with a pointed or cone-shaped punch. In this type of operation, no slug is removed. The hole results from a pushing apart and extrusion of the metal as the punch passes through it. This results in an extruded collar around the hole. The edges of this collar are ragged because of the nature of the break that takes place. The collar is sometimes tapped to provide a means of fastening another part with a bolt.

Pierce-and-extrude operations consist of first punching a hole in the part and then extruding the edges of the hole to provide a circular flange around the enlarged hole. This operation calls for a punch of special design and special features in the die set. The end of the punch is flat, to break out the slug, and a beveled shoulder higher on the punch extrudes the hole edge and enlarges the hole as the punch passes through. Because the hole has been cut larger by the punch end, the edges of the extruded portion are cleaner and freer from cracks than in the case of the pointed punch. However, a certain amount of raggedness exists. Where a very smooth edge is needed, the hole may be reamed before extrusion. Extruded holes of this type may be used to provide spacing between panels; they may be used for strengthening holes in parts (such as hinges) that are mounted on rods or bolts; or they may be used as bosses or locators on which to mount other parts.

Slotting is used to provide holes of special shape (that is, noncircular), such as elliptical holes or rectangular holes. These holes may be required on some parts to allow for adjustment of the fit of a part in an assembly. A slotting punch requires a special design. The cutting portion must be shaped in accordance with the shape of the hole to be produced.

Countersinking is also done with the use of a special punch. This punch is similar to that used in piercing and extruding. In countersinking, however, the distance the punch travels through the stock is less than in extruding. The punch is driven far enough to create a hole and a tapered flange around the hole. This depressed hole is used where the head of an assembly bolt or screw must be flush with or below the surface of the part.

Cutting and Lancing of Tabs. In cutting and lancing of tabs, the tab is usually the item of importance rather than the resulting hole. The tab is created by a punch that cuts all but one side of the opening produced, leaving the tab attached to the part. The operation consists of shearing a U-shaped, V-shaped, or rectangular cut in the metal and pushing the tab downward with an unsharpened edge of the punch. The tab may be used to locate the part for later processing or to facilitate fitting of the part in an assembly.

Tools for Piercing

In a broad sense, the principal components of a set of piercing tools are the punch, the die, and

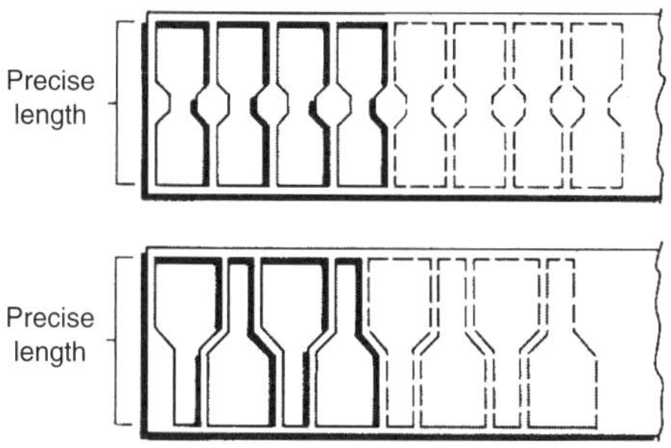

Fig. 15 Layouts for partially developed blanks

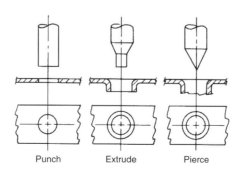

Fig. 16 Types of piercing operations

the stripper. The punch creates a hole as it is forced through the material. The die consists of a steel block in which a hole matching the contours of the punch is provided. The edges of the hole provide the lower cutting edge and cooperate with the punch in shearing the material. The slug from the material usually drops through the hole in the die. The stripper consists of a plate with a suitable hole through which the punch point passes. It has a dual function in that it holds the material from slipping during piercing and keeps on holding it so as to strip the material from the punch as the punch retracts. The tool setup functions in the following manner. The stripper descends to hold the stock against the face of the die. The punch follows through the hole provided in the stripper and is forced through the material into the die hole. On the upstroke, the stripper dwells and holds the material as the punch is withdrawn. The material must be held in this manner because it tends to cling to the punch surface. After the material is stripped from the punch, the stripper plate rises, allowing the part to be removed.

The foregoing description of the makeup and function of the piercing die is couched in the most general terms. Actually, many more factors are involved. The tools are generally much more complicated, the degree of complication depending on the operation involved.

Punch Assembly. Actually, the term *punch* may refer to two different items in considering a piercing die setup. It may refer to a single tool for creating a hole, or it may refer collectively to an assemblage of components that may create several holes and even perform other operations.

The base for building a punch assembly is the upper shoe. This may be made from a mild steel plate, or it may be a casting. All other punch components are fastened to the shoe. Bushings should be provided in all shoes for guide pins to ensure accurate tool travel.

Hardened steel backup plates are attached to the shoe in areas where the individual punches will be placed. This is necessary to prevent the bases of the punches from being driven into the softer metal of the shoe during operation.

Another component of the punch assembly is the punch retainer. This is a device for holding the punch. Standards have been developed for types of retainers to be used with standard punches. A standard retainer is used whenever possible. In some cases, where there is insufficient space, or where several punches are necessarily mounted in a small area, special retainers may have to be used.

The working end of the punch assembly consists of the individual punches. These are held in the punch retainers by several means. The method of holding may depend on the size and shape of the punch and whether it is used singly or in combination with other punches. The service for which the punch is to be used also influences the holding method. For piercing heavy material, a more positive holding method is required than for piercing light material. This is true because of the heavy stripping load that

exists in piercing heavy material. For an operation in which numerous parts are to be pierced, the holding system may be more elaborate and expensive than in an operation where total production quantity is low.

For high-production jobs, the ball-seat holding method may be employed. This system provides for the rapid changing of punches, an item of considerable importance in high-volume operations. In some cases, the stripping pressure may require the use of a shoulder-type holding system. In other cases, bridge-lock or solid-type holding may be best (Fig. 17).

To accommodate the various types of piercing operations, a variety of piercing punches have been developed. Each type has certain special characteristics that make it suitable for the job to which it is applied.

The most widely used type of punch is called the regular-duty punch. It is generally round in shape and is mass produced by a number of tool manufacturing companies. Its simple design and its abundance make it relatively inexpensive. It is of the ball-seat type and thus presents a minimum problem with relation to downtime for changing tools. It is used for making holes up to 25 mm (1 in.) in diameter when the stock thickness is 3 mm (0.12 in.) or less.

For stock that is thicker than 3 mm (0.12 in.), and for holes 25 to 38 mm (1 to 1½ in.) in diameter, a shoulder-type punch is used. The heavier construction of this type of punch provides the added strength necessary to ensure reasonable tool life in this heavier type of work. The shoulder punch also permits a stronger holding method to resist the greater stripping pressure that exists where larger holes or thicker stock is involved.

A solid-type punch is used for piercing holes over 38 mm (1½ in.) in diameter. This type of punch is very heavily constructed and is bolted directly through the backing plate to the upper shoe.

Normally, piercing is confined to production of holes that are greater in diameter than the thickness of the stock involved. This is because piercing of holes with diameters less than stock thickness results in excessive breakage of punches, unless special strengthening provisions are

made in the punch. A punch thin enough to pierce a very small hole will not withstand the forces required to shear the stock on a sustained basis, unless it is sufficiently reinforced. Thin punches tend to deflect or whip on contact with the stock, adding the probability of breakage. The reinforcement provided to enable such piercing to be done consists of several elements. The point of the punch is made as short as possible; the main body of the punch is made large and is brought as close to the point as possible; and a die-cast soft metal sleeve is pressure cast to the punch to absorb the whipping and deflecting forces. This cast sleeve is called a whipsleeve, and punches so constructed are called whipsleeve punches (Fig. 18).

Whipsleeve punches are used for piercing holes where the metal thickness exceeds the punch diameter. They are also used in producing slightly larger holes where whipping and deflection are a problem.

A quill punch (Fig. 18) may also be used to pierce holes smaller than stock thickness. This punch has a very short point and is encased in tool steel for strength. Hard stock may also be pierced with this type of punch.

A large number of special punches are used to create special types and shapes of holes.

A blade-type punch is used for piercing long, narrow slots. It consists of a steel blade mounted in a special retainer.

Another special type of hole is made by the presthole punch (Fig. 18). This punch is designed to pierce out a small amount of stock, countersink the edge of the resulting hole, and slit the countersunk edge. This provides a gripping device into which a screw or bolt can be assembled without using a backing nut.

Another special type of operation is accomplished by the use of a developed punch. This punch is used to create round holes in stock held at an angle, although the punch travel is vertical. To accomplish this, the punch point is ground into an oval shape.

Die Assembly. The die section of the piercing-tool setup is also an assembly of several components. It generally consists of a lower shoe, die sections, die buttons, guide pins, and chutes or trays for the elimination of punched-out slugs.

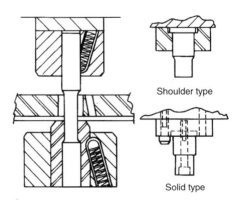

Fig. 17 Punch-retaining methods

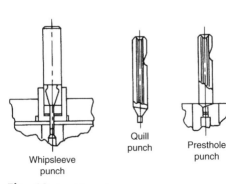

Fig. 18 Special punches

The lower shoe, or die shoe, is a mild steel plate or a casting to which all die components are fastened. Guide pins are fastened to the shoe and are used to ensure accurate travel of the punch. Heel blocks are often added to absorb any tendency of the punch to deflect during the work stroke.

The die sections in a piercing setup usually perform the function of holders for the die buttons. In some cases, the die sections are drilled through and hardened to provide opposing cutting edges for the punch. This practice is usually not followed because of the simpler die maintenance achieved through use of die buttons.

A die button is a hardened steel cylinder with a hole designed to provide opposing cutting edges for a punch. The hole of a die button will conform to the shape of the punch with which it is used. Die buttons, like punches, are available in many standard sizes. Special die buttons are also available for use with punches of special design.

In establishing any piercing operation, consideration should be given to the elimination of slugs from the die. Slug chutes should be provided in the die to move the slugs to convenient accumulation points. In some cases, slug trays should be incorporated in the die to prevent the slugs from piling up on the floor under the press. Where slug trays are not used, adequate receptacles should be provided at the accumulation point.

Stripper. After the punch has penetrated the metal, it must be pulled back. Because of elasticity, the metal compressed at the beginning of piercing tends to expand and grip the sides of the punch. This gripping causes the pierced part to rise with the punch. Thus, a mechanism must be provided to hold the part down while the punch is withdrawn. The holding of the part must be done properly to prevent distortion of the metal. The mechanism used for this task is called the stripper. The hole in the stripper through which the punch passes should be designed with the proper amount of clearance to permit free movement of the punch. In some cases, slender punches may be guided by bushings installed in the stripper plate.

Actually, a stripper may be a component of either the punch assembly or the die assembly. If it is a component of the punch, it is called a spring stripper, because the stripping pressure is supplied by springs that hold the stripper to the stock as the punch begins to ascend. As the punch moves up, the stripper is picked up by the punch after the part has been freed.

If the stripper is mounted on the die, it is called a solid stripper. The solid stripper is firmly mounted and does not move. It is generally used for large stampings or for stampings of heavy-gage material.

Piercing Shear and Clearances

In piercing operations, one of the factors to consider is provision of safeguards against excessive tool breakage. In addition, the holes created must conform to certain quality requirements with respect to size and edge conditions. Edges should generally be free from excessive burrs. The provision of proper clearances and shear can aid considerably in reducing tool breakage and providing desired quality.

Shear. In hole-cutting operations, the metal cut out is the scrap. Hence, any shear provided must be ground on the punch. The main purpose of shear is to reduce the load placed on the punches to accomplish the piercing.

Shear, to be of any advantage, should be ground on punches having diameters that are fairly large in relation to stock thickness. Shear on small punches may cause deflection of the punch and add to the likelihood of tool breakage. However, the benefits of shear may be provided where a number of small punches are being used by stepping the punches (Fig. 19). In this way, the pressure required to effect piercing is reduced by staggering the contact of the punches with the metal. The amount of punch stepping should be held to a minimum to avoid a jerky series of loads and releases as the punches progress through the metal. If the punches must enter the metal by 25% of stock thickness to effect shearing, each punch needs to be shorter than the next by no more than this amount.

Clearances. Tonnage requirements for piercing, and the quality of hole edges, are dependent, in part, on the provisions made for clearance between punch and die. The proper amount of clearance will result in a satisfactorily clean hole. Excessive clearance will result in rough, burred edges, distortion of material surrounding the cut, and pinched-off edge rather than a clean fracture. Insufficient clearance will result in unnecessarily high tonnage requirements because of the secondary and even tertiary shearing that takes place where insufficient clearance is provided. This greater force required under these conditions is apt to shorten tool life. In addition, particles of metal may be released during the secondary shearing. These particles may adhere to the punch surface and result in oversized or irregular holes. This adhering metal, called pickup, also increases tonnage requirements and may result in tool breakage due to deflection caused by a buildup of metal on one side of the punch. In addition to these undesirable effects, secondary or tertiary shearing leaves the internal surfaces of the hole ragged.

Proper clearances between punch and die are based on a percentage of stock thickness. This percentage is ground off the die hole, all around. The clearance varies with the type of material used. Generally, the harder the metal, the less the clearance. The range of clearances for most automotive stampings is considered to lie between 5 and 10% of metal thickness. The proper clearance is usually determined in die tryout.

Special Factors in Piercing

To establish successful piercing operations, thought should be given to certain special factors.

Proximity of Punches. Where a number of holes are punched close together, certain provisions should be made to prevent excessive tool breakage. This breakage may result from the crowding of metal that takes place in piercing. For instance, a small punch adjacent to a large punch may be deflected and broken as the result of the large punch crowding the metal in the direction of the small punch. This problem may be overcome by grinding the smaller punch shorter by an amount equal to the distance the larger punch must travel to shear the stock. In this way, the metal is already crowded before the smaller punch makes contact. Also, the smaller punch retracts first and thus escapes any side pressure resulting from springback when the large punch is withdrawn.

Sharpness of Tools. The sharpness of cutting edges in piercing operations has an important bearing on the quality of work done and the pressure required to do the work. A sharp edge localizes the severest stresses and causes the fracture to occur sooner and with less load. Thus, maintenance of tool edges provides protection against punch breakage. It also helps to ensure clean fractures.

Limitations of Piercing. Theoretically, the size and number of holes that can be pierced, and the thickness of stock that can be used, are limited only by the size and strength of the press. However, for producing specific parts in conformance to specifications, certain practical limitations should be observed.

The number of holes that can be pierced in one die is limited by the space required for assembly of punches and die buttons. The size of the holes to be pierced is another determining factor in the number of these items that can be arranged in a single die setup. While it is usually economical to combine piercing operations in a single die, care should be taken to avoid excessive die-maintenance problems caused by a weakening of the die by inclusion of too many operations.

There is usually no difficulty in piercing any thickness of stock used in automotive stampings. However, in piercing heavier-gage stock, care should be taken to provide punches sturdy enough to withstand the stresses involved. Proper guiding of punches is necessary to avoid excessive tool breakage.

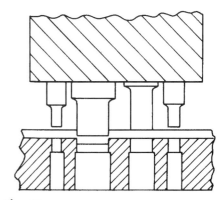

Fig. 19 Punches stepped to avoid breakage

Certain limitations should be observed in piercing close to the edges of stock or flanges. The general rule is to keep the outer edge of the hole at least two stock thicknesses from the edge of the material. With respect to flanges, at least three stock thicknesses should be allowed between the hole edge and the inside surface plane of the flange. The main purpose of this is to preserve adequate strength in the die button. The walls of the button must be thick enough to withstand the force exerted by the punch. If this thickness is reduced too much, in an effort to bring the die button hole close to the edge of the flange, repeated failure of the button may result. If a hole is pierced too close to a flange breakline prior to forming of the flange, undesirable distortion of the hole may take place during flanging. Where part design calls for holes in these limiting areas, a change in design may be necessary.

Most piercing is done with the punch traveling in a direction perpendicular to the horizontal plane of the stock. This provides the optimal conditions for punch strength and clean shearing. However, it is sometimes advantageous to pierce at some other angle (Fig. 20). Such a case may occur when piercing follows forming and the area to be pierced is at an angle to the horizontal. As the punch descends, it first contacts the work on one side of the point. This, of course, tends to deflect the punch, requiring careful guiding. The greater the workpiece angle from horizontal, the more difficult it is to pierce with vertical punch travel. Actually, piercing in such a situation is limited to a part angle of 30° or less. If piercing is attempted at a greater angle, exorbitant costs will be incurred due to punch breakage.

The proximity of holes to each other is another factor that puts a limitation on piercing. This limitation applies only to equipping the punch and die assemblies with standard punches and die buttons. Holes actually can be pierced very close to each other; they may even overlap. However, when such piercing is done, special punches, punch holders, and die buttons are required, because the sides of the punches and the die buttons must be ground down to bring the hole-producing elements close together. The use of special piercing units may be justified on the basis of savings resulting from combining of operations. However, a comparison of tooling and operating costs for special units with costs of other methods should be undertaken to determine the best method.

Cam Piercing

Piercing is a type of press operation that often can be successfully combined with other operations such as blanking, forming, and trimming. The economies that can be realized through combining of operations should always be considered when setting up a piercing operation. In addition, combining piercing with other operations is often a means of carrying out a greater amount of work with a limited number of presses.

In combining piercing operations, it is frequently unfeasible to pierce with a vertically held punch. This is true where the area to be pierced is in an angular position. In such a case, a cam-piercing mechanism may be used to bring the punch into right-angle contact with the part, the cam moving in from the side.

There are several types of cam mechanisms. The type selected for a particular application depends on such factors as space available for mounting the cam, pressure requirements for cam return, and the direction of piercing.

Spring-Return Cam. A spring-return cam (Fig. 21) consists of three main elements: the cam body, the driver, and the spring. The cam body carries the punch and is mounted on a base in such a way that the piercing punch may freely enter the die button, which is mounted in proper position on the die shoe. The driver is a wedge-shaped steel block mounted on the ram of the press. On descent of the ram, the wedge face acts on a similar surface of the cam body, causing the cam body to move in toward the die button. As this occurs, the cam spring is depressed. When the driver moves up, the spring expands to withdraw the cam body and its attached punch.

Air-Cylinder-Return Cam. The components and method of operation of an air-cylinder-return cam are very similar to those of a spring-return cam. The difference lies in the fact that an air cylinder is used instead of a spring to return the cam body to the open position on the upstroke of the press. Air cylinders usually work more satisfactorily than springs as return mechanisms. In addition, the air cylinder is considered safer, because springs sometimes break and may scatter pieces of metal around the work area, endangering employees.

Positive-Motion Cam. Some cam operations are accomplished with the use of a positive-type driver. This driver consists of a mechanism that engages the cam body throughout the work stroke, pressing the cam body inward on the downward stroke and retracting it on the upward stroke. No springs or air cylinders are used. This type of mechanism is called a positive-motion cam (Fig. 22). The cam body may be located on either the punch or the die. If the cam is on the punch, the driver is fastened to the die, and vice versa.

Internal Cams. In some cases, the location of the burr that is created on a part in a piercing operation is highly important. Sometimes, because of limited assembly clearances, a burr may not be tolerated on the interior surface of a part. The part may, however, be so positioned in the die where piercing is to be done that piercing in the normal way would induce burrs on the part interior. To avoid this, it may be necessary to pierce the part from the inside. This may be accomplished through the use of cams that work from inside the part when the part is in the die. These cams are linked to moving press mechanisms and carry the punch. The punch contacts the stock from the inside and forces a slug into a die button mounted outside the part on the punch, or perhaps on another cam mechanism.

Piercing and Extruding

Quite often in press operations, a hole must be produced with an extruded collar all around it. This collar may be designed to strengthen the part or to serve as a locating device or as a bearing for a shaft. To provide such a collar, a special type of die arrangement is necessary. The type of arrangement used will depend on the direction of the extrusion and the gage of the stock.

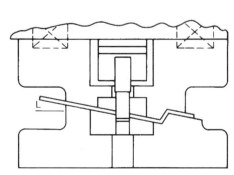

Fig. 20 Piercing at an angle

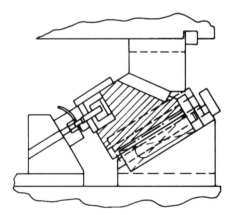

Fig. 21 Piercing with a spring-return cam

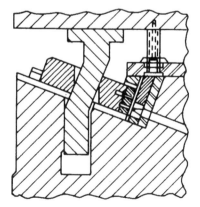

Fig. 22 Piercing with a positive-motion cam

Pierce and Extrude Up. To accommodate other operations, the part may be positioned in a die in such a manner as to require an upward forming of the extrusion (Fig. 23). In such a case, the upper shoe is equipped with a punch and a spring pad to hold the workpiece and keep it from shifting. A spring pad is also fitted to the lower shoe. This spring pad acts as a holding device during extrusion and strips the part when forming is completed. The lower shoe also contains a specially designed die button. In operation, the punch descends and the upper spring pad grips the metal while the punch pierces through the stock into the die button. As the punch continues down, the pressure of the lower spring pad is overcome, and the pad moves downward. As the pad moves down, the free metal around the hole is extruded upward around the die button. The lower spring pad strips the part from the die button on the upstroke.

Pierce and Extrude Down. The position of the part in the die may require that the extrusion be formed downward (Fig. 24). In this type of operation, the piercing punch, is assembled in an extrusion punch, and both are mounted on the upper shoe, along with a spring pad. A lower spring pad is built around the die button, which is mounted on a plate that rides on an air cylinder or on springs. In operation, the upper pad grips the metal as the punch descends. The piercing punch pushes a slug out of the stock and continues down until the extrusion punch contacts the hole edges and wipes them down as the die button is pushed down. On the upstroke, the piercing and extrusion punches are withdrawn, and the die button rises to strip the extruded portion of the panel out of the die opening.

Extruding Heavy-Gage Metal. Heavy stampings are often designed with hole extrusions that can serve as sturdy housings for bushings or other shaft-bearing units. When the material is of very heavy gage, the hole should be drilled rather than pierced. This is necessary because in heavy stock, cracks and fractures occur in extrusions formed from the edges of pierced holes.

Combined Operations

Piercing can be combined with practically any other type of stamping operation. In such com-bined operations, however, it is necessary to provide careful guidance between the punch and die.

Piercing is often combined with blanking when the holes will not be deformed in later operations. Dies for blanking are usually well guided and provide suitable protection for piercing tools. The same is true of trimming operations.

Piercing may also be performed in conjunction with drawing, forming, and restriking operations if hole distortion does not result and if the press tools are properly guided.

Presses Employed in Piercing

In nearly all cases, piercing is done on single-action presses. These presses may be of gap-frame or straight-side design. A triple-action press may also be used for piercing, the piercing being done with the third ram. In rare cases, a double-action press operation may incorporate a piercing punch that cuts through the metal just prior to bottoming to allow relief in an area of deep draw.

Presses selected for piercing should have adequate die area so that provision can be made to allow slugs to fall through the die base. Space for providing slug chutes under the die may be made through the use of risers.

Presses for piercing should also be in good mechanical condition. This will help to avoid tool breakage due to variations or shiftings in the press stroke resulting from worn press parts.

Hydraulic piercing mechanisms are sometimes used for operations in which only piercing is performed. Such mechanisms consist of hydraulic cylinders that are mounted on a fixture base and that carry the punches. The number of cylinders is dependent on the number of holes to be produced. Die buttons are mounted in proper locations on a fixture that positions the part for the operation. In operation, the part moves into the fixture, usually automatically, and the cylinders are activated to effect piercing.

This type of mechanism is used in piercing the various holes in assembled automobile frames. By piercing all of these holes in one fixture, at one time, the physical relationship of the holes may be held to closer tolerances. Performing the same type of operation in a press would be much more costly because of the greater expense of the required equipment.

Hydraulic piercing units are also used for piercing various hole patterns in fenders to accommodate various trim designs. They are also used in piercing rivet holes and air valve holes in automobile wheels.

These mechanisms provide a means of piercing at low cost, are adaptable to quick change-overs, and are versatile in application. However, the use of hydraulic cylinders and hydraulic lines creates maintenance problems that are both costly and troublesome.

Fine-Edge Blanking and Piercing

Fine edge blanking is primarily used to obtain blank or hole edges comparable to machined edges. In fine edge blanking, there is no die break, and the entire wall surface of the cut is burnished. No further finishing or machining operations are necessary. Specially designed single-operation or compound blanking and piercing dies are generally used for the process. Applications of fine-edge blanking include production of components for instruments, watches, and office machines such as levers, gears, fingers, tooth segments, and similar parts (see the article "Fine-Edge Blanking" in this Volume). Fine-edge blanking is also being applied to a wide variety of materials, as well as to thicker stock, in the automotive, farm equipment, ordnance, machine tool, printing machine, household appliance, and textile machine industries. Gears, racks, sprockets, and other toothed forms are easily produced and are common applications.

Fine-edge blanking and piercing are actually more akin to cold extrusion than to a cutting operation. In fine-edge blanking, a V-shaped impingement ring (Fig. 25) is forced into the stock to lock it tightly against the die and force the work metal to flow toward the punch so that the part can be extruded out of the strip without fracture or die break. Die clearance is extremely

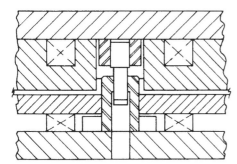

Fig. 23 Pierce and extrude up

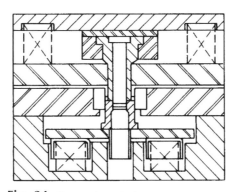

Fig. 24 Pierce and extrude down

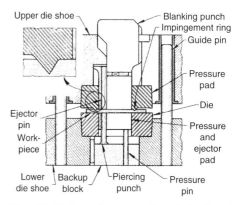

Fig. 25 Typical tooling setup for fine-edge blanking of a simple shape

small and punch speed is much slower than in conventional blanking. Fine-edge piercing can be done either separately or at the same time as fine-edge blanking. In piercing small holes, an impingement ring may not be needed. Specially designed single-operation or compound blanking and piercing dies generally are used for the process.

Process Capabilities. Holes with diameters as small as 50% of stock thickness can be pierced in low-carbon steel. In high-carbon steel, the smallest hole diameter is about 75% of stock thickness. Holes can be spaced as close to each other, or to the edge of the blank, as 50 to 70% of stock thickness. Total tolerances obtainable are 0.013 mm (0.0005 in.) on hole diameter and for accuracy of blank outline; 0.025 mm (0.001 in.) on hole location with respect to a datum surface; and 0.025 mm (0.001 in.) on flatness.

No die break shows on the sheared surface of the hole. Blank edges may be rough for a few thousandths of an inch of thickness on the burr side of the part when the width of the part is about twice the stock thickness or less. Finish on the sheared edge is governed by the condition of the die edge and the land within the die. Parts fine-edge blanked from stainless steel will have a surface finish of 0.8 μm (32 μin.) or better. Smooth edges also are produced on spheroidize-annealed steel parts. Burr formation increases rapidly during a run necessitating frequent grinding of the cutting elements.

Chamfers can be coined around holes and on edges. Forming near the cut edge, or forming offset parts with a bend angle of up to 30°, is possible under restricted conditions.

Metals up to 3.18 mm (0.125 in.) thick having a tensile strength of 585 to 795 MPa (85 to 115 ksi) are easily blanked. Parts up to 13 mm (1/2 in.) thick can be blanked if press capacity is available. Material thicker than 3.18 mm (0.125 in.), especially steel having a carbon content of 0.25% or more, requires an impingement ring on the die so that the corners on the part will not break down. The edges of parts made of 1018 steel work harden as much as 7 to 12 points Rockwell C during blanking. In tests on 0.60% C spring steel with a hardness of 37 to 40 HRC, the surface finish on the sheared edges was 0.8 μm (32 μin.) or better, but punch life was only 6000 pieces.

Blank Design. Limitations on blank size depend on thickness, tensile strength, hardness of work metal, and on available press capacity. For example, perimeters of approximately 635 mm (25 in.) can be blanked in 3.18 mm (0.125 in.) thick low-carbon steel (1008 or 1010). It is possible to blank smaller parts from low-carbon or medium-carbon steel that is about 12.7 mm (1/2 in.) thick.

Sharp corner and fillet radii should be avoided when possible. A radius of 10 to 20% of stock thickness is preferred, particularly on parts over 3.18 mm (0.125 in.) thick or those made of alloy steel. External angles should be at least 90°. The radius should be increased on sharper corners or on hard materials.

Parts with tiny holes or narrow slots to be pierced, or with narrow teeth or projections to be blanked, may not be suited to fine-edge blanking. Hole diameter, slot width, or projection width should be at least 70% of metal thickness for reasonably efficient blanking, although features as small as 50% of stock thickness have been successfully formed.

These limitations have been exceeded. For instance, a 16 mm (5/8 in.) diam hole was pierced in each end of a 1018 steel link 25 mm (1 in.) wide and 7.9 mm (5/16 in.) thick. Because the part had a 13 mm (1/2 in.) radius on each end, the wall thickness was 3.92 mm (3/16 in.). The part was offset 2.54 mm (0.100 in.) in the same die. Holes 3.18 mm (0.125 in.) in diameter were pierced in a part made of 4 mm (0.156 in.) thick aluminum alloy 5052-H34, leaving a wall thickness of 1.0 mm (0.040 in.). A 1.6 mm (0.062 in.) diam hole was pierced in the same part.

The sheared faces of holes pierced during fine-edge blanking are usually vertical, smooth, and free from die break, provided the maximum hole dimensions are not more than a few times the stock thickness. As in conventional piercing, there is a slight radius around the punch side of the hole, but there are no torn edges on the die side of the blank. A rough-sheared surface on the blank may be caused by too great a punch to die clearance, or improper location and height of the impingement ring for the material being blanked. On parts blanked to a small width to thickness ratio, a small, rough surface may be noticeable but may not be detrimental.

Presses. Triple-action hydraulic presses or combination hydraulic and mechanical presses are used for fine-edge blanking. The action is similar to that of a double-action press working against a die cushion. An outer slide holds the stock firmly against the die ring and forces the impingement ring into the metal surrounding the outline of the part. The stock is stripped from the punch during the upstroke of the inner and outer slides. An inner slide carries the blanking punch. A lower slide furnishes the counter action to hold the blank flat and securely against the punch. This slide also ejects the blank.

The stripping and ejection actions are delayed until after the die has opened to at least twice the stock thickness to prevent the blank from being forced into the strip, or slugs from being forced into the blank. Because loads are high and clearance between punch and die is extremely small, the clearance between the gibs and press slides must be so close that they are separated only by an oil film.

Force requirements for fine-edge blanking presses are influenced not only by the work metal and the part dimensions, but also by the special design of the dies and pressure pads used for fine-edge blanking (Fig. 26). Depending on part size and shape, an 890 kN (100 tonf) press can blank stock up to 8 mm (0.315 in.) thick; a 2.2 MN (250 tonf) press can blank stock up to 11.9 mm (0.470 in.) thick; and a 3.6 MN

(400 tonf) press can blank stock up to 12.7 mm (0.500 in.) thick.

Tools. The design of tools for fine-edge blanking is based on the shape of the part, the method of making the die, the required load, and the extremely small punch to die clearance. The considerable loading and required accuracy dictate that the press tools be sturdy and well supported to prevent deflection. The small clearance presupposes precise alignment of the punch and die.

A basic tool consists of three functional components, the die, the punch, and back-pressure components. To produce high-quality blanks, the punch to die clearance must be uniform along the entire profile and must be suitable for the thickness and strength of the work metal. Clearance varies between 0.005 and 0.01 mm (0.0002 and 0.0004 in.).

The components of a typical tooling setup for fine-edge blanking of a part of simple shape are shown in Fig. 25. The profile part of the blanking punch is guided by the pressure pad. A round punch is prevented from rotating by a key fastened to the upper die shoe. The hardened pressure pad is centered by a slightly conical seat in the upper die shoe; this pad contains the V-shaped impingement ring.

Some diemakers put a small radius on the cutting edge of the die. This causes a slight bell-mouth condition that produces a burnishing action as the blank is pushed into the die, improving the edge finish. If holes are to be pierced in the part, the blanking punch will contain the piercing die. The slug is ejected by ejector pins or through holes in the punch.

The die is centered in the lower die shoe by a slightly conical seat, as is the upper pressure pad. Both the die and the upper pressure pad are preloaded to minimize movement caused by compression. The pressure and ejector pad is guided by the die profile and is supported by pressure pins and the lower slide. The backup block for the piercing punch also guides the

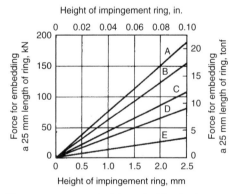

Fig. 26 Force required for embedding impingement rings of various heights into several different work metals. A, stainless steel; B, prehardened alloy steel; C, mild steel, half-hard brass, hard copper, 6xxx aluminum alloys in the H temper; D, soft copper; series 6000 aluminum alloys, half hard; E, commercially pure aluminum, H temper

pressure pins. The die components are mounted in a precision die set with precision guide pins and bushings. Some designers prefer pressing the guide pins into the upper shoe.

Tool Materials. Because of the high loads, close tolerances, and small clearances involved in fine-edge blanking, the die elements are made of high-carbon, high-chromium tool steels, such as AISI D2 or D3, or of A2 tool steel heat treated to about 62 HRC. Punch and die life vary with tool material and hardness, punch-to-die clearance, type of work metal, and workpiece dimensional and surface finish requirements. For most work metals under the usual operating conditions, punch life for fine-edge blanking of 3.2 mm ($1/8$ in.) thick stock is 10,000 to 15,000 blanks between regrinds—assuming that the blanks are of simple shape and that punch wear is such that only 0.05 to 0.13 mm (0.002 to 0.005 in.) of metal must be removed to restore the punch to its original condition. For more information on tool materials and coatings, see the article "Fine-Edge Blanking" in this Volume.

ACKNOWLEDGMENT

This article was adapted from *Handbook of Fabrication Processes,* by O.D. Lascoe, ASM International, 1988.

Blanking of Low-Carbon Steel

BLANKING is the process that uses a die and press to cut or shear a piece of metal from flat or preformed stock. The resulting blank is a piece of metal designed to have a predetermined contour for subsequent forming operations. The blank ordinarily serves as a starting workpiece for forming; less often, it is a desired end product. Making blanks without using a die is referred to as shearing, as discussed in the article "Shearing of Sheet, Strip, and Plate" in this Volume.

The prime objective in planning a blanking-die operation is to design a blank with enough metal to shape the part and, at the same time, prevent excessive scrap loss due to trimming. The types of blanks used in sheet-forming operations fall into four convenient classifications:

- Rectangular blanks
- Rough blanks
- Partially developed blanks
- Fully developed blanks

These types of blanks are described in more detail in the section "Blanking Operations" in the preceding article "Blanking and Piercing" in this Volume. This classification of blanks has permitted the application of some general rules to follow in setting up the blanking operation for a part.

This article discusses the production of blanks from low-carbon steel (such as 1008 and 1010) sheet and strip in dies in a mechanical or hydraulic press. As a point of comparison with higher-strength steel, such as high-carbon steel, the most important difference between the blanking and piercing of high-carbon and of low-carbon steel is that greater clearance between punch and die is required for high-carbon steels. Table 1 gives clearances needed for producing different edge types (Fig. 1) for blanking and piercing high-carbon and low-carbon steels. As shown in Table 1 and in the accompanying Fig. 1, at equal clearance, rollover depth will be smaller and burnish depth greater for high-carbon than for low-carbon steel. For example, 12% clearance per side will produce a type 4 edge on high-carbon steel, and a type 2 edge on low-carbon steel. More information on the forming operations for high-carbon steels and high-strength sheet steels is in the respective articles "Forming of Carbon Steels" and "Forming of Advanced High-Strength Steels" in this Volume.

Methods of Blanking in Presses

Cutting operations that are done by dies in presses to produce blanks include cutoff, parting, blanking, notching, and lancing. The first three of these operations can produce a complete blank in a single press stroke. In progressive dies, two or more of these five operations are done in sequence to develop the complete outline of the blank and to separate it from the sheet, strip, or coil stock.

Trimming is defined as the cutting off of excess material from the periphery of a workpiece. It is usually done in dies and is similar to blanking. It is often the final operation on a formed or drawn part. The quality of blanked edges also can be improved by shaving, as discussed later in this article. Fine-edge blanking is treated in the article "Fine-Edge Blanking" in this Volume.

The applications of these methods are described in examples throughout this article. Other examples of these methods of producing blanks are provided in the articles "Piercing of Low-Carbon Steel," "Blanking and Piercing of Electrical Steel Sheet," "Forming of Carbon Steels," and "Deep Drawing" in this Volume.

Cutoff. This operation consists of cutting along a line to produce blanks without generating any scrap in the cutting operation, most of the part outline having been developed by notching or lancing in preceding stations. The cutoff line can take almost any shape—straight, broken, or curved. After being cut off, the blanks fall onto a conveyor or into a chute or container.

A cutoff die can be used to cut the entire outline of blanks whose shape permits nesting in a layout that uses all of the material (except possibly at the ends of the strip), as shown in Fig. 2. Alternating positions can sometimes be used in nesting (middle, Fig. 2) to avoid producing scrap except at strip ends. Cutoff is also used to cut blanks from strip that has already been notched to separate the blanks along part of their periphery, as described in the following example.

Example 1: Use of Cutoff to Separate Blanks Partly Outlined by Notching. Figure 3 shows the layout for the cutoff of blanks for automobile body-bolt brackets. The brackets were completed by piercing and forming.

The coiled strip of galvanized hot rolled 1006 steel was 1.90 mm (0.075 in.) thick by 495 mm (19.5 in.) wide. This was wide enough for two blanks, as shown in Fig. 3. The blanks were alternated in the layout to facilitate trimming and piercing.

The strip was notched and seminotched in the first stations of a progressive die (making a small amount of scrap). In the following stations, straight cutoff punches made four blanks at each stroke without producing any additional scrap.

The work was done in a 3.6 MN (400 tonf) single-action mechanical press with an air cushion. The press was equipped with a double

Table 1 Punch-to-die clearances for piercing or blanking various metals to produce the five types of edges shown in Fig. 1

For clearances that produce type 1, 2, and 3 edges, it is ordinarily necessary to use ejector punches or other devices to prevent the slug from adhering to the punch.

Work metal	Clearance per side, % of stock thickness				
	Type 1(a)	Type 2	Type 3	Type 4	Type 5
Low-carbon steel	21	11.5–12.5	8–10	5–7	1–2
High-carbon steel	25	17–19	14–16	11–13	2.5–5
Stainless steel	23	12.5–13.5	9–11	3–5	1–2
Aluminum alloys					
Up to 230 MPa (33 ksi) tensile strength	17	8–10	6–8	2–4	0.5–1
Over 230 MPa (33 ksi) tensile strength	20	12.5–14	9–10	5–6	0.5–1
Brass, annealed	21	8–10	6–8	2–3	0.5–1
Brass, half hard	24	9–11	6–8	3–5	0.5–1.5
Phosphor bronze	25	12.5–13.5	10–12	3.5–5	1.5–2.5
Copper, annealed	25	8–9	5–7	2–4	0.5–1
Copper, half hard	25	9–11	6–8	3–5	1–2
Lead	22	8–10	6.5–7.5	4–6	1.5–2.5
Magnesium alloys	16	5–7	3.5–4.5	1.5–2.5	0.5–1

(a) Maximum. Source: Ref 1

roll feed and made 50 strokes (200 blanks) per minute.

Advantages of cutoff in making blanks include:

- The die has few components and is relatively inexpensive.
- Waste of material in blanking is minimized or eliminated.
- The die can be resharpened easily, and maintenance costs are low.

Disadvantages of cutoff include:

- It can be used only to make blanks that nest in the layout without waste.
- Cutting of one edge causes one-way deflection and stress.
- Accuracy may be affected adversely by the method of feeding.

Parting (Fig. 4) is the separation of blanks by cutting away a strip of material between them. Like cutoff, it can be done after most of the part outline has been developed by notching or lancing. It is used to make blanks that do not have mating adjacent surfaces for cutoff (Fig. 4) or to make blanks that must be spaced for ease of handling in order to avoid distortion or to allow room for sturdy tools. Some scrap is produced in making blanks by parting; therefore, this method is less efficient than cutoff in terms of material use.

Blanking (also called punching) is the cutting of the complete outline of a workpiece in a single press stroke. Because a scrap skeleton is usually produced, blanking involves some material waste. However, blanking is usually the fastest and most economical way to make flat parts, particularly in large quantities.

The skeleton left by blanking sometimes has only scrap-metal value, but many shops have organized programs to maximize the use of cutouts and sizable scrap skeletons in making other production parts. Material waste is completely avoided by using the scrap skeleton that remains from certain blanking operations to provide perforated stock for such items as air filters for forced-air furnaces.

Piercing (with a flat-end punch), also called punching or perforating, is similar to blanking except that the punched-out (blanked) slug is the waste and the surrounding metal is the workpiece. Piercing is discussed in the article "Piercing of Low-Carbon Steel" in this Volume.

Notching is an operation in which the individual punch removes a piece of metal from the edge of the blank or strip (Fig. 5).

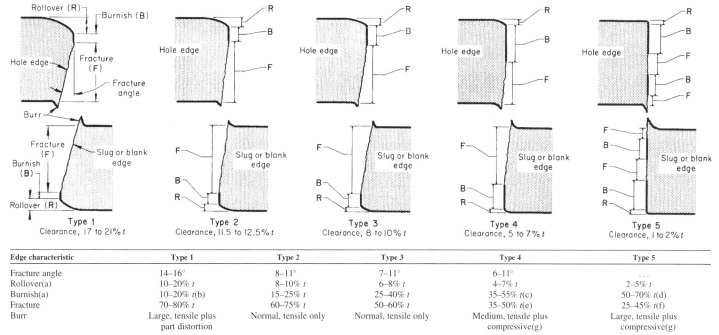

Edge characteristic	Type 1	Type 2	Type 3	Type 4	Type 5
Fracture angle	14–16°	8–11°	7–11°	6–11°	...
Rollover(a)	10–20% t	8–10% t	6–8% t	4–7% t	2–5% t
Burnish(a)	10–20% t(b)	15–25% t	25–40% t	35–55% t(c)	50–70% t(d)
Fracture	70–80% t	60–75% t	50–60% t	35–50% t(e)	25–45% t(f)
Burr	Large, tensile plus part distortion	Normal, tensile only	Normal, tensile only	Medium, tensile plus compressive(g)	Large, tensile plus compressive(g)

(a) Rollover plus burnish approximately equals punch penetration before fracture. (b) Burnish on edge of slug or blank may be small and irregular or even absent. (c) With spotty secondary shear. (d) In two separate portions, alternating with fracture. (e) With rough surface. (f) In two separate portions, alternating with burnish. (g) Amount of compressive burr depends on die sharpness.

Fig. 1 Effect of punch-to-die clearance per side (as a percentage of stock thickness, *t*) on characteristics of edges of holes and slugs (or blanks) produced by piercing or blanking low-carbon steel sheet or strip at a maximum hardness of 75 HRB. Table 1 lists clearances for producing the five types of edges in various metals. See text for additional discussion and for applicability of the five types of edges.

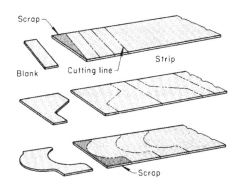

Fig. 2 Nested layouts for making blanks by cutoff

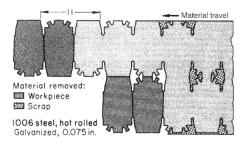

Fig. 3 Layout for the cutoff of four blanks at each press stroke from notched and seminotched strip. Dimensions given in inches

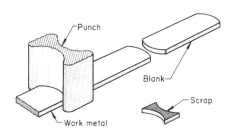

Fig. 4 Use of a parting punch to make blanks not having mating adjacent surfaces

Notching is done for such reasons as the following:

- To free some metal for drawing (Fig. 5a) and for forming (Fig. 5b) while the workpiece remains attached to the strip
- To remove excess metal before forming (Fig. 5c)
- To cut part of the outline of a blank that would be difficult to cut otherwise (Fig. 3 and 29)

The piercing of holes of any shape in a strip to free metal for subsequent forming or to produce surfaces that later coincide with the outline of a blanked part is sometimes called seminotching. The pierced area may outline a portion of one part or of two or more adjacent parts in a strip. Figure 3 illustrates a progressive-die layout incorporating seminotching.

Lancing is a press operation in which a single-line cut or slit is made part way across the strip stock without removing any metal. Lancing is usually done to free metal for forming (Fig. 6). The cut does not have a closed contour and does not release a blank or a piece of scrap. In addition to its use in freeing metal for subsequent forming, lancing is also used to cut partial contours for blanked parts, particularly in progressive dies.

Trimming is an operation for removing excess metal (such as deformed and uneven metal on drawn or formed parts) and metal that was used in a previous operation (such as a blankholding flange for a draw operation). Trimming is done in several ways, depending on the shape of the workpiece, the accuracy required, and the production quantity.

Figure 7 illustrates the tooling for trimming a horizontal flange on a drawn shell in a separate operation. The drawn shell is set on a locating plug for trimming. After scrap from a sufficient number of trimmed shells has accumulated, the

piece of scrap at the bottom is severed at each stroke of the press by the scrap cutters shown in Fig. 7, and falls clear. Except that the die must be constructed to accept and locate the drawn shell, the operation is identical to the blanking of a flat workpiece and produces square edges of the same accuracy and quality. A drawn shell or formed part can be trimmed in a press without leaving a flange on the completed part by using one of three methods: pinch trim, shimmy trim, or trim and wipe-down.

Pinch trimming, shown as a separate operation in a push-through die in Fig. 8, is done only on a part that has at least a narrow flange as-formed. The shell must be free from wrinkles at or near the trimming line. The trimmed edge is not square with the sidewall but has the general shape shown in the lower right corner of Fig. 8. The accuracy of height resulting from pinch trimming is affected by variations in wall thickness and flange radius. To ensure an even pinch-off and to avoid sharp or rough edges, clearance between punch and die must be held to a minimum, and the punch must be kept sharp.

Pinch trimming is also done without a blankholder or holddown, using a die otherwise similar to that shown in Fig. 8. The scrap rings can be blown off the die at each stroke. In another method, the scrap rings climb the punch until they are severed by being compressed against a scrap cutter, after which they are spread apart and allowed to run out along a track for disposal. Pinch trimming without a blankholder is particularly well suited to the high-volume production of eyelets and other small parts.

Pinch trimming is primarily a mass-production method. The production rate is high because only one stroke of the press is required to

complete the trim. The method is often combined with drawing in a compound draw-and-trim die to reduce production costs further. The disadvantages of pinch trimming are excessive burrs, sharp cut edge, and high die maintenance.

Shimmy Trimming. In trimming with a shimmy die (also known as Brehm or model trimming), the drawn shell is held in a close-fitting die of the exact shell height and trimmed in segments by the successive horizontal oscillations of an internal cam-actuated punch toward the outside of the shell. The resulting trimmed edge is square and closely resembles the conventional blanked edge on a flat part. Shell height is more accurate than with pinch trimming. In addition to its application to shells that must have square, accurate edges, shimmy trimming is used on shells that have a wrinkled or otherwise nonuniform top edge as-drawn (cutoff is done below the defects) and on shells that cannot be produced economically with even the narrow flange needed for pinch trimming.

Tooling costs for shimmy trimming are much higher than those for pinch trimming. Shimmy trimming is also slower because it requires four or more oscillations of the punch in one press stroke and cannot be combined with other operations in a compound die. Shimmy dies are inexpensive to maintain because they remain in alignment and therefore are not likely to wear by shearing or chipping.

Trim and Wipe-Down. In this type of trimming, a flange is cut to width using a die such as that shown in Fig. 7 and then wiped or straightened into line with the sidewall of the

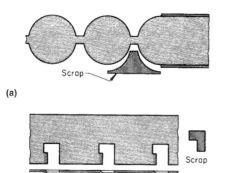

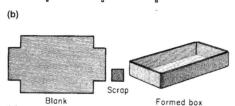

Fig. 5 (a) Notched work illustrating the use of notching for freeing metal before drawing, (b) before forming, and (c) for removing excess metal before forming

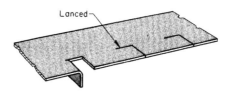

Fig. 6 Strip that was lanced to free metal for forming

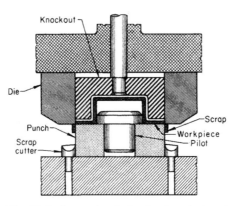

Fig. 7 Single-operation die for trimming a horizontal flange on a drawn shell

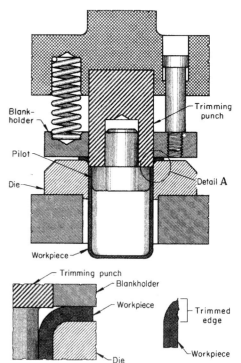

Fig. 8 Pinch trimming a drawn shell in a push-through die

shell or formed part. Because of narrow flange width, trimming and wiping down may be two operations.

The edge is square with the sidewall, but the shell height may be slightly irregular because of the forming characteristics of the metal. In addition, a ring may be visible at the original location of the flange radius.

Trimming, other than shimmy trimming, is frequently combined with one or more other operations in a compound die. Trim stock is often left on a drawn or formed workpiece so that it can be trimmed to size in a second operation. This is done to obtain the most accurate relationship of some other feature, such as a pierced hole, to the trimmed outline of the workpiece.

Characteristics of Blanked Edges

The sheared edges of a blank produced in a conventional die are not smooth and vertical for the entire thickness of the part but exhibit the characteristics represented on an exaggerated scale in Fig. 9. The blank is shown in the position in which it would be cut from the work metal by the downward motion of the punch. A portion of the stock remaining after removal of the blank is shown at the top of the illustration.

Rollover on the lower edges of the blank develops by plastic deformation of the work metal as it is forced into the die by the punch. Compression of the metal above the rollover zone against the walls of the die opening burnishes a portion of the edge of the blank, as shown in Fig. 9. As the punch completes its stroke, the remaining portion of the blank edge is broken away or fractured (resulting in die break), and a tensile burr is formed along the top of the blank edge.

The angle of the fractured portion of the edge is identified in Fig. 9 as the breakout angle. The breakout dimension of the blank and the burnish dimension of the hole in the scrap skeleton are approximately equal to the corresponding punch

dimension, and the burnish dimension of the blank is very close to the corresponding die dimension. Therefore, the punch determines the hole size, and the die governs the blank size.

Penetration depth, or the amount of penetration of the punch into the work metal before fracture occurs, is shown on the edge of the remaining stock or scrap skeleton in Fig. 9. This depth is approximately equal to the sum of the rollover depth and the burnish depth on the blank, except when low die clearance produces secondary burnish. It is usually expressed as a percentage of the work metal thickness.

The percentage of penetration (before fracture) depends on the properties of the work metal, as shown in Table 2, which gives approximate values for various steels and nonferrous metals under typical blanking conditions. The percentage of penetration affects energy consumption and cutting force in blanking, as described in the section "Calculation of Force Requirements" in this article.

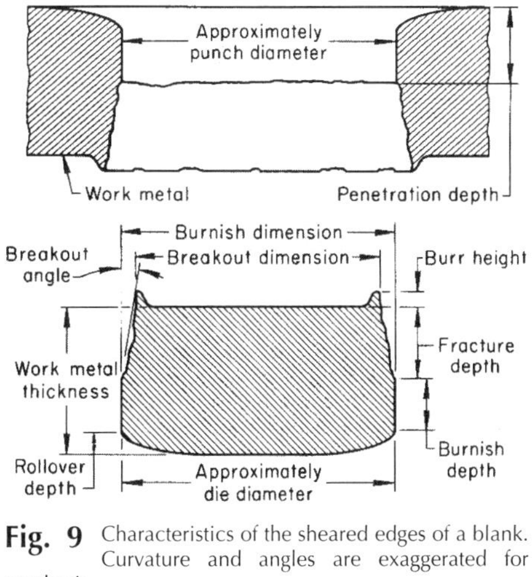

Die Clearance

The terms *clearance, die clearance, and punch-to-die clearance* are used synonymously to refer to the space between punch and die. Clearance is important for the reliable operation of the blanking equipment, the quality and type of cut edges, and the life of the punch and die. In general, the effects of clearance on these factors in blanking are the same as in piercing and are discussed in the article "Piercing of Low-Carbon Steel" in this Volume. The same article also describes the edge characteristics of slugs produced in piercing holes (see Fig. 2 of that article). The data in that illustration can serve as a guide for selecting clearances for blanking. All clearance values given in this article are per side, except where indicated.

Optimal blanking clearance may sometimes be less than optimal piercing clearance. This is partly because the blanked edge is generally

close to the stock edge, and material expansion is therefore less restricted. A piercing tool must move a great deal of material away from its cutting edge, and for longest life, the clearance should be selected to eliminate as much compressive loading on the work metal as possible.

A part blanked using clearance much greater than normal may exhibit double shear, which is ordinarily evident only with extremely small clearance (see edge types 4 and 5 in Fig. 2 in the article "Piercing of Low-Carbon Steel" in this Volume). In addition, a part blanked using large clearances will be smaller than the die opening (except for a deeply dished blank), and it is difficult to correct the tooling to compensate for this. In some applications, retaining the blank becomes almost as great a problem as expelling the slugs into a die cavity after piercing, because of the increased clearance.

Relief in a blanking die (Fig. 10) is the taper provided so that the severed blank can fall free. The relief angle may range from $1/2$ to $2°$ from the vertical wall of the die opening. Relief in a die is sometimes called draft or angular clearance. In some dies, the relief may start at the top of the die surface and have a taper of only 0.002 in./in. (0.002 mm/mm) per side. In other dies, there is a straight, vertical wear land between the top of the die and the relief.

Calculation of Force Requirements

Calculation of the forces and the work involved in blanking gives average figures that are applicable only when the correct shear strength for the material is used and when the die is sharp and the punch is in good condition, has correct clearance, and is functioning properly. The total load on the press, or the press capacity required to do a particular job, is the sum of the cutting force and other forces acting at the same time, such as the blankholding force exerted by a die cushion.

Cutting Force: Square-End Punches and Dies. When punch and die surfaces are flat and at right angles to the motion of the punch, the cutting force can be found by multiplying the area of the cut section by the shear strength of the work material:

$$L = S_s \, tl \qquad \text{(Eq 1)}$$

where L is the load on the press (in pounds) (cutting force), S_s is the shear strength of the stock (in pounds per square inch), t is the

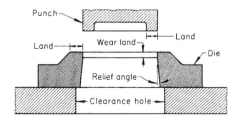

Fig. 9 Characteristics of the sheared edges of a blank. Curvature and angles are exaggerated for emphasis.

Table 2 Approximate penetration of sheet thickness before fracture in blanking

Work metal	Penetration, %
Steels	
Carbon steels(a)	
0.10% C, Ann	50
0.10% C, CR	38
0.20% C, Ann	40
0.20% C, CR	28
0.30% C, Ann	33
0.30% C, CR	22
Silicon steels	30
Nonferrous metals	
Aluminum alloys	60
Brass	50
Bronze	25
Copper	55
Nickel alloys	55
Zinc alloys	50

(a) Ann, annealed; CR, cold rolled

Fig. 10 Relief in a blanking die

stock thickness (in inches), and l is the length or perimeter of the cut (in inches). The shear strengths of various steels and nonferrous metals are given in Table 3.

Cutting Force: Dies with Shear. For cutting large blanks, shear can be applied to the face of the die by grinding it at an angle to the motion of the punch (Fig. 11), but shear is not used in cutting small blanks. Shear reduces shock in the press, as well as blanking noise and blanking force, but the same amount of work is done as with a flat die surface.

The most common type of shear used on the die is convex (Fig. 11a). The apex of the die face is slightly rounded to avoid initiating a crack in the work metal. Concave shear (Fig. 11b) is somewhat more difficult to grind on the die but holds the work metal more securely during blanking. A radius approximately equal to that of the grinding wheel is produced where the ground surfaces meet.

A third type of shear, sometimes used on a die for a large blank, consists of a wavy or scalloped surface around the die opening. This technique uses several convex and concave shear surfaces around the die opening. The punch load is distributed over the entire die surface, thus minimizing punch shift.

The amount of shear in a die can be less than or greater than stock thickness. Shear that is equal in depth or greater than the stock thickness is called full shear.

The cutting force for a die with shear can be calculated after first finding the work done (energy used) in blanking. The work done in blanking equals the force required in blanking (load on the press) multiplied by the distance that the force acted:

$$W = Ls \qquad \text{(Eq 2)}$$

where W is the work done in blanking (in inch-pounds), L is the load (in pounds), and s is the distance the load acts (thickness multiplied by percentage of penetration before fracture) (in inches). To obtain accurate work values,

the percentage penetration must be known accurately.

Cutting or blanking force is reduced by the use of angular shear in the die; the amount of reduction in force depends on the depth of the angular shear. The reduced average cutting force on the press is:

$$L_{sh} = \frac{W}{s + s_1} \qquad \text{(Eq 3)}$$

where L_{sh} is the average cutting force (in pounds), with angular shear, W is the work done in blanking (in inch-pounds), s is the distance (in inches) that the load acts (thickness multiplied by percentage of penetration before fracture), and s_1 is the depth of angular shear (in inches).

In simplified practice, some plants ignore partial shear in calculating cutting force for blanking. When full shear is used, force is calculated as without shear and then reduced by 30%.

Stripping force is the force needed (when drop-through is not used) to free the blank from the die or the strip from the punch when they stick or jam because of springback. Stripping force can be calculated using:

$$L_{st} = kA \qquad \text{(Eq 4)}$$

where L_{st} is the stripping force (in pounds), k is a stripping constant (in pounds per square inch), and A is the area of the cut surface (in square inches) (stock thickness t multiplied by length or perimeter of cut l). Approximate values for the constant k (as determined by experiment for low-carbon steel) are:

- 1500 for sheet metal thinner than 1.57 mm (0.062 in.) when the cut is near an edge or near a preceding cut
- 2100 for other cuts in sheet thinner than 1.57 mm (0.062 in.)
- 3000 for sheet more than 1.57 mm (0.062 in.) thick

Factors That Affect Processing

Factors that affect the processing of blanks include:

- Size and shape of the blank
- Material for blanking

- Form in which the material is supplied
- Thickness of the blank
- Production quantity and schedule
- Quality specifications
- Availability of equipment and tools
- Number and type of subsequent operations required for completing the work

The size and shape of the blank affect the form and handling of the material blanked, the blanking method, and the handling of the completed blank. The thickness of the blank affects the press load required (see the section "Calculation of Force Requirements" in this article), the selection of equipment, and the choice of blanking and handling methods (see the section "Effect of Work Metal Thickness" in this article).

Production quantity and schedule determine the choice of equipment. A total production of fewer than 10,000 pieces is considered a short run; 10,000 to 100,000 pieces, a medium run; and more than 100,000 pieces, a long run.

Quality specifications and tolerances for thickness, camber, width, length, flatness, and finish affect the handling of the material. The availability of single-, double-, or triple-action presses (rated at various force capacities, sizes, speeds, lengths of stroke, strokes per minute, and shut heights) affects the selection of the processing method. The availability and capacity of auxiliary press equipment can have an effect on the selection of a tooling system and on whether a part can be made in-plant.

Operations that follow blanking also affect the choice of equipment, the processing method, and the handling procedures. Such subsequent operations may include piercing, bending, forming, deep drawing, machining, grinding, or finishing. Only rarely is the blank a final product.

Selection of Work Metal Form

Work metal for blanking in presses is usually in the form of flat sheets, strip, or coil stock. Less frequently, steel plate is blanked in presses (see the section "Effect of Work Metal Thickness" in this article). In some applications, the metal is preformed before blanking.

Special preparation of the work metal is usually not required for the blanking operation itself. However, annealing, leveling, or cleaning is often needed because of subsequent forming operations on the blank, as discussed in the article "Forming of Carbon Steels" in this Volume.

Sheet or Strip. Flat sheet is usually the work metal for large blanks, such as automobile roofs. Square-sheared sheet can be used as a blank, or it can be blanked in a die. Small quantities of blanks, regardless of size, are usually made from straight lengths of sheet or strip.

Coil stock is used for mass production, whenever possible. In continuous production, the use of coil stock can save as much as one-third of the time needed for producing an equal quantity

Table 3 Shear strengths of various steels and nonferrous metals at room temperature

Metal	Shear strength	
	MPa	ksi
Steels		
Carbon steels		
0.10% C	241–296	35–43
0.20% C	303–379	44–55
0.30% C	358–462	52–67
High-strength low-alloy steels	310–439	45–63.7
Silicon steels	414–483	60–70
Stainless steels	393–827	57–120
Nonferrous metals		
Aluminum alloys	48–317	7–46
Copper and bronze	152–483	22–70
Lead alloys	13–40	1.83–5.87
Magnesium alloys	117–200	17–29
Nickel alloys	242–800	35–116
Tin alloys	20–77	2.90–11.1
Titanium alloys	414–483	60–70
Zinc alloys	97–262	14–38

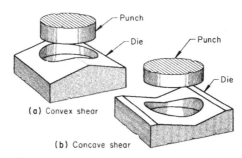

Fig. 11 Convex shear and concave shear on blanking dies. Angle and depth of shear are exaggerated for emphasis. Normally, depth of shear does not greatly exceed stock thickness.

from flat stock. In addition, there are fewer scrap ends when coil stock is used.

Sheet metal is the least expensive when it is supplied in large coils from the mill. For most applications, the coil must be slit to the proper width for blanking, and some edge material must be trimmed off. Parts can sometimes be made most economically in a progressive die by using coil stock that is the width of the developed blank.

Blank Layout

In the medium and high production of medium-sized blanks, the cost of material is 50 to 75% of the total cost of the blank; for large blanks, it may be more than 95% of the total cost of the blank. Substantial savings in net material cost can often be achieved by coordinating blank layout with the selection of stock form and width to minimize the amount of scrap produced.

Several trial layouts may be needed to find the width of stock and the layout that use the material most efficiently while taking into account the possible effects of orientation of parts on subsequent operations. The layout must include the minimum workable scrap allowance between blanks, providing just enough material to support or hold down the strip during blanking. Scrap allowances, based on the use of well-maintained equipment and good shop practice, are given in Table 4.

The percentage of scrap in a strip layout can be calculated as:

$$100\left(1 - \frac{A_B}{A_S}\right) \qquad \text{(Eq 5)}$$

where A_B is the area of blanks produced in one press stroke, and A_S is the area of strip consumed by one press stroke, or strip width times feed length.

Round blanks can be staggered in rows, at the same spacing as for hexagons (Fig. 12), for the most efficient blanking from a long strip. With such a layout, 20 to 40% more blanks can be made from a given amount of material than by blanking each circle from a separate square. With the layout shown in Fig. 12, each press stroke (after the third stroke) produces four blanks—spaced to provide enough room for mounting the punches and dies.

The percentage of scrap loss for the layout of Fig. 12 can be calculated using Eq 5 and the following:

$$A_B = \frac{n\pi D^2}{4}$$

$$A_S = wl$$

$$l = D + s$$

$$w = (d + 2s) + (n - 1)(D + s)\cos\alpha \qquad \text{(Eq 6)}$$

where n is the number of rows of blanks across the strip width, D is the blank diameter (in inches), l is the feed length (one blank made in each row per press stroke) (in inches), w is the strip width (in inches), s is the scrap allowance from the edge of the strip to the blank and between blanks (in inches), and α is the angular displacement between blanks (in degrees). The area of the holes pierced in a blank is not considered in these calculations, because it does not affect the efficiency of the layout.

Rectangular blanks can generally be laid out more easily than other shapes.

Odd-shaped blanks are generally more difficult to lay out for the greatest economy.

Nesting, or the interlocking of blanks in the layout to save material, should be done wherever the shape of the blank permits. Nesting is possible with many irregular blanks.

Figure 13 shows a layout in which irregular blanks are nested so that an appreciable amount of material is saved. A double die can be used with such a layout, blanking two pieces per

stroke. A single die can be used for short runs; the strip is turned around after the first pass and fed through the die again.

Another way of nesting blanks is a layout such as that illustrated in Fig. 14. With this layout, three punches cut four pieces per stroke (after the third press stroke) in a shearing action that produces no scrap except at the ends of the strip. Other strip layouts in which the blanks have been nested are shown in the article "Forming of Carbon Steels" in this Volume.

Use of Full Stock Width. Blanks that have two parallel sides can sometimes be made most economically in a layout that uses the full width of the stock. The remaining outline of the blank is produced by shearing, lancing, notching, parting, or a combination of these.

Effect of Rolling Direction. For blanks that must later be bent or formed, consideration must be given in layout to the orientation of the blanks with respect to the direction of rolling (grain direction). Ideally, blanks should be laid out so that severe bends are made with the bend axis at

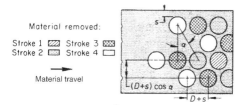

Fig. 12 Strip layout for blanking four circles per stroke with minimal material waste

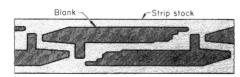

Fig. 13 Nesting of irregular blanks in layout to save material

Table 4 Scrap allowance for blanking

	Scrap allowance when length of skeleton segment between blanks or along edge is:											
	2t or less						Greater than 2t					
	Thickness of stock, t		Edge of stock to blank		Between blanks in row		Thickness of stock, t		Edge of stock to blank		Between blanks in row	
Work metal	mm	in.	mm	in.	mm	in.	mm	in.	mm	in.	mm	in.
Metals in general	Up to 0.53	Up to 0.021	1.27	0.050	1.27	0.050	Up to 1.12	Up to 0.044	1.27	0.050	1.27	0.050
Standard strip stock	0.56–1.40	0.022–0.055	1.02	0.040	1.02	0.040	Over 1.12	Over 0.044	0.9t	0.9t	0.9t	0.9t
	Over 1.40	Over 0.055	0.7t	0.7t	0.7t	0.7t						
Extrawide stock and	Up to 1.07	Up to 0.042	1.52	0.060	1.27	0.050	Up to 0.84	Up to 0.033	1.52	0.060	1.27	0.050
weak scrap skeleton	Over 1.07	Over 0.042	1.4t	1.4t	1.2t	1.2t	Over 0.84	Over 0.033	1.8t	1.8t	1.6t	1.6t
Stock run through twice	Up to 1.07	Up to 0.042	1.52	0.060	1.27	0.050(a)	Up to 0.84	Up to 0.033	1.52	0.060	1.27	0.050(a)
	1.09–1.40	0.043–0.055	1.4t	1.4t	1.02	0.040	0.86–1.12	0.034–0.044	1.8t	1.8t	1.02	0.040
	Over 1.40	Over 0.055	1.4t	1.4t	0.7t	0.7t	Over 1.12	Over 0.044	1.8t	1.8t	0.9t	0.9t
Stock run through twice;	Up to 1.07	Up to 0.042	1.52	0.060	1.27	0.050(b)	Up to 0.84	Up to 0.033	1.52	0.060	1.27	0.050(b)
blanks in rows 1 and	Over 1.40	Over 0.055	1.4t	1.4t	1.4t	1.4t	Over 0.84	Over 0.033	1.8t	1.8t	1.8t	1.8t
2 interlock												
Stainless, silicon and	Up to 1.40	Up to 0.055	1.52 min	0.060 min	1.52 min	0.060 min	Up to 0.84	Up to 0.033	1.52 min	0.060 min	1.52 min	0.060 min
spring steels	Over 1.40	Over 0.055	1.4t	1.4t	1.4t	1.4t	Over 0.84	Over 0.033	1.8t	1.8t	1.8t	1.8t
Nickel-base magnetically	All	All	1.52	0.060	1.52	0.060	All	All	1t	1t(c)	1t	1t(c)
soft alloys												

(a) Allowance between blanks in the same row and also between blanks of the first and second rows. (b) Allow 1.52 mm (0.060 in.) between blanks at first and second rows. (c) When the blank edge is parallel to the edge of the stock or when the length of the skeleton segment between blanks is more than 4t, scrap allowance is 1.8t.

right angles to the direction of rolling or, if this is not practical, with the bend axis at an angle to the direction of rolling. Stretching should be in the direction of rolling, whenever possible. Examples and illustrations of blank layouts are provided in the article "Forming of Carbon Steels" in this Volume.

Welded Blanks

Tailor welded blanks are widely used in automobile manufacturing (see the article "Forming of Steel Tailor-Welded Blanks" in this Volume). Welded blanks also are used when they have advantages over one-piece blanks, as in the following situations:

- The welded blank may cost less than an equivalent one-piece blank if scrap or other low-cost metal can be used to make the welded blank or if tooling and production for the one-piece blank cost more than for the welded blank.
- Stock for a welded blank may be more readily available than stock for an equivalent one-piece blank.
- The blank may have a shape that would waste more material if it were made in one piece instead of being welded. Material can sometimes be saved by welding projecting portions, such as tabs and ears, to simpler shapes.
- The welded blank, when used in subsequent forming operations, may reduce the cost of tooling. Flat or simple shapes are welded in a layout designed to avoid the presence of seams in certain portions of the blank and to permit automatic welding, if possible.

Large blanks that would cost extra because of width or for other reasons if they were made in one piece can sometimes be made at less cost by welding.

Difficult shapes that would waste a considerable amount of material if they were made in one piece can sometimes be made by welding two or more simple blanks together. In the application described in the following example, two developed blanks were welded and then formed into a bent channel.

Example 2: Use of Two Gas Metal Arc Welded Blanks to Form an Automobile-Frame Rail. Figure 15 shows a side rail for an automobile frame that was formed from 5.05 mm (0.199 in.) thick hot rolled commercial-quality 1008 steel. The blank for this rail had a shape that could be made in one piece only with excessive waste of metal in scrap. The two pieces used were blanked from nested layouts with little scrap waste.

Blanking was done in a die that made both pieces in a single press stroke. The two portions were butted and joined by high-speed automatic gas metal arc welding and then formed. Tolerance on all dimensions shown in Fig. 15 was ± 0.51 mm (± 0.020 in.).

The savings in metal exceeded the cost of welding, but if production needs had been fewer than 10,000 pieces, the savings in metal would not have paid for the welding and the blanking dies. The production quantity was 300,000 rails made in lot sizes of 20,000 pieces.

Open Shapes. A blank with a large cutout can sometimes be made at less cost than a one-piece blank by welding simpler pieces together.

Waste metal can sometimes be joined by welding to make a blank that costs less than a one-piece blank.

Welding methods used in making blanks include resistance welding methods (lap-seam, spot, foil butt-seam, mash-seam, flash, and high-frequency butt) and fusion welding.

Lap-seam welding (wheel electrodes) is a frequently used method. Tooling is simple if the components are joined first by tack welding. The disadvantages of lap-seam welding are loose edges, and joints that are double the thickness of the work metal.

Spot welding is fast and needs only simple, inexpensive tooling. The disadvantages of spot welding are loose edges, and joints that are not tight and are not as strong as those made by other methods.

Foil butt-seam welding is a fast method that makes smooth, tight joints with no loose edges. Its disadvantages include the cost of adding foil to one or both sides of the seam and the need to use starting tabs to make strong joints.

Mash-seam welding makes smooth joints that often need no grinding and tight joints that have no loose edges. Disadvantages include the short life of electrodes and the high cost of tooling resulting from the difficulty in maintaining the small overlap.

Flash welding uses simple tooling; the joints are tight and free from loose edges. However, the length of joint that can be produced by flash welding is limited, and the joints are rough and therefore must be ground before the weldment can be worked in a die.

High-frequency butt welding is fast and can be used to join two dissimilar metals. Electrode life is good. However, costly equipment is required, and the technique is generally suitable only for mass production.

Fusion welding methods include gas metal arc welding and gas tungsten arc welding.

Presses

Most blanking is done in single-action mechanical presses. Some dies can be used only with a particular type of press; the die is usually made to suit a specific press. The force capacity rating must be adequate for the work and must be well above the calculated cutting force (see the section "Calculation of Force Requirements" in this article). Press capacities are given in kilonewtons (tons of force) at a certain distance above the bottom of the stroke. This distance must suit the die and the operation. Most blanking is done near the bottom of the stroke where the available force is greatest. In compound dies, blanking may be done near midstroke, where the available force is much lower than that at the bottom.

Size of bed, shut height, stroke length, and speed must all be suitable for the die and the work. Some types of dies can be run at high speeds, and some need moderate or slow speeds, as discussed in the following section in this article.

Construction and Use of Short-Run Dies

Small and medium quantities of blanks are often produced in punch presses by the use of inexpensive short-run dies. These include steel-rule dies, template dies (sometimes called plate dies or continental dies), and subpress dies. Although most applications of such dies are for production quantities of a few hundred to 10,000 pieces, suitably constructed dies of these types have been used for quantities of 100,000 pieces or more.

Short-run dies are used to a limited extent to blank initial quantities of parts that are to be mass produced. Because they can be made and put into operation more quickly than conventional dies, short-run dies make it possible to expedite the delivery of completed parts.

In addition, short-run dies are used to produce trial lots of parts that may be subject to extensive changes in design. If the trial lots show that die design changes are needed, the changes can be

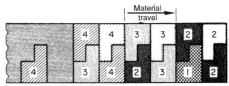

Numbers show order of press strokes.

Fig. 14 Layout for the scrapless blanking of four blanks per press stroke, using three punches. The three shaded shapes for each press stroke denote the blanks produced directly by the three punches. The unshaded shape for each press stroke denotes a blank produced by the action of adjacent punches. Numbers show order of press strokes

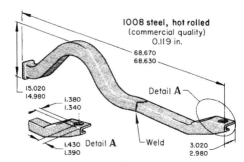

Fig. 15 Automobile-frame side rail that was formed from two blanks joined by gas metal arc welding. Dimensions given in inches

made at less cost before the conventional die is completed. After the conventional die has been set up, either the entire production can be transferred to it or both dies can be used.

For small quantities (<100 pieces), even the most inexpensive short-run blanking die may not be justified. Such small quantities of blanks are generally cut at less cost with standard tools, such as a nibbler, a squaring shear, or a rotary shear. Small quantities of blanks can also be made by contour band sawing, routing, gas cutting, filing, or machining.

Steel-rule dies are simple, inexpensive dies that are made by setting thin, bevel-edged strips of high-carbon tool steel on edge to outline the blank. The rule is set tightly into a slot in plain or impregnated plywood, and the plywood is backed by a steel subplate, as shown in Fig. 16. The die plate or template is attached directly to a steel subplate, and both upper and lower subplates are fastened to master die shoes, which are mounted in a conventional press.

Steel-rule dies are used for blanking, piercing, notching, and shallow forming. For work on flat blanks or on flat, sheared stock of low-carbon steel up to approximately 13 mm ($^1/_2$ in.) thick, a steel-rule die can usually be made more rapidly and at less cost than any other kind of die. Blanks as large as 1.2 by 2.1 m (4 by 7 ft) have been made in steel-rule dies.

Die. The steel rule that is used as the die is made of high-carbon steel or of a tool steel such as W1 or W2, in spring temper and other hardnesses. It is available in stock lengths in several thicknesses, ranging from 0.36 to 4.32 mm (0.014 to 0.170 in.), and in widths of 31.8 mm (1.25 in.) and narrower. Printers' rules of thickness from 1 to 12 points (0.36 to 4.32 mm, or 0.014 to 0.170 in.) are sometimes used. The finished rule usually has a square back edge, and

a cutting edge that is ground to a 45° bevel or to a V-edge.

The back edge is fitted tightly into sawed slots in hard plywood, as shown in Fig. 16, so that it will cut the outline of the blank. The steel subplate is used to back up the steel rule and to support the plywood.

Punch. For blanking low-carbon steel, a die plate (high-carbon or tool steel template) of the same shape as the required blank is used as the punch, opposing the steel rule. Other punch elements and die parts are added as needed to the die for piercing holes and slots at the same time that the blanking is done. Solid steel blocks, instead of a steel rule, can be used in the die to cut sharp corners and notches in the blank. For some work, including the cutting of paper and other soft materials, the punch can be a block of hard wood or a thick sheet of rubber or other soft material with a working surface that extends beyond the area enclosed by the steel rule.

Stripper. For cutting paper and leather, the steel-rule die is stripped by elastic material, such as sponge rubber, which is added to the die. In blanking low-carbon steel, blocks of tougher solid rubber can be used as strippers. Positive spring-loaded steel stripper plates are also used.

The accuracy of the blanks produced in steel-rule dies depends mainly on the skill of the diemaker and the care used in their construction. For noncritical parts blanked in steel-rule dies, the tolerance may be as large as ±0.8 mm (±$^1/_{32}$ in.); for more critical parts, the work can be located accurately to maintain a tolerance of ±0.13 mm (±0.005 in.). Closer tolerances on blanks can be obtained at increased cost by using rotary-head millers or jig boring machines in constructing the dies.

Because holes and slots made by steel-rule blanking are pierced with conventional punch

and die elements that are added to the steel-rule die, they can be produced to the same tolerances as in conventional blanking. Steel-rule dies commonly blank laminations with burrs only 0.05 mm (0.002 in.) high.

Cost. A steel-rule die made to blank low-to-moderate quantities of low-carbon steel generally costs approximately 20% as much as a conventional die made for the mass production of similar work. The following example describes the use of a steel-rule die for stopgap and trial production of an automobile part. A die change was made inexpensively; after it was proved successful in the steel-rule die, the same change was then included in the design of the conventional die for production use.

Example 3: Use of a Steel-Rule Die for Temporary Production. A steel-rule die was used to blank a part for an automobile frame in order to begin production without waiting until the conventional die could be delivered. It was expected that the steel-rule die would have to produce 325 blanks before production could be changed to the conventional die.

The die, shown in Fig. 17, was used in a 2.2 MN (250 tonf) straight-side press to blank annealed cold-rolled low-carbon steel, 3.96 mm (0.156 in.) thick by 229 by 229 mm (9 by 9 in.), with three 14 mm ($^9/_{16}$ in.) diameter holes and three round-end slots 16 by 38 mm ($^5/_8$ by 1$^1/_2$ in.). Tolerances were ±0.81 mm (±$^1/_{32}$ in.) on the blank outline, and ±0.13 mm (±0.005 in.) on the pierced holes and slots. No burr limits were specified.

The steel rule was made of 12-point rule stock (4.32 mm, or 0.170 in., thick by 32 mm, or 1$^1/_4$ in., wide) set full depth into hard plywood 16 mm ($^5/_8$ in.) thick. A die plate of steel, which fit inside the rule, was used as the punch. The die was made by measuring a developed formed blank.

The ease and low cost of making a change in a steel-rule die proved important in this application because it was decided that one of the holes would not be needed. The punch that had been added for that hole was simply removed from the steel-rule die. After tryout, that hole was also eliminated in the conventional die before it was completed. The change, if made after the conventional die had been completed, would have cost much more.

The cost of the steel-rule die was 20% of the cost of the conventional die. Because the steel-rule die was fed and unloaded by hand, production was only a few pieces per minute. More than 1000 blanks were made in this steel-rule die.

Template dies (also called plate dies) are competitive with steel-rule dies in terms of cost and the quantity they can produce. Figure 18 shows an exploded view of the elements of a template die. Punches and die elements can be added, as in steel-rule dies, to combine piercing with blanking.

Punch. The punch or template (Fig. 18) is made to fit the outline of the blank to be produced. The punch is usually made of medium-carbon steel plate (1040, 1050, or 4140) or of

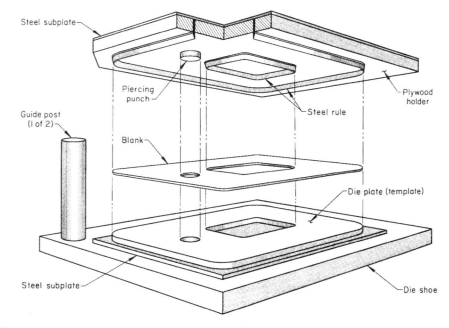

Fig. 16 Exploded view of a steel-rule die. See text for discussion.

ground flat stock such as O2 tool steel, on which the edges may be flame hardened. In more difficult applications or where longer life is needed, the punch can be made of D2 or equivalent tool steel and hardened.

Die. The die is usually assembled of doweled, hardened blocks of steel, ground to fit the punch with proper clearance. The same materials are used as for the punch.

Construction. Typical clearance for template dies is 0.076 mm (0.003 in.). The construction shown in Fig. 18 is satisfactory for blanking low-carbon steel in the same types of applications as steel-rule dies. For severe blanking, a template die can be made stronger by using one-piece construction or adding pins to prevent the die blocks from spreading, and by nesting the die or die segments into a recess in the die backing plate (subplate).

Operation. A continuous strip of stock can be fed into a template die, feeding the stock against a stop for each press stroke and using a side guide for the stock. Blocks of tough rubber are usually used as strippers (Fig. 18), pushing the blank back into the strip so that the blank is removed from the die by the feed motion of the strip. A typical application of template dies is described in the following example.

Example 4: Blanking Laminations in a Template Die. Motor laminations of 1.52 mm (0.060 in.) thick pickled hot rolled low-carbon steel were blanked in a template die. The blank measured 318 by 140 mm (12.5 by 5.5 in.). The punch was made of a one-piece template, except for inserts that were provided where changes may be needed. The die was made with 0.076 mm (0.003 in.) clearance all around punch (template).

The die was made of D2 tool steel. Rubber blocks were used for strippers, which pushed the blank back into the skeleton in the vertical press so that the blank was unloaded by the feeding of the stock. Hand feeding produced fewer than ten pieces per minute.

The burr was approximately 0.051 mm (0.002 in.)—well below the specified limit of

0.102 mm (0.004 in.). Estimated die life was 100,000 pieces.

Subpress dies are short-run die sets that are attached to the press bed but in which the punch shoe is not attached to the ram. They require less setup time than conventional dies but have a lower production rate. The length of stroke of a subpress die is limited because springs are used to raise the punch. Subpress dies are sometimes used to blank the precision parts used in instruments and timepieces.

Self-contained notching tools can be purchased ready to install as subpress tools. Notching units, consisting of both punch and die mounted in individual C-frame units, are available in a variety of standard-corner, V, and square-edged sizes for notching low-carbon steel in thicknesses to 6.4 mm (¼ in.). Special shapes of irregular outline can be incorporated into standard notching units. The notching units can be used singly or in groups, and they can be used in combination with piercing tools of similar construction.

Each unit is self-contained and self-stripping by means of springs. The punches are held in close alignment and are not attached to the ram of the press. Each unit is located, pinned, and bolted to a die plate, template, or T-slot plate and is mounted on the bed of any type of press or press brake of adequate shut height.

This type of subpress tool can be used to make small blanks, but it is more commonly used to notch and pierce precut blanks. The units can be reused to produce parts of different shapes by relocating the tools.

Construction and Use of Conventional Dies

Conventional blanking dies consist basically of one or more mating pairs of rigid punches and dies and are the standard tooling for the production blanking of sheet metal in a press. Mating pairs of metal punches and dies are combined in various ways, and additional com-

ponents are added to make up compound, progressive, transfer, and multiple dies.

Conventional dies are costly, specialized tools that are generally used for only one product, but they are so efficient, accurate, and productive that they are typically the best method of mass production at the lowest cost per piece. They are occasionally used for short-run production when tolerances are exceptionally stringent or when other reasons make the use of short-run dies impractical.

Conventional dies are more accurate than most short-run tooling, and they retain their accuracy for a greater number of pieces. They can also usually be resharpened after wear has affected their action or the quality of the work. Before dies are worn out, they have generally been resharpened many times. Conventional dies commonly produce several million blanks before replacement.

Tool materials used for blanking low-carbon steel sheet in conventional dies include (in order of increasing lot size for which they are recommended) 1020 steel; W1, O1, A2, and D2 tool steels; and, for extremely long runs, carbide. For long runs on steel thicker than approximately 6.4 mm (¼ in.), M2 tool steel is often used instead of carbide because of the limited shock resistance of carbide. Type D2 tool steel is probably the most commonly used and most widely available tool material for the mass-production blanking of steel and other metals.

Cold rolled sheet and hot rolled pickled-and-oiled sheet are far less damaging to tools than gritblasted or hot rolled unpickled surfaces. Tool materials that have a high resistance to abrasion, such as A2 or D2 tool steel, are recommended for use in tools for the blanking of sizable production lots of hot rolled unpickled steel. Detailed information on the selection of tool materials and on tool life is available in the article "Selection of Materials for Shearing, Blanking, and Piercing Tools" in this Volume.

Single-Operation Dies. The simplest conventional blanking dies are single-operation dies.

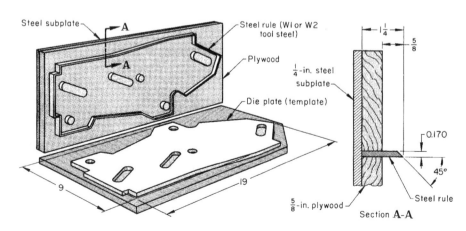

Fig. 17 Steel-rule die that was used as temporary tooling for the blanking of low-carbon steel sheet 3.96 mm (0.156 in.) thick. Dimensions given in inches

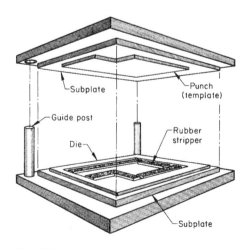

Fig. 18 Exploded view of a template die

They are used as separate units to produce blanks or as parts of more complex dies that perform several operations on a workpiece. The separate stations of a progressive die are similar to single-operation dies (although integrally constructed), and transfer presses use many single-operation dies.

The drop-through die is one of the most common types (Fig. 19). In this die, the severed blank is forced through the die opening by the downward motion of the punch, and it drops through into a chute or container. This type of blanking die has a minimum number of parts and is relatively inexpensive. Another major advantage of drop-through blanking dies is their simple and reliable blank ejection system, which is usually compatible with the use of this type of construction in progressive or transfer dies. In other types of dies, the ejection system may be more complicated than the die itself.

The disadvantages of drop-through dies for blanking include the following:

- Unless parallels are placed between the die and bolster, blanks must be small enough to go through the hole in the bed.
- Blanks may distort by dishing.
- Some shapes make drop-through difficult.
- The die must be on the lower shoe and the punch on the upper shoe of the die set.

Two other types of single-operation blanking dies—inverted and return dies—can be used when, because of size or susceptibility to damage, blanks or workpieces cannot be unloaded by dropping through the die but can be removed between the die and punch faces.

Inverted dies (Fig. 20) have the punch on the lower shoe and the die on the upper shoe. A knockout pin releases the blank from the die, and the blank is removed mechanically or manually from the top of the punch. A scrap cutter is usually included so that the scrap can be blown or knocked away. The scrap is sometimes allowed to stack up in successive collars around the punch and is stripped off manually after a number of pieces have been blanked.

Return dies (Fig. 21) are made in the usual way, with the punch on the upper shoe and the die on the lower shoe. The punch shears the blank and presses it into the die cavity, as with other

types of blanking dies. A spring-loaded pressure plate or die cushion acts as an ejector for the die, returning the blank to the surface of the die, where it can be picked off manually or mechanically. A spring-loaded plate on the upper shoe acts as a blankholder on the downstroke and as a punch stripper on the upstroke. Like many inverted dies, a variation of the return die shown in Fig. 21 includes a scrap cutter (not shown in Fig. 21) that parts the ring of scrap so that it can drop away, can be blown away, or can be removed mechanically.

Inverted and return dies have the advantage of not needing a clearance hole to let the workpiece or blank drop through. The main disadvantages of inverted and return dies are:

- They are more expensive than drop-through dies because they have more parts.
- They may require careful adjustment and synchronization of external ejectors and air blasts, which adds to setup costs.

Of the two types, inverted dies are generally simpler to construct and have less complicated knockout mechanisms than the pressure-plate or die-cushion ejectors in return dies.

Return dies are better suited to continuous strip operation because the strip remains in line and is not pressed down by the die over the punch. If the workpiece must be clamped before blanking, the blankholder or the pressure plate (or both) in a return die holds the workpiece or blank through the entire working stroke.

Instead of being removed by the methods described previously, the severed blank is sometimes pushed back (completely or partly) into the strip to be removed later, as is sometimes done in a progressive die. Pushback can also be used for other purposes, such as to provide knockouts for fitting in electrical panels or junction boxes and in other sheet-metal products. A flattening operation is sometimes added to assist pushback.

Compound dies perform several operations on the same workpiece in the same stroke of the press—such as blank and pierce, or blank, pierce, and form. Figure 22 shows the elements of a compound die for simultaneous blanking and piercing. In this die, the blanking punch is in the bottom; a hole in the punch is used as the piercing die. The piercing punch and the blanking die are in the top.

Compound dies are generally more economical in mass production than a series of one-operation dies, and they are usually more accurate. For example, a compound die that blanks and pierces a workpiece can hold the spacing between pierced holes or the relation of the pierced hole to the edge of the blank more accurately than would be possible if individual operations were done in separate dies. This is because of possible variation in locating the blank for piercing or in locating a prepierced strip for blanking.

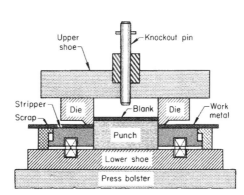

Fig. 20 Inverted blanking die with the punch on the lower shoe and the die on the upper shoe

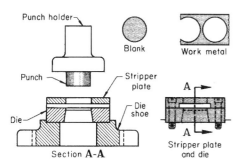

Fig. 19 Elements of a conventional drop-through blanking die

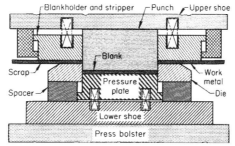

Fig. 21 Return blanking die with spring-loaded pressure plate that acts as a die cushion and an ejector for the die

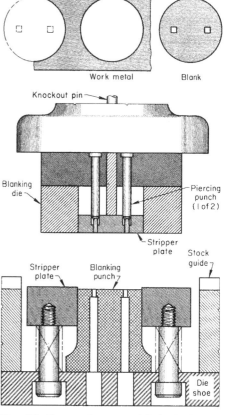

Fig. 22 Elements of a compound die for simultaneous blanking and piercing

Because the complexity of operation causes greater difficulties in unloading the workpiece, compound dies usually run slower than single-operation dies; the maximum speed of a compound die is approximately 250 strokes per minute. Other disadvantages of compound dies in comparison to single-operation dies are that they are more specialized (so that a change in the product is more likely to make the die obsolete) and that initial and maintenance costs are both higher. An advantage of compound dies is that, because of their slower operation, they generally produce more pieces per sharpening than single-operation dies.

A complex compound die can sometimes be more economical than two simpler compound dies in making the same part. This is illustrated in the following example.

Example 5: Replacing Two Two-Operation Dies with a Four-Operation Die. The cup shown in Fig. 23 was originally produced from annealed cold rolled low-carbon steel in two compound dies in separate press operations. The first die was a blank-and-draw die; the second was a pierce-and-pinch-trim die. Two separate dies were used on the assumption that the large pierced hole in the cup would not leave enough tool thickness to sustain all four operations. A 400 kN (45 tonf) open-back inclinable press was used for each of the two operations at a production rate of 500 pieces per hour.

Examination of the procedure led to the design of a die to blank, draw, pierce, and trim—all in one press stroke. The same press was used at the same speed, thus cutting the labor cost per piece in half. Tooling cost also was reduced because the cost of the new die was less than the combined cost of the original two dies.

Production Applications. The following example describes production applications in which compound dies were used for blanking and other operations because of their inherent accuracy.

Example 6: Blanking, Forming, and Piercing a Brake-Drum Back in Compound Dies. An automobile brake-drum back (Fig. 24) was formed from hot rolled 1012 steel in two press operations. A compound die was used to blank and notch the part in a 1.8 MN (200 tonf) press at the rate of 17½ strokes per minute, making one part per stroke. The tolerances shown in Fig. 24 were maintained in production with ordinary shop practice, as was the tolerance on the 14° 24′ angle (±0° 30′). The die, made of vanadium tool steel, required reworking after 50,000 pieces. Chlorinated oils were used as lubricant.

In the second press operation, a compound die in a 3.6 MN (400 tonf) press formed the workpiece, pierced the center hole, and flattened the workpiece at the notches. This die was made of tungsten oil-hardening tool steel. Production rates, lubrication, and die life were the same as in the blank-and-notch operation.

Progressive Dies. In a progressive die, the workpiece, while attached to the strip (or to the scrap skeleton), is fed from station to station at each stroke. At each stroke, the die performs work at some or all of the stations. The workpiece is cut off and unloaded at the last station. Each station can be simple or compound. Figure 25 shows the principal parts of a two-station blank-and-pierce progressive die.

In producing the simple blank shown in Fig. 25 from coiled strip in a progressive die, the round hole is pierced at the first station. The strip is then fed left, to the next station. There, the pilot enters the hole as the blanking punch moves

down to complete the blank. Accurate relation of the hole to the outline of the blank depends on accurate fit of the pilot in the hole. The completed part is not separated from the strip until the last operation, regardless of the number of operations. After the first piece, one piece is completed at each stroke.

A progressive die is expensive, and because it is usually set up in an automatic press with scrap cutter, feeder, straightener, and uncoiler, the total cost of the auxiliary equipment is also high. Other disadvantages of progressive dies are:

- The part cannot be turned over between operations.
- Material may be wasted because the workpiece may not nest well in the strip layout.

Coil stock (or, less often, flat strip stock) is used. Many operations on parts of small and medium size can be done in conjunction with blanking in a progressive die, but the planning may be complicated. Soft or thin stock may be troublesome because the pilot may distort the locating holes. Setup and maintenance may be difficult.

The following example describes an application in which a progressive die was preferred to a compound die because it was too difficult to cut sharp corners in the compound die.

Example 7: Change from a Compound to a Progressive Die to Cut Sharp Corners. Figure 26 shows a fast-idle cam on which the corners had to be sharp. A progressive die was used because it was difficult to make and maintain sharp corners in a compound die. The part was blanked from coiled 1010 steel strip, 3.05 by 60 mm (0.120 by 2⅜ in.), with maximum hardness of 55 HRB. A seven-station progressive die cut the steps in the cam.

The punches were made of M2 tool steel and hardened to 60 to 63 HRC, and the die and the punch holders were made of O1 tool steel and hardened to 56 to 58 HRC. Approximately 10,000 to 15,000 cams were made between sharpenings of the progressive die, using a punch-to-die clearance of 0.10 mm (0.004 in.). This die was mounted in a 530 kN (60 tonf) press that was run at 76 strokes per minute.

In another operation, the steps in the cam were shaved to remove die break so that surfaces were flat across the thickness. Tools were of the same materials as in the progressive die and had a life

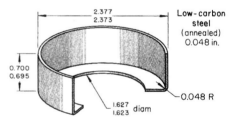

Fig. 23 Cup that was blanked, drawn, pierced, and trimmed in a compound die. This part was previously made in two compound dies: blank and draw, and trim and pierce. Dimensions given in inches

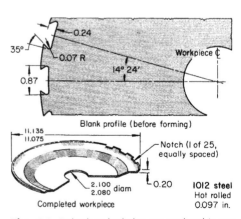

Fig. 24 Brake-drum back that was produced in two press operations, in a blank-and-notch die and in a restrike-and-form die. Dimensions given in inches

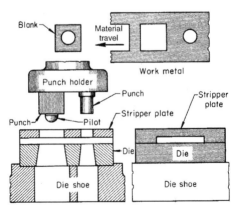

Fig. 25 Principal components of a two-station blank-and-pierce progressive die. Pierced blank and work metal (steel strip) are shown at top. The use of a pilot in the pierced hole ensures accuracy to within a few thousandths of an inch.

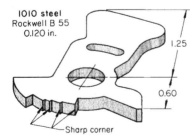

Fig. 26 Fast-idle cam on which sharp corners were cut in a progressive die. Dimensions given in inches

of 5000 to 6000 pieces between sharpenings. The shaving was done at 16 strokes per minute in a 360 kN (40 tonf) press. Finally, the cam was liquid carburized 0.08 to 0.15 mm (0.003 to 0.006 in.) deep and oil quenched to file hardness. Zinc plating (0.005 mm, or 0.0002 in., minimum thickness) and a chromate postplating treatment followed.

Transfer dies, in which separate workpieces are fed from station to station by transfer fingers, are used for blanking only when coil stock is used. Blanking is done at the first station and is followed by other operations.

Transfer dies are typically used for additional operations on precut blanks made in a separate press (blanks that permitted close nesting for best use of the stock). When equally high utilization of stock can be obtained, coil stock can be used in a transfer die, with blanking done at the first station.

Like progressive dies, transfer dies and their related equipment (presses, special attachments, and feeding devices) are expensive and are best suited to mass production. Production rates are high.

Multiple dies (also called multiple-part dies) make two or more workpieces at each stroke of the press. The workpieces can be pairs of right-hand and left-hand parts, duplicate parts, or unrelated parts. Punch height can be staggered to reduce shock and blanking noise. Such dies are used in mass production.

Multiple dies can be multiples of single-operation dies or multiples of compound dies. They generally cost only slightly more than similar dies that make only one part per stroke. A die that makes two parts per stroke may cost only 5% more than a die that makes only one.

Such dies are primarily used for blank-and-form sequences. Draw operations are more difficult to combine with blanking or other operations because of blankholder needs and slower draw operation. Unloading of the work is sometimes difficult.

The use of multiple dies depends on the size and shape of the workpiece, size of the press, production quantity, possible savings in material and labor, and costs for setup and maintenance. Advantages of multiple dies include savings in material by better blank layout, reduced labor costs, and increased production. Disadvantages include increased setup and maintenance costs.

It is often better to increase the production of a single-part die by some simple change, such as putting the die in a faster press, rather than to replace the single-part die with a multiple die. A multiple die may also have to be run slower than the simpler die if a press of greater force capacity must be used to provide ample force. The use of a multiple compound die to blank, pierce, and form three pieces per stroke is illustrated in the following example.

Example 8: Blanking, Piercing, and Forming Three Pieces per Stroke in a Multiple Compound Die. A multiple compound die was used to make a thick dished washer with three flats equally spaced on the edge circle, as shown in Fig. 27. The part was made of hot rolled 1008 or 1010 steel, 2.36 ± 0.18 mm (0.093 ± 0.007 in.) thick.

Operations in the compound die were blank, pierce, and form. The blanks were nested so closely in the stock layout that the pilot holes (half-circles) had to be notched in the edge of the stock. Three parts were made with each stroke of the die, and the parts were pushed partly back into the scrap skeleton so that they were carried out of the press for unloading. Production was 500,000 pieces per month. The die, made of D2 tool steel, needed to be resharpened after 150,000 strokes (450,000 pieces) and required reconditioning (replacement of some parts) after 3.5 million pieces.

Operating Conditions

To achieve high productivity and low unit cost, most blanking is done in high-speed mechanical presses. Speeds as high as 1200 strokes per minute are used. The equipment for high-speed blanking ordinarily includes a short-stroke press, automatic feed devices, and dies designed for bottom ejection.

In most blanking operations, press speed is limited by the length of feed, which is governed by blank size, or by the relationship between the force capacity of the press and the load. The combination of blanking with forming or drawing in compound dies also restricts press speed. Blanking speed may be as low as ten strokes per minute in producing blanks that are extremely large or that present handling problems for other reasons.

Regardless of the number of strokes per minute, the velocity of the punch always approaches zero near the bottom of the stroke. Within the usual range for production work in conventional blanking dies, the speed of the press has little practical effect on the speed of the punch during the blanking portion of the stroke. This effect, however, is critical for fine-edge blanking. Punch speed is usually slower during fine-edge blanking on the order of 7.5 to 1.5 mm/s (0.3 to 0.10 in./s) during the interval while the punch is cutting through the work metal (see also the article "Fine-Edge Blanking" in this Volume).

Lubrication requirements are generally less critical for blanking than for forming or deep drawing; stock to be blanked is often fed into the press with no lubrication other than the residue remaining on the stock from the lubrication at the mill. The stock is sometimes coated with a light mineral oil or a light chlorinated oil. However, lubrication is important in dies that have close clearance between punches and stripper. At speeds of 40 strokes and more per minute, such dies must be lubricated constantly with a spray of light mineral oil to prevent galling of the punches in the stripper. Additional information on lubrication requirements in blanking is available in the article "Selection and Use of Lubricants in Forming of Sheet Metal" in this Volume.

Effect of Work Metal Thickness

Stock thickness affects the selection of material for dies and related components, as well as the selection of die type and design. The amount of shear and relief (angular clearance or draft) built into a blanking die and the amount of clearance between punch and die all depend on blank thickness.

Work metal thickness is also a factor in the selection of blanking method, handling procedure, and handling equipment. Blanking in a punch press is usually the fastest and most economical way of producing blanks thinner than approximately 6.4 mm ($^1/_4$ in.) in medium or large quantities.

Plate stock, in thicknesses of 6.4 to 25 mm ($^1/_4$ to 1 in.), is less frequently blanked in presses than sheet or strip. Blanks of such thick material are often made by gas cutting, sawing, nibbling, or routing instead of by shearing or by press operations; selection of the method depends primarily on plate thickness and production quantity.

Almost all blanks thinner than 3.2 mm ($^1/_8$ in.), except for intricate shapes chemically blanked from foil, are produced with conventional dies in mechanical or hydraulic presses. In only two of the examples of commercial practice presented in this article was work metal thicker than 3.2 mm ($^1/_8$ in.) (example 3: 3.96 mm, or 0.156 in., and example 9: 4.75 mm, or 0.187 in.).

Because of its strength and rigidity, material thicker than 3.2 mm ($^1/_8$ in.) is seldom blanked from coil stock or in a progressive die. On the other hand, because of its lack of strength and extreme flexibility, material thinner than 0.51 mm (0.020 in.) generally requires special handling techniques. The articles "Piercing of Low-Carbon Steel," "Blanking and Piercing of Electrical Steel Sheet," and "Forming of Carbon Steels" in this Volume contain additional information on the effect of work metal thickness on processing.

Distortion is often a problem in blanking complex shapes from thin low-carbon steel sheet by repeated strokes of a notching die. Distortion of such parts can be minimized by the use of hardened stock and by performing the entire blanking operation in one stroke in a single die.

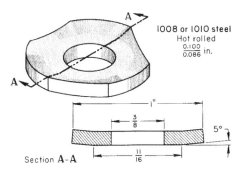

Fig. 27 Dished washer that was blanked, pierced, and formed, three pieces per stroke, in a multiple compound die. Dimensions given in inches

Accuracy

Blanking in conventional dies readily produces parts within a total tolerance of 0.051 to 0.254 mm (0.002 to 0.010 in.), depending on the accuracy of the dies and the condition of the press. The tolerances given in Table 4 in the article "Piercing of Low-Carbon Steel" in this Volume generally also apply to blanks. The total tolerances listed under the column head "Location" apply to the relationship of a point on the periphery of the blank to a hole or other reference feature on the blank; the values listed under "Size" apply to a diameter for round blanks or to a similar control dimension for other blank shapes.

The production of blanks to these tolerances is illustrated by the examples in this article and the article "Piercing of Low-Carbon Steel." The following example describes the use of a compound die to maintain a total (envelope) tolerance of 0.13 mm (0.005 in.) on the relationship of a cam surface to a hole.

Example 9: Blanking and Piercing a Cam to 0.13 mm (0.005 in.) Total Tolerance on Cam Surface in Relationship to Hole Position. A compound die was used to blank and pierce the cam shown in Fig. 28 so that the hole would be in accurate relationship to the cam surface. The die was made of A2 or equivalent tool steel and hardened to 62 HRC, and it had a clearance per side equal to 10% of the stock thickness.

The cam, used in the hinge mechanism of an automobile door, was blanked from hot rolled 1020 steel 4.75 mm (0.187 in.) thick. Samples of the cam were inspected with an optical comparator, which compared the relationship of hole and cam surface to an outline that showed the full 0.13 mm (0.005 in.) tolerance, as illustrated in Fig. 28.

In another operation, the cam surface was machined to remove die break so that the edge would be square. The cam was then case hardened.

The part was produced in a 1.4 MN (160 tonf) open-back inclinable press at 60 strokes per minute. A sulfur-base lubricant was used. Die life was 40,000 pieces per sharpening.

Blanking and Piercing with Compound Dies. Example 6 in this article describes other applications of compound dies in blanking and piercing to conventional tolerances. For parts made in a progressive die, the relationship of the blank outline to features of the part produced in other stations of the die depends on the accurate fit of the pilot in the pilot hole. Transfer dies ordinarily provide better accuracy than progressive dies because the positive location of separate parts can be more precise than the roll feeding of coil stock with pilots.

Holding close tolerances on parts made in a progressive die is particularly difficult on soft or thin material because distortion of the locating holes by the pilot is more likely. Handling problems that may contribute to excessive variation in location or dimensions are not usually encountered in transfer dies, except with blanks of extremely thin material. Shaving (see the

following section in this article) is used to improve the accuracy of blank outlines to meet close tolerances or to improve edge quality.

Short-run dies are generally less accurate than conventional dies. By using more accurate methods of constructing short-run dies, closer tolerances on blanked work can be obtained, but at some increase in die cost (see the section "Construction and Use of Short-Run Dies" in this article). Generally, making blanks by methods other than the use of dies in presses, except for machining or grinding, results in a lower level of accuracy.

Shaving

Shaving is an operation that can be performed after blanking to give a smooth, square edge and greater accuracy than can be achieved in ordinary blanking. Shaving removes only the blanked edge—cutting away the deformed, broken, and burred edge that was left in blanking. The elimination of these irregularities and the removal of locally work-hardened metal minimize breakage of the work metal during subsequent flanging, particularly the flanging of holes. The scrap produced in shaving is so thin that it resembles the chips produced in finish machining, rather than the usual scrap that is produced in a press.

When shaving is planned for, a small amount of extra stock is left on the workpiece to be removed in shaving. Shaving can be done in a separate operation, or it can be included in one station of a progressive die.

The shaving operation produces a straight, square edge, generally to approximately 75% of the metal thickness. Two shaves make a better, straighter edge (to approximately 90% of the metal thickness) than does a single shaving operation. To eliminate rollover from blank edges, which requires the removal of a greater amount of stock, it may be better to consider machining the workpiece rather than shaving.

Punch-to-die clearances range from 0 to 1% of stock thickness per side. Sturdy guideposts in a heavy die set are necessary to maintain the close alignment needed to prevent damage to the punch and die.

Shaving causes more wear on a die than ordinary blanking, so that the die produces fewer parts per grind and needs more frequent maintenance. Slivers of shaving scrap (chips) can jam feeding mechanisms, can become embedded in

workpieces, or can mar the punch and die surfaces if not removed after each press stroke. Because of these problems, special attention must be given to die design when shaving is included among the operations done in a progressive die, as illustrated in the following example.

Example 10: Shaving in a Progressive Die. In evaluating methods for high-volume production of the shaved low-carbon steel part shown in Fig. 29, the use of a single progressive die for all cutting and forming operations was projected as the most economical method, principally because this method would involve fewer operations and less handling than the other methods considered. However, two major problems were anticipated. First, the life of the shaving tools was expected to be much shorter than that of the other tools in the progressive die (which would have resulted in costly interruption of production to sharpen or replace the shaving tools), and, second, it appeared likely that misfeed could occur from jamming of the feeding mechanism by slivers of shaving scrap.

By designing the die and feeding mechanism to eliminate difficulties from these two sources, efficient and economical production was obtained. The tools were of A2 air-hardening tool steel, hardened to 54 to 58 HRC for the forming sections and to 60 to 62 HRC for the cutting sections. The shaving section was made with a replaceable insert to minimize downtime when sharpening was needed. Damage to the progressive die from an accumulation of shaving scrap was prevented by including in the die a stock stop (Fig. 29, station 3) and misfeed and double-thickness protectors.

The shaving was done in station 4, as shown in Fig. 29. The production rate was 80 pieces per minute in a 530 kN (60 tonf) mechanical press with a 50 mm (2 in.) stroke. Tolerance on most sections of the part was ±0.13 mm (±0.005 in.). Additional information on the use of shaving to maintain close tolerances and to improve the quality of hole walls is available

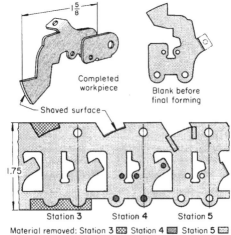

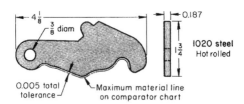

Fig. 28 Cam that was blanked and pierced in a compound die within an envelope tolerance of 0.13 mm (0.005 in.) total indicator reading. Dimensions given in inches

Fig. 29 Formed part on which a blanked edge was made smoother and more accurate by shaving (station 4) after notching (station 3) in a progressive die. Dimensions given in inches

in the article "Piercing of Low-Carbon Steel" in this Volume.

Shaving allowance, or the amount of stock to be removed from the workpiece, depends on the hardness and thickness of the blank. Generally, the smallest amount of stock that will produce the desired result is left for the shaving operation. Table 5 lists shaving allowances recommended by one manufacturer. When shaving only one edge of a blank, shifting of the blank can be reduced by shaving the opposite edge as well, even if not required for function.

Setups for Shaving. Shaving requires that the blank be accurately located over the die or the punch, because only a few thousandths of an inch of metal is removed by the operation (Table 5). Piloting pins, projecting from the punch, can engage holes in the blank to ensure proper location. If the holes are not included in the original design, it may be permissible to add them for locating purposes only.

If adding holes is not permitted, a locating device such as that shown in Fig. 30 can be used. The clamping arms engage the blank at suitable nesting points. When the punch descends, the shaved blank falls through the die. The position of the clamping arms is fixed by the two stop pins.

Operation of this die can be improved by the use of a spring-loaded ejector and pressure pad within the die opening. As the punch ascends, the ejector lifts the shaved workpiece above the die face, thus eliminating the fall through the die

block, which may result in dents or other surface defects.

Burr Removal

The shape, height, and roughness of burrs must be controlled to some degree in nearly all blanking operations. Complete elimination of burrs is not possible, but their formation can be minimized by the use of proper clearance between punch and die and by good maintenance.

Exposed burrs on the finished part can be unsafe and unsightly. Burrs on some blanked work can cause difficulties in forming and can increase the rate of workpiece breakage and die wear. Burrs can be removed by grinding, which generally removes the burr and a portion of the work-hardened edge. Tumbling in a barrel is a common method of deburring small parts. Other deburring methods include chemical and electrolytic deburring, belt grinding, polishing, and ultrasonic methods, as described in *Surface Engineering,* Volume 5 of *ASM Handbook,* 1994. Hand scrapers can be used to remove burrs from irregular shapes or soft metal parts.

Blanking in Presses versus Alternative Methods

Fine-edge blanking is primarily used where die break is unacceptable and would require

removal by subsequent shaving if conventional blanking were used. In fine edge blanking, there is no die break, and the entire wall surface of the cut is burnished (see the article "Fine-Edge Blanking" in this Volume). In fine-edge blanking, the stock is tightly locked against the die and forced to flow toward the punch, so that the part can be extruded out of the strip without fracture or die break. Die clearance is extremely small, and no further finishing or machining operations are necessary to obtain blank or hole edges comparable to machined edges, or to those that are conventionally blanked or pierced and then shaved (see also the section "Fine-Edge Blanking and Piercing" in the article "Blanking and Piercing" in this Volume).

Milling is applicable mainly for cutting stacked parts, for short runs, and for making parts that are subject to frequent design change. It substitutes an inexpensive template for a conventional punch and die.

Chemical blanking may be competitive with blanking in presses for intricate parts that are only a few thousandths of an inch thick. Combs for electric shavers, for example, are more frequently made by the chemical blanking of stainless steel than by mechanical blanking methods.

Contour band sawing and gas cutting may be competitive with blanking for stacked parts and thick material.

Safety

In all blanking operations, as in all press operations, there are hazards to operators, repairmen, and personnel in the vicinity. No press, die, or auxiliary equipment should be considered operable until these hazards are eliminated by installing necessary guards and other safety devices. The operator and all persons working around the blanking operation should be instructed in all precautions for safe operation before work is started. Additional information is available in the article "Presses and Auxiliary Equipment for Forming of Sheet Metal" in this Volume.

REFERENCE

1. L.R. Allingham, ASTME Paper MF64-151, Danley Machine Corporation

Table 5 Shaving allowances recommended by one manufacturer

| Blank thickness | | Allowance per side for steel with HRB hardness of: | | | | | |
| | | 50–66 | | 75–90 | | 90–105 | |
mm	in.	mm	in.	mm	in.	mm	in.
First shave (or a single shave)							
1.19	0.047	0.064	0.0025	0.076	0.003	0.102	0.004
1.57	0.062	0.076	0.003	0.102	0.004	0.127	0.005
1.98	0.078	0.089	0.0035	0.127	0.005	0.152–0.178	0.006–0.007
2.39	0.094	0.102	0.004	0.152	0.006	0.178–0.203	0.007–0.008
2.77	0.109	0.127	0.005	0.178	0.007	0.229–0.279	0.009–0.011
3.18	0.125	0.178	0.007	0.229	0.009	0.305–0.356	0.012–0.014
Second shave (add to first shave)							
1.19	0.047	0.032	0.00125	0.038	0.0015	0.051	0.002
1.57	0.062	0.038	0.0015	0.051	0.002	0.064	0.0025
1.98	0.078	0.044	0.00175	0.064	0.0025	0.076–0.089	0.0030–0.0035
2.39	0.094	0.051	0.002	0.076	0.003	0.089–0.102	0.0035–0.0040
2.77	0.109	0.064	0.0025	0.089	0.0035	0.114–0.140	0.0045–0.0055
3.18	0.125	0.089	0.0035	0.114	0.0045	0.152–0.178	0.006–0.007

Typical shaving of 3.18 mm thick stock (75-90 HRB)

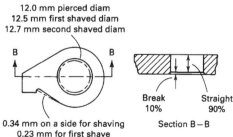

One shave

12.2 mm pierced diam
12.7 mm shaved diam

Break 25%   Straight 75%

Section A–A

0.23 mm on a side for shaving

Two shaves

12.0 mm pierced diam
12.5 mm first shaved diam
12.7 mm second shaved diam

Break 10%   Straight 90%

Section B–B

0.34 mm on a side for shaving
0.23 mm for first shave
0.11 mm for second shave

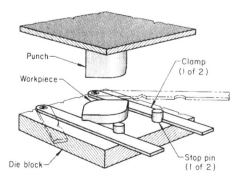

Fig. 30 Shaving die with a device for locating a blank with no holes for piloting

Piercing of Low-Carbon Steel

PIERCING is the cutting of holes in sheet metal, generally by removing a slug of metal, with a punch and die. Piercing is similar to blanking, except that in piercing the work metal that surrounds the piercing punch is the workpiece and the punched-out slug is scrap, while in blanking the workpiece is punched out. Additional information on piercing is available in the articles "Blanking of Low-Carbon Steel," "Fine-Edge Blanking," "Forming of Carbon Steels," and "Blanking and Piercing of Electrical Steel Sheet" in this Volume.

The term *piercing* is used in this article, and in related articles in this Volume, to denote the production of a hole by removing a slug of metal with a punch and die. However, some prefer the terms *punching* or *perforating*, limiting the term *piercing* to the use of a pointed punch that tears and extrudes a hole without cutting a slug of metal. The term *perforating* is also sometimes used in the special sense of cutting many holes in a sheet-metal workpiece by removing slugs with several punches.

Piercing is ordinarily the fastest method of making holes in steel sheet or strip and is generally the most economical method for medium-to-high production. Pierced holes can be almost any size and shape; elongated holes are usually called slots. The accuracy of conventional tool steel or carbide dies provides pierced holes with a degree of quality and accuracy that is satisfactory for a wide variety of applications. Piercing is primarily used when accurate holes are required and when the production lot is large enough to justify the tooling costs. Alternative methods are used for smaller production lots or for holes that have a diameter less than stock thickness. Examples of alternative methods are drilling (including electrochemical machining), milling and sawing, and electric-arc and gas cutting. For increased efficiency, all of these methods can be applied to work that is stacked or nested.

Characteristics of Pierced Holes

Pierced holes are different from through holes that are produced by drilling or other machining methods. A properly drilled or otherwise machined through hole has a sidewall that is straight for the full thickness of the work metal, with a high degree of accuracy in size, roundness, and straightness. The sidewall of a pierced hole is generally straight and smooth for only a portion of the thickness, beginning near the punch end of the hole; the rest of the wall is broken out in an irregular cone beyond the straight portion of the hole, producing fracture, breakout, or die break (Fig. 1).

The piercing operation typically begins as a cut that produces a burnished surface on the hole wall and some rollover (curved surface caused by deformation of the workpiece before cutting begins), as illustrated in Fig. 1. The punch completes its stroke by breaking and tearing away the metal that was not cut during the beginning of the piercing operation.

The combined depth of rollover and burnish is a measure of the penetration depth of the stroke, also shown in Fig. 1. This is the part of the stroke during which the cutting force is exerted, before the metal fractures or breaks away (Fig. 1).

The amount of penetration before fracture is commonly expressed as a percentage of the stock thickness. In general, the percentage of penetration depends more on the material than on other factors, such as punch-to-die clearance. Table 2 in the article "Blanking of Low-Carbon Steel" in this Volume shows the average percentage of penetration (before fracture) in various metals under typical piercing or blanking conditions. The percentage of penetration affects energy consumption and cutting force in blanking or piercing, as described in the article "Blanking of Low-Carbon Steel" in this Volume.

Quality of Hole Wall

If the sidewall of a pierced hole is not smooth or straight enough for the intended application, it can be improved by shaving in a die or by reaming. When done in quantity, shaving is the least expensive method of improving the sidewall of a pierced hole. Shaving in one or two operations generally makes the sidewall of a hole uniform and smooth through 75 to 90% of the stock thickness.

Superior accuracy and smoothness of hole walls can be obtained by fine edge piercing. With this method, one stroke of a triple-action press pierces holes with smooth and precise edges for the entire thickness of the material. Additional information is available in the article "Fine-Edge Blanking" in this Volume.

Burr height is an important element in hole quality, and a maximum burr height is usually specified. For most applications, the limit on burr height is between 5 and 10% of stock thickness. Burr height in piercing a given workpiece is primarily governed by punch-to-die clearance and tool sharpness.

Burr condition and limits usually determine the length of run before the punch and die are resharpened. With good practice, burr height generally ranges from 0.013 to 0.075 mm (0.0005 to 0.003 in.) but may be much greater, depending on workpiece material and thickness, clearance, and tool condition. As an alternative to limiting the length of run to control burr condition, unacceptable burrs can be removed by shaving or deburring.

Selection of Die Clearance

Clearance, or the space between the punch and the sidewall of the die, affects the reliability of operation of piercing (and blanking) equipment, the characteristics of the cut edges, and the life of the punch and die. Published recommendations for clearances have varied widely, with most suggesting a clearance per side of 3 to 12.5% of the stock thickness for steel.

Establishment of the clearance to be used for a given piercing or blanking operation is influenced by the required characteristics of the cut edge of the hole or blank and by the thickness and the properties of the work metal. Larger clearances prolong tool life. An optimal clearance can be defined as the largest clearance that will produce a hole or blank having the required characteristics of the cut edge in a given material

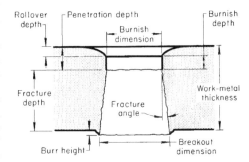

Fig. 1 Characteristics of a pierced hole. Curvature and angles are exaggerated for emphasis.

and thickness. Because of differences in cut-edge requirements and in the effect of tool life on overall cost, clearance practices vary among plants and for different applications.

No single table or formula can specify an optimal clearance for all situations encountered in practice. Starting with general guidelines, trial runs using several different clearances may be needed to establish the most desirable clearance for a specific application. The following general principles are useful in making adjustments:

- Rollover (plastic deformation) and burnish depths are greater in thick material than in thin material and are greater in soft material than in hard material.
- Clearance (in decimal parts of an inch) needed to produce a given type of edge should vary directly with material thickness and hardness, and inversely with ductility.

All clearance values given in this article are for clearance per side, except where indicated.

Edges

More specific guidance in selecting die clearances is provided by considering the types of edges produced with different clearances.

Edge Types. The acceptability of a punched hole or a blank is generally based on the condition of the cut edge and its suitability for the application. Usable holes and blanks can be obtained over a broad range of punch-to-die

clearances, each resulting in a different edge condition. Figure 2 shows five types of edges that result from the use of different clearances in piercing or blanking low-carbon steel at a maximum hardness of 75 HRB. The tabular data accompanying Fig. 2 include approximate ranges of fracture or breakout angles, rollover, burnish and fracture depths, and burr characteristics for the five edge types. Table 1 lists the clearance ranges that will produce these edges when piercing or blanking various metals.

Type 1. This type of edge has a large rollover radius and a large burr that consists of a normal

tensile burr in addition to bending or deformation at the edge. Burnish depth is minimal. Fracture depth is approximately three-fourths of stock thickness, and the fractured surface has a large angle. This edge is satisfactory for noncritical applications in which edge quality and part flatness are not important.

Type 2. This edge, which has a moderate rollover radius, normal tensile burr, and a small fracture angle, provides maximum die life and a hole or blank that is acceptable for general work in which a large burnish depth is not required. Burnish depth plus rollover depth is

Table 1 Punch-to-die clearances for piercing or blanking various metals to produce the five types of edges shown in Fig. 2

For clearances that produce type 1, 2, and 3 edges, it is ordinarily necessary to use ejector punches or other devices to prevent the slug from adhering to the punch.

Work metal	Clearance per side, % of stock thickness				
	Type 1(a)	Type 2	Type 3	Type 4	Type 5
Low-carbon steel	21	11.5–12.5	8–10	5–7	1–2
High-carbon steel	25	17–19	14–16	11–13	2.5–5
Stainless steel	23	12.5–13.5	9–11	3–5	1–2
Aluminum alloys					
Up to 230 MPa (33 ksi) tensile strength	17	8–10	6–8	2–4	0.5–1
Over 230 MPa (33 ksi) tensile strength	20	12.5–14	9–10	5–6	0.5–1
Brass, annealed	21	8–10	6–8	2–3	0.5–1
Brass, half hard	24	9–11	6–8	3–5	0.5–1.5
Phosphor bronze	25	12.5–13.5	10–12	3.5–5	1.5–2.5
Copper, annealed	25	8–9	5–7	2–4	0.5–1
Copper, half hard	25	9–11	6–8	3–5	1–2
Lead	22	8–10	6.5–7.5	4–6	1.5–2.5
Magnesium alloys	16	5–7	3.5–4.5	1.5–2.5	0.5–1

(a) Maximum. Source: Ref 1

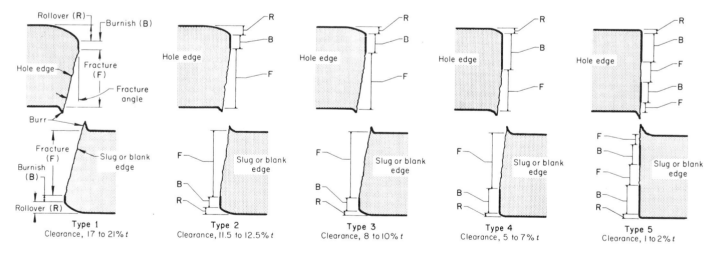

Type 1
Clearance, 17 to 21% t

Type 2
Clearance, 11.5 to 12.5% t

Type 3
Clearance, 8 to 10% t

Type 4
Clearance, 5 to 7% t

Type 5
Clearance, 1 to 2% t

Edge characteristic	Type 1	Type 2	Type 3	Type 4	Type 5
Fracture angle	14–16°	8–11°	7–11°	6–11°	...
Rollover(a)	10–20% t	8–10% t	6–8% t	4–7% t	2–5% t
Burnish(a)	10–20% t(b)	15–25% t	25–40% t	35–55% t(c)	50–70% t(d)
Fracture	70–80% t	60–80% t	50–60% t	35–50% t(e)	25–45% t(f)
Burr	Large, tensile plus part distortion	Normal, tensile only	Normal, tensile only	Medium, tensile plus compressive(g)	Large, tensile plus compressive(g)

(a) Rollover plus burnish approximately equals punch penetration before fracture. (b) Burnish on edge of slug or blank may be small and irregular or even absent. (c) With spotty secondary shear. (d) In two separate portions, alternating with fracture. (e) With rough surface. (f) In two separate portions, alternating with burnish. (g) Amount of compressive burr depends on die sharpness.

Fig. 2 Effect of punch-to-die clearance per side (as a percentage of stock thickness, *t*) on characteristics of edges of holes and slugs (or blanks) produced by piercing or blanking low-carbon steel sheet or strip at a maximum hardness of 75 HRB. Table 1 lists clearances for producing the five types of edges in various metals. See text for additional discussion and for applicability of the five types of edges.

approximately one-third of stock thickness; fracture depth, approximately two-thirds.

Type 3. This edge has a small rollover radius, a normal tensile burr, and a small fracture angle. It has low residual stress and is therefore particularly desirable for use in parts made of work-hardenable material that will undergo severe forming. The clean stress-free edge reduces the possibility of edge cracking during forming. Burnish depth plus rollover depth is one-third to one-half of stock thickness.

Type 4. This is a desirable edge for stampings used for mechanisms or parts that must receive edge finishing such as shaving or machining. The edge has a very small rollover radius, a medium tensile and compressive burr, and a small fracture angle. Burnish depth plus rollover depth is approximately two-thirds of stock thickness. This edge type can be recognized by the spotty appearance of secondary shear on the fractured surface.

Type 5. This edge has a minimum rollover radius and a large tensile and compressive burr, and it can be recognized by the complete secondary shear on the cut surface. It is useful in applications in which edges must have a maximum of straight-wall depth without secondary operations. On steel and other hard metals, die life is extremely short. The edge can be useful on some of the softer metals, which allow a reasonable die life.

Edge Profiles. The exact profile of the edge varies somewhat for different work metals, depending on the properties of the metal. Results are also slightly affected by:

- Face shear on punch or die
- Punch-to-die alignment
- Proximity to adjacent holes
- Distance to adjacent blanked edges
- Orientation of the different portions of the cut edge with respect to the rolling direction of the stock
- Ratio of hole size to stock thickness
- Internal construction of the die cavity
- Lubrication

The edge profiles illustrated in Fig. 2, as well as the estimates of fracture angles and the relative amounts of rollover, burnish, fracture, and burr given in the accompanying table, are intended to represent production conditions, allowing for the normal range of tool sharpness encountered in piercing and blanking low-carbon steel sheet.

The clearance values given in Table 1 for piercing and blanking various metals to produce the five types of edges were obtained in laboratory tests. The cutting edges of the punches were stoned to a radius of 0.05 to 0.15 mm (0.002 to 0.006 in.) to simulate an amount of wear corresponding to the approximate midpoint of a production run. No lubricant was used on the work metal.

As clearance is increased from the low values used for type 5 edges to those used for type 1 edges, several effects are evident. The edge profile deviates more and more from straightness

and perpendicularity as rollover, fracture angle, and fracture depth increase, while burnish depth decreases proportionally. Total burr height initially decreases as its compressive component decreases, leaving only the essentially constant tensile burr on type 2 and 3 edges (usually in the range of 0.013 to 0.076 mm, or 0.0005 to 0.003 in., depending on the work metal and the tool condition).

With further increase in clearance (edge type 1), bending or deformation at and near the edge adds an additional burr component, increasing the total burr height. This part distortion immediately adjacent to the cut edge is usually accompanied by a more gradual curvature, or dishing, on blanks or slugs; the corresponding curvature is much less pronounced on the stock around a hole, which is usually restrained by a stripper (curvature of blanks or stock strip is not shown in Fig. 2). At extremely large clearances (substantially above those shown for type 1 edges), double-shear characteristics are sometimes observed on the cut edge.

Effect of Tool Dulling

The sharpness of punch and die edges has an important effect on cut-edge characteristics in piercing and blanking. At the beginning of a run, with punch and die equally sharp, the hole profile is the same as that of the slug or blank. As the run progresses, dulling of the punch increases the rollover and the burnish depth on the hole wall and increases the burr height on the slug or blank. Dulling of the die increases burnish depth and burr height on the hole edge. The punch dulls faster than the die; therefore, the changes in hole characteristics related to punch dulling proceed more rapidly than those related to die dulling.

On average, the following differences between hole edge and blank edge are observed in production work on sheet metal:

- Rollover is greater on hole edge than on slug or blank edge.
- Burnish depth is greater on hole edge than on slug or blank edge.
- Fracture depth is smaller (and fracture angle greater) on hole edge than on slug or blank edge.
- Burr height on hole edge is less than that on slug or blank edge, and varies with tool sharpness.

These differences are illustrated in Fig. 2.

Use of Small Clearance

Where relatively square edges are required, small clearances can be used to produce holes with type 4 edges. Although tool life is shorter than when larger clearances are used, this may not be an important factor in the overall costs for short- or medium-production runs.

Figure 3 shows the ranges of die clearance per side used by one electronics manufacturer in

piercing and blanking three groups of metals up to 3.18 mm (0.125 in.) thick. The groups and the percentages of stock thickness on which these ranges of clearance were based are listed with Fig. 3. For stainless steel (not included in Fig. 3), nominal clearance per side was 2.5% of the thickness for stock thicknesses up to 4.75 mm (0.187 in.), and 4% for thicknesses between 4.75 and 6.35 mm (0.187 and 0.250 in.).

The data plotted in Fig. 3 were used by the manufacturer to determine whether a given punch and die could be used interchangeably for a metal different in thickness or type from the one for which it had been designed. If the point of intersection of a vertical line for the thickness of the new material with a horizontal line for the existing die clearance was within the range shown for the new material, the punch and die were normally satisfactory for use on the new material.

Use of Large Clearance

Studies of die operation under both laboratory and production conditions have indicated that large clearances can be used to obtain maximum

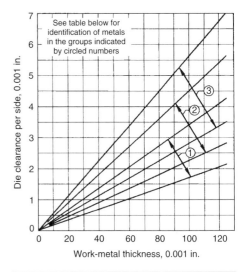

Group	Clearance per side, % of stock thickness(a)	
	Average	Range
1. Aluminum alloys 1100 and 5052, all tempers	2.25	1.7–3.4
2. Aluminum alloys 2024 and 6061, T4 and T6 tempers; brass, all tempers; cold rolled steel, dead soft	3.0	2.25–4.5
3. Cold rolled steel, half hard	3.75	2.8–5.6

Note: Incorrect clearance values twice as large as those shown here have appeared with charts of this type in some publications, apparently because of confusion between clearance per side and total clearance. Also, stainless steel has erroneously been included with the metals in groups 2 and 3 in those publications. (a) Percentages of stock thickness on which the ranges of acceptable clearance in chart are based. See text for clearances used in piercing or blanking of stainless steel.

Fig. 3 Ranges of punch-to-die clearance per side recommended by one manufacturer for piercing and blanking of various metals up to 3.18 mm (0.125 in.) thick

tool life in numerous piercing (and blanking) applications. Table 2 indicates the effect of clearance on force requirements for piercing and stripping, presenting data on the test piercing of 6.53 mm (0.257 in.) diameter holes in cold rolled low-carbon steel of various thicknesses and hardnesses. Although individual results show some inconsistency, as would be expected because of the difficulty in obtaining accurate measurements of this type, the trend toward lower stripping force with an increase in clearance is evident from these data. The amounts of punch penetration into the die needed to release the slug are also given in Table 2. Table 3 lists production data on the effect of increased clearance on tool life for piercing and blanking common metals in various thicknesses.

On the basis of these studies and production experience, clearance per side equal to 12.5% of stock thickness has been recommended by some toolmakers and sheet-metal fabricators for the general-purpose piercing (and blanking) of cold rolled steel 0.51 to 3.18 mm (0.020 to 0.125 in.) thick in all tempers. This practice produces type 2 cut edges (Fig. 2).

The advantages observed when using a clearance of 12.5%, instead of substantially smaller clearances (as for a type 4 edge), include the following:

- Total tool life and tool life between regrinds are considerably increased (Table 3). Punch wear, normally two or three times die wear, is greatly reduced because the hole is larger than the punch size and because stripping wear is minimized.
- Load on the press may be slightly reduced (Table 2).
- Burr height is smaller at the beginning of a run and increases at a slower rate during the run.
- Distortion or waviness of the work surface is reduced, especially with closely spaced holes.
- Stripping force is reduced (Table 2), which partially accounts for the reduced punch wear.

The factors that must be considered in applying the 12.5% clearance include the following:

- A different clearance may be required for steel outside the thickness range of 0.51 to 3.18 mm (0.020 to 0.125 in.), for metals other than steel, or to meet critical edge-quality requirements.
- A spring-loaded stripper should be used instead of the positive or fixed type, and the slug or blank must be prevented from adhering to the end of the punch. These precautions are especially important in transfer dies.
- Hole size is larger than punch size, particularly with hard or thin materials.
- With clearance above 15%, slugs of some materials (for example, 1.3 mm, or 0.050 in., thick type 410 stainless steel at a hardness of 50 HRC) may be ejected from the die at high velocity—possibly requiring special safety precautions.

Effect of Work Metal Hardness. Data on edge characteristics versus punch-to-die clearance in piercing cold rolled low-carbon steel sheet of different hardnesses (tempers) are plotted in Fig. 4. These graphs illustrate the basic characteristics of the edges of pierced holes for the standard AISI tempers at various clearances per side up to 35% of stock thickness.

The data given in Fig. 4 were obtained in laboratory tests on stock 0.58 to 3.18 mm (0.023 to 0.125 in.) thick in the tempers shown. The cutting edges of the punches used were stoned to a radius of 0.1 mm (0.004 in.) to simulate mid-run wear, and no lubricant was used in the tests. Similar trends would be expected for the edge characteristics of slugs or blanks, except that rollover and burnish depth would be smaller and burr height would be greater.

Clearance and Tool Size

In blanking, the die opening is usually made to the desired size of the blank, and the punch size is then equal to the die opening minus twice the specified clearance per side. Conversely, in piercing, the punch is usually made to the desired size of the hole, and the die opening is then equal to the punch size plus twice the specified clearance per side.

Clearance per side for blanking dies is ordinarily calculated from the desired percentage of clearance and the nominal thickness of the stock.

Table 2 Effect of clearance on piercing and stripping force required for piercing 6.53 mm (0.257 in.) diameter holes in cold rolled low-carbon steel
No lubricant was used.

Clearance per side, % of stock thickness	Force required Piercing(a) MPa	ksi	Stripping (total) N	lbf	Punch penetration into die(b) mm	in.	Clearance per side, % of stock thickness	Force required Piercing(a) MPa	ksi	Stripping (total) N	lbf	Punch penetration into die(b) mm	in.
Stock 0.64 mm (0.025 in.) thick, 65 HRB							**Stock 1.27 mm (0.050 in.) thick, 71 HRB**						
6.0	455	66.0	703	158	0.20	0.008	5.0	405	58.7	974	219	0.20	0.008
12.5	462	67.0	480	108	0.20	0.008	12.5	383	55.6	431	97	0.20	0.008
Stock 0.79 mm (0.031 in.) thick, 47 HRB							5.0	404	58.6	956	215	0.51	0.020
5.0	350	50.8	703	158	0.20	0.008	12.5	393	57.0	498	112	0.51	0.020
13.0	341	49.5	503	113	0.20	0.008	**Stock 1.27 mm (0.050 in.) thick, 61 HRB**						
Stock 0.86 mm (0.034 in.) thick, 87 HRB							5.0	367	53.2	1160	260	0.51	0.020
4.5	583	84.5	578	130	0.20	0.008	12.5	374	54.2	418	94	0.51	0.020
13.0	569	82.6	334	75	0.20	0.008	5.0	367	53.2	1312	295	0.51	0.020
Stock 1.07 mm (0.042 in.) thick, 85 HRB							12.5	364	52.7	703	158	0.51	0.020
5.0	551	79.9	1250	282	0.18	0.007	**Stock 1.50 mm (0.059 in.) thick, 74 HRB**						
12.0	527	76.5	783	176	0.18	0.007	5.0	369	53.5	498	112	0.47	0.0185
Stock 1.19 mm (0.047 in.) thick, 47 HRB							6.8	358	52.0	605	136	0.47	0.0185
5.0	341	49.4	1160	260	0.47	0.0185	7.6	350	50.8	383	86	0.47	0.0185
6.5	352	51.0	298	67	0.47	0.0185	8.5	349	50.6	374	84	0.47	0.0185
8.5	339	49.2	267	60	0.47	0.0185	9.8	346	50.2	200	45	0.47	0.0185
9.5	353	51.2	165	37	0.47	0.0185	13.0	355	51.5	89	20	0.47	0.0185
10.5	350	50.8	133	30	0.47	0.0185	**Stock 1.57 mm (0.062 in.) thick, 50 HRB**						
13.0	332	48.2	249	56	0.47	0.0185	5.0	371	53.8	578	130	0.47	0.0185
							6.5	371	53.8	454	102	0.47	0.0185
							7.3	363	52.6	325	73	0.47	0.0185
							8.0	364	52.8	578	130	0.47	0.0185
							9.0	361	52.4	534	120	0.47	0.0185
							12.5	363	52.6	249	56	0.47	0.0185

(a) Pounds per square inch of cross section cut. (b) Penetration required for release of slug.

Table 3 Effect of punch-to-die clearance on tool life in piercing and blanking of ferrous and nonferrous metals of various thicknesses

Stock thickness		Type	Hardness	Initial clearance		Increased clearance		Tool life increase with greater clearance, %
mm	in.			Clearance per side, % of stock thickness	Tool life per grind, holes	Clearance per side, % of stock thickness	Tool life per grind, holes	
Low-carbon steels, cold rolled								
0.41	0.016	Zinc coated	79 HRB	6.3	30,000	12.5	140,000	366
0.51	0.020	1018	22 HRC	2.5	115,000	5.0	230,000	100
0.91	0.036	. . .	(a)	2.8	67,000	12.5	204,000	205
1.19	0.047	1010	. . .	5.0	10,000	12.5	68,000	580
1.52	0.060	. . .	77 HRB	4.5	130,000	12.5	400,000	208
1.78	0.070	Galvanized	32 HRB	5.0	100,000	11.0	300,000	200
Low-carbon steels, hot rolled								
1.35	0.053	. . .	72 HRB	5.0	80,000	12.5	240,000	230
3.23	0.127	. . .	94 HRN	5.0	100,000	12.5	250,000	150
High-carbon steels								
1.52	0.060	1070	. . .	. . .	. . .	15.0	100,000	. . .
1.98	0.078	1090	. . .	. . .	. . .	10.0	835,000	. . .
2.03	0.080	4130	73 HRB	5.0	. . .	7.5	70,000	. . .
3.18	0.125	. . .	9 HRC	2.5	30,000	8.5	240,000	700
Stainless steels								
0.13	0.005	301	45 HRC	20.0	15,000	42.0	125,000	900
0.51	0.020	410	. . .	3.8	5,000	12.5	136,000	2,600
1.14	0.045	304	16 HRC	6.5	12,000	11.0	30,000	150
1.60	0.063	. . .	89 HRB	5.0	175,000	9.0	250,000	60
Co-Cr-Ni-base heat-resistant alloy								
0.91	0.036	HS-25 (L-605)	22 HRC	2.8	1,500	9.5	5,000	230
Aluminum alloys								
0.46	0.018	5086	. . .	16–20(b)	20,000	16–20(c)	70,000	250
1.02	0.040	3003	(d)	5.0	. . .	12.5	(e)	. . .
1.32	0.052	. . .	(f)	5.0	. . .	8.5	. . .	50
Copper alloys								
0.18	0.007	Tin-plated brass	76 HRB	7.0	. . .	14.0	(g)	50
1.14	0.045	Brass	. . .	3.5	15,000	7.0	110,000	(h)
1.19	0.047	Paper-clad brass	81 HRB	5.0	20,000	10.0	25,000	(h)
0.08	0.003	Beryllium copper	95 HRB(j)	8.5	300,000	25.0	600,000	100

(a) No. 4 temper. (b) Punch entered die 1.5 mm (0.060 in.). (c) Punch did not enter die. (d) H12 temper. (e) Higher-quality parts. (f) Soft. (g) Eliminated die breakage. (h) Run completed without regrind. (j) Half hard.

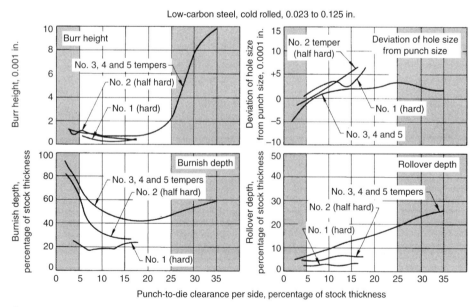

Low-carbon steel, cold rolled, 0.023 to 0.125 in.

Fig. 4 Edge characteristics (burr height, hole-size deviation, burnish depth, and rollover depth) in the piercing of low-carbon steel of different hardnesses with various punch-to-die clearances. Curves are for the AISI tempers shown, corresponding to the following HRB hardness limits: No. 1, 98 and higher; No. 2, 70 to 85; No. 3 to 5, 75 and lower. Results in shaded areas are not closely reproducible and show trends only.

However, to keep the inventory of piercing tools from becoming too large, some manufacturers use a modified practice for stocking piercing punches and die buttons in commonly used diameters. Punches are ordered to size. Work metal thickness is classified into several ranges, and die buttons are ordered to the specified clearance per side for the median stock thickness of the range to which the work metal for the given application belongs.

Hole dimensions are slightly affected as the clearance is changed. When using clearances that produce a type 4 edge, the diameter of the pierced hole is approximately 0.013 mm (0.0005 in.) less than that of the punch used to produce it. By increasing the clearances to those for a type 2 edge, the hole size will be equal to, or approximately 0.013 mm (0.0005 in.) larger than, the punch diameter.

With tight clearances, the slug is wedged into the die cavity. As the clearance is increased, the wedging action decreases; consequently, the slug may be as much as 0.013 mm (0.0005 in.) smaller than the die cavity.

Force Requirements

The force needed to pierce a given material depends on the shear strength of the work metal, the peripheral size of the hole or holes to be pierced, stock thickness, and depth of shear on the punch. The calculation of piercing force is the same as that for cutting force in blanking (see the article "Blanking of Low-Carbon Steel" in this Volume).

Effect of Punch Shear. Shear is the amount of relief ground on the face of a punch (Fig. 5). It is used to reduce the instantaneous total load on the tool and to permit thicker or higher-strength materials to be pierced in the same press. It distributes the total piercing load over a greater portion of the downstroke by introducing

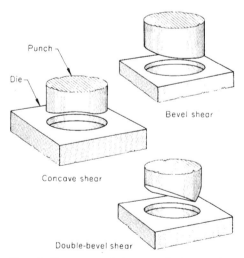

Fig. 5 Three types of shear on piercing punches. Angle and depth of shear are exaggerated for emphasis.

the cutting edge in increments rather than instantaneously.

Piercing force (but not contact edge pressure or total work done) varies with the amount of shear on the punch face. With the bottom of the punch flat and parallel to the face of the die, piercing takes place on the entire periphery at once, requiring maximum force. The load on the press and tools increases rapidly to a maximum after impact and then releases suddenly when piercing is completed. By grinding shear on the punch as shown in Fig. 5, the maximum load is decreased, but the punch travels correspondingly farther to complete the piercing. Load release is also somewhat less sudden.

Shear location is ordinarily selected so as to confine distortion to the scrap metal (slug). Thus, in piercing, shear is ground on the punch because the punched-out metal is to be scrap. Concave shear and double-bevel shear (Fig. 5) provide a balanced load on the punch. Scalloped shear, sometimes ground on round punches, also provides a balanced load on the punch. An unbalanced load may cause deflection and tool breakage or excessive wear.

The amount of shear is determined by trial. However, shear equal to one-third of stock thickness (t/3) will reduce piercing force approximately 25%, and shear equal to stock thickness will reduce piercing force approximately 50%. Shear can be applied to punches for large holes but not to small-diameter punches, because they lack column strength.

When a number of small holes are being pierced and the press load must be reduced, the punches can be ground to different lengths. This enables the punches to start cutting at different times and reduces the maximum load. In selecting a press, it should be noted that the reduction in maximum impact load on the press achieved by staggering punch length or by using shear is not sufficient to enable the use of a press that is significantly lower in tonnage rating, strength, or rigidity.

Presses

Presses used in piercing are the same as those used in other pressworking operations. Open-back gap-frame presses of the fixed upright, fixed inclined, or inclinable type are common. The stock can be fed from the side with minimal interference from the press frame, and the parts can be removed from the front by the operator or ejected out the back by gravity or air jets.

Adjustable-bed or horn presses are used for piercing holes in tubing and in the sides of drawn or formed shells and boxes. Adjustable-bed and gap-frame presses are generally rated at capacities of less than 1.8 MN (200 tonf).

Straight-side presses are commonly used for compound-die and progressive-die operations. Increased accuracy, speed, and stability are required for these operations.

The turret punch press is a special machine in which the punches and dies are mounted in synchronized indexing tables. Several sets of punches and dies are mounted in the table, which can be manually or automatically indexed into operating position. A flat blank is pierced and notched in a turret punch press by positioning it under the operating punch and tripping the punching mechanism. The blank is secured to a free-floating table on which a template containing the hole pattern is also attached. Each hole size and shape is coded so that all such holes can be pierced before indexing a new punch-and-die set under the press ram. The table is moved so that a pin will drop into a hole in the template; this places the blank in the proper position for piercing a hole. After the holes of one size and shape have been pierced, a new punch and die are indexed into operating position, and piercing continues in this manner until the part is finished. Almost any size or shape of hole can be pierced, within the capacity of the machine.

Turret punch presses can be programmed for tape control for increased production. Turret movement can also be controlled semiautomatically or automatically on numerical control or computer numerical control (CNC) punch presses. A closed-loop direct or alternating current drive connected to both the upper and lower turret assemblies provides the automatic turret with either unidirectional or bidirectional movement. Selected CNC presses even offer an optimization feature that automatically determines the most efficient and most cost-effective punching sequence for a specific workpiece.

Tools

A typical piercing die consists of:

- Upper and lower die shoes, to which punch and die retainers are attached
- Punches and die buttons
- A spring-actuated guided stripper (Fig. 6)

Small workpieces are generally pierced in compound dies that blank and pierce in the same stroke. Piercing is also done in the stations of a progressive die or a transfer die.

Any of these dies can be constructed as multiple dies, in which two or more workpieces are pierced at each stroke of the press. Additional information on dies is available in the article "Blanking of Low-Carbon Steel" in this Volume.

Punches. Figure 6 shows three types of punches used for piercing: conventional, standard quill, and telescoping-sleeve quill. Conventional punches are generally available with a standard maximum shank diameter of 25 mm (1 in.), and they can be used to pierce round, square, oblong, or rectangular holes that have a size and shape not exceeding 25 mm (1 in.) in diameter or a size and shape that can be machined or ground on the end of the shank, as shown in Fig. 6(a). Punch shanks are also available in other sizes and shapes. Standard-sized punch retainers, backing

plates, die buttons, and die button retainers are also generally available.

The head-type punch shown in Fig. 6(a) is held by a carefully positioned and ground punch retainer. This type of punch cannot be replaced without removing the retainer. Headless punches are available with locking devices that permit replacement without retainer removal.

A spring-loaded and guided stripper is incorporated into the die design. Conventional punches for small-diameter holes and accurately spaced holes are supported and guided by hardened bushings pressed into the guided stripper plate. Piloting punches can also be guided in the same manner. For larger punches that do not need guidance or support, clearance holes are drilled in the stripper plate, and guide bushings for both the punch and the plate are omitted.

Figures 6(b) and (c) show two types of quill punches that replace conventional punches when piercing conditions are more severe or when more support is required for the punch when piercing small-diameter and accurately spaced holes to close tolerances. The type shown in Fig. 6(b) was designed to hold and align a small-diameter punch. The quill containing a close-fitting

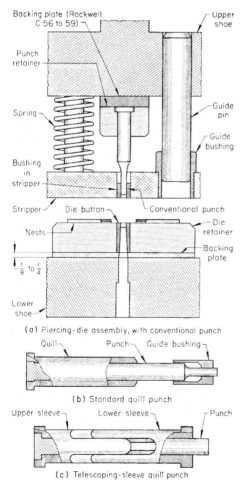

(a) Piercing-die assembly, with conventional punch

(b) Standard quill punch

(c) Telescoping-sleeve quill punch

Fig. 6 Typical piercing die, and three types of punches used. See text for discussion. Dimensions given in inches

punch is pressed into the punch retainer. The punch can easily be changed and still maintain the original alignment and fit in the retainer. A punch guide bushing is used in the stripper plate, and a hardened backup plate supports the head of the punch. Quills are available for punches with body diameters ranging from 1.0 to 9.52 mm (0.040 to 0.375 in.). The nib or end can be ground to a smaller diameter if desired.

The telescoping-sleeve quill punch shown in Fig. 6(c) provides complete support to small-diameter punches and eliminates the weaknesses encountered in other punch-mounting designs. The upper part of the sleeve is press fitted into the punch retainer; the lower part, into the guided stripper plate. The inside diameter of the sleeve will accommodate punch bodies ranging from 0.38 to 9.52 mm (0.015 to 0.375 in.) in diameter. However, a punch body at the larger end of the range and with a ground nib is suggested for best results. When piercing holes in printed circuit boards, a 2° cone-type taper is ground on the bottom surface of the lower sleeve to concentrate the holding and stripping force at the edge of the hole.

Quill punches have been used to pierce holes in low-carbon steel having a thickness up to twice the punch diameter. Supporting sleeves or quills can be used for long, narrow punches of rectangular, oblong, or other shape.

Piercing dies can be ground into a hardened die block, or they can be die buttons that are press fitted into a die retainer, as shown in Fig. 6(a). To coordinate the quill, guide bushing, and die button, the punch retainer, stripper plate, and die retainer can be clamped together and jig bored and ground at the same time. This is possible because the quill, guide bushing, and die button are available with the same body diameters.

Guided Strippers. For all three types of punches, the function of the guided stripper is threefold. On the downstroke, the spring-actuated stripper contacts the work metal ahead of the punches and acts as a holddown. On the upstroke, the metal is stripped from around the punches. The third function is to guide small punches. To ensure that the force at the points of contact and release is sufficient to accomplish the stripping, stripper springs must exert the calculated force in the open-die position (not just in the closed-die position).

Self-Contained Tools. Piercing tools can also be purchased as individual units consisting of frame, punch, die, and spring or hydraulic strippers. The self-contained unit is not attached to the press ram but is located, pinned, and bolted to a die plate, template, or T-slot plate mounted on the bed of any type of press having adequate shut height. The units can be reused by relocating, pinning, and bolting. They can be used singly or in groups and with notching units of the same construction. Punch and die sizes can be replaced as desired, and standard sizes and shapes are available for piercing thicknesses up to 19 mm ($\frac{3}{4}$ in.). The units are available in various styles for the horizontal and vertical piercing of flat or flanged workpieces.

Tool Materials. The materials used for piercing punches and dies are selected to suit the service requirements. In general, the materials used are the same as those for blanking. See the article "Selection of Materials for Shearing, Blanking, and Piercing Tools" in this Volume for information on tool materials and tool life.

Piercing involving unusual shock and high impact may require a shock-resistant tool steel such as S7. As in blanking, M2 high-speed steel is used for long punch life, particularly in piercing thicker steel, or where high abrasion resistance is required. In Example 7 in this article, a cam-actuated punch made of an air-hardening tool steel was replaced by a punch made of low-carbon steel, carburized and hardened. This change increased tool life tenfold.

Use of Single-Operation Dies. Single-operation piercing dies are used:

- When piercing is the only operation to be performed
- When the holes to be pierced are so close to the edge of the work that the incorporation of a piercing operation into a compound die would weaken the die elements
- When the required accuracy or the sequence of operations prevents the inclusion of piercing in compound, progressive, or transfer dies

In large work, such as panels for automobiles, holes are often pierced in a separate operation for numerous reasons. For example, the holes may distort during forming, accuracy of position may be impaired if holes are pierced before forming, or separate operations may provide a more balanced work load and reduce maintenance.

As an example, when a hole is pierced and coined to a bevel with the same punch, coining extrudes an excessive burr between the punch and the die unless a close punch-to-die clearance is maintained. Die maintenance can be reduced by using separate piercing and coining dies.

Use of Compound Dies

Compound dies are used for most piercing operations in which accuracy of position is important. Except for the production of small lots, the use of a compound die is usually the most economical method for making a pierced and formed part to commercial tolerances. The following example describes the use of a compound die to blank and pierce a U-shaped bracket before it was formed.

Example 1: Piercing Five Holes with a Compound Blank-and-Pierce Die. The mounting bracket shown in Fig. 7 was made of 4.55 mm (0.179 in.) thick hot rolled 1010 to 1025 steel strip, 200 mm (7⁷⁄₈ in.) wide, in two operations: blank and pierce, then U-form. A compound die was used to blank the outline and pierce three round holes and two slots. The holes were reamed to size in a secondary operation to hold the 0.025 mm (0.001 in.) tolerance on the diameter. The compound die was made of oil-hardening tool steel (O2) hardened to 58 HRC. Punch-to-die clearance was 5% of stock thickness per side.

The blank was located by pins in the 8.7 by 14 mm ($\frac{11}{32}$ by $\frac{9}{16}$ in.) slots for the forming operation. A spring-loaded pressure pad held the blank firmly against the punch during forming. To overcome springback, the flanges were overbent by 2 to 3°. The forming die was made of air-hardening tool steel (A2).

Total tolerance on location of the in-line holes was 0.25 mm (0.010 in.). Production lots were 50 to 100 pieces. Expected die life was 15,000 to 20,000 pieces before regrinding.

Use of Progressive Dies

Progressive dies perform blanking, piercing, and other operations in successive stations of a die. Each station in a progressive die is similar to a simple die or a compound die. The workpiece in a progressive die remains connected to the strip of work metal until the last station in the die, so that the feeding motion carries the work from station to station.

In a progressive die, piloting holes and notches are pierced in the first station. Other holes can be pierced in any station if they are not affected by subsequent cutting or forming. Holes for which the relative position is critical are pierced in the same station; other holes are distributed among several stations if they are close together or near the edge of a die opening. Tolerances on hole shape, size, or location dictate whether holes are pierced before or after the part is formed.

It is often advisable to add idle stations or to distribute the work over one or two additional stations, so that holes will not be pierced near the edge of a die block. The die block is therefore stronger, and there is less chance of the die cracking in operation or fabrication. Adding stations also allows better support for the piercing punches and increases strength to the strip.

Progressive dies are more expensive than a set of single-operation dies for the same part; therefore, progressive dies are generally used for high production. However, because one part is made at each press stroke, direct labor costs are greatly reduced, and one operator can often

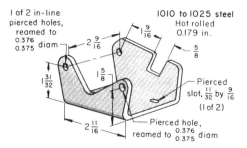

Fig. 7 Bracket that was blanked and pierced in a compound die before being formed. Dimensions given in inches

attend to more than one progressive die. Manufacturing costs can also be reduced by making a pierced and formed part in a progressive die, rather than in two separate dies (one compound and one single-operation die).

The amount of scrap produced in progressive dies is generally high because the nesting of parts is somewhat limited and because material must be provided for connecting tabs and carrier strips. A fully automatic press with cutoff, feed, straightener, and coil cradle or reel is normally used with a progressive die; therefore, press costs are high. Pierced parts can sometimes be made most economically in a progressive die by using coil stock that is the exact width of the developed blank. In the following example, the high-volume production of a bracket in a progressive die is compared with the production of smaller quantities in separate operations in utility tooling.

Example 2: Producing a Bracket in Large Quantities in a Progressive Die and in Small Lots with Utility Tooling. Figure 8 shows a bracket and the strip development for producing it in a five-station progressive die in a 670 kN (75 tonf) mechanical press that had a 102 mm (4 in.) stroke, an air-actuated stock feeder, and an automatic oiler. Material for the bracket was coiled cold rolled low-carbon steel strip, 2.41 mm (0.095 in.) thick by 135 mm (5¼ in.) wide, in the No. 2 (half-hard) temper.

The die was made of D2 tool steel and hardened to 59 to 60 HRC. Setup time was 1.5 h, and the press was stopped for die maintenance at intervals of 15,000 pieces. Production rate was 1200 pieces per hour. A light paraffin oil was the lubricant. The tolerance on the dimensions of all of the pierced holes was +0.051, −0.025 mm (+0.002, −0.001 in.), and the tolerance on the position of the square hole and the two rounded slots was ±0.13 mm (±0.005 in.).

Small quantities of the brackets were made from coil stock 29 mm (1 in.) wide in the following operations, using utility tooling:

- Cut stock into 135 mm (5 in.) lengths (3500 pieces per hour)
- Trim end in a single-operation die, two strokes per piece (1000 pieces per hour)
- Pierce two holes 10.1 by 13.3 mm (0.398 by 0.523 in.), one at each end, in a single-operation die, two strokes per piece (1000 pieces per hour)
- Pierce a 4.88 mm (0.192 in.) wide slot and a 5.11 mm (0.201 in.) square hole, and lance and form two ears, all in a compound die for accurate relative position (2000 pieces per hour)
- Bend the ends down, locating on the ears (1000 pieces per hour)
- Drill four holes 4.90 mm (0.193 in.) in diameter in a multiple-spindle drilling machine (500 pieces per hour)
- Drill small hole next to square hole (500 pieces per hour)

The short-run method required eight times as many man-hours per 1000 pieces as the progressive-die method, and the brackets produced were less accurate.

Use of Transfer Dies

Transfer dies are used for piercing in applications that are similar to those for which progressive dies are used. A number of operations are done in successive stations of the transfer die.

Blanking, cutoff, lancing, notching, forming, and drawing (as well as piercing) can be done in transfer dies. The method differs from progressive-die operation in that the workpiece does not remain attached to the strip for feeding but is fed from station to station by transfer fingers. Production quantities must be large enough to justify the cost of tooling and equipment.

Accuracy in the dimensions between pierced holes is highest for holes that are pierced by the same die in one press stroke. Accuracy in the location of holes relative to an edge or some other feature is highest when the workpiece is blanked and pierced (or pierced and trimmed) in the same stroke in a compound die.

When the aforementioned procedures are used, total tolerances of 0.25 mm (0.010 in.) on hole location and 0.13 mm (0.005 in.) on hole size are readily met in normal production, and closer tolerances can be met with suitable tools, as indicated in Table 4. Tooling cost and per-piece cost usually increase in piercing to closer tolerances. Accuracy is typically somewhat lower for holes pierced by different dies or in different stations of a progressive or transfer die because of piloting and nesting tolerances.

Tolerances smaller than the lowest given in Table 4 can be met with the use of special tooling and gaging and close control over the press operations, but only at increased cost and a lower production rate. The use of shaving to produce holes to a tolerance of less than 0.025 mm (0.001 in.) in size is described in Example 11 in this article. The use of fine-edge blanking for improved accuracy and edge quality is discussed in the article "Fine-Edge Blanking" in this Volume.

Accuracy of hole location is increased by the use of a rigid stripper, precisely aligned on guideposts, to guide the punches. Typical clearance of round punches in the stripper (using drill bushings as guides) is 0.005 to 0.013 mm (0.0002 to 0.0005 in.) total. Lubrication is important in dies that have such close clearance between punches and stripper. At speeds of 40 strokes and more per minute, the die must be lubricated constantly with a spray of light machine oil to prevent galling of the punches in the guides.

Accuracy often requires that holes be pierced after forming. In some cases, it may be necessary to pierce a hole after forming in order to avoid distortion of the hole. In Example 8 in this article, holes on opposing flanges were cam pierced in one stroke after forming for accurate alignment. In Example 7, two slots were pierced in flanges after forming, instead of being machined, for accurate alignment and location.

Noncritical holes, for which there are no close tolerances on size or spacing, can be pierced for venting (to provide for free passage of air or other fluids), for lightening, for improved bond with a molded plastic cover, for increased flexibility of a workpiece, and even to provide

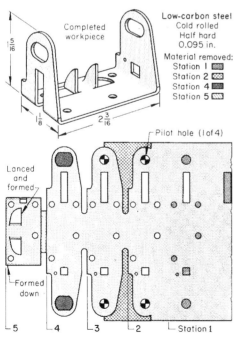

Fig. 8 Bracket that was produced more economically and more accurately in a progressive die (strip layout shown) than by the separate-operation method used for small quantities. Dimensions given in inches

Table 4 Typical accuracy in piercing

For holes that are pierced with a conventional die in the same press stroke. Location will be less accurate for holes that are pierced with different dies, or pierced in different stations of a progressive die or a transfer die.

| Finish on tools | Die retainers used | | Total tolerance on pierced holes | | | |
| | | | Location(a) | | Size(b) | |
	Typical material	Locating holes for tools	mm	in.	mm	in.
Commercial ground	1020 or 4130(c)	Drilled in a drill press	0.25	0.010	0.13	0.005
Commercial ground	1020 or 4130(c)	Jig bored and jig ground	0.10	0.004	0.13	0.005
Precision ground	4130(d)	Jig bored and jig ground	0.05	0.002	0.03	0.001

(a) Relationship between hole centers or between hole center and edge or other reference points on the workpiece. (b) Diameter for round holes, or other control dimension for holes of other shapes. (c) Can be hardened; other grades of steel also can be used. (d) Hardened and tempered before being jig ground

controlled strength. In the following example, noncritical holes were pierced to weaken a part so it would deflect under impact or shock loading.

Example 3: Piercing Noncritical Holes. The workpiece shown in Fig. 9 is a wheel spider that was designed to be enclosed in molded plastic and to deflect under impact load. The spider was made of hot rolled 1008 or 1010 steel, dead soft, pickled and oiled. The steel strip, 3.2 mm ($\frac{1}{8}$ in.) thick by 406 mm (16 in.) wide, was hand fed into a two-stage progressive die in a 4 MN (450 tonf) mechanical blanking press. In the first stage, noncritical holes were pierced to make the spider deflect upon impact and to provide a better bond with the molded plastic. The piece was blanked in the second stage. The pierced blanks were then fed into a 5.3 MN (600 tonf) mechanical forming press, where the part was formed. The forming die was sprayed with soluble oil.

Hole Size

Pierced holes can be of almost any size, ranging from holes as small across as the thickness of the stock to the largest size that can be adapted to the equipment available. Some holes can be pierced that are smaller across than the stock thickness, but such piercing is not common.

Small holes and slots are pierced in much the same manner as large holes, but small holes are more difficult to pierce because slender punches are comparatively weak. To minimize deflection and breakage, punch length is limited to that needed for the operation, and punches are specially stiffened and guided (Fig. 6). The minimum hole size that can be pierced in a specific application is ordinarily found by trial.

Piercing of Thick Stock

The effects of work metal thickness on piercing are generally the same as on blanking. Cutting force increases with thickness; consequently, tool design and material, press selection, and operating conditions are influenced by work metal thickness. The relationship between die clearance and work metal thickness is discussed in the sections on die clearance in this article. The minimum pierced hole size is usually expressed as a function of work metal thickness, as described in the preceding section. The following example illustrates the piercing of unusually thick flat workpieces prior to forming.

Example 4: Piercing a Square Hole in 19 mm ($\frac{3}{4}$ in.) Plate. A hole 17 mm ($\frac{11}{16}$ in.) square was pierced through high-strength low-alloy steel plate 19 mm ($\frac{3}{4}$ in.) thick in a die in a 1.1 MN (120 tonf) open-back inclinable press. The die was made to pierce holes one at a time in any of several flat workpieces. A typical workpiece size was 19 by 305 by 940 mm ($\frac{3}{4}$ by 12 by 37 in.).

The production rate for piercing in D2 tool steel dies was 300 to 360 pieces per hour. Maximum yearly production was 1000 pieces.

After the holes were pierced, the workpiece was heated to 815 °C (1500 °F) and formed into a curved shape. The shape was used as a digger bucket—one of many that were bolted to a large wheel (digger rim) of a machine used for digging trenches for sewers and pipelines.

Radial Piercing of Curved Surfaces. The radial piercing of thick curved workpieces is done in the same manner as the piercing of flat workpieces, except that the die must be designed to accommodate the curved parts. The piercing of a hole through round stock, such as a radial hole through a cylinder, is done by using a die with a heavily loaded stripper, both of which fit the round shape of the workpiece. The round hole is then readily pierced with little bulging of the workpiece, even though the hole can be as large as 40% of the diameter of the workpiece. The center of the hole should be at a distance from the end of the workpiece that is at least equal to the work thickness (diameter of the rod).

Piercing of Thin Stock

Because of its strength and rigidity, material thicker than 3 mm ($\frac{1}{8}$ in.) is seldom blanked or pierced from coil stock or in a progressive die. On the other hand, material thinner than 0.5 mm (0.020 in.), because of its lack of strength and extreme flexibility, generally requires special

handling techniques. The following example describes the blanking of steel shims from thin coil stock in a compound die.

Example 5: Piercing of Shims from 0.25 mm (0.010 in.) Strip in a Compound Die. The shim shown in Fig. 10 was made from coil stock of cold rolled 1008 steel, 19 mm ($\frac{3}{4}$ in.) wide by 0.25 mm (0.010 in.) thick, in a pierce-and-cutoff compound die. Thickness tolerance was +0, -0.025 mm (+0, -0.001 in.). The coiled strip was fed automatically from a stock reel by an air-operated slide feed into a 220 kN (25 tonf) open-back inclinable press that operated at 150 strokes per minute.

The die, made of oil-hardening D3 tool steel, produced 230,000 shims before it needed sharpening. Die life was indefinite because a thickness of 38 mm ($1\frac{1}{2}$ in.) had been provided for grinding allowance in repeated sharpenings in order to restore the cutting edge.

Hole Spacing. Recommended minimum spacings for pierced holes are given in Table 5. These minimum spacings apply when ordinary pressworking practices are followed, without confinement of the workpiece in the die or other special procedures to prevent distortion.

As noted in Table 5, S_5 (the distance from the edge of a hole to the inside of a flange) can be reduced when the metal is relieved near the pierced hole (as by a slot) to prevent distortion of the hole in forming. Accuracy in the shape and location of pierced holes often demands that the holes be pierced after the workpiece has been formed.

When a round hole must be pierced so as to leave almost no metal between the edge of the

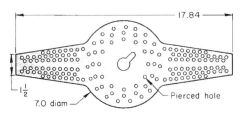

Pierced and blanked workpiece

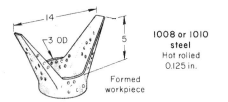

1008 or 1010 steel
Hot rolled
0.125 in.

Formed workpiece

Fig. 9 Wheel spider in which noncritical holes were pierced for deflection under impact and to improve bonding to plastic. Dimensions given in inches

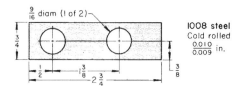

1008 steel
Cold rolled
$\frac{0.010}{0.009}$ in.

Fig. 10 Shim that was pierced and cut off from thin coiled strip in a compound die. Dimensions given in inches

Table 5 Recommended minimum spacings for pierced holes in flat and formed steel and nonferrous metal workpieces

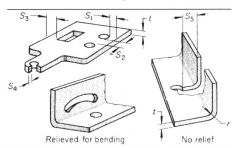

Relieved for bending   No relief

Dimension	Work metal thickness (t), mm (in.)	Minimum distance, mm (in.)
S_1 and S_2	<1.57 (0.062)	3.05 (0.120)
	1.57–9.65 (0.062–0.380)	3.05 (0.120) (but at least 1.5t(a)
S_3 and S_4	<0.81 (0.032)	1.52 (0.060)
	0.81–3.18 (0.032–0.125)	2t
	3.18–9.65 (0.125–0.380)	2.5t
S_5	to 9.65 (0.380)	1.5t + r(b)

(a) For steel. Minimum for nonferrous metals, 2t. (b) For flanges with no relief. Value for S_5 can be reduced when work metal is relieved for bending (as shown at lower left above) to prevent distortion of the pierced hole in forming.

hole and the edge of the part, the hole can be cut through the edge in a keyhole shape that minimizes bulging and does not leave sharp points. In the following example, spacing between the keyhole and the end of the part was only 58% of the recommended spacing given in Table 5, and spacing for the round hole and the slot was only 50% of the recommended spacing. Confinement of the blank in the die prevented distortion at the keyhole and the slot.

Example 6: Piercing Holes at Less Than Recommended Minimum Edge Distance. The flyweight shown in Fig. 11 was part of a centrifugal device used to release pressure. The part was made of hot rolled 1010 steel, pickled and oiled, 6.4 mm (¼ in.) thick. Preliminary design had a wall thickness of 2.4 mm (³/₃₂ in.) between the 11 mm (⁷/₁₆ in.) diameter hole and the edge of the part. The original tooling called for a compound die to pierce and blank the part completely in one press stroke. However, with the fragile punch required, it was impossible to hold the 8.00/7.87 mm (0.315/0.310 in.) dimension on the keyhole opening (detail A, Fig. 11).

Production was successful when the part was made in three separate dies, using the sequence of operations shown in the accompanying table. A keyhole punch was used to pierce the 11 mm (⁷/₁₆ in.) diameter hole. The hole-to-edge spacing was increased from 7.9 to 11.1 mm (⁵/₁₆ to ⁷/₁₆ in.).

Distortion was minimized by confining the blank in a nest during piercing. In the compound die, the hole shown in detail A and the edge of the part were connected with a radius tangent to both. When the keyhole punch was used, the punch surface intersected the outer edge at a 45° angle—a change that did not interfere with the function of the part but avoided difficulty in making a transition radius tangent to the outer edge.

The hole to be reamed was pierced with a 5.74 mm (0.226 in.) diameter punch and a 6.25 mm (0.246 in.) diameter die. The hole was later reamed to 6.29/6.26 mm (0.2475/0.2465 in.) in diameter for nearly its full length and then deburred. This hole was perpendicular to the part surface within +0° 45′.

Effect of Forming Requirements

It is simpler to pierce holes in a flat sheet than in a part that has been formed. Holes near a bend radius (see the illustration in Table 5) are usually distorted when the part is formed. If distorted holes are unacceptable or if an accurate relation of holes to other features in a workpiece is specified, piercing must be done after forming.

Dies for piercing after forming are generally more complex, more expensive, and require more maintenance than dies for flat blanks. These dies often have cam-actuated punches.

In Example 9 in this article, when the part was U-formed after piercing, one hole closed in

during forming, resulting in an elongated hole. This deformation was taken into consideration during product design. The following two examples describe applications in which it was necessary to pierce holes after forming.

Example 7: Piercing of Accurately Located Slots after Forming. Figure 12 shows a formed part that, because of close tolerances, should not be slotted before forming. The original procedure was to drill the holes and mill the slots to hold the alignment of the slots to the end radii and to the tongue within 0.08 mm (0.003 in.). Results from machining were unsatisfactory, and it became necessary to pierce the slots after forming in order to hold tolerances.

The size and shape of the part and the location of the slots precluded piercing from the outside. Piercing was done from the inside by two punches split on the centerline of the part and moved outward by cam action.

The original punches were made of air-hardening tool steel hardened to 58 to 60 HRC. Because of breakage, the maximum punch life was 10,000 pieces. Changing the punch material to carburized and hardened 1025 steel increased the average punch life to 100,000 pieces. Annual production was approximately 240,000 pieces in lots of 50,000 to 60,000.

Example 8: Cam-Piercing Holes in Opposing Flanges of a Formed Part for Accurate Alignment. To maintain alignment of the two opposing holes in the automobile-frame control-arm bracket shown in Fig. 13, the holes were cam pierced in one press stroke after the part was

formed. Tolerance on hole alignment was ±0.13 mm (±0.005 in.).

The bracket was made of commercial-quality 1008 or 1010 steel, as-rolled, 3.78 mm (0.149 in.) thick, in five operations:

- Blank two workpieces per stroke
- Prebend, form, and re-form in three separate dies, side by side
- Form ear
- Trim in two stages
- Restrike and cam pierce

Blanking was done at 1800 pieces per hour in a 4.4 MN (500 tonf) coil-fed automatic blanking press. The remaining operations were done at 135 pieces per hour in a 2.7 MN (300 tonf) or a 3.6 MN (400 tonf) mechanical press.

The use of high-production equipment, including the cam-operated piercing die, was economical for the annual production of 350,000 pieces in 20,000-piece lots. Drilling would have been used for making the holes if 10,000 or fewer pieces had been needed per year to meet production requirements.

Piercing Holes at an Angle to the Surface

For piercing holes that are not perpendicular to the surrounding surface, the workpiece is securely clamped to the die with a pressure pad, and the dies are usually ground to fit the contour of the part. The shape of the punch nose depends

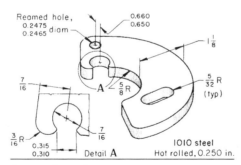

Sequence of operations

1. Shear 1–2 m (4–8 ft) long by 100 mm (3⁷/₈ in.) wide
2. Blank outline; pierce round hole (1000 pieces/h)
3. Ream and deburr round hole (300 pieces/h)
4. Pierce keyhole and oval slot (770 pieces/h)

Operating conditions

Types of press	670 kN (75 tonf) mechanical
Press speed	55 strokes per minute
Die material	A2 tool steel at 58–60 HRC
Lubricant	Sulfur-base, EP type(a)
Production rate(b)	300 pieces per hour
Die life per grind	20,000 pieces
Total die life	1 million pieces

(a) EP, extreme pressure. Applied to strip by roller. (b) Lot size was 2500 pieces; annual production, 10,000 pieces.

Fig. 11 Flyweight in which holes were pierced at less than recommended minimum distances from the edge. Overall length of the flyweight was 90 mm (3.5 in.). Dimensions given in inches

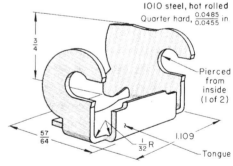

Fig. 12 Part in which accurately aligned slots were pierced (after forming) by two punches that were cam operated from the inside. Dimensions given in inches

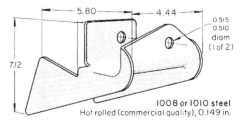

Fig. 13 Bracket in which accurately aligned holes on opposing flanges were cam pierced in one press stroke after forming. Dimensions given in inches

on the angle of contact with the workpiece and on the stock thickness.

In one application, holes 7.8 mm (5/16 in.) in diameter were pierced in 6.35 mm (0.250 in.) thick 1090 steel at an angle of 83° to the surface. The holes were pierced by using sleeved punches and clamping the work tightly to the die with a pressure pad. In the following example, holes were pierced at an angle of 40° 30′ to the surface of a flat workpiece, which was later formed by bending.

Example 9: Piercing Holes at an Angle to the Work Surface. Two 12.8 mm (0.505 in.) diameter holes were pierced at an angle of 40° 30′ to the surface in the 2.67 mm (0.105 in.) thick cold rolled 1010 steel flat blank for the lamp-bracket base shown in Fig. 14. To pierce the compound-angle holes, the blank was placed in a nest, which held it at the proper angle and position for piercing. By repositioning the blank in a second nest, it was possible to pierce both of these holes with one punch and die.

The face of the die button was ground flush with the surface of the nest. The die had a straight land with a minimum length of 9.5 mm (3/8 in.). There was a 16 mm (5/8 in.) diameter relief below the land. The punch was ground with a step, as shown in Fig. 14. This step curled the slug, permitting the use of a long straight land in the die. The punch and die were made of M2 high-speed steel, hardened to 58 to 59 HRC. Punch-to-die clearance was 0.05 mm (0.002 in.) per side. Die life was 15,000 holes per grind.

The punch was mounted in a heavy quill designed for quick changing, which proved unnecessary. A spring-loaded pressure pad guided and added support to the punch and held the blank securely in the nest. The pressure pad was interlocked with the die so that shifting could not take place after piercing started. The die was run in a 90 kN (10 tonf) mechanical press, which produced 300 pieces per hour.

Before the angle holes were pierced as described previously, the outline of the part and seven 90° holes had first been produced in a compound blank-and-pierce die. The blanking die and punch, as well as the punches and die buttons for the 90° holes, were made of O1 tool steel and hardened to 59 to 60 HRC. The die was mounted in a 900 kN (100 tonf) mechanical press producing 300 pieces per hour. Die life was 40,000 pieces per grind.

When the workpiece was bent through 180° after all of the holes had been pierced, the 11.3 mm (0.445 in.) diameter hole changed in shape, assuming final dimensions of 10.4 by 11.3 mm (0.411 by 0.443 in.). This deformation did not affect the function of the hole.

Special Piercing Techniques

Piercing operations that require special tooling and techniques include the piercing and forming of flanged holes, piercing with a fastener and with a pointed punch, and tube piercing.

Flanged holes (sometimes called extruded, countersunk, dimpled, or burred holes) are generally used for assembly purposes, such as providing more thread length for a tapped hole, greater bearing surface, or a recess for a flathead screw or rivet. The flanged hole can be produced by forcing a punch of the desired hole diameter through a smaller prepierced hole or by using a shouldered or pointed punch that both pierces the hole and flanges it.

The depth of flange that is formed depends on the elongation of the metal, and the flange is thinnest at its outer edge. A deeper flange can be made by extruding metal into the flange. Such flanging is done by first piercing a smaller lead hole and then using a punch that extrudes metal around the hole into the die clearance to produce a flange and simultaneously coins or forms a slight chamfer (depressed cone) or other shape of recess into the hole. This kind of extruded flange has uniform wall thickness. Such extrusion causes more metal flow and greater work hardening than ordinary piercing of flanged holes.

Piercing with a fastener (self-piercing) is primarily used as an assembly technique. A rivet, for example, can be used as a punch to pierce a hole through the material that it will join. The following example describes an application in which a square nut with a sharp face served as a piercing tool and then became part of an assembly.

Example 10: Automatic Assembly of Self-Piercing Nut into a Bracket. A square nut was automatically assembled into a body-bolt bracket, as shown in Fig. 15. The nut served as the piercing punch to make a 17.4 mm (0.687 in.) square hole in the embossed portion of the bracket. The bracket was made of galvanized, hot rolled 1006 steel, 1.9 mm (0.075 in.)

thick. After being pierced, the metal sprang back into two grooves in the nut, locking the nut into the pierced square hole. The nut was fed from a special installing head that was loaded from a rotary hopper.

Two assemblies were completed (pierced and installed) at each stroke of a 640 kN (72 tonf) open-back inclinable press running at 45 strokes per minute. Maximum daily production was 12,000 assemblies. The bracket was fed by a gravity-slide feed at the front of the press and was unloaded at the rear with the help of an air blast.

The 20.6 mm (0.812 in.) diameter hole was pierced at the same time the nut was inserted. Commercial punches and die buttons were used.

A standby unit for piercing a square hole was available in case a supply of the square nuts was not on hand when the part was scheduled to run. The nuts were later inserted and clinched using pneumatic equipment.

Holes made with a pointed punch, such as a cone-point, nail-point, or bullet-nose (ogive) punch, have rough or torn flanges. Such a hole is satisfactory for holding a sheet-metal screw, acting as a spacer, or providing a rough surface.

Tube piercing and slotting are done in dies when production lots are large enough to pay for special tooling. Simple dies, as well as more complex tooling such as cam dies, are designed to hold, locate, and pierce tubes, drawn cups, and other round parts.

A mandrel can be used in a horn die for work on tubing and other round parts as well as for piercing, slotting, and notching. One version of a die that uses a mandrel permits piercing two opposing holes in a tube in one stroke, with the slug from one wall going through a hole in the mandrel to act as the punch for the hole in the opposite wall.

Tubes can also be pierced with opposing holes without using a mandrel. The bottom half of the tube is supported by the die, which has a nest with the same diameter as the outside diameter of the tube. A similar nest for the upper half of the tube is in a combined holddown and punch guide. Therefore, the tube is completely surrounded during piercing. Because the bottom side of the tube is supported by the die, the lower hole is pierced without any distortion. The tube will collapse slightly around the hole in the top side of

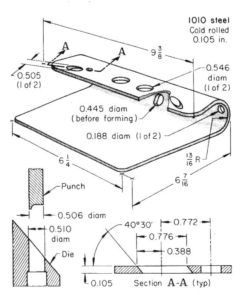

Fig. 14 Lamp-bracket base in which two holes were pierced at an angle to the surface using the punch shown at lower left. All holes were pierced before the part was formed. The 90° hole in the U-bend zone deformed to an elliptical shape during forming. Dimensions given in inches

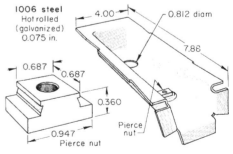

Fig. 15 Bracket with a square nut that pierced a hole for itself. Dimensions given in inches

the tube. The amount of distortion of the top hole varies with the size of the tube and the hole.

Holes on opposite sides of a tube can be pierced simultaneously, using the tool shown in Fig. 16, which does not require a mandrel. The tool consists of identical upper and lower assemblies—one attached to the press ram and the other to the press bed. Each assembly consists of a punch and a spring-loaded combined nest, stripper, and punch guide. When used for multiple-hole piercing, the assemblies are usually mounted to upper and lower plates having tapped holes or T-slots. There will be a slight indentation around each hole, as shown in Fig. 16.

Shaving

Shaving is done in a separate operation or is included in one station of a progressive die (see also the article "Blanking of Low-Carbon Steel" in this Volume). The inclusion of a shaving operation in a progressive die generally increases the need for die maintenance, and the slivers of shaving scrap can jam the feeding mechanism. A replaceable insert can be used in a shaving die for easier maintenance.

Shaving allowance depends on the workpiece material and on its thickness. Shaving plus burnishing is used to produce greater accuracy in a pierced hole than can be obtained by shaving alone, as shown in the next example.

Example 11: Use of Blanking, Piercing, Shaving, and Burnishing in Making Gear Blanks to Close Tolerances. The small gear blank illustrated in Fig. 17 was produced from a 50 mm (2 in.) wide strip of cold rolled 1010 steel of No. 2 temper in a five-station progressive die to the following specifications:

- Critical tolerance of +0.013, −0.010 mm (+0.0005, −0.0004 in.) on the center hole
- Finished blanks flat within 0.05 mm (0.002 in.)
- Surface finish of 0.70 μm (28 μin.) or smoother for 70% of the center-hole surface

These specifications were met by piercing, shaving, and burnishing the center hole in the

sequence of operations indicated by the strip progression. A sulfurized and chlorinated extreme-pressure lubricant was applied to the coil stock by roller coating.

Annual production was 2 million pieces in four lots. The gear blanks were made at 150 pieces per minute in a 530 kN (60 tonf) press. The dies, made of M2 high-speed tool steel, had a life of approximately 100,000 pieces before regrinding and a total life of approximately 10 million pieces.

High-Carbon Steels

The most important difference between the blanking and piercing of high-carbon and low-carbon steel is that greater clearance between punch and die is required for high-carbon steels. The clearances needed for producing each of the five edge types in blanking and piercing high-carbon and low-carbon steels are compared in Table 1 of this article. At equal clearance, roll-

over depth will be smaller and burnish depth greater for high-carbon than for low-carbon steel. For example, 12% clearance per side will produce a type 4 edge on high-carbon steel, and a type 2 edge on low-carbon steel.

Along with the increase in punch-to-die clearance, other tool changes are required for efficiency in the blanking and piercing of high-carbon steel. Dimensional accuracy in the blanking and piercing of high-carbon steel depends largely on the accuracy of the tooling. Initially, the practical accuracy is the same as that for the blanking and piercing of other metals. However, because the rate of tool wear is usually higher in the blanking and piercing of high-carbon steel (especially if pretempered) than for many other work metals, maintenance of tolerances can be more difficult and may require more frequent reconditioning of the tools. The tolerances that can be held and the size and spacing of holes and slots that are practical in press dies are illustrated in the following example.

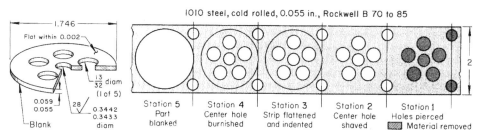

Fig. 17 Gear blank that was produced with an accurate center hole by piercing, shaving, and burnishing in a five-station progressive die. Dimensions given in inches

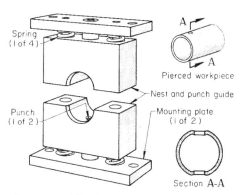

Fig. 16 Tool for tube piercing without a mandrel, and a pierced tube showing the indentations around the holes

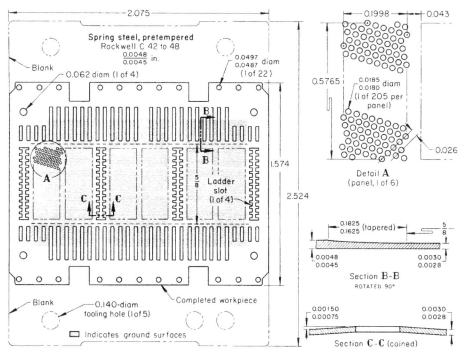

Fig. 18 Comb of an electric shaver made by blanking and piercing. Dimensions given in inches

Example 12: Blanking and Piercing an Intricate Pattern. The comb of an electric shaver (Fig. 18), made of pretempered spring steel, had an intricate pattern of slots and holes. Because more than 7 million of these combs were produced yearly, it was economical to construct the die needed to make them by blanking and piercing. Most of the dimensions on the part were held to a tolerance of ±0.025 mm (±0.001 in.). Similar parts of stainless steel were made by chemical machining.

The stock was 56 mm (2³/₁₆ in.) wide high-carbon spring steel, pretempered to 42 to 48 HRC. Thickness was 0.11 to 0.12 mm (0.0045 to 0.0048 in.). Blanks sheared from the stock were 52.70 mm (2.075 in.) wide by 64.11 mm (2.524 in.) long and included, at each end, 12.1 mm (0.475 in.) that was later trimmed. This trimmed stock contained five tooling holes, one of which was a foolproof hole that prevented incorrect placement of the blank in the die.

In the first piercing operation, the rectangular slots at each end of the cutting areas were pierced, along with the five 3.56 mm (0.140 in.) diameter tooling holes in the trim area and four ladder slots that bordered the three cutting areas. After this operation, the three cutting areas were ground to a thickness of 0.071 to 0.076 mm (0.0028 to 0.0030 in.). This ground surface, 15.9 by 12.6 mm (0.625 by 0.495 in.), tapered to the original stock surface in 4.11 to 4.62 mm (0.162 to 0.182 in.).

In the third operation, 0.46 mm (0.018 in.) diameter holes were pierced in the cutting areas at a centerline spacing of 0.572 mm (0.0225 in.) in one direction and 0.66 mm (0.026 in.) in the other direction. Each of the three cutting areas had 410 holes arranged in two panels of 205 holes each.

After the third operation, workpieces were deburred, and the area around each of the four ladder slots was coined to a depth of 0.019 to 0.038 mm (0.00075 to 0.0015 in.). The surface was buffed before plating, removing 0.008 mm (0.0003 in.) of stock. Finally, the comb was trimmed from the rough blank and four 1.6 mm (0.062 in.) diameter holes for locating studs and 22 rivet holes 1.24 to 1.26 mm (0.0487 to 0.0497 in.) in diameter were pierced. The piece was plated after assembly with the comb support.

Dies were made of D2 tool steel and hardened to 58 to 60 HRC. Some additional wear resistance that could have been attained by hardening to 60 to 62 HRC was sacrificed in order to make the delicate punches more resistant to shock. The presses were 620 and 670 kN (70 and 75 tonf) mechanical presses operated at 5 or 6 strokes per minute to accommodate meticulous hand feeding of the workpieces. Production lots consisted of 50,000 pieces each.

REFERENCE

1. L.R. Allingham, ASTME Paper MF64-151, Danley Machine Corporation

Blanking and Piercing of Electrical Steel Sheet

ELECTRICAL STEELS are used for various static and rotating electrical devices. They are magnetically soft materials; that is, they are not permanent magnets but have properties that make them useful in electrical applications. Most of the parts produced from electrical steels must be laminated. A lamination consists of flat blanked sheets of a particular shape that are stacked to a given height and fastened together by riveting, bolting, or welding. Electrical steel sheet is available in coils or cut-to-length. For most applications, stock thickness ranges from 29 to 24 gage (0.343 to 0.607 mm, or 0.0135 to 0.0239 in.).

Materials

The general category of magnetically soft materials encompasses many types of materials, including iron-nickel, iron-cobalt, and iron-aluminum alloys; ferrites; and austenitic stainless steels. The discussion in this article, however, is limited to the most commonly used magnetically soft materials: low-carbon electrical steels and oriented and nonoriented silicon electrical steels. Table 1 lists some of the characteristics and applications of these materials.

Low-Carbon Steels. For many applications that require less than superior magnetic properties, low-carbon steels (AISI 1010, for example) are used. Higher-than-normal phosphorus and manganese contents are often used to increase electrical resistivity. Such steels are not purchased to magnetic specifications. Although low-carbon steels exhibit power losses higher than those of silicon steels, they have better permeability at high flux density. This combination of magnetic properties, coupled with low price and excellent formability, makes low-carbon steels especially suitable for applications such as fractional-horsepower motors, which are used intermittently.

Nonoriented Silicon Steels. Except for saturation induction, the magnetic properties of iron containing a small amount of silicon are better than those of pure iron. Few commercial steels contain more than 3.5% Si because the steel becomes brittle and difficult to cold roll at silicon levels above 4%.

The commercial grades of silicon steel in common use (0.5 to 3.5% Si) are made primarily in electric or basic-oxygen furnaces. Nonoriented grades are melted with careful control of impurities; better grades have sulfur contents of approximately 0.01% or less. Continuous casting and vacuum degassing can be used. After hot rolling, the hot bands are annealed, pickled, and cold rolled to final thickness as continuous coils.

Semiprocessed grades of strip are not sufficiently decarburized for electrical use; therefore, decarburization and annealing to develop potential magnetic quality must be done by the user. This procedure is practical for small laminations accessible to the annealing atmosphere. Fully processed grades are strand annealed in moist hydrogen at approximately 825 °C (1520 °F) to remove carbon. The final annealing operation is very important and is carried out at a higher temperature (up to 1100 °C, or 2000 °F, for continuous strip) to cause grain growth and the development of magnetic properties. Use of a protective atmosphere is vital. The steel often is coated with organic or inorganic materials after annealing to reduce eddy currents in lamination stacks.

Most finished nonoriented silicon steel is sold in full-width coils (860 to 1220 mm, or 34 to 48 in.) or slit-width coils, but some is sold as sheared sheets. All coils are sampled and tested according to ASTM A 343 and graded as to quality.

Oriented Silicon Steels. Grain size is as important in silicon steel as in iron with regard to core losses and low-flux-density permeability. For high-flux-density permeability, however, crystallographic orientation is the deciding factor. Like iron, silicon steels are more easily magnetized in the direction of the cube edge: ⟨100⟩. For special compositions, rolling and heat treating techniques are used to promote secondary recrystallization in the final anneal at approximately 1175 °C (2150 °F) or higher, which results in a well-developed texture with

Table 1 Silicon contents, densities, and some applications of electrical steel sheet

AISI type	Nominal Si + Al content, %	Assumed density, Mg/m^3	Characteristics and applications
Low-carbon steel			
. . .	0	7.85	High magnetic saturation; magnetic properties may not be guaranteed; intermittent-duty small motors
Nonoriented silicon steels			
M47	1.05	7.80	Ductile, good stamping properties, good permeability at high inductions; small motors, ballasts, relays
M45	1.85	7.75	Good stamping properties, good permeability at moderate and
M43	2.35	7.70	high inductions, good core loss; small generators, high-efficiency continuous-duty rotating machines, ac and dc
M36	2.65	7.70	Good permeability at low and moderate inductions, low
M27	2.80	7.70	core loss; high reactance cores, generators, stators of high-efficiency rotating machines
M22	3.20	7.65	Excellent permeability at low inductions, lowest core loss; small
M19	3.30	7.65	power transformers, high-efficiency rotating machines
M15	3.50	7.65	
Oriented silicon steels			
M6	3.15	7.65	Highly directional magnetic properties with lowest core loss and
M5	3.15	7.65	highest permeability when flux path is parallel to rolling
M4	3.15	7.65	direction; heavier thicknesses used in power transformers,
M3	3.15	7.65	thinner thicknesses generally used in distribution transformers. Energy savings improve with lower core loss.
High-permeability oriented steel			
. . .	2.9–3.15	7.65	Low core loss at high operating inductions

the cube edge parallel to the rolling direction {110} ⟨001⟩. Conventional oriented grades contain approximately 3.15% Si.

In approximately 1970, improved {110} ⟨001⟩ texture was developed in silicon steel through the modification of composition and processing. The high-permeability material usually contains approximately 2.9 to 3.2% Si. Conventional oriented 3.15% Si steel has grains approximately 3 mm (0.12 in.) in diameter. The high-permeability silicon steel tends to have grains approximately 8 mm (0.31 in.) in diameter. Ideally, grain diameter should be less than 3 mm (0.12 in.) to minimize excess eddy-current effects from domain-wall motion. Special coatings provide electrical insulation and induced tensile stresses in the steel substrate. These induced stresses lower core loss and minimize noise in transformers.

Size and Shape. Flat laminations of a wide variety of shapes and sizes are blanked and pierced from electrical sheet. However, most are shaped like those shown in Fig. 1. Laminations similar to those shown in Fig. 1(a) can range in

diameter from less than 25 mm to 1.3 m (1 to 50 in.) or more, and laminations similar to those shown in Fig. 1(b) can range in length from less than 25 to 305 mm (1 to 12 in.) or more.

Punchability. Materials used for electrical sheet can be classified in the following order with respect to decreasing ease of blanking, piercing, and notching:

● Conventional flat-rolled low-carbon steels such as 1008
● Nonoriented silicon steels
● Oriented silicon steels

To a large extent, applications also follow the aforementioned classification (Table 1). Each group has certain distinct characteristics that affect punchability. In addition, differences in composition and hardness within any specific group cause considerable variation in punchability (see the section "Effect of Work Metal Composition and Condition on Blanking and Piercing" in this article).

Presses

A general-purpose punch press in good mechanical condition is acceptable for stamping laminations, but large-volume production of laminations by progressive-die methods requires the use of high-productivity presses (see the article "Presses and Auxiliary Equipment for Forming of Sheet Metal" in this Volume). Most high-productivity presses have heavy bed and crown members to minimize deflection and vibration. Bed deflection for lamination presses should be no more than 0.006 mm/mm (0.006 in./in.) of bed length (measured left-to-right between uprights), with a load equal to the rated capacity of the machine distributed over two-thirds of the bed area between tie rod centers. Deflection of the slide should not exceed 0.006 mm/mm (0.006 in./in.) of the length between the pitman centers, with rated load evenly distributed between those centers. Bending deflection and shear deflection are both considered in these standards. Double-crank presses with two or four points of suspension are preferred for progressive-die applications because of their better resistance to off-center die loads. Parallelism of the bed and slide should be 0.012 mm/mm (0.012 in./in.) of bed dimensions, both left-to-right and front-to-back.

Presses designed for producing laminations have heavy connections, large diameters of the mainshaft and connection bearings, close gib clearances, and thick bolsters. Because of the close gib fits (needed for accurate vertical motion), recirculating oil systems must be used to provide forced-feed lubrication of bearings and slides.

The fact that a die was built with uniform punch-to-die clearance at all cutting edges does not necessarily mean that the clearance is uniform at the instant the punch begins to enter the work metal. The act of applying the load to the

work metal can cause lateral deflections in the die and press, which can change the clearances. To minimize these undesirable deflections, the mechanical condition of the press and die must be maintained at a high level. The total force capacity exerted at each stroke must be in proper relation to the force capacity of the press and to the type of press frame (some types of press frames will deflect laterally more than others). Close-fitting gibs and bearings are essential in minimizing lateral deflection. The die should be built with large guideposts and close-fitting bushings.

A preventive maintenance program must be established to ensure that all presses are kept in top condition. Special attention should be given to bearing clearances, the condition of the counterbalance springs or cylinders, and the parallelism of the slide.

Auxiliary Equipment

When producing motor laminations in individual dies for each operation with upright or inclined presses, blanks can be loaded and unloaded manually. However, when individual dies are used for the simultaneous production of stator and rotor laminations, feeding and stacking equipment is necessary for optimal efficiency. The use of an inclined press is preferred because gravity assists in loading the die and removing the laminations. When progressive dies are used, automatic feeding and scrap-cutting equipment is required.

Stock reels, cradles, and straighteners are required when coil stock is used. Several types and sizes are available (see the article "Presses and Auxiliary Equipment for Forming of Sheet Metal" in this Volume).

Feed Mechanisms. In progressive-die operations, the common types of feed mechanisms, for example, single-roll or double-roll, hitch, grip, and slide, are used to feed strip or coil. Cam feed, which has a fixed feed length, is widely used for large-volume production. This method is accurate at high speeds because it eliminates the slippage that usually occurs in the overriding clutch-and-brake mechanisms of roll feeds.

Magazine feeds have a mechanism that ejects the blank from the bottom of a stack into the die or onto a magnetic belt or a chain feed. In inclined presses, the blank may slide by gravity into the die nest after leaving the magazine.

Stacking. Figure 2 shows a method that can be used for stacking laminations when each operation is done in an individual die. Blanks are fed to the press (inclined 35 to 45° to the rear) from a magazine feeder. Laminations drop from the press into a chute, where they are picked up by a driven elevating belt that conveys them to the stacking chute, from which they fall onto a stacking mast. Stacking masts are usually 380 to 915 mm (15 to 36 in.) high. The tops of the masts are either threaded or have tapped holes so that they can be picked up by handling machines

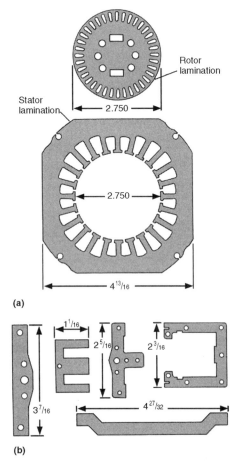

Fig. 1 Typical laminations blanked and pierced from electrical sheet. (a) Laminations for rotating electrical machinery are blanked and pierced in single-station dies (Fig. 3) or progressive dies (Fig. 4). Slots can also be made in precut blanks, one at a time, with notching dies. (b) Typical laminations blanked and pierced from electrical sheet for application in units other than rotating machines. Dimensions given in inches

and moved to subsequent operations. The bases of the stacking masts are large or weighted to prevent the masts from tipping or falling when being loaded or moved to the assembly floor.

Scrap disposal is a major consideration in producing laminations. The removal of scrap from a trimming operation or the removal of slugs from piercing holes and slots requires consideration during die design. Scrap is discharged through holes in the die shoe onto chutes. The chutes convey the scrap into containers or into automatic scrap-conveying systems below the floor. When the die is in the upper shoe, a mechanically operated pan can be used to catch the slugs on the upstroke. The slugs are then ejected into a container on the downstroke.

Dies

Single-station and progressive dies are both used for making laminations.

Single-Station Dies. Each single-station die performs one operation, and a set of dies for a lamination can be mounted in one press or in different presses. Simple laminations such as those shown in Fig. 1(b) are usually produced in

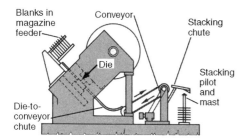

Fig. 2 Method for stacking laminations stamped in individual dies

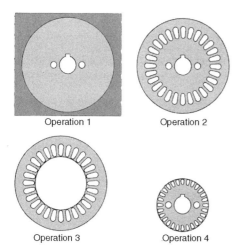

Fig. 3 Sequence of operations for producing stator and rotor laminations using single-station dies. Operation 1, stock blanked and pierced; operation 2, stator lamination notched; operation 3, rotor lamination separated from stator lamination; operation 4, rotor lamination notched. Compare with Fig. 4.

one operation. More complex parts may require several operations. Figure 3 shows a typical sequence for the production of stator and rotor laminations in four operations.

Single-station dies can be used for punching any lamination, regardless of the size, composition, shape, or quality requirements. However, because production with single-station dies is relatively slow, the cost per piece is high for mass production. Laminations such as those shown in Fig. 1 can be produced in large quantities at a lower cost in progressive dies.

The size of the workpiece and the quantity required influence the degree to which operations can be combined and the complexity of any one die. For example, in Fig. 3, the slots made in operations 2 and 4 can be punched in one stroke in a multiple die for each operation if the die sections are strong enough. An alternative method is to punch the slots in 28 strokes with a single-notch die in a high-speed notching press equipped with an indexing mechanism. Single-notch dies are used in the production of laminations for the following reasons:

- Tool costs are lower, and the single-notch die can be used on several different laminations. The cost is sometimes less than 5% of that for a multiple die that can pierce all the holes and slots in one press stroke.
- Laminations more than approximately 380 mm (15 in.) in diameter are sometimes too large to be notched by any other method because of tool and equipment costs. However, many larger-diameter laminations are multiple pierced.
- Limited production does not warrant the cost of a multiple die.
- Available equipment must be used.

Progressive dies perform a series of operations at two or more die stations during each stroke of the press. Each working, or active, station in the die performs one or more operations. The work material progresses through successive stations until a completed part is produced (Fig. 4). Idle stations, in which no work is performed, are added to provide strength to the die, to facilitate material travel through the die, to simplify construction, or to increase flexibility for die changes.

One of the more common progressive dies used in the electric-motor industry is a five-

station die that produces a rotor lamination and a stator lamination with each stroke of the press (Fig. 4). This die can be provided with carbide inserts for the punch and die sections. It has a spring-actuated guided stripper. The die components are mounted on a precision die set with ball bearing guide bushings and hardened guide pins. Slender punches are guided through the stripper by bushings. Such a die usually has four active stations and one idle station.

The progressive die described previously is the blanking or scrap-all-around type. For the most efficient use of material, the cutting-off or parting methods of severing the blank from the strip are used where layout permits.

The principal advantages of progressive dies for blanking laminations are:

- Handling between operations is eliminated; therefore, cost per piece is lower.
- Laminations from progressive dies are generally stacked in chutes that allow the press to be operated at uninterrupted maximum capacity. Stacking chutes fastened to the bottom of the bolster or die shoe keep the laminations oriented in a smooth, uninterrupted flow from the die. Therefore, laminations are better controlled with regard to burr direction and are easier to handle for assembly.

Two disadvantages of progressive dies for blanking laminations are:

- A progressive die is, to a great extent, a single-purpose die. Even a minor change in part design can necessitate an expensive die alteration or can make the die obsolete.
- Progressive dies are more susceptible to damage from accidents than single-station dies. Progressive dies run at high speed and may make many strokes before the press can be stopped. Misfeed detectors built into the die can help prevent damage. Die damage can halt production in progressive operations. In single-station operations, if the inventory of processed material is sufficient to keep other dies running, a breakdown of a die does not interrupt production.

The minimum size of lamination that can be made depends on the slot size and spacing, work metal thickness, and tolerances on the slot dimensions. The die must be strong enough to withstand the blanking pressure. Figure 5 shows

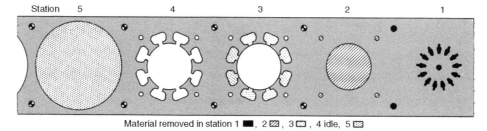

Material removed in station 1 ■, 2 ▨, 3 ▭, 4 idle, 5 ▩

Fig. 4 Blanking and piercing sequence for rotor and stator laminations in a five-station progressive die. Two pilot punches were used at each station. Station 1, pierce pilot holes, rotor slots, and rotor-shaft hole; station 2, pierce stator rivet holes and blank rotor; station 3, pierce stator slots; station 4, idle; station 5, blank stator. Compare with Fig. 3.

some very small laminations that were made in progressive dies. The 6.48 mm (0.255 in.) diameter rotor was made in three stations in order to have the necessary die strength. Part tolerances were such that piercing in three stations produced acceptable parts.

The dies for producing laminations are usually of segmented construction, which provides maximum accuracy. However, electrical discharge or electrochemical machining methods have been used to produce satisfactory dies.

There is no agreement as to the maximum size of lamination that can be efficiently produced in progressive dies. However, progressive dies are seldom used for laminations larger than 380 mm (15 in.) in major dimension. The factors that limit the maximum practical size are as follows:

- Progressive dies for making laminations more than 380 mm (15 in.) in major dimension represent a large investment (and a significant loss if damaged).
- Quantity demands are usually lower for large laminations; therefore, the investment is not warranted.
- Extremely large dies may require a press capacity that is so large as to be impractical.
- Problems from camber and lack of flatness in the stock are magnified in stamping large laminations in progressive dies.

Selection of Die Materials

Almost any hardened tool steel is satisfactory as die material for making a small quantity of

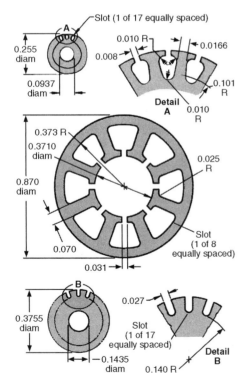

Fig. 5 Small-diameter laminations produced in progressive dies. Dimensions given in inches

laminations. However, for production blanking and piercing, either a high-carbon high-chromium cold work tool steel, such as AISI D2, or carbide must be used to resist the abrasiveness of electrical steels.

The shape or size of the lamination seldom affects the choice of die material. Dies ranging from the smallest to the largest and from the simplest to the most complex have been made from both high-carbon high-chromium tool steel and carbide. In addition, both die materials have been used to blank and pierce all compositions and thicknesses of electrical sheet. The composition and thickness of the stock rarely affect the choice between carbide and tool steel.

Production Quantity. If the dies are of the same design and construction, the total quantity of parts to be produced is the major factor in choosing die material. If the run is so short that it can be made with tool steel dies without sharpening, tool steel is more economical. However, for longer runs, carbide dies have 10 to 20 times as much life per grind as tool steel dies.

Uniform quality of cut edges and minimum burr height will be retained over a much longer run by carbide dies than by tool steel dies. In some cases, the edge condition of the lamination is not critical. However, when automatic stacking and core assembly equipment is used after blanking and piercing, burr height is important. Excessive burr height can cause short circuiting of the laminations in the core.

Cost. Depending on size and design, a die with carbide cutting edges will cost at least $1\frac{1}{2}$ times as much as a steel die. However, in terms of cost per piece, carbide dies may be more economical. Press downtime and die maintenance affect cost per piece; carbide dies can run approximately ten times as long per sharpening as tool steel dies.

Press condition is important in the operation of blanking and piercing dies. To achieve the maximum potential use of carbide dies, press condition must be maintained at a high level. Although tool steel punches and dies can chip and shear because of misalignment, carbide punches and dies are more likely to break. Therefore, the presses used for tool steel dies can be in less than top-level condition and continue to produce quality laminations.

Effect of Stock Thickness

Electrical sheet that is to be blanked and pierced usually ranges in thickness from 29 to 24 gage (0.343 to 0.607 mm, or 0.0135 to 0.0239 in.). Thinner or thicker stock is used for special applications. The blanking and piercing of extremely thin electrical sheet requires close control of equipment and technique. The processing of thick sheet (>1.27 mm, or 0.050 in.) can also cause difficulty, although the force-capacity rating of the press is the major factor that determines the maximum thickness of sheet that can be blanked and pierced.

Punch-to-die clearance for electrical sheet generally ranges from 3 to 7.5% of stock thickness per side, with clearances as large as 20% reported for grain-oriented stock. These values are similar to those used for low-carbon steel, but the stock thicknesses are thinner than those commonly used for the low-carbon steels. This results in close die clearance and requires good diemaking practice and accurate press equipment.

Thin Sheet (≤0.254 mm, or 0.010 in.). Under carefully controlled conditions, laminations can be blanked and pierced from sheet as thin as 0.051 mm (0.002 in.), but the press must be in top condition. Further, the feeding mechanism must be capable of feeding within ±0.076 mm (±0.003 in.) total error per stroke at a feed rate of 23 m/min (75 ft/min).

Punches and dies of hardened tool steel, such as D2, or carbide are satisfactory, although carbide dies and punches will have at least ten times the life of their tool steel counterparts. The punches must be rigidly supported and guided. The entire tool must be made rugged and accurate enough to maintain alignment. To avoid shearing the punch and die during press setup, it is important that the die be handled carefully to prevent the possibility of some of the components moving out of alignment. The press bed, the bottom of the die shoe, the face of the press slide, and the top of the punch holder must be clean and free of any irregularities that would cause a deviation from parallelism.

The punch and the die as a unit should be aligned square with the centerline of the press. The press slide should then be brought down slowly to meet with the top of the punch holder, and the punch holder should be fastened to the face of the slide. The slide should then be adjusted downward so that the punches enter the die cavities. Finally, the die shoe should be fastened to the bolster or press bed.

Dies with this close clearance are often designed as a unit, and the die is not fastened to the press ram. Therefore, the tool is not subject to the inaccuracies of the press.

A back taper of 0.002 mm/mm (0.002 in./in.) per side is commonly used in the die. Because of this angular clearance in the die, total die life is limited by the maximum punch-to-die clearance that can be tolerated. Each time a die having a back taper of 0.002 mm/mm (0.002 in./in.) per side is sharpened, the hole diameter will increase 0.1 μm (4 μin.) for each 0.0254 mm (0.001 in.) ground from the top of the die. After grinding 2.54 mm (0.100 in.) from the die, the punch-to-die clearance will increase 0.005 mm (0.0002 in.) per side.

If this amount of clearance is too great, the original clearance can be restored by installing new die sections or, if the dimensional tolerance permits, by using an oversize punch. The amount that can be removed from the die depends on the amount of back taper in the die, the stock thickness, and the maximum punch-to-die clearance permissible. Dies sometimes have a straight land that is 1.6 to 3.2 mm ($^1/_{16}$ to $^1/_8$ in.) wide before beginning the back taper.

Lubricant is applied to the stock during blanking and piercing to keep wear on the cutting edges of the punch and die at an acceptable level. Use of a water-thin lubricant with rapid evaporation and low residue makes it unnecessary to perform a burn-off operation before hydrogen annealing. Difficulties in producing acceptable laminations from thin sheet are magnified as the plan area of the lamination increases.

Effect of Work Metal Composition and Condition on Blanking and Piercing

Each of the three most widely used electrical sheet materials (low-carbon steels and non-oriented and grain-oriented silicon steels) has distinctive punching characteristics. These characteristics often necessitate specific procedures to produce laminations of the desired quality at the lowest cost.

Low-carbon steels such as 1008 are used as electrical sheet when their electrical properties can meet requirements, primarily because they cost less than silicon steels and the cost is lower for blanking and piercing. More pieces per die sharpening are usually obtained in blanking these steels than in blanking silicon steels. One study of die wear in making stator and rotor laminations similar to those shown in Fig. 1 and ranging in diameter from 92 to 149 mm ($3^5/_8$ to $5^7/_8$ in.) showed that, with tool steel cutting edges, 120,000 to 150,000 pairs were punched per sharpening when stamping 1008.

The condition of the low-carbon steel stock influences power requirements and punching characteristics. When annealed, this steel has a tensile strength of 380 to 414 MPa (55 to 60 ksi), but the strength of full-hard material may be over 690 MPa (100 ksi). Therefore, the material condition must be known before the force-capacity requirements of the presses can be determined. Low-carbon low-silicon steels in the annealed condition are soft, and they are likely to roll at the edges and form excessive burrs. Therefore, punch-to-die clearances must be as close for these steels as for electrical sheet of the same thickness. An annealed product is usually specified, but whether annealed stock is stamped or individual laminations are annealed after stamping is often a matter of convenience, because of press capacity, annealing facilities, or other factors.

Nonoriented silicon steels are available with silicon contents ranging from 0.5 to 3.25%. As silicon content increases, the sheet becomes more brittle and more abrasive. As a result, the edges of higher-silicon steel are less likely to roll and make excessive burrs, but die wear is increased because of abrasion.

Many nonoriented silicon steels are coated with an organic or inorganic material (core plating) to insulate one lamination from another. This organic core plating also improves the punchability of electrical sheet. In one application, carbide dies produced approximately 3.5 million laminations from core-plated M-36 (2.5% Si) between resharpenings. When similar laminations were produced from uncoated M-36, dies required sharpening after each 1.2 million parts. Heating of the coated blanks by welding or die casting may destroy the organic coating. Additional information is available in the section "Core Plating" in this article.

General practice is to use approximately the same punch-to-die clearances for all silicon steels. The tensile strength of the particular steel must be considered in determining press capacity because silicon steels may vary in strength, depending on whether they are fully annealed at the time of stamping.

Oriented silicon steels are relatively high in silicon (3.15% Si + Al) and have most of their grains (crystals) oriented with the cube edges parallel with the rolling direction and face diagonals at 90° to the rolling direction ({110}⟨001⟩). Because of this orientation, these steels have blanking and piercing characteristics that are different from those of nonoriented steels. Tensile strength will vary as much as 20% between the rolling direction and the transverse direction (strength is greater parallel with the rolling direction).

A magnesium hydroxide coating is applied to grain-oriented steel after normalizing. This coating prevents the coiled strip from welding together during annealing. Magnesium hydroxide, in contrast to the organic coatings, is highly abrasive and greatly increases die wear; therefore, it is not recommended for stamped laminations. Tool steel dies wear so rapidly under these conditions, because of the high silicon content of the steel, that carbide cutting edges are almost always used for the blanking and piercing of grain-oriented steel.

Because mechanical properties vary with direction, cutting properties also vary in grain-oriented steels. Cutting across the rolling direction results in a clean break, but the edges are smeared when cutting is parallel with the rolling direction. Therefore, punch-to-die clearance is more critical on the sides parallel with the rolling direction.

Camber and Flatness

Camber in electrical sheet is the deviation (parallel to the stock surface) of a side edge from a straight line that extends to both ends of the side, and it is customarily limited to 6.4 mm ($^1/_4$ in.) for any 2.5 m (96 in.) length or fraction thereof. Flatness, or the degree to which a surface of a flat product approaches a plane, is expressed in terms of the deviation from a plane. Flatness tolerances have not been established for electrical sheet; the operations employed to flatten other steel products cannot be used because of their effect on magnetic quality. Flatness requirements should be specified for a particular application.

Camber and flatness are interrelated; the edge of a 2.5 m (8 ft) section of sheet may come within the 6.4 mm ($^1/_4$ in.) tolerance while lying freely on a flat surface. However, the seemingly flat sheet may have a number of faint waves (sometimes called oil cans). If this sheet is then flattened (as it is in dies), the flattening of these waves causes multidirectional elongation of the sheet, and the edge of the sheet may then be forced into a camber different from that when the sheet is not under flattening pressure. Minimum camber and maximum flatness are desirable for blanking electrical sheet and are especially important in progressive-die operations.

Effect on Progressive-Die Operation. If there is no camber, it is easy to start the stock through the die correctly by aligning the straight edge of the sheet against a straight-edge starting guide, with the end of the material covering the first die stage. Feed rolls are then engaged, and the blanking and piercing can begin.

Even though the edge of the sheet is cambered, it must still be used in the starting alignment. Therefore, the material may be misaligned to some degree as it enters the die. A small degree of misalignment is not readily apparent to the press operator.

Minor misalignment in starting the material through a progressive die may not cause immediate problems. Operating difficulties result from various misalignments, which have a cumulative effect.

In the first stage of the die, pilot holes are pierced into the sheet, often into a portion that will later be scrap. At subsequent die stations, bullet-nose pilots engage the pilot holes as the die closes. The piloting action may cause the sheet to move slightly into true position before the cutting edges of the die meet the sheet. Powered feed rolls move the sheet between press strokes to an approximate position for the next die station. The feed rolls then open, releasing the sheet so that there is no conflict between the locating action of the pilots and the feed rolls.

When the original lineup is not correct or when there are cumulative effects of camber against the stock guides, the sheet may wander from side to side on the die face. Stock guides are provided in progressive dies to limit wandering due to camber and misalignment.

Interference between the pilots and the stock guides may cause the stock to distort and jam in the die, preventing proper flow of the stock. If the stock jams, the press must be stopped at once.

Camber can cause other difficulties. For example, a change in camber or multidirectional elongation of the sheet as the press flattens the waviness may cause the pilots to distort the piloting holes; thus, misaligned rotor and stator laminations are made. This leads to misalignment of the slots in the stacked core. As slots are blanked out, stresses are released that also can change the amount of camber and flatness.

There is no single solution to the problem of camber in the production of laminations in progressive dies, because no two shipments of material are exactly alike. Some manufacturers of laminations use less efficient single-station dies because of difficulties with camber, even though production volume could justify the use of progressive dies. Other manufacturers use progressive dies only for laminations below a certain size.

Burr Height

It is impossible to blank and pierce laminations without producing some burr along the cut edges (Fig. 6). The amount of burr (measured as burr height) depends on the composition and condition of the electrical sheet, the thickness of the sheet, the clearance between punch and die, and the edge condition (sharpness) of the punch and die.

The amount of burr that can be tolerated depends on end use. Burr height influences the stacking factor, which in turn influences magnetic characteristics. Maximum burr height is usually limited to 0.05 to 0.13 mm (0.002 and 0.005 in.).

Length of Die Run. A die is usually run until the maximum allowable burr height is reached, at which time the punch and die are removed for sharpening. Close control is required with this method of determining the length of the die run.

Optimal Die Run. For the greatest economy and convenience of operation, die maintenance requirements and die life (in addition to maximum burr height) should be considered in determining the optimal die run. A die should not be run too long before resharpening; otherwise, excessive stock must be removed from both the punch face and the die face to restore the cutting edges. As a result, fewer laminations can be made during the life of a die.

A common method of determining the optimal die run is to establish an arbitrary number of pieces to be run before die sharpening. At the end of this run, the punch face is sharpened. By knowing the number of pieces run and the amount removed from the die during sharpening, the number of pieces produced per unit of length removed from the die can be established. Using this procedure, the number of pieces run between sharpenings can be varied, and an optimal die life between sharpenings can be determined.

Lubrication

Although uncoated electrical sheet is sometimes blanked and pierced without lubrication, the use of some type of lubricant is preferred. Organic core plate (used on nonoriented silicon steels) serves as a lubricant, and no further lubrication is needed when blanking and piercing sheet that has core plating.

Tool life will be greatly improved by using a lubricant in the blanking and piercing of electrical sheet that has no coating or has been coated with magnesium hydroxide, which acts as an abrasive rather than a lubricant. Oil-type lubricants, such as those used in blanking and forming operations, are not ordinarily used for blanking and piercing electrical sheet, because removal is too expensive.

Some plants purchase nonoriented silicon steel sheet without core plate and then subject the punched laminations to an oxidizing anneal, in which the lubricant is burned off; the only requirement is to select a lubricant that will leave the least residue when it is burned off. Water-soluble oils (1 part oil to approximately 20 parts water) have been used when annealing follows punching. Other low-viscosity low-residue oils, such as the aliphatic petroleums, also burn off with little residue.

Liquid lubricants can be applied in several ways. In low-production operations, the work metal can be dipped into the lubricant just before punching. In high-production operations, in which the stock is continuously fed, the lubricant can be brushed on just before it enters the press, or it can be dripped onto the sheet a few feet from the press. The top of the sheet is then rubbed with a felt wiper that spreads the lubricant over the entire surface. The bottom of the sheet can be coated with lubricant by having a trough under the sheet that catches the excess drip. A piece of felt or similar material in the trough acts as a wick to wet the underside of the moving sheet. A more complete discussion of lubricants is available in the article "Selection and Use of Lubricants in Forming of Sheet Metal" in this Volume.

Molybdenum disulfide is a good lubricant for the blanking and piercing of electrical sheet. The amount used is usually so small that no removal is required. When there is an excess, it can be removed by immersing the sheet for 4 to 5 min in a dilute solution of stripper-type cleaner at 80 °C (180 °F).

Molybdenum disulfide is the basic ingredient of several compounds that are available as dry powder, paste concentrate, and dispersion in liquid. A common method of applying dry molybdenum disulfide is shown in Fig. 7. The sheet passes through a box containing the dry powder, and felt wipers remove the excess. A slurry can be used instead of the dry powder. Sheets can also be coated by a spray timed with the stroke of the press; either powder or a liquid suspension can be sprayed. In low-production operations, a liquid suspension of molybdenum disulfide can be applied to the sheet by brush.

Core Plating

Core plating, or insulation, is a surface coating or treatment applied to electrical steel sheet to reduce interlaminar loss and sometimes to increase punchability. This treatment does not reduce eddy currents within the laminations. Interlamination resistance is usually improved by annealing the laminations under slightly oxidizing conditions and then core plating. Core plating can be classified as organic or inorganic.

Organic insulation generally consists of enamels or varnishes applied to the steel surface. Steels having organic coatings cannot be stress relieved without impairing the insulating value of the coating, but the coating will withstand normal operating temperatures. Coatings are approximately 0.0025 mm (0.0001 in.) thick.

Inorganic insulation usually includes chemical or thermal treatments; it has a high degree of electrical insulation and can withstand stress relieving. Inorganic coatings form a very thin surface layer on the steel and increase lamination thickness only slightly.

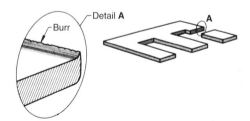

Fig. 6 Burr (exaggerated) produced along the edges of a blanked lamination

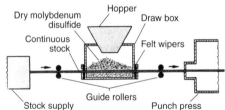

Fig. 7 Setup for applying dry molybdenum disulfide to both sides of electrical sheet

Fine-Edge Blanking

T.C. Lee, C.Y. Tang, and L.C. Chan, The Hong Kong Polytechnic University

THE DEFINITION of fine-edge blanking, commonly known as fine-blanking, is very loose, but the most widely accepted description is that of the International Fine-Blanking Organization: "Fine-blanking is a hybrid metal forming process combining the technologies of stamping and cold extrusion. It produces a part with exceptional edge quality, superior flatness, thin wall sections, and linear spacing held to within 0.001."

Fine-blanking technology was invented by German engineer Scheiss in Switzerland in 1923. By 1959, the technology was being used in industrial applications. In subsequent years, the demand by the office-machinery industry for fine-blanked parts grew considerably. Then, in the 1970s, when office equipment became predominantly electronic, the fine-blanking industry experienced a major setback. This setback proved to be only temporary, however; increasing usage in automobile production and the growing importance of this industrial sector have revitalized fine-blanking.

Initially, fine-blanking dealt mainly with materials 1 to 3 mm (0.04 to 0.12 in.) thick. The material types were limited to soft mild steel such as AISI 1006 (similar to GB St3 or EN 10027) or alloy steels such as AISI 5120 (similar to GB16MnCr5). Considerable technological breakthroughs in tooling and materials have advanced the technology and widened its application. It is now possible to make thicker parts in a variety of shapes and from a host of materials. Today (2006) more than 60% of fine-blanked parts (Fig. 1) are used in the automotive industry with thicknesses up to 19 mm (0.75 in.). Moreover, many alloy and carbon steels with good annealing properties are supplied for fine-blanking. Quality and cost are critical criteria when choosing a manufacturing method. As a cost-effective manufacturing technology, the fine-blanking method has become a necessity in many major industrial sectors, replacing more expensive manufacturing options.

Difference between Fine-Blanking and Conventional Blanking

In conventional blanking, a metal workpiece is removed from the primary metal strip when it is punched. A schematic diagram of the conventional blanking process is shown in Fig. 2.

A typical fine-blanking tool is a single-station compound tool used to produce a finished part in one press stroke. The only additional operation needed is the removal of a slight burr. The process requires a triple-action fine-blanking press that applies clamping force, blanking force, and counterforce to the fine-blanking tool,

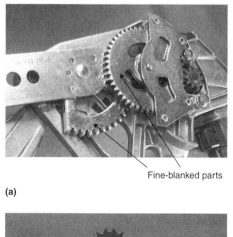

Fig. 1 Fine-blanked parts (a) assembled as the main components of an automatic car window. (b) Other fine-blanked parts

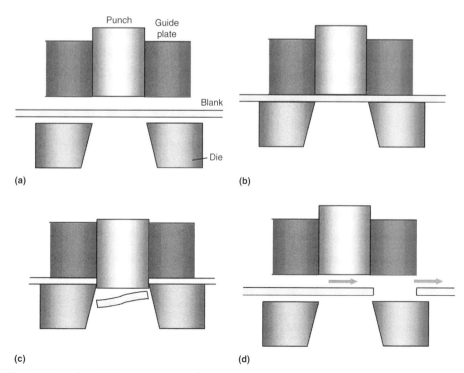

Fig. 2 Conventional blanking process. (a) Steel strip is fed into position when tool is open. (b) Guide plate holds the steel strip and punch drops. (c) Punch continues to pierce into the strip, and the part is blanked. (d) The strip is fed forward, ready for next cycle.

respectively. Figure 3 describes a typical single-station fine-blanking process. For more complex parts, multiple-station and progressive fine-blanking tools may be utilized.

Compared to conventional blanking, fine-blanking operations require more complicated presses (with three drive systems) that can produce high-quality products with good flatness and intricate shapes. Conventional machines are simple and require secondary operations. In conventional blanking, only a single force acts upon the material through the punch: the ram force. Fine-blanking makes use of multiple forces created by a blanking plate, a vee-ring plate, a punch, and an ejector. The ejector maintains pressure against the punch throughout the upward stroke. The three main forces in a fine-blanking cycle—clamping vee-ring force, an ejector/punch counterforce, and a blanking force—transfer through the vee-ring plate, ejector, blanking plate, and punch. At the onset of a typical fine-blanking cycle, a vee-ring force and an ejector counterforce act on the material. The vee-ring plate maintains a vee-ring force against the punch face. The two forces firmly clamp the material, inside and outside the periphery of the cutting edge, prior to shearing. The blanking force, the third force in the fine-blanking cycle, occurs as the punch begins to cut. In a conventional blanking process, the punch tries to push the material through the die. However, the penetration is limited once the deformation exceeds the fracture strength of the material. The material in the clearance zone is subjected to tensile stretching, which, as it reaches the limits of the strength of the material, leads to fracture. The mechanical differences between the conventional blanking process and fine-blanking are listed in the Table 1 and illustrated in Fig. 4.

The distinguishing features of fine-blanking are the vee-ring, the small die clearance, and the counterforce loaded on the material. These generate compressive stresses, prevent horizontal movement of the material during the cycle, and stabilize tool elements.

Design of the Fine-Blanking Process

The three principal design features of the fine-blanking process are the vee-ring, the clearance between punch and die, and the counterforce imposed by the ejector (Fig. 5).

Vee-Ring Dimensions. The vee-ring may be situated on both the guide plate and the die plate, or on only one of these. The dimension and position of the vee-ring depend on material thickness and surface flatness requirements (Fig. 6, 7 and Tables 2, 3). A vee-ring is used on both sides when the material thickness exceeds 4 mm (0.16 in.). Typically, the vee-ring contacts the material in an unbroken line all around the workpiece periphery. However, in special cases, an extra contact line that indents the scrap piece corresponding to the hole can be provided inside this periphery. In some cases, it may be advantageous to have a discontinuous vee-ring contact the periphery.

Determination of Die Clearance. In the fine-blanking process, the clearance should be kept at approximately 0.2 to 0.5% of the working material thickness, which is only one-tenth of that used in conventional blanking. A much smaller die clearance constrains the deformation within a narrow zone, which further restricts the flow of metal outside the deformation zone. Clearance is mainly determined by the material thickness (Fig. 8). However, it is also affected by punch profile and workpiece material.

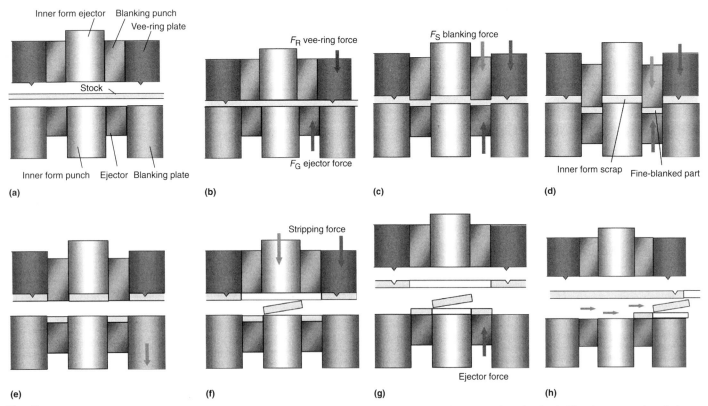

Fig. 3 A complete fine-blanking process. (a) Strip stock is fed into position when the tool is open. (b) Pressure is applied, closing the tool and embedding the vee-ring (also called stinger plate) into the stock. This prevents the material from flowing away from the punch, thus aiding in producing a smooth, exruded edge on the part. For thick parts, a vee-ring is also raised on the face of blanking plate. (c) The blanking force is applied to the stock by the ram, at the same time vee-ring constrains the movement of extra material. During this portion of the cycle, the counterpunch holds the part firmly against the face of the advancing blanking punch, maintaining flatness and preventing the part from moving away from the punch, which could cause die break or edge fracture. (d) The upward-moving ram advances the blanking punch until the part is fully sheared and rests in the upper die opening. In the same action, the punch pierces a hole in the workpiece. The scrap corresponding to the central hole is pushed by the counterforce inside the punch. (e) All three forces are relaxed, and tool starts to open. The ram descends by gravity. (f) The stripping force is 10 to 15% of the blanking force. This force acts to strip the skeleton from the blanking plate and eject the inner-shape scrap. (g) The ejector force pushes the finished fine-blanked part out of the die plate into the tool space. (h) The strip is fed forward, and the part and the scrap are removed mechanically or by an air jet. The system is ready to start the next cycle.

Force Calculation. During a fine-blanking cycle, three forces act upon the material: the vee-ring force, the counterforce, and the blanking force (Table 4). The exact interactions among the three forces have a strong influence on the quality of the part and the performance of the fine-blanking process. Figure 9 and Table 4 show the empirical formulas used to calculate the forces.

Quality of the Cut Edge

A smoothly cut edge requiring no further machining is the most distinctive characteristic of the fine-blanking process. The quality of the cut edge depends on die clearance, lubrication, and blanking force. The cut edge is evaluated by several quality indices: dimensional tolerance, surface quality, and straightness or flatness of the edge (Fig. 10). Dimensional tolerances that can be achieved in fine-blanked parts are listed in Tables 5(a) and (b). Cutting-edge quality consists of surface roughness, surface integrity, and allowable tearing level. The achievable surface roughness of a fine-blanked part is R_a 3.6 to ~0.2 μm; generally, surface roughness is R_a 2.5 to ~0.63 μm. The surface integrity is indicated by the achievable smooth-cut section ratio, which is quantified with five levels (Table 6). The allowable tearing level is 4, according to experience in the fine-blanking field (Table 7).

Advantages and Limitations

As a high-precision manufacturing technology, fine-blanking has some advantages, not only in comparison with conventional stamping but relative to machining as well.

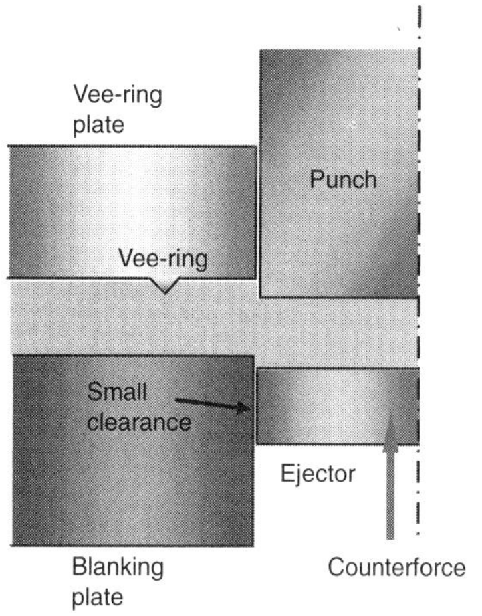

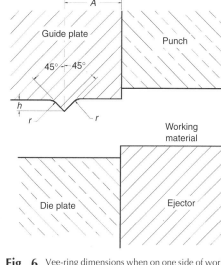

Fig. 5 Three main characteristics of fine blanking

Table 1 Differences between conventional blanking and fine blanking

	Conventional blanking	Fine-blanking
No. of acting forces	One	Three
Force name	Major ram force	Major ram force
		Vee-ring force
		Counter force
Vee-ring	No	Yes
Die clearance	2–5% of material thickness	0.5–0.2% of material thickness
Secondary operation	Many operations are needed	Few or no operations are needed due to the high-quality surface

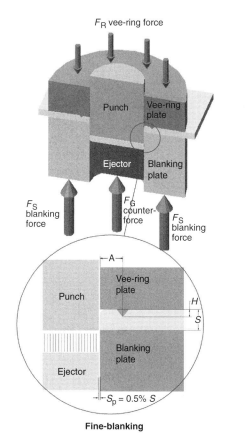

Fig. 4 Main differences between conventional blanking and the fine-blanking process

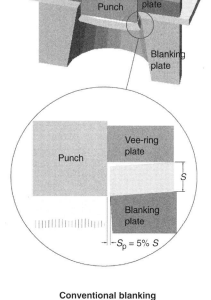

Fig. 6 Vee-ring dimensions when on one side of workpiece. Source: Ref 1, 2

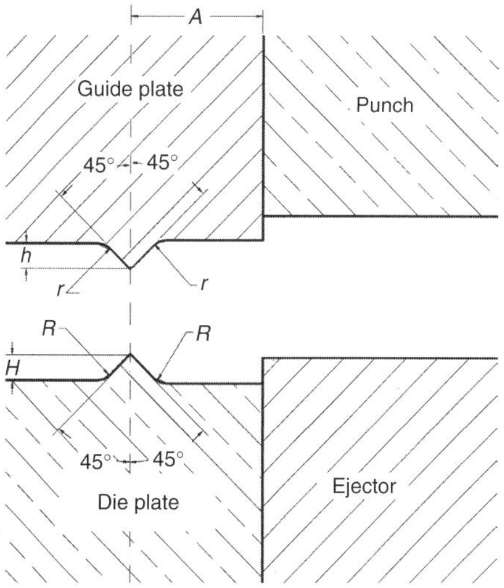

Fig. 7 Vee-ring dimensions when on both sides of the workpiece. Source: Ref 1, 2

Advantages

Edge Quality. The most obvious feature of a fine-blanked part is its completely smooth edge, which, if required, can be free of cracks and tearing. Conventionally stamped parts often have torn, rough, or tapered edges (Fig. 11). Commonly, the taper of the surface from top to bottom is less than 1° in a fine-blanking process and 30° in a conventionally stamped part. At the same time, due to work hardening, the sheared edges have an increased hardness, up to two times that of the base material.

Flatness. Throughout the fine-blanking process, pressure is constantly applied to both sides of the part, which achieves unmatched flatness. In general, fine-blanked part flatness can reach 0.025 mm per 25 mm (0.001 in. per 1 in.) of part length.

Shape Complexity. For a fine-blanked part, hole diameters as small as approximately 40 to 60% of material thickness and holes close to the edge are possible with no loss in structural strength. Webs or slots with widths 60% of material thickness for steel, and 40% for aluminum and copper alloys, can be produced. In conventional stamping, approximately 150 to 200% of material thickness is required.

Finished gears can be fine-blanked with features broached and hobbed at the same time. Fine-blanking also can produce semipierced protrusions firmly affixed to the base metal, unlike conventional stamping, in which fracture occurs. Projections may also extend up to 100% of material thickness (Fig. 12).

Table 4 Calculation of the required forces in a fine-blanking process

Blanking force (F_S), N	Vee-ring force (F_R), N	Counterforce (F_G), N
$F_S = L \cdot s \cdot R_m \cdot f_1$	$F_R = L_R \cdot h \cdot R_m \cdot f_2$	$F_G = A_s \cdot q_G$
L: Developed length of blanked periphery; internal shape plus external shape, mm	L_R: Developed length of vee-ring, mm	A_s: Surface area acted on by ejector, mm^2
s: Material thickness; nominal thickness plus tolerance, mm	h: Height of vee-ring; if two vee-rings, greatest height assumed, mm	q_G: Specific counterforce, N/mm^2; for large, thick parts use 70 N/mm^2 as upper limit; for small, thin parts use 20 N/mm^2 as lower limit
R_m: Maximum tensile strength of material, N/mm^2	f_2: Empirical factor 4, for specified vee-ring shape	
f_1: Empirical factor 0.9		

Table 2 Vee-ring dimensions when material thickness is under 4.5 mm

Material thickness, mm	A, mm	h, mm	r, mm
1.0~1.7	1.0	0.3	0.2
1.8~2.2	1.4	0.4	0.4
2.3~2.7	1.7	0.5	0.5
2.8~3.2	2.1	0.6	0.6
3.3~3.7	2.5	0.7	0.7
3.8 to ~4.5	2.8	0.8	0.8

Source: Ref 1, 2

Table 3 Vee-ring dimensions when material thickness is more than 4.5 mm

Material thickness, mm	A, mm	H, mm	R, mm	h, mm	r, mm
4.5~5.5	2.5	0.8	0.8	0.5	0.2
5.6~7	3	1	1	0.7	0.2
7.1~9	3.5	1.2	1.2	0.8	0.2
9.1~11	4.5	1.5	1.5	1	0.5
11.1~13	5.5	1.8	2	1.2	0.5
13.1 to ~15	7	2.2	3	1.6	0.5

Source: Ref 1, 2

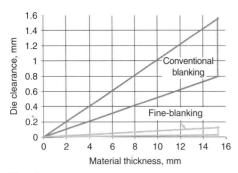

Fig. 8 Die clearance as function of workpiece material thickness. Source: Ref 1

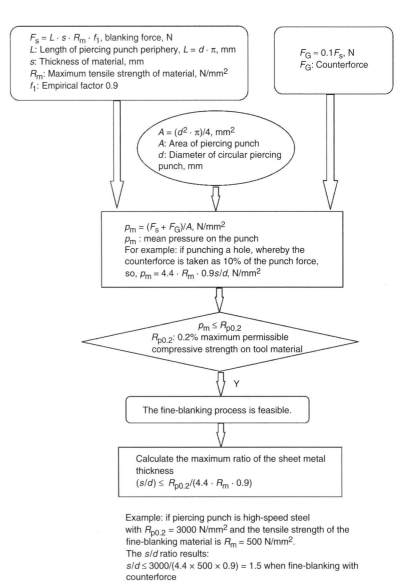

$F_S = L \cdot s \cdot R_m \cdot f_1$, blanking force, N
L: Length of piercing punch periphery, $L = d \cdot \pi$, mm
s: Thickness of material, mm
R_m: Maximum tensile strength of material, N/mm^2
f_1: Empirical factor 0.9

$F_G \approx 0.1 F_s$, N
F_G: Counterforce

$A = (d^2 \cdot \pi)/4$, mm^2
A: Area of piercing punch
d: Diameter of circular piercing punch, mm

$p_m = (F_S + F_G)/A$, N/mm^2
p_m: mean pressure on the punch
For example: if punching a hole, whereby the counterforce is taken as 10% of the punch force, so, $p_m = 4.4 \cdot R_m \cdot 0.9 s/d$, N/mm^2

$p_m \leq R_{p0.2}$
$R_{p0.2}$: 0.2% maximum permissible compressive strength on tool material

Y

The fine-blanking process is feasible.

Calculate the maximum ratio of the sheet metal thickness
$(s/d) \leq R_{p0.2}/(4.4 \cdot R_m \cdot 0.9)$

Example: if piercing punch is high-speed steel with $R_{p0.2} = 3000$ N/mm^2 and the tensile strength of the fine-blanking material is $R_m = 500$ N/mm^2. The s/d ratio results: $s/d \leq 3000/(4.4 \times 500 \times 0.9) = 1.5$ when fine-blanking with counterforce

Fig. 9 Flow diagram to determine the feasibility of a fine-blanking process. Source: Ref 1–3

Two-Dimensional Forming. With the help of the three-drive systems of a fine-blanking press and progressive dies, products with bends and offsets to 70° can be achieved in a single fine-blanking cycle. The maximum offset distance can be four times the sheet thickness (Fig. 13). A complex shape product (Fig. 14) can be finished in one station, saving a considerable amount of time.

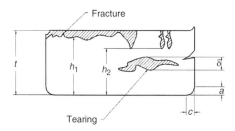

Fig. 10 Evaluation of the cut edge. h_1, smooth cut section length in case of fracture; h_2, the minimum smooth-cut section length in case of shell-shaped fracture; a, die-roll height; c, die-roll width; δ, the maximum width of the tearing band

Table 5(a) Achievable tolerances in fine-blanking

Material thickness, mm	Internal contours ISO quality	External contours ISO quality	Hole distance ISO quality
0.5~1	7	7	7
1~3	8	8	8
3~5	9	9	8
5~6.5	9	9	8
6.5~8	9	9	9
8~10	10	10	9
10~12	10	10	10
12~15	11	11	10

Source: Ref 1–3

Table 5(b) Appendix: ISO tolerances for nominal dimensions

Nominal dimensions, mm	Basic tolerance value, μm					
	IT6	IT7	IT8	IT9	IT10	IT11
≤3	6	10	14	25	40	60
3–6	8	12	18	30	48	75
6–10	9	15	22	36	58	90
10–18	11	18	27	43	70	110
18–30	13	21	33	52	84	130
30–50	16	25	39	62	100	160

Source: Ref 1–3

Table 6 Level of surface smooth-cut ratio

Level	1	2	3	4	5
h_1/t, %	100	100	90	75	50
h_2/t, %	100	90	75	...	...

Table 7 Level of admissible tearing

Level	1	2	3	4
δ, mm		0.3	0.6	1	2

Source: Ref 1, 2

Cost Effectiveness. With fine-blanking, the part is produced to net shape without extensive poststamping machining. Costly machining, such as shaving, milling, reaming, or grinding, is not needed to eliminate die break. The clean edges meet the high standards for parts requiring full bearing or sealing contact on sidewalls or smooth edges for cosmetic reasons. Additionally, small holes and thin web sections can

Fig. 11 Edge quality comparison. Top: Fine-blanking: Full shear, 50 RMS; less than 1° tape. Bottom: Conventional stamping: up to 80% die break; taper as great as 30°

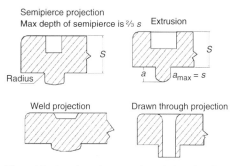

Fig. 12 Complex shapes achieved by the fine-blanking process

Fig. 13 Two-dimensional forming

Fig. 14 A sample two-dimensional part made by fine-blanking

be produced in one stroke of the fine-blanking process; thus, costly secondary drilling and machining operations are unnecessary. Finished parts keep a high tolerance from part to part within a production lot and from the first production lot to the last.

When compared to conventional stamping, the main drawback of the fine-blanking process is the high price of the press system, often several times higher than a conventional system. Generally, if a part can be made to specification by conventional stamping, it will cost somewhat less than if made using the fine-blanking process. However, fine-blanking can eliminate the need for secondary operations in a stamped component and meet dimensional requirements that conventional stamping cannot match.

Limitations

Die Roll and Burr. Two common features, die roll and burr, exist in blanked parts. However, these defects exist to a lesser degree in fine-blanked parts. Die roll is greater on corners than on straight edges and, in general, the harder or thinner the material, the smaller the die roll. The amount of die roll depends primarily on the geometric form, the corner angles, and the radii (Fig. 15). Moreover, as a rule of thumb, the width a of the die roll is usually about 5 times its height b.

Burr, located on the side opposite the die roll, is another inherent feature of a fine-blanked part. Die roll and burr size depend on the part material, blade sharpness, the extent of wear of the die, and the shape of the part. The size of the die roll and burr tend to increase with the number of parts produced, reflecting gradual wear on the tool.

Working Material

Any material suitable for cold forming can be fine-edge blanked, including low- and medium-carbon steels, some alloy and stainless steels, copper and brass, and aluminum alloys. Table 8 (Ref 5) lists typical conditions of fine-blanked materials. Table 9 lists in more

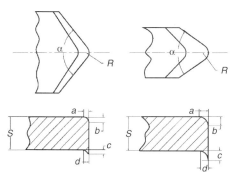

Fig. 15 Die roll and burr in fine-blanked parts. R, corner radius; α, corner angle; a, die roll width; b, die roll height; c, burr height; d, burr width; S, material thickness. Source: Ref 3

Table 8 Materials for fine-edge blanking

Material	Condition
Carbon steels	
1008–1024	Up to fully hard
1025–1095	Spheroidize annealed
Alloy steels	
AISI 4000 series, 8000 series	Spheroidize annealed
Stainless steels	
AISI types 301, 302, 303, 304, 316, 430, 416	Fully annealed; shorter tool life than for plain carbon steels
Aluminum alloys	
1xxx, 3xxx, 5xxx series	H (strain hardened) temper; O (annealed) temper
6061, 7075	T3 or T4 for thinner gages
Copper alloys	
C26000 (catridge brass, 70%), C26800 (yellow brass, 66%)	Annealed, quarter hard, or half hard
C27400 (yellow brass, 63%)	Fully hard in thin gages
C17xxx (beryllium copper alloys)	Annealed only
Other nonferrous metals and alloys	
Monel alloy 400	Annealed
Soft bronzes	Annealed
Silver, gold	Excellent results

Source: Ref 5

Table 9 Types of materials suitable for the fine-blanking process

Material	Most	← Degree of suitability →		Least
Carbon steel				
Low carbon (C1006–C1020)	45–75 HRB, T No. 3 No. 5 (quarter-hard and dead-soft)	70–85 HRB, T No. 2 (half-hard) Annealed, 75–86 HRB, Spheroidized annealed	84–90 HRB, T No. 1 (full-hard) (thin)	T No. 1 (full-hard) (heavier gage)
Medium carbon (C1025–C1035)				
High carbon (C1036–1075)			Spheroidized annealed 80–90 HRB, thin gage	Above 3/8 in.
Spring steel (C1095)				Above 1/8 in.
Alloy steel				
8620, 8630, 4130, 4140, 4340	...	Under 1/8 in., annealed	Above 1/8 in., Spheroidized annealed	Above 7/16 in.
High-strength steel				
040, 050	...	Under 1/8 in.	Above 1/8 in.	Over 1/4 in.
060, 090	...	...	Under 1/8 in.	Over 1/8 in.
Stainless steel				
200, 300 series	...	Annealed, no thickness limit	...	...
Stainless steel (heat treatable)				
400 series	...	Annealed, under 1/8 in.	...	Over 3/16 in.
Aluminum				
1100	0, H2, H4	H6	H8	...
3003	0, H2	H4	H6, H8	...
5052	0	H2, H4	H6, H8	...
Aluminum (heat treatable)				
2024	T0	T3	T4	T6
6061	T0	...	T3, T4	T6, T8
7075	T0	...	...	T6
Brass	1/4 and 1/2 H	3/4 hard	Full hard	...
Bronze	1/4 and 1/2 H	3/4 hard	Full hard	...
Copper	Soft, 45–75 HRB	...	Half hard, 88–95 HRB	Hard > 96 HRB

Table 10 Allowable thickness and tensile strength of the work material

Tensile strength of working material		Maximum allowable thickness of working material	
MPa	ksi	mm	in.
700	102	1.5	0.060
600	87	3.5	0.140
500	73	6.0	0.235
450	65	10.0	0.395
400	58	15.0	0.590

detail suitability of material conditions for fine-blanking.

The mechanical properties of the work material have a direct influence on surface quality, dimensional tolerances of the fine-blanked part, and tool life. The basic requirements of the material are:

- *Good ductility.* The better the ductility, the higher the deformation capacity. The material in the deformation area flows without fracture as the punch moves.
- *Low yield strength.* Low yield strength and tensile strength help to create good lubricating condition, improve the quality of sheared edge, and increase tool life. The allowable thickness of the work material is primarily governed by its strength. Material less than 15 mm thick can be fine blanked. However, work material thickness is usually in the range of 2 to 4 mm (0.08 to 0.16 in.) (Table 10).
- *Homogeneity.* Fine-blanking requires a fine microstructure because different heat treatments result in different microstructures and plasticity which, in turn, affect the quality

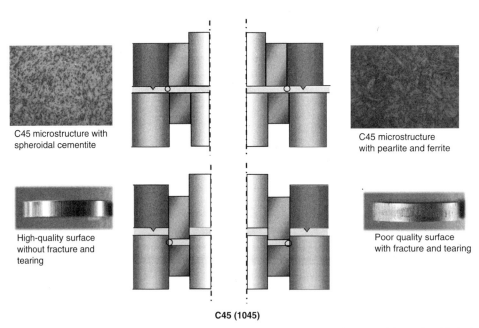

C45 microstructure with spheroidal cementite

C45 microstructure with pearlite and ferrite

High-quality surface without fracture and tearing

Poor quality surface with fracture and tearing

C45 (1045)

Fig. 16 Comparison of suitable and unsuitable 1045 steel microstructure for fine-blanking and the resulting blanked edges. Source: Ref 1

of the fine-blanked part. For carbon steel and alloy steel, the shape and distribution of carbides are critical to the fine-blanking process. Material with perfect spheroidization, even distribution, and good dispersion of fine carbides is most suitable for the fine-blanking process.

The example in Fig. 16 provides various results obtained using the same material but with different microstructures. The material on the left in Fig. 16 with a fine microstructure comprising a ferrite matrix with spheroidal cementite is suitable for fine-blanking. During the fine-blanking process, instead of separation from the matrix, the spheroidal cementite grains are pressed into the soft ferrite matrix, which results in a flat, smooth surface. On the other hand, the material on the right displays an unsuitable microstructure composed of ferrite and pearlite. The hard cementite plates have to be broken when the blanking punch penetrates the material. The result is die fracture.

In addition to material properties, the geometric configuration and the thickness are the prime factors determining the applicability of the fine-blanking process. In order to determine if a part is suitable for fine-blanking, its degree of difficulty must be assessed: S1 (easy), S2 (medium), and S3 (difficult). To gage the feasibility of the fine-blanking process, the various formed elements, such as slot widths, section widths, hole diameters, tooth forms, corner angles, and radii can be evaluated with the help of Fig. 17.

Carbon and Alloy Steels. The easiest materials to fine-edge blank are low-carbon steels (less than approximately 0.25% C). Thin parts, however, have been fine-edge blanked from plain carbon steels containing up to about 0.95% only in the fully annealed, spheroidized condition (Table 8).

Stainless Steels. Austenitic stainless steels, such as AISI types 301 to 304 and 316, can be fine-edge blanked in the fully annealed condition. These materials, as well as the ferritic type 430 and martensitic type 416 (both fully annealed), have good blanked edges, but cause higher tool wear than do plain carbon steels.

Most aluminum alloys, with the exception of alloy 2024, can be fine-edge blanked in the O (annealed) temper. Wrought alloys of the 1*xxx*, 3*xxx*, and 5*xxx* series can be fine-edge blanked in the H (strain hardened) temper with excellent results. In thinner gages, alloys 6061 and 7075 can be fine-edge blanked in the T3 or T4 condition.

Copper alloys are easily fine-edge blanked. The most workable alloys, such as alloy C27400 (yellow brass, 63%) can be fine-edge blanked even in the full hard condition. Other brasses, such as C26000 (cartridge brass, 70%) and C26800 (yellow brass, 66%) can be worked in the annealed, quarter-hard, or half-hard condition. Beryllium-copper alloys can be fine-edge blanked in the annealed condition, as can pure copper, soft bronzes, Monel alloys, nickel-silvers, and silver and gold.

Fine-Blanking Tooling and Lubrication

Types of Dies. Fine-blanking dies are usually classified into "moving punch" and "fixed-punch" systems. The moving-punch system (Fig. 18) is used mainly for the production of small to medium-size parts with few inner forms and, in most cases, on mechanical fine-blanking presses. This tooling system is very compact and economical to make and use. In this moving-punch tool, the moving punch (5) which is connected to the press via the punch base (1), runs in the guide plate or vee-ring plate (19) in the lower die set (2). In the upper die set (7), the blanking die plate (16) is mounted and additionally supported by the base plate (15), piercing punch retaining plate (9), backup plate (10), and supporting plate (12). The ejector (18) is guided into the blanking die plate. The ejector in turn guides

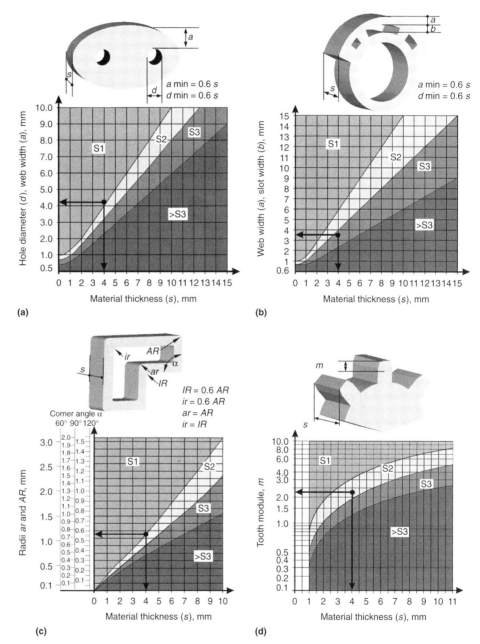

(a)

(b)

(c)

(d)

Different features	Symbol meaning	Requirement
(a) Hole diameter	a, web width	a min = 0.6 s
	d, hole diameter	d min = 0.6 s
(b) Thin slot width	a, web width	a min = 0.6 s
	b, slot width	b min = 0.6 s
(c) Outer and inner radius	IR, ir: inner fillet radius	IR = 0.6 AR
	AR, ar: outer fillet radius	ir = 0.6 AR
		ar = AR
		ir = IR
(d) Tooth modulus	m, tooth modulus	

s, work material thickness. Source: Ref 1, 3

Fig. 17 Difficulty ratings of various features used to evaluate the overall difficulty rating for a part. S1, least difficult; S3, most difficult

the inner forming punch (14). Via the pressure pins (11) and the pressure pad (13), the counterpressure, that is, the ejection force, is applied to the ejector (18). The inner form slugs are ejected by the press hydraulic system via the pressure pins (20) of the ejector bridge (3), and

the inner forming ejector pin (4) is ejected from the punch (5).

For most fine-blanking work, a fixed-punch system (Fig. 19) is preferred and it is suitable for all die types, in particular for the manufacture of thick, large parts. The blanking punch (14) is

positioned on a hardened pressure plate (8) on the upper die set (7), where it is permanently positioned and fastened. The precisely fitted guide plate (6) and the vee-ring plate (17), which bear the latch bolts (15), guarantee the fitting precision of the blanking punch (14) relative

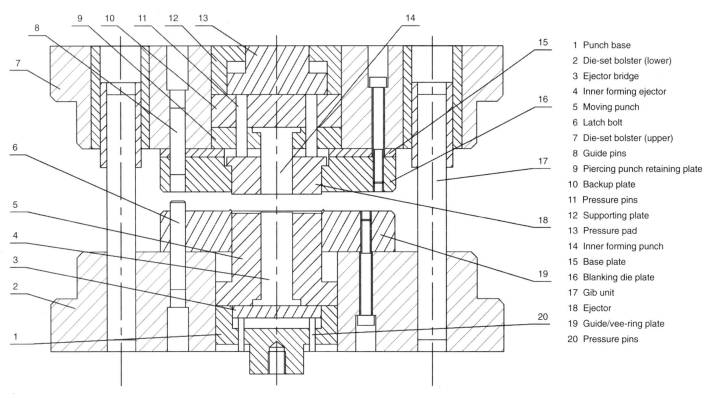

Fig. 18 Moving punch system

1 Punch base
2 Die-set bolster (lower)
3 Ejector bridge
4 Inner forming ejector
5 Moving punch
6 Latch bolt
7 Die-set bolster (upper)
8 Guide pins
9 Piercing punch retaining plate
10 Backup plate
11 Pressure pins
12 Supporting plate
13 Pressure pad
14 Inner forming punch
15 Base plate
16 Blanking die plate
17 Gib unit
18 Ejector
19 Guide/vee-ring plate
20 Pressure pins

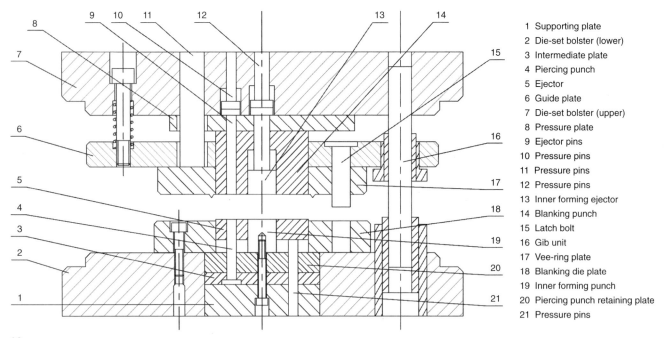

Fig. 19 Fixed-punch system

1 Supporting plate
2 Die-set bolster (lower)
3 Intermediate plate
4 Piercing punch
5 Ejector
6 Guide plate
7 Die-set bolster (upper)
8 Pressure plate
9 Ejector pins
10 Pressure pins
11 Pressure pins
12 Pressure pins
13 Inner forming ejector
14 Blanking punch
15 Latch bolt
16 Gib unit
17 Vee-ring plate
18 Blanking die plate
19 Inner forming punch
20 Piercing punch retaining plate
21 Pressure pins

to the blanking die plate (18) by means of the column gib unit (16) of the frame. The inner form slugs are stripped from the blanking punch (14) by the inner forming ejectors (9) and (13). Pressure is applied on the ejectors (9) and (13) and the vee-ring plate (17) by the vee-ring piston of the press via the pressure pins (11) and (12). The pressure pins (11) transmit the force of the vee-ring during the blanking process and the stripping force during the stripping cycle. The blanking die plate (18) is located, positioned, and fastened on the lower die set (2). The ejector (5), which is guided in the blanking die plate (18), also guides the inner forming punches (19) and (4) and transmits the counterpressure and ejection force from the counterpressure piston of the press via the pressure pins (21). The inner form punches (19) are mounted on the piercing punch retaining plate (20) and supported by the backup plate (1).

Like all conventional stamping tools, both fixed and moving punch systems can be extended to progressive blanking dies, compound progressive dies, and transfer dies (Fig. 20). With progressive and compound dies, an additional coil material guidance system (5), an initial blanking stop (13), pilot pins (21, 14), and positive press-off pins (16) are required. The coil guide ensures not only coil guidance but also stripping of the scrap web from the pilot pins (21, 14). The initial blanking stop (13) ensures precise initial blanking of the sheet metal at the beginning of a coil or strip. The pilot pins (21, 14) position the sheet metal precisely in each die station; that is, they ensure that the prescribed feed step is adhered to precisely. The positive press-off pins (16) compensate for

the counterpressure so that it does not have to be absorbed fully by the individual piercing punches (12) during initial blanking in the first die station.

Tool Material Requirements. The demands placed on the tool materials in service increase in direct proportion to sheet metal thickness and the strength of the workpiece material. If the active elements of the tooling, that is, the punch, undergo severe cyclic impact, they will quickly be destroyed. The challenge posed to the manufacturer of tool materials for fine-blanking is the need to produce tooling grades with better plastic deformation resistance and wear resistance while maintaining toughness. Table 11 gives the tool materials most commonly used today (2006). Most of these are high-speed steels, some of which meet the various requirements of die and periphery components (Table 12). Some tool steel manufactures have developed a powder metallurgy production technique to increase both plastic deformation resistance and wear resistance, as well as to increase the toughness of fine-blanking tools.

Heat Treatment of Tooling. Proper heat treatment is also one of the critical factors for controlling tool performance. The tool properties, including hardness, toughness, compressive yield strength, and dimensional stability, are affected by heat treatment. The two key components of the heat treatment cycle are the quenching rate and the choice of tempering temperature. The minimum quenching rate for a given tooling grade is set by its continuous-cooling-transformation diagram. The ideal cooling rate is sufficient to achieve a fully martensitic microstructure while avoiding the

formation of bainite. Furthermore, pearlite formation should be prevented in all cases, as it is very detrimental to the toughness of a tool. For high-alloy cold-work tool steel grades, minimum quenching rates in excess of 20 °C/min (35 °F/min) are often recommended.

The other crucial aspect of heat treatment is the selection of tempering temperature. Generally, three tempering temperature ranges are available for a specific cold-work grade: low range (approximately 400 to 600 °F, or 200 to 315 °C), intermediate, and high tempering range (980 °F, or 525 °C, and above). The intermediate range is not recommended because tool toughness is severely degraded in this range. The low tempering temperature range yields the best toughness properties for the grade at the selected working hardness. The high range is a sound choice for fine-blanking tooling, although the toughness properties obtained are not as good as those achieved using the low tempering range. With high-temperature tempering, for a high-alloy cold-work grade, some retained austenite transforms to the hardened martensitic structure.

Carbon Steel Performance for Tooling Component. Carbon steel is widely applied because it is inexpensive, possesses high hardness after heat treatment, and can easily be cold worked or hot worked. This material, however, is not without its disadvantages: narrow quenching temperature, poor hot hardness, low wear resistance, and brittleness after the quenching process. Carbon steels are used to produce less critical parts, such as the ejector rod and locator pin. The most commonly used type is AISI W1 (T72301, or GB T8A, T10A). Special forging

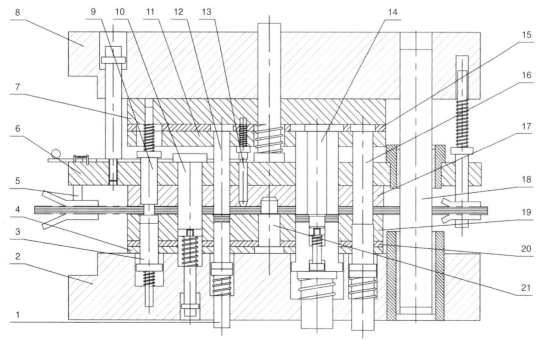

1 Pressure pins
2 Die-set bolster (lower)
3 Ejector
4 Intermediate plate
5 Coil guidance unit
6 Guide plate
7 Back-up plate
8 Die-set bolster (upper)
9 Coil guidance bolt
10 Latch bolt
11 Punch retaining plate
12 Piercing punch
13 Initial blanking stop
14 Feed step pilot/embossing punch
15 Intermediate plate
16 Positive press-off pins
17 Guide/vee-ring plate
18 Gib unit
19 Die plate
20 Retaining plate
21 Feed step pilot/embossing punch

Fig. 20 Compound progressive die. Source: Ref 1, 3

Table 11 Tool materials for fine blanking

Group	Code (ISO)	DIN No.	AISI
	Material types		
Cold-work steels	X 155 CrVMO 12-1	1.2379	...
	X 220 CrVMO 13-4 (PM)	1.2380	...
High-speed steels	S 6-5-2	1.3343	AISI M2
	S 6-5-3 (PM)	...	AISI M3 CL2
	S 6-5-4 (PM)	...	...
	S 11-2-5-8 (PM)	...	...
	S 6-5-3-8	...	...
	ISO range	Grain size, μm	
Cemented carbides	K40 (HiP)	2.5	
	K30-K40 (HiP)	1.0	
	K40 (HiP)	1.0	

Table 12 Materials for different components of fine-blanking tooling

Component of die	Suitable material type Great Britain	Suitable material type	Hardness, HRC
Punch	W18Cr4V	T1 (ASTM)	60~62
	W6Mo5Cr4V2	M2 (ASTM)	
	Cr12MoV	X155Cr (DIN)	
Die	W18Cr4V	T1 (ASTM)	62~64
	Cr12MoV	M2 (ASTM)	
	W6Mo5Cr4V2	X155Cr (DIN)	
Ejector	Cr12MoV	M2 (ASTM)	58~60
	CrWMn	Cr6 (DIN)	
Vee-ring plate	Cr12MoV	M2 (ASTM)	58~60
	CrWMn	Cr6 (DIN)	
Ejector rod or location pin	T10A	C105W1 (DIN)	58~60
	T8A	C85W1 (DIN)	
	9Mn2V	02 (ASTM)	
Pressure pin	CrWMn	Cr6 (DIN)	58~60
	9Mn2V	02 (ASTM)	
	T10A	C105W1 (DIN)	
Supporting plate	Cr12	D3 (ASTM)	56~58
	9Mn2V	02 (ASTM)	
	9SiCr	X100Cr (DIN)	
Guide pillar/guide bush	GCr15	52100 (ASTM)	58 to ~62

Table 13 Properties of the commonly used tool coatings

Coating	Color	Nanohardness, GPa	Coat thickness, μm	Coefficient of friction	Maximum usage temperature, °C	Corrosion resistance
TiN	Golden-yellow	24	1–4	0.4 to ~0.6	600	Good
TiAlN	Black-violet	33	1–4	0.4 to ~0.45	800	Good
CrN	Chrome white	18	1–4	0.5 to ~0.6	700	Excellent
TiCN	Blue-gray	36	1–4	0.25 to ~0.4	400	Good

Source: Ref 6

processes may be used to improve the performance of these materials:

- Heating: 1100 °C (2010 °F)
- Heating duration: approximately 1 to 1.5 min/mm (0.04 to 0.05 min/in.)
- Original forging temperature: approximately 1000 to 1050 °C (1830 to 1920 °F)
- Finished forging temperature: approximately 800 to 850 °C (1470 to 1560 °F)
- Cooling method: air cool

Coatings. Surface treatment is another method used to prolong tool life and increase product dimensional accuracy. Physical vapor deposition (PVD) and chemical vapor deposition (CVD) are two major methods applied to tools and machine elements to improve corrosion, friction, and wear properties. There are four commonly used coating types, titanium nitride (TiN), titanium carbon nitride (TiCN), titanium aluminum nitride (TiAlN), and chromium nitride (CrN), whose properties are listed in Table 13. Fine-blanking tools with PVD/CVD coatings have a longer tool life, ten times greater wear resistance, and high surface smoothness. Examples of coated fine-blanking tooling are shown in Fig. 21.

Lubrication is important in fine-blanking. Without it, the process would lead to cold welding of the active elements (i.e., punch, die) to the blank and rapid wear and blunting of the die. Accordingly, precautions must be taken to ensure that the active elements are supplied with sufficient lubricant at every point of the cutting edge and forming areas in the die. The application of fine-blanking oil to the sheet material can be carried out by using rollers or spray jets. In order to have both the top and the bottom surfaces of the sheet metal evenly wet by a film of lubricant, some wetting substances are added to fine-blanking oils. When a spray system is used, oil mist must be removed using a suitable extraction device. If high-viscosity oils are used for thick, higher-strength materials, it is also possible to apply lubricant through rollers. In order to ensure that the fine-blanking oil applied to the sheet metal surface also reaches the friction pair during the blanking process, parts of the die must be tailor-made to create some lubrication pockets along the cutting edges (Fig. 22).

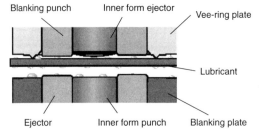

Fig. 21 Examples of coated fine-blanking tooling

Fig. 22 Formation of the lubricant film in the die

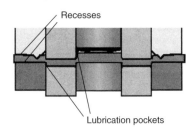

The fine-blanking oil must be capable of withstanding different levels of stress and temperature to the extent that the lubricant film remains intact during the blanking process. No one type of lubricant can satisfy multiple fine-blanking applications. The fine-blanking oil used in each case must be specially adjusted according to the work material, the thickness, and the strength of the part. The viscosity of the oil must match the specific types of stress occurring during fine-blanking. Typical lubricants are listed in Table 14.

Fine-Blanking Press

The fine-blanking process requires a controlled sequence of movements with a precise top dead center. The strict blanking clearance of the dies must not change, even under high levels of stress. Fine-blanking presses, therefore, must meet the stringent precision requirements, including slide gibs, high frame rigidity, and parallelism of die clamping surfaces. Both mechanical and hydraulic systems are used for the main slide drive.

The mechanical fine-blanking press (Fig. 23, 24) is notable for its sturdiness and operational reliability under the tough conditions of daily use. It is equipped with a "combination bed" for the tool system with moving and fixed punches (Fig. 18, 19). Making use of the straight-side principle, the monoblock press frame as a welded construction offers good dimensional rigidity and avoids vibrations. A clearance-free pretensioned slide gib is used. A central support in the upper and lower die clamping plate ensures optimum die support and introduction of force to the die. A controlled, infinitely variable direct-current (d.c.) motor drives the press via the flywheel, consisting of a disk clutch and a worm gear pair on two synchronously running crankshafts of different eccentricities.

The crankshafts drive a double knuckle-joint system that generates the movement sequence of the slide required for fine-blanking. This drive system is particularly well suited for material thicknesses between approximately 1 and 8 mm (0.04 and 0.3 in.) and total press forces up to 2500 kN. The vee-ring force and counterforce are applied by hydraulic systems. The mechanical drive system has:

- A fixed slide-movement sequence with constant stroke and precise position of the top and bottom dead center

- Low energy consumption
- High output (depending on the size of the press) with stroking rates of up to 140/min
- Minimal setting and maintenance input

The high manufacturing precision of fine-blanked parts and the tight clearance of the process demand a rigid press frame and precise ram guidance. In the example below for a mechanical fine-blanking press, ram movement is broken down into four stages: (a) rapid closing, (b) sensing movement and engagement of vee-ring, (c) cutting movement, and (d) rapid opening. During the cutting stage, the travel speed is as slow and as constant as possible, whereas the no-load strokes (a) and (d) are as rapid as possible (Fig. 25).

Figure 26 shows a schematic and photograph of a hydraulic fine-blanking press. Most efficient hydraulic fine-blanking presses with a total force ranging between 2,500 and 14,000 kN are equipped with a pressure accumulator drive. In this kind of press, the body frame is made of a robust welded monoblock straight-side construction with high rigidity. This is achieved by designing large upright cross sections and large ribs. The slide and main pistons together form a unit that is integrated in the press bed. The counterpressure piston is integrated in the slide, and the vee-ring piston is mounted in the press crown. Both pistons are equipped with support bolts to provide support over the entire surface of the dies. The slide is guided in the press body and the gib is a high-precision, multitrack sliding gib. This construction alleviates slide tilting, even under major eccentric loads.

According to the fine-blanking press manufacturer, a hydraulic fine-blanking press has:

- An integrated tool-breakage safety system independent of the tool itself
- All regulation safety systems for personnel protection
- A high degree of operational reliability in the work process

Alternative Tooling and Equipment

Several alternative approaches have been developed for fine blanking. A schematic diagram of a typical hydraulic-tooling die system is shown in Fig. 27. It is composed of a lower and upper die-set bolster, lower and upper piston, lower and upper base plate, gib system, and spring. High precision is guaranteed by the four rigid columns of the gib system. The hydraulic cylinders and pistons provide vee-ring force and counterbalancing force when the die is closing, and unloading force and stripping force when the die rebounds after the oil reservoir in the die bolster is filled with hydraulic fluid.

A conventional hydraulic press system may combine the fine-blanking function with the

Table 14 Lubricants used in the fine-blanking processes

	Composition	Percentage, %
F-1 lubricant	Chloride paraffin	10~15
	Organic acid ester	0.1~1
	S · P additive	5~10
	50# machine oil	75 to ~85
HFF lubricant	Chloride paraffin	66
	Mineral oil	16
	Glycerine	7
	Castor-oil oleate	4
	S · P barium additive	7

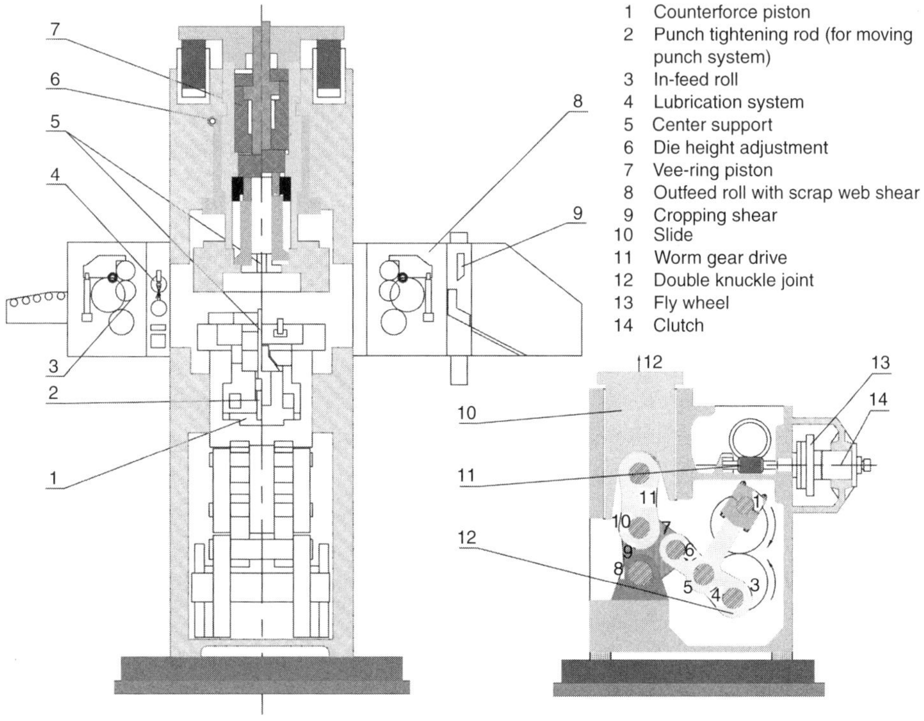

1	Counterforce piston
2	Punch tightening rod (for moving punch system)
3	In-feed roll
4	Lubrication system
5	Center support
6	Die height adjustment
7	Vee-ring piston
8	Outfeed roll with scrap web shear
9	Cropping shear
10	Slide
11	Worm gear drive
12	Double knuckle joint
13	Fly wheel
14	Clutch

Fig. 23 Mechanical fine-blanking press

naddition of a hydraulic tooling-die system. The installation and disassembly is the same as the normal die set without the additional refit to the press. The press can be used for conventional blanking processes when the hydraulic die-set system is removed.

Application and Development in the Future

Current Applications in the Industry. Fine blanking is used not only in the mass-production of precision parts, but also in the batch production of intricate parts. Fine-blanked parts are currently found in many different products, such as automobiles (Fig. 28), airplanes, and office machines. Recent advances in this process have particularly attracted the automotive industry, with its high unit quantities and need to remain competitive.

The bicycle and motorcycle industries have also introduced many fine-blanked parts such as gears, gear levers, gear cams, and cam plates. Chain sprockets and the rotors of centrifugal clutches include a number of fine-blanked parts. Even the perforated disk brake on high-performance motorcycles (Fig. 29) are now made by fine-blanking.

Advances in electronics have virtually eliminated the need for many mechanical control and transmission systems and thus reduced such applications of fine-blanking. However, the electronics and computer sectors, with their high precision requirements and new materials, have created new applications for the process such as computer hard-drive components (Fig. 30). Another important application of fine-blanking is the manufacture of blades, cutlery, hand tools, and surgical instruments.

The potential of the process to manufacture less expensive and better quality parts has not been exhausted. Many new applications in the automotive, electronics, and appliance industries are constantly emerging. By combining cold forming with fine blanking, creative design solutions involving bending, countersinking, semipiercing, coining, cranking, or deep-drawing become possible. In the vast majority of cases, new applications are no longer of simple flat parts, but components in which fine-blanking is used in combination with other technologies (Fig. 31).

Future Trends. With the rapid development of advanced, low-cost computers, the finite-element method (FEM) can be now widely applied for the numerical simulation of sheet-metal forming processes, including the fine-blanking process. With the proper material properties and simplified boundary conditions, FEM can predict the pertinent stress conditions through the whole blanking area to investigate process mechanics. Also, with numerical simulation, working parameters can be optimized without expensive shop trials.

Typically, most engineers use computer-aided design (CAD) programs to create tools with the same geometry as those used in the process simulation or virtual manufacturing. Manufacturers can simulate their production process in a virtual environment to save time and money spent on real pilot production or practical tests at the verification stage. In addition, simulation results provide useful information to

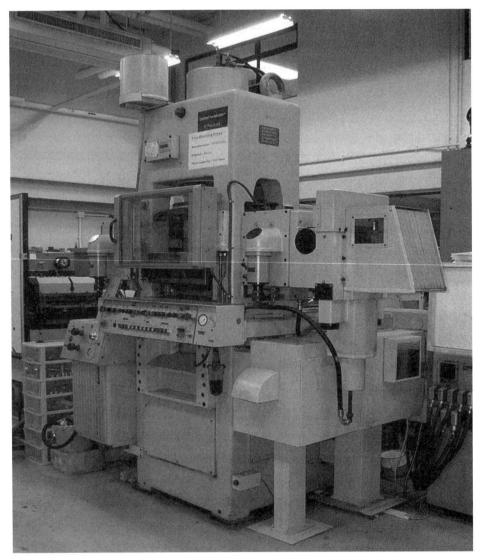

Fig. 24 Mechanical fine-blanking press

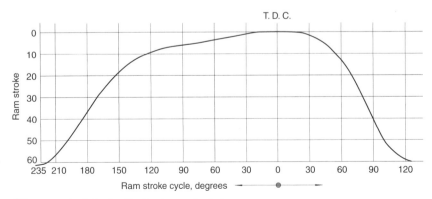

Fig. 25 Ram stroke during fine-blanking in a mechanical press

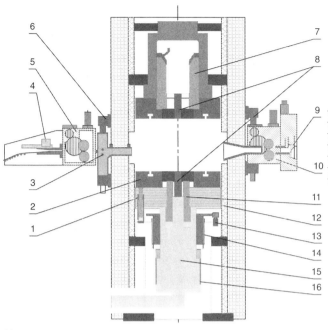

1 Rapid traverse closing piston
2 Die change clamping table
3 Lubrication system
4 Infeed roll basket
5 Infeed roll
6 Feed height adjustment
7 Vee-ring piston
8 Central support
9 Cropping shear
10 Outfeed roll
11 Counterforce piston
12 Slide gib
13 Shut height adjustment motor
14 Fixed stop
15 Main working piston
16 Slide gib

(a)

(b)

Fig. 26 Hydraulic fine-blanking press. (a) Schematic (b) photo of modern hydraulic fine-blanking press

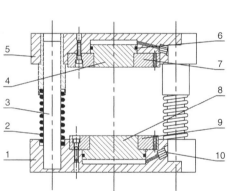

1 Die-set bolster (lower)   6 O-ring seal
2 Spring   7 Upper base plate
3 Gib unit   8 Lower piston
4 Upper piston   9 Lower base plate
5 Die-set bolster (upper)   10 Oil inlet

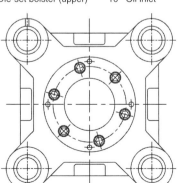

Fig. 27 Hydraulic tooling-die system

Fig. 28 Fine-blanked central gears of car seat adjuster

Fig. 29 Fine-blanked disk brake for motorcycles

Fig. 30 Fine-blanked components for computer hard drives

Fig. 31 Automotive transmission gear-shift fork

address the feasibility of the actual production process. Product quality can also be improved. The risks of tool redesign and modifications are minimized.

REFERENCES

1. *Metal Forming Handbook,* Schuler GmbH/ Springer-Verlag Heidelberg, 1998
2. *Fine-Blanking Technology,* T. Guangqi, Ed., Chinese Mechanical Industrial Publishing House, 1990
3. *Forming and Fineblanking—Cost Effective Manufacture of Accurate Sheetmetal Parts,* Feintool/Verlag Moderne Industrie, 1997
4. L. Shuoben et al., Ed., *Forging Manual Part 2: Stamping,* Chinese Mechanical Industrial Publishing House, 2002; www. fineblanking.com/edges.htm/
5. Fine-Edge Blanking and Piercing, *Forming and Forging,* Vol 14, 9th ed., *Metals Handbook,* ASM International, 1988, p 474
6. B. Janoss, PVD/CVD Tool Coatings Enhance Stamping and Forming of Stainless Steels, *Metal Form.,* Vol 33 (No. 3), March 1999

Equipment for Forming of Sheet Metal

Presses and Auxiliary Equipment for Forming of Sheet Metal

STAMPING PRESSES are built with meticulous attention to accuracy, clearances, parallelism, deflections, and controls. With the use of precision die-making machines and the use of better-grade die materials, precision presses are needed for satisfactory operation in achieving the long life and extremely close clearances of expensive dies. Intricate dies require presses that run with better guiding and deflection characteristics. One example of this tooling is the progressive die, which combines many operations in one press to produce a finished part with each stroke. This tooling, coupled with automatic feeding and ejection equipment, leads to greater press speeds for more production at less cost per piece.

This article describes various types of press construction and the factors that influence the selection of mechanically or hydraulically

powered machines for producing parts from sheet metal. A press (Fig. 1) can represent a substantial capital investment, and proper selection of a press is essential for successful and economical operation. No general-purpose press exists that can provide maximum productivity and economy for all applications. Compromises usually must be made to permit a press to be employed for more different jobs or flexibility for future production requirements.

Types of Presses

There are three basic components in a press: the frame, the drive, and the gibbing. Presses are broadly classified, according to the type of frame used in their construction, into two main groups: gap-frame presses and straight-side presses. Details of construction vary widely in each group, but typical examples are illustrated in Fig. 2 and 3 for a straight-side press and an open-back inclinable (OBI) gap-frame press, respectively. Power presses also can be classified further according to the source of power (mechanical or hydraulic), method of actuation of slides, and the number of slides in action. Presses in any of these classes are available in a

range of capacities (tonnage or bed area), although the range is not necessarily the same for all types of presses. Characteristics of 18 types of presses are summarized in Table 1.

The Joint Industry Conference (JIC), a committee of press builders and large-press users formed some years ago, set guidelines for uniformity with respect to nomenclature, bed and ram sizes, force ranges, and symbols for presses. Although JIC is no longer in existence, most press builders adhere to the standards either completely or in part. Under the JIC press classification system with respect to the number of slides, the first letter in the designation is S for single-action, D for double-action, and T for triple-action presses. Other designations used for visible identification of presses suggested under the JIC classification system are given in Fig. 4. Most press builders place these markings in a prominent position on the fronts of presses.

Source of Power. Power presses can have mechanical or hydraulic drives with principal components (Fig. 5, 6). The performance characteristics and other operational features of hydraulic and mechanical presses are compared in Table 2. Mechanical presses are frequently used, but hydraulic presses are being increasingly

Fig. 1 Stamping press

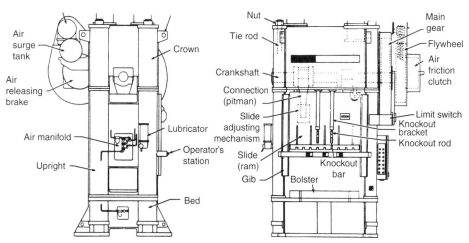

Fig. 2 Typical straight-side press frame

applied. There are applications for which hydraulic presses offer certain advantages and, in some cases, are the only machines that can be used. For example, very high force requirements can be met only with hydraulic presses.

In some respects, the gap between the advantages of selecting a mechanical press over a hydraulic press and vice versa has narrowed due to improvements in design and the addition of features to each type that duplicate the desired actions of the other. An example would be mechanical presses with varidraw drives to improve speed during drawing operations. Hydraulic presses were once the only presses considered for deep draw work, while mechanical presses were used for blanking, forming, and shallow draws. The speed of the press is determined by the speed at which a material can be drawn because the material may tear if the speed is exceeded. In the case of the mechanical press, this then determines the speed of the press through its entire cycle, which is generally slower than the full speed capability of a mechanical press. By installing a varidraw drive, a slower speed can be maintained at the recommended drawing speeds only for that portion of the stroke cycle where the draw occurs, while increasing the speed of the press through the balance of the cycle. Varidraw drives on mechanical presses can as much as double the production of the press in drawing operations.

Another important advantage of hydraulic presses over mechanical presses has been the fact that a hydraulic press cannot be overloaded. This advantage has been eliminated by the introduction of overload units that can be installed in mechanical presses. These units basically consist of a hydraulic cylinder, installed within the connections of the press that maintains a constant and adjustable pressure that, when exceeded, causes the cylinder to collapse and stop the press, preventing any damage to the press proper. After the obstruction (the cause of the overload) has been removed and the overload reset, the unit will automatically build up to the predetermined pressure and be ready to resume its function.

The choice today of a mechanical or hydraulic press becomes more a question of production requirements, economics, versatility, and the type of work to be performed within the press. Hydraulic and mechanical presses can be single-action or double-action, as described in the section "Number of Slides" in this article. Advantages and disadvantages of each are described in the section "Press Selection" in this article.

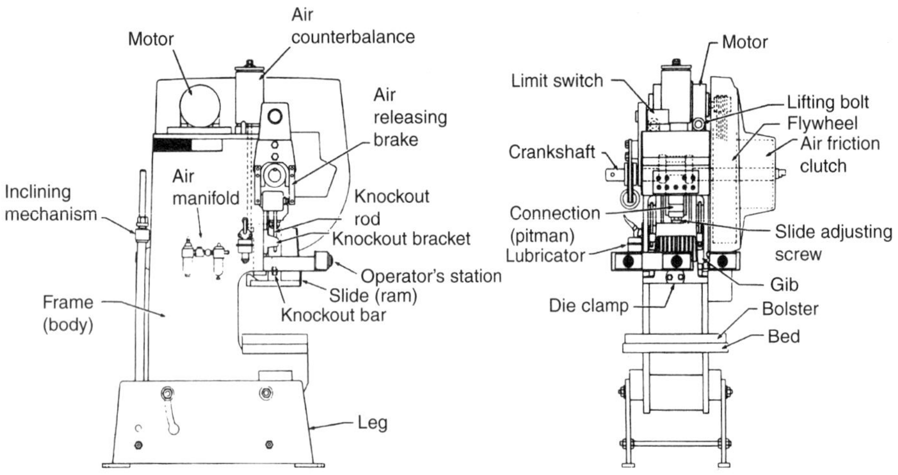

Press Frames

The basic job of the press frame is to contain the load and control the deflection. It follows directly that the bed and slide members of the press, onto which the die shoes are fastened, must exhibit a similar degree of rigidity to back

Fig. 3 Typical gap press frame (open-back inclinable)

Table 1   Characteristics of 18 types of presses

Type of press	Type of frame							Position of frame				Action			Method of actuation								Type of drive				Suspension			Ram		Bed		
	Open-back	Gap	Straight-side	Arch	Pillar	Solid	Tie-rod	Vertical	Horizontal	Inclinable	Inclined	Single	Double	Triple	Crank	Front-to-back crank	Eccentric	Toggle	Screw	Cam	Rack and pinion	Piston	Over direct	Geared, overdrive	Under direct	Geared, underdrive	One-point	Two-point	Four-point	Single	Multiple	Solid	Open	Adjustable
Bench	X	X	...	...	...	X	...	X	...	X	X	X	...	...	X	...	X	...	X	...	X	X	X	...	...	...	X	...	...	X	...	X	X	X
Open-back inclinable	X	X	...	...	...	X	...	X	...	X	...	X	X	...	X	...	X	...	...	...	...	X	X	X	...	...	X	X	...	X	X	...	X	...
Gap-frame	X	X	...	...	...	X	X	X	X	X	X	X	X	...	X	...	X	X	X	...	...	...	X	X	...	...	X	X	X	X	X	X	X	...
Adjustable-bed horn	...	X	...	...	...	X	...	X	...	X	...	X	...	...	X	...	X	...	...	...	...	...	X	X	...	...	X	X	...	X	...	X	X	X
End-wheel	...	X	...	...	...	...	...	X	...	...	...	...	X	X	...	X	X	...	...	...	...	...	X	X	...	...	X	...	...	X	...	...	X	...
Arch-frame	...	...	X	X	...	X	...	X	...	X	...	X	X	...	X	...	X	...	...	...	...	...	X	X	...	...	X	...	...	X	...	...	X	...
Straight-side	...	...	X	X	...	X	...	X	X	X	X	X	X	X	X	X	X	X	...	...	X	X	X	X	X	X	X	X	X	X	X	X	X	...
Reducing	X	X	X	...	...	X	...	X	X	X	X	X	...	...	X	...	X	...	...	...	...	...	X	X	...	...	X	...	...	X	...	...	X	...
Knuckle-lever	...	...	X	...	...	X	...	X	X	...	...	X	X	X	X	X	...	...	...	X	...	...	X	X	...	...	X	X	...	X	X	X	X	...
Toggle-draw	...	...	X	...	...	X	...	X	X	...	...	...	X	X	X	X	X	...	...	...	...	...	X	X	...	...	X	X	X	X	X	...	X	...
Cam-drawing	X	X	X	...	...	X	...	X	X	...	X	...	X	...	X	...	X	...	...	X	...	...	X	X	...	...	X	...	X	X	...	...	X	...
Two-point single-action	...	X	X	...	...	X	...	X	X	...	...	X	...	...	X	...	X	...	...	...	...	...	X	X	...	...	X	...	X	X	...	...	X	...
High-production	...	...	X	...	...	X	X	X	X	...	...	X	X	...	X	...	X	...	...	...	...	...	X	X	...	...	X	X	...	X	...	...	X	...
Die machine	...	...	...	X	...	X	...	X	...	...	...	X	...	...	X	...	X	...	...	...	...	...	X	...	X	...	X	X	...	X	...	...	X	...
Transfer	...	X	X	...	...	X	X	X	X	...	...	X	...	...	X	...	X	X	X	...	...	...	X	X	...	...	X	...	...	X	...	...	X	...
Flat-edge trimming	...	...	...	X	...	X	...	X	...	...	...	X	...	...	X	...	...	...	...	...	X	...	X	...	...	...	X	...	...	X	...	...	X	...
Hydraulic	...	X	X	...	...	X	...	X	X	X	X	X	X	X	...	...	...	...	...	...	...	X	X	...	...	...	X	X	...	X	X	X	X	X
Press brake	X	X	...	...	...	X	...	X	...	X	...	X	...	...	X	...	...	...	...	...	...	X	...	X	...	...	X	X	X	X	...	X	...	...

up the die shoes when under load. If the press members deflect excessively under full load, the die shoes will also deflect, following to some degree the deflection of the press members. When this occurs, both upper and lower die elements are distorted. As a result, the finished stamping may not come up to full expectations because the center of the die has been deflected away from its normal position.

Steel and cast iron are two basic types of materials used in the construction of press frames. The stress design factor used by reputable builders is conservative, and consequently there does not appear to be much difference in the deflection of the component parts, whether they are made of steel or of cast iron. However, one of the big advantages of steel is the straight-line relationship to the yield point (Fig. 7). This provides a big safety factor in the event that the machine is overloaded. Another advantage of steel is that it is basically easier to repair than cast iron. The steel plate used in presses comes as a homogeneous rolled plate, which eliminates the possibility of internal discontinuities such as blowholes and slag inclusions sometimes found in cast frames. Steel fabrications also lend themselves to the utilization of uniform cross sections, which reduces the extent of stress-concentration regions that might lead to trouble later on.

Basic engineering sections can be used in the design of fabricated presses, which allow the use of standard formulae with empirical relations that closely resemble the actual deflection characteristics of the frame. This, of course, is an important point when designing a press for the very-low-deflection characteristics required in many cases with carbide-type tooling. Generally speaking, however, fabricated steel machines do not weigh as much as cast iron machines because with fabricated steel, we can take advantage of standard structural shapes to provide rigidity where it is required. It is not necessary to have the excessive mass found in a cast iron structure.

Fabrication has become increasingly common in recent years, and most press builders will quote on fabricated machines. It is extremely important, however, that the large weldments, which form the basic component of the frame, be thoroughly stress relieved. Also important in conjunction with stress relieving is to have the frame grit blasted to remove objectionable scale and provide a good base for primer paint.

Gap-frame presses (Fig. 8) are sometimes called C-frame presses because the frame resembles the letter C when viewed from the side. Their most distinguishing feature is unobstructed die space accessible from all sides except the rear. In some cases, even the rear is accessible. Because of their extreme versatility, gap-frame presses are used for a great variety of operations such as piercing, blanking, and drawing.

Gap-frame presses feature versatility of application and material handling. The gap makes the die area accessible from either side, as well as from the front, for ease in die setting or for feeding stock. Coil stock often is supplied to gap-frame presses by feeders from stock reels and straighteners. Workpieces usually are ejected through an opening in the press bed, or through the back of an open-back press. (A few gap-frame presses with solid backs are in service, but solid-back presses are coming into increasing disfavor because work cannot be ejected through the back of the press and because the design in general is less convenient than is the open-back type.)

Gap-frame construction has one disadvantage: As the load is imposed, the mouth of the C will have a tendency to "alligator," or open, thus causing angular deflection. (A straight, vertical deflection would be less critical.) Gap deflection resulting from overload causes misalignment of punches and dies, which is a major cause of premature die wear.

Figure 9, which shows a typical right-to-left shaft-inclinable press frame, indicates some significant structural points. Combined bending and direct stress loads are encountered in both the upper and lower cantilever beams (C-C and B-B). These loads extend forward the distance D to support, respectively, the main bearing structure above and the press bed below. It is noteworthy that the upper cantilever beam has a section C-C, which is very deep in relation to the throat, D. Therefore, deflection in this upper cantilever beam is so small that it can be neglected in calculations.

Standard gap-frame presses are built with one-piece frames. Designing these presses with the proper rigidity is, therefore, considerably more complex than for presses with straight-side frames. The total deflection of the gap area is made up of both vertical and angular deflection and is a function of gap height and depth. The angular deflection is the more important of the two as far as the press user is concerned because it may result in misalignment between punch and die.

When a gap-frame press is fabricated from steel plate, the side plates that form the uprights of the frame have a uniform cross-sectional width and thickness. The strength of the frame might be described simply as the depth of this plate times the thickness at the throat, which gives us an areal measure. The thicker and deeper the plate, the stiffer the frame can be. Sometimes on larger gap-frame presses a reinforcing rib is welded around the throat to increase the resistance to deflection.

In designing open-back inclinable (OBI) machines, it is important to have a frame properly designed for "controlled deflection" at maximum tonnage of the machine. This means that, at maximum tonnage, if a perpendicular is

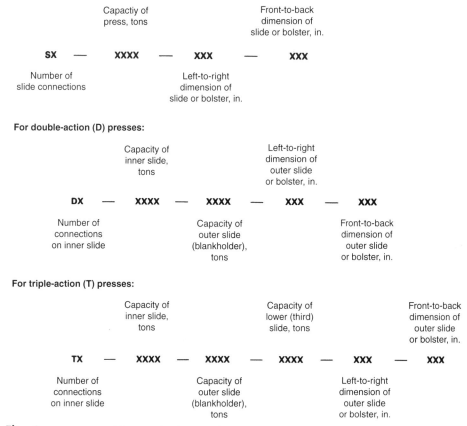

Fig. 4 Designations used for press identification under the JIC classification system

dropped from the slide and from the bed, these lines will intersect at a point approximately at the stock-line height. This then helps to eliminate tooling wear due to the angularity or alligator effect in the machine. It is also interesting to note that in a properly designed OBI machine, its bearings in the frame are set on a 45° angle. This transfers the shaft load directly to the frame and eliminates loading the bolts that attach the bearing caps.

Press builders advocate selection of gap-frame presses of sufficient capacity to perform operations without having to resort to tie-rods for reinforcement. This does not imply that a gap-frame press cannot be used safely up to its rated capacity. However, where angular deflection may be a determining factor for good die life, such as in the use of large-area blanking dies, conservatism is suggested in selecting a press. Tie-rods extending from the top of the frame to the front corners of the bed can be used to minimize deflection. Because the tie-rods close the gap, the accessibility to the die area and the width of parts fed into the die are limited. Tie-rods can be removed for die setup.

Straight-Side Presses. The straight-side frame has the advantage over the gap-frame in that it eliminates angular deflection. Most straight-side frames are four-piece frames, consisting of the crown, two uprights, and the bed. In most straight-side presses, steel tie-rods hold the base and crown against the columns. As with OBI machines, straight-side frames can be made from cast iron or steel.

Straight-side presses have crankshaft, eccentric-shaft, or eccentric-gear drives (see the section "Slide Actuation in Mechanical Presses" in this article). A single-action straight-side mechanical press is shown in Fig. 5. The slide in this illustration is equipped with air counterbalances to assist the drive in lifting the weight of the slide and the upper die to the top of the stroke. Counterbalance cylinders provide a smooth press operation and easy slide adjustment. Die cushions are used in the bed for blank holding and for ejection of the work.

The straight-side design permits the use of an endless variety of bed and slide sizes. Typical presses may range from 180 kN (20 tonf) capacity and a bed of 510×380 mm (20×15 in.) to 36 MN (4000 tonf) capacity with a bed as large as 915 cm (360 in.) left-to-right by 455 cm (180 in.) front-to-back. The size and shape of the slide usually determine the number of points of suspension, or connections between the main shaft and the slide, that are needed.

As noted, the straight-side design can provide less deflection under high pressures or off-center loads than does a gap-frame press. The type of work has a great bearing on how stiff the press bed should be. For instance, when carbide tooling is used with close clearances, it is generally felt that it is important to have a stiffer bed to support the lower die shoe. This in turn extends the life of these very expensive dies.

Basically, the straight-side frame has to resist "oil-canning" and twisting with off-center loads. With progressive dies, the load can be, and quite often is, on one side or the other of the centerline. This then imposes a heavier load on one side of the press. Such off-center loads can create unusual conditions in the frame that must be resisted. The instant the press load is applied, this load has to be transferred from the tooling to the slide and through the gibs to the frame. To eliminate twisting, a triple-box construction is used (Fig. 10). Continuous welds running the full length of the press provide high resistance to twisting under load.

As noted, straight-side four piece frames are held together with tie-rods. The relationship among tie-rod stress, frame stress, and press load in a 200 ton (1.8 MN) press is shown in Fig. 11. Line AB shows the increase in tie-rod stress under increasing press loads. Line AD shows the decrease in frame stress under increasing press loads. Point A indicates the stress in the tie-rods and frames as being equal when there is no workload in the press. Point B indicates the stress in the tie-rods when there is no compression left in the frame. (This is the "liftoff" point. The press workload equals the tie-rod load.) Point D indicates zero compressive stress in the frame at the liftoff point. Point C indicates the tensile stress in the tie-rods at the maximum rated tonnage of the

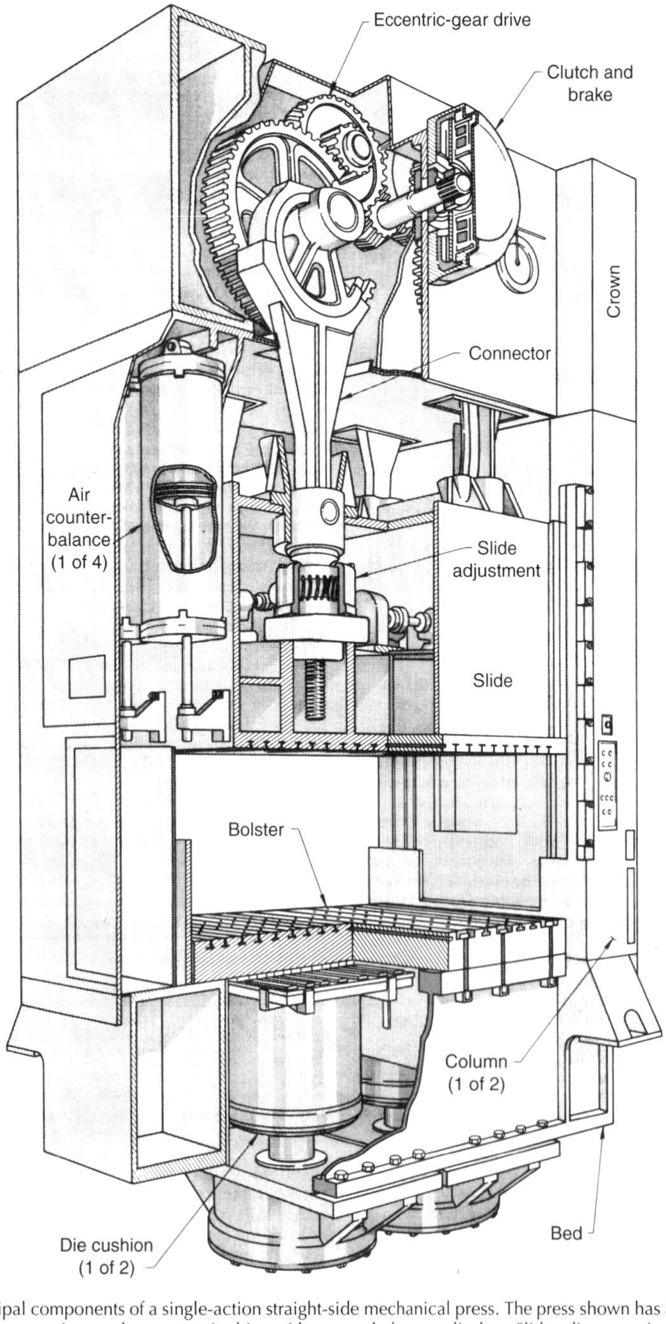

Fig. 5 Principal components of a single-action straight-side mechanical press. The press shown has a large bed, four-point suspension, and an eccentric drive with counterbalance cylinders. Slide adjustment is motorized.

press. The distance CF indicates the press load at 100% capacity. Point F indicates the compressive stress in the frame at the maximum rated tonnage of the press. Line EB shows what the tie-rod loads would be if the tie-rods were not shrunk. BH shows the rate at which tie-rods will yield beyond the liftoff point.

With the rods assembled through the frame parts, the tie-rod nuts are brought into firm con-tact with the crown and bed surfaces. The rod is then heated by any one of several accepted methods. As the temperature of the rod increases, its length increases and the lower tie-rod nut moves away from the lower bed surface. When the rod has been sufficiently lengthened by the application of heat, the lower nut is tightened a predetermined amount. As the rod starts to cool, it tends to shrink to its original length but is prevented from doing so by the semirigid frame parts between the tie-rod nut faces. If the frame parts were absolutely rigid—i.e., if they would not shorten (compress) under the load—the tie rod, when it had cooled off completely, would be stressed in tension to 145 MPa (21 ksi). The frame, of course, would be stressed in compression to a load equal to the tie-rod load.

However, because the frame parts are not absolutely rigid and shorten somewhat under load, the final load equilibrium point between the tie-rods and frame is somewhere between zero and the liftoff point. The exact point of this equilibrium is dependent on the relative stiffness (area and modulus of elasticity) of the tie-rods and the frame parts. In general-purpose press design, this point is targeted to fall somewhere between 90 and 110 MPa (13 and 16 ksi) in the tie-rods. This is, therefore, the residual stress in the tie-rods with a proportionate stress in the frame parts under a no-load press condition. The tie-rods are in tension and the frame parts are in compression. As a workload is developed in the press, the stress in the tie-rods increases while the

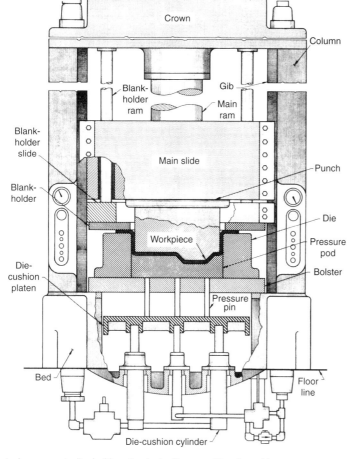

Fig. 6 Principal components of a double-action hydraulic press with a die cushion

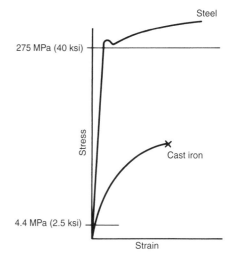

Fig. 7 Stress-strain curves for steel and cast iron

Table 2 Comparison of characteristics of mechanical and hydraulic presses

Force	Capacity	Stroke length	Slide speed	Control	Preferred uses
Mechanical					
Varies depending on slide position	Practical maximum of ~54 MN (6000 tonf)	Limited	Higher than hydraulic, and can be varied. Highest at midstroke	Full stroke is usually required before reversal.	Preferred for operations requiring maximum pressure near the bottom of the stroke. Preferred for cutting operations such as blanking and piercing, and for relatively shallow forming and drawing (depths to about 102 mm or 4 in.). Good for high-production applications and progressive and transfer die operations
Hydraulic					
Relatively constant (does not depend on slide position)	445 MN (50,000 tonf) or more	Capable of long (2.5 m or 100 in.) strokes	Slower pressing speeds, with rapid advance and retraction. Speed is uniform throughout the stroke.	Adjustable; slide can be reversed at any position.	Good for operations requiring steady pressure throughout the stroke. Preferred for deep drawing, die tryout, flexible-die forming, drawing of irregular-shaped parts, straightening, operations requiring high and variable forces, and operations requiring variable or partial strokes.

stress in the frame parts decreases. This change varies directly as the press workload. Figure 11 shows a simple graphical representation of the relationship among tie-rod stress, frame stress, and press load. The distance GA on the diagram is the stress adjustment in the tie-rods, resulting from the shortening of the frame structure under the load developed in shrinking the tie-rods.

Under full-load press conditions, the tie-rods would exhibit a tensile stress of between 117 and 131 MPa (17 and 19 ksi). This stress is made up of two components: a stress between 45 and 59 MPa (6.5 and 8.5 ksi) holding the frame parts in compression, plus a stress of approximately 72 MPa (10.5 ksi) from the press workload. (Refer to points C and F in Fig. 11.)

In order for the crown to lift off the uprights, the frame parts must decompress completely and return to their original height. For this to occur, it would be necessary to stretch the tie-rods to the length they had attained while they were hot prior to their shrinking. It is therefore apparent that at the liftoff point, the load in the tie-rods must equal the workload in the press. This commonly is set by design at 150 to 200% of press capacity. Or, stated in other terms, it would require a 50 to 100% press overload to lift the crown from the top of the uprights.

Although the frame members will not separate at their joint surfaces under normal loads, the distance between the bed and the crown will nevertheless increase slightly from no-load to full-load conditions. Note also in Fig. 11 that the tie-rod stress variation is on the order of 20 MPa (3 ksi): from 100 MPa (15 ksi) at no load to 125 MPa (18 ksi) at full load. Such a stress variation is well within the endurance limits of the types of steels from which tie-rods are made, and long tie-rod life may be anticipated provided the rods are properly shrunk.

Overdesigning of press frames for rigidity should also be cautioned. A frame that is too rigid can be as harmful and troublesome as a too-flexible structure. There are, as a matter of fact, a few cases in which frame structures had to be weakened before satisfactory productivity, press life, and die life were possible. The design of a frame is not wholly dependent on the press tonnage. The type of machine, the class of work for which it is intended, the type and construction of the dies that are contemplated, and even the speed of operation all have some bearing on frame-structure design. A frame that is too unyielding can be the cause of very severe damage to dies or to other parts of the press solely because of the commercial tolerance of the stock being run through the press. The tie-rods on all presses serve as some protection against faulty die setting, doubles, and other

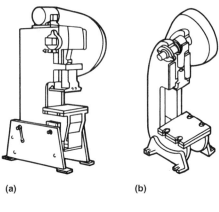

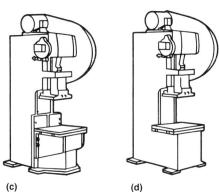

Fig. 8 Four types of gap-frame presses. (a) Open-back inclinable. (b) Bench press. (c) Adjustable-bed stationary. (d) Open-back stationary

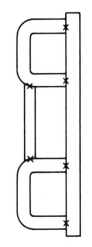

Fig. 10 Triple-box straight-side press frame

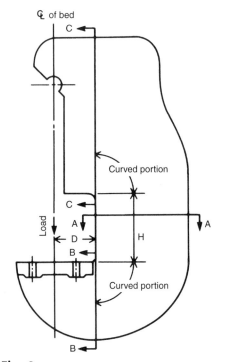

Fig. 9 Typical right-to-left shaft-inclinable press frame

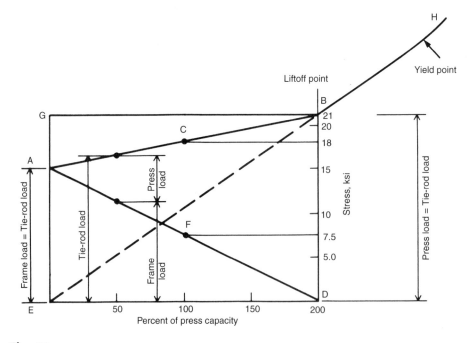

Fig. 11 Relationship among tie-rod stress, frame stress, and press load in a 200 ton press

inaccuracies, but they do not provide safety against breakage of press or die components.

Mechanical Presses

In most mechanical presses, a flywheel is the major source of energy that is applied to the slides by cranks, gears, eccentrics, or linkages during the working part of the stroke. The flywheel runs continuously and is engaged by the clutch only when a press stroke is needed. In some very large mechanical presses, the drive motor is directly connected to the press shaft, thus eliminating the need for a flywheel and a clutch.

Press Drives

The drive on a press is made up of the motor/flywheel combination, clutch and brake, and gearing, if used (Fig. 12, 13). Two basic types of drive, gear and nongeared, are used to transfer the rotational force of the flywheel to the main shaft of the press.

Nongeared Drive. In a nongeared drive (also known as a flywheel drive), the flywheel is on the main shaft (Fig. 14a), and its speed, in revolutions per minute, controls the slide speed. Usually press speeds with this type of drive are high, ranging from 60 to 1000 strokes per minute. The main shaft can have a crankshaft, as shown in Fig. 14(a), or an eccentric.

Energy stored in the flywheel should be sufficient to ensure that the reduction in the speed of the flywheel will be no greater than 10% per press stroke. If the energy in the flywheel is not sufficient to maintain this minimum in speed reduction, a gear-driven press should be used.

Gear drives (Fig. 14b, c, d) have the flywheel on an auxiliary shaft that drives the main shaft through one or more gear reductions. Either single-reduction or multiple-reduction gear drives are used, depending on size and tonnage requirements. In gear-driven presses, there is more flywheel energy available for doing work than there is in the nongeared presses because the speed of the flywheel is higher than that of the main shaft. The flywheel shaft of a gear-driven press often is connected to the main shaft at both ends (Fig. 14c), which results in a more efficient drive.

A single-reduction gear drive develops speeds of 30 to 100 strokes per minute. Speed for a multiple-reduction twin gear drive (Fig. 14d) is usually 10 to 30 strokes per minute, which provides exceptionally steady pressure.

Slide Actuation in Mechanical Presses

Rotary motion of the motor shaft on a mechanical press is converted into reciprocating motion of the slides by a crankshaft, eccentric shaft, eccentric-gear drive, knuckle lever drive, rocker arm drive, or toggle mechanism, each of which is discussed subsequently.

Crankshafts. The most common mechanical drive for presses with capacities up to 2.7 MN (300 tonf) is the crankshaft drive (Fig. 15). A crankshaft is used in both gap-frame and straight-side presses. The crankshaft drive is used most often in the single-suspension design, although some double-crank (two-point suspension) presses, particularly in the 900 to 1800 kN (100 to 200 tonf) range, also have crankshafts.

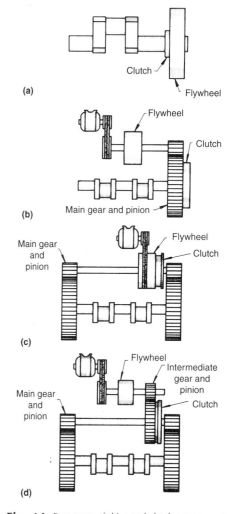

Fig. 14 Four types of drive and clutch arrangements for mechanical presses. (a) Nongeared (or flywheel) drive. (b) Single-reduction, single-gear drive; clutch in gear. (c) Single-reduction, twin-gear drive; clutch on driveshaft. (d) Multiple-reduction, twin-gear drive; clutch on intermediate shaft

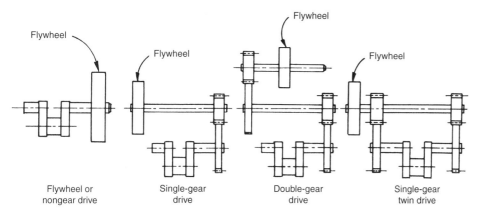

Fig. 12 Types of press drives. The first three press drives, from the left, are called "single-end drives" because the crankshaft is driven from one end only. When the crankshaft is driven from both ends, the drive is termed a "twin drive" (far right).

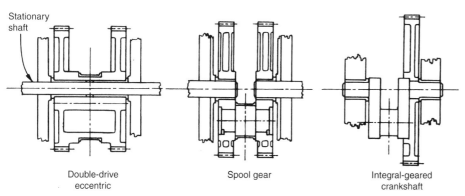

Fig. 13 Eccentric-gear drives. Variations of eccentric-gear drives include the double-drive eccentric, the spool gear, and (at far right) the crankshaft with sides of cheeks keyed directly to the main gear.

The crankshaft imparts a sine-curve speed relation to the press slide. The stroke of a crankshaft-actuated press can be as short as 25 mm (1 in.), in a small gap-frame press, or as much as 760 mm (30 in.), in a straight-side press. However, most mechanical presses with longer strokes are actuated by an eccentric gear because it provides greater strength. Crankshaft drives are usually limited to strokes of 152 to 305 mm (6 to 12 in.).

The main advantage of a crankshaft-driven press is its lower cost, particularly when capacities do not exceed 2.7 MN (300 tonf).

Points of suspension refer to the number of connections between the slide and the actuating mechanism. Presses can have single-point, two-point, or four-point suspension, depending on the number of points at which the slide is pushed or pulled. The simplest mechanical presses have a pitman that connects the eccentric shaft or the crankshaft to the slide at only one point.

Many wide mechanical presses are built with two-point suspension by connecting the slide to the crankshaft (or eccentric) with two pitmans instead of one, for better distribution of force on the slide.

The largest straight-side mechanical presses usually have four-point suspension for more uniform loading of large slides. Four-point suspension is usually accomplished by two interconnected crankshafts or eccentrics and four pitmans; each pitman is connected near a corner of the slide. Hydraulic presses also can have one,

two, or more points of suspension by operating the slide with as many rams as desired.

Eccentric shafts are similar to crankshafts. The eccentric completely fills the space between the supporting bearings of the press crown, thereby eliminating the deflection commonly caused by the unsupported portion where the crank cheeks normally would be. Eccentric drives (Fig. 15) often are used in high-speed, short-stroke, straight-side presses with progressive dies.

The height of the workpiece is the main limitation of the eccentric-shaft drive because the stroke is always equal to twice the eccentricity. When eccentricity is increased, the space available in the press crown determines the maximum stroke that can be used. In most presses of this type, the maximum stroke is usually limited to 152 mm (6 in.). A few presses have been built and used for high-speed operations in which strokes longer than 152 mm (6 in.) were needed. This was accomplished in the press by balancing the eccentric shaft to minimize vibration.

Eccentric-gear drives (Fig. 15) are used almost universally for large, straight-side presses that operate at speeds under 50 strokes per minute. In place of a crankshaft, an eccentric is built as an integral part of the press drive gear. The eccentric gear permits strokes as long as 1.3 m (50 in.); however, with such long strokes, speeds are usually only 8 to 16 strokes per minute. With the eccentric as part of the gear, accuracy of alignment of the slide is determined by accuracy

and alignment of the gears. In a two-point suspension, the parallel condition of the ram is determined by the alignment of the driving gears. The principal advantage of the eccentric gear is that it permits greater torque loads at points above the bottom of the stroke. It also permits multiple-point construction with greater versatility and range of stroke length than is possible with a crankshaft.

The chief limitation of the eccentric-gear design is that it usually requires an overhung flywheel. In addition, a single-gear eccentric press usually costs more than a crankshaft or eccentric-shaft press of equal capacity. A further limitation is that eccentric-gear presses are more likely to stick at the bottom of the stroke than are crankshaft presses. Sticking is caused by greater friction in the connector, which is inherent in the large-diameter eccentrics needed for an equal press stroke. Sticking usually occurs during setup if the press is moved slowly until the bottom of the stroke is reached. At this point, a skilled setup man can usually detect whether the press is likely to stick.

Knuckle lever drives combine the motions of a crank and knuckle lever to drive the press slide (Fig. 15). Their use is limited to operations such as coining or embossing, in which the work is done almost entirely at the bottom of a short stroke. The knuckle lever mechanism permits large capacity in a relatively small press. High mechanical advantage is inherent at the bottom of the stroke. These presses are rated to deliver full capacity at 1.6 to 6.4 mm ($^1/_{16}$ to $^1/_4$ in.) above the bottom of the stroke. The very quick increase in force as the slide nears the bottom of the stroke is the reason its usefulness is limited to operations performed at the bottom of the stroke. Knuckle lever presses usually have capacities of 1.5 to 9 MN (150 to 1000 tonf).

Rocker arm drives apply crank or eccentric motion to a rocker arm that is connected to the press slide (Fig. 15). In this mechanism, the linkage is driven by an eccentric gear and a connecting rod. The rocker arm drive is a variation of the knuckle lever drive. However, a press with rocker arm drive is not limited to coining operations; it can also be used for drawing or forming operations.

The rocker arm drive is used mainly in large-bed underdrive presses. The linkage operates from below the press bed and pulls the slide into the work by a link running up through each of the press columns. In most rocker arm drives, the rocker pin and the connecting eccentric pin do not stop in a vertical plane; thus, the load on the eccentric shaft is relieved at the point of maximum load on the slide, and sticking at the bottom of the stroke is prevented. In addition, a high press capacity is obtained because of the mechanical advantage.

Toggle mechanisms are the most widely used means of providing the second action in double-action mechanical presses. The toggles operate an outer slide, which clamps the blank against the die, while the punch, operated by the inner slide directly from the crankshaft, performs

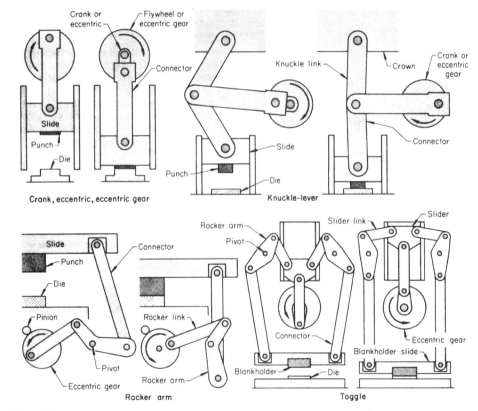

Fig. 15 Operating principles of various drive systems for mechanical presses

the draw operation. Principal components of a toggle mechanism are shown in Fig. 15.

Clutches and Brakes in Mechanical Presses

Both clutches and brakes are essential to the operation of mechanical presses. No other part must work more perfectly if the press is to operate successfully. The clutch must deliver and control the surge of force that is required to shape the work metal. When the press runs continuously, the clutch transmits power from the flywheel to the main shaft. In a single-stroke press, the clutch must accelerate the rotating parts of the drive from stop to full speed at each stroke of the press. Thus, the selection of the proper clutch is dependent on two completely different factors:

- The torque required to deliver adequate force to the slide
- The acceleration of parts that must be started at each stroke of the press

There are two basic types of clutches: positive clutches and friction clutches, as described in this section.

The selection of the brake is dependent on the deceleration of the moving parts that must be stopped at each stroke of the press. The brake must decelerate this moving mass in order to stop the slide at the end of each upstroke. The brake also must be large and efficient enough to stop the press in an emergency, or during inching. Clutches and brakes in presses that are stopped at the end of each stroke need more maintenance than do those in presses operating continuously or that are stopped only a few times a day.

Positive clutches are used on many small and medium-size presses (less than 900 kN, or 100 tonf, capacity) and particularly on gap-frame types. Positive clutches always are on the main shaft and use pins, keys, or jaws to lock the shaft and flywheel together. Because they are shaft mounted, they have a minimum mass to move. All positive clutches engage keys, pins, or multiple jaws.

In comparison with friction clutches, which accelerate the press-drive members by slippage of friction surfaces, the acceleration of the drive members by a positive clutch is very rapid because there is no slip. However, it must be remembered that these clutches are always mounted on the crank or eccentric shaft. Therefore, the inertia of the parts that must be started on each stroke of the press is kept relatively low.

These clutches can be arranged for one-stroke or continuous operation, and usually a throw-out cam disengages the keys, pins, or jaws near the top of the stroke. Positive-engagement clutches are either mechanically or pneumatically actuated. Mechanical actuation is ordinarily done by means of a foot treadle, although the treadle can be replaced by an air cylinder. Mechanically operated units can be engaged and disengaged only once for each stroke. They use some type of

throw-out cam or surface to disengage the keys, pins, or jaws near the top of the stroke. The foot treadle releases the throw-out cam and springs engage the driving members to start the press.

An auxiliary cam may be used to release the foot treadle or air-cylinder linkage at a given point of the stroke. Then the press will single stroke even if the treadle linkage is kept actuated. This single-stroking mechanism can be disconnected for continuous operation. The clutch will then remain in engagement as long as the treadle is depressed. These clutches use a drag friction brake, which is usually on and not released during press operation.

Mechanical positive clutches (Fig. 16) cost less than other types and are compact and easy to operate but are limited in many respects and usually require excessive maintenance. They are not recommended for one-stroke work because wear on the clutch would be severe.

Pneumatic positive clutches are much more efficient than their mechanical counterparts. A pneumatic positive clutch is usually a jaw clutch with 16 or more points of engagement. The jaws are engaged by an air cylinder or a diaphragm and are disengaged by springs. No throw-out cams are used. The clutches have electric controls like those used with friction clutches. A press with a pneumatic positive clutch can be used for a single stroke, operated continuously, jogged in either direction, and stopped for emergencies. The brake used with this type of clutch is usually spring operated and air released as a fail-safe measure.

Friction Clutches and Brakes. Most friction clutches use air pressure to exert force to clamp friction surfaces together, and the brakes use springs to exert this force. Brakes are air-released. The brakes are spring operated rather than air operated so that the press will stop in the event of a power or air failure. This type of clutch provides relatively shock-free acceleration and deceleration of the press-drive parts.

Friction clutches are preferred to positive clutches for most press applications. Friction clutches may be mounted on the crank or eccentric shaft, the intermediate gear shaft, or the driveshaft, as shown in Fig. 14. The location of the clutch brake unit is determined by factors such as press size, press speed, how fast the press is to be single stroked, type of clutch-brake, and inertia of the press drive.

Friction clutches allow the slides to be stopped or started at any point in the stroke. This makes setting and adjusting dies convenient, especially in large presses. With a well-designed air friction clutch, sudden power failure causes the press to stop immediately. Friction clutches using high-pressure oil or magnetic attraction have been employed to some degree.

Most friction clutches are air engaged and spring released. Brakes are spring engaged and air released. Some clutches are combined with integral brakes; others are constructed as individual units for separate mounting. The brakes for presses equipped with friction clutches are a drum type or single- or multiple-disk type and

are operated in unison with the clutches. Some designs use diaphragms or air tubes for actuation instead of pistons with packings.

One of the greatest advantages of friction clutches is their compatibility with electric and electronic controls. In this respect, they are superior to the positive clutch, with the exception of the pneumatic type.

Two basic types of friction clutches are manufactured: the combination clutch-brake unit in which all parts are assembled on one common sleeve mounted on the shaft, and clutch and brake units separately mounted on the shaft with a single valve for synchronization. As the name denotes, the combination clutch-brake is designed as one unit. A common air chamber is used to operate the clutch and release the brake. This means that the clutch and brake are mechanically synchronized. This design inherently means that most of the clutch-brake parts must be started and stopped each time the press is single stroked, and in a high-inertia design.

With the advent of automation and the much faster single-stroking rates of presses, it became necessary to reduce the stored energy of the parts to be started and stopped by using low-inertia clutches and brakes that are separate units.

Torque Control. One important benefit of the modern friction clutch is that it controls torque through air pressure. The torque limitations of a press are determined by the size of the driveshaft, clutch, gears, and crankshaft. A torque load can be imposed when a load is encountered on the upstroke. The press drive has to transmit fly-wheel energy in the form of torque to the slide. Too great a torque load will cause failures of the torque-carrying members, such as shearing of keys, breaking of gear teeth, and excessive wear of clutch members. The clutch is one link in this torque-transmission chain, which can be controlled and which limits torque overload on the rest of the system. If more torque is required than the clutch can deliver at a rated air pressure, it will slip. This is an important feature of the modern-day press and should be understood in order to take advantage of it in the field.

Eddy Current Clutches. A high degree of control over the press is the major advantage of

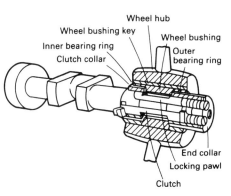

Fig. 16 Mechanically actuated positive clutch mounted on press crankshaft

eddy current clutches, which are, in reality, press drives. Ram speed at any point on the stroke can be programmed.

Eddy current drives consist of a constant-speed flywheel, a variable-speed clutch-and-brake rotor, and a stationary brake field assembly. The clutch-and-brake rotor is directly connected to the press drive shaft.

Clutching and braking are controlled by the current in the coils. In practice, the press drive is usually controlled automatically to speed up the motion of the slide during the idle portion of the stroke and to slow it down just before the tooling contacts the work.

Hydraulic Presses

Hydrostatic pressure against one or more pistons provides the power for a hydraulic press. Most hydraulic presses have a variable-volume, variable-pressure, concentric-piston pump to provide them with a fast slide opening and closing speed. It also provides a slow working speed at high forming pressure.

The principal components of a typical hydraulic press are shown in Fig. 6. A bolster plate is attached to the bed to support the dies and to guide the pressure pins between the die cushion and the pressure pad. Usually all slides are operated by one pumping system. The relation of each action to the others, interaction, and timing all depend on the controls.

The capacity of a hydraulic press depends on the diameter of the hydraulic pistons and on the rated maximum hydraulic pressure, the latter being a function of the pump pressure and related mechanisms. Hydraulic presses with capacities

up to 445 MN (50,000 tonf) have been built, but most have a capacity of less than 133 MN (15,000 tonf). The typical hydraulic press is rated at 900 kN to 9 MN (100 to 1000 tonf). Gap-frame presses are rated at 45 to 450 kN (5 to 50 tonf).

Because of their construction, hydraulic presses can be custom designed at a relatively low cost. They can be designed with a number of slides and motions, or separate hydraulic circuits can be used for various independent actions. In addition, side action can be provided within the frame of the press by means of separate cylinders. Such side action in a mechanical press is usually provided by cams and is complex and expensive. Most hydraulic presses are straight-side models, but small, fast, gap-type presses designed to compete with mechanical open-back inclinable presses have been developed.

Hydraulic press slides, or platens, are actuated by numerous combinations of hydraulic drives. Hydraulic presses usually have a longer stroke than mechanical presses, and force can be constant throughout the stroke. Hydraulic presses have an adjustable stroke for one or more slides. Accumulators or large-volume pumps can provide fast motion for a slide to open and close. High-pressure pumps provide the working force at a slower speed.

The load pressure of a hydraulic press can be set at any required maximum so that if for any reason this maximum is reached, the press will automatically release itself before any damage occurs. This characteristic is important in drawing aluminum alloys, which have a tendency to work harden. As the metal hardens, the pressure required to draw it increases. Spoilage may be avoided by setting the release pressure low

enough to avoid rupture. Should excessive work hardening occur, the press will automatically release, and the partly formed piece can be removed and annealed. Fluid pressure permits a rapid approach of the punch to the work and the pressure and speed at the time of contact, thus avoiding a blow that might harden the metal. Once contact has been made, the speed at which the draw takes place can be controlled, which makes it possible to make a fairly deep draw in one stroke at low operating speed.

The low speed of the hydraulic press permits the use of low-cost punches and dies because these parts receive little or no shock at the time of contact and comparatively little strain during drawing. Also, the stroke is held (dwells) at the bottom position, which gives the metal time to set and prevents springback to a considerable extent.

When the number of pieces to be produced justifies the cost, hardened steel or mild steel dies and punches may be used, but they must be applied carefully because they are apt to foul because of the adhesion of particles of the sheet to the steel. Satisfactory results can generally be obtained by using zinc alloy dies and punches that can readily be replaced at low cost when they become worn.

Number of Slides

Mechanical and hydraulic presses have one, two, or three slides and are referred to as single-action, double-action, or triple-action presses. Each slide can be moved in a separately controlled motion. Tables 3 and 4 compare advantages and disadvantages of single- and

Table 3   Advantages and disadvantages of single-action hydraulic and mechanical presses

Single-action press type	Advantages	Disadvantages
Hydraulic	Press cannot be overloaded because the system is protected with two separately adjusted relief valves. Full tonnage can be developed through the entire stroke. The tonnage of the press is readily adjustable up to the maximum of its rating, allowing for low-tonnage operation with fragile dies. The stroke is adjustable for the work being done. Die setting is easier because it is not necessary to adjust for material thicknesses or variations in stock. The drawing speed remains constant throughout the stroke. In general, for long-stroke presses, a hydraulic press is less expensive than a mechanical press.	Hydraulic press does not require as much electrical current as a comparable mechanical press, but a much larger motor is required as compared with a mechanical press because there is no flywheel in which energy can be stored. Not generally applicable to blanking operations because the shock of breakthrough is detrimental to piping, gaskets, and press connections, but this disadvantage has been lessened somewhat in recent years through the use of better welding techniques, manifolding, and flexible joints. Generally considered more difficult to maintain than mechanical presses, chiefly because breakdowns of mechanical presses are of the visual type and easily detected, whereas tracing of a hydraulic breakdown requires a thorough understanding of the circuit because the source of trouble is seldom visual. Nested or multiple-die setups are not advisable unless pressures are reasonably well balanced. This is seldom possible.
Mechanical	The single-action mechanical press is faster than the conventional hydraulic press. The mechanical press is by far the most suitable for blanking operations because the breakthrough shock is not detrimental to the machine. Does not require as large a motor as the hydraulic press because it can store energy in the flywheel and then dissipate the energy throughout the press stroke. Can be easily adapted for roll and transfer feeds and for progressive dies. In general, mechanical presses with short strokes are more economical than hydraulic presses.	The single-action mechanical press with cushion has a rated tonnage that is full tonnage a certain distance up—usually 13 mm ($^1/_2$ in.) above the bottom of the stroke. Less power is available for work encountered at points above the bottom of the stroke. For instance, a 300-ton (270-tonne) mechanical press with a stroke of 760 mm (30 in.) entering a draw 102 mm (4 in.) above the bottom of the stroke would have an available tonnage of only 114 tons (103 tonnes). At higher points on the stroke, the tonnage factor would be even lower. The mechanical press will not adjust itself for stock variations as does the hydraulic press and thus requires extreme care when setting dies and making allowances for material-thickness variations. If the punch is adjusted too low, the press may stick on dead center. This requires the tedious task of unsticking the press. A single-action mechanical press, when used on a draw that is equal to almost half the stroke, enters the work at the high-midstroke velocity, the drawing speed being reduced as the punch continues downward. This is in direct contrast to the hydraulic press, which slows down before entering the work, then proceeds to draw at a constant rate, giving the metal a chance to flow to its best advantage.

Table 4 Advantages and disadvantages of double-action hydraulic and mechanical presses

Double-action press type	Advantages	Disadvantages
Hydraulic	Double-action hydraulic press has four corner blankholder adjustments at the operator's fingertips. The tonnage set on each corner can readily be seen on the control panel. As with the single-action press neither the blankholder nor the punchholder can be overloaded. Full tonnage is available on both punchholder and blankholder throughout the stroke. Tonnage is adjustable on both the punchholder and the blankholder. Full adjustment of stroke is available on both punchholder and blankholder. Die setting is an easier task because no adjustment need be made for material thicknesses or variations in stock. Drawing speeds are constant throughout the stroke. The double-action hydraulic press can be used to perform single-action operations by tying the blankholder and punchholder together to act as one. This is not possible on a double-action mechanical press.	Disadvantages of double-action hydraulic presses are basically the same as those of the single-action hydraulic press (Table 3)
Mechanical	Double-action mechanical presses are generally faster on long strokes than are double-action hydraulic presses. As in the case of the single-action mechanical press, a smaller motor is required. Although maintenance of the double-action mechanical press is much more complicated than that of the single-action mechanical press, it is much less technical than maintenance of the hydraulic presses, particularly the double-action hydraulic press. This advantage is variable, depending on the number of hydraulic presses in the plant and the relative familiarity of the maintenance men with each type.	In comparison with the double-action hydraulic press, much more time and care is required in the adjustment of the double-action mechanical press (as with the single-action mechanical press), due to variations in stock thickness and other factors. The same variations in drawing speed occur with the double-action mechanical press as with the single-action mechanical press, where the tonnage diminishes the higher up on the stroke the work is performed.

double-action presses with mechanical or hydraulic power drives.

A single-action press has one reciprocating slide (tool carrier) acting against a fixed bed. Presses of this type, which are the most widely used, can be employed for many different metal-stamping operations, including blanking, embossing, coining, and drawing. Depending on the depth of draw, single-action presses often require the use of a die cushion for blankholding. In such applications, a blankholder ring is depressed by the slide (through pins) against the die cushion, usually mounted in the bed of the press (see the section "Die Cushions" in this article).

Single-action presses, either overdrive or underdrive, are the most commonly used because they are well suited for most press operations. Single-action presses are the simpler of the two types, having only one slide, and are generally more versatile. Although the single-action drive has only one slide motion, cushions can be provided to enable single-action presses to perform blankholding, stripping, knockout, and liftout operations. Various operations can be done with single-action hydraulic presses (Fig. 17).

Double-action presses have two slides moving in the same direction against a fixed bed. These slides are generally referred to as the outer (blankholder) slide and the inner (draw) slide. The blankholder slide is a hollow rectangle, while the inner slide is a solid rectangle that reciprocates within the blankholder. With two separate actions provided by two slides—the blankholder for holding the part and the inner slide for drawing—the double-action press is used primarily for deep drawing operations, which would normally cause the part to wrinkle if not held under pressure while being reshaped. Figure 18 illustrates a typical deep-drawing operation with a double-action hydraulic press. Most operations performed on double-action

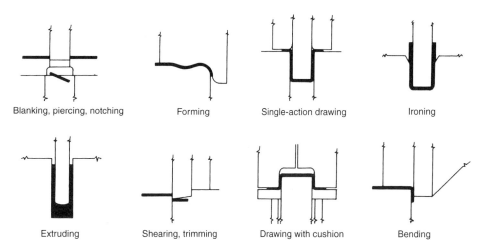

Blanking, piercing, notching   Forming   Single-action drawing   Ironing

Extruding   Shearing, trimming   Drawing with cushion   Bending

Fig. 17 Typical operations of single-action hydraulic presses. In addition to these basic operations, many others can be performed in single-action presses with special adaptions and designs.

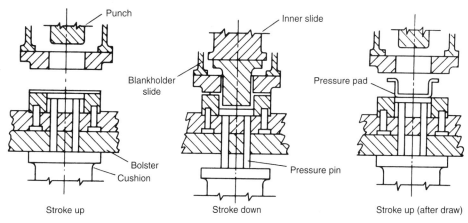

Punch   Inner slide

Blankholder slide   Pressure pad

Bolster   Pressure pin

Cushion

Stroke up   Stroke down   Stroke up (after draw)

Fig. 18 Typical operation (deep drawing) performed in double-action hydraulic press, shown in three positions of stroke

presses require a cushion either for liftout or for reverse drawing of the stamping.

Deep-drawing operations and irregularly shaped stampings generally require the use of a double-action press. Although the double-action drawing operation can sometimes be accomplished in a single-action press using a cushion as a blankholder, this is generally less practical than using a double-action press. In single-action presses, force is required to depress the cushion. In double-action presses, the blankholder slide has a shorter stroke and dwells at the bottom of its stroke before the punch mounted on the inner slide contacts the work. As a result, practically the entire capacity of the press is available for drawing.

A single-action press also requires considerably more energy to depress the cushion than is required by the blankholder in a double-action press. Therefore, a double-action press can perform a deep-drawing operation with a smaller flywheel and motor. Double-action presses also provide more positive blankholder control and more equal blankholder pressure than are possible with single-action presses. The four corners of the blankholder are individually adjustable so that nonuniform forces can be exerted on the work when required. A double-action press equipped with a die having an open bottom permits pushing the stamping through the die to perform other operations, such as ironing, after drawing.

A triple-action press has three moving slides: two slides moving in the same direction as in a double-action press and a third, or lower, slide moving upward through the fixed bed in a direction opposite to the blankholder and inner slides. This action permits reverse drawing, forming, or beading operations against the inner slide while both upper actions are dwelling.

Cycle time for a triple-action press is necessarily longer than it is for a double-action press because of the time required for the third action. Because most drawn stampings require subsequent restriking and/or trimming operations, which are done in faster, single-action presses, most stamping manufacturers consider the triple-action press too slow.

Press Selection

Important factors influencing the selection of a press include size, force, energy, and speed requirements. The press must be capable of exerting force in the amount, location, and direction, as well as for the length of time, needed to perform the specified operation or operations. Other necessary considerations include the size and geometry of the workpieces, the workpiece material, operation or operations to be performed, number of workpieces to be produced, production rate needed, accuracy and finish requirements, equipment costs, and other factors.

Size, Force, and Energy Requirements. Bed and slide areas of the press must be large enough to accommodate the dies to be used and provide adequate space for die changing and maintenance. Space is required around the dies for accessories such as keepers, pads, cam return springs, and gages; space is also needed for attaching the dies to the press. Shut height of the press, with adjustment, must also be suitable for the dies.

Presses with as short a stroke as possible should be selected because they permit higher-speed operation, thus increasing productivity. Stroke requirements, however, depend on the height of the parts to be produced. Blanking can be done with short strokes, but some forming and drawing operations require long strokes, especially for ejection of parts.

Size and type of press to be selected also depend on the method and direction of feeding; the size of sheet, coil stock, blank, or workpiece to be formed; the type of operation; and the material being formed and its strength. Material or workpiece handling and die accessibility generally determine whether the press should be of gap-frame or straight-side construction and whether it should be inclined or inclinable.

Physical size of a press can be misleading with respect to its capacity. Presses having the same force rating can vary considerably in size depending on differences in length of stroke, pressing speed, and number of strokes per minute.

The force required to perform the desired operations determines press capacity, expressed in tons or kilonewtons (kN) (see the section "Press Capacity" in this article). The position on the stroke at which the force is required and the length of stroke must be considered.

Energy or work (force times distance), expressed in inch-tons or joules (J), varies with the operation. Blanking and punching require the force to be exerted over only a short distance; drawing, forming, and other operations require force application over a longer distance. The major source of energy in mechanical presses is the flywheel, the energy varying with the size and speed of the flywheel. The energy available increases with the square of the flywheel speed.

Possible problems are minimized by selecting a press that has the proper frame capacity, drive motor rating, flywheel energy, and clutch torque capacity.

Speed Requirements. Press speed is a relative term that varies with the point of reference. Fast speeds are generally desirable, but they are limited by the operations performed, the distances above stroke bottoms where the forces must be applied, and the stroke lengths. High speed, however, is not necessarily the most efficient or productive. Size and configuration of the workpiece, the material from which it is made, die life, maintenance costs, and other factors must be considered to determine the highest production rate at the lowest cost per workpiece. A lower speed may be more economical because of possible longer production runs with less downtime.

Simple blanking and shallow forming operations, however, can be performed at high speeds. Mechanical presses have been built that operate to 2000 strokes/min with 25 mm (1 in.) stroke, but applications at this maximum speed are rare. Speeds of 600 to 1400 strokes/min are more common for blanking operations, and thick materials are often blanked at much slower speeds. For drawing operations, contact velocities are critical with respect to the workpiece material, and presses are generally operated at slide speeds from 10 to 300 strokes/min, with the slower speed for longer stroke drawing operations.

Press Accuracy

Suggested criteria for the accuracy of a press are:

- Maximum tolerances for parallelism between slide and bed; 0.08 mm/m (0.001 in./ft) at the bottom of the stroke for all slides; 0.24 mm/m (0.003 in./ft) at midstroke for punch slides; and 0.4 mm/m (0.005 in./ft) at midstroke for blankholder slides
- Feed, if used, should be accurate within ± 0.076 mm (± 0.003 in.) at 23 m/min (75 ft/min) (see the section "Press Feeds" in this article).
- Gib clearance should be set as close as required to do the job.

Parallelism. In a single-point press, the slide guides basically determine parallelism of the slide face with respect to the bed or bolster, both in the front-to-back and right-to-left directions. In a two-point press, slide guides determine parallelism only in the front-to-back direction. Out-of-parallel conditions in the right-to-left direction because of faulty adjusting-screw timing, variations in throw between the cranks or eccentrics, or bicycling (periodic tilting of the slide in opposite directions) due to timing errors in the drive cannot be corrected by adjusting the slide guides.

In a four-point press, parallelism is determined strictly by the press-drive and adjusting-screw accuracy and timing. Out-of-parallel conditions at midstroke, however, can be caused by improper centering of the slide guide adjustment.

Gib Clearance. Clearances between press slide guides and gibs are required to compensate for inaccuracies in machining and in the throw and timing of two- and four-point presses, and to allow for expansion and contraction of the guides and gibs. With the bronze-to-cast iron or cast iron-to-cast iron gibbing normally used, some slight clearances must also exist for the oil or grease lubricant.

If sufficient clearance is not provided, excessive pressure on the guides occurs at some point during the stroke, resulting in galling of the mating surfaces. The amount of clearance also affects the repetitive registry of punch with die. If the clearance between punch and die is less than the clearance between the press guides, the punch can mount the die, thereby causing premature wear. Clearance between the press guides also depends on the length of the gibs. A reasonable allowance can be 0.08 to 0.15 mm

(0.003 to 0.006 in.), the exact amount varying with gib length. For high-speed presses, the clearance between a gib-type guide and the press slide at normal operating temperature is generally no more than 0.038 mm (0.0015 in.).

Clearances between the mating surfaces of press guides can be reduced by using self-lubricating reinforced-phenolic liners. The modulus of elasticity of this material is about that of steel. As a result, deformation of the liners due to the inaccuracies discussed does not raise excessive compressive stresses in the liners, provided the liners are of sufficient thickness.

Near-zero clearances can be achieved by using preloaded rolling-contact (ball or roller) bearing guides. When offset or lateral loads are encountered, however, tending to cause the press slide to tilt, only a few rolling-contact members at the top and bottom of the guides take the loads. As a result, stresses exerted on the balls or rollers can be very high.

Press Capacity

Capacity, or rating, of a press is the maximum force that the press can apply. Hydraulic presses can exert maximum force during the full press stroke. Mechanical presses exert maximum force at a specified distance above the bottom of the stroke (usually 1.6 to 13.0 mm, or $1/16$ to $1/2$ in.), and the force decreases to a minimum at midstroke.

The tonnage rating of a press may have little relation to the bed area. This is especially true in the automotive and appliance industries, where presses have large bed areas and die spaces but relatively low tonnage ratings. Coining presses have small bed areas and high tonnage ratings.

Overloading of the press can cause damage to both the die and the press. Several devices based on the strain-gage principle have been developed for accurately measuring the load on a mechanical press with a given die. Misfeeds or double blanks are common causes of press overloading. Detectors built into the die stop the press before overloading occurs.

The capacity of a mechanical press involves consideration of the frame capacity, drive capacity, flywheel energy, and motor size.

Frame Capacity. The press frame must be able to work at its rating without deflecting beyond predetermined standard limits. For general-purpose applications, the bed deflection should not exceed 0.17 mm/m (0.002 in./ft) between tie-rod centers (or per foot of left-to-right bed dimension on presses without tie-rods) when the rated load is evenly distributed over the middle 60% of the distance between tie-rod centers (or of the left-to-right bed length). Slide deflection should not exceed 0.17 mm/m (0.002 in./ft) between pitman centers when the rated load is evenly distributed between pitmans. Both bending and shear deflections are considered. These specifications can be revised to suit more precise applications.

Drive capacity is the capacity a mechanical press develops through the gear train and linkage. The capacity can vary because of the mechanical advantage developed by different types of press linkage and generally is expressed in distance above the bottom of the stroke.

Variation in the capacities of presses with eccentric-gear, crankshaft, or eccentric-shaft drive at any point in the stroke is almost equal. However, the capacity decreases between the bottom of the stroke and midstroke. The capacity of both the knuckle lever drive and the rocker arm drive, however, is much less above the point of rating on the stroke than is that of the crank drives because the knuckle lever drive loses mechanical advantage more rapidly than do crank drives. In addition to loss in capacity, velocity of the slide in the knuckle lever drive and the rocker arm drive is considerably greater at points high above the bottom of the stroke than is the case in the eccentric-gear or crankshaft drives.

Flywheel energy for a given job may be insufficient, although the press frame and shaft may be adequately strong. For a greater working distance or for faster operation, more energy and power must be provided.

Blanking operations are completed in a brief portion of the press cycle. The flywheel instantly supplies practically all of the energy required by its resistance to deceleration. The motor may take the remainder of the press cycle to restore lost energy to the flywheel by bringing it back up to speed. Draw operations may take up to one-fourth of the press cycle.

For intermittent operation, 20% is arbitrarily considered the maximum the flywheel may be slowed when energy from it is being used. For continuous operation, 10% is considered the limit because of the short time available to restore lost energy. The low-speed torque characteristics of the press drive motor greatly affect the amount the flywheel can be safely slowed because the ability of the motor to restore lost kinetic energy is a function of these characteristics.

The amount of energy, E, available at 10% slowdown can be calculated from the following equation:

$$E = \frac{N^2 D^2 W}{5.25 \times 10^9} \qquad \text{(Eq 1)}$$

where E is expressed in inch-tons, N is the rotary speed of the flywheel (rpm), D is flywheel diameter (inches), and W is flywheel weight (pounds). The metric version of Eq 1 is:

$$E = \frac{N^2 D^2 W}{6.8 \times 10^6} \qquad \text{(Eq 2)}$$

where E is given in kilojoules, N remains rpm, D is given in meters, and W is expressed in kilograms.

If calculation indicates that the flywheel will not furnish the necessary energy, it may be necessary to increase the weight, diameter, or speed of the flywheel, or to use a different type of drive or motor.

To change the press speed, devices are used to change the speed of the flywheel. The energy of the flywheel is directly proportional to the square of the flywheel speed of rotation. Therefore, the standard energy of a variable-speed press is calculated at its slowest speed. The intended operating speed should be used in checking the suitability of a press for a specific operation.

Motor Selection. The primary function of the main drive motor on most mechanical presses is to restore energy to the flywheel. During flywheel slowdown, most of the energy is derived from the flywheel, with some contribution from the motor. Following the working stroke, the motor must restore the energy expended by the flywheel while returning the wheel to speed. If the capacity of the press is not adequate to satisfy the operation, slowdown of the flywheel during the working portion of the stroke becomes excessive, resulting in overloading of the motor. Stoppage of the press can occur if there is insufficient time between production strokes to permit recovery of flywheel speed.

When slowdown is rapid and the working stroke is long, it is desirable to use a high starting torque with relatively low starting current, good heat-dissipating capacity, and sufficient slip (slowdown). The slip designation of a motor is the amount of motor slowdown at rated full-load torque. Under conditions of moderate slowdown and short working strokes, general-purpose motors can be used satisfactorily.

Many presses employ alternating current (ac) induction motors having slip ratings of 3 to 5%, 5 to 8%, or 8 to 13%, the choice depending on flywheel design, press speed, and other parameters. For short-stroke presses operating at speeds above 40 strokes/min, general-purpose motors with a slip rating of 3 to 5% are often satisfactory. Motors with a slip rating of 5 to 8% are often used for press speeds of 20 to 40 strokes/min. For long-stroke presses operating at speeds less than 20 strokes/min high-slip (8 to 13%) motors are usually required.

Press Accessories

The setup and operation of a mechanical press are made more versatile through the use of built-in accessories. Included are bolster plates, rolling bolster assemblies, speed change mechanisms, shut height adjusters, and slide counterbalances.

Bolster plates are used in most hydraulic and mechanical presses, between the bed and the die. They provide a flat surface on which to mount the dies and can be remachined to remove nicks and worn areas. T-slots in the top surface facilitate clamping of the die to the bolster plate. Clearance holes in the bolster plate permit pressure pins to extend from the die cushion to the die. Some press beds have a large hole through the top surface for drop through of parts or for mounting of die cushions. For this reason, bolster plates are thick to minimize deflection and support the die properly. When parts are ejected through the die, holes of the proper size and location are cut in the bolster.

The width, length, and thickness dimensions of bolster plates have been standardized for each bed size, as have the size and location of T-slots, pressure pin holes, and holes for fastening to the bed. Standardization facilitates interchangeability of dies between presses. Filler plates can be used either above or below the bolster to reduce the shut height of the press. This is in addition to the normal slide adjustment.

Rolling bolster assemblies are made for some large straight-side presses for fast tooling change. Dies are set up on an assembly outside the press. When a press run is finished, the punch and blankholder are undamped from the slides and the assembly is moved out of the press. Then, another assembly is moved into place, the punch and blankholder are clamped to the sides, adjustments are made, auxiliary equipment is set up, and the press is made ready to run.

Speed change drives are used mostly to change the number of strokes per minute, but some drives also can be used to change speeds during the press stroke for fast approach to the work, slower working stroke, and quick return. Changing speed during the stroke permits an increase in production without increasing the working speed. Press builders commonly supply charts that show the slide speed at any point on the press stroke.

In simple blanking operations, the speed of the working stroke is not critical. In drawing and in some forming operations, the plastic-flow characteristics of the material being formed impose specific speed limitations.

A two-speed drive is combined with clutching and braking in some presses. A clutch can have planetary gears for a two-speed drive. Some two-speed drives have two flywheels with a common brake. With a two-speed gearbox and two speeds from the clutch, a press can have four speeds.

Variable-speed drives may incorporate a speed change belt with adjustable cone pulleys connecting the motor to the flywheel or a steplessly variable electric drive. The eddy current drive, originally developed for inching of mechanical presses, also provides variable speed.

Shut height adjustment is provided in mechanical presses to change the distance between the slide and the bed to fit dies of different sizes. Small, single-point presses have a screw arrangement to provide this adjustment. In heavier presses, a gear drive makes it easier for the operator to move the massive slide. As press size increases, this gear drive is motorized. Motorized slide adjustment is used also in many smaller presses. Air counterbalances on most large presses relieve the load of the slide and die from the adjusting mechanism.

Some presses have dials that indicate the shut height in thousandths of an inch. If the dies to be used in the press are similarly marked, die-setting time will be greatly reduced. Other presses have a motorized adjustment with a dial control. The operator sets the desired shut height, and the slide automatically positions itself.

Counterbalances in press slides provide smooth cycling and reduce backlash and gear wear by:

- Counteracting the moving weight of slides, components, and die members attached to the slides
- Reducing the load on the press brake, thus providing faster stopping
- Taking up clearance on the main bearings, reducing the breakthrough shock for cutting operations
- Reducing backlash in the drive gearing
- Easing the adjustment of slides by reducing the load on the adjusting screws

Excessive counterbalance pressure can prevent the normal breathing of bearings and consequently prevent good lubrication of the bearings.

Most presses manufactured with slide counterbalances use pneumatic cylinders as a means of counterbalance, although springs have been used. To prevent too great an increase of pressure through the full range of press stroke, a surge tank is used in conjunction with the cylinders. The tanks are of such a size that the pressure does not increase more than 20 to 25%. A pressure control valve allows the counterbalance pressure to be adjusted to take care of variation in die weights.

The counterbalance cylinder is attached to the press frame, and the cylinder rod to the press slide. Usually the cylinders are attached to either the crown or the press uprights.

Die Cushions

Die cushions, often referred to as pressure pads, are used to apply pressure to flat blanks for drawing operations. They also serve as knockout or ejection devices to remove stampings from the dies.

Single-action presses do not have an integral means for blankholding and require the use of cushions or other means of applying uniform pressure to the blanks for drawing operations, except for shallow draws in thick stock. The most common means of pressure control for drawing operations on single-action presses are pneumatic and hydropneumatic die cushions. Figure 19(a) shows a single-action press set up for use with a die cushion.

Double- and triple-action presses feature integral blankholders and do not require cushions for drawing operations. Cushions are sometimes employed on double-action presses, however (Fig. 19b), for ejection of triple-action draws, for keeping the bottoms of the stampings

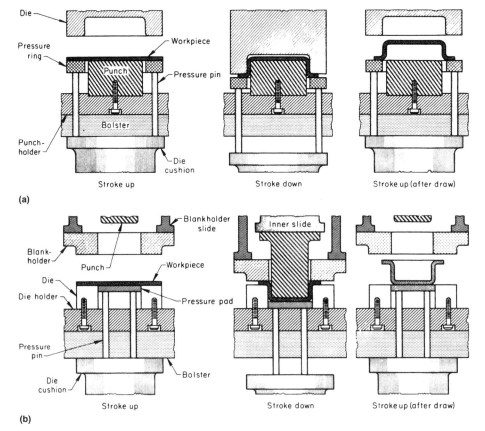

Fig. 19 Setups incorporating die cushions in (a) single-action and (b) double-action presses

flat or ensuring that they hold their shapes, or for preventing slippage while drawing. For such applications, the cushions must be equipped with locking devices to hold the cushions at the bottom of stroke for a predetermined length of the return stroke of the press slide.

Most die cushions are located in the press beds, but there are applications that require installation within or on the press slides. In either case, the functions are similar, and the operations are the same. The recommended capacity of a die cushion (the amount of force it is capable of exerting) is generally about 15 to 20% of the rated press force. Strokes of the cushions are usually one-half the strokes of the press slides, but should not exceed bolster thickness less than 13 mm ($\frac{1}{2}$ in.).

Cushions with different strokes and higher capacities are available, but the size of the press bed opening limits the size, type, and capacity of the cushions. Consideration must be given to the press capacity at the point at which the draw is to begin because the force and energy required to depress the cushion is added to that required to draw the stamping. As a result, the force and energy needed for a high-capacity cushion may not leave enough for the operation to be performed.

In pneumatic cushions, the maximum pressure is controlled by the diameter and number of cylinders and the available air pressure. Shop-line pressure is generally used, but it is possible to use a booster or intensifier to increase the air pressure. Most cushions are normally rated at a pressure of 690 kPa (100 psi), and it is generally recommended that the pressure not exceed 1380 kPa (200 psi). Surge tanks, if required, must conform to local codes and are generally approved for a maximum pressure of 860 kPa (125 psi).

A pneumatic die cushion for a single-point press normally uses one cylinder and one piston. Two or more cushions may be stacked, however, when a high-capacity unit is required in a limited bed area in which vertical space is available. For multiple-point presses, when the pressure pad area requirement is too large for one cushion, multiple cushions can be arranged alongside one another. The cushions may be individually adjustable or tied together. A multiple-die cushion is often preferable to a hydropneumatic die cushion because of the speed restrictions of the latter. Presses to be used with progressive dies can be equipped with a cushion whose position may be changed from right to left in the press bed.

Hydropneumatic cushions are used when higher forces are required or when space does not permit the use of double- or triple-stage cushions. Hydropneumatic cushions are slower acting than the pneumatic cushions; therefore, they are usually used on large presses and on slow presses. They can be adjusted to hold a large, light blank for deep drawing or shallow forming or to grip heavy-gage material as tightly as is required for curved-surface or flat-bottom forming.

A typical hydropneumatic cushion is connected to a surge tank, as shown in Fig. 20. Two individually controlled air lines are required: one connected to the operating valve of the cushion and the other connected to the top of the surge tank. The air pressure supplied to the operating valve determines the capacity of the cushion on the downstroke. The pressure of the air in the surge tank determines the stripping force available on the upstroke. The surge tank may be separate from or integral to the cushion, depending on the space available beneath the press bed.

The pressure of the air in the surge tank is transmitted to the hydraulic fluid, which is free to pass upward through the check valve and force the cushion piston upward. Pressure is also exerted against the face of the operating valve stem, but it is not sufficient to overcome the opposing air pressure working on the operating air piston.

When a downward force is applied to the cushion, the check valve is immediately closed and the pressure of the fluid that is trapped beneath the piston begins to rise. When the pressure against the small face of the operating valve stem reaches a predetermined point, it exceeds the magnitude of the air pressure on the larger area of the air piston and opens the operating valve. As long as the cushion piston continues its downward movement, the fluid beneath it is maintained under constant pressure by the throttling action of the operating valve; the additional fluid replaced by the piston is forced through the valve to the surge tank. Oil pressures are generally limited to approximately 67 MPa (1 ksi).

When the stroke has been completed and the downward force on the cushion piston is removed, the pressure of the fluid beneath the piston is immediately lessened, reducing the air pressure on the air piston, thereby closing the operating valve. Fluid from the surge tank under pressure from the air behind it passes upward through the check valve and raises the cushion piston to top stroke.

Auxiliary Equipment

Most primary press operations are automated so that equipment for feeding and unloading is used even for fairly short runs. Hand-feeding, with its attendant hazards, is often confined to secondary operations on partly completed workpieces. Goals for planning automated operations should include:

- Maximum safety to the operator and to the equipment
- High or nearly continuous production
- Improved quality of the product and minimum scrap
- Reduction in cost of the finished parts

The shape and position of the part before and after each operation must be carefully studied to determine whether design changes, such as adding tabs or extra stock to the blank, will facilitate handling.

Automatic handling equipment can be divided into the following categories: feeding equipment, unloading equipment, and transfer equipment.

Coil-handling equipment moves coiled stock to the press area and uncoils it with a minimum of damage to the stock and danger to the tools and operator. Reliable coil handling is important because coil stock is being used increasingly to supply material to presses.

Other auxiliary equipment discussed in this section includes lubricant applicators, straighteners, and levelers.

Press Feeds

Mechanical feeds are important for high production, combined operations, and operator and press safety. Some feeds supply the presses with stock from a strip or coil; others feed blanks or partly completed workpieces. Either kind, with or without auxiliary hand-feeding, can be used with almost any kind of press. For progressive-die work, the feed length should be accurate and should repeat within ±0.076 mm (±0.003 in.). The stock must advance accurately so that the pilot pin can easily enter the piloting hole and position the strip. Too great a variation in feed length could result in distorted pilot holes and scrap parts.

Feeds for coil stock feed the work metal from a coil to the press. Choosing the optimum type of feed depends mainly on the type of press, strokes per minute, length of feed per stroke, accuracy needed, and the kind of strip (width, thickness, stiffness, and surface condition). The two most common kinds of feeds are slide and roll feeds.

Slide feeds are made in a variety of sizes and capacities. The basic principle of a slide feed is the use of a feed block that is moved between positive stops to advance the material the distance required at each stroke. Slide feeds are very accurate and are particularly suitable for use with coil stock. When strip stock is used, the ends of the strip must be hand-fed into the press.

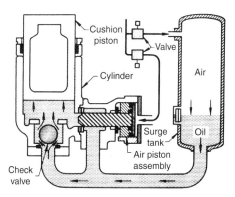

Fig. 20 Components and operating principle of a hydropneumatic die cushion

Some slide feeds are powered by the press through an eccentric mounted on the crankshaft extension (Fig. 21). The eccentric can be a simple one-piece unit keyed to the crankshaft, or it can be adjustable to vary the feed in relation to the rotation of the crankshaft. When changes in feed length are frequent, the adjustable type is usually warranted.

The feed block is mounted on hardened slides and either has a feed-blade holder with an adjustable feed blade (usually carbide tipped) or a pair of eccentric gripping cylinders. The material is gripped during the feed stroke and released on the return stroke. Accurate control of feed length is obtained by the use of adjustable stops.

The direction of feed—left to right, right to left, or front to back—is governed by the location of the crankshaft extension on the press and the arrangement of the die. A mechanical slide feed, feeding from left to right on a press with a front-to-back crankshaft, can be provided by using an appropriate linkage. Slide feeds also can be powered by air cylinders or cams on the press slide.

Roll feeds are available in sizes suitable for use with almost any width and thickness of stock and are used in every type of presswork, from blanking to complex operations in progressive dies.

A roll feed essentially consists of a pair of rolls that can turn in one direction only. The rolls exert pressure on the stock by the use of springs or some other device and are rotated by the motion of the press crankshaft.

Roll feeds are suitable for extremely thin material and material with highly polished surfaces. If hard chromium-plated rolls are substituted for standard ground-steel rolls, polished surfaces will not be scored or marked during feeding. Rubber-coated or plastic-coated rolls can be used on soft finished or prepainted stock.

There are two advantages to using roll feeds for feeding thin stock. With patterned rolls, a flange can be formed on a waste edge of the stock as a stiffener. With a single-roll feed, the stock usually is pulled through the die.

The best method of feeding extra-thin stock is the use of double-roll feeds (Fig. 22), in which roll feeds at each side of the die are set so that the stock between them is always under slight tension. Double-roll feeds eliminate manual feeding of end sections when strip stock is processed and are suitable only when a substantial scrap skeleton remains.

Rack-and-pinion-actuated roll feeds are available in almost all sizes but are used mostly in relatively heavy stamping and drawing operations. In larger presses it is common to use double-roll feeds of the rack-and-pinion type that are attached to the press bolster.

Feeding of blanks or previously formed stampings to presses is accomplished in several ways. Selection of a specific method depends on many factors, including safety considerations, production requirements, and cost.

Manual Feeding. Feeding of blanks or stampings by hand is still a common practice, but this method is generally limited to low-production requirements that do not warrant the cost of automatic or semiautomatic feeds. Manual feeding, however, requires the use of a guard or, if a guard is impossible, hand-feeding tools and a point-of-operation safety device. The use of tools and a safety device eliminates the need for the operator to place hands or fingers within the point of operation and safeguards the operator who inadvertently reaches into the point of operation (see the section "Press Safety" in this article).

Chute Feeds. Simple low-cost chutes are often used for feeding small parts, with the blanks or stampings generally sliding by gravity along skid rails in the bottoms of the chutes. Side members guide the workpieces, and rollers are sometimes added to facilitate sliding. Production rates to 1800 parts per hour are not uncommon for gravity chute feeds.

Blanks or stampings are generally placed in the inclined chutes manually, but the setup can be automated by using hoppers, prestacked magazines, or other means to supply the chutes. Windows are provided at the point at which the workpieces enter the chutes when proper orientation is required.

Push feeds are used when blanks must be oriented in a specific relation to the die, or when irregularly shaped parts are fed that do not slide down a chute and orient themselves properly in the die nest. Workpieces can be manually placed in a nest in a slide, one at a time, and the slide pushed until the piece falls into the die nest. An interlock is generally provided so that the press cannot be operated until the slide has correctly located the part in the die. Slide length should be sufficient to allow placement of workpieces in the pusher slide nest outside a barrier guard enclosure. Strippers, knockouts, or air can be used to eject finished parts from the die. In some cases, holes can be provided in the bottom plates of the slides through which finished pieces fall on the return stroke of the pusher.

Transfer Feeds. In some automated installations, blanks are lifted one at a time from stacks by vacuum or suction cups and moved to the die by transfer units. Separation of the top blank from a stack is usually done magnetically,

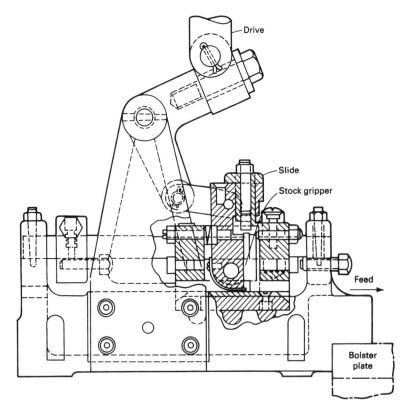

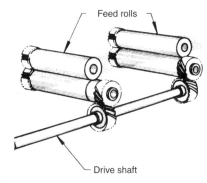

Fig. 21 Press-driven slide feed   **Fig. 22** Components of a double-roll feed

pneumatically, or mechanically. The top level of a stack can be controlled by a height detection system that regulates a stack-elevating cylinder. Two or more stacks can be arranged to be automatically moved into the elevating station when the previous stack has been used up.

Dial feeds are another method of feeding secondary operations that are being increasingly applied because of improved safety provisions and increased productivity. Such feeds consist of rotary indexing tables having nests or fixtures for holding workpieces as they are carried to the press tooling. Parts can be placed in the nests or fixtures at the loading station (away from the point of operation) either manually or by other means, such as with the use of hoppers, chutes, magazines, vibratory feeders, or robots. Dial feeds can be built into or added to presses.

Industrial robots are being used extensively for press loading and other industrial applications. These mechanical arms, manipulators, or universal transfer and positioning units are more sophisticated versions of the mechanical hands or swinging arms long used for press loading and unloading (see the section "Press Unloading" in this article). The main difference between these devices and true robots is that true robots can be programmed to perform different operations. Various types of tooling can be attached to the arms to handle different sizes and shapes of workpieces. Not only do such units increase safety, but they also substantially boost production rates. Robots are particularly suitable for low-volume production requirements and for operations in which there are large differences in the size and geometry of the workpieces to be handled.

Press Unloading

Methods used to unload stampings from presses vary depending on workpiece size, weight, and geometry; production requirements; material from which the stamping is made; press and die design; surface quality requirements; and safety considerations.

Gravity and Air Ejection. Gravity is the simplest and least-expensive method of unloading presses, but it is not applicable for many operations. In some cases, dies can be designed so that the stampings fall through a hole in the press bed. The use of open-back inclinable presses facilitates unloading by means of gravity when there are no holes in the beds; stampings fall out of the open backs of the presses. When press inclination is not practical, chutes are sometimes provided to carry the stampings away. Air ejection is still common for lightweight parts but is expensive and noisy.

Kickers, Lifters, and Shuttle Extractors. Kickers consist of pivoted levers, generally air actuated, that are mounted in the dies and throw stampings out of the dies when the dies open. Lifters are similar devices but simply move vertically and require other means for stamping ejection. Pan shuttle-type extractors swing to and from the die area, catching stamp-

ings as they are stripped from the punches or upper dies and dropping them outside the presses. Actuation of the pans can be from either the press rams or the independent drives.

Mechanical hands, often called iron hands, are actuated by air or electrical mechanisms commonly used to remove stampings from presses. Gripping fingers or jaws are mounted on arms that swing or reciprocate into the die area to lift the stampings and place them on a mechanism for transfer to the next press or operation site. Standard units are available as swing arm or straight-path types.

Interchangeable jaws or fingers are designed to grip the flanges of stampings. Vacuum cups or electromagnetic elements are used in place of jaws or fingers for curved surfaces and fragile or easily damaged workpieces.

Industrial robots are also used for press unloading. An important advantage of robots is their programmability to suit various workpieces and requirements.

Transfer Equipment

Several methods are used to automatically transfer stampings from press to press for high-production requirements. When applicable, the use of chutes on which the stampings slide provides the lowest-cost method. Power-driven slat or belt conveyors are commonly used. Adjustable-speed drives for the conveyors are often desirable to accommodate various cycle times.

Shuttle-type transfer devices are used extensively. With some units, the stampings are pushed by reciprocating fingers that extend and retract as required; other units use the lift-and-carry (walking-beam) method. Shuttle units are driven by hydraulic, pneumatic, or electric power, or they are driven mechanically from the press. Adjustable side rails are often provided to accommodate workpieces having different widths.

Lift-and-Carry Devices. Figure 23 is an example of a lift-and-carry device. Two rails move into slots milled in a die, rise vertically to lift a stamping from the die, retract and lower to deposit the stamping on a set of idle rails, and return to pick up the next stamping. Each time the presses cycle, the stampings are moved progressively from one press to the next. This type of transfer unit maintains full control of the stampings, from unloading them from one die to loading them into the next die.

Turnover or turnaround devices are sometimes added to transfer systems in order to change the positions of the stampings as they pass from one press to another. Turnaround devices generally consist of turntables that lift the stampings, rotate them the required amount, and lower them onto the transfer system. Turnover devices often have one arm and use one or more vacuum cups. In operation, the stamping is transferred to a position above the arm, the arm is raised, the cup or cups engage the stamping, vacuum pulls the stamping against the cups, and

the arm rotates approximately 180°. At the end of the arm movement, the vacuum is released, and the arm returns to its horizontal position. Other devices are of the Ferris wheel type.

Industrial robots, electrically interlocked to two or more presses, are also being used for the automatic unloading, transferring, and loading of stampings. Advantages include increased flexibility, with programmability permitting different stampings to be produced over the same press line.

Computer Numerical Control (CNC). Transfer systems controlled by programmable CNC units are available for automating press lines. Such systems are independent of the presses, can be adapted to stampings of all sizes, and are easily reset. The transfer slides have integral drives, transfer level is programmable in three coordinates, and transfer rates can be varied along certain sections. Grippers or suction cups are used to handle the stampings. Modular construction of the CNC transfer units permits their use with conveyor belts, buffer storage devices, and turnover or turnaround units.

Stackers or conveyor loaders are often provided at the ends of the lines to stack or remove finished stampings that are unloaded from the last press. Low-profile under-the-die conveyors are used for some applications.

Coil-Handling Equipment

Coil cradles, reels, uncoilers, recoilers, and other types of coil-handling equipment are important to the successful operation of a press.

Coil cradles may be either nonpowered or powered. In the nonpowered type, the stock is pulled from the coil by a powered feed, a straightener, or pinch rolls, or by the equipment being fed.

A powered cradle is preferable for coils that weigh more than 900 kg (2000 lb) or when stock is going directly from the reel to the press feed. In a powered cradle, the coil is supported by chain-driven or gear-driven rolls or by a driven conveyor belt. The drive should

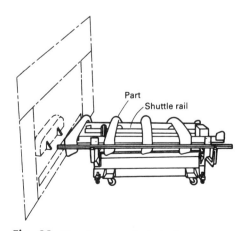

Fig. 23 Lift-and-carry transfer device for stampings

be automatically self-equalizing to prevent skidding of the coil.

Coil cradles should have motors that can stand frequent starting of inert loads. A slack loop is created between the coil and the straightener or feed devices by starting and stopping the motor intermittently on signal from a dancer roll, paddle, or other control device. This intermittent operation may cause a standard motor to fail prematurely. With a variable time delay (electronic or adjustable cam), the motor can overrun to a controllable extent after the control has commanded it to stop.

A variable-speed drive reduces the number of starts and stops, prolongs the life of the motor and drive, and often makes it possible to match the speed of the cradle to that of the machine being fed. A clutch can be used so that the motor will run continuously and the slat conveyor or rolls are driven only when stock is required.

Stock Reels and Uncoilers. Commercial stock reels can accept coils weighing as much as 22,700 kg (50,000 lb). There are reels of the proper size and type for almost any pressworking application.

Selection of a reel should be based on the maximum coil weight and the widths of stock to be unwound. It is better to overestimate future requirements than to underestimate them and find out later that reel capacity limits improvement in equipment and production methods.

Plain or nonpowered reels are usually adequate when the press feed or stock straightener has pinch rolls with enough gripping power to pull the stock from the reel. When stock is going directly from reel to press feed, the reel should be powered so that the feed does not have the job of both feeding the press and unwinding the coil. If the stock becomes taut between the reel and the feed, the feed may start to advance and the stock may slip, resulting in a short feed length. If a straightener is used between the press feed and the reel, a plain reel can be used. However, materials with low tensile strength and lightweight materials should be unwound from a powered reel; otherwise, they might be stretched between the reel and the feeding device.

Powered reels with variable speed and a loop control are preferred for a smooth operation. Noncontact sensor units, such as photoelectric cells or proximity switches, on the loop control should be used for soft metals, polished surfaces, and prepainted stock. These prevent damage inherent with contact-type (rolling or sliding) sensor units. Without powered reels or loop control, a sudden pull can cause the stock feed to slip and mark the work metal.

Other equipment useful for handling coil stock includes recoilers, turnstiles, downlayers, coil cars, coil grabs, and coil ramps.

Recoilers are used for winding coil stock after slitting and for winding the scrap skeleton after pressworking.

Turnstiles (or horns) are two-arm or three-arm devices used to store coils temporarily before processing. In function, a turnstile resembles a coil ramp. Turnstiles may be equipped with hydraulic pushoff devices, which add to their speed and efficiency.

Downlayers, sometimes called up-enders, are turnover devices for rotating the coil from a horizontal to a vertical position.

Coil grabs, for use with cranes, are devices that can handle stock in the horizontal or vertical position. Some similar devices are available for use with forklift trucks. Other devices will pick up a coil and change the position from horizontal to vertical.

Coil ramps are inclined storage units for use with reels or cradles. Most coil ramps operate by gravity.

Other Auxiliary Equipment

Lubricant Applicators. In blanking or forming, a lubricant is usually applied to metal that is fed into the press from coils. The lubricant can be swabbed or brushed onto the metal as it leaves the reel, but this is inefficient and wasteful and produces inconsistent results. An automatic applicator improves efficiency and uniformity. The type of applicator used depends on whether the lubricant is a powder or a liquid and, if a liquid, on its viscosity and flow characteristics. Roller coating, drip feeding, and spraying are common. Information on these application methods, as well as on the types of lubricants used, is available in the article "Selection and Use of Lubricants in Forming of Sheet Metal" in this Volume.

Straighteners have upper and lower rolls mounted alternately in a staggered position. The minimum number of rolls that can be used is three; however, five-roll or seven-roll straighteners are most common for the usual range of stock thickness. Straightening of stock less than 0.50 mm (0.020 in.) thick requires additional rolls; as many as 17 have been used for some thin stock.

Some straighteners have a separate screw adjustment for each of the upper rolls; others have one adjustment for the entire series of upper rolls. A straightener should not be overloaded. When stiff, thick metal is passed through a straightener designed for thin metal; it may deflect the rolls permanently or break their shafts. Stiff, thick stock requires larger, stronger rolls spaced well apart. Thin metal requires more straightening rolls than does thick metal. These rolls are usually smaller in diameter and more closely spaced.

Stock straighteners are available in a wide range of capacities and speeds, with either powered or nonpowered straightening rolls. Either the upper or lower set of rolls, or sometimes both sets, is powered. Nonpowered rolls can be used when there is enough pulling or pushing force to get the stock through the rolls. Powered pinch rolls are used to push or pull the stock through the straightener.

Thin stock requires more working to straighten than does thick stock. For this reason, two sets of pinch rolls are used, and all straight-ening rolls are power driven. The speed of powered straighteners can be adjusted so that the material is delivered by the rolls at the rate it is fed into the press plus 10%. The ideal condition is to have the stock run through the rolls continuously so that there are no breaks or bends in the stock when it is stopped on the rolls. When straighteners are operated intermittently, breaks or bends occur in the stock and are almost impossible to remove.

Roller levelers, like straighteners, have staggered pairs of meshing rolls, but the rolls are smaller and more closely spaced. All of the rolls are powered, and some of the upper and lower working rolls have backup rolls. Levelers with backup rolls can impose strains on the metal to remove stack edges or a crowned center. More information on the use of roller levelers is available in the article "Flattening, Leveling, Slitting, and Shearing of Coiled Product" in this Volume.

Presses for High Production

Mass-produced parts are often formed in presses that are made especially for high-production operation. High speed, or the highest number of strokes per minute, is not the only factor in a high production rate. The capability of a press to run continuously for several hours without full operator attention and with a minimum of wear and vibration contributes more to high productivity than does running at high speed for a short period and then stopping for reconditioning of dies.

The more common types of high-production presses are discussed in this section.

Dieing machines, also known as die presses, are set up with conventional progressive dies for long-run operation. These machines are used extensively for the blanking of laminations; however, drawing and forming can be done. The height of the bed above the floor makes it easy to install stacking chutes for laminations and other parts.

Dieing machines are single-action underdrive presses. The drive mechanism for a dieing machine is located beneath the press bed. Four guide rods from a guided lower crosshead pass through bronze bushings in the bed and are fastened to a platen to which the upper die half is attached. The lower crosshead is reciprocated by a crankshaft through connecting rods. By this action, the die halves are pulled together, rather than pushed together, as in a conventional press.

The size of the guide rods and bushings results in excellent die alignment and long die life. The underdrive construction keeps the center of gravity of the press low. The progressive dies mounted in the machine are near eye level, and there are no columns or side members to obstruct the operator's view. Ejection chutes for finished parts and scrap are comparatively high above the floor so that containers are easy to position. Pneumatic cushions, fastened to the top of the platen for better accessibility for service and adjustment, are used as strippers and blankholders.

Stock is fed through the guide rods with either single- or double-roll feeders. A scrap cutter can be mounted on the end of the machine. Both devices are operated by the upper platen or by a power takeoff on the end of the crankshaft.

Multiple-slide machines are fully automatic machines for mass production of small parts from metal strip or wire in coil form. Detailed information on these machines is given in the article "Forming of Steel Strip in Multiple-Slide Machines" in this Volume.

Transfer Presses. Performing multiple operations on a single press can increase productivity and decrease costs. Transfer presses eliminate the need for secondary operations; annealing requirements between operations; and in-process inspection, storage, and handling of workpieces.

Transfer presses should be considered whenever 4000 or more identical stampings requiring three or more operations are needed daily. A total production run of 30,000 to 50,000 identical parts is generally economical between tooling changes. Most transfer presses are designed to make more than one part, and they are often used for families of parts that are similar in size, shape, and thickness. One press is being used to produce 22 different parts.

Stampings are being produced in a wide range of sizes and shapes. Any configuration that can be grasped by mechanical fingers is suitable, and the parts do not have to be concentric. Practically any operation that can be done in any other press can be performed on transfer presses. Typical operations include blanking, piercing, forming, trimming, drawing, flanging, embossing, and coining.

Major users of transfer presses are the automotive and appliance industries. Automotive parts produced on these presses include wheel covers, taillight assemblies, control and suspension arms, transmission parts, catalytic converters, and timing-gear case covers. Appliance components include refrigerator, freezer, washer, and dryer parts.

Fine Blanking Presses. The fine blanking process is generally performed in special triple-action presses designed specifically for the purpose. The presses are available in a range of sizes varying in capacity from 220 to 22,000 kN (25 to 2500 tonf) or more.

Basic components of most fine blanking presses are the frame, upper and lower tables for supporting the tooling, a power system, a stock feeder and lubricator, a control system, and a tool safety device. The frames are generally of welded plate construction, with four-column or double-frame web design, but some smaller presses have single-casting frames. Most fine blanking presses are designed for vertical operation of the ram, but horizontal presses are available. On vertical presses, ram movement for shearing is usually upward, but some presses have a downward movement.

Flexible-Die Forming Presses. Forming, and sometimes blanking, with flexible dies (rubber pads or diaphragms backed by oil under high pressure) is an economical method because it requires only half a die, and materials of different thicknesses can be formed with the same tool. Also, one pad or diaphragm can be used to produce different workpieces, thereby reducing tooling costs. No scratch marks are produced on the side of the blank facing the flexible die.

Another advantage of flexible-die forming is that localized stress concentrations are avoided because of the uniformly distributed pressure achieved with a rubber pad or diaphragm and the gradual wrapping of the blank around the tool. A limitation is that the process is slower than forming with mating die halves, thus sometimes restricting applications to low-volume requirements. However, depending on workpiece complexity and size, the method may be competitive for part production runs to 20,000.

Flexible-die forming is used extensively by the aircraft and aerospace industries, as well as by other manufacturers with low-volume requirements. The three major types of flexible-die forming are rubber pad, fluid cell, and fluid forming, all of which are performed on either standard or special hydraulic presses. These processes are discussed in the articles "Rubber-Pad Forming and Hydroforming" and "Deep Drawing" in this Volume.

Press Safety

The safest press is one that operates continuously with a stock feeder and part unloader. This type of machine does not require the full attention of an operator, and there is no need for the operator to reach into the danger area. Flywheels, gears, and other moving parts likely to catch an operator or passerby are usually covered.

For hand feeding, shields should be interlocked with press controls so that the press will not run unless the shields are in place. The best practice is to make the guard or shield a part of the die so that protection is automatically in place when the setup is made or installed. Shields also can be attached to the press frame and adjusted for various kinds of work. These guards should suit all the work done in the press, should be easy for the setup man to adjust, and should give the operator an unobstructed view.

Usually, it is more difficult to guard hand-fed secondary operations because the workpiece requires special handling. However, if production rate and quantity warrant the expenditure, standard or special devices can replace hand feeding of presses.

Available safeguards include barriers or interlocking guards that keep the operator away from danger, sweep and pulling devices that push the operator's hands away, and devices that require both hands to trip the press. All safeguards should be inspected and adjusted before and after every press run.

Important considerations in choosing safety devices are: number of operators at the press, size and type of press, size and shape of workpieces, length of press stroke, and number of strokes per minute. Protective devices cannot do the job by themselves; they should be used with a well-planned and strictly enforced safety program.

ACKNOWLEDGMENT

Portions of this article were adapted from the book by O.D. Lascoe, *Handbook of Fabrication Processes,* ASM International, 1988.

Press Brakes

PRESS BRAKES (Fig. 1) are a common and versatile type of equipment for bending metal by delivering an accurate vertical force in a confined longitudinal area. Any metal that can be punched or bent by other processes can be formed on a press brake. Press brakes are manually fed by an operator. The operator holds a metal workpiece between a punch and die and against a gage to apply a bend or multiple bends to the metal workpiece. Press brakes can have one of several types of back gages, depth stops, and pins to engage holes in the workpiece. Gages can be manually placed and adjusted, or computer numerically controlled programmable units can automatically adjust settings after each stroke.

A press brake basically consists of two side housings that are held together at the top by a crosspiece and at the bottom by the bed (Fig. 2). The integral parts of a press brake include the housing, brake, bolster plate, bed, ram, gibs, and drive. A ram slides up and down in front of the side housings and does all its work by pressing against the bed. A gibbing and slide arrangement (Fig. 3) keeps the ram supported to the side housings, and two power sources are mounted on both side housings, above the two ends of the ram to alternately push it down and pull it up during the stroke cycle.

Both the ram and the bed are fitted to accept a wide range of dies for many different operations: bending, flanging, drawing, cutoff, parting, blanking, hemming, curling, staking, notching, coining, piercing, ribbing, lancing, corrugating, beading, seaming, pipe forming, channel forming, embossing, bulging, trimming, perforating, slotting, shearing offsets, seminotching, slitting, louvering, coping, crimping, riveting, and tube forming. Selection of a press brake depends on anticipated production needs in terms of required tonnage (striking force), length of bed, bolster width, stroke length, drive system, and the control system. These choices depend on the material being used, type of operation, and the rate of output. The pressure needed to bend metal on a press brake depends on the hardness and thickness of the metal and the width of the lower die. If the width of the lower die is increased, then less pressure is needed to fill it. Tonnage requirements can rise dramatically with a decrease in the width of the die. The materials generally used on a press brake are low-carbon steel, alloy steel, stainless steel, aluminum alloys, and copper alloys.

Press-brake manufacturers usually offer three alternatives for widening a press brake (Ref 1):

- The press brake can have a wide bolster plate and a removable angle bracket for the ram.
- Both the ram and bolster plate can have angle brakets.
- The ram and bolster plate can have removable angle brackets. This is usually the best choice because it offers greater flexibility.

Widening of a press brake depends on tonnage and the structure of the frame. A large press brake should be widened no more than 915 mm (36 in.) (Ref 1). If work consists of punching and using progressive dies on wide stock in heavier tonnages, then a straight side press with four-point gibbing should be considered instead of a press brake. The straight-side press will have press-brake adaptability with the structural strength of a press. Ram and bed extensions add flexibility to the press brake. Extensions that are

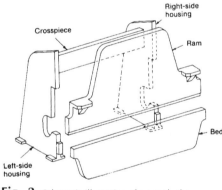

Fig. 2 Schematic illustration of a press brake

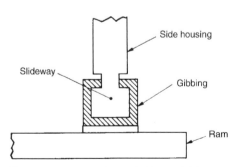

Fig. 3 Cross-sectional view of gibbing and slideway on one end of a press-brake ram

Fig. 1 Press brakes

part of the original equipment are better than add-on extensions, because they have the rigidity of factory installation. A press brake used for horning operations such as the closing of box ends should have extensions, preferably on both sides. It is counterproductive to buy a smaller press brake and add on extensions later. It may cause deflection in areas of the bed and ram outside the housings where die shimming will be a problem (Ref 1).

Mechanical and Hydraulic Press Brakes

Press brakes are divided into two basic categories: mechanical drive systems and hydraulic drive systems. Mechanically driven press brakes have a fixed tonnage and deliver more force at the bottom of its stroke than at the halfway point. Hydraulic press brakes deliver its rated capacity over the entire stroke, and tonnage and ram speed are variable up to the rated limits of a hydraulically driven press brake. A hydraulic drive also allows a longer ram stroke than mechanically driven equipment. Mechanical drives will cycle its ram at more strokes per minute than a hydraulically driven system of the same size. If a user works with dies having a high bottom section, then a hydraulic press brake is recommended. If a machine is needed to exceed its rated tonnage periodically, then a mechanical press brake may be the best choice.

Speed is an important consideration. In the past, only mechanical press brakes were used in high-production shops where product output per hour was important. Later, hydraulic press brakes overcame the slower output rate by adjusting ram speed within a single stroke. Some hydraulic machines offer a third speed that is four times greater than the basic work speed. With this flexibility, the hydraulic press brake can achieve an output rate on long production runs close to comparable mechanically driven press brakes (Ref 1).

Mechanically Driven Press Brakes. Mechanical drive systems range from simple, non-geared belt drives to single- and double-geared drives, with mechanical linkages from motor to flywheel to clutch to gears to crankshaft to crank arm to ram (Fig. 4). They all have one thing in common: a crankshaft action that converts rotary motion into straight, reciprocating motion (Fig. 5). During a stroke cycle, the crank arm drives the ram down to the bottom of the stroke and then back up to the top. An electric motor provides power to a flywheel, which stores energy and provides speed and consistency of motion to the drive shaft on a mechanical system.

Mechanical press brakes have been changed to increase output and for supplying varying ram speed within a single stroke. The ram starts at high speed from the top of the stroke and automatically changes into low speed for the operating position of the stroke. At the bottom of its stroke, the ram again transfers into high speed for its return. A control mechanism provides short, medium, and long periods of time for the ram at slow speeds.

In continuous operation, the ram can be cycled very rapidly for high-speed strokes, due to the mechanical advantage of the crank arm and momentum in the flywheel. The work capability of mechanical drive systems is defined in terms of available tonnage throughout the stroke. Mechanical manufacturers generally cite two reference points for available tonnage: midstroke and bottom-of-stroke (Fig. 6). While the bottom-of-stroke tonnage is the maximum stated tonnage rating for a given press brake, it is not the same as the midstroke, and this difference is due to the nature of the drive system. At midstroke, the crank arm is not in its optimal work position, and the available tonnage is normally approximately two-thirds of the full maximum bottom-of-stroke rating.

As the crank arm continues down from midstroke, the available tonnage rises slowly until near the bottom of the stroke, where the available tonnage rises very sharply. This is because the full mechanical advantage of the crank arm appears near and at the bottom of the stroke. A toggle effect is achieved, and the available tonnage can sometimes exceed three times the maximum published rating for the press brake. This is very useful for certain types of operations that are performed within the last few fractions of an inch from the bottom of the stroke. The full rated tonnage is achieved between 3 and 12 mm (~1/8 and 1/2 in.) from the bottom, depending on the complexity and quality of the drive system. Joint Industry Conference standards establish minimum distance versus tonnage ratings, and, generally, the more complex and expensive drive systems achieve maximum tonnage ratings higher up the stroke. These criteria apply to mechanical presses as well as press brakes. When press capacity is mentioned in terms of tons, a logical question is at what point above bottom stroke is the stated capacity reached.

The reason this is important is because of the danger of overloading the brake. If the operation being performed on the brake requires tonnage at or near the maximum rating of the brake, it should be performed within this area of maximum available rated tonnage at the bottom of the stroke. If it is performed higher up the stroke, the machine will be in danger of being overloaded because of the design limitation of the drive system. Overload can be very serious and can occur when one is least expecting it, even in routine jobs. Commercially available metal varies in hardness and sometimes thickness from

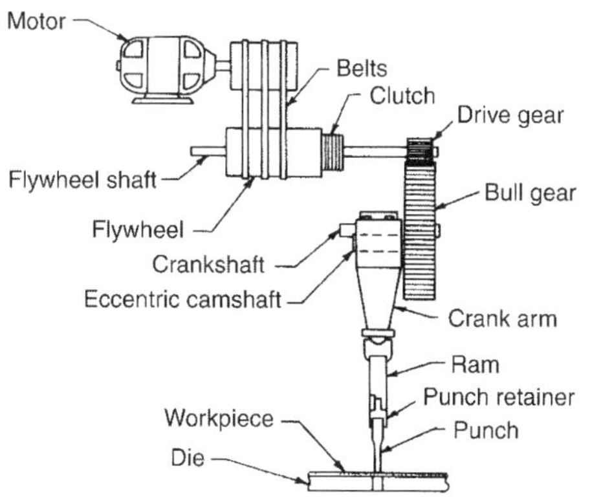

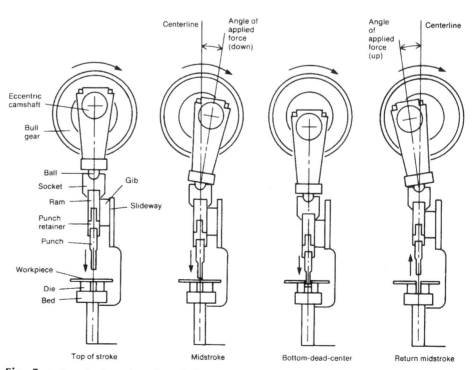

Fig. 4 Drive mechanism of a mechanical press brake

Fig. 5 Stroke cycle of a mechanical press brake

specification among both pieces and sources. These common variations increase tonnage requirements when they occur. In addition, a simple mistake, such as underestimating the tonnage requirement for a job or inserting two thicknesses of metal, can also increase the tonnage requirement.

Mechanical press brakes should not be used under overload conditions, but not because the machine is incapable of delivering extra tonnage. At the bottom of the stroke, a mechanical press brake can deliver as much as three times its maximum rating. The caution against overloading is given because the drive system and even the structure of the press brake can be damaged by the effect of the extra stresses placed on the mechanical linkages, the slide system, and the frame.

When overload occurs at midstroke, it is sometimes called torque overload because the stresses affect the rotating members of the drive system. When it occurs at the bottom of the stroke, it is termed bottoming overload and is a function of the increased mechanical advantage of the toggling action of the crank arm. Between these extremes, an overload demonstrates characteristics of both.

The speed of a mechanical press brake is cited in terms of strokes per minute and is usually stated without a load for continuous or intermittent operation (Fig. 7). The difference lies in the loss of flywheel energy during operation. Intermittent, or stop-start, operation usually involves just one stroke—that is, the ram cycles down and then back up to the top of the stroke and then stops. This action drains up to 35% of the flywheel energy and causes up to a 30% slowdown of the flywheel. If there is a slight pause before the next stroke cycle, the motor will have time to restore the full energy of the flywheel.

In contrast, under continuous operation, the flywheel loses only approximately 20% of its energy and suffers only approximately a 10% slowdown. This lower energy loss is due to the momentum of the flywheel in continuous cycling, which reduces the energy loss and makes energy restoration faster and easier. There are formulas available that can aid in selecting a drive system and motor that will result in the least energy loss under load for a given job.

Stroke control is accomplished through operation of a clutch/flywheel engagement. Hand levers, foot treadles, and air-electric controls are available. Different clutch options are available, from mechanical clutches to air-disk assemblies, and even automatic clutch speed-shifting. Generally, the more complex and precise the control, the more expensive the press brake will be. There are special clutch options that permit successive engagement-disengagement of the clutch during the downstroke to allow inching, and other options permit reversal of the motor to withdraw the ram, but generally, the crank arm and ram must bottom before the ram returns upward.

Press brakes allow operators to choose a high-speed process for full cycle of the ram or a fast advance, low-speed bending, and high-speed response of the ram. The mechanical system increases productivity because once the speed is selected, the machine will cycle automatically. Air-friction clutch systems have a higher degree of consistency. Ram leveling is an aid to the user because it saves time in setting up the job. It is standard on most mechanical press brakes. Ram leveling is accomplished by an independent motor linked through a worm gear drive to a pair of adjusting screws inside the rods joining the drive shaft to the press slide. Tilting adjustments are completed by split couplings and clutches that release the drive on one side of the ram. A calibrated device should be furnished at each end of the ram to show the exact position of the side. The air-friction clutch provides a cushioning effect that extends die life.

Hydraulic press-brake drive systems (Fig. 8) are relatively less complex than mechanical-drive press brakes. They are comprised of a motor, a pump, a valving system, and a hydraulic cylinder with the piston connected to the ram. The cylinder is usually double-acting—that is, oil is pumped under pressure into the top, above the piston, to make the ram move down; to make it move up, oil is pumped into the bottom, under the piston. With oil under pressure both above and beneath the piston, the piston is locked in place anywhere in the stroke cycle. The pressure of the oil under the piston is generally held between 10 and 15% of the pressure of the oil above the piston. This is done to keep the ram from dropping by gravity and also to help control the ram throughout the stroke (Fig. 9). The downward movement of the ram during a stroke cycle means that the pressure of the oil entering the cylinder above the piston is much greater than the pressure of the oil underneath the piston. The upward movement of the ram is caused by a reversal of this action. The valving system ensures the desired control and direction of flow either into or out of the cylinder.

The rotary action of the motor in a hydraulic brake serves to drive a pump, which then forces the oil throughout the system. There is no mechanical linkage to translate rotary action into straight, reciprocating action, as in a mechanical press brake. Because the pressure is a steady, controlled flow, the available tonnage of a hydraulic press brake is the same at all points during the downstroke (Fig. 10). The ram can also be instantly stopped, anywhere during the downstroke, by simply releasing the foot pedal. Stroke control is much more precise than in a mechanical brake and involves much more complex action.

Hydraulic press brakes are available with pressing capacities up to 71 MN (8000 tonf). The tonnage of a hydraulic press brake is a function of the size of its cylinders, pump, and circuit capacity. As noted, hydraulic press brakes deliver its rated capacity over the entire stroke, and the applied load and ram speed are variable up to the machine rated limits. A hydraulic press brake also permits much longer strokes for deep drawing, multiple punch-stepping, or large, complex, multiple-action dies. Extralong cylinders can be installed for extremely long strokes, as long as 1.5 m (5 ft). This is not practical on a

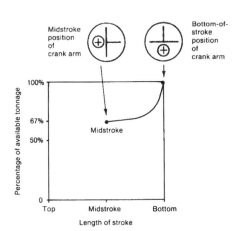

Fig. 6 Available tonnage curve for a typical mechanical press brake

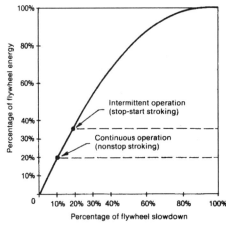

Fig. 7 Flywheel slowdown and loss of energy during operation

Fig. 8 Hydraulic press brake with 1.8 m (6 ft) bed length and 440 kN (50 tonf) load capacity with a 4 kW (5 hp) motor

mechanical press brake, because the crankshaft eccentric limits the length of the stroke. Standard strokes on hydraulic brakes are normally at least three times as long as those on mechanical brakes of equal rating.

The length of the hydraulic stroke can also be varied to suit the job that is being done (Fig. 11). Strokes can be as short as 13 mm ($\frac{1}{2}$ in.). Strokes can be set anywhere throughout the stroke range with limit switches that usually are very easy to set and change. With a shorter stroke, the stroke cycle is faster. For operations requiring very short strokes, such as punching and blanking, a hydraulic press brake can achieve speeds that compare favorably with those of mechanical press brakes, especially when the material is hand fed. Overloading is not

the problem with hydraulic brakes that it is with mechanical brakes. The oil pressure into the cylinders can be set to deliver a given tonnage. Any job requiring more tonnage will cause the ram to stall, but there will be no damage to the press-brake drive system or its structure, because it will not deliver any more than its rated tonnage.

Increasing the pressure of the oil increases the tonnage available on the ram, and hydraulic manufacturers normally build in approximately 10 to 15% reserve as an overload safety feature. However, manufacturers generally do not promote the margins of a safety, so as to curb potential overstressing of the hydraulic system and the structure from too frequent operation in the reserve range.

There are different methods to keep the ram level on a hydraulic press brake. Pressure can be exerted by two cylinders powering the ram to keep it level. There are a wide variety of leveling and tilting controls:

- A servo electric system produces a ram tilt condition by a low-voltage electric signal that is increased and fed back to one of the two variable delivery pumps. This system provides continuous correction to the ram level.
- An electronic system uses an electronic sensing device.
- A proportioning valve system checks the flow of fluid to the cylinders, prevents wavering, and offers continuous correction with high accuracy.
- A limit switch uses a steel tape sensor that drives two highly sensitive limit switches through spring-loaded cams.
- A steel tape system enclosed in a glass tube transmits data to a special level-control unit.

All of the aforementioned systems are used for tilting the ram when required by the type of work.

Bending Capacity

The bending capacity of a hydraulic press brake far exceeds that of a mechanical drive because of its long, full power stroke; it is possible to bend much thicker plate on a hydraulic press brake than on a mechanical brake of the same size. On a hydraulic brake, the chief limitation on bending capacity is the length of the piece to be bent, because of the tonnage requirement. On a mechanical press brake, the limitations include length and thickness, or, in other words, tonnage and stroke (Fig. 12).

As an example, consider the power stroke required to make a 90° air-bend of a given section of mild steel plate in a standard V-die (Fig. 13). The start of the bend requires approximately 80 to 90% of the overall bending tonnage requirement. When the male die achieves a bend close to 20°, most of the work is complete, because the yield point of the metal has been reached. After that, the rest of the bend requires very little increase in tonnage.

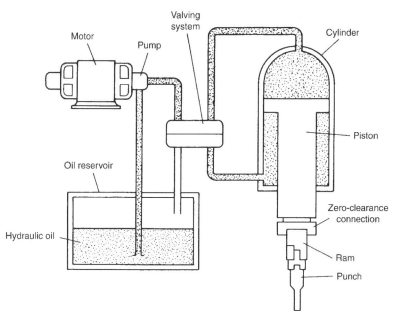

Fig. 9 Drive system of a hydraulic press brake

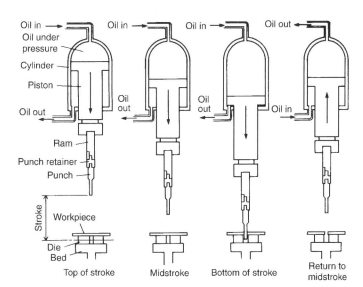

Fig. 10 Stroke cycle of a hydraulic press brake

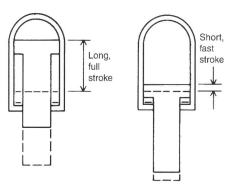

Fig. 11 Cylinder in a hydraulic press brake showing adjustable piston travel

The tonnage required to start the bend should be compared with the available tonnage curve for the mechanical press brake to see if it can deliver the tonnage at that point in the stroke without going into overload. If the beginning of the bend is above midstroke, remember that the available tonnage in this area is even less. For heavy plate, from 13 mm (1/2 in.) thick and up, the recommended width of the required die is ten times the thickness of the metal. This puts the beginning of the bend for heavy plate up in the stroke area of a mechanical press brake where it is at a tonnage disadvantage, compared with a hydraulic. Thus, for example, it is possible to bend short lengths of mild steel plate 25 mm (1 in.) thick on a hydraulic press brake as small as 890 kN (100 tonf), over a 255 mm (10 in.) wide V-die,

requiring a 108 mm (4 1/4 in.) power bending stroke. In contrast, the smallest mechanical press brake that could perform the same bend in one hit without going into overload would have a 9 MN (1000 tonf) rating and a standard 125 mm (5 in.) stroke (Fig. 14).

For thinner workpieces, mechanical press brakes perform very well with light-gage sheet metal and plate up to 6.4 mm (1/4 in.) thick. Their cycling speeds, especially with automated feeding and unloading equipment, are extremely high. They are particularly suited to long runs of the same bending or forming operation. Small hydraulic brakes, however, have several advantages that compare favorably with small mechanical brakes: an inherent versatility in doing the wider range of operations normally found in metal fabricating shops; precise operating control of the stroke; speed in setup and changeover; flexibility in handling both deep-stroke heavy plate bending and short-stroke sheet metal work; and quiet, shockless operation.

In examining both types of press brakes, tonnage-distribution characteristics also must be considered (Fig. 15). The basic construction involves a long ram with two sources of power, one at each end. The power distribution in this design is the full tonnage of both power sources along the entire length of the ram, as long as the load is distributed along the entire length of the ram. If the load is concentrated in the center of the ram, the full tonnage of both power sources will be available for the load. If the load is concentrated under only one of the power sources, only half the rated tonnage will be available to perform the job (Fig. 16). Moving the load toward the center of the ram permits the other power source to participate in the operation, and

so the available tonnage rises as the load is moved until the short load is exactly centered. At that point, the full tonnage of both power sources is concentrated on the load.

Thus, an operator should not count on the full rated tonnage of the press brake if the tonnage on a short job is concentrated at either end of the machine. When horns are used, the available tonnage drops even more, depending on the horn length and the distance of the load concentration from the power source.

The shape of the ram is designed to resist deflection best in the vertical, left-to-right direction. The depth of the center of the ram adds to its stiffness and rigidity as a beam. This characteristic aids in leveling the ram from right to left. From front to back, however, the ram is relatively weak and is dependent on the gibbing and slideways, which are mounted only on the rear of the ram, at both ends. Front-to-back deflection can be a problem even with a stiffener along the rear length of the ram. Operators should therefore center the load in the dies from front-to-back, especially on press brakes fitted with wide platens. Off-center loads from front to back can cause great stress and may damage the gibbing and slideways, because this is their weak direction.

The shape and mounting of a press-brake ram also leads to a unique advantage in permitting press brakes to be connected in tandem (Fig. 17). Both mechanical and hydraulic press brakes can be so connected to accommodate extralong work. The controls can be designed to permit either machine to be operated independently or both together. Tandems are used when designers specify longer beams and structural forms to minimize welding.

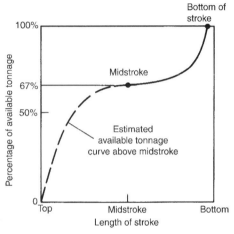

Fig. 12 Available tonnage curve for a mechanical press brake

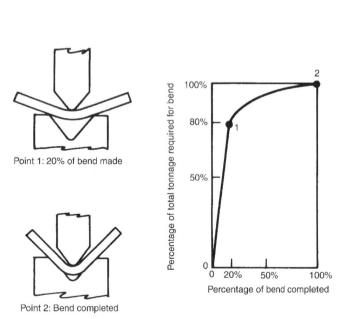

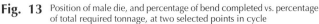

Fig. 13 Position of male die, and percentage of bend completed vs. percentage of total required tonnage, at two selected points in cycle

Point 1: 20% of bend made

Point 2: Bend completed

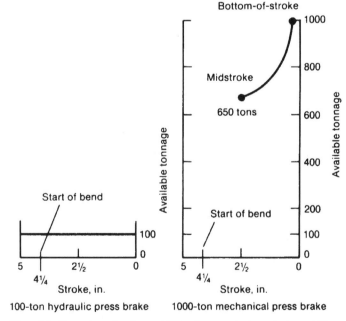

100-ton hydraulic press brake

1000-ton mechanical press brake

Fig. 14 Comparison of stroke capabilities of two standard press brakes: a 100 ton hydraulic brake (left) and a 1000 ton mechanical brake (right). Work metal: mild steel 25 mm (1 in.) thick by 0.46 m (1 1/2 ft) long. Die opening: 255 mm (10 in.) wide. Tonnage requirement: 81 tons. Required bend stroke: 108 mm (4 1/4 in.)

Press-Brake Tooling

Press-brake tooling is mounted on the ram with a holder, which is machined to accept a designated style of tooling. Figure 18 shows examples of American-style tooling. Setup of tooling may involve manual clamping and die shimming. Hydraulic clamping systems (Fig. 18)

reduce setup time with automatic clamping and quicker centering and alignment of the tooling.

The two major types of bending methods are air-bending and bottoming. Bottoming dies (Fig. 19) use pressure to cancel out springback, and they are machined to the exact radius desired in the bend; for example, a 90° bottoming die produces a 90° bend. In contrast, air-bending

dies (Fig. 20) are machined to 88° to produce a 90° bend. Air-bending is commonly used for forming heavy sheets and plates over 10 gage, and bottoming is used only for light-gage forming (approximately 14 gage and lighter) (Fig. 21).

Gooseneck dies (Fig. 22) permit forming of two bends close to each other. The die is designed so that the second bend may be made without the die interfering with the return flange. This design, however, makes it mandatory that the die not be overloaded, and thus it should not be bottomed. Gooseneck dies can also be provided with a radius on the tip for other than sharp-corner bends.

Hemming and seaming dies can consist of a sharp knife die set and a flat die set. The operation is performed in two strokes, with the knife die making the first bend and the flat die closing up the hem or seam. The flattening stroke requires the most tonnage, and there is often no need to bottom the knife die to produce the required bend.

A hem is a fold at the edge of the sheet metal to remove the burred edge and improve the appearance of the edge. The hem also adds slightly to the rigidity of the edge and improves wear resistance. Hems and seams are shown in Fig. 23(a). A hemming die set is shown in Fig. 23(b).

Curling and Wiring. Curling and wiring are used to strengthen the edge of sheet metal. Actually, hems, curls, and wiring may be used on flat parts or on round parts such as cams or drums. During wiring, the metal is curled up and over a length of wire. The edge of the sheet metal is strengthened by both the wire and the curl formed in the metal. When the wire is not used, the operation is known as curling or false wiring (Fig. 24). Curling dies are shown in Fig. 25.

Bead-forming dies are available for forming two main types of beads: the central bead and the tangential bead. To obtain a good bead in thick material or a small bead in thin material, a wire mandrel should be used during the final closing operation. The wire may be left inside the finished bead or be withdrawn, depending on the final use of the product and the tightness of the bead. It is important when forming the bead to keep the burred edges of the metal to the inside of the bend to avoid scoring of the die surface.

Bottoming methods are not recommended because of the die shapes involved. Generally, the pressure required to form beads without

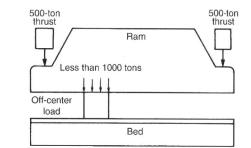

Fig. 15 Effects of load distribution in a 1000 ton press brake

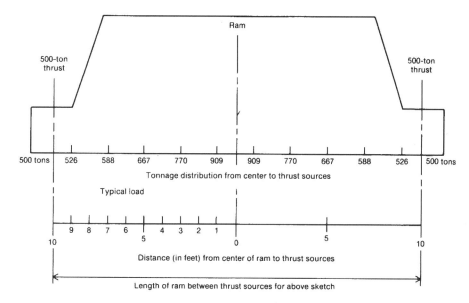

(a)

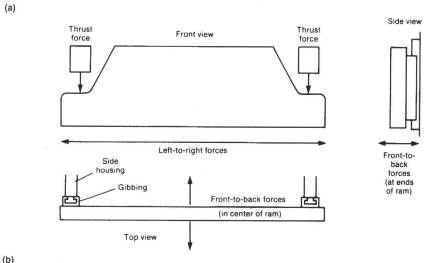

(b)

Fig. 16 Typical load distribution from center to thrust forces (a), and forces that act on the ram (b)

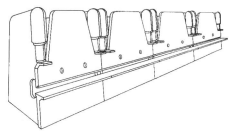

Fig. 17 Extension of the tandem concept

bottoming is approximately four times that of a standard 90° bend.

Pipe-forming dies are not merely scaled-up versions of bead-forming dies. Small tubes or pipes are best formed with three strokes required. Hydraulic press brakes are especially useful in forming pipe because they can be set to hold the pipe closed while the pipe seam is being sealed by tack welding or banding.

Medium-sized pipe can be bumped progressively on larger-radius dies. Each of the edge quadrants should be formed first and the two center quadrants last. Very large pipes may be formed by this method, using larger-radius dies and forming a half-circle at a time. Two half-circles are then welded together to make a complete pipe section. Large-diameter pipe will probably be too large to fit in the press-brake

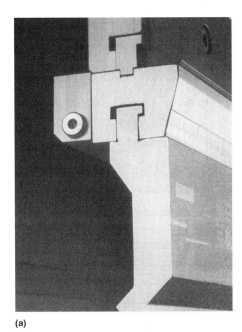

(a)

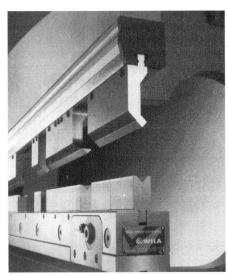

(b)

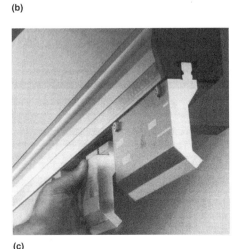

(c)

Fig. 18 American-style press-brake tooling. (a) Closeup. (b) Tooling with hydraulic clamping. (c) Removing tooling from tool clamp. Courtesy of Wila USA

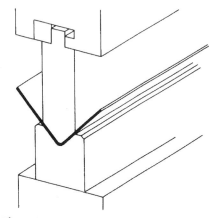

Fig. 19 Principle of a bottoming punch-die set

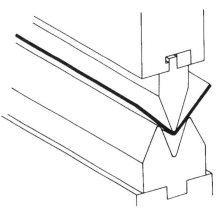

Fig. 20 Principle of an air-bending die on a press brake

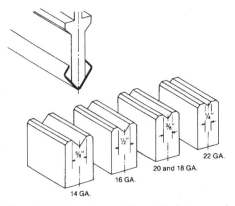

Fig. 21 90° forming punch-die sets for bottoming of light-gage sections

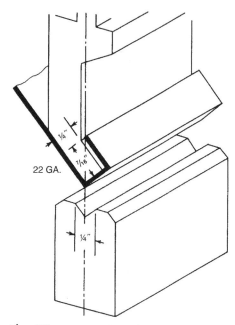

Fig. 22 90° gooseneck punches

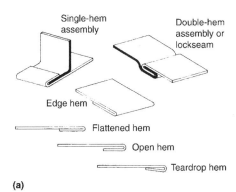

(a)

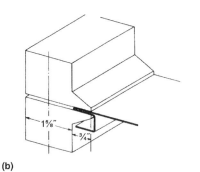

(b)

Fig. 23 Examples of (a) hems and seams and (b) hemming die set

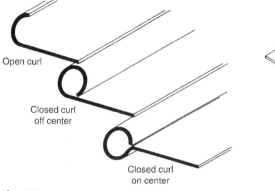

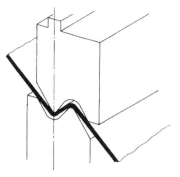

Fig. 24 Curls. (a) Open and closed curls. (b) Wiring and curling

(a)

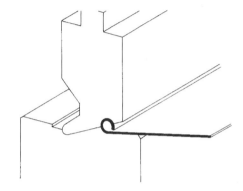

Fig. 25 Curling dies on a press brake

Fig. 26 V-die on a press brake

(b)

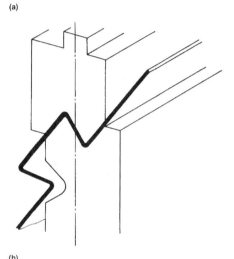

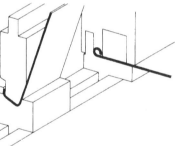

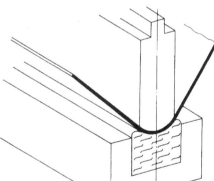

Fig. 27 Radius-forming die

(c)

Fig. 28 Offset dies. (a) Typical. (b) Acute-angle offset die. (c) Z-bend offset die

throat, but it can be accommodated between the side housings or on a horn, depending on its length. An alternative method is to preform the edges in a press brake and form the remainder of the pipe in a roll-former. This method has been successfully used for extremely thick metal that is difficult to get started in the roll-former because of the length of the pipe.

Channel-forming dies can consist of a standard V-die set (Fig. 26) or a gooseneck die (Fig. 22), depending on the size of the channel. Two strokes are needed, and each bend should be accurately gaged. This method reduces die cost, minimizes tonnage, and ensures that the bottom and sides of the channel remain flat.

Channels can also be formed in a single stroke, which is not only faster but also more accurate. The drawback is that channel-forming dies are single-purpose dies, are relatively expensive,

and require at least four times the tonnage of 90° air-bends and thus a larger press brake. Production quantities therefore must justify the added expense.

Deep channels are possible with standard V-dies. A reverse bend is made in the center of the metal to open up the channel so that the straight knife die will fit to form the bottom corners. After both bottom corners have been formed, the bottom is straightened to finish the deep channel. Hydraulic press brakes are especially well adapted to this operation, because a long stroke is required to open the dies far enough to remove the part.

Radius-Forming Dies. A standard knife die and V-die set can be used for small-radius bends on sheet metal. The knife die should have a radius no smaller than the required bend radius. As the metal thickness increases, greater allowance must be made for springback because this is an air-bending operation. Larger radii may require a male die with a machined radius. If this radius is approximately 19 mm (3/4 in.) or less, the metal will follow the male die in air-bending over a channel-type die or a simple V-die. If the radius is larger and a full 90° bend is required, the metal will leave the male die as the die moves down. In this case, a matched-bottom-radius die is needed to force the metal back against the male die contour (Fig. 27). This becomes a bottoming operation with higher tonnage requirements and, of course, added expense for the matched-die set.

Progressive bumping can also be used to form radii in short increments. For best results, the knife die should have a well-rounded nose. To avoid flat sections under the nose of the male die, reducing the V-die opening is recommended. The advantages of this method are low die cost and the ability to compensate for differences in springback. Any desired radius can be formed by adjusting the length of the stroke.

U-Section Dies. U-sections can be formed with simple radius dies by the bottoming method or with springloaded pad dies. In both cases, a second stroke with flat dies may be used to compensate for springback.

Offset dies (Fig. 28) can perform two close bends in one stroke, increasing production up to 500% compared with single bends, but the tonnage requirement can range from 3 to 15 times that of a simple 90° bend. Pressures are highest for Z-bends (Fig. 28c) with sharp right-angle corners. Offset dies are also more expensive.

Offset bends can also be made in two strokes with standard V-dies including a narrow bottom V-die to allow clearance for the sheet to fold down. The sheet must be turned over between strokes, but the tonnage requirements are low, and a single standard die size can be used to form many different shapes.

ACKNOWLEDGMENT

Portions of this article were adapted from Ref 1 and 2.

REFERENCES

1. "How to Buy a Press Brake," http://www.equipmentmls.com/info/howtobuy-press-brakes.html, The Equipment Information Center, EquipmentMLS.com, accessed Nov 2005
2. O.D. Lascoe, Fundamentals of Press Brake Operations, *Handbook of Fabrication Processes*, ASM International, 1988

Die Sets

A DIE ASSEMBLY is a complete punch-press tool that is used to produce large numbers of interchangeable stamped parts. It consists of mating pairs of punches and dies, their retention plates, a stripping device, and a subassembly called a die set. When the press is actuated, the mating components are forcibly brought together to perform the required operations.

The upper die assembly is fastened to the movable press ram, and the lower die assembly is secured to the stationary bolster plate or press bed. The upper member must travel in a precisely controlled path to maintain the alignment and orientation of mating components during the working portion of the press stroke.

Every punch press must permit relative movement between the ram and the press frame in order to satisfy its primary function of delivering energy. The "running" clearance provided, however small, may have an adverse effect on the operational alignment of mating punch-and-die combinations. For example, even when press parts are closely fitted and precisely aligned, the running clearance may actually exceed the required punch-to-die clearances within the die assembly.

The impact nature of the forming process further complicates the alignment problem. It generates deflecting forces that must be effectively opposed so that alignment can be maintained.

Obviously, the alignment system of the press requires that the die assembly provide additional guidance and support at the point of impact and throughout the working cycle. It is in this area that a good die set demonstrates but one facet of its tremendous value.

Caution: Although the die set can be relied on for the ultimate alignment, it cannot be expected to compensate for a press in poor condition or to operate properly if subjected to heavy deflecting loads.

The guide post/guide bushing combinations, properly fitted, ensure precise control of the movable member at every critical point of the stroke to maintain the relative position of each pair of mating components.

Functional Requirements

A good die set is not just a subassembly. It provides the foundation for a basic working system on which a press tool can be assembled, aligned, inspected, and put into operation. It enables the die builder to assemble many separate components into a single integral unit that provides many cost-saving benefits:

- *Ease and accuracy of setup:* The die assembly is installed in the press as a single unit, which minimizes setup time and ensures proper alignment of mating components.
- *Improved and consistent stamped-part quality:* The ensured accuracy of each setup maintains the original degree of punch-to-die clearance uniformity and thereby enhances piece-part quality.
- *Increased die life:* Correct register of punch-to-die reduces the rate of wear on both components and minimizes the risk of component breakage.
- *Ease of die maintenance:* Cutting components can be sharpened in assembly, as units, without removing them from the die set—a distinct advantage over removing the components and sharpening them individually.
- *Simplified die-repair procedures:* Broken or worn components can be removed and replaced without disturbing their relative positions. Alignment and orientation of mating components is virtually ensured.

Die-Set Nomenclature

The die set is a modular unit consisting of a stationary lower plate (die holder), a moveable upper plate (punch holder), and at least two sets of precisely fitted guide posts and guide bushings. Smaller die sets are equipped with a projecting stem (shank) that extends from the top surface of the punch holder (Fig. 1).

Generally, the guide bushings are assembled to the punch holder and the guide posts to the die holder. This arrangement is normally reversed when ball-bearing bushing assemblies are used.

The Shank: Functions and Variations

The shank is used to center the die assembly in the press and to secure the upper die member to the press ram. A large die set may also be equipped with a shank, but it should be used only for centering purposes—not to secure the upper die assembly to the ram.

The shank may be an integral part of the punch holder or it may be a separate unit that is secured by means of a threaded end or individual fasteners. Integrally welded shanks are structurally stronger, cannot loosen in operation, and provide the least interference with knockouts or with adjacent fasteners and components.

Floating shanks are designed to compensate for a ram face that is not perpendicular to the cavity for the die set shank. The top of the punch holder is machined to receive a swivel adapter that contains a separate shank. The shank is permitted to move until it conforms with the ram face. Parallelism between the punch holder and the die holder is not affected.

Alignment Considerations

The die set is expected to maintain the alignment of all mating components during the die-building stage and throughout the productive life of the die assembly. In essence, a good die set must withstand the rigors of use without prematurely losing its ability to control alignment.

Many factors must be considered and controlled in order to satisfy this objective:

- Dimensional specifications and limits
- Allowances for cylindrical fits
- Material types related to the various die-set components
- Geometrical deviations that may be anticipated
- Permissible variations in the alignment of guide posts and guide bushings
- Available styles of die sets and components

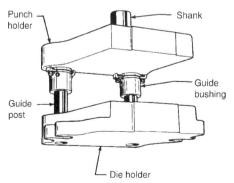

Fig. 1 Die-set nomenclature

ASME B5.25 (B5.25 M), "Punch and Die Sets," provides detailed data for each of the foregoing factors.

Functionally, standards data provide product definition by means of materials specifications, dimensions, and permissible tolerances. In this sense, they tend to establish an area or range from which selections can be made. They fail, however, to provide the basis on which to make a selection.

Application Requirements

The ultimate alignment of mating punches and dies cannot be better than that provided in the die set. Alignment must be established during the building stage when practically no deflecting loads are applied. Alignment must be maintained when the tool is in operation and producing stamped parts.

Under operating conditions, the entire die set and the individual components are continuously subjected to high-impact loads and deflection forces. Consequently, all selections should be predicated on the basis of resistance to shock, deflection, and wear.

Punch and Die Holders

These plates must satisfactorily oppose rapidly applied compression loads. The plate material must have high compressive strength and toughness. Because the die holder may be mounted on parallels or span large bolster-plate openings, it must also exhibit a high degree of stiffness or resistance to bending. The ability to resist these forces is largely dependent on the material selected and the physical proportions of the plates. In general, rigidity increases with plate thickness.

Because of the need for many fastener holes and extensive machining operations, the plate should be free from such internal defects as porosity, hard spots, or shrinkage.

When a die set requires milled pockets, large counterbores, cutouts, or burnouts, the plate should be stress relieved. It is recommended that the die-set manufacturer perform the machining before the plates are bored for guide posts and guide bushings.

Based on these considerations, the most widely used plate materials are:

- Hot-rolled steel containing 0.15 to 0.30% C and 0.30 to 0.90% Mn
- Gray or cast iron and specialized irons such as Meehanite (Meehanite Metal Corp., Mequon, WI)
- A combination wherein one die-set member is made of steel and the other is cast iron
- Tool steels (hardened or partially hardened), aluminum, magnesium, and other special alloys

The steel plate and the specialized irons are the most popular materials. Of the two, the advantages of all steel die sets are clearly established:

- Greatest resistance to impact loads and to loads that are not uniformly distributed
- More uniform density
- Ability to be flame cut and welded

Plate Flatness and Parallelism

The opposite surfaces of the individual plates must be both flat and parallel to avoid an adverse effect on the ultimate alignment of mating components (Fig. 2).

Components mounted on a surface that is not flat may cause the line of action to be canted away from the perpendicular and result in sheared or broken cutting edges. Should a component be assembled over a small bump in the surface, a rocking action may be imparted by the punching load, causing component breakage or premature wear.

Punch-to-die alignment will be impaired if the opposite surfaces of a plate are not parallel even though both may be flat (Fig. 3). The components will be tipped away from the perpendicular, and the lines of action of the upper and lower die assemblies cannot coincide.

Parallelism parameters are applicable to the assembled die set as well as the individual plates (Fig. 4).

Guide Posts and Guide Bushings

Guide posts and guide bushings, in combination, establish and control the path followed by the movable upper die assembly. In effect, alignment uniformity can be maintained only when the relative positions of the upper and lower die members remain unchanged during the working portions of the stroke.

Because deflecting forces are always present and can be minimized but not eliminated, the selection of guide posts and bushings must be based on their ability to resist both wear and deflection (Fig. 5).

Two basic guidance systems enjoy a marked degree of popularity: the ball-bearing bushing assembly and the more widely used conventional system. Each has advantages as well as disadvantages, and selection is usually a function of personal preference.

The ball-bearing bushing assembly, by virtue of its rolling balls, provides unusual ease of assembly. The interference fit or preload on the balls maintains the relative position of the bushing to the post, but the assembly is less

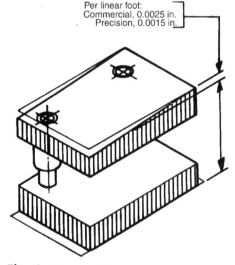

Per linear foot:
Commercial, 0.0025 in.
Precision, 0.0015 in.

Fig. 4 Die-set parallelism

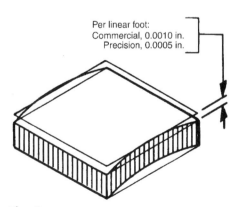

Per linear foot:
Commercial, 0.0010 in.
Precision, 0.0005 in.

Fig. 2 Plate flatness

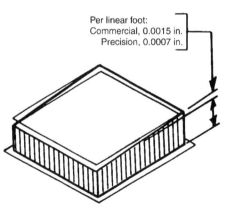

Per linear foot:
Commercial, 0.0015 in.
Precision, 0.0007 in.

Fig. 3 Plate parallelism

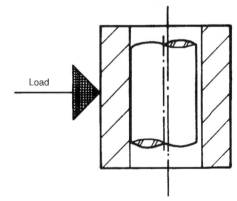

Load

Fig. 5 Guide post/guide bushing deflection

resistant to deflecting loads. For a given bushing size, the post is more likely to bend because its diameter has been reduced in order to provide room for the balls. To compensate for this problem, some die sets maintain the basic guide-post diameter and increase the bushing size. The plate dimensions are correspondingly increased so that the available die space is unchanged. Rolling balls generate less friction than sliding members but are more difficult to lubricate (Fig. 6).

The interference fit with the individual balls eliminates any possibility of a physical barrier being established between the components. Ordinary lubricants will break down due to extremely high pressures that rupture the film. It is necessary to use extreme-pressure additives that react with the various components to form chemical barriers that prevent metal-to-metal point contacts (Fig. 7).

Ball-bearing bushing assemblies are recommended during the building phase, but the post and ball cage can be removed and replaced with the conventional demountable post for optimal performance.

In the conventional system, clearance is provided between the post and the bushing. This space is then filled with lubricant to provide a physical barrier between the moving members. The Die Set Standard establishes the permissible range of clearance without adversely affecting alignment ability and defines the configuration of the lubricant-retention grooves. Guide posts and bushings are prefitted to dimensions, within tolerance, and are furnished as matched units in each die set. By means of color-coded components and selective assembly, three classes of operating fits may be specified.

Guide posts and bushings are subjected primarily to frictional forces, which generate wear. Even deflecting forces are translated into wear factors because they tend to break down the lubricant and thereby increase the coefficient of friction between the sliding members. In selecting the best combination of materials for use in a specific application, the decision is dependent on the anticipated operating conditions:

- *Frequency, amount, and type of lubrication:* Lubrication must be performed on a regular schedule, and the amount must be adequate to replace the lubricant that has been displaced or used. When large deflecting forces are anticipated or encountered, extreme-pressure additives are required.

- *Punch-to-die clearances to be maintained by the die set:* When clearances are extremely close, the guide post and bushing must be selected on the basis of abrasion resistance. Hardened steel posts and bushings are selected for the majority of applications.

- *Condition of the press:* Specially processed bronze-plated bushings should be chosen when considerable ram play is anticipated. These bushings provide a bronze interface with a steel support and are used with hardened steel posts.

- *Velocity of operation:* Press velocity is not merely expressed in terms of strokes per minute. In actuality, it is the product of twice the stroke multiplied by the number of strokes per minute. Bronze-plated bushings and chromium-plated guide posts are the best choice for high-speed operations.

- *Abrasive impurities in the surrounding atmosphere:* When the surrounding atmosphere is charged with abrasive particles that cannot be removed economically, well-lubricated hardened steel guide posts and bushings will provide the best service. The hardness is such that the abrasive particles will have less effect on surfaces.

Surface wear due to friction, either sliding or rolling, is referred to as abrasive wear. Should the abrasive wear condition be permitted to continue without alleviation, the ultimate result will be seizure: the normally moving members will be literally welded together.

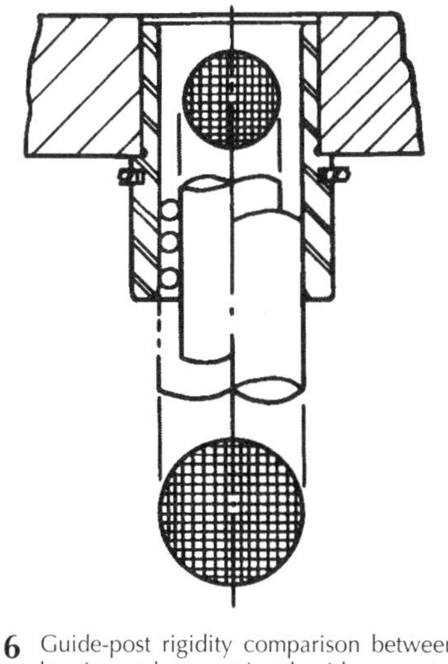

Abrasive Wear in Die Sets

When relative motion exists between two parallel, lubricated surfaces, the protective film is subjected to stretching as well as compressive forces. As a result, there is almost never a complete protective film separating the surfaces. Minute voids in the lubricant film permit metal-to-metal point contacts to occur. These point contacts, along with hard foreign elements in the lubricant, induce an abrasive action on the two surfaces. The result is a removal and/or relocation of surface material caused by a small-scale welding action between contacting points. In its extreme form, this surface relocation action results in seizure.

A periodic visual examination of guide posts and bushings is recommended. In many cases, abrasive wear can be detected by the polished appearance of the surfaces, and corrective action can be initiated to prevent seizure.

Abrasive wear is the result of the frictional forces developed during the relative movement of the two members. It is extremely significant to note that friction cannot be present unless the two surfaces are forced together by the application of an external load. In effect, the conventional post, and bushing combination, are subjected only to deflecting forces; the ball-bearing bushing assembly is subjected to the initial load imparted by the interference fit plus the deflecting forces.

Guide-post and guide-bushing materials must be selected relative to the lowest frictional coefficients and the greatest resistance to deformation. The objective, of course, is to select material combinations that offer the greatest wear resistance with the least rate of wear. To determine which materials best satisfy the objective, a number of mechanical and physical properties must be considered.

Relative Hardness. Where relative sliding action occurs, it is generally agreed that the greater the hardness difference between the two members, the lower is the coefficient of friction. In the presence of hard foreign elements, however, a harder surface on both members will generally provide a lower wear rate. The ability of a material to bury within itself various amounts of foreign particles increases as its hardness decreases. This characteristic is particularly desirable because the abrasive particles are eliminated from the immediate region where the surfaces may be in contact. Die-set operational requirements, however, preclude the use of a very soft material for its embeddability characteristics alone. A layer thick enough to absorb a substantial quantity of particles would soon be plastically and permanently deformed by the lateral pressures imposed by deflecting forces.

Elastic and Plastic Characteristics. Hardened materials exhibit the greatest resistance to deformation. Their relatively high elastic limit permits them to withstand large lateral loads with a minimum of distortion. Soft materials may be permanently distorted when subjected to similar loading conditions. Their relatively low elastic limit may be exceeded, resulting in plastic deformation or displacement. Experiments conducted in a range of low operating pressures and velocities have demonstrated conclusively that hardened steel bushings are more wear resistant than those made of a cast alloy containing zinc, aluminum, and copper. In both instances, the mating surface was a hardened steel guide post.

Presence and Characteristics of Surface Films. Lubricants tend to fail when subjected to

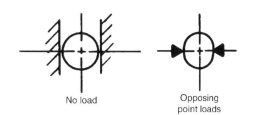

Fig. 6 Guide-post rigidity comparison between ball-bearing and conventional guidance system

Fig. 7 Ball-bearing deformation

No load

Opposing point loads

large pressures and/or high velocities. When the lubricant film is ruptured, metal-to-metal point contacts occur. Extreme-pressure additives such as chlorine and sulfur combine with the base materials to form a chemical film or barrier that can reduce substantially the rate of wear. It is also possible to introduce a thin metallic film having a low coefficient of friction at the interface of the two surfaces. Specially processed bronze-plated bushings exemplify this approach and exhibit an extremely low rate of wear.

Ability to Form a Common Welded Joint. As stated previously, seizure is an extreme form of abrasive wear. It is typified by considerable surface material removal and/or relocation and drastic increases in frictional forces and is usually accompanied by an audible squealing sound. The point at which this action occurs and the extent of damage to the bearing surfaces are largely dependent on the physical properties of the materials involved. Because high local surface temperatures are generated, there is some flow of bearing material, generally resulting in a galled appearance. If the local heating produces a welding effect between minute quantities of the two materials, the strength properties of the resultant substance will influence the degree and direction of the galling. Materials, therefore, should be selected on the basis of least affinity for each other and, in the event of a welding action, the lowest strength properties in the resultant substance. An indication of the properties of the formed alloy will be given by its relative position in the periodic table.

Geometrical Considerations

Guide posts and bushings are secured to the appropriate plates by means of press fitting or the use of external toe clamps and fasteners. Components assembled by the latter method are known as demountable items.

The demountable components provide the best guidance system because all of the desirable geometrical requirements are maintained (Fig. 8). In addition, they can be easily removed and replaced to simplify die-maintenance procedures without disturbing the initial alignment.

The use of a wring or a light tap fit between the plate and the bushing or post, for a length of approximately 4.8 mm (3/16 in.), practically eliminates distortion in any of the components. The guide bushings are machine honed for accurate hole geometry and size. They are then mounted on arbors so that the shoulder seats can be ground flat and perpendicular to the guide posts. This combination of manufacturing accuracy and distortion-free assembly automatically squares up the posts and bushings to the die-set surfaces.

The holes bored in the die-set plates serve only to position the posts and bushings. Retention is accomplished by means of toe clamps and cap screws, which develop greater resistance to stripping than do press-fitted components.

In press-fitted assemblies, the plate holes must provide for the retention of the guide posts and bushings as well as their relative positions. The interference between the bushing and the hole in the plate distorts the bushing hole. The subsequent honing operation may bell-mouth or enlarge the entry end, resulting in loss of proper fit.

Press-hole distortion is an inherent risk in press-fitting operations. Consequently, the posts and bushings may not be square with the die-set surfaces. It may be necessary to tap the components to adjust the grip for squareness.

Abrasion and Seizure Testing Methods

A single unit consisting of a guide post and bushing was installed in a fixture with a variable-speed motor drive that made available speeds of 50 to 430 strokes per minute. The bushing was split into halves to facilitate the application of variable surface loads. Accurately calibrated strain gages were used to measure the magnitudes of the applied loads. Lubrication and cleanness conditions were maintained as uniformly as possible throughout the experiment.

The graph in Fig. 9 illustrates the effects of press speed and variable surface loads on the coefficient of friction between the sliding members.

Endurance Testing

A special die set was installed in a mechanical press continuously operated at 320 strokes per minute. Variable surface loads were applied to the guide post and bushing assembly by means of a preloaded compression spring. Lubrication was introduced between the guide post and bushing at the beginning of each test run, and operation was continued until seizure occurred.

Following each seizure, the specimens were allowed to cool, examined for galling,

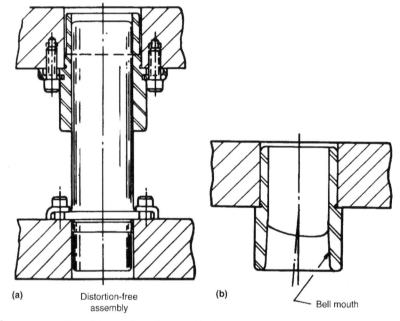

Fig. 8 (a) Demountable and (b) press-fit guide post and bushings

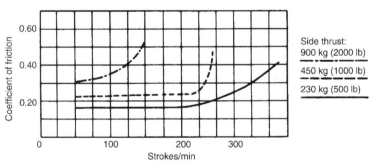

Fig. 9 Hardened steel bushing characteristics (stroke length, 25 mm, or 1 in.)

relubricated if galling was not too severe, and reoperated to determine whether a healing characteristic existed.

This test demonstrated the superior wear resistance of hardened steel posts and bushings. This combination will maintain close tolerances for long periods of time when loads and speeds are below the point of seizure.

Die-Set Recommendations

Catalog die sets that provide up to 635 by 355 mm (25 by 14 in.) of die space are offered in many guide post and bushing arrangements. They include many styles and types of guide post and bushing designs from press-fitted and demountable types to the ball-bearing bushing assembly. Both commercial and precision die sets are available for selection.

The broad areas for selections do not impose any unusual die-design restrictions, and the advantages of die layout for catalog die-set usage are many:

- The punch or die holder can be replaced if necessary.
- Worn or damaged guide posts or bushings can be replaced without impairing the original alignment.

- All steel punch and die holders are stress relieved to maintain dimensional stability during subsequent machining operations. It also helps to soften edges that may have become hard during the flame-cutting operation.

ACKNOWLEDGMENT

This article was adapted from the chapter "Die Sets: Applications and Functional Requirements" by O.D. Lascoe in *Handbook of Fabrication Processes,* ASM International, 1988.

Press Feeders

A STAMPING JOB SHOP generally begins with the installation of a press. Either a reel or a stock cradle is added to handle coiled material. If labor costs are low, the material is pushed into the die by hand, but as labor costs increase, it becomes more and more important to automate the feeding operation.

There are four basic types of feeding arrangements: hand feeding, hitch feeds, roll feeds, and slide feeds. Hand feeding is most useful for short runs and those special cases where secondary sheets and precut lengths can be used to offset the higher cost of coil stock.

Of the automatic feeds, the hitch-type feed is the oldest and least-expensive mechanism. In this type of feed, the material is advanced by a wedging action between a blade and a moving block. These units are best at feeding medium-weight materials at relatively short progressions. Because of the low cost of some commercially available hitch feeds, they are often attached directly to a die where the setup can be of a permanent nature.

A popular type of permanently mounted equipment is the roll feed. In this unit, the material is advanced by the pressure of two opposing rollers, one or both of which are driven from the crankshaft of the press. These units are generally best for feeding rigid materials in moderate to heavy thicknesses. Thinner materials can often be handled by using a double-roll-feed combination in which a set of feeding rollers is attached to each side of a press and synchronized by a connecting link. Roll and mechanical slide feeds generally offer the longest life and highest speeds available today but represent a permanent installation at prices ranging from moderate to high.

Slide feeds fall somewhere between the hitch feed and the roll feed in cost and have the capability of handling materials including the very thin and the very compressible. Air slide feeds are not limited to use in punch presses but can be mounted on special machines in almost any position and can be actuated by mechanical, electrical, or air signals.

Because of the versatility of the air feed, we will examine various feeding concepts for the job shop utilizing air-operated punch-press feeds. The operating principles of air feeds are general and can be applied directly to hitch feeds and roll feeds. The problems that arise in the use of air-operated punch-press feeds are also encountered with other types of feeds. Note that if the proper precautions are followed, almost any type of feed, within its limitations, will give good accuracy, speed, and utility. Before considering a switch from one type of feed to another, always examine the application. The feed at hand is usually capable of doing the job if it is given the proper opportunity.

Types of Material

Automatic feeds are normally designed to feed standard coil stock. If the stock is rigid enough to resist clamping or roller pressure and is self-supporting, it falls into the classification of a standard material. This is the type of material normally found in production shops. In addition to these materials, job shops are often called upon to feed materials not easily accommodated by the production installation.

The simplest special shape is wire. Almost all types of feeds will easily handle wire unless the wire diameter is very small—usually less than 0.38 mm (0.015 in.). The clamping and roller pressures relative to the area of the wire become so high at small diameters that deformation of the wire often occurs. The air feed, with its high clamping area, simplifies this type of feeding job. For small diameters, it is sometimes necessary to use telescoping tube guides to prevent buckling of the wire as it passes through the feed and the die. As the wire diameter becomes smaller, the job becomes more difficult and more delicate, but by no means impossible (Fig. 1).

Thin materials (thicknesses less than about 0.38 mm, or 0.015 in.) have been difficult to feed. Flexible inserts in the clamps can usually prevent scratching of highly polished materials, but again, depending on progression and the self-supporting features of the material, it may be advisable to install antibuckling guides or other types of support for the material as it passes through the feed. Small air feeds excel at feeding thin material.

The ease of grinding special shapes in the clamps makes air feeding of special formed material a simple operation. If the form is rigid, passageways can be machined to accommodate almost any shape. If the form becomes very high, it may be necessary to add special inserts or to machine special clamps—again, a relatively simple operation. Compressible and stretchable materials present a different type of problem. If a material is so flexible that it is distorted and stretched merely by being held in the hands, it will cause considerable feeding problems. Nonetheless, this type of material must be fed, and probably the most practical solution is to bond the material, in a temporary manner, to some sort of carrier strip. This could be, perhaps, a paper carrier approximately the thickness of card stock (but, of course, in coil form).

Compressible materials are best handled by machining a cavity in a clamp slightly smaller in height than the thickness, and slightly larger than

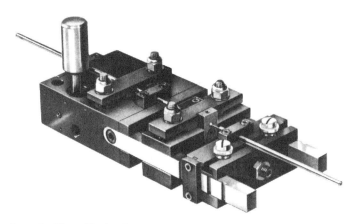

Fig. 1 Telescoping antibuckling guides for wire

the width, of the material. The amount of difference depends on each individual application, but the basic idea is to let the clamp bottom on the feed yet slightly compress the material in order to use the friction of the material to resist the feeding forces. The ease of doing this type of job depends on the resistance of the material to deformation. Proper and imaginative techniques can simplify a difficult job (Fig. 2).

If more than one strip of material is to be fed, it is necessary to grip each piece individually. Again, the air feed makes this a simple job. There are many ways of gripping each individual strip. Among the favorites are use of a flexible insert in the clamp and use of a spring pin to grip each strip (Fig. 3).

Feeds that accommodate strips cut to precut lengths are not as common in the job shop. This type of feed is generally available at a very high price for companies that specialize in high production of special parts. Typical of such parts are the tops for bottles. Usually the press and the feed are available as an integrated package. If the job shop is called upon to handle strips, it must use its ingenuity, and, because hand feeding is often the best method, it should not be overlooked. Encountered in feeding of strips is the problem of what to do with the material between the feed and the die. A simple solution—but one that is relatively expensive and not always applicable—is to use two feeds, one mounted on each side of the die. The strip, of course, must be longer than the distance between the two feeds. Other methods involve either automatic or hand placing of one strip behind the other, perhaps using tape to fasten the strips together as they are running automatically.

Air Feeds

Air feeds offer the utility of use with a wide variety of presses and machines. Their ease of installation and removal expands their versatility because they can be quickly changed from one die set to another. Often the actuating bracket can be left on the die and the feed can be moved from one job to another within a few minutes by removing two bolts and reinstalling the feed on another unit. The many available methods of actuation make air feeds quite simple to use on impact presses, electric presses, hydraulic presses, wire-forming machines, multislide machines, and special machines that bear little relationship to actual presses.

Mounting of Air Feeds. Air feeds typically have almost no mounting limitations. Although they are generally mounted either directly on the die set or on the bolster plate of a press, they can also be attached to the feed vertically, sideways, or upside-down. One common application is to mount two feeds on one die or special machine and process and assemble two different parts. Consider also the possibility of mounting an air feed on a movable platform such as the punch holder of a die set. In one application, two punch-press feeds were mounted one behind the other, thus giving the progression capabilities of two separate increments, depending on which unit was signaled (Fig. 4).

Air-Feed Actuation. There are two basic methods of actuation for air feeds: internal and external. Internal actuation, by means of the mechanical valve built into the feed, is normally used on average-stroke presses for normal conditions. It is the simplest and the preferred method of actuation.

If the stroke of the press becomes extremely short or long, or if special conditions exist, external actuation may be required. This consists of either electric or air actuation in which the mechanical valve is replaced with an electric or air valve mounted as close to the feed as possible. Of course, this external valve must be triggered by a microswitch or air pilot valve generally mounted so as to be operated by a rotary cam on the crankshaft or a linear cam on the ram of the

press. This method is most useful on special machines where a slide may not be adjacent to the feed, or in circumstances where long or short press strokes are being used. Although electric actuation is the most convenient method of external actuation, the cam-operated air valve offers a particular advantage on some special machines, especially where there is danger from electric arcs (Fig. 5).

Most air feeds allow some means of over-travel of the valve mechanism. Normally, the valve switches in the first fraction of an inch of travel, and over-travel is provided to ensure that the feed has time for retraction at the bottom of the stroke. The bottom portion of the stroke is, in most instances, useless for feeding purposes because the punches are perforating the material or forming the part. As the die opens, the feed can be triggered to push the material into the die. A general rule of thumb is to allow 50% of the cycle for feeding and 50% for retraction. This is best accomplished by using the bottom half of the press movement for retraction of the feed. The mechanical valve attached to an air feed accomplishes this condition automatically, but the wide range of variable conditions available with electric actuation calls for greater care in the selection of the actual feeding and retraction signals.

This great versatility in mounting and actuation allows automatic feeding in machinery not normally adaptable to the use of standard feeds.

Types of Jobs

The stamping job shop must have the flexibility to accommodate many different types of jobs. Job shops are called upon to run high-production and low-production jobs and must make a profit on both. In a typical application, two different parts are made from two pieces of material in a single die and are assembled in the final station. Two or more feeds, mounted either at opposite ends of the die or at right angles, are used. The simplified mounting of air feeds again aids in this type of installation. Often a very high-production machine will use literally dozens of air feeds.

Jobs Involving Rewinding. Often the job shop is faced with the realization that the

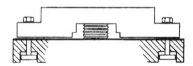

Fig. 2 Clamp design for compressible materials

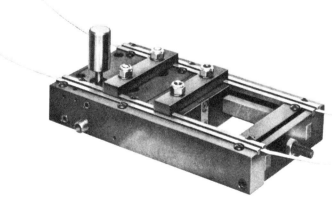

Fig. 3 Simultaneous feeding of two strips of material

Fig. 4 Rapid-air feed mounted on die set

completion of a part in a single die will be expensive for a short run. A solution is to run the material through one die, rewind it on a rewind reel, and then pass it through a second die for the secondary operation. Several good makes of rewind reels are available, but the major problem is generally the axial location of the material from one operation to another. The cumulative error will be enough to scrap the parts; therefore, a pilot is the only solution. The pilot can be placed either in the die or on one of the clamps. Again, the versatility of machining special shapes on the clamps of an air feed simplifies this otherwise rather difficult operation (Fig. 6).

Long progressions present a problem in that long-stroke feeds cost a great deal of money. A relatively simple solution for long progressions is to provide a multistroking device. At the end of the set number of progressions, the feed will stop and provide an electric signal for use in actuating the press. Although this approach is less expensive than buying an extremely long feed, it does have the disadvantage of reducing the available number of strokes per minute. A formula that has been found to be helpful for determining the speed of such a combination states that the number of strokes per minute is equal to $x/2N-1$, where x is the speed of the feed at the maximum progression and N is the number of progressions. For example, if the regular speed for a 300 mm (12 in.) progression is 100 strokes per minute, and it is desired to feed three 300 mm (12 in.) progressions for a total progression of 915 mm (36 in.), the approximate available speed is $100/2(3) - 1$, or 20 strokes per minute. At this speed, the feed is generally operating more slowly than the press, but the job can usually be done at an economic cost.

Another popular use of a counting device is to feed a given number of times before actuating

another secondary operation such as a separate cutoff. This is commonly done in the making of such parts as radiator fins. Instead of a counter, consider the use of a microswitch. This is another example of a simple solution to what could otherwise be a complicated problem.

Separate Progressions. Another somewhat related problem is feeding of two separate and distinct progressions, either alternately or in some sort of fixed sequence. This is difficult with regular hitch feeds or roll feeds, but with a little ingenuity, the air feed, with its reciprocating action between two fixed stops, can provide a solution. Either an air cylinder or a solenoid is used to insert a block of a fixed length between the regular stop on the feed and the sliding block. Generally, two progressions are required, but the solution could be expanded to accommodate a reasonable number of different progressions. The insertion of the fixed block must be timed to occur when the feed is in its forward position, and the actuating means must be fast enough to accomplish the job without interference. The control can be as simple as a push button operated by the press operator or can be electrically programmed. The choice again depends on the length of the job involved.

Continuously Adjustable Progressions. A situation rarely but sometimes encountered in the job shop is the problem of providing a continuously adjustable progression. This is relatively difficult. Although this procedure requires a great deal of work and diligence, it can result, if successful, in a great increase in production. Typical applications are the feeding of pre-printed paper forms to be cut out into individual pieces and the feeding of materials such as labels to be sewn into clothing. The technique begins with an electric eye that detects a light or dark spot on the part. The electric eye operates a

solenoid or air cylinder that adjusts the fine-adjusting screw on the feed to a progression either slightly longer or slightly shorter than necessary. For example, if the electric eye is set to feed long on white material, the overfeeding will eventually allow a control spot to work its way into the range of the electric eye. When the control spot reduces the reflection from the built-in lamp to a minimum level, the electric eye provides a different type of signal, which, through valving or solenoids, adjusts the progression of the feed to a shorter progression. The control spot will gradually recede from the range of the electric eye until the cycle again reverses. The degree of accuracy is relative to the type of electric eye and amplifying system used. This procedure is relatively complicated and, although not beyond the scope of the job shop, should be used only in high-production situations.

Accuracy Control Techniques

The two main reasons for using automatic press feeds are to improve production and to provide uniform, accurate progressions. In some cases, accuracy is not important, but in others it can be the entire reason for using automatic feeds. All press feeds have accuracy limitations. Some are more easily adapted to overcoming these limitations, but accuracy generally can be improved in any feed by using the same techniques.

The simplest accuracy control is a positive or dead stop. When an automatic feed with a positive stop is used, the progression of the feed should be set for a distance slightly greater than the actual progression. This will ensure that the

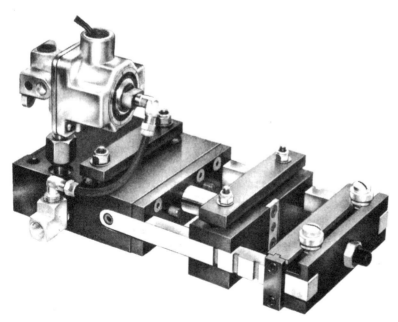

Fig. 5 Electric valve mounted on rapid-air feed

Fig. 6 Hydraulic payout and rewind reel

material always moves the required progression. In order to use this type of stop, the material must be cut off in the last station. The amount of material cut off controls the progression, and thus positive stops can be quite accurate on rigid material. However, this method is ineffective in controlling the progression of thin or compressible materials (Fig. 7).

The most common type of accuracy control is done with pilots. In this operation, a hole, generally round, is pierced in the material, and then, in a subsequent station, a punch with a rounded nose is inserted into the material, aligning it in the proper position. Pilots can be used for both axial and lateral control. In order to avoid uncorrectable axial errors, it is best for the pilot to engage the material in the station immediately following the piercing operation. When a pilot is used with an automatic feed, the usual suggestion is to feed the material slightly short and allow the pilot to pull it into the proper position. Thus, any feeding device should have the capability of pulling material through with just a slight amount of force. Again, pilots are effective on relatively rigid materials, but their performance drops considerably when they are used to feed thin or compressible materials (Fig. 8).

Disappearing stops are related to pilots in that they engage some edge or hole in the strip as it passes through a progressive die. They are generally less expensive than pilots, but do not offer the same accuracy control.

Perhaps the simplest means of controlling accuracy is to rely on the feed itself. Proper attention must be given to control of various factors that affect accuracy, but, with good conditions, repeatability much better than ± 0.025 mm (± 0.001 in.) can be achieved. We will now discuss various factors that must be checked in order to provide maximum accuracy.

The greatest accuracy is achieved when the feed is used only for feeding and is not made to serve some dual purpose. Most readers will be aware of the problems associated with non-powered straighteners, but little attention is generally given to the effects of using non-powered reels. The difficulty that arises when material is pulled directly off a reel comes from the inertia of the coil. At higher speeds, the jerk necessary to start the reel rotating is enough to cause either slipping or stretching of the material. For this reason, it is almost always best to pull the material from a free loop hanging between the feed and the reel or straightener. This loop should contain enough material for three or more progressions. Generally, automatic payoff devices are used to provide this loop, but in simple installations an operator can rotate the reel, or a weighted roller can be hung on the material between the feed and the reel (Fig. 9).

Another cause of apparent inaccuracy in feeding is the shear force generated at the cutoff station in a die. This shear force is generally not present when holes or forms are being pierced through the material but is always present when parts are cut off before some restraining punch or pilot has engaged the material. The feed must then resist this force, and although in many cases it can, the best solution—not only for the sake of the feed, but also for die life—is to provide a holddown clamp in the die that grips the material before the shearing action takes place. The heavier the material, the more pronounced is the shear force (Fig. 10).

In many forming dies, the material is pulled up and down as a part is being formed. This means that the material must be pulled slightly through the feed and then pushed backward. This situation should be avoided in die design, but if it is present, the feed must be mounted as far away from the die as practical so as to maintain the integrity of the geometry to an acceptable level. Somewhat related to this problem is the pulling of material as a part is being drawn. Of course, we are all familiar with the use of cutouts in a strip to reduce this problem (Fig. 11).

Mechanical interference often causes inaccuracy of an elusive nature. It is obvious that camber of the material will cause binding due to failure of the material to pass through the die itself. This type of problem can also occur as a result of improper alignment of the die and feed. Often this alignment is not square because slight differences in the shape or size of the material, or differences existing as the material passes through the feed, can sometimes cause the material to be pushed at a slight angle. The best solution is to observe the results in the die and adjust the feed accordingly. Although a powerful air feed can overcome some of these limitations, it is always advisable to provide the best conditions possible. The possibility of slugs in the die affecting accuracy should not be overlooked. Often a feed will push the slugs aside, but only after the restriction has affected the accuracy. Slugs can be particularly troublesome, and often very difficult to correct, depending on material characteristics. There are commercial devices available to aid in slug removal. Otherwise, the standard die-making techniques should be used to eliminate this problem (Fig. 12, 13).

Hitch feeds and air-operated punch-press feeds require that the impact at the end of the forward stroke be reduced to the minimum. The better air feeds have various devices such

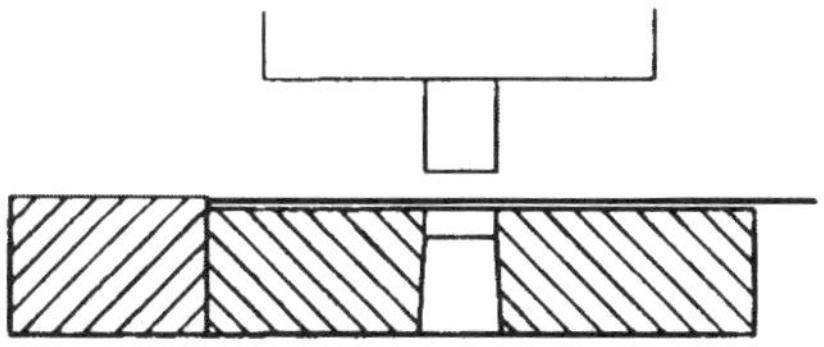

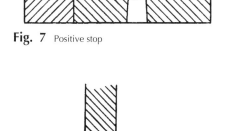

Fig. 7 Positive stop

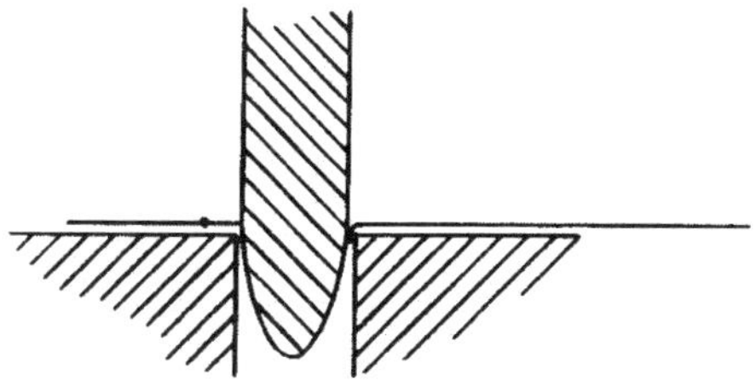

Fig. 9 Powered pedestal-type stock straightener

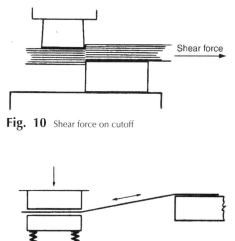

Fig. 10 Shear force on cutoff

Fig. 8 Distortion of thin material by pilot

Fig. 11 Pulling of stock during forming operation

as speed-control valves and cushion pistons designed into the unit. Impact-reducing materials such as rubber and urethane can also be used as cushions. Generally, these techniques are quite simple, but if they are not followed, the material will continue to move forward due to inertia after the feed has stopped. Inertial slippage often will be accentuated by oily material. Although this problem can be reduced by overpowering the system with heavy clamps, this procedure often leads to deformation of the material. A better solution is to provide a material with a high coefficient of friction. Leather is one of the better materials for clamp inserts because it is slightly compressible and will not mark the stock. It provides a high coefficient of friction even under oily conditions.

Some of the most difficult materials to feed by any method are stretchable materials. Often a double-feed combination is the simplest way to accomplish feeding of such materials. One feed is mounted on each side of the press, with the exit feed set at a slightly longer progression. Actuation, of course, must be simultaneous, and a mechanical fink often connects the sliding portions of each feed. Because stretchable materials do not lend themselves to feeding without buckling, special guides can be used. Consider also the possibility of using tension reels to keep the material taut as it passes through the working stations.

Special Speeds

High-speed feeding is popular because it is the most obvious and attractive way to reduce the costs of a part. Air feeds are capable of speeds up to approximately 500 strokes per minute at a 13 mm ($\frac{1}{2}$ in.) progression. Beyond this it is best to rely on the mechanical advantage of a roll feed or on a specialized high-speed feed. These units are usually expensive and should be considered as part of a total package with a good, well-built, high-speed press. The problems encountered with such units are different, and for high production their advantages can be rewarding.

Very low speeds are also sometimes required, typically for handling extremely delicate materials. Air feeds often are called upon to provide extremely slow, uniform feeding, such as feeding of solder into a furnace at a continuous, smooth rate. The air feed provides a very inexpensive solution to this problem. The low speed is usually controlled by a separate speed-control valve.

Air Circuits

Effective use of machinery requires that its operating principles and functions be thoroughly understood. Air feeds usually appear complicated but only because the valves and moving parts are inside the unit and thus not visible.

There are a few simple principles, the understanding of which will greatly increase one's ability to analyze problems and make effective use of air-feed equipment.

The air feed is divided into two separate air circuits. The first circuit is controlled by the main valve, which operates both sets of clamps. The stationary clamp is used to prevent the friction of the slide block from dragging the material backward as the feed retracts for a new progression. The movable clamp provides the greatest force for gripping and pushing the material into the die. The clamp portion of the circuit can be considered as follows. When the main control valve is depressed, the air is exhausted from the clamp circuit. As the pressure falls, the stationary clamp grips the material just before the movable clamp releases. As the main valve is allowed to come up, the action reverses and the movable clamp grips the material just before the stationary clamp opens. The clamps move in sequence by a balance between the areas as determined by the designer of the feed. If it is desired to add motion to this cycle, a pressure-sensing valve is added to the circuit. This auxiliary or pilot-operated valve allows a flow of air into or out of one side of the piston. The area of the pilot-operated valve is adjusted so that the feeding cycle takes place at the appropriate time; retraction occurs while the feed clamp is down. This general principle allows great flexibility in feeding because the two separate circuits ensure that the clamping cycle takes place before the feeding cycle. It also aids in separating the two air circuits so that the pressures necessary for clamping are not greatly reduced by the air flow necessary for feeding.

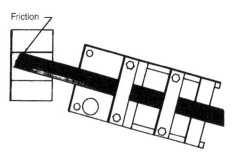

Fig. 12 Binding of stock between feed and die

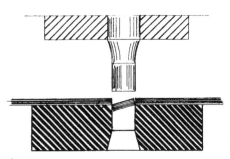

Fig. 13 Slug pickup

Maintenance and Life

Any piece of machinery must be properly maintained for maximum performance and long life. There is no reason why an air feed cannot provide hundreds of millions of cycles. In fact, many feeds provide more than 100 million cycles per year. Needless to say, lubrication and air supply are the key factors to long feed life.

Many shops have a problem with water in the air lines. This is caused by compression of relatively moist air and subsequent condensation of liquid when the air is cooled by being passed through the pipes to its point of use in the shop. Although many extensive systems are installed to eliminate water, generally for the job shop the simplest methods are the best. An after-cooler at the compressor will eliminate most of the water. There is a simple solution to this problem that should not be overlooked: turning the outlet tee in an upward direction and drawing the air from the top of the pipe (Fig. 14).

Most air feeds are built to operate for many millions of cycles before O-rings or seals require replacement. In order to promote the longest life for moving seals, a filter should be used to keep the air as clean as possible. The lubricator should be set to provide a small amount of light machine oil for lubrication. Under good conditions, an O-ring life of at least ten million cycles can be expected. Under excellent conditions, O-ring life is considerably better.

There is no need to use "fancy" oils. One of the least-expensive oils will work best. It should be remembered that motor oils are designed for automobiles and spindle oils are designed for precision spindles. Air feeds use inexpensive No. 10-weight hydraulic or machine oil. A periodic maintenance schedule will provide optimum service. Manufacturers' suggestions

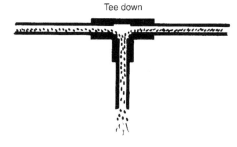

Tee down

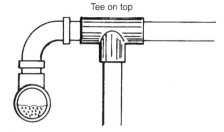

Tee on top

Fig. 14 Air lines with tee down and tee on top

should be consulted for best results from maintenance schedules.

General Job-Shop Problems

The job shop is called upon to make many types of dies. Some shops specialize in high production and others in very short runs. Therefore, the problems of the job shop vary so much that it is hard to specify general rules. There are, however, some universal suggestions that come up often enough to warrant comment—especially because these problems apply to automatic feeds of many different types.

In the straight cutoff-type die, the shear force can push the material backward. In fact, in hand feeding of heavier-gage materials, a considerable shock can be felt in the operator's arm and hand. Pressure pads should be used in the die to prevent this problem.

In dies where bending is performed, careful attention must be given to preventing projections in the strip from catching on cavities on the die. This will lead directly to buckling problems in the die and inaccuracy in feeding. Workers should be cautious of the friction generated by tight bends around forming mandrels. Feeding equipment is usually strong enough to overcome this type of friction, but, depending on the size of the unit and the nature of the stock, inaccuracies can result.

In forming and drawing dies, it is best that the level of material remain constant. Often this is difficult, and the best rule in such cases is to mount the feed as far away from the die as possible. It should be noted that this is exactly opposite to the normal recommendation.

A type of die not to be overlooked in today's production shop is the transfer die. This die is helpful in forming parts around arbors or in performing multiple bends. The part is generally carried in a strip to the next-to-last station, where

a bend is made and the part cut off. The part is then pushed by an air cylinder to the final station, where final forming or assembly is performed.

In any job-shop operation, the length of the run and the setup are of primary importance. It would be convenient if dies and feeds could be mounted on presses permanently. Unfortunately, this would be rather expensive because most jobs are simply not that long. Thus, setup time becomes extremely critical, and air feeds can serve a very useful function. Most air feeds can be completely set up in a matter of minutes and can be moved from press to press or die to die with very little difficulty.

ACKNOWLEDGMENT

This article was adapted from the chapter "Press Feeders for the Job Shop" by O.D. Lascoe in *Handbook of Fabrication Processes,* ASM International, 1988.

Multiple-Slide Machines and Tooling

THE MULTIPLE-SLIDE MACHINE, sometimes called a four-way, four-slide, or multislide machine, is a somewhat specialized item of stamping equipment, although it is very versatile within a limited area of stamping applications. From the design standpoint, a multiple-slide machine may be described as a horizontal stamping press using a system of cam-controlled tools for producing small parts from coil stock.

Numerous small parts are produced by these machines. In addition to stampings from sheet metal, these machines are used to fabricate parts from wire. Brackets, clips, electrical connectors, and clamps are typical of the types of stampings made on such machines.

Machine Construction

The design of a multiple-slide machine is a radical departure from the normal concept of pressworking mechanisms. Nevertheless, the basic principle used in shaping and cutting of stampings is the same as that of any conventional press. Moving slides, fitted with suitable tools, strike the metal while it is supported on tools equivalent to dies, forcing it to assume the desired shape. The base of the multiple-slide machine is essentially a large steel table. A hole is provided near the center of the surface through which completed parts are discharged.

The driving mechanism of the machine consists of an electric motor that is linked by gears or belts to the shafts that operate the tools. The shafting arrangement is the most unusual characteristic of the machine. It consists of four shafts that completely surround the bed of the machine to form a rectangle. The shafts are connected at all four corners of this rectangle by gears. Power is applied at one point of the shaft arrangement and, due to the gearing, gives motion to all four shafts. The slides of the press are connected to cams mounted on the shafts. Split-type cams are often used for this purpose to facilitate conversion to various jobs.

Through this shafting arrangement, slides can be introduced to the die area from all four sides of the press. In addition, several slides may be attached to each shaft, increasing the flexibility of the machine and the number of operations that may be performed. It should be noted that the motion of the slides is in the horizontal plane, another departure from the conventional method of pressworking.

Another important component of the multiple-slide machine is the piercing head. Die sets for piercing, cutting, trimming, blanking, or embossing are mounted in this head. The die components are mounted on edge on a solid backup plate. The punch plate retains the punch components and is mounted on a slide driven by cams on the front shaft. Stock is fed from the left side of the press. Sometimes, two or more piercing heads may be mounted on a single press. The slides for these heads may be mounted on either the front or rear shaft, providing flexibility in tooling. For instance, a burr resulting from the piercing operation may be located on the desired side of the part by locating the tooling on the proper ram. The piercing head is actuated by solid-type cams for reasons of strength. In addition, the cams used in this application are not changed as frequently as are other cams.

After being formed, the part is stripped from the forming tools or the post and dropped from the press.

Advantages of Multiple-Slide Machines

Multiple-slide machines perform in a manner similar to presses equipped for progressive or transfer operations—that is, a number of separate operations, usually all the operations necessary to complete the part, are accomplished in a single machine. In fact, due to its method of operation, this machine may perform the full sequence of operations on a part that could not be completed in a single conventional press, even though transfer or progressive dies were used. This minimizes the number of presses needed to produce a stamping and provides the added advantages of reduced handling costs and conservation of floor space.

The primary advantage of four-slide machines is their versatility, meaningful because it can result in reduced unit part cost. Rapid changes can be made in the combination of press and forming operations, and production can be shifted quickly from one part to another. Forming operations are more flexible than similar operations on power stamping presses because the stroke, shape, and timing of each forming tool can be adjusted independently. For example, changes in the design of a strip part can often be accomplished by single tool adjustments. In a stamping press, an entirely new die may be needed.

In general, press parts require relatively little forming. The parts are quite simple and do not necessarily have a strip shape. Four-slide parts are more complex, involve a higher degree of forming, and usually have the basic strip shape integrated in their design. Despite these differences, four-slide machines and high-speed stamping presses have some common production characteristics:

- High volume
- Low-cost materials (standard mill product strip and wire stock)
- Predictable material strength characteristics
- Uniformity of parts
- Excellent surface finish (surface appearance of original stock is retained)

Other comparisons of four-slide operations with press operations are described next.

Production Speed. Four-slide machines have a wide range of output rates in parts per hour. The production rate depends on the material and the forming operations. Presses are capable of output rates as high as those of four-slides on simple parts, but parts that require complex forming can be produced at higher speeds on four-slides.

Tooling Cost. Because the tool motions in a four-slide are already built into the machine, toolmaking is confined to shaping of the working surfaces of cams, center forms, and slide tools. In progressive-die forming, the entire cam-motion system must be designed and produced. Families of parts, instead of requiring a different die for each size, sometimes need only one or two additional tools; the rest is accomplished with tool and feed adjustments.

Advantages with respect to tooling cost may also be realized through use of multiple-slide machines. The tooling used is generally simpler in construction than the average comparable die sets used in conventional presses. Piercing, shearing, and forming units are of simple design and are relatively inexpensive to build.

Die details are mostly standard parts and can often be made at the plant where the machine is located. Forming units are fastened to the cam-driven rams by simple standard methods. The maintenance cost of this type of tooling is relatively low.

Tool Adjustments. Because the forming motions for a part are built into a fixed set of dies and cams, the initial setup time for equivalent parts is usually greater for a four-slide. However, the time required for tool adjustment and maintenance during production runs is less on four-slides.

Combined Operations. Four-slides perform tapping, welding, and other secondary operations that must be done separately in presswork. Four-slides can also interlock sections of a single part and assemble two or more parts.

Heavy Press Operations. Four-slides cannot do heavy coining or swaging, nor can they blank parts with a large number of die stations or high-tonnage requirements.

Deep Drawing. Draws exceeding 3.2 mm (1/8 in.) are difficult to do on four-slides because of limitations on the lifter motions needed to clear the draws from the die.

Operating Cost. The per-piece cost of operating a multiple-slide machine is comparatively low. This is a direct result of a high rate of production. As many as 20,000 to 70,000 stampings can be made in a 16 hour shift on the fastest models, depending on the length of the stamping and the number of stampings produced per stroke.

Factors in Selecting Multiple-Slide Machines

Multiple-slide machines can be applied to the production of numerous types of small parts. However, there are certain general limitations to observe in assigning work to this type of machine. These limitations are related to part size, design, and production volume.

Size Limitations. Dimensional specifications for stampings produced on multiple-slide machines must be held within the capabilities of the machines themselves. As indicated previously, these machines are applied to the production of relatively small stampings. The largest machines are capable of handling stock up to a maximum of 75 mm (3 in.) in width. The longest feed length possible is 320 mm (12 1/2 in.). Stock thicknesses up to 2.5 mm (3/32 in.) can be handled. This range makes it possible for the largest machines to handle practically any thickness of stock used in automotive body or chassis stampings.

Smaller machines are limited in application to even smaller parts because the opening in the piercing head, which varies proportionately with press size, governs the width of material that may be processed. The smallest machine can accommodate a maximum stock width of 20 mm (3/4 in.) and a feed length of 75 mm (3 in.). Maximum stock thickness on smaller machines is 0.8 mm (1/32 in.).

The stroke of the piercing-head ram varies from 9.5 mm (3/8 in.) on the smallest machine to 19 mm (3/4 in.) on the largest. The throw of the forming cams varies from 22 to 55 mm (7/8 to 2 in.).

Design Limitations. The range of stampings that may be made on multiple-slide machines is somewhat limited in respect to part design. Basically, the use of these machines is limited to parts whose shapes can be achieved by shearing operations and pure bending operations. This confines the use of multiple-slide machines to types of operations such as piercing, notching, slitting, cutting and straight flanging, bending, and sizing. Extrusions are generally impractical because a solid stripper is used in the piercing head. An extrusion usually would prevent or interfere with the movement of the stamping to succeeding stations. Any forming in the piercing head must be done in a manner that will not hinder movement of the part.

Drawing or stretching operations are considered unfeasible both from the standpoint of tonnage requirements and the operational limits of the die setups that may be used. The press and tooling designs do not permit the use of air cushions or blankholding mechanisms required for these operations.

Volume Requirement. The volume requirement for a part is a further consideration in contemplating the use of a multiple-slide machine. Such a machine is a high-volume piece of equipment and as such can be used most profitably for high-volume operations. To avoid excessive downtime and loss of production in changeovers, parts not qualifying as high-production items should generally be produced by some other method.

Tool Planning and Design

The principles followed in planning and designing tools for multiple-slide machines are similar to those used in layout of progressive dies for conventional presses. A strip development is made in which the operational elements are divided and set up in the most effective sequence. This indicates the die stations that must be built and their relative locations (Fig. 1).

Sectional Construction. In the construction of the dies, it is general practice to employ sectional construction. This is done to minimize maintenance problems and reduce downtime. Because a number of slides can be put into operation, the workload can be spread over a considerable distance. When numerous piercing or cutting operations are involved, several piercing heads may be employed. This permits spacing of the work to avoid crowding delicate punches together. It also reduces the necessity of building a long, heavy, and complicated progressive die. Further, versatility and changeability of the tooling setup is increased where multiple piercing heads are used. Making changes in any stage of the die is easier because it is unnecessary to remove and disassemble a large, complex progressive die.

Component Strength. Provisions should be made in designing multiple-slide dies for adequate strength and wearing qualities in all components. Punches, die inserts, and forming tools should be made of materials capable of withstanding rapid repetitive shock. Special hardened inserts should be supplied in all areas where tools are subject to exceptional wearing stresses. Rigid tool mountings are necessary to prevent deflection under load.

Similarities to Presses. Four-slide machines are similar to power presses in many ways. Both have feeding mechanisms that receive

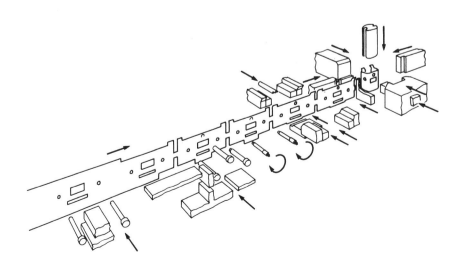

Fig. 1 Hypothetical multislide application showing wide variety of functions that can be performed in production of complete stampings directly from coil stock

strip or wire material from a continuous coil. The material is fed into a die area in which press operations are performed in sequence at one or more stations. There, the similarities end.

Conventional power presses have tool motions in only one direction so that cam systems must be devised to form most complex parts. These systems are limited in what they can accomplish, and parts must be transferred from one machine or area to another for different operations. On the other hand, standard four-slides provide six or more forming slide motions within the forming area. These include four or more motions in the forming plane and two or more motions in a plane perpendicular to the forming plane.

A four-slide machine can perform most press operations within its strip-size capacity. Basic metal-processing operations that can be performed in a four-slide press section are: piercing, notching, coining, swaging, extruding, lancing, embossing, forming, beading, dimpling, stamping, and stippling. These press operations are completed before the forming operation.

Other Operations. Besides performing press operations prior to work the four-slide area, the machine can do other work before and after the forming steps—for instance, small brackets with three holes, one of which is threaded. All three holes are pierced, and one of them extruded, in the first die station in the press area of the four-slide. The extruded hole is then tapped in the press area.

Hose clamps can be formed from strip material that is first stamped to size in the press area. It is then cut off and formed around the center form so that the ends butt at the lugs. The part is pushed farther out on the center form into a welding station where the ends are pressed together and butt welded. By doing the welding independently of forming—so that forming and welding are performed concurrently—the production-cycle time is only that of the slower operation and not the total of the two.

In fabricating a small electrical part, a flat contact strip is dimpled and a hole pierced in the press area. After the strip has been formed on the center form, a threaded stud is fed from a magazine and dropped into the pierced hole. A staking head, located in the position of the right-side tool, finishes the cycle by staking the stud in place.

Materials

The limitations on making parts in four-slide machines are usually related to materials properties, not machine capabilities. The mechanical properties of common strip materials that affect their suitability for four-slide production are tensile strength, yield strength, hardness, and ductility.

Obviously, tensile strength and hardness must be taken into account when planning for production by almost any mechanical method.

However, yield strength and ductility, by themselves, do not directly determine four-slide producibility. Instead, "relative formability," in terms of bending characteristics and forming radii, is used.

Five families of strip materials commonly used in four-slide production are:

- Low-carbon, cold-rolled strip steels
- Spheroidize-annealed, cold-rolled spring steel
- Types 300 and 400 stainless spring steels
- Copper alloys
- Beryllium-copper alloys

High-Formability Steels. Low-carbon, cold-rolled strip steel and spheroidize-annealed spring steel are the most common high-formability materials used in four-slide production (see the article "Forming of Steel Strip in Multiple-Slide Machines" in this Volume). Cold-rolled strip steel (available in five tempers) is the basic ferrous material for four-slide parts other than springs. AISI 1075 spheroidize-annealed spring steel is the main material for spring parts, while AISI 1095 is used for spring parts in thinner sections. These two materials are hardened and tempered after forming to attain desired spring properties.

Tempered Spring Steels. The behavior of tempered spring steels in forming is difficult to predict. To control the mechanical properties, the strip mill should be required to fulfill two of three critical specifications: hardness, tensile strength, and minimum inside-bend radius without fracture. In addition, the allowable limits of reproducibility of mechanical properties should be established with the mill.

Stainless Steels. Type 302 is the most common stainless steel specified for four-slide parts. Other stainless steels that may be used (austenitic types, such as the 200 and 300 series) have mechanical properties similar to those of type 302. Series 400 stainless steels have properties similar to those of AISI 1075 spheroidize-annealed cold-rolled spring steel.

Copper-alloy and beryllium-copper strip are commonly formed on multiple-slide machines. There may be large differences in the hardness and tensile strengths of various alloys, even when they have roughly equivalent formability. It is thus important to evaluate relative formability in terms of bending characteristics and forming radii, even though a wide selection of mechanical properties may be obtained in nonferrous alloys. For example, beryllium-copper alloys are available in various treated conditions or mill-hardened conditions.

Designing Four-Slide Parts

Forming of parts from metal strip on four-slide machines (Fig. 2) is a fast, low-cost production method. Often, parts made in multiple operations in power presses could be produced at much lower cost in four-slides. These machines automatically produce metal parts from wire or strip by the combined action of a power-press section for stamping and cam-actuated sliding tools for forming. With a change in either the tools or the machine adjustments, a wide variety of forms and parts can be produced.

Tolerances

Thickness. If close dimensional tolerances are expected, parts should be designed with mill dimensional standards in mind. Standard thickness tolerances for tempered and untempered cold-rolled carbon spring steel are the AISI mill standards. These standards cover approximately three-quarters of the four-slide parts that are made. Closer thickness tolerances on cold-rolled carbon spring steel are now available from specialty mills.

Edges and Width. Strip edges specified for four-slide strip parts are by far the most common in four-slide parts:

- A prepared edge of a specified contour (round, square, or beveled), which is produced when a very accurate width is required, or when the finish of the edge suitable for electroplating is required, or both
- An approximately square edge produced by slitting
- An approximately square edge produced by rolling or filing for the purpose of eliminating the slitting burr

Because most four-slide strip parts use the full width of the strip, AISI width tolerances on the raw strip material are also significant.

Camber. "Deviation of a side edge from a straight line" is the definition of camber. Standard AISI camber tolerances are 5 mm/m ($1/2$ in./8 ft) for widths between 13 and 38 mm ($1/2$ and $1 1/2$ in.), 2.5 mm/m ($1/4$ in./8 ft) for widths over 38 mm ($1 1/2$ in.).

Finishes

While many types of finishes can be produced in cold-rolled carbon steel strip, three are most commonly used:

- *Dull finish:* A finish without lustre, produced by rolling on rolls roughened by mechanical or chemical means. This finish is especially suitable for lacquer or paint adhesion and is beneficial in aiding drawing operations by reducing the contact friction between the die and the strip.
- *Regular bright finish:* A finish produced by rolling on rolls having a moderately smooth finish. It is suitable for many requirements but not generally applicable to plating.
- *Best bright finish:* Generally of high lustre produced by selective rolling practices, including the use of specially prepared rolls. This is the highest-quality finish produced and is particularly well-suited for electroplating.

Common finishes on hardened-and-tempered cold-rolled carbon spring steel include black, scaleless, bright, polished, and polished-and-colored (blue or straw) finishes.

Design Hints

Design recommendations for producing metal parts in four-slides and power presses are similar to the extent that they are determined by materials characteristics or by part shape. However, there are also special differences.

The following design recommendations will give optimal part quality at maximum production speed.

External Contours. Design recommendations for external contours are (see Fig. 3):

- Lugs or ears (Fig. 3a) should be formed with their bend lines at an angle equal to or greater than 45° to the rolling direction.

- Components requiring rounded ends (Fig. 3b) should have a radius equal to or greater than $3/4 W$, except when otherwise indicated.
- A rounded end with a radius equal to $1/2 W$ may be used if a relief angle of 10° or greater at the point of tangency with the part edge is also used (Fig. 3c).
- Corners along the stock edge should be as near square as practical (Fig. 3d).
- The side that is to be free from burrs should be specified (Fig. 3e).
- All notches (Fig. 3f) should extend inside the stock edge at least $1^1/_2 T$, but not less than 0.5 mm (0.020 in.).
- Tapers should be recessed at least $1T$ from the edge of the part (Fig. 3g).
- Parts should have straight edges on the flat blanks wherever possible (Fig. 3h).
- To form a square corner, the minimum bend allowance should be $1^1/_2 T$ or $R + 1/2 T$ if the corner is on a tapered end, and $2T$ or $R + T$ if on a square end (Fig. 3i).
- Relief slots for tabs, and short flanges whose edges are flush with the external blank outline, should have a depth of at least T plus the bend radius (Fig. 3j).
- When flanges extend over a portion of the part, a notch or circular hole should be used to eliminate tearing. Notch depth should be T plus bend radius (Fig. 3k). A hole should have a diameter of $3T$.

Internal Contours. Design recommendations for internal contours are (see Fig. 4):

- A radius of at least $1/2 T$ should be used at the intersections of all edges that do not lie along the stock edge and should be specified as maximum if radius is not critical (Fig. 4a).
- Punched (pierced) holes should have a diameter of at least $1T$ (Fig. 4b), but not less than 0.76 mm (0.030 in.).
- The minimum distance from the edge of any round, punched hole to the blank edge should be at least $1T$ (Fig. 4c), but not less than 0.76 mm (0.030 in.).
- Edges of adjacent punched holes should be a minimum of $2T$ apart (Fig. 4d), but not less than 1.52 mm (0.060 in.).
- Extruded holes should have a minimum spacing of $6T$ between their edges, and should be at least $4T$ from the blank edge. Depth of the extrusion should be a maximum of 30% of its outside diameter (Fig. 4e).
- Holes should be spaced a minimum of $1^1/_2 T$ from bend tangent lines (Fig. 4f).
- Threaded screw or bolt holes should be at least $1^1/_2$ times the screw diameter from the centerline of the hole to the edge of the part (Fig. 4g).
- Slots that are parallel to the bend should be a minimum of $4T$ from bend tangent lines (Fig. 4h).
- Beads should be a maximum of $2T$ high and have a minimum inner radius of $1T$ (Fig. 4i).

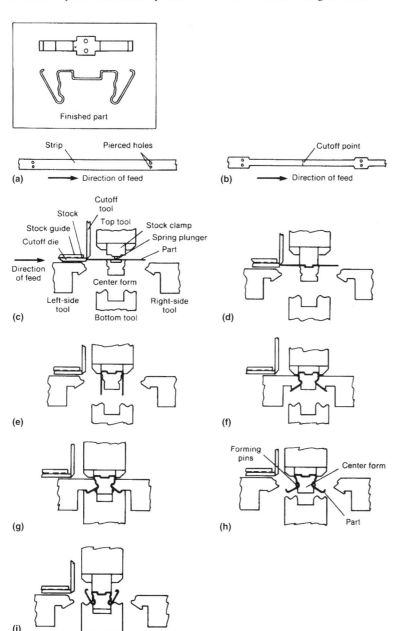

Fig. 2 Forming operations in a four-slide machine. (a) Power-press tools pierce two holes into strip material at first station. (b) Tools at next press station notch strip on both edges. (c) Notched-and-pierced strip is fed in from press area. As strip motion stops, stock clamp descends so that spring plunger holds strip against top of center form. (d) Cutoff tool cuts part off against cutoff die. At the same time, stock clamp descends farther to form workpiece into shaped cavity at top of center form. (e) As top tool descends, its right- and left-hand lobes bend ends of part straight down and hold them there. (f) Left- and right-side tools advance to form ends of part against shaped surfaces on sides of center form. (g) Bottom tool moves up to form tips of part against lower surfaces of left- and right-side tools. (h) Side and bottom tools retract. At the same time, two forming pins move out horizontally (toward the reader) outside the part. (i) Bottom tool advances again and bends ends of part upward around forming pins. Top and bottom tools withdraw vertically, while center form and forming pins withdraw horizontally into machine bed. Finished part falls out of forming area as next section of strip is fed in.

Tabs and Slots. The following recommendations are made for design of tabs and slots (see Fig. 5):

- Widths of tabs and slots should be a minimum of $1\frac{1}{2}T$, or 0.5 mm (0.020 in.). Their length should be a maximum of 5 times their width (Fig. 5a).
- Triangular tabs or slots should have minimum end radii of $1T$ and should form included angles of 60° or more (Fig. 5b).
- Adjacent tabs or slots should be spaced at least $2T$ or 0.76 mm (0.030 in.) apart (Fig. 5c).
- Internal slots should be at least $1\frac{1}{2}T$ or 0.76 mm (0.030 in.) from the edge of the stock (Fig. 5d).

Lugs, Bridges, and Curls. Design recommendations for lugs, bridges, and curls are as follows (see Fig. 6):

- Lanced lugs should have tapered sides (Fig. 6a), except when otherwise indicated.

- A taper is not required when the lug margin is pierced with a clearance prior to forming (Fig. 6b).
- Bridges should have a maximum ratio of lanced length to bridge height of 4 to 1 (Fig. 6c).
- Curls should have inside diameters of $2\frac{1}{2}$ to $8T$ (Fig. 6d).

Cutoff Methods

The four-slide machine is properly credited with built-in versatility in both its press and forming operations. The ability of the four-slide to produce small metal parts—complex or simple, to precise or commercial tolerances—in one machine cycle at high speeds is due also to the flexibility of the cutoff operation. Although the function of the cutoff is simply to separate continuous strip or wire into individual parts, the

ends of the parts can also be shaped and even formed, as needed, at the same time.

The various four-slide cutoff methods are illustrated in Fig. 7 as performed on a vertical-slide (verti-slide) machine. With the exception of the rear-motion cutoff in Fig. 7f, however, all the methods described here apply to any type of four-slide machine. In addition to those shown, there are many other techniques that are being applied to cutting off material in four-slides—and more will undoubtedly be developed in the future by ingenious tool engineers.

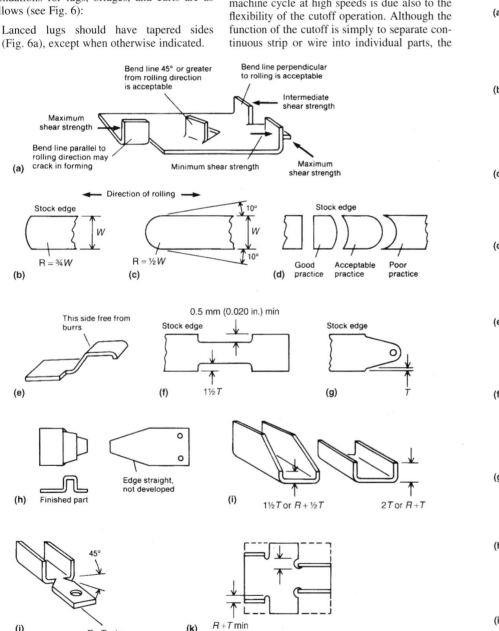

Fig. 3 External contours. *W*, part width; *T*, stock thickness. See text for details.

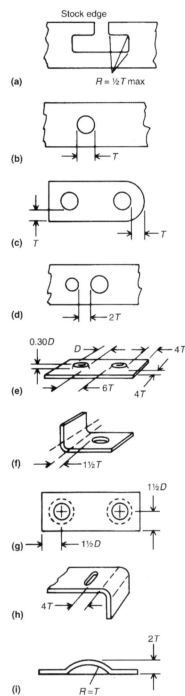

Fig. 4 Internal contours. *T*, stock thickness. See text for details.

The Standard Strip Cutoff Method. A standard cutoff head with a standard cam is usually used with conventional tooling. A straight cutoff (Fig. 7a) is made with a standard cutoff punch and standard die block.

With Limited Space. A straight cutoff can also be made with the standard cutoff die block, but with the cutoff punch or blade mounted on the side of the top forming tool (Fig. 7b). This

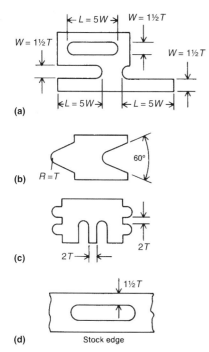

Fig. 5 Tabs and slots. *T*, stock thickness. See text for details.

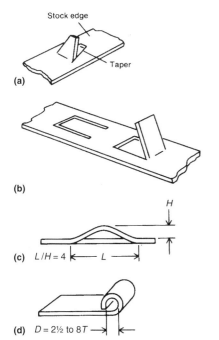

Fig. 6 Lugs, bridges, and curls. *T*, stock thickness. See text for details.

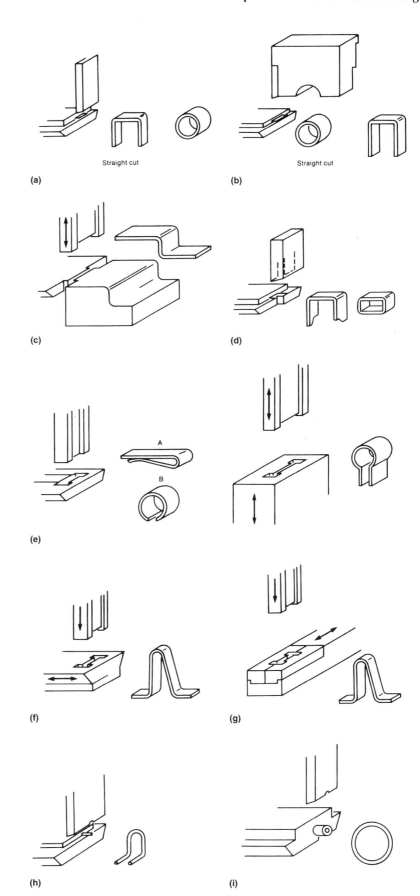

Fig. 7 Four-slide cutoff methods. (a) Standard strip method. (b) Limited space. (c) Stationary forming arbor. (d) Mating or matching cutoff. (e) Double cutoffs. (f) Long legs. (g) Rear-motion cutoff. (h) Wire cutoff. (i) Quill-type cutoff

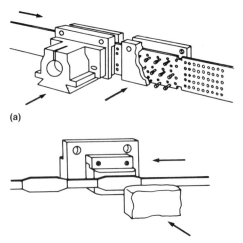

(a)

(b)

Fig. 8 Multislide auxiliary equipment: heavy-duty die heads and toggle press. (a) Heavy-duty die heads. (b) Toggle press, which provides high tonnage at the bottom of the stroke for heavy swaging, coining, or embossing

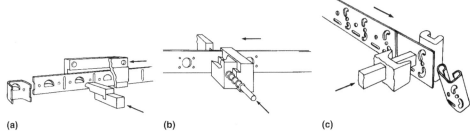

(a) **(b)** **(c)**

Fig. 9 Multislide auxiliary equipment: rear auxiliary slide, rear-position knockout, and positive blankholder. (a) Rear auxiliary slide, for operations requiring motion in direction opposite die head. During die-head dwell, this unit allows a variety of operations, including shearing, forming, extruding, preforming, and lancing. (b) Rear-position knockout, employed to provide positive cam-actuated control, in place of die spring pads, for return of material flush with die surface. (c) Positive blankholder, used in the front forming-tool position to prevent slippage or bulging during cutoff and forming, and providing the ability to pilot and position the blank against the form

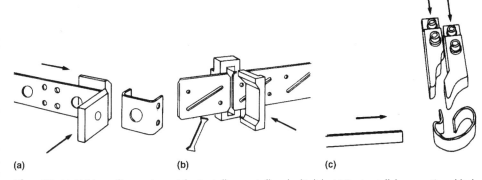

(a) **(b)** **(c)**

Fig. 10 Multislide auxiliary equipment: front cutoff, rear cutoff, and split slide. (a) Front cutoff, for separating a blank from the strip prior to forming when scrap is not removed between blanks. Mounted between dies and forming position. Individual timing of cutoff permits separation at instant blank is held by forming tools or positive blankholder. (b) Rear cutoff, for separating blanks by removal of slugs of straight or shaped configurations. Small back-plate is furnished for cutoff-die mounting, allowing slug-removal-type parting when left-hand forming slide is used. (c) Split slide. A split front, right-hand, or rear slide can be furnished to operate in the standard slide opening, permitting tooling double movements for versatility in forming, locating, and assembly operations.

method is often used when the distance between cutoff die and center form is small.

With Stationary Forming Arbor. The standard cutoff attachment can be combined with a stationary forming arbor (Fig. 7c). The left-hand tail or leg remains in place until forming by the other forming slides has been completed.

For Mating or Matching Cutoffs. Using a standard cutoff attachment, the cutoff die block and cutoff blade can be designed to produce any desired mating or matching cutoff shapes (Fig. 7d).

Double Cutoffs. Unmatched cutoff shapes can be formed with a double cutoff die (Fig. 7e, left).

Another type of double cutoff (Fig. 7e, right) employs the standard cutoff attachment together with the bottom slide. This design usually is used only when the bottom tool is not required for forming operations. If the bottom tool is also needed for forming, special cams can be used.

With Long Legs. Wherever long legs must be formed down past a double cutoff die, a sliding die may be mounted in the standard cutoff attachment (Fig. 7f). This sliding die can be actuated by the left slide cam or by use of a specially designed cam motion. Again, note that independent end shapes are produced.

Rear-Motion Cutoff. The patented rear-motion cutoff (Fig. 7g) allows complete freedom of the verti-slide forming slides for more intricate forming operations.

After the strip has been fed all the way in over the center form, the rear-motion slide containing the cutoff die moves in under the strip. The cutoff punch moves down to cut the strip and then returns to its retracted position. The die then withdraws into the tooling plate, allowing the end of the cutoff piece to be formed downward.

Wire Cutoffs. A typical wire cutoff (Fig. 7h) uses standard cutoff components with regrinding for the particular wire diameter. A cutoff blade

on the side of the top tool may be used. A quill-type cutoff is also available (Fig. 7i).

Multislide Auxiliary Equipment. Four-slide forming machines accept wire or coil stock directly from a reel, straighten, feed, and cut off the required length. Figures 8, 9, and 10 illustrate auxiliary equipment used in typical operations performed by these machines.

ACKNOWLEDGMENT

This article was adapted from the chapter "Multiple-Slide Machines and Tooling" by O.D. Lascoe in *Handbook of Fabrication Processes,* ASM International, 1988.

Tooling and Lubrication for Forming of Sheet, Strip, and Plate

Wear and Lubrication of Sheet-Metal Forming Dies

SHEET METAL WORKING consists of blanking or cutting, forming of shallow parts, and deep drawing. Wear, lubrication, and selection of tool materials used for blanking dies and cutting tools are described in the article "Selection of Materials for Shearing, Blanking, and Piercing Tools" in this Volume. This article introduces the process factors that influence die wear and lubrication for metal forming operations such as bending, spinning, stretching, deep drawing, and ironing. Further details on these processes are provided in separate articles in the Section "Forming Processes for Sheet, Strip, and Plate" in this Volume.

In general, the amount of wear of dies during sheet metal forming is proportional to the distance the sheet metal slides over the die for a given pressure between the surfaces in contact. Thin annealed sheet metal exerts the least pressure and thus causes the least wear; thick hardened sheet metal causes the most rapid wear. However, the rate of wear for different combinations of die/sheet metal pair may vary considerably, depending on material surface characteristics, the speed of forming, and the type of lubrication. Die wear also depends on process characteristics. For example, high localized pressures develop on the tools that produce wrinkles, which then may produce prohibitively high rates of wear and galling in the ironing stage. Wear characteristics also differ for shallow forming and deep drawing, because these two operations impose different stress fields and wear profiles on the forming dies. These general factors are discussed in this article, with some recommendations concerning die materials selection based on the expected die lives. Further information on die materials selection is given in the articles "Selection of Materials for Press-Forming Dies" and "Selection of Materials for Deep-Drawing Dies" in this Volume.

Wear in Shallow Forming Dies

To understand wear in shallow forming dies, it is essential to study the mechanics of the process based on the true shape of the parts to be drawn. For this purpose, six sample part shapes are included in Fig. 1, ranging from the simple shapes, such as those of parts 1 and 2, to moderately severe shapes, such as those of parts 5 and 6.

Recommended die materials for these shapes range from plastics for low-quantity production of simple-to-moderate parts, up to the most wear-resistant (nitrided) tool steels for making severely formed parts. Parts of even greater severity, or those run in quantities larger than one million, may require dies or inserts of cemented carbide.

Effect of Part Shape. The tooling for parts 1 and 2 consists of a punch and an upper and lower die. Because there is little deformation of the sheet metal during forming of such simple parts, little or no sliding of the sheet over the lower die, and little movement over the punch, these components have very low wear rates.

Tooling for part 3 consists of a punch and lower die. In forming, the punch pushes the blank through the lower die, which results in sliding on the lower die, but there is little sliding on the punch. Therefore, the punch generally has ten times the life of the lower die made from the same material. However, wear and galling are expected in the areas of moderate shrinkage of this part, particularly when the part is formed on these single-action dies.

Tooling for part 4 consists of a punch and upper and lower die. Without the upper die, excessive wrinkling would be expected at the shrink flanges. As in part 3, a less wear-resistant material is required for the punch and upper die than for the lower die due to greater sliding occurring on the lower die.

The tooling for parts 5 and 6 consists of a punch and lower die with no upper die required because these parts are produced by stretching rather than by shrinking of the metal along the constricted region. The metal envelops the punch with minimal sliding and consequently produces approximately ten times more wear on the lower die than on the punch. However, the same material can be used for both the punch and

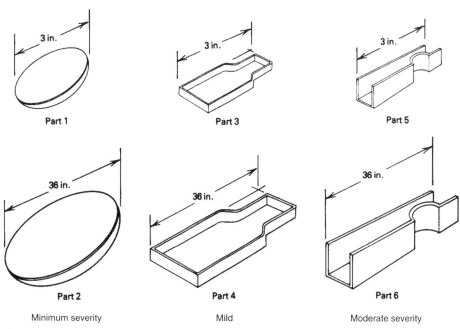

Fig. 1 Parts classified into six classes based on severity of draw

lower die for part 5 because of the smaller part size and minor material cost. In the selection of die material for part 6, the critical locations are the wearing edges of the lower die. The body of the lower die could be made of cast iron with wearing edges of tool steel, and the punch could be of a material less wear resistant than tool steel, for example, low-cost alloy cast iron.

Effect of Sheet Thickness. Thick sheets of any metal will exert greater pressure on the dies than thin sheets of the same metal. Therefore, the effect of galling and wear increases as parts are made from thicker sheets, especially parts with shrink flanges. Consequently, for high-production dies working under severe wear conditions and for production of parts to close tolerances, it is often desirable to use an extremely wear-resistant material such as sintered carbide or nitrided D2 tool steel.

Effect of Tolerance Requirements. Tolerance requirements of the part affect the selection of tool material if the part is to be finished without redrawing. When the part is to be redrawn, the material used in the redrawing die is subjected to less wear than the die that performs the primary operation. A major factor in the choice between a wear-resistant material and a less costly and less wear-resistant material is the necessity for maintenance during the production run. If the production run is large and the tolerance requirements tight, costly wear-resistant materials or coated punches have to be considered.

Effect of Sheet Metal. Sheet stock of a higher hardness will usually wear dies more rapidly, but other factors, such as the presence of scale on the surface of unpickled hot rolled steels, are sometimes of greater importance. However, scales reduce galling, which, on tool materials, may be an even more serious condition than abrasive wear and can result in frequent interruptions for reconditioning the die.

Soft brass or aluminum stock causes less wear and galling than carbon steel, whereas stainless steels and heat-resistant alloys cause more wear and galling. Possible surface treatments include chromium plating of any hardened steel, hardening of alloy cast iron, and nitriding of tool steels such as A2 and D2.

Effect of Lubrication. In low- or medium-production runs, lubricants are often used; in zinc dies, they are a necessity. However, the most effective lubricants are difficult to apply and remove. Efficient application of lubricants is particularly difficult in high-production operations where presses are being fed automatically. In such operations, die materials that are costlier but more resistant to galling (for example, aluminum bronze, nitrided D2 tool steel, and sintered carbides) should be used.

Prevention of Galling. Galling is primarily due to large stretching of sheet metal, poor tool fitting, and rough finish on the surface of tools. For short or medium runs, surface-hardened hot rolled steel dies can be used that are less costly than tool steel dies. Tool steels are recommended for severe reductions or for forming metals that

show a greater tendency to gall, such as austenitic stainless steel.

If galling is encountered, the tool fit and thickness of the stock metal should be checked first to determine whether clearances are adequate. Ironing out wrinkles causes galling. Wrinkles can be prevented by redesign of the tools.

Surface nitriding of dies made from alloy steel or alloy tool steel such as A2 or D2 minimizes galling. However, in die steels containing no nitride-forming elements such as chromium or molybdenum, nitrided surfaces may spall on radii smaller than approximately 3 mm ($^1/_8$ in.), and plated surfaces may spall on radii less than 6 mm ($^1/_4$ in.). Hard chromium plating will usually eliminate galling of mild steel, alloy steel, and tool steel dies; however, to avoid cracking of the hard coating in operations involving high local pressures, hardened steels such as alloy or tool steels should be used as the substrate. Dirt, grit, and shot fragments on the sheet cause greater damage to nitrided and chromium-plated tools than to hardened tool steels; hard particles may cause minute spalling and small pits, especially at radii in areas of high forming pressure.

Galling is less likely if the die and stock materials are dissimilar in hardness, chemical composition, and surface characteristics (for example, aluminum bronze tools for forming carbon steel and stainless tools for aluminum parts).

Galling is also related to the thermal softening resistance of the die material. Galling and pickup are caused by the extremely high frictional heat generated between the die material and the steel being formed, which results in the softening of the die material and increased pickup.

On the basis of tests conducted by an automobile manufacturer, it was concluded that (Ref 1):

- Metal pickup on the faces of flanging dies is due to the frictional heat generated by two metals rubbing together under high pressure.
- Excessive heat is generated if the length of contact is longer than needed to form the flange.
- Correct die lubricants should be chosen. For example, when heavy gages are being formed, an extreme-pressure die lubricant should be used.

Wear and galling are less severe with soft stock material, such as aluminum and copper alloys, than with low-carbon steel; they are more severe with high-strength metals such as stainless steels and heat-resistant alloys.

Die Wear and Die Life Studies. The important factors influencing the wear and life of sheet-metal forming dies are the composition and hardness of the die material and the part, and the thickness, shape, and quantity of the parts to be formed. Effects of these factors are shown graphically in Fig. 2. The data for type 4140 steel show the effect of die hardness: Its die life increases approximately 800% when nitrided

(Fig. 2a). Effects of the tool material composition and hardness are seen in the forming of the label holder (Fig. 2b) and in the forming of the bushing (Fig. 2c). M4 and D7 at 65 HRC gave more than twice the die life of D2 (61 HRC) in flanging the shredder ring of 17-7 PH stainless steel shown in Fig. 2(d). The plots in Fig. 2(e) show the influence of die and stock material on wear of dies for making parts of similar shape. The wear rate gradually decreases for nitrided steels due to the wearing away of the outer layer (white layer).

Wear in Deep-Drawing Dies

This section deals with the wear of material for dies to draw round and square cup-shaped metal parts in a press, primarily by conventional drawing in which each reduction is made in the same direction. Deep drawing is associated with extension and bending of the sheet over the punch nose, together with the gathering and compression of the sheet metal as it is drawn through the die. For economy of manufacture, the drawn part is always produced in the least number of steps using the least number of dies possible. Ironing is used almost universally in multioperation drawing and to increase uniformity of wall thickness. Each operation is designed for maximum practical reduction of the metal being drawn; the examples given in this section involve maximum reductions of approximately 35%.

The performance of a deep-drawing die is determined by the amount of wear or galling during a production run. Wear of a given die is determined largely by the material and thickness of the sheet steel, sharpness of die radii, lubrication, die design, and finish. The amount of wear on die radii can vary by as much as a factor of 20 between the sharpest and most liberal radii.

In drawing square cups, the formation of wrinkles at the corners accompanied by high localized pressures may produce prohibitively high rates of wear. Small corner radii will result in greater wear and shorter life of the dies.

High localized pressures caused by inaccurate fitting of dies during tryout, rough surface on the drawn sheet or on the working surfaces of the die, inadequate lubrication, and poor maintenance (stoning) of dies are typical of uneconomical practice.

Effect of the Thickness of Sheet Metal. In drawing thick sheets of a given metal, the pressure on the dies increases in proportion to the square of the sheet thickness. The die pressure is higher on the draw radius, and increasing sheet thickness will localize the wear in this high-pressure area without a similar effect on other surfaces of the die.

Thick stock will wrinkle less than thin stock; therefore, the pressures required to prevent wrinkling are less than those required for thin stock. However, heavy sheets are often drawn without a pressure pad. Therefore, the wear-resistance requirements for pressure pads used

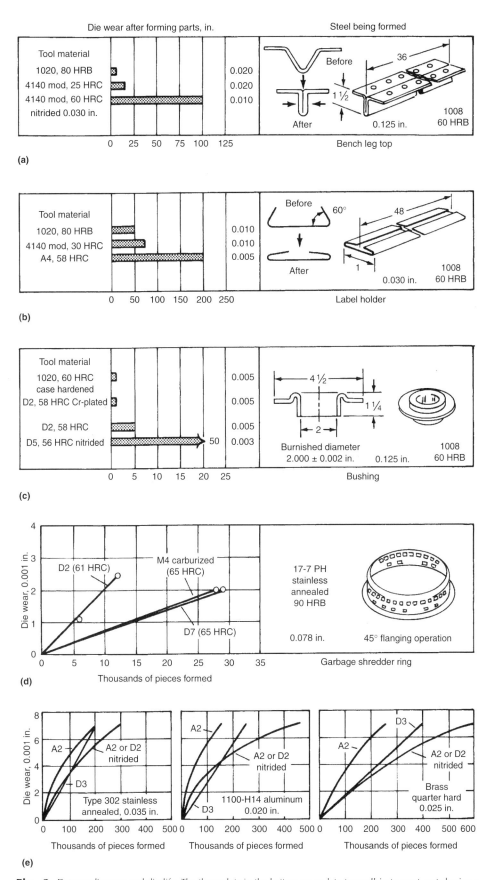

Fig. 2 Data on die wear and die life. The three plots in the bottom row relate to small instrument parts having a maximum area of 19 cm² (3 in.²). Source: Ref 1

on heavy stock are no greater than those for thin material. Alloy cast iron is often flame hardened for long runs and thick sheets; for small quantities and thin sheets, it may be used as-cast. For the mildest applications, unalloyed gray iron may be used.

Prevention of Galling. Common causes of galling are stretching sheet metal beyond practical limits, tool fitting with poor alignment or insufficient die clearance for the given sheet thickness, wrinkles, the use of galling-susceptible tool steel, and rough finish on the surface of tools.

Die materials to resist galling can be selected on the basis of their resistance to softening under heat. For example, T15 steel is the die material that has the highest resistance to softening and thus the best galling resistance; W1 steel has the poorest resistance to softening and galling.

For parts drawn of carbon steel or nonferrous alloy sheet, the die material can be selected without regard to galling; as a finishing operation, the punch and die should be either nitrided or chromium plated. If chromium- and molybdenum-containing tool steels such as A2, D2, D3, or D4 have been selected, the smoothly ground tools should be nitrided and polished or buffed after nitriding. Otherwise, the tools should be hardened to at least 60 HRC, smoothly finished on the wearing surfaces, hard chromium plated, and the plating polished or buffed. Punches should be plated to a thickness of 0.005 to 0.01 mm (0.0002 to 0.0004 in.). To prevent spalling or flaking of the plate, dies should not be plated more than 0.005 mm (0.0002 in.) thick.

When the parts are drawn from stainless steel or from high-nickel alloy steel, the draw ring material with best resistance to galling is aluminum bronze. The second choice is either D2, D3, or D4 (smoothly ground, nitrided, and polished). The third choice is alloy cast iron (quenched and tempered to 400 to 420 HB).

Punch material is selected without regard to galling and is then chromium plated (as with die materials), unless cast iron is chosen.

Sintered carbide, which is only as efficient as the lubricants used, has proved economical for nonferrous alloys, carbon steel, and stainless steel in many long, continuous runs; however, galling will occur as soon as lubrication becomes faulty.

Effect of Lubrication. Correct lubrication of the parts being drawn is essential to reduce friction, wear, and galling. In fact, deep drawing is impossible without lubrication. In actual practice, die materials are selected after trials employing the production lubricants. If excessive wear or galling occurs, a better lubricant is usually applied. For extremely difficult draws, the best lubricants are usually applied at the outset. The effect of process conditions and lubrication is discussed in more detail in the next section, "Lubrication and Wear Process Conditions," in this article. Lubricants are discussed in detail in the article "Selection and Use of Lubricants in Forming of Sheet Metal" in this Volume.

Chromium Plating to Reduce Wear. Chromium plating is used on tool steel draw rings to improve life. On punches, its primary function is to reduce the frictional forces and facilitate removal of the parts from the punch after the sidewalls have been ironed tight to the punch. Usually, chromium plating improves punch life somewhat less than it would be by the use of the next-best tool steel.

For successful performance of tools, chromium plating must always be deposited on a surface harder than 50 HRC, and the thickness of the coating should preferably be 0.005 to 0.01 mm (0.0002 to 0.0004 in.) and never less than 0.0025 mm (0.0001 in.). This gives the required hardness and reduction of friction without excessive spalling or chipping at corners. Chromium-plated dies should be heated to 150 to 205 °C (300 to 400 °F) for a minimum of 3 h immediately after plating to minimize the possibility of hydrogen embrittlement.

The main drawback is environmental, due to the poisonous nature of chromic acid. However, there are several suppliers that have developed the technique further with new additives and closed-circuit systems in order to take care of the waste products. The result is a more environmentally acceptable process as well as the possibility to produce crack-free coatings with superior corrosion-resistant behavior compared to conventional coatings (Ref 2). The price for this type of treatment will increase because of all the environmental issues.

The deposition temperature is approximately 50 °C (120 °F), and the hydrogen outgasing is normally performed at a temperature below or at approximately 200 °C (390 °F).

The deposition rate is typically between 5 and 20 μm/h (195 and 785 μin./h), which results in cycle times of normally less than 1 to 2 h.

Lubrication and Wear Process Conditions

Sheet metal working processes present a wide range of demands on lubricants, from the almost negligible to the most severe. In these processes, as in other non-steady-state processes, whatever lubricant is introduced at the beginning must suffice for the entire course of deformation. Therefore, lubricant breakdown, tool pickup, and scoring of the workpiece are of major concern. For a given process geometry, conditions generally become more severe with increasing sheet thickness.

An important point to note is that in most sheet metal working processes, yielding occurs in a predominantly tensile stress state. Typical conditions range from balanced biaxial tension (point 3 in Fig. 3), through plane-strain tension (point 4 in Fig. 3), to a combination of tension and compression (moving toward point 5 in Fig. 3). The compressive stress often originates in the geometry of the process, and, with the exception of ironing, die pressures remain low, well below the flow stress (σ_f). This means that

contact is limited to asperities, and the Coulomb law of friction holds. Asperities yield at relatively low pressures when the substrate deforms in tension; thus, the real contact area and, with it, frictional stresses increase rapidly, resulting in a relatively high coefficient of friction. Correspondingly, friction plays an often decisive role in the success of the operation, which is measured by the absence of unacceptable necking or fracture of the workpiece. In contrast to bulk deformation, the magnitudes of pressures and forces, and of the effects exerted on them by friction, are of lesser concern.

In terms of lubrication during press forming, the primary function of lubrication is to prevent galling (tool material pickup). A secondary function is to reduce die wear through abrasive and chemical wear mechanisms. Light press forming of most metals requires only light mineral oil or soap lubrication to prevent die wear. More severe conditions, experienced when forming copper-, iron-, and nickel-base alloys, usually require liquid lubrication using either mineral oils with extreme-pressure additives (such as sulfur and chlorine) or zinc stearate soaps in order to prevent galling. Solid-film lubricants in the form of polyethylene, polyvinyl chloride, and polytetrafluoroethylene, which are applied to the workpiece as a powder, coating, or solid sheet, also give adequate protection against adhesive wear. Extreme conditions encountered when press forming iron- and nickel-base alloys require a phosphate-coated product and soap lubrication to prevent excessive die wear. Tin-coated steel, with additional liquid lubrication of mineral oil or emulsions to cool the die, has also been successful in preventing die pickup in severe press-forming applications. For more information on lubricants, see the article "Selection and Use of Lubricants in Forming of Sheet Metal" in this Volume.

Bending. In the basic process of bending, a part is deformed between a punch and a die. The

sheet gradually conforms to the punch nose, at low pressures and with very little sliding; friction effects are negligible here. Meanwhile, the sheet slides over the die radius, and a frictional force is generated. Because interface pressures are low—well below the flow stress of the workpiece material—contact is limited to asperities, and the coefficient of friction can be quite high. If the die/workpiece combination is prone to adhesion, pickup develops. If the workpiece is covered with an abrasive oxide, the die radius wears. In either case, the surface of the part is scored, die wear can become a problem, and lubrication is then essential.

Bending along a straight line is often performed with a steel punch and an elastomeric (polyurethane) pad. The elastomer imposes tensile stresses on the sheet, somewhat as in stretch forming. The stresses are transmitted by a frictional force that can be modeled by a coefficient of friction, the value of which depends on the surface condition of the blank and on the properties of the elastomer. Higher friction reduces springback.

Long lengths of shapes, such as tubes, corrugated sheet, and so on, are produced in large quantities by roll forming. Just as in shape rolling, the workpiece moves at a given constant speed while local velocities around the die profile vary, giving rise to relative sliding. Even though pressures are only fractions of those that occur in shape rolling, adhesive or abrasive wear, tool pickup, and workpiece scoring may be encountered. Lubrication is then helpful, although some friction must be maintained to prevent uncontrolled movement of the strip.

Bending of sections and tubes is widely practiced, and friction on the wiping elements of bending tools requires the use of lubricants. The difficulty of lubricating the inner surfaces of tubes calls for careful selection of the components. In radial bend-drawing of stainless steel, the wiper die and the flexible ball mandrel are made of

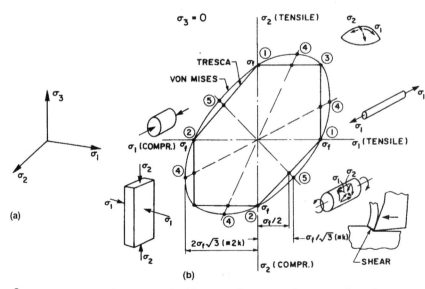

Fig. 3 Directions of principal stresses (a) and yield criteria with some typical stress states (b). σ_f, flow stress

aluminum bronze, and a viscous, compounded lubricant is applied that reduces mandrel tension and minimizes thinning on the tension side.

Spinning is an incremental forming process in which a circular blank is rotated and held between a male die (spinning block) and the tail stock in a machine tool resembling a lathe. The blank is pressed with a hand- or power-actuated tool so that it gradually conforms to the shape of the die (Fig. 4a). Localized pressures on the tool are high, and there is intensive sliding; thus, lubrication of the outer blank surface is essential.

In shear spinning, the wall of a previously formed cup or vessel is reduced. Strictly speaking, this is an incremental bulk deformation process with high surface pressures, and lubrication of the tool is essential. Friction problems are partially circumvented by the use of roller tools (Fig. 4b). Friction on the die (mandrel) hinders the free extension of the wall, but elastic springback helps to relieve the pressures and reduces friction.

In the presence of a liquid lubricant, lubrication is of the mixed-film type. A rougher surface is formed with a more viscous lubricant, higher spinning speeds, or larger rollers. However, it is essential that boundary agents be present, too. In shear spinning of steel, workpieces have been coated with fat and with an emulsion applied during spinning to extract the generated heat.

Stretching. In pure stretching, the sheet is firmly clamped along its circumference while a male die (punch) deforms it (Fig. 5). The shape is developed entirely at the expense of sheet thickness. Therefore, necking and, finally, fracture of the sheet must eventually occur. The depth of draw before fracture and the location of the fracture point are functions of friction and of materials properties.

Thinning of the sheet takes place against frictional restraint over the punch surface. In the absence of friction, a condition that can be obtained by use of a hydraulic fluid as the punch, the sheet thins out gradually toward the apex, and fracture finally occurs in balanced biaxial tension. In the presence of friction, free thinning on the punch nose is hindered, the strain distribution becomes more localized (Fig. 5b), and the fracture point moves toward the die radius, where plane-strain conditions prevail. The total depth, obtainable before fracture sets in, diminishes. A higher strain-hardening exponent (n-value) has the same effect as reduced friction, whereas an increasing strain ratio (r) value has the opposite effect.

At the start of deformation there is normal approach between punch and sheet, and, in the presence of a liquid lubricant, a squeeze film develops. Contact is made at the highest point; in stretching over a hemispherical punch, there is no sliding at this point (Fig. 6). Away from it, sliding distance and velocity increase at a rate proportional to the distance, if friction is low. With high friction, sliding is arrested. Interface pressures are low, because the membrane stress is typically on the order of 0.01 to 0.1 UTS (ultimate tensile strength). Nevertheless, surface deformation is substantial.

Deep Drawing. Tribological conditions vary greatly at various points of a partially drawn cup (Fig. 7). Conditions on the punch bottom

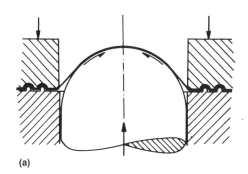

(a)

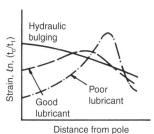

(b)

Fig. 5 Schematic illustration of (a) the limiting dome height test and (b) strain distributions obtained under different lubrication conditions

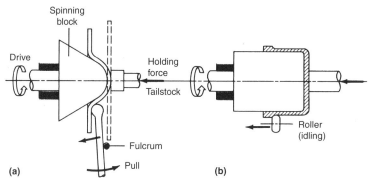

Fig. 4 The processes of (a) spinning and (b) shear spinning

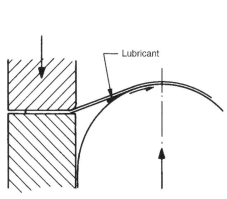

Fig. 6 Negative wedge developing in stretch drawing

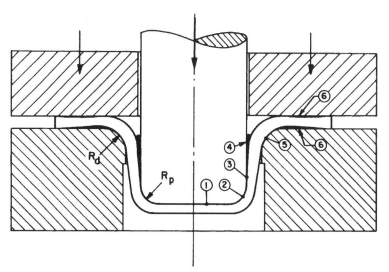

Fig. 7 Zones of differing lubrication conditions in deep drawing (see text for identification)

(zone 1) are the same as in stretching, except that friction around the radius (zone 2) limits the transmission of stresses, and interface pressures may drop close to zero. The sheet roughness is virtually unchanged. On the punch radius (R_p), the stresses are higher, some relative sliding takes place, and the lubricant thins out.

Between the punch flank and the partially drawn cup (zone 3), the lubricant is trapped in a squeeze film, whereas it is drawn in under favorable conditions at the point of transition (zone 4). Pressures are low and can easily drop to zero if elastic springback occurs, or if the clearance is large enough to allow the cup to separate from the punch in the early stages of drawing. The surface roughens as in free deformation, in combined tension/compression, and the presence of a lubricant makes little difference.

Around the die radius (R_d) (zone 5), pressures are somewhat higher—by analogy to a frictionless pulley, on the order of (t/R_d) UTS. Because $R_d = 5t$ to $8t$, the pressure is approximately 0.2 UTS, which is still quite a small value. Sliding conditions are, however, severe. The speed is equal to the drawing speed, and the surface, deformed by circumferential compression and bending and unbending, becomes progressively rougher. There is danger of exposing virgin surfaces, and tool pickup is observed only on the die radius, if at all.

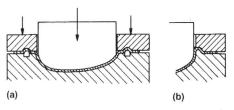

(a)   **(b)**

Fig. 8 Combined stretching/drawing of an irregular part with (a) draw bead and (b) edge bead

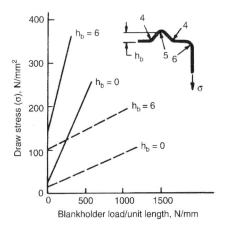

Fig. 9 Draw stresses developed in drawing of 0.9 mm (0.035 in.) steel sheet with a low-viscosity rust-prevention oil ($\mu = 0.15$, solid lines) and with a high-viscosity drawing oil ($\mu = 0.035$, broken lines). μ, coefficient of friction

In the blankholder zone (zone 6), pressures are low—typically 0.002 to 0.02 UTS. A liquid lubricant is trapped by a squeeze-film effect. Thickening of the flange creates a wedge that contributes to lubrication by hydrodynamic action. In materials of planar anisotropy, there is more thickening in the low r-value direction, and thus the film is thicker in the high r-value direction. The surface roughens under the imposed tensile/compressive stresses at a higher rate than in pure tension, but roughness is still proportional to strain if the effects of anisotropy are considered.

When the clearance is such that ironing takes place at the rim, the surface becomes smoother, and roughness depends on the thickness and nature of the lubricant film.

A sheet of directional surface finish or metallurgical structure exhibits visible parabolic markings, which are accentuated in the presence of a liquid lubricant.

Combined Stretching and Drawing. Many parts are made by pressworking processes that incorporate elements of both stretching and drawing. The shapes of these parts range from simple rectangular containers to complex, irregular shapes of automotive components. At the corners, the stress and strain states are similar to those observed in deep drawing, and the wall height is built up by circumferential compression, for which the high r-value direction in a sheet with planar anisotropy is favorable. At the sides, the wall is formed by draw-in of the sheet in plane strain, and a high n-value is favorable. The base of the part is frequently domed (Fig. 8), and therefore, contact between punch and sheet is localized and a large area of the sheet is unsupported, resulting in puckering. Another problem is springback, which distorts the shape of gently curving parts.

Higher pressure on a flat blankholder increases the resistance to draw-in. However, more positive control is obtained with draw beads (Fig. 8a), which function by repeated bending and unbending of the sheet. In principle, they are effective even without friction. The restraining force increases with decreasing bead radius, and a round-cornered square is more effective than a fully radiused bead. Draw beads are usually made as inserts and are more readily adjustable than edge beads (Fig. 8b). Friction on the bend radii, on the blankholder surface, and on the die radius are important contributors to the draw force (Fig. 9). With proper lubrication, the draw beads change the strain history in various parts of the drawn part and through this contribute to an increased limiting depth. Draw beads present the dangers of pickup and sheet damage. Draw beads made of polyurethane are less effective but minimize scoring.

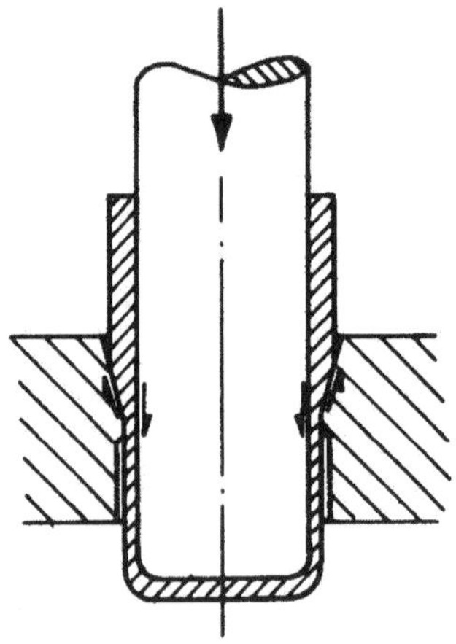

Combined stretching and drawing offers a wide scope for differential lubrication. When the drawing component dominates, use of a dry punch and a lubricated blankholder zone (combined with a high r-value) results in the deepest parts. Overall lubrication provides only a small improvement when stretching is dominant; a high n-value is much more helpful. Punch friction is still important, though. High friction minimizes thinning, whereas low friction promotes deformation in a curved bottom and thus minimizes springback. The punch radius must be lubricated sufficiently to allow plastic deformation over the entire base (for which the stress must exceed the initial flow stress or yield strength) without exceeding the strength of the wall (for which stresses must remain below the UTS).

Ironing and Necking. In ironing, the wall of a cup is reduced over its entire height without materially changing its inner diameter (Fig. 10). Problems in ironing are similar to those encountered in tube drawing on a bar, except that the part is shorter and the lubricant finds relatively easy access to the inside. Interface pressures are on the order of σ_f, and sliding velocity on the die is equal to the punch velocity. Sliding velocities on the punch are much lower.

Ironing can be applied simultaneously with drawing but is generally a separate operation. Long employed for the manufacture of cartridge cases and similar products, it has acquired great significance with the spread of two-piece beverage cans that are drawn and then ironed, usually in three stages, to a much-reduced wall thickness. Substantial expansion of the surface demands a lubricant capable of following surface extension. Friction transfers stresses to the punch and increases the attainable reduction (Fig. 11, Ref 3).

Theoretically, very high reductions can be taken if friction on the punch is high. This conclusion is also confirmed in practice. Reductions close to 60% are feasible with a lubricated punch, and reductions up to 70% can be attained with a dry punch, provided that the die angle is small enough. Even though the frictional force on the die increases with a decrease in half angle, so does the frictional force on the punch. Reductions in excess of 90% are possible in a single pass when the die half angle is less than 6°.

Fig. 10 Ironing of a cup (half arrows indicate directions of frictional stresses acting on cup)

A neutral point develops inside the ironing zone, and the material can emerge at a speed higher than that of the punch. A limit is set by the difficulty of stripping the ironed can, especially when it is a thin-wall can, as in two-piece canmaking. To prevent collapse of the can, the punch is usually smooth and is lubricated together with the die.

Processes of great practical importance are necking (nosing) and expansion of can or tube ends. The maximum necking ratio (ratio of diameters before and after necking) is limited by development of circumferential folds, or by axial collapse or upsetting of the wall. The latter two failure modes are especially sensitive to both friction and die angle. For a given angle, the tube may actually become longer with lower friction, as shown in (Ref 4). The maximum necking ratio is greater for a $30°$ than for a $60°$ included die angle.

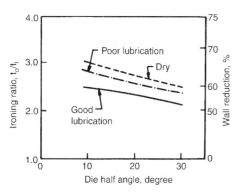

Fig. 11 Effect of lubrication on reduction obtainable in ironing with dies of different half angles.
Source: Ref 3

ACKNOWLEDGMENTS

Portions of this article were adapted from:

- R. Shivpuri and S.L. Semiatin, Friction and Wear of Dies and Die Materials, *Friction, Lubrication, and Wear Technology*, Vol 18, *ASM Handbook,* ASM International, 1992, p 621–648
- J. Schey, Sheet Metalworking, *Tribology in Metalworking,* American Society for Metals, 1983

REFERENCES

1. V.A. Kortesoja, Ed., *Properties and Selection of Tool Materials,* American Society for Metals, 1975, p 131–170
2. G. Håkansson, Experiences of Surface Coated Tools, *Proceedings of the International Conference on Recent Advances in Manufacture and Use of Tools and Dies and Stamping of Steel Sheets,* N. Asnafi, Ed., Volvo, 2004, p 245–254
3. K. Osakada, S. Fujii, R. Narutaki, and S. Sakakura, in *Proc. 18th Int. MTDR Conf.,* Macmillan, London, 1978, p 137–144
4. G.D. Lahoti and T. Altan, in *Proc. Sixth NAMRC,* Society of Manufacturing Engineers, 1978, p 151–157

Selection and Use of Lubricants in Forming of Sheet Metal

LUBRICANTS AND LUBRICATION are relatively low-cost components of a metal-forming system, yet lubricants are important and often indispensable for efficient forming of quality parts. The primary function of lubrication in press forming is to prevent galling (tool material pickup). A secondary function is to reduce die wear through abrasive and chemical wear mechanisms, as briefly discussed in the article "Wear and Lubrication of Sheet-Metal Forming Dies."

This article provides an overview of the interfacial interactions with a lubricant film between a die and metal, lubricant mechanisms, chemistry, qualification testing, application methods, and property test methods. Much of the discussion is relevant to metal-forming operations in general, although it is targeted at sheet-metal forming operations. The final sections deal with lubricant selection as influenced by the metal to be formed and particular sheet-metal forming operations. Some aspects of microbiology and toxicity also are discussed, as are the increasingly important health and economic implications of laws and regulations dealing with metalworking lubrication. Economic and environmental factors are a driver in the development tool coatings that reduce or eliminate the need for lubricants.

Surfaces

In general, the specific characteristics of the surfaces of both the sheet and the dies make up the environment in which sheet metal lubrication is considered. As such they provide the boundaries between which the lubricant must operate. These surfaces are complex. The chemistry of the surface is not the same as the bulk chemistry of the sheet or tools. The lubricant interacts with an oxide layer of varying complexity; various contaminants may infiltrate the oxide layer (oil, gases, acids, rust preventives, and so forth); and the surface may be carburized or decarburized, nitrided or subjected to some other surface treatment, or coated with a polymer or phosphate. Many other surface chemistry variations are possible. Furthermore, the surface may have residual compressive or tensile stresses from prior processing. The fine structure of the metal surface may also vary with prior history and be quite different from that of the bulk material. The surface roughness also varies in many geometrically significant ways. Interposed between these nonhomogeneous surfaces is a lubricating film that may be quite complex in structure, activity, and wetting ability.

The complexity of the lubricant/metal interface fundamentally affects the lubricant performance by preventing galling or wear of the tooling, as well as tearing, scratching, or imperfect forming to desired dimensions of the sheet metal.

Lubrication Mechanisms

In sheet-metal forming, several different lubrication regimes have been identified. More than one of these regimes may occur during sheet-metal forming. Indeed, a single regime often cannot be effective in providing the necessary film integrity between tool and workpiece. The specific regimes that are operational depend on the severity, temperature, and geometry of the deformation mode. Detailed information on lubricating mechanisms is available in Ref 1 and 2.

Thick-film (hydrodynamic) lubrication (Fig. 1a) is the occurrence of a thick film of lubricant between tool and workpiece that completely prevents metal-to-metal contact. In hydrodynamic lubrication, a thick lubricant film is generated by the tangential and normal relative motion between two surfaces. In this regime, the bulk properties of the lubricant (viscosity) and the mechanical system operating during deformation create the necessary conditions. The effectiveness of such lubrication depends directly on relative speed and lubricant viscosity. For very slow contacts, a thick film is not likely to develop unless hydrostatic lubrication (i.e., an externally pressurized lubricant) is introduced into the lubricant film.

Lubricant film thickness is important in hydrodynamic lubrication to prevent possible solid-to-solid contact. It is usually determined from solutions of the flow continuity equation, which calculates the lubricant pressure for a known film thickness. Charts and computer software are available for determining the lubricant film thickness for many common geometrical configurations. Lubricant film thickness increases with the sum of the two surface velocities (v), lubricant dynamic viscosity (η), and bearing size, and decreases with load.

Thin-Film (Quasi-Hydrodynamic) Lubrication. The film between tool and workpiece is thinner, and some metal-to-metal contact takes place (Fig. 1b). In this case, the load is shared by the hydrodynamic (or fluid) pressure and the asperity contact pressure. For this reason, thin-film lubrication is sometimes referred to as mixed lubrication, or partial hydrodynamic lubrication. It occurs in bar and wire drawing and perhaps less frequently in sheet-metal forming. Nonetheless, in most press working, relative sliding speeds are

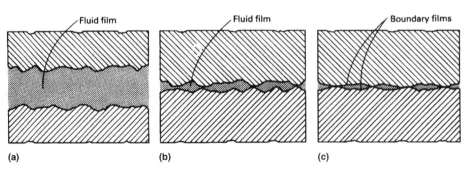

Fig. 1 Three lubrication regimes experienced in sheet metal forming. (a) Thick-film lubrication. (b) Thin-film lubrication. (c) Boundary lubrication

too low and the process geometry is not favorable enough to maintain a full fluid film (Ref 1). This is why solid-film lubricants are often preferred to maintain a full fluid film. Most liquid lubricants operate in the mixed-film regime.

Boundary Lubrication. Physical adherence of the lubricant to the surface occurs, and relatively thin lubricant films may be effective (Fig. 1c). Dependence on viscosity is lower, and chemisorption (chemical adherence of the lubricant to the metal surface) becomes more important. The adherence and strength of the adsorbed film governs lubrication effectiveness.

Extreme-Pressure Lubrication. The metal surfaces are chemically altered because of reaction between the lubricant and the metal surface. These reactions often involve sulfur, chlorine, or phosphorus present in the lubricant reacting with the metal surface to form sulfides, chlorides, or phosphides. These compounds may be very complex in composition. Lubrication is provided because the films formed are low in strength and shear readily under deformation.

Solid Lubricants

When difficult press work operations require film strength in excess of that which is possible with solutions and emulsions, a paste, suspension, or coating may prove to be the best lubricant choice. Pastes may be oil- or water-base, pigmented or nonpigmented. Pigments include talc, mica, or similar inactive but effective high-strength barrier film formers. Sodium or potassium soaps are used widely in the production of all types of pastes. Buildup of lubricant in the tooling and the necessary cleaning of workpieces are the main drawbacks of pastes.

Suspensions are used frequently to provide a desired barrier film. Fine particles of a solid, such as graphite, Teflon (E.I. duPont de Nemours, Wilmington, DE), or clay, are suspended in a fluid carrier. Surfactants of an appropriate chemistry are generally required in order to provide suspension stability. Stability and lubricant buildup on dies are potential problems during use. Coatings applied to sheet metal surfaces—most often at the sheet or strip finish mill—have been used alone or as undercoatings to lubricants applied at the press. Water-soluble polymers, phosphate conversion coatings, soap solutions, and organic resins (either with or without other additives) are used. Generally, coatings are applied by dipping or spraying and then are heated to promote film formation and adhesion. Control of deposit thickness and chemistry, as well as removal, may be operating problems during production.

In solid-film lubrication, separation of tool and workpiece occurs by interposing a solid film. These solids act to provide a hydrodynamic film as long as they maintain surface integrity. The mechanism of solid-film lubrication is essentially the same as that of thick-film lubrication (Fig. 1a), except that the lubricant is a solid substance.

Solid-film lubricants include polymers such as polyethylene (PE), polyvinylchloride (PVC), and polytetrafluoroethylene (PTFE), which are applied to the workpiece as a powder, coating, or solid sheet and also give adequate protection against adhesive wear. Extreme conditions encountered when press forming iron- and nickel-base alloys require a phosphate-coated product and soap lubrication to prevent excessive die wear. Tin-coated steel, with additional liquid lubrication of mineral oil or emulsions to cool the die, has also been successful in preventing die pickup in severe press-forming applications. Other types include molybdenum disulfide, graphite, sodium carbonate, Teflon, nylon, and other solids in a carrier.

Solid films are often preferred so that the film thickness can be controlled by application rather than by process conditions alone. Such films also offer the possibility of differential or selective lubrication.

Dry Soaps. Even though dry soaps may be modeled as viscous substances and have speed-dependent actions, their method of application is a dry-film type. For good adhesion, the surface must be clean, preferably freshly pickled. Soap-film thickness is controlled by the concentration and/or temperature of the aqueous solution. Subsequently, the water is driven off by heating in ovens. These coatings are related to the soaps used in wiredrawing and are based on mixed soaps of selected fatty acids. Borax is often added as a filler and presumably to improve attachment to the surface. Sensitivity to moisture can be reduced by replacing borax with a polymer. The melting point is adjusted according to the severity of operation to avoid melting.

Soaps have been most successful on steel. On stainless steel they are adequate only with coated tools. They are seldom used on nonferrous metals because the slightly alkaline coating etches the metal and develops stiffer, higher-melting soaps on storage. The coating does not protect against corrosion or rough handling, but it has the advantage of water solubility and does not seriously interfere with welding.

Soaps applied to conversion-coated surfaces ensure a high lubricating power that is needed only in severe ironing or in drawing of stainless steel. Their use may be justified also when the part has to be phosphated anyway. Waxes have excellent lubricating properties for all metals but have to be applied from and removed by organic solvents.

Polymer Coatings. Numerous polymers are useful in bearing applications, but for metalworking purposes only thermoplastic polymers of the appropriate glass-transition temperature are of interest (Table 1):

- Polyethylene is readily available in sheet form and possesses substantial elongation (100 to 300%). In combination with mineral oil, sheets 2.5 to 7.5 μm (98 to 295 μin.) thick provide excellent lubrication. Polypropylene has similar properties.

- Polytetrafluoroethylene (PTFE) sheets have limited ductility, but well-adhering films of controlled thickness and good extendability can be deposited from a trichlorotrifluoroethylene dispersion of telomers (which have aliphatic end chains grafted onto the PTFE). Sputtering is possible too.
- Polyvinylchloride (PVC) is available both in sheet form and as a deposited coating.
- Polymethylmethacrylate (PMMA), and acrylics in general, have the advantage that they are readily deposited from either solvent solutions or emulsions, and they can be modified to endow them with boundary-lubrication properties.
- Acetal resins contain the linkage -CH_2O-CH_2O- and have found some limited application.
- Polyimides formed with aromatic radicals have outstanding temperature stability.

Most polymers are used as coatings. Among the earliest were methacrylate copolymers deposited from trichloroethylene solutions or from aqueous dispersions. The value of PTFE, especially in the baked-on form, was discovered early on and PTFE has potential application in drawing of aluminum hollowware. For better adhesion, these films are often deposited onto a ceramic substrate, fired onto the sheet surface. The pressworking of strip precoated in a special facility, often at the rolling mill, has received much attention. These coatings ensure clean shop conditions and protect the part during handling.

Polymer coatings impart greatly increased stretchability, approaching that obtained in hydraulic bulging. In deep drawing and incidental ironing the forces are substantially reduced, thus increasing the limiting draw ratio (LDR). An important requirement is that the film should not wrinkle or stick to the dies. To avoid cracking and peeling in tensile regions, the polymer must have high elongation and good adhesion to the sheet. Coatings that soften at low temperatures may be damaged when worked on high-speed mechanical presses.

Under mild conditions, coated sheets can be formed dry. In more severe situations, additional lubrication is needed. Lubricants applied at the pressworking plant must be chosen for compatibility. For example, PVC is attacked by chlorine-containing lubricants. In general, water-base lubricants, mainly waxes, are preferred; often they can be left on the finished part. It is necessary, however, to test them for possible delayed damage to the coating on the formed part.

Layer Lattice Solid Lubricants. Layer lattice coating is a term used to describe films of atoms with a hexagonal crystal structure (where spacing between basal planes is defined as the *c*-spacing, and spacing between the atoms within the basal planes is defined as the *a*-spacing). Compounds with a high *c/a* ratio have very an-isotropic shear properties with preferred shear parallel to the basal planes or perpendicular to the c-axis of the crystal structure. For example, dichalcogenides (disulfides, diselenides, and ditellurides) of molybdenum and tungsten have this structure. With the possible exception of tungsten diselenide, all of these compounds have lubricating properties, even in a vacuum.

Layer-lattice compounds have been used mostly in pastes, greases, and high-viscosity oils designed for use on heavy-gage material or in pressing at elevated temperatures. Some bonded films can survive severe ironing. For demanding applications, particularly for metals that do not react readily with lubricants, dry films deposited from liquid or volatile carriers have been used. Proprietary or locally made lubricants often contain a variety of other ingredients, sometimes as tackifiers or parting agents and at other times with little clear justification. In general, however, a common problem with layer-lattice compounds and nonorganic fillers is the relative difficulty of their removal.

Metal Coatings. Metal coatings find their most extensive use on sheet products. Most applications involve steel, but the principles are universal. Under mild conditions, tin alone can facilitate drawing.

For corrosion resistance of car bodies, galvanized (zinc-coated) sheet is used. LDR is improved in deep drawing, and higher blankholder forces are needed in press forming. Cracking or flaking of the film may occur, particularly in the compressive zone, and die pickup can be a problem. Polished, hard-chromium coated tools give best results.

Coatings. Soft coatings are used extensively for lubrication or lowered coefficient of friction. In addition, hard coatings are also used for good wear resistance with the accompanying low coefficient of friction. Such coatings are primarily carbides, nitrides, oxides, or borides of simple elements or alloys deposited by physical vapor deposition (PVD) and chemical vapor deposition (CVD) techniques. These include relatively simple compositions such as TiN, TiC, and Al_2O_3; more complex compositions such as (TiAl)N and Ti(C,N); and multilayer coatings, such as $TiC/Al_2O_3/TiN$.

A group of PVD coatings that frequently are used in applications where wear resistance in combination with low friction is required are diamondlike carbide coatings (DLC). These coatings typically are doped with metals, e.g., Cr or metal carbides such as WC. The beneficial low-friction property of DLC coatings makes them increasingly interesting also for different tooling applications such as dry metal forming and dry machining of soft materials such as aluminum. Carbon coatings containing metal (WC/C) and the combination of hard coatings (TiN, TiCN, TiAlN) with WC/C reduce the erosion tendency, and in some cases, they even permit a lubricant-free manufacturing process.

Fluid Lubricants

The three principal fluids (water, oil, and synthetic) that make up the primary ingredients of sheet-metal forming lubricants are combined with additives to achieve the desired operating characteristics. Table 2 lists the relative effectiveness of four types of lubricants for various functions. Some of the more important additives used to achieve particular lubricant characteristics are described in the section "Additives" in this article. However, it is first desirable to discuss the various possible forms of the lubricant (solutions, emulsions, and pastes). Additional information is also described in the section "Lubricant Types."

Solutions. A fluid in which all components are mutually soluble is a solution. All of the fluids mentioned may form solutions, with soluble additives selected for their mutual solubility with the fluid and for their effect on properties. It is common practice in the field to call water-base solutions synthetic fluids or lubricants; oil-base solutions are called compounded oils; synthetic fluids that are compounded are also called synthetic lubricants. This shop language can be misleading if not understood.

Aqueous solutions have the highest cooling capacity and tend to promote operational cleanliness. Resistance to biological contamination is superior except for the possible formation of mold. Corrosion protection and operator skin inflammation (dermatitis) are potential problems.

Compounded oils tend to have superior thick-film lubrication properties and resistance to biological attack. They also provide good corrosion protection. On the other hand, workpiece cleanliness, cleaning of parts, welding, and higher end-use cost may be problems.

Synthetic compounded lubricants have superior high-temperature properties, but much higher costs restrict their use in forming operations.

Emulsions. A mixture in which one immiscible fluid is suspended in another in the form of droplets is called an emulsion. The continuous phase in most sheet-metal forming operations is water. The suspended phase is oil or a synthetic fluid and may contain solid lubricants such as graphite, mica, or sodium carbonate. The continuous phase also may contain additives. In all cases, however, stabilization of an emulsion requires one or more surface-active agents (surfactants), finely divided solids, or special mixing techniques.

Surfactants such as fatty acids exist in an emulsion with the acid radical of the molecule preferentially soluble in the water, and the

Table 1 Thermoplastic polymers and their transition temperatures

Polymer	Repeat unit	Glass transition temperature (T_g), °C (°F)	Melting temperature (T_m), °C (°F)
Polyethylene (PE)	H H │ │ —C—C— │ │ H H	−120 (−185)	137 (280)
Polypropylene (PP)	H CH₃ │ │ —C—C— │ │ H H	−18 (−0.4)	176 (350)
Polytetrafluoroethylene (PTFE)	F F │ │ —C—C— │ │ F F	−50 (−60)	327 (620)
Polyvinylchloride (PVC)	H Cl │ │ —C—C— │ │ H H	87 (190)	212 (415)
Polystyrene	H ◯ │ │ —C—C— │ │ H H	105 (220)	240 (465)
Polymethylmethacrylate (PMMA)	H CH₃ │ │ —C—C— │ │ H COOCH₃	105 (220)	160 (320)
Polyimide	O R O ‖ ‖ —C—N—C—	220 (430) 370 (700)	none

organic, fatty portion of the molecule dissolved in the oil phase. In this way, the surfactant acts as a glue across the water/oil interface, thereby improving stability. This phenomenon is illustrated in Fig. 2, which shows the emulsifier action of three different emulsion systems as droplets of oil are suspended in water. In Fig. 2(a), the surface-active ingredients sodium cetyl sulfate and cholesterol are close-packed at the oil/water interface, thus producing an excellent tight emulsion. In Fig. 2(b), the sodium cetyl sulfate and oleyl alcohol interact at the interface to form a poor, or loose, emulsion. The effect of the surfactants is minimal in reducing the surface free energy of the oil droplets; therefore, emulsion stability is poor. Figure 2(c) illustrates an intermediate case, which may be desirable in some cases because relatively large oil droplets occur, rather than the small droplets that are present in a tight emulsion. The presence of larger oil droplets may benefit the performance of the lubricating film.

A special case of emulsion is used frequently in sheet-metal forming to take advantage of the cooling properties of water and the lubricity of suspended oil. The particle size of the suspended oil is small enough that the lubricant concentrate, even when water is added, remains clear rather than turning white or off-white, as is typically the case with an emulsion.

Emulsions generally are relatively easy to remove from parts and equipment. They provide superior cooling but may be more susceptible to biological contamination and have less-effective lubricating and corrosion-preventive properties than other lubricants.

Lubricant Chemistry

Metalworking lubricant formulations commonly consist of a petroleum-oil carrier (or base) that is compounded with additives of various types. The oil-additive combination is often emulsified in water. Lubricants that are based on synthetic oils, rather than petroleum oils, are also used. Some metalworking lubricants contain no oil at all, but consist of friction-modifying additives and corrosion inhibitors dispersed or solubilized in water.

Tool life and part integrity are directly affected by lubricant choice. The mechanical and chemical properties of the lubricant are influenced by operating conditions (temperature, pressure, geometry), as well as by the surface characteristics of the tool and workpiece. Further, the lubricant chemistry has important implications with respect to cleaning costs, welding integrity, corrosion of equipment and parts, recycling and disposal costs, and operator health and safety.

Carriers (Bases)

The chemical components of the lubricant consist of a combination of a carrier (base) with various additives. Different lubricants have different viscosities: fluids, solids, pastes, or gels may be used. Fluids are currently the most widely used lubricants for sheet-metal forming. Lubricating fluids are based on mineral oils or solvents, synthetic fluids, water, or some combination of these.

Oils. Petroleum oils are naturally occurring materials that are refined from crude oil by fractional distillation processes such as distillation, hydrotreating, solvent extraction, molecular sieving, and dewaxing to give desired properties. Petroleum oils are commonly called mineral oils. The mineral oils or solvents are used as a base for lubricants. Important physical attributes include viscosity, color, and odor, as well as chemical attributes, such as degree of saturation and freedom from undesirable elements such as sulfur.

Petroleum oils are generally classified as paraffinic, naphthenic, or aromatic. Paraffinic oils are classified further as either linear paraffins or isoparaffins. The molecular species associated with each classification is illustrated in Fig. 3. The oils are generally immune to biological

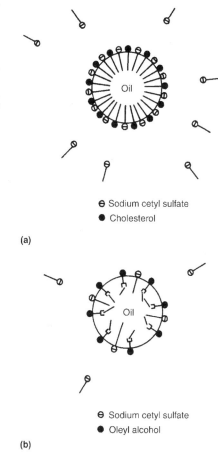

(a)

(b)

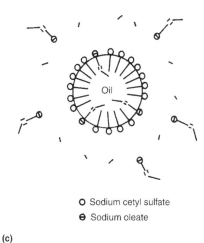

(c)

Fig. 2 Three types of emulsions seen in sheet metal forming lubricants. (a) Excellent (tight) emulsion. (b) Poor (loose) emulsion. (c) Intermediate case with poor emulsion stability

Table 2 Relative effectiveness of different types of lubricants for various functions

Function	Relative effectiveness(a)			
	Compounded oil	Oil-in-water emulsion	Semisynthetic solution	Synthetic solution
Provide lubrication at high pressure (boundary lubrication)	1	2	3	4
Reduce heat from plastic deformation (heat transfer)	5	2	2	1
Provide cushion between workpiece and die to reduce adhesion and pickup (film thickness)	1	2	3	4
Reduce friction between die and workpiece and, thus, heat	1	2	2	1
Reduce wear and galling between tool and workpiece (chemical surface activity)	4	3	2	1
Flushing action to prevent buildup of scale and dirt (fluid flow)	5	4	3	2
Protect surface characteristics; nonstaining	5	2	2	1
Minimize processing costs, welding, cleaning, and painting	5	2	2	1
Minimize environmental effect, air contamination, and water recovery	5	2	2	1

(a) 1, most effective; 5, least effective. Source: After Ref 3 attack, are readily recycled, have inherent lubricity and good metal-wetting characteristics, and afford some metal corrosion protection. The viscosity of selected oil or solvents varies over a wide range, as does the molecular structure.

Linear paraffins consist of straight-chain hydrocarbons, whereas isoparaffins consist of branched-chain hydrocarbons. Naphthenic oils consist of hydrocarbons that contain five- or six-member ring structures that may be unsaturated, but not aromatic. Aromatic oils consist of hydrocarbons that contain totally unsaturated six-member rings known as either benzene or aromatic rings. Petroleum oils generally consist of a mixture of paraffinic, naphthenic, and aromatic molecules and are classified by the species that predominates. For example, a petroleum oil consisting of 70% naphthenic molecules is called naphthenic.

The different classes of oils have different properties. Linear paraffins, for example, have high resistance to oxidative degradation, whereas the naphthenic oils tend to be more reactive oils and are more easily emulsified.

Synthetic Fluids. Synthetic hydrocarbons, such as polyisobutylenes and polyalphaolefins, with a specific molecular structure are used widely as lubricants. Although less commonly used and more expensive than petroleum oils, synthetic fluids are used as bases for metalworking lubricants in cases where their tailored properties can more than make up for their higher cost. Examples of some synthetic fluids are illustrated in Fig. 4. These fluids, which are based on synthesized hydrocarbons (polyalphaolefins) and polybutane derivatives, are often referred to as synthetic oils.

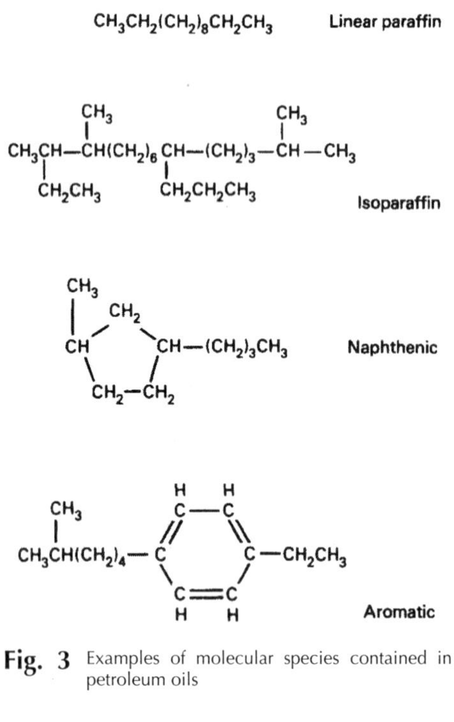

The polyalphaolefins are similar to refined base oils in characteristics but have superior resistance to oxidation. Polyalphaolefins, while not employed as commonly in metalworking operations as are polyglycols and polyisobutylenes, have properties that make them useful for some operations. They can be used over a wide temperature range, and they produce less hydrocarbon emissions than petroleum oils at similar viscosities. They are highly resistant to oxidative and thermal degradation.

The polyglycols, polyolesters, and dibasic acid esters have superior high-temperature stability; however, they are not as readily combined with desired additives for corrosion protection or metalworking ability. Polyglycols are often used in metal removal lubricant formulations. These materials are prepared through polymerization of ethylene oxide and propylene oxide in either random or block fashion. The terminal groups can be either alkyl groups or hydrogen. Properties are controlled by molecular weight, nature of terminal groups, and the ratio of ethylene oxide to propylene oxide. For example, high molecular weight versions have high viscosities. Polyglycols with high ratios of ethylene oxide to propylene oxide tend to be water soluble, whereas those with low ratios tend to be water insoluble. The ratio can be adjusted such that the polyglycol is water soluble at room temperature, but water insoluble at elevated temperatures. This property can be very useful in formulating solutions that are clear in the circulating system but will separate polyglycol and the additive package from solution at the hot tool-workpiece interface.

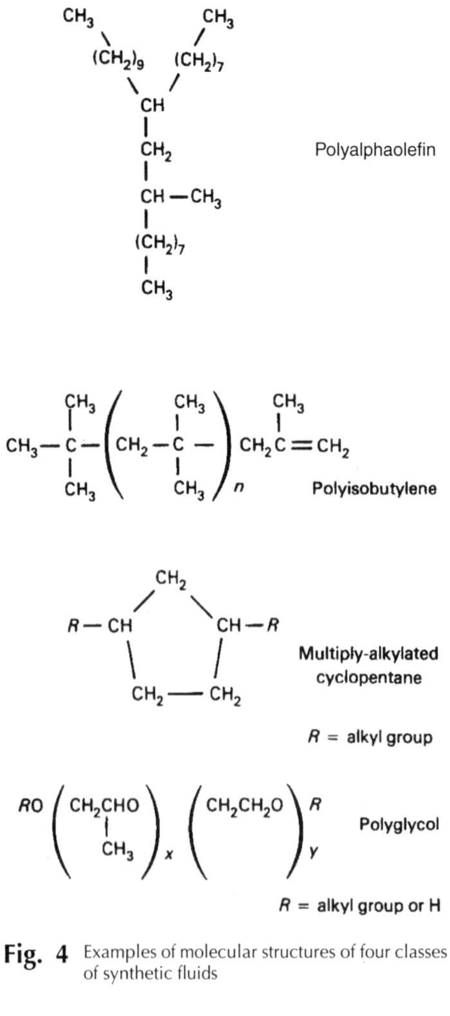

Polyisobutylenes can be made in a wide range of viscosities by controlling molecular weight. They have a unique property of depolymerizing at high temperatures. This makes them useful as rolling and drawing oils for ferrous and nonferrous forming operations where subsequent annealing would produce staining if petroleum oil forming lubricants were used.

Multiply-alkylated cyclopentanes are a new class of synthetic hydrocarbons that are promising in terms of future formulations. Their properties can be varied over a wide range by varying the number and nature of alkyl groups. They have excellent pour points and viscosity indexes, as well as exceptionally low volatility. A number of synthetic fluid categories beyond those discussed previously also exist, but have limited use as base oils in metalworking operations.

Water. Because of the rising cost of refined mineral oil as well as the intrinsically high cost of synthetic fluids, water as the base fluid for sheet metal lubrication has become more widely used. New technology methods use various forms of water-base fluids (solutions, emulsions, gels, and pastes).

Of particular importance in the performance of water-base lubricants is the superior heat transfer capability of water. The sheet-metal forming operation generates heat due to friction and metal deformation; water is the most efficient fluid for dissipating this heat.

Viscosity of Carriers. Viscosity is a measure of the resistance of a fluid to flow and is a very important lubricant property. Kinematic viscosity, which represents the resistance of a fluid to flow under gravity, as measured by the ASTM D 445 test, is the preferred method for describing the viscosity of lubricants. In ASTM D 445, a fixed volume of lubricant is allowed to flow through a calibrated orifice that is held at constant temperature. The kinematic viscosity is calculated by multiplying the flow time in seconds by the calibration constant of the viscometer. The correct SI unit of kinematic viscosity is mm^2/s. A centistoke, which is equivalent to 1 mm^2/s, is also commonly used. The standard temperatures for measuring viscosity are 40 °C (104 °F) and 100 °C (212 °F). The viscosity of fluids decreases as temperature increases. Viscosity index is an empirical, dimensionless number that indicates the rate at which lubricants change kinematic viscosity with temperature. The higher the viscosity index of a lubricant, the less rapidly its viscosity changes with temperature.

In metal forming operations, viscosity is a major factor in determining the lubricant film that separates the tool from the workpiece and

Fig. 3 Examples of molecular species contained in petroleum oils

Fig. 4 Examples of molecular structures of four classes of synthetic fluids

thus is critical in controlling friction and wear. In metal removal operations, optimum lubricant viscosity must be estimated for a particular operation. Factors to be considered in estimating the optimum viscosity of a metal removal fluid include the capability of the lubricant to penetrate and remain in the contact zone, the durability of the lubricant film, the desired rate of spreading, and the cooling capability.

Additives (Ref 4–6)

Although uncompounded fluids may successfully serve as lubricants in light sheet-metal forming, other processing requirements such as cleaning and corrosion protection may be satisfied only by the addition of chemical additives. Further, in more difficult forming operations, the base oil does not totally separate the tool from the workpiece. Thus, additives must be used to provide necessary film characteristics for lubrication. In water-base metalworking lubricants, special materials are employed to disperse or solubilize the oil-additive package in water.

Film-strength additives adsorb on tool-workpiece surfaces and prevent direct metal-to-metal contact and the subsequent welding of asperities and destruction of the workpiece surface. These additives are often called boundary additives, load-bearing additives, oiliness additives, or friction modifiers. Film-strength additives consist of materials with a polar head and a hydrocarbon tail generally containing 10 or more carbon atoms. Compounds that function as film-strength additives include fatty acids, esters, alcohols, amides, amines, and alkyl acid phosphates. Higher-viscosity lubricants give thicker lubricant films and, therefore, less tool-workpiece surface contact and lower friction in metalworking operations.

Emulsifiers, often called surfactants, are materials that have portions within the same molecule that are hydrophilic, or compatible with water, and lipophilic, or compatible with oil. The hydrophile-lipophile balance (HLB) of an emulsifier is a measure of its tendency to be more compatible with water or oil. The HLB scale runs from 0 to more than 30, with oil compatibility decreasing and water compatibility increasing at higher numbers. Emulsifiers with HLB values higher than 13 form clear solutions in water.

Emulsifiers are used to promote stable emulsions and, in some cases, the cleanability of oils. Anionic and nonionic surfactants are the preferred emulsifiers. Typical nonionic emulsifiers are complex esters, fatty acids plus alcohols, monoglycerides, and ethoxylated alcohols. Some anionic surfactants used are soaps of long-chain fatty acids, alkyl aromatic sulfonates, and phosphate esters. Soaps are the most widely used anionic emulsifiers, but they may react with hard water to form compounds that can be difficult to remove from die or part surfaces.

Because emulsifiers have portions of their molecules compatible with water and other portions compatible with oil, they tend to concentrate at the oil-water interface in oil-water mixtures, reduce oil-water interfacial tension, and thereby promote emulsification. The emulsification process is that technique in which oil globules that are larger than colloidal size are dispersed in water. Petroleum oils, animal and vegetable oils, synthetic oils, and waxes each have an HLB value at which they form the most stable emulsion. This is known as the "required HLB" of the material to be emulsified. Required HLB values are different for different materials. For example, the required HLB to emulsify a paraffinic petroleum oil is 10, whereas the required HLB to emulsify castor oil is 14. An emulsifier or emulsifier combination that has the same HLB as that required of the material to be emulsified is chosen if the most stable emulsion is desired.

Emulsifiers are classified as cationic, anionic, amphoteric, and nonionic. The most common cationic emulsifiers are long carbon chain quaternary ammonium halides. Anionic emulsifiers include alkali metal and amine soaps of long carbon chain fatty acids, and long carbon chain sulfates or sulfonates. Examples of amphoteric emulsifiers include long-chain amino acids and alkyl betaines. Examples of nonionic emulsifiers include ethoxylated alcohols, ethoxylated fatty acids, and ethoxylated sorbitol. The most commonly employed emulsifiers in metalworking lubricants are the nonionic and anionic types.

Extreme-pressure (EP) additives are used when the fluid alone does not prevent excessive tool-part friction and wear. The most commonly employed EP additives are compounds containing sulfur, chlorine, phosphorus, or some combination of two or more of these elements. Extreme pressure additives function by forming a reaction layer triggered by the high temperatures reached at the tool-workpiece interface in ferrous metalworking operations. For example, phosphorus-containing compounds produce iron phosphates, iron pyrophosphates, or iron phosphides, depending on the nature of the compound. Sulfur-containing compounds produce iron sulfide, whereas chlorine-containing compounds produce iron chloride. These chemical compounds are relatively weakly bonded and fracture and slide easily under the conditions of forming, providing a separating layer between tool and workpiece. Instability of the lubricant, as well as staining and corrosion, may be encountered during use. Cleaning and welding may also be adversely affected.

Examples of EP additives include sulfurized triglycerides, chlorinated hydrocarbons, chlorinated esters, phosphate esters, and alkyl acid phosphates. Closely associated with EP additives are so-called "antiwear" additives containing the same elements. A commonly used class of antiwear additives are the zinc dialkyl dithiophosphates.

A new class of EP additives, called passive extreme pressure (PEP) additives, was introduced in the 1980s–90s. They function by adsorbing a film of carbonate particles at the tool-workpiece interface in metalworking operations. These films have low shear strengths and high melting points. They reduce friction and minimize metal transfer from the workpiece to the tool. Passive extreme pressure additives do not contain phosphorus, sulfur, or chlorine but are synergistic with sulfur-containing EP additives. They offer advantages over conventional EP additives in that they are less corrosive, are more easily disposed of after use, are low foaming, and are easily cleaned from the workpiece surface. They can be used with both ferrous and nonferrous metals.

Thickeners are used to alter the flow characteristics (viscosity) of the fluid lubricant. Flow characteristics are influenced by temperature and pressure and affect lubricant performance. Organo-clays, polymers, natural gums, and metal hydroxides are used to vary flow properties. Stability of the lubricant system and buildup on tooling may cause operating problems.

Antimisting agents are used to reduce the incidence of airborne particles of the metalworking lubricant. Particularly in the case of solvents or low-viscosity oils, air mists may form that are detrimental to health and may also cause equipment-operating problems. To reduce mist formation, small quantities of polymers such as acrylates or polybutanes are added. When contacted with polymer, the lubricant base builds larger particles so that they do not readily form a mist.

Passivators are added to reduce the activity of the metal surface in order to prevent staining, particularly of nonferrous metals. Organic amines, sulfur, and nitrogen-containing compounds are used as passivators. Frequently, a so-called passivator is used in conjunction with corrosion-prevention additives such as complex borates to improve the performance of each.

Antifoaming agents may be required to prevent foam formation if the lubricant is recirculated. Foaming is undesirable because it may interfere with reservoir cleanup and fluid flow or complicate fluid application. Lubricant foaming can even lead to lubricant starvation if excessive amounts of the lubricant mass are in the form of foam.

Foam is created when air is injected into the lubricant either through spraying at the metalworking operation or through circulation of the lubricant through the lubricant handling system. Nearly all water-base metalworking lubricants contain emulsifiers, which not only lower oil-water interfacial tension but also lower the surface tension of the lubricant compared with water. The surfactants concentrate at the air-fluid interface and form an elastic film that expands, but does not rupture, as air is introduced.

In order to prevent foam formation, the surface free energy of the film must be reduced. This may be achieved with a wide variety of additives, including silicones, amides, glycols, and fine particles of selected solids such as silica. Small additions of these chemical agents, called antifoams or defoamers, can drastically reduce or eliminate foams. The concentration of these agents needed for effective foam control is low;

however, monitoring concentration to maintain control may be a problem.

A good defoamer must have the right combination of dispersibility and surface tension. It should spread throughout the system without dissolving in it, and it should spread over the foam surfaces. When this happens, the defoamer acts at the gas-fluid interface to collapse the elastic film of the fluid, thereby allowing it to release the air and drain. Silicones are very effective defoamers. Their major drawback is that if the workpiece is to be coated or painted in a subsequent operation, adhesion may be adversely affected. Nonsilicone defoamers include long-chain alcohols, certain triglycerides, and water-insoluble polyglycols.

Suspended Solid Lubricants. Solid lubricants are often suspended in oil- or water-base fluids for extremely heavy-duty forming of sheet metal. These solids are graphite, carbonates, mica, polytetrafluoroethylene (Teflon), nylon, metal powders, metal oxides, metal halides, and molybdenum disulfide. The solids are generally in the form of very finely divided powders held in suspension by either mechanical agitation or emulsifiers, or a combination of both. The oil or water carrier often functions solely to coat the die and workpiece in a uniform fashion, and the solid itself functions as a lubricant.

Solid lubricants have the general property of being easily sheared at the tool-workpiece interface. However, buildup on tooling and cleaning of workpieces are potential problems during operations with these suspended-solid lubricants. Suspended solids are not generally used in metal-removal lubricants, although some types of PEP agents are colloidal suspensions.

Corrosion inhibitors are frequently added to both water- and oil-base lubricants to provide parts protection during processing and storage. Parts corrosion may be a particularly difficult problem when water-base lubricants are used; however, sulfonates, carboxylates, borates, and phosphonates have been successful in alleviating corrosion. In oil-base lubricants, organic amines and sulfonates, as well as phosphates and unsaturated fatty acids, have been used successfully.

Prior to the 1980s, a commonly used and very effective corrosion inhibitor was sodium nitrite. In the mid-1970s, it was discovered that nitrosamines, which are carcinogenic, were contained in many commercial metalworking formulations. The nitrosamines were found to arise through a reaction between sodium nitrite and secondary amines, particularly diethanolamine. Therefore, much work has been done to find suitable replacements for sodium nitrite.

Corrosion inhibitors, particularly with water-base lubricants for ferrous metals, protect the workpiece during and after processing. A number of types of compounds have been developed for the prevention of ferrous metal corrosion in modern metalworking lubricants. These include amine-borates, amine carboxylates, amine alkyl acid phosphates, and sulfonates. None of these materials forms nitrosamines and are therefore more acceptable.

Corrosion inhibitors for nonferrous metals are important in some metalworking operations. Benzotriazole is an effective corrosion inhibitor for copper or brass. Cobalt corrosion inhibitors, such as tolyltriazole, are used to prevent cobalt leaching in those operations employing cobalt-cemented tungsten carbide tools. Toluyltriazole has been used successfully to protect copper sheet from staining and corrosion.

Oxidation inhibitors, also called antioxidants, are formulated into metalworking lubricants to minimize oxidative degradation of the lubricant into acidic products that tend to form sludge and corrode metal surfaces. Oxidative degradation involves molecules containing unpaired electrons called free radicals. These radicals are transformed into degradation products while transforming other lubricant molecules into more free radicals in a chain reaction. Small concentrations of oxidation inhibitors functioning as free radical scavengers intercept the unpaired electrons and break the chain. Oxidation inhibitors are not normally used at levels above 0.5%. Commonly used oxidation inhibitors include hindered phenols, such as butylated hydroxytoluene and butylated hydroxy anisole, and secondary aromatic amines, such as phenyl naphthyl amine. Because the rate of oxidation occurs much more rapidly at elevated temperatures, oxidation inhibitors are especially useful for hot metalworking operations, such as the hot rolling of either ferrous or nonferrous metals.

Antimicrobial agents are materials designed to inhibit the growth of bacteria, fungi, and yeast in metalworking lubricants. All water-base metalworking lubricants are vulnerable to attack by one or more of these agents; even oil-base lubricants containing small amounts of water as a contaminant can be degraded by microbes. Thus, antimicrobial agents may be required in order to prevent the growth of anaerobic or aerobic bacteria.

Bacteria frequently cause various operating problems in emulsions or solutions that contain oil and water, and attack of metalworking lubricants by bacteria leads to one or more of the following:

- Buildup of acidic materials
- Buildup of lubricant on workpieces
- Corrosion of machinery and tools
- Objectionable odors
- Destruction of additives
- Loss of emulsion stability

Growth of fungi also can lead to slimy material coating the machinery and tools, as well as the clogging of pumps and filters.

Microbes can generally be controlled at acceptable levels through use of antimicrobial agents known as biocides and fungicides. Care in handling as well as monitoring are important if biocides are to be used safely and effectively. Bacteria, fungi, and yeast are often monitored on a regular basis through commercially available simple culture techniques, and when counts reach a certain level, there is cause for alarm. In solutions formulated with various chemical components, bacteria will generally not be the problem; rather, formation of mold can cause malfunction of pumps and filters, and buildup on tooling and workpiece surfaces. It may also alter the solution chemistry. Biocides are not required when lubricant application does not involve repeated use or recirculation of the fluid.

Standard practice often calls for the addition of two different biocides to the metalworking lubricant at regular intervals in an alternating fashion, in order to guard against microbes developing an immunity to one of them, resulting in an uncontrolled infestation. Although many types of biocides exist, two of the most common are phenolic materials and formaldehyde-release agents. Phenolic materials, such as 2,4,5-trichlorophenol, destroy bacteria directly. Materials such as 1,3-di(hydroxy-methyl)-5,5-dimethyl-2,4-dioxoimidazole, upon being added to water-base metalworking lubricants, release formaldehyde slowly to keep bacteria in check. Materials such as 2,2-dibromo-3-nitrilopropionamide are useful for controlling bacteria, fungi, and yeast.

There are over 50 commercially available antimicrobial agents. In choosing the proper one, parameters such as the required concentration, effect on emulsion stability, and regulations concerning discharge into waste streams must be taken into account. Finally, antimicrobial agents are designed to destroy living organisms. They all display some degree of toxicity toward humans and should be handled with caution.

Lubricant Types (Ref 4)

Four commonly used types of metalworking lubricants are illustrated in Fig. 5. Straight metalworking oils, often simply called straight oils, are given this term because they are not mixed with water prior to use. Emulsions are mixtures of either simple or compounded oils with water, stabilized by the use of emulsifiers. Emulsion droplets are similar in size to, or larger than, the wavelength of visible light; hence, emulsions appear milky white. Microemulsions can be similar to emulsions in composition, but, through emulsifier choice, have oil droplet diameters that are much smaller than the wavelength of visible light and therefore appear transparent. Micellar solutions are similar to microemulsions, except that they contain no oil.

Petroleum oils, as defined previously, are naturally occurring materials that are refined through processes that separate the crude substances into various molecular fractions and remove impurities. In some cases, they are hydrogenated or subjected to reforming catalysts but are not otherwise chemically altered. The chemical reaction of smaller molecules produces the larger molecules of synthetic fluids, such as poly-α-olefins and polyisobutylenes. In metalworking lubricants, the term *synthetic* is often used to describe either transparent micellar solutions (Fig. 5) or true solutions containing no petroleum oils. Unfortunately, this dual definition for synthetic lubricant has led to confusion. It should be obvious that if the definition of synthetic lubricant is one that contains a preponderance of man-made materials, then the straight oil, emulsion, or microemulsion of Fig. 5 is, strictly speaking, synthetic, if the base oil is a synthetic fluid such as poly-α-olefin, rather than a petroleum oil.

Straight oils are generally petroleum oil fractions that are normally formulated with either film-strength additives or EP additives, or a combination of both. They generally provide excellent friction reduction and workpiece surface finish, good corrosion protection, and a long service life. Straight oils containing certain EP additives will stain nonferrous metals, such as copper, and are commonly referred to as staining oils. The major disadvantage of straight oils is their poor capability for heat removal, compared with water. In addition, straight oils with low flash points, coupled with the high temperatures often encountered in metalworking operations, can create fire hazards. Straight oils are commonly used in metalworking operations where lubrication is a major factor and cooling is a minor factor. Examples of such operations include low-to-moderate-speed metal-removal operations where accuracy, tolerance, and workpiece finish are important, and metal forming operations such as aluminum foil rolling where strip surface quality is highly important.

Emulsions. In commonly used metalworking emulsions, oil globules are finely dispersed in water, and this oil-water combination is employed as a lubricant-coolant. In these types of emulsions, oil is said to be the dispersed phase, and water, the continuous phase. Oil-in-water

mixtures are thermodynamically unstable; that is, their state of lowest free energy is total separation. Because of this, oil tends to separate, and emulsifiers are added to stabilize the emulsion. Emulsifiers concentrate at the oil-water interface and inhibit coalescence of oil globules, as illustrated in Fig. 6. The structural formula of sodium oleate, an anionic emulsifier, is shown in Fig. 6(a). A simplified "straight pin" is depicted in Fig. 6(b). An oil-in-water emulsion stabilized by sodium oleate is shown in Fig. 6(c). The hydrocarbon chain of sodium oleate is compatible with the oil globules and penetrates them. The carboxylate head of sodium oleate is compatible with water and lies at the surface of the oil droplet penetrating into the water phase. Because the carboxylate head carries a negative charge, the surface of each oil droplet is negatively charged, and because like charges repel, the oil droplets tend to stay dispersed. Because the state of lowest free energy of the emulsion is still total separation, the emulsion is said to be kinetically stabilized.

Emulsions can vary in stability over a wide range, depending on the nature of the oil phase and the nature and concentration of the emulsifier package. Emulsion oil globule size can vary from approximately 0.2 μm (8 μin.) to as high as 10 μm (400 μin.) or more. Therefore, emulsions appear as off-white to white opaque solutions. The globule sizes within a given emulsion are polydispersed; that is, they vary over some distribution. Stable emulsions have smaller average globule size distributions than do unstable ones.

A major factor in lubricating with emulsions is the availability of the oil phase to lubricate. Two factors control oil availability: the emulsion

stability and the concentration of oil in the emulsion, which is often called "percent oil." In general, the less stable the emulsion and the higher the percent oil, the greater is the availability of oil for lubrication. Unfortunately, the less stable an emulsion is, the higher is the tendency for stability to change, sometimes rapidly, over time. This can lead to undesirable instability in some metalworking operations, such as rolling. Also, as the percent oil in an emulsion increases, cooling capability decreases. Therefore, the stability and percent oil in an emulsion must be carefully balanced to satisfy the lubrication and heat-removal needs of a particular metalworking operation.

The fact that emulsions are kinetically, rather than thermodynamically, stable leads to other factors in their behavior. One of these is called the emulsion "batch life." New emulsions are generally the most stable and have the least oil available for lubrication. Metalworking operations are often not optimal when a new emulsion batch is introduced. Over time, as debris is generated in the emulsion, providing nucleation sites for oil globule coalescence, and as emulsifiers are depleted, the emulsion becomes less stable and performs at its best. At yet a later time, the emulsion becomes so degraded and unstable as to be rendered useless and is discarded. A new batch is introduced, and the process repeats itself. In general, emulsions that are initially less stable have shorter batch lives. A second factor is the care that must be taken upon introducing foreign substances into the emulsion. For example, introduction of a biocide to combat a microbial infestation or contamination by acids or bases can greatly affect emulsion stability and therefore the consistency of the metalworking operation.

Metalworking lubricant emulsions are often complex mixtures of emulsifiers, film-strength additives, oxidation inhibitors, corrosion inhibitors, and coupling agents. Coupling agents are generally lower molecular weight diols and triols that aid in the initial emulsification. Emulsions can also contain various mixtures of EP additives. The formulated oil mixture is called the concentrate and is added to water with agitation to form the emulsion. The quality of the water is extremely important. Distilled or deionized water should be used whenever possible. Metalworking emulsions generally operate at levels between 5 and 10% oil.

Metalworking emulsions are maintained in a variety of ways. The percent oil is determined by breaking the emulsion in a graduated bottle with an acid or salt solution, and it is maintained by adding new concentrate during the life of the batch. Nonemulsifiable tramp oils (those that have leaked into the metalworking lubricant) are skimmed off, and the emulsion is subjected to continuous filtration to remove fine debris. In the case of metal-removal fluids, chips are often removed by mechanical means. Microbe levels are monitored and controlled by the addition of appropriate antimicrobial agents at prescribed intervals. Emulsions have good lubricating and

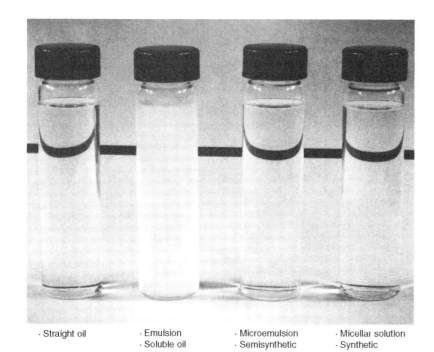

· Straight oil   · Emulsion   · Microemulsion   · Micellar solution
                · Soluble oil   · Semisynthetic   · Synthetic

Fig. 5   Commonly used metalworking lubricants

heat-removal qualities and are used widely in most metal removal operations and many metal forming operations.

Microemulsions are clear-to-translucent solutions containing water; a hydrophobic liquid, that is, an oil phase; and one or more emulsifiers, which are often referred to as surfactants and cosurfactants. Microemulsions in which water is the continuous phase and oil is the dispersed phase are called oil-in-water microemulsions and are the type generally used as metalworking lubricants. Microemulsions employed as lubricants are commonly called semisynthetic fluids.

Several parameters differentiate microemulsions from emulsions. Most importantly, microemulsions are thermodynamically stable; that is, the state of lowest free energy is dispersed rather than separated. Therefore, the stability problems associated with emulsions are nonexistent. Microemulsions remain stable indefinitely, as long as they are maintained in appropriate ranges of pH, oil-to-water ratio, and temperature. These ranges may be very narrow to very broad, depending on the nature of the microemulsion. The diameters of the dispersed oil globules in microemulsions range from approximately 0.01 to 0.2 μm (0.4 to 8 μin.), depending on the nature of the oil and the types and concentrations of emulsifiers. This small oil globule size is the feature that makes them appear clear to translucent. Additionally, the oil globule diameters are much more uniform in microemulsions than they are in emulsions.

Microemulsions are generally produced in the metalworking environment by adding a concentrate to water with agitation. The concentrate generally contains oil, the emulsifier package, film-strength additives, corrosion inhibitors, biocides, and in some cases, EP additives. Dilutions range from about a 10 to 1 ratio of water to concentrate to as high as 60 to 1. Lower dilutions are used in operations were lubrication is more important, whereas higher dilutions are used where cooling is more important. Concentration is commonly determined with a handheld refractometer and a calibration chart that relates the instrument reading to concentration.

Microemulsions offer good resistance to corrosion and to microbial attack, as well as excellent stability and cooling. They suffer from higher initial cost, difficulty of disposal, and a stronger tendency to foam. Microemulsions formulated with fatty acid soap-type emulsifiers tend to degrade rapidly in hard water because of the formation of insoluble calcium and magnesium carboxylates.

Micellar Solutions. When emulsifier molecules are dissolved in water, they tend to aggregate into larger units called micelles. Micelles are formed spontaneously because of the fact that the lipophilic portion of the emulsifier molecule tends to aggregate in the interior of the micelle, whereas the hydrophilic portion tends to penetrate into the water phase. Figure 7 illustrates, in two dimensions, the relation between a molecule of a typical anionic emulsifier, sodium dodecyl sulfate, and the spherical micelle that it forms in water. Micellar solutions used as metalworking lubricants contain neither petroleum oils nor synthetic hydrocarbons. They contain film-strength additives, EP additives as appropriate, and corrosion inhibitors solubilized within the interior of the micelles. Because micelles have diameters typically between approximately 0.005 and 0.015 μm (0.2 and 0.6 μin.), micellar solutions are transparent to the eye and, like microemulsions, are thermodynamically stable. Because virtually all of the components of micellar solutions are obtained by chemical synthesis, they are often referred to as either synthetic lubricants or chemical coolants.

Like emulsions and microemulsions, micellar solutions for metalworking are generally formed by the addition of a concentrate to water with agitation. Dilutions typically range from a water-to-concentrate ratio of 10:1 to about 50:1, depending on the application. As in the case of microemulsions, concentration is determined by refractive index.

Micellar solutions that contain no alkali metal soaps or amine fatty acid soaps show good stability in hard water. They are more resistant to microbial attack than either emulsions or microemulsions. They can be formulated to reject tramp oils, which can then be skimmed and collected for disposal or recycling. They have excellent cooling capability, provide excellent corrosion control, and have a long, useful life.

On the downside, because micellar solutions tend to cost more initially, a total cost-benefit analysis should be performed. Additionally, because they are highly fortified with emulsifiers, foam can be a real problem. Antifoaming agents can be added to control foam. Micellar solutions, in general, have lower lubricating capability than other types of metalworking lubricants. This limits their applications to those

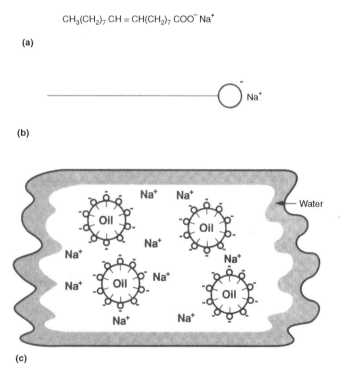

Fig. 6 Sodium oleate. (a) Chemical formula. (b) "Straight pin" depiction. (c) Oil-in-water emulsion stabilized by sodium oleate emulsifier

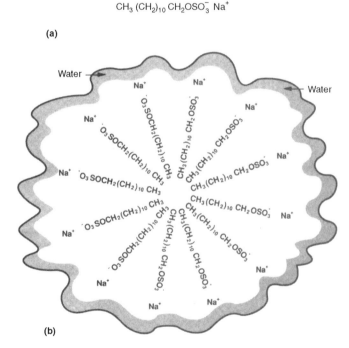

Fig. 7 Sodium dodecyl sulfate. (a) Chemical formula. (b) Sodium dodecyl sulfate micelle in water

metal-removal operations with low tool pressures and high tool speeds where cooling is of paramount importance. In such operations, tool life can be extended as much as 250 times by using micellar solutions, compared with straight oils. One final drawback is waste disposal. It is often very difficult to separate the organic materials from water because of their nature. The organic materials tend to remain soluble over wide ranges of pH, temperature, and salt concentration. It is often necessary to resort to sophisticated techniques, such as reverse osmosis, for disposal.

True solutions differ from micellar solutions in that the molecules of active substances do not form micelles when dissolved in water. Rather, each ion is solvated by water molecules. Because essentially all film-strength additives and EP additives either form micelles or require micelles to dissolve them, true solutions offer lubricating qualities that are little better than those of water.

True solutions are used in cases where cooling is the only consideration. In these cases, their advantages are low cost, very high cooling, stability, low foam, and a very long life. Typical true solutions often contain nothing more than a corrosion inhibitor, such as sodium nitrite, in water.

Solid-Lubricant Suspensions. As previously noted, lubricants for specialized uses often contain solids in the form of finely divided powders suspended in a liquid carrier, such as oil or water. The liquid carriers may also contain soluble additives of the classes previously mentioned. One of the most common suspended substances is colloidal graphite, with specific surface areas that often exceed $100 \ m^2/g$ ($3 \times 10^4 \ ft^2/oz$). It is used extensively in hot forging and extrusion of both ferrous and nonferrous metals. Molybdenum disulfide is another commonly used suspended solid. Both graphite and molybdenum disulfide are compounds that possess layered crystal structures with weak forces bonding the layers together so that they are easily sheared. They function by plating onto tools and workpieces such that the weak shear direction is parallel to the surfaces. As the tool and workpiece surfaces are brought together, they form a solid film, preventing tool-workpiece contact and shear along the weak shear plane, thereby reducing friction.

Other types of solids that are suspended as fine powders in some metalworking lubricants include mica, polymers such as Teflon, certain metal oxides, and glasses. Mica has a layered structure and functions in a way similar to graphite. Polymers mechanically separate metal surfaces, lower friction, and reduce metal transfer. Hard metal oxides, such as aluminum oxide, have good wear resistance but high friction coefficients. Soft oxides, such as lead II oxide, give relatively low friction coefficients that decrease at higher temperatures, where the mechanism of deformation changes from fracture to plastic flow. Glasses are suspended in lubricants for use in metalworking operations at high temperatures, where they soften on hot die and workpiece surfaces and function as parting agents of low shear strength.

Solids can be kept in suspension by using surface active agents, mechanical agitation, or both. Problems can arise if not enough care is taken and the solid is allowed to "settle out." Also, metalworking lubricants containing suspended solids tend to produce buildup on tools and workpieces that are difficult to clean. These problems can be minimized by the appropriate formulation.

Lubricant Effectiveness

Selection of an appropriate lubricant for a specific sheet-metal forming operation or for a series of different operations is often a costly hit-or-miss proposition. There are, however, laboratory test procedures that can often reduce the possibility of error.

Because of the difficulty of translating laboratory test results to an operating condition, several simulation tests have been developed to screen lubricants for sheet-metal forming. Small-capacity presses have been installed as test machines with the ability to control speed of forming, metallurgy of tools and work metal, and lubricant application method.

Strip drawing through a wedge-shaped die on a tensile machine has been developed to measure lubricant effectiveness. One end of the strip is gripped and pulled through the die. The die may also be modified in order to simulate drawing over draw beads. In these tests, resistance to sliding is measured for different lubricants applied to the strip.

Standard deep-drawing tests, such as the Swift cup test, can also be used to measure lubricant effectiveness. Poor surface finish, score marks, or splits are indications of unsatisfactory lubrication. More information on tests for measuring the drawability of metals is available in the article "Formability Testing of Sheet Metals" in this Volume; Ref 7 and 8 include more details on evaluation of lubricant effectiveness. The larger the blanks formed and the deeper the cup, the more effective the lubricant film will be. A wide variety of punch and sheet metal geometry, size, and tool material can be used along with the lubricant coating. The limiting draw ratio (LDR) depends on the draw ratio (DR), which is equal to the ratio of the blank diameter to the punch diameter. For a given punch diameter, the greater the blank diameter, the more severe is the operation. The LDR is the maximum DR that results in a satisfactory cup.

Lubricant Breakdown and Galling (Ref 1). Even under processing conditions that appear mild in comparison with bulk deformation processes, lubricant breakdown is often encountered because the lubricant is wiped off in the course of sliding and asperities are exposed to direct contact with the die. Tool pickup then follows, which in turn leads to scoring or galling of the workpiece surface. Because galling is an event that follows local lubricant failure, it is greatly affected by workpiece and die topography and by adhesion between the contacting materials.

Galling is typical of all processes in which the workpiece slides over the die. Galling is a result of localized adhesion and is, therefore, affected by a number of process variables:

- An asperity of low slope deforms and is pushed into the bulk. As the asperity slope increases, a point is reached where a bulge forms, which is then removed by a shearing (cutting) action.
- Debris formed in the course of sliding may be trapped and rendered harmless in valleys of the sheet surface, may become embedded in other sheet asperities, or may lodge against die scratches. Repeated encounters at many points result in deformation, thinning, and lamination of particles; some of them may even roll up. If adhesion to the die is high, new large asperities form, which lead to scratching, formation of more thick particles, and galling.
- Galling is affected by the macroscopic tool geometry. It usually originates at transition points such as die radii, but its severity greatly increases if the already damaged sheet, with embedded particles, is dragged along a parallel surface such as a die land.
- The incidence of galling is a function also of the stiffness of the system. Damage is localized in a soft (constant-force) system, whereas it is widespread and uniformly distributed in a stiff (constant-gap) setup.

Both workpiece and die roughness have significant influences on the incidence and severity of galling:

- A smoother sheet surface presents a larger bearing area and is more prone to galling. For a given reduction in thickness, smoother sheet galls after shorter sliding or fewer repeat tests. This effect can be quite dramatic; a small change in roughness is sufficient to initiate galling in drawing of materials prone to pickup.
- The size and distribution of asperities is most important. For a given average roughness, larger flat plateaus give longer contact (sliding) distance and more galling. For a given bearing area, a smoother sheet has smaller plateaus but a much greater peak density, and thus it presents more points for sliding contact and is more prone to galling.
- The directionality of the sheet surface profile has a marked effect. Plateaus aligned in the direction of sliding present longer sliding distances and are more likely to allow embedment of the debris. Thus, they are more harmful than transverse ridges, which trap debris in valleys and may even act as rasps to remove incipient pickup from the tooling.
- The depth profile of the sheet surface is important because it determines the ability of the surface to entrap lubricants and accommodate wear debris.

- There has been some success in defining what sheet surface is best, even though there is no general agreement. Invariably, steel sheets with bearing areas of less than 20% at 1.5 μm from the surface prove to be nongalling, whereas those with larger bearing areas (and those whose bearing areas increase more rapidly with depth) are prone to galling.
- A rough die surface is desirable only on the punch because it then increases the LDR and the depth of draw in automotive press-working. If the sheet material has to slide over die elements, roughness leads to higher friction, lubricant breakdown, and galling, although the effect depends also on die composition and hardness. The harder the tool, the more damaging is its roughness. Thus, a cast iron die or a nonhardened steel die can be rougher than a heat treated steel or WC die.

Because of their importance in adhesion, materials composition and hardness have decisive influences. Adhesion between die and workpiece governs whether die pickup will occur and, if it occurs, whether it will be cumulative. Major factors are:

- Compatibility of die and workpiece materials can be quite subtle. For example, increasing the magnesium content of aluminum alloy 2036-T6 reduces adhesion. In steels, segregation of manganese, chromium, and other alloying elements has been suspected to reduce adhesion, perhaps by strengthening the surface layer. Steels precipitation hardened with niobium or titanium additions are more resistant to galling than solid-solution (Si, Mn) hardened steels for the same reason. These effects should be quite general. Martensitic transformation induced in austenitic stainless steel increases friction and draw forces but reduces adhesion.
- Surface composition plays an important role. A sheet subjected to sliding contact galls more readily on subsequent contact. An untouched surface has surface layers that may or may not protect it. For example, steel pickled in HCl

may be more prone to galling than that pickled in H_2SO_4.
- The hardness of the workpiece material plays a role. High-strength steels are not much more adhesive than low-carbon steels but generate higher pressures and are more prone to galling, especially on draw beads. Previously strain-hardened material undergoes less hardening during working, and this is beneficial because it reduces the thickness of wear debris. However, the sheet hardness should be not more than half the tool hardness, otherwise, strain hardening of the asperities makes them strong enough to cause abrasive wear of the die.

Lubricant Application (Ref 9)

The method used to apply the forming lubricant directly influences handling, part quality, and direct costs, as well as indirect costs (cleaning) and the metal-forming operation itself. It is very important to deliver the appropriate volume of lubricant to the right location with the correct velocity at the right temperature in order to achieve optimum lubrication. Some of the more widely used application methods are discussed in this section.

Roller coating (Fig. 8a) is frequently used to coat strip before the strip enters the press. The strip passes through rollers, either top and bottom rollers or one or the other, and is coated with lubricant. The lubricant is fed to the roller either by dripping or by having the roller revolve, in a lubricant reservoir. The roller may be the same width as the strip and may be made of various materials. Pressure on the rollers can be varied to control lubricant thickness.

Air or airless spray devices (Fig. 8b) are used to apply lubricant to preselected areas of the workpiece and/or tooling at the required frequency. Generally, higher-viscosity lubricants are applied by airless spray rather than by air spray units.

Drip application to the workpiece (Fig. 8c) can be used effectively when parts are relatively

small and the cost of alternative application devices is not justified by the production demand. Heavy lubricants cannot be applied readily by this method, nor can the geometry of application be tailored selectively.

Flood application methods (Fig. 8d) require suitable press modifications to provide for recirculation of the fluid. Waterproofing to protect sensing elements must be provided for if a water-base fluid is used, and lubricant flooding must be controlled to confine it to the press work area. Flooding can be very useful when cooling is desirable and debris must be removed continually from the work area.

Control of Lubricant Characteristics

Maintaining the physical and chemical properties of lubricants is central to control of the sheet-metal forming process itself. Table 3 lists some important characteristics of metal-forming lubricants for which tests have been developed and the applicability of testing to various types of lubricants. Control methods developed under the aegis of the American Society for Testing and Materials (ASTM) are widely used. Some of the ASTM test methods for important characteristics are discussed next.

Viscosity (ASTM D 445, D 446, and D 88) is as important as any physical property of fluids used in sheet-metal forming. Viscosity control is based on the measurement of the time required for a lubricant to flow through a fixed orifice at a controlled temperature.

Sulfur and chlorine (ASTM D 129 and D 808) are measured by carrying out a controlled combustion of a lubricant sample in a so-called bomb. ASTM D 129 describes a technique used for sulfur, and ASTM D 808 describes the appropriate technique for chlorine.

Corrosion potential can be very important in both evaluating the appropriateness of a specific lubricant and for purposes of corrosion control during continued use. The methods most commonly used, for example, ASTM D 130 for copper-base metals and ASTM D 1748 for steel

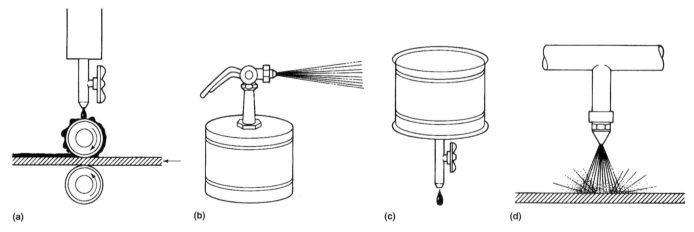

(a) (b) (c) (d)

Fig. 8 Four common methods of applying lubricants. (a) Roller coating. (b) Spraying (simple spray apparatus shown). (c) Drip application. (d) Recirculating flood

corrosion test strips, are based on the potential of water-lubricant interaction with the metal surface in a controlled way. Immersion in the fluid at elevated temperatures or exposure of a lubricant-coated strip in a humidity cabinet are techniques frequently used; they may be varied in many ways to attempt to duplicate as closely as possible particular operating conditions.

Emulsion stability is often of importance in maintaining a reproducible and reliable parts production line. A common method used to evaluate stability is ASTM D 1479, in which water available at the work site is used to make an emulsion. The emulsion is then stored for a specified period of time and evaluated. The concentration of oil in the bottom portion of the sample is a measure of the emulsion stability.

Fatty compounds are frequent components of a metal-forming lubricant. The test method ASTM D 94, using saponification number as a basis for fat determination, may be used as long as the basic fat component is known.

Hydrogen ion concentration (pH) is important in water-base lubricants. Variation of pH can be an important indication of fluid contamination or degradation due to biological or chemical reaction. Monitoring of pH on a regular basis can serve as an important control of fluid properties in a manufacturing environment. The pH can be measured with specially prepared pH paper or a pH meter.

In-Plant Control

Very important test procedures that do not require sophisticated laboratory facilities or highly skilled personnel are used for in-plant control. Some of the test procedures useful for lubricant control in plants are discussed next.

Emulsion concentration is one of the most important characteristics of a lubricant in metalworking. Variation in concentration occurs during use because of water evaporation; selective dragout; and, frequently, lack of emulsion makeup control. A number of control techniques can be used successfully in the plant. These include the use of titration, a refractometer, and acid split.

Titration can be used to measure either emulsifier concentration or, in turn, emulsion concentration. Anionic emulsifiers can be titrated by using a cationic material with an appropriate indicator. Water hardness effects on the titration must be accounted for to improve the reliability of the titration.

A refractometer may be used to measure concentration. The refractive index of the emulsion or water-base solution varies with concentration. Fluid compositions also vary; therefore, the refractive index must be calibrated for each fluid. Fluid contamination can cause a false index reading and may prevent use of this measurement method.

Acid split is used to determine water concentration in an emulsion. The emulsion is split by using a strong salt solution or acid. If oil (tramp or machine oil) enters the emulsion, then the acid split will reflect all the oil present, and thus incorrect concentration values will be obtained.

Corrosion tests have been developed for measuring the possibility of change in this characteristic of a lubricant. A plant procedure is based on taking pieces of sheet metal to be formed, coating them with the lubricant to be used, and placing the specimen in a plant location most likely to encourage corrosion. Experience with a procedure standardized as much as possible with respect to lubricant, metal, and environment can be useful for plant testing. Various tests using metal coupons coated with lubricant and subjected to a high-humidity environment (humidity cabinet) can also be used. Cast iron blocks and chips can be used for accelerated testing, using any of a number of procedures involving contact with lubricant and subsequent aging in a selected room or plant environment.

Biological testing in the plant can be carried out to advantage if reasonable control of test conditions can be maintained. Dipsticks have been developed that consist of a plastic substrate coated with a nutritive gel selected for growth of either bacteria or mold, as desired. The method, though qualitative, is useful as a fluid control procedure in the plant. References 10 and 11 contain more information on control of lubricant characteristics.

Lubricant Toxicity (Ref 12)

Because the environment, as well as the biochemical characteristics of the human body, are extremely complex, effects on health in the workplace are difficult to isolate and measure. In general, exposure to sheet-metal working lubricants may result in dermatitis, acne, tumors, and pigment changes in the skin. All of these may occur also from exposure to other by-products in the workplace, in the home, or in the environment in general.

Skin patch tests of a suspect material may reveal the potential for dermatitis. A small patch is coated with the material in question and taped to the skin.

The Ames test, which measures the effect of a test material on bacteria, has been used as a screen for possible causes of mutagenic effects in humans. Most materials that may cause mutagenic effects are not necessarily carcinogenic. Evaluation of a metalworking fluid for its possible adverse biochemical effects should be a part of the lubricant selection process. The requirement of a completed material safety data sheet on all chemicals and their mixtures can serve as an important source of information for the screening procedure. Appropriate work practices and personal hygiene can significantly reduce adverse biochemical effects.

Microbiology of Lubricants

Microbiological contamination of lubricants can be a serious problem. When water is part of the composition of a metalworking fluid, microbes can grow in the fluid, either deliberately or accidentally as a result of contamination. Bacteria are the main problem in mineral-oil

Table 3   Lubricant characteristics and their applicability to various types of lubricants

Characteristic	Oil-base	Solvent-base	Solids	Water-base emulsions	Solutions (water)	Coatings	Pastes
Particulate concentration	X	...	...	X	X	...	X
Emulsifier level	...	...	...	X	...	...	...
Viscosity	X	...	...	X	X	...	...
Sulfur concentration	X	X	X	X	X	...	...
Chlorine concentration	X	X	X	X	X	...	...
Particle size	...	...	...	X	X	...	...
Solids content	X	X	...	X	X	X	X
Color	X	X	...	X	X	...	...
Saponification number	X	X	...	X	X	...	X
Evaporation	X	X	...	...	...	...	...
Flash point	X	X	...	...	...	...	...
Molybdenum disulfide	...	X	X	...	...	...	...
Graphite	...	...	X	...	...	...	...
Carbonate	...	...	X	...	...	...	...
Stearate	...	...	X	...	...	...	...
Water content	X	X	...	X	X	...	X
Hard-water tolerance	...	...	...	X	X	...	...
Stability	X	X	...	X	X	X	X
Bacteria, mold, fungus	...	...	...	X	X	...	X
Amine concentration	...	...	...	X	X	...	...
Phosphate concentration	...	...	...	X	X	...	...
Corrosion potential	...	...	...	X	X	...	X
pH	...	...	...	X	X	...	...
Conductivity	...	...	...	X	X	...	X
Concentration	...	...	...	X	X	...	...
Odor	X	X	...	X	X	...	...
Coating thickness	X	X	...	X	...	X	...
Foaming characteristics	...	...	...	X	X	...	...

emulsions; in water solutions, molds and yeasts are the principal microbiological contaminants. These microbes feed and grow on one or more of the lubricant constituents or contaminants. If they grow by attacking a single constituent, the constituent can be so altered chemically in short order, frequently as rapidly as in one day, that the lubricant no longer operates effectively as intended. An emulsifier may no longer promote stable emulsification, a rust-preventive additive may no longer prevent rust formation, or a lubricity additive may no longer provide an effective lubricant film.

Anaerobic bacteria, often present in fluids, operate in the absence of oxygen and cause the breakdown of sulfur compounds in the fluid, creating the strong rotten-egg odor that may occur in the fluid reservoir. Aerobic bacteria grow in the presence of oxygen and are the most aggressive in promoting overall chemical breakdown of the lubricant. Molds and yeasts, when present, do not drastically degrade lubricant chemistry; they do, however, block filters, and may build up on tools and generally have a disagreeable feel and appearance.

Bacteria and mold or yeast infestation of the lubricant may be controlled by appropriate formulation of the lubricant, use of an effective biocide, pasteurization, and sometimes effective filtration. Good practices involving clean work habits and avoidance of contamination of the fluids by plant dirt, food, and so forth will contribute to longer life of fluids. References 13 to 15 contain more detailed information on the effects of microbiological contamination.

Environmental and Occupational Health Concerns

Concerns about the environment, health, and safety have led to the enactment and enforcement of a number of laws and regulations. Some of the most important of these as they affect the use and disposal of sheet-metal forming lubricants are discussed here.

The Environmental Protection Agency (EPA) was established by the U.S. Congress to ensure protection of the environment in which we live and work. Over the years, clean air and water have become a subject of increasing concern. Several acts addressing these concerns have been passed, requiring enforcement by EPA. Some of the more important acts are:

- The Toxic Substance Control Act, designed to regulate the manufacture, processing, distribution, use, and disposal of chemicals. Material Safety Data Sheets are required for all lubricants, with information about the toxicity, handling, and safe disposal of such materials.
- The Clean Air Act, as amended in 1977, which specifically regulates the concentration of volatile lubricant components in air. It established permissible concentrations that must not be exceeded in the workplace.

- The Water Pollution Control Act regulates the passage of waste into the water supply. The condition, type, and concentration of waste that may be permitted to enter the water supply is established and regulated. Disposal of lubricants, if it involves entry into a water system, must satisfy the appropriate criteria for disposal.
- The Resource Conservation and Recovery Act (RCRA) regulates hazardous waste management and disposal. Lubricants, if they contain hazardous components, are categorized as hazardous materials and must be stored, transported, and disposed of as prescribed by regulation. Appropriate records must accompany each step, as regulated.

The Occupational Safety and Health Administration (OSHA) is concerned with safe working conditions in the workplace. The impact of occupational safety and health legislation on lubrication as practiced in the workplace concerns prevention of occupational illnesses due to hazardous or toxic chemicals. Companies are required to record all injuries or illnesses as they occur. Relative toxicities of chemicals in the workplace have been surveyed by OSHA.

Product liability is of increasing concern because of the exploding cost of defending a product liability case as well as the bringing of nonmeritorious claims to court. At the same time, increasing attention has been paid to safer, better products as well as a safer, better work and living environment. Lubricant manufacturers and users are faced with claims for improper formulation, usage, and performance. Appropriate steps to anticipate and eliminate possible product liability claims occupy increasing time and attention on the part of both the supplier and user of lubricants.

The laws and regulations described in this article illustrate only a few of the more important ones that govern and regulate the use, disposal, and manufacture of sheet-metal forming lubricants. New developments in lubrication technology will surely be conditioned by this regulatory and liability environment.

Lubricant Selection

Selection of an appropriate lubricant for a specific sheet-metal forming operation depends on a number of diverse criteria that must be successfully satisfied:

- Effect of lubricant on the forming operation (effect on both tool wear and part tolerance and finish)
- Application method and efficiency thereof
- Maintenance of lubricant performance, taking into account problems of recirculation, testing, and disposal
- Corrosion of tooling, machine, and finished parts
- Cleaning
- Welding, painting, or coating of the work material

- Worker response
- Supplier support
- Toxicity and health
- Cost, including cost of lubricant and its effect on overall product cost

A wide variety of operations are commonly used to form parts from sheet metal. Blanking, piercing, ironing and stretching, forming, drawing, fine blanking, and spinning are some of the operations that are used separately or in sequence to form parts, and are discussed in other articles in this Volume. Unlike bulk deformation processes, the temperature increase in sheet-metal forming is generally low unless the complexity of deformation and/or the strength of the material is very high. Generally, except in the case of ironing, sheet-metal forming involves little or no thickness change. Deformation is generally biaxial in the plane of the sheet. However, there can be considerable relative motion between tool and workpiece. Further, the area of tool contact is often large. Surface finish and part tolerance depend on tool geometry and finish, speed and temperature of deformation, the combination of forming steps required, and the choice and application of lubricant. Tables 4 and 5 list some of the types of lubricants used successfully in sheet forming of various metals at room temperature and at elevated temperatures. The type of metal, complexity of part shape, part tolerance, and surface finish specifications all influence the choice of lubricant. The more critical the operation, the more important the need for film strength and adhesion of the lubricant film to the metal substrate. Further, there may be a need for increased film thickness.

The relationship of many process variables, including speed of operation; tool geometry; number and severity of discrete forming operations; worker exposure to lubricant; fluid makeup; circulation, recycling, and disposal; and the effect of the lubricant film on subsequent operations such as welding, cleaning, and plating, influence the choice of lubricant. A simple recommendation based on material, tooling, and process may prove to be far from optimal; many other factors can and should influence lubricant choice.

Nevertheless, in order to give some background information that may provide guidelines in the selection of a lubricant, a number of processes applied to various sheet materials relative to the choice of an appropriate lubricant are discussed subsequently. A summary of commonly used lubricants in sheet forming is in Table 2.

Lubrication of Carbon and Low-Alloy Steels

A wide range of lubricants are successfully used to form steel sheet. The speed of operation and resulting temperature, as well as thickness of stock and complexity of shape, influence lubricant choice. Oil-, solvent-, and water-base fluids,

Table 4 Lubricants for forming of specific sheet materials

| Metal or alloy | Recommended lubricants for: | |
	Cold forming	Hot forming
Aluminum alloys	Synthetic solutions, emulsions, lanolin suspensions, water suspensions, soap solutions, mineral oil plus fatty oils	Graphite suspension
Copper alloys	Emulsions plus fatty oils, mineral oils plus fatty oils, soap suspensions, water suspensions, tallow suspensions, synthetic solutions	Pigmented pastes, graphite suspensions
Magnesium alloys	Solvent plus fatty compounds, mineral oils plus fatty compounds	Graphite plus molybdenum disulfide, soap plus water, tallow plus graphite
Nickel, nickel-base alloys	Emulsions, mineral oils plus EP additives, water plus chlorine additives, conversion coatings plus soap	Graphite suspension, molybdenum disulfide suspension, resin coating plus salts
Refractory metals and alloys	Copper plating	Molybdenum disulfide, graphite suspension
Carbon and low-alloy steels	Emulsions, soap pastes, water, fatty oils plus mineral oils, polymers, conversion coating plus soap, molybdenum disulfide or graphite in grease, synthetic solutions	Graphite suspension
Stainless steels	Fatty oils plus mineral oil, water, polymers, conversion coating plus soap, mineral oil plus EP additives, pigmented soaps	Graphite suspension
Titanium, titanium alloys	Water, pigmented soaps, polymers, conversion coating plus soap	Graphite and molybdenum disulfide suspension

EP, extreme-pressure

Table 5 Commonly Used Sheet Metalworking Lubricants and Typical μm Values

Lubricants are listed according to increasing severity of conditions.

Material	Shearing, bending(a)	Pressworking(a)	μm(b)
Steels	Dry (pickle oil)	Dry (pickle oil)	0.2
	EM (MO + EP)	Soap solution	0.15
	MO + EP	EM (MO + fat) (+EP)	MF
		MO + fat (+EP)	MF
		Fat (tallow)	0.07
		Soap (water-soluble) film	0.05
		Wax (chlorinated)	0.05
		MoS_2 or GR in grease	0.05
		Polymer coating (+MoS_2)	0.05
		Phosphate + soap	0.05
		Metal (Sn) + EM	0.05
Stainless steels and Ni alloys	EM (MO + Cl)	EM (MO + Cl) (or fat)	0.2
	MO + Cl additive	MO + Cl additive	MF
		Chlorinated wax	0.07
		Polymer coating	0.07
		Oxalate + soap	0.07
		Metal (Cu) + MO	0.05
Al and Mg alloys	EM (MO + fatty derivatives)	EM (MO + fatty derivatives)	0.15
	MO + fatty derivatives	MO + fatty derivatives	MF
		Soap or wax (lanolin) coating	0.05
		Polymer coating	0.05
		Warm: GR film	0.1
Cu and Cu alloys	Soap solution	Soap solution	0.1
	EM (MO + fat)	EM (MO + fat) (or fat)	0.1
	MO (+fat) (+EP)	Drawing paste (EM of fat)	0.1
		Fat (tallow)	0.07
		Pigmented tallow (MoS_2, etc.)	0.05
Ti alloys	MO + EP	Wax coat	0.07
		Oxide + soap coating	0.07
		Polymer (+MoS_2)	0.05
		Fluoride-phosphate + soap	0.05
		Metal (Cu or Zn) + MO (or soap)	0.1
Refractory metals	MO + EP	Warm: MoS_2 or graphite	0.2
		Cold: Al-Fe-bronze dies with wax	0.07

(a) EM, emulsion of ingredients shown in parentheses. MO, mineral oil; of higher viscosity for more severe duties, limited by staining. EP, extreme-pressure additive (S, Cl, and/or P; also sulfochlorinated fats). (b) MF, mixed-film lubrication: μm = 0.15 to 0.05. Source: Ref 1

as well as pastes and gels, may be effective, depending on operating conditions.

Light Pressing. Industrial experience shows that almost any lubricant (light oil, dilute soap solution, or emulsion) is satisfactory. The insensitivity of steels to lubrication may well be due to the presence of residual films. Annealed and pickled sheet is coated with a corrosion-prevention lubricant, and cold-rolled sheet carries an oily residue. After degreasing with orthosilicate solutions, a silica film remains, which is likely to have a parting action.

Blanking and Piercing. Straight oil, compound oil, or emulsions with sulfur, chlorine, and/or fatty esters are used for most operations. Solutions may be used when the operation is less severe.

Coining. Because die buildup must be prevented, light oils straight or compounded with fatty additives or chlorine-containing additives are used. Water-base solutions may be effective when formulated with alkanolamides, fatty esters, and/or long-chain derived fatty acid soaps.

Other Operations. Compression, tension, binding, or shearing forces may occur during forming on a particular part; therefore, an appropriate choice of lubricant viscosity and chemistry varies widely from part to part, depending on part configuration. Essentially all classes of lubricants have been successful. Generally, however, the present emphasis is on formulating solutions or emulsions that prevent galling or tearing of the sheet metal.

Multislide Forming. Light compounded oils and water-base emulsions and solutions with chlorine and fatty additives are used successfully. Care must be taken to prevent buildup on the dies and corrosion of machine or formed parts when water-base lubricants are used.

Deep Drawing and Stretching. Progressive- and multiple-die sets are used frequently in deep drawing, and thus carry-through of the lubricant becomes more and more important. Multiple-die lubrication sites that provide die lubrication at succeeding stations may be an alternative method for ensuring adequate lubrication. Compounded oils and emulsions or pastes are being used successfully. Chlorinated compounds and fatty esters are commonly chosen additives. Soap-base lubricants are also applied frequently and successfully in deep drawing.

Severity of duty in deep drawing and stretch drawing is determined by the degree of drawing and sliding, and by sheet thickness.

For light duties, viscous oils, fatty-oil emulsions, or water-soluble polymer formulations are sufficient. When pickup is a problem, a sulfur additive is needed.

For more severe duties, highly viscous oils with fatty oils and EP additives, or drawing pastes based on pigmented soap-fat formulations, can be used. Pigments such as talcum, chalk, and ZnO are good parting agents, but their removal presents difficulties. Fats, particularly tallow, are successful.

Soaps, such as zinc stearate and the specially formulated dry soaps, are excellent, as are waxes. However, on deposition from aqueous solutions, the heating required to drive off the water causes aging in rimming steels. Chlorinated wax is one of the heaviest-duty lubricants.

Graphite and MoS_2 do not necessarily ensure low friction, but when added to a fat such as tallow, or to a drawing paste, they give good results. They are effective under severe conditions such as drawing of high-speed steel. Removal is again a problem, and this can be minimized by applying the layer-lattice compound locally to the die radii.

Polymers such as PE and PTFE have often been found to be useful and, as discussed in conjunction with mill-applied coatings, have found their specific fields of application.

Heavy Drawing and Ironing. High surface pressures and large surface expansions demand a phosphate/soap system. Reactive oils do not give quite the same performance but are suitable for moderately heavy duties. Among the layer-lattice lubricants, bonded MoS_2 coatings can survive. An alternative is to coat with tin 0.25 to 0.75 μm (10 to 30 μin.) thick, as in production of drawn and ironed cans, and to lubricate and cool with the aid of a compounded emulsion.

Spinning. Both lubrication and cooling are important effects provided by appropriate lubricant choices. Water-base emulsions and solutions containing chlorinated or sulfochlorinated additives and/or fatty esters are used. Nonpigmented water-base pastes are also effective.

Roll Forming. Water-base emulsions and solutions have been used to form both bare and coated steel. In the case of galvanized steel, white rust can be a problem if water-base lubricants are used and not properly applied and removed. Solvent-base fluids are often the preferred lubricant, particularly when minimal residue is important. Light oils may also be used, both compounded and uncompounded with chlorinated compounds, soaps, or fatty esters.

Lubrication of Stainless Steels

In general, stainless steels are more difficult to form than carbon or low-alloy steels because of their higher strength. Care must be taken in the choice of chemically reactive lubricants containing sulfur or chlorine. Furnace treatments should be carried out on formed parts after removal of the lubricant film. Sometimes this step may be avoided if high molecular weight polymers or complex fats are used in lubricant formulation. The lack of reactivity coupled with the frequent demand for very high surface quality can present challenges.

With steel dies, a viscous oil with chlorine additives suffices for light duties. Performance improves with increasing viscosity and also when polybutenes are used. Neat chlorinated paraffins give the best performance. Graphited tallow and pigmented oils and pastes are sometimes used, but the graphite causes carburization on annealing and presents difficulties of removal.

Of the separating films, dry soaps and waxes such as beeswax are adequate for medium duties. Chlorinated polymers deposited from aqueous or organic vehicles are more satisfactory, as are polymers in general. Peelable films are indispensable in making parts with high surface quality. However, for the most severe duties, including ironing, an oxalate/soap coating is best. If necessary, MoS_2 may be applied on top of it. When the die is coated with TiC, chlorinated oil suffices. Alternatively, the workpiece can be coated with a metal such as lead, but since the advent of the oxalate coating, this is seldom practiced.

Martensite transformation induced by deformation can lead to cracking of deep drawn, metastable austenitic stainless steel cups. Lubrication seems to play no role in this.

Blanking and Piercing. Solutions and emulsions containing sulfur and/or chlorine are particularly effective for lighter-gage material. Oils compounded with sulfur and/or chlorine are effective in heavier-gage material.

Multiple-Slide Forming. Oil- and water-base fluids are used successfully. Solutions with complex esters and fatty compounds have been used, successfully eliminating cleaning of parts before use.

Deep Drawing. Because of the high work-hardening characteristics of some stainless steels and attendant high die pressure and temperatures, oils with EP additives such as sulfur, chlorine, and phosphorus are most often the lubricants selected. Pigmented and nonpigmented pastes are often used with or without EP additives.

Roll Forming. Thin-gage sheet may be formed by application of solvent-base as well as water-base fluids at the roller. Thicker material and/or more complex shapes, on the other hand, require oils or emulsions compounded with fatty oils, EP additives, and soaps.

Lubrication of Heat-Resistant Alloys

In general, heat-resistant alloys (complex iron-, nickel-, and cobalt-base materials) as a class use lubricants that are suitable for stainless steel. Cobalt-base alloys are particularly difficult to form into sheet because of their high strength and relative inertness. Care must be taken to avoid embrittlement by sulfur diffusion in nickel-base alloys. Pigmented lubricants are difficult to remove after forming, and this may discourage their use. More information on forming of heat-resistant alloy sheet is available in the article "Forming of Nickel and Cobalt Sheet Alloys" in this Volume.

Blanking and Piercing. Soap-base oils and emulsions are used with and without chlorinated additives. Wax and polymer additives may be used if the blanking step is complex.

Spinning. Chlorinated additives in an oil base as well as in emulsions containing fatty esters are effective. Soap-base pastes are also often chosen for application.

Lubrication of Refractory Metals and Alloys

Although warm or hot forming of materials such as molybdenum, tungsten, tantalum, niobium, and their alloys is often required, room-temperature deformation is successfully carried out using metal coatings with auxiliary lubricant application.

The main problem in lubrication of refractory metals usually is brittleness or adhesion. The ductile-to-brittle transition temperature of tungsten sheet is between 100 and 400 °C (212 and 750 °F). Forming is done at elevated temperatures. Molybdenum and its alloys (TZM) are also too brittle to be formed at room temperature. At around 400 °C (750 °F), MoS_2 is the preferred lubricant, although graphite has also been used. Dies of aluminum bronze have been used with castor oil as the lubricant. Bronze dies with beeswax, or chromium-plated steel dies with graphite, are satisfactory in forming of uranium. The basic problem with tantalum is its severe galling on steel dies. One solution relies on bronze dies with beeswax.

Drawing. Fluorinated complex polymers are used to form tantalum; graphite in fat has been used for forming niobium. Both metals may be formed at room temperature.

Spinning. Room-temperature spinning of niobium has been carried out using molybdenum disulfide in oil or graphite in a soap-base paste.

Lubrication of Aluminum and Aluminum Alloys

Aluminum and aluminum alloy sheet may be stained or corrosion may occur through the improper choice of lubricant. Aluminum oxides build up during forming and may necessitate cleaning of both parts and equipment. The lubricant may contribute to undesirable buildup. Lubricant chemistry differs from the chemistry used in formulating lubricants for steel and refractory metals or high-temperature alloys.

For light pressworking, low-viscosity mineral oils, synthetic oils, or oil-base emulsions are suitable, invariably with boundary additives.

For more severe drawing and stretch drawing, the oil viscosity is higher, the emulsion is less diluted, and the boundary additive concentration is higher. Oils with EP additives are also used but are difficult to justify unless it is expected that the tool can develop protective films. When pickup is concentrated at specific areas of the die (such as draw beads), a concentrated lubricant may be applied locally.

Solid-film lubricants such as soaps and waxes are useful—especially if they possess boundary lubricating qualities. Graphite is less favorable because its success depends very much on application and because it leads to corrosion problems. Better is MoS_2, which yielded higher friction but also larger draw ratios in rectangular deep drawing. Except for working at elevated temperatures, there should be no need for layer-lattice lubricants. Polymers are used extensively in the form of precoated (painted) sheet. Peelable films of PVC or PE protect parts with critical surface requirements while providing excellent lubrication when oil is superimposed. Loose PTFE films are useful but are limited by their lack of ductility.

Blanking and Piercing. Oils or compounded oils of light viscosity are used. Fatty esters and chlorinated compounds are the compounding additions frequently chosen. Fatty compounded emulsions as well as soap-base solutions are also effective. Oxide deposits may be a problem with some compounded oils and emulsions.

Roll Forming. Solvent-base lubricants containing fatty esters as well as emulsions and solutions are frequently used lubricants. Light viscosity oils also may be used; however, oxide buildup on the rolls can be a problem. Soap-base lubricants are used if care is taken to prevent soap deposits.

Deep Drawing. Viscosity and lubricity are critical attributes of lubricants. Water-base fluids may be used if appropriate fatty additives and/or soap additives are used to provide the required barrier and lubricity. Oils of varying viscosity or dry soap films may be used for more difficult draws. Pastes that are soap- and fat-base have been applied successfully. Progressive draws involving multiple steps frequently should provide lubrication at more than one station.

Spinning. Waxes and soaps are effective lubricants. Colloidal graphite suspended in an oil carrier is also used successfully. Cleaning can be difficult depending on the severity of the spinning operation.

Lubrication of Copper-Base Alloys

Lubrication difficulties increase from copper (which does not form cumulative pickup) to brass to bronze. For light pressworking, medium-viscosity mineral oils, oil emulsions, or soap solutions are used. For heavier duties, fats are added—sometimes in concentrations up to 50%.

For deep drawing and stretch drawing, straight fatty oils or compounded mineral oils, emulsions, or soap-fat compounds are needed. The limiting draw ratio (LDR) increases with increasing viscosity. Depending on the severity of draw, oil or water-base lubricants in fluid or paste may be effective. Chlorine or sulfur EP additives are often encountered, although sulfur staining is a problem. Staining and/or corrosion of these metals in forming is a particular concern. Lubricants chosen must be formulated to prevent both from taking place. Dry soaps are effective as are, of course, polymers, including methacrylate resins. Layer-lattice compounds have been found to be effective but should not be needed except in working at elevated temperatures. Fillers such as kaolin have been used but, again, should not be needed.

Multislide Forming. Soap- and fat-containing oil and/or water-base lubricants are effective. Solvent-base lubricants containing a fatty ester

or an inhibited chlorine-containing compound may be used.

Roll Forming. All classes of lubricants using fatty compounds and/or soap-base additives are used successfully in water, oil, or solvent fluids.

Deep Drawing. Depending on severity of draw, oil- or water-base lubricants in fluid or paste form may be effective. Soap and fatty compounds, as well as inhibited chlorine-containing compounds, are commonly used additives.

Lubrication of Magnesium Alloys

Magnesium alloys are most often formed warm or hot. Forming at these elevated temperatures is not normally required for other metals of industrial importance. Only moderate deformation of magnesium alloys is possible at low temperatures, and then soaps, waxes, or (in more difficult cases) graphited tallow will suffice.

Stretching and Drawing. At temperatures to approximately 120 °C (250 °F), soap-base lubricants, fatty esters, polymer additives in oil and water, and pastes formulated with chlorinated additives are successful lubricant systems. At temperatures in excess of 120 °C (250 °F), the choice of lubricant is restricted to synthetic fluids formulated with soap, fatty esters, and/or chlorinated compounds. Above 230 °C (450 °F), graphite and/or molybdenum disulfide in various carriers are preferred.

More severe stretching and deep drawing must be done at temperatures from 220 to 370 °C (430 to 700 °F). The lubricant is invariably colloidal graphite in a volatile (organic or water) base. It is applied to the blank prior to heating by spraying or roller coating. To facilitate removal, the blank should not be etched prior to coating so as to retain the oxide, which is then removed together with the graphite coating. Caustic etching is followed by chromic acid-sodium nitrate pickling.

Spinning. At elevated temperatures, the synthetic fluids compounded with graphite and/or molybdenum disulfide are applied. Water carriers for the solid lubricants may be preferred to reduce the occurrence of smoke and the possibility of fire.

Lubrication of Nickel and Nickel-Base Alloys

These metals are difficult to wet with lubricants; thus heavy-duty lubricants with exceptional film-priming characteristics are necessary for effective lubrication. On the other hand, lubricants containing sulfur, chlorine, or solid additives such as zinc oxide or lead carbonate, if not removed from the nickel surface, can cause embrittlement of the metal.

Shearing, Blanking, and Piercing. Oils incorporating sulfur- or chlorine-containing additives may be used. Water-base lubricants of similar composition may be applied if they are

removed as soon as possible after forming to prevent embrittlement. Fatty esters and polymers have grown in application as components of formulated lubricants.

Deep Drawing. Soap-base pastes as well as oils with fatty esters, amides, and/or sulfur and chlorine additives have been used. Emulsions fortified with amides and polymers also have been formulated and applied.

Spinning. Pigmented pastes and chlorinated wax in oils are successful lubricants. Plain waxes and soaps are frequently important components of these lubricants.

Lubrication of Titanium Alloys

Galling of titanium alloys is a particular problem because of the affinity of the metal for die materials. Notch sensitivity and embrittlement may also lead to splitting or cracking of formed parts. Cold and warm forming may be carried out with suitable films designed to prevent metal-to-metal contact. Frequently, overlays of steel sheet or plastic sheet are used with an auxiliary lubricant.

Because of their limited ductility, titanium alloys are usually warm formed, although limited forming and drawing at room temperature are possible.

In cold working, the high adhesion of titanium to die materials necessitates the use of a continuous separating film, such as a polymer or, for stretch forming, even a dry wax. Only for light duties, such as roll forming, will an EP oil suffice. For heavier work, various surface treatments combined with wax or dry soap lubricants can replace the polymers. Oxidizing, fluoride/phosphate coating, and sulfidizing have been found to be effective. Nitriding provides only marginal improvement.

At elevated temperatures, the options are few. Low-melting glasses protect also during heating but are difficult to remove and may require additional graphite lubrication of the die elements. Graphitic lubricants are simpler to apply to the die and blanks but require subsequent removal.

Deep Drawing. Overlays are often used with oil-base lubricants formulated with chlorinated waxes. Oxidized or phosphated coatings are successful in relatively severe drawing operations at elevated temperatures. Graphite and/or molybdenum disulfide in oil may be used.

Roll Forming. Oils compounded with sulfurized or chlorinated fats are used. Oil- or water-soluble polymers may also be added. Chlorinated waxes, high molecular weight waxes, or polymers soluble in oil may be effective for relatively moderate deformation.

Spinning. Colloidal graphite and/or molybdenum disulfide blended in oil may be used at temperatures up to approximately 205 °C (400 °F). Chlorinated wax and/or sulfurized fat in oil also may serve as lubricants. At higher temperatures, fillers such as bentonite or mica with graphite and/or molybdenum disulfide formulated into a grease are used successfully.

Lubrication of Platinum-Group Metals

Surface contamination due to metal contact at surfaces with iron or other metals may adversely affect surface integrity and electrical resistivity. Separation of tool and workpiece by an appropriate lubricant film is critical. Platinum and palladium can be formed by most standard sheet-metal forming operations (blanking, piercing, and deep drawing). Cold welding of the workpiece to the tooling must be avoided, and therefore continuous lubricant films are important in operation. Many of the lubricants used for forming copper alloys may also be used for forming platinum and palladium. Rhodium and iridium are more difficult to form, and ruthenium and iridium are extremely difficult to form.

ACKNOWLEDGMENTS

This article has been adapted from:

- E.S. Nachtman, Selection and Use of Lubricants in Forming of Sheet Metal, *Forming and Forging,* Vol 14, 9th ed., *Metals Handbook,* ASM International, 1988, p 512–520
- J. Schey, *Tribology in Metalworking,* American Society for Metals, 1983, p 27–77

REFERENCES

1. J. Schey, *Tribology in Metalworking,* American Society for Metals, 1983, p 27–77
2. E. Nachtman and S. Kalpakjian, *Lubricants and Lubrication in Metalworking Operations,* Marcel Dekker, 1985, p 49–60
3. *Metalworking Lubrication,* S. Kalpakjian and S.C. Jain, Ed., American Society of Mechanical Engineers, 1980, p 53
4. J.T. Laemmle, Metalworking Lubricants, *Friction, Lubrication, and Wear Technology,* Vol 18, *ASM Handbook,* ASM International, 1992, p 139–149
5. E. Nachtman and S. Kalpakjian, *Lubricants and Lubrication in Metalworking Operations,* Marcel Dekker, 1985, p 63–105
6. J. Schey, *Tribology in Metalworking,* American Society for Metals, 1983, p 131–175
7. E. Nachtman and S. Kalpakjian, *Lubricants and Lubrication in Metalworking Operations,* Marcel Dekker, 1985, p 107–115
8. J. Schey, *Tribology in Metalworking,* American Society for Metals, 1983, p 197–220
9. E. Nachtman and S. Kalpakjian, *Lubricants and Lubrication in Metalworking Operations,* Marcel Dekker, 1985, p 117–123
10. C. Genner and E.C. Hill, Evaluation of the Dip Slide Technique for Cutting Oils, *Tribol. Int.,* Feb 1981, p 11–13
11. E. Nachtman and S. Kalpakjian, *Lubricants and Lubrication in Metalworking Operations,* Marcel Dekker, 1985, p 133–154
12. E.O. Bennett, The Biology of Metalworking Fluids, *Lubr. Eng.,* Vol 28 (No. 7), 1972, p 237–247
13. H.W. Rossmore, Antimicrobial Agents for Water-Based Metalworking Fluids, *Proc. 65th Annual Meeting of the American Occupational Medical Association,* April 1980, p 199–219
14. E. Nachtman and S. Kalpakjian, *Lubricants and Lubrication in Metalworking Operations,* Marcel Dekker, 1985, p 157–171
15. E. Hall, How Bacteria Damage Lubricants, *New Sci.,* Aug 1967, p 17

Selection of Materials for Press-Forming Dies

SHEET METAL is press formed to conform to the contours of a die and punch—largely by bending or moderate stretching, or both—and die material is selected largely by the economics of how many parts that can be produced using a die. Tool materials are usually selected on the basis of providing adequate die life at minimum cost. However, the final choice may depend on availability rather than on a small difference in die life or cost.

The process of selecting the most economical material for a press-forming die is determined by a number of factors that include:

- Requirements for part production, such as quantity of parts required, allowable dimensional tolerances, size, type, thickness, and hardness of the metal being formed
- Equipment capabilities and production practices, such as lubrication methods, die clearance, machine stability, and press speed
- Tool material properties in terms of hardness, hardenability, toughness, and the size and distribution of hard particles such as carbides
- Tool design factors, such as size, thickness, and complexity (fillets, corners, holes)
- Tool manufacturing methods and maintenance (e.g., machining, grinding, regrinding, polishing)
- Tool heat treatments (hardening, stress tempering, surface treatments)

In general, the useful life of a press-forming die is determined by its wear, which is primarily affected by the length of the production run and the severity of the forming operation. The amount of wear on a given die during forming is proportional to the total accumulated distance over which the sheet metal slides against the die at a given pressure between the surfaces in contact. Thin, soft, or weak sheet metals exert the least pressure and thus cause the least wear; thick, moderately hard or strong metals cause the most rapid wear.

The two major mechanisms of die wear in press-forming operations are abrasive wear and adhesive wear (i.e., galling). Galling involves a chemical or physical adhesion of the sheet materials to the tool, and it depends on the surface smoothness and chemical composition of the die and sheet materials. The rate of wear for each combination of work metal and die metal may vary considerably, depending on surface characteristics, speed of forming, and die lubrication. In situations in which wrinkles form in the parts, high localized pressures develop on the tools because of the ironing that takes place at these locations, and prohibitively high rates of abrasive wear and galling are almost always encountered. For more information on the factors of die wear, see the section "Wear in Shallow Forming Dies" in the article "Wear and Lubrication of Sheet-Metal Forming Dies" in this Volume.

In stamping dies, adhesive and abrasive wear are prominent mechanisms in determining tool life, and so both tool material composition and hardness are important die-life factors. Compared to blanking and piercing tools, the danger of spalling, chipping, and cracking is less significant for stamping dies. When the tool material has hard particulates for improved wear resistance, the effective hardness (H_e) depends on the effective surface-area fraction (α) of the hard carbide phase, such that $H_e = H_c + (1-\alpha)H_m$, where H_c is the hardness of the dispersed carbide phase, and H_m is the hardness of the matrix.

This article reviews the production variables that influence the selection of various stamping die materials, such as those listed in Table 1. Details on specific types of die materials are covered in sections on tool steels, cast irons, plastics, aluminum bronze, zinc-aluminum, and steel-bonded carbides. Some of these materials are also used in other tooling components besides the die. For example, guide pins, guide bushings, wear strips and plates, gibs, guide rails, guide blocks, cam-dwell wear plates, and center keys are made from steel, bronze, bronze-plated steel, aluminum bronze, sintered materials, cast iron, polymers, resins, fibers, composites, and other materials (some of which may accommodate graphite plugs for lubricating purposes).

Selection Factors

Recommended die materials depend on a number of variables for part production, such as the required severity of deformation, part size, sheet thickness, quantity, tolerances, and work metal. Selection factors may also depend on the type of forming operations (e.g., simple bending versus press forming). For example, simple bending dies are ordinarily subjected to less shock than other press-working tools; therefore, they can often be made of low-carbon steel, heat treated low-alloy steel such as 4140, or cast iron for low production of low-carbon steel pieces. For moderately high production, they should be made of flame-hardening grades of carbon steel, such as 1045, or flame-hardening cast iron, such as class 40 gray iron. Cam-operated dies, wiping dies, and dies used to make curved flanges (shrink or stretch) must be made of a higher grade of material. Tool steels are used for moderately long production runs.

The following sections describe factors in the selection of materials for press-forming dies. When press forming low-carbon steel, dies are made from a wide range of materials, including plastics, cast irons, tool steels, and cemented carbides. Severity of forming, number of parts to be produced, workpiece shape, work metal hardness, specified surface condition, and tolerances affect selection of the die material. Some typical examples of tool-material selection are listed in Tables 2 through 4 for various workpiece materials, with part size and configurations shown in Fig. 1. These tables are a summary (Ref 1) from observed tool performance and were based on 500 specific forming applications. They include some die materials (e.g., polyesters, phenolics) that have been largely replaced by improved materials for die applications. Details on specific types of die materials are discussed in separate sections of this article.

Effect of Part Shape. In simple parts (parts 1 and 2 in Fig. 1), there is limited deformation of the sheet during forming, and the die components have very low wear rates. The tooling for parts 1 and 2 consists of a punch and an upper and lower die. There is little or no sliding of the sheet over the lower die, and little movement over the punch. Large quantities can be produced in various die materials of convenient size (Table 2).

Tooling for part 3 consists of a punch and lower die (Fig. 2). In forming, the punch pushes the blank through the lower die, which results in sliding on the lower die, but there is little sliding

Table 1 Examples of various materials used for press-forming dies

Material	Nominal composition	Comments
Tool steels		
W1	Fe-1.0C	. . .
S1	Fe-0.50C-1.5Cr-2.5W	. . .
O1	Fe-0.9C-1Mn-0.5Cr-0.5Mo	. . .
A2	Fe-1C-5Cr-1Mo	Commonly used conventional tool steel
A4	Fe-1C-2Mn-1Cr-1Mo	Limited availability
D2	Fe-1.5C-12Cr-1Mo-1V	Commonly used conventional tool steel
D3	Fe-2.25C-12Cr	. . .
D5	Fe-1.5C-12Cr-1Mo-3Co	Limited availability
D7	Fe-2.35C-12Cr-1Mo-4V	Limited availability
M2	Fe-0.8C-4Cr-5Mo-6W-2V	. . .
M4	Fe-1.4C-4Cr-4.5Mo-5.5W-4V	Powder metallurgy tool steel
Vanadis 4	Fe-1.5C-8Cr-4V-1.5Mo	Powder metallurgy tool steel
Vanadis 10	Fe-2.9C-8Cr-9.8V-1.5Mo	Powder metallurgy tool steel
CPM 10V	Fe-2.45C-5Cr-9.75V-1.3Mo	Powder metallurgy tool steel
Other ferrous alloys		
Hot-rolled low-carbon steel	Fe-0.10 to 0.20C	. . .
Unalloyed cast iron, 185 to 225 HB	Fe-3C-1.6Si-0.7Mn	. . .
Alloy cast iron, 200 to 250 HB	Fe-3C-1.6Si-0.4Cr-0.4Mo	. . .
Cast high-carbon steel, 185 to 225 HB	Fe-0.75C	. . .
Cast alloy steel, 200 to 235 HB	Fe-0.45C-1.1Cr-0.4Mo	. . .
4140 alloy steel	Fe-0.4C-0.6Mn-0.3Si-1Cr-0.2Mo	. . .
4140 modified	Fe-0.4C-1.2Cr-0.2Mo-1Al	. . .
Nonferrous alloys		
Zinc alloy (UNS Z35543)	Zn-4Al-3Cu-0.06Mg	See text
Aluminum bronze (UNS C62500), 270 to 300 HB	Cu-13Al-4Fe	See text
Plastics		
Polyester-glass and/or powder filled	Glass filled: 50% polyester, 50% glass in the form of cloth, strand, or chopped fibers	No longer commonly used due to poor dimensional stability
	Powder filled: metal or ceramic powder	
Phenolic	. . .	Used for stretch forming block and some molds, but seldom used anymore
Epoxy-glass and/or powder filled	Glass filled: 50% epoxy, 50% glass (cloth, strand or fibers)	. . .
Nylon-metal filled	Polyamide reinforced with metal powder	Limited use
Polyurethane	. . .	Versatile material for low-cost die pads

on the punch. Therefore, the punch generally has ten times the life of the lower die made from the same material. However, wear and galling are expected in the areas of moderate shrinkage of this part, particularly when the part is formed on these single-action dies. For a small die and punch, the cost of steel is of minor importance, and D2 tool steel may be used for production quantities as low as 10,000. If galling occurs during preproduction trials, the tool can be nitrided. Typical materials for lower dies (Fig. 2) used in press forming small parts similar to that shown in part 3 in Fig. 1 are given in Table 3.

Tooling for part 4 (Fig. 1) consists of a punch and upper and lower die (Fig. 3). Without the upper die, excessive wrinkling would be expected at the shrink flanges. As in part 3, a less wear-resistant material is required for the punch and upper die than for the lower die, due to greater sliding occurring on the lower die. Under conditions for which the tooling was typically made of tool steel (Table 3), the tooling was in the form of inserts in a lower die made of cast iron, as shown in Fig. 3, and the punch was made of a cast tool steel. For example, a cast iron die with tool steel inserts at points of greatest wear was typical for production quantities of 10,000 to 100,000 pieces. When this part must be held to close

tolerances over lengthy production runs, tool steel inserts should be used at all surfaces subject to wear. For quantities of less than 100,000 pieces, the entire lower die was typically made of the material indicated (Table 3), without inserts. The punch was made of a less wear-resistant material, which is usually the same as the lower-die material in the first column to the left of the quantity being considered.

Table 3 may be used to select lower-die materials for parts made of sheet thicker or thinner than the 1.3 mm (0.050 in.) thick sheet used in the tables, or for parts of greater or lesser severity. For parts of greater severity or sheet of greater thickness, use the die material recommended for the next greater production quantity than the quantity actually to be made (the column to the right of the actual production quantity in the table). Similarly, for parts of lesser severity or sheet of lesser thickness, use the die material recommended for the next lower production quantity (the next column to the left of the actual production quantity).

The tooling for parts 5 and 6 (Fig. 1) consists of a punch and lower die with no upper die required, because these parts are produced by stretching rather than by shrinking of the metal along the constricted region. The metal envelops

the punch with minimal sliding and consequently produces approximately ten times more wear on the lower die than on the punch. The body of the lower die could be made of cast iron with wearing edges of tool steel, and the punch could be of a material less wear resistant than tool steel, for example, low-cost alloy cast iron. However, the same material can be used for both the punch and lower die for part 5 because of the smaller part size and minor material cost. Table 4 summarizes tool material selections for a small part of moderate severity (part 5 in Fig. 1).

In the selection of die material for larger parts of moderate severity (part 6 in Fig. 1), the critical locations are the wearing edges of the lower die. The body of the lower die would be made of cast iron with wearing edges of tool steel in instances similar to those where tool steel is indicated (Table 4). The punch for making larger parts of moderate severity could be a material approximately one-tenth as wear resistant as tool steel—for example, cast iron. For a quantity large enough to require a punch material of greater wear resistance than alloy cast iron or cast steel, a tool insert should be considered at the constricted section.

Part Size. For small stampings, cast or plastic dies are uneconomical unless they are made from a model already available and with only minor finishing operations required on the dies. When the cost of patternmaking is included, cast or plastic dies are usually more expensive than are dies machined from other materials. The cost of the die material is usually a small fraction of the total cost of dies for a small part, and the availability of material in such a size that would minimize machining on the dies is usually a greater factor in cost than is any other.

As the size of the part increases, cost savings resulting from minimizing machining by the use of a casting close to final size more than offsets the cost of a pattern. However, tool steel or carbide inserts must be used on high-production dies subject to severe wear and galling. The selection of both the material and the locations of the inserts should be conservative when it is important that production not be interrupted to alter the tooling. If tools can be taken out of production, gray cast iron dies may be used, with the wear surface flame hardened and inserts added later if needed because of wear on the critical surfaces.

Work Metal. High-hardness sheet metals wear dies more rapidly than do softer materials, but other factors, such as the presence of scale on the surface of hot-rolled unpickled steels, cause two to five times more wear. However, scaled surfaces cause less galling, which, on tool materials, may be an even more serious condition than wear, because galling or "pickup" on a die causes frequent interruptions of production forming for reconditioning of the die.

Soft brass and aluminum cause less wear and galling than does carbon steel; stainless steels and heat-resistant alloys cause more wear and galling. When galling is anticipated, it is desirable to use materials that can be treated subsequently, if necessary, to eliminate the difficulty. Possible treatments include chromium plating of any hardened steel, the hardening of alloy cast iron, and the nitriding of or physical vapor deposition coating of tool steels containing alloy elements that provide secondary hardening during heat treatment, thus permitting them to retain their hardness through the temperature exposures in these surface treatments.

Sheet Thickness. Thick sheets of any metal exert greater pressure on the dies than do thin sheets of the same metal. Both abrasive wear and adhesion (galling) increase with increasing sheet thickness.

The selection tables (Tables 2 to 4) from *Properties and Selection of Metals*, Volume 1, 8th edition *Metals Handbook*, 1961 (Ref 1), are based on sheet thickness of 1.3 mm (0.050 in.). Table 5 recommends materials for dies to make parts 3 and 4 (Fig. 1) in three sizes, from sheet of four different thicknesses. The selections in Table 5 illustrate the increasing effect of galling and wear as parts are made from thicker sheets—especially parts with shrink flanges.

Table 5 deals only with die materials for forming steel parts. The forming pressure depends not only on the thickness but also on the strength of the sheet being formed. Wear and

galling are less severe with any thickness of soft metal, such as aluminum and copper alloys, than with low-carbon steel, but more severe with high-strength metals such as stainless steels and heat-resisting alloys.

Quantity. The number of parts to be produced is obviously an important factor. Low production quantities of simple shapes may just require plastic dies, while more severe deformation and higher production quantities need wear-resistant tool steels, nitrided steels, coated tool steel, or perhaps dies or inserts of cemented carbides with parts of extreme severity or production quantities of more than 10^6 parts.

Table 2 Recommended punch material for forming a part of minimum severity

Part (made of 1.3 mm, or 0.050 in., sheet)				Recommended punch material(a)				
	Requirements			Total quantity of parts to be formed				
Metal being formed	Finish	Tolerance, in.	Lubrication(b)	100	1000	10,000	100,000	1,000,000
Small part of minimum severity (part 1 in Fig. 1)								
1100 aluminum, brass, copper(c)	None	None	Yes	Epoxy-metal, mild steel	Epoxy-metal, polyester-metal, mild steel	Epoxy(d), polyester(d), mild steel	Polyester-glass(d), mild steel	O1
1100 aluminum, brass, copper(c)	Best	None	No	Epoxy-metal, mild steel	Epoxy-metal, polyester-metal, mild steel	Epoxy(d), mild steel	Polyester-glass(d), mild steel	O1
Magnesium or titanium(e)	Best	None	Yes	Mild steel	Mild steel	Mild steel	Mild steel	A2
Low-carbon steel, to 1/4 hard	None	None	Yes	Epoxy-metal, polyester-metal, mild steel	Epoxy-glass-metal, polyester-glass-metal, mild steel	Polyester-glass(d), mild steel	Mild steel	O1, A2
Type 300 stainless, to 1/4 hard	None	None	Yes	Epoxy-metal, polyester-metal, mild steel	Epoxy-glass-metal, polyester-glass-metal, mild steel	Polyester-glass(d); mild steel	Mild steel	O1, A2
Low-carbon steel, to 1/4 hard	Best	None	Yes	Epoxy-metal, polyester-metal, mild steel	Epoxy-glass-metal, polyester-glass-metal, mild steel	Mild steel	Mild steel	O1, A2
High-strength aluminum or copper alloys	Best	None	No	Epoxy-metal, polyester-metal, mild steel	Epoxy-glass-metal, polyester-glass-metal, mild steel	Mild steel	Mild steel	O1, A2
Heat-resisting alloys	Best	None	No	Epoxy-metal, polyester-metal, mild steel	Polyester-glass-metal, mild steel	Mild steel	Mild steel	O1, A2
Type 300 stainless, to 1/4 hard	Best	None	No	Epoxy-metal, polyester-metal, mild steel	Epoxy-glass-metal, polyester-glass-metal, mild steel	Mild steel	Mild steel	O1, A2
Large part of minimum severity (part 2 in Fig. 1)								
1100 aluminum, brass, copper(c)	None	None	Yes	Epoxy-metal, zinc alloy	Epoxy-metal, polyester-metal, zinc alloy	Epoxy(d), polyester(d), zinc alloy	Polyester-glass(d), zinc alloy	Cast iron
1100 aluminum, brass, copper(c)	Best	None	No	Epoxy-metal, zinc alloy	Epoxy-metal, polyester-metal, zinc alloy	Epoxy(d), polyester(d), zinc alloy	Zinc alloy	Cast iron
Magnesium or titanium(e)	Best	None	Yes	Zinc alloy	Cast iron	Cast iron	Cast iron	Alloy cast iron
Low-carbon steel, to 1/4 hard	None	None	Yes	Epoxy-metal, polyester-metal, zinc alloy	Epoxy-glass-metal, polyester-glass-metal, zinc alloy	Polyester-glass(d), zinc alloy	Cast iron	Alloy cast iron
Type 300 stainless, to 1/4 hard	None	None	Yes	Epoxy-metal, polyester-metal, zinc alloy	Epoxy-glass-metal, polyester-glass-metal, zinc alloy	Polyester-glass(d), zinc alloy	Cast iron	Alloy cast iron
Low-carbon steel, to 1/4 hard	Best	None	Yes	Epoxy-metal, polyester-metal, zinc alloy	Epoxy-glass-metal, polyester-glass-metal, zinc alloy	Polyester-glass(d), zinc alloy	Cast iron	Alloy cast iron
High-strength aluminum or copper alloys	Best	None	No	Epoxy-metal, polyester-metal, zinc alloy	Epoxy-glass-metal, polyester-glass-metal, zinc alloy	Polyester-glass(d), zinc alloy	Cast iron	Cast iron
Heat-resisting alloys	Best	None	No	Epoxy-metal, polyester-metal, zinc alloy	Epoxy-glass-metal, polyester-glass-metal, zinc alloy	Cast iron	Alloy cast iron	Alloy cast iron
Type 300 stainless, to 1/4 hard	Best	None	No	Epoxy-metal, polyester-metal, zinc alloy	Epoxy-glass-metal, polyester-glass-metal, zinc alloy	Cast iron	Alloy cast iron	Alloy cast iron

(a) Where mild steel is recommended for forming fewer than 10,000 pieces, the dies are not heat treated. For forming 10,000 pieces and more, such dies should be carburized and hardened. (b) Refers to specially applied lubrication rather than mill oil. (c) Soft. (d) With inserts. (e) Heated sheet. Source: Ref 1

Adjustable inserts are often impractical for small dies. Therefore, for high-production dies working under severe wear conditions and producing parts to close tolerances, it is often desirable to use a complete insert or to make the die of wear-resistant material such as carbide, nitrided tool steel, or highly alloyed powder metallurgy (P/M) tool steel.

In materials selection for large dies, the cost of material is equal to or greater than the cost of machining. In smaller dies, the difference between the cost of the most expensive and the cheapest steels is less important than the assurance of long life without the necessity for rebuilding tools if the quantity should be increased above original expectations, or if the die material should prove to be inadequate. However, for large dies, both the choice of tool material and the design of the dies depend on the number of parts to be produced, particularly if it is more than 1000.

Dimensional requirements of a part may have an important effect on the choice of tool material when the part is to be finished without restriking. If the part is to be restruck, the material used in the restriking die is of less importance, because it will usually be subjected to less wear than will the die that performs the primary operation. A major factor in the choice between a wear-resistant material and a less costly and less wear-resistant material is the necessity for maintenance during the production run.

Lubrication Practice. In making parts at low and medium production (up to 10,000 pieces), it is often economical to use lubricants. Lubrication is required when zinc alloy dies are used. However, the most effective lubricants are difficult to apply and remove, and they add significantly to cost. Efficient application of lubricants is particularly difficult in high-production operations in which presses are automatically

Table 3   Recommended lower die material for forming a part of mild severity

Metal being formed	Finish	Tolerance	Lubrication(b)	100	1000	10,000	100,000	1,000,000
Small part of mild severity (part 3 in Fig. 1)								
1100 aluminum, brass, copper(c)	None	None	Yes	Epoxy-metal, mild steel	Polyester-metal, mild and 4140 steel	Polyester-glass(d), mild and 4140 steel	O1, 4140	A2, D2
1100 aluminum, brass, copper(c)	None	±0.005 R	Yes	Epoxy-metal, mild and 4140 steel	Polyester-metal, mild and 4140 steel	Polyester-glass(d), mild and 4140 steel	4140, O1, A2, D2	A2, D2
1100 aluminum, brass, copper(c)	Best	±0.005 R	Yes	Epoxy-metal, mild steel	Polyester-metal, mild and 4140 steel	Polyester-glass(d), mild and 4140 steel	4140, O1, A2	A2, D2
Magnesium or titanium(e)	Best	±0.005 R	Yes	Mild steel	Mild and 4140 steel	A2	A2	A2, D2
Low-carbon steel, to ¼ hard	None	None	Yes	Mild and 4140 steel	Mild and 4140 steel	4140, mild steel chromium plated, D2	A2	D2
Type 300 stainless, to ¼ hard	None	None	Yes	Mild and 4140 steel	Mild and 4140 steel	Mild and 4140 steel	A2, D2	D2
Low-carbon steel	Best	±0.005 R	Yes	Mild and 4140 steel	Mild and 4140 steel	Mild and 4140 steel	A2, D2, nitrided D2	D2, nitrided D2
High-strength aluminum or copper alloys	Best	±0.005 R	No(f)	Mild and 4140 steel	Mild and 4140 steel	Mild steel chromium plated and 4140	Cr plated O1; A2	D2, nitrided D2
Type 300 stainless, to ¼ hard	None	±0.005 R	Yes	Mild and 4140 steel	Mild and 4140 steel	Mild steel and 4140	Cr plated O1; A2	D2
Type 300 stainless, to ¼ hard	Best	±0.005 R	Yes	Mild and 4140 steel	Mild and 4140 steel	Mild steel chromium plated, D2	D2, nitrided D2	D2, nitrided D2
Heat-resisting alloys	Best	±0.005 R	Yes	Mild and 4140 steel	Mild and 4140 steel	Mild steel chromium plated, D2	D2, nitrided D2	D2, nitrided D2
Low-carbon steel	Good	±0.005 R	No(f)	Mild and 4140 steel	Mild and 4140 steel	Mild steel chromium plated	D2, nitrided D2	D2, nitrided D2
Large part of mild severity (part 4 in Fig. 1)								
1100 aluminum, brass, copper(c)	None	None	Yes	Epoxy-metal, polyester-metal, zinc alloy	Polyester-metal, zinc alloy	Epoxy or polyester-glass(d), zinc alloy	Alloy cast iron	Cast iron or A2(g)
1100 aluminum, brass, copper(c)	None	±0.005 R	Yes	Epoxy-metal, polyester-metal, zinc alloy	Polyester-metal, zinc alloy	Alloy cast iron	Alloy cast iron	Alloy cast iron
1100 aluminum, brass, copper(c)	Best	±0.005 R	Yes	Epoxy-metal, polyester-metal, zinc alloy	Polyester-metal, zinc alloy	Alloy cast iron	Alloy cast iron	Alloy cast iron, A2(g)
Magnesium or titanium(e)	Best	±0.005 R	Yes	Cast iron, zinc alloy	Cast iron, zinc alloy	Cast iron	Alloy cast iron	Alloy cast iron, A2(g)
Low-carbon steel, to ¼ hard	None	None	Yes	Epoxy-metal, polyester-metal, zinc alloy	Epoxy-glass, polyester-glass, zinc alloy	Epoxy or polyester-glass(d), cast iron	Alloy cast iron	...
Type 300 stainless, to ¼ hard	None	None	Yes	Epoxy-metal, polyester-metal, zinc alloy	Epoxy-glass, polyester-glass, zinc alloy	Epoxy or polyester-glass(d), cast iron	A2(g)	D2(g)
Low-carbon steel	Best	±0.005 R	Yes	Zinc alloy	Epoxy-glass, polyester-glass, zinc alloy	Alloy cast iron	D2; nitrided A2(g)	D2, nitrided D2(g)
High-strength aluminum or copper alloys	Best	±0.005 R	No(f)	Zinc alloy	polyester-glass, zinc alloy	Alloy cast iron	Alloy cast iron	Nitrided A2(g), nitrided D2(g)
Type 300 stainless to ¼ hard	None	±0.005 R	Yes	Zinc alloy	Zinc alloy	Alloy cast iron	D2; nitrided A2(g)	D2(g), nitrided D2(g)
Type 300 stainless, to ¼ hard	Best	±0.005 R	Yes	Zinc alloy	Zinc alloy	Alloy cast iron	Nitrided D2	Nitrided D2(g)
Heat-resisting alloys	Best	±0.005 R	Yes	Zinc alloy	Zinc alloy	Alloy cast iron	Nitrided D2	Nitrided D2(g)
Low-carbon steel	Good	±0.005 R	No(f)	Zinc alloy	Zinc alloy	Alloy cast iron	Nitrided D2	Nitrided D2(g)

(a) Where mild steel is recommended for forming fewer than 10,000 pieces, the dies are not heat treated. For forming 10,000 pieces and more, such dies should be carburized and hardened. Where 4140 is recommended for fewer than 10,000 pieces, it should be pretreated to a hardness of Rockwell C 28 to 32. Flame hardening of high wear areas is recommended for quantities greater than 10,000 pieces. When more than one material for the same conditions of tooling is given, the materials are listed in order of increasing cost. Under conditions for which tool steel is recommended for making part 4, the lower die should be of cast iron with inserts of wrought tool steel and the punch of a cast tool steel such as D2. For example, for 10,000 to 100,000 pieces, a cast iron die may be used with tool steel inserts at the shrink flanges; over 100,000 pieces, dies would have tool steel inserts at all wear surfaces. (b) Specially applied lubrication, rather than mill oil. (c) Soft. (d) With inserts. (e) Heated sheet. (f) Use lubrication to make 1 to 100 parts. (g) Use as inserts in cast iron body. Source: Ref 1

fed. In such operations, it is often economical to use die metals that are more costly but more resistant to galling in combination with the usually less-effective lubricants that can be applied automatically. Examples of these materials are aluminum bronze, nitrided tool steel, and carbide, which often can be used for forming low-carbon steel with only mill-oil lubrication. More information on lubricants is available in the article "Selection and Use of Lubricants in Forming of Sheet Metal" in this Volume.

Selection of Tool Materials for Galling Resistance. Galling is a process of adhesive wear from the cold welding of the metal being formed to that of the dies. Galling drastically reduces the number of parts that can be made using a particular set of dies. It is caused by attempts to stretch sheet metal beyond practical limits, by inadequate lubrication, by poor tool fitting, or by rough finishes on tool surfaces.

For short and medium runs, surface-hardened hot-rolled steel dies will produce parts equal to those produced from most tool steel dies. Exceptions may be encountered in severe reductions or in forming metals that show a greater tendency to gall, such as austenitic stainless steel. However, tool steel dies may also gall under these conditions.

When galling is encountered, the tool fit and the thickness of the metal being formed should first be checked to determine whether clearance is adequate. If clearance is considered adequate, lubrication practice should be reviewed before considering a change in die materials. Attempts to iron out wrinkles will often cause galling. Whenever possible, wrinkles should be prevented by the design of the tools.

Galling is less likely to occur if the die materials and the metal being formed are dissimilar in hardness, chemical composition, and/or surface characteristics. For example, effective combinations are aluminum bronze tools for forming carbon steel and stainless steel; tool steel tools for forming aluminum and copper alloys; and carbide tools for forming carbon steel, stainless steel, and aluminum. Aluminum bronzes have excellent resistance to galling and are desirable for dies in applications in which the best finish is required on carbon steel or stainless steel parts. However, for medium-to-high production (10,000 to 100,000 parts), the use of inserts permits easy reconditioning of worn tools.

Nitriding minimizes or prevents galling of dies made of alloy steels or alloy tool steels that contain chromium and molybdenum (e.g., D2). It is not recommended for steels that contain no nitride-forming elements, such as chromium or molybdenum. Nitrided surfaces on such steels may spall off on radii smaller than approximately 3.2 mm ($\frac{1}{8}$ in.).

Hard chromium plating usually eliminates galling of mild steel, alloy steel, and tool steel dies, and it is often used for severe duty. For operations involving high local pressures, hardened alloy steels or tool steels are less likely to yield plastically and cause cracking of the hard chromium plating. With dies for complex parts, hard chromium plating may spall off at radii smaller than approximately 6.4 mm ($\frac{1}{4}$ in.).

For some press-forming operations, dies made from tool steels other than those discussed previously may be desirable. For example, shock-resistant tool steels such as S1, S5, and S7 may be used for die components subjected to severe impact in service. H11 and H13, possibly nitrided for greater wear resistance, also may be used for such components. In press-forming operations requiring significantly greater wear life

Table 4 Recommended lower die material for forming a part of moderate severity

Metal being formed	Part (made of 1.3 mm, or 0.050 in., sheet) Requirements Finish	Tolerance	Lubrication(b)	Recommended lower die material(a) Total quantity of parts to be formed 100	1000	10,000	100,000	1,000,000
Small part of moderate severity (part 5 in Fig. 1)								
1100 aluminum, brass, copper(c)	None	None	Yes	Mild steel	Mild steel	Mild steel	O1, A2	A2
1100 aluminum, brass, copper(c)	Best	±0.020 R	Yes	Mild steel	Mild steel	Mild steel	O1, A2	A2
Magnesium or titanium(d)	Best	±0.020 R	Yes	Mild steel	Mild steel	A2	D2	D2
Low-carbon steel, to $\frac{1}{4}$ hard	Best	None	Yes	Mild steel, O1	Mild steel, O1	Mild steel, A2	Mild steel chromium plated, A2	D2
Type 300 stainless, to $\frac{1}{4}$ hard	None	None	Yes	Mild steel, O1	Mild steel, O1, A2 tool steel	Mild steel chromium plated, A2	Mild steel chromium plated, D2	D2
Low-carbon steel	Best	±0.020 R	Yes	Mild steel, O1 or nitrided A2	Mild steel, O1, or nitrided A2	Mild steel chromium plated, D2	Mild steel chromium plated, D2	Nitrided D2
High-strength aluminum or copper alloys	Best	±0.020 R	No(e)	Mild steel, O1	Mild steel, O1	Mild steel chromium plated	A2	D2
Type 300 stainless, to $\frac{1}{4}$ hard	None	±0.020 R	Yes	Mild steel, O1	Mild steel	Mild steel chromium plated, A2	D2	D2
Type 300 stainless, to $\frac{1}{4}$ hard	Best	±0.020 R	Yes	Mild steel, O1	Mild steel, O1	Mild steel chromium plated	Nitrided D2, nitrided A2	Nitrided D2
Low-carbon steel	Good	±0.020 R	No(e)	Mild steel, O1	Mild steel, O1	Mild steel, A2	A2	D2
Large part of moderate severity (part 6 in Fig. 1)								
1100 aluminum, brass, copper(c)	None	None	Yes	Zinc alloy	Zinc alloy	Cast iron	Alloy cast iron	D2, A2
1100 aluminum, brass, copper(c)	Best	±0.031 R	Yes	Zinc alloy	Zinc alloy	Cast iron	Alloy cast iron	A2, D2
Magnesium or titanium(d)	Best	±0.031 R	Yes	Cast iron	Cast iron	Alloy cast iron	A2	D2
Low-carbon steel, to $\frac{1}{4}$ hard	None	None	Yes	Zinc alloy	Cast iron	Cast iron	Alloy cast iron	D2, A2
Type 300 stainless, to $\frac{1}{4}$ hard	None	None	Yes	Zinc alloy	Cast iron	Alloy cast iron	D2, A2	D2, nitrided D2
Low-carbon steel	Best	±0.031 R	Yes	Zinc alloy	Cast iron	Alloy cast iron	Alloy cast iron	Nitrided D2
High-strength aluminum or copper alloys	Best	±0.031 R	Yes	Zinc alloy	Zinc alloy	Alloy cast iron	Alloy cast iron	D2, A2
Type 300 stainless, to $\frac{1}{4}$ hard	None	±0.031 R	Yes	Zinc alloy	Zinc alloy	A2	A2	D2
Type 300 stainless, to $\frac{1}{4}$ hard	Best	±0.031 R	Yes	Zinc alloy	Alloy cast iron	D2; nitrided A2	Nitrided D2	Nitrided D2

(a) Where mild steel is recommended for forming fewer than 10,000 pieces, the dies are not heat treated. For forming 10,000 pieces and more, such dies should be carburized and hardened. When more than one material for the same conditions of tooling is given, the materials are listed in order of increasing cost. Cast iron dies with inserts of tool steel would be needed for making 10,000 parts; however, the same material is recommended for punch and lower die for part 5 because of the small size and minor cost of material. Recommendations for making part 6 are the materials for the wearing edges of the lower die; body of the lower die would be of cast iron with wearing edges of the tool steel recommended; the punch could be of a material approximately one-tenth as wear resistant as tool steel. (b) Refers to specially applied lubrication rather than mill oil. (c) Soft. (d) Heated sheet. (e) Lubrication should be used to make 1 to 1000 parts. Source: Ref 1

than is routinely attained with D2 or nitrided D2, it may be necessary to specify a more wear-resistant cold work tool steel such as a high-speed steel (e.g., M2, M4, or T15) or a more highly alloyed P/M tool steel, particularly those containing high levels of vanadium. Cost generally determines the desirability of changing to an alternative material, although toughness may also be a determining factor. Costs to be considered include not only material costs but also tool fabrication costs and the cost of periodic resharpening.

Dirt, grit, and fragments on the sheet can cause greater damage to nitrided and chromium-plated tools than to hardened tool steels; hard particles may cause minute spalling and small pits, especially at radii in areas of high forming pressure. When these pits in the die cause scratches on the formed sheet, the pits must be stoned out if high finish is required on the part. After such repairs have been made several times in the same area, the soft underlying metal will be exposed, and renewal of the nitrided case or the chromium plate will be necessary.

Effect of Die Finish on Galling. Despite the fact that, in many instances, galling can be minimized or eliminated by careful selection of die materials, die finish is an important factor affecting susceptibility to galling. The probability of galling increases as the severity of forming increases.

The influence of die finish, with other factors constant, is illustrated in the forming of a hat section where the walls of the hat were nearly perpendicular to the original blank, thus representing fairly drastic severity. The material being formed was annealed 1010 steel. The punch portion of the die was made from A6 tool steel at 60 HRC with a finish of 0.75 to 0.9 μm (30 to 35 μin.). The die ring was made from nitrided medium-carbon alloy steel having a hardness of 63 HRC and was also finished to 0.75 to 0.90 μm (30 to 35 μin.).

Both components were copiously lubricated, but after ten hits, there were indications of metal pickup, and after 30 hits the tools were removed because of excessive galling. Both die components were refinished to approximately 0.5 μm (20 μin.), after which 300 parts were produced before galling was excessive. The tools were then refinished to 0.2 to 0.25 μm (8 to 10 μin.), after which 3000 pieces were produced with no sign of galling. However, producing this high finish all over dies is costly and is not always warranted, but it should be considered when galling is experienced in severe forming. Sometimes, the use of lubricants can be minimized by improving the die finish.

Cast Iron

Cast iron is a useful die material for forming parts larger than approximately 300 mm (12 in.). Its performance makes it suitable for use in medium production runs or short runs of large parts. When cast iron is used with inserts, it will produce greater quantities. Cast iron should have a predominantly fine pearlitic matrix with no massive carbides and a minimum of ferrite.

Graphite in gray cast iron should be of type A distribution, with a preferred flake size of 4 to 5. Flaking problems can occur with gray cast iron dies, but flaking problems in gray cast iron tooling can be eliminated by chrome plating or nitriding the tool contact surfaces. Ideally, cast tooling that is to be used to press form a coated steel should be made of cast steel or nodular (ductile) cast iron. Ductile iron has been an attractive material for stamping and drawing dies. To further improve wear resistance, laser surface modification has been used in industrial applications.

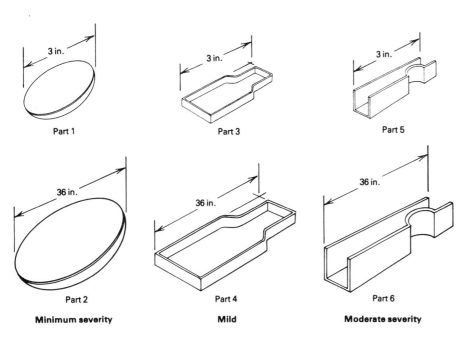

Fig. 1 Parts classified into six classes based on forming severity and part size

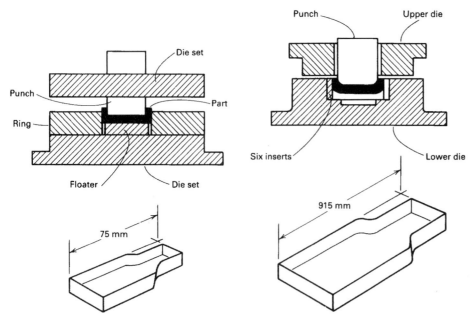

Fig. 2 Cross section of die used to form small part of mild severity

Fig. 3 Cross section of die used to form large part of mild severity

Steels

Hot-rolled mild steel plate with carbon content from 0.10 to 0.20% is in major use as a die material for short-run forming of small parts. This type of steel is not recommended without surface hardening, except for quantities of less than 10,000 parts. In addition, hot-rolled mild steel can be used for dies where straightening facilities are available for correcting distortion induced by heat treatment. Several medium-carbon alloy steels, such as 4140, are available in plate form and are useful for some types of forming dies. Such steels are generally used for lower production runs.

Cast steel is used for forming parts larger than 300 mm (12 in.) long where the shape of the part makes a casting more feasible than wrought stock. Plain carbon steel (0.65 to 0.85% C) and alloy cast steel (Fe-0.45C-1.1Cr-0.4Mo) are two commonly used compositions. Cast steel is more costly and more difficult to machine than cast iron and is more likely to gall. However, it is tougher than cast iron. One important advantage of cast steel, particularly the plain carbon grade, is its weldability. It can be rewelded with steel or hardfaced with tool steel, aluminum bronze, or other wear-resistant hardfacing alloy. Because of its resistance to galling, it is less desirable for dies used for forming carbon and stainless steels than for aluminum and copper alloys. Cast steel is also useful for restrike, flanging, and other types of dies that are less likely to gall or pick up metal from the sheet.

Tool steels. Cold work tool steels for press-work applications contain various amounts of carbon (ranging from 0.6 to 3 wt% C) with alloying of silicon, manganese, chromium, molybdenum, tungsten, and vanadium. Table 6 lists compositions of various press-work tool steels that have been used for press-work dies in automotive stampings. The production methods include:

- Conventional ingot metallurgy
- Conventional ingot metallurgy with electro-slag remelting
- Spray forming
- Powder or particulate metallurgy

Conventional press-work tool steels include the commonly used 5 wt% Cr tool steel (AISI A2) and 12 wt% Cr tool steel (AISI D2) for sheet metal fabrication. However, these materials (cast or wrought form) may not be the most effective choice in more demanding forming or cutting applications. The combination of toughness and hardness of D2 is relatively low compared to other types of tools, such as the martensitic grades (e.g., Carmo, Caldie, Uddeholm Tooling AB; Diemarc, Anchor Lamina Inc.), the spray-formed grades (e.g., Roltec and Weartec, Uddeholm Tooling AB), and the P/M grades (Fig. 4).

Conventional D2 steel also is affected by a wide variation of dimensional changes after heat treatment due to the anisotropy of carbides in the microstructure. The chromium carbides have a stringerlike morphology, with inhomogeneous distribution from the surface to the bar center. The Sleipner grade is an optimized version of the D2 grade, with improved ductility compared to D2 and a smaller amount of chromium carbides (6% versus 13% for D2). This reduces the extent of dimensional change (Fig. 5).

The advantage of the martensitic die steels (such as Carmo, Diemax, and Caldie) is better toughness, with hardness ranging up to approximately 62 HRC (Fig. 4). The Carmo/

Calmax grade (Table 6) was the first-generation version of martensitic press-work tool steels, but it has been largely replaced by new martensitic grades (such as Diemax and Caldie martensitic grades) with better combinations of ductility and hardness (Fig. 4). Electroslag remelting also is used in the production of Diemax and Caldie grades. This results in a significant reduction in the size of second-phase constituents and the improvement of fatigue resistance (Fig. 6).

The most significant advances in tool steels have been made in the development of powder or particulate metallurgy (P/M) tool steels. The powder process allows production with high vanadium content for more effective hardening and wear resistance from carbide formation. This production method results in a very homogeneous distribution of carbides from the surface to the center of the block and thus results in the best dimensional stability (Fig. 5). The size of defects in P/M tool steels is also smaller, which results in higher fatigue limits (Fig. 6).

Powder/particle metallurgy high-speed steels (hot isostatically pressed to full density) offer greater ease of fabrication and significantly improved toughness compared to conventional ingot-cast steels of the same compositions. New grades that could not have been produced economically by conventional steelmaking practices have been introduced through the use of P/M. Several major international manufacturers of tool steels offer a line of P/M-produced high-alloy tool steels. Most of these P/M tool steels feature a high vanadium content, to provide enhanced wear resistance through the presence of a high volume of hard vanadium carbides. The

Table 5 Recommended materials for lower die for forming steel parts of three sizes, each in four thicknesses

Assuming no finish or tolerance requirements

Sheet thickness		Recommended lower die material(a)				
		Total quantity to be produced				
mm	in.	100	1000	10,000	100,000(b)	1,000,000(b)
For part 3, maximum dimension of 76 mm (3 in.)						
0.8	0.031	Mild steel and 4140	Mild steel and 4140	Mild steel and 4140	4140, mild steel chromium plated, D2	D2
1.6	0.062	Mild steel and 4140	Mild steel and 4140	Mild steel and 4140	4140, mild steel chromium plated, D2	D2
3.2	0.125	Mild steel and 4140	Mild steel and 4140	Mild steel, 4140 and A2	D2	D2
6.3	0.250	Mild steel and 4140	Mild steel and 4140	A2 and D2	D2	D2
For a 305 mm (12 in.) part similar to parts 3 and 4						
0.8	0.031	Mild steel, 4140 and zinc alloy	Mild steel, 4140 and zinc alloy	Mild steel, 4140, zinc alloy and alloy cast iron	Alloy cast iron(c), mild steel chromium plated, 4140	D2
1.6	0.062	Mild steel, 4140 and zinc alloy	Mild steel, 4140 and zinc alloy	Mild steel, 4140, zinc alloy and alloy cast iron	Alloy cast iron(c) and A2	D2
3.2	0.125	Mild steel, 4140 and zinc alloy	Mild steel, 4140 and cast iron	Mild(c) and A2 steels	A2 and D2	D2
6.3	0.250	Mild steel, 4140 and cast iron	Mild steel and alloy cast iron	Mild(c) and A2 steels	A2 and D2	D2
For part 4, maximum dimension of 914 mm (36 in.)						
0.8	0.031	Epoxy-metal, polyester-metal, zinc alloy	Epoxy-glass, polyester-glass, zinc alloy	Epoxy or polyester-glass(d), alloy cast iron, zinc alloy(d)	Alloy cast iron(c)	D2
1.6	0.062	Epoxy-metal, polyester-metal, zinc alloy	Epoxy-glass, polyester-glass, zinc alloy	Epoxy or polyester-glass(d), alloy cast iron, zinc alloy(d)	A2	D2
3.2	0.125	Zinc alloy	Zinc alloy, cast iron	Alloy cast iron(c)	A2 and D2	D2
6.3	0.250	Zinc alloy, cast iron	Alloy cast iron	Alloy cast iron(c)	A2 and D2	D2

(a) Materials recommended in above table are for forming parts from low-carbon steel; wear and galling will be less with aluminum and copper alloys, and more with stainless and heat-resisting alloys, than with low-carbon steel.
(b) All A2 or D2 tool steels nitrided after tryout if required. (c) Inserts of A2 tool steel are recommended for wear edges for the shrink flange. (d) At the quantity level indicated, use of inserts with these materials is required.

P/M production process permits the design of tool steels having such an elevated carbide population with minimum detriment to toughness or grindability. One of the first-generation grades is Crucible CPM 10V (Crucible Materials Corp.) (Fe-2.45C-5.0Cr-9.75V-1.25Mo), which is an air-hardened cold work tool steel designed specifically for tooling applications requiring long wear life and good toughness. This material can be a cost-effective alternative to carbide in applications in which breaking or chipping of carbide is a problem or in which the full potential of carbide is either not realized or not required

(see the article "Particle Metallurgy Tool Steels" in *Powder Metal Technologies and Applications*, Volume 7 of *ASM Handbook*, 1998).

Continuing developments in P/M tool steels include the following:

- More wear resistance with ultrahigh vanadium compositions containing 15 to 18% V with up to 30% by volume of primary metal carbide-type carbides
- Low-to-intermediate carbide volume fraction materials moderately alloyed with vanadium and chromium to optimize the toughness

properties while still maintaining good wear resistance
- High-vanadium, high-chromium compositions for wear applications that also require good corrosion resistance
- High-nitrogen P/M steels (e.g., Vancron 40)

These developments address the increasing demands of forming higher-strength sheet materials, especially for the forming of high-strength and advanced high-strength steels in the automotive sector. Higher-strength sheet materials demand tooling with more abrasion resistance, compressive strength, and chipping resistance, along with the ability to be effectively manufactured into multiple-profile sections. In addition, forming of higher-strength steels often means that tool steels have to be coated (and, in some cases, pretreated by nitriding in order to get enough support). The exception may be high-nitrogen P/M steels (such as Vancron 40), which inherently have excellent antigalling characteristics. Table 7 lists examples of tool steels and coatings for various strengths of steel sheet.

Steel-bonded carbides (commonly referred to as Ferro-TiC alloys, Pacific Sintered Metals) have titanium-carbide particles dispersed in a steel matrix. The steel matrix can be heat treated to obtain desirable matrix properties, and the carbides increase effective hardness for improved wear resistance. Steel-bonded carbides do not have suitable high-temperature strength for cutting-tool applications, but they have been useful as a die material. However, application of steel-bonded carbides as a die material has been reduced due to continued innovations with P/M tool steels.

Nonferrous Die Materials

Cemented Carbides. For long runs, inserts of sintered carbide are now widely used, especially for deep-drawing dies. In dies of approximately 200 mm (8 in.) or less for continuous production of over 1,000,000 pieces, carbide has, in many instances, proved to be the most economical die material. Such dies have maintained size in drawing 60% reductions of more than 500,000 pieces and have made as many as 1,000,000 parts with reductions greater than 40% when the steel to be drawn was surface treated with zinc phosphate and soap. However, carbide is not superior to tool steel such as D2 in complex deep-drawing operations, such as those with reductions greater than 40% that combine drawing with coining or stretching.

When maximum resistance to galling and wear is required, cemented carbides have traditionally been recognized as the ultimate tooling materials. However, because of the high cost of these materials and their tendency to be brittle in service, carbides are frequently used only for inserts in critical die areas. These inserts are usually made of a straight grade of tungsten carbide containing approximately 6% Co binder,

Table 6 Nominal compositions of selected cold work tool steels

Steel	AISI designation	Commercial equivalent	Composition, wt%					
			C	Cr	Mo	W	V	N
Powder metallurgy (P/M) cold work tool steels								
PM 3V	...	CPM 3V	0.80	7.50	1.00	...	2.75	...
PM M4	M4	CPM M4HC	1.40	4.00	5.25	5.75	4.00	...
PM 8Cr4V	...	Vanadis 4	1.50	8.00	1.00	...	4.00	...
PM 12Cr4V	D2	K190 PM	2.30	12.00	1.00	...	4.00	...
PM 9V	...	CPM 9V	1.80	5.25	1.30	...	9.00	...
PM 10V	A11	CPM 10V	2.45	5.25	1.30	...	9.75	...
PM8Cr10V	...	Vanadis 10	2.90	8.00	1.50	...	9.80	...
PM 15V	...	CPM 15V	3.50	5.25	1.30	...	14.50	...
PM 18V	...	CPM 18V	3.90	5.25	1.30	...	17.50	...
PM 4V	...	Vanadis 4 Extra	1.4	4.7	3.5	...	3.7	...
PM7Cr5V	...	Vanadis 6	2.1	6.8	1.5	...	5.4	...
...	...	Vancron 40	1.1	4.5	3.2	3.7	8.5	1.9
Conventionally produced (ingot-cast) cold work tool steels								
A2	A2	...	1.00	5.25	1.15	...	0.30	...
D2	D2	...	1.55	11.50	0.80	...	0.90	...
D7	D7	...	2.35	12.00	1.00	...	4.00	...
...	...	Carmo/Calmax	0.6	4.5	0.5	...	0.8	...
...	...	Sleipner	0.9	7.8	2.5	...	0.5	...
Ingot cast with electroslag remelting								
...	...	Diemax	0.5	5.0	2.3	...	0.5	...
...	...	Caldie	0.7	5.0	2.3	...	0.5	...
Spray formed								
...	...	Roltec	1.4	4.6	3.2	...	3.7	...
...	...	Weartec	2.8	7.0	2.3	...	8.9	...

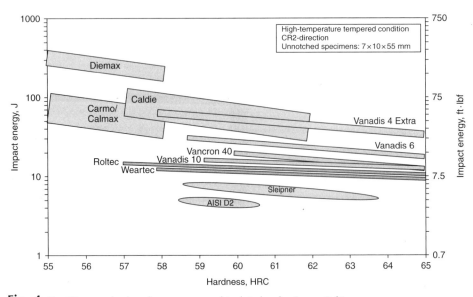

Fig. 4 Ductility versus hardness for some press-work tool steel grades. Source: Ref 2

but higher cobalt contents can be specified to provide greater shock resistance. The aforementioned steel-bonded carbides offer greater ease of fabrication and very often can be demonstrated to be cost-effective substitutes for the more costly cemented carbides with cobalt binder.

Stamping Punches and Dies. The high elastic modulus of carbides combined with their ability to incorporate fine details makes them ideal tool materials for stamping punches and dies. As in many other metal-forming applications, fine-grain carbides are chosen for punches because

of their edge-retention capability and higher abrasion resistance.

Zinc-aluminum die alloy (Kirksite) (Zn-4Al-3Cu-0.06Mg) is used as a forming die alloy and is capable of being sand cast to shape rapidly. It is mainly used in the construction of cast two-piece dies for forming sheet metal parts such as components for use in the transportation and aerospace industries. Two forming-die alloys are included in ASTM B 793 (Table 8).

Kirksite can be advantageous for complicated dies that would require intricate machining and hand tooling if made of cast iron, but which can

be cast more closely with zinc. The alloy has a high pattern fidelity when cast but requires very accurate shrink patterns to minimize hand labor in finishing. It is economical only for die components at least 300 mm (12 in.) long and is most economical in dies approximately 900 mm (36 in.). In tensile strength, compressive strength, and hardness, zinc tooling alloy is inferior to other metals used in die construction.

If compound curves on the face of a stamping have no sharp creases or embossments with sharp corners, it is not unusual for zinc alloy dies to give the length of service indicated in selection Table 2 for part 2 (Fig. 1). Zinc alloy tools are particularly sensitive to surface roughness, thickness, and hardness of the sheet metal being formed. The increase of any one of these multiplies unit pressures over a small area of the die face.

In production use, zinc alloy drawing dies in double-action presses have been found to give the longest life. Contrary to conventional die-making practice, zinc alloy dies, cast to shape, are made of as few components as possible. This not only results in a lower cost, because fasteners are largely eliminated, but it adds to the simplicity and durability of the tool. Otherwise, zinc alloy tools for conventional press production are designed according to the usual practice.

Virtually all the zinc alloy in a tool can be salvaged for reuse simply by melting. For greatest economy, the user must have melting and molding facilities available; the commercial scrap value of the alloy is approximately 65% of the original cost, while remelting costs the user approximately as much as melting purchased metal.

Cycle time or press speed has a great effect on the life of zinc alloy tools. A die component moving more than 63 cm/min (25 in./min) during its working cycle is highly destructive to zinc alloy. The best results have been obtained on double-action presses that average three strokes per minute and involve drawing to a depth of more than 150 mm (6 in.). For this work, hydraulic presses are more desirable than mechanical presses. Therefore, tools made from zinc alloys have a minimum size limit. This is true because even the slowest speed of small presses is greater than is permissible with zinc alloy tooling, and the cost is excessive if a small die is run on a large press merely to suit the speed requirements. Zinc alloy dies cannot be used with unlubricated stock, because wear will be too rapid.

Aluminum bronze (UNS C62500, Cu-13Al-4Fe) is available in hardnesses ranging from 120 to 340 HB. The alloy has excellent resistance to galling and is desirable for dies where best finish is required on carbon and stainless steel parts.

The softer grades (120 to 270 HB) wear rapidly, particularly where wrinkling is likely to occur on the formed part. The harder grades (270 to 340 HB) wear less rapidly but are difficult to machine, drill, and tap. Where elimination of scratches is extremely important, aluminum

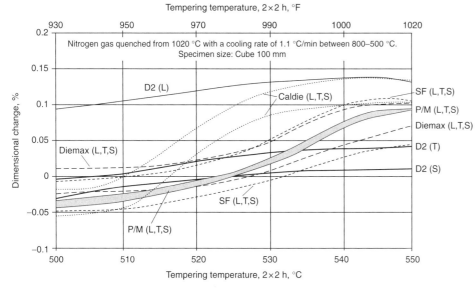

Fig. 5 Dimensional changes after hardening and tempering of some press-work tool steels. For AISI D2, three individual curves are given for dimensional changes in the long (L), transverse (T), and short-transverse (S) directions, respectively. For all other steels, the spread in dimensional changes is narrower and is displayed by an upper and lower boundary line. SF, spray formed; P/M, powder metallurgy. Source: Ref 2

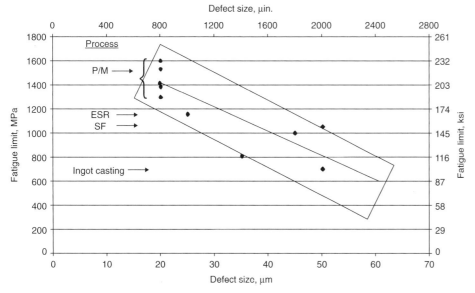

Fig. 6 Fatigue limit at 2 million cycles versus defect size and process route for tool steels. Fatigue limit for various tool steels at 60–62 HRC. R = 0. P/M, powder metallurgy; ESR, electroslag remelt; SF, spray forming. Source: Ref 2

bronzes should be considered, but for medium to high production (10,000 to 100,000), replacement inserts should be available to permit reconditioning of worn tools.

Plastic Die Materials

The major uses of plastics for tooling currently include prototype parts, mockups, model duplications, spotting racks, templates, fixtures, drill jigs, spray masks, foundry patterns, core boxes, and molds and dies for producing prototype plastic and sheet metal parts. Plastic tooling offers several advantages over conventional metal tooling that include:

- Lower cost, often less than half the cost of metal tooling
- Reduced tool manufacturing time, generally a matter of hours or days rather than weeks or months
- Light weight, easily handled and used
- Easy modification for design changes and rework

Polyesters and phenolics were among the first types of plastics used in die applications, but they have been largely replaced by the harder epoxies and improved polyurethane resins. The dimensional instabilities of phenolics and polyesters require frequent checking and modification of tools. Glass-filled polyesters are limited to low-tolerance limits, while phenolics are seldom used anymore for dies. Epoxies and polyurethanes are harder and are used more extensively.

Plastics for dies are typically filled with ceramic or metal powders for wear resistance. The selection tables (Tables 2 and 3) refer to plastics reinforced by the addition of metal powder to the uncured resin. Metal inserts may also be used in plastic tools. Although the hardness of plastics is obviously less than that of metallic die materials, plastic tools can be an economical alternative. Plastic dies are used extensively in forming, primarily for aluminum and other light alloys, but also for forming of low-carbon steels and stainless steels. Other tool applications of plastics include dies for stretch forming and bending. Plastic tooling for ironing and similar operations is not recommended, because sufficient pressure cannot be maintained for the removal of wrinkles in metals.

Urethane is a thermosetting plastic that has the flexibility of rubber and the hardness of structural plastics. In general, urethane offers high load-bearing capacity, high impact resistance, low compression set, high tear strength and cut resistance, and good abrasion resistance. At the pressures used in forming, urethanes are incompressible yet deflectable solids. By controlling movement and displacement and keeping within the established capabilities of the specific formulation of urethane, long life can be maintained. There are many formulations of urethane, each having different characteristics that react quite differently in tooling.

Urethane has become a necessary and useful material in the manufacture of tools for sheet metal fabrication. The use of urethane in metal-forming operations can bring the tool engineer three distinct advantages: cost savings, improved quality, and decreased lead time. More specifically, there are five areas in which urethane may be used advantageously. Applications include:

- Pads that deflect to become female dies
- Male punches
- Wiping blocks
- Springs and pads used to apply pressure
- Wear pads, clamping jaws, and fixtures

The general advantages of urethane are as follows:

- Nonmarking characteristic eliminates tool marks, resulting poor appearance, and need for costly subsequent polishing operations.
- Automatic ability to compensate for metal-thickness variations. Urethane self-adjusts for thickness changes caused by commercial mill tolerances, as well as changes in gage of material.
- Eliminates heat treating of steel components. Many tools can be built with mild steel because of the reduction in material cost and tool fabrication time. Substantial reduction in abrasive wear
- Compensates for other variations. Automatic adjustments for various misalignments, such as ram and bolster alignment, punch and die alignment, and so on. Because part of tool construction is a flexible, self-adjusting material, less time is spent in assembly.
- Eliminates air bending. Complete pressure forming in some tools. Piece part gripped with a blankholding force prior to initial bending. Accurate and consistent flange lengths result. Reduced deformation or fluting in pressure forming throughout the stroke
- Reduces press or die damage. Urethane has the ability to absorb operator and press errors when press is of sufficient tonnage.

In order to realize the full benefit of these advantages, it is important to understand the limitations of urethane materials and design considerations, such as how pressure and deflection are controlled. Undue strain or cutting of the urethane should be avoided. Urethane behaves similarly to a solid fluid with a memory: Under force it will change its shape, but its volume will remain constant. Urethane will produce high, uniform, and continuous

Table 7 Example of tool steel program for car body and structural part forming

Type of sheet	Strength (R_m) MPa	ksi	Steel grade	Base die Bar	Cast	Inserts Bar	Cast	Surface treatment(a)
Soft-mild	<330	<48	Carmo	X	X	X	...	Normally not needed
			Calmax	X	X	X	...	Normally not needed
			Sleipner	X	X	X	X	Normally not needed
High strength	330–570	48–83	Carmo	X	X	X	...	Option needed
			Calmax	X	X	X	...	Option needed
			Caldie	X	X	X	X	Option
			Sleipner	X	X	X	X	Option
			Vanadis 4 Extra(b)	X	...	X	...	Option
			Vanadis 6(b)	X	...	X	...	Option
Extrahigh strength	570–800	83–116	Carmo	X	X	...	...	Nitride + PVD needed
			Calmax	X	X	...	...	Nitride + PVD needed
			Sleipner	X	X	X	X	PVD/CVD needed
			Caldie	X	X	X	X	PVD/CVD needed
			Roltec(c)	X	...	X	...	PVD/CVD needed
			Vanadis 4 Extra(b)	X	...	X	...	PVD/CVD needed
			Vanadis 6(b)	X	...	X	...	PVD/CVD needed
			Vancron 40(b)	X	...	X	...	No coatings needed
Ultrahigh strength	>800	>116	Sleipner	X	X	X	...	PVD/CVD needed
			Caldie	X	X	X	...	PVD/CVD needed
			Roltec(c)	X	...	X	...	PVD/CVD needed
			Weartec(c)	X	...	X	...	PVD/CVD needed
			Vanadis 4 Extra(b)	X	...	X	...	PVD/CVD needed
			Vanadis 6(b)	X	...	X	...	PVD/CVD needed
			Vanadis 10(b)	X	...	X	...	PVD/CVD needed
			Vancron 40(b)	X	...	X	...	No coatings needed

(a) PVD, physical vapor deposition; CVD, chemical vapor deposition. (b) Powder metallurgy grade. (c) Spray-formed grade

Table 8 Nominal compositions of zinc casting alloys used for sheet metal forming dies

Alloy Common designation	UNS No.	Composition, % Al	Cd max	Cu	Fe max	Pb max	Mg	Sn max	Zn
Forming-die alloys (ASTM B 793)									
Alloy A	Z35543	3.5–4.5	0.005	2.5–3.5	0.100	0.007	0.02–0.10	0.005	bal
Alloy B	Z35542	3.9–4.3	0.003	2.5–2.9	0.075	0.003	0.02–0.05	0.001	bal

counterpressure under load and, when the load is removed, quickly return to its original shape.

One of the secrets of good die design is to provide enough blankholding pressure on the downstroke and stripping or shedding pressure on the upstroke. It is frequently impossible with conventional steel springs to generate enough pressure to positively strip, shed, or hold blanks. In blanking and piercing dies, metal holdup or jamming due to double slugs is evidence of insufficient stripping or shedding pressure. In forming dies, a crown next to the bend line or blank slippage is evidence of insufficient blankholding pressure.

Another problem encountered in die design is fitting the required number of steel springs into a limited area, as, for example, in long, narrow press-brake dies. Urethane springs, strippers, shedders, and pressure pads provide more pressure per unit area than conventional springs—an ideal situation for the narrow confines in dies and fixtures.

Urethane springs, strippers, and shedders consist of cylindrical tubes that can be used independently or to which one or two caps can be fitted. This unique combination offers:

- High pressures
- Close center distances
- Positive stripping
- Punch vibration damping
- Simple installation
- Nonmarring

Springs or cylinders are made of a tough, resilient, high-modulus grade of urethane that will withstand more than 100,000 cycles when properly applied. They are quickly installed or removed, requiring neither set screws nor stripper bolts. When fractured under load, these springs will not fly into pieces of shrapnel, thus averting damage to the die and possible injury to the worker.

Epoxy Die Materials. The many advantages of epoxy tools include ease of fabrication, durability, and accuracy. Epoxy resin formulations can be used to make molds and dies for forming prototype parts from sheet metal. Epoxies for tooling applications are available as casting and laminating resins with a wide range of formulations and additives. The compressive strength and wear resistance of epoxy-ceramic systems make their use ideal for hydropress forming of metal sheet. Cast ceramic-filled epoxies also have been successfully used in drop-hammer forming of aluminum, titanium, and steel, with little or no tool wear. Drop-hammer tooling is often cast metal (usually Kirksite or steel). However, the tool-construction cost for the epoxy-ceramic hammer die (not including materials) is approximately one-fourth that of Kirksite.

REFERENCES

1. Selection of Materials for Press-Forming Dies, *Properties and Selection of Metals,* Vol 1, *Metals Handbook,* 8th ed., 1961, p 699–708
2. O. Sandberg and B. Johansson, Tool Steels for Blanking and Forming: New Developments, *Recent Advances in Manufacture and Use of Tools and Dies and Stamping of Steel Sheets,* Volvo, 2004

Selection of Materials for Deep-Drawing Dies

DEEP DRAWING is a process in which sheet metal is formed into round or square cup-shaped parts by making it conform to a punch as it is drawn through a die (see the article "Deep Drawing" in this Volume). In conventional deep drawing, successive draws are made in the same direction. The types of dies and other tooling used for conventional deep drawing are illustrated in Fig. 1.

It is sometimes necessary for redrawn shells to have a wrinkle-free sidewall of uniform thickness or a section in the bottom of the cup that is sharply raised, usually by forming in two operations. Such operations are difficult, impossible, or uneconomical to perform by conventional single-action drawing, but they are easily done by reverse redrawing. Figure 2 shows typical tooling for the reverse redrawing of thin-wall shells.

For economical manufacture, a drawn part should always be produced in the fewest steps possible. Ironing (that is, thinning the walls of the part being drawn by using a reduced clearance between punch and die) is almost universally used in multiple-operation deep drawing. Ironing helps to produce deep draws and uniform wall thickness in the fewest operations. Each operation is designed for maximum practical reduction of the metal being drawn. Accordingly, the information given in this article is based on ironing reductions near the maximum of approximately 35%.

The selection of material for a drawing die is aimed at production of the desired quality and quantity of parts with the least possible tooling cost per part. In small dies (for example, those for making parts up to 75 mm, or 3 in., across), performance is the primary consideration. Material cost is a minor factor, because the cost of even the more highly alloyed tool steels is probably less than 5% of the total die cost. In dies for parts larger than approximately 203 mm (8 in.), material cost is more important, and in a die for a 305 mm (12 in.) part, it may amount to nearly one-half the total die cost, even when the tool consists of a tool steel insert in a flame-hardened alloy cast iron die.

Die Performance

The performance of a drawing die is determined by the total amount of wear (abrasive and adhesive) that occurs during a production run. The wear of a given die material is largely determined by its hardness, type and thickness of the sheet metal being drawn, sharpness of die radii, lubrication, and construction and surface finish of the die. The amount of wear on die radii can vary by a factor of 20 between the sharpest and most liberal radii. In drawing square cups, the formation of wrinkles at the corners, accompanied by high localized pressures, may produce prohibitively high rates of wear.

Lubrication. Correct lubrication of the sheet metal is essential if friction, wear, and galling are to be held to the lowest possible levels during deep drawing. In fact, deep drawing is impossible if the sheet metal is not lubricated. In actual practice, die materials are selected after trials using one or more candidate production lubricants. If excessive wear or galling occurs, a better lubricant is usually applied. For extremely difficult draws, the best lubricants are usually applied at the outset.

Table 1 lists typical lubricants used for different work metals and severities of drawing. Lubricants are marketed under proprietary names, but any supplier of lubricants can recommend commercial compounds fitting the descriptions given in Table 1. More information on lubrication in sheet forming is available in the article "Selection and Use of Lubricants in Forming of Sheet Metal" in this Volume.

Materials for Specific Tools

Draw Rings. Table 2 lists typical materials for draw rings (both dies and backup rings) used in drawing and ironing cups of various diameters and lengths. The data in Table 2 are for round and square cups drawn from stock 1.6 mm (0.062 in.) thick in three typical production quantities. Similar data for a large square cup and a large pan are also provided. Design dimensions for all seven parts referred to in Table 2 are given in Fig. 3. The square parts have liberal corner radii consistent with favorable die life.

Table 3 indicates the effect on material selection of changing the thickness of the sheet metal being drawn. Tool materials of increasingly greater wear resistance are required as the thickness of the work metal or the total quantity of parts is increased.

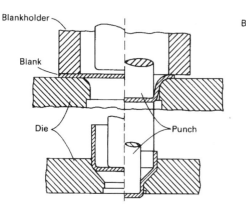

Fig. 1 Schematic showing tools used for the first draw (top) and the first redraw (bottom) in deep drawing

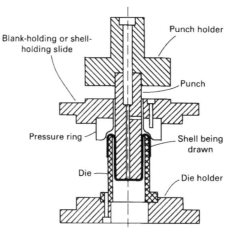

Fig. 2 Typical tooling used for the reverse redrawing of thin-wall shells

Punches and Blankholders. Typical materials for punches and for blankholders or shellholders are listed in Table 4. These materials are for punches and blankholders used in drawing and ironing round and square steel cups similar to parts 2 through 7 in Fig. 3.

More wear-resistant materials are required not only for the tools used in drawing and ironing harder or thicker stock or for those used for longer runs but also for tools used to achieve greater percentages of reduction during ironing. Table 5 lists typical tool steels used in punches and dies for short-, medium-, and long-run production at four levels of reduction in ironing. Typical materials for punches and dies used in the reverse redrawing of steel cups are listed in Table 6. In both cases, for higher production volumes of the more difficult materials or part designs, physical vapor deposition (PVD) coatings and high-alloy powder metallurgy (P/M) tool steels may be considered for longer performance or more severe operating environments than those satisfactorily handled by D2.

Combating Specific Service Problems

Wear (most notably galling) is the most common sign of deterioration in deep-drawing tools. Wear can be reduced by selecting a harder and more wear-resistant material, by applying a surface coating such as chromium plating to the finished tools, or by using a surface treatment such as carburizing or carbonitriding for low-alloy steels or nitriding or PVD coating for tool steels. The following sections in this article are intended to supplement the basic information given in Tables 2 through 6.

Galling. The typical causes of galling of deep-drawing tooling are:

- Attempts to stretch sheet metal beyond practical limits
- Poor tool fit-up, with poor alignment or insufficient die clearance for the sheet thickness
- Excessive wrinkling
- Insufficient or otherwise inadequate lubrication

- Use of tool steels that are susceptible to galling without applying a surface coating to the tools or using a lubricant of superior lubricating qualities
- Rough finishes on tool surfaces

For short runs, dies made of carburized hot rolled steel or hardened alloy steel will often produce parts equal in quality to those drawn over most tool steel dies. Exceptions may be encountered in ironing to severe reductions or in drawing metals that tend to gall, such as austenitic stainless steels. These exceptions may be of little consequence, however, because tool steel dies may also become galled under the same circumstances. The longest die life can be expected when die surfaces have a very fine finish, with final surface scratches parallel to the direction of drawing. Surface treatments that provide lower friction may be beneficial. Die materials can be selected for resistance to galling on the basis of the following two criteria.

First, for parts drawn from carbon steel or nonferrous alloy sheet, the die material can be

Table 1 Typical lubricants for deep drawing

When more than one lubricant is given, they are listed in order of increasing effectiveness.

Metal being drawn	10% or less	Severity of drawing 25% average	50% or more
Aluminum and aluminum alloys	Straight mineral oil, 100 SUS viscosity(a); mineral oil with approximately 10% lard oil	Straight mineral oil, 200–250 SUS viscosity(a); mineral oil with approximately 15% lard oil	Mineral oil with extreme-pressure additives—sulfur and others; coating of soap or wax dried on blanks (or shells) prior to drawing (or redrawing)
Copper and copper alloys	5% soap solution: lard and soap emulsion	10% soap solution with stearic or oleic acid: lard oil and mineral oil with stearic acid	Lard oil blended with 50% mineral oil, coating of soap or wax dried on blanks or draws prior to draw or redraw
Carbon steel	Mineral oil, 250–350 SUS viscosity(a): 5% soap solution	Emulsions of lard oil, mineral oil, and sulfonated oils	Phosphate coating impregnated with dried soap or wax
Stainless steel	Castor oil and soap emulsion	Castor oil with fillers, such as mica or zinc oxide	Boiled linseed oil with mica or lithopone; phosphoric acid etch with dried soap or wax film

(a) Saybolt universal seconds at 40 °C (100 °F)

Table 2 Typical materials for draw rings used in the drawing and ironing of round and square parts

See Fig. 3 for part designs and overall dimensions.

	Total number of parts to be drawn		
Metal to be drawn	10,000	100,000	1,000,000
Cups up to 76 mm (3 in.) across, drawn from 1.6 mm (0.062 in.) sheet (parts 1, 2, and 3)			
Drawing-quality aluminum and copper alloys	W1; O1	O1; A2	A2; D2
Drawing-quality steels	W1; O1	O1; A2	A2; D2
300-series stainless steels	W1 chromium plated; aluminum bronze	Nitrided A2; aluminum bronze	Nitrided D2 or D3; cemented carbide
Cups 305 mm (12 in.) or more across, drawn from 1.6 mm (0.062 in.) sheet (parts 4 and 5)			
Drawing-quality aluminum and copper alloys	Alloy cast iron(a)	Alloy cast iron(a); A2 inserts(b)	A2 or D2 inserts(b)
Drawing-quality steels	Alloy cast iron(a)	Alloy cast iron(c); A2 inserts(b)	A2 or D2 inserts(b)
300-series stainless steels	Alloy cast iron(d); aluminum bronze inserts(b)	A2 or aluminum bronze inserts(b)	Nitrided A2 or D2 inserts(b)
Square cups similar to part 6, drawn from 1.6 mm (0.062 in.) sheet			
Drawing-quality aluminum and copper alloys(e)	W1	O1; A2	A2; D2
Drawing-quality steels(e)	W1	O1; A2	A2; D2; nitrided A2 or D2
300-series stainless steels(f)	W1; aluminum bronze	Nitrided A2; aluminum bronze	Nitrided A2 or D2
Large pans similar to part 7, drawn from 0.8 mm (0.031 in.) sheet			
Drawing-quality aluminum and copper alloys	Alloy cast iron(a)	Alloy cast iron(a); A2 corner inserts(b)	Nitrided A2 or D2 inserts(b)
Drawing-quality steels	Alloy cast iron(a)	Alloy cast iron(a); A2 corner inserts(b)	Nitrided A2 or D2 inserts(b)
300-series stainless steels	Alloy cast iron(d); aluminum bronze	Nitrided A2 or aluminum bronze inserts(b)	Nitrided A2 or D2 inserts(b)

(a) Wearing surfaces flame hardened. (b) In flame-hardened alloy cast iron. (c) Quenched and tempered for part 4; flame hardened for part 5. (d) Flame hardened on wearing surface to not over 420 HB. (e) For drawing aluminum, copper, and steel, the tool material would be used as corner inserts. (f) For drawing stainless steel, inserts would be used for all wear surfaces.

selected without regard to galling, and then, as a finishing operation, the punch and die should be either nitrided, PVD coated, or chromium plated. If a tool steel such as A2, D2, or high-speed steel, which contain chromium, molybdenum, or 2vanadium, has been selected, the smoothly ground tools should be nitrided and then polished or buffed.

Second, for parts drawn from stainless steel or from high-nickel alloy steel, the draw ring material with the best resistance to galling is aluminum bronze. The second choice is D2, high-speed steel, or a P/M tool steel, smoothly ground, nitrided, and polished. The third choice is alloy cast iron, quenched and tempered to 400 to 420 HB.

Chromium plating or PVD coating may be used to extend the service lives of tool steel draw rings. On punches, the primary function of chromium plating is to reduce frictional forces and to facilitate the removal of parts from the punch after the sidewalls have been ironed tight to the punch. The improvement in punch life that results from such surface treatments is usually somewhat less than that attained by changing the punch material to the next-best tool steel.

For successful tool performance, chromium plating must always be deposited on a surface harder than 50 HRC; preferably, plating thickness should be 5 to 10 μm (0.2 to 0.4 mil) and never less than 2.5 μm (0.1 mil). This provides the required hardness and reduction of friction without excessive spalling or chipping at corners. Chromium-plated dies should be heated to 150 to 205 °C (300 to 400 °F) for a minimum of 3 h immediately after plating to minimize the possibility of hydrogen embrittlement.

Combined operations have found increasing use over the past 30 years. The more popular combined operations include one that combines drawing and coining and another that combines successive or tandem drawing (or ironing) operations. This latter combination is called double drawing or double ironing. Advancements in combined operations have paralleled advancements in die materials—for example, better selection of drawing steels as well as improvements in the engineering and construction of tools and especially in surface treatments such as those using zinc phosphate with emulsified soap.

These operations have increased production by doubling reductions and decreasing the number of operations, but at the same time have required capital investment in larger presses. In addition, tool steels of greater resistance to compression and heat have become necessary for drawing and ironing tools.

Double-drawing and double-ironing operations are successive operations in one tooling setup, with two dies placed in tandem so that a punch forces the cup through one die and then directly through the second die while the cup is still warm from deformation heating. The punches are longer than those used in conventional deep drawing and, because of their slenderness, are preferably made of S1 tool steel. Die materials are much the same as in single operations, except that selection is confined to tool steels such as A2, D2, or more highly alloyed tool

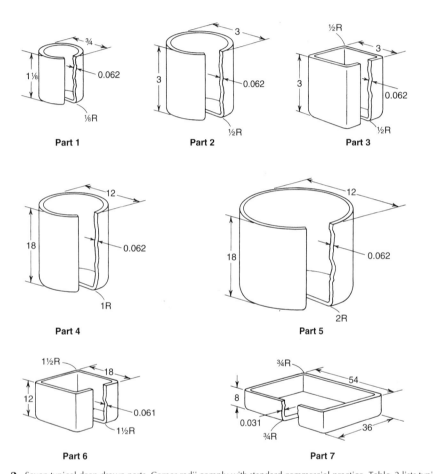

Fig. 3 Seven typical deep-drawn parts. Corner radii comply with standard commercial practice. Table 2 lists typical die materials used for drawing parts of similar configuration. Dimensions given in inches

Table 3 Typical materials for draw rings used in making part 4 from flat-rolled steel of six thicknesses

See Fig. 3 for part design and overall dimensions.

Thickness of steel		Total number of parts to be drawn			
mm	in.	1000	10,000	100,000	1,000,000
0.4	0.015	Alloy cast iron(a)	Alloy cast iron(a)	Alloys cast iron(a)	Alloy cast iron(b), O1, A2
0.8	0.031	Alloy cast iron(a)	Alloy cast iron(a)	Alloy cast iron(b)	A2, D2
1.6	0.062	Alloy cast iron(a)	Alloy cast iron(b)	Alloy cast iron(b), A2	A2, D2
3.2	0.125	Alloy cast iron(b)	Alloy cast iron(b)	A2, D2	D2
6.4	0.250	A2	A2	D2	D2
12.7	0.500	A2(c)	A2(c)	D2(c)	D2(c)

Note: Where tool steels are recommended, they are used as inserts in flame-hardened alloy cast iron. (a) Flame hardening not necessary. (b) Wearing surfaces flame hardened. (c) In drawing 12.7 mm (0.500 in.) plate with A2 or D2 inserts, press speed is slower than for thinner stock, and the plate is phosphate coated.

Table 4 Typical materials for punches and blankholders

See Fig. 3 for part designs and overall dimensions.

Die component	Total number of parts to be drawn		
	10,000	100,000	1,000,000
For round steel cups such as part 2			
Punch(a)	Carburized 4140; W1	W1; carburized S1	A2; D2
Blankholder(b)	W1; O1	W1; O1	W1; O1
For square steel cups such as part 3			
Punch(a)	Carburized 4140; W1	W1; carburized S1	A2; D2
Blankholder(b)	W1; O1	W1; O1	W1; O1
For round steel cups such as parts 4 and 5			
Punch(a)	Alloy cast iron(c)	O1(d)	A2(c); D2(c)
Blankholder(b)	Alloy cast iron(c)	Alloy cast iron(e)	O1; A2
For square steel cups such as parts 6 and 7			
Punch(a)	Carburized 4140(f)	W1; O1(d)	Nitrided A2; D2(d)
Blankholder(b)	Alloy cast iron(c)	W1; O1	O1; A2

(a) Chromium plating is optional on punches to reduce friction between part and punch and therefore facilitate removal of the part. Cast iron, however, should not be plated. (b) Also applies to shellholder and blankholder. (c) Flame hardening not necessary. (d) The punch holder is flame-hardened alloy cast iron with a nose insert of the indicated tool steel. (e) For part 4, this blankholder is quenched and tempered; for part 5, it is flame hardened. (f) The punch holder is alloy cast iron with a nose insert of the indicated steel.

Table 5 Typical tool steels for punches and dies to iron soft steel sheet at various reductions

Ironing reduction, %	Total quantity of shells(a) to be ironed			
	1000	10,000	100,000	1,000,000
Ironing punches(b)				
Up to 25	W1	O1	A2	A2; S1 carburized
25–35	W1	A2	A2; S1 carburized	D2
35–50	A2	A2; S1 carburized	D2	D2
Over 50	D2	D2	D2	D2
Ironing dies				
Up to 25	W1(c)	O1	O1	D2
25–35(d)	W1(c)	O1	D2	D2
35–50(d)	O1	D2	D2	D2
Over 50(d)	D2	D2	D2	D2

(a) Steel sheet up to 75 HRB, or softer metals. (b) All tool steel punches should be plated with chromium 5–10 μm (0.2–0.4 mil) thick for easier removal of the part from the punch. (c) W1 is quenched on the inside and tempered to a minimum of 60 HRC for these applications. (d) Draw rings must be inserted in shrink rings for ironing at reduction greater than 25% and for quantities of more than 10,000 parts.

Table 6 Typical punch and die material for the reverse redrawing of steels

Die component	Total quantity of parts(a) to be redrawn			
	1000	10,000	100,000	1,000,000
Small thick-wall cups				
Die and pressure ring	O1	O1(b)	A2(c)	D2(c)
Punch(d)	4140, 6150	O1, A2	D2	D3
Medium and large thin-wall cups				
Die and pressure ring	1018(e), 4140	4140(f), O1	A2(c)	D2(c)
Punch(d)	W1	A2	D2	D2, D3

(a) No specific finish or tolerance requirements. (b) Dies are polished and chromium plated. (c) A2 and D2 should be nitrided. (d) All punches used for making more than 1000 pieces should be heat treated to 60–62 HRC, polished, and chromium plated. (e) Carburized, hardened, and polished to a fine finish. (f) 4140 or 6150 can be used if carburized and highly polished.

steels when temperatures are high in the second operation. These more temper-resistant steels can better withstand the effects of the higher temperatures developed by increased plastic deformation of the workpiece.

Cemented carbides. For long runs, cemented carbide inserts are widely used in deep-drawing dies. In dies up to 203 mm (8 in.) across for continuous production of over 1 million drawn parts, carbide has often proved to be the most economical die material. Such dies have maintained size in drawing 500,000 parts with 60% reductions and have made as many as 1 million parts with reductions greater than 40% when the steel to be drawn was surface treated with zinc phosphate and soap. However, cemented carbide dies do not provide satisfactory service with inferior lubricants. In addition, carbide dies are not superior to dies made of a tool steel such as D2 in complex deep-drawing operations (for example, those that combine drawing with coining or forming and in which the reduction in drawing is greater than 40%).

The cemented carbides that are most often used for deep-drawing inserts are straight tungsten carbide grades, of normal particle size, that contain approximately 9 to 10% Co or Ni binder. Steel-bonded carbides are also used for deep-drawing tools.

Plastics are the most economical tooling materials for short runs, especially when a prototype part is available as a pattern for the lay-up of plastic-impregnated glass cloth facing, which is backed with chopped glass fibers impregnated with 50% resin. Among resins, polyester, epoxy, phenolic resin, and nylon have been used. The plastic dies that exhibit the longest life are those constructed so that the wearing surface is faced with glass cloth that has had most of the plastic material forced out under pressure before and during curing. Except for very short runs, plastics should not be selected as blankholder materials where burred edges of the blank slide over the plastic surface and produce severe wear or gouging.

Zinc Alloy Tools. Because they are relatively soft, zinc alloy tools should be used only in drawing (without ironing) small quantities of large-diameter thin-wall parts. Zinc alloy tools work best for drawing well-lubricated stock into parts 305 mm (12 in.) or more in diameter under circumstances in which wrinkling is not likely to occur.

Die Manufacture by Electrical Discharge Machining

ELECTRICAL DISCHARGE MACHINING (EDM) is one of several nontraditional methods for machining. In the EDM process, metal is removed by sparks from a shaped electrode (usually graphite or copper), where the result is to make a cavity that is the mirror image of the electrode. There is no direct contact between electrode and workpiece. The sparks travel through a dielectric fluid (typically a light oil) at a controlled distance. Both electrode and work must be electrically conductive. The EDM process has several advantages when machining difficult geometries or materials with poor machinability. Cavities with thin walls and fine features are possible to machine because there is no contact between electrode and work. The removal rate is related to the melting point of the metal being machined, and so the use of EDM is not affected by the high-hardness workpieces, such as cemented tungsten carbide and hardened tool steel.

There are two basic machines for EDM:

- Plunge-type (or ram-type) EDM machine (Fig. 1), where a shaped electrode acts as a plunge in metal removal
- Wire EDM machines (Fig. 2), where the electrode is a continuously spooling conducting wire that moves like a precision jigsaw with respect to the work by a numerically controlled table

The plunge-type machine made its introduction in machining of dies, especially those made of hardened die steel and tungsten carbide in many forms, such as plastic injection molds, extrusion dies, forging dies, and die casting dies. Traveling-wire EDM differs from conventional (shaped electrode) EDM in that a thin-diameter wire acts as the electrode. Electrical discharge wire cutting is used to make stamping dies, tools for lathes, templates for use in tracer lathes, electrodes for vertical EDM, broaches, and extrusion dies. It can also be used for prototype production of parts to be made later by die stamping or Computer numerical controlled (CNC) milling. One of the most important applications of electrical discharge wire cutting is the product of progressive dies (Fig. 3), where a strip of sheet metal is advanced through the die, one station at a time.

Fundamentals of Electrical Discharge Erosion

Electrical discharge machining is a thermo-electric process. Heat from the spark melts the metal, and bulk boiling of superheated metal ejects approximately 1 to 10% of the molten metal. As a result, the hardness or tensile strength of the metal does not have any effect on machining by EDM. This is why EDM has always been successfully applied in machining of high-strength or hardened steels.

The thermal properties of the metal determine its machinability by EDM. Copper, tungsten, and graphite are always applied as electrode materials. This can be explained by the low wear of these materials due to their excellent thermal properties.

Because 90% of the material in the molten pool of metal is not ejected at the end of the pulse, the workpiece is covered with a recast layer 1 to 30 μm (0.04 to 1.2 mils) thick. Sometimes, this layer is desirable, because its hardness and roughness hold lubricating oils. However, in some applications, the recast layer has to be removed in order to avoid surface fatigue.

Setting of an EDM Machine.

The dependent parameters in the EDM process are:

- Metal-removal rate, V_W (mm^3/min)
- Relative electrode wear, θ (%):

$$\theta = \frac{V_E}{V_W} \cdot 100$$

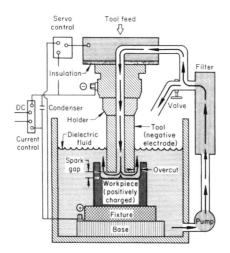

Fig. 1 System concept of plunge electrical discharge machining with a shaped electrode

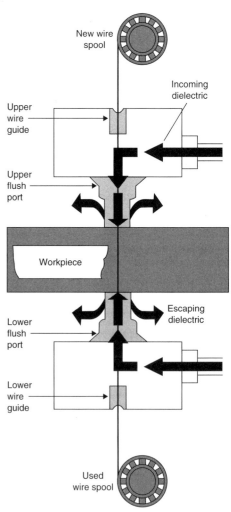

Fig. 2 System concept of wire electrical discharge machining process

where V_E is the electrode-removal rate.

- Surface finish, R_t (peak to valley, μm)
- Thickness of recast layer
- Gap between electrode and workpiece
- Radii of corners and edges

These parameters are dependent on the following independent parameters:

- Electrode material
- Electrode polarity (+ or −)
- Pulse current, $\bar{i}_f$ (A)
- Pulse duration, or on-time, t_j (μs)
- Pulse interval time, or off-time, t_o (μs)
- Average voltage or working voltage, U (V)
- Average current or working current, I (A)
- Working current density, I_d (A/cm^2)
- Open-gap voltage, U_o (V)
- Type of dielectric
- Flushing mode

Dielectric-Fluid Unit. Dielectric fluid is pumped through a filter from a container (100 to 200 L) to the electrode or workpiece. Pressurized flushing through the electrode results in a conical hole in the work because sparks also occur at the sides of the electrode due to the excess of debris in the side gap. When vacuum flushing through the electrode or through the workpiece is used, a straight hole is obtained.

Principles of Electrical Discharge Machining

An electrode, which serves as a cutting tool, and a workpiece are placed face to face with very little clearance (several to several tens of micrometers) in a dielectric fluid (kerosene-based oil specially made for EDM is generally used). Current pulses are continually supplied to the clearance from a pulse power supply (approximately 60 to 300 V) to provide transient arc discharge (discharge retention time: 0.1 μs to 8 ms) at a high frequency so as to remove workpiece metal with a very dense energy provided by the discharge. Figure 4 shows the five-step process from spark discharge, through metal machining, to the original cool state.

It sounds as if it takes a long time for steps 1 to 5, but actually the entire process is completed in $^1/_{50}$ s even for rough machining, which is the slowest among the machining processes. This discharge between the entire surface of the workpiece and that of the electrode is repeated several tens to several hundred thousands of times for machining. If, therefore, electrodes are made only of such easily machinable materials as copper and graphite, workpieces can be machined highly accurately to shapes corresponding to the electrode shapes, regardless of workpiece hardness and machinability.

Surface Roughness, Electrode Wear, and Machining Speed. Because EDM is performed by accumulation of single discharges, the amount machined, machining speed, clearance,

and surface roughness increase as the single-discharge energy increases (Fig. 5). The amount of single-discharge energy depends on the single-discharge peak current (power setting) and time (pulse width). Single-discharge energy is large if the power setting and pulse width are large, and small if both settings are small. To obtain the desired surface roughness, the power setting and pulse width must be elected

from the combinations that have the same single-discharge energy:

- Large power setting and small pulse width
- Small power setting and large pulse width
- Power setting and pulse width intermediate between the previously listed combinations

Figure 6 shows methods of selecting power setting and pulse width for the same surface

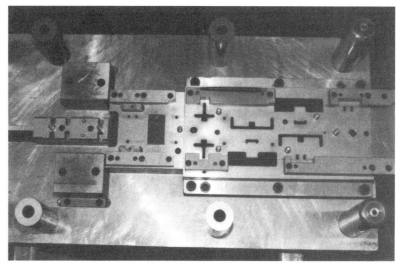

Fig. 3 Progressive die cut by traveling-wire electrical discharge machining. Courtesy of Progressive Tool Company

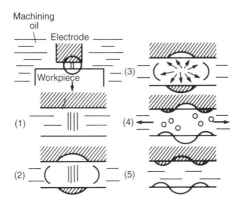

Fig. 4 Process through which metal is machined by spark discharge and oil pressure. (1) When the electrode approaches within several micrometers of the workpiece, a spark is generated at a point of the shortest distance and immediately becomes a fine arc column, or a flow of electrons at high current density, hitting the workpiece at one point. The electron flow generates heat at this point, to a temperature high enough to melt tungsten, which has a high melting point. At the same time, ions, which are generated by collision of the electron current with the dielectric fluid, heat the electrode. (2) The heat vaporizes the dielectric fluid around the flow. (3) The pressure applied to the melted workpiece and electrode is small compared with the entire surface of the workpiece and electrode but is very large per unit area. (4) Molten metal from the workpiece is blown away into the dielectric fluid as small round blocks. The remaining portions on the edge form protrusions on both the workpiece and electrode. The protrusions become discharge points for later discharge. (5) Cool dielectric fluid flows into the hollow where molten metal has been blown off, removing the residual heat from the hollow.

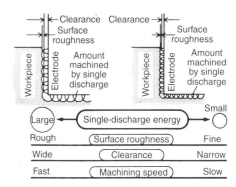

Fig. 5 Relations between single-discharge energy and machining characteristics

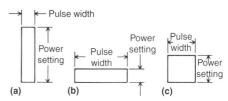

Fig. 6 Method of selecting power setting and pulse width. (a) Fast machining speed, more electrode consumption. (b) Slow machining speed, less electrode consumption. (c) Machining speed and electrode consumption intermediate between (a) and (b)

roughness and the relationships among methods, machining speed, and electrode consumption.

Principles and Features of Orbital-Movement EDM

This machining method provides better performance than that obtainable with conventional EDM, because orbital-movement machining controls the x- and y-axes in addition to the z-axis.

Improvement in Machining Shape Accuracy. Orbital-movement machining prevents loss of accuracy in the side-surface direction (uneven clearance, taper, intermediate protrusion, or uneven surface) so that the entire surface can be machined evenly at any desired clearance (Fig. 7).

Less Electrode Wear. Because orbital-movement machining finishes a shape by machining in the side-surface direction, the electrode wear is not concentrated on the edge, as in conventional machining methods, and consequently, the effect of electrode wear is much smaller than in conventional machining methods (Fig. 8). The result is a finer surface, decreased spark-hardening layer, and increased machining speed.

Improvement in Finish Machining Speed. Discharge area is wide, and sufficient machining current can be supplied (area effect) because machining is performed in both the bottom and side-surface directions at the same time. Orbital movement also helps to remove machining chips more quickly, preventing generation of arcs (Fig. 9).

Reduction of Electrode Manufacturing Man-Hours. Because any values can be selected for machining dimensions by setting the orbital-movement amount, two or more electrodes of different sizes for roughing and finishing are not required. An electrode manufacturing process using only one type of electrode suffices (Fig. 10).

Taper Machining with Straight Electrode. Taper machining can also be performed with a straight electrode, as an orbital-movement machining application.

Examples of taper-machining capabilities include machining of different top and bottom shapes or top and bottom radii, machining punch and die simultaneously, machining sharp edges and corner radii, and angle alteration.

Wire EDM

Wire electrode discharge machining brings a thin wire electrode (made of brass, tungsten, or copper wire) to a workpiece and impresses a voltage between the wire electrode and workpiece, while flooding the cutting areas around the wire electrode with a dielectric fluid. Wire (normally brass) is used once. Servo control of the machine is similar to that of vertical CNC machines. Flushing problems or workpiece distortion sometimes make it necessary for the wire and work to separate. The controller must ensure that the separation takes place along the path previously cut. Work thickness capacity of 150 mm (6 in.) is average, with some machines capable of up to 420 mm (16.5 in.).

Wire EDM machines have at least two controlled axes (x and y), which only allow vertical cuts. Most machines allow simultaneous four-axis control using two auxiliary horizontal axes (called u and v). With four-axis control, the machine can tilt the wire to produce tapered work. Independent four-axis controls (machines with controlled vertical and rotary axes) are also available to allow production of top profiles that are independent of bottom profiles. Constant wire tension is maintained. The tension must be high enough to keep the wire straight in the cutting zone but not high enough to break the wire. The right tension is a function of the wire material and diameter. Wire guides are made of a hard material such as sapphire or diamond to minimize wear. The guides are required to keep the wire vertical, which is necessary to maintain straight sides on the workpiece. Hydraulic servo mechanisms are not used in wire EDM. The servo motion is done with direct-drive alternating current servo motors, direct current servo motors, or stepper motors. The monitored quantity is the voltage between the wire and the workpiece.

Flushing is provided by nozzles at the upper and lower wire guides, with the stream of dielectric being coaxial with the wire. Swarf is removed by filtration and settling. The dielectric fluid is typically deionized water of high insulation resistance, so that the electrode discharges electricity through the deionized water. The discharge forms an arc column at the shortest

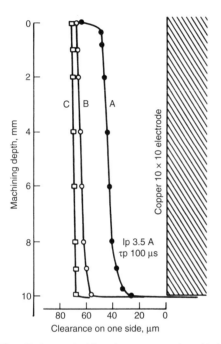

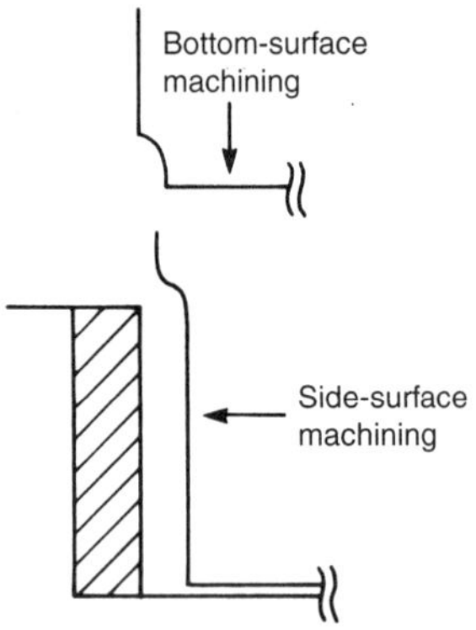

Fig. 7 Improved side-surface accuracy in orbital-movement machining. A: Conventional machining (two electrodes). B: Orbital-movement machining (one electrode), $R = 50$ μm (2 mils). C: Orbital-movement machining (two electrodes), $R = 50$ μm (2 mils)

Fig. 8 Relationship between machining direction and shape of electrode wear

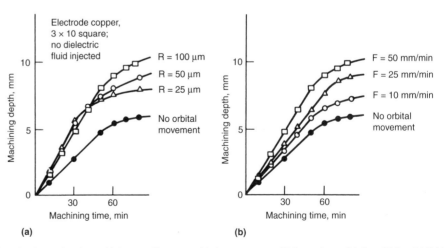

Fig. 9 Comparison in machining speed between orbital-movement machining and nonorbital machining. (a) Orbital circular movement; F = 50 mm/min. (b) Orbital circular movement; radius 100 μmR

distance between the wire electrode and workpiece surfaces and heats both sides locally, melting them with thermal energy, as shown in Fig. 11(a). At the same time, the insulation water around the wire electrode is heated to vapor state and expanded rapidly to create a local explosion (Fig. 11b). The explosion pressure turns molten metal into fine metal particles, which are carried away in the water. The workpiece and wire electrode surfaces are then cooled to leave a dent, as shown in Fig. 11(c). Because the discharge and metal removal are performed repeatedly at a very high frequency, the dent on the workpiece surface grows gradually to form a groove along the wire electrode shape as the wire electrode is fed at increasing speed.

Workpiece Thickness and Machining Speed. Wire electrode discharge machining defines machining speed as sectional area of machining per unit time:

$$\text{Machining speed (mm}^2/\text{min)} =$$
$$\text{Machining feed speed (mm/min)}$$
$$\times \text{Workpiece thickness (mm)}$$

The machining speed of a wire electrode discharge machine is almost proportional to the machining current between the poles, as in general-purpose die-sinking electrical discharge machines. In other words, a higher peak current and shorter rest time increase the machining speed.

A greater workpiece thickness makes the machining surface area larger in the advance direction, and the area effect works effectively to improve the machining speed.

Because the peak current value is selected by machining setting, peak current is higher and thus machining speed is faster if the machining setting is larger. A larger machining setting, however, causes greater surface roughness. Figure 12 shows the relation between machining speed and workpiece thickness.

Machining accuracy in wire electrode discharge machining includes uniformity, straightness, and shape accuracy of machining groove width. Normal accuracy is approximately ±0.013 mm (±0.0005 in.). Special measures

such as multiple passes and precise temperature control are used for a higher accuracy of ±0.005 mm (±0.0002 in.). Power settings may be decreased with each pass for improved surface, as with vertical EDM.

Machining Groove Width and Machining Feed Speed. Dispersion in machining groove width is all equal to that in the final dimensional accuracy because half of the width is considered as an offset value and left in the inside or outside of a size specified in the drawing to obtain the desired dimensions when a workpiece is machined.

Figure 13 shows the relation between machining groove width and machining feed speed. Optimal feed occurs when the electrical condition of the machining power supply is altered greatly using a mean machining voltage constant-control system. As shown in the figure, optimal feed creates an overall change in the machining groove width of approximately 15 μm (0.6 mil) even if the electrical condition

is changed to significantly alter the machining feed speed.

Figure 14 shows the relation between machining groove width and mean machining voltage in a constant-feed/speed system. The relationships at two machining feed speeds, F 1.4 and 2.4 mm/min, are shown with the electrical condition of machining power supply altered for each speed. As illustrated in the figure, the

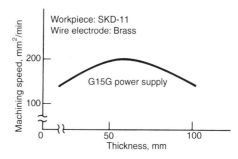

Fig. 12 Relation between wire electrode machining speed and workpiece thickness

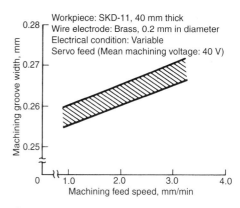

Fig. 13 Relation between machining groove width and machining feed speed in wire electrode discharge machining

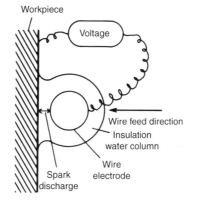

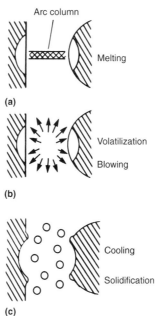

Fig. 11 Schematic showing principles of wire electrode discharge machining. See text for details.

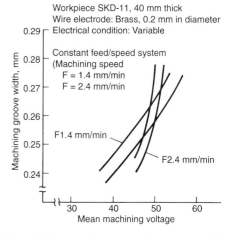

Fig. 14 Relation between machining groove width and mean machining voltage in a constant-feed/speed wire electrode discharge machining system

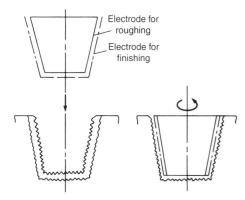

Fig. 10 Comparison of electrodes for conventional and orbital-movement machining processes

machining groove width varies 32 to 35 μm (1.3 to 1.4 mil) with the changes in the electrical condition at either speed in the constant-feed/speed system. In other words, changes in mean power greatly affect changes in groove width.

Consequently, the optimal speed system enables changes in electrical condition and machining feed speed with less effect on machining groove width than the constant-feed/speed system.

Second cut is a finishing process similar to that of a die-sinking EDM. It is performed as follows: the primary machining (first cut) is made on a workpiece, leaving a finish allowance; the electrical condition for the finish allowance is switched to finish condition, decreasing the offset amount gradually so as to perform two or more surface-removal passes for finishing. As in general-purpose EDM, however, increased speed causes greater surface roughness and corner rounding, which results in lower dimensional accuracy. It is the second-cut process that solves these problems; this process provides high accuracy and good surface finish. Figure 15 shows the relation of the number of second-cut passes to machining accuracy and surface roughness. The figure indicates that the required times for one-pass and two-pass machining have a ratio of approximately 1 to 0.3 (the latter being approximately three times shorter than the former). Two-pass machining also gives higher machining accuracy, if, for example, a 40 mm (1.6 in.) thick SKD-11 workpiece is to be finished at a surface roughness of approximately 10 μm R_{max}.

Examples of Applications. Three examples of applications of wire electrode discharge machining, together with the advantages realized from use of the process, are:

- Plastic molding dies, for which a major reduction in machining time can be achieved with the integral machining of sirocco fans.
- Progressive dies for stepping-motor cases, for which the process provides major cost reductions with significantly shortened delivery times.
- Dies for printed-circuit boards, for which the automatic wire-feed option eliminates burrs—an advantage in unattended operation.

Effect of Recast Layer on Surface Integrity and Fatigue Life

Electrical discharge machining is known to have a large detrimental effect on fatigue strength. The nature of the EDM process, which entails removal of metal by melting and vaporization of the metal surface, inherently induces very large residual stresses that are tensile in nature and hence detrimental to fatigue resistance. The rapid cooling of the surface layer by the dielectric also results in surface and subsurface cracks and microcracks. Such cracks also tend to propagate due to the stress-concentration

effect, and, under cyclic loading, fatigue failure occurs prematurely.

Such deterioration in fatigue strength due to EDM has necessitated subsequent surface-layer-removal operations for heavily stressed components or those subjected to cyclic loads. Most users remove the recast layer to restore the fatigue strength.

Figure 16 shows the nomenclature used to define the various layers on the surface of a component machined by EDM. The recast layer is a layer of metal that was in the molten state and that solidified on the surface. The heat-affected zone is a layer, below the recast layer, that has been subjected to elevated temperatures sufficient to alter its mechanical and metallurgical properties. Both layers together are known as the altered metal zone.

EDM Electrodes

Although EDM electrodes can be made of many different types of metals, graphite electrodes have emerged as the best EDM cutting tools. This superiority stems from the excellent electrical and thermal properties of graphite, which together bring about good EDM metal-removal rates and relatively low electrode wear. The importance of electrode wear is evidenced in the fact that the EDM process is frequently referred to as spark erosion. It is inherent in the process that this erosion occurs not only on the workpiece but on the electrode as well (although at a lower rate). If EDM parameters are chosen to keep electrode wear in the minimal range (referred to in the industry as no-wear machining conditions), the ratio of wear between the workpiece and the electrode can still be roughly 100 to 1, that is, machining 25 mm (1 in.) from the workpiece removes 0.25 mm (0.010 in.) from the electrode, and seldom uniformly. Consequently, an electrode can be used for only a

few workpieces (and sometimes not even one completely) before it is worn out of tolerance.

Because electrode wear is one of the major limitations of this process, EDM parameters are set to these no-wear settings to conserve the electrode as much as possible. However, although such machine settings can prolong electrode life, they prolong EDM time as well. To achieve the maximum efficiency from the EDM process and to make it economically justifiable on more machinable materials, the cost and inconvenience of forming the electrodes consumed as tools in EDM is important. At present, the EDM process generally affords the user the simple privilege of conventionally machining the electrode material instead of the workpiece material.

Conventional machining of an electrode is pretty much the same as machining of any other workpiece, except, of course, that the electrode material is probably more machinable. In

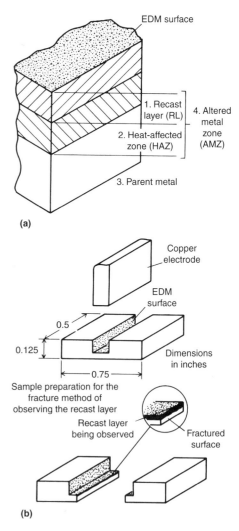

Fig. 16 (a) Surface layers on a component machined by electrical discharge machining (EDM). (b) Fractured samples showing the fractured edges and the recast layer

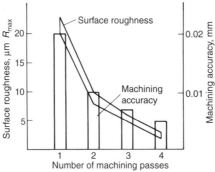

Workpiece: SKD-11, 40 mm thick
Work electrode: Brass, 0.2 mm in diameter
Electrical condition: Standard second-cut condition

Fig. 15 Relation of number of second-cut passes to machining accuracy and surface roughness in wire electrode discharge machining

general, however, conventional machining of new electrodes (or remachining of used electrodes to re-establish dimensions reduced by wear, after some use) involves significant time and cost. The time required, in relation to the time needed for conventional machining of the workpiece itself, depends primarily on the machinability of the workpiece material. Thus, if accurate forming of electrodes were made drastically easier, electrode wear would be less critical and EDM parameters would be less constrained, permitting higher cutting speeds with no sacrifice in accuracy or finish.

Orbital Grinding and Ultrasonic Abrading Techniques of Graphite EDM Electrode Forming. Two systems—orbital abrasive grinding and the proprietary SoneX process (ExOne Co.)—are specifically designed for machining of graphite electrodes. Both of these methods use low-amplitude grinding motions to form the total electrode shape and are a full order of magnitude faster and more convenient than conventional machining.

Orbital abrasive grinding is a method in which two horizontal platens—one holding an abrasive forming tool and the other a graphite electrode block—are oscillated in a precise orbital path with respect to each other. While orbiting, the two platens are brought together, causing the full three-dimensional shape of the abrasive forming tool to be ground into the graphite electrode material. This cutting action occurs simultaneously over the full surface of the abrasive forming tool, producing surface finishes on the electrode that are much finer than those produced by profile milling. The flow of a flushing fluid through the machining gap, combined with frequent vertical reciprocation between the platens, washes away the abraded graphite as fine particles.

The amplitudes used by an orbital abrader range from 0.25 to 6.25 mm (0.010 to 0.250 in.), with 1.5 to 2.5 mm (0.060 to 0.100 in.) being the typical amplitude range at speeds of 1000 to 1200 oscillations per minute.

Orbital abrasion produces an electrode form that is orbitally undersized by the amount of the orbit amplitude. If the abrasive forming tool is the same size as the form to be EDM'd, the resulting electrode is suitable for orbital (sometimes called planetary) EDM. In orbital EDM, rather than following the conventional EDM technique of simply sinking the electrode straight into the metal workpiece, the electrode is instead orbited while machining the workpiece (just as in orbital abrading, the electrode is typically orbited into the abrasive forming tool). When an orbitally abraded forming tool is used, the EDM orbit is normally set to be the same as the abrading orbit was when the electrode was formed, minus the amount per side needed for the EDM overcut (i.e., the spark gap). Orbital EDM can offer dramatic benefits in cutting speed, surface quality, and precision when compared with the conventional EDM technique.

When an orbitally abraded electrode is used in conventional, straight-sinking EDM, an on-size electrode must be produced by using an abrasive forming tool that is oversize by the amount of the orbit amplitude to be used in abrading the electrode.

The actual forming tool for the orbital abrader, when producing electrodes for use in orbital EDM, can simply be an EDM'd form itself, because a fairly rough EDM surface can easily abrade graphite. Alternatively, an abrasive forming tool may be formed by a technique of casting from a reverse model of the form to be EDM'd.

SoneX Process. The second electrode-forming technique, SoneX, also uses a low-amplitude oscillating motion, but as a vertical vibration and not as a horizontal orbital oscillation. The amplitude of tool motion is much lower, typically 0.025 mm (0.001 in.), and the frequency is very much higher, typically approximately 20,000 cycles/s; hence, it is called ultrasonic abrading. In this technique, a non-abrasive forming tool attached to a special toolholder/sonic transducer combination (often called a sonotrode) is vibrated longitudinally at the resonant frequency of the sonotrode/forming tool. At the same time, an abrasive-filled fluid is flushed through the gap between the forming tool and the workpiece. Sometimes, it is desirable to attach the graphite electrode blank to the vibrating sonotrode and then feed it into the stationary forming tool. In either case, a gentle machining action is produced as the sonotrode vibrates the fine abrasive particles flowing throughout the machining gap and propels them against the graphite electrode workpiece material. These shaped vibrations, resonant with the forming tool, cause a shape that is the exact reverse of that of the forming tool to become abraded into the electrode.

SoneX ultrasonic electrode forming relies on a vertical motion—the constant low-amplitude, high-frequency acoustic vibration produced by the sonotrode—to perform the required machining. The forming-tool replica produced by ultrasonic grinding will have an overcut (controlled by the abrasive particle size) that usually can be made to coincide with the expected EDM overcut.

What is particularly dramatic about this technique is its ability to produce intricately detailed electrodes with sharp internal corners and excellent surface finishes. Even very precise or subtle configurations can be produced. Very low machining forces permit manufacture of fragile electrode designs that are preserved by the gentle ultrasonic abrading action—an action in which literally thousands of tiny vibrating abrasive particles serve as the cutting tools. SoneX ultrasonic electrode forming can produce electrodes that once were simply too intricate to be machined or that, if producible at all, could be formed only by laborious hand engraving.

These two new electrode-forming techniques are based on a similar theory of low-amplitude oscillatory abrading action, although they differ somewhat in application, with each approach offering its own strengths. Orbital abrasive grinding uses a low-amplitude horizontal orbiting motion to machine an orbitally undersized reverse impression of the forming tool into the workpiece. This system is suitable even for very large electrodes. One consideration to keep in mind, however, is that, due to the orbital motion, some details (particularly internal corner radii that are smaller than the orbit amplitude) may be lost.

On the other hand, ultrasonic abrading uses an even lower-amplitude and higher-frequency longitudinal vibration to accomplish a similar task. Ultrasonic abrading is often better suited to the forming of small or delicate electrodes. The ultrasonic abrader forming tool imparts its shaped vibrations to the abrasive grains in the fluid flowing throughout the machining gap. These tiny vibrating particles, not the forming tool, perform the actual machining. The SoneX system is limited to electrodes under 100 mm (4 in.) in diameter and has some other limitations in producing certain shapes and depths.

Despite their different mechanisms (and the particular benefits associated with each), these two machining techniques are strikingly similar in their overall effects on the EDM process: the extremely efficient production of graphite electrodes with accurate dimensions and excellent surface finishes allows the use of faster EDM parameters (such as in negative polarity) that have previously been dismissed because of their increased tool wear. Both techniques can use the original EDM'd part itself as the forming tool or, alternatively, can use an epoxy forming tool cast from a model of a production part. Because of their improved machining speed and accuracy, both the orbital abrasive grinding and SoneX systems can machine and re-dress worn electrodes with little of the expense and time normally spent in conventional machining of electrodes. The final result is that these improvements in electrode forming provide for a more economical EDM process, particularly when multiple electrodes are required.

Electrochemical Machining

Another technique that improves the speed and quality of machining of dies is electrochemical machining (ECM). In this process, a low-voltage, high-amperage direct current electrochemically dissolves the die form into the workpiece, regardless of workpiece hardness, in an electrolyte made of a salt and water solution. Compared with EDM, ECM produces surface finishes that are generally much finer and contain no recast layer. There is also little electrode wear.

On the other hand, ECM equipment is expensive, partly because of the special materials that must counteract the corrosive nature of the electrolyte.

Another limitation is the expense of generating a tool (cathode) that will accurately form a cavity to close dimensional tolerances. An

ECM cathode will not produce a form that is its exact geometric opposite, because the overcut across the machining gap will vary due to several factors, including:

- Variations in the flow speed of the electrolyte throughout the gap, particularly at "dead" spots or eddy currents
- Changing electrical conductivity of the electrolyte due to temperature increases (the current flow in the gap heats the electrolyte by electrical resistance) and to changing hydroxide concentrations (the metal removed as the electrolyte flows across the gap becoming, almost instantly, a metal hydroxide that changes the electrolyte conductivity as well)

Because of these factors, tool development for ECM work becomes increasingly expensive as tolerances become closer. However, for high-volume tool and die applications with moderate tolerances, this expense may be justified.

ACKNOWLEDGMENT

This article was adapted from Chapter 5A, "Electrical Discharge Machining," in *Handbook of Fabrication Processes* by O.D. Lascoe, ASM International, 1988.

Progressive Dies

A PROGRESSIVE DIE incorporates several individual work stations where successive operations are performed on a part. Frequently, all the operations required to complete the part are incorporated in a single die.

Progressive dies are also called "cut and carry dies," "follow dies," or "gang dies"—labels descriptive of the operational aspects of such dies. Each workstation performs one or more distinct operations on the part. The part is carried from station to station by the stock strip to which it is left attached throughout the operation. As the part moves forward, it is positively positioned in each workstation until it is completed and cut out of the strip in the final operation.

Factors in Selecting Progressive Dies

The operations performed in a progressive die could be done in a series of dies involving the use of several presses, but this would require a good deal of handling to move the parts from one press to another. Although this multiplicity of machines and materials handling would seem to indicate that progressive dies should always be used, there are a number of factors that must be considered before the feasibility of using a progressive die can be established. Perhaps the most important of these factors is the part itself. Progressive dies are more or less restricted to the production of small stampings, including hubcaps and headlight shells. Handling problems often preclude the use of progressive dies for large parts.

Increase in Productivity. The primary purpose of using a progressive die is to increase productivity—that is, to achieve greater production at the same or less cost in manpower, materials, and machines. Thus, all the factors related to an operation must be considered to determine whether a progressive die will meet these objectives.

The costs involved in the use of progressive dies are frequently a decisive factor in their selection. Progressive dies, due to their complexity, are generally very expensive to build. This expense must be weighed against anticipated savings in manpower, materials, and machines. The progressive die process and alternative processes such as line-die, multislide, or transfer processes must be compared to determine at what piece price each can produce a given part. If there is enough advantage in using a progressive die, the construction cost may be justified by a decreased piece price.

The production volume of the part to be produced is a most important factor in determining whether to use a progressive die. This matter ties in closely with the subject of overall cost. Frequently, the die itself will be the most expensive item in a progressive operation. The cost of the die will ordinarily have to be amortized within the limits of the production run. For this reason, the use of progressive dies is generally restricted to high-volume parts. Except in rare instances, the cost of the die would prohibit progressive operations for low-volume parts. Some low-volume work may be done on progressive dies where only a few operations (two or three) are involved and the die construction is very simple.

Press Availability. Another factor involved in determining the selection of a progressive die is the availability of a press. Presses for progressive dies usually require wide beds to accommodate the multiple work stations of the die. The size of bed is, of course, dependent on the size of the part to be made.

Stock Requirements. Because of certain limitations presently inherent in progressive die operations, the material required to make the part is also a factor in determining whether to use a progressive die.

The progressive die uses a mechanical system to pilot the strip or coil stock into the various workstations. If the stock for a given part is too thin, it will wrinkle or buckle while being fed into the press. The wrinkles formed will result in scrapped production. As a result, progressive dies can be used only with stock that is of heavy enough gage to permit proper piloting.

Restrictions with respect to heavy-gage stock are also encountered in the use of progressive dies. These restrictions are related to the problem of supplying straight, well-leveled stock to the die. Roller levelers or stock straighteners are usually used. If the stock is too heavy; however, proper results may be impossible to achieve. If wavy or uneven stock is introduced to the die, scrapped or damaged parts will generally be made. As a result, there are restrictions on the thickness of stock that can be used in progressive dies.

Because progressive dies depend on a carrier strip to move the part through the workstation, coil stock generally must be used. Some progressive operations may be done with strip stock, but coil stock is preferred for greatest efficiency. Single blanks cannot be accommodated in progressive dies, although they are used for progressive work in transfer-type operations where separate die stations and transfer devices are employed. Therefore, where progressive dies are contemplated, the availability of coil stock should be ascertained well in advance of the beginning of die construction. The proper size of coil and proper gage of steel should be determined and ordered from the steel supplier.

Strip Development for Progressive Dies

To determine the best coil size to use for a progressive die, a strip development should be made. A strip development is a visual representation of the operational steps a part will go through from start to finish in a progressive die. It provides a tentative picture of the position and relationship of each workstation in the die and indicates the distance between each station. From this development, the process engineer determines how wide the strip should be (Fig. 1–7).

Purposes of Strip Development

In addition to providing a means of determining coil size, numerous other important purposes are served by a strip development.

The practicality of using a progressive die to make the part in question is one of the first considerations in using a strip layout. By obtaining a visual representation of the operations involved, it is possible to determine whether it is practical to use the progressive method. In some cases, an excessively complex die design may be required. In others, the operations involved may be too difficult. In these cases, planning would revert to the use of line dies. Layouts that are too long for available presses, are too wide, or require too much tonnage are typical of the unfavorable conditions that may be uncovered in the study of a strip development.

Material Economy. A strip development also provides a means of arriving at the most economical use of material. Economy of material usage must be considered so as to obtain fully the possible reductions of manufacturing costs through use of progressive dies. The strip development should be studied with a view toward using the narrowest possible carrier strips consistent with requirements for producing parts of acceptable quality. It may be discovered that carrier strips outside the part are most feasible. On the other hand, some parts may be carried most effectively by a center strip. In many cases, experimental layouts indicate that the greatest economy can be achieved by running two parts at a time—a procedure that usually provides surer control of the part through the die. This procedure yields more balanced conditions with respect to the thrust of the ram. Often a right-hand and left-hand part can be run together. This

gives a twofold advantage by eliminating separate manufacturing costs while contributing to a better die. The layout should be studied with a view toward reducing scrap production as much as possible. This may call for changes in the original strip layout. Such changes should be discovered early in the die-development process in order to procure the proper size of stock in time for the beginning of production. The objective for every part should be to obtain a "no-scrap" layout. This, of course, is a physical improbability for most parts, but adopting this objective will lead to the maximum scrap reduction.

Part-Design Changes. Study of a strip development may also indicate changes in part design. In progressive die work, some notching

and piercing is usually done in the first stations to allow for forming. When the part is cut out of the strip in the final station, the cutoff punch must line up with one or more of these previously cut notches. To avoid "slivers" and to obtain extreme accuracy in strip location, the cutoff line is usually designed to meet the radius of the notch. To obtain this desirable condition, it may be necessary to change the design of the part to permit the cutoff line to line up with the notch.

Changes in part design may also provide an opportunity to achieve greater material economy through better nesting of parts in the strip. The full effect of any design change should, of course, be studied before such a change is suggested.

Operational Sequences. Sequences of operations that will produce the part most effectively are often determined through analysis of the strip layout. While rigid rules for sequencing operations cannot be formulated, there are principles that should be followed. The requirements for making the part within cost limitations will be the guiding factors in arranging the layout.

All holes that can be pierced without danger of deformation or mislocation should be pierced prior to forming. Some of these holes should be selected as piloting points for locating the part in subsequent operations. Where no suitable holes are available, it may be necessary to pierce pilot holes in the carrier or in the scrap. Holes should be as far from the center axis as the width of the strip will allow. This is particularly important in a strip with a center carrier. Pilot holes designed in this way will prevent mislocation of the part as it advances in the die. They will also prevent shifting of the part in forming operations.

The part should be blanked out following the creation of pilot holes. This should take place in the same station where piercing is done or in the station immediately following.

Forming follows the blanking operation. Where possible, all forming should be done in one station to avoid inaccuracies that are apt to occur from relocating a partially formed part in a subsequent station. All holes that would suffer deformation if made previously should be pierced after forming. Any restrike or additional forming operations should be done just prior to the final station in which the part is severed from the carrier strip.

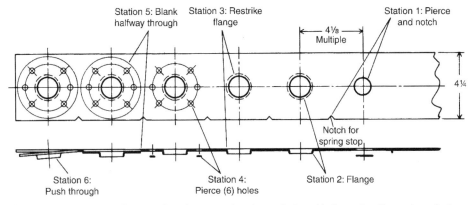

Fig. 1 Strip development for a ring-shaped part. Note locating notch pierced in first station. Flanges formed prior to piercing of nearby holes.

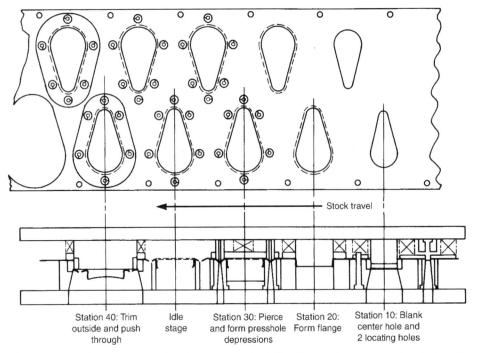

Fig. 2 Strip development for a ring-shaped part (two at a time). Note idle stage for die strength, layout of strip for material economy.

Principles of Strip Development

In making the final development, certain principles should be observed to ensure proper die construction. Some of these principles apply generally to all types of dies. Special considerations must be made where the part design involves drawing or forming.

The following principles apply to all types of parts:

- *Pierce piloting holes and notches in the first station of the die.* An important consideration in progressive die work is the proper gaging and location of the part throughout a sequence of operations. Means to facilitate location should be provided in the first operations. These piloting holes or locating points can then be used throughout the remainder of the operations to ensure that a part will be produced according to specifications.

- *Distribute pierced holes over several stations if they are too close to each other or to the edges of the die openings.* While it is recommended that all holes be pierced in the same station, to better maintain tolerances, holes that are too close to each other are apt to cause damage to punches due to crowding of the metal. This can be compensated for in some instances by stepping the punches to permit larger punches to pierce first followed by the smaller punches. However, if the number of holes to be pierced is great, better results may be achieved by spreading out the piercing operations. Holes that would be distorted by subsequent forming operations should be sequenced to follow forming.

- *Design the shape of blanked-out areas as simply as possible.* The lines along which scrap pieces are cut out should be as simple as possible to reduce the problems involved in cutting them out. The simplest designs may

permit the use of standardized commercial punches, thus obtaining maximum advantages with respect to tooling costs.

- *Consider the use of idle stations for strengthening the die and facilitating strip movement.* Although at first glance it may seem that it would be best to perform work in each and every station of a die, there are many cases in which an idle station offers concrete advantages. At best, a progressive die involves cramming of numerous operations into a short space. The various operating tools are set in the main die body in die sections. The existence of numerous die sections tends to weaken the overall die. By the addition of one or two idle stations, the die can be materially strengthened. Movement of the strip, and control of the metal, also can be facilitated by the inclusion of idle stations. There are often severe forces at work in adjoining work stations that tend to break the carrier strip. If an idle station is placed between two such stations, stresses are exerted over a greater length of strip, providing greater protection.

- *Provide for uniform loading of the press slide.* In a progressive die, there are numerous forces at work. Pressures are mostly vertical, but in many cases sideways thrusts are

encountered. These sideways thrusts should be balanced out as much as possible to prevent damage to punches and dies. Special emphasis should be placed on this factor because of the costly nature of the tools. In addition, damaged tools will result in inferior-quality parts. Balance may frequently be achieved through proper sequencing of operations.

- *Provide for efficient scrap disposal.* The use of a progressive die is aimed primarily at increasing the productivity of a stamping operation. Ineffective scrap removal poses a serious threat to the attainment of this objective. In developing die design, space should be allowed for scrap chutes or other means of efficient scrap removal.

- *Provide efficient means of unloading finished parts.* Parts to be drawn present special problems in the design of progressive dies. Proper provisions must be made to supply the metal required to shape the part. Special care must be taken to provide means of moving the part from station to station.

In the development of progressive dies for drawing, the following principles should be observed:

- *Allow adequate provision for lifting the strip from drawing punches.* Where drawing is involved, the strip must be lifted well above the drawing punch or out of the die cavity before it can advance. Because of this, space must be provided in the die to permit incorporation of mechanical lifters.

- *Provide adequate carrier strips and tabs to allow metal movement.* As drawing takes place, the edges of the blanked-out portion of the strip are pulled in. This in turn exerts tensile stresses on the carrier strip. In fact, a center carrier strip may supply a portion of the metal required for the draw. This requires that the carrier strip be of sufficient size and strength to avoid breakage and of proper design to allow metal to flow into the die cavity.

- *Drawing stages should be properly balanced.* In progressive die design, attempts should not be made to draw too deeply in one station in an effort to reduce the number of draws required. Sound judgment must be used in determining the number of draw stations. It is better to spread the drawing operations out

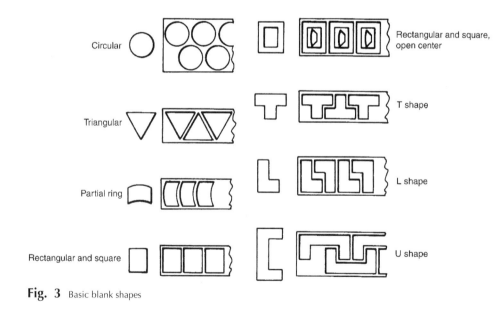

Fig. 3 Basic blank shapes

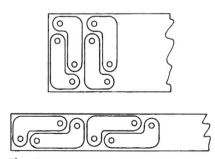

Fig. 4 Single-row blank layouts (shaded areas represent punches)

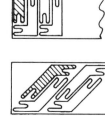

Fig. 5 Double-row blank layouts (shaded areas represent punches)

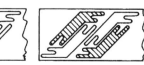

than to work too close to the allowable drawing limits. If excessive draws are attempted, the parts may rupture or tear. The tonnage load on the press may be too great, causing breakdowns. A progressive die with an extra station is less costly than a die of fewer stations that will not produce satisfactory parts.

General Design Features of Progressive Dies

Progressive dies are used for many small parts where tolerances and specifications are relatively severe. Production requirements of parts produced on these dies are high. Because of these factors, quality of design and structure in progressive dies must be of high caliber.

Strength. Progressive dies must be constructed sturdily to withstand sustained high-speed production. This calls for the selection of very strong materials for use in making the die. Areas of greatest stress must be properly built up and reinforced.

The design of progressive dies usually calls for die inserts and retainers and a sectional type of construction. This type of construction is necessary to incorporate the various types of operations that are united in a progressive die. Unless the various die components are properly distributed and strengthened, maintenance costs are apt to be very high.

To facilitate die repair, holes should be supplied in stripper plates and in other areas to facilitate removal of die sections. If such holes are not provided, it may be necessary to dismantle the entire die to remove a single section.

Punch retainers and die-button retainers should be backed up by steel plates to prevent the retainers from penetrating the die material. The die body itself should be made of steel if the demand for die strength is sufficiently high. Where cast iron is used in the die body, special steel wear plates and reinforcements should be added to prevent breakage or excessive wear.

Guiding. The precision requirements, plus the rapid cycling involved, require that the mating portions of the progressive die section be carefully guided. Precision guide pins with steel-backed bronze bushings should be used at all four corners of the die to help ensure exact alignment of mating surfaces. Stop blocks should be built into the die to prevent the punch from descending too far. These blocks should be placed where press operators cannot put their hands on them during operation. Guide pins should be long enough so that they are engaged at least 13 mm (1/2 in.) at the beginning of the press stroke.

All progressive die sets should be adequately heeled. Heeling of all die sets is a practical way of maintaining proper alignment of upper and lower shoes. The heels will offset any tendency of side thrusts to disturb die alignment.

In order to ensure that heels do the job for which they are intended, sufficient bearing surface must be engaged. The heels should be engaged a minimum of 13 mm (1/2 in.) at the bottom of the stroke.

Guide pins, stop blocks, and heel blocks must be provided with suitable safety guards to prevent injury to press operators. They should be shielded wherever there is a possibility of someone getting caught in them. Sheet metal is generally used for shielding. Shielding also provides protection of the press and die by keeping out scrap and other materials that could cause difficulty.

Strippers. Progressive die sets usually employ stripper plates to free the die tools from

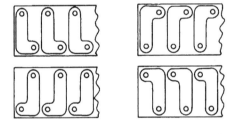

the part being worked on. A stripper prevents distortion of the part and helps to ensure speedy operation. For best operating results, certain factors should be observed in the use of this tool.

The stripper's holes should be carefully designed to allow free movement of punches and other tools through them. Adequate clearances should be provided to prevent binding or excessive wear. Spring strippers should be well regulated with respect to travel. They should be supplied with springs that are sufficiently long to avoid excessive compression. Springs must be strong enough to depress all stock lifters that may be used.

Guiding of strippers is also important. Gibs should be used on two sides of the stripper to provide proper control.

Strippers should be made strong enough, and of a steel that is suitable, for the prevailing working conditions. If parts of a stripper break, they are apt to fall on die surfaces and result in injury to the die.

The lifters used to facilitate movement of the strip through the die must also be constructed so as to withstand operating conditions. Breakage of lifters will often result in scored and damaged dies.

Ejectors. The full effectiveness of progressive die operations cannot be realized if swift, positive means of ejecting the part from the press are not provided. Presses using these dies are usually automatic in operation and run continuously until stopped by the operator. This rapid cycling provides only a short time for removal of the part.

Mechanical ejectors should be provided for removal of parts. Where there is a possibility of the part falling back into the die, air blasts should be incorporated to make ejection speedier and more positive. Air valves should be checked frequently for timing so that the part cannot fall out of the path of the blast.

Auxiliary Equipment

The nature of progressive dies demands the choice of proper auxiliary equipment in order to ensure effective operations.

Coil Feeders. Most progressive die operations use coil stock. Coil feeders are used to pass the stock into the die. These feeders must possess the proper capacity for holding the required stock and should be designed to administer the proper amount of metal to the press at each cycle. Some coil feeders are merely holders for the stock, with automatic feed rolls being used to move the stock. Others are geared to the press cycle and actually feed in the stock.

Scrap Handling. An aspect of some progressive die operations not found in other types of operations is the requirement for handling a carrier strip. This strip issues from the press as finished parts are cut away. Some presses are equipped with devices that roll up this strip as it comes out of the press. In other instances, scrap cutters are mounted at the end of the die to chop

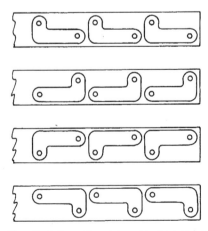

Fig. 6 Triple-row blank layouts (shaded areas represent punches)

Fig. 7 Parts positioned for wide (top) and narrow (bottom) runs

the strip into easily managed lengths. Where the latter method is used, stock boxes or scrap chutes should be used to provide simple means for getting the scrap away from the press.

Presses Used with Progressive Dies

In selecting presses for progressive die work, certain considerations are involved. Some of these considerations include the volume requirements of the part, the accuracy required, the size of the part, pressure requirements, and the relative difficulty of producing the part. Many types of presses are used for this kind of work. The choice will depend on how well the press meets the needs of a given situation plus the inescapable factor of press availability.

Open-Back Inclinable Presses. Numerous progressive operations are performed in open-back inclinable presses. The inclined feature of this press provides an often satisfactory solution to the problem of removing parts because the parts tend to fall easily out of the back of the press. Where a multiple die is used to make two or more parts at a time, a choice must be made as to whether to let the parts running at the uppermost level of the die drop out of the rear or whether to use an air blast to blow them out of the front of the press.

Tie rods usually are used across the gaps of these presses where exceptional accuracy is required. This is done to prevent deflection of the press under load.

Because of pressure limitations, gap-frame presses are usually used on the smaller types of progressively shaped parts.

Four-Column Presses. For progressive work demanding high pressure and exceptional accuracy, straight-side four-column presses are frequently used. This type of press is generally superior in tonnage potential to gap-frame presses. In addition, the four columns absorb sideways thrusts to ensure better alignment of punch and die.

For both gap-frame and straight-side presses, the press bed must be sufficiently wide to provide room for the progressive die used.

Automatic underdrive presses are designed specifically for use with progressive dies. They have wide beds and provisions for easy application of automatic loading and unloading.

ACKNOWLEDGMENT

This article was adapted from the chapter "Progressive Dies" by O.D. Lascoe in *Handbook of Fabrication Processes,* ASM International, 1988.

Forming Processes for Sheet, Strip, and Plate

Bending of Sheet Metal

BENDING is a common metalworking operation to create localized deformation in sheet (or blanks), plates, sections, tubes, and wires. Bending occurs from stressing a workpiece in a localized area, and it results in non-uniform deformation (contradiction and elongation) within the cross-sectional area of bending. Because of these inhomogeneities, analyses of stress strain distributions are of utmost importance when evaluating bending conditions (see the section "Stress-Strain Behavior in Bending" in this article).

Bending is done in various types of operations and machines such as press brakes or four-slide machines. Plates and sections are often bent in three-roll benders into complete circles. Roll forming is another type of bending operation that allows mass production of two-dimensional corrugated sheet, architectural sections, and lock-seam and welded tubes. This article focuses on the bending of sheet metal along with some coverage on flanging. Bending a sheet along a curved line is termed flanging. Circular or other close-shaped flanges (collars) are mass produced in preparation for joining tubes and fasteners to sheet, as in heat exchangers. Limits are set by fracture in a stretch flange and by buckling in a shrink flange. Related processes are flanging and necking of tubes and cans, as on beverage cans.

Bending Theory

The forces applied during bending are in opposite directions, just as in cutting of sheet metal. The bending forces, however, are spread farther apart, resulting in plastic distortion of metal without failure. The simplified sketches in Fig. 1 illustrate the forces applied during bending in V-dies, wiping dies, and U-dies. The latter two are more typical of high-production bending dies. The U-die is often referred to as a channel die. The spread of shear forces in a cutting die is equal to the clearance, which is usually approximately 10% of the sheet-metal thickness. The larger spread of bending forces is accomplished by using a clearance equal to the sheet metal thickness plus the radii used on the punch and die steels.

As the load is applied, the bending is similar to the deflection of a beam under load. The metal on the outside of the bend radius becomes stretched or elongated from tensile stress, while metal on the inside of the bend radius is placed under a compressive stress. If failure or fracturing occurs during bending, it will occur at the outside bend surface. Wrinkling will occur on the inside surface of the bend. When the metal is stressed above its elastic limit, it acquires a permanent set. Some elasticity usually remains, however, so that the metal tends to "spring back" as the pressure of the punch is released. This springback is compensated for by overbending or by bottoming the metal. Bottoming must be done with care, because thickness variations may cause overloading of the press or cause injury to the die.

When the sheet slides over the die radius, a frictional force is generated at the points of contact (Fig. 2). Because the interface pressures are low (well below the flow stress of the workpiece material), contact is limited to the asperities, and the coefficient of friction can be quite high (Ref 1). If the die/workpiece combination is prone to adhesion, pickup develops. If the workpiece is covered with hard abrasives, die wear occurs. In either case, the workpiece becomes scored, and lubrication would be needed.

Neutral Axis. The neutral axis is the line of zero stress and strain in a bend with tension on one surface and compression on the other (Fig. 3). Before bending, the flat blank is of a certain length, and the length of the neutral axis is, of course, exactly equal to this original blank length. During bending, the outside surface of the sheet metal is increased in length, and the inside surface of the sheet metal is decreased in length, but the length of the neutral axis remains the same. Because the neutral axis is a true representation of the original blank length, it is used for blank-development calculations.

When the blank is first being bent, the neutral axis is near the center of the sheet-metal thickness. As bending progresses, the neutral axis shifts toward the inside, or compression side, of the bend. Normally, the neutral axis is measured as a certain distance from the inside surface of the sheet metal at the bend area. Sheet metal thins slightly in the bend area, and the outside surface at the bend radius is not an accurate dimension. The inside surface bends tightly on the die-steel radius and is held to closer tolerances. Therefore, most parts are dimensioned with radii to the inside surface at all bends.

In bending of sheet metal, the distance from the inside surface to the neutral axis is usually approximately 40% of the thickness. Approximate positions of the axis for various thicknesses are shown in Fig. 4. Characteristics of the neutral axis are:

- If the sheet metal thickness is constant, the neutral axis shifts closer to the inside surface as the radius of bend is decreased.
- If the radius of bend is constant, the neutral axis shifts closer to the inside surface as the sheet metal thickness is increased.
- If the radius of bend and sheet metal thickness are constant, the neutral axis shifts closer to the inside surface as the degree of bend is increased.

Because the position of the neutral axis shifts for each variable listed, precise blank-size calculations are frequently difficult. Some alteration of

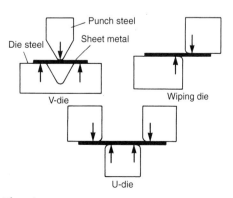

Fig. 1 Forces applied during bending

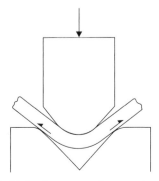

Fig. 2 Frictional stresses acting on sheet during bending

blank dimensions may be necessary after bending the first parts in the die.

Metal Movement. During production bending, one area of the blank is usually held stationary by a pressure plate called a pad. The free blank area is then bent up or down to create the change in contour. The metal forced down or up by the punch moves through space to occupy a new position. This metal movement through space is often called swinging and is a characteristic common only in bending operations. Such metal movement does not occur in embossing, stretch forming, or drawing of sheet metal. Illustrations of bending motions are presented in Fig. 5. In design of a bending die, the swinging action must be predicted so that no obstacles are placed in the way. Because of the metal movement, the larger area is usually held stationary, and the smaller blank area is moved by the punch.

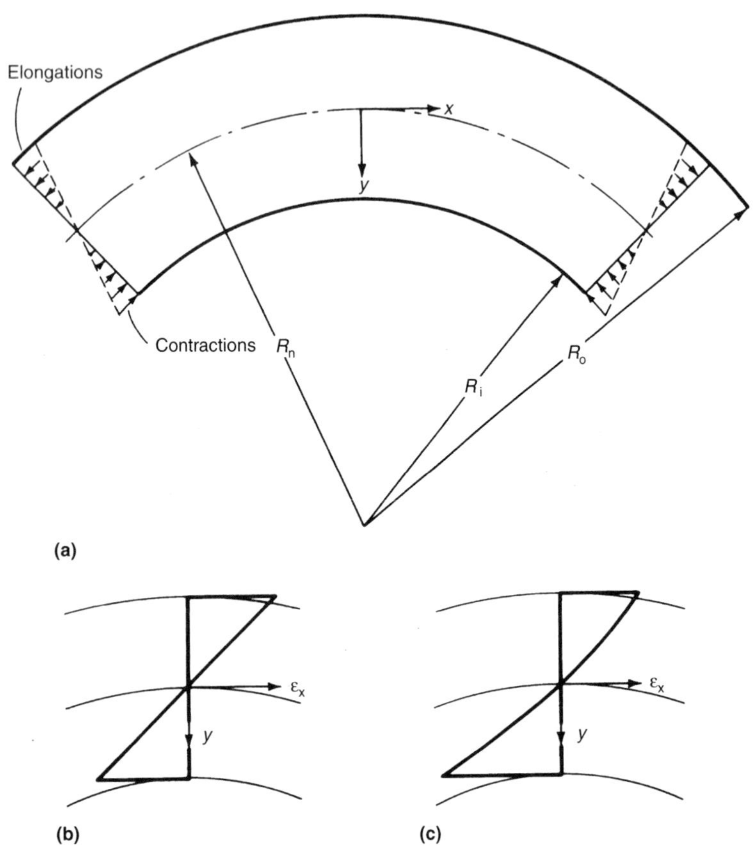

Springback. Variations in bending stresses cause springback after bending. The largest tensile stress occurs in the outside surface metal at the bend. The tensile stress decreases toward the center of the sheet thickness and becomes zero at the neutral axis. The pie-shaped sketch in Fig. 6 depicts the changing tensile and compressive stresses in the bend zone. Because the tensile stresses go from zero at point 0 on the neutral axis to a maximum value at point X on the outside surface, the stress-strain curve developed by the standard tensile test may be used for an analysis of bending. For good bend design, the tensile stress at point X is less than the ultimate tensile strength, as shown. If tensile stress is greater than ultimate strength, the metal may fracture during bending.

The metal nearest the neutral axis has been stressed to values below the elastic limit. This metal creates a narrow elastic band on both sides of the neutral axis, as shown in Fig. 7. The metal farther away from the axis has been stressed beyond the yield strength, however, and has been plastically or permanently deformed. When the die opens, the elastic band tries to return to the original flat condition but cannot, due to restriction by the plastic deformation zones. Some slight return does occur as the elastic and plastic zones reach an equilibrium, and this return is known as springback.

Variables and their effects on springback are as follows:

- Harder sheet metals have greater degrees of springback due to a higher elastic limit.
- A sharper or smaller bend radius could cause tearing, due to higher stresses in the outside surface.
- As metal is bent through greater degrees of bend, the plastic zone is enlarged and springback is reduced for each degree of bend. Total springback is increased, however.

Overcoming springback is a major factor in forming operations and is described in more detail in the article "Springback" in this Volume. Several methods are used to overcome or counteract the effects of springback (Fig. 8). These are:

- Overbending
- Bottoming or setting
- Stretch bending

Sheet metal is often overbent an amount sufficient to produce the desired degree of bend or bend angle after springback.

Fig. 3 Distribution of strain determined by the simple-beam theory. (a) Linear distribution for fiber elongations and contractions. (b) Distribution of engineering strain. (c) Distribution of true strain. R_n, radius of neutral axis; R_i, inner radius; R_o, outer radius

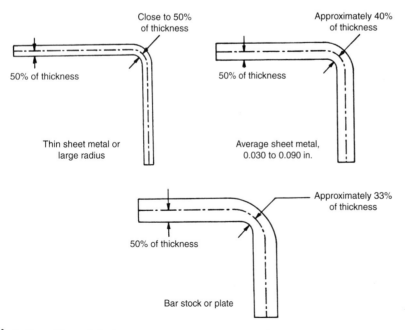

Fig. 4 Positions of the neutral axis

Overbending (Fig. 9) may be accomplished by using cams, by decreasing the die clearance, or by setting the punch and die steels at a smaller angle than required in the case of a V-die. When the clearance is reduced below the sheet metal thickness, the burnishing action wipes the metal against an undersized punch or die steel.

Bottoming or setting consists of striking the metal severely at the radius area. This places the metal under high compressive stresses that set the metal past the yield strength. Bottoming is accomplished by placing a bead on the punch at the bend area. In a wiping die or U-die, the pad must bottom against the shoe or backing plate so that the punch may set the metal at the bend. It would be useless to bottom against the flat areas of sheet metal, because they are not stressed and do not cause springback. Also, bottoming against these larger areas would require extremely high press tonnages. Bottoming must be carefully controlled when adjusting the press ram, or the forces involved will rise at a rapid rate. Also, if two blanks are accidentally placed in a bending die that bottoms, press or die breakage may result.

Stretch bending consists of stretching the blank so that all the metal is stressed past the yield strength. The blank is then forced over the punch to obtain the desired contour. This prestressing before bending results in very little springback. Only relatively large radii are bent by this method, because sharp radii would take the prestressed metal beyond the ultimate tensile strength. The sheet metal must be uniform in strength. Any weak spots or defects will certainly cause failure. Stretch bending is most frequently done with a special hydraulic machine rather than with a die in a press. Hourly production rates are slower for such machines than for presses.

Bending Calculations

Several formulas are available for computing the forces required for bending. These formulas rely on accurate measurement of the values substituted. Calculations are also necessary for predicting blank dimensions.

Blank Development. When metal is bent, the length of the part when measured at the neutral axis is the same as the length of the flat blank. The neutral axis is located midway through the sheet metal thickness from the inside of the bend, which is the compression side.

A common error in determining blank lengths is the failure to add or subtract the sheet metal thickness when necessary. On part prints, the radius at a bend is shown at the inside surface of the bend. Therefore, when radii are on opposite sides, the sheet-metal thickness must often be considered in the calculations.

A general rule in blank development is to first divide the part into straight sections and bends or arcs. Then the length of each section is found. Often, it is necessary to draw in right triangles to connect known to unknown dimensions. Trigonometry is then used to solve for an unknown side or angle. Another rule is that the legs of the triangle should always be drawn parallel to the dimension lines. The hypotenuse then is at the angle of the bend. With this arrangement, the sides of the triangle may be added to or subtracted directly from the dimensions shown on the part print.

Bottoming Forces. To overcome springback, the bend area is often placed under high compressive stress to set the metal. This operation is called bottoming and is accomplished by placing a projection or bead on the punch steel. The force for bottoming is difficult to estimate.

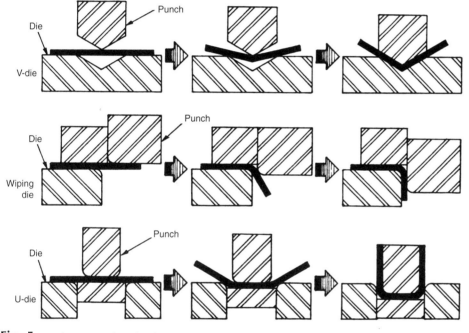

Fig. 5 Metal movement during bending

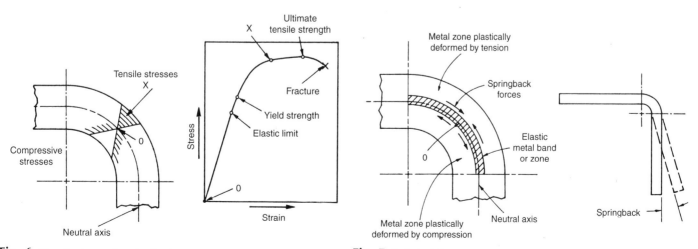

Fig. 6 Changing stress patterns in a bend

Fig. 7 Springback forces

This force will increase due to work hardening of the metal if the yield strength is surpassed. Surpassing of the yield strength is all that is necessary to set the metal. The minimum force for bottoming could be estimated using the sheet metal yield strength. The bottoming force would increase rapidly from this minimum value when press rams were adjusted downward too far.

Pad Forces. Bending along a straight line is often performed with a steel punch and an elastomeric (polyurethane) pad. Most production dies of the wiping and channel styles require a pressure plate called a pad to counteract the bending force. In a wiping die with no pad, the sheet-metal blank would simply flip up vertically in the clearance gap, and no bending would occur. The pad acts as a holddown plate in the die. Besides the holddown function, the pad must hold the metal flat against the tip die surface. The

elastomer imposes tensile stresses on the sheet, somewhat as in stretch forming.

A frequent difficulty in bending is that the sheet metal in the area next to the bend tends to rise or lift from the die-steel surface. To simplify the explanation, consider the pad as a knife-edge blade rather than a plate, as shown in Fig. 10. The lifting action is illustrated; the resulting corner condition at the bend is usually called recoil. Under some lighting conditions, a shadow occurs in the depression. A dark line can be seen to follow the bend and may be called a ghost line. On some parts requiring excellent appearance, such shadow lines are not permitted.

Although recoil cannot be prevented entirely, die designers use many methods to control recoil. The recoiling action causes the pad edge to wear rapidly, and hardened inserts are sometimes used in high-volume dies. The pad can be bottomed

against the shoe, but very high tonnages can cause press damage. Bottoming against a lip on the punch steel is more effective and requires less force.

Several comments concerning pads will aid in the design of wiping dies:

- To restrict recoil, the high pad forces are needed just as bending begins. Springs cannot be fully compressed this early, and space is often limited. Most pads must be powered by air or hydraulic cylinders located in the press bed or lower die shoe.
- Recoil problems are more frequent on long bends or flanges. Not only is the pad force high, but the pad must fit tightly for the entire length of bend. Recoil can occur due to the pad surface not being flat.
- It is vital that pads be well guided so that they do not cock or tip due to the unequal loading caused by recoil. An angular pad would permit some recoil despite high pad forces.
- A third function of the pad is to prevent the blank from slipping toward the bending punch. The moving punch causes a dragging action on the blank. High pad forces create enough static friction to prevent blank slippage.

Pad force is one of the largest variables between different companies doing pressworking of sheet metals. The stresses transmitted by frictional forces depend on the surface condition of the blank and the properties of the elastomers. Schey (Ref 1) cites work that indicates higher friction forces reduce springback (Ref 2).

Bending Operations

Figure 11 illustrates a number of bending terms. The characteristics of bending can be found in many operations for shaping of sheet metal, and each operation has a special name.

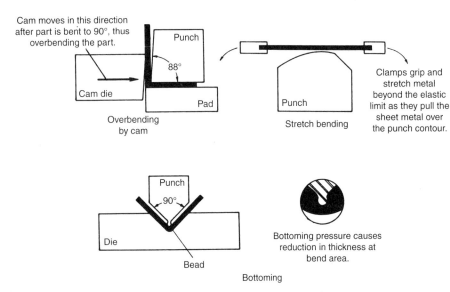

Fig. 8 Methods of overcoming springback

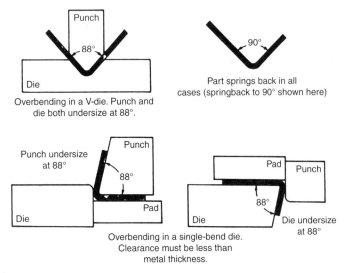

Fig. 9 Methods of overbending

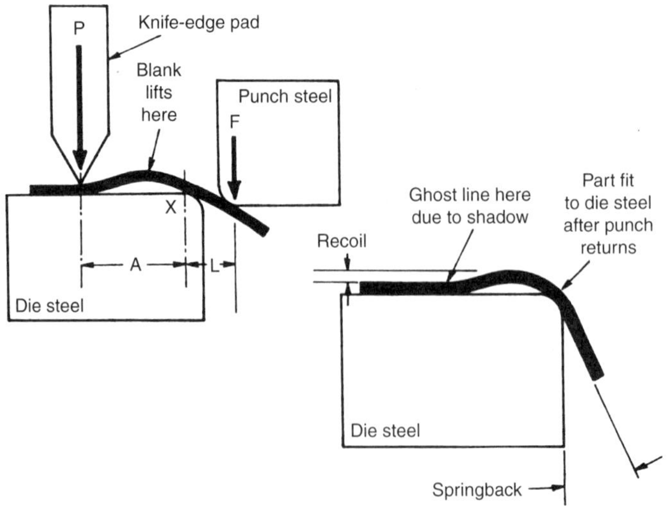

Fig. 10 Metal recoil during bending

Bending in a production plant may be called bending, flanging, hemming, seaming, curling, or corrugating.

Bending. Bends are made in sheet metal to obtain rigidity and to obtain a part of desired shape for performance of a certain function. Bending is commonly used to produce structural stampings, such as braces, brackets, supports, hinges, angles, and channels. One troublesome aspect of bending operations arises from the directionality present in rolled steel. Directional lines are set up in the steel in the direction of rolling, causing points of weakness running parallel to the directional lines. Because of these lines, cracking sometimes occurs when metal is bent parallel to the lines. This is especially true of hard-temper metals. To avoid this, bends may be designed across the grain of the metal.

Flanging is similar to bending of sheet metal except for one factor: during flanging, the metal bent down is short compared with the overall part size. Most forming operations in automobile production have to do with the formation of flanges. Many straight bending operations are used in making flanges. Where straight flanges are made, the operations are fairly uncomplicated.

Contour Flanging. With streamlining and styling of sheet forms, more and more situations have been created that require curved flanges or contour flange. Although contour flanges usually comprise the sides of a part, they are also used extensively in the formation of flanged holes.

The problems involved in making straight flanges are also common in contour flanging. In addition, contour flanging presents a number of unique problems. These problems arise from the stresses placed on metal in shaping of contours and curves.

Contour flanges are subdivided into three distinct types: stretch flanges, shrink flanges, and reverse flanges.

A stretch flange has a concave curvature. It is called a stretch flange because the material forming the flange must be stretched. This stretching is easily seen in forming of a flange for a hole. The metal used in forming the flange must be elongated.

A shrink flange has a convex curvature. In forming a shrink flange, the metal used to form the flange must be reduced in length. This relationship is easily seen in the flanging of a circular cake pan. The diameter of the finished flange is less than the diameter of the disk from which it was formed.

A reverse flange consists of at least one stretched and one shrunk portion.

Each of these types of flanges creates problems in metal control during forming. To obtain parts conforming to design, special blanks are usually developed. These blanks provide relief for the metal in areas of exceptional strain. In some cases, the flanges are formed in rough blanks and then trimmed.

This is usually done when the stresses involved would result in cracking or severe distortion of developed blanks. Where part design permits, relief of the stresses involved in contour forming is frequently provided by notches or cutouts in the developed blank. These notches provide areas into which the stressed metal may flow and prevent buckling or wrinkling.

As in all stamping operations, the sequence of operations for producing a flanged part is very important. A hole pierced in a part prior to flanging may become distorted during flanging; a flange may interfere with tool access for subsequent operations. Each operation must be sequenced for the most effective production arrangement consistent with quality.

Hemming and Seaming. The terms *hemming* and *seaming* are used in a manner similar to their use in the clothing industry. A hem is a fold at the edge of the sheet metal to remove the burred edge and improve the appearance of the edge. The hem also adds slightly to the rigidity of the edge and improves wear resistance. Hems and seams are shown in Fig. 12.

Curling and wiring are used to strengthen the edge of sheet metal. Actually, hems, curls, and wiring may be used on flat parts or on round parts such as cams or drums. During wiring, the metal is curled up and over a length of wire. The edge of the sheet metal is strengthened by both the wire and the curl formed in the metal. When the wire is not used, the operation is known as curling or false wiring (Fig. 13).

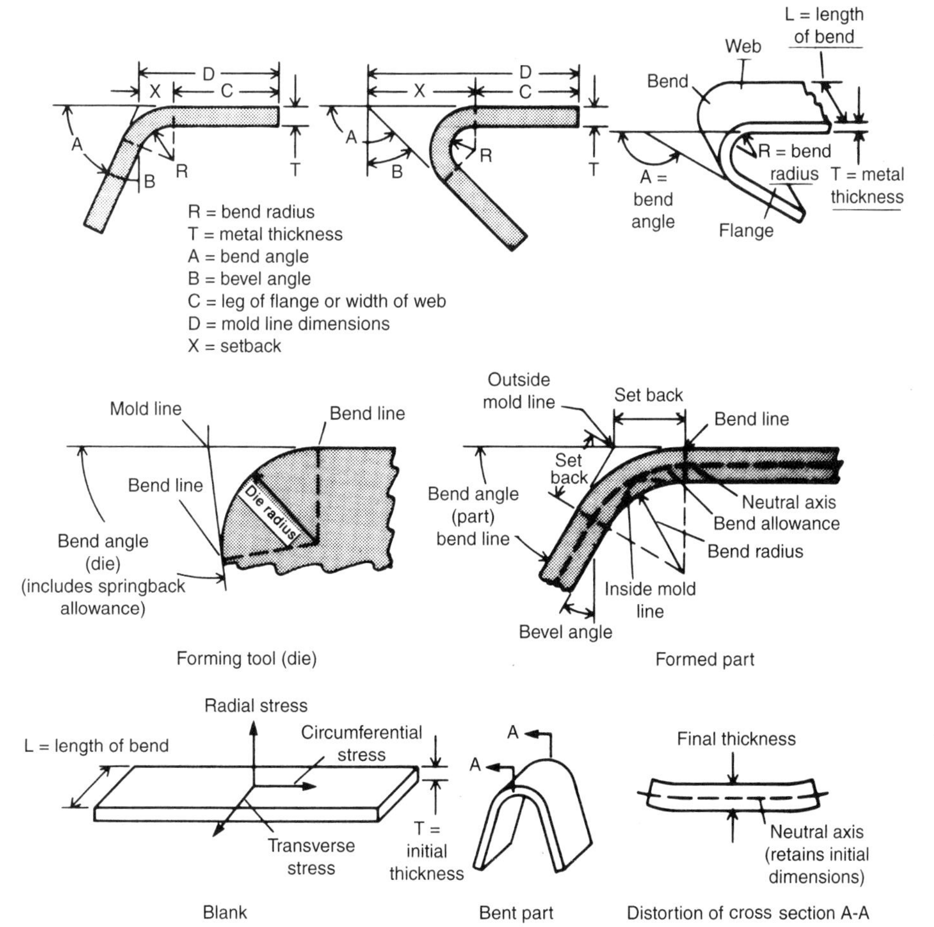

Stress-Strain Behavior in Bending*

Elastic Bending

Elastic analysis of bending deformation can be performed by simple-beam theory (Ref 3), elasticity solutions, and numerical methods such as the finite-difference and finite-element methods (Ref 4). Generally, numerical methods are suitable for bending of specimens that are subjected to complex loading patterns and that have irregular and/or varying cross-sectional areas. Elasticity solutions are useful when accuracies better than ~5% are desired. Simple-beam theory is used in most testing applications in which plates, strips, bars, and rods are bent in three-point or four-point bending modes. The basic assumptions of the simple-beam theory for pure elastic bending (shear force = 0) are: (a) all sections that are initially plane and perpendicular to the axis of the beam remain plane and

R = bend radius
T = metal thickness
A = bend angle
B = bevel angle
C = leg of flange or width of web
D = mold line dimensions
X = setback

L = length of bend
A = bend angle
R = bend radius
T = metal thickness
Web
Bend
Flange

Fig. 11 Bending terms

*Adapted from: P. Dadras, Stress-Strain Behavior in Bending, *Mechanical Testing and Evaluation*, Vol 8, *ASM Handbook*, ASM International, 2000, p 109–114

perpendicular to it after bending; (b) all longitudinal elements (fibers) bend into concentric circular arcs (hence, cylindrical bending); and (c) a one-dimensional stress state is assumed, and the same stress-strain relationship is used for tension and compression.

The first assumption implies a linear distribution for fiber elongations and contractions, as shown in Fig. 3(a). The resulting strain distributions (Fig. 3b, c) are given by (engineering bending strain):

$$\varepsilon_x = -\frac{y}{R_n} \qquad \text{(Eq 1)}$$

and (true bending strain):

$$\varepsilon_x = \ln\left(1 - \frac{y}{R_n}\right) \qquad \text{(Eq 2)}$$

where R_n is the radius of curvature of the neutral axis. For $\varepsilon_x \leq 0.1$, the difference between these two strain definitions is $\leq 5\%$. Therefore, for most elastic bendings, the engineering strain definition is sufficiently accurate and more convenient. As shown in Fig. 3, the true strain description indicates a nonlinear strain distribution and a maximum compressive strain in the concave inner fiber that is greater than the maximum tensile strain in the outermost fiber.

For a linear elastic material:

$$\varepsilon_x = \frac{1}{E}[\sigma_x - \nu(\sigma_y + \sigma_z)] \qquad \text{(Eq 3)}$$

where E is the modulus of elasticity for axial loading (Young's modulus), and ν is Poisson's ratio. From the third assumption, $\sigma_y = \sigma_z = 0$. Therefore:

$$\sigma_x = -E\frac{y}{R_n} \qquad \text{(Eq 4)}$$

When expressions similar to Eq 3 are written for ε_z and ε_y, the following results are obtained:

$$\varepsilon_z = \varepsilon_y = \nu\frac{y}{R_n} \qquad \text{(Eq 5)}$$

Also:

$$\varepsilon_{xy} = \varepsilon_{yz} = \varepsilon_{xz} = 0 \qquad \text{(Eq 6)}$$

Figure 14 illustrates grid deformations in the longitudinal and cross directions. A transverse curvature, called anticlastic curvature (Ref 5), develops with a radius of curvature equal to (R_n/ν). Experimental evidence indicates that the actual radius of anticlastic curvature depends on $(b^2/2R_nh)$, where b is the width, and $2h$ is the thickness of the beam. For $(b^2/2R_nh) \leq 1$ (i.e., narrow beams), the R_n/ν estimate is sufficiently accurate. For plates and wide beams $(b^2/2R_nh > 20)$, the anticlastic deformation is primarily concentrated at the edges.

The location of the neutral axis, which is the line of zero fiber stress in any given section of a member subject to bending, is determined from the condition of zero axial forces acting on the beam. Therefore:

$$\int_A \sigma_x dA = -\frac{E}{R_n}\int_A y\,dA = 0 \qquad \text{(Eq 7)}$$

where A is the cross-sectional area. This equation indicates that the first moment of the cross-sectional area about the neutral axis is zero, which implies that the neutral and the central (centroidal) axes are coincident.

The moment-curvature and moment-stress relationships are found by equating the externally applied bending moment to the internal bending moment at any cross section:

$$M = -\int_A (\sigma_x dA)y \qquad \text{(Eq 8)}$$

where M is the bending moment.

For the linear stress distribution shown in Fig. 15, the results are:

$$R_n = \frac{E}{M}I_z \quad \text{and} \qquad \text{(Eq 9)}$$

$$\sigma_x = -\frac{M_y}{I_z} \qquad \text{(Eq 10)}$$

where I_z is the area moment of inertia of the cross section about the z-axis, which is coincident on the centroidal axis.

The sign (positive or negative) of the bending moment is found from the following relationship: (sign of the bending moment) = (sign of the moment vector)×(sign of the outward normal to the section). For example, the bending moment acting on section $ABCD$ in Fig. 16 is positive, because the moment vector, $\vec{M}$, is in the positive z-direction (right-hand rule), and outward normal of the plane, $\vec{n}$, is also in the positive x-direction.

Noncylindrical Bending

In the previous section, it was stated that the simple-beam theory considers the effect of bending moments alone (shear force = 0) and assumes a bent configuration consisting of concentric circular arcs (cylindrical bending). For cylindrical bending to occur, these conditions must be met:

- Bending must occur under the action of bending moments alone, which implies zero applied shear force.
- The cross-sectional area of the beam must possess at least one axis of symmetry.
- The vector of the applied bending moment must be in the direction of an axis of symmetry.

For asymmetrical beams such as Z-sections and unequal L-sections, the second and third conditions for cylindrical bending are not satisfied; for unsymmetrical bending of symmetrical beams, the third condition is not met. These cases are significant in structural design and are not considered here. Information on unsymmetrical loading of straight beams can be found in Ref 4.

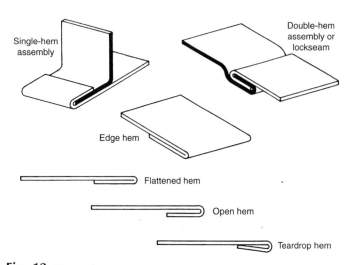

Fig. 12 Hems and seams

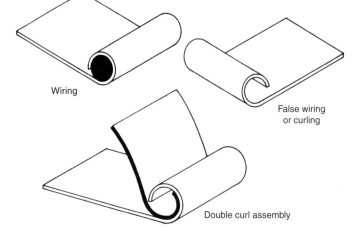

Fig. 13 Wiring and curling

In a majority of testing applications such as three-point bending, roll bending, and press-brake forming, the applied bending moment varies along the length of the specimen. Because shear force $V = (dM/dx)$, such variations in the bending moment imply a nonzero shear force. Therefore, the first condition for cylindrical bending is not met. The resulting shear stress, τ_{xy}, which is determined from the equilibrium considerations at a typical section m-n (Fig. 17a), is:

$$\tau_{xy} = \frac{VQ}{I_z b} \qquad \text{(Eq 11)}$$

where

$$Q = \int_{As} y\, dA \qquad \text{(Eq 12)}$$

is the first moment of the shaded area (A_s) with respect to the neutral axis (Fig. 17b). The first moment of the unshaded area with respect to the neutral axis gives the same Q. The distribution of τ_{xy} for a rectangular cross section is shown in Fig. 17(c).

Elastic-Plastic Bending

The limit for elastic bending, which is the onset of elastic-plastic bending, is reached when the maximum fiber strain $(\varepsilon_x)_{max} = (h/R_n)$ becomes equal to 9 (σ_y/E), where σ_y is the yield strength, and E is Young's modulus of the material. For bending beyond this limit, the beam consists of a central elastic core and two plastically deforming zones remote from the neutral axis. For accurate analysis of elastic-plastic bending, factors such as the shift of the neutral axis from the centroidal axis and the effect of radial (transverse) stresses must be considered. Elastic-perfect plastic (Ref 6) and elastic-linear hardening (Ref 7) analyses are available. However, in view of analytical and computational difficulties, an approximate method, which is an extension of the simple-beam theory, is commonly employed. Therefore, the three assumptions stated for elastic bending will be enforced. The location of the neutral axis is assumed to be fixed at the centroidal axis, as it is for elastic bending.

For a beam with the stress-strain curve shown in Fig. 18(a), the development of longitudinal strain and stress at different stages of deformation is shown in Fig. 18(b) and (c), respectively. In Fig. 18(b), a linear strain distribution given by $\varepsilon_x = (-y/R_n)$ was used. When the strain at the

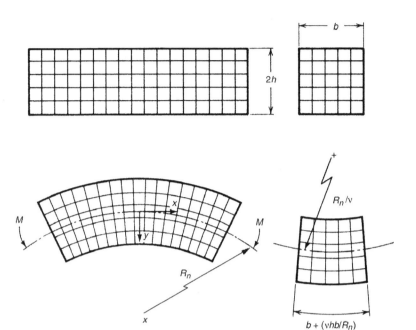

Fig. 14 Grid deformations in the longitudinal and cross directions of a beam

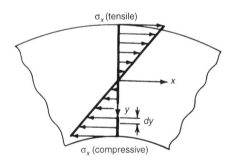

Fig. 15 Linear stress distributions in a beam

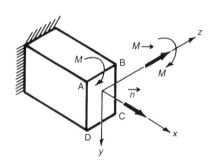

Fig. 16 Sign convention for bending moment

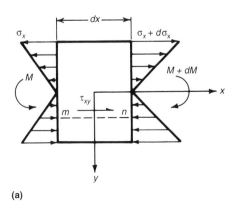

(a)

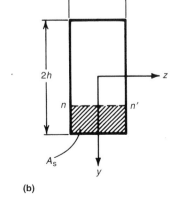

(b)

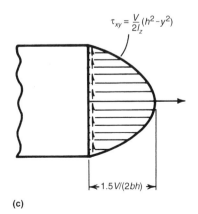

(c)

Fig. 17 Distribution of shear stress, τ_{xy}, for a rectangular specimen. See text for details.

outermost fiber exceeds 0.1, it is suggested that the true strain distribution (Eq 2) be used. For bending to a radius of curvature equal to R_n, the strain distribution and the subsequent stress distribution (from the σ-ε curve or from a known constitutive equation for the σ-ε dependence) can be found. The moment, M, required to produce R_n is:

$$M = -\int_A (\sigma_x dA)y \qquad \text{(Eq 13)}$$

The thickness of the elastic core, C (Fig. 18c), is:

$$C = 2R_n\left(\frac{\sigma_y}{E}\right) \qquad \text{(Eq 14)}$$

Therefore, in bending plates of the same material to the same radius of curvature R_n, the fractional thickness of the elastic core $C/2h$ becomes smaller as the thickness increases.

For elastic-plastic bending, a general equation for the relationship between R_n and M (analogous to Eq 9 for the elastic case) does not exist. For the simple case of an elastic-perfect plastic material (Fig. 18d), the following equation is obtained:

$$R_n = \frac{E}{\sigma_y}\left[3\left(\frac{\sigma_y bh^2 - M}{b\sigma_y}\right)\right]^{1/2} \qquad \text{(Eq 15)}$$

The predictions of this equation at large plastic deformations ($C/2h \leq 0.02$) are not reliable.

Pure Plastic Bending

The fractional thickness of the elastic core decreases as the ratio $R_n/2h$ decreases. For $R_n/2h \leq 10$, the thickness of the elastic core in a hot rolled strip of AISI 1020 steel is ≤3.2%. In such cases, it is possible to ignore the elastic core altogether and use the analysis for pure plastic bending.

Most plastic bending analyses are for sheet-type specimens in which b is much greater than $2h$. The width strain is small for such a geometry, and plane-strain bending deformation ($\varepsilon_z = 0$) is commonly assumed. Also, the first two assumptions of the simple-beam theory are still applied. However, the radial (transverse) stresses and

strains induced by higher curvatures are considered, and a plane-in state, rather than a one-dimensional stress state (the third assumption of the simple-beam theory), is assumed. Also, the neutral axis is not fixed at the centroid, and thickness variations due to bending sometimes are incorporated into the solution. Figure 19 shows the geometry of deformation and the strain and stress states at different zones.

At the onset of the assumed full plastic condition, $R_c = R_n = R_u$. In the current state, the fiber with radius of curvature equal to R_n is being overtaken by the neutral axis and is experiencing unloading from a compressive tangential stress field. Accordingly, all fibers below the neutral axis $r \leq R_n$ have been progressively compressed, while those situated above R_c have been consistently stretched during deformation. All fibers in the interval $R_n \leq r < R_c$ have been overtaken by the neutral axis. Fibers between $R_u < r < R_c$ have now been stretched beyond their original length due to reverse loading, while those located in $R_n \leq r \leq R_u$ have yet to recover their original undeformed length.

As expected, a comprehensive analysis accounting for the described fiber movements is very complicated (Ref 8). A compromise solution (Ref 9), which ignores thickness variations and assumes rigid-perfect plastic material behavior, provides useful approximations for the stress-strain distributions in plastic bending.

When a rigid-perfect plastic material model is used (Fig. 19c), the same stress-strain relationship applies to all fibers with $r \geq R_n$. As a result, the distinction among R_n, R_u, and R_c becomes inconsequential. This eliminates the complicated task of describing the behavior of fibers in reversed loading and the Bauschinger effect.

The state of stress acting on a typical element is shown in Fig. 20, where σ_θ is the circumferential (tangential) stress, and σ_r is the radial (transverse) stress. The equilibrium equation for plane-strain deformation is:

$$\frac{d\sigma_r}{dr} = \frac{\sigma_\theta - \sigma_r}{r} \qquad \text{(Eq 16)}$$

The effective or significant stress, $\bar{\sigma}$, and strain, $\bar{\varepsilon}$, for plane-strain deformation, using von Mises criterion, are, respectively:

$$\bar{\sigma} = \pm\frac{\sqrt{3}}{2}(\sigma_\theta - \sigma_r) \quad \text{(+) for } R_o \geq r \geq R_n$$
$$\text{(−) for } R_n > r > R_i$$
$$\text{(Eq 17)}$$

and

$$\bar{\varepsilon} = \frac{2}{\sqrt{3}}|\varepsilon_\theta| \qquad \text{(Eq 18)}$$

where $\varepsilon_\theta = \ln r/R_n$. Substituting $(\sigma_\theta - \sigma_r)$ from Eq 16 into Eq 15 and putting $\sigma_r = 0$ at $r = R_o$ and $r = R_i$, these equations for the distribution of σ_r are obtained:

$$\sigma_r = -\left(\frac{2\bar{\sigma}}{\sqrt{3}}\right)\ln\left(\frac{R_o}{r}\right) \quad \text{for } R_n \leq r \leq R_o \quad \text{(Eq 19)}$$

$$\sigma_r = -\left(\frac{2\bar{\sigma}}{\sqrt{3}}\right)\ln\left(\frac{r}{R_i}\right) \quad \text{for } R_i \leq r \leq R_n \quad \text{(Eq 20)}$$

From the expressions for σ_r and Eq 16, the following expressions for σ_θ are determined:

$$\sigma_\theta = -\left(\frac{2\bar{\sigma}}{\sqrt{3}}\right)\left[1 - \ln\left(\frac{R_o}{r}\right)\right] \quad \text{for } R_n \leq r \leq R_o$$
$$\text{(Eq 21)}$$

$$\sigma_\theta = -\left(\frac{2\bar{\sigma}}{\sqrt{3}}\right)\left[1 + \ln\left(\frac{r}{R_i}\right)\right] \quad \text{for } R_i \leq r \leq R_n$$
$$\text{(Eq 22)}$$

Because of equilibrium considerations, the radial stress must be continuous at $r = R_n$. Applying this condition to Eq 19 and 20, the location of the neutral axis can be found:

$$R_n = \sqrt{R_i R_o} \qquad \text{(Eq 23)}$$

Figure 21 is a schematic of the distributions of σ_r and σ_θ. In the figure, σ_r is continuous and compressive throughout the plate thickness, while σ_θ changes from tension to compression at the neutral axis. The bending moment, which according to this solution is independent of R_n, becomes:

$$M = \int_{R_i}^{R_o}(\sigma_\theta bdr)r = \left(\frac{2}{\sqrt{3}}\right)\bar{\sigma}bh^2 \qquad \text{(Eq 24)}$$

where h is half-thickness, and b is the plate width.

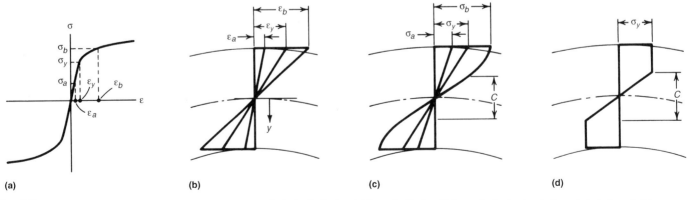

Fig. 18 Stress-strain distributions in a beam. (a) Stress-strain curve. (b) Strain distribution. (c) Stress distribution. (d) Stress distribution for elastic-perfect plastic material

The maximum radial stress occurs at the neutral axis. Its magnitude from Eq 19 and 23 is:

$$(\sigma_r)_{max} = -\left(\frac{\bar{\sigma}}{\sqrt{3}}\right) \ln\left(\frac{R_o}{R_i}\right) \qquad \text{(Eq 25)}$$

The ratio between $(\sigma_r)_{max}$ and the tangential stress at $r = R_o$ for four plates of different thicknesses, all bent to an inside radius of $R_i = 25$ mm (1 in.), is given in Table 1. This table shows that for plastic bending to $(R_n/2h) > 10$, the magnitude of $(\sigma_r)_{max}$ becomes very small. In such cases, the effect of σ_r can be neglected in the analysis, and the elastic-plastic bending solution based on the simple-beam theory can be used.

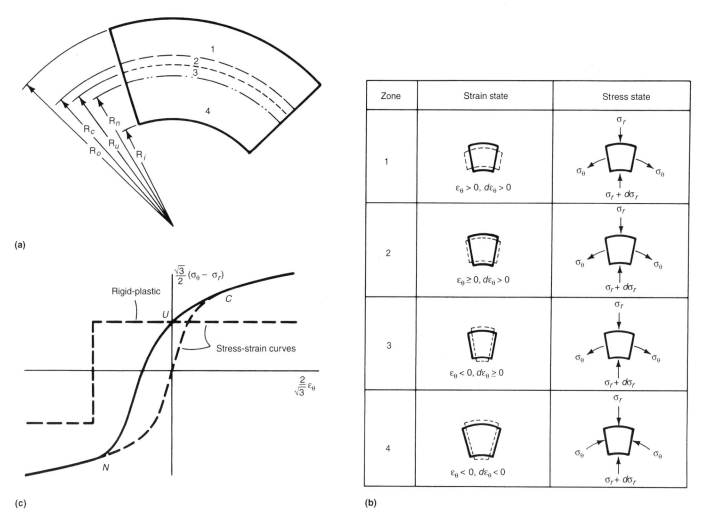

Fig. 19 Pure plastic bending of strip specimen. (a) Geometry of deformation. R_i, inner radius of curvature; R_o, outer radius of curvature; R_n, radius of curvature of the neutral axis; R_u, radius of currently unstretched fiber; R_c, current radius of curvature of original center fiber. (b) Strain and stress states in different zones. $\varepsilon_\theta = \ln r/R_n$ is the circumferential strain. (c) Stress-strain curves and stress-strain states at various locations. N, stress-strain state at R_n; U, stress-strain state at R_u; C, stress-strain state at R_c

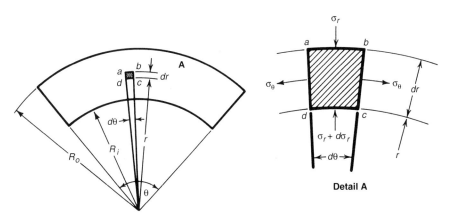

Fig. 20 State of stress acting on a typical element in plane-strain bending

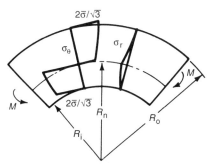

Fig. 21 Schematic of circumferential, σ_θ, and radial, σ_r, stresses in a plate during bending. Source: Ref 9

More elaborate analyses of plastic bending (Ref 8, 10) commonly involve complicated numerical computations. Hence, no equations can be given. A comparison between the results of the analysis in Ref 9 and a solution for rigid work-hardening material behavior (Ref 8) is shown in Fig. 22. In this case, the result from Ref 8 is for a model material with a high rate of strain hardening:

$$\bar{\sigma} = 70 + 300\bar{\varepsilon}^{0.5} \text{ MPa} \qquad \text{(Eq 26)}$$

In using the analysis from Ref 8, $\bar{\sigma} = 120$ MPa (17.4 ksi), which is the approximate average flow stress for bending to $\bar{\varepsilon} = 0.11$, and $\bar{\sigma} = 169.5$ MPa (24.6 ksi), as shown in Fig. 22(b), have been assumed. As expected, some differences in the predicted stress distributions are observed.

However, by using $\bar{\sigma} = 169.5$ MPa (24.6 ksi) in the solution from Ref 8, a close agreement (percent different <6) between the estimates for the fiber stresses at $r = R_i$ and $r = R_o$ is obtained. Because the prediction of maximum fiber stress and strain is of special interest, the following procedure based on the solution in Ref 8 is suggested. First, find the maximum fiber strain:

$$\varepsilon_o = -\varepsilon_i = \ln\sqrt{\frac{R_o}{R_i}} \qquad \text{(Eq 27)}$$

where ε_o and ε_i represent the maximum fiber strain at the outer and inner radii, respectively. Using the stress-strain equation or stress-strain curve for the material, determine $\bar{\sigma}$ as the flow stress at ε_o. The maximum fiber stress is $2\bar{\sigma}/\sqrt{3}$.

Residual Stress and Springback

When a specimen that has been bent beyond the elastic limit is unloaded, the applied moment M becomes zero, and the radius of curvature increases from R_n to R_n'. For a fiber at distance y from the neutral axis, this produces a strain difference:

$$\Delta\varepsilon_x = y\left(\frac{1}{R_n} - \frac{1}{R_n'}\right) \qquad \text{(Eq 28)}$$

The removal of the bending moment, which is an unloading event, is assumed to be elastic. Therefore:

$$\Delta\sigma_x = E\Delta\varepsilon_x = Ey\left(\frac{1}{R_n} - \frac{1}{R_n'}\right) \qquad \text{(Eq 29)}$$

The change in bending moment for complete unloading is $\Delta M = M$, where M is the applied bending moment prior to unloading. Therefore:

$$\int_A \left[Ey\left(\frac{1}{R_n} - \frac{1}{R_n'}\right)dA\right]y = -\int_A (\sigma_x dA)y \qquad \text{(Eq 30)}$$

which reduces to:

$$\frac{1}{R_n} - \frac{1}{R_n'} = -\frac{\int_A \sigma_x y dA}{EI_z} \qquad \text{(Eq 31)}$$

The distribution of the residual stresses can be found from either of the following equations:

$$\sigma_x' = \sigma_x + \Delta\sigma_x = \sigma_x - \left(\frac{y}{I_z}\right)\int_A \sigma_x y dA \qquad \text{(Eq 32)}$$

$$\sigma_x' = \sigma_x + yE\left(\frac{1}{R_n} - \frac{1}{R_n'}\right) \qquad \text{(Eq 32a)}$$

It is important that the correct signs for σ_x and y be used when applying these equations.

As an example, the springback and the residual-stress distribution of a strip of annealed 1095 steel was examined (Ref 11). For this material, yield strength is $\sigma_y = 308$ MPa (44.7 ksi), Poisson's ratio is $v = 0.28$, and the approximate constitutive equation for σ in metric units of measure is:

$$\bar{\sigma} = (2\times10^5 \text{ MPa})\varepsilon \quad \text{for } \varepsilon \leq 0.00154 \qquad \text{(Eq 33)}$$

$$\bar{\sigma} = (896 \text{ MPa})\varepsilon^{0.16} \quad \text{for } \varepsilon \geq 0.00154 \qquad \text{(Eq 34)}$$

In English units of measure, σ is:

$$\bar{\sigma} = (29\times10^6 \text{ psi})\varepsilon \quad \text{for } \varepsilon \leq 0.00154$$

$$\bar{\sigma} = (126 \text{ ksi})^{0.16} \quad \text{for } \varepsilon \geq 0.00154$$

The width of the strip, b, is 50 mm (2 in.), and its thickness, $2h$, is 5 mm (0.2 in.). It is assumed that the strip is bent to $R_n = 100$ mm (4 in.). Because $(b/2h) = 10$, plane-strain deformation prevails. Because of this, the elastic modulus in plane-strain:

$$E' = \frac{E}{1 - v^2} \qquad \text{(Eq 35)}$$

is employed, and the plastic flow stresses (Eq 34) are multiplied by $(2/\sqrt{3})$. These approximate plane-strain adjustments are considered adequate when the simplified elastic-plastic analysis, which was discussed earlier in this article, is used. The thickness of the elastic core in metric units of measure is:

$$C = \frac{2R_n\sigma_y}{E'} = 2(100)\frac{(2\times308)/\sqrt{3}}{2.17\times10^5} \qquad \text{(Eq 36)}$$
$$C = 0.33 \text{ mm}$$

In English units of measure, the thickness of the elastic core is:

$$C = \frac{2R_n\sigma_y}{E'} = \frac{2(3.937)\left(2\times\frac{44.7}{\sqrt{3}}\right)}{(31,465)}$$
$$C = 0.013 \text{ in.}$$

which is 6.6% of the total plate thickness. The final radius of curvature in metric units of measure after springback is found from Eq 31:

$$\frac{1}{100} - \frac{1}{R_n'}$$
$$= -\frac{2}{EI_z}\times\left[\int_0^{C/2} -2.17\times10^5\left(\frac{y}{100}\right)b \, dy \right.$$
$$\left. + \int_{C/2}^h -\left(\frac{2}{\sqrt{3}}\right)869\left(\frac{y}{100}\right)^{0.16}by \, dy\right] \qquad \text{(Eq 37)}$$

Table 1 Ratio between maximum radial stress and tangential stress for plate of various thicknesses

R_i		Thickness			$\sigma_r \text{ max}/(\sigma_\theta$ at $r = R_n)$
mm	in.	mm	in.	$R_n/2h$	
25	1	1.59	0.0625	16.49	0.030
25	1	3.17	0.125	8.48	0.059
25	1	6.35	0.25	4.47	0.112
25	1	12.7	0.5	2.45	0.203

See text for explanation of symbols.

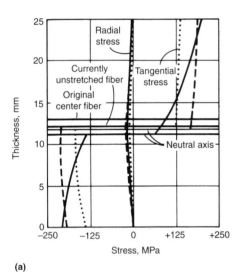

(a)

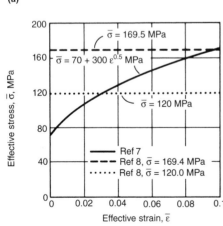

(b)

Fig. 22 Comparison of results for determining plastic bending in a plate. (a) Distribution of tangential and radial stresses for a 25 mm (1 in.) thick plate bent to $R_i = 100$ mm (4 in.). (b) Stress-strain diagrams used in the analyses for (a)

which results in $R'_n = 116.9$ mm (4.60 in.). A more elaborate analysis of springback (Ref 11) for this case predicts $R'_n/2h = 23.41$ (or $R'_n = 117.05$ mm, or 4.608 in.). Also, for bending the same strip to $R_n = 40$ mm (1.6 in.) and $R_n = 500$ mm (19.7 in.), the final radii of curvature from Eq 31 are 42.8 mm (1.68 in.) and 1150 mm (45.3 in.), respectively. The corresponding results from Ref 11 are 42.9 mm (1.69 in.) and 1075 mm (42.3 in.).

The distribution of residual stresses after bending to $R_n = 100$ mm (3.937 in.) is obtained from Eq 32(a). Therefore, in this case, in metric units:

$$\sigma'_x = \sigma_x + yE(0.0014457) \qquad \text{(Eq 38)}$$

In English units:

$$\sigma'_x = \sigma_x + yE(0.03672)$$

At $R = R_i$, $y = h = 2.5$ mm (0.098 in.):

$$\varepsilon_x = -0.025$$

and

$$\sigma_x = -\left(\frac{2}{\sqrt{3}}\right)(869)(0.025)^{0.16} = -556 \text{ MPa}$$

or

$$\sigma_x = -\left(\frac{2}{\sqrt{3}}\right)(126)(0.025)^{0.16} = -80.63 \text{ ksi}$$

For this location:

$$\sigma'_x = -556 + 2.5(2.17 \times 10^5) \times (0.0014457)$$
$$= +228 \text{ MPa}$$

or

$$\sigma'_x = -80.63 + 0.098 \ (31,465)(0.03672)$$
$$= +32.6 \text{ ksi}$$

Similarly, the magnitude of σ_x for other values of y can be determined. Figure 23 shows the distribution of applied and residual stresses.

Bendability

Bendability is the ability of a material to be bend around a specified radius without fracture. It is typically expressed as the minimum bending radius without fracture when bent along a straight line. Sheet metal bendability is a critical factor in many forming operations, and the evaluation of bendability has been addressed in various publications (Ref 12–19). The analytical methods in Ref 12–17 are based on elementary bending theory with the assumption of small curvature bending. In Ref 18, however, an analysis is developed for sheet metal of work hardening materials that are subjected to large curvature plane-strain bending. Bendability criteria for large-curvature bending are suggested in Ref 18.

If small-curvature bending is assumed, two bend-ductility equations can be used to derive minimum bend radii from tensile-test elongation or reduction of area (Ref 13). The derivation of the two relevant bend-ductility equations from Ref 13 (adapted here) assumes the following:

- The fracture strain at the outer fiber of a bending specimen equals the fracture strain on a tensile specimen.
- The materials are homogeneous and isotropic.
- The form (sheet, plate, or bar) bends in plane strain.

To correlate minimum bend radius with reduction in area, it is assumed that the maximum true strain that a tensile specimen can endure is the true strain (ε_f) in the reduced section at the instant of fracture.

$$\text{Then: } \varepsilon_f = \ln \frac{A_0}{A_f} = \frac{w_0 t_0}{w_f t_f}$$

Where, w_0 and t_0 are width and thickness of untested specimen; w_f and t_f are width and thickness of specimen after fracture; A_0 is sectional area of a specimen before test; and A_f is sectional area of a specimen after fracture. By definition, percentage of reduction in area is,

$$(A_r) = \frac{A_0 - A_f}{A_0} \times 100$$

Combining these two equations, we get:

$$\varepsilon_f = \ln \left(\frac{100}{100 - A_r} \right)$$

With reference to the illustrated bending specimen (Fig. 24), the outside fiber is strained in tension; the inside in compression, l_f is length of outside surface; l_0 is length of l_f at the neutral axis; θ is angle of bend; and t is thickness. Assume that the neutral axis of bending lies in the midfiber. Taking the inner bending radius (R) of the beam into consideration, the arc length along the neutral axis subtended by the angle θ is $\left(R + \frac{t}{2}\right) \theta$, and the arc length of the outside fiber is $(R + t) \ \theta$. Therefore, the true strain in the outside fiber (ε_0) is given by:

$$\varepsilon_0 = \ln \left(\frac{l_f}{l_0} \right) = \ln \left[\frac{(R + t) \ \theta}{\left(R + \frac{t}{2}\right) \theta} \right] = \ln \left[\frac{R + t}{R + \frac{t}{2}} \right]$$

Equating the last two equations gives the following:

$$\frac{100}{100 - A_r} = \frac{R + t}{R + \frac{t}{2}} \quad \text{or:}$$

$$\frac{R}{t} = \frac{50}{A_r} - 1 \quad \text{(see Fig. 25).}$$

To correlate minimum bend radius with elongation, we start with the percentage elongation (e_1) of a tensile specimen. By definition,

$$e_1 = \frac{l_f - l_0}{l_0} \times 100 = \left(\frac{l_f}{l_0} - 1 \right) \cdot 100$$

Here, l_f is gage length of a tensile specimen after fracture, and l_0 is initial gage length. ε_f is fracture strain in tension test, as before. Thus $\varepsilon_f = \ln \frac{l_f}{l_0}$. From the first equation $\frac{e_1}{100} = \frac{l_f}{l_0} - 1$. Substituting this equation into the previous one, we get:

$$\varepsilon_f = \ln \left(\frac{e_1}{100} + 1 \right)$$

Equating ε_f to ε_0, we arrive at:

$$\frac{e_1}{100} + 1 = \frac{R + t}{R + \frac{t}{2}}$$

By simplification, the following equations result:

$$\frac{e_1}{100} \left(R + \frac{t}{2} \right) + R + \frac{t}{2} = R + t,$$

$$\text{or } \frac{e_1}{100} \left(R + \frac{t}{2} \right) = \frac{t}{2}$$

Dividing by t, one has:

$$\frac{e_1}{100} \left(\frac{R}{t} + \frac{1}{2} \right) = \frac{1}{2}$$

$$\text{or } \frac{R}{t} + \frac{1}{2} = \frac{50}{e_1}$$

Rearranging gives:

$$\frac{R}{t} = \frac{50}{e_1} - \frac{1}{2} \quad \text{(see Fig. 26).}$$

Testing of this analytical approach was evaluated by comparison with two types of bend tests (Ref 2). Figures 25 and 26 show that experimental data points fall along the theoretical curves. Thus, there is a definite relation between the $\frac{R}{t}$ ratio and the ductility, as given by reduction in arca or elongation. The two equations can indicate minimum bend radii with respect to a particular thickness. Because handbooks usually give elongation and reduction

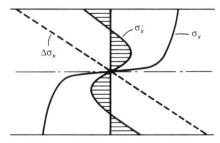

Fig. 23 Distribution of applied and residual stresses

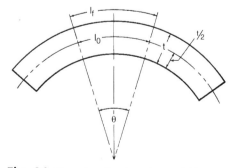

Fig. 24 Illustration of bend-specimen variables

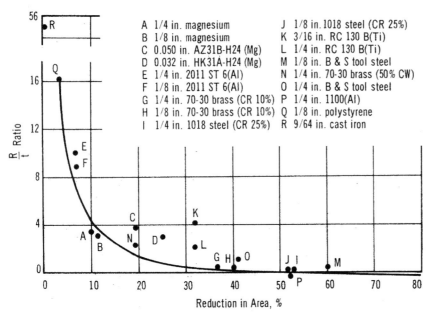

A	1/4 in. magnesium	J	1/8 in. 1018 steel (CR 25%)
B	1/8 in. magnesium	K	3/16 in. RC 130 B(Ti)
C	0.050 in. AZ31B-H24 (Mg)	L	1/4 in. RC 130 B(Ti)
D	0.032 in. HK31A-H24 (Mg)	M	1/8 in. B & S tool steel
E	1/4 in. 2011 ST 6(Al)	N	1/4 in. 70-30 brass (50% CW)
F	1/8 in. 2011 ST 6(Al)	O	1/4 in. B & S tool steel
G	1/4 in. 70-30 brass (CR 10%)	P	1/4 in. 1100(Al)
H	1/8 in. 70-30 brass (CR 10%)	Q	1/8 in. polystyrene
I	1/4 in. 1018 steel (CR 25%)	R	9/64 in. cast iron

Fig. 25 Comparison of bending limit with tensile-test reduction in area. By knowing the reduction in area and thickness (*t*) of a material in a given form, one can calculate the minimum bend radius (*R*) that a sheet, plate, or bar will withstand on being formed. Experimental data points for several materials with varying thicknesses fall on or close to the line calculated with a formula relating to three factors. Initials CR and CW stand for "cold rolled" and "cold worked." Alloy RC 130 B consists of titanium plus 3.5 Mn and 3.0 Al. Source: Ref 13 and 19

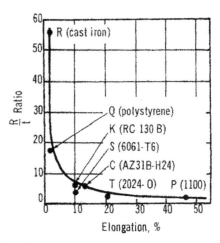

Fig. 26 Correlation of tensile elongations with minimum bend radii (*R*) and thicknesses (*t*). Letters on data points refer to materials listed in Fig. 1 except for S and T, data points for aluminum alloys.

in area values, the equations can be useful in approximating minimum bend radii for small-curvature bends. See also the article "Forming of Carbon Steels" for more information on minimum bend Radii.

ACKNOWLEDGMENT

The sections "Bending Theory," "Bending Calculations," and "Bending Operations" in this article have been adapted from O.D. Lascoe, *Handbook of Fabrication Processes,* ASM International, 1988.

REFERENCES

1. J. Schey, *Tribology in Metalworking,* American Society for Metals, 1983, p 517
2. H.A. Al-Qureshi, *J. Mech. Work. Technol.,* Vol 1, 1977–1978, p 261–275
3. J.M. Gere and S.P. Timoshenko, *Mechanics of Materials,* 4th ed., PWS Publishing Co., 1997
4. A.C. Ugural and S.K. Fenster, *Advanced Strength and Applied Elasticity,* 3rd ed., Prentice Hall, 1995
5. D. Horrocks and W. Johnson, On Anticlastic Curvature with Special Reference to Plastic Bending: A Literature Survey and Some Experimental Investigations, *Int. J. Mech. Sci.,* Vol 9, 1967, p 835–861
6. J. Chakrabarty, *Theory of Plasticity,* McGraw-Hill, 1987
7. P. Dadras, Plane Strain Elastic-Plastic Bending of a Strain-Hardening Curved Beam, *Int. J. Mech. Sci.,* Vol 43 (No. 1), Jan 2001, p 39–56
8. P. Dadras and S.A. Majlessi, Plastic Bending of Work Hardening Materials, *J. Eng. Ind. (Trans. ASME),* Vol 104, 1982, p 224–230
9. R. Hill, *The Mathematical Theory of Plasticity,* Oxford University Press, London, 1950
10. H. Verguts and R. Sowerby, The Pure Plastic Bending of Laminated Sheet Metals, *Int. J. Mech. Sci.,* Vol 17, 1975, p 31
11. O.M. Sidebottom and C.F. Gebhardt, Elastic Springback in Plates and Beams Formed by Bending, *Exp. Mech.,* Vol 19, 1979, p 371–377
12. J. Datsko and C.T. Yang, "Correlation of Bendability of Materials with Their Tensile Properties," *ASME Transactions,* Series B, Vol 82, 1960, p 309–314
13. C.T. Yang, "Calculating Minimum Bend Radii From Ductility Ratings," *Metal Progress,* Vol 98, 1970, p 107–110
14. M. Huang and J.C. Gerdeen, "Springback of Doubly Curved Developable Sheet Metal Surfaces—An Overview," *Analysis of Autobody Stamping Technology* Proceeding of SAE Annual Conference, 1994, p 125–138
15. A. Turno and W. Kunzendorf, "On the Cold Bending Ability of Metal Sheets and Plates," *J. of Mech. Working Technology,* Vol 4, 1980, p 121–131
16. J.L. Duncan, M. Sue-Chu, and X.J. Wang, "Instability in Plastic Bending of Thin Sheet Metal," *Metal Forming Research Report 83-006,* 1983
17. F.N. Mandigo, "Bending Ductility Tests," *Metal Handbook,* 9th ed., Vol 8, 1985, p 125–131
18. M. Huang and J.C. Gerdeen, The Bendability of Sheet Metals, *Sheet Forming Technology,* M.Y. Demeri, Ed., The Minerals, Metals & Materials Society, 1999, p 157–167
19. W.F. Hosford and R.M. Caddell, *Metal Forming—Mechanics and Metallurgy,* 2nd ed., Prentice Hall, 1993

Press-Brake Forming

PRESS-BRAKE FORMING is a process in which the workpiece is placed over an open die and pressed down into the die by a punch that is actuated by the ram portion of a machine called a press brake. The process is most widely used for the forming of relatively long, narrow parts that are not adaptable to press forming and for applications in which production quantities are too small to warrant the tooling cost for contour roll forming.

Simple V-bends or more intricate shapes can be formed in a press brake. Operations such as blanking, piercing, lancing, shearing, straightening, embossing, beading, wiring, flattening, corrugating, and flanging can also be carried out in a press brake. Information on press-brake forming can also be found in the articles "Press Brakes," "Forming of Carbon Steels," "Press Forming of Coated Steel," "Contour Roll Forming," "Forming of Stainless Steel," and "Forming of Aluminum Alloys" in this Volume.

Principles

In press-brake forming, as in other forming processes, when a bend is made, the metal on the inside of the bend is compressed or shrunk, and that on the outside of the bend is stretched. The applied forces create a strain gradient across the thickness of the work metal in the area of die contact. Tensile strain occurs in the outer fiber, and compressive strain in the inner fiber; both decrease in magnitude toward the neutral axis.

The setup and tooling for press-brake forming are relatively simple (Fig. 1). The distance the punch enters the die determines the bend angle and is controlled by the shut height of the machine. The span width of the die, or the width of the die opening, affects the force needed to bend the workpiece. The minimum width is determined by the thickness of the work and sometimes by the punch-nose radius. After the tools have been set up and the shut height has been adjusted, the press brake is cycled, and the work metal is bent to the desired angle around the nose radius of the punch.

Applicability

Press-brake forming is most widely used for producing shapes from ferrous and nonferrous metal sheet and plate. Although sheet or plate 6.4 mm (0.250 in.) thick or less is most commonly formed in a press brake, metals up to 25 mm (1 in.) thick have often been used, as in the following example.

Example 1: Bending 25 mm (1 in.) Thick Steel Plate. A 280 mm (11 in.) long bend was made in a 25 mm (1 in.) thick low-carbon steel plate in a 2700 kN (300 tonf) press brake to make the part shown in Fig. 2. The bend radius was 51 mm (2 in.); the included angle was 130°. The plate was air bent in a V-die with a punch that was made from a steel tube 102 mm (4 in.) in diameter.

After bending, the edges of the part were machined parallel, as shown in Fig. 2. The workpiece became part of a welded assembly—a frame brace on an industrial lift truck.

Workpiece Dimensions. The length of plate or sheet that can be bent is limited only by the size of the press brake. For example, a 5350 kN (600 tonf) press brake can bend a 3 m (10 ft) length of 19 mm (3/4 in.) thick low-carbon steel plate to a 90° angle, with an inside radius of the bend equal to stock thickness. If the included angle of the bend is greater than 90°, if the bend radius is larger than stock thickness, or if the length of bend is less than the bed length, a press of correspondingly lower capacity can be used.

Forming can be done at room or elevated temperature. For elevated-temperature forming in which the punch bottoms, the punch and die should be heated as well as the blank. In air bending, the blank is heated, and the punch is sometimes heated, depending on the area of contact between punch and blank and the metal thickness.

Work Metals. Press-brake forming is applicable to any metal that can be formed by other methods, such as press forming and roll forming. Low-carbon steels, high-strength low-alloy steels, stainless steels, aluminum alloys, and copper alloys are commonly formed in press brakes. High-carbon steels and titanium alloys are less frequently formed in a press brake, because they are more difficult to form.

The formability of all metals decreases as the yield strength increases. Therefore, in press-brake forming, power requirements and springback problems increase and the degree of bending that is practical decreases as the yield strength of the work metal increases.

Press Brakes

The primary advantages of press brakes are versatility, the ease and speed with which they can be changed over to a new setup, and low tooling costs. A press brake is basically a slow-speed punch press that has a long, relatively narrow bed and a ram mounted between end housings (Fig. 3). Rams are actuated mechanically or hydraulically. Additional information on press brakes is provided in the article "Press Brakes" in this Volume.

Mechanical Press Brakes. The ram of a mechanical press brake is actuated by a crank or an eccentric through a gear train in which there is a clutch and a flywheel. The gear train is usually designed to provide fast movement of the ram. Shut height (the distance between ram and bed at the bottom of the stroke) is adjustable by means of a screw (usually powered) in the pitman, or

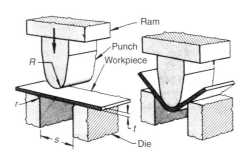

Fig. 1 Typical setup for press-brake forming in a die with a vertical opening. *R*, punch radius; *r*, die radius; *s*, span width; *t*, metal thickness

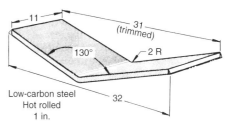

Fig. 2 Frame brace that was air bent from thick plate in a V-die in a 2700 kN (300 tonf) press brake. Dimensions given in inches

link, at each end of the ram. The length of the ram stroke, however, is constant.

One advantage of a mechanical press brake over a hydraulic press brake is that the mechanical type can develop greater-than-rated tonnage, because of the inertia of the flywheel moving the ram and the mechanical advantage of the crank near the bottom of the stroke. As a result, most mechanical press brakes have extra-strong frames to allow for occasional overloading. However, overloading should not be encouraged, because serious damage to the press brake may occur from improper setup. Another advantage is that operating speeds are greater in mechanical press brakes than in hydraulic press brakes. The greater speed is especially useful for long-run production of workpieces that are easily handled. Greater speed also permits instantaneous, high impact forces when the punch contacts the work metal. This impact force is useful in some operations, although it can damage the machine if the setup lacks rigidity.

A disadvantage of a mechanical press brake is that the stroke cannot be adjusted or controlled to the same degree as is possible with the hydraulic type. However, mechanical press brakes are available at additional cost with devices that permit a rapid advance to work and then a slower speed during forming.

Hydraulic Press Brakes. The ram of a hydraulic press brake is actuated by two double-acting cylinders, one at each end of the ram. Force supplied by the hydraulic mechanism will not exceed the press rating; therefore, it is almost impossible to overload a hydraulic press brake. (When thicker metal is inadvertently used, the ram stalls.) Therefore, frames can be lighter and less costly than those for mechanical press brakes, which are subject to overloading.

In hydraulic press brakes, length of stroke and location of the top and bottom of the stroke (within limits of the cylinder length) are adjustable. The point of rapid advance and return of the ram and its speed during contact with the workpiece are also adjustable; this adjustment makes possible a dwell period, which is often helpful in controlling springback. Cycles established by means of the various adjustments are reproduced by switches in the control circuit.

Even though devices are available that permit some control of the stroke of a mechanical press brake, the degree of control that is possible for a hydraulic press brake is considerably greater. For example, the ram on a hydraulic press brake can be reversed or its speed can be changed at any point on the stroke. Because of these features, a hydraulic press brake is often preferred for the segmental forming of stock longer than the dies, for the forming of large sheets that would be likely to whip in a mechanical press brake, and for the forming of difficult-to-form metals.

Hybrid press brakes incorporate both mechanical and hydraulic elements in the ram drive. The hydraulic-mechanical hybrid consists of a mechanical press brake driven by a rotary hydraulic motor. Containing a vane that rotates 270° between stops, the rotary hydraulic motor has replaced the piston used in a hydraulic cylinder. As it moves between the two stops, the motor propels the eccentric shaft through one complete cycle, driving the ram to the stroke bottom and back to the top.

The hybrid press brake combines the best features of both mechanical and hydraulic press brakes. It offers the same accuracy and operating speeds obtainable with the mechanical press brake while providing the adjustable length and controllability of the hydraulic press brake.

Selection of Machine

A mechanical press brake is usually preferred for quantity production because its speed is greater than that of a hydraulic press brake.

Conversely, a hydraulic press brake is generally preferred for varied short-run production because it is more versatile.

Apart from the method of actuating the ram, major factors that must be considered in the selection of a press brake for a given application are the size, length of stroke, and tonnage capacity of the press brake. Table 1 lists capacities and other details for mechanical and hydraulic press brakes.

Size is determined by length of bed and length of stroke (maximum stroke, in a hydraulic press brake). The bed length must be able to accommodate the longest bend required. Bed length can also be dictated by the need to mount more than one die in the press to permit a sequence of related operations in the machine at the same time. Under these conditions, it may be necessary to shim the dies so that the punches will bottom simultaneously. For example, if the parts are 305 mm (12 in.) long and six dies are required, the parts could be mounted on a press brake with a bed length of 1.8 m (72 in.) plus allowance for space between dies.

Standard press brakes are available with a maximum bed length of 7.3 m (24 ft). Still larger press brakes are available on special order. The longer the bed, however, the more massive it must be to provide enough rigidity for holding product dimensions, until a length is finally reached at which cost is prohibitive. Similarly, for a given capacity, the maximum stock thickness that can be accommodated decreases as bed length increases.

Length of stroke is an important consideration in any operation in which the height of the sides of the member after bending (such as a deep channel or box) causes interference between the top edge of the formed section and the ram. In addition, the greater the leg height after forming, the longer the stroke must be to allow the finished part to be withdrawn (unless it can be withdrawn from the end, under which conditions length of stroke is not important). Press brakes having a stroke length as great as 152 mm (6 in.) (mechanical) and 457 mm (18 in.) (hydraulic) are available as standard equipment. Modifications for providing increased stroke length are available at extra cost.

Capacity is stated in tons of force developed by the ram at the midpoint of the stroke. Capacities of commercial press brakes range from 70 kN to 22 MN (8 to 2500 tonf). Required capacity is governed by the size and bending characteristics of the work metal and by the type of bend to be made. A formula for determining the capacity required for 90° bends using V-dies without bottoming is:

$$L = \frac{lt^2 kS}{s} \qquad \text{(Eq 1)}$$

where L is press load (in tonf), l is length of bend (parallel to bend axis) (in inches), t is work metal thickness (in inches), k is a die-opening factor (varying from 1.2 for a die opening of $16t$ to 1.33 for a die opening of $8t$), S is tensile strength of the

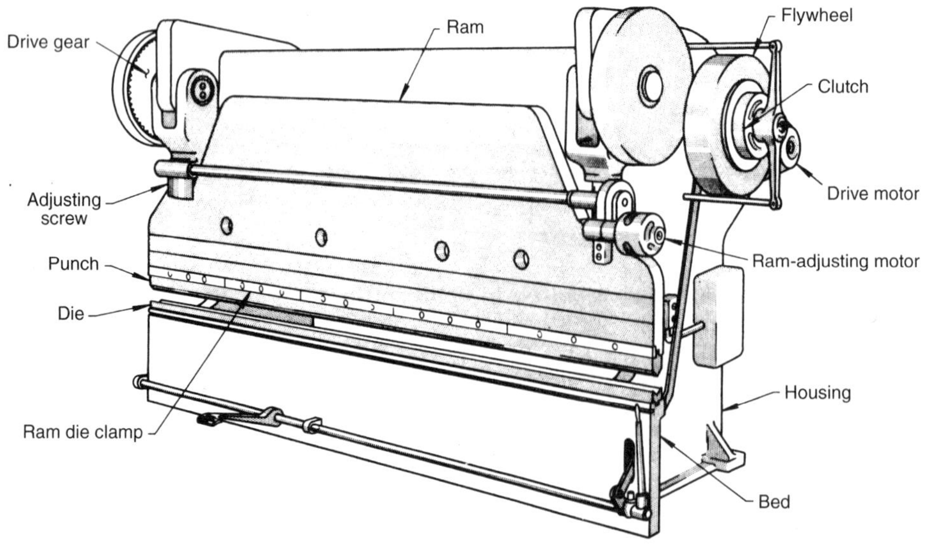

Fig. 3 Principal components of a mechanical press brake

Labels: Drive gear, Ram, Flywheel, Clutch, Drive motor, Ram-adjusting motor, Adjusting screw, Punch, Die, Housing, Ram die clamp, Bed

Table 1 Capacities, sizes, speeds, and ratings for mechanical and hydraulic press brakes

Capacity Midstroke kN	Midstroke tonf	Near bottom of stroke kN	Near bottom of stroke tonf	Bed length m	Bed length ft	Stroke length mm	Stroke length in.	Speed, strokes per min	Bending capacity, m (ft), with standard stroke for low-carbon steel with thickness of: 1.6 (1/16 in.)	4.8 (3/16 in.)	6.4 (1/4 in.)	13 (1/2 in.)	19 (3/4 in.)	25 (1 in.)	Motor, hp
Mechanical press brakes															
...	...	130	15	1.2–3.0	4–10	50	2	20–50	1.2 (4)	0.2 (3/4)	...	...	...	...	3/4–1
...	...	220	25	1.8–3.7	6–12	50	2	20–50	2.0 (6 1/2)	0.5 (1 1/2)	...	...	...	...	1 1/2
320	36	490	55	1.8–3.7	6–12	64	2 1/2	40	3.7 (12)	0.9 (3)	...	...	...	...	3
530	60	800	90	1.8–4.3	6–14	75	3	40	...	1.8 (6)	...	...	...	...	5
800	90	1200	135	1.8–4.3	6–14	75	3	36, 12	...	3.4 (11)	1.8 (6)	...	...	...	7 1/2
1020	115	1560	175	1.8–4.3	6–14	75	3	36, 12	...	...	3.0 (10)	...	...	...	10
1330	150	2000	225	1.8–4.9	6–16	75	3	33, 11	...	...	4.0 (13)	...	...	...	15
1780	200	2670	300	2.4–5.5	8–18	102	4	30, 10	...	...	5.5 (18)	1.8 (6)	...	...	20
2310	260	3560	400	2.6–5.8	8 2/3–18 2/3	102	4	30, 10	...	...	...	2.4 (8)	...	...	20
2980	335	4450	500	2.6–5.8	8 2/3–18 2/3	102	4	30, 10	...	...	...	3.0 (10)	1.5 (5)	...	25
3560	400	5340	600	3.0–7.3	10–24	102	4	30, 10	...	...	...	3.7 (12)	1.5 (5)	...	30
4630	520	6670	750	3.0–7.3	10–24	102	4	23, 7	...	...	...	5.5 (18)	3.0 (10)	...	40
5780	650	8900	1000	3.0–7.3	10–24	127	5	23, 7	...	...	...	7.3 (24)	3.7 (12)	1.8 (6)	40
7340	825	11,100	1250	4.2–6.7	14–22	152	6	20, 6	...	...	...	...	5.2 (17)	3.0 (10)	50
8900	1000	13,300	1500	4.2–7.3	14–24	152	6	20, 6	...	...	...	...	6.4 (21)	3.7 (12)	50
Hydraulic press brakes															
...	...	1780	200	2.6–5.8	8 2/3–18 2/3	305	12	21, 34(a, b)	...	4.3 (14)	3.7 (12)	...	...	...	25
...	...	2670	300	2.6–5.8	8 2/3–18 2/3	305	12	25(a, c)	...	...	4.9 (16)	2.4 (8)	...	...	30
...	...	3560	400	2.6–5.8	8 2/3–18 2/3	305	12	26(a, d)	...	...	...	3.7 (12)	1.8 (6)	...	40
...	...	4450	500	2.6–5.8	8 2/3–18 2/3	305	12	25(a, e)	...	...	...	4.3 (14)	2.7 (9)	...	40
...	...	5340	600	3.0–7.3	10–24	305	12	25(a, f)	...	...	...	4.9 (16)	3.0 (10)	...	50
...	...	6670	750	4.2–7.3	14–24	305	12	21(a, g)	...	...	...	6.7 (22)	4.3 (14)	3.0 (10)	60
...	...	8900	1000	4.2–7.3	14–24	457	18	21(a, h)	...	...	...	...	5.5 (18)	4.3 (14)	75

(a) Normal press speed gives rated capacity. High press speeds, m/min (in./min), together with press tonnage ratings, are as follows: (b) 1.4 and 1.7 m/min (57 and 65 in./min) at 620 kN (70 tonf); (c) 1.1 and 1.6 m/min (44 and 62 in./min) at 1070 kN (120 tonf); (d) 1.3 and 1.6 m/min (51 and 62 in./min) at 1420 kN (160 tonf); (e) 1.4 and 1.5 m/min (54 and 58 in./min) at 1780 kN (200 tonf); (f) 1.4 and 1.3 m/min (56 and 51 in./min) at 2140 kN (240 tonf); (g) 1.2 and 1.2 m/min (48 and 47 in./min) at 2670 kN (300 tonf); (h) 1.5 and 1.1 m/min (58 and 44 in./min) at 3560 kN (400 tonf)

work metal (in tons per square inch), and s is width of die opening (in inches) (Fig. 1).

Sample Calculation. Assume a constant of 1.33, a V-die opening of $8t$, and a bend 305 mm (12 in.) long made in 6.35 mm (0.250 in.) thick plate having a tensile strength of 30 tsi. Substituting these numerical values in Eq 1 yields:

$$L = \frac{12 \cdot 0.250^2 \cdot 1.33 \cdot 30}{2} \qquad \text{(Eq 2)}$$

or approximately a 130 kN (15 tonf) capacity requirement for this 90° bend.

For simple bending, the force required increases proportionally with the length of the workpiece or with the square of the work metal thickness. For example, in Eq 2, if the workpiece were 1220 mm (48 in.) long, a 530 kN (60 tonf) capacity would be needed. For producing offset bends (Fig. 4b), approximately four times as much pressure is required as for simple V-bends.

Dies and Punches

V-bending dies and their corresponding punches (Fig. 4a and d) are the tools most commonly used in press-brake forming. The width of the die opening (s, Fig. 4a) is usually a minimum of $8t$ (eight times the thickness of the work metal).

The nose radius of the punch should not be less than $1t$ for bending low-carbon steel, and it must be increased as the formability of the work metal decreases. The radius of the V-bending die must be greater than the nose radius of the punch by an amount equal to or somewhat greater than the stock thickness in order to allow the punch to bottom. Optimal dimensional control is obtained by bottoming the punch to set the bend.

When producing 90° bends in a bottoming die, the V-die is ordinarily provided with an included angle of 85 to 87°. Several trials are often necessary, and various adjustments must be made on the punch setting before the required 90° bend can be obtained.

Offset Dies. Punch and die combinations such as the one shown in Fig. 4(b) are often used

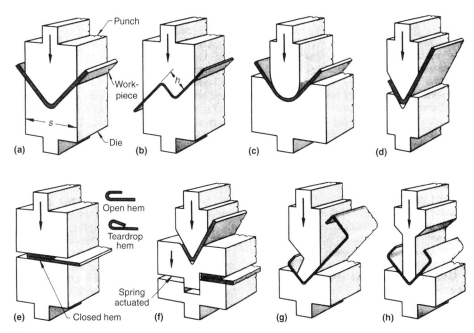

Fig. 4 Dies and punches most commonly used in press-brake forming. (a) 90° V-bending. (b) Offset bending. (c) Radiused 90° bending. (d) Acute-angle bending. (e) Flattening for three types of hems. (f) Combination bending and flattening. (g) Gooseneck punch for multiple bends. (h) Special clearance punch for multiple bends

to produce offset bends. Because an offset bend requires approximately four times as much force as a 90° V-bend, offset bending is usually restricted to relatively light-gage metal (3.2 mm, or 0.125 in., or less). The depth of offset (*h*, Fig. 4b) should be a minimum of six times the work metal thickness to provide stability at the bends.

Radius forming is done with a 90° die and a punch, each having a large radius (Fig. 4c). When the punch is at the bottom position, the inside radius of bend in the workpiece conforms to the radius of the punch over a part of the curve. The harder the punch bottoms, the more closely the work metal wraps around the punch nose, resulting in a smaller radius of bend and less springback. Uniformity of bend angle depends greatly on the uniformity of the work metal thickness.

Acute angles are formed by the die and punch shown in Fig. 4(d). The air-bending technique (see the section "Air Bending" in this article) is often used to produce acute angles. Acute angles are formed as the first step in making a hem. For this purpose, the die is often bottomed to make the bend angle as acute as possible. A disadvantage of bottoming is that the metal becomes work hardened, so that the hem is likely to crack when formed.

Flattening dies, shown in Fig. 4(e), are used to produce three types of hems (also shown in Fig. 4e) after the metal has been formed into an acute angle. The combination die shown in Fig. 4(f) produces an acute angle on one workpiece and a hem on another, so that a piece is started and a piece completed with each stroke of the press brake.

Gooseneck punches (Fig. 4g) and narrow-body, or special clearance, punches (Fig. 4h) are used to form workpieces to shapes that prevent the use of punches having conventional width (two such workpiece shapes are also shown in Fig. 4g and h).

Tongue Design. The punches shown in Fig. 4 as well as in several other illustrations in this article are provided with a simple, straight tongue for securing the punch to the ram. Although this design of tongue is generally accepted, in some shops punches with a hook type of tongue (see Fig. 9, and the punch for operation 4 in Fig. 21) are used exclusively as a safety precaution. A punch mounted with a hooked tongue cannot fall out. In one shop, it was estimated that hooked tongues increased punch cost by approximately 10% over the cost of straight-tongue punches.

Wiping Dies. Another type of bending die is a wiping die (Fig. 5). A pressure pad that is either spring loaded or attached to a fluid cylinder clamps the workpiece to the die before the punch makes contact. The punch descends and wipes one side of the workpiece over the edge of the die. The bend radius is on the edge of the die. To prevent the wiping action from being too severe, there may be a radius or chamfer on the mating face of the punch. When springback must be compensated for, the die is undercut to permit overbending. The flange metal can be put in slight tension by ironing it between the punch and the die. Sharp bends generally cannot be made in one operation in a wiping die without cracking the metal, because a punch or die with a sharp edge will cut the metal rather than bend it.

Special Dies and Punches

Dies that combine two or more operations to increase productivity in press-brake forming are generally more complicated and costly than those illustrated in Fig. 4. Before special dies are designed for a specific application, the increased tooling cost must be balanced against decreased time on the press brake. Generally, the quantity of identical parts to be produced is the major factor in selecting special dies.

Channel Dies. A channel die (Fig. 6a) can form a channel in one stroke of the press brake, while two strokes would be required using a conventional V-die. Because it is necessary to have an ejector in the die to extract the workpiece, channel dies cost more than conventional dies. This higher cost can be justified only on the basis of large-quantity production. It is ordinarily not necessary to have a stripper on the punch, because springback usually causes the part to release. The ejector in the die may be of the spring, hydraulic, or air-return type. The stripper for the punch (if needed) is a release-wedge device or a knockout piece. The use of a channel die, regardless of production quantities, is limited by work metal thickness, corner radii, and required flatness of the web.

A modification of the channel die is the U-bend die (Fig. 6b). Springback is a common problem with this type of die; one means of overcoming it is to perform a secondary operation on flat dies, as shown in Fig. 6(c).

Air Bending. In air bending, the die is deep enough that setting does not take place at the bottom of the stroke. The die can have a V shape (Fig. 7), or the sides can be vertical (Fig. 1). The shape and nose radius of the punch are varied to suit the workpiece. The required angle is

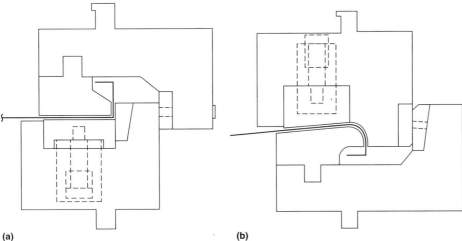

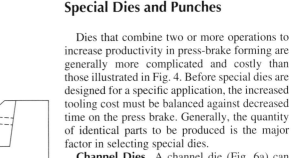

Fig. 5 Wiping dies. (a) Die set for flanging with spring-loaded pressure pad to hold material flat during forming. (b) Die for wiping radius

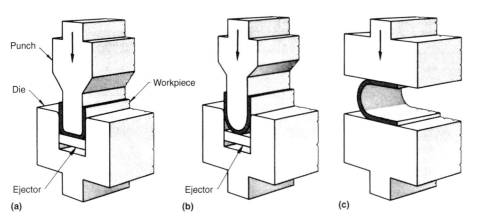

Fig. 6 Three types of special punches and dies for press-brake forming. (a) Forming a channel in one stroke. (b) Forming a U-bend in one stroke. (c) Flattening to remove springback after U-bending

produced on the workpiece by adjusting the depth to which the punch enters the die opening. This permits the operator to overbend the metal sufficiently to produce the required angle after springback.

When metal is bent beyond its yield strength, the radius formed bears a definite relationship to the opening in the die. A small die opening produces a small radius; the use of a large die opening increases the radius but also increases the amount of springback. Springback must be compensated for by overbending. Changing the size of the die opening also changes the amount of force needed to make the bend. As the die opening is increased, less force is required; as the die opening decreases, the bending leverage is less, and more force is therefore required.

For the air bending of metal up to 13 mm (1/2 in.) thick, the die opening is usually equal to eight times the work metal thickness. This keeps the bend radius approximately equal to the metal thickness. For metal thicker than 13 mm (1/2 in.) and for some high-tensile-strength metals, the die opening should be at least ten times the work metal thickness to increase the bend radius and therefore reduce the possibility of fracture at the bend.

The principal advantage of the air-bend method is the variety of forming that can be done with a minimum number of punches and dies. Air bending also requires less force for a given bend, thus preventing excessive strain on the press brake.

The primary disadvantage of air bending is the possible inconsistency in the bends. Because of variations in dimensions and temper of the work metal as it is received from the mill, springback can vary throughout a production run. However, the operator can adjust the ram to compensate for these irregularities. When air bending in a hydraulic press brake, the operator can use a preset pressure, check each part with a gage, and restrike if necessary. With a mechanical press brake, the shut height can be easily adjusted for a restrike and then reset for the next part.

Box-forming dies are similar to standard V-dies, except that the punch is sometimes specially made to clear the sides of the box being formed. For square boxes, the punch length can be the inside length, and the sides of the box can be formed in any sequence. However, for the forming of rectangular boxes, different tools or techniques are required. One approach is the use of a punch that is split vertically (Fig. 8) so that the punch can clear the sides on the long dimension while forming the sides of the short dimension. In most cases, however, a punch long enough for forming the long sides can be used without splitting; this is accomplished by forming the short sides first.

When bottoming dies are used in box forming (to overform and coin the metal to reduce springback), the metal can be work hardened excessively if the force used to form the short sides is the same as that used for the long sides. In some shops, when the short sides are less than two-thirds as long as the long sides, force is reduced for forming the short sides.

Arbor-type punches can be used when the sides of a boxlike workpiece must be folded over (detail A, Fig. 9). The head of the punch extends beyond the punch body so that the formed-over sides can fit over the punch extensions while the remaining folds are made (Fig. 9). The extensions on the punch are approximately triangular in cross section so that the punch can be withdrawn after opposite sides of the workpiece are closed.

Lock-Seam Dies. Lock seams are made in a press brake when quantities are too small to warrant more elaborate equipment. The usual procedure is to form one component with a special punch and die, as shown in Fig. 10(a). The second component of the assembly is formed in a simple V-die. The two components are then locked together in a single stroke of the press brake using another special die (Fig. 10b).

Curling Dies. Curling in a press brake is usually done in two steps using special dies such as those shown in Fig. 11(a).

Tube- and pipe-forming dies resemble curling dies. To ensure that the workpiece rolls up properly, the edges of the metal must initially be bent. Small tubes can be formed by using a two-operation die (Fig. 12a), while larger tubes require the use of a bumping die such as that shown in Fig. 12(b). Where accuracy is required, tubes formed in a bumping die should be sized over a sizing mandrel. If seams are required, they can be formed on the tube edges prior to the rolling operation.

Rocker-Type Dies. Dies that operate with spring-actuated rocker punches can produce bends that would be impossible with the punch operating in a vertical direction only. A typical rocker-type die for U-bending is illustrated in Fig. 11(b).

One-Stroke Hemming Dies. With specially designed spring-actuated dies, it is possible to hem a length of metal sheet in a single stroke. A typical die used for this operation and the movement of die components required for completing the hem are illustrated in Fig. 13. When provided with an adjustable stop, the die

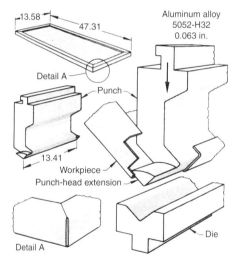

Fig. 9 Folding-over sides of a boxlike workpiece using an arbor-type punch for forming beneath reverse flanges. Dimensions given in inches

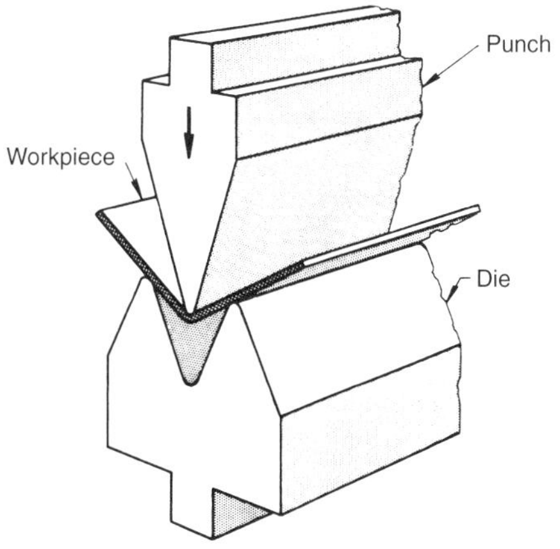

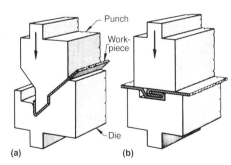

Fig. 10 Special punches and dies for producing lock seams in a press brake

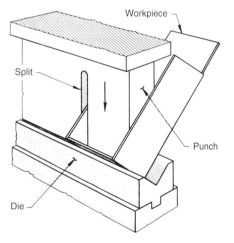

Fig. 7 Setup for air bending with an acute-angle punch and die in a press brake

Fig. 8 Setup for forming a rectangular box using a punch split vertically to clear the long sides during forming of the short sides

shown in Fig. 13 can produce hems of different widths (over a narrow range).

Beading dies form beads (stopped ribs) to add rigidity to flat sheets. The dies form either open beads that extend from edge to edge of the sheet or closed beads that fade out in the sheet. The hemispherical depressions that constitute the open bead are formed by the die shown in Fig. 14. Closed beads, on the other hand, necessitate the use of spring-pressure pads at the ends, which fade out to minimize wrinkling of the metal.

Corrugating dies can be used to produce numerous corrugations, as shown in Fig. 15.

Cam-driven dies, also known as wedge-driven dies, can be used to increase production rates as well as workpiece quality. When using an acute-angle die at normal press speeds, the sheet travels in a large arc, causing a bend at the outside die edge. This problem can be solved by decreasing the speed of the operation, but a better solution is to use a cam-driven die to ensure that the sheet will lay flat while maintaining the press speed at the desired rate.

Dies for Shearing, Lancing, Blanking, Piercing, and Notching

Shearing can be done in a press brake, but holddowns and knife supports must be used to obtain reasonable accuracy. For optimal results, a shearing machine should be used (see the article "Shearing of Sheet, Strip, and Plate" in this Volume). There are, however, applications in which shearing in a press brake is convenient because it can be combined with another operation.

Lancing is often done in a press brake. An example is the production of louvers, which are used in cabinet or locker doors. The punch, die, and die pad used for the simultaneous lancing and forming of sheet metal into louvers are shown in Fig. 16.

Blanking. When long, narrow dies are required for a given application, metal can be blanked in a press brake if adequate support can be provided by the dies or the press brake. The removal of long, narrow workpieces is sometimes a problem. Spring-type or rubber strippers can be added if excessive adherence to the punches is encountered.

Piercing and Notching. Press brakes are extensively used for piercing (punching) and notching. A press brake is more practical than a punch press for the piercing or notching of long, narrow workpieces, such as flats, channels, or other cross-sectional shapes. Press brakes are especially well adapted to piercing holes close to the edge of long panels or to notching the edges. Quick-change punching units help to extend the versatility of a press brake for piercing and notching. These punching units can be quickly changed to accommodate different workpieces by using setup templates.

Rotary Bending

Wiping dies, V-dies, and U-dies have traditionally been used in press-brake forming. Rotary bending is gaining wide acceptance in industry because it significantly reduces the time required to bend materials. Rotary bending is accomplished by using a tool that simultaneously holds and bends the material.

Key components of a rotary bender are the saddle (punch), the adjustable rocker, and the die anvil. The cylindrical rocker features an 88° V-notch cut out along its full length. To minimize marking, the edges of the rocker jaws are flatted and radiused. Figure 17 illustrates the three stages of a rotary-bending operation. Initially (Fig. 17a), the material is clamped, and the rocker rotation begins. Figure 17(b) shows how humping is controlled and confined to the space between the rocker edges. The final step (Fig. 17c) shows how the rocker clamps the workpiece in place and

overbends it sufficiently to compensate for springback.

Rotary benders have been primarily used in conjunction with progressive dies to form Z-bends and short-leg bends in a single operation. Also, dart stiffeners can be rolled into the workpiece simultaneously as it is being bent.

Selection of Tool Material

Selection of tool material for punches and dies used in press brakes depends on the composition of the work metal, the shape of the workpiece (severity of forming), and the quantity to be produced. The tool material used for press-brake bending and forming ranges from hardwood to carbide, although the use of carbide has usually been confined to inserts at high-wear areas. Hardwood and carbide represent the rare extremes; hardwood is suitable only for making a few simple bends in the most formable work metal, and carbide would be considered only for making severe bends in a less-formable work metal (such as high-strength low-alloy steel) in high production.

Simple Bending. Most dies and punches used for simple V-bending operations are made from low-carbon steel (such as 1020) or gray iron. Both of these materials are inexpensive and give acceptable tool life in mild service.

If production runs are long or if the work metal is less formable, some upgrading of the tool material may be desired to retain accuracy over a longer period. Gray iron can be upgraded without adding greatly to tool cost by making both punch and die from a hardenable grade (such as ASTM class 40) and then flame hardening the nose of the punch and the upper edges (high-wear areas) of the V-die to 450 to 550 HB. Low-carbon steel tools can be upgraded by changing to a hardenable grade of steel (such as 1045). High-wear portions of the tools can be flame hardened (usually to 50 to 55 HRC) in the same manner as the gray iron tools.

Severe Bending. As the severity of bending and forming increases, such as in producing channels in a single stroke (Fig. 6a), tool materials should be upgraded when more than low-production quantities are needed. For operations that require severe bending, tool material

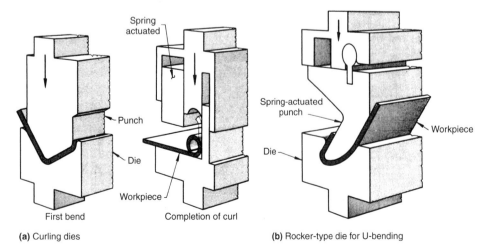

(a) Curling dies

(b) Rocker-type die for U-bending

Fig. 11 Special punches and dies for curling and U-bending in a press brake

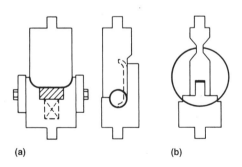

(a)

(b)

Fig. 12 Tube- and pipe-forming dies. (a) Two-operation die for the forming of small tubes. (b) Bumping die used to form large tubes

requirements for press-brake operations parallel those for punch-press operations (see the article "Selection of Materials for Press-Forming Dies" in this Volume).

Blanking and piercing are done in a press brake with tools of the same materials as those used in punch presses (see the article "Selection of Materials for Shearing, Blanking, and Piercing Tools" in this Volume).

Rubber Pads. The use of rubber pads in press-brake dies (Fig. 18) enables the forming of shapes that are difficult or impossible to form without the pads. The pads also minimize damage to work metal surfaces and decrease die cost. Additional information is available in the article "Rubber-Pad Forming and Hydroforming" in this Volume.

Urethane rubber is the type most widely used. Pads inserted into the bottom of the die can be used for forming V and channel sections in various metals ranging from soft aluminum to low-carbon steel up to 12 gage (2.657 mm, or 0.1046 in.) in thickness. When using the urethane-pad technique, the urethane is, in effect,

the die. It is almost impossible to compress urethane; its shape changes but not its volume. With minimum penetration of the punch, the pad begins to deflect, exerting continuous forming pressure around the punch. At the bottom of the stroke, the urethane has assumed the shape of the punch. When the pressure is released, the pad returns to its original shape.

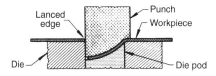

Fig. 16 Punch, die, and die pad used for the simultaneous lancing and forming of louvers in a press brake

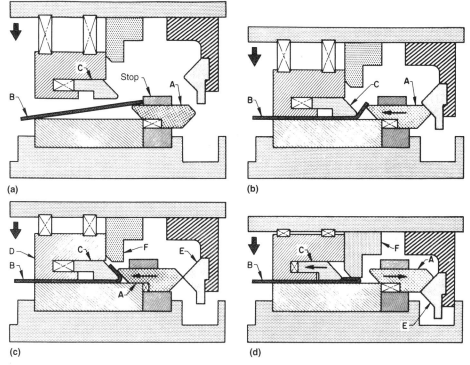

Fig. 13 One-stroke hemming die and movement of die components during hemming. (a) Workpiece, B, is placed over slide A using a component of the die as a stop, and the entire upper die section begins to move down. (b) Slide C contacts the workpiece and makes the first bend for the hem. (c) With slide D holding the workpiece rigidly in place, die component E forces slide A to the left, forming the bent section of the workpiece into an acute angle. (d) As the die continues to move down, die component E permits slide A to retract, providing clearance for F to contact and flatten the workpiece as it forces slide C to the left, thus completing the hem. On the upstroke of the press brake, die components return to original positions.

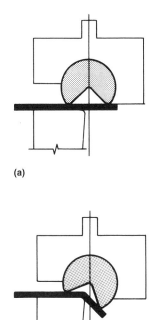

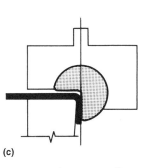

Fig. 17 Sequence of operations illustrating rotary bending in a press brake. (a) Material is clamped, and rocker rotation begins. (b) Humping is controlled and limited to space between edges of rocker. (c) Workpiece is clamped in position by rocker and overbent to allow for springback.

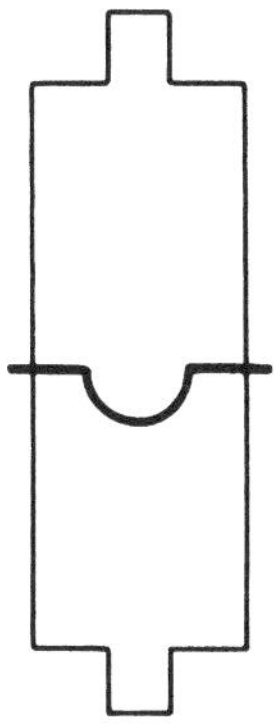

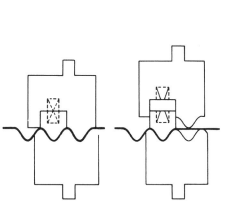

Fig. 14 Die used to form open bead in flat sheet to add stiffness to material

Fig. 15 Corrugating die used to form corrugations in a continuous sheet of material

Urethane pads are generally used for short-run production. However, in one plant, 14,000 boiler-casing channels were formed from 16 gage (1.52 mm, or 0.0598 in.) low-carbon steel in 4.9 m (16 ft) lengths on the same urethane pad before replacement.

Urethane robber is made in several different grades ranging in tensile strength from 18 to 76 MPa (2600 to 11,000 psi) and in hardness from durometer 80A to 79D. (Additional information is available in the article "Miscellaneous Hardness Tests" in *Mechanical Testing and Evaluation,* Volume 8 of *ASM Handbook,* 2000). Selection of grade depends on work metal hardness and thickness and on severity of forming. Experimentation is often needed to determine the optimal grade of urethane for the application.

Work Metal Finish. When preservation of work metal finish is a primary objective, the dies or punches or both are sometimes chromium plated. Other means of preserving the work metal finish include the use of oil-impregnated paper between the tools and the work metal, or spraying the tools with a plastic of the type used to coat metal sheets for deep drawing.

Procedures for Specific Shapes

Procedures and tooling for press-brake operations vary widely and are mainly influenced by workpiece shape. The following examples describe the procedures used for producing several different shapes, including simple boxlike parts, panels, flanged parts, architectural columns, fully closed parts, and semicircular parts.

Example 2: Four-Stroke Forming of Closed-Bottom Boxes from Notched Blanks. Closed-bottom boxes were produced from 1010 steel blanks sheared to 508 by 610 mm (20 by 24 in.)

in a press brake. The four corners were notched to a depth of 102 mm (4 in.) in a standard notching die in the same press brake as that used to make the blanks. The box was then formed in four strokes of the press brake, using a standard 90° V-die but with a deeper-than-normal punch (Fig. 19). By bending the 305 mm (12 in.) sides first, all four bends could be made with a die 406 mm (16 in.) long. Hourly production rates for the three operations were:

Operation	Pieces
Shear blanks	55
Notch corners (four strokes)	57
Bend from sides (four strokes)	44

Example 3: Six-Operation Forming of an Architectural Column. An architectural column 3 m (10 ft) long was produced in six operations in a press brake. Figure 20 shows the sequence of shapes produced. Channel dies were used for operations 1 and 2. Operation 3 required a special punch and die for producing the large-radius bends. A simple channel die was used for operation 4, and a V-die for operation 5. Operation 6 was performed with a gooseneck punch that was necked-in sufficiently to clear the edge flanges as the part closed in.

The major problem in forming this shape was to obtain sharp 90° bends at all corners and to keep the flanges in the same plane. Because the part was 3 m (10 ft) long, considerable shimming of the tools was required to produce satisfactory parts.

Correct shimming is a major factor in maintaining accuracy when producing shapes in a press brake such as that shown in Fig. 20. Shims are required to adjust for the discrepancies between bed plate and bolster. Also, deflections produced by the punch bottoming on all hits will

be greater in the center of the die than at the edges, and shimming is required to equalize the pressure along the entire length of the bend. Optimal shimming is accomplished mainly by trial and error, because of the variations among machines, tools, and workpieces.

Example 4: Producing a Completely Closed Triangular Shape. Figure 21 shows the four separate setups that were used to produce an 865 mm (34 in.) long completely closed triangular part in a press brake. The 540 by 865 mm (21 9/32 by 34 in.) blanks were prepared by shearing on separate equipment. As shown in Fig. 21, the first press-brake operation produced a 90° bend, and the second operation produced a 68° bend; simple straight-sided punches were used for both bends. In the third operation, a special punch 30 mm (1 1/8 in.) thick and having an offset nose was used to produce a 32° bend. By bending only to 32°, sufficient space was allowed for withdrawal of the punch. The punch had an offset nose because of the off-center seam location (a design requirement); if the seam had been centered, the punch would have been symmetrical. In the fourth operation, the part was closed. The part just before the fourth operation

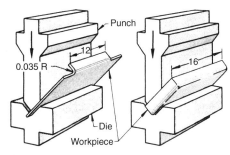

Fig. 19 Die and punch setup for bending sides in the production of a closed-bottom box. 1010 steel, 55–65 HRB. Dimensions given in inches

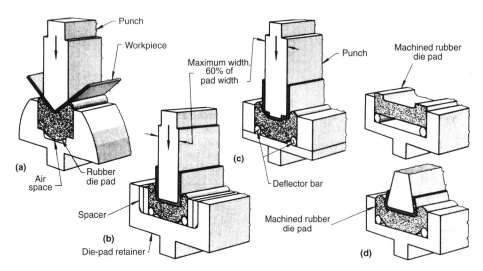

Fig. 18 Setups for rubber pad forming of various shapes in a press brake. (a) Simple 90° V-bend. Air space below die pad permits deep penetration. (b) Simple U-bend or channel. Spacers enable channels of varying widths to be formed in the same die-pad retainer. Deflector bars help to provide uniform distribution of forming pressure. (c) Modified channel, with partial air bending. (d) Acute-angle channel. High side pressures are obtained by using a conforming rubber die pad and deflector bars.

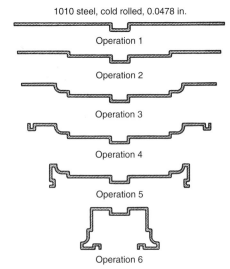

Fig. 20 Shapes progressively produced in six-operation forming of an architectural column in a press brake using 1.21 mm (0.0478 in.) thick steel sheet

is shown at the upper right-hand corner in Fig. 21. Time to complete the four operations was just under 1 min. Total time per part was just over 1 min, with a production rate of 58 parts per hour.

Semicircular Shapes and U-Bends. Flat stock can be formed into semicircular shapes and U-bends in a press brake. If the press capacity is adequate for the work metal thickness and the dimensions required, forming can be done in one operation, as in 90° V-bending. As shown in Fig. 6(b), the radius of the punch nose forms the inside radius of the workpiece.

Air bending is used to form semicircular shapes and U-bends when the work metal thickness and the dimensions exceed press capacity. A typical setup is shown in Fig. 22. Starting from the end of the workpiece at the right, and a distance of half the span of the die opening, pairs of equally spaced center-punch marks are made near each side of the workpiece, progressing to the left in two straight lines at 90° to the bend axis. These two rows of punch marks guide the operator in maintaining the alignment of the bend.

To form the part, the blank is placed across the die opening with the center-punch marks facing up, so that the punch will contact the blank at the first pair of punch marks. A bend is made at each pair of punch marks progressively toward the center of the blank (Fig. 22). When the center is reached, a quarter circle will have been formed. The blank is rotated 180°, and the procedure is repeated until a semicircle is formed. The radius of the semicircle will depend on the amount of bending done with each blow of the press and on the distance between the punch marks.

Bending should always proceed from the end of the blank toward the center in order to avoid interference between the ram and the formed workpiece. After forming, the straight section at each end of the workpiece is sheared off. The following example describes an application of this procedure for producing semicircular parts by air bending.

Example 5: Forming Semicircular Parts by Air Bending. Two semicircular parts were formed from 19 mm (3/4 in.) thick low-carbon

steel plate, and the parts were welded together to produce a 255 mm (10 in.) outside diameter hollow cylinder 255 mm (10 in.) long. Blank size for each semicircular part was 255 by 470 mm (10 by 18½ in.). The bend length required for each blank was 370 mm (14½ in.), but the blanks were cut to 470 mm (18½ in.) to allow for trimming after forming (Fig. 22).

Before forming, two rows of center-punch marks were made on the blanks (Fig. 22). The first mark was 50 mm (2 in.) from the blank edge, with subsequent marks every 13 mm (½ in.) of the 370 mm (14½ in.) bend length.

The parts were formed in an 1800 kN (200 tonf) mechanical press brake using a standard V-die and round-nose punch. Thirty bends were required to form each semicircular part. After several bends, the curve was checked with a template to determine the accuracy of the bend. After forming a quarter of the circle, the workpiece was rotated 180°, and the operation was repeated to complete the half circle. The final bend was made at the center of the workpiece. The 50 mm (2 in.) allowance at each end of the workpiece was then sheared off.

Corrugated Sheet. Special procedures allow bends to be fabricated in corrugated metal that are perpendicular to the corrugations, without flattening the corrugations. This can be achieved by using cast-on plastic blankets.

Effect of Work Metal Variables on Results

Thickness variations, yield strength, and rolling direction are the work metal variables that have the greatest effect on results in press-brake operations. Whenever possible, any metal to be formed should be purchased only to commercial tolerances; special tolerances increase cost. However, when workpiece tolerances are close, it is sometimes necessary to purchase metal with special thickness tolerances, because normal variations can use up a substantial amount of the assigned final tolerance (see the section "Dimensional Accuracy" in this article).

Yield Strength. As the yield strength of the work metal increases, so does the difficulty in bending. This difficulty occurs as cracking at the bends, increased power requirements, or an increase in springback.

For example, bending of stainless steel requires approximately 50 to 60% more power than bending a comparable thickness of low-carbon steel. Because of its resistance to bending, stainless steel often causes difficulty in obtaining acceptable results (see the article "Forming of Stainless Steel" in this Volume).

Springback. In a wiping type of bending operation, in which the metal is bent to position but the corner is not coined to set the bend, the metal attempts to return to its original position. This movement, known as springback, is evident to some extent in all metals, and it increases with the yield strength of the metal. The amount of springback is usually negligible for a soft metal such as 1100 aluminum alloy. However, for aluminum alloys such as 2024, the amount of springback can be significant. In general, low-carbon steels exhibit more springback than aluminum or copper alloys do, and still more springback can be expected for stainless steel.

A common technique for overcoming springback is to overbend by approximately the number of degrees of springback. Several trials in tool development may be needed to obtain the proper angle, because of variations in mechanical properties, work-hardening rate, metal thickness, and die clearances. Springback from one bend can sometimes be used to offset that from another. Tables and graphs for springback have been developed for specific metals. Detailed information on the subject of springback is also addressed in the article "Springback" in this Volume.

Another technique for overcoming springback is the use of specially designed bottoming dies that strike the workpiece severely at the radius of the bend. This action stresses the metal in the bend area beyond the yield point through almost the entire thickness and thus eliminates springback. Bottoming must be carefully controlled, particularly if it is done in a mechanical press brake, because the force developed by this machine can be very high.

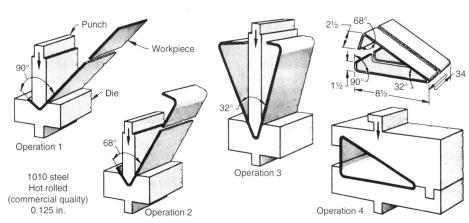

Fig. 21 Sequence of operations for forming a closed triangle in a press brake. Dimensions given in inches

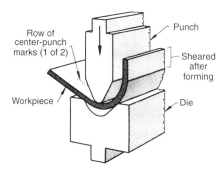

Fig. 22 Air bending to form a semicircular part by progressive strokes of the punch

Restriking in the original dies or special fixtures will reduce springback to a low level. It requires an additional operation but may entail little or no additional equipment. In the following example, a second stroke was used.

Example 6: Correcting Springback in the Forming of a Complex Shape. The shape shown at the lower right-hand corner in Fig. 23 was produced from 0.91 mm (0.036 in.) thick 1010 steel in lengths ranging from 0.9 to 2.4 m (3 to 8 ft). The five operations used in producing the part are shown in Fig. 23(a) to (e). The box section was formed by a wiping action (Fig. 23d) with no force on the outermost portion of the box. Consequently, springback occurred and required another step to correct by overforming (Fig. 23e). The shut height of the die was adjusted to provide the correction.

Rolling Direction. In the press-brake forming of steel, the effect of rolling direction is often a greater problem than in other methods, because long members are usually bent in a press brake and bends are made with axes parallel to the rolling direction, which is the least favorable orientation. However, it is sometimes possible to take advantage of directionality. The most severe bends can be made perpendicular to the direction of rolling, or if several bends are required along axes that are not parallel with each other, the layout can be planned so that all bends run diagonally to the direction of rolling. The difference in behavior of the same steel bent in both directions in a press brake is demonstrated in the following example.

Example 7: Effect of Rolling Direction on Bending. An axle bearing support was produced in four bends (Fig. 24) in a press brake using standard 90° V-dies. Cracks could not be tolerated. No cracks appeared on the flanges formed on the short dimensions, which were bent 90° to the direction of rolling to a 6 mm (1/4 in.) radius. However, in bending the flanges on the long dimensions, parallel with the direction of rolling, open breaks appeared along the length of the bend. To prevent this cracking, it was necessary to increase the bend radius on the long dimensions to 13 mm (1/2 in.) and to prepare the blanks so that the long flanges were formed at a slight angle to the direction of rolling.

Relation Between Bend Angle and Rolling Direction. As the thickness and yield strength of the work metal increase, the relationship between bend angle and grain direction becomes more important. For example, when stock thickness reaches approximately 25 mm (1 in.) and the yield strength is relatively high, as in high-strength low-alloy steels, the bend radius should be at least twice (and preferably three times) the stock thickness, even for bends of no more than 45°, when the bend axis is parallel with the direction of rolling.

In the press-brake forming of long, narrow workpieces, bending at an angle to the direction of rolling is seldom practical. For such work, the use of steel sheet that has been cross rolled or subjected to a pinch pass is a simple but relatively expensive means of minimizing the adverse effects from grain direction.

Dimensional Accuracy

The generally accepted tolerance for dimensions resulting from bending is ±0.4 mm (±0.016 in.) for metals up to and including 3.2 mm (0.125 in.) thick. For thicker metals, the tolerance is increased proportionately. As in many other mechanical operations, obtainable tolerances are influenced by design, stock tolerances, blank preparation, and condition of the machine and tooling. In some cases, close control of variables can provide closer dimensions at no additional cost; in others, cost will be increased.

Design. Bends or holes too close to the workpiece edges make it difficult to maintain an accurate bend line. Notches and cutouts on the bend line make it difficult to hold accurate bend location. Offset bends will shift unless the distance between bends in the offset is at least six times the thickness of the work metal.

Stock tolerances affect the dimensional accuracy of the finished part because they use up a portion of the assigned final tolerance. Commercial tolerances, particularly on thickness to which the specified metal is furnished, should be ascertained. For aluminum, there are minor differences in thickness limits between clad and unclad alloys. For steels, there are significant differences both in thickness tolerances and in cost among hot-rolled sheet, hot-rolled strip, cold-rolled sheet, and cold-rolled strip.

Cold-rolled steel sheet is produced to closer tolerances than hot-rolled sheet, but its cost is higher. Tolerances on steel strip, either hot rolled or cold rolled, are closer than those for corresponding sheet. Established tolerances are closer as the product becomes narrower or thinner.

Thickness tolerances for steel plate are considerably wider than those for hot-rolled steel sheet and strip. When ordered to thickness, the allowable minimum is 0.25 mm (0.010 in.) less than that specified, regardless of thickness, and the allowable maximum for an individual plate is 1 1/3 times the values that are published by the mills and expressed as a percentage of the nominal weight. Therefore, when tolerance requirements are stringent, it should be determined whether the metal can be obtained as strip or sheet rather than as plate.

Blank preparation can have an important effect on the tolerances and the cost of the finished part. If a blank is prepared by merely cutting to length from purchased stock, it will be low in cost, but the width tolerance will be that of the mill product. This may be greater than the tolerance obtainable by shearing. If it is necessary to shear all sides of a blank, the cost will increase, but good shearing can result in greater accuracy.

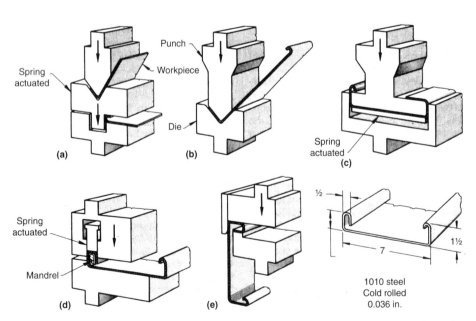

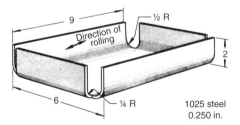

Fig. 23 Setups and sequence of operations for forming a complex shape in a press brake showing use of a restriking operation to eliminate springback. (a) Forming hem in two strokes. (b) Forming of first 90° angle for box section. (c) Forming channel. (d) Closing of box section over a mandrel. Part was moved by sliding it off mandrel. (e) Restriking of box section to eliminate springback. Dimensions given in inches

Fig. 24 Axle bearing support for which blank was prepared so that long flanges were formed at a slight angle to direction of rolling to prevent cracking. Dimensions given in inches

The stock from which blanks are cut must be flat enough for the blanks to be properly inserted into tooling and to remain in position during forming. Stretcher-leveled and resquared sheet costs a little more, but it is usually necessary when tolerances are close.

Blank Size. To determine the size of the blank needed to produce a specified bent part, the blank dimension (usually, the blank width) at 90° to the bend axis can be developed on the basis of the dimension along the neutral axis.

For 90° bends, as shown in Table 2, the developed blank width can be obtained by deducting bend allowances from the theoretical distance along the outside mold line. These allowances take into account the type and thickness of the work metal and the bend radius—each of which can affect the location of the neutral axis and therefore the developed width. The application of these allowances, which are based on shop practice with low-carbon steel and aluminum alloy 5052, is shown in the illustration in Table 2, for parts having one, two, three, or four bends.

For setting the stock stops from the centerline of the punch and die, the distance from the edge of the workpiece to the bend line at the neutral axis (Table 2) must be determined. To establish this value for 90° bends, subtract one-half the bend allowance from the outside flange width.

For bend angles other than 90° or radii other than those listed in Table 2, the width of a strip needed to produce a given shape can be calculated by dividing the shape into its component straight and curved segments and totaling the developed width along the neutral axis. Equation 3 can be used to determine the developed width, w, of a curved segment:

$$w = 0.01745 \; \alpha \; (r + kt) \qquad \text{(Eq 3)}$$

where 0.01745 is a factor to convert degrees to radians; α is the included angle to which the metal is bent (in degrees); r is the inside radius of the bend (in inches); and k is the distance of neutral plane or axis from the inside surface at the bend expressed as the fraction of the metal thickness, t, at the bend. Empirically determined values of k are: $1/3$, for bends of radius less than $2t$; and $1/2$, for bends of greater radius. Sample calculations showing the use of Eq 3 are presented in the article "Contour Roll Forming" in this Volume.

Permissible bend radii depend mainly on the properties of the work metal and on tool design. For most metals, the ratio of minimum bend radius to thickness is approximately constant, because ductility is the primary limitation on minimum bend radius. Another complicating factor is the effect of work hardening during bending, which will vary with metal and heat treatment.

Condition of Machines and Tools. Machines and tools must be kept in the best possible condition for maintaining close dimensions in the finished product. General-purpose tooling is seldom built for precision work and is frequently given hard use, which contributes further to inaccuracy through wear. Uneven wear aggravates the condition. If the press brake has been allowed to become loose and out-of-square and if ram guides and pitman bearings are worn, accurate work cannot be produced. Good maintenance is as essential in successful press-brake operation as in any other mechanical process.

Press-Brake Forming versus Alternative Processes

For many applications, press-brake forming is the only practical method of producing a given shape. In one case, for example, press-brake forming was used for a massive workpiece 3 to 3.7 m (10 to 12 ft) long that required several bends spaced at least 152 mm (6 in.) apart.

Under certain conditions, either a punch press or a contour roll former will compete with a press brake in performance and economy. When a workpiece can be produced by two or all of these methods, the choice will depend mainly on the

Table 2   Bend allowances for 90° bends in low-carbon steel and aluminum alloy 5052

Metal thickness (t)		Bend allowance, mm (in.), for bends with inside radius (r) of:									
		0.8 mm (1/32 in.)		1.6 mm (1/16 in.)		2.4 mm (3/32 in.)		3.2 mm (1/8 in.)		6.4 mm (1/4 in.) steel	13 mm (1/2 in.) steel
mm	in.	Steel	Aluminum	Steel	Aluminum	Steel	Aluminum	Steel	Aluminum		
0.81	0.032	1.50 (0.059)	1.45 (0.057)	1.68 (0.066)	1.73 (0.068)	2.01 (0.079)	2.08 (0.082)	2.36 (0.093)	2.41 (0.095)	3.71 (0.146)	6.45 (0.254)
1.27	0.050	2.21 (0.087)	1.98 (0.078)	2.57 (0.101)	2.31 (0.091)	2.90 (0.114)	2.67 (0.105)	3.28 (0.129)	3.00 (0.118)	4.27 (0.168)	7.01 (0.276)
1.57	0.062	2.67 (0.105)	2.41 (0.095)	3.00 (0.118)	2.74 (0.108)	3.35 (0.132)	3.05 (0.120)	3.68 (0.145)	3.38 (0.133)	4.65 (0.183)	7.37 (0.290)
1.98	0.078	3.25 (0.128)	2.95 (0.116)	3.61 (0.142)	3.33 (0.131)	3.94 (0.155)	3.66 (0.144)	4.29 (0.169)	3.99 (0.157)	5.13 (0.202)	7.87 (0.310)
2.29	0.090	3.71 (0.146)	3.30 (0.130)	4.06 (0.160)	3.66 (0.144)	4.39 (0.173)	3.99 (0.157)	4.75 (0.187)	4.32 (0.170)	5.52 (0.217)	8.23 (0.324)
3.18	0.125	5.03 (0.198)	4.44 (0.175)	5.36 (0.211)	4.80 (0.189)	5.69 (0.224)	5.16 (0.203)	6.17 (0.243)	5.49 (0.216)	6.61 (0.260)	9.32 (0.367)
4.78	0.188	7.34 (0.289)	6.50 (0.256)	7.67 (0.302)	5.51 (0.217)	8.02 (0.316)	7.19 (0.283)	8.36 (0.329)	7.54 (0.297)	9.73 (0.383)	11.3 (0.443)
6.35	0.250	9.71 (0.382)	8.59 (0.338)	10.0 (0.395)	8.92 (0.351)	10.4 (0.409)	9.27 (0.365)	10.8 (0.424)	9.60 (0.378)	12.1 (0.476)	13.2 (0.519)
7.95	0.313	12.0 (0.474)	...	12.4 (0.488)	...	12.7 (0.501)	...	13.1 (0.515)	...	14.5 (0.569)	17.2 (0.676)
9.52	0.375	14.4 (0.566)	...	14.7 (0.580)	...	15.1 (0.593)	...	15.4 (0.607)	...	16.8 (0.661)	19.5 (0.768)
11.1	0.437	16.7 (0.658)	...	17.1 (0.672)	...	17.4 (0.685)	...	17.8 (0.699)	...	19.1 (0.752)	21.8 (0.860)
12.7	0.500	19.0 (0.750)	...	19.4 (0.764)	...	19.7 (0.777)	...	20.1 (0.791)	...	21.5 (0.845)	24.2 (0.952)

$w = a + b -$ bend allowance

$w = a + b + c -$ (2 × bend allowance)

$w = a + b + c + d -$ (3 × bend allowance)

$w = a + b + c + d + e -$ (4 × bend allowance)

Note: w, developed width of blank; t, metal thickness; r, inside radius of bend

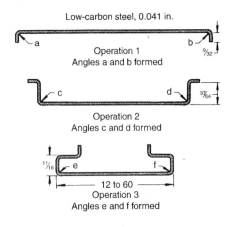

Fig. 25 Workpiece formed in six bends in either a press brake or a ten-station contour roll former. Dimensions given in inches

Item	Time, h	
	Press brake	Roll former
Setup	2.1(a)	9.2(b)
Production of 100 pieces(c)	6.8	0.8

(a) Total for all operations, including dies and gages. (b) Includes dismantling. (c) Pieces are 2.5 m (100 in.) long and 1.2 m (48 in.) wide.

quantity to be produced and the availability of the equipment.

Press Brake versus Punch Press. When a given workpiece can be made to an equal degree of acceptability in either a press brake or a punch press, the punch press is usually more economical, and it is more efficient than the press brake in terms of power requirements for a given force on the ram and number of strokes per unit of time. In addition, air ejection is more readily adapted to a punch press than to a press brake; this is a factor when air is required for ejecting either the workpiece or scrap.

The advantages of a punch press over a press brake are generally greater when production quantities are large and workpieces are relatively small. As workpiece size increases, the advantages of a punch press diminish.

Tooling for a press brake is usually simpler and less costly than counterpart tooling for a punch press—an important consideration for small production quantities. One disadvantage of punch presses is that they are more sensitive to thickness variations of the work metal because they operate at a faster rate.

Press-Brake versus Contour Roll Forming. For many parts usually formed in a press brake, contour roll forming is an acceptable alternative method of production, and the choice between the two processes depends mainly on the quantity to be formed. Press-brake forming is adaptable for quantities ranging from a single piece to a medium-sized production run, while contour roll forming is usually restricted to large-quantity production because of higher tooling costs. An advantage of contour roll forming is that coil stock can be used, while cut-to-length stock must be used in a press brake (see the article "Contour Roll Forming" in this Volume). The following example compares the efficiency of press-brake forming and contour roll forming.

Example 8: Press-Brake Forming versus Contour Roll Forming. Parts were produced to the shape shown in Fig. 25 in lengths up to 3.7 m (12 ft) and widths varying from 0.30 to 1.5 m (12 to 60 in.). The six bends were originally made in a press brake in three operations. When quantity requirements increased, production was changed to contour roll forming in a ten-station machine, from sheared-to-size sheets. Contour roll forming not only decreased the production time (table, Fig. 25) but also resulted in improved surface finish, because less handling of the work metal was required.

Safety

Press-brake operations involve the hazards of other press operations. Proper feeding devices are vital in order to ensure the safety of the press operator. Because more than one operator is often needed, added precautions are necessary to prevent the operation of a press brake without the direct consent of each man.

The article "Presses and Auxiliary Equipment for Forming of Sheet Metal" in this Volume contains information and literature references on safe operation. Some of the precautions noted are discussed as follows.

Barrier guards should be used wherever possible. Hand feeding devices such as vacuum lifters, special pliers, or magnetic pickups should always be used to keep operators' hands clear of dies.

When a large workpiece extends in front of the die, the operator often must use his hands to support the workpiece during forming. If a barrier guard cannot be used, because of the arc of travel of the front leg during forming or because the workpiece is of such shape that a guard would prevent loading or unloading of the workpiece, the sheet should be inserted against back gages. These gages, or stops, are adjusted so that the workpiece cannot slide over them.

The workpiece is supported by hand only if there is no other way to support it, and even then only if the operator's hands are not within reach of the die or any pinch point. A die apron or table should be provided to aid in loading large sheets into the die and to act as a support for sheets that do not require hand support. The formed sheet should be removed from the front of the press; parts that cannot be unloaded from the front of the press are moved to the end for removal. End supports may be required to prevent the workpieces from falling.

For versatility, a press brake is provided with a foot pedal to operate the machine. The foot pedal must have a cover guard so the press cannot be tripped accidentally. A foot-operated press brake should incorporate a single-stroke mechanism and be used as a single-operator machine.

When a press brake is used as a power press for stamping, shearing, and notching operations, the foot pedal should not be used. Instead, the press brake should be equipped with electropneumatic clutch and brake controls and should be provided with a single-stroke device. The foot pedal is replaced with two-hand palm switches, which are spaced so that the operator must use both hands to hold the switches until the die is closed. If a press brake is used exclusively for press work, the foot pedal should be permanently removed.

Deep Drawing

Revised by Mahmoud Y. Demeri, FormSys, Inc.

DEEP DRAWING of sheet metal is used to form parts by a process in which a flat blank is constrained by a blankholder while the central portion of the sheet is pushed into a die opening with a punch to draw the metal into the desired shape without causing wrinkles or splits in the drawn part. This generally requires the use of presses having a double action for hold-down force and punch force. The mechanics of deep drawing of a conical cup are illustrated in Fig. 1, which shows the complexity of the process. Deep drawing involves many types of forces and deformation modes, such as tension in the wall and the bottom, compression and friction in the flange, bending at the die radius, and straightening in the die wall. The process is capable of forming beverage cans, sinks, cooking pots, ammunition shell containers, pressure vessels, and auto body panels and parts. The term *deep drawing* implies that some drawing-in of the flange metal occurs and that the formed parts are deeper than could be obtained by simply

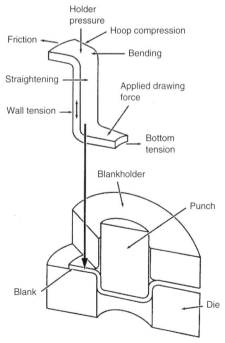

Fig. 1 Mechanics of the deep drawing of a cylindrical cup

stretching the metal over a die. Clearance between the male punch and the female die is closely controlled to minimize the free span so that there is no wrinkling of the sidewall. This clearance is sufficient to prevent ironing of the metal being drawn into the sidewall of the drawn part. If ironing of the walls is to be part of the process, it is done in operations subsequent to deep drawing.

Suitable radii in the punch bottom to side edge, as well as the approach to the die opening, are necessary to allow the sheet metal to be formed without tearing. In most deep-drawing operations, the part has a solid bottom to form a container and a retained flange that is trimmed later in the processing. In some cases, the cup shape is fully drawn into the female die cavity, and a straight-wall cup shape is ejected through the die opening. To control the flange area and to prevent wrinkling, a hold-down force is applied to the blank to keep it in contact with the upper surface of the die. Presses used for deep-drawing operations can be either hydraulic or mechanical, but hydraulic presses are preferred because of better control of the rate of punch travel.

Any metal that can be processed into sheet form by a cold-rolling process should be sufficiently ductile to be capable of deep drawing. Both hot- and cold-rolled sheet products are used in deep-drawing processes. The cold-work effects introduced during processing of the sheet products for deep-drawing applications must be removed (by annealing, for example), and the as-delivered coils should be free of any aging. This would imply that aluminum-killed drawing-quality steel, for example, would be preferred over rimmed steel. After the deep-drawing operation, ductility can be returned to that of the original sheet by in-process annealing, if necessary. In many cases, however, metal that has been deep drawn in a first operation can be further reduced in cup diameter by additional drawing operations, without the need for intermediate annealing.

The properties considered to be important in sheet products designed for deep drawing include:

- Composition, with a minimum amount of inclusions and residual elements contributing to better drawability

- Mechanical properties, of which the elongation as measured in a tension test, the plastic-strain ratio r (see the section "Drawability" in this article), and the strain-hardening exponent n are of primary importance. The strength of the final part as measured by yield strength must also be considered, but this is more a function of the application than forming by deep drawing.

- Physical properties, including dimensions, die geometry, modulus of elasticity, and any special requirements for maintaining shape after forming

Once a metal has been deep drawn into a suitable form, it can be further processed to develop additional shape. The first shape is usually a round cylinder, or a modification of this—a square box with rounded corners, for example. This latter shape is related to the cylinder in that the four corners are essentially quarter segments with straight walls between each segment.

For small cylinders, a relationship between the diameter of a circular blank and the bottom diameter of the cup shape to be formed is sometimes used to measure deep drawability. The most commonly referred to cup test for deep drawing is the Swift cup test, which uses a 2 in. (50 mm) diameter flat-bottomed punch to form test blanks. For ductile low-carbon steel, aluminum, and brass sheets, a 100 mm (4 in.) circular blank can be formed in a single draw. Increased plastic strain ratio r and ductility allow larger blanks to be drawn successfully; the limit is reached when the bottom punches out, rather than forming a cup shape. The blank diameter divided by the punch diameter gives the limiting draw ratio (LDR), which for the previous examples would be 2. More information on the Swift cup test is available in the article "Formability Testing of Sheet Metals" in this Volume.

Of the commonly formed metals, brass and austenitic stainless steels show high limiting draw ratios (up to 2.25). A few samples have reached a value of 2.5 to produce a cup with a sidewall height of nearly 65 mm (2.5 in.). This is possible, although the total length of the cup cross section would be 180 mm (7 in.), which is more than the blank diameter because of the deep-drawing forces. The interstitial-free steels, which have average *r* values of 2.5 or higher, can be deep drawn to limiting draw ratios near 2.5. With these extremely deep draws, sidewall delayed splitting must be prevented by such means as stress-relief annealing immediately following the drawing operation.

The thickness of the work metal does not change appreciably in deep drawing; therefore, the surface area of the final part is approximately the same as that of the initial blank. As the metal in the flange area is drawn into the die opening over the approach radius, it is subjected to radial tension and, concurrently, to circumferential compression. This explains why a 125 mm (5 in.) diameter blank with a surface area of 126 cm^2 (19.6 in.^{2}) can form a cup shape 65 mm (2.5 in.) deep that has a total surface area of approximately 121 cm^2 (18.8 in.^{2}).

With proper balance among the punch force, the hold-down force, and the strength of the sheet metal being formed, a cup shape can be developed. At the start of this process, the metal in the free area between the punch bottom and the flange hold-down is stretched and wrapped over the nose of the punch and the approach radius of the die. During this stretching, strain hardening strengthens the metal. If it is not capable of such strengthening or if its strength at any location is exceeded at any time during the forming, the bottom of the cup shape will break out. Contributing to this strengthening is a high *r* value, which is a measure of resistance to through-thickness changes in the sheet metal. If the metal has a high resistance to thinning and thickening, the bottom radius and the upper sidewall areas remain close to their original thickness, and the radial and circumferential strains obtainable in the drawn-in flange are increased to accommodate the deep-drawing process.

After the bottom has been formed, the clearance between the punch and die is such that the metal in the cup side is free to move without excessive rubbing on the die walls. It has been found that slight roughening of the punch radius and minimizing the lubrication of this area contribute to deeper drawability; however, the die opening should be smooth and well lubricated with a suitable drawing compound.

Fundamentals of Drawing

A flat blank is formed into a cup by forcing a punch against the center portion of a blank that rests on the die ring. The progressive stages of metal flow in drawing a cup from a flat blank are shown schematically in Fig. 2. During the first stage, the punch contacts the blank (Fig. 2a), and metal section 1 is bent and wrapped around the punch nose (Fig. 2b). Simultaneously and in sequence, the outer sections of the blank (2 and 3, Fig. 2) move radially toward the center of the blank until the remainder of the blank has bent around the punch nose and a straight-wall cup is formed (Fig. 2c and d). During drawing, the center of the blank (punch area, Fig. 2a) is

essentially unchanged as it forms the bottom of the drawn cup. The areas that become the sidewall of the cup (1, 2, and 3, Fig. 2) change from the shape of annular segments to longer parallel-side cylindrical elements as they are drawn over the die radius. Metal flow can occur until all the metal has been drawn over the die radius, or a flange can be retained.

A blankholder is used in a draw die to prevent the formation of wrinkles as compressive action rearranges the metal from flange to sidewall. Wrinkling starts because of some lack of uniformity in the movement or because of the resistance to movement in the cross section of the metal. This happens because metal under compression has a great resistance to flow into the die cavity. A blankholder force sufficient to resist or compensate for this nonuniform movement prevents wrinkling. Once a wrinkle starts, the blankholder is raised from the surface of the metal so that other wrinkles can form easily. The force needed to hold the blank flat during drawing of cylindrical shells varies from practically zero for relatively thick blanks to approximately one-third of the drawing load for a blank 0.76 mm (0.030 in.) thick. Thinner blanks often require proportionally greater blankholder force.

Conditions for drawing without a blankholder depend on the ratio of the supported length of the blank to its thickness, the amount of reduction from blank diameter to cup diameter, and the ratio of blank diameter to stock thickness. For thick sheets, the maximum reduction of blank diameter to cup diameter in drawing without a blankholder is approximately 25%. This ratio approaches zero for thin foillike sheet. If a blankholder is employed, the maximum reduction is increased to approximately 50% for metals of maximum drawability and 25 to 30% for metals of marginal drawability in the same equipment. More information on the effect of blank thickness on drawing is in the article "Design for Sheet Forming" in this Volume.

Drawability

In an idealized forming operation, that is, one in which drawing is the only deformation process that occurs, the blankholder force is just sufficient to permit the sheet material to flow radially into the die cavity without wrinkling. Deformation takes place in the flange and over the lip of the die. No deformation occurs over the nose of the punch. The deep-drawing process can be thought of as analogous to wire drawing in that a large cross section is drawn into a smaller cross section of greater length.

The drawability of a metal depends on two factors:

- The ability of the material in the flange region to flow easily in the plane of the sheet under shear
- The ability of the sidewall material to resist deformation in the thickness direction

The punch prevents sidewall material from changing dimensions in the circumferential direction; therefore, the only way the sidewall material can flow is by elongation and thinning. Thus, the ability of the sidewall material to withstand the load imposed by drawing down the flange is determined by its resistance to thinning, and high flow strength in the thickness direction of the sheet is desirable.

Taking both of these factors into account, it is desirable in drawing operations to maximize material flow in the plane of the sheet and to maximize resistance to material flow in a direction perpendicular to the plane of the sheet. Low flow strength in the plane of the sheet is of little value if the work material also has low flow strength in the thickness direction.

The flow strength of sheet metal in the thickness direction is difficult to measure, but the plastic-strain ratio *r* compares strengths in the plane and thickness directions by determining true strains in these directions in a tension test. For a given metal strained in a particular direction, *r* is a constant expressed as:

$$r = \varepsilon_w / \varepsilon_t \qquad \text{(Eq 1)}$$

where ε_w is the true strain in the width direction, and ε_t is the true strain in the thickness direction. Sheet metal is anisotropic; that is, the properties of the sheet are different in different directions. It is, therefore, necessary to use the average of the strain ratios measured parallel to, transverse to, and 45° to the rolling direction of the sheet to obtain an average strain ratio $\bar{r}$, which is expressed as:

$$\bar{r} = r_L + 2r_{45} + r_T / 4 \qquad \text{(Eq 2)}$$

where r_L is the strain ratio in the longitudinal direction, r_{45} is the strain ratio measured at 45° to the rolling direction, and r_T is the strain ratio in the transverse direction.

If flow strength is equal in the plane and thickness directions of the sheet, $\bar{r} = 1$. If strength in the thickness direction is greater than average strength in the directions in the plane of the sheet, $\bar{r} > 1$. In this latter case, the material resists uniform thinning. Generally, the higher

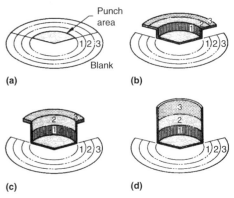

Fig. 2 Progression of metal flow in drawing a cup from a flat blank

the $\bar{r}$ value, the deeper the draw that can be achieved (Fig. 3).

Because the average strain ratio $\bar{r}$ gives the ratio of average flow strength in the plane of the sheet to average flow strength normal to the plane of the sheet, it is a measure of normal anisotropy. Variations of flow strength in the plane of the sheet are termed planar anisotropy. The variation in strain ratio in different directions in the plane of the sheet, $\bar{r} > 1$, is a measure of planar anisotropy, and $\bar{r}$ can be expressed as:

$$\bar{r} = r_L + 2r_{45} = r_T/4 \qquad \text{(Eq 3)}$$

where r_L is the variation in strain ratio, and the other terms are as defined in Eq 2.

A completely isotropic material would have $\bar{r} = 1$ and $\Delta r = 0$. These two parameters are convenient measures of plastic anisotropy in sheet materials. More information on formability is available in the article "Formability Testing of Sheet Metals" in this Volume.

Drawing Ratios. Drawability can also be expressed in terms of a limiting draw ratio or percentage of reduction based on results of Swift cup testing (see the article "Formability Testing of Sheet Metals" in this Volume). The limiting draw ratio is the ratio of the diameter D of the largest blank that can be successfully drawn to the diameter of the punch d:

$$\text{LDR} = \frac{D}{d} \qquad \text{(Eq 4)}$$

Percentage of reduction would then be defined as:

$$\text{Percentage of reduction} = \frac{100(D-d)}{D} \qquad \text{(Eq 5)}$$

Additional information on formability testing and other measures of formability is available in the article "Formability Testing of Sheet Metals" in this Volume.

Defects in Drawing

A number of defects may occur in deep-drawn parts. Figure 4 shows the type of defects that may be found after drawing cylindrical cups. The following is a description of such defects:

- *Earing:* Occurs in deep-drawn parts made from anisotropic materials. Because of planar anisotropy, the sheet metal may be stronger in one direction than in other directions in the plane of the sheet. This causes the formation of ears in the upper edge of a deep-drawn cup even when a circular blank is used. In practice, enough extra metal is left on the drawn cup so that the ears can be trimmed.
- *Wrinkling in the flange:* Wrinkling in deep-drawn parts consists of a series of ridges that form radially in the flange due to compressive forces.
- *Wrinkling in the wall:* Occurs when ridges in the flange are drawn into the vertical wall of the cup

- *Tearing:* Occurs near the base of the drawn cup and results from high stresses in the vertical wall that cause thinning and failure of the metal at that location
- *Surface scratches:* Occur in a drawn part if the punch and die surfaces are not smooth or if lubrication is not enough

Presses

Sheet metal is drawn in either hydraulic or mechanical presses. Double-action presses are required for most deep drawing because a more uniform blankholding force can be maintained for the entire stroke than is possible with a spring-loaded blankholder. Double-action hydraulic presses with a die cushion are often preferred for deep drawing because of their constant drawing speed, stroke adjustment, and uniformity of clamping pressure. Regardless of the source of power for the slides, double-action straight-side presses with die cushions are best for deep drawing. Straight-side presses provide a wide choice of tonnage capacity, bed size, stroke, and shut height.

Factors in Press Selection

Drawing force requirements, die space, and length of stroke are the most important considerations in selecting a press for deep drawing. The condition of the crankshaft, connection bearings, and gibs is also a factor in press selection.

Drawing Force. The required drawing force, as well as its variation along the punch stroke, can be calculated from theoretical equations based on plasticity theory or from empirical equations. The maximum drawing force $F_{d,max}$ required to form a round cup can be expressed by the following empirical relation:

$$F_{d,max} = n\pi dts_u \qquad \text{(Eq 6)}$$

where s_u is the tensile strength of the blank material (in pounds per square inch or megapascals), d is the punch diameter (in inches or millimeters), t is the sheet thickness (in inches or millimeters), and $n = s_D/s_u$, the ratio of drawing stress to tensile strength of the work material. Equation 6 would yield $F_{d,max}$ in either pounds or kilonewtons, depending on the other units used.

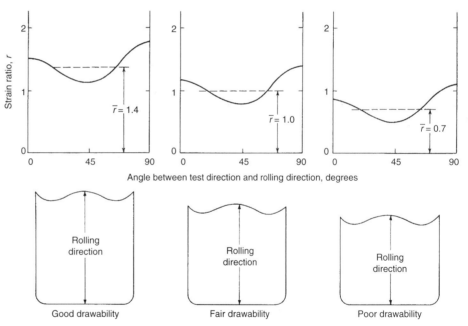

Fig. 3 Variation of strain ratio r with direction in low-carbon steel (top curves) and effect of average strain ratio $\bar{r}$ on drawability of cylindrical cups (bottom). Each cup represents the deepest cup that can be drawn from material with the indicated $\bar{r}$.

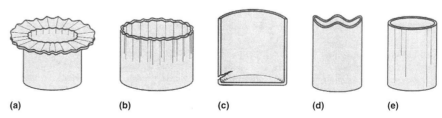

Fig. 4 Defects in deep-drawn cylindrical cups. (a) Flange wrinkling. (b) Wall wrinkling. (c) Tearing. (d) Earing. (e) Surface scratches

The drawing force required to form a round shell can be estimated using Fig. 5. The nomograph shown in Fig. 5 is based on, first, a free draw with sufficient clearance so that there is no ironing and, second, on a maximum reduction of approximately 50% (note also that only English units of measure are used). Figure 5 gives the load required to fracture the cup or the tensile strength of the work metal near the bottom of the shell. An example of its use is the determination of the force required for deep drawing 0.125 in. thick steel stock with a tensile strength of 50,000 psi into a shell 10 in. in diameter:

- Using line 1, connect point 10 on scale 2 to point 0.125 on scale 4.
- Line 1 intersects scale 3 at 4.0, which is the approximate cross-sectional area (in.^{2}) of the shell wall.
- Connect this point using line 2 to point 50,000 on scale 1.
- Project a line to the right to intersect scale 5 at 98 tons, which is the required drawing force.

The force required to draw a rectangular cup can be calculated using Eq 7:

$$F_{d,max} = ts_u \ (2\pi R k_a + L k_b) \qquad \text{(Eq 7)}$$

where R is the corner radius of the cup (in inches), L is the sum of the lengths of straight sections of the sides (in inches), k_a and k_b are constants, and the other quantities are as defined in Eq 6. Values for k_a range from 0.5 for a shallow cup to 2.0 for a cup with a depth five to six times the corner radius; k_b values range from 0.2 (for easy draw radius, ample clearance, and no blankholder force) to a maximum of 1.0 (for metal clamped too tightly to flow).

When blankholder cylinders are mounted on the main slide of the press, the blankholder force must be added to the calculated drawing force. When a die cushion is used to eject workpieces, the main slide works against this force; therefore, such setups require more drawing force than would be calculated using Eq 6 or 7.

In toggle draw presses, the blankholder force is taken on the rocker shaft bearings in the press frame, so that the crankshaft bearings sustain only the drawing load. In other types of presses, both the drawing and blankholding loads are on the crankshaft, and allowances are made when computing press capacity. For round work, the allowance for blankholding should be 30 to 40% of the drawing force. For large rectangular work, the drawing force is relatively lower than that for round work, but the blankholding force may be equal to the drawing force. Where stretching is involved and the blank must be gripped tightly around the edge (and a draw bead is not permissible), the blankholding force may be two or three times the drawing force.

Blank size governs the size of the blankholder surfaces. Some presses with sufficient force cannot be considered for deep drawing, because the bed size and shut height are inadequate. Blank size is determined by calculating the surface area of the part and then converting this area into a flat blank. For a deep-drawn round cup, the surface area of the cup is converted into a flat blank diameter.

Draw Depth. The length of stroke and the force required at the beginning of the working portion of the stroke are both important considerations. Parts that have straight walls can often be drawn through the die cavity and then stripped from the punch and ejected from the bottom of the press. Even under these ideal conditions, the minimum stroke will be equal to the sum of the length of the drawn part, the radius of the draw die, the stock thickness, and the depth of the die to the stripping point, in addition to some clearance for placing the blank in the die.

Workpieces with flanges or tapered walls must be removed from the top of the die. In drawing these workpieces, the minimum press stroke is twice the length of the drawn workpiece, plus clearance for loading the die. In an automatic operation using progressive dies or transfer mechanisms, at least one-half the stroke must be reserved for stock feed, because the tooling must clear the part before feeding begins for the next stroke. For automatic operation, it is common practice to allow a press stroke of four times the length of the drawn workpiece. Therefore, some equipment is not suited to automatic operation, or it is necessary to use manual feed with an automatic unloader, or conversely, because of a shortage of suitable presses.

Slide Velocity. When selecting a press, it is also necessary to check slide velocity through the working portion of the stroke (see the section "Effect of Press Speed" in this article).

Means of Holding the Blank. Double-action presses with a punch slide and a blankholder slide are preferred for deep drawing. Single-action presses with die cushions (pneumatic or hydraulic) can be used but are less suitable for drawing complex parts. Draw beads are incorporated into the blankholder for drawing parts requiring greater restraint of metal flow than can be obtained by using a plain blankholder or for diverting metal flow into or away from specific areas of the part (see the section "Restraint of Metal Flow" in this article).

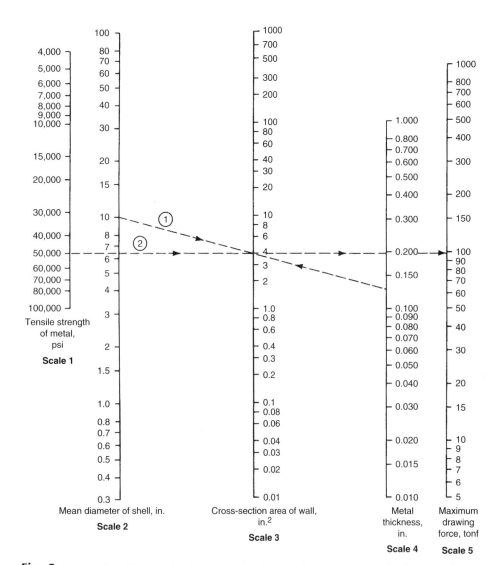

Fig. 5 Nomograph used for estimating drawing force based on several parameters. See text for description of use.

Press Selection versus Availability. The ideal press equipment for a specific job is often not available. This makes it necessary to design tools and to choose product forms of work metal in accordance with available presses and supplementary equipment. For example, if available presses are not adequate for drawing large workpieces, the manufacturing sequence must be completely changed. It may be necessary to draw two sections and weld them together. In addition, operations that could otherwise be combined, such as blanking, piercing, drawing, and trimming, may have to be performed singly in separate presses.

On the other hand, some manufacturers have placed more than one die in a single press because of the availability of a large press and the shortage of smaller presses. This procedure can cause lower production because all blanks must be positioned before the press can be operated. However, storage of partly formed workpieces and additional handling between press operations are eliminated. Where several small dies are used to reduce overall tool cost, there is economic justification for the use of small-capacity presses. If small presses are not available, it is often more economical to use compound dies. This is particularly true if overall part production is likely to exceed original estimates.

The availability of auxiliary equipment may also influence the type of press and tooling used. For example, if equipment is available for handling coils, plans will be made accordingly. However, if coil-handling equipment is not available and straight lengths of sheet or strip are to be processed, a compatible tooling procedure must be used, even though it may not be the most economical procedure.

Dies

Dies used for drawing sheet metal are usually one of the following basic types or some modification of these types:

- Single-action dies
- Double-action dies
- Compound dies
- Progressive dies
- Multiple dies with transfer mechanism

Selection of the die depends largely on part size, severity of draw, and quantity of parts to be produced.

Single-action dies (Fig. 6a) are the simplest of all drawing dies and have only a punch and a die. A nest or locator is provided to position the blank. The drawn part is pushed through the die and is stripped from the punch by the counterbore in the bottom of the die. The rim of the cup expands slightly to make this possible. Single-action dies can be used only when the forming limit permits cupping without the use of a blankholder.

Double-action dies have a blankholder. This permits greater reductions and the drawing of

flanged parts. Figure 6(b) shows a double-action die of the type used in a double-action press. In this design, the die is mounted on the lower shoe, the punch is attached to the inner or punch slide, and the blankholder is attached to the outer slide. The pressure pad is used to hold the blank firmly against the punch nose during the drawing operation and to lift the drawn cup from the die. If a die cushion is not available, springs or air or hydraulic cylinders can be used; however, they are less effective than a die cushion, especially for deep draws.

Figure 6(c) shows an inverted type of double-action die, which is used in single-action presses. In this design, the punch is mounted on the lower shoe; the die on the upper shoe. A die cushion can supply the blankholding force, or springs or air or hydraulic cylinders are incorporated into the die to supply the necessary blankholding force. The drawn cup is removed from the die on the upstroke of the ram, when the pinlike extension of the knockout strikes a stationary knockout bar attached to the press frame.

Compound Dies. When the initial cost is warranted by production demands, it is practical to combine several operations in a single die. Blanking and drawing are two operations commonly placed in compound dies. With compound dies, workpieces can be produced several times as fast as by the simple dies shown in Fig. 6.

Progressive Dies. The initial cost and length of bed needed for progressive dies usually limit

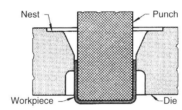

(a) Single-action die

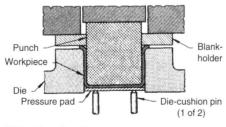

(b) Double-action die

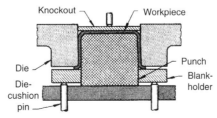

(c) Double-action die, inverted type

Fig. 6 Components of three types of simple dies shown in a setup used for drawing a round cup. See text for discussion.

their application to relatively small workpieces. Figure 7 shows a typical six-station progression for making small shelllike workpieces on a mass-production basis. However, larger parts, such as liners for automobile headlights, have been drawn in progressive dies.

The total number of parts to be produced and the production rate often determine whether or not a progressive die will be used when two or more operations are required. There are, however, some practical considerations that may rule against a progressive die, regardless of quantity:

- The workpiece must remain attached to the scrap skeleton until the final station, without hindering the drawing operations.
- Drawing operations must be completed before the final station is reached.
- In deep drawing, it is sometimes difficult to move the workpiece to the next station.
- If the draw is relatively deep, stripping is often a problem.
- The length of press stroke must be more than twice the depth of draw.

Assuming that a progressive die can be used to make acceptable drawn parts, cost per piece is usually the final consideration. Progressive-die drawing is generally considered to be economical if savings in material and labor can pay for the die in 1 year. Ordinarily, the savings achieved by the use of a progressive die results from decreased labor.

Multiple dies, in conjunction with transfer mechanisms, are often used instead of progressive dies for the mass production of larger parts. Multiple dies and transfer mechanisms are practical for a wider range of workpiece sizes than progressive dies are. Although the eyelet-type transfer method is the most widely used for making parts less than 25 mm (1 in.) in diameter, transfer dies are practical for much larger workpieces. The seven-station operation for making the 165 mm (6.5 in.) outside diameter cylindrical shell shown in Fig. 8 represents a typical sequence for the transfer-die method. The workpiece is mechanically transferred from one die to the next. One advantage of the transfer-die method, as opposed to the progressive-die method, is the greater flexibility permitted in processing procedure, mainly because in transfer dies the workpiece does not remain attached to the scrap skeleton during forming. Because of

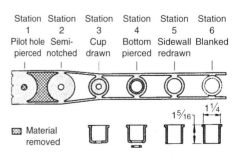

Fig. 7 Production of a small ferrule in a six-station progressive die. Dimensions given in inches

this, precut blanks can be drawn by the transfer method.

Preforms can also be used as blanks. For example, oil pans for automobiles are blanked and partly drawn in a compound die and then finish formed, pierced, and trimmed by the transfer method. Dies for producing a given part usually cost more for the transfer-die method than for a separate-die operation but approximately the same as for a progressive-die operation. The cost of adapting the transfer unit to the part is not included in the die cost. Similarly, the production rate for the transfer method is usually greater than that for a single-die operation but 10 to 25% less than that for drawing in a progressive die. Many parts can be produced equally well by all of these methods. Under these conditions, tool cost, rate of production, and total quantity of parts to be drawn determine the choice of procedure.

Die and Punch Materials. The selection of material for dies and punches for drawing sheet metal depends on work metal composition, workpiece size, severity of the draw, quantity of parts to be drawn, and tolerances and surface finish specified for the drawn workpieces. To meet the wide range of requirements, punch and die materials ranging from polyester, epoxy, phenolic, or nylon resins to highly alloyed tool steels with nitrided surfaces, and even carbide, are used. Detailed information on tool materials is available in the articles "Selection of Materials for Press-Forming Dies" and "Selection of Materials for Deep-Drawing Dies" in this Volume.

Effects of Process Variables in Deep Drawing

The process parameters that affect the success or failure of a deep-drawing operation include punch and die radii, punch-to-die clearance, press speed, lubrication, and type of restraint of metal flow used (if any). Material variables, such as sheet thickness and anisotropy, also affect deep drawing. These are discussed in the section "Effects of Material Variables on Deep Drawing" in this article.

Effect of Punch and Die Radii

As the blank is struck by the punch at the start of drawing, it is wrapped around the punch and die radii; the stress and strain that develop in the workpiece are similar to those developed in bending, with an added stretching component. The bends, once formed, have the radii of the punch and die corners. The bend over the punch is stationary with reference to both punch and shell wall. The bend over the die radius, however, is continuously displaced with reference to both the punch radius and the blank, and it also undergoes a gradual thickening as the shell is drawn. The force required to draw the shell at the

intermediate position has a minimum of three components:

- The force required for bending and unbending the metal flowing from the flange into the sidewall
- The force required for overcoming the frictional resistance of the metal passing under the blankholder and over the die radius
- The force required for circumferential compression and radial stretching of the metal in the flange

Because of the variation in metal volume and in resistance to metal flow, the punch force increases rapidly, passes through a maximum, and gradually decreases to zero as the edges of the flange approach and enter the die opening and pass into the shell wall. With the cup diameter remaining constant, the maximum press load and the length of stroke required to draw the cup depend on the size of the blank. The punch force-stroke relations for drawing blanks of various diameters from brass sheet 1.5 mm (0.060 in.) thick, using a 50 mm (2 in.) diameter punch, are shown in Fig. 9.

Under the conditions shown in Fig. 9, during cupping, the shell bottom is subjected to tensile stress in all directions, while the lower portions

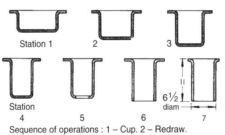

Sequence of operations : 1 – Cup. 2 – Redraw.
3 – Redraw. 4 – Redraw. 5 – Pierce bottom.
6 – Extrude bottom. 7 – Trim flange.

Fig. 8 Seven-station drawing and piercing of a cylindrical part in a multiple die and transfer mechanism. Dimensions given in inches

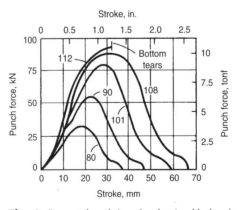

Fig. 9 Force-stroke relations for drawing blanks of various diameters from 1.5 mm (0.060 in.) thick alloy C27400 (yellow brass, 63%) sheet using a 50 mm (2 in.) diameter punch. Numbers indicate blank diameter in millimeters.

of the shell wall, particularly the radiused portion connecting the bottom with the wall, are primarily subjected to longitudinal tension. The stress in the metal being drawn into the shell wall consists of combined compressive and tensile stresses. Separation of the shell bottom from the wall is likely if a reduction is made that requires a force greater than the strength of the shell wall near the bottom (Fig. 9).

The punch and die radii and percentage of reduction determine the load at which the bottom of the shell is torn out. Drawing is promoted by increasing punch and die radii. For a given drawing condition, the punch force needed to move the metal into the die decreases as the die radius increases, as shown in Fig. 10.

The reduction of drawing force in a double-action die by modification of the effective die radius can be accomplished in two convenient ways, as shown in Fig. 11. In the conical lead-in die (Fig. 11a), the cutout is effective in reducing

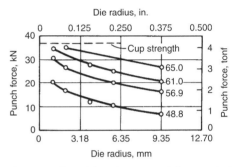

Fig. 10 Effect of die radius on punch force required for cupping various diameters of 1 mm (0.040 in.) thick alloy C27400 (yellow brass, 63%) blanks using a 30.5 mm (1.2 in.) diameter punch with a nose radius of 0.61 mm (0.024 in.). Numbers indicate blank diameter in millimeters.

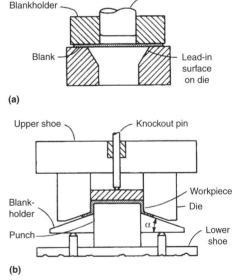

Fig. 11 Two ways of reducing the required drawing forces. (a) Conical lead-in die. (b) Conical blankholder. See text for details.

frictional loads by removal of the portions of the die surface that are usually heavily loaded and increase friction. In Fig. 11(b), the sheet metal is formed into a conical shape before appreciable drawing begins. This has the effect of reducing the area of contact over the die radius by an amount proportional to α/90° (where α is the angle to declination of the hold-down surface to the horizontal, as shown in Fig. 11b).

If the punch nose radius can be increased from one to five times metal thickness, the load in the sidewall of the shell will decrease so that the reduction in blank diameter will increase from 35% to approximately 50% (for steel). The shell can therefore be drawn deeper before the sidewall tears. If the shell bottom radius is less than four times the sheet thickness, it is usually desirable to form it with a larger-radius punch and then to restrike to develop the specified radius. This will minimize bottom failures. However, the bottom corner radius usually cannot be increased beyond ten times the sheet thickness without the likelihood of wrinkling. The metal in dome-shaped parts is likely to pucker in the unconfined area between the punch nose and die radius. High blankholding forces or draw beads are often used to induce combined stretching and drawing of the metal when forming dome shapes.

The deep drawing of stainless steel or high-strength alloy boxes with sides longer than 50 times stock thickness may result in a stability problem called oil canning. The deflection of the sides by snap action can be eliminated by drawing the part in two operations with slightly different punches and an intermediate anneal. The first-draw punch will have a larger nose radius than the second; therefore, in the second drawing operation, the metal can be stretched to eliminate the oil-canning effect. Stretching of the metal in parts with long sidewalls can be improved by gradually increasing the punch nose radius from the corner toward the center. A constant nose radius is used on the second-draw punch.

Effect of Punch-to-Die Clearance

The selection of punch-to-die clearance depends on the requirements of the drawn part and on the work metal. Because there is a decrease and then a gradual increase in the thickness of the metal as it is drawn over the die radius, clearance per side of 7 to 15% greater than stock thickness (1.07 to 1.15t) helps prevent burnishing of the sidewall and punching out of the cup bottom.

The drawing force is minimal when the clearance per side is 15 to 20% greater than stock thickness (1.15 to 1.20t) and the cupped portions of the part are not in contact with the walls of the punch and die. The force increases as the clearance decreases, and a secondary peak occurs on the force-stroke curve where the metal thickness is slightly greater than the clearance and where ironing starts.

Redrawing operations require greater clearance, in relation to blank thickness, than the first draw in order to compensate for the increase in metal thickness during cupping. A sizing redraw is used where the diameter or wall thickness is important or where it is necessary to improve surface finish to reduce finishing costs. The clearance used is less than that for the first draw.

Table 1 lists clearances for cupping, redrawing, and sizing draws of cylindrical parts from metal of various thicknesses. As the tensile strength of the stock decreases, the clearance must be increased.

Clearance between the punch and die for a rectangular shell, at the sidewalls and ends, is approximately the same as, or slightly less than, that for a circular shell. Clearance at the corners may be as much as 50% greater than stock thickness to avoid ironing in these areas and to increase drawability.

Restraint of Metal Flow

Even in the simplest drawing operation, as shown in Fig. 6(a), the thickness of the work metal and the die radius offer some restraint to the flow of metal into the die. For drawing all but the simplest of shapes, some added restraint is generally required in order to control the flow of metal. This additional restraint is usually obtained by the use of a blankholder, as illustrated in Fig. 6(b) and (c). The purpose of the blankholder is to suppress wrinkling and puckering and to control the flow of the work metal into the die.

Drawing without a Blankholder. A blank is not susceptible to wrinkling, and a blankholder need not be used, if the ratio of supported length to sheet thickness is within certain limits. In Fig. 12, the supported length (l) is the length from the edge of the blank to the die cavity (point of tangency). The sheet thickness is denoted as t. The l/t ratio is influenced little by other geometrical conditions, and it differs little for the various metals commonly drawn. When the l/t ratio does not exceed 3 to 1, a cup can be drawn from annealed brass, aluminum (to half-hard), and low-carbon steel without a blankholder. For

slightly harder work metals, such as hard copper or half-hard brass, this ratio should not exceed 2.5 to 1.

An elliptical or conical die opening, such as that shown in Fig. 11(a), can be used where the die radius required to draw the part reduces the length of the blank-supporting surface to less than three times stock thickness. The distance between the die opening and the punch should not exceed ten times stock thickness.

A 30° elliptical radius derived from a circle created by a given draw radius increases the strain on the metal being drawn by 4.2%, but it decreases the metal out of control by 47% of the length of the original draw radius. This shape has been helpful in the drawing of tapered shells from a flat blank. For these draws, it is desirable to increase the strain slightly to prevent puckers and to reduce the metal out of control for the same reason.

A 45° elliptical radius derived as previously reduces the strain on the metal being drawn by 1.03% and reduces the metal out of control by 33% of the length of the original draw radius. The 45° ellipse is useful only when a large radius will draw the part, but produces wrinkles. A smaller radius will not permit the draw.

A 60° elliptical radius does not measurably reduce drawing strain and accounts for only a 9% reduction of metal out of control. Its use on draw dies is not economically feasible when the small gains derived are considered in relation to the cost of producing the contour. The drawing of thick metal without a blankholder is frequently done when the blank diameter is no greater than 20 times stock thickness.

Blankholders. A blankholder, or binder, is used to control metal flow into the die cavity and to prevent wrinkles from forming in the flange of a deep-drawn part. The formation of wrinkles interferes with, or prevents, the compressive action that rearranges the metal from flange to sidewall. Much greater reductions are possible when a blankholder is used.

Blankholders can be used in double- and single-action presses. In a double-action press, the blankholder advances slightly ahead of the punch and dwells at the bottom of its stroke throughout the drawing phase of the punch cycle. The blankholder dwell usually extends to a point on the punch upstroke at which positive stripping of the shell is ensured. By using a die cushion and an inverted die, similar action can be obtained in a single-action press. A die cushion in a double-action press supports the blank and holds it

Table 1   Punch-to-die clearance for drawing operations

Metal thickness, t		Clearance-to-metal-thickness relationship for:		
mm	in.	Cupping	Redrawing	Sizing draws
Up to 0.38	Up to 0.015	1.07–1.09t	1.08–1.10t	1.04–1.05t
0.41–1.27	0.016–0.050	1.08–1.10t	1.09–1.12t	1.05–1.06t
1.29–3.18	0.051–0.125	1.10–1.12t	1.12–1.14t	1.07–1.09t
3.2 and up	0.126 and up	1.12–1.14t	1.15–1.20t	1.08–1.10t

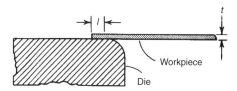

Fig. 12 The ratio of supported length l to sheet thickness t determines whether or not a blankholder is required for deep drawing.

against the punch during the drawing operation; it then lifts the finished part out of the die.

A blankholder must allow the work metal to thicken as the edge of the blank moves inward toward the working edge of the die. The amount of thickening is expressed by:

$$\frac{t_1}{t} = \sqrt{\frac{D}{D_1}} \qquad \text{(Eq 8)}$$

where t is the blank thickness, t_1 is the thickness of the flange at any instant during the drawing operation, D is the blank diameter, and D_1 is the diameter of the flange at any instant during the drawing operation (or the mean diameter of the workpiece without the flange). As the metal flows, paths of least resistance are taken; therefore, the actual value of t_1 will be less than that calculated from the formula.

Types of Blankholders. Blankholders can be flat, or they may have a combination of fixed drawbeads and constant blankholder displacements. In flat binders, restraining forces in the drawn sheet are provided by friction between the binder plates and the sheet metal. In fixed drawbead binders, restraining forces result from a combination of plastic deformation and friction. The simplest type of blankholder is fixed to the die block and has a flat hold-down surface, as shown in Fig. 13(a). A disadvantage of this type of blankholder is that maintenance of the optimal gap between the die surface and the flat hold-down surface requires careful adjustment. As shown in Fig. 13, the blankholder does not quite contact the work metal as drawing begins; restraint begins and increases as the flange portion thickens. A gap that is either too small or too large increases force and reduces drawability. For optimal results, the gap should be slightly smaller than the flange thickness, allowing 50 to 75% of the final thickening before the work metal contacts the blankholder.

The flat controlled-pressure blankholder shown in Fig. 13(b) is generally preferred in production operations because it can be adjusted to a predetermined and closely controlled value by hydraulic or pneumatic pressure. Springs, unless extremely long, are not suitable for supplying pressure to a blankholder during deep drawing, because the force exerted by a spring increases rapidly as it is compressed. The force on hydraulic or pneumatic die cushions will increase approximately 20% when compressed the full stroke length. Some hydraulic systems have pressure control valves that supply a more nearly constant pressure during the entire stroke.

The fixed-type blankholder (Fig. 13a) draws a cup without a flange and ejects it through the bottom of the die. The blankholder shown in Fig. 6(b) and (c) and in Fig. 13(b) can be used for drawing a cup with or without a flange. Cups without a flange can be pushed through the die if a pressure pad is not needed to support the blank.

Flexible Blankholder. Conventional stamping with rigid binders does not allow local control of sheet metal flow into the die cavity, as shown in Fig. 14. This produces a forming system with limited application and leads to an elaborate and costly die tryout process to produce a product. New tooling concepts and an advanced binder control system were developed and implemented in Germany. The new development produces flexible binders with individually controlled hydraulic cylinders, as shown in Fig. 15. Individual control of local binder areas allows the right amount of metal to flow into the die cavity. Unlike rigid binders, a flexible binder produces the right amount of elastic deformation in the binder area, and this produces the necessary friction condition for metal flow.

The cone-shaped segments localize the applied pressure in a specific area without influencing the pressure in adjacent areas. Trials on an automotive front fender were conducted using a ten-point cushion system in a hydraulic press. Results from the trials showed that flexible binder technology improves the quality of fenders by allowing the manipulation of pressure settings in binder areas that control metal flow to the defect location. The effect of binder load control on wrinkling behavior at the nose of the fender is shown in Fig. 16 and 17. Figure 16 shows wrinkles obtained when nonoptimized pin forces

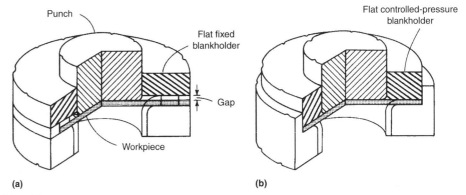

Fig. 13 Setups showing the use of two types of blankholders. See text for details.

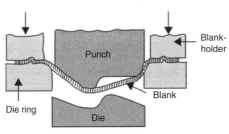

Fig. 14 Schematic of a rigid sheet metal forming system

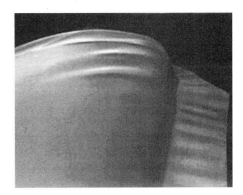

Fig. 16 Wrinkles obtained using nonoptimized pin forces

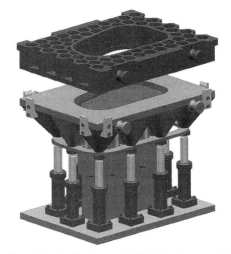

Fig. 15 Flexible binder with individually controlled hydraulic cylinders

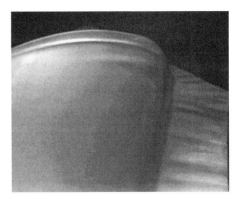

Fig. 17 Reduction in wrinkles after optimizing pin forces

were used. Figure 17 shows that less wrinkles were obtained when the pin forces in the binder area are optimized. Flexible binder control technology is currently being used to successfully deep draw stainless steel double sinks.

Blankholder Force. Compressive forces on the metal in the area beyond the edge of the die cause the work metal to buckle. If this buckled or wrinkled metal is pulled into the die during the drawing operation, it will increase the strain in the area of the punch nose to the point at which the work metal would fracture soon after the beginning of the draw. Blankholder force is used to prevent this buckling and subsequent failure. The amount of blankholder force required is usually approximately one-third that required for drawing. Thickness of the work metal must also be considered when simple shapes are being drawn; the thinner the work metal, the more blankholder force that is required.

There are no absolute rules for calculating blankholder force for a given drawing operation; most blankholder force values are found empirically. Blankholder force should be just sufficient to prevent wrinkling, and it depends on draw reduction, work metal thickness and properties, the type of lubrication used, and other factors. For a particular application, blankholder force is best determined experimentally.

Draw beads help prevent wrinkles and control the flow of metal in the drawing process. The use of draw beads increases the cost of tools, product development, and tool maintenance. However, they are often the only means of controlling metal flow in the drawing of odd shapes. Draw beads are ordinarily used for the first draw only; therefore, production rates are the same as when conventional blankholders are used. For low production, draw beads are often made by laying a weld bead on the die after the optimal location has been determined.

Restraint of the metal flow, to the extreme of locking the flange of the blank to prevent motion, is needed for some draws. A deep shell with sloping walls can be made by drawing, followed by several redraws. This results in a stepped workpiece. The final sizing draw is a stretching operation that is done with the flange secured by a locking bead in the blankholder. This kind of blankholder is also used in making shallow drawn panels. Additional information on the design and use of draw beads is available in the article "Forming of Carbon Steels" in this Volume.

Effect of Press Speed

Drawing speed is usually expressed in meters per minute (m/min) or linear feet per minute (ft/min). Under ideal conditions, press speeds as high as 23 m/min (75 ft/min) are used for the deep drawing of low-carbon steel. However, 6 to 17 m/min (20 to 55 ft/min) is the usual range—up to 17 m/min (55 ft/min) for single-action presses and 11 to 15 m/min (35 to 50 ft/min)

for double-action presses. Ideal conditions include:

- Use of a drawing-quality work metal
- Symmetrical workpieces of relatively mild severity
- Adequate lubrication
- Precision carbide tools
- Carefully controlled blankholding pressure
- Presses that are maintained to a high level of accuracy

When one or more of the aforementioned conditions is less than ideal, some reduction in press speed is required. If all, or nearly all, are substantially less than ideal, press speed may have to be reduced to 6 m/min (20 ft/min). When the operation includes ironing, the drawing speed is usually reduced to approximately 7.5 m/min (25 ft/min) regardless of other factors.

The punch speed in hydraulic presses is relatively constant throughout the stroke. In mechanical presses, punch speed is that at midstroke, because the velocity changes in a characteristic manner throughout the drawing stroke from maximum velocity to zero. The only adjustment in speed that can be made is to decrease flywheel speed or to use a press with a shorter stroke that operates at the same number of strokes per minute. This proportionately decreases maximum punch speed. Speed is of greater significance in drawing stainless steels and heat-resistant alloys than in drawing softer, more ductile metals. Excessive press speeds have caused cracking and excessive wall thinning in drawing these stronger, less ductile metals. At high speeds, the metal thins because it cannot react to the impact speed of the punch. Reducing the speed reduces the stretching and gives the metal enough time to flow plastically. Nominal speeds for drawing various metals are given in Table 2.

Effect of Lubrication

When two metals are in sliding contact under pressure, as with the dies and the work metal in drawing, galling (pressure welding) of the tools and the work metal is likely. When extreme galling occurs, drawing force increases and becomes unevenly distributed, causing fracture of the workpiece.

The likelihood of pressure welding depends on the amount of force and the work metal composition. Some work metals are more "sticky" than others. For example, austenitic stainless steel is more likely to adhere to steel tools than low-carbon steel is.

Lubricants are used in most drawing operations. They range from ordinary machine oil to pigmented compounds.

Selection of lubricant is primarily based on the ability to prevent galling, wrinkling, or tearing during deep drawing. It is also influenced by ease of application and removal, corrosivity, and other factors, as described in the article

"Selection and Use of Lubricants in Forming of Sheet Metal" in this Volume.

If a lubricant cannot be applied uniformly by ordinary shop methods, its purpose is defeated, regardless of its ability to prevent pressure welding. In general, as the effectiveness of a lubricant increases, the difficulty of removing it also increases. For example, grease or oil can be easily removed, but special procedures (frequently including some hand scrubbing) are required for removing lubricants that contain zinc oxide, lithopone, white lead, molybdenum disulfide, or graphite.

A lubricant is sometimes too corrosive for use on certain metals. For example, copper alloys are susceptible to staining by lubricants that contain large amounts of sulfur or chlorine compounds. Lubricants containing lead or zinc compounds are not recommended for drawing stainless steel or heat-resistant alloys, because the compounds, if not thoroughly removed, can cause intergranular attack when the workpieces are heat treated or placed in high-temperature service. Suitable safety precautions are necessary with toxic or flammable lubricants.

Some metals, such as magnesium and titanium, are drawn at elevated temperature, which complicates selection of the lubricant. Most oil-base and soap-base lubricants can be successfully used to 120 °C (250 °F), but above this temperature, the choice narrows rapidly. Some special soap-base lubricants can be used on work metals to 230 °C (450 °F). Molybdenum disulfide and graphite can be used at higher temperatures.

Any lubricant must remain stable, without becoming rancid, when stored for a period of several months at various temperatures. The cost of application and removal of the lubricant, as well as its initial cost, must be considered because all of these items can add substantially to the cost of the drawn workpieces.

In some plants, when a new application is started, a heavily pigmented drawing lubricant is used, regardless of the difficulty of applying and removing it. Lubricant is then downgraded as much as possible to simplify the operation and to reduce costs. In other plants, the reverse of this practice is used; that is, a simple lubricant, such as machine oil, is used at first, and lubricant is then upgraded when necessary.

The difficulty of removing drawing lubricants is an important consideration in production operations. In a number of applications, changes in drawing techniques (such as increasing the

Table 2 Typical drawing speeds for various materials

Material	Drawing speed	
	m/min	ft/min
Aluminum	45.7–53.3	150–175
Brass	53.3–61	175–200
Copper	38.1–45.7	125–150
Steel	5.5–15.2	18–50
Stainless steel	9.1–12.2	30–40
Zinc	38.1–45.7	125–150

number of draws) or in workpiece design (larger radii, for example) have been made solely to permit the use of an easier-to-remove drawing lubricant. Methods of removing lubricants are discussed in the section "Cleaning of Workpieces" in this article.

Typical lubricants used in drawing steel are given in Table 3 according to severity of draw or the percentage of reduction from blank to cup diameter. Zinc phosphate conversion coating of the steel to be drawn is helpful for any drawing operation, and the importance of phosphate coating increases as the severity of the draw increases. Methods of application and other details on the use of phosphate conversion coatings are discussed in the article "Phosphate Coatings" in *Surface Engineering,* Volume 5, *ASM Handbook,* 1994.

Materials for Deep Drawing

Sheet steels and other sheet metals with higher strengths and better formability have recently become available. Developments such as vacuum processing and inclusion shape control have been especially beneficial in increasing the drawability of steels. Other metals and alloys that can be deep drawn include aluminum and aluminum alloys, copper and alloys, some stainless steels, and titanium.

Low-carbon sheet steels are the materials that are most commonly deep drawn and are commonly used, for example, in the automotive industry. Materials such as 1006 and 1008 steel have typical yield strengths in the range of 170 to 240 MPa (25 to 35 ksi) and elongations of 35 to 45% in 50 mm (2 in.). These materials have excellent formability and are available cold or hot finished in various quality levels and a wide range of thicknesses. Table 4 lists mechanical properties of the various qualities of carbon steel sheet.

Other low-carbon steels that are commonly deep drawn are grades 1010 and 1012. These materials are slightly stronger than 1006 and 1008 and are slightly less formable. They are often specified when drawing is not severe and strength of the finished part is of some concern. Grain size affects the drawability of these materials, and it may affect the selection of a grade. Grain sizes of ASTM 5 or coarser may result in excessive surface roughness as well as reduced drawability.

Surface finish also influences drawability. The dull finish normally supplied on drawing steels is designed to hold lubricants and to improve drawability. Brighter finishes may be required if, for example, parts are to be electroplated.

Effects of Material Variables on Deep Drawing

Anisotropy. As mentioned earlier in this article, there are two types of anisotropy that must be considered: planar anisotropy, in which properties vary in the plane of the sheet, and normal anisotropy, in which the properties of the material in the thickness direction differ from those in the plane of the sheet.

Planar anisotropy (variations in normal anisotropy in the plane of the sheet) causes undesirable earing of the work material during drawing.

Table 3 Lubricants commonly used for the drawing of low-carbon steel
Severity is indicated by the percentage of reduction in diameter in drawing a cylindrical shell.

Type or composition of lubricant	Water-base cleaners	Degreasers or solvents	Protection against rusting
Water-base lubricants			
Low severity (10% or less)			
Water emulsion of 5–20% general-purpose soluble oil or wax	Very good	Good	Fair
Moderate severity (11–20%)			
Water solution of 5–20% soap	Very good	Very poor	Fair
Water emulsion of heavy-duty soluble oil (contains sulfurized or chlorinated additives)	Very good	Good	Fair
High severity (21–40%)			
Soap-fat paste, diluted with water (may contain wax)	Fair	Poor	Fair
Water emulsion of heavy-duty soluble oil (contains a high concentration of sulfurized or chlorinated additives)	Very good	Good	Fair to poor
Maximum severity (>40%)			
Pigmented soap-fat paste, diluted with water	Poor	Very poor	Good
Dry soap or wax (applied from water solution or dispersion); may contain soluble filler such as borax	Good	Very poor	Good
Oil-base lubricants			
Low severity (10% or less)			
Mill oil, residual	Good	Very good	Fair
Mineral oil	Good	Very good	Fair
Vanishing oil	Removal not required		...
Moderate severity (11–20%)			
Mineral oil plus 10–30% fatty oil	Good	Very good	Fair
Mineral oil plus 2–20% sulfurized or chlorinated oil (extreme-pressure oil)	Good to fair	Good	Fair to poor
High severity (21–40%)			
Fatty oil	Fair	Fair	Fair
Mineral oil plus 5–50% of:			
(a) Nonemulsifiable chlorinated oil	Poor	Good	Very poor
(b) Emulsifiable chlorinated oil	Good	Good	Very poor
Concentrated phosphated oil	Fair	Fair	Fair
Maximum severity (>40%)			
Blend of pigmented soap-fat paste with mineral oil	Poor	Poor	Fair
Concentrated sulfochlorinated oil (may contain some fatty oil):			
(a) Nonemulsifiable	Very poor	Fair	Poor
(b) Emulsifiable	Good	Fair	Poor
Concentrated chlorinated oil:			
(a) Nonemulsifiable	Very poor	Fair	Very poor
(b) Emulsifiable	Good	Fair	Very poor

Table 4 Typical mechanical properties of low-carbon sheet steels

Quality level	Tensile strength MPa	Tensile strength ksi	Yield strength MPa	Yield strength ksi	Elongation, % in 50 mm (2 in.)	Plastic-strain ratio, r	Strain-hardening exponent, n	Hardness, HRB
Hot rolled								
Commercial quality	358	52	234	34	35	1.0	0.18	58
Drawing quality	345	50	220	32	39	1.0	0.19	52
Drawing quality, aluminum killed	358	52	234	34	38	1.0	0.19	54
Cold rolled, box annealed								
Commercial quality	331	48	234	34	36	1.2	0.20	50
Drawing quality	317	46	207	30	40	1.2	0.21	42
Drawing quality, aluminum killed	303	44	193	28	42	1.5	0.22	42
Interstitial-free	310	45	179	26	45	2.0	0.23	44

Between the ears of the cup are valleys in which the material has thickened under compressive hoop stress rather than elongating under radial tensile stress. This thicker metal sometimes forces the die open against the blankholder pressure, allowing the metal in the relatively thin areas near the ears to wrinkle. Die design, draw reduction, and type of lubricant used all affect earing (see the section "Effects of Process Variables in Deep Drawing" in this article).

Sheet Thickness. In deep drawing, the pressure on the dies increases proportionally to the square of sheet thickness. The pressure involved is concentrated on the draw radius, and increasing sheet thickness will localize wear in this area without similar effect on other surfaces of the die. Thick stock has fewer tendencies to wrinkle than thin stock. As a result, blankholder pressures used for the drawing of thick sheet may be no greater, and may even be less, than those used for thinner blanks.

Redrawing Operations

If the shape change required by the part design is too severe (the drawing ratio is too high), a single drawing operation may not produce complete forming of the part, and more than one drawing step may be required.

Direct Redrawing. In direct redrawing in a single-action die, the drawn cup is slipped over the punch and is loaded in the die, as shown in Fig. 18. At first, the bottom of the cup is wrapped around the punch nose without reducing the diameter of the cylindrical section. The sidewall section then enters the die and is gradually reduced to its final diameter. Metal flow takes place as the cup is drawn into the die so that the wall of the redrawn shell is parallel to, and deeper than, the wall of the cup at the start of the redraw. At the beginning of redrawing, the cup must be supported and guided by a recess in the die or by a blankholder to prevent it from tipping, because tipping would result in an uneven shell.

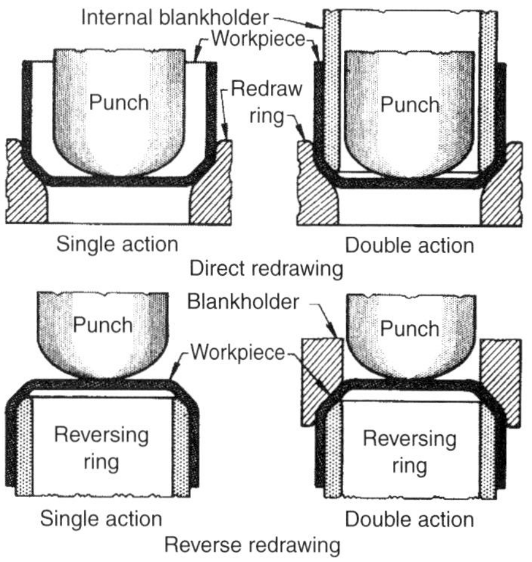

In a single-action redraw, the metal must be thick enough to withstand the compressive forces set up in reducing the cup diameter without wrinkling. Wrinkling can be prevented by the use of an internal blankholder and a double-action press (upper right, Fig. 18), which usually permits a shell to be formed in fewer operations than by single-action drawing without the use of a blankholder.

Internal blankholders (Fig. 19) are slip fitted into drawn shells to provide support and to prevent wrinkling during direct redrawing. The blankholder presses on the drawn shell at the working edge of the die before the punch contacts the bottom of the shell and begins the redraw. It dwells against the shell as the metal is drawn into the die by the punch, preventing wrinkles.

The bottom of the cup to be redrawn can be tapered (Fig. 19a) or radiused (Fig. 19b), with the tip of the blankholder and the mouth of the die designed accordingly. An angle of 30° is used for metal thinner than 0.8 mm (1/32 in.), and 45° is used for thicker work metal. A modification of the aforementioned is a blankholder fitted against an S-curve die (Fig. 19b). The main disadvantage of an S-curve die is that it is more expensive to make and maintain. Near the bottom of a redrawn shell, there is usually a narrow ring, caused by the bottom radius of the preceding shell, which is thinner and harder than the adjacent metal. Redrawing may be required for reasons other than the severity of the drawn shape, for example, to prevent thinning and bulging.

Redrawing can also be done in a progressive die while the part is still attached to the strip. Where space permits the extra stations, the amount of work done in each station will be less than that done in a single die. This reduces the severity of the draw and promotes high-speed operation.

Reverse Redrawing. In reverse redrawing, the cupped workpiece is placed over a reversing ring and redrawn in the direction opposite to that used for drawing the initial cup. As shown in the two lower views in Fig. 18, reverse redrawing can be done with or without a blankholder. The blankholder serves the same purposes as in direct redrawing.

The advantages of reverse redrawing as compared with direct redrawing include:

- Drawing and redrawing can be accomplished in one stroke of a triple-action hydraulic press, or of a double-action mechanical press with a die cushion, which can eliminate the need for a second press.
- Greater reductions per redraw are possible with reverse redrawing.
- One or more intermediate annealing operations can often be eliminated by using the reverse technique.
- Better distribution of metal can be obtained in a complex shape.

In borderline applications, annealing is required between redraws in direct redrawing but is not needed in reverse redrawing.

The disadvantages of reverse redrawing are:

- The technique is not practical for work metal thicker than 6.4 mm (1/4 in.).
- Reverse redrawing requires a longer stroke than direct redrawing.

Usually, metals that can be direct redrawn can be reverse redrawn. All of the carbon and low-alloy steels, austenitic and ferritic stainless steels, aluminum alloys, and copper alloys can be reverse redrawn.

Reverse redrawing requires more closely controlled processing than direct redrawing does. This control must begin with the blanks, which should be free from nicks and scratches, especially at the edges.

The restraint in reverse redrawing must be uniform and low. For low friction, polished dies and effective lubrication of the work are needed. Friction is also affected by hold-down pressure and by the shape of the reversing ring. Radii of tools should be as large as practical, ten times the thickness of the work metal if possible. Reverse redrawing can be done in a progressive die as well as in single-stage dies if the operations are divided to distribute the work and to reduce the severity of each stage.

Tooling for Redrawing. Tooling for redrawing depends mainly on the number of parts to be redrawn and on available equipment. In continuous high production, a complete die is used for each redraw; the workpieces are conveyed from press to press until completed. In low or medium production, it is common practice to use a die with replaceable draw rings and punches. A die of this type used for three redrawing operations is shown in Fig. 20; the three redraws were made by changing to successively smaller draw rings and punches. The cup was drawn in a compound blank-and-draw die from a blank 1.7 mm (0.067 in.) thick and 170 mm (6¾ in.) in diameter.

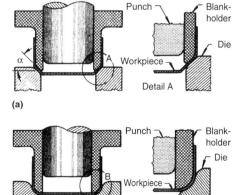

Fig. 18 Direct and reverse redrawing in single-action and double-action dies

Fig. 19 Setups using internal blankholders for restraint of work metal in redrawing shells. See text for details.

Ironing

Ironing is an operation used to increase the length of a tube or cup by reducing wall thickness and outside diameter, while the inner diameter remains unchanged. Wall thickness is reduced by pulling tubes or shells through tight dies. The process of ironing is similar to tube drawing with a moving mandrel, whereby drawing is carried out by using several drawing dies located in tandem. In a typical application, a relatively thick-wall cup is first produced by extrusion or deep drawing. The wall thickness of this cup is then reduced by tandem ironing with a cylindrical punch, while the internal diameter remains unchanged. Hot and cold ironing produce parts with good dimensional accuracy while maintaining or improving concentricity. A very common application of ironing is the production of beverage cans by tandem drawing of steel or aluminum.

Ironing is done to:

- Obtain a wall that is thin compared with the shell bottom
- Obtain a uniform wall
- Obtain a tapered wall (as in cartridge cases), or merely to correct the natural wall thickening that occurs toward the top edge of a drawn shell

The theoretical maximum reduction in wall thickness per operation due to ironing is approximately 50%. In such a case, the cross-sectional area of the (unstrained) metal before ironing is approximately twice the cross-sectional area after ironing. Therefore, the area that is being worked in compression and that must yield is approximately equal to the strain-hardened area that is in tension and that must not yield. This would indicate that the practical limit should apparently be kept below 50%, although slightly higher reductions may be possible.

The chart in Fig. 21 (Ref 1) is arranged to show approximately the natural change in thickness accompanying a change in diameter. The results apply to the upper edges of drawn shells, because the wall thickness tapers from a maximum at that point to a minimum at the bottom corner, where it may be as much as 10 or 15% less than the original metal thickness. Even at the top edge, the metal thickness is likely to be a little less than the theoretical thickness given by the chart because of the thinning effect of bending over the drawing edge. A sharper radius or a deeper draw increases this thinning effect.

An example is indicated by dashed lines in Fig. 21. It is noted that this is based on a draw with sufficient clearance between the punch and die so that there will be no ironing. If the clearance between punch and die is made equal to the metal thickness (0.0625 in., or 1.6 mm, in the example), there will be an ironing load (added to the drawing load) that is sufficient to reduce the wall thickness by 0.45 mm (0.018 in., or 0.0805 in. − 0.0625 in. = 0.018 in.). If a parallel wall is desired, it will be necessary to iron

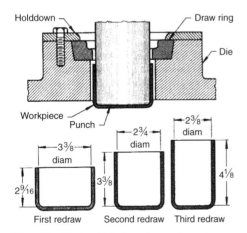

Fig. 20 Die in which draw rings and punches can be replaced for producing three successive redraws. Dimensions given in inches

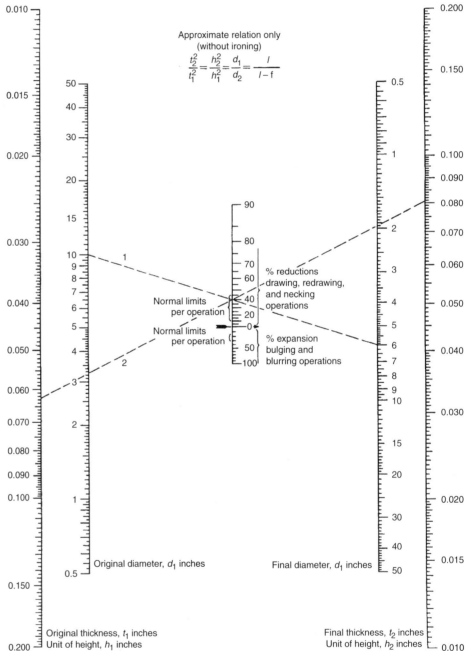

Fig. 21 Chart for determining approximate changes in thickness from changes in outer diameter due to ironing. Read the chart from left to right. (1) Draw a line between the original shell or blank diameter and the final diameter (on the two inner scales) to obtain the maximum percent reduction or expansion on the center scale. (2) Draw a second line through this point, starting from the original metal thickness on the left-hand scale, to indicate the approximate maximum wall thickness at the upper edge of the final shell. Source: Ref 1

down to the thickness of the thinnest part of the wall, which is appreciably less than the original metal thickness, depending on the sharpness of corner radii and the severity of reductions.

The chart in Fig. 22 offers a convenient means of approximating the pressure required in ironing. Referring to the example shown in dashed lines, note that the two inner scales are used first to establish the pivot point on the center scale. Thus, a shell ironed to a finished diameter of 100 mm (4 in.), with a displacement of 0.25 mm (0.010 in.) of the total metal thickness, will require an ironing pressure of approximately 3.8 tons, assuming that the metal is spheroidized steel moderately strain hardened and therefore offering a compressive resistance of approximately 620 MPa (90 ksi). This must be added, of course, to the drawing or redrawing load figured separately. The formula is $P = 1.2 \ \pi d \cdot i \cdot S$, where P is the approximate required maximum pressure (in tons), d is the outside diameter (in inches) after ironing, i is the reduction in wall thickness (in inches), and S is the compressive resistance (in psi) of the metal under existing conditions of strain hardening. There is included a 20% allowance for surface friction in addition to the work of coining. This is an arbitrary figure that should cover well-polished dies and suitable lubrication. Lack of lubricant, toolmark rings on the dies, or surface pickup will increase the friction load considerably.

If the wall of a shell is ironed thinner by the same amount for the entire length of the shell, then the work done is approximately the product of the length of the shell or of the ironed surface and the pressure required for ironing (from the chart). This is expressed as $W = P \cdot l$, where W is work in inch-tons, P is pressure in tons, and l is length in inches. If a reducing operation accompanies ironing, then the drawing pressure should be added to the ironing pressure. If the ironing operation is done merely to correct the natural changes in wall thickness due to drawing and reduces the thickest portion near the top of the shell to equal the thinnest portion near the bottom, the average ironing pressure will equal approximately half the maximum, and the formula will be $W = 0.5 \ P \cdot l$.

Drawing of Boxlike Shells

Square or rectangular shells can be formed by redrawing circular shells when there is no flange. When flanges are required, the difficulty of producing acceptable boxlike shapes by drawing is increased. For deep-drawn square or rectangular shells (for example, where the depth is greater than either length or width), the best approach for forming a narrow flange is to allow sufficient stock and to form the flange after redrawing from a cylindrical shell.

Shallower boxlike shapes (for example, with proportions similar to the box illustrated in Fig. 23) can be drawn with a flange, which is then trimmed to the desired width. Calculations for the area of a blank used for a circular workpiece cannot be used for a square or rectangular box. These require metal in the bottom, ends, sides, and flange, as shown when a box is unfolded (flat pattern, Fig. 23). The excess metal at the corners (shaded areas, Fig. 23) is a problem.

A seamless square or rectangular shell is made by drawing metal into the corners. The metal not needed for the corners is pushed into the walls adjacent to the corner radius and into earlike extensions of the corners. The compressive stresses set up when the metal in the corners is rearranged cause the metal to be thicker in the corners than in the sidewall or in the original blank.

The more difficult draws are made more easily by using a carefully developed blank. There are methods of developing the shape at the corners of a blank for a square or rectangular shell so that there is a minimum of excess metal. However, by cropping the corners as shown in Fig. 23 and by using a blankholder, satisfactory parts can generally be made. Draw beads in the blankholding surface surrounding the die are frequently used.

Drawing of Workpieces with Flanges

Regardless of whether the drawn workpiece is circular, rectangular, or asymmetrical, producing acceptable small-width flanges on workpieces is seldom a problem. Flanged workpieces are usually drawn in two or more operations, frequently with restriking as a final operation.

Cylindrical workpieces with wide flanges are troublesome to draw because of excessive wrinkling or fracturing in the sidewall due to lack of metal flow. Even though the metal is restrained by a blankholder, it is difficult to

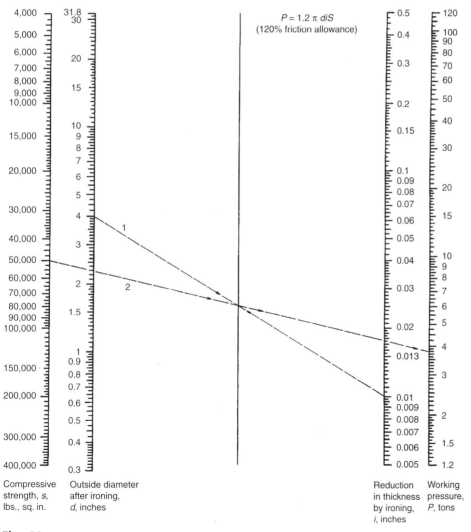

$P = 1.2 \ \pi \ diS$
(120% friction allowance)

| Compressive strength, s, lbs., sq. in. | Outside diameter after ironing, d, inches | | Reduction in thickness by ironing, i, inches | Working pressure, P, tons |

Fig. 22 Chart for determining pressures required for ironing. Source: Ref 1

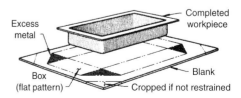

Fig. 23 Flanged rectangular box drawn from a blank with restraint at the corners. See text for details.

obtain acceptable flatness without special procedures. Wide flanges on relatively large workpieces can be made flat by coining after drawing. Another means of dealing with wrinkling, when design permits, is to provide ribs in the flange. This controls the wrinkling by allowing space for the excess metal. Ribs are usually spaced radially around the flange, although circular, concentric ribs also are effective.

Rectangular, boxlike workpieces that have flanges are difficult to redraw in such a manner that the flange is unaffected in redrawing operations. Therefore, it is common practice to draw the part first to a shallower depth and with larger bottom radii than needed for shaping the final contour. The part is then reformed in a final operation.

Asymmetrical workpieces that have flanges are often difficult to draw, particularly when neither draw beads in the die nor ribs in the workpiece can be permitted. Under these conditions, considerable development is usually required to determine the blankholder pressure that will result in the desired metal flow without using a larger blank than necessary.

Drawing of Hemispheres

In the drawing of a hemisphere, metal flow must be closely controlled for balance between excessive thinning in one area and wrinkling in another. In the top illustration in Fig. 24, the punch has begun to stretch the round blank, which is restrained by the blankholder, and the crown section of the hemisphere is being formed. At this stage, the crown section is subjected to biaxial tension, which results in metal thinning. With correct pressure on the blankholder, thinning is in the range of 10 to 15%. More than 15% thinning is likely to result in fracture of the crown section. In the top illustration in Fig. 24, the portion of the blank under the blankholder has not begun to move.

As the drawing operation continues, the metal begins to move from the blankholder, and a different problem develops (center illustration, Fig. 24). Here, the metal has been drawn into a partial hemisphere with unsupported metal in a tangential slope between the punch and the clamped surface. Unlike the drawing of straight-sided shapes, the wide gap (wrinkling area, Fig. 24) prevents the use of the draw ring bore as the means of forcing the metal against the punch surface; therefore, the probability of wrinkling increases. Because the metal cannot be confined between the punch and die, wrinkling is likely to occur in this area.

To prevent wrinkles, the metal must flow from the flange area and, at the same time, must be securely held in tension. This requires an additional stretching force, derived from the portion of the blank that remains clamped. The area of metal between the clamping surfaces is gradually reduced as the punch advances, but the draw radius offers some resistance because the metal

must follow a sharper bend as it moves into the die.

One means of controlling wrinkling is by the use of draw beads, as shown in the bottom illustration in Fig. 24. Another means is by a sharp draw radius. Small radii are susceptible to metal pickup and, depending on sharpness, can produce undesirable circumferential grooves in the hemisphere if the punch does not move at a steady rate.

Reducing Drawn Shells

Necking and nosing are used for reducing the diameter of a drawn cup or shell for a part of its height.

Necking. By the die reduction method, the work metal is forced into compression, resulting in an increase in length and wall thickness. The thicknesses of a shell before and after necking are related by:

$$t_2 = t_1 \sqrt{\frac{d_1}{d_2}} \qquad \text{(Eq 9)}$$

and heights before and after necking by the formula:

$$h_2 = h_1 \sqrt{\frac{d_1}{d_2}} \qquad \text{(Eq 10)}$$

where t_1 is the shell thickness before necking, t_2 is the shell thickness in the necked area after necking, d_1 is the mean diameter of the shell before necking, d_2 is the mean diameter after necking, h_1 is the unit of height before necking, and h_2 is the unit of height after necking.

Figure 25 shows the flow of metal in a necking operation. As the metal flows, paths of least resistance are taken. Therefore, the actual values

for t_2 will be less, and for h_2 greater, than those calculated from Eq 9 and 10.

Necking results are uniformly better if the workpiece has been slightly cold worked. This provides added strength to resist bulging in the column section and buckling in the section being reduced. The entry angle on the necking die is important because the probability that the metal will collapse is decreased as the angle with the vertical becomes smaller. This angle should be less than 45° (Fig. 25). If the angle is greater than 45°, a series of reductions may be necessary, with localized annealing between reducing operations. With a die entry angle less than 45°, thin-wall tubes can be reduced as much as 15% in diameter; thick-wall tubes can be reduced as much as 20%.

Nosing reduces the open end of a shell by tapering or rounding the end (usually by cold reduction) and is primarily used in making ammunition. Shells are often machined before, instead of after, nosing. Shells are usually cold reduced as much as 30% of their original diameter by nosing.

Ironing is the intentional reduction in wall thickness of a shell by confining the metal between the punch and the die wall. When ironing occurs, the force needed to displace the punch often increases to a secondary maximum in the force-displacement curve. The second force maximum can be of such magnitude that the shell will break. However, after ironing has started and metal has been wrapped around the punch, the force is uniform and frequently less than that for redrawing operations.

Ironing is seldom used with redrawing operations unless the amount of wall thinning is relatively small, because it results in excessive die wear, causes workpiece breakage, and increases press tonnage requirements. If a shell with constant wall thickness is needed, however, it can be obtained only by ironing.

Expanding Drawn Workpieces

There are several methods for expanding portions of drawn workpieces in a press. Because the wall thickness is reduced during expansion, it is not advisable to increase the diameter for ductile metal shells (such as low-carbon steel or copper) more than 30%. If a diameter increase of more than 30% is required, the operation should

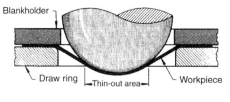
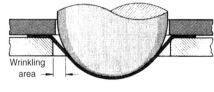

Fig. 24 Drawing of a hemisphere with and without a draw bead. See text for more information.

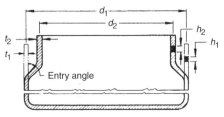

Fig. 25 Flow of metal in the reduction of a drawn shell by necking. See text for details.

be done in two or more stages, with annealing between stages.

Expanding with a Punch. In expanding with a punch, as in Fig. 26, the portion to be expanded is first annealed. Localized annealing, instead of annealing the entire cup, helps retain strength in the remainder of the cup. Regardless of whether or not the strength is required in the finished part, maximum column strength is desirable to prevent buckling as the punch enters the cup.

After the cup has been placed in the die (Fig. 26a), the punch moves downward and expands the top of the cup (Fig. 26b). During the return stroke, the workpiece is stripped from the punch by the stripper ring and is ejected from the die by the ejection pad.

In an expanding operation of this kind, die dimensions are predetermined within reasonably close limits during the design stage. However, the possibility of later design changes must always be considered. Depending on the shape and location of the expanded section, a height reduction of the cup may occur that will require some modification of the die and punch after tryout.

Expanding with segmented dies is often used for forming sidewalls of drawn shells or sections of tubing. The forming segments are contracted by compression springs and expanded radially by a tapered punch. The die is made of two or more segments held apart by compression springs. As the press ram descends, cams move the die segments together. The punch then moves the inner segments outward, thus forming the contours in the sidewall. The presence of gaps between the forming segments is one of the disadvantages of this method and is the reason an alternative method, such as rubber-pad forming, is sometimes selected.

Deep Drawing of Pressure Vessels

Various grades of steel, many of them high-strength alloys, are deep drawn to make cylinders for compressed gases. Joints (when they are made) are around the girth of the vessel, rather than longitudinal. The integrity of the vessel is critical. Commercial-quality hot-rolled steels in the as-rolled condition are generally used. The work metal is usually induction heated or induction annealed to minimize scale.

For propane gas, pressure tanks must have high strength at minimum weight. In one application, the weight of such a tank was reduced from 59 to 32 kg (130 to 70 lb) by changing from 1025 steel to a high-manganese deep-drawing steel (Fe-0.2C-1.6Mn-0.025P-0.3S). Before drawing, the high-manganese steel had a minimum yield strength of 345 MPa (50 ksi) and a minimum tensile strength of 485 MPa (70 ksi).

Bottles for dispensing small quantities of liquefied gases or gases under high pressure are commonly made of drawing-quality low-carbon steel to take advantage of the improved mechanical properties produced by deep draw-

ing. The bottles range in size from 12.7 mm ($\frac{1}{2}$ in.) in diameter by 32 mm ($1\frac{1}{4}$ in.) long to 38 mm ($1\frac{1}{2}$ in.) in diameter by 150 mm (6 in.) long.

Deep Drawing Using Fluid-Forming Presses

Fluid forming (also termed hydroforming) is a deep-drawing process that uses only one solid die half. Forming pressure is applied by the action of hydraulic fluid against a flexible membrane, which forces the blank to assume the shape of the rigid tool.

Fluid forming can be used for deep drawing and, in fact, offers advantages over other forming methods. One of these is that the draw radius can be varied by changing the pressure of the hydraulic fluid during the forming operation. This makes it possible to have, for example, a large draw radius at the start of the operation that decreases as the draw continues. Thus, reductions of up to 70% in a single pass are possible when drawing cylindrical cups; for rectangular-shaped parts, a height of six to eight times the corner radius can be obtained in a single operation.

Presses for fluid forming sometimes use a telescoping ram system. Figure 27 shows a schematic of one type of press used for fluid forming. Figure 28 illustrates the deep-drawing process on a press of this type. More information on fluid-forming equipment and processes is

available in the article "Rubber-Pad Forming and Hydroforming" in this Volume.

Ejection of Workpieces

In drawing operations, the drawn workpiece may adhere to either the punch or the die. Adherence is increased by depth of draw, straightness of workpiece walls, and viscosity of lubricant. The simplest means of ejecting a small workpiece is by compressed air through jets in the punch or the die. Timed air blast is widely used for ejecting relatively small workpieces, for example, where cup diameter is no greater than 102 to 127 mm (4 to 5 in.). In some production drawing operations, the workpiece is ejected by compressed air, and another timed blast of air from the side removes the piece by sending it down a chute or into a container. However, for larger workpieces or for those that are deep, some other means of ejection is required.

Mechanical methods of ejection include:

- Edge stripping by means of a lip on the draw ring (Fig. 29a) or by a spring-actuated stripper (Fig. 29b)
- The use of a blankholder in combination with an upper ejector (Fig. 30)
- The use of a lower ejector in combination with an upper stripper ring (Fig. 31)

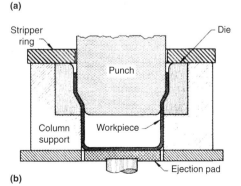

Fig. 26 Expansion of the mouth of a drawn shell with a punch

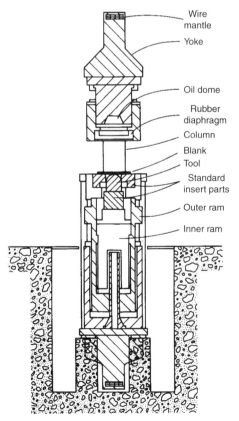

Fig. 27 Schematic of one type of fluid-forming press used for deep drawing

Numerous other mechanical methods using cams or links have been devised to meet specific requirements. These methods are usually modifications of those described previously. For example, thin shells are sometimes stripped from punches near the top of the press stroke by a cam-actuated rod that extends through the punch. This method is often used to avoid damage to the open end of the shell, which can occur when the piece is ejected by other methods. The major factors influencing the method of ejection are workpiece design (especially the presence or absence of a flange), work metal composition and thickness, and the type of equipment available.

Trimming

Trimming in a lathe (using a cutting tool), roll trimming in a lathe, rotary shearing, die trimming (regular and pinch), and trimming on special machines are the methods most commonly used for trimming drawn workpieces.

Methods for Specific Shapes. Cylindrical workpieces, such as the one shown in Fig. 32a, can be trimmed by at least four different methods:

- In a lathe, with a cutting tool, but production cost is high
- By roll trimming in a lathe or in a rotary shear. Production cost is lower than that for trimming with a cutting tool, but the finish of a rolled edge is poor and maintenance cost of the rolls is high.
- By pinch trimming in the press at the bottom of the drawing stroke. This involves almost no increase in production cost but requires a more expensive die. This method produces a thinned edge at the trim line, which may be unacceptable.
- In a shimmy die or trimming machine, but production quantities must be high to warrant the investment

Cylindrical workpieces with flanges, such as the one shown in Fig. 32b, can also be trimmed in a lathe, although shapes such as this are ideal for trimming in a die and can be die trimmed for approximately 5% of the cost of trimming in a lathe. Rotary shearing can also be used for trimming circular drawn parts with flanges if the dimensional tolerance is 0.76 mm (0.030 in.) or more. Drawn workpieces with an irregular trim line, as in Fig. 32c, can be trimmed in a die for

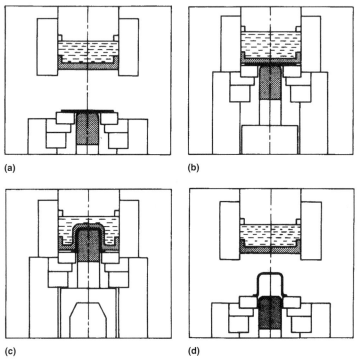

Fig. 28 Deep-drawing process using the fluid-forming press shown in Fig. 27. (a) The blank is placed on the blankholder. (b) The outer ram moves upward, carrying the blank. (c) Oil is pumped into the inner ram system, pressing the punch upward. (d) Outer ram is returned to its initial position, and the punch is retracted to allow removal of the formed part.

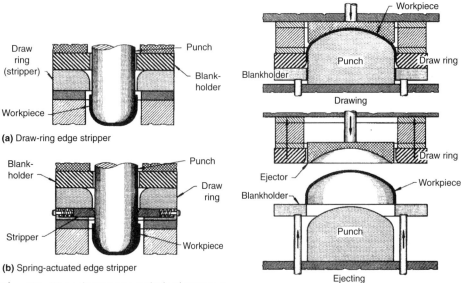

Fig. 29 Setups showing two methods of ejecting a drawn workpiece through the bottom of the press by edge stripping

Fig. 30 Use of a blankholder and ejector for stripping of a drawn workpiece from inverted dies

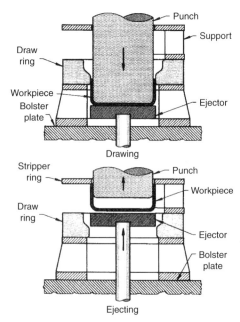

Fig. 31 Setup in which drawn workpiece is lifted from the die by an ejector and is stripped off the punch by a fixed stripper ring

low-production requirements, or with a shimmy die or trimming machine for high-production requirements. The cost of a trimming die is approximately half that for a special trimmer (excluding the cost of the original machine). However, the trimming cost per piece with the special trimmer will be only approximately half the per-piece cost with multiple dies, and the trimmed edges will be better.

Flanged workpieces, such as the one shown in Fig. 32d, can be trimmed in a die for 5% of the cost of trimming in a rotary shear. In low production, drawn workpieces of such a shape are frequently trimmed in a nibbler and filed to conform to a template. This means of trimming costs up to 60 times as much as die trimming.

Developed Blanks versus Final Trimming

The most important factor that influences a choice between using a developed blank or a final trimming operation is whether or not the shape of the drawn edge is acceptable. A semi-developed blank is sometimes necessary to draw an acceptable part, and the edge must be trimmed to meet dimensional tolerances.

The next consideration is the cost of blanking versus final trimming. This would include the adaptability of the process to the available equipment, based on expected production requirements. The principal advantage of a developed blank is that strip or coiled work metal can be used. The use of strip eliminates the need for shearing the work metal to a rough-blank shape, as is sometimes required when final trimming is used. The developed-blank approach is usually more economical than final trimming because a blanking die is frequently less expensive than a final-trimming die.

When using developed blanks, the draw dies are made, and several blanks are drawn to select the optimal developed shape before the blanking die is made. This causes a delay in placing the draw die into production. However, with proper planning and scheduling, this should not be a problem. Another disadvantage of developed blanks occurs when variations in work metal properties and thickness are sufficient to affect the uniformity of the drawn workpiece. Under these conditions, closer tolerances are obtained by final trimming. It is possible to develop blank contours accurately enough so that the outline of the drawn part is within tolerance, thus avoiding a final-trimming operation.

Cleaning of Workpieces

In general, the more effective the lubricant, the more difficult it is to remove. Therefore, an overly effective drawing lubricant should be avoided.

The cleaning method depends on the work metal composition, the lubricant, the degree of cleanliness required, workpiece shape, and sometimes the length of time between application of lubricant and its removal. Some metals will be attacked by cleaners that are not harmful to others. For example, strong alkaline cleaners are suitable for cleaning steel and many other metals, but they are likely to attack aluminum alloys. Detailed information is available in the articles on the surface engineering of specific metals in *Surface Engineering,* Volume 5, *ASM Handbook,* 1994.

Unpigmented oils and greases can be removed from steel workpieces by several simple shop methods, including alkaline dipping, emulsion cleaning, and cold solvent dipping. These methods are usually sufficient for in-process cleaning. However, if the workpieces are to be painted, a more thorough cleaning by emulsion spray or vapor degreasing is required. For plating, electrolytic cleaning plus etching in acid (immediately prior to plating) is required. These latter methods usually follow a rough cleaning operation.

Pigmented drawing lubricants and waxes greatly increase cleaning problems. At a minimum, in-process cleaning usually requires slushing in a hot emulsion or vapor degreasing. If the lubricant is not removed for several days after application, soaking in a hot alkaline cleaner or an emulsion cleaner may be required. Particularly for complex workpiece shapes, some hand or power brush scrubbing may be needed. If the workpieces are to be painted or plated, additional cleaning will be required, as described previously. Detailed information on the choice of

cleaning method is available in the article "Classification and Selection of Cleaning Processes" in *Surface Engineering,* Volume 5, *ASM Handbook,* 1994.

Dimensional Accuracy

Dimensional accuracy in deep drawing is affected by the variation in work metal thickness, variation in work metal condition (chiefly hardness), drawing technique (particularly the number of operations), accuracy of the tools, rate of tool wear, and press condition. Control of dimensions begins with the purchase of sheet to closer-than-commercial thickness tolerance, which adds substantially to the cost. Close control of sheet hardness also costs more. In-process annealing may be required to minimize springback or warpage; it will not be needed if tolerances are more liberal. Annealing, handling, and cleaning operations are costly.

As tolerances become closer, it is often necessary to add more die stations to minimize the amount of drawing in any one station. Close tolerances may demand restriking operations that would not be necessary for parts with more liberal tolerances. Additional operations increase tool costs and decrease productivity, thus increasing the cost per piece.

The initial cost of tools increases as tolerances become closer because of greater cost for precision machining and grinding or more costly tool materials. In addition, tool life before reconditioning and total tool life decrease as tolerances become closer. Maintenance costs and downtime of presses are also greater.

When required, extremely close tolerances can be maintained on some parts (Fig. 33).

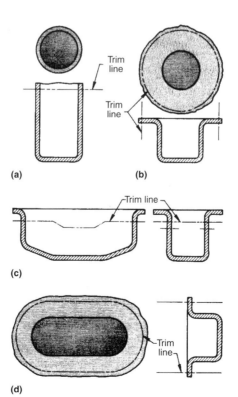

Fig. 32 Typical trim lines on drawn parts. See text for details.

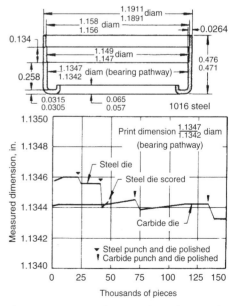

Fig. 33 Variation in bearing pathway diameter of a needle bearing cup drawn using high-speed tool steel and carbide dies

In most deep drawing, the accuracy shown in Fig. 33 is either impossible or impractical. The more usual practice when dimensional accuracy is important is to check critical dimensions at specified intervals during a production run and to plot the variation. Data from this method of quality control show the capabilities of the process under shop conditions and the magnitude of drift during a production run. When results (either initially or during a run) are unacceptable, one or more of the controls discussed at the beginning of this section can be applied.

Safety

Deep drawing, like other press operations, involves potential hazards to operators and other personnel in the work area. No press, die, or auxiliary equipment can be considered ready for operation until these hazards are eliminated by the installation of necessary safety devices. Operators should be properly instructed in safe operation of equipment.

REFERENCE

1. O.D. Lascoe, *Handbook of Fabrication Processes,* ASM International, 1988, p 233

SELECTED REFERENCES

- M. Groover, *Fundamentals of Modern Manufacturing: Materials, Processes, and Systems,* Prentice-Hall, 1996, p 500–542
- W. Hosford and R. Caddell, *Metal Forming: Mechanics and Metallurgy,* 2nd ed., Prentice-Hall, 1993, p 244–358
- *Inform: The Magazine for Metalforming,* Schuler SMG GmbH & Co. KG, Feb 2000, p 32–33
- S. Kalpakjian and S. Schmid, *Manufacturing Engineering and Technology,* 4th ed., Prentice-Hall, 2001, p 392–434
- *Lightweight Vehicle Systems Materials,* Department of Energy Annual Progress Report, 1999, p 12–18
- *Lightweight Vehicle Systems Materials,* Department of Energy Annual Progress Report, 2001, p 13–19
- *Lightweight Vehicle Systems Materials,* Department of Energy Annual Progress Report, 2002, p 9–18
- K. Siegert, S. Wagner, and M. Siegler, Robust Forming Process by Pulsating Blankholder Forces, Multipoint Cushion Systems and Closed Loop Control, *Sheet Metal Forming Technology,* M.Y. Demeri, Ed., TMS, 1999, p 1–21

Stretch Forming

Revised by Mahmoud Y. Demeri, FormSys, Inc.

STRETCH FORMING is the forming of sheet, bars, and rolled or extruded sections over a die or form block of the required shape while the workpiece is held in tension. The workpiece is usually gripped by mechanical jaws on each end and then stretched and simultaneously bent over a die containing the desired shape. The work metal is often stretched just beyond its yield point (generally 2 to 4% total elongation) to retain permanently the contour of the form block. Controlling the amount of stretching is important to avoid splitting. The combination of stretching and bending reduces springback in the formed part. Stretch forming cannot produce parts with sharp contours or depressions. It is primarily used to make aircraft skin panels, window frames, and automotive door panels.

The four methods of stretch forming are:

- Stretch draw forming (Fig. 1a and b)
- Stretch wrapping, also called rotary stretch forming (Fig. 1c)
- Compression forming (Fig. 1d)
- Radial draw forming (Fig. 1e)

These methods are discussed separately in subsequent sections of this article.

Applicability

Almost any shape that can be produced by other sheet-forming methods can be produced by stretch forming. Drawn shapes that involve metal flow, particularly straight cylindrical shells, and details that result from such compression operations as coining and embossing cannot be made. However, some embossing is done by the mating-die method of stretch draw forming (Fig. 1b).

Stretch forming is used to form aerospace parts from steel, nickel, aluminum, and titanium alloys and other heat-resistant and refractory metals. Some of these parts are difficult or impossible to form by other methods, for example, the titanium alloy gas-turbine ring shown in Fig. 2. The procedure for making such a ring is described in Example 5 in this article.

Stretch forming is also used to shape automotive body panels, both inner and outer, and frame members that could be formed by other processes but at higher cost. An example is the automobile roof shown in Fig. 3, which was stretch draw formed using a blank that weighed

2.9 kg (6.4 lb) less than would have been needed for a conventional press-forming process. Architectural shapes and aerospace forms that call for compound curves, reverse bends, twists, and bends in two or more planes are also produced by stretch forming.

Advantages. Stretch forming has the following advantages over conventional press-forming methods:

- Approximately 70% less force is needed than that required for conventional press forming.
- Stretch forming can reduce material costs by as much as 15%. Although allowance must be made on the stock for gripping, it is gripped on two ends only. The allowance for trimming is usually less than that in conventional press forming.
- Because stretch forming is done on the entire area of the workpiece, there is little likelihood of buckles and wrinkles. Tensile strength is increased uniformly by approximately 10%.
- Hardness is increased by approximately 2%.
- Springback is greatly reduced. There is some springback, but it is easily controlled by overforming.

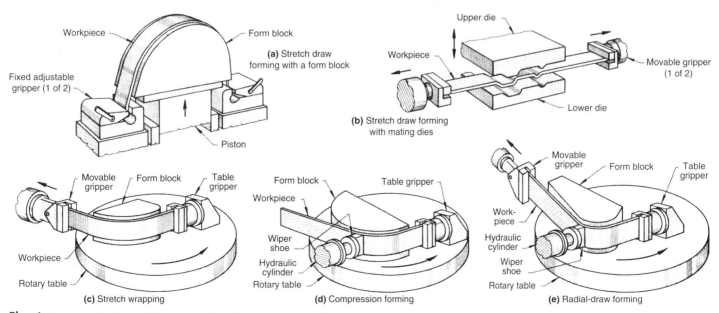

Fig. 1 Fundamentals of the techniques involved in the four methods of stretch forming

- Residual stresses are low in stretch-formed parts.
- Form blocks are made of inexpensive materials, such as wood, plastic, cast iron, or low-carbon steel, and are approximately one-third the cost of conventional forming dies. If the workpiece is formed hot, the dies must be able to withstand the forming temperature. However, most stretch forming is done at room temperature.
- Changeover is simple. Only one form block and two sets of grippers are involved. To make the same part from a different metal or another stock thickness, the same form block and grippers are used, but the tension of the stretch mechanism is adjusted.

Limitations. Stretch forming is subject to the following limitations:

- It is seldom suited to progressive or transfer operations.
- It is limited in its ability to form sharp contours and reentrant angles. It is at its best in forming shallow or nearly flat contours.
- If the piece is not pinched between mating dies, there is no opportunity to coin out or iron out slight irregularities in the surface of the metal.
- In some applications, especially in stretch wrapping, the process is slower than competitive processes, and it is not suited to high-volume production. However, stretch draw forming with mating dies can be done as rapidly and automatically as conventional press operations. In fact, punch presses are used with dies incorporating draw beads or other means of gripping the blank in order to perform some stretch-forming operations.
- Metals with yield strength and tensile strength very nearly the same, such as titanium, necessitate the use of automatic equipment for determining the amount of strain for uniform results.
- Optimal results are achieved with rectangular blanks. The aircraft industry uses trapezoidal blanks but gives greater attention to each piece than is warranted in high-volume production.
- Deep forming in the direction of the free edges is not practical.

Machines and Accessories

Stretch wrapping, compression forming, and radial draw forming use rotary tables (some with sliding leaves) for mounting the form blocks, a ram gripping and tensioning or wiping device, and a mechanically or hydraulically actuated table gripper (Fig. 4). Machines used for these operations have capacities to 8900 kN (1000 tonf).

Stretch draw forming is done in three types of machines. In one type, the form block mounted on a hydraulic cylinder is pushed into the blank, which is held in tension by a pair of pivoting grippers. In another type, the form block is fixed to the table, and the blank is drawn around it by a pair of grippers actuated by slides or a hydraulic cylinder. The third type of machine is a single-action hydraulic press equipped with a means of closing and moving a pair of grippers (see Fig. 7); a mating die is used instead of a form block. The hydraulic presses ordinarily used in stretch draw forming have capacities of 1800 to 7100 kN (200 to 800 tonf).

Accessory Equipment. Grippers and wiping shoes or rollers are made to conform to the rolled or extruded shape that is to be stretch formed. Jaws used for gripping sheet in stretch draw forming can be segmented or contoured to apply equal stretch to all parts of the sheet as it is formed. The vertical adapter shown in Fig. 5 is used with a rotary table; it is fastened to the hydraulic cylinder used for applying tension to the blank. The adapter allows wiper shoes, rollers, and grippers to move up or down as needed in order to accommodate work with bends in two or more planes. Lead screws or hydraulic cylinders position the grippers or wiping devices at the correct position for the forming operation (see Examples 5 and 6).

A yield detector and tension-control device (Fig. 6) provide a means of automatically applying the same amount of stress to every workpiece in a production lot. This is important with metals (for example, titanium) that have yield strength and tensile strength too close for ordinary control of tension for stretch forming (see the article "Forming of Titanium and Titanium Alloys" in this Volume). With this type of control, scrap in the stretch forming of titanium can be reduced to 2%.

The tension control uses two inputs in a null system for its output signal. One input comes from a load cell that gives a signal proportional to the stretch force on the workpiece. The other signal comes from a potentiometer that measures the elongation of the workpiece. As long as the signals are proportional, the metal is not stretched beyond its yield point, and the two inputs balance. When the yield point is reached, the input from the load cell stops increasing, or it increases at a much lower rate, while the potentiometer input continues to rise. This upsets the null balance, and an output signal is given, which can be interpreted as percentage of strain beyond the yield point. Table restretch units are small short-stroke hydraulic cylinders and clamps that can be bolted to the rotary table to give a final stretch set to workpieces that need to be stretched from both ends or restretched after heat treatment. The capacity of a table restretch unit is usually equal to that of the main tensioning gripper.

Stretch Draw Forming

Stretch draw forming is done with either a form block or a mating die.

The Form-Block Method. Bars and structural shapes, although usually radial draw formed, can be stretch draw formed by the form-block method. Also known as drape forming, the form-block method uses either a fixed or a moving form block. A fixed form block is attached to the machine base. Each end of the blank is held by a gripper attached to a hydraulic cylinder. The grippers move to stretch the blank over the form block. Alternatively, the moving

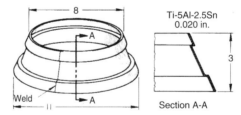

Fig. 2 Titanium alloy gas-turbine ring that was produced by compression forming. Dimensions given in inches

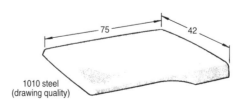

Fig. 3 Stretch draw formed automobile roof. Dimensions given in inches

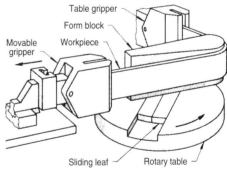

Fig. 4 Stretch-forming machine with rotary table and sliding center leaf

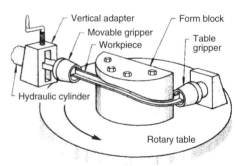

Fig. 5 Use of a vertical adapter on the tension unit to accommodate work with bends in two or more planes

form block is attached to a hydraulic piston. A blank is held by grippers while tension is applied to it, and the form block then moves to form the part, as shown in Fig. 1(a).

The mating-die method uses a two-piece die mounted in a single-action hydraulic press (Fig. 7). This method combines the advantages of stretch forming and conventional press forming. The stretch forming sets the contours of moderately formed workpieces, and the press forming gives definition to sharply formed contours, such as beads or feature lines on automobile body parts.

Grippers preform the blank over the lower die to the curvature of the part (Fig. 7a). There is very little metal flow; the stretching action and the die form the general outline of the part. The upper die then descends to produce the details and to set the contours (Fig. 7b).

Automatic material-handling equipment can be adapted to the machine for production runs. Production rates are comparable to those obtainable for drawing in conventional single- and double-action presses. Stretch draw press tooling for large parts, such as automobile roof panels, weighs only one-third that for a conventional double-action press, as indicated in Table 1.

Exposed parts, such as automobile outer body panels, frequently have a maximum surface roughness specification of 1.1 μm (45 μin.), and stretcher strain marks and other defects, which are still noticeable after painting, must be prevented. One method of avoiding strain marks is the use of segmented or curved grippers, which equalize the amount of stretching. The following example describes the production of a rear-deck lid by stretch draw forming with a mating die.

Example 1: Stretch Draw Forming of an Automobile Rear-Deck Lid. Automobile rear-deck lids were produced in a stretch draw press using mating dies, as shown in Fig. 7. The blanks were commercial-quality cold-rolled 1008 steel 0.91 mm (0.036 in.) thick, 1450 mm (57 in.) wide, and 1600 mm (63 in.) long. Residual mill oil was the only lubricant. The production rate was 360 pieces per hour, and annual production was 400,000 deck lids.

Tension was applied to the sheet by the grippers as they moved apart. (Generally, hydraulic cylinders are used to apply the force in this operation.) The tensioned sheet (still held by the grippers) was then lowered to stretch over the lower die. Finally, the upper die was lowered, pressing the sheet into both dies to form the lid.

The cycle time was 7 s. The finished parts showed uniformly good quality without wrinkles or buckles. Approximately 0.9 kg (2 lb) more sheet steel would have been needed to produce this part in a conventional double-action press.

Lancing. If stretch drawing is used to form severe contours, the stretch limits of the metal may be exceeded in the zones of deep forming, resulting in fracture of the metal. This can be avoided by lancing the metal in areas to be discarded later so that the metal can flow in the severely formed zones. Single-operation production of a truck-cab roof having a combination of gradual curves and sharp contours is described in the following example.

Example 2: Stretch Draw Forming of a Truck-Cab Roof with Reinforcing Beads. By using stretch draw forming with mating dies, the truck-cab roof panel shown in Fig. 8 was produced in one operation. Panels were formed from cold-rolled drawing-quality aluminum-killed 1008 steel. The blanks were 1520 by 813 by 0.89 mm (60 by 32 by 0.035 in.).

With automatic material-handling equipment to load and unload the machine, the production rate was 100 to 150 pieces per hour. Annual production was 25,000 pieces.

The forming dies were of cast iron, with flame-hardened surfaces where severe forming occurred. Radii on the roof beads were 1.0 mm (0.040 in.).

Stretch Wrapping

In stretch wrapping, just enough tension is applied to one end of a workpiece to exceed the yield strength of the material, while the form block revolves into the workpiece with the

Table 1 Comparison of conventional and stretch presses

Press	Capacity							
	Punch		Gripper (blankholder)		Tooling weight		Press height	
	kN	tonf	kN	tonf	kg	tons	mm	in.
Conventional	8000	900	5300	600	20,000	22	7320	288
Stretch	2200	250	760	85	6300	7	5100	200

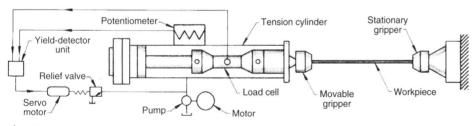

Fig. 6 Components and signal-flow diagram of an automatic tension-control system used in stretch forming

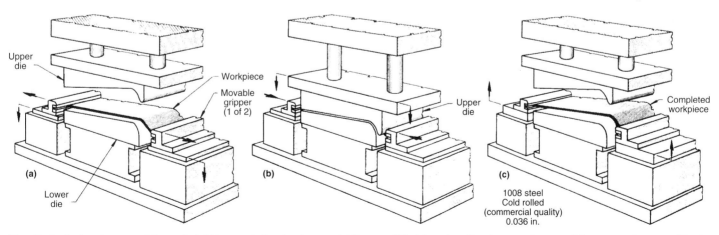

Fig. 7 Production of an automobile rear-deck lid in a stretch draw forming press. (a) Sheet metal blank is tensioned by grippers moving apart. Grippers move down, stretching the workpiece over the lower die. (b) Upper die descends onto the workpiece, pressing the metal into both dies to form the part. (c) After forming, the press opens, and the part is released from the grippers.

turning of the table, as shown in Fig. 1(c). The other end of the workpiece is held in a table gripper or clamped to the end of the form block. The hydraulic cylinder applying tension to the workpiece is free to swivel so that the tension is always tangential to the last point of contact. Thus, the work metal wraps in tension around the form block without the scuffing or friction that occurs with other forming methods. The result is an accurately formed piece with little springback; therefore, form blocks can be made to accurate size.

Because there is no scuffing, form blocks can be made of soft metal, wood, or plastic, although common die materials such as cast iron are often used. Form blocks made of hardwood, masonite, and epoxy have also been used. The contour of the form block can vary throughout the bend, and the workpiece will follow it accurately if there are no concave surfaces on the form block.

Form blocks for the stretch wrapping of rolled and extruded sections are machined to the shape of the section as well as the contour of the finished part. Thus, the shaped form block supports the section during forming. Additional support is sometimes needed for open or hollow sections. A segmented filler, a filler made of low-melting alloy, or a strip of easy-to-form metal can provide this support. Return bends can be made by using additional form blocks on sliding leaves of the turntable and by reversing the table direction to produce the part, as shown in Fig. 9.

Machines for stretch wrapping consist basically of a variable-speed power-driven rotary table and a double-action pressure-controlled hydraulic cylinder. The form block is bolted to the table. Grippers are connected to the hydraulic cylinder so that tension can be applied to the workpiece, as in Fig. 1(c). The fabrication of a typical part by stretch wrapping is described in the following example.

Example 3: Forming of an Aircraft Leading-Edge Wing Panel by Stretch Wrapping. A corrugated leading-edge wing panel of aluminum alloy 6061-O was stretch wrapped in a stretch-forming machine with a vertical-axis turntable. The sheet, with corrugations in the direction of airflow, was gripped at each end with grippers shaped to fit the corrugations. The tension applied was slightly above the yield strength of the work metal. The form block, bolted to the

turntable, rotated slowly into the workpiece, causing it to form smoothly into the shape of the wing without flattening the corrugations. While the form block was moving in the sheet, the hydraulically restrained gripper maintained tension slightly above the yield point. The form block was made to the required final shape without allowance for springback because only a small amount of springback occurred.

Compression Forming

In compression forming, the workpiece is pressed against the rotating form block instead of being wrapped around it. The process is typically used for maintaining or controlling workpiece cross-sectional dimensions throughout the contour, for bending to radii small enough to exceed the elongation limits of the metal if formed by stretch wrapping, and for bending sections too heavy for the capacity of the available stretch wrap machinery.

Compression forming can generally be done in the same machine as stretch wrapping, but the hydraulic cylinder is used to apply pressure instead of tension to the workpiece. The cylinder is locked in place to keep it from swiveling, and the ram head is furnished with a roller or a shoe to press the workpiece against the form block. A clamp or table gripper holds the end of the workpiece against the form block, and as the table rotates, the shoe or roller on the hydraulic cylinder presses the workpiece into the contour of the block, as shown in Fig. 1(d).

Compression forming can often make bends to a smaller radius than stretch wrapping in a part

that has a deep cross section. If the same bend were produced by stretch wrapping, fracturing or overstressing of the outer fibers would result. The total load needed to form large-section pieces, such as cross rails and bumpers, can be as little as 2% of that needed to form them in a punch press. The total energy applied to the workpiece would of course be the same (neglecting efficiency); the smaller compression-forming force is applied for a longer period of time. The wiping shoe or roller can hold the cross-sectional size and shape to close tolerance throughout the contour. Parts that are too heavy in cross section for stretch wrapping can often be compression formed.

Blanks for stretch forming are usually made longer than the finished part so that the surface damaged by the gripper jaws can be trimmed off. However, end details, locating surfaces, and other considerations occasionally necessitate the use of a blank cut to the length of the finished part, and dimensional tolerances still must be met, as in the following example.

Example 4: Use of an Adjustable Form Block to Compression Form a Developed Blank. Because both ends of the piece shown in Fig. 10 had previously produced details, the part could not be trimmed after forming. Instead of a table gripper or clamp, the blank was fastened to the form block by bolts through two 21 mm ($^{13}/_{16}$ in.) diameter holes pierced in one end. The blank was cut slightly shorter than the required length because the length increased from 3.602 to 3.613 m (141.81 to 142.25 in.) during forming.

The 1020 steel structural shape that was being compression formed had considerable springback, which varied with each heat of steel.

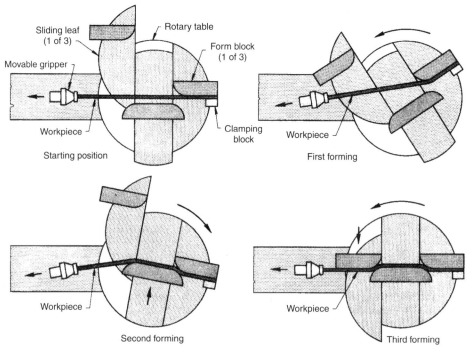

Fig. 9 Stretch wrapping of a part around three form blocks to make two reverse bends

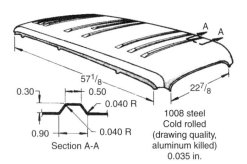

Fig. 8 Beaded truck-cab roof that was stretch draw formed with mating dies. Dimensions given in inches

To obtain uniform results, the form block was made adjustable. As shown in Fig. 10, the wear plate, made of 19 mm (³/₄ in.) thick by 203 mm (8 in.) wide high-carbon steel, was backed up with jackscrews that could be adjusted to change the effective radius. One end of the wear plate was fastened to the base plate, and the other end was free to move in and out of any position, supported by the jackscrews. When a new lot of steel was delivered, an experimental piece was run to determine springback, and the jackscrews were adjusted accordingly.

The work material was a hot-rolled channel 152 mm (6 in.) wide, weighing 15.6 kg/m (10.5 lb/ft), approximating a 1020 steel in composition. The piece was compression formed on a radial draw former into a half circle with a 1145 mm (45 in.) radius. The two holes in the end of the piece were used to connect this section with a fishplate on one end of a mating piece to form an assembled ring. The sequence of operations was as follows:

- Saw ends with a 3° bevel to developed length (3.602 m, or 141.81 in.)
- Deburr
- Pierce two 21 mm (¹³/₁₆ in.) diameter holes
- Form both ends to 1145 mm (45 in.) radius for 152 to 203 mm (6 to 8 in.) of length on a press brake
- Bolt the workpiece to the form block by the two pierced holes. Compression form to 1145 mm (45 in.) radius
- Galvanize after forming
- Flatten as necessary (galvanizing sometimes causes warpage)

A straight mineral oil was used as the lubricant. The overall tolerance on the formed curve was 1.5 mm (0.060 in.) total indicator reading.

Production time was 3 min per piece with two operators, and setup time was 1 h. The production-lot size was 250 pieces.

Two-Plane Curvature. When a part must have curvature in two or more planes, a vertical adapter, either hydraulic powered or screw actuated, is used to permit the ram gripper to move up and down as the work requires. Thus, the work material can be fed into a spiral or other form involving rising and falling curvatures. In the following example, a vertical adapter with a wiper shoe was used to form a low-angle helix that was later welded into a ring.

Example 5: Producing a Ring from a Helix to Counteract Springback. Because springback in the forming of a titanium alloy engine ring made it difficult to weld the workpiece into a true circle after forming, the stock was compression formed into a low-angle helix by using a vertical adapter with a wiper shoe. The form block was smaller in diameter than the finished ring, and when the workpiece was removed from the form block, springback was just sufficient to permit welding into a true ring. The setup used for forming the ring is shown in Fig. 11.

Radial Draw Forming

Radial draw forming is a combination of stretch wrapping and compression forming, as shown in Fig. 1(e). As in stretch wrapping, one end of the workpiece is gripped by stationary jaws attached to the rotary table. The other end is gripped by jaws on the hydraulic cylinder. The cylinder exerts tension on the workpiece as the form block on the rotary table revolves into it. A second hydraulic unit, fitted with a wiper shoe or roller, presses the workpiece into the contour of the form block at the point of tangency. The hydraulic unit applying the compression force can be moved as necessary to keep the wiper shoe in contact with the workpiece. On large machines, an operator sometimes rides a platform on the second unit to observe the point of contact.

Joggles in rolled or extruded sections can be formed after the part has been radial draw formed, without removing the part from the form block. When contour forming is completed, the part is held in tension while the compression unit is repositioned, and the joggle is formed by the wiper shoe. The wiper shoe is sometimes used to apply pressure to a loose joggle block (Fig. 12) if the wiper shoe will not provide the correct shape for the joggle. In either case, the shape of the joggle is also machined into the form block. The vertical adapter shown in Fig. 5 can be used in the radial draw forming of bends in two or more vertical planes.

Architectural sections, extruded shapes, and other sections sometimes have to twist on themselves if they are contour formed through any plane other than the plane of symmetry. In radial draw forming, this can be done by permitting the workpiece to rotate axially as the part follows the twisting contour of the form block. Rotation is obtained by slightly loosening the lock ring on the body of the gripper head, allowing the head to rotate about its own centerline. The following example illustrates the forming of an angle section by this method.

Example 6: Twisting of an Angle Section during Contour Forming. An L-shaped section for the gunwale of an air-sea rescue craft (Fig. 13) had to be twisted as it was radial draw formed. It was made of aluminum alloy 2024-T4. Forming had to be done in several planes. The locking ring of the gripper was loosened, permitting the head to rotate as the part was formed. The usual production-lot size was 500 pieces. Parts were formed at the rate of 10 per hour.

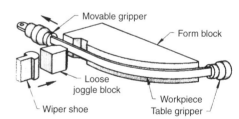

Fig. 12 Forming a joggle with a loose block and a wiper shoe after radial draw forming

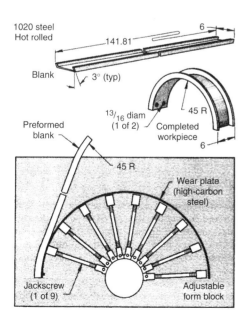

Fig. 10 Steel channel that was compression formed without trim allowance using an adjustable form block. Dimensions given in inches

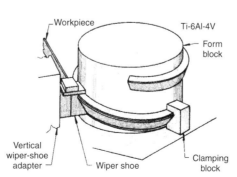

Fig. 11 Use of a vertical wiper-shoe adapter to form a helical shape. After forming, springback of the titanium alloy used brought the formed piece into a circular shape, which could be welded into a ring.

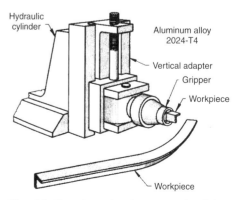

Fig. 13 Gunwale section that was produced from an L-section by a combination of twisting, stretching, and forming in several planes

Stretch-Forming Machines for Large Parts

Drape-forming machines are designed to be used on relatively small lengths and widths. The stretching and bending of greater lengths and widths require the increased capacity of stretch-forming machines. Stretch-forming machines bend the workpiece around the die to elongate the material fibers while simultaneously preventing wrinkles and minimizing springback.

These machines are available in two basic types:

- Moving jaws only, to stretch the blanks around a stationary form block
- Moving jaws combined with a moving-die (form-block) table

Transverse or longitudinal models and combination transverse-longitudinal models are available. Selection of the appropriate machine (jaw width, distance between jaws, and force capacity are key specifications) is determined by the configuration and dimensions of the workpiece.

Transverse Machines. The jaws of transverse machines, on which the workpieces are stretched and bent transversely, must be as long as the workpiece to provide substantial area for gripping. The transverse machine shown in Fig. 14 has a movable die table that can be tilted 15° above or below the horizontal. The jaws can be swiveled 30° in a horizontal plane and 90° in a vertical plane so that the direction of stretching can be aligned with the contour of the die. Rated at 6700 kN (750 tonf), the machine has a 2080 mm (82 in.) stroke, a 25 to 3660 mm (1 to

144 in.) jaw distance, and a 25 to 457 mm/min (1 to 18 in./min) forming speed. This equipment is primarily used for forming large sheets, such as fuselage skins and the leading edges of airplane wings. Large, cumbersome extrusions, such as wing spars, can be formed by changing or adapting the jaws.

Longitudinal Machines. The longitudinal stretch-forming machine shown in Fig. 15 can be swiveled 90° but cannot be raised or lowered. The 2540 mm (100 in.) wide, hydraulically powered, leadscrew-actuated jaws are made in sections for curvature to various radii. These jaws can be swiveled both horizontally and vertically. Each individual jaw develops a 6700 kN (750 tonf) tensile force, and the jaws in tandem are capable of forming 12 by 2.4 m (40 by 8 ft) sheet metal skins. The addition of adapter jaws bolted to the standard jaws allows the machine to form long extruded frame members weighing up to 450 kg (1000 lb).

Dies and Tooling

Traditional dies for stretch forming are made of zinc alloys, steel, plastics, or wood. Nontraditional tools include a recently developed reconfigurable tool system for stretch-forming applications.

Reconfigurable Stretch-Forming Tool. A new stretch-forming tool system for producing large panels with little curvature was recently developed and successfully demonstrated on aircraft parts. The system includes a universal tool that could be reconfigured to produce different part shapes. The working surface of the tool consists of a large number of steel pins with

hemispherical tips. The position of each pin can be changed to accommodate the required shape of the panel. The shape of the die is the envelope of the pin tips, and pin settings are computer controlled to produce the required part shape. A flexible polymer layer is inserted between the tool and the sheet metal during stretch forming to suppress dimpling. The tool can be reconfigured to produce different contoured shapes, such as aircraft wings and fuselage panels.

Figure 16 shows schematic views of the tool design. The overall size of the prototype tool is 1.83 by 2.44 by 1.07 m (6 by 8 by 3.5 ft), and the working volume of the tool surface is 1.22 by 1.83 by 0.31 m (4 by 6 by 1 ft). Individual pins are grouped into modules, and the entire tool is composed of 336 identical modules. Each module can be removed from the tool without disturbing the other modules. A computer with graphical user interface permits configuration of pins to the desired tool shape. Limitations of this technology include minimum radius of curvature and restrictions on the shapes of parts that can be produced. This technology is attractive for low-volume production. It also reduces cost and time by eliminating the need to fabricate as well as store traditional tools and enables transition to a factory with fewer tools.

Accuracy

High-strength alloys, stainless steel, and titanium can be stretch formed to overall tolerances of ±0.25 mm (±0.010 in.) on workpieces bent to approximately a 178 mm (7 in.) radius. With springback allowance, bends can be controlled to ±1/2°. Cross-sectional dimensions have been

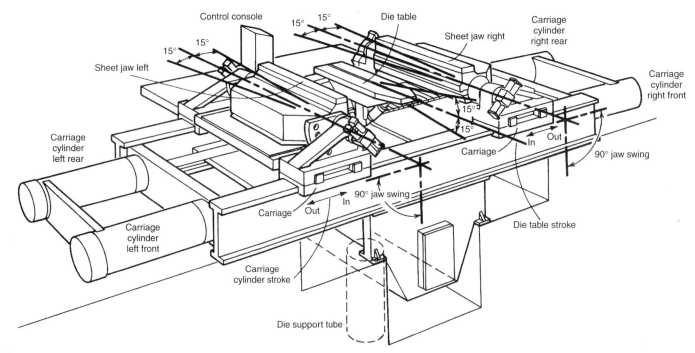

Fig. 14 Transverse stretch-forming machine having both movable and tiltable die table and swiveling, movable jaws

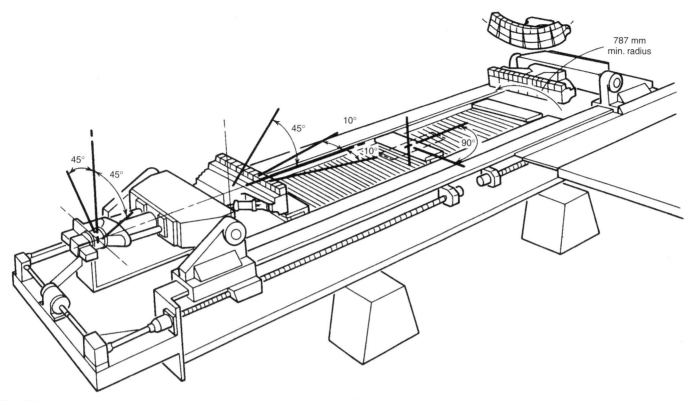

Fig. 15 Longitudinal stretch-forming machine with leadscrew-actuated jaws that can be curved and swiveled both horizontally and vertically

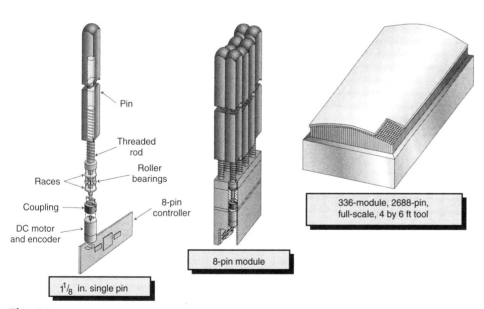

Fig. 16 Schematic views of the modular design of the full-scale reconfigurable stretch-forming tool

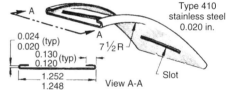

Fig. 17 Guide-vane shroud that was stretch formed without noticeable distortion to the 0.51/0.61 mm (0.020/0.024 in.) groove dimension. Dimensions given in inches

held to ±0.05 mm (±0.002 in.) by close control of raw material, as in the following example.

Example 7: Maintaining Close Tolerances on Grooves While Forming a Guide-Vane Shroud. Because a cover strip had to slide easily, but without play, into the grooves of a stainless steel guide-vane shroud after the vanes were assembled, the width of the grooves had to be held within 0.10 mm (0.004 in.), as shown in Fig. 17. Strip was selected that had thickness variation within ±0.013 mm (±0.0005 in.) because of the close groove tolerances that had to be met. The two U-bends forming the grooves were made in a press brake, and the guide slots were pierced in another operation. The shrouds were then contoured to a 190 mm (7½ in.) radius by stretch forming. The width of the work strip and the width of the grooves were held within the specified tolerances without using filler for support.

The shrouds were produced at a rate of 10 per hour. A typical production run was between 200 and 1000 pieces.

Tolerance Specifications. As shown in the preceding example, it may be necessary to control stock to a small thickness variation if close tolerances on the workpiece must be met. Ordinarily, it is good practice to allow approximately 25% of the finished-part tolerance as the stock thickness tolerance or the tolerance on any preformed dimension that could affect the accuracy of the stretch-formed dimension. In Example 7, the smallest tolerance was ±0.05 mm (±0.002 in.), and the stock thickness variation was controlled to ±0.013 mm (±0.0005 in.) (25% of workpiece tolerance).

Titanium is stretch formed both hot and cold (see the article "Forming of Titanium and Titanium Alloys" in this Volume). In cold stretch forming, shrinkage at right angles to the stretch is ordinarily controlled to ± 0.79 mm ($\pm \frac{1}{32}$ in.) on bends with 229 mm (9 in.) radii. Angular variation on stretch bends in all materials is held to $\pm \frac{1}{2}°$.

In one plant, large rectangular tubing of copper alloy is stretch formed with a filler or with a flexible mandrel similar to those described in the article "Bending and Forming of Tubing" in this Volume. Tubes as large as 102 to 203 mm (4 to 8 in.) square and up to 4.9 m (16 ft) long with 3.2 to 9.5 mm ($\frac{1}{8}$ to $\frac{3}{8}$ in.) walls can be formed. When mandrels are not used, distortion appears as concavity in the face away from the form block and some tapering toward the concave face, as shown in Fig. 18. These tubes are stretch formed to large radii with a ± 0.80 mm ($\pm \frac{1}{32}$ in.) tolerance on the radius of the bend.

In the bending of large tubing, the bend is usually so shallow that the elastic limit of the metal is not reached without gross overbending, unless stretch-forming methods are used. As with conventional methods, overbending leads to unpredictable results. Tolerances on shallow bends can seem deceptively large. To hold a 3.0 m (10 ft) radius within ± 0.80 mm ($\pm \frac{1}{32}$ in.) in a 10° arc means holding an overall tolerance of more than ± 0.13 mm (± 0.005 in.).

Surface Finish

Little can be done in stretch forming to improve surface finish, because tool contact with the surface is incidental. However, some practices can be implemented to help preserve the original finish:

- Avoid overstretching. With most materials, 2 to 4% stretch is sufficient to achieve the results desired in stretch forming. The overstretching of some metals, such as aluminum, simply because they are ductile is a common mistake. This leads to the appearance of stretcher strains or other surface relief effects.
- Plastic wiper shoes can be used in the compression forming or radial draw forming of aluminum alloys to avoid marring the surface. With stainless steel workpieces, well-finished plastic wiper shoes are used with drawing compounds similar to those used for severe deep drawing.

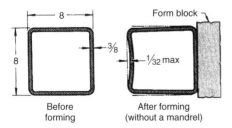

Fig. 18 Typical distortion of square copper alloy tubes in stretch forming. Dimensions given in inches

An extrafine finish is necessary in order to protect the surfaces of aluminum alloys directly in contact with the form block. Special practices used to preserve the finish include cleaning to eliminate abrasive dust particles, the use of polyvinyl chloride instead of a lubricant, and the use of special carrier sheets for protection of the surface during forming.

Stretch Forming Compared with Conventional Drawing

Example 8: Stretch Draw Forming versus Conventional Drawing. This example compares stretch draw forming with drawing for a part manufactured by an automotive supplier. Stretch forming was competitive for quarter-pillar lock panels produced from 0.89 mm (0.035 in.) thick commercial-quality 1008 steel by conventional drawing. In drawing, an 8900 kN (1000 tonf) double-action press with conventional draw dies produced 525 pairs of panels per hour. In stretch draw forming, a 7200 kN (800 tonf) 2.74 by 1.52 m (108 by 60 in.), straight-side, single-action mechanical press with 1.22 m (48 in.) long stretch grippers was used. The production rate was the same as that for the conventional press when automatic loading and unloading were used, and the production cost was less. The process was changed to stretch draw forming.

In high-production forming, the principal disadvantage of stretch forming is the slowness of the hydraulic units used on the grippers, unless pumps of excessively high capacity are used. Mechanical units are available that have rapid response.

Operating Parameters

Size and configuration of the workpiece, material composition, type of forming operation used, machine and tooling used, and production requirements are among the variables that influence stretch forming. Operating parameters, such as force requirements, and the lubricant used must be determined prior to forming.

Force Requirements. The application of excessive tension in stretch forming can cause breakage of the workpiece. On the other hand, too little tension can result in poor contouring, wrinkling, or springback of the formed part. The force capacity of the machine required for stretch forming a part can be calculated by:

$$F = \left(\frac{Y_s + \text{UTS}}{2} \right) A \qquad \text{(Eq 1)}$$

where F is the stretch-forming force (in pounds of force), Y_s is the yield strength of the material (in pounds per square inch), UTS is the ultimate tensile strength of the material (in pounds per square inch), and A is the cross-sectional area of

the workpiece (in square inches). To convert from English units (pounds of force) to metric units (newtons), the force in pounds is multiplied by 4.448.

The estimate of the force required for stretch forming obtained with Eq 1 is generally an average. To compensate for work hardening, friction, more complex contours, and other variables, the force obtained mathematically should be increased by an additional 25% for some applications.

Lubrication. In most stretch forming, little or no lubrication is needed, because the movement between the work metal and the form block is minimal. On sheet steel, residual mill oil is usually sufficient, although some operators spray the stock with light lubricating oil as it enters the forming area. Lubricants are sometimes intentionally avoided because they attract and retain dust particles that could mar the workpiece surface.

In the compression forming of copper alloys, low-carbon steel, and stainless steel, in which a shoe rubs hard against the part or there is considerable movement against the form block, white lead thinned with SAE 30 engine oil can be brushed on the workpiece before forming. In some shops, molybdenum disulfide is similarly used on low-carbon steel. Both lubricants resist heat and pressure and reduce friction. Polyvinyl chloride sheet can be used in place of a lubricant (and to embed dust particles) in the forming of microwave reflectors.

SELECTED REFERENCES

- E.L. Anagnostou, J.M. Papazian, and A.B. Pifko, A Finite Element Simulation System for Sheet Metal Forming Using Reconfigurable Tooling, *Innovations in Processing and Manufacturing of Sheet Materials*, M.Y. Demeri, Ed., TMS, 2001, p 387–403
- M. Groover, *Fundamentals of Modern Manufacturing: Materials, Processes, and Systems*, Prentice-Hall, 1996, p 500–542
- D.E. Hardt, M.C. Boyce, and D.F. Walczyk, A Flexible Forming System for Rapid Response Production of Sheet Metal Parts, *Proceedings of IBEC '93* (Detroit, MI), Society of Automotive Engineers, 1993, p 61–69
- W. Hosford and R. Caddell, *Metal Forming: Mechanics and Metallurgy,* 2nd ed., Prentice-Hall, 1993, p 244–358
- S. Kalpakjian and S. Schmid, *Manufacturing Engineering and Technology,* 4th ed., Prentice-Hall, 2001, p 392–434
- J. Papazian et al., Reconfigurable Tooling for Sheet Metal Forming, *Sheet Metal Forming Technology,* M.Y. Demeri, Ed., TMS, 1999, p 23–38
- J. Papazian et al., Innovative Tooling for Sheet Metal Forming, *Innovations in Processing and Manufacturing of Sheet Materials,* M.Y. Demeri, Ed., TMS, 2001, p 17–31

Superplastic Sheet Forming

A.K. Ghosh, University of Michigan
C.H. Hamilton

SUPERPLASTICITY is a term used to indicate the exceptional ductility that certain metals can exhibit when deformed under proper conditions. The term is most often related to the ductile tensile behavior of the material; however, superplastic deformation has the characteristic of easy deformation under low pressures, and compression deformation characteristics are also described as superplastic. The tensile ductility of superplastic metals typically ranges from 200 to 1000% elongation, but ductilities in excess of 5000% have been reported (Ref 1). Elongations of this magnitude are 1 to 2 orders greater than those observed for conventional metals and alloys, and they are more characteristic of plastics than metals.

Because the capabilities and limitations of sheet metal fabrication are most often determined by the tensile ductility limits, it is clear that there are significant advantages potentially available for forming such materials, provided the high-ductility characteristics observed in the tensile test can be used in production forming processes. This is of course being done, and the number of applications of parts formed by these methods is increasing each year. This article discusses many of the processes and related considerations involved in the forming of superplastic sheet metal parts.

Requirements for Superplasticity

Before discussing the details of the superplastic forming (SPF) processes, it is necessary to review the more important aspects of superplastic material behavior, because some of the specific forming parameters are determined by this behavior. There are several different types of superplasticity in terms of the microstructural mechanisms and deformation conditions, including the following (Ref 2, 3):

- Micrograin superplasticity
- Transformation superplasticity
- Internal stress superplasticity

At this time, only the microgram superplasticity is of importance in the fabrication of parts, and the discussion is limited to this type.

For microgram superplasticity, the high ductilities are observed only under certain conditions, and the basic requirements for this type of superplasticity are:

- Very fine grain size material (of the order of 10 μm, or 400 μin., or finer)
- Relatively high temperature (greater than approximately one-half the absolute melting point), although reduced forming temperature by using finer-grain materials is a goal at this time
- A controlled strain rate, usually 0.0001 to 0.1 s^{-1}

Only a limited number of commercial alloys are superplastic, and these materials are formed using methods and conditions that are different from those used for conventional metals. However, increasingly more and more alloys have been grain-refined to induce superplasticity in them.

Characteristics of Superplastic Metals. For a superplastic metal that is tensile tested under proper conditions of temperature, the observed ductility is seen to vary substantially with strain rate, as shown in Fig. 1 for a zinc-aluminum eutectoid alloy (Ref 4). As shown, there is a maximum in ductility at a specific strain rate, with significant losses in ductility as the strain rate is increased or decreased relative to this maximum. It is well known that the primary factor related to this behavior is the rate of change of flow stress with strain rate, usually measured and reported as m, the strain-rate sensitivity exponent:

$$m = \frac{\partial \ln \sigma}{\partial \ln \dot{\varepsilon}} \qquad \text{(Eq 1)}$$

where σ is the flow stress, and $\dot{\varepsilon}$ is the strain rate.

The aforementioned characteristics of a superplastic alloy indicate that unusual forming capability should be possible with superplastic

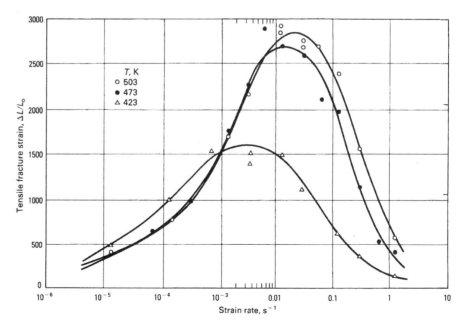

Fig. 1 Tensile fracture strain versus initial strain rate for a Zn-22Al alloy having a grain size of 2.5 μm (100 μin.) tested at temperatures ranging from 423 to 503 K

alloys but that control of the forming process parameters is important to obtain the full potential of this class of material. Such process controls are more demanding than corresponding requirements for conventional forming processes, and the superplastic forming of sheet metals is a technology that is different from the conventional processes. However, superplastic forming offers advantages over other fabrication methods for a number of applications as a result of its unique capability of fabricating complicated components in a single step.

Superplastic Alloys. Because of the stable grain size requirement for a superplastic metal, not all commercially available alloys are superplastic. Many materials have been produced with laboratory or pilot-plant processing, but very few of these have been produced commercially (Ref 3), although greater interest is now seen for their commercial production. As SPF technology develops, alloys are not only being used for aerospace and architectural use but also for automotive application in production volume cars.

A summary of several superplastic alloys is presented in Table 1, along with some of their characteristics. Particularly noteworthy are Ti-6Al-4V, aluminum alloy 7475, and the Supral alloys, which are quite superplastic and are commercially available. Among aluminum alloys, and particularly aluminum-magnesium alloys, alloy 5083 is a favored alloy for superplastic-formed parts in the automotive industry. Optimal superplastic conditions have been determined for fine-grained 5083 aluminum (Ref

5–8); however, the automotive industry generally practices nonoptimal conditions (lower temperature and faster strain rate), to be competitive with mass-production technologies. Low-temperature superplasticity or warm forming of sheet is gaining a great deal of attention for its potential for rapid forming at slightly elevated temperatures. Grain refinement to ultrafine grain sizes (<2 to 3 μm, or 80 to 120 μin.) is becoming more commonplace to permit high ductility. This is currently of interest in relation to magnesium alloys, with potential for lightweight application in the auto industry and space business. Table 2 presents a compilation of superplastic tensile properties of various magnesium alloys, showing that grain refinement is instrumental in making these and other hexagonal close-packed metals highly favorable at low forming temperatures. Titanium alloys are generally superplastic as conventionally produced, although there has not been a need to develop alloy modifications or special mill-processing to make them superplastic. However, this has not been the case with the aluminum alloys, and either special processing (Ref 26) or alloy development (Ref 27) has been necessary to produce superplastic materials. Presently (2006), ultrafine-grain titanium is being produced to reduce the forming temperature of titanium alloys. The Zn-22Al alloy has been the focus of substantial research, because it can be readily processed into the superplastic condition; this alloy has also been made available commercially, but interest in this alloy is much reduced.

Additional information on superplastic alloys is available in the articles "Forming of Aluminum Alloys" and "Forming of Titanium and Titanium Alloys" in this Volume. Superplastic ferrous alloys are discussed in the section "Superplasticity in Iron-Base Alloys" in this article.

Characterization of Superplastic Alloys

The characteristic flow properties of a superplastic metal are exemplified in Fig. 2 for a Ti-6Al-4V alloy tested at 927 °C (1701 °F). The very strong dependence of m on strain rate at various grain sizes (Fig. 3) is typical of superplastic metals, and there is a good correlation between the m-value and superplastic ductility. This relationship is demonstrated most clearly in Ref 28 in a graph of data for a large number of alloys in which the m-value is graphed as a function of elongation. Although the total elongation can also be affected by fracture, the strain-rate sensitivity is a first-order effect. The influence of m on the ductility is understood through mechanics of diffuse neck growth, which slows with increasing value of m, a stabilizing effect of the strain-rate sensitivity of flow stress (Ref 29–31). The strong effect temperature has on superplastic deformation is demonstrated most clearly in Ref 32, in which the ductility of a titanium alloy is graphed as a function of temperature. The elongation of the titanium alloy rises and falls rapidly over a relatively short temperature range, and outside the limits of this temperature span, the ductility is modest in the range of conventional material behavior.

The characterization of superplastic behavior includes the characterization of plastic flow, internal cavitation, and fracture behavior. The processing variables needed for an overall characterization of superplastic behavior are introduced in this section. The parameter that is commonly selected as a measure of superplastic formability is the tensile elongation at the optimal superplastic temperature and strain rate. Because this is a highly strain-rate-sensitive property and because real components can experience significant variations in strain rate during forming, tensile elongation is measured as a function of strain rate. Although this is somewhat time-consuming, an alternative is to determine the strain-rate sensitivity of the flow stress, m, which has been shown to correlate well with tensile elongation for different classes of materials (Ref 28, 29). Measurements of flow stress and strain-rate sensitivity of flow stress can be conducted in a single test and can be used to determine the optimal strain rate for superplastic forming (where m is a maximum). Although strain-rate sensitivity is the dominant parameter in superplastic forming in a great many superplastic alloys, a significant amount of hardening can occur as a function of strain, even at a constant strain rate (Ref 30, 31). This type of

Table 1   Superplastic properties of several aluminum and titanium alloys

Alloy	Test temperature		Strain rate, s^{-1}	Strain-rate sensitivity, m	Elongation, %
	°C	°F			
Aluminum					
Statically recrystallized					
Al-33Cu	400–500	750–930	8×10^{-4}	0.8	400–1000
Al-4.5Zn-4.5Ca	550	1020	8×10^{-3}	0.5	600
Al-6 to 10Zn-1.5Mg-0.2Zr	550	1020	10^{-3}	0.9	1500
Al-5.6Zn-2Mg-1.5Cu-0.2Cr (aluminum alloy 7475)	516	961	2×10^{-4}	0.8–0.9	800–1200
Dynamically recrystallized					
Al-6Cu-0.5Zr (Supral 100)	450	840	10^{-3}	0.3	1000
Al-6Cu-0.35Mg-0.14Si (Supral 220)	450	840	10^{-3}	0.3	900
Al-4Cu-3Li-0.5Zr	450	840	5×10^{-3}	0.5	900
Al-3Cu-2Li-1Mg-0.2Zr	500	930	1.3×10^{-3}	0.4	878
Titanium					
α/β					
Ti-6Al-4V	840–870	1545–1600	1.3×10^{-4} to 10^{-3}	0.75	750–1170
Ti-6Al-5V	850	1560	8×10^{-4}	0.70	700–1100
Ti-6Al-2Sn-4Zr-2Mo	900	1650	2×10^{-4}	0.67	538
Ti-4.5Al-5Mo-1.5Cr	871	1600	2×10^{-4}	0.63–0.81	>510
Ti-6Al-4V-2Ni	815	1499	2×10^{-4}	0.85	720
Ti-6Al-4V-2Co	815	1499	2×10^{-4}	0.53	670
Ti-6Al-4V-2Fe	815	1499	2×10^{-4}	0.54	650
Ti-5Al-2.5Sn	1000	1830	2×10^{-4}	0.49	420
β and near β					
Ti-15V-3Cr-3Sn-3Al	815	1499	2×10^{-4}	0.5	229
Ti-13Cr-11V-3Al	800	1470	...	...	<150
Ti-8Mn	750	1380	...	0.43	150
Ti-15Mo	800	1470	...	0.60	100
α					
CP Ti	850	1560	1.7×10^{-4}	...	115

Table 2 Superplastic properties of selected magnesium alloys

Alloy	Processing method	Grain size		Temperature		Strain rate, s^{-1}	Elongation, %	Gage length	
		μm	μin.	°C	°F			mm	in.
AZ31	Hot rolled	130	5120	325	615	10^{-5}	130	. . .	. . .
	Hot rolled	130	5120	375	710	3×10^{-5}	196	18	0.7
	Hot rolled	53	2090	400	750	1.3×10^{-4}	420	25	1.0
	Hot extruded	15	590	177	351	10^{-5}	120	21	0.8
	Hot rolled	12	472	250–450	480–840	10^{-4} to 2×10^{-4}	360	15	0.6
	Hot extruded	5	200	325	615	10^{-4}	600	. . .	. . .
	Hot rolled	4.5	177	400	750	7×10^{-4}	360	12	0.5
	Hot extruded	2.9	114	300	570	10^{-3}	740	5.5	0.22
AZ61	Hot extruded	17	669	325	615	10^{-5}	390	. . .	. . .
	Hot rolled	17	669	400	750	3×10^{-5}	400	. . .	. . .
	Hot rolled	8.7	343	400	750	2×10^{-4}	580	6	0.24
AZ91	Hot rolled	39.5	1555	300	570	1.5×10^{-3}	604	10	0.4
	Hot rolled	11	433	350	660	10^{-3}	455	10	0.4
	Hot extruded	4.1	161	250	480	3×10^{-4}	425	5	0.2
	ECAP(a)	1.2	47	300	570	5×10^{-4}	270	9	0.35
	ECAP(a)	0.7	28	200	390	6×10^{-5}	660	5	0.2
	ECAP(a)	0.7	28	200	390	3.3×10^{-4}	840	5	0.2
ZK60	Hot extruded	2 to 5	79 to 180	400	750	10^{-3}	550	. . .	. . .
	Hot extruded	2.2(b)	87	177	351	10^{-5}	220	21	0.8
	Hot extruded	2.4	94	300	570	4×10^{-4}	730	5	0.2
ZK61	Hot extruded	1.2	47	300	570	10^{-1}	432	5	0.2
Mg-4Y-3RE	Hot extruded	1.5	59	400	750	4×10^{-3}	1517	10	0.4
	Hot extruded	1.5	59	400	750	4×10^{-1}	358	10	0.4
Mg-8.3Al-8.1Ga	Hot extruded	0.6	24	300	570	10^{-2}	1080	5	0.2

(a) ECAP, equal-channel angular pressed. (b) The microstructure has a bimodal grain size distribution, containing 80% volume fraction of fine grains of size 2 μm (80 μin.) and 20% volume fraction of coarse grains of size 25 μm (985 μin.). Source: Ref 9–25

strain hardening is related primarily to the grain growth that occurs during superplastic forming. At higher strain rates, strain hardening is associated with dislocation cell formation in the classical manner, and grain growth is not observed.

Forming temperature is just as important a variable in superplastic forming as the strain rate. Temperature variation in a forming die is a primary source of localized thinning. Characterization of material behavior must include not only determination of the optimal superplastic temperature but also the sensitivity of flow stress and elongation to temperature. A large temperature sensitivity of flow stress is not desirable, because local hot spots will lead to severe strain localization. When strain localization and necking are the dominant modes of failure, it has been shown that tensile elongation is related to the m-value in a predictable manner (Ref 32). However, when fracture intervenes, the m-value does not provide sufficient quantitative characterization, although within the same alloy system it still provides a qualitative comparison. Fracture is therefore an important consideration in most superplastic materials of engineering application and is preceded by internal cavitation. An exception is materials with anomalously high diffusivity (for example, Ti-6Al-4V alloys), which do not show any evidence of internal cavitation or fracture under typical superplastic forming conditions. Superplastic materials that exhibit cavitation at inclusions, triple points, and second-phase particles generally fail by the interlinking of growing cavities.

Stress-strain-rate behavior is usually characterized by a step strain-rate test, in which strain rate is increased in successive steps and an attempt is made to measure the corresponding steady (or saturated) flow stress. A constant flow stress indicates a negative loading rate, which occurs at a point somewhat beyond the load

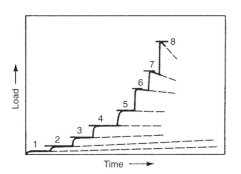

Fig. 2 Schematic plot of load versus time for the tensile deformation of a superplastic Ti-6Al-4V alloy at 927 °C (1701 °F) and at progressively increasing crosshead velocities (step strain-rate test)

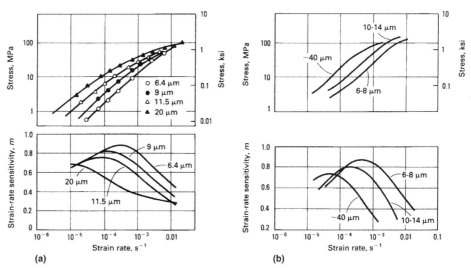

Fig. 3 Stress-strain-rate plots and the corresponding strain-rate sensitivity plots from step strain-rate tests for (a) Ti-6Al-4V at 927 °C (1701 °F) and (b) 7475 aluminum alloy at 516 °C (961 °F), which are both at superplastic temperature

maximum. However, even if the load maximum is used as a criterion for calculating these stresses, it can be shown that the error is negligible, except when considerable strain hardening occurs. In these cases, characterization of strain hardening is carried out separately. For these cases, the knee of the stress-strain curve, past the "yield" point, is the most reasonable representation of isostructural flow stress. Various arguments have been put forward for the proper selection of flow stress from transient loading response (Ref 33–35). However, because of the changing plastic-strain rate during this test, selection of data at the elastic limit from the rapidly rising portion of the load curve is inappropriate.

Figure 2 shows a schematic load versus time plot during a step strain-rate test of a typical superplastic alloy. The interesting features are:

- At the low crosshead speeds, the load does not reach a maximum but continues to show a gradual rise.
- At some intermediate speed, load reaches a constant plateau.
- At higher speeds, it peaks and begins to show a sharp drop.

The load increase at the low strain rates, in spite of a decrease in applied strain rate, indicates hardening of the material with imposed strain. A part of this hardening is due to a rise in plastic-strain rate, which occurs gradually when crosshead speed is low. However, the extent of hardening observed is considerable and does not saturate even after significant plastic strain. This type of hardening has been observed in many superplastic materials, such as aluminum-copper eutectic, aluminum-magnesium, and Al-Zn-Cu-Mg alloys, and is generally attributed to grain growth occurring during deformation (Ref 36, 37).

The criticism against the use of strain-rate jump tests and the selection of maximum load points for obtaining stress values is that the strain automatically becomes a variable along the σ-$\dot{\varepsilon}$ curve. This could be avoided if load relaxation test results were used to derive σ-$\dot{\varepsilon}$ curves. However, more complex transient effects are associated with load relaxation tests, and the results may not be meaningful for a forming application in which strain rate generally increases with accompanying strain. The step strain-rate test is therefore believed to be a logical test method for use in superplastic forming applications, provided the data are obtained with very little strain accumulation.

The σ-$\dot{\varepsilon}$ data for Ti-6Al-4V alloy and 7475 aluminum alloy deformed in the superplastic temperature range are presented in Fig. 3 for a variety of grain sizes. The total accumulated strain is generally less than 0.25 in these tests. At lower strain rates ($<5 \times 10^{-4}$ s^{-1}), strain hardening does not permit the establishment of a load maximum. To characterize flow stress free from grain-growth hardening effects, stresses must be selected soon after the elastic portion. If the load

rises slowly, the plastic-strain rate, $\dot{\varepsilon}_p$, can be obtained from:

$$\dot{\varepsilon}_p = \dot{\varepsilon}_t - \frac{1}{E}\left[\frac{\dot{P}}{A} + \frac{P}{A}\dot{\varepsilon}_t\right] \qquad \text{(Eq 2)}$$

where $\dot{\varepsilon}_t$ is the applied strain rate, E is Young's modulus, P is the load, $\dot{P}$ is the loading rate, and A is the instantaneous area.

The second term within the brackets in Eq 2 can generally be neglected, and even the first term can be neglected for most metals where $\dot{P}/A < 0.015$ MPa s^{-1} (0.0022 ksi s^{-1}). Therefore, for most purposes, $\dot{\varepsilon}_p = \dot{\varepsilon}_t$ is a reasonable assumption as long as the loading rate is low (it need not be zero). The stress versus strain-rate data plotted on the basis of such small strain accumulation serve as the initial behavior of the superplastic material. Deformation produces changes in this behavior, however, and a stress versus strain-rate plot taken after considerable plastic strain exhibits a higher stress level compared with the initial curve (Ref 30).

The proper method for determining m from step strain-rate test results is to obtain the slope of the curve of best fit through the log σ versus log $\dot{\varepsilon}$ data. The determination of m from two consecutive strain-rate jumps assumes a constant m over that strain-rate range and introduces an error that is dependent on the size of the range of jump strain rates. Figure 3 shows that grain size has a strong effect on flow stress and m-value in the superplastic range. Because of the sigmoidal shape of the stress-strain-rate curves, m-values exhibit a maximum at an intermediate strain-rate, and the m-peak shifts to higher strain rates with decreasing grain size. In titanium and aluminum alloys under optimal temperature and strain rate, the value of the m-peak is typically in the range of 0.7 to 0.9 and increases with decreasing grain size. The value of m also exhibits a maximum as a function of temperature. The effects of grain size and temperature are

closely tied to the diffusional creep contribution during superplastic flow (Ref 38, 39). In the power-law creep regime (at higher strain rates), however, the stress values tend to converge for different grain sizes as the grain size dependence decreases.

Grain Size Distribution Effects on Stress-Strain Rate. Grain size has a profound influence on the superplasticity of metals. When the grain size is fine, the flow stress is low, the value of m is generally high, and the tensile elongation is greater. Characterization of grain size is therefore important in the overall characterization of superplasticity. However, because polycrystalline aggregates, in general, possess a distribution of grain size, it is not very meaningful to assign a fixed grain size to metals. The nature of the distribution has also been shown to influence the stress-strain-rate curve (Ref 40, 41). A few coarse grains in an otherwise fine grain structure can control the strain-rate range over which m is high and may, in some cases, cause the appearance of a threshold stress. The important effect of grain size distribution in real materials is to produce a broader strain-rate range for high m, that is, a relatively high m ($m > 0.5$) over almost three decades in strain rate; this is a transition region between power-law creep and diffusional creep (Fig. 4).

Stress-Strain Behavior. Superplastic metals are generally regarded as ideally rate sensitive; that is, no strain hardening occurs during deformation. However, grain-growth-induced strain hardening occurs, and it can be quite significant in some cases (Ref 30, 31, 36, 37). In testing superplastic materials, constant crosshead speed leads to a decreasing strain rate within the specimen gage length, particularly at large tensile strains. An effort to maintain constant strain rate requires that:

- The specimen fillet region be reduced to a minimum to reduce its contribution to the overall extension

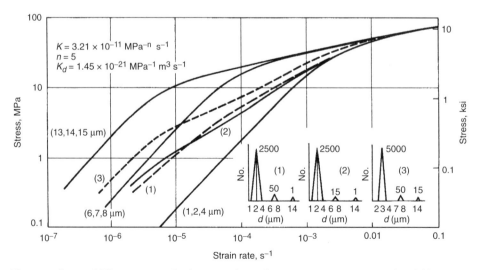

Fig. 4 Influence of different grain size distributions on the steady-state stress-strain-rate curves. The solid lines are for essentially singular grain sizes. Distribution in grain size yields a gradual transition in stress-strain-rate curves.

- The crosshead speed be programmed to increase with specimen elongation in order to maintain a constant strain rate (assuming the elongation to arise uniformly out of the gage length)

An on-line computer serves this function by altering the crosshead speed, v, in the following manner:

$$v = l_0 \dot{\varepsilon}_t \exp(\dot{\varepsilon}_t) \qquad (\text{Eq 3})$$

where l_0 is the initial gage length, $\dot{\varepsilon}_t$ is the applied strain rate, and t is the time from the start of test. Temperature control during the test also must be accurate (within 2 °C, or 4 °F) in order to avoid

any localized deformation. The crosshead speed control process is a function of strain distribution within the specimen, which is now controlled by using a considerably more sophisticated computer program, as illustrated in Ref 42.

Figure 5 shows the stress-strain curves for titanium and aluminum alloys obtained at various constant strain rates. The extent of hardening is quite large and appears to produce a linear stress-strain behavior. In diffusional creep, one can expect $\sigma \sim d^2$; if grain growth kinetics are such that $d \propto t^p$, where t is time, p is the exponent, and d is grain size, then $\sigma \propto t^{2p}$. When p is approximately 0.5, a linear hardening behavior would be expected. If diffusional creep due to

grain-boundary transport is considered, $\sigma \propto d^3$, and p may be as low as 0.33, thus causing the appearance of linear hardening. In reality, the value of the hardening exponent, $2p$ or $3p$, may be somewhat less than unity. A small component of strain-rate-dependent increase of flow stress also influences these experimental results, because strain rate does not remain absolutely constant during the tests. The last portion of the stress-strain curves (shown dashed or with stress drop) is where nonuniformity of deformation within the gage length makes stress measurements incorrect.

The significant strain hardening observed in Fig. 5 is believed to be due to concurrent grain growth, the evidence for which is shown in Fig. 6. In the case of Ti-6Al-4V, the grain growth kinetics appear to some degree to be a function of strain rate as well. It is clear, however, that slower strain rates produce the largest grain sizes, particularly because of the long exposure times, and in all cases, dynamic grain growth is significantly greater than static growth.

Concurrent grain growth effects can also influence the stress-strain-rate data measured at the low strain rates. This can be understood from the consideration of a constitutive equation of the form:

$$\dot{\varepsilon} = \frac{A\Omega D_{\text{eff}}}{kTd^3}\sigma + K\sigma^n \qquad (\text{Eq 4})$$

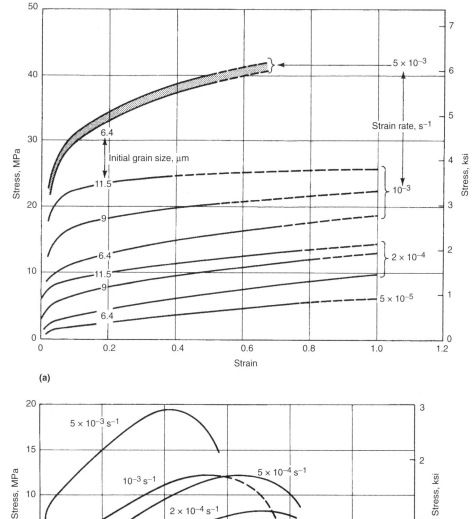

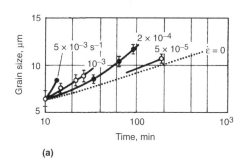

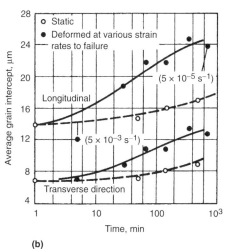

Fig. 5 Stress-strain curves at various constant strain rates for (a) Ti-6Al-4V at 927 °C (1701 °F) and (b) 7475 aluminum alloy at 516 °C (961 °F). Initial grain size: approximately 10 to 14 μm (395 to 550 μin.)

Fig. 6 Grain growth kinetics at four different tensile strain rates compared with static kinetics for (a) Ti-6Al-4V at 927 °C (1701 °F) with 6.4 μm (250 μin.) initial grain size and (b) 7475 aluminum at 515 °C (959 °F)

where A is a constant, Ω is the atomic volume, D_{eff} is the effective diffusion coefficient, k is Boltzmann's constant, T is absolute temperature, K is the constant for power-law creep (containing dependencies on temperature and shear modulus), and n is the power-law creep exponent. Li et al. (Ref 43) provided a detailed analysis of this effect for aluminum-magnesium superplastic alloys. Grain growth increases the proportion of strain contributed by dislocation creep, which reduces the value of m and causes flow hardening. Designating ($A\Omega D_{eff}/kT$ by A', and if dynamic grain growth is given by $d \sim d_o t^p$, the diffusional creep portion can be rewritten as:

$$\dot{\varepsilon}_d = \frac{A'\sigma}{d_o^3 t^{3p}} \qquad \text{(Eq 5)}$$

where d_o refers to the initial grain size. During the step strain-rate test with constant strain-rate segments, if a stress of σ_1 is obtained at a strain rate $\dot{\varepsilon}_1$ and a strain of ε^1, and σ_2 at $\dot{\varepsilon}_2$ and ε_2, from Eq 4, then:

$$\frac{\sigma_2}{\sigma_1} = \left(\frac{\varepsilon_2}{\varepsilon_1}\right)^{3p} \left(\frac{\dot{\varepsilon}_2}{\dot{\varepsilon}_1}\right)^{(1-3p)} \qquad \text{(Eq 6)}$$

because $t = \varepsilon/\dot{\varepsilon}$ for a constant strain-rate test. The strain-rate sensitivity m can be given by:

$$m = \frac{\log\left(\dfrac{\sigma_2}{\sigma_1}\right)}{\log\left(\dfrac{\dot{\varepsilon}_2}{\dot{\varepsilon}_1}\right)} = (1-3p) + 3p\frac{\log\left(\dfrac{\varepsilon_2}{\varepsilon_1}\right)}{\log\left(\dfrac{\dot{\varepsilon}_2}{\dot{\varepsilon}_1}\right)} \quad \text{(Eq 7)}$$

Therefore, the amount of strain accumulated in each strain-rate step will influence the value of m through its influence on dynamic grain growth kinetics. If $p = 0.25$ (Ref 30), then in the limit, m could vary between 0.25 and 1 and may be responsible for the apparent threshold stresslike behavior at low strain rates. The concurrent grain-growth-induced hardening thus makes it difficult to describe superplastic flow behavior accurately.

Step strain-rate tests conducted with minimum strain accumulation are used as the initial stress-strain-rate behavior of the material, and the strain-hardening component is measured and added separately. The rationale for this is seen in Fig. 7, in which two step strain-rate tests con-

ducted with intermediate superplastic deformation clearly show that the hardening effect is not negligible. Although it may not be obvious from the slopes of these plots that m is also influenced by strain, a drop in m occurs due to concurrent grain growth (Fig. 3). The strain dependence of m can also be determined during constant strain-rate tests by making incremental strain-rate changes (Fig. 8a) of a small magnitude. With strain-rate changes of 25 to 40% maintained over a 2 to 3% plastic strain, the microstructure may not be altered significantly after return to the original strain rate. Figure 8(b) shows the decrease in m measured from these tests. Similar results have also been observed in aluminum alloys (Ref 44). The drop in m is usually more rapid at the higher strain rates and may be related to classical hardening due to dislocation buildup, which is also faster at high strain rates. Even though strain hardening has important implications for superplastic flow, the rate sensitivity of strain hardening is usually small, and m is still the parameter of greater interest. The article "Constitutive Equations" in this Volume shows extensive details and data on concurrent grain growth during superplastic deformation and its role in altering strain-rate sensitivity of alloys.

Important considerations in the selection and use of a superplastic alloy are the total elongation capability, the stability of the superplastic microstructure at high temperature, the latitude

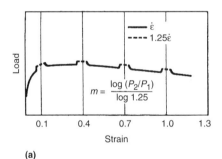

(a)

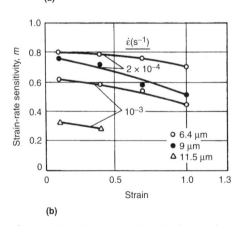

(b)

of the temperature and strain-rate range over which superplasticity is observed, and the rate of development of cavitation during superplastic deformation. All of these factors can change from lot to lot of the material, and it is generally advisable to check each lot for the superplastic properties as well as design properties. Information on superplasticity in specific materials is available in the articles "Forming of Aluminum Alloys" and "Forming of Titanium and Titanium Alloys" in this Volume.

Superplastic Forming Processes

A number of methods and techniques have been reported for forming superplastic materials, each of which has a unique capability and develops a unique set of forming characteristics (Ref 3, 45). The following are forming methods that have been used with superplastic alloys:

- Blow forming
- Vacuum forming
- Thermoforming
- Deep drawing
- Superplastic forming/diffusion bonding (DB)
- Forging
- Extrusion
- Dieless drawing

Only those processes that relate to sheet metal forming are described in this section. Superplasticity as related to bulk forming operations is discussed in the section "Superplasticity in Iron-Base Alloys" in this article and in the article "Isothermal and Hot-Die Forging" in *Metalworking: Bulk Forming*, Volume 14A of *ASM Handbook*, 2005.

Blow forming and vacuum forming are basically the same process (sometimes called stretch forming) in that a gas pressure differential is imposed on the superplastic diaphragm, causing the material to form into the die configuration (Ref 3, 45–47). In vacuum forming, the applied pressure is limited to atmospheric pressure (that is, 100 kPa, or 15 psi), and the forming rate and capability are therefore limited. With blow forming, additional pressure is applied from a gas pressure reservoir, and the only limitations are related to the pressure rating of the system and the pressure of the gas source. A maximum pressure of 690 to 3400 kPa (100 to 500 psi) is typically used in this process.

The blow forming method is illustrated in Fig. 9, which shows a cross section of the dies and forming diaphragm. In this process, the dies and sheet material are normally maintained at the forming temperature, and the gas pressure is imposed over the sheet, causing the sheet to form into the lower die; the gas within the lower die chamber is simply vented to atmosphere. The lower die chamber can also be held under vacuum, or a back pressure can be imposed to suppress cavitation if necessary. The use of back pressure to control or prevent cavitation is discussed in the sections "Pressure Profiling" and

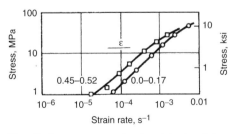

Fig. 7 Stress versus strain-rate plots for 6.4 μm (250 μin.) grain size Ti-6Al-4V at 927 °C (1701 °F) initially (that is, up to ε = 0.17) and after a strain of 0.45 at a rate of 2×10^{-4} s^{-1} showing the hardening contribution due to the deformation

Fig. 8 Schematic representation (a) showing how instantaneous measurements of m can be made at periodic intervals during the tensile test by strain-rate increments of 25%. (b) Corresponding m-value as a function of strain for Ti-6Al-4V at 927 °C (1701 °F)

"Cavitation and Cavitation Control" in this article.

The rate of pressurization is normally established such that the induced strain rates in the forming sheet are maintained in the superplastic range. The rate of pressurization is determined by trial and error or by the application of analytical modeling methods (Ref 48–51). This pressure is generally applied slowly rather than abruptly in order to prevent too rapid a strain rate and consequent rupturing of the part.

The periphery of the sheet is held in a fixed position and does not draw-in, as would be the case in typical deep-drawing processes. It is common to use a raised land (seal bead) machined into the tooling around the periphery, as shown in Fig. 10, to secure the sheet from slippage and draw-in and to form an airtight seal in order to prevent leakage of the forming gas. Therefore, the sheet alloy stretches into the die cavity, and all of the material used to form the part comes from the sheet overlying the die cavity. This results in considerable thinning of the sheet for complex and deep-drawn parts, and it can also result in significant gradients in the thickness in the finished part.

This process is being widely used to fabricate structural and ornamental parts from titanium, aluminum, and other metals. An example of the process applied to the forming of a titanium aircraft nacelle frame is illustrated in Fig. 11 (Ref 52). In this case, the forming is conducted at approximately 900 °C (1650 °F), and inert gas (argon) is used on both sides of the sheet to minimize oxidation and related detrimental surface degradation due to the reactivity of titanium. The use of such protective gases is not usually necessary for aluminum alloys.

Large, complex parts can be readily formed by this method; it has the advantage of no moving die components (that is, no double-acting mechanisms) and does not require mated die components. Multiple parts can be formed in a single process cycle, thus permitting an increase in the production rate for some parts.

Thermoforming methods have been adopted from plastics technology for the forming of superplastic metals, and these methods sometimes use a moving or adjustable die member in conjunction with gas pressure or vacuum (Ref 46, 47, 53). Figure 12 shows two examples of thermoforming methods. In Fig. 12(a), an undersized male die punch is used to stretch form the superplastic sheet, followed by application of gas pressure to force the sheet material against the configurational die to complete the shaping operation. In Fig. 12(b), the first step involves blowing a bubble in the sheet away from the tool.

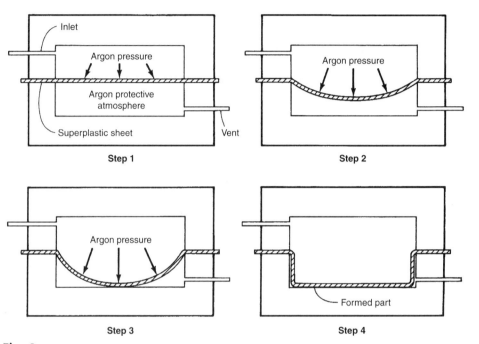

Fig. 9 Schematic of the blow forming technique for superplastic forming

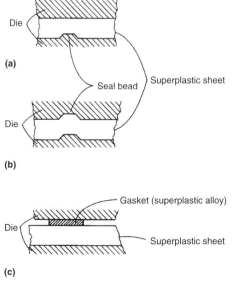

Fig. 10 Various sealing methods that have been used around the sheet to provide a pressure seal suitable for containing the gas pressure during forming. Sections (a) and (b) use seal beads machined into the tooling, and (c) shows the use of a superplastic frame used as a soft gasket.

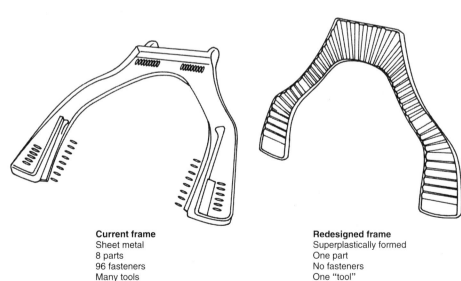

Current frame
Sheet metal
8 parts
96 fasteners
Many tools

Redesigned frame
Superplastically formed
One part
No fasteners
One "tool"

Fig. 11 Ti-6Al-4V aircraft nacelle frame that was redesigned from a conventional configuration to one suitable for superplastic forming having fewer parts and fasteners. The redesigned version of this B-1B aircraft component, having 0.161 m^2 (250 in.^{2}) plan view area, resulted in a 33% weight savings and a 55% cost savings over a conventional multiple-piece assembly.

The male tool is then moved into the bubble, and the pressure is reversed to cause the bubble to conform to the shape of the tool.

Figure 13 illustrates two other methods that employ a movable die member that aids in pre-stretching the sheet material before gas pressure is applied. In this case, the gas pressure is applied from the same side of the sheet as the moving die. These techniques provide ways of producing different shapes of parts and are effective for controlling the thinning characteristics of the finished part.

Deep Drawing. Although deep-drawing studies have been conducted with superplastic metals, this process does not appear to offer many significant advantages in the forming of superplastic materials. Deep drawing depends on strain hardening to achieve the required formability and to prevent thinning and rupture during forming. Superplastic materials do not strain harden to any great extent, but they depend on the high strain-rate hardening for their forming characteristics, and this property seems to offer little aid to deep drawing.

The difficulty is that, in order to draw-in the flange, the material in contact with the punch nose, as well as that in the sidewall, must work harden to carry the increasing stresses required to draw-in increasing amounts of the flange. At superplastic temperatures, no significant work hardening occurs, and the punch typically pierces the blank, or the blank fails in the cup walls if the frictional constraint between the punch and the blank is high. However, in studies on the zinc-aluminum alloy, a maximum draw ratio of 2 to 1 has been developed under optimized conditions (Ref 54).

A technique that tends to improve the drawability of superplastic alloys is discussed in Ref 55. This method (Fig. 14) uses a punch cooled to a temperature below that of the forming blank, while the hold-down tooling is maintained at the forming temperature. It was demonstrated that this differential temperature technique permitted an increase in the limiting draw ratio from less than 2.4 to 1 for isothermal conditions to more than 3.75 to 1 for the differential temperature method. The thinning characteristics for this process are also shown in Fig. 14. Slight thinning can be seen to occur over the (cold) punch nose, and substantial thinning is seen in the material adjacent to the punch. The extent of the thinning in the material adjacent to the punch depends on the blank hold-down load but increases with increasing blank diameter (draw ratio) and decreasing punch speed. Because ironing was not used in these forming tests, thickness increases were observed at the greater distances from the pole of the cup where substantial draw-in of the material occurred.

Another concept evaluated to explore the deep-drawing capability is discussed in Ref 56. This method uses high-pressure oil around a blank periphery to aid in the drawing. It is actually a combined extrusion and drawing process. In this study, a tin-lead eutectic was used, which permitted processing at ambient temperature. Good control of wall thickness was achieved, but the applicability of the process to alloys requiring high temperatures has yet to be demonstrated.

SPF/DB Processes. It is now well established that a number of unique processes are available if joining methods, such as diffusion bonding, can be combined with superplastic forming; these processes are generally referred to as SPF/DB processes (Ref 45, 56, 58). Although diffusion bonding is not a sheet metal process, it complements and enhances superplastic forming to such an extent that the two processes must be discussed together.

The SPF/DB processes evolved as natural combinations of the SPF and DB processes because the process temperature requirements of both are similar. The low flow-stress properties characteristic of the superplastic alloys aid the DB pressure requirements, and it has been found that many superplastic alloys can be diffusion bonded under pressures in the same low range as that used for SPF processing (that is, of the order of 2100 to 3400 kPa, or 300 to 500 psi). The SPF method used with SPF/DB to date is commonly blow forming.

The resulting SPF/DB process consists of the following variations:

- Forming of a single sheet onto preplaced details, followed by diffusion bonding (Fig. 15)
- Diffusion bonding of two sheets at selected locations, followed by the forming of one or both into a die (Fig. 16); the reverse sequence can also be used

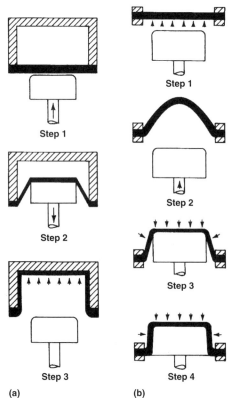

Fig. 12 Examples of thermoforming methods used for superplastic forming. (a) Plug-assisted forming into a female die cavity. (b) Snap-back forming over a male die that is moved up into the sheet

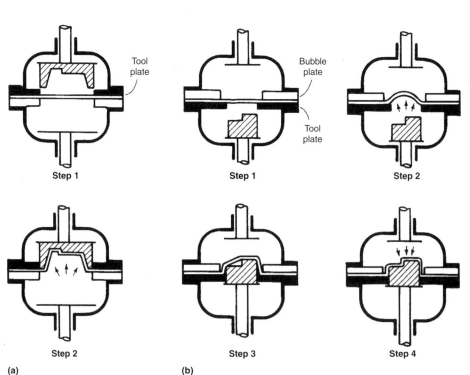

Fig. 13 Thermoforming methods that use gas pressure and movable tools to produce parts from superplastic alloys. (a) Female forming. (b) Male forming

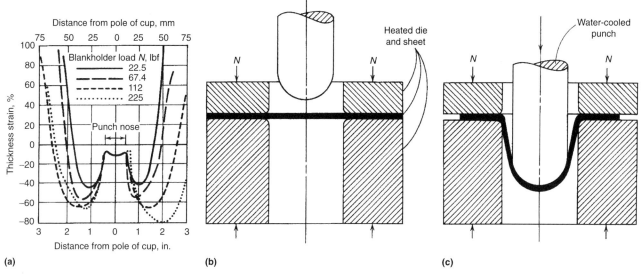

Fig. 14 Punch setup for deep drawing a superplastic sheet. (a) Plot showing thinning characteristics of a 59.9 mm (2.36 in.) diameter Zn-21Al-1Cu-0.1Mg heated sheet that was formed using the 160 mm (6.3 in.) diameter water-cooled punch setup illustrated in (b) and (c). N, blankholder load. Both the die and zinc alloy sheet were heated to 230 °C (445 °F). Punch speed was 33.0 mm/min (1.3 in./min), and maximum punch load was 2150 N (483 lbf). Draw ratio was 3.75. See text for discussion.

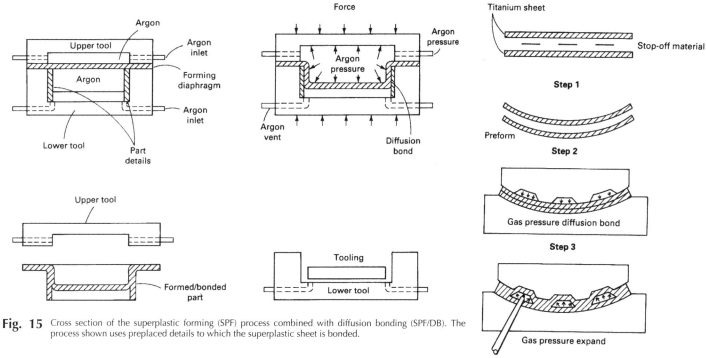

Fig. 15 Cross section of the superplastic forming (SPF) process combined with diffusion bonding (SPF/DB). The process shown uses preplaced details to which the superplastic sheet is bonded.

Fig. 16 Operations required for joining two sheets of superplastic alloy using the superplastic forming/diffusion bonding process

- Diffusion bonding of three or more sheets at selected locations under gas pressure, followed by expansion under internal gas pressure, which forms the outer two sheets into a die; in the process, the center sheet(s) is stretched into a core configuration (Fig. 17)

A few techniques have been used to develop diffusion bonds in predetermined local areas. One of these involves the use of a parting agent, or stop-off material, between the sheets in the local areas where no bonding is desired. Suitable stop-off materials may depend on the alloy being bonded and the temperature being used. For example, yttria or boron nitride has been successfully employed to stop-off titanium alloys processed to temperatures of at least 930 °C (1705 °F). Such stop-off materials can be suspended in an appropriate binder, such as acrylic. After a DB operation, the area of the stop-off pattern is not bonded, and gas can be applied internally along this pattern, thus causing the external sheets to be separated and formed by expanding into a surrounding die.

A variation of the aforementioned method involves the use of a minimum of four sheets to make a sandwich panel. The external (for example, skin) sheets are expanded, the inner two sheets are then bonded (or welded) to define the core structure, and finally the core is expanded to bond with the external sheets and to complete formation of the sandwich structure. This sequence is shown in Fig. 18.

Forming Equipment and Tooling

The forming of superplastic sheet materials involves methods that are generally different from those used in other, more conventional sheet forming processes. The forming environmental conditions are also different. Therefore, the equipment and tooling used are generally different.

Forming Equipment. For the blow forming and vacuum forming methods, there is a need to provide constraint to the forming tools in order to counteract the forming gas pressure. In addition, a seal is generally required at the interface between the sheet and the tool around the periphery in order to prevent leakage of the gas pressure. A press is typically used to meet these requirements. Hydraulic presses and mechanical clamping systems have been used, and each has advantages and disadvantages. The hydraulic press can be loaded and unloaded fairly rapidly, but it requires a significant capital investment. The mechanical clamping systems are much less expensive but are more cumbersome to load and unload. Recently, robotic systems have been coupled with a hydraulic press to aid the loading and unloading, and this type of advanced system is especially beneficial for high-temperature forming operations such as titanium alloy SPF processing.

The hydraulic presses used include both single-action and multiple-action systems (Ref 45, 46, 53). In the single-action press, the press applies the constraining pressure only. In the multiple-action press, the press can also move

dies into the forming sheet and effectively aid in the control of the thinning gradients (Fig. 12, 13).

The heating system used must be tailored to the temperature required and the allowable thermal gradients. The most common heat source is electrical heating, in which resistance heating elements are embedded in ceramic or metal pressure plates placed between the tooling and the press platens. This allows for good control of the temperature and provides a clean source of energy. The heating platens can be arranged in sections of heating elements, and each section can be controlled by independent temperature controllers to minimize thermal gradients in the forming die assembly. Significant thermal gradients can lead to excessive thinning or rupture of the sheet during forming.

Tooling Materials. The tooling used in the SPF process is generally heated to the forming temperature, and it is subjected to internal gas pressure and pressing clamping loads. The internal gas pressure is typically less than approximately 3400 kPa (500 psi), and this is usually not the critical design factor for SPF tools. More important are the clamping loads and thermal stresses encountered during heatup and cooldown and the environmental conditions. The thermal stresses can cause permanent distortions in the die, and this is controlled by selection of a material that has good strength and creep resistance at the forming temperature. Slow heating and cooling of the tooling can reduce the thermal stresses. Materials with a low coefficient of thermal expansion and those that do not undergo a phase transformation during heating and cooling are preferred for the high-temperature SPF processes.

The environmental conditions can be severe for the forming of high-temperature materials, such as the titanium alloys, iron alloys, nickel

alloys, and other high-temperature metals. Oxidation can alter the surface condition of the tooling, thus affecting the surface quality of the SPF part produced and eventually affecting the dimensional characteristics.

Another important environmental factor is the compatibility between the superplastic sheet and the tooling, and the compatibility of these with the stop-off materials that may be used. Interdiffusion at the tooling/sheet interface can result in the degradation of both of these materials. Reactive metals, such as titanium alloys, are especially prone to this type of problem. Tooling materials that have been found to be successful with titanium alloys are the Fe-22Cr-4Ni-9Mn alloy and similar materials. Parting, or stop-off, agents are also helpful in minimizing the interaction, and materials such as boron nitride and yttrium oxide have been successfully used. Generally, materials with a low solid solubility in the sheet are good candidates for compatibility.

A variety of materials have been used for SPF tooling, including metals and alloys, ceramics, and graphite (Ref 59). Metal tools, such as H-13 tool steels, are good for forming aluminum parts and are preferred for large production quantities, such as 100 parts or more. Graphite tools are suitable for approximately 100 parts, and they are readily hand worked, although there is a problem with shop cleanliness with graphite. Ceramics can be cast into the desired shape and are therefore inexpensive for a variety of large parts. Because the ceramic is subject to cracking and rapid degradation, it requires frequent repair. Ceramic tools are considered for small production quantities, usually less than approximately 10 parts.

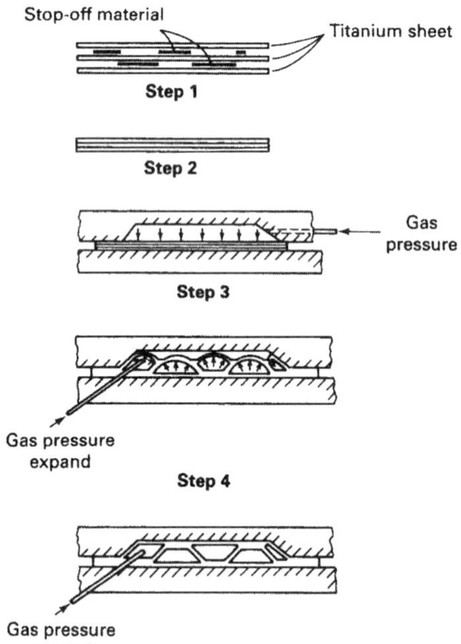

Thinning Characteristics

To take advantage of the very high elongations possible with superplastic metals, it is necessary to accept the accompanying significant thinning in the sheet material. This thinning is a natural consequence of the deformation conditions. For superplastic deformation, elastic strains are negligible; therefore, constancy of volume can be assumed. From this consideration, the sum of the plastic strains is 0, and tensile strain in one direction must be balanced by compressive (negative) strains in another. The strains are:

$$\varepsilon_1 + \varepsilon_2 + \varepsilon_3 = 0 \qquad \text{(Eq 8)}$$

where ε is the strain, and the subscripts indicate the principal directions. For example, in a sheet forming operation under plane-strain conditions, $\varepsilon_2 = 0$ and $\varepsilon_3 = -\varepsilon_1$. In this case, the thinning strain (for example, ε_3) is equal and opposite to the longitudinal tensile strain, and the thinning will therefore match the tensile deformation. For large tensile strains, the thinning will be correspondingly large. Accordingly, as the thinning increases, the tendency to develop thinning gradients also increases.

Stop-off material

Titanium sheet

Step 1

Step 2

Gas pressure

Step 3

Gas pressure expand

Step 4

Gas pressure expand

Step 5

Fig. 17 Operations required for joining three sheets of superplastic alloy using the superplastic forming/diffusion bonding process

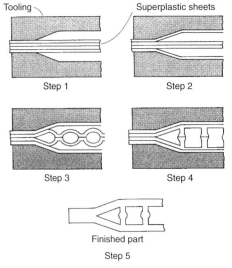

Tooling

Superplastic sheets

Step 1

Step 2

Step 3

Step 4

Finished part

Step 5

Fig. 18 Example of a four-sheet superplastic forming/diffusion bonding process in which the outer sheets are formed first and the center sheets are then formed and bonded to the outer two sheets

Although superplastic materials are effective in resisting the necking process, they nonetheless do neck (in relation to the *m*-value), and thinning gradients do develop. Therefore, in the design and processing of SPF parts, it is important that the thinning be understood and considered.

Uniaxial Tensile Test. It has been shown that superplastic deformation occurs when *m* is large and that under these conditions the deformation process is predominately postuniform, in contrast to conventional metal tensile behavior. In most cases, virtually all deformation is nonuniform, and the issue in the tensile behavior is the extent of this nonuniformity. The thinning in the tensile specimen can be assumed to be the result of a preexisting inhomogeneity, which can grow under the imposed deformation (Ref 60, 61).

The rate of thinning in the tensile specimen is therefore determined not only by the size of the inhomogeneity but also by the *m*-value. This has been demonstrated analytically for an idealized tensile specimen (Fig. 19) containing a geometric inhomogeneity, *f* (for example, a machining defect) (Ref 30). This analysis follows the strain development both inside and outside the inhomogeneity, assuming that the applied load is fully transferred along the length of the specimen and that the material obeys the following constitutive equation:

$$\sigma = K\varepsilon^n\dot{\varepsilon}^m \qquad \text{(Eq 9)}$$

where *n* is the strain-hardening exponent (*n* is small in this case). The results of calculations using Eq 9 are shown in Fig. 20, in which the strain in the inhomogeneity is graphed as a function of the strain outside the inhomogeneity for a number of different *m*-values. The extent of the thinning in the tensile specimen is shown to be strongly related to the *m*-value, although thinning gradients will develop at all *m*-values if the strain is sufficiently large. It will be seen that this is also the case for sheet forming in which the inhomogeneity is caused by stress gradients resulting from the part geometry and tool interactions. The inhomogeneities in tensile specimens have also been found to relate to the *m*-value (Ref 62), as shown in Fig. 21 for the Zn-22Al eutectoid alloy. In Fig. 21, the results are presented for the same alloy tested at different strain rates, for which the *m*-values are known to differ.

Spherical Domes. Although the thinning in superplastic tensile test specimens is the result of geometric inhomogeneities, the corresponding thinning in biaxially formed parts is usually the result of local stress-state differences, which subsequently lead to the development of geometric inhomogeneities. In all of these cases, however, the difference in the local stresses leads to strain-rate gradients, and the strain-rate gradients develop directly into thickness gradients. A major difference between the tensile specimen and the part configuration is that, in the former, the stress gradients can be varied (that is, reduced) by dimensional control during machining. In part forming, however, the configuration determines the stress state, and the stress state is not adjustable without changing the geometry.

The concept of thinning during SPF processing is perhaps best understood in terms of the bulging of a sheet (Ref 48, 49, 53, 63–69). In this geometry, there is a stress-state gradient from the pole of the dome to the edge, as shown in Fig. 22. If the dome is assumed to develop into a part of spherical symmetry, the stress state can be

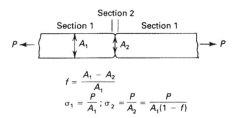

Fig. 19 Geometric inhomogeneity, *f*, in a tensile specimen

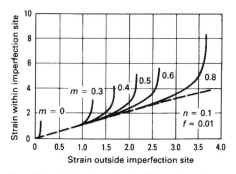

Fig. 20 Calculated strains inside and outside an inhomogeneity in a tensile specimen, such as that shown in Fig. 19, for various *m*-values

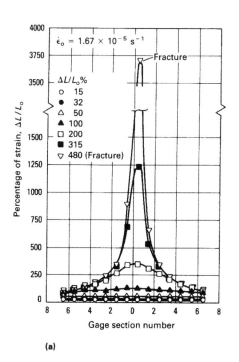

(a)

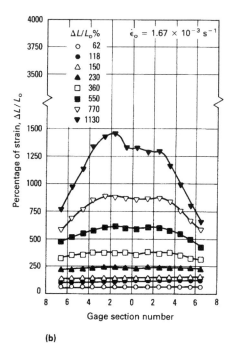

(b)

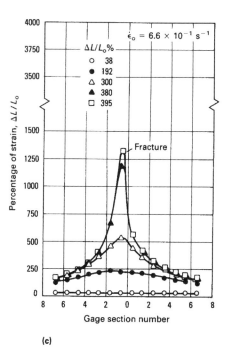

(c)

Fig. 21 Local elongation gradients in tensile specimens of Zn-22Al alloy at *T* = 473 K with initial gage length L_o of 12.7 mm (0.500 in.) that were tested within the superplastic strain-rate range (b) and outside the superplastic strain-rate range (a) and (c). The total percentage of strain at each termination point is given by $\Delta L/L_o$, where ΔL is the overall increase in gage length. The percentage of strain in each of the 14 individual segments of the gage length is given by $\Delta L/L_o$ (%), where ΔL is the increase in length of each small segment of the specimen. The initial strain rate was $\dot{\varepsilon}_o$.

readily described. At the pole, the orthogonal stresses are equal, and the stress state is that of equibiaxial tensile. At the edge of the dome, there is constraint around the periphery, leading to a plane-strain stress state. Because the flow behavior of superplastic metals has been often found to obey the von Mises criterion (Ref 70), it is helpful to examine the effective stress, $\bar{\sigma}$, which will determine the corresponding strain rate:

$$\bar{\sigma} = \frac{1}{\sqrt{2}}[(\sigma_\theta - \sigma_\phi)^2 + (\sigma_\phi - \sigma_t)^2 + (\sigma_t + \sigma_\theta)^2]^{1/2}$$

(Eq 10)

If it is assumed that the through-thickness stress is small with respect to the in-plane stresses, the effective stresses at the pole and the edge can be expressed in terms of the meridional stress, σ_θ, as follows:

$$\bar{\sigma}_p = \sigma_\theta$$

and

$$\bar{\sigma}_e = \frac{\sqrt{3}}{2}\sigma_\theta = 0.87\sigma_\theta$$

(Eq 11)

where the subscripts p and e indicate the pole and the edge, respectively, of the dome. Therefore, the pole is experiencing a 15% higher flow stress than the edge, resulting in a higher strain rate, the initial magnitude of which depends on the m-value.

The stress-state difference between the pole and the edge of the dome is roughly the equivalent of a tensile specimen, which has a local geometric inhomogeneity, f, of 0.13. The initial strain-rate difference between these two areas is dependent on the m-value; the larger the m-value, the smaller the strain-rate difference and the less the tendency to develop a thickness gradient. For example, the ratio of $\dot{\varepsilon}_e/\dot{\varepsilon}_p$ is 0.87 for $m = 1$, and the value is 0.5 for $m = 0.2$, both for the same initial effective stress difference.

Therefore, the stress gradient in a forming dome causes a more rapid thinning rate at the pole, and it may be expected that the thinning difference will accelerate with time, leading to a thickness gradient in the formed dome. There are abundant experimental results to show that this is the case and that the thinning gradient is a function of the m-value. Profiles of thickness

for bulge-formed sheets are shown in Fig. 23 for m-values of 0.57 and 0.23 (Ref 63). The thickness gradient is in agreement with expectations, and the effect of the high m-value in impeding localized thinning at the pole can be seen. Other results for a titanium alloy and a stainless steel for which m-values are 0.75 and 0.4, respectively, are shown in Fig. 24, in which the thickness strain is plotted as a function of the position along the dome cross section (Ref 44). The position along the dome is measured as the fractional height, h/h_o, where h_o is the full height of the dome, and h is the height on the dome at which the thickness measurement is made.

A number of analytical developments have been reported that predict the thinning for the superplastic forming of this type of geometry (Ref 49, 53, 65, 67–69). These models result in relations for thicknesses that are not closed-form but require numerical integration of strain increments. The models predict the thinning characteristics reasonably well, as can be seen by the comparison of experimental and analytical data shown in Fig. 24 and 25.

The theoretical predictions can be used to show the influence of the strain-rate sensitivity of

flow stress on the thinning gradient. For example, the thinning for a hemisphere formed from materials of differing m-values is illustrated in Fig. 25. In Fig. 25, the thinning factor $s/\bar{s}$ is plotted as a function of the fractional height, where s is the local thickness, and $\bar{s}$ is the average dome thickness. The maximum thinning occurs at the pole because of the stress state, as mentioned previously, and the strain-rate sensitivity is a crucial parameter in determining not only the initial strain-rate difference but also the subsequent rate of thinning, as shown in Fig. 26, in which the thinning factor at the pole can be seen to be increasingly influenced by m as the dome height is increased.

The initial stress-state differences and the corresponding strain-rate differences along the meridian of a forming dome lead to a predictable thinning gradient in this type of geometry.

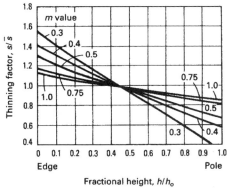

Fig. 25 Theoretical relations for a hemisphere showing the thinning factor as a function of the fractional height for a range of m-values from 0.3 to 1.0

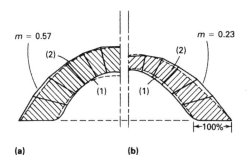

(a)   **(b)**

Fig. 23 Experimentally observed thickness profiles for a hemispherical dome formed from materials with two different m-values (solid lines). The smaller, broken outlines confined mainly within the experimental data silhouette represent bulge profiles (1) and sheet thickness distributions (2) that were calculated using (a) $m = 0.50$ and (b) $m = 0.20$.

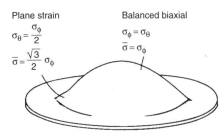

Fig. 22 Illustration of a spherical dome indicating the range of stress states existing between the pole and the edge

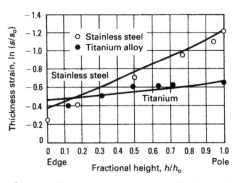

Fig. 24 Thickness strain as a function of the fractional height for dome-shaped parts formed from a stainless steel with an m-value of 0.4 and a titanium alloy with an m-value of 0.75. The values s and s_o in ln (s/s_o) represent dome thickness and initial sheet thickness, respectively.

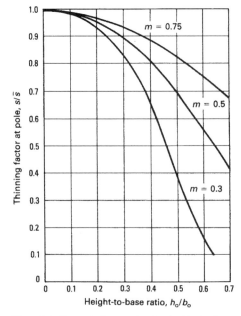

Fig. 26 Theoretical curves showing the thinning factor at the pole as a function of the bulge height-to-base ratio for $m = 0.3$, 0.5, and 0.75

The magnitude of the thinning gradient, however, is determined by the strain-rate sensitivity, *m,* and the height to which the part is formed.

Rectangular Shapes. The factors that contribute to the thinning characteristics in rectangular parts, as well as other shapes, are the same as those for the spherical dome-shaped parts discussed in the previous section in this article. It is the specific geometry that determines the initial stress-state gradients, and different geometries will be expected to develop different stress states in the forming part.

The rectangular shape is one that is common to many parts or sections of parts; therefore, it has been studied by experimental and analytical methods similar to those used for the spherical dome (Ref 45, 50, 70). For the long rectangular shape, there is a plane-stress state throughout the width of the sheet; for the case in which the die entry radius does not cause a significant stress concentration, the sheet will not experience an initial stress-state gradient. This case is very similar to that of the tensile test, in which only thickness variations or material inhomogeneities will cause local stress differences leading to localized thinning. Because these are small in comparison to the magnitude of the stress variation in the forming dome, it may be expected that the thinning gradients would be less pronounced. Experimental results for the free-forming cylindrical section show this to be the case, as illustrated in Fig. 27, and virtually no thinning gradient is seen for hemicylindrical shapes.

Interactions with the tooling do, however, cause local stress variations that can lead to thinning gradients, as shown in Fig. 27. This effect can be considered as two different types resulting from different areas of the die—the die surface at the bottom and sidewall, and the die entry radius.

Die Bottom and Sidewall. Ignoring the die entry effects, the die surface can be considered to restrict deformation in the forming sheet where contact has been made and where friction is nonzero. If the friction is large, the forming characteristic is as illustrated in Fig. 28. When the sheet makes contact with the die wall surface, the deformation in that contact area is restricted, and thinning is localized in the noncontact areas, leading to a greater degree of thinning in the last area to contact the die than in the first areas to make contact (Fig. 28). This results in a thickness gradient, as shown in Fig. 29, for a titanium alloy part formed in a die with no lubricating compounds present.

This type of thinning is readily predicted analytically if it is assumed that the sheet sticks to the die surface after contact is made by using an incremental method (Ref 50). Results of this type of model show that there are a variety of thinning variations corresponding to various width and depth ratios of the rectangular shape, as shown in Fig. 30. It is apparent from Fig. 30 that the narrow and deep parts develop the greatest amount of thinning. In this specific case,

the thickness profiles can be predicted quite well without referring to the strain-rate sensitivity, *m.* This is the result of the dominant effect of the die friction coupled with the uniform initial stress state.

If the interfacial friction is reduced, the thinning gradient will be reduced in the sidewall and bottom areas because continued deformation after die contact is possible. An example of the thinning in a formed rectangular titanium part is shown in Fig. 31, for which forming was conducted with a boron nitride solid lubricant (Ref 45, 50).

A die entry radius causes a local stress concentration in the forming sheet, which then creates a stress-state gradient in the forming sheet, and this can lead to localized thinning, especially if the ratio of die radius to sheet thickness is small and if the surface is lubricated. The source of the

stress concentration is the back pressure exerted by the die radius on the forming sheet and the gas pressure on the opposite side of the sheet from the die. The pressure exerted by the die radius has been shown to be (Ref 50):

$$P_r = \frac{\sigma_w h}{R_1} \qquad \text{(Eq 12)}$$

where P_r is the pressure on the die entry radius, σ_w is the in-sheet stress in the width direction, h is the sheet thickness, and R_1 is the die entry radius.

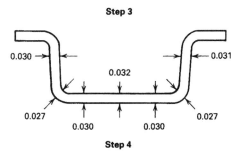

Fig. 27 Thinning development in a 1.37 mm (0.054 in.) thick superplastic formed Ti-6Al-4V part having a rectangular cross section and semi-infinite length. Formed at 870 °C (1600 °F) using a boron nitride lubricant, the sheet required 20 min to fabricate at an average strain rate of $5.8 \times 10^{-4}\,\mathrm{s}^{-1}$. Dimensions given in inches

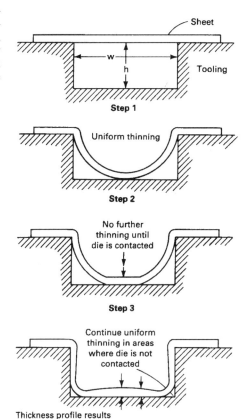

Fig. 28 Illustration of thinning characteristics in the blow forming of an unlubricated part of rectangular cross section and semi-infinite length

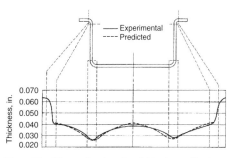

Fig. 29 Observed and predicted thinning profiles in an unlubricated blow-formed Ti-6Al-4V alloy part of rectangular cross section. Forming of the sheet, which had an initial thickness of 1.68 mm (0.066 in.), required 20 min at 925 °C (1700 °F).

This and the applied gas pressure, g, develop an average through-thickness stress σ_h of:

$$\sigma_h = \frac{g + P_r}{2} \qquad \text{(Eq 13)}$$

The magnitude of σ_w will be dependent on the local friction coefficient, μ_r, and the position on the radius, so that the effective stress will vary around the die entry radius. A detailed analytical model of this somewhat complex condition is available in Ref 50, but it has been shown that a local stress increase is developed in this area, causing a tendency toward local thinning. If the friction is sufficiently low, the initially thinned section can continue to thin after die contact is made. Therefore, significant localized thinning can occur, and even rupture may take place if the conditions are sufficiently severe.

Thinning over the die entry radius is the result of stress gradients; therefore, the strain-rate sensitivity m is an important parameter in determining the extent of thinning that will develop. The influence of these variables is illustrated in Fig. 32 for a titanium alloy part formed under the indicated conditions of lubrication and strain rate. The strain-rate variations resulted in corresponding variations in the m-value during the respective forming process. The thinning for the unlubricated part is in agreement with that expected from the discussion in the previous section in this article. For this case where lubricant is used, the strain rate, which determines the corresponding m-value, is a factor in determining the extent of thinning over the die entry radius. The average m-value corresponding to the die entry radius was higher for the forming process that developed the lower average strain rate, resulting in a significantly reduced tendency toward local thinning in that area.

Thinning Control. Because SPF parts are typically stretched to very large elongations, the thickness variations are potentially large for a part. Therefore, it is often important to control

the thickness variations in order to meet part tolerance requirements. Although it is seldom possible to prevent thickness variations, there are techniques that can be used to control this problem. In addition to such methods, the designer can often accommodate variations in thickness if he knows in advance what they may be. This latter approach is an important and viable one but is not addressed in this article because it is a very specialized field of metallurgy.

The methods used to control thinning are:

- Processing of the superplastic material to achieve a high m-value
- Use of surface lubrication, as discussed previously
- Use of thermoforming methods to control the localized deformation
- Modification of the die or part design to minimize local stress concentrations
- Forming a thickness-profiled sheet
- Application of pressure in a controlled and profiled manner to control strain rate to a value corresponding to a high m-value

Because the raw sheet material is generally obtained from a commercial supplier, the material superplastic properties are under the control of the mill. However, it may be judicious for the forming plant to obtain material under the control of an appropriate specification. The effect of lubrication is discussed in the section "Die Entry Radius" in this article, along with the effect of the die entry radius, which may, in some parts, be increased to minimize the thickness gradients.

The thermoforming method has been shown to offer effective techniques that can control the thinning gradients in single-pocketed deep-drawn parts (Ref 47, 53). With these methods, a movable tool is usually used to contact the forming sheet before the finished shape is produced, causing the local friction to minimize deformation in some locations while free-forming sections continue to deform.

The use of thermoforming techniques was demonstrated using apparatuses such as those shown in Fig. 33 and 34 (Ref 53). The test rig used consisted of two cylindrical chambers, each 190 mm (7.5 in.) in inside diameter by 178 mm (7 in.) deep, and a hydraulic ram, positioned in the bottom chamber, that was capable of moving up and down. The material used was Zn-22Al-0.15Cu sheet 1.27 mm (0.050 in.) thick. For the convex upward die, the deformation was restricted in the center of the sheet and concentrated at the outer area, resulting in a strain and thickness profile, as shown in Fig. 35, in which the top center is thicker than the adjacent areas. This thickness profile was substantially modified by the use of a concave

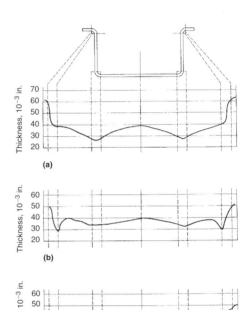

Fig. 31 Observed thickness profiles for 1.68 mm (0.066 in.) thick Ti-6Al-4V blow-formed parts of rectangular cross section formed under different average strain rates, $\dot{\varepsilon}_t$, and with different lubrication conditions. (a) $\dot{\varepsilon}_t = -7 \times 10^{-4}\,\text{s}^{-1}$; no lubricant. (b) $\dot{\varepsilon}_t = -5.6 \times 10^{-4}\,\text{s}^{-1}$; boron nitride lubricant. (c) $\dot{\varepsilon}_t = -5.4 \times 10^{-5}\,\text{s}^{-1}$; boron nitride lubricant

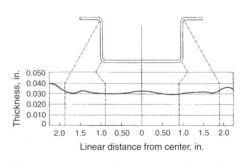

Fig. 32 Observed thickness distribution for Ti-6Al-4V parts with 1.14 mm (0.045 in.) initial thickness at 927 °C (1700 °F) formed at $10^{-3}\,\text{s}^{-1}$ with boron nitride lubrication

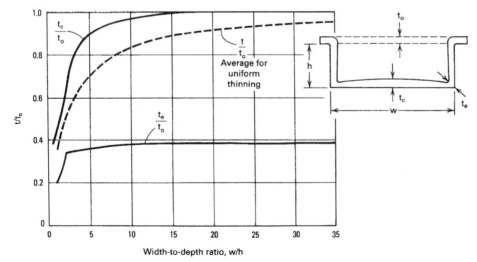

Fig. 30 Predicted minimum thicknesses as function of the width-to-depth (w/h) ratio for unlubricated blow-forming part of rectangular cross section

upward die, as shown in the Fig. 36. In this case, the superplastic diaphragm was formed down into the concave die by gas pressure, and the die was slowly withdrawn until it reached the bottom. The preformed diaphragm was then formed into the upper cylindrical chamber in the same manner as that of the previous figure. The resulting profile is seen to be considerably more uniform across the top of the part.

The use of thickness-profiled sheet has been suggested to control the thickness in the final part (Ref 67). The concept considers that the initial thickness variations can be used to offset the subsequent variations resulting from the stress state and part geometry effects on the thinning, as shown in Fig. 37. If areas that will thin excessively are thicker than surrounding areas, it is possible to develop finished part thickness profiles that are more uniform than those formed of constant-thickness sheet.

Pressure Profiling. It is now well recognized that the m-value for superplastic alloys will vary with strain rate and that it will also often vary with strain. The strain rate imposed during the forming process will therefore determine the m-value, and if the strain rate varies during the forming process, the corresponding m-value and related thinning uniformity will also vary. For example, the simplest pressurization concept for SPF processing is that of constant pressure. The resulting strain-rate variation for a spherical dome part configuration has been shown to be as much as 3 orders of magnitude (Ref 48). A graph of the predicted strain rate for a spherical dome that will be generated under constant pressure for such a part is shown in Fig. 38. The strain rate drops to 0.001 of the initial value when the part becomes a hemisphere. The m-value for a typical superplastic alloy can vary from a maximum m to a value of less than 0.2 over strain-rate ranges of this magnitude. The consequence of forming a part under these conditions is that excessive thinning or part rupture during forming is likely. The initial strain rate corresponds to a large m-value, and good thinning resistance is observed. However, this would be transient; other strain rates would be encountered corresponding to low m-values, and poor resistance to localized thinning would be present.

This condition can be rectified if the strain rate can be maintained at a constant level corresponding to a suitably high m-value. Because the constant pressure is seen to develop a variable strain rate, it is apparent that a variable forming pressure would be required to develop a constant strain rate. Such pressure profiles have been established analytically for the spherical dome (Ref 48, 49) and the rectangular (Ref 50) part configurations. Because most of the analytical models of the SPF process use applied gas pressure to establish the current stress and strain-rate conditions, it is possible to use these same models to adjust the current gas pressure to develop the desired stress and strain rate.

The resulting pressure profiles for the constant strain-rate forming of the spherical dome and the rectangular parts are illustrated in Fig. 39 and 40, respectively. It is typical that the pressure initially rises rapidly, followed by decrease. The rapid initial rise is due to a rapid decrease in the radius of curvature with little change in thickness, and the subsequent decrease is the result of thinning that is more rapid at this stage than the change in radius of curvature. The depth to which a part is formed will also affect the pressure profile for constant strain-rate control, as shown in Fig. 41 for a rectangular part formed with no lubrication. For a deep part ($w = 305$ mm,

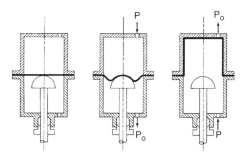

Fig. 33 Apparatus for thermoforming superplastic sheet materials using a convex die member to control thinning in forming of a hat configuration

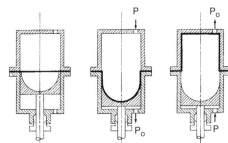

Fig. 34 Apparatus for thermoforming superplastic sheet materials using a concave die member to control thinning in forming of a hat configuration

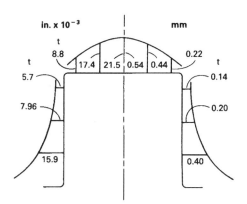

Fig. 35 Thickness profile for hat configuration formed with a convex die member, as shown in Fig. 33. Material formed is Zn-22Al-0.15Cu at a forming temperature of 250 °C (480 °F).

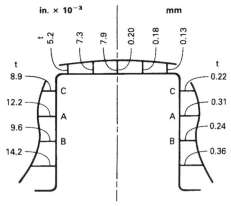

Fig. 36 Thickness profile for hat configuration formed with a concave die member, as shown in Fig. 34. Material formed is Zn-22Al-0.15Cu at a forming temperature of 250 °C (480 °F).

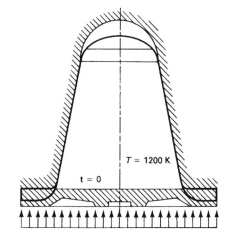

Fig. 37 Schematic of the concept of forming a sheet material that has a thickness profile before forming

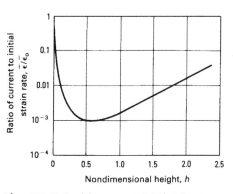

Fig. 38 Ratio of the current to initial strain rate as a function of nondimensional height for a constant pressure application in the forming of a hemispherical configuration

or 12 in.), the applied pressure never reaches a maximum but continues to rise during the forming process. For a shallow part ($w = 50$ mm, or 2 in.), the pressure is decreased for a significant time period before being increased to high levels.

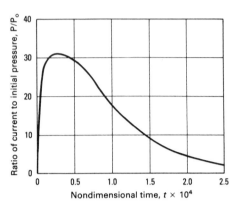

Fig. 39 Ratio of current to initial pressure as a function of a time parameter for forming a spherical configuration under constant strain-rate conditions

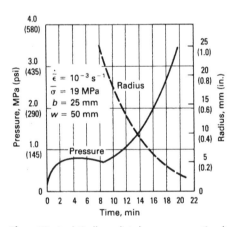

Fig. 40 Analytically predicted pressure versus time for a constant strain rate in the forming of a long, rectangular, 1.27 mm (0.050 in.) thick Ti-6Al-4V part fabricated at 870 °C (1600 °F)

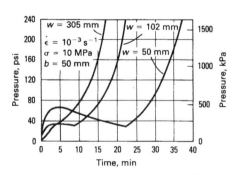

Fig. 41 Analytically predicted pressure profiles for the constant strain-rate forming of long rectangular parts of different cross section width, b, and height, w, dimensions

Cavitation and Cavitation Control

Most superplastic alloys tend to form voids, or cavities, at intergranular locations during the superplastic deformation. This process is termed cavitation. Cavitation can lead to the degradation of strength and other design properties, and it is dealt with in one of two ways:

- Establishing reduced design properties
- Using a back-pressure technique to control cavitation

Typical cavitation as a function of strain is shown for an aluminum alloy in Fig. 42. It can be seen that the absolute amount of cavitation in terms of the volume fraction is not large but depends on the strain imposed. The use of the back-pressure concept imposes a hydrostatic pressure on the sheet during forming, and if this pressure is of the order of the flow stress, cavitation can be reduced or completely suppressed. An example of the effect of back pressure on the rate of development of cavitation is shown in Fig. 43.

In practice, the back pressure is achieved by imposing a pressure on the back side of the sheet to oppose the forming pressure and by sustaining this pressure during the forming cycle. The forming pressure must be higher than the back

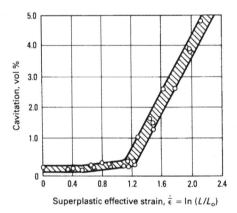

Fig. 42 Development of cavitation with uniaxial tensile strain in a 7475 aluminum alloy specimen of 0.8 cm^2 (1/8 in.^{2}) cross-sectional area deformed at 516 °C (961 °F) under a constant strain rate of 2×10^{-4} s^{-1}

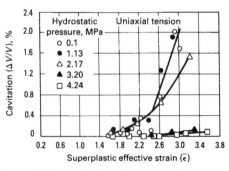

Fig. 43 Effect of hydrostatic pressure on the suppression of cavitation in a 7475 aluminum alloy specimen superplastically deformed at 516 °C (961 °F)

pressure, and the same forming rates can be achieved with or without back pressure if the pressure differential is the same. For example, if a pressure profile is desired, such as that shown in Fig. 39 or 40, the pressure profile is simply raised in magnitude by an amount equal to the back pressure. Because the back pressure is normally of the order of the material flow properties, pressures of approximately 690 to 3400 kPa (100 to 500 psi) are generally suitable for suppressing cavitation.

Manufacturing Practice

The SPF process is unique in terms of the complexity of parts that can be produced and the methods that can be used to shape such a material. A number of processing methods are currently being used, most of which involve significant stretching of sheet material. The high ductility that can be achieved with these types of materials also has a consequence that must be understood and dealt with, namely, thinning gradients. The thinning gradients are a natural consequence of stress gradients that develop in the various die configurations, and the superplastic property of the strain-rate sensitivity of the flow stress then determines the subsequent thinning gradient that will result in the part. Control of the forming process, die configuration, and material characteristics are all factors that can affect the thinning.

The SPF processes are being increasingly used for a wide range of structural and nonstructural applications. The aerospace industry depends greatly on the structural integrity of the parts, but the automotive industry uses body sheet metal parts that do not require very high structure strength, and cavitation can be tolerated. Under rapid forming conditions at low forming temperatures, cavitation in aluminum alloys is generally found (Ref 71–73). However, rapid forming technology is in great demand for improving the efficiency of the superplastic forming process. Faster forming rates are obtained by mated punch-die technology as in sheet stamping, rather than by gas pressure forming. This has led to the development of warm forming technologies, which use embedded heating elements in the punch and die systems. Figure 44 shows an arrangement for a heated die system for warm stamping of aluminum alloys. This has been developed with integrally embedded heating elements within punch-die systems rather than gas pressure forming systems. Part forming with 5083-type aluminum alloys has been demonstrated in less than a minute per part, and hundreds of parts have been produced (Ref 74, 75). Using warm forming technology, full-scale parts such as the Chrysler Neon inner door panel were successfully formed from fine-grain 5182 aluminum alloy containing 1% Mn at the rate of one minute per part (shown in Fig. 44c). This development, originally spearheaded by the U.S. Department of Energy in 2000, has spawned other rapid gas-pressure

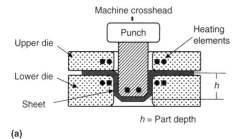

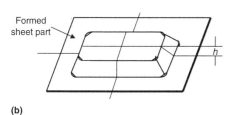

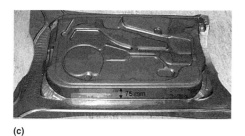

Fig. 44 Heated die system for warm stamping of aluminum alloys. (a) Schematic diagram of warm forming dies. (b) Schematic of formed sheet part. (c) Chrysler Neon door inner panel successfully formed at 350 °C (660 °F) using aluminum 5182 plus manganese sheet

forming technologies discussed in the next section.

Quick Plastic Forming Cells

In terms of gas pressure forming, General Motors has concentrated on the rapid transfer of sheet between tooling, blank preheating, and the use of robotics in systems referred to as quick plastic forming cells. The quick plastic forming (QPF) system (Ref 76) was developed by General Motors to address the major deficiencies of the automated SPF system:

- Inefficient press utilization
- Inability to achieve the desired pressure/time cycle in the tool
- Inability to maintain the desired process temperature for short forming cycles
- Tool temperature variation
- Forming cell inflexibility

A forming cell configured with these characteristics would run with the process steps shown in Fig. 45. A forming cell capable of maintaining the desired forming temperature for very short cycles must incorporate tools of very high thermal conductivity, elaborate control systems, or tools with heating sources very close to the working surfaces. Further, improved

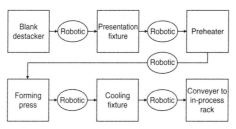

Fig. 45 Operating process for purpose-built quick plastic forming cell

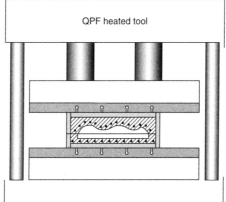

Fig. 46 The quick plastic forming (QPF) process can be run with integrally heated dies as for warm forming applications.

methods to model the forming process were needed to facilitate application to products within the normal GM vehicle development timing. An integrated QPF cell was developed and demonstrated for a decklid inner panel.

Purpose-Built QPF Cell Technology Elements

Blank preheating is done outside the press in a purpose-built oven. The preheating system has at least two additional advantages over the ability to support a faster process time. First, productivity of the press doubles over that of the in-press heating system, because the press space and time previously used for preheating can now be used for forming. Further, the faster heating rate may produce a material microstructure with better inherent formability.

Improved Pressure/Time Control. In typical QPF cycles, the gas flow rate increases substantially between the initial and final stages of panel forming. To allow for good pressure control at both high and low gas delivery volumes, QPF gas management systems are based on proportional control valves capable of adequate pressure resolution at low pressures but also of large-volume gas delivery at high pressures.

Improved Thermal Control. The heated die-tool system shown in Fig. 46 offers many advantages in performance and flexibility in terms of minimizing heat loss. One downside to

moving to heated and insulated tools is the increased cost and complexity of the tool system. However, the increased cost can be amortized over a larger number of panels due to the increase in system productivity. The higher tool cost is also partially offset in that the heating system can be run at a much lower temperature than the press heating system for a given tool temperature. Further, the convective and radiation losses of the heated tool system are much lower than the heated press system, and energy savings of 50 to 70% have been demonstrated in the QPF process.

Improved Forming Cell Flexibility. By developing the heated tool as a self-contained system fully enclosed in insulation, the tool package can be handled using ordinary means (cranes, lift trucks) without risk to the tool-change operators. The tool insulation system was developed to maintain an exterior surface temperature of < 55 °C (130 °F). That temperature facilitates fast tool-change capability. Further, only very minor modifications are needed to allow the heated/insulated tool to be mounted to the press with existing automated clamping systems. A quick tool-change is possible with the heated-tool systems, and the newly installed tool is immediately available for forming because it can be preheated off-line.

The use of fully insulated tools and the accompanying fast tool-change capability allow a QPF press line to be run in a manner similar to a conventional press line; that is, a single line could support more than one product without significant downtime during tooling changeover.

Superplasticity in Iron-Base Alloys*

Superplasticity is the ability of certain polycrystalline metallic materials to extend plastically to large strains when deformed in tension. Strains to failure in superplastic materials range from several hundred to several thousand percent. In general, superplastic materials also exhibit low resistance to plastic flow in specific temperature and strain-rate regions. These characteristics of high plasticity and low strength are ideal for the manufacturer who needs to fabricate a material into a complex but sound body with a minimum expenditure of energy.

The phenomenon of superplasticity was first observed over 70 years ago (Ref 77), but active research in this field did not begin until 1962 (Ref 78). Since that time, superplastic effects have been reported in over 100 alloy systems (Ref 78). The micromechanisms that permit extraordinary elongations in these materials are still under investigation (Ref 79), but it is generally accepted that a major microstructural requirement for

*Reprinted from "Appendix: Superplasticity in Iron-Base Alloys" by Oleg D. Sherby, Stanford University; Jeffrey Wadsworth, Lockheed Missiles and Space Company; Robert D. Caligiuri, Failure Analysis Associates, in *Forming and Forging*, Volume 14, *Metals Handbook*, 9th ed., ASM International, 1988, p 868–872.

superplasticity is the development of a stable, very fine grain size (typically <10 μm, or 400 μin.) (Ref 80). This requirement also usually (but not always) leads to the necessity for a uniform distribution of a fine, second phase to inhibit matrix grain growth. Furthermore, the grain boundaries should be high angle (that is, disordered), equiaxed, mobile, and must resist separation under tensile stresses (Ref 81). To prevent cavitation around the second-phase particles, the strengths of the two phases should be similar at the temperature of deformation (Ref 81). These microstructural characteristics can lead to a flow stress that is highly sensitive to changes in strain rate, which in turn resists the formation of instabilities during tensile deformation.

Even though superplasticity has been observed in many metallic alloy systems, only a relatively small number of them are iron-base. This is because of the difficulty in generating microstructures in these systems with the aforementioned characteristics. As a result, there have been relatively few industrial applications involving the superplastic bulk forming of ferrous alloys. Those iron-base systems in which the microstructural requirements have been met include:

- Hypoeutectoid and eutectoid plain carbon steels (ferrite-carbide two-phase system)
- Hypereutectoid plain carbon steel (ferrite-carbide and austenite-carbide two-phase systems) and white cast iron (ferrite-carbide two-phase system)
- Low-to-medium alloy steels (ferrite-austenite two-phase system)
- Microduplex stainless steels (ferrite-austenite two-phase system)

Of these, only the hypereutectoid steels have been extensively investigated for applications involving superplastic forming.

The development and existence of superplasticity in all of the aforementioned categories of iron-base alloys have been reviewed in detail (Ref 82, 83). This article describes the key characteristics of superplastic behavior in each of these iron-base systems, with emphasis on hypereutectoid plain carbon steels, which currently have the most industrial potential.

Hypoeutectoid and Eutectoid Plain Carbon Steels. Early attempts at making superplastic plain carbon steels were not particularly successful (Ref 82). In these early efforts, extensive heat treatments were used to produce microstructures of fine ferrite grains pinned by spheroidized cementite particles in compositions containing between 0.2 and 1.0% C. Even though the overall grain structure was correct, elongations of only approximately 130% were observed, primarily because the grain boundaries were low-angle dislocation boundaries. The flow stress of materials with low-angle grain boundaries is usually not sensitive to changes in strain rate; therefore, superplasticity was not observed. However, later work showed that thermally cycling such materials across the eutectoid transformation temperature after thermomechanical processing changes the grain boundaries from low angle to high angle; tensile elongations of the order of 1000% have been achieved in thermally cycled, hypoeutectoid plain carbon steels (Ref 80).

Hypereutectoid Plain Carbon Steels and White Cast Irons. Hypereutectoid plain carbon steels, also known as ultrahigh-carbon (UHC) steels, will exhibit a microstructure consisting of a continuous network of proeutectoid carbide (at prior-austenite grain boundaries) surrounding pearlite colonies when slowly cooled from a temperature in the single-phase austenite region. This carbide network in turn imparts poor mechanical properties to these steels. However, work conducted in the 1970s showed that, with proper thermomechanical processing, microstructures can be achieved in UHC steels that consist of ultrafine, equiaxed grains of ferrite and a uniform distribution of fine, spherical, discontinuous proeutectoid carbide particles (Ref 84). Such microstructures in UHC steels have led to superplastic behavior; tensile elongations up to 1500% have been reported (Ref 85).

The requisite thermomechanical processing of UHC steels to obtain the desired fine spheroidized microstructure usually involves two steps. In the first step, the UHC ingot or billet is solution heat treated at typically 1150 °C (2100 °F), followed by hot and warm working during cooling down to 750 °C (1380 °F). This step refines the austenite grains, uniformly distributes the proeutectoid carbides at prior-austenite grain and subgrain boundaries, and forms pearlite in the matrix of the prior-austenite grains. The second step incorporates a divorced eutectoid transformation in which the UHC steel is heated to just above the eutectoid transformation temperature and air cooled (with or without accompanying deformation). This step transforms the fine pearlite from step 1 into a fully spheroidized fine structure. The divorced eutectoid transformation is discussed in more detail in Ref 86.

Although not yet fully commercialized, the industrial potential of UHC steels has been thoroughly investigated. One promising application is in the press forging of gears. Figure 47 illustrates a bevel gear that was warm forged from a fine-grain 1.25% C UHC steel in a single operation. An additional advantage of using superplastic UHC steels is that, because of their high carbon content, the carburization step in normal gear production is eliminated. A second example of the superplastic press forming of fine-grain UHC steels is the aft closure for a guided missile, shown in Fig. 48. In this case, a 1.6% C steel casting was liquid atomized, and the resulting fine-grain powders were warm pressed into a billet at 800 °C (1470 °F). A third application for superplastically formed fine-grain UHC steels is in the manufacture of die components by superplastic hobbing operations (Ref 87).

The original work on UHC steels demonstrated that, given the proper microstructure, superplasticity could be developed over the composition range of approximately 0.8 to 2.1% C and the temperature range of approximately 650 to 800 °C (1200 to 1470 °F). Therefore, superplastic UHC steels include both austenite-carbide as well as ferrite-carbide two-phase structures. Ultrahigh-carbon steels are generally not susceptible to cavitation, because the ferrite, austenite, and carbide all have approximately the same strength in the intermediate temperature range in which superplasticity occurs.

Furthermore, in the original work, superplasticity in UHC steels was limited to intermediate strain rates (that is, forming rates): 10^{-5}

Fig. 47 Warm precision forging of a 1.25% C ultrahigh-carbon steel billet into a bevel gear. Forging temperature was 650 °C (1200 °F).

(a)   (b)   (c)

Fig. 48 Aft closure, for a guided missile, that was superplastically formed at 815 °C (1500 °F) from a 1.6% C ultrahigh-carbon steel. The processing procedure consisted of warm pressing (800 °C, or 1470 °F) liquid-atomized powders into a billet and forging the resulting billet into plate. The plate was then superplastically formed to the final shape (a) and (b). (c) Cross section of the part

to 10^{-3} s^{-1}. The upper limits on strain rate and temperature are primarily related to the destruction of the fine uniform carbide structure. Commercialization of superplastic UHC steels has in fact been hampered by these somewhat limited strain-rate and temperature ranges. However, work has shown that these ranges could be substantially extended by careful alloying additions of silicon or aluminum (Ref 88). These elements enhance superplasticity in UHC steels because they:

● Increase the eutectoid transformation temperature (increase the stability of ferrite)
● Inhibit carbide coarsening at high temperatures (increase the activity of carbon in ferrite)
● Increase the volume fraction of proeutectoid carbides
● Do not form active sites for cavitation to occur

Ultrahigh-carbon steels containing 3% Si or 1.6% Al have exhibited superplasticity at strain rates to 10^{-2} s^{-1} at a temperature of 800 °C (1470 °F). Superplasticity has been projected to occur in a UHC 12% Al alloy at strain rates to 3×10^{-1} s^{-1} at 950 °C (1740 °F); this would make superplastic forming of UHC steels both economical and feasible for many operations.

An additional benefit of adding aluminum or silicon to UHC steels is a low, fairly constant (14 ± 1.5 MPa, or 2.00 ± 0.22 ksi) flow stress over the temperature range of 750 to 925 °C (1380 to 1700 °F). This suggests that thin-sheet UHC steels containing the proper amounts of aluminum or silicon can be readily blow formed over a wide temperature range.

Hypereutectoid White Cast Irons. Application of rapid solidification processing to UHC steels has also helped to extend the compositional range for superplasticity well into the white cast irons. Such techniques can create very high-carbon-content powders that, upon annealing at intermediate temperatures (595 to 700 °C, or 1100 to 1290 °F), exhibit the requisite fine microstructure (Ref 89). These powders are then readily consolidated into fully dense compacts at temperatures below the subcritical annealing temperature (A_1) (Ref 90). White cast irons processed in this manner have been shown to exhibit superplasticity at intermediate temperatures; a maximum tensile elongation of 1410% has been observed in a Fe-3.0C-1.5Cr (Ref 91). The shaded region on the iron-carbon phase diagram shown in Fig. 49 illustrates the range of carbon content and temperature over which superplasticity has been documented in fine-grain UHC steels and cast irons.

Hypereutectoid Steel Properties. Because of their carbon content, UHC steels, after superplastic forming, can be heat treated to very high hardness levels (65 to 68 HRC). This can be important in a wide variety of applications, including gears, tool bits, abrasion-resistant surfaces, and military vehicle armor. However, because of their very fine grain size, plain carbon UHC steels also exhibit poor hardenability. This hardenability problem can be alleviated by dilute

alloying additions of hardenability-enhancing elements, such as chromium, manganese, and molybdenum. The dramatic effect of alloying additions on the hardenability of UHC steels is shown in Table 3.

The room-temperature tensile properties of fine-grain UHC steels have been extensively studied (Ref 92). As expected, the properties are very sensitive to heat treatment. Figure 50 compares the tensile properties of UHC steels in two different heat treatment conditions with the properties of other plain carbon and low-alloy structural steels. In addition, if properly heat treated to achieve an extremely fine (optically unresolvable) martensite, fully hardened UHC steels will exhibit a surprisingly high strain to failure in compression. As shown in Fig. 51, a 1.3% C steel that was water quenched from

770 °C (1420 °F) (steel A) will exhibit a compression fracture strength of 4.5 GPa (650 ksi) and a strain to failure of 10%. However, austenitizing the same steel at 1100 °C (2010 °F) prior to water quenching from 770 °C (1420 °F) (steel B) will coarsen the resulting martensite and reduce the compression strain to failure to less than 2%.

Because they contain a high volume fraction of carbide, UHC steels exhibit only moderate impact resistance (<25 J, or 18 ft·lbf, Charpy V-notch at 25 °C, or 80 °F). The impact resistance of fine-grain UHC steels can be enhanced by laminating them to tougher (but weaker) materials. Fine-grain UHC steels can be readily solid-state bonded to themselves or other ferrous materials, partly because of their superplastic nature (Ref 93). The lamination procedure can

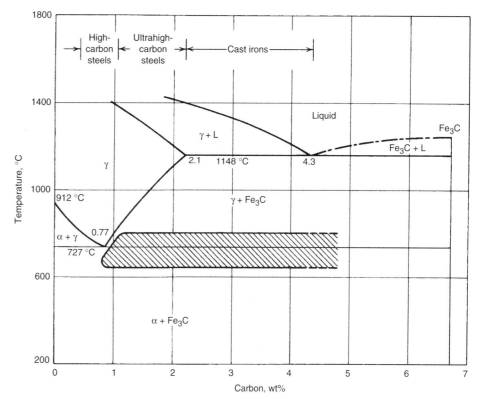

Fig. 49 Iron-carbon phase diagram. The shaded area illustrates the temperature and composition over which ultra-high-carbon steels and white cast irons have been made superplastic.

Table 3   Hardenability of ultrahigh-carbon (UHC) steels as a function of alloying additions

Material composition	Austenitizing temperature ($A_1 + 50$)		Critical brine temperature to achieve 62 HRC, T_c		Critical cylinder diameter, D_c		M.F.(a)
	°C	°F	°C	°F	mm	in.	
1.25C-0.5Mn	773	1425	66	151	6.9	0.27	1.0
1.25C-0.5Mn-0.2P	795	1465	72	162	10.9	0.43	1.6
1.25C-1Cr-0.5Mn-0.25Mo	790	1455	73	163	11.4	0.45	1.7
1.25C-3Si-1.4Cr-0.5Mn	825	1520	78	172	15.5	0.61	2.2
1.25C-1.6Al-1.5Cr-0.5Mn	860	1580	93	199	22.4	0.88	3.2
1.25C-2Mn-1Cr	795	1465	>100	>212	>23.1	>0.91	>3.3

(a) M.F., multiplying factor = $D_{c\,\text{UHC steel alloy}}/D_{c\,\text{UHC steel}}$

consist of conventional roll bonding, press bonding, or explosive techniques as long as a good metallurgical bond results. The effect of such lamination on the impact properties of UHC steels is illustrated in Fig. 52. The impact properties of a UHC steel/1020 steel laminated composite are clearly superior to those of either the monolithic UHC steel or the monolithic

1020 low-carbon steel. Most UHC steel-base laminates will also exhibit superplasticity over limited temperature and strain-rate ranges and therefore can be superplastically formed (Ref 94, 95). The room-temperature tensile strength of a UHC steel laminate will of course be reduced relative to that of a monolithic UHC steel by an amount governed by the rule of

mixtures; however, this loss in strength can be partially compensated for by heat treating the laminate to selectively harden the UHC steel component to very high strength levels.

Low-to-Medium Alloy Steels. Superplastic studies have been conducted on several alloy steels (Ref 82). Compositions and thermo-mechanical processing procedures are generally chosen to generate fine-grain (1 to 2 μm, or 40 to 80 μin.) microstructures consisting of roughly equal amounts of ferrite and austenite. These phases will coexist only in a narrow temperature range, but within that range each phase will inhibit the other from growing, and superplasticity will occur. Elongations of 300 to 500% have been achieved in steels containing 1 to 2% Mn and 0.1 to 0.4% C, and elongations of up to 600% have been achieved in Fe-4Ni-3Mo-1-2Ti alloys when tested at temperatures in the dual-phase region. However, the potential of superplastic forming these alloys has not been exploited, because of the narrow temperature range over which superplastic flow occurs and because of the rapid growth that can occur even in the two-phase region at high temperatures.

Microduplex stainless steels are so called because their stable microstructure consists of both ferrite and austenite in fine grain size form—approximately 2 to 3 μm (80 to 120 min). They typically contain 18 to 26% Cr and 5 to 8% Ni and can also contain molybdenum, titanium, copper, silicon, manganese, carbon, and diatomic nitrogen. Elongations to failure in these alloys are commonly in excess of 500% at strain rates between 10^{-3} and 10^{-2} s^{-1} and at temperatures between 900 and 1000 °C (1650 and 1830 °F) (Ref 95). Although microduplex stainless steels have found a number of commercial applications in the chemical industry because of their high strength and corrosion resistance, their superplastic forming potential has not been exploited. This is because

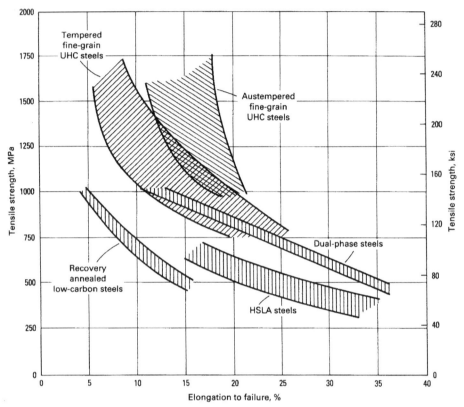

Fig. 50 Tensile strength versus elongation to failure of heat treated fine-grain ultrahigh-carbon (UHC) steels compared to low-carbon steel, high-strength low-alloy (HSLA) steels, and dual-phase steels

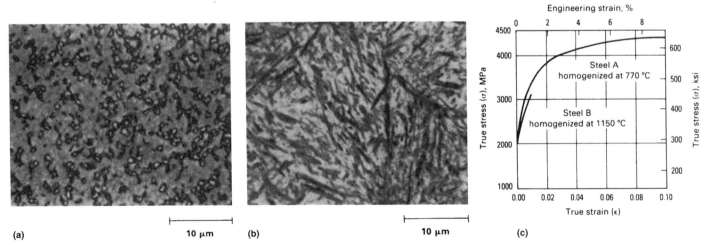

Fig. 51 Influence of prior heat treatment on (a) and (b) the microstructure and (c) compression stress-strain behavior of a 1.3% C ultrahigh-carbon steel quenched from 770 °C (1420 °F)

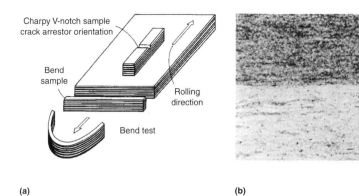

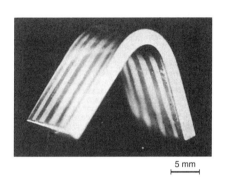

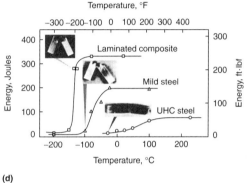

Fig. 52 Ultrahigh-carbon (UHC) steel/1020 steel laminated composite to improve impact resistance of fine-grain UHC steels. (a) Orientation of mechanical test samples taken from a laminated composite of UHC steel and 1020 steel. (b) Optical micrograph of interface in laminated composite of UHC steel and 1020 showing an excellent metallurgical bond. (c) Bend test sample of the laminated composite. (d) Impact properties of UHC steel, 1020 steel, and UHC steel/1020 steel laminated composite, including photographs of the tested samples

microduplex stainless steels are susceptible to cavitation during superplastic flow due to the strength differential between the ferrite and the austenite at these high temperatures (Ref 96).

Outlook for Superplastic Ferrous Alloys. The number of iron-base alloys exhibiting superplasticity is relatively limited. This is primarily because of the difficulty in satisfying all of the microstructural requirements generally required for superplasticity in ferrous materials. Of the alloys identified to date for possible superplastic forming, the hypereutectoid (or UHC) steels show the most promise. These steels can be stretched without cavitation over 1000% at reasonable forming rates and intermediate temperatures. Ultrahigh-carbon steels are heat treatable to high hardnesses, exhibit good room-temperature mechanical properties, and can be readily laminated to other ferrous materials to enhance toughness and impact resistance. Future efforts in the superplastic forming of iron-base alloys will undoubtedly focus on this class of steels.

REFERENCES

1. M.M.I. Ahmed and T.G. Langdon, *Metall. Trans. A,* Vol 8, 1977, p 1832
2. J.W. Edington, K.N. Melton, and C.P. Cutter, *Prog. Mater. Sci.,* Vol 21, 1976, p 61
3. K.A. Padmanabhan and G.T. Davies, *Superplasticity,* Springer-Verlag, 1980
4. F.A. Mohamed, M.M.I. Ahmed, and T.G. Langdon, *Metall. Trans. A,* Vol 8, 1977, p 933
5. R. Verma, A.K. Ghosh, S. Kim, and C. Kim, Grain Refinement and Superplasticity in 5083 Al, *J. Mater. Sci. Eng.,* Vol A191, 1995, p 143–150
6. T. Chanda, A.K. Ghosh, and C. Lavender, The Effect of Dispersoid Particle Size on the Superplasticity of Al-Mg Alloy, *Superplasticity and Superplastic Forming,* A. Ghosh and T. Bieler, Ed., TMS, 1995, p 41–48
7. R. Verma, P.A. Friedman, A.K. Ghosh, S. Kim, and C. Kim, Characterization of Superplastic Deformation Behavior of a Fine Grain 5083 Al Alloy Sheet, *Metall. Mater. Trans. A,* Vol 27, May 1996, p 1889
8. R. Verma, P. Friedman, and A.K. Ghosh, Superplastic Gas Forming Characteristics of Fine Grain 5083 Al, *J. Mater. Eng. Perform.,* Sept 1995–1997
9. T. Mukai, H. Watanabe, and K. Higashi, *Mater. Sci. Technol.,* Vol 16, 2000, p 1314
10. H. Watanabe, H. Tsutsui, T. Mukai, M. Kohzu, S. Tanabe, and K. Higashi, *Plasticity,* Vol 17, 2001, p 387
11. A. Jager, P. Lukai, V. Gartnerova, J. Bohlen, and K.U. Kainer, *J. Alloy. Compd.,* Vol 378, 2004, p 184
12. A. Bussiba, A. Ben Artzy, A. Shtechman, S. Ifergan, and M. Kupiec, *Mater. Sci. Eng. A,* Vol 302, 2001, p 56
13. J.C. Tan and M.J. Tan, *Mater. Sci. Eng. A,* Vol 339, 2003, p 81
14. D.L. Yin, K.F. Zhang, G.F. Wang, and W.B. Han, *Mater. Lett.,* Vol 59, 2005, p 1714
15. C.J. Lee and J.C. Huang, *Acta Mater.,* Vol 52, 2004, p 3111
16. T. Mukai, H. Tsutsui, H. Watanabe, K. Ishikawa, Y. Okanda, M. Kohzu, S. Tanabe, and K. Higashi, in *Proceedings of the Eighth International Conference on Creep and Fracture of Engineering Materials and Structures, Key Engineering Materials,* Vol 171–174, T. Sakuma and K. Yagi, Ed., Trans Tech Publications, Tsukuba, Japan, 2000, p 337
17. W.J. Kim, S.W. Chung, C.S. Chung, and D. Kum, *Acta Mater.,* Vol 49, 2001, p 3337
18. T. Mohri, M. Mabuchi, M. Nakamura, T. Asahina, H. Iwasaki, T. Aizawa, and K. Higashi, *Mater. Sci. Eng. A,* Vol 290, 2000, p 139
19. Y.H. Wei, Q.D. Wang, Y.P. Zhu, H.T. Zhou, W.J. Ding, Y. Chino, and M. Mabuchi, *Mater. Sci. Eng. A,* Vol 360, 2003, p 107
20. M. Mabuchi, T. Asahina, and K. Higashi, *Mater. Sci. Technol.,* Vol 13, 1997, p 825
21. K. Mathis, J. Gubicza, and N.H. Nam, *J. Alloy. Compd.,* Vol 394, 2005, p 194
22. M. Mabuchi, K. Ameyama, H. Iwasaki, and K. Higashi, *Acta Mater.,* Vol 47, 1999, p 2047
23. K. Matsubara, Y. Miyahara, Z. Horita, and T.G. Langdon, *Acta Mater.,* Vol 51, 2003, p 3073
24. T. Mohri, M. Mabuchi, N. Saito, and M. Nakamura, *Mater. Sci. Eng. A,* Vol 257, 1998, p 287
25. A. Uoya, T. Shibata, K. Higashi, A. Inoue, and T. Masumoto, *J. Mater. Res.,* Vol 11, 1996, p 2731
26. J.A. Wert, N.E. Paton, C.H. Hamilton, and M.W. Mahoney, *Metall. Trans. A,* Vol 12, 1981, p 1265
27. B.M. Watts, M.J. Stowell, B.L. Baikie, and D.G.E. Owen, *Met. Sci.,* Vol 10, 1976, p 198
28. D.A. Woodford, *Trans. ASM,* Vol 62, 1969, p 291
29. D. Lee and W.A. Backofen, *Trans. TMS-AIME,* Vol 239, 1967, p 1034
30. A.K. Ghosh and C.H. Hamilton, *Metall. Trans. A,* Vol 10, 1979, p 699
31. A.K. Ghosh, Deformation of Polycrystals: Mechanisms and Microstructures, *Proceedings of the Second Riso International Symposium on Metallurgy and Materials Science,* N. Hansen et al., Ed., 1981, p 277
32. A.K. Ghosh and R.A. Ayres, *Metall. Trans. A,* Vol 7, 1976, p 1589
33. J. Hedworth and M.J. Stowell, *J. Mater. Sci.,* Vol 6, 1971, p 1061
34. C.P. Cutler, Ph.D. thesis, University of Cambridge, 1971
35. A. Arieli and A. Rosen, *Metall. Trans. A,* Vol 8, 1977, p 1591

36. B.M. Watts and M.J. Stowell, *J. Mater. Sci.,* Vol 6, 1971, p 228
37. M. Surey and B. Baudelet, *J. Mater. Sci.,* Vol 8, 1973, p 363
38. C. Herring, *J. Appl. Phys.,* Vol 21, 1950, p 437
39. R.L. Coble, *J. Appl. Phys.,* Vol 34, 1963, p 1679
40. N.E. Paton and C.H. Hamilton, *Metall. Trans. A,* Vol 10, 1979, p 241
41. A.K. Ghosh and R. Raj, *Acta Metall.,* Vol 29, 1981, p 607
42. P.A. Friedman and A.K. Ghosh, Control of Superplastic Deformation Rate During Uniaxial Tensile Tests, *Metall. Mater. Trans. A,* Vol 27, 1996, p 3030
43. F. Li and A.K. Ghosh, *Acta Mater.,* Vol 45, 1997, p 3887
44. A.K. Ghosh and C.H. Hamilton, in *Proceedings of the Fifth International Conference on the Strength of Metals and Alloys,* P. Haasen, Ed., 1979, p 905
45. C.H. Hamilton, *Formability: Analysis, Modeling, and Experimentation,* S.S. Hecker, A.K. Ghosh, and H.L. Gegel, Ed., The Metallurgical Society, 1977
46. O.W. Davis, C.F. Osborne, and G.J. Brooks, *Met. Eng. Q.,* Nov 1973, p 5
47. R. Pearce, *Mod. Met.,* April 1986, p 188
48. F. Jovane, *Int. J. Mech. Sci.,* Vol 10, 1968, p 403
49. G.C. Cornfield and R.H. Johnson, *Int. J. Mech. Sci.,* Vol 12, 1970, p 479
50. A.K. Ghosh and C.H. Hamilton, *Process Modeling: Fundamentals and Applications to Metals,* American Society for Metals, 1979, p 303
51. C.D. Ingelbrecht, *J. Mater. Sci. Lett.,* Vol 4, 1985, p 1021
52. C.H. Hamilton and G.W. Stacher, *Met. Prog.,* March 1976, p 34
53. W. Johnson, T.Y.M. Al-Naib, and J.L. Duncan, *J. Inst. Met.,* Vol 100, 1972, p 45
54. T. Oshita and H. Takei, *J. Jpn. Inst. Met.,* Vol 36, 1973, p 1081
55. R. Hawkins and J.A. Belk, *Met. Technol.,* Nov 1976, p 516
56. T.Y.M. Al-Naib and J.L. Duncan, *Int. J. Mech. Sci.,* Vol 12, 1970, p 463
57. B.W. Kim and A. Arieli, Paper 8201-076, presented at the ASM Metals Congress (St. Louis, MO), ASM Metals/Materials Technology Series, American Society for Metals, 1982
58. J.R. Williamson, Diffusion Bonding in SPF/DB, *Proceedings of the American Welding Society* (Las Vegas, NV), 1982
59. J.F. Hubert, *Tool. Prod.,* Vol 42, 1977, p 74

60. A.K. Ghosh, *Acta Metall.,* Vol 25, 1977, p 1413
61. F.A. Nichols, *Acta Metall.,* Vol 28, 1980, p 663
62. F.A. Mohamed and T.G. Langdon, *Acta Metall.,* Vol 29, 1981, p 911
63. D.L. Holt, *Int. J. Mech. Sci.,* Vol 12, 1970, p 491
64. G.G.W. Clemas, S.T.S. Al-Hassani, and W. Johnson, *Int. J. Mech. Sci.,* Vol 17, 1975, p 711
65. J. Belk, *Int. J. Mech. Sci.,* Vol 17, 1975, p 505
66. G.J. Cocks, C. Rowbottom, and D.M.R. Taplin, *Met. Technol.,* July 1976, p 332
67. J. Argyris and J. St. Doltsinis, *Plasticity Today,* A. Sawczuk and G. Bianchi, Ed., Elsevier, 1985, p 715
68. K.S.K. Chockalingam, M. Neelakantan, S. Devaraj, and K.A. Padmanabhan, *J. Mater. Sci.,* Vol 20, 1985, p 1310
69. S. Yu-Quan, *Mater. Sci. Eng.,* Vol 84, 1986, p 111
70. A.K. Ghosh and C.H. Hamilton, *Metall. Trans. A,* Vol 11, 1980, p 1915
71. D.H. Bae and A.K. Ghosh, Cavitation in a Uniaxially Deformed Superplastic Al-Mg Alloy, *Proc. MRS Fall Meeting,* Vol 601 (Boston, MA), MRS, 2000
72. A.K. Ghosh, D.H. Bae, and S.L. Semiatin, Initiation and Early Stages of Cavity Growth During Superplastic and Hot Deformation, *Mater. Sci. Forum,* Vol 04-306, 1999, p 609–616
73. A.K. Ghosh and D.-H. Bae, *Mater. Sci. Forum,* Vol 89, 1997, p 243–245
74. D. Li and A. Ghosh, Tensile Deformation Behavior of Aluminum Alloys at Warm Forming Temperatures, *Mater. Sci. Eng. A,* Vol 352 (No. 1–2), July 15, 2003, p 279–286
75. D. Li and A.K. Ghosh, Effects of Temperature and Blank Holding Force on Biaxial Forming Behavior of Aluminum Sheet Alloys, *J. Mater. Eng. Perform.,* Vol 13 (No. 3), June 2004, p 348–360
76. J.G. Schroth, General Motors' Quick Plastic Forming Process, *Advances in Superplasticity and Superplastic Forming,* 2004 TMS Annual Meeting (Charlotte, NC), March 14–18, 2004, p 9–20
77. W. Rosenhain and D. Ewen, *J. Inst. Met.,* Vol 8, 1912, p 149
78. E.E. Underwood, *J. Met.,* Vol 914, 1962, p 919
79. A.K. Mukherjee, in *Annual Reviews of Materials Science,* Vol 9, R.A. Huggins, Ed., Annual Reviews, Inc., 1979

80. O.D. Sherby and J. Wadsworth, in *Deformation, Processing, and Structure,* G. Krauss, Ed., American Society for Metals, 1984, p 355
81. O.D. Sherby and R.D. Caligiuri, in *Superplasticity: AGARD(NATO) Lecture Series No. 154,* Advisory Group for Aerospace Research and Development, 1987, p 3-1
82. N. Ridley, in *Superplastic Forming of Structural Alloys,* N.E. Paton and C.H. Hamilton, Ed., The Metallurgical Society, 1982, p 191
83. B. Walser and U. Ritter, in *International Conference on Superplasticity,* B. Baudelet and M. Suery, Ed., Editions du Centre Nationale de la Recherche Scientifique, 1985, p 15.1
84. O.D. Sherby, B. Walser, C.M. Young, and E.M. Cady, *Scr. Metall.,* Vol 9, 1975, p 569
85. T. Oyama, J. Wadsworth, M. Korchynsky, and O.D. Sherby, in *Proceedings of the Fifth International Conference on the Strength of Metals and Alloys, International Series on the Strength and Fracture of Materials and Structures,* Pergamon Press, 1980, p 381
86. T. Oyama, O.D. Sherby, J. Wadsworth, and B. Walser, *Scr. Metall.,* Vol 18, 1984, p 799
87. R. Pearce and E.W.J. Miller, in *Superplastic Forming of Structural Alloys,* N.E. Paton and C.H. Hamilton, Ed., The Metallurgical Society, 1982, p 191
88. O.D. Sherby, T. Oyama, D.W. Kum, B. Walser, and J. Wadsworth, *J. Met.,* Vol 37 (No. 6), 1985, p 50
89. L.E. Eiselstein, O.A. Ruano, and O.D. Sherby, *J. Mater. Sci.,* Vol 18, 1983, p 483
90. R.D. Caligiuri, R.T. Whalen, and O.D. Sherby, *Int. J. Powder Metall. Powder Technol.,* Vol 3, 1976, p 154
91. O.A. Ruano, L.E. Eiselstein, and O.D. Sherby, *Metall. Trans. A,* Vol 13, 1982, p 1785
92. H. Sunada, J. Wadsworth, J. Lin, and O.D. Sherby, *Mater. Sci. Eng.,* Vol 38, 1979, p 35
93. O.D. Sherby, J. Wadsworth, R.D. Caligiuri, L.E. Eiselstein, B.C. Snyder, and R.T. Whalen, *Scr. Metall.,* Vol 13, 1979, p 941
94. B.C. Snyder, J. Wadsworth, and O.D. Sherby, *Acta Metall.,* Vol 32, 1984, p 919
95. G.S. Daehn, D.W. Kum, and O.D. Sherby, *Metall. Trans. A,* Vol 17, 1986, p 2295
96. G.W. Hayden, R.C. Gibson, H.P. Merrick, and J.H. Brophy, *Trans. ASM,* Vol 60, 1967, p 3

Spinning

B.P. Bewlay, General Electric Global Research
D.U. Furrer, Ladish Company

METAL SPINNING is a term used to describe the forming of metal into seamless, axisymmetric shapes by a combination of rotational motion and force (Ref 1–4). Metal spinning typically involves the forming of axisymmetric components over a rotating mandrel using rigid tools or rollers. There are three types of metal-spinning techniques that are practiced: manual (conventional) spinning (Ref 1, 2), power spinning (Ref 4–11), and tube spinning (Ref 7, 8). The first two of these techniques are described in this article. Tube-spinning technology is described in the articles "Flow Forming" and "Roll Forming of Axially Symmetric Components" in *Metalworking: Bulk Forming,* Vol 14A of *ASM Handbook,* 2005.

Figure 1 shows examples of products from metal spinning. The range of components include:

- Bases, baskets, basins, and bowls
- Bottoms for tanks, hoppers, and kettles
- Canopies, caps, and canisters
- Housings for blowers, fans, filters, and fly-wheels
- Ladles, nozzles, orifices, and tank outlets
- Pails, pans, and pontoons
- Cones, covers, and cups
- Cylinders and drums
- Funnels and horns
- Domes, hemispheres, and shells
- Rings, spun tubing, and seamless shapes
- Vents, venturis, and fan wheels

The equipment for metal spinning is based on lathe technology, with appropriate modifications for the components that are being formed. Typically, sheet preforms are employed to allow relatively low forming stresses. Metal spinning can be used to cost-effectively produce single or a small number of parts out of expensive materials, such as platinum, or large quantities of components of low-cost materials, such as aluminum reflectors. In this article, the term *preform* is used to describe the component both before and during metal spinning; other terms that are sometimes used include *workpiece* and *starting blank.*

In manual spinning, a circular blank of a flat sheet, or preform, is pressed against a rotating mandrel using a rigid tool (Ref 1, 2). The tool is moved either manually or hydraulically over the mandrel to form the component, as shown in Fig. 2. The forming operation can be performed using several passes.

Manual metal spinning is typically performed at room temperature. However, elevated-temperature metal spinning is performed for components with thick sections or for alloys with low ductility. Typical shapes that can be formed using manual metal spinning are shown in Fig. 3 and 4; these shapes are difficult to form economically using other techniques. Manual spinning is only economical for low-volume production. Manual metal spinning is

Fig. 1 Various components produced by metal spinning. Courtesy of Leifeld USA Metal Spinning, Inc.

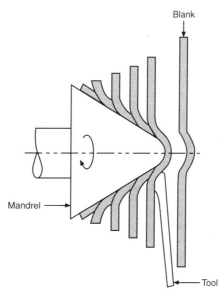

Fig. 2 Schematic diagram of the manual metal-spinning process, showing the deformation of a metal disk over a mandrel to form a cone

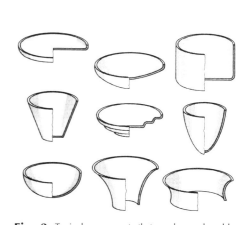

Fig. 3 Typical components that can be produced by manual metal spinning. Conical, cylindrical, and dome shapes are shown. Some product examples include bells, tank ends, funnels, caps, aluminum kitchen utensils, and light reflectors.

extensively used for prototypes or for production runs of less than ~1000 pieces, because of the low tooling costs. Larger volumes can usually be produced at lower cost by power spinning or press forming.

Power spinning of metals and alloys is also known as shear spinning (Ref 4–11), because this forming method employs high shear forces. There are two large-volume applications of power spinning: metal cone spinning and metal tube spinning. Power spinning can also be employed to produce hemispheres, provided a preform of the appropriate design is employed. In cone spinning, a flat blank or metal sheet is deformed over a mandrel at high speed.

Almost all ductile metals can be shaped using power spinning (provided they have a minimum ductility corresponding to an elongation of ~2%). Some of the alloys used to make components by metal spinning include:

- 300- and 400-series stainless steels
- Precipitation-hardening stainless steels (17-4PH, 17-7PH)
- Iron-nickel superalloys (A-286)
- Nickel-base superalloys such as Hastelloys (X, C, B, S) and Inconels (600, 625, 718, X-750), N-155, Nimonic 263, and Waspaloy
- Cobalt-base alloys (Haynes 188, Haynes 230)
- Other nonferrous alloys such as aluminum, brass, copper, and platinum

A wide range of shapes can be produced with relatively simple tooling. Power spinning can be used to form large conical parts (up to 3 m, or 10 ft, in diameter) to close tolerances (<±0.5 mm, or 0.020 in.); there is little scrap material, and the forming operation can be completed quickly. The high material utilization is highly desirable, because the preforms for power spinning can be expensive, particularly for aerospace components.

Products that can be produced by metal spinning range from small hardware items made in large quantities (such as metal tumblers and automotive components) to household articles to large components for aerospace applications in low-volume production. Some examples of metal components that are spun include trophies, kettles, kettle drums, cymbals, tank ends, centrifuge parts, pressure bottles, venturis, radar

reflectors, parabolic dishes, wheel discs, and wheel rims.

The use of elevated metal temperatures is sometimes required during metal spinning to reduce the flow stress and increase the ductility of the component, particularly if the machine capacity is insufficient for cold forming the component or if the alloy ductility is too low. Although power spinning can be performed with only one roller, two rollers are generally preferred to balance the radial forces acting on the mandrel and the bearings within the lathe. Power spinning involves large plastic strains and relatively high strain rates, and as a result, significant heat can be generated during forming; the heat is generally dissipated using a coolant/lubricant during power spinning.

The surface finish of spun components is usually of sufficient quality that no additional machining is required after spinning. The surface finish of spun components is typically approximately 1.5 μm (0.06 mil), although finishes as smooth as 0.5 μm (0.02 mil) have been produced by power spinning. Preforms are generally used for cone spinning when the included angle of the component is less than 35° or when the percentage of wall reduction is high, as is described in more detail subsequently. The preforms are typically cold formed in a die, although hot forging or machining, or a combination of both, can also be used.

In this article, manual spinning is described first, and power spinning is described second. For each of these techniques, the process technology, equipment, and tooling is described.

Manual Spinning of Metallic Components

General Description. Manual metal spinning is practiced by pressing a tool against a circular metal preform that is rotated using a lathe-type spinning machine. The tool typically has a work face that is rounded and hardened. Some of the traditional tools are given curious names that describe their shape, such as "sheep's nose" and "duck's bill." The first manual spinning machine was developed in the 1930s. Manual metal spinning involves no significant thinning of the work metal; it is essentially a shaping technique. Metal spinning can be performed with or without a forming mandrel. The sheet preform is usually deformed over a mandrel of a predetermined shape, but simple shapes can be spun without a mandrel. Various mechanical devices and/or levers are typically used to increase the force that can be applied to the preform. Most ductile metals and alloys can be formed using metal spinning. Manual metal spinning is generally performed without heating the workpiece; the preform can also be preheated to increase ductility and/or reduce the flow stress and thereby allow thicker sections to be formed.

Manual metal spinning is used to form cups, cones, flanges, rolled rims, and double-curved surfaces of revolution (such as bells). Typical

shapes that can be formed by manual metal spinning are shown in Fig. 3 and 4; these shapes include components such as light reflectors, tank ends, covers, housings, shields, and components for musical instruments. The maximum practical component diameter is often limited by the size of the available equipment. The upper limit of component thickness increases as preform ductility increases or as flow stress decreases. For example, the manual spinning of aluminum as thick as ~6 mm (0.24 in.) is possible. The practical maximum thickness of low-carbon steel that can be deformed by spinning without mechanical assistance is ~3 mm (0.12 in.).

Manual metal spinning has several advantages and several disadvantages over alternate processes such as press forming or forging (Ref 1, 2). There are three advantages of manual metal spinning. First, the tooling costs and investment in capital equipment are relatively small (typically, at least an order of magnitude less than a typical forging press that can effect the same operation). Second, the setup time is shorter than for forging. Third, the design changes in the workpiece can be made at relatively low cost. However, there are several disadvantages of manual metal spinning. First, highly skilled operators are required, because the uniformity of the formed part depends to a large degree on the skill of the operator. Second, manual metal spinning is usually significantly slower than press forming. Third, the deformation loads available are much lower in manual metal spinning than in press forming. Manual metal spinning and power spinning are generally in competition with pressing and deep drawing.

Equipment for Manual Spinning—Lathes and Tooling. A simple tool and sheet preform setup for manual metal spinning is shown in Fig. 5(a). The forming mandrel is mounted on the headstock of a lathe. The circular preform is clamped to the mandrel by the follower. Pressure is applied at the tailstock by means of an anti-friction center and suitable pressure to form the component. The tool rest and pedestal permit the support pin (fulcrum) to be moved to various positions. Metal spinning is performed by manually applying the friction-type spinning tool as the preform is rotated. Figure 5(b) shows a more complex setup for manual metal spinning. In this arrangement, the spinning rollers are mounted in the fork sections of long levers, and the tool support has a series of holes to adjust the tool position. The roller is manipulated by moving two scissorlike handles around the preform/workpiece.

Horizontal metal-spinning lathes that can spin preforms with diameters in the range of ~6 mm to 1.8 m (0.24 in. to 5.9 ft) have been built. For large-diameter parts that may be formed at high speeds, special pit lathes (per the safety requirements) that permit the spinning of blanks as large as ~5 m (16 ft) in diameter have been built. Standard lathes can also be fitted with special tooling for making ovular parts. Tooling costs are generally low for manual spinning. However, manual spinning is generally

Fig. 4 Photograph of conical components that were produced by metal spinning. Courtesy of Leifeld USA Metal Spinning, Inc.

performed using multiple passes, and tool life can be low. Hydraulic lathes were introduced after about 1945 for forming components with either a thicker section or higher-strength alloys. A reliable spinning machine must possess a significant mass in order to ensure stability; the mass provides vibration-free operation when producing components to tight and repeatable tolerances at high speed. The high speeds associated with metal-spinning processes require considerations for all safety aspects of the process.

Mandrel technology plays a very important role in metal spinning. The mandrels are also sometimes referred to as form blocks or spin blocks for manual spinning. The mandrels can be made of seasoned hard-maple wood or metals or combinations of the two. Most hardwood mandrels are constructed by gluing strips of 25 to 50 mm (1.0 to 2.0 in.) thick maple into the main block to create a cross-laminated structure to increase strength. Such mandrels are stronger and more durable than mandrels machined from a solid block. Some wooden mandrels are reinforced with steel at the ends and at small radii, to ensure maintenance of radii in the final part. Minimum inside radii of 1.6 mm (0.06 in.) are possible using mandrels of appropriate construction; corners with radii of smaller than 1.6 mm (0.06 in.) are not desirable. Corners with radii of greater than 3 mm (0.12 in.) are preferred where possible.

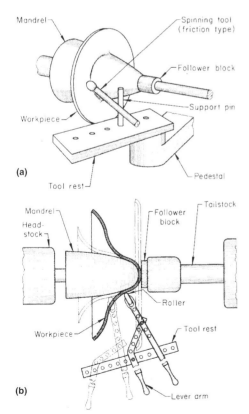

Fig. 5 Typical arrangements for manual spinning using a lathe. (a) Simple arrangement with a friction-type spinning tool. (b) More complex setup using levers and a spinning roller

Other mandrel materials include steel, cast iron, aluminum, magnesium, and plastic-coated wood. When it is necessary to produce parts to close tolerance, the mandrels are typically made entirely of steel and cast iron. Cored castings of steel or cast iron are preferred in order to reduce the rotating weight. Mandrels must be statically balanced, and, when used at high speed, the mandrels should also be dynamically balanced.

Simple metal-spinning tools can be made by forging carbon or low-alloy tool steels to the desired shape and hardening the working surfaces to a hardness of ~60 HRC. The rollers also need to be polished when surface finish of the final part is important. Typically, the rollers are made of hardened tool steel or aluminum bronze.

Process Technology for Manual Spinning. Manual metal spinning is extensively used for prototypes or for production runs of less than ~1000 pieces, because of the low tooling costs. Larger volumes can usually be produced at lower cost by power spinning or press forming. For large-quantity production, power spinning can generally be conducted at lower cost than manual metal spinning. For example, a stainless steel cover for a food-processing machine (with a shape similar to that shown in the middle of Fig. 3) can be produced economically at the rate of 100 per year using manual metal spinning.

Conical parts (such as the shapes in Fig. 4) are ideally suited for metal spinning because only one tool is required; drawing in dies would require four or five operations. Many such cones can be spun in one operation at a moderate production rate, depending on their included angle. For large-quantity production, power spinning is generally less expensive than manual metal spinning. In cone spinning, the deformation of the metal from the flat blank is performed in accordance with the sine law, as is described in more detail in the section "Mechanics of Cone Spinning" in this article.

The rotational speeds that are best suited to manual metal spinning depend mainly on work metal composition and thickness. For example, a given blank of stainless steel can be spun at a surface speed of 60 m/min (200 surface feet per minute, or sfm). Under otherwise identical conditions, changing to an aluminum blank will permit speeds of 120 to 180 m/min (400 to 600 sfm). Selection of optimal speed depends largely on operator skill. In many metal-spinning operations, speed is changed (usually increased) during the operation by means of a variable-speed drive on the headstock. The dimensional tolerances that can be achieved by manual spinning increase as the diameter of the component decreases. For components up to 300 mm (12 in.) in diameter, tolerances of ±0.20 mm (0.008 in.) can be achieved. For larger-diameter components, the tolerances are worse. For example, parts ~4 m (13 ft) in diameter can only be produced to tolerances of approximately ±1.0 mm (0.04 in.), but this is machine dependent.

Lubricants generally need to be used in all metal-spinning operations, regardless of the preform composition or shape or the type of metal-spinning tools that are used. Lubricants are typically required both before and during forming. The need for lubrication during spinning depends on the tenacity of the lubricant used and on the rotational speed of the preform. The lubricant must continue to adhere to the rotating preform during spinning. Ordinary cup grease is often used. It can be heated to reduce its viscosity, for ease of application. Other lubricants used for metal spinning include soaps, waxes and tallows, and pigmented drawing compounds; in the selection of the most suitable lubricant, the ease of removal of the lubricant after forming has to be considered.

Power Spinning

General Description. Power spinning of metals and alloys is also known as shear spinning, because in this method metal is deformed using high shear forces (up to 3.5 MN, or 800,000 lbf). There are two broad applications of power spinning: metal cone spinning and metal tube spinning. In cone spinning, the deformation of the metal from the flat blank is performed in accordance with the sine law, as is described subsequently. Almost all ductile metals can be shaped using power spinning (provided they have a minimum ductility of ~2%). Products range from small hardware items made in large quantities (metal tumblers, for example) to large components for aerospace applications in low-volume production. A classic low-production-volume example of power spinning involved the Concorde engine compressor shaft (Ref 4), which was formed by a combination of forging and spinning, because this was more efficient than forging alone.

Metal blanks as large as 6 m (20 ft) in diameter have been successfully formed using power spinning. Conical and curvilinear shapes are most commonly produced from flat (preformed) blanks by power spinning. Plate stock up to 25 mm (1.0 in.) thick can be power spun at room temperature. Blanks as thick as 140 mm (5.5 in.) have been successfully spun at elevated temperature.

Mechanics of Cone Spinning. The most common application of power spinning is for conical shapes. In this variant, the metal is volumetrically displaced in the axial direction. The metal deformation occurs in accordance with the sine law, which relates the wall thickness of the starting blank, t_1, and the wall thickness of the finished workpiece, t_2, as $t_2 = t_1 (\sin\alpha)$, where α is half the apex angle of the cone (assuming uniform wall thickness in the conical section). The diameter of the finished component is the same as that of the starting blank. When metal spinning is performed in accordance with the so-called sine law (Ref 1, 2, 4), the thickness of the component in the axial direction is the same as the thickness of the starting blank

(Fig. 6). The arrangement shown in Fig. 6 is for cone spinning using a single pass.

When spinning metal cones to small singles (<35° included angle), it is generally easier to use multiple spinning passes with different cone angles for each pass, as illustrated in Fig. 7; typically, the component is annealed or stress relieved between passes. The practice of multiple passes with intermediate anneals permits a high total reduction while maintaining a practical reduction limit of 50 to 75% between process anneals. The reduction between successive annealing operations is determined by the maximum deformation limit for the metal being spun, as is described subsequently.

Deformation limits are shown for a range of alloys in Table 1; the deformation limit is obtained by multiplying the thickness of the starting preform, t_1, by the maximum reduction factor and then dividing the result by t_1 to obtain the sine of the half-angle required for the conical mandrel.

In power spinning of small-angle cones (as shown in Fig. 7), even when multiple-pass spinning is used, the original blank diameter is retained, and the exact volume of material is used in the final part. At any diameter of either the preform or the completed workpiece, the axial thickness equals the thickness of the original blank. For example, if a flat plate has a diameter of 190 mm (7.5 in.) and a thickness of 12.5 mm (0.5 in.), the spun preform has the same 12.5 mm (0.5 in.) axial thickness, but the wall thickness is only 6.25 mm (0.25 in.) (t_2 in Fig. 7), thus satisfying the sine law. Similarly, the final workpiece has an axial thickness of 12.5 mm (0.5 in.), but in accordance with the sine law, it has a wall thickness of only 3.1 mm (0.125 in.) (t_3 in Fig. 7).

Deviations from the sine law that can occur are usually expressed in terms of overreduction or underreduction. In overreduction, the final thickness of the workpiece is less than that indicated by the sine law; in underreduction, the thickness is greater. In overreduction, the flange on the cone will lean forward; in underreduction, the flange on the cone will lean backward. If a thin blank is spun with severe underreduction, the flange can also wrinkle.

Machines for Power Spinning—Lathes and Tooling. Power spinning is generally performed using special-purpose machines. The significant components of a power-spinning machine are shown schematically in Fig. 8.

Although Fig. 8 illustrates power spinning of a conical shape, similar machines can be used for power spinning of tubes. Figure 9 shows a horizontal lathe for spinning large-diameter cone and dish-shaped components, and Fig. 10 is a photograph of some remarkably large-diameter (~2 m, or 6.5 ft) cone- and dish-shaped components that were produced by metal spinning. Spinning machines can also be configured to

Table 1 Maximum preform thickness reductions (approximate), or deformation limits for single-pass power spinning of a range of metals and alloys

Typically, the maximum reduction that can be employed to form a hemisphere is less than can be used for forming a cone.

Material	Maximum reduction for a cone, %	Maximum reduction for a hemisphere, %
Aluminum alloys		
2014	50	40
2024	50	...
3000	60	50
5086	65	50
5256	50	35
6061	75	50
7075	65	50
Beryllium	35	...
Copper	75	...
Nickel alloys		
Waspaloy	40	35
René 41	40	35
Steels		
4130	75	50
4340	70	50
6434	70	50
D6ac	70	50
H11	50	35
Stainless steels		
321	75	50
347	75	50
410	60	50
17-7PH	65	45
A-286	70	55
Titanium		
Commercially pure titanium	45	...
Ti-6Al-4V	55	...
Ti-3Al-13V-11Cr	30	...
Ti-6Al-6V-2.5Sn	50	...
Tungsten	45	...

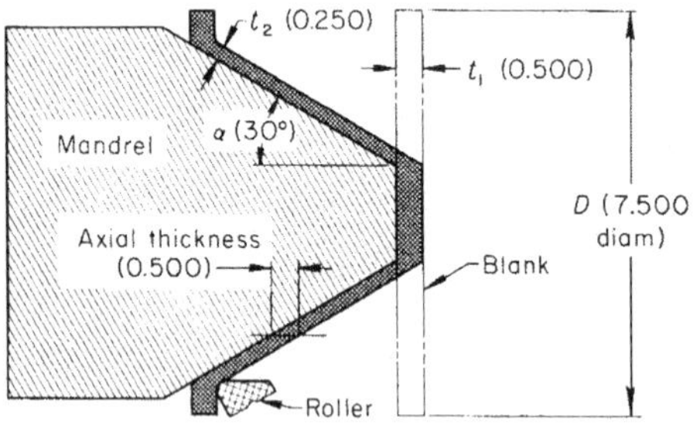

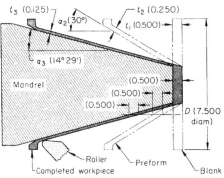

Fig. 6 Typical arrangement for power spinning a cone in a single operation. The mandrel diameter is 188 mm (7.5 in.), t_1 is the thickness of the preform, and t_2 is the wall thickness of the final conical component. The included angle of the cone is α. For the case of power spinning, the diameter of the final component is the same as the starting sheet preform. Dimensions given in inches

Fig. 7 Typical arrangement for power spinning a cone in two stages. The two-step approach is used for small included cone angles (35° in this figure). Dimensions given in inches

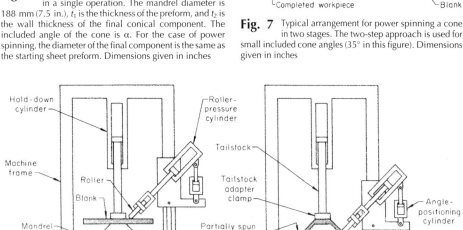

Fig. 8 Schematic diagrams of a vertical arrangement employed for power spinning of large-diameter cones. The diagram shows the preform, clamping cylinder, and the positioning cylinders that are used to control the axial, radial, and angular positions of the roller shed for the forming scheme used to generate the cone.

Fig. 9 Photograph of a horizontal lathe, workpiece, and mandrel arrangement for spinning large-diameter (~2 m, or 6.5 ft) cone- and dish-shaped components. Courtesy of Leifeld USA Metal Spinning, Inc.

accommodate several different rollers with quick switchovers.

Machines for power spinning are generally specified by the diameter and length of the largest component that can be spun and the maximum load that can be applied to the work. Metal-spinning machines can be vertical or horizontal. Machines used for spinning large-diameter and large-mass preforms, such as 1.8 m (6 ft) or more in diameter, are usually vertical because they are better suited to handling large components. A broad range of power-spinning machines has been built. The capacity of spinning machines ranges from 455 mm (18 in.) diameter and 380 mm (15 in.) length (maximum component dimensions) to machines capable of spinning workpieces as large as 6 m (20 ft) in diameter and 6 m (20 ft) long. The load on the work can be as great as 3.5 MN (800,000 lbf). Figure 11 shows an automated metal-spinning machine that can be configured for forming cone-shaped components. The lathe, roller, mandrel, and controls work station can be seen.

Machines for power spinning can be automated in a variety of ways. Contemporary metal-spinning machines use template guides that control the shape and accuracy of the workpiece. Most modern machines used for production spinning are at least semiautomatic; that is, they are loaded and unloaded by the operator, but the entire spinning cycle is controlled automatically. Machines can also be equipped with automatic loading and unloading devices for high-volume production. The most recent machines employ computer numerically controlled (CNC)-based techniques, with both playback and CNC controls. In playback mode, the first component is run in manual mode by the operator, typically with joystick control; the program that has been acquired can be modified and optimized for high-speed production.

During power spinning, the tooling is subjected to more severe service conditions than during manual spinning, and as a result, design and manufacture of the tooling must be performed in a more rigorous manner. The tooling that is used for both the rollers and the mandrels is described in the following paragraphs. A typical mandrel profile for cone spinning is shown in Fig. 12. The flange diameter, dimension A, and the diameter of the nose of the cone, dimension B, and angle α can be adjusted as required. The usual practice is to have an integral flange to permit the mandrel to be bolted to the headstock. The radius R can vary from a minimum of 0.8 mm (0.03 in.) to a round nose. Mandrel wear or failure can be a problem in the power spinning of cones. The mandrels used for production spinning of cones must be hard in order to resist wear, and they must have a high fatigue strength in order to resist the fatigue loading due to the normal eccentric loading during power spinning. Failure is typically caused by spallation of regions from the surface.

The materials used for the mandrels for cone spinning are selected primarily on the basis of the desired mandrel life. The most commonly used materials are cast irons and tool steels; the actual mandrel material selection depends on the part design, part material, and desired life. For example, gray cast iron can be used for the low-volume (10 to 100 pieces) spinning of soft metals, and alloy cast iron for spinning 100 to 250 pieces; the mandrels can be hardened in areas of high wear. For high-production volumes (250 to 750 pieces), 4150 or 52100 steel hardened to approximately 60 HRC can be used. Tool steels such as O6, A2, D2, or D4 hardened to 60 HRC or slightly higher are more suitable for high-volume production. The surface finish of the mandrels should be at least 1.5 μm (0.06 mil). The mandrel dimensions should be machined so that they are within ±0.025 mm (0.0010 in.) of being concentric with each other.

Three types of rollers are used in power spinning; these are shown in Fig. 13. The roller

Fig. 11 Automated metal-spinning machine for forming cone-shaped components. The lathe, roller, mandrel, and controls work station can be seen. Courtesy of Leifeld USA Metal Spinning, Inc.

Fig. 10 Photograph of very large-diameter (~2 m, or 6.5 ft) cone- and dish-shaped components produced by the Leifeld Company. Courtesy of Leifeld USA Metal Spinning, Inc.

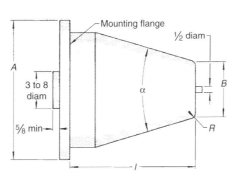

Fig. 12 Typical mandrel used for power spinning of cones. Generally, there are small bosses on the nose and tail for clamping in the tailstock and headstock, respectively. Dimensions given in inches

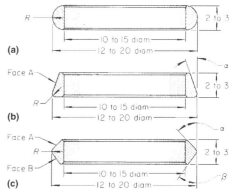

Fig. 13 Diagrams of typical forming rollers used for spinning of cones and hemispheres. (a) Full-radius roller. (b) Roller profiled for forming corners. (c) Roller used for reducing the wall thickness. Dimensions given in inches

designs shown in Fig. 13 typically have outside diameters in the range 305 to 510 mm (12 to 20 in.), depending on the type and size of the spinning machine and the part to be formed. Roller widths are usually 50 to 75 mm (2 to 3 in.). The design of the roller employed depends principally on the shape of the component that is to be formed. The full-radius roller design shown in Fig. 13(a) (two axes of symmetry) is generally used to produce curvilinear shapes, and the designs with the radii of curvature shown in Fig. 13(b) and (c) are preferred for the spinning of cones. The design of the rollers, and the alloy used for the rollers, play a critical role in ensuring efficient power spinning.

The roller angle α shown in Fig. 13(b) and (c) is adjusted to suit the geometry of the component that is being spun (the included angle of the cone has a significant effect on selection of roller design). This roller angle is selected to provide clearance such that the work metal does not contact the faces of the roller where the metal is being deformed (surfaces A and B, shown in Fig. 13(c). The radius R should not be less than the final wall thickness.

The roller design illustrated in Fig. 13(b) has been widely used for cone spinning. A typical arrangement for cone spinning, using two opposed rollers, is shown in Fig. 14.

When two rollers are used to spin a part from flat plate, the rollers are positioned at equivalent/symmetric conditions with respect to the preform. However, when metal spinning is performed from a preform, a lead roller is often used, and it is set ahead of the other by 1.5 to 3 mm (0.06 to 0.12 in.). The angle between the axis of rotation of the rollers and the surface of rotation of the workpiece (angle β in Fig. 14) is typically ~10°. The angle between the axis of rotation of the roller and the peripheral face of the roller (angle γ in Fig. 14) can be adjusted for different shapes, and it is also often adjusted during the forming operation; this angle is shown in Fig. 14 as approximately 30°.

Rollers for power spinning are typically made from tool steel or tungsten carbide. A variety of tool steels have been employed, including W2, O6, D2, and D4. The roller material is selected on the basis of the number of parts that are to be formed. D2 and D4 tool steels are preferred for high-production quantities (they should be hardened to 60 to 65 HRC). Tungsten carbide is only used for specialized applications when the high cost can be justified. The rollers should be polished to a maximum surface roughness of 0.25 μm (0.010 mil).

Process Technology for Power Spinning. An important factor in power spinning is the deformation limit, or so-called spinnability, of the metal; the spinnability is the smallest section thickness (or the maximum reduction in thickness) to which a component can be formed by metal spinning without failure of the component. A simple test has been established (Ref 1, 4, 6–8) to determine deformation limit (or spinnability) of a metal, as shown in Fig. 15.

The deformation limit, or spinnability, test is performed by spinning a circular blank over an ellipsoidal mandrel, and spinning is performed so that the outside diameter of the final component is the same as the initial blank. Because the thickness is eventually reduced to zero for the ellipsoidal mandrel, all metals will eventually fail at some thickness, t_f. The deformation-limit data on a range of materials with different tensile strengths and different formabilities are shown in Fig. 16 and Table 1 (Ref 1, 4).

The deformation limit is defined as:

$$\text{Maximum thickness reduction} = (t_0 - t_f) \times 100/t_0$$

The maximum reduction is plotted against the tensile reduction in area of the material in Fig. 16. It can be seen that if the metal possesses a tensile reduction in area of 50% or greater, the metal can be reduced by power spinning to a thickness of up to 80% in one pass for spinning a cone. The maximum reduction that can be employed to form a hemisphere is less than can be employed for forming a cone. Also, any increase in the material tensile ductility (as described by the reduction in area) above 50% reduction in area does not increase formability or spinnability. For materials with low ductility, if the ductility can be increased by increasing temperature, then the formability can be improved. Process parameters, such as the feed rate and the rotational speed, have a less significant effect on the spinnability.

The best quality for most components is achieved when spinning at high speeds. The minimum surface speed considered to be

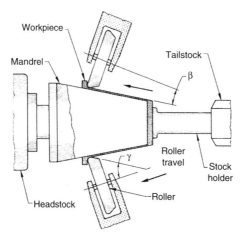

Fig. 14 Schematic diagram showing the relative positions of the preform and two forming rollers used for spinning a cone. One roller can be positioned to follow the second roller, if appropriate for the forming scheme that is being employed.

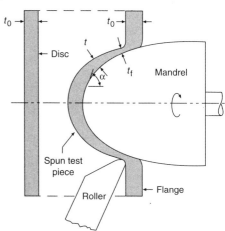

Fig. 15 Schematic diagram of the spinnability test. The thickness of the test coupon is reduced as the roller is advanced; the reduction in thickness at the point of fracture is taken as the maximum spinning reduction per pass. Source: Ref 6

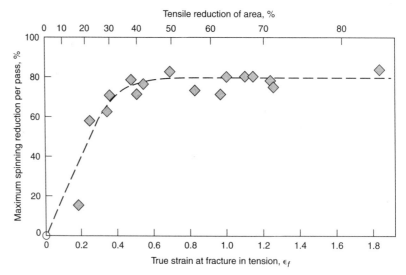

Fig. 16 Maximum spinning reduction per pass as a function of tensile fracture strain for materials of a range of tensile strengths. For materials with tensile ductilities of greater than 50%, there is no further increase in the spinnability. Source: Ref 1, 4

practical for metal spinning is approximately 120 m/min (400 sfm), and this is only used for spinning small-diameter workpieces. Surface speeds of 300 to 600 m/min (1000 to 2000 sfm) are typically used; this speed range is suitable for a range of metal compositions, preform shapes, and process conditions (such as reduction per pass, roller design, roller position, and forming temperature).

Most cone-spinning operations are performed at linear feed rates of 0.25 to 2 mm/rev (0.010 to 0.08 in./rev); for typical spinning machines, this equates to linear feed rates in the range 38 to 380 mm/min (1.5 to 15 in./min) (feed rates are usually measured in millimeters per minute). Most machines used in cone spinning are equipped with devices that continuously change the rate of feed with the diameter on which the rollers are working.

The feed rate controls the workpiece finish and the material properties and the fit of the workpiece to the mandrel. With all other factors constant, a decrease in the feed rate will improve surface finish. An increase in feed rate will make the workpiece fit tighter on the mandrel, and the finish of the workpiece will become coarser. The use of preforms can influence surface finish and is common in cone spinning when the included angle of the cone is less than 35° or when the percentage of wall reduction is high. Preforms are usually prepared by cold forming in a die, although hot forging or machining or a combination of both can be used.

The surface finish of a spun component is usually of sufficient quality that no additional machining is required after spinning. The surface finish of spun components can typically be approximately 1.5 μm (0.06 mil), although surface finishes as smooth as 0.5 μm (0.02 mil) have been produced by power spinning, when using appropriate tooling and surface finish of the tooling.

A lubricant is almost always used during power spinning. The fluid used serves as both a lubricant and a coolant. A water-based coolant, such as an emulsion of soluble oil in water, is most commonly used, and in large quantities because of the large amount of heat generated. When spinning aluminum, stainless steel, or titanium, the workpieces or mandrels or both are sometimes coated with the lubricant before spinning. An increase in the forming temperature can lead to a reduction in the flow stress and an increase in the ductility of the preform; this is sometimes required if the load capacity of the spinning machine is not sufficient for cold forming the preform or if the room-temperature ductility of the work metal is too low. When operating at elevated temperatures, great diligence must be exercised in the selection and use of an appropriate lubricant.

Power Spinning of Hemispheres. Spinning of hemispheres is more complicated than spinning of cones. However, in order to spin hemispheres, preforms of specially designed geometries can be used to adjust the percentage of reduction as a function of radial position, as is described in the following section. This approach has enabled power spinning to be applied to the forming of hemispheres, ellipses, ogives, and almost any curvilinear surface of revolution. However, the design of the preform for curvilinear shapes is more complicated than that for conical shapes. For the case of spinning of conical shapes, it is possible to determine an axial thickness of the spun part that corresponds to the thickness of the blank (Fig. 6, 7). However, the same relationship does not exist for a curvilinear surface; this problem is illustrated in Fig. 17. In the path from the pole to the equator of a hemisphere, the axial thickness of the metal on a hemisphere changes from stock thickness at the pole to infinity (the inverse of sin 0°) at the equator (the wall thickness, in the normal direction, goes to zero). The blank thickness must therefore be back-tapered to compensate for the change in thickness that takes place during spinning of hemispheres. Figure 17 shows a preform for a ~1.5 m (5 ft) diameter hemisphere; the machined taper started at 3.8 mm (0.15 in.) in thickness in the center of the preform and ended at a thickness of 7.5 mm (0.30 in.) in thickness at the circle where the 30° radial line of the sphere was projected to the blank. At the corresponding 45° line, the blank thickness was 5.4 mm (0.21 in.), and the final part thickness was 0.71 times the original thickness. For the region of the hemisphere below the 30° line, the reduction of the preform was greater than permissible for spinning aluminum alloy 6061 (according to the previous description), and the forming operation was performed as if spinning a cylinder; the preform for this region had a flange with a thickness proportional to the designed percentage of reduction.

As an example, a suitable preform for spinning a hemisphere was designed by first finding in Table 1 the maximum allowable reduction for the material that was used in order to obtain the minimum part thickness associated with the deformation limit and minimum angle of the cone. A beginning stock thickness was selected that, with the maximum allowable reduction, gave the thickness desired in the final hemisphere sphere. The ratio of finished stock thickness to original stock thickness was then taken as the sine of an angle, which was the angle of the surface at the latitude at which forming was started. Beyond this point, the reduction required to make the hemisphere was greater than is permissible for the 6061 aluminum alloy. At 45° from the pole, final part thickness was 0.71 times the original thickness. Forming started at the circle corresponding to the latitude associated with the forming limit (the point where the maximum permissible reduction has taken place). In a cross-sectional view, the circles resulting from the aforementioned approach become points, and the thickness of the stock at these points can be determined. The correct roller locus can be programmed with state-of-the-art CNC-based techniques.

The following example describes forming a 1.5 m (5 ft) diameter hemisphere by power spinning. Large hemispheres (Fig. 17) have been power spun from a solution-treated aluminum alloy 6061 using the following calculations. From Table 1 it was determined that a 50% reduction could be used with this alloy. Preliminary calculations for the thickness of the starting preform indicated a thickness of 7.6 mm (0.30 in.) was required (preform thickness = final wall thickness/maximum reduction = 3.8 mm/2).

In calculating the blank thickness profile for various points on the sphere, it was found that at the pole, or 90° point, the thickness had to be reduced to 3.8 mm (0.15 in.) and that a linear reduction was required out to a point directly above the 30° tangency on the hemisphere, where the thickness of the starting blank had to be 7.6 mm (0.30 in.). Beyond this point, a flange was incorporated; the preform thickness was increased in the region by 30% to allow for this flange, and the initial blank thickness was established at 9.9 mm (0.39 in.). Final spinning was accomplished in one pass of the rollers.

The type of procedure described in the previous example has also been successfully used to form both hemispheres and ellipses with diameters in the range of 150 mm to 1.8 m (6 in. to 6 ft). Hemispheres from the following alloys have also been formed: 17-7PH and type 410 stainless steels, alloy steels such as 4130 and 4140, and from aluminum alloys 5086, 2014, 2024, and 6061.

Effects of Power Spinning on Component Properties. Power spinning is a severe cold working operation, and it therefore can have a very significant effect on the mechanical properties of the component. Typically, well-defined flow patterns are generated in the grain structure by power spinning. In many applications, the increase in strength caused by spinning is highly desirable, because it eliminates the need for subsequent heat treating. In those applications where the change in mechanical properties is not desired, the component must be annealed after

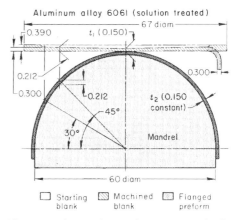

Fig. 17 Schematic diagram showing an example of a forming scheme used to spin a large-diameter hemisphere from a preformed blank metal sheet, such as aluminum alloy 6061. The preform is tapered to allow spinning of a hemisphere with a uniform wall thickness. Dimensions given in inches

metal spinning. The effect of power spinning on mechanical properties, such as fatigue performance and creep resistance, is similar to that of other cold working operations.

Conclusions

This article has described two forming techniques, manual spinning and power spinning, for forming seamless metal components. The equipment for both of these two spinning techniques is based on lathe technology, with appropriate modifications for the components that are being formed. A wide range of components can be produced using these two metal-spinning techniques with relatively simple tooling. Metal spinning is very competitive with other forming processes, such as pressing and deep drawing; it is a highly flexible forming technique.

Metal spinning can be operated economically to produce complicated parts for single applications, low-volume production, and mass production. Manual-metal-spinning and power-spinning processes are very flexible and lend themselves to broad automation. Machine design changes, innovations in control systems, and process developments have led to improvements in all aspects of metal-spinning technology since the 1980s. Manual spinning is generally suitable for low-volume production of components.

This article has described process technology, equipment, and tooling for both manual spinning and power spinning. Power spinning can be used to form large parts (up to 6 m, or 20 ft, in diameter); there is little scrap material, and the forming operation can be completed quickly. A wide range of shapes can be produced with relatively simple tooling. Power spinning is particularly suited to cones and hemispheres. Other components that can be produced by metal spinning range from small hardware items made in large quantities (such as metal tumblers and automotive components) to large components for high-performance aerospace applications in low-volume production (such as rocket engine casings and missile nose cones).

Other examples of metal components that are spun include trophies, kettles, kettle drums, cymbals, tank ends, centrifuge parts, pressure bottles, venturis, radar reflectors, parabolic dishes, wheel discs, and wheel rims. For these types of complex geometries, manual-metal-spinning and power-spinning techniques are generally preferred over pressing and deep drawing; the advantages of spinning include flexible production, relatively low tooling costs, and short setup times.

ACKNOWLEDGMENTS

This article is adapted from Spinning, *Forming and Forging*, Volume 14, *Metals Handbook*, 9th ed., 1988. The authors would also like to acknowledge Dirk Palton and Leifeld USA Metal Spinning, Inc. for the use of several photographs.

REFERENCES

1. Spinning, *Metals Handbook Desk Edition*, American Society for Metals, 1985
2. K. Lange, Metal Spinning, *Handbook of Metal Forming*, Society of Manufacturing Engineers, 1985, p 21.1–21.14
3. H. Palten and D. Palten, *Met. Form.*, Sept 2002
4. S. Kalpakjian, *Manufacturing Processes for Engineering Materials*, 2nd ed., Addison-Wesley, 1991, p 436–438
5. *Roll Forming*, Leico Machine Company, Germany
6. R.L. Kegg, A New Method for Determination of Spinnability of Metals, *J. Eng. Ind. (Trans. ASME)*, Vol 83, 1961, p 119–124
7. S. Kalpakjian, A Study of Shear-Spinnability of Metals, *J. Eng. Ind. (Trans. ASME)*, Vol 83, 1961, p 478–484
8. S. Kalpakjian, *J. Eng. Ind. (Trans. ASME)*, Vol 86, 1964, p 49–54
9. *Flow Forming*, Dynamic Machine Works, Billerica, MA
10. D. Furrer, *Adv. Mater. Process.*, Vol 155 (No. 3), 1999, p 33–36
11. C.T. Olofson, T.G. Byrer, and F.W. Boulger, Ladish Company, Battelle DMIC Review, May 29, 1969

Rubber-Pad Forming and Hydroforming

RUBBER-PAD FORMING, also known as flexible-die forming, employs a rubber pad or a flexible diaphragm as one tool half, requiring only one solid tool half to form a part to final shape. The solid tool half is usually similar to the punch in a conventional die, but it can be the die cavity. The rubber acts somewhat like hydraulic fluid in exerting nearly equal pressure on all workpiece surfaces as it is pressed around the form block.

Rubber-pad forming is designed to be used on moderately shallow, recessed parts having simple flanges and relatively simple configurations. Form block height is usually less than 100 mm (4 in.). The production rates are relatively high, with cycle times averaging 1 min or less.

The advantages of the rubber-pad forming processes compared to conventional forming processes are:

- Only a single rigid tool half is required to form a part.
- One rubber pad or diaphragm takes the place of many different die shapes, returning to its original shape when the pressure is released.
- Tools can be made of low-cost, easy-to-machine materials due to the hydrostatic pressure exerted on the tools.
- The forming radius decreases progressively during the forming stroke, unlike the fixed radius on conventional dies.
- Thinning of the work metal, as occurs in conventional deep drawing, is reduced considerably.
- Different metals and thicknesses can be formed in the same tool.
- Parts with excellent surface finish can be formed, because no tool marks are created.
- Set-up time is considerably shorter, because no lining-up of tools is necessary.

The disadvantages are:

- The pad or diaphragm has a limited lifetime that depends on the severity of the forming in combination with the pressure level.
- Lack of sufficient forming pressure results in parts with less sharpness or with wrinkles, which may require subsequent hand work.
- The production rate is relatively slow, making the process suitable primarily for prototype and low-volume production work.

Equipment. The hydraulic presses used in most flexible-die forming are similar to those described in the article "Presses and Auxiliary Equipment for Forming of Sheet Metal" in this Volume. Some processes use special machines, which are described in this article in the discussions of the specific processes. In most applications, only one solid tool half is specially made. The tool half can be made of epoxy resin, zinc alloys, hardwood, or other inexpensive material, as well as aluminum, cast iron, or steel.

Equipment is available with cycling rates as high as 1500 per hour. Some flexible-die forming methods have been applied to high-volume production, such as the forming of deeply recessed taillight reflectors for automobiles, and the deep drawing of toaster shells (example 3).

The application of rubber pads in press-brake dies is discussed in the article "Press-Brake Forming" in this Volume. In the past, flexible-die forming methods were designated by specific processes: Guerin process, Verson-Wheelon process, trapped-rubber process, Marform process, Hydroform process, SAAB process, and Demarest process. Modern technology has reduced this list, categorizing the methods into three basic groups: rubber pad, fluid cell, and fluid forming. Detailed applications of these rubber-die forming processes to specific metals are available in the articles "Forming of Stainless Steel," "Forming of Aluminum Alloys," "Forming of Copper and Copper Alloys," and "Forming of Titanium and Titanium Alloys" in this Volume.

Rubber-Pad Forming

The Guerin process is synonymous with the term *rubber-pad forming*. An improvement over the Guerin process is the Marform process, which features the addition of a blankholder and die cushion to make this process suitable for deeper draws and to alleviate the wrinkling problems common to the Guerin process. Another variation of the Guerin process is the trapped-rubber process, in which the forming force is provided by a hammer instead of a hydraulic press. Like the Marform process, the trapped-rubber process can be used for deeper draws and results in less scrap due to wrinkling than the basic Guerin process. The design and construction of the ASEA Quintus rubber-pad presses are further refinements of the Guerin and Marform processes.

Guerin Process

The Guerin process is the oldest and most basic of the production rubber-pad forming processes. Its advantages are simplicity of equipment, adaptation to small-lot production, and ease of changeover.

Some metals that are commonly formed by the Guerin process are listed in Table 1. Titanium can be formed only if the workpiece and the form block are both heated. The resulting deterioration of the rubber pad often makes the process too costly, as compared to forming by conventional dies.

Presses. For maximum forming capability, the force capacity of the press and the area of the rubber pad must be suitable for the operation under consideration. The rubber pad is generally approximately the same size as the press ram, but it can be smaller (example 2).

Tools. The principal tools are the rubber pad and the form block, or punch (Fig. 1). The rubber pad is relatively soft (approximately Durometer A 60 to 75) and is usually three times as deep as the part to be formed. The pad can consist of a solid block of rubber, or of laminated slabs cemented together and held in a retainer, as shown in Fig. 1. The slabs can also be held loose in a flanged retainer. The retainer is generally made of steel or cast iron, and it is approximately 25 mm (1 in.) deeper than the rubber pad. It is

Table 1 Metals commonly formed by the Guerin process

Metal	Maximum thickness(a)	
	mm	in.
Mild forming		
Aluminum alloys		
2024-O, 7075-W	4.7	0.187
2024-T4	1.6	0.064
Austenitic stainless steels		
Annealed	1.3	0.050(b)
Quarter hard	0.8	0.032(c)
Titanium alloys	1.0	0.040(d)
Stretch flanging		
Aluminum alloy 2024-T4	1.6	0.064
Austenitic stainless steels		
Annealed	1.3	0.050
Quarter hard	0.8	0.030

(a) Typical; varies with type of equipment and part design. (b) Up to 2.0 mm (0.078 in.) when compression dams are used (see the section "Accessory Equipment" and Fig. 2 in this article). (c) Only very mild forming. (d) When heated to 315 °C (600 °F)

also strong enough to withstand the forming pressures generated (up to 140 MPa, or 20 ksi, in some applications, although an upper limit of 14 MPa, or 2 ksi, is more common).

The minimum pad thickness is $1\frac{1}{3}$ times the height of the form block, as shown in Fig. 1. Pad thicknesses generally vary from 150 to 300 mm (6 to 12 in.), and the most commonly used thickness is 200 to 230 mm (8 or 9 in.).

Form blocks are made of wood, plastic, masonite, richlite, cast iron, steel, or alloys of aluminum, magnesium, zinc, kirksite, or bismuth. The softer materials are used in making prototypes or experimental models or in small production runs. The life of a wood, plastic, or soft-metal form block can be extended by facing it with steel. Form blocks are fitted with locating pins to hold the blanks in position while they are being formed.

The form block is placed on a platen, or pressing block (Fig. 1), which fits closely into the rubber-pad retainer to avoid extrusion of the rubber during the forming process. Several form blocks are often positioned on one platen so that several parts can be formed simultaneously with one stroke of the press. Two or three platens can be used with each press; they can be slid or rotated from under the press ram for loading and unloading.

Cavity Tools. With the high pressure attainable on the new hydropresses, more complex parts are able to be formed. The use of traps and dams in conjunction with cavity dies makes it possible to form parts that could only be hammer formed in the past. These parts are often formed in stages, with one or two anneals in between forming steps. When doing short-production or spares work, cavity dies can offer large cost reductions over other conventional processes.

Accessory equipment includes draw clips, cover plates, wiping plates, forming rings and forming bars, and dams and wedge blocks. These are used to increase the pressure on the work piece in specific locations and to aid in the forming of difficult shapes.

Draw clips are fastened to the edge of a blank to equalize the drawing force on a flange and to keep it from wrinkling. Wiping plates, usually hinged to the pressing block, mechanically transfer the pressure of the rubber pad to hard-to-form flanges. Forming rings and forming bars work in the same way, except that they encircle the part and therefore are not hinged. Dams are shaped and positioned so that with the sidewall of the form block they form a trap. The dam has a face sloping toward this trap so that rubber is cammed against the sidewall as the rubber pad moves down, thus increasing the pressure in that area. Wedge blocks use the same kind of camming action to apply mechanical pressure to the side of a workpiece. Examples 1 and 4 in this article illustrate the use of dams. Cover plates are used to hold blanks flat during forming or to protect previously formed areas from distortion.

Procedure. The rubber-pad retainer is fixed to the upper ram of the press, and the platen, containing the form block, is placed on the bed of the press. A blank is placed on the form block and is held in position by two or more locating pins. The pins must be rigidly mounted in the form block so that the rubber will not drive them down into the pinhole or push them out of position; and they must be no higher than necessary to hold the blank, or they will puncture the rubber pad. In some applications, nests can be used to locate the blank during forming.

As the ram descends, the rubber presses the blank around the form block, thus forming the workpiece. The rubber-pad retainer fits closely around the platen, forming an enclosure that traps the rubber as pressure is applied. The pressure produced in the Guerin process is ordinarily between 6.9 and 48 MPa (1 and 7 ksi). The pressure can be increased by reducing the size of the platen. Pressures as high as 140 MPa (20 ksi) have been developed through the use of small platens in high-capacity presses (see example 1).

The pressure is not a function of the number of parts being formed but of the platen area. To obtain maximum production with each stroke of the press, therefore, as many form blocks as possible are mounted on a single platen. The depth of the finished parts formed by this process seldom exceeds 38 mm ($1\frac{1}{2}$ in.). However, deeper parts can be formed by using a press with a high force capacity and a rubber pad with a small surface area. In one application, such a setup produced 140 MPa (20 ksi) of pressure and was able to form a flange 70 mm ($2\frac{3}{4}$ in.) deep.

Straight flanges can be easily bent by the Guerin process if they are wide enough to develop adequate forming force. If the flanges are not wide enough, accessory tools must be used.

Minimum widths for flanges of stainless steel and aluminum alloys that can be bent by rubber-pad forming are listed in Table 2. Angles on flanges in soft metal can generally be held to a maximum variation of $\pm 1°$. In hard metals, such as half-hard stainless steels, which have more springback than annealed stainless steels, a $\pm 5°$ tolerance can be met only with special care. An envelope (all-around) tolerance of ± 0.38 mm (± 0.015 in.) is possible on the contour of soft-metal pieces, but on hard metal, the tolerance must be increased to ± 0.51 mm (± 0.020 in.).

Stretch flanges and shrink flanges can be formed around curves and holes if the deformation is slight to moderate. If forming is severe, auxiliary tools must be used to support the work and to prevent wrinkling.

Shallow Drawing. Cover plates are often used to hold webs flat while flanges are being formed. In the following example, a pressure of 140 MPa (20 ksi) was used to form a part so well that only minimal hand reworking was required. To obtain the 140 MPa (20 ksi) pressure, a rubber pad 508 mm (20 in.) in diameter (surface area: ~0.19 m^2, or 300 $in.^2$) was mounted in a 27 MN (3000 tonf) hydraulic press. The bed size of the press was 1520 mm (60 in.) front-to-back and 1570 (62 in.) left-to-right. Stroke length was 457 mm (18 in.), and shut height was 1270 mm (50 in.). The press had a turntable that held two form blocks; therefore, one block could be unloaded and loaded while the other was under the press ram. The rubber pad was 203 mm (8 in.) thick and made in two pieces. One piece was 178 mm (7 in.) thick and had a hardness of Durometer A 80 to 85. The second piece, which was replaceable, was 25 mm (1 in.) thick and had a hardness of Durometer A 70 to 75.

Example 1: Shallow Drawing of a Fuselage Tail Cap by the Guerin Process. The fuselage tail cap shown in Fig. 2 was rubber-pad formed at 140 MPa (20 ksi) using the 27 MN (3000 tonf) hydraulic press and the 508 mm (20 in.) diameter pad described previously. The cap was

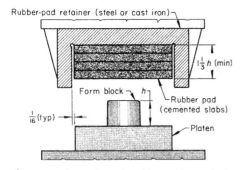

Fig. 1 Tooling and setup for rubber-pad forming by the Guerin process. Dimensions given in inches

Table 2 Minimum formable flange widths for the rubber-pad forming of stainless steels and aluminum alloys

Alloy and/or temper	Minimum flange width(a)	
	mm	in.
Stainless steels		
Annealed	4.8 + 4.5t	$\frac{3}{16}$ + 4.5t(b)
Quarter hard	16	$\frac{5}{8}$
Aluminum alloys		
2024-O, 7075-O	1.6 + 2.5t	$\frac{1}{16}$ + 2.5t(b)
2024-T3, 2024-T4	3.2 + 4t	$\frac{1}{8}$ + 4t(b)

(a) Using minimum permissible bend radius; a larger bend radius requires a wider flange. (b) t, sheet thickness

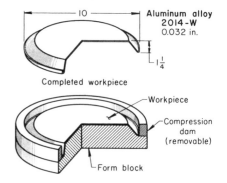

Fig. 2 Fuselage tail cap that was formed by the Guerin process in a high-pressure setup. Dimensions given in inches

originally made by spinning, but at a rate of only one piece per hour. Changing to high-pressure rubber-pad forming by the Guerin process increased the production rate to 12 pieces per hour.

A blank of aluminum alloy 2014-O, 0.81 mm (0.032 in.) thick, was solution heat treated to the W temper and was lubricated with heavy-duty floor wax. The part was formed before age hardening was complete. A compression dam surrounded the form block (Fig. 2) and was used to concentrate pressure on the flange.

Matched Laminations. In the following example, accurately matched laminations were made by forming one over the other on a form block by the Guerin process.

Example 2: Forming of a Two-Piece Cockpit Rail on a Single Form Block. By using a rubber pad instead of a conventional die, it was possible to form the two mating parts of a cockpit rail section on a single form block. One of the two parts is shown in Fig. 3. The second part was formed over the first after it had been formed.

A fully developed blank, 2.54 mm (0.100 in.) thick, was cut from aluminum alloy 2014-O and solution heat treated. Forming was done by the Guerin process with a minimum pressure of 52 MPa (7.5 ksi). No lubricant was used. The form block (Fig. 3) was made of masonite and was plastic faced. The production rate was 20 pieces per hour.

Blanking. With the Guerin process, rubber pads can be used for blanking and piercing as well as for forming. Rubber pads produce better edges on the workpiece than band sawing, and almost as good as those made by routing. An edge radius up to the thickness of the metal can be produced on some heavy-gage metals. The rubber-pad method can blank aluminum alloy 2024-O up to 0.81 mm (0.032 in.) thick and, for some shapes, up to 1.0 mm (0.040 in.) thick. The minimum hole diameter or width of cutout is 50 mm (2 in.). A minimum of 38 mm (1½ in.) trim is needed for external cuts.

The form block has a sharp cutting edge where the blank is to be sheared. In hard-metal blocks, this edge can be cut into the form block, as shown in Fig. 4(a) and (b). Form blocks of soft metal, plastic, or wood need a steel shear plate for the cutting edge (Fig. 4c); the shearing edge should be undercut 3 to 6°.

The trim metal beyond the line of shear must be clamped firmly so that the work metal will break over the sharp edge instead of forming around it. This clamping is done by a lock ring (Fig. 4a), a grip plate (Fig. 4b), or a raised extension of the form block (Fig. 4c).

These clamping devices also localize pressure at or near the cutting line. A rounded edge on the finished blank can be produced by locating the lock ring or grip plate a small distance from the shear edge (Fig. 4a and b). The metal droops in the unsupported area and forms around the sharp corner before it shears. The result is a smooth, rounded edge.

Drawing of shallow parts is often done by a modification of the Guerin process in which the contour is recessed (for example, a die cavity) into the form block rather than being raised on it. The blank is securely clamped between the rubber pad and the surface around the recess before forming begins.

Clamping the work metal before drawing and the amount of pressure used are both important for successful drawing. The work metal must be securely clamped to prevent it from flowing irregularly and subsequently forming wrinkles, but not so tightly that the metal cannot flow at all, which will cause thinning, or even tearing, of the work metal. To avoid this, either the edges can be lubricated or a protecting block with an undercut slot to accommodate the flange (Fig. 5) can be placed over the edges of the workpiece. The width of the block and undercut must provide the correct balance between clamping force and drawing force. The undercut should be 0.08 to 0.15 mm (0.003 to 0.006 in.) higher than the thickness of the work metal.

Marform Process

The Marform process was developed to apply the inexpensive tooling of the Guerin and Verson-Wheelon processes (see the section "Verson-Wheelon Process" in this article) to the deep drawing and forming of wrinkle-free shrink flanges. A blankholder plate and a hydraulic cylinder with a pressure-regulating valve are used with a thick rubber pad and a form block similar to those used in the Guerin process. The blank is gripped between the blankholder and the rubber pad. The pressure-regulating valve controls the pressure applied to the blank while it is being drawn over the form block.

While forming a soft aluminum alloy blank, the diameter can usually be reduced 57%, and reductions as high as 72% have been obtained. A shell depth equal to the shell diameter is normal when the minimum stock thickness is 1% of the cup diameter. Depths up to three times shell diameter have been reached with multiple-operation forming. The minimum cup diameter is 38 mm (1½ in.).

Foil as thin as 0.038 mm (0.0015 in.) can be formed by placing the blank between two aluminum blanks approximately 0.76 mm (0.030 in.) thick and forming the three pieces as a unit. The inner and outer shells are discarded.

Presses. The Marform process is best suited to a single-action hydraulic press in which pressure and speed of operation can be varied and controlled. A Marform unit comes as a package that can be installed in a hydraulic press having ample stroke length and shut height. However, a press that incorporates a hydraulic cushion system into its bed has been designed specifically for Marforming.

The rubber pressures used depend on the force capacity of the press and the surface area of the rubber pad. Recent installations range from 34 to 69 MPa (5 to 10 ksi).

Tools. The rubber pad used in Marforming is similar to that used in the Guerin process. It is normally 1½ to 2 times as thick as the total depth of the part, including trim allowance. The rubber pad can be protected from scoring by the use of a throw sheet, which is either cemented to the pad or thrown over the blank.

Well-polished steel form blocks are used for long runs and deep draws. Aluminum alloy form blocks must be hard coated to prevent galling for draws deeper than 38 mm (1½ in.). Masonite form blocks can be used if they can withstand the abuse and wear of forming a particular part in a given quantity. When a cast shape is more economical, aluminum or zinc alloy form blocks can be used.

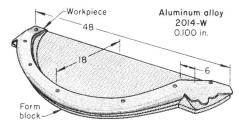

Fig. 3 Cockpit rail section that was formed on a single form block by the Guerin process. Dimensions given in inches

Aluminum alloy 2014-W 0.100 in.

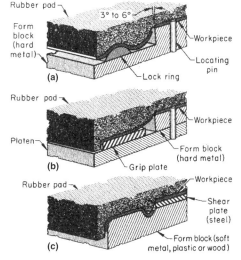

Fig. 4 Three techniques for blanking by the Guerin process categorized by clamping method. (a) Lock ring. (b) Grip plate. (c) Raised extension of the form block

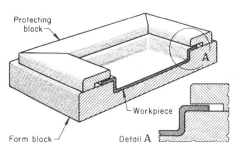

Fig. 5 Use of a protecting block to prevent work metal irregularities in shallow drawing by the Guerin process

Blankholder plates are usually made of low-carbon steel. The contact surface is ground flat and polished to avoid scratching of the blank. Clearance between the form block and the mating hole in the blankholder is 0.75 to 1.5 mm (0.030 to 0.060 in.) per side. The edge should have a 1.6 mm (1/16 in.) radius.

A radius plate is necessary when the machine pressure is insufficient for forming the flange radius within tolerance. The part is drawn first without the plate, then redrawn using the plate to form the exact radius. The radius plate is usually 13 mm (1/2 in.) thick and 25 mm (1 in.) wider than the workpiece. A sealing ring is used to prevent the rubber pad from extruding out of the container.

Procedure. The blank rests on the blankholder plate above the form block. The rods supporting the seal ring and blankholder plate (Fig. 6) are supported on a variable-pressure hydraulic cushion. As the press ram is lowered, the blank is clamped between the rubber pad and the blankholder before forming begins. As the rubber pad continues to descend, the blank is drawn over the form block while the pressure control valve in the hydraulic cushion releases fluid at a controlled rate. The pressure in the hydraulic cushion must be adjusted to prevent wrinkles from forming in the flange but to permit the blank to be drawn into a smooth shell. The part is stripped from the form block by the blankholder. The following example describes an application of the process.

Example 3: Deep Drawing of Toaster Shells by the Marform Process. The toaster shell shown in Fig. 7 was deep drawn in large quantities (80,000 pieces) from 0.76 mm (0.030 in.) thick deep-drawing-quality 1010 steel. The blanks were lubricated by brushing with a soap compound. Available pressure was 41 MPa (6 ksi). The depth of the trimmed shell was 127 mm (5 in.). The reduction time per piece was 22 s.

Drop Hammer Forming with Trapped Rubber

A process similar to the Guerin process, for forming shallow workpieces, is a trapped-rubber process, which uses a drop hammer in place of the hydraulic press; the primary differences are the faster forming speed and the impact force of

the hammer. The use of rubber pads in drop hammer forming is illustrated in the article "Drop Hammer Forming" in this Volume.

Figure 8 shows the effects of forming flanges on aluminum alloys 5052-O and 2024-O by the drop hammer (trapped-rubber) and Guerin processes. When flanges deeper than 32 mm (1 1/4 in.) are made by the Guerin process, stretch flanges can tear and shrink flanges can wrinkle. However, when the drop hammer process is used, fewer deformities occur (Fig. 8).

ASEA Quintus Rubber-Pad Press

ASEA presses, generally designed with force capacities of 50 to 500 MN (5600 to 56,000 tonf), are constructed of wire-wound frames and have separate guiding columns (Fig. 9). By winding the press frames with pre-stressed wire, only compressional stresses are present in the large castings or forgings of the yokes and columns, even when subjected to maximum forming pressure. Therefore, when the press is loaded, the frame remains in slight compression, and the major structural components never operate in the tensile mode.

The press is equipped with a forged-steel rubber-pad retainer that has a replaceable insert to allow for forming at even higher pressures. Although the maximum tool height is sacrificed by using these high-pressure inserts, cutting the

work area in half doubles the maximum forming pressure of the press when needed.

Standard table sizes range from 0.7 by 1.0 m (28 by 39 in.) to 2 by 3 m (79 by 118 in.). These presses provide forming pressures to 100 MPa (15 ksi). Hard and brittle materials such as titanium, along with the die, can be heated outside the press by an infrared heater; blanks can also be heated by conduction from the table through the heat transferred through the die (Fig. 10).

Fluid-Cell Forming

Initially developed as the Verson-Wheelon process, the fluid-cell process uses a fluid cell (a flexible bladder) backed up by hydraulic fluid to exert a uniform pressure directly on the form block positioned on the press table. This process can be classified in terms of the presses used (Verson-Wheelon and ASEA Quintus) as well as

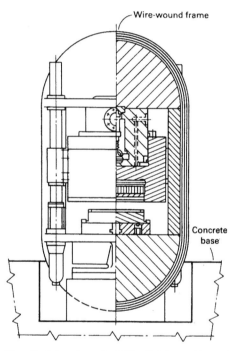

Fig. 9 Schematic of ASEA Quintus rubber-pad hydraulic press with wire-wound frame

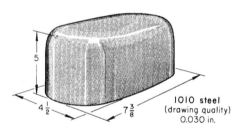

Fig. 7 Toaster shell that was deep drawn by the Marform process. Dimensions given in inches

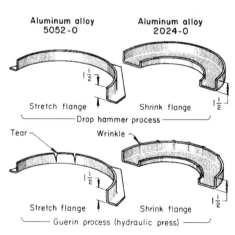

Fig. 8 Effect of impact in forming stretch and shrink flanges by the drop hammer (or trapped-rubber) and Guerin processes. Dimensions given in inches

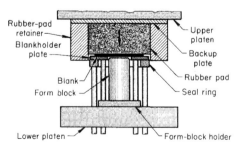

Fig. 6 Tooling and setup for rubber-pad forming by the Marform process

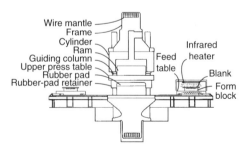

Fig. 10 Schematic of ASEA Quintus rubber-pad press with provision for heating hard and brittle materials using infrared heater or heating elements contained in feed table

a specialized method (Demarest process) for producing cylindrical and conical parts. The Verson-Wheelon press has cylindrical press housings of laminated, prestressed steel that serve as pressure chambers. The ASEA Quintus press has a forged steel cylinder that is wound with high-strength steel wire to create a pre-stressed press frame with extremely good fatigue properties.

Fluid-cell forming can be used for recessed parts that are beyond the capabilities of rubber-pad forming, for all flange configurations (including C-shaped flanges), and for complex parts with reentrant features and intricate joggles. Maximum form block height is 425 mm (16.7 in.), and typical cycle time is 1 to 2 min.

Verson-Wheelon Process

The Verson-Wheelon process was developed from the Guerin process. It uses higher pressure and is primarily designed for forming shallow parts, using a rubber pad as either the die or punch. A flexible hydraulic fluid cell forces an auxiliary rubber pad to follow the contour of the form block and to exert a nearly uniform pressure at all points.

The distribution of pressure on the sides of the form block permits the forming of wider flanges than with the Guerin process. In addition, shrink flanges, joggles, and beads and ribs in flanges and web surfaces can be formed in one operation to rather sharp detail in aluminum, low-carbon steel, stainless steel, heat-resistant alloys, and titanium.

Presses. The Verson-Wheelon press has a horizontal cylindrical steel housing, the roof of which contains a hydraulic fluid cell (Fig. 11). Fluid-cell bladders can be of neoprene or polyurethane composition. Hydraulic fluid is pumped into the cell, causing it to inflate or expand. The expansion creates the force needed to flow the rubber of the work pad downward, over and around the form block and the metal to be formed.

Below the chamber containing the rubber pad and the hydraulic fluid cell is a passage, extending the length of the press, that is wide and high enough to accommodate a sliding table containing form blocks. At each end of the passage is a sliding table that is moved into position for forming.

The rubber pad used in the Verson-Wheelon process has a hardness of approximately Durometer A 35. It is usually protected from sharp corners on the form block and blank by a throw sheet or work pad that is harder and tougher than the pad itself. The throw sheet is much less costly to replace than the rubber pad below the fluid cell.

Verson-Wheelon presses are available with forming pressures ranging from 35 to 140 MPa (5 to 20 ksi) and force capacities of 22 to 730 MN (2500 to 82,000 tonf). Sliding tables range in size from 508 by 1270 mm (20 by 50 in.) to 1270 by 4170 mm (50 by 164 in.). The

larger machine can form parts having flange widths to 238 mm (9⅜ in.).

Heating elements can be used with the sliding tables for producing parts made of magnesium. The maximum temperature is 315 °C (600 °F), and special heat-resistant throw pads are used to protect the hydraulic fluid cell.

Tools. The form blocks for the Verson-Wheelon process are made in much the same way as those for the Guerin process. Compression dams or deflector bars can be used to direct pressure into local areas for forming shrink flanges or return bends, as described in example 4.

Aluminum alloy or zinc alloy form blocks are recommended. Because of the high pressures, masonite or wood form blocks may break down from repeated use. More than one form block can be used at a time; the quantity depends on the size and shape of the form block.

Damage to the rubber pad can be reduced by removing all burrs and sharp edges from the blank. Form blocks should be smooth; all sharp corners and projecting edges should be well rounded; high tooling pins should be eliminated; deep, narrow crevices or gaps between parts of form blocks should be eliminated; and holes in blocks should be plugged during forming.

Procedure. The sliding table containing the form blocks is loaded and slid into the press. Hydraulic fluid is pumped into the cell, expanding it and driving the rubber pad down against the workpiece and around the form blocks. The pressure is released, and the table of formed pieces is slid out, unloaded, and reloaded for another cycle.

Repositioning the form blocks after a few cycles will distribute the wear on the rubber pad and lengthen its life. The use of a hard-rubber (or

occasionally leather) pad slightly larger than the blank assists in the uniform forming of flanges and prevents wrinkles.

The cycle time for the Verson-Wheelon process is longer than that of conventional presses, such as those used with the Guerin process. To reduce cell filling and draining time, it is good practice to load the table to capacity or to have dummy blocks on the sliding table when only one part is being formed.

In the following example, the time to form a part in a Verson-Wheelon press was less than when the part was made by the Guerin process. The higher forming pressure completely formed the part in one operation (whereas, the Guerin process had required two operations) and reduced hand work after machine forming.

Example 4: Verson-Wheelon versus Guerin Process for the Forming of a Complex Part. The complex part shown in Fig. 12 was originally formed by the Guerin process in a 40 MN (4500 tonf) hydraulic press from 1.0 mm (0.040 in.) thick alclad aluminum alloy 7075-W. The improved method used a Verson-Wheelon press that could exert a pressure of 69 MPa (10 ksi) and had a capacity of 360 MN (41,000 tonf). A pressure of 48 MPa (7 ksi) was needed to form the part. The same tool was used for both processes.

In the Guerin process, forming was done in two operations. Joggles and stringer tabs were set by hand after forming. In the first press operation, the outer flange was formed, and the inner and return flanges were partly formed. In the second press operation, the forming was completed with rubber strips confined by the dams.

In the Verson-Wheelon process, the outer, inner, and return flanges and the joggles were formed in one operation. This translated into a 30% labor-cost advantage for the Verson-Wheelon process over the Guerin process for this workpiece.

The heat treated aluminum alloy 6061 form block was mounted to a baseplate, with the inner and outer rims acting as dams. Because of the return flange on the inside radius of the part, the aluminum alloy form block was split longitudinally, and the outer half was fastened to the

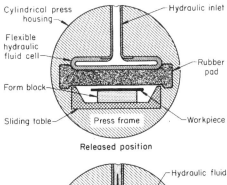

Released position

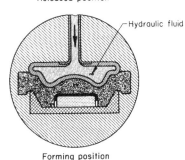

Forming position

Fig. 11 Principal components of the Verson-Wheelon process

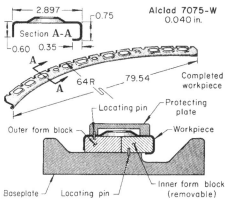

Fig. 12 Complex part that was formed in a Verson-Wheelon press. Dimensions given in inches

baseplate (see tooling setup in Fig. 12). The inner half, bushed and located on pins projecting from the baseplate, was removed from the base with the finished part. Locating holes in the blank and the outer form block matched locating pins in the cover plate.

The 127 by 2240 mm (5 by 88 in.) blank was routed, and lightening holes were individually pierced and flanged, in a punch press in both processes before rubber-pad forming. After forming, the part was aged to the T6 temper. The production lot was 20 pieces. Several thousand pieces were produced on the form blocks.

Secondary Operations. Even though higher forming pressures are used, many pieces made by the Verson-Wheelon process (as well as by the Guerin process) need hand work to remove wrinkles and to add definition to details. A further refinement in the use of throw pads is a shaped rubber pad. Pad laminations are built up around a cavity that approximates the shape of the part, so that flow of the rubber is less severe and forming pressure is more evenly distributed than with the conventional flat rubber pad. The shape of the cavity is only approximate and can be used for similar parts.

Demarest Process

Cylindrical and conical parts can also be formed by a modified rubber bulging punch. The punch, equipped with a hydraulic cell, is placed inside the workpiece, which is in turn placed inside the die. Hydraulic pressure expands the punch. Forming with an expanding punch using the Demarest process is described in the following example.

Example 5: Use of Expanding Punches to Form Aircraft Fuel-Tank Sections. Aluminum alloy workpieces, rolled and welded into cones (Fig. 13a), were formed into aircraft fuel-tank sections with expanding rubber punches

(Fig. 13b). The cones were lowered into cast iron dies, which weighed 1600 kg (3500 lb) each and were designed to withstand 10 MPa (1.5 ksi) of forming pressure.

The rubber punch was lowered into the workpiece, and a steel cover was clamped over the whole assembly (Fig. 13c). The punch was expanded under 2800 kPa (400 psi) of hydraulic pressure, which formed the work metal into the curved shape of the die (Fig. 13d).

The time taken for the entire process, including dismantling of the die and unloading of the workpiece, was 3 min. In contrast, spinning requires 15 to 20 min.

ASEA Quintus Fluid-Cell Process

The ASEA Quintus fluid-cell process is a further development of the Guerin process, allowing deeper and more complex parts to be formed. It uses a flexible rubber diaphragm backed up by oil as either the male or female tool half. The pressurized diaphragm forces the blanks to assume the shape of the solid-tool halves. The high uniform hydrostatic pressure permits the forming of shallow- to medium-depth parts with complex shapes to final shape, practically eliminating the subsequent hand forming usually required to apply the Guerin process.

Presses. The ASEA Quintus fluid-cell press (Fig. 14) consists of a horizontally placed cylindrical press frame, two rectangular, independently operated press tables on each side of the press, and all electric and hydraulic equipment necessary to operate the press placed on top of and alongside the press. The press frame is a forged steel cylinder that has been wound under prestress with high-strength steel wire, so that only compressional stresses can occur in the steel forgings, even at maximum pressure. Therefore,

the fatigue properties of the Quintus wire-wound frame are excellent and the risk of fracture virtually eliminated.

A diaphragm (bladder) of natural rubber is located inside the press and covers the entire surface of the press table. Hydraulic fluid is pumped into the cell created by the frame and the diaphragm, and the diaphragm is forced by the increased pressure to expand into the table, which forces the blank to assume the shape of the form block, thus forming the part. Once forming is done, the table is shuttled out to one side of the cylindrical press frame, and the table on the other side can be shuttled inside the press to allow forming of parts placed on that table.

ASEA Quintus fluid-cell presses are available with maximum forming pressures ranging from 100 to 200 MPa (14 to 29 ksi) and force capacities up to 1400 MN (157,000 tonf). The forming tables range in size from 700 by 2000 mm (27.5 by 78.7 in.) to 2000 by 5000 mm (78.7 by 196.8 in.). The large presses can accommodate tools as high as 425 mm (16.7 in.) and consequently form parts as deep or flanges as wide as 425 mm (16.7 in.).

Personal Presses. A new generation of Guerin-type press has been developed, called the personal hydropress. They are small presses and are well suited for small cells using lean-manufacturing principles. They generate pressures in the range of 14 to 21 MPa (2 to 3 ksi) and work well on less-complex parts. As the gage increases, the definition of the formed part is reduced. Aluminum parts with thicknesses up to 1.6 mm (0.063 in.) can be run on this new generation of press. The definition on joggles begins to deteriorate with thicknesses of approximately 1.27 mm (0.050 in.).

Tools. Because of the uniform hydrostatic pressure exerted by the press on the tools, they can be made of low-strength and low-cost tool materials such as hardwood, bakelite, epoxy,

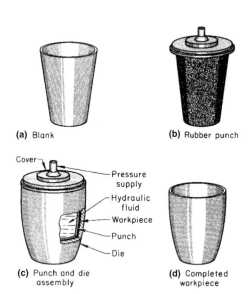

Fig. 13 Forming of a fuel-tank section from a blank using the Demarest process

(a) Blank
(b) Rubber punch
(c) Punch and die assembly
(d) Completed workpiece

Cover — Pressure supply — Hydraulic fluid — Workpiece — Punch — Die

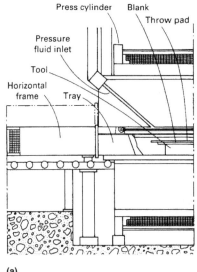

Press cylinder   Blank   Throw pad
Pressure fluid inlet
Tool
Horizontal frame   Tray

(a)

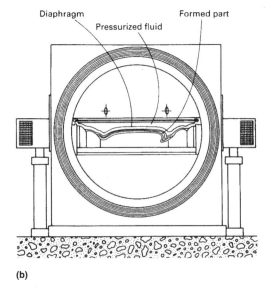

Diaphragm   Formed part   Pressurized fluid

(b)

Fig. 14 Illustration of the principal components of a new-generation ASEA Quintus fluid-cell press. (a) Tray containing blanks inserted into press prior to pressurizing. (b) Pressurized fluid-cell diaphragm forming a blank

zinc alloy, and so on, as well as of stronger materials such as aluminum, cast iron, and steel. Depending on the shape of the part, and tolerance and surface finish requirements, parts can be formed over a male die (form block), in a female die, or in an expansion die. The side of a part requiring a high surface finish should face the diaphragm, while the side of a part requiring close tolerances should face the tool. Parts with complex shapes, tight radii or return flanges or made of high-strength materials, demanding high pressures, and parts required in large quantities may require tools of zinc or aluminum alloys.

Procedure. The ASEA Quintus fluid-cell press has two independently operated press tables. This allows parts to be unloaded and new blanks loaded on tools on one press table while parts are being formed on the other table inside the press. A press table can be loaded with one large single tool half occupying the full size of the table or several smaller tool halves; the number is restricted only by the size of the table.

Once blanks are loaded on the tools, the table is shuttled into the cylindrical press frame to a position where the flexible diaphragm, located in the upper half of the press, covers the entire table area. Oil is then pumped into the cell, causing the diaphragm to expand, which forces the blanks to assume the shapes of the tools. When the parts are formed, the pressure is relieved and the oil is evacuated from the cell, allowing the table to be shuttled out from the press so that parts can be unloaded from the tools. The cycle time for a Quintus fluid cell is usually 1 to 3 min, depending on the press size and forming pressure selected.

Fluid Forming

In contrast to conventional two-die forming, which produces local stress concentrations in a workpiece, fluid forming (previously classified as rubber-diaphragm forming) employs a flexible-die technique that inhibits thinning and crack initiation due to the uniformly distributed pressure. In the fluid forming process, a rubber diaphragm serves as both the blankholder and a flexible-die member. Fluid forming differs from the rubber-pad and fluid-cell processes in that the

forming pressure can be controlled as a function of the draw depth of the part.

Fluid forming was initially known as the Hydroform process. The process, as originally conceived, is incorporated into the Verson Hydroform press. After a hydraulic pump delivers fluid under pressure into the pressure-dome cavity, the punch containing the die is driven upward into the cavity against the resistance provided by the fluid, and the workpiece is formed.

In the SAAB rubber-diaphragm method, hydraulic fluid being compressed by the press piston alone (no moving die is involved) forces the workpiece to assume the contour of the die. Air vents incorporated into the die facilitate the removal of trapped air and eliminate blisters on the surface of the workpiece. The ASEA Quintus fluid forming press was developed through slight modification of the SAAB process and optimization of press specifications with easily changeable pressure domes.

Another adaption of the fluid forming method is the ASEA Quintus technique used in the ASEA Quintus deep-drawing fluid forming press. Dome pressure and punch draw are both controlled from below the blank by two concentric rams, each of which governs one of these two variables. The units have interchangeable pressure domes.

Fluid forming is intended for punch, cavity, hydroblock, or expansion forming of deep-recessed parts (Fig. 15). Cycle time is 15 to 20 s for most parts.

Verson Hydroform Process with Rubber Diaphragm

This process differs from those previously described in that the die cavity is not completely filled with rubber but with hydraulic fluid retained by a 65 mm (2½ in.) thick cup-shaped rubber diaphragm. This cavity is termed the pressure dome (Fig. 16). A replaceable wear sheet is cemented to the lower surface of the diaphragm, as shown in Fig. 16. More severe draws can be made by this method than in conventional draw dies because the oil pressure against the diaphragm causes the metal to be held tightly against the sides as well as against the top of the punch.

Reductions in blank diameter of 60 to 70% are common for a first draw. When redrawing is necessary, reductions can reach 40%. Low-carbon steel, stainless steel, and aluminum in thicknesses from 0.25 to 1.65 mm (0.010 to 0.065 in.) are commonly formed. Parts made of heat-resistant alloys and copper alloys are also formed by this process.

Presses. A special press, called a Hydroform press, is used for this process. A lower hydraulic ram drives the punch upward; the upper ram is basically a positioning device. A hydraulic pump delivers fluid under pressure to the pressure dome. The blankholder is supported by a solid bolster and does not move during the operation.

Dome pressures range from 40 to 100 MPa (6 to 15 ksi), and punch force capacities vary from 3 to 19 MN (370 to 2090 tonf). Special guide pins and a platen adaptor convert a standard Hydroform press into a single-action conventional hydraulic press with a force capacity of 6 to 40 MN (700 to 4470 tonf). This variation of the process has the punch stationary and the blankholder actuated by the die cushion of a single-action hydraulic press, as shown in Fig. 20.

Maximum blank diameters are 300 to 1020 mm (12 to 40 in.), maximum punch diameters are 255 to 865 mm (10 to 34 in.), and maximum draw depths are 180 to 300 mm (7 to 12 in.). The maximum rating is 1500 cycles per hour. The practical production rate in cycles per hour is usually approximately two-thirds the machine rating. However, the operation often takes the place of two or three conventional press operations.

Tools. Punches can be made of tool steel, cold rolled steel, cast iron, zinc alloy, plastic, brass, aluminum, or hardwood. Choice of material depends largely on the work metal to be formed, number of parts to be made, shape of the part, and severity of the draw.

Blankholders are usually made of cast iron or steel and are hardened if necessary. Clearance between punch and blankholder is not critical; it may be 50% or more of the thickness of the metal being drawn.

For short runs, an auxiliary blankholding plate can be placed on a blankholder that is already in place. The auxiliary blankholder plate should not overhang in the punch clearance more than its thickness, and it should not be larger than the blankholder.

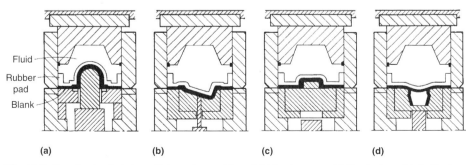

Fig. 15 Four forming techniques that can be used in a fluid forming press. (a) Punch draw. (b) Cavity draw. (c) Hydroblock draw (male-die forming). (d) Expansion draw

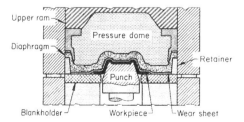

Fig. 16 Fluid-cell forming in a Hydroform press with rubber diaphragm

Rubber strips are placed on the blank to break the vacuum caused by dome action during drawing. Blankholders can be contoured to match the shape of a preformed blank or to preform the blank as an aid in forming.

Procedure. The blank to be formed is placed on the blankholder. The pressure dome, filled with the hydraulic fluid and covered by the rubber diaphragm, is lowered over the blank, and preliminary pressure is applied through a pump in the hydraulic supply line. The preliminary pressure can range from 14 to 70 MPa (2 to 10 ksi), depending on the part to be formed.

The punch is raised and pushed into the blank from underneath. As the form in the blank rises into the hydraulic chamber, the pressure in the chamber increases sharply, reaching as high as 100 MPa (15 ksi). A pressure control valve keeps the pressure within programmed limits. When parts are formed of thin metal, a vacuum release valve can be built into the punch to aid stripping after forming.

Three cams are programmed to control the operation of the machine. The first controls the height of rise of the punch, the second controls edging or sharpening of the corner radii, and the third returns the punch at the end of the stroke while the blankholder strips the finished part from the punch. The forming of a complex part by the rubber-diaphragm process is described in the following example.

Example 6: Rubber-Diaphragm Forming of a Complex Jet-Engine Part. Fuel-nozzle swirl cups for high-performance turbojet engines were originally produced by welding six press-formed sections of type 310 stainless steel, AMS 5521 (Fig. 17a). Forming the six sections was difficult, and the finished parts were expensive. The rejection rate was also high.

Rubber-diaphragm forming in a Hydroform press was tried. This press formed the part from one blank 1.1 mm (0.043 in.) thick by 325 mm (12¾ in.) in diameter. Less press force was used, and costs were reduced 50%.

Before forming, the blank was rough drawn (Fig. 17b) in a 1330 kN (150 tonf) hydraulic press to a depth of 35.6 mm (1.40 in.), and its thickness was reduced to 0.99 mm (0.039 in.). After degreasing and annealing, the partly formed blank was drawn in a 305 mm (12 in.) Hydroform press, using the punch shown in Fig. 17(c). The blank rested on a blankholder mounted on a subbolster. Diametral clearance between punch and blankholder was a minimum of 50% of the work metal thickness. The production rate was 30 pieces per hour.

After forming, six equally spaced 36.50 mm (1.437 in.) diameter holes and a 63.50 mm (2.500 in.) diameter center hole were pierced in a 490 kN (55 tonf) mechanical press. The outside diameter was trimmed in a lathe after the part had been pierced, annealed, and restruck. The completed workpiece is shown in Fig. 17(d).

Lubricants. The following example shows the importance of the lubricant and its application when the depth of draw is near the limit for the rubber-diaphragm process.

Example 7: Use of Lubricant to Eliminate Tearing and Wrinkling in Severe Rubber-Diaphragm Drawing. The stepped cover shown in Fig. 18 represented the limit of forming severity for the rubber-diaphragm equipment that was available. The material was 1.0 mm (0.040 in.) thick cold rolled drawing-quality 1008 steel. The shell was 102 mm (4 in.) deep and had a step in its outer contour. Attempts to draw the stepped shell in one operation in a Hydroform press were not successful. Subsequently, two Hydroforming operations were developed in which the larger width of the cover was drawn first, and then the narrower portion

above the step was produced in a redrawing operation to complete the part.

In the first operation, the blankholding pressure had to be carefully adjusted. When the pressure was too low, the metal moved freely and wrinkles appeared at the corners. Too high a blankholding pressure caused tears along the narrow end. Tears and wrinkles damaged the wear sheet and, in extreme cases, the diaphragm itself.

A lubrication program was developed that prevented wrinkling or tearing. After the first draw, the workpiece was cleaned, annealed, and phosphate coated. The phosphate made it possible to use a lighter oil and to apply it more effectively, with heavy applications in some areas and little or none in others. With experience, the operators became expert at judging the location and thickness of the lubricant. Mechanical application of lubricant could not be made selective enough or controlled closely enough for consistent results.

Because the part was nearly impossible to produce by conventional deep-drawing techniques, rubber-diaphragm forming was used. The tools, consisting of two punches and a blankholder, cost considerably less than the several sets of draw dies that would otherwise have been needed, and with the lubricating technique that was developed, there was less danger of tearing or wrinkling than by other processes.

Surface Finish. A major reason for using any rubber-pad process is to preserve the surface finish of the work metal, which would be scuffed or marked by ordinary press-forming tools. In the following example, appearance was an important consideration. The part was to be plated with copper-nickel-chromium. Forming by the rubber-diaphragm method prevented marks that would have been difficult to buff out before plating.

Example 8: Use of a Rubber-Diaphragm Process to Preserve Surface Finish on a Flatiron Shell. Because a mechanical draw press caused an impact line on the workpiece that was difficult to remove by buffing, production of the flatiron shell shown in Fig. 19 was

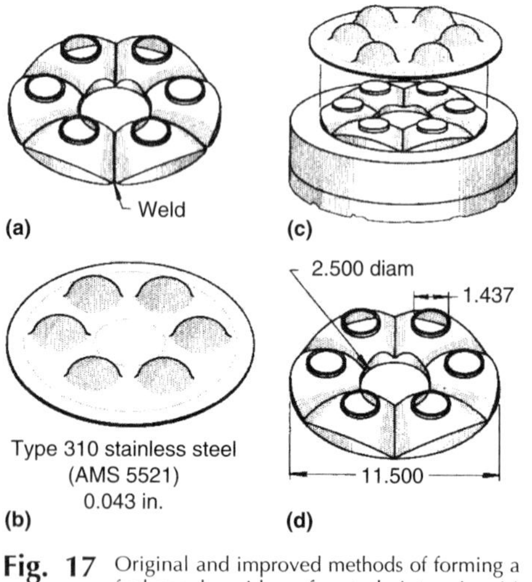

Fig. 17 Original and improved methods of forming a fuel-nozzle swirl cup for a turbojet engine. (a) Part formed by original method; six press-formed sections welded together. (b) Partly drawn blank ready for rubber-diaphragm forming. (c) Punch of six similar wedge-shaped segments doweled into bottom plate, used for rubber-diaphragm forming. (d) Swirl cup as formed by the rubber-diaphragm method and subsequently pierced and trimmed. Dimensions given in inches

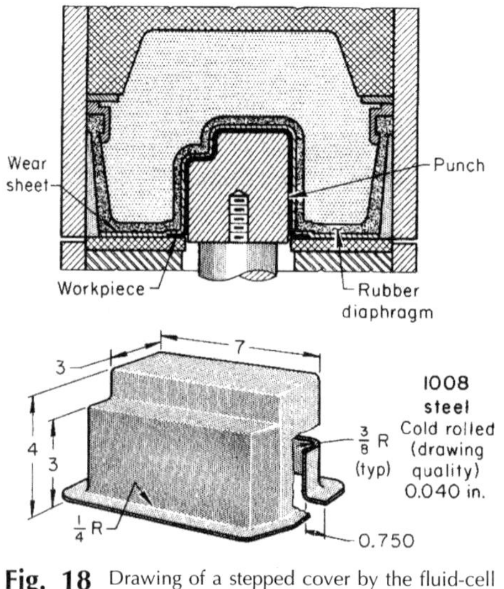

Fig. 18 Drawing of a stepped cover by the fluid-cell process. Dimensions given in inches

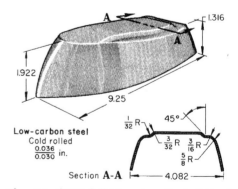

Fig. 19 Flatiron shell that was formed by the fluid-cell process in a Hydroform press to preserve the surface finish. When this shell was drawn in conventional dies, an impact line was caused below the radius that was difficult to remove by buffing. Dimensions given in inches

changed to a rubber-diaphragm process, using a 3.6 MN (400 tonf) Hydroform press. A rubber draw ring of Durometer A 92 hardness helped adjust holddown pressure so that wrinkles were avoided in the finished product. Two rubber pads were used on the rubber diaphragm. One covered the diaphragm as a reinforcement and protector; the other was a 9.5 mm (3/8 in.) thick ring molded to the shell outline. The blank was located in a nest on the blankholder.

Previously, the part had been drawn on a single-action mechanical press of 890 kN (100 tonf) capacity. In this press, the tools had been made of D2 tool steel. The stock was treated with soap and wiped with hydraulic oil near the point of the shell to minimize tearing.

The Hydroform press cycled at 450 strokes per minute. The production-lot size was 50,000 pieces, and yearly production was 850,000 pieces. Life of the rubber pads was as high as 20,000 pieces, and the finish of the part was good enough for subsequent plating with a minimum of buffing.

The sequence of operations was as follows: cut off blank, draw in Hydroform press, trim, pierce, copper plate, buff, nickel-chromium plate. The stock was 0.84 ± 0.08 mm (0.033 ± 0.003 in.) thick cold rolled low-carbon steel sheet slit to width. Two different qualities of steel were used:

- Aluminum-killed drawing-quality special-surface steel with a commercial finish, dry, maximum hardness HRB 60
- Cold rolled aluminum-killed steel strip with a No. 2 finish, dry, dead soft, maximum hardness HRB 55

The tolerance on important dimensions was ± 0.08 mm (± 0.003 in.); on angles, ± 1/2°.

Single-Draw Operation. In the following example, a pressure dome was mounted on the ram of a single-action hydraulic press. The punch was fixed to a shoe mounted on the bolster plate. A die cushion provided the blankholding force. This setup functioned much like a conventional draw die except that the oil-filled pressure dome and rubber diaphragm replaced the draw ring and die cavity.

Example 9: Forming of an Automotive Tail-Lamp Housing in One Drawing Operation in a Rubber-Diaphragm Press. An automotive tail-lamp housing was drawn in one operation from an aluminum alloy 5457-O blank 1.2 mm (0.048 in.) thick and 311 mm (12 1/4 in.) in diameter, in a rubber-diaphragm press rated at 69 MPa (10 ksi), as shown in Fig. 20. A water-soluble low-foaming lubricant was used. The production rate was 425 to 450 pieces per hour.

To produce this part with conventional tooling, two drawing operations would have been needed to form the sharp radii at the top and bottom of the part. Tooling costs for the rubber-diaphragm press were less than one-third the cost for conventional press tooling.

The blanks were moved from a stack adjacent to the press to an automatic feeder by a pneu-matic suction transfer device. A photoelectric cell prevented more than one blank being transferred. The blank passed between lubricating rollers before being fed automatically into the die. In subsequent operations, the housing was trimmed, flanged, and pierced in a mechanical press, using two conventional dies.

SAAB Rubber-Diaphragm Method

For some applications, the male member of a die set is made of rubber, and the female member is made of a hard material. In the Guerin process, shallow draws are made by recessing the form block and using the rubber pad as a punch to form the part (see the section "Shallow Drawing" in this article). The advantage of this method is that the flange is clamped before drawing, thus preventing wrinkling.

In the SAAB rubber-diaphragm method, hydraulic fluid is used behind a comparatively thin rubber pad or diaphragm. A hydraulic piston compresses the fluid against the rubber and forces the blank into the die, as shown in Fig. 21.

In all rubber-punch forming processes, air vents are provided in the die to allow the air trapped between workpiece and die to escape (Fig. 21). Without air vents, the trapped air would prevent the workpiece from reaching the full contours of the die, and the workpiece would have to be removed after partial forming to release the compressed air and then replaced in the same die to complete the forming.

Bulging Punches

Rubber punches can be used to make tubular parts that must be expanded or beaded somewhere along their lengths. If such parts were made with solid punches, the punches would have to be collapsible so that they could be withdrawn.

Hollow shapes can be bulged into suitable mating dies by applying a vertical force to the punch. The dies must be segmented so that the resulting bulged product can be removed, as shown in the following example.

Example 10: Forming of a Mushroom Shape in a Segmented Bulging Die. Figure 22 shows the process used in forming a mushroom-shaped frying-pan cover. The workpiece was a rectangular drawn shell of stainless steel, which was placed over a rubber punch of the same shape.

The two dies, which contained between them a cavity of the shape required, were closed until the rubber punch bulged the workpiece. The amount of bulge was determined by the depth of stroke. When the dies were opened, the punch returned to its original shape and was easily extracted from the finished part.

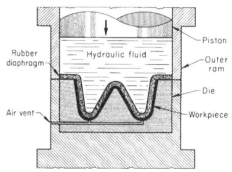

Fig. 21 Principal components of SAAB rubber-diaphragm (fluid forming) method. The air vents keep trapped air from causing blisters on the workpiece.

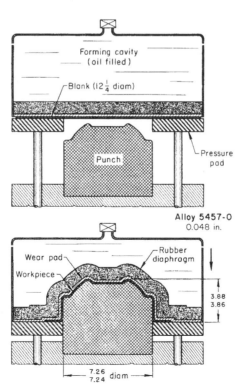

Fig. 20 Forming of an automotive tail-lamp housing in one draw in a fluid forming press. Dimensions given in inches

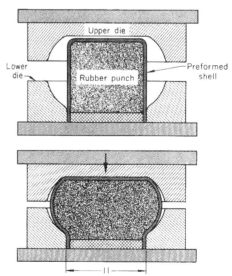

Fig. 22 Bulging of a mushroom shape from a preformed shell in a two-piece die with a rubber punch

ASEA Quintus Fluid Forming Press

The ASEA Quintus fluid forming press is a vertical press having a circular fluid form unit that contains the rubber diaphragm and pressure medium (Fig. 23). These modular fluid form units serve the same function as the units used in the ASEA Quintus deep-drawing fluid forming press (see the section "ASEA Quintus Deep-Drawing Technique" in this article). The rigid tool half may be a male block, a cavity die, or an expansion die that is situated in a movable tool holder. Blanks are loaded onto the tool holder prior to shuttling the holder into the press for the 10 to 50 s forming cycle required to produce the part.

For a 56 MN (6300 tonf) fluid forming press, two fluid form units designed for maximum 315 mm (12.4 in.) draw depth can be used. One unit, with a 63 MPa (9100 psi) forming pressure, has a 1090 mm (43 in.) blank diameter capacity. The other unit, providing a 160 MPa (23.2 ksi) forming pressure, has a 690 mm (27 in.) blank diameter capacity.

ASEA Quintus Deep-Drawing Technique

Fluid forming has been optimized using the ASEA Quintus deep-drawing fluid forming technique, a variation of the SAAB rubber-diaphragm method. The ASEA Quintus fluid forming method incorporates two telescopic rams: an outer ram to control dome pressure, and an inner ram to regulate the length of the punch draw (Fig. 24). Interchangeable domes, which require 20 min to change, allow the user to select a dome of optimal size and to avoid uneconomical use of an oversized dome when manufacturing small parts. Maximum forging pressure can also be increased by installing a smaller dome.

Each dome is a completely self-contained unit, with rubber diaphragms sealing off the pressure medium. Fluid form units for various pressure levels between 50 and 200 MPa (7.3 and 29 ksi) are available for each press size. Two or three such units should enable the user to make the most cost-effective use of the deep-drawing press. As an example, a set of interchangeable fluid form units for a 27 MN (3000 tonf) press ranges from an 800 mm (31.5 in.) diameter unit with a maximum 60 MPa (8.7 ksi) forming pressure to a 450 mm (17.7 in.) diameter unit with a maximum 190 MPa (27 ksi) forming pressure. Cycle time ranges from 10 to 60 s, depending on draw depth, part configuration, installed power, and selected pressure.

Complex shapes require accurate press control. As a result of 96 photocells that monitor the continuously varying pressure in the dome, greater accuracy and reliability are attained using a computer program to regulate the position and velocity of the draw depth. A paper cam cut with scissors gives dome pressure versus draw depth. The network of photocells reads the cam and governs an electrohydraulic control valve that controls the oil pressure in the outer cylinder, which is proportional to the dome pressure but much lower. By controlling the low counteracting pressure instead of the dome pressure, increased accuracy and reliability are achieved.

Diameters to 2000 mm (79 in.) can be formed from blanks ranging from 0.1 to 16 mm (0.004 to 0.63 in.) thick. Draw ratios to 3:1 can be produced, making it possible to form a complex part in one operation.

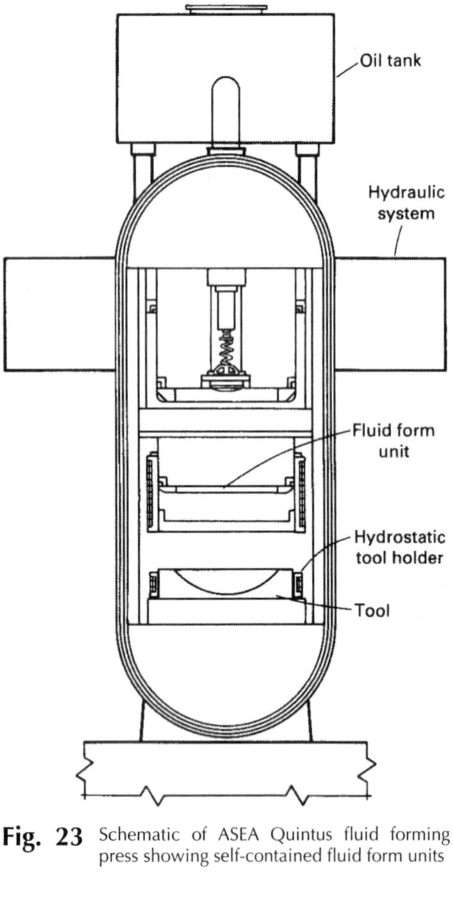

Other Methods of Hydraulic Forming (Ref 1)

Other methods for hydraulic forming of thin metal parts have used special machines such as diaphragm-forming hydrobuckling machines, or traditional presses such as gasket-draw and pressure-controlled machines, or more simple machines that are without gasket or control. The principles of different processes are discussed in the following.

Hydraulic Forming with Diaphragm. During the 1950s, this process was proposed on the market of special machines where oil pressure is applied by a synthetic rubber diaphragm on the metal to be formed (Fig. 25). Contrary to rubber stamping, where the elastic bulk is crushed by greatly increased local pressures (to the detriment of the sunken areas of the part), the pressure here is uniformly distributed, and very compact, concave, rounded forms result.

Hydrobuckling. The hydrobuckling process, patented by P. Cuq, can be included here because it incorporates metal-deformation control when there is minimum friction and an absence of thickness reduction. The part is formed by two combined actions, from a blank obtained by traditional drawing (Fig. 26). Hydraulic pressure (P), created by a pump inside the part, flattens the metal against an exterior die or mold from collapsible load action (F).

The pressure, P, is adjusted in ratio to the upstroke of the piston to fit the form, so that the metal areas that flatten the sunken parts of the mold are not obstructed by friction and therefore undergo no expansion—no thickness reduction. The word *buckling* is not quite suitable, because it implies a brutal, more-or-less controlled deformation.

This technique has been used for several years in the production of pulleys for automobile trapezoidal belts. In relation to the traditional

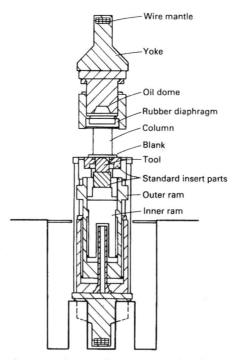

Fig. 23 Schematic of ASEA Quintus fluid forming press showing self-contained fluid form units

Fig. 24 Schematic of ASEA Quintus deep-drawing press, a fluid forming press with a telescopic ram system

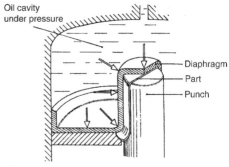

Fig. 25 Hydraulic forming with diaphragm

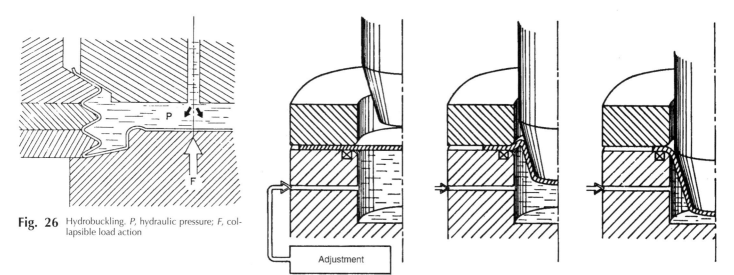

Fig. 26 Hydrobuckling. *P*, hydraulic pressure; *F*, collapsible load action

Fig. 27 Hydraulic drawing with joints and pressure control

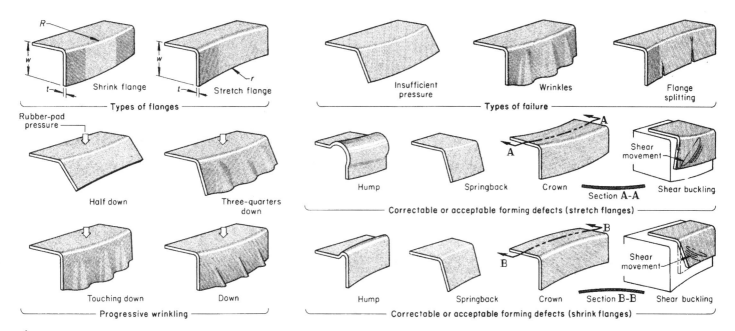

Fig. 28 Principal types of failure in curved flanges made by rubber-die forming

process (roller), the support facings of the belts show not even the slightest bulge.

Hydraulic Forming with Gasket and Pressure Control. This technique requires, in principle, nothing but a traditional machine—a double-effect hydraulic press (Fig. 27). Impermeability is ensured by means of a gasket between the die plate and the blank (or the flange), against which the blank grip leans. Pressure starts up when the punch goes into action, and its value is controlled during the entire operation in the leakproof chamber of the die cavity.

This process is most useful in making conical or vertical-wall draws, if there is enough play between the punch and the die to allow sufficient pressure to form metal beads around the punch

at the start of the operation. This pressure should be high: 30 to 50 MPa (300 to 500 bars) for soft steel, and 40 to 80 MPa (400 to 800 bars) for stainless steel. Here, the path of deformation becomes most important. A sheet has important deformation possibilities if it has been previously subjected to a restraining effort, but these possibilities are reduced if it has been subjected to expansion. The formation of beads around the punch can be compared with a beginning of return, or, again, to a first-draw blow corresponding to a less-severe reduction ratio. The metal starts deforming in the direction of restraint. In practice, it has been found that it is not always easy to define pressure control or the height of the beads.

Failures in Rubber-Die Flanging

The rubber-pad forming of flanges can be performed within certain limits. The flange must be wide enough to develop sufficient bending force (Table 2) but not so wide as to exceed the permissible depth of the part. Figure 28 shows some typical flanging failures.

REFERENCE

1. O.D. Lascoe, *Handbook of Fabrication Processes*, ASM International, 1988

Three-Roll Forming

THREE-ROLL FORMING is a process for forming plate, sheet, bars, beams, angles, or pipe into various shapes by passing the work metal between three properly spaced rolls. This article discusses sheet and plate, the mill products most often formed by the three-roll process.

Shapes Produced. Figure 1 illustrates some of the shapes commonly produced from flat stock by three-roll forming.

The plain round cylinder shown in Fig. 1 is used for pressure tanks, boilers, and related containers, and it represents a large portion of the shapes produced. The corrugated cylinder is produced in quantity for culvert pipe and is formed from flat stock corrugated at the mill. To retain the corrugations in the workpiece, the forming rolls also must be corrugated.

The flattened cylinder (obround) is primarily used for oil-supply tanks for heating systems and transformer cases. The elliptical cylinder is used for tank trucks hauling liquid food products, petroleum products, and chemicals.

Symmetrical and asymmetrical cones are both used in a wide variety of hoppers, bins, vertical storage tanks, concrete mixers, and vessels for chemical and food processing, as well as in piping and ductwork. In addition to the shapes produced for commercial use, three-roll forming is also used to produce various regular and irregular shapes for structural sections of submarines, aircraft, and nuclear reactors.

Metals Formed. Any metal ductile enough to be cold formed by other processes can be formed in a three-roll machine. Steels with a maximum carbon content of 0.25% constitute a major portion of the total tonnage used in three-roll forming. Steel sheet or plate in the 1010 to 1020 category is sometimes used, but most of the steels formed by this process conform to one of the plate specifications: either plain carbon or low-alloy steels, such as ASTM A515 grade 60, A515 grade 70, A516 grade 70, A285, A441, A283, A306, and A36. For the most successful three-roll forming, steels with a minimum elongation of 18% are preferred. Stainless steels, heat-resistant alloys, and aluminum and copper alloys can also be successfully formed by the three-roll process.

Metal thicknesses commonly used range from 1.5 mm (0.06 in.) sheet (16 gage) to 255 mm (10 in.) plate. In a few applications, 300 mm (12 in.) plate has been successfully formed. The principal factors limiting maximum thickness are the size and power of the rolling machine. Minimum thickness is typically limited only by handling equipment. Any sheet that can be handled without damage can usually be rolled.

It is impractical to roll thicknesses ranging from 1.5 to 255 mm (0.06 to 10 in.) on the same machine, although any machine can handle a relatively wide range of work metal thicknesses. For example, a machine capable of rolling 9.5 mm ($^3/_8$ in.) plate (maximum or near maximum) can generally roll sheet as thin as 1.5 mm (0.06 in.), while a machine with a maximum capability for rolling 150 mm (6 in.) plate can successfully roll plate as thin as 13 mm ($^1/_2$ in.)—even less on some machines.

Diameter and Width. The minimum diameter of a workpiece that can be successfully formed in a given machine is governed by the diameter of the top roll on either of the two types of machines used in three-roll forming—pinch type or pyramid type. In general, the smallest cylinder that can be rolled under optimal conditions is 50 mm (2 in.) larger in diameter than the top roll of a pinch-type machine. On a pyramid-type machine, the minimum workpiece diameter is rarely less than 150 mm (6 in.) greater than the top roll. However, more power is required to form sheet or plate into cylinders of minimum diameter than to form cylinders substantially larger than the top roll.

The maximum workpiece diameter that can be rolled is primarily limited by the space available above the machine to accommodate extremely large circles. Thin-gage metal rolled to a large diameter on horizontal rolls becomes less self-supporting as the workpiece diameter increases, and out-of-round cylinders will result if supports are not used. However, by using supports, almost any diameter can be rolled from thin metal. In general, 1.5 mm (0.06 in.) thick low-carbon steel sheet can be formed into cylinders as large as 120 cm (47 in.) in diameter without support, while 6.4 mm ($^1/_4$ in.) thick low-carbon steel can be formed into cylinders as large as 210 cm (83 in.) in diameter without support.

The width (dimension of the work metal parallel with the axes of the rolls, designated as length in the formed cylinder) of sheet or plate that can be rolled is limited by the size of the equipment; machines with rolls as long as 12 m (40 ft) have been built. The width-to-diameter relationship for workpieces that are extremely large in both directions is limited by problems in handling.

Machines

There are two basic types of three-roll forming machines: the pinch-roll type and the pyramid-roll type. The rolls on most three-roll machines are positioned horizontally; a few vertical machines are used, primarily in shipyards. Vertical machines have one advantage over horizontal machines in forming scaly plate: Loose scale is less likely to become embedded in the work metal. With vertical rolls, however, it is difficult to handle wide sections that require careful support to avoid skewness in rolling. Most vertical machines have short rolls for fast unloading and are used for bending narrow plate, bars, and structural sections.

Conventional pinch-type machines have the roll arrangement shown in Fig. 2. For rolling flat stock up to approximately 25 mm (1 in.) thick, each roll is of the same diameter.

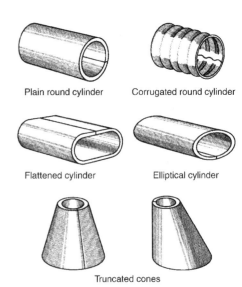

Plain round cylinder   Corrugated round cylinder

Flattened cylinder   Elliptical cylinder

Truncated cones

Fig. 1 Typical shapes produced from flat stock by three-roll forming

However, on larger machines, the top rolls are sometimes smaller in diameter to maintain approximately the same surface speed on both the inside and outside surfaces of the plate being formed. These heavier machines are also supplied with a slip-friction drive on the front roll to permit slip, because of the differential in surface speed of the rolls. Therefore, as work metal thickness increases, the diameter of the top roll is decreased in relation to the diameter of the lower rolls.

The position of the top roll is fixed, while the lower front roll is adjustable vertically to suit the thickness of the blank. Optimal adjustment of the lower roll is important for gripping the stock and for minimizing the length of the flat areas on the workpiece. The rear, or bending, roll is adjustable angularly (usually 30° off vertical), as shown in Fig. 2. Angular movement of this roll determines the diameter of the cylinder to be formed.

All of the rolls are powered in most pinch-type machines. On some machines, however, only the two front rolls are powered, and the bending roll is rotated by friction between the roll and the work metal (Fig. 2). This arrangement is usually satisfactory in forming medium-to-heavy stock to large diameters. However, when forming sheet or plate that is thin or soft (or both) or when the diameter is large, the amount of friction is sometimes insufficient to rotate the bending roll. This condition can result in a marred surface if the work metal is soft or has a bright mill finish (aluminum sheet, for example).

A pinch-type machine can produce a more nearly true cylindrical shape than a pyramid-type machine because the work metal is held more firmly. This results in smaller flat areas on the leading and trailing ends of the workpiece.

As shown in Fig. 2, the work metal is fed to the powered pinch rolls (front), which grip the plate and move it through the machine. Forming begins when the work metal contacts the bending roll (rear) and is forced upward. As the forward motion of the workpiece continues, a cylindrical shape is produced, except for the unformed flat area along the leading end and a small flat area at the trailing end of the workpiece (Fig. 2). The width of the flat area on the trailing end usually ranges from $\frac{1}{2}t$ to $2t$ (t, work metal thickness), depending on the design of the machine.

In most pinch-roll forming, one of two procedures is used to minimize flat areas. The most common method is to preform both ends of the work metal in the machine. This is done by reversing the rotation of the rolls and feeding a short section of the work metal from the rear, thus preforming one end. The work metal is then removed from the machine; the formed section can be fed into the machine from the front, or it can be turned to the opposite unformed end and fed through from the rear of the machine. This procedure eliminates most of the flat areas.

Another method is to preform the leading and trailing ends of the work metal in a press brake, hydraulic press, or joggling press. However, this technique is seldom used, because it is usually more convenient to preform in the pinch-roll machine.

On the other hand, preforming in a press brake or in hydraulic or joggling presses can sometimes save time in the rolling machine, thus increasing the productivity of the machine. Additional advantages of a pinch-roll machine, as compared to a pyramid-roll machine, are:

- When all rolls are power driven, thinner sheets can be rolled, and cylinders can be formed to within approximately 50 mm (2 in.) of the diameter of the top roll.
- A given size of a pinch-type machine can roll a greater range of metal thicknesses because of the method of feed.
- Greater dimensional accuracy can be obtained in one pass in a pinch-type machine than in a pyramid-type machine.

The principal disadvantage of a pinch-roll machine is its unsuitability for rolling workpieces from angles, channels, and other structural forms.

Shoe-Type Pinch-Roll Machines. One important modification of the conventional three-roll pinch-type machine is the shoe-type machine, which uses the pinch principle and incorporates a forming shoe, as shown in Fig. 3. Because of the relationship of the two front rolls and the forming shoe to the workpiece, the flat area becomes barely discernible compared with the length of flat area obtained when rolling in a conventional machine (without preforming).

The shoe-type machine is often used to manufacture transformer cases and small tanks, such as jackets for hot-water tanks. This type of machine can be completely automated; therefore, the work metal can be positioned on the table and fed into the machine automatically. During the work cycle, the cylinder is formed and ejected by means of an ejector mechanism and an automatically controlled drop end. Thus, a shoe-type machine is primarily a production machine that is used where large quantities of identical workpieces are to be rolled. For this reason and because of the limitations listed subsequently, shoe-type machines seldom compete directly with conventional pinch-type machines:

- Thickness of the work metal is limited to 12 gage (2.657 mm, or 0.1046 in.).
- Width of the sheet is limited to 183 cm (72 in.).
- Shoe-type machines are best adapted to the rolling of round cylinders; the rolling of ovals or obrounds is impractical.
- Shoe-type machines are applicable only to cold forming.

Within their range of applicability, however, shoe-type machines can produce a rolled cylinder in approximately half the time required in a conventional machine, primarily because preforming is not required with a shoe-type machine.

Pyramid-Type Machines. Figure 4 illustrates the arrangement of the rolls in a pyramid-type machine. The bottom rolls are of equal diameter but are approximately 50% smaller in diameter than the top roll. The bottom rolls are gear-driven and are normally fixed; each roll is supported by two smaller rolls (Fig. 4). The top roll is adjustable vertically to control the diameter of the cylinder formed. The top roll, which rotates freely, depends on friction with the work metal for rotation. Backup rolls are not used on the top roll.

As shown in Fig. 4, the work metal is placed on the bottom rolls while the top roll is in a raised position. The top roll is then lowered to contact and bend the work metal a predetermined amount, depending on the diameter of the

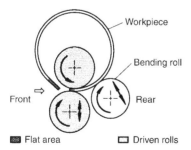

Fig. 2 End view of a cylindrical workpiece being rolled in a conventional pinch-type machine. Note the large flat area on the leading end and the smaller flat area on the trailing end.

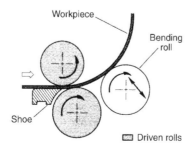

Fig. 3 End view of a cylindrical workpiece being rolled in a shoe-type machine with two powered rolls

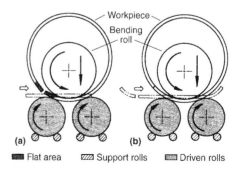

Fig. 4 Arrangement of rolls in a pyramid-type machine. (a) Entrance of flat workpiece and shape of a nearly finished workpiece, including the flat areas on the leading and trailing ends. (b) Similar, except that the workpiece was prebent to minimize the flat areas on the ends

workpiece to be formed. Machines are usually equipped with a device that indicates the amount of initial bend. Some machines use an ammeter, which shows the amount of current used in forcing the roll downward. However, this device measures force only; variables in the work metal can cause differences in the amount of bend for a given force.

As required bending force increases, machines are designed with rolls of larger diameter, and the distance between centers of the bottom rolls increases. Bending forces are applied midway between the bottom rolls; therefore, less force is needed for a given deflection, but less curvature is produced.

Because the top roll is adjustable, pyramid rolls can be used for forming irregular shapes by bolting dies to the top roll—a technique that is not adaptable to pinch-type machines. In addition, plate, beams, angles, and other structural forms can be straightened with greater ease because the bottom rolls are on the same elevation.

The top roll is an idler; therefore, there are definite limitations on the minimum thickness of work metal that can be rolled (especially when forming large diameters). Adequate stiffness in the work metal is essential to provide enough friction to rotate the top roll. The minimum thickness that can be rolled varies, depending on the specific machine and the work metal composition.

Another disadvantage of the pyramid machine is the large flat areas that remain on both the leading and trailing ends of the work metal. Because the workpiece must remain supported by the bottom rolls at all times, the ends of the work can never get closer to the top roll than the distance between the points of tangency of the workpiece and the rolls. Therefore, it is impossible to eliminate these flat areas by rolling (Fig. 4a).

To minimize flat areas when using pyramid machines, the usual procedure is to preform the ends to the desired radius in a press brake or to roll an oversize blank, then trim the flat ends after. The shell can sometimes be returned to the rolls for truing after the seam has been joined. Occasionally, a narrow shim is placed at the ends to increase the bend radius, but care must be taken to avoid machine overload. The techniques used in forming with pyramid rolls make it more difficult to achieve the accuracy that is obtainable with pinch-type rolls.

Capacity. Three-roll forming machines are rated by the manufacturer according to the maximum thickness and width of low-carbon steel plate the machine can form at room temperature. Values are usually given for single-pass rolling, and allowances are then made for multiple-pass rolling. For example, a machine rated at 19 by 3660 mm ($3/4$ by 144 in.) (thickness and width of plate, respectively) for work metal with a maximum tensile strength of 414 MPa (60 ksi) and capable of rolling plate to a diameter of 244 cm (96 in.) in a single pass can roll to a final diameter of 58 cm (23 in.) in multiple passes if

the top roll is no larger than approximately 37 cm ($14^{1}/_{2}$ in.) in diameter.

If plate thickness is increased to 25 mm (1 in.), the same diameter restrictions would apply, but the allowable plate width would be reduced from 366 to 142 cm (144 to 56 in.) because of the additional power required for the increased thickness of work metal. At this point, another limitation may be encountered because the load imposed on the shorter surface area can become excessive as the plate becomes narrower and thicker. On the other hand, assuming all other factors remain constant, if plate thickness is reduced to 16 mm ($5/_8$ in.), the allowable width would revert to full capacity of the machine (366 cm, or 144 in.), but the rolled diameter could be reduced to 42 cm ($16^{1}/_2$ in.).

The maximum plate thickness that can be handled by this machine depends on the pinch opening and is rated by the manufacturer of the machine. For example, some machines rated as described previously can accommodate work 38 mm ($1^{1}/_2$ in.) thick, but for forming this thickness in a machine having the indicated capacity, the allowable plate width would be reduced to 53 cm (21 in.) because of the aforementioned factors. All of the previously mentioned calculations also take into consideration the limiting factor of roll deflection.

With all other conditions constant, power requirements increase according to the square of metal thickness. Therefore, the power required for forming plate 50 mm (20 in.) thick is four times as great as that required for forming 25 mm (1 in.) thick plate of the same width.

Selection of Machine

Selection between pinch-type and pyramid-type machines depends mainly on the shape of the starting form and of the finished workpiece, the number of formed parts to be produced, accuracy requirements, and the cost. The pinch-type machine produces more accurate workpieces, and it can be loaded and unloaded much faster than the pyramid-type machine. Although both machines can produce shapes other than plain cylinders, the pinch type is capable of rolling a wider range of thicknesses. However, the pyramid-type machine is often preferred for small quantities of varied work, as in a job shop. Because of the wide space that can be obtained between the upper roll and the two lower rolls in a pyramid machine, various types of dies and fixtures can be fastened to the upper roll, thus permitting channels, angles, and various other structural shapes to be rolled or bent, either hot or cold.

Rolls

Rolls used in three-roll forming machines are machined from steel forgings having a carbon content of 0.40 to 0.50% and a hardness of 160

to 210 HB. Plain carbon steel such as 1045 has often been used; when greater strength is needed, rolls are forged from an alloy steel such as 4340. Because the modulus of elasticity is the same for all carbon and low-alloy steels of medium carbon content, roll deflection for a given force will be the same.

Although the hardness range of 160 to 210 HB can be obtained by annealing, rolls with a microstructure obtained by quenching and tempering or by normalizing and tempering are less subject to surface deterioration from spalling. Therefore, the forged rolls are heat treated before being machined.

Roll diameter varies with the length and thickness of plate to be rolled. A typical top roll in a pinch-type machine, rated for forming steel plate up to 65 mm ($2^{1}/_2$ in.) thick and 366 cm (144 in.) wide, would have a minimum diameter of 76 cm (30 in.). Journals for rolls of this diameter are approximately 43 cm (17 in.) in diameter.

Crowning of rolls to compensate for deflection is common practice. The amount of crowning is not necessarily the same for all rolls in a given machine. For example, in some machines, the rolls are not all of the same diameter; under these conditions, a roll that is smaller in diameter requires more crowning than a larger roll because the stress on all rolls is the same. When a machine is used for both light and heavy work, it is usual to crown the rolls for average conditions and then to use strips either at the center of the rolls to compensate for extreme deflection or at the ends to compensate for a lack of deflection (see the section "Roll Deflection" in this article).

Roll Maintenance. The extreme pressures to which rolls are subjected cause them to work harden. Rolls used in continuous production under high pressure sometimes elongate and reduce slightly in diameter. The amount of elongation or reduction in diameter is seldom significant, although the ends of rolls may require trimming after long periods of use.

There is no standard practice for reconditioning rolls. In some plants, rolls that have been subjected to long periods of severe service are trued by removing some or all of the work-hardened layer by turning. When required, the diameter is built up by welding an overlay on the rolls and then finish turning them. On the other hand, some manufacturers recommend that roll surfaces should never be turned. If the surfaces are spalled or otherwise damaged, any protruding metal should be removed by grinding. Although indentations in the rolls are less likely to be harmful, they may mark polished or clad surfaces. When rolling scaly plate, blowing away loose scale with an air lance is helpful in preventing scale from indenting the rolls or the work metal.

Bearings and Lubricants. Bronze has been successfully used for main bearings and is sometimes specified by the user. However, tin-base Babbitt is superior to bronze for most applications and is used in most machines.

Tin-base bearings are more compatible with the relatively soft steel journals at the pressures and speeds involved, and their ability to absorb particles of scale minimizes the possibility of scoring journals or bearings.

Extreme-pressure lubricants are recommended for the main bearings on all rolls, and a grade containing molybdenum disulfide is especially desirable. Because environmental conditions are likely to vary considerably where three-roll forming is done, the lubricant should have good pumpability over a range of temperatures. Extreme-pressure lubricants are satisfactory for both cold and hot forming.

Preparation of Blanks

Blanks are usually cut to the desired size before forming. The length of plate (dimension of the work metal perpendicular to the axes of the rolls) required to form a given shape is determined by measuring the mean circumference (or perimeter, if the shape is other than a cylinder), which is the circumference taken at one-half the distance between the inside diameter and the outside diameter of the shape to be formed. This method of calculation is the one most generally used in both the cold and hot forming of plate.

Allowance for Shift in Neutral Axis. When greater accuracy is required, the more exact location of the neutral axis is considered when computing the blank, particularly if heavy plate thicknesses are involved. The neutral axis is the boundary between metal in tension and in compression and is usually one-quarter to one-half the thickness of the metal being bent, as measured from the inside of the bend. The exact location of this axis varies to some extent with the bend radius and the mechanical properties of the metal.

During cold forming, the neutral axis shifts inward from the mean by approximately 26% of the plate thickness. Therefore, for a 46 cm (18 in.) inside diameter (ID) cylinder, rolled from 13 mm ($^1/_2$ in.) thick plate, the mean circumference is approximately 147.6 cm (58.12 in.), and for a 26% shift, the circumference at the neutral axis becomes 145.5 cm (57.30 in.). For a 46 cm (18 in.) ID cylinder of 6.4 mm ($^1/_4$ in.) thick plate, the mean circumference is 145.6 cm (57.33 in.), and with a 26% shift, it changes to 144.6 cm (56.93 in.); the amount of shift is approximately one-half that for the 13 mm ($^1/_2$ in.) thick plate. Therefore, the shift of neutral axis is usually disregarded for plate thicknesses less than 13 mm ($^1/_2$ in.) except where greater accuracy is required.

In cold forming, the length of the blank is calculated by using a radius that is determined by subtracting 26% of the plate thickness from the mean radius or by adding 24% to the inside radius. When dimensional requirements are stringent, similar allowances are made for shift of the neutral axis and for thermal expansion in the hot forming of thick

plates (75 to 150 mm, or 3 to 6 in.) at 870 °C (1600 °F).

Cutting of blanks can be done by shearing, if shearing equipment is available for the width and thickness of the work metal. Gas cutting is commonly used for preparing blanks that are too thick for shearing. Additional information is available in the article "Shearing of Sheet, Strip, and Plate" in this Volume.

Edge Preparation. The cut edges of any plate (high-strength steel, in particular) can be a serious problem because of cracking during cold forming, which is cause for rejection of the workpiece. When plate is sheared, the edges are rough and often have surface cracks. Gas cutting usually produces smoother edges, but the edges of gas-cut steel plate will frequently be hardened in cooling from the cutting temperature. Therefore, nucleation sites for cracks are likely to be present from either method of cutting.

The danger from cracking caused by rough edges increases as plate thickness increases and the finished diameter of the cylinder decreases. Because the plate surface that forms the outside diameter of the cylinder is in tension during forming, cracks propagate from edges, which are in tension.

On plate 25 mm (1 in.) or more in thickness, the edges indicated in Fig. 5 should be removed before cold forming. This is not required before hot forming. Usual practice is to employ a chipping hammer and then a portable grinder to smooth the edges. The amount of metal removed is usually negligible, and a slight bevel on the critical edges is sufficient. If a substantial amount of metal is to be removed, allowance must be made for it when calculating the dimensions of the blank.

Cold versus Hot Forming

Because cold forming involves fewer problems and is less costly than hot forming, it is preferred practice to form workpieces at room temperature. In hot forming, dimensional accuracy is more difficult to control, and cost is significantly increased by:

- Heating the blank
- Handling both the blank and the workpiece while the metal is hot

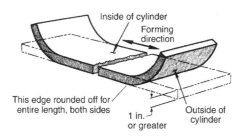

Fig. 5 Sheared or gas-cut blank, showing where metal should be removed from edges before cold forming, to reduce susceptibility to cracking

- The necessity for restoring acceptable surfaces by pickling, blast cleaning, or other surface treatment
- The accelerated rate of deterioration of rolls and other equipment because of contact with hot metal

Forming Capacity. When carbon or low-alloy steel is heated, tensile strength decreases and formability increases. Heating the work metal therefore extends the usefulness of a roll-forming machine. For example, a 3 m (10 ft) long pinch-type machine with 495 mm (19$^1/_2$ in.) diameter rolls can form a 366 cm (144 in.) diameter cylinder from 64 mm (2$^1/_2$ in.) thick by 62 cm (24$^1/_2$ in.) wide plate at room temperature in one pass (assuming the work metal has a tensile strength of 410 MPa, or 60 ksi, at room temperature). With all other conditions remaining constant, by heating the work metal to a temperature high enough to reduce the tensile strength to 70 MPa (10 ksi) or lower, the width of the plate (measured parallel to the roll axes) can be increased to 210 cm (82 in.) and rolled in one pass, using the same amount of power needed for rolling plate 62 cm (24$^1/_2$ in.) wide at room temperature. To reduce the tensile strength of low-carbon steel and to obtain optimal formability, the usual practice is to heat the steel to 870 °C (1600 °F).

Similarly, the size of the machine described previously is capable of cold forming a 366 cm (144 in.) diameter cylinder from 44 mm (1$^3/_4$ in.) thick by 150 cm (60 in.) wide low-carbon steel plate in one pass. Under the same conditions, except for heating to 870 °C (1600 °F), the thickness of the plate can be increased to 70 mm (2$^3/_4$ in.).

One method of evaluating the difference in formability between cold and hot rolling is to measure the force required to form a given plate. This is done on pyramid-type rolls by measuring, in terms of amperage, the downward force on the upper roll and the force required to rotate the powered rolls (Fig. 6). The following example

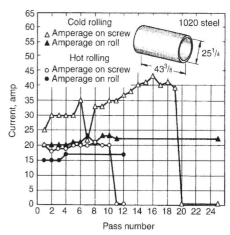

Fig. 6 Comparison of current flow (proportional to force) measured on screw and rolls during cold rolling and during the rolling of 44 mm (1$^3/_4$ in.) thick plate preheated to 870 °C (1600 °F). Dimensions given in inches

demonstrates the difference in current flow, number of passes, and time needed for the cold and hot forming of similar cylinders.

Example 1: Cold versus Hot Forming of 44 mm (³/₄ in.) Thick Steel Plate. A pyramid-type machine was used to produce 64 cm (25¼ in.) ID by 110 cm (43⅜ in.) long cylinders from 1020 steel blanks 44 mm (1¾ in.) thick by 110 cm (43⅜ in.) wide by 215 cm (84½ in.) long. When forming was done at room temperature, 25 passes were required with current flow on the upper roll (screw) and power rolls as shown in Fig. 6. Rolling time was 40 min per cylinder. When blanks were heated to 870 °C (1600 °F) and finished at 565 °C (1050 °F), the number of passes was reduced to 12 and rolling time to 11 min per cylinder. Current flow was also reduced, as shown in Fig. 6.

Hot Forming Formability Problems. The technique described in the preceding example for increasing the effective capacity of equipment by hot forming does not apply to all metals, and the amount of decrease in tensile strength varies considerably among carbon and low-alloy steels. For example, heat-resistant alloys, by definition, resist the softening effect of heat, and many of these alloys precipitation harden in the temperature range that would be used for the hot forming of carbon steel (see the article "Forming of Nickel and Cobalt Sheet Alloys" in this Volume). Magnesium and titanium alloys are usually formed at the same elevated temperature used to form the same alloy by other methods (see the articles "Forming of Magnesium Alloys" and "Forming of Titanium and Titanium Alloys" in this Volume). Copper and aluminum alloys are usually formed at room temperature. However, alloys 7075 and 7079, as well as some other precipitation-hardening alloys, must be formed within 24 h after solution treatment. If forming cannot be done within this length of time, the work metal must be stored at −12 °C (−10 °F) to prevent precipitation hardening (see the articles "Forming of Aluminum Alloys" and "Forming of Copper and Copper Alloys" in this Volume).

Maximum Elongation. In many applications with steel, hot forming is mandatory regardless of the capacity of available forming equipment. Common practice is to compute a maximum elongation in the outer surface for cold forming, as determined by:

$$E = \left(\frac{t}{d+t}\right)100 \qquad \text{(Eq 1)}$$

where E is the percentage of elongation in the outer surface of the cylinder, d is the inside diameter of the cylinder (in inches), and t is the plate thickness (in inches). For example, for a cylinder having an inside diameter of 145 cm (57 in.) to be formed from 75 mm (3 in.) thick steel plate:

$$E = \left(\frac{3}{57+3}\right)100 = 5\% \qquad \text{(Eq 2)}$$

Elongation of more than 5% (determined by Eq 2) is seldom permitted for cold forming, and the maximum is often 3.5%. If the maximum permissible elongation was 3.5% in the aforementioned example, either the minimum cylinder diameter would be near 210 cm (83 in.) or the plate thickness would have to be reduced before cold forming would be permitted. Maximum elongation is established by the user of the formed product, and hot forming is used when specifications cannot be met by cold forming.

Combination Hot and Cold Forming. A combination of hot and cold forming is sometimes advantageous and permits elongation requirements to be met. For example, in forming the 145 cm (57 in.) diameter cylinder described previously, one procedure is first to hot form the 75 mm (3 in.) thick plate to a circular segment of approximately 230 cm (90 in.), allow the workpiece to cool to room temperature, and then clean and finish form at room temperature. This procedure makes it possible to meet more severe elongation requirements and to retain some of the advantages of cold forming, such as greater accuracy.

Hot Forming Temperatures for Steel

Carbon or low-alloy steel plate is commonly heated to 870 °C (1600 °F) for hot forming after normalizing at the mill. However, plate in the as-rolled condition is less costly. The as-rolled steel is normalized while it is being heated for forming and cooled during forming. In such an operation, the steel is heated to 900 to 925 °C (1650 to 1700 °F), instead of to 870 °C (1600 °F), before forming.

Finishing temperature is critical for some steels, especially the plain carbon grades, because of the blue-brittle temperature range. It is generally recommended that the finish-rolling temperature should be 565 °C (1050 °F) or higher. If the workpiece cannot be completely formed before it cools to 565 °C (1050 °F), it should be removed from the machine and reheated.

Warm forming is often used when forming requirements are too severe for room temperature and when heating to the conventional hot forming temperature cannot be permitted because the mechanical properties of the steel would be impaired. A notable example is the forming of quenched-and-tempered grades of high-strength low-alloy steel. Common practice is to heat these steels no higher than the temperatures at which they were tempered and then form at once.

Power Requirements

The power required to form a given cylinder on three-roll equipment depends on the strength of the work metal, plate thickness, plate width, finished diameter of the cylinder, number of passes used, and temperature (hot or cold forming).

Strength of the work metal, plate thickness, and plate width (measured parallel to the roll axes) determine the diameter of a cylinder that can be formed in a given machine. The curves in Fig. 7, for a 3 m (10 ft) long pinch-type roll machine with 483 mm (19½ in.) diameter rolls, represent combinations of maximum plate thickness and width that can be rolled at room temperature into cylinders of 365 cm (144 in.) in diameter or larger in a single pass. The significant influence of the strength of the steel is evident from Fig. 7. For example, 38 mm (1.5 in.) thick low-carbon steel plate of 410 MPa (60 ksi) tensile strength can be rolled into 365 cm (144 in.) diameter cylinders in one pass in widths up to 305 cm (120 in.) in this machine. Under similar conditions, the width of 38 mm (1.5 in.) thick high-strength low-alloy steel that can be formed to the same cylinder diameter is restricted to approximately 56 cm (22 in.) (Fig. 7). The machine used was rated at 65 mm (2.5 in.) by 300 cm (120 in.).

Work Hardening. Most metals are susceptible to strengthening by cold work (work hardening), although the extent to which metals are affected varies widely among the various compositions. Of the steels commonly processed by three-roll forming, those of low carbon content, such as 1010, are least susceptible to work hardening and seldom present any serious problems. As carbon or alloy content increases, the rate of work hardening increases.

For metals that work harden rapidly, power consumption increases as forming proceeds. Eventually, the machine is overloaded, or the work metal fractures.

Intermediate annealing must be used when work-hardenable metal is severely formed. For

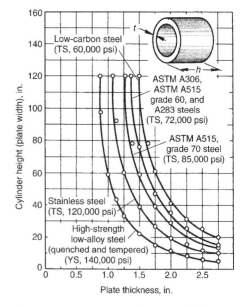

Fig. 7 Interrelationship of cylinder height (h), thickness (t), and work metal strength for forming in a single pass in a pinch-type roll machine at room temperature. TS, tensile strength; YS, yield strength

steel, full annealing is usually recommended. Process (subcritical) annealing is sometimes used. However, process annealing alternated with cold work is likely to result in excessive grain growth and subsequent poor formability, despite the hardness.

Cylinder Diameter. With other conditions constant, power requirements increase as cylinder diameter decreases when the cylinder is completely formed in one pass. However, the use of two or more passes (up to 12 is not uncommon) permits the rolling of smaller-diameter cylinders without increasing machine size. For example, Fig. 7 shows that 38 mm (1.5 in.) thick by 305 cm (120 in.) wide plate of low-carbon steel can be formed into a cylinder 366 cm (144 in.) in diameter in one pass on a given machine. By using 10 or 12 passes, cylinders as small as 65 cm (25½ in.) in diameter have been formed from the maximum thicknesses shown in Fig. 7. For plate thickness of 30 mm ($1^{3}/_{16}$ in.) or less, cylinders as small as 55 cm (21½ in.) in diameter can be formed in multiple passes.

Similarly, for carbon steel having a tensile strength to 590 MPa (85 ksi) and stainless steel to 830 MPa (120 ksi), any of the plate thicknesses shown in Fig. 7 can be rolled to cylinders as small as 70 cm (27½ in.) in diameter. For steels having a tensile strength of approximately 500 MPa (72 ksi), plate thicknesses of 28 mm ($1^{3}/_{32}$ in.) or less can be rolled to cylinders 56 cm (22 in.) in diameter, using the multiple-pass procedure. The plate thickness limitations for rolling 56 cm (22 in.) diameter cylinders then decrease to 26 mm ($1^{1}/_{64}$ in.) (maximum) for steel with a tensile strength of 590 MPa (85 ksi) and to 21 mm ($^{13}/_{16}$ in.) for stainless steel with a tensile strength of approximately 830 MPa (120 ksi).

Power requirements for quenched-and-tempered high-strength low-alloy steel (970 MPa, or 140 ksi, in Fig. 7) are high, and there is an increased probability of cracking. Suggested limits for maximum plate thickness and minimum cylinder diameter for multiple-pass rolling, regardless of power, are:

Maximum plate thickness		Minimum cylinder diameter	
mm	in.	mm	in.
25	1	1170	46
19	3/4	890	35
13	1/2	825	32½
9.5	3/8	775	30½
6.4	1/4	710	28

Temperature. The limitations imposed by steel composition and other factors shown in Fig. 7 are markedly changed when the work metal is heated (see the section "Cold versus Hot Forming" and example 1 in this article). However, it is not always possible to use hot forming—for example, for quenched-and-tempered high-strength low-alloy steel (see the section "Warm Forming" in this article).

Forming Small Cylinders

The cold forming of small cylinders by the three-roll process requires extra care, especially when the diameter of the cylinder to be formed is near that of the rolls. The rolling of cylinders having an inside diameter of less than 50 mm (2 in.) more than the outside diameter of the roll is not generally recommended. However, a skilled operator, using special care, can form cylinders within 38 mm (1½ in.) in diameter in a pinch-type machine.

Forming Large Cylinders

Cylinders that are large in diameter or in length can be formed by the three-roll process. Some special procedures may be required, especially when the sheet is so thin that the cylinder cannot retain a round shape without support. Under these conditions, overhead cranes or temporary braces or both can be used for support.

When flat blanks of the required length are unavailable, two or more sections can be welded together to obtain the required length. The following example describes this practice in the forming of large cylinders.

Example 2: Forming 472 cm (186 in.) Outside Diameter Cylinders in a Pinch-Type Machine. Cylinders 472 cm (186 in.) in diameter and 284 cm (112 in.) long were formed from blanks prepared by welding together three sections of copper-clad low-carbon steel. The fabricated blanks were 1480 cm (582¾ in.) long by 284 cm (112 in.) wide by 14 mm ($^{9}/_{16}$ in.) thick. The cylinders were formed cold on a 19 mm ($^{3}/_{4}$ in.) by 3.7 m (12 ft) pinch-type machine. The major problem in this operation was support because the work was not thick enough to provide natural support. Overhead cranes were used to hold the formed section of the plate during rolling. The ends of the cylinder were tack welded before the workpiece was removed from the machine. Temporary braces were then used to hold the cylinders in a near-round condition while the seam was welded. After welding, outside rounding rings made from 38 by 125 mm (1½ by 5 in.) rectangular steel bars were used to hold the cylinder shape for subsequent attachment to other cylinders. In use, the welded internal parts stiffened the assembly.

Forming Truncated Cones

Pyramid-type machines are generally used in the forming of truncated cones. There are two basic limitations on the shape of conical configurations that can be formed by the three-roll process. First, the smaller diameter (A, Fig. 8) must be large enough to maintain the established workpiece-to-roll relationship on minimum obtainable diameter, and second, the plate must

be thick enough to be formed by the holdback method shown in Fig. 8 so that the pressure buildup on the holdback pin does not upset the plate edge or damage the holdback attachment or both.

The size of cone that can be formed depends largely on the pitch, the spacing and length of rolls, and the diameter of the upper roll. As the cone height increases, difficulties in getting the large diameter to follow the small diameter accurately also increase. If the lower roll spacing is too wide, the size of the smaller diameter will be severely limited, because as the work metal passes the holdback pin (Fig. 8), it will curve into the housing that supports the rolls and rolling will be impossible.

Procedure. There are several significant differences between forming conical shapes and forming cylinders by the pyramid-type three-roll process. In rolling cylinders, the blank is rectangular and is rolled in a direction perpendicular to the rolls. In contrast, curved blanks are used for conical shapes, and they are rolled on a curve. In addition, no pin is needed in rolling cylinders, and the top roll is pitched for forming cones but is straight and level when forming cylinders.

Before the developed blank is placed in the pyramid-type three-roll machine for forming a cone, the ends of the blank are preformed in a pinch-type roll from the rear or in a press brake. For forming the cone, the top roll is pitched as shown in Fig. 8. As the rolls drive the blank, its edge drags around the pin, and the various diameters of the cone are formed. The blank is then rolled in multiple passes, dragging around the pin until the ends of the blank are closed. If the pitch of the rolls matches the pitch of the entire length of the cone, a nearly true cone will result.

Truncated cones are sometimes produced by forming two semicircular half-cones and welding them together; the completed workpiece has two longitudinal seams instead of one. In another method, two or more circular tapered sections are formed and welded together. When produced by this method, the finished cone has

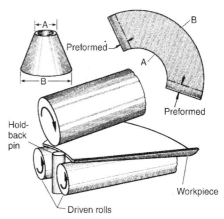

Fig. 8 Rolling a truncated cone in a three-roll pyramid-type machine from a blank with preformed ends

one longitudinal seam and one or more circumferential seams.

Forming Bars and Shapes

Pyramid rolls and machines with overhanging rolls are used to form bars, bar sections, and structural shapes into circles. Shapes that can be processed by this method include rounds, squares, flats (on the edge or on the flat), I-beams, L-shaped structurals, and channels. Hardened roll sections that are adjustable to the thickness or cross section of the shape are used (Fig. 9). Some bars or shapes can be formed with plain, flat rolls, but more often rolls conforming to the shape of the unformed workpiece are required.

The rolls are adjusted to produce the required workpiece diameter in the same manner as in rolling cylinders from sheet or plate. Prebending the ends and rolling the workpiece back and forth are also employed in the three-roll forming of bars and shapes. A significant difference in the rolling of bars and shapes is the frequent use of guide rollers or guide fingers or both, mounted so as to contact the sides of the workpiece and prevent it from twisting during forming.

One of the most difficult shapes to form by the three-roll process is an angle with one leg inside the circle (Fig. 9). When forming this shape, spiraling, twisting, buckling of the inside leg, and reduction of the angle between the two legs are likely to occur. These difficulties can be minimized by using a hardened upper roll having a removable end plate and a spacer that is not more than 0.81 mm ($^1/_{32}$ in.) thicker than one leg of the angle (Fig. 9). The radius on the upper roll (R, Fig. 9) must conform to the fillet radius of the angle to be rolled. The other end of this roll (R_1, Fig. 9) can have the same radius or a radius conforming to another angle to be rolled. This end of the upper roll can be used by reversing the roll end for end. In production rolling of the same angle, common practice is to have the same values for R and R_1 (Fig. 9), thus permitting double the roll life by reversing the roll. Guide rollers and fingers (not required for the section shown in Fig. 9) also help to produce accurate circles from bar sections.

Out-of-Roundness

Out-of-roundness, or ovality, of cylinders produced by three-roll forming is caused by one or more of the following variables:

- Varying thickness of the flat blank
- Varying hardness within the blank
- Overforming or underforming of the ends in the preforming operation
- Springback of the work metal
- Temperature of the metal being formed
- Number of passes
- Condition of the equipment
- Operator skill

The most important of the aforementioned factors is operator skill; condition of the equipment is also a major factor.

Variations in the mill product (thickness and hardness within a single sheet or plate) are seldom great enough to warrant the extra cost that would be necessary for closer-than-normal control of the work metal. Out-of-roundness and other dimensional variations in the product increase as workpiece diameter increases. Plate thickness above or below the actual crown thickness of the rolls can also cause dimensional variations. For large or small workpieces, much out-of-roundness is caused by variations in preforming the ends, regardless of whether pinch rolls or press brakes are used in preforming.

Springback is overcome by forming to a circle smaller than that required for the finished cylinder. However, the overforming of high-springback material must be done with caution; as the elastic limit is exceeded, metals will take a permanent set, and too much overforming can result.

Hot forming may contribute to out-of-roundness because considerable plastic flow can take place when a steel workpiece is heated to 870 °C (1600 °F) or higher. The amount of plastic flow varies as the bending load is applied during forming, and variations are increased by uneven cooling of the work metal. Resulting variations in plate thickness and curvature contribute to out-of-roundness.

Production Example. Without the use of special techniques or secondary operations, there is likely to be considerable variation in out-of-roundness among workpieces that are intended to be identical and are produced under the same conditions. This is demonstrated in the following example.

Example 3: Out-of-Roundness Variations in 44 Cylinders, 432 mm (17 in.) in Diameter. Cylinders 432 mm (17 in.) in diameter were produced on a pinch-type machine in one

pass. Leading edges of the blanks were prebent 15° for a distance of 50 mm (2 in.) with a 102 mm (4 in.) radius. After the cylinders were formed and tack welded, measurements were taken. Out-of-roundness variation in 44 cylinders is plotted in Fig. 10.

The number of passes used to form a given cylinder can have a significant effect on out-of-roundness. In most applications, two or more passes (sometimes as many as 12) will produce cylinders that are more nearly true round than those formed in a single pass.

Methods of Correction. The most effective method for correcting out-of-roundness is to reroll the cylinder carefully before welding. This operation causes a one-third greater load on the machine than the original rolling.

For cylinders having wall thickness no greater than approximately 9.5 mm ($^3/_8$ in.), a draw-down type of expander applied after rolling and welding is an effective means of correcting out-of-roundness.

If design permits, beads or flanges can be rolled into the cylinder wall to reinforce it and to help maintain roundness. Oil drums are examples of the effective use of this technique. When dimensional requirements are plus or minus a few thousandths of an inch, stock must be allowed for machine boring the cylinder to the specified diameter.

Forming Speed

The speed of forming is a critical factor in product quality. Low-carbon steel plate up to 8 mm ($^5/_{16}$ in.) thick is sometimes rolled at speeds to 18 m/min (60 ft/min). For a speed this high, however, workpiece diameter is necessarily medium to large, because it is impractical to control the machine for rolling small diameters at high speed.

The most commonly used speeds for cold forming (particularly for thick plate) range from 3.6 to 6 m/min (12 to 20 ft/min). This range is usually maintained in both cold and hot forming; however, to complete hot forming with a minimum decrease in temperature of the work metal, it is sometimes necessary to increase the speed of the bending roll.

Roll Deflection

Roll deflection can be calculated by standard formulas, considering the roll as a simple beam

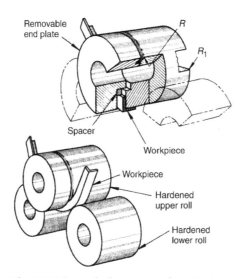

Fig. 9 Roll setup for forming an angle section into a circle. Guide rolls and guide fingers are not required for this application.

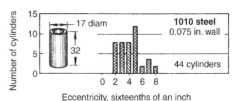

Fig. 10 Out-of-roundness variations in 44 cylinders of the same diameter produced in a pinch-type roll machine. Dimensions given in inches

supported at both ends. On pyramid rolls, deflection is often minimized by support rollers applied to the lower rolls. These rolls act as backup rolls.

In forming heavy plate, pressures are high and all three rolls are crowned (made larger at the centers than at the ends). Crowning is necessary because the rolls deflect under the bending load; if they were straight, all formed cylinders would bulge somewhat at the center. Because the amount of deflection depends on the bending load, usual practice is to crown the rolls enough to compensate for the average job in the plant. When forming plate thicker than the actual crown deflection, the rolls are shimmed by running strips of thin metal (16, 14, 12, or 10 gage) between the rolls and the inside diameter of the workpiece at the center of the rolls. This shimming compensates for excessive deflection.

When forming metal that is too thin to cause deflection of the rolls, crowning will cause the formed cylinder to be larger in diameter at the ends than in the center. Correction can be made by shimming the ends of the rolls in a manner similar to that described previously for shimming the centers.

Alternative Processes

Three-roll forming is the most practical method of producing large cylinders and truncated cones from heavy plate.

Deep drawing is often the most economical method of producing small cylinders from sheet no thicker than 3.18 mm (0.125 in.). Seamless cylinders or cones can be produced by deep drawing, piercing, and trimming (see the article "Deep Drawing" in this Volume). However, as cylinder size or wall thickness increases, forming by deep drawing becomes impracticable. In some applications, three-roll forming and welding are preferred for producing a hollow shape from stock that is substantially thinner than 3.18 mm (0.125 in.).

Contour Roll Forming. Theoretically, the diameter and length of straight cylinders producible by contour roll forming are almost unlimited. In practice, however, diameter and wall thickness are limited by the size of available equipment. Contour roll forming is rarely used for rolling metal thicker than 6.35 mm (0.250 in.) and is most often used for thicknesses less than 3.18 mm (0.125 in.). Therefore, contour roll forming is impractical for producing large heavy-wall cylinders (see the article "Contour Roll Forming" in this Volume).

Forming two halves (semicircles) between dies in a press and then welding the two half-cylinders is sometimes practical. However, even when presses of sufficient size are available for forming large plate, die cost is likely to be prohibitive. For limited production, a press brake is often used to produce semicircles that can subsequently be joined by welding into cylinders.

Safety

Forming rolls move relatively slowly but require protection for the operator. The most positive method of protection is to cover the nip point between the feed rolls. One effective guarding device is a solid metal plate covering the nip point between the feed table and the rolls for the full length of the rolls. This plate, with a stud welded to each end, is attached to slotted vertical brackets by nuts and washers so that it is adjustable vertically. The brackets are securely fastened to the feed table.

The height of the feed table can be made adjustable by welding a nut to the lower end of each tubular leg. A long bolt, with a large washer welded to the top of the head, is screwed into the nut to achieve the desired height. A locknut can be used to prevent the bolt from turning because of vibration.

Emergency tripping bars connected to electric cutoff switches or, preferably, to reverse electric switches can be used to stop the rolls. The bars may be at knee level in front of the operator, or directly in front of the bottom feed roll and far enough below the feed point to avoid accidental tripping.

Feeding guides for narrow workpieces can be made of bar stock or angles that are bolted to the feed table. The guides should be slotted for ease of adjustment to various widths of workpieces.

Contour Roll Forming

CONTOUR ROLL FORMING (also known as roll forming or cold roll forming) is a continuous process for forming metal from sheet, strip, or coiled stock into desired shapes of uniform cross section by feeding the stock through a series of roll stations equipped with contoured rolls (sometimes called roller dies). There are two or more rolls per station. Most contour roll forming is done by working the stock progressively in two or more stations until the finished shape is produced.

Only bending takes place in contour roll forming; the stock thickness is unchanged except for a slight thinning at bend radii. The process is particularly suited to the production of large quantities and long lengths to close tolerances and involves a minimum of handling. Auxiliary operations, such as notching, slotting, punching, embossing, curving, and coiling, can easily be combined with contour roll forming.

Contour roll forming is used in many diverse industries to produce a variety of shapes and products. The process is also used for parts that were previously manufactured by extrusion processes. This use is limited, however, to parts that can be redesigned to have a constant wall thickness. Industries that use roll-formed products include the automotive; building; office furniture; home appliance and home product; medical; railcar; aircraft; and heating, ventilation, and air conditioning industries.

Contour roll forming can be divided into two broad categories: a process using precut lengths of work metal (precut or cut-to-length method) and a process that uses coil stock that is trimmed to size after forming (postcut method).

In precut operations, the work metal is cut to length before entering the forming machine. The precut process usually employs a stacking and feeding system to move blanks into the machine, a contour roll-forming machine operating at a fixed speed of about 15 to 75 m/min (50 to 250 ft/min), an exit conveyor, and a stacking system. The precut method is primarily used for low-volume operations and when notching cannot be easily accomplished in a postcut line. Often, the material is run from a coil to a shear or blanking press and then fed mechanically to the contour roll former.

Tooling for the precut method is relatively inexpensive because cutting requires only a flat shear die or an end notch die. End flare is more pronounced than it is with the postcut method, however, and side roll tooling is required to obtain a good finished shape.

The most efficient, productive, and consistent contour roll forming process is the postcut method. This method requires an uncoiler, a roll-forming machine, a cutoff machine, and a runout table (Fig. 1). Postcut contour roll forming can be augmented by various auxiliary operations, such as prenotching, punching, embossing, marking, trimming, welding, curving, coiling, and die forming. These auxiliary operations can be used to eliminate the need for subsequent operations, resulting in production of a finished product.

Tooling cost and tooling changeover time are greater for the postcut method than is the case with precut operations, but the increased efficiency of the postcut process balances this limitation.

Materials

Any material that can withstand bending to the desired radius can be contour roll formed. Thicknesses of 0.13 to 19 mm (0.005 to 3/4 in.) and material widths of 3.2 to 1830 mm (1/8 to 72 in.) can be used. Length of the formed part is limited only by the length that can be conveniently handled after forming.

In some cases, multiple sections can be formed from a single strip; in other cases, several strips can be fed into the machine simultaneously and combined after forming to produce a composite section. Contour roll forming is almost always performed at room temperature; however, some materials, such as certain titanium alloys, must be formed at elevated temperatures. This is done on specially designed machines.

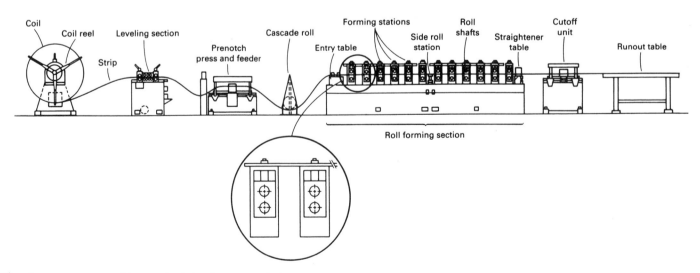

Fig. 1 Setup for contour roll forming of coiled stock (postcut method)

Influence of Work Metal Composition and Condition. The effect of work metal formability on procedures and results in contour roll forming is generally the same as it is in other forming methods.

Initial yield strength and rate of strain hardening of the work metal affect contour roll forming. One measure for predicting the formability of a work metal is the Olsen cup test (see the article "Formability Testing of Sheet Metals" in this Volume).

Work metals of equal thickness, with differing compositions, initial yield strengths, and rates of strain hardening, could require alterations in the contour rolling operation. Factors to be considered are: power requirements, number of stations, roll material, lubrication, and speed.

In contour roll forming, it is not uncommon to change the work metal but retain the same shape. In such instances, changes may be necessary in equipment and tooling.

When work metal thickness and tooling remain unchanged, it is seldom a problem to change to a more formable work metal. For example, in changing from low-carbon steel to aluminum, major changes would be unlikely. However, it may be necessary to regrind the finish-station rolls to avoid overbending the section; this would usually be determined by a trial run.

When the change is to a metal of higher strength, one or more changes in procedure may be required to achieve desired results. In forming higher-strength metals such as stainless steels, some overforming is usually required to allow for springback. Residual stress in highly cold-worked metal often causes straightening problems, particularly when forming asymmetrical shapes. A common remedy is to add roll stations to decrease the amount of forming in a given station.

In most instances, difficulty in maintaining size and angle tolerance increases as the yield strength of the work metal increases. Optimum rolling speed decreases as yield strength or hardness increases. For example, in changing from carbon steel to stainless, speed is usually decreased 10 to 25%, mainly to prevent rolls from galling when there is an appreciable amount of roll sweep. One method of combating roll galling without greatly reducing speed is to use extreme-pressure (EP) additives in the lubricant; for extreme conditions, pigmented drawing compounds can be added to the lubricant. The main disadvantage of special lubricants is the difficulty and expense involved in removing them from finished workpieces.

Aluminum bronze rolls are often an advantage in shaping the difficult-to-form metals because they resist galling; however, bronze rolls are softer and wear faster than do tool steel rolls. Rolls made from D2 tool steel, hardened and chromium plated, are usually best for contour roll forming of high-strength metals (see the section "Roll Materials" in this article).

Process Variables

In contour roll forming, material is progressively formed as it passes from one station to another. The variable parameters in a roll-forming operation include power requirement, forming speed, and type of lubricant. These parameters are determined by width, thickness, and type of material; complexity of the cross section to be formed; coating (if any) on the material; and accuracy required.

The power required by a roll-forming machine depends on the torque loss through the drive gearing and the friction between the material and the rolls as the material is being formed. The particular alloy and its thickness must be taken into consideration when looking at the effect of material on power requirements. Generally, contour roll-forming machines have motors ranging from 10 to 50 hp on small machines and from 50 to 125 hp on larger machines.

Forming Speed. Speeds used in contour roll forming can range from 0.5 to 245 m/min (1.5 to 800 ft/min), although this speed range represents unusual extremes. Speeds between 25 and 30 m/min (80 and 100 ft/min) are most widely used. One or more of the following can influence optimum forming speed:

- Composition of the work metal
- Yield strength or hardness of the work metal
- Thickness of the work metal
- Severity of the forming operation
- Cutting of finished shapes to length
- Number of roll stations
- Required auxiliary operations
- Use of lubricant (coolant)

Lower speeds in the range indicated previously (near 0.5 m/min, or 1.5 ft/min) are required for contour roll forming titanium into a relatively complex shape. At the other extreme, a speed of 245 m/min (800 ft/min) has been used in production operations in which conditions were nearly ideal, that is, for contour rolling low yield strength metal, such as aluminum or annealed low-carbon steel, in thicknesses less than 0.91 mm (0.0359 in.), in an operation having mild forming severity and requiring cutoff into relatively long lengths (about 25 m, or 80 ft). To use such high speeds effectively, even though forming is not severe, more stations are usually required to minimize the amount of forming in any one station. High forming speed usually precludes auxiliary operations, such as punching, notching, or welding, and requires a flood of lubricant at each station.

The preceding first four factors are closely related and influence permissible forming speed. In addition, one or more of the last four factors may dictate a lower speed regardless of the otherwise permissible speed.

Lubricants prevent metal pickup by the rolls (thus improving finish on the work metal and prolonging roll life) and also prevent overheating of rolls and work metal. When rolls become overheated, their life is shortened. If work metal is overheated, it may warp and require straightening. When lubricants can be tolerated, rolling efficiency is usually increased by their use.

Soluble oils (in a 1-to-12 mixture with water) are the most commonly used lubricants. They are usually applied by a pumping action from a self-contained sump in the machine base through a manifold having flexible tubes and nozzles that direct the fluid to the required locations. Gutters are arranged around the top of the machine to catch the fluid and return it to the sump.

Other lubricants have been used satisfactorily for specific applications. In addition to lubricating and cooling, however, a lubricant must be nontoxic, noncorrosive to the metal being formed (as well as to rolls and other machine components), and removable by available shop cleaning facilities. For instance, some silicone-base fluids are excellent lubricants for roll forming, but they are extremely difficult to remove from metal surfaces. This poses problems in obtaining satisfactory plating or adherence of organic coatings or adhesives. Extreme-pressure lubricants are sometimes used in severe roll forming.

For some applications, no lubricant is permitted (for example, for the forming of painted or otherwise coated metals or for the forming of complex shapes that would entrap lubricants). The result could be a reduction of rolling speed or lower-quality finish, or both. However, in some instances, even though flooding with lubricant cannot be tolerated, other means can be used to supply some lubricant to the rolls. One method is to mount cellulose sponges in constant contact with the rolls and to keep them wetted with lubricant by hand or by drip applicators.

Despite the fact that lubrication is helpful and often necessary in contour rolling, application and subsequent removal of lubricants are significant cost items. When roll forming steel, however, the selection of hot-rolled, pickled, and oiled grades of work metal has often eliminated the need for an additional lubricant. More information on the role of lubrication and the types of lubricants used in sheet forming is available in the article "Selection and Use of Lubricants in Forming of Sheet Metal" in this Volume.

Machines

The contour roll-forming machine most commonly used has a number of individual units, each of which is actually a dual-spindle roll-forming machine, mounted on a suitable baseplate to make a multiple-unit machine. The flexibility of this construction permits the user to purchase enough units for immediate needs only. The purchase of additional length of baseplate on the machine allows the addition of units at any time for future needs. Some of these machines are provided with machined ends on the baseplates, making it possible to couple several machines together, in tandem, to provide additional units as required.

Screws for making vertical adjustments of the top rolls are designed with dials and scales to provide micrometer adjustment and a means of recording the position of the top shaft for each roll pass and each shape being formed. Shaft diameter on most machines ranges from 25 to 102 mm (1 to 4 in.).

Several types of roll-forming machines or roll formers are used. They can be classified according to spindle support, station configuration, and drive system.

Spindle Support. Roll-forming machines can be classified according to the method by which the spindles are supported in the unit. Generally, two types exist: inboard or overhung spindle machines and outboard machines.

Inboard-type machines (Fig. 2a) have spindle shafts supported on one end that are 25 to 38 mm (1 to 1.5 in.) in diameter and up to 102 mm (4 in.) in length. They are used for forming light-gage moldings, weather strips, and other simple shapes. Material thickness is limited to about 1 mm (0.040 in.), and the top roll shaft is generally geared directly to the bottom shaft. This direct-mesh gearing permits only a small amount of roll redressing (no more than the thickness of the material being formed) on the top and bottom rolls. Tooling changeover is faster on this machine than on the outboard type of machine.

Outboard-type machines (Fig. 2b) have housings supporting both ends of the spindle shafts. The outboard housing is generally adjustable along the spindles, permitting shortening of the distance between the supports to accommodate the roll forming of small shapes of heavy-gage material. This adjustment also permits the machine to be used as an inboard type of machine when desired. Outboard machines can be readily designed to accommodate any width of material by making the spindle lengths suit the material width and then mounting the individual units and spindles on a baseplate of suitable width. This type of machine is built with spindle sizes ranging from 38 to 102 mm (1.5 to 4 in.) diameter and with width capacities up to 1830 mm (72 in.).

Generally, for roll-forming material more than 5 mm (^{3}/₁₆ in.) thick, machines are constructed so that both top and bottom shafts can be removed by lifting them vertically from the housings after the housing caps have been removed. This permits rolls to be mounted on the shafts away from the machine, an important consideration when heavy rolls are being handled. This type of machine is built in spindle sizes ranging from 50 to 380 mm (2 to 15 in.) in diameter.

Station Configuration. As mentioned previously, a typical contour roll-forming machine consists of several individual forming units mounted on a common baseplate. The manner in which the forming units are mounted determines

to a great extent the type of shapes that are formed on the machine.

Single-duty machines are built and designed for a one-purpose profile or for one particular set of roll tooling and are not normally designed for convenient roll changing. This machine is generally used for long production runs, and its cost is low in comparison with the other styles.

Conventional (standard) machines are more versatile than single-duty machines because the outboard supports are easily removed. This facilitates roll changes, making conventional machines suitable for a variety of production requirements.

To change the tooling, the top and bottom spindle lock nuts are removed and the outboard

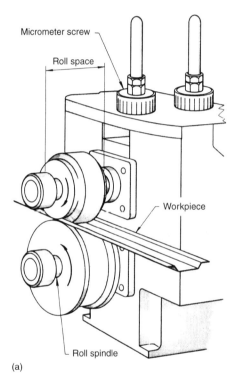

(a)

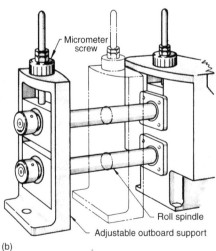

(b)

Fig. 2 Two basic machine concepts for contour roll forming. (a) Inboard-type machine. (b) Outboard-type machine

housing is pulled off the spindles. The tooling can then be removed and replaced with the desired profile.

Side-by-side machines (Fig. 3) are designed for multiple-profiled tooling and provide the flexibility of having more than one set of roll tooling mounted on the spindle shaft at the same time. Generally, this type of machine is limited to two sets of rolls at a given time, but there can be up to three or four sets of rolls when small profiles are being run in production. Changeover from one production profile to another is accomplished by shifting the machine bed to the desired profile. The main advantages of the side-by-side configuration are low initial investments, fast tooling change, and reduced floor space requirement. Roll wear, however, can create problems because one set cannot be reground without regrinding the others at the same time. Adjusting for material variations can also be a problem.

The double-high machine configuration consists of one set of roll tooling mounted on its own roll shafts and housings at one level on the bed frame, and a second complete set of roll tooling and housings mounted at a different level on the same frame. This particular type of machine is used in the metal building industry for forming building panels up to 1520 mm (60 in.) wide.

The rafted machine configuration resembles the single-duty and conventional configurations because each configuration has housings and spindle shafts with one particular set of roll tooling mounted on it. However, the rafted configuration has several roll-forming units mounted on rafts or subplates that are removable from the roll-former base. During tool changeover, the individual rafts are removed from the base, and the replacement rafts with the roll-forming units and tooling are installed. On a typical 16-stand roll-forming machine, there are four sets of rafts containing four forming units each.

Double-head machines are designed and constructed with two separate sets of housings and roll shafts mounted so that they face one another. Each housing is mounted on an

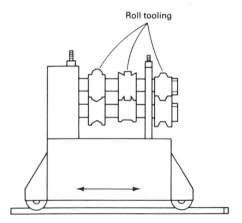

Fig. 3 A side-by-side contour roll-forming machine, which allows forming of several different profiles on the same machine

adjustable plate mechanism to allow the housing to be shifted for a change in overall width while at the same time maintaining the same profile for the edge formation.

This type of machine is very popular in the shelving industry, in which large, flat panels are rolled on a production basis and the panel widths change regularly. A disadvantage of this type of machine is that it does not lend itself to forming the center of the panel. Two of these machines, connected by an automatic transfer mechanism, are used to form the four edges of a shelf; the first machine forms the two long edges, and the second machine forms the two end configurations.

Drive Systems. The five basic methods used to drive roll-forming units are chain drive, spur gear drive, worm gear drive, square gearing, and universal drive.

A *chain drive* consists of a sprocket attached to the individual roll-forming unit and connected to the main drive by means of a roller chain. This is accomplished using a continuous roller chain, with one long chain driving each unit, or a shorter chain connected to each individual unit. This drive system is inexpensive and allows flexibility in the construction of the machine.

A *spur gear drive* consists of a continuous train of spur gears mounted at the rear end of each spindle shaft. Idler gears are positioned between each unit to transfer the drive equally to all the units.

A *worm gear drive* is very similar to the spur gear drive. However, instead of using the idler gear to transfer the drive to each unit, an individual worm gearbox is mounted on the bottom spindle of each unit. The worm gearboxes are coupled in line, which permits the machine designer to spread out the horizontal centers of each roll-forming station without being concerned about properly meshing the gear train to the idler gear.

Square gearing also incorporates both spur gears and a worm gear. This type of gearing permits a vertical adjustment of the upper spindle and allows use of a wide range of roll diameters.

Universal drive eliminates the need for any spur gearing or roller chain and sprocket drives. It consists of a series of worm-driven gear boxes with top and bottom outputs that transfer the power source to the individual shafts through a double-jointed universal coupling. On certain applications, only the bottom spindle is driven. This drive system is generally used with rafted-style machines to permit quick tool changeover. Simplicity of design and minimal maintenance are two important advantages of this drive system.

Machine Selection. Several factors must be taken into consideration when selecting a machine to be used in a roll-forming operation. These include load capacity, section size and shape, and roll changeover.

Load Capacity. The type and thickness of the material being formed determine to a large extent the load capacity that a given forming machine is required to produce. If material type

and thickness change, it is best to select a machine that can provide the additional capacity.

In forming material up to 1.5 mm (0.060 in.) thick, a machine with 38 mm (1.5 in.) diameter spindles should be used as long as the part is not too wide. A machine with 50 mm (2 in.) diameter spindles can be used for forming material thicknesses up to 2 mm (0.080 in.). As the material thickness increases, the diameter of the spindles must also increase to provide strength to create the pressure required to do the forming. Center distances must be increased as the size of the part shape and the movement of material between forming stations increases. These distances are usually determined by the machine builder.

Section Size and Shape. Wide roll-formed sections require wide roll spaces. To support the pressure of the rolls, the spindle shafts must be large enough in diameter to prevent shaft deflection during forming. The distance between the centerline of the bottom spindle and the machine bed determines the maximum roll diameter and hence the maximum section depth.

The more complex the shape of the section being formed, the greater the number of passes (pairs of rolls) required to roll form the section. It is best to select a machine that provides the flexibility of adding or subtracting pairs of rolls in accordance with the part design.

Roll changeover can be costly and time consuming because material variations may require different roll pressure settings. When several part configurations must be run on one machine, it is best to select a machine that permits the tooling to be changed quickly. If only two or three profiles are run, the side-by-side and the double-high machines are possible selections; the changeover can be performed quickly without losing valuable production time. Another machine to consider is the rafted machine.

Auxiliary Equipment

In addition to the machines that do the roll forming, several other pieces of equipment are usually required for production operation. Stock for roll forming is usually received in coils; thus, an electric hoist on an overhead track is needed to lift coils from skids and transfer them to a cradle or reel (another piece of auxiliary equipment). Also, equipment for welding the end of an expended coil to the lead of the next one, an entrance guide, intermediate guides, a straightening device, and cutoff equipment may be needed.

Stock reels should be equipped with an expandable arbor to fit the inside diameter of the coil, and with a friction drag, Stock reels incorporating these features are commercially available in a wide range of coil capacities (see the article "Presses and Auxiliary Equipment for Forming of Sheet Metal" in this Volume).

The friction drag is necessary to prevent the coil stock from overrunning onto the floor in the event of a sudden stoppage of the

roll-forming equipment. In the simplest type of motor-driven stock reel, a dancer arm and roll ride the stock in a loop-detector arrangement, which starts and stops the motor as required, supplying stock at the average rate used by the roll former. Stock speed is matched approximately by adjusting a variable-pitch sheave to prevent too-frequent stopping and starting of the alternating current drive motor. This type of control on the stock reel provides acceptable results for most applications. More elaborate controls can be used, such as a direct current motor drive with feedback control to match stock speed with machine speed. Elaborate controls are expensive and should not be considered unless they are needed to meet special workpiece requirements.

Stock reels are available with a swivel base and two arbors. A coil may be positioned on one arbor while the first coil is being used, thus reducing change time. This arrangement is advantageous when coils are relatively small and production requirements are high because time consumed in changing coils can become a substantial portion of the total production time.

Welding Equipment. Thread-up time can be eliminated by manually welding the end of each expended coil to the leading end of the next one. For stock thickness of 1.6 mm ($\frac{1}{16}$ in.) or more, a semiautomatic welder can be placed in the line. Regardless of the welding method used, any appreciable flash must be removed before the welded joint reaches the first roll station. Provision can be made to remove flash by installing a grinder similar to a band saw blade grinder. More information on combined roll-forming and welding operations is available in the section "Tube and Pipe Rolling" in this article.

Entrance guides positioned in front of the first forming station ensure correct alignment of the work metal entering the starting rolls. This is particularly desirable when the part being formed is asymmetrical in the first station because the stock could climb or shift to one side without guides. The simplest form of entrance guide consists of a flat plate with a channel milled to the proper width and depth to accept the strip at its maximum tolerance, plus a simple, removable lid to hold the stock in place. The mounting for this guide should permit adjustment vertically and laterally.

When wide variations in stock width are encountered, a self-centering, parallel-rule entrance guide is preferred. This guide is constructed like a navigator's parallel rule, with the crossbars pivoted and mounted at their centers with a spring, causing the side bars or rules to close on the stock under spring-load.

Stock drags are occasionally used to place a slight tension on the stock and to cause it to feed more uniformly through the first few stations. The simplest form consists of two pieces of hardwood. The stock is clamped between the wooden members, which butt against the entrance guide, thus providing enough friction to keep the stock under tension. The amount of tension can be regulated by the clamping force on the wooden members.

Guides between roll stations facilitate entrance of the partly formed stock into the next station. In theory, if rolls are properly designed, guides between stations are unnecessary because each set of rolls should accept the cross section from the preceding set of rolls. In practice, however, because of such factors as cost, lead time, and availability of space or equipment, the number of roll stations is often fewer than the ideal number; this necessitates more forming in each station than is consistent with the best practice. Therefore, the use of guides between stations helps to compensate for this lack of additional stations and to minimize springback.

Generally, guides between stations are required to contact only the critical points of the work metal, not the entire contour. Regardless of their shape, guides should be designed with removable top portions to facilitate threading.

Various metals may be used for guides, depending on the end-use of the workpiece. For the areas contacting the moving workpiece, hardened steel (usually, case-hardened low-carbon steel) is preferred from the standpoint of guide life. However, when workpiece finish is critical and hardened steel guides are likely to scratch the surface, bronze or aluminum guides are used. For some work, hard chromium plating of guides minimizes damage to workpiece surfaces and still provides acceptable guide life. Guides can be replated when the plating becomes worn.

Straightening Equipment. Usually it is necessary to straighten the workpiece after it leaves the final roll station. This is done by standard straightening guides attached to the machine beyond the last set of rolls, or by special devices designed for individual applications.

Straightening guides are usually adjustable vertically and laterally; the most versatile types can be swiveled in either elevation or azimuth and can also be rotated about an axis. Most straighteners employed for contour roll forming are either of the roll type or the shoe type.

A roll straightener consists of multiple rolls (individually adjustable) arranged to contact the stock in selected areas. A shoe straightener consists of one or more shoes, usually made of bronze, properly fitted to the contour and adjustable in at least one direction that will crimp the stock to correct for sweep or twist.

There are also applications in which a sweep (curve) is deliberate and desired. A sweep guide is similar to a straightening guide, and a straightening guide often can be adjusted to give the required sweep in a constant radius. For more detailed information on straightening equipment, see the articles "Straightening of Bars, Shapes, and Long Parts" and "Straightening of Tubing" in this Volume.

Cutoff Equipment. Because most contour rolled products are made from coil stock, a system of cutting the formed shapes to length must be provided. There are several types and sizes of flying-shear cutoff machines.

The sliding-die cutoff machine is most commonly used. The action of this machine is similar to that of a punch press, although construction of the machine differs. The flywheel and clutch are placed below the bed, with the ram posts passing through the bed. Gibs are provided in the bed and ram to accept a gibbed die and a punch holder that permits linear movement of the die to match work-metal speed during the cutoff cycle.

Tooling

Tooling used in roll forming includes the forming rolls and the dies for punching and cutting off the material. Tube mills require some additional tooling to weld, size, and straighten the tubes as they are produced on the machine; the needed tooling is discussed in the section "Tube and Pipe Rolling" in this article.

Forming Rolls

The rolls are the tools that do the actual forming of the material as it moves through the roll-forming machine. Several factors must be considered when designing the rolls to form a particular part. These include the number of required passes, the material width, the "flower" design, the roll design parameters, and the roll material. *Flower* is the name given to the progressive section contours, starting with the flat material and ending with the desired section profile.

Number of Passes. The roll forming of material into a desired final shape is a progressive operation in which small amounts of forming are performed at each pass or pair of rolls. The amount of change of shape or contour in each pass must be restricted so that the required bends can be formed without elongating the material. Too few passes can cause distortion and loss of tolerances; too many passes increase the initial tooling cost.

Generally, the number of passes depends on the properties of the material and the complexity of the shape. Other areas to consider are part width, horizontal center distance between the individual stations, and part tolerances. The number of passes must be increased as the tolerances of the shape become tighter.

Material. Material thickness, hardness, and composition all affect the number of passes required to achieve a desired shape. As the thickness of the metal increases, the number of passes required to form the material increases. Steel that has a high yield strength should be overformed approximately 2° and then brought back to finish size on the final pass. Overforming compensates for springback that is encountered when materials having high yield strengths are formed. Material that is coated or that has a polished surface generally requires more passes than does uncoated material. Precut material may also require more passes so that

the rolls can pick up the leading end of each section.

Shape complexity is determined by the number of bends and the total number of degrees that the formed part must be bent. It is also influenced by the symmetry of the part design. The forming angle method is a rule of thumb that roll designers use to determine the approximate number of passes.

On simple shapes, a forming angle of 1 to 2° is recommended. This forming angle is based on the amount of bending performed for every inch of distance between station centers (horizontal center distances). The minimum forming length for a single bend is determined by multiplying the height of the desired section by the cotangent of the forming angle. This length is then divided by the distance between station centers on a given machine to determine the approximate number of passes. For multiple bends, the number of passes must be determined for each bend and then, after the formation of bends have been combined where possible, the approximate number of passes can be determined.

Horizontal Center Distance. If the machine on which the section must be run is predetermined, the specifications and limitations of the machine will have a bearing on the number of passes required. The distance between stations (horizontal center distance) may dictate more stations if that distance is too short. The total distance from flat material to finished section is more critical than the number of stations because undue stresses are created by forming too fast.

Strip Width. The width of strip required to produce a given shape is determined by making a large-scale layout, dividing it into its component straight and curved segments, and totaling the developed width along the neutral axis. The outside profile and the neutral axis of each curved segment can usually be treated as circular arcs. Also, for bends having an inside radius of up to about twice the stock thickness in low-carbon steel, the neutral plane or axis is located approximately one-third of the distance from the inside surface to the outside surface at the bend.

Developed width, sometimes termed *bend allowance,* is the amount of material required to form a curved section of a particular shape properly. The two methods for calculating developed width described in this section use general equations and can be employed for all shapes. The equations given are applicable when low-carbon steel is formed; for less-formable materials, values should be increased.

Method One. Using this method, developed width w is calculated as:

$$w = r\frac{\alpha}{57.3} \qquad \text{(Eq 1)}$$

If the inside bend radius is less than two times the material thickness, then:

$$r = r_i + 0.4t \qquad \text{(Eq 2)}$$

where w is in millimeters (inches), r is the bend radius in millimeters (inches), α is the angle (in degrees) through which the material is bent, r_i is the inside bend radius in millimeters (inches), and t is the metal thickness in millimeters (inches).

If the inside bend radius is greater than $2t$, then:

$$r = r_i + 0.5t \qquad (Eq\ 3)$$

If the material is bent through a 90° zero radius or a 180° zero radius, w is $1/3\,t$ or $2/3\,t$, respectively.

Method Two. Another method used to determine developed width for a roll-formed part is with the empirical equation:

$$w = (t \times p + r_i)0.01745\,\alpha \qquad (Eq\ 4)$$

where w is in millimeters (inches), p is a bend factor based on the ratio of inside bend radius to material thickness expressed as a percentage, and the other quantities are as previously described.

The bend factor p is obtained by first dividing the inside bend radius by the material thickness. After this ratio is obtained, p may be determined using either the nomograph shown in Fig. 4 or the following calculations. For a ratio less than one:

$$p = r_A \times 0.04 + 0.3 \qquad (Eq\ 5)$$

For a ratio greater than one or equal to one:

$$p = (r_A - 1.0)0.6 + 0.34 \qquad (Eq\ 6)$$

where r_A is the inside bend radius divided by the material thickness, r_i/t. If p is calculated to be greater than 45%, the value is 0.45.

Flower Design. The development of the flower—the station-by-station overlay of progressive section contours, starting with the flat strip width before forming and ending with the final desired section profile—is the first step in the design of tooling for contour roll forming. The intermediate profiles between flat material and finished profile are graduated at a rate that enables the section to be completed in the fewest number of stations of passes without compromising general roll-forming parameters. The flower graphically shows the number of passes required to roll form the given profile (Fig. 5).

The two prime considerations in designing the flower are: a smooth flow of material from first to last pass and maximum control over fixed dimensions while roll forming. Other factors to be considered include forming position (a section is usually formed in an upward position), vertical reference line with respect to the number and severity of bends, and drive line (the optimal placement in the rolls for equal top and bottom surface speeds).

Roll Design Parameters. After the flower is completed to the satisfaction of the designer, rolls may be drawn around each overlay. For small sections, the roll material should contact the section material as much as possible. It is possible, however, to go too far and overdesign rolls, thereby creating too much roll contact, which can be detrimental. Each roll pass must

be examined not only by itself but as part of the total job to determine where to make contact with the section material, where to exaggerate pressures or dimensions, where to clear out rolls so that material flows without restriction from one pass to another, and how to accept the section material from the previous configuration.

Rolls are usually made progressively larger in diameter from one pass to the next to permit the surface speed of each succeeding station to increase. The diameter increase is called step-up. The speed differential between passes creates a tension in the section material and eliminates the possibility of an overfeed between passes. Overfeed is created by an excessive amount of work being done in a single pass, which stretches the section material. Normal step-up is approximately 0.8 mm (1/32 in.) per pass on diameter, but it varies depending on the gage of the section and the particular amount of forming being done in those passes.

Rolls may be solid or split (segmented) depending on the complexity of the section. Simple rolls are usually of a one-piece design, but as complexity of the workpiece increases, the use of split rolls should be considered. There are seldom any marked disadvantages in using split rolls, and one or more of the following advantages can often be gained:

- Turning, grinding, or other machining operations are usually easier to perform on the separate sections of split rolls than on one-piece rolls having a complex contour.
- Sections of split rolls are less susceptible to cracking in heat treatment than are single complex rolls.
- Handling problems are simplified, particularly for large roll sections.
- Because sections of rolls subject to excessive wear or breakage can be replaced separately, split rolls can be more economical than one-piece rolls.
- Split rolls permit the use of different roll materials, as needed, for areas of high and low wear.
- Split rolls allow flexibility in making different widths of the same section by the use of spacers or additional roll sections.

- Split rolls allow for minor adjustments that cannot be made with one-piece rolls.

Figure 6 illustrates some of the advantages of split rolls. In Fig. 6(a), the upper roll is composed of five separate sections. In addition to being easier to manufacture, split rolls of this type can accommodate minor adjustments (by shims or similar means) after the initial trial and before the first production run.

Figure 6(b) shows an upper forming roll made up of three sections—a narrow center section flanked by two wider sections. In use, the center section is subjected to a higher rate of wear than are the adjacent flanking sections. The center section of this forming roll can be replaced without changing the flanking sections, or can be made from a more wear-resistant metal.

Roll Materials. The materials that are most commonly used for contour rolls are:

- Low-carbon steel, turned and polished but not hardened
- Gray iron (such as class 30), turned and polished but not hardened
- Low-alloy tool steel (such as O1 or L6), hardened to 60 to 63 HRC and sometimes chromium plated
- High-carbon high-chromium tool steel (such as D2), hardened to 60 to 63 HRC and sometimes chromium plated
- Bronze (usually aluminum bronze)

The quantity of parts to be rolled is usually the major factor in choosing the most appropriate roll material, although other factors, as noted subsequently, also affect selection to some extent, and one or more of them may become definitive in particular applications.

Short-Run Production. For rolling small quantities of a specific shape, or when repeat orders are not expected, rolls made of either

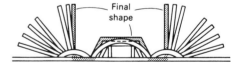

Fig. 5 Development of a flower for design of roll-forming rolls. See text for details.

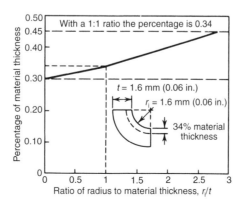

Fig. 4 Nomograph used to determine percentage of material thickness required when calculating bend allowance

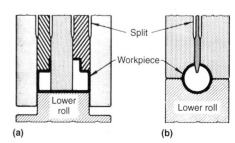

Fig. 6 Two types of split rolls used in contour roll forming. (a) Upper roll constructed in five sections allows for minor adjustments. (b) Three-section upper roll allows replacement of center section.

low-carbon steel or gray iron are commonly used. When the work metal is soft and corner radii are generous, low-carbon steel or gray iron rolls can be used for medium- or high-production runs because rolls are easily made from these materials and are not expensive.

Medium-Run Production. As the size of the production run increases, rolls made of hardened, ground, polished tool steel, such as O1 or L6, are usually more economical and are widely used.

Not only are these materials relatively inexpensive, they machine more easily than more highly alloyed tool steels and they can be heat treated by simple procedures. Rolls made of these steels can be plated with up to 25 μm (1 mil) of chromium to reduce galling or scratching of the work metal, or to extend tool life between grinds by decreasing roll wear or minimizing corrosion or pitting.

Long-Run Production. For long production runs (>5 million ft) or continuous high production, it is usually more economical to make rolls from one of the high-carbon high-chromium tool steels, such as D2. These highly alloyed grades cost nearly twice as much as O1, are more difficult to machine and grind, and require more complex heat treatments, but because of their longer life between regrinds, are usually more economical for long runs.

Factors Other Than Quantity. Three other conditions, under any one of which rolls made of steel such as D2 are often preferred (quantity becoming a secondary consideration), are: excessively hard work metal (low formability), excessively sharp radii or other severe forming conditions, and work metal surfaces that are abrasive (such as unpickled hot-rolled steel) and cause excessive roll wear. Rolls made of highly alloyed tool steel may also be plated with chromium for the reasons mentioned in the section "Medium-Run Production."

Special Finish Requirements. In many applications, such as the rolling of light-gage stainless steel, aluminum alloys, or coated stock, preservation of surface finish is of primary concern. With these work metals, softer rolls are used to avoid damaging the work metal surface, even though there may be a substantial reduction in roll life.

As workpiece surface finish requirements become more rigorous, softer rolls are required. Rolls made of bronze are often used. In some applications, the plating of hardened steel rolls is sufficient to prevent the marring of work metal surface finish.

Tube and Pipe Rolling

Welded-seam pipe and tube are contour roll formed by three methods: edge forming, center forming, and true-radius forming. Edge and center forming (Fig. 7a and b) require complete sets of rolls for each size of tube because the stations that precede welding have the final radii in the roll contours. In true-radius forming

(Fig. 7c), the breakdown rolls can be used for a range of sizes, which reduces tooling cost and setup time.

For all three methods of producing welded tubing, the forming rolls are termed as breakdown and finishing rolls. Breakdown is usually accomplished in the first three or four roll stations, as indicated in Fig. 7.

Optimum distance between the edges of the stock in the final finishing station (stations 6, 7, and 5 in Fig. 7a, b, and c, respectively) is affected by the method of welding and tube size. For resistance welding, however, the following relations of tube size to distance are typical:

Tube diameter		Distance	
mm	in.	mm	in.
9.5	3/8	1.57	0.062
25	1	3.15	0.124
51	2	4.37	0.172
102	4	6.35	0.250

Machine size is determined by material thickness and tubing size. Large pipe sometimes requires a machine having spindles 305 to 355 mm (12 to 14 in.) in diameter.

Speed of production is controlled by thickness and type of work metal. Aluminum can be formed and welded as fast as 75 m/min (250 ft/min), whereas titanium tubing is produced at a rate of only 455 mm/min (18 in./min). The welding operation is often the main limitation on the speed at which tubing can be produced by roll forming.

Soluble oils are the most practical lubricant for forming tube and should be used to prevent galling. A mixture of 25 to 40 parts water to 1 part oil is commonly used.

Long roll life can be expected in producing round tubing. Three to four million feet between regrinds is considered normal.

Lock-seam tube (Fig. 7d) has two edges folded over to form a lock. Production of lock-seam tubing is restricted to relatively thin material (usually less than 0.91 mm, or 0.0359 in.)

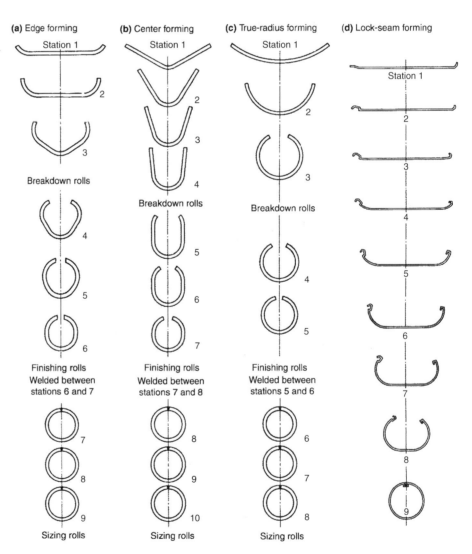

(a) Edge forming   **(b)** Center forming   **(c)** True-radius forming   **(d)** Lock-seam forming

Fig. 7 Typical sequences for forming round pipes and tubes. (a to c) With a butt-welded longitudinal seam. (d) With a lock-seam joint.

because heavier stock is too difficult to lock. The lock-seam method is used extensively for thicknesses that are impractical for welding because they are too thin. Maximum thickness is also restricted by tube diameter. As a general rule, stock thickness should not be greater than 3% of the tube diameter. For instance, 0.76 mm (0.030 in.) is about the maximum thickness of strip that should be used to produce 25 mm (1 in.) outside diameter (OD) tubing.

Minimum width of the lock should be five times the material thickness. The lock-seam method is also applicable to square tubing.

The various stages of lock-seam forming are shown in Fig. 7(d). Idlers are used between stations 4 to 8. Between stations 8 and 9, a lock housing unit that is equivalent to two stations is used. Both side rolls and vertical rolls are used in the locking stand, the main purpose of which is to lock the two edges of work metal in the groove. A mandrel is placed inside the tube to help form the small lock. This mandrel is in the locking stand and extends beyond station 9. In the last station, two small rolls ride opposite each other on the mandrel, pushing the work metal to form a tight lock between the top and bottom rolls in station 9. It is necessary to have close control over the stock thickness, because a 0.025 mm (0.001 in.) difference in thickness will result in a difference of about 0.13 mm (0.005 in.) in the outside diameter of the tube.

Roll Design for Tube Rolling

A number of designs and methods are used for forming strip into a tubular shape suitable for welding, with many factors involved in choosing the proper roll design for producing a particular tube.

Figure 8 illustrates one of the most commonly used designs for rolls used for forming the tube before welding. The rolls are designed with a single forming radius in each roll pass. This radius decreases progressively in each roll pass until the final pass. The radius of the final roll pass is slightly larger than the finished tube size to permit the insertion of a thin fin in the top roll to act as a guide for the two edges of the material. Generally, for the last two or three roll passes, there are fin rolls in the top rolls to guide the two edges of the strip, prevent twisting of the tube, and ensure accurate positioning of the seam entering the welder. Idler rolls mounted on vertical spindles between the driven-roll passes are positioned to prevent excessive rubbing and scuffing of the side of the tube as it passes through the succeeding driven-roll pass. The number of driven-roll passes can vary, increasing as the tube diameter increases, but five driven passes are considered a minimum.

For forming tube with either a very thin wall or a very thick wall (<3% of the OD of the tube, or >10% of the OD of the tube), a modification of the design shown in Fig. 8 is used. This modification is obtained by forming the portion of the strip adjacent to the edges to the finished radius of the tube in the first forming pass, instead of depending on the fin passes to finish form at the edges. This finished form at the edges of the strip helps to prevent buckles and waviness at the edges of the strip as it passes through the forming rolls when very thin material is being formed, and it helps to avoid the necessity of extreme pressures at the fin rolls when extra-heavy-gage material is being formed. Figure 9 illustrates the first four sets of rolls used in this method of tube forming. The remaining rolls are similar to those shown in Fig. 8.

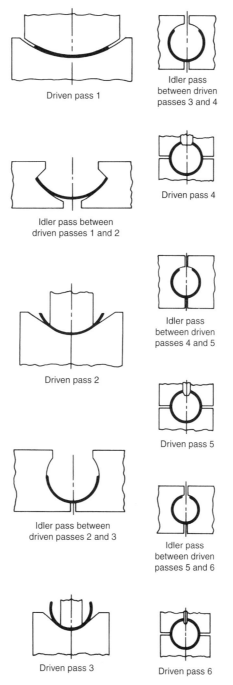

Fig. 8 Typical sequences for roll forming tubing from strip material.

In forming material of high tensile strength when springback of the metal is a factor, a third method of forming is sometimes used. In this design, a part of the strip is formed to the finished tube radius in each roll pass, progressing from the two edges toward the middle until the fin passes are reached. The forming radius on the rolls is less than the finished radius of the tube to compensate for the springback of the material.

Welding

Seam welding of pipe and tubing is generally performed using either low-frequency rotary-electrode welding or high-frequency welding; laser welding is beginning to be used (see the section "Laser Welding"). High-frequency

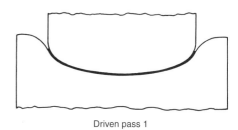

Driven pass 1

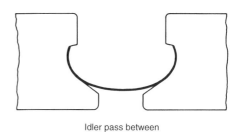

Idler pass between driven passes 1 and 2

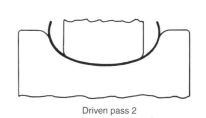

Driven pass 2

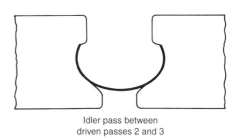

Idler pass between driven passes 2 and 3

Fig. 9 First group of rolls for forming tubing from thin strip material from the edges inward.

resistance welding and high-frequency induction welding for tube diameters less than 25 mm (1 in.) (Fig. 10) have become more predominant in recent years in the production of welded tubing (see the article "High Frequency Welding" in *Welding, Brazing, and Soldering*, Volume 6 of *ASM Handbook*). With rotary-electrode welding, the maximum wall thickness that is economically feasible is 4.5 mm (0.180 in.). Using high-frequency welders, wall thicknesses as thick as 19 mm (0.75 in.) and as thin as 0.13 mm (0.005 in.) are obtainable. After the welding operation, the piece is usually sized and then straightened before being cut to length.

Laser Welding. Powerful carbon dioxide (CO_2) lasers also have been used to make longitudinal welds in contour roll-formed stainless steel tube. In this process autogenous (no filler metal) welds were made in types 304 and 430 stainless steel tubes using a 5 kW CO_2 laser at a speed of 5 m/min (16.5 ft/min). Tubes had a wall thickness of 1.5 mm (0.060 in.) and a diameter of 48.5 mm (1.91 in.). Laser welding resulted in joints that were tougher and more ductile than joints made by other welding processes.

Sizing and Straightening. Tube and pipe are welded to an OD slightly larger than the finished diameter. The sizing rolls can then produce round, accurately dimensioned, straight, finished tube. A typical set of sizing and straightening rolls consists of three driven-roll

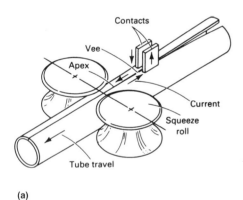

(a)

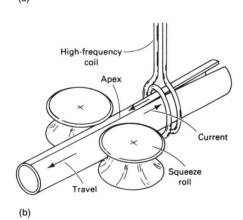

(b)

Fig. 10 Two methods of high-frequency welding of longitudinal seams in tubing. (a) Sliding contacts introduce current to the tube edges. (b) Multiturn induction coil induces current to the tube edges.

passes, vertically mounted idler rolls between each driven pass, and finally, a set of idle cluster rolls that are adjustable both vertically and horizontally for final straightening of the tube. The roll radius for each driven roll successively decreases to size the tube down to its proper diameter.

Reshaping of Round Tubing

Several cross sections that can be feasibly produced by reshaping round tubing are shown in Fig. 11. Reshaping can be done either continuously in sequence with the production of round tubing or in a separate operation.

On square and rectangular tubing, the flatness of the sides will vary with work metal thickness and hardness. On light gages and in the harder tempers, springback results in a crown effect on the sides. This condition can be corrected by overforming in the final station. In forming rectangles, the longer sides may become convex, and the short sides may become concave. Flatness can be controlled to a minor extent by roll adjustments in the finish station, but a major correction must be made by using concave or convex contours in final Turk's-head rolls.

Rolling speeds used in reshaping tubing of light-gage metal are often equal to those used in producing the basic round tube. For reshaping heavier-gage tubing (for example, over 0.89 mm, or 0.035 in.), speed should be reduced by as much as 25% of that used for forming the round tube, depending mainly on the capacity of the equipment.

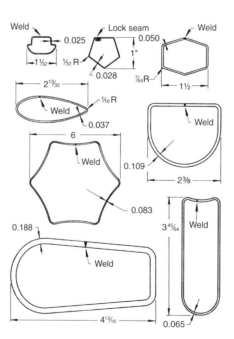

Fig. 11 Cross sections showing typical contours that can be produced by reshaping welded or lock-seam round tubing. In addition to these contours, square, triangular, and rectangular shapes can be formed from round tubing in one or more roll stations. Dimensions given in inches

Tooling requirements for reshaping round tubing vary with the gage, size, and complexity of the final shape. It is often possible to form simple squares from thin-gage round tubing in one roll station and one Turk's-head station. However, as stock thickness or complexity of shape, or both, increase, more stations are required.

Tolerances

Cross-sectional tolerances on part dimensions are a result of variations in material width and thickness, physical properties of the material, quality of the tooling, conditions of the machine, and operator skill. Dimensional cross-sectional tolerances of ±0.25 to ±0.78 mm (0.010 to 0.030 in.) and angular tolerances of ±1° are common. Tolerances are slightly greater when wide building panels and deep sections are being formed. If a closer tolerance is required, material with a controlled-thickness tolerance of ±0.05 mm (0.002 in.) should be used.

Length tolerances are dependent on material thickness, part length, line speed, equipment quality and condition, and type of measuring and cutoff system used. For thin material (0.38 to 0.64 mm, or 0.015 to 0.025 in., thick), tolerances of ±0.51 to ±2.36 mm (0.020 to 0.093 in.) are obtainable. For material more than 0.64 mm (0.025 in.) thick, tolerances of ±0.38 to ±1.52 mm (0.015 to 0.060 in.) are obtainable. The minimum tolerances are based on part lengths up to 915 mm (36 in.), and the maximum tolerances are based on lengths up to 3.66 m (12 ft). Tolerances would generally be greater on parts longer than those specified.

In roll forming, it is generally advisable to order the material to be formed with somewhat tighter than commercial quality tolerances. If this is done, a great many dimensional problems can be eliminated. Failure to consider material quality results in needless problems and frustrations.

Straightness

In addition to cross-sectional, angular, and length tolerances, another tolerance to consider is the straightness of the material and the formed section. Some of the parameters that determine straightness include camber, curve or sweep, bow, and twist. The terms camber, curve, and bow are often used synonymously when describing straightness. The horizontal and vertical planes of the formed part are determined by the position in which the part is formed.

Camber (Fig. 12a) is the deviation of a side edge from a straight line. Measured prior to roll forming, the maximum allowable camber is 3.2 mm/m (3/8 in. in 10 ft.). Excessive camber contributes to curve, bow, and twist in the finished part.

Curve or sweep (Fig. 12b) is the deviation from a straight line in the horizontal plane measured after the part has formed. The curve in a formed part can be held to within ± 1 mm/m ($\pm 1/8$ in. in 10 ft). Curve or sweep can result from incorrect horizontal roll alignment and uneven forming pressure in a pair of rolls.

Bow (Fig. 12c) is the deviation from a straight line in the vertical plane and can be either cross bow or longitudinal bow. Bow results from uneven vertical gaps on symmetrical sections and from uneven forming areas on unsymmetrical sections. Generally, bow can be held to within ± 1 mm/m ($\pm 1/8$ in. in 10 ft).

Twist in a formed part resembles a corkscrew effect and often results from excessive forming pressure. Twist is generally held to less than $5°$ in 3 m (10 ft).

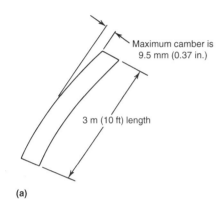

(a)

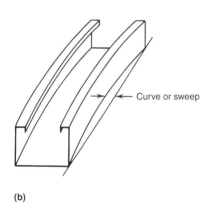

(b)

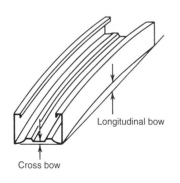

(c)

Fig. 12 Straightness parameters for roll formed parts. (a) Camber. (b) Curve or sweep. (c) Bow. See text for details.

Quality and Accuracy

Two factors that can affect the quality and accuracy of a roll-formed section are springback and end flare.

Springback is a distortion that becomes evident after the straining of the part has been discontinued. The amount of springback varies with different metals properties such as yield and elastic modulus. Springback can be compensated for in the tool design by overforming. Overforming forms the material past its expected final shape.

End flare is the distortion that appears at the ends of a roll-formed part. The internal stresses incurred in roll forming are much more complex than in other types of bending. These stresses are usually higher in the edges of the material being formed and are released when the part is cut off.

Control of End Flare. End flare occurs to some degree in nearly all roll-formed shapes. It can extend back from the end as little as 50 or 75 mm (2 or 3 in.) or as much as 305 mm (12 in.). End flare can range from a few thousandths of an inch (this small amount is usually ignored) to 12.7 mm ($1/2$ in.) or more. The possibilities for flare can pose difficult problems. Both ends may flare outward, or one end may flare outward and the other inward. Excessive flare is caused by hard work metal, workpiece shape, too few roll stations, unsuitable roll design, or a combination of all of these.

There are several means of keeping end flare within acceptable limits. Replacing a hard work metal with one that is more formable will lessen, but not always eliminate, end flare. Minor changes in shape or design of the workpiece are often feasible and should be considered when end flare is a problem. It is much better to recognize the possibility of excessive flare before designing the rolls than to attempt correction after the rolls are made. The best way to minimize end flare is to provide enough roll stations to prevent the sides or edges of the work metal strip from elongating.

Surface Finish

Contour roll forming seldom improves the initial finish of the work metal. One exception is the roll forming of unpickled hot-finished steel, from which much scale is removed. It is usually possible to preserve the existing finish on unformed areas of the work metal. Sheet and strip ranging from unpickled hot-rolled steel to highly polished stainless steel are contour roll formed with a minimum of damage to the surface finish. In addition, work metals having almost every known type of coating are contour roll formed in high production with no damage to the coatings.

This does not mean that no damage to the work metal will occur if the specific applications are not carefully considered in planning the processing technique. In addition to normal precautionary measures, such as keeping the work metal clean before rolling and maintaining the equipment properly, one or more of the following must be considered and possibly adjusted when minimum damage to the surface finish is a primary requirement:

- Roll design or number of rolls in a given station
- Number of stations
- Roll material and finish
- Lubricant
- Rolling speed

As severity of forming increases, the possibility of damage to the work metal surface increases and may require alteration of the rolls within a station. For example, in forming deep channels, the shape can be produced using top and bottom rolls. Side rolls can be added to improve dimensional tolerance. When maintenance of surface finish is a problem, the use of side rolls is helpful because it minimizes roll sweep, which is inevitable when the shape is produced solely by top and bottom rolls. Excessive roll sweep is likely to damage both the work metal and the rolls.

Although sliding friction caused by roll sweep may damage surface finish, there is even greater likelihood of damage from excessive forming pressure in a given roll station. Therefore, as severity of forming increases, the possibility of damage to the work metal finish can be lessened by adding stations, thus decreasing the amount of forming done by a given set of rolls and reducing forming pressures.

Roll material and roll finish also contribute to the surface finish obtained in contour roll forming. Chromium-plated steel rolls or aluminum bronze rolls are best for preserving work metal finish.

Lubrication is preferred in contour roll forming and has a significant effect on work metal finish. When lubricants cannot be tolerated, as in the roll forming of coated metals, more attention must be given to roll design, additional stations, roll materials, and possibly lower rolling speeds than would be used if copious amounts of lubricant were permitted.

Each metal presents a different problem in maintaining surface finish.

Hot-rolled unpickled steel seldom offers any problem in maintaining surface finish. Rolling removes much of the scale and usually improves the finish, provided a flood of lubricant is used to flush away the scale. Otherwise, this scale will be trapped between work metal and rolls, resulting in damage to both surfaces. Rolls made from abrasion-resistant tool steel such as D2 are especially recommended for roll forming of hot-rolled unpickled steel.

Cold-finished carbon steel, aluminum, and brass are usually rolled with a minimum of damage to work metal finish. One or more of the foregoing conditions may require special attention, depending mainly on severity of forming. A flood of lubricant is desirable in roll forming cold-finished metals.

Highly polished stainless steel or aluminum can also be contour roll formed without damage to surfaces. However, each step of the procedure becomes more critical than is the case with roll forming lower-quality finishes. Greater attention must be given to roll design, fitting, and maintenance. Chromium-plated rolls are usually preferred when work metal finishes are critical. Maximum cleanliness in all phases of the operation (including the use of freshly cleaned work metal) is mandatory for achieving desired results. Special lubricants are preferred for roll forming stainless steel and may be essential when forming is severe and quality of finish is critical.

Galvanized Steel. Success in roll forming hot dip galvanized steel depends mainly on the quality of the zinc coating, maintenance of the rolls, and lubrication. Inferior galvanizing or severe bends, or both, cause the coating to loosen and stick to the rolls. Wipers that contact the working surfaces of the rolls will aid in preventing surface damage. Chromium-plated rolls are also helpful in minimizing damage to galvanized work metal.

Precoated metals (vinyl and other organic coatings) must be rolled without lubricant and sometimes pose problems, although by paying careful attention to the conditions listed previously, precoated metals can be contour roll formed without damage to the coating surface. One of the most common applications is the forming of aluminum siding for buildings. Complete preservation of finish depends mainly on severity of forming. Sometimes it is necessary to increase radii if the particular operation is to be successful.

Embossed metals are also roll formed without lubricants. Shapes are designed to avoid excessive forming pressure, and bend radii not less than twice the metal thickness are used to prevent distortion of the embossing. Forming of aluminum eaves troughs is an example of this operation.

Use of Computers

Computers are becoming an important aid in the design of roll-forming tooling. Consistency, accuracy, and speed enable the designer to determine the optimal design for each roll pass in less time than is required when the calculations are performed by hand. The capability to display the profile of the part enables the designer to see how the material flows through each pass. This profile enables the designer to determine whether too much work is being performed at a particular pass.

The numerical information compiled by the computer can be employed in numerically controlled machining operations to ensure that rolls are accurately produced. The computer also aids in the setup of the rolls on the machine by specifying the size and locations of the required shims and spacers. All this information and data can be stored for future use and reproduced whenever necessary.

To design the rolls, information about the roll forming machine, the cross section of the final shape, and the initial forming sequence are entered in the computer program. The computer numerically defines the coordinates of each corner and displays on the computer terminal the profile of the part at each pass as well as various perspectives of the flower diagram. Input changes can be made to vary the material flow through the roll forming machine so that optimum flow is achieved. Computer output includes flower diagrams, drawings of the cross-sectional shape, drawings of the rolls, and tabular data defining the material and rolls. A computer can also produce the tapes used in the manufacturing of the rolls on numerically controlled machines.

High-Velocity Metal Forming

Glenn S. Daehn, The Ohio State University

HAMMERS were the first tool used to shape billets into useful shapes, but fixed tools, such as stamping dies, became the standard for metal forming during the industrial revolution. Indeed, fixed tools are highly productive, with forces and displacements remotely controlled. However, fixed tools do not offer the local control that was provided by hammer forming. Even today, difficult components in prototype production and aerospace manufacture are commonly produced with the aid of manually operated hammers. High-velocity metal forming techniques are more akin to hammers than presswork, because the impulse is controlled rather than forces or displacements. Also like hammer work, when properly implemented, one can control the spatial distribution of the impulse.

High-velocity forming methods include techniques such as explosive forming and electromagnetic forming. These techniques are distinct from most other metal forming methods in that the explosive or electromagnetic force first accelerates the workpiece to a high velocity, and the kinetic energy of the workpiece is significant. The sheet metal workpiece then changes shape, either as it strikes a die or as it is decelerated by plastic deformation. These methods generally provide robust methods of performing metal forming operations that are quite difficult conventionally. Although these methods have been known for over 100 years and saw significant development (particularly in the 1960s), these methods have not been developed or documented so that they can be routinely used to their potential. However, because these methods work well on even hard-to-form materials, and because manufacturing systems are generally very simple and can be established quickly, there is a recent resurgence in interest in these methods.

This article first provides some general background and emphasis on the traits that are common to high-velocity forming operations (considered here to be operations where the workpiece velocity is above approximately 50 m/s, or 165 ft/s). This overview includes a description of general principles on how metal forming is accomplished and analyzed when inertial forces are large. Second, the principal methods of high-velocity forming are described. This includes the methods of explosive forming, electrohydraulic forming, and electromagnetic forming. These are all similar in the respect that the workpiece is given kinetic energy early in the process and where forming is largely inertial because kinetic energy is dissipated as plastic deformation. For each of the forming methods, the physics is quite different in determining what the as-launched velocity profile of the workpiece will be. Finally, this article gives examples showing how these techniques can be practically applied. The article concludes with some comments on the status and development potential for this technology.

Background

High-velocity forming methods were discovered in the late 1800s and saw some application in forming thick plates in the 1930s. Between approximately 1950 and the early 1970s, the United States government funded numerous studies on the development and application of high-velocity forming. The U.S. Defense Advanced Research Projects Agency (DARPA) invested heavily in explosive forming technology in the mid-1960s. DARPA (Ref 1) cites explosive forming as "a cost-effective process for forming a variety of metals and alloys that results in remarkably high reproducibility (~0.5%) for complex, large metal structures. Used extensively in Department of Defense projects, the applications include making afterburner rings for the SR-71, jet engine diffusers, Titan 'manhole' covers, rocket engine seals, P-3 Orion aircraft skin, tactical missile domes, jet engine sound suppressors, and heat shields for turbine engines." Also in the same period, significant research investments were made in the allied technologies of electromagnetic forming and electrohydraulic forming. These methods are explained later and are based on storing the forming energy in a capacitor bank instead of as chemical energy. Electrohydraulic forming is, in operation, quite similar to explosive forming but limited to lower energies. Electromagnetic forming is only directly applicable to materials with high electrical conductivity, and this has been well developed as an assembly technique. Explosive, electrohydraulic, and electro-

magnetic forming have all seen continued use since their development in the 1960s. However, there has been only sporadic research or development in these areas since the early 1970s. It is rather unfortunate that while the old literature shows that these techniques have significant potential, in many cases the information available is not detailed enough to design new processes.

Fundamentals of High-Velocity Deformation

High-velocity forming is characterized by first imparting a high velocity to a workpiece, and this energy is turned to plastic deformation by constraint of the part or impact with a die. The velocity distribution for the part is determined by the pressure distribution or developed from the explosive, electromagnetic pressure distribution (determined by coil shape) or shock wave profile. Changing the amount of energy available will change the absolute values of the velocities. Once the velocity is imparted, the shape can be developed either by free forming or die forming. Examples of these are shown in Fig. 1. Free forming can be used to create bulges that take on roughly spherical section geometries. Because there is no direct die contact, these methods do not have great dimensional precision. This method has been used, for example, to form the ends of pressure vessels from a monolithic metal plate. When better dimensional precision is required, forming into a die is used. In die forming, one must often pull a rough vacuum between the sheet and die before forming, or the rapidly compressed air will cause the part to rebound. An appropriate array of holes can also permit air movement in forming. Also, if the workpiece strikes the die at too high a velocity, it may rebound (i.e., bounce off). This will also impair dimensional tolerance.

Note that these techniques are distinctly different from conventional stamping in that only a single tool is typically used. In this respect, the techniques can be compared to methods such as superplastic forming or hydroforming. However, high-velocity techniques can have significant advantages over both. Unlike superplastic forming, the process is typically carried out at room temperature, and it is a robust forming technique for almost any alloy, whereas superplastic forming can only be applied to a very limited set of materials at low strain rates. In comparison to hydroforming, because large forces and pressures in high-velocity forming only act for a short period of time, the inertia of tools is often more than sufficient to constrain them, and very light tooling systems can be used. This avoids the high die, equipment, and setup cost associated with hydroforming. However, high-velocity forming can be well suited to forming large and complex parts, resulting in large cost savings versus assemblies.

Basic Elements. There are a number of basic technical elements that distinguish high-velocity forming from conventional stamping. These must be appreciated if the techniques are to be used effectively. Important issues include understanding how to account for inertia, impact, and the new loading modes available. These are described briefly, and their direct application to metal forming is shown later.

Inertia—Resistance to Further Velocity Change. Newton's law, $\vec{F} = m\,\vec{a}$, can be summarized as saying a body (or a material point in a deforming body) prefers to maintain its present velocity. If one piece of material is to change its velocity by some amount, the force required to do so is inversely proportional to the time available to effect that change. That is to say, after

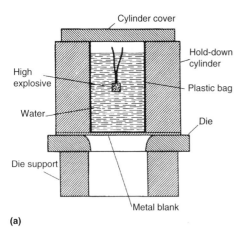

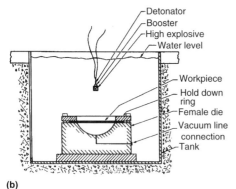

Fig. 1 Example of (a) free forming and (b) die forming as accomplished with explosive forming. Source: Ref 2

a workpiece has been launched in short time periods (or at high velocity), bodies very much prefer to maintain their launch velocity profile. A basic difference between conventional forming and high-velocity forming is that in the former, use inertial forces can be ignored, while they are dominant in the latter case. As is described later, this has important implications to both formability and wrinkling.

Impact—High Pressures Easily Created. Something fairly unique to high-velocity forming is that very high pressures are created when two solid bodies impact with significant velocity. A detailed understanding of this requires the study of wave behavior in solids and is beyond the scope of this article, but Ref 3 gives an excellent treatment of the background. One useful result is the impact pressure, P, that is developed when two semi-infinite elastic bodies, labeled "1" and "2," collide at an impact velocity, V_i:

$$P = \frac{\rho_1 \rho_2 C_1 C_2}{\rho_1 C_1 + \rho_2 C_2} V_i \qquad \text{(Eq 1)}$$

where for each material, ρ represents density, and C is the longitudinal wave speed, which can be expressed as:

$$C = \sqrt{\frac{3K(1-\nu)}{\rho(1+\nu)}} \qquad \text{(Eq 2)}$$

where K represents bulk modulus, and ν is Poisson's ratio. Longitudinal wave speeds are on the order of approximately 7000 m/s (23,000 ft/s) for most structural metals. Equation 1 shows that for aluminum-steel and steel-steel couples, impact pressures of 500 MPa (73 ksi) and 1.4 GPa (0.20×10^6 psi) are generated for a 50 m/s (165 ft/s) impact. This analysis shows that it is very easy to develop pressures large enough to produce plastic deformation at the interface. More sophisticated analyses are required to study the deformation of the interface once plastic deformation takes place. However, it is clear that even modest impact speeds can produce significant plastic deformation at the interface. Such pressures can be used to put great surface detail into material or to perform embossing or coining-like operations. Also, explosives themselves can provide enormous pressures that can be used practically. Common explosives can produce pressures on the order of 20 GPa (3×10^6 psi). Here, impact is not required to generate huge pressures.

Changes in Constitutive Behavior. High-velocity deformation may or may not involve high strain rates. It is well known that the fundamental constitutive behavior (stress, strain, strain-rate relations) for most metals changes qualitatively at strain rates above approximately $10^4\ \text{s}^{-1}$. Above these strain rates, the apparent strain-rate sensitivity of the material increases markedly. This may actually be related to the increase of the material rate of strain hardening instead. However, many high-velocity forming operations do not result in high strain-rate effects. For example, if one is to consider tube

expansion at 200 m/s (655 ft/s) a tube diameter of 4 cm (1.6 in.) will sustain a strain rate of $10^4\ \text{s}^{-1}$ in the hoop direction. However, a 1 m (3.3 ft) diameter tube will only have a hoop strain rate of $400\ \text{s}^{-1}$ at the same expansion velocity. Thus, while these changes in constitutive behavior can be important, they are not universal to all forming operations, even if they are carried out at high velocity.

New Loading Modes Possible. The methods used to impart velocity to a sheet are quite distinct from those traditionally used, in that launch does not usually involve contact with a hard punch or tool. Specifically, in the case of electromagnetic forming, body forces are produced by the interactions of magnetic fields without any contact. As is shown later, these forming devices can be integrated with conventional tooling, greatly expanding the capabilities of traditional stamping. Also, these unusual modes of imparting energy to a body, along with the short loading duration, offer the ability to use very lightweight and simple forming systems.

Characteristics of High-Velocity Metal Forming

The basic elements (large inertial forces, impact forces, changes in constitutive behavior, and new loading modes) offer many useful attributes to metal forming. These attributes can be designed into a number of new or enhanced manufacturing processes. The characteristics of these elements are described in this section, and examples follow at the end of this article.

Improved Formability. In conventional stamping, formability, as described by the set of available strain states that can be attained, is described by the forming limit diagram (see the article "Formability Testing of Sheet Metals" in this Volume). This forming limit diagram does not include inertial effects and therefore cannot describe formability in high-velocity forming; inertia and specific boundary conditions can help or hurt formability. Different issues control formability, and the forming limit diagram no longer approximately describes formability, as a few brief examples illustrate.

First, consider a simple tensile test. As is standard, one end of the sample is held fixed and the other is moved with some velocity. W.W. Wood (Ref 4) carried out a series of experiments where the end velocity was varied by both conventional testing and impact experiments. He studied several materials and consistently found that there were three distinct regions in the ductility-velocity plot. Below a first critical velocity, strain to failure was independent of endpoint velocity. Between a first and second critical velocity, strain to failure was significantly increased, and beyond this second critical velocity, strain to failure dropped to nearly zero. These results are summarized in Fig. 2.

Below the first critical velocity, the sample behaves in the ordinary way, where ductility is not a strong function of strain rate. Above the second critical velocity, the velocity imposed to the endpoint is greater than the effective plastic wave speed. Therefore, significant deformation is deposited in the driven end of the sample before the stress is fully transmitted to the fixed end of the sample. This results in the sample pinching off near the driven end. This second critical velocity is known as the von Karman velocity, and this is related to the plastic wave speed in the material (which is much lower than the elastic wave speed, because both are related to the slope of the stress-strain curve in the respective elastic and plastic regimes). This second critical velocity is between approximately 30 and 150 m/s (100 and 490 ft/s) for most important engineering alloys. Between these two critical velocities, a period of enhanced ductility is expected and observed.

The reduced ductility at high velocities observed in uniaxial tensile deformation can be attributed to the strongly nonuniform velocity distribution initially delivered to the sample. Initially, one end moves at a high velocity while the other is fixed. This problem can be addressed by using a different launch profile or different boundary conditions. In Fig. 3, electromagnetic forming was used to launch an aluminum ring in an axisymmetric radially outward manner. Now, if only inertial forces were active, every material point along the ring would move along the radial outward path that it was launched along. However, in simple tension, the usual instability present in tensile test eventually takes over, and

the material will eventually fail. However, at high velocity, the ductility in a slender ring can be doubled versus its quasi-static value, and upon failure, it will break into a number of fragments. Increasing the height of a launched tube or stopping it from high velocity by impact can be used to develop even greater ductility, as detailed elsewhere (Ref 7).

Figure 4 shows a comparison between forming 2024-T4 aluminum into a conical die using quasi-static fluid pressure versus at high velocity using a shock wave developed through electrohydraulic forming. In quasi-static forming, the strains achieved are in good accord with the conventional forming limit diagram. However, when electrohydraulic forming was used, extension strains in excess of 100% were seen (near plane strain and without signs of failure) in

the upper section of the cone. Similar experiments were carried out with copper and iron alloys with similar results. In the case seen in Fig. 4, the strain to failure of the part was increased by a factor greater than 4. Such dramatic increases in formability cannot be explained using the inertial stabilization models used in the discussion of Fig. 2 and 3. The plastic deformation generated when the sheet strikes the die is also important. This will develop a through-thickness compression that deforms the sheet in a manner similar to ironing. This technique is more general than ironing, however, because it is not very restricted with respect to configurations that can be used. As a general summary, in high-velocity forming, one is not constrained by the usual limits prescribed by the forming limit diagram. Instead, if launch and

Quasi-static

Peak ductility

High velocity

(a)

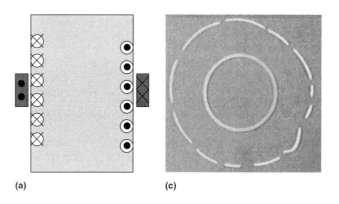

(b)

Fig. 2 (a) Schematic of the tensile test geometry and failure morphology. Right side of sample is driven and failure produced at driven end at high velocity. (b) Plot of experimentally observed strain to failure as a function of velocity with predicted behavior. Source: Ref 5

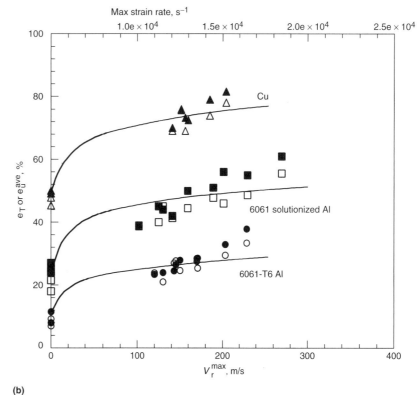

(b)

Fig. 3 Electromagnetic forming of an aluminum ring. (a) Schematic of ring launch. (b) Ductility as a function of launch velocity. (c) Photograph of the original ring geometry and fragmentation after a total strain of 45%. Source: Ref 6

Fig. 4 Comparison of two methods of forming 2024-T4 aluminum into a conical die. (a) Schematic of the setup used in hydroforming and electrohydraulic forming. (b) Comparison of forming 2024-T4 sheet aluminum into a conical die using a hydroforming process (left) and using high-velocity deformation developed by electrohydraulic forming (right). Plane-strain extensions over 100% are seen in the electrohydraulically formed sample. Source: Ref 8

boundary conditions are properly chosen, ductility far beyond typical quasi-static ductility can be achieved.

More Uniform Strain Distribution in a Single Operation. In a practical forming operation, the ability to make a part is not only governed by the formability of a material (which is related to the magnitudes of the strains available without failure), but the ability to uniformly distribute strain is also crucial. The total increase in part surface area is related to the total average amount of stretching over the part. In conventional stamping, friction causes the strain distribution to become quite nonuniform. This problem is particularly important in high-strength materials that have little strain hardening. In punchless forming operations (such as hydroforming, gas pressure forming, or high-velocity forming), the strain distributions can be much more uniform. This will facilitate forming many common component geometries, because strains are more uniformly distributed.

Reduced Wrinkling. Inertial forces can also be used to effectively resist wrinkling. Consider a simple example of a ring being compressed radially inward. If this is done at low velocity, the tube will collapse similar to a drinking straw under suction. Under conditions where inertial forces dominate and the tube is launched uniformly, each material point will want to travel radially inward, keeping the circular cross section at each time. Figure 5 demonstrates this schematically and with actual images from ring compression. These results show that inertia strongly suppresses wrinkling in ring compression. Experiments in tube and sheet flanging (Ref 9) show that inertial forces can be similarly effective in inhibiting wrinkling in sheet forming.

Coining and Detonography. The very large pressures available from impact or directly from explosives make possible several things that are difficult or impossible with conventional forming. First, by producing a sufficiently high impact velocity, large pressures (Eq 1) can be developed over a large surface. Usually, if fine surface detail is to be captured in a metal surface, it is done with the high-pressure forging process known as coining. Coining pressures usually operate only on small areas (such as coins).

High-velocity forming techniques offer the ability to perform coining-like operations over large areas. An example of this effect appears in the next section.

Lightweight Tooling and Equipment. Although it may seem initially counterintuitive, very light tooling and equipment can be used in high-velocity forming operations. This can be demonstrated with the following simple but quantitative example. Imagine one would like to emboss a 1 m (3.3 ft) square aluminum sheet with exceptional surface detail. The traditional solution to this problem would be to use a coining technique. Coining simply uses a large static pressure to embed the shape in the plate. Typical coining pressures for moderate-strength aluminum are approximately 300 MPa (45 ksi). Thus, a press force of approximately $300{\times}10^6$ (67.5 million pounds) would be required to do this. This requires a very large press, and this is a key reason coining is only carried out on small items. As discussed earlier (Eq 1 and 2), if the aluminum sheet were accelerated to 50 m/s (165 ft/s) and impacted into a steel die, appropriate pressures can be developed to coin the sheet. Now, imagine specifically, a 1 mm (0.04 in.) thick sheet of aluminum is to be coined, and the steel die is 5 cm (2 in.) thick. By conservation of momentum, if the steel die was completely unrestrained, after impact it would develop a velocity of only 0.34 m/s (1.1 ft/s). A very light framework can easily contain this and dissipate the kinetic energy. In general, because with dynamic and impact events the force will dissipate quickly, much lighter equipment can be used than will be required for static forces.

This potential to reduce equipment size by using dynamic phenomena is true for both simple sheet forming and fluid-pressure forming, such as hydroforming, where again, large pressures acting over large areas require the use of very large presses. Shock wave forming can permit the use of much smaller systems, because the inertia of the forming system itself is primarily responsible for limiting the motion of the machine.

Natural Interference Fits and Reduced Springback. The large impact pressure typically generated in high-velocity deformation generally causes sheet metal formed at high

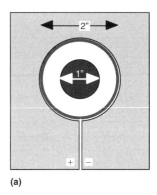

(a)

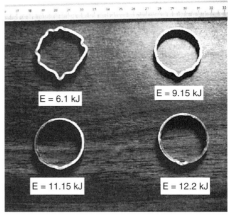

(b)

Fig. 5 Example of electromagnetic high-velocity crimping of thin aluminum tube onto a mandrel. (a) The coil is made from a 1.32 cm (0.52 in.) thick 6061-T6 plate. (b) Behavior of initial 1.651 mm (0.065 in.) thick, 5.08 cm (2.0 in.) outside diameter 6061-T6 rings compressed onto a 3.175 cm (1.25 in.) diameter mandrel at varied indicated launch energies. Source: Ref 9

velocity to closely conform to the shape of the die. One of the primary reasons for this is that on impact with a rigid tool, a plastic wave front will propagate through the sheet thickness; this will cause the residual elastic strains in the sheet to be minimized (these elastic strains cause distortion and springback). Second, on impact, the large pressures will cause the struck surface to displace, and the rebound wave will tend to lock, producing an interference fit between the sheet component and the die.

There are two important practical implications of this. First, it has been demonstrated that high-velocity forming, which induces impact with a die, can be used for calibration, meaning it can be used to make a sheet workpiece very nearly dimensionally conform with a die surface, even if the contour is so shallow that that shape would normally only introduce elastic strains into the part (and very significant springback) (Ref 10). Second, this can be used to join a tube to a cylinder (Ref 11). This can be accomplished either by expanding into a hollow cylinder or by collapsing onto a cylinder. In either case, natural interference fits between the driven and static tube can be routinely achieved. A simple example of this is shown in Fig. 6. Here, a simple 32 mm (1.25 in.) diameter tube is electromagnetically compressed onto two ordinary hex nuts using two electromagnetic impulses. If the crimping energy is chosen appropriately, a simple crimped joint (i.e., with no metallurgical or chemical bonding) has a strength that is approximately 90% of that of the parent tube. With appropriate fixturing, such joints can have excellent dimensional and alignment precision. They also do not suffer from heat-affected zones or thermal distortion. A more advanced example of this is shown in the examples portion of this article.

Impact Welding. The large impact pressures possible in high-velocity forming can also cause extensive plastic deformation near the interface between two mating materials. This can cause solid-state welding to take place between the impacting materials. This type of joining can have several significant advantages relative to traditional fusion welding. Most notably, this is a solid-state process, which is very quick and takes place at relatively low temperatures; dissimilar metals can be joined without the formation of deleterious intermetallic phases. This allows

Fig. 6 Example of a 6061-T6 32 mm (1.25 in.) diameter, 1 mm (0.04 in.) wall thickness tube electromagnetically crimped onto four nuts using four electromagnetic discharges and a coil similar to that shown in Fig. 5(a). The nuts are tight after forming and after destructive testing.

joining pairs such as aluminum and steel that cannot be joined by fusion techniques, because the iron aluminides that would form are unacceptably brittle.

Several studies have shown that impact welding can be carried out for a range of geometries, including concentric tube cladding, lap welding, and welding tube ends into plates. By far, the most common practical implementation of this is in plate cladding. A detailed discussion of these techniques is beyond the scope of this article, but the basic techniques have been discussed in the literature (Ref 2, 12). Generally, to create a good weld, the surfaces must impact at a relatively high velocity (typically greater that approximately 300 m/s, or 985 ft/s), and they must impact at an appropriate angle. Under these conditions, there is a jetting phenomenon that takes place that cleans nascent oxides off the mating surfaces, and this allows solid-state bonding without extensive chemical reaction. Designing an impact welding process typically involves selecting the standoff between the mating surfaces so that impact takes place, meeting the critical impact velocity and angle. Recently, a technique has been demonstrated for impact welding aluminum, using a projectile and a shaped die, where standoff between the mating sheets is not required (Ref 13). Again, pressure with extensive plastic deformation allows clean metal from the two sheets to meet, because the oxide is disrupted.

Enhanced Powder Consolidation. There is fairly extensive literature on the shock wave compaction of both metal and ceramic powders; good summaries are provided in Ref 14 and 15. The fundamentals of the process are quite different than quasi-static compaction. First, the compact is not in static equilibrium; instead, a shock wave moves through the body, producing a distinct boundary between the consolidated and unconsolidated regions. The mass and force distribution that results can often make adding lubricants to the powders less important or unnecessary. Also, the rapid increase in pressure and rapid energy dissipation can cause local heating at particle boundaries. This can soften or often melt a small volume of the powder and can aid consolidation. Lastly, like the examples discussed previously, fairly lightweight compaction systems can produce very high pressures. For all of these reasons, higher density can be developed in dynamic compaction than can easily be achieved with traditional quasi-static compaction. Therefore, sintering steps are sometimes not required to reach target densities. This enables the fabrication of materials such as metal-polymer composites, which are useful as magnetic materials, because they do not effectively transmit eddy currents. The traditional challenge in dynamic powder compaction is avoiding cracks that can be the result of tensile waves, which are the natural consequence of impact operations. Some recent examples show, however, that dynamic powder compaction can be quite effective on relatively complex parts, such as large gears (Ref 16).

High-Velocity Methods

The old literature on high-velocity forming (Ref 17) documents that a wide range of techniques were used to perform several variants of high-velocity forming or high-energy-rate forming. These included pneumatically driven hammers, exploding gases, and burning of propellants, among others. Of these, there are two that are still used fairly regularly: explosive forming and electromagnetic forming. A third, electrohydraulic forming, has some use but only sporadically and in a few locations. This method essentially combines aspects of explosive and electromagnetic forming. This review focuses on explosive and electromagnetic forming but also briefly discusses electrohydraulic forming and other techniques.

Explosive Forming

There are a couple of fairly detailed books that have been written on explosive forming and the industrial applications of explosives (Ref 2, 18). These are now fairly old, and while the essential technology of explosive forming has changed little, many analytical tools are now available that were not at the time those books were written. There is also newer literature in impact engineering (Ref 19) and dynamic materials behavior (Ref 20) that is useful but less directly applicable to explosive metal forming. A review of the technology was also recently written (Ref 21).

Technology Overview. In order to understand explosive forming, one must first understand some of the essentials of explosive technology. Detailed information on explosives technology can be obtained through the International Society of Explosives Engineers (Ref 22) and their comprehensive background materials, such as *Blasters' Handbook* (Ref 23). There are two general categories of explosives: high and low explosives. Low explosives are substances such as gunpowder that basically burn, generating heat and hot gas. The velocity of the reaction front is much less than the speed of sound. Thus, unless these substances are contained, a shock wave impulse is not obtained. However, they contain enough energy that small masses of gunpowder can do significant amounts of useful work (for example, propelling a bullet down a rifle barrel). High explosives have detonation velocities on the order of 3500 to 8000 m/s (11,500 to 26,000 ft/s) and energies on the order of 1 MJ/kg, as summarized in Table 1. High explosives are available in a wide variety of product forms and compositions, such as granules, powders, shapeable putty, hot-castable compounds, and detonation cord. Detonation cord is especially useful in metal forming, because it allows one to easily prescribe a line source for the explosive event, and it allows the fairly precise handling of the quantities of explosives that are used in metal forming (which are usually much less than those used in typical mining or military applications). Detonation cord is usually specified in terms of grains (one grain is 1/7000 pound) of explosive per foot of cord, which usually ranges between 10 and 500 grains per foot or 2 to 100 grams per meter.

There are also two types of high explosives: primary and secondary. Primary explosives are relatively easy to detonate and are used to initiate the explosion, usually by means of a blasting-cap configuration. Secondary explosives provide the vast majority of the energy. These are engineered such that they are very difficult to detonate with incidental impact, spark, or flame. Ideally, only a sharp shock wave from something like a blasting cap can detonate them. By properly engineering detonators and secondary high explosives, the explosives industry has developed an excellent safety record, and there are no fundamental reasons explosive metal forming cannot be carried out in a safe and routine way.

In principle, explosive metalworking can be quite inexpensive. A few cents' worth of explosives can perform all but the biggest explosive forming experiments. However, finding a facility where explosive forming work can be carried out is often problematic. The U.S. Department of Transportation governs the transportation of explosive materials. The driver must have a commercial driver's license with a hazardous materials certification. The other main requirement is that explosives be stored in a safe and secure manner. The Bureau of Alcohol, Tobacco, and Firearms (ATF) governs this, and their regulations are summarized in ATF pamphlet 5400.7. Two trade organizations, the Institute of Makers of Explosives (Ref 24) and the International Society of Explosives Engineers (Ref 22), can also be useful sources of information on the use and regulation of explosives.

Methods of Explosive Forming. As the voluminous older literature on explosive forming shows, there are a variety of ways that explosives can be used to form or shape metal. There are two types of classifications that can be used to broadly distinguish one kind of process from another. The first is defined by the spatial relationship between the explosive and the workpiece. When very high velocities or pressures are needed, the explosive may be placed directly (or with a thin protective layer) on the workpiece. This is classified as a contact operation. One of the more common contact operations is in the explosive cladding of materials via solid-state impact welding. Standoff operations

are more common in explosive sheet metal forming. Here, the explosive charge is separated from the workpiece by a substantial distance (on the order of the part size). To improve the system efficiency, a transfer medium couples the charge and workpiece. Water is typically used; however, sand and other substances have been used in special cases. The other way to broadly classify forming operations is related to the way the explosion is contained. Most commonly, explosive systems are open. The workpiece, die, and explosive are positioned in what resembles a modified swimming pool. The major modification that is typically made versus a swimming pool is that an array of air bubbles may be generated to blanket the wall of the pool. The purpose of this is to keep the explosive shock from damaging the tank wall. This open approach is quite general, so that almost any explosive/die and workpiece arrangement can be immersed in the tank and formed explosively. One shortcoming of this open technique, however, is that significant energy may be lost in kinetic agitation of the water bath. This can be remedied by using a confined or closed system. Here, the water is largely contained in a strong and rigid enclosure. While this approach has better efficiency and produces less noise, it must be designed carefully in order to achieve adequate die life. Reference 18 has example calculations on sizing dies for confined explosive forming. Lastly, very simple setups can also be used. For example, a plastic garbage bag affixed to a simple support and filled with water lying over the metal to be formed can also act as a bath to contain the explosive charge and comply to the workpiece. This versatility and simplicity in setup is one of the most attractive features of explosive forming.

Typical explosive forming operations are summarized here.

Sheet and Tube Forming. Explosive forming has been commonly used to form relatively thin sheet metal in standoff operations. If explosives are to be used in contact, something similar to a rubber sheet is often required to protect the sheet. The sheet can either be deformed into a cavity or formed over a male die. Figures 1 and 7 show some examples of sheet forming operations carried out by explosive forming. The main advantage of this technique over more conventional techniques is that one-sided dies are used. This reduces the cost of the forming system as well as the lead time needed to create it. This can

be especially important for large components. Also, extended formability can be observed relative to conventional forming methods. In addition, unlike methods such as superplastic forming, ordinary materials can be used.

Another attractive feature of explosive sheet metal forming is the versatility in die-making procedures. For modest pressures and for relatively small numbers of parts, castable die materials (such as kirksite, fiberglass, urethane, and concrete) can be used. The great advantage of this is that models cut from wood or Styrofoam (Dow Chemical Co.) or created by solid free-form fabrication can be used as the pattern, and a working die can be created very quickly. As with all other die technologies, for higher pressures and longer runs of parts, harder die materials, such as hardened steel, are required.

Explosive forming can be used to develop very high forming pressures (again, often high enough to form tight external radii or emboss the surface). This could possibly be used to augment hydroforming, or, in short run production, this could replace tube hydroforming. Also, because the shock wave strikes the entire surface to be formed at essentially the same time, a single impulse can be used to cut out a series of holes or cutouts in a single operation. Reference 18 provides an analysis and experimental data on such operations. Figure 8 shows an example of a tube that received many perforations as the result of a single explosive impulse.

Plate Forming. Explosive forming represents one of the only affordable methods of fabricating large sections from thick plates. Thus, it has often been used to create components such as bulkheads for large pressure vessels and sections of ships and large vehicles. The method has also been used to form several large nuclear reactor components. Contact forming is often used in these situations, because even the availability of an appropriate pool may be problematic. Concrete dies are often used to set the stage.

Dimensional Tolerance. Like any manufacturing process, the dimensional tolerance or reproducibility one can obtain from the process is very dependent on the care used in the operation. However, in general, explosive forming operations can have significantly reduced springback as compared to conventional operations (Ref 25). There are a couple of prominent reasons for this. First, the relatively high forming pressures can cause dies to deform somewhat elastically in a manner that opposes springback. Also, the high pressures developed on impact can produce plastic deformation through the thickness of the sheet. This can permit relaxation of the elastic strains that are the root cause of springback distortion. Generally, tolerances on the order of ± 0.025 mm (0.001 in.) have been obtained in sheet and plate forming of fairly large structures. Small-diameter tubes have also been explosively formed to tolerances of ± 0.025 mm (0.001 in.).

Joining and Welding. Explosives are commonly used to drive joining applications based either on explosive welding or cladding. This is

Table 1 Common explosives and their typical properties

Explosive	Form of charge	Deformation velocity		Energy, J/g	Storage life	Pressure	
		m/s	ft/s			GPa	10^6 psi
2,4,6-trinitrotoluene (TNT)	Cast	7010.4	23,000	780	Moderate	16.536	2.40
Cyclotrimethylene trinitramine (RDX)	Pressure granules	8382	27,500	1265	Very good	23.426	3.40
Pentaerythrite (PETN)	Pressure granules	8290.56	27,200	1300	Excellent	22.048	3.20
Pentolite (50/50)	Cast	7620	25,000	945	Good	19.92	2.89
Smokeless powder	Powder	<1	<3.3	300	Excellent	0.35	0.05

treated well elsewhere (Ref 12). Also, the natural interference fits formed on impact can be useful in joining operations. One example where welding and/or mechanical crimping can be used to form joints is in the explosive expansion of tubes into tubeplates, as takes place in heat exchangers. The British company TEI Ltd. (Ref 26) advertises the use of explosives for bonding of tubes to tubeplates in heat exchangers, expanding tubes into structural supports, as well as expanding sleeves into corroded or worn tubes for life extension. They also use the same basic explosive technology for the removal of adherent scales in power plant applications.

Electromagnetic Forming

Basic Physics. Electromagnetic forming is a noncontact technique where large forces can be imparted to any electrically conductive workpiece by a pure electromagnetic interaction. The physics of this interaction are covered fairly well in several sources (Ref 14, 17, 27, 28); only a short description is included here. A schematic diagram is provided in Fig. 9. The process is driven by the primary circuit, labeled "1." A significant amount of energy (usually between 5 and 200 kJ) is stored in a large capacitor or bank of capacitors by charging to a high voltage (usually between 3000 and 30,000 V). The charge is switched over low-inductance conductive buswork through a coil or actuator. Large currents run through the coil. The currents take the form of a damped sine wave and can be understood as a ringing inductance-resistance-capacitance (LRC) circuit. The peak current is typically between approximately 10^4 to 10^6 amperes, and the time to peak current is on the order of tens of microseconds. This creates an extremely strong transient magnetic field in the vicinity of the coil. The magnetic field induces eddy currents in any conductive materials nearby in much the way the primary circuit of a transformer induces voltage and current in the secondary. Hence, any metallic workpiece nearby will have currents induced, and these will generally be opposite in direction to the primary current. The opposed fields in the coil and workpiece set up an electromagnetic repulsion between the coil and workpiece. That electromagnetic force can produce stresses in the workpiece that are several times larger than the material flow stress. Ultimately, this can cause the workpiece to deform plastically and to be accelerated at velocities exceeding 100 m/s (330 ft/s). The key governing equations for this process are also summarized in Fig. 9.

Developing a relatively high system ringing frequency is often quite important. If the electrical oscillation frequency is too low, intense eddy currents are not induced in the workpiece, and the force developed is low. The ringing frequency is directly related to the circuit LRC characteristics, and low capacitance and low inductance favor a high ringing frequency. Materials of lower conductivity demand higher ringing frequency for effective forming. For these reasons, metals with high conductivity, such as aluminum and copper, are very well

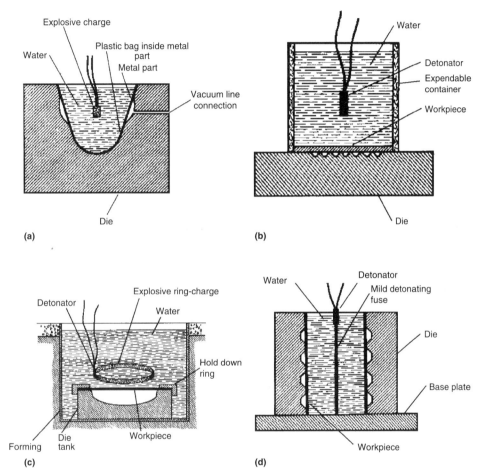

Fig. 7 Schematic examples of typical explosive forming operations. (a) Sizing with a water-filled die cavity. (b) Method for forming a flat panel. (c) Use of detonation cord to prescribe the pressure distribution in an open forming system. (d) Use of detonation cord to form a cylinder. Open- or closed-die systems can be used. Source: Ref 2

Fig. 8 Example of a 4130 cylinder given multiple perforations in a single explosive operation. Source: Ref 17

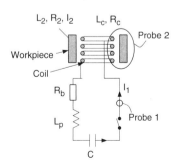

$$\frac{d}{dt}(L_1I_1 + MI_2) + R_1I_1 + \frac{Q_1}{C_1} = 0$$

$$\frac{d}{dt}(L_2I_2 + MI_1) + R_2I_2 = 0$$

$$P_m = \frac{1}{A}\frac{dM}{dh}I_1I_2$$

where $L_1 = L_p + L_c$, $R_1 = R_b + R_c$

Fig. 9 Schematic of the general electromagnetic forming process with key equations

suited to electromagnetic forming. Carbon steel can be formed, but one should pay attention to the system ringing frequency. Metals with relatively low electrical conductivity, such as titanium and austenitic stainless steels, are almost impossible to directly form by electromagnetic forming (but they can be formed with the aid of a more conductive driver plate).

Historically, the workpiece has usually been a tubular metal piece that is either expanded or compressed using a helical electromagnetic coil. These applications are particularly efficient and easy to model. However, the process is actually quite versatile, and the workpiece may be cylindrical, a flat sheet, or an arbitrarily curved surface. In making an actuator that would act against an arbitrarily curved surface, one would like to place a forming coil in close proximity to the sheet. A continuous sheet will generally find a way to support eddy currents to develop mutual repulsion. A complete calculation of the spatial and temporal distribution of force is difficult, particularly when the sheet is moving and deforming while the electromagnetic force is being applied. However, it is useful to develop a quantitative estimate of the electromagnetic pressures present. An upper bound on force on the sheet can be roughly calculated with the following assumptions and principles:

- The electromagnetic force will always act normal to the plane of the sheet. It can be seen as being driven by the magnetic flux between the actuator and workpiece wanting to expand to fill space.
- If both elements are highly conductive and close enough to each other to be electromagnetically well coupled, and the actuator width is large relative to its gap, it is appropriate to make the assumption that the current in the workpiece is equal and opposite to the current in the actuator.

Under this upper bound assumption, the magnetic pressure, P_m, in the region between the actuator and conductor can be approximated as:

$$P_m = \frac{\mu_o H^2}{2} \qquad \text{(Eq 3)}$$

where H is the electromagnetic field intensity, expressed in amperes per width. Thus, decreasing the width of a conductor or increasing the number of turns can both powerfully increase the magnetic pressure. A version of this equation that is often simpler to use states that magnetic pressure in megapascal units can be related to the current density in the conductor (in units of KA/cm or KA·turns/cm) as:

$$P_m = 0.0063 \ H^2 \qquad \text{(Eq 4)}$$

Actually, if the current densities in the primary and secondary conductor are known by estimate or measurement, H^2 can be replaced with the actual product of these current densities. The equation will be accurate so long as the gap between the current-carrying elements is small with respect to their in-plane dimensions. The

peak current generated by a capacitor bank discharge, I_{max}, can be estimated from standard LRC equations for the primary circuit. As long as the circuit resistance is low, this can be estimated as:

$$I_{max} = V_o \sqrt{\frac{C}{L}} \qquad \text{(Eq 5)}$$

where V_o is the voltage the capacitor bank is initially charged to, and L is the total system inductance. At a fixed primary circuit current, the magnetic pressure (electromagnetic field intensity) is increased by increasing the number of windings in the forming coil or solenoid, but this will increase the inductance of the system. This in turn reduces the peak current. The result is that for any situation, there are an optimal number of coil windings. As a general rule for efficient forming, most of the total system inductance should reside in the forming coil. There is an engineering trade-off, however; as the number of turns increases, the inductance of the system rises, reducing ringing frequency. This reduces the peak current and will allow more flux to leak through the workpiece.

The correspondence between electromagnetic field intensity and magnetic pressure can be exploited in the design of electromagnetic forming coils. For example, Fig. 10 shows a simple coil created by machining a thick plate. In regions where the current paths are closely packed, there will be a much higher pressure than in regions where the coil has thick, sparse conductors. This ultimately determines the launch velocity distribution for the sheet and the free-form shape the sheet metal shape would take.

Equipment. An electromagnetic forming system has two essential electrical components—a capacitor bank and a forming coil—that do work on a workpiece to be formed. Also, some tool or die system that produces the final part shape may also be important. There are two opposing philosophies that one may follow in developing an electromagnetic forming system. One could develop a general-purpose setup that is useful, but not optimal, for a range of metal forming activities. Or, a system may be designed that is optimal for a particular operation. In this latter case, one can be much more efficient (developing equivalent metal forming with less stored electrical energy). These approaches become clear when system details are discussed.

Capacitor Banks. The heart of an electromagnetic (or electrohydraulic) forming system is a capacitor bank. The essential components are a number of capacitors that store energy, E (typically in the range of 5 to 200 kJ), by storing electrical charge at voltages, V, between 3 and 30 kV. The associated bank capacitance, C, can be determined from:

$$E = \frac{CV^2}{2} \qquad \text{(Eq 6)}$$

The energy is provided to the capacitors by a charging system. A transformer steps the voltage up from line voltage to that required by the

capacitor bank. The capacity of the transformer will largely control the time required to charge the bank between discharge events. In repeated use, active cooling of the elements that dissipate energy will limit the ultimate cycle period for such a bank.

The capacitors are tied to the current output with buswork. Two considerations are key in the design of the buswork: it must have low inductance (in general, the majority of the system inductance should be at the forming coil), and long-lasting spark-free electrical contacts are required.

In use, the capacitors are relatively slowly charged to the forming voltage, and once this is reached, a fast-action switch is used to provide the current pulse to the coil. Several types of switches have been used, including rat-trap-type switches and fairly simple spark gaps. Commercial capacitor banks typically have either mercury-filled ignitron-type switches or solid-state silicon controlled rectifier switches. The former perform well but can be temperamental and have a somewhat limited life, while the latter can be inefficient if not designed well, and it can be expensive to equip the bank with the number of switches required to handle the large currents seen in electromagnetic forming.

In a simple LRC-type circuit, the current will ring with a damped sinusoidal current-time profile. However, a high-capacity diode is often put in parallel with the circuit. This allows only the first half-cycle of the sine wave to be output. Because most of the useful work is done in this first half-cycle, this has little downside and has the benefit of adding life to the capacitors as well as limiting damage and sparking at contacts. The disadvantage of a clamped (diode-including) capacitor bank circuit is that if too large a current runs through the diodes, they will burn out. Such events may be rare in a steady-state manufacturing environment, but shorts or low-inductance coils may be common in research and development applications.

Fig. 10 Example of a coil designed for forming flat sheet where the electromagnetic field intensity is intentionally varied from place to place. In this case, the coil is simply machined from a high-strength aluminum plate. Much higher magnetic pressure is developed in the bottom half of the coil than in the top.

Coils. There are several different philosophies on how electromagnetic forming actuators can be fabricated, and different approaches have different optimal capacitor bank system characteristics associated with them.

Single Turn and Machined Actuators. The simplest way to make an electromagnetic compression coil is simply to cut a slot and hole in a conductive plate, as shown in Fig. 5(a). High material conductivity will improve system efficiency, and high material yield strength provides a strong, robust, long-life coil. High-strength aluminum alloys, copper-beryllium alloys, brasses, and oxide-dispersion-strengthened materials such as Glidcop (OMG Americas Corp.) are all good candidate coil materials. Polyamide films such as Kapton (E.I. du Pont de Nemours & Co.) can provide insulation in a convenient way.

This basic approach of machining coils from relatively thick, flat plate can also be used to form effective actuators for the electromagnetic forming of flat or curved sheet. Figure 10 shows an example of such a coil. As discussed previously, the shape of the coil determines the pressure distribution and launch velocity distribution for the sheet.

Multiturn Coils. Often, electromagnetic forming actuators are fabricated by the simple coiling of conductive wire (often common copper magnet wire) over a strong nonconductive mandrel such as grade G-10 laminated phenolic composite (or a single-use coil can be created by using only the wire itself). Reinforced epoxy or urethane is often used as a structural and insulating overlayer. Compression, expansion, or flat coils can all be made following this same general approach. There are two important limitations of this type of coil. First, the forces on the workpiece and coil are equal and opposite (in addition to other magnetic interactions within the coil). Therefore, the coil construction materials determine the pressure the coil can withstand. For this reason, coils of this construction are usually limited to pressures on the order of 48 MPa (7000 psi), if they are to be used for many

operations. Second, it is usually difficult to use a very fine winding pitch to increase local field intensity. This also limits the local pressures that can be generated from this kind of coil. Both of these issues can be partly treated in short-run production by using inexpensive coils that are essentially disposable.

Field Shaper. In general, the highest electromagnetic pressures can be generated when the working surface of the coil is made from a monolithic block of a high-strength (and high-conductivity) metal. Hence, large pressures can often be created with single-turn coils, but because single-turn coils typically have quite low inductance, they are often quite inefficient. Field-shaper-based coils can be used to develop high electromagnetic pressure while being able to increase and tailor the inductance of the coil. The principle of a field shaper coil is illustrated in Fig. 11(a) (b) and (c). A typical multiturn coil is used to create a magnetic field. This is coupled to a secondary inductor or field shaper. As schematically described in Fig. 11, if properly designed, essentially the entire current flux (amp · turns) that is created in the primary current can be transferred to the bore of the shaper. The advantage of a coil of this type is that it can concentrate large magnetic pressures in a desired area. This is done at the cost of being less efficient than a well-designed single-stage coil. Figure 11(c) shows a production coil with field shaper. Notice that the primary conductor, which is connected to the high-voltage side of the capacitor bank, is embedded in a secondary coil core. To enhance safety in a production environment, the core is electrically isolated from the primary winding. It is only electromagnetically coupled to it. It is also liquid cooled. The current is then transferred from the primary to the coil core secondary and then to the field shaper. This type of coil will accept field shapers with varied forming diameter, which makes it extremely versatile. Thus, a single coil with a number of varied inserts can be applied to a wide range of compressive forming operations. This basic approach has been creatively applied to making

coils of many types for many operations (Ref 29). Another common application of the field-shaper concept is shown in Fig. 12, which shows a common wafer coil. Coils of this type can generate very high electromagnetic pressures (up to approximately 350 MPa, or 50 ksi) repeatedly without failure.

Typical Applications. While electromagnetic force can be used in very general and creative ways to form conductive sheet metal, to date it has been almost exclusively used to form crimped assemblies that resemble those shown in Fig. 6. Such assemblies can be designed to optimize axial or torsional strength, and in either case, the joint strength can often exceed the strength of the parent tube. Despite this, most of the assemblies that have been produced do not require high mechanical performance. Instead, the high reproducibility and dimensional tolerance afforded by the process are more typically the reasons for choosing electromagnetic crimping. The technology has also found application in the aircraft maintenance industry. Magnetic hammers are used (Ref 30), and a somewhat more complex application of this has been developed to create electromagnetic dent-pullers (Ref 31).

Electrohydraulic Forming and Other Methods

In many regards, electrohydraulic forming is a hybrid between explosive forming and electromagnetic forming. Here, an intense liquid-based shock wave is produced by an intense electrical current vaporizing a small volume of liquid in a spark gap between two electrodes. The current pulse is produced using essentially the same equipment (capacitor banks) that produces the currents in electromagnetic forming. However, the liquid-based shock wave that results is very similar to what would be produced by an explosive point or line charge. The energy efficiency of this technique is similar to that for electromagnetic forming.

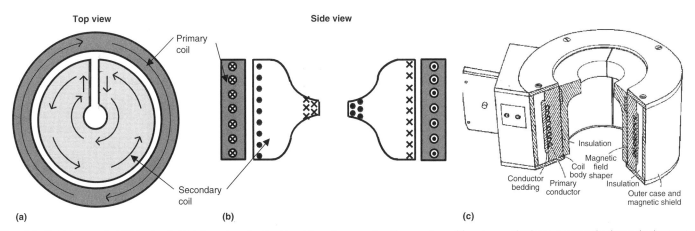

Fig. 11 Field shaper coils. (a) Behavior of a field-shaper-based coil looking at the coil cross section. The many turns of the primary coil induce a current in the shaper. (b) The situation looking down the axis of the shaper. A slot in the coil forces the current to the bore of the coil, concentrating current there. (c) Practical construction of a field-shaper-based coil

Electrohydraulic forming is not widely practiced; however, a few companies do use it for short-run production (Ref 32). It is better controlled than explosive forming and therefore more amenable for operation in usual factory conditions. However, it is difficult to access the same quantities of forming energy that explosive forming can. Reference 17 contains a good review of this technique. The use of electrohydraulic forming is described in Example 4 in this article.

In addition to electrohydraulic and explosive forming, other methods for developing a pressure pulse in water have been considered. For example, Ref 33 examines methods of water-hammer forming. Also, for forming small features, water-pulse methods and electromagnetic methods may be quite inefficient. In these cases, it is possible to use a deformable projectile traveling at a high velocity to accomplish metal forming. This can quite easily develop high strain rates and pressures over a small area. This technique has been demonstrated for solid-state welding (Ref 13) and is an excellent candidate technique so long as the area to be formed is smaller than the projectile.

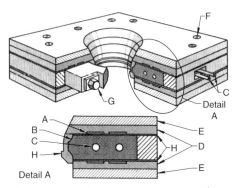

Fig. 12 Heavy-duty wafer-type compression forming coil, which is also based on complied primary and secondary coils. A, primary conductor; B, beryllium-copper field shaper; C, water passage; D, fiberglass insulation; E, steel backup plates; F, press bolt; G, shaper press bolt; H, shaper insulation

Hybrid Methods

There are a number of ways that the impulse-based metal forming methods discussed in this article can be used on their own to accomplish many metalworking tasks. These methods are not commonly practiced, but they are effective and robust. Traditional displacement-controlled forming is usually chosen because these methods are well developed, understood, and (except for springback) the part shape is definitely determined by the forced conformation with a hard tool. Impulse-based methods can offer improved strain distribution and formability, springback reduction, and the ability to add crisper part detail. In many instances, it may be best to use impulse-based techniques only in regions where these special advantages are required. The following brief discussion is broken into two subsections. The first deals with augmenting the matched-die systems that are at the heart of mass production stamping. The second section addresses using one-sided dies as are used in hydroforming, which dominates aircraft production and is used in other low-volume processes such as prototyping. In both cases, the general goal is to get the best from established methods while treating difficult part features by using impulse-based metal forming methods where needed.

Hard Tool. Electromagnetic forming coils can be rather simply embedded into traditional stamping tools (Ref 34). These actuators essentially add new degrees of freedom to the forming process, allowing operations to be completed in a single press tool that would typically require multistage forming operations. Figure 13 shows two different approaches for integrating electromagnetic forming tools with stamping systems. Figure 13(a) shows a system where traditional tooling is used to make the bulk of the shape, but discrete features and corner radii, for example, require greater formability than the material can offer. These features can be created using a single electromagnetic pulse that can be applied at the end of the forming cycle. This will create the relatively small but difficult feature. This basic approach has been demonstrated in the context of forming an automotive door inner from aluminum (Ref 35) and is briefly summarized in the examples part of this article. Typically, aluminum does not have sufficient formability to form a door inner, and an electromagnetic actuator was used to put a sharp corner in a panel that had a large radius of curvature put in by traditional forming.

The approach of using a single electromagnetic discharge to form a feature (Fig. 13a) has two drawbacks. First, quite high magnetic flux densities are required to give sufficient electromagnetic force to create the feature in a single discharge. This makes the development of long-life actuators a challenging issue. Second, in making features by a single high-velocity impulse, die bounce-off at areas of high strike velocity can make it difficult to maintain dimensional tolerance. These drawbacks can be avoided by using a number of electromagnetic pulses in a hybrid tool of the type shown in Fig. 13(b). The key idea here is that the capacitor bank is optimized for quickly producing a large number of small electromagnetic pulses, and it is coupled to coils in regions of the tool where increased strains are required in order to make a part. The setup is shown schematically in Fig. 13(b). Typically, in forming such a part, the sheet will neck and tear just outside the punch perimeter. The sheet can be stretched in regions 2 and 3 to add strains that can provide increased line length to provide further punch displacement. Also, actuators 1b and 1a can be sequentially discharged in order to facilitate sheet draw-in into the die. This can be efficiently carried out by using many (5 to 50) small discharges to strain the sheet just a few percent with each discharge and by moving the tool with each cycle to re-establish contact between the sheet and actuator. This avoids the challenging issues associated with handling the large and potentially destructive energies associated with forming with a single discharge designed to develop large strains. The hard die set controls part shape, and

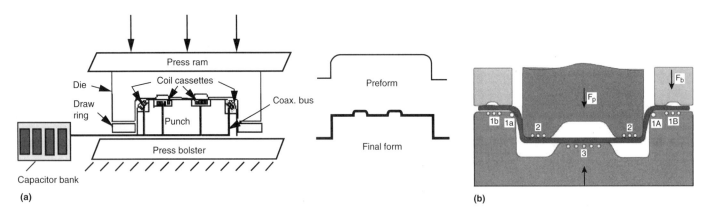

Fig. 13 Two concepts of how electromagnetic forming coils can be used profitably in press systems. (a) Single significant discharges can be used to sharpen or add features that would cause failure of the part if they were produced in the initial stamping operation. Increases in formability and improved strain distributions aid the situation. (b) Schematic demonstration of a press where a large number of small discharges can be used to aid draw-in (region 1) and to produce strain in desirable areas (region 2). After strain or displacement, the punch is moved to re-establish close proximity between the sheet and coil.

the relatively small pulses required to produce only a few percent strain with each impulse are not damaging on the electromagnetic coils.

Single-Sided Tool. In applications where on the order of 100 or fewer parts of a particular shape are required each year, the design and manufacture of the matching tools that are used in stamping operations is cost-prohibitive. As a result, in the aircraft and aerospace industries, single-sided (or so-called soft) tools are commonly used. A large fraction of doubly-curved parts used in aircraft production are created by rubber pad hydroforming. In this process, a large rubber mat is subjected to fluid pressure on one side, and this is used to drive sheet metal into a single-sided die. Often, these dies can be made from materials that are inexpensive to shape, such as masonite and kirksite. There are two persistent problems associated with hydroforming. First, although the strain distribution can be improved relative to double-sided hard tooling (because frictional forces are generally reduced), material forming limits often limit the ability to make a given part in a single operation. The usual approach for treating the problem associated with the material forming limit is to only partially form the part (by using a pressure that is only a fraction of that needed to completely form the part), remove the part from the die, wash off all lubricants, anneal the part to soften it, re-lubricate it, and place it in the die where a larger pressure is now applied. Many such cycles may be used to fabricate the part. The other problem in forming with soft tooling is springback. Usually, the tool is cut to the required contour, and springback is treated by manual deformation back to shape with soft hammers.

Two hybrid concepts that can aid the situation with soft tools are shown in Fig. 14. Figure 14(a) shows the use of an electrohydraulically or explosively generated shock wave that can be used to form features that may need improved definition (such as sharper corners) or improved formability (such as recessed regions in otherwise flat panels). There are significant advantages of such an approach. First, all of the issues discussed with respect to formability, reduced springback, improved part definition, and blanking of features can be accessed with such a method. Second, the approach may ultimately be less expensive with respect to both equipment cost and cost per article. This is because the peak pressure in a hydroforming process determines the sharpness of features that can be formed. High static pressures require large equipment and longer cycle times. If the high pressure is obtained by an explosive or electrohydraulic impulse, the cycle time could be much shorter and delivered from smaller and less expensive pumps and enclosures.

Figure 14(b) demonstrates another concept that could be useful with soft tooling. It is closely related to the approach discussed in Fig. 13(b). Electromagnetic forming coils can be embedded into a hydroforming die, as shown in Fig. 14(b). The electromagnetic repulsive pressure between the coil and sheet can be several times that of the fluid pressure. This would cause the sheet in the proximity of the coil to move away and stretch in this region. Once the magnetic pressure naturally dissipates, the fluid pressure will naturally move the sheet back to the die, but it will have stretched in the area of the electromagnetic coil. This will cause controlled stretching and motion of the sheet. The important element in designing systems of this type is to place the coils either in regions that see insufficient stretch (such as regions 1 to 3 in Fig. 14b) or in areas that can encourage the part to draw in at a fixed fluid pressure. If uniform strain can be encouraged in this way, the multiple cycles of forming, cleaning, annealing, and reforming could be reduced or avoided.

Forming Examples

Example 1: Electromagnetic Forming and Piercing. Figure 15 shows an example of how versatile the process is and how, with some ingenuity, very simple forming operations can be developed to form complex parts. The tubular part shown in Fig. 15 was produced in one electromagnetic forming operation using an expansion coil to form, flange, and pierce a length of tubing against a single-piece die. The workpiece was a 100 mm (4 in.) length of 6061-O tubing with an 83 mm (3.25 in.) outside diameter and an 0.89 mm (0.035 in.) wall thickness.

The die was turned from 4340 steel tubing with a 12.7 mm (0.5 in.) wall thickness. Two drilled holes are used for the piercing operation. The formed part was removed from the die by means of a simple ejector. The parts used a 6 kJ impulse for forming, and production rates of 240 pieces per hour could be achieved using only manual loading and unloading.

Example 2: Aluminum Automotive Door Inner Panel. In collaboration with the automakers, The Ohio State University carried out a demonstration of electromagnetically assisted stamping that conforms generally to the hard tool hybrid forming approach discussed earlier and shown in Fig. 13(a). The experimental approach and results (detailed in Ref 35) involved a trial of electromagnetically assisted stamping to allow forming of aluminum door inners. Aluminum door inners can provide weight savings (about 6 kg per door), but the formability of aluminum is poor relative to steel. Without substantial changes in the product geometry, the door inner is not even close to fabricable using traditional press operations.

The hybrid electromagnetic-assisted stamping process was used to treat the GM J-Car hinge face. The concept was to soften the corner (Fig. 16) on the hinge face of the door in such a manner that the door was formable and then use an electromagnetic actuator to reform the door panel into the required shape. This was carried out in two discrete operations, but the concept can be carried out in one press stand.

Despite the successes, this experiment points to reasons that the approach of using many small impulses to form the material may be superior. In this example, a forming energy of approximately 36 kJ was required to change the shape of the preform to the final shape. This produced currents near 500 kA in the coil. It is presently difficult to design, build, and predict remaining

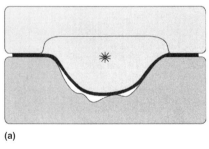

(a)

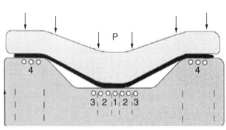

(b)

Fig. 14 Hybrid methods for use with soft tools. (a) Use of an electrohydraulic impulse in conjunction with sheet hydroforming. Hydroform forms the basic shape, while the shock wave imparts detail. (b) Use of electromagnetic coils to influence the strain distribution in electromagnetic augmented hydroforming

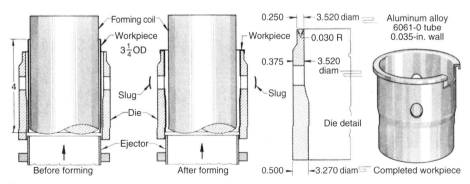

Fig. 15 Forming and piercing of a tubular part in one operation

lifetime for coils that are so heavily mechanically and electrically stressed. Another problem with the approach is that the final shape of the part is not directly controlled. In the experiment, the final shape was very close to that required. However, a restrike operation may be required to meet required tolerance.

Example 3: Hybrid Incremental Forming. The concept of using a series of small electromagnetic discharges to alter the strain distribution in sheet metal parts has been tested (Ref 34). Figure 17 shows a series of images related to this. Figure 17(a) shows a coil path that was inserted into a die. An electromagnetic pressure acts along this line when the actuator is energized by a capacitor bank discharge. During the stamping process, the clamp load is fixed, and the punch is advanced a prescribed amount between discharges. This is repeated with each discharge until failure. A result of hybrid incremental forming is shown in Fig. 17(b). It shows that the draw depth of the parts can be dramatically increased without any reliance on lubricants. The draw depth increased by 44% (from 4.4 to 6.35 cm, or 1.7 to 2.5 in.). This experiment shows that the approach of placing forming energy where required can significantly increase the ability to make aggressive sheet metal parts. The effect of the embedded electromagnetic coil is to produce tensile strain across the top surface of the part, and the tooling system defines the part shape in the usual way. Figure 17(c) shows the punch force versus stroke profile for the standard process and the electromagnetically assisted bump forming process. In the electromagnetically assisted process, with each cycle the strain across the top reduces the punch force. This reduction in punch force ultimately is responsible for the delay in tearing. Figure 17 represents one of the first experiments of this kind to be carried out. It simply shows that formability can be significantly enhanced and electromagnetic actuators can be embedded in forming dies. Better results will certainly follow as this type of process is further developed and improved.

Example 4: Auto Body Prototype Forming. Explosive and electrohydraulic forming represents cost-effective alternatives to hydroforming in the prototyping of auto body panels. The pressure pulse in water holds several advantages

to hydroforming: formability can be increased, peak pressures attained can be much higher, and springback can be reduced. Both approaches have been used for the low-volume production of aircraft parts as well as for prototype automotive parts, primarily in the former Soviet Union.

Joining Examples

Example 5: Solid-State Impact Welding. If two metals strike at an appropriate velocity (typically above 300 m/s, or 985 ft/s) and angle (typically approximately 10°), they will often form a solid-state metallurgical bond. Dissimilar metals can often be bonded without the deleterious effects of intermetallic formation that limit the use of fusion welding for dissimilar metals. Explosively clad plates can be purchased that are based on this solid-state welding approach. Basically, explosives are placed in contact with the cladding metal that is given an appropriate standoff from the base metal plate. When the explosive is detonated, the contact angle between the cladding and base metal is determined by a combination of the explosive detonation velocity, the clad metal velocity, and the original standoff distance. Reference 12 summarizes both the theory and practice in this area.

In recent years, there has been considerable interest in using electromagnetic pulses to provide the impulse required for solid-state welding. Such techniques have the advantage of being better adapted to joining small articles (such as tubing, driveshafts, or automotive frames). They are more controllable and reproducible. Also, they do not require the use of high explosives. While electromagnetic pulse welding techniques

have been demonstrated at least as early as 1978 (Ref 37), they still have not been widely commercialized. This is in part because the process is somewhat difficult due to the high impact velocities that are required for robust joining. This makes it difficult to develop long-life electromagnetic coils that can repeatedly produce the required high pressures.

Example 6: Lightweight Torque-Carrying Shafts. For many demanding structural applications, crimping can be used instead of the formation of metallurgical welds to carry force or torque. Structural crimping does not require the large impact velocities that are necessary in impact welding. One of the most impressive applications of electromagnetic crimping is in high-lift torque tubes, which Boeing uses in its 777, 737, and 747 aircraft. Here, 0.325 cm (0.128 in.) wall thickness aluminum alloy tubes are crimped onto steel yoke fittings. Peak electromagnetic pressures of approximately 205 MPa (30,000 psi) are created from a 60 kJ energy storage bank. Figure 18 shows a typical manufacturing system for torque tube assembly. Figure 19 shows the detail on a torque tube joint, including a tube after destructive torsion testing. The tube does not fail in the joint but in the parent material. These tubes have been shown to have a fatigue life equivalent to several times the life of the airframe structure. This electromagnetic crimp replaces a design that was troublesome with respect to fatigue.

Technology Status

High-velocity forming methods have a long and significant history. However, these

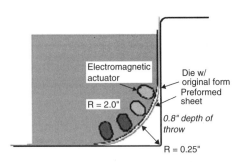

Fig. 16 Hybrid electromagnetic-assisted stamping process used to form hinge faces on aluminum inner door

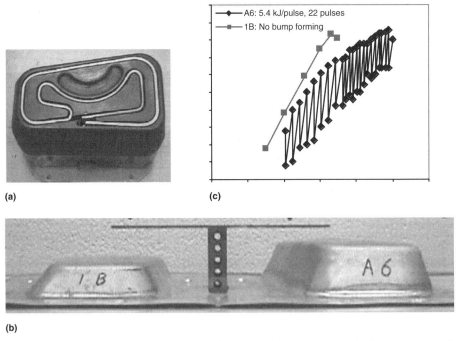

Fig. 17 Illustrations of the bump forming process. (a) Actuator path in an electromagnetically augmented stamping punch. (b) Increase in draw depth available using such a hybrid process. (c) Force-displacement trace for the traditional forming process as well as when periodic electromagnetic impulses are used

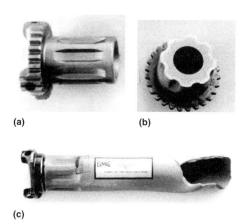

Fig. 19 Torque tube details. (a) Steel yoke used in torque tubes. (b) Cross section of the torque-carrying joint. (c) Behavior of a tube subjected to torque overload testing. Failure is outside the joint region.

Fig. 18 Photograph of the type of system used to assemble torque tubes, such as those shown in Fig. 19. The coil in this system is a wafer-type coil.

techniques have never quite achieved a clear place among mainstream manufacturing methods. This may be attributed to the fact that there is little intersection between the communities that understand capacitor bank discharge or explosive engineering with the metal forming community. As a result, neither toolmakers nor engineers have become comfortable with or trained in these methods, and hence, they are not used.

Precisely because these methods are very different than conventional metalworking by stamping, they may be very well suited for treating many issues that are difficult with conventional technology (formability, strain distribution, controlling springback, etc.). Although these techniques have fundamentally been known for a long while, there remain significant opportunities in their development and application.

REFERENCES

1. "DARPA Technology Transition Report 2002," www.darpa.mil.body/pdf/transition.pdf, Defense Advanced Research Projects Agency, 2002
2. J.S. Rinehart and J. Pearson, *Explosive Working of Metals,* Macmillan, 1963
3. W. Johnson, *Impact Strength of Materials,* Edward Arnold, London, 1970
4. W.W. Wood, *Exp. Mech.,* Vol 19, 1967, p 441
5. X. Hu and G.S. Daehn, Effect of Velocity on Flow Localization in Tension, *Acta Mater.,* Vol 44, 1996, p 1021–1033
6. M.M. Altynova, X. Hu, and G.S. Daehn, Increased Ductility in High-Velocity Electromagnetic Ring Expansion, *Metall. Mater. Trans. A,* Vol 27, 1996, p 1837–1844
7. A.A. Tamhane, M.M. Altynova, and G.S. Daehn, Effect of Sample Size on Ductility

in Electromagnetic Ring Expansion, *Scr. Mater.,* Vol 34 (No. 8), 1996, p 1345
8. V.S. Balanethiram and G.S Daehn, Increased Forming Limits at High Workpiece Velocities, *Scr. Metall,* Vol 31, 1994, p 515–520
9. M. Padmanabhan, Wrinkling and Spring-back in Electromagnetic Sheet Forming and Electromagnetic Ring Compression, *Materials Science and Engineering,* Ohio State University, 1997
10. M. Stuivinga et al., Explosive Forming in Enabling Technology, *Sixth International Conference on Sheet Metal,* University of Twente, 1998
11. S. Golovaschenko, Methodology of Pulsed Electromagnetic Joining of Tubes, *Innovations in Processing and Manufacturing of Sheet Materials,* TMS, 2001
12. B. Crossland, *Explosive Welding of Metals and Its Application,* Claredon Press, Oxford, 1982
13. A. Turner et al., Spot Impact Welding of Sheet Aluminum, *Aluminum Alloys 2002: Their Physical and Mechanical Properties Pts 1-3,* Trans Tech Publications Ltd., 2002 p 1573
14. W.H. Gourdin, Dynamic Consolidation of Metal Powders, *Prog. Mater. Sci.,* Vol 30 (No. 1), 1986, p 39
15. R. Prummer, Powder Compaction, *Explosive Welding Forming and Compaction,* T.Z. Blazynski, Ed., Applied Science Publishers, London, 1983
16. www.iap.com, IAP Research, Inc., Dayton, OH, 2003
17. E.J., Bruno, Ed., *American Society of Tool and Manufacturing Engineers: High Velocity Forming of Metals,* Prentice Hall, Inc., 1968
18. A.A. Ezra, *Principles and Practices of Explosive Metallurgy, Metal Working,* Industrial Newspapers, Ltd., London, 1973

19. Impact, *Int. J. Impact Eng.,* Pergamon Press, Oxford, 1983–present
20. M.A. Meyers, *Dynamic Behavior of Materials,* John Wiley and Sons, 1994
21. D.J. Mynors and B. Zhang, Application and Capabilities of Explosive Forming, *J. Mater. Process. Technol.,* Vol 125–126, 2002, p 1–25
22. "ISEE 2003 International Society of Explosives Engineers," www.isee.org, International Society of Explosives Engineers, 2003
23. *Blasters' Handbook,* International Society of Explosives Engineers, Cleveland, OH, 2003
24. www.ime.org, Institute of Makers of Explosives, Washington, DC, 2003
25. C.P. Williams, *J. Met.,* Vol 1, 1960, p 12
26. www.tei.co.uk, TEI Company, Wakefield, England, 2005
27. J. Jablonski and R. Winkler, Analysis of the Electromagnetic Forming Process, *Int. J. Mech. Sci.,* Vol 20, 1978, p 315–325
28. F.C. Moon, *Magneto-Solid Mechanics,* John Wiley and Sons, 1984
29. I.V. Belyy, S.M. Fertik, and L.T. Khimenko, *Electromagnetic Forming Handbook,* M.M. Altynova, Trans., Kharkov, Ukraine, 1977; www.osu.edu/hyperplasticity
30. "The Magnetic Hammer," in NASA SP-5034, National Aeronautics and Space Administration, 1965, p 1465
31. "Electroimpact," www.electroimpact.com/edr.asp, Electroimpact Corp., Seattle, WA, 2003
32. www.miller-company.com, Miller Company, Burleson, TX, 2003
33. J.C. Duncan, W. Johnson, and J. Miller, *Conference on Electrical Methods of Forming and Coating,* Institution of Electrical Engineers, 1975
34. G.S. Daehn, J. Shang, and V.J Vohnout. Electromagnetically Assisted

Sheet Forming: Enabling Difficult Shapes and Materials by Controlled Energy Distribution, *The MPMD Fourth Global Innovation Symposium,* TMS, 2003

35. V.J. Vohnout, A Hybrid Quasi-Static Dynamic Process for Forming Large Sheet Metal Parts from Aluminum Alloys, *Materials Science and Engineering,* Ohio State University, 1998

36. M.K. Knyazyev, *EM Forming Handbook,* Internal reports of the Zhukovsky National Aerospace University, KhAI, Kharkiv, Ukraine, 2003

37. W.F. Brown, J. Bandas, and N.T. Olson, Pulsed Magnetic Welding of Breeder Reactor Fuel Pin End Closures, *Weld. J.,* June 1978

Drop Hammer Forming

DROP HAMMER FORMING is a process for producing shapes by the progressive deformation of sheet metal in matched dies under the repetitive blows of a gravity-drop hammer or a power-drop hammer. The configurations most commonly formed by the process include shallow, smoothly contoured double-curvature parts; shallow-beaded parts; and parts with irregular and comparatively deep recesses. Small quantities of cup-shaped and box-shaped parts, curved sections, and contoured flanged parts are also formed.

Advantages and Limitations. The main advantages of drop hammer forming are:

- Low cost for limited production
- Relatively low tooling costs
- Dies that can be cast from low-melting alloys and that are relatively simple to make
- Short delivery time of product because of simplicity of toolmaking
- The possibility of combining coining with forming

These advantages must be weighed against the following limitations:

- Probability of forming wrinkles
- Need for skilled, specially trained operators
- Restriction to relatively shallow parts with generous radii
- Restriction to relatively thin sheet (approximately 0.61 to 1.63 mm, or 0.024 to 0.064 in.; thicker sheet can be formed only if the parts are shallow and have generous radii)

Drop hammer forming is not a precision forming method; tolerances of less than 0.8 to 1.6 mm ($\frac{1}{32}$ to $\frac{1}{16}$ in.) are not practical. Nevertheless, the process is often used for sheet metal parts, such as aircraft components, that undergo frequent design changes or for which there is a short run expectancy.

Process Description (Ref 1)

The drop hammer forming process methods can be classified into three categories: bare punch, blow down, and blow up. Bare punch is the simplest procedure, using a single hammer blow and no forming aids. Blow down and blow up procedures require the use of forming aids and multiple punch blows (Fig. 1). The most commonly used forming aids are made of rubber, because this has the ability to flow like liquid when squeezed or impacted. The rubber sheets are placed on (blow down) or under (blow up) the part to form a pyramid. During the forming process, the pyramid causes the impact force to flow in the direction of the least rubber buildup, thus preventing wrinkles and folding. After each punch blow, one or two sheets of rubber will be removed before the next punch blow is applied. The distance of the punch blow may vary depending on the part and die configuration.

Key factors in determining a process plan include:

- Type of part material
- Type of tool (male or female)
- Maximum draw depth of the tool
- Number of corners of the part

Part Material Type. Three types of part material are used frequently, namely, aluminum, steel, and titanium. The material of the part can determine the selection of different types of rubber as forming aids. For example, because titanium is very hard, usually the part is heat treated (above 600 °C, or 1100 °F, for 10 min) before the forming process is applied. As a result, harder and heat-resistant rubbers are needed as forming aids.

Type of Tool. Generally, there are two types of tool according to their geometric configuration, namely, male and female. The geometric configuration determines where forming aids (rubber sheets) can be placed, that is, on or under the part. The determination of tool type is not always straightforward, because it depends not only on the tool geometry but also on the part geometry. For example, a male tool may be considered when the part is formed using the rib located in the middle of the tool. A female tool is considered when the part will be drawn into the cavity during forming.

Maximum Draw Depth. The maximum draw depth, to some extent, determines the number of hits the operator should apply during the drop hammer forming process. Softer materials, such as aluminum and mild steel, can be drawn in one stage to approximately 50 mm (2 in.). However, hard material, such as titanium, can only be drawn in one stage to approximately 13 mm (0.5 in.).

Number of Corners. The severity of permissible deformation in drop hammer forming is limited both by geometrical considerations and by the properties of the work metal. The number of corners is a major geometrical consideration that influences the determination of process conditions.

Hammers for Forming

Gravity-drop hammers and power-drop hammers are comparable to a single-action press. However, they can be used to perform the work of a press equipped with double-action dies through the use of rubber pads, beads in the die surfaces, draw rings, and other auxiliary equipment.

Because they can be controlled more accurately and because their blows can be varied in intensity and speed, power-drop hammers, particularly the air-actuated types, have virtually replaced gravity-drop hammers. A typical air-drop hammer, equipped for drop hammer forming, is shown in Fig. 2.

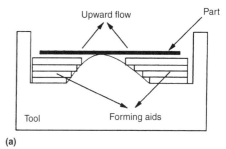

(a)

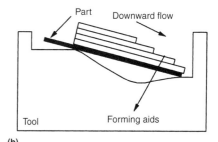

(b)

Fig. 1 Process methods with forming aids. (a) Blow up. (b) Blow down. Source: Ref 1

Power-drop hammers are rated from 4.5 to 155 kN (1,000 to 35,000 lbf), representing energies from 15 to 575 kJ (11,000 to 425,000 ft · lbf). Air-drop hammers range in size (ram area) from 762 by 610 mm (30 by 24 in.) to 3.05 by 3.05 m (120 by 120 in.), with impact energies ranging from 8.9 to 134 kJ (6,600 to 99,000 ft · lbf). Ram dimensions and other pertinent details concerning these hammers can be found in the article "Hammers and Presses for Forging" in *Metalworking: Bulk Forming*, Volume 14A, *ASM Handbook*, 2005.

Planishing hammers are used to supplement drop hammer forming. These are fast-operating air-driven or motor-driven machines that are generally used for low-production operations to form dual-curvature surfaces. They are also used to planish welds and to smooth out wrinkles or other imperfections in drawn or drop hammer formed parts.

Tooling

Dies for drop hammer forming are usually made by casting metals such as kirksite, which is a zinc-aluminum alloy, for rapid production of press tooling. These dies can be rapidly produced; are more economical than permanent dies; can be melted and recast; and can be reinforced at selected points of wear by facing with harder material, such as tool steel, for long production runs (Ref 2).

Normally, drop hammer forming is accomplished without the benefit of holddown. The metal is slowly forced into shape by controlling the impact of the blows. In many instances, it is necessary to use two- or three-stage dies, supplemental equipment, and hard forming, such as a bumping hammer or wooden mallet, to remove wrinkles, and so on.

To successfully complete forming operations, another aid that may be necessary is to anneal material between die stages and intermediately for single-stage die forming.

Parts should be cleaned prior to annealing to protect the finish. Care should be taken to remove all traces of zinc that may be picked up from kirksite forming dies, because failure to remove the zinc will result in penetration of the steel (stainless) when treated and will cause cracking (Ref 2).

Tool Materials. Dies are cast from zinc alloy (Zn-4Al-3.5Cu-0.04 Mg), aluminum alloy, beryllium copper, ductile iron, or steel. The wide use of zinc alloy as a die material stems from the ease of casting it close to the final shape desired. Its low melting point (380.5 °C, or 717 °F) is also advantageous. All dies, regardless of die material, are polished.

Punches are usually made of lead or a low-melting alloy, although zinc or a reinforced plastic can also be used. The sharpness of the contours to be formed, the production quantity, and the desired accuracy primarily govern the choice of punch material. Lead has the advantage of not having to be cast accurately to shape, because it deforms to assume the shape of the die during the first forming trial with a blank.

Rubber Padding. In some drop hammer forming, both a working (roughing) punch and a coining (finishing) punch are used. When the working punch becomes excessively worn, it is replaced by the coining punch, and a new coining punch is prepared. Another method of achieving the same results with one punch is to use 3.2 to 25 mm (1/8 to 1 in.) thick rubber pads. Rubber that is suitable for this purpose should have a hardness of durometer A 60 to 90. In some cases, soft semicured rubber is used. In the

positioning of pads for a particular part (Fig. 3), the maximum thickness of rubber is situated where the greatest amount of pressure is to be applied in the initial forming. As forming progresses, the thickness of the rubber is reduced by removing one or more of the pads after each impact.

Trapped Rubber Forming. Drop hammer forming using trapped rubber dies is a process derived from the Guerin process, which is today (2005) synonymous with the term *rubber-pad* forming (see the article "Rubber-Pad Forming and Hydroforming" in this Volume). Both the Guerin and Marform (which is a refinement of the Guerin process) processes use hydraulic pressure instead of a drop hammer as the forming force. The trapped-rubber process using a drop hammer is used extensively by the aerospace and aircraft industries to fabricate sheet metal parts such as instrument panels, tank sections, air frames, stabilizer tips, air ducts, and doors made of aluminum alloys, titanium alloys, and stainless steels.

The drop hammer used in the trapped-rubber process offers several advantages over the hydraulic press used in the Guerin and Marform processes:

- Develops greater force (up to 44 MN, or 5000 tonf) per unit of applied energy in shorter time and at less cost
- Minimizes springback as a result of dynamic application of force
- Reduces tendency of material to wrinkle due to better material flow
- Reduces amount of hand finishing required due to dynamic application of pressure

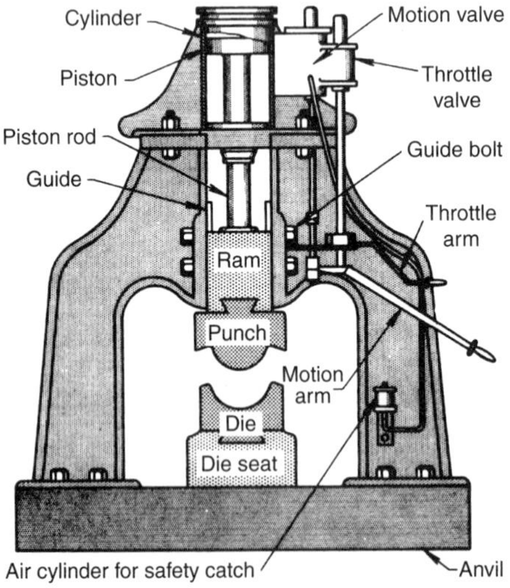

Fig. 2 Schematic of an air-actuated power-drop hammer equipped for drop hammer forming

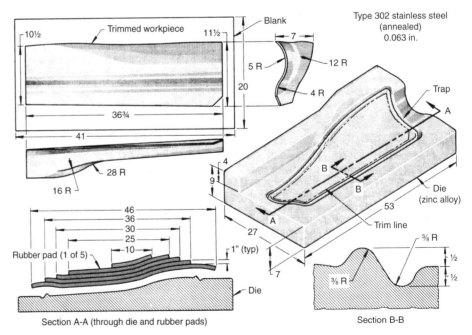

Fig. 3 Smoothly contoured stainless steel part that was hammer formed in a die with a peripheral trap for holddown. Dimensions given in inches

Lubricants

Lubricants are used in drop hammer forming to facilitate deformation by reducing friction and minimizing galling and sticking, and to preserve or improve surface finish. Selection of a lubricant depends primarily on the type of work metal, the forming temperature, the severity of forming, and the subsequent processing. Recommendations for lubricants used with steels and with aluminum, magnesium, and titanium alloys are given in the sections of this article that deal with the processing of those metals.

Blank Preparation

The blanks for drop hammer forming are generally rectangular and are prepared by shearing. The blank should be large enough to yield a part with a flange 50 to 75 mm (2 to 3 in.) wide in order to facilitate drawing of the metal during forming. When multistage forming is used, the part can be trimmed to provide a flange not less than 13 mm ($^{1}/_{2}$ in.) wide for the final forming stage.

Sheared edges are generally satisfactory for drop hammer forming, because the wide flange permits some cracking in the flange area without harming the part. The blank should be deburred to avoid possible damage to the tooling.

Drop Hammer Coining

Tableware, coins, and a variety of decorative items produced in copper alloys, stainless steel, sterling silver, and other metals are commonly coined by the drop hammer process. Processing details and production examples of drop hammer coining are given in the article "Coining" in *Metalworking: Bulk Forming,* Volume 14A, *ASM Handbook,* 2005.

Multistage Forming

With the drop hammer process, complicated parts can be formed by means of a single die and punch. However, when a large quantity of a particular part is required, it is common practice to adopt a multistage forming technique, employing several sets of dies and punches. The forming operation in any one stage is less severe than if the part were formed in a single operation. Metals that work harden appreciably are usually annealed after each operation, unless a suitable die sequence eliminates the need for annealing.

Unless contact of the work metal with lead is undesirable, the number of stage dies can be kept at a minimum by using lead pads to reduce the depth of the die cavity. Lead can be poured into the die, or lead sheet can be laid in the bottom of the recess, to the desired height; the lead is then formed to contour by a heavy blow of the punch. After the part has been preformed with the padded die, the pad is removed, and the part is then formed to the full depth of the die.

Control of Buckling

Forming a deeply recessed part in thin sheet by any conventional method usually requires a high holddown force in order to prevent buckling. In drop hammer forming, the holddown action is restricted to the end of the stroke. Therefore, buckles are free to form during most of the stroke. The holddown pressure takes effect when the punch contacts the top of the wrinkles formed in the flange and rapidly increases until the die is bottomed. The wrinkles can be removed only at the end of the stroke and only if they are not too deep.

Multistage Processing. To avoid the formation of wrinkles that are too deep to be removed in the late stages of the stroke, the forming process is divided into several forming stages. The wrinkles formed in each stage are slight and can be eliminated at the end of the stroke or, if necessary, by manual hammering. To provide adequate holddown action at the end of each operation, it is common practice to use stage dies with surfaces that extend slightly beyond the trim line and to use punches equipped with suitable beads or traps.

Holddown for Deep Parts. To form deep parts, a series of plywood or metal holddown rings can be used; one ring is removed after each blow. These rings usually vary in thickness from 6.4 to 25 mm ($^{1}/_{4}$ to 1 in.). The desired metal flow can be approached during the early stages of forming by using a blank that is considerably larger than that required to form the part. The excess metal is trimmed off after several blows. The stiffness of the oversized blank prevents draw-in and therefore induces stretching, which counteracts the tendency toward buckling. The die for a part with a deep recess must be designed with a horizontal surface to accommodate beading.

Processing of Steels

Carbon and low-alloy steels containing less than 0.30% C are the most easily formed by the drop hammer process. Higher carbon content decreases formability and promotes cracking. Although lead additions do not adversely affect the formability of steel, the sulfur additions that are characteristic of resulfurized free-machining steels promote susceptibility to cracking. All carbon and low-alloy steels require full annealing for satisfactory drop hammer forming.

Stainless steels that are extensively formed by the drop hammer process include AISI types 302, 304, 305, 321, and 347. The common types of corrosion-resistant steel used for drop hammer forming include 301, 302, 304, 305, and stabilized grades 321 and 347 (Ref 2). Type 301 work hardens more rapidly and is subject to strain cracking. It is possible to form some types (301 and 302) in quarter- and half-hard condition. However, the severity of the forming operation must be reduced to compensate for the prehardened material. For severe drop hammer forming, grades containing not less than 10% Ni (and preferably approximately 12%) should be selected in order to minimize cracking. The condition of material for the best forming should be the fully annealed (solution-treated) condition.

Sheet Thickness. The drop hammer forming of steel sheet (particularly stainless steel sheet) less than 0.46 mm (0.018 in.) thick is impractical, because of wrinkling and the difficulties encountered in attempting to planish the wrinkles. The most common range of steel sheet thickness for drop hammer formed parts is 0.61 to 1.6 mm (0.024 to 0.063 in.). Thicknesses up to 1.98 mm (0.078 in.) have been hammer formed.

Tool Materials. Cast zinc alloy is the most widely used die material for the drop hammer forming of carbon, low-alloy, and stainless steels. Alloy cast iron dies are substituted when a large quantity of parts is required. Inserts of an air-hardening tool steel can be used in order to increase the life of dies with sharp fillets and corners.

Punches are made of either zinc alloy or cast iron, and are ground to size. A zinc alloy punch, cast directly into the die, can be used for shallow parts, but it may be undersized (because of shrinkage upon cooling) if the part contains large cavities. Lead punches are also extensively used, although they are easily distorted when used to form steels. Punch life can be considerably increased by facing the punch with an untrimmed finish-formed steel part.

Lubricants. When zinc alloy dies and zinc alloy or lead punches are employed, many steel parts can be formed without a lubricant. Harder tool materials or more severe forming may require the use of a light lubricant, such as SAE 10 or SAE 30 mineral oil.

Precautions for Stainless Steels. Stainless steels, especially the austenitic grades, work harden more than the carbon and low-alloy steels that are suitable for hammer forming. In the cold forming of stainless steel, it is necessary to stretch the metal, rather than allow it to draw into the die. Stretching prevents the formation of wrinkles that are difficult to eliminate. By means of stretching, quarter-hard and even half-hard types 301 and 302 can be drop hammer formed, although only to a very limited extent. Part configuration must be simple and of only moderate depth; otherwise, wrinkles (in a shallow part) or distortion (in a complex shape) will occur. Although it is preferable that the part be made in a single die with a single blow, some commercial quarter-hard parts have required as many as three or four blows for successful forming.

When moderately complex parts are formed in a drop hammer in several stages, it is advisable to consider intermediate annealing in order to offset the effects of work hardening. It is not necessary to pickle after each annealing treatment (provided scaling is not too heavy), except before the finish-forming operation and after the final annealing treatment. If the part is formed in zinc alloy dies, any adhering zinc particles must be removed by pickling or by treatment in a fused salt bath (caustic soda) before annealing treatments (both intermediate and final). This requirement is most important for parts that are to be welded or that will be exposed to elevated-temperature service. Failure to remove the zinc may result in cracking.

Springback. Carbon and alloy steel parts, and especially stainless steel parts, having large radii and smooth contours are more difficult to maintain in desired shape than parts with relatively sharp radii, because of the greater springback under these conditions. Common practice is to compensate for this springback from the desired contour by trial and error. If this method is not successful, the part must be distorted elastically upon assembly, that is, sprung into the final shape.

Parts with reverse contours (saddleback parts) are extremely difficult to form without excessive wrinkling.

Limits in Deep Recessing. When deeply recessed parts are to be formed in a drop hammer, the recesses are limited in both depth and contour. With a single die, a cup-shaped or dome-shaped part can be formed to a limiting depth of 60 to 70% of that obtainable by means of double-action dies. Square and rectangular steel boxes (even shallow ones) require a minimum corner radius of 6.4 mm ($\frac{1}{4}$ in.) or five times the metal thickness, whichever is larger. For deeper boxes, progressively larger corner radii are necessary, and these minimum radii apply to boxes of any width.

Processing of Aluminum Alloys

The drop hammer forming of aluminum alloys is most suitable for limited production runs that do not warrant expensive tooling. The process is often used for parts, such as aircraft components, that undergo frequent design changes. Some forming applications also involve coining and embossing. The article "Forming of Aluminum Alloys" in this Volume contains more information on the forming of aluminum alloy sheet.

The drop hammer can be used to form deep pan-shaped and beaded-type parts. Kirksite with a plastic surface insert is satisfactory for male and female dies (Ref 2). The surface of kirksite dies used without a plastic insert should be smooth to prevent galling and scratching of the aluminum surface. When forming deep pans and complicated shaped parts, it is often necessary to use drawings rings, pads, or two- to three-stage dies. An intermediate anneal is

sometimes used to relieve the hardened condition (cold work) resulting from the forming operation.

Work Metal. Annealed tempers of all aluminum alloys are the most suitable for hammer forming. Intermediate work-hardened tempers of the non-heat-treatable alloys are often used for channel shapes and shallow embossed panels.

Heat-treatable alloys are often partly formed in the annealed condition. The part is then solution heat treated, quenched, restruck to size, and artificially aged. Restriking is also necessary to remove distortion caused by quenching. Drop hammer forming can be done on freshly quenched alloys immediately after quenching, or it can be done later if the alloys are refrigerated to prevent aging.

Sheet Thickness. Under comparable conditions, with the same equipment and with the same thickness of sheet, aluminum wrinkles more easily than steel under a drop hammer. To obtain results comparable to those obtained with steel, aluminum alloy sheet should be at least 40% thicker than the steel, or preferably in the approximate thickness range of 0.86 to 3.18 mm (0.034 to 0.125 in.).

Equipment and Tool Materials. Aluminum alloys are drop hammer formed in gravity-drop, power-drop, and planishing hammers. Dies are cast from aluminum, zinc alloy, iron, or steel. Dies for high production are usually cast in iron or steel. All dies are polished. Most punches are made of lead or a low-melting alloy, although zinc alloy or reinforced plastic can also be used. The softer punch materials have the advantage of deforming readily to assume the shape of the die during forming trials. When planishing hammers are used, the preferred tool material is hardened, polished tool steel.

Forming Characteristics. Annealed aluminum alloys are readily formed under the drop hammer. Simple components can often be produced by a single blow. Deep shapes require extreme care in blank development and die design. Blankholders are not used; therefore, wrinkles are difficult to avoid, especially when thin sheet is being formed.

Processing of Magnesium Alloys

The drop hammer forming of magnesium alloys is performed on preheated sheet in heated dies. This procedure is suited to the production of formed parts having shallow depths and asymmetrical shapes and to parts for which special springback control is required.

Work Metal. Magnesium sheet alloys in the annealed condition are preferred for drop hammer forming. The ideal sheet thickness for forming is 3.2 mm ($\frac{1}{8}$ in.), and the part should be designed so as to be formable in six stages or fewer. For sheet thinner than 3.2 mm ($\frac{1}{8}$ in.), ten stages or fewer are recommended.

Equipment and Tool Materials. Both gravity-drop and power-drop hammers are sui-

table for the forming of magnesium alloys. Zinc alloy is the preferred punch and die material, although lead punches are sometimes used for production runs of not more than 50 pieces. When lead comes in contact with magnesium sheet, there is danger of lead pickup, which can cause corrosion of the sheet. Although lead pickup may occur at room temperature, it is more likely to occur at the elevated temperatures at which magnesium alloys are formed. Therefore, if lead pickup cannot be tolerated or if the production run exceeds 50 pieces, either zinc alloy or cast iron can be substituted for lead.

Lubricants. Vegetable-lecithin oils provide good lubrication at temperatures to 260 °C (500 °F). Suspensions of colloidal graphite may have to be used if temperatures are to exceed 260 °C (500 °F). However, these suspensions are more difficult to remove when parts are cleaned after forming.

Preheating. Magnesium alloy parts are usually formed at temperatures of 230 to 260 °C (450 to 500 °F), depending on the alloy (see the article "Forming of Magnesium Alloys" in this Volume). Heating times are 5 min per stage for sheet up to approximately 1.29 mm (0.051 in.) thick, and up to 9 min per stage for thicker sheet (up to 3.2 mm, or 0.125 in.).

The oven used to heat the parts between stages should be situated near the drop hammer; the decrease in temperature during transfer from the oven to the hammer will range from 17 to 25 °C (30 to 45 °F) in 5 s. The dies can be heated by placing them in an oven located near the hammer, and they can then be kept at temperature with ring burners or torches during the forming of the part.

Small dies can be anchored to an electrically heated cast iron platen installed on the hammer bed, but this method is impractical for large dies. The punch and die can also be heated by electric elements or by a heat-transfer fluid. The working temperatures should not exceed those recommended.

Rubber pads can be used in the initial forming operation. At 230 °C (450 °F), the reduction obtainable with rubber staging is approximately 10%. Special types of rubber are available for forming at temperatures to 315 °C (600 °F). The rubber pads are removed before the final blow is delivered to set the material.

Springback. One advantage of the elevated temperatures used in the drop hammer forming of magnesium is the marked reduction or total elimination of springback, provided the maximum practical temperatures are always employed. The rate of the deformation is important in drop hammer forming and must be carefully controlled upon severe drawing or when material in the hard (H24) temper is being formed. The rate of deformation can be controlled by the operator, although not to close limits. Parts that require relatively severe forming can be started by allowing the punch to descend slowly into the die and by using subsequent strikes to set the material.

Dimensional Tolerances. Tolerances of ±0.76 mm (±0.03 in.) have been held in the production of magnesium parts. When close tolerances are important, press forming is usually the preferred method for the forming of magnesium parts.

Processing of Titanium Alloys

Various titanium sheet alloys have been formed by the drop hammer process, including Ti-13V-11Cr-3Al, Ti-8Al-1Mo-1V, Ti-6Al-4V, and Ti-5Al-2.5Sn. In general, the alloys containing aluminum as the principal alloying element are the most difficult to form. The minimum thickness of titanium sheet for hammer-formed parts is approximately 0.64 mm (0.025 in.).

Drop hammer forming of titanium has been very successful and has been accomplished at both room and elevated temperatures. Kirksite is satisfactory for male and female dies where only a few parts are required. If long runs are to be made, ductile iron or laminated steel dies are usually necessary. In drop hammer forming, the best results are usually by warming the female die to a temperature of 425 to 540 °C (800 to 1000 °F) for 10 to 15 min (Ref 2). The part is then struck and set in the die. Usually, a stress-relief operation at 540 °C (1000 °F) for 20 min is necessary, then a restrike operation. In most instances, a finished part requiring no handwork is obtained.

Parts should be cleaned prior to annealing to protect the finish. Care should be taken to remove all traces of zinc that may be picked up from kirksite forming dies, because failure to remove the zinc will result in penetration of the steel (stainless) when treated and will cause cracking (Ref 2).

Tool Materials. Contact between titanium and low-melting tool materials, such as zinc alloy or lead, should be avoided—particularly when the titanium is formed at elevated temperature or must be heat treated after forming. When these tool materials are used, contact with the workpiece can be avoided by capping the

punch and die with sheet steel, stainless steel, or a nickel-base alloy. The choice of capping material depends on the tool life desired. The longest tool life is obtained by capping with nickel alloy sheet such as alloy 600 (UNS N06600) in thicknesses of 0.64 to 0.81 mm (0.025 to 0.032 in.).

In general, steel and ductile iron dies are used when the tooling must be heated above 205 °C (400 °F). Preheating of both the work metal and the tooling is not uncommon.

Rubber Pads. High-temperature rubber pads are used both in preforming operations before the final strike and as electrical insulators to prevent current loss to the tooling when the work metal blank is heated by the electrical-resistance method.

Lubricants used in the drop hammer forming of titanium should be nonchlorinated. Extreme-pressure oils and both pigmented and nonpigmented drawing compounds are used in most operations.

Preheating Tools and Blanks. Difficult titanium parts are formed at elevated temperature (see the article "Forming of Titanium and Titanium Alloys" in this Volume for recommendations and precautions). Thermal expansion of the blank and the tooling must be considered. If the tooling is not preheated, the amount it expands will depend on the length of time it is in contact

with the blank. The allowance for thermal expansion used in the design of tooling for titanium is 0.006 mm/mm (0.006 in./in.) for a forming temperature of 540 °C (1000 °F). The allowance for expansion of circular or elliptical parts should be made radially, not peripherally. When hot sizing is to follow forming, the drop hammer tooling is usually made to net dimensions without consideration of thermal expansion.

Formability versus Temperature. Variation of the drop hammer formability index for two titanium alloys with temperature is given in Fig. 4. It is evident from the curves that significant increases in formability can be achieved at temperatures above 540 °C (1000 °F).

Drop Hammer Forming Limits

The severity of permissible deformation in drop hammer forming is limited both by geometrical considerations and by the properties of the work metal. The forming limits can be predicted by considering parts of interest as variations of beaded panels. For parts characterized in this way, the critical geometrical factors are the bead radius, r, the spacing between beads, s, and the thickness of the work metal, t (Fig. 5).

Two of the forming limits depend entirely on dimensional relations and are the same for all materials; the ratio of the bead radius, r, to bead spacing must lie between 0.35 and 0.06. The lower formability limit is controlled by the necessity of producing uniform stretching and avoiding excessive springback. If the r/s ratio is too small, there will be greater localized stretching at the nose of the punch.

Within the limits set for all materials by the r/s ratio, success or failure in forming beaded panels depends on the ratio of the bead radius to the sheet thickness (r/t) and on the ductility of the work metal. The part will split if the necessary amount of stretching exceeds the ductility available in the material. The splitting limit can be predicted from elongation in a 12.7 mm (0.5 in.) gage length in tension tests at the temperature of interest.

Formability limits for two titanium alloys are plotted in Fig. 5. Both charts show the marked improvements in formability resulting from the better elongation values at elevated temperature.

REFERENCES

1. S.H. Huang, H Xing, and G. Wang, Intelligent Classification of the Drop Hammer Forming Process Method, *Int. J. Adv. Manuf. Technol.*, Vol 18, 2001, p 89–97
2. *Aerospace Metals—General Data and Usage Factors,* technical manual, *Engineering Series for Aircraft Repair,* To 1-1A-9, NAVAIR 01-1A-9, Feb 26, 1999

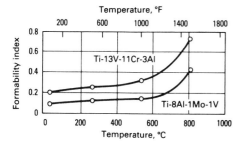

Fig. 4 Effect of forming temperature on the drop hammer formability of two titanium alloys

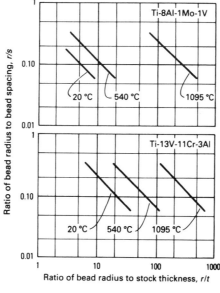

Fig. 5 Formability limits of beaded titanium alloy panels at room temperature and at elevated temperature

Thermal Forming of Sheet and Plate

Alan Male, University of Kentucky

THERMAL FORMING OF METALS refers to the development of permanent deformation without the application of external mechanical forces. It is based on the fundamental physical principle that most metals expand when heated and contract when cooled. If the expansion is limited in two dimensions, compressive stresses build up and result in localized plastic deformation if the temperature is sufficiently high. This plastic deformation then occurs in the third dimension. Upon cooling, the deformed material does not return to its original dimension, thus imparting a shape change upon the whole component. It is important, of course, that no melting is allowed to take place.

Historically, thermal forming has been used for flame straightening of structures and weldments, and for bending of plates for ship construction. As its name implies, a gas flame was used as the means of heating. More recently, interest has been revived in such processing because of the development of laser technology for heating.

Until recently, industrial application of thermal forming techniques using gas flames has been based largely on empiricism, and the ensuing results have depended principally on the skill and experience of the operators performing the process. Considerable research has been conducted with the aim of providing greater understanding and quantification of the technology. An excellent detailed status of this research may be found in the proceedings of an international workshop (Ref 1). Today, thermal forming is developing as a highly sophisticated potential manufacturing technology. Well-defined and closely controllable heat sources such as laser beams and plasma jets allow for the precise and reproducible heating of a very localized area of material and thus allow the possibility of automation. Additionally, the process is contributing to the development of rapid prototyping technologies. Furthermore, it provides other nontraditional possibilities, such as in-use repair or modification of sheet metal components.

Energy Sources

Thermal forming relies on the availability of a localized and controllable energy source to produce local heating of the material to be formed. In fact, four energy sources have been used for thermal forming: an electric induction coil, a gas flame, a plasma torch, and a laser beam.

The gas flame is usually from a handheld oxyacetylene torch and is obviously the least controllable and reproducible method of heating. Equipment for such heating is very inexpensive. Details of such torches can be found in many welding manuals, including Volume 6 of *ASM Handbook,* and is not further elaborated here.

A high-frequency electric induction coil is probably the most efficient of the available energy sources because the heat is created directly in the metal by the induced eddy currents. The higher the frequency used, the shallower the depth of heating occurring. As will be apparent later, this may be an advantage in promoting faster forming.

Light amplification by stimulated emission of radiation (laser) provides a beam of visible light of significant energy. It may be used for rapidly heating metallic materials provided the energy of the beam can be transferred to the material by "coupling" rather than being reflected away. Coupling is usually accomplished by use of a spray coating of graphite on the workpiece surface to promote heat absorption. This must be periodically reapplied because of its gradual oxidation during multiple exposures. It is generally accepted that the overall energy transfer ratio of laser systems into the workpiece material is rather low (<10%).

Lasers are available in a variety of types and power levels. Low power lasers of the neodymium-yttrium (Nd-YAG) type may use fiber-optic beam delivery systems. High-power CO_2 lasers require water-cooled copper mirror systems and are expensive. All laser systems require extensive safety precautions because of the potential for damaging stray reflections. Further details of laser systems may be found in Volume 6 of *ASM Handbook.*

Plasma arc systems are less expensive than most high-power laser systems (usually <10%) and are readily available. To avoid the need for making electrical connection to the workpiece, a nontransferred plasma arc torch system is preferred. Although normal safety precautions must be strictly adhered to, they are normally safer than laser systems. The overall energy transfer ratio is generally >85%. Details of plasma arc torches and systems may be found in Volume 6 of *ASM Handbook* and in the article "Electric Arc Cutting" in this Volume.

The Basic Process

Modern thermal processing using a laser has been pioneered in Poland (Ref 2, 3) and in Germany (Ref 4, 5). More recently, the concept of using a plasma arc heat source was introduced in the United States (Ref 6, 7). With the exception of the heat source details, the processes are similar in concept.

The basic process for a plasma arc heat source is illustrated in Fig. 1. A cooling jet (cooling stream 1), usually of CO_2 gas, is positioned on the back of the workpiece directly opposite the heat source. Occasionally, and more normally with a laser system, an alternate cooling jet (cooling stream 2) is located behind the heat source.

The amount of deformation achieved with each thermal scan is limited. Thus, a multiple of repeated scans are required to develop significant amounts of bending or of change in shape.

Mechanisms of Forming

Bending toward the heat source is a term used to describe a metal sheet bending process in which there is a concave hinge on the bending straight line. During the heating process, the heat input and cooling stream 1 (Fig. 1) are combined to make the upper part of the sheet metal section (zone S_1, Fig. 2) reach its plastic state and allow the lower part (zone S_2) to maintain its elastic

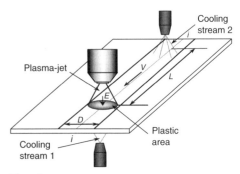

Fig. 1 Schematic of thermal forming of metal sheet

state. The material in zone S_1 expands because of being heated. However, because the zone S_2 retains its stiffness, it inhibits free expansion within S_2. The partial outflow will occur in zone S_1, as shown in Fig. 2. Stresses in the S_1 zone are compressive because the expansion of these areas is restrained by surrounding metal that is at lower temperature. The compressive stresses will reach the yield level of the metal at that temperature, and plastic deformation will be produced in zone S_1. Because zone S_2 is in an elastic state, the workpiece becomes bent away from the heat source by an angle α_1. The combination of the heat source input and cooling stream 1 from the back side can control the temperature and size of zones S_1 and S_2. The cooling stream 2 that follows the heat source decreases the temperature of zone S_1 and produces a contraction in the previously heated and compressed zone S_1. As the metal in the heated and cooled zone tries to shrink, tensile stresses are induced in zone S_1. This manifests itself by a bend toward the heat source at an angle α_2. Because the yield stress at low temperature is higher than that at high temperature, the tensile stress produced during cooling will result in a relatively greater shrinkage. As a result of these two processes (the heating and the cooling), a permanent deformation appears so as to produce a concave hinge on the heated side and a permanent bending angle of $\alpha_3 = \alpha_2 - \alpha_1$, as shown in Fig. 3.

Bending Away from the Heat Source refers to metal sheet bending such that the surface being heated exhibits a convex hinge. The heating and cooling process along *i-i* is very close to that shown in Fig. 1. There is, however, the difference that the energy beam is wider when compared with the sheet thickness and the total thickness is heated thoroughly (Fig. 4a). The cooling stream 2 following the heat source is at a greater distance behind (or does not use a front cooling stream) and the flow rate of cooling stream 1 is much lower. Because the heated surface of the sheet is away from its neutral axis, a bending moment is produced. The metal near the heat source is heated to higher temperatures than the metal away from the heat source. The hotter metal expands and the sheet bends away from the plasma. Additionally, the total thickness of the sheet is heated thoroughly, a relatively great plastic strain is produced by the thermal cycle and after cooling, the convex shaped is maintained, as shown in Fig. 4b.

A hollow box component formed using both of these bending methods is shown in Fig. 5.

Other Deformation Modes. Combination of the aforementioned two concepts can be used together with particular workpiece geometries to achieve different modes of deformation. For example, the local diameters of tubular workpieces can be either increased or decreased, or the wall thickness increased or decreased at will (Ref 8), as depicted in Fig. 6 and 7.

Effect of Process Parameters on Forming

Any process parameter that is likely to raise the workpiece surface temperature to the highest level, without causing melting or other adverse metallurgical effect, and to do it as rapidly as possible will generate the maximum amount of bending for each thermal scan (Ref 9). With a laser beam, this refers to the laser power, beam spot size, rate of travel over the workpiece surface, and the efficacy of the coupling medium on the surface.

For plasma forming, the diameter of the plasma jet increases with distance from the orifice. Thus, as the distance between a plasma torch nozzle and the workpiece surface increases, for a given travel speed, the temperature developed in the workpiece surface is reduced, thus resulting in a lower rate of forming (Ref 10), as shown in Fig. 8. Similarly, as the travel speed is increased, the workpiece surface temperature is reduced, also resulting in a lower rate of forming (Fig. 9). Other parameters will also control the temperature of the workpiece surface. A smaller plasma nozzle diameter and/or an increased plasma gas flow rate will increase

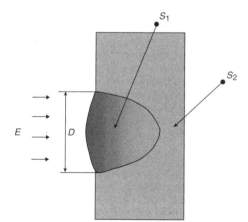

Fig. 2 Sectional view of a heating phase of the plate to be bent

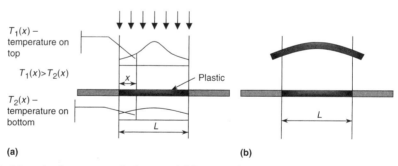

Fig. 4 Schematic of temperature distribution and deformation of plate bent away from heat source. (a) Heating differential. (b) Final convex form

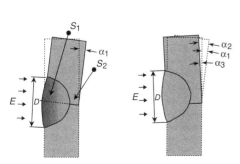

Fig. 3 The deformation of metal sheet during heating and cooling processes

Fig. 5 Photograph of a box formed with plasma arc (Ref 7). Blank: stainless steel, 0.8 mm (0.0312 in.) thick

Fig. 6 Thermally formed components showing change in diameter and wall thickness (Ref 8)

work surface temperatures and promote higher forming rates.

Effect of Material Parameters on Forming

As can be seen from the mechanisms of bending outlined earlier, any material parameter that affects the amount of plastic upsetting occurring during the heating cycle will control the amount of shape change (bending) upon cooling. These parameters are likely to be material thickness, thermal expansion coefficient, thermal conductivity, and detailed effects of temperature on the plasticity characteristics of the material. In general terms, any factors likely to affect the temperature gradient between the top (heated) surface of the workpiece and the bottom (cooled) surface, will affect the degree of thermal forming achieved.

Bending trials were reported (Ref 11) using four different materials of thickness 0.8 mm (0.04 in.): stainless steel, mild steel, copper, and aluminum. The results are given in Fig. 10 and show that, in general, under the forming condi-

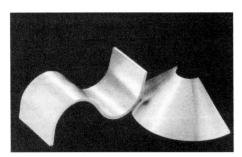

Fig. 7 Three-dimensional thermally formed components (Ref 8)

tions used, the materials behave somewhat similarly with the bend angle increasing with the cumulative increase in the number of plasma scans accomplished. Note the small degree of bending that can be induced with the copper material as compared with the other materials. Figure 11 shows the bend angle cumulatively generated by 12 plasma scans at various travel speeds for the four different materials, but plotted as a function of the material thermal conductivity. This indicates the large effect of that physical parameter on the bending behavior of the sheet material.

This observation is in accord with the concept that the achievement of a large thermal gradient between the front (heated) side and the back (cooled) side of the sheet material would promote more localized thermal expansion and associated plastic deformation, which would enhance the force available to cause bending upon subsequent cooling. It is easier to develop large thermal gradients in materials with relatively low thermal conductivities such as stainless steel. Indeed, one must be careful with such materials because it is possible to induce surface melting under conditions where higher thermal conductivity, but lower melting temperature, materials do not suffer the same degradation.

As noted previously, the basic mechanism of bending sheet metal using a heat source is the plastic compression of the heated zone during heating and the support of the compressed zone during cooling by the surrounding material. Thus, the forming behavior should be similar regardless of the type of heat source, whether it be a laser beam or a plasma torch. For example, the work using plasma heating (Ref 10) strongly indicates that the thermal conductivity of the material has a significant effect on the bending behavior, depending on the particular process parameters used. On the other hand, the findings of a researcher (Ref 12) who conducted bending

experiments using a laser suggested that the thermal conductivity of the material had only a secondary effect on the bend angle. The apparent difference in findings may be due to the two widely differing regimes used in the two sets of experiments. The work in Ref 11 used a sheet thickness of 0.8 mm (0.04 in.) and heat source travel speeds of 2 to 8 mm/s (0.08 to 0.3 in./s), while Vollertsen (Ref 12) used sheet thicknesses of 1 to 7 mm (0.04 to 0.28 in.) and heat source travel speeds of 25 to 583 mm/s (1 to 23 in./s). Thus, according to the data shown in Fig. 11, Vollertsen's results could be compressed toward relatively low heat input rates producing low bend angles and therefore mask the thermal conductivity effect.

Structure and Properties

It must be kept in mind that thermal forming is also a heat treatment operation by virtue of the thermal exposure. Simple or complex metallurgical changes are likely to occur, depending on the particular workpiece material. Mechanical property changes will occur as a result of these metallurgical changes. Limited increases were reported (Ref 13) in hardness in the bend region of AISI 304 stainless steel formed using a plasma torch. However, these increases were significantly less than what was found with conventional mechanical bending at room temperature. Other work by (Ref 10) evaluated the effect of plasma forming on the microstructure of both 304 stainless steel and mild steel. The 304 stainless steel has an austenitic structure over a wide temperature range, and no significant change in microstructure was observed after thermal forming. The mild steel, on the other hand, undergoes a phase transformation from ferrite to austenite on heating and the reverse on cooling. Thermal forming of this

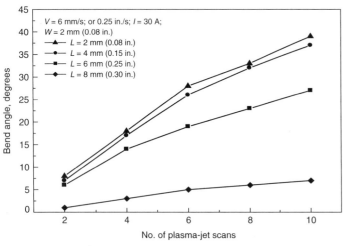
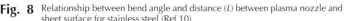

Fig. 8 Relationship between bend angle and distance (L) between plasma nozzle and sheet surface for stainless steel (Ref 10)

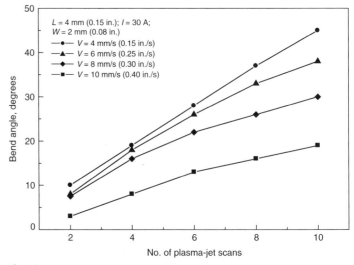

Fig. 9 Relationship between bend angle and plasma jet travel speed (V) for stainless steel (Ref 10)

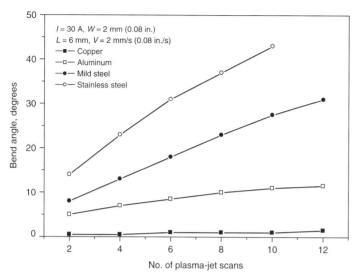

Fig. 10 Bend characteristics of various metals of the same thickness (0.8 mm, or 0.03 in.) under identical process conditions (Ref 11)

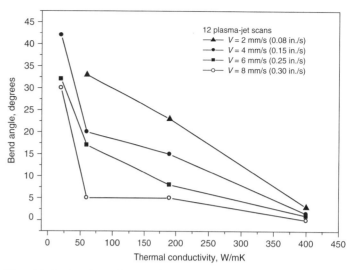

Fig. 11 Bend angle as a function of material thermal conductivity at various travel speeds of a plasma torch (Ref 11)

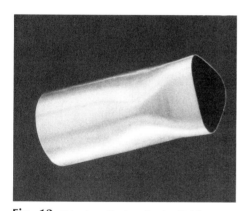

Fig. 12 Tube shaped by laser forming (Ref 8)

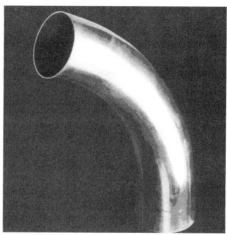

Fig. 13 Tube bend manufactured by laser forming (Ref 8)

Fig. 14 Sheet contoured by multiple laser passes (Ref 21)

material was found to cause considerable grain refinement in the areas most affected by the plasma heating. No associated data were reported for any mechanical property changes.

Other researchers (Ref 14) performed a microstructural investigation of laser forming of aluminum and aluminum alloys using optical and electron microscopy. Zones with different substructures were observed depending on processing parameters. The degree of substructural development could not be explained entirely by laser parameters and material characteristics. Hardness tests of laser-formed material indicated the existence of different zones depending on the specific material and heat treatment.

Applications

For a number of years "flame straightening" has been used to both shape and straighten steel beams and girders. Indeed, it has been used for the in-situ repair and straightening of

damaged bridge structures (Ref 15). Line heating using gas flame torches has also been used for bending and for weld distortion removal in shipbuilding (Ref 16). Such applications rely exclusively on the experience and expertise of the operator.

Because of the potential for automation in producing three-dimensional sheet profiles, much research has been oriented toward analysis and control of the process—for example, the works of various researchers in Britain (Ref 17–20). However, functioning systems are not known to be available at the present time.

It must be stressed that this is an evolving area of technology with no known applications reduced to manufacturing practice. Some potential applications are shown in Fig. 12 to 14. On the other hand, Geiger, Vollertsen, and coworkers in Germany are developing applications in the electronics manufacturing area.

REFERENCES

1. Thermal Forming, *Proc. IWOTE'05: 1st International Workshop on Thermal Forming* (Bremen, Germany), F. Vollertsen and T. Seefeld, Ed., April 13–14, 2005
2. H. Frackiewicz, Z. Mucha, W. Trampoczynski, A. Baranowski, and A. Cybulski, Method of Bending Metal Objects, U.S. Patent 5,228,324, July 20, 1993
3. H. Frackiewicz, Laser Metal Forming Technology, *Proc. Conference Fabtech International '93* (Rosemont, IL), Oct 18–21, 1993
4. M. Gieger and F. Vollertson, Mechanism of Laser Forming, *Ann. CIRP,* Vol 42 (No. 1), 1993, p 301–304
5. F. Vollertsen, I. Komel, and R. Kals, Laser Bending of Steel Foils for Microparts by the Buckling Mechanism—a Model, Modelling and Simulation, *Mater. Sci. Eng.,* 1995, Vol 3 (No. 1), p 107–119
6. A.T. Male, P.J. Li, Y.W. Chen, and Y.M. Zhang, Flexible Forming of Sheet Metal Using Plasma Arc, *Sheet Metal 1999, Proc.*

International Conference (Erlangen, Germany), p 555–560

7. P.J. Li, Y.W. Chen, A.T. Male, and Y.M. Zhang, Flexible Forming of Sheet Metal Using a Plasma Arc, *Proc. Inst. Mech. Eng.*, Vol 214 B, 2000, p 117–125

8. H. Frackiewicz, *Laser Shaping Technology*, publicity brochure of Institute of Fundamental Technological Research, Polish Academy of Sciences

9. A.T. Male, Y.W. Chen, P.J. Li, and Y.M. Zhang, Modeling and Control of Plasma-Jet Forming Process, *Transactions of 28th North American Manufacturing Research Institution of SME*, Vol. XXVIII, 1988, p 39–44

10. A.T. Male, C. Pan, Y.W. Chen, P.J. Li, and Y.M. Zhang, *Processing Effects in Plasma Forming of Sheet Metal, Ann. CIRP*, 2000, Vol 49 (No. 1), p 213–216

11. A.T. Male, Y.W. Chen, C. Pan, and Y.M. Zhang, Influence of Material Physical Properties in Plasma-Jet Forming of Sheet Metals, *Sheet Metal 2001, Proc.*

International Conference (Leuven, Belgium), p 183–190

12. F. Vollertsen, An Analytical Model for Laser Bending, *Lasers in Eng.*, 1994, Vol 2, p 261–267

13. A.T. Male, Y.W. Chen, P.J. Li, and Y.M. Zhang, Material Properties in Plasma-Jet Forming Process, *Sheet Metal 2000, Proc. International Conference* (Birmingham, U.K.), p 371–380

14. M. Merklein, T. Hennige, and M. Geiger, Laser Forming of Aluminum and Aluminum Alloys—Microstructural Investigation, *Sheet Metal 1999, Proc. International Conference* (Erlangen, Germany), p 285–294

15. J.W. Post, Getting the Kinks Out, *Claims Magazine*, Dec 2001

16. "Line Heating," A National Shipbuilding Research Program report, U.S. Department of Commerce, Nov 1982

17. G. Thomson and M. Pridham, A Feedback Control System for Laser Forming, *Mechatronics*, Vol 7 (No. 5), 1997, p 429–441

18. G. Dearden and S.P. Edwardson, "Some Recent Developments in Two- and Three-Dimensional Laser Forming for 'Macro' and 'Micro' Applications," *J. Optics: Pure Appl. Opt.*, Vol 5, 2003, p S8–S15

19. G. Dearden and S.P. Edwardson, Laser Assisted Forming for Ship Building, *SAIL Conference* (Williamsburg, VA), June 2–4, 2003

20. S.P. Edwardson, A.J. Moore, E. Abed, R. McBride, P. French, D.P. Hand, G. Dearden, J.D.C. Jones, and K.G. Watkins, Iterative 3D Laser Forming of Continuous Surfaces, *Proc. 23rd International Congress on Applications of Lasers and Electro-Optics*, Laser Institute of America, 2004

21. L. Sovilla, K. Carcy, and J.R. Dydo, "An Analysis Tool for Forming Complex Shapes in Ship Plate Structures by Line Heating," presented at Ship Production Symposium and Exposition, Society of Naval Architects and Marine Engineers, 2001

Peen Forming

R. Kopp and J. Schulz, Institute of Metal Forming, Aachen University, Germany

SHOT PEEN FORMING is a manufacturing process in which local compressive residual stresses form thin sheet metals and structural components in one or more dimensions. It was developed out of blasting technologies that are better known, such as cleaning, finishing, and hardening blasting. During these blasting technologies, the elastic-plastic impacts of the shot medium onto the thin workpiece often unintentionally cause macroscopic deformations. Shot peen forming uses this normally undesirable phenomenon to specifically form sheet metal components.

The process is primarily used to manufacture large sheet metals and single components or components with a relatively small lot size. Such components are mainly used in the transportation industry, particularly in the aircraft and aerospace industries and in shipbuilding and carriage building, for example (Ref 1, 2).

The forming tool (the balls) can be used universally; that is, the manufacturing of individually formed dies is not necessary. This substantial tool cost reduction leads to high profitability in spite of the relatively long forming cycles according to the size of the component and the forming parameters. The cost advantage increases with decreasing lot size.

that can be divided into eight influence zones (Fig. 1), according to Ref 3. The deformation mainly depends on the tool-workpiece contact area, the deformation zone, the tool, and the shot peening facility. The deformation zone of a single shot depends on the flow resistance of the material, that is, the resistance against being formed, and on the ball energy resulting from the ball mass and the ball velocity. Each shot causes a scallop-shaped deformation on the tool-workpiece contact surface.

Shot peen forming is based on the principle of changing the material state by locally effective inhomogeneous elastic-plastic deformation. With hardening peening, a deformation only occurs near the surface up to a depth of 200 μm, depending on the shot medium, the material, and the parameters selected. Deformation occurring during shot peening can, however, affect thicker surface layers up to the entire cross section of the material. Because the hits of the shot medium or the individual balls onto the workpiece are partly plastic, the kinetic energy involved is transformed into elastic and plastic deformation of the workpiece.

Depending on the local plastic deformation depth of the material beneath each impact, it is possible to generate either a convex or a concave curvature.

Convex Curvature. During shot peening, the tool (the ball) transfers a part of its impulse onto the workpiece during the elastic-plastic impact. When the stress induced is higher than the flow stress of the workpiece material, the contact area is compressed and displaced perpendicular to the force direction. These deformations describe a radius around the ball impression zone, as illustrated in Fig. 2 for peening with and without the influence of friction.

A three-dimensional residual-stress and deformation state beneath the shot impact is produced. In the case of peening under frictional influences, the maximum radial strain is located at a certain depth, because the material flow, and thus the plastic deformation, is obstructed by friction between the ball and the surface of the workpiece.

When a large number of balls hits the workpiece, the material areas lying near the surface undergo a plastic elongation (Fig. 3). This plastic deformation reaches a maximum on the surface, while it is reduced with increasing depth (Ref 4).

The plastic elongation of the component surface implies elastic strains in the other layers of the material, leading to a convex bending of the peened material. The residual-stress distribution

Principle of the Process

Fundamental Mechanisms. Shot peen forming is a flexible pressure-forming process

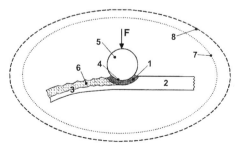

Fig. 1 Zones of influence in shot peen forming. F, force direction; 1, forming zone; 2, properties prior to forming; 3, properties after forming; 4, effective contact area; 5, tool (shot); 6, surface reactions; 7, shot peen forming machine; 8, factory. Source: Ref 3

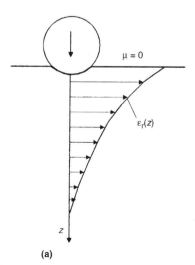 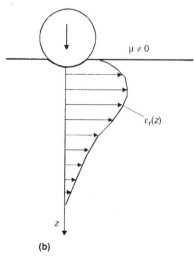

Fig. 2 Radial strains (ε_r) underneath a ball mark with and without friction (μ). (a) Without friction. (b) With friction.

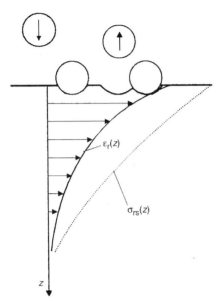

Fig. 3 Plastic deformation and elongation of the near-surface layers. ε_r, radial strain; σ_{rs}, radial stress. Source: Ref 4

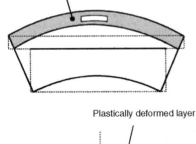

The plastically deformed layer is being elongated.

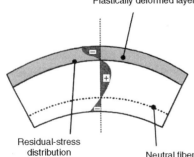

Fig. 4 Principle of convex shot peen forming. Source: Ref 1

produced by this convex bending ensures a balance of forces and moments.

The residual-stress distribution shows residual compression stresses at the surface and residual tensile stresses at greater depths in the workpiece (Fig. 4).

The mechanism of convex forming takes place when shot peening causes only plastic deformation and elongation of a thin surface layer compared to the cross section of the peened workpiece. In this case, the workpiece is bent in the direction of the shot. An elementary model to describe this residual-stress distribution is contained in Ref 1.

Concave Curvature. Plastic deformation of the entire cross section generates a curvature in the direction of the force; the formed sheet metal will thus emerge concave. The depth of the

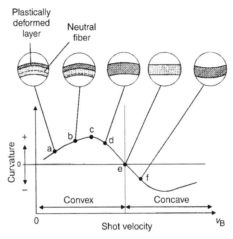

Fig. 5 Strains in the cross section dependent on the shot velocity (v_B). See text for description. Source: Ref 5

plastically deformed layer depends directly on the kinetic energy of the shot medium, which can be modified simply by changing the shot velocity. Figure 5 illustrates the relation between convex or concave curvature and the shot velocity (Ref 5).

At low shot velocities, a convex curvature is produced. The fiber on the top of the workpiece is plastically elongated. The balance of forces and moments induced leads to elastic compression at the bottom fiber. As the shot velocity increases, the neutral fiber initially located in the middle of the cross section moves to the bottom (a, b in Fig. 5). The maximum convex curvature is reached when the neutral fiber is on the bottom surface of the sheet metal (c). When the shot velocity is increased further, the bottom fiber will also be plastically elongated (d), but the plastic elongation is smaller below than above this fiber. Because all fibers of the cross section are now plastically elongated, there is no longer neutral fiber. At the inflection point of the curve, the plastic elongations on the top are as high as on the bottom (e). Despite being elongated, the workpiece remains flat. With even higher shot velocity, the plastic elongations are higher on the bottom than on the top surface (f); the curvature thus turns into a concave form.

To confirm this relation between curvature and shot velocity, tests were conducted with sample strips of the aluminum alloy AlMg$_3$ in the dimensions 250 by 100 mm^2 (Fig. 6). Moreover, the sheet metal thickness was varied (2, 3, and 4 mm). Shot peening tests were carried out using balls with a 6.4 mm diameter (injector-gravitation peening system). The distance between the nozzle and the sample was 200 mm, with a constant mass flow of 20 kg/min. Sample strips were shot peened sinusoidally, with a distance of 10 mm between peening traverses.

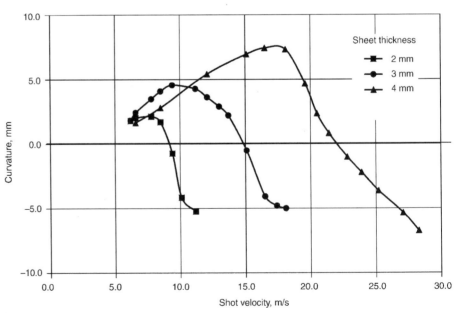

Fig. 6 Curvature dependence on shot velocity. Alloy: AlMg$_3$; workpiece geometry: 250×100 mm^2. Source: Ref 5

Strain rates during shot peen forming are very high. For a rough estimate of local strain rates, the deformation zone beneath the shot impact is simplified to form an axisymmetric cylinder. Instead of homogeneous deformation resulting from a linear velocity distribution, however, the velocity field $v_z(z)$ is assumed to be a squared function, according to Fig. 7, which must be consistent with the shot velocity conditions on the one hand and with the demand for constant volume on the other (Ref 6). The z-direction velocity, $v_z(z)$, is assumed to be a function of the distance z above the depth h_0 (Fig. 7) such that:

$$v_z(z) = f(z^2) = -v_B \frac{z^2}{h^2}$$

where v_B is the velocity of the peen ball. Consequently, the strain rate for axisymmetric deformation is:

$$\dot{\varepsilon}_z = \frac{\partial v_z}{\partial z} = -v_B \frac{2z}{h^2}$$

and, with constant volume:

$$\dot{\varepsilon}_r = -\frac{\dot{\varepsilon}_z}{2} = -v_B \frac{z}{h^2}$$

As an example, using the parameters $v_B = 18$ m/s and $h_0 = 2.1$ mm, a ball impact would produce the following strain rates:

$$\dot{\varepsilon}_z(z = h_0) = 17,140 \text{ s}^{-1}$$

$$\dot{\varepsilon}_r(z = h_0) = 8570 \text{ s}^{-1}$$

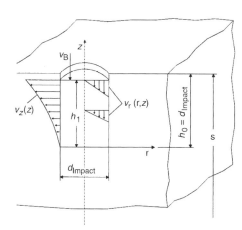

Fig. 7 Velocity field for the determination of strain rates during shot peen forming. Source: Ref 6

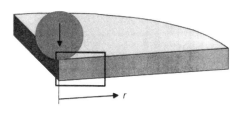

Side A: Ball diameter, 6.4 mm

Fig. 8 Finite element model for single impact

Simulation of Shot Peen Forming

Basic Considerations. The number of balls used for shot peen forming of large sheet metal components is 10^5 to 10^6. Interactions occur between the balls by direct contact and scatter. Modeling is difficult because of the fact that the actual forming process can take several minutes or even hours, whereas the individual ball hits occur within a period of 10^{-4} to 10^{-5} s. Due to the large dimensions of the components, a very rough mesh is required. To be able to describe the deformation zone beneath each shot impact up to a depth of approximately 10^{-1} mm and in order to determine the stress gradients in the cross section, however, the length of the element edges should be no longer than 10^{-2} to 10^{-1} mm. Simplified material laws, with the material considered as a continuum, are usually used to describe the material behavior. In addition, the fact that the initial material may contain inhomogeneities (due to the forming history of the material) may also have an influence on the residual-stress distribution induced by shot peen forming. Moreover, elastic stress caused by clamping or preloading influences the results to a great extent. Also, the flow curve description has to take into account the high strain rates of 10^3 to 10^4 s^{-1}, which makes it difficult to determine suitable material laws. Nevertheless, modern

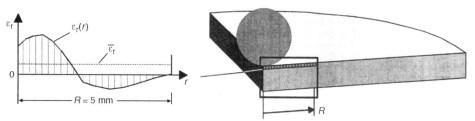

Fig. 9 Determination of average radial strains

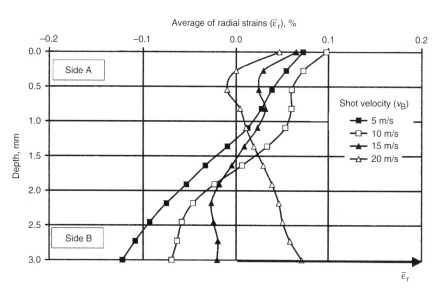

Fig. 10 Average radial strains in the cross section after a single impact. Source: Ref 5

finite element modeling simulation techniques can contribute to a better understanding of the mechanisms operating during shot peening.

Single Impact. Simulations were carried out with different shot velocities and ball dimensions according to Fig. 8 (Ref 5). For the simulations, flow curves for the material AlMg$_3$ were used, which are valid for a range of strain rates from 0.001 s^{-1} to 10,000 s^{-1}. Figure 9 shows the average radial elongation for each layer as a typical measure for a larger area in the radial direction. The average value of the radial strain ($\bar{\varepsilon}_r$) in the radial direction (Ref 7) over a distance R (Fig. 9) is:

$$\bar{\varepsilon}_r = \frac{1}{R} \int_0^R \varepsilon_r dr = \frac{\sum_i \varepsilon_{r,i}}{i}$$

Figure 10 shows the average radial strain distribution over the cross section.

After a shot impact, a specific area on the top surface is elongated. As mentioned previously, at low shot velocities, the fibers at the bottom are compressed. While being high at low velocities, these compressions are reduced and become *elongations* with increasing shot velocity. At the highest shot-velocity of 20 m/s, the bottom side of the component is stretched. This is due to the fact that the short-term stretching of the fibers is higher in lower layers of the cross section than in upper layers. This elongation on the bottom side produced by the shot impact dynamics leads to a concave curvature of the component.

Multiple Impacts. Multiple-impact peening simulations are limited by local geometry and time discretization. When simulating a single impact whose deformation is very local, the mesh chosen for the impact zone has to be very fine in order to simulate the real processes occurring at the contact surface and in the deformation zone.

Because ball impacts have an effect not only in one location but can influence the entire workpiece surface, it is often necessary to have a fine mesh for the entire component. Also, process time discretization has to be fine, with the deformation during each impact calculated in several steps due to the nonlinearity of the process. The chronological order of the impacts has to be simulated as well, so that it is impossible to calculate all impacts simultaneously. That is the reason why correct simulation requires a large number of steps for a large model, entailing a great deal of computation and memory. Figure 11 represents a simulation using 210 balls.

Equipment/Tooling

Peen forming facilities were developed as a spin-off of the first sandblasting facilities designed for cleaning. The first peening facilities were simple manual blasting facilities, still widely used today, followed by the first automatic peen forming machines. Nowadays, peen forming

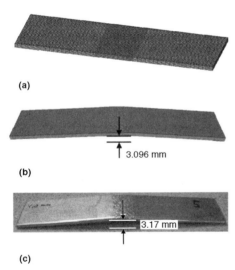

(a)

(b) 3.096 mm

(c) 3.17 mm

Fig. 11 Example simulation of shot peen forming for a metal plate. The conditions for this simulation were: explicit calculation with finite element software LS-DYNA3D (Livermore Software Technology Corp.); geometry: $100 \times 20 \times 2$ mm^3; material: AlMg$_3$; simultaneous double-sided peen forming, upper side: 70 balls, 6.4 mm diam, $v_B = 17.0$ m/s; lower side: 140 balls, 4.0 mm diam, $v_B = 6.1$ m/s; stochastically distributed shot impacts; control of kinetic energy of single impacts with calculation of the contact, three-dimensional mesh; fine discretization in the middle of the sample; elastic-plastic material behaviour. (a) Before the simulation. (b) After the simulation. (c) Real part after shot peening. Source: Ref 5

facilities can be distinguished by the way the shot medium is accelerated (Ref 8). The most important machines are illustrated schematically in Fig. 12.

Centrifugal wheel spinning systems (Fig. 12a) do not use pressure to accelerate the shot medium. The balls are accelerated directly, that is, without any other medium, by the centrifugal force of the spinner. This type is used for a high mass per area and large surfaces to be peened.

The air pressure system (Fig. 12b) uses compressed air to accelerate the shot medium. The compressed air and the shot medium are merged below the reservoir, then transported to the nozzle through a common hose. Shot velocities may be much higher than with spinner peening facilities. The shot medium is more concentrated so that it can be used more effectively.

The injector peening system (Fig. 12c) uses two separate hoses leading to the peening nozzle, one for the compressed air and one for the shot medium. Acceleration of the shot medium does not take place before it is merged with the compressed air in the peening nozzle. The shot medium is sucked into the nozzle by the low pressure generated. Injector peening machines are not very expensive and can be easily automated because the shot medium does not need any transport or conveying facilities.

The injector-gravitation peening process (Fig. 12d) is slightly more complex. It also contains two separate hoses designed for the compressed air and the shot medium. However, the shot medium is not sucked from the reservoir below but transported through a conveying facility to a shot medium supply tank above the peening cubicle. The shot medium falls under gravitational forces through a pipe into the peening nozzle, where acceleration takes place by compressed air. The main advantage of this design is the continuous, smooth exit of the shot medium with relatively low peening pressure.

Shot Properties. In contrast to hardening blasting, the most suitable shot medium for shot peen forming consists of round balls as tools. Therefore, hardened roller bearing balls are generally used, mostly of 100Cr6 (52100) steel with a hardness of approximately 57 to 60 HRC. The shot medium is used in a closed system with return. Ball diameters are usually from 2 to 4 mm, in some cases up to 10 mm.

Quality Control of the Shot. Shot peen forming facilities usually control the quality of the shot in the process (Ref 9) to sort out deformed or broken balls. Starting with a sieving process, the shot medium is sorted into different sizes. The shot is fed perpendicularly, continuously, and at constant velocity onto a rotating sieve. A sorting machine checks the roundness of the balls, using the principle of the inclined plane. The shot medium is fed at a defined velocity onto a transverse continuous conveyor belt. Depending on the quality of the balls, they reach the opposite conveyor edge earlier (round balls) or later (balls out of round). An adjustable withdrawal facility classifies the balls according to their quality into separate boxes. Balls of very poor quality stay on the conveyor, drop at the conveyor edge, and fall into the scrap box.

The facility can be designed to be suitable for any shot medium by changing the conveyor inclination, the conveyor speed, the flow, or the positioning of the withdrawal facility.

Shot Distribution. The shot impacts on a peened component are statistically distributed, with the shot medium usually spread corresponding to a Gaussian distribution (Fig. 13) (Ref 10).

Fixtures. Continuous clamping of the component avoids undesirable inhomogeneities caused by preloading or induced stress. Therefore, the component should not only be fixed at certain points but, if possible, all along its edge. Fixturing on a rubber layer is recommended to allow small movements of the material without inducing additional stresses.

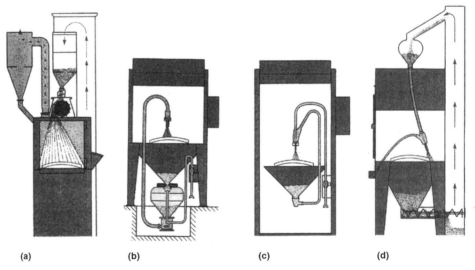

(a) **(b)** **(c)** **(d)**

Fig. 12 Types of peening facilities. (a) Centrifugal wheel peening system. (b) Air pressure peening system. (c) Injector peening system. (d) Injector-gravitation peening system. Source: Ref 8

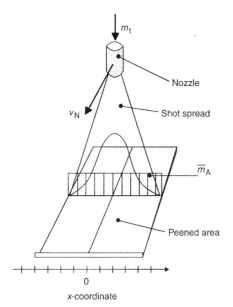

Fig. 13 Gaussian distribution of mass per area in the shot spread. v_N, nozzle velocity; m_t, total mass; m_A, average mass per unit area

Process Parameters and Controls

Influence of Process Parameters on Shot Peen Forming. The success of shot peen forming of sheet metal components depends on a large number of parameters. To a large extent, they can be determined and used effectively in order to achieve a specific effect, but partly they are determined by the facility itself or the process and cannot be influenced by the operator. Figure 14 gives an overview of all of the process parameters for shot peen forming.

It is important to keep in mind that most of the parameters described previously have a direct or indirect influence on each other. A mass flow increase, for instance, leads to a reduction of the shot velocity with constant peening pressure. The shot velocity and the shot distribution depend highly on the nozzle geometry. Therefore, only the use of different inner diameters of the nozzle leads to greatly different shot intensities and thus different forming results, while all other parameters are unchanged.

The most important process parameters in shot peen forming are the shot velocity and the degree of coverage.

Shot Velocity. The shot velocity has to be squared when calculating the kinetic energy during shot peen forming. The kinetic energy (E_{kin}) of a single impact (*i*) is:

$$E_{kin} = \tfrac{1}{2}\dot{m} \cdot t \cdot \bar{v}^2 = \sum_i \tfrac{1}{2}m_i \cdot v_i^2$$

Figure 15 is a schematic overview describing the influence of the shot velocity and other material and the process parameters on the residual stresses induced in the cross section (Ref 11).

Degree of Coverage. Coverage describes the surface covered by peening impacts compared to

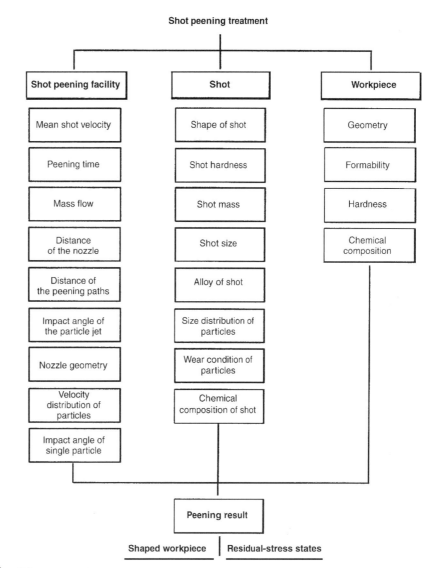

Fig. 14 Variables influencing peen forming

the total surface of the component. It can also be interpreted as the probability of at least one peening impact on an infinitely small surface element of the component. This interpretation is used to formulate the relation between the shot distribution on the component and the coverage as:

$$A^* = 1 - \exp\left(-\frac{m_A}{m_B}A_{id}\right)$$

where id is impact diameter. To calculate the degree of coverage (A^*), it is necessary to know the shot medium quantity per surface area (m_A), the mass of a single ball (m_B), and the surface of a single peening impact (A_{id}). The shot medium quantity per surface is simplified to:

$$m_A = \frac{1}{z} \cdot \frac{1}{v_n} \cdot \frac{dm}{dt}$$

when the shot spread of the peening path is $s > 2\sigma$ of the shot distribution. z is the distance between

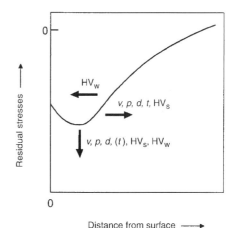

Fig. 15 Effect of alloy and process parameters on the generation of residual-stress states. v, shot velocity; p, peening pressure; t, peening time; d, ball diameter; HV_s, ball hardness; HV_w, workpiece hardness. Source: Ref 11

the peening paths (v_n is nozzle velocity; dm/dt is mass flow).

Material Anisotropy. Sheet materials with a texture may exhibit anisotropic deformation behavior during peen forming. When being peened, such materials tend to develop larger curvatures transverse to the rolling direction or texture orientation than in the longitudinal direction.

Workpiece Geometry. Many components, especially integrated components with a complex geometry and reinforcements, such as stringer and ribs, have different geometrical moments of inertia related to direction. The material flow occurs preferably in the direction of the smallest geometrical moment of inertia, because the lowest forces and moments are needed for material bending.

Peening Tracks. The curvature of a component can also be influenced by selecting or determining the peening tracks (Ref 12). For instance, a pressure peening facility can be used to pass over the entire component in one peening process, thus ensuring 100% coverage (Fig. 16a). This leads to a "wrap-up" of the workpiece from one side by generating a curvature perpendicular to the direction of the peening path or track. The stiffening of the sheet metal produced in this way makes it difficult to form a curvature parallel to the direction of the peening path.

The same degree of coverage can be achieved by several peening processes with partial coverage as well, with increasing distance between the peening paths, for example, while all other parameters are unchanged (Fig. 16b). Curvatures after the first peening are not very pronounced, so that bending parallel to the peening path direction is less obstructed.

Process Controls. Initially, shot peen forming was performed using a trial-and-error method. Technological progress in facilities and measuring techniques now yield better forming reproducibility. All parameters have to be controlled, especially the positioning, the mass flow, and the peening pressure or the speed. Recent developments deal with on-line measuring of the shot velocity, one of the most important peening parameters. The principle of this measuring method is the scanning of the values by a sensor (Ref 13).

Modern shot peening facilities contain process control equipment, enabling the measurement of the shape of the component by contact scanning or by noncontact optic methods.

Process automation of the entire forming process is difficult due to the tolerances in terms of residual stresses, material characteristics,

geometries, and so on, which can have a greater or lesser important influence on the forming behavior. Moreover, operational problems may occur during the process without being registered at once.

Modern process automation follows the principle of adaptive control (Ref 12, 14). With this method, the final degree of coverage required for the desired curvature is achieved in several steps. After each step, the curvature is measured, and the result is used to calculate the forming parameters for the next peening step. Thus, this method allows automatic on-line registering and documentation of all the main process parameters as well as the measurement and subsequent generation of the parameters for the next peening step.

Peen Forming with Prestresses. When peen forming is performed together with prestressing of the component by elastic bending or via a method analogous to stretch drawing, it is possible to reduce the flow resistance of the component considerably. Larger deformations, and thus higher curvatures, can thus be developed to strengthen or, in some cases, prevent specific curvatures in a certain direction.

Because the desired curvature of the component is achieved earlier while the peening intensity remains the same, prestressing can increase the effectiveness and thus the profitability of the shot peening process. This is especially valid for simple shapes for which simple prestressing facilities can be used (Fig. 17).

Materials

In principle, all materials having an elastic-plastic material behavior are suitable for shot peen forming. The process is mainly used to produce large and thin components, as they are used in the aircraft and aerospace industries. Following the need to reduce weight, these are predominantly light metals, above all aluminum alloys. For single-sided concave shot peen forming, the kinetic energy of the balls necessary for plastic deformation on the bottom side of the component has to be high enough, depending on the sheet metal thickness. That is why the sheet

metal thickness suitable for peen forming is from 1 to approximately 8 mm, depending on the specific material. Steel sheet materials can also be used but only with small thickness due to their higher strength.

In general, it has to be considered that the difference between the tool and the workpiece hardness should be high enough to allow high transformation of kinetic energy into deformation energy.

Specifications

Figure 18 shows some examples of simple and more complicated structural components produced with shot peen forming, demonstrating the wide applicability of this process. It is possible to generate geometries curved in one or more dimensions, according to Ref 15.

Curvatures and Accuracy. Due to the large number of different applications, it is impossible to indicate reliable figures. The curvatures that can be achieved as well as the accuracy of the shape depend on the material, the component geometry, and the sheet metal thickness. In general, single-sided convex shot peen forming can be used to obtain radii of curvature of only a few hundred millimeters. On the other hand, there is virtually no limit for the maximum radius of curvatures; that is, almost flat sheet with very large radii of 30 m and even more can be produced. Tolerances for convex shot peen forming are in tenths of millimeters. Single-sided concave shot peen forming can be used to produce curvatures with radii of approximately 1000 to 3000 mm, depending on the component geometry and the material used. Due to higher kinetic energy generated and process variations occurring during concave shot peen forming, compared to convex forming, the accuracy of concave forming with large panels is usually between 1 and 3 mm.

Surface Characteristics. Every shot peening process changes the topography of the peened surface. The surface roughness is increased when the shot velocity is increased or the ball diameter reduced. The harder the shot compared to the workpiece, the lesser the plastic deformation of

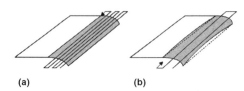

(a)   (b)

Fig. 16 Influence of the nozzle guidance. Source: Ref 12

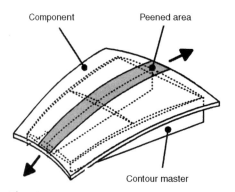

Fig. 17 Examples of clamping fixtures for pretensioning. Source: Ref 12

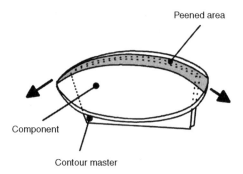

the shot and thus the higher the energy transfer into the workpiece. According to Ref 16, the shot hardness should be at least twice the workpiece hardness to avoid plastic deformation.

When peened, the surface of a component is characterized by circular impressions that are statistically distributed and can thus be overlapping. To give an example, some surface topographies are illustrated in Fig. 19, 14 times magnified, for two ball diameters (6.4 and 4.0 mm diameter) at different shot velocities.

Ball impressions become larger and deeper with increasing velocity, whereas larger ball diameters make local deformation more difficult. As a consequence, larger ball diameters generally lead to lower surface roughness.

Material Removal. Because the shot used for shot peen forming has smooth, round surfaces, there is almost no material removal from the workpiece.

Work Hardening. The impacts of the balls onto the workpiece surface cause work hardening or partial even smoothening. Which of the two effects occurs depends mainly on the material hardness. Smooth materials will harden; hard materials can become smoother (Ref 4). This behavior can be explained by an increase of the dislocation density within smooth materials and the rearrangement of dislocations within hard materials.

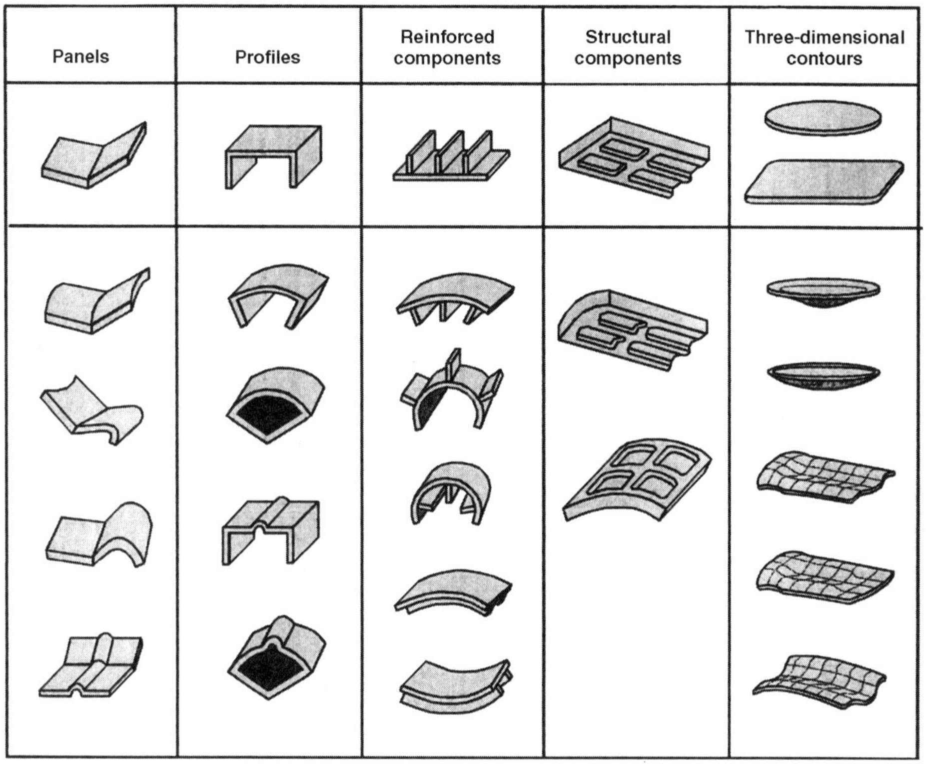

Typical Applications

Aerospace Industry. In the aircraft and aerospace industries, shot peen forming has been successfully used for many years to form numerically controlled milled components, such as airplane wings, stringer-strengthened fuselages (Alpha Jet, Airbus), or structural segments of spacecraft (Ref 12, 17, 18). The first industrial application of the shot peen forming process was as early as 1952 when Lockheed applied the method to form stringer-stiffened wing panels for the Super Constellation (Ref 19).

Table 1 gives an overview of shot peen forming applications in the aircraft and aero-

Panels	Profiles	Reinforced components	Structural components	Three-dimensional contours

Fig. 18 Possible shapes and contours producible by shot peen forming. Source: Ref 15

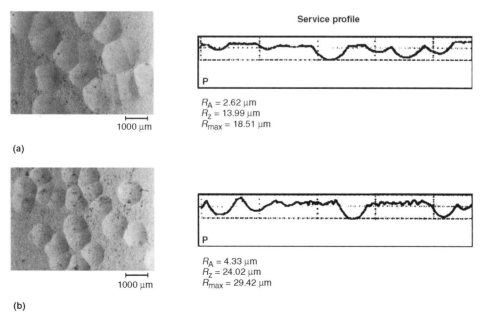

Service profile

(a)
$R_A = 2.62 \ \mu m$
$R_z = 13.99 \ \mu m$
$R_{max} = 18.51 \ \mu m$

1000 μm

(b)
$R_A = 4.33 \ \mu m$
$R_z = 24.02 \ \mu m$
$R_{max} = 29.42 \ \mu m$

1000 μm

Fig. 19 Surface topography of shot peened workpieces with (a) 6.4 mm ball diameter and 15 m/s ball velocity, and (b) 4.0 mm ball diameter and 12.8 m/s ball velocity. R_A, roughness average; R_z, ten-point height (roughness average); R_{max}, maximum peak-to-valley roughness height

Table 1 Typical peen forming applications in the aircraft and aerospace industries

Commercial	• Airbus: A310, A320, A321, A330, A340, A380
	• Boeing: 707, 727, 737, 747, 757, 767, 777
	• Canadair: Regional Jet
	• Ehaviland: DASH 7, DASH 8
	• Lockheed: L1011
	• McDonnell Douglas: MD11, MD80, MD90, MD95, DC10
Business	• Canadair: Challenger, Regional Jet, Global Express
	• Cessna: Citation X
	• Gulfstream: 2, 3, 4, 5
	• Israel Aircraft: Astra, Galaxy
	• Lear: 30, 45, 60
	• Raytheon: Premier One
	• Sino Swearingen: SJ-30
Military	• Embraer: EM120
	• Fairchild: A10
	• Grumman: A6, EA6
	• LTV: S3A
	• Lockheed: P3, C5, C130, C141
	• McDonnell Douglas: F15
	• Northrop: F5E
	• Rockwell: B1
Missiles	• Ariane: 4, 5
	• Atlas II

Source: Ref 20

Fig. 20 Typical application of shot peen forming: Ariane 5 power module frame. Source: Ref 21

(a)

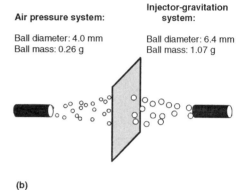

Air pressure system:

Ball diameter: 4.0 mm
Ball mass: 0.26 g

Injector-gravitation system:

Ball diameter: 6.4 mm
Ball mass: 1.07 g

(b)

Fig. 22 Equipment layout for double-sided simultaneous shot peen forming. Source: Ref 5

Fig. 21 Ariane 4 structural tank segment. Source: Ref 22

Fig. 23 Straightening of stringer-reinforced fuselage shells for the Airbus Deutschland GmbH at Kugelstrahlzentrum Aachen GmbH. Source: Ref 21

space industries (Ref 20). Some typical components used in the aircraft and aerospace industries are described in Ref 21 to 24. Examples are shown in Fig. 20 and 21.

Double-sided simultaneous shot peen forming (Ref 25) is an optimized process that enables three-dimensional components to be formed much more efficiently. On the one hand, process productivity is enhanced by conducting shot peening on both sides at the same time; on the other hand, this technique can be used to achieve smaller radii and to produce superposing curvatures. With double-sided simultaneous shot peen forming, the plastic deformation of the entire cross section can be guaranteed despite considerably lower shot intensity.

All parameters used in double-sided shot peening have to be adapted for each side due to the influence they have on each other. Therefore, pretests run for the peening of each side separately cannot be used to determine the optimal parameters for each side by itself. Figure 22 shows the configuration of a facility designed for double-sided simultaneous peen forming.

Straightening of Stringer Reinforcements (Ref 21). Modern aircraft contain fuselages with stringer reinforcements fixed by laser welding. The heat generated during the welding of the stringers onto the fuselage may lead to an unintended deformation of the component. With shot peen forming, it is possible to straighten the fuselages and restore the original shape. This is done by specifically peening local areas of the stringer reinforcements (Fig. 23).

ACKNOWLEDGMENTS

Parts of the presented research results were supported by the Deutsche Forschungsgemeinschaft DFG in Bonn, Germany. The authors also thank the Kugelstrahlzentrum Aachen GmbH for their cooperation.

REFERENCES

1. R. Kopp, Ein Analytischer Beitrag zum Kugelstrahlumformen, *Bänder Bleche Rohre,* Vol 12, 1974, p 512–522
2. W. Köhler, R. Kopp, P. Hornauer, F. Wüstefeld, and K.-R. Baldner, Verfahren zum Umformen von ebenen, plattenförmigen Bauteilen in eine sphärische gekrümmte Form mit definierten Krümmungsradien, German patent DE 3842064 C2
3. K. Lange, *Lehrbuch der Umformtechnik,* Band 1, Springer-Verlag, Berlin/Heidelberg, 1974, p 30, 226
4. B. Scholtes and O. Vöhringer, Grundlagen der mechanischen Oberflächenbehandlungen, *Mechanische Oberflächenbehandlungen,* E. Broszeit and H. Steindorf, Ed., DGM Informationsgesellschaft, Oberursel, 1989, p 3ff
5. J. Schulz, "Geschwindigkeitskontrolliertes Kugelstrahlen und Kugelstrahlumformen," Ph.D. thesis, Aachen University, 2003
6. R. Kopp, K.-P. Hornauer, and H.-W. Ball, "Kugelstrahlumformen, neuere technologische und theoretische Entwicklungen," Second International Conference on Shot Peening (Chicago, IL), 1984
7. R. Kopp and H. Wiegels, *Einführung in die Umformtechnik,* Verlag der Augustinus Buchhandlung, Aachen, 1998, p 18/19
8. F.P. Auer, "Strahlen weltweit," technical documentation, Germany, 2002
9. "Strahlverfahrenstechnik," technical documentation, A.G. Baiker, Switzerland
10. F. Wüstefeld, "Modelle zur quantitativen Abschätzung der Strahlmittelwirkung beim Kugelstrahlumformen," Ph.D. thesis, Aachen University, Germany, 1993
11. O. Vöhringer, *Changes in the State of the Material by Shot Peening,* Third International Conference on Shot Peening (Garmisch-Partenkirchen, Germany), 1987, p 191
12. H.-W. Ball, "Beitrag zur Theorie und Praxis des Kugelstrahlumformens," Ph.D. thesis, Aachen University, 1989
13. R. Kopp, W. Linnemann, and F. Wüstefeld, *Shot Velocity Measurement,* Sixth International Conference on Shot Peening (San Francisco, CA), 1996, p 118–129
14. S. Kittel, W. Linnemann, F. Wüstefeld, and R. Kopp, "Tight Tolerance Peen Forming with On-Line Shape Control," Seventh International Conference on Shot Peening (Warsaw, Poland), 1999

15. R. Kopp and K.-P. Hornauer, "Kugel-strahlumformen, ein flexibles Umformver-fahren," First International Conference on Shot Peening (Paris), 1981

16. H. Breckel, "Kenngrößen und Verschleiß beim Stoß metallischer Werkstoffe," Ph.D. thesis, Stuttgart University, 1967

17. R. Meyer, H. Reccius, et al., *Shot Peen Forming of NC-Machined Parts with Inte-grated Stringers Using Large Balls,* Third International Conference on Shot Peening (Garmisch-Partenkirchen, Germany), 1987, p 327–334

18. R. Kopp and H.-W. Ball, *Recent Develop-ments in Shot Peen Forming,* Third Inter-national Conference on Shot Peening (Garmisch-Partenkirchen, Germany), 1987, p 297–308

19. T.C. Simmons, *Integrally Stiffened Wing Panels Formed by Shot Peening Method,* Western Metals, Jan. 1952

20. "Homepage of Curtiss-Wright Corpora-tion," http://www.curtisswright.com/seg-ments/metal_treatment/shot_peen_forming. asp, Curtiss-Wright Corporation, 2005

21. A. Friese, J. Lohmar, and F. Wüstefeld, "Current Applications of Advanced Peen Forming Implementation," Eighth Interna-tional Conference on Shot Peening (Garmisch-Partenkirchen, Germany), 2002

22. F. Wüstefeld, W. Linnemann, and S. Kittel, "Towards Peen Forming Process Automa-tion," Eighth International Conference on Shot Peening (Garmisch-Partenkirchen, Germany), 2002

23. "Homepage of Metal Improvement Com-pany, Inc.," http://www.metalimprove-ment.com, Metal Improvement Company, Inc., 2005

24. S. Ramati, S. Kennerknecht, and G. Levas-seur, "Single Piece Wing Skin Utilization via Advanced Peen Forming Technologies," Seventh International Conference on Shot Peening (Warsaw, Poland), 1999

25. R. Kopp and J. Schulz, Flexible Sheet Forming Technology by Double-Sided Simultaneous Shot Peen Forming, *CIRP Ann.,* Vol 51 (No. 1), 2002

Age Forming

Jacob A. Kallivayalil, Alcoa Technical Center

AGE FORMING is a shaping process for heat treatable aluminum alloys that is gaining popularity. In this process, parts are given an aging treatment while simultaneously being subjected to mechanical shaping loads. It is used mainly for large parts that have to be imparted a curvature. The part shape is obtained due to the creep that occurs at the aging temperatures (120 to 190 °C, or 250 to 375 °F). Because creep is the phenomenon responsible for achieving the part shape, age forming is sometimes referred to as creep forming. The most common application for age forming is the shaping of upper wing skins in the aerospace industry. Recent investigations (Ref 1) have looked into the possibility of using age forming for damage-tolerant lower wing applications. Age forming is also used for shaping extrusions for the transportation industry. An example of extrusions shaped by age forming is shown in Fig. 1. This method offers several advantages over the conventional process path where the material is typically solution heat treated and then cold formed by shot peening or roll forming.

During conventional cold forming, plastic deformation is imparted to the surface layers of the part such that after the forming loads are released, the part springs back to the desired shape. This results in nonuniform microstructure, because the surface layer has a significantly larger plastic deformation than the bulk. During age forming, the forming loads are often lower than the yield stress of the material, and the part shape is obtained due to the low-temperature creep that occurs during the aging process; hence, there is less nonuniformity in the microstructure. Compared to cold-formed parts, age-formed parts have lower residual stresses and consequently better stress corrosion resistance.

Technical Issues

Creep in aluminum alloys at slightly elevated temperatures was first investigated as part of the effort to develop alloys for supersonic flight (Ref 2, 3), where wing tip heating is a concern. Most investigations (Ref 4–6) of creep in precipitate-strengthened materials have, however, dealt with fairly stable precipitate microstructures. The situation where the precipitate microstructure is unstable has not been investigated thoroughly.

Rösler and Arzt (Ref 7–9) have investigated the interaction between precipitation and creep. They suggested that the creep resistance may be due to an attractive interaction between dislocation segments and dispersed particles. They have also suggested relations for the dependence of the creep rate on the particle size and particle spacing for dispersion-hardened material. According to them, for a fixed volume fraction of particles, the creep resistance increases with increasing particle size up to an optimal size, beyond which the creep resistance falls. The optimal size is determined as a trade-off between the probability of detachment of dislocation segments from a precipitate and the actual athermal stress required for detachment. This optimal precipitate size is not the same as the precipitate size where the peak aged strength is observed. The peak aged strength is determined by the precipitate size where the precipitate cutting and bypassing strengths are equal during dislocation glide.

The variation in age formability with aging may be observed in Fig. 2. Here, the stress relaxation curves for alloy 7055 given four

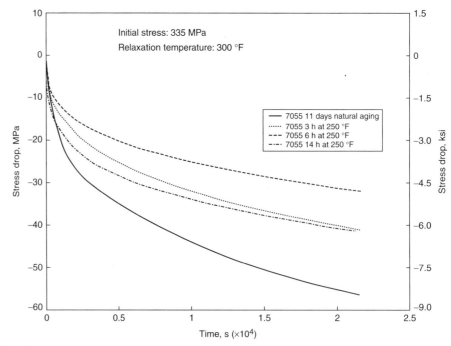

Fig. 1 Age-formed extrusion profiles

Fig. 2 Impact of initial state on relaxation behavior

different initial treatments prior to a stress relaxation test are presented. The stress drop shown in Fig. 2 is the difference between the initial applied stress and the stress measured at any instant. This stress drop is proportional to the retained plastic strain and hence the age formability of the specimen. The creep resistance increases up to 6 h of aging at 120 °C (250 °F) but subsequently decreases for an initial treatment of 14 h at 120 °C (250 °F). This observation is consistent with the mechanism suggested by Rösler and Arzt.

In addition to the effect of the evolving precipitate microstructure on creep, one must also consider what effect the applied stress has on the aging process. The impact of applied stress on the aging process has been addressed in a series of investigations on aluminum-copper alloys by Zhu and Starke (Ref 10–12). They addressed the issues of stress affecting the aging kinetics and the changes to the precipitate morphology as a result of applied stress. From hardness measurements they concluded that the overall aging kinetics were unaffected by

the applied stress, at least for stress magnitudes normally encountered during the age-forming process. The main effect of applied stress was on the orientation of the platelike precipitates that form in aluminum-copper alloys. These effects of applied stress on the aging process have not been assessed in other alloy systems.

Prediction of Final Part Shape

The ability to predict the final part shape is crucial for a successful age-forming operation, because parts that are out of tolerance have to be mechanically formed to get the final part shape, voiding many of the advantages of the age-forming process. The general program for predicting final part shape is shown in Fig. 3. The different steps involved in this program are discussed in further detail during the rest of this article. The basic steps involve developing mechanical tests to study creep at low temperatures and low stresses, describing low-temperature creep in terms of a constitutive model, and then using the constitutive model in a process model or finite element analysis to predict final part shape. For simple cross-sectional shapes such as plates, a process model is sufficient to predict the final part shape. For more complicated cross-sectional shapes such as extrusions, finite element analyses that incorporate the appropriate constitutive models have to be employed.

Testing Methods for Studying Creep. The strain rates encountered during age forming are in the range of 10^{-8} to 10^{-4} s^{-1}. To study constitutive behavior in this regime, stress relaxation tests or creep tests can be used. During stress relaxation tests, a fixed displacement is applied to the specimen, and the variation in load with time is studied. During creep tests, a fixed load is applied to the specimen, and the variation of strain with time is studied. Both types of tests generate equivalent information.

A typical stress relaxation curve is shown in Fig. 4. During a fixed displacement test, the plastic strain rate can be related to the stress rate through the elastic modulus, according to Eq 1. The elastic modulus, $E(T)$, used in determining the plastic strain rate is temperature dependent. The elastic modulus at the testing temperature can be estimated during the unloading part of the stress relaxation test:

$$\varepsilon^{Total} = \varepsilon^{Plastic} + \varepsilon^{Elastic} = constant;$$

$$\frac{d(\varepsilon^{Plastic})}{dt} = \frac{-d(\varepsilon^{Elastic})}{dt} = \frac{-1}{E(T)}\frac{d\sigma}{dt} \quad (Eq\ 1)$$

The differentiation of stress/strain histories presents numerical difficulties due to the noise that is inherent in experimental data. Hence, the stress/strain history is first fit to an algebraic expression and then differentiated. The fitting of algebraic expressions to the experimental data is

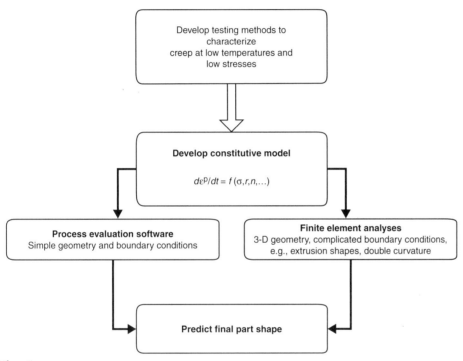

Fig. 3 General program for predicting final part shape during age forming

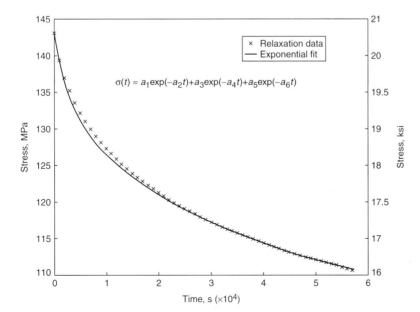

Fig. 4 Typical stress relaxation curve

an ill-posed problem, and proper regularization techniques (Ref 13) should be employed to obtain meaningful fits. A fit of the load history to a sum of exponential terms is shown in Fig. 4. The sum of exponential terms can be differentiated to obtain the creep rate as a function of stress.

Constitutive Model Development

To develop constitutive equations, simple uniaxial tension or compression tests are performed, because the stress state is easy to determine. The constitutive equation predicts the creep rate as a function of the applied stress (σ), temperature (T), and other parameters representing the microstructure:

$$\frac{d\varepsilon^p}{dt} = f(\sigma, T, r, n, \ldots) \qquad \text{(Eq 2)}$$

During aging, the parameters representing the microstructure could be the average radius (r) and number density (n) of precipitates. These parameters evolve, and hence, one needs evolution equations for them to complete the constitutive equation. In the event that explicit evolution equations for microstructural parameters cannot be determined, approximate phenomenological constitutive relations can be obtained by scaling experimentally determined stress-versus-strain-rate curves. Guidelines for scaling data obtained during creep or stress relaxation tests have been provided by Povolo and Rubiolo (Ref 14) and also by Fortes and Rosa (Ref 15).

Once a constitutive model has been developed, it can be verified by applying a four-point bend loading and comparing the final measured curvature to the curvature predicted by the constitutive model. The final curvature can be predicted once the constitutive model is determined, using a general process model described later. The oven and loading fixtures used for these tests are shown in Fig. 5 and 6.

Process Model for Age Forming

The necessary pieces of information required for an age-forming process model are shown in Fig. 7. In Fig. 7, the arrows on the top represent inputs that are material constants or thermal and mechanical boundary conditions. The arrows to the left are inputs representing the evolving microstructure. These inputs have to be initialized prior to running the model. The arrows below represent the quantities that the model tries to predict.

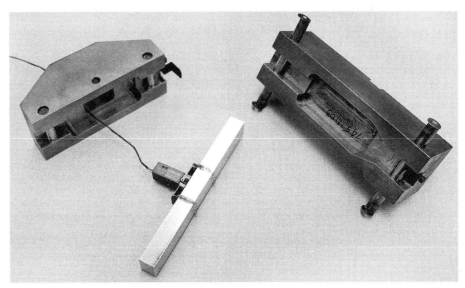

Fig. 6 Bending fixture and specimen with extensometer

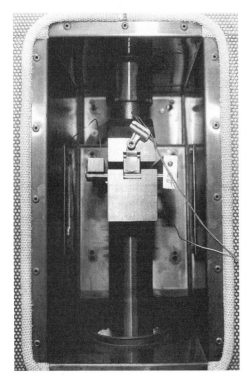

Fig. 5 Bending-loading fixture in temperature-controlled oven

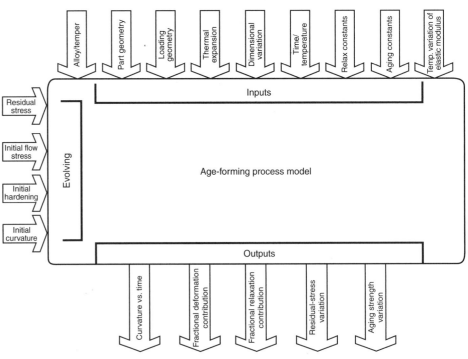

Fig. 7 General process model for age forming

For extrusion and plate products, an initial residual-stress state may be present as a result of heat treating and stretching. These stresses are superposed on those applied by the loading fixture. The residual stresses may be comparable in magnitude to the applied loads and should not be ignored.

Because the loading fixtures try to impart a curvature, there is a variation in applied stress through the cross section of the part. Thus, depending on the initial yield strength of the part and the curvature imposed, it is possible that the outer regions of the part yield plastically. This will result in a curvature that should not be mistaken for age formability. In order to determine the initial stress state after yielding, it is necessary to know the hardening behavior of the material. Once the initial stresses are known, it is possible to determine the impact of the initial plastic deformation on the final curvature of the part by the moment release method described later.

Different means of load application can be used. Dies are typically used for plate products and clamps for extrusions. Occasionally, dies that impart double curvatures are used. Depending on how the part is clamped, the differential thermal expansion between the clamping material and the part could introduce additional thermal stresses, which are superposed on the applied stresses. Further, there is a volume change associated with precipitation that can change the stress state. The volume shrinkage can be significant for large parts such as wing skins.

The final shape after elastic springback can be estimated for simple shapes by the moment release method. For circular dies, a formula for estimating the curvature in terms of the constitutive model is presented in Fig. 8. Sallah, Peddieson, and Foroudastan (Ref 16) have suggested ways of dealing with dies that are not circular in shape. The constitutive model allows one to predict the stress, $\sigma(z,t)$, at any point in the cross section at any time. From this, the retained radius at any time, $r(t)$, can be estimated.

Conclusion

Age forming is a shaping process that offers significant advantages, but it also presents numerous challenges. It is anticipated that with increasing research activity in the area of creep at intermediate temperatures, age forming will become the method of choice for the shaping of large aluminum parts. The issues addressed in this article will help develop alloys and tempers that are particularly suited for the age-forming process.

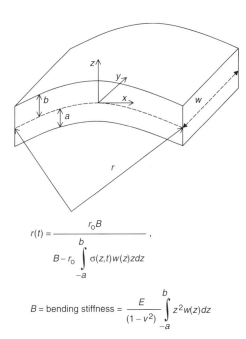

$$r(t) = \frac{r_0 B}{B - r_0 \int_{-a}^{b} \sigma(z,t)w(z)z\,dz},$$

$$B = \text{bending stiffness} = \frac{E}{(1-v^2)} \int_{-a}^{b} z^2 w(z)\,dz$$

Fig. 8 Retained radius calculation using moment release method, where $\sigma(z,t)$ is the stress at position z and time t, r_0 is the die radius, neutral axis $Z = 0$, $w(z) \gg (a+b) \rightarrow$ plane strain, $r \gg (a+b) \rightarrow$ small strain, and $\varepsilon = z/r$.

REFERENCES

1. M.J. Starink, I. Sinclair, N. Gao, N. Kamp, P.J. Gregson, P.D. Pitcher, A. Levers, and S. Gardiner, Development of New Damage Tolerant Alloys for Age Forming, *Mater. Sci. Forum*, Vol 396–402, 2002, p 601–606
2. W.M. Doyle, The Development of Hiduminium-RR58 Aluminum Alloy for Use in the Concorde, *Metallurgia*, Vol 80, Sept 1969, p 101
3. R. Singer and W. Blum, The Influence of Thermomechanical Treatments on the Creep Resistance of RR58 at Elevated Temperatures, *Z. Metallkd.*, Vol 68, 1977, p 328
4. G.S. Ansell and J. Weertman, Creep of a Dispersion-Hardened Aluminum Alloy, *Trans. Metall. Soc. AIME*, Vol 215, Oct 1959, p 838–843
5. R. Lagneborg, Bypassing of Dislocations Past Particles by a Climb Mechanism, *Scr. Metall.*, Vol 7, 1973, p 605–614
6. E. Arzt and M.F. Ashby, Threshold Stresses in Materials Containing Dispersed Particles, *Scr. Metall.*, Vol 16, 1982, p 1285–1290
7. J. Rösler and E. Arzt, The Kinetics of Dislocation Climb over Hard Particles—I. Climb without Attractive Particle-Dislocation Interaction, *Acta Metall.*, Vol 36, (No. 4), 1988, p 1043–1051
8. J. Rösler and E. Arzt, The Kinetics of Dislocation Climb over Hard Particles—II. Effects of an Attractive Particle-Dislocation Interaction, *Acta Metall.*, Vol 36 (No. 4), 1988, p 1053–1060
9. J. Rösler and E. Arzt, A New Model-Based Creep Equation for Dispersion Strengthened Materials, *Acta Metall.*, Vol 38 (No. 4), 1990, p 671–683
10. A.W. Zhu, J. Chen, and E.A. Starke, Jr., Precipitation Strengthening of Stress-Aged Al-xCu Alloys, *Acta Mater.*, Vol 48 (No. 9), 2000, p 2239–2246
11. A.W. Zhu and E.A. Starke, Jr., Stress Aging of Al-Cu Alloys: Computer Modeling, *Acta Mater.*, Vol 49 (No. 15), 2001, p 3063–3069
12. A.W. Zhu and E.A. Starke, Jr., Materials Aspects of Age-Forming of Al-Cu Alloys, *J. Mater. Process. Technol.*, Vol 117 (No. 3), Nov 2001, p 354–358
13. A.N. Tikhnov and V.Y. Arsenin, *Solutions of Ill-Posed Problems*, V.H. Winston & Sons, 1977
14. F. Povolo and G.H. Rubiolo, Scaling Relationship in the Log σ-Log ε Creep and Stress-Relaxation Curves and the Plastic Equation of State, *J. Mater. Sci.*, Vol 18, 1983, p 821–826
15. M.A. Fortes and M.E. Rosa, The Form of a Constitutive Equation of Plastic Deformation Compatible with Stress Relaxation Data, *Acta Metall.*, Vol 32 (No. 5), 1984, p 663–670
16. M. Sallah, J. Peddieson, Jr., and S. Foroudastan, A Mathematical Model of Autoclave Age Forming, *J. Mater. Process. Technol.*, Vol 28, 1991, p 211–219

Forming of Bar, Tube, and Wire

Shearing of Bars and Bar Sections

BARS AND BAR SECTIONS are sheared between the lower and upper blades of a machine in which only the upper blade is movable. As the upper blade is forced down, the work metal is distorted and caused to fracture. There are also shears, such as impact cutoff machines, that use a horizontal knife movement to shear the bar sections. Figure 1 shows the appearance of a sheared round bar. The burnished area, or depth of shear action by the blade, is usually one-fifth to one-fourth the diameter of the bar. In visual examination of a sheared edge, the burnished portion appears smooth, while the fractured portion is comparatively rough. Shearing, in some cases, may be associated with crude cuts and distorted parts—generally when parts are cut only on inexpensive universal-type shears. However, with proper tooling and machines, nondistorted, burr-free cuts are possible.

Applicability

In structural and bar shearing, the cost of cutoff is a small percentage of the overall cost. Materials handling and setup, along with die changes, are most important from a cost standpoint.

The cost of cutting off a bar of steel (Fig. 2), regardless of how it is cut, is determined by the following formula:

$$C = \frac{S+T}{N} + R$$

where C is total cost per part, S is setup cost, T is tool cost, N is number of pieces to be cut, and R is running cost (feed, cut, discharge).

In general, any metal that can be machined can be sheared, but power requirements increase as the strength of the work metal increases. Further, blade design is more critical and blade life decreases as the strength of the work metal increases. Equipment is available for shearing round, hexagonal, or octagonal bars up to 152 mm (6 in.) in diameter or thickness, rectangular bar and billets up to 75 by 305 mm (3 by 12 in.) in cross section, and angles up to 203 by 203 by 38 mm (8 by 8 by 1½ in.).

Straight blades can be used to shear bars and bar sections, although a considerable amount of distortion occurs, as shown in Fig. 1. In addition, the concentration of shock on the blades is high when shearing with straight blades (particularly when shearing round bars). Preferred practice is to use blades that conform to the shape of the work metal, as discussed in the section "Blade Design and Production Practice" in this article. Angles are usually sheared in a special machine or in a special setup with conforming blades.

Because of the conditions illustrated in Fig. 1, the edges sheared with straight blades are not as high quality as edges that are sawed or otherwise machined. However, when blades are sharp and accurately adjusted, sheared edges that are acceptable for a wide range of applications can be obtained. The quality of sheared edges usually increases as thickness of the work metal decreases.

Accuracy of Cut. Workpieces properly supported on both sides of the shear blades by a roller conveyor table and placed squarely against a gage stop securely bolted to the exit side of the machine can ordinarily be cut to lengths accurate to +3.2, −0 mm (+⅛, −0 in.) on shears that can cut bars up to 102 mm (4 in.) in diameter. When larger shears are used, the breakaway of the metal can cause a variation of ±4.8 mm (³⁄₁₆ in.). Fairly consistent accuracy in the shearing of slugs can be obtained by careful adjustment of the gage setting, especially if the slugs are produced on a weight-per-piece basis. Supporting the free end of the material on a spring-supported table will minimize bending during the shearing operation, thus providing better control over the length of cut.

Selection of Cutoff Method. The method of cutting off bars can be determined by the edge condition required for subsequent operations. Sawing usually produces a uniform cut edge with little or no damage to the microstructure in the immediate area. Gas cutting produces an edge that resembles a sawed edge in smoothness and squareness. However, the cut edge of some steels becomes hardened during gas cutting, thus making subsequent machining difficult.

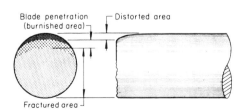

Fig. 1 Effects of shearing a round bar with a straight blade

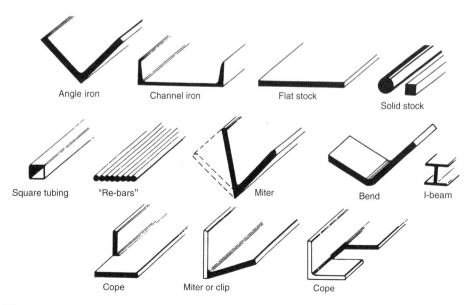

Fig. 2 Typical shapes that can be sheared

A sheared edge is usually easy to machine but can make the fit-up of parts of a weldment more difficult and can increase warpage because of wider gaps.

Power Requirements. The net horsepower (hp) required for shearing can be estimated from the following formula:

$$hp = A \cdot V \cdot S / 33,000 \qquad \text{(Eq 1)}$$

where A is the cross-sectional area of the workpiece (in square inches), V is the speed of the shear blade (in feet per minute), and S is the shear strength of the work metal (in pounds per square inch). The 33,000 is foot-pounds per minute per horsepower. For metric use, the power in English units (hp) should be multiplied by 0.746 to obtain kilowatts. It may be necessary to increase the calculated value as much as 25% to compensate for machine inefficiency.

Although Eq 1 is used for estimating, it is of limited value because it does not consider the ductility of the metal. The formula is based on shearing low-carbon steel. Copper, for example, is more ductile than steel; therefore, the distance of blade penetration before fracture in copper will be greater than that in steel. Conversely, when shearing metals that are less ductile than low-carbon steel, the distance of blade penetration before breakaway will be less. Power requirements are also affected by the ductility of the work metal.

Cutting Speed. The speed at which material is sheared without adverse effect can range from almost zero to 21 or 24 m (70 or 80 ft) per min. However, as speed increases above 6.1 or 7.6 m (20 to 25 ft) per min, problems are encountered in holding the workpiece securely at the blade without the far end whipping, especially with material 6.4 mm ($\frac{1}{4}$ in.) thick or more. When bars harder than 30 HRC are cut at speeds of 12 to 15 m (40 to 50 ft) per min or higher, chipping of the blade is common (see the section "Example 1: Reduction in Speed for Prolonged Blade Life" in this article).

Alternative Cutoff Methods. Sawing with a hacksaw, a radial saw, or a band saw is the most widely used cutting method, but it is also the most costly. Sawing must be used for cutting certain shapes and sizes for economic or quality reasons, for example, cutting of thick-wall round tubing. However, the average plant does not have to saw everything, and consequently, vast savings can be realized on parts that can be sheared.

Table 1 is a chart compiling the times required to cut various types of bars using a hacksaw, a band saw, a high-speed steel band saw, and a hydraulic shear. The cost of deburring the sawed ends is not included, but in most cases, this operation is required after shearing. Another cost not included is the blade cost per cut, which is much higher for sawing than for shearing.

Abrasive and friction cutting offer the advantages of cutting various shapes without changeover costs, production of flat nondistorted ends, and rapid cutting. Their main drawbacks are the high blade cost per cut and the cost of deburring both ends of the cut part. A very sharp razor edge is usually produced and is dangerous if not completely removed.

Punching and Shear Machines

Punching and shearing machines are deep-throat C-frame machines for the punching, shearing, notching, or coping of plates, bars, and structural sections. The production shearing of bars and bar sections is usually done in machines with a throat opening designed for large, bulky workpieces. High-quality cut slugs or blanks of precise length from round, square, or specially shaped bars or coils can also be produced with impact cutoff machines (see the section "Impact Cutoff Machines" in this article).

Iron workers (Fig. 3) are heavy-duty shears used primarily for preparing structural shapes such as bars, angles, and T-bars. Like the shears used for plate and flat sheet, iron workers leave a square butt edge that may require further preparation for heavier sections. These machines include alligator and guillotine shears and multipurpose (combination) machines. The multipurpose machines feature interchangeable punches and dies for shearing, punching, and coping. Squaring shears, normally used for sheet and plate, can also be used for cutting bar stock to length.

Punch presses and press brakes can be provided with appropriate tooling for cutting operations. Shoes, into which punches, dies, and shear blades can readily be inserted, are mounted on the bed and ram. The shoes either rest on the bed or are overhung; the overhanging type is designed so that structural shapes can be punched in both web and flange. The plain type of shoe is primarily used for plate work; however, plates can be worked with the overhanging shoes. Both types of die blocks are fitted with die sockets that hold dies of different inside diameters. The punch holder is adjustable to suit the location of the dies. Table 2 lists the capacities for punching and shearing with this type of machine.

Hydraulically powered shears help fill the need for fast-cycling shear cutting of structural and bar-shaped parts with little or no distortion or burring. Hydraulic power can apply high forces throughout a relatively long stroke, due to the rake angle of the shear blade. Hydraulic power is noted for its maintenance-free performance, shock-free operation, and the ability to operate more than one machine from a single power source. The stroke of the ram can also be controlled for quicker cycle times. For example, if the cycle time required for cutting 150 by 150 by 13 mm (6 by 6 by $\frac{1}{2}$ in.) angle iron is 7 s, the cutting of 50 by 50 by 6 mm (2 by 2 by $\frac{1}{4}$ in.) angle iron can be done with a cycle time of only 3 s when the stroke length is reduced by more than half. Several different stroke lengths may be

Fig. 3 Hydraulic iron worker with 590 kN (66 tonf) punching capacity

Table 1 Time comparison between sawing and hydraulic shearing

Size		Material or shape	Time required for cutting by:				Time savings, %, realized by shearing compared with:		
mm	in.		Hacksawing	Band sawing	High-speed steel band sawing	Shearing	Hacksawing	Band sawing	High-speed steel band sawing
32	1$\frac{1}{4}$	1017 steel	30 s	1 min, 2 s	7 s	2 s	93	97	71
5×127	$\frac{3}{16}$×5	1018 steel	27 s	1 min, 15 s	5 s	3 s	89	96	40
13×228	$\frac{1}{2}$×9	1018 steel	3 min, 54 s	6 min, 15 s	18 s	3 s	99	99	83
150	6	Channel	45 s	2 min, 54 s	23 s	6 s	89	96	74
13×100×100	$\frac{1}{2}$×4×4	Angle iron	54 s	3 min, 38 s	21 s	6 s	88	97	71
13×150×150	$\frac{1}{2}$×6×6	Angle iron	1 min, 59 s	5 min, 50 s	34 s	6 s	94	98	82
5×75×75	Miter $\frac{3}{16}$×3×3	Angle iron	...	1 min, 10 s	12 s	6 s	...	91	50
13×100×100	Miter $\frac{1}{2}$×4×4	Angle iron	...	4 min	30 s	6 s	...	97	80

available on hydraulic machines for selection by the operator.

Alligator Shears

In alligator shears (also known as pivot or nutcracker shears), the lower blade is stationary, and the upper blade, held securely in an arm, moves in an arc around a fulcrum pin (Fig. 4). The shearing action is similar to that of a pair of scissors. A crankshaft transmits power to the shearing arm, and the leverage applied produces the force for shearing. Maximum shearing force is obtained closest to the fulcrum, and the mechanical advantage decreases as the distance between the point of shearing and the fulcrum increases. Therefore, with the maximum opening (largest rake angle), capacity is maximum for any shear. As the blade begins downward travel and rotation about its fulcrum, the cross-sectional area engaged by the blade is increased; therefore, more energy is expended, and the upper blade is slowed down until the breakaway point is reached. At this point, the mechanical forces have overcome the resistance of the metal, and the remainder of the cross-sectional area breaks off.

The capacity of an alligator shear is designated as the maximum cross-sectional area of flat plate or round bar that can be sheared, based on work metal having a tensile strength of 275 MPa (40 ksi). An alligator shear can be used not only for bars and bar sections but also for plate, sheet, and strip within the limitations of the blade length of the machine. Alligator shears are extensively used for preparing scrap because the single pivot point allows the blades to open wide enough to accept bulky objects. For example, on a machine with 300 mm (12 in.) long blades, the opening between upper and lower shear blades is approximately 135 mm (5¼ in.); on a machine with 915 mm (36 in.) long blades, the opening can be as great as 280 mm (11 in.).

The two types of shearing arms used on alligator shears are low-knife and high-knife (Fig. 4). In the low-knife type, the cutting edge of the lower blade is in line with the center of its fulcrum pin; in high-knife shear, the cutting edge of the lower blade is on a plane above the centerline of the fulcrum pin. In general, the low-knife shear is preferred for cutting bars and bar sections. The high-knife shear is preferred for cutting flat stock and for use in scrap yards. A high-knife shear, by making successive cuts, can shear flat stock that is wider than the length of the shear blades. An alligator shear is further classified as either right-hand or left-hand (each is shown in Fig. 4), depending on the side from which it is fed.

The weight of alligator shears ranges from approximately 1,100 to 19,500 kg (2,500 to 43,000 lb). The lighter shears can be made portable on wheels, on a sled-type skid, or on blocks. The heavier machines must be anchored in concrete.

The speed of alligator shears ranges from approximately 50 strokes per minute for the smallest power-driven machine to approximately 18 strokes per minute for the largest heavy-duty shear. Except for small hand-operated types, alligator shears are usually mechanically driven; a flywheel provides uniform, sustained power during cutting.

The number of strokes per minute can be reduced by changing either the motor speed or the diameter of the flywheel pulley. An increase in the number of strokes may affect the stored energy of the flywheel and reduce the shearing capacity. On continuously cutting shears, greater speed will reduce workpiece-positioning time for the operator and therefore may reduce output rather than increase it. On single-stroke machines, cutting time is minimal compared to the time required for accurate positioning of the stock.

Guillotine Shears

Guillotine shears are designed for cutting bars and bar sections to desired lengths from mill stock. They are extensively used throughout the fabricating industry. Guillotine shears for bars and angles are available in capacities to 2700 kN (300 tonf). Either intermittent or continuous operation is possible. Guillotine shears can be equipped with simple straight blades (Fig. 5) or with two or more short blades having specific shapes.

Two general types are available: open end and closed end. An open-end shear has a C-frame construction (Fig. 5) with one end open and unsupported. Open-end shears are either single end, for one operator, or double end, for two operators. On double-end machines, both ends can be right-hand or left-hand, or one end can be right-hand and the other end left-hand, depending on the type of shearing to be done.

An open-end shear has the advantage of giving the operator a clear view of the blades. However, because one end is open, a heavier frame and more floor space are required than for a closed-end shear of equal capacity. A closed-end guillotine shear, on the other hand, is basically the same as the one shown in (Fig. 5) except that it has frame supports on both sides.

Guillotine shears are actuated mechanically, hydraulically, or pneumatically. Hydraulic and pneumatic machines are lighter in weight for a specific shearing power, are more economical,

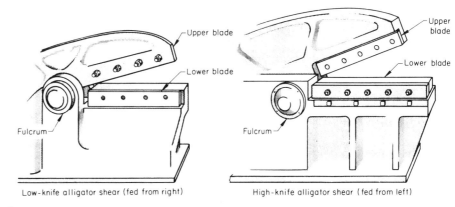

Fig. 4 Low-knife and high-knife alligator shears. See text for discussion.

Table 2   Maximum workpiece dimensions that can be accommodated in vertical open-cap punching and shearing machines of various tonnage ratings

Machine rating, kN (tonf)	Punching		Shearing					Size of angle	
	Hole diam, mm (in.)	Plate thickness, mm (in.)	Plate thickness, mm (in.)	Size of bar section or plate, mm (in.)	Diam of round, mm (in.)	Size of square, mm (in.)		Cutting on square, mm (in.)	Cutting on angle, mm (in.)
445 (50)	21 (¹³/₁₆)	19 (³/₄)	13 (½)	127×13 (5×½)	38 (1½)	32×32 (1¼×1¼)		75×75×9.5 (3×3×³/₈)	102×102×9.5 (4×4×³/₈)
870 (97½)	32 (1¼)	25 (1)	22 (⁷/₈)	152×25 (6×1)	50 (2)	38×38 (1½×1½)		102×102×13 (4×4×½)	152×152×9.5 (6×6×³/₈)
1390 (156)	50 (2)	25 (1)	29 (1¹/₈)	203×25 (8×1)	64 (2½)	50×50 (2×2)		152×152×13 (6×6×½)	. . .
2780 (312½)	64 (2½)	38 (1½)	38 (1½)	254×38 (10×1½)	89 (3½)	75×75 (3×3)		203×203×19 (8×8×³/₄)	. . .
4450 (500)	102 (4)	38 (1½)	50 (2)	254×64 (10×2½)	127 (5)	89×89 (3½×3½)		203×203×32 (8×8×1¼)	. . .
6230 (700)	152 (6)	38 (1½)	64 (2½)	305×75 (12×3)	152 (6)	108×108 (4¼×4¼)		203×203×38 (8×8×1½)	. . .

and operate with less vibration than mechanical shears. Therefore, hydraulic or pneumatic machines up to 90 kN (10 tonf) capacity are completely portable, and units of 1800 to 2700 kN (200 to 300 tonf) capacity are semiportable (need not be solidly mounted in concrete).

Combination Machines

Combination machines are multipurpose machines used primarily in metal-fabricating shops where there is a constant need for shearing small quantities (often two or three pieces) of a variety of shapes and sizes from bars or bar sections. Some combination machines can also be used for punching, slotting, and notching.

Many combination machines incorporate several devices within the frame for performing different operations; therefore, a new setup is not required for each. A holder for an interchangeable punch and die is located in an area with a deep throat. This facilitates the punching of holes, slots, or notches in plates and bars, and in webs or legs of structural members.

A slide moving at 45° from vertical carries a blade for shearing angles. The support bed is on a swivel so that the ends of the angle section can be varied as desired from 45 to 90°. In two strokes of the machine, angles can be sheared to produce miter joints for subsequent welding (see the section "Shearing of Angles" in this article).

Combination machines can be used for cutting square and rectangular notches in the leg of an angle. These machines can be set up to cut a 90° V-shaped notch in angles that subsequently will be bent into frames. Other shapes, such as beams and channels, can be notched in a similar manner if the machine can accommodate the vertical height between the upper and lower legs of the workpiece. Provision is also made for shearing bars with guillotine-type blades or special blades (see the section "Conforming Blades" in this article).

Combination machines are available in capacities ranging from 110 to 890 kN (12 to 100 tonf). The 110 kN (12 tonf) machine can punch a 14 mm (9/16 in.) diameter hole through a 6.5 mm (1/4 in.) thick section and can shear 75 by 75 by 6.4 mm (3 by 3 by 1/4 in.) angle sections, 22 mm (7/8 in.) diameter rounds, 19 mm (3/4 in.) squares, and 100 by 6.5 mm (4 by 1/4 in.) flats. The 890 kN (100 tonf) machine can shear 150 by 150 by 16 mm (6 by 6 by 5/8 in.) rounds, 50 by 50 mm (2 in.) squares, and 200 by 19 mm (8 by 3/4 in.) flats.

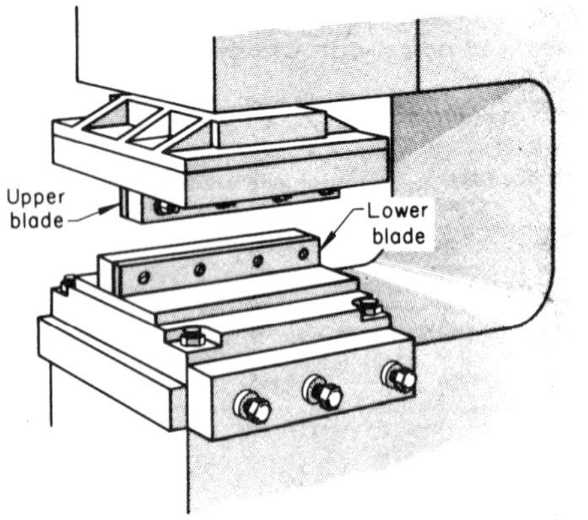

Shear Blades

Shock-resistant tool steels such as S2 or S5 are most commonly used as blade materials for cold shearing, although L6 has also been successfully used for some applications. Some plants use shear blades made from carbon or alloy steel with hardfaced shearing edges.

The hardness of tool steel blades usually ranges from 45 to 55 HRC (sometimes as high as 58 HRC). The upper end of this range can be used when work metal thickness does not exceed 13 mm (1/2 in.). As work metal thickness increases, the hardness of the blade should be decreased to the lower end of this range, but not below 45 HRC unless experience with previous applications warrants it. The practice in most plants is to start with blades near the lower end of the range, increasing their hardness only after experience proves it safe. Excessive blade wear is usually preferable to blade breakage.

Blade Material for Hot Shearing. Blades for the hot shearing of bar stock are usually made of H11, H12, or H13 hot work tool steel. Tool steel for blades can also be made of compositions manufactured by powder metallurgy processes. There are no data to prove the superiority of one of these steels over the others. Grades H21 and H25 are sometimes used, but they are more costly and are recommended only when H11 has been tried and found to be inadequate.

The hardness of blades for hot shearing varies considerably with the thickness and temperature of the metal to be sheared and with the type and condition of the shearing equipment. However, hardness is usually maintained at 38 to 48 HRC.

For high-alloy metals to be sheared at high temperatures, higher-alloy blades may be needed. High-temperature engine-valve alloys have been sheared with T1 high-speed steel blades. Hardfaced blades are satisfactory for hot shearing and are used exclusively in some plants. The material for the blade body is usually 1030 or 1045 steel.

Blade Profile. The cross section of an alligator shear blade for cutting bars and shapes is normally rectangular. Light-duty blades are approximately 32 mm (1 1/4 in.) wide by 100 mm (4 in.) deep by 300 mm (12 in.) long. Blades for machines of approximately maximum size are commonly about 50 by 125 by 915 mm (2 by 5 by 36 in.). For mounting, blades are provided with countersunk holes, as shown in Fig. 6, that allow bolt heads to be sunk sufficiently to prevent interference between blades.

Blade clearance for shearing bars and bar sections ranges from 0.13 to 0.38 mm (0.005 to 0.015 in.). The smaller clearance is used for shearing clean work metal; the larger clearance is preferred for shearing scaly products to prevent scale or other foreign material from lodging between blades and scoring the surfaces.

Blades for alligator shears are available with grooves across the width to prevent forward movement of the work metal when the upper blade descends, so that more of the cutting length of the blade can be used. Most blades have four cutting edges that are identically ground (Fig. 6); therefore, by inverting the blade and reversing its direction, all four cutting edges can be used before the blade is returned for sharpening. Resharpening any of the four edges requires grinding of one or both faces of an edge. Consequently, a blade that shows severe damage, such as breakout of a section, must be ground to a new, clean, and sharp edge. To avoid such major regrinding, blades should be kept free from large nicks and mushrooming.

Most blades are ground to a slight negative rake, as shown in Fig. 6. The intent is to cause the work metal to begin to flow from a slight bending action before actual shearing takes place. Blades provided with a negative rake of 5 to 10° are often less susceptible to chipping at the cutting edge than those ground with a 90° edge (zero rake).

Shear Blade Life. Service data on shear blade life are scarce because maintenance programs in most high-production mills call for removal of blades and redressing during scheduled shutdowns, regardless of the condition of the blades at the time. Blade life in number of cuts before regrinding has been variously reported at 5000 to more than 2 million. Even when an attempt is made to make blade material and cutting conditions as nearly identical as possible, variations in blade life of 100% or more have been reported.

Blade life depends to a great extent on the composition and hardness of the work metal (see the article "Selection of Materials for Shearing, Blanking, and Piercing Tools" in this Volume. The angle of the cutting edge often affects blade life, and in some cases, shearing speed has a marked effect on blade life. For example, harder work metal usually requires a lower shearing speed in order to avoid blade chipping and premature dulling. When work metal hardness is

Fig. 5 Open-end guillotine shear

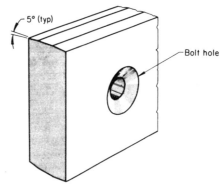

Fig. 6 Straight shear blade ground with negative rake on all four cutting edges

30 HRC or higher, speeds greater than 15 m (50 ft) per minute are not recommended, and much slower speeds may be required for acceptable blade life. The following example demonstrates the effects of blade angle and speed on blade life.

Example 1: Reduction in Speed for Prolonged Blade Life. Blades had to be replaced every 2 to 3 weeks (200 to 500 cuts) because of chipping when shearing 1085 flat spring steel with hardness ranging from 30 to 34 HRC. The stock was 3.2 and 4.8 mm (1/8 and 3/16 in.) thick and 50 mm (2 in.) wide, and it was sheared to lengths of 1.2 and 1.5 m (4 and 5 ft) at a blade speed of 21 m (70 ft) per minute in a punch press. In addition, regrinding was required during this period. When the procedure was changed to shearing at 3 m (10 ft) per minute in a C-frame shear, blade life was increased to an average of 10,000 cuts before regrinding was required, and chipping was eliminated. Blades for both the punch press and the shear were made from S1 tool steel and ranged in hardness from 54 to 56 HRC.

Cost. Because of the relatively small amount of machining needed to make a shear blade, compared with the machining needed to make an intricate impression in a die block or a forming die, the cost of the material is an important part of the total cost of a blade. A blade made from S2 tool steel costs 1.8 times as much as one made from W2 tool steel, and 0.7 times as much as one made from D2 tool steel.

Blade Design and Production Practice

The straight-edge blades (Fig. 6) described in the preceding section can shear almost any bar or shape that is within the capacity of the machine. However, unacceptable distortion may result in some shapes of workpieces when they are sheared with blades that are not designed for cutting specific shapes.

Conforming Blades. One method of minimizing distortion in sheared bars employs two hardened blades mounted face-to-face, with identical holes through each blade. The holes should conform to the shape of the work metal and should be large enough to allow easy passage through the blades (Fig. 7a). One blade is movable vertically and one is stationary. Relatively little movement of the machine is required when blades of this type are used. In addition, because the blades completely encircle the work metal, holddowns are not needed. However, these blades are usually limited for use on specially built or combination machines.

The shearing of round and square bars is more frequently done with the open-type blades illustrated in Fig. 7(b). Each blade is contoured to accommodate one-half the cross section of the work metal. The upper blade moves in a vertical direction, while the lower blade remains stationary. When using this technique, some type of holddown is needed. Because of the stiffness of the work metal, the holddown for bars should permit slight movement of the work metal in the axial direction to avoid double shearing. The holddown can be a simple set screw (to permit adjustment) in a bracket or can be a more elaborate unit, such as a handwheel assembly using an Acme thread.

For shearing square bars with any type of blade, the work metal should be placed so that the movement of the blade is across the diagonal of the square. With this technique, the shearing force is applied to four sides instead of two, resulting in a smoother sheared surface. Shearing across the diagonal provides support on two sides of the square shape, which minimizes distortion, and permits more than one size of bar stock to be sheared in a given hole.

Best practice for shearing round bars is to use blades with holes for each size of stock to be cut. Blade holes appreciably larger than the stock size cause excessive distortion of the workpiece.

Shearing of angles is done either in a combination machine or by double cutting. In a combination machine—the more common method—two blades such as those shown in Fig. 8(a) are used. One blade, usually the one that is stationary, is L-shaped and is positioned as shown. The movable blade is square or rectangular and is mounted with its two cutting edges parallel to those of the stationary blade. Figure 8(a) also shows that the space between the blades in the loading position is the same shape as the workpiece.

The movable blade travels at 45° toward the stationary blade, and both blades contact the work metal uniformly. Shearing by this technique is essentially a blanking cut, and distortion of the workpiece is minimal. One disadvantage of the method is that all cutting occurs at once, resulting in a high shear load. This condition is not important when small angle sections are cut; however, for work metal larger than 100 by 100 by 13 mm (4 by 4 by 1/2 in.), the movable blade should be provided with a rake to prevent excessive loading.

To provide a rake angle between the movable and stationary blades, the included angle between the cutting edges of the upper or movable blade is increased to 95°, as shown in Fig. 8(b). Shearing begins at the extremity of each leg and progresses toward the root of the angle. The increase in the included angle of the movable blade results in some distortion of the drop-cut piece; the amount of distortion is approximately equal to the difference in angle between the movable and stationary blades (5° is normal). The part remaining on the table or stationary blade is not distorted.

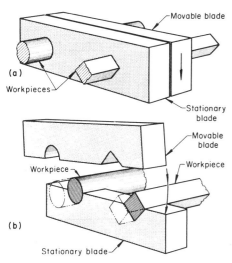

Fig. 7 Two types of blades for the shearing of bars. See text for discussion.

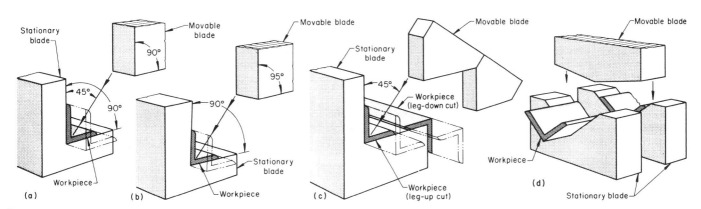

Fig. 8 Four types of blades for the shearing of angle sections. (a) to (c) Shearing in a combination machine. (d) Double-cutting method

Most combination shearing machines use a more versatile blade arrangement than those shown in Fig. 8(a) and (b). The setup shown in Fig. 8(c) is used to shear angle sections in both the leg-up and the leg-down positions. A swiveling table locates and holds the workpiece during shearing. With the swiveling table and two positions for the workpiece, the flanges can be easily mitered to any specific angle. For example, when shearing angle sections for a frame having the leg on the inside, the table would be set and locked at 45°. One end is mitered by placing the section in a leg-down position on the table and shearing off enough to make a clean cut. The other end is mitered by placing the section leg-up on the table and shearing to the proper length. The opposite positions are applicable when angle sections for a frame having the leg on the outside are being cut.

Shearing at a 45° angle reduces the capacity of the machine because a greater length of metal is cut at one time when a 90° cut is made. For example, a machine with a capacity of 200 by 200 by 32 mm (8 by 8 by 1¼ in.) when making a 90° cut has a capacity of only 200 by 200 by 25 mm (8 by 8 by 1 in.) when cutting at 45°.

Double cutting of angle sections, also called slugging, is used less frequently than shearing in a combination-type machine. This technique uses two stationary blades, spaced 13 mm (½ in.) apart, and one movable V-shaped blade arranged as shown in Fig. 8(d). The movable blade has a shallow V-shape that does not conform to the shape of the workpiece. Shearing starts at the extremity of each leg and progresses to the root of the angle, producing a 13 mm (½ in.) wide slug that is pushed out the bottom between the stationary blades.

In the double-cutting method of shearing, distortion occurs only in the slug, because the work metal is supported by the two stationary blades. There are two disadvantages of using the double-cutting method. First, increased power is required for making two cuts at the same time, and second, some metal is lost in the slug. The two stationary blades must be firmly supported to prevent their spreading during the cutting operation.

A similar tool can be used for shearing a channel section. The stationary blades should closely fit the contour of the channel section. Double cutting is adaptable to a guillotine shear, a combination machine, or a press.

Multiple Setups. Fabricating shops often must shear small quantities to a variety of shapes. To handle such work, many shops use a machine with a multiple setup, such as the one shown in Fig. 9. Without changing blades, the following operations can be performed:

- Double shearing of angle sections (Fig. 9, left)
- Straight-blade shearing (Fig. 9, center)
- Shearing of round and square bars and single shearing of L-sections (Fig. 9, right)

In this type of setup, all movable blades are attached to a single ram.

Fixturing is an important consideration in the shearing of bars. For safety and the proper functioning of open-end machines (Fig. 5) and many other iron-working machines, holddown fixtures are essential. Guide pins are also helpful, especially when cutting with conforming blades such as those shown in the center and at the right in Fig. 9.

Double-Cut versus Single-Cut Shearing

Two types of dies are used for shearing of structural and bar shapes: double-cut dies and single-cut dies.

Double-Cut Shearing. In shearing with double-cut dies, the shear blade acts as a punch and has a die on either side of it. A thin slug is sheared or punched out of the bar, its width depending on the size of the shear roughly as follows:

Shears with capacities of:		Slug (blade) thicknesses of:	
kN	tonf	mm	in.
160	18	10	3/8
445	50	13	1/2
1300	150	20	3/4

Double-cut shearing has three advantages. First, it produces distortion-free cuts on shapes that can be closely matched to the contour of the dies and where the surface in contact with the blade varies in shape. For example, angle iron and channel iron have square outside corners that can be closely matched by the dies, but the inside fillets vary with size. Respective fillet radii for various angle-iron dimensions are as follows:

Angle-iron dimensions		Fillet radius	
mm	in.	mm	in.
50×50×6	2×2×1/4	5	3/16
100×100×6	4×4×1/4	10	3/8
150×150×13	6×6×1/2	13	1/2
200×200×20	8×8×3/4	16	5/8

Both ends of a sheared angle or channel have the same square, burr-free quality after being double cut. Double cutting does not improve the quality of flat-stock, round-stock, or square-stock shearing.

Second, shearing forces are greatly reduced due to the rake angle on the blades. The angle-iron shear blade has a 120° included angle, which results in a 15° effective rake angle per side. Table 3 gives a comparison of the forces required for double-cut and single-cut shearing. The rake angle on the blade badly distorts the slug, which is thrown away. Flat stock is double-cut sheared to increase the capacity of the shears.

Capacities of double-cut dies are determined primarily by stock thickness, but there is a limit on the bar width or flange size. With the rake angle, only a small area is being sheared at a time; therefore, the only way to increase that area is to increase the bar thickness. For example, a 160 kN (18 tonf) shear will cut 100 by 100 by 6 mm (4 by 4 by ¼ in.) angle iron but will not cut 50 by 50 by 10 mm (2 by 2 by 3/8 in.) angle iron even though its total area is smaller.

The third advantage of double-cut shearing is that no holddown is required to keep the material from "kicking" up. The blade holds the material firmly against the dies, which creates a balanced condition. The material must span the two dies, and a minimum length equal to the stock thickness must be cut off to prevent damage to the blade or dies.

Single-Cut Shearing. In shearing with single-cut dies, the material is sheared by two opposed cutting surfaces that closely pass by one another—as in, for example, a pair of scissors or a typical mechanically driven universal shear. Nondistorted cuts are possible only where both the upper blade and lower die match the contour of the bar. For example, flat bars are cut free of distortion when the upper blade closely parallels the lower die. If a rake angle were applied to the blade, it would distort the part in contact with the blade; however, the other side of the cut in contact with the die would be free of distortion. Round and square bars are always single cut. Flat bars are usually single cut but may be double cut for increased size capacity. Angle iron must be double cut for best nondistorted cuts but can be single cut if distortion is not a problem.

Single-cut shearing causes distortion of angle iron mainly because of the mismatch between the radius on the blade and the fillet size in the angle. The fillet size varies with the angle size, as noted previously. Single-cut shearing requires a hold down ahead of the die and, for best results, a support on the opposite side. Figure 10 illustrates

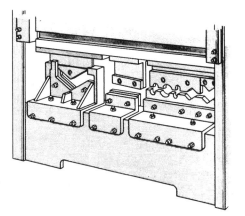

Fig. 9 Multiple setup for various types of shearing—singly or in combination

Table 3 Forces required for single-cut and double-cut shearing of angle iron

Size of angle iron(a)		Required force			
		Single cut		Double cut	
mm	in.	kN	tonf	kN	tonf
100×100×6	4×4×1/4	535	60	130	15
150×150×13	6×6×1/2	1560	175	355	40
200×200×19	8×8×3/4	3070	345	1245	140

(a) A-36; shear strength, 415 MPa (60,000 psi)

the importance of a holddown and a support. The support is not required on single-cut angle-iron dies due to the shape of the bar, which prevents it from bending. A fixed support, as shown, is satisfactory, provided that the part being sheared is sufficiently long (approximately 150 mm, or 6 in., on 445 kN, or 50 tonf, shears). Shorter parts can be sheared if a spring-loaded support pad is used. The new style of round-stock dies used in four-post injector-type shears do not require a support due to their unique design, which is explained later.

Nondistorted, Burr-Free Cuts

Figure 11 illustrates what is meant by the phrase "nondistorted, burr-free cut." Normal deformation on the opposite side contacting the die still exists due to the inherent shearing characteristics of materials. The amount of deformation depends on the hardness of the material: The harder the material, the less severe the deformation. The amount of "break" depends on the thickness and type of material.

Shearing of Specific Forms

Shearing of Angle Iron. Angle iron can be sheared with double-cut or single-cut dies. Figure 12 shows a double-cut die and how it can be used to miter angle iron up to 45° and cut square tubing and flat stock. This die will also cut small bar-sized channels where flange height does not exceed 14 mm (9/16 in.). The versatility of this assembly, plus the fact that a holddown is not required, makes it extremely popular.

Square tubing can be cut with little distortion, provided that the wall thickness in proportion to the tube size is great enough to withstand the shearing force. The angle-iron blades normally have a large radius on the point, but in shearing of square tubing, it is best to use a pointed blade.

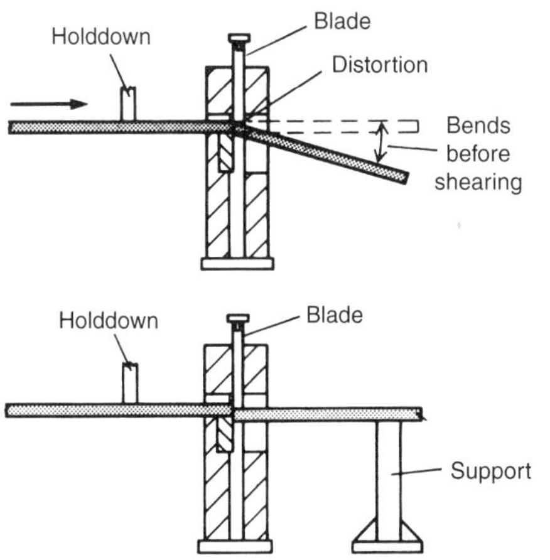

The type of material has a large effect on the quality of cutting.

Shearing of flat stock in a double-cut angle die creates a side thrust that is taken by blade guides. For this reason, it is best to cut two bars at once, one on each side; in high production, it is best to use a flat-stock die designed for that purpose.

The clearance between the blade and the dies must be adjusted to suit the angle-iron thickness. This is accomplished by inserting spacers behind the die inserts or holders. No spacers are required on 160 kN (18 tonf) shears, one set on 400 and 445 kN (45 and 50 tonf) models, and two sets on 1330 kN (150 tonf) models.

Single-cut angle dies are limited to angle-iron cutting. Figure 13 shows an injector-type die with a manually adjusted holddown on the feed side of the machine. The holddown must be adjusted so that the material clears by only 1.6 mm (1/16 in.). An air-operated holddown is used with feed systems that have a lift under the conveyor for lifting the material off the die during the feed cycle. The radius on the single-cut angle shear blade is of an intermediate size so that it can be used to cut a range of angle sizes with minimum distortion.

Shearing of Flat Stock. Figure 14 illustrates the double-cut flat-stock tooling assembly for a 1330 kN (150 tonf) shear. Note the concave rake angle on the leading face of the blade. Its concave shape balances the material through the center of the die.

The single-cut flat-stock die looks similar to the double-cut die except that the blade does not have a severe rake but a mere 31.2 mm to the meter (3/8 in. to the foot). This rake goes straight across the blade and causes little or no distortion. If absolute flatness on the part is required, the blade should not have a rake angle. The 160 and 400 kN (18 and 45 tonf) shear models have no rake.

Shearing of Round, Hexagonal, and Square Bars. Figure 15 illustrates the type of die used in the 160 and 400 kN (18 and 45 tonf) shear models. Five die cavities are provided—three for round bars, one for hexagonal bars, and one for squares. The die cavity is made to fit the bars very closely for minimum distortion, and therefore, this die is limited to shearing of five bar sizes in

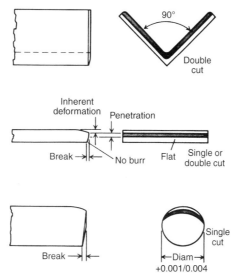

Fig. 11 Typical quality of cuts made by hydraulic shears. Dimensions given in inches

Fig. 13 Single-cut angle die with adjustable holddown

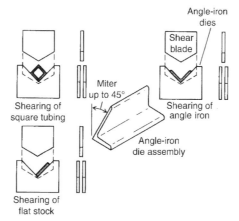

Fig. 10 Single-cut shearing of flat, round, and square bars, showing holddown and support required for nondistorted cutting

Fig. 12 Double-cut angle die for cutting various shapes without changeover

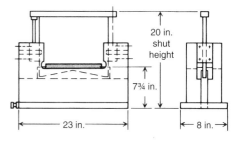

Fig. 14 Double-cut tooling assembly for cutting flat stock and bar stock

the shapes noted. The die cavity or hole for cold-finished round bars is 0.05 to 0.10 mm (0.002 to 0.004 in.) larger than the bar. After the bar has been sheared, it will still freely slip back through the hole, which points out how little the bar is distorted. Hexagonal and square bars can be sheared with the same accuracy. Smaller squares can be sheared with good success using a larger die cavity, provided that a holddown is added to the front of the machine. Normally, the die itself acts as a holddown, because it completely confines the bar.

The single-cut blade shown in Fig. 15 has only a half-hole to fit over the bar and must be used with a rear support. This blade is simple in design and inexpensive. Figure 16 shows a round-stock die for use in injector-type shears. This type of die features interchangeable bushings to suit the bar being sheared. Two identical bushings are used: One is for the fixed die, and the other fits into a moveable slide, forming what is normally called the blade. Three bushing sizes (outside diameter) are used to cover a range of bar diameters up to 63 mm (2½ in.) in the 1330 kN (150 tonf) shear. The reason that there are three sizes is cost, because they are perishable. The 197 mm (7¾ in.) die height (feed-conveyor height) is maintained only when the bar or hole size for each of the three bushings is at a maximum. Therefore, the height increases slightly when going to smaller bar.

The round-stock die in Fig. 16 has a spring-loaded slide that can be set in the open position to very precisely align the front and rear die bushings for feeding. Bars are sheared in less than a second as the ram forces the slide down, off-setting the bushings. Hexagonal and square bushings can be made for those shapes. A holddown is used on the larger bars due to the high amount of torque involved.

As cited previously, cold-finished round bars are closely fit to the die hole within 0.05 to 0.10 mm (0.002 to 0.004 in.), but hot-rolled bars need more clearance due to their oversized, out-of-round, scaly condition. Usually 0.8 mm (1/32 in.) is allowed for bars up to 32 mm (1¼ in.) in diameter, and 1.6 mm (1/16 in.) for bar diameters up to 63 mm (2½ in.).

Shearing of Channel and I-Sections. Channel iron and I-sections are difficult to shear without distortion, mainly due to the mill toler-

ance that allows their width to vary, depending on size, by as much as 5 mm (3/16 in.). On the other hand, Fig. 17 shows how a channel is cut in a double-cut die, which supports it on three sides. The "W"-shaped blade shears the flanges outward against the inserts. This die automatically, through hydraulics, causes the side inserts to move in and clamp the channel prior to contact by the blade. After the cut, when the blade starts to ascend, the inserts unclamp and move approximately 5.5 mm (7/32 in.) to allow the channel to be fed through the die.

One channel die assembly cuts a range of sizes. For example, a 1330 kN (150 tonf) shearing die cuts American Standard tapered flanges and Ship and Car channels in sizes from 75 to 250 mm (3 to 10 in.). To adjust from one size to the next, the clamping cylinder located on the side of the shear (Fig. 18) is disengaged, and a lever is turned to move the inserts in or out. Right- and left-hand screws are rotated simultaneously through a line shaft that keeps the inserts always centered about the blade.

Two shear blades are used to cut the 75 through 250 mm (3 through 10 in.) American Standard channel sizes. One cuts 75 to 125 mm (3 to 5 in.) channels, and the other cuts 150 to 250 mm (6 to 10 in.) channels. Changing the blades takes only 30 s.

I-beams are cut in the channel die by changing the lower die inserts and sometimes the blade. Figure 17 illustrates how the inserts are made higher to fit up into the "I" for supporting the web sections. A set of two lower inserts is required for each size of I-section, but in some cases, one shear blade will cut more than one size.

The 160 and 400 kN (18 and 45 tonf) shears use fixed dies for cutting channels, as shown in

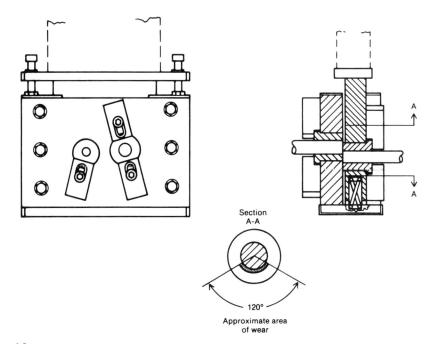

Fig. 16 Round-stock die assembly for four-post shear, featuring interchangeable bushings

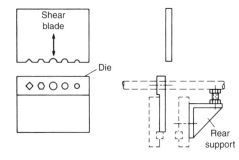

Fig. 15 Single-cut shearing die for cutting round, square, and hexagonal stock in two-post shears

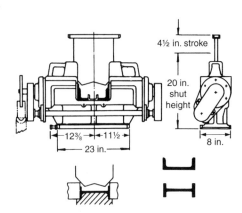

Fig. 17 Double-cut tooling assembly with adjustable die for nondistorted shearing of channels and I-beams

Fig. 18 Shearing die with hydraulic cylinder that clamps channel for distortion-free shearing

Fig. 19. These allow the flanges to spread, causing distortion, and the stock cannot be fed as easily as in the dies described previously. Fixed channel dies are not recommended for high-production shearing.

Shearing of T-Bars. T-bars are sheared in double-cut angle-iron dies by using a special set of side inserts, as shown in Fig. 20. The angle-iron shear blade is used. Stroke is the limiting factor when determining shearing capacity:

- A 160 kN (18 tonf) shear will cut T-bars up to 38 by 38 by 5 mm ($1\frac{1}{2}$ by $1\frac{1}{2}$ by $\frac{3}{16}$ in.).
- A 400 kN (45 tonf) shear will cut bars up to 63 by 63 by 6 mm ($2\frac{1}{2}$ by $2\frac{1}{2}$ by $\frac{1}{4}$ in.).
- A 445 kN (50 tonf) shear will cut bars up to 75 by 75 by 10 mm (3 by 3 by $\frac{3}{8}$ in.).
- A 1330 kN (150 tonf) shear will cut bars up to 100 by 100 by 13 mm (4 by 4 by $\frac{1}{2}$ in.).

A slight amount of distortion can be expected in the web due to shearing.

Shearing of Z-Bars. Structural Z-bars are single cut with a special die designed for that purpose. The blade and insert must be made to match the Z-bar contour; however, due to the mill tolerance, some distortion can be expected. Only the 445 and 1330 kN (50 and 150 tonf) shears are adaptable for shearing of Z-bars, and a special holddown is required (Fig. 21).

Structural Z-bars cannot be double cut due to their great thickness. Light-gage roll-formed Z-purlins, used in the metal building trade, have to be double-cut sheared.

Materials Handling

Materials handling and gaging have presented problems with many of the commonly used universal shears. Materials handling on some shears constitutes 95% of overall cost. Thus, inexpensive shear cutting can be very costly if feeding and gaging of the material are not done efficiently. If the shear machine has different die locations and feed and stop heights, it may be very difficult to equip this shear with a feed and stop system that will result in fast cutting to length. With hydraulic bar shears, a common feed conveyor and stop (that readily adapt to the various bar shapes) reduce the materials-handling costs and the overall shearing cost to a minimum.

Because four-post injector-type shears are normally used for production cutting, their feed systems are discussed; however, much is applicable to other models.

Two die heights are used on four-post shears: one for angle iron; and the other for flats, channels, rounds, and other shapes. Figure 22 shows how a dual-purpose conveyor is made to adapt to the two die heights and provide positive alignment for the bars. The Y-rolls are permanently mounted to the conveyor base for guiding of angle iron. The other dies, for flats and channels, are 38 mm ($1\frac{1}{2}$ in.) higher, which makes it convenient simply to raise the flat rolls to allow the material to pass over the V-rolls, as shown at left in Fig. 22. The right-hand view shows the flat roll in the lowered position.

A cut can be no squarer than the guiding system. Thus, the V-rolls for angles maintain positive alignment for both single- and double-cut dies. Adjustable side guides must be set for guiding the other shapes, such as flats and channels. Whether powered or manually fed, the cut-to-length system greatly reduces shearing costs and increases production.

Hydraulic shears lend themselves to many nonstandard applications. One such case is illustrated in Fig. 23, which shows how a 160 kN (18 tonf) shear for cutting angle iron up to 100 by 100 by 6 mm (4 by 4 by $\frac{1}{4}$ in.) is used in line with a hydraulic modular gang punching machine. This approach to metal fabrication is a prime example of how operations can be grouped to reduce material handling.

Another example of a nonstandard application, which required only a special die assembly, is shown in Fig. 24. Here, a 150 mm (6 in.) wide flat bar is sheared, leaving one end with a large radius and the other end with a slot. Three operations were combined into one, producing significant savings in labor.

These are only two of many applications that prove the versatility of hydraulic shears. These shears can also be used for forming, notching, punching, riveting, and many other operations. The four-post shear has a large tooling area and is adaptable to many different applications, with the main limitation being imagination.

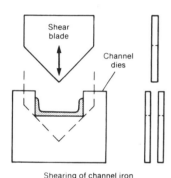

Fig. 19 Fixed channel die for two-post shears

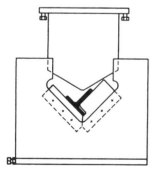

Fig. 20 Double-cut angle-iron die assembly with side inserts for shearing of T-bars

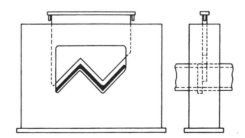

Fig. 21 Single-cut die assembly for shearing of Z-bars

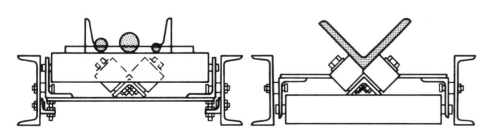

Fig. 22 Dual-height system with fixed V-rolls and movable flat rolls

Fig. 23 A 160 kN (18 tonf) shear mounted at 45° angle for use with gang punching machine

The advantages of hydraulic bar and structural shears are:

- Reduced setup, tooling, and running costs: quick-change tooling for all shapes; long tool life—no shock; fast, efficient feeding; only one operator required
- Ease of materials handling: common feed line for all shapes; powered feed available; back gage and conveyor provide efficient takeaway
- Improved quality of cut: little or no distortion; no burrs; square cuts gaged to close length tolerances
- Elimination of secondary grinding operation to remove burrs
- Reduced physical strain
- Increased production
- Increased versatility (shearing, forming, notching, punching, riveting, etc.)

Impact Cutoff Machines

Impact cutoff machines produce high-quality cut slugs or blanks of precise length from round, square, or specially shaped bars or coils. Two precision cutoff dies are engaged in opposite directions within the impact block, with short, simultaneous strokes fracturing the metal. The result (Fig. 25) is a clean cut at the interface of the two dies, producing slugs or blanks with length tolerances of well within ± 0.13 mm $(+0.005$ in.) and virtually no deformities at reduction speeds of 300 cuts per minute.

The quality of cut obtained surpasses that produced by any other known shear-type cutoff device. Squareness is held to close tolerances, and the cut pieces are virtually free of burrs, distortion, and edge rollover.

On materials such as carbon steel of 415 MPa (60 ksi) tensile strength, impact cutoff machines are capable of cutting 65 mm ($2\frac{1}{2}$ in.) diameter stock up to 915 mm (36 in.) long. Doubling the tensile strength of the material to 820 MPa (120 ksi) reduces the maximum diameter capacity to 45 mm ($1\frac{3}{4}$ in.).

Principle and Machine Construction. Impact cutoff machines use a unique double-impact cutting principle (Fig. 26). Stock is fed into a pair of precision cutoff dies that have a cavity shaped to the same configuration as the stock. These opposed dies are actuated with short, simultaneous strokes by two flywheel-cam assemblies. This double impact fractures the

metal, cleanly cutting the confined stock at the interface of the two dies.

The cutoff dies are located within an impact block. The flywheels rotate at a constant speed, and a cut takes place only when a pair of air-operated wedges are brought into position to close the gap. Wedge elevation is triggered by a positive stop that controls the length of cut. The

positive stop features a micrometer adjustment for precise control of the blank length.

Either a short or a long target—stock-stop or length gage—is used, depending on the length of cut. In actual operation, the target moves out of the way when the leading edge of the workpiece engages it. This motion activates the circuit that controls the wedges to make the cut. Stock is fed

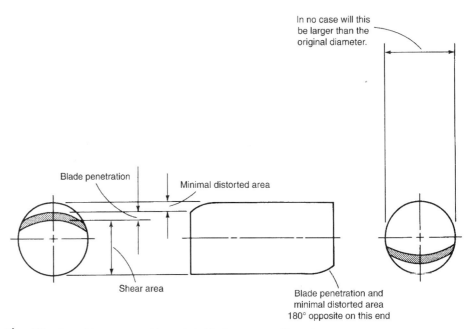

Fig. 25 Effects of shearing a round bar with double-die impact cutoff-type shear

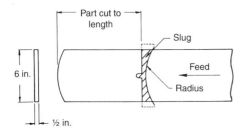

Fig. 24 Shaped blade and die that produce radiused and slotted ends with each stroke

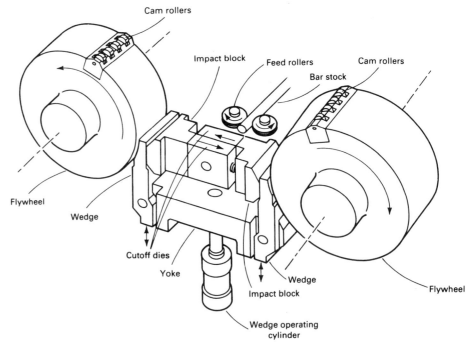

Fig. 26 Impact cutoff machine shearing bar stock with twin cutoff dies that are actuated by cam rollers on identical flywheel-cam assemblies

into the dies and up to the positive stop by a hydraulically driven roll feed.

Die life is excellent in impact cutoff machines and ranges from tens of thousands to hundreds of thousands of cuts before resharpening is required. A number of variables affect this, with the most important being the amount of clearance between the workpiece and the die.

The dies are usually of the inserted type. The longest service life has been obtained with inserts made of M2 high-speed steel hardened to 60 to 62 HRC.

When resharpening of the cutoff die is required because of wear, the insert is first removed and then replaced in the holder with a 0.5 mm (0.020 in.) shim behind it. The protruding portion of the insert is then sharpened by a simple surface grinding operation. Removing the die blocks for sharpening or replacement is fast and simple. The die-retaining plate unbolts, allowing the cutting dies to be easily lifted out.

ACKNOWLEDGMENT

This article was adapted from *Forming and Forging,* Volume 14, *Metals Handbook,* 9th ed., ASM International, 1988, p 714, and from O.D. Lascoe, *Handbook of Fabrication Processes,* ASM International, 1988.

Bending of Bars and Bar Sections

ROD AND BAR may be bent in the same manner as tubing. The possibility of collapsing or wrinkling is eliminated, because the solid section provides its own internal support. Generally, rod and bar in the annealed condition are preferred for cold forming. Material in other tempers may be required for some forming operations or when properties that cannot be obtained by heat treatment after forming are desired in the end product.

Bending Methods

Bending of bar and other nontubular sections involves similar techniques to that of tubing, although tube bending may require additional supports inside the tube (see the article "Bending and Forming of Tubing" in this Volume). Methods common to the bending of bar or tubes include draw bending, compression bending, roll bending, stretch bending, and ram-and-press bending. Bending methods unique to pipe or tube include wrinkle bending and roll-extrusion bending.

In draw bending, the workpiece is clamped to a rotating form and drawn by the form against a pressure die (Fig. 1a). The pressure die can be either fixed or movable along its longitudinal axis. A fixed pressure die must be able to withstand abrasion caused by the sliding of the work metal over its surface. A movable pressure die, because it moves forward with the workpiece as it is bent, is less subject to such abrasion. It provides better guidance and more uniform restraint of the work material. Draw bends are made when more precise dimensional tolerances are needed, or when tight bends of thin-walled tubing are required (see the article "Bending and Forming of Tubing" in this Volume). Bends up to 180° can be made. Draw bending is more common than any other bending method when power bending machines are used.

In compression bending, the workpiece is clamped to a fixed form, and a wiper shoe revolves around the form to bend the workpiece (Fig. 1b). Compression bending is most useful in bending rolled and extruded shapes. A bend can be made close to another bend in the workpiece without the need for the compound dies required in draw bending. Although compression bending does not control the flow of metal as well as draw bending, it is widely used in bending presses and in rotary bending machines.

The fixed form may have a flat surface or be contoured to match the diameter of the workpiece. When the radius block is contoured, a stop plug may be used instead of a clamp. Clamping is often necessary when bar stock is being bent against a flat radius block. The action of the roller on the stock is sometimes through a wiper shoe. The wiper shoe converts the single fixed load of the roller into a uniformly distributed load on the stock.

Roll bending uses three or more parallel rolls. In an arrangement using three rolls (Fig. 2a), the axes of the two bottom rolls are fixed in a horizontal plane. The top roll (bending roll) is lowered toward the plane of the bottom rolls to make the bend. The three rolls are power driven; the top roll is moved up or down by a hydraulic cylinder. Another roll arrangement is four-roll bending (Fig. 2b). The bar enters between the two powered rolls on the left. The lower bending roll is then adjusted in two directions according to the thickness of the bar and the desired angle of bend.

Rings, arcs of any length, and helical coils are easily fabricated in a roll bender. The bend radius usually must be at least six times the bar diameter or the section thickness in the direction of the bend. To limit distortion in the roll bending of asymmetrical sections, a double section can be made and split in two after bending. Rings are sometimes made by roll bending coils and cutting them into rings for welding. A typical roll bender with three driven rolls is shown in Fig. 3. Roll benders may also be manually powered (Fig. 4).

Roll bending is impractical for making more than one bend in a bar. It is difficult to control springback in a roll bender, and it may take several passes through the rolls to make the needed bend. Therefore, this method of making bends is slower than other methods. Another disadvantage of roll bending is that a short section of each end of the bar is left straight. For three-roll bending, the ends can be preformed in a press before bending, or the straight parts can be trimmed off. The following example describes a production application of three-roll bending. Other examples of three-roll bending are covered in the article "Three-Roll Forming" in this Volume.

Example 1: Three-Roll Bending of a Structural Section. A 7.6 kW (10 hp) three-roll bender was used to bend a steel angle 75 by 75 by 9.5 mm (3 by 3 by 3/8 in.) into a circular reinforcing flange 1520 mm (60 in.) in diameter. The angle was of hot-rolled ASTM A107 steel. The top roll of the bender was a plain cylinder; each of the two bottom rolls consisted of two cylindrical sections held apart by a spacer to provide a recess for the edge-bent flange.

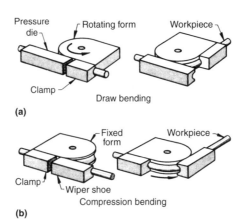

Fig. 1 Essential components and mechanics of (a) draw bending and (b) compression bending of bars and bar sections

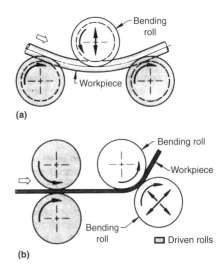

Fig. 2 Operating essentials in (a) one method of three-roll bending and (b) four-roll bending

The operations were performed in the following sequence:

- Cut angle to developed length plus 254 mm (10 in.)
- Set rolls to bend correct radius; roll 360°
- Cut off ends of the rolled bar
- Weld the bar into a ring
- Grind the weld flush
- Roll the ring in the three-roll bender to make it a true circle

Stretch bending is used for bending large, irregular curves. The workpiece is gripped at its ends and bent as it is stretched around a form (Fig. 5). The gripped ends are customarily trimmed off. This method can accomplish in one operation what may otherwise take several operations for parts with compound curvatures or large parts with shallow curvatures. The result is a possible savings in time and labor, even though stretch bending is a slow process. Tooling and machines may also be more costly, because stretch forming is a relatively sophisticated operation.

One benefit of stretch forming is that it elongates the metal throughout the section (Fig. 6a). This differs from the other methods (draw bending, compression bending, press bending, and roll bending) that result in some compression in various sections of the bend (Fig. 6b). Thus, less springback occurs when the work is stretched formed, due to the lack of compressive strain. Wall thickness is generally reduced but with greater uniformity.

Ram-and-Press Bending. With this method, a workpiece is placed between two supports, and a rounded form die (ram or punch) is pressed against it (Fig. 7). The two supports pivot as the ram moves forward, maintaining support of the workpiece. The two supports may be fixed or allowed to pivot. This method is rapid but provides less control over metal flow. It is used in production applications on heavy sections, depending on the capacity of the available press. Most ram benders are driven hydraulically, and they are commonly used to bend tube but are also used to bend solid sections.

Bending Machines

The machines used for the bending of bars include the following: devices and fixtures for manual bending, press brakes, conventional mechanical and hydraulic presses, horizontal bending machines, rotary benders, and bending

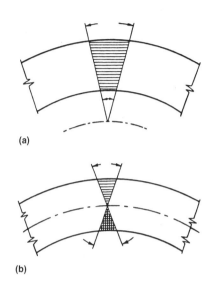

Fig. 6 Strain distribution for section formed by (a) stretching and (b) bending. Draw bending, compression bending, press bending, and roll bending compress the metal in various sections of the bend. Stretch bending imparts elongation throughout the bend and thus minimizes springback.

Fig. 3 Roll bender with three driven rolls and hydraulic positioning of the top roll. Courtesy of Baileigh Industrial Inc.

Fig. 4 Manually powered three-roll bender. Courtesy of Baileigh Industrial Inc.

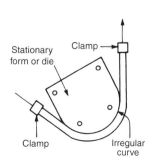

Fig. 5 Principle of stretch bending

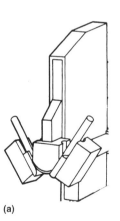

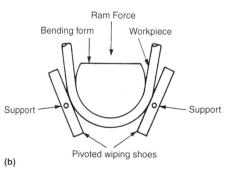

Fig. 7 Ram-and-press bending. (a) General configuration. (b) Tool area

presses. Shapers have also been used to perform specific bending operations.

Manual Bending. Hand-powered machines or fixtures are used in many shops for making bends that do not require much energy to form. This equipment is supplied with ratchets, levers, or gears to give the operator mechanical advantage. Different types of fixtures are used for manual draw bending, stretch bending, or compression bending. Roll bending is seldom done by hand. The tools used in manual bending are the same as those used on some power bending machines. The maximum sizes of low-carbon steel bars that can be manually cold bent are given in Table 1.

Press brakes are used for all types of bending, especially in small-lot production (25 to 500 pieces), when standard tooling or low-cost special tooling can be used. Often, the punch is not bottomed in the die, but the stroke is controlled, and the bar is bent "in air" (Fig. 8). With this technique, various bend angles can be made with the same die (see also the article "Press-Brake Forming" in this Volume).

Mechanical presses are generally used only for mass production, because only large production lots can justify the cost of tooling, which is more than that for most standard bending tools. Figure 9 shows a round bar being bent into a U-bolt in a press. The bar is first cut to length and pointed at both ends (preliminary to a later threading operation). The bar is then loaded into the press and held in a grooved die that bends the bar into a "U" in one stroke. In the setup shown in Fig. 9, more than one workpiece can be bent at a time. The following example describes an application of a mechanical press in bending bars.

Example 2: Bending a Welded Assembly in a Mechanical Press. The wheel spider shown in Fig. 10 had three 255 mm (10 in.) long spokes of 9.52 mm (0.375 in.) diameter low-carbon steel. The spider was assembled by welding the three spokes to a 13 mm (1/2 in.) thick steel hub. The assembly was loaded into a 670 kN (75 tonf) mechanical press, a double bend (joggle) was made in the spokes, and the short straight surface between the two bends was flattened to 6.4 mm (1/4 in.) thick. The wheel rim was then welded to the spokes, as shown in Fig. 10(d). Next, the assembly was loaded into another press, in which the legs were sheared flush with the outer edge of the rim, and a 3.2 mm (1/8 in.) diameter hole was pierced in the flattened area of each spoke. The production rate was 25 per minute.

Hydraulic presses are often used to bend bars in much the same manner as mechanical presses. Although hydraulic presses are usually slower than mechanical presses, they have the advantage of exerting full force over a long stroke. Therefore, deep bends can often be made on a hydraulic press much smaller than the mechanical press that would be required. In the following example, a hydraulic press needed so little head room that a closed shape could be bent over it.

Example 3: Bending a Double-Bar Structure in a Hydraulic Press. A double-bar structure was constructed of two 11 mm (7/16 in.) diameter bars that were connected by welded cross members to form a ladderlike structure. A rectangular shape was formed by making four 90° bends having 16 mm (5/8 in.) inside radii. The two bars (sides of the ladderlike structure) were bent simultaneously, using a punch that forced the bars between rollers. By using a small (27 kN, or 3 tonf) vertical hydraulic press, the four bends could be made consecutively, allowing the workpiece to encircle the press ram as bending was completed. The overhead clearance

would not have been available with a mechanical press. This technique permitted the fabrication of 360 double bends (90 frames) per hour.

Horizontal bending machines for bending bars consist of a horizontal bed with a powered crosshead that is driven along the bed through connecting rods, crankshaft, clutch, and gear train. Dies are mounted on the bed, and forward motion of the crosshead pushes the bar through the die. The long stroke and generous die space make this machine useful for a variety of cold and hot bending operations, although speeds are lower than those for mechanical presses of similar capacity. Horizontal benders are available in capacities from 89 to 2700 kN (10 to over 300 tonf).

Rotary benders, either vertical or horizontal, are used for the draw, compression, or stretch bending of bars. Such machines consist of a rotary table in either a horizontal or vertical position on which the form block or die is mounted (Fig. 1). Suitable hydraulic or mechanical clamping, tensioning, or compressing devices are provided to hold the workpiece while the die rotates to the required position, or while the workpiece is bent about the central forming die. Some machines can make bends by two, or all three, methods. This type of bender is often built for manual bending of tube or pipe. A programmable rotary bender with a universal bending plate is shown in Fig. 11.

Bending presses are most widely used for bending tubing because of the wide variety of techniques and tooling options. However, bending presses are occasionally used for bending bars, as in the following example.

Example 4: Making a Double Crank in a Bending Press. Double cranks, such as the one shown in Fig. 12, were made in a bending press from round bars 8 or 9.5 mm (5/16 or 3/8 in.) in diameter. The bars were cut to length and fed into the press to flatten the ends and pierce the holes. The two sharp bends were made one at a time in

Table 1 Maximum sizes of low-carbon steel bars for manual bending

Shape	Size mm (in.)
Rounds	25 (1) (diam)
Squares	19 (3/4) (per side)
Flats bent on flat	9.5 × 102 (3/8 × 4)
Flats bent on edge	6.4 × 25 (1/4 × 1)
Angles	4.8 × 25 × 25 (3/16 × 1 × 1)
Channels	4.8 × 13 × 25 (3/16 × 1/2 × 1)

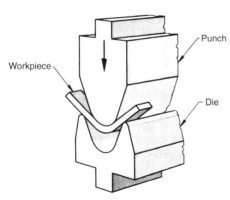

Fig. 8 Air bending of a bar in a press brake

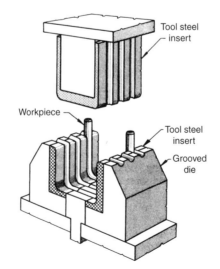

Fig. 9 Use of a grooved die in a mechanical press for bending a round bar into a U-bolt in one stroke

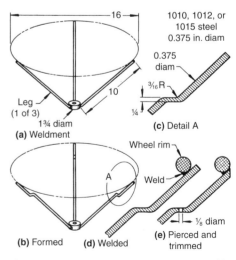

Fig. 10 Welded bar assembly that was formed by bending in a mechanical press. Dimensions given in inches

the same press with a V-die. A bearing bracket was assembled on the crank, followed by a double-staking operation in the same press. The greatest demand on the press was in the end-flattening operation, which required a press capacity of 890 to 1330 kN (100 to 150 tonf).

Shapers can be tooled for bending operations. One method is to have the fixed die, or anvil, held on the knee of the shaper and the punch mounted on the ram (Fig. 13). The shaper must make one stroke only. The stroke can be adjusted to allow for springback in the workpiece; therefore, the workpiece can be bent to fairly close limits.

The setup shown in Fig. 13 is used to correct for springback in formed parts such as J-bolts, U-bolts, and rings. A V-punch and die mounted in the shaper can make bends of various kinds.

A shaper can also be provided with a rack and pinion to produce rotary motion for bending (Fig. 14). Bend radius can be varied by the use of center pins of different diameters. A typical use of a shaper in the bending of bars is described in the following example.

Example 5: Shaper versus Press for the Bending of J-bolts. Originally, J-bolts were press bent cold in a die from sheared lengths of 13 mm ($1/2$ in.) diameter hot-rolled merchant bars of 1025 steel. The press made the J-bolts by bending the bar into a half circle with one long arm. Because the unbalanced support caused variations in bending, some of the parts were unacceptable.

The job was put on a shaper. For this operation, the stock was sheared to double lengths, which the shaper bent into a circular loop with two long arms (the ends). The looped bar was then sheared in half. The cost of tooling for the shaper was much less than that for the press.

A rack-and-pinion-actuated fixture similar to that shown in Fig. 14 was used to bend the bar into a circular loop. The bar stock was placed in the fixture at a slight angle to allow one leg to pass the other during forming. After the bar was cut in two, the J-bolt was finished in a setup such as that shown in Fig. 13.

Tools

Tools for draw and compression bending are shown in Fig. 1. The form used in both processes is shaped to the contour of the bend. It is usually grooved to fit the work. Often, the form is part of a right cylinder whose straight portion (frequently, an insert) provides the surface against which the work is clamped. Hydraulic or mechanical pressure holds the clamp against the workpiece. Annular grooves or roughened surfaces grip the bar or bar section.

To allow for springback, the bend radius is made smaller on the form than is required for the workpiece. The form is also designed for a greater angle of bend than is needed. These two adjustments permit overbending the piece to allow for springback. Such adjustments are made by trial and error. The form is tested and corrections are made before it is heat treated.

The finish on the form (rotating or fixed) should be just good enough to avoid marring the workpiece. For most bar bending, a machined finish is sufficient. For decorative stainless steel and polished aluminum, grinding or polishing of the form surfaces may be needed. However, the clamping area should not be ground or polished unless necessary. The smoother the finish in the clamping area, the greater the danger that the workpiece will slip through the clamp. The

pressure die and wiper shoe require a good finish (usually ground) because the work metal must slide along them.

When air bending bars in a press brake, simple V-blocks will suffice for the female dies. The opening of the V-blocks should be eight times stock thickness for standard sections, and ten times stock thickness for heavy sections. The upper die (punch) is shaped to the inside radius of the bend, and the angle of bend is controlled by the length of stroke. The same die set can be used for making various bends, as well as for various stock thicknesses, by adjusting the stroke.

Press brakes can use rubber pads for tooling on bends that need support (see the article "Press-Brake Forming" in this Volume). Completely shaped dies for bottoming can also be used in a press brake, as shown in Fig. 15.

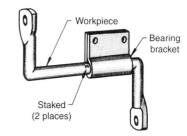

Fig. 12 Double crank produced in a bending press

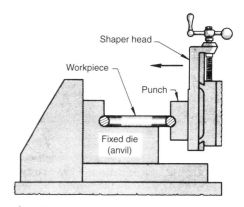

Fig. 13 Use of a shaper for correction of springback in U-bolts, J-bolts, or rings

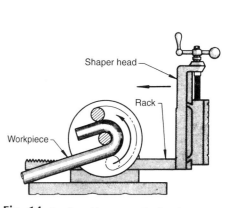

Fig. 14 Bending with rotary motion in a shaper

Fig. 11 Rotary bender with programmable controls and universal bending plate capable of 360° bends. Courtesy of Baileigh Industrial Inc.

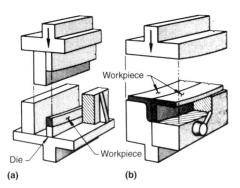

Fig. 15 Die setups in a press brake for (a) edge bending a bar and for (b) bending two structural angles

(a) **(b)**

Die   Workpiece   Workpiece

Most dies used in conventional mechanical or hydraulic presses are completely shaped bottoming dies, which makes them more expensive than tooling for other bending machines. Tooling for bending presses is specially designed to fit the needs of the machine and the work to be done. Dies for bending presses are simple to construct and relatively inexpensive. Tools for stretch bending are covered in the article "Stretch Forming" in this Volume.

Die Materials. Dies are usually made of hardened steel for the production of thousands of pieces per month. Tool steel is used for small one-piece dies. Larger dies are made of low-carbon steel and then carburized and hardened. Clamping inserts are made separately.

For moderate production of a few hundred pieces per month, unhardened carbon steel is often used. If only a few parts are needed, wood or an aluminum alloy may be strong enough for dies.

For bottoming dies in presses, hardened tool steel is always used. For cold bending, A2 tool steel hardened to 58 to 62 HRC is most often selected. A hot work tool steel, such as H11,

Table 2 Typical lubricants for bending various metals

Work metal	Lubricant
Low-carbon steel	Water-soluble, vegetable-oil based drawing oil(a)
Stainless steel and other high-alloy iron-base alloys	Mineral-oil-based drawing oil(a)
Aluminum alloys and copper alloys	Mineral oil
Brass (severe bends)	Soap solution(b)
Hot bending of carbon, alloy, and stainless steels	Molybdenum disulfide

(a) Available as proprietary material. (b) Creamy mixture of laundry soap and water

hardened to 45 to 50 HRC is usually the choice for hot bending.

Bend Allowance

The stock consumed in a bend (that is, overall length of a bend) can be computed from the radius of curvature at the neutral axis and from the angle of the bend. A formula often used for this computation is:

$$W = 0.01745\,\alpha\,(r + \delta)$$

where W is the bend allowance, α is the angle of bend (in degrees), r is the radius of bend to inner stock surface, and δ is the distance from the inner surface to the neutral layer (a commonly used approximation when this figure is not known is one-third to one-half stock thickness). The constant 0.01745 is a conversion factor changing degrees to radians.

When bar stock for a workpiece whose ends are within, or very near, the bend area is cut square to the neutral axis, the ends, after forming, will not be square to the neutral axis. The basic reason for this is the difference in circumference of the outer and the inner surfaces of the bend. Additional deviation from squareness can be expected because all of the material toward the outside of the bend from the neutral axis has undergone a tensile load, and the material inside

this line has been under compressive load. Unless compensation can be made for these variations, the ends of the formed part must be trimmed if they are to be square to the axis. However, when end details, locating surfaces, and other considerations necessitate such action, it is still possible to cut the blanks to size in such a way that trimming after forming is eliminated.

Lubrication

Successful bending depends to a large extent on the type of lubricant used. No one lubricant works equally well on all materials. Selection of a lubricant varies among different shops. Typical lubricants for bending specific metals are listed in Table 2.

Overlubrication, in either quantity or type of lubricant, must be avoided. Not only is excessive lubrication likely to cause wrinkling, but the cost of removal must be considered. It is never good practice to use a pigmented compound if successful results can be obtained with an unpigmented compound, because pigmented compounds are more difficult to remove.

Wiper dies are lubricated with a very small quantity of high-grade drawing lubricant. It is important not to overlubricate pressure dies and wiper shoes.

Bending and Forming of Tubing

THE PRINCIPLES for bending tubing are much the same as those for bending bars (see the article "Bending of Bars and Bar Sections" in this Volume). Two important additional features in the bending of tubes are that internal support is often needed and that support is sometimes needed on the inner side of a tube bend.

The wall thickness of the tubing affects the distribution of tensile and compressive stresses in bending. A thick-wall tube will usually bend more readily to a small radius than a thin-wall tube. Table 1 lists the minimum practical inside radii for the cold draw bending of round steel or copper tubing, with and without various supports against flattening and wrinkling.

Selection of Bending Method

The four most common methods of bending tubing are basically the same as those used in the bending of bars: compression bending, stretch bending, draw bending, and roll bending. The method selected for a particular application depends on the equipment available, the number of parts required, the size and wall thickness of the tubing, the work metal, the bend radius, the number of bends in the workpiece, the accuracy required, and the amount of flattening that can be tolerated. Two additional methods for pipe bending are wrinkle bending and roll-extrusion bending.

Compression bending of tube is similar to that for bar bending. One end of the workpiece is fixed in place relative to a stationary bending form, while a moveable shoe or roller compresses the section against the form. Figure 1 illustrates an example, where one end is fixed by a stop plug and the tube is bent around a fixed bending form. Compression bending requires much less clamping area than for draw bending because the work material is pressed against the stationary form by the pressure tool. The various types of pressure tools include: a static wiper shoe, and follower block, or a roller.

Compression bending of tubing is often performed on manually operated machines. Mandrels cannot be used in compression bending because the mandrel would have to move with the workpiece when it is wrapped around the form. Thin-walled tubing is not usually bent by this method. However, bend radii as small as 2.5 times diameter of thin-walled tubing can be obtained with internal support to prevent wrinkling. The normal minimum centerline radius for compression bends is 4 times the tubing diameter (Ref 1). Bends up to 170° can be made.

Compression bending results in less stretching on the outer surface of the bend than with draw bending. In compression bending, the neutral axis lies in the outer third of the bend so that less stretching occurs along the outer surface of the bend. In contrast, the neutral axis in draw bending occurs at the inner third of the bend section. Because less stretching occurs, compression bending is effective in the forming of coated tubing.

Draw bending of tube involves a rotating bending form that pulls the tube through a pressure and wraps the tube around the bending die (Fig. 2). The tube is fastened to the bending die by a clamp with sufficient pressure to allow pulling the tube around the die. Flattening during draw bending tends to be more severe than with compression bending, and so typical machines for draw bending pull the tube over a stationary mandrel. The pressure tool can be a roller, a sliding shoe, or a static shoe.

Draw bending is used widely for tube forming. It is a versatile method for bending to tight radii with better dimensional control than compression bending. Bends up to 180° can be made with tube diameters ranging from about 13 to 250 mm ($\frac{1}{2}$ to 10 in.). When power machines are required, the draw bending method is more commonly employed than by powered compression bending. Draw bending can be done on

Table 1 Minimum practical inside radii for the cold draw bending of annealed steel or copper round tubing to 180°

Radii can be slightly less for a 90° bend, but must be slightly larger for 360°.

Tubing outside diam		Minimum practical inside radius					
		Grooved bending tools				Cylindrical bending block without mandrel; ratio, <30(a) (poor conditions)	
		With mandrel; ratio, <15(a) (best conditions)		With mandrel or filler; ratio, <50(a) (normal conditions)			
mm	in.	mm	in.	mm	in.	mm	in.
3.2	$\frac{1}{8}$	1.6	$\frac{1}{16}$	6.4	$\frac{1}{4}$	13	$\frac{1}{2}$
6.4	$\frac{1}{4}$	3.2	$\frac{1}{8}$	7.9	$\frac{5}{16}$	25	1
9.5	$\frac{3}{8}$	4.8	$\frac{3}{16}$	9.5	$\frac{3}{8}$	50	2
12	$\frac{1}{2}$	6.4	$\frac{1}{4}$	11	$\frac{7}{16}$	75	3
16	$\frac{5}{8}$	7.9	$\frac{5}{16}$	14	$\frac{9}{16}$	102	4
19	$\frac{3}{4}$	11	$\frac{7}{16}$	17	$\frac{11}{16}$	152	6
22	$\frac{7}{8}$	13	$\frac{1}{2}$	19	$\frac{3}{4}$	203	8
25	1	14	$\frac{9}{16}$	22	$\frac{7}{8}$	254	10
32	$1\frac{1}{4}$	17	$\frac{11}{16}$	25	1	381	15
38	$1\frac{1}{2}$	21	$\frac{13}{16}$	29	$1\frac{1}{8}$	508	20
44	$1\frac{3}{4}$	24	$\frac{15}{16}$	32	$1\frac{1}{4}$	686	27
50	2	27	$1\frac{1}{16}$	35	$1\frac{3}{8}$	889	35
64	$2\frac{1}{2}$	35	$1\frac{3}{8}$	41	$1\frac{5}{8}$	...	...
75	3	41	$1\frac{5}{8}$	48	$1\frac{7}{8}$	...	...
89	$3\frac{1}{2}$	48	$1\frac{7}{8}$	54	$2\frac{1}{8}$	...	...
102	4	54	$2\frac{1}{8}$	60	$2\frac{3}{8}$	...	...

(a) Ratio of outside diameter to wall thickness of tubing

Fig. 1 Tube bending around a stationary form with outside of tube held against a bearing block

manually operated machines (Fig. 3) or powered units for heavy pipe (Fig. 4). Close-fitting mandrels, dies, and shoes can be used to achieve bend radii as close as 1 diameter (Ref 1).

Roll bending of tube is limited to thick-walled product, but it is used to produce full circles of helical coils (see the article "Three-Roll Forming" in this Volume).

Stretch bending of tube may or may not involve the use of mandrel. Like stretch forming of bar, the advantage of this method is reduced springback, as the metal is stretched without compression in the inner part of the bend section.

Ram-and-Press Bending. Like solid section, ram-and-press bending of tube is rapid for bends up to about 165° when close tolerances or low distortion are not critical. The minimum center-line bending radius is about 3 times diameter, but 4 to 6 times is preferable (Ref 1).

Wrinkle Bending. In this method, wrinkles are made on one side of the tube by heating with a gas torch while applying a compressive load on the tube section. Wrinkling shortens one side and thus results in a bend. The method is used in the field and is applied to heavy-walled steel pipe up to 650 mm (26 in.) with bend radii of about 2 times diameter (Ref 1).

Roll extrusion bending is a method used to create a large-radius bend in large, heavy-walled pipe by internal swaging on one side. Commercial equipment has capacities from 125 to 300 mm (5 to 12 in.) diameters to produce distortion-free bends with radii of about 3 times diameter (Ref 1).

Hand versus Power Bending. Depending on the material and workpiece size, the bending methods may be manual or with power machines. Steel tubing as large as 40 mm (1.50 in.) in outside diameter (OD) with a 1.65 mm (0.065 in.) wall thickness can be bent by hand, but the process is slow and repeatability is questionable.

Manual bending is commonly done by a machine for compression or draw bending, and these manual machines operate on the same basic principles as do powered types. Hand bending techniques include simple form tool for bending ductile material like aluminum tubing (Fig. 5) or machines (Fig. 3). Some hand benders use an adjustable friction device, a kind of sliding brake, to prevent sliding of the tubing. The friction prevents wrinkles and other defects in bending.

The following two examples illustrate several of the factors that must be taken into consideration in selecting either hand bending or power bending to fabricate tubing.

Example 1: Hand Bending of a U-Shaped Furniture Part Having Two 90° Bends. Bending equipment was needed to produce a U-shaped furniture part with two 90° bends from 19 mm (3/4 in.) OD 1010 steel welded tubing with a 1.25 mm (0.049 in.) wall thickness. The two bends were made to a 50 mm (2 in.) radius as measured on the tube centerline. A small amount of flattening was tolerated. Production rate was 500 pieces per month.

Because no great accuracy was needed and because the production volume did not warrant more than a minimum investment, a bending fixture for compression bending by hand was selected, along with tooling to allow for both bends to be made in one setup. If production volume had been larger, a power-driven bender or a bending press might have been selected.

Example 2: Power Bending of Machine Tool Hydraulic Lines. In the production of hydraulic lines for machine tools, from 9.5 and 13 mm (3/8 and 1/2 in.) OD steel tubing and from 6.4 mm (1/4 in.) OD copper tubing, hand bending required 6400 man-hours per year for 44,000 bends. A change to power bending reduced the man-hours needed to 450.

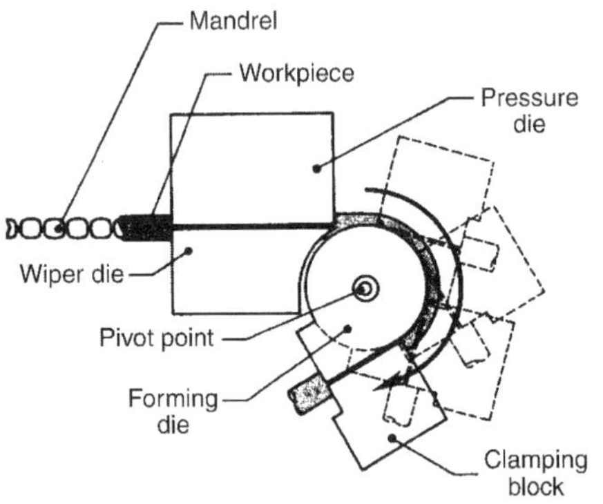

Tools

Tools used for the bending of tubes are similar to those for the bending of bars (see the article "Bending of Bars and Bar Sections" in this Volume). One important difference is that tools for tubes need carefully shaped guide grooves to support the sidewalls and to preserve the cross section during the bend.

Form blocks, or bending dies, resemble those for bending of bars. They either rotate or are fixed, depending on the arrangement of the machine in which they are used. One end of the tube is clamped at the end of the groove in the form block, and the tube is bent by being forced around the block and into the groove. For round tubes, the depth of the groove in the form block should be one-half the OD of the tube to provide sufficient sidewall support.

The block becomes the template for holding the shape of the bend. Form blocks can be made of wood, plastic, or hardboard; if they are to be used for an extensive production run, they can be made of tool steel and hardened.

Clamping blocks hold the end of the tube to the form block and maintain the holding force necessary to make the bending action effective. Although the groove in the clamping block should be well formed, the finish should not be so fine that the tube will slip. Ordinarily, the as-machined finish is adequate, but sometimes

Fig. 2 Sequence of movement of rotating die in table drawing

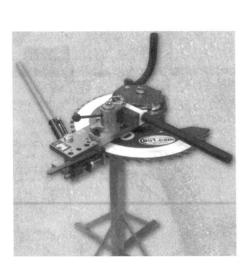

Fig. 3 Manual rotary draw tube bender. Courtesy of Baileigh Industrial Inc.

Fig. 4 Programmable rotary draw bender with capacity to bend 50 mm (2 in.). Schedule 40 pipe. Courtesy of Baileigh Industrial Inc.

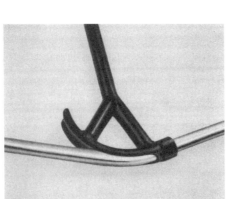

Fig. 5 Bending of aluminum tube with a bending form

ridges or serrations are machined into the clamp to increase the holding force. Rosin can be applied to the tube to prevent it from slipping in the clamp.

If the clamped area is to be part of the finished piece, care must be taken to prevent scratches or mars. If the clamping groove has to be ground or polished to provide a good surface, the portion of the tube to be clamped will have to be longer to distribute the higher clamping force better. When the clamping length is short, the end of the tube is sometimes plugged to prevent it from deforming from high clamping forces. Table 2 lists typical clamping lengths for bending steel tubing.

Pressure dies are used in the draw bending of tubing to press the workpiece into the groove in the form block and to support the outer half of the tube. The most commonly used pressure die is as long as the developed length of the bend plus some allowance for holding, and it does not slide over the tube but travels with it as it moves toward the bend area (see the article "Bending of Bars and Bar Sections" in this Volume). In one face, it has a groove with a depth that is slightly less than one-half the OD of the tube.

A stationary pressure die or even a roller can be used on noncritical work. Either unit has a tube-forming groove machined in its face. Most stationary pressure dies are made of low-carbon steel, which can be case hardened to resist wear. Tool steel such as O1, A2, or D2, hardened to 55 to 60 HRC, or aluminum bronze is commonly used for sliding dies.

In compression bending equipment, where the tube is clamped to a nonrotating form block, a wiper shoe replaces the pressure die. Its relationship to the workpiece is similar to that of the stationary die described previously in that the wiper shoe slides over the workpiece. However, instead of being fixed, the wiper shoe revolves around the stationary form block, progressively pressing the tube into the form block groove. For most applications, the length of the wiper shoe is from three to five times the OD of the tube. The wiper shoe is made of tool steel and hardened to 55 to 60 HRC, or of a bearing bronze.

Wiper dies are stationary straight-groove dies (not to be confused with the wiper shoes already described) that are sometimes needed in draw bending to support the tube on the side opposite the pressure die as the tube is about to be drawn into the contour of the form block. Metal that will form the inside of the bend undergoes severe compression that is transmitted back toward the as yet unbent end of the tube, where it could cause wrinkles if not for the support of the wiper die.

The wiper die has a groove that is machined and ground to conform to the tube being bent and to fit the groove and lips of the form block, ending in a featheredge pointing toward the tangent point of the bend and extending to within 3.2 to 13 mm ($\frac{1}{8}$ to $\frac{1}{2}$ in.) of the tangent point. Although it is difficult to maintain this distance without deflection, it must be done meticulously if the wiper die is to prevent compression wrin-

kles. Wiper dies are machined from AISI 52100 (or L2 tool steel) for low-carbon steel tubing or from aluminum bronze for stainless steel tubing. Wiper dies are never hardened.

Mandrels, which are described in detail in the following section of this article, are of three general types—rigid, flexible, and articulated—and are made to support the inside of the tube during bending. Rigid mandrels fit the interior of the tube and are sometimes shaped to conform to the start of the bend. However, because they are rigid, they support the entire circumference of the tube only as far as the point of bending and not beyond the tangent of the bend. Plug mandrels and formed mandrels are included in this category.

Flexible mandrels bend with the tube. They are generally built up of shims or laminae. This type of mandrel is sometimes used with square tubes and box sections where only a few bends are needed. Inserting and removing flexible mandrels is usually difficult. Articulated mandrels include ball mandrels (discussed subsequently) and various other shaped mandrels that are used in much the same way as ball mandrels.

Loose fillers such as sand and various low-melting alloys also serve as mandrels for low-production applications.

Dies used in press-type bending machines are similar to those described in the article "Bending of Bars and Bar Sections" in this Volume. Dies, including wing dies for bending presses, may have grooves for one to six tubes to be bent in one press stroke.

Formed rolls are used in the roll bending of tubes. Grooves corresponding to the outer surfaces of the tubes to be bent are cut or ground into the outer surfaces of the rolls so that they fit the surface of the tube as it is bent. A more complete description of the rolls used in roll bending is available in the article "Three-Roll Forming" in this Volume.

Bending Tubing with a Mandrel

Mandrels are sometimes used in bending to prevent collapse of the tubing or uncontrolled flattening in the bend. A mandrel can neither correct failure in bending after the failure initiates, nor remove wrinkles.

Table 2 Typical clamping lengths for bending steel tubing

Radius of bend centerline	Wall thickness of tube, mm (in.)	Typical length clamped
1 × OD	<0.89 (0.035)	4–5 × OD
	0.89 – 1.65 (0.035 – 0.065)	3–4 × OD
	>1.65 (0.065)	2–3 × OD
2 × OD	<0.89 (0.035)	3–4 × OD
	0.89 – 1.65 (0.035 – 0.065)	2–3 × OD
	>1.65 (0.065)	$1\frac{1}{2}$–$2\frac{1}{2}$ × OD
3 × OD	<1.65 (0.065)	2–3 × OD
	≥1.65 (0.065)	1–2 × OD

OD, outside diameter

Figure 6 shows five types of mandrels used in the bending of tubing. The plug mandrel and the formed mandrel are rigid, but the three other types shown are flexible or jointed to reach farther into the bend.

The largest diameter of the rigid portion of the mandrel should reach a short distance into the bend; the distance that it extends past the tangent straight portion depends on the type of mandrel and the size of the tube and is usually established by trial. If the mandrel extends too far, it can cause a bulge in the bend. Conversely, if the mandrel does not extend far enough, wrinkles may form, or the outer tube surface may flatten in the bend area.

The need for a mandrel depends on the tube and bend ratios. The tube ratio is D/t, where D is the OD and t is the wall thickness. The bend ratio is R/D, where R is the radius of bend measured to the centerline.

Table 3 can be used to determine whether a mandrel is needed for bending steel tubing. Figure 7 shows the usual conditions that require the use of a mandrel and what type of a mandrel is needed.

The nomographs given in Fig. 7 are used in two steps to determine if and when a mandrel is required and which specific type will suffice. In nomograph A, the first step is to find the tube ratio and the bend ratio in the left-hand and center scales and to lay a straightedge across them. The zone on the right-hand scale where the straightedge falls shows whether a mandrel is required, and which type. Bends for which D/t is more than 40 always require a multiball mandrel.

If a multiball mandrel is indicated, step two requires the user to refer to nomograph B in Fig. 7. As before, the tube ratio and the bend ratio are located in the left-hand and center scales and a straightedge is laid across them. The number of balls needed in the multiball mandrel will be indicated on the right-hand scale.

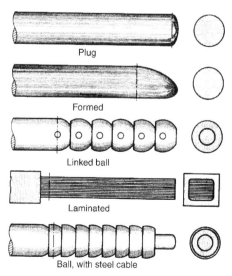

Fig. 6 Five types of mandrels used in the bending of tubing. Broken vertical lines are points at which bends should be tangential to mandrel centerlines.

Plug and formed mandrels are fixed, and the tube is drawn over the tip of the mandrel in forming. This action work hardens the tube so that it resists flattening during bending. Clearance between the mandrel and the inside of the tube should not be more than 20% of the wall thickness. If the mandrel is too tight, the tube is likely to fail in the bend. Mandrels are necessary in tubes other than round in order to avoid distortion of the cross section. The use of plug and formed mandrels is shown in the following two examples.

Example 3: Use of a Plug Mandrel in Bending Welded Low-Carbon Steel Pipe. A short 20 mm ($3/4$ in.) section of thick-walled (2.9 mm, or 0.113 in., thick) welded low-carbon steel pipe with an OD of 27 mm (1.05 in.) was bent 90° to a 50 mm (2 in.) radius. Despite the thick wall, the small bend radius made it necessary to use a plug mandrel to support the pipe against flattening. Other tools used were a form block, a clamp, and a pressure die.

The bending machine was a power-driven rotary draw bender rated for a maximum of 25 mm (1 in.) extra-strong seamless low-carbon steel pipe (33.40 mm, or 1.315 in., OD by 4.55 mm, or 0.179 in., wall). The bends were made at the rate of 300 per hour.

Example 4: Bending Oval Tubing with a Formed Mandrel. Oval tubing measuring 25×110 mm (1×$43/8$ in.) in outside dimensions and with a wall thickness of 1.65 mm (0.065 in.) was bent on edge to make a U-shape with two 90° bends at 230 mm (9 in.) radius. The tubing was welded hot-rolled low-carbon steel. Wrinkles, shear marks, or other visible defects were not permitted.

The bends were made in a draw bender rated for 90 mm ($31/2$ in.) OD by 2.10 mm (0.083 in.) wall thickness steel tubing with 275 MPa (40 ksi) yield strength. This piece, bent with a formed mandrel, form block, clamp, and pressure die, demanded the full rated torque of the machine. The mandrel was lubricated. The bends were made at a rate of 250 per hour.

Ball mandrels with one or more balls are used for many bends. During bending, the metal is stretched tightly over the mandrel, making withdrawal difficult. Withdrawal mechanisms are needed. In thin-wall tubing of softer metals, as the mandrel is withdrawn, it sizes the bend somewhat, smoothing the stretched metal and correcting the shape of the cross section.

The bodies and balls of one-ball mandrels used on most tubing are commonly made of carburized low-carbon steel, hardened, ground, and polished. For the bending of stainless steels, they are made of polished aluminum bronze.

One-ball mandrels used in the bending of tubes up to 32 mm ($11/4$ in.) in OD generally have a body 0.13 to 0.18 mm (0.005 to 0.007 in.) smaller and a ball 0.25 to 0.36 mm (0.010 to 0.014 in.) smaller than the inside diameter of the tube. Square or shaped tubes require a mandrel that fits closer. If the bends are in one plane, the body and the ball of the mandrel can be grooved to clear weld flash or seams. More commonly, a mandrel is made undersize to clear the obstruction. When the workpiece must be bent in several planes, it can be reinserted with the seam in the groove, but it is usually better to specify tubing with a controlled weld flash.

Ball mandrels are often made with several balls, as shown in Fig. 6. The balls or segments are always smaller than the body, and they can be jointed by links and pins, ball joints, or steel cable. A linked or jointed mandrel is usually stronger than a comparable mandrel joined by steel cable. The linked mandrel bends in only one plane and is easier to load than one that is less rigid. The ball-jointed mandrel is also in wide use, and it has the advantage of having rotating balls to equalize wear. Ball-jointed ball mandrels are made in many sizes—down to one for tubes as small as 5.64 mm (0.222 in.) in inside diameter. Ball-jointed and steel cable ball mandrels cannot be grooved to clear weld flash and seams because the mandrel segments rotate.

Many multiball mandrels make bends with a centerline radius that equals the outside diameter of the tubing. Ball mandrels can be used on bends that are not possible with formed mandrels.

Example 5: Multiball versus Formed Mandrel in Forming a U-Shaped Bend. A U-shape was produced by making two 90° bends in 32 mm ($11/4$ in.) OD welded tubing of low-carbon steel with wall thicknesses of 1.25 and 1.65 mm (0.049 and 0.065 in.). The bend radius for both types of tube was 60 mm ($23/8$ in.). Wrinkles, shear marks, or other visible defects were not permitted.

For the 1.65 mm (0.065 in.) wall thickness tubing, the bend was made with a formed mandrel, which adequately supported tubing of this wall thickness. However, a formed mandrel could not be used for the 1.25 mm (0.049 in.) wall thickness tubing. For the thinner-wall tubing, the D/t ratio was so large that it was necessary to use a well-lubricated three-ball mandrel and wiper die. The ball mandrel was required to support the outer wall in the bend area, and the wiper die was required to prevent wrinkling caused by compression in the inner wall of the bend.

The machine used was a draw bender rated for steel tubing with 90 mm ($31/2$ in.) OD, 2.10 mm

Table 3 Minimum centerline radii for bending steel tubing without a mandrel

Tubing outside diam		Minimum centerline radius for tubing with wall thickness of:											
mm	in.	0.89 mm	0.035 in.	1.24 mm	0.049 in.	1.65 mm	0.065 in.	2.11 mm	0.083 in.	2.36 mm	0.093 in.	3.05 mm	0.120 in.
4.8	$3/16$	7.9	$5/16$	6.4	$1/4$	4.8	$3/16$	...	...	...	...	...	...
6.4	$1/4$	13	$1/2$	9.5	$3/8$	7.9	$5/16$	...	...	...	...	...	...
7.9	$5/16$	22	$7/8$	19	$3/4$	16	$5/8$	...	...	...	...	...	...
9.5	$3/8$	38	$11/2$	32	$11/4$	29	$11/8$	25	1	...	...	...	...
13	$1/2$	57	$21/4$	50	2	44	$13/4$	38	$11/2$	...	...	...	...
19	$3/4$	102	4	75	3	64	$21/2$	50	2	...	...	...	...
25	1	203	8	152	6	102	4	75	3	50	2	50	2
38	$11/2$	...	...	...	...	305	12	254	10	203	8	152	6
50	2	...	...	...	...	...	...	610	24	508	20	406	16
64	$21/2$	...	...	...	...	...	...	...	...	610	24	508	20
75	3	...	...	...	...	...	...	...	...	...	...	635	25

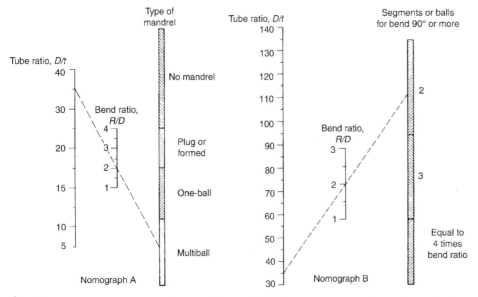

Fig. 7 Nomographs for determining when a mandrel is needed and the correct type to use. See text for explanation

(0.083 in.) wall, and 275 MPa (40 ksi) yield strength.

If the formed mandrel had been used in bending the 1.25 mm (0.049 in.) wall tube, the tube would have deformed excessively from inadequate support. A formed mandrel would have to be advanced farther into the bend area than is normally done. This would cause a hump and greater thinning of the tube wall where the outside of the bend was stretched over the end of the mandrel. When wall thickness is less than approximately 1.25 mm (0.049 in.), this technique should not be used because there is not enough metal to take the stretch without breaking.

Clearances for mandrels vary from 0 (to produce some bends, the mandrel is forced into thin-wall tubing) to 2.4 mm (0.095 in.) or more. The clearance needed depends on stock material, wall thickness, bend radius, and quality of the bend. The better the bend, the more closely the mandrel must fit.

Mandrels are even more necessary in bending nonround tubing than in bending round tubing (see Example 4). Segmented mandrels must be used in almost all bending of square, hexagonal, and octagonal tubing. The number of segments that are needed in the mandrel usually depends on the wall thickness of the tubing.

The bending of a fragile thin-wall tube may require the careful use of a multiball mandrel and a wiper die. If the bend is changed to a larger radius or if a stronger or thicker-wall tube is substituted, the mandrel can be changed to a less complicated one, and the wiper die may not be needed. A plug or form mandrel can sometimes be used instead of a multiball mandrel, even when the bend is made to the minimum practical radius.

Dimensional Accuracy. Regardless of other conditions, when accuracy is important, the use of a mandrel is mandatory. The following example describes an application that illustrates the degree of accuracy that can be achieved by using a mandrel.

Example 6: Use of a Plug Mandrel to Hold Close Tolerance on Four Bends. The tube shown in Fig. 8 was used in a return-line manifold of a high-pressure hydraulic system on a large tractor. The radius of each of the four bends was required to be within ±1°. Overall length was required to be within ±0.38 mm (±0.015 in.).

To achieve this degree of accuracy, a considerable amount of tool adjustment was required during each setup. From two to six tubes were bent before acceptable tubes were produced. To produce a lot of 40 tubes, 2½ man-hours were required, including setup.

Bending was done in a powered draw bender, using a round-end plug mandrel. The mandrel had a drilled hole to deliver a constant flow of lubricant inside the tube during bending. The lubricant was a mixture of lard oil and low-viscosity mineral oil.

Mandrel Materials. Most mandrels are made of tool steel (W1, O1, A2, and F1 are typical selections) and hardened to 55 to 60 HRC. Polished mandrels usually work best, but ground or machine surfaces are satisfactory when slight markings on the inside of the tube is acceptable. Possible surface treatments include chromium plating of any hardened steel and nitriding of tool steels such as A2 and D2.

Although chromium is now subject to environmental regulation, chromium plating has been used to extend the life of some mandrels. Platings were no more than 0.008 to 0.013 mm (0.0003 to 0.0005 in.) thick because thicker platings may flake off. Sometimes platings were renewed for further use. Plated mandrels would be stripped and replated when the plating was worn through at any point.

Bending Tubing without a Mandrel

It is less expensive to bend tubing without a mandrel. Trial bending is generally necessary to determine which bends can be made. Tubing with thick walls is more likely to be bendable without a mandrel than thin-wall tubing. Bends with large radii are more likely to be formable without a mandrel than those with small radii. Slight bends are more feasible than acute bends. Wide tolerances on permissible flattening make a bend easier to form without a mandrel. Springback is greater without a mandrel, but it can be compensated for by overbending or lessened by increasing force on the pressure die.

When bending nickel alloys with no internal support, the dies should be slightly smaller than those used for bending with a mandrel or filler. Bending without use of a mandrel or filler is suitable only for tube and pipe that have a wall thickness greater than 7% of the OD, or for large-radius bends. Within this ratio, nickel alloy tube can be bent through 180° with no mandrel or filler to a minimum mean radius of three times the OD of the tube (3D).

Machines

The machines used in the bending of tubes are essentially the same as those used in the bending of bars. In general, bending machines fall into three categories: rotary benders (stretch, compression, and draw bending), press benders (stretch and compression bending), and roll benders.

Powered rotary benders are commonly used to bend tubing as large as 200 mm (8 in.) in OD. At least one machine can bend tubing as large as 300 mm (12 in.) OD with a 6.5 mm (¼ in.) wall, and a few special power benders can bend 460 mm (18 in.) pipe.

Boilermakers normally use power benders that can bend 75 mm (3 in.) OD steel tubing with 13 mm (½ in.) wall to a centerline radius as small as 75 mm (3 in.). The following example describes the use of rotary benders.

Example 7: Bending Steel Tubing for Automobile Seat Frames. Steel tubing with 25 mm (1 in.) OD and 1.24 mm (0.049 in.) wall was bent 90° to make the frame for automobile seats. The inside of the bend was collapsed into a dimple for clearance, but most of the column strength was maintained by holding the shape of the outside of the bend. This was done in part by collapsing the inner wall of the tube over a convex punch instead of a conventional groove. The centerline radius of the 90° bend was 16 mm (⅝ in.). The bends were made two at a time in a fast rotary bender.

Bending presses are hydraulic machines that are made especially for bending both bars and tubes, but most often for tubes. The ram of a bending press can be stopped at any point in the stroke. Wing dies and a cushioning device help to wrap the work around the ram die, as shown in Fig. 9. When the ram moves down, it causes the wing dies to pivot by a camming action and to wrap the workpiece around the ram die. The wing dies wipe the work to control the flow of metal; a compression bend is made on each side of the ram die, without wrinkles or distortion.

A bending press can usually make bends much faster than machines that are not made especially for bending. The open design of the press makes possible the bending of complex shapes in one setup. Single bends can be made sequentially, or the press can make several bends simultaneously. Bends can be made to various angles and in various planes. The tube or bar is usually passed through the press in one direction, and the press makes a sequence of bends automatically. The work is held against stops to locate each bend.

When several bends are made in one or more workpieces at each stroke of the press, all bends are in the same plane. Different angles and bend radii can also be made in the same workpiece, and the angles and spacing of bends can be

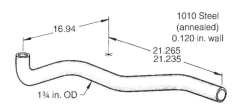

Fig. 8 Manifold tube that was bent to an accuracy of ±1° on each of four bends. Dimensions given in inches

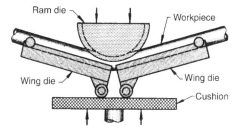

Fig. 9 Essential components and mechanics of a bending press

adjusted. One disadvantage of the bending press is that it causes a slight reduction in the thickness of the workpiece at the bend.

Automatic bending presses are used for production bending. Capacities are 26 to 360 kN (3 to 40 tonf) for bends that generally do not exceed 165°. Bends in the same plane should be separated by a distance equal to twice the OD of the tube. Bends in different planes should be separated by a distance equal to at least three times the OD of the tube. If the bent portion is to be joined with another bend into an accurate circle, the length of the straight legs on each end of the bend should be at least twice the OD of the tube. Bends can be made beyond these limits, but at greater cost. Bends are usually made in dies that are slightly tight on the tubes to prevent flattening and wrinkling. Exhaust pipes for automobiles, with bends in various planes, are made in automatic bending presses, as in the following example.

Example 8: Forming Automobile Exhaust Pipes from 1010 Steel Welded Tubing. Figure 10 shows an exhaust pipe 210 cm (82 in.) long with bends to bypass the obstructions on the underside of a vehicle. These pipes were made of hot-rolled or cold-rolled 1010 steel welded tubing with 44 mm (1¾ in.) OD and 1.52 mm (0.060 in.) wall. The hardness was 63 to 78 HRB.

Bending was done in an automatic 180 kN (20 tonf) hydraulic bending press, using a punch with a 125 mm (5 in.) centerline radius in the groove and two wing-die shoes. Locating fixtures and a checking fixture (off the press) were also used.

The press made different angles of bend by using turret-mounted stops to control the stroke. The location of the bend was set by backstops at the right side of the press. Counting the plane of the first bend as 0°, the radial position (roll of the tube) for the plane of each bend in turn was located by stops along a bar on the left of the machine. In loading the tubes for the first bend, it was important to keep all weld seams in the same place so that all tubes would bend alike.

The tube was nested in the wing-die shoes for the first bend, and the work was then moved by hand for each of the other bends. All the bends were standard. Tolerances were:

Angles of bend	±0° 15′
Angles of tube ends (all planes)	±0° 30′
Maximum depth of depressions and wrinkles	3.0 mm (0.12 in.)
Linear dimensions	±1.5 mm (±0.06 in.)

All the bends were made cold. The punch was set to clear the work by 13 mm (½ in.) for loading, and the total machine time to make five bends was less than 8 seconds. Production rate was 130 pieces per hour on a 10,000 piece order.

Press Bending Nickel Alloy Tubing (Ref 2). Press or ram bending can be used for nickel alloy tubing, although it does not provide close tolerances and is applicable only to large-radius bends. Bends from press bending are limited to 120°. If a smooth bend is desired, the radius of the bend should not be less than six times the OD of the tube (6D). A filler material should be used for bends of radii less than 6D.

Pressure dies used in press bending should be at least two times longer than the OD of the tube. Press bending with wing dies is used for unfilled, thin-wall, large-diameter tube.

Annealed nickel alloy tubing is not always preferred for press bending. Annealed tubing of low base hardness is not stiff enough to withstand deformation without excessive flattening. Consequently, nickel and nickel-copper alloys are usually press-bent in the stress-relieved temper. Nickel-chromium alloys have higher mechanical properties in the annealed condition than do nickel and nickel-copper alloys and should be press-bent in the annealed temper. Ideally, the choice of temper for a specific bend should be determined from the results of several trial bends.

Roll benders for bending tubes are similar to those used for bending bars, as described in the article "Bending of Bars and Bar Sections" in this Volume, but tolerances are more critical on the rolls and spacing. Roll benders are discussed in the article "Three-Roll Forming" in this Volume. The contour of the rolls must match that of the tube to minimize wrinkling or flattening. Tubes of sizes up to 203 mm (8 in.) OD by 6.10 mm (0.240 in.) wall can be bent into arcs, circles, or helixes. Rings are easily made on three-roll benders. Helix coiling is described in the following example.

Example 9: Coiling a Helix of Round Steel Tubing in a Three-Roll Bender. Steel tubing of 50 mm (2 in.) OD and 6.10 mm (0.240 in.) wall was coiled in a three-roll bender into a helix 610 mm (24 in.) in mean diameter with 65 mm (2½ in.) pitch and 20 turns, as shown in Fig. 11. The tubing, 38.4 m (126 ft) long, was made by welding together random lengths 3.0 to 7.3 m (10 to 24 ft) long. The welds were then ground flush. The form rolls were contoured to fit the 50 mm (2 in.) OD of the tubing. The tubing was started into the rolls so that it coiled without stop until the helix was completed. If the coiling had been stopped, the diameter of the helix might not have been constant. After coiling, the helix was removed and trimmed to length.

Roll bending is the principal method of producing helical coils, spirals, and circular configurations from nickel alloy tubing. Bending may be done on either unfilled or filled tube. The

minimum bend radius that can be attained on unfilled tube is approximately six times the OD of the tube.

Hot Bending

Most of the bends described so far in this article have been made by cold bending (workpiece at room temperature). There are obvious advantages to cold bending:

- Heating equipment is not needed.
- The benefits of previous heat treatment are not destroyed.
- Subsequent cleaning or descaling is less likely to be needed.
- Workpiece finish is better.
- Thermal distortion is avoided.

On the other hand, cold bending demands more energy than hot bending for the same bend. There is more springback after a cold bend and more residual stress in the tube. Bends cannot be made to as small a radius cold as they can hot.

Tubes are bent hot to make bends of small radius, adjacent bends with little or no straight tube between them, bends in material with little cold ductility, bends that take too much power to bend cold, and bends in fragile assemblies where the force of cold bending might cause damage. Temperatures and procedures in the hot bending of carbon, low-alloy, medium-alloy, and stainless steel tubes are summarized in Table 4. The disadvantages of bending tubes hot include high cost, slow production, and poor finish on bends.

Tubes of carbon steel and most alloy steels can be bent to a much smaller radius by hot bending than by cold bending. A bend radius of 0.7 to 1.5 times the OD of the tube can usually be made by hot bending. The wall thickness of the tube affects this range. The tube wall must not be so thin that it will distort or thin excessively in the outer wall. If the tube ratio (OD divided by wall thickness) is more than 10, the tube probably needs internal support in bending,

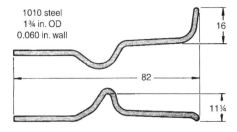

Fig. 10 Automobile exhaust pipe produced in an automatic hydraulic bending press. Dimensions given in inches

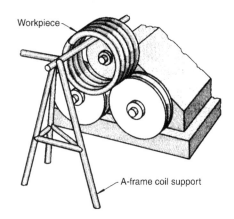

Fig. 11 Coiling a helix from round steel tubing by three-roll bending

unless there is some upsetting of the tube that thickens the wall.

Figure 12 shows the tube-geometry characteristics that can usually be bent hot in a die without a mandrel. Figure 13 shows the tube configurations that can usually be bent hot with a mandrel or filler. Figure 12 applies to tubes 38 to 75 mm (1.5 to 3 in.) in diameter; Fig. 13 is for tubes of all diameters.

If a mandrel is not practical for internal support of the tube, the tube can be packed with dry sand. First, a plate is welded to one end of the tube to block the end. The tube is filled with sand and is tamped or vibrated to make sure the sand is packed tightly. Another plate is welded to the open end, the tube is heated, and the bend is made. Finally, the plates are cut off and the sand is emptied out. Various specialized techniques for bending tubes hot are described in Examples 10, 11, and 12.

Example 10: Five-Step Bending of a Tube. Return bends for the coils of a boiler were made by bending 50 mm (2 in.) OD carbon steel tube 180° on a 38 mm (1½ in.) centerline radius.

The tube had a 6.60 mm (0.260 in.) wall. The sequence of operations was:

- Bend 180° on 114 mm (4½ in.) centerline radius in a conventional rotary bender.
- Heat the bend area to 980 to 1095 °C (1800 to 2000 °F) in a furnace.
- Reshape the bend to 50 mm (2 in.) radius in a bending press.
- Close legs to 75 mm (3 in.) between centers in a vertical press.
- Restrike in the vertical press to make bend radius 38 mm (1½ in.).

The last three operations were performed in rapid sequence so that all bends were made before the tube cooled below 870 °C (1600 °F). Production rate was 30 pieces per hour.

Thinning of the outer wall can sometimes be controlled better in hot bending by heating only the part of the workpiece that will be the inner wall of the bend. This reduces its compressive strength so that the bend causes very little stretch of the outer wall. This method makes good bends with centerline radii of 1.3 to

1.5 times the OD of the tube with the usual tooling in a rotary bender.

Boiler tubes are bent by heating one side for one-third of the way around the tube. The heated portion becomes the inside of the bend, and because it yields more easily in compression, thinning of the outer wall is limited. The tubes are bent to a U-shape in a bending machine. The U can then be reheated on the inside as before, and the bend can be squeezed in a press in one or more steps to make a narrower U, as in the following example.

Example 11: Localized Heating for a Compression Bend. Return bends were made for use in the economizer of a boiler. The tube was made of carbon steel, 50 mm (2 in.) OD by 7.21 mm (0.284 in.) wall. The operations were:

- Heat the tube to 705 to 730 °C (1300 to 1350 °F) in a special burner that heated the bottom 120° of the tube for a 457 mm (18 in.) length where the bend was to be made (Fig. 14).
- Bend the tube in a rotary bender 180° to 190 mm (7½ in.) between centers with the heated portion on the inside of the bend.
- Squeeze the tube between dies in a hydraulic press to 150 mm (6 in.) between centers.
- Squeeze, as in the third step, to 120 mm (4¾ in.) between centers.
- Squeeze, as in the third step, to 90 mm (3½ in.) between centers.
- Reheat the entire surface of the bend for 200 mm (8 in.) on each side.
- Size in a die.
- Squeeze, as in the third step, to 64 mm (2½ in.) between centers.
- Normalize.

The production rate, with three men working, was 20 bends per hour.

Dry Sand as Filler. Large tubes are commonly bent hot while they are filled with dry sand. The bending is done on a bending table made of cast iron plates. The plates have a continuous pattern of cored holes. A bending form for the desired radius is bolted to the bending table. Steel pins or stops can be placed in the holes in the table to keep the work in line.

Table 4 Temperatures and procedures for hot bending of steel tubes

Steel	Temperature, °C (°F)	Procedures
Carbon steels		
ASTM A106, A178, A192, A210	980–1095 (1800–2000) or <730–(1350)	Do not heat beyond 1095 °C (2000 °F) and do not bend between 730 and 870 °C (1350 and 1600 °F)(a).
Low-alloy steels		
ASTM A209, A213 (grade T11 and T22), A335 (grade P2)	980–1095 (1800–2000) or <730–(1350)	Do not heat beyond 1095 °C (2000 °F) and do not bend between 730 and 870 °C (1350 and 1600 °F)(a).
Alloy steels		
ASTM A213 (grade T5 and T9)	980–1095 (1800–2000)	Do not heat beyond 1095 °C (2000 °F) and do not bend between 730 and 870 °C (1350 and 1600 °F)(a). Heat treat after bending 730–745 °C (1350–1375 °F).
Stainless steels		
Types 304, 310, 321	>1150 (2100)	After bending, heat treat to 1095–1120 °C (2000–2050 °F). Furnace cool to 315 °C (600 °F); air cool(b).
Type 446	>1150 (2100)	Do not bend less than 870 °C (1600 °F). Heat treat at 790–870 °C (1450–1600 °F); water quench.

(a) Ductility is sometimes low in this range; therefore, the range should be avoided for hot bending. (b) This treatment has proved best for maximum strength in service at elevated temperature.

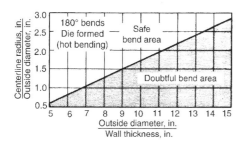

Fig. 12 Chart for determining conditions for the successful hot bending of tubes 38 to 75 mm (1.5 to 3 in.) in diameter without the use of mandrel

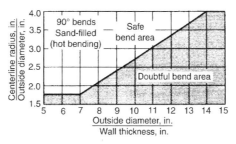

Fig. 13 Chart for determining conditions for the successful hot bending of tubes of all diameters with the use of a mandrel or filler

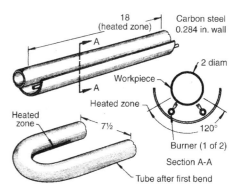

Fig. 14 Compression bend produced in tubing, and special burner used for localized heating of the workpiece before bending. Dimensions given in inches

A clamshell furnace in sections is the usual source of heat. Winches, jacks, and hoists supply the bending force. A typical application is described in the following example.

Example 12: Hot Bending a Large Tube on a Cast Iron Bending Table. A 180° bend with a 380 mm (15 in.) radius was made near the center of a 460 cm (15 ft) length of steel tube, 185 mm (7¼ in.) OD by 25 mm (1 in.) wall. A plate was welded to one end of the tube to close it. The tube was dropped, closed-end down, into a pit and was filled with sand. The tube was vibrated and tamped to make sure the sand was well compacted, and the open end was closed with another welded-on plate.

The bending table was set up for the bend. The proper form block was bolted to the table, and lines tangential to the bend were laid out for proper location of stop pins (Fig. 15).

The tube was heated to 980 to 1095 °C (1800 to 2000 °F) in a clamshell furnace and set on the bending table. One end was clamped down to the table. The other end was pulled by a cable, which, guided by a series of strategically located pulleys fastened to the table, bent the tube incrementally around the form block to the final position shown in Fig. 15.

When the bend passed inspection, the ends of the tube were cut off and the sand was poured out. The entire process, including preparation and handling, took 45 min.

Hot Bending of Nickel Alloy Tubing (Ref 2). When possible, nickel alloy tube and pipe should be formed by cold bending. If hot bending is necessary, it can be performed by standard hot bending methods. Nickel alloy pipe should be cleaned thoroughly prior to heating and bending. The metal should be worked as soon as possible after removal from the furnace to avoid cooling before bending is completed.

Hot bending is normally limited to tube and pipe larger than 2 in. schedule 80 (60.5 mm, or 2.375 in., OD, and 5.54 mm, or 0.218 in., wall thickness). Thin-wall tubing should not be bent hot because it is difficult to retain sufficient heat to make the bend.

Hot bending should be done on filled tube only. As mentioned previously, sand is a suitable filler material. For nickel alloys, the sand must be free of sulfur because contamination of nickel alloys by sulfur causes cracking during bending. Sulfur can be removed from sand by heating to about 1150 °C (2100 °F) in an oxidizing atmosphere. In addition, the tubing must be thoroughly cleaned prior to filling.

Sand-filled tube and pipe in small sizes (60.2 to 72.9 mm, or 2.37 to 2.87 in., OD) can be bent hot to a minimum mean radius of two times the OD of the tube. Larger sizes require greater bend radii.

Because of the wide variety of nickel alloys and the multitude of uses, alloy producers should be consulted regarding cooling rates for specific alloys after hot forming operations. Improper cooling rates can lead to sensitization to corrosion, lower strength, and reduced ductility.

Local heating by induction heating can rapidly heat a specific section of tubing or pipe, which can then be bent. The pipe is clamped at both ends and the circumferential section where the pipe is to be bent is inductively heated. Once the appropriate temperature is attained, the pipe is pushed through the induction coil, and pipe is bent by the action of the bending arm clamped to the front end of the tube. Bends of up to 180° can be made with this arrangement. Bending speeds can range from 13 to 150 mm/minute (0.5 to 0.6 in./minute) (Ref 3–5). If large bend radii are required, bending rolls can be used.

Induction heating is an efficient and cost-effective method of hot bending pipe or tubing. The benefits of using induction heating for hot bending of pipe or tube include (Ref 3–5):

- Reduced power consumption required for heating
- Predictable shape (ovality)
- Minimal and uniform wall thinning across bend
- Little change in surface finish
- Creation of small radii
- Creation of multiradius curves in one tube

Hot versus Cold Bending. The selection of hot or cold bending can depend on available equipment, the cost of new equipment, labor costs, the urgency of the job order, and the size of the production lot.

Tube Stock

Tubes are classed as seamless, welded, lock-seam, butt-seam, and jacketed.

Steel tubing is available both seamed and seamless. Seamed tubing with internal flash demands special consideration when a mandrel is to be used in bending (see the section "Bending Tubing with a Mandrel" in this article).

Lock-seam tubing can be bent if the seams are tight. A test is to twist a 914 mm (3 ft) long section of tubing in the hands; any grating or slipping indicates a seam too loose to make good bends.

Butt-seam tubing is similar to welded tubing but with no weld at the joint; it is seldom used. To make good bends, it must be accurate in dimensions, have no scale, and be bent with the seam in the plane of the bend. A mandrel must always be used. It is more economical to use welded tubing.

Stainless steel jacketed tubing is made by roll forming a sheet of stainless steel onto a butt-seam tube of low-carbon steel. The stainless steel jacket is rolled into a lock seam in the open seam of the inner tubing. For best bending, the stainless steel jacket should be at least 0.51 mm (0.020 in.) thick, and the two layers should be rolled tightly together. Tools for bending such jacketed tubing cost more than tools for bending plain tubing; an aluminum bronze wiper die, a hardened steel mandrel, and unusually high pressure on all tools are required, and all of these add to the cost of the tooling. In addition, the seam must be in the plane of the bend, either inside or outside.

Galvanized steel tubing can be bent to a radius as small as four times the OD. For tube to be bent to smaller radii, galvanizing should be done after bending because the galvanized coating is likely to flake if galvanized stock is used.

Aluminum-coated tubing (hot dipped) can be bent by essentially the same techniques used for uncoated tubing of the same diameter and wall thickness, using slightly higher clamping pressures to avoid slipping. Additional information is available in the article "Press Forming of Coated Steel" in this Volume.

Seamless tubing should be free of scale or rust. Wall thickness, concentricity, and hardness vary in seamless steel tubing. These variations are likely to cause variable springback, wrinkles, and excessive flattening. Common pipe in all sizes and thicknesses is easily bent if it is clean and free of rust or scale, inside and out.

Stainless steel tubing can be bent to a greater angle, at a given radius, than low-carbon steel. Austenitic types in the 300 series are most commonly bent because they are strong and ductile. Tubing in a stabilized condition at a temper no higher than quarter hard will make good bends with low scrap rates. Both welded and seamless tubing are available. Thin-wall tubes should have the exact diameter and wall thickness specified. Annealing is usually recommended after bending operations.

Copper alloy tubing is usually extruded. It is easily bent in the annealed condition, and it has little springback. Copper and some brasses may not need to be annealed. Copper-nickel alloys, however, are more difficult to bend and have greater springback.

When copper alloys are annealed, as most of them are, oxides should be removed by pickling before the tube is bent to protect the tooling. Oxides increase friction and wear in bending.

Aluminum alloy tubing, like copper alloy tubing, is usually extruded, or extruded and drawn. Soft aluminum may tear or collapse upon bending. The oxide coating that forms on

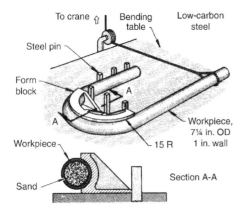

Fig. 15 Final position in the hot bending of a large sand-filled tube

exposed surfaces of aluminum alloys is abrasive to tooling. Lubrication prolongs tool life.

Anodized aluminum and decorated aluminum can usually be bent without damaging the finish. Aluminum pipe is bent by hand to a radius usually not less than four times the OD. The bending of aluminum tubing is discussed in the article "Forming of Aluminum Alloys" in this Volume.

Nickel alloy tubing and pipe can be formed by all common forming operations. Bending, coiling, and expanding can be performed readily on nickel alloy tube and pipe, using the same type of equipment as is used for other metals. In general, material in the annealed condition is recommended. Alloys 400, 200, and 201 can be formed in the stress-relieved temper; however, the amount of deformation will be limited by the higher tensile strength and lower ductility. In bending, the minimum radius to which stress-relieved tubing can be bent is 25 to 50% greater than it is for annealed tubing of the same size.

The minimum radii to which nickel alloy tubing can be bent by various methods are given in Table 5. Depending on equipment design, tube size, and quality of the finished bend, it is possible to bend to smaller radii than those listed; trial bends should be made to determine whether the smaller radii are practical.

Heat treatment should be considered if cold worked of nickel alloys intended for use at elevated temperatures. Although cold working generally increases strength, it can also increase precipitation rates, reducing corrosion resistance and ductility in materials exposed to elevated temperatures. Consult with materials producers for the appropriate heat treatment after cold formed parts intended for elevated temperature service.

Draw bending is the most common bending process and the preferred method for bending nickel alloy tube. The process is similar to compression bending, except that the bending form revolves and the pressure die either remains stationary or slides along a straight line. The sliding pressure die is preferred because it distributes the applied stresses more evenly.

Bends of up to 180° with a minimum radius of 2*D* can be produced by draw bending. Bending can be done with or without a mandrel. In general, a mandrel is preferred and must be used when the ratio of tube diameter to wall thickness is above the limit suitable for bending without tube wrinkling or collapsing. Various types of mandrels are used, including ball and plug types.

Compression bending uses a stationary bending form and a movable wiper shoe. This method is unsuitable for thin-wall tubing and is generally used with no mandrel support. Compression bending can produce bend radii down to 2.5*D* but is normally used only for large-radius bends. The maximum included angle that can be produced is 180°.

Bending Thin-Wall Tubes

The techniques used to bend thin-wall tubes are the same as those used to bend standard tube and pipe, but they are more carefully applied. A tube can be classified as thin wall if the ratio of OD to wall thickness (*D/t*) is greater than 30 to 1. The wall thickness, if not related to the tube diameter, is a meaningless measure. For example, a tube wall 0.51 mm (0.020 in.) thick would be a standard wall thickness for a tube 3.2 mm (¹/₈ in.) in OD, but for a 152 mm (6 in.) tube it would be a very thin wall. The centerline radii given in Table 6 for bending tubing of various *D/t* ratios with a ball mandrel and a wiper die are conservative and are often exceeded.

Machines used to bend thin-wall tubing have a greater capacity than necessary so that they will be stable and rigid. Their bending action must be smooth and steady. Runout on machine spindles should not exceed 0.013 mm (0.0005 in.). Mandrel rods must be heavy enough so that they do not stretch or buckle when the slip-fitting mandrel is inserted into the tubing.

Auxiliary equipment includes:

- A means for pressurizing the tubes with air or hydraulic oil (hydrostatic mandrel) to keep them from necking after they are drawn past the last mandrel ball
- Hydraulic feed on pressure dies to hold tubes in compression
- Mandrel oscillators that move the mandrel back and forth to keep the tubes from necking down

The amplitude of mandrel oscillation can be adjusted from 3.2 to 25 mm (¹/₈ to 1 in.); frequency, from 1 to 500 cpm.

Tools for bending thin-wall tubes must be more accurately made than those for standard tubes. The form block or bending die should have a runout at the bottom of the groove of not more than 0.025 mm (0.001 in.). The depth of the groove should equal 55% of the OD of the tube. The width of the groove should equal the OD of the tube plus 10% of the wall thickness. The width of the clamping groove on the bending die should equal the OD of the tube minus 10% of the wall thickness. The length of the clamping groove should be five to six times the OD of the tube, unless special clamping provisions such as flaring or clamping cleats are included. The clamp and the bending die can be keyed or doweled for perfect alignment. Clamping plugs are sometimes used; these should either be slip fitted in the tube or expandable.

The pressure die should have a groove wider than the tube OD by an amount equal to 15% of the wall thickness. The width should not vary from end to end by more than 0.013 mm (0.0005 in.). Variation in the groove will cause a pinching or relieving effect. If all tools are properly adjusted, only light pressure is needed on the pressure die, which can be adjusted against a solid bar with the same diameter as the OD of the tube.

The wiper die has a groove whose width is equal to the OD of the tube plus 10% of the wall thickness. The groove should be highly polished and have a thin coat of light oil. Too much or too heavy oil will cause wrinkles. The groove must be a full half-circle in cross section to support the entire inner half of the tube. The groove must also fit closely to the form die for at least 15° back of the point of bend so that it cannot be forced away by the pressure buildup of the compressed inner wall of the bend. Failure to maintain the position of the wiper die can cause wrinkles.

Multiball mandrels are generally used with thin-wall tubes. They have a clearance no greater than 10% of the wall thickness of the tube. The mandrel must be positioned very carefully so that the full diameter of the body is just at the start

Table 5 Minimum bend radii for nickel-base alloy tubes

Bending method	Minimum mean bend radius(a)	Maximum included angle of bend, degrees
Press bending, unfilled tube	6*D*	120
Roll bending, filled tube	4*D*	360
Compression bending		
Unfilled tube	2.5*D*	180
Filled tube(b)	2*D*	180
Draw bending		
Unfilled tube	3*D*	180
Filled tube(b)	2*D*	180

(a) *D*, tube outside diameter. (b) Or using mandrel

Table 6 Average practical centerline radii for bending thin-wall steel tubing with a ball mandrel and wiper die

Tubing outside diam		Average centerline radii for tubing with wall thickness of:											
		0.89 mm	0.035 in.	1.24 mm	0.049 in.	1.65 mm	0.065 in.	2.11 mm	0.083 in.	2.36 mm	0.093 in.	3.05 mm	0.120 in.
mm	in.	mm	in.	mm	in.	mm	in.	mm	in.	mm	in.	mm	in.
13	¹/₂	13	¹/₂(a)	13	¹/₂ (a)	...	...	...	...	...	...	...	...
16	⁵/₈	16	⁵/₈(a)	16	⁵/₈(a)	...	...	...	...	...	...	...	...
19	³/₄	19	³/₄	19	³/₄(a)	19	³/₄(a)	...	...	...	...	...	...
22	⁷/₈	32	1¹/₄	29	1¹/₈	25	1(a)	...	...	...	...	...	...
25	1	44	1³/₄	38	1¹/₂	32	1¹/₄	29	1¹/₈(a)	...	...	...	...
29	1¹/₈	64	2¹/₂	50	2	44	1³/₄	38	1¹/₂	...	...	...	...
32	1¹/₄	98	3⁷/₈	89	3¹/₂	75	3	64	2¹/₂	50	2	...	...
38	1¹/₂	127	5	108	4¹/₄	95	3³/₄	83	3¹/₄	70	2³/₄	57	2¹/₄
50	2	229	9	203	8	178	7	152	6	127	5	89	3¹/₂
64	2¹/₂	305	12	267	10¹/₂	235	9¹/₄	203	8	165	6¹/₂	127	5
75	3	381	15	330	13	279	11	254	10	229	9	203	8

(a) No wiper die required

of the bend (first ball of cable or ball-socket mandrels at the bend tangent). A template should be used to set the mandrel. If auxiliary oil or air pressure in the tube is not used, there must be enough balls to reach completely around the bend.

Interlock tooling is sometimes used for bending thin-wall tubes. The clamp is keyed to the form block, the wiper die is locked to the pressure die, and the pressure die is locked into alignment with the form block. Interlock tooling was developed specifically for automatic bending, but it has some advantages for general bending. The tools will not crush or mark the work, and setup time and scrap can be reduced.

Material should be especially uniform in thin-wall tubing that is to be bent and should all be from the same source—preferably the same heat. Because tooling dimensions are held closely, close-tolerance tubing is recommended despite its added cost.

Thin-wall tubing of nickel alloys can be mandrel-bent through 180° to a minimum mean radius of 2D. To minimize galling of the inside surface of the tube, mandrels should be made of hard alloy bronze rather than of steel. If steel mandrels are used, they should be chromium plated to reduce galling (Ref 2).

Mandrels must be lubricated before use; chlorinated oils with extreme-pressure additives are recommended for severe bending. For less severe bending or for ease of removal, water-soluble lubricants can be used.

If a mandrel is not practical, any conventional filler material, such as sand, resin, and low-melting alloys, can be used. Sand is the least desirable because it is difficult to pack tightly and thus can lead to the formation of wrinkles or kinks during bending.

Low-melting alloy fillers produce the best bends. The expansion characteristics of these fillers ensure that voids are eliminated and a sound carrier is created. Alloy fillers are removed by heating the bent tube in steam or hot water. Metallic fillers must not be removed by direct torch heating because they contain elements such as lead, tin, and bismuth that will embrittle nickel alloys at elevated temperatures. All traces of metallic fillers must be removed if the tube is to be subjected to elevated temperatures during subsequent fabrication or during service.

Production Example. Thin-wall tubing is frequently bent to elbows that have a centerline radius equal to the diameter, and it is not uncommon for the diameter to be as much as 90 times greater than the wall thickness—for example, a 152 mm (6 in.) diameter tube with a 1.65 mm (0.065 in.) thick wall. Many such elbows are used in vacuum-line service, an application in which no wrinkles are permitted. They are commonly made from 1020 steel tubing in the as-received condition. Bends are made with ball mandrels, wiper dies, and an oil-base lubricant. Some manufacturers of elbows use chromium-plated tools to minimize tool wear.

It is often difficult to prevent thin-wall tubing from slipping during bending. Methods used to provide adequate clamping are described in the following example.

Example 13: Procedures to Prevent Slipping of Tubes during Bending. The tube shown in Fig. 16 was used in a high-pressure hydraulic circuit of an earthmover. Five bends, ranging from approximately 20 to 86°, were made in a powered compression bender. All bends were made on a 152 mm (6 in.) centerline radius.

The 86° bend posed a problem because only 127 mm (5 in.) of tube was available for clamping; adjacent bends were in other planes. Surfaces of the clamps were rough, but the tube slipped during bending, causing unacceptable bends. The first approach was to line the clamps with emery cloth, but this did not add enough friction to prevent slipping. A second approach was to increase the force on the clamp. Until a hydraulic cylinder could be installed to provide this force, a factory lift truck was used. Acceptable bends were produced by increasing the force on the pressure die.

The tubing had a phosphate coating. A three-ball mandrel with an oil hole provided a constant supply of lubricant (a mixture of lard oil and mineral oil) to the inside of the tube. Tools were of W1 or W2 tool steel, hardened.

Lubrication for Tube Bending

Where a mandrel is used, both the mandrel and the interior of the tube are heavily coated with a thick lubricant. Pigmented lubricants are useful for adding body between the mandrel and the tube. Thick lubricants are sometimes heated to 120 °C (250 °F) and sprayed onto the inner surface of the tube. An oil hole in a mandrel can be used to lubricate the inside of a tube during bending (Example 13).

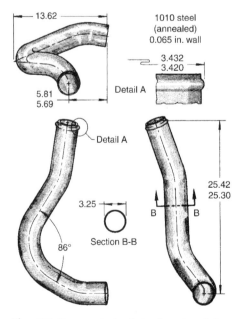

Fig. 16 Component of a hydraulic system that required five bends. Dimensions given in inches

The wiper die, on the other hand, needs only a very light lubricant, applied sparingly, if at all. Nothing must interfere with the close fit between wiper die and tube, which prevents compression wrinkles.

Not all metals react to lubricants in the same way. In general, mineral oils are always acceptable, as are organic fats. Certain sulfur and chlorine additives can stain or corrode stainless steel or copper and should be used with caution. For aluminum, special additives have been developed for use with light or medium mineral oil; the same formulations work well on copper and brass.

Tube Forming

Tubular sections are converted into a variety of products. One or more forming processes can be used, including press forming, contour roll forming, tube spinning, rotary swaging, hydraulic bulging, explosive forming, electromagnetic forming, and electrohydraulic forming.

Press Methods. Expanding with segmented dies (Fig. 17) is a frequently used method for forming configurations in the sidewalls of drawn shells or sections of tubing. With this method, the forming segments are contracted by compression springs and are expanded radially by a tapered punch. The backup, or outer, die is made of two segments, held apart by compression springs. In operation, as the press ram descends, cams move the two segments of the backup die together. The punch then moves the inner segments outward, forming the shapes in the sidewall. As the press ram is raised, springs return the die parts to their original positions, and the workpiece is rotated 45° for a restrike operation. Restriking is required to remove flat spots resulting from the stretch across the gaps between the forming segments in the expanded position. The size of the gap can be decreased by increasing the number of segments. The presence of the gaps

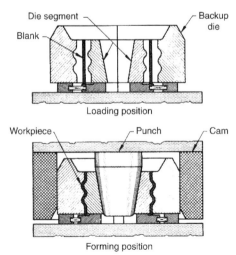

Fig. 17 Use of segmented dies for expanding a section of tubing in a press

is one of the disadvantages of this method and is the reason that an alternative method, such as rubber-pad forming, is sometimes used (see the article "Rubber-Pad Forming and Hydroforming" in this Volume). The following example describes the forming of a tube section by the rubber-pad method.

Example 14: Bulging by Rubber-Pad Forming in a Segmented Die. The hose-coupling tube shown in Fig. 18 was formed by bulging two beads with a rubber punch in a

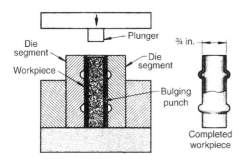

Fig. 18 Bulge-formed beads on a coupling tube using a rubber punch in a segmented die

segmented die. The workpiece, with the rubber punch in it, was held in the die on the bed of the press while a closely fitting plunger was lowered into the tube by the upper ram. The compression of the rubber punch caused outward expansion of the tube walls into the two grooves in the die, thus forming the part.

Nosing of Tubing. Tubing is often used for making mortar shells by nosing. Shells are machined before instead of after nosing and can be reduced by as much as 30% of their original diameter by nosing.

Contour roll forming is used not only to make seam-welded tubing but also to produce a variety of cross sections by reshaping the tubing. This process is discussed in the article "Contour Roll Forming" in this Volume.

Tube spinning is an established method for altering the shape of tubing, most commonly to produce a tubular part with two or more wall thicknesses. Tube spinning often precedes some other forming operation.

Other tube-forming methods include rotary swaging, hydraulic bulging, explosive forming, and electromagnetic forming. These processes are discussed in numerous articles in this Volume.

Combination Procedures. A tubular blank can often be effectively used to produce a specific shape by several different operations, which may include piercing, reducing, expanding, and upsetting. The rear-axle housings described in the following example were made in seven steps from tubular blanks. Several types of forming were involved.

Example 15: Rear-Axle Housing Formed from a Tubular Bank. The rear-axle housing shown in Fig. 19 was produced from 1035 steel tubing in seven manufacturing operations. The 2020 mm (79$\frac{1}{2}$ in.) long tube (Fig. 19a) was first pierced to form the slot (Fig. 19b). Each end was then reduced (at room temperature) from 194 to 132 mm (7$\frac{5}{8}$ to 5$\frac{3}{16}$ in.) in diameter, as shown in Fig. 19(c). This operation increased total length to 2290 mm (90$\frac{1}{4}$ in.) and increased wall thickness in the reduced sections from 7.70 to 9.27 mm (0.303 to 0.365 in.). The workpiece was then heated to approximately 760 °C (1400 °F) in a gas-fired conveyor-type furnace. In the heated condition, the center section of the workpiece was rough formed (Fig. 19d) by means of an expanding bar that was inserted through the slot, rotated 90°, and expanded with a tapered shaft. While the workpiece was still hot, the tapered transition sections were formed by pulling a shaped plug into each throat area.

The circular section was formed by cold rolling (Fig. 19e). At the same time, the flanges were flattened. The flanges were machined, and reinforcing rings were welded on, as shown in Fig. 19(e). The wheel ends were tapered, reduced in diameter, and increased in wall thickness by hot upsetting (Fig. 19f). The housings were machined to locate the brake flanges and the spring pads, which were welded on (Fig. 19g). Other machining followed. The machined housing is shown in Fig. 19(h).

REFERENCES

1. J.G. Bralla, *Handbook of Product Design for Manufacture,* McGraw-Hill, 1986
2. R.W. Breitzig, Forming of Nickel-Base Alloys, *Forming and Forging,* Vol 14, *ASM Metals Handbook,* 9th ed., ASM International, 1988, p 831–837
3. V. Rudnev, A Fresh Look at Induction Heating of Tubular Products: Part 2, *Heat Treat. Progress,* Vol 4 (No. 4), July/Aug 2004, p 23–25
4. Recommended Standards for Induction Bending of Pipe and Tube," TPA-IBS-98, Tube and Pipe Association International, Rockford, IL, 1998
5. B. Gover, Exploring Applications, Bending Methods for Structural Tubing, *Tube and Pipe J.,* June 2000

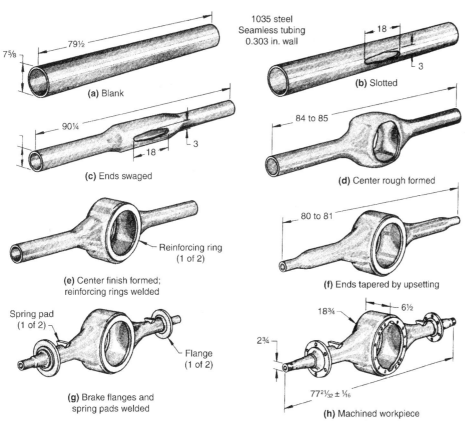

Fig. 19 Steps in the production of a rear-axle housing from a tubular blank. Dimensions given in inches

Straightening of Bars, Shapes, and Long Parts

BARS, bar sections, structural shapes, and long parts are straightened by bending, twisting, or stretching. Deviation from straightness in round bars can be expressed either as camber (deviation from a straight line) or as total indicator reading (TIR) per unit of length. Total indicator reading, which is twice the camber, is measured by rotating a round bar on its axis on rollers or centers and recording the needle travel on a dial gage placed in contact with the bar surface, generally midway between the supports. The indicator reading divided by the distance between the supports gives the straightness in TIR per unit of length. Alternatively, the deviation is expressed in terms of the distance between the supports. The effect that changing the distance between supports has on the reading is illustrated in Fig. 1; the difference in readings illustrates the importance of including support distance and location of indicators in a straightness specification.

Sections other than round are usually checked for camber by placing a straightedge against the bar and measuring with suitable gages the distance between the straightedge and the bar at the midpoint of its length. In flat bars and structural members, camber is sometimes referred to as the deviation from straightness parallel to the width, and bow as the deviation parallel to the thickness (Fig. 2).

Runout is minimal at the nodes of curvature. If a bar has compound curvature, it can be misleading to check for camber or TIR only at the midpoint of the bar length. Serious errors may result unless dial-gage readings are taken at short intervals over the entire length of the bar, or unless gangs of indicators are used at short intervals along the bar.

Straightness Tolerances. Federal Specification 48 ("Tolerances for Steel and Iron Wrought Products") establishes the straightness tolerances on some steels as:

- Hot-rolled carbon steel bars: 6.4 mm deviation per 1.5 m ($1/4$ in. deviation per 5 ft) or 4.2 mm per m (0.050 in. per ft)
- Hot-rolled alloy steel and high-strength low-alloy steel bars: 3.2 mm deviation per 1.5 m ($1/8$ in. deviation per 5 ft) or 2.1 mm per m (0.025 in. per ft)
- Hot-rolled stainless steel and heat-resisting steel bars for machining: 3.2 mm deviation per 1.5 m ($1/8$ in. deviation per 5 ft) but not to exceed 2.1 mm per m (0.025 in. per ft)
- Cold-finished carbon steel bars; turned, ground, and polished bars, or drawn, ground, and polished bars: machine straightened within limits (1.6 mm deviation per 1.5 m, or $1/16$ in. deviation per 5 ft) reasonable for satisfactory machining in an automatic bar machine
- Cold-finished stainless steel and heat-resistant steel bars for machining: 1.6 mm deviation per 1.5 m ($1/16$ in. deviation per 5 ft) but not to exceed 1.0 mm per m (0.0125 in. per ft)

- Carbon steel, stainless steel, and heat-resistant steel structural shapes (except wide flange sections): 2.1 mm deviation per m (0.025 in. deviation per ft)
- Wide flange sections used as beams: 1.0 mm deviation per m (0.0125 in. deviation per ft)
- Wide flange sections used as columns: up to 14 m (45 ft) long, 1.0 mm deviation per m (0.0125 in. deviation per ft) but not over 9.5 mm ($3/8$ in.); over 14 m (45 ft) long, 9.5 mm ($3/8$ in.) plus 1.0 mm per m (0.0125 in. per ft) beyond the 14 m (45 ft) length

Material Displacement Straightening

Manual Straightening. The original method of hand straightening is still used extensively when accuracy and precision are required or when the shape of the bar or part makes machine straightening impractical. The tools used in manual straightening include hammers and mallets, anvils, surface tables, vises, levers, grooved blocks, grooved rolls, twisting devices, various fixtures, and heating torches. The use of a grooved block (Fig. 3a) illustrates the basic principle of manual straightening by bending.

Shafts for centrifugal irrigation pumps are an example of parts that usually must be manually straightened because of the accuracy required and the necessity of doing the work at the installation site. Most of these shafts are 3.0 to 6.1 m (10 to 20 ft) long, have diameters of 19 to 50 mm ($3/4$ to 2 in.), and are of cold-drawn 1045 steel. The steel supplier generally straightens the cold-drawn stock within 0.13 or 0.25 mm (0.005 or 0.010 in.) TIR in 3.0 m (10 ft), but the shafts are often bent slightly in transport and in handling. It is common to hand straighten the shafts at installation—within 0.13 mm (0.005 in.) TIR in 6.1 m (20 ft). The shafts are rotated on supports and are deflected with a lever.

Special cold-drawn sections as long as 3.7 m (12 ft) are commonly straightened manually. Many special sections are similar enough to standard flats that they can be straightened in standard two-direction roll straighteners, but quantities are often too small to warrant the cost

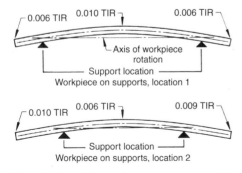

Fig. 1 Effect of distance between supports on straightness readings for round bars

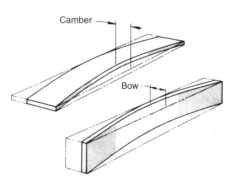

Fig. 2 Camber and bow in flat bars

of special rolls. Other special sections may be too complex in shape for machine straightening.

Special sections almost always have twist after cold drawing. The twist must be removed before the section can be straightened. One end of the section is held in a vise or in a special fixture, while the other end is twisted with a wrench or special handle. When the twist is corrected, the bar can be straightened to remove camber and bow.

Manual Peening. Straightening of a workpiece by peening (material displacement) is accomplished manually by placing the workpiece on heavy, flat plates and rapidly hammering on the concave side of the distorted portion of the workpiece surface. Elimination of the distortion allows the workpiece to lie flat. This method is applicable when straightening round parts with a high degree of hardness, such as drill bit blanks. Unfortunately, the technique is time consuming and does require highly skilled operators.

High-Production Peening. In peen straightening, also known as pulse straightening, the workpiece is pulsed on the low side opposite the

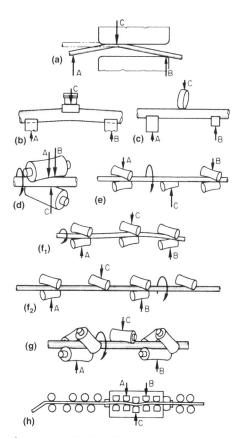

Fig. 3 Principle of straightening by bending. (a) Manual straightening with a grooved block. (b) Straightening in a press. (c) Simplest form of rotary straightening. (d) Two-roll straightening. (e) Five-roll straightening. (f1 and f2) Two arrangements of rolls for six-roll straightening. (g) Seven-roll straightening. (h) Wire straightening. In all methods shown, the bar is supported at points A and B, and force at C on the convex side causes straightening. See text and subsequent illustrations for details of the straightening methods shown here.

high point of the bend (Fig. 4). The pulsating tooling compresses the workpiece material and burnishes its surface. To counteract existing stresses in the workpiece, the material expands to straighten the part. The metal structure is stabilized, the part retains its straight configuration in storage, and the material is unaffected by the shock and vibration encountered during subsequent machining operations. This method is gaining wide acceptance in high-volume production operations in which cast and forged steel camshafts, hardened steel transmission shafts, crankshafts, and disk plates require straightening.

Stretch Straightening. Many bars and shapes can be easily straightened by stretching. However, this technique is usually confined to the straightening of shapes that are uniform in cross section and length. The advantages of stretch straightening include low costs for tooling and for maintenance, simplicity of operation, and (usually) completion of straightening in one operation. The disadvantages include waste by trimming 152 to 467 mm (6 to 18 in.) from the ends of bars damaged by gripping, the need (usually) for two men to do the straightening, the need for cutoff equipment, and production of only 30 to 40 bars per hour.

A stretch straightener has two heads with grips that clamp on the ends of the bar. One head can be adjusted to suit the workpiece length. The other head (tailstock) is powered for stretching and for rotation to correct twist in the workpiece.

Stretching machines are made in sizes to exert stretching forces of 135 to 4450 kN (15 to 500 tonf) to workpieces that may be from 6 to 30 m (20 to 100 ft) long. Stretching must stress the work beyond its yield strength. For complete straightening, the bar should be stretched 2%. To accomplish this and to overcome the greater strength caused by work hardening, the stretching machine needs a capacity of 10 to 15% beyond the yield strength of the bar.

Some straightening can be done by stretching only to the yield strength of the work, but this would not be sufficient to remove completely some sharp bends and twists. A stretcher with a capacity of 1340 kN (150 tonf) can straighten, to some extent, a low-carbon steel bar with a cross section of 9700 mm² (15 in.²), 6450 mm² (10 in.²), of austenitic stainless steel, or 4550 mm² (7 in.²) of ferritic stainless and can do complete straightening on 8400 mm² (13 in.²) of low-carbon steel, 5150 mm² (8 in.²) of austenitic stainless, or 3900 mm² (6 in.²) of ferritic stainless steel. Figure 5 shows the relationship between stretcher force and the cross-sectional area of the bar. The deviation that remains after

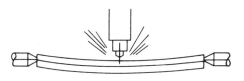

Fig. 4 Schematic of principles involved in straightening by automatic peening

stretch straightening can sometimes be corrected by manual straightening.

Most hot-rolled bars can be straightened by stretching. Straightening by stretching also works well on rolled and extruded bars of aluminum and austenitic and ferritic stainless steels, but not on martensitic stainless steels unless they are first annealed. Low-carbon steels are easy to stretch straighten, but annealing before stretching becomes more necessary as the carbon content increases. Because it is slow and limited in effectiveness, stretch straightening is not widely used.

Straightening by Heating

Alloy steel bars and shapes with a hardness exceeding 50 HRC, as well as fabricated stainless steel parts, frequently warp because of the stress set up during fabrication, machining, or heat treatment. These items can usually be straightened by the application of heat and, in most cases, force. The heat can be localized in the area to be straightened, or the entire piece can be heated—either to the tempering temperature or to about 30 °C (50 °F) below it. (Heating to a temperature above that required for tempering will reduce the hardness, as will prolonged heating at the tempering temperature itself.) Low-carbon steel bars can also be straightened by heating.

Localized Heating. Torches are used to apply heat to the convex side of warped parts. A small area is heated to a dull red. The localized heating causes the workpiece to expand, but some straightening occurs during cooling. Skillful heating, cooling, and gaging of the workpiece can result in reasonable straightness.

Torch heating causes soft spots in hardened steel workpieces. Localized heating with a torch can also cause localized residual tensile stress that can be undesirable even in an unhardened workpiece if it is subjected to cyclic loading.

In press straightening with the use of localized heat, the workpiece is supported at each end with

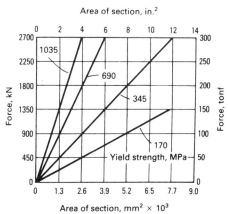

Fig. 5 Relationship between stretcher force and area of bar cross section in the stretch straightening of steel bars of various yield strengths

suitable blocks. A stop block is placed directly under the ram to limit the amount of deflection. With the high points of its curvature up, the workpiece is pressed down until it rests lightly on the stop block; heat is then applied. For a heat-treated workpiece, the amount of heat is usually governed by the original tempering temperature, and the distance the workpiece can be deflected and released without fracture depends on the type and hardness of the steel, the heat treatment, and the shape of the workpiece. Another method of controlling deflection without breakage is to place the workpiece on shims while it rests on a flat surface, apply pressure to the surface of the workpiece, then heat and release. If the workpiece is still not straight, it will be necessary to use more shims and reheat or to allow the workpiece to cool longer before releasing the pressure.

Where a flame cannot be effectively directed or may damage the metal, a small weld bead can sometimes be used as the source of heat. Weld beads are applied to the convex area, allowed to cool, and machined off if necessary.

Heating below Tempering Temperature. Heating and press straightening are generally not applicable to steel at high hardness levels. The force required to cause permanent set is close to the rupture strength of the steel, and even with extreme care, failure is probable. At medium and lower hardness levels, heating to a temperature about 30 °C (50 °F) below the tempering temperature will permit press straightening to be done successfully. Straightening becomes more difficult as the part cools, and only slight corrective straightening should be attempted at the lower temperature levels. Considerable skill is required to perform such operations and to hold tolerances within 0.08 to 0.25 mm (0.003 to 0.010 in.) over a length of 0.45 to 1.22 m (18 to 48 in.).

After the workpiece has been straightened, it is tempered to the required hardness. Tempering relieves stress set up during straightening and during the hardening cycle. This stress will often deform the workpiece; consequently, work-pieces straightened by heating and pressing should be clamped in restraining fixtures during tempering. Fixturing can correct a slight distortion and prevent distortion during tempering.

Temper straightening is used to correct the distortion caused by heat treatment. The work-piece is first tempered to a hardness somewhat higher than required, then clamped in a straight-ening fixture and tempered to the required hardness. The greater the hardness difference between the first and the corrective tempering operations, the more accurate the dimensions will be. Temper straightening is most successful at hardness levels of 55 HRC and lower.

Deep-hardening alloy and tool steels that are being martempered to minimize distortion should be held straight during the cooling period after austenitizing and until the completion of martempering. If straightness is not maintained throughout martempering, the workpiece will warp as martensite continues to form. Straight-

ening should be done below 480 °C (900 °F). Cold bars or chills contacting the high side will more rapidly extract the heat from the workpiece and aid in straightening.

Straightening in Presses

Round bars up to 50 mm (2 in.) in diameter and from 0.6 to 3.0 m (2 to 10 ft) in length are often straightened in an arbor press. Larger workpieces are similarly straightened in power presses, which may have power rolls and hoists to move the work.

The principle of press straightening is illustrated in Fig. 3(b). The bar to be straightened is supported at points A and B with the convex side of the bow or kink toward point C. Sufficient force is applied at C to cause the bar to become bowed in the opposite direction. The force must be great enough to exceed the elastic limit of the material, but it must set up just enough strain in the bar to allow it to return to the straight position (but no farther) when the pressure is released. The greater the bow in the bar and the higher its elastic limit, the greater the force required to produce the correct amount of strain. To straighten a bar by press straightening, the metal must be capable of cold deformation, and it must strain harden.

In press straightening, the operator usually locates kinks or bows in round bars by holding a piece of chalk close to the surface of the bar and then rotating the bar so that any high spots will be marked by the chalk. The high spot is then brought under the ram of the straightening press, and sufficient force is applied to remove the kink or bow. In shapes other than rounds, the out-of-straight condition must be detected visually or with the aid of a straightedge. This type of straightening requires considerable skill on the part of the operator. A straightening press is sometimes referred to as a gag press and the straightening operation as gagging.

Hydraulic or mechanical presses are used to straighten bars, shapes, and shaftlike parts before, between, and after heat-treating operations. Some bars that are roll straightened do not meet straightness requirements and must receive a final press straightening. Presses are also used to straighten large diameter bars in preparation for turning or grinding. Finish-ground or turned products with highly finished surfaces are press straightened to avoid the spiral marks produced by rotary straightening. Press straightening does not change the size of the bar, but rotary straightening—because of the rolling action or the alteration in residual stress caused by bending or both—may cause a change in bar size.

Some high-strength steels and stainless steels are too hard to be straightened in any way except in a press, and some metals are too hard to be straightened without heat, unless the bar is first annealed. The cold straightening of bars of these metals may cause the bars to break.

Press straightening is easier when the bar has a hardness of less than 40 HRC. The bar can be

retempered to relieve the stress introduced during straightening.

In a straightening press, a round bar is usually set on spring-loaded rollers near the ends of the bar. Thus, the bar can be rotated on its axis while a dial gage shows any deviations from straightness. As the press ram moves down, it presses the bar into V-blocks that support the straightening pressure. This action is repeated, sometimes with the bar shifted or the roller supports moved, until the bar is straight enough to meet specifications. The V-blocks and roller supports can be moved to change the leverage and to adjust the application of the force. A straightening press is better suited to the correction of short bends and kinks than to the correction of long bends. A straightening press can hold the distortion of heat treated bars and shafts to 0.25 mm (0.010 in.) or less, as shown in examples 1 and 2.

Example 1: Straightening of a Heat Treated Shaft in a Press. A D2 tool steel shaft with a hardness of 63 to 65 HRC and a straightness specification of 0.25 mm (0.010 in.) TIR is shown in Fig. 6. The shaft was hung in a vertical furnace and preheated to 205 °C (400 °F), 540 °C (1000 °F), and 815 °C (1500 °F) before heating at 540 °C (1000 °F). The part was air cooled to 260 °C (500 °F). Because inspection at that temperature showed distortion of 0.38 to 0.64 mm (0.015 to 0.025 in.) TIR, the bar was straightened in a manual hydraulic press to 0.18 mm (0.007 in.) TIR before it cooled to 205 °C (400 °F). At 65 °C (150 °F), the shaft was again straightened to within 0.18 mm (0.007 in.) TIR.

The shaft was clamped in a V-block for tempering at 150 °C (300 °F) for 6 h, then given a subzero treatment and straightened to within 0.20 mm (0.008 in.) TIR. The shaft was again clamped in the V-block, using shims for straightening, while being retempered at 150 °C (300 °F) for 6 h. Final straightening in the press was to 0.18 mm (0.007 in.) TIR, which was better than was required.

Example 2: Straightening of a Heat Treated Bar in a Press. A flat bar of O1 tool steel having

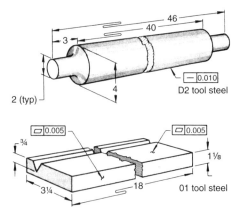

Fig. 6 Shaft and grooved bar that were press straightened after heat treatment. Shaft hardness: 63 to 65 HRC. Bar hardness: 61 to 63 HRC. Dimensions given in inches

a 19 mm (³/₄ in.) deep V-groove along one edge is shown in Fig. 6. The specified hardness was 61 to 63 HRC. The top and sides had to be flat within 0.13 mm (0.005 in.).

The bar was preheated to 595 °C (1100 °F), heated at 805 °C (1480 °F), and marquenched at 190 °C (375 °F). The bar was then clamped in a fixture and cooled to 38 °C (100 °F). Inspection showed 0.15 mm (0.006 in.) maximum variation for the top and side. The bar was reclamped in the fixture and tempered at 150 °C (300 °F) to a hardness of 63 to 64 HRC. The bar was reclamped with shims to straighten it and was retempered at 165 °C (325 °F) to a hardness of 61 to 62 HRC. The bar was then straightened in a press within 0.13 mm (0.005 in.) at the top and side.

Clamping of Workpiece. If the workpieces in examples 1 and 2 had been heat treated without being clamped, they would have been free to distort and would have needed more press straightening. The straightening of bars and shafts during transformation and during tempering is more efficient and costs less, and it is sometimes the only way in which straightness specifications can be met. Structural parts for aircraft are commonly straightened by a combination of methods, as shown in examples 3, 4, and 5. Example 6 describes a procedure for heat treating and straightening a long, thin rectangular bar in a press. Round shafts are often straightened before they are ground, as in examples 7 and 8.

Straightening presses are used as accessories to other equipment, such as blooming mills that roll blooms or billets 127 to 178 mm (5 to 7 in.) thick. A 1.8 MN (200 tonf) press with a bed 0.6 by 1.2 m (2 by 4 ft) can straighten such blooms or billets within 6.4 mm per 1.8 m (¹/₄ in. per 6 ft) in lengths as great as 4.9 m (16 ft) without the use of spacers or shims. Many blooms or billets do not need straightening—for example, if they are to be cut into pieces for forging stock.

Example 6 describes a procedure for heat treating and straightening a long, thin rectangular bar in a press. Round shafts are often straightened before they are ground, as in examples 7 and 8.

Example 3: A structural aircraft part called a cap was made from modified 4330 steel bar 4.7 mm (³/₁₆ in.) thick by 1.7 m (66 in.) long

(Fig. 7). The part was channel shaped with one flange removed for a portion of its length to produce an angle section. Holes 3.2 mm (¹/₈ in.) in diameter were made in both sides of the angle section and in the channel section. Heat treatment consisted of suspending the workpiece by the angle end in a salt bath at 845 °C (1550 °F), quenching in salt at 245 °C (470 °F), and air cooling. After being heat treated, the parts were cleaned and checked for hardness (aim was 46 to 49 HRC). With hardness less than 50 HRC, a subzero treatment was given prior to tempering; with hardness of 50 HRC or harder, the workpiece was clamped in a fixture, tempered at 315 °C (600 °F) for 5 h, and finally air cooled. The workpiece warped about 25 mm in 1.7 m (1 in. in 66 in.) after quenching, and a warp of about 6.4 mm (¹/₄ in.) remained after fixture tempering. A camber of 6.4 mm (¹/₄ in.) was easily removed manually, but a workpiece with a camber of more than 6.4 mm (¹/₄ in.) was retempered in the fixture before straightening. Shape and straightness were checked in a fixture using a 0.51 mm (0.020 in.) feeler gage. After being tempered at 315 °C (600 °F), the workpiece had a hardness of 46 to 49 HRC.

Rotation of about 60° was required to correct for 0.76 mm (0.030 in.) twist. An 89 kN (10 tonf) hydraulic press was used to correct short bends. The workpiece was supported on slotted blocks, placed about 360 mm (14 in.) apart, which gave support to the flanges while pressure was applied. A force of about 13 kN (3000 lbf) deflected the channel section 9.5 to 13 mm (³/₈ to ¹/₂ in.) for a camber correction of 3.3 mm/m (0.020 in. per 6 in.). A force of 4.4 to 6.7 kN (1000 to 1500 lbf) on the angle section produced about the same deflection and correction.

Stretching was used to maintain spacing and alignment for the 3.2 mm (¹/₈ in.) diam holes. Because of the thin sections, the part shrank 0.76 to 1.0 mm (0.030 to 0.040 in.) after quenching in salt at 245 °C (470 °F). To correct this shrinkage, the part was preheated at 290 °C (550 °F) for 30 min, then clamped in the tempering fixture and heated at 315 °C (600 °F) for 5 h. As the fixture expanded from the heat, the part was stretched. After slowly cooling in the fixture, the part had a permanent stretch of 0.51 to 0.76 mm (0.020 to 0.030 in.), which corrected hole

alignment and spacing. About 8 min was required for clamping the part in the tempering fixture, and 22 min for hand straightening.

Example 4: A welded double-channel structural member made of 4340 steel 1.8 m (72 in.) long and weighing 5.2 kg (11.5 lb) (Fig. 8) was austenitized at 830 °C (1525 °F) for 40 min, martempered at 245 °C (470 °F), stress relieved at 205 °C (400 °F), and cleaned. The unusual feature of this operation was the simultaneous straightening, bending, and tempering.

The workpiece was machined on a straight plane and then fixture bent 7° on one end during tempering. The fixture was constructed with various gibs and filler blocks milled to fit the contours of the workpiece, including allowance for the 7° bend, which extended for approximately 305 mm (12 in.) (Fig. 8). The workpiece was then placed on the fixture, and all holding clamps were placed in position. The fixture-and-workpiece assembly was then heated to 315 °C (600 °F). At this time, all clamps were tightened, thus making the 7° bend at one end. The assembly was then heated (tempered) to 540 °C (1000 °F) for 4 h.

After tempering the channel was straightened in a hydraulic press to achieve final alignment within 0.76 mm (0.030 in.). A special gage was used to inspect alignment and the bend. Time for straightening and gaging was 30 min.

Example 5: A structural bar with a tapered-width channel 4.8 mm (³/₁₆ in.) thick and 1.7 m (66 in.) long was made of 17-4 PH stainless steel (Fig. 9). The part was solution treated to 39 to 42 HRC, finish machined, then aged for 1 h at 480 °C (900 °F). It was straightened to within 0.38 mm (0.015 in.) camber in a hydraulic press upon removal from the aging treatment. The part could be straightened only until it cooled to 370 °C (700 °F), which took 10 min. Because 20 min was required to straighten the part, reheating to 480 °C (900 °F) was necessary.

The channel was supported on two blocks 406 mm (16 in.) apart while force was applied by a pressure ram through a block fitted to the inside contour (upper view, Fig. 9). A force of 13.3 kN (3000 lbf) deflected the part about 25 mm (1 in.) for a correction of 0.51 mm (0.020 in.). Force was applied at 50 mm (2 in.) increments along the bar.

Two 44 kN (5 tonf) hydraulic presses mounted on a large steel table were used to remove the twist. One ram held one end of the part against a block on the table while the second ram untwisted the part (lower view, Fig. 9). A slotted

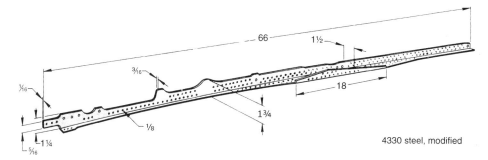

Fig. 7 Aircraft part that was straightened by a combination of methods. Hardness: 46 to 49 HRC. Dimensions given in inches

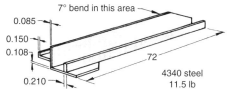

Fig. 8 Welded structural member that was straightened during and after heat treatment. Dimensions given in inches

bar served as a lever to twist the channel 30° for a permanent correction of 0.83 mm per 1 m (0.020 in. per 24 in.) or 1.3 mm (0.050 in.) for the entire length.

Example 6: Straightening a Long, Thin Rectangular Bar in a Hydraulic Press. A rectangular bar of 17-4 PH stainless steel was 75 mm (3 in.) wide, 2.1 m (84 in.) long, and 6.4 mm (1/4 in.) thick (except for 75 mm, or 3 in., at each end, where it was 25 mm, or 1 in., thick). The bar was solution treated and finish machined, which caused a bowing of 6.4 mm (1/4 in.). The bar was clamped in a fixture and aged at 480 °C (900 °F), which reduced the bowing to 3.2 mm (1/8 in.). After aging, the bar was removed from the fixture, reheated to 425 °C (800 °F), and straightened in an 89 kN (10 tonf) hydraulic press, using 13.3 kN (3000 lbf) of force between support blocks 406 mm (16 in.) apart. This caused deflection of 19 mm (3/4 in.) for a correction of 0.76 mm (0.030 in.). It took 20 min and two to three ram strokes per 406 mm (16 in.) setup, as well as five setups per bar, to straighten the bar within 1.5 mm (0.060 in.).

Example 7: A shaft of medium-carbon steel, 102 mm (4 in.) in diameter by 6.1 m (20 ft) long, was heat treated to 269 to 321 HB, straightened for turning, turned in a lathe, and then straightened for centerless grinding. The shaft lay on rollers beneath the ram of the press, which permitted it to be rotated and to be moved along its axis. Spring-loaded blocks supported the rollers so that straightening pressure would first deflect the springs, letting the shaft down on movable V-block anvils for straightening. The springs pushed the rollers up, lifting the shaft off the anvils when the ram moved up. The shaft was rotated under a dial indicator to find the high spots. The spots were marked with chalk so that they could be moved beneath the ram. The shaft was straightened within tolerance by pressing, moving the shaft, and pressing again. In an 8 h day, 5 to 15 shafts were straightened.

Example 8: After being heated to 1010 °C (1850 °F) and air cooled, a shaft of D2 tool steel, 50 mm (2 in.) in diameter by 1.7 m (66 in.) long,

was straightened within 0.51 mm (0.020 in.). The straightening began when the shaft had cooled to 480 °C (900 °F) from 1010 °C (1850 °F). An 89 kN (10 tonf) hydraulic press applied 8.9 kN (2000 lbf) of force to the shaft, which was supported on anvil blocks 457 mm (18 in.) apart. The shaft was continuously gaged with dial indicators as it was rotated on its axis in order to check the location of the high and low points.

Between 480 and 260 °C (900 and 500 °F), the shaft deflected easily under a load of 910 kg (2000 lb), causing a deflection of 3.2 mm per 508 mm (1/8 in. per 20 in.). When the shaft cooled further, straightening was more difficult and required increased force and a longer holding time.

Gaging and straightening continued until the shaft had cooled to 65 °C (150 °F); the shaft was then tempered at 480 °C (900 °F) for 2 h, resulting in a hardness of 59 to 60 HRC. A difficult shaft was sometimes clamped to a 76 by 76 by 1830 mm (3 by 3 by 72 in.) bar with shims and retempered. If the straightness error was 2.0 mm (0.080 in.), a 1.5 mm (0.060 in.) shim was used on each end for a deflection of 3.5 mm (2.0+1.5 mm), or 0.140 in. (0.060+0.080 in.), and the shaft was retempered at 495 °C (925 °F). The higher tempering temperature made the shaft slightly softer but generally straightened it within the specified 0.51 mm (0.020 in.).

Parallel-Roll Straightening

Roll straightening is a cold-finishing mill process by which bars and structural shapes are provided with straightness adequate for most applications. For bars and shapes on which close tolerances must be maintained, roll straightening can be followed by press straightening.

In one type of roll straightening, square, flat, hexagonal, and other flat-sided bars are continuously passed between sets of parallel-axis rolls (Fig. 10, 11). Uniform bends are introduced such that the bar is straight when it leaves the rolls. By varying the distance between roll centers and the amount of offset, the degree of

bend can be adjusted according to the section size and yield strength of the metal being straightened.

Round bars can be straightened on parallel-axis roll straighteners, but there is no way to prevent a round workpiece from turning on its axis as it passes through the machine. In rotary straighteners, round bars rotate and advance through the rolls so that the bars are bent uniformly in all planes; the rolls are adjustable so that the bars emerge straightened. Rotary straighteners for round bars usually consist of rolls that can be set at variable angles to each other (see the section "Rotary Straighteners" in this article).

Parallel-Roll Straighteners. Square, flat, hexagonal, and other flat-sided bars can be straightened in both directions in one pass through a straightener having two sets of parallel rolls in planes 90° to each other (Fig. 10), or in two passes through a straightener having a set of parallel rolls in only one plane, by turning the bar 90° on its axis between passes.

In machines with a single straightening plane, the rolls are mounted on horizontal shafts as shown in unit 2 of Fig. 10. If the horizontal rolls are grooved (like the vertical rolls of unit 1, Fig. 10), the straightening in one plane also produces some straightening in a plane 90° to the first plane.

For high accuracy of straightness and high production rate, a second unit is added in a plane perpendicular to the first unit (Fig. 10). Unit 1 has vertical shafts for straightening curvature in the horizontal plane; unit 2 has horizontal shafts for straightening curvature in the vertical plane. The two driven rolls rotate but are otherwise stationary; the three idler rolls are adjustable away from and toward the workpiece.

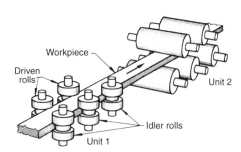

Fig. 10 Arrangement of vertical-shaft and horizontal-shaft rolls in a roll straightener for straightening a rectangular-section bar

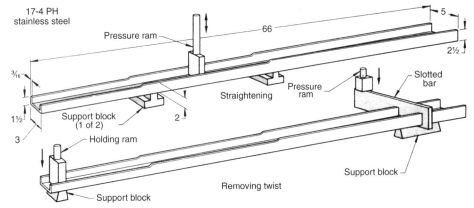

Fig. 9 Setups for straightening and removing twist from a stepped channel. Dimensions given in inches

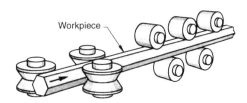

Fig. 11 Straightening of a hexagonal bar in a two-plane roll straightener

The first driven roll contacted by the bar is set with enough space between the first two idler rolls to curve the bar uniformly with the concavity toward the first driven roll. As the bar passes over the second idler roll and is held in position by the second driven roll, the concave side of the bar is reversed. The amount of reversal can be controlled by the position of the second idler roll, and with that roll properly positioned, the bar will emerge straight from the third idler roll. With a greater number of rolls, the most severe curvatures are reduced at the entry end of the machine. This results in less work for the remaining rolls and provides for better straightening of small curvatures. The number of rolls in a set of straightening rolls ranges from 4 to as many as 13; the most common number is eight or nine.

The amount of adjustment in roll spacing is determined by:

- The resistance of the work metal to deflection beyond its elastic limit. The greater the resistance, the farther apart the rolls must be spaced to provide sufficient shaft and bearing capacities.
- The distance must be short enough to produce a permanent set in the smallest bar and to straighten it.
- If sections with high width-to-thickness ratios are being straightened, the distance must be great enough so that pressure from the rolls does not upset the edges of the bar.

Machines are built with both fixed and adjustable roll-center distances. In adjustable center distance machines, a separate housing carries the roll assembly to permit positioning along the straightener bed. These housings either reduce the space for the shafts or limit the minimum distance between the rolls.

Most sections can be straightened adequately in two planes. A flat bar is shown in Fig. 10 as it passes between grooved straightening rolls on vertical shafts and then through plain-face rolls on horizontal shafts. The bar passes first through the vertical-shaft unit because it is natural for the bar to enter with the flat side lying against the feed table. It is also natural to straighten in the

grooved rolls first because the grooves that guide the bar also produce some straightening in the second plane and ensure proper entry of the bar into the second unit. If the machine were reversed, it would be desirable to groove the rolls on the horizontal shafts to make certain that the bar would not "walk off" the rolls.

Grooving the rolls for thin flat bars helps to reduce upsetting of the edge and twisting of the bar. A hexagonal bar is shown in Fig. 11 as it passes between grooved straightening rolls on vertical shafts and through flat-face rolls on horizontal shafts. This roll arrangement follows the same principles as those for flat bars.

Sections that are symmetrical in both planes are easier to roll straighten than nonsymmetrical sections. Sections that are symmetrical in one plane but nonsymmetrical in the plane at 90° are usually best straightened in the symmetrical plane. Axial adjustment of some of the rolls to produce additional straightening in the nonsymmetrical plane is necessary in this method, as in the straightening of the angle section shown in Fig. 12.

Angles are best straightened in horizontal-shaft rolls with the apex of the angle up and roll adjustment made in the vertical plane. The angle lies naturally on the feed table and is straightened in both directions. To take out bends 90° to the plane in which the angle is straightened, one or two top rolls are adjusted axially to deflect the angle in this direction.

A structural channel can be passed through horizontal-shaft rolls with the flanges of the channel up or down. The upper rolls are staggered vertically and horizontally to remove camber in both planes.

Square or nearly square bars can be straightened on the diagonal in a single-plane machine using V-shaped upper and lower rolls (Fig. 13a) similar to the upper rolls used for straightening the angle shown in Fig. 12. In this method of straightening, only the shaded portion of the cross section shown in Fig. 13(b) is stressed beyond the elastic limit; therefore, the results may not be satisfactory. A square is better straightened in a two-plane machine, in which the areas stressed beyond the elastic limit are greater and more nearly symmetrical, as shown

in Fig. 13(c). Square and hexagonal bars as large as 102 mm (4 in.) and flat bars as wide as 305 mm (12 in.) are straightened in roll straighteners. Larger bars are usually straightened in presses.

Bars of low-carbon steel seldom change size in straightening. Steel having 0.30% C or more may enlarge slightly in section because straightening redistributes the stress that remains from previous operations. For example, a bar of 1045 steel 30 mm ($1^3/_{16}$ in.) square may enlarge 0.05 mm (0.002 in.) in one pass through a roll straightener, and 0.10 mm (0.004 in.) in two passes. The bars shorten as they enlarge in section in accordance with the Poisson's ratio of the material.

Square, hexagonal, and flat bars to 19 mm ($3/_4$ in.) in cross section are sometimes cold drawn from coils of hot-rolled stock. The drawn bars are straightened in roll straighteners and sheared to length. To correct the curvature resulting from coiling, rotary or two-plane straighteners with sets of six to eight rolls in each plane are used.

Rotary Straighteners

Round bars or shaftlike parts of all types of metal are straightened in rotary straightening machines of two basic types: crossed-axis-roll machines and rotary-arbor machines. The basic principle of rotary straightening is that the workpiece is fed forward and deflected beyond its elastic limit by crossed-axis rolls that also impart the rotary motion. The surface of the bar is alternately subjected to tensile and compressive stresses as it rotates in the straightener. Rotary straighteners are available with two to nine rolls.

A two-roll rotary straightener consists of two rolls that are directly opposed and positively driven. One of the rolls is concave, and the other has a relatively straight face (Fig. 14). The angularity adjustment of the rolls at opposite inclinations rotates and feeds the bar through the machine. Straightening is accomplished by flexing the workpiece into the throat of the concave roll by the modified straight-face roll (Fig. 3d). The bar is positioned vertically by means of a bottom guide or top and bottom guides (not shown in Fig. 14) so that the axis of the bar coincides with the centerline of its path between the rolls.

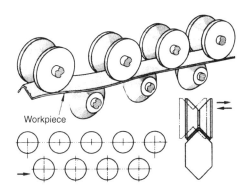

Fig. 12 Roll straightening of a structural angle. The top rolls can be adjusted horizontally and vertically.

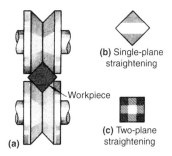

(b) Single-plane straightening

Workpiece

(c) Two-plane straightening

(a)

Fig. 13 Roll straightening of square bars. (a) Setup for single-plane roll straightening. (b) and (c) Stress patterns that result from single-plane and two-plane straightening

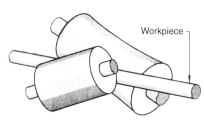

Workpiece

Fig. 14 Arrangement of rolls in a two-roll rotary straightener (top and bottom guides not shown)

The amount of bend given a bar as it passes through the machine depends on two adjustments made by the operator:

- The angle of the rolls to the axis of the bar
- The roll pressure, which is selected by adjusting one of the rolls toward or away from the other

Roll-angle and pressure adjustments depend on the size of the bar being straightened and its mechanical properties. In general, the larger the bar, the greater the roll angle, provided the mechanical properties are about the same. A heat treated bar (tensile strength: 862 to 1030 MPa, or 125 to 150 ksi) will require a smaller roll angle and more pressure than a bar of the same size and grade that has been annealed, or annealed and cold drawn.

In two-roll rotary straightening, the workpiece is subjected to a continuous straightening action from the point of entrance to the work rolls to the end of the workpiece as it leaves the rolls. Therefore, there is no variation in size within the bar, as is sometimes encountered with multiroll straighteners. Two-roll straighteners can be used for short workpieces, such as rocker-arm shafts and chain-link pins, because all of the flexing is contained within the cavity of one roll. Two-roll straighteners are also used for sizing or for correcting out-of-roundness in hot-rolled bars. Extremely soft metal may be reduced in diameter if too much pressure or too large a roll angle is used. Two-roll straighteners can be used to remove end kinks and to round out squashed ends, both of which sometimes occur when bars are cold sheared to length prior to straightening.

Two-roll rotary straighteners inherently have a lower through speed than do multiroll straighteners. The roll inclination must be kept lower (about 20°) in two-roll straighteners; therefore, the rotational speed of the bar is much higher in relationship to the forward speed.

The span over which bending takes place is considerably shorter in two-roll machines than in multiroll rotary straighteners because in two-roll machines, all bending takes place within the length of the rolls and not from roll to roll. With such a short span, much more force must be applied to the bar by the bending equipment than with multiroll machines. Bars from 1.6 to 255 mm ($\frac{1}{16}$ to 10 in.) in diameter can be straightened in two-roll rotary machines.

In addition to finish straightening, the two-roll rotary machine can be used to rough straighten hot-rolled round bars, which may be very crooked and may have sharp hooks and round and scaly surfaces; to straighten and size cold-drawn round bars, which may be bowed but have no sharp bends; and to polish or burnish to improve surface finish after grinding. Additional rolls should be kept for straightening only, sizing only, and polishing only.

Multiroll Rotary Straighteners. Another type of machine used in the straightening of bars is the multiroll rotary straightener. Figure 15 shows a five-roll rotary straightener, which consists of two driven rolls and three idler rolls.

The two end idlers oppose the driven rolls, and between them is the middle or pressure roll. All rolls are concave, and the roll inclination is adjustable in order to obtain the maximum length of contact between the roll surface and the workpiece (see also Fig. 3e). Bottom cast iron guide shoes are located at the entry and exit ends between the driven rolls and their respective opposing idlers to position the bar properly.

A six-roll rotary straightener has a roll arrangement similar to that of a five-roll machine; the sixth roll is placed either opposite the middle roll or outboard of the exit-end powered rolls (see these roll arrangements in Fig. 3f). Seven-roll arrangements consist of two three-roll clusters with a middle idler roll (Fig. 3g). Small cluster-roll straighteners have been used extensively for specialty work on small workpieces, such as valve push rods (~7.9 mm, or $\frac{5}{16}$ in., in diam) and rocker-arm shafts; however, these straighteners are used most often for straightening large tubing (60 to 610 mm, or $2\frac{3}{8}$ to 24 in., in diameter).

In operation, the rolls are adjusted angularly to accommodate various bar sizes. With the average angle selected as 30°, the adjustment may vary from approximately 28 to 30°, depending on the size of the bar being straightened.

In a five-roll straightener, the middle idler roll is adjusted to put enough bend in the bar to exceed the elastic limit of the metal. As the bar is fed through the straightener and rotated by the entrance and exit rolls, the adjustment of the pressure roll causes the bar to bend beyond its elastic limit in all directions perpendicular to its longitudinal axis. This action produces a straight bar with symmetrical stresses.

Optimum settings of roll angle vary somewhat with bar size. Typical settings are given in Table 1.

Cold-drawn bars that are straightened in a multiroll rotary straightener usually increase in diameter during the straightening operation. Low-carbon steel bars with up to about 0.15% C show a negligible increase in diameter. However, as the carbon content increases, the amount of change increases. It is not uncommon for 50 mm (2 in.) diameter cold-drawn bars of 1050 steel to increase as much as 0.1 mm (0.004 in.) in diameter. These bars will decrease in length by approximately 13 mm per 3.7 m ($\frac{1}{2}$ in. per 12 ft) as a result of the increase in diameter. This shortening must be considered when bars are cut to exact lengths before straightening.

When cold-drawn bars are to be straightened in a multiroll rotary straightener, selection of the cold-drawing die size is important if the bars are to be held within standard size tolerances. Most

grades, particularly those having high carbon content, should be drawn to the low side of the diameter tolerance to compensate for the increase during straightening. The extreme ends of the bars, which do not get the full effect of the bend by the pressure roll, do not increase in diameter. After straightening, the bar ends will remain the same size as when cold drawn.

Straightening in a multiroll rotary straightener does not work harden the bar stock to any appreciable extent. This is desirable when the bars are to be cold headed or cold extruded.

The basic five-roll straightener has been modified so that four of the rolls are driven and only the middle pressure roll is an idler. The heavier feeding pressure obtainable with driven entry and exit pressure rolls is advantageous in that a polishing effect can be obtained on products such as cold-drawn steel. In addition, the driven rolls provide more traction so that a heavier deflection can be exerted by the middle straightening roll.

In a further modification of the five-roll straightener, all five rolls are driven. This eliminates the need for guides between the rolls, but roll speed synchronization becomes important.

Hot-rolled round steel bars are generally straightened commercially in two-roll or multiroll rotary straighteners. Bars as large as 255 mm (10 in.) in diameter and having yield strengths up to 690 MPa (100 ksi) have been straightened in these machines. Some machines are modified by adding hydraulic loading to the straightening mechanisms and by adding a shear pin to provide for any shock loading that might be encountered because of the extreme out-of-roundness of large hot-finished bars.

Rotary-arbor straighteners are used to straighten coiled rod or wire up to 32 mm ($1\frac{1}{4}$ in.) in diameter. The straightening is done by an arbor rotating around the wire as it passes through the machine, as shown in Fig. 16 (see also Fig. 3h). The arbor encloses five pairs of cast iron straightening dies. The dies are equally

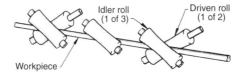

Fig. 15 Arrangement of rolls in a five-roll rotary straightener

Table 1 Typical settings of roll angle for five-roll rotary straighteners for use on bars of various diameters

Diameter of bar		Setting of roll angle, degree
mm	in.	
19	$\frac{3}{4}$	$26\frac{5}{8}$
25	1	$26\frac{7}{8}$
32	$1\frac{1}{4}$	27
38	$1\frac{1}{2}$	$27\frac{1}{4}$
44	$1\frac{3}{4}$	$27\frac{1}{2}$
50	2	$27\frac{3}{4}$
57	$2\frac{1}{4}$	28
64	$2\frac{1}{2}$	$28\frac{1}{4}$
70	$2\frac{3}{4}$	$28\frac{1}{2}$
75	3	$28\frac{3}{4}$
83	$3\frac{1}{4}$	29
89	$3\frac{1}{2}$	$29\frac{1}{4}$
95	$3\frac{3}{4}$	$29\frac{1}{2}$
102	4	$29\frac{5}{8}$
108	$4\frac{1}{4}$	$29\frac{7}{8}$
114	$4\frac{1}{2}$	30

Recommended settings for the starting setup; these will vary slightly in the actual setup used.

spaced in a fixed spacing that relates to the size capacity of the machine. The greater the capacity, the greater the fixed spacing of the dies will be.

The dies, bell-mouthed for easier entrance of the wire, are locked in place by adjusting screws. The pairs of dies at the two ends of the straightener arbor are set so that the wire is always at the center of the arbor in these dies. The middle die is called the pressure die because it is set to bend the wire slightly, as shown in Fig. 16. The dies on either side bend the wire slightly in the opposite direction.

The machine is set by trial. If the wire comes out bowed, usually the pressure die has not been set to bend the wire enough. If the wire comes out wavy, the pressure die may be bending the wire too far. Helical marks may be caused by lack of lubrication, imperfect dies, or an embedded sliver of metal.

Growth. During straightening in a rotary-arbor machine, most grades of cold-drawn carbon steel and of alloy steel with more than 0.15% C will increase in diameter (as much as 0.15 mm, or 0.006 in.) unless the wire has been stress relieved before straightening.

Speed of straightening in rotary-arbor straighteners is usually 23 to 61 m (75 to 200 ft) per min, depending on the size capacity of the machine and on the type of wire.

Cut lengths of bars are straightened in mechanisms such as the one shown in Fig. 17. A helical motion is imparted to the bars by pairs of rotating, offset friction disks, which burnish the bar as they feed it through three straightening bushings that turn freely in bearings. The middle bushing is adjusted to deflect the bar, just enough for good straightening action. Such a machine made in various size capacities can straighten bars 2 to 32 mm (5/64 to 1 1/4 in.) in diameter at speeds of 28 to 50 m (92 to 164 ft) per min.

Straightening Stainless Steel for Cold Heading. Coiled stainless steel wire (series 300 and 400) 1.6 to 15.5 mm (0.062 to 0.610 in.) in diameter requires moderate straightness while being fed into cold-heading machines. Wire 1.0 to 3.2 mm (0.040 to 0.125 in.) in diameter can be hand straightened sufficiently for entering the feed rolls. The feed rolls then pull the wire with enough tension to remove the coil radius as the wire leaves the coil reel. For parts having a length-to-diameter ratio of 4 to 1 to 8 to 1, no further straightening is necessary. The feed rolls provide sufficient straightness to permit the blank to be cut to length and transferred to the die station. After cold heading, the part has the straightness obtained in the heading operation.

Parts up to 152 mm (6 in.) long cold headed in open-die headers require a straightness of 0.1 mm in 102 mm (0.004 in. in 4 in.). A single-plane five- or six-roll straightener placed 90° to the feed roll is usually used. Single-plane and two-plane straighteners mounted on portable pedestals are available as machine accessories.

Automatic Press Roll Straightening

One of the fastest straightening processes available is the automatic press roll-straightening method. It is capable of straightening small concentric parts at up to 1200 pieces per hour; larger parts with greater ovality, such as cold extrusion axles and transmission shafts, can be made at 225 pieces per hour.

Roll straightening features support and straightening roll assemblies that resemble those used in a roller V-block, but it includes a headstock unit equipped with a drive mechanism to rotate the workpiece. This process uses a press frame with a hydraulically powered ram (Fig. 18). Two or more lower support roll assemblies are mounted on the bed of the device, while one or more upper support rolls are mounted on the ram. Additional roll assemblies are needed to straighten a series of bows (known as snaking). The headstock consists of a chuck or driver used to rotate the workpiece. The equipment is particularly suited to the straightening of cylindrical solid parts as well as tubular parts having walls that are thick enough to withstand the pressure of the rolls without being deformed (see the article "Straightening of Tubing" in this Volume).

The process is initiated by placing the workpiece on the lower rolls and having the driver move forward to engage the part and to rotate the workpiece. The overhead rolls situated on the ram are then applied to deflect the part in a bowlike arc while the workpiece rotates. This procedure subjects the material to a plastic strain. When the material exceeds the yield point, the ram action is released to complete the cycle.

The advantages of this method include a stretch-relieving action that accompanies the straightening as well as the ability to handle a wide range of types and degrees of out-of-straightness conditions. Concentric parts with no ovality can be held to a close tolerance and require minimal operator skill.

The disadvantages of this technique are that it cannot be used on nonround parts, thin-wall tubes, or parts having variable diameters. Only hardened shafts with a hardness of 38 HRC or less should be straightened using this method because the depth of hardness limits its effectiveness.

Moving-Insert Straightening

Designed for use on linear, flat, or irregularly shaped parts, moving-insert straightening is accomplished by reciprocal strokes transmitted to tooling inserts by a rotary-cam action. When positioned between two rows of movable inserts situated on a tool base, a part is subjected to a series of reciprocal strokes that overbend the workpiece by a preset amount (Fig. 19). The amplitude of the movement is progressively reduced during the cycle until it approaches a straight line, at which point the workpiece is also straight. The degree of bending movement and

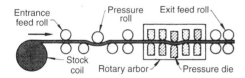

Fig. 16 Arrangement of rolls and dies in a rotary-arbor straightener used for the straightening of coiled rod or wire

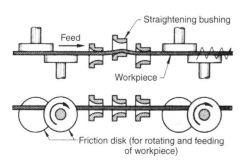

Fig. 17 Plane and side views of a mechanism for straightening cut lengths of bars

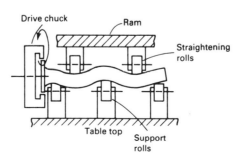

Fig. 18 Illustration of equipment used in automatic press roll straightening

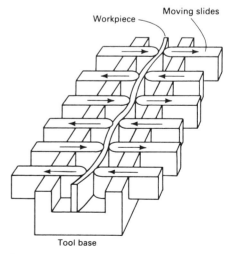

Fig. 19 Moving-insert straightening for linear, flat, or irregularly shaped objects

the number of bending cycles are adjustable, and varying insert spacing is available to accommodate a wide range of soft or heat treated components. The primary advantages of this straightening technique are its ability to straighten flat or irregularly shaped parts, with or without projections, bends, and so on; the ability to produce straightness or slight curves when needed; a tolerance as close as 0.03 mm (0.001 in.) throughout the length of the part, depending on its configuration; and minimal skill requirements imposed on the operator.

Parallel-Rail Straightening

Parts that feature multiple diameters and heads, such as bolts and spindles, require a straightening device that adjusts to the contours of the workpiece. The parallel-rail method performs this function by employing a series of parallel rails. One group of rails is located on a slide, and the other group is located on a ram or head above and between the bottom rails. As shown in Fig. 20, the cylindrical or thick-wall tubular part is positioned between the rails, and the ram is lowered by adjustable hydraulic pressure to overbend the workpiece. Simultaneously, the lower slide moves forward and rotates the workpiece. At the end of the stroke cycle, the pressure decreases to zero, and the part is now straightened. The configuration of the workpiece determines the number and position of the adjustable rails needed to complete the straightening operation.

With this technique, production rates of 500 to 600 pieces per hour can be obtained with manual loading and unloading methods. Production rates of 1000 pieces or more per hour can be obtained with automatic loading and unloading equipment.

Among the advantages of parallel-rail straightening are:

- Minimal operator skill required
- Easy adjustment to make rails conform to workpiece configuration
- Complex parts featuring headed areas and multiple diameters can be straightened
- High production rates combined with low tooling costs

The primary disadvantage of parallel-rail straightening is that workpiece dimensions are limited by the machine width of 610 mm (24 in.) as well as maximum diameters of 20 mm (0.8 in.).

Epicyclic Straightening

Epicyclic straightening is suited specifically to straightening linear parts, tubing, and solid cylindrical parts featuring a variety of cross-sectional shapes. After the workpiece is securely supported by locating fixtures at either end, a straightening arm secures the part in the approximate center of its length (additional straightening arms may be necessary on some parts due to length and configuration considerations). The straightening arm is programmed to move in a circular or elliptical path about the neutral axis of the part. The cross section of the part is taken through its elastic limit as the amplitude of the arm motion is increased. At the elastic limit, the motion decreases, and the arm moves to a stationary position at the neutral axis of the workpiece to produce a straight and stable part.

The motion of the straightening arm is a circular path for round parts, while parts with varying cross sections, such as an I-beam section, require an elliptical path (Fig. 21). The feed rate should be maximized to reach the yield point as quickly as possible. The degree of straightness required determines the feed rate necessary to return the part to its neutral axis.

Epicyclic straightening can be used on such parts as I-beam axles, trailer axles, asymmetrical forgings, and thin-wall propeller shaft or drive

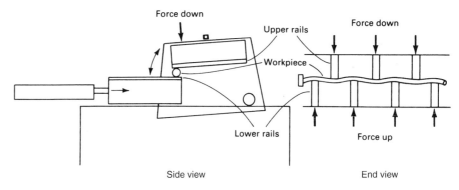

Fig. 20 Parallel-rail straightening for cylindrical or thick-wall tubular parts, either symmetrical or with variations in diameter

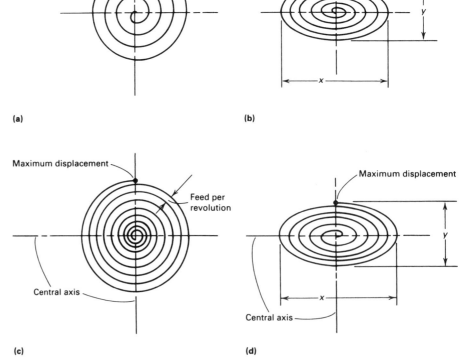

Fig. 21 Circular and elliptical paths of the motion arm on an epicyclic straightener. (a) Displacement circular. (b) Displacement elliptical. Ratio $x:y = 2:1$. (c) Feed to central axis circular. (d) Feed to central axis elliptical. Ratio $x:y = 2:1$

shaft tubes. Production rates range from 80 parts per hour (trailer axles) to 300 parts per hour (propeller shaft tubes).

Straightening in Bar Production (Ref 1)

Coil Straightening on Spinner Type Machine. One way coil is straightened is on a spinner type machine having a rotary spinner with generally five dies or more that are offset. The spinner rotates around the bar as the bar is driven through the machine by driven feed rolls. The material from the spinner dies is normally somewhat bright with a condition of straightness that is acceptable to the trade if done properly with tolerances of 0.3 to 0.55 mm/m (0.004 to 0.005 in. per foot) total indicator reading (TIR) as a minimum achievement without special care and operator attention. The reduction in scrap here is the proper size machine for the grade of material to be run. Drawn material from a spinner type machine may be marginal because the material is either at the large or the small end of the range of the machine. If the material resists being straightened, roping (twisted material looking like a rope) can occur. The roping can be very severe; even after centerless grinding there may be low or high spots resembling a wire rope braid.

To reduce scrap from the coil as described it is necessary to use the proper size machine. If a machine has a range of 6 to 19 mm ($^1/_4$ to $^3/_4$ in.), then the smallest and largest material, as it varies from a mild-steel center state, should be reduced in diameter. For example, a mild steel machine with a 6 to 19 mm ($^1/_4$ to $^3/_4$ in.) overall diameter would be reduced to 9 to 16 mm ($^3/_8$ to $^5/_8$ in.) when used with titanium, 1045 steel, or stainless material. This will assure that there is a much better chance of scrap reduction in a rotary center type of machine. There are also variables that can be used between the throughput of a machine and the feed of the spinner. Too high a spinner speed with too slow a throughput speed on extremely difficult to straighten materials on either end of the range of the machine produce roping or, even worse, surface defects. Scrap reduction method, therefore, is to use the optimum speeds and feeds. The selection of the spinner dies is quite important so they are compatible with the material to be straightened. Die materials can range from cast iron through hardened grades of the bronze alloys, stainless and the exotics. The total surface of the coiled material being straightened is contacted by these spinner dies and lubrication is necessary. It is still possible with good lubrication and proper die to get scratches from an improper spinner setting. As the carbon increases in this coiled stock, it is difficult for the spinner to remove all of the stress that is in the coil after it has been drawn through the die around the bull block drum. If the material is 1045 steel or thereabout, a well-straightened scrap-free product must have these stresses completely relieved.

Spinner die type machines are frequently followed by two roll straighteners, either operating in connection with them or off to the side as recovery machines. These machines will remove the wire roping and have the ability, through the action of the concave and straight rolls on the bar, which are restrained by the upper and lower guides, to slow the material to relieve the stresses. Material above 0.30 wt% carbon must be expanded on to be straightened properly, 0.30 wt% carbon about the same; material below 0.30 wt% carbon is compressed when straightened and can lose size. This is an example of one straightening process overcoming the difficulties of another straightening process and recovering what would possibly be scrap or marginal product.

Straightening Coils from Continuous Drawing Machines. Another method of producing material from coils is to use a continuous drawing machine. This has a distinct advantage over the coil to coil followed by the spinner type of straightening. The material is in the same coil form but the continuous drawing machine uses reciprocating, nonmarking, gripping guides. The material is drawn through a die in a straight line. Then the material, when cut off, goes to a straightening machine automatically. Unfortunately, the normal straightening machine furnished with the continuous drawing machine is limited to straightness of approximately 0.5 mm TIR/m (0.006 in. TIR/ft). This straightening machine has a series of rotating nozzles—generally five, sometimes as low as three, sometimes as many as seven—and the material is spun up by two attrition disks on either side of these nozzles. The nozzles are then offset, which produces the straightening. The major cause of scrap from straightening by this process is there cannot be a complete end-to-end straightening operation because of the fixed distance between the nozzles. There is generally a scrap problem with the ends of the bars as they are twisted and scarred in the attrition disks. For this reason, they generally are sold in lengths of + 13 mm ($^1/_2$ in.).

The surface finish from this type of nozzle straightener is generally bright but does have a pattern of loose particles caused by the attrition wheels, and they cling to the bar. The scrap reduction in this straightening is to carefully select the range of materials to be run in the continuous draw machine with only the nozzle type straightener. After passing 30 carbon, this straightening method becomes somewhat erratic and the scrap does go up, especially if trying to maintain a TIR rating of less than 0.5 mm/m (0.006 in./ft).

Again, the scrap reduction method is to put a two-roll straightening machine inline or offline through which to run all or a select group of bars. A direct inline machine eliminates the straightening problems that are attendant with hooked ends that occur in either the shearing and/or the gripping of the continuous drawn machine. The ends of the bars are, for all practical purposes, perfect, having only a 1.6 mm ($^1/_{16}$ in.) or, at the

most, 3 mm ($^1/_8$ in.) slight roll down due to the rotational speed. The higher carbon bars are then straightened as precisely as the lower carbon bars because of the ability of the two-roll machine to exert such straightening forces as blow the bars up in diameter by stress relief.

Nonetheless, a continuous draw machine can have an inherent drawback with this operation. It can draw material so fast that a heavy draft (e.g., a $^1/_8$ in. reduction in 1 in.) will produce a phenomena that causes erratic straightening at higher speeds, and as the severity of the reduction drops off, the phenomena also disappears. Also, the hooks and bends in the incoming coil of material play an important part in the final straightness. Scrap reduction starts with the incoming material and the method of straightening it prior to the drawing, even going back further to handling the coils so they do not become battered or have wild loops. The addition of a two-roll continuous straightening machine to a continuous drawing line provides a method of scrap reduction in that the material can now be produced to tolerances of 0.15 mm TIR/m (0.002 in. TIR/ft) and less on certain materials, or an average of 0.25 mm TIR/m (0.003 in. TIR/ft) on most materials. The surface finish, while not as bright as that on the attrition wheel, actually is smoother because of the burnishing action of the concave and straight rolls of the two-roll machine, and there are less metal particles on the surface.

Guide scratching in the straightening line is reduced to a very low level with a heavy flow of coolant and the use of the proper grade of guide material, such as cast iron or bronze. Also, an excellent means of scrap reduction is an addition to the two-roll straightening machine of a digital readout that reads the position of the straight roll, i.e., distance, down to 0.25 mm (0.001 in.). The operator then sets the preload, that is, the amount of load on the two rolls, by reducing the opening between the rolls. Customarily, the higher the carbon, the more preload, the lower the carbon, the less preload there will be. For example, if 1045 steel is being run and it is in the 25 mm (1 in.) diameter range with the normal draft of 1.5 mm (0.060 in.), the preload may be only 1 mm (0.040 in.) less than the diameter of the bar. However, if this is increased to a draft of 2 mm (0.090 in.), then the preload might go up to 1.5 mm (0.060 in.). Scrap reduction, therefore, involves the proper use of instrumentation on the straightening machine to inform the operator that the straightening conditions are optimal. It is also exceedingly important that the operator be trained in the proper preparation of the grip jaws as the action of the two-roll machine will provide a very nice polished background; an improperly ground jaw will emboss the surface of the bar and be easily visible normally as four almost continuous lines.

The straightening of round bars coming from a draw bench uses the same principles as coil drawing. However, the setup of the two-roll straightening machine, which is always used in this process, is less severe for a given diameter than the same size from a coil. The product of the fixed draw line straightener tends to have higher yields and less scrap than that from a coil. There are certain economic factors that rule out the production of material on a fixed length draw bench when it can be made from coil at much higher speeds. The operation on larger sizes of material is just as important, and scrap reduction by the two-roll straightener saves a significant amount of material because the weight per foot increases. A slight imperfection caused by the straightening machine process on the surface of an 80 mm (3½ in.) bar is far more costly than it would be to a 20 mm (¾ in.) bar. Again, the straightening machine scrap reduction program starts back at the draw bench, or even before the draw bench. Overblasting or lining can cause difficulties in straightening if they are severe enough. However, most of the difficulty is surface finish. The biggest detriment to the scrap-reduction program in the straightening of drawn bars of fixed length is the proper setup of the drawing die. The drawing die affects not only the size of the bar but also the curvature. The dies should be universally adjustable by means of four bolts in the corners, operating on a spherical surface to cant the die one way or another. The second or third bar being drawn should be as straight as possible. The scrap reduction from the straightener is then an easy matter because the straightener is not required to overstraighten the product and therefore can straighten every bar.

In the draw line, the shear is important, and especially in that the shear blades be set close enough to give a good cut without a lip. Many a bar is poorly straightened, and scrap results because a bad shear is used or an improper setting that puts a lip on the end of the bar. The operator then cannot feed the lip into the straightener properly and therefore is obliged to misset the straightener to get the lip through. Scrap reduction in the straightening machine is dependent on the improvement of the shear cut and proper straightener setup. The shear table is another scrap producer if it does not drop with the shear blade in the same articulated motion; a hooked end under the shear blade occurs. A good holddown is required to do the same action on the incoming bars. Often, bars that are improperly straightened can be traced to bad hooks that are produced by the shears.

The proper feed and handling equipment on the input side of the machine with the proper entry device, as well as the proper handling equipment on the exit side of the machine, is required to restrain the bar from high rotation. This almost automatically occurs in the inline drawing but is often neglected in the drawing of bars from a draw bench.

Improperly set up input or discharge tables will cause straightening scrap. Scrap reduction in the straightening of this material requires not only proper drawing but also proper setup of the input table, straightening machine, and the proper discharge table.

REFERENCE

1. W.L. Siegerist, Straightening Processes— Scrap Reduction Methods, *12th Technical Conference on the Methods of Reducing Scrap in the Production of Tubes, Bars, and Shapes,* ASM International, 1989, p 145–165

Straightening of Tubing

TUBING of any cross-sectional shape can be straightened by using equipment and techniques that are basically the same as those discussed in the article "Straightening of Bars, Shapes, and Long Parts" in this Volume. In general, a round tube that has warped in annealing or other heat treatment is given a rough-straightening pass in a press or a roll straightener, followed by one or more passes in a rotary straightener and, if required, a finish pass in a press straightener.

Rough straightening eliminates excessive bow that would cause whipping in the inlet trough. Because the tube does not rotate during rough straightening, almost any amount of bow can be accepted; therefore, both roll and press straightening can be used. If the tube is only slightly bowed, initial rough straightening can be omitted.

A rotary straightening pass is necessary for finish straightening but need not be used when the tube requires another cold drawing operation. Two or more rotary passes may be required when the tube has developed excessive out-of-roundness, bow, or multiple kinks in previous processing.

If the tube is considerably out-of-round, a normal straightening pass, with only a middle roll offset, will not be effective. A preliminary pass with no offset in the middle rolls and all rolls set to ovalize may be required before the normal straightening pass.

Most tubing, especially the smaller sizes (32 mm, or 1¼ in., in outside diameter and smaller), requires only one or two rotary straightening passes. For larger-diameter tubing (38 mm, or 1½ in., in outside diameter and larger), or when straightness better than standard is required, additional work can be done in a press straightener to remove short end hooks or to secure precision straightness. Press straightening converts long, gentle bows into a series of short bows ending at each point where the punch hits.

Straightening of tubing having shapes other than round is done in roll or press straighteners with much the same considerations governing the choice as those for similar solid shapes. Asymmetrical sections, such as airfoil or teardrop designs, require press straightening. Stretch straighteners are helpful in detwisting and straightening asymmetrical shapes.

Effect of Tubing Material

Procedures and tooling are influenced by the tubing material. The principal factors are the composition and condition of the work metal, wall thickness, type of tubing (seamless or welded), and type and extent of distortion. Greater force and more rugged equipment are required for straightening tubes with thick walls and high elastic limit.

Wall Thickness. The amount of distortion that occurs in the hollow section where it bears against the straightening tools is relatively small in thick-wall tubes, which approach solids in their behavior. Distortion becomes significant as tube walls become thinner, necessitating precautions against permanent distortion. Damage to thin-wall tubing appears as spot dinges or ovality from punch straightening, or as spiral dinges (called rings) or ovality from rotary straightening.

Type of Tubing. Seamless tubes often have a nonuniform wall thickness, which makes the straightening operations more difficult. In a welded tube, the weld and adjacent material may have mechanical properties that differ considerably from those of the rest of the tube. Residual stress from welding may also affect straightening.

Type and Extent of Distortion. Crookedness in tubing can result from distortion during heat treatment or cooling, from distortion during cold drawing as the result of eccentric wall, from hooked ends on hot-finished tubes (usually caused by misalignment or roll wear in the hot sinking mill), or from accidental bending during processing. End hooks are pronounced bends that appear close (usually within 455 mm, or 18 in.) to the leading or trailing ends of some tubular products.

Control of Straightening Pressure

Pressure exerted on the workpiece by the straightening rolls must be carefully controlled to prevent permanent damage, especially to thin-wall tubing. This can be done by increasing the distance between the points at which the tool contacts the workpiece, thus reducing the total force, or by increasing the contact area between tool and tubing, thus reducing the unit pressure on the workpiece.

In the rotary straightening of thin-wall tubing, the length of contact between the roll and the tube and the distance between sets of rolls should both be maximum, and care should be exercised in the adjustment of opposed rolls to avoid excessive ovalizing and pressure on a short section of tubing. In a two-roll machine, the length of curve is limited to the length of the roll, and tubing with very thin walls may be subject to ringing at the roll shoulders where the pressure is greatest. This limits the two-roll machine to a maximum diameter-to-wall thickness ratio of approximately 15 to 1 for full straightening.

Press Straightening

The straightening press can rough straighten tubing prior to roll straightening or rotary straightening, or it can completely straighten tubing by removing end hooks that cannot be removed easily by any other method. End hooks and large cambers sometimes must be removed to permit the tubing to enter the rotary straightener and to prevent dangerous whipping of the work. A press can also be used to straighten welded tubing before the bead is reduced. It is best to reduce the bead before straightening the welded tubing in a rotary straightener.

Tooling. In the press straightening of thin-wall tubing, there should be a wide spacing between saddles. If there are short kinks in the tube, the spacing is dictated by their length, and operator skill must be relied on to obtain the best possible straightening.

Rolls with a semicircular groove for holding the tubing are sometimes used for end supports. The spacing of these rolls can easily be adjusted to suit the length of the bow. The pressure shoe attached to the press ram is flat, with a semicircular groove to distribute force over a greater area. The diameter of the groove should approximate the diameter of the tube to avoid flattening of the tube under pressure.

Applications. Straightening presses are used on tubes to remove kinks, camber, and other distortions caused by mishandling, cutting off, and heat treatment. Straightening presses can also be used to reduce the out-of-roundness of tubing.

Parallel-Roll Straightening

Single-plane and two-plane roll straighteners for tubular products are basically the same as those for solid bars. Machines for tubular products may have somewhat longer center distances than bar straighteners and at the same time may not require shafts and bearings that are as large. The roll-center distance is generally a function of the outside dimensions of the tube, and it increases approximately as the outside dimensions of the tube increase. Less force is required for straightening tubes than solid bars of equal outside dimensions.

Round tubing is best straightened on machines equipped with semicircular grooved rolls. The grooves must conform closely to the tube size, and each size must have its own set of rolls, or set of grooves in a multigrooved roll. Round tubing will twist slightly in the rolls, thus avoiding the effect of straightening and resulting only in rough straightening.

Tubing that is badly warped can be rough straightened as easily in a roll straightener as in a press. The roll straightener operates faster than the press but takes longer to set up. When straightening 50 mm (2 in.) diameter, 6 m (20 ft) long tubes, for example, the roll straightener can process 250 to 400 pieces per hour as compared to 100 to 120 pieces per hour in a press. However, changeover time for a roll straightener is 16 min; a press setup can be changed in as little as 6 min.

The cost of roll straightening increases greatly for the larger tube sizes. Therefore, a large tonnage of tubes must be processed to justify the cost of machines for straightening tubes that are larger in outside diameter than approximately 75 mm (3 in.).

Square and rectangular tubing is commonly straightened in either single-plane or two-plane machines (see the article "Straightening of Bars, Shapes, and Long Parts" in this Volume). If the rolls of a single-plane machine are grooved to fit the tube, an appreciable straightening effect in a plane spaced at 90° is provided. When additional straightening is required, square tubes are turned 90° for a second pass through the rolls. A two-plane roll straightener is better suited to rectangular tubing than a single-plane roller because the rolls need not be reset for the second pass.

Two-Roll Rotary Straightening

The principle employed in the rotary straightening of round tubes is basically the same as that for solid round bars. The driven rolls, set at a predetermined angle, rotate the tube while conveying it in a lineal direction. The crest of the bow is stressed to, or beyond, the elastic limit once during each revolution, and the maximum stress point is repeated spirally along the length of the tube. The distance between each stress point depends on the lineal travel for each revolution of the tube. Approximate values for lineal travel can be determined by multiplying the tube circumference by the tangent of the angle of the rolls. Special contoured rolls can be used to eliminate stress points, yielding tubes with uniform stresses.

Two-roll rotary straighteners (see the article "Straightening of Bars, Shapes, and Long Parts" in this Volume) are used primarily on tubes having a diameter-to-wall thickness ratio of no more than 15 to 1. The machine is equipped with two skewed rolls, between which two guide shoes are mounted. One roll has a concave contour; the other is straight or convex. The rolls can be arranged in a horizontal or a vertical plane. Machines with rolls arranged in the vertical plane are of relatively new design.

The tube is held between the guide shoes while the straight or convex roll bends the tube between the ends of the concave roll. The maximum deflection depends on the depth and the skew angle of the concave roll.

Two concave rolls have also been used to straighten tubes. The concave rolls are set to make full-length contact along the surface of the tube, and the crest of the bow is ovalized between the rolls several times before the tube emerges from the machine.

Straightening of End Hooks. The two-roll machine can remove most of the sharp bend at the end of tubing if the rolls are ground to suit the deflection requirements of the tube material and are set at an angle to suit the size of tube and length of end hook. The resulting curve in the rolls is suitable for a specific range of tube sizes with a specific elastic limit. Wider variations in tube size and grade of material can be processed by changing the angle setting of one or both rolls to produce the required deflection.

The design of the two-roll machine permits the rolls to be set at a small angle with the centerline of the stock. The smaller the angle, the less through feed per revolution of the tube, and because of the small helix angle created by feed per revolution, a large portion of the tube is subject to maximum bending stress.

Short bends and end hooks can be straightened in a two-roll machine, because all bending takes place within the length of the rolls and not from roll to roll, as is the case with multiple-roll machines. The short span greatly increases the load necessary for straightening, which partly explains why a two-roll machine is unsuccessful in straightening thin-wall tubing.

Roll Angle. Contoured rolls are designed for a specific range of tube sizes and materials. This is approximately the same range of conditions that can be handled by the equivalent multiple-roll straighteners. For most applications, the rolls are set at an angle of 15 to 25°. The contour of the rolls can be varied to suit specific applications. For example, a roll of shallow concavity is used for materials having low elastic limits, while a roll of deeper concavity can be used to straighten materials having higher elastic limits.

Limitations. The two-roll machine is not ordinarily used to finish straighten tubing if the ratio of outside diameter to wall thickness is greater than 15 to 1. The crushing strength of a thin-wall tube in the short span of a two-roll straightener is such that the tube will crush, or ring, before it bends if the rolls are set to remove the maximum bend. However, if the amount of bend to be removed is reduced by a preliminary rough-straightening operation, the machine can be used to finish straighten tubing having a diameter-to-wall thickness ratio considerably greater than 15 to 1, depending on the amount of straightening done in the preliminary operation.

Polishing of the tube surface can be either beneficial or detrimental. The hourglass shape of the rolls presents different diameters to the surface of the tube, resulting in some slipping. This burnishing action improves surface finish, although excessive slipping can produce a burnished spiral on the work surface.

Scratches may result when foreign material becomes embedded in the guide shoes. The use of nylon shoes and a soluble oil as a lubricant will reduce scratches.

Sizing after Derodding. Producers of cold, drawn tubing use internal mandrels to control the inside diameter, and two-roll crossed-axis machines are used to extract the mandrel (derod) after the drawing process. The identical rolls apply heavy pressure on the workpiece, thus expanding the tubing to allow removal of the mandrel.

The expanded tubing does not always return to its drawn diameter. The external dimensions of tubing can be corrected by drawing the tubing through dies that are undersize by an amount equal to the amount by which the tubing expands during derodding.

Multiple-Roll Rotary Straightening

Rotary straighteners with five, six, seven, or even more rolls are also used to straighten tubing. The five-roll machine consists of two two-roll clusters and a middle deflecting roll. This machine has two large rolls on one side that are opposed or nearly opposed by three small rolls on the other. Two of the three small rolls and the two large rolls function as entry and exit feed rolls. The third small roll located between the other two small rolls functions as a deflecting roll. The rolls can be arranged in the horizontal or in the vertical plane.

In some machines, only the two large rolls are driven; in others, all rolls except the deflecting roll are driven, and sometimes all five rolls are driven. When more than two rolls are driven, the speed matching of roll surfaces becomes important. Matching can be obtained by maintaining the correct relationship between roll diameters or, more easily and accurately, by a differential drive between the two roll banks or, in some machines, by driving the rolls with individual motors having relay, continuous-feedback, or similar controls.

The six-roll machine has two middle deflecting rolls opposed to one another, similar to the entry and exit rolls in the five-roll straightener, but it differs from the five-roll straightener in that all rolls are of equal diameter. Normally, four or all six of the rolls are powered. Another type of six-roll straightener has a roll arrangement similar to that of a five-roll straightener with an additional outboard roll.

A seven-roll rotary straightener has two three-roll clusters—one at the entry end and one at the exit end of the straightener—and a middle deflecting roll (Fig. 1). Normally, the two bottom rolls are driven and the five others are idlers. The middle roll (deflecting roll) moves vertically, and the four end idler rolls move in a circular path about pivot points in the base and apply pressure to the tube for feeding and straightening. This seven-roll arrangement has the greatest effect at the middle roll because the tube is held perfectly in the pass.

Various other multiple-roll straighteners have been built for specific applications. The general principles that apply to the use of multiple-roll machines are the same as those already described for two-roll straighteners.

Middle-roll offset straightening bends a straight tube by an amount that stresses the outer fibers of the tube to the elastic limit. When the straightening load is released, the tube springs back to its initially straight position. The outer fibers of an initially bowed tube are stressed beyond the elastic limit, and the tube springs back to a straight position.

Straightening of tubing in this manner can be termed single-pass straightening. It is possible, however, to arrange the rolls in the multiple-roll machine to double straighten each tube as it is conveyed through the machine. The middle roll stand is offset to deflect a bowed tube and to stress its outer fibers beyond the yield point, and in a similar manner, the tube is deflected over the third roll stand by an auxiliary roll or by the discharge table. The tube is stressed at both the middle and third roll stands—or, in effect, double straightened—with each pass through the machine.

Bending the tube, however, creates an area of strain or cold work directly under the bending load. The rotation of the tube as it goes through the machine generates a spiral of strained area around the periphery of the tube. The centerline

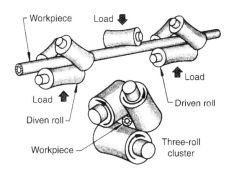

Fig. 1 Arrangement and principles of operation of three-roll clusters in a seven-roll rotary straightener

of the third roll stand should be spaced such that no one point on the tube duplicates itself at the same location on both the middle and third rolls, thus eliminating the possibility of reworking the strained area.

Tube Deflection. The amount, δ, by which the middle roll or rolls must be offset from the entry and delivery rolls varies inversely with the outside diameter of the tube to be processed, as demonstrated by the following equation:

$$\delta = Sl^2/6ED$$

where S is the yield strength of the metal, l is the distance between the outboard rolls, E is the modulus of elasticity, and D is the outside diameter of the tube.

As the diameter of the tube decreases, the deflection requirements increase over a given span. Therefore, the factor that determines the smallest-diameter tube that a machine can process is the deflection requirement of the tube over the span of the rolls. Approximately 19 mm ($^3/_4$ in.) is the maximum deflection that can be applied to most tubes without adversely affecting travel of the tube through the rolls of a rotary straightening machine.

Applications. Multiple-roll machines are advantageous in processing thin-wall tubes, and thick-wall tubes having a high ratio of diameter-to-wall thickness. This is because of the lower unit loading on the workpiece that is applied across the longer bending spans between adjacent rolls. The multiple-roll straightener has an additional advantage in higher throughput speeds than other rotary straighteners, because of the higher angularity settings of the rolls. Therefore, multiple-roll straighteners are widely used in tube mill production lines.

Multiple-roll machines are ordinarily used for applications in which the primary purpose is the sizing or burnishing of the workpiece. They do not straighten as accurately as two-roll machines, nor do they remove end hooks effectively.

Long bows in medium- and heavy-wall tubes are removed by closing each roll pass and bending the tube with the middle roll. Closed-pass rolling provides full-length support within each roll stand. Straightening in this manner requires less deflection and spreads the bending load over the full length of each roll.

Long bows can sometimes be removed from extra-heavy-wall tubes by opening each roll stand and bending the tube between the two bottom rolls with the top middle roll. The object is to exert a minimum load on the tube.

Thin-wall tubes should be processed with low unit loads applied to the surface of the tube and with minimum deflection. High loads and excessive deflection will ring the interior of the tube. The bending load can be spread across the surface of the tube by angling the middle roll toward the centerline of the pass. Ovalizing and spreading the roll contact lower the required amount of deflection to produce a straight tube. Deflecting a tube over a long span requires a large offset that can make it difficult to convey

the tube through the machine. However, proper roll design and roll spacing make it possible to process a wide range of outside diameters in the multiple-roll rotary straightening machine.

End hook can be removed from thin- and medium-wall tubes by ovalizing and bending the tube in the middle roll pass. Removing end hook from thick-wall tubing on multiple-roll straighteners is difficult because the center-to-center distance between the No. 1 and 2 roll stands is usually greater than the length of the end hook.

The roll angle of a straightening machine depends on the outside diameter of the tube to be straightened. In general, a machine used for small-diameter tubing in the range of 6.4 to 19 mm ($^1/_4$ to $^3/_4$ in.) may have a roll angle of 40 to 45°. In some special high-speed straighteners, when the tubing is almost straight to begin with and is easy to straighten, a 40° angle can be used for tubing as large as 50 mm (2 in.). For machines in the range of 50 to 152 mm (2 to 6 in.), a common angle is 30°. For very large tubing 455 to 610 mm (18 to 24 in.) in diameter, the angle can be as low as $17^1/_2$°.

To fit a large tube within the size range of the machine to the roll contour, the angle between the centerline of the tube and the roll must be greater than that required for a small-diameter tube. The angle can be adjusted from 2 to 3° on either side of the nominal angle for a given machine—the lowest angle for the smallest size and the largest angle for the largest size.

Entry and Delivery Tables. Well-designed entry and delivery tables are important in the rotary straightening of tubular products. The rotary straightening machine is designed to support, guide, and straighten a tube within the length between the first and last roll stands. It is not designed to feed, support, or confine a long tube over its full length.

Entry-table guides confine the portion of the tube that is beyond the limits controlled by the rotary straightening machine in order to minimize rotary whipping of the unstraightened portion. The table guides must be designed to suit the size range, grade of material, and travel rate of tube that is to be processed. A combination of entry and delivery tables and guides keeps the amount of straightener deflection, or offset, to a minimum, increases the range of sizes that can be straightened in any one machine, and decreases the possibility of damaging the tube.

The entry table should also feed the tube into the rotary straightener with skewed rolls, so that the tube will be rotating and moving lineally when it enters the straightener. When in-line straighteners are used at the mill, the straightener and tables must be designed to keep up with the speed of the mill.

Ovalizing in Rotary Straighteners

Some straightening effect can be produced in certain tubular products not only by bending but also by squeezing them elastically between the opposed straightening rolls as the tubes are processed through a rotary straightening machine. The rolls have a concave curvature through which a line of contact is produced between the roll and the workpiece for almost the full length of the roll.

Machines having sets of two opposed rolls are used to combine bending and ovalizing in the straightening of round tubes. Squeezing results in high residual stresses that can be removed by subsequent bending operations.

Collapse Strength. The cold working of a tubular product by bending and ovalizing in a rotary straightening machine can reduce the collapse strength or external-pressure resistance of the tube. Therefore, cold working should be kept to a minimum. The least amount of bending and/or ovalizing that will still yield a straight tube should be used. It is believed that a round tube obtained by squeezing will result in good collapse resistance.

Straightening in Tube Production (Ref 1)

The simple rule for scrap reduction due to straightness in the tubing industry is to use the proper straightening machine. The range of tolerances of 0.8 mm of total indicator reading (TIR) per meter (0.010 in. TIR/ft) and above can be handled by the normal multicycle, five-, six-, or seven-roll rotary straightening machine. The machine, because it does not have any restriction to the path of the tube, cannot achieve greater tolerances than this unless the incoming material itself is almost to these tolerances prior to being

straightened, and thus, the scrap reduction in the straightener is not normally great because of the loose tolerances.

However, anytime a tube has a ratio of thickness to diameter of less than 30 to 1 and required tolerances better than commercial, a two-roll type of machine is used. The scrap reduction method is to treat this type of tubing as though it were a precision-drawn solid bar. The same comments concerning the ends can be attributed to the method of cutoff. The same tolerances can be lost if the tube is not held against rotation, which would cause it to throw itself out of straightness before going into the straightener, and surface condition on the tube is generally, except for the food-handling type of tubing, important but not necessarily bright.

One interesting example is a machine that straightens tapered golf shafts. When the straightening machine is able to straighten a multidiameter drawn shaft, it is an important cost savings, because it eliminates hand straightening. The material to be straightened has to fit within certain categories, and the diameter of the small end to the large end is approximately a 2.5 to 1 ratio. Precise electronic controls are used to set the machine for each diameter.

The reduction in scrap in some of the drawn-with-mandrel tubing operations is unique. After drawing the mandrel with the tube, it is passed through the standard two-roll straightening machine. The mandrel, being a harder material, is unaffected, and the tube is expanded due to deflection in the straightening machine. The deflection is caused by the straight roll pressing the tube against the concave roll to a measured amount, as indicated by the digital readout on the

straight roll. In this manner, there is a great deal less scrap because of the predictability of the operation and because the straightening itself is done so that the mandrel will be easily withdrawn from the tube. Therefore, this is an indirect scrap reduction; however, it has been proven that it is much more gentle on the tube wall than the so-called cross-rolled derodder, where the straight roll is rammed into the center of the concave roll, and plain, raw pressure ovals the tube and pressure-irons it out with stresses against the hardened mandrel. This savings results later on when the material is straightened by other means, generally, six-roll machines if it is not to go into screw machines or other applications.

The reduction of scrap in the straightening process of tubing depends on the method of manufacture of the tubing. The simple straightening of large-diameter tubes (approximately 127 mm, or 5 in., with a heavy wall of approximately 16 mm, or $5/8$ in.) that are used in trailer axles, for example, was done in a multiroll or on a gag press. Experiments indicate that the rotary straightening of these in a two-roll machine will increase their fatigue life more than five times. Thus, scrap reduction by straightening is enhanced by the selection of the proper straightening method.

REFERENCE

1. W.L. Siegerist, Straightening Processes—Scrap Reduction Methods, *12th Technical Conference on the Methods of Reducing Scrap in the Production of Tubes, Bars and Shapes,* ASM International, 1989, p 145–165

Forming of Wire

Revised by David Berardis, Fenn Technologies

WIRE FORMS are used to give a high strength-to-weight ratio, an open construction (as in fan guards or baskets), resilience to absorb shock, and the economy of automated production of formed parts. When production quantities are small or the size of the finished article is large, the wire may be straightened and cut to length as a preliminary operation before the individual pieces are fed into hand benders, kick presses, power presses equipped with appropriate dies, or coiling devices. For large quantities, the wire is straightened as it comes from the coil and is fed directly and continuously into power presses, automatic forming or spring-coiling machines, multiple-slide machines, or special machines actuated by cams, air, or hydraulic cylinders. Wire drawing is discussed in the article "Wire, Rod, and Tube Drawing" in Volume 14A of the *ASM Handbook,* 2005.

Operations other than bending that are performed on wire include:

- *Threading* with single-head or multiple-head chasers, or with flat-die or rotary-die roll threaders. Roll dies can also be used for knurling, pointing, and chamfering.
- *Heading* in open-die rod headers, to make a variety of heads such as flat, round, slotted, indented hexagon, tee, and ball
- *Swaging or extruding* of long points or reduced-diameter sections on rotary-die swagers, stationary-die swagers, or long-stroke headers. While rotary swagers are used for pointing and reducing the diameter, stationary-die swagers are used to produce flattened ends, such as for screwdrivers and chisels.
- *Welding* with resistance, arc, or gas

Speed of Forming. Increasing the speed of forming can result in out-of-tolerance parts, increased springback, and wear on the tools and machine caused by increased force and torque. With older machines in which some of the tools are air-actuated and some are not, enough time must be allowed for the air-actuated tools to cycle to prevent the machine from going out of phase. For example, a machine with a mechanical drive and air-actuated tools was constantly out of phase when the production rate was increased. By reducing the speed slightly, the air valves and cylinders had time to complete their cycles and were in phase with the mechanical devices. Higher speeds can be achieved using servo motors and drives rather than the more antiquated pneumatics.

Tools used for forming wire should be made of tool steel hardened to 56 to 61 HRC. Water-hardening tool steels such as AISI W1 are usually adequate. For more severe forming and longer tool life, D2 tool steel is recommended. For forming wire at higher temperatures, that is, over 540 °C (1000 °F), H-13 tool steel is recommended. Surfaces contacting the wire should be polished to prevent marking. They can usually be hardened after tryout in the soft state. For operations that induce shock loading, such as repetitious swaging, an S-7 tool steel is recommended to prevent cracks from prematurely developing.

Springback is variable and difficult to control in the forming of wire, as it is in most press-working operations. Springback varies with the type and temper of wire and may be different for each lot of a specific type and temper. The most practical way to determine springback is to make trial bends either before the tools have been hardened or on temporary tools. The necessary final correction for springback is usually made at the final tool setup, after the tools have been hardened.

Effect of Material Condition

Most wire forming is done at room temperature. Wire made of low-carbon steel is usually formed in the as-drawn condition. Medium-carbon steel wire (1035 to 1060) is usually annealed before severe forming and heat treated after forming.

Surface Finish. A rough surface on the wire may cause short tool life. Plated wire is as easily formed as is bare wire, except that if the plating loosens or peels, it may damage the tools. Platings of gold, tin, solder, or other soft metals may show marks readily; however, soft plating may act as a lubricant during the forming of wire. Whether the wire can be plated before forming may depend on the severity of forming and the subsequent fabricating operations. Welding, for instance, may require that plating be done after forming.

Properties. The strength of wire is important in forming, especially when making steel springs. The required tensile strength is developed in spring wire either by cold drawing through a series of dies with up to 85% reduction in cross section, or by heat treating steel containing 0.60 to 0.70% C, quenching in oil, and tempering the wire. The elastic limit in torsion of spring wire is more important to its use in a spring than is its tensile strength. Information on the mechanical properties of steel spring wire is available in the article "Steel Springs" in *Properties and Selection: Irons, Steels, and High-Performance Alloys,* Volume 1 of *ASM Handbook,* 1990.

Rolling of Wire in a Turks Head Machine

A Turks Head machine (Fig. 1) generally has four rolls that will accommodate wire of one cross section (generally, round) and cold roll it to another shape. A Turks Head may be compared to an adjustable draw die but is actually infinitely adjustable within its limiting dimensions. The Turks Head operates on the rolling mill principle and imparts the same qualities to the metal as the rolling mill does, namely superior surface finish, accurate size and shape, and improved grain structure of the metal. Uses of the machine are:

- To make accurate square and rectangular wire directly from round wire
- To finish special shapes from round or preformed rough shapes
- To put edge contours on flat metal ribbon or rectangular wire

Operation. The machine has a cluster of four rolls with the four axes in the same plane and at right angles to each other, as shown in Fig. 1(a) and (b). In operation, a coil of wire is supported in a pay-off reel; the wire is pulled through the rolls by a capstan and then recoiled. A draw-bench can be used for pulling short lengths (up to 30 m, or 100 ft) through the rolls, while a capstan or drawing block is used to pull through continuous lengths of wire. Larger Turks Heads can be powered by driving the upper and lower roll assemblies. The power-driven Turks Heads eliminate the need for a pull-through device.

This is advantageous for forming irregular shapes that distort when coiled, for materials of low tensile strength, or precious metal wire where the leader cannot be wasted in threading.

Depending on the shape to be formed, the narrow rolls can be centered (opposed), as in Fig. 1(a), or they can be offset, as in Fig. 1(b). Although the rolls shown are plain cylinders, these can be replaced with rolls ground to any shape that will form the desired cross section. Some sections require several passes through the machine. It may take two passes to roll an accurate sharp-cornered square wire, and three or more passes may be needed to make a complex section fill properly.

Simple or complex shapes may be drawn through a standard Turks Head machine as fast as 180 m/min (600 ft/min), depending on the force and speed available in the capstan or drawbench and the amount of heating in the operation. A standard Turks Head contains needle bearings in its roll assemblies. High-speed Turks Heads use tapered roller bearings in the roll assemblies, which allows for higher speeds with the trade-off of lighter loading. Wire can be drawn through a high-speed Turks Head at speeds of up to 450 m/min (1500 ft/min), although these Turks Heads are limited to lighter reductions than standard Turks Heads. In general, the sections that can be formed depend on the ductility of the wire and are limited to shapes that can be ground into the rolls and shapes that are suited to the roll design, symmetrical, and no wider than twice the round wire thickness (unless preformed wire is used).

Some Turks Head machines have three rolls to make triangular shapes and other shapes suited to a three-roll design.

Accuracy of forming in a Turks Head machine depends on:

- Accuracy and uniformity of the initial round wire in size, shape, smoothness, hardness, and ductility
- Tensile strength of the incoming wire
- Dissipation of heat caused by cold working
- Smooth operation of drive in accelerating, running, and decelerating

- The amount of reduction in area or change in section in one pass
- Amount and type of lubrication used on the wire as it is formed

Any variation in size of the round wire pulled through the rolls may cause changes in size and shape of the product. If the round wire is oversize in a portion of its length, it may cause a sharper corner in the shape and thus a longer cross-corner dimension, or it may form a fin. If the round wire is undersize, the shape will not be well filled, and the cross-corner dimension will be decreased. Similarly, too heavy of a reduction will cause a fin in the corners, while too light of a reduction will lead to corners that are not well filled.

Variations in the hardness and ductility of round wire can also cause variations in the cross-corner dimension of a shape. Hard spots increase the cross-corner dimension; soft spots decrease it. A rough or unlubricated surface increases cross-corner size; a smooth, oiled surface makes it smaller. Heating of the rolls caused by cold working may enlarge the rolls, making the product smaller. A coolant is frequently used to remove heat from the rolls. Mineral oil is typically used in this application, because it is superior to water-soluble oil for lubrication.

Nonuniform acceleration, running and deceleration of the capstan, and changes in tension on the wire may also cause variations in the formed shape. The greater the reduction in section size of the wire as it is drawn through the rolls, the greater the chances of variations in the formed shape.

Tolerances in wire formed in a Turks Head machine in ordinary production are ±0.05 mm (±0.002 in.), but ±0.013 mm (±0.0005 in.) is a reasonable tolerance if all important factors are controlled.

Spring Coiling

Production coiling is done in automatic spring coilers. In a standard spring-coiling machine, a pair of feed rolls pushes a calculated length of straightened wire through restricting guides against a coiling point and around a fixed arbor into a coil. At the end of the coiling cycle, the feed rolls stop, and a cutoff mechanism actuates a knife, which severs the completed spring against the arbor. A flying knife separates the completed spring from the wire strand.

Modern spring coilers are computer numerical controlled (CNC) and use high-speed servo motors, servo drives, and computers to facilitate easy programming and easy operation. While traditional gear and cam mechanical spring coilers are limited in the types and sizes of springs they can produce, an automated CNC spring coiler can produce almost any kind of spring as well as extremely complex wire forms. A modern computer-controlled Torin spring coiler is shown in Fig. 2.

Standard spring-coiling machines range in size from those that can coil only fine wire to those that can form 19 mm (3/4 in.) diameter cold-drawn or 16 mm (5/8 in.) diameter pre-tempered wire. Each coiler can process a range of wire diameters, depending on the number and size of half-round grooves in the feed rolls. A set of feed rolls usually has grooves of three or four different sizes. For example, a machine may coil wire 2.3 to 5.36 mm (0.09 to 0.21 in.) in diameter and make a spring with an index (ratio of mean spring diameter to wire diameter) ranging from 3 to 18. The length of wire fed is controlled by the feed rolls.

A coiler equipped with a variety of attachments and cams or servo motors can produce almost any type of spring, including tight-wound extension springs, common compression springs with either open ends or ends closed for grinding, barrel-type springs of various contours, tapered springs, single-coil springs, variable-pitch compression springs, and torsion springs. A modern computer-controlled spring coiler can also produce an infinitely wide variety of bent wire forms.

Setup time is significantly reduced with modern CNC camless spring coilers, because these machines are very easy to set up. What used to take hours to set up with mechanical machines with cams and levers is now reduced to minutes on new CNC spring coilers. Now, small production runs can be profitable, because a

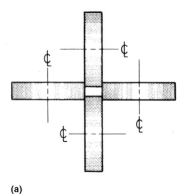

(a)

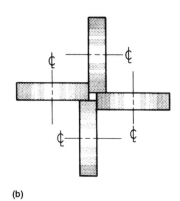

(b)

(c)

Fig. 1 Turks Head rolls. (a) Positioned in line to form a rectangular cross section. (b) Offset to form a square section. (c) Machine section of a Turks Head roll

recalled spring can be quickly and accurately produced. Preset production count automatically stops the machine when the desired production run is reached.

New five-axis CNC springmakers are designed for a full range of traditional wire sizes but differ from the mechanical counterparts by being fully servo controlled. The feed, diameter, pitch, torsion, and coiling-point axes (Fig. 3) are directly driven with externally commutated alternating current brushless servo motors. There are no cams or associated levers and linkages used, and there are no motor brushes to wear out. Torque transmission on the diameter, pitch, torsion, and coiling point is done through preloaded ball screws, thus eliminating backlash effects. The servo motors have a very high torque-to-inertia ratio to maximize responsiveness.

A tight-wound extension spring is wound in a standard coiler, but the end loops are formed in one of three ways:

- One or more of the end loops are opened up or pulled out in a secondary operation to form the required end hooks.
- End loops are automatically formed by deflecting the wire into shape as a part of the coiling operation.
- Automatic handling equipment and attachments are incorporated as additions to a regular coiler to make the loops.

The second method is less complicated to set up and operate, but both it and the third method are limited to wire less than 1.25 mm (0.050 in.) in diameter and, because of setup time, to production runs of not less than 10,000 pieces.

A machine equipped with a torsion-spring attachment forms straight extended arms; these arms can be formed and looped as desired in a second operation.

Compression springs are wound in standard coilers equipped with a pitch tool located under the first formed coil. This tool, controlled by cams, regulates the spacing between the coils, which may be either uniform or variable. The ends of compression springs may be plain, plain and ground, square, or square and ground.

The use of round wire predominates in making compression springs, although square, rectangular, or special-section wire is necessary in some applications. For example, die springs are formed with rectangular wire. Square or rectangular wire is used to obtain the maximum load capacity for a given space. Wire with square corners before coiling will upset at the inside of the coil and become trapezoidal in section after coiling. This limits the deflection per coil, especially with small ratios of mean diameter to wire thickness.

Accuracy. With older mechanical spring coilers, the cams, gears, and other parts of a coiling machine become worn as the machine is used, resulting in a less accurate product. Some product inaccuracies can be reduced by control of the speed of the machine. With newer CNC spring coilers with servo drives, there are no mechanical parts such as cams to wear out, so more accurate parts are produced in the long run. Additionally, servo motors and precision, hardened, preloaded ball screws on diameter, pitch, and torsion provide a zero backlash operation. Accuracy such as 0.013 mm (0.0005 in.) on feed and 0.0013 mm (0.00005 in.) on pitch, diameter, and torsion can be achieved.

Dimensional variations for different materials are caused by variations in springback and by distortion during heat treatment. Variations in pitch and diameter depend on the speed of coiling. Dimensional variation depends also on the ratio of wire size to diameter and the ratio of pitch to spring diameter. By increasing the limits slightly, coiling speed and production rate may be increased.

Many springs are acceptable with inaccuracies or with a wide tolerance in dimensions and performance, but some springs (valve springs, for instance) must be more accurate. Variations in mechanical properties of wire will result in nonuniform springs. Standard spring wire has a permissible tensile-strength variation of 170 to 240 MPa (25 to 35 ksi); wire for valve springs has a limit of variation of 140 MPa (20 ksi). The range for any one coil seldom exceeds 35 MPa (5 ksi).

Coiling must be fast enough to produce an even flow of wire from the pay-off reel. If the wire flow is jerky or nonuniform, the dimensions of the spring may vary excessively.

Fig. 2 A modern computer-controlled spring coiler (camless computer numerical controlled spring coiler with three-axis control). Courtesy of Fenn-Torin, Inc.

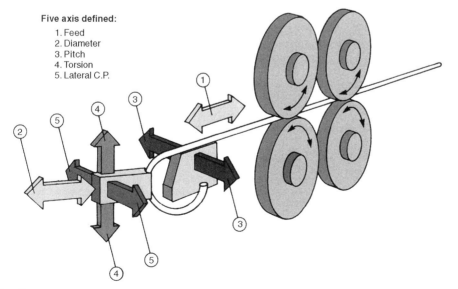

Five axis defined:
1. Feed
2. Diameter
3. Pitch
4. Torsion
5. Lateral C.P.

Fig. 3 Directional controls in a five-axis spring coiler. 1. feed; 2. diameter; 3. pitch; 4. torsion; 5. lateral coiling point

Manual and Power Bending

Manual and power bending are done in rotary benders, pneumatic or hydraulic formers, and special fixtures to locate the bend and hold the part. The blanks are straightened and cut to length before forming.

A rotary bender can be either manual or power operated. The precut blank is placed between a center pin or form block and a stop pin. The arm supporting the wiper block is rotated clockwise, thus forming the wire around the center pin. An adjustable stop controls rotation of the arm so that uniformity is maintained from piece to piece. The eyebolt shown in Fig. 4 was formed in a rotary bender from 13 mm (1/2 in.) diameter cold-drawn stock at a production rate of 300 pieces/h.

The center pin can be changed to suit the bend radius. Form blocks can be used for larger radii. The stop pin is movable to accommodate different bend radii and stock thicknesses. Round, square, or rectangular wire can be bent in this equipment.

A more complex wire form, such as the scroll shown in Fig. 5, can be formed around blocks mounted in a rotary bender. The compact shape at each end could be formed by holding the wire taut while it is wrapped around a rotating form block, or in a machine with a spring-loaded or air-loaded wiper block. Another form block could be used for the long curves. The form blocks in these two operations must be designed to compensate for springback. The sharp bends at the outer ends can be made in a rotary bender around a center pin. Many types of complex wire forms can also easily be formed by a modern computer-controlled spring coiler.

The CNC servo hydraulic three-axis wire bending machines can automatically take wire directly from the coil, straighten it, make any number of bends through any angle and curvature and in any plane, and cut the finished product. The three-axis CNC wire bender can handle mild steel, spring steel, or aluminum wire from 2 to 7 mm (0.08 to 0.27 in.) in diameter. Bending is by a single head that rotates to give the third dimension. When the wire diameter is unaltered, changeover from one bend to another can take as little as 5 s, because bending programs are stored in a built-in memory. When the wire diameter varies, changeovers can be accomplished in approximately 15 min.

Forming in Multiple-Slide Machines

Multiple-slide machines are automatic mass-production machines that make completed wire (or sheet metal) products from coiled stock. They can straighten, feed, cut, stamp, and form the wire, all in one continuous operation. Attachments are used for additional operations on wire up to 13 mm (1/2 in.) in diameter. Feed lengths of wire used to make one part may be as much as 1 m (3 ft).

Most multiple-slide machines are horizontal, but some are vertical and some are inclinable. Production lots of 10,000 pieces usually justify the use of a multiple-slide machine, but smaller lot sizes may also be economical.

A multiple-slide machine usually includes a wire straightener, feed mechanism, and stock clamp and has a bed with four or more forming slides, a center post, and a stripper. Traditional machines known as four-slide machines in fact have four separate forming slides. New modern multislide machines may have as many as eleven independent forming slides. Some of these slides may be replaced with welding heads, stamping heads, drilling and tapping heads, or others as needed, depending on the part to be made. More information on forming in multiple-slide machines is available in the article "Forming of Steel Strip in Multiple-Slide Machines" in this Volume.

Production Problems and Solutions

Examples of several wire-forming production problems and solutions are given in Table 1.

Lubricants

Requirements of lubricants for wire-forming operations are more severe than for most other metalworking operations. The exceptionally high working pressures that may be reached require special lubricants to prevent galling, seizure, or fracture of the wire, as well as excessive tool wear. Improper lubricating oils or compounds interfere with close-tolerance work and cause variations in the finished parts. The lubricant varies with the type of wire. Aluminum, copper alloys, basic steel wire, and steel spring wire each require a different lubricant. Lubricants for wire forming can generally be classed in three groups: inorganic fillers, soluble oils, and boundary lubricants.

Inorganic fillers include solids such as white lead, talc, graphite, and molybdenum disulfide in a vehicle such as a neutral oil or paraffin oil.

Soluble oils include mineral oils to which agents such as sodium sulfonates have been added to make the oil emulsifiable in water. Soluble oils are good for cooling and corrosion prevention.

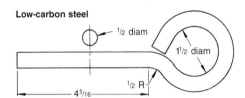

Fig. 4 Eyebolt formed in a rotary bender. Dimensions given in inches

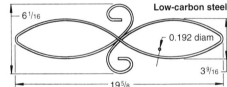

Fig. 5 Complex wire scroll formed in a rotary bender. Dimensions given in inches

Table 1 Wire-forming production problems and solutions

Problem	Solution
A small garment eye was produced at the rate of 300 parts/min, but desired flatness was not obtained. The eye was to be formed in one plane.	The tooling was changed to form the eye on a plate, confining the wire. Forming sections were supported in the plate and were retracted below the surface of the plate before ejection.
Broaching of a flat on a small formed part for a business machine caused fine chips to collect in the forming tools, interfering with the work.	Forming was done on the upper level, and broaching was accomplished on the lower level of the machine. A small jet of air blew the chips down and away from the forming tools.
A small electronics part was formed around a fragile center post that broke after a few thousand pieces were formed.	The tooling was changed so that the part was made in two stages. A heavy center post was used for the first form, where the greatest pressure was exerted. A center post having the shape of the smaller section controlled final closing of the part.
Ordinary forming tools could not form brass wire of 1.0 mm (0.040 in.) diameter into a 9.5 mm (0.375 in.) outside diameter ring.	The formed ring was sized by being pushed through a die below the forming level, by means of a ring-setting attachment.
Steel wire 5.7 mm (0.225 in.) in diameter was formed into seat wire in a large multiple-slide machine, using an 860 mm (34 in.) feed length. The front tool was 510 mm (20 in.) wide. The bearings of the front shaft became hot, and the cam roller of the front forming slide had to be replaced often.	The addition of auxiliary front slides at each side of the standard front slide provided more direct application of the forming pressure, with three motions from the front position, reducing the load on each tool. Forming loads were exerted at different points in the machine cycle, reducing bearing pressure and allowing faster machine speed.
A specially arranged multiple-slide machine for forming and welding handles from low-carbon steel wire 2.4 mm (3/32 in.) in diameter made imperfect welds. There were also variations in forming. The wire varied in diameter and tensile strength, which affected the feed length and forming. Operators had not been trained to adjust and maintain the machine.	The wire was specified to closer tolerances and inspected before use. Operators were trained, and a program of preventive maintenance was initiated. Production efficiency increased to 75% from as low as 25%.

Boundary lubricants are thin, absorbed films and are usually subjected to high unit pressures. Thin-film lubricants are of two basic types:

- *Polar lubricants:* lubricants, or constituents of lubricants, capable of either physical or chemical adsorption on a solid surface to form a thin film that resists mechanical removal and provides lubrication under high unit pressures
- *Extreme-pressure lubricants:* lubricants capable of reacting chemically with solid surfaces under rubbing conditions, to prevent welding and provide lubricant reaction products on the surface. Extreme-pressure lubricants permit high unit loading with a minimum of surface wear and damage.

Chemically active constituents of typical boundary lubricants are sulfur, chlorine, and phosphorus compounds.

Applications. Certain types of oil, wax, and tallow are used to lubricate aluminum and scaly steel wire. Mixtures of lard oil or of paraffin oil in kerosene, or of oil and a soap solution, have been used as lubricants for wire forming.

Often, the lubricant used in drawing wire is expected to stay on the drawn wire in a quantity that is adequate for subsequent forming operations. Many severe operations, such as upsetting and spring coiling, may be done without additional lubrication, but additional lubricant may be used in some press and rolling operations.

The lubricant remaining after wiredrawing should be enough to lubricate a wire formed over a form tool or a mandrel. The lubricant should be a hard, dry coating, such as a mixture of lime and metal soaps. This will protect the wire from damage in forming and will extend the life of the tools without sticking to them.

Zinc phosphate is often used to coat wire before it is redrawn into smaller sizes. It is also used in coiling thick, high-tensile spring wire into a closed helix. In other difficult forming, such as a large upset, a zinc phosphate coating is a good lubricant for all tools.

Wire for forming into products to be electroplated is usually drawn with a lubricant that can be easily removed and that does not contain small particles that could become embedded in the surface of the wire. After drawing, the wire may be sprayed or dipped in thin oil. The oil protects the wire from corrosion and serves as a lubricant in the forming operations.

Steel wire with a metal coating, such as zinc, tin, copper, brass, or lead, is often used in forming. In some operations, the metal coating provides all the lubrication necessary.

When lubricant must be added to the wire, it can be applied at the uncoiler or at the tools. Soluble oil or wax in water is most practical and is easy to remove in cleaning. Some formed wire must be completely clean.

Spring wire is supplied with a coating that acts as a lubricant. The coating may be a mixture of soap and lime or of borax or phosphate, or it may be a plating (or displacement coating) of cadmium, zinc, tin, or copper. When the coil is to be electrically normalized, a borax coating is specified; other coatings insulate the wire from good electrical contact.

The most unusual lubricant may be the one that comes on the oil-tempered grade of valve spring wire. During heat treatment, oxidation of the surface is permitted under carefully controlled conditions. The scale thus formed acts as a lubricant during coiling. Its characteristics must be carefully controlled with respect to thickness, adherence, and flakiness, because not only must it supply the required lubrication during coiling, but it should detach from the surface at the same time.

Sheet Forming of Specific Metals

Forming of Carbon Steels

Revised by Mahmoud Y. Demeri, FormSys Inc., and Steve Lampman, ASM International

FABRICATION of sheet products includes various forming operations such as press forming, press bending, press-brake forming, stretch forming, and deep drawing. Fabrication of sheet products may also include cutting operations, such as blanking, trimming, and piercing (for example, see the articles "Blanking of Low-Carbon Steel" and "Piercing of Low-Carbon Steel" in this Volume).

The choice of forming method depends on the desired size, configuration, and production quantities of a part. For example, press forming (where a punch presses the workpiece into a die) is widely used and is often the most effective method when production quantities are large and workpieces are relatively small. In some applications, however, other forming methods may be desirable or needed. For example, the production of hollow shells from flat blanks requires deep drawing, where sheet is pressed into a die opening to draw the metal into the desired shape without folding of the corners (see the article "Deep Drawing" in this Volume).

Another example is press-brake forming, which sometimes is the only practical method of producing a given shape. Compared to press forming, press-brake forming is more effective when only a few pieces are needed or when the workpiece is large. Tooling for a press brake (see the article "Press Brakes" in this Volume) is usually simpler and less costly than counterpart tooling for a punch press, and so it can be effective for small production quantities. Press brakes are also more effective as the size of the workpiece increases. Presses are usually used for workpieces less than 610 mm (2 ft) long, while press brakes are used for pieces longer than 610 mm (2 ft)—except perhaps in the automotive industry, where very large sheet metal parts are mass produced on large presses.

In general, press forming is often more economical than the press-brake forming method, because of high production rates and the efficient use of metal. When a given workpiece can be made to an equal degree of acceptability in either a press brake or a punch press, the punch press is usually more economical. The punch press is more efficient than the press brake in terms of power requirements for a given force on the ram and number of strokes per unit of time. Air ejection (of either the workpiece or scrap) also is

more readily adapted to a punch press than to a press brake. In addition, punch presses can be applied to produce various surface profiles, with forming conditions ranging from pure bending (press bending) to more complex forming conditions involving stretching, bending, and drawing.

This article reviews the selection and formability characteristics of steels, with emphasis on low-carbon steels and some coverage on the forming of high-carbon steels (which, despite the higher strength and lower ductility compared to other carbon steels, are formed into products requiring high strength, stiffness, and hardness). Low-carbon steels are the most commonly used sheet metal because of their low cost and good formability. However, ordinary low-strength, low-carbon sheet steel has been replaced by a number of higher-strength sheet steels requiring new process technology. These new steels include the high-strength, precipitation-strengthened steels, the dual-phase and tri-phase steels, and the bake-hardenable steels. Also, new coating techniques have been developed to protect these new steels from corrosion. These topics are discussed in more detail in the articles "Forming of Advanced High-Strength Steels" and "Press Forming of Coated Steel" in this Volume.

Presses and Dies

The characteristics of the various types of presses used in forming sheet metal parts are discussed in the article "Presses and Auxiliary Equipment for Forming of Sheet Metal" in this Volume. Single- and double-action presses are available in approximately the same ranges of bed size and force capacity. Restriking, coining, and embossing are usually done in presses with more available force capacity than that needed for the simple forming of similarly sized areas, because in these operations the metal is confined while being forced into plastic flow. Progressive dies are used in presses with enough force capacity to meet the total demands of the various stations and with enough dimensional capacity for the long multiple-station dies. Although some progressive dies are hand fed, most have auxiliary equipment, such as stock feeders, scrap

choppers, coil reels, and chutes, to carry the finished parts to containers. The use of blank feeders and piece ejectors or extractors depends on the production rate and safety requirements. Springs, cams, fluid pressure, or press knockouts are used for piece ejection.

Shallow forming can be done in single-action presses using die cushions or springs to provide the blankholder pressure. Deeper forming and the forming of large, irregular shapes generally must be done in double-action presses with die cushions. The trend in stamping technology is to use single-action and transfer presses with hydraulic multipoint cushion systems in the press table (Ref 1). Multipoint cushion systems (Fig. 1) make it possible to control material flow between the binders of draw dies, due to closed-loop controls that automatically adjust blankholder forces according to changes in parameters affecting friction (e.g., lubrication, blank surface, die roughness, etc.).

The hydraulic four-point cushion (Fig. 1b) system is mainly a passive system, in that the pressure in the four cylinders increases by compression when the lower binder is forced downward. It is also possible to separately control pressure for each cylinder. It has also been shown that single-action presses with hydraulic four-point cushions give better and more reproducible draw results than mechanical double-action presses (Ref 1). Whether or not a press has a die cushion has some effect on die design and construction costs. Special die designs are needed to take advantage of hydraulic cushion technology.

Speed of forming has little effect on the formability of steels used for simple bending or flanging or for moderate stretching. The maximum velocity of the punch when it contacts the blank in such conventional press forming is usually not greater than approximately 1 m/s (200 ft/min). However, the steels used for most parts that involve local stretching of more than 20% in forming move considerably over the face of the punch or flow appreciably over the blankholder. The flow of the metal in such operations is controlled by frictional forces, and so, stretching failures can be sensitive to speed. The critical value differs for each steel and die combination, but a maximum punch velocity of 0.2 m/s (40 ft/min) is recommended; high punch speeds also shorten tool life.

Dies

Low-carbon steel can be formed by any of several types of dies. Bending dies include V-dies, wiping dies, U-bending dies, rotary bending dies, cam-actuated flanging dies, wing dies, and compound flanging dies (see the section "Bending Dies" in this article). Dies may also be designed for single-operation function, or multiple forming functions can be implemented more efficiently with the design of compound dies, progressive dies, and transfer dies (see the subsequent information). Dies for press brakes are discussed in more detail in the article "Press-Brake Forming" in this Volume.

Workpiece size and shape, production volume, tolerances, and available presses are the major factors that determine the most suitable type of die for a specific application. Dies for the press forming of low-carbon steel are made from a wide range of materials, including plastics, cast irons, tool steels, and cemented carbides. Severity of forming, number of parts to be produced, workpiece shape, work metal hardness, specified surface condition, and tolerances affect selection of the die material (see the article "Selection of Materials for Press-Forming Dies" in this Volume).

Simple bending dies are ordinarily subjected to less shock than other press-working tools; therefore, they can often be made of low-carbon steel, heat treated low-alloy steel such as 4140, or cast iron for low production of low-carbon steel pieces. For moderately high production, they should be made of flame-hardening grades of carbon steel, such as 1045, or flame-hardening cast iron, such as class 40 gray iron. Cam-operated dies, wiping dies, and dies used to make curved flanges (shrink or stretch) must be made of a higher grade of material. Tool steels such as O1 or A2 are used for moderately long production runs. For a total tool life of 1 million or more pieces, D2 tool steel is used.

Bending dies used in press brakes can generally be used in presses. However, because presses ordinarily are not long and narrow like press brakes, more consideration must be given to clearance for removing the finished workpiece when the press is open, as well as clearance for the legs of the bend when the piece is being formed. The bed dimensions of a press also limit the size of workpiece that can be bent.

Presses also cycle more rapidly than press brakes, and the shut height is not as easy to change; therefore, fewer pieces are bent in air. More frequently, pieces are formed by bottoming the dies. This has the advantage of decreasing springback. One disadvantage of punch presses is that they are more sensitive to thickness variations of the work metal, because they operate at a faster rate than press brakes.

Die Development and Press Setting. To help identify problems during computer modeling and tool development and die tryouts, it is helpful to use steel near maximum in hardness and near minimum in formability. Tools developed with tryouts of below-average-quality steels are less troublesome for production runs and are less sensitive to pressure adjustments, variations in sheet thickness, or normal variations in steel properties.

After the dies have been manufactured, finished to a high polish, and mounted in the press, the first job is to "bed-in" the blankholder with the die plate. It is most important that neither the blankholder nor die plate have any high spots that may locally restrain the blank and prevent even flow of material during the draw.

Secondly, all the radii should be checked to ensure they blend in smoothly, then a blank, cut to the shape as predicted by the rules for blank size and shape, should be tried in the press. A lubricant should be selected to suit the operation. For deep drawing a good-quality drawing compound or for tapered shells where restraint by the blankholder is needed, a lubricant that prevents galling of the sheet steel to the press tools but that has a relatively high coefficient of friction will be more suitable. After a few blanks have been drawn and the trouble spots identified, a series of blanks may be marked with circles for grid analysis of formability.

Single-operation dies perform one operation at a time and are individually loaded and unloaded. Single-operation dies are used for low production or on pieces that are so difficult to bend that only one operation at a time is feasible. Some single-operation dies are general-purpose dies that can make bends in simple workpieces of many different designs. Others are single-purpose dies for a particular piece.

Single-operation dies are usually set up in a press, and the operation is performed on a specific lot size. The die is then removed from the press, and the next die in the sequence is set up. For continuous production, a line of presses, each operating a single die, can produce finished pieces from raw stock without interruption for change in setup. Occasionally, more than one die is set up in a press at a time, and the parts are moved manually from one die to the next. With this type of tooling, more than one operation is done in each stroke of the press.

Single-operation press forming dies are used when:

- The operations are so interrelated that they cannot be done in a compound die.
- The amount of work done on a part is approaching press capacity, and more work would overload the press.
- Production quantity is low, and two or more single-operation dies would be less costly than a die combining operations.

Forming presses with single-operation dies do not necessarily have a low production rate. They often can be run at high speed with automated materials handling (e.g., coil stock fed automatically into blanking dies) or inclined beds (that permit high-speed loading and unloading). When a higher rate of production is needed, it is sometimes more practical to increase the speed of the press than to use an additional die, provided the flywheel, bearings, gibs, and gears can withstand the additional speed.

Compound dies are one-station dies in which more than one operation is done without relocating the workpiece in the die. One or more of the operations done in the die can be bending. The operations must be such that their inclusion does not weaken the die elements or restrict other operations.

Although the operations are generally done in succession in the course of the press stroke (rather than simultaneously), the operations may sometimes be done in one press stroke, so that a finished piece is produced with each stroke of the press. The press speed (strokes per minute) for compound dies is generally only slightly lower than for single-operation dies; therefore, production time and labor cost per piece are decreased almost in proportion to the number of operations done in the compound die.

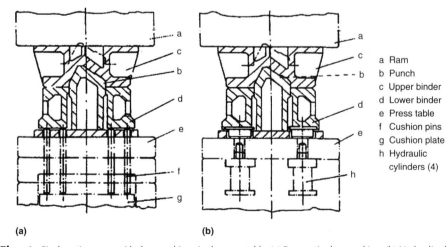

a Ram
b Punch
c Upper binder
d Lower binder
e Press table
f Cushion pins
g Cushion plate
h Hydraulic cylinders (4)

(a)   (b)

Fig. 1 Single-action press with draw cushions in the press table. (a) Pneumatic draw cushion. (b) Hydraulic draw cushion. Source: Ref 1

Typical combinations of operations include:

- Cutting a blank from a strip and then forming
- Lancing and forming a tab or louver
- Forming a flange and embossing a stiffening bead

When a die is used for blanking and forming a part, holes can often be pierced in the bottom with the same die. When pierced holes are required in a flange, piercing should be done after the flange has been formed; otherwise, the hole (and perhaps the edge of the flange) can be distorted. The combination of lancing and forming is common. Continued travel of the lancing punch does the forming. Flanging can be combined with forming or embossing if no metal flow is necessary after the flange has been formed.

A compound die may or may not cost more than a set of single-operation dies. Loading and unloading can be automated or manual. Compound dies are generally operated at slower speed than single-operation dies. In the automotive industry, single-operation and compound dies are both set up in a press line. Coil stock or blanks are automatically fed into the first press, and the workpiece is automatically removed and transferred to the next press, where the cycle is repeated until the workpiece is completed. Typical parts are front grills, hoods, roof panels, and deck lids.

Several operations can be performed successively on a workpiece in a press, using two or more compound or single-operation dies. The parts can be manually transferred from die to die, eliminating storage and transfer between presses. The capacity of a large-bed press may be more fully used by performing several operations during each press stroke.

In most applications, the cost of a compound die does not differ greatly from the combined cost for equivalent separate dies, and is sometimes less. Operations must be judiciously combined in making a compound die in order not to have die sections that are too thin to be heat treated without distortion, or that will break down under the cyclic loading of ordinary die operation. If these precautions are observed, die life and die maintenance costs should be nearly the same as the costs for equivalent simple dies. Compound dies can be fed individual blanks, or they can be fed strip stock.

Progressive dies are similar in function to compound dies in that they combine in one die set several operations that are performed with one stroke of the press. In a progressive die, however, the operations are separated and distributed among a number of stations. The stock progresses through these stations in the strip form until the finished workpiece is cut from the strip at the last station. Consequently, at the start of a strip, there are several press strokes before a piece is produced. Thereafter, a finished workpiece is produced with each stroke of the press, up to the end of the strip (Fig. 2). Tandem part production is shown in Fig. 3.

Although a progressive die runs more slowly than a single-operation or a compound die for similar work, overall production is usually higher because the die is operated more continuously. The initial cost of a progressive die is generally greater than that of a series of individual dies for the same workpiece. However, unless the production quantity is low, the lower setup, maintenance, and direct-labor costs for the progressive die will often outweigh its higher initial cost.

The choice between a progressive die and two or more separate dies (single-operation or compound) is not always clear-cut. Various considerations can influence the decision; perhaps the most important is the size of the production order. Other considerations are the rate of obsolescence of the product and the rate at which tool cost must be amortized. Example 2 (which follows) shows how these considerations can affect choice of type of die. In this example, increased annual production requirements justified the use of a progressive die.

A set of individual dies is sometimes used for making a complex part prior to the designing and building of a progressive die. This is done for two reasons. First, a set of individual dies can usually be made in less time than a progressive die, thus permitting earlier production startup, and second, the experience gained in producing the parts in individual dies can be used in designing the progressive die. From this experience, it can be determined:

- How the metal flows and reacts in the die
- How much work can be done in each operation
- What is the best sequence of operations
- What the size and shape of the developed blank should be

Progressive dies are used to perform an almost endless variety of operations on one piece. Operations that can be combined in a progressive die include notching, piercing, coining, embossing, lancing, forming, cupping, drawing, and trimming. As the outline of the workpiece is developed in the trimming or forming stations, connecting tabs link the workpiece to the strip until the workpiece reaches the last station, where it is cut off and ejected from the die. Pilot holes that are engaged by pilot pins in the die keep the workpieces aligned and properly spaced as they progress through the die.

If the strips are short, sheared from sheets, they can be fed into the die so that each strip is butted against the end of the preceding strip for continuous production. Otherwise, starting stops are used for each new strip. Coil stock is fed into the die at one end. A hole or notch is usually make in the strip at the first station, and subsequent stations use this as a pilot to keep the strip properly aligned and positioned while the operations are performed. When the workpiece is complete, it is cut from the strip, which has acted as a holder to carry the piece from station to station. Intricate pieces, often needing no further work, can be make in one press.

In planning the strip layout for progressive-die operations, consideration must be given to development of the part outline, to provision for piloting, to distribution of press load and strength of die elements, and to ensuring minimum metal waste. Some compromise among these factors is usually necessary in developing a sequence of operations and designing a progressive die to accomplish these operations. The strip layout in the following example (example 1) used the metal efficiently by having the developed flanges surround the body of the preceding part.

Example 1: Nesting of Workpieces to Minimize Stock Waste. The bracket shown in the lower part of Fig. 4 was formed from hot-rolled 1010 steel strip in a progressive die with three working and two idle stations. As shown in the upper part of Fig. 4, pieces were nested on the strip to minimize scrap. The operations in the three working stations were as follows: pierce one pilot hole and two flange holes and notch the contour, bend two tabs upward, and flange and cut off from the center connecting tab.

The die was used in a 1330 kN (150 tonf) mechanical press that could make 50 strokes per minute. Allowing for setup and downtime, production was 2800 pieces per hour. A light mineral oil was the lubricant.

Fig. 3 Tandem part production on progressive die machine. Courtesy of Müller Weingarten

Fig. 2 Combined sequential production of a structural part with progressive die technology

The die was made of W1 tool steel hardened to 58 and 60 HRC. The punch-to-die clearance for cutting elements was 6% of stock thickness per side. Annual production was 100,000 brackets.

Example 2: Change from Three Separate Dies to a Progressive Die. The part shown in Fig. 5(b) was produced from 1.6 mm (0.062 in.) thick cold-rolled 1010 steel that had a hardness of 58 HRB maximum. A progressive die, which made a finished part at each press stroke, replaced three dies in which 12 press strokes were required for producing one piece. The separate dies were a standard piercing and notching die, an embossing die, and a con-

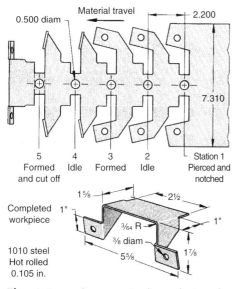

Fig. 4 Layout for progressive-die production of a bracket, with blanks nested to save stock. Dimensions given in inches

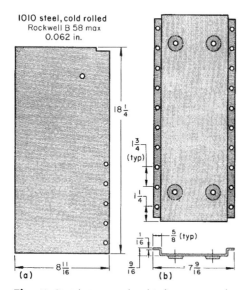

Fig. 5 Strut that was produced in fewer press strokes with a progressive die than with standard press-brake tooling. (a) Blank after first stroke with standard tooling. (b) Finished part. Dimensions given in inches

ventional V-die. Setup time for each die was 30 min.

The cost of the progressive die was justified for the annual production of 60,000 parts (break-even point was 34,000 parts). Both methods made parts that were within the maximum dimensional tolerance of ±0.4 mm (±1/64 in.), and to the maximum bend radius of 0.8 mm (1/32 in.).

The original dies had been used in an 890 kN (100 tonf) press brake that operated at 600 strokes per hour on sheared blanks 220 mm (811/16 in.) wide by 464 mm (181/4 in.) long. The press-brake operation consisted of piercing five holes in the flange and one in the center and notching one corner (Fig. 5a). The other holes and the three other corner notches were produced with three more strokes, with the workpiece being turned after each stroke. Four more strokes were needed for embossing four holes (Fig. 5b). Bending required four strokes, bringing the total to 12 strokes.

The progressive die, made of O1 tool steel, pierced, embossed, notched, and cut off the strut from 220 mm (811/16 in.) wide coil stock at a rate of 225 per hour. The die was set up in a 2700 kN (300 tonf) mechanical press. Mineral oil was used as the lubricant for both methods.

Example 3: Bending a Cylindrical Part in a Progressive Die. The part shown in Fig. 6 was made in a six-station progressive die from 1.2 mm (0.048 in.) thick cold-rolled 1010 or 1020 steel having a hardness of 65 to 75 HRB. The part was a valve used to adjust the size of the air intake in a gas burner.

The operations for making the part included piercing the 2.84/2.77 mm (0.112/0.109 in.) diameter hole, blanking the rectangular cutout, notching the outside of the blank (leaving a 13 mm, or 0.5 in., wide center carrier tab), bending the circle in three steps starting at the outside edge, and cutoff. The 2.84/2.77 mm (0.112/0.109 in.) diameter hole was tapped with 6–32 UNC threads in an automatic tapper at 12 pieces per minute. The two 1.6 mm (1/16 in.) deep circular notches were to ensure that the air intake could not be closed completely.

The cutting and forming elements of the die were made of O1 tool steel hardened to 58 to 60 HRC. Die life was 80,000 pieces per regrind. The die was run in a 400 kN (45 tonf) press at the rate of 40 pieces per minute to produce lots averaging

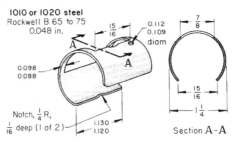

Fig. 6 Air-intake valve that was bent to circular form in three stages in a six-station progressive die. Dimensions given in inches

10,000 pieces. The part later was zinc plated and bright dipped.

Transfer dies are similar to progressive dies except that the workpieces being processed are not attached to a strip but are mechanically moved from station to station. In a progressive die, the workpieces remain fastened to the stock strip. In a transfer die, workpieces are separate and are transferred from station to station within the die between press strokes by mechanical fingers, levers, or cams. A blank is automatically fed into the first station and moved to the next at each press stroke. The first station can be a blanking die, which cuts a blank from manually or automatically fed stock during each press stroke.

Transfer dies are particularly suited to the fabrication of parts that would be difficult to connect to the stock skeleton with carrier tabs. Bends that cannot be made in a single step are often made in several stages in transfer dies, and bending is often combined with cutting or other forming operations in transfer dies. Transfer dies are well suited to the bending of small rings, cups, and cylinders. Transfer dies are used to make rings with one joint.

The advantages of transfer dies for bending include high production rate, greater versatility than progressive dies, and more efficient use of stock. The last advantage is ordinarily achieved by blanking in a separate press, which permits close nesting of parts. The disadvantages include high equipment cost (for dies, press attachments, and feeding devices), high setup and tool maintenance cost, difficulty in handling thin work metal, and poor applicability to large or oddly shaped parts that need variations in blankholder pressure and contour.

Transfer Presses. In transfer machines (eyelet machines), the mechanism for moving the workpiece from station to station is a part of the machine to which suitable transfer fingers are attached. This is the classic gripper transfer mode (Fig. 7). Electronic transfer technology also enables a cross-bar mode (Fig. 8) in transfer presses. Transfer presses are generally long-bed straight-side presses (Fig. 9). The transfer mechanism as a part of the press is actuated by the main press drive or is powered separately. A dial feed is a type of transfer mechanism that moves the workpiece from die to die in a circular path rather than in a straight line. Transfer press technology is used in forming large automotive outer-body panels (Fig. 10).

Multiple-Slide Machines

Multiple-slide machines are designed for automatic, complete production of a variety of small formed parts. Flat stock is fed into a straightener, into a feed mechanism, and then through one or more presses incorporated in the multiple-slide machine for operations such as piercing, notching, and bending—often in a progressive die. The feed mechanism then moves the metal into the multiple-slide forming area, where it is first severed by a cutoff mechanism to predetermined lengths. The piece is usually formed around a center post by four sets of tools mounted 90° apart around the forming post.

Finally, the part is stripped off the center post and dropped through a hole in the bed.

Small parts that are used in large quantities and require considerable forming are often produced on multiple-slide machines. In general, more severe forming can be done in the forming station of a multiple-slide machine than in a progressive die. The following example illustrates high-carbon steel parts produced in a multiple-slide forming machine. Additional information on multiple-slide forming is available in the article "Forming of Steel Strip in Multiple-Slide Machines" in this Volume.

Example 4: Blanking and Bending a Bracket in a Multiple-Slide Machine. The mounting bracket shown in Fig. 11 was embossed, pierced, and notched in a progressive die in the press station of a multiple-slide machine. It was then cut off and bent. Production was at the rate of 100 pieces per minute. The work metal was coiled cold-rolled 1050 steel strip, 0.91 mm (0.036 in.) thick by 32 mm (1¼ in.) wide.

In the press station, two holes were pierced, two weld projections were embossed, one hole was flanged, the outline was notched, and the top surface was flanged to improve stiffness. The strip then moved to the forming station, where the part was cut off and bent to the final form shown in Fig. 11. All slides in this station moved in a horizontal plane. A front slide tool held the part against the center post and preformed the flanges. A cam action on the right and left slides formed the flanges against the post and the rear slide tool.

Lubrication

Lubrication in Bending. Lubrication is less important for most bending operations than for other types of forming. In many bending operations, no lubricant is used; in others, the mill oil remaining on the stock or a light mineral oil applied before forming is sufficient to prevent galling. Exceptions to this practice are hole flanging, compression and stretch flanging, and severe bending in which wiping, ironing, or drawing of the work metal may call for more effective lubrication. An application in which lubricants were used because of the nature of the bending operations is in example 5, in which accuracy was important and sharp-radius bends were made parallel with the direction of rolling; the workpiece was lubricated with mineral oil before bending. Another example may be lubricant applied to a workpiece before hemming or flattening a bend to 180°.

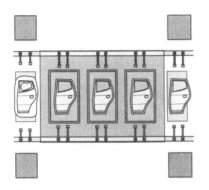

Fig. 7 Classic gripper system in a transfer press

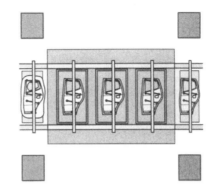

Fig. 8 Cross-bar mode in a transfer press

Fig. 9 Classic in-line arrangement of automatic blanking and forming machine

(a)

(b)

Fig. 10 Example of automotive panels formed in a transfer press. (a) Press space in transfer press. (b) Door elements on the outfeed conveyor

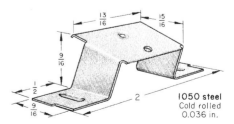

Fig. 11 Mounting bracket that was blanked and bent in a multiple-slide machine. Dimensions given in inches

Example 5: Bending Parallel with Rolling Direction. Conflicting demands called for a choice in the orientation of the blank for the can-opener blade shown in Fig. 12. Because an orientation that would favor the bends would have meant a cross-grain surface on the cutting edge, with consequent poor wearing quality, the blank was oriented so that the grain favored the cutting edge, and the bends were made nearly parallel with the direction of the rolling.

To ensure that the stock would withstand the three sharp-radius 90° bends, the steel specification called for stock that would withstand a 180° bend both parallel with and across the direction of the rolling. Steel that could be oil hardened was specified in order to limit distortion during subsequent heat treatment. A modified 1023 steel (with 0.85 to 1.15% Mn) met all of the requirements. The stock was 50 mm (2 in.) wide cold-rolled strip 1.1 mm (0.045 in.) thick. A No. 2 finish was specified to minimize the amount of polishing or burnishing before plating.

The blade was made in a 12-station progressive die, run in a 670 kN (75 tonf) mechanical press. Operations performed in the die included:

- Pierce four holes (one in the scrap area served as a pilot hole)
- Notch outline of part
- Coin cutting edge
- Emboss center hole
- Form bends
- Cut off

The die was made of D2 tool steel, except for the bending and cutoff station, for which the die material was C-5 carbide at 71 HRC. The die life per grind was 300,000 pieces. Mineral oil was the lubricant.

The production rate was 4500 pieces per hour. Annual production was 8 million pieces, in 700,000-piece lots. After forming, the part was oil hardened, barrel finished in oil and sawdust, and bright nickel plated.

Lubrication in Press Forming. The type of lubricant used usually has little effect on the grade of steel selected to form a given part. The main purposes of a lubricant are to prevent die galling and die wear and to reduce the friction over critical areas, thus allowing proper flow of metal and possibly a reduction in severity class. The selection of the optimal lubricant for a given part is a complex problem that depends on part geometry and the forming process used.

In progressive dies, a light oil sprayed on the strip as it enters the die is often enough to keep the stock lubricated through all stages. The oil is generally applied to the stock between the feeding device and the die. Applying oil to the stock ahead of the feeder may cause variation in the feed length, depending on the type of feeder.

For some applications, residual mill oil or the residue from emulsion cleaning provides enough lubrication for forming. When this is not adequate, a spray or mist lubricant can be applied to the work metal as it enters the die. More information on lubricants is available in the article "Selection and Use of Lubricants in Forming of Sheet Metal" in this Volume.

Sheet Steels

Steel sheet selection should be based on an understanding of available grades of sheet and forming requirements. Other factors that should be considered when selecting a material for forming into a particular part include:

- Purpose of the part and its service requirements
- Thickness of the sheet metal and allowable tolerances
- Size and shape of blanks for the forming operation
- Equipment available for forming
- Quantities required
- Available handling equipment for sheets or coils
- Local availability of sheet products
- Surface characteristics of the steel sheet
- Special finishes or coatings for appearance or for corrosion resistance
- Aging propensity and its relation to time before use
- Strength of the steel sheets as-delivered
- Strength requirements in the formed part

Because these factors may be interdependent, technical representatives of suppliers or the steel producer should be consulted, especially if steel selection is for either large quantities or special requirements. Some parts require specialized low-carbon steel that has been processed to enhance a given mechanical property and/or reduce production costs and forming problems.

Low-carbon steels, coated and uncoated, are generally supplied as commercial-quality, drawing-quality, and drawing-quality special-killed grades. Typical mechanical properties of low-carbon sheet steels are given in Table 1. Some steel mills also offer specialized grades, such as interstitial-free deep-drawing steels and enameling steels. Grade designations of common formable grades include:

- Commercial-quality (CQ) steel
- Drawing steel (DS)
- Extra-deep-drawing steel (EDDS)
- Extra-deep-drawing steel plus (EDDS+)
- Structural steel (SS)
- High-strength low-alloy (HSLA) steel
- Dent-resistant (DR) steel
- Bake-hardenable (BH) steel
- Inclusion-shape-controlled steel

General forming characteristics of the more commonly used formable grades are:

- *Commercial-quality (CQ) steel:* Available in hot-rolled, cold-rolled, and coated grades. The least expensive grade of sheet steel. Subject to aging (mechanical properties may deteriorate with time). Not intended for difficult-to-form shapes
- *Drawing steel (DS):* Available in hot-rolled, cold-rolled, and coated grades. Exhibits better ductility than CQ-grade steels, but has low plastic-strain ratio (*r*) values. Subject to aging (mechanical properties may deteriorate with time). Has excellent base metal surface quality
- *Drawing-quality special-killed (DQSK) steel:* Available in hot-rolled, cold-rolled, and coated grades, with good forming capabilities. Not subject to aging (mechanical properties do not change with time)

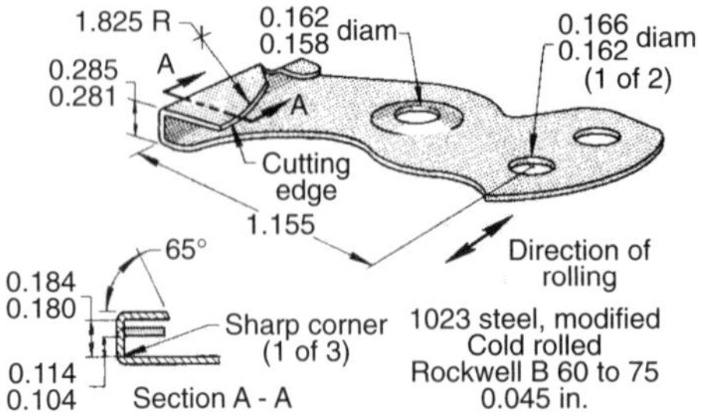

Fig. 12 Can-opener blade that was bent parallel with the direction of rolling in order to promote long service life of the cutting edge. Dimensions given in inches

Table 1 Typical mechanical properties of low-carbon sheet steels

Quality level	Tensile strength MPa	ksi	Yield strength MPa	ksi	Elongation in 50 mm (2 in.), %	Plastic-strain ratio, *r*	Strain-hardening exponent, *n*	Hardness, HRB
Hot rolled								
Commercial quality	358	52	234	34	35	1.0	0.18	58
Drawing quality	345	50	220	32	39	1.0	0.19	52
Drawing quality, aluminum killed	358	52	234	34	38	1.0	0.19	54
Cold rolled, box annealed								
Commercial quality	331	48	234	34	36	1.2	0.20	50
Drawing quality	317	46	207	30	40	1.2	0.21	42
Drawing quality, aluminum killed	303	44	193	28	42	1.5	0.22	42
Interstitial-free	310	45	179	26	45	2.0	0.23	44

- *Interstitial-free steels:* Available in cold-rolled and coated grades, with excellent forming capabilities for deep drawing (as EDDS or EDDS+ grades)
- *Enameling steels:* Available in cold-rolled grades. Various types of processing are used to make a product that is satisfactory for porcelain enameling. All grades have good forming capabilities.
- *Structural steels:* When higher strength is required, structural-quality sheet, also called physical-quality sheet, can be specified, although at some sacrifice in ductility.
- *Higher-strength steel sheets:* Available in hot-rolled, cold-rolled, and coated grades. Various types of processing are used to obtain the desired strength levels. In general, the formability of these grades decreases as yield strength increases. Springback may be a problem at lower sheet thicknesses.
- *Inclusion-shape-controlled steels:* Cold formability has been substantially improved by inclusion-shape-controlled steels, which enables steel to be formed to nearly the same extent in both the longitudinal and transverse directions. Any grade produced with inclusion shape control can be more severely formed than a grade of the same strength level produced without inclusion shape control. Inclusion-shape-controlled steels are responsible for the moderately good formability of the higher-strength HSLA steels, such as the grades having 550 MPa (80 ksi) yield strengths.

Low-carbon steel sheet and strip are used for low-cost production of various products having good dimensional tolerance and appearance. The steels used for these products are supplied over a wide range of chemical compositions; however, the vast majority are unalloyed, low-carbon steels selected for stamping applications, such as automobile bodies and appliances. For these major applications, typical compositions are 0.03 to 0.10% C, 0.15 to 0.50% Mn, 0.035% P (max), and 0.04% S (max).

In the past, rimmed (or capped) ingot cast steel has been used because of its lower price. More recently, however, rimmed steels have been largely replaced by killed steels produced by the continuous casting process. Continuous casting is inherently suited to the production of killed steels, but killed steels are also produced by ingot metallurgy. Regardless of the method of casting or manufacture, killed steels are preferred because they have better formability and are not subject to aging or strain aging (that is, mechanical properties do not change with time).

The width differentiation between sheet and strip made of plain carbon steel depends on the rolling process. It should be noted that both sheet and strip can be purchased as either cut lengths or coils. The standard dimensional tolerances for plain carbon steel strip are more restrictive than those for sheet. Standard size ranges of plain carbon steel sheet and strip are given in Table 2. Typical characteristics of the various qualities of these products are listed in Tables 3(a) and 3(b).

Modified Low-Carbon Steel Sheet and Strip. In addition to the low-carbon steel sheet and strip products, there are numerous additional products available that are designed to satisfy specific customer requirements. These products are often made with low-carbon steels having chemical compositions slightly modified from those discussed earlier. For example, in structural-quality (SQ) steels, alloying additions of manganese and phosphorus are used to increase strength by substitutional solid-solution strengthening: approximately 3 MPa (0.4 ksi) per 0.1% Mn, and 7 MPa (1 ksi) per 0.0 1% P. Hot-rolled SQ steels contain from 0.90 to 1.35% max Mn and 0.035% max P. Cold-rolled SQ steels contain 0.60 to 0.90% max Mn and 0.035 to 0.20% P. Carbon contents for SQ steels are generally 0.20 to 0.25%.

Interstitial-Free Steels. In interstitial-free (IF) steels, which are also referred to as extra-deep-drawing quality, the elimination of interstitials (carbon and nitrogen) is accomplished by adding sufficient amounts of carbide/nitride-forming elements (generally titanium and/or niobium) to tie up carbon and nitrogen completely, the levels of which can be reduced to less than 50 ppm by modern steelmaking/casting practices, including vacuum degassing.

Steels with very low interstitial content exhibit excellent formability with low yield strength (138 to 165 MPa, or 20 to 24 ksi), high elongation (41 to 45%), and good deep drawability. With the addition of carbonitride-forming elements, the deep drawability and the nonaging properties are further improved.

Bake-hardening steels are characterized by their ability to exhibit an increase in yield strength due to carbon strain aging during paint-baking operations at moderate temperature (125 to 180 °C, or 260 to 355 °F). Bake hardening has little effect on tensile strength. Bake-hardening steels are finding increased use in automotive outer-body applications (hoods, doors, fenders) to achieve an improvement in dent resistance and, in some cases, a sheet thickness reduction as well.

The bake-hardening behavior is dependent on steel chemistry and processing, in addition to the amount of forming strain and paint-baking conditions (temperature and time). Steels that exhibit bake-hardening behavior include plain low-carbon steels (continuously annealed or batch annealed), IF steels (continuously annealed), and dual-phase steels (continuously annealed).

Automotive specifications for BH steels can be categorized according to those that specify a minimum yield strength level or a minimum bake-hardening increment, in the formed (strained) plus baked condition. The conventional test for determining bake-hardenability characteristics involves a 2% tensile prestrain, followed by baking at 175 ± 5 °C (345 ± 10 °F). The resulting increase in yield strength measures the bake hardenability of the material.

While all the specifications call for a minimum yield strength level in the as-received (that is, prior to forming) condition, some also require a minimum yield strength after baking the as-received material in the absence of any tensile prestrain. The as-received yield strength is in the range of 210 to 310 MPa (30 to 45 ksi) (compared with approximately 175 MPa, or 25 ksi, for DQSK), while the final yield strength, that is, after 2% prestrain plus bake, ranges between 280 and 365 MPa (40 to 53 ksi)

Table 2 Standard sizes of low-carbon sheet and strip

	Thickness		Width			Specification symbol (ASTM No.)	
Product	mm	in.	mm	in.	Other limitations	Metric units	English units
Hot-rolled sheet	1.2–6.0	0.045–0.230 incl	> 300–1200 incl	> 12–48 incl	Coils and cut lengths	A 569M, A 621M, or A 622M	A 569, A 621, or A 622
	1.2–4.5	0.45–0.180 incl	>1200	>48	Coils and cut lengths	A 569M, A 621M, or A 622M	A 569, A 621, or A 622
	6.0–12.5	0.230–0.500 incl	> 300–1200 incl	> 12–48 incl	Coils only	A 635M	A 635
	4.5–12.5	0.180–0.500 incl	>1200–1800 incl	>48–72 incl	Coils only	A 635M	A 635
Hot-rolled strip	1.2–5.0	0.45–0.203 incl	≤200	≤8	Coils and cut lengths	A 569M, A 621M, or A 622M	A 569, A 621, or A 622
	1.2–6.0	0.045–0.230 incl	> 200–300 incl	> 8–12 incl	Coils and cut lengths	A 569M, A 621M, or A 622M	A 569, A 621, or A 622
	6.0–12.5	0.230–0.500 incl	> 200–300 incl	> 8–12 incl	Coils only	A 635M	A 635M
Cold-rolled sheet	0.35–2.0	0.014–0.082 incl	> 50–300 incl	> 2–12 incl	(a)	A 366M, A 619M, or A 620M	A 366, A 619, or A 620
	≥0.35	>0.014	> 300	>12	(b)	A 366M, A 619M, or A 620M	A 366, A 619, or A 620
Cold-rolled strip	≤6.0	≤0.230	> 12–600 incl	> 0.50–23.9 incl	(c)	A 109M	A 109

Note: Incl, inclusive. (a) Cold-rolled sheet, coils, and cut lengths, slit from wider coils with cut edge (only) thickness 0.356–2.08 mm (0.014–0.082 in.) and 0.25% C (max) by cost analysis. (b) When no special edge or finish (other than matte, commercial bright, or luster) is required and/or single-strand rolling of widths under 610 mm (24 in.) is not required. (c) Width 51–305 mm (2–12 in.) with thickness of 0.356–2.08 mm (0.014–0.082 in.) are classified as sheet when slit from wider coils, have a cut edge only, and contain 0.25% C (max) by cost analysis.

(compared with approximately 225 MPa, or 33 ksi, for DQSK).

The HSLA steels are generally formed at room temperature using conventional equipment. Cold forming should not be done at temperatures below 10 °C (50 °F). As a class, high-strength steels are inherently less formable than low-carbon steels because of their greater strength and lower ductility. This reduces their ability to distribute strain. The greater strength makes it necessary to use greater forming pressure and to allow for more springback compared to low-carbon steels. However, high-strength steels have good formability, and straight bends can be made to relatively tight bend radii, especially with the grades having lower strengths and greater ductility. Further, high-strength steels can be stamped to relatively severe shapes, such as automotive bumper facings, wheel spiders, and engine-mounting brackets.

High-strength low-alloy steels can be hot formed. However, hot forming usually alters mechanical properties, and a particular problem that arises in many applications is that some of the more recent thermomechanical processing techniques (such as controlled rolling), used for plates in particular, are not suitable where hot forming is used during fabrication. This problem can be circumvented by the use of a rolling finishing temperature that coincides with the hot forming temperature (900 to 930 °C, or 1650 to 1700 °F). Subsequent hot forming therefore simply repeats this operation, and deterioration in properties is then small or even absent, provided that grain growth does not occur. Producers should be consulted for recommendations of specific hot forming temperatures and for comments on their effects on mechanical properties.

High-Carbon Steels. High-carbon steel strip (including spring steel and tool steel) is blanked, pierced, and formed to make a variety of parts. The practices, precautions, presses, and tools used in making high-carbon steel parts are comparable to those used for producing similar parts of low-carbon steel. The key differences in blanking, piercing, and forming of high-carbon steels compared to low-carbon steels include the following:

- Higher yield strength and lower ductility results in less bendability and, therefore, requires greater bend radii than plain carbon steels.
- Greater allowance is required for springback.
- More force is required for high-carbon steel because of its higher strength.
- Greater clearance between the punch and die is necessary in blanking and piercing.
- A more wear-resistant tool material may be required before acceptable tool life can be obtained.

In blanking and piercing of high-carbon steels, the most important difference with that of low-carbon steel is the need for greater clearance between punch and die (see the article "Blanking of Low-Carbon Steel" in this Volume). Mold forming of high-carbon steel in the quenched-and-tempered (pretempered) condition (usually 47 to 55 HRC) is common practice. The severity of forming that can be done without cracking of the work metal depends mainly on thickness. When metal thickness is no more than approximately 0.38 mm (0.015 in.),

Table 3(a) Summary of available types of hot-rolled and cold-rolled plain carbon steel sheet and strip

			Surface finish			
	Applicable basic specification number	SAE-AISE grade designation	Temper-rolled; for exposed parts(a)		Annealed last; for unexposed parts(a)	
Quality or temper			Description	Symbol	Description	Symbol
Hot-rolled sheet						
Commercial quality	A 569, A 635	1008–1012	As-rolled (black)	A	As-rolled (black)	A
			Pickled—dry	P	Pickled—dry	P
			Pickled and oiled	O	Pickled and oiled	O
Drawing quality	A 621	1006–1008	As-rolled (black)	A	As-rolled (black)	A
			Pickled—dry	P	Pickled—dry	P
			Pickled and oiled	O	Pickled and oiled	O
Drawing quality, special killed	A622	1006–1008	As-rolled (black)	A	As-rolled (black)	A
			Pickled—dry	P	Pickled—dry	P
			Pickled and oiled	O	Pickled and oiled	O
Hot-rolled strip						
Commercial quality	A 569	1008–1012	As-rolled (black)	A	As-rolled (black)	A
			Pickled—dry	P	Pickled—dry	P
			Pickled and oiled	O	Pickled and oiled	O
Drawing quality	A 621	1006–1008	As-rolled (black)	A	As-rolled (black)	A
			Pickled—dry	P	Pickled—dry	P
			Pickled and oiled	O	Pickled and oiled	O
Drawing quality, special killed	A622	1006–1008	As-rolled (black)	A	As-rolled (black)	A
			Pickled—dry	P	Pickled—dry	P
			Pickled and oiled	O	Pickled and oiled	O
Cold-rolled sheet						
Commercial quality	A 366	1008–1012	Matte	E	Matte	U
			Commercial bright	B	...	...
			Luster	L	...	...
Drawing quality	A 619	1006–1008	Matte	E	Matte	U
			Commercial bright	B	...	...
			Luster	L	...	...
Drawing quality, special killed	A620	1006–1008	Matte	E	Matte	U
			Commercial bright	B	...	...
			Luster	L	...	...
Cold-rolled strip						
Temper description numbers 1, 2, 3, 4, 5	A 109	(b)	Matte	1	Matte	1
			Regular bright	2	Regular bright	2
			Best bright	3	Best bright	3

(a) See Table 3(b) for a description of the surface finish listed. (b) Produced in five tempers with specific hardness and bend test limits; composition subordinate to mechanical properties

Table 3(b) Selection and specification of surface condition for plain carbon steel sheet

Specification symbol	Description of surface	Surface described applicable to
U(a)	Surface finish as normally used for unexposed automotive parts. Matte appearance. Normally annealed last	Cold-rolled sheet
E(b)	Surface finish as normally used for exposed automotive parts that require a good painted surface. Free from strain markings and fluting. Matte appearance. Temper rolled	Cold-rolled sheet
B	Same as above, except commercial bright appearance	Cold-rolled sheet
L	Same as above, except luster appearance	Cold-rolled sheet
1	No. 1 or dull finish (no luster). Especially suitable for lacquer or paint adhesion. Facilitates drawing by reducing the contact friction between the die and the metal	Cold-rolled strip
2	No. 2 or regular bright finish (moderately smooth). Suitable for many applications but not generally applicable for parts to be plated, unless polished and buffed	Cold-rolled strip
3	No. 3 or best bright finish (relatively high luster). Particularly suitable for parts to be plated	Cold-rolled strip
A	As-rolled or black (oxide or scale not removed)	Hot-rolled sheet and strip
P	Pickled (scale removed), not oiled	Hot-rolled sheet and strip
O	Same as above, except oiled	Hot-rolled sheet and strip

(a) U, unexposed; also designated as class 2, cold-rolled sheet. (b) E, exposed; also designated as class 1, cold-rolled sheet

Table 4   Typical effects of carbon content and sheet thickness on minimum bend radius of annealed steels

Thickness of sheet		Minimum bend radius					
		Steels 1020 to 1025		Steels 4130 and 8630		Steels 1070 and 1095	
mm	in.	mm	in.	mm	in.	mm	in.
0.41	0.016	0.8	0.03	0.8	0.03	1.5	0.06
0.51	0.020	0.8	0.03	0.8	0.03	1.5	0.06
0.64	0.025	0.8	0.03	0.8	0.03	1.5	0.06
0.76	0.030	0.8	0.03	1.5	0.06	2.3	0.09
0.89	0.035	1.5	0.06	1.5	0.06	2.3	0.09
1.07	0.042	1.5	0.06	1.5	0.06	3.3	0.13
1.27	0.050	1.5	0.06	2.3	0.09	3.3	0.13
1.57	0.062	1.5	0.06	2.3	0.09	4.1	0.16
1.98	0.078	2.3	0.09	3.3	0.13	4.8	0.19
2.36	0.093	2.3	0.09	4.1	0.16	6.4	0.25
2.77	0.109	3.3	0.13	4.1	0.16	7.9	0.31
3.18	0.125	3.3	0.13	4.8	0.19	7.9	0.31
3.96	0.156	4.1	0.16	6.4	0.25	9.7	0.38
4.75	0.187	4.8	0.19	7.9	0.31	12.7	0.50

Source: Ref 2

Table 5   Typical values for strength coefficient, K, and strain-hardening exponent, n, at room temperature

Material	K		n
	MPa	ksi	
Aluminum			
1100-O	180	26	0.20
2024-T4	690	100	0.16
5052-O	210	30	0.13
6061-O	205	30	0.20
6061-T6	410	59	0.05
7075-O	400	58	0.17
Brass			
60-39-1 Pb, annealed	800	116	0.33
70-30, annealed	895	130	0.49
85-15, cold rolled	580	84	0.34
Bronze (phosphor), annealed	720	104	0.46
Cobalt-base alloy, heat treated	2070	300	0.50
Copper, annealed	315	46	0.54
Molybdenum, annealed	725	105	0.13
Steel			
Low-carbon annealed	530	77	0.26
1045, hot rolled	965	140	0.14
1112, annealed	760	110	0.19
1112, cold rolled	760	110	0.08
4135, annealed	1015	147	0.17
4135, cold rolled	1100	160	0.14
4340, annealed	640	93	0.15
17-4 P-H, annealed	1200	174	0.05
52100, annealed	1450	210	0.07
302 stainless, annealed	1300	188	0.30
304 stainless, annealed	1275	185	0.45
410 stainless, annealed	960	139	0.10

Source: Ref 3

it is possible to make relatively severe bends without fracturing the work metal. However, as metal thickness increases, the amount of forming that can be done on pretempered steel decreases rapidly.

Moderately severe forming can be done on cold-rolled stock that has not been quenched and tempered and on high-carbon steel that has been spheroidize-annealed. Such materials are usually hardened and tempered after forming to improve spring properties. Table 4 shows the effect of the carbon content of steel on bendability and demonstrates the major importance of sheet thickness.

Formability of Steels

Formability is the ability of sheet metal to be deformed into a desired shape while maintaining structural integrity without tearing, buckling and wrinkling, excessive thinning, and so on. Formability depends greatly on the nature of the forming operation, and various criteria of formability have been developed, depending on the nature of the forming operation and the applied forces (which may be tensile, compressive, bending, shearing, or various combinations of these).

Forming applications may range from simple bending to stamping and deep drawing of complex shapes. Appreciable stretching of the metal also may occur in some operations. The emphasis of this article is on the pressworking of carbon steels, where presses are employed with sets of tools to produce sheet metal products or stampings. This includes press forming, press bending, and press-brake forming. However, this section also addresses formability of sheet steel in drawing and stretch-forming operations, with additional information given in the articles "Deep Drawing" and "Stretch Forming" in this Volume.

In general, one key aspect of formability is the distribution of strain during forming and the evaluation of mechanical properties in terms

of the following (determined from tensile testing):

- Strain-hardening exponent (n)
- Strain-rate sensitivity factor (m)
- Plastic-strain ratio, or normal anisotropy factor (r, or r_m)

These three properties are used to evaluate the behavior of strain distribution under various forming conditions, and their relevance depends on the nature of the forming operation. Most sheet forming operations usually involve stretching and some shallow drawing, in which case the product of the strain-hardening exponent (n) and the normal anisotropy (r) of the sheet has been shown to be a significant parameter. In simple stretching operations, however, a forming limit may be determined by the uniform elongation of the metal as it is related to the strain-hardening exponent, n. Strain-hardening exponents are listed in Tables 5 through 7 for various steels, along with strength coefficients (K) for the power-law relation $\sigma = K\varepsilon^n$.

The r-value is a measure of the ability of the material to resist thinning while undergoing width reduction. A high r-value is the measure of a material with good deep-drawing characteristics. High values of n and m lead to good formability in stretching operations, because they promote uniform strain distribution, but they have little effect on drawability. Table 8 lists m-values for various steels and nonferrous alloys. These and other factors influencing formability are covered in more detail in the article "Formability Testing of Sheet Metals" in this Volume.

Bendability

Bendability commonly is defined as the minimum bend radius, R (measured to the inner surface of the bent part), to which a sheet metal can be bent without cracking of its outer surface. It is usually given as the minimum R/T ratio, where T is the sheet thickness. The minimum R/T

ratio also has been shown to be related to the tensile reduction of the area (RA) of the sheet metal, as obtained from a tension test specimen and cut in the direction of bending.* This relationship is:

$$\text{Minimum } R/T = (50/\text{RA}) - 1$$

This equation has been derived by equating the true strain at which the outer fiber in bending begins to crack to the true fracture strain of the sheet specimen in simple tension, and it is in reasonably good agreement with experimental results (see the section "Bendability" in the article "Bending of Sheet Metal" in this Volume). A curve-fitting modification is also made by increasing the numerator in the equation from 50 to 60 (Fig. 13), which indicates that a sheet with a tensile reduction of area of 60% can be bent completely over itself (hemming) without cracking. However, bendability depends not only on the property of the metal but also on the state of stress (geometric factors) and edge quality, as described in the following sections.

Factors Affecting Bendability. Besides ductility of the material, minimum bend radii are affected by angle of bend, length of bend, condition of the cut edge perpendicular to the bend line, and orientation of bend with respect to

---

*Because of planar anisotropy of cold rolled sheets (with higher ductility in the rolling direction than in the transverse direction), it is important to prepare the specimens accordingly.

Table 6 Strength coefficient, K, and strain-hardening exponent, n, values for flow stress-strain relation, $\bar{\sigma} = K(\bar{\varepsilon})^n$ of various steels

Steel	Composition(a), %												Material history(b)	Temperature		Strain rate l/s	Strain range	K, ksi	n
	C	Mn	P	S	Si	N	Al	V	Ni	Cr	Mo	W		°C	°F				
Armco iron	0.02	0.03	0.021	0.010	Tr	...	...	...	...	...	...	...	A	20	68	(c)	0.1–0.7	88.2	0.25
1006	0.06	0.29	0.02	0.042	Tr	0.004	...	...	...	...	...	...	A	20	68	(c)	0.1–0.7	89.6	0.31
1008	0.08	0.36	0.023	0.031	0.06	0.007	...	...	...	...	...	...	A	20	68	(c)	0.1–0.7	95.3	0.24
...	0.07	0.28	...	...	0.27	...	...	...	...	...	...	...	A	20	68	(c)	0.1–0.7	95.3	0.17
1010	0.13	0.31	0.010	0.022	0.23	0.004	...	...	...	...	...	...	A	20	68	(c)	0.1–0.7	103.8	0.22
1015	0.15	0.40	0.01	0.016	Tr	...	...	...	...	...	...	...	F, A	0	32	30	0.2–0.7	91.4	0.116
1015	0.15	0.40	0.01	0.016	Tr	...	...	...	...	...	...	...	F, A	200	390	30	0.2–0.6	73.7	0.140
1015(d)	0.15	0.40	0.045	0.045	0.25	...	...	...	...	...	...	...	A	20	68	1.6	...	113.8	0.10
1015(d)	0.15	0.40	0.045	0.045	0.25	...	...	...	...	...	...	...	A	300	572	1.6	...	115.2	0.11
1020	0.22	0.44	0.017	0.043	Tr	0.005	...	...	...	...	...	...	A	20	68	(c)	0.1–0.7	108.1	0.20
1035	0.36	0.69	0.025	0.032	0.27	0.004	...	...	...	...	...	...	A	20	68	(c)	0.1–0.7	130.8	0.17
...	...	...	...	...	...	...	...	...	...	...	...	...	A	20	68	1.6	...	139.4	0.11
...	...	...	...	...	...	...	...	...	...	...	...	...	A	300	572	1.6	...	122.3	0.16
1045(d)	0.45	0.65	0.045	0.045	0.25	...	...	...	...	...	...	...	A	20	68	1.6	...	147.9	0.11
...	...	...	...	...	...	...	...	...	...	...	...	...	A	20	68	1.5	...	137.9	0.14
...	...	...	...	...	...	...	...	...	...	...	...	...	A	300	572	1.6	...	126.6	0.15
1050(e)	0.51	0.55	0.016	0.041	0.28	0.0062	0.03	...	...	...	...	...	A	20	68	(c)	0.1–0.7	140.8	0.16
1060	...	...	...	...	...	...	...	...	...	...	...	...	A	20	68	1.6	...	163.5	0.09
...	...	...	...	...	...	...	...	...	...	...	...	...	A	20	68	1.5	...	157.8	0.12
2817(e)	0.19	0.55	0.057	0.023	0.26	0.016	...	...	...	...	...	...	A	20	68	(c)	0.2–1.0	111.2	0.170
5115	0.14	0.53	0.028	0.027	0.37	...	...	...	...	0.71	...	...	A	20	68	(c)	0.1–0.7	115.2	0.18
...	...	...	...	...	...	...	...	...	...	...	...	...	A	20	68	1.6	...	123.7	0.09
...	...	...	...	...	...	...	...	...	...	...	...	...	A	300	572	1.6	...	102.4	0.15
5120(e)	0.18	1.13	0.019	0.023	0.27	...	...	...	...	0.86	...	...	A	20	68	(c)	0.1–0.7	126.6	0.18
...	...	...	...	...	...	...	...	...	...	...	...	...	A	20	68	1.6	...	116.6	0.09
...	...	...	...	...	...	...	...	...	...	...	...	...	A	300	572	1.6	...	98.1	0.16
5140	0.41	0.67	0.04	0.019	0.35	...	...	...	...	1.07	...	...	A	20	68	(c)	0.1–0.7	125.1	0.15
...	...	...	...	...	...	...	...	...	...	...	...	...	A	20	68	1.6	...	133.7	0.09
...	...	...	...	...	...	...	...	...	...	...	...	...	A	300	572	1.6	...	112.3	0.12
D2 tool steel(e)	1.60	0.45	...	...	0.24	...	...	0.46	...	11.70	0.75	0.59	A	20	68	(c)	0.2–1.0	191.0	0.157
L6 tool steel	0.56	...	...	...	...	...	...	0.14	1.60	1.21	0.47	...	A	20	68	(c)	0.2–1.0	170.2	0.128
W1-1.0C special	1.05	0.21	...	...	0.16	...	...	...	...	...	...	...	A	20	68	(c)	0.2–1.0	135.6	0.179
302 SS	0.08	1.06	0.037	0.005	0.49	...	...	...	9.16	18.37	...	...	HR, A	0	32	10	0.25–0.7	185.7	0.295
...	...	...	...	...	...	...	...	...	...	...	...	...	HR, A	200	390	30	0.25–0.7	120.8	0.278
...	...	...	...	...	...	...	...	...	...	...	...	...	HR, A	400	750	30	0.25–0.7	92.7	0.279
302 SS	0.053	1.08	0.027	0.015	0.27	...	...	...	10.2	17.8	...	...	A	20	68	(c)	0.1–0.7	210.5	0.6
304 SS(e)	0.030	1.05	0.023	0.014	0.47	...	...	...	10.6	18.7	...	...	A	20	68	(c)	0.1–0.7	210.5	0.6
316 SS	0.055	0.92	0.030	0.008	0.49	...	...	...	12.9	18.1	2.05	...	A	20	68	(c)	0.1–0.7	182.0	0.59
410 SS	0.093	0.31	0.026	0.012	0.33	...	...	...	...	13.8	...	...	A	20	68	(c)	0.1–0.7	119.4	0.2
...	...	...	...	...	...	...	...	...	...	...	...	...	A	20	68	1.6	...	137.9	0.09
431 SS	0.23	0.38	0.020	0.006	0.42	...	...	...	1.72	16.32	...	...	A	20	68	(c)	0.1–0.7	189.1	0.11

(a) Tr, trace. (b) A, annealed; F, forged; HR, hot rolled, (c) Low-speed testing machine; no specific rate given. (d) Composition given is nominal (analysis not given in original reference). (e) Approximate composition. Source: Ref 4

Table 7 Monotonic stress-strain properties of selected steels

Alloy	Condition	Monotonic properties							
		Elastic modulus, 10 psi	Yield strength, ksi	Ultimate tensile strength, ksi	Strength coefficient, K, ksi	Strain-hardening exponent, n	Reduction in area, %	σ_f, ksi	ε_f
A136	As-rec'd	30	46.5	80.6	144	0.21	67	143.6	1.06
A136	150 HB	30	46.0	81.9	...	0.21	69	145.0	1.19
SAE950X	As-rec'd 137 HB	30	62.6	75.8	94.9	0.11	54	...	...
SAE950X	As-rec'd 146 HB	30	56.7	74.0	116.0	0.15	74	141.8	1.34
SAE980X	Prestrained 225 HB	28	83.5	100.8	143.9	0.13	68	176.8	1.15
1006	Hot rolled 85 HB	30	36.0	46.1	60.0	0.14	73	...	...
1020	Annealed 108 HB	27	36.8	56.9	57.9	0.07	64	95.9	1.02
1045	225 HB	29	74.8	108.9	151.8	0.12	44	144.7	...
1045	Q&T 390 HB	29	184.8	194.8	...	0.04	59	269.8	0.89
1045	Q&T 500 HB	29	250.6	283.7	341.0	0.04	38	334.4	...
1045	Q&T 705 HB	29	264.7	299.8	...	0.19	2	309.6	0.02
10B21	Q&T 320 HB	29	144.9	152.0	187.7	0.05	67	217.4	1.13
1080	Q&T 421 HB	30	141.8	195.6	323.0	0.15	32	238.6	...
4340	Q&T 350 HB	29	170.8	179.8	229.2	0.07	57	239.7	0.84
4340	Q&T 410 HB	30	198.8	212.8	...	...	38	225.8	0.48
5160	Q&T 440 HB	30	215.7	230.0	281.4	0.05	39	280.0	0.51
8630	Q&T 254 HB	30	102.8	113.9	153.9	0.08	16	121.8	0.17

σ_f, fracture stress; ε_f, strain to fracture

Table 8 Approximate range of parameters in equation for strain-rate sensitivity, $\sigma = C\dot{\varepsilon}^m$

Material	Temperature		C		
	°C	°F	MPa	ksi	m(a)
Aluminum	200–500	390–930	82–14	12–2	0.07–0.23
Aluminum alloys	200–500	390–930	310–35	45–5	0–0.20
Copper	300–900	570–1650	240–20	35–3	0.06–0.17
Copper alloys (brasses)	200–800	390–1470	415–14	60–2	0.02–0.3
Lead	100–300	212–570	11–2	1.6–0.3	0.1–0.2
Magnesium	200–400	390–750	140–14	20–2	0.07–0.43
Steel					
Low-carbon	900–1200	1650–2190	165–48	24–7	0.08–0.22
Medium-carbon	900–1200	1650–2190	160–48	23–7	0.07–0.24
Stainless	600–1200	1110–2190	415–35	60–5	0.02–0.4
Titanium	200–1000	390–1830	930–14	135–2	0.04–0.3
Titanium alloys	200–1000	390–1830	900–35	130–5	0.02–0.3
Ti-6Al-4V(b)	815–930	1500–1700	65–11	9.5–1.6	0.05–0.80
Zirconium	200–1000	390–1830	830–27	120–4	0.04–0.4

Note: C, Strength coefficient; $\dot{\varepsilon}$, true strain rate; m, strain-rate sensitivity exponent.

(a) As temperature increases, C decreases and m increases. As strain increases, C increases, and m may increase or decrease. Also, m may become negative within certain ranges of temperature and strain. (b) At a strain rate of 2×10^{-4} s^{-1}. Source: Ref 3

direction of rolling. These factors are described subsequently, followed by typical values of minimum bend radii reported for various types of steels.

Effect of Bend Angle and Length. Minimum bend radii are larger for a larger angle of bend. Parts in which the length of the bend (direction parallel with the bend axis) exceeds eight times metal thickness have a fairly constant minimum bend radius. When the bend length is less than eight times metal thickness, the bend radius generally must be greater.

Influence of Stress State. The effects of notch sensitivity, surface finish of the sheet metal and its lay, and rate of deformation are factors that should be taken into consideration. The beneficial effect of hydrostatic pressure has also been observed in bending. Although specimen size is limited, bending of metals with limited ductility has been carried out successfully in a pressurized chamber, and major increases in bendability have been observed. It has also been shown that as the sheet width-to-thickness ratio increases (thus changing the deformation condition from one of plane stress to plane strain), bendability decreases (Fig. 14).

Effect of Edge Condition. Edge condition of the sheet is also significant; the rougher the edge, the greater is the tendency for edge cracking. Bendability thus decreases. When bending low-carbon steel, the condition of the edge perpendicular to the bend axis has little effect on the minimum bend radius. Steels that are susceptible to work hardening or hardening by heating during gas or electric-arc cutting may crack during bending because of edge condition. For these steels, it is often necessary to remove burrs and hardened edge metal in the bend area to prevent fracture. Edges can be prepared for bending by grinding parallel with the surface of the sheet and removing sharp corners in the bend area by radiusing or chamfering.

If the burr side is on the inside of the bend, cracking is less likely to occur during bending. This is important on parts with small bend radii in comparison with the metal thickness and on parts with metal thickness greater than 1.6 mm ($^1/_{16}$ in.), because fractures are likely to start from stress-raising irregularities in the burr edge if it is on the outside of the bend.

Effect of Metal Thickness. Sheet thickness is, of course, a major factor, because minimum bend radii are generally expressed in multiples or fractions of the thickness of the work metal. On parts that require a minimum flange width or a minimum width of flat on the flange, stock thickness will limit both of these dimensions. If thickness is not critical in the design, the use of thinner stock can make the bending of small radii and narrow flanges feasible.

Orientation of Bend. With improved steelmaking practices (and the reduction of inclusions), the bend orientation (with respect to the rolling direction of the sheet, Fig. 15) is less critical than in the past. However, it is better to orient a part on the stock so that bends are made across the rolling direction. Sharper bends can be made across than can be made parallel with the rolling direction, without increasing the probability of cracking the work metal. When bends are to be made in two or more directions, the piece can sometimes be oriented in the layout such that none of the bends is parallel with the rolling direction.

In some applications, there is no practical way to avoid making bends parallel with the rolling direction (see example 5 in the section "Lubrication" in this article). A choice must be made in orientation to favor one or another consideration. For example, a blank can be oriented in a strip for economy and the least possible scrap. It can be oriented so that the grain direction will reinforce the metal that receives maximum stress in service. Alternatively, it can be oriented so there is no end-grain runout on a wear surface. In any of

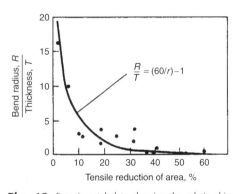

Fig. 13 Experimental data showing the relationship between bend radius to sheet thickness ratio and the tensile reduction of area for various sheet metals. Source: Ref 3

$\frac{R}{T} = (60/r) - 1$

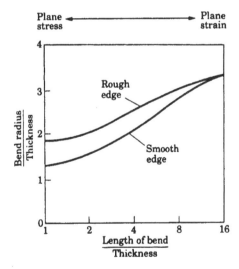

Fig. 14 The effect of length of bend (strip width) and sheared-edge condition on bend radius to sheet thickness ratio for 7075 aluminum. Source: Ref 3

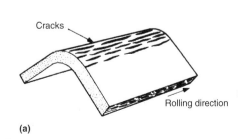

(a)

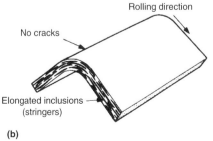

(b)

Fig. 15 Effect of orientation on bendability

these cases, orientation may not be optimal for bending.

Minimum Bend Radii of Steels. As noted, the minimum bend radii correlates with the ductility (elongation or reduction of area) determined from tensile test. Typical values of minimum bend radii are listed in Table 9 for various materials. Table 10 lists minimum bend radii for 1008 or 1010 hot- and cold-rolled steel sheet in various conditions. The quality of steel has a major influence on the minimum bend radius, especially for hot-rolled steel (Table 10). Table 11 indicates typical commercial-quality steels suitable for 90 and 180° bends.

The types of bends suitable for different hardness levels (tempers) of cold-rolled carbon steel strip are illustrated in Fig. 16. Steel in the higher tempers (low hardness and high ductility) can be bent 180° to a sharp radius without cracks or tears. Bend radii may be smaller for bends made across the rolling direction than for bends made parallel to it. Full-hard strip (84 HRB min) is not recommended for bending, except to large radii. Half-hard tempers (70 to 85 HRB) can be bent 90° over a radius equal to strip thickness, perpendicular to the rolling direction. Quarter-hard tempers (60 to 75 HRB) can be bent 90° over a radius equal to strip thickness, parallel to the rolling direction; it can also be bent 180° around a strip of the same thickness when the bend is perpendicular to the rolling direction.

Skin-rolled (65 HRB max) steel and dead-soft carbon steel (55 HRB max) steel can be bent 180° flat on itself in any direction.

Table 12 shows the effect of carbon content of some grades of carbon steel strip and sheet on bend radius in standard bend tests. Bend radii for higher-carbon steels and two low-alloy steels are given in Table 4. Low-carbon steel sheet also includes various types of higher-strength steels, which may include carbon steels with some manganese or microalloying. This includes various types of HSLA steels, such as those in SAE J410c. Bend radii are listed in Table 13. Examples of bend radii for cold-rolled and hot-rolled HSLA steel are given in Tables 14 and 15, respectively. Table 16 lists bend radii for a sheet steel with inclusion shape control.

Formability Diagrams

Sheet metal formability under biaxial stresses is evaluated with the construction of forming-limit diagrams (see the article "Formability Testing of Sheet Metals" in this Volume). By observing and recording strain along both the major and minor axes of sheet deformation, it is possible to more accurately identify the region of maximum deformation (which would eventually lead to thinning and tearing). Major strains are always positive, because the sheet forming operation always involves some stretching (hence positive strain) in at least one direction. Minor strains may be positive or negative (Fig. 17). The broken straight lines show the states of strain in Fig. 17. The pure shear line indicates that the tensile and compressive strains on the sheet are equal (therefore, the negative 45° slope). The simple tension line indicates a slope that relates to Poisson's ratio in plastic deformation (i.e., 0.5). The vertical line shows the plane-strain condition, where there is no width change (minor strain is zero) as the sheet is stretched over the punch. Finally, the 45° line on the right side of the diagram indicates equal biaxial tension, meaning the major and the minor strains are both tensile and equal.

Table 9 Minimum bend radii for various materials at room temperature

Material	Condition(a) Soft	Condition(a) Hard
Aluminum alloys	0	6*t*
Beryllium copper	0	4*t*
Brass, low-leaded	0	2*t*
Magnesium	5*t*	13*t*
Steels		
Austenitic stainless	0.5*t*	6*t*
Low-carbon, low-alloy, and HSLA(b)	0.5*t*	4*t*
Titanium	0.7*t*	3*t*
Titanium alloys	2.6*t*	4*t*

(a) *t*, sheet thickness. (b) HSLA, high-strength low-alloy

Table 10 Minimum bend radii for 1008 or 1010 steel sheet

Quality or temper	Minimum bend radius, mm (in.) Parallel to rolling direction	Minimum bend radius, mm (in.) Across rolling direction
Cold rolled, special properties		
Quarter hard(a)	1*t*(b)	¹/₂*t*
Half hard(c)	NR(d)	1*t*
Full hard(e)	NR	NR
Hot rolled		
Commercial, mm (in.)		
Up to 2.3 (0.090)	³/₄*t*	¹/₂*t*
Over 2.3 (0.090)	1¹/₂*t*	1*t*
Drawing, mm (in.)		
Up to 2.3 (0.090)	¹/₂*t*	¹/₄*t*
Over 2.3 (0.090)	³/₄*t*	¹/₂*t*

(a) 60–75 HRB. (b) *t*, sheet thickness. (c) 70–85 HRB. (d) NR, not recommended. (e) 84 HRB minimum

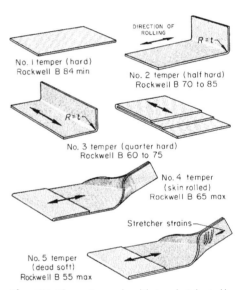

Fig. 16 The most severe bend that can be tolerated by each of the standard tempers of cold-rolled carbon steel strip. Stock of No. 1 (hard) temper is sometimes used for bending to large radii; each lot should be checked for suitability, unless furnished for specified end use by prior agreement. Hardnesses shown are for steel containing 0.25% C (max) in the three hardest tempers and 0.15% C (max) in the No. 4 and 5 tempers. Hardness for No. 1 temper applies to thicknesses of 1.8 mm (0.070 in.) and greater; for thinner sheet, hardness would be a minimum of 90 HRB.

Table 11 Suitability of commercial-quality low-carbon steel sheet for bending

Class and hardness of steel, and bending conditions	90° bends		180° bends	
Cold-rolled steel up to 1.6 mm (0.062 in.) thick				
Suitable commercial-quality steels	1008 or 1010, rimmed, temper passed		1008 or 1010, rimmed, annealed	
Maximum HRB hardness	80(a)		65	
Minimum bend radius	1*t*		0.25 mm (0.01 in.)	
Hot-rolled steel up to 6.4 mm (0.250 in.) thick				
Suitable commercial-quality steels(b)	1008, 1010	Up to 1030	1008, 1010	Up to 1015
Maximum HRB hardness(c)	68	80	68	72
Minimum bend radius				
Sheet up to 2.3 mm (0.090 in.) thick	³/₄*t*	1¹/₂*t*	1*t*	1¹/₂*t*
Sheet 2.3–6.4 mm (0.090–0.250 in.) thick	1*t*	2*t*	1¹/₂*t*	2*t*

(a) For 90° bends made across the direction of rolling. The acceptable maximum hardness for 90° bends parallel with the direction of rolling is 70 HRB. (b) Rimmed or capped. (c) Can be met on hot-rolled unpickled steel or steel pickled in sheet form. Hardness values will be higher on mill-pickled hot-rolled coil. With higher hardness values, somewhat larger bend radii will sometimes be required.

Table 12 Bending limits for hot-rolled commercial-quality carbon steel strip and cold- or hot-rolled carbon steel sheet

If greater ductility is needed, drawing-quality or physical-quality steel can be used.

Carbon, %	Bending limit
0.15 or less	180° bend flat on itself, in any direction
0.15–0.25	180° bend around one thickness of the material, in any direction

Because of volume constancy in plastic deformation, the thickness change in a particular location can also be calculated by comparing the surface areas of the original and distorted circles. If the area of the ellipse is larger than the area of the original circle, the sheet has undergone thinning. If the areas are the same, there is no thickness change. It is thus possible to predict thinning and tearing of the sheet by observing the shape changes of the original circles during forming (see the article "Troubleshooting Formability Problems Using Strain Analysis" in this Volume).

As expected, friction between the punch and the sheet affects the test results; hence, lubrication of the punch-sheet interface is an important parameter. As for the effect of sheet thickness, experimental results indicate that the boundaries in Fig. 17 rise with increasing sheet thickness; that is, the safe zone is expanded. Also note

that the left side of the forming-limit diagram has a larger safe zone than the right side, indicating the desirability of encouraging the development of a compressive strain during the forming of the sheet. This phenomenon has been used successfully in practice, such as in bulging of tubular workpieces or bending of tubes by applying a compressive force through the axis of the tube.

Experimentally derived forming-limit curves for various steels are given in Fig. 18 to 20. The experimental process of developing forming-limit curves is labor-intensive, and so there is interest to develop computer-based models of formability and/or to develop techniques of deriving formability limits from tensile-test data. For example, Ref 6 describes prediction of forming-limit curves from a single limit yield stress obtained from tensile testing. Hill's anisotropic yield criteria are used to obtain a continuous limit yield locus, which can be termed as forming-limit stress curve. The methodology was tested using literature data on some deep-drawing-quality low-carbon steels, with a band of results shown in Fig. 21. Tensile tests also were conducted on sheet specimens of brass that were selected in three important directions with reference to the original sheet rolling direction. The mechanical properties indicated a significant variation in the three different directions.

Drawability

Drawability of a metal depends on two factors: the ability of the material in the flange region to flow easily in the plane of the sheet under shear, and the ability of the sidewall material to resist deformation in the thickness direction. Taking

both of these factors into account, it is desirable in drawing operations to maximize material flow in the plane of the sheet and to maximize resistance to material flow in a direction perpendicular to the plane of the sheet. Low flow strength in the plane of the sheet is of little value if the work material also has low flow strength in the thickness direction.

The flow strength of sheet metal in the thickness direction is difficult to measure, but the plastic-strain ratio r compares strengths in the plane and thickness direction by determining true strains in these directions in a tension test. As noted, a high r-value is the measure of a material with good deep-drawing characteristics. Rather than average r over a plane, another way to examine the effect of planar anisotropy is by Δr, which indicates the variation in r with direction in the plane of the sheet. For values of Δr less than or greater than 1, nonuniform straining during deep drawing will occur in directions parallel and normal to the rolling direction, resulting in the formation of ears.

Earing in deep-drawn parts is related to planar anisotropy (i.e., the sheet metal is stronger in one direction than another). This causes ears to form on the drawn part even when a circular blank is used. Ears are shown in Fig. 22(a), while 22(b) shows the orientation dependence on the ears with rolling direction for positive and negative Δr-values. It is important for the process designer to know whether or not the sheet metal is subject to earing, because an extra step is needed to trim away the ears.

Effects of Alloying

Low-carbon sheet steels are generally preferred for forming. These steels typically contain less than 0.10% C and less than 1% total intentional and residual alloying elements. The amount of manganese, the principal alloying addition, normally ranges from 0.15 to 0.35%. Controlled amounts of silicon, niobium, titanium, or aluminum may be added either as deoxidizers or to develop certain properties. Residual elements, such as sulfur, chromium, nickel, molybdenum, copper, nitrogen, and phosphorus, are usually limited as much as possible. In steelmaking shops, these amounts are based on the quality of sheet being produced. Alloy sheet steels (including high-strength low-alloy grades), however, contain specified amounts of one or more of these elements.

Table 13   Minimum bend radii for cold bending of SAE 410c high-strength low-alloy steel sheet, strip, and plate

| SAE grade | Ratio of bend radius to material thickness of: | | |
	To 4.57 mm (0.180 in.)	4.60–6.35 mm (0.181–0.250 in.)	6.38–12.7 mm (0.251–0.500 in.)
942X	. . .	1	2
945A, C	1	2	2.5
945X	1	1	2
950A, B, C, D	1	2	3
950X	1.5	2.5	2.5
955X	2	3	3
960X	2.5	3.5	3.5
965X	3	4	4
970X	3.5	4.5	4.5
980X	3.5	4.5	4.5(a)

(a) Available to 9.53 mm (0.375 in.) inclusive

Table 14   Minimum and typical minimum bend radii of a cold-rolled high-strength low-alloy steel

| Typical hardness | Tensile strength(a) | | Yield strength(a) | | Elongation in 50 mm (2 in.)(a),% | Minimum bend radius | Typical bend radius |
	MPa	ksi	MPa	ksi			
77 HRB(b)	448–490	65–71	345–379	50–55	22–26	$1/2 t$	0
81 HRB(b)	517–545	75–79	414–455	60–66	18–23	$1/2 t$	0
89 HRB(b)	552–607	80–88	483–524	70–76	14–19	$1/2 t$	0
91 HRB(b)	621–669	90–97	552–614	80–89	12–17	$1/2 t$	0
99 HRB(c)	724–800	105–116	689–745	100–108	8–12	$1 t$	$1/2 t$
25 HRC(c)	827–903	120–131	827–883	120–128	4–10	$2 t$	$1 t$
32 HRC(c)	903–1048	140–152	965–1041	140–151	2–5	$4 t$	$3 t$

Note: t, sheet thickness. (a) Listed values are minimum tensile properties followed by typical tensile properties. (b) Grades with typical composition of 0.07C, 0.45Mn, 0.045Al, and 0.19Ti. (c) Grades with typical composition of 0.07C, 0.90Mn, 0.045Al, and 0.24Ti

Table 15   Minimum bend radii of a hot-rolled high-strength low-alloy steel sheet and strip

| Composition(a), max wt% | Tensile strength, min | | Yield strength, min | | Elongation, min, in 50 mm (2 in.), % | Minimum bend radius for sheet thickness(t) of: | | |
	MPa	ksi	MPA	ksi		Less than 4.8 mm (3/16 in.)	4.8 to 6.4 mm (3/16 to 1/4 in.)	Over 6.4 mm (1/4 in.) up to 13 mm (1/2 in.)(b)
0.22C, 1.35Mn	415	60	310	45	25	$1 t$	$1 t$	$2 t$
0.23C, 1.35Mn	450	65	345	50	22	$1 t$	$1 t$	$2 t$
0.25C, 1.35Mn	480	70	380	55	20	$11/2 t$	$2 t$	$21/2 t$
0.26C, 1.50Mn	520	75	415	60	18	$2 t$	$3 t$	$31/2 t$

(a) Maximum of 0.4 wt% P and 0.05 wt% S; minimum of 0.005 wt% Nb and 0.01 wt% V. (b) Thicknesses above 13 mm (1/2 in.) should be hot formed.

Carbon content is particularly significant in steels that are intended for complex forming applications. An increase in the carbon content of steel increases the strength of the steel and reduces its formability. These effects are caused by the formation of carbide particles in the ferrite matrix and by the resulting small grain size. The amount of carbon in steel sheet is generally limited to 0.10% or less to maximize the formability of the sheet.

Manganese enhances the hot working characteristics of the steel and facilitates the development of the desired grain size. Some manganese is also necessary to neutralize the detrimental effects of sulfur, particularly for hot workability. Typical manganese contents for low-carbon steel sheet range from 0.15 to 0.35%; manganese contents up to 2.0% may be specified in high-strength low-alloy steels. When the sulfur content of the steel is very low, the manganese content also can be low, which allows the steel to be processed to develop high *r*-values.

Phosphorus and sulfur are considered undesirable in steel sheet intended for forming,

drawing, or bending because their presence increases the likelihood of cracking or splitting. Allowable levels of phosphorus and sulfur depend on the desired quality level. For example, commercial-quality cold-rolled sheet must contain less than 0.035% P and 0.040% S. For more applications, phosphorus may be added to the steel to increase the strength. Sulfur usually appears as manganese sulfide stringers in the microstructure.. These stringers can promote splitting, particularly whenever an unrestrained edge is deformed.

Silicon content in low-carbon steel varies according to the deoxidation practice employed during production. In rimmed steels (so called because of the rimming action caused by outgassing during solidification from the molten state), the silicon content is generally less than 0.10%. When silicon rather than aluminum is

used to kill the rimming action, the silicon content may be as high as 0.40%. Silicon may cause silicate inclusions, which increase the likelihood of cracking during bending. Silicon also increases the strength of the steel and thus decreases its formability.

Chromium, nickel, molybdenum, vanadium, and other alloying elements are present in low-carbon steel only as residual elements. With proper scrap selection and control of steelmaking operations, these elements are generally held to

Table 16 Specified bend test radii for inclusion-shape-controlled ASTM A 715 steel sheet

| Grade | Bend test radius for: | |
	Transverse bends(a)	Longitudinal bends(a)
50	0.5t	0
60	0.5t	0
70	0.75t	0.5t
80	0.75t	0.5t

(a) For sheet thicknesses (*t*) up to 5.84 mm (0.2299 in.)

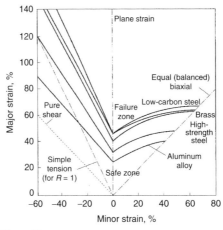

Fig. 17 Forming-limit diagram for various sheet metals

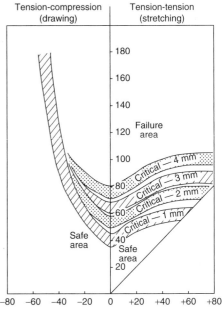

Fig. 18 Forming-limit curves for drawing-quality and mild sheet steels of various thicknesses. Source: Ref 5

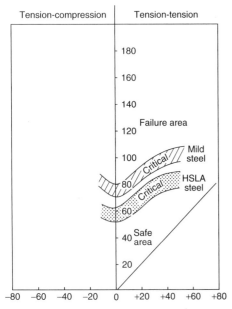

Fig. 19 Forming-limit curves for high-strength low-alloy (HSLA) steel compared with mild steel for sheet thicknesses of 4 mm (0.16 in.). Source: Ref 5

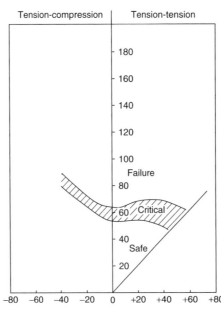

Fig. 20 Forming-limit curves for 301 austenitic stainless steel

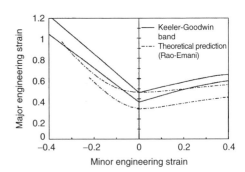

| Steel | Work-hardening exponent, *n* | Strength coefficient, *K* | | Strain ratio, *r* |
		MPa	ksi	
1	0.255	516	75	1.10
2	0.227	557	81	1.47
3	0.244	521	76	1.65

Fig. 21 Comparison of Keeler-Goodwin forming-limit band with a predicted band by Rao and Emani (Ref 6) from tensile-test data of selected low-carbon steels. Tensile properties of steels are given in the table.

minimum amounts. Each of these elements increases the strength and decreases the formability of steel sheet. High-strength low-alloy steels may contain specified amounts of one or more of these elements.

Copper is generally considered an innocuous residual element in steel sheet. The strengthening effect of copper is almost negligible in typical residual amounts of less than 0.10%. However, copper is added to steel in amounts exceeding 0.20% to improve resistance to atmospheric corrosion.

Niobium strengthens high-strength low-alloy steel through the formation of niobium carbides and nitrides. It can also be used either alone or in combination with titanium to develop high r-values in interstitial-free steels. These alloying elements remove the interstitial elements carbon and nitrogen from solid solution. Consequently, the steel shows no yield-point elongation.

Titanium is a strong carbide and nitride former. It helps develop high r-values and eliminates yield-point elongation and the aging of cold-rolled annealed steel sheet. Titanium streaks may be a problem in some grades, especially in the form of surface defects in exposed applications.

Aluminum is added to steel to kill the rimming action and thus produce a very clean steel known as an aluminum-killed, or special-killed, steel. Aluminum combines with both the oxygen and nitrogen to stop the outgassing of the molten steel when it is added to the ladle or mold. Aluminum also aids the development of preferred grain orientations to attain high r-values in cold-rolled and annealed steel sheet. Elongated grains of an approximate ASTM 7 size are found in most well-processed aluminum-killed steels. Because the aluminum combines with the nitrogen, the steel is not subject to strain aging.

Nitrogen can significantly strengthen low-carbon steel. It also causes strain aging of the steel. The effects of nitrogen can be controlled by deoxidizing the melt with aluminum.

Cerium and other rare earth elements may be added to steel to change the shape of manganese sulfide inclusions from being needle-like or ribbonlike to being globular. Globular inclusions reduce the likelihood of cracking if the sheet is formed without restraining the edges.

Oxygen content of molten steel determines its solidification characteristics in the ingot. Excessive amounts of oxygen impede nitride formation and thus negate the effects of alloying elements added to minimize strain aging. Deoxidizers such as silicon, aluminum, and titanium will control the oxygen content. When oxygen combines with these deoxidants, complex nonmetallics are formed. Although most nonmetallics dissolve in the slag, some may become trapped in the steel, causing the surface defects of seams and slivers.

Effects on Formability

This section describes some key factors that affect the formability of steels in terms of steelmaking practices, surface finish, and metal thickness. From the standpoint of formability, an ideal material should:

- Distribute the strain in the sheet uniformly (high m)
- Achieve high strain levels without necking or fracture (high n)
- Withstand in-plane compressive stresses without wrinkling
- Withstand in-plane shear stresses without fracturing
- Retain part shape on removal from the die
- Retain a smooth surface and resist surface damage

For tentative severity classification of a part, steel near maximum in hardness and near minimum in formability may help identify problems during computer modeling and tool development and tryout runs. Tools developed with steel of below-average quality are seldom troublesome and run with minimum tool breakage and steel rejection when the production run begins. They are also less sensitive to pressure adjustments, variations in sheet thickness, or normal variations in steel properties, and maintenance costs are usually less. Conversely, tools developed with steel of above-average quality are often unsatisfactory when forming regular production shipments of steel.

Effects of Steelmaking Practices

The formability of steel sheet is determined to a great extent by the steelmaking practices employed in manufacturing. The user of steel sheet normally specifies certain characteristics for the sheet, thus ensuring that the material can be formed in a predictable manner. Adherence to these specifications also implies that the producer of the sheet has observed whatever steelmaking practices are necessary to enable the product to perform as indicated. The user can specify either hot-rolled or cold-rolled sheet, and he must select an appropriate quality designation.

Rimmed steels refer to the relatively pure iron layer on the surface of the ingot caused by outgassing during solidification. As previously noted, rimmed (or capped) ingot cast steel has been largely replaced by continuous casting, which is more suited to the production of killed (deoxidized) steels. However, rimmed steels may still be available in some circumstances. Rimmed steel generally has a better surface finish than killed steel but is subject to aging.

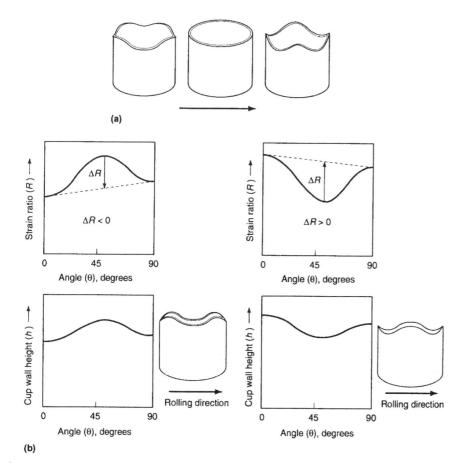

Fig. 22 Ear formation. (a) Drawn copper cups showing earing. The rolling direction is indicated by the arrow. Source: Ref 7. (b) The relation of ear formation to the direction of rolling. Source: Ref 8

Rimmed steels have been available as both hot-rolled and cold-rolled products. Hot-rolled commercial-quality rimmed steel is suitable for many forming applications and has the advantage of minimum cost. Cold-rolled, drawing-quality, special-killed, and temper-passed steel has maximum formability and yields parts with the best appearance and finishing characteristics, but it is more expensive. Annealing treatments are used to regain ductility in the product following cold reduction to final thickness. After annealing, rimmed steels must be temper rolled to prevent the formation of Lüders lines during forming.

If rimmed steels are being selected for forming, attention must be paid to deoxidation practice. Killed steel is preferred where sheets must be free of significant changes in mechanical properties (strain aging) for a long time, where neither stretcher strains (Lüders lines) nor roller leveling is permitted, or where better mechanical properties are desired for severe forming applications. Rimmed steels are more suited to stretch-type deformation than to deep drawing, for which aluminum-killed steels are generally recommended. Rimmed steels will age after a period of time following temper rolling. Quality levels of rimmed steel are achieved by controlling chemical composition and annealing practice. Commercial quality is standard, whereas drawing quality is produced under stricter tolerance levels for impurities and is given a longer anneal to ensure uniformity throughout the coil, as well as good formability.

Strain Aging of Rimmed Steels. The effect of aging of rimmed steel on formability is difficult to predict on the basis of tests. A certain rimmed steel may not age at all, while another may make the most difficult draws when received and, after aging 30 days, may not make minimum draws.

Strain aging is more pronounced after steel undergoes blanking, forming, or finishing operations. It is therefore advisable to complete the sequence of operations on a part without intervening storage unless artificial aging tests positively indicate the absence of aging.

Artificial aging tests give an approximate measure of the strain-aging characteristics of the steel but do not predict the time at which definite changes in mechanical properties will occur. Artificial aging does not change the tensile strength appreciably; however, yield strength and hardness will always increase, while elongation and uniform elongation will always decrease.

Adding a die operation may reduce the severity of forming enough so that a rimmed steel can be used instead of a killed steel. The overall cost of producing a part is usually the criterion for determining whether to use a more expensive grade of steel or a more expensive die system. Depending on blank size, die complexity, and number of pieces to be produced, the savings in material may offset the additional die costs.

Stretcher strains, or Lüders lines, and the Piobert effect, also known as worms, are char-acteristic markings that appear on the surface of low-carbon steel that has been annealed as a final mill operation. These lines appear during the early stages of stretching and almost disappear as the stretch exceeds 5 to 10%. In tension, the lines are depressions in the surface; in compression, they are raised; and in bending, the same phenomenon causes flutes or kinks. Stretcher strains have no harmful effect on strength. In stampings that are visible in service, stretcher strains are generally unacceptable, because they show clearly through paint.

Stretcher strains can be avoided by a temper pass of approximately 1% cold reduction after final annealing. The correction is normally permanent in killed steel, but stretcher strains frequently recur in rimmed steel unless it is formed in 1 week or less, depending on the amount of temper pass, the temperature, steelmaking practice, and the amount of forming in the stamping.

The probability of stretcher straining can be eliminated from temper-passed rimmed steel and from insufficiently temper-passed killed steel by roller leveling through a machine that flexes the sheet sufficiently in bending to remove the sharp yield point and the yield-point elongation that cause stretcher strains. This amount of cold work does not reduce drawing quality (in some steels, the quality may be improved by reducing the yield point), but additional strain aging is induced, which reduces formability if the steel is stored after roller leveling. Roller-leveled steel should be used within 24 to 72 h after leveling. The sheet should be passed through the roller leveler once in each direction, because approximately 455 mm (18 in.) of the entering end of the sheet is not flexed.

Occasionally, a lift, or even a shipment, of steel does not respond to roller leveling. If such material is unsatisfactory after two passes through the roller leveler, it should not be used for parts that will be exposed in service. However, the performance of annealed steel used for a very difficult unexposed part can be improved by a single pass through the roller leveler.

Annealed sheet cannot be roller leveled for an exposed part, because the flex roll kinks the sheet so severely that, after forming, the deformation will not disappear. In addition, small stretcher strains will occur between the kinks.

Coil breaks and stickers have the appearance of stretcher strains, but both are distinctly different. Coil breaks are regularly spaced, and stickers are spotty. Roller leveling has no effect on these defects.

Killed Steel. Generally, killed steel has mechanical properties that are superior to those of drawing-quality rimmed steel (particularly, it has low yield point); killed steel also has better formability and performance, less tendency to form buckles, and is usually free from aging. However, inferior surface properties and more surface defects can be expected from killed steel than from rimmed steel, with consequent higher scrap or repair loss because of these defects. In addition, panels produced from killed steel are usually less resistant to handling damage and oil canning, because of the lower yield strength of this steel.

Most major producers of stampings restrict the use of killed steel to the most severe draws, to low-volume parts when the steel inventory cannot be used before aging begins in rimmed steel, and to small or irregularly shaped parts for which sheet cannot be roller leveled successfully. To ensure optimal performance, killed steel should have a fine, flat, elongated grain; ASTM grain size 7 to 8 is preferred. Stretcher strains may often be removed by roller leveling if the size and shape of the blank permit. However, this is the responsibility of the supplier, because killed steel is expected to be usable without roller leveling.

Aluminum-killed steels are deoxidized with aluminum and, possibly, with silicon. As already mentioned, use of aluminum results in a very clean steel, known as aluminum-killed or drawing-quality special-killed steel. Exceptional resistance to thinning through the sheet thickness (as measured by the plastic-strain ratio, r) can be developed through the controlled processing of these steels. Because the pure iron skin characteristic of rimmed steel does not exist in aluminum-killed steel, surface imperfections may occasionally be encountered on aluminum-killed sheet. Both class 1 and class 2 drawing-quality aluminum-killed steels are produced. It should be noted that some aluminum-killed steels that cannot meet the formability requirements for drawing-quality sheet are sold as commercial-quality steel.

Hot-Rolled Steel. In the as-hot-rolled condition, the steel has a dark-gray oxide coating on its surface, which offers limited corrosion protection as long as it is undisturbed. However, the oxide flakes off during forming and may be undesirable around the press. Because the oxide coating also interferes with steel surface lubricants, it should be removed before the final finishing of most formed parts. Hot-rolled steel may be ordered pickled (using either hot sulfuric or hydrochloric acid to remove the oxide) and oiled to inhibit in-transit rusting. Rimmed hot-rolled steel in the as-pickled condition will show stretcher strains or Lüders lines on the surface after forming. Whenever surface appearance is important, the steel should be ordered with a temper-rolled surface (skin pass of less than 2% cold reduction) to reduce this tendency. If aging is a problem because of storage requirements, special-killed hot-rolled steel should be ordered.

There is no preferred grain orientation providing high r-values in hot-rolled steel, but improved grain size and resistance to longitudinal splitting may be attained by closely controlling chemical composition, which differs between commercial- and drawing-quality hot-rolled steel. Higher strength levels (when necessary for the part being formed) are obtained by alloy additions and processing controls to develop improved structure. Because higher strength is associated with forming problems such as lower ductility, increased springback, and longitudinal bend failures, only high-strength low-alloy steels designed for improved formability should be used in structural parts made by press operations. Mechanical properties of several types of hot-rolled sheet are given in Table 17.

Cold-rolled steel sheet forming is produced by the cold reduction of hot-rolled pickled coils, followed by annealing and possibly additional processing, such as temper rolling. Class 1 (E, exposed) should be ordered when a controlled surface finish is required. Class 2 (U, unexposed) is intended for applications in which surface appearance is not of primary importance. Both classes are available as commercial-quality, drawing-quality, or drawing-quality special-killed cold-rolled steel. Mechanical properties of cold-rolled steel sheet are given in Table 18.

Cold-rolled steels in the as-annealed condition may exhibit Lüders lines or stretcher strains from yield-point elongation, if the steel is subject to aging or is not properly killed. The yield-point elongation may be removed by temper rolling the annealed coils. Because temper coiling strengthens the steel and reduces its ductility, it is usually limited to 0.5 to 1.5% elongation of the strip. Temper rolling under tension is more effective than flex rolling or roller leveling for eliminating yield-point elongation, because the steel is more uniformly strained through the thickness. These latter methods are sometimes used in the plants of fabricators because the equipment is less expensive and because it permits the use of aged coils of rimmed steels that may show strain on the surface of formed parts.

In addition to many as-processed surface finishes, cold-rolled sheet may be ordered with a metallic coating that provides corrosion protection or a decorative finish that reduces the manufacturing costs of parts such as appliances or building panels.

Interstitial-free steel is vacuum degassed to reduce the amounts of the interstitial elements carbon, nitrogen, and oxygen. It is usually processed to achieve high r-values ($r_m \sim 2.00$). This type of steel is not subject to strain aging at any stage of processing or manufacture; it exhibits no yield-point elongation. Interstitial-free steel can withstand deeper draws with less breakage than other grades of steel sheet, and coated products made from it generally retain excellent formability.

Effect of Surface Finish

The surface roughness of sheet steel has an effect on the finishing cost and the appearance of the formed product as well as on processing in dies and on other operations. Dull or slightly roughened surfaces are used especially in parts with the deepest draws in order to retain lubricant through the operations for minimum scoring of the dies and for better flow of metal over pressure pads. Sheet with a surface roughness of approximately 0.75 to 1.3 μm (30 to 50 μin.) draws well and is smooth enough for most painted parts, such as hood tops and fenders, which require average paint finish.

The need for uniformity among parts that must have matching surface finishes (such as automobile fenders and hoods), even when made from different materials, often dictates the sheet finish. A surface roughness of 0.8 to 1.5 μm (30 to 60 μin.) for average peak height and two to six peaks per millimeter is considered standard for cold-rolled steel sheet. The surface finish is determined by the finish applied to the cold-mill rolls and the temper-mill rolls. Roll finishes are obtained by shot blasting or electroetching a ground roll surface so that the roll is roughened sufficiently to transfer the pattern to the sheet. As these rolls are used, their finish tends to become smoother; there may be a consequent change in appearance among coils, and press

performance may vary slightly. A rougher sheet surface tends to hold lubricant better and resists galling and cold welding to die surfaces during forming. For parts requiring little forming, a smoother and often preferred finish can be attained when roughness is minimized.

Single-dip or painted parts intended for trim and interior moldings require a smoother surface of approximately 0.25 to 0.5 μm (10 to 20 μin.). Sheet for average decorative chromium-plated parts should have surface roughness no greater than 0.25 μm (10 μin.) where the surface is to have no preparation except a light polishing to remove die marks. Parts with surface roughness as high as 0.4 μm (15 μin.) require additional surface preparation, for example, buffed copper plate applied before another plating.

The selection of fine-grain steel (ASTM grain size of 9) with minimum surface roughness for forming usually sacrifices some ductility and latitude in die design. With fine grain, the steel will be somewhat harder, higher in yield point and elastic ratio (yield point/tensile strength), lower in elongation and uniform elongation, and more likely to strain age.

Surface defects in unexposed parts may be acceptable if the function or the strength of the part is not affected. The following example describes the use of two different grades of steel for forming a concealed panel on which surface defects were acceptable and for forming a panel that required a smooth surface for painting.

Example 6: Effect of Grade of Steel on Surface Finish of Severely Formed Parts. Figure 23 shows an inner panel for a glove-compartment door that was made of 0.81 mm (0.032 in.) thick cold-rolled drawing-quality rimmed 1008 steel. On some parts, stretcher strains appeared in the severely stretched flat surface, as shown in Fig. 23. When the stock strained, the parts were used for a part number where the surface was covered by another detail. Parts that had severe stretcher

Table 17 Typical mechanical properties of hot-rolled steel sheet

Type or quality	Special feature	Yield strength		Tensile strength		Elongation in 50 mm (2 in.), %	Hardness, HRB	Strain-hardening exponent, n	Plastic-strain ratio, r_m
		MPa	ksi	MPa	ksi				
Commercial	Standard properties	262	38	359	52	30	55	0.15	0.9
Drawing (rimmed)	Improved properties	241	35	345	50	35	50	0.18	1.0
Drawing (special killed)	Nonaging	241	35	345	50	40	50	0.20	1.0
Medium strength	Inclusion shape control	345	50	414	60	25	70	0.15	0.9
High strength	Inclusion shape control	552	80	620	90	15	90	...	...

Table 18 Typical mechanical properties of cold-rolled steel sheet

Type or quality	Special feature	Yield strength		Tensile strength		Elongation in 50 mm (2 in.), %	Hardness, HRB	Strain-hardening exponent, n	Plastic-strain ratio, r_m
		MPa	ksi	MPa	ksi				
Commercial	Standard properties	234	34	317	46	35	45	0.18	1.0
Drawing (rimmed)	Stretchable	207	30	310	45	42	40	0.22	1.2
Drawing (special killed)	Deep drawing	172	25	296	43	42	40	0.22	1.6
Interstitial-free	Extra-deep drawing	152	22	317	46	42	45	0.24	2.0
Medium strength	Formable	414	60	483	70	25	85	0.20	1.2
High strength	Moderately formable	689	100	724	105	10	25(a)	...	...

(a) HRC

strains also cracked in the areas shown in Fig. 23 and were not acceptable.

The outer panel for the same door, however, was visible and had to have a maximum surface roughness of 1.15 μm (45 μin.) before it was painted. Killed or flex-rolled steel strip was used for the outer panel to minimize the stretcher strains.

The transfer die for the door panel was set up in a 7 MN (800 tonf) straight-side mechanical press operating at 500 strokes per hour. The die was cleaned after each shift, and it was resharpened after making 40,000 pieces. Lubrication was a chlorinated oil applied to the stock by rollers.

Effect of Cleaning

The procedure for cleaning steel sheet before forming depends on the type and amount of soil present and on the finish specified for the formed surface. Removing soil before forming improves the surface finish, prevents marking of the formed piece, and prolongs die life. Large particles can be removed by wiping, which allows oil to remain and act as a lubricant. Coil stock can be cleaned by feeding it between wiping pads at the press. Another method is to feed the coil through a vat of emulsion cleaner or a light-duty drawing lubricant before it passes between the wiper pads. This procedure is an economical means of simultaneously removing foreign material and providing lubrication.

Cleaning the steel in this manner normally does not completely remove the smudge from the surface. This condition is generally desirable because the smudge acts as filler in the lubricant. Many parts are formed using only mill oil with its smudge for lubrication. If the smudge must be removed, a rotating brush can be used between the emulsion cleaner and the press to scrub the work metal surfaces.

Scale is often removed by abrasive blasting or pickling before the hot-rolled sheet is formed. Both methods remove the residual lubricant. Workpieces that have been in-process annealed without a protective atmosphere are usually

cleaned by abrasive blasting before final forming. The need to clean parts after forming is determined by the subsequent operations and by the necessary finish. The necessity for cleaning immediately after forming can be minimized by choosing lubricants that are compatible with welding, painting, plating cycles, and handling.

Effect of Work Metal Thickness

Thickness variations in sheet steel can cause parts made on the same tooling to be of different shapes. This is because of springback or because the pressure applied is either insufficient or excessive at sharp corners or at sides that are to be held at a predetermined angle.

If the sheet is too thick, a die or roll adjusted to a certain thickness may pinch the steel and localize the stretching, thus causing fracturing; or it may work harden the steel and cause excessive springback in a subsequent operation. Thickness greater than the die clearance may cause undesirable marring of the surface of the part or galling and scoring on the surface of the tools, and, in some cases, it may be the reason for tool breakage.

Bending

Bending is one of the most common sheet metal forming operations used to form flanges, seams, curls and hems. Figure 24(a) shows the parameters involved in bending of sheet metal. During bending, the outer fibers of the bent metal are in tension, while the inner fibers are in compression (Fig. 24b). The metal undergoes plastic deformation with little change in its thickness. Bendability is typically expressed in terms of minimum bend radii, as previously described in the section "Bendability" in this article.

Bending operations are performed using presses, punches, dies, and fixtures. In addition to simple V-bending and edge bending, the industry employs other bending operations, such as flanging, hemming, seaming, and curling (Fig. 25). Additional bending operations, shown in Fig. 26, include channel bending, U-bending, air bending, offset bending, corrugating, and tube forming (see the articles "Bending of Sheet Metal" and "Press-Brake Forming" in this Volume).

Punch presses are used for bending, flanging, and hemming carbon steel when production quantities are large, when close tolerances must be met, or when the parts are relatively small. Press brakes are ordinarily used for small lots, uncritical work, and long parts. As workpiece size increases, the advantages of a punch press diminish.

The punch press is usually more economical, and it is more efficient than the press brake in terms of power requirements for a given force on the ram and number of strokes per unit of time. To estimate the press capacity needed for bending in V-dies, the bending load in tons can be computed from:

$$L = lt^2 kS/s$$

where L is press load (in tons of force), l is length of bend (parallel to bend axis) (in inches), t is work metal thickness (in inches), k is a die-opening factor (varying from 1.2 for a die opening of $16t$ to 1.33 for a die opening of $8t$), S is tensile strength of the work metal (in tons of force per square inch), and s is width of die opening (in inches). For U-dies, the constant k should be twice the values given previously.

The generally accepted tolerance for dimensions resulting from bending is ± 0.41 mm (± 0.016 in.) for metals up to and including 3.2 mm (0.125 in.) thick. For thicker metals, the tolerance is increased proportionately. As in many other mechanical operations, obtainable tolerances are influenced by design, stock tolerances, blank preparation, and condition of the machine and tooling (e.g., see the section "Dimensional Accuracy" in the article "Press-Brake Forming" in this Volume).

Bending Dies

Typically, bending dies used in press brakes can generally be used in presses, except that the die design must give more consideration to clearance for removing the finished workpiece, because presses ordinarily are not long and narrow like press brakes. Secondly, presses cycle more rapidly than press brakes, and the shut height is not as easy to change; therefore, fewer pieces are bent in air (see the article "Press-Brake Forming" in this Volume). More frequently, pieces are formed by bottoming the dies. This has the advantage of decreasing springback.

V-dies are composed of a V-block for a die and a wedge-shaped punch (Fig. 27a). The width of the opening in the V is ordinarily at least eight times the stock thickness. In bending, the workpiece is laid over the V in the die, and the punch descends to press the workpiece into the V to form the bend.

The included angle of a V-bend can be changed by adjusting the distance that the punch forces the sheet metal into the V-die. When the piece must be overbent (to allow for springback), the angle of the punch is smaller than the included angle on the part. Bottoming the punch

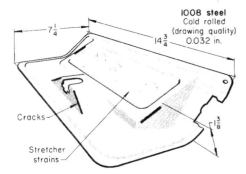

Fig. 23 Formed panel on which stretcher strains and cracks sometimes occurred. Some of the strained parts could be used in applications in which they were concealed in service; cracked parts were unacceptable. Dimensions given in inches

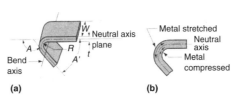

Fig. 24 Bending. (a) Bending of sheet metal. (b) Tension and compression in bending

and striking the metal severely at the bend is a means of reducing springback.

In V-die bending of a flange along the edge of a wide sheet, distortion is likely to occur. Most of the sheet overhangs the die and lifts up as bending takes place. If the punch strikes too fast, the workpiece will distort and will have irregular break lines. However, if the press ram is slowed down just before the punch hits the work, distortion is minimized.

For this type of V-die work, presses are available in which the ram advances rapidly, slows down just above the work, proceeds slowly through the bottom of the stroke, and returns rapidly. In addition, there are presses in which the rate of ram advance can be controlled somewhat by the operator.

Wiping Dies. Another type of bending die is the wiping die (Fig. 27b), where a pressure pad clamps the workpiece to the die before the punch makes contact. The punch descends and wipes one side of the workpiece over the edge of the die. The bend radius is on the edge of the die. To prevent the wiping action from being too severe, there may be a radius or chamfer on the mating face of the punch. When springback must be compensated for, the die is undercut to permit overbending. The flange metal can be put in slight tension by ironing it between the punch and the die. Sharp bends generally cannot be made in one operation in a wiping die without cracking the metal, because a punch or die with a sharp edge will cut the metal rather than bend it.

Interchangeable V-Dies. Figure 28 shows equipment for making various sizes of V-bends in a punch press. Four different sizes of punches can be mounted into the punch holder, which is attached to the press slide. In operation, the groove in the die that gives the needed bend is aligned with the punch, and the die is then fastened to the bolster plate on the press bed. Adjustable side and end stops can be used to position the blanks for bending.

With equipment such as that illustrated in Fig. 28, rectangular boxes in a range of shapes and sizes are often produced more economically

by bending flat blanks into folded-end shapes than by deep drawing. Making folded-end pans by bending in adjustable wing dies and cam dies with interchangeable punches is described in the article "Press Forming of Coated Steel" in this Volume.

U-Bending Dies. U-shaped pieces can be bent in a die such as the one shown in Fig. 29. The width of the U is adjustable by means of spacers and by changing the width of the knockout. Punches can be mounted in the press with a holder and can be provided to the proper width and shape to make either a U or a channel shape. Side clearance should be 10% more than stock thickness.

Rotary-bending dies (Fig. 30) are used to make bends or twists in bars or strip. These dies use cam action to rotate the workpiece. As shown in Fig. 30, a 90° twist is given to strip metal to make a connecting link. The punch is made in two major parts: a hollow cylinder that is solidly mounted to the ram, and inside it a solid cylinder that is free to rotate. The inner cylinder has a 90° helical cam groove in its cylindrical surface that engages a hardened pin in the outer cylinder.

When the ram descends, a slot in the face of the inner cylinder engages the end of the workpiece. After the inner cylinder bottoms, as the ram continues to move down, the spring compresses, letting the outer cylinder move down over the inner cylinder. The action of the pin in the groove makes the inner cylinder rotate, giving the workpiece a 90° twist.

An auxiliary cam keeps the inner cylinder from rotating back in the return stroke, until it

has cleared the workpiece. Near the top of the stroke, the auxiliary cam is released by a stop, allowing the inner cylinder to return to its starting position.

Cam-Actuated Flanging Dies. Horizontal motion is often needed to form, or partially form, a flange on a workpiece. One of the most commonly used methods of producing this motion at right angles to the motion of the main press ram is with an inclined surface, or cam, in the die mechanism.

As shown in Fig. 31, a blankholder contacts the work first and holds it in position. Resiliency, either in the form of pressure pins leading to a die cushion or in the form of a spring, allows the ram to continue to descend. A cam actuates the sliding punch, which either forms or completes the forming of the flange and is then retracted. The blank is placed on the pressure pad, where it is held by punch A and wiped past the cam-actuated sliding punch to form the flange. Near the bottom of the stroke, punch B contacts the cam head, which moves the sliding punch to set the flange to the 10.3 mm (0.406 in.) dimension and a 90° angle. Cam-actuated sliding punches on each side of the forming punch can be used for setting flanges on channel and U-shaped parts.

Cam-actuated dies are often used in combination with other tooling to produce complicated parts. When used in tandem with a progressive die, a cam-actuated die can significantly reduce the number of operations needed to produce a part. A drawer front that originally required nine separate operations (two shearing, two notching, one piercing, one box-forming, and three

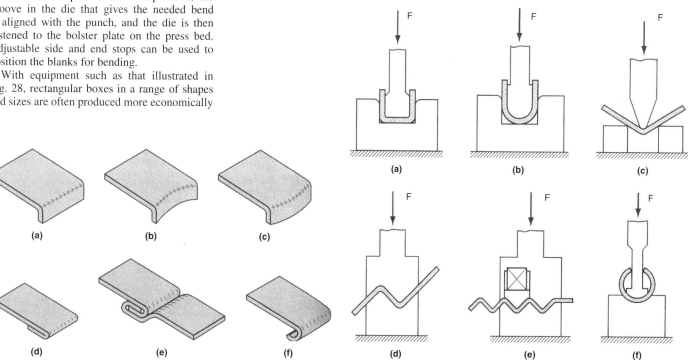

Fig. 25 Sheet metal bending operations. (a) Straight flanging. (b) Stretch flanging. (c) Shrink flanging. (d) Hemming. (e) Seaming. (f) Curling

Fig. 26 Various bending operations. (a) Channel bending. (b) U-bending. (c) Air bending. (d) Offset bending. (e) Corrugating. (f) Tube forming

flanging) when using a press brake was produced in only four separate operations (slitting coil to width; notching, piercing, and cutting off in a progressive die; box-forming in a second die; and flanging the sides, top, and bottom in a cam-actuated flanging die) with the incorporation of a cam-actuated die.

Compound Flanging and Hemming Die. The compound flanging and hemming die shown in Fig. 32 is unusual in having no horizontal motion of punches or dies. There are two cushions: a spring-loaded holddown plate

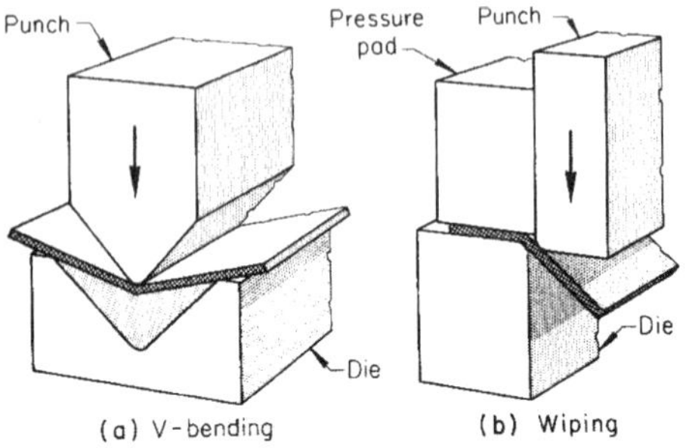

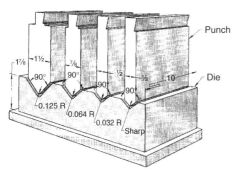

Fig. 27 Bending in (a) a V-die and (b) a wiping die. The wiping die can be of inverted design. The workpiece would move downward by the action of the punch against the pressure pad, and the flange would bend upward.

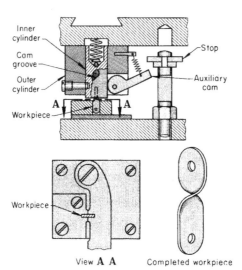

Fig. 30 Rotary-bending die used for 90° twisting of strip metal. Die is shown in closed position; inner cylinder has rotated to give workpiece a 90° twist. The auxiliary cam prevents rotation of the inner cylinder until it is free of workpiece.

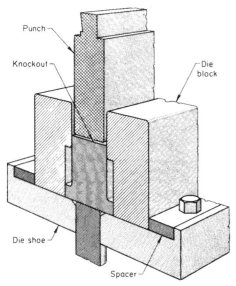

Fig. 28 V-bending and die for making a variety of bends in a punch press. Dimensions given in inches

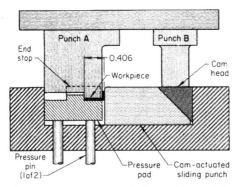

Fig. 31 Cam-actuated single flanging die used for producing a multiflanged part. See text for description of operation. Dimensions given in inches

Fig. 29 Adjustable U-bending die for use in a punch press

and an air cushion for the die plate. As the ram descends, the holddown makes first contact, clamping the piece securely to the die. As the ram continues to descend, the springs compress, and the angled flange is formed between the angled face of the bending punch and the die plate.

The angle of the bend is set and slightly coined between the chamfer on the edge of the die plate and the angled face of the punch. At this point, the holddown plate bottoms against the bending-punch holder, then the air cushion begins to yield, lowering the die plate. The edge of the workpiece bumps the corner of the lower die member and folds up into a hem that is completed between the angled face of the bending punch and the mating face of the lower die member.

Wing Dies. Special dies for making U-shaped bends have wings that turn up on each side as the punch descends. These dies are described in the article "Bending and Forming of Tubing" in this Volume.

Single-Curved Bends

Bending Cylindrical Parts. Generally, as the bend radius becomes larger, the allowance for springback must be more generous, because less of the bent metal has been stressed beyond its yield strength. Very large radii cannot be easily formed by ordinary bending but must be stretch formed (see the article "Stretch Forming" in this Volume). For large-radius bends, in which the workpiece is formed to half a circle or more, the bend is often made in several stages, as in the following example.

Edge bending is the bending of parts with the bend radius perpendicular to the width rather than to the thickness, as is usually done. It is sometimes possible to save material by producing fairly large blanks in simple rectangular form (square-sheared) and then edge bending them into shape. Finish blanking (trimming) is frequently done after edge bending. The size of the offset radii, the length and depth of the offset, and the location of the offset with respect to the flanges may preclude the edge bending of a rectangular blank. The potential cracking and thinning of the outer edge and the wrinkling of the inner edge make the use of a blanked shape more practical.

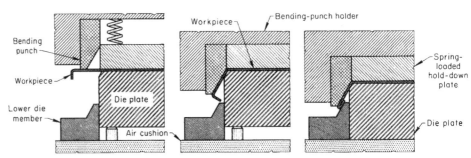

Fig. 32 Compound flanging and hemming die without horizontal motion

Before edge bending can be selected in preference to blanking from sheet, it is necessary to consider the effect of cold work from the bending operation on subsequent forming. High breakage during edge bending can eliminate the cost advantage of the savings in material. Another consideration is the higher cost of labor and tooling for edge bending, which can also offset the savings in material. The following example describes edge bending that was done to save material and to strengthen the part.

Example 7: Edge Bending. Figure 33 shows a curved-end automotive foot-pedal lever that was produced by edge bending a blank sheared from a bar of hot-rolled 1010 steel. The bar, 6.4 mm ($^1/_4$ in.) thick, 60 mm ($2^3/_8$ in.) wide, and 368 mm ($14^1/_2$ in.) long, was sheared lengthwise to produce two blanks 41 mm ($1^5/_8$ in.) wide at one end and 20 mm ($^3/_4$ in.) wide at the other. Edge bending resulted in substantial material savings over blanking and also a stronger part.

Each blank was edge bent to a 50 mm (2 in.) radius at the narrow end. Other operations included bending the offset and making the two other bends, piercing the 13 mm ($^1/_2$ in.) diameter hole, extruding four weld projections (section A-A, Fig. 33), and trimming the radius on the small end. The cutting and forming sections of the dies were made of air-hardened tool steel.

The stock was cut to length in an 1100 kN (120 tonf) end-wheel press (mainshaft extending front-to-back) operating at 60 strokes per minute. Edge bending, forming, trimming, and piercing were done by separate dies in a 1700 kN (190 tonf) open-back inclinable press at 30 pieces per minute.

Straight Flanging. Flange bending (flanging) in a wiping die is similar to the cantilever loading of a beam. To prevent movement during bending, the workpiece is clamped to the die by a pressure pad before the punch contacts the workpiece. The bend axis is parallel with the edge of the die. Flanging dies are often cam actuated, with an accompanying loss of efficiency. Holddown pads must be used, adding further to the press capacity requirement.

Considering all factors, the press capacity for flanging in a wiping die may be up to ten times that for forming a similar length of bend in a V-die with a spacing of at least eight times the thickness of the work metal. In some operations, only single flanges are bent. More often, more than one flange is bent at a time. Dies can be simple V-dies, U-dies, wiping dies, or complex flanging dies, such as shown in Fig. 31.

Even when fairly close tolerances must be held, simple V-dies can be used to make a complex part if production is low. Flanging dies are more expensive than ordinary press-brake dies, but considering the time and labor saved in making simple flanged pieces in flanging dies, they often pay for themselves quickly.

Hemming is an operation in which flanges are flattened against the workpiece in 180° bends to make a finished or reinforced edge. If the flange to be hemmed has been bent somewhat more than 90°, the hemming die can be a simple flat bed or anvil and a simple flat punch. Flanging and hemming can both be done in one press in a compound die, as shown in Fig. 32.

Bending of Curved Flanges

When a flange has concave curvature, the metal in the flange is in tension, and the flange is called a stretch flange (Fig. 34a). When the curvature is convex, the metal in the flange is in compression, and the flange is called a shrink flange (Fig. 34b). The amount of tension or compression in either type of flange increases from the bend radius to the edge of the flange.

Excessive tension in a concave (stretch) flange causes cracks and tears; excessive compression in the convex (shrink) flange causes wrinkles. Stretch and shrink flanges are commonly formed adjacent to each other, producing a reverse flange (as shown in Fig. 36, example 9). See the section "Blank Size" in the article "Press-Brake Forming" in this Volume for bend allowances when using a flat blank.

Flanging Limits. Flange radii, flange width, and angle of bend for curved flanges are primarily limited by the amount of deformation that can be tolerated by the flange edges, which depends on the type, thickness, and hardness of the metal and the method of forming. Greater fineness of detail can be achieved in conventional dies than by rubber-pad forming, because of the limited pressures in rubber-pad forming. However, rubber pads provide a uniform pressure over the entire surface of the workpiece, and they can be used to advantage where conventional dies would shear or tear the material.

The approximate percentage of deformation of the free edge is equal to $100 \left[(R_2/R_1) - 1 \right]$, where R_1 is the edge radius before forming (flat-pattern radius), and R_2 is the edge radius after forming, as shown in Fig. 34. For 90° flanges, R_2 is the same as R_m. Positive values of percentage of deformation indicate elongation (stretch); negative values indicate compression (shrink).

The permissible limits for the conventional die forming of curved flanges in three common

steels 1.0 mm (0.040 in.) thick or more are as follows:

Steel	Stretch, %	Shrink, %
1010	38	10
1020	22	10
8630	17	8

These limits should be reduced slightly for thin stock, particularly for shrink flanges. The limits for stretch flanges can be increased if there is an adjoining shrink flange. The compression in the shrink flange helps to relieve the stress in the stretch flange (and vice versa). In addition, the limits of permissible stretch may be increased by filing or grinding the edges of the blank lengthwise; flanges having a calculated stretch of approximately 100% can be formed by this procedure, as in example 9. Shrink flanges are more easily formed if the motion of the die causes some ironing of the metal in the direction of the edge of the flange.

Severe Contour Flanging. In small pieces, when enough press force is available, shrink flanges can be formed with sufficient metal flow that the flanges resemble drawn shapes, without using bend relief. Such severe flanging is shown in the following example.

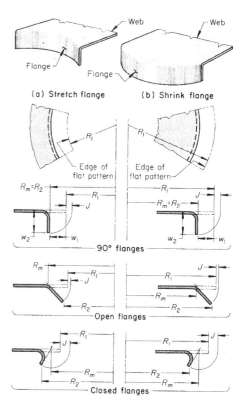

R_1 = edge radius before forming
R_2 = edge radius after forming
R_m = mold-line radius
W_1 = width of flange before forming
W_2 = width of formed flange
J = bend allowance (difference between flange width measured from the mold line after bending and the length of the neutral axis of the metal before bending)

Fig. 34 Dimensional relationships for three types of stretch and shrink flanges

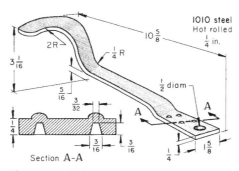

Fig. 33 Pedal lever that was edge bent to save metal and to increase the strength of the part. Dimensions given in inches

Example 8: Severe Contour Flanging of Thick Stock. The hinge half shown in Fig. 35 was formed in one stroke of an 8000 kN (900 tonf) press. By close control of clearances, it was possible to produce the contour, the sharp inside corner on the bend of the flange, and close matching of the outside planes of the flanges and the edges of the part. The metal was drawing-quality hot-rolled 1010 steel 6.07 mm (0.239 in.) thick.

A slight puckering of the metal at the corners between the flanges and the back indicated the severity of the forming. Relief cutouts were not used in this piece, but because of the severity of the forming, the five holes were pierced after forming to hold the location of the two in-line holes in the flanges within ±0.25 mm (±0.010 in.) of the three other holes. This practice is quite different from that described in example 13, in which holes were pierced before the part was flanged, and yet hole locations were held within ±0.25 mm (±0.010 in.).

The forming die was made of W1 tool steel and was hardened to 58 to 60 HRC. The die was reconditioned after each production lot of 20,000 to 25,000 pieces. The production rate was 600 hinge halves per hour.

Edge Grinding. If they are rough, the edges of stretch flanges may crack or tear at critical points. In the following example, a severe bend was made in a difficult-to-form flange, without cracking, after the rough edges had been ground to smooth the surface.

Example 9: Use of an Edge-Ground Blank to Minimize Cracking of a Severe Stretch Flange. The cross member of a truck chassis was made from hot-rolled low-carbon steel that had been pickled and oiled. It was bent from a developed blank into the channel shape shown in Fig. 36. In bending the blank, the edges of the stretch flanges cracked and tore. This problem was solved by filing the edges smooth in the portion of the blank where the stretch and shrink flanges would be formed. Bends were made parallel to the rolling direction.

Because hand filing of the rough edges was costly, grinding was tried, with good results so long as the direction of grinding was along the edge rather than across it. Cross grinding left grooves that increased stress to the same degree as did the original roughness.

Forming was done in a 3800 kN (425 tonf) hydraulic press. The die was a single-action pressure-pad type made of tool steel and had an estimated life of 300,000 pieces. The production rate was two or three channels per minute, in 800-piece production lots.

Hole Flanging. Flanges are formed around holes to increase bearing surface or to increase the number or threads that will fit in a tapped hole. A flange formed around a pierced hole is a continuous stretch flange, and Fig. 37 illustrates flange dimensions (in relation to stock thickness, hole size, and flange height) in a particular example for flange holes to be tapped in low-carbon steel. In thick stock, the length of the flange around a hole can be greater than that shown in Fig. 32, but the flange thickness will

taper rather than be relatively uniform. In the following example, flanges were formed around holes that were large compared to the flange width in order to provide a bearing surface.

Example 10: Swivel Washer with Flanged Holes. Annealed 1070 spring steel 0.25 mm (0.010 in.) thick was used to make the swivel washer shown in Fig. 38. After forming, the parts were heat treated to 46 to 48 HRC.

The strip layout for the six-station progressive die is also shown in Fig. 38. In the first station, two rectangular holes, one 6.35 mm (0.250 in.) diameter pilot hole, and two 18.0 mm (0.710 in.) diameter holes were pierced. The larger round holes were flanged to 20.3 mm (0.798 in.) in diameter by 1.6 mm ($^1/_{16}$ in.) in depth in the second station. The washer was lanced in the third station and formed in the fifth. The part was cut off in the sixth station. Station 4 was idle.

The die was made of A2 tool steel and was hardened to 60 to 61 HRC. It ran in a 130 kN (15 tonf) open-back inclinable press at 2000 strokes per hour. To maintain a minimum burr height, the die was sharpened after making 60,000 pieces. Total die life was more than 3 million pieces.

Control of Springback

Springback has little effect in the bending of low-carbon steel. It is considered only when close dimensional control is needed. Springback ordinarily ranges from $^1/_2$ to $1^1/_2°$ and can be controlled by overbending or by restriking the bend area. Factors that affect springback include ratio of bend radius to stock thickness, angle of bend (degrees of bend from flat), method of bending (V-bending or wiping), and amount of compression in the bend zone. The amount of springback with a carbon steel and a high-strength low-alloy steel is compared in Fig. 39.

When the bend radius is several times the stock thickness, the metal will need more overbending to stress it beyond the yield point than when the radius is $2t$ or less. A greater amount of overbending is needed to correct for springback on small bend angles than on large bend angles.

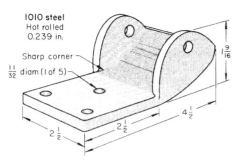

Fig. 35 Hinge half on which flanges were formed flush with adjacent surfaces. Dimensions given in inches

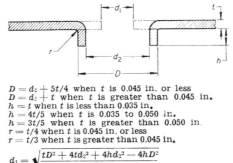

$D = d_2 + 5t/4$ when t is 0.045 in. or less
$D = d_2 + t$ when t is greater than 0.045 in.
$h = t$ when t is less than 0.035 in.
$h = 4t/5$ when t is 0.035 to 0.050 in.
$h = 3t/5$ when t is greater than 0.050 in.
$r = t/4$ when t is 0.045 in. or less
$r = t/3$ when t is greater than 0.045 in.

$$d_1 = \sqrt{\frac{tD^2 + 4td_2^2 + 4hd_2^2 - 4hD^2}{9t}}$$

Fig. 37 Dimensions of flanged holes to be tapped, as a function of thickness, for low-carbon steel

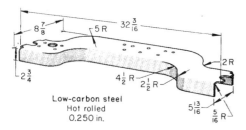

Fig. 36 Truck-frame cross member that was bent from edge-ground blanks to prevent cracking and tearing of the stretch flanges. Over the length of the two stretch flanges, the 7.9 mm ($^5/_{16}$ in.) inside bend radius was increased, varying to a maximum of 25 mm (1 in.). Dimensions given in inches

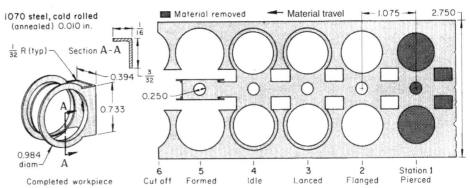

Fig. 38 Swivel washer, with flanged holes, that was made in a progressive die. Dimensions given in inches

In curved flanges, the radius of curvature and the flange length have an effect on the tension or compression in the flange metal, which in turn affects springback. Springback can vary in a production run of a given part because of variation in stock thickness, variation in stock hardness or temper, tool wear, variation in tool adjustment, and variation in power input (line surges).

Multiple Bends. When more than one bend is made in a part, the effect of springback is ordinarily cumulative and may necessitate closer control of the operation than would be needed for just one bend. The following example demonstrates the variation in springback in a part with more than one bend.

Example 11: Variation in Springback in a Part with Eight Bends. The part shown in Fig. 40 was produced from 1008 steel in three operations: blank, bend, and trim. In three of the eight bends, springback reduced the 54.74 mm (2.155 in.) dimension, and in five, it increased that dimension.

To find the net magnitude of the springback and how much variation could be expected, 100 random samples were measured from a production lot of 10,000 pieces. As expected, the net effect of the springback was to enlarge the 54.74 mm (2.155 in.) dimension by amounts ranging from 0.64 to 1.52 mm (0.025 to 0.060 in.), as shown in Fig. 40.

The variation in stock thickness was ±0.05 mm (±0.002 in.). The inside radius on all bends was equal to stock thickness.

Restriking and overbending can be used to set flanges and to offset the effects of springback, as in the following example.

Example 12: Use of Restriking and Close-Fitting Tools to Control Springback in a Flanged Part. The column-support bracket shown in Fig. 41 was made of 3.35 mm (0.132 in.) thick drawing-quality cold-rolled 1008 or 1010 steel having a hardness of 48 to 51 HRB. The use of close-fitting punches and dies for certain areas and restriking other areas made it possible to obtain the dimensional accuracy necessary for hole positions, cutouts, and flanges.

All of the round holes were pierced in the flat strip before notching, trimming, punching, and forming. The positions of the small holes with respect to the center were important. The spacing of each set of four small holes was also critical. The flanges along the outer edges had to be uniform in height and set square with the adjacent surface. The open center design and the unbalanced form made it difficult to hold the hole distances and the uniform flange heights.

The forming was done in a progressive die, which was made of O2 tool steel. The cutting punch-to-die clearance was 10% of stock thickness per side. To hold the burr to a minimum height, the die was resharpened after each 50,000 strokes.

The die operated at 400 to 500 strokes per hour. The die was expected to produce parts at this rate for 2 years, and it would then be used for making replacement parts, for which the tolerances were less critical.

Accurate Location and Form of Holes

Hole position sometimes cannot be held to tolerance when a workpiece is pierced before bending. When this is so, holes must be pierced after bending, which may require the use of cam-actuated punches, specially shaped dies to support overhanging formed flanges, or punches and dies that are shaped to pierce at an angle.

Holes made before bending are likely to be displaced during bending. Whether or not this displacement can be tolerated must be carefully considered in planning the sequence of operations.

The following example typifies the kind of variation in hole location that can be expected in formed pieces. In this example, if the tolerance on the spacing of the holes had been ±0.025 mm (±0.001 in.) instead of ±0.25 mm (±0.010 in.), the holes would have been pierced after forming, as they were in example 8.

Example 13: Variation in Distance between Prepierced Holes in a Formed Part. The bracket shown in Fig. 42 was formed in three operations: blanking and piercing, bending the two small flanges, and bending the two large flanges. The 130.0 mm (5.120 in.) spacing between hole centerlines in opposite flanges had to be held within ±0.25 mm (±0.010 in.).

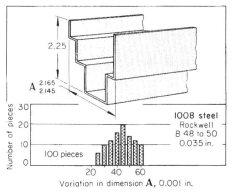

Fig. 40 Variation in flange spacing caused by springback in a part with eight bends. Dimensions given in inches

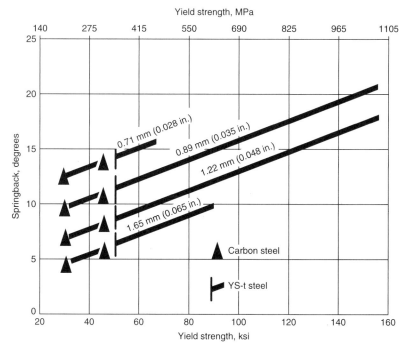

Fig. 39 Springback of a carbon steel and a high-strength low-alloy steel

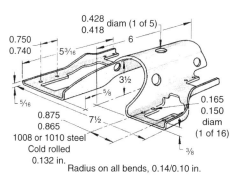

Fig. 41 Bracket with unbalanced shape that was held within bending tolerances by restriking and by extra-close tolerances on some parts of bending dies. Dimensions given in inches

A reference dimension of 117 mm (4.620 in.) between the inner edges of the holes was computed for the purpose of checking hole spacing with vernier calipers. Caliper measurements of that dimension on 100 randomly selected pieces (from a run of 10,000 pieces) showed variations of −0.025 to +0.18 mm (−0.001 to +0.007 in.) (Fig. 42).

Free bending of a workpiece is likely to displace holes from true position and to distort and elongate them. In many press-bent workpieces, this deformation is slight enough to be negligible, but it sometimes causes serious difficulties in assembly or fitting. The likelihood of distortion is minimized when holes are located at least one stock thickness away from the beginning curve of a bend. Other precautions include trapping the hole with a holddown pad that is heavily loaded or relieving the area around the hole with a crescent-shaped cutout. Additional information on the design and protection of pierced holes during subsequent bending is available in the section "Effect of Forming Requirements" in the article "Piercing of Low-Carbon Steel" in this Volume. Examples in that article describe applications in which holes were pierced after bending to avoid hole distortion and displacement during the bending operation.

An instance of acceptable distortion of pierced holes by subsequent bending is described in the section "Piercing Holes at an Angle to the Surface" in the article "Piercing of Low-Carbon Steel" in this Volume. That article also contains other examples on piercing holes before and after bending.

Accurate Spacing of Flanges

The distance between flanges is another dimension that depends on accuracy of bending. For assembly purposes, this distance may need to be held very closely. Ordinarily, the dimensions at the bases of the flanges are fairly uniform, and variation in the distance between flanges is

greater near the free edges. If a flanged part is to fit over a mating piece, oversized dimensions may be preferable to undersized dimensions, as in the following example.

Example 14: Flange Spacing That Was Obtained by Maintaining Tight Gibs and a True Ram. The bracket shown in Fig. 43 was produced from 5.54 mm (0.218 in.) thick commercial-quality 1010 steel by shearing, blanking, forming, and piercing. The flanges and other bends in the bracket were produced in a five-station transfer die mounted in a 3600 kN (400 tonf) mechanical press. The opposing holes in the flanges were either 16.0/15.9 or 12.8/12.7 mm (0.630/0.626 or 0.505/0.501 in.) in diameter and were pierced after forming. Bend radii were 5.6 mm (0.22 in.). Workmetal hardness was 55 HRB maximum.

For proper fit of the bracket with its mating part, the dimension between the two opposing flanges had to be held between 71.12 and 72.64 mm (2.800 and 2.860 in.); the basic dimension was 71.12 mm (2.800 in.). Figure 43 plots the distance between flanges as measured on 74 pieces during a total production run of 6345 brackets. As these data show, all samples were well within tolerance.

To obtain this degree of accuracy on the heavy-gage bracket, gib clearance in the press was kept between 0.15 and 0.20 mm (0.006 and 0.008 in.), and parallelism of the face of the ram with the top of the bolster was maintained within 0.51 mm (0.020 in.). In addition, before forming, the blanks were checked for thickness, and the forming tool was adjusted to compensate for variations. The dies were made of A2 tool steel and were hardened to 60 to 62 HRC.

Press Forming

Press forming is a metalworking process in which the workpiece takes the shape imposed by the punch and die. The applied forces may be tensile, compressive, bending, shearing, or various combinations of these. As in bending operations, fabrication may also include various cutting operations, such as blanking, trimming, and piercing. Presses are used to fasten parts together by rivets or by plastically deforming

mating areas in either or both of the parts. In operations such as staking, folding, crimping, curling, or press assembly, projections of various sizes and shapes are deformed so that the assembly is reasonably strong.

This section describes the accuracy in press forming of steels, effect of workpiece shape, and the press forming of ribs, beads, and bosses. If a part can be made by forming sheet metal in a press, this is ordinarily the least expensive method of manufacture, except when only a few pieces are needed. The low cost per piece results from the high production rates and the efficient use of metal usually obtainable in the press forming of sheet metal. The accuracy inherent in press forming is satisfactory for most requirements. Greater accuracy can be achieved by the use of precision tooling and by maintaining close control over press conditions.

In some applications, the metal requires appreciable stretching in order to retain the shape of the formed part. Speed of forming has little effect on the formability of steels used for simple bending or flanging or for moderate stretching. The maximum velocity of the punch when it contacts the blank in such conventional press forming is usually not greater than approximately 1 m/s (200 ft/min). When local stretching is more than 20%, workpiece movement or flow over the face of the punch or the blankholder may be considerable. The flow of the metal in such operations is controlled by frictional forces, so stretching failures are sensitive to speed. In this case, a maximum punch velocity of 0.2 m/s (40 ft/min) is recommended, although the critical value differs for each steel and die combination. High punch speeds also shorten tool life.

Accuracy in Press Forming

Tolerances that can be maintained in the press forming of steel depend on press condition, accuracy of the tools, workpiece shape, and hardness and variation in the thickness of the work metal.

Flange Width. In forming 90° flanges, if the break lines in the plan view and side elevation are straight or nearly straight, there should be no difficulty in holding a tolerance of ±0.76 mm (±0.030 in.) on flange width. When flange break lines are curved significantly, tolerance usually must be increased to ±1.52 mm (±0.060 in.).

The overall length and width of large formed parts such as automobile hoods and deck lids can usually be held to a tolerance of ±0.76 mm (±0.030 in.). For small formed parts, closer tolerances can be met in production.

Bend Angle. Tolerances that can be held on bend angles depend greatly on the thickness and hardness of the work metal and on the flange height, because these factors affect springback. A 90° bend in low-carbon steel will spring back approximately 3°; therefore, either the dies are built to overform by this amount or the bend region is compressed to set the corner. To provide enough metal for gripping in the die, the

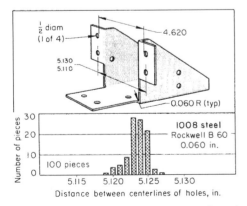

Fig. 42 Variation in distance between centerlines of holes prepierced in opposite flanges of press-bent workpieces. The plotted distances from centerline to centerline were determined by adding 13 mm (0.500 in.) to the gage readings of distances between the inner edges of the holes. Dimensions given in inches

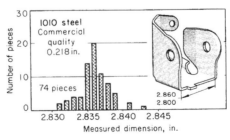

Fig. 43 Accuracy of flange spacing obtained by maintenance of small gib clearance and close ram-to-bolster parallelism in press bending of control-arm brackets for passenger-car frames. Dimensions given in inches

minimum inside flange height should be two times the stock thickness, plus the bend radius.

Distance between Holes. Distances between the centerlines of pierced holes that lie in a common surface of a formed part are commonly held within a total tolerance of 0.25 mm (0.010 in.) or less (see the article "Piercing of Low-Carbon Steel" in this Volume). The relation of hole positions in parallel or in-line surfaces of formed parts can be held to ±0.38 mm (±0.015 in.); holes with a smaller tolerance on the centerline distance should be pierced after forming (Fig. 44).

Distance between Offset Surfaces. In ordinary commercial practice, the distance between two offset flat surfaces in parallel planes can be maintained within ±0.38 mm (±0.015 in.). For closer tolerances, extreme accuracy must be built into the tools, and the stock thickness must be held closer than normal mill tolerances. The following example describes the techniques used to maintain a tolerance of ±0.25 mm (±0.010 in.) on the distance between surfaces.

Example 15: Variation of a Critical Distance between Offset Parallel Planes. On the brake-assembly backing plate shown in Fig. 40, the distance between the tops of the six pads (A to F, Fig. 45) and the bottom of the center mounting face had to be within ±0.25 mm (±0.010 in.). The ranges of dimensions for all six pads as measured on 22 pieces are plotted in the graph in Fig. 45. Of the 132 individual measurements, only 7 exceeded the limits. The spread in dimensions was the result of a combination of variables, including stock thickness (±0.18 mm, or ±0.007 in.), springback after forming and coining, and condition of the press.

Springback was unpredictable and varied among different lots of steel. Therefore, the die was made extra rigid, and the inserts for coining the pads (inserts 1 and 3, section B-B, Fig. 45) were made adjustable to compensate for variations in stock thickness. Coining reduced stock thickness at the 1600 mm² (2.5 in.²) pads to 2 to 2.3 mm (0.080 to 0.090 in.)—a 15% reduction.

The backing plate was made in two operations, both in a 7.1 MN (800 tonf) mechanical knuckle press with a 125 mm (5 in.) stroke, which was operated at 20 strokes per minute. In the first operation, coil stock was fed into a progressive die, where the inner portions of the part were formed, the six pads were partly completed, the center hole was pierced, and the outside diameter was blanked. The outside flange was formed, and the six pads were coined to height and parallelism in a second die (section B-B, Fig. 45).

The press slide was parallel to the bolster within 0.25 mm/m (0.003 in./ft). The slide and the press bed were large enough to keep the die set centered, thus minimizing the possibility of error. The die set had four heavy guideposts in long bushings. The main parts of the punch and die were made of 6145 steel hardened to 45 to 50 HRC and ground after hardening, and the inserts for coining (inserts 1 through 4, section B-B, Fig. 45) were made of O1 tool steel hardened to 60 to 61 HRC and ground.

Holes Close to a Bend. Whether a hole that is close to a bend is pierced before or after forming depends on the function of the hole, its closeness to the bend, the bend radius, and the stock thickness. When the distance from the edge of the hole to the inside surface of the other leg of the bend is less than 1½ times the stock thickness, plus the bend radius, the outer portion of the hole is likely to deform as a result of stretching of the metal. If the deformation is acceptable, the hole can be pierced before forming; otherwise, it must be pierced after forming. When the offset angle on formed parts is unimportant, stating the minimum flat and maximum radius dimensions,

as in Fig. 44, allows the maximum practical tolerance in producing acceptable parts.

Trimmed edges can be held to close tolerances when they are trimmed by one punch and when trimming is done after any forming that would cause distortion, as in the following example.

Example 16: Forming and Accurate Trimming in a Progressive Die. Figure 46 shows an eggbeater side frame that was produced in a four-station progressive die from 1 mm (0.040 in.) thick cold-rolled 1008 to 1010 steel strip 19 mm (3/4 in.) wide. For accurate fit in assembly, the end of the part was trimmed to the close tolerances indicated in Fig. 46 after the U-shaped bead flanked by the vertical tabs had been formed. The bead was formed by drawing metal from the surrounding area and by stretching.

The stock had a No. 6 edge, a No. 3 finish, and a hardness of 55 to 65 HRB. The No. 6 edge (a square edge produced by edge rolling the natural edge of hot-rolled strip or slit-edge strip) was relatively burr free. Because the edges of the strip were also the edges of the completed part, the use of a No. 6 edge eliminated a deburring operation. The No. 3 finish was particularly suited to bright nickel plating without prior polishing or buffing. Stock was purchased to average in the lower part of the standard thickness range of ±0.05 mm (±0.002 in.) and within the standard width tolerance of ±0.38 mm (±0.015 in.). The sequence of operations in the four die stations was as follows:

- Lance and form semicircular indent, and notch-trim
- Pierce rectangular slot at one end, pierce round hole, and emboss weld projections
- Lance tabs, form bead, and form flange around round hole
- Cut off (trim) and form contour

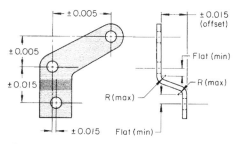

Fig. 44 Tolerances on hole position and distance between offset parallel surfaces on formed parts, and recommended dimensions to be specified for formed offsets. The tolerances shown apply to holes pierced before forming; holes in offset parallel surfaces with position tolerances less than ±0.38 mm (±0.015 in.) should be pierced after forming. Dimensions given in inches

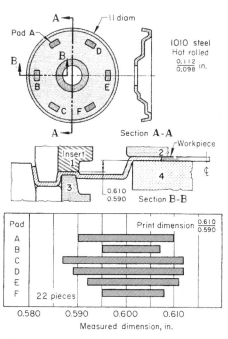

Fig. 45 Variation of offset distance in a formed and coined backing plate for a brake assembly. Dimensions given in inches

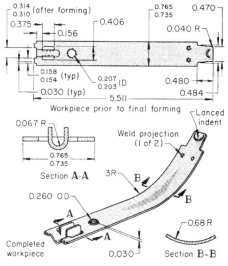

Fig. 46 Eggbeater side frame that was formed and accurately trimmed in a progressive die. Dimensions given in inches

The die, made of D2 tool steel, was used in a 665 kN (75 tonf) press; die life between regrinds was 65,000 pieces. Mineral oil was used as the lubricant. The production rate was 360 pieces per hour; annual production quantity was 600,000 pieces.

Close Tolerances. Closer-than-conventional accuracy can be attained in sheet metal parts with accurate dies, precise location of the parts in the dies, and handling equipment designed to avoid damage to semifinished or finished workpieces. Press condition is also an important factor. The following example describes the techniques used for close-tolerance forming in progressive dies.

Example 17: Use of Progressive Die to Meet Close Tolerances on a Lamp Bracket. The lamp bracket shown in Fig. 47 had several tolerances that were closer than normal. Three holes had to be pierced to a tolerance of +0.13, −0 mm (+0.005, −0 in.) on diameter, and two of the holes had to be in line within 0.25 mm (0.010 in.) after forming. In addition, when the 104° bend was made with the preformed flanges out, the radius of bend on the flanges had to be held within 0.38 mm (0.015 in.) total indicator reading, and the 10° angle of the flanges had to be held within ±5°. The bracket was produced in a seven-station progressive die with two piloting stations that could have been used for auxiliary operations such as shaving or restriking, if necessary. The operations were as follows:

- Notch for stop, pierce one 10 mm (0.395 in.) hole, pierce two 6.7 mm (0.265 in.) holes
- Blank the contour of the ears
- Pilot
- Blank the partial contour
- Form the two long flanges
- Pilot
- Form the 104° angle and cut off

The material was cold-rolled 1010 steel strip in No. 4 temper. The strip was 2.4 mm (0.095 in.) thick by 86 mm (3⅜ in.) wide. The progressive die was mounted in a 665 kN (75 tonf) straight-side mechanical press with a flywheel drive, a 152 mm (6 in.) stroke, and a maximum rate of 60 strokes per minute. The press was equipped with an air clutch. Ball-lock punches, die bushings, and easily reproducible die sections were used throughout.

The die was made of D2 tool steel hardened to 58 to 60 HRC and had a life of 55,000 pieces per regrind. Production was discontinued after 2.5 million lamp brackets had been made.

Flatness. The longitudinal and transverse stresses set up during the forming of a part can cause it to warp, particularly if the part is made of steel that is not uniform. Depending on the size and shape of the part, either smooth flattening or embossing can be used to maintain flatness.

Press Condition. The accuracy and condition of presses must be maintained within close limits when tolerances on formed parts are critical, regardless of the type of operation. Press slides that are not parallel with the bed at the bottom of the stroke can cause uneven stock thicknesses when the punch bottoms against the die surface. Unbalanced forming forces can shift the punch, producing out-of-tolerance workpieces. In some forming applications, shifting of the punch can be minimized by tipping the workpiece to balance the forces. Heel blocks and other means of positively maintaining the punch-to-die relationship are used to overcome unbalanced forming forces and shifting of the punch slide.

Workpiece Shape

The size and shape of the workpiece and the number of operations needed to make it must be considered in determining press capacity (both force capacity and bed size) and the type of tooling used. Open-end parts, or parts with one or more open edges, can be formed two or more at a time from a single blank. Sheet metal elbows that cannot be made from tubing are formed four halves at a time, then separated and assembled into two elbows. Small flanged parts with a low ratio of flange width to stock thickness are often difficult to form because of slippage and unbalanced forces. Forming two parts at a time can balance the forces and reduce scrap. More information on the factors that affect press selection is available in the article "Presses and Auxiliary Equipment for Forming of Sheet Metal" in this Volume.

Recessed parts require special precautions in forming to avoid wrinkling in the flat area surrounding the recesses and to prevent cracking the corners of the recess. In the forming of a deep recess in a large panel, for example, adjustment of the blankholder pressure may be necessary to prevent such failures. In this application, the deep recess is formed first, with the material being drawn into the recess from the end and two sides of the blank; stretching done in the final forming removes wrinkles.

When the cross section of the recess is a circular arc, acceptable percentages of stretch for recesses with various height-to-diameter, h/d, ratios are:

h/d ratio	0.10	0.15	0.20	0.25	0.30
Stretch, %	3	6	10	15	22

Dish-shaped parts having only one recess, of regular or irregular shape, are commonly formed by stretching the metal over a punch. The punch nose should be smoothly contoured so as not to trap the metal and should have as large a radius as possible. Parts with a straight sloping surface can be free formed by stretching between a clamp ring and a small-diameter punch. On large parts, stretch is most severe near the punch, and the work metal elongates and drapes from its own weight, forming an undesirable concavity in the wall. Both of these defects, uneven stretch and concavity, can be minimized by using a stepped punch.

Cone-shaped parts that are formed by a combined stretch-and-draw operation can be made in a progressive die without first cutting the contour. This reduces the number of die stations that are required, and more than one part can often be formed at a time.

Shapes with Locked-In Metal. In the forming of some shapes, metal may become locked-in (formed so that metal flow is stopped) before enough of it has been drawn into the cavity to form the part completely, and the metal sometimes fractures before the punch has reached the bottom of the stroke. In some cases, the strain that causes fracturing can be relieved by piercing a hole or lancing the metal in a noncritical area.

Severely Formed Shapes. Developed blanks are sometimes used to provide sufficient metal in critical areas of severely formed parts. Preforming helps to distribute the metal before final forming and restriking operations, thus reducing the severity of these operations. Edge condition and ductility greatly influence the success of severe forming.

Offset Parts (Edge Bending). Scrap loss can sometimes be reduced when blanks can be produced in simple rectangular form and subsequently edge bent in the final shape. Whether or not the severity of the edge-bending operation will adversely affect subsequent forming operations must be considered. A potential savings of work metal can be quickly lost if edge bending work hardens the pieces to the extent that many are broken in forming. In some applications, the tensile and compressive stresses set up in edge-bent blanks can be counteracted by forming stretch flanges around the inner contours of the bends and compression flanges on the outer contours. The same flanges made on a cut blank instead of an edge-bent blank could fracture during flanging. On the other hand, reverse flanges made on an edge-bent blank may cause more severe stress than they would on a cut blank and may possibly lead to a high scrap rate.

The edge-bending operation and its tooling are an added expense in the fabrication of a part.

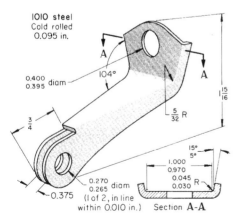

Fig. 47 Lamp bracket that was press formed to close tolerances in a progressive die. Dimensions given in inches

This expense must be compared to the savings of work metal when one method is to be selected over the other.

Large Irregular Shapes

Large parts are usually formed in single-operation or compound dies mounted in large-bed presses. If the part is irregular in shape, care must be taken in shaping the blankholder surfaces, in planning the orientation of the workpiece in the die, and in planning the die operation to obtain approximately even stressing of all areas of the workpiece and to form the piece in the fewest possible operations.

To make shallow draws, sometimes the metal must be locked in place by the blankholder in order to stretch the sheet metal over the punch enough to set the contours. Draw beads or selective local applications of drawing compound are used to restrict and control metal flow.

Large, irregularly shaped parts are often made by preshaping blanks before forming. This method reduces scrap when making the blank and neutralizes the tensile and compressive stresses in the flanges. The blanks for automobile fenders and other parts are preshaped by bending and tack welding them into a cone shape.

Use of Shaped Blankholders. Some parts that are made from preformed blanks are formed with the aid of shaped blankholders to avoid distortion of the previously formed contours. In addition, parts that are not completely formed in one operation may have to be clamped in a shaped blankholder in a second-operation die. The part in the following example had a peripheral offset that served as a locating surface in the second-operation die, but the part had to be clamped to prevent distortion during final forming.

Example 18: Holding a Previously Formed Workpiece in a Shaped Blankholder to Prevent Distortion during Subsequent Forming. After the first forming operation, the blank for the fan shown in Fig. 48 had a 4.78 mm (0.188 in.) offset outer rim and a 22 mm ($7/8$ in.) deep recessed inner web. The surfaces were used as locators in a shaped blankholder for the subsequent trimming and forming operations.

Spring-loaded pressure pads were used to hold the rim and the web while the blades were twisted. The pressure pads bottomed at the end of the stroke so that any deformation in the rim or the web could be flattened. The 4.78 mm (0.188 in.) offset was placed over a close-fitting plug to prevent radial distortion of the rim.

Three operations were needed to produce the part. In the first operation, the 4.78 mm (0.188 in.) offset was formed to 471 mm (18.550 in.) in diameter, and the 22 mm ($7/8$ in.) deep recess was formed at the same time. In the second operation, the 485.7 mm (19.121 in.) outside diameter was trimmed, and the 44.6 mm (1.755 in.) diameter hole and 24 T-shaped cutouts at 15° spacing were pierced. The T-slots

gave shape to the fan blades prior to twisting at 45°, which was done in the third operation.

The cutting elements in the dies were made of A2 tool steel. The forming dies for the offset rim and recessed web were of low-carbon high-strength alloy cast iron. The blade formers were made of prehardened low-alloy tool steel. The workpiece was made of a 0.89 mm (0.035 in.) thick cold-rolled 1010 steel blank 510 mm (20 in.) square.

Recesses in Large Panels. The forming of recesses in large panels at some distance from the edge can be difficult. Because the sheet is large, the recesses appear to be shallow, but forming may actually be severe in terms of localized deformation.

If the recess is fairly small and far enough from the edges of the blank so that there is a large resistance to metal flow, the metal can be overstressed at the punch nose and can fracture or tear. To prevent tearing, blankholder pressure is kept as low as possible to allow almost unrestricted metal flow, but this may produce wrinkles radiating from the lip of the recess. These wrinkles can be removed by stretching the sidewalls of the recess in a second forming operation.

Large cup-shaped parts that must be stronger in the bottom surface than in the wall are often designed with tapered wall sections. A tapered section can be made by machining the part before or after drawing.

High-strength low-carbon steels of thinner gage than conventional low-carbon steels can be used without tapering, but they are more difficult

to draw and are more expensive. The use of a tapered blank for a cup-shaped part is described in the following example.

Example 19: Forming a Large Disk from a Tapered Circular Blank. The truck-wheel disk illustrated in Fig. 49 was formed from a blank 610 mm (24 in.) in diameter that had been roller tapered from a 500 mm (19 in.) diameter blank with a 90.5 mm ($3\frac{1}{2}$ in.) diameter center hole. Taper was approximately 0.04 mm/mm (0.04 in./in.) of radius. The material was annealed hot-rolled 1012 or 1015 steel, 9.52 mm (0.375 in.) thick, abrasive blasted to remove scale and oxide.

The blanks were tapered back to back, two at a time, giving each disk only one rolled surface. The tapering rolls were stopped before reaching the edge of the blank, in order to maintain an even taper. Otherwise, lack of resisting stock at the edge would have caused a torn edge. Tapering started at a diameter of 260 mm (10.25 in.), just beyond the circle describing the outer edges of the six round holes; this gave a stock thickness of 8.38 mm (0.330 in.) at the 17.5/12.7 mm (0.69/0.50 in.) radius.

The combined forming and trimming operation was done in a 13.3 MN (1500 tonf) hydraulic press, with the rolled surface of the work face down in the die. The blank was trimmed to 584 mm (23 in.) in diameter in a compound blank-and-draw die at a rate of 375 per hour. The die was made of O1 tool steel and was reworked after making 40,000 pieces.

A second operation sized the 498.2 mm (19.616 in.) outside diameter, enlarged the center hole to 164.4 ± 0.05 mm (6.471 ± 0.002 in.) in diameter, and pierced six 32.5 mm (1.28 in.) diameter holes, using a 7.1 MN (800 tonf) mechanical press. In the third operation, six hand holes were punched, one at a time. To attach a rim to a disk, 16 rivet holes evenly spaced around the circumference of the flange were pierced in both the disk and the rim at the same time.

The optional 1015 steel had the highest carbon content that could be used in this part. Work hardening increased its strength and hardness. If the hardness of the workpieces exceeded 91 HRB, hairline cracks radiated from the rivet holes, causing rejection, as observed when 1020 steel was used.

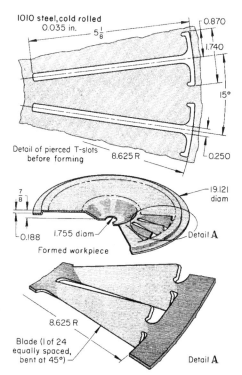

Fig. 48 Large sheet metal fan that was clamped in a shaped blankholder during the second forming operation. Dimensions given in inches

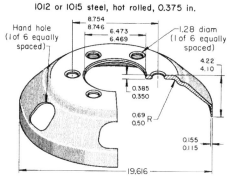

Fig. 49 Truck-wheel disk that was formed from a tapered blank. Dimensions given in inches

Localized severe forming is encountered in making many large irregular shapes by press forming. This imposes stringent demands on process planning; quality of work metal; lubrication; and the design, material, and maintenance of dies, as demonstrated in the following example.

Example 20: Severe Embossing and Hole Flanging in Press Forming a Control Arm. The severe embossing and hole flanging demanded on the automobile control arm shown in Fig. 50 required the use of drawing-quality steel for the workpiece and high-quality tool steel for the dies, close attention to tool maintenance, and the use of a heavy-duty lubricant. The stock was hot-rolled drawing-quality rimmed 1008 or 1010 steel, pickled and oiled, with a hardness of 55 HRB. Commercial-quality steel had been tried but was unsatisfactory for the severe forming. Stock thickness was 3.96 or 4.17 mm (0.156 or 0.164 in.). Wall thickness of the hole flanges had to be at least 2.67 mm (0.105 in.), and flange width was 7.62 mm (0.30 in.). The parts were formed with dies of hardened W2 tool steel. A developed blank was used, making a final trimming operation unnecessary. The operations were as follows:

- Blank to developed outline, pierce two locating holes, form center boss—done in a 1.8 MN (200 tonf) mechanical press with W2 tool steel die inserts hardened to 57 to 60 HRC
- Form in a 5.3 MN (600 tonf) mechanical press using W2 tool steel dies hardened to 61 to 64 HRC
- Pierce two holes in flanges, enlarge one locating hole, pierce oval hole in center boss—done in a 620 kN (70 tonf) inclined-bed mechanical press
- Restrike to size and sharpen radii, form dimple, flange round hole and oval hole in a 5.3 MN (600 tonf) mechanical press

- Pierce remaining holes in a 620 kN (70 tonf) inclined-bed mechanical press
- Outward flange two side holes to 35.5/35.4 mm (1.398/1.394 in.) in diameter, 2.67 mm (0.105 in.) minimum wall thickness, and 7.62 mm (0.30 in.) minimum flange height, using hydraulic equipment designed for this part

A dry-film lubricant consisting of soap and borax was applied to the blank. Where further lubrication was needed, an oil-based compound was added. Production rate ranged from 200 to 325 pieces per hour, production-lot size was 10,000 pieces, and annual demand was more than 200,000 pieces. Except for minor repairs, the dies were good for 1 year of production.

Blanks That Cannot Be Nested. Irregularly shaped parts frequently must be made from developed blanks with a contour that makes close nesting of the blanks impossible. Channel-shaped parts with flanges or web of varying width use blanks that are cut with more scrap than parts with flanges or web of unvarying width, and an excessive amount of scrap may be generated in producing them.

In some applications, material that would otherwise be wasted can be moved into a useful location after notching or lancing. This was done in the following example, in which the web was notched at the end and then spread into a V-shape to increase flange height at the ends and to reduce blanking scrap.

Example 21: Use of a Notched Blank to Increase Flange Width and Reduce Scrap in Forming an Irregular Channel-Shaped Part. A flange of varying width was needed on the channel-shaped truck-frame member shown in Fig. 51. Instead of using a contoured blank, which would have meant considerable waste in blanking, the additional flange width was gained by notching each end of a nearly rectangular blank and spreading the notch into a V-shape during the forming operation. The end of each

notch was radiused to minimize cracking during forming. The stock saved by notching the blank was 1.4 kg (3 lb) per piece.

The workpiece was made of 4.54 mm (0.179 in.) thick hot-rolled, rimmed 1008 steel, pickled and oiled. Each of the two notches in the blank was 9.1 mm (0.36 in.) wide by 181 mm (7 in.) long. The notches were spread to make V-shaped openings 56.9 mm (2.24 in.) wide at the ends. Flange widths varied from 50 mm (2 in.) at the center of the part to 127 mm (5 in.) at each end of the channel. The operations were:

- Cut blank to developed outline, and pierce large center hole. A 3.1 MN (350 tonf) mechanical press with a mechanical unloader produced 300 to 400 blanks per hour.
- Form completely, including flange around center hole, in a 16 MN (1800 tonf) hydraulic press
- Pierce all holes in the web, and trim 18.8 mm (0.74 in.) radius at two places on each end in a 2.9 MN (325 tonf) mechanical press
- Pierce holes in top and bottom flanges, and restrike flange ends in a 2.9 MN (325 tonf) mechanical press

Presses for the aforementioned second, third, and fourth operations, operating at 250 to 310 strokes per hour, were set up in a line with transfer equipment between them. In the second operation, a spreader was incorporated into the die to assist in opening the notch to a V-shape. The 181 mm (7 in.) length of the notch had been carefully developed so that the notch would spread to the required maximum width without causing the work metal to split. After forming, a 15.87 mm (0.625 in.) diameter hole was pierced at the end of each notch to remove any fractured material or other stress raisers aggravated by the severe edge forming.

The forming punches were made of 1045 steel. The wear surfaces on the punch and die were W2 tool steel hardened to 61 to 64 HRC. As a lubricant, a soap solution was dried on the stock. Some heats of steel were difficult to form; for these, an oil-based compound was used as additional lubricant.

The truck-frame member was made in lots of 27,000 pieces for an annual production of 270,000 pieces. Except for minor repairs, such as replacing small punches, the dies were good for 1 year of production.

Use of Draw Beads. A draw bead in a blankholder controls the movement of metal into the die cavity by providing additional resistance to metal flow. Draw-bead geometry and placement have major influence on the strains in sheet forming. The location of the beads is usually determined in die tryout; dies for producing similar parts can be used as a guide. The restraining action must be sufficiently high to provide the force necessary to prevent wrinkling and surface distortions in both the flange and unsupported areas, but not so high to result in excessive thinning or splitting in the sheet metal. Draw bead restraining force is caused, to a large

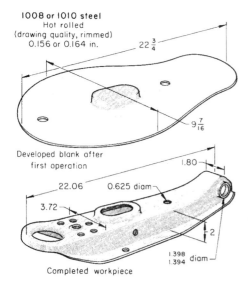

Fig. 50 Press-formed control arm on which embossing and hole flanging were of near-maximum severity. Dimensions given in inches

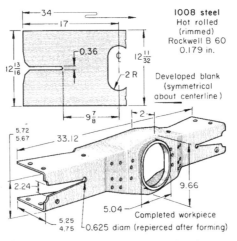

Fig. 51 Press-formed truck-frame member that was notched and spread to increase flange width locally and therefore save material in blanking. Dimensions given in inches

extent, by deformation resulting from bending and unbending of the sheet metal in the draw bead area, and to a lesser extent, by the frictional forces resulting from the sliding contact between the binder and the sheet metal. The bending force was found to be inversely proportional with the radius of curvature of the bead, whereas the friction force was found to be directly proportional with the coefficient of friction and the binder force.

A single bead is generally placed around the cavity, and additional beads are placed in areas where more control is needed. Conditions may indicate that the bead size should be reduced or that the whole bead omitted in some places. Short beads can be placed at an angle to deflect metal into or away from local areas.

Whether the bead is placed in the draw ring or in the blankholder is determined by the die construction. Placing the groove in the upper member has the advantage that it will not catch dirt. However, the groove should be put in the member that is to be altered during spotting for mating of opposing surfaces. For convenience in making alterations, this is usually the lower member.

Unless they are part of the product design, draw beads are placed outside the trim line, as shown in Fig. 52(a). The trim line can be on the punch or the blankholder. A locking draw bead, such as that shown in Fig. 52(b), is used to provide maximum restriction to metal flow. Locking beads are used when forming to shape is done primarily by stretching the metal under the punch, rather than by moving metal into the cavity.

Figure 52(c) shows the use of conventional and locking beads. Here, conventional beads control metal flow into the die cavity until the last portion of the punch movement, when the locking bead gradually engages the metal to restrict its flow. The last fraction of a millimeter of punch travel causes stretch in the metal under the punch.

Beads in the concave surfaces of a blankholder are usually 3 mm (0.12 in.) deeper than beads on the top or straight surfaces. This eliminates locking on the top surfaces during preforming of the blank to the shape of the concave blankholder surface.

Sometimes, draw beads need not be used for the full depth of draw, or need to be used only in certain locations, such as the corners of regular or irregular polygon-shaped parts. In such cases, some of the material may be allowed to slip through the draw bead to be retained at the end of the stroke only by blankholder pressure. More information on the design and construction of draw beads is available in Ref 9 to 11 and the article "Deep Drawing" in this Volume.

Use of Flat Binders. In a flat binder, sheet metal restraining force is caused by the friction force resulting from the sliding contact between the binder and the sheet metal. The friction force was found to be directly proportional with the coefficient of friction and the binder load. Restraining forces in flat binders are largely determined by the binder load. Hydraulic or pneumatic cushions with constant binder load during forming have been used to provide the blankholding action in many press systems.

During stamping, the flange tends to thicken in specific areas, and this thickening tendency is highly dependent on the normal anisotropy of the sheet metal. Flange thickening results in the development of separating forces that exceed the binder load and therefore increase the gap between the binder surfaces. This tendency to open up the binder can be a problem in stamping because the increased gap reduces the restraining force needed to form the desired part. To account for the nonuniform thickening in the flange, binders undergo "spotting," which consists of local grinding of a specific area, by trial and error, to produce a more uniform distribution of binder load.

A uniform pressure in the binder area can be achieved without "spotting" by using die cushion pins to flex the binder. The cushion pins extend through the bolster such that the binder load can be concentrated in certain areas. The height of the pins can affect the distribution of the binder load. Other methods for locally controlling binder load distribution involve segmented binders, shown in Fig. 53. Different segments can be independently actuated by hydraulic cylinders to give spatial and temporal binder load control.

Forming of Ribs, Beads, and Bosses

Unsupported sheet metal surfaces that may buckle or oil can are often stiffened by the addition of long, thin bosses called beads or ribs. Round or nearly round bosses are sometimes called buttons. Dimples are occasionally used as a recess for a rivet or screw head. The forming of ribs, beads, and bosses is a combination of bending, stretching, and drawing and involves high shear forces.

Design of Bosses. Bosses are usually approximately one stock thickness in depth. The radius of curvature on the inside of bends is approximately one stock thickness. A typical bend appears to be approximately four stock thicknesses wide on the convex side and approximately three stock thicknesses wide on the concave side. Some recesses for screw heads have sloping sides with very little curvature.

Bosses are typically produced at the same time that other forming operations are done and in the same dies, although they can be formed at separate stations in a progressive die. Die clearances can be critical in the forming of decorative bosses and embossed lettering, especially when definition must be sharp and detail must be accurate. On the other hand, reinforcing beads usually do not require great accuracy and can be produced in dies with the clearances typical of ordinary forming.

Use of Bosses. Bosses are often used to provide flatness, stiffness, and reinforcement to formed parts. They can also serve as locators for subsequent operations. In example 20, an oval boss was used as a hole-locating surface and as the hole flange. The most frequent use of embossed beads is to lighten the weight of products by making it possible to use thinner metal than would be feasible without bosses, as described in the following example.

Example 22: Use of a Bead for Stiffening a Corner Bracket. The corner bracket shown in Fig. 54 had a bead on the flat surface to produce stiffness. The work material was 1.09 mm (0.043 in.) thick cold-rolled 1008 or 1010 steel. The bead made the use of heavier stock unnecessary. A short flange around the periphery of the bracket also added to the stiffness.

The bracket was produced in a progressive die in the following operations: trim to shape, pierce, form bead, form flange, and pinch trim from the carrier strip. The die was run in a 4.45 MN (500 tonf) press operating at 600 strokes per hour. Forming of the bead caused some distortion of the pilot holes and resulted in stretcher strains that extended from the inside radius of the bead to the edge.

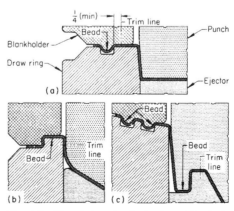

Fig. 52 Use of draw beads. (a) Conventional. (b) Locking. (c) Combined conventional and locking. Dimensions given in inches

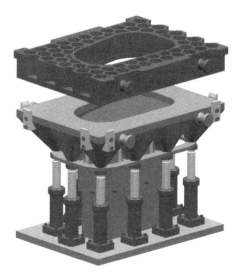

Fig. 53 Flexible flat binder with individually controlled hydraulic cylinders

The progressive die was made of D2 tool steel hardened to 62 to 64 HRC. One hour per week was needed to clean the die and to make minor repairs. Die lubricant was chlorinated oil.

Shells

Shells are produced in various ways, as described in this section adapted from Ref 12. For shells, a special type of press forming is deep drawing, where flat blank is pressed into a die opening to draw the metal into the desired shape without folding of the corners (see the article "Deep Drawing" in this Volume).

Doubly Contoured Shells. In forming doubly contoured shells, the major problem is in controlling wrinkling in the outer portion of the blank. In thin materials, which are prone to wrinkling, it is usually necessary to incorporate a blankholder to constrain the outer edge. Alternatively, for lower-volume production, the use of rubber pad, urethane, or fluid forming techniques may be used. When a blankholder is used, the material is stretch formed over the punch, and stretching limits can be determined by measuring the initial blank length, R_1, and the final perimeter length of several cross sections through the center of the part, R_2, and using the maximum R_2/R_1 values given in Table 19 for stretch flanging.

Deep Recessed Shells with Vertical Walls. In these shells, the walls are produced by drawing in the outer portion of the original blank, then bending the material through 90° and restraightening it. Due to the reduction in diameter of the blank, compressive circumferential stresses are developed, and wrinkles will tend to form in the outer flange (analogous to the buckling of a thin plate under compression)—the thinner the material, the more prone to wrinkling.

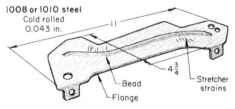

Fig. 54 Corner bracket that was stiffened by beading and flanging. Dimensions given in inches

To prevent this, pressure is applied by means of the blankholding plate—for mild steel sheet, the interfacial pressure is typically 2 to 3.5 MPa (0.3 to 0.5 ksi), and for austenitic stainless steel 10 to 14 MPa (1.5 to 2.0 ksi).*

In the initial stage of the draw, the blank is stretched across the punch (Fig. 55a). As the punch is pushed through the draw ring, the metal is bent over the die radius and straightened into the cup wall (Fig. 55b). The outer portion of the blank is reduced in diameter as the metal is drawn in (Fig. 55c). The load required to shrink the blank and to bend and unbend the metal over the die radius is carried by the wall. Therefore, the strength of the wall controls the maximum size of shell that can be drawn. An analysis of wall thickness shows that there are two thin areas at which fracture may occur. The most common fracture is at the nose of the punch, where the material was stretched between punch and die in the initial stage of the draw—this is related to the punch radius, as shown in Fig. 56. The punch radius should be above $6t$ for drawing-grade mild steel and $4t$ for austenitic stainless steel (where t is the thickness of the material) to avoid severe thinning.

The second region occurs further up the wall, above the die impact line, that is, in the drawn region where the strain history is complex. The material has been reduced in diameter, bent and unbent under tension while passing over the draw radius, and then carries the high tensile load to draw in the remainder of the blank. This thin area can be related to the draw radius (Fig. 57), which should be greater than $8t$ for drawing-grade mild steel and $10t$ for austenitic stainless steels to prevent severe thinning. However, the die radius should not be too great, because the conventional

Table 19 Limiting radii for contour flanging

Material	Thickness		Maximum R_2/R_1	
	mm	in.	Stretch flanging	Shrink flanging
Deep-drawing grade	1	0.04	1.44	1.10
Aluminum-stabilized steel	2	0.08	1.44	1.20
Drawing-grade rimmed steel	1	0.04	1.38	1.10
Cold-rolled mild steel	1	0.04	1.22	1.10
Hot-rolled mild steel	2	0.08	1.18	1.20

R_1, initial blank length; R_2, final perimeter length

flat blankholder does not contact the blank over the draw radius, and control of wrinkling may be difficult, especially near the end of the draw.

A third form of fracture may be encountered in deep draws when the die-profile radius is too small. This takes the form of a vertical crack in the wall and is found most often in thick-walled cups when $t/d > 0.1$ (Fig. 58).

As mentioned previously, it is usual to subdivide this category into shallow circular shells, deep circular shells, wide-flanged circular shells, and rectangular and noncircular shells.

Shallow Circular Shells with Vertical Walls. These shells are produced in a one-stage draw. The maximum height of flat-bottomed shell that can be achieved in typical commercial practice for drawing-quality rimmed steel is generally assumed to be $h/d \sim 0.75$. This assumes that the optimal die-profile radius and punch-profile radius, a moderate punch speed (200 mm/s, or 8 in./s), and a good drawing-compound-type lubricant are used. Deeper shells are produced with improvement in steel grade, and the effect is summarized in Table 20.

The effect of altering punch-profile radius and die-profile radius is shown in Fig. 59 and 60 for drawing-grade rimmed steel and austenitic stainless steel, respectively.

The use of better lubricant and slower press speeds should improve the limits for rimmed or capped steels. In the case of continuously cast aluminum stabilized steel, there is evidence to suggest that higher press speeds (5000 mm/s, or 200 in./s) will improve the drawability. Much more significant improvements are obtained by the use of fluid forming, in which the female die is a rubber diaphragm backed by hydraulic fluid at high pressure. During the draw, the water becomes pressurized and is squeezed out between the die and the shell, providing lubrication over the die radius and, more importantly, forcing the material against the punch, which reduces thinning over the punch nose. The limiting draw ratio, D/d, may be improved from 2.0 to 2.5 and the height-to-diameter ratio, h/d, from 0.75 to 1.3 for drawing-quality rimmed steel by this process.

The limits given previously have been for flat-bottomed shells but also apply to contour-bottomed shells where h is the height of the vertical portion of the wall, provided the material is not too thin, $t > d/100$.

To determine the blank diameter, D, for a given shell diameter, d, and height, h, the following formula is used:

$$D = \sqrt{d^2 + 4d \cdot h}$$

This is shown in terms of D/d versus h/d in Fig. 61.

The punch force to draw the shell is given by

$$F = \pi dtS(D/d - k)$$

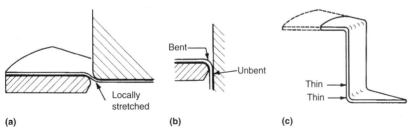

Fig. 55 Deformation areas. (a) Stretching as punch contacts blank. (b) Bending and unbending over die radius. (c) Local thinning produced by stretching and bending

*These interfacial pressures are for controlling wrinkling only. Higher pressures are used when it is necessary to hold back the blank, as in tapered shells.

where S (MPa) is the ultimate tensile strength of the material, k is usually 0.6 to 0.7, and F is in Newtons, provided d, D, and t are in mm.

Deep Circular Shells. For deep cups, multistage drawing must be used, as shown in Fig. 62. The reduction of diameter in each stage is lower with succeeding stages. Table 21 gives practical maximum values for reduction in diameter and height-to-diameter ratios achieved in four successive stages using the optimal die-and punch-profile radii, a good drawing compound, and drawing-quality rimmed steel sheet. Figure 61 also shows the number of stages necessary as a function of height-to-diameter ratio.

Substantially deeper draws, better control, and less likelihood of wrinkling can be obtained by reverse redrawing instead of conventional redrawing. In this process, the shell is turned inside out while reducing its diameter (Fig. 63). The flexure of the metal as it passes over the die radius is then all in one direction rather than first flexing one way then the other, as in conventional redrawing (Fig. 64). Reverse redrawing also has the advantage that it can be accomplished in one stroke of a triple-action press or double-action press with die cushion. Also, greater interstage reductions are possible (Table 22).

Wide-Flanged Shells. The additional flange to the shell is, of course, drawn in by the wall and therefore raises the load carried by the wall. Thus, the maximum height that can be drawn in one stage is reduced. Figure 61 gives a plot of the blank size necessary and the number of stages required to draw a flanged shell as a function of height to diameter and flange width, f. Again, the optimal die- and punch-profile radii and a good lubricant are assumed.

Rectangular Shell. The depth of rectangular shells drawn in one operation is basically dictated by corner radii. Table 23 gives the maximum heigh-to-width ratio for square or rectangular shells as a function of corner radius to shell width. Die-profile radius and punch-profile radius are assumed to be the optimum ($10t$ and $8t$, respectively) for drawing-quality rimmed sheet steel.

As in the case of all deep-drawn shells, it is important to use the correct blank size and shape in order to decrease the stresses in the walls and reduce the probability of fracture. If a

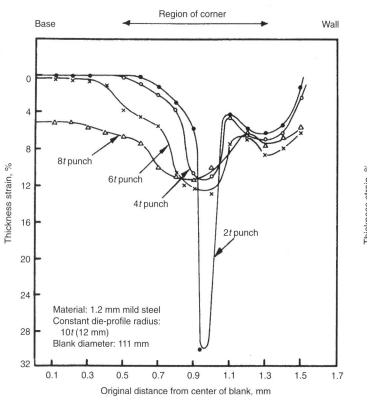

Fig. 56 Effect of increasing punch-profile radius on thickness strain (using constant blank diameter). t, material thickness

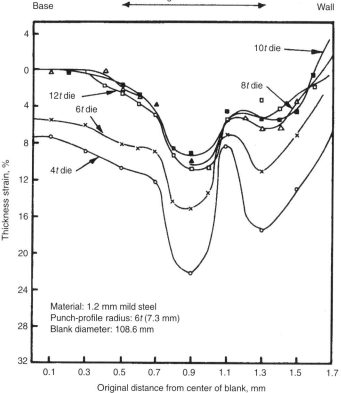

Fig. 57 Effect of increasing die-profile radius on thickness strain (using constant blank diameter). t, material thickness

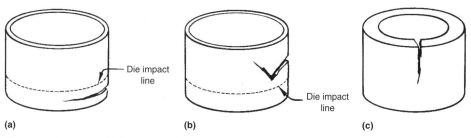

Fig. 58 Three types of failure in deep drawing. (a) Fracture over punch nose; punch nose radius is too sharp. (b) Chevron fracture in wall; die-profile radius is too sharp. (c) Vertical crack in thick-walled cups; die-profile radius may be too sharp, and blank edge may be poor.

Table 20 Maximum height and draw ratio of the single-drawn shallow flat-bottomed circular shell

Material grade	Maximum h/d	Maximum D/d
Drawing-quality rimmed steel	0.75	2.0
Deep-drawing-quality aluminum-stabilized steel	0.85	2.1
Low-carbon, niobium-stabilized, cold-rolled steel	1.1	2.25
Hot-rolled drawing-quality steel	0.7	1.9
Austenitic stainless steel grade 302	0.75	2.0

h, height; d, shell diameter; D, blank diameter

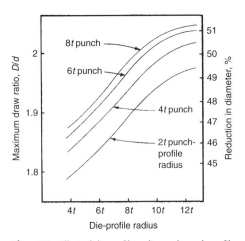

Fig. 59 Effect of die-profile radius and punch-profile radius on maximum draw ratio of single-stage cup drawn in drawing-quality mild steel. *D*, blank diameter; *d*, shell diameter; *t*, material thickness

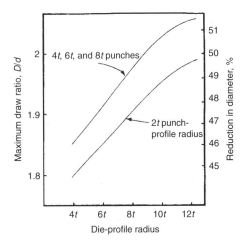

Fig. 60 Effect of die-profile radius and punch-profile radius on the maximum draw ratio of a single-stage cup drawn in austenitic stainless steel. *D*, blank diameter; *d*, shell diameter; *t*, material thickness

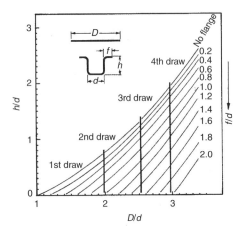

Fig. 61 Relationship of blank diameter (*D*) to cup height (*h*) and flange width (*f*)

rectangular blank is used for pressing a rectangular shell, then large ears will remain at the corners after pressing (Fig. 22).

It is general practice to cut the corners of the blank in order to avoid the ears and even up the flow. Figure 65 shows a worked example for a rectangular pressing 355 mm by 89 mm by 76 mm (14 by 3.5 by 3 in.) deep with a 19 mm (0.75 in.) flange. First, the rectangular blank dimensions are calculated by measuring the total perimeter along the major and minor center lines. Then, the amount of corner cutting is determined by reference to Fig. 66. In cases that may be critical, the blank is developed further by

rounding the cut corners, as shown in Fig. 67 (there are three examples, depending on whether $R = 0.65$, $R > 0.65$ or $R < 0.65$).

Table 21 Maximum reductions in diameter and cup height/diameter (*h/d*) for successive redraws of drawing-quality rimmed steel

Draw stage	$\dfrac{d_i}{d_{i-1}}$	$\dfrac{D_{blank}}{d_i}$	$\dfrac{h_i}{d_i}$
First draw	0.53	1.90	0.75
Second draw	0.75	2.50	1.30
Third draw	0.84	3.00	~2.0
Fourth draw	0.86	3.50	~2.8

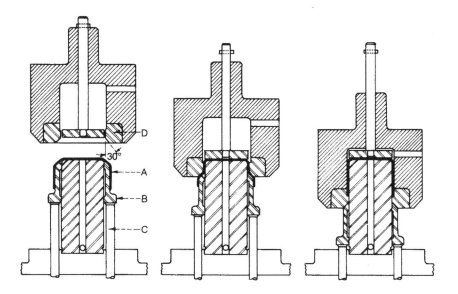

Fig. 62 Redrawing operation. The optimal die throat angle for redrawing thin sheet steel is 30°. When a drawn shell A is to be redrawn, it is placed over the draw collar B (with dies open). Partially redrawn shell illustrates how metal is caused to flow into a shell of smaller diameter and greater length. Metal flow is caused by descent of die ring D against the resistance of the draw collar B, supported by the pressure pins C. The draw collar will strip the shell from the punch during the press upstroke.

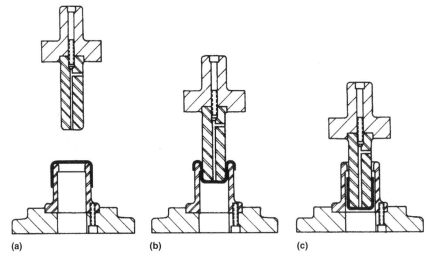

Fig. 63 The reverse redrawing operation. Three steps in reverse drawing a small thick walled shell. (a) The shell is placed over the die member. (b) The punch turns the shell inside out. (c) The reduced shell will be stripped by the sharp edge in the die cavity when the punch ascends.

Another approach that has been used is to leave the corners of the blank uncut, so that fracture high in the corners is eliminated, and deeper shells can be obtained. The difficulty is in controlling the flow of material in the sidewalls, leading to slackness and oil canning effects.

Blank orientation can play a major part, because commercial capped and stabilized drawing-quality steels develop the highest r-values (i.e., best resistance to thinning) parallel and at right angles to the rolling direction. If the blank can be oriented with these directions running

into the corners, then deeper shells will be obtained. However, unless the developed blank has substantial corner cutting, this will lead to extra scrap in cutting blanks from coil (Fig. 68). Improvements in shell depth of 30 to 40% can be achieved using the low-carbon steels with high r-values at 45° to the rolling direction, especially for shells with tight corners, $r/w < 0.2$.

Another variable used to control blank flow is the flexing of the blankholder plate by inserting shims between the plate and blankholder bolster to provide higher local pressures on the blank.

This can be very effective and completely alter the flow pattern. For a rectangular shell, the shims are usually placed to hold the sides back and encourage the corner to flow in. Draw beads can be used in a similar manner and provide more effective control; however, they tend to leave galling marks on the part.

Very deep rectangular pans should be drawn in two operations, following the procedure shown in Fig. 69. The width and length of the first-stage shell should be larger than the final shell by approximately three times the final corner radius.

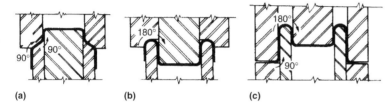

Fig. 64 Flexure in redrawing. (a) Reducing a shell supported by a collar causes the metal to flow inward 90° then upward 90°. (b) Reverse drawing also causes 180° flexure, but in one direction. (c) Overstressing can occur if the metal is pulled through more than 180°.

Table 22 Maximum reductions in diameter for successive reverse redraws

Draw stage	$\frac{d_i}{d_{i-1}}$	$\frac{D_{blank}}{d_i}$	$\frac{h_i}{d_i}$
First (conventional) draw	0.53	1.90	0.75
Second (reverse) redraw	0.70	2.70	1.50
Third (reverse) redraw	0.75	3.60	~2.9

Table 23 Maximum height-to-width ratios (h/w) for square or rectangular shells drawn in one operation as a function of corner radius-to-width ratio (r/w)

r/w	h/w
0.4	1.0
0.3	0.9
0.2	0.76
0.15	0.65
0.10	0.50
0.05	0.30
0.02	0.12

Note: These figures are conservative, especially for $r/w < 0.15$; however, considerable blank and die development will be necessary if they are exceeded.

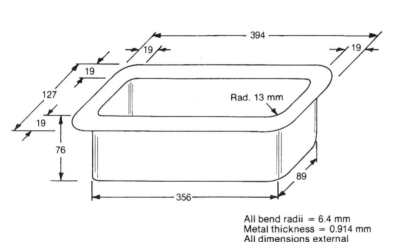

All bend radii = 6.4 mm
Metal thickness = 0.914 mm
All dimensions external

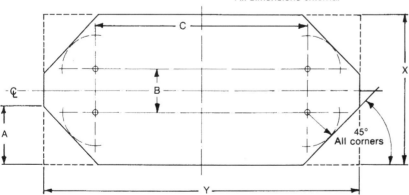

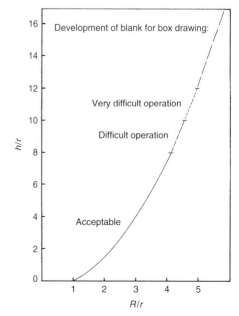

Fig. 65 Laying out the blank. The dimensions of the rectangular blank are established as X = 264 mm (10 in.) and Y = 531 mm (21 in.) by measuring the perimeter along the major and minor center lines. Mark out the positions of the center lines and then the positions of the centers of the corner radii. This establishes B = 64 mm (2.5 in.) and C = 330 mm (13 in.). From the four center points, scribe a radius selected from figures for the corner radii of 13 mm (0.5 in.) and depth of 76 mm (3 in.). This gives R = 47 mm (1.9 in.), with the note that this will be a difficult pressing and will probably require further blank development. Add ≤ 17 mm (0.7 in.) to R to allow for the flange to give R = 64 mm (2.5 in.).

Fig. 66 Plot for determining amount of corner cutting, R, as a function of corner radius, r, and height, h

The corner radius in the first stage should be four to five times the final corner radius, and the punch-profile radius should be approximately 1.5 times the final corner radius.

Deep Recessed Shells with Sloping Walls. Shells with sloping walls are more difficult to produce than shells with vertical sides because of the unsupported material in the sidewalls, which will tend to buckle or pucker. As explained previously, the blank will attempt to buckle as it is reduced in diameter. The blankholder prevents this, but after the material has passed the die orifice and moves into the shell wall, it is not constrained by a blankholder, and with further reduction in diameter, it will tend to pucker. Puckering can only be prevented by stretching the material. Failure modes are therefore two. The first failure occurs in the stretch region over or near the punch nose and is related to the work-hardening ability of the material, punch-profile radius, and coefficient of friction between punch and blank. The second failure is puckering in the sloping wall and is related to the angle of taper and the material-thickness-to-shell-diameter ratio. Obviously, the thicker the material, the greater will be its resistance to puckering. For thickness, $t > d/100$, where d is the die orifice diameter, puckering does not occur.

Because puckering is prevented by stretching the material in the shell wall, the drawing process is controlled by the blankholder force. At low blankholder force, the walls will pucker at low shell depths; as the blankholder force is increased, deeper shells will be drawn before puckering occurs. At high blankholder force, the metal will fracture over the nose of the punch, and as blankholder force is increased further, the shell height will reduce. A plot of shell height against blankholder force (Fig. 70) illustrates this effect.

There is an optimal blankholder force that produced the deepest shell. At this force, both wall puckering and splitting over the punch nose occur simultaneously at the failure point. The difficulty of drawing a shell of a given height can be determined from this plot; for example, a shell drawn to half maximum height can be produced over a wider range of blankholder force than a shell drawn to three-quarter maximum height, and therefore it will be a less critical draw.

The curves in Fig. 71 illustrate the importance of metal thickness in drawing tapered shells; doubling the metal thickness allows the shell to be drawn 50% deeper through a die throat diameter of 100 mm (4 in.) with a punch diameter of 75 mm (3 in.). The effect of sheet steel grade is seen in Fig. 72; changing from rimmed drawing-quality to aluminum-stabilized deep-drawing-quality steel improved the depth of shell by 20% through a 203 mm (8 in.) die throat diameter.

General rules for tapered shells are:

- Shallow tapered shells with height less than one-quarter the diameter at the open end, d, and with thickness $t > d/200$ can be drawn in one operation.

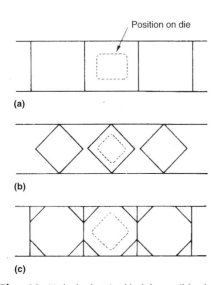

Fig. 67 Development of blank corners for rectangular shells

Corner blank development
(1) Mark out four centers, A
(2) Mark out sides and ends, S
(3) Mark out radius R (see Fig. 66)
(4) Mark out $L/2$
(5) Draw lines cd, from points O
(6) Draw radii $R_1 = R$

Shapes of corners

For $R = 0.6S$

For $R < 0.6S$

For $R > 0.6S$

Fig. 68 Methods of cutting blank from coil for drawing-quality rimmed and aluminum-stabilized steels. (a) Blanks cut across rolling direction. (b) Blanks cut at 45° to rolling direction; produces high scrap but deeper draws. (c) Blanks cut at 45° to rolling direction for deepest draws. Scrap is reduced when corner cutting is used.

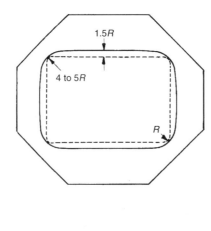

Fig. 69 Profiles of first and second stages in drawing a very deep rectangular shell with tight corner radius, R

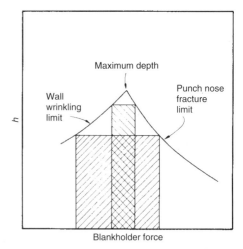

Fig. 70 Drawing of a tapered shell—a typical plot of depth of shell against blankholder force. The shallower the shell, the greater the latitude in blankholder force to produce a successful shell.

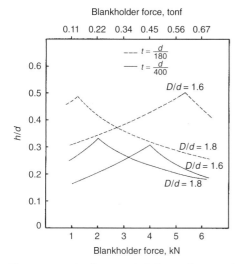

Fig. 71 Depth of a tapered shell as a function of blankholder depth. The effect of metal thickness and blank size is shown. Die orifice diameter is 100 mm (4 in.), and punch orifice diameter is 75 mm (3 in.).

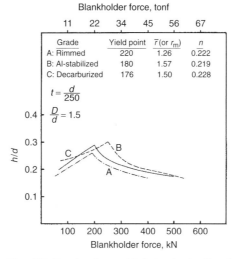

Fig. 72 Drawing of tapered shell, showing the effect of steel grade on depth of shell. Die orifice diameter is 203 mm (8 in.), and punch diameter is 150 mm (6 in.).

- Tapered shells with $1/4 < h/d < 1/2$ should be drawn into a similar shape having vertical or a small taper in a preliminary operation.
- Deep tapered shells with $h/d > 1/2$ must be drawn first into stepped shells having vertical walls and then stretched to shape in a final operation.
- where $t < d/100$, tapered shells can be drawn to between $0.4d$ and $0.5d$, provided the punch-profile radius is $10t$ or greater. However, the sides will tend to bow out at the top, and control of the shape will be difficult unless matching male and female dies are used and the part is coined in the dies at the bottom of the stroke.

Hemispherical Shells. As with tapered shells, the main problem encountered in drawing spherical shells is puckering. To avoid this, blankholder forces are higher than in drawing vertical-sided shells, in order to control flow and promote stretching.

The important parameters in the drawing of spherical shells are height of crown, h; diameter of blank, d; thickness of material, t; and flange width, f.

A full hemisphere can be drawn to a diameter equal to $100t$, and this diameter can be pro-portionately increased as the height of the shell is decreased.

Shallow shells give problems with springback, and to counteract this, the flanges must be drawn to a larger diameter, and the shell to a greater depth.

When controlling flow, it is sometimes found that the blankholder force is insufficient, and recourse is then made to:

- Oversized blanks, giving an increase in the area acted on by the blankholder
- Draw beads, which provide localized resistance to flow in the required areas

When a shell proves impossible to produce in a single operation, either:

- A two-piece punch is used. This is a method of reducing the pull exerted by a domed punch by dividing the punch into two parts that act separately.
- The shell is drawn in two operations, where the final shape is either stretched or redrawn from a straight-walled shell.

Special Shells. A special shell is one in which the body is not concentric with other parts of the shell, or the flange at the open end is not parallel with the bottom, or the shell is made up of a combination of straight, tapered, and spherical surfaces. Most special shells have sections shaped like the common shells covered previously, and it is possible to make judicious use of the rules for the parts in combination to produce the die design for the special shell.

REFERENCES

1. K. Siegert, S. Wagner, and M. Ziegler, Robust Forming Process by Pulsating Blankholder Forces, Multipoint Cushion Systems and Closed Loop Control, *Sheet Metal Forming Technology,* M.Y. Demeri, Ed., TMS, 1999, p 1–21
2. *Die Design Handbook,* 2nd ed., McGraw-Hill, 1964
3. S. Kalpakjian, *Manufacturing Processes for Engineering Materials,* 3rd ed., Addison-Wesley, 1997
4. T. Altan, S.-I. Oh, and H.L. Gegel, *Metal Forming: Fundamentals and Applications,* American Society for Metals, 1983, p 56–57
5. P. Lee and J.R. Hiam, *Factors Influencing the Forming-Limit Curves for Press Forming,* Dofasco, Canada, 1973
6. K.P. Rao and M. Emani, Simple Prediction of Sheet Metal Forming Limits Stresses and Strains, *Sheet Metal Forming Technology,* M. Demeri, Ed., TMS, 1999, p 205–231
7. D.V. Wilson and R.D. Butler, *J. Inst. Met.,* 1961–1962, p 473–483
8. W.F. Hosford and R.M. Caddell, *Metal Forming: Mechanics and Metallurgy,* 2nd ed., Prentice-Hall, 1993, p 268
9. M.Y. Demeri, Drawbeads in Sheet Metal Forming, *J. Mater. Eng. Perform.,* Vol 2 (No. 6), Dec 1993, p 863–866
10. M.Y. Demeri, Drawbeads in Sheet Metal Forming Simulation, *Computer Applications in Shaping and Forming of Materials,* M.Y. Demeri, Ed., TMS, 1993, p 301–310
11. H.D. Nine, New Drawbead Concepts for Sheet Metal Forming, *J. Appl. Metalwork.,* Vol 2 (No. 3), July 1982, p 185
12. R.M. Hobbs, Prediction and Analysis of Press Performance for Sheet Steels, *Source Book on Forming of Steel Sheet,* American Society for Metals, 1976, p 1–11

Forming of Advanced High-Strength Steels

Mahmoud Y. Demeri, FormSys Inc.

THE STEEL INDUSTRY has responded to competition from alternative materials for lightweighting and performance enhancements by developing new steel grades with superior product attributes to leverage steel as the optimal automotive material. New technologies such as continuous casting and thermomechanical processing have resulted in the development of several new grades of steel. High-strength low-alloy (HSLA) steels achieve their good combination of formability and weldability through microalloying. Advanced high-strength steels (AHSS) such as dual-phase (DP), complex-phase (CP), and transformation-induced plasticity (TRIP) steels show superior strength compared to the HSLA grades with the same formability. The AHSS derive their superior mechanical and forming properties from their final micro-constituents (ferrite, bainite, martensite, and retained austenite). For example, DP steels derive their strength from the martensite phase and their ductility from the ferrite phase. Automotive engineers are specifying increased amounts of AHSS to achieve greater strength without a corresponding increase in weight.

Interest in the automotive applications of high-strength steels (HSS) and AHSS is based on the many advantages that such materials offer over mild steel, aluminum, and magnesium alloys. Some of these advantages are:

- Weight reductions through reduced sheet thickness
- Safety improvement through high crash resistance
- Better appearance through elevated dent resistance
- Better performance through increased fatigue strength
- Cost reduction through reduced material use due to down-gaging
- Cost savings through material cost compared to aluminum and magnesium
- Fuel economy improvement through weight reduction

Classification of HSS and AHSS

The HSS and AHSS have been developed to improve the strength-to-weight ratio of low-carbon steels. Higher strengths are achieved by microalloying, melting practice, heat treatment, or a combination of these. The HSS and AHSS cover a wide range of properties and strength levels. A number of arbitrary definitions of the classes of such materials have been developed. The American Iron and Steel Institute (AISI) designation of HSS and AHSS is shown in Table 1.

The International Ultralight Steel Auto Body (ULSAB) Consortium (Ref 1) defines the strength levels of mild steel, HSS, and AHSS as shown in Table 2.

It is important to realize that some HSS may be designated by minimum tensile strength and not by minimum yield strength. To provide a consistent nomenclature, the ULSAB Consortium adopted a standard practice that defines both yield strength and ultimate tensile strength. In this classification system, HSS and AHSS steels are identified as:

$$XX \ aaa/bbb$$

where XX is the type of steel, aaa is the minimum yield strength in MPa, and bbb is the minimum ultimate tensile strength in MPa.

The steel type designator uses the classification shown in Table 3 (Ref 1).

As an example, a classification of DP 500/800 refers to dual-phase steel with 500 MPa (73 ksi) minimum yield strength and 800 MPa (116 ksi) minimum ultimate tensile strength.

The strength levels of HSS are achieved by using steel grades such as rephosphorized, high-strength interstitial-free (HS-IF), bake-hardenable (BH), microalloyed (HSLA), and phosphor-alloyed steels. The strength levels of AHSS are achieved by using steel grades such as dual-phase (DP), transformation-induced plasticity (TRIP), complex-phase (CP), and martensitic (MART) steels. Such steels are available in hot-rolled, cold-rolled, and zinc-coated products. Coatings may be electrogalvanized or hot dipped.

The principal differences between HSS and AHSS are due to their microstructures. The HSS have small amounts of alloying elements to attain high strength levels. The AHSS are multiphase steels that contain martensite, bainite, and/or retained austenite in quantities sufficient to produce distinct mechanical properties. They exhibit a superior combination of high strength with good formability and high strain hardening capacity. Typical values of mechanical and forming properties of some HSS and AHSS steel grades are given in Table 4 (Ref 1).

Figure 1 shows the relationship between total elongation and ultimate tensile strength for some of the HSS and AHSS grades. The figure is adapted from a study commissioned by AISI and the United States Department of Energy (DOE) to advance the use of HSS and AHSS in the North American automotive industry (Ref 2).

Figure 1 shows that ductility, as measured by percent total elongation, decreases with increasing the ultimate tensile strength of steels.

Table 1 American Iron and Steel Institute designation of high-strength steel and advanced high-strength steel

Steel type	Yield strength	
	MPa	ksi
High-strength steel	255–550	37–80
Advanced high-strength steel	>550	>80

Table 2 Ultralight Steel Auto Body definition of mild, high-strength steel, and advanced high-strength steel

Steel type	Yield strength	
	MPa	ksi
Mild steel	140–<210	20–<30
High-strength steel	210–550	30–80
Advanced high-strength steel	>550	>80

Table 3 Type classifications for high-strength steels and advanced high-strength steels

Designation	Type
High-strength steels	
BH	Bake hardenable
HSLA	High strength low alloy
Advanced high-strength steels	
DP	Dual phase
TRIP	Transformation-induced plasticity
MART	Martensitic
CP	Complex phase

The objective of the AISI/DOE formability project was to conduct tests on HSS and AHSS grades to characterize mechanical behavior and material limits under different forming conditions. Table 5 summarizes the results of the AISI/DOE project (Ref 2).

Strengthening in HSS and AHSS includes one or more of the following mechanisms (Ref 3):

- Solid-solution hardening (alloying)
- Precipitation hardening
- Grain-refinement hardening
- Phase-transformation hardening (heat treating and deformation-induced austenite-to-martensite transformation)

Many HSS and AHSS grades, such as BH, HSLA, and DP steels, are commercially available and are used in the current production of automotive components. Other grades, such as TRIP and CP, are under development and available only in laboratory quantities.

Substituting HSS for mild steel in an automotive body structure may reduce its weight by approximately 40% without compromising its performance (Ref 4). In general, a weight reduction of 10% can increase the fuel economy of a vehicle by approximately 8%.

Review of HSS and AHSS Grades (Ref 3)

The following is a review of the major grades of HSS and AHSS that are currently used or will potentially be used in industrial applications.

Bake-Hardening Grades. The BH steels have a good combination of strength and formability. This combination makes them ideal for dent-resistance auto applications such as hoods, doors, and fenders. The increase in yield strength in this class of steel results from strain hardening and strain aging during paint baking after the stamping process. Bake hardening produces an increase in yield strength ranging from 28 to 56 MPa (4 to 8 ksi) after a low-temperature heat treatment cycle similar to a paint-bake schedule (175 °C, or 345 °F, for 20 to 30 min). The bake-hardening effect is possible only if the steel has undergone low plastic deformation of <5% and has been exposed to a baking temperature of at least 150 °C (300 °F). These steels are appropriate for some low-deformation body panels.

High-Strength Low-Alloy Grades. The HSLA steels have small amounts of alloying elements, such as phosphorus, manganese, or silicon, added to low-carbon (0.02 to 0.13% C) steels to attain high strength levels. In these steels, higher strength is achieved by rapid cooling to produce a very fine ferrite grain size; by solid-solution strengthening with phosphorus, nitrogen, silicon, and manganese; and by formation of carbides or carbonitrides with vanadium, nickel, and titanium. These steels have better mechanical properties, corrosion resistance, and weldability than mild steels. The HSLA steels are produced as hot-rolled products and cold-rolled sheet. These steels can achieve yield strengths up to 485 MPa (70 ksi) without heat treatment. This class of steel has low formability and is used for shallow drawn parts.

Dual-Phase Grades. The duplex microstructure of DP steels is comprised of a soft ferrite matrix and between 20 and 70% volume fraction of martensite. The volume fraction of martensite determines the strength level of this steel. Special heat treating practices that involve quenching and tempering are used to generate the martensite phase. The DP structure is produced by quenching low-carbon steels from the alpha + gamma phase region to form a microstructure of martensite islands in a ferrite matrix. This microstructure gives DP steels their high strain hardening capability and better formability compared to HSLA grades. For a given yield strength, the tensile strength of DP steels is also higher than HSLA steel (Fig. 2).

Table 4 Mechanical properties of some high-strength steel and advanced high-strength steel grades

| Steel grade(a) | Minimum yield strength | | Minimum ultimate tensile strength | | Total elongation(b), % | n-value (5–15%)(b) | Average strain ratio (r_m or $\bar{r}$)(b) | K-value(b) | |
	MPa	ksi	MPa	ksi				MPa	ksi
BH 210/340	210	30	340	49	34–39	0.18	1.8	582	84
BH 260/370	260	38	370	54	29–34	0.13	1.6	550	80
DP 280/600	280	41	600	87	30–34	0.21	1.0	1082	157
DP 300/500	300	44	500	73	30–34	0.16	1.0	762	110
HSLA 350/450	350	51	450	65	23–27	0.14	1.1	807	117
DP 350/600	350	51	600	87	24–30	0.14	1.0	976	142
DP 400/700	400	58	700	102	19–25	0.14	1.0	1028	149
TRIP 450/800	450	65	800	116	26–32	0.24	0.9	1690	245
DP 500/800	500	73	800	116	14–20	0.14	1.0	1303	189
CP 700/800	700	102	800	116	10–15	0.13	1.0	1380	200
DP 700/1000	700	102	1000	145	12–17	0.09	0.9	1521	220
MART 950/1200	950	138	1200	174	5–7	0.07	0.9	1678	243
MART 1250/1520	1250	181	1520	220	4–6	0.065	0.9	2021	293

(a) BH, bake hardenable; DP, dual phase; HSLA, high strength low alloy; TRIP, transformation-induced plasticity; CP, complex phase; MART, martensitic. (b) Typical

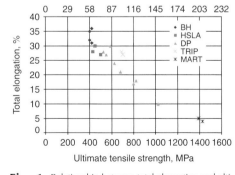

Fig. 1 Relationship between total elongation and ultimate tensile strength for some high-strength steel and advanced high-strength steel grades. BH, bake hardenable; HSLA, high strength low alloy; DP, dual phase; TRIP, transformation-induced plasticity; MART, martensitic

Table 5 Summary of American Iron and Steel Institute/Department of Energy formability project

Formability attribute	Test method	Formability parameters characterized	Factors influencing formability
Forming limits	Full dome test	Forming-limit diagram	n-value, thickness
Sheared-edge stretching limits	Hole extrusion test	% hole expansion	Ultimate tensile strength (UTS), r-bar value
Bending under tension limits	Angular stretch-bend test	Height at failure, stretch-bendability index	UTS, r-bar value
Springback and curl	Channel draw test	Springback opening angle, radius of sidewall curl	Yield strength, tool radii, draw bead restraining force, tool gap
Stretch formability	Pan forming (fully locked conditions)	Height at failure	n-value, thickness
Stretch drawability	Square draw test	1. Height at failure 2. Binder control 3. Strain measurement	Uniform elongation, r-bar, blank size, coating, lubrication

The DP steels are designed to provide ultimate tensile strengths of 600 to 1000 MPa (87 to 145 ksi). The DP steels can also exhibit a bake-hardening effect, which is the increase in yield strength resulting from prestraining and elevated-temperature aging. The extent of the bake-hardening effect in AHSS depends on the specific chemistry and thermal histories of the steels. Dual-phase steels are usually bake hardened.

TRIP Grades. The TRIP steels consist of two phases: a ferrite/bainite matrix and a 5 to 20% volume fraction of metastable retained austenite, which progressively transforms to martensite during plastic deformation. This combination of phases gives TRIP steels the high formability of austenite during the initial stages of the stamping process, followed by the high strength of martensite at the end of the forming process.

The TRIP steels are characterized by strain hardening rates higher than DP steels (Fig. 3), high tensile strengths, and ductility. The high strain hardening rate of TRIP steels is one reason for improved formability. The progressive transformation of retained austenite to martensite in TRIP steels also leads to microstructural volume and shape changes that accommodate strain and increase ductility. The transformation thus provides strengthening and extends strain beyond that of DP steels (Fig. 2).

The excellent formability of TRIP steels combined with their high strength make them attractive candidate materials for lightweight applications. Also, because the martensite formation occurs at extremely fast rates, the TRIP phenomenon occurs even at high deformation rates, as encountered during vehicle crash conditions. Therefore, TRIP steels are also being considered for applications requiring maximization of crash energy absorption.

Complex-Phase Grades. The CP steels consist of a very fine microstructure of ferrite and a higher volume fraction of hard phases that are further strengthened by fine precipitates. They use many of the same alloy elements found in DP and TRIP steels but additionally have small quantities of niobium, titanium, and/or vanadium to form fine strengthening precipitates. Complex-phase steels provide ultimate tensile strengths of 800 MPa (116 ksi) and greater. They are characterized by high deformability and high energy absorption, which makes

them ideal candidates for automotive crash applications, such as bumper and B-pillar reinforcements.

Martensitic Grades. The MART steels contain a high fraction of martensite. The microstructure is low-carbon martensite, with the carbon content determining the strength level of the material. The austenite that exists during hot rolling or annealing of the material is transformed to martensite during quenching and/or postforming heat treatment. Martensitic steels are often subjected to postquench tempering to improve ductility and can provide good formability even at extremely high ultimate tensile strengths of up to 1500 MPa (217 ksi). Carbon is added to martensitic steels to increase hardenability and also to strengthen the martensite. Elements such as manganese, silicon, chromium, molybdenum, boron, vanadium, and nickel are also used in various combinations to increase hardenability.

High-Hole-Expansion (HHE) Steels. The AHSS microstructures result in improved total elongations, but these same microstructures also reduce local elongations or ductility (measured by hole-expansion factor, λ) that affect hole expansion, stretch flanging, and bending. In responses, HHE steel is an area of development for applications that require a high degree of sheared-edge elongation (hole flanging). The key to improved sheared-edge stretchability is a homogeneous microstructure, such as a single phase of bainite or multiple phases with bainite

(Fig. 4), and the removal of large martensitic constituents. The multiple-phase microstructure is primarily ferrite and bainite with some retained austenite. Parts stamped from these grades are replacing cast and forged parts (Ref 7).

Properties of HSS and AHSS (Ref 3)

Plastic Flow. A comparison between typical engineering stress-strain curves of various grades of cold-rolled, hot dip galvanized HSS and AHSS (HSLA 340, DP 600, TRIP 700) is shown in Fig. 5.

The figure shows a significant increase in the tensile strengths of these materials compared to mild steel. The materials also shows superior crash performance due to the energy-absorption capability that results from their high toughness. The true stress-strain curves of various HSS and AHSS grades (HSLA 300, DP 600, TRIP 800, CP 1000) are shown in Fig. 6. Mild steel is included for reference.

Strain Hardening. A high strain hardening capacity positively influences formability by resisting local necking during stamping and is especially important in the stretch-forming

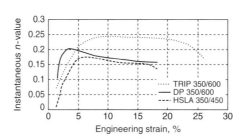

Fig. 3 Instantaneous *n*-values versus strain for transformation-induced plasticity (TRIP), dual phase (DP), and high-strength low-alloy (HSLA) steels. Source: Ref 5

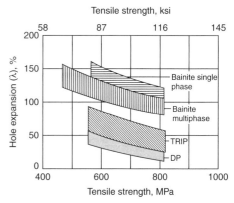

Fig. 4 Improvements in hole expansion by modification of microstructure. TRIP, transformation-induced plasticity; DP, dual phase. Source: Ref 6

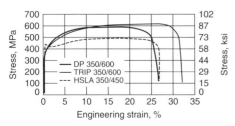

Fig. 2 Stress-strain curves for high-strength low-alloy (HSLA), dual-phase (DP), and transformation-induced plasticity (TRIP) steels with 350 MPa (50 ksi) yield strength. Source: Ref 5

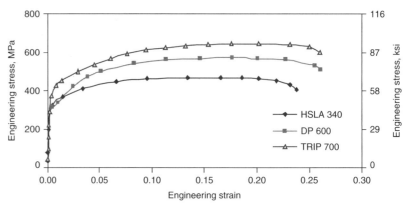

Fig. 5 Engineering stress-strain curves of various grades of high-strength steel and advanced high-strength steel

deformation modes typically encountered in the manufacture of many automotive body components. High strain hardening capacities also result in higher ultimate tensile strengths in the manufactured component, which enhances crash energy absorption and fatigue performance.

Strength. A comparison between typical strength levels of various grades of cold-rolled, hot dip galvanized HSS and AHSS (BH 180, HSLA 220, IF 260, DP 600, TRIP 700, CP 1000) is shown in Fig. 7.

Ductility and Bendability. A comparison between the ductility of the same grades of HSS and AHSS is shown in Fig. 8. The HSS and AHSS are more difficult to bend than plain carbon steels because of their higher yield strength and lower ductility. This requires more power, greater bend radius, more die clearance, and greater allowance for springback. It may be necessary to remove shear burrs and to smooth corners in the area of the bend. Whenever possible, the axis of the bend should be perpendicular to the rolling direction. If the bend axis must be parallel with the rolling direction, it may be necessary to use cross-rolled material, depending on the severity of the bend. All HSLA steels are not equal in formability. However, for the more readily formable grades and the quenched-and-tempered grades, the minimum bend radii in Table 6 are recommended.

These minimum bend radii are for bending with the bend axis across the rolling direction. The use of smaller bend radii increases the probability of cracking. Hot bending is recommended for thicknesses greater than 13 mm (1/2 in.).

Hot bending is necessary when the product shapes are too complex or when bend radii are too small for cold forming. The HSLA steels can be successfully hot bent at temperatures as low as 650 °C (1200 °F); however, when maximum bendability is needed, temperatures of 845 to 900 °C (1550 to 1650 °F) are recommended. Cooling in still air from these temperatures returns the material nearly to the as-rolled mechanical properties. Tables in the article "Forming of Carbon Steels" in this Volume show the effect of composition on minimum bend radius by comparing the minimum radii for common grades of carbon and low-alloy steels.

Springback is the change in dimension in the shape of the part from that of the shape of the die after forming and unloading the part. It may involve angular change (typically referred to as springback) or sidewall curl of a section. In general, springback depends on the ratio of yield strength to elastic modulus. A higher yield strength or a lower modulus increases springback for a given section thickness and tooling arrangement.

Due to the high ratio of yield strength to elastic modulus of HSS and AHSS, a larger amount of springback, compared to mild steel, develops in the formed part. Springback is of great concern to sheet metal forming tool designers, because it can cause serious problems in the assembly of parts and can lead to expensive modifications of

the forming tools. Springback and sidewall curl after unloading from stamping dies is a very important technical barrier to the widespread use of HSS. Springback can be reduced through proper tooling design and by controlling the binder forces acting on the drawn sheet.

Recent studies (Ref 8) of springback in U-channel forming of HSLA 350, DP 600, and

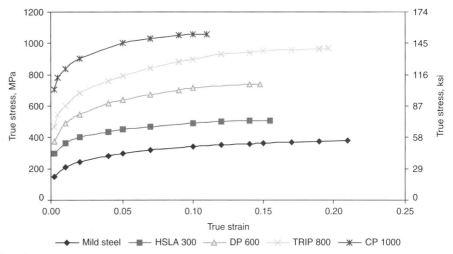

Fig. 6 True stress-strain curves of various high-strength steels, advanced high-strength steels, and mild steel

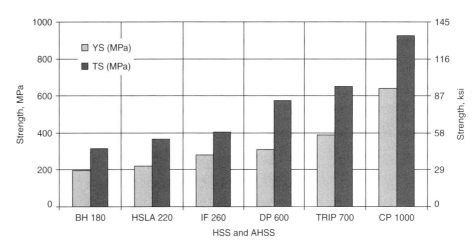

Fig. 7 Strength levels of various grades of high-strength steel (HSS) and advanced high-strength steel (AHSS). YS, yield strength; TS, tensile strength

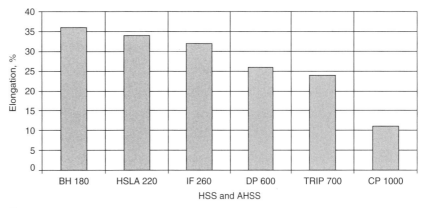

Fig. 8 Percent elongation of various grades of high-strength steel (HSS) and advanced high-strength steel (AHSS)

TRIP 700 steels showed that springback depends on the strength level of the stamped part (that is, the yield strength after forming). Materials with a high *n*-value (DP 600, TRIP 700) show increased strength after forming, which results in higher springback. The study also showed that a reduction of sheet thickness, and therefore an increase in the die gap, increases springback. The amount of springback increases with increase in the strength level of the sheet materials. It was found that the most influential factors controlling springback and curl were the tooling radius and draw bead penetration (see the section "Stamping Issues and Forming Guidelines for HSS and AHSS" in this article).

Formability. Forming limits of sheet steel are described by a forming-limit diagram that shows the limit strains of the steel when it is formed through a linear strain path. The formability of various grades of HSS, AHSS, and mild steel (for reference), as determined by the forming-limit diagram, is shown in Fig. 9. The figure shows that formability follows the general trend of decreasing with increase in the strength of the steel.

Hot Forming of HSLA. In hot forming, the work metal is heated above its recrystallization temperature. A lower press capacity is needed to hot form a given shape than to cold form it. However, the press, dies, and related equipment

must be designed to withstand high temperature. Meeting this requirement is sometimes more difficult than obtaining a higher-capacity press for cold forming.

In some applications, a steel workpiece can be quenched directly from the forming temperature. In one example, a 25 mm (1 in.) thick HSLA steel blank was hot formed into a bucket blade for earthmoving equipment. The blank was cut by oxyfuel torch to 305 by 1700 mm (12 by 67 in.) from a flat plate, then furnace heated to 815 °C (1500 °F). The hot workpiece was formed with one stroke of a 1.8 MN (200 tonf) hydraulic press. No lubricant was used, because production quantities were small and the surface condition of the part was not critical. After being formed, the part was quickly removed from the die and allowed to cool on the floor. Time from furnace to press to floor was 4 min per piece. Heating the blank to 815 °C (1500 °F) permitted forming in a much smaller press and with greater accuracy than could have been done if the blank had been formed at room temperature. Tolerances of ±1.6 mm (±1/16 in.) were maintained on the overall dimensions of the formed part. The forming die was a weldment made of hot-rolled 1045 steel, flame hardened at the critical wear points. Production was 400 buckets per year in lots of 60. Die life was 2400 pieces before reworking.

HSS and AHSS Automotive Applications (Ref 3)

Applications of HSS in the automotive industry revolve around two major areas:

- Improving crash performance (front, rear, side) via increased strength levels
- Reducing weight via gage reduction

Other important applications criteria include stiffness, fatigue life, corrosion resistance, formability, and weldability. Typical applications of HSS include panels, chassis, and structural components (door beams, wheels, bumpers, seats, suspensions, etc.).

The ULSAB project used 91% of the vehicle weight from HSS and AHSS grades, as shown in Table 7 (Ref 1). The Ultralight Steel Auto Closures (ULSAC) project selected BH 210, BH 260, and DP 600 to develop new concepts for automotive steel closures that can achieve 10% mass reduction while maintaining structural performance with no cost penalty (Ref 9).

Recently, the Auto Steel Partnership (ASP) commissioned a world-wide survey of the automotive applications of HSS (Ref 10). The report concluded that weight reduction and improved crash resistance are the driving forces for using more HSS and AHSS in vehicles. The report also concluded that Europe and Japan are more aggressive in increasing the HSS and AHSS content in their cars. There is a general agreement that the HSS and AHSS content will increase in the Body-in-White (BIW) from its current 10 to 40% to >50% as the availability of DP, TRIP, and CP steels increases.

Benefits Analysis. The benefits of using HSS in car body structures and components can be analyzed by using the performance indices that were developed for materials selection. Table 8 shows the performance indices that must be considered for selecting the best materials for car body applications. Performance indices are groupings of material properties developed by Ashby to maximize some aspect of the performance of a component (see the article "Performance Indices" in *Materials Selection and Design*, Volume 20 of *ASM Handbook*, 1997).

Table 6 Minimum bend radii for high-strength low-alloy steels

Steel thickness(t)		Minimum bend radii for minimum yield strength of:	
		310 MPa (45 ksi)	345 MPa (50 ksi)
mm	in.		
<1.6	1/16	1/2*t*	1*t*
1.6–6.4	1/16–1/4	1*t*	2*t*
6.4–13	1/4–1/2	2*t*	3*t*

Table 7 High-strength steel (HSS) and advanced high-strength steel (AHSS) grades used in the Ultralight Steel Auto Body project

Steel type	Yield strength		Amount used, %
	MPa	ksi	
Mild steel	<210	<30	8
HSS	210	30	27
HSS	280	41	13
HSS	350	51	45
HSS	420	61	3
AHSS	>550	>80	3

Table 8 Performance indices for car body applications

Application requirements	Performance index(a)
Stiff and light panel	E/ρ
Dent and crash resistance of light panel	$(\sigma_y)^{1/2}/\rho$
Tensile/compressive strength of light panel	σ_y/ρ
Bending stiffness of a light panel	$(E)^{1/3}/\rho$
Buckling stiffness of a light panel	$(E)^{1/2}/\rho$

(a) E, elastic modulus; σ_y, yield strength; ρ, density of material

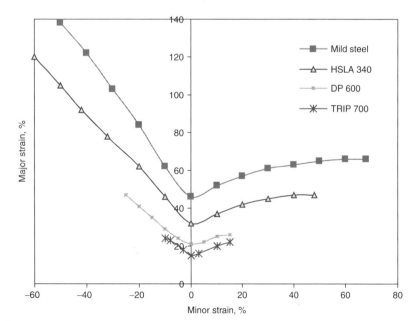

Fig. 9 Forming-limit diagrams for various grades of high-strength steel, advanced high-strength steel, and mild steel

Some of the performance indices are stiffness-specified (specific stiffness) and others are strength-specified (specific strength). Materials with large values of performance indices are the best materials for the intended application.

For example, the table shows that if the critical buckling stress in impact is considered, increasing the yield strength by a factor of 2 raises the critical buckling stress by 50%. This can be achieved with high-strength grades.

Outer Panels. High-strength steels are used in outer-body panels for weight reduction, increasing panel stiffness, and improving dent resistance. Most outer panels are made of steel sheets with tensile strength of 260 to 340 MPa (38 to 49 ksi). This may include BH and HSLA grades. The BH steels are often used to make doors, hoods, fenders, deck lids, and truck beds. In such applications, dent resistance is important. Dent resistance (DR) increases with increase in yield strength (YS) and sheet thickness (t) in accordance with:

$$DR = K(YS)(t^m) \tag{Eq 1}$$

where $m = 1$ to 3, and K is a constant depending on the shape of the panel.

In HSS, the increase in strength can compensate for the decrease in dent resistance due to thickness reduction. This is not the case for panel stiffness (PS), which increases with increase in elastic modulus (E) and sheet thickness (t) in accordance with:

$$PS = K(E)(t^m) \tag{Eq 2}$$

Reducing sheet thickness for weight savings lowers panel stiffness, and this leads to surface deflection. Because of the negative effect of thickness reduction on panel stiffness, limits on the amount of down-gaging are established. Because outer panels are subjected to bending and buckling, stiffness is more important than strength for this application. In such cases, substitution of HSS and AHSS for mild steel cannot be rationalized.

The ULSAC program demonstrated that mass reduction of approximately 10% could be achieved by using BH 210, BH 260, and DP 600 grades (Ref 9). The mass reduction was achieved while maintaining structural performances and without cost penalty.

Inner panels have complicated shapes and are formed by deep drawing. High-strength steels used for this application have tensile strength of 260 to 390 MPa (38 to 57 ksi). This may include BH and HSLA grades. However, because of the limited formability of HSS, complicated inner panels cannot be formed successfully using traditional forming methods.

Structural and Chassis Components. Structural parts in automobiles require high crash resistance, which, for a specific part, depends on material fracture toughness and sheet thickness. Most structural components in current production are made of steel sheets with tensile strength of 270 to 540 MPa (39 to 78 ksi). This may include thicker gages of BH, HSLA, and DP 500 grades. Higher-strength steels offer an opportunity for reducing weight by reducing component thickness. A major problem in using higher-strength steels for automotive applications is the increase of springback of the formed parts. For a constant elastic modulus, springback increases with increase in the flow strength of the sheet metal. For structural applications, the yield strength is the major consideration in materials selection. In such cases, HSS and AHSS can be used, and weight savings of 40 to 60% are possible when substituting these materials for mild steel.

The BIW is the heaviest and largest structure in vehicles. It comprises approximately 20 to 30% of the total weight of a medium-sized passenger car. Reducing the mass of BIW by using high-strength can produce significant weight savings and improved crash resistance. Examples of such applications are impact beams, B-pillar reinforcements, and bumper bars. It is anticipated that the HSS and AHSS content in the BIW may increase to over 50% as more of these

materials become commercially available and as new designs and manufacturing methods are implemented.

Applications of DP steels include seat parts, bumpers, door beams, frame components, rails, and pillars. Typical automotive parts for improving crashworthiness include front rails, center pillars, and rear-side members. Side impact resistance has been improved and weight reduction has been achieved by incorporating many DP steel parts in body structures. A 13% weight reduction has been accomplished by using TRIP steel in several suspension parts.

Table 9 shows some typical properties and application examples of various grades of HSS and AHSS (Ref 3).

Stamping Issues and Forming Guidelines for HSS and AHSS

The ASP has issued a guide for using HSS for BIW applications (Ref 11). The guide explains the steel characteristics and provides guidelines for die design, die construction, and die tryout. It also discusses product design and advanced die-process concepts for forming HSS. Because the properties of HSS are characterized by low elongation and high resistance to compression, splitting and buckling tendencies increase with increase in tensile strength. Parts are designed with minimum draw depth and reduced stretch and compression of the sheet metal to avoid splits. Also, gentle shape transitions are used to avoid wrinkles, and large radii are specified to facilitate metal flow.

Press Loads. The energy required to plastically deform steel is proportional to the area under the stress-strain curve. Some of the HSS and AHSS exhibit high strain hardening behavior, which can also be used to estimate the press capacity needed for stamping a part. Higher-than-normal binder pressure and press tonnage is necessary with HSS and AHSS to maintain process control and to minimize

Table 9 Typical properties and applications of high-strength steels and advanced high-strength steels

Steel type(a)	Gage		Yield strength		Tensile strength		Elongation,%	Application examples
	mm	in.	MPa	ksi	MPa	ksi		
HSLA (HR)	1.6	0.063	340	49	470	68	34	Roof side rail, reinforced front pillar, front floor cross member
HSLA (CR)	1.2	0.047	410	59	465	67	35	Center pillar, inner-side sill inner, rear floor cross member
IF (CR)	1.4	0.055	325	47	455	66	39	Dash panel
DP 500 (CR)	2.0	0.079	315	46	480	70	38	Frame, pillar
HSLA (GA, HR)	2.0	0.079	372	54	453	66	35	Side sill outer reinforced front-side member, front-side member reinforcement
DP 500 (GA, CR)	1.8	0.071	290	42	462	67	39	Inside sill
DP 600 (CR)	1.4	0.055	350	51	635	92	29	Roof-side rail inner, sill front side, front floor cross member
DP 600 (HR)	2.0	0.079	437	63	618	90	28	Reinforced center pillar, front pillar outer, external front-side member
DP 600 (GA, CR)	1.4	0.055	363	53	621	90	29	Front-side member, extension sill inner, center frame
DP 800 (HR)	3.2	0.126	630	91	800	116	23	Seat rail wheel rim, wheel disk
DP 800 (CR)	1.6	0.063	430	62	820	119	21	Center pillar inner, roof-side rail
DP 1000 (CR)	1.6	0.063	635	92	1040	151	14	Reinforced front bumper, center pillar reinforcement

(a) HSLA, high strength low alloy; IF, interstitial-free; DP, dual phase; CR, cold rolled; HR, hot rolled; GA, galvanized and annealed

buckles on the binder. A double-action press or hydraulic press cushions may be required to achieve the necessary binder forces required to control sheet metal flow into the die cavity. Air cushions or nitrogen cylinders may not provide the required force for setting the draw beads or maintaining binder closure.

Edge Cracking, Trimming, and Flanging. The part and process thus needs to be designed to reduce stretch flangeability, because martensite and other hard phases reduce stretch flangeability (as noted in the section "High-Hole-Expansion (HHE) Steels" in this article). Local ductility, which is influenced by microstructure, affects stretch flanging and hole expansion. The edge condition after trimming is influenced by die clearance and wear. A clean edge cut is especially important for blanked or sheared edges and punched holes on AHSS material (Ref 12). A consistent and appropriate clearance is vital to minimize burr height. Cutting tools also need to be sufficiently sharp. Compared to sharp tools, worn tools result in a 20% reduction in hole expansion (stretch flangeability) in mild steels but a reduction of 50% or more in DP and TRIP grades (Ref 13).

Panels should be supported during trimming to eliminate the occurrence of bad burrs and to avoid edge cracking and tearing. This reduces the tendency for a bad burr, which in turn minimizes the tendency for edge cracking. Trim steels also need to be engineered with a higher strength in mind; the tensile strength of the AHSS grades can be substantially higher than that of conventional HSS (Ref 12). Abrupt changes in flange length act as local stress raisers that lead to edge cracks. Sharp notch features should be avoided in curved flanges. Flange die bend radii must be small in order to minimize springback.

In general, sheared-edge stretching limits are strongly related to the tensile strength, showing a significant decrease with increasing tensile strength for steels of tensile strength less than 700 MPa (102 ksi) and then saturating for steels of higher strength. Furthermore, during tool development, the blank die sometimes is the last one to be completed, necessitating the use of laser-cut blanks during prototype and even at the beginning of hard tool tryout. Laser-cut edges have much higher stretch flangeability than the sheared edge obtained from a conventional blank die, and as a result, tryout performance may be different (Ref 12).

Springback. As noted, springback is a function of the yield-strength-to-modulus ratio. Springback of AHSS grades is greater than that of mild steel or HSS, because of higher yield strength. The degree of springback correlates with yield strength after forming (rather than yield strength of the flat sheet). Thus, the higher strain hardening rate of DP and TRIP steels (Fig. 3) results in more springback than HSLA grades, even though the initial yield strength of the unformed blank may be similar.

Addressing springback requires careful assessment of both part and process design. Springback computer simulations should be considered whenever possible to test process and design solutions (e.g., see the article "Computer-Aided Engineering in Sheet Metal Forming" in this Volume). Part and tool design should try to minimize springback effect rather than compensate for it. For example, springback can be reduced with sharper punch radii (Fig. 10). However, sharp radii may also increase forming challenges and press-load requirements. Sharp radii can contribute to excessive thinning, and stretch bending will be more difficult as yield strength increases.

Springback is a complicated problem. Its value depends on the nature of the deformation. Under certain conditions, the final bend angle can be smaller than the original angle, to give negative springback! Figure 10 is for simple bends, where the bar is held at one end and a bending force is applied to the other end. This figure is a plot of springback, defined as the percent change of curvature such that:

Springback (% change in curvature)

$$= \left[\frac{\frac{1}{R} - \frac{1}{r}}{\frac{1}{R}} \right] 100 \qquad \text{(Eq 3)}$$

For a material that is perfectly plastic (i.e., it does not work harden), the previous equation under conditions of rigid springback (Eq 8 and 9a in the article "Springback" in this Volume) can be expressed as:

Rigid-perfectly plastic springback

$$= \frac{3\sigma_0 R}{E \cdot t} \times 100 \qquad \text{(Eq 4)}$$

where σ_0 is the tensile yield stress, E is elastic modulus, and t is thickness. A higher-order term occurs for elastic-perfectly plastic springback (Eq 12 and 13 in the article "Springback" in this Volume), but this additional term is not a significant factor in the plot of Fig. 10 for a given yield strength.

However, it is very important to understand that this depends on how springback is defined. Springback does increase with R/t (Fig. 10) when springback is defined as a fractional curvature change. However, if springback is defined as a curvature change (not some fractional representation thereof), the opposite result is obtained, at least for the elastic-perfectly plastic case. In the article "Springback" in this Volume, the definition of springback is based on curvature change throughout.

Tool geometry also plays an important role in determining the resultant springback, because this affects the conditions of deformation. For example, if a very small amount of stretching is superimposed during bending, springback can be reduced or even eliminated. Under pure bending, springback is proportional to R/t for cylindrical tooling (Ref 15–17), U-bending/channel bending (Ref 18–20), V-bending (Ref 19–21), and flanging (Ref 22). However, under stretch-bending situations, it has been reported that springback decreased with increasing R/t in flanged channel (Ref 23) and draw bending (Ref 24).

In body panels, springback can be reduced by compensation through overbending of the flange. Guidelines for reducing springback in stamped components include avoiding right or acute angles, using larger open-wall angles, and avoiding large transition radii between two walls. Springback allowance must be increased as material strength increases: 3° for mild steels but 6° or more for HSS and AHSS.

Wrinkling (Ref 12). Although the benefit of the AHSS grades is the ability to reduce the sheet steel thickness, the decreased thickness has the potential to increase wrinkling if die clearances are not adjusted to reflect the reduced gage. Controlling the wrinkling requires higher press

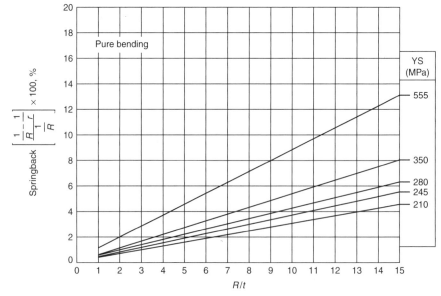

Fig. 10 Fractional change in curvature (springback) increased by yield strength (YS) and bend radius (R) to sheet thickness (t) ratio. Source: Ref 14

forces, which may lead to the need for higher-tonnage presses. The wrinkling, combined with the sheet steel overall strength, increases the potential for die wear. The high press forces can result in higher temperatures, which can cause lubricants to break down or burn, resulting in extreme wear conditions.

Residual stresses are induced in parts during their manufacture. Such stresses do not disappear when parts relax and all external constraints are removed. Some stamped parts produce significant residual stresses that create large springback when allowed to relax. Springback in a part results from stretch-bend forming. The stretch portion varies with the amount of the clamp force applied by the binder. Residual stress can cause distortion in stamped parts. It can be reduced in HSS stampings by proper part design and die-process planning. An effective way of reducing residual stresses is to use a die process where some form of post-stretch or shape-set may be required in order to induce a minimum of 2% stretch in the part near the bottom of the press stroke.

Draw beads control metal flow into the die cavity. Draw beads for HSS and AHSS should not extend around corners of the draw die, because this will result in locking out the metal flow and cause splitting in the corners of the part. Draw beads should "run out" at the tangent of the corner radius to minimize metal compression in corners, as shown in Fig 11.

Tool Design. The HSS and AHSS require a stiffer die to resist flexing that may lead to part distortions, especially for channel or hat-section parts. Tool geometry plays an important role in determining the final value of springback. As previously noted in the section "Springback" in this article, the influence of punch radii is different depending on the nature of the forming operation.

In pure bending, for example, springback decreases with increase in die radius because of diminishing bending forces in larger radii. In pure bending, the bending moment is inversely proportional to the radius of curvature, so that springback thus decreases with an increase in radius of curvature (see, for example, Eq 6 in the article "Springback" in this Volume). This is true for fairly gentle radii (i.e., $R/t > 5$), where simple rules can be derived. Most sheet forming falls in this region. However, the trend reverses with very tight radii. The mechanics of very tight bending is complicated, so these results are mainly from finite element simulations using solid elements, which include three-dimensional states of stress and strain. Shell elements are not accurate in this region, and the simple rules do not apply.

In stretch-bend forming, springback (expressed as sidewall curl) is reduced by tighter tool radii (Fig. 10). Punch radii must be fairly sharp, approximately $1t$ for HSS and greater than $1t$ for AHSS. Many experiments and simulations show that the sidewall curl goes to nearly zero for R/t approximately equal to 2. In a series of stretch-bend tests of metal strips, the curl in DP 90 was removed when the specimen was drawn over a small radius ($r/t < 1.5$) under high tension (Ref 26). The small radius is to pin the material and allow high tension to reduce springback.

Die Processes. The quality of stamped parts is critical in avoiding problems in assembly and in the final product performance. Quality problems may result from formability issues such as wrinkling caused by excessive compression or splitting caused by excessive tension, and from dimensional accuracy issues resulting from springback caused by elastic recovery. Controlling metal flow into the die cavity has been found to be crucial to both part quality and dimensional accuracy. Several alternate die processes for HSS are being evaluated. These are primarily focused on better process control and various means of inducing shape-set stretch in the stamping. Among these processes are programmable hydraulic blankholders for varying the force-stroke trajectory as the sheet is drawn, active drawbeads for increasing the restraining force as the press approaches bottom dead center, and flexible binder technology for local control of the binder area. In-process adjustment of the binder force can lead to better quality and improved part consistency. Mechanical presses

are being retrofitted with hydraulic multicylinder cushion systems to provide more control of the blankholder force during forming (Ref 25).

Die wear. High binder pressure is needed for stamping HSS and AHSS to maintain process control and to minimize buckles on the binder. Local contact loads can be very high for features with small radii. The higher compressive forces on the binder will require better die material, surface treatments, and lubricants. Tools must have sufficient toughness to avoid fracturing and good wear resistance for long-life operation and production reliability. Hard material coatings and nitriding have been used to improve the tribological properties of die surfaces. New developments in hard material coatings include advances in chemical vapor deposition, physical vapor deposition, and plasma-assisted chemical vapor deposition coatings. A new chemical vapor deposition coating has been developed to reduce abrasive wear during stamping of high-strength materials. It employs selective combinations of individual coating layers of titanium carbide (TiC), titanium carbonitride (TiCN), and titanium nitride (TiN) (Ref 27).

Advantages and Disadvantages of Using HSS and AHSS (Ref 3)

The HSS and AHSS content in vehicles is expected to increase as more grades become commercially available and as new designs and manufacturing methods are developed.

Considerations should be taken as to whether a structure is designed to maximize strength or stiffness. For a constant modulus, component stiffness increases with increasing yield strength and sheet thickness. In a down-gaging strategy, HSS and AHSS will offer no advantage in improving stiffness, because all steels have the same modulus of elasticity.

Major advantages of using HSS and AHSS in automotive applications include potential for reducing vehicle weight, increased tensile strength, improved fatigue and impact strengths, and reduced material and transportation cost from weight reduction.

Major disadvantages of using HSS and AHSS include reduced stiffness due to thickness reduction, increased springback due to increase of flow strength, and limited formability due to low ductility. Also included are increased notch sensitivity, reduced dent resistance (when increase in strength cannot compensate for thickness reduction), and increased tool wear on the draw-die binder surfaces due to the high strength of these materials. However, tool wear can be reduced by upgrading tool materials and by applying protective coatings on the surface of the forming tools.

The implications of using thinner gages on the elastic stiffness and corrosion resistance need to be considered. Because the modulus of elasticity is the same for all steels, down-gaging reduces component stiffness. Also, because corrosion

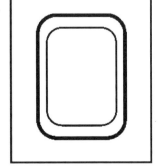

Conventional draw beads | Run-out draw beads for high-strength steels | Lock beads for stretch-form die

Fig. 11 Draw bead types showing conventional, run-out, and lock beads. Source: Ref 25

resistance of HSS and AHSS is similar to that of mild steel, thinner gages rust through faster under the same corrosion conditions.

Other concerns include weldability and paintability of such materials. Also, the effect of austenite-to-martensite transformation in TRIP grades on formability and process stability is yet to be determined.

REFERENCES

1. "Advanced Vehicle Concepts," Technical Transfer Dispatch 6, ULSAB-AVC Consortium, May 2001
2. "Formability Characterization of a New Generation of High Strength Steels," final report, AISI/DOE Technology Roadmap Program, March 2003
3. M.Y. Demeri, High Strength and Stainless Steels for Automotive Applications, *Innovations in Processing and Manufacturing of Sheet Materials,* M.Y. Demeri, Ed., TMS Proceedings, Feb 11–15, 2001 (New Orleans, LA), p 83–95
4. "Weight Reduction for Safer, Affordable Passenger Cars by Using Extra Formable High Strength Austenitic Steels," Research and Technological Development Project Proposal, European Consortium, March 28, 2000
5. A. Konieczny, "Advanced High-Strength Steels—Formability," Great Designs in Steel Seminar, American Iron and Steel Institute, Feb 2003; www.autosteel.org (accessed April 14, 2006)
6. *Advanced High-Strength Steel (AHSS) Application Guidelines,* International Iron and Steel Institute, March 2005, p 2–16; www.worldautosteel.org (accessed April 14, 2006)
7. D.J. Schaeffler, "Introduction to Advanced High Strength Steels, Part I: Grade Overview," http://www.thefabricator.com, FMA Communications, Inc., Aug 2005 (accessed April 14, 2006)
8. *Automotive Steel Design Manual,* Auto Steel Partnership, April 1998
9. "ULSAC Engineering Report," Auto Steel Partnership, April 2000
10. P.R. Mould, "World Wide Survey of the Automotive Application of Advanced High Strength Steels," Automotive Steel Technologies, Inc., 2000
11. *High Strength Steel Stamping Design Manual,* Auto Steel Partnership, 2000
12. D.J. Schaeffler, "Introduction to Advanced High Strength Steels, Part II: Processing Considerations," http://www.thefabricator.com/Articles/Stamping_Exclusive.cfm?ID=1158, FMA Communications, Inc., Sept 2005 (accessed April 14, 2006)
13. B. Carlsson, P. Bustard, and D. Eriksson, "Formability of High Strength Dual Phase Steels," Paper F2004F454, SSAB Tunnplåt AB, Borlänge, Sweden, 2004
14. M. Shi, "Springback and Springback Variation Design Guidelines and Literature Review," internal report, National Steel Corporation, 1994
15. T.X. Yu and W. Johnson, Plastic Bending, Theory and Applications, *Mater. Technol.,* Vol 10, 1983, p 439–447
16. W.P.D. Yuen, Springback in the Stretch-Bending of Sheet Metal with Non-Uniform Deformation, *J. Mater. Process. Technol.,* Vol 2, 1990, p 1–20
17. L. Sanchez, R.D. Robertson, and J.C. Gerdeen, "Springback of Sheet Metal Bent to Small Radius/Thickness Ratios," SAE Paper 960595, Society of Automotive Engineers, 1996
18. C. Sudo, M. Kojima, and T. Matsuoka, Some Investigations on Elastic Recovery of Press Formed Parts of High Strength Steel Sheets, *Eighth Biennial Congress of IDDRG,* 1974, p 192–202
19. M.L. Chakhari and J.N. Jalinier, Springback of Complex Parts, *13th Biennial Congress of IDDRG,* 1984, p 148–159
20. R. Hino, Y. Goto, and M. Shirashi, Springback of Sheet Metal Laminates Subjected to Stretch Bending and the Subsequent Unbending, *Advanced Technology of Plasticity,* Vol II, 1999, p 1077–1082
21. Z.T. Zhang and S.J. Hu, "Mathematical Modeling in Plane Strain Bending, SAE Technical Publication 970439, Society of Automotive Engineers, 1997
22. N.-M. Wang, Predicting the Effect of Die Gap on Flange Springback, *Efficiency in Sheet Metal Forming,* IDDRG 13th Biennial Congress, Feb 20–24, 1984 (Melbourne, Australia), 1984, p 133–147
23. Y.C. Liu, The Effect of Restraining Forces on Shape Deviations in Flanged Channels, *J. Eng. Mater. Technol., (Trans. ASME),* Vol 110, 1988, p 389–394
24. T.S. Kuwabara, K. Takahashi, and K. Ito, Springback Analysis of Sheet Metal Subject to Bending-Unbending under Tension, Part II (Experimental Verification), *Advanced Technology of Plasticity,* Vol II, T. Altan, Ed., Proc. Fifth ICTP (Columbus, OH), 1996, p 747–750
25. M.Y. Demeri, C.-W. Hsu, and A.G. Ulsoy, Application of Real Time Process Control in Sheet Metal Forming, *New Developments in Sheet Metal Forming,* K. Siegert, Ed., Institute for Metal Forming Technology, University of Stuttgart, Stuttgart, Germany, May 23, 2000, p 213–228
26. A.A. Konieczny, M.F. Shi, and C. Du, "An Experimental Study of Springback for Dual Phase Steel and Conventional High Strength Steel," SAE 2001-01-3106, Society of Automotive Engineers, 2001
27. C. Escher and T. Henke, Tool Materials for Processing of High Strength Steel Sheets, *New Developments in Sheet Metal Forming,* K. Siegert, Ed., Institute for Metal Forming Technology Proceedings, University of Stuttgart, Stuttgart, Germany, 2004, p 103–122

Forming of Steel Tailor-Welded Blanks*

TAILOR-WELDED BLANKS consist of different types of sheet stock (with different thicknesses, grades/strengths, and sometimes coatings) that are welded together prior to forming. With variable-thickness blanks produced by welding together different sheet-stock combinations in specific regions, it is possible to make finished parts with a desirable variation in properties such as strength, corrosion resistance, and so on. In the automotive industry, for example, steel tailor-welded blanks are used to provide optimum strength using the least amount of material possible for specific points on a vehicle. By combining various steels into a single welded blank, product and manufacturing engineers can "tailor" the best properties of different steels so that they are located precisely within the part where they are needed.

Tailor-welded blanks are made from prime stock as nested blanks or from collectible offal. Although the tailor-welded-blank process was first developed as a method for utilizing collectible offal and improving blank nesting possibilities, its greatest potential lies in the area of blanks with different thicknesses, coatings, and material grades. Tailor-welded blanks give the product designer the opportunity to eliminate reinforcements while improving structural and dimensional characteristics.

Although tailor-welded-blank (TWB) technology with aluminum alloy sheet is an area of development, the main application is with steel TWB sheet for automotive components such as body side frames, door inner panels, motor compartment rails, center pillar inner panels, and wheelhouse/shock tower panels. The main benefit is weight reduction, which is derived from optimization of material and gage reductions in specified areas of the TWB and the finished part. In addition, the TWB process allows for part integration, eliminating the need for reinforcements and stiffeners (e.g., in situations when weight is reduced by using lighter-gage high-strength steel in place of heavier gage steels, leading to reduced structural stiffness). Material yield is also improved both by making use of collectible offal and by reducing offal as a result of a product design, thereby optimizing the material for the specific requirements of a given part. Overall, the benefits of steel TWB methods in automotive vehicles include:

- Fewer parts
- Fewer dies
- Fewer spot welds
- Reduced design and development time
- Lower manufacturing costs
- Reduced material use
- Optimization of the use of steel properties
- Weight reduction
- Improved dimensional accuracy
- Improved structural integrity
- Stiffness
- Improved safety
- Increased offal utilization and reduced scrap

This article briefly reviews the forming of steel tailor-welded blanks with some discussion of the effects of welding on forming. As with all stamping operations, control of the stamping operation for tailor-welded blanks requires monitoring parameters such as blankholder pressure, material uniformity, die gaging, lubrication, press tonnage, and overall process monitoring. This article is adapted from Ref 1 and covers weld factors (such as orientation of weld relative to metal movement in dies), the formability of TWB materials, die and press considerations, and specific factors for the drawing, stretching, and bending of steel tailor-welded blanks. Other product factors and process considerations are discussed in more detail in Ref 1.

Welding Methods for Tailored Blanks

The types of welds that are used or have been considered for tailor-welded blanks include:

- Laser
- Resistance mash seam
- High-frequency induction
- Electron beam (nonvacuum)
- Friction-stir welding (more recently considered for aluminum TWB)

The type of weld joint selected for a specific application depends on a number of factors such as aesthetic and structural requirements, cost, and availability. Performance and the cost of welding operations are influenced by material handling and preparation, material gage, welding methods, overall blank dimensions, length of weld, and number and orientation of welds.

Currently, the most common methods of joining tailor-welded blanks are resistance mash-seam welding and laser welding. One of the earliest European applications of tailor-welded blanks for automotive applications was by Swedish carmaker Volvo using the resistance mash-seam process in 1979. An early use of laser-welded blanks in Europe was the Audi floor pan made by Thyssen A.G. in Germany since 1985. In North America, the initial use of tailor-welded blanks was on structural members such as frames, center pillar inners, and motor compartment upper rails. In the late 1960s, electron-beam welding was used by A.O. Smith to weld thicker gage (3.0 to 5.0 mm, or 0.12 to 0.20 in.) metal blanks for automotive frame members.

In Japan, welded blanks have been produced by laser methods both with and without filler wire by Toyota since 1986. Filler wire is used for applications that have an exposed weld in the finished product, such as body side frames. Filler wire welds are ground flush to improve surface appearance after welding. Welds that do not require a flush surface for aesthetic or sealing purposes are welded with Toyota's beam-weaving process (U.S. patent No. 5,245,156). Although beam weaving can reduce the concavity of the weld, it does not replace the filler wire process for body side frames, inasmuch as it does not satisfy the requirements for outer surfaces. Both filler wire and beam weaving are used to reduce the need for precision shearing and to allow laser welding of multipiece, multiweld blanks in a one fixture setup. By using die-cut blanks without precision shearing before each weld, Toyota eliminates the repeated processing steps of shearing and welding and the inefficient batch processing that is typical for multipiece, multiweld blanks being made on other welding systems.

Selection of Welding Method

Tailored-blank welding processes have different capabilities that must be considered when designing a tailor-welded blank or selecting a welding process. For example, limits on length of weld and gage ratios are critical and influence the selection of an optimal welding process for

*Adapted with permission from *Tailor Welded Blank Design and Manufacturing Manual,* Auto Steel Partnership Inc., July 1995; available from the Auto-Steel Partnership website: a-sp.org

a specific tailor-welded blank. Some processes have only straight-line capabilities, and edge-preparation is more critical with some processes than with others. The capabilities of various processes are summarized in Table 1. Figure 1 illustrates typical welding geometries, welds, and heat-affected zones for the various welding techniques.

Laser beam butt welding is a full-penetration fusion-welding process that results in a high depth-to-width ratio and therefore generally produces a narrow weld seam. Due to the extremely high cooling rate of the molten metal and narrow weld zone, the laser weld has a higher hardness than a resistance mash-seam weld. For laser butt welding, good edge preparation and beam alignment are critical to produce an acceptable weld.

For the laser-welding process, weld quality depends on parameters such as weld height, weld width, angle of surface alignment, mismatch, centering, and undercut or concavity (as defined from earlier testing of specimens). Similar tests and process controls can be applied to high-frequency induction and electron-beam welding processes. Postweld testing of laser or high-frequency-induction butt welds usually involves cup tests, dome tests, or hydraulic-bulge testing performed on randomly selected samples.

The laser beam tailor-welded blank has a butt-welded joint with a slight concavity at the joint with a narrower (about 1.0 mm, or 0.04 in.) heat-affected zone. Minimizing concavity of the weld joint is critical to good formability and mechanical properties. Minimum concavity in the weld joint is usually achieved by precision shearing the blank edges to be welded. A good fit-up prior to welding and proper laser-beam alignment with respect to the joint are important for a good laser weld. Other methods used to minimize concavity and reduce fit-up problems include the use of filler wire, beam weaving, beam orbiting, or beam spinning during the welding process.

Laser welding (Fig. 2, 3) using filler wire was developed by Toyota in order to increase the allowable gap of butt welding and to improve the weld appearance. This type of welding system does not require a precision sheared edge, but edge preparation is still a critical factor in producing a good weld. Cutting dies must be carefully maintained to produce an edge that is flat and straight, with minimum breakaway in order to obtain a secure weld with minimum concavity. The use of filler wire also requires additional process controls for the wire feed and a postweld grinding operation of the excess filler material whenever needed by appearance or functional requirements.

Beam weaving also reduces the sharpness of the offset when welding different metal thicknesses (Fig. 3). The beam-weaving process

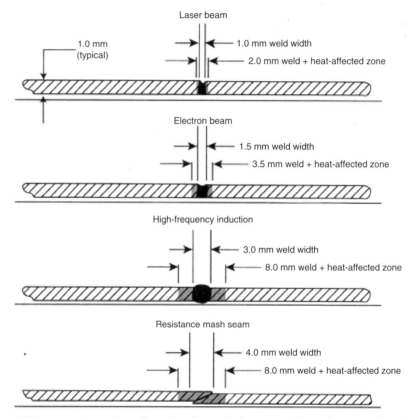

Fig. 1 Welding geometries, welds, and heat-affected zones for the various welding techniques used to make tailor-welded blanks. (Values are typical for 1 mm, or 0.04 in., thick sheet and may vary depending on welding speed and power input.)

Table 1 Capabilities of blank-welding methods

Capability	Process			
	Resistance mash-seam welding	Laser welding	High-frequency induction welding	Electron-beam welding
Straight-line capabilities	50 mm (2 in.) minimum and 2500 mm (100 in.) maximum weld length in a straight line only	About 3800 mm (150 in.) maximum weld length. The laser welding process can weld nonparallel lines using a multiaxis gantry robot system.	1 m (40 in.) maximum weld length	Weld length is limited only by size of blank-handling equipment. The electron-beam welding process can also make nonparallel welds. Because the beam location is fixed in the plan view, however, the X and Y axes must be on the table. For large blanks, this will result in a large shielding enclosure to allow room for the table movement.
Edge preparation	Edge preparation is unnecessary unless welding multiple-piece tailor-welded blanks, such as a body side ring.	Maintaining edge alignment becomes more difficult with longer welds (over 500 mm, or 20 in.). Therefore, edge preparation becomes more critical.	Edge preparation is not as critical as with the laser welding process.	Edge preparation is almost as critical as with the laser welding process.
Gages	0.7 mm (0.03 in.) minimum thickness 3 mm (0.12 in.) maximum thickness 5 mm (0.20 in.) total thickness ($T1 + T2$) 3-to-1 maximum material ratio	0.7 mm (0.03 in.) minimum thickness 3 mm (0.12 in.) is maximum thickness(a) 3-to-1 maximum material ratio	0.7 mm (0.03 in.) minimum thickness 2.5 mm (0.10 in.) maximum thickness 2.5-to-1 maximum material ratio	Minimum thickness is the same as for the laser case. 5 to 6 mm (0.20 to 0.24 in.) maximum thicknesses are possible(a).

(a) Depending on the maximum power capability with a 35 kW maximum for electron beam versus 8 kW maximum for laser. Electron-beam welding is not only faster than laser welding, but can also weld a thicker gage.

typically produces a wider laser weld (1.5 to 2.5 mm, or 0.06 to 0.10 in.), but reduces fit-up problems and concavity without precision shearing (for shorted welds) or the postweld grinding associated with filler wire. Use of the beam-weaving process makes beam alignment less critical in addition to reducing fit-up problems. As shown in Fig. 3, blanks can be aligned center-to-center or one-side flush. In either case, beam weaving will reduce the sharpness of the offset.

Blanks of different thicknesses are typically welded center-to-center to eliminate the need for right- or left-handed tools during the blank welding process. Blank stacks must be inverted for left-handed blanks. When appearance is important and weather-seal surface and wind-noise suppression are critical, the laser welding process may have advantages over the resistance mash-seam process due to surface geometry. Hem flange requirements may also favor laser welding.

Laser Welding with Fixed Optics. Laser systems with fixed optics are best for maintaining beam alignment and focus. This is a good process for two-piece tailored blanks with long welds, such as door inners and two-piece floor pans, but can require inefficient batch processing for multipieces moved under the fixed laser beam. Using a laser with fixed optics requires precision edge preparation before each weld. This requires a very large, multistage, dedicated system or an inefficient series of flexible systems for batch processing.

Laser Welding with Moving Optics. Laser systems with moving optics can be built in single-axis or multiaxis configurations. This process can be used for multipiece, multiweld tailor-welded blanks in order to weld several pieces in one setup. A gantry robot can be coupled with the laser system to move the laser beam over a fixed workpiece. In addition to movement in the robot axes, the moving-optic system may include a reciprocating condenser lens that produces a beam-weaving effect along the weld path. This reduces the sharp offsets of heavier gages, slightly increases gap tolerances, and eliminates the need for filler wire or precision shearing before each welding operation.

Resistance mash-seam welding is a roller seam resistance-welding process that requires a small overlap (approximately 3 mm, or 0.12 in., depending on gage) with forging ("mashing") of the weld seam as the current is applied. Resistance mash-seam welding is thus a forging process that produces a solid-state weld. It is a relatively wide weld compared to the laser weld. Use of a resistance mash seam requires in-line welds for most efficient operation. For those welds that are not in-line, a retrim operation before the next weld may be required to ensure a consistent edge for the weld rollers. The result can be delays or make batch processing necessary.

Resistance mash-seam tailor-welded blanks have an overlap seam joint that is 10 to 50% thicker than the thicker of the joined materials and a heat-affected zone approximately twice the width of the weld. The reduction of thickness at the joint is limited by the necessity to avoid weld current shunting, depending on the gage of the materials. If the lap joint is deformed excessively during welding, shunts occur where the weld current can flow through a single thickness of material.

The thickened joint can be reduced to ~10% more than the thickness of the heavier-gage material by postweld planishing. The planishing process involves compressing the weld joint between steel rollers. This can be done immediately after welding at a temperature above the recrystallization temperature (~550 °C, or 1000 °F, for sheet steel) as in hot planishing or after the weld has cooled via cold planishing. Hot planishing has better formability characteristics, but coated sheet steels can be a problem for this process due to coating pickup on the planishing rollers.

During the mash-seam welding operation, parameters such as voltage, amperage, temperature, and weld speed are monitored and recorded to ensure optimum weld conditions. Postweld testing of resistance mash-seam welds involves both cup testing for formability and peel testing for weld integrity. These are both destructive tests and are done only on random samples.

High-frequency induction welding is a butt-upset welding process. Volvo (with Elva Induction) developed this process for joining blanks. This type of weld has a thickening at the joint and a heat-affected zone similar in width to the resistance mash-seam weld. Grinding the joint can remove the thickened area to improve fit-up in assemblies. High-frequency-induction welding has a 1 m (40 in.) maximum weld length and ~50% thickening at the weld seam.

During the welding operation, the two blanks are pressed together edge-to-edge. They are held together by a clamp on each side of the joint over the full length of the weld. One clamp presses down on the fixed frame, while the other is pressed against the movable frame and follows its movement. The parts are in firm contact under fairly high pressure when the welding current is switched on. During welding, the edges are compressed by a certain amount. The material fuses without the use of filler material to a weld about 1 to 2 mm (0.04 to 0.08 in.) wide, depending on the thickness of the material. According to the developers, induction welding is considered to have a faster cycle time and lower operating costs than laser or resistance-mash-seam welding.

Electron-beam (nonvacuum) welding is also a fusion process. Power is typically in the

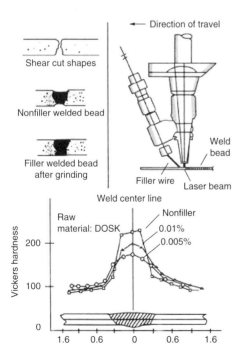

Fig. 2 Laser welding using filler wire to make tailor-welded blanks and the resulting hardness distribution

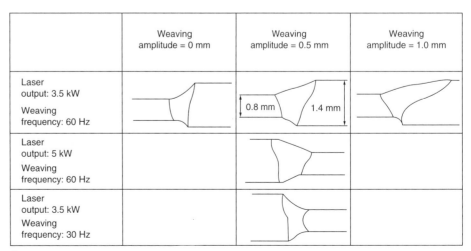

Fig. 3 Effect of laser power and beam-weaving amplitude and frequency on laser welds

25 to 35 kW range, making it a viable process for heavier-gage material. Compared to the laser welding process, electron-beam welding can tolerate a slightly larger gap between the edges to be welded. It is faster, and the beam is easier to focus. Although electron-beam welding offers some advantages over the laser-welding process, its use has been limited by radiation-shielding requirements to contain the x-rays generated by the process. The electron-beam welding process has been used by A.O. Smith to make tailor-welded blanks for heavy automotive frames, but is not being used in production currently. This type of weld is similar in appearance to a laser weld, but is about 50% wider.

Special Considerations for Welding and Weld Design

Blank Preparation Prior to Welding. Most resistance mash-seam tailor-welded blanks have a die-cut edge at the weld seam. Butt-welding processes, on the other hand, require more accurate edge treatment. Long laser welds (>500 mm, or 20 in.) require a precision shearing operation prior to welding. The concavity of the laser weld is increased by excessive metal breakage at the trim edge. Some metal processors use long sections and "pocket" these sections during blanking (Fig. 4) in order to eliminate movement.

Short laser welds, electron-beam welds, and induction welds can be made with die-cut blanks if the dies are constructed to a standard similar to fine blanking. These standards demand more stringent tolerances than normal blanking dies. Die clearances that are normally between 6 and 10% of the sheet metal thickness must be reduced to a minimum, and holding forces on the pressure pads must be increased sufficiently to avoid recoil during blanking. This will reduce metal breakage and improve the fit-up at the weld joint. Also, the die shear must be increased to at least four times the metal thickness in order to reduce metal breakage.

Once the blank is trimmed, extreme care must be taken to not damage the edge to be welded. The blank must not be allowed to drop through the die nor down metal chutes or conveyors, and blank edges should not impact against stacker pins. Rather, each blank should be removed from the top of the die by vacuum cups or magnetic belts and placed on the stack with a vertical motion in order to avoid edge damage.

Die and press guidance is also critical. With the reduced die clearance, it is important to guide the die with oversize pins and bushings and run the die in a press with accurate slide guidance and parallelism.

Pre- or Postweld Blank Washing and Lubrication. Mill oils, dry lubricants, and other coatings can pose a problem for some welding systems, especially those that move the material with skew rolls. In such cases, blanks should be prewashed before welding in order to avoid slippage. It may also be necessary to reapply lubricants or to restore mill-applied prelube conditions to the finished blank after welding for formability and corrosion-resistance performance.

Location of Welds Relative to Capabilities of Welding Equipment and Processes. Straight-line welds for two-piece blanks can be made with most job-shop welding equipment, whether laser or resistance mash-seam. The main considerations for these types of welds concern limitations on length, thickness, formability, or aesthetics. Multipiece, multiweld blanks require careful planning to achieve economical production. These blanks often cannot be run in job-shop equipment without inefficient batch processing. Also, some weld orientations can cause efficiency problems in certain types of welding equipment (See Fig. 5).

High-volume blanks with multiple nonparallel welds and blanks with circular welds require more sophisticated welding equipment and processes. A gantry-mounted laser with three-axis capabilities and good edge preparation and clamping are necessary to accomplish this with a minimum amount of process steps and in a cost-effective manner.

Location of Welds in Blanks and Formed Parts. Careful consideration should be given to metal movement in the forming dies when planning orientation of welds in the blank. The weld area and heat-affected zone reduce formability limits, and deformation transverse to the weld can result in failure in the base metal. Areas of excessive deformation in line with the weld are undesirable. For welds made from materials with similar thickness and strength, deformation transverse to the weld should not cause formability problems unless the weld is poor.

For welds made from materials with different thicknesses and/or strengths, failure normally occurs on the side for which the material is thinner or of lower strength because of the natural tendency for plastic flow to localize in weaker areas. For tailor-welded blanks made from materials of dissimilar thickness or strength, particular care therefore needs to be taken to minimize strength on the side with the greater thickness or higher strength. This can be accomplished by allowing greater material flow from the side of the weld with the greater thickness or higher strength by using less severe draw beads, reduced blankholder pressure, or modified blank outlines. Furthermore, blanks with significant differences in thickness or strength that could cause metal flow perpendicular to the weld seam may lead to excessive die wear and, if possible, should be relocated. Other measures to prevent die wear include the use of more wear-resistant die materials and die surface treatments and special die lubrications.

Areas that are subjected to bending and straight areas with minimal elongation are good choices for welds in line with metal movement. On the other hand, stretch flange areas (e.g., the inside corners of a body side frame such as corners of window or door openings) and deep-drawn areas with large elongation in the direction of the weld are poor places to locate a weld

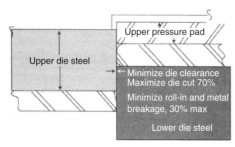

Fig. 4 Blanking-tooling design to ensure proper edge preparation for laser welding

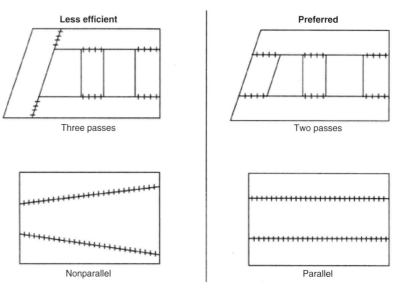

Fig. 5 Preferred blank designs for resistance mash-seam or single-axis laser-welding systems

joint. The flange edge adjacent to the weld is particularly susceptible to failure during stretch flanging (see Fig. 6). The reason for this precaution is because of the reduction in stretch flange capability in the area of the weld. Thicker/high-strength metal will tend to pull thinner/lower-strength metal during this type of deformation. Splitting usually occurs in the base metal or heat-affected zone beside the weld if weld lines are located in the stretch flange areas at the corners of these openings. Welds should not be located within 25 mm (1.0 in.) of an area subject to the biaxial strain of stretch flanging (see Fig. 7) without first determining the formability potential of the material through computer modeling.

Postweld Blanking and Notching. Use of square-sheared blanks for component blanks rather than pattern blanks usually requires postweld blanking for door and window parts. Square-sheared blanks allow the use of standard equipment such as square shears, trap shear lines, and oscillating cutoff dies. This type of standard blanking equipment is normally available at outside job shops. Such equipment, however, decreases material utilization because of inefficient nesting. It also increases the length of the weld.

Blanking dies for pattern blanks will not only improve material utilization but will also eliminate postweld blanking and shorten the length of welds because of the smaller blank size.

Postweld notching is used to remove a small area of imperfect weld at the end of some weld lines and to improve stress relief in stretch flange areas, as for the inside corners of body side frames. Postweld notching can be accomplished with a small hydraulic unit in line with the welder, instead of dies and presses. These units are similar to the units used for dimpling and embossing.

Formability Considerations

Formability of tailor-welded blanks is an important consideration particularly in the weld area because of the heat-affected zone and hardening in the weld-joint region. Some of the important factors related to formability are summarized in this section. Additional information (for both steel and aluminum TWB formability) can be found in the Selected References cited at the end of this article.

Formability tests for tailor-welded blanks can be performed via tension testing (in line with

the weld and transverse to the weld) and by simulative forming tests. For example, a limiting dome-height test machine can be used to perform:

- Spherical-punch stretching tests (to simulate stretch forming)
- Hole-expansion tests (to simulate stretch flanging)
- Cylindrical contraction or cup-drawing tests (to simulate deep drawing)

These tests have shown that the most deleterious deformation modes for welds in tailor-welded blanks are plane strain (in line with the weld) and stretch flanging, especially with different metal thicknesses or strengths. With tailor-welded blanks that have different strengths or sheet thicknesses, the forming limit is further reduced as the strength ratio increases.

General Formability Guidelines. Early research on the formability of tailor-welded blanks (Ref 2) revealed important formability considerations for three standard forming operations: stretch forming, stretch flanging, and deep drawing. In stretch forming, it was found that the weld bead ductility is the limiting factor when the weld is oriented parallel to the major stretch axis, while the formability of the weaker base material is the limiting factor when the weld is normal to the stretch axis. It was also shown that stretch flangeability drops by 25 to 30% compared to the base material because of the presence of the weld, while deep drawability is unaffected.

Several subsequent investigations (Ref 3, 4) shed further insight into the formability of steel tailor-welded blanks. This work revealed that there is almost no difference in the formability of blanks made by resistance mash-seam and laser welding processes. Specifically, the tensile elongation in line with the weld is significantly

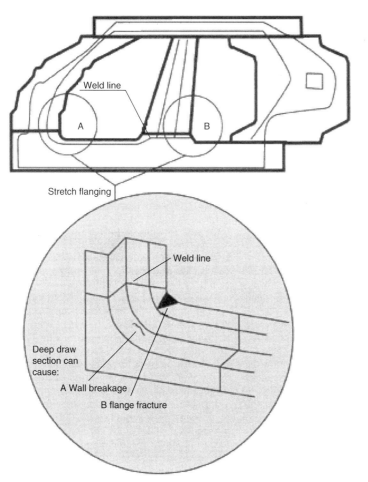

Fig. 6 Factors affecting deformation and failure during stretch flanging of tailor-welded blanks

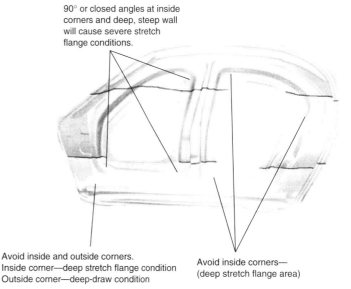

Fig. 7 Difficult forming areas with respect to the placement of welded joints on tailor-welded blanks

reduced for blanks made by both techniques. Correspondingly, the limit strain can be reduced by more than 50%. For resistance mash-seam welds, the formability for as-welded blanks and as-welded blanks with hot planishing can be equivalent to laser-welded blanks of the same material combination. Cold planishing after resistance mash-seam welding, however, significantly reduces formability of the blank in the area of the weld.

It is important to note that tailor-weld blanks should only undergo deformation parallel to the weld line regardless of the global deformation mode. If strains perpendicular to the weld line exceed the thinning limit of the material, the result is failure parallel to the weld line of the thinner or lower-strength material. Thus, in the case of dissimilar blanks, the influence of geometry (i.e., thickness dissimilarity) generally outweighs differences in welding characteristics (see Fig. 8).

Additional formability studies on tailor-welded blanks performed at the Institute for Metal Forming at the University of Stuttgart (Ref 3) have shown that certain guidelines should be followed in the forming of welded blanks of different thicknesses. Because of the variation in the thickness of sheet steel, offset blankholder surfaces should be used in order to ensure a uniform surface pressure on the binder area of the drawn stamping. Furthermore, weld movement during the drawing process requires that the blankholder be designed with appropriate clearances in the region of the welds so that they can move through the area during the drawing process. These clearances, though necessary for movement of the heavier gage metal and the welds, can also be a source of wrinkling with lighter gages due to the lack of hold-down force in this area.

Another challenge with tailor-welded blanks is that due to the welding together of sheet segments with different thicknesses; different thickness variations within permitted tolerance ranges may have unfavorable effects on die performance. This is especially true in multigage blanks when the thinner material is at the lower gage tolerance and the thicker material is at the upper gage tolerance.

The formability of tailor-welded blanks has also been investigated at Ohio State University (OSU) (Ref 4). Failure of tailor-welded blanks is highly dependent on the amount the weld line moves. If the weld movement is too severe, the base material parallel to the weld line will tear. If the weld movement is not severe, the blank can either fail in the weld bead itself or in some other part of the blank, depending on the loading conditions. To evaluate the sensitivity of the weld movement to material and processing conditions, OSU conducted computer modeling. It was found that two conditions must be present for weld movement: a difference in the load-bearing capacity of the joined materials and stretching perpendicular to the weld. This combination forces the weaker material to deform, and the weld moves in the direction of the stronger material to compensate for it. If either a force normal to the weld is absent or if there is no difference in strength of the parent materials, no weld movement is observed. Hence, the severity of the weld movement is dependent on the strength ratio of the materials and the magnitude of the boundary forces. Computer modeling also indicated that weld movement is influenced to a much lesser extent by the global friction conditions, the local friction conditions in the weld area (especially for resistance mash-seam welds), the weld geometry, and the mechanical properties of the weld. When modeling the forming of a laser-welded blank, for example, the mechanical properties of the weld bead itself can be neglected with no significant loss of accuracy in the results. In order to predict the forming behavior of a tailor-welded blank, however, it is essential to have a clear understanding of the restraining forces around the edge of the blank. There were greater errors in model predictions due to discrepancies in the boundary forces than from the friction coefficient.

Draw, Stretch, and Bend Considerations

In general, the performance of tailor-welded blanks during deep drawing is similar to the base metal unless stretching along the weld during drawing exceeds the forming limit of the weld. In this case, the stretchability of tailor-welded blanks is significantly less than the base metal. Stretching perpendicular to the weld normally does not cause any problems as long as the thinning limit of the weld material is not exceeded. On the other hand, a large amount of stretch in the direction of the weld may result in tearing across the weld line due to the poorer mechanical properties of the weld relative to those of the base material.

The formability of tailor-welded blanks depends on the type of welding, welding schedules, material, and thickness combinations. For a typical aluminum-killed drawing-quality (AKDQ) steel, for instance, the plane-strain forming limit can be between 15 and 20% for a

typical laser-welded blank and a resistance mash-seam blank with hot planish. To avoid splitting in the weld, the strain component along the weld line should be below the forming limit of the welded blanks.

In general, two types of splits may occur when forming tailor-welded blanks. One is across the weld, which can occur in blanks with similar-or-dissimilar thickness-or-strength material combinations. The other is parallel and immediately adjacent to the weld, which usually occurs in blanks made from materials with dissimilar thicknesses or strengths. To avoid splits in the weld, the weld should be placed away from the major strain direction and away from areas with high strains along the weld direction. The best orientation is placing the weld perpendicular to the major-strain direction. To avoid splits adjacent to the weld in the thinner gage or lower-strength material, the shift of the weld line from its initial position should be minimized. This may be accomplished by allowing more metal to flow from the binder surface on the thicker or higher-strength side of the draw die by reducing the severity of the draw beads, by reducing blankholder pressure, or by modifying the blank outline on the higher-strength or thicker gage side of the die. Another technique is to add a local nitrogen pad (such as a Teledyne DRAC system) along the weld line in order to minimize the weld-line shift.

Bending perpendicular to the weld line is normally not a problem. When bending on the weld and in-line with the weld, sharp radii should be avoided since severe bending will result in cracking or failure on the tensile surface of the weld in the bending area.

Die and Press Considerations

Stamping presses must be equipped with tonnage monitors and strain gages in order to measure and record blankholder pressure and press tonnage. This is required for all stamping operations that bottom, such as draw dies and form dies. Pressure in each corner of the press can be optimized to suit the formability characteristics of the material. This tonnage must be determined and recorded at die tryout and maintained in each production run.

Drawing dies. Laser- and electron-beam-welded blanks of the same grade and thickness are not a major problem for drawing operations when the product planning and draw die development is done properly. As noted previously, it is important to minimize elongation in line with the weld and to avoid a stretch-flange condition perpendicular to the weld.

Resistance mash-seam and high-frequency induction tailor-welded blanks of the same grade require additional die clearance due to increased thickness at the weld line.

In all instances, careful planning for the weld-joint locations during product design and drawing-die development is necessary. Product design engineers and die engineers must work in

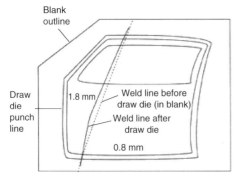

Fig. 8 Deformation of weld line during drawing operation using blanks with dissimilar thickness

close collaboration during the development phase in order to achieve maximum cost effectiveness with the tailor-welded blanks. CAD/CAE software (such as ABAQUS, PMASTAMP, or DYNA3D) should be used to determine local deformations during forming.

Blanks of different thicknesses and strengths made with any welding process can pose challenges during die tryout and production because of the tendency for the elongation to be concentrated in the thinner or lower-strength material when deformation occurs perpendicular to the weld line. Also, when the metal-thickness difference is significant, press tonnage and balance can be a problem. Press-tonnage monitors are essential for this type of part, not only for protection of the press equipment, but also for good process control. If press balance is a problem, the tonnage monitors will indicate where the problem is. Supplemental balance devices, such as nitrogen cylinders, may be required in the die, or, in extreme cases, a larger press may be required.

Another consideration when planning blanks of mixed strength is that the die process for high-strength steel is often quite different than that for mild steel. High-strength steel can exhibit side-wall curl, springback, and drawability characteristics different from mild steel. Mild steel can usually be formed with high-strength-steel die processes, but high-strength steel will not always form satisfactorily with conventional-die processes. The use of developmental soft tooling is often advisable in such cases.

Forming dies require planning similar to draw dies. Elongation in line with and perpendicular to the weld must be considered before construction of hard tooling begins. Effects of different strengths and gages on formability and press tonnage and balance must also be evaluated. High-strength steel components in the welded blank require die technology and process planning suitable for this material.

Trim Dies. Trimming of stampings with major differences in metal thickness or strength can cause press imbalance and excessive shock loads as the press energy is released when metal breakage occurs. Increasing the shear on the upper trim steel to four times the metal thickness per linear foot for high-strength steel will greatly reduce the shock loads. In extreme cases, nitrogen cylinders or urethane blocks may be necessary in the lower die to contain this sudden release of energy and reduce noise levels. Care must be taken during die construction to provide proper die clearance and shear for the different metal thicknesses in the stamping. As for forming dies, high-strength steel components in the welded blank require die technology and process planning suitable for this material (e.g., Fig. 9).

Flanging Dies. When flanging material of different strengths, springback allowance must be made for the higher-strength material. This means that a panel setup that allows a wiping action (where the flange surface is vertical or perpendicular to the die surface) on flanges of the higher-strength material will result in a bumping action (where the flange surface is struck home to an open angle when the press is on bottom dead center) for the lower-strength material, which requires less overbend for springback. Usually, this difference will be no more than three degrees, but could be more for some higher-strength steels.

Stretch flanging operations must be designed carefully with respect to the placement of the weld joint and when using high-strength steels. Weld placements that should be avoided are illustrated in Fig. 9; the flange length for high-strength steels in these areas should be limited.

Hemming Dies and Hemming Machines. Hemming over different metal thicknesses on an inner panel can present a challenging quality problem for outer panel surface with respect to hem radius uniformity. In such cases, a simple yet effective course of action for door assemblies is to make the metal thickness transition as gradual as possible by coining the thicker portion of the hem surface on the inner door panel adjacent to the weld in one of the form dies.

This will reduce the abrupt transition from the thicker material to the thinner gage.

Another consideration is the type of hemming operation and its effect on surface and hem-radius quality. Hemming can be done in mechanical or hydraulic presses with separate prehem and finish-hem dies or a two-stage hem die. Also, hemming machines with hydraulic, pneumatic, or electric drives have been used. In order to ensure the best surface and hem radius quality, a process with a controlled pressure is necessary. This requirement excludes any hem operation in a mechanical press that will hem by hitting solid at the bottom of the press stroke. A process that flattens the hem with a squeezing action is preferred. In addition, the prehem and finish hem should be completed in one tool to avoid mismatches in the outline. Hemming fixtures or dies in hydraulic presses can produce the desired results if constructed with proper gaging and cam designs or mechanical linkages for effective prehem and final-hem actions.

REFERENCES

1. *Tailor Welded Blank Design and Manufacturing Manual,* Auto Steel Partnership, July 1995, www.a-sp.org
2. K. Azuma et al., "Press Formability of Laser Welded Blanks," Toyota, 1990
3. K. Siegert and E. Knabe, Fundamental Research and Draw Die Concepts for Deep Drawing of Tailored Blanks, *Automotive Stamping Technology* (Detroit, MI), Feb 27 to March 2, 1995, Society of Automotive Engineers SAE, 1995, p 159–169 and also *SAE Trans.: J. Mater. Manuf.,* Vol 104, 1995, p 866–876
4. F.I. Saunders and R.H. Wagoner, "A Better Sheet Formability Test," Ohio State University, Department of Materials Science and Engineering

SELECTED REFERENCES

Steel Tailor-Welded Blanks

- M.A. Ahmetoglu, D. Brouwers, L. Shulkin, L. Taupin, G.L. Kinzel, and T. Altan, Deep Drawing of Round Cups from Tailor-Welded Blanks, *J. Mater. Process. Technol. (Switzerland),* Vol 53 (No. 3–4), Sept 1, 1995, p 684–694
- V.V. Bhaskar and K. Narasimhan, Formability Studies on Transverse Tailor Welded Blanks, *NUMISHEET 2005: Proc. Sixth International Conference and Workshop on Numerical Simulation of 3D Sheet Metal Forming Processes,* Part A, Vol 778, 2005, p 699–704
- V.V. Bhaskar, R.G. Narayanan, and K. Narasimhan, Effect of Thickness Ratio on Formability of Tailor Welded Blanks (TWB), *Proc. Eighth International Conf. on Numerical Methods in Industrial Forming*

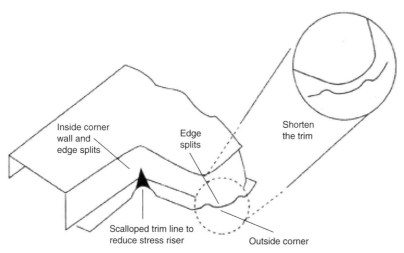

Fig. 9 Trim line changes required to avoid edge splitting during trimming of formed tailor-welded blanks. It is important that the trim quality be maintained to prevent edge splitting.

Processes (Columbus, OH), June 13–17, 2003, 2004, p 863–868

- J. Cao, B. Kinsey, and Z. Liu, A Novel Forming Technology for Tailor-Welded Blanks, *J. Mater. Process. Technol. (Netherlands),* Vol 99 (No. 1–3), March 1, 2000, p 145–153
- S.M. Chan, L.C. Chan, and T.C. Lee, Thickness Ratio Effects on Forming Limit Diagram of Tailor-Welded Blanks, *Fourth International ESAFORM Conference on Material Forming* (Liege, Belgium), April 23–25, 2001, p 329–332
- S.M. Chan, L.C. Chan, and T.C. Lee, Tailor-Welded Blanks of Different Thickness Ratios Effects on Forming Limit Diagrams, *J. Mater. Process. Technol. (Netherlands),* Vol 132 (No. 1–3), Jan 10, 2003, p 95–101
- Deep Drawing Tailor-Welded Blanks, *Stamp. J.,* Vol 9 (No. 2), March–April 1997, p 26–32
- P. Dong, Y.P. Yang, Z. Cao, and J. Zhang, "A Finite Element Study on Effects of Welding Procedures on Formability for Tailor-Welded Blanks," paper No.1999-01-0680, Society of Automotive Engineers, 1999
- Y. Heo, Y. Choi, H.Y. Kim, and D. Seo, Characteristics of Weld Line Movements for the Deep Drawing with Drawbeads of Tailor-Welded Blanks, *J. Mater. Process. Technol. (Netherlands),* Vol 111 (No. 1–3), April 25, 2001, p 164–169
- Y.M. Heo, S.H. Wang, H.Y. Kim, and D.G. Seo, The Effect of the Drawbead Dimensions on the Weld-Line Movements in the Deep Drawing of Tailor-Welded Blanks, *J. Mater. Process. Technol. (Netherlands),* Vol 113 (No. 1–3), June 15, 2001, p 686–691
- H.M. Jiang, S.H. Li, H. Wu, and X.P. Chen, Numerical Simulation and Experimental Verification in the Use of Tailor-Welded Blanks in the Multi-Stage Stamping Process, *J. Mater. Process. Technol.,* Vol 151 (No. 1–3), Sept 1, 2004, p 316–320
- S.S. Kang, K.B. Min, and K.S. Kim, A Study on Resistance Welding in Steel Sheets Using a Tailor-Welded Blank. I. Evaluation of Upset Weldability and Formability, *J. Mater. Process. Technol. (Netherlands),* Vol 101 (No. 1–3), April 14, 2000, p 186–192
- B. Kinsey, Z. Liu, and J. Cao, New Apparatus and Method for Forming Tailor Welded Blanks, *SAE Trans.: J. Mater. Manuf. (USA),* Vol 108, 1999, p 653–660
- C.H. Lee, H. Huh, S.S. Han, and O. Kwon, Optimum Design of Tailor Welded Blanks in Sheet Metal Forming Processes by Inverse Finite Element Analysis, *Met. Mater. (South Korea),* Vol 4 (No. 3), 1998, p 458–463
- K.B. Min and S.S. Kang, A Study on Resistance Welding in Steel Sheets for Tailor Welded Blank. Evaluation of Flash Weldability and Formability. II, *J. Mater. Process. Technol. (Netherlands),* Vol 103 (No. 2), June 15, 2000, p 218–224
- R.G. Narayanan, V.V. Bhaskar, and K. Narasimhan, Effect of Weld Conditions on the Deformation Behaviour of Tailor Welded Blanks (TWB), *Proc. Eighth International Conference on Numerical Methods in Industrial Forming Processes* (Columbus, OH), June 13–17, 2003, 2004, p 856–862
- A. Piela and J. Rojek, Validation of the Results of Numerical Simulation of Deep Drawing of Tailor Welded Blanks, *Arch. Metall.,* Vol 48 (No. 1), 2003, p 37–51
- F.I. Saunders, Forming of Tailor-Welded Blanks, *Diss. Abstr. Int.,* Vol 55 (No. 10), April 1995, p 217
- F.I. Saunders and R.H. Wagoner, Forming of Tailor-Welded Blanks, *Metall. Mater. Trans. A,* Vol 27A (No. 9), Sept 1996, p 2605–2616
- D. Seo, Y. Choi, Y. Heo, and H.Y. Kim, Investigations of Weld-Line Movements for the Deep Drawing Process of Tailor Welded Blanks, *J. Mater. Process. Technol. (Netherlands),* Vol 108 (No. 1), Dec 1, 2000, p 1–7
- M.F. Shi, K.M. Pickett, and K.K. Bhatt, Formability Issues in the Application of Tailor Welded Blank Sheets, paper 930278, SAE SP-944, *Sheet Metal and Stamping,* Proceedings, Symposium, SAE International, 1993, p 27–37
- K.M. Zhao, B.K. Chun, and J.K. Lee, Finite Element Analysis of Tailor-Welded Blanks, *Finite Elements Analysis Design (Netherlands),* Vol 37 (No. 2), Feb 2001, p 117–130

Aluminum Tailor-Welded Blanks

- A.V. Bhagwas, T. Kridli, and P.A. Friedman, Influence of Weld Characteristics on Numerically Predicted Deformation Behavior of Aluminum Tailor Welded Blanks, "Magnesium Technologies for the Automotive Industry" and "Advances in Aluminum Technology for Automobile Applications," Sessions at the SAE 2002 World Congress (Detroit, MI), March 4–7, 2002, p 79–85
- A. Buste, X. Lalbin, M.J. Worswick, J.A. Clarke, B. Altshuller, M.J. Finn, and M. Jain, Prediction of Strain Distribution in Aluminum Tailor Welded Blanks for Different Welding Techniques, International Symposium on Light Metals as held at the *38th Annual Conference of Metallurgists of CIM* (Quebec City, Quebec, Canada), Aug 22–26, 1999, p 485–500
- S. Das, Aluminium Tailor Welded Blanks, *Adv. Mater. Process.,* Vol 157 (No. 3), March 2000, p 41–42
- R.W. Davies, H.E. Oliver, M.T. Smith, and G.J. Grant, Characterizing Al Tailor-Welded Blanks for Automotive Applications, *JOM,* Vol 51 (No. 11), Nov 1999, p 46–50
- R. Davies, G. Grant, M. Smith, and E. Oliver, Formability and Fatigue of Aluminium Tailor Welded Blanks, *IBEC 2000, Proc., International Body Engineering Conference* (Detroit, MI), paper 2664, Oct 3–5, 2000, 10 pages
- R.W. Davies, J.S. Vetrano, M.T. Smith, and S.G. Pitman, Mechanical Properties of Aluminum Tailor Welded Blanks at Superplastic Temperatures, *J. Mater. Process. Technol. (Netherlands),* Vol 128 (No. 1–3), Oct 6, 2002, p 38–47
- R. Davies, G. Grant, M. Smith, and S. McCleary, Describing the Formability of Tailor Welded Blanks, *SAE Trans.: J. Mater. Manuf.,* Vol 111, 2003, p 935–941
- P.A. Friedman and G.T. Kridli, Microstructural and Mechanical Investigation of Aluminum Tailor-Welded Blanks, *J. Mater. Eng. Perform.,* Vol 9 (No. 5), Oct 2000, p 541–551
- B. Kinsey, V. Viswanathan, and J. Cao, Forming of Aluminum Tailor Welded Blanks, *SAE Trans.: J. Mater. Manuf.,* Vol 110, 2001, p 673–679
- G. Kridli, P. Friedman, and A. Sherman, "Formability of Aluminium Tailor-Welded Blanks," paper 2000-01-0772, Society of Automotive Engineers, 2000
- Y. Lee, M.J. Worswick, H. Shakeri, S. Truttmann, M. Finn, M. Jain, F. Feng, and W. Christy, Formability of Large-Scale AA 5182 Tailor Welded Blanks, *Fourth International ESAFORM Conference on Material Forming* (Liege, Belgium), April 23–25, 2001, p 325–328
- Y. Lee, M. Worswick, S. Truttmann, H. Shaken, M. Finn, F. Feng, B. Christy, and D. Green, Evaluation of Small Scale Formability Results on Large Scale Parts: Aluminum Alloy Tailor Welded Blanks, *SAE Trans.: J. Mater. Manuf.,* Vol 110, 2001, p 680–686
- P. Martin, R. Zhang, B. Altshuller, M.J. Finn, and M.J. Worswick, Formability of Aluminum Multi-Gage Tailor Welded Blanks, *International Symposium on Light Metals 1998 Metaux Legers* as held at the *37th Annual Conference of Metallurgists of CIM* (Calgary, Alberta, Canada), Aug 16–19, 1998, p 409–423
- M.J. Saran, E.R. Pickering, M.A. Glagola, and R.T. Vandyke, Manufacturing and Forming of Aluminium Tailor Welded Blanks, Materials and Body Testing IBEC 1995, *Proc., International Body Engineering Conference and Exposition* (Detroit, MI), Oct 31 to Nov 2, 1995, p 120–124
- H.R. Shakeri, Y. Lee, M.J. Worswick, F. Feng, W. Christy, and J.A. Clarke, Weld Failure in Formability Testing of Aluminum Tailor Welded Blanks, *SAE Trans.: J. Mater. Manuf.,* Vol 110, 2001, p 101–110
- H. Wang, P. Martin, J.A. Clarke, B. Altshuller, M.J. Finn, and F. Feng, Aspects of Weld Metallurgical Characteristics and Fracture Mechanism in Formability Tests of Non-Vacuum Electron Beam Tailor-Welded AA5754-O Temper Aluminium Blanks, *IBEC 2000, Proc., International Body Engineering Conference* (Detroit, MI), paper 2663, Oct 3–5, 2000, 17 pages

Press Forming of Coated Steel

Revised by Brian Allen, AK Steel Corporation

COATED FLAT-ROLLED STEELS are formed using the same general equipment, tooling, and lubrication used to form uncoated steels. While the properties of the base steel remain the primary determinants of the formability of a coated product, coatings do have an effect on the forming process and must be taken into account when designing parts, dies, and forming strategies. Not only do the surface coatings affect the lubricity at the die/part interface, the forming process must be carried out in such a way as to maintain the coating integrity and preserve the original purpose of the coating.

Most coatings commonly used on sheet steel substrates have corrosion resistance as their primary function, although appearance is often important as well, especially for the organic coatings. Coatings can have a different coefficient of friction with common die materials than the steel substrate and may be more ductile or more brittle than the base metal. Various coatings also react differently with different lubricants and die materials than bare steels. Given the friction and formability differences between steel substrates and common coating materials, some adjustments to the forming process are usually necessary to obtain optimal results when switching from bare to coated steels or when changing coatings. While coated steels can usually be successfully formed on tooling designed for bare steel, part cost and quality improvements can frequently be realized if tooling, lubrication, forming parameters, and part geometry are adjusted to suit the coating/substrate combination being used.

Because most coatings are softer than the steel substrate, steps to protect the coating from damage are sometimes necessary. Coating damage can take place in the die but also occurs during blank or part handling, especially if sharp edges or burrs on blanks or formed parts are allowed to rub or scrape against adjacent pieces during stacking or handling. The brittle nature of some coatings may impose forming path or bend radii limits on coated steels that are much more stringent than would be the case with bare steels. These limits must be observed not only in the finished part but also in all areas of the part, including drawbeads and other areas outside the final part shape. Even if coating damage to the material outside the part trim line is acceptable from a part quality perspective, coating scraped or flaked off the part scrap will build up in the tooling and eventually result in dents, defects, or increased downtime and labor for die cleaning.

This article provides a general overview of some of the more common sheet steel coatings available. It also discusses formability differences between coated and bare steel and provides some general guidelines about the forming of coated steels. Information on the forming of uncoated steels can be found in other articles in this Volume.

Coated Steels

Coated steels are classified according to the nature of the substrate, the type of coating, and the method used to apply the coating. Nearly every type of sheet steel commonly used in the automotive, appliance, or building industries is available coated, from high-strength steels used for automotive structural members, steel roofing, and architectural framing, to vacuum degassed interstitial-free steels used in applications demanding the highest available formability. The types of substrate steels available are discussed in *Properties and Selection: Irons, Steels, and High-Performance Alloys,* Volume 1, *ASM Handbook,* 1990.

Common steel coatings include zinc, aluminum, tin, lead, nickel, and various alloys of these metals, as well as a range of organic coatings. Coatings such as porcelain enamels and electroplated copper or chromium are also commonly found on sheet steel parts, but these are typically applied after forming and are not discussed here. Most sheet steel coatings are applied for purposes of corrosion protection, but some coatings, particularly the organics, are also applied for aesthetic reasons. Application methods vary widely and include dipping in a molten bath (zinc, aluminum, tin, and terne—a lead-tin alloy), electroplating (zinc and tin), and spraying or roll coating (paints or organics).

Die Interaction Effects. The interaction between die materials and coatings is very complicated and should be considered when switching from bare to coated steels, when switching lubricants, or when developing new parts using prototype tooling. Both electro-galvanized and galvannealed steels have been shown to be sensitive to changes in tooling material between gray cast iron, nitrided gray cast iron, torch-hardened cast iron, bare steel, TiN-coated steel, and CrN-coated steel (Ref 1).

Galvannealed steel performed best with gray cast iron (GCI) tooling, followed by torch-hardened GCI and nitrided GCI. Electrogalvanized steel performed best in TiN- and CrN-coated dies, followed by bare steel and nitrided GCI dies. Cold rolled steel was only slightly sensitive to tooling material changes, and hot dip galvanized material was relatively insensitive to tool material. Coated steels were also observed to have a higher sensitivity to lubricant changes than were bare steels. For the two lubricants used in the referenced study, electrogalvanized sheet was found to be the most sensitive to lubricant change, followed by galvannealed, hot dip galvanized, and uncoated steel.

Coated steels may also show different behavior than cold rolled steels when switching from zinc-base Kirksite or other types of soft tooling prototype dies to hard production dies (Ref 2). Higher punch loads have been noted on a Kirksite prototype die as compared to the hardened-steel production die for cold rolled steel, while electrogalvanized steel showed the reverse trend. The punch load was also found to be more sensitive to punch speed when forming electrogalvanized steel as compared with uncoated steel for both die materials (Ref 3).

Zinc-Coated Steels

By far, the most widely used steel coating applied prior to press forming is zinc. Zinc coating is usually referred to as galvanizing, after the galvanic corrosion protection provided to the steel substrate by the zinc coating. Zinc coatings are applied to the strip via electroplating or by immersion of the strip in a bath of molten zinc. Both methods are common, but the hot dip coating is more prevalent.

Hot Dip Galvanizing

Hot dip galvanizing is conducted using one of two processes. Historically, most hot dip galvanized material was batch annealed in coil form, then processed on a flux line, where it was unwound and passed through a series of cleaning and fluxing or pickling tanks for removal of any oxides or other surface contamination prior to entry into a molten zinc bath. Today, most material is processed on continuous annealing lines, where coils of cold rolled material are fed through an atmosphere-controlled, in-line annealing furnace designed to both anneal the strip and burn off or reduce any oils, contaminants, or surface oxides that may be present. The strip passes from the continuous annealing furnace through a controlled cooling section, which reduces the strip temperature to molten zinc levels, and enters the bath through an inert-gas-filled snout submerged in the liquid zinc.

On emergence from the zinc bath, the strip passes through two blowoffs, or air knives, where the thickness of the still-molten zinc is controlled by varying the air pressure fed to the knives. On-line measurements of the coating thickness provide feedback control to the wiping system, allowing continuous monitoring and control of the amount of zinc being deposited on each side of the strip. Coating thicknesses on each side of the strip are typically kept equal but may be varied independently to some extent. One-side-only coatings are not feasible via the standard hot dipping process. Procedures to remove the semisolidified zinc on one side of the sheet via brushing or other physical methods have been developed but are not widely used.

Steel processing is typically complete after cooling and a temper rolling operation designed to eliminate discontinuous yielding, which would otherwise result in objectionable stretcher strain lines (Lüders bands) in the formed parts. Most modern hot dip galvanizing lines incorporate inline temper rolling, but off-line temper rolling or stretch leveling may also be used. In some specialized instances, the coated steel coils can be low-temperature batch annealed after hot dip coating to further alter the mechanical properties of the strip. However, due to the relative cost of the additional annealing operation, steels are designed to meet the final specifications directly off the galvanizing line, if possible.

Formability Considerations. At molten zinc temperatures, zinc and iron react, forming a series of intermetallic compounds with varying iron-to-zinc ratios. This reaction results in an alloy layer between the steel surface and the molten zinc, the thickness of which depends on the time and temperature profile of the coating process as well as the alloy composition of the steel substrate. The rate of the reaction is highly dependent on the presence of intentional alloy additions and unavoidable impurities, particularly lead and aluminum, which are present in the zinc coating bath. Because this alloy layer is significantly more brittle than either the zinc coating or the steel substrate, the degree of forming a coated steel can withstand prior to coating flaking is reduced as the thickness of the alloy layer increases. Steel producers strive to

minimize the thickness and alter the phases present in the alloy layer to mitigate the effects that the brittle iron-zinc intermetallics have on the subsequent formability and coating adhesion of the material.

Given proper tooling and press setup, the layer of free zinc on the surface of the steel substrate can result in an increased formability of hot dip galvanized steel over bare sheet, particularly in deep-drawing operations. The soft zinc on the tool surface prevents direct contact between the steel substrate and the punch or die, thus eliminating galling between the two steel surfaces and resulting in increased draw depths.

Typical Products. Hot dip galvanized coatings are readily available on all standard and most proprietary grades of sheet steels. Common ASTM and SAE steel grade designations include:

- Commercial steel (QS)
- Drawing steel (DS)
- Extradeep-drawing steel (EDDS)
- Extradeep-drawing steel plus (EDDS+)
- Structural steel (SS)
- High-strength low-alloy steel (HSLAS)
- Dent-resistant steel (DR)
- Bake-hardenable steel (BH)

Thickness of the zinc coating is typically specified by coating weight per surface area in the units of grams per square meter, with typical commercially available thickness ranges starting at 20 g/m^2 (0.07 oz/ft^2) and ranging up to 600 g/m^2 (2.0 oz/ft^2). Zinc coating weight is also occasionally discussed in terms of coating thickness, with the aforementioned coating weights correlating to coating thicknesses of 2.8 to 84.0 μm (0.0001 to 0.0034 in.).

Good coating thickness control is particularly important in parts that will be assembled or affixed by welding, because small variations in zinc thickness can have an appreciable impact on the welding parameters needed to achieve high-quality welds.

Tool Design. Tools for forming galvanized steels are manufactured from the same materials as tools used for bare steel. Springback compensation is based on the strength and properties of the base metal and is not appreciably affected by the zinc coating. As hot dip coating is applied to a wide range of steels, from the most formable interstitial-free grades to very high-strength dual-phase and transformation-induced plasticity steels, the primary consideration when designing forming tools for a galvanized steel is the base metal. The effects of the zinc layer on formability are small compared to the effects of the base metal and stem primarily from the soft nature of the coating and the presence of the brittle intermetallic layer that may exist in varying levels of severity between the coating and the substrate.

Galvanized coatings will typically stand tensile strains that result in failure of the base metal without flaking, but compressive strains are much more damaging to the brittle iron-zinc intermetallic alloy layer. Under moderate tensile

strains, the thin intermetallic layer cracks but does not lose adherence with the steel substrate. The ductile zinc layer bridges these cracks in the alloy layer, and no loss of coating integrity results. During compression, the alloy layer not only cracks, but also the compressive stresses can cause portions of the coating to wedge against each other. Even in this state, coating adhesion is typically adequate, but if the compressive strains are followed by further tensile strains, the fractured alloy layer is unable to maintain cohesion with the substrate, and large particles of the coating may flake off. Figure 1 schematically illustrates the process.

Excessive coating loss typically requires a compressive strain, followed by a tensile strain in the same direction. Most zinc adhesion issues in hot dip galvanized steels occur when material is pulled over a sharp radius, then stretched in tension. As the material flows over a sharp radius, a high compressive bending strain is imposed on the coating on the inside diameter (ID) of the bend. As the material flows off the bend, it is straightened and placed in tension. This sequence of events occurs on the outside diameter (OD) of a drawn cup, explaining why coating adhesion problems are typically worse on the OD than the ID of a drawn cup. As the blank flows into the die, the material on the cup OD is compressed as it bends to flow over the die radius, then is straightened and stretched as it moves into the sidewall of the cup. The material

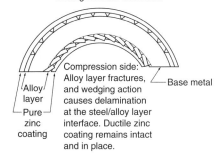

Initial bending

Tension side: Alloy layer cracks but maintains adhesion with base metal. Ductile zinc coating remains unbroken.

Compression side: Alloy layer fractures, and wedging action causes delamination at the steel/alloy layer interface. Ductile zinc coating remains intact and in place.

Alloy layer
Pure zinc coating
Base metal

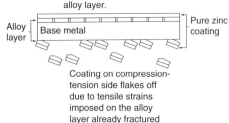

Subsequent tension

Coating on tension-tension side maintains adhesion. Zinc coating maintains integrity over the fractured alloy layer.

Alloy layer
Base metal
Pure zinc coating

Coating on compression-tension side flakes off due to tensile strains imposed on the alloy layer already fractured under compression.

Fig. 1 The effects of tensile and compressive strains on the alloy layer of hot dip galvanized sheet steels

on the cup ID remains in tension, both when being drawn in the die and when being stretched in the cup sidewall. Adhesion can be improved by reducing the compressive stresses via an increased die radius or blankholder force or by decreasing the subsequent tensile stresses by limiting the draw depth of the cup.

Pulling material through an excessively small radius or across a sharp drawbead not only affects coating adherence but can result in scoring of the soft zinc or, in extreme cases, the shaving of flakes of zinc from the surface. This not only results in a cosmetic defect but also reduces the corrosion protection of the finished part. Zinc particles removed from the steel surface can build up in the press or die and cause dents, buildup, and numerous other problems.

The soft zinc coating can also be easily damaged during forming and handling operations. Thus, dies used for galvanized material must be kept free of damage or rough spots that could impart scratches and other cosmetic marks to the soft zinc coating. Sharp burrs on sheared edges can easily cause objectionable coating marks, especially if the blanks or formed parts are slid against one another during stacking, handling, or assembly. The degree of effort necessary to prevent these defects depends heavily on the nature of the part, the manufacturing process, and any secondary finishing or coating operations.

If there are no severe radii in the part or excessively sharp corners over which metal is forced to flow, most tools designed for bare steel can be used for coated sheet. Due to variances in coatings, thicknesses, and lubrication, there are no definitive radius limit guidelines, but in general, radii over which material is being drawn should be made as generous as possible. Simple bending is less of an issue, because the compressive-to-tensile strain reversal that is most damaging to the coating is not encountered in a simple bend. Table 1 shows radii minimums for simple bending. Radii over which material is pulled should be significantly more generous than radii for simple bending. Few problems should be encountered as long as dies are kept polished enough to prevent damage to the soft

zinc surface and radii are kept large enough to avoid severe fracturing of the brittle iron-zinc intermetallic layer due to compressive bending stresses.

Electrogalvanizing

Electrogalvanized steels are produced by batch annealing the cold rolled steel in coil form, temper rolling, then electroplating a layer of essentially pure zinc crystals on the surface of the sheet. As with hot dip galvanized products, the zinc coating is much softer than the base steel and can help provide some lubricity in forming, especially in preventing galling during deep-drawing operations. Because the electrocoating process is not conducted at elevated temperatures, the brittle iron-zinc intermetallic layer that sometimes imposes forming limits on hot dip galvanized steels does not form between the base steel and the coating. Therefore, the forming limits on electrogalvanized material are very similar to those of the bare substrate. Similar care must be taken to avoid rough spots in dies and burrs on blanks or finished parts, which could damage the soft zinc layer during forming or part handling.

Electrogalvanized coatings are typically much thinner than hot dip coatings, generally ranging from approximately 20 to 100 g/m^2 (0.07 to 0.33 oz/ft^2). These coating weights correlate to coating thicknesses of 2.8 to 13.5 μm (0.0001 to 0.0005 in.).

Like hot dip galvanized material, electrogalvanized material can be supplied with different coating weights on each side of the material. Unlike the hot dip process, electrogalvanized material is commonly supplied "one side only," where one side of the material remains bare. Because of the galvanic nature of the corrosion protection, the zinc coating on one side of the sheet will still provide corrosion protection for the bare side.

Zinc-Nickel

A 10 to 14% Ni addition can be made to the zinc electroplated on the steel substrate to provide an electrogalvanized zinc-nickel alloy coating. The nickel addition is used primarily to improve the corrosion resistance of the coating, but it also dramatically increases the coating hardness. Typical applications include fuel tanks and other automotive components. The increased hardness results in a reduced coefficient of friction at higher coating nickel contents. Nickel contents above 15% are not used because the corrosion resistance of the coating begins to decrease at higher nickel levels, and the interfacial shear strength of the coating to the base steel falls below acceptable levels. Weight loss due to coating exfoliation on a drawn cup has been shown to increase dramatically for coating nickel contents over 12% (Ref 5). The higher coating hardness helps limit coating damage, and the lower coefficient of friction may help

improve formability by helping to distribute strain more evenly. However, at higher nickel levels, part radii may be limited by coating adherence.

Electrogalvanized zinc-nickel coatings can be applied to any steel that can be electrogalvanized, but availability is limited due to a limited number of suppliers and a limited demand.

Galvannealed Steels

Galvannealed steels are hot dip galvanized steels that are given a secondary heat treatment after zinc plating to alloy the zinc coating with the base steel. The resulting iron-zinc alloy coating typically consists of a thin layer of FeZn$_4$ (gamma) at the interface, with the bulk of the coating consisting of FeZn$_7$ (delta). Crystals of FeZn$_{13}$ (zeta) frequently exist at the surface of the coating. Producers generally try to avoid the formation of excessive amounts of zeta phase, because the softer, blocky zeta crystals have been shown to increase friction, with the associated reduction of formability (Ref 6).

The intermetallic galvanneal coatings are specified because of their advantages with respect to welding and paint adhesion. The coating is harder than pure zinc, so it is more resistant to scratching and handling damage, but it is brittle and therefore much more prone to exfoliation during forming.

Coating exfoliation in galvannealed sheet steels takes place by one of two mechanisms: powdering or flaking. Powdering occurs due to cracking in the delta layer of the coating, causing the coating to exfoliate in small, generally equiaxed particles with diameters less than the coating thickness. Flaking is primarily caused by fracture of the gamma layer at the steel/coating interface and results in flakes of material with a thickness equal to the coating thickness and a diameter much larger than the coating thickness.

Powdering has been shown to result from excessively tight radii in the part itself or at the draw bead. The brittle coating cannot deform in tandem with the base steel and cracks to relieve the stress. The level of this cracking is related to the amount of surface strain in the part; thus, tighter radii result in increased coating cracking and increased powdering (Ref 7, 8). The ID radii are more prone to powdering than the OD radii, because the compressive stresses force the fractured coating pieces together and tend to wedge many of them up and break them free from the coating. Powdering is not generally severe enough to cause corrosion or cosmetic problems on the sheet itself, but the buildup of coating powder in the dies can quickly result in dents and other problems on formed parts, increasing scrap rates and necessitating frequent line shutdowns and increased die cleaning.

Flaking is more predominant where coated sheets are slid against a die under high speed and/or high pressure, such as would occur during an ironing operation. Higher pressures and forming

Table 1 Minimum simple bend radii for hot dip galvanized sheet steel

Coating thickness: minimum triple-spot coating weight		Sheet thickness (t) range		
g/m^2	oz/ft^2	Through 1.0 mm (0.04 in.)	Over 1.0 through 2.0 mm (0.04 through 0.08 in.)	Over 2.0 mm (0.08 in.)
700	2.35	2t	3t	3t
600	2.10	2t	2t	2t
550	1.85	2t	2t	2t
500	1.65	2t	2t	2t
425	1.40	t	t	2t
350	1.15	0	0	t
275	0.90	0	0	t
180	0.60	0	0	0

Radii value is the minimum diameter rod, in multiples of the sheet thickness, around which the galvanized sheet can be bent 180° in any direction at room temperature without flaking of the coating on the tensile side of the bend. Source: Ref 4

speeds increase flaking, presumably due to frictional and heat effects that create shear forces exceeding the shear strength of the gamma layer at the steel/coating interface (Ref 9). Figure 2 shows a schematic of this process.

Electrogalvannealed Steel

Galvannealed steels are also produced by electroplating the zinc-iron alloy directly onto the surface of the strip, similar to the way zinc-nickel electrogalvanized coatings are produced. The electroplated galvanneal coating contains the same phases as a hot-dipped and furnace-alloyed galvanneal coating, but the phases in the electroplated coating are uniformly and homogeneously distributed as opposed to the layered structure that develops in the hot dip galvannealed process. This coating homogeneity, combined with the lack of a continuous gamma layer at the steel/coating interface, results in a significantly higher resistance to powdering and flaking in the electroplated coating as compared to the furnace-alloyed coating (Ref 10).

Aluminum-Coated Steels

Hot dip aluminum-coated steels (aluminized steels) are used in applications requiring greater

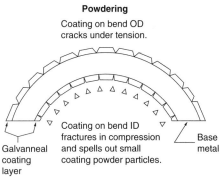

Powdering

Coating on bend OD cracks under tension.

Coating on bend ID fractures in compression and spells out small coating powder particles.

Galvanneal coating layer

Base metal

Flaking

Tool pressure

Die

Coating layer

Base metal

Forming stress

Shear stress imposed by tool pressure and forming strain fractures the brittle gamma layer at the steel/coating interface, resulting in coating flaking.

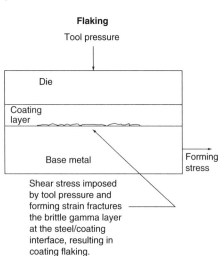

Fig. 2 Schematic of the powdering and flaking of galvannealed coatings on sheet steels. OD, outside diameter; ID, inside diameter

salt spray or atmospheric corrosion protection than is offered by zinc-coated steels. The aluminized coating also provides corrosion protection at maximum temperatures that are significantly higher than those available with zinc coatings, making aluminized steel the material of choice for mufflers, heat exchangers, and many other parts that must exhibit both good corrosion resistance and high-temperature stability of the coating.

Aluminized steels are formed using the same general practices as are used for forming bare and zinc-galvanized sheet. There are two types of aluminized coatings commonly used on carbon and ferritic stainless steels, referred to as type 1 and type 2. Type 1 is an aluminum-silicon alloy with a silicon content from 5 to 11% used primarily for high-temperature applications. Type 2 is a commercially pure aluminum coating used for resistance to atmospheric corrosion. The

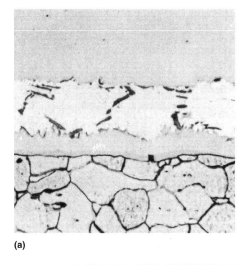

(a)

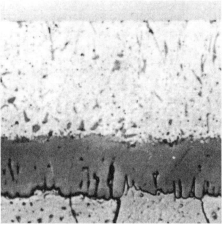

(b)

Fig. 3 Microstructure of aluminum coatings on steel. (a) Type 1 coating from top: a nickel filler, aluminum-silicon alloy, aluminum-silicon-iron alloy, and steel base metal. (b) Type 2 coating forms a layer of essentially pure aluminum (top) with scattered gray particles of aluminum-iron; the light-gray center layer is aluminum-iron, and the bottom layer is the base steel. Original magnification for both 1000×

rapid reaction of iron with molten aluminum at hot dip coating temperatures results in a relatively thick Fe_2Al_5 iron-aluminum intermetallic layer at the steel/coating interface. This brittle adhesion layer is typically the limiting factor in the formability of both types of coatings, with excessive forming severely damaging the alloy layer and resulting in coating exfoliation.

The addition of silicon to the molten aluminum coating bath reduces the required bath temperature and inhibits the growth rate of the intermetallic layer. Both of these effects retard the formation of Fe_2Al_5 at the steel/coating interface, resulting in a thinner and smoother intermetallic layer. This alloy layer difference causes the type 1 coating to have better adhesion and higher forming limits than the type 2 coating. Micrographs showing the alloy layer differences between typical type 1 and type 2 coatings are displayed in Fig. 3. Precise formability limits of the coated steel are difficult to characterize, because alloy layer thickness and properties vary from product to product, being highly dependent on substrate chemistry, overall coating thickness, bath composition, bath temperature, and several coating line parameters. Because adhesion of aluminized coatings is primarily dependent on alloy layer thickness, thicker coatings, which have a thicker alloy layer, will exhibit lower coating adhesion than thinner coatings.

Formability limits are sometimes controlled by the required corrosion resistance of the part being formed. Hairline cracks that develop in the coating during forming can limit the corrosion resistance of the coating, even if no visible coating damage develops. Table 2 shows the effects of coating thickness on the atmospheric corrosion resistance of type 1 and type 2 aluminized coatings of varying thicknesses.

While sufficient tensile strains will cause coating cracking that may impact corrosion resistance, tensile stresses alone will rarely result in extensive coating exfoliation. Even coatings with excessively thick alloy layers will typically not exhibit any coating peeling in tension, even if strained up to the point of base metal fracture, as long as no compressive stresses have been previously applied. Tensile forces will cause the

Table 2 Minimum simple bend radii for corrosion-resistant 180° bends in various thicknesses of aluminized sheet steels

| Steel sheet thickness (t) | | Minimum bend radius | |
mm	in.	Type 1 coating(a)	Type 2 coating(b)
1.61	0.0635	1.5t	2.5t
1.31	0.0516	1.5t	2t
1.00	0.0396	1.5t	1.5t
0.85	0.0336	...	1t
0.70	0.0276	0.5t	0.5t

(a) Coating containing approximately 9% Si and weighing 150 g/m² (0.5 oz/ft²). Minimum radii are for no rust on the outside of the bend after exposure to air for 25 cycles consisting of 30 min at 595 °C (1100 °F) and 30 min of cooling. (b) Coating of commercially pure aluminum weighing 350 g/m² (1.15 oz/ft²). Minimum bend radii are for no rusting at the outside of the bend after 1 year of exposure to a mild industrial atmosphere.

brittle alloy layer to crack, but the high ductility of the aluminum coating allows the outer layer of the coating to elongate and provide unbroken coverage to the underlying steel.

Compressive stresses cause the alloy layer to fracture and wedge up but do not typically result in coating exfoliation unless a tensile strain is subsequently applied. Compressive strains followed by tensile strains, such as occurs on the OD of a drawn cup, are most damaging to the coating. In cup drawing, as the material flows into the deep-drawing die, the material on the OD side of the cup is placed in compression as it bends over the die radius. As the material flows into the wall of the cup, it is unbent and stretched in tension, sometimes resulting in coating loss. The process is identical to that illustrated in Fig. 1 for hot dip galvanized steels. The extent of coating damage is dependent on the levels of compressive and tensile strain applied. Coating exfoliation has been shown to increase dramatically on the OD of deep-drawn cups as coating compressive stresses are increased by reducing the die radius. Coating damage was prevented by providing a large enough die radii to reduce coating compressive stresses to acceptable levels (Ref 11).

In most stamped parts, the majority of the part is in tension, and therefore, coating peeling is not a danger. The potential problems come from areas where material is first pulled through a drawbead, forcing one side of the coating into compression, then subsequently strained in tension. Eliminating drawbeads wherever possible may improve coating adhesion, but the increased binder force required will increase the potential for coating scratches or gouging and may require higher levels of binder maintenance. Type 1 coatings are harder than type 2, but all aluminum coatings, such as zinc, are softer than the underlying steel substrate so require the same additional care to avoid cosmetic damage to the surface of the coating.

Aluminum-Zinc Alloy Coatings. Two types of aluminum-zinc alloy coatings are generally available: one consisting of approximately 45% Zn and 55% Al, the other consisting of 5% Al with the remainder zinc. Both types of coatings frequently contain other alloy additions designed to improve strip wettability during coating and to help limit the growth of any brittle intermetallic layers at the steel/coating interface.

The 55% Al coating has a superior atmospheric corrosion resistance to pure zinc coatings while retaining enough galvanic corrosion protection to eliminate rust staining at sheared edges, screw holes, or other minor coating breeches. The primary use of this material is in roofing and other architectural panels, but it also finds applications in the appliance and automotive markets. The heat resistance of the alloy coating is intermediate to the pure zinc or aluminum coatings, while the formability is similar to a hot dip zinc-galvanized product.

The 5% Al coating has a slightly higher atmospheric corrosion resistance than zinc-galvanized material. Its primary advantage is a slightly improved resistance to cracking during forming (Ref 12), which makes it attractive for the manufacture of deep-drawn parts.

Tin-Coated Steels

Tin-coated steels are extensively used in the food packaging industry but also find applications in the hardware, appliance, automotive, and kitchen equipment industries. Almost all tin-coated steel is produced using an electrocoating process, although tin can be and sometimes is applied by hot dipping. Electroplated tin is typically applied in a thickness range of 0.4 to 2.3 μm (0.02 to 0.1 mil); hot-dipped product is typically 1.5 to 2 times thicker. As with the zinc- and aluminum- coated steels discussed previously, the soft, ductile tin coating helps to increase die life and reduce lubrication requirements as compared to bare steel. The main difference between tin and hot-dipped zinc or aluminum is that tin, even hot dipped, usually does not develop an alloy layer of any appreciable thickness between the coating and the substrate. Thus, the formability of tin-coated steels is limited only by the formability of the base metal. Coating powdering, flaking, and decohesion are not typically factors of concern. The high lubricity of the tin coating results in a significantly improved drawability of tin-coated steels as compared to bare steels.

The tin coating is much softer and more easily damaged than the surface of the steel substrate, so all of the warnings against cosmetic coating damage from rough spots in dies, burrs, or sharp edges of adjacent parts or blanks typically given for galvanized or aluminized steels also apply to tin-coated steels.

Terne-Coated Steels

Terne coat is a lead-tin alloy coating that typically contains 10 to 25% Sn by weight. Terne coating is used where low cost and good atmospheric corrosion resistance are necessary. It finds applications in roofing, architectural items, and automotive fuel tanks. The toxicity of the lead component of the terne coating is an environmental concern, and the use of terne coatings has significantly declined in recent years. Terne coatings will probably be totally phased out in the near future.

The properties of terne-coated steels are very similar to those of tin-coated steels. Like tin, terne coatings do not develop any significant brittle alloy layer between the base steel and the ductile coating. Thus, the coating does not limit the formability of the base steel, and the lubricity increase over bare steel may even result in increased formability of terne-coated steels. Like most other coatings, terne is much softer than the underlying steel, and some precautions must be taken to prevent coating damage.

Nickel, Copper, and Chromium Plating

Decorative items are sometimes made from steel that has been electroplated with nickel, copper, or chromium, but typically these coatings are applied to finished parts after forming. Due to the limited amount of forming done on steels coated with these materials, they are not covered in depth here.

Organic-Coated Steels

Typical organic coatings applied to steels prior to forming include several classes of paints and plastic coatings. While most organic-coated parts are coated after forming, parts can often be successfully formed from steel coated in the coil form at a significant cost advantage over parts painted after forming. Precoated sheet steels are formed using processes and equipment similar to those used in forming cold rolled steel. Most standard grades and gages of flat-rolled steel can be purchased with any of a variety of organic coatings available from multiple suppliers.

Tooling used for cold rolled steel is usually suitable for forming organically coated steels, although the less abrasive surface presented to the tool by the organic coating frequently allows forming tools to be manufactured of lower-strength, softer, and less shock-resistant materials. In appearance-critical applications, rubber or plastic die inserts are sometimes used in select areas of the die to help prevent coating damage during forming.

Organic coatings are much more susceptible to scratching and other surface damage than cold rolled or galvanized steels. To avoid scoring or scratching the organic coatings, die surfaces must be highly polished and kept free of burrs, or rough spots. Blank and part handling is critical to avoid damage to the coating caused by foreign particles, burrs, or the edges of adjacent blanks and parts.

Lubricants. Any lubricants used with organic-coated steels must be compatible with the coating composition and must be removable in a cleaning system that is also compatible with the coating. Plastic sheets or adhesive-backed paper are sometimes used instead of, or in conjunction with, more conventional lubricants to avoid problems with compatibility or lubricant removal.

Formability. The wide variety of organic coatings available prevents any detailed discussion of the formability of each separate class of organic material used to coat coated sheet steels. Some organic coatings are flexible enough to withstand any forming operations that will not cause fracture in the steel substrate. The formability of other coatings is much more limited, depending on the particular coating, application method, and the appearance and protection requirements of the specific application. Coating thickness, metal pretreatment, and coating curing parameters all strongly affect the properties and adherence of the organic coating and, subsequently, the formability of the coated sheet.

Table 3 lists several common types of organic coatings and gives the typically applied coating thicknesses, formability ratings, adhesion ratings, and minimum acceptable bend radii for each type. The coating thickness values presented in Table 3 represent the maximum and minimum thickness ranges; typical coating thickness windows will be appreciably narrower. The thickness of any primer coating is not included in the coating thickness values.

Base metal surface roughness can influence the formability of organically coated steels, especially for the thinnest coatings. Lower surface roughness values (smoother surfaces) give a more uniform distribution of stresses during the forming process, reducing the peak strains experienced by the coating and generally resulting in better surface appearance. Base metal with a rougher surface finish subjects the coating over the high spots on the substrate to higher forming stresses and greater wear during forming. This generally results in more cosmetic damage to the surface, but the lower heat generated by the reduced friction can partially or totally reduce this effect in certain applications.

Adhesion and Flexibility. The adhesion of commonly used coatings is rated in Table 3. The minimum bend radii and formability ratings shown are a measure of the combined effect of coating adhesion and flexibility. Vinyl plastisol coatings have such outstanding flexibility that they can bridge cracks produced in the metal base by severe forming.

To provide the adhesion generally required for forming, chemical conversion coatings, selected for the metal surface to be painted or plastic coated, are applied to coil stock or blanks by spray or immersion treatment. This treatment is usually followed by the application of a prime coat compatible with the final coating material. Bare steel is given an iron phosphate coating with a weight of 375 to 480 mg/m^2 (0.013 to 0.016 oz/ft^2). Zinc-coated steel receives a zinc phosphate coating—1.6 to 2.7 g/m^2 (0.005 to 0.009 oz/ft^2) on hot dip coatings and 1.08 to 1.9 g/m^2 (0.0035 to 0.006 oz/ft^2) on electroplated coatings. Aluminum-coated steel is treated with a chromate coating weighing 215 to 270 mg/m^2 (0.0007 to 0.0008 oz/ft^2). These coating weights refer to area of stock and must be doubled when treatment of both sides of the stock is considered.

Organic coatings that are applied to untreated and unprimed metal surfaces have better flexibility than those applied to pretreated surfaces. However, without pretreatment or the use of a suitable primer, the degree of adhesion necessary to withstand forming forces or to have adequate service life and corrosion resistance usually cannot be assured. As indicated in Table 3, forming characteristics are generally improved by the use of a primer. An exception is tin-coated steel.

The thicker films of organosols, plastisols, and vinyl-film laminates have better flexibility than

films of solution coatings in the 0.025 mm (1 mil) dry-film range. The stress of forming is absorbed within the thicker films and is not transferred to the interface with the metal, at which adhesion is established.

Solution coatings, because of increasing film strength, have less flexibility and adhesion near the upper thickness limit. The most severe forming can be done near the lower thickness limit, but such thin coatings may not fulfill service requirements. Although solution coatings of silicone-polyesters and thermosetting acrylics can be applied as thin as 2.5 μm (0.1 mil), as exterior coatings, they are used almost exclusively in the heavier thicknesses shown in Table 3. Laminated plastic coatings may show excessive film strength and therefore less flexibility and adhesion at the upper end of the thickness range, and low tensile strength at the lower end. The color and gloss of an organic coating also affect coating flexibility, which decreases with increased pigment loading.

The hardness of organic coatings is in the range of HB to 3H pencil hardness for most of the commonly used paints and approximately Durometer A 85 to 90 for plastisol coatings and laminated plastics. Softer coatings are more likely to be damaged by scoring in the forming die. Die pressure transmitted through an organic coating to the interface can destroy adhesion. In heavy films and vinyl-film laminates, the elastic, compressible finish coating normally yields under die pressure, but the relatively brittle adhesive primer layer can be damaged by localized high die pressure.

Shearing, Blanking, and Piercing. Sharpness of cutting tools and direction and speed of cut affect the performance of coatings in the cut area. Dull cutting tools or high impact speeds cause high-energy impact on the coating surface and may shatter the bond in the surrounding areas, particularly in coatings of borderline adhesion strength. Flaking or lifting of the coating may result.

Bending. Minimum bend radii for organic-coated metals are given in Table 3. Slow bending will prevent breakage of the coating more effectively than rapid bending. When bending with contour forming rolls, the finish of the organic coating will be preserved, and less stress will be imposed on the steel base if the radii are bent over several rolls instead of one or two rolls.

Bending short flanges close to a cut edge or where the coating is scored by the bending tool at the peak of the bend can cause the coating to lift off the steel base. In both cases, the cohesive strength of the film has been weakened, and coatings with high film strength will attempt to return to the shape in which they were applied.

Deep Drawing. Suitability of the organic-coated metals for deep drawing (or severe forming) is also rated in Table 3. The effect of speed of drawing or forming is generally the same as that described for bending. The more

Table 3 Forming characteristics of organic precoated sheet steels

For coiled bare, hot dip or electrogalvanized, and aluminized steels. Applied also to aluminum, copper, and brass substrates. For tin-coated steel, ratings apply only for epoxy coatings; other coatings have lower ratings. For copper and brass, ratings apply only when primers based on epoxy or phenolic resins are used. Data are based on the use of suitable chemical conversion treatments and primers; results may vary for different substrates. Conversion coatings and primers improve results with most coating-substrate combinations.

Type of coating	Coating thickness		Coating adhesion(a)	Minimum 180° bend radius(b)	Suitability for severe forming(a)
	mm	mils			
Solution paints					
Alkyd-amino	0.0025–0.03	0.1–1.2	G	3t	F
Vinyl-alkyd	0.0025–0.03	0.1–1.2	G	2t	F
Silicone-polyester	0.018–0.03	0.7–1.2	G	2t–3t	F
Thermoset acrylic	0.023–0.03	0.9–1.2	E	1t–3t	F–G(c)(d)
Solution epoxy	0.0025–0.025	0.1–1.0	E	0t	E
Ester epoxy	0.0025–0.025	0.1–1.0	E	0t	E
Polyester	0.0025–0.03	0.1–1.2	G	2t	G
Solution vinyl	0.0025–0.03	0.1–1.2	E	0t	E
Dispersion paints					
Organosol vinyl	0.018–0.10	0.7–4.0	E	0t	E
Plastisol vinyl	0.10–0.50	4.0–20.0	E	0t	E(e)
Polyvinyl fluoride	0.013–0.05	0.5–2.0	G	0t	G
Polyvinylidene fluoride	0.013–0.05	0.5–2.0	G	0t	G
Laminated plastics					
Polyvinyl fluoride	0.04–0.05	1.5–2.0	G	3.2 mm (1/8 in.)	G(f)
Polyvinyl chloride	0.10–0.64	4.0–25.0	E	0t	E(g)
Polyester	0.013–0.36	0.5–14.0	F	0t	F(c)(h)
Tetrafluoroethylene	0.025–0.50	0.1–20.0	G	0t	E(j)
Acrylic	0.075–0.15	3.0–6.0	G	0t	G

(a) Ratings: E, excellent; G, good; F, fair. (b) t, thickness of sheet. (c) Results are greatly affected by coating thickness. (d) Coating of medium thickness is good for deep drawing. (e) Coating can bride cracks in the metal produced by severe forming; compressibility of the coating must be considered in forming to close tolerances. (f) Results are greatly affected by substrate material and thickness. (g) Bonds may be destroyed in extreme draws or sharp bends. (h) Bond strength may be seriously reduced after slight deformation. (i) Particularly susceptible to damage by scoring of coating during forming

steps used, the more severely a part can be drawn or formed without damaging the organic coating. However, the ductility and work-hardening behavior of the steel, as well as the flexibility of the coating system, must be considered in the design of a forming or drawing die.

Forming Temperature. Depending on the effect of temperature on the properties of the organic coating, heating up to 50 °C (120 °F) before forming will reduce the likelihood of coating fracture. Some coatings, such as silicone-polyester coatings with a high silicone content, can be formed at a temperature as high as 65 °C (150 °F). Heating can be done with infrared radiant heaters, hot air, or an open gas flame or by storing the coil stock in a heated room until fabrication. Overheating must be avoided. Organic coatings, especially the thermosetting types, can be softened enough to make them subject to surface damage from die action and handling.

The design and surface finish of tools and the selection of handling procedures and equipment are important in forming organic-coated metals. Depending on the accuracy requirements and the coating thickness, allowance may have to be made for the coating thickness in dimensioning the forming tools. When accuracy requirements and coating thicknesses permit, the same dies can often be used for painted and plated metals.

To prevent scratching and scoring of prepainted surfaces, the die surfaces must be polished and sharp corners rounded off. In addition, damage during ejection and subsequent handling of the parts must be prevented.

REFERENCES

1. C. Magny, "Tool Materials and Coatings: Optimization in Relation to the Stamping Behavior," SAE Technical Paper Series 2002-01-1060, Society of Automotive Engineers, 2002, p 1–9
2. D. Cartwright, P.R. Drake, and M.J. Godwin, Effect of Low Cost Press Tool Materials on Formability of Sheet Steel, *Ironmaking Steelmaking,* Vol 25 (No. 2), 1998, p 131–135
3. I. Kim and D. Lee, The Effect of Die Materials on the Strain Distribution in Electro-Galvanized Sheet Steel Forming, *Metals and Mater.,* Vol 4 (No. 4), 1998, p 695–701
4. "Standard Specification for Steel Sheet, Zinc-Coated (Galvanized) or Zinc-Iron Alloy-Coated (Galvannealed) by the Hot-Dip Process," A 653/A 653M-05, *Annual Book of ASTM Standards,* ASTM International, 1994
5. C.S. Lin, H.B. Lee, and S.H. Hsieh, Microstructure and Formability of ZnNi Alloy Electrodeposited Sheet Steel, *Metall. Mater. Trans. A,* Vol 31, Feb 2000, p 475–485
6. D.A. Haynes and P.D. Hodgson, Influence of Galvanneal Coating Structure on Press Formability, *Proceedings of the 44th Mechanical Working and Steel Processing Conference of the Iron and Steel Society,* Vol XL, 2002, p 1271–1279
7. V. Rangarajan, V. Jagannathan, and K.S. Raghavan, "Influence of Strain State on Powdering of Galvannealed Sheet Steel," SAE Technical Paper 960026, Society of Automotive Engineers, 1996
8. G.M. Michal and D.J. Piak, The Effects of Forming Velocity and Strain Path on the Performance of Galvannealed Sheet Steels, *Galvatech 2001 Proceedings,* Stahleisen Verlag, Düsseldorf, 2001
9. M. Urai, J. Iwaya, and M. Iwai, Effects of Press Forming Conditions and Coating Structure on Flaking Phenomenon in Galvannealed Steel Sheets, *Kobelco Technol. Rev.,* No. 22, April 1999, p 48–51
10. K. De Wit, A. De Boeck, and B.C. De Cooman, Study of the Influence of Phase Composition and Iron Content on the Formability Characteristics of Zinc-Iron Electroplated Sheet Steel, *J Mater. Eng. Perform.,* Vol 8 (No. 5), Oct 1999, p 531–537
11. H. Kawase and A. Takezoe, Peeling of Coated Metal due to Severe Press-Forming in Aluminized Sheet Steel, *ISIJ Trans.,* Vol 22, 1982, p 371–376
12. R.F. Lynch and F.E. Goodwin, "Galfan Coated Steel for Automotive Applications," SAE Technical Paper 860658, Society of Automotive Engineers, 1986

Forming of Steel Strip in Multiple-Slide Machines

MULTIPLE-SLIDE FORMING is a process in which the workpiece is progressively formed in a combination of units that can be used in various ways for the automated fabrication of a large variety of simple and intricately shaped parts from coil stock or wire. Operations such as straightening, feeding, trimming, blanking, embossing, coining, lettering, forming to shape, and ejecting can be done in one cycle of a multiple-slide machine. Forming is generally limited to bending operations, but the four slides and center post permit the fabrication of very complex parts, as described in more detail in the article "Multiple-Slide Machines and Tooling" in this Volume. Deep drawing is generally not done in the forming or press stations of a multiple-slide machine.

Applicability

Multiple-slide forming is used to produce shapes from coiled strip or wire. The maximum size of workpiece that can be formed from strip metal in a multiple-slide machine is 203 mm (8 in.) wide by 685 mm (27 in.) long. Parts made from wire up to 1015 mm (40 in.) long (or longer if a special machine is used) and up to 9.5 mm (^{3}/8 in.) in diameter can be formed automatically from coil stock. This article deals with the forming of strip stock; the fabrication of wire forms is discussed in the article "Forming of Wire" in this Volume.

If the work metal is comparatively thin and the bending is not severe, tempered strip material can be formed. Plated or coated materials can be formed, but it is usually better to coat after forming because it is difficult to avoid marring coated surfaces during forming. However, nonmetallic inserts at appropriate points in the straightener, feeder, and forming tools can be used to reduce tool marks.

Springback must be considered in bending such materials as stainless steel, phosphor bronze, certain grades of brass and beryllium copper, or high-carbon steel. Adjustments can be made in the forming tools to provide the amount of overbending required for the accuracy of the finished work.

More than one piece can be made in each cycle of a multiple-slide machine. For example, a part that had been made in seven conventional press operations was replanned for the multiple-slide production of four pieces per cycle at 200 cycles per minute.

Multiple-Slide Machines

Multiple-slide machines are made in a range of sizes, all similar in construction and principle. The larger machines have a longer die space, which enables more die stations to be used for the manufacture of complicated components. Generally, the number of strokes per minute decreases and the horsepower increases as the machine size increases.

The four forming slides of a typical multiple-slide machine are generally sufficient for ordinary part-forming needs. However, complex parts can be formed at two or three levels around the center post, thus doubling or tripling the number of forming positions available.

Figure 1 shows a plan view of the main units of a medium-size multiple-slide machine that uses a floor space of 3.7 by 1.5 m (12 by 5 ft), including the stock reel. Four shafts (A, B, C, and D), mounted to a flat-top bedplate, are driven at equal speed through spur gearing (E) by an electric motor. Each of the four shafts is fitted with a positive-action cam (F) that drives a slide (G) (only two of four are identified) on which the forming tools can be secured. In the center of the machine is a vertical post (H) into which the center post or former is fixed and around which the work material is bent. The formed workpiece is removed from the center post by a stripper mechanism, which usually consists of a hardened steel plate surrounding the center post and secured to a vertical rod operated by a cam (F) on the right-hand shaft to give up-and-down motion to the stripper. All of these parts constitute the forming station of the machine.

To the left of the machine proper is a stock straightener (L), shown in working position with strip stock passing through it. Intermittent feeding of the work metal is accomplished by an automatic gripper in the feed slide (M) and an adjustable crank (O), which is attached to a

shaft (C). A separate gripper (P) is provided with cam-operated jaws, which grip the strip on the return stroke of the feed slide to prevent backward motion of the strip.

The work metal strip, fed through the machine in a vertical plane (on edge), passes horizontally through dies in a horizontal press (Q). A short, powerful stroke is given to the horizontal press slide by a cam on the front shaft (D) (Fig. 2).

Stock straighteners used on multiple-slide machines are similar to those on stamping presses. The primary difference is that in a multiple-slide machine the rolls are mounted vertically to straighten the work metal as it passes through the machine on edge, instead of horizontally as in a conventional press (see the article "Multiple-Slide Machines and Tooling" in this Volume).

The stock-feed mechanism of a multiple-slide machine is made of two separate units constituting the forward gripping and transporting device (Fig. 1, M) and a stationary gripping unit (P) to hold the strip when it is released on the return stroke of the feed. The stock-feed slide is reciprocated through a system of links from a crank disk keyed to the left-hand camshaft of the machine (Fig. 1, M, O, and C).

Press Station. The die used for piercing, trimming, embossing, and minor forming of the stock is mounted in the press station, which consists essentially of a horizontal press operated by a cam on the front shaft (Fig. 1, Q and D). It may consist of a single unit, as shown in Fig. 2, or additional units can be placed side by side, especially in the larger machines.

Die head units can also be operated from front, rear, or both in machines that are constructed symmetrically. Burr direction can be controlled at designer option.

Cams provide a positive movement to the press slide in both directions so that the tools are withdrawn from the strip at the end of the working stroke. A dwell at completion of the in-stroke provides time for opposing motions. This permits working cleanly from each side of the stock. A means of adjusting the shut height of the die is provided. The entire unit can also be moved longitudinally along the bedplate to the desired position established by the feed length. A bolster provides support for the die shoe, and a

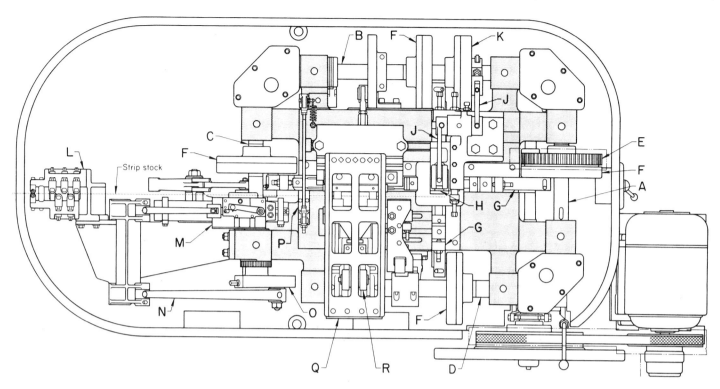

Fig. 1 Plan view of a multiple-slide machine showing major components. A to D, integrated shafts; E, spur gearing; F, positive-action cam; G, slide; H, vertical post; J, bell crank; K and R, cams; L, stock straightener; M, automatic gripper in feed slide; N, links; O, adjustable crank; P, stationary gripper with cam-operated jaws; Q, horizontal press with dies; R, cam. See text for description of operation.

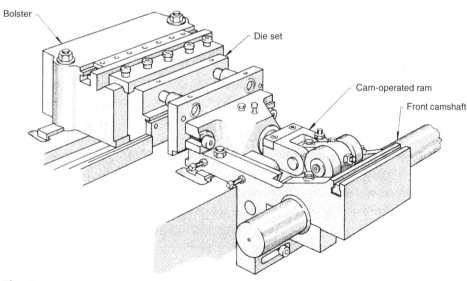

Fig. 2 Press station of a multiple-slide forming machine. See text for details.

cam-actuated ram provides support and motion for the punch holder.

The cutoff unit, placed between the press and the front forming slide, is used for cutting work metal into blanks before they are bent to shape. The ends can be cut off straight, or the cut can be curved.

Various types of cutoff units are described in the article "Multiple-Slide Machines and Tooling" in this Volume. The cutoff unit consists basically of a horizontal slide that is operated through a lever from a cam on the front shaft. The unit can be adjusted along the bedplate in order to cut off the blank at the required distance from the center post. A positive cam action returns the slide so that the cutoff tool will not interfere with forming. Generally, the cutting die and stripper, through which the work metal is fed, are stationary, while the punch moves to cut off the blank.

It is sometimes desirable to install a second cutoff unit on the right-hand side of the center post. The two units trim the two ends of the blank to an accurate length. Because both ends of the blank are trimmed, slightly more stock is required when this method is used, but the inaccuracies of a long feed length are corrected.

The forming station consists of four slides, a center post, and a stripper mechanism. Shaped tools in the four slides progressively bend the workpiece around the center post. In most applications, the center post controls the shape of the workpiece. The first forming tool, usually on the front slide, holds the blank against the center post during cutoff.

Each of the four forming slides runs in its own slideway machined in the bedplate, and each slide is operated positively in both directions by cams. The front, rear, and right-hand slides are all similar in design; the left-hand slide is different because it must pass underneath the press unit.

The tool holder and slide have a mating key and keyway machined parallel to the slide motion so that the forming tool can be adjusted in this direction. A keyway in the top of the tool holder at right angles to the slide movement and a mating key on the end of the forming tool allow sidewise adjustment. The cam that operates each forming slide is made in two halves and has radially disposed slots so that it can be easily exchanged or timing can be adjusted.

The unit shown in Fig. 3 holds the center post rigidly in position for the forming tools to operate around and provides the means for stripping the completed parts off the center post. The center post is fitted and rigidly secured to a king post, which generally consists of a square or rectangular steel bar. The king post fits into a recess in an overhead horizontal slide and is clamped in position by a steel plate and screws. This arrangement holds the king post and center tool rigid and allows for easy adjustment of the vertical position of the center post. The center tool is held square with the forming tools at all times.

The overhead slide, carrying the center post, moves parallel with the front and rear slides and is pushed backward by the action of the front tool holding the strip against the center post until the slide reaches a stop, when the metal is bent to shape. When the front tool retracts, the slide containing the center post is returned by spring pressure until a stop is reached. If desired, the slide can be held in a fixed position by adjustment of the two stops. The sliding center post arrangement minimizes interference between the cutoff die and the left-hand slide tools.

Workpieces are ejected from the center post by the downward movement of a suitably shaped stripper plate. The stripper plate is held on a vertical rod, which derives its motion through a system of levers from a positive-action cam on the rear shaft. During forming, the stripper plate is positioned above the forming tools. The vertical rod is guided by a bushing in the overhead slide.

The stripper-cam lever has provision for the roller to be in one of two positions. With the roller as shown in position Fig. 3, A, the cam acts as a normal ejector, moving quickly down and returning the stripper rod almost immediately. With the roller in position B, the rod is held down for a longer period and remains in the upward position only for a short time. This arrangement is used when a retractable mandrel is attached to the mechanism in place of a stripper plate. The mandrel can be held down for a long time while the forming takes place and can then be quickly retracted, as may be required in a further forming or closing operation.

Blanking

The blank that is bent to shape in the forming station has all of the prior operations done in the press station. Trimming the blank outline; piercing holes; embossing ribs, weld projections, and hole flanges; and stamping letters and numerals are done before the blank is cut off and formed. The blank is severed from the strip by cutting off, parting, or blanking methods.

Dies used in multiple-slide machines are either single-stage or progressive, depending on the complexity of the part being formed. Generally, a progressive die is used, even though it may be only a simple two-station pierce-and-pilot die.

The dies for producing the blank must be made as accurately for a multiple-slide machine as they are for a conventional press. However, the multiple-slide dies are of a simpler design and are less expensive than conventional press dies because most of the forming is done in a separate station and around the center post.

The strip layout and die construction are similar to those for progressive dies used in conventional presses. However, as noted previously in this article, the bending operations are usually done in the forming station and cutting off is usually done in the cutoff unit, unless a transfer unit is used to transport the blank from the press station to the forming station. Air jets are used where possible to eject the pierced slugs from the die.

The forming of lanced detents and flanged holes toward the punch side of the blank can be done by actuating the punches with cams on the rear camshaft. Additional movements of

die units can be obtained from any of the three other camshafts.

Extended dies are progressive dies that are mounted in the press station and extend into the forming station. After the usual piercing, notching, and piloting operations, bending or forming is done by a combination of elements in the progressive die and those actuated by the front and rear camshafts. The stripper mechanism can be used to advance and retract a mandrel around which the part is bent. The moving or positioning of punches and dies by the camshafts can make it possible to use tools that are much simpler in design than those made for operations in conventional presses. The parts made in extended dies are generally those that can be retained on the strip until all operations are complete before the part is cut off.

Cutting off the blank is usually done with a blade and die mounted in the cutoff unit. The die is fixed to the housing across which the strip of work metal passes. The blade is secured to the slide so that when it is given a forward movement, the blank is sheared from the strip.

Basically, the cutoff tool consists of two flat pieces of metal: the cutoff die (A) and the blade (B), as shown in Fig. 4(a). A fixed stripper (D) holds the work metal (C) against the die, and when the blade moves across the face of the die, the metal is severed. Most of the parts illustrated in this article were cut from the strip by this method.

The die and blade can be modified to give a shaped cut, as shown in Fig. 4(b) to (d). The shapes illustrated can be used when forming ringlike parts that have a smooth-fitting joint.

Shaped ends on a workpiece can be formed by two methods. The first method consists of cutting the full contour of the ends of two adjacent parts in one machine stroke (Fig. 5a). The second method consists of partly developing the contour in the press station and then using the cutoff station to complete the contour by parting the blank from the strip by shearing the connecting tab, as shown in Fig. 5(b).

Blending of the parting punch with the previously cut contour may necessitate the use of a punch with sharp corners, which could shorten punch life. If not properly blended, flats or other evidence of mismatch will appear on the severed workpiece.

As shown in Fig. 5, the parting punch is mounted on the plate at the rear of the strip.

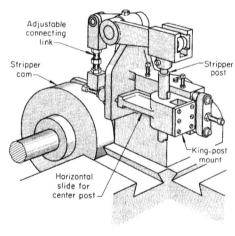

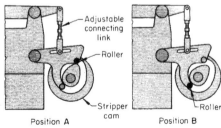

Fig. 3 Unit containing cam-operated stripping mechanism and horizontal slide for mounting of the center post assembly. See text for details.

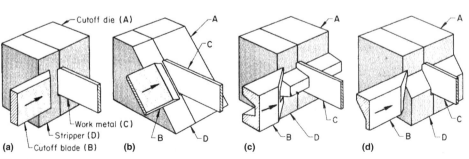

Fig. 4 Four arrangements of cutoff blades and dies for multiple-slide forming machines. See text for details.

The die is fitted with a stripper into which the metal is guided across the front of the die. The die and the stripper are both mounted on the slide of the cutoff unit and move horizontally toward the punch, carrying the strip metal with them. When the work metal contacts the punch, the ends are sheared. The scrap is ejected through the die.

Normal press practice is for the die to remain stationary and the punch to move. This practice is reversed when parting in the cutoff unit so that the end of the metal can swing clear, which would not be possible if the die were at the back of the strip.

Another way in which shaped ends can be produced is to provide a cutoff unit at each side of the front forming tool, each unit being fitted with a die to cut one end. When pilots cannot be used in the die or when a long blank is being cut, the second unit helps in solving tooling problems.

Short Blanks. When the length of the blank is short and the end is very close to the forming slides, the cutoff blade can be fastened to the front tool. The blade is fitted and secured to the front tool; an adjustment screw provides for positioning of the blade. The cutoff die is positioned to mate with the blade.

Transfer of blanks from the cutoff unit to the forming station is generally unnecessary because the blank can be moved into the forming station before it is cut from the strip. However, it is sometimes desirable to cut the blank from the strip at the last station in the progressive die, as in ordinary press blanking. In such a cycle, a transfer unit such as that shown in Fig. 6(a) is used, which transfers the blank to the forming station. The transfer head is actuated by a stub shaft located below the left-hand camshaft (Fig. 1, C).

After the blanking punch has retracted, the blank is held in the die cavity. The blank is then pushed out of the die cavity with a plunger that is moved forward by a cam fixed to the rear shaft and into the waiting pickup finger. The blank is transported horizontally to the forming station, where the first forming tool pushes the blank out of the pickup finger and against the center post.

While the forming tools are bending the blank around the center post, the transfer head moves back to the press station for another blank.

A blank can be picked up in the blanking station in one position and indexed to the forming position during the transfer motion (Fig. 6b). This permits the strip layout to be designed for optimal stock use. Bends can also be positioned favorably with respect to the rolling direction. The transfer of blanks from the blanking position to the forming position without reorientation is shown in Fig. 6(c). The pickup finger is mounted in the transfer head so that no swiveling or indexing takes place.

Timing of the transfer-head motion is important. The finger must be in position for loading or unloading but must not interfere with the blanking punch or the forming slides.

Forming

As the work metal strip leaves the press station, it passes through the cutoff unit and between the center post and the first forming tool (usually the front tool). Simultaneously with the cutting off of the blank, the front forming tool moves forward and holds the blank firmly against the center post. By continued movement of the front tool, the blank is bent around the center post. Tools on the two side slides move in to make further bends, and these can be followed by a fourth tool on the rear slide to complete the forming. The sequence in which the slides move is not fixed, nor is it always necessary to use all four slides when forming a part.

Holding blanks firmly in position against the center post is necessary while they are being severed from the stock and also during the initial forming operation in order to prevent slipping and to ensure against premature bending or kinking across a weak section. The work metal is likely to bulge when a U-shaped part is being

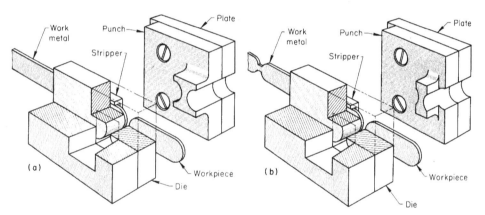

Fig. 5 Two types of parting dies used to shape the ends of blanks before forming. (a) Die for cutting the contour in one stroke. (b) Die for completing the contour by parting a partially developed blank from the strip

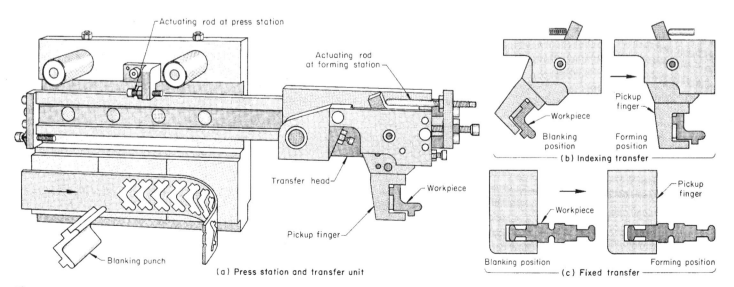

Fig. 6 Mechanism for indexing and fixed transfer of blanks from the press station to the forming station of a multiple-slide machine

formed. Such a bulge is not removed when the tools are fully closed because the surplus metal cannot flow while the blank is held firmly by the two corners of the bending tool. In some applications, a spring-loaded plunger in the first forming tool is positioned slightly in advance of the bending portions and prevents the blank from bulging when the bends are started. A positive blankholder is essential for holding forms having a weak section between the bends or for preventing slipping or dragging of a blank around the corner when bending takes place. A partly formed blank also must be held while being moved to the lower level and until a form tool can grip it.

There are two general types of mechanically operated blankholders. One type makes use of the standard stripping mechanism, and the other is operated by a cam and lever from one of the shafts.

Part Ejection. After all tools are retracted, the formed part is ready for ejection by the stripper plate. During forming, the stripper plate is positioned at the top and clear of the forming tools. After forming, the stripper plate moves down and ejects the part from the center post. The stripper can also move a partly formed component into a second, or even a third, forming level on the center post.

Multiple-Part Forming. Forming of more than one part per cycle of the machine should be considered when planning the tools for production, but the increased loading must be within the capacity of the equipment. The way in which the

outline of the part is developed depends largely on its finished shape. If a part can be made of strip stock by piercing, cutting off, and forming, then more than one part can be made in each cycle by slitting the strip into two or more ribbons. For parts having shaped ends (for example, semicircular), the strip can be slotted and the blank then cut off. Slotting the strip may eliminate the need for sharp corners on the punch and die, thus increasing their service lives. Parts with more complex outlines require more complex trimming punches. The blank is severed from the strip by either cutoff or parting methods.

Some parts, such as those having the basic shape of an L, are easier to form into a U-shape and then part after forming. The bottom forming level of a multiple-slide machine can be used for parting. The parting punch is positioned on the rear tool slide, and the die on the front tool. The center post has a clearance hole for the punch.

Forming Level. Forming can be done around the center post at the same level at which the blank enters the forming station (single-level forming) or at one or two positions below that level (two- or three-level forming). Parts can be finish formed at the lower level, resistance welded into ringlike parts, or finish formed and then cut into two or more pieces. The partly completed workpieces are usually moved to the lower level by the stripper mechanism.

Single-level forming is used when all of the bends can be made with one set of forming tools, usually with one forward stroke per machine cycle. Sometimes, a bend can be made by partly

forming, retracting the tool, doing some work by another tool, and then advancing the first tool to finish the bend. Auxiliary forming tools actuated by separate cams or lever arrangements can be used to do more work on a piece. Wide blanks are usually formed at one level because of the length of center post needed to support more than one part and because of the limited space available for mounting tools.

During two-level forming, two components are on the center post—one at the top level and the other at the bottom level. Work is done simultaneously on each piece by the forming tools. Forming is often done at both levels simultaneously by providing each slide with a tool that is shaped to perform an operation on each piece. The workpiece at the lower level is pushed off of the center post by the downward movement of the partly completed piece from the top level.

One of the simplest applications of two-level forming is the manufacture of bushings and ferrules by forming the center and the ends of the blank at the top level and finish forming to final diameter at the bottom level. A more complicated application of two-level forming is described in the following example.

Example 1: Two-Level Forming of a Hose Clamp. The hose clamp shown in Fig. 7 was made from cadmium-plated steel 9.5 mm (3/8 in.) wide at two levels on the center post in a multiple-slide machine. At the press station, the three holes were pierced and a score mark or slight kink was formed between two holes to

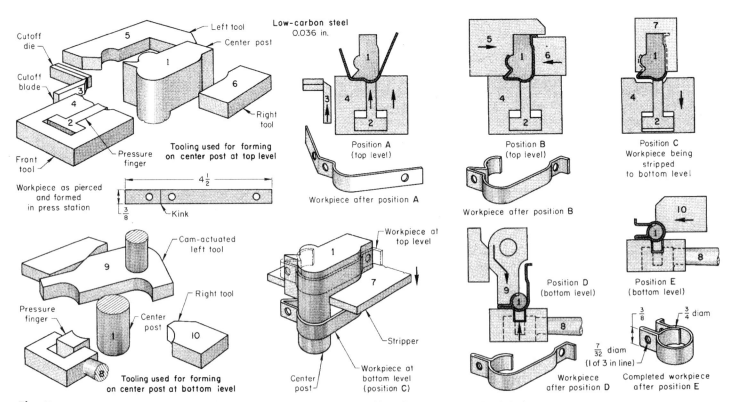

Fig. 7 Forming tools used and sequence of operation in the forming of a low-carbon steel hose clamp. Dimensions given in inches

ensure that the two holes would coincide after the metal had been folded back on itself. Pilots in the press station ensured accurate location of the strip for cutoff and forming. The forming tools used for the two levels are shown at left in Fig. 7.

At position A in Fig. 7, the cutoff blade (3) sheared the blank from the strip while it was held against the center post (1) by the pressure finger (2). The front tool (4) formed the blank as shown. The side tools (5 and 6) are shown pressing the stock around the center post in position B. With the front and side tools retracted in position C, the pressure finger held the formed part against the center post while the stripper (7) moved the part to the lower level, where the spring-loaded lever arm and pressure finger (8) held the part.

In position D at the bottom level, the pivoting rear tool (9) formed the part around the center post (circular in shape at this level). The rear tool also set the folded-over metal against an extension of the front tool (4). The final forming operation by the right tool (10) was as shown in position E.

In the next cycle, the stripper pushed a new workpiece down to the bottom level, as the completed hose clamp was pushed off of the center post. The cycle time was about 60 components per minute.

Lockseaming. Metal strip is used to make large quantities and varieties of small open-end boxes, tubes, and cylinders to be used as ferrules for paint brushes, radio shields, and many other common products. All of these are made with some locking method for joining the ends of the metal.

In lockseaming or can seaming, one end of the metal is formed over the other and flattened into a tight joint. Other methods of locking two ends of metal together in a multiple-slide machine include interlocking dovetails, raised tabs inserted through slots and then flattened or twisted, a tongue inserted into a lanced slot and then swaged, and lancing and forming through two stock thicknesses.

Lockseaming can be done at either one or two levels around the center post. Single-level forming is used where the tubes are long or where projecting lugs would be too weak to push the finished part off of the center post. In this method, the rear and side tools must be advanced two or three times during each cycle of the machine.

Two-level forming with an attachment that gives a positive movement to the seaming tools is a more efficient method of making lockseams. The sliding head containing the center post can be used to make several sizes of lockseam tubes by setting the rear and side tools a fixed distance from the strip line, then changing only the front tool and the center post. The following example describes a two-level forming operation for external lockseaming to make a short cylindrical tube.

Example 2: Two-Level External Lockseaming to Produce a Round Tube. The external lockseam of the tube shown in Fig. 8 was formed at two levels in a multiple-slide machine. At position A, the strip had been fed between the center post (1) and the pressure finger (2) and sheared by the cutoff die (3) and blade (4). At the same time, the right-hand end of the

blank had a small flange formed by the fixed tool (5) and the blade (6). Position B shows the blank formed into a U-shape by the front tool (7). Forming of the cylinder by the side tools (8 and 9) is shown in position C. The two ends extend at right angles to the surface, one end being longer than the other. The right tool (10), moved by the cam (11), supported the previously bent flange as the left tool (8) formed the longer end. At position D, the left tool (12), actuated by the cam (13), bent the long end over the shorter end.

At position E, the cams (11 and 13) retracted, allowing the side tools (10 and 12) to be pushed aside by the advancing rear tool (14), thus permitting the end of the seam to be bent. The rear tool retracted, and the cam (11) advanced to move the side tool (10) to flatten the seam against the left tool (8), as shown in position F. The part then was transferred to the lower level by the stripper (not shown).

At the bottom level, the front tool (15) and the side tools (16 and 17) held the cylinder on the center post. The front tool also backed up the center post during the final forming operations. As shown in position G, the anvil (18) supported the seam while the swinging tool (19) moved forward to bend the seam at an angle. The anvil is moved by the cam (20). Final flattening of the seam is shown in position H. As the rear tool (21) advanced, the cam projection (22) contacted the anvil, pushing it clear as the swinging tool (19) closed the seam.

Two workpieces were on the center post at this stage, and both were formed at the same time. One, formed through position F, was at the top

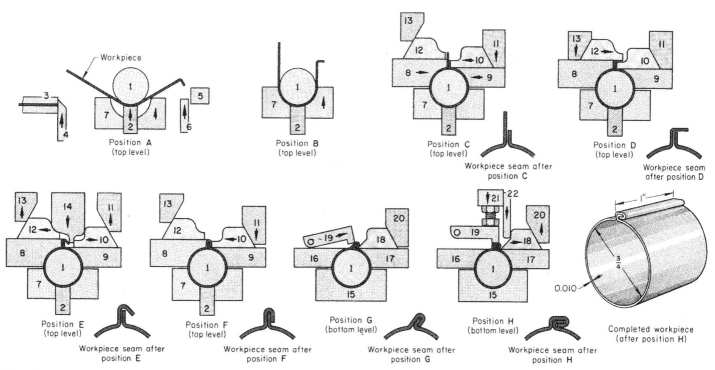

Fig. 8 Two-level forming of an external lockseam to produce a cylindrical part. Dimensions given in inches

level; the other was at the bottom level. Both were held tightly on the center post. The completed piece was pushed off of the center post and into a container.

Internal lockseams can be made by forming an external seam, as shown in Fig. 8, and then reforming the part to make the seam flush with the outer surface. The part is moved to the lower level, where a vertical slot in the center post provides space for the seam as it is pushed inward.

Assembly Operations

By means of auxiliary feeding attachments, one or more components can be assembled to a part being formed during the same machine cycle. Hopper feeding units can sort and orient machined or previously formed parts for feeding into the part being formed in the multiple-slide machine. These devices can be arranged to feed horizontally adjacent to any forming slide, or can be operated vertically.

Hopper feeding involves the automatic sorting and transfer of components placed randomly in a hopper of suitable design and shape for delivery by gravity to a track, correctly positioned and in ordered sequence to a machine.

Production Practice. Figure 9 shows the assembly of a flat steel spring into a small steel bracket at the lower level of a two-level multiple-slide machine operation. The bracket was made from 12.7 mm (1/2 in.) wide stock 0.76 mm (0.030 in.) thick. In the press station, the two

holes were pierced, and the lug was slit and raised. The bracket was bent to shape around the center post at the top level by the action of the front, rear, and right forming tools. It was then moved to the bottom forming level by the stripper and held in position against the center post by a spring-loaded retainer. A sliding pin mounted in the retainer was used to flatten the lug against the spring steel strip.

The spring steel strip was fed into the recess formed by the lug. The front tool then moved in, contacted the sliding pin, flattened the lug, secured the strip to the bracket, and at the same time severed the spring steel strip by a blade mounted to the front tool.

On the next cycle, the assembly was ejected from the center post by the downward movement of the succeeding component. The lower part of

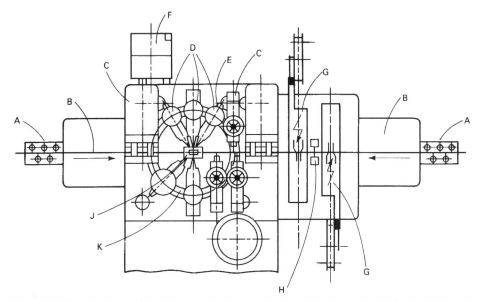

Fig. 10 Schematic of a multiple-slide rotary forming machine. A, straightening stations; B, feed mechanism; C, stamping stations; D, forming tools; E, thread tapping unit; F, station for feeding and assembling; G, welding stations; H, sizing unit; J, standard forming tool guide; K, central gear wheel. See text for description of operation.

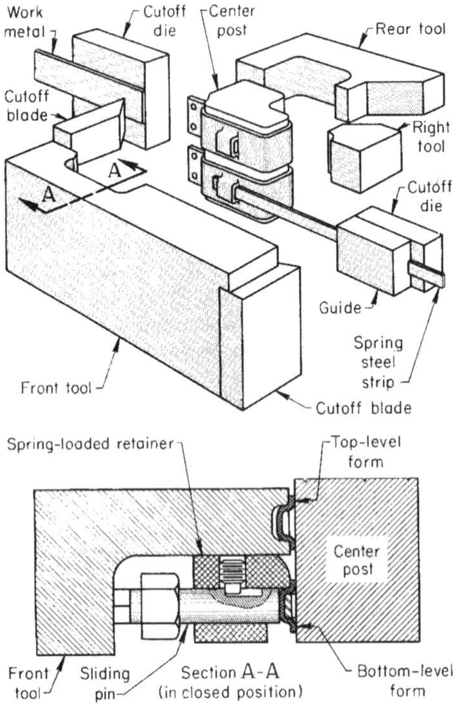

Fig. 9 Assembly of spring steel strip into a bracket at the bottom level of a two-level multiple-slide forming operation

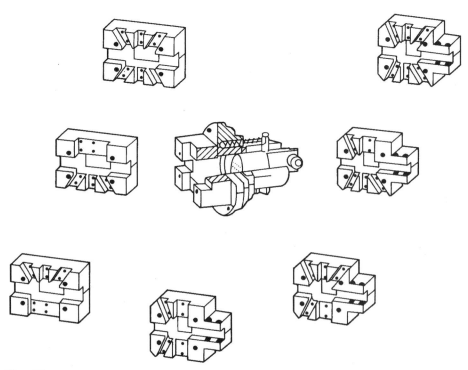

Fig. 11 Standard forming tool guides for multiple-slide rotary forming machines

the spring-loaded retainer was cut away to clear the strip as it was fed into the lug and to permit ejection.

Multiple-Slide Rotary Forming Machines

The multiple-slide rotary forming machine with open planetary gearing is simple in concept and design. In this type of machine, the slides are driven by a central gear wheel, and the central forming motions are controlled by cams.

Figure 10 shows a schematic of a multiple-slide rotary forming machine. The material is first uncoiled from the reel, then pulled through the straightening station (A) by the material feeder (B). It is then fed to the press (C) and forming slides (D), where the workpieces are formed and ejected. Recessed rails guide the material between the feeder and the stamping station, as well as between the cutting and forming stations. Here, the blanks are either completely stamped out in the press and then pushed into the bending station by the following material, or they remain attached to the strip by means of a web that is not completely separated until it reaches the final forming station. Material feed, stamping presses, forming slides, and operation of the central tool are synchronously coupled to each other.

Auxiliary Equipment. A second material feed and press, a part-feeding mechanism, three thread-tapping units, and two welding units can be added to a multiple-slide rotary forming machine (Fig. 10). It is possible to have additional material feeds from the front and rear sides. Additional cam motions allow forming slides and standardized tools to be mounted on the rear side of the bed in the forming area. Machined assembly surfaces on the bed and cover plate offer further possibilities for fastening attachments.

Standardized Tooling. The forming tool guide (the basic tool) consists of the tool plate with a central slider for one or more tools (Fig. 11). Forming tools are suspended in the tool holders mounted on the forming slides, which can be easily inserted or removed. In operation, the tool guide is enclosed by a cover plate, with an opening at its center for ejection of the working piece. Control cams drive the forming slides and consist of two complementary disk cams.

Forming of Stainless Steel

Revised by Joseph A. Douthett, AK Steel Corporation, Inc.

STAINLESS STEELS are blanked, pierced, formed, and drawn in basically the same press tools and machines as those used for other metals. However, because stainless steels have higher strength and are more prone to galling than low-carbon steels and because they have a surface finish that often must be preserved, the techniques used in the fabrication of sheet metal parts from stainless steels are more exacting than those used for low-carbon steels. In general, stainless steels have the following characteristics, as compared with those of carbon steels:

- Greater strength
- Greater susceptibility to work hardening
- Higher propensity to weld or gall to tooling
- Lower heat conductivity

Selection of Alloy

General ratings of the relative suitability of the commonly used austenitic, martensitic, and ferritic types of stainless steels to various methods of forming are given in Table 1. These ratings are based on formability and on the power required for forming.

As Table 1 shows, the austenitic and ferritic steels are, almost without exception, well suited to all of the forming methods listed. Of the martensitic steels, only types 403, 410, and 414 (the lower-chromium and -carbon alloys) are generally recommended for cold forming applications. The higher carbon content of the remaining martensitic alloys and higher hardness of the martensitic precipitation-hardenable alloys 17-4PH or 15-5PH act to limit the formability of these alloys. Moderate warming can be used to improve the formability of these alloys, as in general with all the classes of stainless steels. Most martensitic stainless alloys are mill supplied with an in-process anneal generally between 650 and 760 °C (1200 and 1400 °F). This subcritical temperature cycle is designed to transform the martensite present into chromium carbides and ferrite. Hardness is reduced to 80 to 90 HRB, thereby improving formability. With carbides present, the alloy resistance to corrosion is reduced. After forming, a resolution-hardening heat treatment is required to return the alloys to maximum corrosion resistance.

The forming of higher-carbon martensitic alloys is normally restricted to blanking, piercing, and mild bending. These steels are sometimes formed warm. Warm forming can also be used to advantage with other stainless steels in difficult applications.

Formability. The characteristics of stainless steel that affect its formability include yield strength, tensile strength, ductility (and the effect of work hardening on these properties), and the r-value. The composition of stainless steel is also an important factor in formability. Figure 1 compares the effect of cold work on the tensile strength and yield strength of type 301 (an austenitic alloy), types 409 and 430 (both ferritic alloys), and 1008 low-carbon steel sheet.

In Fig. 1, it is noteworthy that the three stainless alloys all have higher tensile strengths than 1008 carbon steel after any given amount of cold reduction. One reason for this finding is that

Table 1 Relative suitability of stainless steels for various methods of forming

Suitability ratings are based on comparison of the steels within any one class; therefore, it should not be inferred that a ferritic steel with an A rating is more formable than an austenitic steel with a C rating for a particular method. A, excellent; B, good; C, fair; D, not generally recommended

Steel	0.2% yield strength, 6.89 MPa (1 ksi)	Blanking	Piercing	Press-brake forming	Deep drawing	Spinning	Roll forming	Coining	Embossing
Austenitic steels									
201	55	B	C	B	A–B	C–D	B	B–C	B–C
202	55	B	B	A	A	B–C	A	B	B
301	40	B	C	B	A–B	C–D	A	B–C	B–C
302	37	B	B	A	A	B–C	A	B	B
302B	40	B	B	B	B–C	C	...	C	B–C
303, 303(Se)	35	B	B	D(a)	D	D	D	C–D	C
304	35	B	B	A	A	B	A	B	B
304L	30	B	B	A	A	B	A	B	B
305	37	B	B	A	B	A	A	A–B	A–B
308	35	B	...	B(a)	D	D	...	D	D
309, 309S	40	B	B	A(a)	B	C	B	B	B
310, 310S	40	B	B	A(a)	B	B	A	B	B
314	50	B	B	A(a)	B–C	C	B	B	B–C
316	35	B	B	A(a)	B	B	A	B	B
316L	30	B	B	A(a)	B	B	A	B	B
317	40	B	B	A(a)	B	B–C	B	B	B
321, 347, 348	35	B	B	A	B	B–C	B	B	B
Martensitic steels									
403, 410	40	A	A–B	A	A	A	A	A	A
414	95	A	B	A(a)	B	C	C	B	C
416, 416(Se)	40	B	A–B	C(a)	D	D	D	D	C
420	50	B	B–C	C(a)	C–D	D	C–D	C–D	C
431	95	C–D	C–D	C(a)	C–D	D	C–D	C–D	C–D
440A	60	B–C	B	C(a)	C–D	D	C–D	D	C
440B	62	...	...	...	...	D	...	D	D
440C	65	...	...	...	...	D	...	D	D
Ferritic steels									
405	40	A	A–B	A(a)	A	A	A	A	A
409	38	A	A–B	A	A	A	A	A	A
430	45	A	A–B	A(a)	A–B	A	A	A	A
430F, 430F(Se)	55	B	A–B	B–C(a)	D	D	D	C–D	C
442	...	A	A–B	A(a)	B	B–C	A	B	B
446	50	A	B	A(a)	B–C	C	B	B	B

(a) Severe sharp bends should be avoided.

the initial annealed tensile strength of the stainless alloys is greater than that of 1008. The rate of work hardening (how fast strength levels increase with strain or cold work) is measured by determining the *n*-value. The *n*-value is the slope of the true stress-strain curve determined by the uniaxial tension test. The measurement is made between predetermined strain levels usually 10% to the ultimate. Steel values normally range from 0.15 to 0.50, with annealed carbon steels, ferritic stainless, and martensitic alloys at 0.15 to 0.25, and austenitic stainless at 0.35 to 0.6. The higher work-hardening rate of 300-series austenitics, as can be seen in Fig. 1, is a key factor in needing better lubricants when drawing stainless alloys, particularly when forming is carried out in multiple stages.

Formability of Austenitic Types. Type 301 stainless steel has the lowest nickel and chromium contents of the standard austenitic types. It also has the highest tensile strength in the annealed condition. Among the 200-series Cr-Mn-Ni-N austenitic alloys, type 201 would be considered an alloy equivalent to type 301, although the typical yield strength of type 201 is higher and tensile strength lower. The extremely high rate of work hardening of type 301 results in appreciable increases in tensile strength and yield strength with each increase in the amount of cold working, as measured by cold reduction (Fig. 1). The *n*-value for this alloy can typically

be measured at 0.45 or higher. This response to work hardening is particularly important for structural parts, including angles and channel sections, that, after fabrication, are expected to have additional strength and stiffness. On the other hand, for deep-drawing applications, a lower rate of work hardening is usually preferable (*n*-values of 0.38 to 0.42) and can be obtained in the austenitic alloys that have higher nickel contents, notably, types 304, 304L, and 305.

In general, the austenitic alloys are more difficult to form as the nickel content or both the nickel and the chromium contents are lowered, as in type 301. Such alloys show increased work-hardening rates (higher *n*-values) and are less suitable for deep-drawing or multiple forming operations. The presence of the stabilizing elements niobium, titanium, and vanadium, as well as higher carbon contents, also exerts an adverse effect on the forming characteristics of the austenitic stainless steels. Therefore, the forming properties of types 321 and 347 stainless steel are less favorable than those of types 304 and 305.

In general, an austenitic alloy work-hardening rate (*n*-value) decreases with increasing alloy content. The more solid-solution (chromium, nickel, manganese, copper) and interstitial (carbon, nitrogen) alloying elements added to a standard 300-series alloy, the lower the resultant work-hardening rate and the more deep drawable

the alloy. In recent years, copper additions of 0.5 to 4.0% have been added to effect such a reduction.

In large part, the higher work-hardening rate of 301 and lower-nickel 304 alloys is attributable to the formation of strain-induced or deformation martensite, a magnetic body-centered tetragonal crystalline phase, in the austenitic grain structure. While this phase has the same composition as the face-centered cubic austenite from which it transformed, it is stronger, harder, and somewhat needlelike in microstructural appearance. The volume of this phase that forms during cold work is dependent on the alloy composition, amount of cold deformation, and temperature of forming. The lower the alloy content (lower the nickel, chromium, manganese, copper, carbon, nitrogen) of the austenitic alloy being formed and the greater the degree of cold work, the more volume percent deformation martensite that will form.

Temperature of forming is a third key determinant of the amount of deformation martensite that forms during cold deformation of an austenitic alloy. The lower the temperature, the greater and more rapid the transformation. Conversely, heating a leaner alloyed 300-series stainless steel reduces the transformation rate of austenite to martensite. For each alloy composition, a critical temperature called the martensite deformation (MD) temperature exists where warming above this temperature suppresses the shear transformation to martensite. Cooling below this MD temperature will conversely induce the transformation, with the lower the temperature, the greater the degree of formed martensite. Figure 2 visualizes the deformation martensite tendencies discussed previously. While dependent on alloy content, the MD temperature will vary from well below room temperature for high-nickel austenitic stainless steels such as 305, 309, and 316 to somewhere between room temperature up to perhaps 93 °C (200 °F) for leaner 304 versions or 301/201 stainless alloys. When forming metastable austenitic alloys (leaner alloys that more readily form deformation martensite), adiabatic heating due to deformation can raise the part temperature above the MD temperature and slow down the strengthening or work-hardening effect due to

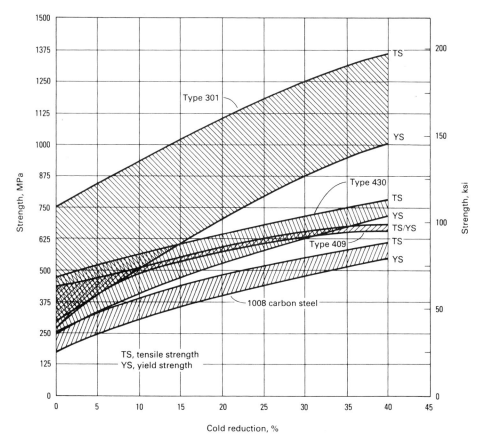

Fig. 1 Comparison of work-hardening qualities of type 301 austenitic stainless steel, types 409 and 430 ferritic stainless steels, and 1008 low-carbon steel

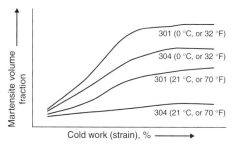

Fig. 2 Comparison of deformation martensite formation tendencies versus temperature, amount of cold work, and alloy content. Type 301 is a lower-alloy 17%Cr-7%Ni 300-series austenitic alloy versus a more highly alloyed 18%Cr-8%Ni type 304 alloy.

the formation of martensite. In multiple forming operations, such as progressive die setups, interrupting the stamping cycle of a 301 or 304 part can mean that with press restart, partially formed pieces will stamp differently. The cooler part will form a greater volume of martensite than one kept warm without a forming cycle interruption.

Formability of Ferritic Types. The range between yield strength and tensile strength of types 409 and 430 narrows markedly as cold work increases, as shown in Fig. 1. This response is typical of the ferritic alloys and limits their formability (ductility) (in comparison with the austenitic alloys). Nevertheless, types 409 and 430, although lacking the formability of type 304, are widely used in applications that require forming by blanking, bending, drawing, or spinning. One of the most important applications for type 430 stainless steel is in automotive trim or molding. Type 409 stainless steel has found wide acceptance as the material of choice in automotive exhaust systems.

The ferritic stainless alloys, in many respects, act like stronger, somewhat less ductile low-carbon steels. Their work-hardening rate (*n*-value) is similar to or slightly less than 1004/1008 carbon steel. The plastic strain ratio (*r*-value) for 409 or 430 stainless at 1.1 to 1.8 is somewhat less than the 1.5 to 2 value typical of drawable low-carbon cold-rolled steels.

In addition, the ferritic stainless alloys can present some unusual forming problems in the cold work forming of thicker sections. At cooler temperatures and higher strain rates, ferritic stainless steels can fail in a brittle manner due to a lack of toughness. Toughness is defined as the ability of a material to absorb a large volume of impact energy and deform in a ductile manner. Carbon steels and ferritic stainless alloys have a characteristic temperature or range of temperatures above which alloy specimens absorb a large volume of energy and fail in a ductile manner when impact loaded and below which impact loads cause brittle fracturing at low levels of absorbed energy. This alloy characteristic temperature is known as the ductile-to-brittle transition temperature (DBTT). Figure 3 shows plots of absorbed impact energy versus test temperature for two alloys subjected to impact testing.

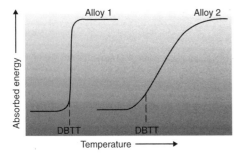

Fig. 3 Ductile-to-brittle transition temperature (DBTT) curves for two ferritic alloys. Alloy 1 has a well-defined transition temperature. Alloy 2 has a higher and less well-defined DBTT.

Alloy 1 shows a transition from ductile (high absorbed energy levels) to brittle behavior at a specific temperature, while alloy 2 shows the same behavior change over a range of temperatures. With materials such as alloy 2, the transition temperature is more difficult to pinpoint. Frequently, a specified amount of absorbed energy per unit area of sample cross section (W/ A of 1000 ft-lb/in.^{2} in Charpy tests) is chosen as the minimum value to exceed to be classified as ductile behavior.

With ferritic stainless steels, the DBTT depends on alloy content, thickness, grain size, the presence of cold work, and strain rate. For higher-alloyed (18% Cr and above) ferritic stainless steels, the DBTT can range from below −18 °C (0 °F) for 1 mm (0.040 in.) thick material to above 40 °C (100 °F) for 2.5 mm (0.100 in.) and thicker. As ferritic alloy content increases, so does the DBTT for materials of the same thickness. Larger grain sizes also lead to higher DBTTs, as does the presence of a weld. The presence of embrittling second phases (sigma or carbides due to sensitization) can lower toughness. Finally, the DBTT of a well-annealed base alloy is lower than for the same material after cold working or forming. Stored strain energy reduces the amount of impact energy that can be absorbed if the partially cold-worked material is impact loaded.

In forming ferritic stainless steels, the concern for unexpected DBTT cracking, particularly in colder winter months or in northern climates, has given rise to the following precautions:

- Do not attempt to form ferritic stainless alloys when cold. Store raw material in heated areas, or intentionally warm just prior to fabrication.
- If brittle fracturing occurs, consider moderate warming (20 to 60 °C, or 70 to 140 °F).
- If warming still leads to brittle-appearing fractures, consider reducing the rate of forming.
- Take the greatest cold-weather precaution with higher-alloyed ferritic stainless alloys (superferritics, 25 to 30% Cr, more than 439 and 439 more than 409).
- Take greater precaution with thicker materials (above 2.5 mm, or 0.100 in., thick).
- Coarser-grained materials and welds should be given care.
- The presence of cold work should dictate added care.

This last precaution can come into play in forming parts in multiple stages. If the forming sequence is done in-line (transfer press) or rapid sequence (cell operation), the alloy is warmed by deformation and, in spite of being cold worked, may pass through the subsequent forming steps above the DBTT. Interrupting the sequence of forming can allow partially produced parts to cool down and lead to brittle fractures when the cycle again commences. Originally, the annealed incoming ferritic alloy had a DBTT below room temperature, but initial forming imparted cold work, boosting the DBTT to room temperature or above. Interrupting the forming sequence per-

mitted components to then cool below the transition temperature, setting the stage for fracture on restart.

Comparison with Carbon Steel. The curves for 1008 low-carbon steel are included in Fig. 1 as a reference for the evaluation of stainless steels. The decrease in formability of 1008 steel with cold work appears to fall between that of types 409/430 and that of the more formable type 301. Figure 1 also shows that cold work does not increase the strength of 1008 as rapidly as it does that of type 301 and the ferritic alloys.

Stress-Strain Relations. Figure 4 shows tensile test load-elongation curves for six types of stainless steel: four austenitic (202, 301, 302, and 304), one martensitic (410), and one ferritic (430). The figure also shows that the type of failure in cup drawing of the austenitic types was different from that of types 410 and 430. The austenitic types broke in a fairly clean line near the punch nose radius, almost as if the bottom of the drawn cup were blanked out; types 410 and 430 broke in the sidewall in sharp jagged lines.

As suggested by the data in Fig. 4, the power required to form type 301 exceeds that required by the other austenitic alloys. In addition, type 301 will develop maximum elongation before failing. Much of the additional power required to form 301 and the greater elongation of the alloy in tensile tests can be attributed to the alloy tendency to form deformation martensite (α′). The martensite that forms due to shear deformation is harder and stronger than the austenite from which it transformed. The martensite hardness, as with carbon steels, is proportional to the weight percent carbon and nitrogen present. The 201 and 301 melt compositions with higher carbon and nitrogen contents would be expected to form a higher-hardness deformation martensite, assuming the overall composition is lean enough and the temperature is at or below the earlier described MD so that martensite forms with cold work. As the 301 tensile specimen of Fig. 4 is pulled to form the load/elongation curve, specimen strain with accompanying and thinning would trigger martensite formation and thereby strengthen the initially necked region. Tensile strain and necking would then be transferred to specimen regions adjacent to the transformed area. Once more, strain and necking would initiate martensite formation, strengthening, and additional transfer of the strain to an adjacent region along the tensile blank. With such a sequence of events, the 301 tensile blank develops a greater amount of overall elongation before the transformation sequences cease and the martensite-containing tensile dogbone section necks and fractures. Types 410 and 430 require considerably less power to form but fail at comparatively low elongation levels.

Power requirements for the forming of stainless steel, because of the high yield strength, are greater than those for low-carbon steel; generally, twice as much power is used in forming stainless steel. Because the austenitic steels work harden rapidly in cold forming operations, the need for added power after the

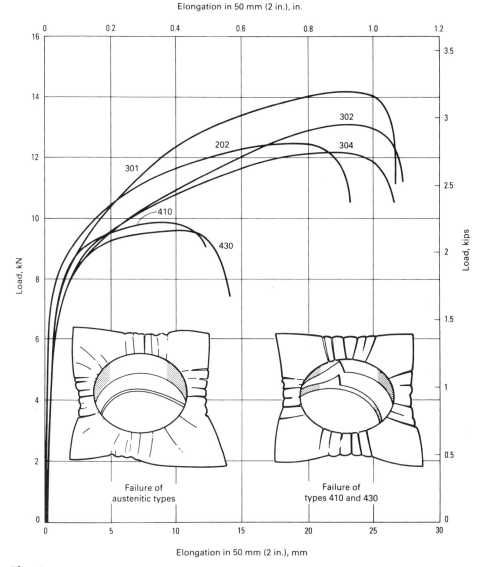

Fig. 4 Comparison of ductility of six stainless steels and of the types of failure resulting from deep drawing

start of initial deformation is greater than that for the ferritic steels. The ferritic steels behave much like plain carbon steels once deformation begins, although higher power is also needed to start plastic deformation.

Lubrication

Lubrication requirements are more critical in forming stainless steels than in forming carbon and alloy steels, because it usually is necessary to preserve the high-quality surface on stainless steels and because stainless steels have higher strength, greater hardness, lower thermal conductivity, and higher coefficient of friction. In forming stainless steels, galling and spalling occur more readily, and higher temperatures are reached in a larger volume of the workpiece. Local or general overheating can change the properties of the work metal and lubricant.

Table 2 lists the lubricants ordinarily used in forming stainless steel by various processes. Except for the special-purpose lubricants graphite and molybdenum disulfide, the lubricants are listed in the approximate order of increasing ability to reduce galling and friction. The ratings in Table 2 also consider other suitability factors, such as cleanliness and ease of removal. The desirable characteristics of a lubricant include the ability to reduce friction and wear, the dissipation of heat, durability, nonreactiveness with the base steel, and ease of removal. The higher temperatures generated when forming stainless alloys, particularly the austenitic varieties, frequently lead to breakdown of the polar lubricant molecules.

In recent years, a family of organic waterborne polymer dry-film lubricants has been introduced that can be roll coated onto stainless steel coils and moderately warmed (95 °C, or 200 °F) to dry or cure in-line. These dry-film lubricants

applied in uniform film thicknesses measuring 1 to 3 g/m^2 have proven to have very low (0.05) coefficients of friction, resist buildup on tooling, can be welded without removal, and can be readily removed, depending on formulation, with either warm water or alkaline detergents. Dry-film lubricants are not recommended for use with other wet lubricants.

Among stainless steel lubricants for severe drawing applications, the extreme-pressure (EP) additive types are the most desirable. Additives of chlorine or sulfur tend to react chemically with the steel surface at higher temperatures and form a readily shearable compound. Chlorine is the more popular EP additive, because sulfur tends to react with some steel tooling.

Mineral oils, soap solutions, and water emulsions of general-purpose soluble oils are omitted because they are ineffective in most forming of stainless. The recommended lubricants are discussed further in the sections that deal with the individual forming processes in the remainder of this article.

As a precaution, all lubricants should be removed, and the parts thoroughly dried, after completion of the sheet metalworking operation. Most lubricants must be removed before the formed parts are heat treated; this applies particularly to those containing insoluble solids, sulfur, or chlorine. In addition, certain metastable austenitic stainless alloys can react with their lubricant and cause delayed cracking in heavily strained areas. Rapid removal of the lubricant is therefore desirable.

Chlorinated lubricants, if left on formed parts and put into service applications, could lead to pitting or stress cracking if moisture and humidity are present. Conversely, the roll coated dry-film polymer lubricants painted onto stainless coils do not seem to chemically react with the stainless surface and have been found to serve as a barrier coating, reducing corrosion potential for coils seeing moisture in transit.

Blanking and Piercing

The shear strength of stainless steel is approximately twice that of low-carbon steel. Therefore, the available force for the blanking or piercing of stainless steel should be 50 to 100% higher than that for equivalent work on carbon steel.

Tools and power can be saved if the stock can be blanked at approximately 175 °C (350 °F). The finish will be better as well. Power requirements can also be reduced by using angular shear on the punch or the die.

Die Materials. Cutting edges must be of a hard, strong material. Recommended die materials, in order of suitability for increasing quantities, include O1, A2, D2, and D4 tool steels and carbide. Additional information is available in the article "Selection of Materials for Shearing, Blanking and Piercing Dies" in this Volume. The use of carbide for high-volume production in applications that do not require the impact resistance of tool steels is illustrated in the following example.

Example 1: Use of a Carbide Die to Form a Miniature Piece. The cathode shown in Fig. 5 was produced in a three-stage progressive die made of carbide by piercing, blanking, and forming. The piece was trough-shaped, 6.4 mm (1/4 in.) long, of type 304 stainless steel, 0.08 mm (0.003 in.) thick. One end was rounded, and the other was V-shaped. The difference in contour of the two ends kept the pieces from stacking. Before forming, the blank was 9.5 mm (0.374 in.) wide. The piece was pierced with 68 holes, each 0.31 mm (0.012 in.) square. In this operation, the material was displaced by a pointed punch, rather than removed by a flat-nose punch. The pieces were cut from 152 mm (6 in.) wide strip, producing 16 pieces at a stroke. The press was a 130 kN (15 tonf) mechanical press that ran at 240 strokes per minute.

Clearance between punch and die should be approximately the same as that for the blanking and piercing of cold-rolled low-carbon steel. Some manufacturers use less than 0.03 mm (0.001 in.) per side; others specify 5 to 10% of stock thickness per side for sheet and 10 to 15% of stock thickness for plates and bars. Studies have shown, however, that larger clearances (12.5 to 13.5%, and even up to 42%, of stock thickness) have resulted in increased die life (see the article "Piercing of Low-Carbon Steel" in this Volume).

Cutting edges should be carefully aligned, sharp, clean and free of burrs. The importance of sharpness of cutting edges cannot be over-emphasized.

Deburring. Generally, stainless steel does not shear clean but leaves a rough work-hardened edge that is dangerous to handle and may adversely affect subsequent operations. Flat pieces can be rolled or pressed between dies adjusted exactly to the thickness of the stock, or the burrs can be removed by grinding, stoning, or filing.

Lubrication. The blanking and piercing of stainless is often done dry, but the lubricants indicated in Table 2 are sometimes used to prolong die life. Lubricants containing sulfur or chlorine are the most effective for this purpose. Emulsions are used for high-speed work.

Dimensions. Pierced holes should not be smaller than the thickness of the stock. Holes larger than 3.18 mm (0.125 in.) should be spaced so that the distance between centers is not less than 1 1/2 times the hole diameter. Small holes should have a distance between centers of at least 1 3/4 times the diameter of the holes. Holes should never be closer together than one stock thickness, nor should the edge of blanks be less than one stock thickness from the edge of the stock. For progressive-die operation, edge distances should be between 1 1/2 and 2 times stock thickness.

Nibbling. In some applications, an irregular contour is cut out by punching a series of overlapping holes along the contour. This process is called nibbling. A variety of unusual shapes can be cut at 300 to 900 strokes per minute by a press equipped with either a round or a rectangular punch.

Press-Brake Forming

All of the austenitic stainless steels in the soft condition can be bent 180° over one stock thickness but need up to 50% more power to form than that required by low-carbon steel. Springback is more severe with austenitic stainless steels than with low-carbon steel, and it must be allowed for. Work-hardened austenitic steel can be press-brake formed only to a very limited degree. If austenitic stainless steel is heated to approximately 65 °C (150 °F), it can be formed with appreciably less power than that required when it is cold and yet can be handled easily. A temperature of 65 °C (150 °F) is normally high enough to exceed an austenitic stainless steel MD temperature and eliminate the strengthening contribution attributed to the formation of deformation martensite.

The straight-chromium grades of stainless steels vary in their response to press-brake forming. The low-carbon stainless steels containing 12 to 17% Cr bend readily but, like the austenitic steels, need more power for bending than that required for low-carbon steel. High-chromium low-carbon types, such as 446, bend better when heated to 175 to 205 °C (350 to 400 °F). The heating of these high-chromium low-carbon grades tends to lower the yield strength but can simultaneously aid in allowing the forming to be done above the brittle-to-ductile transition temperature. For these alloys, that temperature can be at or above room temperature, depending on thickness. In room-temperature forming, the highly alloyed ferritic stainless steels have been known to benefit from slower bend speeds, which minimize the possibility of an impactlike load and resultant brittle fracture. High-carbon heat-treatable stainless steels are not recommended for press-brake forming, even if in the annealed condition.

Typical bending limits for the major stainless steels are shown in Table 3. A completely flat bend can generally be made in the 18-8 and similar alloys.

Dies. Press brakes can use dies with cross sections such as those shown in Fig. 6 for forming stainless steel in sheets up to 0.89 mm (0.035 in.) thick. Adjustable dies, such as that shown in Fig. 7, can be used for forming 180° bends in stainless steel sheet 0.30 to 0.46 mm (0.012 to 0.018 in.) thick.

Springback is a function of the strength of the material, the radius and angle of bend, and the thickness of the stock; the thicker the stock, the less severe the problem. Table 4 shows the relationship between radius of bend and springback for three austenitic stainless steels. Ferritic steels usually exhibit less springback than austenitic steels, because the rate of work hardening of ferritic steels is lower. As a practical guide, the amount of springback is normally proportional to $(0.2YS + UTS)/2$, where YS is yield strength, and UTS is ultimate tensile strength.

Table 2 Suitability of various lubricants for use in the forming of stainless steel

Ratings consider effectiveness, cleanliness, ease of removal, and other suitability factors.
A, excellent; B, good; C, acceptable; NR, not recommended

Lubricant	Blanking and piercing	Press-brake forming	Press forming	Multiple-slide forming	Deep drawing	Spinning	Drop-hammer forming	Contour roll forming	Embossing
Fatty oils and blends(a)	C	B	C	A	C	A	C	B	B
Soap-fat pastes(b)	NR	NR	C	A	B	B	C	B	C
Wax-based pastes(b)	B	B	B	A	B	B	C	B	A
Heavy-duty emulsions(c)	B	NR	B	A	B	B	NR	A	B
Dry-film (wax, or soap plus borax)	B	B	B	NR	B	A	B	NR	A
Pigmented pastes(b)(d)	B	NR	A	B	A	C	NR	NR	NR
Sulfurized or sulfochlorinated oils(e)	A	A	B+	A	C	NR	A	B	A
Chlorinated oils or waxes(f)									
High-viscosity types(g)	A(h)	NR	A	NR	A	NR	A(j)	A	NR
Low-viscosity types(k)	B+	A	A	A	B	NR	A(j)	A	A
Graphite or molybdenum disulfide(m)	NR	(n)	(n)	NR	(n)	NR	(n)	NR	NR

(a) Vegetable or animal types; mineral oil is used for blending. (b) May be diluted with water. (c) Water emulsions of soluble oils; contain a high concentration of extreme-pressure sulfur or chlorine compounds. (d) Chalk (whiting) is commonest pigment; others sometimes used. (e) Extreme-pressure types; may contain some mineral or fatty oil. (f) Extreme-pressure chlorinated mineral oils or waxes; may contain emulsifiers for ease of removal in water-based cleaners. (g) Viscosity of 4000 to 20,000 SUS (Saybolt Universal seconds, see ASTM D 2161 for more detailed information). (h) For heavy plate. (j) For cold forming only. (k) Viscosity (200 to 1000 SUS) is influenced by base oil or wax, degree of chlorination, and additions of mineral oil. (m) Solid lubricant applied from dispersions in oil, solvent, or water. (n) For hot forming applications only

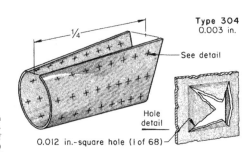

Fig. 5 Cathode produced in a progressive die with carbide tools. Dimensions given in inches

Springback can be controlled by reducing the punch radius, by coining the line of bend (if the shape of the die is such that bottoming is feasible), and by overbending. For overbending, it is sometimes necessary only to make the punch angle smaller than the desired final angle of the workpiece, as in the following example.

Example 2: Setting a Flange Angle in a Press Brake. The bracket shown in Fig. 8 was preformed in a U-die from a developed blank of type 302 stainless steel, half-hard, 1.0 mm (0.040 in.) thick. Only the punch angle needed to be reduced to set the angle on the flange.

As the bracket came from the U-die, the springback in each flange was 15°. To correct this spread, the piece was put in a restrike die in a press brake, which set each angle separately. The restrike die angle was 90° with a 3.2 mm (0.126 in.) radius. The restrike punch was made to an angle of 86° with a 2.4 mm (0.094 in.) radius to coin the bend, so that the flanges would form to 90° ± 1°.

The lubricant was a water-soluble pigmented drawing compound. The workpiece was degreased after forming.

Lubricants. For ordinary press-brake operations (chiefly, bending and simple forming), lubricants are not used as frequently as with higher-speed press operations. Convenience of use is a major factor in selecting lubricants for this type of press-brake forming. Pigmented lubricants are not favored, and cooling effectiveness is of little significance at low production rates. For severe forming and for operations that would ordinarily be done in a press, if available, the recommendations in the "Press forming" column in Table 2 apply.

Applications of press-break forming are described in the following examples. Repetitive bends, as in corrugated stock, are frequently made one at a time in a press brake if the quantity of production is not sufficient to warrant a special die, as in the subsequent example.

Example 3: Press-Brake Forming of Corrugations. The corrugated sheet shown in Fig. 9 was formed from 0.41 mm (0.016 in.) thick full-hard type 302 stainless steel. The finished sheets, after bending, were 419 mm (16½ in.) long, as shown, but the width, w, varied according to the use of the piece.

The corrugations were made one at a time by air bending in the tooling shown at lower right in Fig. 9. Pilot holes in the workpiece and locating pins in the punch helped to keep the workpiece aligned. Deviation from flatness in the pieces was corrected by restriking some of the bends.

Irregular contours on long, narrow parts are conveniently produced by bending in a press brake. Because of the strength of stainless steels, the forming often must be divided among several successive operations, as in the subsequent example.

Example 4: Forming of Stainless Steel Handrails in a Press Brake. Figure 10 illustrates the shapes produced in five successive operations that were required for forming a handrail from 1.57 mm (0.062 in.) thick type 304 stainless steel. Because of flatness requirements and the resistance of the metal to bending, a 3600 kN (400 tonf) press brake was used.

Forming the 1.6 mm (0.063 in.) radius beads (operation 1, Fig. 10) was particularly troublesome because of the difficulty in retaining flatness. A force of 5300 kN (600 tonf), which

Table 3 Typical bending limits for six commonly formed stainless steels

	Minimum bend radius		
	Annealed to 4.75 mm (0.187 in.) thick (180° bend)	Quarter hard, cold rolled	
		To 1.27 mm (0.050 in.) thick (180° bend)	1.30–4.75 mm (0.051–0.187 in.) thick (90° bend)
Type			
301, 302, 304	½t	½t	1t
316	½t	1t	1t
410, 430	1t	...	...

t, stock thickness

Table 4 Springback of three austenitic stainless steels bent 90° to various radii

	Springback for bend radius of:		
Steel and temper	1t	6t	20t
302 and 304, annealed	2°	4°	15°
301, half-hard	4°	13°	43°

t, stock thickness

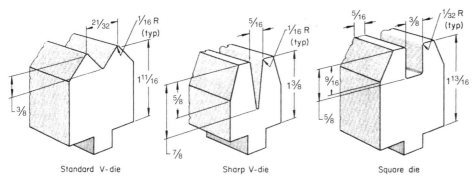

Fig. 6 Typical dies for the press-brake forming of stainless steel sheet up to 0.9 mm (0.035 in.) thick. Dimensions given in inches

Standard V-die   Sharp V-die   Square die

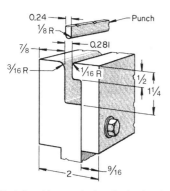

Fig. 7 Adjustable press-brake die for forming 180° bends in stainless steel sheet. Setup can be used for forming bends to 3.2 mm (0.125 in.) inside radius in sheet 0.30 to 0.46 mm (0.012 to 0.018 in.) thick, and it will produce 4.0 mm (0.157 in.) radius bends in half-hard stainless steel. Detachable side of die can be shimmed for bending thicker sheet or for bending with larger-radius punches. Dimensions given in inches

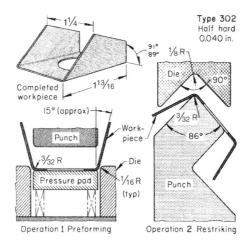

Fig. 8 Production of a U-shaped bracket from a developed blank by preforming, and restriking to set flange angles, in a press brake. Dimensions given in inches

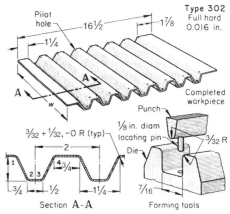

Fig. 9 Corrugated sheet in which corrugations were formed one at a time in a press brake, using tools shown. Dimensions given in inches

exceeded the rating of the press brake, was used to form the beads.

The second and third operations presented no problems, but the fourth operation was difficult because the workpiece had to be held without marring the polished surface. Similar parts were produced from low-carbon steel without difficulty.

Press Forming

Stainless steels are press formed with the same kind of equipment as that used in the forming of low-carbon steel. However, although all stainless steels are not the same in strength or ductility, they all need more power to form than carbon steels do. In general, presses should have the capacity for 100% more ram force than that needed for equivalent work in low-carbon steel, and frames should have the rigidity and bulk necessary to withstand this greater force.

Dies. In addition to wearing out faster, dies may fracture more readily when used with stainless steel than when used with low-carbon or medium-carbon steel. This is because of the greater forces needed for the working of stainless steel.

For the longest service in mass production, the wearing parts of the dies should be made of carbide, D2 tool steel, or high-strength aluminum bronze. Carbide can last ten times as long as most tool steels, but it is more expensive and does not have the shock resistance of tool steels and aluminum bronze. Tool steels such as D2 are preferred when resistance to both shock and wear is required.

Aluminum bronze offers the most protection against galling and scuffing of the workpiece. An oil-hardening tool steel such as O2 can be used for short production runs.

Austenitic Alloys. Workpieces can be stretched by applying high blankholder pressures to the flange areas to prevent metal from flowing into the die. This causes severe thinning, but work hardening may cause the thinned metal to be as strong as or stronger than the thicker

unworked sections. Figure 11 shows a section of an automobile wheel cover made of type 301 stainless steel; the central portion was purposely thinned and work hardened by stretching. In the following example, one of the principal reasons for stressing the workpiece to the limits of formability was to work harden it for increased strength.

Example 5: Severe Forming for Intentional Work Hardening. The material for a muffler header (Fig. 12) for a small aircraft engine was intentionally stressed nearly to the limits of formability to increase rigidity and to impart the necessary fatigue strength. The headers were made in two operations in a 530 kN (60 tonf) open-back-inclineable mechanical press having a 127 mm (5 in.) stroke. Each operation used a tool steel die hardened to 59 to 62 HRC. Production was 400 pieces per month.

The first die (Fig. 12a) was a compound die that formed the dish of the part, formed the bead in the dish, and blanked the inside and outside diameters. The blankholder at the outer edge of the workpiece was spring loaded, and a rubber pad supported the inner surface of the workpiece against the center blanking punch. The sequence was programmed so that the forming was completed before the outside and inside diameters were blanked, thus making the flange dimensions more accurate and concentric than would have otherwise been possible. Die life was approximately 20,000 pieces.

The second die (Fig. 12b) formed both the inner (stretch) flange and the outer (compression) flange. A spring-loaded pressure pad maintained the correct gripping pressure against the muffler-header body during this operation.

The blank was annealed type 321 stainless steel 0.81 mm (0.032 in.) thick, sheared to 216 mm (8½ in.) square. The bead formed in the first die was used as a locating surface in the second die. The dies were brushed with oil between pieces.

The production rate for both operations was seven pieces per minute. Setup time for the first operation was 0.17 h; for the second operation, 0.31 h.

Stretching. Stainless steel has high ductility but wrinkles easily in compression. Therefore, if there is a choice in the direction of metal flow

during forming, a better part is likely to be produced by stretching than by compression, as in the following example.

Example 6: Use of Clamping Plates and Bead to Control Metal Flow. The dome section shown in Fig. 13 was formed from a tapered blank of annealed type 302 stainless steel in a 2200 kN (250 tonf) double-action hydraulic press. It was desirable to maximize metal flow from the narrow end of the blank in order to cause stretching rather than contraction in the metal and thus avoid wrinkles.

The dies could not be oriented to let the blank lie flat, because of the necessity for forming the reentrant angle next to the lower clamping plates. Both the upper and the lower edges of the blank were held between steel plates during forming. The clamping force on each pair of plates was 320 kN (36 tonf). Because the upper plates were twice as long as the lower and two-thirds as wide, the clamping force was distributed over a larger area and could have permitted most of the metal flow from the larger end, with attendant wrinkling of the work metal. The addition of a bead to the upper clamping plates improved holding at that end and caused most of the metal flow to occur at the small end of the blank. The application of a fatty-acid-type, nonpigmented drawing compound to the lower plates further encouraged metal flow from the small end of the blank. Scrap loss because of tearing over the relatively sharp lower die radius was 3%.

Ferritic Alloys. The formability of ferritic stainless steels, particularly the higher-chromium types, can be improved by warm

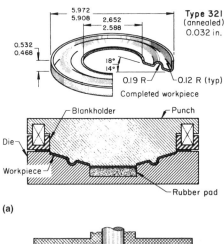

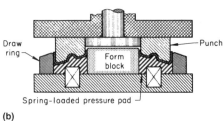

Fig. 12 Severe forming of an austenitic stainless steel aircraft muffler header to produce work hardening that would increase the rigidity and fatigue strength of the part. Dimensions given in inches

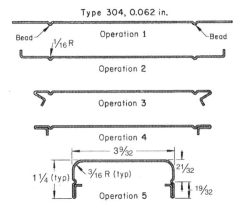

Fig. 10 Shapes progressively produced in the five-operation forming of a handrail in a 3600 kN (400 tonf) press brake. Dimensions given in inches

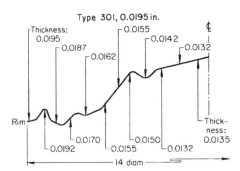

Fig. 11 Profile of a press-formed automobile wheel cover showing thinning purposely produced by severe stretching. Dimensions given in inches

forming at 120 to 200 °C (250 to 400 °F), rather than cold forming. The metal is more ductile at the higher temperatures, and less power is needed in forming. Some pieces that cannot be made by cold forming can be successfully made by warm forming. As discussed earlier, with ferritic stainless steels the presence of cold work raises the alloy DBTT. Such cold work can promote brittle fractures in multiple forming operations, particularly if the forming sequence is interrupted and the partially formed part is allowed to cool off. Keeping heat in the part is important in multiple forming operations of thicker and higher-alloyed ferritic stainless blanks.

Lubrication. The lubricant used most often in the press forming of stainless steel is the chlorinated type. It has unexcelled chemical EP activity, and the ability to adjust this activity and viscosity independently over an extremely wide range makes it the most versatile lubricant for this purpose. All chlorinated lubricants are readily removable in degreasers or solvents, and emulsifiers can be added to them for easy removal in water-based cleaners.

As shown in Table 2, pigmented pastes, sulfurized or sulfochlorinated oils, and dry wax or soap-borax films are also highly effective lubricants for press forming but are less convenient to use. Heavy-duty emulsions, because of their superior characteristics as coolants, are preferred for high-speed operations. In the following example, high chlorine content and high viscosity were needed to produce acceptable parts (see also example 15, in which a low-viscosity chlorine-base lubricant replaced a viscous mineral oil).

Example 7: Increase in Chlorine Content and Viscosity of Lubricant that Improved Results in Forming. A wheel cover was made from a type 302 stainless steel blank, 457 mm (18 in.) in diameter by 0.71 mm (0.028 in.) thick, in two operations: draw, then trim and pierce. At first, a lightly chlorinated oil (10% Cl) of medium viscosity (1500 SUS, Saybolt Universal seconds, at 40 °C, or 100 °F) was used in drawing. Even though the draw was shallow, 12% of the wheel covers were rejected for splits and scratches.

A change was made to a highly chlorinated oil (36% Cl) of much higher viscosity (4000 SUS at 40 °C, or 100 °F). As a result, the rejection rate decreased to less than 1%. After forming, the wheel covers were vapor degreased.

Combined Operations in Compound and Progressive Dies

The use of compound and progressive dies for the mass production of parts that require many operations or for an operation that is too severe to be done economically in a single-operation die is discussed in the article "Forming of Carbon Steels" in this Volume. The same principles apply to their use on stainless steel for blanking, piercing, bending, forming, drawing, coining, embossing, or combinations of these operations.

Both compound and progressive dies must be made of die materials that are hard enough to withstand the most severe demands of blanking and are tough enough for the most severe forming or coining operations. The lubricant must have enough body for the most severe draw, yet must be light enough not to interfere with the production of coined or embossed details or to gum up cutting edges. In a compound die in a double-action press, two draws can be made in stainless steel if the press capacity is not exceeded. The following example demonstrates the near-maximum severity of forming that can be achieved in a blank-and-draw compound die.

Example 8: Blanking and Severe Drawing in One Operation in a Compound Die. The shell illustrated in Table 5 was blanked and drawn in a severe forming operation in a compound die at the rate of 16,000 pieces per year. The die was used in a 400 kN (45 tonf) mechanical press with an air cushion. The formed piece was restruck in the same die to sharpen the draw radius and to flatten the flange within 0.15 mm (0.006 in.). The die was made of A2 tool steel and had a life of 50,000 pieces per grind. An emulsified chlorinated concentrate was used as a lubricant.

After forming, the piece was then moved to a 200 kN (22 tonf) mechanical press, in which the 2.4 mm (0.093 in.) diameter hole was pierced and the flange was trimmed to an oval shape. A second piercing operation, in a horn die in a 200 kN (22 tonf) mechanical press, pierced two 1.6 mm (0.062 in.) diameter holes in the side of the shell. Air ejection was used in all operations except the final piercing, where the piece was picked off.

The material was type 302 stainless steel, 0.94 mm (0.037 in.) thick and 57 mm (2¼ in.) wide, which had been annealed. Table 5 lists the production rate and labor time for each of the four operations.

Small, complex parts that must be made in large quantities are well suited to production in progressive or transfer dies. A transfer die uses a minimum of material and can accept coil stock, loose blanks, or partially formed parts. Scrap removal problems are lessened. A progressive die is preferred when the piece can remain attached to the strip.

In the following example, piercing, blanking, and forming were combined in a seven-stage progressive die. Although progressive, it was hand fed—a rather unusual combination.

Example 9: Producing a Small Bracket in a Progressive Die with Hand Feeding. The small bracket shown in Fig. 14 was made in a seven-station progressive die from 9.5 mm (0.374 in.) wide stock that was hand fed into the 50 kN (6 tonf) press. Hand feeding was done because close operator attention was required to prevent jamming, which would have damaged the frail dies. The sequence of operations was as follows:

- Feed strip to finger stop; pierce
- Feed to notch-die opening; pierce

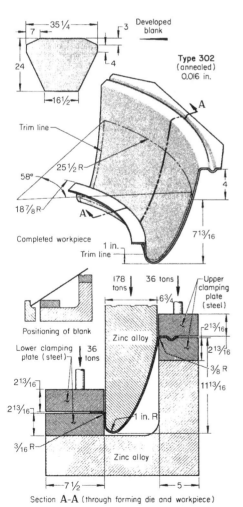

Fig. 13 Tools and clamping plates for controlling metal flow in press forming the part shown. Dimensions given in inches

Table 5 Production rates and labor time for making a severely drawn shell

Operation	Production, pieces/h	Labor/100 pieces, h
Blank and draw(a)	922	0.108
Restrike(a)	1429	0.070
Pierce center hole; trim(b)	845	0.118
Pierce side holes(c)	786	0.127

(a) In a compound die, in a 400 kN (45 tonf) mechanical press with an air cushion. (b) In a 200 kN (22 tonf) mechanical press. (c) In a horn die in a 200 kN (22 tonf) mechanical press

- Notch and trim lugs
- Form lugs
- Form and cut off
- Unload by blast of air

The parts were barrel finished to remove burrs and to provide a smooth finish and high luster. The production rate was 2175 pieces per hour. Annual production was 2 million pieces. A chlorinated and inhibited oil was used as a lubricant.

Progressive Dies versus Simple Dies. There is often a choice as to whether a stainless steel piece is to be made in a progressive die or in a series of single-operation (simple) dies. Deep forming usually presents difficulties in designing and constructing efficient and long-life progressive dies. The cost and delay involved in developing progressive tooling was justified for producing the frame described in the following example in quantities of 100,000 or more per year.

Example 10: Use of Progressive Dies for High-Quantity Production of Frames. The frame shown in Fig. 15 was made of 0.56 mm (0.022 in.) thick type 430 stainless steel coil

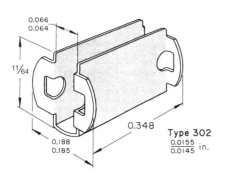

Fig. 14 Bracket that was made by hand feeding stock into a progressive die to avoid the jamming that would have been likely from automatic feeding. Dimensions given in inches

stock, 95.3 mm (3¾ in.) wide. The maximum hardness was 83 HRB.

A nine-station progressive die was used to pierce and flange the holes, to emboss the stiffening beads on the two legs, to trim and form the tabs, to coin identification data, and to blank the part from the strip. Stops were then lanced and formed, and bottom flanges were formed in a forming die. A final forming die was used for the deep side flanges.

The progressive die was run in a 670 kN (75 tonf) mechanical press at a rate of 5000 pieces per hour. The first and second forming dies were run in a 270 kN (30 tonf) press at speeds of 984 and 936 pieces per hour, respectively.

Annual production was 90,000 frames, and demand was expected to increase. This, in addition to the short press time (0.2284 h per 100 pieces, as against an estimated 0.6665 h per 100 pieces if produced in eight separate dies) and the greater accuracy obtainable in the progressive die, justified the higher tooling cost for the progressive-die method (60% higher when compared to separate dies).

The dies were made of A2 tool steel and had a life of 50,000 to 75,000 pieces between regrinds. The lubricant was an emulsifiable chlorinated oil concentrate.

Multiple-Slide Forming

Small high-production stainless steel parts can sometimes be formed in multiple-slide machines with the same kinds of tools as those used for the forming of low-carbon steel. Additional information is available in the article "Forming of Steel Strip in Multiple-Slide Machines" in this Volume. The following example describes the forming of a link for a flexible expanding wristband.

Example 11: Multiple-Slide Forming of a Wristband Link. The workpiece shown in Fig. 16, a link for an expanding wristband, was formed in a multiple-slide machine from stainless steel strip 0.25 mm (0.010 in.) thick by 8.99 mm (0.354 in.) wide, and it was locked in shape by bent lugs. The production rate was 6000 pieces per hour.

The blank for the link was made in a five-station progressive die mounted in the press station of the machine. As shown in the upper right corner of Fig. 16, the strip was notched in stations 1, 2, and 3 by four small heeled punches. For support against side thrust, the heels entered the die before engaging the stock. An air blast entering through holes in the punches removed the scrap in order to protect the die and the feed mechanisms. In the fifth die station, two lugs on the blank were bent 60°. Spring-actuated lifters stripped the blank from the bending section of the die after the lugs were bent. The workpieces were held together by a narrow strip of stock that was left to index the workpiece through the stations of the progressive die.

The blank was then fed to the forming station so that it was edge-up between the center post (7) and the front tool (6), as shown in position A in Fig. 16. As the blank entered the forming station, the center post moved upward into the forming position. The shear blade (1) then moved forward against the fixed die (2) to trim off the joining strip. The shear blade (1) also bent the end of the blank against the auxiliary rear tool (3), which then retracted. The other end of the blank was cut off by the shear blade (4) against the die (5).

After the blank was cut off by the shear blade (4), the front tool (6) bent the workpiece around the center post (7), as shown in position B in Fig. 16.

In position C in Fig. 16, the workpiece was formed on the center post by the side tools (8 and 9) while still being held by the front tool (6). The front tool was wide enough to form the full width of the workpiece, including the lugs, but the side

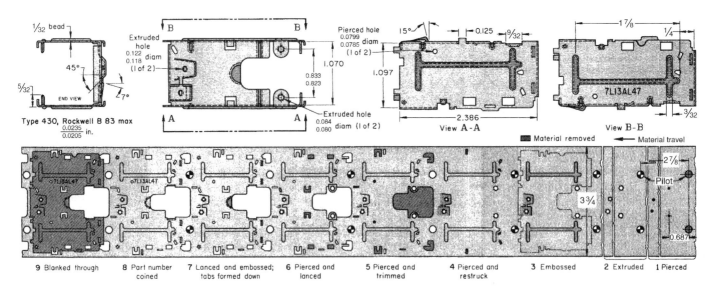

Fig. 15 Frame produced in a nine-station progressive die in the sequence of operations indicated on the strip development shown. Final forming was done in separate dies. Dimensions given in inches

tools (8 and 9) were narrower, leaving exposed the top and bottom lugs that had been formed in the last press-die station.

In position D in Fig. 16, the front and side tools (6, 8, and 9) held the part against the center post (7), while the rear tools (11) flattened the top and bottom lugs against the center post. The center post was then lowered from the workpiece. The side tool (9), which was spring loaded, slid between the top and bottom tools (10), permitting them to advance to form the top and bottom lugs into a U-shape. The side tool (9) held the workpiece against the front and side tools (6 and 8), while the top tools (10) tucked in the lugs.

With all the other tools holding the closed position against the workpiece, the rear tools (11) moved slightly to press the lugs closed against the top tools (10). As the tools opened, the completed link was then ejected by an air jet.

Deep Drawing

The percentages of reduction obtainable in deep drawing range from 40 to 60% for the chromium-nickel (austenitic) stainless steels of best drawability and from 40 to 55% for the straight-chromium (ferritic) grades (percentage of reduction = $[(D-d)/D] \times 100$, where D is the diameter of the blank, and d is the inside diameter of the drawn piece). The amount of reduction obtainable varies greatly with the radius of the die and, to a lesser extent, with the radius of the punch nose. As the die radius decreases, the drawability decreases, as shown in Table 6 for austenitic stainless steel. Typically

used punch and die radii are five to ten times metal thickness. With the ferritic grades, the drawability and ductility usually decrease with increasing chromium content. To offset this, steels with high chromium content are often warmed moderately before drawing.

Presses used for the deep drawing of stainless steel differ only in power and rigidity from those used for low-carbon steel. Because of the higher work-hardening rate of stainless steel and its inherent higher strength, presses used for the deep drawing of stainless steel often need 100% more ram force and the necessary frame stiffness to support this greater force.

Dies for drawing stainless steel must be able to withstand the high force and resist galling. For ordinary service, D2 tool steel dies give a good combination of hardness and toughness. On long runs, carbide draw rings have exceptionally long life. Where friction and galling are the principal problems, draw rings are sometimes made of high-strength aluminum bronze. The following example describes an application in which the selection of tool material was critical in order to avoid scoring of the workpiece and to obtain acceptable die life in drawing.

Example 12: Use of a Carbide Blank-and-Draw Ring. An orifice cup, 25 mm (1 in.) in diameter by 11 mm ($^{7}/_{16}$ in.) deep, was blanked and drawn in one operation. A 1.35 mm (0.053 in.) diameter orifice was pierced in the cup in a second operation. The specifications called for the sides of the cup to be free of score marks from the die. The blank was 40.0 mm (1.575 in.) in diameter, cut from 0.97 mm (0.038 in.) thick type 302 stainless steel strip 50 mm (2 in.) wide.

The blank-and-draw tooling shown in Fig. 17 was originally made of tool steel of a grade no longer used. It produced fewer than 50 pieces without scoring the workpieces. The combination blanking punch and draw ring was chromium plated in an attempt to increase its durability. Adhesion of the plating was not satisfactory; the chromium started to peel after 180 pieces had been produced. A draw ring of graphitic tool steel was then tried, but this also scored the workpieces.

Finally, a new draw ring was made of sintered carbide consisting of 81% tungsten carbide, 15%Co, and 4%Ta—a composition especially recommended for draw dies. The new ring, used with a chlorinated oil-based lubricant, withstood the heat and pressure generated by the severe blank-and-draw operation and produced mar-free parts. Maintenance was negligible, and after 3 years the carbide draw ring had produced 180,000 pieces, with little evidence of wear.

Table 6   Effect of die radius on percentage of reduction obtainable in the deep drawing of austenitic stainless steel

Percentage of reduction = $[(D-d)/D] \times 100$, where D is the diameter of blank, and d is the inside diameter of the drawn piece

Die radius(a)	Reduction in drawing, %
15t	50–60
10t	40–50
5t	30–40
2t	0–10

(a) t, stock thickness

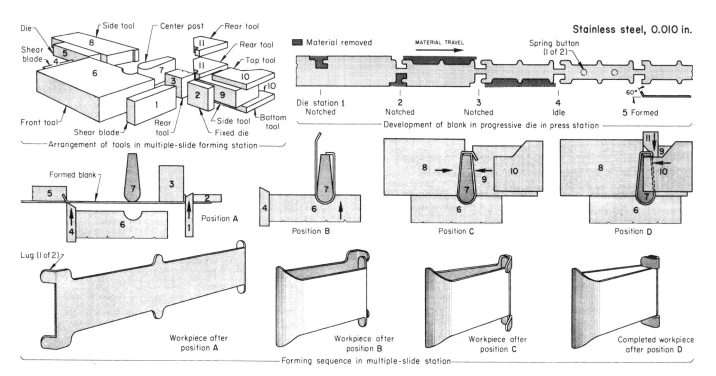

Fig. 16   Operations in the forming of a wristband link in a multiple-slide machine

The blanking punch-to-die clearance was 0.05 mm (0.002 in.) per side. Drawing punch-to-die clearance was 0.851 mm (0.0335 in.) plus 3° taper per side on the draw punch. The punch nose radius was 0.38 mm (0.015 in.), and the draw radius was 2.4 mm (0.093 in.).

Effect of Worn Draw Rings. The following example shows how the gradual wear of carbide draw rings in severe drawing affected the outside diameter of drawn shells.

Example 13: Effect of Wear of a Carbide Draw Ring on the Diameter of a Drawn Shell. The carbide draw ring used in deep drawing a shell for pens and pencils made more than 225,000 pieces before it was replaced. Measurements of the pieces were made at production intervals, as shown in Fig. 18.

Shortly after 225,000 pieces had been drawn, shells began to be produced that would no longer enter the "go" ring gage freely, because of wear on the draw ring. The worn draw ring, which permitted excessive springback, was replaced before the beginning of the next production run.

The shell was drawn from a blank of type 317 stainless steel 48.4 mm (1.906 in.) in diameter and 0.18 mm (0.007 in.) thick to a finished depth of 69.0 mm (2.718 in.) using chromium-plated punches. The shell was made in eight single-station dies, seven drawing and one end forming, at a rate of 600 per hour. The punches had a 2.29 mm (0.090 in.) nose radius, and the draw dies had a 90° conical entrance angle with a 1.52 mm (0.060 in.) radius blending the corners. A mixture of three parts inhibited hydraulic oil

and one part chlorinated oil was used as lubricant.

Die clearance for heavy draws is 35 to 40% greater than the original metal thickness for austenitic alloys. For the ferritic alloys, which thicken less, 10 to 15% is generally adequate.

Figure 19 shows a profile of an austenitic stainless steel drawn part that illustrates the thickening pattern observed in drawing a cup from this material. If the process is one of stretching more than of drawing, the clearances do not have to compensate for natural thickening.

Clearances of less than the metal thickness are generally not used with stainless steel, because they result in ironing (squeezing of the metal between the punch and die). The austenitic stainless steels are not suited to ironing, because their high rate of work hardening promotes scoring and rapid wear of the dies. In addition, any substantial ironing in the drawing of austenitic stainless steels greatly increases the likelihood of fracturing the workpiece.

The subsequent example describes an application in which the work metal was changed from galvanized carbon steel to a thinner ferritic stainless steel without a revision of die clearance. The resultant problems were solved by substituting an austenitic stainless steel that was better suited to the original clearance even though it had the same thickness as the ferritic steel.

Example 14: Matching Work Metal to Die Clearance. Using the tooling shown in Fig. 20, basins were made at the rate of 10,000 to 15,000 pieces per year from galvanized carbon steel, 1.27 mm (0.050 in.) thick. The press was an 8900 kN (1000 tonf) hydraulic press with an air-over-oil pressure pad and a draw rate of 152 mm (6 in.) in 5 s. The punch, draw ring, and pressure pad for the drawing die (Fig. 20a) were hardened cast iron. Carbide inserts were used as cutting edges on the trimming punch and die (Fig. 20b). The locator on the trimming die was molded plastic, and the die plate was cast iron. Both dies were used side-by-side in the press at the same time because it had enough capacity to draw and trim in one stroke. Therefore, a finished piece was produced with each stroke of the press, using manual transfer.

To produce a more corrosion-resistant basin, type 430 stainless steel was substituted for the galvanized carbon steel. The type 430 was only 0.79 mm (0.031 in.) thick in order to minimize the increase in material costs; however, the same tooling was used because the relatively low annual quantity did not warrant the cost of retooling. Because the holddown forces were not suitable for the ferritic stainless steel, several hundred pieces out of the first run were fractured in drawing.

When the holddown pressure was adjusted to a level suitable for a ferritic stainless steel, contraction wrinkles formed where the material entered the throat of the die (Fig. 20c) because

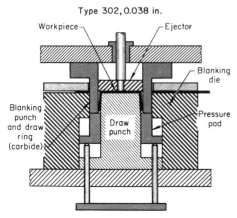

Type 302, 0.038 in.

Fig. 17 Forming an orifice cup in a blank-and-draw die with a carbide punch and draw ring. Orifice was pierced in a second operation. Annual production was 60,000 pieces. Rate of blanking and drawing was 670 pieces per hour. Rate of piercing was 153 pieces per hour. Dimensions given in inches

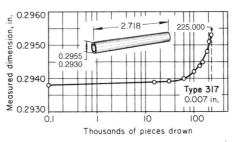

Fig. 18 Variation in diameter of a deep-drawn shell that resulted from wear of the carbide draw ring used. Dimensions given in inches

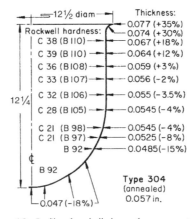

Fig. 19 Profile of a shell drawn from an austenitic stainless steel showing variations in hardness and thickness produced by drawing. Dimensions given in inches

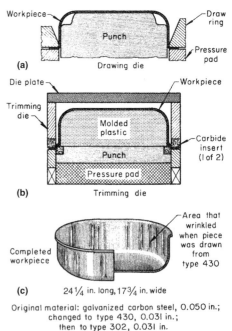

Original material: galvanized carbon steel, 0.050 in.; changed to type 430, 0.031 in.; then to type 302, 0.031 in.

Fig. 20 Setups for drawing and trimming a basin. Die clearance (1.40 mm, or 0.055 in., per side) and drawing radius (6.4 mm, or 1/4 in.) were not changed when 0.79 mm (0.031 in.) thick type 430 was substituted for 1.27 mm (0.050 in.) thick galvanized carbon steel as the work metal, and wrinkles resulted in drawing. Dimensions given in inches

Table 7 Effect of die clearance and draw ring radius on noncylindrical draws of stainless steel

Stock thickness, t		Die clearance per side					
		Carbon steel		Type 430		Types 302 and 304	
mm	in.	mm	in.	mm	in.	mm	in.
1.27	0.050	1.40	0.055	1.40	0.055	2.29	0.090
0.76	0.030	0.84	0.033	0.84	0.033	1.37	0.054

Stock thickness, t		Draw ring radius				
		Carbon steel		Type 430		
mm	in.	mm	in.	mm	in.	Types 302 and 304
1.27	0.050	6.4–9.5	$^1/_4$–$^3/_8$	6.4–9.5	$^1/_4$–$^3/_8$	4t min
0.76	0.030	4.8–7.9	$^3/_{16}$–$^5/_{16}$	4.8–7.9	$^3/_{16}$–$^5/_{16}$	4t min

the die clearance was too great. The corners of the blank were cropped, the viscosity of the lubricant was changed, and the holddown pressures were more closely adjusted in an effort to control wrinkling.

The data given in Table 7 for noncylindrical draws provide an explanation for the difficulties encountered in changing the work metal as well as guidance for the selection of a suitable stainless steel. According to these data, the thickness of type 430 stock that could best be formed by the die would be the same as that of the carbon steel previously formed. In addition, if the stock thickness were reduced, an austenitic steel such as type 302 could be used. The die clearance was 1.4 mm (0.055 in.) per side, and the draw ring radius was 6.4 mm (0.25 in.). Therefore, the die was suited for the 1.27 mm (0.050 in.) thick carbon steel but not for the 0.79 mm (0.031 in.) thick type 430. However, 0.79 mm (0.031 in.) thick type 302 would be closely matched to the die capacity.

A change was made to type 302 stainless steel, 0.79 mm (0.031 in.) thick, with no further difficulty. A change to 1.27 mm (0.050 in.) thick stock of type 430 may have been successful.

Speed of drawing has an important bearing on the success of the draw. A rate of 6 to 7.5 m (20 to 25 ft) per minute is a good compromise between the rate of work hardening and the uniform distribution of stress. With proper forming techniques, the rate of fracture at this speed is often less than 2%.

Lubricants. Ordinarily, both sides of the workpiece need to be lubricated for each draw. If too little lubricant is used, tools may accumulate enough heat during a production run to cause the work metal to fracture because of galling. In tests with a minimum of lubricant, failures occurred after 25 draws.

The chemical type and the viscosity of the lubricant are both important. Either chemical or mechanical EP activity (see the article "Selection and Use of Lubricants in Forming of Sheet Metal" in this Volume) is needed for the severe deep drawing of stainless steel.

Viscosity or pigment loading must not be too high or too low. Too thick a lubricant can cause wrinkling of compressed metal; too thin, seizing or galling. The ability to remove a lubricant readily is also important. In general, the higher

the viscosity, the more difficult the lubricant is to apply and remove.

The same characteristics that make chlorinated oils and waxes useful for the press forming of stainless steel (see the section "Press-Brake Forming" in this article) also make them useful for the deep drawing of these alloys. Table 2 lists other lubricants used in the deep drawing of stainless steel. Pigmented pastes and dry-films are also effective (and, in some cases, superior) in deep drawing.

In Example 15, changing from a viscous mineral oil to a low-viscosity mineral oil blend of a chlorinated wax eliminated wrinkling and galling. Sometimes, however, there is no substitute for the physical separation and equalization of pressure provided by pigments, as in Example 16.

Example 15: Effect of Reducing Viscosity and Adding Chlorinated Wax to Mineral Oil Lubricant in Deep Drawing. A coffeepot was deep drawn from a type 302 stainless steel blank, 355 mm (14 in.) in diameter by 0.81 mm (0.032 in.) thick, in two deep draws and one bulging operation. At first, the blanks were lubricated by brushing both sides with mineral oil having a viscosity of 6000 SUS at 40 °C (100 °F). The workpiece wrinkled in the first draw, and it galled in the second draw and in bulging.

The lubricant was replaced with a thinner mineral oil (viscosity: 500 SUS at 40 °C, or 100 °F) that was fortified with a chlorinated wax. The lubricant was brushed on, as before. Not only did the use of the modified lubricant eliminate the wrinkles in the first draw, but enough lubricant remained on the surface to prevent galling in the two other operations. Even though a fluid of much lower viscosity was used, the tenacity imparted by the chlorinated wax permitted the retention of sufficient lubricant for the subsequent bulging and deep-drawing operations.

Example 16: Pigmented Paste versus Chlorinated Oil for Deep Drawing. For easy cleaning in a vapor degreaser, highly fortified oils were specified for the deep drawing of a rectangular shell from 0.89 mm (0.035 in.) thick type 304 stainless steel. Chlorinated and sulfochlorinated oils with viscosities of 4000 to 20,000 SUS at 40 °C (100 °F) failed to eliminate

welding to the dies and splitting of the workpiece at the corners. The shell, a well for a steam table, was deep drawn from a rectangular blank measuring 760 by 585 mm (30 by 23 in.) with corners trimmed at 45°. The shell was drawn in one operation, and the flange was then trimmed. Interior dimensions of the drawn shell were 510 by 305 by 150 mm (20 by 12 by 6 in.). Bottom corners had 16 mm (0.63 in.) radii; vertical corners, 29 mm (1.14 in.) radii; and the flange, a 6.4 mm (0.25 in.) radius. The shell had approximately 3° taper on each side. The clearance between the punch and die was equal to the stock thickness.

The oil-type lubricant was replaced with a highly pigmented water-miscible fatty paste, diluted with two parts of water, which was applied to both sides of each blank by rollers. This lubricant eliminated the welding and allowed enough metal flow to prevent splitting. The drawn parts were cleaned with hot alkaline solution in a soak tank.

Lubricant Location. The location of the lubricant on the blank is also critical in the successful fabrication of a drawn part. Because all draws are made up of a combination of stretching and deep drawing, the lubricant location often depends on which type of forming is dominant. In a stretch condition, lubricant should especially be applied on the steel surface contacting the punch so that friction is minimized and the steel slips over the punch surface during stretching and thinning. Under deep-draw conditions, the steel surface contacting the die is definitely lubricated in order to allow ease of movement into the die cavity. However, whether stretch or deep drawing dominates, some lubricant is necessary on both steel surfaces to minimize the galling tendencies of stainless alloys.

Drawing Cylindrical Parts. When a part is made in several drawing operations, the amount of reduction in redrawing is related to the condition of the metal in the first drawing operation (cupping). If the material is highly stressed because of excessive blankholder pressure or because of small die radius, very little reduction can be made in the second operation.

General practice on the more formable grades of austenitic stainless steel is to allow 40 to 45% reduction in the first operation, followed by a maximum of 30% in the second operation, if the workpiece is not annealed between draws. With an anneal, the second reduction is usually 30 to 40%. On some parts, it may be preferable to spread the reduction over four draws before annealing—for example, successive reductions of 35, 30, 20, and 10%.

There is usually a decrease in drawability upon redrawing, and the greatest total reduction in a two-draw operation is most often produced by having the first-stage reduction as large as possible. During redrawing, it is advisable to use a tapered or rounded-end internal blankholder or sleeve to allow easy flow of metal into the die, as indicated in the article "Deep Drawing" in this Volume. An internal blankholder with small-radius 90° corners causes the metal to be bent severely through two 90° bends before flowing into the die.

Optimal drawability is available at ram speeds of not more than 6 to 9 m (20 to 30 ft) per minute. Because of the strain-rate sensitivity of most stainless steels, work hardening of these alloys is minimized by slow forming.

The following example describes an application in which small shells were deep drawn in several steps to reduce the amount of work done in a single operation. Because production quantities were small, individual dies were more economical than a transfer die.

Example 17: Seven-Step Deep Drawing of a Fountain-Pen Cap. Fountain-pen caps of various closely related designs were made on the same production line by one blanking and cupping operation and six redraws. A flat blank of type 302 stainless steel having a hardness of 83 to 88 HR15-T was used. The first five draws were usually the same for any of the caps made on the line; therefore, to set up for a different size of cap, only the compound blank-and-cup die and the last die (or, for some caps, the last two dies) needed to be changed. As a result, the changeover time was only approximately 45 minutes.

In the first operation, which was done in a 160 kN (18 tonf) mechanical press, a compound blank-and-cup die equipped with a rubber die cushion was used to cut circular blanks from 0.267 to 0.279 mm (0.0105 to 0.0110 in.) thick strip and to draw them into a cup. To make a typical cap 90 mm (3½ in.) long by 8.55 to 8.57 mm (0.3365 to 0.3375 in.) in outside diameter, a blank 55.9 mm (2.200 in.) in diameter was cut from stock 57 mm (2¼ in.) wide and was drawn into a cup 19 mm (0.75 in.) deep by 31.8 mm (1.250 in.) in diameter (a 43% reduction in diameter). Reductions in the subsequent redraws were 27, 22, 18, 18, 16, and 15%, respectively. All except the last redraw were done in 35 kN (4 tonf) hydraulic presses with 152 mm (6 in.) strokes. The final redraw was made in a 55 kN (6 tonf) hydraulic press with a 305 mm (12 in.) stroke.

The draw dies were carbide inserts 13 to 16 mm (0.51 to 0.63 in.) thick. The die openings had a 4.8 mm (0.19 in.) radius blending with a 1.6 mm (0.06 in.) wide land. There was a 2° relief per side below the land. The high-speed steel punches had a 2.4 mm (0.094 in.) nose radius and were chromium plated for smoothness and wear characteristics. The workpiece was pushed through the die and stripped from the punch by a split stripper plate under the draw die. The strippers were closed by cam action from the press stripper rod.

Because production quantities of any one part were small, this technique was preferable to making a transfer die for each of the several caps produced on this line. Operations were set up in machines in the line as they were needed and as the machines became available.

The final draw, which was the deepest, governed the final production rate of 575 pieces per hour. However, when there was a backlog of pieces, this operation was set up on two machines at the same time.

The blank-and-cup die made approximately 45,000 pieces before resharpening. The draw rings were used for 150,000 to 200,000 pieces before wear was too great. Dies in the first few draws were allowed to wear over a fairly wide range. As the die opening increased, clearance was maintained by increasing the thickness of the chromium plating on the punch. When the die openings were 0.10 to 0.13 mm (0.004 to 0.005 in.) oversized, the dies were replaced, and punches were returned to the original size by stripping, polishing, and replating.

The lubricant was a mixture of one part sulfur-free chlorinated oil with three parts inhibited hydraulic oil having a viscosity of 250 SUS at 40 °C (100 °F). This lubricant was furnished to all presses through a central pumping system.

Critical tolerances on these fountain-pen caps were ±0.02 mm (±0.0008 in.) on outside diameter and ±0.01 mm (±0.0004 in.) on inside diameter. Holding the clearance between the draw die and punch to 10% greater than stock thickness helped to maintain these tolerances.

Steel Drawing Forces. Estimates of the maximum drawing forces necessary to form cups from an austenitic stainless steel, a ferritic stainless steel, and low-carbon steel are compared in Table 8. These drawing forces, in tons of force, are based on the formula $S\pi Dt$, where S is the tensile strength of the metal in tons of force per square inch, D is the cup diameter in inches, and t is the metal thickness in inches.

Blankholding pressures for the austenitic alloys must be much higher than those for the ferritic types or low-carbon steels. For austenitic alloys, the pressure, P, on the metal under the blankholder is usually approximately 6.9 MPa (1.0 ksi); for the ferritic alloys, 1.4 to 3.4 MPa (0.2 to 0.5 ksi). Thinner material and larger flange areas generally require greater pressure.

Drawing Hemispherical Parts. The drawing of hemispherical, or dome-shaped, parts demands special attention to blankholder pressure to prevent wrinkling because so much of the metal surface is not in contact with any die surface for most of the draw. Only the very tip of the punch is in contact with the work at the start of the stroke, and the surface between the tip and the blankholder draws or stretches free until the punch descends far enough to contact it. An undersized punch can sometimes be used to draw or stretch the blank into a preform before the dome-shaped punch makes the final draw, as in the following example.

Example 18: Two-Stage Drawing of a Stepped-Diameter Hemisphere. One of the critical points in the production of the vacuum-bottle top shown in Fig. 21 was the forming of the shoulder at the large end of the dome-shaped top. The stepped inside diameter of this shoulder had to be an exact fit with the body of the vacuum-bottle jacket. The pierced hole at the small end also had to be accurately formed to conform to the mouth of the inner container.

The stock was 289 mm (11.375 in.) wide annealed type 304 stainless steel strip, 1.1 mm (0.042 in.) thick. A single-action mechanical press with a spring-loaded pressure pad was used to cut 280 mm (11⅛ in.) diameter blanks from the strip, leaving 3.2 mm (0.13 in.) minimum scrap on each side of the strip.

The first draw was made in a 2200 kN (250 tonf) double-action mechanical press. The punch was 83 mm (3¼ in.) in diameter; therefore, much of the surface of the dome was drawn free (Fig. 21, operation 2). This required careful control of the blankholder pressure to prevent puckers and wrinkles. Blankholder pressure had to be adjusted for every lot of steel; it varied from

Table 8 Force required for drawing two stainless steels and low-carbon steel of 1.27 mm (0.050 in.) thickness to various diameters

Diameter of piece		Approximate drawing force required					
		Austenitic stainless steel, type 18-8		Ferritic stainless steel, 17% Cr		Low-carbon steel	
mm	in.	kN	tonf	kN	tonf	kN	tonf
125	5	350	39	180	20	160	18
255	10	700	78	520	59	350	39
510	20	1400	157	1040	117	700	78

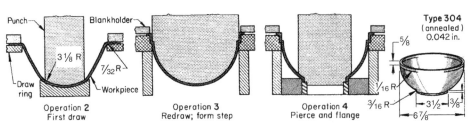

Fig. 21 Production of a stepped-diameter flanged hemisphere, in which a narrow punch was used in predrawing the dome. The piece was drawn from a 280 mm (11⅛ in.) diameter blank produced in operation 1 (not shown). Dimensions given in inches

5.5 to 6.9 MPa (0.8 to 1.0 ksi). The die radius also had to be held closely (5.2 times the stock thickness). The first draw produced a cup 175 mm (6$^7/_8$ in.) in diameter with a 235 mm (9$^1/_4$ in.) diameter flange.

The second draw was also made in the 2200 kN (250 tonf) double-action press. The punch for the second draw was shaped to the required inner contour of the part, including the step at the base of the dome, which was formed as the press bottomed at the end of the second draw stroke (Fig. 21, operation 3). This operation formed the dome shape of the bottle top by reshaping (mostly by stretching) the cup formed in the first draw. The metal for the cylindrical area above the step was drawn from the flange metal remaining after the first draw.

In the fourth operation, the hole in the top of the dome was pierced, and an internal stretch flange was formed around the hole. This was done with a spring-loaded piercing die, which gave sufficient resistance to let the piercing punch shear the material and then retreat under pressure from the flange-forming part of the punch. Both ends of the part were later trimmed in a lathe.

Drawing Rectangular Parts. During the deep drawing of a box-shaped part, the metal in the corners of the part and in the flange around the corner undergoes a change much like that which takes place when a round shell is drawn from a circular blank. Metal is compressed at the corners, and significant thickening occurs where the metal flows into the corners. The sides of the box undergo essentially no thickening, because there is no compression of the metal in the flange areas as it flows or bends over the die radius.

Clearances in the sides between the punch and die are ordinarily approximately 10% greater than the metal thickness to compensate for gage variations and to allow for metal flow. At the corners, punch-to-die clearances are similar to those used for cylindrical parts to allow for thickening.

Blankholding devices are almost always used in producing deeply recessed box-shaped parts in order to control the metal movement, particularly in the corners. The corners are under severe strain because of the intense compression of the flange metal, and most fracturing, if it does occur, takes place in the lower wall corner sections.

Punch and die radii are generally the same for rectangular draws as for circular draws. Some fabricators prefer to make the punch and die radii at the corners larger than along the sides in order to equalize the stress in the metal at the corners. The top surface of the draw die and the draw radii should be polished smooth (free of grind marks and well blended) to prevent localized retardation of metal flow with resultant uneven drawing of the metal. Burrs and bent edges on the blank often restrict metal flow or movement along the blankholder surface to such a degree that vertical wall fractures can occur.

Semideveloped blanks usually produce better results than rectangular ones. There are a number of patterns for trimming the corners, ranging from a simple 45° trim to patterns with a carefully developed area containing the optimal volume and area of metal.

The economic success of the run is related to tool wear and scrap rate. The following example describes a combination of tool materials that has given satisfactory performance in terms of parts or draws per regrind and redress. The same tooling can be used for both drawing operations, with the draw ring reversed to present a different radius for the second draw, as in the following example.

Example 19: Use of a Reversible Two-Radius Draw Ring for Drawing and Redrawing of a Flanged Rectangular Shell. The flat-flanged single-sump kitchen sink shown in Fig. 22 was formed in four operations: blank, draw, redraw, and trim. Forming of the part was a combined draw-and-stretch operation. Because several different models, with drain holes in various locations, were made from the same drawn part, the drain hole was not pierced in the trimming operation but was made separately. The production rate was 50,000 to 100,000 pieces per year.

The material was annealed type 302 or 304 stainless steel coil stock 735 mm (29 in.) wide and 1.27 mm (0.050 in.) thick, with a No. 2D sheet finish. Blanks 635 mm (25 in.) long were sheared from the coil at the rate of 40 per minute in a single-action mechanical press. Corners of the blanks were trimmed at 45°, removing 50 mm (2 in.) from each edge of the blank at each corner. Clearance for this trimming was kept at less than 5% of metal thickness to minimize edge distortion and burrs.

The draws were made in a 3600 kN (400 tonf) double-action mechanical press with 2200 kN (250 tonf) available for blank holding. The draw punch was made of alloy tool steel, and the blankholder was made of alloy cast iron. The reversible draw ring (Fig. 22) was made of hard aluminum bronze and had a 19 mm (0.75 in.) draw radius on one side for the first draw and a 13 mm (0.51 in.) draw radius on the other side for the redraw. The workpiece was annealed in an inert atmosphere at 1065 °C (1950 °F) between the first and second draws and then air cooled rapidly to room temperature.

The depth of the sink after the first draw was 127 mm (5 in.); after the second draw it was 170 mm (6$^3/_4$ in.). Draws were made at a punch speed of approximately 6.4 m (21 ft) per minute, with less than 2% of the workpieces fracturing.

A similar 3600 kN (400 tonf) press was used to trim the piece. Carbide inserts provided shearing edges for the trimming operation. The sink was held on a form block of molded plastic or cast iron for trimming.

The second draw operation sharpened the bottom and flange corner radii and stretched the bottom surface and the side walls to remove any loose metal. Little or no metal was drawn into the part from the flange during the second draw.

Spinning

Stainless steel parts such as cups, cones, and dished heads can be readily formed by manual or power spinning, although more power is required than that needed for the spinning of low-carbon steel. Equipment and techniques for these processes are described in the article "Spinning" in this Volume.

Manual Spinning. The amount of thinning that occurs during manual spinning is related to the severity of the formed shape. A cross section is shown in Fig. 23 of a manually spun piece that thinned out to such a degree that it often fractured. This piece was excessively worked, and the midcenter area was work hardened beyond the capacity of the material, causing the workpiece to fracture. The piece was later made by press drawing the dome-shaped cup and spinning the broad flared flange.

The approximate limits of stretch in manual spinning are given in Table 9. These are for 1.57 mm (0.062 in.) thick fully annealed stock. The second stretch after annealing is approximately 8% less than the first. The amount of stretch is not necessarily uniform over the entire part; it varies with the severity of the form.

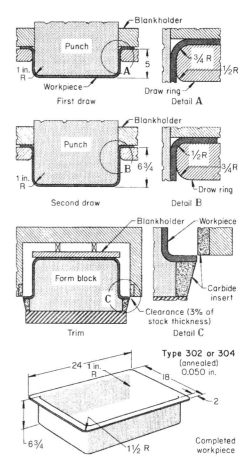

Fig. 22 Production of a flat-flanged sink basin by drawing and redrawing (using a two-radius reversible draw ring) and trimming. Dimensions given in inches

Although 300-series stainless steels can be formed by spinning, 302, 304, and 305 can be spun to greater reductions than other stainless steels before intermediate annealing becomes necessary. All anneals must be followed by pickling to remove oxides, thus restoring the clean, smooth surface. The No. 1 strip or 2D sheet finish is best for severe applications because the metal is in the softest stress-free condition and will take the greatest amount of working. The following example demonstrates the spinability of type 305 stainless steel.

Example 20: Four-Pass Manual Spinning of a Cone from Type 305 Stainless Steel. The 355 mm (14 in.) diameter cone shown in Fig. 24 was produced in eight operations, including four manual spinning passes, from a 405 mm (16 in.) diameter blank of 0.94 mm (0.037 in.) thick annealed type 305 stainless steel that had a No. 2D sheet finish or a No. 1 strip finish. Other types of austenitic stainless steel could have been used, but the reduction per pass would have been lower, in proportion to the increase in rate of work hardening.

As shown in Fig. 24, the mandrels for spinning were made of wood or steel. The spinning roller was made of hardened steel. Pressure was applied to the entire blank in the first spinning pass. In the three other passes, the outer 25 mm (1 in.) of the blank was not spun. This caused the edge to thicken to 1.78 mm (0.070 in.) and helped hold the outer shape. Thinning was greatest at the middle of the cone, to approximately 0.69 mm (0.027 in.) wall thickness (28% reduction). The surface area of the piece was increased 40%. The drastic working that accompanied the thinning and the increase in area made two anneals necessary (see sequence of operations, Fig. 24). Annual production quantity was 500 pieces.

The 400-series stainless steels, because of their relatively low ductility, do not adapt readily to manual spinning, especially when the deformation is severe. The high pressure of the forming tool causes wear of the work metal, resulting in early thinning and fracturing. Figure 25 shows forming speeds used for manual spinning of 400-series stainless steels.

The surface of severely spun parts is often very rough because of the action of the tools on the metal, and the production of a buffed or highly polished finish on a part spun from a 400-series stainless steel can be expensive. It is generally necessary to rough grind the material to smooth out the irregularities before polishing and buffing.

Typical stock thicknesses of stainless steel for manual spinning are 0.30 to 3.18 mm (0.012 to 0.125 in.), although stainless steels as thin as 0.13 mm (0.005 in.) and as thick as 6.35 mm (0.250 in.) have been spun by hand. The corner radius should be at least five times the thickness of the work metal. Allowance must be made in the size and shape of mandrels for springback and for heat-induced dimensional changes.

Power spinning is used for severe reductions and for work that cannot be done by hand. Stainless steels in both the 300 and 400 series are readily formed by power spinning, but the low-work-hardening types 302 and 305 are superior. Much larger reductions of type 430 can be made by power spinning than by manual spinning.

Spinning can be done hot or cold, although the severe reduction accompanying power spinning may cause so much heat that the spinning that began cold becomes warm spinning. Hot spinning, done only above 790 °C (1450 °F), is commonly used for work 4.8 to 13 mm (0.19 to 0.51 in.) thick. The need for careful control of the temperature makes it difficult to hot spin metal that is less than 6.4 mm (0.25 in.) thick. Thicker stainless steel can be hot spun as easily as low-carbon steel.

Cracking at the edge is the main problem in the power spinning of austenitic stainless steels. The edge of the blank may need to be ground smooth to prevent cracking. A generous trim allowance is helpful so that the cracked edge can be cut off. Cracking and distortion can be prevented by keeping a narrow flange on the work. If the size of the spun piece is not correct after it cools (because of springback and heat expansion), the piece can be annealed and spun to size while it is still above 150 °C (300 °F).

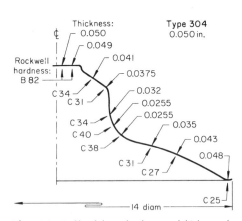

Fig. 23 Profile of shape, hardness, and thickness of a manually spun part that often fractured in its thinnest section. Dimensions given in inches

Table 9   Approximate limits of stretch in the manual spinning of stainless steels 1.57 mm (0.062 in.) thick

Type	Stretch (max), %	Type	Stretch (max), %	Type	Stretch (max), %
305	45	321	35	202	25
302	40	309	30	301	25
304	40	310	30	405	25
302B	35	317	30	446	25
316	35	430	30	403	20
316L	35	201	25	410	20

These limits are for stretching during one spinning pass, after being annealed, the metal can be respun to 8% less than the first stretch.

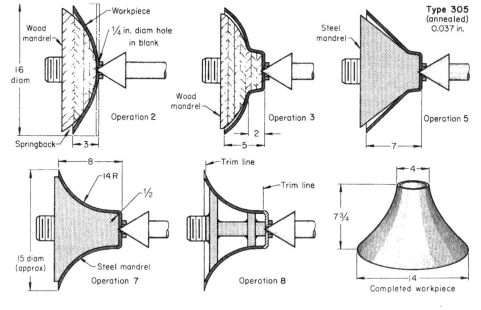

Sequence of operations
1. Drill a 6.4 mm (¼ in.) diam center hole in a 405 mm (16 in.) diam blank 0.94 mm (0.037 in.) thick.
2. Spin to 75 mm (3 in.) depth on a laminated hardwood mandrel at 300 rpm, applying manual pressure on lever and roller.
3. Spin to 125 mm (5 in.) depth on a second laminated hardwood mandrel to within 25 mm (1 in.) of edge.
4. Anneal in hydrogen atmosphere at 1040 °C (1900 °F); air cool.
5. Spin to 178 mm (7 in.) depth on a steel mandrel to within 25 mm (1 in.) of edge.
6. Anneal as in operation 4.
7. Spin to 205 mm (8 in.) depth and final shape on a steel mandrel.
8. Lathe-trim top and bottom ends to 195 mm (7¾ in.) final height of cone.

Fig. 24   Production of a stainless steel cone by four-pass manual spinning. Dimensions given in inches

Considerable thinning can be produced by power spinning, as indicated by the cross section of a deeply spun vessel shown in Fig. 26. The thickness of the vessel was reduced from 1.90 to 0.66 mm (0.075 to 0.026 in.) in one spinning operation. A preformed cup 152 mm (6 in.) in diameter and 75 mm (3 in.) deep, drawn on a conventional press, was used as the starting shape. The top of the vessel is much thicker than the wall. Thickening of the rim occurred during drawing, and it remained thick because there was essentially no deformation in this region during spinning.

The surfaces of power-spun pieces are rough, and extensive finishing is required to make them smooth and bright. The spun surface is rough because the roller usually imparts a spiral or helical groove to the surface as the roller is fed into the metal while it rotates. Except for this disadvantage, power spinning is an excellent way of forming pieces from stainless steel.

Lubricants (see Table 2) are used to reduce friction, to minimize galling and tool drag, and to provide cooling. For manual spinning, firmly adherent lubricants are preferred; for power spinning, coolant action is more important. Lubricants containing sulfur or chlorine are usually avoided; they are difficult to remove completely and have harmful effects on heated stainless surfaces.

Rubber-Pad Forming

Annealed austenitic stainless steels—types 301, 302, 304, 305, 321, and 347—are rubber-pad formed in thicknesses to 1.3 mm (0.050 in.). Most of the operations are straight flanging, especially in thicker workpieces. With auxiliary devices, such as wedges or rollers, pieces up to 2.0 mm (0.078 in.) thick can be formed. Flanges must be wide enough to develop adequate forming force from the unit pressure on their surface. For annealed stainless steels, the following minimum flange widths beyond the bend radius are recommended for successful forming:

Thickness		Flange width	
mm	in.	mm	in.
0.41	0.016	6.35	0.250
0.51	0.020	6.86	0.270
0.64	0.025	7.37	0.290
0.81	0.032	8.38	0.330
1.02	0.040	9.14	0.360
1.30	0.051	10.0	0.410
1.63	0.064	12.2	0.480
1.83	0.072	13.0	0.510

In quarter-hard temper, types 301 and 302 up to 0.81 mm (0.032 in.) thick can be flanged if the flange is at least 16 mm (0.63 in.) wide.

The rubber-pad forming of contoured flanges in stainless steel requires more powerful equipment than that used for flat flanges. Most forming of contoured flanges is done on annealed stainless steel, but a limited amount is done on quarter-hard stock.

Stretch flanges are readily formed on annealed stainless steel up to 1.3 mm (0.050 in.) thick. Rubber-pad-formed stretch flanges of thin metal are generally smoother and more accurately formed than those formed by single-action dies. Die-formed flanges often curl outward, requiring considerable hand work for correction.

The hydraulic presses used in the Guerin process develop forming pressures to 34.5 MPa (5 ksi). Narrow stretch flanges that require pressures greater than 34.5 MPa (5 ksi) are formed with the aid of auxiliary devices, such as traps and wedge blocks, that raise the forming pressure locally (see the article "Rubber-Pad Forming and Hydroforming" in this Volume).

Thin metal can be formed by means of a simple form block, but if the web is narrow, the workpiece should be protected by a cover plate to avoid distortion. The following example demonstrates the limits of rubber-pad forming of stretch flanges in stainless steel:

- The stock was quarter-hard type 302.
- The workpiece had a narrow web and therefore required the use of a cover plate in forming.

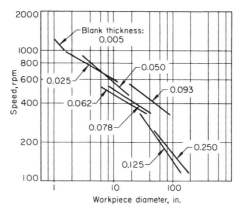

Fig. 25 Effect of workpiece diameter and blank thickness on rotational speed for the manual spinning of austenitic stainless steel. Dimensions given in inches

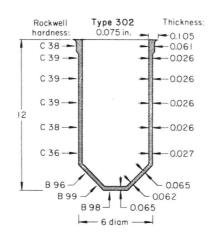

Fig. 26 Variations in hardness and thickness of a shell that was power spun from a preform drawn from stainless steel 1.9 mm (0.075 in.) thick. Dimensions given in inches

- The workpiece had external hole flanges.
- The stretch flange was only 7.9 mm (0.31 in.) wide.
- The curved workpiece was nearly 965 mm (38 in.) long.

If a stainless steel part of this shape is more than 610 mm (24 in.) long, it is almost impossible to prevent the springback of the flange material from bowing the part unless curved dies are used.

Example 21: Use of a Curved Die with Cover Plates in Rubber-Pad Forming. The strut shown in Fig. 27 had a 7.9 mm (0.31 in.) wide stretch flange and external 65° flanges on two lightening holes at the large end. It was rubber-pad formed from a quarter-hard type 302 stainless steel blank 0.41 mm (0.016 in.) thick.

The zinc alloy die used was made with a curve to offset the springback of the flange (Fig. 27). Right-hand and left-hand pieces were flanged at the same time in the same die. A steel cover plate protected the thin web of each piece from distortion during forming. No lubricant was used. The pressure developed by the rubber was 10.3 MPa (1.5 ksi).

Deep Drawing. By the rubber-pad and rubber-diaphragm processes, stainless steels in both the 300 and the 400 series can be deep drawn to greater reductions than can be achieved with conventional methods. For extremely deep sections, the lower-work-hardening austenitic types 302 and 305 are recommended.

Two characteristics of rubber-pad methods make this great depth of draw possible. The first is controlled, continuously adjustable pressure on the blankholder or holddown mechanism, and

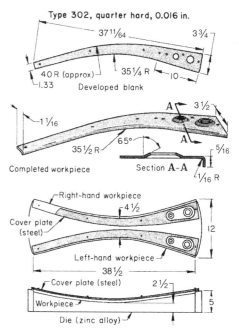

Fig. 27 Long narrow strut with a contoured stretch flange that was made by rubber-pad forming in a curved die with cover plates to prevent springback. Dimensions given in inches

the second is the continuously variable draw ring radius. There is no draw ring as such, but the rubber that forms around the workpiece functions as a draw ring and conforms to the radius that will apply equal pressure to the entire surface of the workpiece. This minimizes both thinning at the punch radius and work hardening as the flange metal is drawn into the cup.

Figure 28 illustrates the relatively uniform wall thickness that can be produced by drawing using the rubber-pad method. For comparison, Fig. 19 shows the much greater variation in wall thickness produced by conventional deep drawing. Additional information on deep drawing by rubber-pad techniques is available in the article "Rubber-Pad Forming and Hydroforming" in this Volume.

Drop Hammer Forming

A wide variety of sizes and shapes can be formed in thin stainless steel by drop hammer forming. The advantages of this method include high impact energy (which often means that a piece can be formed by one blow, as compared to four or five by other processes) and suitability to low-volume and experimental production.

Dies. Die material for drop hammer forming is less critical than for press forming. The dies are made of steel, plastic, zinc alloy, and lead. Zinc alloy is widely used.

Punches are often made of lead because it can be cast directly on the lower die and because its weight adds energy to the stroke of the drop hammer. Although the lead is reusable, the number of pieces that can be made from each cast punch is small—approximately 200. Plastic punches and dies impart a finish to formed parts that would otherwise be difficult to obtain. Steel dies are used for high production and for coining and sizing.

Die designs are generally similar to those for press forming, with the same punch and die radii to reduce stress on the work metal. Die design for the forming of beads and methods of relieving entrapment to ensure good metal flow in drop hammer forming are also similar to those used in press forming. A trapped-rubber technique somewhat similar to the Guerin process is described in the article "Rubber-Pad Forming and Hydroforming" in this Volume.

Quality of Product. The dimensions of workpieces formed in a drop hammer are less consistent than those made by other processes, because the degree of impact is subject to operator skill and because the punch can shift under localized high loads. However, springback is less pronounced in drop hammer forming than in other forming methods because of the high impact and forming speed.

Lubrication. The lubricants that can be used in drop hammer forming are listed in Table 2. If working is severe enough to require annealing between stages, contaminants such as graphite or sulfur (from the lubricant) or zinc or lead (from the die) must be removed from the work surface. If these contaminants are left on the surface of the stainless steel when it is heated, they can cause serious surface deterioration.

Comparison with Press Forming. Press forming, although done rapidly, is inherently an operation in which ram speed and holding pressures can be closely controlled; however, in drop hammer forming, the only way to form a part is by sudden impact. In some applications, production difficulties are overcome by the high rate of energy release in a drop hammer. In others, especially those in which blankholder pressure is critical, press forming produces better parts more economically if the die is properly made, as in the following example.

Example 22: Change from Drop Hammer to Press Forming that Eliminated Wrinkling and Reduced Cost. The tailpipe half shown at the top in Fig. 29 was originally produced in a drop hammer, using the tooling setup shown at the lower left in Fig. 29. The operation was unsatisfactory, however, because wrinkles occurred at the intersection of the 30° risers, and six operations totaling nearly 2 min per piece were required to complete each piece.

The tools were redesigned for use in a 4400 kN (500 tonf) hydraulic press (lower right, Fig. 29). The zinc alloy die used in the drop hammer was reused in the press; to make it resistant to the abrasion of press forming with stainless steel, the die was faced off, and a low-carbon steel wear plate was installed. The 43 mm (1¹¹/₁₆ in.) radius had formed well in the drop hammer with very little springback, but springback in the press made it necessary to deepen the die. This was done by inserting shims between the die and the wear plate.

The press produced pieces that were completely free of wrinkles at the rate of two pieces per minute. This was a 1¹/₂ min savings per piece.

The blank for both methods was annealed type 321 stainless steel measuring 510 by 610 mm (20 by 24 in.) and 0.71 mm (0.028 in.) thick. No lubricant was required for drop hammer forming; a wax emulsion was used for the press operation. Trimming after forming was done in a second press.

A drop hammer is ordinarily used for prototypes, and a press, using the prototype or improved dies, is used for mass production. If the quality of the prototype die is good, the drop hammer can be used for low production.

Three-Roll Forming

The three-roll forming of stainless steel is, in general, similar to the three-roll forming of other metals (see the article "Three-Roll Forming" in this Volume). Springback is a major problem with austenitic stainless steels, primarily because of the large radii involved and work hardening. It is important that the equipment be set up so that the desired curvature can be made in one pass. Because of the high rate of work hardening of austenitic stainless steels, subsequent passes are sometimes difficult to accomplish and control unless heavy equipment is used. The response of annealed ferritic stainless steel to three-roll forming is quite similar to that of hot-rolled low-carbon steel.

Three-roll and two-roll formers can be put in sequence with contour roll formers to make a cross-sectional shape and to bend or coil it, all in one production line. The following example describes an application in which three-roll forming was combined with press forming and hydraulic expansion forming.

Example 23: Use of Three-Roll Forming in the Production of a Container for Liquid. Figure 30 shows eight of the 14 operations entailed in the production of a container for liquids by press forming and hydraulic expansion forming of a welded cylinder made from a radiused flat blank by three-roll forming in pyramid-type rolls. The six other operations are identified in the table that accompanies Fig. 30. These containers were produced in annual

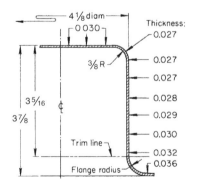

Fig. 28 Profile of a shell that was deep drawn from 0.76 mm (0.030 in.) thick stainless steel by the rubber-pad method showing the relatively uniform wall thickness obtained. Dimensions given in inches

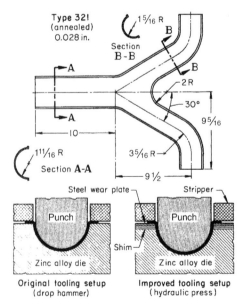

Fig. 29 Aircraft tailpipe half that was formed by the drop hammer and hydraulic press tooling setups shown. Dimensions given in inches

quantities of 10,000 to 100,000 pieces from annealed type 304 stainless steel coil stock 0.79 mm (0.031 in.) thick and 585.8 mm (23.06 in.) wide.

Blanking the rectangular sheets for three-roll forming gave the workpiece the uniform square edges needed for maintaining the welded seam of the tube in axial alignment. The blanking tools were hardened high-carbon high-chromium tool steel; clearance was 0.08 mm (0.003 in.) per side. The stock was lubricated for blanking and edge radiusing, but the blanks were vapor degreased before three-roll forming.

Contour Roll Forming

Stainless steel is ordinarily contour roll formed in the annealed condition. Types 410 and 430 are usually roll formed on equipment similar to that used for carbon steel, with a No. 2 finish generally specified. Speeds are usually in the range of 7.6 to 30 m (25 to 100 ft) per minute, with the heavier gages and more difficult sections being roll formed at the slower speeds.

Stainless steels in hard tempers, such as quarter-hard and half-hard type 301, are also frequently roll formed. Increased power over that used for forming the same steels in the annealed condition is necessary because of the higher initial strength of the strip. Springback must be compensated for by adequate overbending.

Longitudinal cracking can be a problem with the hard tempers if adequate radii are not included in the design of the part.

Distortion or warpage of straight sections causes the greatest problem in roll forming the 300-series steels, particularly when the steel is thick. The distortion can be minimized by using more sets of rolls, or more passes, for greater control during each stage of bending. However, the skill of the operator is all-important in controlling distortion. Various straightening devices are usually attached or used on the last pass as the section emerges from the machine. In some applications, sections are deliberately curved.

With the chromium-nickel stainless steels, pickup on the rolls and galling of the strip sometimes occur. Highly polished rolls or bronze rolls are used with lubrication to minimize this problem when high pressure is needed. Heavy-duty emulsions containing chlorine offer the best combination of chemical EP and coolant activity (Table 2). Chlorinated oils or waxes are easy to use but are less effective as coolants. For severe forming, the cushioning effect of pigments is sometimes needed (as in the next example), as well as efficient cooling.

Example 24: Nine-Station Contour Roll Forming of Annealed 304 Stainless Steel. Figure 31 shows the sectional shapes progressively produced in the nine-station roll forming of a sheave track from annealed type 304 stainless steel strip 67.3 mm (2.648 in.) wide by

0.79 mm (0.031 in.) thick, with a No. 2 finish. In a tenth station, the formed track was straightened. As the track left the tenth station, it was clamped to a moving table that conveyed it to an abrasive wheel for cutoff into lengths of 3 to 6 m (10 to 20 ft). The material weighed 0.414 kg/m (0.278 lb/ft); annual production was 180 Mg (400,000 lb).

During forming, the developed width of the section measured along the neutral axis increased only 1.02 mm (0.040 in.) to 68.2 mm (2.688 in.), corresponding to only 1.5% stretch. The stretch was limited because the metal was restrained by the six pinch beads that were rolled into the strip before it was bent (stations 1, 2, and 3, Fig. 31). Each bead, 1.6 mm ($^1/_{16}$ in.) wide by 0.8 mm (0.030 in.) deep, permitted a sharp bend at that point without tearing or breaking the steel. The 50 to 55% elongation property of austenitic 304 stainless steel made it unlikely that the metal would break in bending. The strip was rolled with the slitting burr down so that the burr was flattened by the shoulders of the bottom roll of station 2.

The forming rolls were made of hardened steel, and the straightening rolls of hard bronze, for a good finish. Rolling speed was 17 m (55 ft) per minute. The lubricant was a pigmented water-soluble oil.

Plastic protective coatings are sometimes applied to the strip to minimize or prevent scratches and scuffing when high pressures are used and surface finish requirements are critical. On light-gage material (especially type 430), such protection is generally unnecessary if the fabricator is experienced in processing stainless steel. A detailed discussion of the equipment and techniques employed in contour roll forming is available in the article "Contour Roll Forming" in this Volume.

Stretch Forming

The tools and techniques for stretch forming described in the article "Stretch Forming" in this Volume are applicable to stainless steel. Machines used for the stretch forming of stainless steel require 60 to 100% more power than that needed for similar operations on low-carbon steel of the same thickness.

Because of the abrasiveness of stainless steels, forming tools must be especially abrasion resistant. Wiping dies, wiping shoes, mandrels, and wear plates must be made of wear-resistant tool steel, carbide, or a bearing grade of bronze in order to avoid galling and welding.

Although the 300-series stainless steels are especially suitable for stretch forming because of their high work-hardening rate and ability to take large elongations, the 400-series steels are usable only for shallow stretched shapes. Type 301 is the austenitic steel that is best suited to stretch forming. Because of its high rate of work hardening, forming should be done slowly to derive maximum benefit from the ductility of type 301.

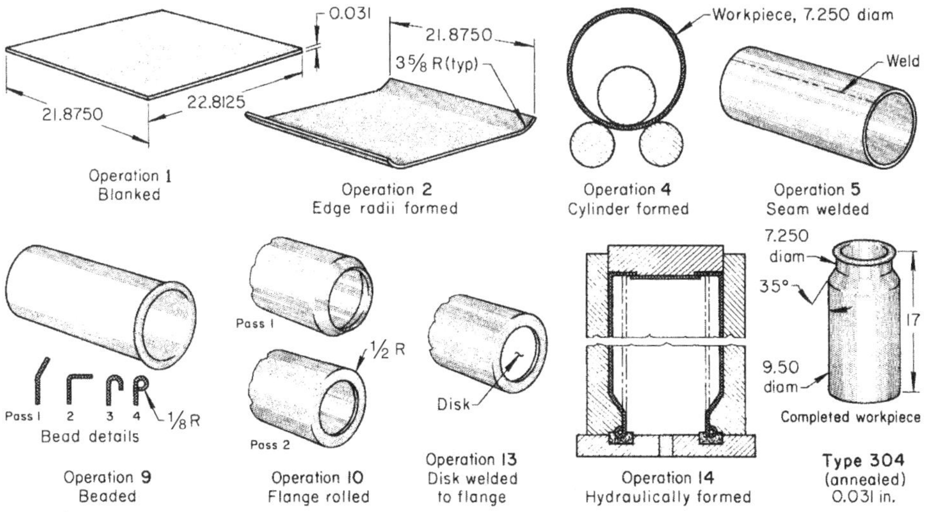

Sequence of operations

1. Blank in die, in single-action press.
2. Form edge radii on blank, in a press brake.
3. Vapor degrease, to remove lubricant used in operations 1 and 2.
4. Roll cylinder, in three-roll former.
5. Weld cylinder seam, in automatic Heliarc setup using starting and stop-off tabs.
6. Trim tabs.
7. Hammer weld to induce compressive stress, using an air hammer at 310 kPa (45 psi).
8. Restore roundness of cylinder by rerolling several times in three-roll former.
9. Form bead on one end of cylinder, in four passes in an edger.
10. Roll flange on opposite end of cylinder, in two passes.
11. Trim flange.
12. Vapor degrease.
13. Weld (Heliarc) disk to inside of flange.
14. Expand and form to final shape (30% reduction in wall thickness), in a hydraulic expansion die (final pressure: 4800 kPa, or 700 psi).

Fig. 30 Use of three-roll forming in conjunction with press forming and hydraulic expansion forming, in the 14-operation production of a container for liquids. Dimensions given in inches

Maximum percentages of stretch for one-directional forming of various kinds of austenitic stainless steels are as follows:

- Annealed types 301, 302, 304, 305, 316, 321, and 347: 20% typical; possibly 30% on symmetrical and solid sections
- Quarter-hard types 301 and 302: 15% typical; possibly 20% on optimal sections
- Half-hard types 301 and 302: 5% typical; possibly 10% on optimal sections
- Full-hard type 301: possibly 2% on optimal sections

These figures should not be confused with permissible stretch in bending, nor are they the limits to which these stainless steels will stretch (which are considerably greater). Instead, these percentages, which determine the possible curvature of stretch-formed sectional shapes of stainless steel, are based on the distortion susceptibility of severely stretched stainless steel.

The upper limits can be extended by very slow stretching and forming, especially with hardened metal. In addition, to obtain maximum stretch from the harder tempers, workpieces should be carefully deburred. Automatic programming is valuable in applying continuously increasing tension to overcome the continuously increasing strength as work hardening takes place during stretch forming.

Lubricants. If there is little or no movement after contact between workpiece and form block, as in stretch wrapping or single-die draw forming, little or no lubricant need be used except when deformation is severe. A low-viscosity chlorinated oil or wax provides excellent chemical EP action and convenience of use. If there is considerable movement of the work metal against the dies (such as against the wiper shoe in radial-draw forming), pigmented lubricants are sometimes used. The following example describes an application in which no lubricant

was used in the stretch forming of a sharply contoured part.

Example 25: Dry Stretch Forming of an Airfoil Leading Edge. The leading edge of an airfoil was stretch formed dry from a type 302 stainless steel blank, 0.20 mm (0.008 in.) thick, 115 mm (4$^1/_2$ in.) wide, and 5.5 to 6.7 m (18 to 22 ft) long, that had been roll formed to the airfoil contour shown in section A-A in Fig. 32. The blank had been annealed before roll forming, and it was stretch formed, without further annealing, to a 7.6 m (25 ft) radius with the heel of the contour pointing out (Fig. 32).

The airfoil was stretch formed in a radial-draw former over a hard-maple form block with the airfoil contour carved into its surface (Fig. 32). Lubricant was not used, because it had previously caused local variations in friction. Time for forming was 10 min per piece with three men working. Setup time was 2 h. A typical production lot was 100 pieces.

The rolled contour had to be held within ±0.1 mm (±0.004 in.) after stretch forming. The envelope tolerance on the stretch-formed shape was 0.76 mm (0.030 in.).

Springback. In sharply contoured pieces that have a relatively deep, wide cross section, some springback cannot be avoided, even in annealed metal. During severe stretch forming, considerably higher strength, and therefore appreciably higher elastic recovery, is developed in the more highly stressed convex surface.

Springback in regular, symmetrical sections can usually be offset by overbending the piece. Dimensional variations in workpieces are primarily caused by variations in springback, which are in turn caused by variations in mechanical properties from sheet to sheet.

If the workpiece is irregular in cross section, if preformed flanges are to be held to a certain angular position, or if the curve of the form varies in severity, springback may cause twist or irregular distortion of the workpiece. Various methods of blocking, pretwisting, or overforming are used to prevent or correct this distortion. In the following example, an asymmetrical cross section was twisted during forming to offset the twist caused by springback.

Example 26: Use of Twisting to Compensate for Springback in Stretch Forming. The curved channel section shown in Fig. 33 was

stretch formed from quarter-hard type 302 stainless steel strip, 1.07 mm (0.042 in.) thick, that had been preformed in a press brake. Although the channel fit closely in the groove of the form block, springback caused considerable twist in the finished piece.

Elastic recovery of the outer flange and the metal near the outer edge of the web caused buckling and twisting in the part as forming tension was released. To overcome this, the part was canted by the form block, and tension on the part was gradually increased during forming.

To establish a compensating initial reverse twist in the workpiece, spacers were added to the built-up form block to wedge the section to a 5° angle, as shown in Fig. 33. At the same time, a fiber filler strip with maple filler blocks was closely fitted into the channel to hold the cross-sectional contour. Details of the tooling are shown in Fig. 33.

The applied tension during stretch forming was 83.2 kN (18,700 lbf) at the start, 87.0 kN (19,550 lbf) at 45° bend, 90.7 kN (20,400 lbf) at 90°, 94.5 kN (21,250 lbf) at 135°, and 98.3 kN (22,100 lbf) on completion of the bend. A non-pigmented fatty acid was used as the forming lubricant. After forming, the workpiece was trimmed to a 145° arc with a band saw.

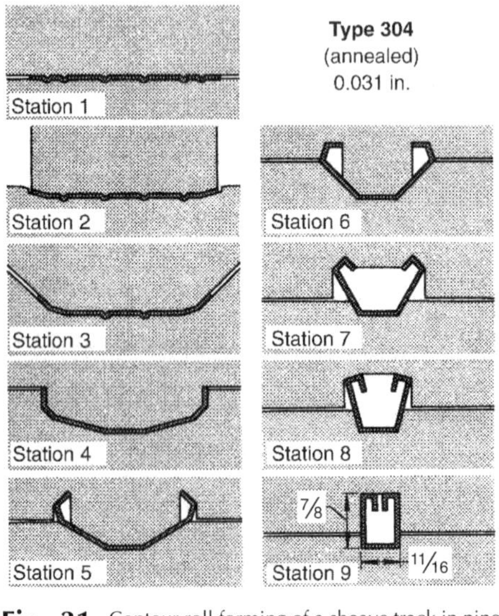

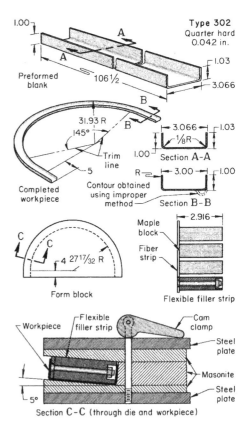

Fig. 33 Channel section that was stretch formed from a preform produced in a press brake, and details of tooling used in stretch forming, which provided reverse twist to compensate for springback. Dimensions given in inches

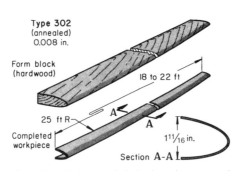

Fig. 31 Contour roll forming of a sheave track in nine stations. Dimensions given in inches

Fig. 32 Airfoil on which the leading edge was stretch formed to a long convex shape without lubricant in a radial-draw former

Equalizing Stretch. In the stretch forming of sheets to a curvature in two directions (especially in stretching tempered material when the limits of stretch are very close), the quality of the product can be controlled much better if the stretch is uniform across the workpiece. One means of obtaining uniform stretch is to provide compensating contours (which are later trimmed off) at the end of the form block.

Bending of Tubing

Austenitic stainless steel tubing can be bent to a centerline radius of $1\frac{1}{2}$ times tube diameter. As the ratio of tube diameter to wall thickness, D/t, increases, it becomes increasingly necessary to provide both internal and external support to keep the tube from collapsing as it is bent. When D/t is greater than 30, the tube is classed as a thin-wall tube. Interlocked tooling, as well as bending machines of a greater capacity than that required for thick-wall tubes, is strongly recommended for thin-wall tubing (see the article "Bending and Forming of Tubing" in this Volume).

For the bending of stainless steel tubing, wiper dies and mandrels are often made of aluminum bronze or a chromium-plated tool steel. Lubricants for the mandrel should be fairly heavy. Viscous or pigmented oil-based lubricants containing emulsifiers for ease of removal are used. Only the very lightest of lubricants should be used between the wiper die and the tube. A thin application of very light chlorinated mineral oil can be used in some bending operations without causing wrinkling. The following example describes techniques used in the bending of stainless steel tubing.

Example 27: Bending Difficult-to-Form Tubing into an Aerospace Component. The bent tube shown in Fig. 34, used in an aerospace assembly, was difficult to form within the specified tolerances (dimensions within 0.25 mm, or $\pm$0.010 in.; angles within $\pm\frac{1}{2}°$; and flattening of the tube at bends not more than 0.05 mm, or 0.002 in.). The piece was produced from type 304 stainless steel tubing in nine operations in the following sequence (times shown are for the production of 100-piece lots):

- Cut tubing into lengths of 160 mm ($6\frac{1}{4}$ in.) with an abrasive cutoff wheel; deburr roughly (3 h)
- Fill each workpiece with low-melting alloy (8 h)
- Make 160° bend in powered draw bender; gage the bend (5 h)
- Make 24° bend in hand bender; gage the bend (5 h)
- Trim ends to length in a cutoff fixture using an abrasive wheel (3 h)
- Melt out the filler (6 h)
- Deburr by hand, using a grinder and a drill (3 h)
- Passivate in a chemical dip (1 h)
- Inspect 100% with gage and by rolling an accurate ball through the completed part (2 h)

Springback in bending, approximately 5°, was corrected by overbending to a degree established in trial bends.

Other Forming Operations on Tubing

Stainless steel tubing can be easily flared to increase the diameter 25 to 30% if it is annealed. The diameter can be reduced by rotary swaging, or it can be increased by bulging or beading. Rubber punches are often used for this purpose, as described in the article "Rubber-Pad Forming and Hydroforming" in this Volume.

Tubing of austenitic stainless steel can be hot formed by heating to 1175 to 1260 °C (2150 to 2300 °F). Work should be halted when the tube has cooled to 925 °C (1700 °F), and the tube should then be cooled rapidly to minimize the precipitation of carbides.

Because austenitic stainless steel tubing is stronger than carbon steel tubing and work hardens rapidly, warm forming (below the recrystallization temperature) is also used on this material. The temperature for warm forming should be kept below 425 °C (800 °F) to prevent the formation of carbides.

Tubing of ferritic stainless steels, such as types 430 and 446, is less easily formed than similar tubing of austenitic stainless steels. Ferritic tubing is hot formed at 1035 to 1095 °C (1900 to 2000 °F), and forming is stopped when the tubing cools to 815 °C (1500 °F). For best results, the range from 815 to 980 °C (1500 to 1800 °F) should be avoided, because ductility and notch toughness are progressively impaired as the tube cools through that range. Hot shortness may be encountered in the upper part of the range. Tubing of ferritic stainless steels is warm formed at 120 to 205 °C (250 to 400 °F).

Steel producers have studied the cold formability of 11% Cr (409) and 17% Cr (439 or 18% Cr-Nb) tubing materials, primarily because of requests from the automotive industry to use titanium- or titanium-plus-niobium-stabilized ferritic alloys in exhaust systems. Such alloys are normally used in high-frequency welded, laser welded, or gas tungsten arc (GTA) welded (autogenous) and annealed tubing. Traditionally, the GTA welded and annealed tubes had more

formability because of the elimination of the 8 to 15% cold work induced in forming the tube.

As these ferritic alloys were subjected to the demands of high-speed vector bending, particularly in making tubular exhaust manifolds, breakage rates increased to over 50%. In response, stainless steel producers borrowed technology from low-carbon steel production practices and developed a line of high-performance ferritic alloys with improved elongations and higher r-values (1.4 or above). Additional information on the determination of r-values is available in the article "Formability Testing of Sheet Metals" in this Volume.

Such alloys have permitted the greater use of high-frequency welded and unannealed tubes for thin-wall bends with a centerline bend radius less than twice the tube diameter. Furthermore, such bends can be made at room temperature, although care should be exercised in cold weather not to fabricate sub-room-temperature tubing. Finally, through tighter control of both melt chemistries and processing parameters, ferritic tube alloys with excellent welding and bending reproducibility from heat to heat have been developed.

Forming versus Machining

Although forming ordinarily requires expensive tooling and bulky equipment, it is a high-speed process, and for most parts that can be formed from sheet, it is more economical than machining for mass production. The following example shows how production techniques can vary with the size of the production lot to make the best use of each technique.

Example 28: Influence of Change in Quantity of Production Method and Product Design. A threaded cap was made of type 347 stainless steel by three different methods. Each method involved a change in design, as illustrated in Fig. 35.

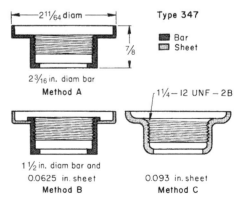

Fig. 35 Machining versus press forming for producing a cap. In method A, the cap (illustrated above) was completely machined from bar stock. In method B, the cap (redesigned) combined components that were press formed from sheet and machined from bar stock. In method C, the entire cap (again redesigned) was press formed from sheet and then partly machined. Dimensions given in inches

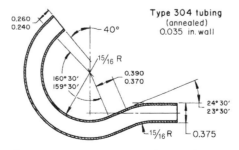

Fig. 34 Aerospace component that was bent from stainless steel tubing with the use of a low-melting alloy as a filler during bending. Dimensions given in inches

The original order was for 100 caps, with an anticipated design change on future orders. The quickest and most economical production method was to machine the cap in one piece from bar stock (Fig. 35, method A).

The next order was for 1000 caps. The design and manufacturing methods were revised so that the cap was produced as an assembly of two components, one press formed from sheet and the other machined from bar stock (Fig. 35, method B).

When requirements increased to 5000 caps, a cost reduction was essential to obtain the order against a competitor's bid. The part was redesigned for production entirely from sheet by press forming and partial machining (Fig. 35, method C). Overall cost was reduced nearly 50% as compared to methods A and B.

The press-formed part of method B was made in a 400 kN (45 tonf) open-back-inclinable mechanical press at a rate of 200 to 250 pieces per hour. The die was made of oil-hardening tool steel. Method C used an air-hardening tool steel die and a 530 kN (60 tonf) open-back-inclineable mechanical press that made 300 to 350 pieces per hour. Mineral oil was used as a lubricant in both methods.

Forming of Aluminum Alloys

Revised by Pawel Kazanowski, Hydro Aluminum Cedar Tools

ALUMINUM and its alloys are among the more formable materials of commonly fabricated metals. There are, of course, differences between aluminum alloys and other metals in the amount of permissible deformation, in some aspects of tool design, and in details of procedure. These differences stem primarily from the lower tensile and yield strengths of aluminum alloys, their lower modulus of elasticity, and their comparatively low rate of work hardening as compared with typical draw-quality steels. Some aluminum alloys also have a low and even negative strain-rate hardening factor, making them significantly more sensitive to problems related to careless die design. Formability is also a function of the wide range of compositions and tempers available in aluminum alloys. This article emphasizes those aspects of commercial forming processes and equipment that apply specifically to aluminum alloys. More general information on the forming of metals is given in other articles in this Volume.

General Formability Considerations

One aspect of the formability of a material is the extent to which it can be deformed in a particular process before the onset of cracking, splitting, or buckling. Another major aspect of formability relates to the consistency of producing nominally identical parts in sufficient quantity. Variation in yield strength, strain hardening, and gage or thickness can all contribute to poor formability in a given coil or lot of material. Considerable advances have been made in the development of alloys with good formability, but, in general, an alloy cannot be optimized on this basis alone. The function of the formed part must also be considered, and improvements in functional characteristics, such as strength and ease of machining, often tend to reduce the formability of the alloy.

Aluminum alloy sheet can fail by cracking or splitting during forming by ductile fracture, with either localized or diffuse necking. Necking is governed largely by material properties such as strain hardening and strain-rate hardening and depends critically on the strain path followed by the forming process. In dilute alloys, the extent of necking or limit strain is reduced by cold work, age hardening, gross defects, large grain size, and the presence of alloying elements in solid solution. Ductile fracture occurs as a result of the nucleation and linking of microscopic voids at particles and the concentration of strain in narrow shear bands. Fracture occurs very soon after the onset of local necking, but fracture usually occurs at larger strains than does localized necking and therefore is usually important only when necking is suppressed due to stamping geometry or tooling issues. Common examples where fracture is encountered are at small-radius bends and at severe drawing, ironing, and stretching near notches or sheared edges (Ref 1, 2).

Most sheet metal elements may also undergo complicated deformation processes. In most sheet forming operations, aluminum and aluminum alloys are susceptible to the Bauschinger effect. The effect involves stress-strain asymmetry that results in a stress-strain hysteresis loop. From the experimental results (Ref 3), it is apparent that two identical sheet metal specimens with 6111-T4 alloy can have the same final total strains but have distinctly different springback amounts. The reason is that their deformation histories in the space are different.

For intricate forming operations, it is necessary to use annealed (condition O) material and the final strength developed by heat treating after the forming has been accomplished. Heat treated alloys can also be formed at room temperature immediately after quenching (W temper), which is much more formable than the fully heat treated temper. The part is then aged to develop full strength. The forming operation should be performed as soon after quenching as possible, in view of the natural aging that occurs at room temperature on all the heat treatable alloys. The natural aging can be delayed to a certain extent by placing the part in a cold storage area of 0 °C (32 °F) or lower. The lower the temperature, the longer the delay to a point where maximum delay is obtained.

Effects of Alloying Elements. The principal alloys that are strengthened by the presence of alloying elements in solid solution (often coupled with cold work) are those in the aluminum-magnesium (5xxx) series. Aluminum and magnesium form solid solution over a wide range of compositions, and wrought alloys containing from 0.8% to slightly more than 5% Mg are widely used. Strength values in the annealed condition range from 40 MPa (6 ksi) yield and 125 MPa (18 ksi) tensile for alloy 5005 (Al-0.8Mg) to 160 MPa (23 ksi) yield and 310 MPa (45 ksi) tensile for the strongest alloy, 5456 (Al-5.0Mg). These alloys often contain small additions of transition elements such as chromium or manganese, and less frequently zirconium, to control the grain or subgrain structure and iron and silicon impurities that are usually present in the form of intermetallic particles. Small additions of chromium and manganese raise the recrystallization temperatures and increase the tensile properties for a given magnesium content. Figure 1 illustrates the effect of magnesium in solid solution on the yield strength and tensile elongation for most of the common aluminum-magnesium commercial alloys. Note the large initial reduction in tensile elongation with the addition of small amounts of magnesium.

The reductions in the forming limit produced by additions of magnesium and copper appear to be related to the tendency of the solute atoms to migrate to dislocations (strain age). This tends to increase strain (or work) hardening at low

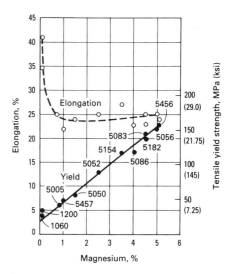

Fig. 1 Correlation between tensile yield strength, elongation, and magnesium content for some commercial aluminum alloys in the annealed temper. Source: Ref 4

strains, where dislocations are pinned by solute atoms, but it also decreases strain hardening at large strains. Small amounts of magnesium or copper also reduce the strain-rate hardening, which in turn reduces the amount of useful diffuse necking that occurs after the uniform elongation. Zinc in dilute alloys has little effect on work hardening (i.e., necking), and it does not cause strain aging.

Elements that have low solid solubilities at typical processing temperatures, such as iron, silicon, and manganese, are present in the form of second-phase particles and have little influence on either strain hardening or strain-rate hardening and thus a relatively minor influence on necking behavior. Second-phase particles do, however, have a large influence on strain at fracture. The addition of magnesium promotes an additional reduction in fracture strain, because the higher flow stresses aid in the formation and growth of voids at the intermetallic particles. Magnesium in solid solution also promotes the localization of strain into shear bands, which concentrates the voids in a thin plane of highly localized strain.

Precipitation-strengthened alloys are usually formed in the naturally aged (T4) condition or in the annealed (O) condition but rarely in the peak strength (T6) condition where both the necking and fracture limits are low. Figure 2 shows the effect of a wide range of precipitate structures on some of the forming properties of alloy 2036 (Al-2.5Cu-0.5Mg). Curves similar in shape can be drawn for most of the precipitation-strengthened alloys in the 2xxx and 6xxx series.

The properties in Fig. 2 were obtained from sheet tensile specimens first solution heat treated, then aged at temperatures ranging from room temperature to 350 °C (660 °F). This produced a full range of structures from solid solution (as-quenched) through T4 and T6 tempers to various degrees of overaging and precipitate agglomeration.

Deformation of aluminum and its alloys proceeds by crystallographic slip that normally occurs on the {111} planes in the ⟨110⟩ directions. Large amounts of deformation at ambient temperatures lead to some strengthening through the development of textures, the nature of which depends in part on the mode of working. In rolled sheet, the texture that is developed may be described as a tube of preferred orientation. Preferred orientations in the plane of a sheet that are associated with textures may cause a problem known as earing. Earing describes the phenomenon of small undulations that may appear on the top of drawn cups and is wasteful of material, because this uneven end of the cup must be trimmed off. Moreover, it may lead to production problems due to difficulties in ejecting parts after a processing operation. Four ears usually form because of nonuniform plastic deformation along the rim of a deep-drawn product. If the rolling texture is predominant, then ears will appear at 45° to the rolling direction of the sheet, whereas in the presence of the annealing (or cube) texture, they form in the direction of rolling and at right angles to it. If there is a desirable balance between the two textures, there will be either eight small ears or none at all (Ref 3).

Formability Properties and Testing

To conduct a complete analysis of a formed part, the required mechanical properties, as determined by several standard tests, must be considered. These properties include those determined by tension testing and by other tests designed to simulate various production forming processes, including cup tests and bend tests. More information on the test methods briefly described here is available in the article "Formability Testing of Sheet Metals" in this Volume.

Tension testing is used to determine the commonly reported properties—(ultimate) tensile strength, (tensile) yield strength, and (total) elongation—as well as two properties especially important in forming (Ref 1), that is, the strain-hardening exponent n and the plastic-strain ratio r.

The strain-hardening exponent n of a material is determined from the true stress-true strain curve for that material using the formula:

$$\sigma = K\varepsilon^n \qquad \text{(Eq 1)}$$

where σ is true stress, ε is true strain, and K is a strength coefficient. Equation 1 is the Holloman strain-hardening relation and is commonly used in modeling. However, strain-hardening behavior is also described by other relations (see the article "Constitutive Equations" in this Volume). For example, the Voce relation [$\bar{\sigma} = A - B\exp(-c\bar{\varepsilon})$] describes aluminum sheet very well (e.g., see the article "Flanging of Aluminum" in this Volume).

The plastic-strain ratio r describes the resistance of the material to thinning during forming operations and is the ratio of the true strain in the width direction (ε_w) to the true strain in the thickness direction (ε_t) of plastically strained sheet:

$$r = \varepsilon_w / \varepsilon_t \qquad \text{(Eq 2)}$$

A standard method for determining r using a tension specimen is given in ASTM E 517. The strain ratio (or anisotropy factor, r) depends on orientation relative to the rolling direction of the sheet. Thus, the average value ($\bar{r}$ or r_m) is commonly use in modeling and practice where:

$$r_m = \bar{r} = \frac{(r_0 + r_{90} + 2r_{45})}{4}$$

where r_0, r_{90}, and r_{45} are strain ratios parallel, perpendicular, and at 45° to the rolling direction.

The tensile properties (as well as other mechanical properties) of many aluminum sheet alloys in medium and hard tempers exhibit directional sensitivity. The test direction should be reported along with the test results. Directional sensitivity is important in the analysis of forming operations that involve bending, flange stretching, or plane straining, all of which are encountered in the forming of ribs and troughs. Orientation of the rolling direction of the sheet relative to the direction of critical strain in the part often can mean the difference between producing a good part and producing scrap.

Simulative Tests. For many forming operations, tests that simulate the operation are more useful and relevant than fundamental intrinsic property measurement tests. These tests subject the work material to deformation that closely approximates the production operation, including the effects of factors not present in the intrinsic tests, such as bending and unbending and friction between the work materials and die surfaces. Because these additional factors are present, simulative tests tend to be less reproducible than intrinsic tests and must be performed under carefully controlled conditions to minimize variability in the results. Simulative tests can be classified on the basis of the predominant forming operation involved: bending, stretching, drawing, and stretch-drawing. In addition, tests have been developed to measure wrinkling and the springback that occurs after bending or another forming operation.

The Olsen cup test is a test used to predict resistance to local necking in biaxial-stretch-forming operations. The Olsen cup test measures the maximum penetration of a 22 mm (0.875 in.) diameter hemispherical punch into a clamped flat blank (see the section "Stretching Tests" in the article "Formability Testing of Sheet Metals" in this Volume). The punch depth at failure is the Olsen cup value. The test has been used for various materials, including aluminum alloys (Ref 5).

In Europe, several methods similar to the Olsen cup test are accepted. In eastern and northern Europe, the Erichsen cup test is widely used. The Erichsen test is similar to the Olsen cup test, except that the punch diameter is 20 mm (0.790 in.). The value reported is the ratio of cup height to cup diameter when the specimen is stretched over the 20 mm (0.790 in.) diameter ball. In Great Britain, the Avery test is very common, while in France the Guillery test is employed. Both methods are very similar to the Erichsen cup test, with slight modifications to the tooling and its setup.

Historically, ball punch tests, such as the Olsen cup test and Erichsen cup test, have been

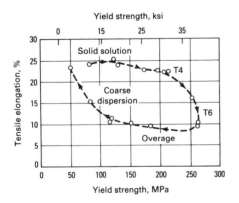

Fig. 2 Effect of precipitation on yield strength and elongation of aluminum alloy 2036. Source: Ref 4

used to determine the properties of sheet metals in stretching. However, the correlation between the test results and the steel performance in many sheet metal forming operations has not been good. These tests stretch a specimen over a hardened steel ball and measure the height of the cup produced. Tests that stretch the specimen over a much larger hemispherical dome have been developed, including the limiting dome height test, which uses specimens of different widths to control the strain ratio at fracture.

In the Swift cup test, a deep-drawn cup is used to determine the limiting draw ratio (LDR) of blank size to cup diameter. It is obtained with a 51 mm (2.0 in.) diameter flat-bottom punch and a draw die appropriate for the thickness of the specimen. A circular blank is cut to a diameter smaller than the expected draw limit. Lubrication is provided by two oiled polyethylene disks, one on each side of the blank. The blank is drawn to maximum punch load, which occurs before the cup is fully formed. Successively larger blanks are drawn until one fractures before being drawn completely through the die. The diameter of the largest blank that can be drawn without fracturing, divided by cup diameter, determines the LDR.

In bend tests, strips of material are bent around mandrels having different tip radii. Often, mandrels are in the form of pins or rods. The mandrel is forced against one side of the specimen strip, and the other side is simply supported at the end points. The value reported is the minimum radius of mandrel, in multiples of the material thickness t around which the material can be bent 180° without cracking. The direction of the bend relative to the rolling (or extrusion) direction should be recorded with the test results. Bend test results may vary significantly, depending on the test method. Bend-test methods include a guided-bend method, several semiguided methods, free-bend methods, and a bend-flatten method (ASTM E 290). Correlation of minimum bend radii with tensile-test data is discussed in the section "Bendability" in the article "Bending of Sheet Metal" in this Volume.

Correlation of Simulative Test Results with Tensile-Test Data. In some cases, results of simulative forming tests correlate with results of tension tests. Results of cup ductility tests, such as the Olsen and Swift tests, show good correlation with values of tensile elongation, strain-hardening exponent, and plastic-strain ratio. Olsen cup values correlate well with tensile elongation, and Swift cup values correlate with plastic-strain ratio. Neither of these two cup-ductility tests correlate with bend tests.

Forming-limit diagrams, also known as forming-limit curves, are direct and useful representations of the formability of aluminum sheet. These diagrams illustrate the biaxial combinations of strain that can occur without splitting. These diagrams are commonly derived from experimental testing, although computer-based modeling is an area of interest (Ref 6) due to the labor-intensive process of the experimental method.

To construct forming-limit diagrams experimentally, an array of circles, which often are 2.5 mm (0.1 in.) in diameter, is first imprinted by photoprinting, photoetching, or electroetching on the surface of the sheet metal before forming. After deformation, the imprinted circles either remain as circles in areas where pure biaxial stretching occurs or become ellipses wherever deformation is not purely biaxial stretching. The major and minor axes of the ellipses are compared with the circles of the original grid to determine the major and minor strains at each location. The areas immediately adjacent to splits are of particular concern in evaluating the forming capabilities of the metal. Failure can be defined by several criteria, but the onset of visible necking is the most widely used. The loci of strain combinations that produce failures define the forming-limit curve. The area below this curve encompasses all the combinations of strain that the metal can withstand. Forming-limit diagrams for a variety of aluminum alloys and tempers are shown in Fig. 3 to 5.

Stress-Based Forming-Limit Curves. A relatively new approach to refine formability theory is the stress-based forming-limit curve (FLC) developed by T. Stoughton of the General Motors Research Laboratories (Ref 8). It has been shown that the stress-based FLC is independent of strain path and is unique for a given material. Conventional FLCs are strain-based, created by stretching various-width strips of material over a smooth ball. The ball is large enough so that relatively little stress is introduced through the thickness of the material. The result is that the material is strained proportionately throughout the test.

The strain-based FLC is relevant for forming processes with a linear strain path, which is representative for most first-draw forming operations. However, a nonlinear strain path may be found in secondary forming processes, such as some restrike operations, and even in some areas of the first-draw die, such as re-entrant surfaces such as door handles on doors that have been subjected to high prestrain. In these cases, the strain path can significantly alter the FLC (see Fig. 9 in the article "Modeling and Simulation of the Forming of Aluminum Sheet Alloys" in

this Volume). Stress-based FLCs (Ref 9, 10) resolve this discrepancy that results from non-uniform strain paths.

Equipment and Tools

Most of the equipment used in the forming of steels and other metals is suitable for use with aluminum alloys. Because of the generally lower yield strengths of aluminum alloys, however, required press capacities are usually lower than for comparable operations on steel, and higher press speeds can be used. Similarly, equipment for roll forming, spinning, stretch forming, and other forming operations on aluminum need not be as massive or rated for such heavy loading as for comparable operations on steel. The lower elastic modulus leads to greater springback, which thus requires different die designs than for steel forming.

The forming of aluminum (1100) is relatively easy, using approximately the same procedures as those used for common steel, except that care must be taken to prevent scratching. Do not mark on any metal surface to be used as a structural component with a graphite pencil or any type of sharp, pointed instrument. All shop equipment, tools, and work area should be kept smooth, clean, and free of rust and other foreign matter. Alloyed aluminum (2024, 7075, 7178, etc.) is

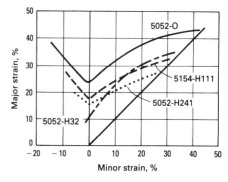

Fig. 4 Forming-limit diagrams for four 5*xxx*-series aluminum alloys. Source: Ref 7

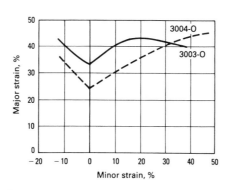

Fig. 3 Forming-limit diagrams for two 3*xxx*-series aluminum alloys. Source: Ref 7

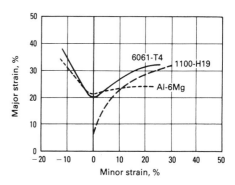

Fig. 5 Forming-limit diagrams for aluminum alloys 1100-H19 and 6061-T4 and for Al-6Mg. Source: Ref 7

more difficult to form, and extensive control is required to prevent scratching and radii cracking. Scratching will make forming more difficult, plus it provides an easy path for corrosion attack, especially on clad materials.

Tools. Total wear in forming aluminum is somewhat less than that of steel. This results in part from the lower force levels involved and in part from the smoother surface condition that is characteristic of aluminum alloys. Accordingly, tools can sometimes be made from less expensive materials, even for relatively long runs.

However, a higher-quality surface finish is generally required on tools used with aluminum alloys, to avoid marking. The oxide film on the surface of aluminum alloys is highly abrasive, and for this reason many forming tools are made of hardened tool steels. As a rule, these tools, even if otherwise suitable, should not be used interchangeably to form steel parts, because this could destroy the high-quality finish on the tools.

Most aluminum alloys require smaller clearances between punches and dies in blanking and piercing than do steels. On drawing tools, they require larger clearances but approximately the same radii, to allow the free flow of metal and avoid excessive stretching.

The amount of springback in forming aluminum alloys, which is generally more than that of steel, must be considered in tool design (see the article "Springback" in this Volume). The amount of springback is roughly proportional to the ratio of yield strength to elastic modulus of a metal. Additionally, the lower rate of work hardening of aluminum alloys permits a greater number of successive draws than is usually possible with steel.

Lubricants

Lubricants must be specifically selected for their compatibility with aluminum alloys and their suitability for the particular forming operation in question. A lubricant suitable for use on a steel part will not necessarily be suitable for use in the forming of a similar aluminum alloy part.

The proper formulation of lubricants for the forming of aluminum alloys must take into account the special requirements of regulation of moisture content in nonaqueous systems, corrosion inhibitors, and pH control, in order to prevent staining or corrosion and to make duration of contact with the workpiece less critical.

The lubricants most widely used in the forming of aluminum alloys are listed in approximate order of increasing effectiveness:

- Kerosene
- Mineral oil (viscosity of 40 to 300 SUS at 40 °C, or 100 °F)
- Petroleum jelly
- Mineral oil plus 10 to 20% fatty oil
- Tallow plus 50% paraffin
- Tallow plus 70% paraffin
- Mineral oil plus 10 to 15% sulfurized fatty oil and 10% fatty oil

- Dried soap films or wax films
- Fat emulsions in aqueous soap solutions with finely divided fillers
- Mineral oil with sulfurized fatty oil, fatty oil, and finely divided fillers

The use of various special-purpose lubricants is discussed in sections of this article that deal with individual forming processes. The article "Selection and Use of Lubricants in Forming of Sheet Metal" in this Volume contains more information on lubricants for sheet forming.

Blanking and Piercing

Blanking and piercing of aluminum alloy flat stock are ordinarily done in punch presses because of their high production rates and ability to maintain close tolerances. Press brakes are sometimes used, particularly for experimental or short-run production.

The generally lower shear strength of aluminum alloys usually permits the use of presses or press brakes of lower capacity than those used for comparable operations with steel. A conservative (high) calculation of the required shearing force is the product of shear strength times the total length of cut and metal thickness. This calculation will give high force requirements, because the shear action is not typically applied to the whole circumference simultaneously. Allowance also must be made for different alloys, for dulling of the cutting edges of punches and dies, and for variation in clearance between punch and die. The shear strengths of the commonly used aluminum alloys range from 62 to 338 MPa (9 to 49 ksi), whereas the shear strength of a typical low-carbon steel is 240 to 460 MPa (35 to 67 ksi).

Tool Materials. A discussion of materials for blanking and piercing dies is given in the article "Selection of Materials for Press-Forming Dies" in this Volume. Aluminum alloys are classed with other soft materials, such as copper and magnesium alloys. In general, for a given tool material, tool life is longer for blanking and piercing aluminum alloys than for blanking and piercing steel.

In some applications, a less expensive die can be used than is true for steel parts, particularly for relatively short runs. Cast zinc dies, which cost only approximately one-fifth as much as tool steel dies, are used for runs of up to approximately 2000 parts (as in forming steels). Steel-rule dies and template dies also reduce tooling costs for short runs or moderate-length runs. For example, an aluminum alloy blank 495 by 305 mm (19.5 by 12 in.) by 1 mm (0.040 in.) in thickness was made in a steel-rule die having an expected life of 150 pieces. For the production quantity, burr height did not exceed 0.127 mm (0.005 in.). Punches and die buttons for seven pierced holes of 3.9, 4.8, and 6.4 mm ($^5/_{32}$, $^3/_{16}$, and $^1/_4$ in.) diameter were incorporated in the die.

Low-carbon steel or cast iron dies sometimes replace hardened tool steel dies, even for long

runs. Punches are usually made from annealed or hardened tool steel, depending on the size and complexity of the part and the length of the run. Cemented carbide tools are seldom required, even for extremely long runs.

Tolerances. A tolerance of ±0.127 mm (0.005 in.) is normal in the blanking and piercing of aluminum alloy parts in a punch press. Using a press brake, it is possible to blank and pierce to a location tolerance of ±0.25 mm (0.010 in.) or less, although tolerances for general press brake operations usually range from ±0.51 to ±0.76 mm (0.020 to 0.030 in.).

For economy in tool cost, specified tolerance should be no less than is actually necessary for the particular part. A tolerance of ±0.127 mm (0.005 in.) would probably require that the punch and die be jig ground, adding 30 to 40% to their cost. A tolerance of ±0.05 mm (0.002 in.) may require the addition of a shaving operation. In addition to the cost of an extra die, labor costs would be increased by the added operation.

For extremely accurate work, an allowance must be made for the shrinkage of holes and expansion of blanks resulting from the elasticity of the stock. This allowance, made to both punch and die, does not change the clearance between them, and is primarily a function of stock thickness. For large sizes and normal tolerances, this correction is not very important.

Clearance between punch and die must be controlled in blanking and piercing in order to obtain a uniform shearing action. Clearance is usually expressed as the distance between mating surfaces of punch and die (per side) in percentage of work thickness.

Correct clearance between punch and die depends on the alloy as well as the sheet thickness. Suggested punch-to-die clearances in terms of percentage of sheet thickness t for blanking and piercing aluminum alloys in various tempers are listed in Table 1.

The character of the shearing action also depends on the sharpness of the tools. Dull cutting edges on punch and die have effects similar to those of excessive clearance, with the effect on burr size being particularly pronounced.

With proper clearance, the fractures proceeding from the punch surface and from the die surface of the work meet cleanly without secondary shearing and excessive plastic deformation. Secondary shearing indicates that the clearance is too small; a large radius or dished contour at the sheared edge and a stringy burr indicate that the clearance is too large.

Blanking tools are generally assigned with less clearance than piercing tools, so that the burnished area of the piece part (slug) is greater. For typical aluminum alloy and sheet thickness greater than 5 mm (0.2 in.), piercing clearance should be 5% higher than blanking clearance. This leads to a higher-quality piece part; however, due to the smaller clearances, blanking tools become dull more quickly.

Die Taper. The walls of die openings in blanking or piercing dies are often tapered $^1/_2$° from the vertical, to minimize sticking of the blank or slug in the die. A straight, vertical section of at least 3.2 mm (1/8 in.) or equal to the metal thickness for stock thicker than 3.2 mm (1/8 in.) is usually left at the upper end of the die opening, to provide for sharpening without changing the clearance. Tapered die relief is usually more suitable for piercing aluminum than is counterbore design relief.

Stripping force of 3 to 20% of the total capacity needed for blanking and piercing is used for aluminum alloys. The force needed depends on the alloy, temper, and stock thickness. Sharpness of cutting edges on punch and die,

lubrication, and uniformity of application of stripper-plate pressure also affect stripping force.

Lubricants are normally used in blanking or piercing aluminum alloy parts to reduce sticking of slugs or blanks in the die opening and to facilitate clean stripping from the punch without buckling. Lower tool maintenance costs and smoother edges on blanks or holes can be obtained with suitable lubrication.

Forming Using Tailor-Welded Blanks

A promising approach to reduce manufacturing costs, decrease vehicle weight, and improve the quality of automotive body components is through the use of tailor-welded blanks. This term refers to blanks where multiple sheets of material are welded together to create a single blank prior to the forming process. The welding process creates formability concerns in a traditional forming process due to material property changes in the weld and in the heat-affected zone adjacent to the weld. Tailored blanks are also produced by roll forming of blanks with tailored variation of section thickness in desired locations.

In the conventional fabrication of automobile body component assemblies, several stampings are formed individually and subsequently spot welded together in order to obtain the material and strength requirements at various locations in the assembly. Alternatively, the various materials can be welded together prior to the forming process to produce what are known as tailor-welded blanks (TWBs). This term is derived from the notion that the automobile designer will be able to tailor the location in the stamping where specific material properties are desired. These differences can be in the material grade, gage thickness, strength, or coating, for example, galvanized versus ungalvanized (Ref 11).

Forming of TWB steel is described in more detail in the article "Forming of Steel Tailor-Welded Blanks" in this Volume. The production of aluminum TWBs involves:

- Blanking aluminum alloy sheet into desired sizes and orientations
- Cleaning the material
- Precisely positioning the materials relative to each other
- Welding the sheets together into a single flat sheet

Each step in the process is considered critical to avoid excessive gap between sheets prior to welding and to prevent or eliminate contamination on the sheet that may result in a weld imperfection (Ref 12). The TWBs began appearing in Europe and Japan in the mid-1980s, and their use has continued to increase. Forty to sixty million TWBs were produced in 2000 (Ref 13).

The TWBs have generated enormous interest in the automotive industry as of late, due to the substantial benefits they produce. These include:

- Reduced manufacturing costs due to fewer forming dies
- Elimination of downstream spot-welding operations
- Reduced scrap; weight reductions due to the combining of parts into a single component; improved dimensional part consistency from the reduction of inaccurate spot-welding processes
- Improved corrosion resistance through the elimination of lap joints by integration of reinforcements (Ref 11)
- Improved crash test results due to the increased stiffness of laser and mash-seam welds in comparison to traditionally used spot welds

However, because the material properties in the heat-affected zone (HAZ) adjacent to the weld and in the weld itself are significantly less desirable from a forming standpoint compared to those of the base metal, creating a failure-free deep-drawing process for TWBs is difficult. While there are not specific published studies on the formability of aluminum in TWB applications, the effects on the mechanical properties of welding aluminum alloys has been studied (Ref 14).

Aluminum has several chemical and physical properties that need to be understood when using the various joining processes. The specific properties that affect welding are:

- Aluminum oxide characteristic
- Solubility of hydrogen in molten aluminum
- Aluminum thermal, electrical, and nonmagnetic characteristics
- Wide range of mechanical properties and melting temperatures that results from alloying with other metals (Ref 14)

Significant reductions in the potential elongation, both transverse (i.e., perpendicular) and longitudinal (i.e., parallel), to the weld line have been reported. In addition, unlike steel, the strength of aluminum in the HAZ decreases due to welding (Ref 14). For a 6xxx-series aluminum alloy, the tensile strength in the HAZ can be reduced by as much as 40% of the base metal value (Ref 14). However, this reduction varies greatly, based on the filler alloy used in the welding, postweld heat treatment or aging, and other welding process parameters, such as the amount and duration of weld heating.

When choosing the optimal filler alloy, the end use of the weldment and its desired performance must be the prime consideration. Many alloys and alloy combinations can be joined using one of several filler alloys, but only one filler may be optimal for a specific application. Table 2 (Ref 15) lists the chemical composition and melting range of standard aluminum filler alloys.

The primary factors commonly considered when selecting a welding filler alloy are:

- Ease of welding freedom from cracking
- Tensile or shear strength of the weld
- Weld ductility
- Service temperature

Table 1 Punch-to-die clearances for blanking and piercing aluminum alloys

Alloy and temper	Clearance per side(a), % t
1100	
O	5.0
H12, H14	6.0
H16, H18	7.0
2014	
O	6.5
T4, T6	8.0
2024	
O	6.5
T3, T36, T4	8.0
3003	
O	5.0
H12, H14	6.0
H16, H18	7.0
3004	
O	6.5
H32, H34	7.0
H36, H38	7.5
5005	
O	5.0
H12, H14, H32, H34	6.0
H36, H38	7.0
5050	
O	5.0
H32, H34	6.0
H36, H38	7.0
5052	
O	6.5
H32, H34	7.0
H36, H38	7.5
5083	
O	7.0
H323, H343	7.5
5086	
O, H112	7.0
H32, H34, H36	7.5
5154	
O, H112	7.0
H32, H34, H36, H38	7.5
5257(b)	
O	5.0
H25	6.0
H28	7.0
5454	
O, H112	7.0
H32, H34	7.5
6061	
O	5.5
T4	6.0
T6	7.0
7075	
O	6.5
W, T6	8.0
7178	
O	6.5
W, T6	8.0

(a) t, sheet thickness. (b) Also alloys 5357, 5457, 5557, and 5657

- Corrosion resistance
- Color match between the weld and base alloy after anodizing

Another concern in TWBs is weld-line movement. Whether produced with steel or aluminum, the stronger material in the TWB will resist deformation more than the weaker material, causing the weld line to move in the stamped part. This effect limits the ability of the designer to position the specific material properties in the final stamping where desired and may create other forming problems, such as wrinkling, tearing, and uncontrollable springback.

Despite these formability concerns, TWBs incorporated into the automobile body were prudent to reap the potential rewards they offer. Some of the components where TWBs are currently being used are body side panels, motor compartment rails, center pillar inner panels, door inner panels, and wheel-house/shock tower panels (Ref 16, 17).

Bending

The most common problems encountered in practice are springback and cracking within the bend area. Problems associated with bend cracking are usually a result of improper bend radii, rough edges of material being formed or the forming equipment, and bending parallel to the direction of grain flow. For the approximate bend radius to use in bending various thicknesses and types of aluminum, see Table 3. Actual practice may reveal that a larger or a smaller radius may be used in some instances. If tighter bend radii are required, then fabricators should proceed with additional caution and, if needed, should seek the assistance of engineering or laboratory metallurgists.

Difficulties encountered with springback are most commonly associated with bending of the stronger alloys, especially those having high yield strength. Springback problems associated with this material can be overcome to a certain degree by overforming. The amount of overforming used will depend on the temper and the alloy; the softer the material, the less springback compensation required. Other means of reducing springback are to bend the material in the soft condition (condition O) or immediately after quenching, and to reduce the thickness or the radius if allowed. Avoid reducing radii to the point that grain separation or bend cracking results.

Table 2 Nominal composition and melting range of standard aluminum filler alloys

Aluminum alloy	Nominal composition, wt%								Approximate melting range	
	Si	Cu	Mn	Mg	Cr	Ti	Al	Others	°C	°F
1100	...	0.12	...	...	...	...	≥99.00	...	643–657	1190–1215
1188	...	...	...	...	...	...	≥99.88	...	657–660	1215–1220
2319	...	6.3	0.30	...	...	0.15	bal	0.18 Zr; 0.10 V	543–643	1010–1190
4009(a)	5.0	1.25	...	0.50	...	...	bal	...	546–621	1015–1150
4010(b)	7.0	...	...	0.35	...	...	bal	...	557–613	1035–1135
4011(c)	7.0	...	...	0.58	...	0.12	bal	0.55 Be	557–613	1035–1135
4043	5.25	...	...	...	...	...	bal	...	574–632	1065–1170
4047	12.0	...	...	...	...	...	bal	...	577–582	1070–1080
4145	10.0	4.0	...	...	...	...	bal	...	521–585	970–1085
4643	4.1	...	...	0.20	...	...	bal	...	574–635	1065–1175
5183	...	...	0.75	4.75	0.15	...	bal	...	579–638	1075–1180
5356	...	...	0.12	5.0	0.12	0.13	bal	...	571–635	1060–1175
5554	...	...	0.75	2.7	0.12	0.12	bal	...	602–646	1115–1195
5556	...	...	0.75	5.1	0.12	0.12	bal	...	568–635	1055–1175
5654	...	...	...	3.5	0.25	0.10	bal	...	593–643	1100–1190
C355.0	5.0	1.25	...	0.50	...	...	bal	...	546–621	1015–1150
A356.0	7.0	...	...	0.35	...	...	bal	...	557–613	1035–1135
A357.0	7.0	...	...	0.58	...	0.12	bal	0.55 Be	557–613	1035–1135

(a) Wrought alloy with composition identical to cast alloy C355.0. (b) Wrought alloy with composition identical to cast alloy A356.0. (c) Wrought alloy with composition identical to cast alloy A357.0. Source: Ref 15

Press-Brake Forming

The press-brake forming techniques used with aluminum alloys are similar to those used with steel and other metals, differing only in some details of tool design (see the article "Press-Brake Forming" in this Volume).

Tolerances in press-brake forming are larger than those in punch-press operations. For simple shapes that are relatively long and narrow, a tolerance of ±0.8 mm (1/32 in.) can usually be maintained. On larger parts of more complex cross section, the tolerance may be as much as ±1.6 mm (1/16 in.).

Springback, or partial return to the original shape upon removal of the bending forces, occurs in most forming operations. The amount of springback depends on the yield strength, strain hardening, the rigidity of the section (which depends on the elastic modulus and the thickness of the material), and on the bend radius. Springback also depends strongly on concentration of strain and the presence of additional stress, such as in stretch bending.

Table 4 shows the effects of these variables, giving springback allowances in degrees of overbending that have been used for high-strength aluminum alloys 2024 and 7075. The springback allowance, or number of degrees of overbending required, ranges from 1 to 12° for 2024-O and 7075-O (yield strength of 76 MPa, or 11 ksi, min), and from $7\frac{1}{4}$ to $33\frac{1}{2}°$ for

Table 3 Cold bend radii (inside) for general applications

Alloy and temper	Bend radii in mm(in.) or as a fraction of sheet thickness (t)													
	mm	in.	mm	in.	mm	in.	mm	in.	mm	in.	mm	in.	mm	in.
Sheet Thickness:	**0.4**	**0.16**	**0.8**	**0.032**	**1.0**	**0.4**	**1.6**	**0.63**	**3.2**	**0.125**	**4.76**	**0.1875**	**6.35**	**0.250**
1100-O	0.5	0.02	0.8	0.03	0.8	0.03	1.5	0.06	3.2	0.125	4.75	0.187	6.35	0.250
3003-O	0.8	0.03	0.8	0.03	1.5	0.06	1.5	0.06	4.1	0.160	4.75	0.187	6.35	0.250
5052-O	0.8	0.03	0.8	0.03	1.5	0.06	1.5	0.06	4.1	0.160	4.75	0.187	6.35	0.250
6061-O	0.8	0.03	0.8	0.03	1.5	0.06	1.5	0.06	4.1	0.16	4.76	0.1875	6.35	0.250
2014-O	0.8	0.03	1.5	0.06	2.3	0.09	2.3	0.09	4.8	0.19	7.92	0.312	11.2	0.44
2219-O	0	0	0	0	...	...	0.5–1.5t	0.5–1.5t	0.5–1.5t	0.5–1.5t	...	0.5–1.5t	...	1–2t
7075-O	0.8	0.03	...	2t	...	2t	...	2t	...	2t	...	2.5t	...	3t
7178-O	0.8	0.03	...	2t	...	2t	...	2t	...	2t	...	2.5t	...	3t
1100-H12	0.5	0.02	0.8	0.03	0.8	0.03	...	1t	...	1t	...	1.5t	...	1.5t
3003-H12	0.8	0.03	0.8	0.03	0.8	0.03	...	1t	...	1t	...	1.5t	...	1.5t
5052-H32	0.8	0.03	1.5	0.06	1.5	0.06	...	1.5t	...	2t	...	2.5t	...	2t
1100-H16	0.8	0.03	1.5	0.06	...	2t	...	2t	...	2t	...	2.5t	...	3t
3003-H16	0.8	0.03	1.5	0.06	...	2t	...	2t	...	2.5t	...	4t	...	5t
5052-H36	0.8	0.03	1.5	0.06	...	2t	...	2t	...	2.5t	...	4t	...	5t
Sheet Thickness:	**0.41**	**0.016**	**0.81**	**0.032**	**1.62**	**0.064**	**3.2**	**0.125**	**4.76**	**0.1875**	**6.35**	**0.250**	**>6.35 mm**	
1100-H18	0.8	0.03	...	2t	...	2t	...	2.5t	...	3t	...	3.5t	...	...
3003-H18	0.8	0.03	...	2t	...	2t	...	2.5t	...	3t	...	4.5t	...	...
5052-H38	0.8	0.03	...	2t	...	3t	...	4t	...	5t	...	6t	...	...
6061-T4	0.8	0.03	...	2t	...	2t	...	2t	...	3t	...	4t	...	...
6061-T6	0.8	0.03	...	2t	...	2t	...	2t	...	3t	...	4t	...	...
2219-T4	...	0–1t	...	0–1t	...	1–2t	...	1–2t	...	1.5–2.5t	...	1.5–2.5t	...	...
2219-T62, T81	...	2–3.5t	...	2.5–4t	...	3.5t	...	4–6t	...	4–6t	...	5–7t	...	...
2024-T4	1.5	0.06	...	4t	...	4t	...	5t	...	6t	...	6t	...	...
2024-T3	1.5	0.06	...	4t	...	4t	...	5t	...	6t	...	6t	...	...
2014-T3	1.5	0.06	...	3t	...	4t	...	5t	...	6t	...	6t	...	...
7075-T6	1.5	0.06	...	5t	...	6t	...	6t	...	6–8t	...	9–10t	...	...
7178-T6	1.5	0.06	...	5t	...	6t	...	6t	...	6–8t	...	9–10t	...	...
7050-T7	...	...	...	...	...	...	...	...	...	8t	...	9t	...	9.5t
7475	...	...	...	...	...	...	...	...	...	...	...	...	...	...

Table 4 Springback allowances for 90° bends in 2024 and 7075 aluminum alloy sheet

Sheet thickness		Springback allowance, in degrees, for bend radius, mm (in.) of:							
mm	in.	2.4 (3/32)	3.2 (1/8)	4.8 (3/16)	6.4 (1/4)	7.9 (5/16)	9.5 (3/8)	11.1 (7/16)	12.7 (1/2)
2024-O and 7075-O									
0.51	0.020	3	4	5 1/2	7 1/2	8 1/2	9	9 1/2	12
0.64	0.025	2 3/4	3 3/4	5 1/2	6 1/2	8	8 1/4	8 3/4	10 3/4
0.81	0.032	2 1/4	3	4 3/4	6	6 3/4	7	7 1/2	9 1/2
1.02	0.040	2	3	4	5	6	6 1/4	6 3/4	8 3/4
1.29	0.051	2	2 1/2	3 1/2	4	5	5 1/4	5 3/4	7 1/2
1.63	0.064	1 1/2	2	2 3/4	3 3/4	4 1/2	5	5 1/2	6 3/4
2.06	0.081	1	1 1/2	2	2 1/2	3 1/4	3 1/2	4	4 3/4
2.39	0.094	...	...	1 3/4	2 1/2	3	3 1/4	3 3/4	4 1/2
3.18	0.125	...	...	1 1/2	2	2 1/4	2 3/4	3	3 3/4
2024-T3									
0.51	0.200	10	12	15 1/2	19	22 1/2	24	27 1/4	33 1/2
0.64	0.025	8 3/4	10 1/2	14	16 3/4	17 3/4	21	23	28 1/2
0.81	0.032	7 3/4	8 3/4	12	14 1/2	16 3/4	17 3/4	19 1/4	24
1.02	0.040	7 1/4	8 1/4	10 3/4	12 3/4	14 1/2	15 1/4	17	20 1/2
1.29	0.051	...	...	9	10 1/2	12 1/4	13	14 1/2	16 3/4
1.63	0.064	...	...	8	9 3/4	11 1/4	12	12 3/4	15
2.06	0.081	...	...	...	...	9 1/2	10 1/2	11 1/4	13
2.39	0.094	...	...	...	...	8 3/4	9 3/4	10 1/2	12

2024-T3 (yield strength of 345 MPa, or 50 ksi). The allowance increases with increasing yield strength and bend radius but varies inversely with stock thickness. The allowance for bends of other than 90° can be estimated on a proportional basis. For bend angles of less than 90°, the springback may be greater unless the bend radius is decreased, because the metal in the bend area may not have been stressed beyond its yield point.

Radii to which bends can be made depend on the properties of the metal and the design, dimensions, and condition of the tools. For most metals, the ratio of minimum bend radius to sheet thickness is approximately constant, because ductility is the primary limiting factor on minimum bend radius. This is not true for aluminum alloys, for which the ratio of bend radius to sheet thickness increases with the thickness.

With special tooling, aluminum alloys can be bent to smaller radii than those indicated in standard tables. Bottoming dies and dies that combine bottoming with air bending are used for this purpose. Hydraulic forming, forming with rubber-pad dies, and high-energy-rate forming also produce good small-radius bends.

Sometimes, it is possible to take advantage of the grain direction in the work metal: The most severe bends can be made across the direction of rolling. If similar bends are made in two or more directions, it is recommended that, if possible, all bends be made at an angle to the direction of rolling. Local heating along the bend lines can sometimes be used to produce small bend radii without fracture; this is particularly useful in bending plate.

The maximum temperature that can be used without serious loss in mechanical properties is 150 to 205 °C (300 to 400 °F) for cold-worked material. Reheating of naturally aged aluminum alloys 2014 and 2024 is not recommended unless the part is to be artificially aged. Generally, any reheating sufficient to improve formability will lower the resistance to corrosion to an undesirable degree, except with alclad sheet.

Blank Development. For relatively simple parts, particularly those for which close tolerances are not required, the blank layout can be developed directly by using bend-allowance tables or equations. As a rule, the initial calculated blank layout and die design are developed into final form by successive trial and modification.

Lubricants are needed for nearly all press-brake forming of aluminum alloys. The light protective film of oil sometimes present on mill stock is often adequate for mild bending operations, but when this is not sufficient, a lubricant is usually applied to the working surfaces of the tools and the bend area of the workpiece to prevent scoring and metal pickup.

Tools. The bending, forming, piercing, and notching dies used in press brakes for aluminum alloys are much the same as those used for low-carbon steel. To prevent marring or scratching of the workpiece, tools used for bending steel should be carefully cleaned and polished before being used for aluminum alloys. Rubber pads used in press-brake dies, when clean, will not scratch the surface of an aluminum sheet.

Because of the differences in tensile strength and springback, shut height settings for aluminum alloys may be different from those for low-carbon steel.

Contour Roll Forming

Aluminum alloys are readily shaped by contour roll forming, using equipment and techniques similar to those used for steel (see the article "Contour Roll Forming" in this Volume). Operating speeds can be higher for the more ductile aluminum alloys than for most other metals. Speeds as high as 245 m/min (800 ft/min) have been used in mild roll forming of 0.8 mm (1/32 in.) thick alloy 1100-O sections 15 to 30 m (50 to 100 ft) long. Power requirements for roll forming of aluminum alloys are generally lower than is the case for comparable operations on steel, because of the lower yield strength of most aluminum alloys.

Tooling. The design of rolls and related equipment, as well as the selection of tool materials, is discussed in the article "Contour Roll Forming" in this Volume. The most commonly used material is L6 tool steel, a low-alloy nickel-chromium grade with excellent toughness, wear resistance, and hardenability. For extremely severe forming operations or exceptionally long runs, a high-carbon high-chromium grade such as D2 is preferred because it has superior resistance to galling and wear. These tool steels are hardened to 60 to 63 HRC. The tools are highly polished and are sometimes chromium plated to prevent scratching and to minimize the pickup of chips when surface finish of the work is critical.

For short runs and mild forming operations, rolls can be made of turned and polished gray cast iron (class 30 or better) or low-carbon steel. For light-gage metals, tools made of plastics reinforced with metal powder, or of specially treated hardwood, have occasionally been used. For some applications in the roll forming of light-gage alloys when quality of surface finish is the primary concern, use has been made of cast zinc tools, at the cost of shorter tool life.

Extremely close tolerances are required on tool dimensions. Allowance for springback must be varied with alloy and temper, as well as with material thickness and radius of forming, as indicated in Table 4. Final adjustments must be made on the basis of production trials.

Tolerances of ±0.127 mm (0.005 in.) are common in contour roll forming, and ±0.05 mm (0.002 in.) can be maintained on small, simple shapes formed from light-gage metals. One or two final sizing stations may be required for intricate contours or when springback effects are great.

Lubricants are required in nearly all contour roll forming of aluminum alloys. For high-speed or severe forming operations, the rolls and workpiece may be flooded with a liquid that functions as both a lubricant and a coolant. A soluble oil in water is preferred for this type of operation. When a more effective lubricant is required, a 10% soap solution or an extreme-pressure compound may be used. These are better suited for minimizing tool wear and producing a high-quality finish, but are more difficult to remove.

Applications. Roll-formed aluminum alloy parts made from sheet or coiled strip include furniture parts, architectural moldings, window and door frames, gutters and downspouts, automotive trim, roofing and siding panels, and shelving.

Tubing in sizes ranging from 20 to 200 mm (3/4 to 8 in.) in outside diameter and from 0.64 to 3.9 mm (0.025 to 0.156 in.) in wall thickness is made in a combined roll-forming and welding operation (see the article "Contour Roll Forming" in this Volume). Linear speeds of 9 to

60 m/min (30 to 200 ft/min) are used in this process. Applications include irrigation pipe, condenser tubing, and furniture parts.

Other applications of contour roll forming include the forming of patterned, anodized, or pre-enameled material. Such applications impose stringent requirements on tool design and maintenance, and lubrication sometimes cannot be used because of the nature of the coating or because of end-use requirements.

Deep Drawing

Equipment, tools, and techniques used for deep drawing of aluminum and aluminum alloys are similar to those used for other metals and are described in more detail in the article "Deep Drawing" in this Volume. This section deals with those aspects of deep drawing that are specific to aluminum alloys and is restricted to procedures using a rigid punch and die. Other procedures are described in subsequent sections of this article.

Equipment. Punch presses are used for nearly all deep drawing; press brakes are sometimes used for experimental or very short runs. Presses used for steel are also suitable for aluminum.

Capacity requirements, determined by the same method used for steel, are generally lower for comparable operations because of the lower tensile strength of aluminum alloys.

Press speeds are ordinarily higher than they are for steel. For mild draws, single-action presses are usually operated at 27 to 43 m/min (90 to 140 ft/min). Double-action presses are operated at 12 to 30 m/min (40 to 100 ft/min) for mild draws, and at less than 15 m/min (50 ft/min) for deeper draws with low- and medium-strength alloys. Drawing speeds on double-action presses are approximately 6 to 12 m/min (20 to 40 ft/min) with high-strength alloys.

Tool Design. Tools for deep drawing have the same general construction as those used with steel, but there are some significant differences. Aluminum alloy stock must be allowed to flow without undue restraint or excessive stretching. The original thickness of the metal is changed very little. This differs from the deep drawing of stainless steel and brass sheet, each of which may be reduced by as much as 25% in thickness in a single draw.

Clearances between punch and die are usually equal to the metal thickness plus approximately 10% per side for drawing alloys of low or intermediate strength. An additional 5 to 10% clearance may be needed for the higher-strength alloys and harder tempers.

With circular shells, metal thickening occurs with each draw; therefore, clearance is usually increased with each successive draw. The restrictions imposed on the drawing of rectangular shells by metal flow at the corners make equal clearances for each draw satisfactory. The final operation with tapered or rectangular shells serves primarily to straighten walls, sharpen radii, and size the part accurately. Therefore, the clearance for these operations is equal to the thickness of the stock.

Insufficient clearance burnishes the sidewalls and increases the force required for drawing. Excessive clearance may result in wrinkling of the sidewalls of the drawn shell.

Wrinkling is a wavy condition obtained in deep drawing of sheet metal, in the area of the metal between the edge of the flange and the draw radius. Wrinkling tendency may also be due to material factors such as variations in thickness and anisotropy, or wrinkles may occur if the blankholder pressure is not sufficiently high.

Radii on Tools. Tools used for drawing aluminum alloys are ordinarily provided with draw radii equal to four to eight times the stock thickness. A punch nose radius is sometimes as large as ten times the stock thickness.

A die radius that is too large may lead to wrinkling. A punch nose radius that is too sharp increases the probability of fracture or of residual circular shock lines that can be removed only by polishing.

Nonetheless, failure by fracture can sometimes be eliminated by increasing the die radius, or by making the drawing edge an elliptic form instead of a circular arc.

Surface Finish on Tools. Draw dies and punches should have a surface finish of 0.4 μm (16 μin.) or smoother for most applications. A finish of 0.08 to 0.1 μm (3 to 4 μin.) is often specified on high-production tooling for drawing light-gage or precoated stock. Chromium plating may also be specified to minimize friction and prevent pickup of dirt or other particles that could damage the finish on the part.

Tool Materials. The selection of materials for deep-drawing tools is discussed in the article "Selection of Materials for Deep-Drawing Dies" in this Volume. Materials for small dies are chosen almost entirely on the basis of performance, but cost becomes a significant factor for large dies. Local variation in wear on tools is an important factor in tool life. A twentyfold variation in rate of wear can be observed on the die radius.

Lubricants for deep drawing of aluminum alloys must allow the blank to slip readily and uniformly between the blankholder and the die and must prevent stretching and galling while this movement takes place.

The drawing compounds can be applied only to the areas that will be subjected to a significant amount of cold working, unless local application interferes with the requirements of high-speed operation. Uniformity of application is critical, especially to enable the maintenance of correct blankholder pressure around the periphery of the die.

Drawing Limits. The reduction in diameter that is possible in a single operation with aluminum alloys is approximately the same as that obtainable with drawing-quality steel. For deep-drawn cylindrical shells, reductions in diameter of approximately 40% for the first draw, 20% for the second draw, and 15% for the third and subsequent draws can be obtained with good practice. The part can usually be completely formed without intermediate annealing. Four or more successive draws without annealing can be performed, with proper die design and effective lubrication, on such alloys as 1100, 3003, and 5005. The amount of reduction decreases in successive draws because of the loss in workability due to strain hardening. The total depth of draw thus obtainable without intermediate annealing exceeds that obtainable from steel, copper, brass, or other common materials.

For high-strength aluminum alloys, the approximate amount of permissible reduction is 30% for the first draw, 15% for the second draw, and 10% for the third draw.

Local or complete annealing is usually necessary after the third draw on alloys such as 2014 and 2024. Alloys 3004, 5052, and 6061 are intermediate in behavior.

The rate of strain hardening is greatest for the high-strength alloys and least for the low-strength alloys. Table 5 shows the changes in mechanical properties that result from successive draws with alloys 3003 and 5052. The major portion of the change is accomplished in the first draw. The rate of strain hardening is more rapid with high-strength heat treatable alloys such as 2014 and 2024.

Practical limits for single-operation deep drawing of cylindrical cups and rectangular boxes have been expressed in terms of dimensional ratios, as shown in Fig. 6. (Reverse redrawing can be used to obtain a deeper shell than indicated by the limits in Fig. 6 for conventional drawing methods.)

The relation of the metal thickness t to the blank diameter D is an important factor in determining the percentage reduction for each drawing operation. As this ratio decreases, the probability of wrinkling increases, requiring more blankholding pressure to control metal flow and prevent wrinkles from starting. Figure 7 shows the effect of this ratio on percentage reduction of successive draws, without

Table 5 Effect of drawing on mechanical properties of aluminum alloys 3003 and 5052

Number of draws	Tensile strength		Yield strength		Elongation in 50 mm (2 in.), %
	MPa	ksi	MPa	ksi	
Alloy 3003					
0	110	16	41	6	30
1	131	19	117	17	11
2	152	22	145	21	9
3	162	23.5	152	22	8
4	169	24.5	155	22.5	8
(a)	(200)	(29)	(186)	(27)	(4)
Alloy 5052					
0	193	28	90	13	25
1	238	34.5	221	32	6
2	272	39.5	248	36	6
3	296	43	255	37	6
4	303	44	262	38	6
(a)	(290)	(42)	(255)	(37)	(7)

(a) Values in parentheses are typical values for these alloys in the full hard condition.

intermediate annealing, for low-strength alloys such as 3003-O.

Blank development is of particular importance in the deep drawing of large rectangular and irregular shapes. Excessive stock at the corners must be avoided, because it hinders the uniform flow of metal under the blankholder and thus leads to wrinkles or fractures.

With suitable tooling and careful blank development, large rectangular and irregular shapes can often be produced economically in large quantities by deep drawing. Smaller quantities are made in sections with inexpensive tooling and then assembled by welding. Both the welding operation and the subsequent grinding and polishing of the weld areas are time-consuming and costly.

Warping. The nonuniformity of stress distribution in the drawing of rectangular or irregular shapes increases the tendency toward warping. Bowing or oilcan effects on the major surfaces become more pronounced with increasing size of the part. Changes can sometimes be made in dimensional details of the drawing tools to eliminate these defects without the need for extra forming operations.

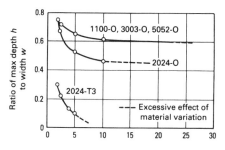

Fig. 6 Drawing limits for single-operation drawing of cylindrical cups or rectangular boxes from aluminum alloy sheet 0.66 to 1.63 mm (0.026 to 0.064 in.) thick. For cylindrical cups, width w equals diameter, and vertical corner radius r equals half the diameter. For rectangular boxes, width w equals the square root of the projected bottom area. If length is more than three times width, drawing limits will be more severe than those shown above. For flanged boxes, the flange width must be included in depth h.

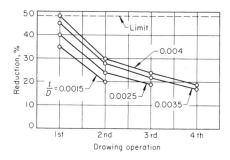

Fig. 7 Effect of thickness-to-diameter ratio on percentage of reduction for successive drawing operations without intermediate annealing for low-strength aluminum alloys such as 3003-O. t, metal thickness; D, blank diameter

Miscellaneous Shapes. Other shapes often produced by deep drawing (besides cylindrical and rectangular shells) include hemispherical shells, flat-bottom hemispherical shells, and tapered shells.

Hemispherical shells with a final inside diameter of less than approximately 150 times the original metal thickness can be drawn in one operation. For inside diameters of more than 150 times thickness, two draws are usually required, to avoid wrinkles. Local thinning in the first draw must be avoided if the second draw is to be successful.

Flat-bottom hemispherical shells, unless very shallow, require at least two draws. The first draw produces a rounded shape, with a larger radius in the bottom area than on the side areas. The final draw flattens the bottom and gives the sides a uniform curvature of the radius required.

Tapered shells require more drawing operations for a given depth of draw than do most other symmetrical shapes. The number of steps required increases with the taper angle.

The bottom edges, except for the final operation, do not have the contour of a circular arc. The profile consists of essentially flat sections at an angle of approximately 40 to 50° from the horizontal. Stepwise reductions are made along the line of final contour, as shown in Fig. 8, and the final draw straightens out the sidewalls to the desired shape.

Each operation after the first is restricted to a shallow draw to minimize strain hardening. With alloys of low and intermediate strength, this procedure makes it possible to complete the

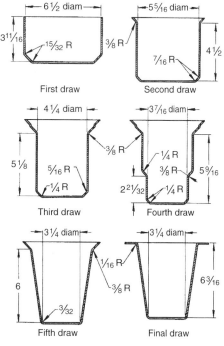

Fig. 8 Typical progression of shapes in multiple-draw forming of a tapered shell from an aluminum alloy blank 1.63 mm (0.064 in.) thick and 292 mm (11½ in.) in diameter. Dimensions given in inches

series of draws without annealing. Contrary to normal practice, the amount of reduction per draw need not be lowered after the second draw. However, polishing or burnishing is often required on the completed shell to obtain a good-quality finish on the sidewalls.

Ironing is avoided in some deep-drawing applications with aluminum alloys but can be used to produce a shell with a heavy bottom and thin sidewalls. The shell is first drawn to approximately the final diameter. The drawing lubricant is then removed, and the shell is annealed, bringing it to temperature rapidly to minimize the formation of coarse grains in areas that have been only slightly cold worked.

The sidewalls can then be reduced in thickness by 30 to 40% in an ironing operation. By repeating the cleaning, annealing, and ironing steps, an additional reduction of 20 to 25% can be obtained, with good control over wall thickness.

A typical use of ironing is shown in Fig. 9. Here, a cylindrical shell is produced with a thick bottom and thin sidewalls by a single deep draw and two successive ironing operations. The approximate final diameter and approximately half the final depth are obtained in the drawing operation. Wall thickness is reduced 33% in the first ironing step and 19% in the second.

Hot Drawing. Severe drawing operations are often impossible to perform at room temperature on large and relatively thick shapes made from high-strength aluminum alloys. However, the lower strength and increased ductility at temperatures above the recrystallization point of the alloy make it possible to produce large and relatively thick shapes by hot drawing. There is little or no advantage when stock is less than 3.2 mm (0.125 in.) thick. Alloys frequently used in applications of this type include non-heat-treatable alloys 5083, 5086, and 5456, and heat treatable alloys 2024, 2219, 6061, 7075, and 7178.

Heavy-duty presses and related equipment are required. Drawing temperatures range from 175

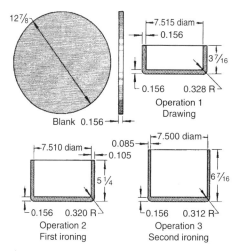

Fig. 9 Progression of shapes in production of a shell with a thick bottom and thin sides in one draw and two ironing operations. Dimensions given in inches

to 315 °C (350 to 600 °F). The length of time the workpiece is held at temperature is controlled to avoid excessive grain growth in areas with little strain hardening. Ordinary drawing compounds break down or burn at elevated temperature and are not suitable for hot drawing operations.

Graphite-containing tallow and hard yellow naphtha soap have sometimes been used as lubricants at intermediate elevated temperature. Lubricants that remain stable above 260 °C (500 °F) include graphite and molybdenum disulfide. These materials can be used in the colloidal form with a volatile vehicle, mixed with other lubricants, or applied to the die as powders.

Spinning

Spinning is often used for the forming of aluminum alloy shapes that are surfaces of revolutions. The manual lathes, automatic spinning machines, chucks, and tools used for aluminum alloys are essentially the same as those used for steel and the other metals commonly formed by spinning (see the article "Spinning" in this Volume).

Hand-spinning lathes and simple tools are suitable for forming aluminum alloy blanks 0.51 to 2.05 mm (0.020 to 0.081 in.) thick; with proper care, stock as thin as 0.10 mm (0.004 in.) can be spun. For thicker and larger blanks, auxiliary equipment is used to apply pressure to the workpiece. This equipment varies from a simple scissors arrangement to feed screws for controlling tool advance; pressure against the work is provided by air or hydraulic cylinders.

Blanks up to 6.4 mm (¼ in.) thick can usually be spun at room temperature. For greater thicknesses, semimechanical to fully mechanical equipment is used, and the work metal is heated. Work metal 25 mm (1 in.) or more in thickness requires special heavy-duty machines and hot spinning.

Aluminum alloy parts 75 mm (3 in.) thick have been spun experimentally. Equipment is available for the spinning of parts as large as 5 m (16 ft) in diameter.

Tolerances for the spinning of aluminum alloys are essentially the same as those for other common metals.

Alloys. A number of aluminum alloys are widely used in spinning applications. Desirable properties are ductility, relatively low ratio of yield strength to ultimate strength, low rate of work hardening, and small grain size.

The alloys of low and intermediate strength that are spun most frequently include 1100, 2219, 3003, 3004, 5052, 5086, and 5154. Annealed blanks are generally used for severe forming; however, a harder temper is sometimes preferred, if it is sufficiently formable, to avoid a tendency to ball up ahead of the tool. A harder temper also may be used when forming is not severe enough to give the product its necessary strength by work hardening.

Heat treatable alloys used for high strength in the finished part are 2014, 2024, and 6061. If the forming is extensive, these alloys often must be annealed several times during spinning, or they may be spun hot.

One method used frequently for spinning heat treatable alloys is:

- Spin annealed blank to approximate form
- Solution heat treat and quench
- Spin to final form at once, before appreciable age hardening

If spinning to the final form cannot be done after solution heat treating and quenching, the quenched parts should be placed in a refrigerator, or packed in dry ice, and held as close to −20 °C (0 °F) as possible until they can be spun. The parts are aged to the T6 temper after spinning has been completed.

Typical spindle speeds for spinning flat blanks and drawn shells of various diameters are listed in Table 6. Rotational speed is decreased as blank diameter increases, so that peripheral speed is maintained in the same range regardless of the size of the workpiece. Peripheral speed ordinarily averages approximately 915 m/min (3000 ft/min) for aluminum alloys. This is somewhat faster than the speeds normally used in spinning copper, brass, stainless steel, and low-carbon steel.

Lubricants are needed in nearly all spinning operations. Beeswax, tallow, and petroleum jelly are suitable for most small parts. Hard yellow naphtha soap is an effective lubricant for larger workpieces. Colloidal graphite in kerosene, or compounds containing molybdenum disulfide, are used in hot spinning. Lubricating compounds used must be easily removable from the finished part without costly treatments.

Applications. Parts produced from aluminum alloys by spinning include tumblers, pitchers, bowls, cooking utensils, ring molds, milk cans, processing kettles, reflectors, aircraft and aerospace parts, architectural sections, tank heads, and streetlight standards.

Spinning is often selected in preference to drawing when quick delivery of small quantities is important, because the spun parts can usually be delivered before drawing tools have been made. Cones, hemispheres, tapered shapes, and parts with complex or reentrant contours (if surfaces of revolution) are often more readily formed by spinning than by other methods. Spinning is also used for very large parts when suitable press equipment and tools are not readily available or are too costly.

Spinning is not usually economical for quantities of more than 5000 to 10,000 pieces because of comparatively low production rates and resulting high unit labor costs. There are exceptions, especially in the power spinning of truncated cone-shaped parts having included angles of 40° or more. Spinning is capable of producing such parts at lower cost than deep drawing, it gives a uniform wall thickness and a surface free from wrinkles, and it increases the tensile strength of the work metal by as much as 100%.

Stretch Forming

Almost all of the aluminum alloys can be shaped by stretch forming. In this process, the work metal is stretched over a form and stressed beyond its yield point to produce the desired contour (for a detailed description, see the article "Stretch Forming" in this Volume).

Typical shapes produced by stretch forming are shown in Fig. 10. These include large shapes with compound curvature formed by longitudinal and transverse stretching of sheet, and compound bends or long, sweeping bends formed from extrusions.

Stretch forming is normally restricted to relatively large parts with large radii of curvature and shallow depth, such as contoured skin. The advantage is uniform contoured parts at faster speed than can be obtained by hand forming with

Table 6 Typical spindle speeds for the spinning of aluminum alloy flat blanks

Blank diameter		Spindle speed, rpm
m	in.	
Flat blanks		
Up to 0.3	Up to 12	600–1100
0.3–0.6	12–24	400–700
0.6–0.9	24–36	250–550
0.9–1.8	36–72	50–250
1.8–3.0	72–120	25–50
3.0–4.5	120–180	12–25
4.5–5.3	180–210	12
Drawn shells		
0.25–0.35	10–14	1000–1200
0.35–0.50	14–20	650–800
0.50–0.75	20–30	475–550
0.75–1.0	30–40	325–375
1.0–1.3	40–50	250–300
1.3–1.8	50–70	200–210
1.8–2.3	70–90	150–175

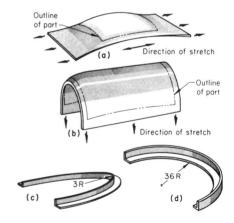

Fig. 10 Typical stretch-formed shapes. (a) Longitudinal stretching. (b) Transverse stretching. (c) Compound bend from extrusion. (d) Long, sweeping bend from extrusion. Dimensions given in inches

a yoder hammer or other means. Also, the condition of the material is more uniform than that obtained by hand forming. The disadvantage is the high cost of initial equipment.

Material used for stretch forming should be limited to alloys with fairly high elongation and good spread between yield and tensile strength. Most of the common alloys are formed in the annealed condition. It is possible to stretch form the heat treatable alloys in tempers T4 or T6, where the shape is not too deep or where narrow width material is used. For the deeper curved shapes, the material is formed in the annealed O temper, heat treated, and reformed, to eliminate distortion resulting from heat treatment. As previously stated, the material should be reformed as fast as possible after heat treatment. In some instances, the material is formed immediately after heat treating and quenching.

Alloys. Properties desirable for stretch forming are high elongation, wide forming range (spread between yield strength and tensile strength), toughness, and fine grain structure. Table 7 shows the effect of elongation and forming range on stretchability ratings for the alloys most commonly used in stretch forming. The stretchability rating varies directly with the forming range, except for 6061-W (which has somewhat higher elongation than adjacent alloys) and 7075-T6 (which has by far the lowest elongation listed). Alloys 1100-O and 3003-O, with the highest elongation shown, nevertheless are less desirable for stretch forming than are the alloys above them in the list. Their low strength and the narrow spread between yield strength and tensile strength make them particularly susceptible to local necking and premature failure in stretch forming.

Tools. The materials used for the form block or die depend on the production quantities required, the severity of local stress and wear on the die, and the thickness and wear properties of the alloy to be formed. Materials include wood, plastics, faced concrete, cast zinc alloys, aluminum tool and jig plate, cast iron, and (rarely) steel or chromium-plated steel.

Lubricants are recommended in the stretch forming of aluminum alloys. Water-soluble oils are commonly used, with viscosity dependent on the severity of forming. Calcium-base greases, paraffin, beeswax, and commercial waxes also are used. The application of too much lubricant can result in buckling of the workpiece.

Sometimes, a layer of sheet rubber, glass cloth, or plastic between die and workpiece serves as a lubricant. Because of their inherent lubricity, zinc alloy dies require only a minimum of lubrication. Smooth-surface plastic dies may require no lubrication, because of their low coefficient of friction against aluminum.

Applications. The various stretch-forming techniques (including stretch drawing, stretch wrapping, and compression and radial drawing) are used extensively in the aerospace industry. Typical parts produced include wing-skin and fuselage panels, engine cowlings, window and door frames, and trim panels used in aerospace, automotive, architectural, and appliance industries.

Stretch draw forming of aluminum is done using both the matched-die and form block techniques. The matched-die method uses a single-action hydraulic press equipped with a means of closing and moving the jaws that grip each end of the blank. The punch is attached to the bed of the press, and the die is attached to the ram.

The alternate method uses a form block that is attached to a stationary bed or a hydraulic cylinder. With this method, the blank is gripped with jaws that hold it in tension or draw it over the form block.

Stretch wrapping uses a form block that is bolted to a rotary table. One end of the blank is clamped to the form block or to a table-mounted gripper. A hydraulic cylinder or a gripper applies tension to the other end of the blank while the form block revolves into it with the turning of the table.

Shaped form blocks that match the contour of extruded or rolled sections are used for support during forming. Filler strips, either segmented or made of low-melting alloys or strips of aluminum, are used to prevent the collapse of sections.

Radial draw forming is a combination of stretch wrapping and compression forming. The workpiece is pressed against the form block by a roller or shoe while being wrapped around the turning form block. This method can be used, for example, to form a flange to a compound curvature while forming a leg, as in the part shown in Fig. 11.

Rubber-Pad Forming

Aluminum alloys are formed by several techniques that can be classified as rubber-pad forming. A general description of processes, equipment, tools, and applications is given in the article "Rubber-Pad Forming and Hydroforming" in this Volume.

Alloys for rubber-pad forming are selected on the same basis as they are selected for similar bending or deep-drawing operations. With non-heat-treatable aluminum alloys, the temper that will meet the forming requirements and give the maximum strength in unworked areas is usually chosen.

Heat treatable aluminum alloys ordinarily are either formed in the annealed temper and then solution heat treated or formed in the freshly quenched W temper.

Tool materials are usually masonite for short runs, and aluminum alloy, zinc alloy, or steel for longer runs. Several types of rubber have been used as the pad material. Certain grades of rubber have particularly good resistance to oils and forming lubricants and are available in a range of hardness, tensile strength, and deflection characteristics to meet different forming requirements.

Capabilities. A given alloy and temper can sometimes be formed more severely by rubber-pad forming than with conventional tools because of the multidirectional nature of the force exerted against the workpiece. Also, the variable radius of the forming pad assists in producing a more uniform elongation of the workpiece than in conventional forming operations.

Forming the shallow part shown in Fig. 12 with a rubber pad and a rigid female die used the variable radius to advantage. The development of wrinkles was almost eliminated, because the rubber acted as a blankholder and kept the work in contact with the flat and contoured die surfaces

Table 7 Mechanical properties and stretchability ratings for aluminum alloys most commonly used in stretch forming

Alloy	Tensile strength		Yield strength		Forming range(a)		Elongation in 50 mm (2 in.), %	Stretchability rating(b)
	MPa	ksi	MPa	ksi	MPa	ksi		
7075-W(c)	331	48	138	20	193	28	19	100
2024-W(c)	317	46	124	18	193	28	20	98
2024-T3	441	64	303	44	138	20	18	95
6061-W(c)	241	35	145	21	97	14	22	90
7075-O	221	32	97	14	124	18	17	80
2024-O	186	27	76	11	110	16	19	80
6061-O	124	18	55	8	69	10	22	75
3003-O	110	16	41	6	69	10	30	75
1100-O	90	13	35	5	55	8	35	70
7075-T6	524	76	462	67	62	9	11	10

(a) Tensile strength minus yield strength. (b) Relative amount of stretch permissible in stretch forming, based on 7075-W as 100. (c) Freshly quenched after solution heat treatment

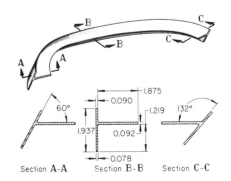

Fig. 11 Aluminum alloy 7075-O radial draw formed T-section with radical changes in angle between leg and flange. Dimensions given in inches

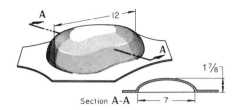

Fig. 12 Shallow part that was drawn from aluminum alloy 6061-O with a rubber pad and a rigid female steel die, in one operation. Dimensions given in inches

as the drawing progressed. A drawing compound was used on the blank.

Limitations. The simpler types of rubber-pad forming have relatively low production rates and correspondingly high unit labor costs compared with punch-press operations. However, the hydroforming (fluid forming) process is adaptable to automatic loading equipment and thus has fairly high production rates.

Applications. Rubber-pad forming is widely used in the aerospace industry, especially for structural parts and skin components. Products made in other industries include appliance parts, license plates, numerals, lighting reflectors, skin panels for buildings, moldings, utensils, and parts drawn from prefinished sheet.

Most rubber-pad forming is done on material 1.6 mm ($^1/_{16}$ in.) or less in thickness, with only a small percentage being thicker than 6.4 mm ($^1/_4$ in.). However, aluminum alloy parts 15.8 mm ($^5/_8$ in.) thick have been formed in special heavy-duty equipment of the rubber-diaphragm type.

Some bulkheads and brackets have both straight and curved flanges with joggles at both ends. The form blocks for such parts are sometimes interchangeable between the Guerin and Verson-Wheelon rubber-pad processes (see the article "Rubber-Pad Forming and Hydroforming" in this Volume). Handwork is usually necessary to set the joggles and to smooth minor buckling in the shrink flanges.

The simultaneous blanking and piercing of flat stock can also be done with rubber-pad tooling. This type of operation is limited to aluminum alloy sheet no thicker than approximately 1.63 mm (0.064 in.).

The control of metal movement that can be obtained with rubber-pad forming not only permits more severe forming than do conventional tools but also is applicable to beading operations. Beads are frequently used to obtain rigidity on large surfaces without increasing the metal thickness.

With a conventional steel punch, die, and blankholder, metal is moved from the edges of the workpiece toward the bead, making the edges somewhat concave and sometimes producing warpage or oilcan effects. Some movement of metal toward the formed area is usually desirable, in order to prevent excessive thinning or cracking of the beads. In the forming of some parts, however, it may be necessary to restrict

metal movement to the immediate vicinity of the beads.

The deep-drawing capabilities of rubber-pad processes vary with the different types of equipment. The severity of drawing possible with heavy-duty rubber-pad drawing by the Marform process (see the article "Rubber-Pad Forming and Hydroforming" in this Volume) is compared as follows with that possible in conventional drawing. The comparison is based on the drawing of alloys 1100-O and 3003-O:

Drawing severity	Reduction in diameter, %	Ratio of depth to diameter
Rubber-pad drawing		
Typical	57	1.1
Maximum	72	3.0
Conventional drawing		
Maximum	40	0.45

Warm Forming of Aluminum

Warm forming is a method for improving the formability of aluminum. Interest in warm forming of lightweight materials began in the 1970s, when it was discovered that an aluminum alloy with 6% Mg content could give a 300% total elongation at approximately 250 °C (480 °F) (Ref 18).

In warm forming of aluminum, the die and the blankholder usually are heated to a temperature in the range of 200 to 300 °C (390 to 570 °F). Many studies show a significant increase in formability of 5xxx and 6xxx series when warm forming is used.

The dies and blankholder are heated with electrical heating rods that are located in the dies. It is necessary to heat up the corners of the dies, because the corners are critical in controlling the metal flow. So, in most cases, it is not necessary to heat up the entire die. Straight sides can be cooled by using water or oil. This reduces material flow, similar to the effect of a draw bead.

Many studies have been conducted in warm forming of aluminum alloys. For example, experimental analyses of rectangular conical cups from an aluminum alloy (5754-O) drawn at room temperature (20, 100, 175, and 250 °C, or 70, 212, 350, and 480 °F) were conducted (Ref 19). Maximum cup heights, obtained without fracture, were compared. These cup heights were 35, 38, 38, and 60 mm (1.4, 1.5, 1.5, and 2.4 in.) for temperatures of 20, 100, 175, and 250 °C (70, 212, 350, and 480 °F), respectively (Ref 19).

Uniaxial tensile deformation and biaxial forming behavior of three aluminum sheet alloys, 5182 + 1% Mn, 5754, and 6111-T4, were studied in the warm forming temperature range of 200 to 350 °C (390 to 660 °F) and in the strain rate range of 0.015 to 1.5 s^{-1} (Ref 20). Attempts have been made to process the selected aluminum sheet alloys so that the microstructural change during warm forming provides adequate recovery favorable to formability but does not

deteriorate the postforming properties. The total elongation in uniaxial tension was found to increase with increasing temperature and to decrease with increasing strain rate. The enhanced ductility at elevated temperatures is contributed primarily from the postuniform elongation that becomes dominant at elevated temperatures and/or slow strain rates. The results of the biaxial forming tests also indicate that, as temperature is increased, part depth, stretchability, and limiting strain level before failure are all increased. The enhancement of strain-rate sensitivity (m-value) with increasing temperature accounts for the plasticity improvement at elevated temperatures under both the uniaxial tension and the biaxial forming conditions. The uniaxial tensile test was identified to serve as a screening test for ranking relative formability among different sheet alloys. The strain-hardened 5xxx alloys (5182 + Mn and 5754) have shown better formabilities than the precipitation-hardened alloy (6111-T4). Furthermore, warm forming limits for the aluminum alloys are superior to those for aluminum-killed steels at room temperature.

Superplastic Forming (Ref 21)

Superplastic behavior has been demonstrated in several aluminum alloys, including the high-strength alloy 7475. The prime material requirement for superplasticity—a fine, stable grain size—can be achieved in aluminum alloys by either static or dynamic recrystallization. In static recrystallization, a deformed microstructure is allowed to undergo discontinuous recrystallization during static annealing, leading to a fine-grain microstructure at the start of superplastic forming.

In dynamic recrystallization, a deformed microstructure undergoes gradual, continuous recrystallization and grain refinement in the course of superplastic forming.

The microstructures of superplastic aluminum alloys can be either dual-phase or essentially a single phase with very small amounts of second phase present. Some amount of second phase is always necessary to develop and stabilize a fine-grain structure.

Table 8 lists the nominal compositions of several superplastic aluminum alloys, their typical grain sizes, and selected mechanical properties. For comparison, the elongations and yield strengths of aluminum alloys 1100-O and 2024-T3 are also shown.

Superplasticity in Aluminum Alloy 7475. The high strength and high fracture toughness of alloy 7475 are the main reasons for examining superplasticity in this material. A number of grain refinement methods have been developed for 7xxx-series aluminum alloys; of these, the Rockwell method (Ref 30) has been relatively easy to implement. A schematic of the Rockwell grain refinement process is shown in Fig. 13. A critical aspect of the process is the heating rate in recrystallization, which must be extremely rapid in order to activate simultaneously as many nuclei as possible.

Stress versus strain-rate plots and corresponding values of strain-rate sensitivity index m for alloy 7475 are shown in Fig. 14. The effect of grain size on flow stress and m value is apparent: Flow stress increases with grain size, while m decreases. At a grain size of 10 to 14 μm (400 to 560 μin.) and a strain rate of 2×10^{-4} s^{-1}, flow stresses are very low (~690 kPa, or 100 psi). Peak m values are very high (~0.8). Flow stress at the same strain rate but a grain size of 40 μm (1600 μin.) is more than doubled to approximately 1380 kPa (200 psi); peak m value has decreased to approximately 0.7.

Superplastic Forming Processes. A number of processes are used for superplastic forming, including blow forming, vacuum forming, thermoforming, deep drawing, and dieless drawing. All of these processes are discussed in detail in the article "Superplastic Sheet Forming" in this Volume. Information on superplastic forming of titanium alloys also is discussed in the article "Forming of Titanium and Titanium Alloys" in this Volume.

Cavitation (formation of internal microvoids during superplastic forming) is a problem in most superplastic aluminum alloys. Many factors, including alloy cleanliness, grain size, flow stress, strain rate, forming temperature, and hydrostatic pressure, influence cavitation in aluminum alloys. Factors that increase the creep-flow stress of the alloy increase the tendency toward cavitation. Creep-flow stress increases with high applied strain rate, low forming temperature, or with large grain size or excessive grain growth (when above the homologous temperature for creep deformation).

Cavitation can be reduced by imposing a pressure on the back side of the sheet during forming (Fig. 15). The forming pressure must be higher than this back pressure. The same forming rates can be achieved with or without back pressure. Back pressures of 690 to 3450 kPa (100 to 500 psi) are generally suitable for suppressing cavitation. A die apparatus used to provide back pressure during forming is illustrated in Fig. 15.

Applications. The use of superplastically formed aluminum components in the aircraft industry is increasing (Ref 31). Figure 16 illustrates the cost and weight savings possible when conventionally fabricated components (in this case, an airframe member) are replaced by superplastically formed parts.

Table 8 Nominal compositions, typical grain sizes, and selected mechanical properties of several superplastic aluminum alloys

Elongation and yield strength of aluminum alloys 2024-T3 and 1100-O are shown for comparison.

Alloy	Nominal composition, %	Grain size μm	Grain size μin.	Tensile elongation, %	Room-temperature yield strength(a) MPa	Room-temperature yield strength(a) ksi	Reference
Al-33Cu	Al-33Cu	3–4	120–160	400–1000(b)	186	27	22
08050	Al-5Cu-5Zn	1–2	40–80	600(b)	152	22	23
Al-8.5Zn-1.25Mg-0.3Zr	Al-8.5Zn-1.25Mg-0.3Zr	8	320	1500(b)	...	...	24
7475	Al-5.8Zn-1.6Cu-2.3Mg-0.22Cr	10–14	400–560	800–1200(b)	483	70	25
Supral 100	Al-6Cu-0.4Zr	2–3(c)	80–120	1000(b)	283	41	26
Supral 220	Al-6Cu-0.35Mg-0.1Ge-0.1Si	2–3(c)	80–120	900(b)	448	65	27
Aluminum-lithium	Al-2.5Li-1.2Cu-0.6Mg-0.13Zr	2–5(c)	80–200	800(b)	469	68	28, 29
2024-T3	Al-4.4Cu-1.5Mg-0.6Mn	...	...	18	345	50	...
1100-O	99.00 min Al	...	...	35	90	13	...

(a) In aged condition whenever applicable. (b) Determined at optimal strain rate and temperature for the specific material. (c) In dynamically recrystallized condition. Source: Compiled from Ref 21

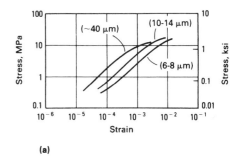

(a)

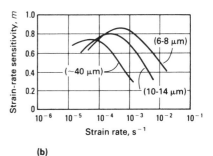

(b)

Fig. 14 (a) Stress versus strain and (b) corresponding strain-rate sensitivity m for superplastic aluminum alloy 7475 in three different grain sizes. Tests were performed at 516 °C (960 °F), the optimal forming temperature for alloy 7475. Source: Ref 21

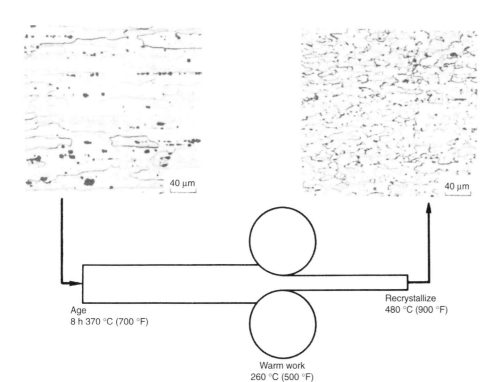

Fig. 13 Steps involved in thermal and mechanical processing to produce superplastic aluminum alloy 7475. Source: Ref 30

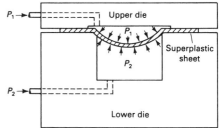

Fig. 15 Die apparatus for providing back pressure during superplastic forming to suppress cavitation. P_1, forming pressure; P_2, back pressure. Source: Ref 21

Explosive Forming

Explosive forming is one of the high-energy-rate forming methods that are employed in the production of aluminum alloy parts, mainly in the aerospace industries. It is often used to produce parts whose size exceeds the limits of conventional equipment or whose thickness requires pressures not obtainable with conventional equipment. It is also used to form small quantities of complex parts that would be more costly to produce by conventional techniques.

Deformation velocities are several hundred feet per second, compared with 0.15 to approximately 6 m/s (0.5 to approximately 20 ft/s) for conventional forming processes. The time required for the workpiece to deform to its final shape is a few milliseconds, with working pressures of several thousand to several hundred thousand pounds per square inch. Water usually serves as the pressure medium.

Details of equipment, tools, and procedures used in explosive forming are available in the article "High-Velocity Metal Forming" in this Volume.

Capabilities. Types of operations possible in explosive forming include panel forming (bending), piercing, flanging, shallow dishing, deep drawing, and cylindrical bulging. Part dimensions range from 25 mm (1 in.) to approximately 15 m (50 ft); work metal thicknesses range from several thousandths of an inch to approximately 152 mm (6 in.).

Alloys. The explosive forming process can be used with any aluminum alloy. Formability may be roughly compared in terms of the ordinary tensile elongation values, but the function is different for each alloy, because of different strain-rate behavior. Alloy 1100-O is rated the most formable of all common metals by explosive forming.

Effect on Mechanical Properties. Changes in mechanical properties as a result of explosive free-forming operations are essentially the same as those observed with conventional forming techniques to produce the same part. Explosive

forming in a die, however, often causes the metal to strike the die at extremely high velocity. The resulting high interface pressures can increase the yield and tensile strengths substantially. Forming capability is increased when critical forming velocities are exceeded.

Dies. Only a forming die or cavity is needed for explosive forming, because the shock wave acts as a punch. Some direction and concentration of the shock wave is obtained with suitably shaped and positioned reflectors.

Cast iron and cast steel are the most frequently used die materials. A variety of other materials and combinations of materials are used, depending on the impact of the shock wave workpiece against the die, the size of the die, dimensional tolerances on the part, and quantity of parts. These materials include low-melting cast alloys and plastics, reinforced concrete, concrete faced with plastic-glass composites, and high-impact steel.

The air between the workpiece and die cavity must be evacuated before forming, because the forming speed is so great that the air will be trapped between the workpiece and die rather than displaced, as in conventional press forming. Trapped air and excessive lubrication cause malformed areas. The vent holes for evacuating the air must be placed in noncritical areas; otherwise, marks will appear on the formed parts. In thinner parts, the forming force will pierce holes in the parts, with the vent hole acting as a piercing die. The surface finish of the die cavity is also important because it is reproduced in mirror image on the workpiece.

Lubricants, if used, are usually extreme-pressure types. Because of the high velocity and the extreme pressures of forming, excessive lubrication must be avoided. Dies of low-melting alloys or those with smooth surfaces require little or no lubrication.

Springback is of importance in die design. Increasing the explosive charge or reducing the standoff distance reduces springback. However, die wear is thereby increased, and the more brittle die materials may fracture. A compromise is often required. Compensation is sometimes made for die wear by reducing charge size or increasing standoff distance to produce a controlled amount of springback and maintain dimensional tolerances.

Studies on alloy 2219 have shown springback to increase when sheet thickness decreases between 6.35 and 0.81 mm (0.250 and

0.032 in.), and also to increase substantially when a lubricant is applied. Incremental forming, on the other hand, has been observed to reduce the extent of springback. Draw radius, draw depth, and die material have shown no significant effect on springback behavior.

Examples of Applications. The forming of flat and moderately curved shapes has been one of the most useful applications of explosive forming. These have included parts ranging from small, detailed items a few square inches in area to large panels with areas in excess of 2.8 m² (30 ft²).

The curved, corrugated panel shown in Fig. 17 was formed from alloy 2014 in the O, T4, and T6 tempers in a laminated epoxy-fiberglass die. The panel was formed in a single shot, using a detonating fuse as a source of energy.

Another example of an explosively formed aluminum part is the alloy 6061 instrument container shown in Fig. 18. The part was produced in a closed die using a hydraulic clamping system; tolerances were ±0.076 mm (0.003 in.).

Tubular parts also are readily shaped by explosive forming, using a length of detonating cord suspended along the axis of the tube.

Electrohydraulic Forming

Another high-energy-rate forming (HERF) method used in the fabrication of aluminum alloy parts is electrohydraulic forming (EHF). In this process, either a spark gap or an exploding bridgewire is employed to discharge

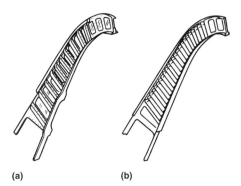

Fig. 16 Example of cost and weight savings obtainable using superplastic forming in the aircraft industry. Conventionally fabricated part (a) had 15 pieces and required 212 fasteners; the superplastically formed part (b) consists of 3 parts and requires 45 fasteners. This results in a 56% cost savings and a 13% weight savings.

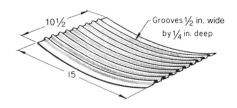

Fig. 17 Curved, corrugated panel produced by explosive forming from aluminum alloy 2014 0.51 mm (0.020 in.) thick. Dimensions given in inches

Fig. 18 Aluminum alloy 6061 instrument container fabricated from a blank by explosive forming. Courtesy of Explosive Fabricators, Inc.

electrical energy in water or another liquid. This generates an extremely high pressure and a shock wave similar to those produced in explosive forming.

Once the energy is released in the transfer medium, the remainder of the operation is essentially the same as it is for explosive forming.

Capabilities of EHF differ somewhat from those of explosive forming. The spark-gap method can apply programmed repetitive shock waves of varying magnitude without removal of the workpiece from the die.

The exploding-bridgewire method is less readily automated, but the shock wave can be localized and directed by the shape and placement of the wire.

Dimensional tolerances can be held to lower limits than with explosive forming, because the discharge of energy is more closely controlled. For this reason, EHF is sometimes used for a restrike or sizing operation after preliminary explosive forming to an approximate contour.

Commercial equipment is available that can produce approximately 3000 small- or medium-sized pieces per week.

Examples of Applications. Electrohydraulic forming is well suited to the production of parts such as those shown in Fig. 19 and to the production of other transitional shapes in tubing. Both of the parts shown in Fig. 19 were originally fabricated by welding two drawn pieces, but the use of EHF resulted in considerable cost savings as well as in parts with closer tolerances and better surface finish.

Electromagnetic Forming

Operations generally similar to those described for the preceding two HERF methods also can be carried out by electromagnetic forming (EMF). In this process, the discharge of a capacitor through a coil generates an intense magnetic field. This field interacts with the electric currents induced in a conductive workpiece to produce a force perpendicular to the workpiece surface.

Details of the process and of equipment, tools, and procedures are described in the article "High-Velocity Metal Forming" in this Volume. The method is suitable for aluminum alloys because of their formability and high electrical conductivity. Pressure-tight joints, electrically or thermally conductive joints, torque joints, and structural joints between metals can be produced by EMF techniques in a variety of shapes.

Examples of Applications. Electromagnetic forming is being used to attach an aluminum skirt to a machined bulkhead as part of an engine inlet mounting assembly for an aircraft (Fig. 20). The skirt is positioned over the bulkhead, and EMF is used to compress the skirt locally into a premachined configuration in the bulkhead.

Welding and mechanical fasteners also were considered for this application. Welding was eliminated because of the large difference in section thicknesses being joined and the distortion that would accompany the welding operation. Mechanical fasteners were eliminated because of the increased stresses that would be caused by adding holes to the bulkhead. Joint integrity is of paramount importance because the completed assembly is used in the very front of an aircraft engine, and any failure could result in ingestion of debris into the engine itself. The EMF joint meets all design requirements for the application. Another example of the use of EMF is in the joining of both ends to a tubular aluminum alloy 6063-T832 drive shaft (Fig. 21).

Hydraulic Forming

True hydraulic forming by direct oil pressure against the surface of the workpiece has been applied to aluminum alloy flat stock. The process has been used mainly for the drawing of multiple beads on small quantities of large, flat sheets of thin material for aerospace applications. As shown in Fig. 22, a form block attached to the ram of the press holds the workpiece tightly against a selector plate, through which oil is introduced into channels at the bead locations.

In typical applications, up to 20 beads have been drawn in parts 510 to 760 mm (20 to 30 in.) wide by 1525 to 2030 mm (60 to 80 in.) long and approximately 0.30 mm (0.012 in.) thick, made from alloy 2024. Clamping force required is approximately 2.7 MN (300 tonf), and necessary forming oil pressure is approximately 6.9 MPa (1 ksi). Vents are provided in the form block to allow the escape of air from each bead cavity. The oil film left on the form block after each operation provides sufficient lubrication to draw the next part.

Forming by Shot Peening

The major application of shot peening is to increase the fatigue life of metal parts by producing a uniform compressive stress in the surface layers. Shot peening is sometimes used as a metal-forming process and is especially useful in the forming of large, irregularly shaped parts from aluminum alloy sheet stock.

Shot. When steel shot is used to peen form aluminum alloy parts, the parts are usually treated chemically after forming to remove

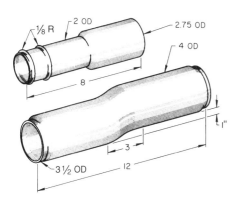

Fig. 19 Alloy 5052-O stepped tube and offset tube produced as one-piece units by electrohydraulic forming. The parts were originally produced as welded assemblies. Dimensions given in inches

Fig. 20 Aluminum alloy engine inlet mounting assembly for an aircraft before (right) and after (left) assembly by electromagnetic forming. Courtesy of Grumman Aircraft Systems

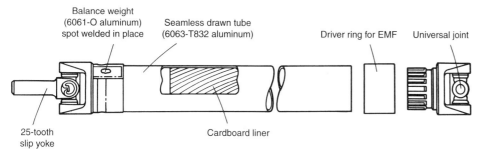

Fig. 21 Schematic of aluminum alloy 6063-T832 drive shaft with ends attached to drawn aluminum shaft by electromagnetic forming (EMF). Source: Ref 32

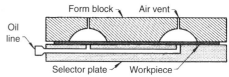

Fig. 22 Typical tooling setup for hydraulic forming of multiple beads in flat stock

particles of iron or iron oxides that may be embedded in the surface.

Slugs cut from stainless steel or aluminum alloy wire are sometimes used. When peening with aluminum alloy slugs, no subsequent chemical treatment is needed, and the danger of overpeening and high localized residual stress (which sometimes occurs with steel or iron shot) is also eliminated.

Automatic or semiautomatic devices are available for the separation and removal of fines and undersized shot, and for the addition of new shot. Manual handling of shot and batch replacement may be more feasible for small-scale operations. The proportion of full-sized shot in the system is usually maintained at a minimum of 85%.

Control. The effectiveness of shot peening depends on the size, shape, material, and velocity of the shot, and on the quantity of shot striking a unit area per unit time. The combined effect of these variables is known as peening intensity.

The angle at which the shot strikes the work also affects the peening intensity, which is proportional to the sine of the angle of impingement. The amount of breakdown of the shot will, of course, also affect peening intensity. The extent of surface coverage as measured by visual or instrumental techniques is often used, together with Almen test strips, to control peening operations.

Applications. One of the earliest forming applications of shot peening was the contour forming of integrally stiffened aircraft wing panels. Because of their extreme length and variable thickness, these parts are ill suited for forming by mechanical processes. Other parts formed by shot peening include honeycomb panels and large tubular shapes. Large, irregularly shaped parts are conveniently formed by this method.

If a part is deformed beyond the specified amount, the contour can be corrected by peening the reverse side. Also, peening can be used as a salvage procedure to correct the contours of bent or distorted parts.

The process is usually carried out as a free-forming technique, without dies or form blocks. Contour is checked against a template. The peening intensities and the number of passes are varied depending on the material and the severity of forming required. Local areas can be subjected to the required treatment.

Drop Hammer Forming

Drop hammer forming is of value for limited production runs that do not warrant expensive tooling. For example, it is often used in experimental work to make trial parts and parts that are expected to undergo frequent design changes.

Tooling costs are low, and finished parts can be produced quickly. However, only relatively shallow parts with liberal radii can be drop hammer formed, and material thickness must be in the range of approximately 0.61 to 1.63 mm (0.024 to 0.064 in.). Also, wrinkling occurs frequently, and a high degree of operator skill is required.

Equipment and Tools. Air-operated hammers with sensitive and accurate control are usually preferred to hammers operated by gravity or steam.

The material is formed in a sequence of small steps. In a typical setup, several plywood or rubber spacers are stacked on the die face, and one or more are removed after each stroke to form the workpiece progressively.

In a variation of this procedure, a series of dies can be used to accomplish the progressive forming. Only the last of these dies requires close tolerances. A rubber pad several inches thick is sometimes used between workpiece and punch in all but the final step.

Dies are simple and inexpensive. Bottom dies are cast from zinc alloy. Punches can also be made from zinc alloy, but if requirements on sharpness of radii and accuracy of contour are not stringent, punches cast from lead are used for short runs. These need not be cast accurately because they deform to the shape of the bottom die in a few strokes. For longer runs, tools can be made of cast iron or cast steel. Lubrication requirements are similar to those for drawing operations.

Alloys used most frequently are 1100, 3003, 2024, 5052, 6061, and 7075. Annealed tempers permit the greatest severity of forming. Intermediate tempers of the non-heat-treatable alloys are often used for channel shapes and shallow, embossed panels. Heat treatable alloys can be partly formed in the annealed condition and given a restrike operation after heat treatment, or they can be formed in the fresh W temper.

All processing conditions being equal, aluminum alloy stock will wrinkle more readily than the same thickness of steel sheet. For comparable results in forming, aluminum alloys must be approximately 40% thicker than steel. More information on this process is available in the article "Drop Hammer Forming" in this Volume.

Other Forming Methods

A number of additional conventional forming processes are applied to aluminum alloy sheet, including embossing, coining, stamping, curling, expanding or bulging, contracting or necking, hole flanging, and beading or ribbing.

Embossing, Coining, and Stamping. These three closely related methods for making shallow impressions and patterns by compression between a punch and a die are frequently combined with drawing. In these operations, the material must yield under impact and compression, and it must be ductile to avoid fracture in tension.

Uniform thickness in all areas of the workpiece generally is maintained in embossing; however, some stretching occurs. Simple designs are produced with light pressure, using a punch of the desired shape and an open female die. Complex patterns require high pressure and a closed matching female die or a rubber female die.

Coining differs from embossing in that the metal is made to flow, thus producing local differences in metal thickness. The design on the top and bottom surfaces may be different. Very high pressure is required.

Stamping produces cut lines of lettering or patterns in one side of the workpiece, to a depth of 0.51 to 1.0 mm (0.020 to 0.040 in.). The depth of penetration must be carefully controlled to minimize distortion and to prevent the design from appearing on the opposite side. Outline or open-face stamps are preferred.

Curling or false wiring can be done in a variety of machines, such as press brakes, single-action punch presses, lathes, roll-forming machines, or special beading machines. The selection of machine depends on the shape and the number of parts required. Circular parts are usually curled on spinning lathes, and rectangular parts are curled in presses. Long, relatively narrow parts can be curled in press brakes or roll-forming machines. Various types of machines have been built specifically for curling in high production quantities.

The edge to be curled should be of uniform height and free from roughness on the outside of the curl. Preferably, it should be rounded slightly before beginning the operation. The minimum radius for curling should be $1\frac{1}{2}$ to 4 times the metal thickness, depending on alloy and temper.

Expanding or bulging of aluminum alloy parts can be carried out by several means, including segmented mechanical dies, rubber punches, or hydraulic pressure.

Segmented mechanical expanding dies are relatively inexpensive and are capable of high production rates but are limited to certain shapes and may produce marks on thin stock or low-strength alloys.

Rubber punches are widely used and are applicable to extremely difficult operations or those impossible to do by other means. Rubber is selected at a hardness, tensile strength, and deflection most suitable for the workpiece shape. The rubber punch or pad must be correctly shaped and located to apply pressure to the shell wall at the required points; it must be kept free from oil; and it should be lubricated with talc, pumice, or other powder-type lubricant.

Water and oil can also be used to exert pressure directly against the workpiece, but this technique requires expensive tooling and controls and is often messy.

Contracting or necking operations reduce the diameter of a shell, usually at the open end. This entails reductions ranging in severity from the forming of a shallow circumferential groove to the forming of a bottleneck shape.

Reduction in diameter in a single operation should not exceed 8 to 15%, depending on alloy, temper, and extent of prior work hardening. The angle from the body to the necked diameter should be less than 45°, to prevent collapse of the shell. It may be necessary to anneal the workpiece locally.

Hole flanging, the forming of a flange or collar around a hole in sheet stock, can be a critical operation. The hole should be punched from the side opposite the intended flange. This prevents splitting of the severely stretched outer edge of the flange. Splitting could be initiated by the burred edge of the hole.

Shallow-flanged holes can be produced in a single pierce-and-flange operation with a stepped punch. The edges of the pierced hole should also be as smooth as possible. Low-strength ductile alloys in the annealed temper will permit forming the deepest flanges and the sharpest bend radii.

Beading or ribbing is usually the most economical way to provide stiffness and avoid oil-can or buckling effects in large panels. Beads that extend from edge to edge of the workpiece are conveniently formed either by bending in a press brake or with corrugating rolls.

Beads that do not extend all the way across the part require a stretching or forming operation either with a rubber-pad die or in a punch press with a rigid punch and a rigid die. A double-acting die and a blankholder can be used to prevent wrinkling at the ends of the beads, and deep, parallel beads are often made one at a time. Rubber-pad forming can also be used, as can drop hammer forming for small quantities.

REFERENCES

1. S.S. Hecker, Forming Limit Diagrams, *Met. Eng. Q.,* Vol 14, 1974, p 30–36
2. I.J. Polmear, *Light Alloys,* Edward Arnold, 1981
3. J.T. Gau and G.L. Kinzel, *J. Mater. Process. Technol.,* Vol 108, 2001, p 369
4. L.R. Morris et al., Formability of Aluminum Sheet Alloys, *Aluminum Transformation Technology and Applications,* C.A. Pampillo, Ed., American Society for Metals, 1982, p 549–582
5. "Comparison of Olsen Cup Values on Aluminum Alloys," Publication T13, Aluminum Association, Feb 1975
6. K. Rao and E. Mohan, Simple Prediction of Sheet Metal Forming Limit Stresses and Strains, *Sheet Metal Forming Technology,* M. Demeri, Ed., TMS, 1999, p 205
7. S.S. Hecker, "A Simple Forming Limit Curve Technique and Results on Aluminum Alloys," International Deep Drawing Research Group Congress, Oct 1972
8. T.B. Stoughton, Stress-Based Forming Limits in Sheet Metal Forming, *J. Eng. Mater. Technol. (Trans. ASME),* Vol 123, 2001, p 123
9. T.B. Stoughton and J.W. Yoon, Sheet Metal Formability Analysis for Anisotropic Materials under Non-Proportional Loading, *Int. J. Mech. Sci.,* in press
10. T. Kuwabara, K. Yoshida, K. Narihara, and S. Takahashi, Forming Limits of Aluminum Alloy Tubes under Axial Load and Internal Pressure, *Proc. of Plasticity '03,* NEAT Press, 2003, p 388–390
11. E. Kubel, *Manuf. Eng.,* Vol 119, 1997, p 38
12. R.W. Davis, H.E. Oliver, M.T. Smith, and G.J. Grant, *JOM,* Nov 1999, p 46
13. S. Das, *Adv. Mater. Process.,* Vol 157 (No. 3), 2000, p 41
14. J.R. Davis, *Aluminum and Aluminum Alloys, ASM Specialty Handbook,* ASM International, 1993
15. J.R. Davis, *Aluminum and Aluminum Alloys, ASM Specialty Handbook,* ASM International, 1993, p 378
16. B. Kinsey, V. Viswanathan, and J. Cao, *J. Mater. Manuf.,* Vol 110, 2001, p 673
17. B. Kinsey and J. Cao, *J. Manuf. Sci. Eng.,* Vol 125, 2003, p 344
18. T. Altan et al., *Fabricator,* Jan 2002
19. P.J. Bolt, N.A.M.P. Lamboo, and P.J.C.M. Rozier, *J. Mater. Process. Technol.,* Vol 115, 2001, p 118
20. D. Li and A. Ghosh, *Mater. Sci. Eng. A,* Vol 352, 2003, p 279
21. A.K. Ghosh, Superplasticity in Aluminum Alloys, *Superplastic Forming,* S.P. Agrawal, Ed., American Society for Metals, 1985, p 23–31
22. D.L. Holt and W. Backofen, *Trans. ASM,* Vol 59, 1966, p 755
23. D.J. Lloyd and D.M. Moore, in *Superplastic Forming of Structural Alloys,* N.E. Paton and C.H. Hamilton, Ed., American Institute of Mining, Metallurgical, and Petroleum Engineers, 1982, p 147
24. K. Matsuki, H. Morita, M. Yamada, and Y. Murakami, *Met. Sci.,* Vol 11, 1977, p 156
25. A.K. Ghosh, in *Superplastic Forming of Structural Alloys,* N.E. Paton and C.H. Hamilton, Ed., American Institute of Mining, Metallurgical, and Petroleum Engineers, 1982, p 85
26. B.M. Watts, M.J. Stowell, B.L. Baikie, and D.G.E. Owen, *Met. Sci.,* Vol 10, 1976, p 189
27. A.J. Barnes, Paper presented to the Society of Automotive Engineers, Detroit, MI, Feb 1984
28. A.K. Ghosh, Rockwell International Science Center, unpublished research, 1984
29. J. Wadsworth, The Development of Superplasticity in Aluminum-Lithium Base Alloys, *Superplastic Forming,* S.P. Agrawal, Ed., American Society for Metals, 1985, p 43–57
30. N.E. Paton and C.H. Hamilton, U.S. Patent, 1978
31. C. Bampton, F. McQuilkin, and G. Stacher, Superplastic Forming Applications to Bomber Aircraft, *Superplastic Forming,* S.P. Agrawal, Ed., American Society for Metals, 1985, p 76–83
32. S.B. Carl and C.M. Foster, "Aerostar Aluminum Driveshaft," Technical Paper 841697, Society of Automotive Engineers, 1984

Flanging of Aluminum

Jian Cao, Northwestern University

FLANGING is a process used to form a projecting rim or edge on a part. The process definition and the hardware needed in the operations of bending, flanging, and hemming are provided in the Section "Forming Processes for Sheet, Strip, and Plate" in this Volume. In this article, emphasis is given to how to determine the flanging limits in terms of fracture, wrinkling, and springback, and their influencing material and process parameters. Similar to the flanging process, hemming involves severe bending operations as well. Interested readers can refer to Ref 1 for an experimental study of hemming of aluminum 1050.

Fracture

Bendability is often defined as the ratio of the thickness of the material to the minimum bend radius that a material can achieve before failure (Ref 2). Figure 1 shows a fractured aluminum autobody sheet resulting from a flat-hemming operation (Ref 3). It can be seen from Fig. 2, a micrograph of the outer surface of a bend specimen of 1.0 mm (0.04 in.) thick 6111-T4 sheet for a bend radius of 0.406 mm (0.016 in.), that the fiber at the outer surface is elongated quite significantly. It has been widely accepted that the well-known forming limit diagram is not suitable for predicting failure in bending, flanging, and hemming. The concept of the forming

limit diagram is explained in the article "Formability Testing of Sheet Metals" in this Volume. In general, 5xxx alloys have significantly better bendability than 6xxx alloys; however, they experience surface roughening on the outer surface of the bend sample that is similar to the 6xxx alloys (Ref 2).

In general, hemmability and bendability are primarily influenced by four factors: work hardening, constituent particles, grain-boundary precipitates, and surface roughening. Hemmability can be enhanced by more work hardening, less volume fraction, more uniformly distributed constituent particles, cleaner grain boundaries, smaller size of grain-boundary precipitates (especially avoiding plate-shaped grain-boundary precipitates), smoother as-rolled surface, and suppressed surface roughening in deformation (Ref 3).

Example 1: Bendability Cannot be Measured from the Forming Limit Diagram. In Fig. 2, a sheet of 1.0 mm (0.04 in.) thick 6111-T4 was bent to an arc with an inner radius of 0.406 mm (0.016 in.). A rough estimation of the strain along the circumferential direction accounting for the shift of neutral axis is:

$$\varepsilon = \ln \frac{r+t}{\sqrt{r(r+t)}} = \frac{1}{2} \ln \frac{r+t}{r}$$
$$= \frac{1}{2} \ln \frac{0.406+1.0}{0.406} = 0.62 \qquad \text{(Eq 1)}$$

where r is bend radius, and t is thickness. This is way beyond the elongation limit of this material

obtained from a tensile test—23% recorded in the same study. The cause of this extra stretchability in bending is believed to be due to the high strain gradient across the thickness in this kind of tight bend. A closed-form solution for predicting this bendability is not yet available, because it highly depends on the microstructure of the material, for example, the size and distribution of precipitate particles. Dao and Li (Ref 3) attempted a crystal-plasticity-based computational approach to study the localization and fracture initiation modes in bending of aluminum sheets, and they showed some success.

In 1992, Mnif studied the bendability of aluminum sheet AlMg0.4Si1.2 during a 90° straight flanging process (Ref 4). Figure 3 shows the results of his study when 1.25 mm (0.05 in.) sheets were subjected to uniaxial prestretching of 5, 10, and 15% in the bending direction and then bending with bend radii of 1, 2, 3, and 4 mm (0.04, 0.08, 0.12, and 0.16 in.) and a bend gap of 1.35 mm (0.053 in.).

Example 2: Bend Performance of 6xxx Alloys. Friedman and Luckey (Ref 2) experimentally investigated the bendability and surface roughness of several aluminum alloys, including 6111-T4 (Al-0.82 Si-0.26 Fe-0.59 Cu-0.20 Mn-0.58 mg) and 6022-T4 (Al-0.82 Si-0.13 Fe-0.07 Cu-0.06 Mn-0.60 mg). Figure 4 shows the initial yield stresses and the ultimate stresses along the longitudinal and transverse directions for these

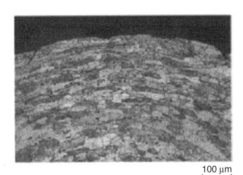

Fig. 1 Example of an aluminum alloy fractured in a flat-hemming operation. Source: Ref 3

Fig. 2 Micrograph of the outer surface of an aluminum alloy 6111-T4 bend specimen. Source: Ref 2

100 μm

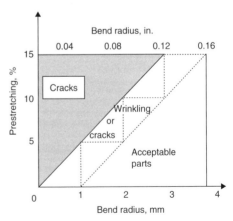

Fig. 3 Fracture limit of an aluminum sheet (AlMg0.4Si1.2-ka; sheet thickness, 1.25 mm or 0.05 in.) in straight flanging. Source: Ref 4

two alloys. As can be seen from Fig. 5, interestingly, the bendability (a good bendability is indicated by a low value of r/t in the figure) of 6111-T4 is better than that of 6022-T4 when the prestrain is lower than 5%. Once the prestrain exceeds 5%, the trend is reversed. This phenomenon is expected to be related to alloying of the two different aluminum materials. More research is needed to find out the exact cause.

Wrinkling

Wrinkling in shrink or convex flanging can be problematic in the flanging area as the outer edge is compressed. Figure 6 shows the setup and a before-and-after flanged sheet. These wrinkles can significantly reduce the assembly accuracy. Furthermore, wrinkles are sometimes pressed (or ironed) out when the clearance between the punch and the die is small. This complicates the deformation history and generates an uneven stress distribution in the sheet, which makes the accurate prediction of springback (or distortion) extremely difficult.

Theoretical prediction of the wrinkling limit in shrink flanging has been specifically studied (Ref 5, 6). In Ref 5, the critical wrinkling strain was derived as:

$$\varepsilon_\theta^{cr} = \left\{ \frac{1}{3} \left(\frac{t}{R_2} \sin \alpha \right)^2 \frac{E}{K} n \right\}^{\frac{1}{n+1}} - \varepsilon_0 \qquad \text{(Eq 2)}$$

where K and n are the material parameters in the power law to describe the material hardening

behavior ($\bar{\sigma} = K\bar{\varepsilon}^n$), t is the sheet thickness, ε_0 is the prestretching amount in the sheet before the flanging process, R_2 is shown in Fig. 7, and E is Young's modulus.

Wang et al. (Ref 6) investigated the wrinkling limit of aluminum shrink flanging and compared the results to the experimental study conducted by Li at Alcoa, which is summarized here. More details can be found in their paper. The approach used in the analysis was based on the energy method. The material hardening behavior was modeled with Voce's law:

$$\bar{\sigma} = A - B \exp(-c\bar{\varepsilon}) \qquad \text{(Eq 3)}$$

which describes the aluminum sheet very well. The stress, $\sigma_{applied}$, corresponds to the maximum compressive hoop stress in the flange sheet and can be calculated as:

$$\sigma_{applied} = A - B$$
$$\times \exp\left(-c\sqrt{\frac{2(1+R)}{1+2R}} \left| \ln\left(\frac{R_f + f_l \sin \beta}{R_0} \right) \right| \right)$$
$$\text{(Eq 4)}$$

where A, B, and c are material parameters in Voce's law (Eq 3), R is the planar anisotropic factor, R_f, f_l, and R_0 are geometric parameters (Fig. 6a), and β is the folding angle shown in Fig. 8.

This applied stress, $\sigma_{applied}$, should be compared with the critical buckling stress, σ_{cr}. Wrinkling is assumed to occur when $\sigma_{applied}$ is

greater than σ_{cr}. The calculation of σ_{cr} is more complicated and out of the scope of this work, but an executable program is available from the author. Note that the program is suitable for large panels, that is, when the planar radius is significantly larger than the sheet thickness.

Example 3: Shrink Flanging of 6111-T4. The example here was presented in Ref 6. Table 1 shows the material properties for inputting to the aforementioned executable program. Table 2 shows the comparison of wave number (i.e., number of wrinkling peaks) obtained from simulations and experiments, where plane view radius is the R_f in Fig. 6(a), and f_l is the flange length.

Springback

Earlier experimental and analytical work on springback in the straight flanging process was

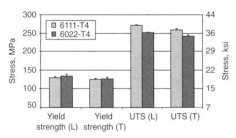

Fig. 4 Mechanical properties of 6111-T4 and 6022-T4. L, longitudinal; T, transverse; UTS, ultimate tensile strength

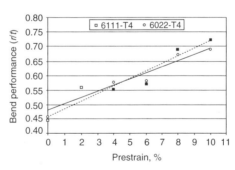

Fig. 5 Bend performance (r/t) for both 6xxx alloys in the longitudinal orientation as a function of prestrain

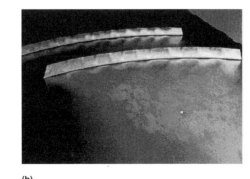

(a)

(b)

Fig. 6 Schematic of a shrink flanging and its key geometry parameters. (a) Unflanged blank. (b) Flanged sheet with wrinkles

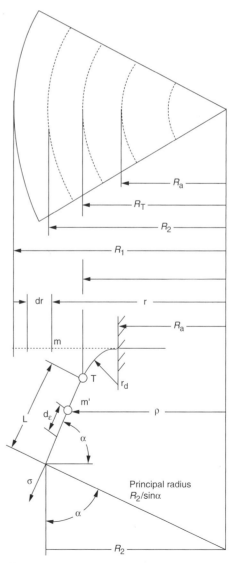

Fig. 7 Schematic of flanging process parameter definitions used in Eq 2. Source: Ref 6

mainly concentrated on steel alloys, such as the study in Ref 7. Recently, as aluminum alloys are being increasingly used in automotive applications, experimental study of springback in flanging aluminum has been reported in literature, including straight flanging of aluminum AlMg0.4Si1.2 (Ref 4), shrink flanging of 1015 (Ref 1), straight flanging of 6111 (Ref 8), and straight flanging of 5182 (Ref 9). In addition, the shape distortion of aluminum 6111-T4 in shrink/stretch flanging has been reported (Ref 10). Figures 9 to 14 collectively show the springback values reported in the aforemen-tioned studies on straight flanging. In general, the springback angle increases as the gap between the punch and the die increases and as the bend radius increases, except for the case when 15% prestretching was imposed (Ref 4) (Fig. 9).

In Ref 9, a straight flanging experiment with alloy 5182 was conducted. The relevant material properties of this material are listed in Table 3. The dimension of the 1 mm (0.04 in.) thick samples was 150 by 150 mm (6 by 6 in.) (width and length), and the flange length of the samples was 20 mm (0.8 in.). The die corner radius was 3 mm (0.12 in.). The binder force was set to approximately 460 kN (103,000 lbf).

The effect of gap on springback is presented in Fig. 13. Each isolated symbol represents one experimental datum, and various finite element models and material models were compared.

A trend of major parameters to springback is summarized in Fig. 14.

Prediction of springback in flanging is not an easy task, because sheet material is under large deformation of bending and unbending. The fundamental recovery mechanism has not been

---

Table 1 Material properties of 6111-T4 for shrink flanging analysis

See Example 3.

Yield strength, MPa	191.0
Material parameters (Voce's law, Eq 3)	
A, MPa	417.5
B, MPa	230.3
c	8.624
Planar anisotropic factor (R)	0.679

Source: Ref 6

Table 2 Comparisons of wave number between predictions and experiments for 6111-T4

Plane view radius, mm	Wave number for indicted flange length (f_1)					
	5 mm	8 mm	10 mm	12 mm	15 mm	20 mm
500						
Predicted	No	16	13	11	9	7
Experiment	No	No	10–11	13–14	9–10	7–8
1000						
Predicted	No	No	13	11	9	6
Experiment	No	No	Just seen	10–11	8–9	6
1500						
Predicted	No	No	No	11	9	5
Experiment	No	No	No	Just seen	8	5
2000						
Predicted	No	No	No	11	9	5
Experiment	No	No	No	No	8–10	5–6

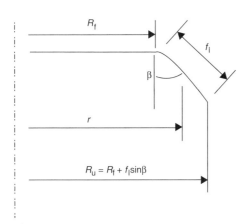

Fig. 8 Side-section view of a flanged sheet

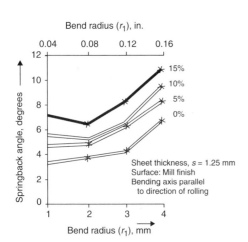

Fig. 9 Dependence of springback angle on bend radius and prestretching for alloy AlMg5Mn-w. Source: Ref 4

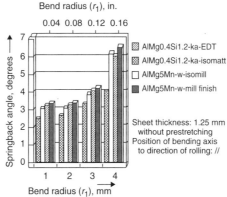

Fig. 10 Dependence of springback angle on surface finish and bend radius. Source: Ref 4

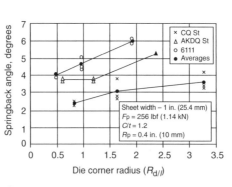

Fig. 11 Dependence of springback angle on die corner radius for aluminum alloy 6111. F_p, pad force; C/t, clearance ratio; R_p, punch radius. Source: Ref 8

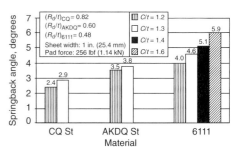

Fig. 12 Dependence of springback angle on gap for aluminum alloy 6111. C/t, clearance ratio. Source: Ref 8

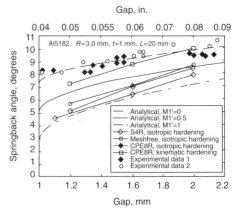

Fig. 13 Dependence of springback angle on gap for aluminum alloy 5182. Source: Ref 9

well understood. For straight flanging processes, a number of analytical springback-prediction models have been proposed (Ref 9, 11, 12). The key mechanism assumed in all those models was that the work done during the forming process is reversed elastically during the springback step. An analytical model for springback prediction in Ref 12 is available (Ref 13) with the following main assumptions: plane strain, isotropic material, power hardening law ($\sigma = K\varepsilon^n$), and the same tensile and compressive behavior. It is important to point out that common engineering sense needs to be applied when interpreting the result obtained from the program, especially when the thickness or panel geometry is significantly different from those used in the illustrations.

In Ref 1, shrink flanging of alloy 1050 was investigated. Relevant properties are given in Table 3. The experimental setup is presented in Fig. 15. No effect of binder force was studied in the experiment.

Varying the shrinking radius (x-axis: R_{sh}) and flange length (h), different springback angles are obtained and shown in Fig. 16. One can observe that if the flange length (h) is large enough, less springback variation is developed. Additional information is provided in the article "Springback" in this Volume.

Conclusions

This article reviews several common concerns with designs involving aluminum flanging, that is, fracture, wrinkling in shrink flanging, and springback in straight and shrink flanging. Each of these phenomena depends on material conditions and process parameters, such as the surface finish of sheet metal, tooling radii, gap between the punch and the die, and so forth, as illustrated previously. Readers are encouraged to explore more details in the references and in recent conference publications.

REFERENCES

1. A. Muderrisoglu, M. Murata, M.A. Ahmetoglu, G. Kinzel, and T. Altan, Bending, Flanging and Hemming of Aluminum Sheet—An Experimental Study, *J. Mater. Process. Technol.*, Vol 59 (No. 1–2), 1996, p 10–17
2. P.A. Friedman and S.G. Luckey, Bendability of Al-Mg-Si Sheet Alloys for Automotive Closure Applications, *TMS Annual Meeting*, 2001, p 3–15
3. M. Dao and M. Li, A Micromechanics Study on Strain-Localization-Induced Fracture Initiation in Bending Using Crystal Plasticity Models, *Philos. Mag. A: Phys. Conden. Matter, Struct., Defects Mech. Prop.*, Vol 81 (No. 8), Aug 2001, p 1997–2020
4. J. Mnif, "90° Bending of Sheets Made out of Aluminum Alloys," Institute for Metal Working, Stuttgart, Germany
5. C.T. Wang, G. Kinzel, and T. Altan, Failure and Wrinkling Criteria and Mathematical Modeling of Shrink and Stretch Flanging Operations in Sheet-Metal Forming, *J. Mater. Process. Technol.*, Vol 53, 1995, p 759–780
6. X. Wang, J. Cao, and M. Li, Wrinkling Analysis in Shrink Flanging, *J. Manuf. Sci. Eng.*, Vol 123, Aug 2001, p 426–432
7. N.M. Wang and M.L. Wenner, An Analytical and Experimental Study of Stretch Flanging, *Int. J. Mech. Sci.*, Vol 16 (No. 2), 1974, p 135–136
8. H. Livatyali and T. Altan, Prediction and Elimination of Springback in Straight

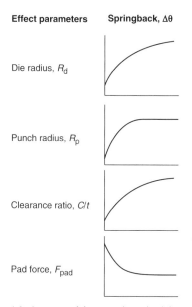

Fig. 14 Summary of the general trends of die corner radius, punch-nose corner radius, gap, and binder force on springback angle. Source: Ref 8

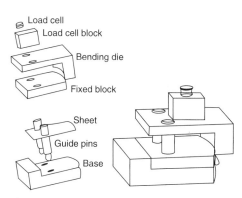

Fig. 15 Experimental setup for an aluminum shrink flanging study. Source: Ref 1

Table 3 Material properties of alloys studied in flanging springback experiments

Property	Aluminum alloy	
	5182	**1050**
Yield strength, mpa	146	139
Young's modulus, GPa	70	69
Strength coefficient (K), MPa	592.0	158.8
Strain-hardening exponent (n)	0.306	0.295
Poisson's ratio	0.3	0.3

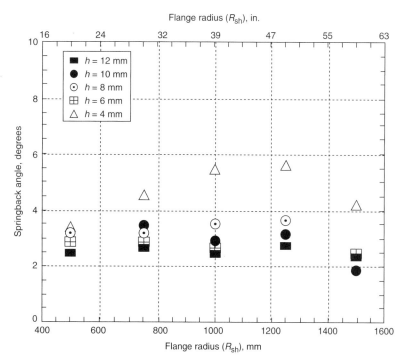

Fig. 16 Variation of springback on the aluminum shrink flanging due to flange length (h) and shrinking radius (R_{sh}). Source: Ref 1

Flanging Using Computer-Aided Design Methods, Part 1: Experimental Investigations, *J. Mater. Process. Technol.,* Vol 117, 2001, p 262–268

9. N. Song, D. Qian, J. Cao, W. Liu, V. Viswanathan, and S.F. Li, Effective Models for Prediction of Springback in Flanging, *J. Eng. Mater. Technol. (Trans. ASME)*, Vol 123, Oct 2001, p 456–461

10. K.N. Shah, M. Li, and E. Chu, "Effects of Part and Tool Design on Surface Distortion after Down Flanging of Aluminum Panels," Numisheet, 1999

11. N.M. Wang, Predicting the Effect of Die Gap of Flange Springback, *Proceedings of 13th Biennial IDDRG Congress* (Melbourne, Australia), International Deep Drawing Research Group, 1984, p 133–147

12. T. Buranathiti and J. Cao, An Effective Analytical Model for Springback Prediction in Straight Flanging Processes, *Int. J. Mater. Prod. Technol.,* Vol 21 (No. 1/2/3), 2004, p 137–153

13. www.mech.northwestern.edu/ampl/programs, Advanced Materials Processing Laboratory, Department of Mechanical Engineering, Northwestern University, accessed Nov 2005

Forming of Beryllium

BERYLLIUM has been successfully formed by most common sheet metal forming operations. The following are required:

- Equipment that can be controlled at slow speeds and that can withstand the use of heated dies
- Dies that can withstand the temperatures at which beryllium is commonly formed
- Facilities for preheating and controlling the temperature of dies and workpieces
- In some applications, facilities for stress relieving the work at 705 to 790 °C (1300 to 1450 °F)
- Special lubrication
- Safety precautions when grit blasting is required for cleaning after forming

Almost all beryllium currently used is produced by consolidating beryllium powder into a block by vacuum hot pressing. The powder is obtained by chipping and then mechanically or pneumatically pulverizing a vacuum-cast ingot. The hot pressed block can be warm rolled to the desired sheet thickness.

Unalloyed beryllium is available in two grades, I (instrument grade) and S (structural grade). Typical applications for instrument-grade beryllium include gyroscopes, components in inertial guidance systems, and precision satellite and airborne optical components. Structural grades find application as satellite superstructures, antenna booms, and optical support structures. Table 1 lists the compositions of four grades of vacuum hot pressed beryllium.

Formability

The formability of beryllium is low compared with that of most other metals. Beryllium has a hexagonal close-packed crystal structure; thus, there are relatively few slip planes, and plastic deformation is limited. For this reason, all beryllium products should be formed at elevated temperature (generally 540 to 815 °C, or 1000 to 1500 °F) and at slow speeds.

Temperature, composition, strain rate, and previous fabrication history have marked effects on the results obtained in the forming of beryllium.

Effect of temperature on formability (in terms of bend angle at fracture) of two grades of powder sheet is shown in Fig. 1. Although these data show the effect of temperature on bendability, maximum strain on a 2t bend radius is not achieved at less than 90°. Therefore, it should not be assumed that the quantitative results shown in Fig. 1 can always be applied directly in practice.

It should be noted that Fig. 1 was generated using beryllium sheet with a guaranteed elongation of only 5%. Current beryllium sheet products have guaranteed room-temperature elongations of 10%; typical values of 15 to 20% indicate that, if the test illustrated in Fig. 1 were repeated today (2005), improvement in results would be significant. In one case, a 90° bend with a 2t radius was achieved in 0.5 mm (0.020 in.) thick beryllium sheet.

Effect of Composition. The oxide content of ingot and powder sheet has a significant effect on formability, as shown by the curves in Fig. 1. As the oxide content increases, yield strength increases and ductility decreases.

Effect of Strain Rate. Strain rate greatly influences the formability of beryllium. For instance, the stroke of a press brake is too fast for making sharp bends in hot beryllium. Slow bending, by means of equipment such as a hydraulic or air-operated press, is usually used. Minimum bend limits for the press-brake method and the slower-press method are compared in Fig. 2 for bending of cross-rolled powder sheet.

In a laboratory, a radius of $2\frac{1}{2}t$ was bent in beryllium sheet at the rate of 50 mm/min (2 in./min), and a radius of $5\frac{1}{2}t$ at 305 mm/min (12 in./min). Forming temperature in both cases was 745 °C (1375 °F).

Effect of Fabrication History. Beryllium products consolidated by vacuum hot pressing have low ductility, even at a theoretical density of 100%. The ductility of hot pressed beryllium can be increased by hot mechanical working.

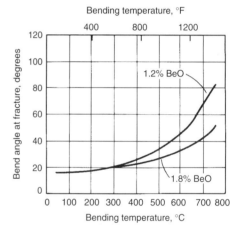

Fig. 1 Bend angle to fracture versus temperature of beryllium sheet using a 2t bend radius

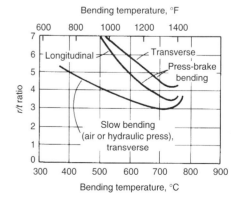

Fig. 2 Minimum bending limits for press-brake versus slower (hydraulic) bending of beryllium sheet in transverse and longitudinal directions. r, bend radius; t, sheet thickness

Table 1 Compositions of four grades of vacuum hot pressed beryllium

Grade	Composition(a)							
	Be, %	BeO, %	Al, ppm	C, ppm	Fe, ppm	Mg, ppm	Si, ppm	Other, ppm
S-65B	99.0 min	1.0	600	1000	800	600	600	400
S-200F	98.5 min	1.5	1000	1500	1300	800	600	400
I-220A	98.0 min	2.2	1000	1500	1500	800	800	400
I-400	94.0 min	4.2 min	1600	2500	2500	800	800	1000

(a) Maximum, unless otherwise indicated

Equipment and Tooling

Presses operated by air or hydraulic systems are usually used for forming beryllium, because of the slow speeds required. Standard mechanical presses or other fast forming presses are not suitable.

Critical components of the equipment must be protected against damage by the heat of forming. This protection usually is achieved by means of simple insulation.

Tooling. Because the tools used for forming beryllium will be heated, allowances must be made for thermal expansion, high-temperature strength, and oxidation when selecting tool material and designing tools. Tooling requirements for forming beryllium are similar to those for hot forming titanium (see the article "Forming of Titanium and Titanium Alloys" in this Volume). When only a few pieces are required, mild steel is usually used for dies. However, mild steel oxidizes rapidly at elevated temperatures, and when more than a few identical pieces are to be formed, the best practice is to make dies from hot work die steels, stainless steel, or one of the nickel-base or cobalt-base heat-resistant alloys.

Heating Dies and Workpieces

In most forming applications, both the die and the workpiece must be preheated. Dies are specially constructed to permit heating; heat may be supplied by either electrical elements or gas burners. Although sometimes torches are satisfactory for heating the work (as when heating sheet for spinning), usually a furnace is preferred. No specially prepared atmosphere is needed.

At the maximum temperature used for forming beryllium, surface oxidation is usually negligible. However, if desired, to prevent surface discoloration (hard oxide layer), the workpiece can be coated with a film of commercial heat-resistant oil. After forming, the film of oil can be removed by wet blasting, or by degreasing with an agent such as trichloroethylene.

In the forming of thin sheet (less than ~1 mm, or 0.040 in., thick), cooling of the work between the furnace and the forming equipment is often a problem. Overheating to compensate for this heat loss is not recommended. One satisfactory solution is to "sandwich" thin sheets of beryllium between two sheets of low-carbon steel. This sandwich is retained throughout heating and forming.

Stress Relieving

Stress relieving between stages of forming, or after forming is completed, is needed only in the forming of relatively thick sheet or in severe forming. For some finish-formed parts, stress relieving has proved an effective means of counteracting "oil canning" or excessive warpage. When stress relieving is used, regardless of whether it is an intermediate step or a final operation, holding at 705 to 760 °C (1300 to 1400 °F) for 30 min is recommended. No specially prepared atmosphere is needed.

Lubrication

Lubrication or coating of some type is needed in most beryllium forming operations. For less severe operations, such as bending, powdered mica has been used.

For operations such as joggling, forming in matched dies, or deep drawing, colloidal graphite in oil is commonly used. The role of lubrication is especially critical in deep drawing and is discussed in more detail in the section "Deep Drawing" in this article.

Safety Practice

No special precautions or safety measures are required in forming of beryllium because no fines or oxide dust is created in forming; and the maximum temperature (815 °C, or 1500 °F) used for preheating causes the formation of only a thin film of hard oxide, which under normal operating conditions will not harm personnel. Extreme caution should be used, however, and safety equipment should be available in the event of a furnace overrun.

However, if parts require cleaning after forming and if grit blasting is used, the wet method is recommended. Wet blasting minimizes the possibility that beryllium oxide dust will contaminate the surrounding atmosphere. Adequate ventilation must be provided if parts are processed by chemical etching after forming.

The usual precautions observed in working with beryllium must be taken. Details on protection can be obtained from the publication "Health Protection in Beryllium Facilities," which is available from the U.S. Atomic Energy Commission. Also, a video tape, "Beryllium: Safe Handling," is available through Brush Wellman Inc.

Deep Drawing (Ref 1)

Deep drawing is the forming of deeply recessed (cuplike) parts by means of plastic flow of the material (see the article "Deep Drawing" in this Volume). Tooling consists of a punch and a suitable die or draw ring. Normally, the deformation in deep drawing is actually a combination of deep drawing and stretching.

There are two parameters that must be under control during any successful deep-drawing operation: friction and holddown pressure. Both can be controlled by proper die design and lubricant selection, as discussed subsequently.

Lubrication is required to prevent galling between the beryllium workpiece and the die. A lubricant film must be maintained over that portion of the blank surface making contact with the drawing surfaces of the die throughout the entire draw. Because elevated temperatures (595 to 675 °C, or 1100 to 1250 °F, for the workpiece; 400 to 500 °C, or 750 to 930 °F, for the dies) are required to deep draw beryllium, conventional lubricants applied directly to the blank and die will burn off, causing galling between workpiece and die at high-pressure areas such as the draw ring. The solution to this problem is best achieved by using die materials that are self-lubricating, such as graphite or an overlay of colloidal suspension of graphite on an asbestos paper carrier.

The technique of using consolidated graphite as a self-lubricating die material was initially developed for forming small, thin-walled parts to finished size. This technique has evolved to the point that very deep drawing of 6.35 mm (0.25 in.) thick blanks over a graphite draw ring is routine. The disadvantage is that such draw rings have short service life.

Organic emulsified suspensions of powdered graphite, aluminum, and copper have all been used successfully to lubricate punches to facilitate part stripping. These materials can also be applied to the draw ring to improve lubricity of the drawing surface under the graphite-impregnated paper.

Blank development for deep drawing of beryllium generally follows the same rules as for other metals. Blanks too thin to support themselves during the early stages of drawing will buckle or wrinkle. A restraining force is required to prevent this.

There are numerous factors involved in determining whether blank restraint is required during any drawing operation. The two most important are the ratio of blank diameter d to blank thickness t and the percentage of reduction from one draw to the next.

The relationship between reduction R and d/t is shown in Fig. 3 for cylindrical parts, whether they are flat-bottomed or hemispherical cups. The areas under the curves were determined experimentally, with the curves themselves being the normal limit of formability for a given reduction at a given d/t ratio.

The curves in Fig. 3 describe formability limits; therefore, some consideration should be given during design to avoid borderline cases. Reductions of more than 50% are possible but will require partial drawing followed by several anneals, usually with a high failure rate. Several stages of tooling requiring smaller reductions is a more practical approach.

Tool Design. There are many different tool designs for deep drawing sheet metal parts. Two general types are described here. One type does not apply blank restraint to prevent wrinkling and is referred to as single action. The other type does apply blank restraint and is referred to as double action.

Single-action tooling should be used to form parts that fall in the no-restraint section of Fig. 3. Double-action tooling, or tooling that applies blank restraint to avoid wrinkling, was developed in two forms. In one system, the lower cushion ram in a hydraulic press is used as the second action for blank restraint (Fig. 4). The other type of double-action tooling used to deep draw beryllium is described in detail in Ref 1.

Materials used for beryllium deep-drawing tooling need not be exotic. Gray cast iron is satisfactory for most punch and die applications. Drawing surfaces are usually made from a free-machining tool steel or, in the case of very large dies, low-carbon steel that has been carburized after machining.

Strain rates during deep drawing of beryllium may vary widely, depending on the severity of the draw. For deep drawing simple hemispherical shells, punch speeds of 760 to 1270 mm/min (30 to 50 in./min) are commonly used. With optimal die clearance and lubrication, strain rates in excess of 2500 mm/min (100 in./min) have been observed in successful deep draws.

Applications. Numerous shapes have been deep drawn from beryllium. Considerable material savings may be achieved by deep drawing rather than machining thin-walled parts. The process lends itself to cup-shaped parts that have a slightly thicker wall at the equator than at the pole because of thickening in this area during forming.

Three-Roll Bending (Ref 1)

Three-roll bending is a process for shaping smoothly contoured, large-radius parts by applying three-point bending forces progressively along the part surface (see the article "Three-Roll Forming" in this Volume). Usually, one or more of the forming rolls is driven. The process has been used to form curved panel sections and full cylinders from beryllium. As in all forming operations for beryllium, it is necessary to heat the blank to achieve the necessary ductility to avoid cracking.

Applications. Three-roll bending has been used to form precision beryllium cylinders. The cylinders were joined by an electron beam fusion weld and are round within 0.5 mm (0.02 in.) total indicator reading on the diameter.

Panels for the Agena spacecraft also have been formed to a 762 mm (30 in.) radius of curvature. There were two sizes of panels formed, 635 by 635 mm (25 by 25 in.) and 559 by 355 mm (22 by 14 in.), at thicknesses of 1.4 and 1.88 mm (0.055 and 0.074 in.), respectively, from cross-rolled beryllium powder sheet. The flat beryllium sheet was heated to approximately 427 °C (800 °F), placed on a stainless steel sheet somewhat longer than the beryllium, and manually rolled to contour. The stainless steel sheet was used to "lead in" the beryllium and reduce the flat end inherent to roll forming. The rolled panels were stress relieved at 732 °C (1350 °F) for 20 min.

Stretch Forming (Ref 1)

Stretch forming is the shaping of a sheet or part, usually of uniform cross section, by first applying suitable tension or stretch, then wrapping it around a die of desired shape (see the article "Stretch Forming" in this Volume). When applying this technique to beryllium, the wrapping operation usually takes place quite slowly.

Tooling. Two commonly used types of tooling to stretch form beryllium are generally described as open-die and closed-die tooling. Open die, the most common, consists of a male die with the desired contour and some means of forcing the blank to assume that contour. Tension is not normally required for beryllium because the high modulus resists buckling and wrinkling.

Closed-die tooling has male and female counterparts. The male die is used to force the blank into the female die, thereby causing the blank to assume the contour of the male die. This type of tooling lends itself well to beryllium forming because both portions of the die may be heated to facilitate maintenance of the heat necessary in the blank to avoid cracking. Friction forces on the female die can help to restrain the part and cause stretching.

Spinning

Beryllium sheets up to 5.1 mm (0.200 in.) thick have been successfully formed by spinning. For sheets less than approximately 1 mm (0.040 in.) thick, a common practice is to sandwich the beryllium between two 1.5 mm (0.060 in.) sheets of low-carbon steel and heat the sandwich to 620 °C (1150 °F) for spinning. The steel sheets not only help to maintain temperature but also help to prevent buckling. Beryllium sheets more than approximately 1 mm (0.040 in.) thick usually are not sandwiched between steel sheets for spinning and are heated to 730 to 815 °C (1350 to 1500 °F).

Hemispherical shapes have been spun in as many as nine stages with no adverse effect on the properties of the beryllium. The part and mandrel often are torch-heated during spinning.

Lubrication is especially important in spinning. Colloidal graphite or glass is usually used. Wet blasting is the recommended means of cleaning the workpiece after spinning.

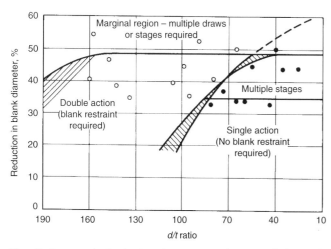

Fig. 3 Percent reduction in deep-drawing versus diameter-to-thickness (d/t) ratio for deep drawing of cylindrical beryllium shells. Datapoints are experimental observations (double action or single action) used to derive the curve limits; d, blank diameter; t, blank thickness; shaded areas, marginal. Source: Ref 1

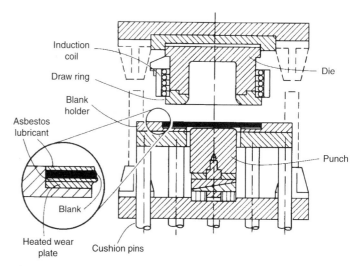

Fig. 4 A double-action tool for deep drawing of beryllium that uses the action of the lower press action for blank restraint. Lubrication with this type of tooling is best achieved using asbestos paper impregnated with colloidal graphite (see inset). Source: Ref 2

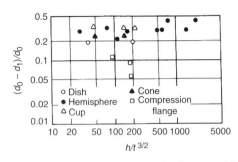

Fig. 5 Dimensional combinations for the successful spinning of beryllium sheet. d_0, blank diameter; d_1, diameter of spun part; t, blank thickness; h, height of spun part

Figure 5 plots combinations of conditions under which parts of a variety of shapes have been successfully produced by spinning cross-rolled beryllium powder sheet. The points plotted, however, represent only limited data, and many more points would have to be established before it would be safe to designate dimensional limitations for spinning specific shapes. More information on the spinning process is available in the article "Spinning" in this Volume.

ACKNOWLEDGMENT

This article is adapted from L.A. Grant, Forming of Beryllium, Forming and Forging, Vol 14, *Metals Handbook,* 9th ed., ASM International, 1988, p 805–808.

REFERENCES

1. J.J. Blakeslee, Chapter 7, Metalworking IV: Forming, *Beryllium Science and Technology,* Vol 2, D.R. Floyd and J.N. Lowe, Ed., Plenum Press, 1979, p 107–124
2. J.L. Frankeny and D.R. Floyd, "Ingot Sheet Beryllium Fabrication," RFP-910, Rocky Flats Division, Dow Chemical Company, Feb 1968

Forming of Copper and Copper Alloys

Frank Mandigo and Derek Tyler, Olin Corporation

COPPER AND MOST COPPER ALLOYS are readily formed at all sheet gages. The copper alloys commonly formed are characterized by strength and work-hardening rates between those of steel and aluminum alloys. This article reviews the general characteristics of copper and copper alloys and how these characteristics affect the behavior of strip in different types of forming operations. The attempt is to provide an understanding of copper alloy formability coupled with illustrative data rather than to offer a complete single-source alloy data reference.

General Considerations

The combination of moderate-to-high strength, high electrical and thermal conductivity, stress-relaxation resistance, modest cost, good corrosion and stress-corrosion resistance, and ease of plating and joining, coupled with good formability, accounts for the use of copper and copper alloys in a wide range of applications. A list of forming operations typically used to fabricate various parts from copper and copper alloy strips is as follows:

Application	Forming operations
Electrical terminals and connectors	Bending, stretch forming, blanking, coining, drawing
Electronic leadframes	Bending, coining, blanking
Hollow ware, flatware	Roll forming, blanking
Builder's hardware	Shallow and deep drawing, and stretch forming operations
Heat exchangers	Roll forming, bending, sinking, blanking
Coinage	Blanking, coining, embossing
Bellows, flexible hose	Cupping, deep drawing, bending
Musical instruments	Blanking, drawing, coining, bending, spinning
Ammunition	Blanking, deep drawing

These applications are illustrated in Fig. 1 to 8.

The forming of any part involves the interaction of material, tooling, and lubrication. Tooling and lubrication are discussed in the articles "Presses and Auxiliary Equipment for Forming of Sheet Metal" and "Selection and Use of Lubricants in Forming of Sheet Metal" in this Volume and are noted in this article only where there are unique or specific requirements for a copper alloy or a given forming operation. The forming characteristics of copper and copper alloys are discussed at length. All of the various major forming operations considered in this article—blanking, bending, stretch forming, drawing, and coining—depend on some optimal combination of strength, ductility, and work-hardening behavior of the sheet metal to provide the most cost-effective part. Therefore, much of this article is devoted to understanding the trade-offs in strength, work hardening, and ductility available by selection of material composition and temper. Strain-rate sensitivity (m) is also a factor in some forming operations. However, m is of practical significance only at elevated temperature. A more comprehensive treatment of the relationships between these materials characteristics and formability is available in the article "Formability Testing of Sheet Metals" in this Volume and in the article "Bend Testing" in *Mechanical Testing and Evaluation*, Volume 8 of *ASM Handbook*, 2000.

Other test methods used to assess sheet metal forming characteristics are the plastic-strain ratio (r), which is a measure of sheet anisotropy; the limiting draw ratio (LDR); bulge height; and minimum bend-forming radius. Comparison of test results can provide a relative ranking of the deep draw and/or stretch forming capabilities of various alloy compositions in different temper conditions.

Effects of Composition, Cold Work, and Heat Treatment on Formability.

Copper alloys are primarily strengthened by cold work or by alloying additions that solid-solution strengthen and enhance strain hardening. A finely dispersed second phase is sometimes used as a grain refiner to increase strength/ductility combinations and/or to minimize or prevent surface roughening ("orange peel") during forming.

Precipitation hardening is important to a small but important class of alloys, most notably, the beryllium-copper and copper-nickel-silicon alloy systems. Spinodal and/or precipitation hardening is available in the copper-nickel-tin, copper-nickel-chromium, and copper-titanium systems. Hardening by martensite transformation is available in the copper-aluminum system but is rarely used commercially.

Copper alloys are classified using the Unified Numbering System (UNS). The designations of the Copper Development Association are also used and correspond closely to UNS designations. Wrought copper alloys are divided in the UNS system into the following groups:

Alloy type	UNS No.
Copper and high-copper alloys	C1xxxx
Zinc brasses	C2xxxx
Zinc-lead brasses	C3xxxx
Zinc-tin brasses	C4xxxx
Tin bronzes	C5xxxx
Aluminum, manganese, and silicon bronzes	C6xxxx
Copper-nickel and copper-nickel-zinc alloys	C7xxxx

UNS numbers, common alloy names, and the nominal alloy chemistries of the principal copper alloys are listed in Table 1.

The temper designations for solid-solution-strengthened and precipitation-hardened copper and copper alloys are reviewed in the sections "Solid-Solution Strengthening and Cold Working" and "Precipitation Hardening and Cold Working" in this article.

Solid-Solution Strengthening and Cold Working. Solute elements provide a major means of strengthening copper, and the magnitude of strengthening depends on the type and level of addition. Temper designations for solid-solution-strengthened copper and copper alloys are listed in Table 2. Table 3 lists mechanical properties resulting from various alloying additions to copper in the annealed condition. These data indicate strength higher than that of pure copper (alloy C11000) can be acquired with limited or no loss of tensile ductility by solid-solution alloying. A listing of properties that can be typically expected is available in the article "Properties of Wrought Coppers and Copper Alloys" in *Properties and Selection: Nonferrous Alloys and Special-Purpose Materials*, Volume 2 of *ASM Handbook*, 1990. Suppliers of copper and copper alloy strip should be consulted for specific property/temper characteristics.

Figure 9 shows the work-hardening behavior of copper (C11000) and several copper alloys in terms of strength and ductility versus cold reduction. The relative work-hardening effects of various alloying elements are evident; the strong effect of aluminum is contrasted with the weak effect of nickel, with zinc and tin being intermediate. Ductility, as indicated by tensile elongation, decreases with cold reduction. Again, however, the combination of strength and

(a)

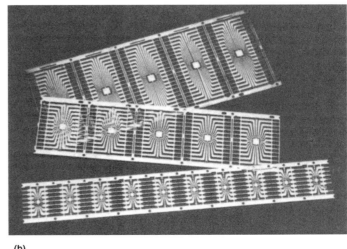

(b)

Fig. 1 Electrical and electronic applications for formed copper alloy parts. (a) Connectors used in home appliances and automotive electrical systems. (b) Copper alloy leadframe for a semiconductor device

Fig. 2 Typical household flatware utensils formed from copper alloys

ductility is enhanced by solid-solution additives even after cold working.

Precipitation Hardening and Cold Working. Alloys containing beryllium (0.15 to 2.0% Be) with nickel (0.3 to 2.7% Ni) or cobalt (1.0 to 2.2% Co) or alloys with nickel (1.0 to 3.5% Ni) and silicon (0.25 to 0.80% Si) can be strengthened by solid-state precipitation. Alloys with high beryllium content (1.8 to 2.0%) can be cold worked and precipitation hardened to produce material with tensile strength above 1380 MPa (200 ksi) at 15 to 28% International Annealed Copper Standard (IACS). In 1913, values of electrical conductivity were established and

expressed as a percent of a standard. The standard chosen was an annealed copper wire with a density of 8.89 g/cm^3, a length of 1 m (3 ft), a weight of 1 g, with a resistance of 0.1532 Ω at 20 °C (70 °F). The 100% IACS value was assigned with a corresponding resistivity of 0.017241 Ωmm^2/m. The percent IACS for any material can be calculated by:

$$\%\text{IACS} = \frac{0.017241 \ \Omega \ \text{mm}^2/\text{m} \cdot 100}{\rho_v}$$

where ρ_v is volume resistivity.

Alloys with lower beryllium contents provide tensile strengths from 655 to 965 MPa (95 to 140 ksi) with better thermal and electrical conductivities (45 to 60% IACS). Alloys with nickel and silicon additions can provide tensile properties from 585 to 860 MPa (85 to 125 ksi) with electrical conductivities from 35 to 60% IACS. A listing of properties that can be typically expected for precipitation-hardened copper alloys is available in the articles "Properties of Wrought Coppers and Copper Alloys" and "Beryllium-Copper and Other Beryllium-Containing Alloys" in *Properties and Selection: Nonferrous Alloys and Special-Purpose Materials,* Volume 2 of *ASM Handbook,* 1990. Suppliers of precipitation-hardened copper alloy strip should be consulted for specific property/temper characteristics.

Precipitation-hardenable alloys offer the opportunity to form parts in the maximum-ductility (as-solution heat treated) conditions prior to final precipitation hardening to develop maximum strength. Part geometry, platings or coatings on the surfaces of copper alloy strip, volume changes that may occur during precipitation-hardening treatments, overall cost, and/or fabrication requirements are factors that determine whether forming precedes or follows the final precipitation-hardening treatment of precipitation-hardened copper alloys. Typically, only parts that require sharp bends or have very severe formability requirements are fabricated

from strip in the annealed or rolled tempers prior to final precipitation hardening. Electrical connector parts are typically fabricated from precipitation-hardened copper alloys in mill-hardened tempers to minimize cost (customer heat treatment, cleaning, and/or surface plating or coating steps). Temper designations for precipitation-hardenable copper alloys are provided in Table 4.

The mechanical properties of exemplary precipitation-hardenable alloys in the solution-annealed condition are given in Table 5. The work-hardening behavior of several precipitation-hardening systems in the solution-annealed condition is shown in Fig. 10. The strong effect of beryllium content on work hardening is evident in Fig. 10 for alloy C17200. Table 6 lists the mechanical properties of selected tempers of mill-hardened alloys.

Postforming Heat Treatment. Heat treatments, aside from those employed to precipitation harden, are used after forming to reduce susceptibility to stress corrosion (primarily the brasses) or to increase the stiffness or stress-relaxation resistance of electrical or electronic springs (mainly the brasses, aluminum bronzes, and copper-silicon alloys). These postforming treatments are performed at low temperatures.

Formability of Copper Alloys versus Other Metals

In forming a given part, no single materials property completely defines formability. As previously noted, formability can best be rationalized in terms of the strength, work hardening, and ductility of a copper alloy, but these parameters do not directly correlate with formability. The problem becomes even more difficult when comparing different alloy systems, for example, ferrous and nonferrous.

Figure 11 shows the annealed ultimate tensile and yield strengths and response to cold rolling

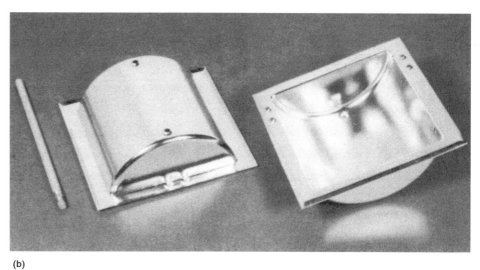

(a)   **(b)**

Fig. 3 Builder's hardware formed from copper alloys. (a) Doorknob fabricated by deep drawing. (b) Recessed fixture for kitchen and bathroom accessories

Fig. 5 Copper alloy U.S. currency with heavy coining and embossing

Fig. 4 Automotive radiator fabricated from several formed copper alloy components, including a deep-drawn water tank, roll-formed cooling tubes, and formed cooling fins

for type 304 stainless steel, 1045 steel, aluminum alloy 1100, copper alloy C11000, and other selected copper alloys. The high work-hardening rate and strength of the austenitic stainless steel are evident. The copper alloys range from above aluminum to above medium-carbon steel in strength and work-hardening rate. A comparison of limiting draw ratio with the plastic-strain ratio (r) for ferrous and nonferrous alloys is shown in Fig. 12. Increasing values of r and LDR reflect increasing drawability (see the section "Drawing and Stretch Forming" in this article).

In general, copper alloys typically offer better combinations of strength, formability, electrical conductivity, corrosion, ease of joining and plating, and/or stress-relaxation performance when compared to most other alloy systems.

Accordingly, copper alloy strips are replaced by other materials (aluminum, steel, stainless steel, or plastic) only when economics (material and/or fabrication costs) dictate or if the stamped part does not require a unique property (such as moderate electrical conductivity or ease of joining and plating) offered by copper alloy strip.

Blanking and Piercing

Nature of the Operation. Blanking, piercing, and related cutting operations (trimming, notching, parting, and so on) are often used to provide parts that are subsequently formed to final shape by such operations as bending, drawing, coining, and spinning. Cutting oper-

ations are frequently conducted in the same press tooling used to form and shape the final part geometry. The principal objective of any cutting operation is to produce a workpiece that has the correct geometric shape, is free of distortion, and possesses sheared edges that are of sufficient quality to allow subsequent forming, finishing, and/or handling operations.

Materials Considerations. Copper and copper alloys can be readily blanked and pierced. The quality of the final part produced and the ease of fabrication of the part are dependent on the flatness and dimensional tolerances (width, thickness, and so on) of the initial strip as well as the quality of the blanked or cut edge of the strip. The flatness and dimensional tolerances of copper alloy strip are dependent on the equipment and manufacturing expertise of the strip supplier. The quality of the blanked or cut edges—shear-to-break, rollover, breakout angle, burr height, and so on—is determined by die clearance, die sharpness, and material characteristics.

Effects of Alloy Composition and Temper. Burr-free and distortion-free parts can be cut from annealed copper alloy strip at die clearances to approximately 5% of strip thickness. Unalloyed coppers, such as C10100 and C10200, require smaller clearances (usually <5%) and less latitude in actual values to produce burr-free edges, even in rolled tempers. Copper alloys that contain second-phase particles (for example, C19400) that have high solute additions (such as C26000 or C51000) and/or that are cold rolled more than 50% generally exhibit high-quality blanked edges at die clearances in the range of 3 to 12%. Low-lead additions to brass and other

copper alloys will decrease burrs and the shear-to-break ratio in blanking operations, but at some cost to formability in almost all types of forming operations.

Bending

Nature of the Operation. Many connectors, terminals, and springlike components are fabricated by simple bending operations. Bending is a forming method in which a blanked coupon is wrapped, wiped, or stretched over a die to a specified radius and bend angle. Bend formability is usually expressed as minimum bend radius R in terms of strip thickness t (R/t). Minimum bend radius is defined as the smallest radius around which a specimen can be bent without cracks being observed on the exterior bend (tension) surface. Bend deformation is highly localized and is confined to the region of the workpiece in contact with the bending die. Workpiece thickness is not substantially reduced unless the bend radius is less than $1.0t$ or the part is coined or stretched during bending. A detailed review of bend testing is provided in the article "Bend Testing" in *Mechanical Testing and Evaluation,* Volume 8 of *ASM Handbook,* 2000.

Materials Characteristics. Ductility is the principal materials factor that determines bend formability. The ductility factor of first-order importance is the ability of a material to distribute strain in a highly localized region, that is, necking strain. The necking strain available depends on alloy composition and temper. As strength is increased by cold work, the ability of

an alloy to distribute necking strain decreases. The extent to which bend formability (necking strain) is decreased with increasing strength is dependent on the alloy composition and the strengthening mechanism. Conventional tensile elongation cannot be used to predict bend formability, because it does not adequately account for the contribution of necking strain. However, if the tensile specimen gage length were decreased to define an area of deformation equal to that deformed during bending, comparable ductility values would be obtained.

Effect of Alloy Composition, Temper, and Orientation. Bend data for a wide range of copper alloys are summarized in Table 7. Strength-to-bend formability characteristics are dependent on alloy composition, temper, and orientation. The principal strengthening mechanism is through solute additions to increase the work-hardening rate. For example, additions of 15 and 30% Zn to copper increase the tensile-strength-to-bend properties by 220 and 290 MPa (32 and 42 ksi), respectively, for 0.25 mm (0.010 in.) thick good-way bends at a bend radius of 0.4 mm ($1/64$ in.). Precipitation strengthening is also an important mechanism employed to improve the strength-to-bend performance of copper alloy strip, particularly if the part is bent in a softer temper and subsequently precipitation age hardened to a higher strength.

The practice of cold rolling to increase strip temper degrades bend formability. However, it is often used because most alloys still exhibit useful bend formability at modest cold rolling reductions. Product applications that require both high strength and good bend performance are usually satisfied by selecting copper alloys that are precipitation and/or solute strengthened with additions that greatly increase the work-hardening rate and thus minimize cold rolling requirements to achieve the desired strength.

Bend formability is typically dependent on bend direction with respect to strip-rolling direction (Fig. 13 and Table 7). All cold-rolled materials exhibit directionality. The extent of bend directionality varies from alloy to alloy but always increases with increasing cold reduction.

Fig. 6 Deep-drawn and corrugated copper alloy bellows

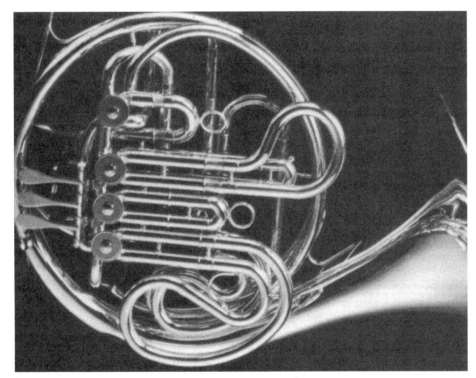

Fig. 7 French horn fabricated from copper alloys using complex bending and spinning operations

Fig. 8 Ammunition using a deep-drawn copper alloy cartridge case

Table 1 Nominal compositions of copper and copper alloys

UNS No.	Common name	Nominal composition, wt%
Wrought copper		
C10100	Oxygen-free electronic copper (OFE)	99.99 (min) Cu
C10200	Oxygen-free copper (OF)	99.95 (min) Cu
C11000	Electrolytic tough pitch copper (ETP)	99.90 (min) Cu
C15100	Zirconium copper	98.5 (min) Cu, 0.10 Zr
Wrought high-copper alloys		
C17200	Beryllium copper	bal Cu, 1.90 Be, 0.40 Co
C17410	Beryllium copper	bal Cu, 0.45 Co, 0.35 Be
C17460	Beryllium copper	bal Cu, 0.35 Be, 1.2 Ni, 0.1 Zr
C17510	Beryllium copper	bal Cu, 1.8 Ni, 0.4 Be
C18080	. . .	bal Cu, 0.3 Cr, 0.15 Ag, 0.1 Fe, 0.1 Ti, 0.05 Si
C19400	Iron-bearing copper	97 (min) Cu, 2.35 Fe, 0.125 Zn, 0.05 P
C19500	Iron-cobalt-bearing copper	97 (min) Cu, 1.5 Fe, 0.18 P, 0.8 Co, 0.6 Sn
C19700	. . .	99 (min) Cu, 0.6 Fe, 0.2 P, 0.05 Mg
Wrought brass alloys		
C21000	Gilding metal, 95%	bal Cu, 5 Zn
C22000	Commercial bronze, 90%	bal Cu, 10 Zn
C23000	Red brass, 85%	bal Cu, 15 Zn
C26000	Cartridge brass, 70%	bal Cu, 30 Zn
C28000	Muntz metal, 60%	bal Cu, 40 Zn
C35300	High leaded brass, 62%	bal Cu, 36 Zn, 2 Pb
C41100	Lubaloy	bal Cu, 8 Zn, 0.5 Sn
C42200	Lubronze	bal Cu, 11.5 Zn, 1 Sn
C42500	Lubaloy X	bal Cu, 9.5 Zn, 2 Sn
C46400	Uninhibited naval brass	bal Cu, 38 Zn, 1 Sn
Wrought bronze alloys		
C50500	Phosphor bronze, 1.25% E	bal Cu, 1.25 Sn, 0.1 P
C51000	Phosphor bronze, 5% A	bal Cu, 5.0 Sn, 0.2 P
C52100	Phosphor bronze, 8% C	bal Cu, 8 Sn, 0.1 P
C60800	Aluminum bronze, 5%	bal Cu, 5.8 Al, 0.2 As
C63800	. . .	bal Cu, 2.8 Al, 1.8 Si, 0.4 Co
C65500	High silicon bronze A	bal Cu, 3.3 Si, 0.9 Mn
C66400	. . .	bal Cu, 11.5 Zn, 1.5 Fe, 0.5 Co
C68800	. . .	bal Cu, 22.7 Zn, 3.4 Al, 0.4 Co
Wrought copper-nickel alloys and nickel silvers		
C70600	Copper nickel, 10%	bal Cu, 10 Ni, 1.4 Fe
C70250	. . .	bal Cu, 3.0 Ni, 0.65 Si 0.15 Mg
C71300	. . .	bal Cu, 25 Ni
C71500	Copper nickel, 30%	bal Cu, 31 Ni, 0.50 Fe
C72500	. . .	bal Cu, 9.5 Ni, 2.3 Sn
C73500	. . .	bal Cu, 18 Ni, 10 Zn
C74000	. . .	bal Cu, 10 Ni, 20 Zn
C74300	. . .	bal Cu, 8 Ni, 27 Zn
C75200	Nickel silver, 65-18	bal Cu, 18 Ni, 17 Zn
C75400	Nickel silver, 65-15	bal Cu, 15 Ni, 20 Zn
C77000	Nickel silver, 55-18	bal Cu, 18 Ni, 27 Zn

Table 2 ASTM B 601 temper designations for copper and copper alloys

Temper designation	Temper name or material condition
Annealed tempers	
025	Hot rolled and annealed
050	Light annealed
060	Soft annealed
061	Annealed
065	Drawing annealed
068	Deep-drawing annealed
070	Dead soft annealed
080	Annealed to temper—1/8 hard
081	Annealed to temper—1/4 hard
082	Annealed lo temper—1/2 hard
OS005	Average grain size 0.005 mm
OS010	Average grain size 0.010 mm
OS015	Average grain size 0.015 mm
OS025	Average gram size 0.025 mm
OS035	Average grain size 0.035 mm
OS050	Average grain size 0.050 mm
OS070	Average grain size 0.070 mm
OS100	Average grain size 0.100 mm
OS120	Average grain size 0.120 mm
OS150	Average grain size 0.150 mm
OS200	Average grain size 0.200 mm
Cold-worked tempers	
H00	1/8 hard
H01	1/4 hard
H02	1/2 hard
H03	3/4 hard
H04	Hard
H06	Extra hard
H08	Spring
H10	Extra spring
HI2	Special spring
H13	Ultra spring
H14	Super spring
Cold-worked and stress-relieved tempers	
HR01	H01 and stress relieved
HR02	H02 and stress relieved
HR04	H04 and stress relieved
HR06	H06 and stress relieved
HR08	H08 and stress relieved
HR10	H10 and stress relieved
HR50	Drawn and stress relieved
Cold-worked and order-strengthened tempers	
HT04	H04 and order heat treated
HT06	H06 and order heat treated
HT08	H08 and order heat treated

Bend directionality results from the development of strong textures during rolling. Copper alloys with low stacking fault energy, such as alloy C26000 (cartridge brass), develop strong {110}⟨112⟩ textures during rolling and can exhibit bend directionality even at approximately 30% cold rolling reduction. Dilute copper alloys and copper-nickel alloys do not develop well-defined rolling textures, and they show less bend directionality even at high (70%) cold rolling reductions. In general, sharper bends can be made in the good-way than in the bad-way orientations for alloys that are cold rolled and/or solute strengthened. Bend anisotropy in precipitation-hardening systems is strongly process dependent.

The bend performance of copper alloy strip is also dependent on bend angle, bend radius-to-strip-thickness ratio, and the width-to-thickness ratio of the bend regions of a stamped part. A 180° bend is more demanding than a 90° bend when the R/t ratio of a strip is less than approximately 2t. Typically, thin-gage strip, $t < 0.15$ mm (0.006 in.), provides slightly better bend formability than strip at thicker gages, $t > 0.38$ mm (0.015 in.). The bend performance of strip can be improved by as much as a factor of 3 or more in both good-way and bad-way orientations by reducing the width-to-thickness

Table 3 Mechanical properties of selected solid-solution copper alloys

Grain sizes of all materials listed ranged from 0.010 to 0.025 mm (0.0004 to 0.001 in.).

Alloy designation and common name	Nominal composition, %	0.2% offset yield strength MPa	ksi	Tensile strength MPa	ksi	Elongation, %
C11000 (Electrolytic tough pitch)	99.90 min Cu	83	12	241	35	48
C21000 (Gilding, 95%)	Cu-5Zn	97	14	262	38	45
C23000 (Red brass, 85%)	Cu-15Zn	110	16	290	42	45
C26000 (Cartridge brass, 70%)	Cu-30Zn	179	26	379	55	48
C50500 (Phosphor bronze, 1.25% E)	Cu-1.4Sn	124	18	290	42	47
C51000 (Phosphor bronze, 5% A)	Cu-5Sn	165	24	345	50	50
C61000(a) (. . .)	Cu-8Al	207	30	483	70	65
C70600 (Copper nickel, 10%)	Cu-10Ni	124	18	317	46	38
C71500 (Copper nickel, 30%)	Cu-30Ni	172	25	400	58	32
C75200 (Nickel silver, 65-18)	Cu-18Ni-18Zn	179	26	414	60	37

(a) Available only as tube, but properties are illustrative of copper-aluminum alloy strip properties.

ratio of the bend region of the part to less than 8 to 1.

Figure 14 shows the effects of bend directionality on part layout. This part includes both good-way and bad-way bends. If the part were fabricated from an alloy with strong bend directionality, for example, phosphor bronze (alloy C51000) in spring temper, the part layout would be restricted to avoid failure at bad-way bends. With alloys such as C68800 or C72500, which exhibit significantly less bend directionality, the part layout is not as restricted. It is not always possible to orient parts to minimize web scrap, regardless of the alloy selected, because tool design and part-handling and transfer costs may override the cost penalty of poor strip utilization.

Special Considerations. The minimum bend radii values listed in Table 7 for various alloys as a function of temper are approximate; actual results are dependent on the tooling and stamping methods used. For example, bends can be made in one step or in multiple stamping steps with decreasing bend radii or by different methods (wipe or V-block) or with (different types) or without lubricant or with different tool materials (steel, carbide, coated tooling, etc.). Other examples of special considerations are parts to be plated after forming or subjected to other finishing operations after bending (no orange peel or surface roughening allowed) or plated and/or coated parts that have the requirement that no copper be visible at the exterior bend apex (no cracks in surface plating and/or coatings). Often, more than one alloy and

fabrication method is available that will meet product bend requirements. In the absence of other limitations, bend formability may be the deciding factor in alloy selection.

Drawing and Stretch Forming

In drawing and stretch forming, a suitably shaped blank of sheet metal is drawn or formed into a die cavity to produce a part. A clamping ring, draw beads, and/or other restraints are usually applied at the periphery of the blank to prevent wrinkling and/or tearing of the blank as it is drawn or formed into the die cavity. The complexity of the required restraint at the periphery of the blank is usually directly proportional to the complexity of the formed part.

A deep-drawn part is characterized by having a depth greater than the minimum part width. A deep-drawn part can be fabricated in a single

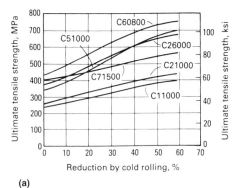

(a)

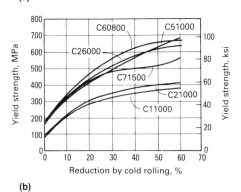

(b)

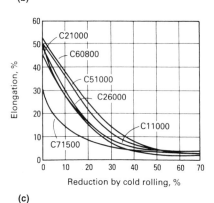

(c)

Fig. 9 Work-hardening behavior of copper and some solid-solution copper alloys. (a) Effect of cold work by rolling reduction on ultimate tensile strength. (b) Effect of cold work on yield strength. (c) Effect of cold work on elongation

Table 4   ASTM B 601 temper designations for precipitation-hardened copper alloys

Temper designation	Temper name or material condition
Solution-treated temper	
TB00	Solution heat treated
Solution-treated and cold-worked tempers	
TD00	TB00 cold worked to $1/8$ hard
TD01	TB00 cold worked to $1/4$ hard
TD02	TB00 cold worked to $1/2$ hard
TD03	TB00 cold worked to $3/4$ hard
TD04	TB00 cold worked to full hard
Precipitation-hardened temper	
TF00	TB00 and precipitation hardened
Cold-worked and precipitation-hardened tempers	
TH01	TD01 and precipitation hardened
TH02	TD02 and precipitation hardened
TH03	TD03 and precipitation hardened
TH04	TD04 and precipitation hardened
Precipitation-hardened and cold-worked tempers	
TL00	TF00 cold worked to $1/8$ hard
TL01	TF00 cold worked to $1/4$ hard
TL02	TF00 cold worked to $1/2$ hard
TL04	TF00 cold worked to full hard
TL08	TF00 cold worked to spring
TL10	TF00 cold worked to extra spring
TR01	TL01 and stress relieved
TR02	TL02 and stress relieved
TR04	TL04 and stress relieved
Mill-hardened tempers	
TM00	AM
TM01	$1/4$ HM
TM02	$1/2$ HM
TM04	HM
TM06	XHM
TM08	XHMS

Table 5   Mechanical properties of precipitation-hardenable copper alloys in the annealed condition

UNS designation	0.2% offset yield strength		Tensile strength		Elongation, %
	MPa	ksi	MPa	ksi	
C17200	290	42	476	69	40
C17500/C17510	207	30	310	45	27
C70250	138	20	338	49	37
C72400	276	40	483	70	27
C72900	241	35	517	75	40

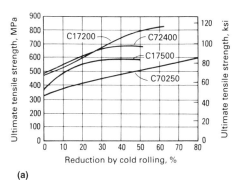

(a)

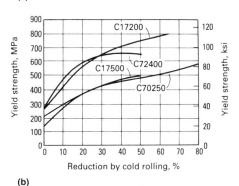

(b)

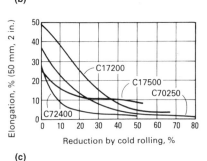

(c)

Fig. 10 Work-hardening behavior of four precipitation-hardening copper alloys in the solution-annealed condition. (a) Effect of cold work by rolling reduction on ultimate tensile strength. (b) Effect of cold work on yield strength. (c) Effect of cold work on elongation

Table 6 Mechanical properties of mill-hardened copper alloys

UNS designation	Temper	0.2% offset yield strength		Tensile strength		Elongation, %
		MPa	ksi	MPa	ksi	
C17200	TM02	690–862	100–125	827–931	120–135	12–18
	TM04	793–931	115–135	931–1034	135–150	9–15
C17410	TM04	690–827	100–120	758–896	110–130	7–17
C17460	3/4 HT	655–790	95–115	790–930	115–135	11 min
	HT	720–860	105–125	825–965	120–140	10 min
	TM06	931–1055	135–153	951–1089	138–158	1 min
C17510/C17500	HT	655–827	95–120	758–931	110–135	8–20
C19900	TM00–TM08	820–1115	119–162	980–1200	142–174	2–15 min
C70250	TM00	552 min	80 min	620 min	88 min	6 min
	TR04	690 min	100 min	731 min	106 min	2 min
C72900	TM00–TM08	515–1170	75–170	655–1255	95–180	2–22 min

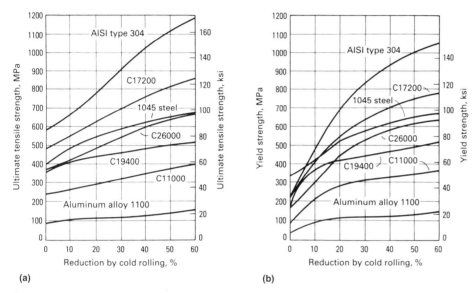

Fig. 11 Work-hardening behavior of copper alloys versus that of low-carbon steel, austenitic stainless steel, and aluminum. (a) Effect of cold work by rolling reduction on ultimate tensile strength. (b) Effect of cold work on yield strength

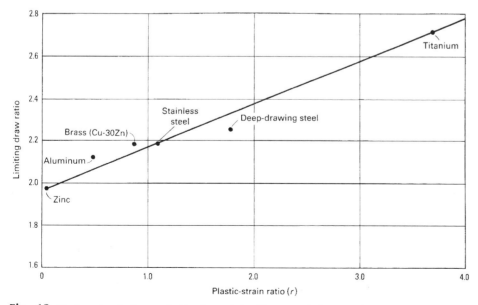

Fig. 12 Plastic-strain ratio (*r*) versus limiting draw ratio for different metals. Source: Ref 1

drawing step or in multiple steps by redrawing preforms developed by deep-draw, shallow-draw, and/or stretch-forming methods. Ironing can be used during redrawing to control the wall thickness of the final part. Additional anneals may be required between redrawing steps. Successful deep drawing is highly dependent on the control of the frictional forces between the tools and the metal. The friction between the metal and the holddown pad and between the metal and the punch provides the tension that draws the metal to the deep-drawn or stretch-formed shape. The gradual and uniform release of the metal under the holddown pad allows it to flow into the die without developing wrinkles and puckers. The thickness of the metal, clearance between tools, condition of the metal surfaces, and lubrication all help to determine how high and how uniform the frictional forces will be (Ref 3).

A shallow-drawn part has a depth less than the minimum part width and is usually formed in one process step. It can be a final part or the preform for deep drawing.

A stretch-formed part is fabricated by pressing a punch into a blank that is fully or partially restrained at its periphery to develop positive biaxial strain on the part surface. A stretch-formed part can be a final part or the preform for drawing operations. Additional information on drawing and stretch forming is available in the articles "Deep Drawing" and "Stretch Forming" in this Volume.

Materials Characteristics and Effects of Alloy Composition and Temper

Single-Step Drawing. Copper alloys that have high *r* values will provide the largest limiting draw ratio in a single deep-draw step. The *r* value is defined as the ratio of true width strain to true thickness strain in the region of uniform uniaxial elongation during a tensile test. It measures the resistance of a material to thinning. The *r* value correlates with deep-drawing performance because it reflects the difference between the load-carrying capability of the cup sidewall and the load required to draw in the flange of the cup or blank during a deep-drawing operation.

The deepest single-step draws (highest LDR) can be made with alloy C52100, followed by the brasses (in order of decreasing zinc level) and by copper. The LDR of cartridge brass (alloy C26000) increases as its grain size increases.

Multiple-Step Deep Drawing. The number of redrawing steps and the frequency of intermediate annealing treatments depend on the initial preform geometry, the extent of ironing required, and the work-hardening rate of the particular alloy. Fewer redrawing steps are required if the preform geometry closely matches that of the final part. The trade-offs involved in selecting a fabrication procedure for the initial preform (for example, deep drawing, shallow drawing, or stretch forming) are complex.

In contrast to single-step deep drawing, in which alloys with high work-hardening rates give the highest LDR, copper alloys with lower work-hardening rates can be redrawn and ironed to larger height/diameter ratios (a higher number of deep-draw steps) without intermediate annealing. The curves shown in Fig. 15 suggest that alloy C11000 (electrolytic tough pitch copper) will possess better redrawing and ironing characteristics and will require lower press forces than copper alloys with solute additions of zinc, tin, and/or silicon.

In general, the reduction at each redrawing step is successively decreased to ensure that the punch forces required to decrease the flange circumference do not exceed the load-carrying capability of the part sidewall. The magnitude of the incremental steps of redrawing is decreased if the part sidewall is to be ironed. Ironing increases the strength of the sidewall and flange proportionally to the distance from the cup bottom. In some applications, redrawing capacity can be improved by increasing the temper of the initial strip to enhance the load-carrying capability at the junction of the part sidewall and the cup bottom. The alternative is to use an alloy with a lower work-hardening behavior (Fig. 9 to 11) or an alloy with a higher r value.

Stretch Forming. The stretch formability of copper alloys correlates with the total elongation measured in a tension test. Annealed alloys that show high work-hardening rates offer the best stretch-forming characteristics. Improved combinations of strength and stretch formability are achieved by solute elements that greatly increase the work-hardening rate. Cold rolling to increase strip temper (strength) significantly reduces stretch formability.

The variation of tensile elongation with cold rolling reduction for copper alloys is shown in Fig. 16. These data indicate that high-tin and high-zinc alloys offer the best combinations of strength and stretch formability.

Specific Characteristics of Copper Alloys. The higher-zinc brasses, such as alloy C24000 (low brass), alloy C26000 (cartridge brass), and alloy C26200 (high brass), have strengths comparable to those of low-carbon steels. They are outstanding materials for deep drawing and stretch forming.

Many other families of copper alloys also have good deep-drawing and stretch-forming properties. Phosphor bronze A (alloy C51000) has an excellent combination of high strength and high ductility and is used to form deep-drawn thin-wall shells that are then annealed and corrugated

to produce bellows with high fatigue strength, corrosion resistance, and excellent flexibility.

The nickel silvers (copper-nickel-zinc) are copper alloys with a white or gray color that also have excellent deep-drawing characteristics similar to those of the high-zinc brasses. However, they have somewhat higher work-hardening rates and require more frequent anneal treatments during redrawing when compared to

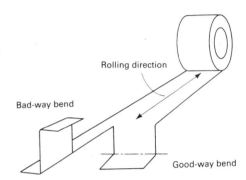

Fig. 13 Bend formability of copper alloys as a function of rolling direction. Bends with the axis transverse to the rolling direction are termed good-way bends; bends with the axis parallel to the rolling direction are bad-way bends. See also Table 7. Source: Ref 3

Table 7   Maximum tensile strengths available to make the indicated bends (1.6 R/t) in various copper alloys

	Maximum tensile strength available to make bend of the indicated radius in material of indicated thickness, MPa (ksi)					
	Good-way bend(s)(a)			Bad-way bend(s)(a)		
	0.25 (0.010)(b)	0.50 (0.020)(b)	0.76 (0.030)(b)	0.25 (0.010)(b)	0.50 (0.020)(b)	0.76 (0.030)(b)
UNS designation	0.4 (1/64)(c)	0.8 (1/32)(c)	1.2 (3/64)(c)	0.4 (1/64)(c)	0.8 (1/32)(c)	1.2 (3/64)(c)
C11000	372 (54)	352 (51)	352 (51)	365 (53)	331 (48)	345 (50)
C15100	428 (62)	400 (58)	400 (58)	407 (59)	400 (58)	400 (58)
C17200(d)	896 (130)	896 (130)	896 (130)	896 (130)	896 (130)	896 (130)
C17200(e)	1413 (205)	1413 (205)	1413 (205)	1379 (200)	1379 (200)	1379 (200)
C17460(d)	896 (130)	896 (130)	...	896 (130)	896 (130)	...
C17510, C17500(d)	724 (105)	724 (105)	724 (105)	724 (105)	724 (105)	724 (105)
C18080(d)	585 (85)	585 (85)	...	585 (85)	858 (85)	...
C19400	538 (78)	510 (74)	496 (72)	517 (75)	496 (72)	490 (71)
C19500	614 (89)	572 (83)	572 (83)	592 (86)	572 (83)	558 (81)
C19700	538 (78)	510 (74)	496 (72)	517 (75)	496 (72)	490 (71)
C19900(d)	1034 (150)	1034 (150)	...	1034 (150)	1034 (150)	...
C23000	593 (86)	593 (86)	593 (86)	572 (83)	552 (80)	538 (78)
C26000	662 (96)	662 (96)	662 (96)	627 (91)	524 (76)	524 (76)
C35300	641 (93)	572 (83)	572 (83)	496 (72)	483 (70)	469 (68)
C41100	517 (75)	496 (72)	496 (72)	468 (68)	448 (65)	434 (63)
C42500	621 (90)	621 (90)	621 (90)	552 (80)	475 (69)	462 (67)
C50500	490 (71)	469 (68)	469 (68)	490 (71)	468 (68)	469 (68)
C51000	710 (103)	662 (96)	648 (94)	621 (90)	572 (83)	538 (78)
C52100	765 (111)	745 (108)	731 (106)	614 (89)	558 (81)	552 (80)
C63800	827 (120)	807 (117)	793 (115)	724 (105)	696 (101)	696 (101)
C65400	745 (108)	731 (106)	731 (106)	627 (91)	627 (91)	627 (91)
C66600	669 (97)	655 (95)	641 (93)	613 (89)	586 (85)	579 (84)
C68800	786 (114)	744 (108)	745 (108)	786 (114)	745 (108)	731 (106)
C70250(d)	690 (100)	655 (95)	...	552 (80)	517 (75)	...
C70600	524 (76)	496 (72)	496 (72)	489 (71)	483 (70)	483 (70)
C72500	572 (83)	517 (75)	517 (75)	531 (77)	504 (73)	503 (73)
C72900(d)	896 (130)	896 (130)	...	862 (125)	862 (125)	...
C72900(e)	1275 (185)	1275 (185)	1275 (185)	1207 (175)	1207 (175)	1207 (175)
C73500	579 (84)	579 (84)	579 (84)	525 (76)	518 (75)	517 (75)
C74000	648 (94)	600 (87)	586 (85)	593 (86)	565 (82)	552 (80)
C75200	579 (84)	579 (84)	579 (84)	558 (81)	558 (81)	558 (81)
C77000	807 (117)	751 (109)	717 (104)	758 (110)	696 (101)	676 (98)

(a) "Good way" and "Bad way" refer to the orientation of the bend with respect to the sheet or strip rolling direction (see Fig. 13). (b) Sheet thickness, mm (in.). (c) Bend radius, mm (in.). (d) Mill hardened to strength shown, then formed. (e) Formed from rolled temper, then precipitation hardened to achieve strength shown. Source: Ref 2

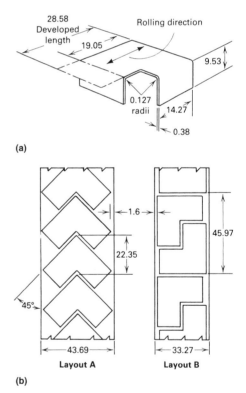

Fig. 14 Impact of bend anisotropy on part layout. (a) Hypothetical part, which has equal-radius bends at 90° orientations in the plane of the strip. Selection of the appropriate copper strip alloy for this application depends on the material strength and the bend properties in the relevant orientations. (b) Potential nesting of blanks for the part shown in (a). Layout A is required for directional alloys such as C51000 and results in 38% scrap; a nondirectional alloy such as C68800 would allow the more efficient layout B, with 23% scrap. Dimensions given in millimeters (1 in. = 25.4 mm). Source: Ref 3

cartridge brass. In the fully annealed condition, alloy C63800 (Cu-3Al-2Si-0.4Co) also exhibits good deep drawability (similar to that of the nickel silvers). Annealed high-zinc leaded brasses are suitable for shallow-drawn parts, such as garden-hose coupling nuts.

Copper-zinc-tin alloys such as C40500, C41100, C42200, and C425000 respond well to drawing and redrawing operations. With regard to deep-drawing properties, C40500 and C41100 are similar to the high-copper brasses, and C42200 and C42500 are similar to C24000.

Beryllium coppers can be drawn in the solution-annealed temper and then age hardened. For example, annealed alloy C17200 has been deep drawn to 80% reduction before annealing. Parts drawn from beryllium-copper alloys can subsequently be heat treated to produce tensile strengths to 1275 to 1380 MPa (185 to 200 ksi). Beryllium-copper alloys are often used in parts that require high strength and high fatigue performance, for example, bellows applications.

There are many other special-purpose coppers and copper alloys. By examining their compositions and mechanical properties carefully and by comparing them with standard alloys, the user can estimate how they will respond in deep-drawing applications.

Grain Size Effects. For the coppers and single-phase alloys in annealed or light cold-rolled tempers, grain size is the basic criterion by which deep drawability and stretch forming are measured. In general, for a given alloy and sheet thickness, ductility increases with grain size, and strength decreases. However, when grain size is so large that there are only a few grains through the thickness of the sheet or strip, both ductility and strength, as measured by tensile testing, decrease. Figure 17 illustrates how elongation changes with grain size for three different thicknesses of alloy C26000 (cartridge brass). General recommendations for the grain size of annealed strip for drawing and stretch-forming operations are provided in Table 8, along with the expected surface characteristics.

For optimal deep-drawing and stretch-forming properties, the grain size chosen should provide maximum elongation. With reference to Fig. 17, peak elongation for 0.15 mm (0.006 in.) thick

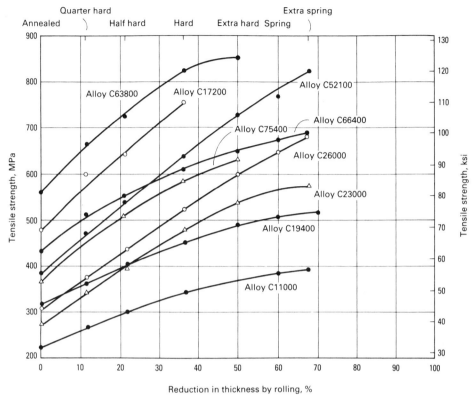

Fig. 15 Redrawing characteristics of 1.0 mm (0.040 in.) thick sheets of several copper alloys. Curves of lower slope indicate a lower rate of work hardening and therefore a higher capacity for redrawing. Source: Ref 3

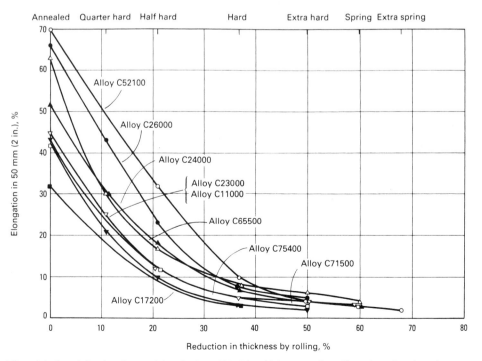

Fig. 16 Stretch-forming characteristics of 1.0 mm (0.040 in.) thick copper alloys. Elongation values for a given percentage of cold reduction indicate the remaining capacity for stretch forming in a single operation. Source: Ref 3

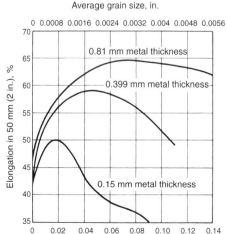

Fig. 17 Elongation versus grain size for alloy C26000 sheets of various thicknesses. Source: Ref 3

strip occurs at an average grain size of 0.020 mm (0.0008 in.). For 0.40 mm (0.0157 in.) thick brass, a range of 0.038 to 0.061 mm (0.0015 to 0.0024 in.) average grain size would provide maximum drawability. For 0.81 mm (0.032 in.) thick material, a range of 0.060 to 0.090 mm (0.0024 to 0.0035 in.) average grain size would give optimal performance.

The surface finish required on the final part is an important consideration when selecting the grain size to be used. When metal with a coarse grain size is drawn or stretch formed, the surface roughens and develops an appearance resembling orange peel. Such a surface is more difficult and costly to polish and buff. Therefore, when a part has a buffed surface requirement, much effort is often expended to design tools and fabrication sequences that allow the use of brass strip with a fine grain size.

A classic example of this situation is the one-piece brass or bronze doorknob (Fig. 3a). Such useful and decorative articles are made by the millions, and these types of shapes are difficult to produce on draw presses. These parts are usually produced in transfer presses, and the process can include 15 to 20 operations with one intermediate anneal or partial anneal. The alloy C26000 or C22000 strip from which these parts are made is usually approximately 0.76 mm (0.030 in.) thick, and the grain size is usually 0.020 to 0.035 mm (0.0008 to 0.0014 in.) or 0.015 to 0.030 mm (0.0006 to 0.0012 in.) to provide sufficient ductility for the part to be drawn without surface roughening.

Friction Effects (Ref 3). Each copper alloy has inherently different coefficients of friction against tool steels. For example, the coefficient of friction of copper against tool steels is higher than that of brass. The high-zinc brasses have coefficients of friction lower than the copper-rich brasses, and these values vary with zinc content. The phosphor bronzes have high coefficients of friction when compared to high-zinc brasses. The nickel silvers have higher friction coefficients than brasses, and cupro-nickels higher

values than the nickel silvers. These general guidelines regarding friction coefficients are from experience; actual values are dependent on tooling, lubricant, strip characteristics (gage, gage tolerance, tensile properties, etc.) and the deep-draw part geometry.

Other material factors that can directly influence deep-draw characteristics are surface topography produced by cold rolling, surface cleanliness (annealing atmospheres and/or cleaning practices during strip processing), mechanical surface damage (scratches, rolled-in defects, flakes, etc.), residual fluids from cold rolling or slitting, and/or moisture stains from shipping or storage.

Special Considerations (Ref 4). A common concern in all drawing operations is the formation of ears at the top of the cup sidewall. Ears occur in preferred directions (usually 45° or 0° and 90°) relative to the strip-rolling direction and indicate a difference in the cup sidewall thickness (thinner in earing orientations and thicker in valley orientations) in parts deep drawn or redrawn without ironing. Earing reflects the crystallographic texture of the strip. In part manufacture, parts with ears may be difficult to transfer station-to-station in stamping presses. The ears trimmed from finished parts represent excess scrap, and parts with ears will have varying sidewall thickness. Accordingly, non-earing grades of copper and copper alloy strip are almost always specified for drawn parts.

For copper alloys, the reduction in diameter in a single draw (cupping) usually ranges from 35 to 50%, with a 50% reduction corresponding to ideal conditions. Drawing procedures vary widely in commercial practice. Reductions for successive draws of the commonly formed brasses, under favorable operating conditions and without intermediate annealing, are usually 45% for cupping; 25% for the first redraw; and 20, 16, 13, and 10% for subsequent redraws. Greater reductions are usually obtained with blank thicknesses greater than approximately 1.62 mm (0.064 in.); for blank thicknesses

less than approximately 0.38 mm (0.015 in.), reductions are usually approximately 80% of the percentages given previously. With an annealing operation before each redraw, a reduction of 35 to 45% in each successive redraw can be obtained under favorable operating conditions, assuming that the accompanying reduction in wall thickness is acceptable.

Die radius usually varies from approximately 20 times the metal thickness for material 0.127 mm (0.005 in.) thick to approximately 5 times the metal thickness for material 3.18 mm (0.125 in.) thick. Radii of this size prevent high stress concentrations at the die opening, which can lead to tearing in subsequent draws. Sharper radii are needed for flanged shells and for meeting special design requirements.

The nose radius of the punch, except for the final stages of drawing, is usually less than one-third of the punch diameter, or four to ten times the metal thickness. Clearance between punch and die is usually maintained at values to produce a slight amount of ironing of the sidewalls.

Coining

Nature of the Operation. Coining is a cold forming process in which the work metal is compressed between two dies to fill the depression of both dies in relief or to reduce the strip thickness. The most familiar coining operation is the minting of coins. However, one of the most common uses of coining is in reducing the thickness or width of localized regions of electrical and electronic connectors and leadframe leads.

Materials Characteristics. The ability of a material to be coined is determined by its strength and work-hardening rate. In general, copper, the lower-zinc brasses, the lower-alloy nickel silvers, and the copper nickels, which all exhibit low work-hardening rates, exhibit good coinability (Fig. 15).

Spinning

Nature of the Operation. Spinning is a method of forming sheet metal or tubing into seamless hollow cylinders, cones, hemispheres, or other circular shapes by a combination of rotation and force. Manual and power-automated equipment is used for spinning copper alloys. More information is available in the article "Spinning" in this Volume.

Materials Characteristics. The principal materials factors that determine the spinnability of copper alloys are plastic-strain ratio (r), total available elongation, and work-hardening rate. In general, alloys with high r values, high tensile elongation, and low work-hardening rates exhibit the highest spinnability.

Effects of Alloy Composition and Temper. Tough pitch copper (alloy C11000) is the easiest copper material to spin and usually does not require intermediate annealing. Brasses, except

Table 8 Available grain size ranges and recommended applications

Average grain size		
mm	in.	Type of operation and surface characteristics
0.005–0.015	0.0002–0.0006	Shallow forming or stamping. Parts will have good strength and very smooth surface. Also used for very thin metal
0.010–0.025	0.0004–0.001	Stampings and shallow-drawn parts. Parts will have high strength and smooth surface. General use for metal thinner than 0.25 mm (0.010 in.)
0.015–0.030	0.0006–0.0012	Shallow-drawn parts, stampings, and deep-drawn parts that require buffable surfaces. General use for thicknesses under 0.3 mm (0.012 in.)
0.020–0.035	0.0008–0.0014	This grain size range includes the largest average grain that will produce parts essentially free of orange peel. Therefore, it is used for all types of drawn parts produced from brass up to 0.8 mm (0.032 in.) thick.
0.010–0.040	0.0004–0.0016	Begins to show some roughening of the surface when severely stretched. Good deep-drawing quality in 0.4–0.5 mm (0.015–0.020 in.) thickness range
0.030–0.050	0.0012–0.002	Drawn parts from 0.4–0.64 mm (0.015–0.025 in.) thick brass requiring relatively good surface, or stamped parts requiring no polishing or buffing
0.040–0.060	0.0016–0.0024	Commonly used for general applications for the deep and shallow drawing of parts from brass in 0.5–1.0 mm (0.020–0.040 in.) thicknesses. Moderate orange peel may develop on drawn surfaces.
0.050–0.119	0.002–0.0047	Large average grain sizes are used for the deep drawing of difficult shapes or deep-drawing parts for gages 1.0 mm (0.040 in.) and thicker. Drawn parts will have rough surfaces with orange peel except where smoothed by ironing.

for the multiphase alloy Muntz metal (C28000), are readily spun, although the higher-zinc brasses sometimes require intermediate annealing. Tin brasses containing at least 87% Cu require higher spinning pressure and more frequent annealing than brasses. Nickel silvers that contain at least 65% Cu, as well as the copper nickels, are also well suited for spinning. Phosphor bronzes, aluminum bronzes, and silicon bronzes are difficult to spin but can be spun into shallow shapes under favorable conditions. Copper alloys that are difficult to spin include Muntz metal, nickel silvers containing 55% Cu or less, beryllium coppers, alloys containing more than approximately 0.5% Pb, naval brass (C46400), and other multiphase alloys.

The single-phase high-strength copper alloys can be heated for spinning to reduce the force required to permit the spinning of thicker material or to permit more severe deformation, provided the increased cost for heating is justified. The forming characteristics of Muntz metal, extra-high-leaded brass, and naval brass are also improved at elevated temperature, but special precautions must be taken to avoid even the unintentional heating of the workpiece in spinning brasses that contain 0.5% Pb or more and more than 64% Cu.

Annealed tempers are almost always used in spinning copper alloys. Larger grain sizes (lower hardnesses) are easier to spin; finer grain sizes may be needed to meet surface finish requirements.

Although stock as thin as 0.1 mm (0.004 in.) can be manually spun under special conditions, manual spinning is usually restricted to thicknesses of 0.51 to 6.35 mm (0.020 to 0.250 in.). Powered equipment is used in the upper part of this range, and stock thicknesses in excess of 25 mm (1 in.) can be shaped by hot power spinning.

Applications. Typical products that are spun from copper alloys include bell-mouth shapes for musical instruments, lighting fixture components, vases, tumblers, decorative articles, pressure vessel parts, and other circular parts with bulged or recessed contours.

Contour Roll Forming

Nature of the Operation. Contour roll forming is an automated high-speed production process that is capable of producing tubular, box, angular, and folded parts of varied and complex shapes (see the article "Contour Roll Forming" in this Volume). Auxiliary operations such as notching, slotting, punching, and embossing can be combined with contour roll forming.

The materials characteristics that determine the roll-forming capability of copper alloy strip are the same as those that govern bend and stretch formability (see the sections "Bending" and "Drawing and Stretch Forming" in this article).

Alloy and Temper Effects. The bend properties given in Table 7 provide an indication of the relative suitability of copper alloys for contour roll forming. Annealed tempers are needed for complicated shapes and parts with extremely sharp bends or for severe stretch forming.

Applications. Contour roll forming is used less extensively with copper alloys than with steel and aluminum alloys because there are fewer copper alloy parts that are made in sufficient volume to be produced economically by this type of forming operation. Applications are primarily in the automotive and architectural industries.

Rubber-Pad Forming

Nature of the Operation. In this process, the rubber pad usually serves as the female die, in conjunction with an inexpensive male punch. The pad is practically incompressible, and it transmits pressure in all directions in the same manner as hydraulic fluid. Rubber-diaphragm forming uses hydraulic fluid behind the rubber pad. The most important reasons for using rubber-pad forming in preference to conventional press techniques or other production methods are improved formability, low tooling costs, and freedom from marking of workpiece surfaces. This is the most cost-effective method of fabricating one-piece doorknobs.

Deep drawing by rubber-diaphragm or Marforming techniques often permits a 65% reduction of diameter in a single draw, without producing wrinkles or surface defects that could require expensive finishing operations. More information on the Marform process and other rubber-pad forming techniques is available in the article "Rubber-Pad Forming and Hydroforming" in this Volume.

Materials Characteristics. The materials properties of greatest importance in rubber-pad forming are the same as those that control strip performance in metal dies; that is, deep drawing is dependent on the plastic-strain ratio (r), stretch forming is dependent on tensile elongation, bending is determined by strip ductility, and so on.

Effects of Alloy Composition and Temper. The same principles can be used to select the appropriate alloy composition and temper for rubber-pad die forming that are used for parts formed with conventional metal dies.

Specialized Forming Operations

Hydraulic Forming. Copper alloys are sometimes formed by applying direct hydraulic pressure to the surface of the workpiece in order to shape the workpiece against a rigid die. This procedure can be used to form grooves on large, thin, flat sheets and to shape small parts to irregular contours. Tool cost is low, but the method is ordinarily applicable only to small-lot production because of comparatively low production rates.

Embossing and swaging, which are closely related to coining (being compressive or deformation operations), are also frequently used in the cold forming of copper alloys. The principles of alloy selection described for coining apply equally to embossing and swaging. However, embossing (impressing letters, numerals, or designs into a surface by displacing metal to either side) can be done on any copper alloy, with special attention to tooling and selection of temper on the less formable alloys. Swaging is often used for the production of complicated electrical contacts from copper or brass.

High-velocity metal forming, also known as electromagnetic forming or magnetic pulse forming, is a process for forming metal by the direct application of an intense, transient magnetic field. The workpiece is formed without mechanical contact by the passage of a pulse of electric current through a forming coil (see the article "High-Velocity Metal Forming" in this Volume).

Electromagnetic forming can be used on copper and some brasses because of their high electrical conductivity and excellent formability. Metals with a resistivity greater than approximately 16 m$\Omega \cdot$cm are formed by the use of a copper or aluminum electromagnetic driver that is one to three times the thickness of the work metal. Thermally or electrically conductive joints and structural joints are produced in a single forming operation. Field shapers are frequently used to concentrate the forming force.

Electrical connections are made by electromagnetically swaging a copper band onto the end of stranded electrical conductor wire before insertion into a brass terminal. Optimal conductivity with 100% mechanical strength and long life under severe service conditions are obtained by using swaging forces great enough to compact the strands of the conductor so that a cross section of the joint appears to be essentially solid copper.

Special Forming Considerations for Conductive Spring Materials. Increasingly, contact designers are developing parts that rely on stepped or tapered beam thickness for optimal deflection or normal force characteristics. Some designs involve complex geometries requiring high formability in some regions (as for crimp connections) coupled with high strength in other regions (to resist permanent set in spring connections). Stepped or tapered contact beam thicknesses can be achieved by coining heavier-gage strip in progressive dies. This practice, however, rapidly work hardens copper alloys and reduces their formability. Die progressions that include the forming of contacts after a coining operation must incorporate more generous minimum bend radii than those suggested in the product literature of the supplier.

Figure 18 shows this change in formability for a mill-hardened temper of alloy C17200 that was subjected to coining up to 50% reduction in area and simulated by cold rolling after mill hardening. To avoid this formability problem, strip can be purchased with variable gage across the slit width, which is produced by profile milling or skiving or by the longitudinal electron beam

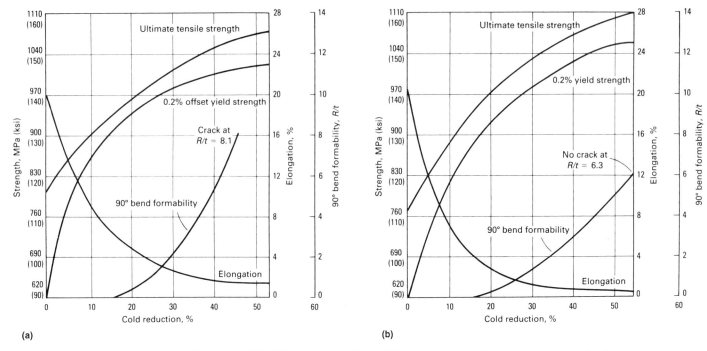

Fig. 18 Change in formability as a function of the coining of alloy C17200 in (a) longitudinal and (b) transverse directions. The effect of coining is simulated by cold reduction. Original strip thickness in both cases was 0.41 mm (0.016 in.). Bend formability is measured as the ratio of bend radius (R) to strip thickness (t)

welding of dissimilar thicknesses of strip. The need for localized high formability can also be met by the longitudinal electron beam welding of dissimilar metals, combining, for example, ductile C19500 with high-strength mill-hardened C17200.

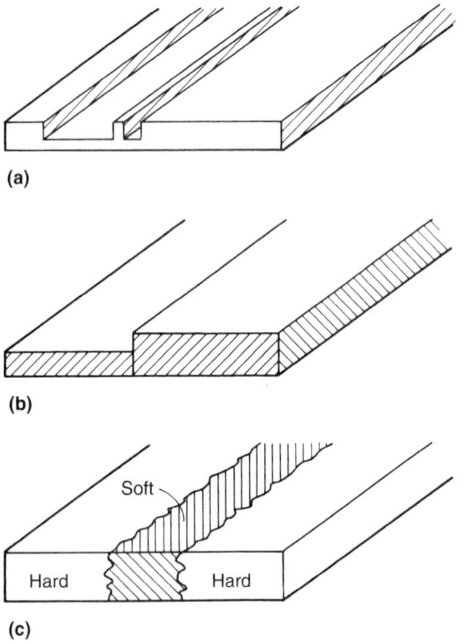

An emerging electron beam application is the localized thermal softening of mill-hardened copper alloy strip to provide increased formability with no sacrifice in strength in the remainder of a contact. Examples of these unique copper alloy strip forms are shown in Fig. 19.

Springback

Springback is the elastic recovery that occurs in a plastically deformed part when it is released from tooling. It causes the final part to have a geometry different from that of the press tooling. The springback that occurs in a bending operation is shown schematically in Fig. 20. Springback must be taken into account in design of tooling and materials selection.

Springback depends on alloy, temper, thickness, bend radius, and the angle of bend. For fixed tooling and press conditions, springback increases as the strength of the copper alloy strip increases. Therefore, springback is increased by cold rolling to increase strip temper and/or by alloy additions that increase strength. The springback behavior of three copper alloys (C21000, C26000, and C35300) is shown in Fig. 21. These data indicate that springback increases with increasing bend radius and decreasing strip gage. Springback values for tempers or bend radii not shown can be interpolated from Fig. 21. Some strip suppliers will provide springback data for selective copper alloys on request.

Three techniques are commonly used to compensate for springback: overbending, restriking, and the use of special dies. Overbending simply deforms the part to a larger bend angle so that it is at the desired value after springback. Restriking in original dies reduces springback in much the same manner as overbending, that is, by the introduction of additional plastic deformation. Special dies often use coining action at bend radii to deform the metal plastically in the bend area beyond the elastic limit. In other die modifications, the metal is pinched slightly at the bend region. When special dies are used, careful control must be exercised because excessive thinning can cause part failure during bending or can make the part susceptible to early failure in service.

Forming of Larger Parts from Copper and Copper Alloy Strips

Forming limit analysis provides the means to assess sheet metal formability over a wide range

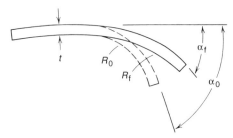

Fig. 19 Special treatment of copper alloy strip for optimized combinations of formability and spring characteristics. (a) Profile milled strip. (b) Dissimilar thicknesses longitudinally welded; this method can also be used to join dissimilar alloys. (c) Localized heat treatment (electron beam softening)

Fig. 20 Schematic of springback in a bending operation. t is sheet thickness, R_0 and, α_0 are the die radius and bend angle, and R_f and α_f are the part radius and bend angle after springback.

of forming conditions, including drawing, bending, and stretching. The amount of deformation that occurs during sheet forming, that is, the strain state, is given in terms of, or related to, major and minor strains (e_1 and e_2, respectively) measured from fiducial markings (typically contacting circles with 2.5 mm, or 0.1 in., diameters) printed or etched onto strip surfaces prior to fabrication. The analysis requires two curves:

- A forming limit curve (FLC), which indicates the ability of the material to distribute localized strain
- A limiting dome height (LDH) curve, which indicates the overall ductility for forming of the material

These empirically determined curves show the biaxial strain or deformation limits beyond which failure may occur in sheet metal forming. More information on the development and use of these curves is available in the article "Formability Testing of Sheet Metals" in this Volume.

Forming limit and limiting dome height curves for 13 copper alloys are shown in Fig. 22 and 23. Table 9 lists UNS designations, common names, alloy compositions, and tempers for the alloys tested. These data indicate that, in annealed tempers, high-copper and copper-zinc alloys exhibit the highest FLC values, followed closely by alloys C72500, C51000, and C74300; these materials in turn are slightly better than alloys C19400, C75200, and C70600. Increasing the temper by cold rolling decreases forming capability, as shown in Fig. 22. The LDH data essentially follow the trend shown in FLC behavior.

Solving Forming Problems. In addition to displaying the relative formability of one material versus that of another, FLCs and LDH curves are valuable for identifying the cause of a sudden production problem that may arise from changes in tooling, lubrication, or material suppliers. This permits the forming process to be modified to maximize formability and productivity.

The most direct approach for determining the cause for an unexpected forming problem is to compare the LDH curve of the suspect lot with that of the control lot of known good material. If only one region of the part is subject to critical strains, it may be necessary only to test the blank width that will produce that critical value of minor strain. If the LDH curve of the new material is the same as that of the control lot, then tooling or lubrication is suspect. If the LDH curve of the new material is below that of the control lot, the material is the problem.

The best way to determine whether tooling or lubrication conditions have changed is to form a gridded sample under current tool conditions from a control lot held in inventory. Strain distribution and critical grid strains measured on this sample can be placed on the established FLC and compared with those before the problem arose in order to establish their relationship to known, safe strain levels. If changes are detected, they can often be remedied by adjusting press

conditions to change the magnitude of stretch or draw components.

This is illustrated in Fig. 24. Point A on this FLC represents the strains in the critical region of a part when the part was being formed satisfactorily. Point B represents the critical strains when

forming became a problem because the major and minor strains were too high. Draw beads, blank holddown pressure, blank size, and/or lubrication can be modified to change the amount of major and minor strains. The effects on critical strain can be compared on the FLC to ensure that

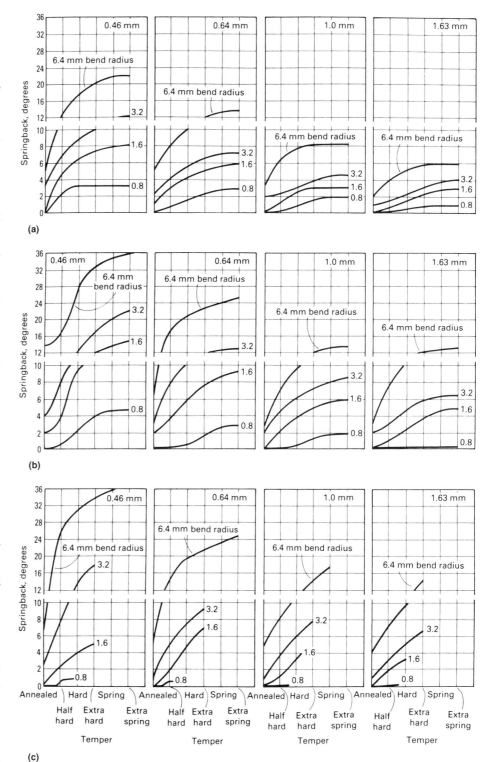

Fig. 21 Springback behavior of copper alloys as a function of temper, sheet thickness, and bend radius (90° bends). (a) Alloy C21000. (b) Alloy C26000. (c) Alloy 35300

the adjustments will indeed enable the part to be formed.

A similar approach can be used to adjust the forming operation so that a less ductile material can be formed. Figure 25, for example, shows FLCs for two materials (A and B) and the critical strain combination (point X) measured on a formed part. Material A forms successfully, but material B fractures during forming, as indicated by the location of point X relative to the FLC of each material.

Because of the shape of the FLCs, it is possible to maintain approximately the same e_1 value for point X but to fall in the safe region by changing e_2, as indicated by points X′ and X″. In this case, moving toward X′ requires that the draw component be increased during forming; moving toward X″ requires that the stretch component be increased. Either can be accomplished by altering lubrication, tooling, and/or blank holddown pressure, thus enabling the part to be formed in material B.

Forming of Smaller Parts from Copper and Copper Alloy Strips (Ref 7–10)

With the recent advances in software, computing speed, and storage capacity, parts are often developed or designed by finite-element analysis (FEA) techniques. Simplistically, the FEA approach uses strip property data, the geometry of the part, and the boundary and loading conditions to calculate the strain and stress distributions of a formed part. By matching the calculated strains to a material forming limit diagram (or a similar failure criterion), an assessment can be made as to whether the part can be formed from a specific copper alloy in a specific temper condition.

For larger parts, it is possible to experimentally verify (by fudicial grid measurement or by analytical calculation) that the strains calculated by FEA match the actual strain in the part. This verification is not possible for typical electronic terminals, connectors, leadframes, and/or bellows parts, which frequently have part dimensions that are less than the 2.5 mm (0.1 in.), or the plastic deformation in these parts is localized to regions that are less than the 2.5 mm (0.1 in.), the diameter of contacting circles typically used to measure forming strains in larger parts.

A measurement technology first published in the early 1990s that could address this shortfall is digital image correlation (DIC), a computer vision deformation mapping technique. This methodology can map local surface strains to an accuracy of approximately 0.05% when strip is plastically deformed in two dimensions (a tensile test or coining operation, for example) and approximately 0.5% when strip is deformed in three dimensions (a part with bends, stretch, draw, or redraw forming, for example). The minimum effective gage length over which strain and strain gradients can be measured is on the order of 0.05 mm (0.002 in.) (dependent on pixel density of the digital image and reference grid dimensions). The accuracy of the DIC method is sufficient to measure the diffuse and local necking strains during tensile testing of copper alloy strip in cold-rolled or precipitation-hardened tempers to verify the accuracy of constitutive equations used in the FEA. This accuracy of strain measurement is also sufficient to verify whether FEA-calculated strains match actual strain values in formed parts. Cooperation is required between suppliers and users of copper alloy strips with high strength and moderate electrical conductivity to determine if application of DIC methodology can increase the accuracy of FEA to provide connector parts with improved performance.

Property Requirements for Various Formed Products (Ref 11)

Because the selection of an alloy begins with performance requirements even before formability is considered, this section offers a brief discussion of alloy effects relative to the performance of copper alloys. When these factors are matched with the best formability considerations, optimal materials selection is attained.

Electrical Conductors. High conductivity is a primary requisite for many conductors, but not all conductors require high conductivity. Strength and resistance to creep or softening can be traded for some loss of conductivity. Thermal conductivity goes hand-in-hand with electrical conductivity and is usually required in all conductors operating in the high-amperage range.

Terminals and connectors are used in both electronic and electrical applications. Electrical circuits require greater current-carrying ability than electronic circuits, for which low-to-moderate conductivity will often suffice. Good formability at the required strength level, well-defined load deflection characteristics, and resistance to stress relaxation are usually required for both electrical and electronic connectors. Corrosion and stress-corrosion resistance and solderability are also often demanded.

Other Electronic Applications. The primary electronic application for copper and copper alloys, other than connectors and printed circuits, is as leadframes for semiconductor devices. Transistors, diodes, and integrated circuits can be fabricated on a semiconductor chip often less than 5 mm (0.2 in.) square. The chip is bonded to a substrate or leadframe, which serves both structurally and electrically to connect it to the outside world. Where copper alloys can be used, the materials requirements are expressed in terms of cost, conductivity, strength, softening resistance, formability, low-cycle fatigue resistance, and surface characteristics (such as plating, wire bonding, solderability, and plastic molding compound adherence).

Hollow Ware, Flatware, and Decorative Applications. The production of hollow ware products requires materials with good drawability. In addition, the material must have good solderability, corrosion resistance, sufficient strength to resist denting during manufacture or use, and good buffing and plating characteristics; most hollow ware products are silver or gold plated. The least expensive alloys that meet these

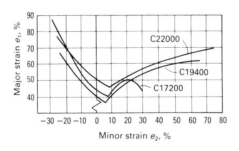

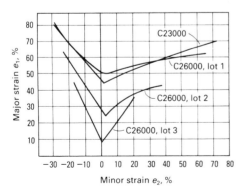

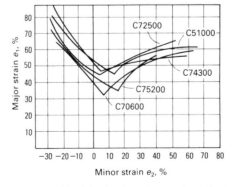

Fig. 22 Forming limit curves (FLCs) for selected copper alloys. FLCs reveal local ductility during forming. See Table 9 for material designations, thicknesses, and tempers. Source: Ref 5, 6

criteria are the 10 to 30% Zn brasses. The lower zinc levels are used for multiple-redrawing applications, and the higher zinc levels are for parts demanding higher strength and/or deep-drawing capability. Copper and phosphorus-deoxidized copper are significantly softer and are primarily used for decorative applications without plating, where the red color of the metal is considered appealing. They have adequate drawability for parts that do not require side-wall ironing and a high work-hardening rate. Phosphorus-deoxidized copper is required where brazing is needed.

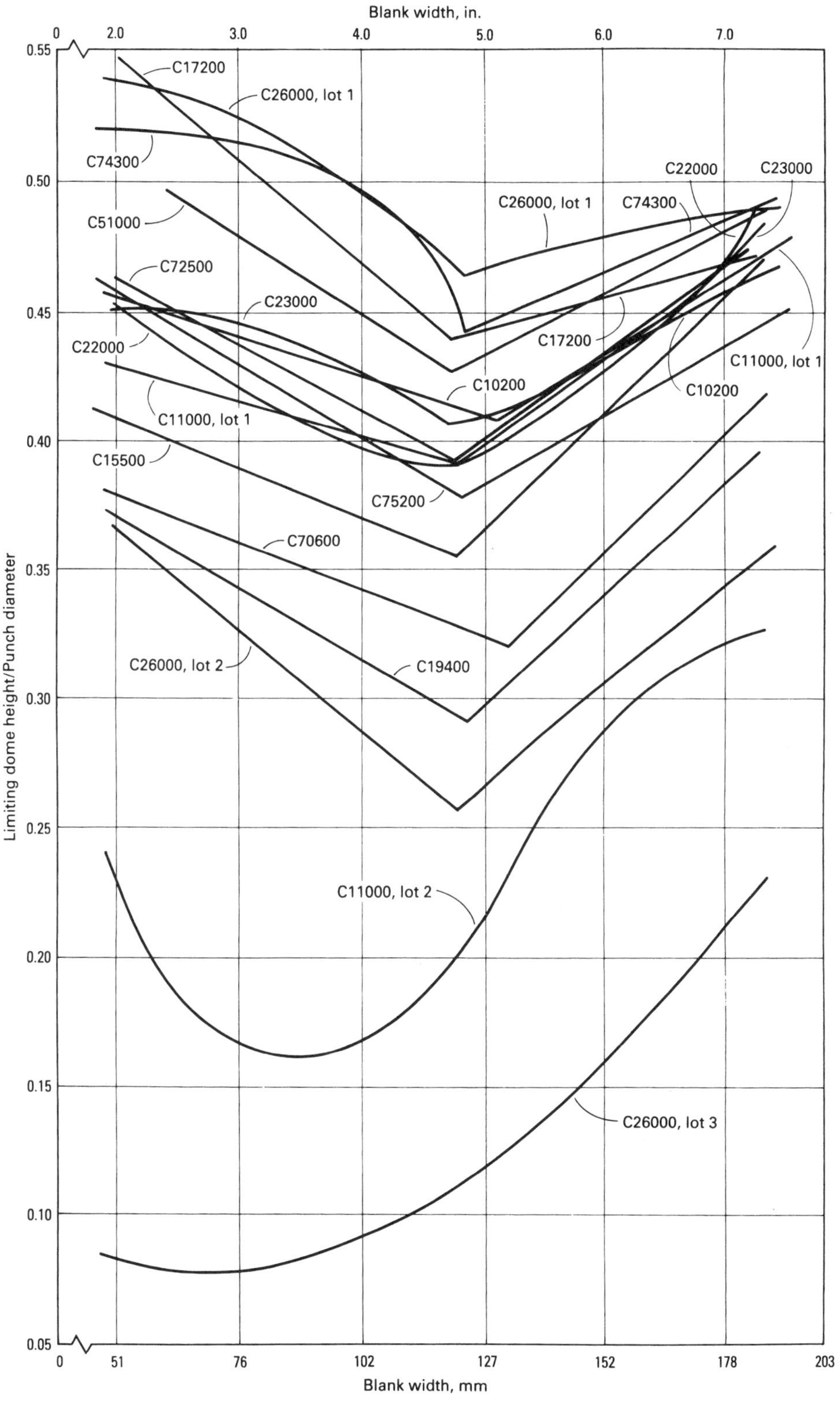

Flatware items are generally produced by roll forming. In addition to good formability, flatwork alloys must have good solderability, corrosion resistance, good buffing and plating characteristics, and low cost. Embossed items use annealed tempers of materials with sufficiently low work-hardening rates to give faithful reproduction of detailed patterns. Copper-zinc and copper-nickel-zinc alloys offer the required combination of properties; the zinc content is varied to suit the work hardening needed. A copper-nickel-zinc alloy that has a silvery color is generally used for silver-plated flatware, and Cu-30Zn is used for gold-plated flatware to minimize the color contrast between base metal and plate if damage to the plate occurs.

Heat exchangers require good thermal conductivity, corrosion and stress-corrosion resistance, joinability, and strength at modest cost. These requirements vary in importance for each application. Copper and copper alloys offer good combinations of these properties. The two major heat exchanger applications are steam condenser tubing and automotive radiators.

Condenser tubing must withstand potentially corrosive cooling water as well as the volatile components carried by the steam, which condense on the tubing. Corrosion requirements are paramount, but strength at elevated temperature and thermal conductivity are also required.

The copper alloys commonly used in power utility condensers include arsenic-, antimony-, or phosphorus-inhibited brasses; aluminum bronzes; or copper-nickels, depending on corrosion and stress-corrosion requirements. For automotive radiators, corrosion resistance, thermal conductivity, and fabricability are the primary requisites. Certain applications require strength at elevated temperature. Fabricability demands the ability to solder and to braze. Resistance to both atmospheric corrosion and corrosion by heat-transfer media and their decomposition products is required. Cooling fins are made of pure copper or high-copper copper alloys.

Coinage. General requirements include low cost, attractive appearance and high density for high denominations (to give the impression of intrinsic value), tarnish and corrosion resistance, modest strength, and ability to be coined easily. Specific additional requirements for vending machine use include control of conductivity, density, magnetic permeability, and eddy-current response. Apart from those few coins that are fabricated from stainless steel, most non-precious metal coins are made of copper alloys, which can meet these requirements.

Ammunition. The cartridge case that contains the explosive powder and primer for ammunition is made by a cup-and-draw process; a blanked disk is cupped, drawn to extend the sidewall, and redrawn for the same purpose. The drawing process is repeated until the wall is

Fig. 23 Limiting dome height (LDH) curves for copper and copper alloys. LDH curves illustrate the overall ductility of the coppers and copper alloys evaluated. See Table 9 for material designations, thicknesses, and tempers. Source: Ref 5, 6

sufficiently thinned and extended. The case can be annealed between a series of draws to permit sufficient extension without cracking.

The primary materials requirements are related to fabricability, but stress-corrosion resistance and strength are also needed.

Cartridge brass (Cu-30Zn) is the most widely used copper alloy for shells and other ammunition; hence its name. Cartridge brass offers the best materials compromise for such applications, having excellent deep drawability, moderate redrawability, and sufficient strength. In press operations, its low coefficient of friction and absence of refractory oxides contribute toward low press forces, low tool maintenance, and good surfaces on the parts being formed. It is also low in cost. Formed cartridge cases must be stress-relief annealed to minimize the possibility of stress-corrosion cracking.

Table 9 Coppers and copper alloys evaluated using forming limit analysis

See Fig. 22 and 23 for results of analysis.

UNS designation	Common name	Material conditions applicable to forming limit curves and limiting dome height (LDH) curves
C10200	Oxygen-free copper	Annealed, 0.66 mm (0.026 in.) thick, 0.014 mm (0.0006 in.) grain, 234 MPa (34 ksi) UTS(a)
C11000, lot 1	Electrolytic tough pitch copper	Annealed, 0.74 mm (0.029 in.) thick, 0.016 mm (0.00063 in.) grain, 224 MPa (32.5 ksi) UTS(b)
C11000, lot 2	Electrolytic tough pitch copper	Half hard, 0.69 mm (0.027 in.) thick, 268 MPa (38.8 ksi) UTS, 20% tensile elongation(c)
C15500	Silver copper	Annealed, 0.71 mm (0.028 in.) thick, 0.009 mm (0.00035 in.) grain, 288 MPa (41.8 ksi) UTS
C17200	Beryllium copper	Annealed, 0.25 mm (0.010 in.) thick, 0.019 mm (0.00075 in.) grain, 491 MPa (71.2 ksi) UTS
C19400	HSM copper	Annealed, 0.69 mm (0.027 in.) thick, 319 MPa (46.3 ksi) UTS, 29% tensile elongation(d)
C22000	Commercial bronze	Annealed, 0.69 mm (0.027 in.) thick, 0.017 mm (0.00067 in.) grain, 234 MPa (34 ksi) UTS(d)
C23000	Red brass	Annealed, 0.69 mm (0.027 in.) thick, 0.024 mm (0.00094 in.) grain, 293 MPa (42.5 ksi) UTS(e)
C26000, lot 1	Cartridge brass	Annealed, 0.64 mm (0.025 in.) thick, 0.025 mm (0.00098 in.) grain, 345 MPa (50 ksi) UTS(f)
C26000, lot 2	Cartridge brass	Half hard, 0.69 mm (0.027 in.) thick, 407 MPa (59 ksi) UTS, 28% tensile elongation(e)
C26000, lot 3	Cartridge brass	Full hard, 0.51 mm (0.020 in.) thick, 531 MPa (77 ksi) tensile strength
C51000	Phosphor bronze A	Annealed, 0.69 mm (0.027 in.) thick, 0.014 mm (0.0006 in.) grain, 374 MPa (54.3 ksi) UTS
C70600	Copper nickel, 10%	Annealed, 0.81 mm (0.032 in.) thick, 0.016 mm (0.00063 in.) grain, 361 MPa (52.4 ksi) UTS
C72500	Copper-nickel-tin alloy	Annealed, 0.69 mm (0.027 in.) thick, 0.023 mm (0.0009 in.) grain, 356 MPa (51.6 ksi) UTS
C74300	Nickel silver	Annealed, 0.69 mm (0.027 in.) thick, 0.035 mm (0.0014 in.) grain, 387 MPa (56.1 ksi) UTS
C75200	Nickel silver	Annealed, 0.69 mm (0.027 in.) thick, 0.020 mm (0.0008 in.) grain, 405 MPa (58.7 ksi) UTS

(a) UTS, ultimate tensile strength. (b) LDH curves are medians based on 0.69, 0.74, and 0.79 mm (0.027, 0.029, and 0.031 in.) thickness data. (c) LDH curves are medians based on 0.64, 0.69, and 0.79 mm (0.025, 0.027, and 0.031 in.) data. (d) LDH curves are medians based on 0.69 and 0.74 mm (0.027 and 0.029 in.) thickness data. (e) LDH curves are medians based on 0.69, 0.79, and 0.81 mm (0.027, 0.031, and 0.032 in.) data. (f) LDH curves are medians based on 0.66 and 0.69 mm (0.026 and 0.027 in.) data

ACKNOWLEDGMENTS

The authors would like to acknowledge the information and assistance provided by Peter W. Robinson, Manager of Connector Products, Olin Corporation; Michael J. Gedeon, Market Applications Development Engineer, Brush Wellman Inc.; and W. Raymond Cribb, Director of Technology, Brush Wellman Inc.

REFERENCES

1. *Sheet Metal Industries—Yearbook,* Fuel and Metallurgical Journals Ltd., 1972/1973
2. T.E. Bersett, Back to Basics: Properties of Copper Alloy Strip for Contacts and Terminals, *Proceedings of the 14th Annual Connector Symposium,* Electronic Connector Study Group, 1981
3. J.H. Mendenhall, Ed., *Understanding Copper Alloys,* Olin Corporation, 1977
4. "Advanced Sheet Metal Forming Course," Metals Engineering Institute Home Study and Extension Course, American Society for Metals, 1979
5. Forming Limits Set for Copper Metals, *Am. Mach.,* April 1983, p 99
6. "Forming Limit Analysis for Enhanced Fabrication," Report 310A, International Copper Research Association, Dec 1981
7. W. Tong, Detection of Plastic Deformation Patterns in a Binary Aluminum Alloy, *Exp. Mech.,* Vol 37 (No. 4), 1997, p 452–459
8. B.W. Smith, X. Li, and W. Tong, Error Assessment for Strain Mapping by Digital Image Correlation, *Exp. Tech.,* Vol 22 (No. 4), 1998, p 19–21
9. W. Tong and X. Li, Evaluation of Two DIC Methods for Plastic Deformation Mapping, *Proceedings of the SEM Annual Conference on Theoretical, Experimental, and Computational Mechanics* (Cincinnati, OH), 1999, p 23–26
10. W. Tong, Plastic Surface Strain Mapping of Bent Sheets by Image Correlation, *Exp. Mech.,* Vol 44 (No. 5), Oct 2004, p 502–511
11. M.B. Bever, Ed., Copper: Selection of Wrought Alloys, *Encyclopedia of Materials Science and Engineering,* Vol 2, Pergamon Press and The MIT Press, 1986, p 866

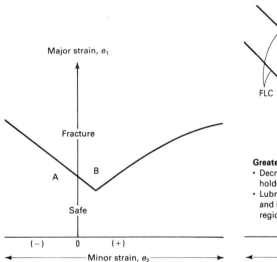

Fig. 24 Effect of tooling or lubrication on critical strain. Inadvertent changes in tooling or lubrication can shift strain from point A to point B, causing parts that previously were readily formed to fail during forming. Source: Ref 5, 6

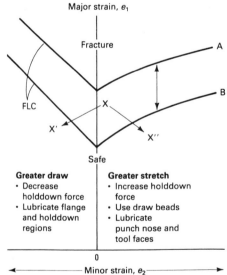

Fig. 25 Effect of changes in forming operation on critical strain. Shifting strain from X to X′ by increasing draw or from X to X″ by increasing stretch permits material B to be used in place of material A despite its lower ductility. FLC, forming limit curve. Source: Ref 5, 6

Forming of Magnesium Alloys

E. Doege*, B.-A. Behrens, G. Kurz, and O. Vogt, University of Hanover

MAGNESIUM ALLOYS are an important category of lightweight structural material with favorable specific strengths. For example, an important goal of the automotive industry is fuel efficiency, and one area of potential savings is weight reduction offered by magnesium alloy components, such as sheet metal body shell panels that are exposed to bending stresses (Ref 1). Currently, the weight of the car body is reduced by using high-strength steel with a decreased sheet thickness or conventional lightweight materials such as aluminum. In the future, aluminum and steel sheets may be replaced by magnesium sheets in some applications.

Currently, structural magnesium alloys are mainly processed by die casting in the automotive industry. Die casting technology allows the manufacture of parts with a complex geometry (Ref 2). The mechanical properties of these parts often do not meet the requirements concerning endurance, strength, ductility, and so on. A promising alternative for thin, large-surfaced parts, such as automotive body constructions, are components manufactured by sheet metal forming. In comparison to die-casted parts, sheet metal formed parts are characterized by advantageous mechanical properties and a good surface without pores. Therefore, the substitution of conventional sheet materials by magnesium sheets could lead to essential weight savings.

In terms of forming, magnesium alloys are much more workable at elevated temperatures due to their hexagonal crystal structures (as described in the following section, "Deformation Mechanisms of Magnesium," in this article). Magnesium alloys are usually formed at elevated temperatures, although some forming of magnesium alloys can be done at room temperature (see the section "Cold Forming" in this article). The methods and equipment used in forming magnesium alloys are the same as those commonly employed in forming other metals, except for differences in tooling and technique that are required when forming is done at elevated temperatures. Heat for manufacturing sheet metal components was first applied by the American aerospace industry. At first, cold-formed sheet metal parts were only finished by use of heated straightening machines. Afterward, heat was applied during the forming process itself, using heatable tools and presses. Today (2006), deep drawing at elevated temperatures is only used in a few special applications.

Deformation Mechanisms of Magnesium

Conventional magnesium alloys have a hexagonal lattice structure and possess a low formability at room temperature. Previous research projects thoroughly investigated the temperature-dependent processes in the lattice structure that increase the formability of magnesium alloys. The increase in plasticity that takes place in the temperature range from 200 to 225 °C (390 to 435 °F) has been described (Ref 3). The temperature at which plasticity increases depends on the metal alloy being used (Ref 3). The plasticity of pure magnesium increases at a temperature of 225 °C (435 °F). Studies report that slip happens basically in the basal plane of the hexagonal lattice structure at temperatures up to 225 °C (435 °F) (Fig. 1) (Ref 4–6).

Slip occurs in the direction in which the atoms are most closely packed. Another deformation mechanism occurring at room temperature is twinning. Twinning is a shear deformation process where parts of one crystal are reflected across a mirror plane. The mirror plane is called a twinning plane (Fig. 2). The main part of the twinning process happens in pyramidal planes (second-order type 1) (Fig. 3).

When the temperature exceeds 225 °C (435 °F), the mobility of atoms in the matrix increases as well, so that additional pyramidal slip planes are activated (Fig. 4). The additional slip planes cause a sudden increase in plastic deformation. In monocrystals, these slip systems are already activated at room temperature when the angle between crystal slip plane and stress direction is 6°, as demonstrated in Ref 7 and 8. Dislocation slip of prismatic planes is another forming mechanism of magnesium crystals (Ref 9). At low strain rates and a temperature of approximately 260 °C (500 °F), it is assumed that alternating slip of both pyramidal and prismatic slip planes leads to wavy slip bands on the material surface (Ref 10).

The Dow Chemical Company gives advice on the tool geometry of the MgAl3Zn1 (AZ31) magnesium alloy when this material is bent and deep drawn (Ref 11). Choosing large tool radii leads to slight deformations even at room temperature. Various authors state the minimum radii for bending magnesium sheets in a temperature range of 21 to 371 °C (70 to 700 °F).

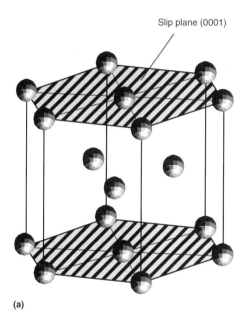

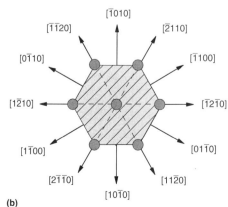

Fig. 1 Slip. (a) Slip plane in the hexagonal lattice structure. (b) Directions for the slip plane

*Deceased

Allowable bend radii at room and elevated temperature from various sources are given in Tables 1 and 2 (Ref 12–14).

Advice on tool radii for deep drawing rectangular components with a heated tool is given in Ref 12 (Table 3). The temperatures of the die (T_d), the blankholder (T_{bh}), and the punch (T_p) are $T_d = 316\ °C$ (600 °F), $T_{bh} = 343\ °C$ (650 °F), and $T_p = 232\ °C$ (450 °F). Punch edge radius values and die radius values also apply to deep drawing round components.

In order to avoid stress-corrosion cracking, cold-formed magnesium-aluminum-zinc alloy components and magnesium-thorium-manganese/zirconium alloy components need to be annealed after forming. Advice for choosing the anneal temperature and the necessary holding period is given in Ref 12.

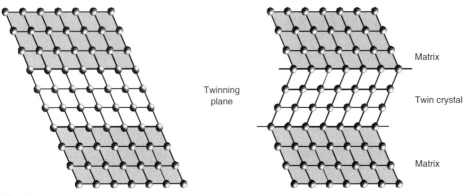

Fig. 2 Principle of twinning

Influences on Formability

Influence of Temperature on Flow Stress. A key parameter in the forming of magnesium alloys is the forming temperature. Figure 5 displays the flow curves of the magnesium sheet material MgAl3Zn1 (AZ31) at different temperatures, determined in the uniaxial tensile test according to EN 10130 and EN 10002, part V. It is obvious that the stresses and possible strains depend strongly on the forming temperature. At higher temperatures, the flow stress decreases with increasing strain, because of the thermally activated work softening (Ref 15).

Influence of Temperature on Deep Drawing. Experiments with magnesium alloy MgAl3Zn1 (AZ31) at room temperature show that the ductility of the material is sufficient to form inner and outer flat vaulted car body parts. A partial heating could be used

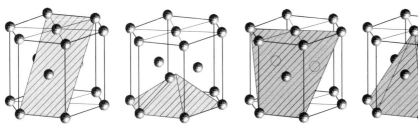

Fig. 3 Pyramidal slip planes, second-order type 1   **Fig. 4** Pyramidal slip planes, first-order type 1

Table 1 Smallest possible bending radii for magnesium sheets

Temperature		Minimum bending radius as multiple of the base sheet thickness						
°C	°F	MgAl3Zn1 (AZ31)	MgAl3Zn1 (AZ31B-H24)	MgZn1SE (ZE10A-O)	MgZn1SE (ZE10A-H24)	MgTh3Zr1 (HK31A-O)	MgTh3Zr1 (HK31A-H24)	MgTh2Mn1 (HM21A-T8)
21	70	5.5	8.0	5.5	8.0	6.0	13.0	9.0
149	300	4.0	6.0	4.0	6.0	6.0	13.0	9.0
204	400	3.0	3.0	3.0	...	5.0	9.0	9.0
260	500	2.0	2.0	2.0	...	4.0	8.0	9.0
316	600	...	...	...	...	3.0	5.0	8.0
371	700	...	...	...	...	2.0	3.0	6.0

Source: Ref 12, 13

Table 3 Tool radii for drawing rectangular magnesium sheet components with a heated tool

Tool geometry	Minimum value(a)	Proposed value(a)
Punch edge radius	$3s_0$	$4s_0$
Die radius	$4s_0$	$6s_0$
Edge radius (h, drawing depth)	h/20	h/12

(a) s_0, initial sheet thickness. Source: Ref 12

Table 2 Recommended minimum bend radii for the fast forming of magnesium alloys at room temperature

Alloy and temper	Minimum bend radius(a) in terms of workpiece thickness, t									
	At 20 °C (70 °F)	At 100 °C (212 °F)	At 150 °C (300 °F)	At 200 °C (400 °F)	At 230 °C (450 °F)	At 260 °C (500 °F)	At 290 °C (550 °F)	At 315 °C (600 °F)	At 370 °C (700 °F)	At 425 °C (800 °F)
Sheet 0.51–6.3 mm (0.020–0.249 in.) thick										
AZ31B-O	5.5	5.5	4.0	3.0	...	2.0	...	...	...	...
AZ31B-H24	8.0	8.0	6.0	...	...	...	...	...	...	...
Extruded flat strip 22.2 × 2.3 mm (0.875 × 0.090 in.) thick										
AZ31C-F	2.4	...	...	...	...	...	1.5	...	...	...
AZ31B-F	2.4	...	...	...	...	...	1.5	...	...	...
AZ61A-F	1.9	...	...	...	...	...	1.0	...	...	...
AZ80A-F	2.4	...	...	...	...	...	0.7	...	...	...
AZ80A-T5	8.3	...	...	...	1.7	...	...	...	...	...
ZK21A-F	15.0	...	...	...	...	...	...	5.0	...	...
ZK60A-F	12.0	...	...	...	...	2.0	...	...	...	...
ZK60A-T5	12.0	...	...	...	...	6.6	...	...	...	...

(a) Values based on bending a 150 mm (6 in.) wide specimen through 90° (99% success rate).

in areas of higher true strains. At higher drawing ratios, deep drawing with process annealing or warm deep drawing should be used (Ref 16). For the production of parts at high true strains, it is necessary to heat the deep drawing process.

Figure 6 shows the results of deep drawing tests at different temperatures. These tests confirm the low formability of magnesium alloys at low temperatures and a good formability at elevated temperatures. The work-softening behavior of magnesium at higher temperatures, as shown in the uniaxial tensile tests, influences the heated deep drawing process, too. Consequently, the limit drawing ratio reaches a maximum in the temperature range from 200 to 250 °C (390 to 480 °F), which further activates the slip planes. Above this temperature level, the work-softening behavior of the material

prevails, and the limit drawing ratio decreases (Ref 15, 17).

Influence of Temperature on Stretch Forming. The manufacturing of flat vaulted parts is especially characterized by high loading under stretch forming conditions (biaxial tension). These stress combinations often result in a local fracture of the formed components, so that process modifications, for example, of the tool geometry and lubrication, become necessary.

The maximum stretching height (h_{max}) until fracture of the sheet material was used to evaluate the material behavior under stretch forming conditions. Figure 7 shows the values of h_{max} that have been determined in stretch forming tests at three different temperatures (150, 200, and 250 °C, or 300, 390, and 480 °F). This figure shows that the maximum stretching heights of the magnesium alloys increase as temperature

increases (Ref 15). Figure 8 shows a stretch-formed specimen.

Influence of Strain Rate. Figure 9 shows the influence of testing velocity on flow stress for the sheet material MgAl3Zn1 (AZ31) at a temperature of 200 °C (390 °F). Raising the elongation rate from $\dot{\varepsilon} = 0.004$ s^{-1} (EN 10002) to $\dot{\varepsilon} = 3.0$ s^{-1} leads to a considerable increase in flow stress. This effect is less significant if the experiments are carried out at lower temperatures. Furthermore, the tensile tests show that increasing velocity results in a lower maximum elongation of the material at any temperature (Ref 15).

The results gained by the tensile tests showed the significant influence of strain rate on the mechanical properties of magnesium alloys. Figure 10 displays the dependence of the limit drawing ratio on punch velocity during deep drawing at a temperature of 200 °C (390 °F). Due to the higher required stresses and the lower maximum strain at higher velocities, the limit drawing ratio shows a considerable decrease at higher punch velocities (Ref 15).

Influence of Alloy Composition on Deep Drawing. Forming properties are vitally

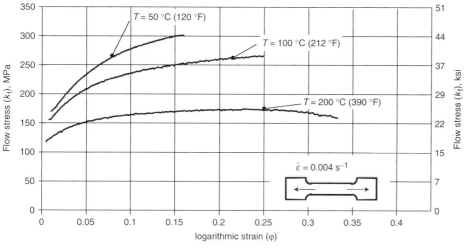

Fig. 5 Temperature-dependent flow curves of MgAl3Zn1 (AZ31) (initial sheet thickness, $s_0 = 1.0$ mm, or 0.039 in.) determined in the uniaxial tensile test. $\dot{\varepsilon}$, strain rate. Source: Ref 15

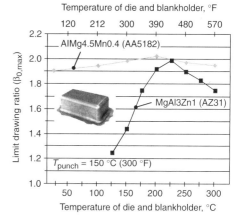

Fig. 6 Temperature-dependent limit drawing ratio for MgAl3Zn1 (AZ31) and AlMg4.5Mn0.4 (AA5182). Source: Ref 15

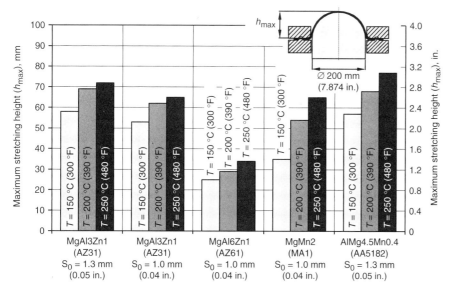

Fig. 7 Maximum stretching height (h_{max}) for different sheet materials and temperatures. s_0, initial sheet thickness. Source: Ref 15

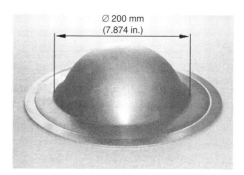

Fig. 8 Stretch-formed specimen. Material: MgAl3Zn1 (AZ31); initial sheet thickness, s_0: 1.3 mm (0.051 in.); forming temperature: 250 °C (480 °F). Source: Ref 15

important for producing magnesium sheet components. The limit drawing ratio ($\beta_{0,max}$) can be used to assess formability under deep drawing conditions. Figure 11 shows the limit drawing ratio of MgAl2RE1 (AE21) at temperatures between 200 and 325 °C (390 and 620 °F). It can be seen that at 300 °C (570 °F), $\beta_{0,max}$ is optimized ($\beta_{0,max} = 2.2$) for this alloy. In contrast to the results of the tensile tests, the limit drawing ratio for MgZnREZr (ZEK100) at 225 °C (440 °F) is significantly higher ($\beta_{0,max} = 3.0$) than for MgAl2RE1 (AE21) (Ref 18).

Influence of As-Rolled Conditions on Flow Stress. Figure 12 illustrates the flow curves of the magnesium sheet material AZ31 in different as-rolled conditions at room temperature and at 200 °C (390 °F). The flow curves show the dependence of the stresses on the as-rolled conditions, especially at low temperatures. The influence on the flow stress decreases significantly at elevated temperatures (above 200 °C, or 390 °F). At room temperature, the curves demonstrate that the strain to failure of the rolled and heat treated sheet material is higher than the strain to failure of the sheet material under other as-rolled conditions. These results point out that the choice of the optimal as-rolled condition influences the formability of the magnesium sheet material, especially at low temperatures (Ref 19).

Influence of As-Rolled Conditions on Deep Drawing. Figure 13 displays the limit drawing ratio of the magnesium alloy MgAl2RE1 (AE21) under different as-rolled conditions at a forming temperature of 300 °C (570 °F). The limit drawing ratio ($\beta_{0,max}$) is 2.0 for the rolled, heat treated, and temper-rolled sheet in comparison to the rolled and heat treated sheet material ($\beta_{0,max} = 2.2$). This result corresponds to the outcomes of the uniaxial tensile tests, where the heat treated and temper-rolled sheets showed higher flow stresses and lower strains in comparison to the rolled and heat treated sheet material (Ref 20).

Cold Forming

Cold forming of magnesium alloys is restricted to mild deformation with a generous bend radius. Cylinders and cones can be formed from magnesium alloys at room temperature by using standard power rolls. Simple flanges can be press formed at room temperature. Table 2 (Ref 14) gives minimum radii for fast bending at room temperature, as in a press brake. Slightly smaller bend radii than those given in Table 2 may be used when forming speeds are slower, as in a hydraulic press or when proved by trial and error. Also, the minimum radius changes with the angle. Little or no change occurs as the angle becomes larger than 90°, but the radius decreases markedly as the angle becomes smaller than 90°.

Surface Protection. In low-production cold forming, a common method of preventing surface damage is to apply tape to critical areas of the work metal and, if feasible, to tool surfaces. It is especially important to keep the die clean and free from particles of foreign metal that can become embedded in the surface of the workpiece and impair its corrosion resistance.

Hard rubber inserts in dies are sometimes helpful in preventing damage to the work metal and in forming large radii. However, this practice is not recommended for high production, because the inserts wear rapidly and cause nonuniform workpieces.

Reworking of bends by straightening and rebending the same portion should not be done in cold forming, because of the possibility of failure.

Springback can be as much as 30° for a 90° bend in cold forming of magnesium alloys. The amount of springback increases as the angle of bend decreases and as the radius of bend curvature increases. Springback decreases as temperature is raised. The suggested springback allowances for right-angle bends in AZ31B sheet at room temperature are given in Table 4. Because springback is also affected by the materials being bent, the tooling used, bending speed, and dwell time, the correct amount of overbend (springback allowance) depends on the specific application. See the article "Springback" in this Volume for more details.

Effect of Bending on Length. Unlike aluminum alloys and steel, which lengthen in bending, magnesium alloys shorten, because the

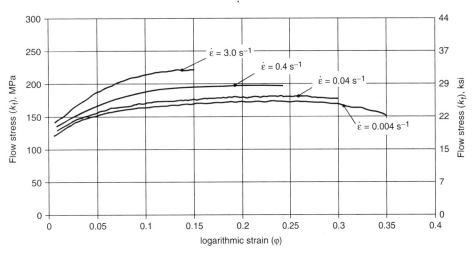

Fig. 9 Strain-rate-dependent flow curves of MgAl3Zn1 (AZ31) (initial sheet thickness, $s_0 = 1.0$ mm, or 0.04 in.) determined in the uniaxial tensile test at a temperature of 200 °C (390 °F). $\dot{\varepsilon}$, strain rate. Source: Ref 15

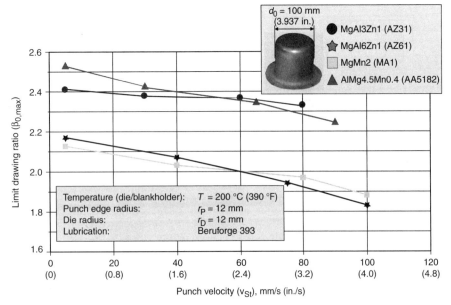

Fig. 10 Limit drawing ratios in dependence on drawing velocity. $d_0 = 100$ mm (4 in.). Source: Ref 15

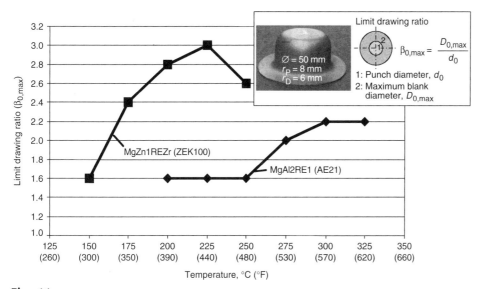

Fig. 11 Limit drawing ratios in dependence on alloy composition. $d_0 = 50$ mm (2 in.). Source: Ref 18

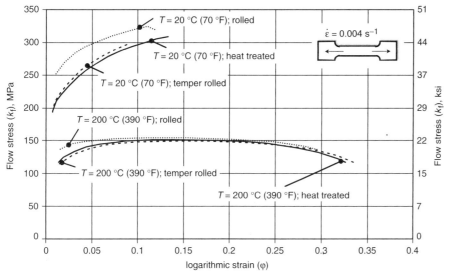

Fig. 12 As-rolled-dependent flow curves of MgAl3Zn1 (AZ31) (initial sheet thickness, $s_0 = 1.0$ mm, or 0.039 in.) determined in the uniaxial tensile test at temperatures of 20 and 200 °C (70 and 390 °F). $\dot{\varepsilon}$, strain rate

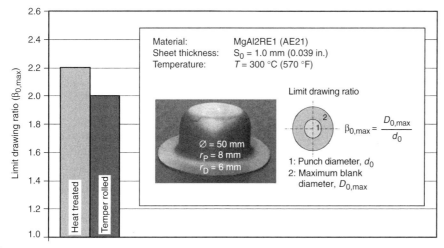

Fig. 13 Limit drawing ratios in dependence on as-rolled conditions. $d_0 = 50$ mm (2 in.). Source: Ref 20

neutral axis moves slightly toward the tension side of the bend. For thin sheet, the extent of this shortening is small, because the axis shifts only 5 to 10%. However, in thicker sheet when several bends are made, the amount of shortening can be significant and must be allowed for in the development of the blank. A nomograph (Fig. 14) aids in this calculation.

Stress Relieving. Magnesium-aluminum-zinc (AZ) alloys should be stress relieved after cold forming to prevent stress corrosion. It may be desirable to stress relieve workpieces formed from the magnesium-thorium (HM, HK) alloys, particularly if they require straightening in fixtures. Recommended temperatures and times for stress relieving the magnesium alloys that are most commonly cold formed are given in Table 5.

Hot Forming

The heating of the blank represents an additional process stage in sheet metal forming of magnesium alloys, which does not exist in the conventional sheet metal forming at room temperature. However, hot forming has several advantages over cold forming. Magnesium alloy parts usually are drawn at elevated temperature in one operation without repeated annealing and redrawing, thus reducing the time involved for making the part and also eliminating the necessity for additional die equipment for extra stages. Hardened dies are unnecessary for most types of forming. Hot-formed parts can be made to closer dimensional tolerances than can cold-formed parts because of less springback. Table 6 lists suggested maximum forming temperatures and times for various wrought magnesium alloys. The times given indicate the maximum time the alloy can be held at temperature without adversely affecting mechanical properties.

Table 4 Suggested springback allowances for right-angle bends (0.4–1.6 mm, or 0.016–0.064 in., thick sheet)

Temperature		Bend, R/t	AZ31B-O, degrees	AZ31B-H24, degrees
°C	°F			
20	70	4	8	10
		5	11	13
		10	17	21
		15	25	29
100	212	3	4	5
		5	5	7
		10	8	12
		15	13	17
150	300	2	1	2
		5	3	4
		10	5	7
		15	8	11
230	450	2	0	0
		5	1	1
		10	2	2
		15	4	4
290	550	Up to 15	0	0

Heating of the workpiece (or blank) can be done in an oven prior to placement in a die (Fig. 15, top), or the blank can be heated inside the deep drawing tool as a result of heat conduction (Fig. 15, bottom). Heating equipment includes ovens, platen heaters, ring burners, electric heating elements, heat-transfer liquids, induction heaters, and lamps and other types of infrared heating (see the section "Heating Methods" in this article). For small lots, tools and work may be heated by hand torches. Small dies that can be handled rapidly can be heated in ovens adjacent to the forming equipment.

External heating of the blank in an oven guarantees homogeneous blank temperature, but the disadvantage is a loss of temperature during transportation from the oven to the tool. For industrial processes, an automated transportation system with insulated boxes could be applied (Ref 21). When heating the blank inside the tool, the blank must be clamped between the blankholder and the die for a short time before drawing. Investigations show that only a few seconds of flat contact between the blank and the die are required for heating the blank to temperatures of approximately 200 °C (390 °F), due to the high coefficient of thermal conduction and the low heat capacity of magnesium alloys.

An advantage of heating inside the tool is the opportunity to achieve an optimized temperature distribution over the blank area by choosing different temperatures in individual tool areas. Experimental results showed that the limit drawing ratio of magnesium sheet can be increased significantly if the punch has a lower temperature than the die and the blankholder. In order to maintain constant temperature conditions, additional cooling of the punch could be necessary (Ref 22). Similar results have been achieved for the deep drawing of aluminum sheets at elevated temperatures (Ref 23, 24).

Another possibility for carrying out the deep drawing process at higher temperatures is the forming of an externally heated blank in an unheated tool. However, exact forming temperatures cannot be guaranteed in this case, due to the rapid heat transmission from the blank into the tool. For deep drawing of magnesium sheets, this process variant cannot be recommended because of the strong dependence of forming performance on temperature (Ref 25). In most hot forming methods, both the tools and the magnesium alloy work metal must be heated.

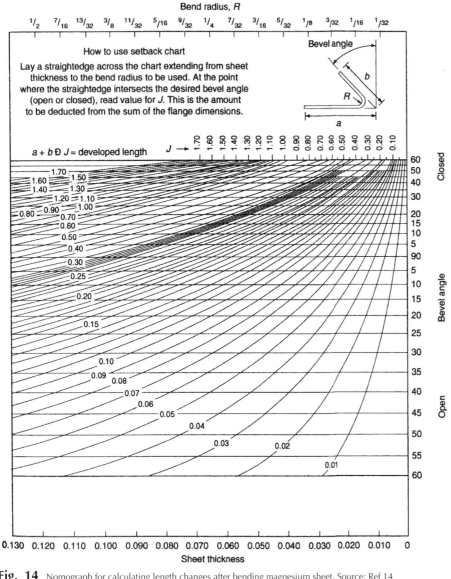

Fig. 14 Nomograph for calculating length changes after bending magnesium sheet. Source: Ref 14

Table 5 Stress-relief treatments for the magnesium alloys most commonly cold formed

Alloy and temper	Temperature °C	°F	Time at temperature, min
Sheet			
AZ31B-O	260	500	15
AZ31B-H24	150	300	60
Extruded flat strip			
AZ31B-F	260	500	15
AZ61A-F, AZ80A-F	260	500	15
AZ80A-T5	205	400	60

Table 6 Maximum forming temperatures and times for wrought magnesium alloys

Alloy and temper	Temperature, °C (°F), at exposure times of: Minimum(a)	1 min	3 min	10 min	30 min	60 min
Sheet						
AZ31B-O	120 (250)	...	...	...	...	290 (555)
AZ31B-H24	120 (250)	225 (435)	200 (395)	180 (360)	175 (345)	150 (300)
AZ31B-H26	120 (250)	225 (435)	200 (395)	180 (360)	175 (345)	150 (300)
Extrusions						
AZ31B-F	120 (250)	...	...	...	...	290 (555)
AZ61A-F	200 (395)	...	...	...	...	290 (555)
AZ80A-F	140 (285)	...	...	...	290 (555)	...
AZ80A-T5	140 (285)	...	...	...	...	195 (385)
ZK60A-F	150 (300)	...	...	...	...	290 (555)
ZK60A-T5	150 (300)	...	...	...	200 (395)	...

(a) If material is deformed below the minimum temperature shown, stress relief may be required.

Figure 16 shows the principal design of a deep drawing tool for use at elevated temperatures (Ref 26). Today (2006), deep drawing at elevated temperatures is only used in a few special applications. Therefore, the knowledge of tool and process design is limited.

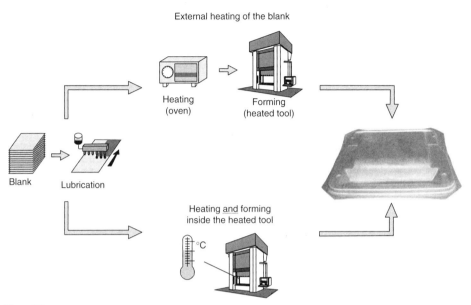

External heating of the blank

Blank

Lubrication

Heating (oven)

Forming (heated tool)

Heating and forming inside the heated tool

°C

Fig. 15 Process chain for deep drawing at elevated temperatures. Source: Ref 15

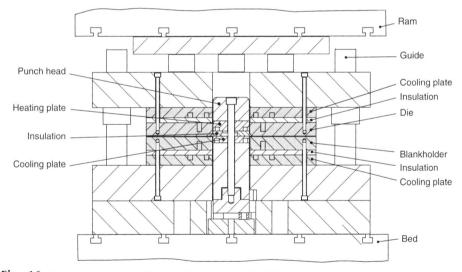

Ram

Guide

Punch head

Heating plate

Insulation

Cooling plate

Cooling plate

Insulation

Die

Blankholder

Insulation

Cooling plate

Bed

Fig. 16 Design of a heatable tool for deep drawing. Source: Ref 26

Evaluation of Hot Forming Conditions

Evaluation of Appropriate Heating Methods for Deep Drawing Tools. Appropriate heating methods for deep drawing tools have been determined. Heating of tools can be achieved by means of resistance heating, induction heating, and heating with fluid. To determine the best solution, the three methods were evaluated according to the criteria of transferable heat quantity, control accuracy, thermal efficiency, and initial and operational costs. Table 7 shows that resistance heating provides the best solution to heat deep drawing tools, from a thermal and economic point of view (Ref 27).

To avoid heating of the tool frame and thus failure during operation, it is required that heated tool elements and the tool frame be thermally separated. This can be done by using thermal insulations, fluid-cooled plates between tool and tool frame, or by cooling the surfaces with air. When using thermal insulations and cooling plates together, a sufficient thermal separation is possible. Therefore, low flow rates of cooling water are sufficient (Fig. 17). The combined use of insulation and air is not recommended, because even if the flow velocity is 50 m/s (2000 in./s) (compressed air), the cooling of the forming machine is insufficient (Ref 27).

The limit drawing ratio reaches a maximum in the temperature range from 200 to 250 °C (390 to 480 °F), which activates further slip planes. Above this temperature level, the work-softening behavior of the material prevails, and the limit drawing ratio decreases (Ref 15, 17).

Deep drawing of components with complex and/or irregular geometries is accompanied by nonuniform conditions of stress and strain in the flange area. While the straight and slightly bent sides of the drawn part mainly undergo radial stretching (less true strain), the corners are dominated by an overlapping of radial stretching and tangential upsetting (high true strain) (Ref 28, 29). From the resulting differences in equivalent strain, it is advantageous to subject the flange area to locally differentiated heating. The corner areas with higher true strains are more heated than the side areas with less true strains. The advantage of this partial heating is the increase of the forming limits by a better load transmission in the material during the deep drawing process, because of the lower work softening in the less heated areas. Furthermore,

Table 7 Results of investigations concerning appropriate heating principles

| Criterion | Heating by transfer fluid | Resistance heating | | Inductive heating |
		Heating cartridge	Tubular radiator	
Transferable amount of heat	Max. allowed oil temperature: 350 °C (660 °F)	Max. allowed oil temperature: 750 °C (1380 °F)		Max. allowed inductor temperature: 70 °C (160 °F) Tool temperature: 250–260 °C (480–500 °F)
Control accuracy	Max. difference of temperature: ±4 K	Max. difference of temperature: ±0.5 K		Max. difference of temperature ±0.5 K
Costs of the heating system	500% (oil preheater, pump, flow control)	100% (temperature-contol unit)		1000% (HF-generator, inductor cooler)
Costs of the tool		Similar for all heating principles (construction with milled channels necessary)		
Efficiency	30–58%	94–96%		78–88%

Source: Ref 27

in order to reduce the energy consumption, it is advantageous to heat the side areas of the part less. Figure 18 displays the construction of a partially heated tool. Investigations have proven that a distinct extension of the forming limits is possible by increasing the temperature in the more severely formed areas of the formed part than in the less severely formed zones (Fig. 19).

Influence of Temperature on Flow Stress. Figure 20 displays the flow curves of the magnesium alloy MgAl3Zn1 (AZ31) at the temperature range of 250 to 400 °C (480 to 750 °F) and at two strain rates, 1/s and 10/s.

The flow curves decrease at all tested temperatures. The gradient of the curve quantifies the work softening. The decrease of the flow curves is stronger at all temperatures under recrystallization temperature (Ref 30). In these tests, partial microcracks or microporosities appear, which lead to strong work softening or to the crack of the specimen. At temperatures above 300 °C (570 °F), higher strain rates can be obtained without shear fracture of the specimen under 45° to the deformation direction (Fig. 21). Above the recrystallization temperature, the decrease of the flow curves can be explained by an overlapping of work-hardening mechanisms and a recovery of the material. The recovery of the material has a strong influence on the decrease of the flow curve (Ref 31).

Influence of Strain Rate on Flow Stress. Figure 22 shows flow curves for MgAl3Zn1 (AZ31) at the temperature of 300 °C (570 °F) and at different strain rates. The yield stresses were determined by means of the uniaxial upsetting test.

The stresses increase at a higher deformation rate and decrease at a higher forming tempera-ture. Due to the temperature increase of the specimens during forming, the yield stress decreases at higher degrees of deformation. These temperature differences may amount to 70 °C (160 °F). The higher the inner resistance to a deformation within a specimen, the higher the temperature increase (Ref 31).

Hot Forming Practices

Sheet and Plate. Rolled magnesium alloy products include flat sheet and plate, coiled sheet, circles, tooling plate, and tread plate. These products are supplied in a variety of sizes.

The ability to use increased section thickness without weight penalty is of particular importance in designs that employ magnesium sheet. Thick-sheet construction provides the rigidity necessary in a structure, without the need for costly assembly of ribs and similar reinforcing members.

Rolled magnesium alloy products can be worked by most conventional methods. For severe forming, sheet in the annealed (O temper) condition is preferred. However, sheet in the partially annealed (H24 temper) condition can be formed to a considerable extent. Because heat has significant effects on properties of hard-rolled magnesium, properties of the metal after exposure to elevated temperature must be considered in forming. The design curves shown in Fig. 23 give minimum values suitable for design use. Although the curves are based primarily on tests of sheet 1.63 mm (0.064 in.) thick or less, check tests indicate reasonable applicability for thicknesses up to 6.35 mm (0.250 in.).

Figure 23 shows how the properties of AZ31B-H24 change with exposure time at various temperatures. The curves have been extrapolated above the typical property levels of AZ31B-H24 sheet. Thus, if the value selected from a curve exceeds the actual property level of the material before exposure, the actual figure

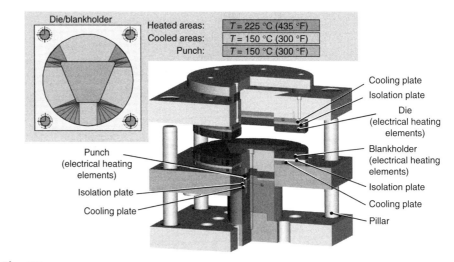

Fig. 18 Construction of a partially heated tool. Source: Ref 27

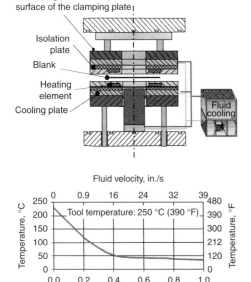

Fig. 17 Temperature of the clamping plate if cooling plates and thermal insulations are used together. Source: Ref 27

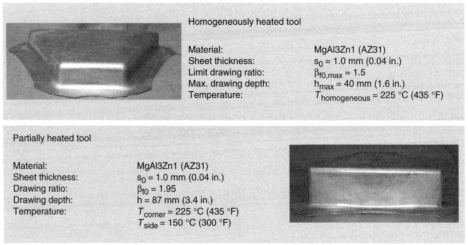

Fig. 19 Effect of temperature distribution on forming limits. Source: Ref 27

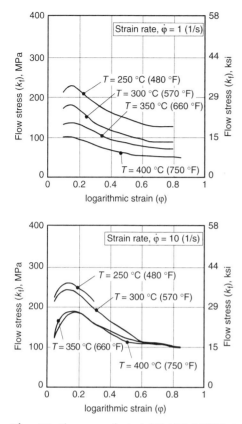

Fig. 20 Flow curves of extruded MgAl3Zn1 (AZ31) at different temperatures and strain rates determined in the uniaxial upsetting test

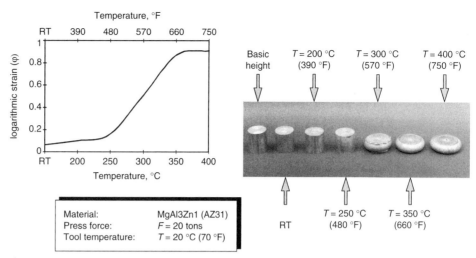

Fig. 21 Formability of MgAl3Zn1 (AZ31) with dependence on temperature. RT, room temperature

Material: MgAl3Zn1 (AZ31)
Press force: F = 20 tons
Tool temperature: T = 20 °C (70 °F)

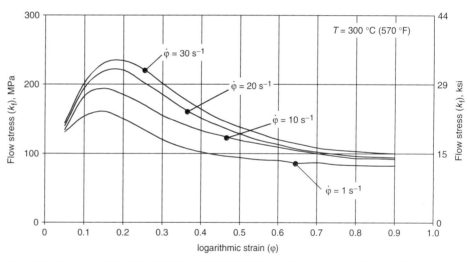

Fig. 22 Flow curves of MgAl3Zn1 (AZ31) at 300 °C (570 °F) and different strain rates determined in the uniaxial upsetting test. Source: Ref 32

must be used. Also, it should be kept in mind that the effects of multiple exposures at elevated temperature are cumulative.

AZ31B-H24 sheet is commonly hot formed at temperatures below 160 °C (320 °F) to avoid annealing it to room-temperature property levels lower than the specified minimums. Annealing is a function of both time of exposure and temperature; thus, temperatures higher than 160 °C (320 °F) can be tolerated if exposure is carefully controlled.

Thermal Expansion. Magnesium alloys have a very high rate of thermal expansion. At 260 °C (500 °F), for example, the thermal expansion of magnesium is more than twice that of steel. Therefore, when parts made of magnesium alloys are hot formed in tool steel or cast iron dies, the difference in the thermal expansion of the tool material and the work metal must be considered.

Figure 24 shows the relation between the size of a magnesium part and the size of a steel die at 20 °C (70 °F) and at 205 °C (400 °F). The dimensional factor need not be applied when the dies are made of zinc or aluminum alloys, because the coefficients of expansion of these alloys are similar to those of magnesium alloys.

Precautions. Magnesium alloy stock to be hot formed must be clean. Protective coatings,

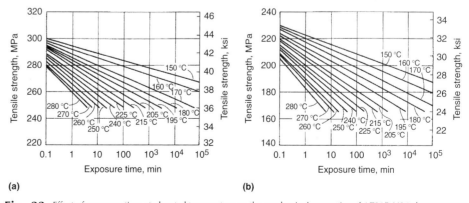

(a)   **(b)**

Fig. 23 Effect of exposure time at elevated temperature on the mechanical properties of AZ31B-H24 sheet at room temperature. (a) Tensile strength. (b) Tensile yield strength. Data are based on tests of sheet 1.63 mm (0.064 in.) thick, and have reasonable applicability for thicknesses up to 6.35 mm (0.250 in.).

oil, dirt, moisture, or other foreign matter must be removed.

Dies, punches, and form blocks should be clean and free of scratches; tooling should be cleaned with solvent. Rust, scratches, and minor imperfections can be removed by light polishing with fine-grit abrasive cloth. However, polishing must not alter the dimensions of the tool.

In forming, the possibility of fire due to ignition of magnesium is remote. However, an ample supply of suitable fire-extinguishing material, such as dry sand or commercially available powder, must be kept in the work area.

Heating Methods. Electric heating elements often are used for heating dies and other forming tools. Electric-resistance heating can be used for heating some dies. Low-voltage high-amperage current is passed into the die through conducting grips or clamps.

Radiant heating with electricity and gas is useful for heating dies and workpieces in some applications. Radiant heating is particularly useful for rapid heating of the workpiece and for use in rapid-action presses. Also, with this heating method, cloth covers can be used on the workpiece to minimize heat loss.

Infrared heating is also commonly used. A bank of infrared lamps is the most common method, but gas-fired units are also used. The principal advantage of infrared heating is that only the die and workpiece are heated and not the surrounding area. Also, the cost of heating is less, and working conditions are cooler and less hazardous.

Gas heating often is advantageous, because the installation of equipment is simple and fuel cost is generally low. Burners up to 1.9 m (75 in.) long can be formed and welded from 19 mm (3/4 in.) black iron pipe; 25 mm (1 in.) pipe is suggested for burners more than 1.9 m (75 in.) long. Burners are attached to the dies so that the flames touch the die surface. Hollow punches can be heated by a burner inside the punch.

Four gas-mixing systems are used for heating tools and dies:

- Simple venturis, in which gas flows through a mixer that draws in air
- Proportional mixers, which use compressed air flowing through a venturi (or an air injector) that pulls the gas into the burner at atmospheric pressure
- A gas-air carburetor system, which uses a low-pressure turbocompressor to compress the gas-air mixture. The carburetor holds a constant fuel-to-air ratio regardless of the volume of flow.
- A venturi mixing system combined with turbocompressors, which gives accurate temperature control with minimum overshoot

Heat-transfer fluids are used for heating platens, form blocks, drop hammer dies, and other forming tools large enough to have passages in the die. Heating by this method is rapid and permits good temperature control. Heat-transfer fluids with a working-temperature range of 150 to 400 °C (300 to 750 °F) are available. Hot oils, natural and synthetic, that can withstand temperatures up to 345 °C (650 °F) are commonly circulated in passages in the dies. Steam is readily available and is circulated through ducts in the tools or dies, but its maximum temperature is usually approximately 175 °C (350 °F). Commercially available equipment for use with heat-transfer fluids includes vapor generators, circulating mechanisms, and means for temperature control.

Temperature control is important. For forming a few pieces, contact pyrometers or temperature-sensitive crayons are satisfactory for determining temperature. Blue carpenter's chalk can sometimes be used. A streak of this chalk on a metal surface will turn white at approximately 315 °C (600 °F).

Automatic temperature controls are essential for most magnesium-forming operations. Radiant and infrared heat are more difficult to control than other kinds of heat. One type of infrared lamp has a control that extends or retracts the lamp when the tool or workpiece has reached the desired temperature. Another temperature control for infrared heating consists of a special radiometer that senses only the heat radiated by the surfaces being heated.

To maintain the desired temperature, controls used in gas-heating systems usually operate by adjusting a solenoid valve in the line to lower or raise the flame. Electric heating by elements, resistance heating, and heating by means of heat-transfer fluids usually provide good temperature control.

Lubricants (Ref 33)

Generally, lubrication is more important in hot than in cold forming of magnesium alloys, because the likelihood of galling increases with the increase of temperature.

Lubricants used in forming magnesium alloys include mineral oil, grease, tallow, soap, wax, molybdenum disulfide, colloidal graphite in a volatile vehicle, colloidal graphite in tallow, and thin sheets of paper or fiberglass.

Selection of a lubricant depends primarily on forming temperature. For temperatures up to 120 °C (250 °F), oil, grease, tallow, soap, and wax are generally used.

In spinning, it is essential that the lubricant cling to the work metal; otherwise the lubricant will be thrown off by centrifugal force. This is not a problem when drawing in a die or bending in a press brake.

Frequently, a lubricant that is used for other operations in the plant can be used, up to a forming temperature of 120 °C (250 °F). It is common practice to use the lubricant that can be most easily removed after forming and to apply it by roller coating or swabbing. Sometimes, the lubricant is applied to both the work metal and the tools.

When forming is done at temperatures above 120 °C (250 °F), the selection of a lubricant is more limited; ordinary oil, grease, and wax are eliminated. Although colloidal graphite can be applied at any temperature that is used for forming magnesium alloys, its use is usually avoided, because graphite is difficult to remove and interferes with subsequent surface treatments.

A soap lubricant is acceptable for temperatures as high as 230 °C (450 °F). This compound is an aqueous solution and is applied to the work metal by dipping, brushing, or roller coating. After coating, the work metal blanks are dried in still or forced air. After drying, the blanks can be stored for an indefinite period for future processing because the dried lubricant is stable. Lubricant that remains after forming can be completely removed by cleaning in hot water.

When forming temperatures are higher than 230 °C (450 °F), the choice of lubricant is restricted to colloidal graphite or molybdenum disulfide. Graphite in a vehicle such as spirits (2% graphite) is widely used; for spinning, the graphite is mixed with tallow to improve adherence.

Procedures. Lubricants should be cleaned from parts as soon as possible after forming, to prevent corrosion and to avoid difficulty in their removal. Colloidal graphite is particularly difficult to remove if allowed to remain on parts for any length of time.

Because some work lubricants cannot be tolerated at any forming temperature, thin sheets of paper or fiberglass (depending on temperature) are placed between the work metal and the tools. More information on lubricants for sheet forming is available in the article "Selection

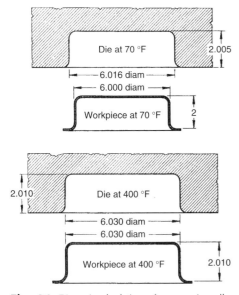

Fig. 24 Dimensional relations of a magnesium alloy workpiece and a steel die at room temperature and at a forming temperature of 205 °C (400 °F). Room-temperature dimensions of the steel die are determined by multiplying the design dimensions of the magnesium part by 1.00270. Dimensions given in inches

and Use of Lubricants in Forming of Sheet Metal" in this Volume.

Press-Brake Forming

The press-brake forming of magnesium alloys is the same as it is for other metals, except that the work metal and the dies are usually heated. Top and bottom dies can be made of steel, or if the workpiece permits cold forming, the steel punch can be bottomed in a rubber die held in a retaining box. Metal punches and dies should be highly polished to prevent marking of workpiece surfaces.

A preferred method of heating the punch and die for hot forming is shown in Fig. 25. When only a few workpieces are to be formed, heating with a gas torch is satisfactory; however, care must be taken to ensure that the area of the workpiece to be formed is uniformly heated. If the press-brake die is not heated, the workpiece should be heated to the maximum allowable temperature and formed quickly, before the tools can cool the work metal too much.

Deep Drawing (Ref 14)

Hydraulic presses are usually preferred for deep drawing magnesium alloys because they can operate at slower and more uniform speeds. Mechanical presses are used, however, for making draws that are not too severe. The force required for deep drawing magnesium to a cylinder can be estimated by the following empirical formula:

$$L = \pi dtS \ (D/d - k)$$

where L is the press load in Newtons, d is the cup diameter in mm, D is the blank diameter in mm, t is the blank thickness in mm, S is the tensile strength at temperature in MPa, and k is a constant, which is between 0.8 and 1.1 for magnesium conversion factor: Newtons $\times$ 0.225 = lbf.

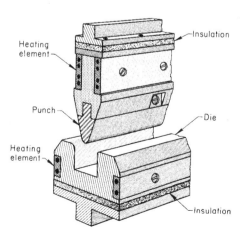

Fig. 25 Preferred method of heating a punch and die for hot forming magnesium alloys on a press brake

The values for the tensile strength of a particular magnesium alloy at elevated temperatures can be used to estimate the force that will be needed to make any given draw. The holddown force needed to prevent wrinkling depends on blank thickness and temperature but is between 1 and 10% of the drawing force.

Blankholder pressures for magnesium alloys vary from the lowest obtainable to as much as 5 MPa (700 psi). In average draws, the blankholder pressure is usually 345 to 1380 kPa (50 to 200 psi). To secure proper wall thinning, blankholder pressures are obtained by trial and error.

Die Heating. Electric heating elements are usually used for heating the tools, although when intricate parts are drawn, gas-ring burners provide more flexibility, particularly if the work metal is likely to pucker. The base of the punch sometimes can be heated.

Another type of differential heating is used in the production of workpieces that have a slight crown. When the crown must be held to close tolerance, burners with separate controls are provided inside and outside the die. For more crown, the outside of the mating die is heated to a higher temperature than the inside. When the workpiece reaches die temperature and the outside of the sheet is hotter than the inside, the work metal retracts upon removal of the die, thus causing the crown to form. Alternatively, to eliminate crowning and obtain flatness, the inside of the die is heated to a higher temperature than the outside.

Drawability. Annealed magnesium alloy sheet can be cold drawn to a maximum reduction (blank diameter to cup diameter) of 15 to 25%. With heat, drawability is greatly improved, and up to 70% reduction is possible. The drawability, or percentage of reduction in blank diameter, is calculated by the formula:

$$\text{Percentage reduction} = \frac{D - d}{D} \times 100$$

where D is the blank diameter before drawing, and d is the diameter of the punch.

While multiple draws can be made at room temperature to arrive at a final deep draw, in practice all deep drawing of magnesium is done at elevated temperatures. The effect of temperature on lubrication practice and the modifications of tool design required to account for thermal expansion must therefore be considered. At high temperatures, magnesium alloys can

be deep drawn in one operation to a cylinder by a reduction of as much as 75%.

In general, as the temperature is raised and the speed of the draw is reduced, the greater will be the amount draw possible (Table 8). The speed of drawing can vary from approximately 4 to as much as 2500 cm/min (1.5 in./min to 82 ft/min); the speed selected will depend on the severity of the draw and any compromise required between using single and multiple draws, the latter requiring additional tooling.

Drawability is also influenced by the shape of the workpiece. A maximum drawability of 70% is for drawing a cylindrical cup. Square and rectangular boxes, for example, seldom are drawn as severely.

In addition, the die radius of curvature, which is a measure of the severity of the bend required as the blank is pushed through the die, is another consideration of severity of a draw, along with the radius of curvature between two side walls for boxlike shells, and the radius of curvature of the punch, which forms the radius between the side wall and the bottom. All of these parameters must be kept within workable limits if the draw, or series of draws, is to be successful.

As a round blank is drawn into a cup, the metal that will be forced through the die must first be compressed from its radius to the radius of the die opening. Because the volume occupied by the metal does not change, the blank becomes thicker as it reaches the die and remains thicker as it bends to go through the die. After being bent over the radius and stretched into the wall of the cup, thinning occurs. Therefore, die design must provide sufficient clearance to allow for the thickening that occurs as the blank is drawn to the die. A guide for the required clearances is:

Blank thickness		Clearance
mm	in.	
0.40	<0.016	1.07–1.09t
0.40–1.3	0.016–0.050	1.08–1.10t
1.3–3.2	0.050–0.125	1.10–1.12t
>3.2	>0.125	1.12–1.14t

t, blank thickness

Because, during the draw, the metal that has reached the wall is thinner than the metal that has just passed the die radius, the metal in the wall must support a higher stress than the metal at the radius in order to continue stretching the entire blank into the wall. This is possible when the draw is done at room or low temperatures

Table 8 Drawability of AZ31B-O at various temperatures and speeds

Drawing temperature		% reduction		
°C	°F	At 600 cm/min (235 in./min)	At 1500 cm/min (590 in./min)	At 2400 cm/min (945 in./min)
20	70	14.3	14.3	...
120	250	30.3	30.3	...
200	390	58.6	57.1	53.8
260	500	62.5	61.3	53.8

76 mm (3 in.) diam cylindrical cup; sheet thickness, 1.65 mm (0.064 in.); draw-ring radius, 6t; punch radius, 10t

because the metal that has been stretched has been strengthened by work hardening. There is added flexibility, however, when the draw is at elevated temperatures; the punch can be kept at a lower temperature than the die, resulting in higher strength in the cooler wall metal than in the warmer metal at the die.

Effect of Speed on Drawability. Drawability at any temperature varies with the speed of drawing, which ranges from 0.6 to 405 mm/s (1½ to 960 in./min). Large reductions (70%, for example) require slower speeds than do moderate reductions (up to 55%).

Reductions up to approximately 55% can often be made on high-speed hydraulic or mechanical presses. Also, milder draws permit lower forming temperatures, and costs can be reduced because strip feeding, blanking, lubrication, trimming, and cleaning can be simplified.

For most parts, however, depth of draw is not a primary consideration, and usually no trouble is experienced in drawing to the depth required. More trouble is encountered in keeping the metal free from puckers in parts with rounded corners or contours. Temperatures above those required for maximum drawability are often necessary to eliminate these puckers. On unusual or difficult jobs, it may be necessary to vary the procedure to obtain minimum scrap.

Redrawing. The possibilities inherent in two-step draws are illustrated by the following parts: In the first operation, 610 mm (24 in.) blanks of 0.64 mm (0.025 in.) annealed sheet were drawn to a cup 200 mm (8 in.) in diameter by 400 mm (16 in.) in depth; they were redrawn to a cup 140 mm (5½ in.) in diameter by 585 mm (23 in.) in depth. Starting with a rectangular blank of 1.3 mm (0.051 in.) AZ31B-O, 455 by 485 mm (18 by 19 in.), a rectangular box 111 by 273 by 165 mm (4⅜ by 10¾ by 6½ in.) in depth was drawn in the first operation. This box was then redrawn into a rectangular box 89 by 254 by 171 mm (3½ by 10 by 6¾ in.) in depth having 5.6 mm (7/32 in.) corner radii.

Choice of die materials is chiefly influenced by the severity of the operation and the number of parts to be produced. For most applications, unhardened low-carbon steel boiler plate or cast iron is satisfactory. For runs of 10,000 parts or more, for maximum surface smoothness, or for close tolerances where no significant die wear can be allowed, hardened tool steels are recommended. Tool steels W1 or O1 are satisfactory for extremely long runs (1 million parts). For the most severe draws, however, the more abrasion-resistant tool steels, such as A2 or D2, will probably be more satisfactory and economical. For room-temperature drawing, it is usually desirable for die steels to be heat treated to obtain near-maximum hardness in service. However, for elevated-temperature drawing, the maximum temperature to which the dies will be exposed in drawing must also be considered. In this situation, the dies must be tempered slightly above the maximum service temperature, even though some hardness may be sacrificed.

Manual Spinning

Various conical and hemispherical shapes can be produced from magnesium alloys by manual spinning. Because tooling is inexpensive, manual spinning of small quantities is often more economical than press forming. When press tooling would be complex, manual spinning may be used for medium to large production quantities.

Equipment and tooling for manual spinning of magnesium alloys are essentially the same as those used for other metals (see the article "Spinning" in this Volume), except that when magnesium alloys are to be heated, the mandrels (spin blocks) should be made of metal, with provision for controlled heating of the work metal.

For spinning a few pieces, it is common practice to heat the blanks with a hand torch, using temperature-sensitive crayons to indicate the temperature. For production spinning, however, the use of a thermostatically controlled burner on the lathe is preferred.

Procedure. Annealed sheet is usually used in spinning. Manual spinning depends to a large extent on operator skill, especially when spinning magnesium alloys, which are more temperature sensitive than most metals.

Many shapes can be spun from unheated blanks by a skilled operator, especially when thin sheet is used, because friction between the spinning tool and workpiece generates a substantial amount of heat. As severity of sheet thickness increases, the work metal must be heated. Temperatures of 260 to 315 °C (500 to 600 °F) are common.

Whether spinning is done hot or cold, a lubricant should be used (see the section "Lubricants" in this article). Spindle speed should be such that the speed of the edge of the blank is approximately 610 m/min, or 2000 surface feet per minute (sfm), when spinning begins.

Tolerances. Typical tolerances that can be maintained in manual spinning of magnesium alloys are as follows:

Workpiece diameter		Tolerance	
mm	in.	mm	in.
<455	<18	±0.8	±1/32
455–915	18–36	±1.6	±1/16
>915	>36	±3.2	±1/8

Spinning Extrusions. Manual spinning can be used either to close or flare the ends of extruded round tubing.

Closing the ends of tubes is done by slowly forcing a rotating hemispherical cup over the end of the tube until the end is closed and has assumed the hemispherical shape of the cup (spinning tool). This can be done on almost any machine that can hold the workpiece and rotate the cuplike tool. A drill press is frequently used. The use of grease or soap as a lubricant is help-ful in producing a better workpiece finish and prolonging the life of spinning tools. In most applications, tube ends can be closed without the use of heat.

Tube flaring is done either by inserting a stationary mandrel into the tube and pushing it out against a stationary die on the outside, or by spinning or rolling the flare against a stationary outside die with a conical, rotating, inside mandrel. The shape required for the flare usually determines the preferred procedure.

Tube-flaring machines have a stationary outside die and conical, rotating, inside spindle with adjustable eccentricity, the axis of which rotates off center at approximately 1600 rpm. The eccentric spindle forces the tube against the outer die to form the flare. In the flaring of magnesium alloy tubes, the outer die should be heated, preferably by electric heating elements in the die holder, to a preferred temperature of approximately 260 °C (500 °F). The tube to be flared is preheated to the same temperature. Lubrication may be required during tube flaring. The lubricants recommended for spinning can be used.

Power Spinning

Power spinning (shear spinning) can be used for magnesium alloys. Both cone spinning (spinning in accordance with the sine law) and tube spinning (spinning in which metal displacement is strictly volumetric) are used for magnesium alloys.

Equipment and Tooling. Special machines are used in the cone and tube spinning of magnesium alloys. However, the equipment used in power spinning is the same as that used for other metals (see the article "Spinning" in this Volume), except when hot spinning is done; then, torches or other heating equipment must be added to the machine. Consequently, the mandrels and rollers must be made from an alloy tool steel that will not be softened by heat. Tool steels such as H12 or H13 hardened to 54 to 58 HRC are used in many applications.

Procedure. Magnesium alloys are sometimes power spun without heat, but more often the major portion of the reduction is performed hot and finished cold, or is rough worked hot and then finished at a somewhat lower temperature (warm). For the most successful results, a definite procedure of alternate spinning and heating should be followed, whether the metal is finished cold or warm. Table 9 gives recommended procedures for the two methods often used for alloys HK31A and HM21A. The use of the procedures outlined in Table 9 has resulted in total wall thickness reductions as high as 80%.

Rubber-Pad Forming

Hydraulic presses are generally used for the rubber-pad forming of magnesium alloys.

Tooling is simple because only a form block is used (Fig. 26). A conventional die is not needed.

For forming at room temperature, particularly for limited use, form blocks can be made of wood or masonite; or for higher production runs, they can be made of aluminum, zinc, or magnesium, which are more durable than wood or masonite. However, large radii must be used in cold forming.

When rubber-pad forming at elevated temperature, form blocks must be made from metal that will not creep excessively at the working temperature and pressure; magnesium, aluminum, or zinc can be used up to approximately 230 °C (450 °F). However, forming at temperatures higher than 230 °C (450 °F) requires steel form blocks.

Specially compounded grades of solid rubber or laminated sheets are used for the rubber pad when forming at temperatures up to 315 °C (600 °F). Hardness of the rubber is important; Durometer A 40 to 70 is the common range.

Heating. As shown in Fig. 26, the heating elements heat the steel platen, and the heat is transferred to the form block, which is not fastened to the platen, by conduction. Alternatively, the form block can be heated separately in an oven and then placed on the platen. With this method, a fireproof blanket often is placed between the heated form block and the cold platen for insulation. Usually, blanks are heated in ovens situated near the press to minimize loss of heat.

Forming Pressure. Pressure for rubber-pad forming is a function of sheet thickness and forming temperatures; 6200 kPa (900 psi) is adequate for most work.

Rubber-pad forming is generally done by shaping the blank around a form block with pressure from the rubber pad. However, when pressure must be concentrated at one point, or metal flow must start before general pressure is applied, deflector bars are used (see the article "Rubber-Pad Forming and Hydroforming" in this Volume).

Some severe forming is done in two operations: The workpiece is partly formed, is removed from the press for hand smoothing of wrinkles, and then is returned to the press for final forming under full pressure. In one-operation forming, thin throw sheets of heat-resistant rubber can be placed over the blank, or attached to the pad, to protect the rubber.

Shrink flanges that are wrinkle-free can be made with a higher percentage of compression from magnesium alloys than from other metals, such as aluminum, of the same gage. Minor wrinkles can be hand corrected after flanging. If the part is likely to wrinkle severely, scalloped cutouts or recesses in the form blocks may correct this condition. A draw plate to iron out wrinkles during forming is often helpful.

Stretch flanges of up to 40% stretch in hard-rolled magnesium alloy sheet (H24 temper) and 70% in annealed sheet (O temper) can be made by rubber-pad forming. The ranges of stretch flange limits for various thicknesses of AZ31B-O and AZ31B-H24 alloy sheet rubber pad formed at 150 °C (300 °F) are shown in Fig. 27.

A minimum flange radius of $5t$ is suggested for alloy AZ31B-H24 at 160 °C (320 °F). The radius of the die should be approximately $1/2 t$ less to compensate for springback.

Beads. In rubber-pad forming, both internal and external beads can be formed in magnesium alloy sheet. Usually, external beads are easier to produce, although wrinkling is slight in both types.

Severity of forming internal and external beads is expressed as the ratio w/h, where w is the width of the bead, and h is the height of the bead. Beading is essentially a stretching operation, and this w/h ratio is related to the maximum percentage of stretch obtained in a given bead.

External beads can be made to equal or more severe ratios; the minimum bead margin should be 9.6 mm (0.38 in.), and beads should be separated by a minimum centerpoint distance of 19.3 mm (0.76 in.).

Hand forming can slightly improve the definition of a part after rubber-pad forming, while the form block and workpiece are still hot. A leather or plastic forming tool can be used to correct minor irregularities or improve flange angles. To avoid damage to the form block, hard tools should not be used.

Stretch Forming

The stretch forming of magnesium alloys is the same as it is for other metals, except that magnesium alloys are generally stretch formed at elevated temperatures. The fundamentals of stretch forming, compression forming, and radial draw forming are described in the article "Stretch Forming" in this Volume.

Dies, or form blocks, made from magnesium, aluminum, or zinc alloys are suitable for forming at temperatures up to 230 °C (450 °F). Concrete form blocks containing wire mesh heated by electrical resistance may also be used

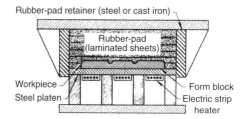

Fig. 26 Tooling and heating setup for rubber-pad forming of magnesium alloys at elevated temperature

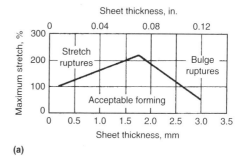

(a)

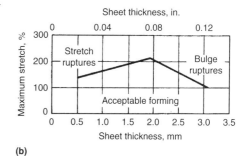

(b)

Fig. 27 Stretch flange limits for various thicknesses of magnesium alloys AZ31B-O and AZ31B-H24 rubber-pad formed at 150 °C (300 °F). (a) Alloy AZ31B-O. (b) Alloy AZ31B-H24

Table 9 Procedures for power spinning two magnesium alloys to obtain acceptable properties

Alloy	Procedure
Cold finishing	
HK31A	Hot work roughly to shape at 425 ± 30 °C (800 ± 54 °F). Heat treat for 30–60 min at 455–480 °C (850–900 °F)(a). Cold work to a total reduction in thickness of at least 25% using low reductions per pass. Heat 1 h at 315–330 °C (600–625 °F)
HM21A	Hot work roughly to shape at 455 ± 10 °C (850 ± 18 °F). Heat treat for 30–60 min at 480–510 °C (900–950 °F)(b). Cold work to a total reduction in thickness of 15–25% using low reductions per pass. Heat 1 h at 370 ± 16 °C (700 ± 29 °F)
Warm finishing	
HK31A	Hot work roughly to shape at 425 ± 30 °C (800 ± 54 °F) if necessary. Heat treat for 30–60 min at 455–480 °C (850–900 °F)(a). Warm work at 315–370 °C (600–700 °F) to a total reduction in thickness >50% with a minimum number of passes. Heat 16 h at 205 °C (400 °F)
HM21A	Hot work roughly to shape at 455 ± 30 °C (850 ± 54 °F) if necessary. Heat treat for 30–60 min at 480–510 °C (900–950 °F)(b). Warm work at 315–370 °C (600–700 °F) to a total reduction in thickness >50% with a minimum number of passes. Heat 16 h at 230 °C (450 °F)

Note: Properties obtained will approach those of the H24 temper for HK31A and the T8 temper for HM21A. (a) Fairly rapid cooling is desirable but less critical than for HM21A. (b) Should be cooled from the heat treating temperature to 315 °C (600 °F) or below within 5 min at temperatures up to 230 °C (450 °F). For temperatures higher than 230 °C (450 °F), cast iron form blocks are used.

To prevent tearing of the work metals, grippers used in the forming of magnesium alloys should not have serrated jaws. Coarse emery paper or cloth can be placed between the work metal and the jaws to preclude tearing.

Tools and work metal can be heated by electric heating elements or by radiant heat. Proper distribution of heat is important, and units should be placed at critical forming areas.

For differential stretching of sheet over forms of low curvature, the practical maximum stretch is approximately 15%. The maximum is 12% if allowance (overstretch) is made for springback; however, normally little springback is encountered at elevated temperatures; therefore, an addition of 1% to the total stretch usually compensates for any springback that may occur.

Although freedom from wrinkles is an advantage of stretch forming in most applications, wrinkles can be a problem when making asymmetrical low-curvature parts. Wrinkles can be controlled by including proper restraints in the die. The skill of the operator largely determines where such restraints are needed.

Drop Hammer Forming

Drop hammer forming is used for producing shallow depths and asymmetrical shapes in magnesium alloys when quantities are small and for applications requiring minimum springback. Successful results depend on operator skill. Except for heating, drop hammer forming of magnesium alloys is the same as it is for other metals (see the article "Drop Hammer Forming" in this Volume).

Zinc alloy can be used for both punch and die. Lead punches are sometimes used, but lead pickup can cause corrosion of the sheet. For production quantities greater than approximately 50 pieces, however, cast iron punches and dies are recommended, because zinc alloy tools lose their shape at these quantities.

Annealed sheet is preferred for drop hammer forming. Blanks should be heated near the hammer, because the work cools rapidly—usually 16 to 25 °C (30 to 45 °F) in 5 s. Ten blows may be needed to form a part, with reheating between blows—5 min for metal thicknesses up to 1.3 mm (0.051 in.) and 9 min for thicknesses of 1.3 to 3.18 mm (0.051 to 0.125 in.).

Heat-resistant rubber pads are often used in the dies for preliminary forming and are removed before final forming.

Dies can be heated in an oven near the hammer, or by torches or ring burners during operation. Small dies can be used on an electrically heated cast iron platen on the hammer bed, but this method is not practical for large dies. Heating of the punch and die by electric heating elements or by a heat-transfer fluid is also used.

The elevated temperatures used in drop hammer forming of magnesium alloys can reduce or eliminate springback; therefore, the maximum practical temperature should always be used. The rate of deformation must be carefully controlled, especially when deformation of the work metal is severe, or when the metal is in the H24 temper. For workpieces that require severe forming, the punch is lowered slowly, and forming is completed with subsequent blows. Tolerances of ±0.76 mm (±0.030 in.) can be maintained in production.

Precision Forging

Precision forging for the near-net shape production of high-quality components is a relatively new method of die forging (see the article "Precision Hot Forging" in *Metalworking: Bulk Forming,* Volume 14A of *ASM Handbook,* 2005). With steel and aluminum, the procedure has been successfully used in the automotive and airplane industry. Besides simple precision-forged workpieces, no new developments for magnesium have been made in recent years.

Precision forging, which was developed as a bspecial method of die forging, is realized in a closed die without room for compensation, so that a flash is almost nonexistent. Workpieces can be produced with tolerances, where machining can be reduced to a minimum. Forging processes of magnesium alloys, as near-net shape forming technologies, have the potential to become more and more relevant in the production of lightweight applications. In particular, aerospace and automotive components, with their high requirements in view of safety and weight, provide a range of applications for forging processes. Not only aluminum or steel parts but even plastic or fiberglass components could be replaced by magnesium alloys (Ref 34). Precision forging has the following advantages compared to conventional die forging methods:

- Production of near-net shape workpieces
- More economic use of material because of flashless forging
- Reduction of machining processes

Precision Forging and Forging Simulation of Test Specimens. To investigate the temperature dependence of the forgeability of magnesium alloys, the spike forging test can be used. The shape of the test specimens was defined in a way in which the relevant material flow processes could be investigated. The height of the cone and the broadening of the flange could be used to assess the formability of the material under the particular forging condition. The tool temperatures (upper and lower die) and the heating temperature are varied in relevant ranges.

Based on the material data of the fundamental investigations, the forging process is simulated.

The simulated material flow is in good accordance with the experiment (Fig. 28). Forging processes for the considered alloy can be simulated by verifying the material description on which the simulation is based and which was obtained by experimental material data.

During the forging operation, the microstructure recrystallizes. Due to the finer grain, the hardness (Brinell) of the forged specimen is higher than the hardness of the initial material. It decreases with increasing tool temperature and seems to be more sensitive to tool temperatures than the initial workpiece temperature. Although the highest values can be achieved at low initial temperatures, under these conditions, high punch forces, which are normally limited by the press, are necessary for complete form filling (Ref 32).

Precision Forging of a Sprocket. Based on information from the test specimens, a tool system for the precision-forging process of a sprocket was developed (Fig. 29) (Ref 32).

The sprocket was forged in one hit from a cylindrical billet on a screw press with a high punch speed of 365 mm/s (14 in./s). The investigations show that complete die filling for components with slim variant structures can be realized by thermally controlled tools. An optimized die filling was reached at billet temperatures between 300 and 380 °C (570 and 715 °F),

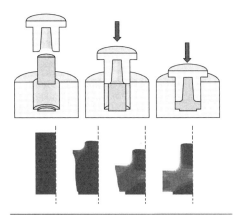

Fig. 28 Test specimens by experience and simulation. Source: Ref 32

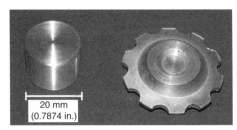

Fig. 29 Forged sprocket of MgAl3Zn1 (Az31). Source: Ref 32

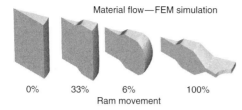

Fig. 30 Finite element method (FEM) simulation of material flow and form filling. Source: Ref 32

with a die temperature of 250 °C (480 °F) for the upper die and 300 °C (570 °F) for the lower die. Some graphite and graphite-free lubricants were tested, and the best results were achieved for Molykote (Dow Corning Corp.), a graphite-based MoS$_2$ oil. Numerical thermomechanical-coupled finite element simulation, based on the elaborated material properties for the considered magnesium alloys, enables the layout of forging processes. The temperature distribution, die filling, punch forces, as well as the tool loading can be predicted. The die filling starts with the compression of the initial cylindrical geometry, followed by a radial material flow. The shape of the teeth is realized at the end of the forging operation (Fig. 30).

ACKNOWLEDGMENTS

Portions of this article were adapted from Ref 14 and 33:

- Forming, *Magnesium and Magnesium Alloys, ASM Specialty Handbook,* ASM International, 1999
- Forming of Magnesium Alloys, *Forming and Forging,* Vol 14, *Metals Handbook,* 9th ed., ASM International, 1988, p 825–830

REFERENCES

1. S. Schumann and H. Friedrich, The Use of Magnesium in Cars—Today and in the Future, *Magnesium Alloys and Their Applications,* Wolfsburg, April 28–30, 1998, p 3–13
2. A. Metz, "Structural Parts for Car Bodies Made of Extruded Profiles and Pressure Die-Casted Magnesium," Progress with Magnesium in Automotive Manufacture, Automotive Manufacture Working Group, Bad Nauheim, Feb 17–18, 2000 (in German)
3. G. Siebel, *Technology of Magnesium and Its Alloys,* A. Beck, Ed., F.A. Hughes & Co. Ltd., London, 1940
4. J.A. Chapman, Ph.D. thesis, University of Birmingham, 1963
5. C.S. Roberts, *Magnesium and Its Alloys,* Wiley, New York, 1960
6. G.V. Raynor, *The Physical Metallurgy of Magnesium and Its Alloys,* Pergamon Press, London, 1959
7. R.E. Reed-Hill and W.D. Robertson, Deformation of Magnesium Single Crystals by Nonbasal Slip, *JOM,* Vol 9 (No. 4), 1957, p 496–502
8. R.E. Reed-Hill and W.D. Robertson, Pyramidal Slip in Magnesium, *Trans. AIME,* Vol 212, 1958, p 256
9. F.E. Hauser, P.R. Landon, and J.E. Dorn, *Trans. ASM,* 1958, p 50
10. A.R. Chaudhuri, H.C. Chang, and N.J. Grant, Creep Deformation of Magnesium at Elevated Temperatures by Nonbasal Slip, *JOM,* May 1955, p 682
11. *Fabricating Magnesium,* The Dow Chemical Company, 1984
12. L. Taylor, H.E. Boyer, E.A. Durand, et al., Ed., *Forming,* Vol 4, *Metals Handbook,* 8th ed., American Society for Metals, 1969
13. M. Bauccio et al., Ed., *ASM Metals Reference Book,* 3rd ed., ASM International, 1993
14. Forming, *Magnesium and Magnesium Alloys, ASM Specialty Handbook,* ASM International, 1999
15. K. Dröder, "Investigations of Forming of Magnesium Sheet Metal," Ph.D. dissertation, Hanover University, 1999 (in German)
16. H.-W. Wagener and J. Hosse-Hartmann, Deep Drawing of Magnesium Sheet Metal *UTF Sci. IV,* Quartal 2001, p 28–34 (in German)
17. L.R. Morris and R.A. George, Warm Forming High-Strength Aluminium Automotive Parts, *International Automotive Engineering Congress and Exposition,* Feb 28 to March 4, 1977 (Detroit, MI), p 1–9
18. H. Haferkamp, E. Doege, M. Rodman, C. Jaschik, and G. Kurz, "Heated Hydromechanical Deep Drawing of Optimized Rolled Magnesium Sheet Metal," Kolloquium Wirkmedien Blechumformung—Sheet Metal Hydroforming—2002, Dec 10, 2002 (Dortmund, Germany) (in German)
19. E. Doege and G. Kurz, Development of a Formulation to Describe the Worksoftening Behaviour of Magnesium Sheets for Heated Deep Drawing Processes, *Ann. CIRP,* Vol 50 (No. 1), 2001, p 177–180
20. H. Haferkamp, E. Doege, M. Schaper, M. Rodman, C. Jaschik, and G. Kurz, Influence of Rolling Parameters on the Formability of Magnesium Sheet Metal, *Metal,* Vol 56 (No. 12), 2002, p 801–805 (in German)
21. K. Siegert et al., Superplastic Aluminum Sheet Metal—Processing with Numerical Controlled Presses, *Metall,* Vol 45 (No. 4), 1991 (in German)
22. E.F. Emley, *Principles of Magnesium Technology,* Pergamon Press, Oxford, London, 1966
23. D. Schmoeckel, "Temperature Guided Process Control for Forming of Aluminum Sheet Metal," EFB report 55, 1994 (in German)
24. H. Beißwänger, "Warm Drawing of Lightweight Materials," Report 27, Research Society for Sheet Metal Forming, 1950 (in German)
25. E.G. Donaldson, Cast High-Duty Iron Tooling for Sheet Metal Work, *Sheet Met. Ind.,* Vol 50, 1973
26. E. Doege and K. Droeder, Sheet Metal Forming of Magnesium Wrought Alloys—Formability and Process Technology, *J. Mater. Process. Technol.,* Vol 115, 2001, p 14–19
27. E. Doege, G. Walter, G. Kurz, and T. Meyer, "Forming of Magnesium Sheet Metal with Heated Deep Drawing Tools," EFB report 195 (in German)
28. K. Yoshida, K. Miyauchi, H. Hayashi, and S. Kuriyama, Flange Behaviour Analysis of Complex Geometries, *Tenth Biennial Congress of the IDDRG,* 1972, p 1–27
29. W. Strackerjahn, "Prediction of Failure at Deep Drawing of Rectangular Parts," Ph.D. dissertation, Hanover University, 1982 (in German)
30. M. Papke, "Powder and Precision Forging of Ultra Lightweight Magnesium-Lithium Alloys," Ph.D. dissertation, Hanover University, 1996 (in German)
31. J. Becker, G. Fischer, and K. Schemme, Production and Properties of Extruded and Forged Magnesium Parts, *Metall,* Vol 52 (No. 9), 1998, p 528–536 (in German)
32. S. Janssen, "Closed Die Forging of Aluminum-Zinc Based Magnesium Wrought Alloys," Ph.D. dissertation, Hanover University, 2000 (in German)
33. Forming of Magnesium Alloys, *Forming and Forging,* Vol 14, *Metals Handbook,* 9th ed., ASM International, 1988, p 825–830
34. P. Aroule, Magnesium Demand and Supply, *Automot. Sourc.,* Vol 6 (No. 2 Supplement), 1999, p 30–33

Forming of Nickel and Cobalt Sheet Alloys*

Revised by Howard W. Sizek, Air Force Research Laboratory

NICKEL AND COBALT ALLOYS are used in a wide range of products and processes where they must resist corrosion and elevated temperatures. These alloys have a wide range of compositions and are often alloyed for specific applications. Table 1 lists the nominal compositions for the nickel and cobalt alloys discussed in this article. Included in this list are some iron alloys with significant levels of nickel, chromium, and/or cobalt. These alloys are often grouped with nickel and cobalt alloys because of their capabilities in engineering applications. These iron alloys also have forming characteristics similar to the nickel and cobalt alloys.

Depending on the specific alloy, one or more of the following strengthening mechanisms is used in the nickel and cobalt alloys: solid-solution strengthening, precipitation hardening, and dispersion strengthening. These strengthening mechanisms result from wide-ranging microstructural, compositional, and processing variations. Each of these variations contributes to the response of the material during forming into the final product.

Cold forming is preferred for nickel and cobalt alloys, especially in thin sheets, because most of

Table 1 Compositions of selected nickel and cobalt alloys

Alloy	UNS number	Composition(a), wt%						
		C	Fe	Co	Cr	Ti	Mo	Others
Nickel 200	N02200	0.10 M	0.4 M	...	...	0.10 M	...	0.25 Cu (M), 0.15 Si (M)
Nickel 201	N02201	0.02 M	0.4 M	...	...	0.10 M	...	0.25 Cu (M), 0.15 Si (M)
Alloy 400	N04400	0.15 M	1.25	...	...	...	...	31.5 Cu, 0.5 Si
Alloy K500	N05500	0.25 M	2.0 M	...	...	0.6	...	31.5 Cu, 2.7 Al, 0.5 Si (M)
Alloy X	N06002	0.10	18.0	1.5	22.0	...	9.0	0.6 W, 1 Mn (M), 1 Si (M), 0.008 B (M)
Alloy C-22	N06022	0.01 M	3.0	2.5 M	22.0	...	13.0	3.0 W, 0.08 Si (M), 0.5 Mn (M), 0.35 V (M), 0.5 Cu (M)
Alloy 75	N06075	0.15 M	5.0 M	...	19.5	0.6 M	...	0.5 Cu (M), 1.0 Mn (M), 1.0 Si (M)
Alloy C-2000	N06200	0.01 M	3.0 M	2.0 M	23.0	...	16.0	1.6 Cu, 0.08 Si (M), 0.5 Mn (M), 0.5 Al (M)
Alloy 230	N06230	0.10	3 M	5 M	22.0	...	2.0	14 W, 0.4 Si, 0.3 Al, 0.02 La, 0.5 Mn, 0.015 B (M)
Alloy 600	N06600	0.1 M	8.0	...	15.5	...	...	...
Alloy 617	N06617	0.10	3 M	12.5	22.0	0.6 M	9.0	1.15 Al
Alloy 625	N06625	0.10 M	5.0 M	1.0 M	21.0	0.4 M	9.0	3.7 Nb + Ta, 0.5 Si (M), 0.5 Mn (M), 0.4 Al (M)
Alloy 686	N06686	0.01 M	1.0	...	20.5	0.15	16.0	3.9 W
Alloy 690	N06690	0.05 M	9.0	...	29.0	...	...	...
Alloy 693	N06693	0.15 M	4.25	...	29.0	1.0 M	...	3.25 Al, 1.5 Nb
Waspaloy	N07001	0.08	2.0 M	13.5	19.0	3.0	4.25	1.4 Al, 0.07 Zr, 0.005 B
Alloy R-41	N07041	0.09	5 M	11.0	19.0	3.15	10	Al = 1.6
Alloy 80	N07080	0.10 M	5.0 M	2.0 M	19.5	2.25	...	1.15 Al
Alloy 214	N07214	0.04	3.0	...	16.0	...	...	4.5 Al, 0.01 Y, 0.5 Mn (M)
Alloy 263	N07623	0.06	...	20.0	20.0	2.15	6	Ti + Al = 2.6
Alloy 718	N07718	0.05 M	19.0	1.0 M	18.0	1.0	3.0	5 Nb, 0.6 Al
Alloy X-750	N07750	0.04	7.0	...	15.5	2.5	...	0.95 Nb + Ta, 0.7 Al
Alloy HR-120	N08120	0.05	33.0	3.0 M	25.0	...	1.0 M	0.5 W (M), 0.6 Si, 0.7 Mn, 0.7 Nb, 0.1 Al, 0.2 N, 0.004 B
Alloy 800	N08800	0.1 M	44.0	...	21.0	0.38	...	0.35 Al, 0.75 Cu (M), 1.0 Si (M)
Alloy 825	N08825	0.05 M	30.0	...	21.5	0.9	3.0	2.25 Cu
Alloy 25-6Mo	N08926	0.02 M	46.0	...	20.0	...	6.5	1 Cu, 0.2 N
Alloy W	N10002	0.12 M	6	2.5 M	5.0	...	24	0.6 V (M)
Alloy 242	N10242	0.03 M	2.0 M	2.5 M	8.0	...	25.0	0.8 Mn (M), 0.8 Si (M), 0.5 Al (M), 0.006 B (M), 0.5 Cu (M)
Alloy C276	N10276	0.01 M	5.0	2.5 M	15.5	...	16	4.0 W, 0.35 V (M)
Alloy B-3	N10675	0.01 M	2.0 M	1.0 M	1.0 M	0.2 M	28.5	3.0 W (M), 0.1 Si (M), 3.0 Mn (M), 0.5 Al (M)
N-155	R30155	0.12	30.0	20.0	21.0	...	3.0	2.5 W, 1.5 Mn, 1 (Nb + Ta), 0.15 N, 1 Si (M)
Alloy 188	R30188	0.1	3 M	39.0	22.0	...	...	14 W, 0.35 Si, 0.03 La, 1.25 Mn (M)
Alloy 25	R30605	0.1	3 M	51.0	20.0	...	...	0.4 Si (M), 15 W, 1.5 Mn
Alloy 27-7Mo	S31277	0.02 M	40.0	...	21.75	...	7.25	1 Cu, 0.35 N
Alloy 864	S35135	0.08 M	37.0	...	22.5	0.7	4.4	0.8 Si
Alloy A286	S66286	0.08 M	51.0	...	14.75	2.13	1.25	Al = 0.35
PE16	...	0.06	34.0	20 M	16.5	1.2	3.3	Ni + Co = 43.5, Al = 1.2

M, maximum. (a) Balance nickel

*This article was adapted from Ref 1 and 2. For additional credits, see "Acknowledgments" at the end of this article.

these alloys can be hot formed effectively only in a narrow temperature range between approximately 925 and 1260 °C (1700 and 2300 °F). Annealing between cold forming operations is usually preferred to hot forming.

This article discusses forming techniques for these alloys as well as provides several examples of the forming techniques.

Alloying and Strain Hardening

Strain hardening can be related to the solid-solution strengthening afforded by alloying elements (Ref 3–5). In general, the strain-hardening rates increase with the complexity of the alloy. Accordingly, strain-hardening rates range from moderately low for nickel and nickel-copper alloys to moderately high for the nickel-chromium, nickel-chromium-cobalt, and nickel-iron-chromium alloys. Figures 1 and 2 compare the strain-hardening rates of a number of alloys in terms of the increase in hardness with increasing cold reduction. Note that the strain-hardening rates of the nickel and cobalt alloys are greater than that of 1020 steel, but some are less than that of AISI type 304 stainless steel. Based on this comparison, most nickel and cobalt alloys can be formed by techniques similar to those employed for the forming of AISI 300-series stainless steels. Table 2 lists the Vickers hardness versus percent cold work for a number of nickel and cobalt alloys.

The differences in work hardening and strength between solid-solution-strengthened alloys can be attributed to compositional differences (Table 1). Alloys with the greatest amount of cobalt, such as alloy 25 (UNS R30605) and alloy 188 (UNS R30188), nominally 51 and 39 wt% Co, respectively, require a greater

magnitude of force to form than iron or nickel alloys. These two alloys have the highest hardening rates of the alloys discussed.

Alloys containing substantial amounts of molybdenum or tungsten for strengthening, such as alloy 230 (UNS N06230) or alloy 41 (UNS N07041), are harder to form than alloys with lesser amounts of these elements. These alloying elements not only increase the strength of the alloy but also increase the strain-hardening rate (Ref 5).

Similarly, the age-hardenable alloys have higher strain-hardening rates than their solid-solution equivalents. Aluminum and titanium are added to nickel alloys to provide strengthening by the precipitation of the γ' phase. The volume fraction γ' is a function of the amount of aluminum and titanium present as well as the overall composition. Examples of γ'-containing alloys include alloy 80A (UNS N07080), Waspaloy alloy (UNS N07011), and alloy 214 (UNS N07214), typically containing 15, 20, and 33% γ', respectively. Although parts from precipitation-hardened alloys are used primarily in the aged state, aluminum and titanium in these alloys increase the strength in the annealed state, increasing the difficulty of forming parts. The hardening curves for alloys 600 and X-750, its precipitation-hardened counterpart, illustrate the contribution of aluminum and titanium to work hardening (Fig. 1). Alloys 690 and 693, an alloy 690 variant with additional aluminum and niobium, illustrate similar hardening effects in data presented in Table 2. As a result of the high strain-hardening rate, many precipitation-hardened alloys require complex production steps to produce satisfactory components.

Because the modulus of elasticity of the nickel alloys is relatively high (similar to that of steel), a

small amount of springback in cold forming operations may be expected. However, springback is also a function of proportional limit, which can increase greatly during cold working of strain-hardenable materials. For instance, a yield strength of 170 MPa (25 ksi) of an alloy in the annealed condition may increase to 520 MPa (75 ksi) during a drawing operation. Therefore, the amount of springback for this alloy must be computed from the 520 MPa (75 ksi) flow stress, rather than from the initial value of the yield strength. A number of nickel and cobalt alloy stress-strain curves through fracture can be found in Ref 6.

Forming Practice for Age-Hardenable Alloys

To soften the age-hardenable nickel alloys for forming, two types of annealing treatments are used based on the ductility needed for forming and, if subsequent welding is required, the avoidance of adverse metallurgical effects during and after welding. A high-temperature anneal is used to obtain maximum ductility when the formed part will not be welded. A lower-temperature anneal is used when welding is required, but at the expense of forming ductility.

For example, solution annealing of alloy 41 at 1175 °C (2150 °F) followed by quenching in water gives maximum ductility. However, parts formed from solution-annealed and quenched sheet should not be welded; during welding or subsequent heat treatment, they are likely to crack at the brittle carbide network developed in the grain boundaries. A lower annealing temperature, preferably 1065 to 1080 °C (1950 to 1975 °F), results in less sensitization during welding and decreases the likelihood of grain-boundary cracking. Formability is reduced by 10 to 20% but is adequate for most forming operations.

Cold-Formed Parts for High-Temperature Service. A part that is highly stressed during cold forming may require heat treatment to avoid excessive creep in service above its recrystallization temperature. Recrystallization of a specific alloy is determined largely by the extent of cold work. In general, increasing the amount of cold work reduces the recrystallization temperature. The grain size and exact composition of the material also complicate prediction of the recrystallization temperature. Alloys used for high-temperature service are frequently used in the grain-coarsened condition at service temperatures above 595 °C (1100 °F).

Generally, cold-formed nickel alloys should be heat treated if they have been strained in either tension or compression by more than 10% and will be in service at temperatures above 650 °C (1200 °F). Sheet producers should be consulted for the proper thermal treatments.

In certain alloys, heavy cold working (for example, highly restrained bending) followed by exposure at moderate to high temperatures (for example, stress relieving or age hardening) can

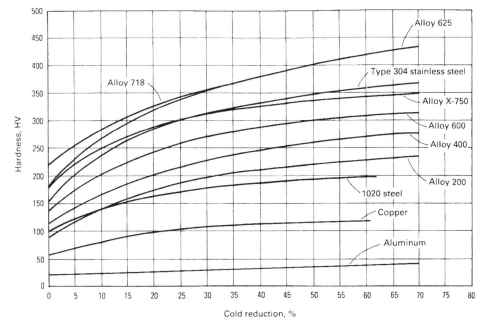

Fig. 1 Effect of cold reduction on the hardness of various nickel alloy sheet materials and, for comparison, aluminum, copper, type 304 stainless steel, and low-carbon ferritic steel

lead to cracking. In age-hardenable alloys, the combination of high residual tensile stress and the stress associated with the aging response may exceed the stress-rupture strength of the material. In non-age-hardenable alloys, excessive cold working of coarse-grain material (grain size of ASTM No. 5 or coarser) without the recommended intermediate annealing can cause cracking during subsequent exposure at stress-relieving or annealing temperatures. Testing the material under actual conditions of forming and heating will determine its susceptibility to cracking. Springs can be cold formed from age-hardenable alloys in the annealed or cold-drawn temper. For service at temperatures above 315 °C (600 °F), springs should be solution annealed before aging to prevent loss of strength from relaxation.

Temper

Most cold forming operations require the use of annealed material. However, the softer alloys, such as nickel 200, the NILO alloys, and alloy 400, are frequently used in $1/8$-hard and $1/4$-hard tempers for improved shearing and piercing. To obtain the fine grain structure that is best for cold forming, the nickel and cobalt alloys must be cold worked (reduced) beyond a critical percentage reduction and then annealed. The critical amount of cold work varies with the alloy and with the annealing temperature but is usually 8 to 10%. Reheating metal that is only slightly cold worked can result in critical grain growth, which can cause orange-peel or alligator-hide effects in subsequent forming operations.

For example, an alloy X (UNS N06002) workpiece, partly formed, stress relieved, and then given the final form, had severe orange peel on much of its surface. The partial forming resulted in approximately 5% cold working, and during stress relief, an abnormally coarse grain structure developed. The problem was resolved by making certain that the metal was stretched 10% or more before it was stress relieved. In addition, stress relieving was done at the lowest temperature and shortest time that could be used, because higher temperatures and longer times increased grain growth. Optimal time and temperature were determined by hardness testing.

Severely cold-formed parts should be fully annealed after final forming. If annealing causes distortion, the work can be formed within 10% of the intended shape, annealed, pickled, and then given the final forming.

Solution-annealed products are usually soft enough to permit mild forming. If the solution-annealed alloy is not soft enough for the forming operation, an annealing treatment must be used that will remove the effects of cold work and dissolve the age-hardening and other secondary phases. Some control of grain size is sacrificed, but if cooling from the annealing temperature is very rapid, the age-hardening elements will remain in solution. Further annealing after forming can be done at a lower temperature to decrease the risk of critical grain growth. Several process anneals may be required in severe forming, but the high-temperature anneal need not be repeated. Annealing should be performed at a temperature that produces optimal ductility for the specific metal, as shown in the following example.

Example 1: Change in Heat Treatment to Eliminate Cracking. A large manifold was made by welding together two drawn halves into

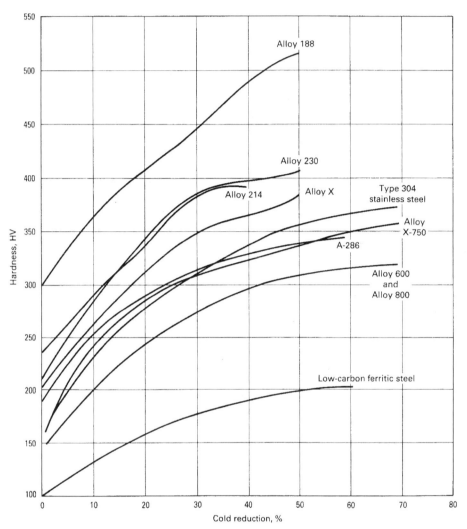

Fig. 2 Effect of cold reduction on the hardness of several heat-resistant nickel and cobalt alloys and, for comparison, a type 304 stainless steel and low-carbon ferritic steel

Table 2 Effect of cold work on hardness for selected nickel and cobalt alloys

Alloy	UNS number	Hardness, HV, at indicated cold work percentage					
		0	10	25	40	50	75
Alloy X	N06002	180	226	327	363	382	...
Alloy C-22	N06022	200	260	340	392	402	...
Alloy 230	N06230	210	286	363	392	412	...
Alloy 617	N06617	163	227	313	376	418	519(a)
Alloy 625	N06625	222	318	377	412	446	...
Alloy 686	N06686	205	240	300	365	382	...
Alloy 690	N06690	145	218	276	310	326	351(a)
Alloy 693	N06693	220	269	347	386	408	...
Alloy HR-120	N08120	200	279	327	345	354	...
Alloy 825	N08825	145	227	310	348	373	385
Alloy 25-6Mo	N08926	190	240	307	341	363	400
Alloy C-276	N10276	195	205	327	345	372	...
Alloy B-3	N10675	228	302	382	434	458	...
Alloy 27-7Mo	S31277	225	274	326	368	390	436
Alloy 864	S35135	160	215	273	304	317	382(a)

(a) Interpolated from 70% cold work

a doughnut shape. Each half was drawn to a depth of 125 mm (5 in.) from 6.35 mm (0.25 in.) thick alloy 41 that had been solution treated at 1175 °C (2150 °F) and water quenched. Drawing of the plate stock on a 31,000 N (7000 tonf) drop hammer produced severe work hardening, and cracking occurred frequently. To eliminate the cracking, forming was done in three steps, and the parts were annealed at 1080 °C (1975 °F) before the second and the third step.

The forming characteristics of the alloy 41 plate were greatly improved by modifying the solution treatment. The revised treatment consisted of first soaking the alloy at 540 °C (1000 °F), transferring it to a gantry furnace, and holding it at 1080 °C (1975 °F) for 30 min. The work was then lowered rapidly through the bottom of the furnace into a salt bath at 205 to 260 °C (400 to 500 °F). Thus, the elapsed time between leaving the high-temperature zone and entering the quench was kept to 4 or 5 s, the alloy was in the precipitation range, 595 to 1010 °C (1100 to 1850 °F) for a minimum time, and minimum hardness (16 to 21 HRC) was obtained. The salt bath provided a more uniform quench and a more ductile alloy than the original water quench. The improved ductility of the alloy allowed forming of the manifold halves in two operations.

Formability

Formability refers to the ease with which sheet metal can be formed. Material formability is difficult to measure because there is no single index for predicting specific material formability for all processing conditions. The deformation modes in the forming of most sheet metal components are complex and consist of bending, unbending, stretching, and deep drawing. Forming technology depends a great deal on practical experience. Material characteristics such as tensile ductility, strain-hardening exponent, and anisotropy parameters can act as guides to the nature of formability and are useful for comparing materials.

In any forming operation, the useful ductility of material is that amount up to the point of necking. Greater ductility at peak load and a large separation between yield and tensile strengths are desirable. A measure of stretchability is provided by the strain-hardening exponent (n-value). Plastic deformation in a tensile test is most commonly related to true stress in the following manner:

$$\sigma = K\varepsilon^n \qquad \text{(Eq 1)}$$

where K is the strength constant, σ is true stress, ε is true strain, and n is the strain-hardening coefficient.

Most nickel and cobalt alloys possess n-values between 0.35 and 0.5; a high strain-hardening capacity results in spreading the strain away from any local region in the presence of a stress gradient.

The rolling and rerolling of a metal during its manufacture may align individual grains in the sheet. This crystallographic texture imparts anisotropic plastic properties to the sheet. Depending on the alloy and processing temperatures, recrystallization during annealing will tend to restore the random grain orientation and resultant plastic isotropy. The plastic strain ratio r (the ratio of the width strain to thickness strain in a uniaxial tensile test) is a measure of normal anisotropy, that is, the variation of properties in the plane of the sheet relative to those perpendicular to the sheet surface. The average r-value is given by:

$$r = r_L + 2r_D + r_T/4 \qquad \text{(Eq 2)}$$

with the subscripts referring to the tensile test measurements made in the longitudinal (L), diagonal (D), and transverse (T) orientations of rolled sheet. The variation of the properties in the plane of the sheet is termed planar anisotropy (Δr) and is given by:

$$\Delta r = r_L + r_T - 2r_D/2 \qquad \text{(Eq 3)}$$

For an isotropic material $r = 1$ and $\Delta r = 0$. A material with a high r-value resists localized necking in the thickness direction; therefore, deep drawability is high. There are various correlations between deep drawability and r-value. Planar anisotropy causes uneven flow of metal, resulting in earing of drawn cups. Some typical forming characteristics of several heat-resistant alloys are given in Table 3.

Forming limit curves are often used to predict the formability of materials. The forming limit curve is experimentally constructed for combinations of strain paths to describe strain to necking (or fracture). The forming limit diagrams for three nickel sheet alloys are given in Ref 9. The effect of small changes in the carbon and niobium content on the formability of alloy 718 is also presented in this article. More information on formability testing is available in the article "Formability Testing of Sheet Metals" in this Volume.

Effect of Rolling Direction on Formability. Depending on the size, amount, and dispersion of secondary phases, the age-hardenable alloys show greater directional effects than non-age-hardenable alloys. However, vacuum melting and solution annealing serve to reduce directional effects (anisotropy). Vacuum melting increases the cleanliness of the metal, reducing the number of nitride and carbide precipitates that can reduce ductility and formability. Solution annealing dissolves the precipitated phases and also recrystallizes the cold-worked structure. As shown by data for press-brake bending in Fig. 3, directional effects contribute erratically to cracking and surface defects. The following example shows how directionality seriously affected the forming characteristics of iron-base alloy A-286 (UNS S66286).

Example 2: Effect of Directionality in Bulging A-286. A contoured exhaust cone (Fig. 4a) was made by cutting a flat blank from mill-annealed A-286 sheet, rolling and welding a cone from the blank, and then bulging the cone into final shape. Developed blanks for two cones were cut from one sheared rectangle (Fig. 4b) with little waste of stock.

Several lots of A-286 produced good parts, but one lot of material cracked in bulging. As shown in Fig. 4(a), cracks occurred in the cone adjacent to the weld at the location where the forming stresses were perpendicular to the rolling direction (which was also the direction of minimum elongation). The tensile elongations of the good and inferior lots of A-286 were compared parallel and perpendicular to the rolling direction (Table 4). The inferior lot showed substantially greater difference in elongation between the two test directions.

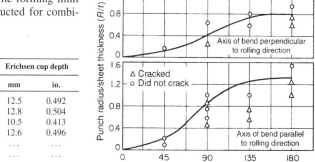

Fig. 3 Effect of forming direction relative to rolling direction on the formability of alloy 41 sheet 0.5 to 4.75 mm (0.02 to 0.187 in.) thick in press-brake bending

Table 3 Forming characteristics of selected heat-resistant alloys

Alloy	UNS number	Thickness mm	Thickness in.	Anisotropy(a) r	Anisotropy(a) Δr	Olsen cup depth mm	Olsen cup depth in.	Erichsen cup depth mm	Erichsen cup depth in.
Alloy 80A	N07080	0.90	0.035	0.91	−0.020	...	...	12.5	0.492
Alloy 263	N07263	0.90	0.035	0.86	0.100	...	...	12.8	0.504
Alloy PE16	...	0.90	0.035	0.98	−0.400	...	...	10.5	0.413
Alloy 188	R30188	1.20	0.047	0.94	0.130	...	...	12.6	0.496
		0.63	0.025	0.95	−0.024	12.5	0.492	...	...
Alloy 230	N06230	0.76	0.030	0.93	−0.059	11.0	0.433	...	...
Alloy 625	N06625	0.61	0.024	0.97	−0.139	11.7	0.461	...	...
Alloy X	N06002	0.61	0.024	0.95	−0.105	10.2	0.402	...	...
Alloy 600	N06600	1.57	0.062	...	...	11.9	0.47	...	...

(a) r, plastic strain ratio; Δr, planar anisotropy. Source: Ref 7, 8

Annealing the welded cones before bulging reduced the number of cracked cones, but not by a satisfactory percentage. A higher percentage of acceptable cones resulted when the blanks were cut with their edges oriented to the rolling direction, as shown in Fig. 4(c). Cones made from these blanks had less abrupt change in the forming direction relative to rolling direction on each side of the weld, and the forming stresses were never perpendicular to the rolling direction; however, there was more scrap material from cutting the blank. When a revision of production techniques at the mill reduced the elongation difference in the two directions of stress, it was possible to use the more economical blank layout shown in Fig. 4(b).

Speed of Forming. The speed at which a metal is deformed affects its formability. In general, each metal has a critical speed of forming. In some cases, the ductility increases until this critical speed is reached, after which it decreases sharply with increasing speed, as indicated by the curve in Fig. 5. This curve has a plateau of maximum strain where ductility is greatest. This plateau seems to be broad for most nickel and cobalt alloys. The breadth of the plateau depends on the use of biaxial or triaxial loading of the material during forming. Table 5 recommends forming speeds for three alloys and three forming operations.

Galling

Nickel Alloys. Because nickel alloys do not readily develop an oxide film that presents a barrier to diffusion bonding, they cold weld

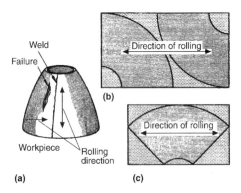

Fig. 4 Alloy A-286 exhaust cone that cracked in bulge forming, and the two layouts used in cutting the cone blank from 1.0 to 1.3 mm (0.04 to 0.05 in.) thick sheet

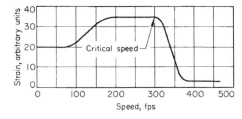

Fig. 5 Effect of forming speed on ductility. fps, feet per second. Source: Ref 10

(gall) easily to materials of similar atomic diameter. When a cold weld is formed, the high shear strength and ductility of the alloys prevent the weld from being easily broken. For these reasons, the coefficient of friction between nickel alloys and other metals, including most die materials, is usually high.

Alloying with highly reactive elements that readily form oxide films, such as chromium, reduces the galling, or cold welding, propensity of nickel alloys. Accordingly, the nickel-chromium and nickel-iron-chromium alloys are less likely to gall than are the nickel and nickel-copper alloys. However, chromium oxide films are thin and brittle and provide only limited protection because they are easily broken when the substrate is deformed. The use of heavy-duty lubricants minimizes galling in most cold forming processes.

Cobalt Alloys. In situations where oxides cannot prevent intimate metal-to-metal contact, the cobalt alloys are well known for their resistance to galling, especially when self-mated. This attribute is related to their response to mechanical stress. It is believed that their low stacking fault energies, and their tendencies to transform (from face-centered cubic to hexagonal close-packed) and to twin, limit the depths affected and allow easy shear of surface layers.

Lubricants

Heavy-duty lubricants are required on nickel alloys and cobalt alloys for most cold forming operations. Although zinc-, sulfur-, and chlorine-containing additives can improve lubricants, they can also have harmful effects if not completely removed after forming. Sulfur and zinc will embrittle nickel alloys at elevated temperatures such as may be encountered in annealing or age hardening, and chlorine can cause pitting of the alloys after long exposure. Although sulfurized and chlorinated lubricants can be used if the work is carefully cleaned with a

Table 4 Elongation differences in alloy A-286 sheet

Test direction with respect to rolling	Elongation, %	
	Good	Inferior
Perpendicular	41	43.5
Parallel	38.5	37.2
Difference	2.5	6.3

degreaser or an alkaline cleaner, they should not be used if any difficulty is anticipated in cleaning the formed part. Molybdenum disulfide is seldom recommended for use with nickel alloys because removal is difficult. Work that has been formed in zinc alloy dies should be flash-pickled in nitric acid before heat treatment to prevent the possibility of zinc embrittlement.

Pigmented oils and greases should be selected with care, because the pigment may be white lead (lead carbonate), zinc oxide, or similar metallic compounds with low melting points. These elements can embrittle nickel alloys if the compounds are left on the metal during heat treatment. Inert fillers such as talc can be used safely.

Metallic coatings such as copper offer the maximum film strength for a lubricant. Because application and removal are expensive, metallic coatings are used as lubricants only in severe cold forming operations, and then only when they can be properly removed.

Ordinary petroleum greases are seldom used in forming nickel alloys, because they have neither the film strength indicated by their viscosity nor a strong polar attraction for metals to keep them from breaking down during the forming process.

Phosphates do not form usable surface compounds on nickel alloys and cannot be used as lubricant carriers.

Mild forming operations requiring less than a 10% reduction can usually be accomplished successfully with unpigmented mineral oils and greases. Polar lubricants, such as lard oil and castor oil, are preferred for mild forming. They will usually produce acceptable results and are easily removed.

Some lubrication is usually required for optimal results in drawing, stretch forming, or spinning. Lubrication is seldom needed for the press-brake forming of V-bends but will greatly improve results if a square punch is used. For more severe forming, metallic soaps or extreme-pressure (EP) lubricants, such as chlorinated, sulfochlorinated, or sulfurized oils or waxes, are recommended. They can be pigmented with a material such as mica for extremely severe forming.

Lubricants used for spinning operations must cling tenaciously; otherwise, they will be thrown off the workpiece by centrifugal force. Metallic soap or wax applied to the workpiece before spinning is usually satisfactory. Sulfurized and chlorinated lubricants are not recommended for

Table 5 Recommended forming speeds for selected heat-resistant alloys

Alloy	UNS No.	Forming speed for:					
		Tensile forming(a)		Bulge forming(a)		Draw forming	
		m/s	ft/s	m/s	ft/s	m/s	ft/s
A-286	S66286	20 to 130	65 to 425	0 to 213	0 to 700	0 to 236	0 to 775
Alloy 41	N07041	0 to 107	0 to 350	0 to 213	0 to 700	0 to 229	0 to 750
Alloy 25	R30605	15 to 85	50 to 275	. . .	. . .	198 to 251	650 to 825

(a) Maximum possible tensile and bulge forming rates may be greater than stated here. Source: Ref 1, 10

use in spinning, because this operation may burnish the lubricant into the surface of the metal. In power spinning, a coolant should also be used during the process (see the article "Spinning" in this Volume).

Occasionally, it is advantageous to use two kinds of lubricant in the same operation. In one stretch forming application, the strain at the middle of the work was 3 to 4%, but near the ends, where the metal pulled tangentially to the die, the strain was 10 to 12%. A light coat of thin oil was adequate for most of the work, but an EP lubricant was used at the ends. More information on lubricants for sheet forming is available in the article "Selection and Use of Lubricants in Forming of Sheet Metal" in this Volume.

Tools and Equipment

Nickel and cobalt alloys do not require special equipment for cold forming. However, the physical and mechanical properties of these materials frequently necessitate modification of tools and dies used for cold forming other metals. These modifications are discussed in this section. Information applying to specific forming operations is presented in the sections covering those operations.

Few nickel and cobalt alloys are used in applications requiring large quantities that warrant the use of high-production methods and tools. Because only a few to a few hundred parts are needed, methods requiring minimal tooling, such as press-brake forming, drop-hammer forming, spinning, and explosive forming, have been used more than other methods. Presses or other machines are the same as those used for forming steel, but because nickel and cobalt alloys have higher yield strengths and strain-hardening rates, they require stronger and harder dies and more powerful forming equipment than do low-carbon steels. Generally, 30 to 50% more power is required for nickel alloys than is needed for low-carbon steel. High-cobalt or -tungsten alloys may require 50 to 100% more power than low-carbon steels.

Die materials used in forming austenitic stainless steels (see the article "Forming of Stainless Steel" in this Volume) are suitable for similar operations on nickel alloys. Soft die materials such as aluminum bronze, nickel-aluminum bronze, and zinc alloys are used when superior surface finishes are desired. These materials, however, have a relatively short service life. Parts formed with zinc alloy dies should be flash-pickled in dilute nitric acid to remove any traces of zinc picked up from the dies during forming. Zinc can cause embrittlement of nickel alloys during heat treatment or high-temperature service. For similar reasons, parts formed with brass or bronze dies should be pickled if the dies impart a bronze color to the workpiece.

Cast iron has proved adequate and nongalling for many low-production forming tools. If a heat-treatable grade of iron is used, anticipated areas of high wear can be locally hardened. Steel dies,

punches, or mandrels can be plated with approximately 5 to 13 μm (0.2 to 0.5 mil) of chromium in order to minimize adherence.

Tool Design. Because nickel alloys are likely to gall, and because of the high pressures developed in forming, tooling should be designed with liberal radii, fillets, and clearances. The radii and clearances used in the cold forming of nickel alloys are usually larger than those for brass and low-carbon steel and approximately equal to those for the austenitic stainless steels.

Equipment Operation. The strain-rate sensitivity and frictional characteristics of nickel alloys dictate that all forming operations be performed at relatively slow speeds. For instance, the slide speed in shearing, deep drawing, and press-brake bending is usually 9 to 15 m/min (30 to 50 ft/min). Cold heading, piercing, and similar operations are normally done at speeds of 60 to 100 strokes per minute.

Shearing, Blanking, and Piercing

The optimal temper of nickel alloys for shearing, blanking, and piercing varies from skin hard to full hard, depending on the alloy and thickness. For instance, thin strip nickel 200 and the NILO alloys should be blanked in full-hard temper for maximum die life and minimum edge burr, but alloy 600 (UNS N06600) usually gives best results in skin-hard temper. Annealed temper is usually suitable for blanking of the precipitation-hardenable alloys such as alloy X-750 (UNS N07750).

Punch-to-die clearance per side should be 3 to 5% of stock thickness for thin material and 5 to 10% of stock thickness for thick (≥3.2 mm, or ⅛ in.) material. The clearance between the punch and the stripper plate should be as small as is practical.

Shears should have a low-carbon steel rating of 50% greater than the size of the nickel alloy material to be sheared. For example, a shear with a low-carbon steel rating of 9.5 mm (⅜ in.) should be used to cut 6.4 mm (¼ in.) thick alloy 400 (UNS N04400) plate.

Lubricants are not usually used in shearing but should be used in blanking and piercing. A light mineral oil fortified with lard oil can be used for material less than 3.2 mm (⅛ in.) thick. Heavier sulfurized oil should be used for material that is thicker than 3.2 mm (⅛ in.). If a sulfurized lubricant is used during blanking or piercing, the part should be cleaned thoroughly prior to heat treatment or exposure to elevated temperatures.

Procedure. In piercing, the minimum hole diameter is usually equal to or greater than the thickness of the material, depending on the thickness, temper, and specific alloy. Minimum hole diameters for given thicknesses of alloys 200, 400, and 600 are given in Table 6. Hole diameters equal to the thickness of the sheet have been produced in material as thin as 0.46 mm (0.018 in.), but only after considerable experience and with proper equipment.

The softer alloys, such as nickel 200, have greater impact strength than do the harder, chromium-containing alloys. Consequently, the softer alloys are more sensitive to the condition of dies and equipment. Shear knives may penetrate 65 to 75% of the material thickness before separation occurs in shearing nickel 200, whereas penetration may be only 20 to 30% in shearing the harder alloys.

Laboratory tests have indicated that the shear strength of nickel alloys in double shear averages approximately 65% of the tensile strength. However, these values were obtained under essentially ideal conditions using laboratory testing equipment with sharp edges and controlled clearances. Shear loads on nickel alloys ranged from 113 to 131% of those for low-carbon steel in production shearing, based on tests using a power shear with a rake of 31 mm/m (⅜ in./ft) of blade length. More information on shearing is available in the article "Shearing of Sheet, Strip, and Plate" in this Volume.

Example 3: Forming and Slotting of Alloy X. The 88 flutes in the workpiece shown in Fig. 6 were finish formed and slotted one at a time with hand indexing in a 450 kN (50 tonf) mechanical press at the rate of one piece every 15 min, including setup. Slots were required to be within 0.5 mm (0.02 in.) of true position. The work metal was 1 to 1.1 mm (0.04 to 0.044 in.) alloy X sheet. Before the mechanical press operations, the sheet had been formed by a rubber-diaphragm process, electrolytically cleaned, annealed to 74.5 to 81.5 HR30T hardness, pickled, restruck in the forming press, and trimmed. The flutes were partially formed in this series of operations.

In choosing a method of finish forming, it was decided that the only way to form the flutes to the required shape was with a solid tool. The rubber-diaphragm forming process, however, was the best way to form the main contours of the part. The flutes could not be fully formed by a conventional die alone, because the percentage elongation exceeded the limits for alloy X (38 to 42% elongation in 50 mm, or 2 in.). By making use of the natural tendency of the blank to form wrinkles, the flutes were preformed during rubber-diaphragm forming. Although pressures were only enough to form them 75% complete, the amount of elongation needed in the final die-forming operation was lowered, and definite locations for flutes were provided. As a result, each flute could be produced in one stroke of the mechanical press. The tooling (Fig. 6, right side) consisted of a die and a cam-actuated punch of high-carbon, high-chromium tool steel hardened

Table 6 Minimum hole diameter for piercing selected nickel alloys

Sheet thickness, t		Minimum hole
mm	in.	diameter
0.46–0.86	0.018–0.034	$1.5t$
0.94–1.78	0.037–0.070	$1.3t$
1.98–3.56	0.078–0.140	$1.2t$
3.97	0.156	$1t$

to 58 to 60 HRC, as well as die inserts, stripper, and cam sections of lower-alloy, air-hardening tool steel. The punch pierced the slot and flattened the bulge above the flute. The stripper formed the flute when struck by the punch holder.

Deep Drawing

Nickel alloys can be drawn into any shape that is feasible with deep drawing steel. The physical characteristics of nickel alloys differ from those of deep drawing steel but not so much as to require different manipulation of dies for the average deep drawing operation.

Most simple shapes can be deep drawn in nickel alloys using dies and tools designed for use on steel or copper alloys. However, when intricate shapes with accurate finished dimensions are required, minor die alterations are necessary. These alterations usually involve increasing clearances and enlarging the radius of the draw ring or of the punch nose.

Double-Action Drawing. In drawing and redrawing of thin stock (≤1.6 mm, or 1/16 in.) into cylindrical shells with no ironing, the diameter reduction should be 35 to 40% on the first operation and 15 to 25% on redraws. If the walls are held to size, the first and second operations may be the same as suggested previously, but the amount of reduction should be diminished by approximately 5% on each successive redraw.

Although reductions of up to 50% can be made in one operation, this is not advisable because of the possibility of excessive shell breakage. Also, large reductions may open the surface of the metal and cause finishing difficulty.

The number of redraws that can be made before annealing depends on the alloy being drawn. Alloys with lower work-hardening rates (Fig. 1 and 2 and Table 2) can often be redrawn more than once without an intermediate anneal. Trial runs may be needed to determine when annealing is necessary.

Single-Action Drawing. As with all metals, the depth to which nickel alloys can be drawn in single-action presses without some means of blank restraint is controlled by the blank-thickness-to-diameter ratio. For single-action drawing without holddown pressure, the blank thickness should be at least 2% of the blank or workpiece diameter for reductions of up to 35%. With properly designed dies and sufficiently thick material, the reduction on the first (cupping) operation with a single-action setup may be made equal to those recommended for double-action dies—that is, 35 to 40%. Redraws should not exceed a 20% reduction.

If the shell wall is to be ironed, the increased pressure on the bottom of the shell usually necessitates a decrease in the amount of reduction to prevent shell breakage. With reductions of 5% or less, the shell wall may be thinned by as much as 30% in one draw. With medium reductions of approximately 12%, the thickness of the shell wall can be decreased by approximately 15%. If the wall is to be reduced by a large amount, the shell should first be drawn to the approximate size with little or no wall thinning and the ironing done last. If a good surface finish is desired, the final operation should have a burnishing effect with only a slight change in wall thickness.

Clearances. Because nickel alloys have higher strengths than do low-carbon steel of drawing quality, nickel alloys have greater resistance to the wall thinning caused by the pressure of the punch on the bottom of the shell. Consequently, greater die clearances are required than for steel if the natural flow of the metal is not to be resisted. However, the clearances required for nickel alloys are only slightly greater than those required for steel, and if dies used for steel have greater-than-minimum clearances, they are usually satisfactory for drawing nickel alloys, depending primarily on the mechanical properties of the alloy.

For ordinary deep drawing of cylindrical shells, a punch-die clearance per side of 120 to 125% of the blank thickness is sufficient and will prevent the formation of wrinkles. In the drawing of sheet thicker than 1.6 mm (1/16 in.), the general practice is to have the inside diameter of the draw ring larger than the diameter of the

punch by three times the thickness of the blank (150% of stock thickness per side).

Draw-Ring and Punch Radii. Because nickel alloys work harden rapidly, relatively large draw-ring and punch radii should be used, especially for the early operations in a series of draws. Nickel alloys require more power to draw than do steels; consequently, the punch imposes a greater stress on the bottom corner of the shell. Small punch radii cause thinning of the shell at the line of contact, and if such a shell is further reduced, the thinned areas will appear farther up the shell wall and may result in visible necking or rupture. Also, buffing a shell having thinned areas will cause the shell wall to have a wavy appearance. For redraws, it is preferable to draw over a beveled edge and to avoid round-edged punches except for the final draw.

The draw-ring radius for a circular die is principally governed by the thickness of the material to be drawn and the amount of reduction to be made. A general rule for light-gage material is to have the draw-ring radius from 5 to 12 times the thickness of the metal. Insufficient draw-ring radius may result in galling and excessive thinning of the wall.

Drawing Rectangular Shells. As with other materials, the depth to which rectangular shapes can be drawn in nickel alloys in one operation is principally governed by the corner radius. To permit drawing to substantial depths, the corner radii should be as large as possible. Even with large corner radii, the depth of draw should be limited to from two to five times the corner radius for alloys 400, 200, and 201, and to four times the corner radius for alloy 600 and alloy 75 (UNS N06075). The permissible depth also depends on the dimensions of the shape and on whether the shape has straight or tapered sides. The depth of draw for sheet less than 0.64 mm (0.025 in.) thick should not exceed an amount equal to three times the corner radius for alloys 400, 200, and 201 and should be less than that for alloy 600.

The corner radius on the drawing edge of the die should be as large as possible—approximately four to ten times the thickness of

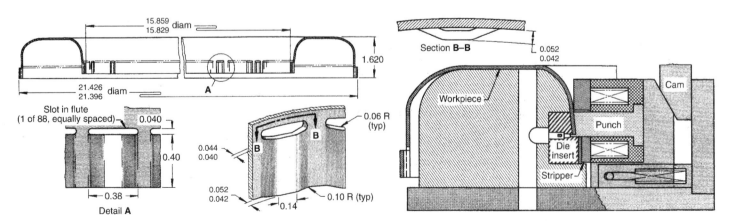

Fig. 6 Finish forming of flutes and piercing of slots one at a time in an alloy X workpiece using a mechanical press. Hardness of the workpiece was 74.5 to 81.5 HR30T (186 to 247 HV). Dimensions given in inches

the material. To avoid wrinkles around the top corner of the shape, it is essential that the blank not be released prematurely.

In redrawing for the purpose of sharpening the corners or smoothing out wrinkles along the sides, only a small amount of metal should be left in the corners.

Frequently, it is necessary to draw shapes on dies designed to make a deeper single draw than is practical for nickel alloys. With such dies, the general practice is to draw approximately two-thirds of the full depth, anneal the shape after this draw, and then complete the draw to full depth on the same dies. This same practice can be used to avoid wrinkling in drawing to lesser depths. More information on the deep drawing process is available in the article "Deep Drawing" in this Volume.

Hydroforming is also used to deep draw parts (Ref 11). A study on hydroforming of a nonsymmetric part of alloy 718 is described in Ref 12. In this study, a viscous material was used as the pressure medium. Viscous pressure forming can form deep-drawn parts with more uniform wall and less thinning than is obtainable with deep drawing dies, because the forming is done with near frictionless contact between the forming fluid and the workpiece. Hydroforming is also used to create bellows or flexible coupling out of nickel alloys. For additional information, see the article "Rubber-pad Forming and Hydroforming" in this Volume.

Spinning

Power spinning is preferred over manual spinning for nickel alloys (see the article "Spinning" in this Volume). However, thin material, particularly alloys 200 and 400, can be manually spun with no difficulty. Table 7 gives practical limits on blank thickness for manual spinning of six nickel alloys.

Tools. Except for small, light shapes, the required pressure cannot be exerted with the ordinary bar or hand tool pivoted on a fixed pin. Most shapes require the use of a tool that is mechanically adapted for the application of greater force, such as a compound-lever tool or roller tools that are operated by a screw. For small jobs, a ball-bearing assembly on the end of a compound lever can be used to make a good roller tool. Roller tools should be used whenever practical in order to keep friction at a minimum and exert maximum pressure. Roller tools should

Table 7   Minimum blank hardness and thickness for manual spinning of nickel alloys

| Alloy | Minimum hardness, HRB | Minimum thickness | |
		mm	in.
Nickel 200	64	1.57	0.062
Nickel 201	55	1.98	0.078
Alloy 400	68	1.27	0.050
Alloy 600	80	0.94	0.037
Alloy X-750	94	0.94	0.037

also be used to perfect contours in the spinning of press-drawn shapes.

When possible, tools used for spinning nickel alloys should be broader and flatter than those used for softer materials. The broader tool distributes plastic flow over a greater area and reduces overstraining. Except for this consideration, bar and roller tools should be designed the same as those used for spinning copper or brass.

Correct tool materials are essential for successful spinning. The most suitable material for bar tools is a highly polished, hard alloy bronze. Hardened tool steels are preferred for roller and beading tools. Chromium-plated hardened tool steel is recommended, because it decreases metal pickup by the tool. Tools of common brass and carbon steel, which are used for spinning softer materials, are unsatisfactory for use with nickel alloys.

Rotary cutting shears are preferred for edge trimming. If rotary shears are not available, hand trimming bars hard faced with a cobalt alloy may be used, but the trimming speed must be reduced. Hand trimming bars should be ground so that they have a back-rake angle of 15° to 20° from the cutting edge, and the edge must be kept sharp. A tool shaped like a thread-cutting lathe tool can be used for trimming. This tool also has a back rake from the cutting edge. With this type of tool, the material is not sheared off the edge; instead, the tool is fed into the side of the workpiece, and a narrow ring is cut from the edge. The workpiece should be supported at the back during all trimming operation.

Mandrels. Hardened cast iron and steel mandrels give longer life and better results than softer materials such as wood. Hard maple or birch mandrels may be used for intermediate operations if production quantities are small and tolerances are liberal.

Spinning nickel alloys over mandrels that are the same as those used for copper alloys will not necessarily result in spun shapes of exactly the same dimensions as those of the softer metal. Because of the greater springback, most nickel alloy shapes will have slightly larger peripheries than those of softer metals spun over the same mandrel.

Lubricants. Heavy-bodied, solid lubricants, such as yellow laundry soap, beeswax, and tallow, are recommended for spinning. These lubricants can be manually applied to the blank as it rotates. Blanks can be electroplated with 5 to 18 μm (0.2 to 0.7 mil) of copper to improve lubrication on difficult shapes.

Procedure. The spinning procedures for nickel alloys are essentially the same as that used for other metals (see the article "Spinning" in this Volume). Generally, when planning a spinning sequence for alloy 400, an increase in height of 25 to 38 mm (1 to 1½ in.) on the article being spun constitutes an operation if spinning is being done with a bar tool. Approximately twice that depth per operation may be obtained with a compound-lever or roller tool. The workpiece

should be trimmed and annealed before it is spun to greater depths.

A hard-surfaced mandrel should be provided for each operation so that the metal can always be pushed firmly against the surface of the mandrel. This procedure keeps the workpiece surface smooth and dense and ensures the best results in annealing. With an insufficient number of intermediate mandrels, the material is subjected to an excessive amount of cold working. This may result in either spinning a buckle into the material or formation of a pebbled surface. It is virtually impossible to smooth out the former by additional cold work, or to correct the latter by annealing.

Figure 7 illustrates the number of mandrels and annealing operations necessary for spinning deep cups from 0.94 mm (0.037 in.) thick alloy 200, 400, and 600 blanks using hand tools. Figure 7 also shows the amount of forming that can be done before annealing and between intermediate anneals. The spinnability of other alloys can be estimated from their relative work-hardening rates (Fig. 1 and 2) and from their tensile properties.

In spinning, the optimal speed of the rotating blank is governed by its diameter and thickness. Small, thin blanks can be spun at greater speeds than larger or thicker pieces. Nickel alloys are often spun at speeds of one-half to three-fourths those normally used in spinning the same shape from softer metals. Lathe speeds of 250 to 1000 rpm are usually satisfactory. Trimming speeds, by necessity, must be slow and are ordinarily done at the minimum speed of the lathe.

Example 4: Forming Alloy A-286 Tube by Spinning. The tube shown at the top of Fig. 8 was backward spun from a roll forging that had been solution annealed at 980 °C (1800 °F). A starting groove had been machined into the tube in a previous operation. Spinning was performed in three passes on a machine capable of spinning a part 1065 mm (42 in.) in diameter and 1270 mm (50 in.) in length. Backward spinning was used in preference to forward spinning because:

- The finished workpiece was longer than the mandrel.
- Forward spinning would have required a change in workpiece design to permit hooking over the mandrel.
- Backward spinning is faster than forward spinning.

For convenience, the flanges were left at both ends and trimmed off later. The flanges prevented bell-mouthing and permitted trimming of the portions likely to have small radial cracks.

Explosive Forming

Explosive forming has been used extensively to form sheet metal parts. Part tolerances

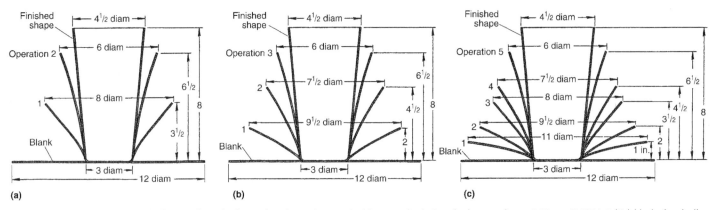

Fig. 7 Spinnability of three nickel alloys, as shown by the number of operations required for manual spinning of a deep cup from a 0.94 mm (0.037 in.) thick blank of each alloy. Workpieces were annealed between all operations. Dimensions given in inches. (a) Alloy 200. (b) Alloy 400. (c) Alloy 600

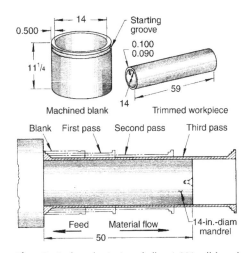

Fig. 8 Backward spinning of alloy A-286 roll-forged tube (hardness, 203 HV max). Dimensions given in inches

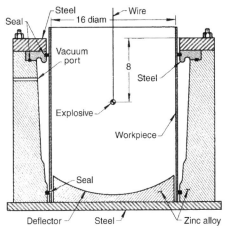

Fig. 9 Explosive forming of a case from 1.5 mm (0.060 in.) thick alloy A-286 sheet. Dimensions given in inches

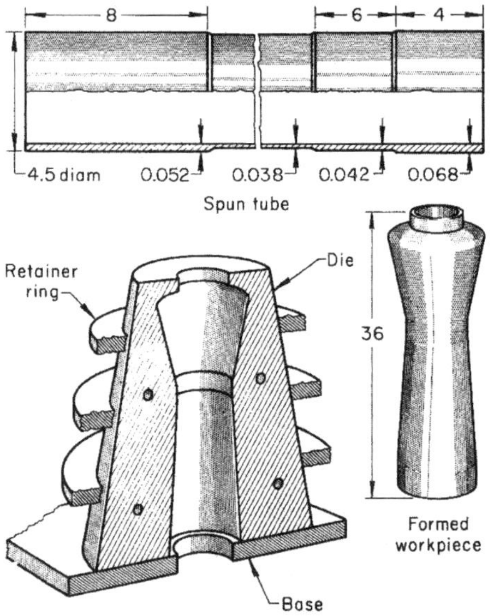

Fig. 10 Alloy 718 flame deflector formed from sheet 1.8 mm (0.072 in.) thick by explosive forming in three successive charges. Dimensions given in inches

obtainable from explosive forming can range from ±0.025 mm (0.001 in.) to ±0.25 mm (0.01 in.) depending on the size and complexity of the part. The article "High-Velocity Metal Forming" in this Volume describes the technique and capabilities of this method. Some examples of explosive forming to create parts from nickel and cobalt alloys are given as follows.

Safety in Explosive Forming. Operations involving explosives and pressure vessels are governed by state, county, and municipal regulations. The requirements and restrictions of these regulations should be taken into account in tool design and operational setup for explosive forming.

Example 5: Explosive Forming of Alloy A-286. A tubular workpiece was explosively formed inside a die (Fig. 9) to produce a part with an internal flange. If this part had been produced by other methods, such a flange would have had to be welded on. The A-286 sheet was rolled into a round cylinder 405 mm (16 in.) in diameter, welded, solution treated, and descaled. Tolerance on the diameter was ±0.75 mm (±0.03 in.).

Gel dynamite was the explosive used. Six shots were used to form the workpiece. For shots 1 and 2, 15 g of explosive was used; for shot 3, 18 g; for shot 4, 20 g; and for shots 5 and 6, 25 g. After three shots, the workpiece was solution annealed and descaled.

Example 6: Explosive Forming of Alloy 718. Fully annealed alloy 718 (UNS N07718) sheet was used to make the flame deflector shown in Fig. 10. The sheet was rolled onto a cylinder, with the grain direction at right angles to the long axis of the cylinder. A 115 mm (4.5 in.) outside diameter by 815 mm (32 in.) long tube was gas tungsten arc welded from the cylinder using alloy 41 filler rod. The weld was made flush on the inside, and the outside was ground flush to +0.13 mm (+0.005 in.). The tube was spun to the dimensions shown in Fig. 10, fully annealed at 955 °C (1750 °F), and grit blasted. An outstanding characteristic of this alloy is its slow response to age hardening, which enables it to be welded and annealed with no spontaneous hardening unless cooled slowly. Explosive forming of the flame deflector was accomplished by three successive charges in a

split die, and the workpiece was fully annealed after explosive forming.

Example 7: Explosive Forming of Alloy 25. Figure 11 shows the setup used for the explosive forming of a tail-pipe ball from alloy 25 sheet. The sheet was gas tungsten arc welded (butt) into a cylinder, and the shape was formed by three explosive charges. No annealing was done between welding and the first two shots of explosive forming, but after the first two shots (50 g of dynamite for each), the workpiece was withdrawn from the die, annealed at 1175 °C (2150 °F), and descaled. The workpiece was returned to the die for further forming. The third explosive charge used 62 g of dynamite. Tolerance on diameters was maintained within ±0.25 mm (±0.01 in.).

Explosive forming was preferred over forming on an expanding mandrel, because the mandrel left flats on the wall of the workpiece, and explosive forming did not.

Example 8: Alloy N-155 Exit Nozzle Produced by Tube Spinning and Explosive Forming. The exit nozzle shown in Fig. 12 was produced from fully annealed 3.4 mm (0.135 in.) thick alloy N-155 sheet. The sheet was rolled into a cylinder, with grain direction at right angles to the long axis, and was gas tungsten arc welded. The weld was ground flush on both the inside and outside, after which the cylinder was tube spun to the various wall thicknesses shown in Fig. 12. The workpiece was then placed in a die and explosively formed to the shape shown at right in Fig. 12.

The underwater explosive-forming technique was used, with a vacuum of 3 kPa (0.03 atm) between the workpiece and the die. The explosive charge was equal to 620 g of TNT and was placed at an average distance of 190 mm (7½ in.) from the workpiece walls. The first shot produced approximately 90% of the final shape. A second shot, using the same size charge, completed the workpiece, after which it was fully annealed.

Bending Sheet, Strip, and Plate

Table 8 lists minimum bend diameters for hot-rolled and annealed nickel alloy sheet, plate, and strip. In compiling these data, a sample was

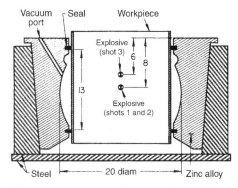

Fig. 11 Alloy 25 welded cylinder (sheet thickness, 1.7 mm, or 0.066 in.) in position for explosive forming. Dimensions given in inches

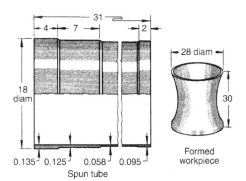

Fig. 12 Alloy N-155 exit nozzle produced by tube spinning and explosive forming. Dimensions given in inches

judged to have passed the 180° bend test if its surface showed no ductile fracturing. Because of the effects of various surface conditions and heat treatments on bending, the data in Table 8 should be regarded as general guidelines. Many of the materials can be bent in stages to tighter radii than those suggested, provided the initial bend is not too severe.

The importance of surface condition is demonstrated by the alloys from which scale or oxides must be removed to ensure successful bending. As indicated in Table 8, scale can be removed either by chemical or mechanical means, depending on the alloy.

Expanding/Tube Forming

Nickel alloy tubing can be expanded into tube sheets for heat exchanger applications by any conventional method. The oversize allowance on tube sheet holes to the nominal outside diameter of the tube should be kept to a minimum. The tube sheet hole should be 0.10 to 0.20 mm (0.004 to 0.008 in.) larger than the nominal outside diameter of the tube for tubing less than 38 mm (1½ in.) in outside diameter. For tubing 38 mm (1½ in.) or larger in outside diameter, the oversize allowance should be 0.23 to 0.25 mm (0.009 to 0.010 in.).

Procedure. Expanding may be done by drifting with sectional expanders or by rolling with three-roll expanders. Three-roll expanders are preferred. The ends of rolled-in tubing are flared in the conventional manner.

The tube sheet hole and both the outside and inside surfaces of the tube must be free of all foreign matter such as oxides, dirt, and oil. The ends of the tube should also be deburred before rolling. Lubrication should be provided between the rollers of the tool and the inside surface of the tube. Any sulfur-free mineral oil or lard oil,

either diluted or straight, can be used. Lubricants that contain embrittling or contaminating elements such as sulfur, zinc, or lead should be avoided, because of the difficulty in cleaning the finished assembly.

Controlled rolling equipment should be used to prevent overexpanding, which may distort the tube sheet and deform the tube sheet ligaments, causing loose-fitting tubes. This is particularly true when the tube has a higher hardness than the tube sheet or a significantly higher rate of work hardening.

Temper. The tube sheet should be harder than the tube being rolled into it. Otherwise, springback in the tube may be greater than in the tube sheet, causing a gap between the two when the expanding tool is removed. For this reason, tube sheets are usually supplied in the as-rolled or as-forged temper, and tube is supplied in the annealed temper. The need for the tube sheet to be harder than the tube is greatest when the thickness of the tube sheet is less than the outside diameter of the tube, and when the center-to-center spacing of the tubes (tube pitch) is less than 1¼ times the outside diameter of the tube or the outside diameter plus 6.4 mm (¼ in.), whichever is greater.

Stress-relieved tubing may be slightly harder than the tube sheet but can be expanded to form a satisfactory connection if greater care is exercised in expanding. For greater assurance of pressure tightness, a seal weld may be placed around the end of the tube after expanding. The stress-relieved temper is suitable for either welding or silver brazing.

Tubing in the annealed condition is used when optimal rolling or expanding characteristics are desired or for severe cold bending and flaring.

Example 9: Manufacture of Bellows from Nickel Alloys. Bellows manufacturing is a good illustration of the excellent fabricability of

Table 8 Minimum bend diameters for annealed sheet and strip and hot-rolled annealed plate

Alloys were bent 180°(a). Minimum bend diameters are given in terms of material thickness.

Alloy	UNS number	Product form(b)	Thickness, t mm	Thickness, t in.	Minimum bend diameter
Nickel 200	N02200	Sheet/strip	0.30–6.35	0.012–0.250	1t
		Plate	4.75–6.35	0.187–0.250	2t
Alloy 400	N04400	Sheet/strip	0.30–2.77(c)	0.012–0.109(c)	1t
			2.79–6.35	0.110–0.250	2t
		Plate	4.75–6.35(d)	0.187–0.250(d)	2t
Alloy 600	N06600	Sheet/strip	0.30–6.35	0.012–0.250	1t
		Plate	4.75–6.35	0.187–0.250	2t
Alloy 625	N06625	Sheet/strip	0.30–6.35	0.012–0.250	2t
		Plate	4.75–6.35	0.187–0.250	2t
Alloy 718	N07718	Sheet/strip	0.30–1.34	0.012–0.049	1t
			1.27–6.35	0.050–0.250	2t
Alloy 230	N06230	Strip	<1.37	<0.050	1t
		Sheet	1.27–4.75	0.050–0.187	1.5t
		Plate	4.75–6.35	0.187–0.250	2t
Alloy X-750	N07750	Sheet/strip	0.30–1.24	0.012–0.049	1t
			1.27–6.35	0.050–0.250	2t
Alloy 800	N08800	Sheet/strip	0.30–6.35(c)	0.012–0.250(c)	1t
		Plate	4.75–6.35(c)	0.187–0.250(c)	2t
Alloy 825	N08825	Sheet/strip	0.30–6.35(c)	0.012–0.250(c)	2t
		Plate	4.75–6.35(c)	0.187–0.250(c)	2t

(a) Bend test performed according to ASTM E 290 with a guided bend jig as described in ASTM E 190. (b) For plate, sheared edges were ground or machined. (c) Successful bending depended on surface condition of the samples, in particular to freedom from oxidation. (d) Samples were descaled.

many corrosion-resistant and high-temperature-resistant nickel alloys. Nickel alloy bellows are used in expansion joints and flexible couplings to accommodate mechanical and thermal stresses found in many industrial, aerospace, and automotive components requiring corrosion resistance from the operating environment. Fatigue-resistant high-nickel alloys are readily cold formed into intricate bellows assemblies. Typical sizes range from 6.5 cm (2.5 in.) diameter in automotive and aerospace applications to 100 cm (3 ft) in diameter in chemical and refining industry applications.

Coiled strip is the typical starting material for bellows. The strip is fed into a roll former to make the proper-diameter tube, which is then welded longitudinally with an automatic gas tungsten arc weld without a filler metal. During this autogenous weld, inert gas is supplied continuously to both sides of the weld to reduce the formation of deleterious oxides, nitrides, and carbides. After the welding is complete, the weld is planished between hard rollers to flatten the weld profile and increase the strength of the weld joint. Heat treatment of nickel alloys is typically not required prior to or after bellows forming. Once the welding and planishing are complete, the tube is trimmed to length for forming.

The most common methods for forming bellows are punch forming, roll forming, and hydraulic forming. In expanding mandrel or punch forming (Fig. 13), the welded tube is located around a flat circular die set that consists of several pie-shaped segments. The segments are forced radially outward with a cone-shaped ram, thereby expanding the tube to form a convolution.

In roll forming (Fig. 14), the welded tube is located over a flat circular inner roller that is forced radially outward between two outer rollers while the tube is rotating. The convolutions are formed one at a time in both punch forming and roll forming.

In hydraulic forming (Fig. 15), the welded tube is located inside of external dies or forming rings. The tube is internally pressurized, forcing the nickel alloy material to expand outward between the rings. As forming progresses, the rings move together. In hydraulic forming, several of the convolutions are formed at one time.

Once all of the convolutions are formed, excess material is trimmed from the bellows tangents, and the bellows are nondestructively tested to ensure that the part is free of defects. After inspection is complete, the parts are ready for installation.

Example 10: Superplastic Forming of Alloy 718. *Superplasticity* is a term used to describe large elongations in polycrystalline material, typically several hundred percent, obtained during tensile loading (see the article "Superplastic Sheet Forming" in this Volume). Generally, materials exhibiting superplasticity exhibit low resistance to plastic flow in specific temperature and strain-rate regions. The cost savings in producing superplastic parts can be considerable; the complex parts formed using superplastic

forming technology enable the elimination of extensive welding or other joining methods in addition to the lowering of inventory of parts and a simplification of the manufacturing process.

The high-strength nickel alloys used in the construction of commercial and military aircraft components are superplastically formed on a commercial basis. One noteworthy example is

alloy 718 (UNS N07718) sheet produced to a grain size of ASTM 10 or finer using rigorous controls of composition, melt practice, and rolling conditions. The resulting sheet product is provided as UNS N07719 to AMS 5950. The controlled production practice provides a sheet product that enables the product to have the two essential characteristics that make the alloy

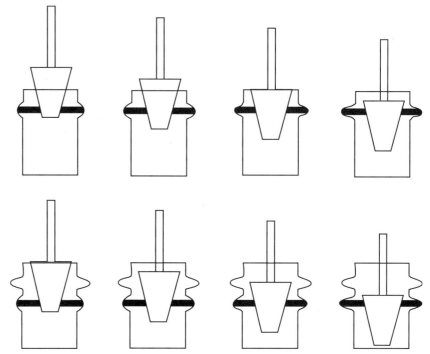

Fig. 13 Schematic of expanding mandrel or punch forming of bellows

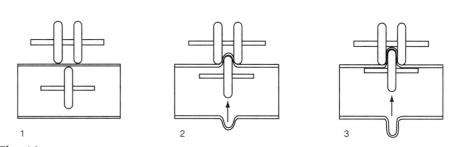

Fig. 14 Schematic for roll forming of bellows

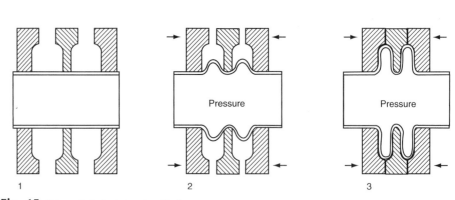

Fig. 15 Schematic for hydroforming of bellows

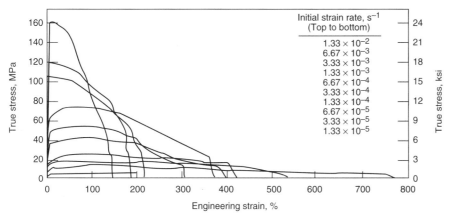

Fig. 16 Plot of true stress vs. engineering strain for ten strain rates at 954 °C (1750 °F) for alloy 718 (UNS N07719)

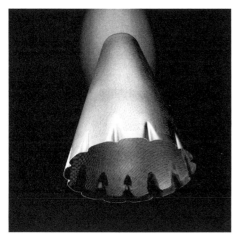

Fig. 17 Alloy 718 (UNS N07719) perforated sheet superplastically formed for an aircraft gas turbine tail cone skin. Courtesy of Special Metals Corporation

amenable to superplastic forming: grain-size stability over the time and temperature regime employed for superplastic forming, and a combination of low flow stress and significant ductility using typical superplastic forming process conditions (Fig. 16) (Ref 13–15). The superplastic forming process for alloy 718 produced to UNS N07719 properties involves the shaping of sheet between ceramic or metallic platens at temperatures near 954 °C (1750 °F) at an inert gas pressure of approximately 2.1 MPa (300 psi) for periods up to 4 h.

Figure 17 shows a UNS N07719 perforated sheet superplastically formed for an aircraft gas turbine tail cone skin. The sheet was 360° superplastically formed in a cold-wall furnace and illustrates the complex shapes that can be formed with this technique.

ACKNOWLEDGMENTS

This article was adapted from Ref 1 and 2. New material submitted by Don Tillack, Tillack Metallurgical Consulting, and H. Lee Flower, Haynes International Inc., was incorporated into this article. Additional guidance and information was received from Brian Baker of Special Metals Corporation and from Dwaine Klarstrom, Krishna Srivastava, Paul Crook, and Steve Matthews of Haynes International Inc.

REFERENCES

1. S.K. Srivastava and E.W. Kelley, Forming of Heat-Resistant Alloys, *Forming and Forging*, Vol 14, *Metals Handbook*, 9th ed., ASM International, 1988, p 779–784
2. R.W. Breitzig, Forming of Nickel-Base Alloys, *Forming and Forging*, Vol 14, *Metals Handbook*, 9th ed., ASM International, 1988, p 831–837
3. U.F. Kocks, Kinetics of Solution Hardening, *Metall. Trans. A*, Vol 16, 1985, p 2109–2129
4. P.S. Follansbee, J.C. Huang, and G.T. Gray, Low-Temperature and High-Strain-Rate Deformation of Nickel and Nickel-Carbon Alloys and Analysis of the Constitutive Behavior According to an Internal Stat Variable Model, *Acta Metall. Mater.*, Vol 38, 1990, p 1241–1254
5. T.A. Bloom, U.F. Kocks, and P. Nash, Deformation Behavior of Ni-Mo Alloys, *Acta Metall.*, Vol 33, 1985, p 265–272
6. *Atlas of Stress-Strain Curves,* 2nd ed., ASM International, 2002
7. B. Hicks, *The Development of Gas Turbine Materials,* Applied Science, 1981, p 229–258
8. S.K. Srivastava, Haynes International, unpublished research, 1985
9. P. Roamer, C.J. Van Tyne, D.K. Matlock, A.M. Meier, H.H. Ruble, and F.S. Suarez, Room Temperature Formability of Alloys 625LCF, 718 and 718SPF, *Superalloys 718, 625, 706 and Various Derivatives* (Pittsburgh, PA), June 15–18, 1997, TMS, p 315–329
10. W.W. Wood et al., Report AFML 64-411, Project 8-143, U.S. Air Force Materials Laboratory, Jan 1965
11. H.-U. Lücke, C. Hartl, and T. Abbey, Hydroforming, *J. Mater. Process. Technol.,* Vol 115, 2001, p 87–91
12. M. Ahmetoglu, J. Hua, S. Kulukuru, and T. Altan, Hydroforming of Sheet Metal Using a Viscous Pressure Medium, *J. Mater. Process. Technol.,* Vol 146, 2004, p 97–107
13. G.D. Smith and H.L. Flower, Superplastic Forming of Alloy 718, *Adv. Mater. Process.,* Vol 145 (No. 4), April 1994, p 32–34
14. G.D. Smith, S.R. Gregory, Y. Ma, Y. Li, and T.G. Langdon, Process Modeling the Superplastic Forming Behaviour of Inconel Alloy 718SPF, *Superalloys 718, 625, 706 and Various Derivatives,* E.A. Loria, Ed., TMS, 1997, p 303–314
15. Y. Huang and P.L. Blackwell, Microstructure Development and Superplasticity in Inconel 718 Sheet, *Mater. Sci. Technol.,* Vol 19, 2003, p 461–466

Forming of Refractory Metals

Revised by Louis E. Huber, Jr., Cabot Supermetals Corporation
Christopher A. Michaluk, Michaluk and Associates

REFRACTORY METALS are generally worked in small quantities. Production rates are low, each piece is handled separately, and the forming process is closely controlled.

Table 1 shows the composition of refractory alloys available as sheet. Typical conditions for bending 0.5 to 1.3 mm (0.020 to 0.050 in.) thick sheet are given in Table 2. Tensile-forming parameters for sheet materials are summarized in Table 3.

Formability

Niobium and tantalum alloys are usually formed at room temperature in the annealed (recrystallized) condition, although the stress-relieved alloys are sufficiently ductile for most forming operations. Work hardening, especially of the stronger alloys, often necessitates annealing after severe forming.

Strong alloys of niobium and tantalum, which are made in limited quantity, are not listed in Table 1. These alloys have varying degrees of brittleness at low temperature but can be formed by the same procedures used for molybdenum.

Molybdenum and tungsten are more difficult to form than niobium and tantalum, but if they are heated and certain precautions are taken, even complex parts can be formed. The greatest difficulty in forming these metals is their tendency toward brittle fracture (cracks and ruptures that occur with little or no plastic deformation) and delamination (a type of brittle behavior that produces cracks or ruptures parallel to the plane of the sheet). Tungsten can be hot formed only; it is brittle at room temperature.

At slow strain rates in tension and in bending, molybdenum and titanium-zirconium-molybdenum (TZM) alloys are ductile at room temperature, becoming brittle at lower temperatures. However, because of the high variable strain rates and triaxial stresses produced in the usual forming processes, these metals are usually hot formed in order to decrease the probability of brittle fracture. Molybdenum and tungsten blanks must have prepared edges to prevent cracking and splitting during forming.

Molybdenum and tungsten are generally supplied in the stress-relieved condition. Recrystallization increases the ductile-to-brittle transition temperature.

Effects of Composition on Embrittlement. Niobium and tantalum are severely embrittled by oxygen, nitrogen, and hydrogen on exposures at temperatures above approximately 300 °C (570 °F). However, the usual melting and processing techniques keep the metals pure enough for good formability.

Some niobium alloys are more resistant to grain growth at high temperature than high-purity niobium. Alloys such as Nb-1Zr and C-103 are dispersion-strengthened materials that resist grain growth at high temperature. These alloys have fine grain structure and elongate uniformly for forming and drawing operations.

Surface Contamination. The most common causes of surface contamination are failure to clean the surface properly and failure to provide the proper atmosphere in heat treatment. Niobium and tantalum are usually acid pickled, and they are heat treated in a vacuum or an inert gas atmosphere. Additional information on the heat treating of refractory metals is available in the article "Heat Treating of Refractory Metals and Alloys" in *Heat Treating,* Volume 4 of *ASM Handbook,* 1991.

Generally, the high-strength alloys are more severely embrittled by surface contamination than the lower-strength alloys. Molybdenum and tungsten are much less susceptible to surface

Table 1 Nominal compositions of refractory alloys available as sheet

Commercially pure niobium, tantalum, molybdenum, and tungsten sheets are also available.

Alloy	Composition, %				
	Zr	Ti	Hf	W	Other
Niobium alloys					
Nb-1Zr	1.0	...	...	...	...
FS-85	1.0	...	...	11.0	28.0Ta
C-103	...	1.0	10.0	...	...
C-129Y	...	...	10.0	10.0	0.10Y
Nb-752	2.5	...	...	10.0	...
Tantalum alloys					
Ta-2.5W	...	...	...	2.5	...
Ta-10W	...	...	...	10.0	...
Ta-8W-2Hf	...	...	2.0	8.0	...
Molybdenum alloys					
Mo-0.5Ti	...	0.5	...	...	0.03C
TZM(a)	0.1	0.5	...	...	0.03C

TZM, titanium-zirconium-molybdenum

Table 2 Conditions for the press-brake forming of refractory metal sheet 0.5 to 1.3 mm (0.020 to 0.050 in.) thick

Formed to a 120° bend angle in a 60° V-die at a ram speed of 254 to 3050 mm/min (10 to 120 in./min)

Metal or alloy	Forming temperature		Minimum bend radius(a)		Springback, degrees
	°C	°F	Test data	Preferred	
Niobium alloys (annealed)					
C-103, C-129Y	Room		<1t	1t	2–6
Tantalum alloys (annealed)					
Tantalum	Room		<1t	1t	...
Ta-10W	Room		<1t	2t	1–5
Molybdenum alloys (stress relieved)					
Mo-0.5Ti, TZM(b)	150	300	2t–5t	5t	3–8
Tungsten (stress relieved)					
Tungsten	315	600	2t–5t	5t	2–8

(a) *t*, sheet thickness. (b) TZM, titanium-zirconium-molybdenum

contamination by oxygen and nitrogen than niobium and tantalum.

Factors That Affect Mechanical Properties

The major variables that affect mechanical properties and formability are working temperature, temperature of anneals between operations, percentage of reduction after the final anneal, and temperature of final heat treatment.

Rolling. Refractory metal sheet is generally made by hot forging or extruding billets to make sheet bars, which are rolled to sheet at high temperature. The final rolling of niobium and tantalum alloys is done below 540 °C (1000 °F), often at room temperature. The cold-worked sheet is given a final recrystallization anneal to improve formability and ductility. The finish rolling of molybdenum and tungsten is done at high temperature, and final heat treatment is usually for stress relief only. Cross-rolled sheet is generally more formable, because cross rolling promotes planar isotropy desired for uniform deformation.

Heat Treatment. In the annealed (recrystallized) condition and in the ductile range, refractory metals behave much like low-carbon or interstitial-free steels. For example, recrystallized tungsten, although brittle at low temperature, has 35% uniform elongation and 50% total elongation at 400 °C (750 °F). Cold working strengthens molybdenum and tungsten and makes them less formable. Although molybdenum and tungsten are given a final stress relief, they retain their cold-worked structure.

Figure 1 shows the effect of heat treatment and strain hardening on the ductility and the ductile-to-brittle transition temperature range of unalloyed molybdenum. The curves in Fig. 1 show that the ductile-to-brittle transition for unalloyed molybdenum is between −18 and −45 °C (0 and −50 °F) for the stress-relieved condition, between 25 and −25 °C (80 and −10 °F) for the stretch-strained condition, and at approximately 25 °C (80 °F) for the recrystallized condition.

Transition Temperature. Niobium, tantalum, and their most frequently used alloys are readily formable, and they are ductile at temperatures as low as −195 °C (−320 °F). Molybdenum and TZM, in the stress-relieved condition, have transition temperatures just below room temperature. Molybdenum may fracture or delaminate at room temperature under the high deformation rates and stresses generally encountered in forming practice. Therefore, molybdenum alloys are generally formed at moderate-to-high temperatures. Stress-relieved tungsten has a transition temperature of 150 to 315 °C (300 to 600 °F), so that all forming of tungsten must be done at high temperatures.

Figure 2 shows how temperature changes the strength and elongation of a typical high-strength niobium alloy with good formability. A slight increase in temperature reduces yield strength but also reduces ductility. The ductility is

lowest at approximately 650 °C (1200 °F) and then increases with temperature. This reduced ductility is caused by strain aging, which is characteristic of body-centered cubic metals.

Figure 3 shows how temperature changes the ductility of four typical refractory metals. The ductility minimums lie between 540 and 1095 °C (1000 and 2000 °F). The tantalum and niobium alloys in Fig. 3 were annealed (recrystallized), and the molybdenum alloy and tungsten were stress relieved. Tests above 260 °C (500 °F) were conducted in a vacuum.

Effect of Temperature on Formability

Annealed niobium and tantalum alloys are formed at room temperature. Heating these alloys would reduce their formability because of strain aging and would cause oxidation and possible surface contamination.

Tungsten is brittle at room temperature. Therefore, thin tungsten sheet is formed at 315 to 540 °C (600 to 1000 °F), and thicker sheet or complex shapes are formed at 540 to 815 °C (1000 to 1500 °F) after stress relief.

Table 3 Elongation and true strain in forming refractory metal sheet of various thicknesses and grain directions

Results are based on testing of one heat of material for each alloy.

| Alloy | Condition(a) | Thickness | | Forming temperature | | Grain direction(b) | Elongation, % in | | True strain(c) | |
		mm	in.	°C	°F		25 mm (1 in.)	50 mm (2 in.)	ε_m	ε_c
Niobium alloys										
C-103	SR	0.76	0.030	Room		L	18.0	14.0	0.122	0.145
						T	6.0	4.0	0.041	0.046
	A	0.76	0.030	Room		L	30.0	24.0	0.152	0.232
						T	26.0	21.0	0.150	0.197
Tantalum alloys										
Ta-10W	A	1.0	0.040	Room		L	39.0	30.5	0.180	0.283
						T	38.0	30.0	0.197	0.282
Molybdenum alloys										
Mo-0.5Ti	SR	0.51	0.020	Room		L	19.0	15.0	0.102	0.164
						T	11.0	9.0	0.052	0.089
TZM(d)	SR	0.89	0.035	Room		L	19.0	15.0	0.074	0.130
				Room		L	16.0	11.5	0.060	0.075
Tungsten										
Tungsten	SR	0.89	0.035	595	1100	T	4.0	2.5	0.019	0.022
				980	1800	T	3.5	...	0.021	0.023

(a) A, annealed; SR, stress relieved. (b) L, longitudinal; T, transverse (c) ε_m, true strain at maximum load; ε_c, maximum true strain at maximum true stress. (d) TZM, titanium-zirconium-molybdenum

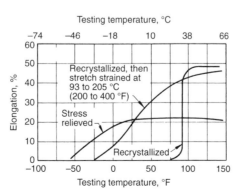

Fig. 1 Effect of heat treatment and strain hardening on the ductility and ductile-to-brittle transition temperature range of unalloyed molybdenum sheet as determined in tensile tests. The ductile-to-brittle transition occurs in the temperature range in the steep portion of the ductility curves.

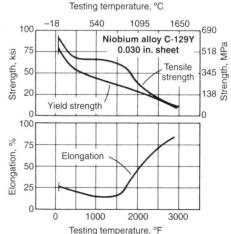

Fig. 2 Effect of temperature on strength and elongation of vacuum-annealed (recrystallized) niobium alloy sheet

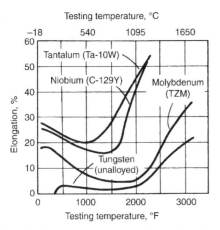

Fig. 3 Effect of temperature on the ductility of four refractory metals

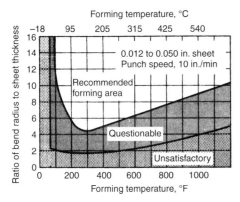

Fig. 4 Effect of temperature on the formability of Mo-0.5Ti sheet as indicated by the ratio of bend radius to sheet thickness

Molybdenum and molybdenum alloys, in thin sheets, can be cold formed to some extent, but heating helps to prevent fracture and delamination. As shown in Fig. 3, the TZM alloy is most ductile at 95 °C (200 °F). Further increases in temperature lessen ductility, because of strain aging. Most molybdenum is formed at 95 to 315 °C (200 to 600 °F), but thicker metal or complex shapes are formed at 315 to 650 °C (600 to 1200 °F).

Figure 4 shows the effect of heating on the bending of TZM sheet. This material is most formable at 95 to 205 °C (200 to 400 °F). Forming at lower temperatures may crack it. In one case, two sheets of TZM required bending. One sheet shattered when formed at room temperature, but the second sheet formed well in severe forming at 150 °C (300 °F). Erickson cup ductility tests indicated the two sheets to be of equal formability at room temperature.

Forming of Sheet

All of the common sheet-forming methods are used for refractory metals. However, the necessity of using elevated temperature in forming molybdenum and tungsten usually precludes the stretch forming and rubber-pad forming of these two metals.

Niobium and Tantalum. Almost all forming of niobium and tantalum is done at room temperature using conventional tools. A backup sheet is frequently used in a press brake to reduce galling or to provide support, so that the part will more closely follow the punch radius.

Niobium alloy C-103 is more ductile than type 310 stainless steel; it can be power spun to 60% reduction and deflects the rolls only half as much as type 310 stainless. Tantalum can be spun in thicknesses as great as 15.75 mm (0.620 in.).

Tantalum, niobium, and Nb-1Zr sheet can be readily drawn into cups, tubes, or other shapes amenable to drawing methods. However, these materials exhibit a serious tendency toward galling to tool surfaces at contact pressures that are almost always exceeded in the drawing process. The tendency toward galling increases with each redraw but can be significantly reduced or eliminated through careful attention to tool geometry, material surface condition, and workpiece lubrication.

The following rules should be observed when designing parts for deep drawing in these materials:

- Parts should have uniform wall thickness.
- A 25% thinning allowance should be made on tight corners or extreme reductions in diameter; thinning can be effectively controlled through careful tool design.
- The overall length should be less than nine times the smallest diameter in most cases, unless intermediate annealing is considered.
- Bends of up to 90° can be made on inside radii of one-half the material thickness; bends over 90° should have an inside radius of at least one wall thickness.
- Resultant surface finish is a function of grain size, severity of cold working, and original surface finish; it is difficult to "iron" to improve the surface finish because of galling.

One method used to reduce galling consists of oxidizing the material surface by heating in an open furnace to temperatures as high as 650 °C (1200 °F) for tantalum or as high as 625 °C (1155 °F) for niobium or Nb-1Zr. The thickness of the oxide produced is related to the length of time at temperature and the surface condition of the material. A soak of 1 to 2 min produces a surface with greatly reduced tendency to gall. In most cases, this oxide must be removed from completed workpieces by acid etching or other means. The oxide is quite stable and is strongly abrasive to draw tooling. Serious reduction in tool life can be a problem when oxides are used.

Standard chlorinated drawing compounds are appropriate lubricants for the drawing of these materials. Spray or flood-type lubrication systems help to ensure adequate lubrication for parts requiring multiple redraws. Particular care must be taken to lubricate die surfaces, because dry spots will initiate galling.

Trimming or blanking operations should be conducted with a minimum punch-to-die clearance. This reduces metal pickup on tool surfaces. Burr-free or nearly burr-free results can be achieved.

Molybdenum and Tungsten. All of the common sheet-metal forming methods except rubber-pad forming and stretch forming are used for molybdenum and tungsten. These metals are formed at high temperature to prevent the cracking and delamination that occur when forming at room temperature. Stretch forming has not been successful, because of the difficulties in adapting high temperatures to the process.

Proper preparation of the edges of blanks is necessary in the forming of molybdenum and tungsten. All edges in tension during forming must be rounded or polished to prevent fracture. Shearing and sawing may cause edge cracking and delamination, which must be removed before forming.

Power and manual spinning are used extensively to work tungsten sheet. Tungsten can be power spun in machines that are capable of power spinning steel. Complex contoured or deeply recessed parts are often produced by drop hammer forming. All work is done with heated tools and with work metal temperatures ranging from 595 to 1095 °C (1100 to 2000 °F). Many failures in forming tungsten are caused by the stressing of edge defects in the starting blanks. These defects originate in sawing, shearing, or blanking and are difficult to detect visually.

Tools and workpiece blanks are heated by electrical-resistance elements, heat lamps, and gas torches. An allowance for the difference in thermal expansion between steel dies and a tungsten or molybdenum workpiece is required for all parts whose shape or dimensions will be out of tolerance because of forming at high temperature. Hot work tool steels are satisfactory die materials. A bronze facing is recommended for steel dies if galling becomes a problem. Aluminum tooling is not recommended, because of its high thermal expansion.

Localized deformation and wrinkling of complex parts formed from molybdenum and tungsten generally result from poor die design and operation. These problems can be avoided by proper die clearance, staging and contours, and mechanical support. A steel backup sheet is sometimes used to ensure that the part more closely follows the punch or to reduce galling.

Forming of Preformed Blanks

Refractory metals that are preformed and welded into shaped blanks, cones, or cylinders can be formed by the same process used for unwelded blanks of common sheet metals. The welds must be of high quality to avoid defects or embrittlement. Chemical blanking, electrical discharge machining, abrasive cutting, and milling are preferred for making blanks. The following sequence of operations is generally used in preforming:

- Form intermediate shape
- Weld by the gas tungsten arc method
- Grind weld flush and inspect
- Stress relieve or anneal
- Form to final shape

Weldments of molybdenum and tungsten are generally formed at temperatures 95 to 150 °C (200 to 300 °F) higher than unwelded sheet of the same metals, and the weldments are usually stress relieved before forming. In some extreme applications, parts are stress relieved before and after welding, and after forming.

Lubricants

The types of lubricants used in the forming of refractory metals include oils, extreme-pressure lubricants, soaps, waxes, silicones, graphite, molybdenum disulfide, copper plating, and an acrylic enamel coating made by suspending powdered copper in acrylic resin.

Ordinary oils and greases are commonly used in the forming of niobium and tantalum, because these metals are generally formed at room temperature. Petrolatum is frequently used for severe forming operations. Solid lubricants and suspensions of suitable pigments, such as molybdenum disulfide with or without colloidal graphite, are used in the hot forming of molybdenum and tungsten. Chlorinated lubricants and others that decompose upon heating to form toxic or noxious fumes must not be used without proper safety precautions.

Forming of Titanium and Titanium Alloys

Joseph D. Beal, Rodney Boyer, and Daniel Sanders, The Boeing Company

TITANIUM AND ITS ALLOYS can be formed in standard machines to tolerances similar to those obtained in the forming of stainless steel. However, to reduce the effect of springback variation, improve accuracy, and to gain the advantage of increased ductility, the great majority of formed titanium parts are made by hot forming or by cold preforming and then hot sizing.

The following characteristics of titanium and titanium alloy sheet materials must be considered in forming:

- Notch sensitivity, which may cause cracking and tearing, especially in cold forming
- Galling (more severe than with stainless steel)
- Relatively poor ability to shrink (a disadvantage in some flanging operations)
- Potential embrittlement from overheating and from absorption of gases, principally hydrogen and oxygen (scale and the surface layer adversely affected by the slower penetration of oxygen can be removed readily)
- Limited workability—varies with the alloy
- Higher springback than that encountered in ferrous alloys at the same strength level

However, as long as these limitations are recognized and established guidelines for hot and cold forming are followed, titanium and titanium alloys can be successfully formed into complex parts.

The mechanical properties, and therefore the formabilities, of titanium and its alloys vary widely. For example, the tensile strength of different grades of commercially pure (CP) titanium ranges from 240 to 550 MPa (35 to 80 ksi); correspondingly large differences in formability are obtainable at room temperature. The tensile strength and ductility of CP titanium are largely dependent on its oxygen content. Table 1 lists the common designations, compositions, and selected mechanical properties of some titanium alloys.

Titanium Alloys

Alloy Ti-6Al-4V is the most widely used titanium alloy, accounting for approximately 60% of total titanium production. Unalloyed grades constitute approximately 20% of production, and all other alloys make up the remaining 20%. Selection of an unalloyed grade of titanium, an α or near-α alloy, an α-β alloy, or β metastable alloy depends on desired mechanical properties, forming method, service requirements, cost, and the other factors that enter into any materials selection process. The high solubility of the interstitial elements oxygen and nitrogen makes titanium rather unique among metals and creates problems that are not of concern in most other metals. For example, heating titanium in air at high temperature results not only in oxidation but also in solid-solution hardening of the surface as a result of inward diffusion of oxygen. A surface-hardened zone (alpha case) is formed. This layer is usually removed by machining or chemical milling prior to placing a part in service. The presence of this layer reduces fatigue strength and ductility.

Commercially pure titanium is usually selected for its excellent corrosion resistance, especially in applications in which high strength is not required. The yield strengths of CP grades (Table 1) vary from less than 170 to more than 480 MPa (25 to 70 ksi) as a result of variation in the interstitial, grain size, and impurity levels. Oxygen and iron are the primary variants in these grades; strength increases with increasing oxygen and iron contents and decreases with grain size. Grades of higher purity (lower interstitial content) are lower in strength, hardness, and transformation temperature than those higher in interstitial content.

Alpha and Near-Alpha Alloys. Alpha alloys that contain aluminum, tin, and/or zirconium are preferred for high-temperature and cryogenic applications. Alpha-rich alloys are generally more resistant to creep at high temperature than α-β or metastable β alloys. The extra-low-interstitial α alloys (ELI grades) retain ductility and toughness at cryogenic temperatures, and Ti-5Al-2.5Sn-ELI has been extensively used in such applications. Unlike α-β and metastable β alloys, α alloys cannot be strengthened by heat treatment. Generally, α alloys are annealed or recrystallized to remove residual stresses induced by cold working. Alpha alloys that contain small additions of β stabilizers (for example, Ti-8Al-1V-1Mo or Ti-6Al-2Nb-1Ta-0.8Mo) are sometimes classed as near-α alloys.

Although they contain some retained β phase, these alloys consist primarily of α and behave more like conventional α alloys than α-β alloys. They can, however, be strengthened by grain size.

Alpha-beta alloys contain one or more α stabilizers or α-soluble element plus one or more β stabilizers. These alloys retain more β phase after final heat treatment than near-α alloys; the specific amount depends on the amount of β stabilizers present and on the solution heat treating temperature and time.

Alpha-beta alloys can be strengthened by solution treating and aging. Solution treating is usually done at a temperature high in the two-phase α-β field and is followed by quenching in water, oil, or other suitable quenchant. The β phase present at the solution-treating temperature may be retained or may be partly transformed during cooling by either martensitic transformation or nucleation and growth. The specific response depends on alloy composition, solution-treating temperature (β-phase composition at the solution temperature), cooling rate, and section size. Solution treatment is followed by aging, usually in the 480 to 650 °C (900 to 1200 °F) range.

Solution treating and aging can increase the strength of α-β alloys 20 to 50%, or more, over the annealed or overage condition. Response to solution treating and aging depends on section size; alloys relatively low in stabilizers (Ti-6Al-4V, for example) have poor hardenability and must be quenched rapidly to achieve significant strengthening. For Ti-6Al-4V, the cooling rate of a water quench is not rapid enough to cause significant hardening of sections thicker than approximately 25 mm (1 in.). Hardenability increases as the content of β stabilizers increases. It should be noted that distortion can also be experienced during the solution-treating operation. The thinner the material, the greater the distortion when using water quench. It is best to use sheet material in the annealed condition to eliminate this problem.

Metastable beta alloys are richer in β-phase stabilizers and leaner in α stabilizers than α-β alloys. They are characterized by high hardenability, with β phase completely retained upon the air cooling of thin sections or the water quenching of thick sections. Beta alloys in sheet

Table 1 Designations, nominal compositions, and selected mechanical properties of selected titanium alloys

ASTM	MIL-T-9046F	MIL-T-9046H	MIL-T-9046J/ AMS-T-9046A	Minimum ultimate tensile strength		Minimum 0.20% yield strength		Elongation, %
				MPa	ksi	MPa	ksi	
Type I: Commercially pure titanium								
ASTM grade 2	Comp. A: Unalloyed (275 MPa, or 40 ksi, yield)	Comp. A: Unalloyed (275 MPa, or 40 ksi, yield)	CP-3	345	50	280–450	40–65	20
ASTM grade 4	Comp. B: Unalloyed (480 MPa, or 70 ksi, yield)	Comp. B: Unalloyed (480 MPa, or 70 ksi, yield)	CP-1	550	80	480–655	70–95	15
ASTM grade 3	Comp. C: Unalloyed (380 MPa, or 55 ksi, yield)	Comp. C: Unalloyed (380 MPa, or 55 ksi, yield)	CP-2	450	65	380–550	55–80	18
ASTM grade 1	...	...	CP-4	240	35	170–310	25–45	24
Type II: Alpha titanium alloy								
...	Comp. A: 5Al-2.5Sn	Comp. A: 5Al-2.5Sn	A-1	790	115	760	110	10
...	Comp. B: 5Al-2.5Sn (ELI)(a)	Comp. B: 5Al-2.5Sn (ELI)(a)	A-2	690	100	657	95	6
...	Comp. F: 8Al-1Mo-1V	Comp. F: 8Al-1Mo-1V	A-4	828	120	760	110	6
...	Comp. GT: 6Al-2Cb-1Ta-0.8Mo	Comp. GT: 6Al-2Cb-1Ta-0.8Mo	A-3	711	103	657	95	10
Type III: Alpha-beta titanium								
...	Comp. A: 8Mn	...	AB-6	863	125	761	110	10
ASTM grade 5	Comp. C: 6Al-4V	Comp. C: 6Al-4V	AB-1	897	130	830	120	8
...	Comp. D: 6Al-4V (ELI)(a)	Comp. D: 6Al-4V (ELI)(a)	AB-2	863	125	934	135	6
...	Comp. E: 6Al-6V-2Sn	Comp. E: 6Al-6V-2Sn	AB-3	1001	145	934	135	8
...	Comp. G: 6Al-2Sn-4Zr-2Mo	Comp. G: 6Al-2Sn-4Zr-2Mo	AB-4	897	130	830	120	8
...	...	Comp. H: 6Al-4V-SPL	AB-5	621	90	519	75	15
Type IV: Beta titanium								
...	Comp. A: 13V-11Cr-3Al	Comp. A: 13V-11Cr-3Al	B-1	911	132	872	126	8
...	...	Comp. B: 11.5Mo-6Zr-4.5Sn	B-2	690	100	623	90	10
...	...	Comp. C: 3Al-8V-6Cr-4Mo-4Zr	B-3	828	120	796	115	6
...	...	Comp. D: 8Mo-8V-2Fe-3Al	B-4	828	120	796	115	8
Ti-15V-3Al-3Cr-3Sn	...	...	...	790	115	770	112	20–25

(a) ELI, extra-low interstitial

form can be cold formed more readily than high-strength α-β or α alloys. An example of this is the Ti-15V-3Sn-3Cr-3Al alloy, which is formed almost exclusively at room temperature. After solution treating, metastable β alloys are aged at temperatures of 450 to 650 °C (850 to 1200 °F) to partially transform the β phase to α. The α forms as finely dispersed particles in the retained β and gives strength levels comparable to or superior to those of aged α-β alloys.

In the solution-treated condition (100% retained β), metastable β alloys have good ductility and toughness, relatively low strength, and excellent uniaxial formability. However, their formability is less for biaxial forming. Solution-treated metastable β alloys begin to precipitate α phase at slightly elevated temperatures and are therefore generally unsuitable for elevated-temperature service without prior stabilization or overaging treatment.

Table 2 Superplastic characteristics of titanium alloys

Alloy	Test temperature		Strain rate, s^{-1}	Strain-rate sensitivity factor, m	Elongation, %
	°C	°F			
Commercially pure titanium	850	1560	17×10^{-4}	...	115
α-β alloys					
Ti-6Al-4V	840–870	1545–1600	1.3×10^{-4} to 10^{-3}	0.75	750–1170
Ti-6Al-5V	850	1560	8×10^{-4}	0.70	700–1100
Ti-6Al-2Sn-4Zr-2Mo	900	1650	2×10^{-4}	0.67	538
Ti-4.5AL-5Mo-1.5Cr	870	1600	2×10^{-4}	0.63–0.81	>510
Ti-6Al-4V-2Ni	815	1500	2×10^{-4}	0.85	720
Ti-6Al-4V-2Co	815	1500	2×10^{-4}	0.53	670
Ti-6Al-4V-2Fe	815	1500	2×10^{-4}	0.54	650
Ti-5Al-2.5Sn	1000	1830	2×10^{-4}	0.49	420
Near-β and β alloys					
Ti-15V-3Sn-3Cr-3Al	815	1500	2×10^{-4}	0.50	229
Ti-13Cr-11V-3Al	800	1470	...	...	<150
Ti-8Mn	750	1380	...	0.43	150
Ti-15Mo	800	1470	...	0.60	100

Source: Ref 1

Superplastic Alloys

The workhorse superplastic titanium alloy is Ti-6Al-4V, and the state-of-the-art in titanium superplastic forming is largely based on this alloy. However, a number of titanium alloys, especially the α-β alloys, exhibit superplastic behavior. Many of these materials, such as Ti-6Al-4V, are superplastic without special processing. Table 2 gives data concerning the superplastic behavior of some titanium alloys and lists the characteristics used to describe superplastic properties in engineering alloys: strain-rate sensitivity factor, m, and tensile elongation. The m-value is a measure of the rate of change of flow stress with strain rate; the higher the m value of an alloy, the greater its superplasticity. Titanium alloys that have exhibited superplasticity but are not listed in Table 2 include Ti-3Al-2.5V (ASTM grade 9), Ti-4.5Al-1.5Cr-5Mo (Corona 5), and Ti-0.3Mo-0.8Ni (ASTM grade 12).

Metallurgical variables that affect superplastic behavior in titanium alloys include grain size, grain size distribution, grain growth kinetics, diffusivity, phase ratio in α-β alloys, and texture (Ref 1). Alloy composition is also significant, because it has a pronounced effect on α-β phase ratio and on diffusivity.

Grain size is known to have a strong influence on the superplastic behavior of Ti-6Al-4V (Ref 2, 3). This is illustrated in Fig. 1, which shows flow stress and strain-rate sensitivity factor, m, as a function of strain rate for Ti-6Al-4V materials with four different grain sizes. Increasing grain size increases the flow stress, reduces maximum m-value, and reduces the strain rate at which maximum m is observed.

Grain Size Distribution. Figure 2 shows flow stress versus strain rate for Ti-6Al-4V alloys with two different grain size ranges. The material with the smaller grain size distribution (lot A) exhibits significantly lower flow stresses than the material with the larger grain size distribution (lot B). Maximum m-value is also higher for the lot A material.

Grain growth kinetics affect superplastic behavior in direct relation to the grain size developed in the material. A study of grain growth effects on Ti-6Al-4V found that the flow hardening observed during constant strain-rate superplastic flow was the direct result of grain growth (Ref 3). It was also observed that grain growth accelerated with increasing strain rate. This grain growth causes an increase in flow stress and a decrease in maximum m-value.

Diffusivity is an important quantity in the superplastic flow of titanium alloys (and other engineering materials). The best indicator of diffusivity is usually activation energy, Q, which can be determined from the change in strain rate with temperature (Ref 1). Values of Q have been determined for several titanium alloys and for the α and β phases of titanium alloys. As indicated in Table 3, the activation energies determined from superplastic data are consistently higher than those for self-diffusion. It has been suggested that the higher Q values seen in superplastic alloys are due to the fact that the volume fraction of β phase in the alloys investigated increases with temperature, exaggerating the strain-rate increase and resulting in falsely high Q values. This complicates efforts to establish specific deformation mechanisms.

Phase Ratio Effects. Figure 3 shows that the two-phase (α-β) titanium alloys seem to exhibit greater superplasticity than other titanium alloys. The α and β phases are quite different in terms of crystal structure (hexagonal close-packed for α, and body-centered cubic for β) and diffusion kinetics. Beta phase exhibits a diffusivity approximately 2 orders of magnitude greater than that of α phase. For this reason alone it should be expected that the amount of β phase present in a titanium alloy would have an effect on superplastic behavior.

Figure 3 shows elongations and m-values for several titanium alloys as a function of the volume fraction of β phase present in the alloys. It can be readily seen that elongation values reach a peak at approximately 20 to 30 vol% β phase (Fig. 3a), while m-values peak at β contents of approximately 40 to 50 vol% (Fig. 3b). Because m is usually considered to be a good indicator of superplasticity, this discrepancy in the location of maximal of the curves in Fig. 3 may be surprising. It is believed that the difference stems from a grain growth effect during superplastic deformation. Beta phase is known to exhibit more rapid grain coarsening than α, and the maximum ductility may be the result of a balance between moderated grain growth (due to the presence of α phase) and enhanced diffusivity (due to the presence of β).

General Formability

Titanium metals exhibit a high degree of springback in cold forming. To overcome this characteristic, titanium must be extensively overformed or hot sized after cold forming. Aging or stress-relief operations are usually conducted on titanium alloys that are cold formed. Straightening can be done during the aging or stress-relief cycle with proper tools.

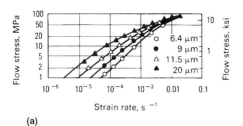

(a)

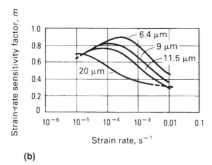

(b)

Fig. 1 (a) Flow stress and (b) strain-rate sensitivity factor, m, versus strain rate for Ti-6Al-4V materials with four different grain sizes. Test temperature: 927 °C (1700 °F). Source: Ref 3

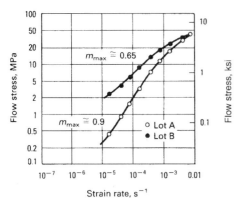

Fig. 2 Effect of grain size distribution on flow stress versus strain-rate data for Ti-6Al-4V at 927 °C (1700 °F). Lot A, average grain size of 4 μm and grain size range of 1 to 10 μm; lot B, average grain size of 4.6 μm but grain size range of 1 to >20 μm. Source: Ref 4

Table 3   Activation energies for superplastic deformation and self-diffusion in titanium alloys

Alloy	Temperature range		Activation energy (Q), kcal/mol	Ref
	°C	°F		
Ti-5Al-2.5Sn	800–950	1470–1740	50–65	2
Ti-6Al-4V	800–950	1470–1740	45	5
Ti-6Al-4V	850–910	1560–1670	45–99	6
Ti-6Al-4V	815–927	1500–1700	45–52	7
Ti-6Al-2Sn-4Zr-2Mo	843–900	1550–1650	38–58	8
Self-diffusion, α phase	...	...	40.4	9
Self-diffusion, β phase	...	...	36.5	10
Self-diffusion, β phase	...	...	31.3	11

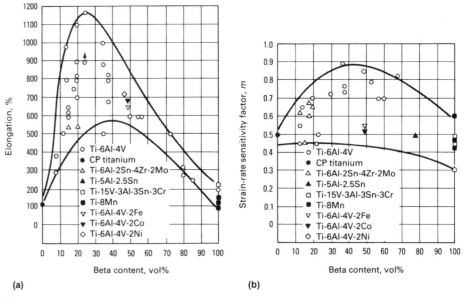

(a)   (b)

Fig. 3 (a) Elongation and (b) m-value as a function of β-phase content for several titanium alloys

Hot forming does not greatly affect final properties. Forming at temperatures ranging from 595 to 815 °C (1100 to 1500 °F) allows the material to deform more readily and simultaneously stress relieves the deformed material; it also minimizes springback. The net effect in any forming operation depends on total deformation and actual temperature during forming. Because titanium metals tend to creep at elevated temperature, holding under load at the forming temperature (creep forming) is another alternative for achieving the desired shape without the need to compensate for extensive springback.

The Bauschinger Effect. In all forming operations, titanium and its alloys are susceptible to the Bauschinger effect, which is a drop in compressive yield strength subsequent to tensile straining and vice versa. The Bauschinger effect, unlike the strain-hardening behavior observed in other metals, involves stress-strain asymmetry that results in hysteretis stress-strain loops such as those shown schematically in Fig. 4. The Bauschinger effect is most pronounced at room temperature; plastic deformation (1 to 5% tensile elongation) at room temperature always introduces a significant loss in compressive yield strength, regardless of the initial heat treatment or strength of the alloys. At 2% tensile strain, for example, the compressive yield strength of Ti-6Al-4V drops to less than one-half the value for solution-treated material. Increasing the temperature reduces the Bauschinger effect; subsequent full thermal stress relieving completely removes it.

The Bauschinger effect can be removed at temperatures as low as the aging temperature in solution-treated titanium alloys. Heating or plastic deformation at temperatures above the normal aging temperature for solution-treated Ti-6Al-4V will cause overaging; as a result, all mechanical properties will decrease.

Sheet Preparation

Before titanium sheet is formed, it should be inspected for flatness, uniformity, and thickness. Some manufacturers test incoming material for hardness, strength, and bending behavior.

Critical regions of titanium sheet should not be nicked, scratched, or marred by tool or grinding marks, because the metal is notch sensitive. All scratches deeper than the finish produced by 180-grit emery should be removed by sanding the surface. Edges of the workpieces should be smooth, and scratches, if any, should be parallel to the edge of the blank to prevent stress concentration that could cause the workpiece to break. To aid in forming, surface oxide or scale should be removed before forming.

Cleaning. Grease, oil, stencils, fingerprints, dirt, and all chemicals or residues that contain halogen compounds must be removed from titanium before any heating operation. Salt residues on the surface of the workpiece can cause hot-salt cracking in service or in heat treating; even the salt from a fingerprint can cause problems. Therefore, titanium is often handled with clean cotton gloves after cleaning and before hot forming, hot sizing, or heat treatment.

Ordinary cleaners and solvents such as isopropyl alcohol and acetone are used on titanium. Halogen compounds, such as trichlorethylene, should not be used, unless the titanium is pickled in acid after cleaning. Titanium that has been straightened or formed with tools made of lead or low-melting alloy should be cleaned in nitric acid. Detailed information on the cleaning of titanium is given in the article "Surface Engineering of Titanium and Titanium Alloys" in *Surface Engineering,* Volume 5 of *ASM Handbook,* 1994.

Grinding the sheet to final thickness leaves marks in titanium that can be moderated in an aqueous acid bath containing (by volume) 30% concentrated nitric acid and not more than 3% hydrofluoric acid. Failure to keep the ratio of nitric to hydrofluoric acid at 10 to 1 or greater (to suppress the formation of hydrogen gas during pickling), or the use of any pickling bath that produces hydrogen, can result in excessive hydrogen pickup. The acid bath should remove 0.025 to 0.075 mm (0.001 to 0.003 in.) of thickness from each surface to eliminate the marks made by abrasives. Titanium should be washed or cleaned before it is immersed in acid. The material left on the surface may protect the surface from the acid.

After thermal exposure, thin oxides that form at temperatures below 540 °C (1000 °F) can be removed by acid pickling. Very tenacious oxides may require grit blasting prior to pickling.

Exposure above 540 °C (1000 °F) forms an oxygen-rich surface layer on titanium. This surface is made up of a scale and an alpha case layer. The scale is normally reduced by the use of scale-inhibiting coatings put on prior to the thermal exposure. Scale can also be removed by abrasive blasting; however, this may cause distortion in thin parts. The removal of this scale prior to metal removal in the cleaning process improves the surface appearance. The alpha case layer is a brittle layer that must be removed to restore the base metal properties. Chemical milling, machining, or other similar methods accomplish the surface removal. Another method of limiting the formation of the oxygen-rich surface layer is to use a protective atmosphere or vacuum. Argon gas is a commonly used atmosphere for superplastic forming. Argon is applied to one side, with the other being exposed to air. This can cause an alpha case thickness difference from one side to the other.

Trimming

Blanking of titanium alloy sheet and plate 6.4 mm (0.25 in.) thick or less is done in a punch press. As with other metals, maximum blank size depends on stock thickness, shear strength, and available press capacity. Dies must be rigid and sharp to prevent cracking of the work metal. Hardened tool steel must be used for adequate die life.

In one application, holes 6.4 mm (0.25 in.) in diameter were punched in 1.02 to 3.56 mm (0.040 to 0.140 in.) thick annealed alloy Ti-6Al-4V sheet to within ±0.051 mm (±0.002 in.) of diameter and with surface roughness of less than 1.3 mm (0.05 in.). The best holes were produced with flat-point punches having 0.025 mm (0.001 in.) die clearance.

Shearing of titanium sheet up to 3.56 mm (0.140 in.) thick can generally be done without difficulty; with extra care, titanium sheet as thick as 6.0 mm (0.25 in.) can be sheared. Shears intended for low-carbon steel may not have enough holddown force to prevent titanium sheets from slipping. A sharp shear blade in good condition with a capacity for cutting 4.8 mm (0.188 in.) thick low-carbon steel can cut 3.2 mm (0.125 in.) thick titanium sheet. Cutters should be kept sharp to prevent edge cracking of the blank.

Sheared edges, especially on thicker sheet metal, can have straightness deviations of 0.25 to 5 mm (0.01 to 0.20 in.), usually because the shear blade is not stiff enough. Shearing can cause cracks at the edges of some titanium sheet thicker than 2.0 mm (0.080 in.). If cracks or other irregularities develop in a critical portion of the workpiece, an alternative method of cutting should be used, such as band sawing, abrasive waterjet cutting, or laser cutting (see the articles "Abrasive Waterjet Cutting" and "Laser Cutting" in this Volume).

Slitting of titanium alloy sheet can be done with conventional slitting equipment and with

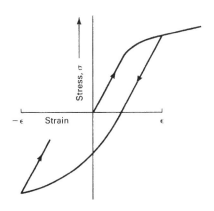

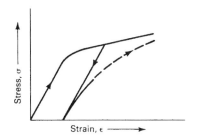

Fig. 4 Schematics showing two types of hysteresis stress-strain loops resulting from the Bauschinger effect in titanium alloys. Source: Ref 12

draw-bench equipment. Slitting shears are capable of straight cuts only; rotary shears can cut gentle contours (minimum radii: ~250 mm, or 10 in.). The process can be used for sheet thickness to 2.54 mm (0.100 in.).

Band sawing prevents cracking at the edges of titanium sheet but causes large burrs. This is usually followed by an edge-sanding operation to remove burrs.

Nibbling can be used to cut irregular blanks of titanium, but most blanks need filing or sanding after nibbling to produce a uniform edge.

All visual evidence of a sheared or broken edge on a part should be removed by machining, sanding, or filing before final deburring or polishing. All rough projections, scratches, and nicks must be removed. Extra material must be allowed at the edges of titanium blanks so that shear cracks and other defects can be removed. On sheared parts, a minimum of 0.25 mm (0.010 in.) must be removed from the edge; on punched holes, 0.35 mm (0.014 in.). On parts cut by friction band sawing or abrasive sawing, 6.35 mm (0.25 in.) or one thickness of sheet should be removed, whichever is the smaller.

The lay of the finish on the edges of sheet metal parts should be parallel to the edge surface of the blank, and sharp edges should be removed. Edges of shrink flanges and stretch flanges must be polished before forming. To prevent scratching the forming dies, edges of holes and cutouts should be deburred on both sides and should be polished where they are likely to stretch during forming.

Tool Materials and Lubricants

Tool materials for forming titanium are chosen to suit the forming operation, forming temperature, and expected quantity of production. The cost of tool material is generally a small fraction of the cost of tools, unless forming temperature is such that heat-resistant alloy tooling is required. Cold forming can be done with epoxy-faced aluminum, steel, or zinc tools. Hot forming tools are fabricated from ceramic, cast iron, tool steel, stainless steel, and nickel-base alloys. Materials selection is based on the service temperature, hardness of the material and tool at the forming temperature, and the number of parts being formed. Tool steel does not work well when temperatures are above the tempering temperature of the steel or above the temperatures where oxides form on the tool.

Titanium alloys are often formed in heated dies and presses that have a slow, controlled motion and that can dwell in the position needed during the press cycle. Hot forming is sometimes done in dies that include heating elements or in dies that are heated by the press platens. Press platens heated to 650 °C (1200 °F) can transmit enough heat to keep the working faces of the die at 425 to 480 °C (800 to 900 °F). Other methods of heating include electrical-resistance heating and the use of quartz lamps and portable furnaces.

Tool materials for the superplastic forming of titanium alloys are a special case (see the section "Superplastic Forming" in this article). They must be able to withstand the high temperatures (870 to 925 °C, or 1600 to 1700 °F) required for superplastic forming. Cast ceramics, 22-4-9 stainless steel (Fe-0.5C-22Cr-9Mn-4Ni), and 49M steel are used for this purpose. Figure 5 shows a typical tool used at elevated temperatures, in the 870 to 925 °C (1600 to 1700 °F) range. The heavy scale and oxide layer is due to the long exposure at elevated temperatures. Dies are usually cleaned between production runs to keep the surface smooth and ensure an acceptable finish on the parts.

Lubricants. Galling is the most severe problem to be overcome in hot forming. Lubricants may react unfavorably with titanium when it is heated, although molybdenum disulfide suspended in a volatile carrier, colloidal graphite, and graphite-molybdenum disulfide mixtures have been successfully used. Boron nitride slurries also are used. If the lubricant reacts with oxidation products to produce a tenacious surface coating, it must be removed by sandblasting with garnet grit or 120-mesh aluminum oxide, followed by acid pickling. Parts can be preformed cold and hot sized to minimize the galling effects seen in hot forming parts; however, this is not always practical.

Boron nitride is the preferred temperature-resistant lubricant because of its higher lubricity, as well as ease of application and removal. Other lubricants used for hot forming have a graphite, molybdenum disulfide, or Y_2O_3 base.

Lubricants for the cold forming of titanium are generally similar to those used for the severe forming of aluminum alloys (see the articles "Forming of Aluminum Alloys" and "Selection and Use of Lubricants in Forming of Sheet Metal" in this Volume). Tool materials and lubricants for the cold and hot forming of titanium alloys are given in Table 4.

Cold Forming

Commercially pure titanium and the most ductile metastable β titanium alloys, such as Ti-15V-3Sn-3Cr-3Al and Ti-3Al-8V-6Cr-4Zr-4Mo, can be formed cold to a limited extent. Alloy Ti-8Al-1Mo-1V sheet can be cold formed to shallow shapes by standard methods, but the bends must be of larger radii than in hot forming and must have shallower stretch flanges. The cold forming of other alloys generally results in

Fig. 5 Typical tool finish after being exposed to elevated temperatures. This hot forming die has a heavy scale and oxide layer caused by long exposure to high temperatures. Dies must be cleaned using abrasives in between production runs in order to avoid mark-off being transferred to the part surface. The light area is alpha case that has been migrated to the die from the titanium parts.

excessive springback, requires stress relieving between operations, and requires more power.

Titanium and titanium alloys are commonly stretch formed without being heated, although the die is sometimes warmed to 150 °C (300 °F). For the cold forming of all titanium alloys, formability is best at low forming speeds.

Hot sizing and stress relieving are ordinarily needed to improve part contour, reduce stress, and avoid delayed cracking and stress corrosion. Stress relief is also needed to restore compressive yield strength after cold forming. Hot sizing is often combined with stress relieving, by holding the workpiece in fixtures or form dies to prevent distortion. Stress-relief treatments for CP titanium and some titanium alloys are given

in Table 5. Hot sizing for shorter times than reflected in the table will remove the springback on some materials. This would indicate that shorter times may be acceptable. Detailed information on the heat treatment of titanium alloys is available in the article "Heat Treating of Titanium and Titanium Alloys" in *Heat Treating,* Volume 4 of *ASM Handbook,* 1991.

The only true cold-formable titanium alloy is Ti-15V-3Sn-3Cr-3Al. Hot sizing is usually not used for this alloy; however, properties must be developed with an aging treatment (8 h at 540 °C, or 1000 °F, is typical). Because of the high springback rates encountered with this alloy, more elaborate tooling must be used. Hot sizing can be used at the solution-treatment

temperature, followed by air cooling. This helps to solve the springback problems seen during the aging process. A restraint fixture can also be used to straighten during aging.

Hot Forming

Heating titanium increases formability, reduces springback, takes advantage of a lesser variation in yield strength, and allows for maximum deformation with minimum annealing between forming operations. Severe forming must be done in hot dies, generally with preheated stock. Figure 6 shows the removal of a set of curved channels after being hot formed from a flat sheet. The flat sheet is located in the die and allowed to heat up to temperature. Pressure is slowly applied, bringing down the matching punch and holding for 10 min under pressure prior to removal. Hot forming is ordinarily done at 730 °C (1350 °F) for Ti-6Al-4V material.

The greatest improvement in the ductility and uniformity of properties for most titanium alloys is at temperatures above 540 °C (1000 °F). At still higher temperatures, some alloys exhibit superplasticity (see the section "Superplastic Forming" in this article). However, contamination is also more severe at the higher temperatures. Above approximately 870 °C (1600 °F), forming should be done in vacuum or under a protective atmosphere, such as argon, to minimize oxidation. When done in air, metal removal is required to remove the oxygen-rich layer that forms on the surface of the titanium.

As indicated in Table 6, most hot forming operations are done at temperatures above 540 °C (1000 °F). For applications in which the utmost in ductility is required, temperatures below 315 to 425 °C (600 to 800 °F) are usually avoided. Alpha-beta alloys should not be formed above the β-transus temperature.

Temperatures generally must be kept below 815 °C (1500 °F) to avoid marked deterioration in mechanical properties. Superplastic forming, however, is performed at 870 to 925 °C (1600 to 1700 °F) for some alloys, such as Ti-6Al-4V. At these temperatures, care must be taken not to exceed the β-transus temperature of Ti-6Al-4V. Heating temperature and time at temperature must be controlled so that the titanium is hot for the shortest time practical and the metal temperature is in the correct range.

Reference 14 gives details about forming and the tolerance that can be expected, as well as some of the strength effects. The equipment used is described in detail. The information was generated in 1968 and reflects much of the technology of the day. Some of the forming tools have been improved, and some have not changed.

Hot sizing is used to correct inaccuracies in shape and dimensions in preformed parts. Hot forming takes a flat blank and forms it to the final shape. Hot sizing uses the creep-forming principle to force irregularly shaped parts to assume

Table 4 Tool materials and lubricants used for forming titanium alloys

Operation(s)	Tool materials	Lubricants
Cold forming		
Press forming, drawing, drop hammer forming	Cast zinc die or lead punch with stainless steel caps	Graphite suspension in a suitable solvent
Press-brake forming	4340 steel (36–40 HRC)	Graphite suspension in a suitable solvent
Contour roll forming, three-roll forming	AISI A2 tool steel	SAE 60 oil
Stretch forming	Epoxy-faced cast aluminum, cast zinc, cast bronze	Grease-oil mixtures, wax; 10 : 1 wax-graphite mixture
Hot forming		
Press forming, drawing, drop hammer forming	High-silicon cast iron, stainless steels, heat-resistant alloys	Graphite suspension, boron nitride
Sizing	Low-carbon steel, high-silicon gray or ductile iron, AISI H13 tool steel, stainless steels, heat-resistant alloys	Graphite suspension, boron nitride
Press-brake forming	AISI H11 or H13 tool steel, heat-resistant alloys	Graphite suspension, boron nitride
Contour roll forming, three-roll forming	AISI H11 or H13 tool steel	Graphite suspension, boron nitride
Stretch forming	Cast ceramics, AISI H11 or H13 tool steel, high-silicon gray iron	Graphite suspension, 10:1 wax-graphite mixture, boron nitride
Superplastic forming	Ceramics, 22-4-9 stainless steel, 49M heat-resistant steel	Boron nitride

Table 5 Stress-relief schedule for titanium and titanium alloys

Alloy	Stress-relief temperature		Time, min
	°C	°F	
Commercially pure titanium (all grades)	480–595	900–1100	15–240
Alpha alloys			
5Al-2.5Sn	540–650	1000–1200	15–360
5Al-2.5Sn (ELI)(a)	540–650	1000–1200	15–360
6Al-2Cb-1Ta-0.8Mo	540–650	1000–1200	15–60
8Al-1Mo-1V	595–760	1100–1400	15–75
11Sn-5Zr-2Al-1Mo	480–540	900–1000	120–480
Alpha-beta alloys			
3Al-2.5V	370–595	700–1100	15–240
6Al-4V	480–650	900–1200	60–240
6Al-4V (ELI)(a)	480–650	900–1200	60–240
6Al-6V-2Sn	480–650	900–1200	60–240
6Al-2Sn-4Zr-2Mo	480–650	900–1200	60–240
5Al-2Sn-2Zr-4Mo-4Cr	480–650	900–1200	60–240
6Al-2Sn-2Zr-2Mo-2Cr-0.25Si	480–650	900–1200	60–240
Metastable beta alloys			
13V-11Cr-3Al	705–730	1300–1350	30–60
3Al-8V-6Cr-4Mo-4Zr	705–760	1300–1400	30–60
15V-3Al-3Cr-3Sn	790–815	1450–1500	30–60
10V-2Fe-3Al	675–705	1250–1300	30–60

(a) ELI, extra-low interstitial. Source: Ref 13

the correct shape against a heated die by the controlled application of horizontal and vertical forces over a period of time. Some buckles and wrinkles can be removed from preformed parts in this way. A combination of creep and compression forming is used when reducing bend radii by hot sizing. The effect of temperature on the properties of the metal may limit the maximum useful temperature. Figure 7 shows the setup for creep forming a B-737 part using a hot ceramic die in a conventional furnace. The die is heated up in the furnace, then rolled out and separated. The blank is placed in the tool, the die is then closed up, and weight is added to the top. The tool and weights are then rolled back into the furnace and held at temperature for a period of time sufficient to allow for the creep forming of the part to contour, prior to rolling the tool out,

removing the formed part, and placing another blank into the tool.

Hot platen presses are commonly used for the hot sizing and forming of titanium. The tooling is designed for holding the workpiece to the required shape for the necessary time at temperature. Hot forming/sizing in hot platen presses is done in the following sequence of operations:

- Parts are usually cleaned and coated with a scale inhibitor.
- Parts are loaded on hot forming tools, the press closed, and the parts allowed to heat up prior to applying the forming force.
- Force is applied through the platens and auxiliary side rams as required and held to complete the forming/annealing cycle.

- Parts are removed and cooled in a uniform manner. Hot parts are very susceptible to handling distortion.

Some hot forming temperatures are high enough to age a titanium alloy. Heat-treatable α-β alloys generally must be resolution heat treated after hot forming. Some of the metastable β alloys have solution temperatures in the hot forming range and can be resolution heat treated during the hot forming operation. Solution heat treating thin-gage alloys that require water quench is risky because of distortion.

Hot forming has the advantage of improved uniformity in yield strength, especially when the forming or sizing temperature is above 540 °C (1000 °F). However, care must be taken to limit the accumulation of dimensional errors resulting from:

- Differences in thermal expansion
- Variations in temperature
- Dimensional changes from scale formation
- Changes in dimensions of tools
- Reduction in thickness from chemical pickling operations

Fig. 6 Hot-formed parts being removed from a hot press

Table 6 Temperatures for the hot forming and annealing of titanium alloys

Material	Annealing/forming temperature		Soak time, min
	°C	°F	
Commercially pure titanium			
All grades	650–815	1200–1500	15–120
Alpha alloys			
5Al-2.5Sn	705–845	1300–1550	10–120
5Al-2.5Sn (ELI)(a)	705–900	1300–1650	10–120
6Al-2Cb-1Ta-0.8Mo	790–900	1450–1650	30–120
8Al-1Mo-1V	760–815	1400–1500	60–480
Alpha-beta alloys			
3Al-2.5V	650–790	1200–1450	30–120
6Al-4V	705–870	1300–1600	15–60
6Al-4V (ELI)(a)	705–870	1300–1600	15–60
6Al-6V-2Sn	705–815	1300–1500	10–120
6Al-2Sn-4Zr-2Mo	870–925	1600–1700	10–60
6Al-2Sn-2Zr-2Cr-2Mo	690–870	1275–1600	15–360
Metastable beta alloys			
13V-11Cr-3Al	760–815	1400–1500	10–60
3Al-8V-6Cr-4Mo-4Zr	760–925	1400–1700	10–60
15V-3Al-3Cr-3Sn	760–815	1400–1500	3–30

(a) ELI, extra-low interstitial. Source: Ref 13

Superplastic Forming

The superplastic forming of titanium is currently being used to fabricate a number of sheet metal components for a range of aircraft and aerospace systems. Hundreds of parts are in production, and significant cost savings are being realized through the use of superplastic forming. Additional advantages of superplastic forming over other forming processes include the following:

- Very complex part configurations are readily formed.
- Lighter, more efficient structures are possible.
- It is performed in a single operation, reducing fabrication labor time.
- Depending on part size, more than one piece can be produced per machine cycle.
- The force needed for forming is supplied by a gas, resulting in the application of equal amounts of pressure to all areas of the workpiece.

Superplastic forming is similar to vacuum forming of plastics. A computer system is used to control the gas pressure so that the part forms into the cavity at a constant strain rate. The Ti-6Al-4V material is generally used for this process; however, there are other alloys that work. The material needs to have fine, equiaxed grains and high elongation at the elevated temperature.

The superplastic forming process puts the part on top of a cavity die, as shown in Fig. 8(a). With the tool and blank heated up to forming temperatures in the superplastic range, gas pressure is applied at a predetermined rate to keep a constant strain rate. As the part is formed down into smaller features, the pressure is increased to maintain the strain rate, and the thickness decreases. The resultant process produces a part that has thinned out based on the geometry of the die (Fig. 8b). There are variations to this process to improve thickness distributions in die design and how the parts are formed.

Figure 9 shows a part blank being loaded onto a hot die in a shuttle press. The lower platen shuttles out on a track to make loading and unloading much easier. The operators wear reflective suits to protect them from the thermal exposure. The tool only needs to reflect one side of the part, whereas, in hot sizing the tool matches both sides of the part.

The limitations of the process include:

- Heat-resistant tool materials are required.
- Equipment that can provide high temperatures and tonnage to balance forming pressures is necessary.
- Long preheat times are necessary to reach the forming temperature.
- A protective atmosphere, such as argon, is helpful.

Several forming processes are used in the superplastic forming of titanium alloys. Among these are blow forming, vacuum forming, thermoforming, deep drawing, and superplastic forming/diffusion bonding (see the section "Superplastic Forming/Diffusion Bonding" in this article). All of these processes are discussed in more detail in the article "Superplastic Sheet Forming" in this Volume.

Superplastic Forming/Diffusion Bonding

The superplastic forming process can be enhanced with diffusion bonding (solid-state joining). Both processes require similar conditions, such as heat, pressure, clean surfaces, and an inert environment. The combined process is referred to as superplastic forming/diffusion bonding (SPF/DB). Diffusion bonding can be carried out as the first part of the superplastic forming cycle, thus eliminating the need for welding or brazing for complex parts.

The SPF/DB process has greatly extended the applicability of superplastic forming. Using SPF/DB, a sheet can be diffusion bonded and formed onto preplaced details, or two or more sheets can be bonded and formed at selected locations. Figure 10 illustrates the SPF/DB process for three-sheet parts.

Diffusion bonding can be applied only to selected areas of a part by using a stop-off material (Fig. 10 step 1, and Fig. 11) that is placed between the sheets at locations where no bonding is desired. Suitable stop-off materials depend on the alloy being bonded and the temperatures employed; yttria and boron nitride have been successfully used. The powder is mixed and applied to the part in selected areas using the silk screening process (Fig. 11). Figure 10, step 2 shows the sheets sealed into an airtight pack. In step 3, the pack is bonded

Fig. 7 Creep forming a part for the B-737 in a ceramic die using a conventional furnace. The titanium is pushed into the correct shape through the application of heat, weight, and time.

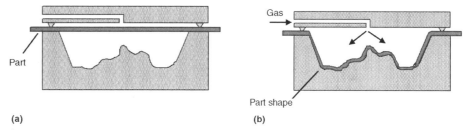

(a)   (b)

Fig. 8 Superplastic forming of titanium. (a) Setup at the start of the forming cycle. (b) After forming is completed

Fig. 9 Loading a sheet of titanium into a superplastic forming die

together under pressure and temperature. In step 4, the pack is inflated to start forming areas where the stop-off material was applied. Step 5 completes the forming cycle, fully forming the part prior to removal.

Diffusion bonding in combination with superplastic forming can produce lightweight panel structures, as shown in Fig. 12.

Superplastic forming and SPF/DB are gaining acceptance in the aircraft/aerospace industry. Figure 13 shows the increase in applications for superplastically formed titanium parts in four military aircraft since 1980; applications for commercial aircraft and in the aerospace industry also are increasing. Inspection of the bond is usually done with ultrasonic inspection methods.

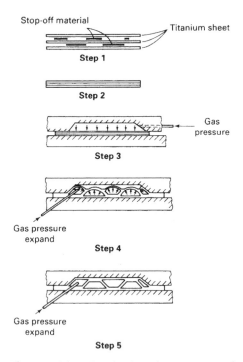

Fig. 10 Schematic showing the sequence of operations for superplastic forming/diffusion bonding of three-sheet titanium parts

Mechanical analysis of the joint and bond is based on part configuration.

Applications range from simple clips and brackets to airframe components and other load-bearing structures. Figures 14 and 15 show current applications for superplastically formed parts and illustrate the cost and weight savings that can be realized by using superplastic forming. Reducing the part count and assembly costs and making a more structurally efficient part result in cost and weight savings.

Press-Brake Forming

Titanium alloys cold formed in a press brake behave like work-hardened stainless steel, except that springback is considerably greater (see the article "Forming of Stainless Steel" in this Volume). If bend radii are large enough, forming can be done cold. However, if bend radii are small enough to cause cracking in cold forming, either hot forming or the process of cold forming followed by hot sizing must be used.

The setup and tooling for press-brake air bending are relatively simple because the ram stroke determines the bend angle. The only tooling adjustments are the span width of the die and the radii of the punch. The span width of the

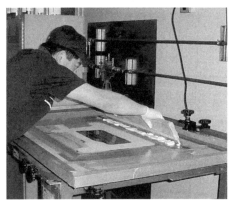

Fig. 11 Typical silk screen application of the stop-off material

die affects the formability of bend specimens and is determined by the punch radius and the work metal thickness, as shown in Fig. 16. Acceptable conditions for dies in press-brake forming are shown as the shaded area between the upper and lower limits in Fig. 16.

The minimum bend radius obtainable in press-brake forming depends on the alloy, work metal thickness, and forming temperature (Table 7). Springback in press-brake forming depends on the ratio of punch radius (bend radius) to stock thickness and on forming temperature, as shown in Fig. 17 for alloy Ti-6Al-4V (Fig. 17 is not to be used for minimum bend radii).

Hot-brake forming puts linear bends in a sheet by heating up the blank, then forming in a cold press brake. When this technique is used, a stress relief follows because the forming takes place quicker than the stress-relief time required to prevent springback of the part. Springback appears to be approximately the same with the blank heated as with a cold blank when the forming takes place very quickly. This process appears to work well when the bend radii is three times the thickness or larger in Ti-6Al-4V material. It is difficult to determine the temperature at which the forming takes place.

The deep drawing of titanium alloys is limited to the more formable alloys, such as CP titanium in the lower-strength grades. Superplastic forming can also be used for deep drawing; however, a draft angle is usually required because it is difficult to remove parts from a shape that has vertical sides, unless there is a guided removal system that keeps the part aligned.

However, general guidelines for the deep drawing of titanium alloy into dome shapes at room temperature are:

- The edges and surface of the blank should be smooth to prevent cracking during forming.
- The flange radius should be at least 9.5 to 12.7 mm (0.375 to 0.500 in.).
- The workpiece should be clean before each forming operation.
- An overlay or pressure cap can be used to prevent wrinkles.

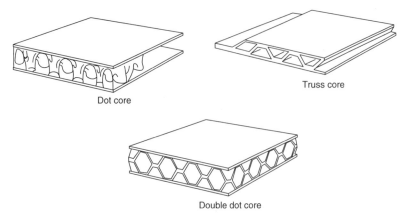

Fig. 12 Typical lightweight panels produced with diffusion bonding and superplastic forming

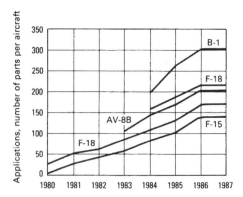

Fig. 13 Applications of superplastically formed titanium parts in military aircraft. Source: Ref 15

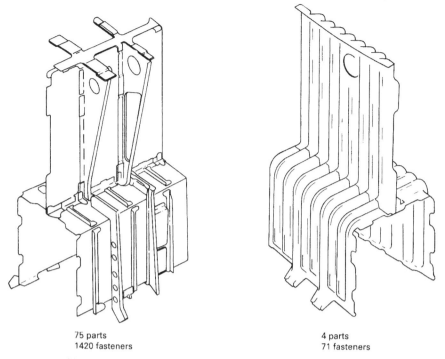

Fig. 14 Original keel design (left) and superplastically formed titanium keel section (right) for F-15 fighter aircraft. The change to the superplastically formed part resulted in a 58% cost savings and a 31% weight savings. Source: Ref 15

75 parts
1420 fasteners

4 parts
71 fasteners

(a)

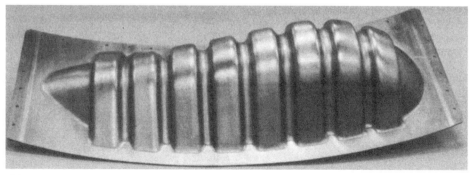

(b)

Fig. 15 Ti-6Al-4V engine nacelle component for the Boeing 757 aircraft. (a) Part as previously fabricated required 41 detail parts and more than 200 fasteners. (b) Superplastically formed part is formed from a single sheet.

- Severe forming and localized deformation should be avoided; forming pressure should be applied slowly.
- The punch should be polished to prevent galling, regardless of lubrication. Often, it is preferred to weld a layer of hard bronze on top of more conventional tooling steels to minimize galling and damage to the part, the tool, or both.

The deep drawing of dome and hemisphere shapes has also been accomplished at room temperature in a rubber-diaphragm press. A detailed description of rubber-diaphragm forming is available in the article "Rubber-Pad Forming and Hydroforming" in this Volume. Deep drawing is discussed in more detail in the article "Deep Drawing" in this Volume.

Hot Drawing. At temperatures of approximately 675 °C (1250 °F), titanium can be drawn deeper, with more difficult forming than at room temperature. Generally, depth of draw depends on material, workpiece shape, required radii, forming temperature, die design, die material, and lubricant. The setup becomes more critical than in hot sizing because the sides become more vertical. A setup that resembles a cold forming die works best to maintain alignment of the tools, and the tool is only heated where it contacts the part. Normal hot sizing presses usually have distorted heated platens that make alignment difficult when making vertical draws.

Power (Shear) Spinning

Most titanium alloys are difficult to form by power spinning. Alloys Ti-6Al-4V and some grades of CP titanium are the most responsive to forming by this method.

Most tools for the power spinning of titanium are made of high-speed steel and hardened to 60 HRC. Mandrels are heated for hot spinning, though. It may be advantageous to heat the workpiece also. Tube preforms can be heated by radiation. The hot power spinning of titanium is done at 205 to 980 °C (400 to 1800 °F), depending on the alloy and the operation.

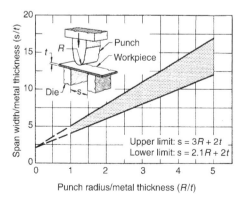

Fig. 16 Optimal relationships among span width of die, punch radius, and work metal thickness in the press-brake forming of titanium alloys. Shaded area indicates acceptable forming limits.

Lubricants for the power spinning of titanium depend on the forming temperature used. At temperatures up to 205 °C (400 °F), heavy drawing oils, graphite-containing greases, and colloidal graphite are used. Colloidal graphite and molybdenum disulfide are employed at temperatures to 425 °C (800 °F); above this temperature, colloidal graphite, powdered mica, and boron nitride are used. More information on power spinning is available in the article "Spinning" in this Volume.

Rubber-Pad Forming

The cold forming of titanium in a press with tooling that includes a rubber pad is used mostly for flanging thin stock and for forming beads and shallow recesses. The capacity of the press controls the range in size, strength, and thickness of blanks that can be formed. Within this range, however, additional limits may be set by buckling and splitting. Auxiliary devices, such as overlays, wiper rings, and sandwiches, are usually needed in rubber-pad forming to improve the forming and to reduce the amount of wrinkling and buckling. Rubber-pad forming is generally done at room temperature or with only moderate heat. Forming is almost always followed by hot sizing to remove springback, to sharpen radii, to smooth out wrinkles and

buckles, to stress relieve, and to complete the forming. Handwork is sometimes needed to complete the forming.

Sharp bends can be made at higher forming pressures. Figure 18 shows the effect of pad pressure on bend radius for two titanium alloys.

Springback behavior of titanium and its alloys in rubber-pad forming differs somewhat from that observed in other methods of forming. In general, springback in forming titanium varies directly with the ratio of bend radius to work metal thickness, and inversely with forming temperature. Springback is also inversely proportional to forming pressure.

Beads can be formed to a limited extent in titanium alloy sheet by rubber-pad forming. However, beads are readily formed by superplastic forming, and this process is preferred. Additional information on rubber-pad forming is available in the article "Rubber-Pad Forming and Hydroforming" in this Volume.

Stretch Forming

Tooling that is used for the stretch forming of stainless steel is generally suitable for the cold stretch forming of titanium, when used with a high clamping force that will prevent slipping and tearing. Particular attention should be paid to the tooth or serration pattern for the jaws to preclude slipping and/or breaking. Titanium may exhibit irregular incremental stretch under tension loads; therefore, optimal results are obtained when titanium is stretch formed at slow strain

rates. The rate of wrapping around a die should be no more than 205 mm/min (8 in./min). Poststretching of Ti-6Al-4V material does not work well because it is very notch sensitive and may break. Lower-strength CP titanium stretch forms well. In general, material preparation is critical for cold stretching of titanium alloys due to the notch sensitivity.

In the stretch forming of angles, channels, and hat-shaped sections, deformation occurs mainly by bending at the fulcrum point of the die surface; compression buckling is avoided by applying enough tensile load to produce approximately 1% elongation in the inner fibers. The outer fibers elongate more; the extent depends on the curvature of the die and on the shape of the workpiece. It is sometimes preferable or required (especially if sufficient forming power is not available) to stretch wrap at elevated temperature. Again, the wrapping speed must be slow to prevent local overheating or necking.

Formability limits can be extended by permitting small compression buckles to occur at the inner fibers and removing them later by hot sizing. The buckled region represents a condition of overforming and should be limited to the amount that can be effectively removed by hot sizing. Care must be taken to only permit buckling that can be removed. Small, sharp wrinkles may indent the hot sizing tool rather than be removed.

Compression buckling is not a problem when sheet is stretch formed to produce single or compound curves. The ductility of sheet varies with orientation and is generally better when the direction of rolling concedes with the direction of stretching. In the stretch forming of compound curves, the stretching force should be applied in the direction of the smaller radius. The rate of wrapping around the die should be no more than 205 mm/min (8 in./min).

Stretch forming is being replaced in many applications by superplastic forming. Additional information on the stretch forming process is available in the article "Stretch Forming" in this Volume.

Contour Roll Forming

Titanium sheet can be contour roll formed like any other sheet metal, but with special consideration for allowable bend radius and for the greater springback that is characteristic of titanium. Springback is affected to some extent by roll pressure. Often, hot rolling must be done on heated work metal with heated rolls. Additional information is available in the article "Contour Roll Forming" in this Volume.

Roll forming is an economical method of forming titanium alloy sheet into aircraft skins, cylinders, or parts of cylinders. The sheet should be flat within 0.15 mm (0.006 in.) for each 51 mm (2 in.) of length. The corners of the sheet should be chamfered to prevent marking of the rolls.

Table 7 Minimum bend radii obtainable in the cold press-brake bending of annealed or solution-treated titanium alloys

Alloy	Minimum bend radius as a function of sheet thickness, t	
	$t < 1.75$ mm (0.069 in.)	1.75 mm (0.069 in.) $< t < 4.76$ mm (0.1875 in.)
Commercially pure titanium		
ASTM grade 1	2.5	3.0
ASTM grade 2	2.0	2.5
ASTM grade 3	2.0	2.5
ASTM grade 4	1.5	2.0
α alloys		
Ti-5Al-2.5Sn	4.0	4.5
Ti-5Al-2.5Sn ELI	4.0	4.5
Ti-6Al-2Nb-1Ta-0.8Mo	. . .	. . .
Ti-8Al-1Mo-1V	4.5(a)	5.0(b)
α-β alloys		
Ti-6Al-4V	4.5	5.0
Ti-6Al-4V ELI	4.5	5.0
Ti-6V-2Sn	4.0	4.5
Ti-6Al-2Sn-4Zr-2Mo	4.5	5.0
Ti-3Al-2.5V	2.5	3.0
Ti-8Mn	6.0	7.0
β alloys		
Ti-13V-11Cr-3Al	3.0	3.5
Ti-11.5Mo-6Zr-4.5Sn	3.0	3.0
Ti-3Al-8V-6Cr-4Mo-4Zr	3.5	4.0
Ti-8Mo-8V-2Fe-3Al	3.5	3.5
Ti-15V-3Cr-3Sn-3Al(c)	2.0	2.0

ELI, extra-low interstitial. Source: Ref 16. (a) 4.0 in transverse direction. (b) 4.5 in transverse direction. (c) Source: Ref 17

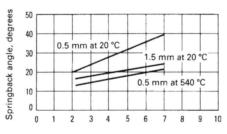

Fig. 17 Effect of ratio of punch radius to work metal thickness on springback in the press-brake bending of Ti-6Al-4V at two temperatures

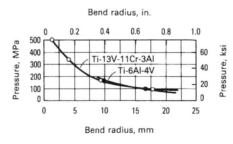

Fig. 18 Effect of pad pressure on radii formed in 1.60 mm (0.063 in.) thick titanium alloy sheets at room temperature

The upper roll of the three-roll assembly can be adjusted vertically. The radius of the bend is controlled by the roll adjustment. Premature failure will occur if the contour radius is decreased too rapidly; however, too many passes through the rolls may cause excessive work hardening of the work metal. Several trial parts must sometimes be made in a new material or shape to establish suitable operating conditions.

Three-roll forming is also used to form curves in channels that have flanges of 38 mm (1.5 in.) or less. Figure 19 shows the use of the process for curving a channel with the heel in. Transverse buckling and wrinkling are common failures in the forming of channels. The article "Three-Roll Forming" in this Volume contains more information on this process.

Creep Forming

In creep forming, heat and pressure are combined to cause the slow forming of titanium sheet into various shapes, such as double-curve panels, channel sections, Z-sections, large rings, and small joggles. The metal flows plastically at a stress below its yield strength. At low temperature, creep rates are ordinarily very low (for example, 0.1% elongation in 1000 h), but the creep rate of titanium accelerates sharply with increasing temperature.

Creep forming/hot straightening is done by applying a force on the part over a period of time while the part is at temperature. The desired effect is to force the part into the correct configuration while stress relieving. This will ensure that the part stays in the correct configuration when cooled down. Methods may include the following:

- A blank is clamped at the edges, as for stretch forming, and a heated male tool is loaded to press against the unsupported portion of the blank; the metal yields under the combination of heat and pressure and slowly creeps to fit the tool.
- The part is located on a tool with weights and run through a stress-relief cycle. During this cycle, the part will deform into the correct configuration.
- The part is forced into the correct configuration, then a stress-relief cycle is run.

- A heated die and vacuum bag is used. The blank is placed in the tool under heat and a vacuum is applied to produce the necessary forming force, then the part is run through a stress-relief cycle. Additional time and/or pressure may be required to obtain the desired contour.

Temperatures for creep forming are the same as those used in hot forming (Table 6). Generally, titanium must be held at the creep-forming temperature for 3 to 20 min per operation; creep forming sometimes takes as long as 2 h.

Vacuum Forming

Large panels (some as much as 18 m, or 60 ft, long) for aircraft are sometimes vacuum formed from titanium alloy sheet. Vacuum forming, however, can be by superplastic forming for smaller panels. There are some advantages to developing stand-alone vacuum forming tools, because they tend to be simpler to maintain and do not require a large press to create the forces. For vacuum forming, the blank is laid on a die of heated concrete, ceramic, or metal, and a somewhat larger flexible diaphragm is laid on top of the blank to provide a seal around its edges. Usually, insulation is placed between the part and the flexible diaphragm. This helps to hold in the heat as well as keep the heat off the diaphragm. It should be noted that the water needs to be removed from concrete and ceramic material through a proper curing cycle prior to heating, or it will come out in a most unsatisfactory way. Ceramic material is normally preferred because it has a very low coefficient of expansion and does not flake off under a temperature gradient. After the blank has been heated to forming

temperature, the air is pumped out from between the blank and the die so that atmospheric pressure is used to form the work. This method, a kind of creep forming, cannot bend the work to sharp radii. Finite element analysis, similar to that done for superplastic forming, can be used to determine forming capabilities at a given pressure, temperature, and time.

Drop Hammer Forming

Forming of titanium using drop hammers is becoming a lost art and perhaps the method of last resort. The tooling is quick, and the method does provide a preform. Hot sizing is required to obtain the desired contour. As shown in Fig. 20, a heat source is normally used to preheat the blank prior to forming. The alloys used in hammer tools contain lead, zinc, and other low-melting metals that contaminate titanium. These need to be removed from the titanium prior to heating. This can be done in a couple of ways. One is to not permit lead, zinc, or other low-melting metals that contaminate titanium to come in contact with the titanium. To do this, the drop hammer tools can be capped with sheet steel, stainless steel, or nickel alloy, depending on the expected tool life. Nickel-base alloys, in thicknesses of 0.635 to 0.813 mm (0.025 to 0.032 in.), have the longest life. The other way is to chemically remove the contamination from the titanium prior to reheating.

As indicated in Table 6, severe forming of most titanium alloys, which includes drop hammer forming, is done at approximately 500 to 800 °C (900 to 1500 °F). Thermal expansion of the dies must be considered in the design. The approximate rate of expansion for steel dies is 0.006 mm/mm (0.006 in./in.). The expansion

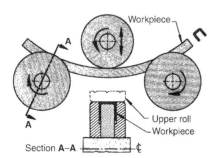

Fig. 19 Use of three-roll forming to produce a curve in a U-section channel

Fig. 20 Drop hammer forming showing oven next to the drop hammer

rate will be different for different alloys and temperatures.

Multistage tools can be used if the part shape is complex and cannot be formed in one blow. The minimum thickness of titanium sheet for drop hammer forming is 0.635 mm (0.025 in.); thicker sheet is used for complex shapes. Minimum thickness is determined from a number of variables. Contour will cause buckling in the thinner gages. Surface damage seems to occur more on the thinner gages. Total tolerance on parts formed in drop hammers is usually 1.6 mm (0.06 in.). Typically, hammer forming is used as a preform to hot sizing. More information on the drop hammer forming process is available in the article "Drop Hammer Forming" in this Volume.

Joggling

Joggling is frequently done on titanium alloy sheet. A joggle is an offset in a flat plane, consisting of two parallel bends in opposite directions at the same angle (Fig. 21). Generally, the joggle angle is less than 45°.

Depending on joggle depth, joggles can be either formed completely at room temperature or at elevated temperature in press brakes and mechanical or hydraulic presses. Room-temperature joggle limits are given in Table 8. The practice is to preform at room temperature and then hot size ("set" the joggle) in a heated die. The sizing operation is usually done under conditions that result in stress relieving or aging.

Joggles with radii smaller than the minimum bend radii (Table 7) at room temperature, or joggles with length-to-depth ratios of less than approximately 6 to 1, are more successfully formed at elevated temperature. Forming temperature varies between 315 and 650 °C (600 and 1200 °F), depending on the alloy and its heat treated condition. Annealed alloys are joggled at 315 to 425 °C (600 to 800 °F). Heat treated or partly heat treated alloys are joggled at, or near, their aging temperature.

Dimpling

Dimpling produces a small conical flange around a hole in sheet metal parts that are to be assembled with flush or flathead fasteners. Dimpling is most commonly applied to sheets that are too thin for countersinking. Sheets are always dimpled in the condition in which they are to be used, because subsequent heat treatment may cause distortion of the holes or dimensional changes in the sheet.

The hot ram-coin dimpling process is generally used. In hot ram-coin dimpling, force in excess of that required for forming is applied to coin the dimpled area and to reduce the amount of springback.

Titanium is dimpled at up to 650 °C (1200 °F) with tool steel dies. If higher temperatures are required, heat-resistant alloy or ceramic tooling is needed in order to prevent deformation of the dies during dimpling. The work metal is usually heated by conduction from the dimpling tools, which are automated to complete the dimpling stroke at a predetermined temperature.

Pilot holes must be drilled, rather than punched, and must be smooth, round, cylindrical, and free of burrs. Because of the notch sensitivity of titanium, care must be taken in deburring the holes.

The amount of stretch required to form a dimple varies with the head and body diameters of the fastener and the bend angle. If the metal is not ductile enough to withstand forming to the required shape, cracks will occur radially in the edge of the stretch flange, or circumferentially at the bend radius. Circumferential cracks are more common in thin sheet; radial cracks are more common in thick stock.

Explosive Forming

Within the limits set by its mechanical properties, titanium can be explosive formed like other metals. Explosive forming is most commonly used for cladding titanium to other metals. Titanium is explosive formed using techniques similar to those used for other metals and alloys (see the article "High-Velocity Metal Forming" in this Volume).

Bending of Tubing

Round tubing of CP titanium and alloy Ti-3Al-2.5V can be formed at room temperature in ordinary draw bending machines. When hot bending is required, the equipment is modified

Table 8 Room-temperature joggle limits for several annealed titanium alloys

See Fig. 21 for definitions of joggle dimensions given here, and Table 7 for minimum bend radii.

Alloy	Sheet thickness, t mm	Sheet thickness, t in.	A, minimum	D, maximum	L, minimum
Commercially pure titanium(a)	Up to 4.75	Up to 0.187	6D	3t	5D
Commercially pure titanium(b)	Up to 4.75	Up to 0.187	4D	4t	5D
Ti-8Al-1Mo-1V	Up to 2.29	Up to 0.090	8D	2.5t	6D
Ti-6Al-4V	Up to 2.29	Up to 0.090	8D	2.5t	6D
Ti-6Al-6V-2Sn	Up to 2.29	Up to 0.090	8D	2t	6D
Ti-5Al-2.5Sn	Up to 3.18	Up to 0.125	6D	3t	6D
Ti-13V-11Cr-3Al	Up to 4.75	Up to 0.187	6D	3t	6D
Ti-15V-3Cr-3Sn-3Al	Up to 2.29	Up to 0.090	4D	4t	5D

(a) Minimum yield strength: 483 MPa (70 ksi). (b) Minimum yield strength: <483 MPa (70 ksi). Source: Ref 18

Table 9 Limits on radii and angles in bending of commercially pure titanium

Tube outside diameter mm	Tube outside diameter in.	Wall thickness mm	Wall thickness in.	Minimum bend radius mm	Minimum bend radius in.	Bending conditions Maximum angle(a), degrees	Bending conditions Preferred minimum bend radius mm	Bending conditions Preferred minimum bend radius in.	Bending conditions Preferred maximum angle(a), degrees
Room-temperature bending									
38.1	1.5	0.41	0.016	57.2	2.25	90	75	3	120
		0.51	0.020	57.2	2.25	100	75	3	160
50.8	2.0	0.41	0.016	76.2	3.00	80	100	4	110
		0.51	0.020	76.2	3.00	100	100	4	150
63.5	2.5	0.41	0.016	95.3	3.75	70	127	5	100
		0.89	0.035	95.3	3.75	110	127	5	180
Elevated-temperature bending (175 to 205 °C, or 350 to 400 °F)									
76.2	3.0	0.41	0.016	114.3	4.50	90	150	6	120
		0.89	0.035	114.3	4.50	130	150	6	180
88.9	3.5	0.41	0.016	133.4	5.25	90	178	7	120
		0.89	0.035	133.4	5.25	130	178	7	180
101.6	4.0	0.41	0.016	152.4	6.00	110	203	8	160
		0.89	0.035	152.4	6.00	120	203	8	180
114.3	4.5	0.41	0.016	171.5	6.75	130	229	9	140
		0.89	0.035	171.5	6.75	140	229	9	140
127.0	5.0	0.51	0.020	254.0	10.00	...	254	10	110
152.4	6.0	0.51	0.020	304.8	12.00	...	305	12	100

(a) Maximum bend angles are based on the use of a clamp section three times as long as the diameter of the tubing and on maximum mandrel-ball support of the tubing.

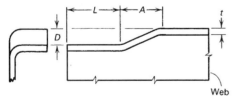

Fig. 21 Details of a joggle. See Table 8 for room-temperature joggle limits of several titanium alloys. t, sheet thickness; D, joggle height; L, joggle length; A, joggle allowance. Source: Ref 18

by adding heat to the tools. Minimum and preferred conditions for bending tubing of CP titanium at room temperature and at elevated temperatures are given in Table 9. As indicated in the table, tubing up to 63.5 mm (2.5 in.) in diameter ordinarily is bent at room temperature, while larger sizes are bent at temperatures of 175 to 205 °C (350 to 400 °F). In either case, bend radius is limited chiefly by tubing diameter, but maximum bend angle is affected by both diameter and wall thickness.

Commercially pure titanium deforms locally if tension is not applied evenly. Bending should be slow; rates of $1/4°$ to 4° per minute are suitable. A lubricant should be used.

Tools used in bending titanium and titanium alloy tubing are shown in Fig. 22. In this type of apparatus, the tubing is gripped between the clamp and the straight portion of the rotating form block tightly enough to prevent axial slip-

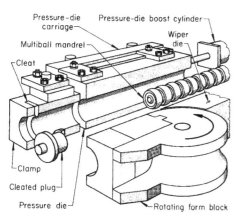

Fig. 22 Tools used for bending titanium tubing. The cleats on the clamp and plug are used only for bending of large-diameter tubing with thin walls. For hot bending, the pressure die and mandrel are integrally heated.

ping during bending. The clamped end of the tubing is supported by a plug. The cleat insert in the clamp and that attached to the end of the plug (Fig. 22) are used only in bending the larger sizes of tubing that have thin walls, for which greater gripping power is needed.

Computers are also being applied to titanium tube bending, especially at large aircraft and aerospace companies. Computer measurement systems are used during bending, and software packages are available that can design bend geometries. Completely automated precision bending can be performed using computers and numerically controlled bending equipment. More information on automated tube bending is available in the article "Bending and Forming of Tubing" in this Volume.

Drawing oils are used as lubricants for forming CP titanium tubing at room temperature. Grease with graphite is used as a lubricant for the hot bending of CP titanium tubing but is not recommended for temperatures above 315 °C (600 °F). Phosphate conversion coatings are sometimes used for hot bending of titanium tubing.

REFERENCES

1. C.H. Hamilton, Superplasticity in Titanium Alloys, *Superplastic Forming*, S.P. Agrawal, Ed., American Society for Metals, 1985, p 13–22
2. D. Lee and W. Backofen, *Trans. TMS-AIME*, Vol 239, 1967, p 1034
3. A.K. Ghosh and C.H. Hamilton, *Metall. Trans. A*, Vol 10, 1979, p 699
4. N.E. Paton and C.H. Hamilton, *Metall. Trans. A*, Vol 10, 1979, p 241
5. A. Arieli and A. Rosen, *Metall. Trans. A*, Vol 8, 1977, p 1591
6. T.L. Mackay, S.M.L. Sastry, and C.F. Yolton, Report AFWAL-TR-80-4038, Air Force Wright Aeronautical Laboratories, Sept 1980
7. J.A. Wert and N.E. Paton, *Metall. Trans. A*, Vol 14, 1983, p 2535
8. C.H. Hamilton and L.F. Nevarez, Rockwell International Science Center, unpublished research
9. F. Dyment, Self and Solute Diffusion in Titanium and Titanium Alloys, *Titanium '80: Science and Technology*, Vol 1, H. Kimura and O. Izumi, Ed., The Metallurgical Society, 1980, p 519
10. N.E.W. DeReca and C.M. Libanat, *Acta Metall.*, Vol 16, 1968, p 1297
11. A. Pontau and D. Lazarus, *Phys. Rev. B*, Vol 19, 1979, p 4027
12. E.W. Collings, *The Physical Metallurgy of Titanium Alloys*, American Society for Metals, 1984, p 151
13. Military Standard MIL-H-81200B, U.S. Government Printing Office
14. J.S. Newman and J.S. Caramanica, "Optimum Forming Processes and Equipment Necessary to Produce High Quality, Close Tolerance Titanium Alloy Parts," AFMR-TR-68-257, final technical report, 1968
15. J.R. Williamson, *Superplastic Forming/Diffusion Bonding of Titanium: An Air Force Overview*, Air Force Wright Aeronautical Laboratories, 1986
16. Military Standard MIL-T-9046J, U.S. Government Printing Office
17. G.A. Lenning, J.A. Hall, M.E. Rosenblum, and W.B. Trepel, "Cold Formable Titanium Sheet Material Ti-15-3-3-3," Report AFWAL-TR-82-4174, Air Force Wright Aeronautical Laboratories, Dec 1982
18. "Fabrication Practices for Titanium and Titanium Alloys," Lockheed Corporate Process Specification LCP70-1099, Revision B, Lockheed-California Company, Oct 1983

Formability Analysis

Formability Testing of Sheet Metals

Revised by Michael Miles, Brigham Young University

SHEET METAL FORMING is the process of converting a flat sheet of metal into a part of desired shape without fracture or excessive localized thinning. The process may be simple, such as a bending operation, or a sequence of very complex operations such as those performed in high-volume stamping plants. In the manufacture of most large stampings, a sheet metal blank is held on its edges by a blankholder ring and is deformed by means of a punch and die. The movement of the blank into the die cavity is controlled by pressure between the upper and lower parts of the blankholder ring.

This control is usually increased by means of one or more sets of drawbeads. These consist of an almost semicylindrical ridge on the upper part of the blankholder and a corresponding groove in the lower part (the positions are sometimes reversed). The drawbeads force the periphery of the blank to bend and unbend as it is pulled into the die; this increases the restraining force considerably. Presses with capacities to 17.8 MN (2000 tonf) are commonly used for the manufacture of large stampings, and presses to 26.7 MN (3000 tonf) are used for heavy-gage parts.

Sheet metal forming operations are so diverse in type, extent, and rate that no single test provides an accurate indication of the formability of a material in all situations. However, knowledge of materials properties and careful analysis of the various types of forming involved in making a particular part are indispensable in determining the probability of successful part production and in developing the most efficient process.

Types of Forming

Many forming operations are complex, but all consist of combinations or sequences of the basic forming operations—bending, stretching, drawing, and coining (see the section "Forming Processes for Sheet, Strip, and Plate" in this Volume).

Bending is the most common type of deformation, and it occurs in almost all forming operations. Bending around small radii can lead to splitting in the early stages of a forming process because it localizes strain and prevents its distribution throughout the part. Ideally, strain should be distributed as uniformly as possible to maximize the amount of deformation that can be obtained. Even a slight increase in the radius in a given location can sometimes significantly improve strain distribution. Frequently, designs specify smaller radii than necessary, which results in manufacturing problems and increased costs.

Lubrication is not recommended when bending over a sharp radius because die friction reduces strain localization by restricting metal movement away from the radius. The orientation of the sheet in relation to the rolling direction can also be important in a bending operation. During rolling, inclusions and other defects become elongated in the rolling direction, producing lines of weakness. When the axis of bending is in this direction, there is a tendency toward splitting along the lines of weakness. This lowers resistance to fracture compared with when the axis is inclined to the rolling direction. Detailed information on sheet metal lubrication is available in the article "Selection and Use of Lubricants in Forming of Sheet Metal" in this Volume.

The outer and inner panels of a part are frequently assembled by bending (hemming) the edges of the outer panel around the inner panel. This requires material that can be easily bent over very small radii. In the absence of other types of deformation, bending produces tensile stresses on the outside surface. These decrease to zero at an interior level known as the neutral axis. These stresses then become compressive on the inside of the bend. They can cause springback (shape distortion) upon removal of the applied forces. If tensile deformation is also present, the compressive stresses may be reversed, but through-thickness stress and strain gradients generally will still exist.

Many forming operations involve pulling metal over a die radius so that it is initially bent and subsequently straightened. The net strain resulting from this process may be quite small, depending on the size of the die radius and the tensile forces involved. However, the bending and straightening process cold works the metal, particularly at the surfaces, and reduces its subsequent formability.

Stretching is caused by tensile stresses in excess of the yield stress. These forces produce biaxial stretching when they are applied in perpendicular directions in the plane of the sheet. Balanced biaxial stretching occurs when the perpendicular forces are equal. Much higher levels of deformation, as measured by an increase in area, can be reached in balanced biaxial stretching than in any other forming mode.

Many forming operations involve stretching of some areas within the stamping. Automotive outer body panels are typical examples of parts formed primarily by stretching. Parts with regions containing domes, ribs, and embossments also involve stretching.

Plane-strain stretching results in elongation in one direction and no dimensional change in the perpendicular direction. It frequently occurs when a wide, flat area of sheet metal is stretched longitudinally—for example, in the sidewall of a stamping. In this case, strain in the transverse direction is prevented by the adjacent metal. Plane-strain stretching is an important type of deformation because most materials fracture at a lower level of strain in plane strain than in any other condition. Many of the fractures that occur in stamping operations are in the plane-strain region.

Drawing of sheet metal causes elongation in one direction and compression in the perpendicular direction. The simplest example is the drawing of a flat-bottomed cylindrical cup. In this process, a circular disk is held between two flat annular dies and impacted in the center by a flat-bottomed punch. This draws (pulls) the edges of the disk inward to form the wall of the cup. The metal is stretched radially by the tensile forces produced by the punch, but it is compressed circumferentially as its diameter decreases. Many other forming operations involve substantial drawing.

Coining occurs when metal is compressed between two die surfaces. It is used extensively for making coins and parts with similar surface features, for flattening, and for reducing springback upon removal of parts from a die. In many stretching and drawing operations, coining is undesirable because it restricts metal movement, localizes strain, and produces surface damage. Much of the die preparation for these operations concentrates on locating and eliminating coining.

Combinations of Types of Forming. Most forming operations involve combinations of different types of forming. Figure 1 illustrates a part design that requires drawing in the corners; biaxial stretching in the dome; bending, straightening, and plane-strain stretching in the walls; and bending and plane-strain stretching at the tops and bottoms of the walls.

Formability Problems

The major problems encountered in sheet metal forming are fracturing, buckling and wrinkling, shape distortion, loose metal, and undesirable surface textures. The occurrence of any one or a combination of these conditions can render the sheet metal part unusable. The effects of these problems are discussed next.

Fracturing occurs when a sheet metal blank is subjected to stretching or shearing (drawing) forces that exceed the failure limits of the material for a given strain history, strain state, strain rate, and temperature. In stretching, the sheet initially thins uniformly, at least in a local area. Eventually, a point is reached at which deformation concentrates and causes a band of localized thinning known as a neck, which ultimately fractures. The formation of a neck is generally regarded as failure because it produces a visible defect and a structural weakness. Most current formability tests are concerned with fracture occurring in stretching operations.

In shearing, fracture can take place without prior thinning. The most common examples of this type of fracture occur in slitting, blanking, and trimming. In these operations, sheets are sheared by knife edges that apply forces normal to the plane of the sheet. Shearing failures are sometimes produced in stamping operations by shearing forces in the plane of the sheet, but they are much less common than stretching failures.

Buckling and Wrinkling. In a typical stamping operation, the punch contacts the blank, stretches it, and starts to pull it through the blankholder ring. The edges of the blank are pulled into regions with progressively smaller perimeters. This produces compressive stresses in the circumferential direction. If these stresses reach a critical level characteristic of the material and its thickness, they cause slight undulations known as buckles. Buckles may develop into more pronounced undulations or waves known as wrinkles if the blankholder pressure is not sufficiently high.

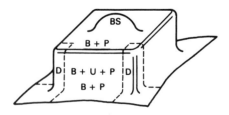

Fig. 1 Part design requiring a combination of types of forming. B, bending; BS, biaxial stretching; D, drawing; P, plane-strain stretching; U, unbending (straightening). Source: Ref 1

This effect can also cause wrinkles in other locations, particularly in regions with abrupt changes in section and in regions where the metal is unsupported or contacted on one side only. In extreme cases, folds and double or triple metal may develop. These may in turn lead to splitting in another location by preventing metal flow or by locking the metal out. Therefore, increasing the blankholder pressure often corrects a splitting problem.

Shape Distortion. In forming operations, metal is deformed elastically and plastically by applied forces. Upon removal of the external forces, the internal elastic stresses relax. In some locations, they can relax completely, with only a very slight change in the dimensions of the part. However, in areas subjected to bending, through-thickness gradients in the elastic stresses will occur; that is, the stresses on the outer surfaces will be different from those on the inner surfaces.

If these stresses are not constrained or "locked in" by the geometry of the part, relaxation will cause a change in the part shape known as shape distortion or springback. Springback can be compensated for in die design for a specific set of materials properties but may still be a problem if there are large material property or process variations from blank to blank.

Loose metal occurs in undeformed regions and is undesirable because it can be easily deflected. A phenomenon usually referred to as oil canning, in which a local area can be either concave or convex, can also be encountered. In stampings with two or more sharp bends of the same sign in roughly the same direction, such as a pair of feature lines, a tendency exists for the metal between them to be loose because of the difficulty involved in pulling metal across a sharp radius.

It is sometimes possible to avoid the problem by ensuring that the metal is not contacted by both lines at the same time; thus, some stretching can occur before the second line is contacted (see the article "Press Forming of Coated Steel" in this Volume). There is a tendency for loose metal to occur toward the center of large, flat, or slightly curved parts. Increasing the restraining forces on the blank edges usually improves this condition.

Undesirable Surface Textures. Heavily deformed sheet metal, particularly if it is coarse grained, often develops a rough surface texture commonly known as orange peel (see the article "Press Forming of Coated Steel" in this Volume). This is usually unacceptable in parts that are visible in service.

Another source of surface problems occurs in metals that have a pronounced yield point elongation, that is, materials that stretch several percent without an increase in load after yielding. In these metals, deformation at low strain levels is concentrated in irregular bands known as Lüders lines (or bands), or stretcher strains. These defects disappear at moderate and high strain levels. However, almost all parts have some low-strain regions. These defects

are unsightly and are not concealed by painting. Aged rimmed steels and some aluminum-magnesium alloys develop severe Lüders lines.

In some cases, zinc-coated steels exhibit surface defects known as spangles. This phenomenon occurs only in hot-dipped products and is caused by the development of a coarse grain size in the galvanic coating, which makes the individual grains clearly visible. This problem can be corrected in the coating process. In addition to the aforementioned occurrences, handling damage, dents caused by dirt or slivers in the die, and scoring or galling caused by a rough die surface or inadequate lubrication sometimes produce unacceptable surfaces.

Measurement of Deformation

The principal methods of measuring deformation are gage marks, strain gages, optical extensometers, and thickness and shape measurements.

Gage Marks. The most widely used method for measuring deformation is to mark the sheet by etching or scribing or by means of ink, dye, or paint and to measure the changes in the separations of the marks caused by the deformation. Rectangular grid markings and arrays of small-diameter circles (for example, 2.5 mm, or 0.1 in.) are frequently used.

In most production forming operations and in the later stages of tension testing, deformation varies rapidly with location, which can lead to large differences in strain measurements made over different gage lengths. Therefore, small gage lengths, such as the diameters of small, closely spaced circles, are commonly used. Circular markings provide an additional advantage in that it is easy to identify the directions of the maximum (or major) and minimum (or minor) strains and thus to measure their values. Upon deforming, the circles change into ellipses with their major axes in the direction of the maximum strain and their minor axes in the direction of the minimum strain.

This information is essential in determining how close the local strain state is to the maximum the material can withstand without fracturing, which depends on the ratio of the strains. It is also useful in determining how the geometry of a die must be modified when the formability limits of the work material are exceeded.

Strain Gages and Extensometers. In some cases, a strain gage or a strain gage extensometer is attached to the sheet or test sample. Accurate strain measurements are thus obtained continuously during a forming operation or test. Optical extensometers, which are particularly effective at high strain rates, can also be used.

Thickness and Shape Measurements. Thickness measurements, which can be made rapidly by ultrasonic methods, can sometimes be used to determine strains. In practice, this method is limited to situations in which the ratio of the major and minor strains is known from previous measurements because many different combinations of strains can lead to the same change in thickness.

Part shape is measured by using templates, checking fixtures, or shadowgraphs or by using a profile meter that employs a stylus to contact the surface. Profile meters may give two- or three-dimensional digital representations of the part. Noncontacting surface digitizers and systems for measuring deformation by locating grid markings in three dimensions are also used.

Representation of Strain

The most common method of representing strain defines the engineering strain, e, as the ratio of the change in length, ΔL, to the original length, L_0:

$$e = \frac{\Delta L}{L_0} = \frac{L - L_0}{L_0} = \frac{L}{L_0} - 1 \qquad \text{(Eq 1)}$$

The second method defines the true strain, ε, in the region of uniform elongation as the integral of the incremental change in length, dL, divided by the actual (instantaneous) length, L:

$$\varepsilon = \int_{L_0}^{L} \frac{dL}{L} = \ln\left(\frac{L}{L_0}\right) = \ln(1 + e) \qquad \text{(Eq 2)}$$

The engineering strain is easier to calculate and is satisfactory for many applications. The true strain is used in the theoretical analysis of formability and is advantageous in that successive strains can be added to give the cumulative strain.

The strain state of a deformed sheet is frequently represented graphically by plotting the maximum or major strain, e_1, on the vertical axis and the minimum or minor strain, e_2, which can be positive or negative, on the horizontal axis. This is illustrated in Fig. 2, which shows five strain paths, each leading to the same major strain of 40% but with minor strains ranging from -40 to $+40\%$. The ellipses shown

were originally circles (shown dashed) in the undeformed sheet.

On the right side of Fig. 2, the circles have transformed into ellipses that are larger in all directions than the original circles. This is the region of biaxial stretching and, in the diagonal ($45°$) direction, of balanced biaxial stretching. In this direction, the circles have expanded without changing shape.

On the left side of Fig. 2, the circles have transformed into ellipses, which are larger in one direction but smaller in the perpendicular direction than the original circles. This is the region of drawing and is the strain state developed in the tension test. On the vertical axis, the ellipses are larger in one direction but unchanged dimensionally from the original circles in the perpendicular direction. This is the region of plane strain.

Effect of Materials Properties on Formability

The properties of sheet metals vary considerably, depending on the base metal (steel, aluminum, copper, and so on), alloying elements present, processing, heat treatment, gage, and level of cold work. In selecting material for a particular application, a compromise usually must be made between the functional properties required in the part and the forming properties of the available materials. For optimal formability in a wide range of applications, the work material should:

- Distribute strain uniformly
- Reach high strain levels without necking or fracturing
- Withstand in-plane compressive stresses without wrinkling
- Withstand in-plane shear stresses without fracturing
- Retain part shape upon removal from the die
- Retain a smooth surface and resist surface damage

Some production processes can be successfully operated only when the forming properties of the work material are within a narrow range. More frequently, the process can be adjusted to accommodate shifts in work materials properties from one range to another, although sometimes at the cost of lower production and higher materials waste. Some processes can be successfully operated using work material that has a wide range of properties. In general, consistency in the forming properties of the work material is an important factor in producing a high output of dimensionally accurate parts.

Strain Distribution

Three materials properties determine the strain distribution in a forming operation:

- The strain-hardening coefficient (also known as the work-hardening coefficient or exponent), or n value
- The strain rate sensitivity, or m value
- The plastic strain ratio (anisotropy factor), or r value

The ability to distribute strain evenly depends on the n value and the m value. The ability to reach high overall strain levels depends on many factors, such as the base material, alloying elements, temper, n value, m value, r value, thickness, uniformity, and freedom from defects and inclusions.

The n value, or strain-hardening coefficient, is determined by the dependence of the flow (yield) stress on the level of strain. In materials with a high n value, the flow stress increases rapidly with strain. This tends to distribute further strain to regions of lower strain and flow stress. A high n value is also an indication of good formability in a stretching operation.

In the region of uniform elongation, the n value is defined as:

$$n = \frac{d \ln \sigma_T}{d \ln \varepsilon} \qquad \text{(Eq 3)}$$

where σ_T is the true stress (load/instantaneous area) and ε is true strain, as defined by Eq 2. This relationship implies that the true stress-strain curve of the material can be approximated by a power law constitutive equation proposed in Ref 2:

$$\sigma_T = k\varepsilon^n \qquad \text{(Eq 4)}$$

where k is a constant known as the strength coefficient.

Equation 4 provides a good approximation for most steels but is not very accurate for dual-phase steels and some aluminum alloys. For these materials, two or three n values may need to be calculated for the low, intermediate, and high strain regions.

When Eq 4 is an accurate representation of material behavior, $n = \ln(1 + e_u)$, where e_u is the uniform elongation, or elongation at maximum load in a tension test. By definition, $\ln(1 + e_u)$ is

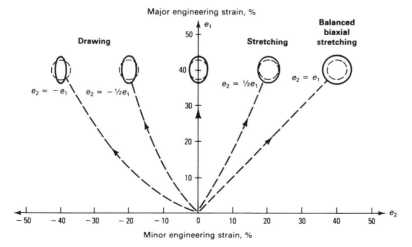

Fig. 2 Schematic of several major strain/minor strain combinations

identical to ε_u, which is the true strain at uniform elongation.

Most steels with yield strengths below 345 MPa (50 ksi) and many aluminum alloys have n values ranging from 0.2 to 0.3. For many higher-yield-strength steels, n is given by the relationship (Ref 3):

$$n = \frac{70}{(\text{yield strength in MPa})} \qquad (\text{Eq 5})$$

A high n value leads to a large difference between yield strength and ultimate tensile strength (engineering stress at maximum load in a tension test). The ratio of these properties therefore provides another measure of formability.

The m value, or strain rate sensitivity, is defined by:

$$m = \frac{d \ln \sigma}{d \ln \dot{\varepsilon}} \qquad (\text{Eq 6})$$

A constitutive law that incorporates both strain hardening and strain rate sensitivity is:

$$\sigma_T = k\varepsilon^n \dot{\varepsilon}^m \qquad (\text{Eq 7})$$

A positive strain rate sensitivity indicates that the flow stress increases with the rate of deformation. This has two consequences. First, higher stresses are required to form parts at higher rates. Second, at a given forming rate, the material resists further deformation in regions that are being strained more rapidly than adjacent regions by increasing the flow stress in these regions. This helps to distribute the strain more uniformly.

The need for higher stresses in a forming operation is usually not a major consideration, but the ability to distribute strains can be crucial. This becomes particularly important in the post-uniform elongation region, where necking and high strain concentrations occur. An approximately linear relationship has been reported between the m value and the post-uniform elongation for a variety of steels and nonferrous alloys (Ref 4). As m increases from -0.01 to $+0.06$, the post-uniform elongation increases from 2 to 40%.

Metals in the superplastic range have high m values, usually greater than 0.5, which are one to two orders of magnitude higher than typical values for steel. At ambient temperatures, some metals, such as aluminum alloys and brass, have low or slightly negative m values, which explains their low post-uniform elongation.

High n and m values lead to good formability in stretching operations but have little effect on drawability. In a drawing operation, metal in the flange must be drawn in without causing fracture in the wall. In this case, high n and m values strengthen the wall, which is beneficial, but they also strengthen the flange and make it harder to draw in, which is detrimental.

The r value, or plastic strain ratio, relates to drawability and is known as the anisotropy factor. This is defined as the ratio of the true width strain to the true thickness strain in the uniform elongation region of a tension test:

$$r = \frac{\varepsilon_w}{\varepsilon_t} = \frac{\ln\left(\frac{w}{w_0}\right)}{\ln\left(\frac{t}{t_0}\right)} \qquad (\text{Eq 8})$$

The r value is a measure of the ability of a material to resist thinning. In drawing, material in the flange is stretched in one direction (radially) and compressed in the perpendicular direction (circumferentially). A high r value indicates a material with good drawing properties.

The r value frequently changes with direction in the sheet. In a cylindrical cup drawing operation, this variation leads to a cup with a wall that varies in height, a phenomenon known as earing (Fig. 3). It is therefore common to measure the average r value, or average normal anisotropy, r_m, and the planar anisotropy, Δr.

The property r_m is defined as $(r_0 + 2r_{45} + r_{90})/4$, where the subscripts refer to the angle between the tensile specimen axis and the rolling direction. The value Δr is defined as $(r_0 - 2r_{45} + r_{90})/2$. It is a measure of the variation of r with direction in the plane of a sheet. The value r_m determines the average depth (that is, the wall height) of the deepest draw possible. The value Δr determines the extent of earing. A combination of a high r_m value and a low Δr value provides optimal drawability.

Hot-rolled low-carbon steels have r_m values ranging from 0.8 to 1.0; cold-rolled rimmed steels range from 1.0 to 1.4, and cold-rolled aluminum-killed (deoxidized) steels range from 1.4 to 2.0. Interstitial-free steels have values ranging from 1.8 to 2.5, and aluminum alloys range from 0.6 to 0.8. The theoretical maximum r_m value for a ferritic steel is 3.0; a measured value of 2.8 has been reported (Ref 5).

Maximum Strain Levels: The Forming Limit Diagram

Each type of steel, aluminum, brass, or other sheet metal can be deformed only to a certain level before local thinning (necking) and fracture occur. This level depends principally on the combination of strains imposed, that is, the ratio of major and minor strains. The lowest level occurs at or near plane strain, that is, when the minor strain is zero.

This information was first represented graphically as the forming limit diagram, which is a graph of the major strain at the onset of necking for all values of the minor strain that can be realized (Ref 6, 7). Figure 4 shows a typical forming limit diagram for steel. The diagram is used in combination with strain measurements, usually obtained from circle grids, to determine how close to failure (necking) a forming operation is or whether a particular failure is due to inferior work material or to a poor die design (Ref 8).

For most low-carbon steels, the forming limit diagram has the same shape as the one shown in Fig. 4, but the vertical position of the curve depends on the sheet thickness and the n value. The intercept of the curve with the vertical axis, which represents plane strain and is also the minimum point on the curve, has a value equal to n in the (extrapolated) zero thickness limit. The intercept increases linearly with thickness to a thickness of about 3 mm (0.12 in.).

The rate of increase is proportional to the n value (Fig. 5). The level of the forming limits also increases with the m value (Ref 4).

The shape of the curve for aluminum alloys, brass, and other materials differs from that in Fig. 4 and varies from alloy to alloy within a system. The position of the curve also varies and rises with an increase in the thickness, n value, or m value, but at rates that are generally not the same as those for low-carbon steel.

The forming limit diagram is also dependent on the strain path. The standard diagram is based on an approximately uniform strain path. Diagrams generated by uniaxial straining followed by biaxial straining, or the reverse, differ considerably from the standard diagram. Therefore, the effect of the strain path must be taken into account when using the diagram to analyze a forming problem.

Materials Properties and Wrinkling

The effect of materials properties on the formation of buckles or wrinkles is the subject of extensive research. In drawing operations, there

Fig. 3 Drawn cup with ears in the directions of high r value

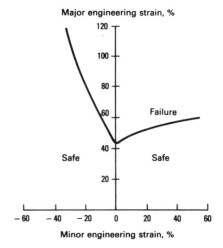

Fig. 4 Typical forming limit diagram for steel

is general agreement, based primarily on experiments with conical and cylindrical cups, that a high r_m value and a low Δr value reduce buckling in both flanges and walls (Ref 10–12). In addition to the aforementioned correlations, a low flow-stress-to-elastic-modulus ratio (σ_F/E) decreases wall wrinkling (Ref 13). The n value has an indirect effect. When the binder force is kept constant, the n value has no effect. However, high n values enable higher binder forces to be used, which reduces buckling.

In stretching operations, the situation appears to be different. A close correlation between the formation of buckles at low strain levels and the yield-strength-to-tensile-strength ratio (YS/TS) has been reported, as well as an inverse correlation with the low strain n value and an absence of correlation with the r_m value and uniform elongation (Ref 14). Some of the differences between these results may be attributed to the fact that the experiments with cups involved high strains and high compressive stresses, while the stretching experiments were conducted at low strain and low compressive stress levels. In both situations, the problem becomes significantly more severe as the sheet thickness decreases.

Materials Properties and Shear Fracture

Shear fractures due to in-plane shear stresses are more prevalent in high-strength cold-worked materials, particularly when internal defects such as inclusions are present. Typical strain combinations that cause shear fracture are shown on the forming limit diagram in Fig. 6. For this material, Fig. 6 shows that, at high strain levels in the regions close to $\varepsilon_2 = \pm\varepsilon_1$, failure occurs by shearing before the initiation of necking.

The position and shape of the shear fracture curve depends on the material, its temper, and the

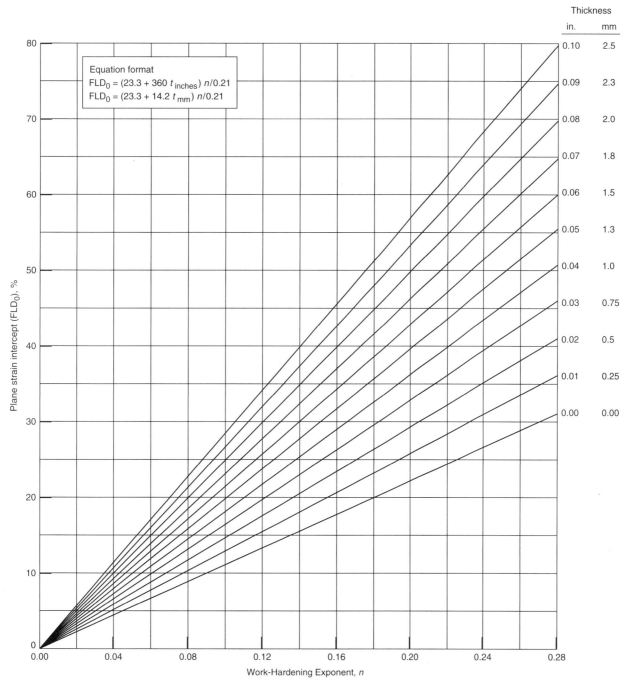

Equation format
$$FLD_0 = (23.3 + 360\, t_{inches})\, n/0.21$$
$$FLD_0 = (23.3 + 14.2\, t_{mm})\, n/0.21$$

Fig. 5 Effect of thickness and n value on the plane-strain intercept of a forming limit diagram for low-carbon steel. Source: Ref 9

type and degree of prestrain or cold work (Ref 15–17). Limited data are available on shear fracture.

Materials Properties and Springback

Materials properties that control the amount of springback that occurs after a forming operation are:

- Elastic modulus, E
- Yield stress, σ_y
- Slope of the true stress/strain curve, or tangent modulus, $d\sigma_T/d\varepsilon$

Springback is best described by means of three examples involving a rectangular beam: elastic bending below the yield stress, simple bending with the yield stress exceeded in the outer layers of the beam, and combined stretching and bending. In an actual part, springback is determined by the complex interaction of the residual internal elastic stresses, subject to the constraints of the part geometry.

Elastic Bending below the Yield Stress. Tensile elastic stresses are generated on the outside of the bend. These stresses decrease linearly from a maximum at the surface to zero at the center (neutral axis). They then become compressive and increase linearly to a maximum at the inner surface. Upon removal of the externally applied bending forces, the internal elastic forces cause the beam to unbend as they decrease to zero throughout the cross section (Fig. 7a).

The maximum amount of elastic deflection that can be produced without entering the plastic range is proportional to the yield stress divided by the elastic modulus. The strain at the yield point is equal to σ_y/E ($E = \sigma/\varepsilon$). The springback moment for a given deflection is therefore proportional to the elastic modulus ($\sigma = E\varepsilon$).

Simple Bending. In this example, the yield stress is exceeded in the outer layers of the beam. The outer layers deform plastically, and their stored elastic stresses continue to increase, but at a much lower rate that is proportional to the slope of the true stress-strain curve, or tangent modulus, $d\sigma_T/d\varepsilon$, instead of the elastic modulus. Figure 7(b) illustrates this condition for a beam bent so that 50% of its volume is in the plastic range.

Upon removal of the externally applied bending forces, the stored elastic stresses cause the beam to unbend until their combined bending moment is zero. This produces compressive stresses at the outer surface and tensile stresses at the inner surface.

The springback in this case is less than for a material whose yield strength is not exceeded at the same strain level. This can result from either a higher yield stress or a lower elastic modulus. It is also apparent that higher values of the tangent modulus cause greater springback when the yield strength is exceeded.

In actual conditions, the neutral axis moves inward upon bending because the outer part of the beam is stretched and becomes thinner and because the inner part is compressed and

becomes thicker. This effect is analyzed in detail in Ref 18.

Combined Stretching and Bending. In this case, the entire beam can be plastically deformed in tension by as little as 0.5% stretching. However, a stress gradient still exists from the outer to the inner surface (Fig. 7c). Upon removing the external forces, the internal elastic stresses recover.

This causes unbending, but to a lesser extent than in the previous cases. As the level of stretching is increased, the amount of springback decreases because the tangent modulus and therefore the stress gradient through the beam decrease at higher strains. The yield strength ceases to be a factor in springback once all regions are plastically deformed in tension.

In the bending of wide sheets, the metal is deformed in plane strain, and the plane strain properties (elastic modulus, yield stress, and tangent modulus) should be used. The effects of a low elastic modulus and a high yield stress and

tangent modulus in increasing springback have been experienced in forming operations. Springback is more severe with aluminum alloys than with low-carbon steel (1 to 3 modulus ratio). High-strength steels exhibit more springback than low-carbon steels (~2 to 1 yield strength ratio), and dual-phase steels spring back more than high-strength steels of the same yield strength (higher tangent modulus).

The effect of stretching in reducing springback to very low levels has also been reported (Ref 19). Springback is also greatly influenced by geometrical factors, and it increases as the bend angle and ratio of bend radius to sheet thickness increase.

Effect of Temperature on Formability

A change in the overall temperature alters the properties of the material, which thus affects

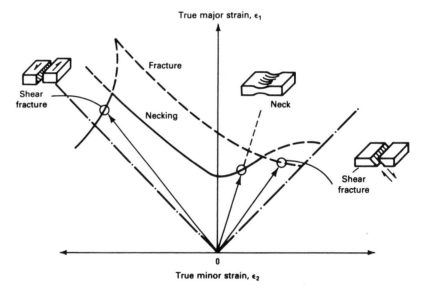

Fig. 6 Forming limit diagram including shear fracture. Source: Ref 15

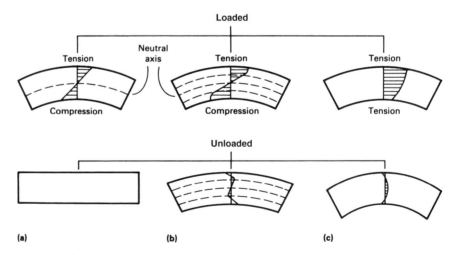

Fig. 7 Springback of a beam in simple bending. (a) Elastic bending. (b) Elastic and plastic bending. (c) Bending and stretching

formability. In addition, local temperature differences within a deforming blank lead to local differences in properties that affect formability.

At high temperatures, above one-half of the melting point on the absolute temperature scale, extremely fine-grain aluminum, copper, magnesium, nickel, stainless steel, steel, titanium, zinc, and other alloys become superplastic. Superplasticity is characterized by extremely high elongation, ranging from several hundred to more than 1000%, but only at low strain rates (usually below approximately $10^{-2}/s^{-1}$) at high temperatures.

The requirements of high temperatures and low forming rates have limited superplastic forming to low-volume production. In the aerospace industry, titanium is formed in this manner. The process is particularly attractive for zinc alloys because they require comparatively low temperatures (~270 °C, or ~520 °F).

At intermediate elevated temperatures, steels and many other alloys have less ductility than at room temperature (Ref 20, 21). Aluminum and magnesium alloys are exceptions and have minimum ductility near room temperature. Alloys of these metals have been formed commercially at slightly elevated temperatures (~250 °C, or ~480 °F). The strain rate sensitivity (m value) and post-uniform elongation for aluminum-magnesium alloys have been found to increase significantly in this temperature range (Ref 22).

Low-temperature forming has potential advantages for some materials, based on their tensile properties, but practical problems have limited application. Local increases in temperature occur during forming because of the surface friction and internal heating produced by the deformation. Generally, this is detrimental because it lowers the flow stress in the area of greatest strain and tends to localize deformation.

A method of improving drawability by creating local temperature differences has been developed and is being used commercially (Ref 23). It involves water cooling the punch in a deep-drawing operation. This lowers the temperature of the blank where it contacts the punch, which is the principal failure zone, and increases the local flow stress. Heating the die in order to lower the flow stress in the deformation zone at the top of the draw wall has also been found to be beneficial. The combination of these procedures has produced an increase of over 20% in the drawability of an austenitic stainless steel.

Types of Formability Tests

Formability tests are of two basic types: intrinsic and simulative. Intrinsic tests measure the basic characteristic properties of materials that can be related to their formability. Simulative tests subject the material to deformation that closely resembles the deformation that occurs in a particular forming operation.

Intrinsic tests provide comprehensive information that is insensitive to the thickness and surface condition of the material. The most important and extensively used intrinsic test is the uniaxial tension test, which provides the values of many materials properties for a wide range of forming operations. Other commercially important intrinsic tests are the plane-strain tension test, the Marciniak stretching and sheet torsion tests, the hydraulic bulge test, the Miyauchi shear test, and hardness tests.

Simulative tests provide limited and specific information that is usually sensitive to thickness, surface condition, lubrication, and geometry and type of tooling. This information usually relates to only one type of forming operation. Many simulative tests, such as the Olsen and Swift cup test, have been used extensively for many years with good correlation to production in specific cases. Several simulative tests are described later in this article.

Uniaxial Tension Testing

The most widely used intrinsic test of sheet metal formability is the uniaxial tension test. A specimen such as that illustrated in Fig. 8 is used; its sides are accurately parallel over the gage length, which is usually 50.8 mm (2.00 in.) long and 12.7 mm (0.50 in.) wide. The specimen is gripped at each end and stretched at a constant rate in a tension machine until it fractures, as described in ASTM E 8. The applied load and extension are measured by means of a load cell and strain gage extensometer.

The load extension data can be plotted directly. However, data are usually converted into engineering (conventional) stress, σ_E (load/original cross section), and engineering strain, e (elongation/original length), or to true stress, σ_T (load/instantaneous cross section), and true strain, ε (natural logarithm of strained length/original length).

In addition, for formability testing, it is common practice to measure the width of the specimen during the test. This is done either intermittently by interrupting the test at preselected elongations to make measurements manually or continuously by means of width extensometers. From these measurements, the plastic strain ratio (anisotropy factor), or r value, can be determined.

During the rolling process used to produce metals in sheet form and the subsequent annealing, the grains and any inclusions present become elongated in the rolling direction, and a preferred crystallographic orientation develops. This causes a variation of properties with direction. Therefore, it is common practice to test specimens cut parallel to the rolling direction and at 45 and 90° to this direction. These are known as longitudinal, diagonal, and transverse specimens, respectively. This also enables the values of r_m and Δr to be calculated. Because the mechanical properties and elongation tend to be lower in the transverse direction, tests in this direction are often used as the basis for specifications.

The rate at which the test is performed can have a significant effect on the end results. Two methods are commonly used to determine this effect. In the first method, replicate samples are tested at different rates, and the results are influenced by variations between the samples. In the second method, the test rate is alternated between two levels. This approach avoids the problem of variation between samples, but it cannot be used at very high rates and is complicated by transients, which occur each time the rate is changed. The strain rate sensitivity, or m value, can be calculated from these tests.

Figure 9 shows a typical engineering stress-strain curve and the corresponding true stress-strain curve for a material that has a smooth transition between the very low strain (elastic) and the higher strain (plastic) regions of the curve. When the load is removed in the elastic

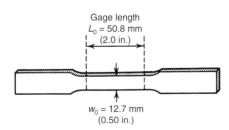

Fig. 8 Sheet tensile test specimen

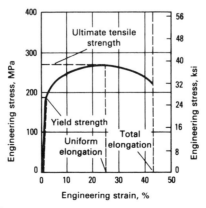

Fig. 9 Typical engineering and true stress-strain curves

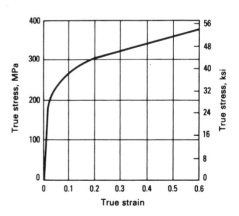

region, the sample returns to its original dimensions. When this is done in the plastic region, the sample retains permanent deformation.

In the tension test, the load increases to a maximum value and then decreases prior to fracture. The decrease is due to the localization of the deformation, which causes a reduction in cross section. This reduction has a greater effect than the opposing increase in flow stress due to strain hardening.

Test Procedure

For accurate and reproducible results, uniaxial tension testing must be performed in a carefully controlled manner. The main steps in the procedure are discussed in detail in the article "Uniaxial Tension Testing" in *Mechanical Testing and Evaluation,* Volume 8 of *ASM Handbook.* These procedures are summarized next.

Specimen Preparation. The surfaces of the specimen should be free from scratches or other damage that can act as stress raisers and cause early failure. The edges should be smooth and free from irregularities. Care should be taken not to cold work the edges, or to ensure that any cold work introduced is removed in a subsequent operation, because this changes mechanical properties and lowers ductility.

It is common practice to mill and grind the edges, but other procedures such as fine milling, nibbling, and laser cutting are also used. When a new method is used, initial tests should be performed to compare the results with those obtained by conventional methods.

The width of a nominally 12.7 mm (0.50 in.) wide specimen should be measured to the nearest 0.025 mm (0.001 in.), and the thickness for specimens in the range of 0.5 to 2.5 mm (0.02 to 0.1 in.) should be measured to the nearest 0.0025 mm (0.0001 in.). If this is impractical because of surface roughness, the thickness should be measured to the nearest 0.025 mm (0.001 in.).

The tension test is sensitive to variations in the width of the specimen, which should be accurately controlled. For a specimen 12.7 mm (0.50 in.) wide, the width of the reduced section should not deviate by more than ± 0.25 mm (± 0.01 in.) from the nominal value and should not differ by more than ± 0.05 mm (± 0.002 in.) from end to end.

Some investigators intentionally taper the reduced section slightly toward the center to increase the probability that fracture will occur within the gage length. In this case, the center should not be narrower than the ends by more than 0.10 mm (0.004 in.).

Alignment of Specimens. The specimen should be accurately aligned with the centerline of the grips. The effect of small displacements (10% of the specimen width) of one or both ends from the centerline has been calculated (Ref 24). It has been determined that the latter case is the more serious, but both strongly affect the strain in the outermost fibers. It has also been concluded that the calculated stress-strain curve is not significantly affected at strains above 0.3%.

Measurement of Load and Elongation. The applied load is measured by means of a load cell in the test machine, for which the usual calibration procedures must be followed (ASTM E 4). Elongation is usually determined by using a clip-on strain gage extensometer (ASTM E 83). In addition, small scratches are often scribed across the specimen at the ends of the gage length so that the total elongation can be determined from the broken specimen.

Circle grids are sometimes etched or printed on the specimen. These can be used to measure the strain distribution and width strain as well as the overall strain. This can be done continuously by means of a video camera and data processing system if required. Optical extensometers are used for some applications, particularly high-speed testing. These units require well-illuminated boundaries that are clearly delineated by means of high-contrast coatings, such as black-and-white paint.

An approximate measure of elongation can be obtained from the crosshead travel. This involves errors due to elongation of the specimen outside the gage length and elastic strain in the grips, which can be compensated for to some extent. This method is used when the specimen is inaccessible, such as in nonambient testing.

The signals from the load cell and extensometer can be plotted on a chart recorder or processed by a data processing system to the required form, such as plots of stress versus strain or tables of mechanical and forming properties.

Measurement of Width and Thickness. In addition to the initial measurements of specimen width and thickness, which are required to calculate the stress, measurements can be made at intervals during the test to determine the r value (ASTM E 517) and to determine the reduction in area and true strain. The r value is measured at a specified strain level between the yield point and the uniform elongation (for example, at 15% elongation). It can be measured by stopping the test at this strain level and then measuring the width accurately (± 0.013 mm, or ± 0.0005 in.) at a minimum of three equally spaced points in the gage length (for a 50.8 mm, or 2.0 in., gage length). In practice, the thickness is calculated from the specimen width and length, assuming no change in volume.

Alternatively, width measurements can be made during the test using width extensometers, although this is a more complicated procedure. Attempts are underway to develop combined width and length extensometers to simplify this method.

Reduction in area is the ratio $(A_0 - A)/A_0$, where A is the instantaneous cross-sectional area and A_0 is the original cross-sectional area. It is used to calculate the true strain in the region of post-uniform elongation. A large reduction in area at fracture correlates with a small minimum bend radius and high energy absorption. To calculate the reduction in area, the width and thickness must be measured in the narrowest part of the necked region.

Effect of Gage Length on Elongation. In post-uniform elongation, part of the specimen is elongated uniformly, and the remainder is narrowed into a necked region of higher strain level. A change in the gage length alters the ratio of these two regions and has a significant effect on the total elongation measurement. This phenomenon is discussed in detail in Ref 25.

To obtain results that are comparable for different gage lengths, the ratio of the square root of the cross-sectional area to the length, $\sqrt{A/L}$, should be the same. When comparing samples of different thickness, this implies that the gage length or the width should be adjusted to maintain this ratio.

Rate of Testing. Most tension tests are performed on screw-driven or hydraulic testing machines at strain rates of 10^{-5} to 10^{-2} s^{-1}. The strain rate is defined as the increase in length per unit length per second. These tests are known as low strain rate or static tests.

Table 1 Typical tensile properties of selected sheet metals

Material	Young's modulus, E		Yield strength		Tensile strength		Uniform elongation, %	Total elongation, %	Strain-hardening exponent, n	Average normal anisotropy, r_m	Planar anisotropy, Δr	Strain rate sensitivity, m
	GPa	10^6 psi	MPa	ksi	MPa	ksi						
Aluminum-killed drawing quality steel	207	30	193	28	296	43	24	43	0.22	1.8	0.7	0.013
Interstitial-free steel	207	30	165	24	317	46	25	45	0.23	1.9	0.5	0.015
High-strength low-alloy steel	207	30	345	50	448	65	20	31	0.18	1.2	0.2	0.007
Dual-phase steel	207	30	414	60	621	90	14	20	0.16	1.0	0.1	0.008
301 stainless steel	193	28	276	40	690	100	58	60	0.48	1.0	0.0	0.012
409 stainless steel	207	30	262	38	469	68	23	30	0.20	1.2	0.1	0.012
3003-O aluminum	69	10	48	7	110	16	23	33	0.24	0.6	0.2	0.005
6009-T4 aluminum	69	10	131	19	234	34	21	26	0.23	0.6	0.1	−0.002
70-30 brass	110	16	110	16	331	48	54	61	0.56	0.9	0.2	0.001

Most high-volume production forming operations are performed at considerably higher strain rates—in the range of 1 to 10^2 s^{-1}. To determine the tensile properties in this range, dynamic test machines, which operate at rates of 10^{-1} to 10^2 s^{-1}, are used (Ref 25). As mentioned previously, steels have higher tensile properties and lower elongations at high strain rates. The properties of aluminum alloys have little sensitivity to the strain rate.

Materials Properties

The stress-strain curve determined by uniaxial tension testing provides values of many formability related materials properties. Several of these properties and methods for measurement are discussed subsequently. Table 1 lists typical values of properties measured in tension tests on thin (0.5 to 1.0 mm, or 0.02 to 0.04 in.) sheet materials.

Young's Modulus. The initial slope of the stress-strain curve, that is, the ratio of the stress to the strain in the elastic region before any plastic deformation has occurred, is the Young's modulus, E, of the material. This property affects springback and shape distortion at low strains. For accurate measurement of Young's modulus, a low strain rate and a high data acquisition rate should be used in the elastic region (below approximately 0.5% elongation), and a very stiff tension-testing machine should be used if strain is inferred from crosshead displacement.

Yield Strength. The stress at which the stress-strain curve deviates in elongation from the initial elastic slope by a specified amount, commonly 0.2%, is known as the yield strength (YS). The yield strength determines the load necessary to initiate deformation in a forming operation, which is usually a high percentage (40 to 90%) of the maximum load required.

For accurate measurement of yield strength, a rate of loading of less than 690 MPa/min (100 ksi/min) is specified. Beyond this point, the strain rate should not exceed 0.08 s^{-1}. Some materials elongate without an increase in load, or at a decreased load, at the transition between the elastic and plastic regions. The point at which this initiates is known as the yield point.

With a decrease in load, the material has an upper yield point and a lower yield point. The upper yield point is difficult to measure reproducibly. The lower yield stress usually fluctuates, and the minimum value is used.

The elongation that occurs after yielding before the load starts to increase monotonically is known as the yield point elongation. Yield point elongation leads to nonuniform deformation at low strains in forming operations. If it exceeds approximately 1.5%, irregular surface markings known as Lüders lines or stretcher strains may occur to an extent that is unacceptable in visible parts.

Tensile Strength. The maximum stress observed in the test is known as the tensile strength (TS), or ultimate tensile strength (UTS).

Tensile strength determines the maximum load that can be usefully applied in a forming operation.

Uniform Elongation. The engineering strain at the maximum engineering stress is known as the uniform elongation, e_u. Prior to this point, the sample deforms uniformly. Subsequently, deformation concentrates—initially in a fairly large region known as a diffuse neck, and ultimately in a localized region of sharply reduced cross section known as a local neck. Deformation continues to concentrate in this region until fracture occurs.

Total Elongation. Elongation at the point of fracture is known as total elongation, e_T. It has been used extensively as an approximate indication of sheet metal formability. However, no single property is a reliable indicator of formability under all conditions.

Reduction in area $(A_0 - A)/A_0$ is calculated from measurements of actual specimen width and thickness in the narrowest part of the necked region. The true strain, which cannot be determined from length measurements in the post-uniform elongation region, is also calculated from these values.

The true strain in the necked region is equal to $\ln(dL/dL_0)$, where dL is a small element of length in this region, whose original length was dL_0. Equating the original and final volumes of this element of length gives:

$$V_0 = A_0 dL_0 = V = A dL$$

or

$$\varepsilon = \ln\left(\frac{dL}{dL_0}\right) = \ln\left(\frac{A_0}{A}\right) \quad \text{(Eq 9)}$$

The relationship between the reduction in area at fracture and the minimum bend radius is described as follows (Ref 25): for values of reduction in area at fracture, q, below 0.2, the ratio of the minimum bend radius, R_m, to sheet thickness, t, is given by:

$$\frac{R_m}{t} = \frac{1}{2q} - 1 \quad \text{(Eq 10)}$$

For values of q greater than 0.2:

$$\frac{R_m}{t} = \frac{(1-q)^2}{(2q-q^2)} \quad \text{(Eq 11)}$$

Strain-Hardening Exponent. The n value, $d\ln\sigma_T/d\ln\varepsilon$, is given by the slope of a graph of the logarithm of the true stress versus the logarithm of the true strain in the region of uniform elongation. For materials that closely follow the Holloman constitutive equation (Eq 4), an approximate n value can be obtained from two points on the stress-strain curve by the Nelson-Winlock procedure (Ref 26). The two points commonly used are at 10% strain and at the maximum load. The ratio of the loads or stresses at these two points is calculated, and the n value and uniform elongation can then be determined from a table or graph. The accuracy of the n value determined in this way is ± 0.02.

The n value can be determined more accurately by linear regression analysis, as in ASTM E 646. For some materials, n is not constant, and initial (low strain), terminal (high strain), and sometimes intermediate n values are determined. The initial n value relates to the low deformation region, in which springback is often a problem. The terminal n value relates to the high deformation region, in which fracture may occur.

Plastic Strain Ratio. The r value, or anisotropy factor, is defined as the ratio of the true width strain to the true thickness strain in a tension test. Generally, its value depends on the elongation at which it is measured. It is usually measured at 10, 15, or 20% elongation.

The r value is calculated approximately from the measured width and length as:

$$\varepsilon_w = \ln\left(\frac{w}{w_0}\right)$$

$$\varepsilon_t = \ln\left(\frac{t}{t_0}\right) = \ln\left(\frac{L_0 w_0}{L w}\right) \quad \text{(Eq 12)}$$

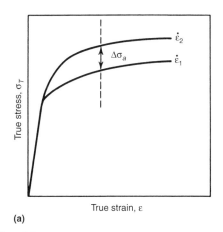

(a)

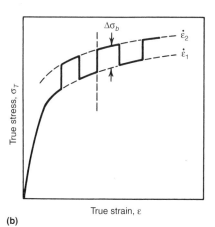

(b)

Fig. 10 Methods for determining strain-rate sensitivity (m value). (a) Duplicate test method. (b) Changing rate method

where constancy of volume ($Lwt = L_0w_0t_0$) has been used and

$$r = \frac{\varepsilon_w}{\varepsilon_t} = \frac{\ln\left(\dfrac{w}{w_0}\right)}{\ln\left(\dfrac{L_0w_0}{Lw}\right)} \qquad \text{(Eq 13)}$$

The average r value, or normal, anisotropy (r_m), and the planar anisotropy, or Δr value, can be calculated from the values of r in different directions using Eq 1, 2, 8, and 12.

Strain Rate Sensitivity. The m value, $d \ln \sigma/d \ln \dot{\varepsilon}$, is determined either from duplicate tension tests performed at different strain rates or from a single test in which the rate is alternated between two levels during the test. These methods are shown schematically in Fig. 10. The m value can be determined at various strain levels in the region of uniform elongation:

$$m = \frac{\ln\left(\dfrac{\sigma_1}{\sigma_2}\right)}{\ln\left(\dfrac{\dot{\varepsilon}_1}{\dot{\varepsilon}_2}\right)} \qquad \text{(Eq 14)}$$

In some materials, m is insensitive to strain (Ref 4, 27). In other materials, however, m is sensitive to strain and strain rate (Ref 28). In many materials, m increases and n decreases with an increase in temperature (Ref 29), sometimes to the extent that superplastic properties develop.

Determining n and r Values

The time and facilities required for sample preparation and for performing the uniaxial tension test make it difficult to use for online process control. The following simplified tests for determining n and r are more suitable for this purpose. The circle arc elongation test and the rapid-n test use tension specimens with two sections that differ in width by approximately 5% to determine n values. Fracture almost always occurs in the narrow section, but the final measurements are made on the wide section, which elongates uniformly. The r value can be obtained from these tests, but for ferritic steels, the Modul-r test is faster and easier to perform. This test actually measures the elastic modulus of the specimen and uses an empirically determined correlation between the modulus and the r value.

The circle arc elongation test does not require measurement of the applied loads (Ref 30). It uses a rectangular tension specimen with a reduced width section produced by milling a pair of small circular arc notches on opposite sides. The gage length is marked in the full-width section, and the specimen is pulled to fracture, which usually occurs in the narrow section. The uniform elongation is measured in the full-width section. The value is slightly lower than that obtained in the conventional tension test and gives a slightly lower n value. However, it is suitable for production control. The r value can be determined by the additional measurement of the change in width in the full-width section.

The rapid-n test provides rapid and fairly accurate measurements of yield and tensile strengths, elongation, and n and r values (Ref 31). It requires relatively simple equipment and can be performed in less than 5 min, including specimen preparation. The test is suitable for sheet metals whose properties are represented accurately by the Holloman equation, $\sigma_T = k\varepsilon^n$. It has been used successfully on low-carbon and stainless steels and on a variety of nonferrous alloys.

The test specimen, which is punched directly from the sheet sample, has the dimensions shown in Fig. 11. Generally, 25 mm (1.0 in.) gage lengths are marked on both the wide and narrow sections, and the specimen is strained to fracture in a manual or motorized load frame or in a tension-testing machine. The yield load is measured if there is discontinuous yielding, and the maximum load is measured. Yield and tensile strengths are calculated from the measured loads and the initial dimensions of the narrow section. If the yielding is continuous, the yield strength can be calculated from the tensile strength and n value, as indicated subsequently.

An empirically determined correction is applied to compensate for the effect of the sheared edges of the specimens. For steel, this correction reduces the measured yield and tensile strengths by 13.6 MPa/mm (50.0 ksi/in.) of initial sample thickness. The n value is calculated from:

$$n = \frac{\ln\left(\dfrac{w_{02}t_0}{w_{f1}t_f}\right) - n}{\left[\dfrac{\ln(w_{01}t_0/w_{f1}t_f)}{n}\right]} \qquad \text{(Eq 15)}$$

where w_{01} and w_{02} are the initial widths of the wide and narrow sections, respectively; w_f is the final width of the wide section; and t_0 and t_f are the initial and final thicknesses of the wide section, respectively. Equation 15 can be solved iteratively in four steps, or by means of a simple computer program, beginning with a trial value of 0.24 for n.

For materials that do not have a discontinuous yield point, the yield strength can be calculated as:

$$YS = TS\left(\frac{C}{n}\right)^n \qquad \text{(Eq 16)}$$

where TS is the tensile strength and C is a constant. For low-carbon steels, $C \approx 0.02$. For other materials, C must be determined empirically. The r value can be computed from the initial and

final dimensions of the wide section as described previously.

The Modul-r test measures the elastic modulus (Young's modulus) of low-carbon steel samples by determining their resonant frequencies by exciting them using an oscillating magnetic field (Ref 32). The elastic modulus is directly proportional to the square of the resonant frequency, and simple empirical relationships exist between the directionally averaged elastic modulus and the r_m value and between the planar variation of the modulus and the Δr value.

This test uses a flat 102×6.35 mm (4.0×0.25 in.) punched specimen and a specially designed, commercially available magnetostrictive oscillator. The specimen is placed inside drive and pickup coils in the oscillator. An alternating current passed through the drive coil produces an alternating magnetic field, which causes magnetostrictive oscillations in the sample. The oscillations induce an alternating current in the pickup coil. This current is used to change the frequency of the current in the drive coil to maximize the amplitude of the oscillations, that is, to obtain the resonant frequency, which is displayed digitally.

The relationships used to determine r_m and Δr from the resonant frequency, f, are:

$$E = 4_\rho L^2 f^2 \qquad \text{(Eq 17)}$$

$$r = \frac{4822.6}{(E_m - 267.7)^2} - 0.564 \qquad \text{(Eq 18)}$$

$$\Delta r = 0.031 - 0.0468\Delta E \qquad \text{(Eq 19)}$$

where E is the elastic modulus in gigapascals; ρ and L are the density and length of the specimen, respectively; and E_m and ΔE are defined analogously to r_m and Δr.

This test provides a more reproducible measure of r_m and Δr than the conventional tension-testing method and is less sensitive to differences between operators. It can be performed in 5 min, including specimen preparation. When alloy steels or ferritic stainless steels are tested, a different correlation between the modulus and r value must be used. This must be determined experimentally. When the test is used on coated products, the coating must be removed chemically prior to testing.

Plane-Strain Tension Testing

In conventional uniaxial tension testing, the sample is strained in the region of drawing; that

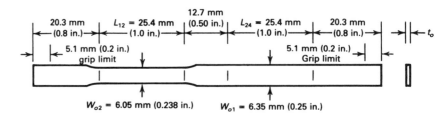

Fig. 11 Rapid-n test specimen. Source: Ref 31

is, the minor or width strain is negative. The test does not provide information on the response of sheet materials in the plane-strain state, in which the minor strain is zero. However, it can be modified to produce this strain state in part of the sample. This modification involves the use of a very wide, short sample or the use of knife edges to prevent transverse (width) strain in part of the sample.

Wide Sample Methods. Increasing the width of the sample and decreasing the gage length changes the strain state from one with a large negative minor strain component toward the plane-strain state, in which the minor strain component is zero. In the rectangular sheet tension test, samples with length-to-width ratios of 1 to 1, 1 to 2, and 1 to 4 are used to approach the plane-strain conditions (Ref 33). Gage lengths are constrained further by reinforcements welded onto each side of the sample at both ends, thus making the samples three layers thick except in the gage length.

The minimum minor strain obtained with the 1 to 4 length-to-width ratio is −0.05 times the major strain, which is close to the plane-strain condition of zero minor strain. The in-plane strains are measured by means of grid markings on the samples, and through-thickness deformations can be observed by holographic interferometry.

A similar approach was used in testing many wide specimen designs to determine the effect of edge profile and length-to-width ratio on strain state (Ref 34–36). The specimen geometry that yielded the highest center strain at failure with a large region of plane strain is shown in Fig. 12. The plane-strain region, which is arbitrarily taken as the region where $|e_2/e_1|$ is less than 0.2, occupies approximately 80% of the specimen width. The outer part of the specimen deforms in a similar manner to a standard tension test specimen.

Special grips were developed that exert a high clamping force at the inner contact lines. This minimizes distortion and slippage in these regions, giving the test well-defined boundary conditions. The results of both aforementioned types of wide specimen tension tests correlated well with stress-strain predictions obtained by finite element modeling using materials properties obtained in the standard tension test (Ref 34, 37).

Width Constraint Method. In the width constraint method, a rectangular sample is used that has a central gage section reduced in width by circular notches (Ref 38). The gage section is clamped between two pairs of opposing parallel knife edges (stingers) aligned with the sample axis. The knife edges prevent transverse (width) strain in this region. The sample is pulled to fracture in a tension-testing machine, and the plane-strain limit (necking) and fracture strains are determined from thickness measurements made on the fractured sample. This procedure is described in detail in Ref 38. The use of a spring-loaded clamp around the knife edges makes adjustment of the clamp during testing unnecessary.

Biaxial Stretch Testing

Two tests that determine the properties of sheet metals in biaxial stretching without involving surface friction effects are the Marciniak biaxial stretching test (Ref 39) and the hydraulic bulge test (Ref 40). The Marciniak test subjects the sample to in-plane biaxial stretching, but does not determine the stresses. In the hydraulic bulge test, the stresses can be determined, but the sample is deformed into a dome, which involves out-of-plane stresses and strains.

Marciniak Biaxial Stretching Test. A disk of the test material is stretched over a flat-bottomed punch of cylindrical or elliptical cross section. This creates uniform in-plane biaxial strain in the center of the sample, with a strain ratio that is determined by the ratio of the major and minor diameters of the punch. Most testing has been performed with a cylindrical punch, which produces balanced biaxial stretching.

The center of the punch is hollowed out to eliminate friction in this area, and a spacer is placed between the sample and the punch. The spacer is a disk of material similar to that under test—with the same diameter, but with a hole at the center. The experimental arrangement is shown in Fig. 13. As the disk and spacer are stretched over the punch, the hole in the spacer enlarges, and the central part of the test sample is deformed in uniform in-plane biaxial stretching.

The function of the spacer is to reverse the direction of the surface friction experienced by the sample. In the absence of the spacer, the surface friction opposes the movement of the sample over the punch and reduces the maximum strain level attainable. The spacer deforms more easily than the test sample because of the hole in the center, and it exerts a frictional force on the sample directed outward over the punch radius.

For the material to stretch freely, the punch and die radii must be adequate for the thickness of the material under test. A ratio of spacer hole diameter to punch diameter of 1 to 3 has been used successfully. The strains can be measured by using grid circles, squares, or other suitable markings. The test has the following applications:

- Determination of the limiting strains of materials in uniform in-plane biaxial stretching without surface friction
- Application of a carefully controlled level of uniform in-plane biaxial strain to samples with large areas to be used in other tests—for example, tests to determine the effect of different strain paths on the limiting strain levels
- Detection of defects, such as inclusions, by straining a sample of large area to a uniformly high level; defects will cause early localized fracture, usually parallel to the rolling direction

Hydraulic Bulge Test. The periphery of a sheet metal sample is clamped between circular or elliptical die rings, and hydraulic pressure is applied on one side of the sample to deform it into a dome, as shown in Fig. 14. The edge of the sample is prevented from slipping by a lock bead placed in the die rings. This consists of a ridge with small radii on one ring and a matching groove on the other.

With circular die rings, the center of the dome has been found to be nearly spherical (Ref 40). The stress and strain states in this region can be determined from the curvature and extension and the fluid pressure. A biaxial test extensometer has been developed that measures the extension and curvature by means of a spherometer and an extensometer that are in direct contact with the dome (Ref 41).

A system for controlling the strain rate in this test was developed (Ref 42) because significantly different test results have been obtained (Ref 43) under conditions of constant strain rate and constant fluid flow. The system

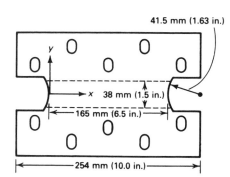

Fig. 12 Plane-strain tensile test specimen. Source: Ref 36

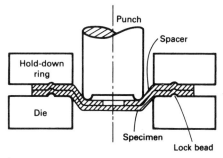

Fig. 13 Schematic of Marciniak biaxial stretching test

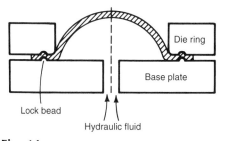

Fig. 14 Schematic of hydraulic bulge test

uses feedback from the extensometer signal to operate a servovalve, which controls the flow of hydraulic oil to the bulge. This system was used to determine the strain rate sensitivity of aluminum alloys to a much higher strain level than is possible in the tension test.

A computerized system is available that uses electronic vision to measure the principal strains and closed-loop feedback to control the strain rate (Ref 44). This system monitors the relative positions of the centers of three closely spaced white dots painted on a black background at the center of the sample. Initially, the dots form a right-angled isosceles triangle. The principal strains are computed from the change in the spacings of the dots and changes in the angle they subtend. This information is recorded and also used to maintain a constant strain rate by controlling the hydraulic pressure. Strains can be computed only once per second, which limits the maximum controllable strain rate.

A camera is mounted so that it maintains a constant distance from the top of the dome and the dots, except for the effect of the curvature of the dome, which is negligible. The stress state is determined by measuring the curvature of the dome with a contacting spherometer and by measuring the hydraulic pressure with a strain gage pressure transducer.

For thin samples, bending stresses can be neglected, and the radial (meridional) stress, σ_r, is given by:

$$\sigma_r = \frac{pR}{2t} \qquad \text{(Eq 20)}$$

where p is the hydraulic pressure, R is the radius of curvature, and t is the instantaneous thickness. The thickness is calculated from the measured strains using constancy of volume (Fig. 15). At the top of the dome, the sample is in balanced biaxial stretching, and the circumferential stress, σ_c, is equal to the radial stress.

For convenience, it is customary to express the results of hydraulic bulge tests in terms of the true thickness stress and strain. This is done by theoretically superimposing a hydrostatic compressive stress that does not influence deformation and that has in-plane components equal to the actual radial and circumferential tensile stresses. This converts the stress state to a simple uniaxial thickness compressive stress, as shown in Fig. 16.

The true thickness strain, ε_t, can be obtained from the radial strain, ε_r, and circumferential strain, ε_c, by using the constancy of volume condition:

$$\varepsilon_t = -\varepsilon_r - \varepsilon_c \qquad \text{(Eq 21)}$$

This enables the results to be represented by a true compressive stress-strain curve in the thickness direction. The hydraulic bulge test has the following applications:

- Intrinsic materials characterization in biaxial stretching, which is a very common strain state in production stampings

- Testing to much higher strain levels than those achievable in tension testing (in some cases, by as much as a factor of ten), particularly for heavily cold-worked materials

- Checking the validity of plasticity theories that attempt to predict the yielding behavior of metals in all stress states from properties measured in uniaxial and plane-strain tension testing

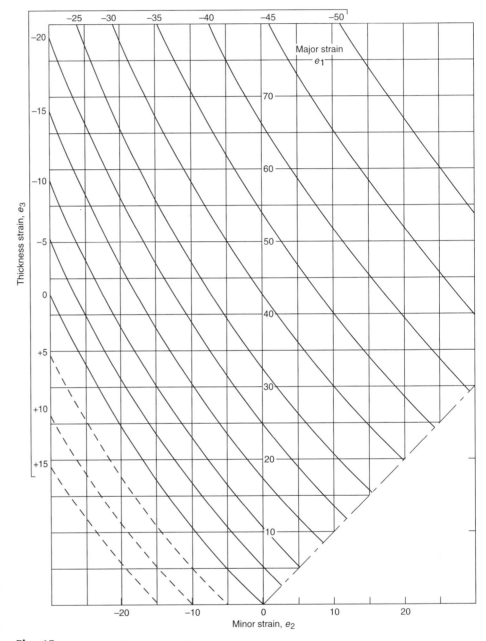

Fig. 15 Constancy of volume nomograph

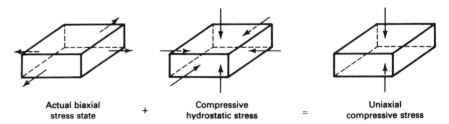

Fig. 16 Addition of compressive hydrostatic stress to biaxial tensile stress

Shear Testing

Two tests have been developed to determine the properties of sheet metals subjected to planar shear deformation: the Marciniak in-plane sheet torsion test (Ref 45) and the Miyauchi shear test (Ref 46).

Marciniak In-Plane Sheet Torsion Test. A flat 50 mm (1.97 in.) square sample is effectively divided into three zones: an inner circular zone, which is clamped; a ring-shaped middle zone, which surrounds the inner zone and is free to deform; and an outer ring-shaped zone, which is clamped. The inner zone is rotated in its plane relative to the outer zone, which deforms the middle zone in shear. The sample is deformed to fracture, and the angular rotations at two radii in the middle zone are measured by means of calibrated drums that rotate with the sheet. This is shown schematically in Fig. 17.

For materials that follow the power law shear strain-hardening relationship:

$$\tau = C\gamma^n \tag{Eq 22}$$

where τ is the shear stress, C is a constant, and γ is the shear strain; it can be shown that (Ref 45):

$$n = \frac{2\log\left(\frac{R_a}{R_b}\right)}{\log\left(\frac{\alpha_a}{\alpha_b}\right)} \tag{Eq 23}$$

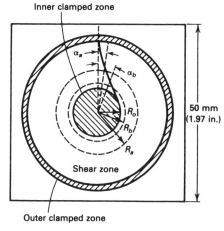

Fig. 17 Deformed Marciniak in-plane torsion test specimen

where α_a and α_b are the angular displacements at radii R_a and R_b.

Shear fracture occurs where the shear stress is greatest—at the inner radius, R_0, of the deforming ring. The shear fracture strain, γ_f, is given by:

$$\gamma_f = \frac{2\alpha_a\left(\frac{R_a}{R_0}\right)^{2/n}}{n} \tag{Eq 24}$$

This test enables the forming properties of sheet metals to be determined at much higher strain levels than is possible in the uniaxial tension test and in a different strain state, that is, in shear.

The Miyauchi shear test determines the properties of sheet metals in planar shear deformation by means of a modified tension technique (Ref 46). The test uses flat, rectangular specimens whose ends are divided into three equal sections by parallel longitudinal slits, as shown in Fig. 18(a).

The specimen is clamped in a fixture that prevents out-of-plane deformation. The inner and outer sections are then pulled in opposite directions in a tension machine. This produces a shear stress in the regions between the inner and outer sections and deforms the specimen, as shown in Fig. 18(b). The deformation in these regions is uniform, except at the ends.

The shear strain, γ, is the tangent of angle θ, which is the change in direction of lines scribed across the specimen as they pass through the shear zone, as shown in Fig. 18(b). The strain can also be determined from the displacement of the inner section once a relationship has been established between the displacement and θ, which must be done for each type of sheet metal tested. Shear stress-strain curves are given in Ref 46 for three different steels. These curves show differences in the strain dependence of the work-hardening coefficient from that in the tension test.

Hardness Testing

Hardness, or the resistance to indentation by a concentrated load applied by a suitable indenter, has been used in many stamping plants as a measure of formability. Generally, formability decreases with increasing hardness, but the fine-

scale correlation between these properties has not been reliable. The test can be used effectively to monitor changes in a particular grade of material caused by changes in processing that may affect formability.

For steels, hardness measurements correlate well with yield strength values (Ref 47). Therefore, hardness testing is useful in quality control to ensure that the material in use is the specified grade and has the required strength level.

The Rockwell hardness test (ASTM E 18), which is described in detail in the article "Macroindentation Hardness Testing" in Volume 8 of *ASM Handbook*, is typically used to determine the hardness of such materials. The load and indenter must be selected for the gage and hardness range of the material according to the test specification to ensure that the indentation is the appropriate size. If the indentation is too deep in a sheet sample, the reading will be artificially high because of the influence of the supporting anvil. This becomes a more serious consideration with the use of thinner gage sheet metal in many industries. For thicker and harder materials, the Rockwell B scale is commonly used, and for thinner or softer materials, the Rockwell 30T superficial scale is used. Hardness readings are influenced by the degree of flatness and surface conditions. In addition, the presence of cold-worked surface layers can cause high hardness readings that suggest a lower formability level than actually exists.

Simulative Tests

For many forming operations, tests that simulate the operation are more useful and relevant than fundamental intrinsic property measurement tests. These tests subject the work material to deformation that closely approximates the production operation, including the effects of factors not present in the intrinsic tests, such as bending and unbending and friction between the work materials and die surfaces. Because these additional factors are present, simulative tests tend to be less reproducible than intrinsic tests and must be performed under carefully controlled conditions to minimize variability in the results. Simulative tests can be classified on the basis of the predominant forming operation involved: bending, stretching, drawing, and stretch-drawing. In addition, tests have been developed to measure wrinkling and the springback that occurs after bending or another forming operation.

Bending Tests

Two types of bending tests relate to sheet metal forming: simple bending and stretch-bending tests. Simple bending tests are useful in predicting how the sheet metal will perform when bent without tension, as in a hemming operation. Stretch-bending tests relate the

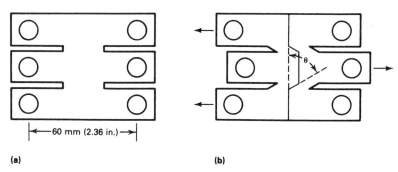

Fig. 18 Miyauchi shear test specimen. (a) Undeformed. (b) Deformed. Source: Ref 46

response to combined bending and stretching, as when sheet metal is pulled over a punch or die radius.

Simple bending tests can be performed in various ways (ASTM E 290). The simplest method for thin sheet material is to clamp a specimen and a bending die in a vise, as shown in Fig. 19, and to bend the specimen over the die manually or with a nonmetallic mallet.

If the specimen bends through 180° without fracturing or cracking, the experiment is repeated using a bending die of smaller radius. A modified test is performed for highly ductile metals that have extremely small bend radii. The specimen is initially bent at its midpoint, through less than 90°, over a small radius. The test is then completed by pressing the ends of the specimen together between flat platens without a bending die placed between the platens.

The ratio of specimen width to thickness should be greater than 8 to 1, and sheared edges should be machined, filed, or sanded to remove the heavily cold-worked metal present. The orientation of the specimen with respect to the rolling direction may be important because it affects the resistance of the specimen to fracture. Specimens cut perpendicular to the rolling direction usually require a larger bend radius and therefore provide a more conservative measure of this property.

For low-carbon sheet steels, the minimum bend radius is usually not a limiting factor. For high-strength steels and aluminum alloys, it sometimes is, and methods such as rope hemming, which increase the bend radius, have been developed to prevent cracking during the hemming of these materials.

Stretch-Bending Tests. A rectangular strip of sheet metal is clamped at its ends in lock beads and deformed in the center by a punch, as shown in Fig. 20. There are two types of stretch-bending tests: the hemispherical test, in which a hemispherical-tipped punch and a concentric circular lock bead are used, and the angular test, in which a wedge-shaped punch and straight parallel lock beads are used. The hemispherical test involves a range of strain states. The angular test produces the plane-strain state.

The punch travel between initial contact and specimen fracture is measured. The conditions are chosen so that fracture occurs in the region of punch contact. When fracture occurs in the unsupported region, which tends to happen with narrow thin-gage specimens and large punch radii, the test effectively becomes a tension test.

The results of several hemispherical and angular stretch-bending tests on three types of steels and an aluminum alloy have been reported (Ref 48). For the hemispherical test, the effects of variations in punch tip radii ranging from 3.2 to 51 mm (0.13 to 2.0 in.), in sheet thicknesses ranging from 0.5 to 3.3 mm (0.02 to 0.13 in.), and in specimen widths ranging from 25 to 203 mm (1.0 to 8.0 in.) were investigated in the dry and lubricated conditions. The tests showed that the height at fracture increased with increasing punch radius and sheet thickness and with the use of lubricants. It decreased with increasing specimen width in the range of 102 to 203 mm (4.0 to 8.0 in.), in which fracture occurred in the region of punch contact. The ranking of two of the steels was found to be dependent on specimen thickness.

Fewer conditions were investigated in the angular test. The results for a 76 mm (3.0 in.) wide specimen and punch radii ranging from 1.6 to 6.4 mm (0.06 to 0.25 in.) showed much greater heights than for the same conditions in the hemispherical test. Increases in height with increasing punch radius were also evident, but in contrast to the hemispherical case, a decrease with increasing thickness was observed. Correlation between the results of these tests and production experience is reported to be fairly good.

Data from the angular stretch-bending test have been analyzed and indicate that fracture occurs at a constant limit strain that is independent of sheet thickness and punch radius (Ref 49). Stretch-bending tests are useful for materials selection and for predicting the effects of materials substitution and gage reduction in many forming operations.

Stretching Tests

Historically, ball punch tests, such as the Olsen cup test and Erichsen cup test, have been used to determine the properties of sheet metals in stretching. These tests stretch a specimen over a hardened steel ball and measure the height of the cup produced. Tests that stretch the specimen over a much larger hemispherical dome have been developed, including the limiting dome height test (LDH), which uses specimens of different widths to control the strain ratio at fracture.

Many forming operations involve stretching an edge of a part or a cutout (hole) in a part. For example, when a concavely contoured edge is flanged, the metal is stretched. The ability of the material to undergo this type of forming operation can be measured by the hole expansion test. In this test, a cylindrical, hemispherical, or conical punch is pushed through a circular hole of smaller diameter in the specimen. This initially increases the diameter of the hole and then forms a rim of stretched metal. The edge ductility of the material is indicated by the amount of hole expansion that occurs without edge cracking.

Ball Punch Tests. The Olsen and Erichsen cup tests are similar, differing principally in the dimensions of the tooling used. The Olsen test (ASTM E 643) uses a 22.2 mm (0.875 in.) diameter hardened steel ball and a die with a 25.4 mm (1.0 in.) internal diameter (28.6 mm, or 1.125 in., for gages over 1.5 mm, or 0.06 in.) and a 0.81 mm (0.032 in.) die profile radius, as shown in Fig. 21. The Erichsen test, which is extensively used in Europe, uses a 20 mm (0.79 in.) diameter ball and a die with a 27 mm (1.06 in.) internal diameter and a 0.75 mm (0.03 in.) die profile radius.

In both tests, the cup height at fracture is used as the measure of stretchability. The preferred criterion for determining this point is the maximum load. When this cannot be determined, the onset of a visible neck or fracture can be used, but this yields a slightly different value. The cup height measured by means of a visible fracture is 0.3 to 0.5 mm (0.012 to 0.020 in.) greater than the height measured at the maximum load.

These tests, as indicators of stretchability, should correlate with the n value, but the correlation is not satisfactory. Improved correlations with the total elongation (Ref 50) and reduction in area (Ref 51) have been reported. Some investigators have reported poor reproducibility of results in the Olsen and Erichsen tests and poor correlation with production experience (Ref 52, 53). Satisfactory reproducibility and correlation in specific cases have been reported when

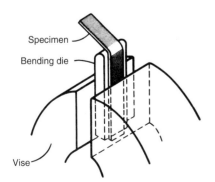

Fig. 19 Schematic of simple bending test

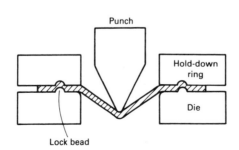

Fig. 20 Schematic of stretch-bending test

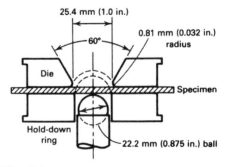

Fig. 21 Schematic of Olsen cup test

experimental conditions were carefully controlled (Ref 50).

The variability in tests has been attributed to the small size of the penetrator, uncontrolled drawing-in of the flange, and inconsistent lubrication (Ref 52, 53). The small size of the penetrator leads to excessive bending, particularly in thicker sheet, and is generally unrepresentative of production conditions. Drawing-in can be controlled somewhat by standardizing the specimen size and by using a high (~71 kN, or ~8 tonf) clamping force. Even greater control can be achieved by using lock beads or serrated dies (dies with concentric circular ridges of triangular cross section that dig into the specimen and prevent slippage).

Consistent lubrication can be achieved by using oiled polyethylene between the specimen and penetrator. The problems with the Olsen and Erichsen tests have led to the development of stretching tests that use a much larger diameter punch and a lock bead to prevent drawing-in.

Hemispherical dome tests using 50.8, 76.2, and 101.6 mm (2.0, 3.0, and 4.0 in.) punches have been reported (Ref 52, 53). A 100 mm (3.94 in.) test is the most widely used. Typical tooling designed for this test is shown in Fig. 22. The lock bead, in combination with a hold-down force of about 222 kN (25 tonf), completely prevents drawing-in of the flanges.

The specimens fracture circumferentially at a distance (for lightly lubricated low-carbon steel) of 35 to 40 mm (1.38 to 1.57 in.) from the pole, at which point the radial strain peaks sharply. The circumferential strain varies gradually from a maximum of 10 to 20% at the pole to zero at the lock bead.

The hemispherical dome test yields more reproducible results than the Olsen and Erichsen cup tests. For low-carbon steels, the dome height, which is measured at the point of maximum load, increases linearly with the n value. For a wide range of material (including brasses, aluminum alloys, and zinc), optimal correlation is found between the dome height and the total elongation, which incorporates the effects of strain rate hardening and limiting strains.

Overall, the use of lubrication in hemispherical dome tests is beneficial. A thin layer of a standard lubricant, applied in a consistent manner, reduces scatter in test results, simulates production conditions more closely, reduces damage to the tooling, and simplifies specimen preparation. The improved sensitivity obtained in the dry condition is negated by the increased scatter in the results.

The use of lubrication makes the strain ratio at fracture more biaxial. This is undesirable for production simulation because most production failures occur in the region of plane strain, that is, in a less biaxial manner. To control the strain ratio at fracture, specimens of different widths were used (Ref 54). This technique has been developed further into the limiting dome height test (Ref 55, 56).

Limiting Dome Height (LDH) Test. Specimens of various widths are held in a circular lock bead and stretched over a 100 mm (3.94 in.) dome using tooling of the type shown in Fig. 22. In principle, this test can be used to duplicate a large range of production failure strain states and to select the most suitable material for each particular operation. In practice, most production failures occur close to plane strain, which is generally the strain state at the minimum on a plot of dome height versus specimen width. Consequently, attention has concentrated on this minimum value.

When testing a new material, initial tests should be performed to determine the specimen width that yields the minimum dome height, or LDH_0 value, and the corresponding minor strain. Once this has been established, tests can be conducted at this width only. For low-carbon steels, the minimum dome height occurs at a width of approximately 124 mm (4.9 in.). This can also be used as an approximation for other materials. Increments in test specimen width of ±3 mm (±0.12 in.) are sufficiently close.

It has been found that for specimens lubricated lightly with a wash oil, the dome height increases with decreasing hold-down force below about 250 kN (28 tonf). This is attributed to the drawing-in of the flange. Therefore, a hold-down force of at least 250 kN (28 tonf) should be used. The limiting dome height is taken as the height at which the maximum load occurs.

Testing has shown a correlation between the limiting dome height test and production stamping performance (Ref 57). Some problems have been encountered with test reproducibility over a period of time and among different test facilities (Ref 58). Numerous attempts have been made to determine a correlation between the limiting dome height test and mechanical and forming property measurements. The dome height depends on the ability of the material to distribute strain and on the limiting strain level and would therefore be expected to correlate with the total elongation. Correlation for a range of different materials has been reported (Ref 59).

The specimens used in the limiting dome height test can be sheared or blanked from the sheet sample, and the test can be performed rapidly on equipment that automatically measures the dome height at the maximum punch load. The test has been widely used for both production control and research applications.

OSU Test

The LDH test has been widely used, but as mentioned in the prior section, has some problems with reproducibility, which can make interpretation of test results difficult. The primary reason for this lack of consistency is that the hemispherical geometry of the tooling causes both spatial and temporal variation in strain during the test (Ref 60–62). Only a narrow window of opportunity exists within which to obtain a plane-strain fracture. For example, changing friction conditions slightly will result in different degrees of draw-in of the unclamped sides of the blank, causing different strain states at failure.

After considering both the positive and negative attributes of the LDH test, a new formability test was developed at The Ohio State University (Ref 63–65). This test, named the OSU test, uses a cylindrical punch and dies with a lockbead to stretch a wide blank to failure (Fig. 23).

The punch height at failure is an index of formability, as in other stretch formability tests. The test is quick and simple to perform because only one specimen width is required (typically about 5 in.) to obtain a plane-strain failure. The strain path remains at or near plane strain throughout the test, which improves test consistency significantly compared with LDH (Ref 63, 65). This contrasts with the LDH test geometry, which can produce a large range of strains, from near uniaxial tension to balanced biaxial tension (see the section "Forming Limit Diagrams" in this article). In this sense, the LDH tooling is quite versatile; however, for rapid and consistent evaluation of sheet metal stretch formability, as needed for quality assurance in a factory environment, the OSU test is a good option.

Hole Expansion Test. A flat sheet specimen with a circular hole in the center is clamped between annular die plates and deformed by a punch, which expands and ultimately cracks the

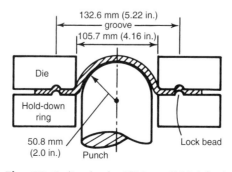

Fig. 22 Tooling for the 101.6 mm (4.0 in.) hemispherical dome test. Source: Ref 52

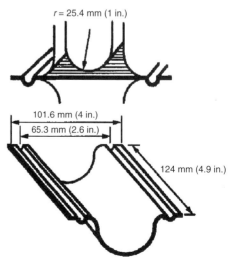

Fig. 23 Schematic of the OSU formability test. Source: Ref 65

edge of the hole. Flat-bottomed hemispherical and conical punches have been used, and in some cases die plates have been equipped with lock beads to prevent drawing-in of the flange. The punch should be well lubricated and should have a large profile radius. A spacer can be used between the punch and the sample, as in the Marciniak test. Figure 24 illustrates the hole expansion test using a flat-bottomed punch.

The test is terminated when a visible crack is observed, and the hole expansion is expressed as the percentage of increase in hole diameter:

$$\text{Hole expansion (\%)} = \frac{100(D_f - D_0)}{D_0} \quad \text{(Eq 25)}$$

where D_0 and D_f are the initial and final hole diameters, respectively.

The results of several hole expansion tests on eight different types of steel are reported in Ref 66. Square specimens measuring 203 mm (8.0 in.) on each side with a 25 mm (1.0 in.) diam punched hole, a 101.6 mm (4.0 in.) diam hemispherical punch, and die plates with a 2 mm (0.08 in.) radius lock bead were used. The measured hole expansion ranged from 24 to 82% for steels with yield strengths ranging from 253 to 537 MPa (36.7 to 77.9 ksi).

In most cases, removing the burr and cold-worked metal from the edge of the punched hole increased the hole expansion considerably. The hole expansion also increased with increasing total elongation and r_m value and decreased with increasing tensile strength (which was anticipated because total elongation decreases with increasing tensile strength). Inclusions were observed in crack locations, and inclusion shape control improved hole expansion performance.

Drawing Test

Swift Cup Test. The most commonly used test for deep drawability is the Swift cup test. Circular blanks of various diameters are clamped in a die ring and deep drawn into cups by a flat-bottomed cylindrical punch. The standard tooling for this test is shown in Fig. 25. Drawability is expressed as either the limiting draw ratio (LDR) or the percentage of reduction. The limiting draw ratio is the ratio of the diameter, D, of the largest blank that can be successfully drawn to the diameter, d, of the punch:

$$\text{LDR} = \frac{\text{maximum blank diameter}}{\text{punch diameter}} = \frac{D}{d} \quad \text{(Eq 26)}$$

Percentage of reduction is defined as:

$$\text{Percentage of reduction} = \frac{100(D-d)}{D}(\%) \quad \text{(Eq 27)}$$

Cup height, h, is approximately (Ref 67):

$$h = \frac{(D^2 - d^2)}{4d} \quad \text{(Eq 28)}$$

An alternative method for determining the limiting draw ratio uses blanks of a single diameter, which is less than the critical diameter in the standard test (Ref 68). The blanks are drawn to the maximum load, which usually takes place before 50% of the draw has occurred. The clamping force is then increased to prevent further drawing-in of the flange, and the load is increased to the point of fracture. The limiting blank diameter (LBD) is defined by:

$$\text{LBD} = \left[\frac{\text{fracture load} \cdot (\text{blank diam} - \text{die diam})}{\text{maximum drawing load}} \right] + \text{die diam} \quad \text{(Eq 29)}$$

The limiting draw ratio is given by:

$$\text{LDR} = \frac{\text{LBD}}{\text{punch diameter}} \quad \text{(Eq 30)}$$

This method has been shown to correlate well with the standard test for a range of materials of widely different drawability (Ref 68).

The limiting draw ratio increases with normal anisotropy (r_m) and thickness, particularly at the low ends of the ranges for these variables, but is not sensitive to the n value (Ref 69). The limiting draw ratio also increases as the punch profile radius increases up to about eight times the sheet metal thickness, as the die profile radius increases up to about 12 times the metal thickness, and as the punch speed increases. The height of the ears formed in this test is proportional to the Δr value.

Too low a blankholder force may cause wrinkling, and too high a blankholder force may cause fracture at the punch profile radius. The die rings should be well lubricated, but the punch should not be lubricated. By not lubricating the punch, the amount of stretching that occurs over the punch profile radius and the tendency for splitting to occur at this location are reduced.

Stretch-Drawing Tests

Many forming operations involve stretching and drawing; for example, square cups have drawn corners and stretched sides. The ratio of stretching to drawing in an actual part can be measured by a shape analysis technique (Ref 70).

A line is drawn from a reference point (for example, the center of the blank) to the edge of the blank, through the critical forming area. After forming, the ratio of the increases in length of this line inside and outside the initial die contact line is taken as the ratio of stretching to drawing. Two tests are commonly used for stretch-drawing: the Swift round-bottomed cup test and the Fukui conical cup test.

The Swift round-bottomed cup test resembles the Swift flat-bottomed cup test described previously. However, the top of the punch is hemispherical, which causes stretching in the center of the specimen in addition to the drawing-in of the flange to produce the wall of the cup.

This test was used to evaluate 50 different steels with a 50 mm (1.97 in.) diameter punch and 127 mm (5.0 in.) diameter specimens and with a 65 mm (2.56 in.) diameter punch and 165 mm (6.5 in.) diameter specimens (Ref 71). Hold-down forces of 490 and 981 N (110 and 220 lbf), respectively, were used at a test speed of 1 mm/s (0.04 in./s). Both sides of the specimens were lubricated with thin polyethylene sheet.

The end point of the test is determined by observing fracture visually or by detecting a drop in the punch load. Multiple regression analysis of the test results showed that the cup height at fracture increased linearly with increases in the r_m value, n value, and metal thickness.

To determine the correlation between performance of the steels in the stretch-drawing test and in actual parts production, 4 automotive stampings were made, using 12 different steels for each. The stampings had stretch-to-draw ratios ranging from approximately 1 to 5 to 2 to 1, and minor-to-major strain ratios in critical areas ranging from -0.3 to $+0.45$. The correlation coefficients between the test and stamping results had an average value of 0.92 and ranged from 0.89 to 0.94 (a value of 1.00 indicates perfect correlation). In another trial on a stamping with a stretch-to-draw ratio of 4.5 to 1, the test results did not correlate. These tests indicate that for parts that involve both stretching and

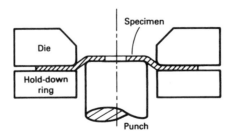

Fig. 24 Schematic of the hole expansion test with a flat-bottomed punch

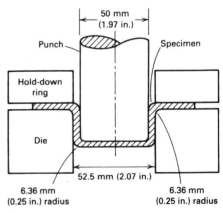

Fig. 25 Standard tooling for the Swift flat-bottomed cup test

drawing, without excessive stretching, the Swift flat-bottomed cup test is useful as a quality control tool.

Fukui Conical Cup Test. In the Fukui conical cup test, circular specimens punched from a sample of sheet metal are deformed into conical cups by means of a 12.5 to 27 mm (0.5 to 1.1 in.) diameter ball and tooling of the type shown in Fig. 26 (Ref 69, 72, JIS Z 2249). The ball size depends on the sheet thickness. The specimens are lubricated on the die side only. Lubrication on the punch side leads to tilting of the specimens. Specimens are centered and held in place by the hold-down ring and deformed to fracture by the punch.

The diameter of the base of the conical cup formed is measured and divided by the diameter of the original specimen to give the Fukui conical cup value. The end point of the test is not critical because the diameter of the cone does not change after fracture. A constant punch travel is usually used. When the test material has a high level of planar anisotropy (a high Δr value), the conical cup is asymmetric, and an average diameter must be determined. A high correlation between the Fukui conical cup value and the product of the average n value and the average r value has been reported for low-carbon steels (Ref 69).

An alternative method has been developed for performing this test (Ref 50). The punch travel between the initial contact with the specimen and the onset of a drop in the punch load, which coincides with the formation of a visible neck, is measured and used instead of the ratio of the diameters. This value, known as the formability index, correlates with the uniform elongation and therefore with the n value for low-carbon steels.

Wrinkling and Buckling Tests

Two principal types of tests are used for wrinkling and buckling: the conical cup wrinkling test and the Yoshida buckling test. The conical cup wrinkling test is similar to the Swift flat-bottomed cup test but uses a punch that is much smaller than the die opening. Consequently, the cup wall is conical and is not in contact with the punch. Under some conditions,

wrinkles form in the cup wall. In the Yoshida buckling test, a flat, square specimen is stretched slightly in the diagonal direction, and the height of the buckle that is formed is measured (Ref 73).

Conical Cup Wrinkling Test. A circular blank is clamped between annular dies and deformed by a flat-bottomed punch with a diameter that is typically about 75% of the internal diameter of the die. This procedure is illustrated in Fig. 27. At very low levels of hold-down force, wrinkling occurs in the flange. At higher levels, flange wrinkling is suppressed, but wrinkling occurs in the unsupported wall. This is caused by compressive stresses in the circumferential direction (hoop stresses) that are due to the local reduction in diameter as drawing progresses. For example, with a 75 mm (2.96 in.) diameter punch and a 100 mm (3.94 in.) diameter die, the top of the wall has a diameter of 100 mm (3.94 in.). If the cup depth is doubled, the original top of the wall becomes the new midpoint and must decrease in diameter to 87.5 mm (3.44 in.).

At high levels of blankholder force, the tensile stresses in the radial direction in the wall prevent the formation of wrinkles, and fracture at the punch or die radius becomes the limiting factor. The maximum cup height occurs at the intersection of the wall wrinkling and fracture limits, as shown in Fig. 28.

The results of experiments on several types of steel with different thicknesses and tooling of various dimensions have been reported in Ref 13 and 74. Wrinkling occurred in the unsupported

wall when the true compressive hoop strain exceeded a certain value for each level of the tensile radial strain for all tooling geometries and forming conditions. The critical wrinkling strains were plotted on the forming limit diagram, as shown in Fig. 29.

Attaining the critical wrinkling strain is strongly influenced by the dimensions of the specimen and tooling, lubrication, and the hold-down force. Changes in these variables that reduce the radial stress (that is, an increase in the die radius, improved lubrication, or a reduction in the blank diameter or the hold-down force) increase the tendency toward wrinkle formation.

Materials properties that affect wrinkling in the conical cup test are the r_m, Δr, and n values and the ratio of the flow stress to the elastic modulus. A high r_m value and low Δr value reduce wrinkling, which initiates in the directions of lowest r value. A high n value enables the hold-down force to be increased, which increases the radial force and reduces wrinkling. A low flow-stress-to-elastic-modulus ratio also reduces wrinkling.

Yoshida Buckling Test. A flat, square specimen is gripped at opposite corners and pulled in tension in the diagonal direction, as shown in Fig. 30 (Ref 73, 75). The standard

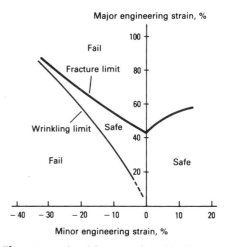

Fig. 29 Combined forming and wrinkling limit diagram. Source: Ref 68

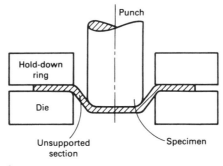

Fig. 27 Schematic of the conical cup wrinkling test

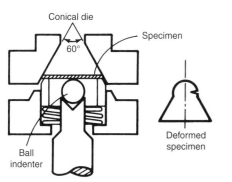

Fig. 26 Schematic of the Fukui conical cup test. Source: Ref 62

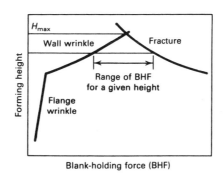

Fig. 28 Wrinkling and fracture limits in conical cup drawing. Source: Ref 67

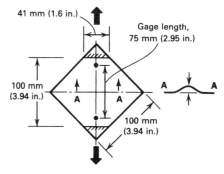

Fig. 30 Schematic of the Yoshida buckling test. Source: Ref 67

specimen is 100 mm (3.94 in.) square with 41 mm (1.6 in.) wide grips and a gage length of 75 mm (2.95 in.). The buckle height is measured over a 25.4 mm (1.0 in.) width at the center of the specimen.

Nonuniform stresses are generated in the specimen, and these stresses cause a buckle to form in the center along the direction of loading. The height of the buckle at a given elongation, for example 2%, is used as the measure of buckling.

Several investigations have been conducted on the correlation between buckle height and test material properties. The Yoshida buckling test and a conical cone wrinkling test (using a hemispherical punch) were performed on several ferrous and nonferrous materials in different tempers (Ref 76). A direct correlation for both tests between the buckling or wrinkling height and the yield strength, an inverse correlation with the work-hardening exponent, and a lack of correlation with the normal anisotropy were reported. The Yoshida test was not successful for aluminum because the specimens fractured before buckling.

The Yoshida test was performed on 31 steels of different types and thicknesses, and correlations between the slope of the buckle height versus elongation curve, which is easier to determine than the height at a particular elongation, and the yield strength and the ratio of yield strength to tensile strength were obtained (Ref 14). An inverse correlation with the instantaneous (2%) strain-hardening exponent and a lack of correlation with the uniform elongation and normal anisotropy were also noted.

Correlation between Simulative Tests and Materials Properties

Quantitative correlations between the results of simulative tests and select tensile properties have been determined for the Olsen cup, Swift flat-bottomed cup, and Fukui conical cup tests (Ref 50) and for the Swift round-bottomed cup test (Ref 71). In the first correlation, 48 materials were tested, including aluminum-killed, rimmed, and stainless steels and aluminum alloys in various tempers (Ref 50). Tensile properties included the directionally averaged percentage of total elongation, $\bar{e}_T$, to indicate the stretchability, and the normal anisotropy, r_m, to indicate drawability. The following relationships and correlation coefficients were obtained:

Test parameter	Relationship	Correlation coefficient
Olsen cup height/ punch diameter	$0.217 + 0.00474\ \bar{e}_T + 0.00392\ r_m$	0.925
Limiting draw ratio (Swift)	$1.93 + 0.00216\ \bar{e}_T + 0.226\ r_m$	0.835
Formability index (Fukui)	$0.525 + 0.0134\ \bar{e}_T + 0.207\ r_m$	0.757

The Olsen test involves a much greater ratio of stretching to drawing than the Fukui test, which in turn involves a much greater ratio than the Swift flat-bottomed cup test.

In the second correlation, 50 different steels were tested (Ref 71), and the results were correlated with the average n and r values and thickness, t, as:

$$\frac{\text{Swift round-bottomed cup height}}{\text{blank diameter}} =$$

$$0.0830t + 0.679\bar{n} + 0.0594r_m - 0.036$$

(Eq 31)

Forming Limit Diagrams

A traditional forming limit diagram (FLD) indicates the limiting strains that sheet metals can sustain over a range of major-to-minor strain ratios. Two main types of laboratory tests are used to determine these limiting strains. The first type of test involves stretching test specimens over a punch or by means of hydraulic pressure—for example, the hemispherical punch method. This produces some out-of-plane deformation and, when a punch is used, surface friction effects. The second test produces only in-plane deformation and does not involve any contact with the sample within the gage length.

The first type of test has been used much more extensively (Ref 6, 77–79) than the second and provides slightly different results (Ref 33, 80). Good correlation has been obtained between FLDs determined in the laboratory and production experience.

The hemispherical punch method for determining forming limit diagrams uses circle-gridded strips of the test material ranging in width from 25.4 to 203 mm (1.0 to 8.0 in.) that are clamped in a die ring and stretched to incipient fracture by a 102 mm (4.0 in.) diam steel punch (Ref 77, 79). The narrowest strip fractures at a minor-to-major strain ratio of about −0.5, which is comparable to that obtained in a tension test. As the strip width is increased, the strain ratio passes through plane strain toward positive values for a full-width specimen. Further increases in the ratio to a maximum value of +1.0 (balanced biaxial stretching) are achieved by using progressively improved punch lubrication (oiled polyethylene, oiled neoprene) and by increasing thicknesses of polyurethane rubber between the sheet and the punch.

The strains are measured in and around regions of visible necking and fracture. The forming limit curve is drawn above the strains measured outside the necked regions and below those measured in the necked and fractured regions, as shown in Fig. 31.

In-plane determination of the forming limit diagram can be achieved by using the uniaxial tension test, rectangular sheet tension test, or Marciniak biaxial stretching test with elliptical and circular punches, as described earlier in this article. The forming limit curve can be determined over the full range of strain ratios, without introducing any out-of-plane deformation. A comparison of the in-plane and punch methods showed close agreement for negative strain ratios and slightly higher values in the punch test at plane strain and for positive strain ratios (Ref 33).

Circle Grid Analysis and Use of FLDs

Circle grid analysis is a useful technique for ensuring that a die is adequately prepared for production and for diagnosing the causes of necking and splitting failures in production (Ref 81, 82). The forming limit diagram for the type and gage of work material selected must first be obtained. Arrays of small diameter (2.5 mm, or 0.1 in.), evenly spaced circles are printed or etched on several blanks in the critical strain regions, preferably in the same location on each blank. Some of the blanks are formed into parts, and the major and minor axes of the deformed circles are measured in the critical locations. The critical strain regions of the part are identified by visual observation of necking or splitting, or by previous experience with similar parts. The local strains are then calculated from the measured dimensions and plotted on the forming limit diagram.

If the maximum strains measured are close to or above the forming limit curve (FLC), problems with the tooling, lubrication, blank size or positioning, or press variables are indicated, whether necking or splitting actually occurs. Fluctuations that occur in operating conditions and in the properties of the work material over a production run will eventually cause failure if the material is strained to its full capacity. If the greatest levels of strain in the stamping are more than 10% below the FLC, the dies are considered "green," or safe. Many companies require this type of analysis, proving that the new stamping dies are capable of producing a part with strain levels in the "green" zone of the FLD, before buying off on the dies (Ref 83). When new dies cause levels of strain between 0 and 10% below the FLC, this is considered a "yellow" zone, where caution requires further development of either the dies themselves, the drawbeads, or other process conditions including lubrication. Any strain readings that exceed the FLC indicate a "red" condition, where the likelihood of failure is high if the dies are not modified prior to being used for production.

The material used in die tryout should have typical, or slightly lower, forming properties than the production material. The use of superior material may indicate an adequate forming safety margin that will disappear when a more typical or lower formability material is used. It is good practice to form a few gridded blanks of a standard (nonaging) reference material periodically during a production run to determine the trends in the maximum strains. If the strains are approaching the maximum limits, corrective measures can be taken before any actual failures occur.

A nonexhaustive list of industrial applications of the FLD includes (Ref 84):

Visual display of the cause of a problem	Tight radii Restrictions Insufficient punch-die clearance Blank too big
Strain comparisons	Mirror image stampings Before and after changes Last one shutdown, first one startup
Setting dies	Press lines, types and speeds Die location Ram adjustments
Guide modifications	Detect changes Rapid 2-blank method Sensitivity to changes Shop records
Strain history	Source of problem Breakdown sequence Status at shutdown
Monitor production runs	Die wear Sudden breakage Status at breakdown
Initial specifications	Material Lubrication Die release
Experimental changes	Material Lubrication Die modifications Press adjustments
Education	Apprentices New material New processes

Circle Grids. Many types of circle grid patterns have been used, such as square arrays of contacting or closely spaced noncontacting circles and arrays of overlapping circles. The contacting and overlapping circles provide improved coverage but are more difficult to measure manually. With small, closely spaced circles, it is possible to determine strain gradients accurately, provided the circles are not too small for accurate measurement. Circles with 2.5 mm (0.1 in.) diameters have been found to be a good size. Both open and solid circles have been used successfully, and automatic systems have been developed for measuring both types.

Applying Circle Grids to the Blanks. The circle grids can be applied to the blanks by a printing or photographic technique or by electrochemical etching. Printed and photographically applied circles are easily damaged and tend to rub off in areas contacted by the dies. This has led to general acceptance of etched circles.

In the electrochemical etching process, an electric stencil with the required grid pattern is placed on the blank and covered with a felt pad soaked in an etching solution. An electrode is placed on the pad, and low-voltage (up to 14 V) current is passed between the electrode and the blank for a short time, usually less than 1 min. This produces a lightly etched and oxidized pattern on the surface of the blank. The stencils, etching solutions, and power supplies for this process are commercially available. Different metals require different solutions, levels and types of voltage, and etching times.

Measuring Strains from Deformed Circles. Deformed circles can be measured manually by means of dividers and a ruler, graduated transparent tapes, or a low-power microscope with a graduated stage. Automatic systems, known as grid circle analyzers, have also been developed for measuring the dimensions of the circles and calculating and displaying the major and minor strains (Ref 85–87).

In regions of high curvature, the most accurate method of measurement is use of the transparent tape because it follows the contour of the part and measures the arc length, while the other methods measure the chord length. The tapes have a pair of diverging lines graduated to give direct readings of the strain, as shown in Fig. 32.

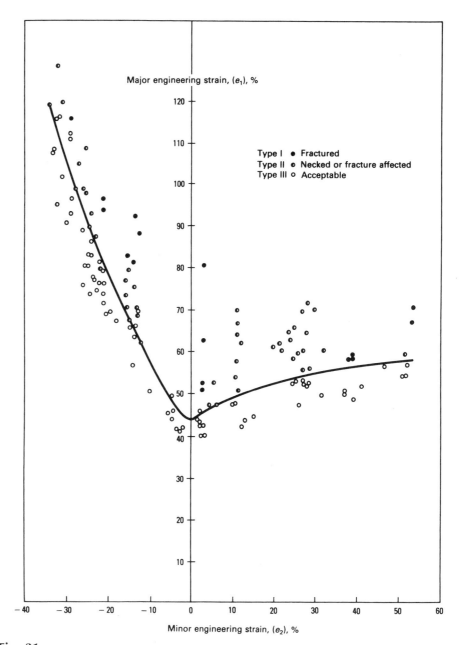

Fig. 31 Strain measurements and forming limit diagram for aluminum-killed steel. Source: Ref 76

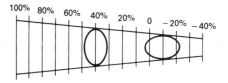

Fig. 32 Transparent tape measurement of deformed circles. Source: Ref 78

Grid circle analyzers use a solid-state digital array camera with a built-in light source, a minicomputer, keyboard, cathode-ray tube (CRT) display, and printer. An image of a given deformed circle is displayed on the CRT, and a least squares curve fitting program selects the most suitable ellipse, which is displayed simultaneously. The major and minor strains, computed from the equation for the ellipse and the diameter of the original circle, are displayed on the screen and printed. A typical layout for the equipment is shown in Fig. 33.

Drawbead Forces

It is common practice in production stamping operations to control the movement of the edges of the blank into the die cavity by means of drawbeads placed in the blankholder. These consist of a semicylindrical ridge in the upper part of the blankholder and a corresponding groove with rounded shoulders in the lower part, or a similar but opposite configuration. The drawbeads cause the periphery of the blank to bend and straighten three times as it passes through each bead, as shown in Fig. 34.

The repeated bending and straightening produces a restraining force in addition to that caused by surface friction. A method has been devised for measuring the restraining force due to deformation, independently of the effects of friction, using a drawbead simulator with low-friction rollers instead of a fixed bead and groove (Ref 88, 89). A second drawbead simulator with nonrotating parts can be used to measure the combined effects of friction and deformation. Figure 35 shows both types of simulators.

Strips of 0.75 to 1.00 mm (0.03 to 0.04 in.) thick and 50 mm (1.97 in.) wide rimmed and aluminum-killed steels and two aluminum alloys were tested using simulators and a universal testing machine. The contribution from deformation to the total restraining force depended on the lubricant used and ranged from an average of 60% with poor lubrication to 85% with very good lubrication. The required clamping forces, surface strains in the workpiece at various locations in the drawbead simulators, effect of drawbead radius, and effect of rate of testing were also investigated.

Improvements to Forming Limit Diagram Technology

Enhanced Forming Limit Curve Effect

The bending and unbending that occurs when a sheet metal is pulled through a drawbead has been shown to result in measured strain values that are misleading in terms of material formability. Experiments conducted by the Auto/Steel Partnership have shown that thinning strains caused by bending and unbending deformation, either through drawbeads or over tight punch and die radii, are only 40% as damaging as uniform thinning caused by pure stretching (Ref 83). This work resulted in a strain correction that is used to determine a new, correct forming limit. The strain correction is called "bead correction factor" (BCF) and is given by the simple

relationship:

$$BCF = 60\varepsilon_t \qquad (Eq\ 32)$$

where ε_t is the true thinning strain that occurs as sheet metal passes through a drawbead, or through a drawbead and die radius combination. This corrective factor is subtracted from the measured major strain so that only 40% of this strain should be used to determine forming severity. The practical advantage of using this correction lies in being able to quantify the true forming severity of material undergoing bending and unbending so that the results of die tryout experiments can be interpreted properly. In the past, strain levels in a sheet stamping may have exceeded the FLC for the material being tried but did not result in a failure. This result was problematic because all areas of the stamping needed to be in the "green" zone of the FLD in order for the tryout to be successful. Now this correction provides the press shop with a proven method of interpreting the strains produced at locations in a sheet stamping where material has passed through a drawbead or where material has passed over a sharp die radius. Additional

information on the enhanced FLC effect can be found in Ref 83.

Stress-Based Forming Limit Curves

One of the difficulties with the traditional FLC is that it is generated by straining sheet metal specimens along essentially linear strain paths. As a result, a traditional FLC may not capture the formability of complex forming operations, including many first-stage operations, but especially when two or three forming steps are required to make a part. The inaccuracy that results from making the FLC path-dependant may be overcome by using a stress-based forming limit criterion, as proposed in Ref 90. The traditional strain-based FLC is equivalent to the stress-based FLC when loading paths are linear, but not when they are nonlinear.

This new approach involves plotting the forming limit curve in stress space, rather than in strain space. An appropriate constitutive model must be used to convert strain values, which may be easily measured experimentally, into values of stress for the material being studied. While

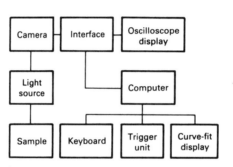

Fig. 33 Layout of a grid circle analyzer. Source: Ref 79

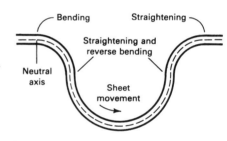

Fig. 34 Repeated bending and straightening of a blank edge in a drawbead

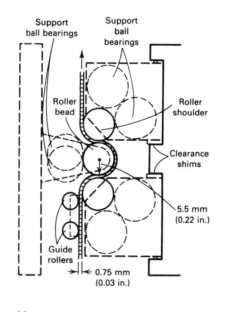

(a)

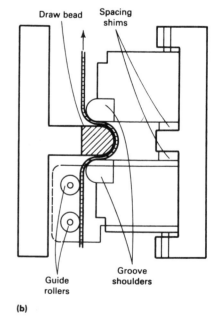

(b)

Fig. 35 Drawbead simulators. (a) Frictionless simulator. (b) Standard simulator. Source: Ref 83

concerns about the dependence of this method on the choice of constitutive model have been raised, it was shown that this approach produces robust results that converge to the same values for a given material and for a given model (Ref 91, 92). An example of a stress-based FLC for 2008-T4 aluminum is shown in Fig. 36, using a von Mises yield surface with a power law stress-strain relationship.

The path-independence of this approach is demonstrated in Fig. 36, where the FLC in stress space for the as-received 2008-T4 material is nearly the same as that for an equibiaxial prestrained 2008-T4. The increased measure of

accuracy gained by this new approach can be used to improve the predictions of finite-element simulations of forming processes, which up to now have employed mostly the traditional strain-based FLC as a criterion for necking failure.

Lubricants

The use of lubricants is essential in most forming, operations (see the article "Selection and Use of Lubricants in Forming of Sheet Metal" in this Volume). An effective lubricant provides the following advantages:

- Reduction or elimination of direct sheet metal to die contact and the associated wear and galling
- Control of friction
- More uniform distribution of strain and therefore an increase in the overall level of deformation
- Reduction of heating

Lubricants must meet many requirements to be used in a production operation, such as:

- Suitable viscosity over the ranges of temperatures and pressures encountered
- Chemical and physical compatibility with work and die materials
- Ease of application, removal, and disposal
- Compatibility with welding operations, sealants, and paint systems

No single lubricant is optimal for all types and rates of forming and all combinations of work and die materials. Rankings of lubricants change considerably for different types of operations and material combinations; this necessitates evaluation on an individual basis. Differences in the performance of lubricants are to be anticipated in view of the differences in surface composition, roughness, and texture of work and die materials and the different strain paths and rates of different forming operations. For example, some stretching operations involve local increases in area in excess of 100%, but in many drawing operations, a negligible or negative change in area occurs. In addition, the rate at which the work material slides over the dies varies widely. A simple test for evaluating lubricants measures the frictional force exerted on a lubricated strip of sheet metal when it is pulled between two rectangular blocks of die material. The force used between the blocks and the test rate can be varied. The number of strips that can be tested before the onset of galling provides an additional measure of the effectiveness of the lubricant. However, this test does not include some of the important aspects of actual forming operations, such as plastic deformation of the work material, the ranges of sliding speeds and rates of straining involved, die geometry (which influences the amount of residual lubricant in various locations as the operation progresses), and the heating that occurs in high-volume production operations.

The drawbead simulator described earlier is more realistic and has been used to measure friction forces with various lubricants (Ref 88, 89). Under most of the conditions tested, friction was described by Coulomb's law, which states that the friction force, F, is directly proportional to the normal force, N, between the contacting surfaces:

$$F = \mu N \qquad \text{(Eq 33)}$$

where μ is the coefficient of friction. Coulomb's law did not apply at the highest contact pressure tested. The value of the coefficient of friction for mill oil, a poor lubricant, was found to be 0.17;

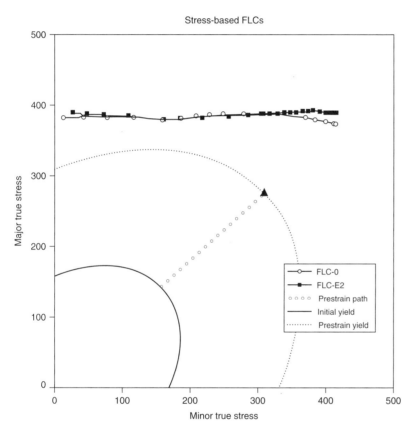

Fig. 36 Stress-based forming limit curve for as-received 2008-T4 and prestrained 2008-T4. Source: Ref 90

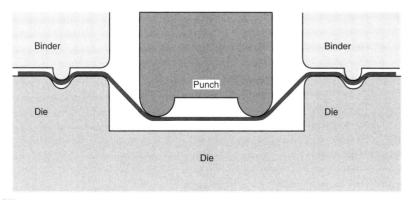

Fig. 37 Schematic of OSU friction test. Source: Ref 94

for the best lubricant tested—a soap-based lubricant—it was 0.06.

In addition, simulative tests can be used as lubricant evaluation tests. The production operation should be characterized in terms of the principal types of forming operations involved and their relative severities, and the appropriate simulative test selected. The Swift flat-bottomed cup test has been used extensively for the evaluation of lubricants for deep drawing, and the Swift round-bottomed and Fukui conical cup tests have been used for stretch-drawing operations. The 100 mm (3.94 in.) hemispherical dome test can be used for stretching operations. A more recent test developed at The Ohio State University (OSU friction test) is capable of ranking lubricants through experimental determination of the friction coefficient (Ref 64, 93–96). A gridded strip of metal is clamped and then stretched by a punch, as shown in Fig. 37.

Measurement of the grids after the experiment allows for calculating ε_1 and ε_2, which may be used to determine the forces if the constitutive behavior of the material is known. Combining this information with the contact angle (which can be obtained from the punch height) allows for calculating the friction coefficient, which may then be used as an index to rank lubricants.

The various regimes of lubrication—that is, thick film, thin film, mixed, and boundary lubrication—and their characteristics are reviewed in Ref 95, which also discusses the limited validity of Coulomb's law, various methods for lubricant evaluation, and some of the current limitations in this area.

ACKNOWLEDGMENT

This article has been revised and updated from B. Taylor, Formability Testing of Sheet Metals, *Forming and Forging,* Vol 14, *Metals Handbook,* 9th ed., ASM International, 1988, p 877–899.

REFERENCES

1. S.P. Keeler, "Understanding Sheet Metal Formability, Vol 2," paper 350A, Metal Fabricating Institute, 1970
2. J.H. Holloman, Tensile Deformation, *Trans. AIME,* Vol 162, 1945, p 268–290
3. W.A. Backofen, Massachusetts Institute of Technology Industrial Liaison Symposium, Chicago, March 1974
4. A.K. Ghosh, The Influence of Strain Hardening and Strain-Rate Sensitivity on Sheet Metal Forming, *Trans. ASME,* Vol 99, July 1977, p 264–274
5. I.S. Brammar and D.A. Harris, Production and Properties of Sheet Steel and Aluminum Alloys for Forming Applications, *J. Austral. Inst. Met.,* Vol 20 (No. 2), 1975, p 85–100
6. S.P. Keeler and W.A. Backofen, Plastic Instability and Fracture in Sheets Stretched over Rigid Punches, *Trans. ASM,* Vol 56 (No. 1), 1963, p 25–48
7. G.M. Goodwin, "Application of Strain Analysis to Sheet Metal Forming Problems in the Press Shop," Paper 680093, Society of Automotive Engineers, 1968
8. S.P. Keeler, Determination of Forming Limits in Automotive Stampings, *Sheet Met. Ind.,* Vol 42, Sept 1965, p 683–691
9. S.P. Keeler and W.G. Brazier, Relationship between Laboratory Material Characterization and Press-Shop Formability, *Microalloying 75 Proceedings,* Union Carbide Corporation, 1977, p 517–530
10. H. Naziri and R. Pearce, The Effect of Plastic Anisotropy on Flange-Wrinkling Behavior during Sheet Metal Forming, *Int. J. Mech. Sci.,* Vol 10, 1968, p 681–694
11. K. Yoshida and K. Miyauchi, Experimental Studies of Material Behavior as Related to Sheet Metal Forming, *Mechanics of Sheet Metal Forming,* Plenum Press, 1978, p 19–49
12. W.F. Hosford and R.M. Caddell, *Metal Forming, Mechanics and Metallurgy,* Prentice-Hall, 1983, p 273, 309
13. J. Havranek, The Effect of Mechanical Properties of Sheet Steels on the Wrinkling Behavior during Deep Drawing of Conical Shells, *Sheet Metal Forming and Energy Conservation, Proc. 9th Biennial Congress of the International Deep Drawing Research Group* (Ann Arbor, MI), American Society for Metals, 1976, p 245–263
14. J.S.H. Lake, The Yoshida Test—A Critical Evaluation and Correlation with Low-Strain Tensile Parameters, *Efficiency in Sheet Metal Forming, Proc. 13th Biennial Congress* (Melbourne, Australia), International Deep Drawing Research Group, Feb 1984, p 555–564
15. J.L. Duncan, R. Sowerby, and M.P. Sklad, "Failure Modes in Aluminum Sheet in Deep Drawing Square Cups," Paper presented at the Conference on Sheet Forming, University of Aston, Birmingham, England, Sept 1981
16. G. Glover, J.L. Duncan, and J.D. Embury, Failure Maps for Sheet Metal, *Met. Technol.,* March 1977, p 153–159
17. J.D. Embury and J.L. Duncan, Formability Maps, *Ann. Rev. Mater. Sci.,* Vol 11, 1981, p 505–521
18. J. Datsko, Materials in Design and Manufacturing, J. Datsko Consultants, Ann Arbor, MI, 1977, p 7–16
19. N. Kuhn, On the Springback Behavior of Low-Carbon, Steel Sheet after Stretch Bending, *J. Austral. Inst. Met.,* Vol 12 (No. 1), Feb 1967, p 71–76
20. G.V. Smith, Elevated Temperature Static Properties of Wrought Carbon Steel, *Special Technical Publication on Temperature Effects,* STP 503, American Society for Testing and Materials, 1972
21. F.N. Rhines and P.J. Wray, Investigation of the Intermediate Temperature Ductility Minimum in Metals, *Trans. ASM,* Vol 54, 1961, p 117–128
22. B. Taylor, R.A. Heimbuch, and S.G. Babcock, Warm Forming of Aluminum, *Proc. Second International Conference on Mechanical Behavior of Materials,* American Society for Metals, 1976, p 2004–2008
23. W.G. Granzow, The Influence of Tooling Temperature on the Formability of Stainless Sheet Steel, *Formability of Metallic Materials-2000 A.D.,* STP 753, J.R. Newby and B.A. Niemeier, Ed., American Society for Testing and Materials, 1981, p 137–146
24. H.C. Wu and D.R. Rummler, Analysis of Misalignment in the Tension Test, *Trans. ASME,* Vol 101, Jan 1979, p 68–74
25. G.E. Dieter, *Mechanical Metallurgy,* 2nd ed., McGraw-Hill, 1976, p 347, 349, 681
26. R.L. Whitely, "Correlation of Deep Drawing Press Performance with Tensile Properties," STP 390, American Society for Testing and Materials, 1965
27. S.J. Green, J.J. Langan, J.D. Leasia, and W.H. Yang, Material Properties, Including Strain-Rate Effects, as Related to Sheet Metal Forming, *Metall. Trans. A,* Vol 2A, 1971, p 1813–1820
28. G. Rai and N.J. Grant, On the Measurements of Superplasticity in an Al-Cu Alloy, *Metall. Trans A.,* Vol 6A, 1975, p 385–390
29. W.J. McGregor Tegart, in *Elements of Mechanical Metallurgy,* Macmillan, 1966, p 29–38
30. R.H. Heyer and J.R. Newby, Measurement of Strain Hardening and Plastic Strain Ratio Using the Circle-Arc Specimen, *Sheet Met. Ind.,* Vol 43, Dec 1966, p 910–914
31. D.C. Ludwigson, A Rapid Test for the Assessment of Steel Sheet and Tinplate Properties, *J. Test. Eval.,* Vol 7 (No. 6), 1979, p 301–309
32. P.R. Mould and T.E. Johnson, Rapid Assessment of Drawability at Cold-Rolled Low-Carbon Steel Sheets, *Sheet Met. Ind.,* Vol 50, June 1973, p 328–348
33. M.L. Devenpeck and O. Richmond, Limiting Strain Tests for In-Plane Sheet Stretching, *Novel Techniques in Metal Deformation Testing,* The Metallurgical Society, 1983, p 79–88
34. R.H. Wagoner and N.M. Wang, An Experimental and Analytical Investigation of In-Plane Deformation of 2036-T4 Aluminum, *Int. J. Mech. Sci.,* Vol 21, 1979, p 255–264
35. R.H. Wagoner, Measurement and Analysis of Plane-Strain Work Hardening, *Metall. Trans. A,* Vol 11A, Jan 1980, p 165–175
36. R.H. Wagoner, Plane-Strain and Tensile Hardening Behavior of Three Automotive Sheet Alloys, *Experimental Verification of Process Models,* Symposium Proc., Cincinnati, OH, Sept 1981, American Society for Metals, 1983, p 236
37. E.J. Appleby, M.L. Devenpeck, L.M. O'Hara, and O. Richmond, Finite Element Analysis and Experimental Examination of

the Rectangular-Sheet Tension Test, *Applications of Numerical Methods to Forming Processes,* Vol 28, *Proc. ASME Winter Annual Meeting* (San Francisco, CA), Applied Mechanics Division, American Society of Mechanical Engineers, Dec 1978, p 95–105

38. H. Sang and Y. Nishikawa, A Plane Strain Tensile Apparatus, *J. Met.,* Feb 1983, p 30–33

39. Z. Marciniak and K. Kuczynski, Limit Strains in the Processes of Stretch-Forming Sheet Metal, *Int. J. Mech. Sci.,* Vol 9, 1967, p 609–620

40. J.L. Duncan, J. Kolodziejski, and G. Glover, Bulge Testing as an Aid to Formability Assessment, *Sheet Metal Forming and Energy Conservation, Proc. 9th Biennial Congress of the International Deep Drawing Research Group* (Ann Arbor, MI), American Society for Metals, 1976, p 131–150

41. W. Johnson and J.L. Duncan, The Use of the Biaxial Test Extensometer, *Sheet Met. Ind.,* Vol 42, April 1965, p 271–275

42. J.E. Bird, Hydraulic Bulge Testing at Controlled Strain Rate, *Novel Techniques in Metal Deformation Testing,* The Metallurgical Society, 1983, p 403–416

43. A.J. Ranta-Eskola, Use of the Hydraulic Bulge Test in Biaxial Tensile Testing, *Int. J. Mech. Sci.,* Vol 21, 1979, p 457–465

44. D.N. Harvey, Electronic Vision as Input to Calculated Variable Control of the Hydraulic Bulge Test, *Proc. International Symposium on Automotive Technology and Automation* (Wolfsburg, West Germany), Sept 1982

45. Z. Marciniak, *Aspects of Material Formability,* Metalworking Research Group, McMaster University, Hamilton, Ontario, Canada, 1973, p 84–91

46. K. Miyauchi, Stress-Strain Relationship in Simple Shear of In-Plane Deformation for Various Steel Sheets, *Efficiency in Sheet Metal Forming, Proc. 13th Biennial Congress* (Melbourne, Australia), International Deep Drawing Research Group, Feb 1984, p 360–371

47. R.A. George, S. Dinda, and A.S. Kasper, Estimating Yield Strength from Hardness Data, *Met. Prog.,* May 1976, p 30–35

48. M.Y. Demeri, The Stretch-Bend Forming of Sheet Metal, *J. Appl. Metalwork.,* Vol 2 (No. 1), 1981, p 3–10

49. O.S. Narayanaswamy and M.Y. Demeri, Analysis of the Angular Stretch Bend Test, *Novel Techniques in Metal Deformation Testing,* The Metallurgical Society, 1983, p 99–112

50. A.S. Kasper, "Forming Sheet Metal Parts," Paper MF 69-516, Society of Manufacturing Engineers, 1969

51. T.R. Thompson, G. Glover, and R. Jackson, "The Formability of Sheet Steels," Technical Bulletin 18, No. 2, Broken Hill Proprietary Company, Melbourne, Australia, 1974, p 15–19

52. S.S. Hecker, A Cup Test for Assessing Stretchability, *Met. Eng. Quart.,* Nov 1974, p 30–36

53. J.R. Newby, A Practical Look at Biaxial Stretch Cupping Tests, *Sheet Met. Ind.,* Vol 54, March 1977, p 240–252

54. K. Nakazima, T. Kikuma, and K. Hasuka, "Study on the Formability of Sheet Steels," Technical Report 264, Yawata Iron & Steel Company, Japan, Sept 1968, p 141

55. A.K. Ghosh, How to Rate Stretch Formability of Sheet Metals, *Met. Prog.,* May 1975, p 52–54

56. A.K. Ghosh, The Effect of Lateral Drawing-In on Stretch Formability, *Met. Eng. Quart.,* Aug 1975, p 53–64

57. R.A. Ayres, W.G. Brazier, and V.F. Sajewski, Evaluating the Limiting Dome Height Test as a New Measure of Press Formability, *J. Appl. Metalwork.,* Vol 1 (No. 1), 1979, p 41–49

58. LDH Committee/NADDRG, "Round Robin Evaluation of LDH Test Variability," Final Report of the LDH Committee, Ann Arbor, MI, May 1990

59. R. Stevenson, Correlation of Tensile Properties with Plane-Strain Limiting Dome Height, *J. Appl. Metalwork.,* Vol 13 (No. 3), July 1984, p 272–280

60. S. Sadagopan and R.H. Wagoner, "Simulating the LDH Test," NUMISHEET 96, J.K. Lee, G.L. Kinzel, and R.H. Wagoner, Ed., The Ohio State University, 1996, p 128–135

61. D.J. Meuleman, J.L. Siles, and J.J. Zoldak, SAE Paper 85005, 1985, p 1–9

62. J.A. Shey, *J. Mater. Shaping Technol.,* 1988, Vol 6 (No. 2), p 103

63. M.P. Miles, J.L. Siles, R.H. Wagoner, and K. Narasimhan, A Better Sheet Metal Formability Test, *Metall. Trans. A,* Vol 24A, 1993, p 1143–1151

64. Y.S. Suh and R.H. Wagoner, Application of the Finite-Element Method to a Design of Optimized Tool Geometry for the O.S.U. Formability Test, *J. Mater. Eng. Perform.* 1996, Vol 5, p 489–499

65. R.H. Wagoner, W. Wang, and S. Sriram, Development of the OSU Formability Test and the OSU Friction Test, *J. Mater. Process. Technol.,* Vol 45, 1994, p 13–18

66. R.D. Adamczyk, D.W. Dickinson, and R.P. Krupitzer, "The Edge Formability of High-Strength Cold-Rolled Steel," Paper 830237, Society of Automotive Engineers, 1983

67. D.F. Eary and E.A. Reed, *Techniques of Pressworking Sheet Metal,* 2nd ed., Prentice-Hall, 1974, p 136–172

68. D.V. Wilson, B.J. Sunter, and D.F. Martin, A Single-Blank Test for Drawability, *Sheet Met. Ind.,* Vol 43, June 1966, p 465–476

69. K.M. Frommann, The Prediction of Metal Stamping Behavior, *Proc. American Metal Stamping Association Conference* (Detroit, MI), April 1968, p 1–55

70. A.S. Kasper and P.J. VanderVeen, "A New Method of Predicting the Formability of Materials," Paper 720019, Society of Automotive Engineers, 1972

71. M.W. Boyles and H.S. Chilcott, Recent Developments in the Use of the Stretch-Draw Test, *Sheet Met. Ind.,* Vol 59, Feb 1982, p 149–156

72. G.M. Goodwin, "Formability Index," Paper MF 71-165, Society of Manufacturing Engineers, 1971

73. K. Yoshida, H. Hayashi, K. Miyauchi, M. Hirata, T. Hira, and S. Ujihara, Yoshida Buckling Test, *Proc. International Symposium on New Aspects of Sheet Metal Forming,* Iron and Steel Institute of Japan, 1981, p 125–148

74. J. Havranek, Wrinkling Limit of Tapered Pressings, *J. Austral. Inst. Met.,* Vol 20, 1975, p 114–119

75. H. Abe, K. Nakagawa, and S. Sato, Proposal of Conical Cup Buckling Test, *Proc. Congress of the International Deep Drawing Research Group* (Kyoto, Japan), May 1981, p 1–19

76. A.M. Szacinski and P.F. Thompson, The Effect of Mechanical Properties on the Wrinkling Behavior of Sheet Material in the Yoshida Test and in a Cone Forming Test, *Efficiency in Sheet Metal Forming, Proc. 13th Biennial Congress* (Melbourne, Australia), International Deep Drawing Research Group, Feb 1984, p 532–542

77. C.C. Veerman, L. Hartman, J.J. Peels, and P.F. Neve, Determination of Appearing and Admissible Strain in Cold Reduced Sheets, *Sheet Met. Ind.,* Vol 48, Sept 1971, p 678–694

78. A.B. Haberfield and M.W. Boyles, Laboratory Determined Forming Limit Diagrams, *Sheet Met. Ind.,* Vol 50, July 1973, p 400–411

79. S.S. Hecker, Simple Technique for Determining Forming Limit Curves, *Sheet Met. Ind.,* Vol 52, Nov 1975, p 671–676

80. A.K. Ghosh and S.S. Hecker, Stretching Limits in Sheet Metals: In-Plane Versus Out-of-Plane Deformation, *Metall. Trans.,* Vol 5, Oct 1974, p 2161–2164

81. S. Dinda, K.F. James, S.P. Keeler, and P.A. Stine, How to Use Circle Grid Analysis for Die Tryout, American Society for Metals, 1981

82. Z. Buchar, Circle Grid Analysis Applied to the Production Problems of the Car Body Panel, *J. Mater. Process. Technol.,* Vol 60 (No. 1-4), June 1996, p 205–208

83. Stuart Keeler, The Enhanced FLC Effect, The Auto/Steel Partnership—Enhanced Forming Limit Diagram Project Team, Southfield, MI, Jan 2003

84. Stuart Keeler, private communication, May 2004

85. R.A. Ayres, E.G. Brewer, and S.W. Holland, Grid Circle Analyzer: Computer Aided Measurement of Deformation, *Trans. SAE,* Vol 88 (No. 3), 1979, p 2630–2634 (SAE Paper 790741)

86. D.N. Harvey, Optimising Patterns and Computational Algorithms for Automated, Optical Strain Measurement in Sheet Metal, *Efficiency in Sheet Metal Forming, Proc. 13th Biennial Congress* (Melbourne, Australia), International Deep Drawing Research Group, Feb 1984, p 403–414

87. D.W. Manthey, R.M. Pearce, and D. Lee, The Need for Surface Strain Measurement, *Met. Form.,* Vol 30 (No. 5), 1996, p 48–54

88. H.D. Nine, Drawbead Forces in Sheet Metal Forming, *Mechanics of Sheet Metal Forming,* Plenum Press, 1978, p 179–211

89. H.D. Nine, The Applicability of Coulomb's Friction Law to Drawbeads in Sheet Metal Forming, *J. Appl. Metalwork.,* Vol 2 (No. 3), July 1982, p 200–210

90. T.B. Stoughton, A General Forming Limit Criterion for Sheet Metal Forming, *Int. J. Mech. Sci.,* Vol 42, 2000, p 1–27

91. T.B. Stoughton, The Influence of the Material Model on the Stress-Based Forming Limit Criterion, SAE Paper 2002-01-0157, 2002

92. T.B. Stoughton and X. Zhu, Review of the Theoretical Models of the Strain-Based FLD and Their Relevance to the Stress-Based FLD, *Int. J. Plast.,* Vol 20, 2004, p 1463–1486

93. W. Wang and R.H. Wagoner, A Realistic Friction Test for Sheet Forming Operations, (SAE Paper 930807), SAE Transactions, *J. Mat. Manuf.,* Vol 2, 1993, p 915–922

94. W. Wang, R.H. Wagoner, and X.-J. Wang, Measurement of Punch Friction under Realistic Sheet Forming Conditions, *Metall. Mater. Trans. A,* Vol 27A, 1996, p 3971–3981

95. R.H. Wagoner and J.L. Chenot, *Fundamentals of Metalforming,* John Wiley & Sons, 1996, p 324–328

96. W.R.D. Wilson, Friction and Lubrication in Sheet Metal Forming, *Mechanics of Sheet Metal Forming,* Plenum Press, 1978, p 157–177

SELECTED REFERENCES

- A.K. Ghosh, S.S. Hecker, and S.P. Keeler, Sheet Metal Forming and Testing, *Workability Testing Techniques,* American Society for Metals, 1984, p 135–195

- S.S. Hecker and A.K. Ghosh, The Forming of Sheet Metal, *Sci. Am.,* Vol 235, Nov 1976, p 100–108

- W.F. Hosford and R.M. Caddell, *Metal Forming: Mechanics and Metallurgy,* 2nd ed., Prentice Hall, 1993

- J. Hu, Z Marciniak, and J Duncan, *Mechanics of Sheet Metal Forming,* 2nd ed., Elsevier, 2002

- V. Karthik, R.J. Comstock, D.L. Hershberger, and R.H. Wagoner, Variability of Sheet Formability and Formability Testing, *J. Mater. Process. Technol.,* Vol 121 (No. 2-3), Feb 28, 2002, p 350–362

- S.P. Keeler, Sheet Metal Forming in the 80's, *Met. Prog.,* July 1980, p 25–29

- "Sheet Formability Testing," Auto-Steel Partnership, www.a-sp.org, accessed Jan 2006

- R.H. Wagoner and J.-L. Chenot, *Fundamentals of Metal Forming,* John Wiley and Sons, 1996

- R.H. Wagoner and J.-L. Chenot, *Metal Forming Analysis,* Cambridge University Press, 2001

Troubleshooting Formability Problems Using Strain Analysis

Daniel J. Schaeffler and Evan J. Vineberg, Engineering Quality Solutions, Inc.

STRAIN ANALYSIS is a technique to assess the degree of stamping process robustness as it relates to the sheet properties of the incoming blank used to form the part in question. The analysis is typically done on parts formed from known sheet properties, but it is possible to extrapolate the forming behavior when the steel properties are at the low end of the ordered grade and thickness. Strain analysis is also useful during tool development to help limit problems during production. Although parts on day one may "hold water," strain analysis can highlight those areas that may need some tooling work to ensure that the ordered steel or aluminum grade and the forming process, together, are sufficiently robust to produce a split-free panel over the entire property range associated with the particular sheet material grade. Strain analysis can help determine if a change to a more formable grade or heavier gage is warranted to reduce the potential for splitting. Similarly, strain analysis can help determine if a less formable grade (which is typically less expensive) or lighter gage (also a cost-savings opportunity) can be used.

This article describes techniques that can help with troubleshooting production process discrepancies. If supposedly symmetrical areas on a given part are not behaving the same way (e.g., the left side splits but the right side works), strain analysis can help determine if the locations in question are actually forming the same way and undergoing the same deformation. These differences in strain distribution can then be used to determine the root cause of the observed performance differences.

Once a robust condition has been achieved at the beginning of production stamping (or during the production stamping life cycle), a "reference panel" can be established with the aid of strain analysis techniques. This reference panel documents all properties, settings, and conditions that produced a good panel, such that these parameters can be replicated if issues are encountered during production. Using the strain distribution information from the reference panel, strains/thicknesses at selected areas can be checked intermittently during ongoing production. Tracking these data as a function of time can be used as a statistical process control and quality-monitoring tool, providing indication of part/process consistency as well as raising a "red flag" if something with the material, tooling, and/or process setup is going out of control.

The same strain-analysis techniques described in this article thus provide a useful and versatile tool that can be used in troubleshooting formability and process discrepancies throughout the entire tooling development and production stamping cycle: initial die tryout, tooling buyoff, home pressline die tryout, and ongoing production stamping.

Strain Calculations

Strain calculations can be visualized using the circle grid method by picturing a circle on a surface of a flat blank. After the blank is formed, the circle from the flat blank will become an ellipse (unless deformation is not pure biaxial stretching). The longest dimension of the ellipse is the major axis, and the dimension perpendicular to the major axis is the minor axis. Knowing the exact dimension of the starting circle as well as obtaining an accurate measurement of the length of the major and minor axes

(Fig. 1) allows for the direct calculation of the engineering major and minor strains, where:

$$e_{ma} = \text{Major strain (\%)} = 100 \times \left(\frac{L_{ma} - L_i}{L_i}\right)$$

$$e_{mi} = \text{Minor strain (\%)} = 100 \times \left(\frac{L_{mi} - L_i}{L_i}\right)$$

The major strain is always positive and always greater than the minor strain. In the sketches shown in Fig. 2, the shape of the initial circle is shown as a dashed line. After forming, there are three possible descriptions of the shape of the resulting ellipse.

In this process of metal deformation, a unit of material maintains a constant volume before, during, and after forming. This allows for calculation of a thickness strain (e_t), because the major, minor, and thickness strains in that unit of material multiplied together must equal 1, such that:

$$(e_{ma}+1) \times (e_{mi}+1) \times (e_t+1) = 1, \text{ or}$$

$$e_t = \left[\frac{1}{(e_{ma}+1) \times (e_{mi}+1)}\right] - 1$$

A nomograph of this is shown in Fig. 15 of the article "Formability Testing of Sheet Metals" in this Volume. As an example, consider a 0.100 in. diameter circle that was deformed into an ellipse with the longest dimension of 0.125 in., and

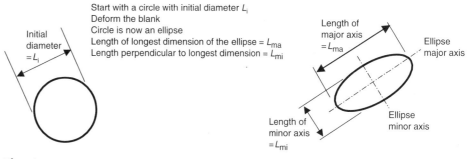

Start with a circle with initial diameter L_i
Deform the blank
Circle is now an ellipse
Length of longest dimension of the ellipse = L_{ma}
Length perpendicular to longest dimension = L_{mi}

Initial diameter = L_i

Length of major axis = L_{ma}

Ellipse major axis

Length of minor axis = L_{mi}

Ellipse minor axis

Fig. 1 Definition of major strain and minor strain

perpendicular to that, the ellipse measured 0.090 in.:

$$e_{ma} = \text{Major strain (\%)}$$

$$= 100 \times \left(\frac{0.125 \text{ in.} - 0.1 \text{ in.}}{0.1 \text{ in.}} \right)$$

$$= 100 \times \left(\frac{0.025 \text{ in.}}{0.1 \text{ in.}} \right) = +25\%$$

$$e_{mi} = \text{Minor strain (\%)}$$

$$= 100 \times \left(\frac{0.09 \text{ in.} - 0.1 \text{ in.}}{0.1 \text{ in.}} \right)$$

$$= 100 \times \left(\frac{-0.01 \text{ in.}}{0.1 \text{ in.}} \right) = -10\%$$

$$e_t = \text{Thickness strain (\%)}$$

$$= \frac{1}{(e_{ma}+1) \times (e_{mi}+1)} - 1$$

$$= \frac{1}{(0.25+1) \times (-0.10+1)} - 1$$

$$= \frac{1}{(1.25) \times (0.9)} - 1 = \frac{1}{1.125} - 1 = 0.89 - 1$$

$$= -0.111 = -11.1\%$$

Forming Limit Curves

Every sheet product has a limit to which it can be deformed before it splits. The maximum amount to which the material can be deformed before splitting is a function of not just the maximum major strain on a formed part but also the associated minor strain at each location as well. In experiments conducted since the late 1960s, numerous grades and thicknesses of sheet products have been deformed over a range of minor strain conditions (varying degrees of draw, plane strain, and stretch) to just before and just beyond failure. The strains on these parts were measured and plotted on a graph of major strain versus minor strain. The resulting curve defining the boundary between split-free and split conditions is called the forming limit curve (FLC).

Interestingly, the shape of the FLC is essentially the same for all sheet steel products. The only thing that changes is where the FLC is plotted on the major strain versus minor strain graph. The location of the FLC is determined by the value of FLC_0, which is defined as the lowest point on the FLC and typically occurs in plane-strain deformation (when the minor strain is zero).

The FLC_0 is a function only of initial sheet metal thickness (before forming) and n-value (strain-hardening exponent). As thickness and n-value increase, the FLC_0 is greater and is therefore plotted higher on the graph.

Based on empirical data, FLC_0 is calculated as:

$$FLC_0 = (23.3 + 360 \times t) \times (n/0.21)$$

where t is the sheet thickness in inches, and n is the n-value, and:

$$FLC_0 = (23.3 + 14.2 \times t) \times (n/0.21)$$

where t is the sheet thickness in millimeters, and n is the n-value.

Note that some versions of these equations show the coefficients to one or two additional decimal places. The added precision does not appear to affect the results significantly, and the equations in the aforementioned format are consistent with what is used within American Iron and Steel Institute and International Iron and Steel Institute documents written by Dr. Stuart Keeler, who first developed the FLC concept and strain analysis techniques in the 1960s and 1970s.

In the example in Fig. 3, the top FLC was generated using an n-value of 0.24 and a thickness of 0.90 mm (0.035 in.), for an FLC_0 of 41.2%. The bottom FLC was generated using an n-value of 0.20 and a thickness of 0.80 mm (0.032 in.), for an FLC_0 of 33.0%. Note that n-value has a much greater impact on the value of FLC_0 than does thickness.

The shape and placement of the FLC for aluminum alloys, stainless steels, and some Advanced High Strength Steels like Transformation Induced Plasticity (TRIP) steels, cannot be represented by a simple curve or a universally applicable equation. For these products, a curve needs to be established for each grade and thickness combination. Once this is done however, the same principles described in this article are applicable.

Forming Limit Diagrams

A flat blank is deformed during stamping. The strains in this deformation occur both on the surface of the sheet (major and minor strain) as well as the through the thickness (thinning strain). These strains are primarily a function of the part design and the forming process. It is possible to measure these strains using relatively simple techniques described in this article.

Plotting these measured strains along with the FLC plot on a graph of major strain versus minor strain creates a forming limit diagram (FLD). An FLD provides a graphical indication of the stamping performance that can be expected when using the ordered grade of steel and when the stamping conditions (press, die, lubricant, etc.) that existed during the forming of the gridded blank are also present.

Measured strains that plot higher than the FLC are said to be in the failure zone (also known as the necking zone or red zone). These locations indicate areas that likely will exhibit localized thinning, necking, or splitting.

Locations having strains that plot close to, but not in, the failure zone are also of concern, because a slight change in a host of factors affecting forming (such as the normal day-to-day variability in stamping or lubricant conditions, or variations in the surface properties of the steel) can increase the strains just enough to push them into the red zone.

To reduce the likelihood that these subtle changes have a detrimental effect on the forming process, it is typically recommended that all measured areas on an FLD plot at least 10% lower than the FLC on the major strain axis. Locations that meet this criterion are said to be in the safe zone (or green zone) of the FLD. Locations that plot between the failure zone and the safe zone are said to be in the marginal zone (or

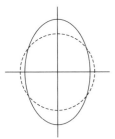

Draw deformation

Negative minor strain
Circle diameter is greater
than the ellipse minor axis

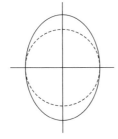

Plane-strain deformation

Zero minor strain
Circle diameter is equal
to the ellipse minor axis

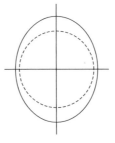

Stretch deformation

Positive minor strain
Circle diameter is less
than the ellipse minor axis

Fig. 2 Deformation mode as a function of minor strain

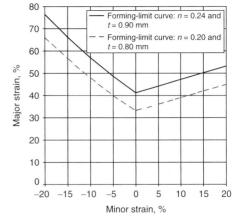

Fig. 3 Basic shape of the forming-limit curve, with FLC_0 defined by the strain-hardening exponent (n-value) and sheet metal thickness (t)

yellow zone). Offsetting the FLC 10% on the major strain axis from the appropriate failure curve creates the marginal zone boundary with the safe zone.

Safety Margin or Safety Factor

The vertical distance between the strain measured at a particular location and the strain on the FLC defines the major strain safety margin at that location. The lowest safety margin of all the locations is considered to be the overall safety margin of the part.

By definition, locations with major and minor strains that plot exactly on the FLC are said to have 0% safety margin (no room for error), and locations having safety margins between 0 and 10% are said to be in the marginal zone. A robust stamping process typically occurs when the measured strains plot in the safe zone, where all locations are more than 10% away from the FLC.

Greater values of FLC_0 lead to plotting the FLC higher on the graph, which increases the vertical distance to the measured strains. This is the primary reason why greater safety margins are associated with steels of higher n-values. Increased n-value also lowers the peak strain severity and provides for a more even strain distribution, thereby also increasing the minimum strain safety margin.

More generous radii or better lubrication during forming can reduce the forming strains, and as such, these locations will plot lower on the FLD. This leads to increased safety margins as well.

In the example shown in Fig. 4, at a minor strain value of −20%, the FLC major strain value is 77%. The vertical distance between point A, which plots as +50% major and −20% minor, and the FLC is 27%. This means that location A is in the safe zone with a 27% major strain safety margin.

At plane strain, where the minor strain is 0%, the FLC is at its lowest point, the FLC_0. In this example, for the material used ($n = 0.24$,

$t = 0.90$ mm, or 0.035 in.), the FLC_0 is calculated as 41%. Point B plots as +50% major and 0% minor, which falls above the FLC and, as such, is in the failure zone with a major strain safety margin of −9%.

At a minor strain value of +20%, the FLC major strain value is 53%. The vertical distance between point C, which plots as +50% major and +20% minor, and the FLC is 3%. This means that location C is in the marginal zone with a 3% major strain safety margin.

Types of Strain Analysis

There are two general types of strain analysis, known as thinning strain analysis and circle grid (or surface) strain analysis.

Thinning Strain Analysis

Thinning strain analysis (TSA) is easier to perform and requires very little operator training. TSA can be done on any formed part, without special preparation of the blank or tools before stamping. TSA does not require as much stamping plant involvement and assistance as circle grid analysis (CGA). Production stamping schedules are less likely to be disrupted when a TSA is needed. TSA will also be used when it is not practical to place circle grids on a blank. Examples include parts stamped on a progressive die or in the central portion of a very large blank.

In addition, during die development of new parts, TSA is a useful tool to assess the effect of tooling changes when the minor strain has not changed from the previous CGA. An example of this would be when a radius is increased along a sidewall that is in plane strain (minor strain = 0). Knowing the minor strain and measuring the thickness allows for calculation of the major strain.

The concern is that this approach may indicate that this location is not safe (suggesting that the time and cost of tooling work may be warranted). In reality, the metal deformation may be in stretch mode, with an associated greater safety margin, reducing the need for tooling work because the part may actually be safe.

TSA Technique. The tools of the TSA technique include an ultrasonic thickness (UT) gage and a micrometer or some other measuring tool to determine a reference thickness. If it is possible to obtain a reading before forming, measure the thickness of the flat blank, using the reference micrometer, and calibrate the UT gage at this location. It is important to remember that the UT gage is sensitive to the type of base metal, coating, and nominal thickness. Therefore, every time there is a change in any of these conditions, it is important to perform this procedure to ensure the UT gage is properly calibrated for the particular subject material.

If the blank is not available before forming, use the micrometer to obtain a known reference thickness on the flat edge of the formed draw

panel (or some other flat stock of the same nominal thickness made of the same base steel and coating as the formed part). Calibrate the UT gage at this location.

After forming, measure the thickness at various regions on the formed part, using the UT gage. Focus on the critical areas, and determine the exact location of the thinnest area within a given region. If finite element analysis (FEA) has been performed on the part design/tooling development and the FEA package is available, it will generally include a map of potential "hot spots" that should be the first locations checked for thinning. Other locations should also be checked to confirm that the FEA package is an accurate representation of the current state of the part. During or after measurement, mark the locations on the panel where measurements were made, and photograph the part so it is clear as to which specific areas were analyzed.

After measurements, then calculate the percent difference between the formed part thickness reading and the initial blank thickness to determine the thickness reduction percentage, which is the thinning strain:

$$\text{Thinning strain (\%)} = e_t$$
$$= \frac{\begin{array}{c}\text{Formed part} \\ \text{thickness}\end{array} - \begin{array}{c}\text{Initial blank} \\ \text{thickness}\end{array}}{\begin{array}{c}\text{Initial blank} \\ \text{thickness}\end{array}} \times 100\%$$

Note that the initial blank thickness is not necessarily the same as what was measured in the previous calibration stage. Even though the edge of a formed draw panel may be flat, there may have been some thinning. Also note that the initial blank thickness is not necessarily the gage reported on the steel certification. The material certification may list only the minimum or nominal ordered thickness. Even if an actual thickness is reported on the certification, this was likely determined from the coil end at the quarter- or center-width position and may not be exactly representative of the blank used to form the analyzed part. Ideally, the initial blank thickness is determined from the specific blank used to form the analyzed part.

Thinning Strain Calculation Example. This example shows how thinning strains can be calculated and also illustrates the importance of knowing the actual flat blank material thickness in order to avoid discrepancies in interpretation of thinning strain results as being either "Acceptable" versus "Unacceptable" thinning conditions. Assumptions are:

• The specified nominal gage for a given part is 0.80 mm (0.032 in.) +/−0.02 mm (0.0008 in.).
• The thickness measured at a particular location is 0.64 mm (0.025 in.).
• For this particular application, thinning strains below −20% are considered "Acceptable" (or "Safe/pass"), and thinning strains exceeding −20% are considered "Unacceptable" (or "Fail"). Thinning strains of

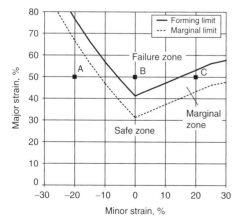

Fig. 4 A Representative forming limit diagram and 10% marginal zone

exactly −20% are considered "Marginal" (or questionable).

If the initial blank thickness is known, the actual thinning strain can be directly calculated:

$$e_t(\%) = [(t_{measured} - t_{initial})/t_{initial}] \times 100\%$$

If the actual initial thickness is not known, logical options for estimating the thinning strain include the nominal gage, the minimum specified gage, or the maximum specified gage. In the example described here:

- Initial thickness assumed to be the minimum of the allowed tolerance (gage = 0.80 − 0.02 = 0.78 mm), such that: $e_t(\%)$ = [(0.64 − 0.78)/0.78] × 100% = −17.9%. Compared to the 20% maximum thinning criterion, this area is "Acceptable."
- Initial thickness assumed to be "Nominal" (gage = 0.80 mm), such that: $e_t(\%)$ = [(0.64 − 0.80)/0.80] × 100% = −20.0%. Compared to the 20% maximum thinning criterion, this area is "Marginal."
- Initial thickness assumed to be the maximum of the allowed tolerance (gage = 0.80 + 0.02 = 0.82 mm), such that: $e_t(\%)$=[(0.64 − 0.82)/0.82]×100% = −22.0%. Compared to the 20% maximum thinning criterion, this area is "Unacceptable."

If the maximum allowable gage is used as the basis for assessing part-thinning status, areas appearing to exhibit "Unacceptable" thinning conditions may actually be "Acceptable" if the original initial blank thickness is within specification but substantially less than the allowed maximum. The risk is that some "Acceptable" areas may appear as "Unacceptable," which could lead to more tooling work to bring everything to compliance.

Conversely, if the minimum allowable gage is used as the basis for assessing part-thinning status, areas appearing as "Acceptable" could actually exhibit "Unacceptable" thinning conditions if the initial blank thickness is again within specification but substantially greater than the allowed minimum. The risk here is that some "Unacceptable" areas may appear as "Acceptable," which could lead to problems after the tooling is bought off and the part is in production.

Circle Grid Application and Measurement

Circle grid strain analysis (CGSA or CGA) takes more operator training and requires application of the grid pattern on the flat blank before it is formed. CGA will provide a great deal of information on the forming of the part and can be used to project panel performance with a change in sheet metal properties.

Thinning strain analysis by itself is a useful quality-assurance tool to assess panel performance over time. A CGA is useful as an up-front analysis, done at hard tool buyoff and/or home line commissioning, to assess overall process robustness as it relates to sheet metal properties. When there are forming problems in production, CGA can provide insight on the forming modes across the panel.

Circle Grid Strain Analysis Technique. The tools for CGA are:

- Ultrasonic thickness gage
- Micrometer or some other measuring tool to determine a reference thickness
- Electrolyte
- Stencil containing the grid pattern
- Felt pad
- Power source with grid-application device (rocker, roller, or grid wand)
- "Cleaner" (which, in reality, is a neutralizer)
- Calibrated Mylar (E.I. du Pont de Nemours and Company) strip with divergent "railroad" tracks
- Point dividers

In order to perform a CGA, one must have access to the flat blank before stamping. Measure and record the thickness of the blank as well as other descriptive information, such as dimensions and configuration (rectangular, trapezoid, developed, etc.).

Application of Circle Grids to Blank. Wipe off any oils that are on the blank (i.e., mill oil or prelube). A rag is often sufficient, and solvents are not necessarily needed. Do *not* use the aforementioned "cleaner" at this time. It is used after gridding to neutralize the acid contained within the electrolyte.

Next, determine which surface will be electrochemically etched with a grid of circles of known diameter on the blank. If it is an exposed part, the circles typically are on the exposed surface. For an unexposed part, the circles typically are on the surface that will be easiest to read once the part is formed (generally, the cavity side of the blank rather than the side that comes in contact with the punch).

Application steps are as follows:

- Put a light coating of electrolyte in the area that will be etched with a grid.
- Lay the stencil on top of that area. Smooth out any air pockets. This is crucial, because an air pocket will prevent transfer of a perfect circle.
- Lay a felt pad on top of the stencil. This felt pad should be saturated with the electrolyte.
- Connect the gridding device lead(s) to the power source. Connect the ground clip to the blank. If needed, connect the other end of the cable to the grid-application device.
- Set the amperage to approximately one-half to three-fourths of the way to the 100% setting.
- Pick up the grid-application device off the blank (or wherever it is placed). Turn the power on. Place the grid-application device on a given area on top of the felt pad (in the areas on top of the stencil) and gently rub it across that area a few times, then move to a different area until the area of the stencil has been covered. *Note:* Excessive pressure on the grid-application device is neither needed nor

desired. Too much force may have the tendency to shift the stencil.

- Turn the power off, and then put the grid-application device down.
- Gently lift one corner of the stencil to determine if the transfer is sufficient. If the transfer is lighter than desired, increase the amperage slightly, pick up the grid-application device, turn on the power, and repeat the grid-application process until an acceptable grid is produced. *Note:* Care must be taken not to shift the stencil when the operator "peeks" under one corner. If the stencil does shift, then a series of double circles will be produced that will be difficult to analyze. If the stencil does not shift, then the grid pattern should be quite readable, even though one corner has been lifted.
- Turn the power off, and then put the grid-application device down.

In this process, it is necessary to always turn the power on as the last step before gridding and turn the power off as the first step after gridding. If the grid-application device touches a metal surface with the power on, it will produce a spark. The resulting weld nugget may be large enough to interfere with forming. Also, the operator may receive an unpleasant shock if the metal edges of the grid-application device are touched when hands are covered by electrolyte. Wearing rubber gloves will help avoid the shock as well as protect the operator's skin from the electrolyte and cleaner.

The process is repeated until the entire blank is covered with the grid pattern. The panel is then wiped with the cleaner (neutralizer) to prevent further etching from the electrolyte.

There is some concern that the electrochemically etched circle pattern may change the frictional characteristics of material pulling in across draw beads, thereby changing the formability characteristics of the overall forming process. Therefore, when gridding, try to estimate the location(s) of draw beads around the binder, and apply the grid pattern only inside the perimeter of the draw beads. Depending on the size of the blank/part and the nature of the binder development, the grid pattern should generally be applied to within 25 to 50 mm (1.0 to 2.0 in.) of the blank edges.

After application of the circle-grid pattern on the appropriate surface, relubricate the blank, using the correct lubricant specified for the part. Attempt to relubricate to a level similar to that seen on production coils (light, uniform coating; no large puddles). This protects the blank from rusting between the time it is gridded and the time it is actually formed, and also provides lubrication in the forming operation similar to that experienced with the specified production material under prescribed forming conditions.

Forming of Blanks with Circle Grids. Form the gridded blank using production-representative techniques, if possible. Ideally, the press should be warmed up, and several satisfactory nongridded blanks have just been formed. Wipe any excess oils off the panel.

In the areas of interest, measure each circle (which are now ellipses). To narrow down the number of circles that need to be measured, use the UT gage to find the thinnest area. Measure the circle associated with this thinnest area. Each general section on a panel should be documented. Photograph the part so it is clear which areas were measured.

Plot the major and minor strains on the appropriate FLD. Different customers have different acceptance criteria. Examples include:

- All points are more than 10% safe on the FLD created for minimum *n*-value and minimum allowable thickness.
- All points are more than 10% safe on the FLD created for minimum *n*-value and actual thickness.
- All points are more than 10% safe on the FLD created for actual *n*-value and actual thickness.

Measuring Strains Using a Calibrated Mylar Strip Ruler. Accurate measurements of

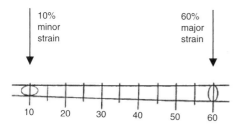

Fig. 5 Use of the Mylar strip to measure major and minor strains

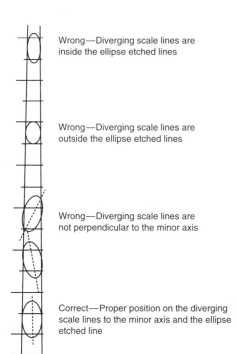

Fig. 6 Schematic representation of correct versus incorrect use of the Mylar strip

major and minor strains are crucial to obtaining the correct information on which to base forming-severity conclusions. A flexible Mylar strip ruler is the most common tool used to measure the deformation on electrochemically etched circles/ellipses. These Mylar strips are calibrated to provide direct readings of the percent strains for initial circles of a given diameter.

To measure the major strain, position the Mylar strip so the wedge-shaped scale is perpendicular to the ellipse major axis. Then, slide the scale until the wedge scale lines are centered on the ellipse etched lines. Read the engineering major strain directly from the scale (Fig. 5 and 6) with the following in mind.

- It is very important to have the diverging lines perpendicular to the major axis in making a correct measurement. Also, the diverging lines should be over the ellipse etch line, not outside or inside, for a correct measurement.
- Circles/ellipses to be read should be uniform in shape. A distorted ellipse (looking like an egg) is too close to the neck/fracture and will not provide the correct major and minor strain readings.
- The minor strain is measured in the same manner, rotating the scale accordingly.

Measuring Strains Using Automated Grid-Reading Techniques. In addition to the conventional Mylar strip approach to read circles/ellipses, there are a number of video-camera noncontact grid-reading systems available that offer data-logging capability, direct hookup to a laptop computer, and varying degrees of portability.

This approach does take some of the operator subjectivity out of the measurements, but the readings can be influenced by the available lighting, the clarity/crispness of the lines of the ellipse, the angle of the equipment relative to the surface of the part, and other system-specific issues. Still, under controlled conditions, this approach can be much faster than the Mylar strip method, especially when reading numerous data

points, such as what would be necessary to generate an FLC.

Whereas the more portable units can be used in field applications, for example, in a stamping plant setting, even the most rugged of these camera systems is still somewhat delicate and may be more appropriate for laboratory or benchtop applications rather than the typical environment encountered on the stamping plant floor.

Circle Grid Analysis

Example: Plotting Strains on a Forming Limit Diagram. Surface strains measured at three locations on a part are shown in Fig. 7. These strains are then plotted on an FLD, which is the same one shown in Fig. 4.

Observe that the corner of the part in draw (location A: negative minor strain) is not near failure, because the measured strain plotted on the FLD is in the safe zone with a 27% major strain safety margin.

The sidewall of the part in plane-strain deformation (FLD center, location B: zero minor strain) exhibits a strain level located in the failure zone. That part location should exhibit either a neck or a split.

The location of the part in stretch (location C: positive minor strain), plotted on the right side of the FLD, is located in the marginal zone with a 3% major strain safety margin. The marginal zone signals that the part is near the onset of necking.

Also note that to go from the draw location, A, to the plane-strain location, B, (which changes the part from being very safe and robust to a location that will probably split) involves a measurement difference of just 0.5 mm (0.02 in.) (from 2.0 to 2.5 mm, or 0.08 to 0.10 in.) in the minor strain reading. This should reinforce how critical it is to accurately measure the strains.

Failure Thinning Limit Curve. The failure thinning strain is the through-thickness strain

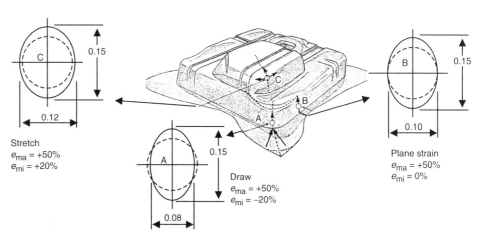

Fig. 7 Example of measured strains on a complex stamping. These strains are plotted on the forming limit diagram in Fig. 4, which was drawn assuming an *n*-value of 0.24 and a thickness of 0.90 mm (0.035 in.). e_{ma}, major strain; e_{mi}, minor strain. Dimensions given in inches

before the onset of a necking failure. A failure thinning strain can be calculated for each failure major and minor strain combination on the FLC using the equation:

$$e_t = \frac{1}{(e_{ma}+1) \times (e_{mi}+1)} - 1$$

Plotting these results against the minor strain values yields the thinning-limit curve (TLC) shown as the dashed curve in Fig. 8. The failure thinning strain under draw and plane-strain deformation is calculated using FLC_0 as the major strain value and a minor strain value of zero:

$$TLC_0 = e_{tf} = \frac{1}{(FLC_0+1) \times (0+1)} - 1$$

The TLC_0 (which is the border of the marginal and failure thinning strains) is based on FLC_0, as shown in the previous equation. Similarly, the border of the safe and marginal thinning strains of the TLC (TLC_M, where "M" stands for marginal) is based on $FLC_0 - 10\%$:

$$TLC_M = e_{tM} = \frac{1}{([FLC_0 - 10\%]+1) \times (0+1)} - 1$$

Note that although a 10% safety margin is used for the FLC (10% on the major strain axis separates the marginal and failure zones on the FLC), this does not translate into 10% on the thinning strain axis separating the marginal and failure zones of the thinning limit diagram. This value is closer to 7% on the thinning strain axis and can be shown by $TLC_0 - TLC_M$ in the previous equations.

From here, it is possible to calculate a failure thickness and a marginal thickness:

- *Failure thickness* = Initial blank thickness × (1 + Failure thinning strain), or

$$t_f = t_0 \times (1 + e_{tf})$$

- *Marginal thickness* = Initial blank thickness × (1 + Marginal thinning strain), or

$$t_M = t_0 \times (1 + e_{tM})$$

Areas thicker than the marginal thickness are said to be in the safe zone of the thinning limit diagram, while areas thinner than the failure thickness are in the failure zone of the thinning limit diagram. Areas in between the marginal and failure thickness values are said to be in the marginal zone of the thinning limit diagram.

It is possible to determine the formed part thickness at all locations, without needing a UT gage, by measuring the major and minor strains and using a similar equation to the previously mentioned one. Using the contour thinning strain calculated from the major and minor strain values measured at any given location, the corresponding part contour thickness is calculated as:

Part contour thickness
= Initial blank thickness
× (1 + Contour thinning strain)

Example: Failure and Marginal Thickness Determination. When it is not possible to do a circle grid, the following information could be useful (albeit conservative) for monitoring performance. Knowing the minimum specified thickness and either the minimum specified or expected n-value, one can calculate a worst-case FLC_0. Then, the thinning limit under draw and plane strain conditions can be calculated using the FLC. A similar calculation can be done for the marginal zone by substituting ($FLC_0 - 10$).

As an example, assume that the minimum ordered gage is 0.80 mm (0.032 in.) and the minimum allowable n-value is 0.20:

$$FLC_0 = (23.3 + 14.2 \times 0.80) \\ \times (0.20/0.21) = 33.0\%$$

The failure thinning strain is therefore:

$$TLC_0 = e_{tf} = \frac{1}{(0.33+1) \times (1)} - 1 \\ = \frac{1}{1.33} - 1 = -0.248 = -24.8\%$$

If the initial blank thickness is 0.81 mm (0.032 in.), the thickness value where the part enters the failure zone in plane-strain and draw deformation is:

Failure thickness = 0.81 mm (1 + [−0.248])
= 0.81 mm (0.752) = 0.61 mm

The marginal-zone boundary is determined in a similar manner. With the $FLC_0 = 33.0\%$, the marginal zone is at a major strain value of 23.0%. As such, the marginal thinning strain is:

$$e_{tM} = \frac{1}{(0.23+1) \times (1)} - 1 = -0.187 = -18.7\%$$

The thickness value where the part enters the marginal zone in plane-strain and draw deformation is:

Marginal thickness
= 0.81 mm (1 + [−0.187])
= 0.81 mm (0.813) = 0.66 mm

This can be interpreted as meaning the part will be in the safe zone as long as all areas measure greater than 0.66 mm (0.026 in.).

It is important to remember that these values for marginal and failure thickness of 0.66 and 0.61 mm (0.026 and 0.024 in.), respectively, were generated based on a thickness of 0.81 mm (0.032 in.) for the initial blank. The marginal and failure thickness values will increase as the initial blank thickness increases.

Also note that these values are for plane-strain and draw deformation only. The part can actually be thinner than these values in stretch deformation and still be safe. However, if it is not possible to do a circle grid to determine the minor strain value to confirm if the part is in stretch mode, then a conservative approach is recommended, where the values calculated from the aforementioned methodology are used.

Bead Correction Factor

Over the years, it has been observed that in areas where steel is bent and unbent around draw beads, split-free parts can be produced even though measured strains are deep in the failure zone. With these sections in the failure zone, unnecessary die work or material grade changes may have been done to bring these areas into the safe zone, whereas in actuality, these areas were sufficiently robust. In 2003, the Auto/Steel Partnership (A/SP) published the results of an extensive study on this phenomenon, along with a procedure to adjust the measured strain readings to bring them closer to the field observations (Ref 1). This discussion and the following example are adapted from the A/SP training manual.

Example of Bead Correction Factor. The bead correction factor (BCF) is computed from the amount of sheet metal thinning caused by bending and unbending. Ultrasonic thickness measurements of sheet metal entering and exiting the draw bead (or draw bead and die radius combination) easily permit computation of the BCF. The magnitude of BCF can be large enough to reduce the FLC severity rating of a die from red (failure) to green (safe). The procedure is shown as follows, with example values.

Data required:

- n-value (0.218)
- Thickness at area of concern (0.0213 in.)
- Initial blank thickness (0.0321 in.)
- Thickness at entry side of draw bead (0.0305 in.)

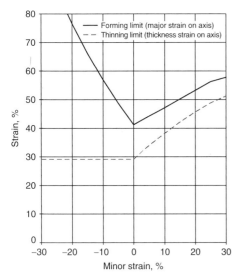

Fig. 8 Representative thinning limit curve, as derived from the corresponding forming limit curve

Compute FLC$_0$:

$$FLC_0 = (23.3 + 360 \times t) \times (n/0.21) = 36.2\%$$

Determine Major Strain. Use either the major strain measurement on a gridded sample or convert from the UT value under plane-strain assumptions:

Thickness strain (e_t)

$$= \frac{\text{Formed part thickness} - \text{Blank thickness}}{\text{Blank thickness}} \times 100$$

So that $e_t = \dfrac{0.0213 - 0.0321}{0.0321} \times 100 = -33.64\%$

Major strain $(\%) = e_{ma} = \dfrac{1}{(e_t + 1) \times (0 + 1)} - 1$

So that $e_{ma} = \dfrac{1}{([-0.3364] + 1)} - 1 = 50.7\%$

Compute as-measured safety margin:

- $FLC_0 = 36.2\%$
- Subtract as-measured major strain = $(-)50.7\%$
- As-measured safety margin = -14.5% (red zone)

Compute BCF:

- $BCF = 60 \times \ln$ (t of exit bead/t of enter bead) $= 60 \times \ln (0.0213/0.0305)$
- $BCF = -21.5\%$

Compute corrected engineering major strain:

- As-measured major strain = $+50.7\%$
- Add BCF = -21.5%
- Corrected major strain = $+29.2\%$

Compute corrected safety margin:

- FLC_0 (step 1) = 36.2%
- Subtract corrected major strain = $(-)29.2\%$
- Corrected safety margin = $+7.0\%$ (yellow zone)

Techniques for Measuring Small Strains Distributed Over Large Areas. For such parts as automotive outer skin panels (e.g., hoods, roofs, doors, quarter panels), performance characteristics such as strength for the finished part, dent resistance, and oil canning resistance are often key design concerns. These characteristics result from the cold work imparted in the forming process through the relatively small biaxial surface strains (typically on the order of 0 to 5% major strain by -2 to 5% minor strain) induced over the large and fairly flat portions of the formed panels.

To quantify these effects, it is necessary to accurately measure these generally small strains. The ability to do so by using the conventional CGSA techniques described previously (using the pattern of electrochemically etched small circles of typically 2.5 or 5.0 mm, or 0.10 or 0.20 in., diameter) is quite limited, because the very small change in dimension of an already small circle is difficult to measure with any reliable degree of accuracy or precision.

Instead, a pattern of larger circles (typically 100 mm diameter) is mechanically scribed on those portions of the flat blank corresponding to the areas of interest/concern on the formed parts. With forming, the originally round circles become elliptical in shape (similar in nature to what happens in conventional CGSA). By determining the directions of the major and minor strain axes and measuring the extent of deformation along those axes, the distributed biaxial strain state across the face of these panels can be established.

There are two basic approaches to achieving careful and accurate readings of the small strains distributed over large areas. Key to both is how the circles are initially scribed on the flat blank. A circle template of the precise circle diameter can be used to scribe the circles, but this does not provide an indication of the circle center; the lack of the circle center location makes it extremely difficult to precisely determine the major and minor strain axes. A preferred circle-scribing technique involves (light) center-punching of each circle location, from which a needle-tipped compass is used to "swing" the circle of desired diameter.

Templates have been developed that can be used to measure the described biaxial surface strains. The template is essentially a calibrated spiral printed on a Mylar sheet that, when centered over the originally round circle and appropriately rotated, provides an instant readout of the respective major and minor strain values. One drawback to this technique is that it is necessary to ensure that the spiral template being used is either an original (created to accurate dimensions and indeed printed on Mylar) or has been checked against an original template. For purposes of convenience, these templates are often reproduced on clear transparency material on a copying machine, and this results in two possible sources of inaccuracy:

- The image generally increases or decreases in the process of copying. The difference is generally small but significant enough to compromise the accuracy of the measured strain reading.
- The extent of possible expansion or shrinkage of the template (depending on the nature of the transparency material used) can significantly compromise the measured results.

An alternative approach is to begin with the same center-punched scribed circles on the flat blanks. After forming, "reswing" a circle of the same initial diameter from each of the center-punched locations. Comparing the round, rescribed circle to the shape of the formed ellipse provides a graphic indication of the major and minor strain axes.

It is important to note that the major and minor strain directions do not necessarily need to align with the edges of a rectangular blank. The rescribing of the round circles after forming will make the true alignment of the major and minor strain axes immediately apparent. To merely assume the major and minor strain axes direc-

tions based on the blank shape will, in many cases, lead to a misleading and inaccurate indication of the true biaxial strain distribution across the face of a formed panel.

Once the major and minor strain axes have been identified, the respective major and minor strain values are determined by calculating the difference between the final respective ellipse axis measurement and the original circle diameter. The easiest way to do this is to use 100 mm diameter circles, a metric scale (marked off to the nearest half of a millimeter), and a magnifying visor (with approximately a $5\times$ magnification lens) to measure the ellipse axes. Doing so provides an all-but-direct readout of the respective major and minor strain values.

For example, with an initial circle diameter of 100 mm:

- Ellipse major axis = 104.5 mm, so major strain = 4.5%
- Ellipse minor axis = 98.3 mm, so major strain = -1.7%

Circles of 5 in. diameter can similarly be used, but even by using an English-unit scale marked off in decimal divisions (0.10 and 0.01 in. divisions), some level of calculation will be required to convert the measured ellipse dimensions to corresponding strain values.

TSA Application Examples

Production Process Robustness and Tracking with the Reference Panel. Once a robust condition has been achieved at the beginning of production stamping (or during the production stamping life cycle), a reference panel can be established, using strain analysis techniques. This reference panel documents as a "snapshot in time" all of the material and part characteristics, peak strain and strain distribution conditions, the tooling conditions, and the press and process operating conditions and parameters used in forming the sound part under robust processing conditions. The reference panel is particularly useful if forming difficulties are experienced at any point during ongoing production stamping. If the part now experiences increased breakage rates, wrinkles/buckles in the formed parts, reduced productivity rates, compromised panel integrity characteristics (strength, dent resistance, oil canning resistance, etc.), or surface issues (scratches, scoring, powdering/flaking of surface coatings, etc.), there is a set of known conditions that was able to produce a satisfactory part, and the settings can be reverted to those conditions.

The required documentation falls into three basic categories:

- Material
  a. Complete mechanical properties (actual thickness, yield strength, tensile strength, total elongation, n-value, r-value, K-value)
  b. Surface coating—type and coating weight

c. Lubrication—type and amount (film thickness)
d. Blank type (rectangular, trapezoid, developed) and dimensions
e. Surface roughness characteristics
f. Detailed strain and thinning analyses of the draw panel
g. Documentation of breakdown panel characteristics (if available)

- Tooling conditions
  a. Tooling surface characteristics—polishing, hardness, coating, and so on
  b. Bead conditions—geometry, locations, dimensions, surface conditions, and so on
  c. Engineering change level of the current die configuration

- Press and process parameters
  a. Press number and type
  b. Binder and punch tonnage conditions
  c. Shims—locations and thickness
  d. Detailed binder set draw-in measurements at key locations
  e. Shut height
  f. Press speed

If problems are experienced in the course of ongoing downstream production, the reference panel documentation is there as a benchmark to compare the current/problematic conditions to those known to yield a combination of sound and robust part and process. With this knowledge, it should now be possible to both identify the source of the difficulty and return to the favorable/desired material/tooling/process conditions. The onset of forming/process difficulties can be either sudden/abrupt (as, for instance, if wrong material is inadvertently run, or tonnage settings are grossly off) or more gradual in nature (as, for instance, the deterioration of panel quality/integrity over time with the slow wearing of draw beads with ongoing production). In either case, the reference panel documentation should provide significant help both in determining the source of the change and in restoring the favorable forming and processing conditions.

Thickness Strain Analysis as a Statistical Process Control Tool. Another useful and nondestructive application of thinning strain analysis is in statistical process control. Critical locations can be defined from structural requirements (determined from finite-element analysis, or FEA, or physical crash testing, for example) or from formability requirements (hot spots from FEA or hands-on strain analyses). A draw panel is then marked with the critical locations, and small circles are cut out of these critical locations. These cut circles should be big enough to accommodate the probe tip of a UT gage. The draw panel with cut holes is now a template for measuring critical locations and can be laid on top of any subsequent panel to ensure that the same areas are measured each time. If the part in question is made from a relatively large draw panel that is hard to maneuver, then the full draw panel template can be cut into smaller pieces, with one or more critical locations on each minitemplate.

Using the strain data for critical locations on a given draw panel (as established, for instance, in documenting the reference panel described previously), areas are cut from a typical draw panel to be used as templates overlaying the particular key locations of the draw panel.

On a prescribed schedule (once an hour, once a shift, beginning and end of each production run, etc.), a draw panel is taken off-line, and UT measurements are made, using the templates for each of the key locations. After the measurements are made and recorded, the panel can be returned to the rack and used for regular production requirements. Although it is possible to use a hand-held micrometer instead of a UT gage, it is more time-consuming, because the panel needs to be cut up into small enough pieces to measure, and as such, the panel is now scrap after the measurements are made and recorded.

Logging these data points and plotting them as a function of time provides an indication as to process stability. If, for whatever reason, something in the process is drifting out of control, the effects should show up as changes in the noted-versus-expected strain levels and distributions. Once such changes are observed, the reference panel can again be referred to in order to identify the source/nature of the material/tooling/process discrepancy, and to return to the prescribed/favorable conditions known to be capable of robustly producing quality parts.

How Circle Grid Analysis Techniques Can Influence the Part Safety Assessment. Based on the principle of volume constancy, from any given pair of major and minor surface strain measurements, it is possible to calculate a corresponding thinning strain. Also, from the determined thinning strain and the initial blank thickness, the remaining metal thickness at the location in question can be calculated. If all of the measurements were perfect, the calculated thickness would be identical to that determined using a (properly calibrated) UT gage. However, due to surface-strain circle measurement error combined with the inevitable slight variation in UT gage readings, the measured and calculated thickness values are rarely in absolute agreement. For all practical purposes, if these values are within $\pm 3\%$ of each other, they are typically considered to be close enough.

The following is an account of an actual situation in which CGSA practices came into direct question, the outcome of the ensuing investigation having a direct and significant impact on the ultimate determination of the strain-state safety assessment of the respective panels.

Example: Incorrect Method Used in Circle Grid Analysis. Two steel suppliers had been sourced on opposite sides of an otherwise virtually symmetric major automotive body outer panel. Draw dies for the two panels were in similar stages of tooling development and die tryout when representatives from the two steel companies were asked to perform surface and thinning strain analysis of their respective parts.

The results of one company were called into question, with the tool shop asking why—with two symmetric but essentially identical dies and both in the same state of tooling development—the supplier would be reporting only a few areas as exhibiting "Marginal" strain conditions. At the same time, the other supplier, analyzing basically symmetrically located areas on the other side of the part, was reporting "Failure" and "Marginal" conditions at nearly every location analyzed.

Before even rechecking their own circle readings or checking the readings provided by the other source, representatives from the first steel company noted that the calculated thicknesses they had reported agreed closely with the area-by-area UT gage data they had recorded and reported. Upon checking the results reported by the other steel company, they noted widely varying agreement (or lack thereof) between their calculated and measured thickness results, with some differing by as much as ~20%. This immediately suggested the likelihood that there was something discrepant in either the circle readings or UT gage results reported by the other steel company.

After verifying their own findings, representatives of the first steel company spot-checked readings reported by the other supplier. They found that, based on their reading and plotting of areas that the other supplier had documented earlier, areas that the other supplier had reported as "Failure" now typically appeared as "Marginal," and areas reported earlier as "Marginal" now typically appeared as "Safe." In order to reproduce the strain measurements and safety status indications reported by the other steel company, representatives of the first supplier determined that the earlier results had to have been made by determining both major and minor strains by measuring from outside edge to outside edge of the respective circles—a practice that is both incorrect and inaccurate. This is shown in the top schematic sketch in Fig. 9. Below that sketch are several alternative options for correct reading techniques.

In the schematic representation of this hypothetical example, it is seen that outside-to-outside edge yields strain readings of 40% major strain by −5% minor strain, which plots as a "Marginal" condition. Reading the circle using any of the correct options yields major and minor strain readings of 30 by −10%, which plots as "Safe." Also shown in each case are the circle dimensions that would be measured if a very precise scale were used rather than the calibrated Mylar strip usually used to directly determine surface-strain values. The dimensional difference of the 0.100 in. diameter circle between the correct major strain reading of 30% and the erroneous 40% indication is seen to be just 0.010 in. (ten thousandths of an inch!). Similarly, the difference in circle-dimension measurements between the correct minor strain determination of −10% and the incorrect indication of −5% is just 0.005 in. (five thousandths of an inch!). These dimensional numbers,

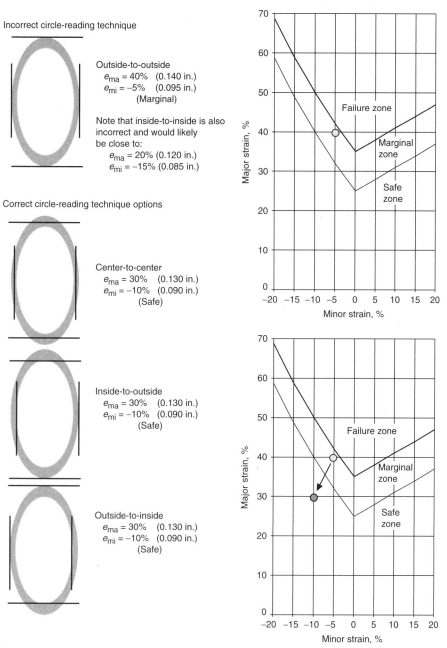

Fig. 9 Schematic drawings of correct and incorrect circle-reading techniques. e_{ma}, major strain; e_{mi}, minor strain

Table 1 Thinning percentages from UT gage measurements

	Thickness by UTG(a)		Percentage
Area	mm	in.	thinning, %
A	0.644	0.0253	19.5
B	0.576	0.0227	28.0
C	0.530	0.0209	33.8
D	0.654	0.0257	18.3

while generally not measured or reported, are highlighted here to illustrate that even the slightest error in circle reading can quite easily result in significant misrepresentation of the actual/correct strain safety status of a part being analyzed.

To further illustrate this situation, data as measured and reported by the second supplier and rechecked by the first supplier for four representative areas are summarized in Tables 1, 2 and 3. The actual thinning percentages, calculated by comparing the point-by-point UT gage readings to the initial blank thickness, are indicated in the last column of Table 1. For each of the supplier's reported respective circle readings, the corresponding thinning strains also can be calculated. These values are in turn compared to the actual thinning strains (Table 2). When based on the incorrect circle readings, wide variation in agreement between the actual and calculated thinning strains is noted (differences ranging between 3.4 and 17.7%). For the correctly measured circles, the actual and calculated thinning strains are in much closer agreement (differences ranging from −2.2 to +2.4%).

The respective circle-reading data for each of these four representative locations are plotted in the FLD presented in Fig. 10. Here, the initial results, based on the incorrect circle-reading practices, are shown with the letter identifications A through D. For each point, the correct/corrected results are indicated as A′ through D′, respectively. In each case, a consistent trend is noted: The major strain is reduced, and the minor strain shifts toward the left on the FLD. This, in turn, results in a significant increase in the indicated strain safety margin in each case (Table 3). With location A, for example, going from a "Failure" condition with strain safety margin of −19% to a "Safe" condition for A′ with a strain safety margin of +17%. For location C, the indicated strain safety status does not change, even with the correction for circle-reading technique, and the strain safety margin is seen to go from −13% for the (incorrect) location C indication to −4% for the (corrected) indication for location C′—a 9% strain improvement in the apparent strain safety status. However, even this would indicate significant work remains to be done to achieve a "Safe" configuration at this particular location.

Conversely, if circles are measured by likewise incorrectly reading from inside edge to inside edge, the results can be similarly misleading, but in the opposite direction: Areas that

Table 2 Comparison of calculated to measured thinning strains based on major and minor surface strains as determined by incorrect (outside edge to outside edge) versus correct circle-reading techniques

Area	Major strain, %	Minor strain, %	Calculated thickness mm	Calculated thickness in.	Percentage thinning, %	Difference in % thinning, calculated versus UTG
Incorrect circle measurements						
A	75	−9	0.502	0.0198	37.2	17.7
B	35	8	0.549	0.0216	31.4	3.4
C	58	6	0.478	0.0188	40.3	6.5
D	43	−1	0.565	0.0222	29.4	11.1
Correct circle measurements						
A′	60	−20	0.625	0.0246	21.9	2.4
B′	31	3	0.593	0.0233	25.9	−2.1
C′	49	−2	0.548	0.0216	31.5	−2.2
D′	36	−11	0.661	0.0260	17.4	−0.9

(a) UTG, ultrasonic thickness gage

Table 3 Safety margins of calculated thickness strains (Table 2, Fig. 10)

Location	Major strain, %	Minor strain, %	Calculated thickness strain		Actual properties	
			Gage, mm	Thinning, %	Safety, %	Status
A	75	−9	0.502	37	−19	Failure
A′	60	−20	0.625	22	17	Safe
B	35	8	0.549	31	12	Safe
B′	31	3	0.593	26	13	Safe
C	58	6	0.478	40	−13	Failure
C′	49	−2	0.548	32	−4	Failure
D	43	−1	0.565	29	0	Marginal
D′	36	−11	0.661	17	23	Safe

Parameter	Actual properties
n-value	0.253
Initial thickness, mm	0.800
FLC_0, %	41.8
Failure thinning, %	29.5
Failure thinout limit, mm	0.564
Marginal FLC_0, %	31.8
Marginal thinning, %	24.1
Marginal thinout limit, %	0.607

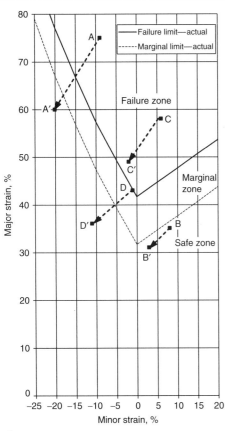

Fig. 10 Forming limit diagram (FLD) comparing correct and incorrect circle-reading techniques

Table 4 Summary of illustrative inside-to-inside circle-reading results

Location	Major strain, %	Minor strain, %	Strain safety margin, %	Status
E	30	−10	21	Safe
E′	40	−5	3	Marginal
F	35	−5	8	Marginal
F′	45	0	−9	Failure
G	25	5	14	Safe
G′	35	10	7	Marginal

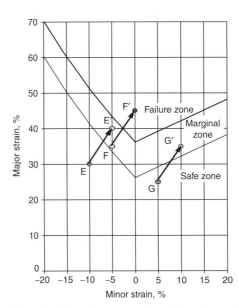

Fig. 11 Schematic representation of the effects of incorrectly measuring circles by reading from inside edge to inside edge

are actually "Marginal" could appear as "Safe," and areas that are actually "Failure" could appear as "Marginal." Unlike the previous situation where unnecessary die work could result, here areas legitimately requiring further die work

could be overlooked. Data for several examples of the potential problems resulting from incorrectly measuring circles by reading from inside edge to inside edge are summarized in Table 4 and shown schematically in Fig. 11. In this example, the strains measured by reading from inside edge to inside edge are shown as the unprimed locations (E, F, and G). The strains for circles measured using the correct reading technique (e.g., center-to-center) are shown as the respective primed locations (E′, F′, and G′).

Inferences. When performing CGSA, be sure to follow proper procedures, and be careful to ensure accurate circle measurement/strain data. Using incorrect reading techniques can result in either passing areas appearing as failure situations and/or failure conditions appearing as safe. Either case results in a grossly inaccurate indication of the actual panel strain safety status. In addition, the following practices will ensure accuracy:

- Use a UT gage, specifically calibrated for the particular material in question, to determine the actual remaining sheet metal thickness at each specific point being analyzed. Also, wherever and whenever possible, make every effort to ascertain the initial blank thickness of the actual material being used in the specific tryout activity.
- In order to ensure accuracy and consistency, cross-check the thickness values determined by UT gage measurement versus those calculated from the measured major and minor strain data. If there is significant discrepancy between the two, it is an indication that either the UT gage reading or the strain data are in error, and in either case, one should recheck both values.

REFERENCE

1. "Enhanced Forming Limit Diagrams," Auto/ Steel Partnership, http://www.a-sp.org/ publications.htm (accessed March 27, 2006)

Process Design for Sheet Forming

Constitutive Equations

Amit Ghosh, University of Michigan

CONSTITUTIVE RELATIONS for metal-working include elements of behavior at ambient temperature as well as high-temperature response. Because bulk forming is conducted over a wide range of temperatures and strain rates, it is best to review the low-temperature behavior first and gradually build in the effects of high-temperature response. In the following sections, equations are presented first for strain hardening and then for strain-rate-sensitive flow, with alternate sections on empirically determined properties, followed by models of constitutive behavior.

Strain Hardening (Ref 1)*

Several phenomenological models used to describe hardening are summarized in Table 1. The first three predict continued hardening, whereas the Voce models predict saturation as a stress, σ_s. Kocks' mechanistic model also predicts saturation (Ref 2). In fact, a plot of θ versus σ presents a good graphical picture of the saturation stress. Kocks' model is based on dislocation interactions and predicts an early steady-state saturation behavior. Figure 1 (Ref 3) demonstrates the hardening behavior of 1100 aluminum deformed under several different modes. The hardening does not saturate but persists to high stresses at low hardening rate.

Strain Rate Effects (Ref 4)**

Usually, flow stress increases with strain rate, and the effect at constant strain can be approximated by:

$$\sigma = C\dot{\varepsilon}^{m} \qquad \text{(Eq 1)}$$

where C is a strength constant that depends on strain, temperature, and material; and m is the strain-rate sensitivity of the flow stress. For most metals at room temperature, the magnitude of m

*Adapted from: S.S. Hecker and M.G. Stout, Strain Hardening of Heavily Cold Worked Metals, *Deformation, Processing, and Structure*, G. Krauss, Ed., American Society for Metals, 1984, p 1–46

**Adapted with permission from: W.F. Hosford and R.M. Caddell, *Metal Forming: Mechanics and Metallurgy*, Prentice Hall, 1983, p 80–98

is quite low (between 0 and 0.03). If the flow stresses, σ_2 and σ_1, at two strain rates, $\dot{\varepsilon}_2$ and $\dot{\varepsilon}_1$, are compared at the same strain:

$$\frac{\sigma_2}{\sigma_1} = \left(\frac{\dot{\varepsilon}_2}{\dot{\varepsilon}_1}\right)^{m} \qquad \text{(Eq 2)}$$

or $\ln(\sigma_2/\sigma_1) = m \ln(\dot{\varepsilon}_2/\dot{\varepsilon}_1)$. If, as is likely at low temperatures, σ_2 is not much greater than σ_1, Eq 2 can be simplified to:

$$\frac{\Delta\sigma}{\sigma} \simeq m \ln\frac{\dot{\varepsilon}_2}{\dot{\varepsilon}_1} = 2.3\, m \log\frac{\dot{\varepsilon}_2}{\dot{\varepsilon}_1} \qquad \text{(Eq 3)}$$

For example, if $m = 0.01$, increasing the strain rate by a factor of 10 would raise the flow stress by only $0.01 \times 2.3 \simeq 2\%$, which illustrates why rate effects are often ignored.

However, rate effects can be important in certain cases. If, for example, one wishes to predict forming loads in wire drawing or sheet rolling (where the strain rates may be as high as 10^4/s) from data obtained in a laboratory tension test, in which the strain rates may be as low as 10^{-4}/s, the flow stress should be corrected unless m is very small.

At hot-working temperatures, m typically rises to 0.10 or 0.20, so rate effects are much larger than at room temperature. Under certain circumstances, m values of 0.5 or higher have been observed in various metals. Ratios of (σ_2/σ_1) calculated form Eq 2 for various levels of $(\dot{\varepsilon}_2/\dot{\varepsilon}_1)$ and m are shown in Fig. 2.

There are two commonly used methods of measuring m. One is to obtain continuous stress-strain curves at several different strain rates and compare the levels of stress at a fixed strain using Eq 2. The other is to make abrupt changes of strain rate during a tension test and use the corresponding level of $\Delta\sigma$ in Eq 3. These are illustrated in Fig. 3. Increased strain rates cause somewhat greater strain hardening, so the use of continuous stress-strain curves yields larger values of m than the second method, which compares the flow stresses for the same structure. The second method has the advantage that several strain-rate changes can be made on one specimen, whereas continuous stress-strain curves require a specimen for each strain rate.

Strain-rate sensitivity is also temperature dependent; Fig. 4 (Ref 5) shows data for a number

of metals obtained from continuous constant strain-rate tests. Below $T/T_M = \frac{1}{2}$ (T/T_M is the ratio of testing temperature to melting point on an absolute scale), the rate sensitivity is low, but it climbs rapidly for $T > T_M/2$.

More detailed data for aluminum alloys are given in Fig. 5 (Ref 6). Although the definition of m in this figure is based on shear stress and strain rate, it is equivalent to the definition derived from Eq 1. For these and many other alloys, there is a minimum in m near room temperature, and, as indicated, negative m values are sometimes found.

Alternative Description of Strain-Rate Dependence. For steels, Eq 1 may not be the best description of the strain-rate dependence of flow stress. There is considerable evidence that the strain-rate exponent, m, decreases as the steel is strain hardened and that it is lower for high-strength steels than for weaker steels. The data of Fig. 6 (Ref 7) show that m is inversely proportional to the flow stress, σ. This suggests that a better description is:

$$\frac{d\sigma}{d(\ln\dot{\varepsilon})} = m' \qquad \text{(Eq 4)}$$

or

$$\sigma = m' \ln\frac{\dot{\varepsilon}}{\dot{\varepsilon}_0} + C \qquad \text{(Eq 5)}$$

where m' is the new rate-sensitivity constant, and C is the value of σ at $\dot{\varepsilon} = \dot{\varepsilon}_0$ (C and $\dot{\varepsilon}_0$ are not

Table 1 Stress-strain relations based on empirical (phenomenological) and theoretical (Kocks) considerations

The Voce and Kocks relations predict saturation.

Phenomenological models

Holloman (parabolic)	$\sigma = K\varepsilon^{n}$
Ludwik	$\sigma = \sigma_0 + K'\varepsilon^{n'}$
Swift	$\sigma = K_2(\varepsilon + \varepsilon_0)^{n_2}$
Voce	$\sigma = \sigma_s - (\sigma_s - \sigma_0)\exp(-N\varepsilon)$
Modified Voce (Hockett-Sherby)	$\sigma = \sigma_s - (\sigma_s - \sigma_0)\exp(-N'\varepsilon^P)$
Kocks Model	$\theta = \theta_0\left(1 - \dfrac{\sigma}{\sigma_s}\right)$
	where $\theta = \dfrac{d\sigma}{d\varepsilon}\bigg|_{\dot{\varepsilon},T}$

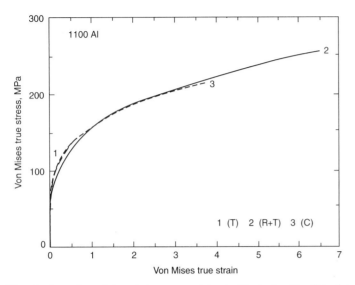

Fig. 1 Comparison of stress-strain curves as determined by tension (T), rolling plus tension (R+T), and compression (C) of annealed 1100 aluminum. Source: Ref 3

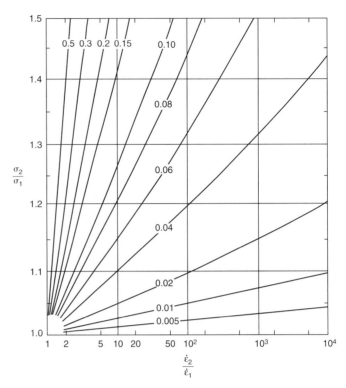

Fig. 2 Influence of strain rate on flow stress for various levels of strain-rate sensitivity, *m*, indicated on the curves

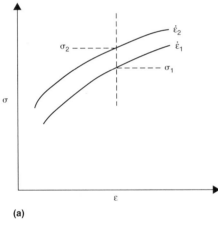

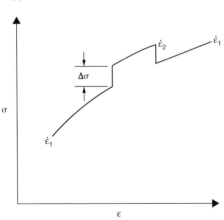

Fig. 3 Two methods of determining *m*. (a) Two continuous stress-strain curves at different strain rates are compared at the same strain, and $m = \ln(\sigma_2/\sigma_1)/\ln(\dot{\varepsilon}_2/\dot{\varepsilon}_1)$. (b) Abrupt strain-rate changes are made during a tension test, and $m = (\Delta\sigma/\sigma)/\ln(\dot{\varepsilon}_2/\dot{\varepsilon}_1)$.

independent constants; if $\dot{\varepsilon}_0$ is taken as unity, $C = \sigma$ for $\dot{\varepsilon} = 1$).

Because C must depend on strain, Eq 5 indicates that the contributions of strain and strain rate to flow stress are additive rather than multiplicative. Another indication supporting this postulate is the observation that the strain-hardening exponent, *n*, for steels decreases at high strain rates (Fig. 7).

When C in Eq 5 is $K'\varepsilon^{n'}$:

$$\sigma = m' \ln \frac{\dot{\varepsilon}}{\dot{\varepsilon}_0} + K'\varepsilon^{n'} \tag{Eq 6}$$

The usual exponent *n* in $\sigma = K\varepsilon^n$ can be expressed as:

$$n = \frac{\varepsilon \, d\sigma}{\sigma \, d\varepsilon} \tag{Eq 7}$$

Then, evaluating *n* in terms of Eq 6:

$$n = n' \bigg/ \left[\frac{m' \ln(\dot{\varepsilon}/\dot{\varepsilon}_0)}{K'\varepsilon^{n'}} + 1 \right] \tag{Eq 8}$$

so, if *m'* and *n'* are truly constant, *n* would appear to decrease with strain rate.

Temperature Dependence of Flow Stress. At elevated temperatures, the rate of strain hardening falls rapidly in most metals with an increase in temperature, as shown in Fig. 8 (Ref 8). The flow stress and tensile strength, measured at constant strain and strain rate, also drop with increasing temperature, as illustrated in Fig. 9. However, the drop is not always continuous; often, there is a temperature range over which the flow stress is only slightly temperature dependent or, in some cases, even increases

slightly with temperature. The temperature dependence of flow stress is closely related to its strain-rate dependence. Decreasing the strain rate has the same effect on flow stress as raising the temperature, as indicated schematically in Fig. 10. Here, it is clear that at a given temperature, the strain-rate dependence is related to the slope of the σ versus T curve; where σ increases with T, *m* must be negative.

The simplest quantitative treatment of temperature dependence is that of Zener and Hollomon, who argued that plastic straining could be treated as a rate process using the Arrhenius rate law, rate $\propto \exp(-Q/RT)$, which has been successfully applied to many rate processes. They proposed that:

$$\dot{\varepsilon} = Ae^{-Q/RT} \tag{Eq 9}$$

where Q is an activation energy, T the absolute temperature, and R the gas constant. Here, the constant of proportionality, A, is both stress and strain dependent. At constant strain, A is a function of stress alone [$A = A(\sigma)$], so Eq 9 can be written as:

$$A(\sigma) = \dot{\varepsilon}e^{Q/RT} \tag{Eq 10}$$

or more simply as:

$$\sigma = f(Z) \tag{Eq 11}$$

where the Zener-Hollomon parameter $Z = \dot{\varepsilon}e^{Q/RT}$. This development predicts that if the strain rate to produce a given stress at a given temperature is plotted on a logarithmic scale against $1/T$, a straight line should result with a slope of $-Q/R$. Figure 11 shows such a plot.

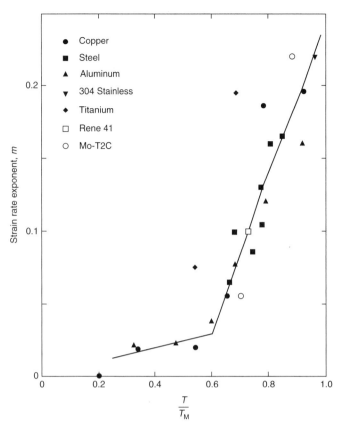

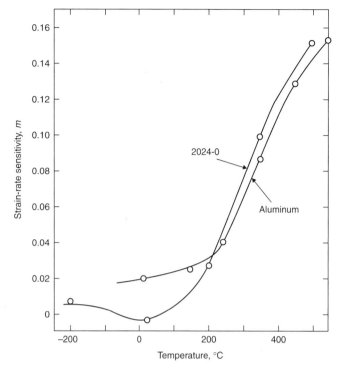

Fig. 5 Temperature dependence of the strain-rate sensitivities of 2024 and pure aluminum. Source: Ref 6

Fig. 4 Variation of the strain-rate sensitivity of different materials with homologous temperature, T/T_M. Adapted from Ref 5

Correlations of this type are very useful in relating temperature and strain-rate effects, particularly in the high-temperature range. However, such correlations may break down if applied over too large a range of temperatures, strains, or strain rates. One reason is that the rate-controlling process, and hence Q, may change with temperature or strain. Another is connected with the original formulation of the Arrhenius rate law in which it was supposed that thermal fluctuations alone overcome an activation barrier, whereas in plastic deformation, the applied stress acts together with thermal fluctuations in overcoming the barriers, as indicated in the following development.

Consider an activation barrier for the rate-controlling process as in Fig. 12. The process may be cross slip, dislocation climb, and so on. Ignoring the details, assume that the dislocation moves from left to right. In the absence of applied stress, the activation barrier has a height Q, and the rate of overcoming this barrier would be proportional to $\exp(-Q/RT)$. However, unless the position at the right is more stable, that is, has a lower energy than the position on the left, the rate of overcoming the barrier from right to left would be exactly equal to that in overcoming it from left to right, so there would be no net dislocation movement. With an applied stress, σ, the energy on the left is raised by σV, where V is a constant with dimensions of volume, and on the right the energy is lowered by σV. Thus,

the rate from left to right is proportional to $\exp[-(Q-\sigma V)/RT]$, and from right to left the rate is proportional to $\exp[-(Q+\sigma V)/RT]$. The net strain rate then is:

$$\begin{aligned}\dot{\varepsilon} &= C\{\exp[-(Q-\sigma V)/RT] \\ &\quad - \exp[-(Q+\sigma V)/RT]\} \\ &= C\exp(-Q/RT)[\exp(\sigma V/RT) \\ &\quad - \exp(-\sigma V/RT)] \\ &= 2C\exp(-Q/RT)\sinh(\sigma V/RT)\end{aligned} \quad \text{(Eq 12)}$$

To accommodate data better, and for some theoretical reasons, a modification of Eq 12 has been suggested (Ref 9, 10). It is:

$$\dot{\varepsilon} = A[\sinh(\alpha\sigma)]^{1/m}\exp(-Q/RT) \quad \text{(Eq 13)}$$

Steady-state creep data over many orders of magnitude of strain rate correlate very well with Eq 13, as shown in Fig. 13 (Ref 11).

It should be noted that if $\alpha\sigma \ll 1$, $\sinh(\alpha\sigma) \approx \alpha\sigma$, so Eq 13 reduces to:

$$\dot{\varepsilon} = A\exp(-Q/RT)\cdot(\alpha\sigma)^{1/m}$$

or

$$\sigma = A'\dot{\varepsilon}^m\exp(mQ/RT)$$
$$\sigma = A'Z^m \quad \text{(Eq 14)}$$

which is consistent with both the Zener-Hollomon development, Eq 11, and the power-law expression, Eq 1.

Because $\sinh(x) \to e^x/2$ for $x \gg 1$, at low temperatures and high stresses, Eq 13 reduces to:

$$\dot{\varepsilon} = C\exp(\alpha'\sigma - Q/RT) \quad \text{(Eq 15)}$$

but now strain hardening becomes important, so C and α' are both strain and temperature dependent. Equation 15 reduces to:

$$\sigma = C + m'\ln\dot{\varepsilon} \quad \text{(Eq 16)}$$

which is consistent with Eq 5 and explains the often-observed breakdown in the power-law strain-rate dependence at low temperatures and high strain rates.

Isothermal Constitutive Model (Ref 12)*

The temperature dependence of the constituent processes is essential in a description of metal deformation, but it is convenient to first develop the model for a constant temperature. The grain deformation consists of elastic, microplastic, and grossly plastic (macroplastic) parts, as detailed in Fig. 14. The elastic part is assumed Hookean, and stress rate, $\dot{\sigma}$, is given in terms of elastic strain rate, $\dot{\varepsilon}_e$, by:

$$\dot{\sigma} = E\dot{\varepsilon}_e \quad \text{(Eq 17)}$$

where E is the modulus of elasticity. This strain is instantaneously recoverable during unloading.

*Adapted with permission from: A.K. Ghosh, A Physically Based Constitutive Model for Metal Deformation, *Acta Metallurgica,* Vol 28 (No. 11), Nov 1980, p 1443–1465

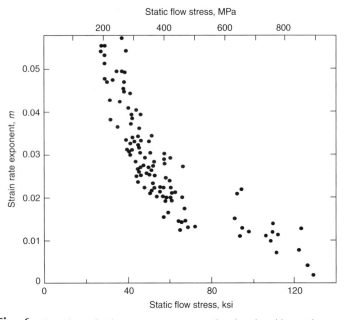

Fig. 6 Effect of stress level on strain-rate sensitivity of steels. Adapted from Ref 7

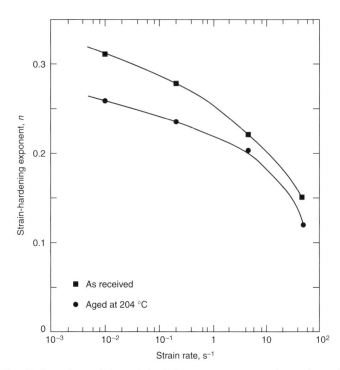

Fig. 7 Dependence of the strain-hardening exponent, n, on strain rate for steels. Adapted from Ref 7

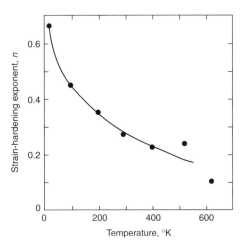

Fig. 8 Decrease of the strain-hardening exponent, n, of pure aluminum with temperature. Adapted from Ref 8

Microplastic strain involves dislocation motion over extremely short distances, such as that between obstacles. The term *microplastic* is used in a broad sense and includes anelastic strain (completely recoverable) and other microplastic components that may not be fully recoverable. Such strain can occur during the piling up of dislocations against barriers, which requires increased stress for their motion. As the stress rises, new dislocations of the appropriate sign are activated and those of the opposite sign are annihilated. Thus, the density of mobile dislocations increases during uploading without their release from the barriers. Conversely, their density is decreased due to annihilation and dislocation runback during unloading. Upon reverse reloading, there is again a rise in the dislocation density of the reverse sign. Thus,

stress depends on mobile dislocation density, ρ_m, through (Ref 13):

$$\sigma \propto \sqrt{\rho_m} \qquad (Eq\ 18)$$

where ρ_m in turn is proportional to the microplastic strain, ε_a, through the relation (Ref 14):

$$\rho_m \propto \varepsilon_a \qquad (Eq\ 19)$$

Combining Eq 18 and 19, $s \propto \sqrt{\varepsilon_a}$, a convenient form being $s = K'\varepsilon_a^{0.5}$, where K' is the microplastic strength constant, and s is the internal back stress. This kind of parabolic dependence is in agreement with the nonlinear nature of microplastic deformation commonly observed and contrasts with the choice of a single anelastic modulus that has been attempted by some investigators. To indicate the nonlinear microplastic part, the curve in Fig. 14 is extrapolated beyond the grain yield stress.

In addition to the strain dependency, there is also a velocity dependence of stress even for such small degrees of dislocation movement, and the effective stress, σ_e, that drives the dislocation velocity is given, following Johnston and Gilman's work (Ref 15), by:

$$\sigma_e = L\dot{\varepsilon}_a^m \qquad (Eq\ 20)$$

where L is a constant, m is the strain-rate sensitivity index, and σ_e is given by $\sigma - (K'\varepsilon_a^{0.5})$. Therefore, stress is given in terms of anelastic strain by:

$$\sigma = K'\varepsilon_a^{0.5} + L\dot{\varepsilon}_a^m \qquad (Eq\ 21)$$

Microplastic strain is essentially plastic in the sense of its origin; on unloading, however,

it is recoverable either fully or partly in a time-dependent manner, because there is no dislocation intersections and defect structure development to cause its storage. Thus, stress decreases during and after unloading, releasing ε_a. Equation 21 suggests that upon rapid loading in the anelastic regime, the rate-dependent part of the stress, $L\dot{\varepsilon}_a^m$, could initially be large. However, when stress is held constant, $\dot{\varepsilon}_a$ drops, and the strain-dependent part increases at the expense of the rate-dependent part. In addition to the dislocation-based anelasticity discussed so far, anelastic strain is also contributed by grain-boundary sliding at extremely low stresses, which is excluded from the present model.

Plastic. As the mobile dislocation density builds up to a significant level, leakage occurs through the barriers, aided by thermal activation. Consequently, yielding begins, starting with the grains most favorably aligned to the applied stress direction (highest critical resolved shear stress). This occurs as σ (given by Eq 21) exceeds the grain yield stress, y. The deformation is truly plastic (nonreversible), because dislocations start intersecting and defect structure begins to become stored. Total dislocation density increases and leads to strain hardening, that is, greater internal back stress (or glide resistance) has to be overcome in order to keep dislocations moving at the desired rate.

Strain rate is, again, related to the effective stress through its dependence on dislocation velocity, as in the Gilman-Johnston relationship. Thus, the grain contribution to the plastic strain rate is given by:

$$\dot{\varepsilon}_g \propto (\sigma - g)^p \qquad (Eq\ 22)$$

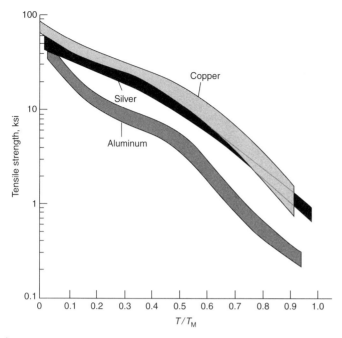

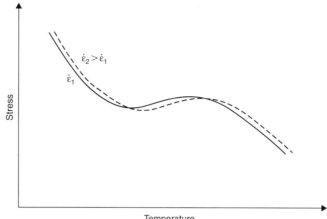

Fig. 10 Schematic plot showing the temperature dependence of flow stress for some alloys. In the temperature region where flow stress increases with temperature, the strain-rate sensitivity is negative.

Fig. 9 Decrease of tensile strength of pure copper, silver, and aluminum with homologous temperature. Source: Ref 8

where p is a constant exponent, and g is the internal strength due to defect structure, which equals the values of back stress, s, during monotonic loading in the plastic range. From Eq 22, the applied stress can be written as:

$$\sigma = g + L\dot{\varepsilon}_g^m \qquad \text{(Eq 23)}$$

where L is the proportionality constant in Eq 22, and $m(= 1/p)$ is the strain-rate sensitivity index.

Internal strength is a function of the current defect structure and is determined by the net storage of dislocations that occurs through the simultaneous operation of work hardening (dislocation multiplication, tangle formation, etc.) and thermally assisted recovery processes (annihilation, rearrangement, etc.) (Fig. 15). Thus, g is a scalar quantity unlike applied stress, internal stress, or strain rate, which can change sign. While g can decrease as a result of recovery during unloading or relaxation, the maximum value of g remains fixed as g^* and represents the strength state of the material. Further plastic deformation is possible only if $g \geq g^*$ (the equality indicates a steady state). During plastic deformation, as g increases, microplastic strain also increases, because $|s|$ must maintain equilibrium with g.

The functional form for the strain-hardening part alone for polycrystalline aggregates is not adequately known, because measured stress-strain curves already contain elements of recovery. However, multiple slip is known to start at relatively low strains in polycrystals. For single crystals, this corresponds to stage II hardening, which is approximated by a linear behavior. The accuracy of the linearity may, however, be argued in many instances. In fact, the change of curvature from stage I to stage III

invariably imposes a linear region, thereby questioning the existence of a real linear hardening regime.

The parabolic nature of polycrystalline hardening curves is often attributed to the action of cross slip, which continuously reduces the hardening rate. Although this is frequently referred to as dynamic recovery and is aided by thermal fluctuation, it is distinctly different from a thermally assisted, time-dependent rearrangement of dislocations (considered in the next section). Because cross slip is stress assisted as well as time dependent, even during infinitely fast tests cross slip would occur as the stress rises above that required to overcome the Cottrell-Lomar barriers. Because more of these barriers are present at higher levels of internal stress, cross slip at higher stresses would release greater bursts of strain and would lead to a hardening rate decaying with stress. Thus, while there is an additional time dependency of cross slip, the stress dependency alone can be responsible for the nonlinear hardening rate (concave toward the strain axis) in metals. Polycrystals have an additional requirement of grain rotation during deformation, which changes the slip direction orientation and is likely to cause a further departure from linear hardening rate. Thus, it seems reasonable to represent the strain hardening component by an equation of the form:

$$g = K\varepsilon_g^n \qquad \text{(Eq 24)}$$

where K is the strength constant, and n, the strain-hardening exponent, varies between 0 and 1 and thus can include both linear and parabolic hardening behaviors (Fig. 15). Both K and n may be functions of temperature. Again, the exact functional form of the strain-hardening equation

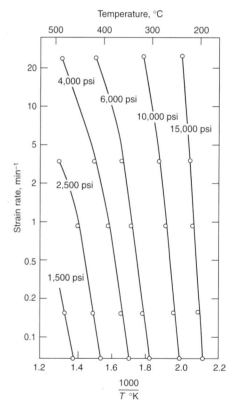

Fig. 11 Strain-rate and temperature combination for various levels of flow stress. Source: Ref 6

is not emphasized here but is left to be determined experimentally; this simple form is chosen only for analytic convenience. Because the internal strength also depends on recovery effects, the proper form for expressing strain-hardening effects is through the differential of Eq 24 and only in terms of the state parameter, g, as follows:

$$\partial g / \partial \varepsilon_g = nK^{1/n}/g^{1/n-1} \qquad \text{(Eq 25)}$$

In addition to the parabolic nature of the strain-hardening part, one reason why experimental stress-strain curves are significantly concave toward the strain axis is the simultaneous operation of the time-dependent thermal recovery process, which reduces internal stress by annihilation of loose dislocations and rearranging them in stabler networks and cell walls. This can occur by cross slip, climb, and similar processes. Further recovery can occur by coarsening of the dislocation network structure. The rate of recovery has been shown to be strongly dependent on the dislocation density (Ref 16), which in turn determines the internal strength. In general, the recovery process cannot reduce the internal strength to the original yield strength, y. The residual stress after very long times of recovery, x, is usually greater than y and is dependent on the value of the internal strength experienced by the material prior to start of a long-time recovery (Ref 16, 17). Thus, the time rate of recovery, $\partial g/\partial t$, is assumed to be proportional to the recoverable part of the internal strength ($g - x$). In fact, Friedel's network coarsening model (Ref 18) suggests:

$$-\partial g/\partial t = \alpha_d (g - x)^r \qquad \text{(Eq 26)}$$

where the exponent $r \sim 3$. The rate constant for dynamic recovery, α_d, needs to be determined experimentally. The value of x is assumed to be given by:

$$g^* - x = f(g^* - y) \qquad \text{(Eq 27)}$$

where f is the recoverable fraction, and g^* is the internal strength to which the material was raised prior to recovery. During monotonic loading, $g^* = g = |s|$, whereas during unloading or load relaxation, $g^* \geq g$. Substituting g in place of g^* in Eq 27 for monotonic loading, recovery rate in Eq 26 is given by:

$$-\partial g/\partial t = \alpha_d [f(g - y)]^r \qquad \text{(Eq 28)}$$

More is said later about recovery during unloading and load relaxation tests.

In resemblance to the Bailey-Orowan model (Ref 19, 20), the deformation resistance of the stored dislocation structure is given by the incremental integration of strain-hardening and dynamic recovery components. Thus, the rate of increment in internal stress, $\dot{g} = (\partial g/\partial \varepsilon_g)\dot{\varepsilon}_g + (\partial g/\partial t)_{recov}$, is given by combining Eq 25 and 28 as:

$$\dot{g} = (nK^{1/n}/g^{1/n-1})\dot{\varepsilon}_g - \alpha_d[f(g - y)]^r \qquad \text{(Eq 29)}$$

This, however, is quite different from Bailey-Orowan's formulation in that the constituent terms are now dependent on internal stress. Furthermore, the hardening Eq 25 is temperature dependent and is not simply the same as that at absolute zero. The change in it is caused by an increase in stress-dependent cross slip as well as a decrease in shear modulus with increasing temperature. The value of internal strength can be obtained by integrating Eq 29 over time and then combining that with Eq 23 to obtain the flow stress.

Unloading and Reverse Loading. The unloading curve from a stress of σ^* is shown diagrammatically in Fig. 16. As stress drops from σ^* to g^* (internal strength corresponding to σ^*), it actually accompanies a small forward plastic strain component; this is why the slope of the unloading curve in this region is steeper than elastic slope. Below g^*, all forward plastic strain stops, and reverse anelastic strain from dislocation runback starts. Thus, below the point g^*, an equation similar to Eq 21 is valid for the reverse anelastic strain. A general equation for this that incorporates the direction of stress and strain is given by:

$$\sigma - g^* \operatorname{sgn}(\sigma^*) = [K'|\varepsilon_a|^{0.5} + L|\dot{\varepsilon}_a|^m]\operatorname{sgn}(\dot{\varepsilon}_a) \qquad \text{(Eq 30)}$$

where reverse ε_a is measured from the point g^* on the diagram. When zero stress is attained, ε_e becomes zero; however, reverse $\dot{\varepsilon}_a$ cannot become zero instantaneously but decreases slowly as the reverse ε_a increases, thereby maintaining zero stress. After very long times, all of the forward anelastic strain is recovered. This recovery of microplastic strain is known as elastic aftereffect.

If no recovery of internal strength occurs during unloading and reverse loading, reverse plastic flow would begin at a stress of $-g^*$, as shown in Fig. 16. Once $-g^*$ is reached, internal and applied stress curves are essentially the continuations of the forward loading curves, with reverse sign for stress and strain. However, because this is preceded by a fairly large amount of anelastic strain (corresponding nearly to a stress change of $2 g^*$), both internal and applied stress curves in the plastic regime are below the corresponding curves had there been no reverse anelastic strain. Thus, Bauschinger effect is incorporated into the model through microplastic strain. In precipitation-hardening materials, the dislocation loops formed around precipitates provide a larger number of dislocations of the

opposite sign during reverse loading. The effect can be thought to produce either a lower microplastic strength constant, K', or a reduced g (for reverse flow), leading to an enhanced Bauschinger effect. Which one of these two possibilities is more plausible needs be determined from experimental data.

Furthermore, when there is thermal recovery or internal strength during unloading and reverse loading, reverse flow begins at a stress whose absolute magnitude is less than g^*. A more pronounced Bauschinger effect results from this.

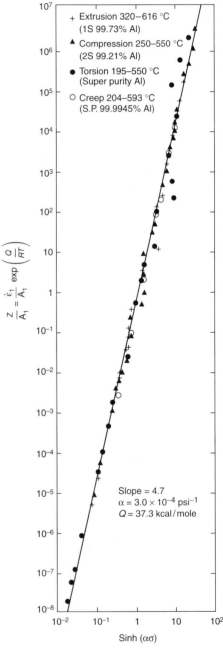

Fig. 13 Plot of Zener-Hollomon parameter versus flow-stress data for aluminum showing the validity of the hyperbolic sine relationship, Eq 13. Source: Ref 11

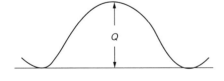

Fig. 12 Schematic illustration of an activation barrier for slip and the effect of applied stress on skewing the barrier

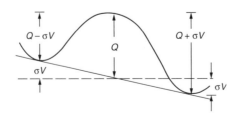

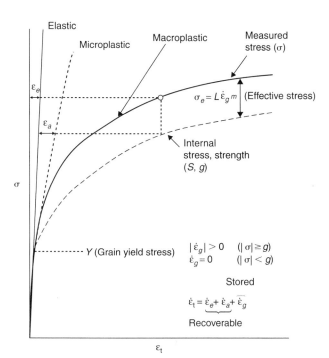

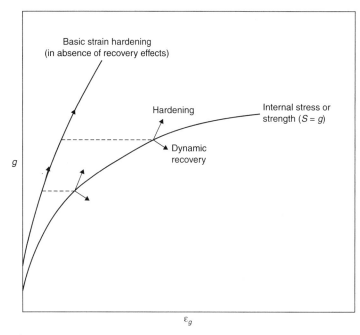

Fig. 14 Schematic stress-strain diagram illustrating elastic (ε_e), microplastic (ε_a), and macroplastic (ε_g) strains. Note that in the macroplastic range, ε_a is associated with internal back stress (s), while $\dot{\varepsilon}_g$ is associated with effective stress.

Fig. 15 Internal strength is the sum of the net increments from strain hardening and dynamic recovery components. The rate of the former is obtained from the basic hardening curve (same for the same level of g) and decreases with increasing g, while that of the latter increases. The basic hardening curve is a function of temperature and not the one for absolute zero.

It should be pointed out that Eq 30 is a simplistic form for monotonic loading within the microplastic region, with no change in the direction of strain rate. A differential form is preferred, however, for internal stress under arbitrary strain-rate history. This is given by:

$$\dot{s} = \left(\frac{0.5K'^2}{s - s^*}\right)\dot{\varepsilon}_a \qquad \text{(Eq 31)}$$

A comparison between constant load and constant stress results on steady-state creep rate versus stress may be made from the following considerations: If $\dot{\varepsilon}_b$ is ignored for the moment, minimum or steady-state strain rate corresponds to $\ddot{\varepsilon}_g = 0$. It can be shown by considering $g = g_0$ that for minimum creep rate (steady state):

$$\dot{\sigma} = \dot{g} \qquad \text{(Eq 32)}$$

For a constant stress test ($\dot{\sigma} = 0$), Eq 32 with the help of Eq 29 leads to:

$$\dot{\varepsilon}_s = \frac{\alpha_d[f(g - y)]^r}{nK^{1/n}/g^{1/n-1}} \qquad \text{(Eq 33)}$$

where $\dot{\varepsilon}_s$ is the steady-state $\dot{\varepsilon}_g$. If $y \ll s$, $\dot{\varepsilon}_s$ may further be simplified into the expression:

$$\dot{\varepsilon}_s = A'g^{(1/n+r-1)} = A'(\sigma - L\dot{\varepsilon}_s^m)^{(1/n+r-1)} \qquad \text{(Eq 34)}$$

where $A' = [\alpha_d f^r/nK^{1/n}]$. Rearranging, stress can be expressed as:

$$\sigma = A\dot{\varepsilon}_s^M + L\dot{\varepsilon}_s^m \qquad \text{(Eq 35)}$$

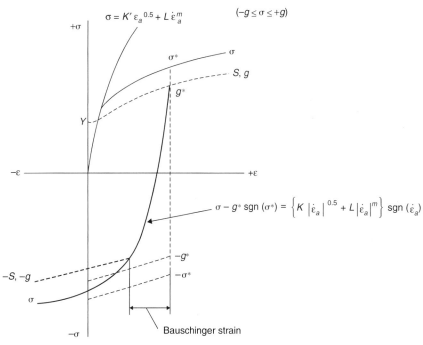

Fig. 16 Schematic loading and reverse loading diagram in the plastic range

where $A = 1/A'$ and $M = (1/n + r - 1)^{-1}$. At low temperatures, A is very large, and the hardening behavior predominantly determines flow stress. The value of M at low temperature is difficult to determine experimentally, because steady state occurs at extremely large strains for which the measurements of stress in a monotonic uniaxial test is complicated by necking, fracture, or barreling. The instantaneous rate sensitivity resulting from the effective stress may be determined by the step strain-rate test and is usually found to be small

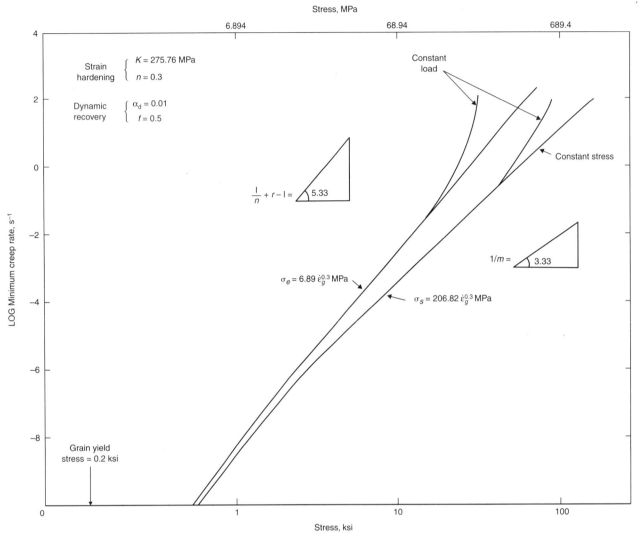

Fig. 17 Minimum creep rate versus stress curves for two different effective stress levels, showing the importance of its exponent, *m*, in comparison to that of the internal stress, *M*

at low temperature, because both L and m are small.

As temperature is raised, A decreases while L and m increase. The value of m can approach 0.3, and when A is small, the effective stress may become the dominant term (as in class I alloys). Conversely, when A still remains large, the creep exponent may be approximately 5 (or more), that is, $M = 0.2$ or less, as in the case of class II alloys or pure metals. In the case of diffusional creep, which has not been modeled here, M could approach the value of 1.

Thus, depending on whether the first term (saturated internal strength) or the second term (effective stress) on the right side of Eq 35 is dominant, σ versus $\dot{\varepsilon}_s$ data can exhibit either a rate-sensitivity index of M or m or somewhere in between (creep exponent is inverse of this index). This is shown in Fig. 17, where constant stress test yields a nearly linear behavior (at least in a double-logarithmic scale), even though no attempt has been made to simulate the absolute positions of the curves for any particular

material. The low rate portions of the relaxation curves can be nearly superimposed by translation along a slope of M. Thus, while the low strain-rate behavior is governed by the rate sensitivity of internal strength, the higher rate behavior is controlled by that of effective stress, with an intermediate behavior persisting in between. For creep at low strain rates (or stress), the influence of grain yield stress (y) becomes important as the curves in Fig. 17 bend toward a lower rate sensitivity (or higher stress exponent). At low enough strain rates, a higher rate sensitivity may again develop because of the dominance of grain-boundary processes ($\dot{\varepsilon}_b$), depending on the relative magnitudes of σ_0 and y.

If load is maintained constant instead of stress, $\dot{\sigma}$ is no longer 0 but equal to $\sigma\dot{\varepsilon}_t$. Again because $\ddot{\varepsilon}_g = 0$, $\dot{g} = \sigma\dot{\varepsilon}_t$. Thus, in place of Eq 33 and 34, one obtains:

$$\dot{\varepsilon}_s = \frac{\alpha_d[f(g-y)]^r}{nK^{1/n}/g^{1/n-1} - \sigma} \qquad \text{(Eq 36)}$$

This equation can be solved for $\dot{\varepsilon}_s$ for given values of σ and the constants.

Dynamic Recovery (Ref 21)*

Mechanisms for Dynamic Recovery. Very few efforts have been reported pertaining to the operative mechanisms during restoration via dynamic recovery at hot-working strain rates. However, it is a reasonable assumption that the basic softening events, characterizing constant-rate testing at low and intermediate temperatures and also creep, are relevant even under the conditions of hot deformation.

Mecking and Lücke (Ref 22, 23) introduced a formal description of the superposition of

---

*Adapted from: William Roberts, "Dynamic Changes That Occur during Hot Working and Their Significance Regarding Microstructural Development and Hot Workability," in *Deformation, Processing, and Structure*, G. Krauss, Ed., American Society for Metals, 1984, p 109–184

hardening and softening during plastic flow, where it is assumed that the individual contributions of these two processes to stress and strain are additive, that is:

$$d\varepsilon = d\varepsilon_h + d\varepsilon_{sf}; \quad d\sigma = d\sigma_h - d\sigma_{sf} \quad \text{(Eq 37)}$$

In terms of the work-hardening rate, θ,

$$\theta = \theta_{II} - \frac{1}{\dot{\varepsilon}}(\theta_{II}\dot{\varepsilon}_{sf} + \dot{\sigma}_{sf}) \quad \text{(Eq 38)}$$

where θ_{II} is a constant athermal hardening rate equivalent to that in stage II single-crystal hardening; even at high temperatures, it is presumed that θ approaches θ_{II} asymptotically at small strains. If the softening (dynamic recovery) processes are considered as single dislocation events with frequency $\dot{N}$, then both $\dot{\varepsilon}_{sf}$ and $\dot{\sigma}_{sf}$ are proportional to $\dot{N}$, and Eq 38 is replaced by:

$$\theta = \theta_{II} - q\left(\frac{\dot{N}}{\dot{\varepsilon}}\right) \quad \text{(Eq 39)}$$

where the proportionality constant q is related to the strain increase and stress decrease associated with a single softening event. Equation 39 should be compared with that derivable from the assumptions of the Bailey-Orowan (Ref 19, 20) theory of creep, in which the unit softening events are considered to be activated by thermal vibrations only, that is, exactly as for static recovery. On this basis, $\dot{N}$ is constant, and Eq 39 can be modified ($\dot{N}$ proportional to r = rate of recovery) to yield:

$$\theta = \theta_{II} - \frac{r}{\dot{\varepsilon}} \quad \text{(Eq 40)}$$

In the steady-state limit, $\theta = 0$, and $\dot{\varepsilon}_s = r/\theta_{II}$, which is the Bailey-Orowan relation for recovery-controlled creep. By applying Eq 48, one can write for r during creep at low stresses:

$$r = r_0\sigma^n\exp(-Q/RT) \quad \text{(Eq 41)}$$

where $r_0 = B\theta_{II}$ (θ_{II} independent of T, $\dot{\varepsilon}$ except via the temperature dependence of the shear modulus). However, because Eq 40 should also apply to the situation away from the steady state, then this relationship together with Eq 41 predicts that the work-hardening rate will exhibit a strong dependence on both stress and strain rate that is much more pronounced than the experimental evidence would indicate.

The physical basis of the Bailey-Orowan approach has been disputed in a number of papers (Ref 2, 24–27). The crux of the problem lies in the assumption that, at a given dislocation structure, the time frequency of recovery events is constant, and that each event only contributes to an elementary reduction in stress but not to a corresponding increase in strain. It would appear that this presumption is questionable. In terms of the formalism given by Mecking and Lücke (Ref 22, 23), the proportionality constant, q, can contain terms due to both the elementary stress decrease and the elementary strain increase from a single softening event, that is:

$$q = \theta_{II} \cdot d\varepsilon_{el} + d\sigma_{el}$$

Hence, $\dot{N}$ in Eq 39 can depend on the current level of applied stress or strain rate, that is, the restoration takes place via thermally assisted strain softening rather than the time-dependent, static-type recovery of the Bailey-Orowan approach and, as such, is dynamic recovery in its true sense. The interdependence of strain rate and stress at a given structure makes it difficult to decide, from experimental information, whether the stress level (the driving force) or the strain rate (the motion of dislocations) is the quantity that determines the frequency of softening events. Mecking and Lücke (Ref 22, 23) considered the $\sigma(\varepsilon)$ behavior in terms of a spectrum of softening centers, each with its specific activation stress. The number of active centers then increases with increasing degree of hardening, and the $\sigma(\varepsilon)$ curve flattens continuously. Assistance from thermal vibrations causes the work-hardening rate at a given strain to decrease with increasing temperature and decreasing strain rate. An alternative approach, in which the softening events are considered to be triggered by moving dislocations, has been presented by Kocks (Ref 2).

In the detailed models for dislocation creep presented by Öström and Lagneborg (Ref 28, 29), the dynamic recovery mechanism presumed to operate is one proposed originally by Freidel (Ref 18). He considered the mesh growth of a Frank network (which may exist uniformly throughout the crystal or locally in subgrain walls) via diffusion-controlled climb. The rate of growth of the average link size ($\bar{l}$) is:

$$\frac{d\bar{l}}{dt} = \frac{M\tau}{\bar{l}} \quad \text{(Eq 42)}$$

where M is a mobility that, for diffusion-controlled climb, is related to the self-diffusion coefficient. For growth of an individual link of size 1, Öström and Lagneborg (Ref 28) draw an analogy with grain growth and propose the expression:

$$\frac{dl}{dt} = M\tau\left(\frac{1}{l_{cr}} - \frac{1}{l}\right) \quad \text{(Eq 43)}$$

where l_{cr} is a critical value above which links increase in size and below which they shrink. In a simple treatment, the dislocation density is functionally related to $\bar{l}$, that is, $\rho = \bar{l}^{-2}$, and the rate of decrease of dislocation density due to recovery is then easily obtained via integration of Eq 42:

$$\left.\frac{d\rho}{dt}\right|_{rec} = -2M\tau\rho^2 \quad \text{(Eq 44)}$$

Öström and Lagneborg (Ref 28, 29) have adopted a more advanced formulation based on Eq 43, because in their model they consider a distribution of link sizes that is continuously modified via the accumulation of links due to glide/storage and their disappearance through shrinkage of the smallest meshes and participation in glide (largest links for which $l > \alpha' \mu b/\sigma$). In its most refined form (Ref 29), the model

offers a satisfactory description of the primary and secondary stages of creep in austenitic stainless steels. The steady-state dislocation density and its stress dependence are determined principally by $l^* = 2\mu b/\sigma$. In this sense, the simple argument culminating in Eq 44 and the more advanced theory lead to similar results.

Öström and Lagneborg's treatment has been criticized by Kocks and Mecking (Ref 26) on a number of points, the principal one being that the Friedel theory, for the growth of meshes in a Frank network, is essentially a model of time-dependent static recovery. However, in a more recent paper (Ref 27), these latter authors have attempted to develop a unified treatment of static (i.e., time dependent) and dynamic recovery. In this treatment, the problem associated with the simple Bailey-Orowan formalism for recovery creep, that is, that the wrong dependence of the work-hardening rate on stress and strain rate is predicted (Eq 40 and 41), is avoided because the recovery rate, r, is not controlled by a constant activation energy but rather by one that depends on the local forward internal stress experienced by dislocation segments in tangles. The substance of this model is a distribution of forward internal stresses on dislocation segments. Above a critical forward stress, σ_s, the athermal storage rate (defined by θ_{II}) is balanced by recovery; the rate of recovery under these conditions may then be written as:

$$-\left.\frac{df}{dt}\right|_{rec} = f(\sigma_s)v_0\exp\left\{-\frac{\Delta G(\sigma_s/\sigma_m)}{RT}\right\} \quad \text{(Eq 45)}$$

where f is the distribution function for forward stresses, and σ_m is the mechanical collapse stress at which dislocations would break free of the tangles even in the absence of thermal activation. On this basis, the net change in dislocation density for applied stress levels below σ_s ($\equiv$ steady-stage stress) is:

$$\frac{d\rho}{d\varepsilon} = h(\sigma, \sigma_s)(\sigma_s - \sigma) \quad \text{(Eq 46)}$$

This applies because, above σ_s, $df/d\varepsilon = 0$. The function h depends on the average segment length being stored, $\bar{l}$, and on $df/d\varepsilon$. Kocks and Mecking argue that $h \propto \sigma/\sigma_s$ and so, assuming a proportionality between σ and $\sqrt{\rho}$:

$$\frac{d\sigma}{d\varepsilon} = \text{const.} (1 - \sigma/\sigma_s) \quad \text{(Eq 47)}$$

which is the formalism for the $\sigma(\varepsilon)$ curve proposed originally by Voce (Ref 30). Equation 47 is characterized by a weak dependence of $d\sigma/d\varepsilon$ on σ and $\dot{\varepsilon}$ (through that of σ_s) and is thus consistent with experimental observations pertaining to the dependence both of the strain-hardening rate prior to σ_s and of the steady-state stress on $\dot{\varepsilon}$. However, the authors point out that this model, which involves only time-dependent recovery, will be invalidated if stress- or strain-rate-activated recovery events are rate controlling.

Steady-State Stress. In general, creep and high-temperature deformation at a given rate should, under equivalent conditions, lead to the same value for σ_s; this has been confirmed for pure aluminum by Mecking and Gottstein (Ref 31). However, such a situation is not likely to hold in a comparison of constant-rate tests under hot-working conditions and creep, which are often characterized by different activation energies. Based on the so-called temperature-compensated strain rate, as originally proposed by Zener and Hollomon, where $\sigma = f(\dot{\varepsilon} \exp (Q/RT)) = f(Z)$, the temperature and strain-rate dependence of σ_s is defined by the following equations:

$$Z = B(\sigma^n) \qquad \text{(Eq 48)}$$

or

$$Z = B' \exp(\beta\sigma) \qquad \text{(Eq 49)}$$

or if such simplification is not always possible, in the unified form:

$$Z = B''[\sinh\,(\alpha\sigma)]^n \qquad \text{(Eq 50)}$$

Kocks (Ref 2) has proposed that, at low temperatures, n in Eq 48 can be identified with ζ/kT, where ζ is a constant for a given material (face-centered cubic, or fcc) and k is Boltzmann's constant; the former can be derived from a so-called τ_{III} analysis of single-crystal stress-strain curves.

Kocks (Ref 2) presented a treatment for $\sigma(\varepsilon)$ in which recovery events are considered to be controlled by strain rate (moving dislocations). This argument leads to a formula:

$$\theta = \theta_0 (1 - \sigma/\sigma_s) \qquad \text{(Eq 51)}$$

where θ_0 is an athermal hardening rate to which all $\sigma(\varepsilon)$ curves are asymptotic at low strains (related to, but not necessarily equal to, θ_{II}). Apart from minor deviations at small strains, Eq 51 was found to describe accurately the stress dependence of the work-hardening rate during the tensile testing of aluminum, copper, and austenitic stainless steel (all polycrystalline). Because necking intervened before a steady state could be established, σ_s was evaluated via the extrapolation of $\theta - \sigma$ plots to $\theta = 0$.

Equation 51 is rationalized by Kocks in terms of dislocation storage at a rate, with respect to strain, that is proportional to $\sqrt{\rho}$; in fact:

$$\left.\frac{d\rho}{d\varepsilon}\right|_{stor} = \frac{k_1\sqrt{\rho}}{b}$$

where k_1 is a proportionality constant between mean-free path and $\sqrt{\rho}$. In evaluating $(d\rho/d\varepsilon)|_{rec}$, it is assumed that the probability of a recovery event is proportional to the number of times a potential recovery site is contacted by a moving dislocation. If a length of dislocation l_r is annihilated per recovery event, then for a unit area of slip plane $dl_{rec} = l_r\rho$, because the number of potential recovery sites is ρ. The shear-strain increment for unit area of slip plane is b, and so:

$$\frac{d\rho}{d\varepsilon} = \left.\frac{d\rho}{d\varepsilon}\right|_{stor} - \left.\frac{d\rho}{d\varepsilon}\right|_{rec} = \frac{1}{b}(k_1\sqrt{\rho} - k_2 l_r\rho)$$

which, in terms of hardening rates, becomes:

$$\theta = \frac{d\sigma}{d\varepsilon} = \frac{\alpha\mu k_1}{2} - \frac{k_2 l_r}{2b} \cdot \sigma \qquad \text{(Eq 52)}$$

This expression is equivalent to Eq 51. Kocks and Mecking (Ref 27) have also derived the Voce law from fundamental arguments using a time-dependent (i.e., static) model for recovery.

In order to check whether Eq 51 applies all the way up to σ_s, and thereby to examine the validity of the extrapolations made by Kocks on the basis of his tensile data, $\sigma(\varepsilon)$ was determined for polycrystalline superpure aluminum (200 μm grain size) using compression testing between 450 and 600 K and at $\dot{\varepsilon} = 0.01$ and $1\ \mathrm{s}^{-1}$. Lubrication with polytetrafluoroethylene permitted friction-free compression up to a true strain of 1, which is sufficient to attain steady state for all the conditions examined. Over the relevant strain range, the compressive $\sigma(\varepsilon)$ curves are in excellent agreement with the corresponding tensile data reported by Kocks (Ref 2).

Figure 18 shows typical θ-σ($\theta \equiv d\sigma/d\varepsilon$) plots derived from the $\sigma(\varepsilon)$ curves for aluminum. These conform accurately to Eq 51 at low stresses, but a systematic deviation from this law is found as σ_s is approached. Extrapolation of the low-stress behavior will clearly lead to an underestimation of σ_s. For testing at 500 and 600 K and at 0.01 s^{-1}, an extrapolation after the fashion of that performed by Kocks gives $\sigma_s = 23$ and 15 MPa, respectively; the corresponding values reported by Kocks for $\dot{\varepsilon} = 1.6 \times 10^{-2}\ \mathrm{s}^{-1}$ are 23 and 14 MPa. However, the correct, experimentally determined levels of σ_s are 34.5 and 16 MPa.

A careful examination of the θ values suggests that, with the exception of small strains (less than 0.05), θ is linearly related to $1/\sigma$. Some minor deviations from the law are found, especially as σ_s is approached, but the overall conformity must be regarded as acceptable. The slope of the $1/\sigma$-versus-θ plot increases systematically with increasing temperature and decreasing strain rate, or, more specifically, as $1/\sigma_s$. Data are reasonably well described by a straight line, which passes through the origin. Accordingly, the observed $\sigma(\varepsilon)$ behavior does not conform to Eq 51 but rather follows the law:

$$\theta = P\left(\frac{\sigma_s}{\sigma} - 1\right) \qquad \text{(Eq 53)}$$

where P is the slope of $1/\sigma_s$ versus $d(1/\sigma)/d\theta$ (116 MPa) and is very close to being constant for the range of temperatures and strain rates investigated.

Phenomenologically, Eq 53 is easily shown to be concomitant with the following relationship for the rate of increase of dislocation density with strain:

$$\frac{d\rho}{d\varepsilon} = k_1 - k_2\sqrt{\rho} \qquad \text{(Eq 54)}$$

that is, dislocation accumulation at a constant rate, which is realistic if a cell or subgrain structure is established early in the deformation process, combined with dynamic recovery at a rate proportional to $\sqrt{\rho}$. The $\sigma(\varepsilon)$ law derived from Eq 53 is:

$$\sigma_s \ln\left(\frac{\sigma_s}{\sigma_s - \sigma}\right) - \sigma = P\varepsilon \qquad \text{(Eq 55)}$$

The constants in Eq 54 are given by:

$$\frac{k_1}{k_2} = \frac{\sigma_s}{\alpha'\mu b}\ ;\ k_2 = \frac{2P}{\alpha'\mu b} \qquad \text{(Eq 56)}$$

Examples illustrating the degree of accord between Eq 55 and the experimentally determined $\sigma(\varepsilon)$ behavior are given in Fig. 19. In a more accurate appraisal, it is necessary to take into account the temperature dependence of elastic modulus in the evaluation of P. The general measure of agreement between the two-parameter formalism (Eq 55) and the experimental $\sigma(\varepsilon)$ is quite good but hardly perfect, especially at low strains. This reflects the inability of Eq 53 to describe the behavior under these circumstances; such is not to be expected either, in view of the fact that some initial strain is required before the equilibrium cell size (defining a constant rate of dislocation storage) is established. However, Eq 55 represents the best two-parameter description of

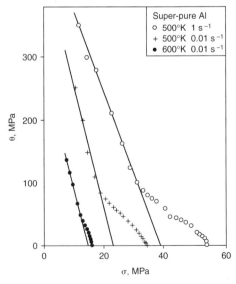

Fig. 18 Plot of work-hardening rate versus flow stress for superpure aluminum, illustrating the type of extrapolation performed by Kocks (Ref 2) on his tensile data

$\sigma(\varepsilon)$ under the range of experimental conditions studied.

From Eq 56:

$$k_1 = \frac{2P\sigma_s}{(\alpha'\mu b)^2} = \frac{1}{\omega b d_{sub}} \quad \text{(Eq 57)}$$

where ω is a factor for converting from shear to normal quantities. Because P is virtually independent of T and $\dot{\varepsilon}$, then σ_s should be proportional to the inverse of subgrain size, which is approximately the situation found experimentally. In actual fact, the predicted exponent for d_{sub} must deviate somewhat from -1 because of the temperature dependence of μ (assuming, of course, that σ_s and the subgrain size are varied via changes in test temperature). Turning to the recovery term, it is clear that the proportionality constant, k_2, is approximately

independent of T and $\dot{\varepsilon}$ (Eq 56); taking $\mu(500\ K) = 2.24 \times 10^4$ MPa, $\alpha' \doteq 1$, and $b = 0.286$ nm, then $k_2 \doteq 4 \times 10^7\ m^{-1}$. Following Kocks (Ref 2), it is assumed that recovery events are triggered by mobile dislocations; on this basis, the rate of decrease of ρ with respect to strain (see previous text) is:

$$\left.\frac{d\rho}{d\varepsilon}\right|_{rec} = \text{const.}\frac{l_r\rho}{b}$$

where l_r is the length of dislocation lost per recovery event. The constant in the previous expression can be identified with $\omega \cdot g$, g being the fraction of encounters between mobile and stationary dislocations that leads to recovery events. Kocks takes l_r to be fixed and independent of total dislcoation density. However, because the unit recovery process is likely to involve indi-

vidual dislocation links, then l_r can be expected to decrease as ρ increases. For both a uniformly distributed network and a subgrain structure, one can thus anticipate that l_r is proportional to $1/\sqrt{\rho}$ and that:

$$\left.\frac{d\rho}{d\varepsilon}\right|_{rec} = \text{const.}\frac{\sqrt{\rho}}{b} \quad \text{(Eq 58)}$$

in agreement with the formalism derived from experimental observations (Eq 54). Making the rough approximation that $l_r = 1/\sqrt{\rho}$, one has $k_2 = \omega g/b$, and so, with the previous value of k_2 plus $\omega = 3.1$ (fcc polycrystals), g works out to be 4×10^{-3}, that is, approximately one encounter in 200 results in a recovery event.

The two-parameter description of $\sigma(\varepsilon)$ embodied in Eq 55 is a very attractive one. Within the range of temperatures and strain rates investigated, the stress-strain behavior up to $\varepsilon = 1$ can be described quite accurately in terms of a material constant, P, and the steady-state stress. Furthermore, the experimental data for aluminum indicate that $\sigma_s(\dot{\varepsilon},T)$ conforms well with $Q = 130$ kJ·mol^{-1}, which is very close to Q_{SD} (138 kJ·mol^{-1}). Hence, the description of $\sigma(\varepsilon,\dot{\varepsilon},T)$ is simplified further because σ_s can be evaluated for any temperature and strain rate if Q is known. The entire formalism is thus based on just two quantities, which are, at least to a first approximation, independent of temperature or strain rate.

It is of interest to investigate whether or not Eq 55 can be applied to a typical commercial material, characterized by restoration during hot deformation via dynamic recovery alone. Figure 20 shows $\sigma(\varepsilon)$ curves for a ferritic stainless steel (19Cr-0.6Ti) tested in compression at temperatures between 750 and 1150 °C (1380 and 2100 °F) and at $\dot{\varepsilon} = 1\ s^{-1}$. The agreement between the experimental points and the theoretical curves is acceptable. However, in this case, P is not constant but decreases systematically with increasing temperature. The probable explanation lies in the existence of a strain-independent flow-stress component in this commercial steel, that is, a friction stress, σ_0. This being the case, Eq 53 must be changed to:

$$\theta = P\left(\frac{\sigma_s - \sigma}{\sigma - \sigma_0}\right) \quad \text{(Eq 59)}$$

and Eq 55 must be changed to:

$$\sigma_0 - \sigma + (\sigma_s - \sigma_0)\ln\left(\frac{\sigma_s - \sigma_0}{\sigma_s - \sigma}\right) = P\varepsilon \quad \text{(Eq 60)}$$

It is readily shown that the application of Eq 53 and 55 to $\sigma(\varepsilon)$ behavior characterized by a non-zero σ_0 will result in a P value that increases with increasing σ_0. Hence, the data for P listed in Fig. 20 are consistent with a diminishing σ_0 as the temperature is raised, which seems plausible. For a more accurate correlation with $\sigma(\varepsilon)$ from commercial materials, Eq 59 and 60 should

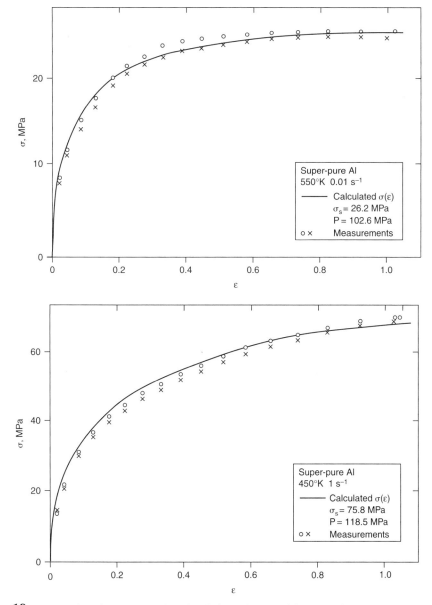

Fig. 19 Correspondence between $\sigma(\varepsilon)$ evaluated on the basis of Eq 55 and the measured curve

be used; as a first approximation, σ_0 can be set equal to the yield stress under the conditions of interest.

Diffusional Flow Mechanisms

Creep (Ref 32)*

Nabarro-Herring Creep. Dislocation glide creep involves no atomic diffusional flow. Nabarro-Herring creep is the opposite of dislocation glide creep in that Nabarro-Herring creep is accomplished solely by diffusional mass transport. Nabarro-Herring creep dominates creep processes at much lower stress levels and higher temperatures than those at which creep is controlled by dislocation glide. Because it does not devolve on dislocation glide, Nabarro-Herring creep is also observed in amorphous materials. However, discussion of Nabarro-Herring creep is facilitated by considering first how it is accomplished in a crystalline material subjected to the stress state shown in Fig. 21.

The grain illustrated in Fig. 21 may be considered either an isolated single crystal or an individual grain within a polycrystal. As indicated, the lateral sides of the crystal are assumed to be subjected to a compressive stress, and the horizontal sides to a tensile stress. The stresses alter the atomic volume in these regions; it is increased in regions experiencing a tensile stress and decreased in the volume under compression. As a result, the effective activation energy for vacancy formation is altered by $\pm \sigma\Omega$, where Ω is the atomic volume, and the $\pm$ signs refer to compressive and tension regions, respectively. Thus, the fractional vacancy concentration in the tensile and compressively stressed regions is given as:

$$N_v(\text{tension}) \simeq \exp\left(-\frac{Q_f}{kT}\right)\exp\left(\frac{\sigma\Omega}{kT}\right)$$

(Eq 61)

and

$$N_v(\text{compression}) \cong \exp\left(-\frac{Q_f}{kT}\right)\exp\left(-\frac{\sigma\Omega}{kT}\right)$$

(Eq 62)

where Q_f is the vacancy-formation energy. Provided the grain boundary is an ideal source or sink for vacancies if the grain of Fig. 21 is a polycrystal or, if it is a single crystal, if the surface of it behaves likewise, the vacancy concentrations given by Eq 61 and 62 are maintained at the horizontal and lateral surfaces. The different concentrations at the surfaces drive a net flux of vacancies from the tensile to the compressively stressed regions, and this is equivalent

to a net mass flux in the opposite direction. As illustrated in Fig. 21(b), this produces a change in grain shape. The grain elongates in one direction and contracts in the other; that is, creep deformation occurs.

The creep rate resulting from this process is estimated as follows. The vacancy flux, J,

through the crystal volume is given by:

$$J_v = -D_v\left(\frac{\delta N_v}{\delta x}\right)$$

(Eq 63)

where D_v is the vacancy diffusivity [$= D_{0v}$ $\exp(-Q_m/kT)$, where Q_m is the vacancy motion

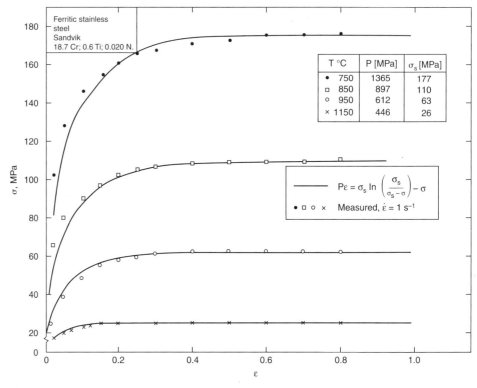

Fig. 20 $\sigma(\varepsilon)$ curves for a commercial ferritic stainless steel at various temperatures; experimental measurements compared with curves evaluated from Eq 55

Ferritic stainless steel Sandvik 18.7 Cr; 0.6 Ti; 0.020 N.

T °C	P [MPa]	σ_s [MPa]
• 750	1365	177
□ 850	897	110
○ 950	612	63
× 1150	446	26

$$P\varepsilon = \sigma_s \ln\left(\frac{\sigma_s}{\sigma_s - \sigma}\right) - \sigma$$

• □ ○ ×   Measured, $\dot{\varepsilon} = 1\ s^{-1}$

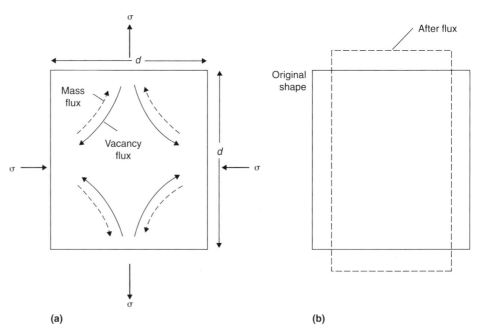

(a) (b)

Fig. 21 Nabarro-Herring creep results from a higher vacancy concentration in regions of a material experiencing a tensile stress vis-à-vis regions subject to a compressive stress. (a) This results in a vacancy flux from the former to the latter areas, and a mass flux in the opposite direction. (b) The resulting change in grain dimensions is equivalent to a creep strain.

*Adapted with permission from: Thomas H. Courtney, *Mechanical Behavior of Materials*, McGraw-Hill, 1990, p 271–279, 285, 304, 307

energy] and $\delta N_v/\delta x$ is the vacancy gradient. The term δx can be taken as a characteristic diffusion distance that is proportional to the grain size, d (compare with Fig. 21a), whereas δN_v is given as the difference of Eq 61 and 62. Multiplication of Eq 63 by the diffusion area (proportional to d^2) gives the volumetric flow rate $\delta V/\delta t$; $\delta V/\delta t$ represents the volume transferred per unit time from the lateral to the horizontal sides of the crystal. According to this reasoning, it is given by:

$$\frac{\delta V}{\delta t} \cong -D_{0v}d \exp\left[-\frac{Q_f+Q_m}{kT}\right]$$
$$\times \left[\exp\left(\frac{\sigma\Omega}{kT}\right) - \exp\left(-\frac{\sigma\Omega}{kT}\right)\right] \quad \text{(Eq 64)}$$

The change in length (δd) of the crystal along the tensile axis is related to δV by $\delta V \cong d^2\,\delta d$. The corresponding Nabarro-Herring creep rate ($\dot\varepsilon_{NH}$) is expressed as $1/d(\delta d/\delta t)$; thus:

$$\dot\varepsilon_{NH} = \left(\frac{D_{0v}}{d^2}\right)\exp\left[-\frac{Q_f+Q_m}{kT}\right]$$
$$\times \left[\exp\left(\frac{\sigma\Omega}{kT}\right) - \exp\left(-\frac{\sigma\Omega}{kT}\right)\right] \quad \text{(Eq 65)}$$

The term $D_{0v}\exp\left[-(Q_f+Q_m)/kT\right]$ of Eq 65 is identically equal to the lattice self-diffusion coefficient, D_L. Moreover, at the high temperatures and low stresses at which Nabarro-Herring creep is important, $\sigma\Omega$ is much less than kT, so that $\exp\left[\pm\sigma\Omega/kT\right] = 1 \pm \sigma\Omega/kT$. Using these relations, and letting a constant A_{NH} represent geometrical factors that are only incompletely considered, $\dot\varepsilon_{NH}$ can be written as:

$$\dot\varepsilon_{NH} = A_{NH}\left(\frac{D_L}{d^2}\right)\left(\frac{\sigma\Omega}{kT}\right) \quad \text{(Eq 66)}$$

As mentioned, Nabarro-Herring creep is important at high temperatures and low stresses, that is, in the temperature-stress regime where dislocation glide is not important. It is more important in creep of ceramic materials than in metals. This is the case because glide mechanisms of creep can be considered competitive with Nabarro-Herring creep, and dislocation glide is generally more difficult to effect in ceramics than in metals.

Coble creep is closely related to Nabarro-Herring creep. Coble creep is driven by the same vacancy concentration gradient that causes Nabarro-Herring creep. However, in Coble creep, mass transport occurs by diffusion along grain boundaries in a polycrystal or along the surface of a single crystal. For polycrystals, the diffusion area is thus proportional to $\delta'd$, where δ' is an appropriate grain-boundary thickness. Analysis similar to that employed previously yields an expression for Coble creep:

$$\dot\varepsilon_C = A_C \exp\left(-\frac{Q_f}{kT}\right)D_{0GB}\left[\exp\left(-\frac{Q_m}{kT}\right)\right]$$
$$\times \left(\frac{\delta'}{d^3}\right)\left(\frac{\sigma\Omega}{kT}\right) = A_C\left(\frac{D_{GB}\delta'}{d^3}\right)\left(\frac{\sigma\Omega}{kT}\right)$$
$$\text{(Eq 67)}$$

In Eq 67, Q_f represents, as it did previously, the vacancy formation energy, but Q_m represents the energy of atomic motion along the boundary. Thus, the exponentials containing these terms have been incorporated into D_{GB}, which represents an effective grain-boundary diffusivity (or surface diffusivity if a single crystal is considered). As indicated by Eq 67, Coble creep is more sensitive to grain size than is Nabarro-Herring creep. Thus, even though both forms of creep are favored by high temperature and low stress, it is expected that Coble creep will dominate the creep rate in very fine-grained materials. In the general case, the creep rate due to diffusional flow should be considered a sum of $\dot\varepsilon_{NH}$ and $\dot\varepsilon_C$, because the mechanisms operate in tandem; that is, they are parallel creep processes.

Creep Mechanisms Involving Dislocation and Diffusional Flow (Ref 32)

The linear dependence of creep on stress predicted by the previously mentioned mechanisms is not observed for many materials under conditions of moderate applied stress and temperature. Instead, the value of the stress exponent m' is found to range from approximately 3 to 7 (with $m' = 4.5$ being observed as often as not).

Under these conditions, creep involves dislocation-recovery processes. Numerous dislocation mechanisms for creep have been postulated, and although it has been difficult to experimentally verify their applicability in individual cases, several of the mechanisms discussed in this section are physically appealing, and they, or similar processes, undoubtedly occur during dislocation creep. The mechanisms described are useful, too, for illustrating the strong stress dependence of dislocation creep.

Nabarro-Herring Creep of Subgrains. If a subgrain structure is developed, as it often is, during creep, then Nabarro-Herring creep of subgrains can occur. Provided each dislocation in the subboundary is an ideal source and sink for vacancies, the subgrain size, d', substitutes for the grain size in Eq 66 and 67. It has been observed experimentally that d' is inversely related to stress, that is, $d' \cong K/\sigma$, where K is a material constant. Thus, when mass transfer occurs by volume diffusion, the subgrain creep rate ($\dot\varepsilon_{sg}$) is given as:

$$\dot\varepsilon_{sg} = A_{sg}D_L\left(\frac{\sigma^2}{K^2}\right)\left(\frac{\sigma\Omega}{kT}\right) \quad \text{(Eq 68)}$$

and m' for this type of creep is 3.

Dislocation Glide-Climb Mechanisms. As mentioned, creep can conveniently be viewed as a manifestation of competitive work hardening and recovery. The work-hardening aspects of creep are related to factors involving dislocation glide. Recovery aspects are related to nonconservative dislocation motion, for exam-

ple, dislocation climb in which obstacles to dislocation motion are circumvented and/or by which dislocations are annihilated—that is, removed from the structure.

Because the processes are sequential, the creep rate in a coupled glide-climb process is determined by the lesser of the glide-climb rates. In some cases, glide controls creep rate. For example, dislocation glide creep is controlled solely by glide, because no atomic mass transport is involved in this type of creep. Even at higher temperatures, creep may be limited by glide. At certain stress-temperature combinations, for example, solute atoms are able to diffuse at velocities comparable to those of moving dislocations. In these circumstances, the solute concentration in the dislocation vicinity far exceeds the average concentration, and the resulting solute drag markedly increases lattice resistance to dislocation motion. As a consequence, the glide velocity is less than the climb velocity.

Most coupled creep mechanisms, however, are climb controlled; that is, the dislocation climb velocity is less than the glide velocity. One potential mechanism is illustrated in Fig. 22, where X dislocation sources per unit volume emit dislocations that glide a distance L in their slip plane. These interact with similar dislocations emitted from sources on parallel slip planes separated vertically by the distance h. Consistent with the constant dislocation structure associated with creep, each source is assumed to have a fixed number of loops around it. Thus, continued emission of dislocations from the sources (i.e., continued glide) is dependent on the annihilation rate of the outermost dislocations. Annihilation is effected by climb over the distance h. Dislocation pairs of the type $[\frac{\perp}{\top}]$ climb and annihilate each other by addition of solute atoms (i.e., removal of vacancies) to the atomic planes separating them. Conversely, dislocation pairs of the type $[\frac{\top}{\perp}]$ on the opposite side of the loop are removed by addition of vacancies (i.e., removal of solute atoms) between their respective planes. Thus, the climb process involves mass transfer from one side of the loop to the other.

The strain rate associated with this process can be written in the conventional form:

$$\dot\varepsilon = \rho b v_g \quad \text{(Eq 69)}$$

where ρ is the dislocation density, and v_g is the dislocation glide velocity, which is less than the climb velocity v_c. Climb and glide are coupled, and it can be shown that v_g and v_c are related through the geometrical ratio h/L as $v_g = \frac{L}{h}v_c$, where it is assumed that $L > h$. The dislocation density is obtained by multiplication of X by the average loop diameter ($\cong L$) and the number of loops per source, which can be shown to be proportional to L/h; thus:

$$\dot\varepsilon \sim \frac{XL^3}{h^2}v_c \quad \text{(Eq 70)}$$

where X is the number of dislocation sources per unit volume. Moreover, X multiplied by the

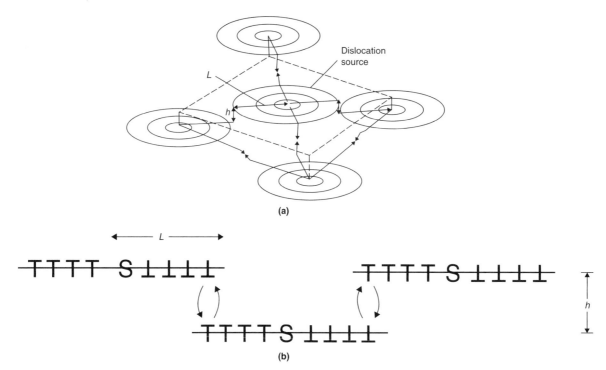

Fig. 22 (a) X dislocation sources per unit volume emit dislocations over a radius L. Continued emission (and hence strain) requires that the outermost dislocations in each loop be annihilated by climb processes between nearby loops separated by the distance h. (b) A two-dimensional view of the loop array illustrating how the climb process results in dislocation annihilation. Adapted from Ref 33

volume per source ($\cong \pi L^2 h$) is constant, or $L \cong (Xh)^{-1/2}$. Hence:

$$\dot{\varepsilon} \sim \frac{v_c}{h^{3.5} X^{1/2}} \qquad \text{(Eq 71)}$$

Dislocation climb is driven by stress fields similar to those driving diffusional creep. Thus, if mass transfer occurs by volume diffusion, v_c is given by:

$$v_c \sim D_L \left[\exp\left(\frac{\sigma\Omega}{kT}\right) - \exp\left(-\frac{\sigma\Omega}{kT}\right) \right] \cong 2D_L \frac{\sigma\Omega}{kT}$$
$$\text{(Eq 72)}$$

and the glide-climb creep rate, $\dot{\varepsilon}_{GC}$, can finally be written as:

$$\dot{\varepsilon}_{GC} = \frac{A_{gc} D_L}{h^{3.5} X^{1/2}} \left(\frac{\sigma\Omega}{kT}\right) \qquad \text{(Eq 73)}$$

The distance h is thought to scale inversely with stress. Thus, if X is assumed independent of stress, the exponent m' for this type of dislocation creep is 4.5, and this is in accord with considerable experimental evidence.

General Equation Form

In spite of the apparent diversity of the formulations for the several creep rates presented in this section, all of them can be expressed in a similar form, that is:

$$\dot{\varepsilon}_i = A_i D_i \left(\frac{\sigma}{\mu}\right)^{m''} \left(\frac{\sigma\Omega}{kT}\right) \left(\frac{b}{d}\right)^{n'} \qquad \text{(Eq 74)}$$

Table 2 Values of the parameters m'' and n' and approximate values of the constant A_i in the expression for the steady-state creep rate $A_i D_i \, (\sigma/\mu)^{m''} (\sigma\Omega/kT)(b/d)^{n'}$

Mechanism	Favored by	$A_i(m^{-2})$	m''	n'
Nabarro-Herring (NH) creep	High temperature, low stress, and large grain sizes	$7(\Omega)^{-2/3}$	0	2
Coble creep	Low stress, fine grain sizes, and temperatures less than those for which N-H creep dominates	$50(\Omega)^{-2/3}$	0	3
Nabarro-Herring creep of subgrains	High temperature and stresses such that the subgrain size is less than the grain size. (Subgrain size, d'_s, scales with stress approximately as $d'_s = 20(\mu b/\sigma)$	$0.01(\Omega)^{-2/3}$	2	0
Generalized power-law creep	High stress, lower temperatures in comparison to Coble creep, and large grain sizes	(Several to several million) $\times (\Omega)^{-2/3}$(a)	2–6(a)	0

(a) The terms A_i and m'' are strongly dependent on the mechanism controlling power-law creep at the substructural level. Adapted from Ref 34

In Eq 74, the last three parameters on the right-hand side are dimensionless. As mentioned, the ratio $\sigma\Omega/KT$ is the ratio of a mechanical to a thermal energy. The parameter $(b/d)^{n'}$, where b is the Burgers vector, represents a grain-size dependence of creep; for example, $n' = 2$ for Nabarro-Herring creep and $n' = 0$ for dislocation climb-glide creep. The term σ/μ is the ratio of the applied stress to the shear modulus; this ratio is important in determining the creep rate when dislocations are involved in creep (note that $m'' = 0$ for diffusional creep).

The diffusion coefficient, D_i of Eq 74, is usually the volume diffusion coefficient (Coble creep is an exception), and A_i has dimensions of m^{-2}. Thus, A^{-1} can be considered a measure of the diffusion area. The coefficients m'' ($= m' - 1$), n', and approximate values of A_i are listed in Table 2 for the mechanisms described in this section. Comments in the table also indicate the approximate stress, temperature, and, if appropriate, grain-size regimes in which a particular mechanism may be expected to dominate the creep rate.

Although Table 2 is useful for clarifying to some extent the conditions under which a particular creep mechanism is most important, a graphical description is usually preferable, such as by representation in so-called deformation mechanism maps.

Grain-boundary sliding accommodated by diffusional flow is the superplastic analog of Nabarro-Herring and Coble creep. As proposed originally by Ashby and Verrall (Ref 35), grain shape is preserved during superplastic deformation by a grain-switching mechanism that also

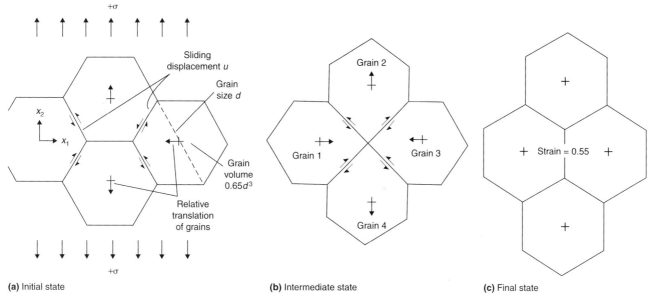

Fig. 23 The grain-switching mechanism of Ashby and Verrall. Relative grain-boundary sliding produces a strain (c) without a change in shape of the grains (compare a with c). However, the intermediate step (b) of the process is associated with an increased grain-boundary area. Source: Ref 35

provides for the resulting strain. The grain-switching mechanism is illustrated in Fig. 23, which shows a two-dimensional configuration of four grains before (Fig. 23a), after (Fig. 23c), and at an intermediate state (Fig. 23b) of the grain-switching process. For the geometry illustrated, a true tensile strain of 0.55 is effected by the grain-switching event. The intermediate stage of the process illustrates several significant microstructural features of the switching mechanism. First, there is an increase in grain-boundary area in the intermediate state as compared to the initial and final states. This results in a threshold stress below which grain switching cannot occur. In effect, the applied stress must perform a component of irreversible work associated with the formation of the increased grain-boundary area, and this stress must be exceeded in order for additional stress to drive diffusional flow.

Second, diffusional flow provides for the shape accommodation necessitated by the intermediate state (Fig. 24). In the Ashby-Verrall model, the flow can be either within grains (analogous to Nabarro-Herring creep) or along the grain boundaries (analogous to Coble creep). Provided the applied stress is considerably in excess of the threshold stress, the strain rate for the grain-switching mechanisms exceeds considerably that due to conventional creep mechanisms. This is related to several geometrical features of grain switching. First, the volume of material that must be transported to effect a given strain via grain switching is approximately 1/7 that required for diffusional creep. Additionally, the grain-switching diffusion distance is reduced by a factor of approximately 3 vis-à-vis the diffusional-creep distance, and there are six such paths for grain switching as opposed to four for diffusional creep.

Although these factors are mitigated to a degree by the fact that some of the grain

boundaries are at angles of neither 0 nor 90° to the tensile axis (thus reducing the effective driving stress for diffusional flow), the net result is that the strain rate for the grain-switching mechanism is approximately an order of magnitude higher than it is for diffusional creep. Ashby and Verrall, considering both volumetric and grain-boundary mass transport, developed the following constitutive equation to describe the grain-switching creep rate:

$$\dot{\varepsilon}_{GS} \cong \frac{100\Omega}{kTd^2} \left(\sigma - \frac{0.72\gamma}{d} \right) D_L$$
$$\times \left(1 + \frac{3.3\delta' D_{GB}}{d D_L} \right) \quad \text{(Eq 75)}$$

The term $0.72\,\gamma/d$ represents the threshold stress for the grain-switching event. If only boundary transport is important (i.e., if $3.3\,\delta' D_{GB} \gg d D_L$), Eq 75 reduces to:

$$\dot{\varepsilon}_{GS} \cong \frac{330\Omega}{kT} \frac{\delta' D_{GB}}{d^3} \left(\sigma - \frac{0.72\gamma}{d} \right) \quad \text{(Eq 76)}$$

and the grain-switching mechanism can be considered competitive with ordinary Coble creep. Grain switching dominates Coble creep at stresses large in comparison to $0.72\,\gamma/d$ and vice versa.

Grain-switching creep, as described by Eq 76, is also competitive with ordinary dislocation creep, with the latter dominating at high stress levels. This leads to stress-strain-rate behavior very similar to that shown in Fig. 25. In region I, grain switching dominates, and $\sigma \cong \sigma_0 + A\dot{\varepsilon}$. Likewise, dislocation creep dominates region III, and the transition region II is characterized by a rapid increase in stress with strain rate and associated superplastic behavior. The overall scheme as envisaged by Ashby and Verrall is summarized in Ref 35.

Grain-switching events have been observed in emulsions and in thin films of the superplastic zinc-aluminum metallic alloy. The mechanism is also conceptually appealing, because it predicts the generally observed stress-strain-rate relationship. Additionally, it does not, except as a transition event, invoke dislocation mechanisms, and, in view of the absence of observed concentrated dislocation activity in superplastic materials, this was for some time considered additional support for the grain-switching mechanism. It was also realized that dislocation and diffusional flow accommodation could occur concurrently. However, the grain-switching model is not without its shortcomings. For example, it generally does not predict accurately the stress level of region I. Nor, for that matter, does it consider in detail the previously mentioned dislocation accommodation, and recent studies have revealed dislocation activity, especially in the grain-boundary vicinity, in

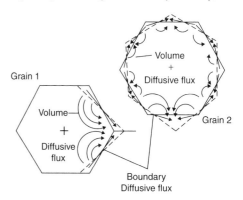

Fig. 24 During the intermediate stage of grain switching, grains 1 and 2 (compare with Fig. 23) change their shape from that indicated by the solid (initial state) lines to that of the dotted lines. The shape change is provided for by diffusional flow, which can take place by volumetric or boundary diffusion. Source: Ref 35

superplastically deformed materials. A qualitative description of accommodation provided by dislocation flow is therefore provided in the next section.

Grain-Boundary Sliding Accommodated by Dislocation Flow. Models of superplasticity that invoked accommodation by dislocation flow were developed even before microscopic observations confirmed dislocation activity in superplastic materials. This was a natural result of the realization that creep strains can also be accommodated by dislocation flow. The accommodation mechanisms advanced for superplasticity are similar in some respects to those expected in dislocation climb-glide creep processes. They must, however, provide for a means by which grain shape remains unaltered during straining. It is significant, too, that in contrast to dislocation creep mechanisms, grain size is explicitly recognized as important in relating strain rate to stress. This arises because, at the low stress levels associated with superplastic deformation, the grain size of a superplastic material is expected to be less than the cell or subgrain size.

The absence of a grain substructure does not, per se, eliminate dislocation activity during superplastic deformation. It only means that the characteristic diffusion distance of dislocation accommodation is that of the grain dimension.

As a consequence, several of the dislocation models for superplasticity arrive at a constitutive equation of the form:

$$\dot{\varepsilon}_{DS} = C_1 D \frac{\sigma\Omega}{kT} \frac{\sigma}{\mu} \left(\frac{b}{d}\right)^2 \qquad \text{(Eq 77)}$$

where the constant C_1 depends on the details of the model, and the diffusivity D is usually a grain-boundary diffusion coefficient. Such descriptions are reasonable with respect to the magnitude of the stress required to produce a given strain rate, the m value of 0.5, and the grain-size dependence of strain rate. In common with the Ashby and Verrall model, dislocation models for superplasticity consider region III to be controlled by conventional dislocation creep. In contrast to the grain-switching model, however, region II is viewed as controlled by an independent mechanism and is not considered a transition stage from region I to region III.

According to the dislocation models of superplasticity, the transition from region II to region III happens when the subgrain size becomes less than the grain size. For several materials, the inverse relationship between subgrain size, d', and stress is approximated by:

$$\frac{d'}{b} = 10 \frac{\mu}{\tau} \qquad \text{(Eq 78)}$$

and Eq 78 therefore defines the stress for the transition from superplasticity to dislocation creep. That this reasonably describes observed transitions is confirmed by Fig. 26(a) and (b) (Ref 36), which are deformation mechanism maps (in terms of d/b versus τ/μ for the superplastic aluminum-zinc and lead-tin alloys. The dotted lines in these diagrams represent the transition predicted by Eq 78, and it is seen that it describes well the experimental results. Figures 26(a) and (b) also show that superplastic flow is competitive with diffusional creep and dominates it at higher stress levels. Although region I is shown as separate areas in these figures, there is disagreement among the dislocation advocates of superplasticity as to whether it represents a separate flow mechanism or is merely an obscure manifestation of Coble creep.

Although no model of superplasticity is able to describe accurately the phenomenon in all of its quantitative and qualitative aspects, the models discussed are nonetheless useful from a technological viewpoint. In effect, they can predict approximately the grain size, temperature, and strain-rate regimes where superplasticity is likely to be observed, and this has proven useful in engineering applications. Indeed, superplastic forming of commercial aluminum-, titanium-, and nickel-base high-temperature alloys is done routinely, and this process could not be considered without appropriate constitutive equations. One drawback of superplastic forming, though, is the (frequently) low strain rates used in the process. These result in low production rates and correspondingly higher costs.

Grain Growth Effects (Ref 37)

As shown by Eq 77, grain growth during superplastic flow can have a significant effect on the strain rate developed under a given imposed stress. A detailed investigation of such an effect was conducted by Ghosh and Hamilton (Ref 37) for Ti-6Al-4V. Hot tension tests were conducted on material with three different starting alpha-grain (particle) sizes. The grain growth during deformation at constant strain rate was determined by quantitative metallography. To obtain a unified view of grain-growth behavior, the results for each initial grain size (Fig. 27) were translated along the time abscissa such that the initial grain size on these plots came to rest on the corresponding plots in Fig. 28 (i.e., for the same strain rate in each case). The resulting log (grain size) versus log (time) plots are shown in Fig. 29. The approximation to linear behavior in these plots is reasonable, thus suggesting the relationship:

$$d = d_0 \, (t/10)^n \qquad \text{(Eq 79)}$$

where d is the current grain size, d_0 is the initial grain size, both in micrometers, t is time in minutes, and n is a parameter that increases with increasing strain rate. An approximate value for

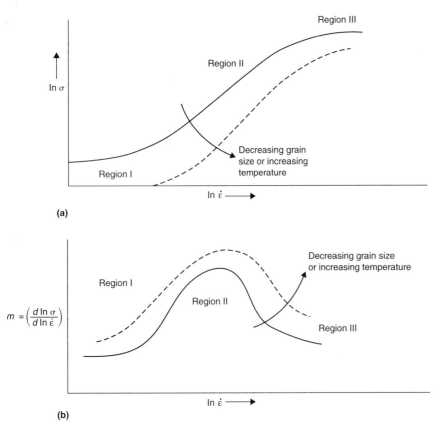

Fig. 25 (a) The low stress-strain-rate behavior of a material manifesting superplasticity. In regions I and III, the strain-rate sensitivity (b) is fairly small, whereas it is high in region II where superplasticity is observed. As indicated in (a), increases in temperature or decreases in grain size shift the σ-ε̇ curve downward and to the right. The same changes produce a somewhat higher value of m, as shown in (b).

this parameter for Ti-6Al-4V at 927 °C (1700 °F) is found to be:

$$n = 1.8 \, (\dot{\varepsilon}_t + 0.00005)^{0.237} \qquad \text{(Eq 80)}$$

where $\dot{\varepsilon}_t$ has the dimension of s^{-1}. Equations 79 and 80 are used subsequently in developing a constitutive equation for superplastic flow in this material.

With evidences of hardening as well as grain growth during deformation available, an attempt is now made to relate the two in Fig. 30. This was done by cross plotting flow stress from Fig. 31 against grain size during deformation, obtainable from the grain growth kinetics plots of Fig. 27 and 28 (for each applied strain rate). The solid curves in Fig. 32 are thus grain-growth hardening curves for initial grain sizes of 6.4, 9, and 11.5 μm for the various applied strain rates. The flow stress corresponding to each initial grain size is indicated by a data point taken from the "knee" between the elastic and grossly plastic parts of the load-extension plots and represents as much as 3% total strain. While there is some subjectivity in determining these points, the errors are no more than ±1 MPa (0.15 ksi).

Physical Model for Superplastic Flow (Ref 38)*

Research on the mechanism of superplastic flow in fine-grain metals has encompassed many ideas, such as diffusional creep (Ref 39–41), dislocation creep (Ref 42, 43) with diffusional accommodation at grain boundaries (Ref 35), concepts of grain-mantle deformation (Ref 44, 45), and so on. While these have broadened the view of the underlying physics of micrograin superplasticity, controversy still exists on the details of the microscopic process of deformation. A few known facts are:

- Sliding of grains along grain boundaries is observed (Ref 46)
- Both grain stretching and dynamic grain growth accompany superplastic deformation (Ref 37, 47)

With respect to the mechanical response under optimal superplastic conditions, the peak value of m ($d \log \sigma / \log \dot{\varepsilon}$) can be 0.7 to 0.8 for fully recrystallized single-phase alloys (typical grain size ~6 to 15 μm), and can be as low as 0.4 to 0.6 for very fine-grain (and subgrain-containing) alloys with a large amount of dispersoid particles (grain size ~0.5 to 4 μm), with a bell-shaped distribution of m as a function of strain rate (Ref 48, 49). The existing superplastic deformation models are unsatisfactory in terms of explaining values of $m > 0.5$. These steady-state models ignore microstructural evolution, which is integral to the deformation process. Furthermore, for most alloys, the m value is usually high, over several decades in strain rate, which does not agree with predominantly diffusion-based models, for example, the Ashby-Verrall model (suitable for $m > 0.5$).

While grain-boundary shear and sliding are important to the superplastic deformation process, the grain neighbor-exchange process (Ref 43, 46) is regarded primarily as a surface phenomenon due to lower constraints at the free surface. Any interior neighbor-exchange effect is small and results from a combined effect of concurrent grain growth and grain-boundary sliding, which causes smaller grains to shrink and move out of the path of larger growing grains. Fine-grain alloys whose grain boundaries are typically pinned by dispersoid particles often show grain-boundary migration during superplastic deformation, resulting in unpinning of boundaries from the particles. The mechanical effects and the microstructural evolution processes mentioned here are intrinsic to the superplastic deformation process and must be incorporated into a general theory of deformation of polycrystalline metals.

A new model to address this from the viewpoint of dislocation micromechanics is described here based on the following. At elevated temperature, the grain-boundary bonds are weakened, and yet shear and normal stresses must be transferred across them. Thus, inelastic displacement in grain-boundary regions must be larger than those within the grain interior, which may even remain elastic. This view is essentially the same as the grain mantle versus core deformation model proposed by Gifkins (Ref 44).

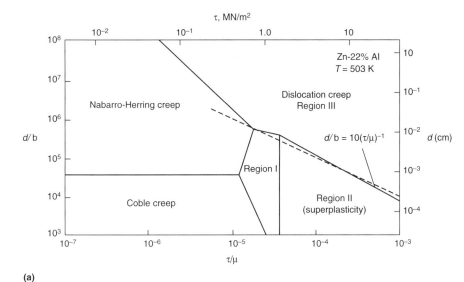

(a)

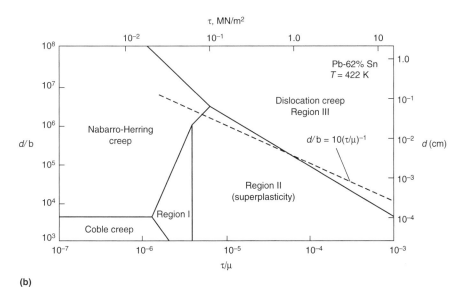

(b)

Fig. 26 Deformation mechanism maps for (a) zinc-aluminum and (b) lead-tin in terms of grain size versus stress. According to the dislocation mechanisms of superplasticity, a transition from superplasticity to dislocation creep should occur when $d/b \cong 10 \, \mu/\tau$. This is approximately observed for these materials. Source: Ref 36

*Adapted with permission from: A.K. Ghosh, A New Physical Model for Superplastic Flow, *Materials Science Forum*, 1994, p 39–46

While diffusion of atoms is critical to the deformation process, it is not believed that large-scale atom transport (Ref 35, 39, 41) occurs to any appreciable extent in metallic alloys. Figures 33(a) and (b) schematically show this new concept of mantle and core regions. Due to the higher defect density and weakened bonds, the diffusivity in the mantle region can be very high (close to grain-boundary diffusivity). Arrows within the mantle region in Fig. 33(a) indicate the direction of the tensile component parallel to the boundary. Local inelastic shear within the mantle, and atom transport, are also in the direction of these arrows, and marker line

offsets are created (from *ab* to *a'b'*) during superplastic deformation. The details of the atom transport, however, involve both glide and climb, as explained subsequently.

Figure 33(b) shows how grain-boundary sliding may be resisted at grain-boundary steps, ledges, and particles (not shown). Stress concentration leads to initation of glide along slip planes in the favored orientations. Thus, discontinuities such as ledges, nondeformable particles, and dispersoids on the grain boundaries act as dislocation sources. At low stresses, glide velocities of the dislocations are low, and therefore, these dislocations can climb out

of their planes toward the grain boundaries (lower chemical potential) before they have a chance to glide into the grain core. The rows of atoms, as they arrive at the grain boundary, can plate on the boundary as well as travel along the boundary away from the ledge that produced them. The rise of the subsurface atoms to the grain-boundary surface helps them attach preferentially to the grain steps and ledges existing on a concave inward boundary, thus leading to boundary migration and enhanced concurrent grain growth. Grain-boundary sliding and grain growth are thus tied to the same mantle deformation process. The sliding strain is directly tied to extension of the grains due to atom flow, not additive to it. When applied stresses are high, some dislocations can still climb to the nearest grain boundaries, but many more travel across the grain to cause grain core deformation via conventional climb-guide creep. Kinetics of these processes are given as follows.

Low-Stress Behavior. Stress concentration at grain-boundary discontinuities is a function of their size and spacing. The largest discontinuities initiate slip first at the lowest stresses. A popular view of source activation is by bowing of grain-boundary dislocation segments between their pinning points (particles, jogs, etc.) A source is activated when the local stress, τ, exceeds $\mu b/\lambda$, where μ is shear modulus, b is Burgers vector, and λ is spacing between pinning points. For a distribution in the value of λ, τ for source initiation ranges from a minimum value of τ_0 (for the largest λ) to the largest value corresponding to the smallest λ. The number of activated sources, N, per grain boundary as a function of $(\tau - \tau_0)$ will have a decaying slope approximated by a parabolic function:

$$N \propto (\tau - \tau_0)^q \qquad \text{(Eq 81)}$$

where q is the source activation exponent (0.1 to 1 depending on the alloy).

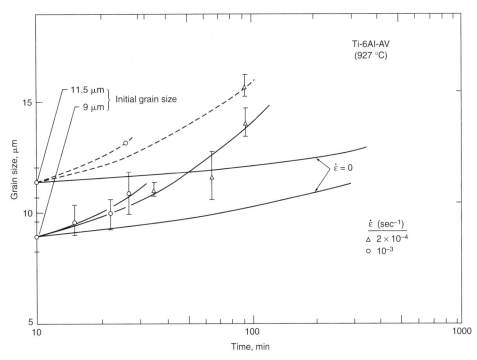

Fig. 27 Grain-growth kinetics at two different tensile strain rates compared with static kinetics for initial grain sizes 9.0 and 11.5 μm, respectively

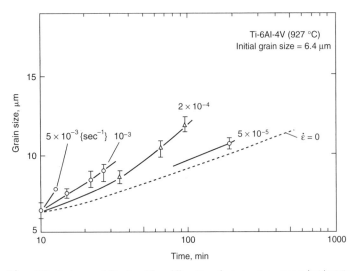

Fig. 28 Grain-growth kinetics at four different tensile strain rates compared with static kinetics for an initial grain size of 6.4 μm

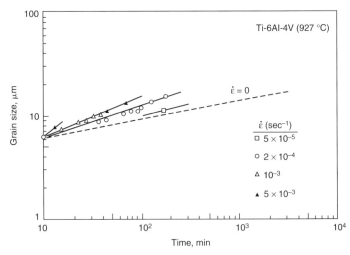

Fig. 29 Grain-growth kinetics data from Fig. 27 and 28 have been reassembled in a log-log plot.

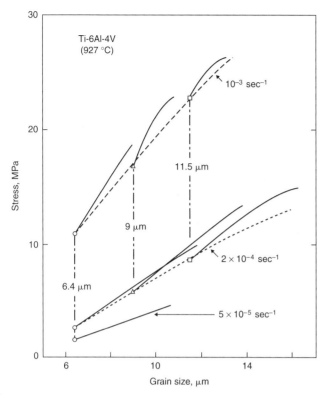

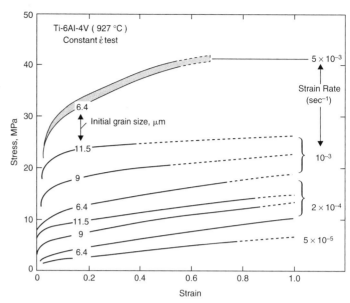

Fig. 31 True stress-strain curves for three initial grain sizes of 6.4, 9, and 11.5 μm, respectively. The bottommost curve shown is for a strain rate of 5×10^{-5}/s, and the topmost one for 5×10^{-3}/s, both for the 6.4 μm grain size material. The dotted part indicates a significant strain gradient developing in the specimen.

Fig. 30 The change in flow stress as a function of grain size (changing during tensile test), indicated by solid curves. The data points indicate the three initial grain sizes from which tests were started. The appropriate strain rates are shown on the plot.

The shear displacement rate, $\dot{x}_m$, in the mantle may be expressed as:

$$\dot{x}_m = b \cdot N \cdot v \qquad \text{(Eq 82)}$$

where b is the Burgers vector, and v is the glide velocity. As commonly assumed, v is proportional to climb velocity, which in turn is related to the volume transport rate of atoms per unit area of the grain boundary driven by the effective stress. It can be shown then that:

$$v \propto \frac{1}{d^2} \left[\frac{D_m(\tau - \tau_0)\Omega}{kT} \right] \cdot \delta d \qquad \text{(Eq 83)}$$

where D_m is the effective diffusivity in the mantle region (similar to grain-boundary diffusivity), d is grain size, δ is mantle width, Ω is atomic volume, k is Boltzmann's constant, T is absolute temperature, and $(\tau - \tau_0)\Omega$ represents chemical potential driving diffusional climb. Using Eq 81 to 83, the mantle shear strain rate, $\dot{\gamma}_m$, can be expressed as:

$$\dot{\gamma}_m \propto \frac{\dot{x}_m}{d} \propto \frac{b}{d^2} \cdot (\tau - \tau_0)^q \left(\frac{\delta D_m \Omega}{kT} \right) (\tau - \tau_0) \qquad \text{(Eq 84)}$$

Because this strain rate directly leads to tensile strain rate, the mantle contribution to the tensile strain rate can be written as:

$$\dot{\varepsilon}_m = (A/d^2(\sigma - \sigma_0)^{1+q} \qquad \text{(Eq 85)}$$

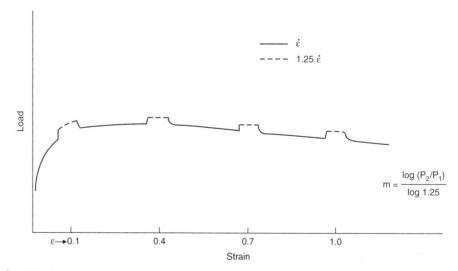

Fig. 32 A schematic representation showing how instantaneous measurements of m were made at periodic intervals during the tensile test, by strain-rate increments of 25%

where $A \propto (b \, \delta \Omega D_m / kT)$ is constant for a fixed temperature, σ is tensile stress, and σ_0 is the tensile equivalent of τ_0. The value of q may vary between 0.1 and 0.4 for a well-recrystallized grain structure containing a small volume fraction of dispersoids, but for a recovered subgrain containing material with a high volume of fine dispersoids, the stress dependence of source activation can be greater ($0.3 \le q \le 1$).

Concurrent Grain Growth. Dynamic grain growth in this model is directly related to the transport occurring from strain contribution due to the mantle region. This occurs in addition to the surface-energy-driven static grain growth. Thus, instantaneous grain size, d, may be given by:

$$d = d_0 + at^p + \beta \varepsilon_m \qquad \text{(Eq 86)}$$

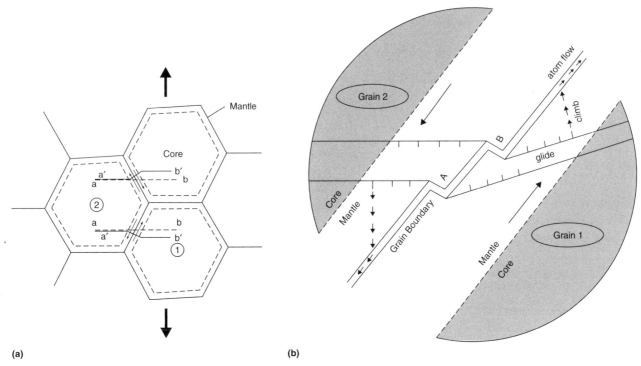

Fig. 33 Schematic illustrations of mantle-core deformation model. (a) Atom transport along the direction of arrow leads to offsets in the marker line *ab*. New position of marker *a′b′* lies on deformed grains (not shown). (b) A portion of the mantle magnified to show local glide-and-climb effect initiated from grain-boundary steps A and B

where d_0 is the initial grain size, t is time, a is a constant for static grain growth, p is a static grain growth exponent (typically 0.5), ε_m is the strain contributed from mantle deformation, and β is a proportionality constant dependent on temperature, alloy chemistry, and dispersoid volume fraction.

High-Stress Behavior. At high stresses, climb transport is too slow to fully relieve stress concentration at grain boundaries. Dislocations, injected through the grain core, participate in the normal glide-climb creep process with a creep law similar to:

$$\dot{\varepsilon} = K_1 \sigma^n e^{-Q/kT} \qquad \text{(Eq 87)}$$

where K_1 is the standard constant for dislocation creep, and Q is the activation energy for dislocation creep. Stress concentration at grain corners cannot, however, be relieved by the general dislocation creep alone, and additional local deformation must occur at all grain corners where slip incompatibility develops. An estimate of this effect has been made in a previous paper (Ref 50), which suggests:

$$\dot{\varepsilon}_c = (K + A_1/d^3)\sigma^n \qquad \text{(Eq 88)}$$

where $\dot{\varepsilon}_c$ is creep rate contributed by grain core, including accelerated deformation at grain corners, K is $K_1 e^{-Q/kT}$, and A_1 is a constant relating to stress concentration and enhanced dislocation creep at grain corners. The term d^3 arises due to the number of grain corners present per unit volume. The overall superplastic creep rate is then obtained by assuming that both mantle and

core deformation contribute to the overall deformation; that is:

$$\dot{\varepsilon} = \dot{\varepsilon}_m + \dot{\varepsilon}_c \qquad \text{(Eq 89)}$$

If all relevant constants are known, Eq 85 and 88 can be combined with Eq 86 to obtain instantaneous stress-strain response as a function of applied strain rate. A simple computer program was developed to calculate stress, strain, and grain size incrementally in time for a given strain-rate history and material property parameters.

Results of Simulation Experiments. The model presented previously is used to stimulate mechanical and microstructural response of two aluminum-base alloys, similar to 7475 Al and a 7000-series alloy containing a large volume of chromium and zirconium intermetallic dispersoids (e.g., Al_3Zr, Cr_2Al_9) (Ref 48, 51). Their compositions are given in Table 3.

The addition of a large volume of dispersoids suppresses nucleation of recrystallized grains in alloy 2, and a particle pinned subgrain structure forms. Thus, finer structures are typically associated with higher dispersoid volume, and the simulation is carried out with this realistic feature in mind. The experimental stress versus strain-rate data of these two materials in uniaxial tension are shown in Fig. 34. Alloy 2 has lower flow stresses than alloy 1, and the strain rate for its peak m (~0.5) is almost 2 orders of magnitude higher than that of alloy 1 (peak m ~0.8). Note that alloy 3 in Fig. 34 is for a similar alloy containing only 0.2% Zr and no chromium. It has a

grain size of 6 μm and appears to be more fully recrystallized than alloy 2.

Predictions of the Ashby-Verrall model for similar grain sizes (shown for comparison) do not provide an adequate match in slope with these. Their microstructure, stress-strain-rate data, and dynamic grain growth results were examined in detail to determine relevant parameters to be used to stimulate their behavior. Table 4 lists parameters representative of this general class of alloys without any attempt to exactly duplicate them for a specific alloy. The source activation exponent q is expected to increase with increased dispersoid content. While not directly measurable, its value for the two alloys is assumed to be 0.25 and 0.65, respectively. Also assumed is the value of A, which is maintained constant for the same type of alloy. All other parameters can be assessed from experimental data on these materials. Both static and dynamic grain growth for finer-grain alloy 2 are lower.

Figure 35 shows calculated stress-strain curves for alloy 1 for a variety of strain rates. Considerable strain hardening is observed at lower strain rates and very little hardening at the higher strain rates. It has been recognized recently that experimental strain measurements based on crosshead displacement are incorrect because of considerable stretching of material in the grip and specimen shoulder. Many existing data in the literature suffer from this problem. The reported strain-hardening rates are considerably higher than true hardening rates. This general behavior is in good agreement with results on superplastic materials (Ref 37, 48).

Table 3 Chemical compositions of alloys for study

| Material | Composition, wt% | | | | | Condition |
	Zn	Mg	Cu	Zr	Cr	
Alloy 1 (7475 Al)	5.7	2.3	1.5	. . .	0.14	Fully recrystallized microstructure ($d_0 = 12$ μm)
Alloy 2 (7000+dispersoids)	7.2	2.5	2.0	0.3	0.3	Recovered subgrain structure ($d_0 = 1.5$ μm)

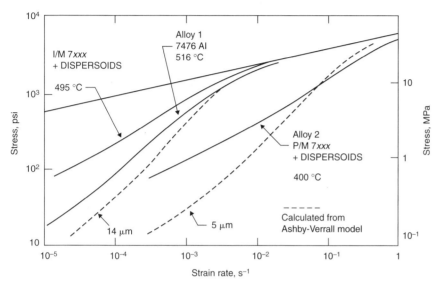

Fig. 34 Experimental stress-strain-rate plots for two aluminum alloys (Table 3) having similar matrix chemistry. Calculated results for similar grain sizes are shown for comparison. The top solid line represents dislocation creep.

Table 4 Parameters for core-mantle model used in simulating behavior of two 7000-series aluminum alloys

Material parameters	Alloy 1 — Low density of dispersoid practicles; fully recrystallized grain structure	Alloy 2 — High density of dispersoid particles; recovered subgrain structure
Tensile threshold glide resistance, σ_0 (MPa)	0.08	0.12
Source activation exponent, q	0.25	0.65
Pre-exponential constant for mantle strain rate, A (m^{-12} $MPa^{-(1+q)}s^{-1}$)	0.0075	0.0075
Constant for dislocation creep, K ($MPa^{-4}s^{-1}$)	1.18×10^{-7}	1.18×10^{-7}
Stress exponent for dislocation creep, n	4	4
Constant for grain corner creep, A_1 ($m^{-18}MPa^{-4}s^{-1}$)	0.5×10^{-4}	0.5×10^{-4}
Initial grain size, d_0 (μm)	10	3
Static grain growth, $\alpha(p)$, $\alpha(μm/\sqrt{s})$	0.025 (0.5)	0.015 (0.5)
Dynamic grain growth, $\beta(μm)$	5	3
Possible examples within the alloy category	7475 Al (Al-Zn-Mg-Cu) 5083 Al (Al-Mg-Mn) Al-33%Cu Zn-22%Al Ti-6Al-4V	7000 Al (Al-Zn-Mg-Cu)+dispersoids Al-Li-Zr alloys 2000 Al (Al-Cu)+dispersoids 5000 Al (Al-Mg)+dispersoids

The stress values from the early part of the stress-strain curves are plotted in Fig. 36(b) as a function of strain rate for both alloys 1 and 2. Note that the curve for alloy 2 is less steep. It also shows a peak m of ~0.6 (Fig. 36a) at a strain rate of $2 \times 10^{-3} s^{-1}$, which is approximately 20 times faster than that for alloy 1. The calculated peak m for alloy 1 is approximately 0.8. These values as well as the general trend of data are in excellent agreement with experimental observations. The wide range of strain rates over which alloy 2 shows a high m agree well with many results on superplastic alloys. Figure 37 shows the mantle strain rate ($\dot{\varepsilon}_m$) as a function of total applied strain rate for alloys 1 and 2. Below $10^{-4}s^{-1}$, deformation is entirely via mantle deformation. Above this rate, core deformation begins, more quickly for alloy 1 with gradual

exhausted source activation capability. The finer-subgrain-containing alloy does exhibit mantle deformation continuing to significantly higher strain rates.

The concurrent grain-growth behavior has been studied as a function of imposed strain rate and plotted in Fig. 38 and 39. As experimentally observed, the growth kinetics are accelerated by strain rate (Fig. 38a) and show the correct functional dependence. When plotted as a function of strain, the slower strain rates show larger grain size (Fig. 38b). Grain-growth kinetics normalized with respect to the initial grain size, d_0, $(\Delta d/\Delta t)d_0$, are shown in Fig. 39 as a function of applied strain rate. The simulated results match data from several investigators well (Ref 37, 47, 52), showing that normalized grain-growth rate rises sharply at intermediate strain rates (10^{-4} to $10^{-3}s^{-1}$). Comparing with Fig. 36, in this region of strain rates, m value begins to drop off rapidly.

Examination of flow stability was conducted by determining periodic m values from small step strain-rate (30% change) tests conducted during simulations from constant strain-rate tensile tests, and also by determining normalized slopes, $(1/\sigma)(d\sigma/d\varepsilon)$, of stress-strain curves. These results are shown in Fig. 40. Figure 40(a) shows that m drops only slightly with increasing superplastic strain when m is high (~0.7). At higher strain rates, when m is lower (0.3 to 0.4) and mantle strain is smaller, the change in m is negligible. Figure 40(b) shows that $(1/\sigma)(d\sigma/d\varepsilon)$ at a strain rate of $10^{-3}s^{-1}$ drops sharply as a function of strain and falls below 1 (maximum load point) at low strain; however, at a strain rate of $10^{-6}s^{-1}$, it decreases slowly, and maximum load is not reached before a strain of 0.4. Thus, flow stability is a result of concurrent grain growth and contributes to higher tensile elongation.

Summary of Mantle-Core Model. The new concept in the aforementioned mantle-core model of superplasticity is that dislocation sources are activated at the grain boundaries, and glide and climb in the grain-mantle region at low stresses control both grain-boundary sliding and concurrent grain growth. The model accurately predicts stress-strain-rate characteristics of metals with fully and partially recrystallized microstructures and strain-hardening characteristics. The details of grain-growth kinetics and flow stability are also simulated. This suggests that microstructural evolution is integrally connected with the superplastic deformation process.

ACKNOWLEDGMENTS

This article is adapted from the following sources with permissions:

- T.H. Courtney, *Mechanical Behavior of Materials,* McGraw-Hill, 1990, p 271–279, 285, 304, 307
- A.K. Ghosh, A Physically Based Constitutive Model for Metal Deformation, *Acta*

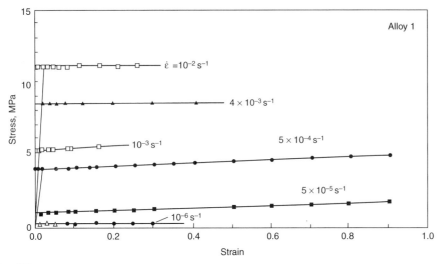

Fig. 35 Calculated stress-strain curves for alloy 1 at various strain rates, based on parameters listed in Table 4

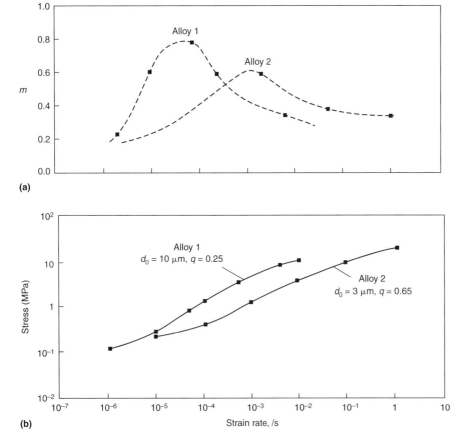

Fig. 36 Simulated stress versus strain-rate results (b) for alloy 1, taken from data similar to that in Fig. 34. *m* values or slopes of curves in (b) are shown in (a).

Metall., Vol 28 (No. 11), Nov 1980, p 1443–1465

- A.K. Ghosh, A New Physical Model for Superplastic Flow, Superplasticity in Advanced Materials ICSAM-94, *Mater. Sci. Forum,* Vol 170–172, 1994, p 39–46

- A.K. Ghosh and C.H. Hamilton, Mechanical Behavior and Hardening Characteristics of a Superplastic Ti-6Al-4V Alloy, *Met. Trans.,* Vol 10A, June 1979, p 699–706

- S.S. Hecker and M.G. Stout, Strain Hardening of Heavily Cold Worked Metals, *Deforma-*

tion, Processing, and Structure, G. Krauss, Ed., American Society for Metals, 1984, p 1–46

- W.F. Hosford and R.M. Caddell, *Metal Forming: Mechanics and Metallurgy,* Prentice-Hall, 1983

- W. Roberts, Dynamic Changes That Occur during Hot Working and Their Significance Regarding Microstructural Development and Hot Workability, *Deformation, Processing, and Structure,* G. Krauss, Ed., American Society for Metals, 1984, p 109–184

REFERENCES

1. S.S. Hecker and M.G. Stout, Strain Hardening of Heavily Cold Worked Metals, *Deformation, Processing, and Structure,* G. Krauss, Ed., American Society for Metals, 1984, p 1–46
2. U.F. Kocks, Laws for Work-Hardening and Low Temperature Creep, *J. Eng. Mater. Tech. (Trans. ASME),* Vol 98, 1976, p 76–85
3. P.E. Armstrong, J.E. Hockett, and O.D. Sherby, Large Strain Multidirectional Deformation of 1100 Aluminum at 300 K, *J. Mech. Phys. Solids,* Vol 30, 1982, p 37–58
4. W.F. Hosford and R.M. Caddell, *Metal Forming: Mechanics and Metallurgy,* Prentice Hall, 1983, p 80–98
5. F.W. Boulger, DMIC Report 226, Battelle Memorial Institute, 1966, p 13–37
6. D.S. Fields and W.A. Backofen, *Trans. ASM,* Vol 51, 1959, p 946–960
7. A. Saxena and D.A. Chatfield, SAE paper 760209, 1976
8. R.P. Carreker and W.R. Hibbard, Jr., *Trans. TMS-AIME,* Vol 209, 1957, p 1157–1163
9. F. Garofalo, *TMS-AIME,* Vol 227, 1963, p 251
10. J.J. Jonas, C.M. Sellars, and W.J. McG Tegart, *Met. Rev.,* Vol 14, 1969, p 1
11. J.J. Jonas, *Trans Q. ASM,* Vol 62, 1969, p 300–303
12. A.K. Ghosh, A Physically Based Constitutive Model for Metal Deformation, *Acta Metall.,* Vol 28 (No. 11), Nov 1980, p 1443–1465
13. F.R.N. Nabarro, Z.S. Basinski, and D.B. Holt, *Adv. Phys.,* Vol 13, 1964, p 193
14. F.W. Young, *J. Phys. Soc. Jpn.,* Vol 18, 1963, Suppl. 1
15. W.G. Johnston and J.J. Gilman, *J. Appl. Phys.,* Vol 30, 1959, p 129
16. J. Hausselt and W. Blum, *Acta Metall.,* Vol 24, 1976, p 1027
17. J.G. Byrne, *Recovery, Recrystallization and Grain Growth,* McMillan, 1965, p 44
18. J. Friedel, *Dislocations,* Pergamon Press, 1964, p 239, 278
19. R.W. Bailey, Note on the Softening of Strain-Hardened Metals and Its Relation to Creep, *J. Inst. Met.,* Vol 35, 1926, p 27
20. E. Orowan, The Creep of Metals, *J. West Scotland Iron Steel Inst.,* Vol 54, 1946–1947, p 45

21. W. Roberts, Dynamic Changes That Occur during Hot Working and Their Significance Regarding Microstructural Development and Hot Workability, *Deformation, Processing, and Structure,* G. Krauss, Ed., American Society for Metals, 1984, p 109–184

22. H. Mecking and K. Lücke, Quantitative Analyse der Bereich III-Verfestigung von Silber-Einkristallen, *Acta Metall.,* Vol 17, 1969, p 279

23. K. Lücke and H. Mecking, Dynamic Recovery, *Inhomogeneity of Plastic Deformation,* R.E. Reed-Hill, Ed., American Society for Metals, 1973, p 223–250

24. H. Mecking, U.F. Kocks, and H. Fischer, Hardening, Recovery and Creep in f.c.c. Mono- and Polycrystals, *Proc. Fourth Int. Conf. on Strength of Metals and Alloys* (Nancy), 1976, Vol 1, p 334

25. H. Mecking, Description of Hardening Curves of f.c.c. Single and Polycrystals, *Work Hardening in Tension and Fatigue,* A.W. Thompson, Ed., AIME, 1977, p 67–88

26. U.F. Knocks and H. Mecking, Discussion of Ref 30, *J. Eng. Mater. Tech. (ASME H),* Vol 98, 1976, p 121

27. U.F. Kocks and H. Mecking, A Mechanism for Static and Dynamic Recovery, *Proc. Fifth Int. Conf. on Strength of Metals and Alloys* (Aachen), Vol 1, 1979, p 345

28. P. Ostrom and R. Lagneborg, A Recovery-Athermal Glide Creep Model, *J. Eng. Mater. Tech. (ASME H),* Vol 98, 1976, p 114

29. P. Ostrom and R. Lagneborg, A Dislocation Link-Length Model for Creep, *Res. Mech.,* Vol 1, 1980, p 159

30. E. Voce, The Relationship Between Stress and Strain for Homogeneous Deformation, *J. Inst. Met.,* Vol 74, 1948, p 537

31. H. Mecking and G. Gottstein, Recovery and Recrystallization during Deformation, *Recrystallization of Metallic Materials,* F. Haessner, Ed., Riederer Verlag, Stuttgart, 1979

32. T.H. Courtney, *Mechanical Behavior of Materials,* McGraw-Hill, 1990, p 271–279, 285, 304, 307

33. J. Weertman, *Trans. ASM,* Vol 61, 1968, p 681

34. A.K. Mukherjee, *Treatise Mater. Sci. Technol.,* R.J. Arsenault, Ed., Vol 6, 1975, p 163

35. M.F. Ashby and R.A. Verall, *Acta Metall.,* Vol 21, 1973, p 149

36. F.A. Mohammed and T.G. Langdon, *Scr. Metall.,* Vol 10, 1976, p 759

37. A.K. Ghosh and C.H. Hamilton, Mechanical Behavior and Hardening Characteristics of a Superplastic Ti-6Al-4V Alloy, *Metall. Trans. A,* Vol 10, June 1979, p 699–706

38. A.K. Ghosh, A New Physical Model for Superplastic Flow, Superplasticity in Advanced Materials, ICSAM-94 *Mater. Sci. Forum,* Vol 170–172, 1994, p 39–46

39. R.L. Coble, *J. Appl. Phys.,* Vol 34, 1963, p 1679

40. W.A. Backofen, F.J. Azzarto, G.S. Murty, and S.W. Zehr, *Ductility,* American Society for Metals, 1968, p 279

41. J.R. Spingarn and W.D. Nix, *Acta Metall.,* Vol 27, 1979, p 171

42. A. Ball and M.M. Hutchinson, *Met. Sci.,* Vol 3, 1969, p 3

43. A.K. Mukherjee, *Mater. Sci. Eng.,* Vol 8, 1971, p 83

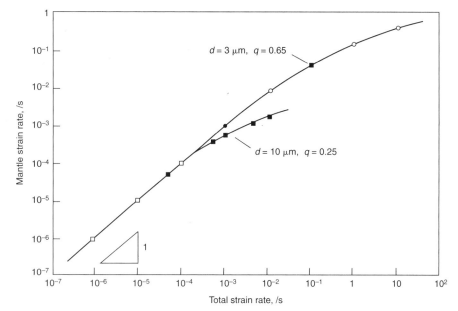

Fig. 37 Mantle strain rate as function of total strain rate for the two alloys from Table 4. Note the finer-grain alloy (alloy 2) also has higher dispersoid density (q is higher).

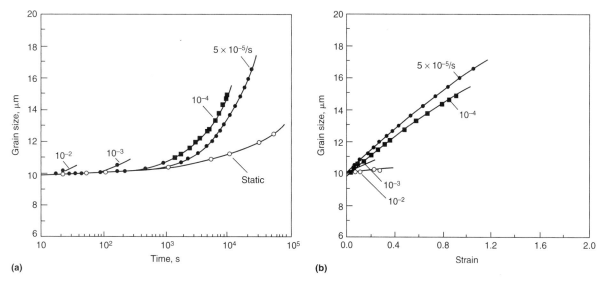

Fig. 38 Concurrent grain growth during superplastic deformation simulated for alloy 1 for a variety of strain rates plotted as a function of (a) time and (b) strain

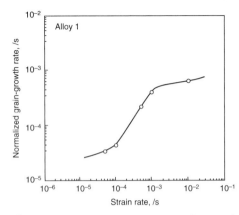

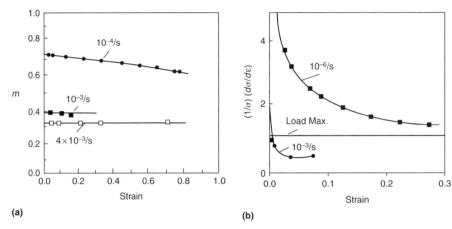

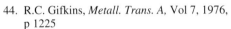

(a)

(b)

Fig. 39 Simulated normalized grain-growth rate, $1/d_0$ $(\Delta d/\Delta t)$, for alloy 1 as a function of applied strain rate

Fig. 40 (a) Instantaneous values of m determined from simulated step changes in strain rate at periodic intervals during a tensile test. (b) Normalized strain-hardening rate as a function of superplastic strain

44. R.C. Gifkins, *Metall. Trans. A,* Vol 7, 1976, p 1225
45. R.C. Gifkins, *Metall. Trans. A,* Vol 8, 1977, p 1507
46. J.W. Edington, K.N. Melton, and C.P. Cutler, *Prog. Mater. Sci.,* Vol 21, 1970, p 61
47. M.A. Clark and T.H. Alden, *Acta. Metall.,* Vol 21, 1973, p 1195
48. A.K. Ghosh, in *Deformation of Polycrystals: Mechanics and Microstructures,* Proc. Second Riso Intl. Symp. (Roskilde, Denmark), 1981, p 277; also A.K. Ghosh and C.H. Hamilton, *Metall. Trans. A,* Vol 13, 1982, p 1955
49. B.M. Watts, M.J. Stowell, B.L. Baikie, and D.G.E. Owen, *Met. Sci.,* Vol 10, 1976, p 198
50. A.K. Ghosh and A. Basu, in *Critical Issues in the Development of High Temperature Structural Materials,* N.S. Stoloff, D.J. Duquette, and A.F. Giamei, Ed., TMS, 1993, p 291
51. A.K. Ghosh and C. Gandhi, U.S. Patent 4,770,848, Sept 1988
52. D.S. Wilkinson and C.A. Caceres, *Acta Metall.,* Vol 32, 1984, p 1335

Springback

R.H. Wagoner, J.F. Wang, and M. Li, The Ohio State University

This article is intended as an introduction to the concepts of springback simulation as well as recommendations for its practice in a metal forming setting. Most of the developments focus on thin beams or sheets, where springback is most pronounced. The underlying mechanics of large-strain, elastic-plastic deformation are treated in a simplified, intuitive way, with numerous references provided for those wishing to delve into the theoretical underpinnings in more detail. Simple bending is first considered, along with a discussion of approximations, then bending with tension and finally, more complex numerical treatments. Compensation of die design to account for springback is also presented briefly.

This treatment is intended for practitioners with widely differing backgrounds and needs. The early treatments are suited to a limited class of problems but are best suited for understanding the direction of the effects of various material properties and process parameters. The roles and effects of various simplifying assumptions are also treated naturally with these closed-form solutions. The later treatments are intended to augment the practice of applied sheet forming analysis (almost always finite element based) to include postforming springback analysis. As is shown, the choices of numerical parameters can be quite different for springback, so these aspects are emphasized.

As used throughout this article, *springback* refers to the elastically driven change of shape that occurs after deforming a body and then releasing it. The concept is understood by anyone who has manually bent a metal wire or strip. For a sufficiently small bend radius, some part of the bending remains after unloading and some part is recovered during unloading (or has sprung back). For bend radii larger than some critical value, the initial shape of the body is recovered. The recovered portion of the deformation is referred to as springback. As such, the definition inherently refers to a difference in geometry between the loaded state and the unloaded state.

The word *springback* as a single, unhyphenated word appears in virtually no standard dictionaries but has been in technical use since at least the early 1940s. A search of the internet in April 2005 found more than 26,800 occurrences of the word, and a contemporaneous search of the ISI Web of Science (Thomson Scientific) of published technical papers located 334 such references appearing since 1980. These numbers represent increases of 460 and 27%, respectively, over similar searches performed 20 months earlier, in April 2003. Two inferences may be drawn: the technical meaning of springback is well established, although formal definitions appearing in dictionaries lag, and interest in springback is growing rapidly.

The definition of springback can be broad, applying to the action of springs, for example, but the principal technical intent of the word and interest in the phenomenon refers to the undesirable shape change that occurs after forming a part. The change is undesirable because it creates a difference of part shape from the tool shapes that were used to carry out the forming operation. If this difference is not predicted accurately and compensated for in the design of the tools, the part may not meet specifications.

Consideration of springback is of prime interest for bodies that have high aspect ratios; that is, at least one dimension is much larger than, or smaller than, the other dimensions. Examples include slender beams and thin sheets (Fig. 1, 2). For these cases, the overall geometric

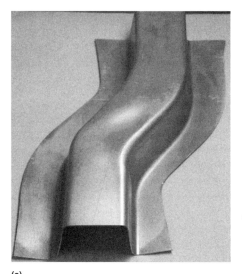

(a)

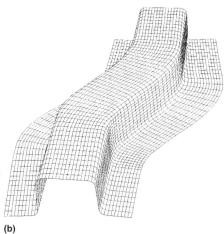

(b)

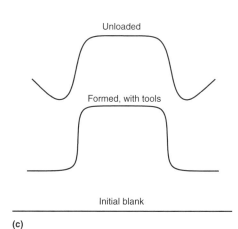

(c)

Fig. 1 Typical automotive sheet-formed part, the S-rail. (a) Formed part. (b) Finite element representation, as formed. (c) Cross-sectional schematics at three forming stages. Source: Ref 1

changes caused by springback can be very significant even though the elastic strains driving the springback can be tiny.

To introduce an applied example, Fig. 1(a) shows a representative automotive formed part referred to as the S-rail (Ref 1). Figure 1(b) depicts a corresponding finite element mesh, and Fig. 1(c) focuses on a schematic cross section of this part at three stages: initial (flat), as-formed (with tools in place), and unloaded (after springback). Inspection of the operation (and ignoring slight stretching at the top web of the part) reveals that the upper corners of the cross section are essentially bent to conform to the punch radius (the punch in this orientation lies below the sheet). When the tooling is removed, these radii open up to larger radii. This is typical of an idealized bending-with-tension operation. The sidewall regions of the formed rail or channel are drawn over the die radii (the die lies above the sheet in this orientation) over large distances, such that each element undergoes bending and unbending sequentially, also under the action of tension. When loaded, the sidewalls are flat. The final shape of the sidewalls incorporates what is known as sidewall curl. The level of tension for each location is related to the binder force, the friction with the tooling, and the work required to bend, unbend, and draw. If a draw bead were involved, this would add yet another element to the sheet tension determination.

The primary focus in this article is on sheet metal forming operations, such as the one shown in Fig. 1. This focus allows conclusions to be drawn with reference to a relatively narrow range of thicknesses and bend radii, both of which are small relative to the width of the body. The equations and results are nonetheless applicable to other geometries, with restrictions specified as necessary.

This article is organized in sections. The subject of springback is first addressed for the simplest, most easily understood cases, that is, pure bending of slender beams or sheets. While such treatments are applicable to few problems of applied interest, their study reveals the principles governing the problem and addresses the limitations of the various assumptions. To these treatments is then added the effect of superimposed tension, which is shown to be a critical variable for accurate prediction of springback. From these generally closed-form treatments, a leap is made to the much more general and practical prediction of springback for real forming operations, using either experience or finite element modeling. Finally, the design of dies and tooling using an assumed springback prediction capability is addressed.

Pure Bending—Classical Results

In order to understand the phenomenon of springback, it is instructive to begin with the simplest case and the most restrictive assumptions. In this section, the case of pure, or simple, bending is considered, that is, bending under the action of an applied moment without applied sheet tension. The springback consists of assumed elastic unbending on removal of the applied moment.

Assumptions. The assumptions that apply to this case in the simplest treatment may be listed as:

1. Plane sections remain planar.
2. No change in sheet thickness
3. Two-dimensional geometry, either plane strain or plane stress in width direction
4. Constant curvature (i.e., no instability of shape)
5. No stress in the radial, or through-thickness, direction
6. The neutral (stress-free) axis is the center fiber and is the zero-extension fiber.
7. No distinction between engineering and true strain
8. Isotropic, homogeneous material behavior
9. Elastic straining only during springback

The validity of these assumptions is discussed in the next section, but the simple results for springback under these conditions are first presented here.

Within these assumptions, the primary differences among treatments appearing in the literature relate to the assumed material constitutive behavior. There are two basic choices to be made: whether to treat the problem as purely plastic or elastic-plastic, and what form of stress-strain law to adopt in the plastic range. Results have been presented for nearly all choices: perfectly plastic (no hardening), linear hardening (Ref 2), power-law hardening (Ref 3–7), or a general approach (requiring graphical or other numerical integration) (Ref 8, 9).

Basic Equations and Approach. The approach is illustrated in Fig. 2. An initially flat sheet or beam is envisioned. For these purposes, a sheet denotes a part that is very wide relative to its thickness and bend radius and implies that the deformation is nearly plane strain; that is, the strain in width direction is zero. A beam denotes a part that is very narrow relative to thickness and bend radius and implies that the deformation is nearly plane stress; that is, the stress in the width direction is zero. The part is bent to a starting radius (R) under the action of a moment (M). The value of M acting on the sheet or beam is obtained by integrating the stress distribution as:

$$M = \int_{-t/2}^{t/2} \sigma_x(\varepsilon_x) z w(z) dz \qquad \text{(Eq 1)}$$

where ε_x is the circumferential strain, σ_x the circumferential stress (Fig. 2b), t is the sheet thickness, and $w(z)$ is the width of the sheet, which in general may vary with the z-coordinate (that is, the cross section need not be rectangular). Assuming a rectangular cross section, taking advantage of the symmetry of the problem (assumptions 6 and 8 in the section "Approximations in Classical Bending Theory" in this article), and substituting into Eq 1 obtains the moment per unit width (M/w), which may be expressed more simply:

$$\frac{M}{w} = \int_{-t/2}^{t/2} \sigma_x(\varepsilon_x) z \, dz = 2 \int_0^{t/2} \sigma_x(\varepsilon_x) z \, dz \qquad \text{(Eq 2)}$$

The strain shown in Eq 1 and 2, ε_x, depends on z. Within the given assumptions, the circumfer-

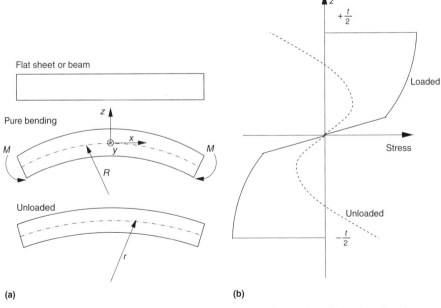

(a)

Flat sheet or beam

Pure bending

Unloaded

(b)

Fig. 2 Schematics of pure bending. (a) Configurations with coordinates defined. (b) Through-thickness stress distribution

ential strains (ε_x) are linearly related to the distance from the center of the sheet (z) and inversely to the bend radius (of the center fiber of the body) R:

$$\varepsilon_x \approx e_x = \frac{z}{R} \qquad \text{(Eq 3)}$$

where it is assumed that the true strain (ε_x) and the engineering strain (e_x) are small enough to be used indistinguishably. As is shown explicitly, the bending moment can be calculated using Eq 2 and 3, along with a constitutive relationship between stress and strain.

Note: In order to simplify the notation, the subscript x is dropped from the terms σ_x and ε_x, with the understanding that these represent the principal components of stress and strain normal to the beam or sheet cross section (as labeled in Fig. 2).

In order to compute the springback after bending, the moment per unit width of sheet, M/w, is removed from the sheet or beam while the material responds elastically. Because elastic stresses and strains can be superimposed, an alternative view of this operation is obtained by applying a moment (M) to the stress-free body in the configuration of the bent beam or sheet. An isotropic linear elastic beam or sheet has a constitutive response of $\sigma_x = E' \varepsilon_x$, where E' is the effective modulus for the beam (plane-stress case) or $E' = E/(1-v^2)$, where v is Poisson's ratio, for the sheet (plane-strain case).

For elastic recovery from an initially curved configuration (radius = R) to a final configuration (radius = r), the relationship for a body of general cross section is:

$$\frac{1}{R} - \frac{1}{r} = \frac{M}{E'I} \qquad \text{(Eq 4)}$$

where I is the moment of inertia of the cross section.

Note: The springback results for plane strain and plane stress do not differ greatly for most materials. Assuming that the bending moment is proportional to the operative flow stress for an isotropic, nonhardening, von Mises material, the plane-strain bending moment is $2/\sqrt{3}$ (1.15) times the plane-stress moment. Assuming a typical Poisson's ratio of $1/3$, the plane-strain elastic modulus is 1.12 times the plane-stress (i.e., uniaxial tension) one. Thus, the differences between Eq 4 interpreted for plane stress or plane strain is only approximately 1.15 versus 1.12, or approximately a 3% differential. Elastic and plastic anisotropy may change this value.

Moments of inertia may be readily calculated for complex shapes by integration and have been tabulated for a variety of standard structural shapes (Ref 10). For the case of a rectangular cross section, which is assumed in the remainder of this article, the moment of inertia is taken as:

$$I = \frac{wt^3}{12} \qquad \text{(Eq 5)}$$

where t is the sheet thickness. Equation 4 may be rewritten for a rectangular cross section in a per-width format as:

$$\frac{1}{R} - \frac{1}{r} = \frac{M}{E'I} = \frac{12M/w}{E't^3} \qquad \text{(Eq 6)}$$

which may be readily rewritten in the alternate form:

$$\frac{r}{R} = \left[1 - R\frac{M}{E'I} \right]^{-1} = \left[1 - R\frac{12M/w}{E't^3} \right]^{-1} \qquad \text{(Eq 7)}$$

The form $\left[\frac{1}{R} - \frac{1}{r} \right]$ is called springback in this article, while the second form, r/R, is called the springback ratio. In general, the springback is positive ($r > R$), and the springback ratio is thus greater than unity. The relationship between the two measures is as Springback ratio = $1/(1 - R \cdot \text{Springback})$. For most applications, springback as defined previously is the quantity of interest. For small curvature changes, the shape change displacements are proportional to springback. The springback ratio is occasionally used with some analytical procedures, so a few results in this article are presented using it.

Note that a fractional error associated with the evaluation of springback may be quite different from the fractional error associated with the springback ratio, depending on how large the second term of Eq 7 is relative to 1. That is, when the second term is small, errors of R/r will appear to be small even though the fractional errors on moment can be significant.

Equations 6 and 7 represent the fundamental springback result for pure bending with the assumptions listed. To apply Eq 6, it is necessary to first choose the plane-stress or plane-strain approximation based on width with respect to bend radius and thickness. The bending moment is computed using Eq 2 and 3 and an explicit material stress-strain law (and known stress state). This approach is used to reproduce some classical springback results in the remainder of this section.

Rigid, Perfectly Plastic Result. The simplest springback result for pure bending makes use of a rigid (i.e., no elastic strains), perfectly plastic (no strain hardening) material model (Ref 4–6, 8, 11–13). Under these assumptions, the bending moment (and thus the springback) is independent of the original bend radius:

$$\frac{M}{w} = 2 \int_0^{t/2} \sigma_0' z\, dz = \frac{\sigma_0' t^2}{4} \quad \text{(rigid, perfectly plastic)} \qquad \text{(Eq 8)}$$

where σ_0' is the yield stress (also the flow stress) of the material in plane stress or plane strain. The springback, defined here, is obtained using Eq 4:

$$\frac{1}{R} - \frac{1}{r} = \frac{3\sigma_0'}{E't} \quad \text{or, alternatively,}$$

$$\frac{R}{r} = 1 - \frac{3\sigma_0' R}{E't} \quad \text{(rigid, perfectly plastic)}$$

$$\text{(Eq 9)}$$

This result is often sufficient for springback prediction, and it reveals the importance of the principal material properties as they affect springback:

- Springback is proportional to strength/stiffness, that is, σ_0/E.
- Springback is inversely proportional to sheet thickness.

More detailed analysis alters the exact form of these dependencies, but the conclusion remains the same: materials that are strong relative to their elastic modulus are more susceptible to large springback, as are thinner materials. Thus, aluminum sheet of comparable strength to a steel alloy exhibits springback approximately three times greater, because its elastic modulus is approximately $1/3$ as large as that of steel.

Elastic, Perfectly Plastic Result. The first refinement of Eq 9 is by the inclusion of elastic, perfectly plastic bending behavior (Ref 14, 15). That is, there will be an elastic core near the neutral axis. The location of the elastic-plastic transition has a z-coordinate of z^*, which is found by setting the yield strain (σ_0'/E') equal to the bending strain (z/R):

$$z^* = \begin{cases} \dfrac{R\sigma_0'}{E'} & \text{for} \quad \dfrac{R\sigma_0'}{E'} \leq \dfrac{t}{2} \\[2mm] \text{(elastic-plastic case)} & \text{(Eq 10a)} \\[2mm] \dfrac{t}{2} & \text{for} \quad \dfrac{R\sigma_0'}{E'} > \dfrac{t}{2} \\[2mm] \text{(elastic only, no springback)} \end{cases}$$

$$\text{(Eq 10b)}$$

Note that the extent of the elastic core is proportional to the bend radius (i.e., inversely proportional to curvature), proportional to the yield stress, and inversely proportional to the elastic modulus. Thus, inclusion of the elastic part of the material response becomes progressively more important for gentle bending of high-specific-strength materials. As is shown later, for typical sheet metal press forming, the bend radius is typically small enough that the elastic region may be neglected without significant loss of accuracy.

Evaluation of the required integral to obtain the moment for elastic-plastic cases may be conveniently split into two terms, the second identical to the rigid, perfectly plastic case outside of the elastic core:

$$\frac{M}{w} = 2 \int_0^{z^*} \frac{E'z}{R} z\, dz + 2 \int_{z^*}^{t/2} \sigma_0' z\, dz \qquad \text{(Eq 11)}$$

$$= \frac{2E'z^{*3}}{3R} + \frac{\sigma_0' t^2}{4} - \sigma_0' z^{*2}$$

Substituting the relationship for the elastic-plastic location, z^* (Eq 10a), obtains the elastic-plastic moment and springback results:

$$\frac{M}{w} = \frac{\sigma_0' t^2}{4} - \frac{\sigma_0'^3 R^2}{3E'^2} \quad \text{(elastic, perfectly plastic)}$$

$$\text{(Eq 12)}$$

$$\frac{1}{R} - \frac{1}{r} = \frac{12M/w}{E't^3} = \frac{12}{E't^3}\left[\frac{\sigma_0't^2}{4} - \frac{\sigma_0'^3R^2}{3E'^2}\right]$$

(elastic, perfectly plastic)   (Eq13)

The springback equation may be rewritten in an alternate form presented by Gardiner (Ref 14) using Eq 7:

$$\frac{R}{r} = 1 - \frac{3\sigma_0'R}{E't} + \frac{4\sigma_0'^3R^3}{E'^3t^3} = 1 - 3x + 4x^3,$$

$$\text{where } x = \frac{\sigma_0'R}{E't} \qquad \text{(Eq 14)}$$

Note that the left side of Eq 14 is the reciprocal of the springback ratio as defined in this article. The error introduced by ignoring the elastic core in springback calculations may be evaluated by comparing Eq 9 and 13 or, equivalently, Eq 8 and 11. Table 1 presents R/t ratios where the moment error is limited to 1, 2, 5, and 10%. For a given desired level of accuracy, the R/t ratio is the largest one that can be safely considered.

Typical R/t ratios for automotive press forming lie in the range of 5 to 25, although many examples outside of that range may be found, particularly with general three-dimensional shapes that are not amenable to simple analysis. The results in Table 1 show that ignoring the elastic core leads to very small errors in this range for normal materials (aluminum alloys with a yield stress of 500 MPa, or 73 ksi, are seldom suitable for complex press forming).

Rigid, Strain-Hardening Results. In addition to the results presented previously, bending moments and springback relationships for strain-hardening material models have been presented in various forms, including the following selections:

- Empirical forms (Ref 16)
- Rigid, arbitrary hardening (Ref 8)
- Rigid, power-law hardening (Ref 3–7)
- Rigid, linear hardening (Ref 2)

Power-law hardening models are frequently used for sheet forming analysis. The hardening law, often attributed to Hollomon (Ref 17), may be written as follows, in uniaxial stress and other fixed stress- or strain-ratio forms:

$$\sigma = K\varepsilon^n \text{ (uniaxial stress) or}$$

$$\sigma' = K'\varepsilon'^n \text{ (general stress state)} \qquad \text{(Eq 15)}$$

where K is the strength parameter, K' is the effective strength parameter, n is the strain-hardening index, and the primes indicate that the strains and stresses to be considered must take into account the stress-strain state and the form of the yield function (anisotropic, quadratic, etc.). (Because elasticity is ignored, ε is the total strain, equal to the plastic strain.) Typical results for such hardening may be summarized as (Ref 6):

$$\frac{M}{w} = \left(\frac{2}{n+2}\right)\frac{K'}{R^n}\left(\frac{t}{2}\right)^{n+2} \qquad \text{(Eq 16)}$$

$$\frac{1}{R} - \frac{1}{r} = \left(\frac{6}{n+2}\right)\frac{K'}{E'}\left(\frac{t}{2R}\right)^n\frac{1}{t} \qquad \text{(Eq 17)}$$

Elastic-Plastic Result. Bending moments and springback relationships for elastoplastic, strain-hardening material models have also been presented in various forms, including:

- Elastic, power-law hardening (Ref 18, 19)
- Rigid, linear hardening (Ref 18)

In order to assess the importance of strain hardening in pure bending results, moment and springback formulas were derived based on a hardening law of the following form:

$$\sigma = \sigma_0 + K\varepsilon_p^n \text{ (uniaxial stress) or}$$

$$\sigma = \sigma_0' + K'\varepsilon_p'^n \text{ (general case)} \qquad \text{(Eq 18)}$$

For the plane-stress case, ε_p signifies the approximate plastic strain, that is, the total strain less the elastic yield strain:

$$\varepsilon_p = \varepsilon - \varepsilon_e \approx \frac{z - z^*}{R} \qquad \text{(Eq 19)}$$

The second equality of Eq 19 is approximate because the elastic strain (ε_e) is treated as a constant corresponding to the value at first yield, rather than as evolving with hardening. This approximation, adopted for simplicity, has little effect on the result.

The moment consists of three terms, the first two identical to the elastic, perfectly plastic result, that is, Eq 12 (with yield stress, σ_0), and the third an integral corresponding to the additional moment caused by the hardening beyond the yield stress. This third term may be evaluated as:

$$\frac{\Delta M}{w} = 2\int_{z^*}^{t/2} K\varepsilon_p^n z\,dz = \frac{2K}{R^n}\int_{z^*}^{t/2}(z-z^*)^n z\,dz$$

(uniaxial)   (Eq 20)

where, as before, $z^* = R\sigma_0/E$. Equation 20 may be evaluated to obtain the explicit form of the incremental moment:

$$\frac{\Delta M}{w} = \frac{2K}{R^n(n+2)}\left(\frac{t}{2} - z^*\right)^{n+2} + \frac{2Kz^*}{R^n(n+1)}\left(\frac{t}{2} - z^*\right)^{n+1} \qquad \text{(Eq 21)}$$

The full moment for the elastic, hardening plastic case is thus:

$$\frac{M}{w} = \frac{\sigma_0 t^2}{4} - \frac{\sigma_0^3 R^2}{3E^2} + \frac{2K}{R^n(n+2)}\left(\frac{t}{2} - z^*\right)^{n+2} + \frac{2Kz^*}{R^n(n+1)}\left(\frac{t}{2} - z^*\right)^{n+1} \qquad \text{(Eq 22)}$$

and the springback is then:

$$\frac{1}{R} - \frac{1}{r} = \frac{12M/w}{Et^3} \qquad \text{(Eq 23)}$$

where M is given by Eq 22.

Springback based on Eq 23 is compared in Fig. 3 with springback computed analytically using elastic, perfectly plastic material behavior (Eq 13), rigid, perfectly plastic material behavior (Eq 9), and elastic-plastic finite element (FE) simulations of four-point bending (Fig. 4). (Finite element simulations are introduced later in this article, but are included here for completeness. The FE results presented in Fig. 3 were verified by refining meshes and number of integration points until no significant changes were observed.) The FE simulations make use of either plane-stress quadratic solid elements (labeled CPS8) or plane-stress beam elements with shear terms (labeled B21). The element labels correspond to the ABAQUS (Ref 20) elements used.

For purposes of Fig. 3 and subsequent use in this article, two material models corresponding to ratios of extremes of yield stress (σ_y) to Young's modulus for typical forming materials were defined based on Eq 18. The soft material, low-strength steel, is based on properties appearing in the literature (Ref 21) for interstitial-free steel: yield stress is 150 MPa (22 ksi), ultimate tensile strength is 310 MPa (45 ksi), and uniform elongation is 28.5%. The hard material, high-strength aluminum, is based on the properties appearing in the literature (Ref 22) for 7075-T6: yield stress is 500 MPa (73 ksi), ultimate tensile strength is 572 MPa (83 ksi), and uniform elongation is 11%.

Parameters of Eq 18 can be determined from the yield stress, ultimate tensile strength, and uniform elongation by noting two conditions: 1) the engineering stress at the uniform elongation is equal to the ultimate tensile strength (σ_{UTS}), and 2) the derivative of the true stress/true plastic strain equation is equal to the true stress at the uniform elongation (σ_f) (or corresponding plastic true strain, ε_f^p). Use of these two conditions, and associating σ_0 of Eq 18 with the yield stress, σ_y, allows determination of the parameters of Eq 18

Table 1 Maximum ratios of bending radius to thickness (R/t) for specified moment errors by neglecting the elastic core in bending (perfectly plastic)

| Material | Yield stress | | Young's modulus | | Maximum R/t for moment error of: | | | |
	MPa	ksi	GPa	10^6 psi	1%	2%	5%	10%
Low-strength steel	150	22	210	30	85	120	190	270
Low-strength aluminum	150	22	70	10	28	40	64	90
High-strength steel	500	73	210	30	26	36	57	81
High-strength aluminum	500	73	70	10	9	12	19	27

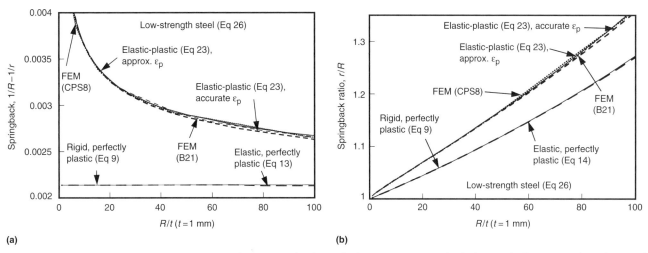

Fig. 3 Effect of various approximations (Eq 9, 13, 14, 23, and 26) on simulated springback quantities. (a) Springback. (b) Springback ratio. R, radius of primary bending curvature; r, radius of curvature after springback; ε_p, plastic strain; FEM, finite element modeling; t, thickness. CPS8 and B21 are ABAQUS 6.2 element designations. Source: Ref 20

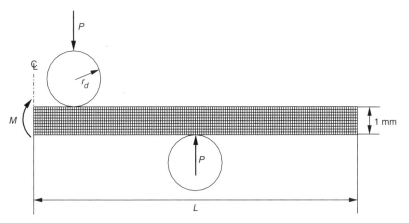

Fig. 4 Finite element modeling (FEM) mesh (CPS8 plane-stress solid elements shown) and tools for simulation of pure bending. M, bending moment; P, applied load in four-point bend FEM; r_d, tool radius in four-point bend FEM; L, half-sample length. Source: Ref 20

as follows:

$$n = \frac{\sigma_f}{\sigma_f - \sigma_y}\,\varepsilon_f^p \quad \text{(Eq 24)}$$

$$K = \frac{\sigma_f - \sigma_y}{(\varepsilon_f^p)^n} \quad \text{(Eq 25)}$$

For the two chosen materials, explicit elastic-plastic constitutive equations are:

Low-strength steel:

$$\sigma = 150\,\text{MPa} + 425\,\text{MPa}\ \varepsilon_p^{0.4},\ E = 210\,\text{GPa} \quad \text{(Eq 26)}$$

High-strength aluminum:

$$\sigma = 500\,\text{MPa} + 400\,\text{MPa}\ \varepsilon_p^{0.45},\ E = 70\ \text{GPa} \quad \text{(Eq 27)}$$

As illustrated in Fig. 3(a) neglecting strain hardening leads to large errors in springback throughout the range of R/t tested, varying from approximately 20 to 50%. (The smaller fractional error for springback ratio is illustrated in Fig. 3b.) For pure bending, Eq 23 is in good agreement with the FE results, whether or not the strain distribution through the thickness of the sheet is approximated.

Approximations in Classical Bending Theory

In this section, the assumptions introduced previously are discussed and, in some cases, evaluated semiquantitatively. As is shown throughout this article, the most important aspects for accuracy in springback prediction for typical sheet forming operations (R/t assumed to be in the range of 5 to 25) involve:

- Sheet tension (most critical aspect, presented in the next section)

- The hardening law (discussed in the previous and last sections)
- Presence of anticlastic curvature (this section and the last)

The basic assumptions of the previous section for pure bending have fairly small errors associated with them. However, these errors can grow when elastic-plastic laws are considered and when bending and unbending occur. It is useful to consider first the effect of the various approximations made within the foregoing pure-bending analysis.

Assumption 1: Plane Sections Remain Planar. For bending and bending under tension, this assumption is very nearly satisfied under most circumstances. For $R/t \geq 5$, shell finite elements (which incorporate this assumption, among others) agree well with full solid elements, which allow general deformation patterns (see later section of this article). Therefore, for $R/t > 5$, the assumption degrades the accuracy little. Another indicator is the accuracy of more complicated closed-form solutions for pure bending. These solutions (Ref 4, 5, 19, 23) retain the planar section assumption but allow through-thickness stresses to develop. These solutions are in good agreement with experiments for small R/t, thus indicating again that the assumption of planar sections has little effect on pure bending over a wide range of R/t.

There are two circumstances where this assumption may be significantly violated: when the frictional stress of sliding on the inner surface of the part is significant relative to bending and stretching forces, and when the hardening law is such that instabilities can occur (assumption 4 in this section). For pure bending and bending under tension (i.e., without frictional contact), the assumption of plane sections remaining plane is reasonable for most situations, probably down to R/t ratios as small as 1.

Assumption 2: No Change in Sheet Thickness. This assumption is related intimately with pure bending, for which it is very accurate, even to small R/t ratios. Thick shell results (Ref 23, 24) show this directly, although for R/t ratios less than 1, some thickness changes can occur (Ref 25). The use of the phrase "thick shell" in this article refers to a relaxation of some thin-shell approximations. This is distinct from the specialized use of this phrase in the mechanics literature to refer to a particular, systematic development of the kinematics of shell theory. For bending under tension, as presented in the next section, the thickness change is marked.

Assumption 3: Two-Dimensional Geometry, Either Plane Strain or Plane Stress in Width Direction. This is not a good assumption for many bending operations. As illustrated in Fig. 5, as bending occurs in the principal axis, a curvature develops across the width of the specimen. The effect is well known (Ref 26, 27).

The origin of this anticlastic curvature is easily understood: the principal bending causes lengthening of fibers above the neutral axis and shortening of those below it. For lengthened fibers there are Poisson contractions in the width and thickness directions, while for the shortened fibers there are expansions. Across the entire thickness, for pure bending, these very nearly balance each other, hence assumption 2 (no change in sheet thickness) is very accurate under most circumstances.

When the width changes are considered, the tendency to develop a secondary curvature is clear. The outer fibers tend to contract laterally and the inner ones to expand, so a concave-up curvature is favored. For very wide geometries (relative to thickness and bend radius), the plane-strain assumption becomes the limiting case (although there will always be some anticlastic curvature near the sheet edges). For narrow geometries, the anticlastic curvature is unimpeded by shear terms, and the cross section adopts a circular shape with radius of curvature $R_a = \nu/R$ (Ref 9). The plane-stress assumption is the limiting case when there are no stresses resisting the adoption of this shape.

For pure-elastic bending, the shape of the cross section has been found analytically (Ref 28), and a literature review of the subject has appeared (Ref 29). In spite of the limited accuracy of the result for small R/t (the inaccuracy arises by considering the bent configuration to be parabolic), the results are illuminating for typical sheet-forming cases.

The most important result is that the configuration of the bent part is determined by a single dimensionless parameter (β) describing the normalized width (w) of the specimen, sometimes called Searle's parameter (Ref 30, 31):

$$\beta = \frac{w^2}{Rt} = \frac{w}{R} \cdot \frac{w}{t} \quad \text{(Searle's parameter)}$$

(Eq 28)

along with the Poisson's ratio, ν. The actual shape of the cross section for various values of β is illustrated in Fig. 6, where the bend radii

are also shown assuming a fixed width of 50 and unit thickness of 1. The results in Fig. 6 correspond to the analytic solution as follows, where C_1 and C_2 are numerical constants:

$$\left(\frac{z}{t}\right) = C_1 \cosh\left(\frac{y\alpha}{w}\right)\cos\left(\frac{y\alpha}{w}\right) + C_2 \sinh\left(\frac{y\alpha}{w}\right)\sin\left(\frac{y\alpha}{w}\right)$$

(Eq 29)

where $C_{2,1} = \dfrac{\nu}{\sqrt{3(1-\nu^2)}}$

$$\times \frac{\sinh\left(\frac{\alpha}{2}\right)\cos\left(\frac{\alpha}{2}\right) \pm \cosh\left(\frac{\alpha}{2}\right)\sin\left(\frac{\alpha}{2}\right)}{\sinh(\alpha) + \sin(\alpha)}$$

with $\alpha = \sqrt[4]{3(1-\nu^2)}\sqrt{\beta}$.

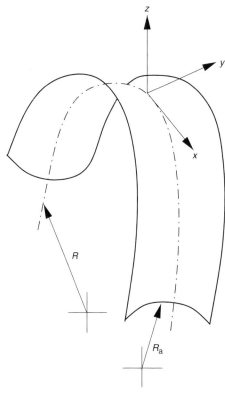

Fig. 5 Anticlastic surface with two orthogonal curvatures. R, radius of primary bending curvature; R_a, radius of anticlastic curvature

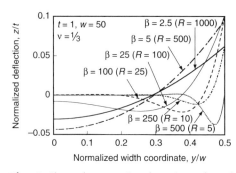

Fig. 6 Shape of cross sections for various values of Searle's parameter, $\beta = (w^2/Rt)$. z, thickness coordinate; t, sheet thickness; y, width coordinate; w, sheet width; ν, Poisson's ratio; R, radius of primary bending curvature; β, Searle's parameter

For β greater than approximately 100, the edge regions look similar and the center region is essentially flat, thus implying that plane-strain conditions are accurate except at the local edge regions. For small β, the cross-sectional shape is essentially circular, which implies that the stresses resisting this curvature may be safely ignored (i.e., plane stress).

The transition from the plane-stress limit to the plane-strain limit is a smooth function of β, as shown in Fig. 7. The limiting values of β for specified errors of the effective moment of inertia are shown in Table 2.

For springback application, the most important effect of anticlastic curvature is two-fold: on the bending moment-curvature relationship, and on the elastic unloading response. The latter will depend greatly on the degree to which the anticlastic shape persists, that is, how much of the anticlastic shape is retained after unloading. For the case of pure-elastic bending and unloading, there is no difference in springback, that is, zero springback, for cases with or without anticlastic effects. Persistent anticlastic curvature is particularly important for the typical sheet-forming case bending and unbending, as is discussed in the last section of this article. For parts curved in three dimensions, the anticlastic curvature may manifest itself as wrinkling, twisting, or generalized distortion. See, for example, the wrinkled area of the S-rail (Fig. 1a).

Assumption 4: Constant Curvature (i.e., No Instability in Shape). This is closely related to assumption 1. For bending and bending under tension, instabilities can occur because of shear

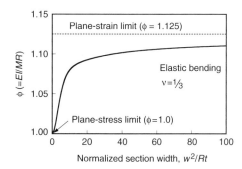

Fig. 7 The transition from the plane stress to plane strain, as a function of $\beta = (w^2/Rt)$. ϕ, anticlastic factor; E, Young's modulus; I, moment of inertia; M, bending moment; R, radius of primary bending curvature; w, width; t, thickness; ν, Poisson's ratio

Table 2 Limiting values of β (w^2/Rt) for specified accuracy limits using plane-stress or plane-strain bending formulas

β limit	Limiting value of β for an accuracy limit of:			
	1%	2%	5%	10%
Plane stress ($\beta <$)	2	3	5	34
Plane strain ($\beta >$)	170	42	7	2

banding, for example. These instabilities are expected to be of a similar magnitude and importance and in the same range of strains as for tension or compression. Therefore, when a stable hardening law is obtained (without serrated yielding, Lüder's banding, or yield-point phenomenon), bending may be assumed to behave similarly.

Assumptions 5, 6, and 7: No Stress in the Radial, or Through-Thickness, Direction. The Neutral (Stress-Free) Axis is the Center Fiber and is the Zero-Extension Fiber. No Distinction between Engineering and True Strain. These three assumptions are closely related and may be considered together naturally. To consider these effects profitably, it is simpler to ignore, for the moment, anticlastic curvature. Each of these assumptions is closely related to the R/t ratio to which a flat beam or plate is bent initially. The effect of bending to R/t less than approximately 5 produces significant through-thickness stresses in the interior of the body and causes the stress-free axis to vary significantly from the zero-extension fiber. Bending to small R/t also produces large strains at the outer fibers, thus making the distinction between true and engineering strains more significant (Ref 32).

The last of these effects is the simplest to quantify. In order to do so, consider the elastic-plastic result of Eq 20, but this time evaluate the strains in terms of true strains, that is:

$$\varepsilon_p = \varepsilon - \varepsilon_e \approx \ln\left(1 + \frac{z}{R}\right) - \ln\left(1 + \frac{z^*}{R}\right) \quad \text{(Eq 30)}$$

Note that again the additional elastic strain that accrues with strain hardening after the elastic-plastic transition point has been ignored. This allows Eq 20 to be rewritten as:

$$\frac{M'}{w} = 2\int_{z^*}^{t/2} K\varepsilon_p^n z\, dz$$

$$= 2K\int_{z^*}^{t/2}\left[\ln\left(1 + \frac{z}{R}\right) - \ln\left(1 + \frac{z^*}{R}\right)\right]^n z\, dz$$

$$\text{(Eq 31)}$$

where, except for the use of true strains, all the other assumptions remain the same. Evaluation of Eq 31 shows that the error on M introduced by neglecting true strain is 3.8% for $R/t = 1$ and less than 0.64% for $R/t \geq 5$. Therefore, other assumptions introduce larger errors than this one. (Note that a true kinematical description of bending, whether finite or infinitesimal, is not used for any derivations here. Only the evaluation of strain differs.)

The remaining assumptions related to small R/t can be assessed by the thick-shell solutions for rigid, perfectly plastic behavior presented first by Hill (Ref 33) and later in simplified forms (Ref 4, 5) and less restrictive forms for elastic-plastic cases (Ref 19). The conditions found to hold even during bending to small R/t are plane strain (assumed) and no thickness change.

Maintaining these conditions during bending to small R/t requires consideration of the quadratic terms in the value of the circular arc. The results show that significant through-thickness stresses develop and the stress-free fiber is no longer the zero-extension fiber. The details of the derivations may be found in the references provided, but the general result is that the total true strain for the more precise form is given by:

$$\varepsilon = \ln\left[1 + \left(\frac{t}{2R}\right)^2 + \left(\frac{2z}{R}\right)\right]^{1/2} \quad \text{(Eq 32)}$$

and the plastic problem must be solved incrementally in order to determine the stress distribution throughout the plastic deformation.

Figure 8 illustrates the relative magnitude of the various approximations for pure bending. The deviations are quite small for the pure bending case.

Assumption 8: Isotropic, Homogeneous Material Behavior. For fixed stress state (plane stress, for example) or strain state (plane strain, for example), the introduction of anisotropy, elastic and plastic, makes no fundamental change in the treatment of two-dimensional bending and springback results. While an extensive treatment of anisotropy is beyond the scope of this article, a simple result can illustrate the general procedure. For more complicated cases, FE analysis is usually required, and FE programs usually have capabilities incorporating material anisotropy.

It is important to note that anisotropy does not affect the basic equations for the simple bending case. For the plastic bending, the relationship between tensile strain at a given fiber and the fiber location (Eq 3) is independent of anisotropy. The relationship between the normal principal stress (σ_x) and the bending moment (Eq 2) is also unchanged. For the elastic unloading, the relationship governing the change of curvature remains the same, except that the modulus is the effective one relating tensile strain (ε_x) to normal principal stress (σ_x), taking anisotropy into account.

The role of anisotropy may be reflected sufficiently in a generalized elastic relationship:

$$\sigma_x = E'\varepsilon_x = f_E E\varepsilon_x \quad \text{(elastic)} \quad \text{(Eq 33)}$$

where E' is the effective modulus for the given strain-stress state, and f_E is a constant factor equal to E'/E.

For the plastic relationship, it should be noted that strain hardening is usually specified in terms of effective strain ($\bar{\varepsilon}$) and stress ($\bar{\sigma}$) based on a tensile test:

$$\bar{\sigma} = f(\bar{\varepsilon}) \text{ for example } \bar{\sigma} = K\bar{\varepsilon}^n \quad \text{(Eq 34)}$$

Using similar notation, the constant factor reflecting plastic anisotropy and stress-strain state may be defined (Ref 34, 35):

$$\bar{\varepsilon} = f_\varepsilon \varepsilon_x, \ \bar{\sigma} = f_\sigma \sigma_x \quad \text{(Eq 35)}$$

For any fixed anisotropy and strain-stress state, the values of f_E, f_σ, and f_ε may be found and

used to complete the basic bending and springback equation, such as Eq 2 to 4.

As an example, consider a sheet with normal plastic anisotropy according to Hill's quadratic yield function (Ref 33). The factors f_ε and f_σ can be derived for a given plastic anisotropy parameter (Ref 34), $\bar{r}$, as:

$$f_\varepsilon = \frac{1 + \bar{r}}{\sqrt{1 + 2\bar{r}}}, f_\sigma = \frac{\sqrt{1 + 2\bar{r}}}{1 + \bar{r}} \quad \text{(Eq 36)}$$

The necessary strain-hardening relationship, $\sigma_x(\varepsilon_x)$ in Eq 2, may then be found in terms of a measured uniaxial hardening law, $\bar{\sigma} = f(\bar{\varepsilon})$:

$$\sigma_x(\varepsilon_x) = \frac{\bar{\sigma}}{f_\sigma} = \frac{1}{f_\sigma}f(\bar{\varepsilon}) = \frac{1}{f_\sigma}f(f_\varepsilon \varepsilon_x) = \frac{1}{f_\sigma}f\left(f_\varepsilon\frac{z}{R}\right)$$

$$\text{(Eq 37)}$$

The substitution may be illustrated conveniently by taking a particular hardening law, say $\bar{\sigma} = K\bar{\varepsilon}^n$:

$$\sigma_x = \frac{K}{f_\sigma}\left(f_\varepsilon\frac{z}{R}\right)^n = \frac{K(1 + \bar{r})}{\sqrt{1 + 2\bar{r}}}\left[\frac{1 + \bar{r}}{\sqrt{1 + 2\bar{r}}}\frac{z}{R}\right]^n$$

$$= \frac{(1 + \bar{r})^{n+1}}{(1 + 2\bar{r})^{\frac{n+1}{2}}}K\left(\frac{z}{R}\right)^n = K'\left(\frac{z}{R}\right)^n \quad \text{(Eq 38)}$$

where K' represents all of the needed changes. The same procedure may be applied to any strain state, tensile hardening law, and fixed plastic anisotropy.

It should also be noted that an assumption of symmetric yielding in tension and compression has been made. For as-received sheet material, this is usually a reasonably accurate picture of the stress-strain behavior. However, as is discussed in the section, "Applied Analysis of Simple Forming Operations" in this article, the hardening behavior can become complex in reverse bending, which is common in many sheet-forming situations. Under these conditions, the Bauschinger effect on strain reversal must be considered. (Strictly speaking, some strain increment reversal can take place in single bending, because the neutral surfaces move relative to the midplane. This effect appears not to have been analyzed and is likely very small in practical cases.)

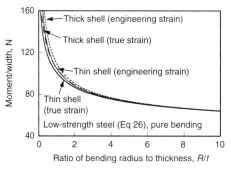

Fig. 8 Comparison of various approximations for elastoplastic pure bending

Finally, initial material properties are assumed to be the same at each point in the body.

Assumption 9: Springback Occurs Elastically. For all pure bending and nearly all bending under tension cases, this is very accurate. However, contrary to assertions in the literature (Ref 36), elastic-plastic springback can occur for bending under high sheet tensions (approaching and beyond the yield stress) (Ref 37–40) and when the material behavior is time-dependent (i.e., via creep or anelasticity) (Ref 41, 42). A few examples of such situations are mentioned in the section "Applied Analysis of Simple Forming Operations" in this article.

For most situations, these effects can be ignored without greatly affecting the result. However, it should be noted that unloading itself may involve inelastic effects (Ref 43) that produce changes in the observed modulus (Ref 44). There has been no clear approach on how such effects can be incorporated in springback analysis except for adjusting the effective elastic modulus.

Bending with Tension

The effect of superimposed tension during bending plays a dominant role in determining springback, as is demonstrated with simple analyses. Nearly all sheet-forming operations involve at least some sheet tension, whether introduced by remote sections of the part undergoing deformation, local friction conditions, or the intentional action of a draw bead or other restraint. Increasing sheet tension to reduce springback and its variability has been the principal industrial solution to the problem of inadequate shape fixability.

Analyses similar to those for pure bending can be carried out by relaxing just one of the assumptions listed in the first section, namely the sixth one, that is, that the neutral (stress-free) axis is the center fiber and is the zero-extension fiber. The sheet thickness may change substantially if the tension is high during bending (Ref 45) (and particularly for bending and unbending, which is not considered until the next section), but this effect is often ignored for simplicity. (For FE simulation, in the next section, shell elements usually assume zero thickness change in one time step, but the thickness is updated at the end of each step.)

Springback solutions for bending with tension have been presented with various levels of complexity, including elastic, perfectly plastic (Ref 6–13); elastic, power-law hardening (Ref 18, 19, 46–48); elastic, linear hardening (Ref 18, 49, 50); and rigid, power-law hardening (Ref 51–54). Extension to more complex cases includes: biaxially loaded plates (Ref 36), bending to small radii with tension (Ref 19), the effect on nonsimultaneous tension and bending (i.e., prebending or postbending) (Ref 51), taking into account section changes in narrow strips (Ref 7), the role of nonuniform deformation (Ref 55),

results for laminated sheets (Ref 56), and the effect on formability and residual-stress distribution (Ref 57).

Elastic, Perfectly Plastic Result. The simplest elastic, perfectly plastic solution for bending with tension is sufficient to reveal the dominating importance of tension relative to other variables. Initially considering the thickness constant, the strain distribution through the sheet thickness is a simple superposition of a tensile or membrane strain (ε_m) and the bending strain (ε_b) as before:

$$\varepsilon = \varepsilon_m + \varepsilon_b = \varepsilon_m + \frac{z}{R} \quad \text{(Eq 39)}$$

At the neutral axis (assumed to be the zero-extension axis), located at z_0, the strain is zero, so an expression relating the membrane strain and the neutral axis location is obtained:

$$\varepsilon_m = -\frac{z_0}{R} \quad \text{(Eq 40)}$$

The stress distribution is similar to the one shown in Fig. 9, which may be integrated to obtain the overall sheet tension, T, (per unit width, w) operating:

$$\frac{T}{w} = \int_{-t/2}^{t/2} \sigma_0' \, dz = \int_{-t/2}^{z_0} -\sigma_0' \, dz + \int_{z_0}^{t/2} \sigma_0' \, dz = -2z_0\sigma_0'$$
$$\text{(Eq 41)}$$

It is convenient to rewrite Eq 41 in terms of normalized quantities: z_0/t, the fractional location of the neutral axis, and $\overline{T}$, the average sheet tension stress (T divided by sheet width and thickness) divided by stress to yield the sheet (σ_0'), yielding $\left(\overline{T} = \frac{T}{wt\sigma_0'} \right)$. In terms of these reduced variables, Eq 41 becomes:

$$\left(\frac{z_0}{t} \right) = -\frac{\overline{T}}{2} \quad \text{(Eq 42)}$$

With the location of the neutral axis known explicitly in terms of the sheet tension, the moment may be evaluated in closed form:

$$\frac{M}{w} = \frac{\sigma_0' t^2}{4} \left[1 - \overline{T}^2 \right], \text{ or} \quad \text{(Eq 43)}$$

$$\left(\frac{M/w}{t^2} \right) = \frac{\sigma_0'}{4} \left[1 - \overline{T}^2 \right] \quad \text{(Eq 44)}$$

where Eq 44 emphasizes the proper normalization with thickness. The springback may then be presented in standard and normalized closed forms with the help of Eq 6 and 14 as:

$$\frac{1}{R} - \frac{1}{r} = \frac{12M/w}{E't^3} = \frac{3\sigma_0'}{E't} \left[1 - \overline{T}^2 \right] \quad \text{(Eq 45)}$$

$$\frac{R}{r} = 1 - \frac{3\sigma_0'}{E'} \left(\frac{R}{t} \right) \left[1 - \overline{T}^2 \right] \quad \text{(Eq 46)}$$

Equations 45 and 46 ignore the thickness change that occurs by the action of the sheet

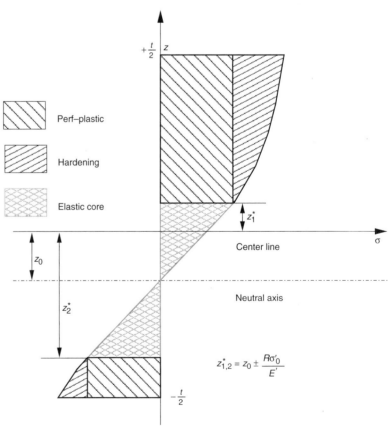

Fig. 9 Schematic of the stress distribution in a beam or sheet, bent to radius R, with definition of various coordinates used in the analysis of springback

tension; that is, final thickness t is assumed to be equal to original thickness t_0. The expressions may be approximately corrected for thickness changes by assuming that the final thickness is related to the original thickness such that the volume is maintained using the membrane strain (a linear approximation for the definition of strain is used for simplicity); that is:

$$\frac{t-t_0}{t_0} = -\varepsilon_m = \frac{z_0}{R} = -\frac{t\overline{T}}{2R} \qquad \text{(Eq 47)}$$

where t_0 is the initial thickness, and the final thickness, t, is given by:

$$t = \left(\frac{1}{t_0} + \frac{\overline{T}}{2R}\right)^{-1} \quad \text{or} \quad \frac{t}{R} = \frac{1}{(R/t_0)+\overline{T}/2} \qquad \text{(Eq 48)}$$

which gives an expression for the final thickness that may be substituted into Eq 45 and 46 to obtain thickness-corrected versions:

$$\frac{1}{R} - \frac{1}{r} = \frac{12M/w}{E't^3} = \frac{3\sigma_0'}{E'R}\left(\frac{R}{t_0}+\frac{\overline{T}}{2}\right)(1-\overline{T}^2) \qquad \text{(Eq 49)}$$

$$\frac{R}{r} = 1 - \frac{RM/w}{E'I} = 1 - \frac{3\sigma_0'}{E'}\left(\frac{R}{t_0}+\frac{\overline{T}}{2}\right)(1-\overline{T}^2) \qquad \text{(Eq 50)}$$

The results represented by Eq 46 and 50 are shown graphically in Fig. 10. As can be seen readily, the application of sheet tension substantially reduces springback. For the perfectly plastic case, springback disappears when the normalized sheet tension approaches unity, that is, when the average tensile stress approaches the appropriate yield stress. By setting $\overline{T}=0$, the pure bending result (Eq 9, for example) is obtained.

Rigid, Power-Law Hardening Result. Using the approach followed previously, the springback predicted for bending with tension can be derived. Unfortunately, it is not in a form as convenient as for the perfectly plastic case. The

only additional complexity is that t cannot be found explicitly in terms of $\overline{T}$, so that M cannot be written as an explicit function of $\overline{T}$. It is simplest to proceed by choosing R/t_0 and z_0/t_0 as independent variables, then evaluating the sheet tension, bending moment, and thus springback. In this way, springback may be obtained as a function of sheet tension but not in a closed equation form.

A power-law hardening law with a yield stress as in Eq 15 is adopted, and similar assumptions to the ones mentioned previously are made. Because total strain is represented by Eq 39, the stress throughout the sheet thickness is:

$$\sigma = \begin{cases} \sigma_0' + K'\varepsilon^n, \ \varepsilon > 0 \quad \text{or} \\ \sigma_0' + K'\left(\frac{z-z_0}{R}\right)^n, \ z > z_0 \\ -(\sigma_0' + K'|\varepsilon|^n), \ \varepsilon < 0 \quad \text{or} \\ -\sigma_0' - K'\left(\frac{z_0-z}{R}\right)^n, \ z < z_0 \end{cases} \qquad \text{(Eq 51)}$$

where, as was previously done, the plastic strain is approximated by the total strain (evaluated using the linear, small-strain definition) minus the elastic strain at yield. The normalized sheet tension and bending moment increment (beyond perfectly plastic) may then be obtained as:

$$\overline{T} = -2\left(\frac{z_0}{t}\right) + \frac{K'}{(n+1)\sigma_0'}\left(\frac{t}{R}\right)^n$$
$$\times \left[\left(\frac{1}{2}-\frac{z_0}{t}\right)^{n+1} - \left(\frac{1}{2}+\frac{z_0}{t}\right)^{n+1}\right] \qquad \text{(Eq 52)}$$

$$\frac{\Delta M/w}{t^2} = K\left(\frac{t}{R}\right)^n\left[\frac{\left(0.5+\frac{z_0}{t}\right)^{n+2}+\left(0.5-\frac{z_0}{t}\right)^{n+2}}{n+2}\right.$$
$$\left. - \left(\frac{z_0}{t}\right) \times \frac{\left(0.5-\frac{z_0}{t}\right)^{n+1}+\left(0.5+\frac{z_0}{t}\right)^{n+1}}{n+1}\right] \qquad \text{(Eq 53)}$$

where the current thickness must be evaluated in terms of z_0 using either the true-strain definition

or the small-strain approximation:

$$\frac{t}{t_0} = \exp\left(\frac{z_0}{R}\right) \approx \left(1+\frac{z_0}{R}\right) \qquad \text{(Eq 54)}$$

Equation 53 represents the additional bending moment caused by strain hardening that must be added to the perfectly plastic moment (Eq 43). The springback ratio is then evaluated using Eq 7. The final springback ratio is shown in Fig. 11.

Corrections to the Simple Power-Law Hardening Result. It is possible to obtain more accurate solutions for this case; however, the equations become rather bulky, will usually require numerical evaluation of integrals, and they will differ for each kind of hardening law considered. (In the truly arbitrary case, a numerical integration can be carried out to obtain the appropriate solution.) Nonetheless, it is useful to illustrate the additional terms that can be considered for completeness and estimation of importance (still adopting assumptions 1 and 3).

For large $\overline{T}$ or small R/t (i.e., large strain), the true-strain definition should be used such that the strains no longer vary linearly through the thickness, except approximately:

$$\varepsilon = \varepsilon_m + \varepsilon_b = \ln\left(1-\frac{z_0}{R}\right) + \ln\left(1+\frac{z}{R}\right)$$
$$= \ln\left(1-\frac{z_0}{R}\right)\left(1+\frac{z}{R}\right)$$
$$= \ln\left(1-\frac{z_0}{R}+\frac{z}{R}-\frac{z_0z}{R^2}\right)$$
$$\cong -\frac{z_0}{R}+\frac{z}{R}\left(1-\frac{z_0}{R}\right) \qquad \text{(Eq 55)}$$

Inclusion of the large strain formula via Eq 55 will usually require numerical evaluation of the integrals to obtain $\overline{T}$ and M. Furthermore, bending to large curvatures (R/t less than approximately 5) introduces errors in the other approximations that are more significant than the small strain form (see the section "Approximations in Classical Bending Theory" in this article). Forms equivalent to the last approximate

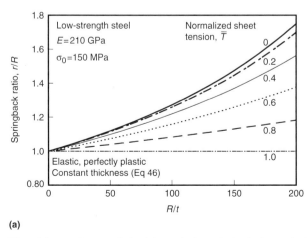

(a)

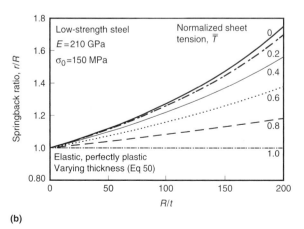

(b)

Fig. 10 Effect of sheet tension on springback for an elastic, perfectly plastic constitutive equation for low-strength steel. (a) Constant thickness. (b) Changing thickness. r, radius of curvature after springback; R, radius of primary bending curvature; t, sheet thickness; E, Young's modulus; σ_0, initial yield stress

one shown in Eq 55 have been used in the derivations already presented.

Figure 12 shows that the use of true or engineering strain (ε_t) has little effect on either the thin-shell or thick-shell solutions. Even at $R/t = 2$, the error is a few percent, and it is inconsequential for $R/t \geq 5$, where the overall approach applies.

The proper plastic strain can be found by subtracting the elastic strain, which depends on the current flow stress of the material:

$$\varepsilon_p = \varepsilon_t - \frac{\sigma_f'}{E'} = \varepsilon_t - \frac{1}{E'}\left(\sigma_0 + K\varepsilon_p^n\right)$$

$$= \varepsilon_t - \frac{\sigma_0}{E'} + \frac{K\varepsilon_p^n}{E'} \qquad \text{(Eq 56)}$$

Equation 56 cannot be rewritten to find ε_p explicitly, as required for substitution in the hardening law, but the result may be used nonetheless in evaluating the required integral numerically. Note that the third term on the right side of Eq 56 has been ignored throughout the previous derivations, thus enabling explicit evaluation of ε_p from ε_t. This has little effect for materials with typical hardening laws (Fig. 13).

For large R/t (small bending strain), the elastic core may make an appreciable contribution to the evaluation of the bending moment. (This was previously illustrated for the nonhardening case.) In this case, the integrals for $\overline{T}$ and M are carried out over only the part of the thickness subjected to plastic strain. The results for hardening can be added to the elastic, perfectly plastic result (e.g., Eq 12) to obtain the full solution, as was illustrated in the previous section for pure bending.

The result for the plastic part of these quantities, say $\overline{T}$ and ΔM, is the same as Eq 52 and 53, with z_0 replaced by z_1^* and z_2^*, where z_1^* and z_2^* are the location of the transitions from elastic to plastic behavior above and below the neutral axis. That is:

$$\overline{T} = -\frac{2z_0}{t} + \frac{K}{(n+1)\sigma_0}\left(\frac{t}{R}\right)^n$$

$$\times \left[\left(\frac{1}{2} - \frac{z_1^*}{t}\right)^{n+1} - \left(\frac{1}{2} + \frac{z_2^*}{t}\right)^{n+1}\right]$$

$$\qquad \text{(Eq 57)}$$

$$\frac{\Delta M}{w} = K\left(\frac{1}{R}\right)^n$$

$$\times \left[\frac{(0.5t - z_1^*)^{n+2} + (0.5t + z_2^*)^{n+2}}{n+2}\right.$$

$$\left. + \frac{z_1^*(0.5t - z_1^*)^{n+1} - z_2^*(0.5t + z_2^*)^{n+1}}{n+1}\right]$$

$$\qquad \text{(Eq 58)}$$

$$\frac{M}{w} = \frac{2}{3}E'R^2\left(\frac{\sigma_0'}{E'}\right)^3 + \frac{\sigma_0'}{4}$$

$$\times \left[t^2 - 2\left(z_1^{*2} + z_2^{*2}\right)\right] \qquad \text{(Eq 59)}$$

Care must be taken using Eq 57 and 58 that z_0, z_1^*, and z_2^* are not assigned nonphysical values outside of the sheet thickness. That is, the following rules apply to all of the calculations illustrated so far:

$$z_0 = \begin{cases} -\varepsilon_m R & -\dfrac{t}{2} \leq -\varepsilon_m R \leq \dfrac{t}{2} \\[2mm] \dfrac{t}{2} & -\varepsilon_m R > \dfrac{t}{2} \\[2mm] -\dfrac{t}{2} & -\varepsilon_m R < -\dfrac{t}{2} \end{cases} \qquad \text{(Eq 60)}$$

$$z_1^* = \begin{cases} z_0 + \dfrac{R\sigma_0'}{E'}z_0 & \dfrac{R\sigma_0'}{E'} \leq \dfrac{t}{2} \\[2mm] \dfrac{t}{2} & z_0 + \dfrac{R\sigma_0'}{E'} > \dfrac{t}{2} \end{cases} \qquad \text{(Eq 61)}$$

$$z_2^* = \begin{cases} z_0 - \dfrac{R\sigma_0'}{E'} & z_0 - \dfrac{R\sigma_0'}{E'} \geq -\dfrac{t}{2} \\[2mm] -\dfrac{t}{2} & z_0 - \dfrac{R\sigma_0'}{E'} < -\dfrac{t}{2} \end{cases} \qquad \text{(Eq 62)}$$

Evaluation of Eq 58 and 59 (Table 3), shows that for $0 < \overline{T} < 1$, the maximum error caused by neglecting the elastic core is 0.17% ($R/t = 25$) in sheet-forming range and grows to 4.6 and 13.8% for R/t of 100 and 200, respectively.

The foregoing results and discussions show that for the typical sheet-forming regions ($5 < R/t < 25$), the elastic response of the material may be safely ignored, along with more complicated treatments of the strain and thick shells. For large R/t, the elastic strains become significant, and for smaller R/t, the thick-shell approach and proper plastic-strain measures become significant.

Applied Analysis of Simple Forming Operations

For a typical industrial sheet-forming operation, the sheet workpiece is pressed between nearly rigid tools with draw-in constraints enforced, usually via draw beads. The general operation may be arbitrary in three dimensions, and conformance to the tools is by no means assured (Ref 58), thus making it impossible to know, a priori, the bend radius of the sheet. Many material elements undergo bending and unbending with superimposed tension, whereas the closed-form analyses usually assume a flat starting configuration in both in-plane directions. Determining sheet tension, which was shown in the last section to be critically important in springback, is complex, depending on friction, bending and unbending, and boundary constraints. All of these variables may change throughout the part and the forming stroke, over small distances and times.

For arbitrary, three-dimensional (3-D) forming operations, FE analysis (or a similar numerical method) is required throughout the forming operation to obtain a final, as-formed state. This configuration (with tool contact forces) may then be used as a basis for a general springback

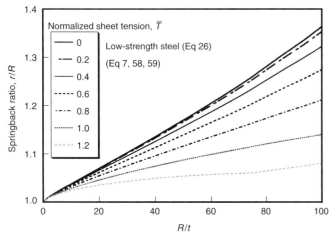

Fig. 11 Role of normalized sheet tension and bend radius on springback for elastoplastic, hardening behavior (Eq 58 and 59) of low-strength steel (Eq 26). See Eq 7, 58, and 59. r, radius of curvature after springback; R, radius of primary bending curvature; t, sheet thickness

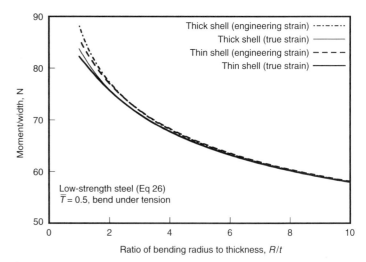

Fig. 12 Comparison of bending moment using thin and thick shells, with either engineering strain or true strain. $\overline{T}$, normalized sheet tension

analysis using the same, or a different, FE model. Such analysis is discussed subsequently; however, the application of two-dimensional (2-D) closed-form methods is possible and profitable for some classes of forming operations that are first mentioned.

In spite of the difficulties of applied springback analysis discussed previously, certain applied problems have sufficiently restrictive characteristics to provide a basis for closed-form or empirical analysis. Pure bending by dies in two dimensions, for example, may be analyzed using closed-form bending solutions if the workpiece is assumed to conform to the punch surface. Results have been presented for U-bending and V-bending (Fig. 14) using such analysis (Ref 32, 59–63) and empirical approaches (Ref 36, 64–66). A closely related application in sheet metal forming is flanging, for which analysis (Ref 67) and empirical approaches (Ref 68) have been presented.

Closely related operations involving significant tension are often called stretch-bend or draw-bend problems. These operations involve the bending and unbending of sheet as it is progressively drawn over a die. The typical application is often referred to as a top-hat section (Fig. 1c, 15a) and may be called channel forming, among other common names. For large draw-in, the principal springback typically occurs in the form of sidewall curl, which is the curling of the material that was drawn over the die radius (and which was flat while the workpiece was held in the dies during the operation itself). Various analyses based on the methods presented in the last section have appeared (Ref 23, 47–54, 57, 69, 70) and empirical methods have been applied (Ref 71, 72). Such analysis has been extended to consider the differing roles of postbending tension versus prebending tension (Ref 47), laminated materials (Ref 56), and nonuniform bending (Ref 55).

Much of the experimental work appearing in the literature for draw bending must be examined

critically because the tensile stress or load is often not carefully controlled or measured. In a few exceptional cases (Ref 53, 73, 74), direct control was imposed to obtain draw-bend results. For other work, experiments rely on indirect control of tension via friction, draw beads, or die clearances to establish the essentially unknown value of sheet tension. As shown in the last section, the tensile stress has a dominant effect on springback, particularly for values approaching the yield tension, thus leading to large uncertainties in measured results unless the tensile stress is known accurately.

A wide range of experimental data for a draw-bend problem from various sources appears as part of the NUMISHEET '93 *U-Channel Benchmark* (Ref 75). Geometry, material, lubrication, and forming parameters were fully specified by the conference organizers, and numerous laboratories were asked to carry out independent measurements and simulations. The results, shown in Fig. 15(b), illustrate the wide scatter that was obtained. In general, the experimental scatter was greater than or equal to the simulation scatter, illustrating the difficulty in carrying out such experiments with normal industrial forming machinery. It appears that the scatter of experimental results is typical for experiments employing indirect control of sheet tension. The sources of error for the FE simulations are considered in more detail

Table 3 Maximum error in moment caused by neglecting elastic core for $0 < \overline{T} < 1$, low-strength steel (Eq 26)

R/t	Maximum error in moment, %				
	$\overline{T}=0$	$\overline{T}=0.2$	$\overline{T}=0.5$	$\overline{T}=0.7$	$\overline{T}=1.0$
5	0.002	0.002	0.002	0.002	0.002
25	0.06	0.06	0.06	0.07	0.09
100	1.09	1.12	1.27	1.5	2.4
200	4.8	4.93	5.73	7.04	10.4

R/t, ratio of bending radius to thickness; $\overline{T}$, normalized sheet tension

subsequently, along with a summary of draw-bend results for which the sheet tension is carefully controlled.

Finite Element Analysis

It is only through a complete analysis of the forming operation that the critical variables in springback analysis may be obtained reliably, notably sheet tension, prespringback part shape, distribution of internal properties (such as yield strength), and external loading/internal stresses prior to springback. The geometric complexity of general bodies with curves in three dimensions (such as typical autobody parts, for example) requires discretized treatment. However, finite

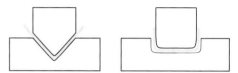

Fig. 14 Schematics of V-bend and U-bend forming operations

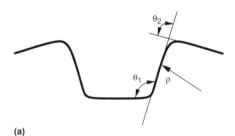

(a)

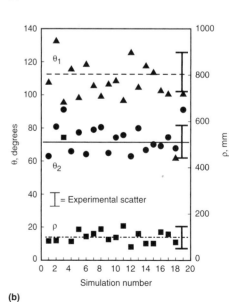

(b)

Fig. 15 U-channel forming and benchmark results. (a) Cross-sectional geometry, after springback. (b) Simulation results (points) and superimposed estimated experimental scatter (bars) for results reported by various laboratories. θ_1 and θ_2, angles characterizing springback in top-hat samples; ρ, radius of curvature of sidewall curl

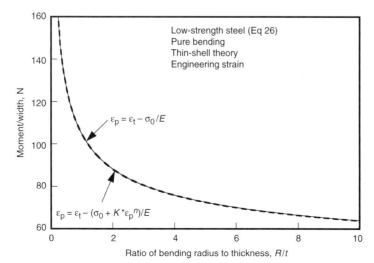

Fig. 13 Effect of strain approximation on springback calculation. See text for definition of variables.

Plot labels:
- Low-strength steel (Eq 26)
- Pure bending
- Thin-shell theory
- Engineering strain
- $\varepsilon_p = \varepsilon_t - \sigma_0/E$
- $\varepsilon_p = \varepsilon_t - (\sigma_0 + K^*\varepsilon_p{}^n)/E$
- y-axis: Moment/width, N
- x-axis: Ratio of bending radius to thickness, R/t

element analysis (FEA) offers several other advantages as well. Most FE programs readily accept complex laws of material behavior, including anisotropy, elastic-plastic behavior, rate sensitivity, complex hardening, and so on. The more sophisticated programs handle large strain and rotation properly, and arbitrary geometry is treated naturally.

The FEA of sheet forming is well established and now routine (see, for example, various benchmark tests on the subject) (Ref 1, 75–78). The FEA of springback, while appearing to be a simpler problem, requires higher accuracy of both the forming solution and throughout the springback simulation (Ref 39, 41, 79, 80). The choice of element (Ref 38), unloading procedure (Ref 39, 55, 81), and integration scheme (Ref 38, 39, 62, 80, 82, 83) must all faithfully reproduce the stresses and part configuration.

In view of the need for high precision of stress results, implicit forming simulation and implicit springback schemes seem the most likely to succeed (Ref 82). However, claims of success have appeared for nearly every possible combination of procedure: implicit/implicit (Ref 38, 39, 41, 84, 85), explicit/implicit (Ref 79, 82, 83, 86, 87), explicit/explicit (Ref 88, 89), and one-step approaches (Ref 90).

In the section "Draw-Bend Experience" in this article, these general observations are probed with tests and simulations corresponding to draw-in over a die radius in a press-forming operation. In order to understand those results, a brief introduction to the FE method is first presented.

A presentation of the FE method is beyond the ambit of this article. Nonetheless, an understanding of the basic method is helpful in understanding the particular constraints presented by springback analysis following forming analysis.

The following is based on a presentation of FEA particularly aimed at metal forming (Ref 91). Numerous books on the subject of general-purpose FEA have appeared. References 92 to 97 may be of interest for those seeking additional information. They are presented in approximate order of increasing difficulty.

The FE method applied to forming analysis consists of the following steps:

1. Establish the governing equations: equilibrium (or momentum for dynamic cases), elasticity and plasticity rules, and so on.
2. Discretize the spatially continuous structure by choosing a mesh and element type.
3. Convert the partial differential equations representing the continuum motion into sequential sets of linear equations representing nodal displacements.
4. Solve the sets of linear equations sequentially, step forward and repeat.

Items 1 and 2 of this list are of particular importance for springback analysis. For item 1, many choices of material model may be used,

but most forming simulations rely on two basic governing equations: either static equilibrium is imposed in a discrete sense (i.e., at nodes rather than continuous material points), or else a momentum equation in the form of $F = MA$ is satisfied for dynamic approaches.* For nearly all commercial codes used for metal forming, the static equilibrium solutions are obtained with implicit methods that solve for equilibrium at each time step by iteration, starting from a trial solution. Thus, such programs are often referred to as static implicit. Examples of static implicit include ABAQUS Standard (Ref 98) and ANSYS (Ref 99). Forming programs that solve a momentum equation typically use explicit methods that convert unbalanced forces at each time increment into accelerations but do not iterate to find an assured solution. Examples include ABAQUS Explicit (Ref 100) and LS-DYNA (Ref 101).

The choice of element refers to the number of nodes per element, the number of degrees of freedom at each node, and the relationship with the assumed interior configuration (among other things) that define the element type. The nodal displacements are the primary variables to be solved for. A fixed relationship between the displacements of points within the finite element to the displacements of the nodes is assumed. In this way, the continuous nature of the deformation within the element is related to a small number of variables. Of course, the distribution in the element may be quite different from the continuum solution, but this difference can often be progressively reduced by refining the mesh, that is, by choosing finer and finer elements. By comparing the solutions, an adequate mesh size can be determined.

The essence of the FE method, as opposed to other discrete treatments such as the finite difference method, lies in an equivalent work principle. As described previously, the continuous displacements within an element are represented by a small number of nodal displacements (and possibly other variables) for that element. Similarly, the work done by the deformation throughout the element is equated to the work done by the displacements of the nodes and thus are defined equivalent internal forces at these nodes.** The internal work is computed by integrating the stress-strain relation over the volume of the element. This frequently cannot be accomplished in closed form, so certain sampling points, or integration points, inside the element are chosen to simplify this integration. The number and location of integration points may be selected to provide the desired balance between efficiency and accuracy, or between locking and hourglassing, as mentioned subsequently.

For forming and springback analysis, the procedure consists of applying boundary conditions (i.e., the motion of a punch or die, the action of draw beads, frictional constraints, and so on), stopping at the end of the forming operation, replacing the various contact forces

by fixed external forces (without changing the shape of the workpiece), and then relaxing the external forces until they disappear. The last step (or steps) produces the springback shape.

Because the choices of program, element, and procedure usually apply to both the forming and springback steps, it is difficult to separate discussions of accuracy between the two stages. The deformation history established in the forming operation is used in the springback simulation via the final shape, loads, internal stresses, and material properties.

Two choices are of particular importance in forming and springback analysis: the type of solution algorithm/governing equation, and the type of element. These two aspects are discussed as follows.

Explicit and Implicit Programs. The first choice facing one wishing to do forming/springback analysis is the type of program. As noted previously, the two standard choices are a dynamic explicit program or a static implicit program, although several companies have both options available, sometimes even during a single simulation. (Also, there are other variations available, such as dynamic solutions solved implicitly.) Table 4 lists general advantages and disadvantages of the two methods.

Most applied sheet-forming analysis in industry currently uses dynamic explicit methods. The complicated die shapes and contact conditions that occur in complex industrial forming are more easily handled by the very small steps required by the dynamic explicit methods. The stress solutions are of little importance. Often, even simulations that are inaccurate in an absolute sense can be used by experienced die designers to guide sequential modifications leading to improved dies. On the other hand, if a certain set of tools cannot be simulated successfully, that is, with an implicit method that does not converge, the die improvement process is stymied completely.

As is shown in the next sections, springback simulation is much more sensitive to numerical procedure than forming analysis. The reason is simple: springback simulation relies on accurate knowledge of stresses throughout the part at the end of the forming operation. Conversely, forming analysis is primarily concerned with

---

*It should be noted that nearly all commercial forming operations may properly be considered static. That is, the inertial forces are orders of magnitude smaller than the deformation forces and thus may be safely ignored. The use of dynamic solutions to solve such quasi-static problems is for numerical convenience, with a corresponding loss of accuracy whenever the inertial forces are magnified (by mass scaling, for example). Thus, mass scaling must always be examined to quantify the errors introduced.

**Equilibrium in a weak sense is imposed by requiring that the sum of all such internal forces is zero at each node. Compatibility in a weak sense is automatically satisfied because the common nodes of adjacent elements have a single displacement. These are called weak forms because they do not ensure equilibrium or compatibility throughout the entire body, only at the nodes.

the distribution of strain within the shape of the part. The shape of the final part is largely determined by the shape of the dies because, near the end of the forming operation, contact occurs over a large fraction of the workpiece. Therefore, the oscillatory nature of the stresses obtained at the end of a dynamic explicit analysis may be unsuited for accurate springback analysis. Poor and uncertain results have been reported (Ref 82).

Developments are currently proceeding in attempts to artificially smooth or damp dynamic explicit forming solutions in the hope of providing a stable base for springback calculations. It is too early to be confident that these approaches will be successful. Certain isolated results can appear promising; however, as is shown later in this section, it is not unusual to obtain fortuitously accurate results in springback analysis. For this reason, great caution should be used in drawing conclusions from a small number of apparently accurate predictions.

The foregoing refers to the drawbacks of a dynamic explicit simulation of a forming operation prior to a springback analysis. The springback simulation itself is also much better suited to implicit methods because the operation is dominated by quasi-static elastic deformation that is computed very inefficiently by dynamic explicit methods. For this reason, implicit springback analysis is often favored even after explicit forming analysis.

It is for these two reasons that static implicit methods are better suited to forming analysis where accurate springback predictions are required. The obvious drawback is the uncertain convergence of current versions of such methods.

Choice of Element. For sheet-forming analysis, two principal kinds of elements are popular, although nearly limitless variations are found within each category. The major choices are solid elements and thin-shell elements.

The simplest to understand is the standard eight-node, trilinear solid element, sometimes called a brick element. (When an element is described as linear or quadratic, it refers to the polynomial order of the shape function, that is,

the mapping equation that relates the motion of the nodes to the motion of the continuous interior points in the element. In a linear element, the displacements of interior points vary linearly with their original positions. Trilinear refers to a shape function that is linear in all three dimensions.) The incompressibility constraint of plastic deformation is readily adapted with this element, and it is used for a variety of structural applications, including bulk forming. The major disadvantage is simple: because many such elements are required through the thickness of the sheet, particularly to accommodate springback analysis, the final number of degrees of freedom for applied problems is enormous and the resulting computation time far beyond today's (2005) computers. Such elements are also typically very stiff in bending, thus making them very poor for springback. One improvement is the use of quadratic or higher-order elements, which have better bending behavior, but which add even more degrees of freedom and further aggravate the overwhelming computational intensity.

Many variations of the standard brick element have been introduced with the goal of enabling use of coarse meshes without locking (nonphysically high stiffness in certain modes of deformation). Unfortunately, these numerical corrections often introduce the converse phenomenon of hourglassing (nonphysically low stiffness in certain modes of deformation). New solid elements can balance these problems well, but often at the cost of complexity (Ref 102, 103). Unfortunately, even with elements optimized for coarse meshes, the numbers still overwhelm today's (2005) computers for applied sheet-forming analysis, because very small features must be simulated.

By far, the predominant element used for sheet-forming analysis is the thin-shell element, or simple shell element. One can imagine generating such an element by starting with solid elements and shrinking one dimension to small thickness, a strategy that has been pursued (Ref 104, 105). Unfortunately, this procedure leads to locking, and special integration methods (reduced integration, selective reduced

integration, assumed strain methods, and so on) are then necessary to recover reasonable behavior.

Typical shell elements for sheet forming are based on some version of thin-shell theory, which itself makes assumptions about the strain and stress state throughout the body. They may be triangles or quadrilaterals, although quadrilaterals are more common. There is no real thickness to the elements, so clearances between die faces can be a problem in FE simulations where that aspect is important. Usually, the strain is assumed to vary linearly through the "thickness" of the shell for purposes of evaluating the stresses and work of deformation.* Likewise, through-thickness stresses are usually ignored. The usual procedure introduces two new degrees of freedom at each node corresponding to the slope or local rotation of the material plane at that location.

The advantages of shells for applied sheet forming and springback analysis are so persuasive in today's (2005) computing environment that solids are seldom used except for research or when certain conditions are present. These conditions include critical die clearances, two-sided contact, significant through-thickness stresses, and R/t ratios less than 5 to 6. Outside of these cases, shell elements are many times more efficient, and they capture the necessary phenomena in most sheet-forming operations, which are dominated by bending and stretching.

Draw-Bend Experience

The advantages and pitfalls of FEA of springback for many sheet-forming operations are revealed by the draw-bend test (Fig. 16), which closely represents the situation in channel forming (Fig. 14b, 15a) and many other press-forming operations. The advantage of the test is that sheet tension may be closely controlled.

The material in the test is drawn over a round tool under the action of a pulling displacement (and corresponding front force) and resisting force (back force). The workpiece may or may not conform completely to the tool surface when under load; it undergoes bending and then unbending under tension, then rebending under the final unloading when it is released from the fixtures. When released, the drawn length of the strip specimen adopts a final radius of curvature (r'). This is precisely analogous to the channel-forming operation, where the pulling displacement is provided by the punch displacement, the back force is provided by a draw bead or frictional resistance over a binder

Table 4 Advantages and disadvantages of static implicit and dynamic explicit finite element programs

Program type	Advantages	Disadvantages
Static implicit	• Known accuracy • Equilibrium satisfied • Smooth stress variation • Elastic solutions are possible • Unconditionally stable	• Solution not always assured • Complex contact difficult to enforce • Long computer processing times for complex contact
Dynamic explicit	• Solution always obtained • Simple contact • Short computer processing times with mass scaling	• Uncertain accuracy • Equilibrium not satisfied in general • Mass scaling introduces error in static problems • Oscillatory stress variation • Elastic solutions are difficult and slow • Conditionally stable

*In the limit of no thickness, and consequently no bending stiffness, a thin-shell element becomes a membrane element. These elements are efficient because they do not require additional degrees of freedom at the nodes, but they are unstable in bending, a common mode for many sheet-forming operations.

surface, and the final radius of curvature of the drawn section is referred to as sidewall curl. When the drawn distance is sufficiently large, the final springback changes are dominated by the sidewall curl (radius r') rather than the changes in the small region in contact with the tool at the end of the test.

Results of draw-bend tests and parallel FE simulations (Ref 40, 74) form the basis for the following observations.

Numerical Parameters. Finite element sensitivity studies (Ref 38, 39, 106) of the draw-bend test revealed that accurate springback prediction requires much tighter tolerances and closer attention to numerical parameters than does forming analysis. Furthermore, the tighter tolerances must be maintained throughout the forming operation; that is, it is not sufficient to do a coarse forming simulation followed by a precise springback simulation (Ref 82).

Using meshes, tolerances, and numerical parameters typical for forming analysis to analyze the draw-bend test gave nonphysical predictions, including, under some conditions, simulated springback opposite to the direction observed. Figure 17(a) (Ref 38) shows the initial simulations and the final ones (i.e., with appropriate choices of model parameters). A mesh size four times finer along the draw direction was required, combined with a number of integration points ten times larger than normal.

The FE sensitivity results based on the draw-bend test may be summarized as:

- The finite elements in contact with tooling should be limited in size to approximately 5 to 10° of turning angle. This is approximately 2 to 4 times the refinement typically recommended for simulation of forming operations.

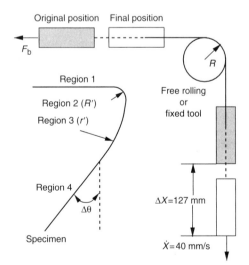

Fig. 16 Schematics of the draw-bend test and final configuration of the unloaded specimen. F_b, normalized back force; R, tool radius; R', radius of curvature of region in contact with tool, after unloading; r', radius of curvature in curl region, after springback; $\Delta\theta$, springback angle; ΔX, displacement, the distance between the original and final positions; $\dot{X}$, displacement rate

- The convergence tolerance and contact tolerance must typically be set tighter than for forming analysis. There are a variety of ways to define such measures, depending on the programs used, so again, the best policy is to refine the measure until the differences become insignificant.
- While most applied sheet-forming analyses use shell elements with three to seven integration points through the thickness, up to 51 integration points are required to assure simulated springback results within 1% of the "converged" solutions. (Converged solutions were obtained by using very large numbers of integration points, until no appreciable change in springback was observed.) More typically, 25 integration points were found to be sufficient for many simulations.

The last of these conclusions represents a dramatic divergence from current practice and remains surprisingly controversial, with researchers continuing to recommend using numbers of integration points ranging from five to nine (Ref 107–111). For this reason, non-FEA numerical studies (Ref 112, 113) were undertaken to explore the errors associated with numerical integration for finite numbers of integration points. Results from the FE sensitivity studies are presented first, then those from the non-FEA numerical studies. The two approaches serve to confirm the main conclusions.

Figure 17(b) (Ref 39) shows a typical result of the FE sensitivity analysis for tests of 6022-T4 aluminum with R/t of 10. In one case, for a normalized back force of 0.9 (relative to the force to yield the strip in tension), 35 integration points were required to meet the 1% tolerance requirement, while for a back force of 0.5, only 21 integration points were required. It is important to note the following points with respect to these results:

- The choice of R/t, back force, and other process and material parameters changes the number of required integration points. However, there is currently no good way to predict the exact number before a simulation is carried out, because many of these quantities are available only after solving the boundary-value problem. As shown subsequently, the effect of process variables on the required number of through-thickness integration points can be determined. Therefore, 25 to 51 integration points are recommended for general cases (if 1% accuracy of springback prediction is desired). As with all FEA, the best policy is to refine the parameters (number of integration points, in this case) to verify that no significant changes take place with continued refinement.
- It is possible, by chance, to obtain accurate results for a given forming problem with a small number of integration points. Note that the results in Fig. 17(b) cross the converged solution several times, with as few as three to five integration points. However, the result

is fortuitous and cannot be assured unless many more integration points are employed. Extrapolating from this result may explain why seemingly rough simulation techniques (such as dynamic explicit solutions, which show oscillatory stress behavior) can produce, with carefully selected (postsimulation) parameters (or sufficient luck!), accurate springback results. The best way to verify the robustness and predictivity of such solutions is by changes of the critical parameters to test whether the solution is a stable one.

- Many 3-D forming operations are much stiffer than the 2-D draw-bend geometry because of the final form of the part. It is unknown what effect this has on the need for numerical accuracy. However, it is clear that for smaller springback, a larger percentage error may be acceptable in terms of the overall geometry changes. Thus, it may not be necessary to demand a 1% springback accuracy; instead, perhaps 10% is adequate. In such cases, the number of integration points may be reduced.

The FEA sensitivity results for the draw-bend test can be understood with the aid of a related but simpler problem: the springback of a beam subjected to an applied R/t and tension force. That is, the state after forming is known analytically, and only the springback is computed, both analytically and numerically. The springback is proportional to the applied moment (Eq 4) so that any fractional numerical error that occurs in evaluating the moment produces a corresponding fractional error in the computed springback. The analytical moment (and springback) may be computed exactly for this problem, and thus, the error induced by the numerical integration scheme during the springback simulation alone can be evaluated separately from other effects. This error is less than the combined errors that can occur during both the forming and springback stages of a simulation.

The following results are extracted from more detailed publications that should be consulted by those seeking more complete information (Ref 107, 108). Equations 57 to 62 represent the analytical tension and moment calculations. Material models were adopted representing a low-strength steel and high-strength aluminum (Eqs 26, 27). Numerical integrations were carried out with three common choices of integration rules: trapezoidal (Ref 114), Simpson (Ref 115), or Gauss (Ref 116). Figure 18 presents the fractional moment error for the low-strength steel bent to $R/t = 5$ using a trapezoidal integration rule with 51 integration points through the thickness (N_{IP}) for various normalized sheet tension forces. (The normalized sheet tension force is defined by dividing the applied tension force by the force to yield the sheet, which is equal to the yield stress times width times thickness of the sheet.) The range of sheet tensions considered is from zero (pure bending case) to $\overline{T}_{max}$, where $\overline{T}_{max}$ is the normalized sheet tension force where the entire thickness of the sheet is plastically deforming; that is, the elastic core has moved to the edge of the sheet. For the low-strength steel at $R/t = 5$, $\overline{T}_{max}$ is 2.06.

Two conclusions can be drawn from the set of computations shown in Fig. 18. First, the errors fall into a range depending on the exact choice of parameters. For some particular cases, the error may be zero, while at nearby adjacent states the error may be maximum. (For Fig. 18, this effect can be observed by small changes of the normalized tension force.) The behavior is oscillatory (as most easily seen in Fig. 18 at smaller tension forces), depending on where the integration points fall relative to the actual through-thickness stress distribution. The smooth lines drawn on Fig. 18 represent the assured error limit, that is, the limit of error in the vicinity of a

given set of conditions. For any set of conditions and numerical integration choices, the numerical integration error will always be less than or equal to this assured error limit, but the actual error in the nearby vicinity of those conditions may be anywhere from zero to this value. In this manner, the bounding error limit can be defined.

The second conclusion illustrated by Fig. 18 is that the assured fractional moment error (or, more succinctly, limiting error) increases for increasing sheet tension. This can be understood in terms of the decreasing moment with increasing sheet tension. Thus, a given absolute error in reproducing the analytical stress distribution by numerical integration technique represents a larger fractional error of moment (or springback). While assured fractional moment error is increased for larger sheet tensions, note that the

oscillatory behavior still exists, so it is possible in the vicinity of any sheet tension to obtain a particular result with nearly zero error. Such an isolated result is fortuitous and should not be relied on to estimate future performance.

A comparison of the three tested integration methods (Fig. 19) shows that no single integration method works best throughout the range of parameters. In view of the oscillatory nature of the error, for any given set of conditions, any one of these may outperform the other two. In Fig. 19, the unsigned assured error limit is presented. This error limit is smaller at low back forces for trapezoidal integration, while Gauss integration is better at higher normalized tension forces. This may be simply understood by noting that the stress distribution is smoothest in this region, most like a polynomial, and thus

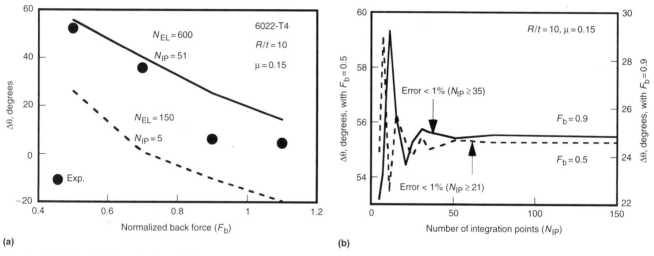

Fig. 17 Sensitivity of simulated draw-bend springback to mesh size (N_{EL}) and number of through-thickness integration points (N_{IP}). (a) Nonphysical springback predictions obtained using typical sheet-forming simulation parameters. (b) Accuracy of selected springback solutions depending on the number of through-thickness integration points. $\Delta\theta$, springback angle; R/t, ratio of bending radius to thickness; μ, friction coefficient

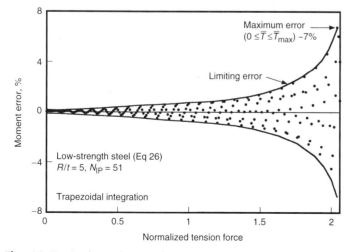

Fig. 18 Fractional error of computed moment (or springback) using trapezoidal integration and 51 integration points (N_{IP}) through the thickness. The limiting error (i.e., the assured maximum error in the vicinity of a set of conditions) is shown as smooth curves, and the maximum error occurs for a normalized tension force of 2.06 ($\overline{T}_{max}$). R/t, ratio of bending radius to thickness

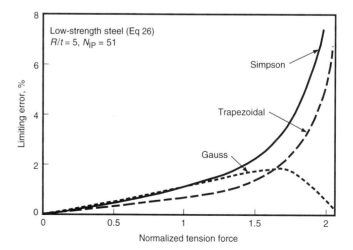

Fig. 19 Comparison of the variation of limiting error (unsigned) with normalized tension for three common integration schemes using 51 integration points (N_{IP}) through the thickness. R/t, ratio of bending radius to thickness

is reproduced better by Gauss integration in general. Because the largest possible fractional errors occur at larger tension forces, the Gauss integration scheme is a good choice (but by no means the best for all cases).

The role of R/t in affecting integration error is illustrated in Fig. 20. A maximum error is defined, as shown in Fig. 18 and as plotted in Fig. 20. That is, this quantity represents the assured fractional error limit for a range of normalized tension forces between 0 (pure bending) and $\overline{T}_{max}$ (where the entire thickness yields plastically in tension). This maximum error is of interest because for most forming operations, the value of the tension force is unknown except after the forming simulation is completed. Furthermore, it varies with time and location in the workpiece, as does R/t. Therefore, the maximum error must be considered as the numerical tolerance when performing a forming/springback simulation without *a priori* knowledge of the sheet tension. The maximum error is highest for small R/t values and lowest for larger R/t. Note the different scales on Fig. 20(a) and (b), illustrating the larger fractional errors for the low-strength steel.

A summary of many results such as those in Fig. 18 to 20 is presented in Fig. 21 and Table 5. Both may be used to determine the number of integration points needed to assure a limiting fractional error based only on numerical integration errors. (Of course, other sources of error, for example, incurred during the forming simulation or by other numerical aspects, can contribute to larger overall errors.) Note that Gauss integration is usually more efficient because, in the regime where the fractional springback errors are largest, it reproduces the analytical result best. The required number of integration points for a specified accuracy varies widely. For an assured 1% error tolerance, Gauss integration requires between 17 and 68 integration points through the thickness (depending on R/t and

material properties), while Simpson integration requires between 35 and 139 integration points through the thickness. For 10% tolerance, Gauss integration requires 5 to 16, depending on R/t and material properties, while Simpson requires 9 to 41.

Choice of Element, R/t, Anticlastic Curvature. Shell elements are preferred for springback simulation of sheet metal forming, because they can capture the bending behavior accurately while being computationally efficient as compared with solid elements, as long at the material thickness is small relative to the radius of bending. Membrane elements exclude bending effects and thus miss the major part of springback, while solid elements are very time-consuming for use with sheets because numerous layers must be used through the thickness (for the same reasons that many integration points are required for shells).

Draw-bend tests and simulations for drawing-quality special killed steel (Fig. 22a) showed that shell elements are accurate for R/t ratios as small as approximately 5 or 6, while solid elements are needed for smaller values, at much increased cost. Solid elements capture the through-thickness stresses that become significant for small R/t, but the typical brick elements with linear shape functions are very stiff in bending and provide poor results (Fig. 22b). For this reason, higher-order solid elements are preferred. For larger R/t ratios, shell elements normally provide better accuracy at much lower computation time.

One surprising result from the draw-bend simulations is that 3-D elements are required for the nominally 2-D problem (Fig. 22b). The answer lies in anticlastic curvature, discussed more fully in the section "Approximations in Classical Bending Theory" in this article. This is the secondary curvature that develops orthogonal to bending because of differential lateral contraction at the inner and outer

surfaces. As shown in Fig. 23, 3-D shell elements and 20-node quadratic solid elements capture the anticlastic curvature well, whereas 2-D elements and linear solid elements do not. Poor treatment of anticlastic curvature is the principal source of error in springback prediction for the draw-bend tests with back forces near the yield force. For arbitrary 3-D parts, the analog of anticlastic curvature causes distortion or wrinkling out of the plane of bending, as can be seen in the dimpling of the S-rail (Fig. 1a).

Unloading Scheme. Another surprising result from the draw-bend simulations is the importance, under some conditions, of plastic deformation during unloading. A comparison of 2-D simulation results is shown in Fig. 24, where the unloading was carried out purely elastically and elastic-plastically. Significant differences of the springback angle and residual stress are evident. However, it should be noted that in spite of the small plastic contribution to unloading behavior under some circumstances, the choice of path taken during unloading seems to have no significant effect on the final configuration (Ref 39). That is, the various unloading schemes and sequences for removing the tool constraints seem to give nearly identical results. It should also be noted that for many springback problems, particularly those with fairly small tension forces relative to yielding, the unloading will be purely elastic.

Plastic Constitutive Equation. At least two aspects of the plasticity law are important for springback prediction: plastic anisotropy (yield stress and strain ratios) and strain hardening. In particular, strain hardening must be suitable for a path reversal in draw-bend or channel-type forming, as is encountered when a material element undergoes sequential bending and unbending with superimposed tension (Ref 117). The yield surface anisotropy affects not only

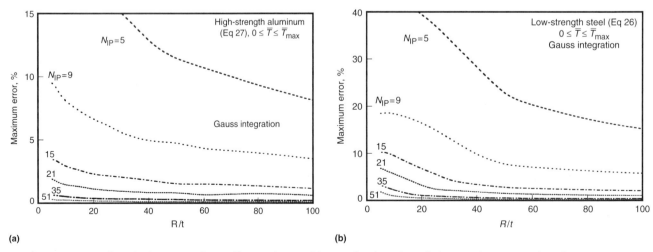

Fig. 20 The maximum error (that is, for sheet tensions from 0 to $\overline{T}_{max}$) as a function of the ratio of bending radius to thickness (R/t) for various numbers of integration points (N_{IP}). (a) For material properties corresponding to a typical high-strength aluminum (Eq 27, $E = 70$ GPa, where E is Young's modulus). (b) For properties corresponding to typical low-strength steel (Eq 26, $E = 210$ GPa)

the loaded bending moment but also the anticlastic curvature via the lateral strain ratios (related to the plastic anisotropy parameter, r, in various directions).

Table 6 compares the standard errors of fit for the simulated draw-bend springback angles for 6022-T4 aluminum alloy with $R/t = 10$ and normalized back forces ranging from 0.5 to 1.05.

The results illustrate that choosing an anisotropic yield function gives no guarantee of improved results. The Hill 48 quadratic yield function (Ref 23) gives significantly poorer fit than the von Mises isotropic yield function. The reason for this result is shown in Fig. 25, where the variation of yield stress and strain ratio with direction is poorly matched by the Hill 48 yield function and well represented by the Barlat YLD 96 (Ref 118) yield function. These results are presented and discussed in more detail in the literature (Ref 119), including excellent final prediction of anticlastic curvature. There are likely compensating effects of the yield stress variance and strain-ratio variance for the von Mises simulations that are not easily deconvoluted in predicting the final springback angle.

For all choices of yield function, taking into account the Bauschinger effect improves the prediction, as would be expected based on the reverse straining that takes place for most material elements throughout the draw-bend process (Ref 119).

Summary. The relative errors induced by the various factors discussed previously are illustrated in Fig. 26. Using normal forming simulation parameters (mesh size, integration points) and standard plastic laws (von Mises, isotropic hardening) for 2-D springback simulation leads to very poor results, including springback of the wrong sign (Fig. 17a, for example). The overall standard error of fit to the measured values is 26°. The 2-D simulations using adequately selected numerical parameters reduce the error to 19° degrees (plane strain) and 11° (plane stress). Use of 3-D shell elements for simulation (i.e., taking anticlastic curvature into account) reduces the error to 5.7°. Incorporation of the proper yield function form, Barlat YLD 96 (Ref 118), and a treatment of the Bauschinger effect (Geng-Wagoner hardening, Ref 119) reduce the final error to 1.2°, approximately the same magnitude as the experimental scatter of the measurements.

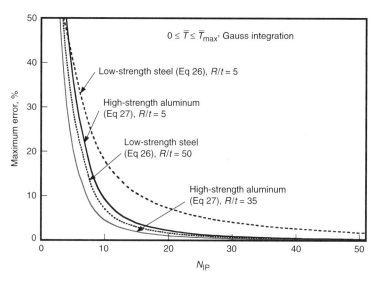

Fig. 21 Summary of the relationship between the maximum fractional error (for normalized sheet tension between 0 and $\overline{T}_{max}$) in moment (or springback) as a function of the number of integration points (N_{IP}) used with Gauss quadrature for several typical process and material parameters. R/t, ratio of bending radius to sheet thickness

Table 5 Number of through-thickness integration points (N_{IP}) required for a specified springback accuracy. Numbers without parentheses refer to Gauss integration; numbers in parentheses refer to Simpson integration

Maximum error	1%	5%	10%	50%
Low-strength steel ($0 \leq \overline{T} \leq \overline{T}_{max}$)				
$R/t = 5$	68 (139)	26 (69)	16 (41)	4 (9)
$R/t = 20$	38 (91)	18 (37)	13 (29)	4 (9)
$R/t = 100$	22 (57)	10 (23)	6 (15)	3 (7)
High-strength aluminum ($0 \leq \overline{T} \leq \overline{T}_{max}$)				
$R/t = 5$	30 (79)	13 (33)	9 (21)	4 (7)
$R/t = 20$	22 (55)	11 (25)	7 (15)	4 (7)
$R/t = 100$	17 (35)	8 (15)	5 (9)	3 (5)

$\overline{T}$, normalized sheet tension; R/t, ratio of bending radius to sheet thickness

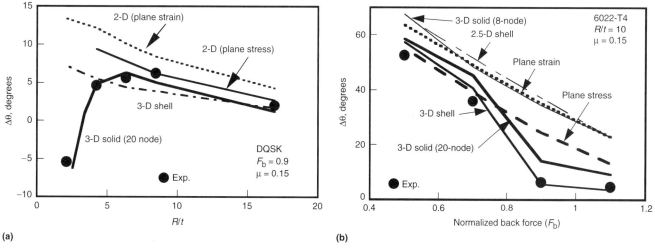

(a)   **(b)**

Fig. 22 The role of finite element type on draw-bend springback prediction. (a) Results for various R/t (bend radius/sheet thickness). (b) Results for various F_b (normalized back force). $\Delta\theta$, springback angle; DQSK, drawing quality special killed; μ, friction coefficient

Therefore, while it is essential to perform springback simulations carefully and to properly take into account the material behavior, it is possible to predict springback accurately for general cases. The biggest challenge for large-scale industrial applications lies with the convergence of implicit solutions during forming and the computational intensity of such simulations.

Springback Control and Compensation

The prediction of springback has been discussed in some detail at various levels of complexity in the foregoing sections. However, practical mitigation of springback in industry relies on two principal strategies: control (reduction) of springback via forming operation changes, and compensation for springback via changes of die shape. These two approaches are discussed briefly, with references to original works provided.

Control of Springback. As illustrated in Fig. 10, application of sheet tension, particularly near the tension to yield of the sheet, drastically reduces springback by reducing the stress gradient through the thickness and hence the bending moment. Most industrial schemes for reducing springback rely on this principle. However, increasing sheet tension moves a forming operation closer to failure by splitting (Ref 47), such that many optimized forming operations walk a fine line between splitting and excessive springback.

Most springback control methods focus on increasing the sheet tension while mitigating the negative effect on formability. The tension can be applied during the drawing part of the forming operation or subsequently (Ref 48, 71). By the use of a variable blank-holder force throughout the punch stroke, the sheet tension can be varied arbitrarily (Ref 120–122). Alternatively, special reverse-bending tooling may be devised either for pure bending (Ref 123, 124) or in the context of a more general channel forming (Ref 73, 125).

More general empirical approaches based on observations of numerous forming operations (Ref 126–129) involve a range of options including altering R/t, die clearances, punch bottoming, coining, and off-axis/in-plane compression.

Compensation of Springback. Instead of trying to reduce springback, which invokes penalties in formability, an alternative approach is to design dies that produce the desired final part shape after springback. (Usually, the die face for a sheet-formed part is very close to the desired part shape; therefore, springback moves the final part configuration away from the desired one.)

If springback prediction is considered a forward analysis, then a backward analysis is needed to use such results to modify a die design in order to achieve a given final part shape. For simple bending operations involving constant radii of curvature, dies can be designed to account for springback using handbook tables or closed-form solutions, which can be inverted to specify tool radii for desired final-part radii. For new or unusual materials, varying radii of curvature (i.e., arbitrary curves), or arbitrary or

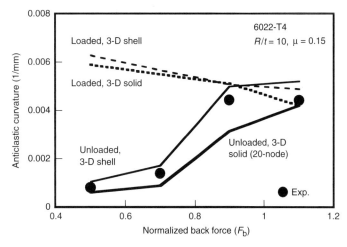

Fig. 23 Measured and simulated anticlastic curvature in the loaded and unloaded draw-bend specimens. R/t, ratio of bending radius to sheet thickness; μ, friction coefficient

Table 6 Standard errors of fit of simulated springback angle compared with measured ones for 6022-T4, $R/t=10$, $F_b=0.5$ to 1.05

	Standard error of fit, degrees, for indicated yield function		
Hardening law	von Mises	Hill 1948 (Ref 23)	Barlat YLD 96 (Ref 118)
Isotropic hardening	5.7	11.1	2.0
Geng-Wagoner hardening (Ref 95)	2.7	8.7	1.2

R/t, ratio of bending radius to sheet thickness; F_b, back force

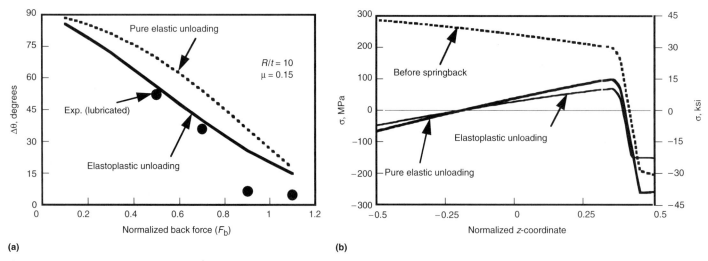

(a)   **(b)**

Fig. 24 Simulated role of plasticity in springback for a draw-bend test. (a) Difference of springback angle ($\Delta\theta$) for pure-elastic and elastoplastic springback simulations. (b) Differences in through-thickness stress distribution following pure-elastic and elastoplastic springback. R/t, ratio of bending radius to sheet thickness; μ, friction coefficient; σ, stress

3-D shapes (with compound curvature), springback compensation has traditionally been carried out by simple trial and error or some variation thereof (Ref 130–132). Unfortunately, this procedure often produces unsatisfactory results and depends intimately on the skill and experience of the user. The process can take many months, thus extending critical tooling lead times. The trial-and-error method can be applied using FE forward analysis trials instead of experiments, but the backward design steps are equally inefficient and may take a similar amount of time.

Schemes for guiding the forward and backward analyses may be found in the literature based on various optimization strategies (Ref 89, 133, 134). These methods typically involve a gradient calculation and sensitivity analyses. Considerable complexity in formulation and implementation is involved, and usually special programming within special-purpose FE programs is required.

A promising approach integrating forward (FE) and backward analyses in an iterative scheme was reported by Karafillis and Boyce

(Ref 135–137). This method, denoted the springforward method, may in principle be used with any FE program. As discussed subsequently, however, its application suffers from lack of convergence (Ref 138, 139) unless the forming operation is symmetric or has very limited geometric change during springback (Ref 138).

Figure 27 outlines the steps of the springforward method. A flat sheet is first deformed into the target shape, and the external forces are recorded. At step 3, the target shape is elastically

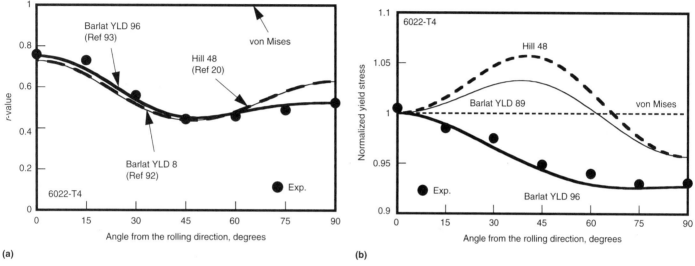

Fig. 25 Comparison of plasticity predictions for various in-plane sheet directions for several yield functions fit to a 6022-T4 aluminum alloy sheet. (a) Plastic anisotropy parameter (*r*-value). (b) Yield stresses normalized to yield stress in the rolling direction (0°)

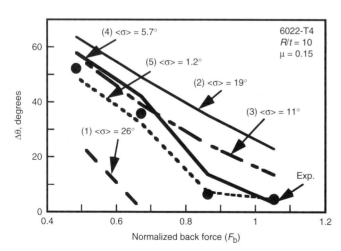

Fig. 26 The role of numerical procedures and constitutive modeling in springback accuracy as represented by standard error of fit, <σ>, to measurement. Springback angles (Δθ): (1) Plane stress, 5 integration points (N_{IP}), 600 elements along length (N_{EL}), von Mises yield, and isotropic hardening. (2) Plane strain, $N_{IP} = 51$, $N_{EL} = 600$, von Mises yield, and isotropic hardening. (3) Plane stress, $N_{IP} = 51$, $N_{EL} = 600$, von Mises yield, and isotropic hardening. (4) 3-D shell, $N_{IP} = 51$, $N_{EL} = 600$, von Mises yield, and isotropic hardening. (5) 3-D shell, $N_{IP} = 51$, $N_{EL} = 600$, Barlat 96 yield, and Geng-Wagoner anisotropic hardening. *R/t*, ratio of bending radius to sheet thickness; μ, friction coefficient

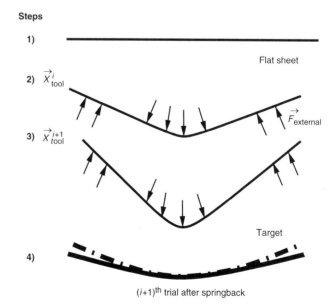

Fig. 27 Schematic representation of the steps undertaken in the springforward method of springback compensation. Step 1: flat sheet before deformation; Step 2: form to the tool shape and record external force field; Step 3: apply the recorded force field to the previous tooling shape and obtain the new tooling shape; Step 4: evaluate the $(i + 1)^{th}$ tooling shape by comparing the part shape, after forming and springback, with the target shape. x^i_{tool}, forming tool position at i^{th} iteration; *F*, force. Source: Ref 135–137

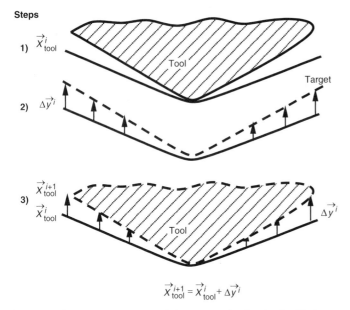

Steps

1) $\vec{X}^i_{tool}$

Tool

Target

2) $\Delta\vec{y}^i$

3) $\vec{X}^{i+1}_{tool}$ $\vec{X}^i_{tool}$

Tool

$\Delta\vec{y}^i$

$$\vec{X}^{i+1}_{tool} = \vec{X}^i_{tool} + \Delta\vec{y}^i$$

Fig. 28 Schematic representation of the steps undertaken in the Displacement Adjustment method of springback compensation. Step 1: form the part to the ith tooling shape; Step 2: compare the part shape after springback with the target shape; Step 3: generate the $(i+1)^{th}$ tooling shape by adding the displacement error field to the ith tooling. x^i_{tool}, forming tool position at i^{th} iteration; Δy^i, the ith displacement correction. Source: Ref 138

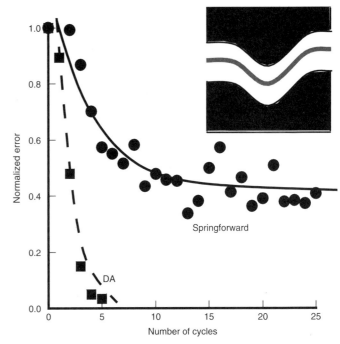

Fig. 29 Convergence and accuracy comparison of springforward and displacement adjustment (DA) springback compensation

loaded by the recorded external forces, and the next trial die shape is obtained (the same shape as the deformed blank at the end of this step). This is the springforward step. The accuracy of the trial die shape is next checked by doing a forming and springback simulation. If the resulting springback shape is not the same as the target, another cycle will be carried out from step 1. Now, a flat sheet is deformed to the trial die shape just obtained, instead of being deformed to the target shape as in the first cycle. External forces are recorded and applied to the target shape. A new trial die shape will be obtained. The new trial die shape will be checked again at step 4, and iterations will continue until the target part shape is attained within a specified tolerance. Variations of this approach have been presented making use of internal forces instead of contact forces (Ref 137).

An alternate iterative design method that avoids many of the limitations of springforward while maintaining its generality and ease of implementation may be designated the displacement adjustment method (Ref 138). Instead of a springforward step using simulated contact forces, the simulated forming and springback displacements are used to predict the next die design iteration. A similar approach has been used in one-step simulation versions (Ref 140) and via experimental iteration (Ref 67, 141). The displacement adjustment method appears to offer several advantages over the springforward method, including excellent convergence rate, ease of implementation, and considerable generality. However, it has only been tested for rather simple 2-D, bending-dominated problems as of this writing.

The displacement adjustment method is outlined in Fig. 28. First, a flat sheet of metal is deformed to a trial die shape (for the first cycle, the trial die shape is the target shape). After springback, the springback shape is compared with its target. The shape error is defined as $\Delta\vec{y}^i$, which is the vector difference of coordinates of a FE node in the springback shape and in the target shape, at the ith iteration. At step 3, the $\Delta\vec{y}^i$ is added to the current die shape nodal positions, obtaining a new tooling shape of $\vec{X}^{i+1}_{tool}$. For the next cycle, a flat sheet is deformed to this new tooling shape. If the springback shape is not within a specified tolerance of the target (checked at step 2), another iteration will be conducted.

Comparison of springforward and displacement adjustment methods for a simple bending-dominated forming operation have been presented (Ref 138). For an arbitrary, nonsymmetric, nonconstant-radius part shape (Ref 138), both methods were applied, with iterative results shown in Fig. 29. The normalized error for the nth iteration cycle in Fig. 29 is defined as:

$$\frac{rms\ error^{(n)}}{rms\ error^{(1)}} \qquad (Eq\ 63)$$

where:

$$rms\ error = \left[\frac{\sum \Delta\vec{y}_k^2}{N}\right]^{1/2} \qquad (Eq\ 64)$$

where rms is root mean square, K is a counting variable that progresses from 1 to N, and N is the total number of nodes of displacement.

Analysis of simpler, symmetric operations showed that the obstacle to applying the springforward method lies with the unknown constraints that must be enforced during the springforward step to restrict rigid body motion. Unless such conditions are applied at exactly the right location, which usually is unknown for a general problem, the springforward forces distort the workpiece and lead to nonconvergence of the technique.

ACKNOWLEDGMENTS

The authors would like to thank the funding support of the National Science Foundation (DMR 0139045) and the Center for Advanced Materials and Manufacturing of Automotive Components at The Ohio State University. Computer simulations were conducted with the help of the Ohio Supercomputer Center (PAS080). Special thanks to Christine Putnam, who helped prepare the manuscript.

REFERENCES

1. J.K. Lee, G.L. Kinzel, and R.H. Wagoner, Ed., *Proceedings of NUMISHEET '96,* The Ohio State University (Columbus, OH), 1996
2. R. Böklen, *Z. Metallkd.,* Vol 44, 1953, p 382–386
3. C.A. Queener and R.J. DeAngelis, *Trans. ASME,* Vol 61, 1968, p 757–768
4. Z. Marciniak and J.L. Duncan, *Mechanics of Sheet Metal Forming,* 1st ed., Edward Arnold, London, England, 1992

5. Z. Marciniak, J.L. Duncan, and S.J. Hu, *Mechanics of Sheet Metal Forming,* 2nd ed., Butterworth-Heinemann, 2002

6. W.F. Hosford and R.M. Caddell, *Metal Forming: Mechanics and Metallurgy,* 2nd ed., Prentice Hall, 1993

7. K.C. Chan and S.H. Wang, *J. Mater. Process. Technol.,* Vol 91, 1999, p 111–115

8. W. Schroeder, *J. Appl. Mech. (Trans. ASME),* Vol 65, 1943, p 817–827

9. T.X. Yu and L.C. Zhang, *Plastic Bending, Theory and Applications,* World Scientific, 1996

10. W.C. Young, Table I, Case 20, *Roark's Formulas for Stress and Strain,* 6th ed., McGraw-Hill, 1989, p 68

11. B.W. Shaffer and E.E. Ungar, *J. Appl. Mech. (Trans. ASME),* Vol 82, 1960, p 34–40

12. S.H. Crandall and N.C. Dahl, *Introduction to the Mechanics of Solids,* McGraw-Hill, 1961, p 327–336

13. J.D. Lubahn and G. Sachs, *Trans. ASME,* Vol 72, 1950, p 201–208

14. F.J. Gardiner, *J. Appl. Mech. (Trans. ASME),* Vol 79, 1957, p 1–9

15. J.M. Alexander, An Analysis of the Plastic Bending of Wide Plate, and the Effect of Stretching on Transverse Residual Stress, *Proc. Inst. Mech. Eng.,* Vol 173 (No. 1), 1959, p 73–96

16. G. Sachs, *Principles and Methods of Sheet-Metal Fabricating,* Reinhold, 1951

17. J.H. Hollomon, *A.I.M.E. Trans.,* Vol 162, 1945, p 268

18. D.M. Woo and J. Marshall, *The Engineer,* Vol 208, 1959, p 135–136

19. Z.T. Zhang and S.J. Hu, "Mathematical Bending in Plane Strain Bending," Paper 970439, Society of Automotive Engineers, 1997

20. User Manual, ABAQUS Standard, Version 6.2, ABAQUS, Inc., Pawtucket, RI, 2002

21. A.K. Sachdev and R.H. Wagoner, Uniaxial Strain Hardening at Large Strain in Several Sheet Steels, *J. Appl. Metalwork.,* Vol 3 (No. 1), 1983, p 32

22. *Properties of Some Metals and Alloys,* 3rd ed., The International Nickel Company, Inc., 1968, p 14

23. R. Hill, Theory of Yielding and Plastic Flow of Anisotropic Metals, *Proc. R. Soc. (London),* 1948, p 193–281

24. Z.T. Zhang and D. Lee, "Effect of Process Variables and Material Properties on the Springback Behavior of 2-D Draw Bending Parts," Paper 950692, Society of Automotive Engineers, 1995

25. E.A. Popov, *Russ. Eng. J.,* Vol 43 (No. 10), 1963, p 39–40

26. H. Lamb, *Philos. Mag.,* Vol 31, 1891, p 182–195

27. H.D. Conway and W.E. Nickols, *Exp. Mech.,* Vol 5, 1965, p 115–119

28. D.G. Ashwell, *J. R. Aero. Soc.,* Vol 54, 1950, p 708–715

29. D. Horrocks and W. Johnson, *Int. J. Mech. Sci.,* Vol 9, 1967, p 835–861

30. G.F.C. Searle, *Experimental Elasticity: A Manual for the Laboratory,* Cambridge University Press, Cambridge, England, 1908

31. J. Case, *Strength of Materials,* 3rd ed., Arnold, 1938, p 645–654

32. C.T. Wang, G. Kinzel, and T. Altan, *J. Mater. Process. Technol.,* Vol 39, 1993, p 279–304

33. R. Hill, *The Mathematical Theory of Plasticity,* Oxford University Press, London, England, 1950

34. R.H. Wagoner, *Metall. Trans. A,* Vol 11, 1980, p 165–175

35. R.H. Wagoner, *Metall. Trans. A,* Vol 12, 1981, p 2142–2145

36. T.X. Yu and W. Johnson, *J. Mech. Work. Technol.,* Vol 6, 1982, p 5–21

37. K.P. Li and R.H. Wagoner, Simulation of Springback, *Simulation of Materials Processing, Proceedings of NUMIFORM '98* (Enschede, Netherlands), 1998, p 21–31

38. K.P. Li, L.M. Geng, and R.H. Wagoner, Simulation of Springback: Choice of Element, *Advanced Technology of Plasticity, Proceedings of the Sixth ICTP* (Nuremberg, Germany), Vol III, 1999a, p 2091–2098

39. K.P. Li, L.M. Geng, and R.H. Wagoner, Simulation of Springback with the Draw/Bend Test, *Proceedings of IPMM '99* (Vancouver, B.C., Canada), 1999b, p 1

40. K.P. Li, W.P. Carden, and R.H. Wagoner, *Int. J. Mech. Sci.,* Vol 44, 2002, p 103–122

41. R.H. Wagoner, W.D. Carden, W.P. Carden, and D.K. Matlock, Springback after Drawing and Bending of Metal Sheets, *Proceedings, IPMM '97—Intelligent Processing and Manufacturing of Materials,* University of Wollongong, Vol 1, 1997, p 1–10

42. J.F. Wang, R.H. Wagoner, and D.K. Matlock, Creep following Springback, *Proceedings, Tenth Conference of Plasticity* (Quebec City, Canada), 2003, p 211–213

43. R.M. Cleveland and A.K. Ghosh, *Int. J. Plast.,* Vol 28, 2002, p 769–785

44. F. Morestin and M. Boivin, *Nucl. Eng. Des.,* Vol 162, 1996, p 107–116

45. H.W. Swift, *Engineering,* Oct 1948, p 333–359

46. M.L. Wenner, Report MA-233, General Motors Research Laboratories, May 1982

47. M.L. Wenner, *J. Appl. Metalwork.,* Vol 2 (No. 4), 1983, p 277–287

48. M.K. Mickalich and M.L. Wenner, Report GMR-6108, General Motors Research Laboratories, Feb 1988

49. J.L. Duncan and J.E. Bird, Approximate calculations for Draw Die Forming and Their Application to Aluminium Alloy Sheet, *Proceedings of the Tenth Biennial Cong. IDDRG,* (Warwick, England), Vol 1, The International Deep Drawing Research Group, April 1978, p 45–52

50. J.L. Duncan and J.E. Bird, *Sheet Met. Ind.,* Vol 55 (No. 9), Sept 1978, p 1015–1023

51. T. Kuwabara, S. Takahashi, K. Akiyama, and Y. Miyashita, "2-D Springback Analysis for Stretch-Bending Processes Based on Total Strain Theory," Paper 950691, Society of Automotive Engineers, 1995, p 1–10

52. T. Kuwabara, S. Takahashi, and K. Ito, Springback Analysis of Sheet Metal Subjected to Bending-Unbending under Tension I, *Advanced Technology of Plasticity, Proceedings of the Fifth ICTP* (Columbus, OH), Vol 2, 1996, p 743–746

53. S. Takahashi, T. Kuwabara, and K. Ito, Springback Analysis of Sheet Metal Subjected to Bending-Unbending under Tension II, *Advanced Technology of Plasticity, Proceedings of the Fifth ICTP* (Columbus, OH), Vol 2, 1996, p 747–750

54. T. Kuwabara, N. Sekiand, and S. Takahashi, A Rigorous Numerical Analysis of Residual Curvature of Sheet Metals Subjected to Bending-Unbending under Tension, *Advanced Technology of Plasticity, Proceedings of the Sixth ICTP* (Nuremberg, Germany), Vol 2, 1999, p 1071–1076

55. W.Y.D. Yuen, *J. Mater. Process. Technol.,* Vol 22, 1990, p 1–20

56. R. Hino, Y. Goto, and M. Shiraishi, Springback of Sheet Metal Laminates Subjected to Stretch Bending and the Subsequent Unbending, *Advanced Technology of Plasticity, Proceedings of the Sixth ICTP* (Nuremberg, Germany), Vol 2, Sept 19–24, 1999, p 1077–1082

57. A.A. El-Domiaty, M.A.N. Shabara, and M.D. Al-Ansary, *Int. J. Mach. Tools Manuf.,* Vol 36 (No. 5), 1996, p 635–650

58. W. Johnson and T.X. Yu, *Int. J. Mech. Sci.,* Vol 23, 1981, p 687–696

59. C. Sudo, M. Kojima, and T. Matsuoka, Some Investigations on Elastic Recovery of Press Formed Parts of High Strength Steel Sheets, *Proceedings of the Eighth Biennial Congress of IDDRG* (Gothenburg, Sweden), Vol 2, The International Deep Drawing Research Group, Sept 1974, p 192–202

60. M.L. Chakhari and J.N. Jalinier, Springback of Complex Parts, *Proceedings of the 13th Biennial Congress of IDDRG* (Melbourne, Australia), The International Deep Drawing Research Group, Feb 1984, p 148–159

61. L.C. Zhang and Z. Lin, *J. Mater. Process. Technol.,* Vol 63, 1997, p 49–54

62. A. Focellese, L. Fratini, F. Gabrielli, and F. Micari, *J. Mater. Process. Technol.,* Vol 80–81, 1998, p 108–112

63. H. Ogawa and A. Makinouchi, Small Radius Bending of Sheet Metal by

Indentation with V-Shape Punch, *Advanced Technology of Plasticity, Proceedings of the Sixth ICTP* (Nuremberg, Germany), Vol 2, Sept 19–24, 1999, p 1059–1064

64. B.S. Levy, *J. Appl. Metalwork.,* Vol 3 (No. 2), 1984, p 135–141

65. S.S. Han and K.C. Park, An Investigation of the Factors Influencing Springback by Empirical and Simulative Techniques, *Proceedings of NUMISHEET '99* (Besancon, France), 1999, p 53–57

66. L.R. Sanchez, D. Robertson, and J.C. Gerdeen, "Springback of Sheet Metal Bent to Small Radius/Thickness Ratios," Paper 960595, Society of Automotive Engineers, 1996, p 650–656

67. C. Wang, An Industrial Outlook for Springback Predictability, Measurement Reliability, and Compensation Technology, *Proceedings of NUMISHEET 2002* (Jeju Island, Korea), 2002, p 597–604

68. R.G. Davies, *J. Appl. Metalwork.,* Vol 1, (No. 4), 1981, p 45–52

69. F. Pourboghrat and E. Chu, *Int. J. Mech. Sci.,* Vol 36 (No. 3), 1995, p 327–341

70. F. Pourboghrat and E. Chu, *J. Mater. Process. Technol.,* Vol 50, 1995, p 361–374

71. R.G. Davies, *J. Appl. Metalwork.,* Vol 3 (No. 2), 1984, p 120–126

72. N.E. Thompson and C.H. Ellen, *J. Appl. Metalwork.,* Vol 4 (No. 1), 1985, p 39–42

73. Y.C. Liu, *J. Eng. Mater. Technol. (Trans. ASME),* Vol 110, 1988, p 389–394

74. W.D. Carden, L.M. Geng, D.K. Matlock, and R.H. Wagoner, *Int. J. Mech. Sci.,* Vol 44, 2002, p 79–101

75. A. Makinouchi, E. Nakamachi, E. Onate, and R.H. Wagoner, Ed., *Proceedings of NUMISHEET '93* (Isehara, Japan), The Institute of Physical and Chemical Research, 1993

76. S.F. Shen and P.R. Dawson, Ed., *Simulation of Materials Processing: Theory, Methods and Applications,* A.A. Balkema, Rotterdam, Holland, 1995

77. J.C. Gelin and P. Picart, Ed., *Proceedings of NUMISHEET '99* (Besancon, France), Burs Publications, 1999

78. D.-Y. Yang, S.I. Oh, H. Huh, and Y.H. Kim, Ed., *NUMISHEET 2002—Proc. Fifth Int. Conf. Workshop Num. Simul. 3D Sheet Forming Processes* (Jeju, Korea), Vol 1, 2002

79. K. Mattiasson, A. Strange, P. Thilderkvist, and A. Samuelsson, Simulation of Springback in Sheet Metal Forming, *Proceedings of the Fifth International Conference on Numerical Methods in Industrial Forming Process* (New York), 1995, p 115–124

80. S.W. Lee and D.Y. Yang, *J. Mater. Process. Technol.,* Vol 80–81, 1998, p 60–67

81. S.C. Tang, Analysis of Springback in Sheet Forming Operation, *Advanced Technology of Plasticity, Proceedings, Second ICTP* (Stuttgart, Germany), Vol 1, 1987, p 193–197

82. N. He and R.H. Wagoner, Springback Simulation in Sheet Metal Forming, *Proceedings of NUMISHEET '96* (Columbus, OH), 1996, p 308–315

83. N. Narasimhan and M. Lovell, *Finite Elem. Anal. Des.,* Vol 33, 1999, p 29–42

84. Y. Hu and C. Du, Quasi Static Finite Element Algorithms for Sheet Metal Stamping Springback Simulation, *Proceedings of NUMISHEET '99* (Besancon, France), 1999, p 71–76

85. L.M. Geng and R.H. Wagoner, Springback Analysis with a Modified Hardening Model, Sheet Metal Forming, *Sing Tang 65th Anniversary Volume,* SP-1536, SAE Publication 2000-01-0768, Society of Automotive Engineers, 2000, p 21–32

86. D.W. Park, J.J. Kang, J.P. Hong, and S.I. Oh, Springback Simulation by Combined Method of Explicit and Implicit FEM, *Proceedings of NUMISHEET '99* (Besancon, France), 1999, p 35–40

87. F. Valente and D. Traversa, Springback Calculation of Sheet Metal Parts after Trimming and Flanging, *Proceedings of NUMISHEET '99* (Besancon, France), 1999, p 59–64

88. N. Montmayeur and C. Staub, Springback Prediction with OPTRIS, *Proceedings of NUMISHEET '99* (Besancon, France), 1999, p 41–46

89. G.Y. Li, M.J. Tan, and K.M. Liew, *Int. J. Solids Struct.,* Vol 36 (No. 30), Oct 1999, p 4653–4668

90. U. Abdelsalam, A. Sikorski, and M. Karima, Application of One Step Springback for Product and Early Process Feasibility on Sheet Metal Stampings, *Proceedings of NUMISHEET '99* (Besancon, France), 1999, p 47–52

91. R.H. Wagoner and J.-L. Chenot, *Metal Forming Analysis,* Cambridge University Press, Cambridge, U.K., 2001

92. R.D. Cook, D.R. Malkus, M.E. Plesha, R.J. Witt, *Concepts and Applications of Finite Element Analysis,* 4th ed., Wiley, 2002

93. J.N. Reddy, *Introduction to the Finite Element Method,* 2nd ed., McGraw-Hill, 1993

94. O.C. Zienkiewicz and R.L. Taylor, *The Finite Element Method: Volume 1, The Basis,* 5th ed., Butterworth-Heinemann, Oxford, U.K., 2000

95. O.C. Zienkiewicz and R.L. Taylor, *The Finite Element Method: Volume 2, Solid Mechanics,* 5th ed., Butterworth-Heinemann, Oxford, U.K., 2000

96. T.J.R. Hughes, *The Finite Element Method: Linear Static and Dynamic Analysis,* Prentice-Hall, 1987

97. K.-J. Bathe, *Finite Element Procedures,* Prentice-Hall, 1995

98. http://www.abaqus.com/products/products_standard.html, ABAQUS, Inc., 2005

99. http://www.ansys.com/products/mechanical.asp, ANSYS, Inc., 2005

100. http://www.abaqus.com/products/products_explicit.html, ABAQUS, Inc., 2005

101. http://www.lstc.com, Livermore Software Technology Corp., 2005

102. J. Wang and R.H. Wagoner, A New Hexahedral Solid Element for 3D FEM Simulation of Sheet Metal Forming, *Proc. of NUMIFORM 2004, Materials Processing and Design* (Columbus, OH), S. Ghosh, J.M. Casto, and J.K. Lee, Ed., June 2004, p 2181–2186

103. J. Wang and R.H. Wagoner, A Practical Large Strain Solid Finite Element for Sheet Forming, *Int. J. Numer. Methods Eng.,* in press

104. T. Shimizu, E. Massoni, N. Soyris, and J.-L. Chenot, A Modified 3-Dimensional Finite Element Model for Deep-Drawing Analysis, *Advances in Finite Deformation Problems in Material Processing and Structure,* N. Chandra et al., Ed., AMD Vol 125, ASME, 1991, p 113–118

105. M. Kawka and A. Makinouchi, Finite Element Simulation of Sheet Metal Forming Processes by Simultaneous Use of Membrane, Shell and Solid Elements, *Numerical Methods in Industrial Forming Processes,* J.-L. Chenot et al., Ed., Balkema, Rotterdam, 1992, p 491–496

106. K.P. Li and R.H. Wagoner, Simulation of Deep Drawing with Various Elements, *Proceedings of NUMISHEET '99* (Besancon, France), 1999, p 151–156

107. A. Andersson and S. Holmberg, Simulation and Verification of Different Parameters Effect on Springback Results, *Proceedings of NUMISHEET 2002* (Jeju Island, Korea), 2002, p 201–206

108. V.T. Nguyen, Z. Chen, et al., Prediction of Spring-Back in Anisotropic Sheet Metals, *Proc. Inst. of Mech. Eng. C, J. Mech. Eng. Sci.,* Vol 218 (No. 6), 2004, p 651–661

109. W.L. Xu, C.H. Ma, et al., Sensitive Factors in Springback Simulation for Sheet Metal Forming, *J. Mater. Process. Technol.,* Vol 151 (No. 1–3), 2004, p 217–222

110. N. Yamamura, T. Kuwabara, et al., Springback Simulations for Stretch-Bending and Draw-Bending Processes Using the Static Explicit FEM Code, with an Algorithm for Canceling Non-equilibrated Forces, *Proceedings of NUMISHEET 2002* (Jeju Island, Korea), 2002, p 25–30

111. H. Yao, S.-D. Liu, et al., Techniques to Improve Springback Prediction Accuracy Using Dynamic Explicit FEA Codes, *SAE Trans. J. Mater. Manuf.,* Vol 111, 2002, p 100–106

112. R.H. Wagoner and M. Li, Advances in Understanding Springback Simulations, *Proceedings of NUMISHEET 2005,* 2005

113. R.H. Wagoner and M. Li, Simulation of Springback: Through-Thickness Integration, *Int. J. Plasticity,* submitted

114. E.W. Weisstein, Trapezoidal Rule, *Math-World—A Wolfram Web Resource,* http://

mathworld.wolfram.com/TrapezoidalRule. html, 2005

115. E.W. Weisstein, Simpson's Rule, *Math-World—A Wolfram Web Resource,* http:// mathworld.wolfram.com/SimpsonsRule. html, 2005

116. E.W. Weisstein, Legendre-Gauss Quadrature, *MathWorld—A Wolfram Web Resource,* http://mathworld.wolfram.com/ Legendre-GaussQuadrature.html, 2005

117. S.C. Tang, Application of an Anisotropic Hardening Rule to Springback Prediction, *Advanced Technology of Plasticity, Proceedings, Fifth ICTP* (Columbus, OH), Vol 2, 1996, p 719–722

118. F. Barlat, Y. Maeda, K. Chung, M. Yanagawa, J.C. Brem, Y. Hayashida, D.J. Legfe, S.J. Matsui, S.J. Murtha, S. Hattori, R.C. Becker, and S. Makosey, *J. Mech. Phys. Solids.* Vol 45 (No. 11/12), 1997, p 1727–1763

119. L.M. Geng and R.H. Wagoner, *Int. J. Mech. Sci.,* Vol 44, 2002, p 123–148

120. Y. Hishida and R.H. Wagoner, SAE paper 930285, *J. Mater. Manuf.,* Vol 102, 1993, p 409–415

121. D. Schmoeckel and M. Beth, *Ann. CIRP,* Vol 42 (No. 1), 1993, p 339–342

122. M. Sunseri, J. Cao, A.P. Karafillis, and M.C. Boyce, *Trans. ASME,* Vol 118, 1996, p 426–435

123. C.C. Chu, *Int. J. Solids Struct.,* Vol 22 (No. 10), 1986, p 1071–1081

124. Y. Nagai, *Jpn. Soc. Technol. Plast.,* Vol 28, 1987, p 143–149

125. R.A. Ayres, *J. Appl. Metalwork.,* Vol 3 (No. 2), 1984, p 127–134

126. Y. Hayashi, Analysis of Surface Defects and Side Wall Curl in Press Forming, *Proceedings of the 13th Biennial Congress of IDDRG* (Melbourne, Australia), The International Deep Drawing Research Group, Feb 1984, p 565–580

127. Y. Tozawa, *J. Mater. Process. Technol.,* Vol 22, 1990, p 343–351

128. Y. Umehara, *J. Mater. Process. Technol.,* Vol 22, 1990, p 239–256

129. M. Ueda and K. Ueno, *J. Mech. Work. Technol.,* Vol 5, 1981, p 163–179

130. R.D. Webb, "Spatial Frequency Based Closed-Loop Control of Sheet Metal Forming," Ph.D. thesis, Massachusetts Institute of Technology, 1987

131. R.D. Webb and D.E. Hardt, *J. Eng. Ind. (Trans. ASME),* Vol 113, 1991, p 44–52

132. C. Hindman and K.B. Ousterhout, *J. Mater. Process. Technol.,* Vol 99, 2000, p 38–48

133. O. Ghouati, D. Joannic, and J.C. Gelin, Optimisation of Process Parameters for the Control of Springback in Deep Drawing, *Proceedings of NUMIFORM '98* (Netherlands), 1998, p 819–824

134. I. Chou and C. Hung, *Int. J. Mach. Tools Manuf.,* Vol 39, 1999, p 517–536

135. A.P. Karafillis and M.C. Boyce, *Int. J. Mech. Sci.,* Vol 34 (No. 2), 1992, p 113–131

136. A.P. Karafillis and M.C. Boyce, *J. Mater. Process. Technol.,* Vol 32, 1992, p 499–508

137. A.P. Karafillis and M.C. Boyce, *Int. J. Mach. Tools Manuf.,* Vol 36 (No. 4), 1996, p 503–526

138. W. Gan and R.H. Wagoner, Die Design Method for Sheet Springback, *Int. J. Mech. Sci.,* Vol 46 (No. 7), 2004, p 1097–1113

139. R. Lingbeek, J. Huétink, S. Ohnimus, M. Petzoldt, and J. Weiher, The Development of a Finite Elements Based Springback Compensation Tool for Sheet Metal Products, *J. Manuf. Prod. Technol.,* in press

140. F. Valente, X.P. Li, and A. Messina, in *Proceedings of Fourth COMPLAS* (Barcelona, Spain), Vol 2, 1997, p 1431–1488

141. E. Chu, L. Zhang, S. Wang, X. Zhu, and B. Maker, Validation of Springback Predictability with Experimental Measurements and Die Compensation for Automotive Panels, *Proceedings of NUMISHEET 2002* (Jeju Island, Korea), 2002, p 313–318

CAD/CAM and Die Face Design in Sheet Metal Forming

Ed Herman, Creative Concepts Company, Inc.
Daniel J. Schaeffler and Evan J. Vineberg, Engineering Quality Solutions, Inc.

SINCE THE EARLY 1980s, ever-evolving computer technology and processing speed have brought about revolutionary changes to the sheet metal forming industry, beginning with the development of computer-aided design (CAD) and computer-aided manufacturing (CAM) capabilities and continuing currently with the development of computer-aided engineering (CAE) tools to facilitate the application of physics and mathematics in the process of part and die design. These changes have taken the industry from a purely art form to the art/science form that it is today (2006). The sheet metal forming industry is now poised for the next step of going to a truly engineering form, where the meaning of CAE will incorporate true syntheses of design as well as analysis (as it now does) and optimization (as is starting to show up in limited forms). (Computer-aided engineering is discussed in more detail in the article "Process Modeling and Simulation for Sheet Forming of Steel" in this Volume.)

Critical in providing the infrastructure needed for the computerized "science" in use today 2006 and for the computerized "engineering" of the future is the need for better descriptions of material properties—not better materials, but a better/different approach to materials characterization based on the traditional mechanical properties. Furthermore, an expanded understanding of the role of friction is needed, along with the proper techniques to apply this knowledge to the die development and simulation process.

With a significant portion of the die development process occurring with the aid of computers, it is imperative that the files used are complete and that appropriate information can be exchanged between them. The CAD data for product and manufacturing should include files that contain the geometry of the sheet metal in all stages of production and use, plus files that contain the geometry of the actual die faces (addendum, binder shape, and die wrap). Creating these die face files is called functionalizing the die face. Once functionalized in the CAD system, the die construction and tryout costs are significantly reduced.

This article includes the following sections:

- Grade designations
- Materials specifications
- Forming limit curve
- Friction (draw bead, binder, and internal die radii)
- Product development process
- Direct engineering for formability
- CAD data sets

Grade Designations

Sheet steels for draw forming are typically sold under grade designations that are generally indicative of their relative strengths and formability or how they achieve their strength. Typical grades include:

- Lower-strength steels, including commercial steel, forming steel, drawing steel, deep drawing steel, and extra-deep-drawing steel (also known as interstitial-free steel)
- Medium-strength steels, including bake-hardenable steel, interstitial free-rephosphorized, and carbon-manganese steel
- Higher-strength steels, including high-strength low-alloy steel, dual-phase (DP) steel, transformation-induced plasticity (TRIP) steel, and others. DP and TRIP steels are part of a family of steels collectively known as Advanced high strength steels (AHSS). For more information on AHSS, see the article "Forming of Advanced High-Strength Steels" in this Volume and Ref 1 to 5.

Such specifications as SAE 1008 and 1010 cover chemistry requirements only and do not have any minimum mechanical property requirements. Some minimum mechanical property requirements are listed in SAE J2329 (lower-strength steels) and SAE J2340 (for some medium- and higher-strength steels). The Society of Automotive Engineers is also developing a specification to cover the AHSS grades (tentatively designated as SAE J2745). However, these specifications typically list only minimum and maximum values for yield strength, with minimums only on tensile strength, elongation, n-value, and r-value.

The stress-strain curves and the range of variation of those curves for the actual material produced by any one mill will vary somewhat from those of other steel mills. To some extent, this is true even for different mills within one steel company. The industry practice for making new dies work is to tune them in to the material from the mill that will be supplying the steel. Naturally, the tuning process ends when the die just works, that is, when the first "safe" part is achieved. It is common for dies to not work well with material from any other mill.

To make the dies robust to the normal and reasonable variation of the material is to design the dies to be robust to that variation rather than to "just get them to work." Such designs were impossible in the past, but now that computer simulation and direct engineering are available, such design objectives are at least technically feasible. The practicality of creating such designs is still not perceived by the industry as realistic but must be made so if the full utility of the materials and the stamping processes are to be realized. A major reason why the robust die designs are not being created today (2006) is that the designers and formability engineers do not have the data of the range of variation of materials (as illustrated in Fig. 1), reiterating the need for this information.

Material characteristics that are used in defining a sheet metal grade include but are not limited to:

- Tensile stress-strain curve
- Forming limit curve
- Minimum and maximum yield strengths
- Minimum tensile strengths
- Minimum, or typical, total elongations
- Minimum, or typical, n-values
- Minimum, or typical, r-bar values, as well as their directional components (r_0, r_{45}, and r_{90})
- Thickness tolerances (Present standard tolerances should be revisited.)
- Crown limit
- Surface roughness

Materials Specifications

Historically, the stamping industry has pressured the materials suppliers for "better" materials. "Better steel" has different meaning to the product engineer than to the manufacturing engineer. The product engineer is usually trying to get higher strength (e.g., yield strength) at low cost so he can reduce the gage and save weight. At the same time, he needs good formability so his designs can be formed into the shapes he needs. The manufacturing engineer is looking for better formability and more consistency in forming performance. The manufacturing people are periodically faced with situations where one batch of steel works in the dies and another batch will not work. Yet, both meet the specifications for the steel and satisfies the requirements on the purchase order. The reason one coil will work and another will not work is the normal and reasonable variation in even sequential coils of sheet metal, as illustrated in Fig. 1, and the die creation process. The die creation process often does not allow time, money, or even the possibility of testing the die for the entire range of material variation during the die design simulation or the physical die try-out stages.

Exactly how the material will perform in the dies is not known until the part design is completed, a die face is designed, and the operation of the die on the material is simulated. Even then, designs that work in the simulation may not work with every coil of steel supplied to the specification. Unless simulations are made to study the behavior for the total range of reasonable and normal variation of the material, the design may not be robust to that total range of properties, and some coils of material will not work in the dies. The same can be said for the variation in friction between the sheet metal and the die. Today (2006), it is possible to explore the robustness of the part/die designs throughout the range of variation with computer simulations. Such sensitivity studies are not yet common practice but are starting to be used. Not only do such explorations take more time and computing costs, but the necessary information for the material

and friction is generally not available. At best, the analysis uses typical values (or what the analyst considers to be worst-case values) for both the material behavior and the friction and then adjusts to be (hopefully) conservative. By being conservative, failures are minimized, but optimal use of the material is not necessarily realized. There is a tendency to err on the side of requiring unnecessary design changes rather than risk production failures.

There is considerable opportunity for the materials supplier industry to better support the available engineering calculations that are in use today (2006). (Future advances in product and die engineering calculations will also need the materials performance information being suggested here.) Such improved support is not just in creating "better" materials but in providing more descriptive data and standards for existing materials. One way would be to provide the bounding stress-strain curves defining the six-sigma normal and reasonable variation of the material falling within each grade designation, as illustrated in Fig. 2, in addition to the specified properties listed previously. These bounding curves may not be actual data for any specific piece of material but rather composite curves created to bound the variation. These bounding curves are in addition to the average and standard deviation of the yield and tensile strength, elongation, r-value, and n-values. Another improvement would be to specify the maximum and minimum values for each of the properties listed previously instead of just minimum or typical values.

The stress-strain curve for most sheet draw forming materials is commonly represented by what is called the power-law relation:

$$\sigma = K(\varepsilon_0 + \varepsilon_i)^n$$

where σ is the true stress, K is the extrapolated true stress at a true strain of 1.00, ε_i is the instantaneous true strain, ε_0 is the strain offset to get the 0.002 strain offset yield stress to hit the power-law stress-strain curve (The ε_0 can be considered the equivalent strain induced by the steel processing to achieve a specified yield stress.), and n is the the strain-hardening exponent.

Some argue that the power law is insufficient to represent all materials well enough and they fit the stress-strain curve with other equation forms and other coefficients. This argument is particularly relevant for sheet aluminum and the advanced high-strength steels such as DP and TRIP steels. Although these discrepancies exist, the error is insignificant when compared to the noise of variation from coil to coil of material. Because a design must be robust to the total spectrum of normal and reasonable variation, the power law is therefore quite adequate. However, a practical resolution may be for the suppliers to provide both the stress-strain curves and the coefficients that give the best power-law fit. From the power-law coefficients, the user can recreate the stress-strain curves. Or, if the user prefers, he/she can fit another equation that he/she feels is more appropriate to the supplied curves. The user can also simply digitize a number of points on the curve and interpolate between the points.

To use the engineering computational capability available and practical today (2006) for both product and die engineering, the engineer must be able to reconstruct the stress-strain curve and compute the stress that results from any strain, and vise versa. Therefore, the material behavior must be provided to him.

Forming Limit Curve

Another very useful representation of the material is the forming limit curve. The stress-strain curve from a tensile test, described previously and in Fig. 1 and 2, represents the behavior of the material in a single mode of deformation. It is pulled in one direction only, and the thickness and width of the pulled specimen are unrestrained. As the material is being formed into a real part in a real die, the material is deformed by external forces pulling or pushing on the material in all three directions. The forming limit curve, shown in Fig. 3, captures the necking failure limit of the material through all the common deformation modes. Fortunately, for low-carbon steels, the FLC_0 point (the lowest point on the forming limit curve, or FLC) can be estimated from the n-value and the thickness of the material, and the shape of the curve remains the same. However, that simple relationship does not hold true for other materials, such as aluminum and stainless steel, and it may not hold true for all steels in the emerging range of new products, such as TRIP steels.

The conventional strain-based FLC is created by stretching various-width strips of material over a smooth ball. The ball is large enough so that relatively little stress is introduced through the thickness of the material. The result is that the material is strained proportionately throughout the test. What is meant by a proportional, or linear, strain path is that halfway through the test, the strip will be stretched half of its final lengthwise strain and will have half of its final

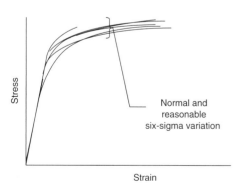

Fig. 1 The stress-strain curves for consecutive coils of steel for a single grade of steel, illustrating that no two are alike. The parts and the dies must be designed to be sufficiently robust to handle such normal and reasonable variation.

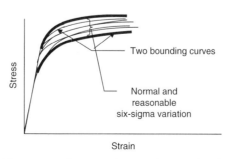

Fig. 2 Bounding curves for a specific material grade. Any specific grade of material should be defined by two stress-strain curves that describe the plus/minus three-sigma boundaries of normal and reasonable variation.

width strain. This strain-based FLC is relevant only to those forming processes that result in a linear strain path. Fortunately, most first-draw forming operations impart nearly proportional strain paths, so the discrepancy between the simulation result and the failure criteria is not considered to be a serious issue. Most computer simulation programs actually track the strain path to determine/predict the final strain conditions throughout the formed part. Those final strains are then compared to the strain-based FLC to show where the material may split, where it is dangerously close to splitting, and where it may wrinkle under certain part geometries and material flow conditions.

However, a nonlinear strain path may be found in secondary forming processes, such as some restrike operations, and even in some areas of the first-draw die, like re-entrant surfaces such as door handles on doors that have been subjected to high prestrain. In these cases, because the strain paths significantly differ from the ones used to form each point on the linear strain path strain-based FLC, that FLC is not applicable. Recent work (Ref 6–10) has shown that there is a corresponding stress-based FLC, and this stress-based FLC is independent of the strain path and is unique for a given material, such as steel or aluminum (Ref 11). Using a stress-based FLC resolves the discrepancy of calculating strains that result from nonuniform strain paths and then evaluating the risk of failure (i.e., splitting/breaking) by comparing those strains to proportional strain path failure data (i.e., the strain-based FLC).

The stress-based FLC is a relatively new approach to refine formability theory. When a new material or grade is made available to the manufacturing industry, the FLC for that material must also be provided. Conventionally, the strain-based FLC was generated, but the merits of stress-based FLCs are helping this approach to gain acceptance.

The FLC is also extensively used to assess the strain safety status ("healthiness" or robustness) of an actual formed sheet metal panel. Before forming in the die, the blank is gridded with a pattern of etched and stained circles of a precise and known diameter. After the gridded panel is formed in the die, the circles have been deformed into ellipses (Fig. 3). These ellipses are measured and the applied strains plotted on the forming limit diagram. Any plotted points that are above the FLC indicate conditions that are likely to split. Those locations that plot too close to the FLC (i.e., in the marginal zone) indicate conditions that are likely to split should the material or process change. Only those locations that plot sufficiently below the FLC are considered safe in terms of formability. More details on how to perform a strain analysis and relate the data to a forming limit diagram can be found in the articles "Formability Testing of Sheet Metals" and "Troubleshooting Formability Problems Using Strain Analysis" in this Volume.

Once the bounding curves, as illustrated in Fig. 2, are generated for a particular grade from a given supplier, they can then be widened to encompass the normal and reasonable variation of the materials as available in the industry, as illustrated in Fig. 4. Such standards would not be more restrictive than what is produced today (2006) but would simply quantify what is being produced and supplied in a way that the user can optimize the use of the material.

Friction

One of the least understood and most significant factors in sheet metal forming is the role of friction between the sheet metal and the die as the metal slides across the die face as the part is being formed. The advent of computer simulation and the future direct-engineering methods have provided the ability to put the effects of friction into the part and die designs mathematically. The problem (circa 2006) is that the effects are not well quantified. The practical approach has been to adjust a single friction coefficient used in the analysis until the analysis matches well a known condition, and then use that coefficient for future analyses. As with material properties, analysts have tended to err on the side of conservative friction values.

There are three separate frictional conditions acting in a draw die. These are (1) the metal passing through a draw bead, (2) the metal clamped in the binder, and (3) the metal sliding across a die radius while it is simultaneously changing the wrap angle on the radius and is being increasingly plastically strained.

The draw bead, illustrated in Fig. 5, has been the object of the greatest investigative effort. Several physical draw bead simulators exist, and many data have been extracted from them and many papers written on those investigations. However, the draw bead performance of specific materials with specific lubricants has not been published as any type of accepted, and greatly needed, industry standard.

The amount of restraining force generated by a draw bead is a function of the sheet material being pulled through it, the lubricant, the die condition (e.g., material roughness, etc.), and the dimensions, depth (D) and width (W), of the bead. For any specific set of material, die condition, dimension W, and lubricant, the restraining force as a function of dimension D can be established in the laboratory by experimentation with a draw bead simulator (DBS). The results are illustrated in Fig. 6.

The bead restraining force is not used in the direct engineering of the die design. This

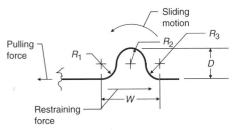

Fig. 5 Draw bead. The draw bead creates a restraining force on the metal as it slides through the binder into the die cavity by bending, unbending, and friction as it is pulled through three (or four) radii (R). D, depth; W, width

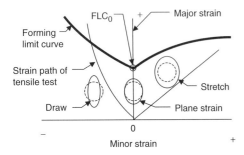

Fig. 3 The forming limit diagram. In real forming operation, the material experiences stretching in both directions, as illustrated by the "Stretch" circle; stretching in only one direction, as illustrated by the "Plane strain" circle; or stretching in one direction while reducing in the other direction, as illustrated by the "Draw" circle. The forming limit shows the conditions at which the material will start to neck, after which it will soon fracture. For most low-carbon sheet steels, the FLC_0 point (the lowest point on the forming limit curve, or FLC) can be calculated from the sheet thickness and n-value. The shape of the forming limit curve remains the same. It is moved up or down on the graph to fit to the FLC_0 point.

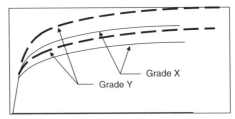

Fig. 4 Bounding curves for different grades. Each grade of draw forming sheet metal (e.g., steel or aluminum) should have the stress-strain bounding curves. The properties of grades probably will overlap, as illustrated here. Any coil with properties outside the bounding curves could not be considered to be that grade.

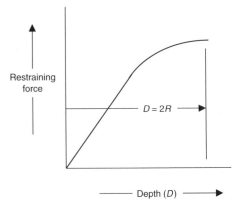

Fig. 6 Effect of depth on draw bead restraining force. The draw bead restraining force is nearly proportional to the depth until the maximum depth of twice the bead radius (R) is approached. As the bead depth exceeds twice the radius, no additional restraining force is generated.

analytical technique inputs (or calculates) the restraining force. The bead force is used after the design has been finalized to determine the bead geometry that will create the calculated required restraining force. That geometry is then built into the CAD data files for the die face.

There are several draw bead computer models in use today (2006). Some of these are commercially available, and some are proprietary. These models allow interpolation between, and extrapolation beyond, actual test data to provide bead performance data on a wide range of conditions. For these models to work correctly, they must be supplied with the bead geometry, the material performance data (as described previously), and the friction for the die, material, and lubricant combination. The friction values are calculated from the DBS test data. These friction values represent a complex and low-pressure condition, and it can be argued that the values calculated by the normal methods are not true friction coefficients but rather some combined friction coefficient and error correction factor. The use of the friction coefficients is valid as long as the coefficients are used for the friction models and are not used for other conditions (described subsequently) that exist in the die.

The binder is that section of the die outside the part that first clamps the sheet metal before the punch enters and stretches the sheet metal into the part shape. The binder serves several functions. The first is to preform the sheet into a developable shape that places, to the best possible extent, the right amounts of material into the right areas for subsequent forming. The second function is to keep the sheet metal within the binder from wrinkling. The third function is to create the necessary backforce for properly forming the part as the punch stretches it into the desired shape. A part of that third function is to be the mounting platform for the draw bead.

As the sheet metal is stretched in the generally horizontal part area to set the part shape into the sheet metal, some of the extra metal (from stretching) will move (displace) into the more generally vertical stretch or draw wall between the part area and the binder, as indicated by the diverging arrows in the part area shown in Fig. 7. Similarly, any metal being pulled (drawn) from the binder into the stretch or draw wall will result in hoop compression, as indicated by the converging arrows in the binder area shown in Fig. 7. Two elements of sheet metal in the binder area located at points A and B, respectively, will move to points A′ and B′, respectively, causing them to become closer together. The material between the two elements must do some combination of:

- Be compressed and thicken
- Become elongated in the direction of the arrows by having sufficient pulling force in the direction of the arrows to plastically deform the material to the uniaxial tension condition or greater
- Wrinkle

One purpose of the binder is to control the wrinkling to within some allowable amount (what is allowable can vary from application to application) and thereby force some combination of the other two possibilities into the sheet metal.

The binder should not actually clamp the sheet metal. Instead, the binder should be closed onto blocks outside of the sheet metal and those blocks hold the binder open from the sheet metal by approximately one-fourth of the metal thickness. The binder should be held onto the blocks with sufficient force that the metal can not push it off the blocks as the metal starts to wrinkle. Although the metal will wrinkle slightly, it can not wrinkle as much as the displacement arrows imply. Hence, the material must plastically deform.

If that deformation can be represented by a circle that is deformed into an ellipse, the largest diameter of the ellipse (the major strain axis) is aligned in the direction of the displacement arrow. The plastic deformation of the material (one can think of this as rearranging the surface area) requires a pulling force on the metal in the direction of the displacement arrow that is equal to a force in the opposite direction restraining the sliding motion. The slight wrinkling of the metal creates a normal force on the binder that is a function of the material internal resistance to deformation, the material thickness, and the height of the wrinkle. The friction between the metal and the binder is small but can still generate a significant force acting to restrain (but not stop) the sliding motion of the material through the binder and toward the part being formed. The friction-induced restraining force is additive to the plastic deformation restraining force. Figure 8 is a view of the section from Fig. 7 and illustrates the wrinkling that occurs as a result of the converging material displacements in the binder. When more force is needed to adequately restrain the sliding motion of the material, draw beads are added.

The friction coefficients representative of what is happening in the binder can be measured by pulling strips of material through flat plates. Such friction tests are available and should be used to generate the data. These conditions are very low-pressure situations, and the data should not be used for the internal areas of the die where the pressures and the metal straining conditions are much more severe.

Internal Die Radii. The third and most complex friction condition is where material is sliding across the curved punch and/or die faces and simultaneously being stretched in plastic deformation. The condition is illustrated in Fig. 9. The pressure of the sheet metal against the die face is a function of the radius and the stress in the metal and is calculated with the pressure vessel equation:

Pressure = Instantaneous stress in the metal

× (Metal thickness/Die radius)

The equation shows frictional restraining force to be small when the die radius-to-metal-thickness ratio is large (e.g., >15) but can be quite large when the radius-to-thickness ratio is small (e.g., <5).

The actual value of the restraining force is a function of:

- Die radius
- Included arc angle
- Amount of straining in the sheet metal

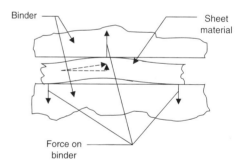

Fig. 8 Detail of binder effect. As the material wrinkles, it bumps against the binder in this view of the section cut from Fig. 7. The binder limits the height of the wrinkle and withstands the force of the material pushing against it. The force vector diagram within the material illustrates the force component pushing on the die.

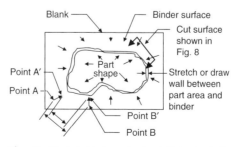

Fig. 7 Effect of a binder on sheet metal behavior. As the sheet metal blank is drawn into a part shape, the material that forms the part shape is generally stretched outward, and the material within the binder slides inward, as indicated by the arrows. The converging pattern of the inwardly sliding material in the blank implies that the material will try to wrinkle.

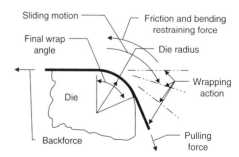

Fig. 9 Force acting to restrain the sliding motion generated by friction and bending as sheet metal slides across a die face. The magnitude of that force is strongly affected by the radius and wrap angle.

- Stress in the sheet metal in the direction of motion
- Velocity of the sheet metal relative to the die
- Displacement distance (separate from velocity)
- Die material and coating (e.g., chrome plating)
- Die surface finish
- Lubricant type
- Lubricant amount

The value also depends on interactions between these factors. In the late 1990s, a team at the General Motors Metal Fabricating Division created a friction test that captured the global effects of the conditions by having the die radius on a separate block resting on load cells. This work showed the variability of friction and the ability to accurately measure it at actual press speeds and with large specimens. The test equipment, procedure, and results of that work are included in the final report of the National Institute of Standards and Technology (NIST) Springback Predictability Project (Ref 12). Such a test measures the combined effects of friction and the forces necessary to accomplish the bending and unbending. To reduce the data to just friction, several data-processing steps are required. First, the test must be modeled by the simulation program in which the test data are going to be used. That simulation modeling of the test is done with the friction value set to zero and the simulation post processed to output the same data curves as derived from the actual test. These curves simulate the zero friction condition, so they represent what the test data would be if there was no friction. Thus, these simulations show what the simulation code would predict for only the effects of bending and unbending. Then, the difference between the simulation data curves and the actual data collected can be mathematically analyzed to deduce the friction values. The difference between the simulation and the test data is a value that includes the friction and is also a correction for any errors in how the simulation software computes the bending contributions. Because the various simulation software programs have different methods of computing bending and different interpretations of how friction is applied, the adaptation of the test data must be done for each simulation package. The results of the test data must be reduced by regression analysis to coefficients of a basic equation. The form of the basic equation can be standardized, and a set of coefficients can be created for each sheet material, die material, die coating, and lubricant combination. Such data should be created and made available as an industry standard.

The NIST Springback Predictability Project, completed in 2000, also enhanced the LS-DYNA (Livermore Software Technology Corp.) software package to enable the program to work with the user's friction models at every node and time step, to ensure the use of the correct friction factor at all times and at all locations during the simulation.

Product Development Process

As an automobile is conceived and the engineering data created, the data forms go though many stages. The automobile example is used here because it represents one of the most complex situations. Other situations will usually be some subset of what is described here. The following descriptions of vehicle development only apply to those portions of the vehicle development process relating to the stamped sheet metal components. The basic steps in the vehicle development process are shown in Fig. 10.

The objective is to design the part, the die, and the stamping process to fully use the capability of the material. Both simulation and direct engineering are used in the binder development step, shown in Fig. 10, to achieve that objective.

The die engineering community advises the product community as to the feasibility of the product geometry. For the most part, that advice is still experience based, as illustrated in Fig. 11(a). Even when the advice is in formally packaged standards or design guidelines, these standards and/or guidelines are created based on experience. As the industry moves into the future, such advice will be based more on actual calculation of the material's capability for each situation.

The processing steps and the draw die binder developments are also created from experience, as illustrated in Fig. 11(b). The shapes in the binder development are transferred into the die as the die faces, and therefore act to collect and focus the force and motion from the stamping press onto the sheet metal to change it into the formed part.

In the past, a die was built and then tried out (see the "Tryout and buyoff" step in Fig. 10) to determine if the experience-based design would actually form the material properly. The die was then physically modified until the part was free of splits and wrinkles. Today (2006), both the part and the binder development are created in CAD systems with the same experience-based inputs, which results in the same experience-based designs but in a math form. The performance of the die is now simulated in a computer (instead of being built), as illustrated in Fig. 12,

to determine if it will perform as desired. Any undesirable condition is modified, based on experience, and the new design is then simulated again. It must be noted at this point that the designer of the die binder now has a much faster and accurate learning experience, because of the computer simulations. The simulations give the designer much faster feedback on the performance of his design; he actually revises the design (instead of the try-out die maker), and the simulations give him much more information as to how the material failed.

Today's (2006) practice of computer simulation of the design performance introduces science and mathematics into the formability aspects of the die creation process and has

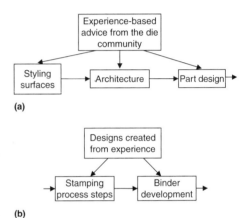

(a)

(b)

Fig. 11 Experience-based product and die design. (a) Advice as to how the part geometry should be designed to accommodate the material formability is generally supplied to the product community by the die community but is experience based, and as such is neither scientific nor mathematical. (b) The die community creates the binder development (which defines the die faces and focuses energy onto the sheet metal to change it from what it is to what one wants it to be) by experienced judgment.

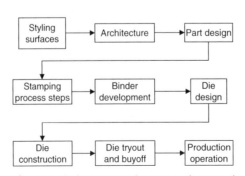

Fig. 10 The basic steps in the creation of a stamped sheet metal part for an automobile. The steps shown are those where formability issues must be addressed in one form or another.

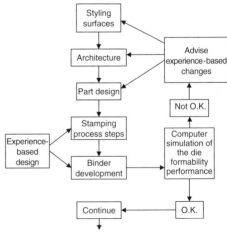

Fig. 12 Use of computer simulations to test experience-based designs. Today (2006), the experience-based product and die face (i.e., binder development) designs are tried by computer simulation. Any undesirable condition discovered by the simulation is modified by experience and then tried again.

resulted in untold improvements in the cost, performance, and lead time of that process.

The future direct-engineering methods alluded to previously would facilitate the application of science- and math-based advice and design, as illustrated in Fig. 13. By applying in a direct-engineering (i.e., synthesis) approach, using the same formability science and math now used in the computer simulation codes, an optimized design will be created directly as it is being drafted. Such practice will result in better use of the materials and far fewer simulation-design iterations. Cost, quality, and lead time will be further reduced.

Direct Engineering for Formability

As mentioned previously, the present method of creating product and die geometries and the sizing of those geometries is done by experience. The result is that the part and die designs are much the same as those done in prior times with manual designs and physical models, except they are now being done electronically with a CAD system in a computer. Those same CAD systems and computers now enable the application of the same science and math used in the finite element simulation programs to be used as the design (either product or die face*) unfolds. The science and math can be used in a synthesis mode rather than just a postmortem analysis mode. The

following is a simplified example to demonstrate the basic direct-engineering method. What makes the approach possible is the way the problem is structured.

Direct-Engineering Method Example. The example is a stylized engine hood for an automobile, as shown in Fig. 14. The design of the engine hood is provided, and the design to be completed is the die face, which consists of the addendum, binder shape, and binder wrap. For simplicity, only one section—through the hood—is used in the example, as shown in Fig. 15. An actual design of the die face (i.e., the binder development) for such an engine hood would require six to eight such sections plus a reconciliation of the binder wrap to all the "X" points. The objective here is to present the thought process for direct engineering (synthesis) but not to actually design the die face.

Step 1. The first step is to clearly define the design objectives. For a part such as the example, a normal set of objectives would be:

- No splitting of the sheet metal
- No wrinkling of the sheet metal
- No impact lines or skid marks
- No highs or lows
- No springback, flattening, crowning, or twist

Some objectives that may not be so common but could have significant economic consequences are:

- Robustness to all normal and reasonable material and processing variation
- Short die-change time (where the first part made after a die set is a good part)
- Work-hardening drawing steel (180 MPa, or 26 ksi, yield strength) to 350 MPa (51 ksi) yield strength on the vehicle
- Size consistency (to fit the panel to the subsequent dies in the process)

Step 2. This step is unique to all net shape processes. One characterizing aspect of the net shape processes is that there is not a direct cause-and-effect relationship between the processing variables and the product features. A ramification of that aspect of the processes is that there are no die features or press operating settings that

relate directly to the aforementioned list of objectives. There are no splitting or wrinkling control knobs. There are controls for shut height, tonnage, lubrication, blank location, and so on. The controls have different physical meanings and different metrics than do the objective requirements of the part.

To work through the aforementioned dilemma, the product objectives must be redefined as changed conditions of one or more transformation characteristics. A transformation characteristic is a characteristic of the material that can be transformed into some different condition through the application of energy. The mathematical relationship between the application of energy and the change in condition is known, as is the mathematical relationship between the new condition and the product objective conditions. For sheet metal forming, the transformation characteristics are strains and displacements. Thus, critical and target strains must be specified and the displacement pattern and critical displacements identified.

The critical strains are the forming limit of the material (Fig. 3). Conventional use typically requires that all major and minor strain combinations must plot at least 10% lower on the major strain axis than the forming limit curve. This approach typically ensures that the first requirement (i.e., no splitting) will be met. The target strains are usually the smallest strains that must be achieved and the location of those strains. In the selected example, the target strains are driven by the need to work harden the material up to a yield strength of 350 MPa (51 ksi), and these strains must be obtained or slightly exceeded throughout the expanse of the engine hood.

First, it is necessary to calculate the strain required to work harden the formed panel to the desired yield strength of 350 MPa (51 ksi). This is done using a power-law approximation of the stress-strain curve, with the following inputs:

- $K = 560$ MPa (81 ksi)
- $\varepsilon_0 = 0.006$
- $n = 0.18$
- $R = 1.5$
- $t = 0.8$ mm (0.03 in.)

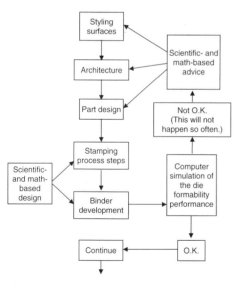

Fig. 13 Direct engineering. The future direct-engineering methods will change the experience-based advice-and-design practice shown in Fig. 12 to a scientific- and math-based practice, as shown here.

---

*The phrase *die face* is used here to differentiate between the creation of the die surfaces that actually contact the sheet metal and the design of the rest of the die. Once the shape of the die face, the forces acting on it, and the relative motions of those surfaces have been determined, the metal thickness is the only property of the material that is of any concern to the design of the structural and mechanical die. The design of the rest of the die happens after the die face is created and entails considerably more work by a different group or person.

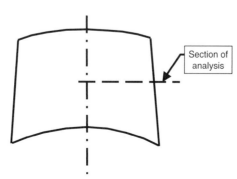

Fig. 14 Plan view of a stylized automobile engine hood used to illustrate the direct-engineering design method. The design developed will be for the addendum, binder wrap, and binder for the section of analysis identified here and shown in Fig. 15.

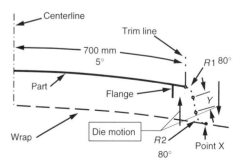

Fig. 15 Section of the automobile hood. The product shape (heavy line) is the section of analysis from Fig. 14 and is given as a simple 700 mm (28 in.) long arc through an angle of 5°. The addendum is shown in dotted lines as a parametric form consisting of R1, Y, and R2, ending at point X, which must be sized. The length of the wrap (dashed line) must be determined and then curved as necessary to fit between the centerline and point X.

The material behavior is defined by the stress-strain curve and the R-value derived from tensile testing. The tensile test imposes a very special strain relationship, because there is a force applied to the testpiece in only one direction. In a real part, pulling forces are applied to each small local region of the sheet in two directions. Also, as shown previously for the friction test, there can be a compressive force applied through the thickness when the metal is sliding over a curved die or punch surface. These real-situation forces must be mathematically related back to the tensile stress-strain curve. The equations for that relationship are well established in the literature and are used in all simulation and synthesis codes. As an example, when the aforementioned material properties are used in the power law to calculate the pulling stress (all other stresses are zero) for 5% strain in the pulling direction, one obtains 325.15 MPa (47.16 ksi). That is also called the effective stress (which is related to measurements after biaxial deformation). The yield stress would compute to 183 MPa (26.5 ksi). The effective stress is what would be measured as the yield stress from a tensile test specimen extracted from the formed part. The following table shows several strain pairs that could happen on a real part to result in the same 325.15 MPa (47.16 ksi) of effective stress:

Minor strain, %	Major strain, %
−2.5	+4.37
−2.0	+4.32
−1.0	+4.01
0.0	+3.48
+1.0	+2.77
+1.5	+2.34
+1.94	+1.94

Although several strain pairs will work, equal strains in both directions were chosen for the example because the engine hood is relatively square, and it simplifies the example. Straining the panel by +3.5% in both directions (hood centerline and normal) results in a formed panel yield strength of 357.7 MPa (51.9 ksi),which is only slightly more than the 350 MPa (51 ksi) desired. Note that the line of analysis of the hood centerline and its normal direction may or may not be identical to the major/minor strain axis determined from actual circle grid measurements or the output of a simulation.

The target strains must be achieved throughout the panel, but the location at which they are assigned by the designer to be the boundary condition for his calculations is at the neutral line. For the engine hood example, the neutral line is the front-to-rear centerline. The definition of a neutral line is the location on the panel where the designer wants no sliding motion of the sheet metal across the line as the die forms the part. The stretching of the sheet metal will result in sliding motion away from the neutral line, and the resulting displacements increase with distance from the neutral line. For the engine hood example, the 3.5% stretching of the 700 mm (28 in.) width of half the hood will be 24.5 mm

(0.96 in.) (3.5%×700 mm, or 28 in.) at the fender edge of the hood. For this example, those displacements will cause no problem with impact lines or skid marks as long as the centerline is maintained as a neutral line, hence satisfying two more objectives.

Highs, lows, flattening, overcrowning, twist, and size variation are all aspects of springback resulting from the elastic straining (i.e., shrinkage) that results from the relaxation of the forming stresses. The designer should review the forming stresses as the design progresses and ensure that the die face geometry and strain assignments do not result in stresses that will twist the part out of shape when they relax as the die opens.

True stresses in the direction along the line of analysis and normal (i.e., into and out of the drawing) to the line of analysis are the stresses that relax when the die opens. As these stresses relax, the part gets smaller with elastic strain. The amount of size change is very nearly equal to the ratio of these stresses to Young's modulus of the material times the length of the region of the part affected. If the forming stresses are not close to equal when they relax as the die opens, the part will warp out of shape. Such warping out of shape is generally referred to as springback. Technically, springback is the elastic relaxation of the forming stresses when the die releases the part. However, when that relaxation is proportional in all directions, it does not normally attract much attention. Sometimes, when the elastic relaxation is not proportional, the conflicting elastic strains balance forces before completely relaxing, and there will be residual stresses; thus, the full dimensional size change will not be realized. Adjacent forming stresses usually should be within 10% to avoid shape distortion of the stamping. A uniform shrinkage is readily accommodated by a simple global expansion of the die (similar to the shrinkage allowance built into casting molds) and is easy to calculate, but twisting and bending are difficult to predict and equally difficult to accommodate with compensating changes to the shape of the die. (See the article "Springback" in this Volume.)

The robustness of the design to normal and reasonable material and processing variation is maximized when the design can be formulated to provide lock beads instead of draw beads in the binder. When no metal slides through the beads, there is a minimum of variation in the resulting part. The design method should always incorporate a lock bead if at all possible.

Step 3. The designer must now determine if the sheet metal can be pulled on at the part perimeter with sufficient force to make the target strains happen at the neutral line and without having the material strain more than the critical strains anywhere. That pulling force is applied mathematically. The mechanics of how it will be accomplished come later. The section line of analysis for the product, as illustrated in Fig. 15, is approximated by a series of true arcs connected tangentially from the neutral line to the edge of the part. The illustration in Fig. 15 for the

engine hood example consists of a single such arc of 700 mm (28 in.) arc length and through 5° of arc angle.

The forming problem of the sheet metal being stretched over that arc of the die face is illustrated in Fig. 9. The forces to achieve bending and unbending plus the force to overcome friction must be added to the force to achieve the target strain to calculate the required pulling force at the part edge. This can be done by defining any two dimensions of the arc (i.e., radius, angle, or arc length), the coefficient of friction, and the normal strain (target strain) at the outside end of the arc (+3.5%, as previously determined). The bending and through-thickness stresses and the friction force values are then calculated to help the designer determine where the geometry is causing excessive restriction to the movement of the sheet metal across the die face, should there be a problem with strains approaching the critical strain limits. All of these calculations can be done with a spreadsheet application once the proper equations, material properties, and cell relationships are set up. In this example, such a spreadsheet program shows a 3.5% target strain, which is also the peak strain, and this should be small enough to cause no concern.

Note that most parts will have more than one arc representing the shape of the part between the neutral line and the edge of the part. The output of each arc must be used as the input to the next so that the part analysis can proceed to the edge of the part.

In summary, the designer can enter the problem as radii, angles, and/or arcs. Then, forces are calculated that will accomplish the bending and overcome friction. Finally, the amount of stretch in that segment is calculated, and the amount of wrap length that is needed to start with is determined.

Step 4. The task is now to determine the dimensions for the addendum shape, shown as $R1$, $R2$, and Y in Fig. 15. The parametric addendum shape would usually have 11 segments for unfolding flanges and trimming reasons, but the three shown in Fig. 15 are essential for forming the part and adequate for showing how the design process works.

In the example, there is an accumulated displacement of 25.97 mm (1.02 in.) determined from the spreadsheet calculations, which means that approximately 26 mm (1 in.) of sheet metal will be stretched off the punch and out of the part. That material must be put somewhere. However, the 26 mm (1 in.) of material is sliding generally horizontally while the press is closing the die vertically. Hence, the 26 mm (1 in.) of material must be dragged across $R1$ like a rope over a pulley. The $R1$ is simply another arc in the sequence, so the output of the calculation for the 700 mm (28 in.) arc of the product is the input for the calculations of what will happen to the metal as it is pulled across $R1$ and further stretched. The 80° of arc angle shown in Fig. 15 is the maximum arc angle for geometric and mechanical reasons. If there is a forming problem, that angle can be made smaller.

As an example, the designer can select a radius of 8 mm (0.32 in.) and an arc angle of 70°, which the spreadsheet calculations show to result in the maximum (i.e., peak) strain being only 15.55%. That peak strain is approximately halfway to the forming limit for the material, which looks perfectly safe. However, a 7 mm (0.28 in.) radius with the 70° angle is shown by the spreadsheet calculations to make the material split, as will a 75° angle with the 8 mm (0.32 in.) radius. The decision as to what the angle and radius must be depends on the designer's understanding of the precision of the die-making process and the ability to maintain the initial geometry of the die during the production life of the die. A general rule is to make the angle 5° smaller than the largest that will work, and to make the radii 1 mm (0.04 in.) larger than the smallest that will work. The maximum angle and the minimum radius that work should be those that work for all conditions within the range of reasonable and normal material and processing variation.

Using an 8 mm (0.32 in.) radius, a 70° arc angle, and a 15.55% peak strain, the spreadsheet calculation shows that there is an accumulated 29.7 mm (1.17 in.) of material sliding out of $R1$ into Y. Of that, 3.7 mm (0.15 in.) is due to $R1$ alone, added to the 26 mm (1.02 in.) stretched off the punch. That 29.7 mm (1.17 in.) of material will be subject to additional stretching once in Y. The opposing punch and die motions (shown by the die motion arrows in Fig. 15) cause stretching of the sheet metal in element Y. That stretching acts like a spring, generating the force to do all the work of stretching, bending, and overcoming friction throughout the sheet metal. Therefore, that element is called the stretch wall. All of the accumulated stretching of the arcs between the neutral line and the beginning of the stretch wall is used to make the stretch wall. (The total length of the line of analysis except for the stretch wall is the summation of the original individual arc lengths, each of which is stretched some delta length. The summation of these delta lengths is what is used to make the stretch wall.) The length of the stretch wall is not added into the accumulated wrap, and the actual length of the element in the die is subtracted from the accumulated displacement. If the arc length of the stretch wall is made to exactly equal the accumulated displacements, the accumulated displacement out of the stretch wall will be zero. If the stretch wall is longer than the accumulated displacements, the accumulated displacement out of the stretch wall will be the difference and will be negative. A negative displacement is draw-in.

To achieve the small negative displacement out of the stretch wall, the designer must specify the arc length to be greater than the total stretched length of the metal sliding into it. In the ongoing example, it can be calculated with the spreadsheet that the 29.7 mm (1.17 in.) of metal sliding into element Y would stretch to 34.73 mm (1.37 in.) while in element Y. As such, the designer can specify the element to be 35 mm (1.38 in.) long to obtain a −0.27 mm

(−0.01 in.) of displacement out. By building the die to these dimensions, the designer is assured that a locking bead can be used and that he has achieved the objective of maximum robustness to material and processing variation.

Note that a small negative displacement (usually between −0.2 and −0.5 mm, or −0.008 and −0.02 in.) is required to ensure that a stretch die construction will work. To test if the negative displacement could be excessive, one can divide the amount (in the previous example, 0.27 mm, or 0.01 in.) by the total length of the analysis line, including the stretch wall (which is 744.77 mm, or 29.32 in., for the previous example), to get the extra average strain in the line, should the negative displacement not occur. For the previous example, that would be 0.27/744.77 (0.01/29.32) or 0.036%. This error strain can then be added to the "along line" strain assignment to determine if it could possibly cause failure at any segment.

Step 5. The last step is to establish $R2$ and the binder shape. The parameter $R2$ is just another arc in series with all the others. However, because the displacement entering it is negative, the forces required to achieve bending and to overcome friction are subtracted from the accumulated forces. In the example, the friction force is 51 N/mm (7.4 ksi). With a lock bead at point X, a force of 289.8 N/mm (42.03 ksi) will be generated at this location, which in turn will generate the desired centerline characteristics outlined in step 1 (e.g., work hardening to 350 MPa, or 51 ksi; no skid marks off the centerline; and no splitting). Because the dimensions for the shapes in the addendum are now calculated, point X (Fig. 15) can now be positioned in space on the die design.

Finally, it is necessary to calculate the length of the material that must be contained by the binder between point X and the neutral line, as indicated by the dashed wrap in Fig. 15, which is determined to be 715.36 mm (28.19 in.). As such, the binder must be shaped (curved) so the length of the sheet metal when the binder has closed on it is 715.36 mm (28.19 in.) between the neutral line and point X.

Similar calculations will also give the amount of draw-in and the required draw bead restraining stress if a lock bead proves not to be feasible.

Direct engineering to establish the dimensions and shapes of the die addendum and binder has been illustrated. That work is done before any investment in drafting on the CAD system. These dimensions are then drafted into a die face on the CAD system, and the resulting surfaces are then analyzed with a computer simulation. The simulation must be set to evaluate the blank size, draw-in amounts, and the bead restraining stresses determined using the direct-engineering methodology outlined here, all of which can be facilitated by a spreadsheet to automate the calculations. Simple dies, such as the engine hood example shown, should not require computer simulation.

Spreadsheet Calculation Method. All of the processes and calculations described

previously are based on a fundamental understanding of the mechanics of metal flow. No one calculation is overly challenging, but there are numerous steps requiring a detailed setup. This lends itself to using a spreadsheet program to facilitate the calculations.

One such program has been developed by one of the authors as part of a binder development training program. The product analysis and binder synthesis (PABS) program is a Microsoft Excel-based spreadsheet that makes these calculations easy. PABS is based on power-law material behavior and is organized to support problems as described in this section. It performs the mathematical calculations for converting between engineering and true stresses and strains, computing effective strains and stresses, determining the force necessary on a hypothetical strip of material (with user-specified thickness) in the direction of the section of interest, calculating the frictional restraining force with a pulley friction formulation, converting bending energy into a required pulling force, and decomposing the results into strains on the output side of the subject die radius. Then, the force output of one radius is the force input to the next. PABS also keeps track of the accumulated stretching of the material. The equations used are evident to the user, and the user can change the formulations if other equations better describe the conditions being evaluated. The values quoted throughout the preceding example were generated with the PABS program for illustrating the engineering method.

Other Forming Situations. Many forming issues/questions can be resolved by even simpler techniques. Almost any section through a sheet metal stamping can be divided into a series of S-shaped inflections, as illustrated in Fig. 16. The metal to the left of the inflection sits on and slides along the lower die face (left horizontal arrow), and the metal to the right of the inflection sits on and slides along the upper die face (right horizontal arrow). The metal passing through the inflection is being stretched by the opposing motion of the two die halves.

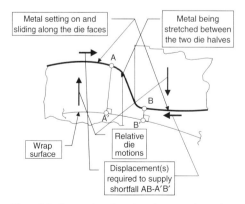

Fig. 16 Any section through a sheet metal part is a series of S-shaped inflections, where the metal is sitting on and sliding across the die faces on each side of the inflection and being stretched through the inflection by the opposing motions of the two die halves.

The length of material (length AB) through the inflection is often too long to be stretched from the available adjacent material (length A′B′) in the wrap. The difference (D = AB − A′B′) must be supplied from having material slide off one or both of the adjacent die faces, as indicated by the horizontal arrows. Because the lengths can be measured exactly in the CAD system, the amount of required displacement can be calculated exactly. In practice, both the left and the right horizontal arrows would then be drawn to the total length of the required displacement. Only one (or part of one and part of the other) would actually be used. The designer must decide which displacement best achieves the design intent. To decide, the displacement arrows of adjacent sections must be plotted to scale in the flat, and the resulting straining calculated, as illustrated in Fig. 17.

The resulting strains can then be plotted on a forming limit diagram (Fig. 3) and evaluated to determine if the locations are sufficiently safe. Then, a design decision can be made as to which displacement the die design should impose onto the material. The displacement thus determined becomes a target displacement for use in subsequent PABS calculations.

CAD Data Sets

The CAD data must be created and maintained during the design process. The content and controls on each and the interrelationships between them must be clearly defined. Likewise, the hierarchy of the files, how the files are generated, and how they are used must be carefully established and understood by those using them. To maximize the benefits, industry standards on the nomenclature and content of such files are likely needed.

To begin with, the concept of CAD/CAM is inadequate for net shape parts. The concept was created when computer automation was focused on the machining of machined parts. For stamping, the concept must be expanded to CAD-CAD/CAM-CAM. The first CAD is the computer-aided design of the actual piece part—the stamping. The second CAD is the computer-aided design of the dies that make the part. The first CAM is the computer-aided machining of the dies that make the part. The second CAM is the computer-aided manufacturing (e.g., robot programming, automation programming, weld schedules) and inspection (i.e., coordinate-measuring machines) of the parts.

The hierarchy of CAD files may be as illustrated in Fig. 18. Data must flow from one data set to the next, as indicated by the arrows in the figure. Within each box, some, but not necessarily all, of the data arriving is used, and additional information is added. When the data leave the last CAD function, they go to a CAM function, such as computer-controlled machining, robot programming, or materials handling design. The nature of the data in some of the function boxes is discussed in the following sections.

When data from one CAD file are to be forwarded to another CAD function, the data should be filtered so that only those data needed by the receiving function are sent to it. Otherwise, the receiving function must sort through the data. Such sorting by the receiving function is usually inefficient, because the data are new to the receiving function, and that receiving function must study and understand the data before deciding what to use. Much of the reading and understanding is redundant work. The sending function is familiar with the data and can sort them much more efficiently. Also, much of the sorting could be automated within the CAD system.

Requirements. The requirements CAD file defines what the piece part must do and what it must not do. The requirements include but are not limited to:

- Forces that will act on the part
- The location at which the forces will act
- Limits on deflections resulting from the forces
- Vibration frequencies
- Styling surfaces to be incorporated
- Mechanical connections to other parts (e.g., bolting patterns)
- Surface finishes
- Other objects that the part must not interfere with
- Corrosion resistance
- Mass targets
- Other requirements

Many of the requirements are drawn to scale, while others are simply listed.

Piece Part in Assembly. Most parts are assembled together with other parts to make a consumer product. The stamping is one of the piece parts and must usually be shown as assembled with the other components of the consumer product.

Piece Part in Service and in Free State. In many instances, the subject piece part is put under some loading when assembled into the consumer product or when used as a component of the consumer product. These loadings cause the piece part (the stamping, in this case) to deform elastically, as illustrated in Fig. 19 for an automotive engine hood.

The design of the engine hood must be in the correct position relative to the rest of the vehicle when it has been deflected by the load imposed by the holddown latch and resisted by the outboard downstops. However, for it to end up in that position when the loading is imposed on it, it must be made to some specific overcrowned shape, as indicated by the phantom lines in Fig. 19. The overcrowned shape must be established by the product engineer, and that shape is what is manufactured. Except for springs, the design of the free-state part is skipped. In most instances, the part is sufficiently stiff that it can be considered a rigid body, and the consumer product will still work. However, the need to reduce mass and to minimize costs will force the consideration of the free versus stressed conditions.

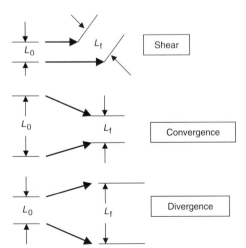

Fig. 17 When adjacent arrows are plotted to scale in the flat, the necessary strains can be calculated by the measured distances between the heads and the tails of the arrows, where the strain is $(L_f − L_0)/L_0$.

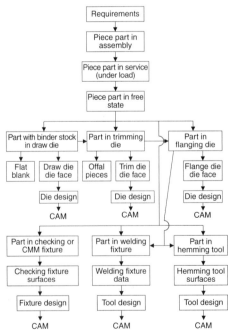

Fig. 18 Possible hierarchy of computer-aided design data sets for a typical automobile stamping. CMM, coordinate-measuring machine; CAM, computer-aided manufacturing

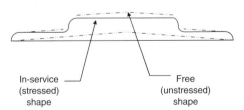

Fig. 19 Example of deformation in an automobile hood. The holddown latch force causes the engine hood to deflect when closed.

Part with Binder Stock in Draw Die. The piece of sheet metal from which the part is made is cut out of a coil. That blank can be rectangular, trapezoidal, or some developed shape, but it will be larger than the final part. Some of the extra material is the result of cutting from a coil that has parallel sides, and at least part of that exterior material comprises the binder area that is used to generate the pulling forces around the edge of the part that are necessary to stretch the sheet metal into the desired shape. The size and shape of that blank and the shape of the die faces outside of the actual part edge must be designed to impart those forces onto the sheet metal. The die shapes outside the part (and sometimes inside the part as well) are referred to as the binder development. Some of the final product shape may be out of position as it sits in the draw die when the die is closed on it. Those out-of-position regions are finish formed in subsequent dies. So, a CAD file must be made to define all of these geometries of the sheet metal.

Also, the draw forming operation imposes severe stresses into the sheet metal to form it. Then, when the die opens and releases the sheet metal, those stresses relax, and the sheet metal changes size and sometimes shape. It is the size and shape of the part defined by the free-state CAD files that must be achieved after this relaxation has occurred. Therefore, to compensate for the size and shape changes that result from this stress relaxation, the die (even in the product regions) must be designed and machined to a configuration that will be other than that defined in the free-state CAD file.

Flat Blank. A CAD file of the size and shape of the flat blank is required to build a blanking die, if one is needed, to calculate the material cost for the part and to specify the coil size to be purchased. (The blank may not be flat when the die has closed on it. That shape, whatever it is, is unfolded to the flat condition to determine the blank size and shape.)

Die Face of Draw Die. The previous CAD file was of the part as it is when the draw die is closed on it. However, that does not reflect what the shape of the draw die die faces should be. The die faces may not look much like the sheet metal. The die faces should only touch the sheet metal on one side at any location, but the surface that touches will change from side to side of the sheet metal, as illustrated in Fig. 20. Also, the die faces

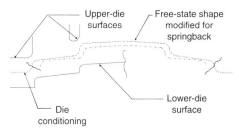

Fig. 20 Die conditioning in the forming of an automobile hood. The shape of the die faces (solid lines) may not follow the shape of the sheet metal (dashed lines), because the die should only touch the sheet metal on one side at any location.

must have additional surface features for runoff as the die face falls away from the sheet metal.

The die faces of the trimming and flanging dies must have offsets for trimming steel clearances, flanging material clearances, shear, trimming and flanging steel-to-nest bypass, and scrap cutters. These features are not reflected in the CAD file of the sheet metal sitting in the die.

Offal Pieces. It is easy to forget that the pieces cut off from the part in the trimming die (i.e., the offal pieces) must be managed after trimming. They must go somewhere and get away from the die. So, accurate CAD files of these pieces must be provided so that the proper materials handling of the offal can be designed and built.

Die Design. Once the die faces have been defined with CAD files, the die structure and mechanism can be designed. Each piece must be designed. Modern CAD systems use solid modeling so that the CAD files can drive cutter path programming directly. The die design must accommodate the fitting of the die to the press and integrate with the materials-handling automation.

Other Tools and Fixtures. Similar arguments and needs apply to all of the other tooling, such as the hemming tools, welding fixtures, and inspection fixtures. Welding CAD files are somewhat different in that only selected and small areas are extracted from the product surfaces for the locating and clamping points. Fixtures for spot welding need the x,y,z coordinates of each weld spot identified as a position in space, along with the surface normal vector of the weld gun alignment. Arc welding CAD files need the line to be welded and the surface normal for aligning the welding rod. The actual design of the fixture from this file is similar in functional concept to that for the dies, even though there is not similarity in appearance.

Computer-Aided Manufacturing. When a tool design CAD file arrives at the Computer Numerical Control machining station, the cutter path programmer should have no decisions as to what is to be made. He should only have to decide how to make it.

Significance. Significant efficiencies can be realized by having a proper hierarchy of CAD files and a clear understanding throughout the organization (and supplier community) as to what each file is called, what it contains, who is responsible for its content, and how it is to be used. As soon as these relationships between the data and the physical functionality are mixed up, enormous inefficiencies materialize.

This article places considerable emphasis on the need for the designer to clearly define the die/tooling faces in the CAD system before the data are passed on to the construction functions. This approach differs from what has typically been done. It has been the general practice to pass on to the tooling construction source only the design of the part as it sits in the closed die or tool. The tool or die builder (e.g., cutter path programmer) then must make many of the die/tool face functionalizing decisions and modify the CAD

data prior to programming the cutter paths. The suggested approach provides for a more controlled and consistent implementation of tooling and processing standards, eliminates a learning step, eliminates some sorting through of the data and the learning processes, and, in general, places the decision-making functions into engineering activities, where both the computer tools and the people's mind sets are more in line with the job functions.

REFERENCES

Advanced high-strength steels

1. D.J. Schaeffler, Advanced High-Strength Steels: Enhanced Formability Adds to Vehicle Strength, *Stamp. J.*, Nov 2004
2. D.J. Schaeffler, "Introduction to Advanced High Strength Steels—Part I: Grade Overview," www.thefabricator.com/Articles/Stamping_Exclusive.cfm?ID=1139, FMA Communications, Inc., Aug 2005
3. D.J. Schaeffler, "Introduction to Advanced High Strength Steels—Part II: Processing Considerations," www.thefabricator.com/Articles/Stamping_Exclusive.cfm?ID=1158, FMA Communications, Inc., Sept 2005
4. www.autosteel.org
5. www.worldautosteel.org

Stress-based forming limit curves

6. T.B. Stoughton, A General Forming Limit Criterion for Sheet Metal Forming, *Int. J. Mech. Sci.*, Vol 42, 2000, p 1–41
7. T.B. Stoughton, Stress-Based Forming Limits in Sheet Metal Forming, *J. Eng. Mater. Technol.*, Vol 123, 2001, p 123
8. T.B. Stoughton, "The Influence of Material Model on the Stress-Based Forming Limit Criterion," SAE 2002-01-0157, Society of Automotive Engineers, 2002
9. T.B. Stoughton and X. Zhu, Review of Theoretical Models of the Strain-Based FLD and Their Relevance to the Stress-Based FLC, *Int. J. Plast.*, Vol 20, 2004, p 1463–1486
10. T.B. Stoughton and J.W. Yoon, Sheet Metal Formability Analysis for Anisotropic Materials under Non-Proportional Loading, *Int. J. Mech. Sci.*, Vol 47, Issue 12, Dec 2005, p 1972–2002
11. T. Kuwabara, K. Yoshida, K. Narihara, and S. Takahashi, Forming Limits of Aluminum Alloy Tubes under Axial Load and Internal Pressure, *Proceedings of Plasticity '03*, NEAT Press, 2003, p 388–390

Springback predictability

12. Springback Predictability Project (SPP), NIST-ATP Final Technical Report, NIST Cooperative Agreement 70NANB5H1149, Dec 2000

Computer-Aided Engineering in Sheet Metal Forming

Chung-Yeh Sa, General Motors

SHEET METAL FORMING is defined as deforming thin sheet metal (or tubes) into a useful shape by means of plastic deformation using a set of dies. It was a black art in real production until the 1980s. Many fundamental theories in plasticity had been published prior to this time. However, most application research work at the time was limited to simple shapes or idealized conditions. Hence, the direct impact of that work to practical applications in the industry was not very significant.

Computer-aided engineering (CAE) analysis using the finite element method was very successful in many different application areas in the 1980s. Unfortunately, the technology was not mature enough to be useful for daily production work until the mid-1990s. Since then, however, great progress has been made in this field. This article focuses mainly on the technology breakthroughs that make forming simulation a routine work throughout the industry.

Background

Industry Trends. Since the dawn of the 21st century, the tool and die industry in developed regions such as North America has experienced several challenges that affect its ability to remain competitive in the marketplace. Among the most significant challenges are overcapacity, growing international competition, technology improvements, decreasing demand from automotive customers, and increasing pressure from customers to lower prices and expand services. Original equipment manufacturers (OEMs) and first-tier sheet metal parts suppliers have begun sourcing to low-cost producers around the world in recent years, which adds strong negative impact to the traditional tooling industry in developed countries. These challenges have resulted in many tool and die shops going out of business during the first few years of the new millennium.

To survive and remain competitive in the marketplace, tool and die makers have to adopt new practices, such as the lean manufacturing process, new technologies, and collaboration. These practices have been shown to result in cost reduction and increased manufacturing performance. One innovative practice is the adoption of forming CAE to enable early evaluation of the product and die-face development. Proper implementation of this technology will help provide better cost estimation, lead time, and cost reduction, as well as help to meet quality and fast-to-market requirements. The CAE technology essentially replaces the traditional trial-and-error black art with scientific engineering calculations. Furthermore, tool and die makers can employ highly efficient and process-oriented tools to enable designers to use stamping CAE tools.

The aesthetics of automobile design have evolved dramatically, especially in the past 5 years. Modern vehicle designs embrace a more aerodynamic shape, with large sweeping curves, sharp features with tight radii, and retro styling with round but deep panels. The unveiling of these new designs turns what were once just futuristic concept cars into mainstream production vehicles. This trend brings many unknowns and challenges to the sheet metal industry.

Another challenge comes from an increased market demand for "green" (i.e., fuel efficient) vehicles with stringent safety criteria. This trend requires manufacturers to evaluate and use new materials for structural components and body panels. As a result, many new aluminum alloys and new advanced high-strength steel have been developed by metal suppliers and used by OEMs. The use of dual-phase, complex-phase, transformation-induced plasticity steels, and even ultrahigh-strength steel in the automotive industry has been increasing significantly in the last 10 years. These new materials are much stronger than the conventional mild steel (Fig. 1), thus allowing gage reduction while providing better crashworthiness.

Inevitably, the use of higher-strength steel leads to more challenges for traditional formability issues and dimensional accuracy due to springback. Due to the lack of experience with these new materials, the conventional wisdom for formability and die development may fall short and sometimes will be erroneous.

Using CAE technology is a perfect solution for both challenges. Engineers can test out the new materials in a virtual environment before committing to build the tool. For example, simulation allows CAE analysts to adjust boundary conditions (e.g., binder tonnage) to quickly identify the optimal value. It is much faster and more reliable than trial-and-error, especially for complex parts. The CAE technology also allows the investigation of several alternative forming options, such as tailor-welded blanks, sheet hydroforming, and superplastic forming, before making a feasible and cost-effective decision.

Sheet Metal Forming Processes. Sheet metal forming is one of the most important manufacturing processes in the automotive industry. Several different processes fall into this category. Among them, the conventional stamping process is the most widely used method because of its high speed and low cost. In this article, the term *sheet metal forming* implies primarily the stamping process and some variations listed as follows. The CAE technologies needed for simulating these forming processes are very similar:

- Conventional stamping
- Sheet hydroforming
- Tube hydroforming
- Superplastic forming

To limit the scope of this article, unusual forming processes, such as warm forming, electromagnetic forming, and explosive forming, are not covered.

One of the prerequisites for obtaining good results with CAE is to simulate the forming process as closely as possible to the physical process. The CAE by itself is not a magic tool. It will only give results of the process being modeled. The analysts must understand the details of the physical process and model it accordingly.

In a conventional stamping process, sheet metal parts are generally made in four stages, not including the blanking operation:

- Stage 10: (draw) forming
- Stage 20: trimming
- Stage 30: flanging
- Stage 40: final trimming and flanging

The stages are often called operation 10 through 40. The first die is the most important operation, in that it makes the majority of the final product shape. (Note: The term *die* usually implies the entire die, although sometimes this term refers to the female die, or die cavity). The first die has four pieces (punch, die, upper binder, and lower binder). This design allows die makers more flexibility to control the operation and to make good parts. However, it is rarely used at the present time due to the high cost. The other three stages often perform combined functions for complex parts. For example, the flanging die may perform a forming or restriking function in some areas while performing the flanging operation around the perimeter.

The majority of the deformation in the first die in a stamping operation is draw forming; hence, it is often called the draw die. Generally

speaking, there are two types of three-piece die: toggle draw (Fig. 2) and air draw (Fig. 3). The schematic drawings in the figures show the two different concepts. Each one has its advantages and disadvantages. The toggle draw requires a double-action press so that the punch and the binder can be controlled independently. In this process, the sheet metal blank rests on the binder portion of the lower die, which is stationary throughout the entire forming process. The upper ring moves down to force the blank to conform to the binder and hold the sheet metal in that position. The punch then follows to deform the sheet metal into the die cavity. The binder force may increase as the blank becomes thicker as it is being pushed into the cavity.

The air draw is the opposite of the toggle draw. The punch is stationary and sits at the bottom. The lower ring, supported by air pins or nitrogen

cylinders, is generally set above the punch so that the blank rests on the ring initially. The female die is on top of the blank and is attached to the ram. As the die moves down, it initially forces the blank to conform to the binder. The die continues to wrap the blank over the lower punch as the blank is being held by the upper and lower binder. The maximum binder force is dictated by the nitrogen cylinders. Insufficient binder force may force the binder open during the process.

Formability and Sheet Metal Forming. Before using any CAE methodology to simulate a process, one must understand the goal of the analysis. The goal for metal forming is to solve formability problems. Formability is defined as "the ability of the material (sheet metal) to be formed (under the specific process conditions) into the designed shape without failures." This simple statement covers five important issues. First of all, failure must be adequately and accurately defined. The other important concept is that formability is governed by four groups of factors:

- Product design (shape of the part)
- Blank used (the mechanical and physical properties of the blank)
- Die-face design (the addendum and binder)
- Forming process and operations

All four groups play an important role in affecting the formability in sheet metal forming (Table 1).

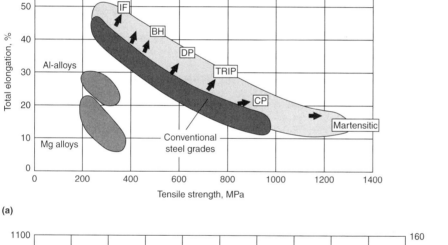

(a)

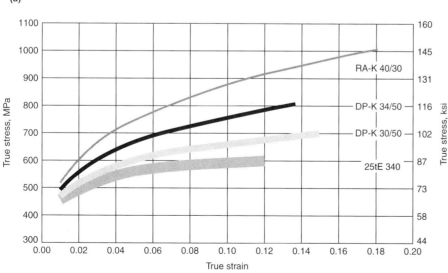

(b)

Fig. 1 Trends in development of steel sheet properties. (a) Elongation and tensile strength of sheet steels. (b) Stress-strain curves for different cold-rolled steel grades (without predeformation, aged 170 °C, or 340 °F, for 20 min). IF, interstitial free; BH, bake hardened; DP, dual phase; TRIP, transformation-induced plasticity; CP, complex phase

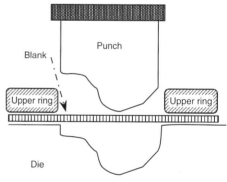

Fig. 2 Section view of a die in a double-action toggle press

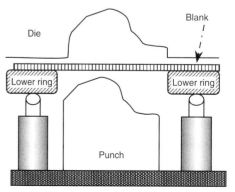

Fig. 3 Section view of a three-piece air-draw die

Although there are many factors, not all factors have the same impact on overall formability. Generally speaking, the following list gives the primary factors that a CAE model should focus on:

- Product design (shape of the part)
- Die-face development
- Blank shape and location
- Boundary conditions
- Material properties
- Process development

The most challenging and sometimes confusing part is that there is a strong interaction among these factors. The importance of each factor often changes, depending on other factors. For example, boundary conditions and blank shape/location may become more critical as the blank material is changed from steel to aluminum. A well-designed product and die-face development will be less sensitive to friction variations.

The formability concept discussed previously represents only a narrow view from the manufacturing perspective. A more complete definition of formability should be broadened so that "the formed part must meet all intended design functions and performance criteria." It is essential to embed the product requirements in the formability guideline. If a product design is completely dominated by formability issues, many decisions made will lead to higher cost or an inferior design not meeting performance criteria.

For example, aluminum-killed deep-drawing-quality mild steel (AKDDQ or DDS) offers excellent formability. It can be used to make complex or challenging parts. However, this grade of steel is soft and has relatively low yield strength. Structural parts made of this grade will have less desirable structural performance (including crash/safety) unless the gage of the steel is increased to compensate for the lack of strength. This countermeasure increases part weight and may not be an acceptable solution in today's (2006) mass-conscious environment.

The narrow view of formability may also drive the design to be rounder (large radii), shallower, and more open angle (relative to the vertical wall). Those changes, while sometimes necessary, will compromise the structural performance. An optimal product development needs to carefully balance all requirements, including other manufacturing concerns, mass, and cost. The CAE offers opportunities to evaluate many of those issues before the die and other hardware are built.

Stamping simulation technology has matured over the past decade. Finer and more detailed models have drastically improved the forming analysis capability from investigation level to tuning and validation, with proven productivity gain and cost-efficient solution. However, during recent years, new requirements have surged, such as quality, springback, and tolerance control, as well as integration of the manufacturing effects (including assembly) on product performance analysis, such as fatigue, crashworthiness, and so on. These requirements have become even more persistent with the introduction of new material grades such as aluminum alloys, ultrahigh-strength steels, dual-phase steels, and so on in order to control energy consumption in the transport industry in particular.

Current available stamping simulation software, which was designed more than 10 years ago to support formability issues, has started to reach its limits. Consequently, new-generation software has become a necessity to handle new industrial requirements. In-depth solver review, including element formulation, contact algorithm, parametric material modeling, and so on, was fundamental in ensuring the accuracy improvement of the forming results regarding stress and strain within the required tolerance. Moreover, the strong time-to-market pressure left a very narrow time window to perform forming simulations in order to impact the design decision. Therefore, the overall simulation process has been upgraded and streamlined. Quicker die run-off design and a quicker incremental solver fill the gap between die design and simulation. The objective is to create the first draft of a die addendum in just a few minutes. This allows earlier simulation involvement, process optimization, and eliminates costly downstream problems. Furthermore, massive parallel processing (MPP) represents the optimal solution for the speedup of the validation phase with the hardware evolution. The industrial MPP version needs new innovative solver architecture. This new architecture also allows strong interaction with the customized user interface. Advanced object concept and new data model assess sophisticated simulation data management and transfer for active process/performance coupling analysis, in a collaborative and concurrent engineering environment.

Evolution of Computer-Aided Engineering for Sheet Metal Forming—History Overview. The use of the finite element method (FEM) for sheet metal forming applications created strong interest for researchers and industry users in the fourth quarter of the 20th century. Throughout the 1970s and early 1980s, most studies were limited to either axisymmetric or plane-strain problems or to three-dimensional (3D) problems with simple geometry, such as cups. However, these studies only provided general information on a variety of important issues on sheet metal forming and were of little use for the complicated 3D sheet metal forming problems. Except for can making, the technology did not offer practical industry applications.

The technology continued to grow in the ensuing 10 years, with an emphasis on real applications in the automotive industry. People began identifying and studying various issues related to CAE methodology for sheet metal forming. Better material models, such as Barlat's 1989 and Hill's 1990 yield loci, were published (Ref 1, 2). More practical contact treatment was developed. Some automotive applications, such as roofs, fenders, and hoods, were studied. At this stage, different attempts were also made with non-FEM-based approaches, such as expert system and knowledge-based methods.

In the 1990s, more vigorous research and development was being carried out to improve the reliability and accuracy of sheet metal forming simulation. Many valuable papers were published in the NUMISHEET and NUMIFORM series of international conferences (see the Selected References), although a good number of papers also appeared in various conferences (such as the SAE Congress) and publications. NUMISHEET has become the most recognized conference for sheet metal forming simulation. It is exclusively dedicated to sheet metal forming (including hydroforming) simulations and has been organized by volunteers without the support of any permanent organization since the first conference was held in 1991 in Zurich. The conference traditionally covers the following subjects:

- Theoretical development in all aspects
- Experimental works
- Industrial applications
- Benchmarks

Since 1993, the technology has made significant and rapid progress in many different areas. Considerable efforts were made in improving the challenges for dealing with complex industrial parts so that more process details and bigger

Table 1   Variables that affect formability

Variable	Property affected
Product	Shape (depth, radii, angle)
	Flange
Blank physical characteristics	Shape
	Location
	Thickness
	Edge condition
	Tailor-welded blank weld line
Blank material mechanical properties	All properties related to the constitutive model/equation
Die	Face design (tip, addendum, binder)
	Surface finish and coating
	Material
	Surface hardness
	Rigidity
	Guidance
	Alignment
	Temperature
Die lubrication	Friction
	Lubricant type
	Film thickness
	Wetness (distribution)
Process	Forming type (stamping, hydroforming, etc.)
	Forming process (trimming, flanging, etc.)
	Ram guidance and speed
	Counterbalance/rigidity of the press
	Shim blocks
	Ambient temperature and humidity
	Prebending
Process control	Binder pressure/force
	Pad pressure/force
	Draw bead

model sizes can be analyzed. Several critical technology breakthroughs occurred in the mid-1990s. Most notable were the successful development of automatic meshing for dies and the adaptivity for the deformable sheet metal. The main emphasis during that period was focused on formability evaluation from blanking through trimming operations. Users were looking for answers to straining, thinning, wrinkling, and failure predictions.

There were two major breakthroughs in the next period, beginning in 1999. One was the improvement of springback prediction, as a result of several global multiyear consortia. The most important of these was the Springback Predictability Project, funded by the National Institute of Standards and Technology Advanced Technology Program (NIST-ATP). Due to the work of this consortium, the industry finally had a code that could provide decent springback prediction for some parts. Another major development was the emphasis of a user-friendly metal-forming system that integrated all needed functions, including some computer-aided design (CAD) work that used to be done outside of the system.

The rapid developments of software technology, together with faster and lower-cost computer hardware, have enabled many metal-forming operations to be modeled cost-effectively. Simulation lead time has been decreasing significantly since 1990. Figure 4 shows that the simulation lead time for total analysis (including meshing, setup, calculation, and visualizing the results) of one panel (based on a front fender from Mazda) was reduced from 50 days in 1990 to approximately 10 days in 2002.

The total turnaround time for all tasks, including die-face generation, formability analysis, and springback prediction, has been reduced to less than one week since 2004. These data are very representative throughout the global industry. This dramatic reduction in total simulation time was also influenced by standardized setup and more experienced CAE analysts to produce a rapid and qualitative forming evaluation.

Proper use of forming simulations can effectively reduce physical testing and costly downstream problems by solving the problems upfront in the early development stage. Over the years, successful stamping simulation has helped to considerably reduce the costs and lead time for the automotive industry, partially due to the technology growth. The industry applications of sheet metal forming simulation have grown enormously in recent years as the benefits of troubleshooting dies on the computer, rather than through extensive shop trials, have been realized.

There are countless examples of these savings. For example, Hyundai was able to cut total turnaround time by half, while improving the quality in the development of a full-body side panel with the use of stamping simulation. This led to a large increase in the need for CAE analysts for this field, and a huge increase in further research and development work supporting the technology.

Over the last decade, more drastic lead time and cost reduction have come from overall process optimization with the introduction of concurrent engineering, parallel tasking, and the optimization of different tasks using the CAE technology. Further speedup in this area is expected with the emergance of new computer technologies and the industrialization of MPP technology using modern configuration such as clusters and so on.

Methodology

The term *computer-aided engineering* simply means performing engineering analysis as the product is being designed and developed. It does not imply finite element analysis (FEA) using FEM. However, many people use all three terms interchangeably, mainly because FEM has become more powerful and accurate in the last 10 years. Supporting technology such as automatic mesh generation has also improved significantly. Hence, other simpler forms of analysis have become nearly obsolete.

For the completeness of this article, many forms of CAE methodology are discussed.

However, the emphasis is given to software tools using 3D FEM technology.

Simplistic Approaches. There are many different approaches to analyzing formability. Length-of-line analysis is at one end of the spectrum, and 3D FEA is at the other. The length-of-line analysis is unbelievably crude in today's (2006) standard. However, it was used well into the 1990s, before any CAE methodology became truly useful. The method simply calculates the ratio between the final length of a cut section and its original length (Fig. 5).

Before 3D FEM became practical, some 2D analysis techniques were explored. The earlier approach used either bar or beam elements to model a section line of a forming die. One good example of such a code is LINEFORM, developed at General Motors. These 2D solutions could take care of the challenging sliding contact problems in an ideal condition. A more advanced approach was to use shell elements with additional boundary conditions to force a 3D model to behave like a 2D problem. The first approach is much faster but has very limited applications. The second approach offers more realistic contact, especially for a more complex section line. However, it is also much slower compared to the previous method.

There were two major reasons to use the 2D approach. First, 3D FEM was not very practical until the mid-1990s. The 2D approach was much faster and more stable. Secondly, developing the die face on the computer was a very slow process in those days. The 2D solution provided some guidelines to the initial die-face design. Although the technology was relatively crude compared to a 3D analysis, it was better than experience-based die-design methodology. With the advancement of CAD and CAE technology, the need for 2D FEM simulation has diminished tremendously.

3D FEM Approaches. The earliest form of FEM was first introduced in the 1940s. The initial general application was for structural stiffness analysis in the 1960s. However, simulation of general sheet metal forming problems came much later. The work done by Wang and Budiansky was probably one of the earliest successful cases to model it in 3D formulations (Ref 3).

There are many technical reasons that 3D FEM simulation for metal forming did not become practical for production use compared to other application areas. In order to obtain reasonably reliable results for real production parts, the following formulation issues for FEM technology needed to be resolved:

- Large displacement
- Large deformation
- Large rotation
- Large sliding with nonlinear contact
- Efficient and reliable contact treatment
- Element formulation to handle through-thickness variation in order to model combined stretching and multibending and unbending

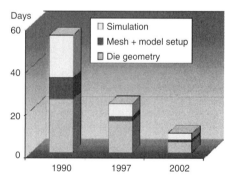

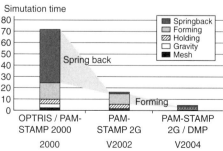

Fig. 4 Reduction in lead time for simulating the stamping of an automotive panel

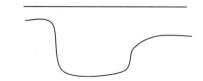

Fig. 5 Schematic view of a section

- Advanced material model to handle anisotropy and cyclic loading
- Advanced friction model to capture the surface traction
- Intelligent adaptivity algorithm to refine the mesh as needed

A significant advancement in this area began in the early 1990s. The majority of the issues listed have been overcome with new technology that was developed and enhanced between 1990 and 2000. Although most of the early work was done by academia, codes developed by some commercial software companies have been the primary workhorses in real-world applications. The current market is primarily dominated by AutoForm of AutoForm Engineering, LS-DYNA of Livermore Software Technology Corporation, and PAM-STAMP of ESI Group. They successfully built their own codes based on early codes from academia. Over the years, they continuously enhanced their codes with in-house development or implementation of published work. Few people in the industry use codes other than these three codes.

One of the most important technologies in a 3D FEM code is the time-integration scheme. Generally speaking, this can be divided into two major families: explicit time integration and implicit time integration. Both algorithms can be applied to either static or dynamic governing equations. Hence, there are four possible fundamental formulations used by the FEM codes:

- Static Implicit (SI)
- Static Explicit (SE)
- Dynamic Implicit (DI)
- Dynamic Explicit (DE)

Among them, the static implicit formulation is widely used for static or quasi-static problems such as structural analyses. The dynamic explicit formulation is widely used for dynamic problems such as drop tests and vehicle crash simulations. However, the choice of integration algorithm for metal-forming applications was not that obvious in the early years as FEM started to be used in this field. The majority of researchers thought static implicit formulation was more rigorous and suitable for accurate results for metal-forming simulations.

By 1990, almost all successful sheet metal forming simulations published in major conferences and technical journals used the static implicit approach, with a few exceptions using the static explicit formulation. As papers using dynamic explicit formulations began to increase

throughout the 1990s, the debate about this issue continued for many years. One major reason fueling the debate was that the majority, if not all, of the codes used a single time integration method. Most debates seemed to focus on justifying the superiority of their own codes rather than a fair comparison of the advantages and disadvantages of each method. Table 2 gives some well-known codes representing each method.

Note that the codes listed in Table 2 were those popular during that period. Some of them may not be available today. SHEET-x represents several codes from The Ohio State University. The dynamic implicit method was available as an option in some dynamic-explicit-based codes.

Many comparison works were published by static implicit proponents using unreasonable parameters to illustrate why dynamic explicit was not appropriate for metal-forming simulations. Those conclusions were obviously very misleading. There is no single ideal method that can be used for all types of forming problems or all stages of the forming process. The choice of method hinges on a good balance of many critical issues, depending on the specific application being simulated.

The static implicit formulation used for metal-forming simulation is very similar to the one used for structural analysis. It transforms all externally applied force (such as the moving punch) into internal energy, due to the deformation of the sheet metal. The equilibrium state is maintained at every state. This method has been demonstrated to be far superior to other methods simulating small-scale lab-type parts such as cups. The drawbacks of the static implicit method are its storage and central processing unit (CPU) requirements that grow nearly as a cubic function of the number of elements. As the size or the complexity of the geometry grows, the static implicit approach becomes less practical. As computer random access memory price decreases and CPU speed increases as Moore's law predicts, this may become more possible but still less efficient compared to dynamic explicit codes.

There are other issues that limit the widespread use of static implicit codes in a production environment. Sheet metal forming is not really a "pseudostatic" problem. In a typical stamping operation, the actual forming completes the total travel of 100 to 150 mm (4 to 6 in.) in 2 to 3 s, with the peak speed close to 500 mm/s, (20 in./s). The sheet metal flows in various

directions and changes in the process of forming. Most areas (nodes in FEA) of the sheet metal have large displacement and large rotation in addition to large deformation (in elements). This limits the step size of static implicit codes in order to capture the physical behavior of the metal.

The other challenge is that the contact between the metal and the dies is highly nonlinear. Correctly capturing the contact and calculating the friction force is not an easy task for static implicit codes. Capturing the onset of wrinkling and correctly predicting the actual formation of wrinkles are also difficult challenges for static implicit codes. One fundamental characteristic of the static implicit algorithm is the need to invert the stiffness matrix. As the complexity and the size of the problem grow, convergence slows down tremendously or cannot be accomplished at all.

The basic governing equation of dynamic explicit codes has an inertia term similar to the following equation. Different codes may use different formulations, but the concept of the momentum equation is the same:

$$f^{int}(u) + c\dot{u} + m\ddot{u} = f^{ext} \qquad \text{(Eq 1)}$$

where f^{int} is the internal force, f^{ext} is the external force, c is damping, m is mass, u is displacement, $\dot{u}$ is velocity, and $\ddot{u}$ is acceleration.

Critics of dynamic explicit codes argue that equilibrium is never maintained. This statement is a little misleading. In dynamic explicit codes, equilibrium is dynamically balanced. It is true that the inertia force may be exaggerated, mostly due to user error. In that case, the quality of the solution will deteriorate. If a user chooses a proper setting so that the inertia is no more than 5%, the solution quality should be good.

Static explicit codes were presented as taking the advantages of both static implicit and dynamic explicit codes. It was actively promoted in the 1990s. The results seem to indicate that static explicit codes have the weaknesses of both codes. The popularity of this group greatly diminished in the late 1990s after the third NUMISHEET conference in 1996.

AutoForm, from AutoForm Engineering, is a unique code. Some people classify it as a *fast implicit* code. It uses the same governing equation as the static implicit codes. The major reasons for its fast performance are that it uses membrane element and some decoupled proprietary techniques. Among them, the matrix is decoupled and simplified. This allows the matrix to remain small with a high convergence rate. Although many experts doubt the technical soundness of some of the techniques used by AutoForm, the code seems to generate relatively good results in most cases.

Adaptivity. Proper mesh definition is always important for all FEM simulation. However, it is not practical to use fine mesh for the entire blank from the beginning of a forming operation due to the high CPU hours needed. Adaptivity can intelligently refine the blank mesh only in the

Table 2 Summary of metal stamping codes by integration algorithm

Method	Code names
Static implicit	ABAQUS (standard), ADINA, MARC, INDEED, NIKE, Metalform
Dynamic explicit	LS-DYNA, PAM-STAMP, OPTRIS, RADIOSS, LLNL-DYNA, ABAQUS/Explicit, DYTRAN, DYNAMIC
Static explicit	ITAS, ROBUST, SHEET-x, Panelform

Note: ABAQUS, ABAQUS, Inc.; ADINA, ADINA R&D, Inc,; NIKE, Lawrence Livermore National Laboratory; LS-DYNA, Livermore Software Technology Corp.; PAM-STAMP, ESI Group; OPTRIS, Dynamic Software; RADIOSS, Mecalog Group; LLNL-DYNA, Lawrence Livermore National Laboratory; DYTRAN, MSC Software; SHEET-x, The Ohio State University

areas needed. Users do not have to figure out a "smart" mesh pattern to capture the deformation gradient correctly. It is especially essential for the deep drawing operation, when the sheet metal often has large displacement. Adaptivity has been around for many years but was not popular until 1995, due to some limitations of earlier adaptive algorithms.

Since 1997, it has become standard practice among the majority of CAE analysts to use adaptivity regardless of the software used, because the technology provides fast and reliable results. By the late 1990s, all metal-forming software also developed intelligent algorithms that can foresee contact problems and refine the mesh before it is needed. Although the details of the algorithms may be different, they are all based on the element normal differences between adjacent elements in the approaching tooling mesh.

The algorithms work well in most stamping cases. However, a different criterion may be needed in some special cases. It does not work at all in other types of simulation, such as crashworthiness. It is desirable to modify the algorithm to use a different index other than angle differences. Those indexes may include thinning, effective stress, or effective strain.

The purpose of adaptivity is to refine the mesh. There are three fundamentally different approaches: *h*-adaptivity, *p*-adaptivity, and *r*-adaptivity. The most common method used by sheet metal forming software is *h*-adaptivity. This method splits the elements when the criterion is met; hence, it is dubbed as "fission." Although there are some variations in *h*-adaptivity, often each element is split into

four elements. Adaptive constraints are applied to middle nodes to take care of connectivity requirements. This method is relatively easier to implement, especially for a complex code so as to avoid incompatibility issues. Figure 6 is an example of a uniform blank mesh and a refined mesh at the end of forming, with four levels of refinement.

It is potentially possible to merge previously split elements into a bigger element when desired. This is often dubbed as "fusion." The main purpose of fusion is to speed up the calculation and reduce the memory required. However, fusion provides some but not significant speed-up benefit for dynamic explicit codes. On-the-fly fusion, shown in Fig. 7, often causes more perturbation and creates more stress noise in dynamic explicit codes. Hence, it may not be a preferred practice. However, it can be applied between operations to save time for the subsequent calculation. This is also called coarsening by some codes.

Material model is very important to get good results in a metal-forming simulation. There are many available material yield criteria to describe different material properties. One of the most popular models used in the early years of stamping simulation was a planar-isotropic, which was based on Hill's material model published in 1948 (Ref 4). This yield criterion is an improvement over the von Mises yield criterion by accounting for normal anisotropy. This planar-isotropic yield criterion for plane stress is:

$$\bar{\sigma}^2 = \sigma_{11}^2 - \frac{2R}{(R+1)}\sigma_{11}\sigma_{22} + \sigma_{22}^2 + \frac{2(2R+1)}{(R+1)}\sigma_{12}^2 \quad \text{(Eq 2)}$$

where σ is stress, and R is the anisotropy parameter.

Considering only the principal stresses, this equation can be simplified as:

$$\bar{\sigma}^2 = \sigma_1^2 + \sigma_2^2 + R(\sigma_1 - \sigma_2)^2 \quad \text{(Eq 3)}$$

This yield surface is an improvement over the Tresca yield surface (an irregular hexagon) and the von Mises yield surface (an ellipse). However, it seems to exaggerate the influence of the anisotropy, especially in the biaxial stretching quadrant. Hosford proposed a yield criterion in

1979 that is mathematically similar but with a high-order exponent (Ref 5):

$$\bar{\sigma}^m = \sigma_1^m + \sigma_2^m + R(\sigma_1 - \sigma_2)^m \quad \text{(Eq 4)}$$

where m is a mathetical exponent.

Exponents of $m = 6$ and $m = 8$ were proposed for body-centered cubic materials such as ferritic steel and face-centered cubic metals such as aluminum, respectively. This locus lies between the Tresca criterion and Hill's quadratic form and matches experiments better, compared to the other three criteria. However, this minimizes the influence of anisotropy due to its high exponent, and it does not address the planar anisotropy issue. Deep-drawn cups such as beverage cans made of certain metals have a strong earing phenomenon. This is due to the directional anisotropy in the plane of certain metals, such as the extradeep-draw-quality steel. The following equation is a general form of the aforementioned yield function that considers the anisotropy in both the rolling and transverse directions:

$$2\bar{\sigma}^m = \frac{2}{(1+r_0)}[\sigma_1^m + (r_0/r_{90})\sigma_2^m + r_0(\sigma_1 - \sigma_2)^m] \quad \text{(Eq 5)}$$

where r is the anisotropy parameter.

When $m = 2$, Eq 5 is the same as the von Mises yield criterion if $r_0 = r_{90} = 1$, and it represents Hill's 1948 quadratic yield criterion if $r_0 = r_{90} = R$. It coincides with Hosford's 1979 yield criterion given in Eq 4 when $r_0 = r_{90} = R$.

While the thinning calculation in dynamic explicit codes is less dependent on the stress and material model, it is crucial to have a good material model and stress calculation for springback analysis. As the use of aluminum and advanced high-strength steel increases, springback becomes a pronounced concern. Accordingly, the accuracy of springback prediction is vital before it can be minimized. The first step in solving the problems mathematically is to obtain accurate stress distribution. To this end, the choice of the right material model and the yield criterion is very important.

Barlat and Lian proposed a model in 1989 that was developed with aluminum alloys in mind (Ref 1). This model offers improved agreement

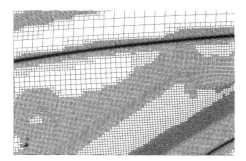

(a)

(b)

Fig. 6 Example of (a) uniform mesh of blank and (b) refined meshes after forming

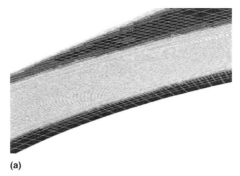

(a)

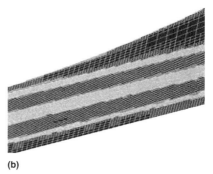

(b)

Fig. 7 Example of a mesh pattern difference between (a) regular adaptivity and (b) adaptivity with fusion

compared to models with Hill's 1948 model with planar isotropy. It is less complex compared to Barlat's later models; hence, it is easier for implementation:

$$2\bar{\sigma}^m = a\{|K_1 + K_2|^m + |K_1 - K_2|^m\} + c \cdot |2K_2|^m$$

(Eq 6)

where K_1, K_2, a, c, and h are defined as:

$$K_1 = \frac{1}{2}(\sigma_1 + h\sigma_2)$$

$$K_2 = \sqrt{\left(\frac{\sigma_1 - h\sigma_2}{2}\right)^2 + p^2\sigma_{12}^2}$$

$$a = 2\left(1 - \sqrt{\frac{r_0}{1+r_0}\frac{r_{90}}{1+r_{90}}}\right)$$

$$c = 2\sqrt{\frac{r_0}{1+r_0}\frac{r_{90}}{1+r_{90}}}$$

$$h = \sqrt{\frac{r_0}{1+r_0}\frac{1+r_{90}}{r_{90}}}$$

(Eq 7)

The variable p does not have a close form solution, and it is found by an iterative search to solve the $g(p)$:

$$g(p) = [2m\bar{\sigma}^m/(\partial\Phi/\partial\sigma_1 + \partial\Phi/\partial\sigma_2) \cdot \sigma_{45}]$$
$$- 1 - r_{45}$$

(Eq 8)

The Corus-Vegter model is a refined case of the original Vegter model, and it is essentially a discrete yield description, derived from several experimental measurements. This yield model has the potential to be more accurate in representing the yield surface and effective in improving the results of simulations, in both formability and springback predictions.

The Corus-Vegter model may have some advantages over the commonly used yield models today (2006). A comparison between Hill 1948, Hill 1990, and Vegter yield loci is shown in Fig. 8. The Vegter model is constructed through four points on a quarter of the yield ellipse; both the point and the slope (or tangent) are either measured or are known. Bezier interpolation between the points and two symmetry planes are then used to construct the planar ellipse. Cosine interpolation between the 0, 45, and 90° measurement sets is used to construct the entire yield surface description.

The general consensus is that isotropic hardening does not accurately describe the sheet metal behavior and always notably under-predicts springback. This helped encourage a substantial amount of work on the kinematic hardening model. However, the results do not support the theory that kinematic hardening models are generally superior to isotropic hardening models. Springback calculations using a kinematic hardening model often (but not always) overpredict the springback amounts. The Chaboche model was implemented by some metal-forming codes. However, the results in

terms of springback prediction are inconclusive for various reasons. A combination of isotropic and kinematic hardening models that account for anisotropic behavior may be a potential material model to use.

There are several important criteria for choosing a proper model. First, the parameters needed for the model can be easily acquired through relatively simple tests. Second, the model should not create too great a cost penalty in using it. Third, it should not cause too many numerical complications that compromise the accuracy due to approximation or round-off error. Fourth, it should not involve parameters that require adjustment from case to case.

Many more advanced models and yield criteria have been proposed since 1989. However, these advanced material models, although theoretically sound, often do not give better springback prediction. The confusing results may be due, in part, to the stress noise commonly seen in analyses with dynamic explicit codes. The stress noise is the result of stress wave propagation and penalty-method-based contact algorithm. More studies are needed to better understand the root cause.

Failure Criteria for Sheet Metal Forming

In FEM simulation, the sheet metal does not necessarily fail unless a proper failure criterion is used. Besides, the code cannot always simulate the failure precisely at the right location and at the right time. The use of different material criteria and numerical noise may have an effect on the initialization of necking, and it is necessary for the user to specify a failure criterion in order to be more accurate in evaluating sheet metal formability. Most users determine failure by postprocessing the results, which can be done by comparing thickness distribution, for example, against a predetermined value. Any areas with values above the limit are considered failures. This approach is simple and efficient. However, it may not accurately show

the problem for several reasons. The limiting thickness is not a constant value but a strain-path-dependent value. To better predict the failure of sheet metal, several failure criteria have been used, as discussed subsequently.

Experimentally Based Forming-Limit Curves and Forming-Limit Diagrams. Before CAE technology became mature enough for use in metal-forming analysis, empirical methods were used to provide the relationship between acceptable formability and strain state. Perhaps the best known of these is circle grid analysis (CGA). The CGA technique was used with controlled tests forming domes, cylindrical cups, and so on and special tensile tests to determine limiting strain states. First, grid patterns of circles are imprinted or etched on the surface of a sheet metal blank. The marked blank is then formed into a part. During forming, as the sheet metal is deformed to the final shape, the circles are deformed into ellipses (Fig. 9). The change of shape then gives the state of strain at each point, as shown in the diagram in Fig. 9. The data can be used to construct a forming-limit curve (FLC).

Although it was initially developed to be used in the field, CGA concepts still provide a useful way to visualize and comprehend the changes that occur during the forming process. The strain induced from the deformation, whether from measurement or from analysis, can be plotted on a forming-limit diagram (FLD). The FLD is a diagram with all of the strain data, compared against the FLC that is based on the material being used.

The FLC developed by Dr. Stuart Keeler is the most popularly used failure criterion for sheet metal stamping, and sometimes for hydroforming. The concept and data were first published by Dr. Keeler in the 1960s (Ref 6, 7). Dr. Keeler later refined his original publication with more details based on his work at National Steel and at Budd Company. However, the later information

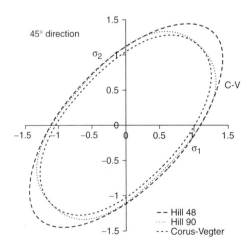

Fig. 8 Comparison of Hill 1948, Hill 1990, and Vegter yield ellipses in the 45° direction

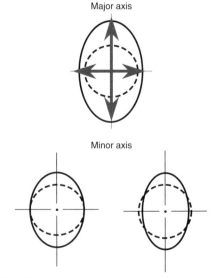

Fig. 9 Deformation of an imaginary circle on the blank

was released with reference to the earlier publications without complete details. Some information was given in graphic form without equations. Hence, it is necessary to summarize this standard FLC in equation form for anyone who is interested in doing his/her own calculation or implementing it in computer codes.

The fundamental methodology (data and guidelines) that Keeler developed was for low-carbon steel (including mild steel and high-strength low-alloy steel). It can be summarized as follows:

- In sheet metal forming, any deformation can be described using the primary principal surface strain (major) and secondary principal surface strain (minor).
- The shape of the FLC remains constant.
- The FLC moves vertically on a major-minor strain chart based on the position of the FLD_0 that is influenced *only* by the n-value and the thickness of the steel.
- The marginal curve is generated by subtracting 10% (engineering strain) from the FLC (e.g., 25% for FLD_0 of 35%).

The previous statements are not perfectly accurate but are good enough for practical use. This simplified version makes implementation of CGA and FLD on the shop floor much easier. The equations and the guidelines can be used on low-carbon steel only. No comprehensive study for other types of materials is available.

Empirical FLD_0. The FLD_0, in percent engineering strain, is defined as:

$$FLD_0 = N \cdot T \qquad (Eq\ 9)$$

with:

$$N = \mathbf{min}\ (n/0.2116,\ 1.0)$$

$$T = 23.26 + 14.02 \cdot \mathbf{min}\ (t,\ 3.0)$$

where N represents the material effect due to work hardening (n-value), and T represents the thickness effect based on the initial thickness of the sheet metal in millimeters. There are some minor variations of the formula and parameters throughout the industry, but the difference is so minute that it is not of concern. The stamping community slightly modified Dr. Keeler's theory in the late 1980s (according to his advice). The revised guidelines set no limitation on the n-value effect and reduced the thickness effect as follows:

$$Let\ N = n/0.2116 \qquad (Eq\ 10)$$

Then:

For $t < 2.5$ mm	$T = 23.26 + 14.02 \cdot t$	(Eq 11)
For 2.5 mm $< t < 5$ mm	$T = 20.0 + 20.2 \cdot t - 1.95 \cdot t^2$	(Reduced thickness effect)
For $t > 5$ mm	$T = 72.25$	(No more thickness effect)

The new equations affect only heavy-gage steel (thickness greater than 2.5 mm, or 0.10 in.). The results of the two sets of equations are shown in Fig. 10. The revised equations better describe the low-carbon steel characteristics and should be used. However, many people are still using the original equations for various reasons.

Empirical FLC. The shape of the FLC for low-carbon steel can be described with the following equations:

$$e_{maj} = FLD_0 + e_{min}(0.027254 e_{min} - 1.1965) \qquad (Eq\ 12)$$

$$e_{maj} = FLD_0 + e_{min}(-0.008565 e_{min} + 0.784854) \qquad (Eq\ 13)$$

Equation 12 is for the left side ($e_{min} < 0$) of the curve, and Eq 13 is for the right side ($e_{min} > 0$). All of the strains in the preceding equations are percent engineering strain. In other words, the value of 5% engineering strain in the equation is

5. For a material with the n-value equal to 0.212 and 0.75 mm (0.03 in.) thickness (t), the FLD_0 is 33.78% (engineering strain). This standard Keeler FLD in engineering strain is shown in Fig. 11.

For FEM codes using true strain, all strain data must be converted using the engineering strain value (i.e., 0.05, not 5 for 5% strain) to the true strain value. In the true strain domain, the left side of the FLC is a 45° straight line (Fig. 12). Hence, only one data point is needed on the far left in a true strain domain using the following equation:

$$e_{maj} = FLD_0 - e_{min} \qquad (Eq\ 14)$$

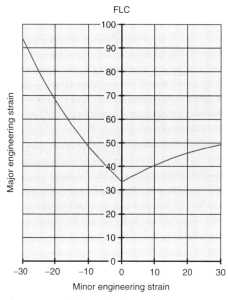

Fig. 11 Standard Keeler forming-limit curve (FLC) ($n = 0.21$; $t = 0.75$ mm, or 0.03 in.)

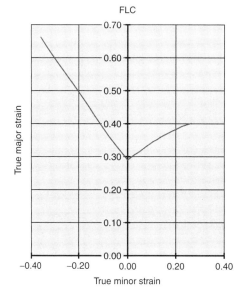

Fig. 12 Forming-limit curve (FLC) in true strain ($n = 0.21$; $t = 0.75$ mm, or 0.03 in.)

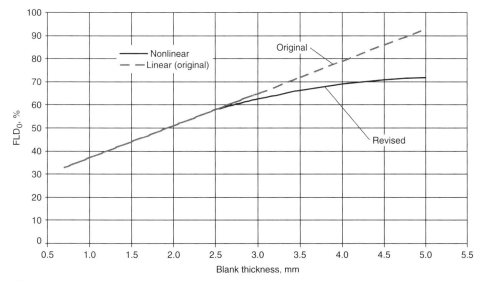

Fig. 10 Difference between the original and revised formula for the thickness effect

All numbers in this equation are true strain. For example, 35% engineering strain becomes 0.30 (= ln [1 + 0.35]) in true strain, not 35 or 0.35. Figure 12 shows an FLD in true strain units.

Although Keeler's standard FLD is quite simple, it may still be used incorrectly due to misunderstandings. The most common mistake is using the percent engineering strain value on an FLD in true strain units. Then, the mistake could be as big as 50% (engineering strain). The preceding equations have been adopted and used by several commercial codes.

Generally speaking, most European interpretations of the FLD_0 have little or no thickness effect. The marginal curve is generated differently in many other countries (especially in Europe). Instead of shifting the FLC down by a fixed amount, (the major strain value of) the marginal curve is a percentage (e.g., 80% or 0.80) of (the major strain value of) the FLC, hence creating a variable bandwidth for the marginal zone.

Thickness-Based Forming Limit. Some analysts prefer the thickness-based forming limit for various reasons. First, it is easy to use and easy to compare against the measurement. Second, the Keeler FLC previously described was based on a nearly linear strain path, without significant bending and unbending deformation. It becomes very questionable to use the same failure criterion for multistage forming analysis. Thickness limit is an option to address this issue. One popular method is to simply convert the FLD_0 to limiting thickness based on constancy of volume. Some engineers choose to use a safety factor in conjunction with the limiting thickness reduction. Some simply choose a conservative number, such as 20%, as the failure criterion. Although these methods work fine, more study is needed to offer a more definitive criterion.

Theoretical Forming Limit. Due to the limitation of Keeler's FLC, numerous researchers have been working on theoretical forming-limit prediction for many years. That work is very useful because metal normally does not follow a simple linear strain path in a typical forming operation. It is more crucial to have a reliable way to predict failure in a multistage forming operation. Most theoretical forming-limit prediction assumes that the failure is caused by localized necking on the left side of the FLD and diffuse necking on the right. The classical theory used simple power law for the constitutive equation and isotropic hardening. The prediction was much lower than experimental results.

The most common method used for predicting localized necking for the right side of the FLD is the Marciniak-Kuczynski (M-K) method (Ref 8, 9). It is often used to generate FLCs for various reasons, especially in the absence of reliable or sufficient experimental data. The M-K method requires a pre-existing defect in the material. This defect can be caused by any combination of geometric and material nonuniformity. The most common approach is to characterize the defect

by a reduction in the material thickness. The thinner material will actually experience higher strain than the neighboring material and eventually reaches the forming limit.

The M-K method is a strain-based approach that is easy to understand and seems to be a sound approach; however, comparisons between the predictions and the experimental data are inconsistent. It is influenced by the material model used and generally predicts a very high forming limit unless a large pre-existing defect value is used. Many technical papers based on the M-K model have been published to improve the correlation between the prediction and measurements for various materials.

Another approach toward predictive FLDs is to apply a damage mechanics model to predict the failure controlled by fracture. This method takes into account the effects of initiation, growth, and coalescence of the microdefects in "real-life" material up to its final rupture. It can predict diffuse necking, localized necking, and rupture of metal sheets under both proportional and nonproportional loading conditions. The advantage of this approach, according to Ref 10 and 11, is its potential broader application to cover deformation of any strain path. The results seem to be better than the conventional methods in predicting the formability of the sheets under biaxial stretching. The theoretical failure model can be incorporated into stamping finite element codes for more realistic prediction, and its results can be incorporated into other applications such as crash simulations to improve the correlation. However, this model has not been adopted widely, possibly due to its intense calculation.

Strain path effects on conventional strain-based FLDs have been published by many researchers working on theoretical FLC prediction. It has also been reported by many experimental works. Figure 13, from the work done by Graf and Hosford on 2008-T4 aluminum (Ref 12, 13), is an excellent example of the path dependency of the forming limit. First, legs of the dashed lines represent the prestrain state for each test condition. The heavy black line is the FLC for the as-received material. Each additional FLC represents the new forming limit following the indicated prestrain.

The huge variation of the FLC in a two-stage deformation makes one wonder whether the material truly has significant forming limit. Experimental evidence for a path-independent stress-based FLD has been reported in the literature, suggesting that the path dependency of the strain-based approach arises from the path-dependent constitutive laws governing the relationship between the stress and strain tensors. The stress-based FLD can now be used equally well for all forming processes, without concern for path effects.

In 1977, Kleemola and Pelkkikangas first proposed using a stress-based FLD as an alternative for FLC prediction (Ref 15). They gave some experimental results that supported the path independence of the forming limit in stress

space for these materials. Arrieux et al. (Ref 16) reaffirmed this phenomenon and proposed using a stress-based criterion for all secondary forming operations. Despite these earlier works, the significance of a path-independent stress-based FLD went largely unnoticed until 2000. Stoughton (Ref 17) rediscovered the effect and proposed that it is necessary to use the stress-based criterion in all forming operations, including the first draw die, in order to get a robust measure of forming severity.

Stress-based FLCs are very useful, not only for multistage deformation. Many areas in a panel with complex product shapes, especially around the depressions with re-entrant surfaces, often go through strain-path changes in the first forming operation. A simple and consistent forming-limit algorithm that can be used throughout the entire forming operation without prior knowledge of the strain path is very beneficial and accurate in predicting the failure of the material.

The use of stress-based FLDs still faces several challenges before it can be fully accepted in production use. First, in any FEA calculation (especially with dynamic explicit codes), noise in stress calculation is unavoidable. If those stress states are put into FLDs for evaluation, stress point will change dramatically from one step to the next. As a result, it will generate obvious error in failure evaluations. Secondly, unloading is another issue that needs to be addressed in stress-based FLDs. When the material encounters unloading in the subsequent forming operation after exceeding the limiting stress, this location will appear safe on the FLDs. Thirdly, it is difficult to evaluate safety margin. From Fig. 13, it is seen that after initial yielding, the band between the limit stress and the initial yield stress is very narrow. Hence, it will be very difficult to estimate the remaining formability.

Numerical Procedure for Sheet Metal Forming

The most important concept in sheet metal forming CAE is simulation. To obtain

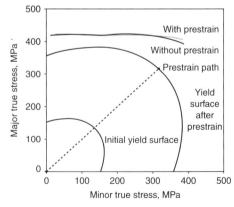

Fig. 13 Prestrain path dependency for the forming limit of 2008-T4 aluminum. Source: Ref 14

meaningful results, the analysis must be set up to emulate the physical process as closely as possible; otherwise, the results may not be meaningful. For example, gravity may not be important in the draw-forming operation. It could make a difference in blank setting and binder wrap.

A typical complete stamping operation for the entire line of dies has five stages:

- Blanking
- Forming
- Trimming
- Flanging
- Trimming, flanging, and restriking

However, a complete stamping simulation will require more steps:

1. Blanking
2. Gravity loading
3. Binder closure
4. Forming
5. Springback
6. Trimming
7. Springback
8. Flanging
9. Springback
10. Trimming, flanging, and restriking
11. Springback

Although some software allows the user to set up the entire operation in one job setup, those functions still need to be activated sequentially. It is not common to do the entire line die analysis for various reasons. These include lack of manpower, insufficient computer capacity, lack of knowledge, and limitation of the software. On the other hand, it is not necessary to perform all functions.

Blanking simulation is rarely done with a "solver" in a production environment. Blank definition and meshing is generally done in the preprocessor. The analyst simply generates mesh for a predefined blank boundary. Alternatively, a mesh for a rectangular blank is created first. Then, the blank mesh is trimmed using the blank boundary line. Most users prefer the first method because it is simpler. The second method, however, gives the analyst a mesh pattern lining up with some reference points that could be useful for validation and correlation.

Springback is the result of stress relaxation. The springback analysis in step 5 is often skipped to save time. The reality is that springback occurs automatically as soon as the die opens. In some areas of a complex automotive part, the stresses cannot be totally released, or the stresses will release to a different pattern due to the constraint of the addendum. Further springback will occur when the addendum (working like a rubber band providing extra constraint) is trimmed away. The importance of this extra step may be case dependent.

Besides the functions listed previously for a complete line die simulation, it is necessary to perform additional numerical work for accuracy and performance. A trimmed mesh often leaves "bad" elements around the trimmed boundary. Those elements can be classified into three categories (Fig. 14) that may lead to problems in the subsequent analysis:

- Improperly defined elements
- Triangular elements with bad aspect ratio
- Small elements

The first category can be easily identified. When a corner of a quad element is cut away, the remaining element has five nodes (Fig. 14a). Bad triangular elements (Fig. 14b) are too stiff and often influence other elements surrounding them in the subsequent analysis. Small elements (Fig. 14c) will slow down the calculation efficiency for dynamic explicit codes. It will also cause the stiffness matrix to be ill-behaved, leading to slow convergence or even divergence for static implicit codes. All three cases are undesirable and must be remeshed for good results.

In a real die lineup, the line dies (the second stage through the last stage) often do not line up with the first draw die operation for various reasons. Hence, it is necessary to reposition the panel from stage to stage so it will fit into the die properly. This creates two issues that analysts should be aware of. First, the residual tensors cannot be rotated like a rigid body. Second, the sprung part may cause some initial interference with the prepositioned die members or components. Those die members must be repositioned, and the travel must be adjusted accordingly.

Alternatively, the job can be set up by repositioning die members of the subsequent operations without changing the sheet metal. This may take care of the tensor rotation problem, but it is not necessarily simpler. The motion of the moving piece will move in a user-defined local coordinate system, not in the global z-direction.

Job Setup. Setting up a stamping simulation today (2006) is quite different from what was done ten years ago. Most codes have an integrated process definition tool built inside the preprocessor. The tool does much more than a generic preprocessor such as Hypermesh (Altair Engineering, Inc.).

A complete input deck is very complicated, with many types of keyword control cards. It may be overwhelming for new users. It will be helpful to group the keywords into several categories.

General Job Definition. In this section, job title, comments, and external files are defined. In addition, the global job controls, including termination criteria, time-step control, and adaptivity, are also defined.

Parts Definition. All parts used in the simulation are defined, including material model and data, metal thickness, rigid body and constraints, and element type for the sheet metal.

Interface Definition. All contacts between the sheet metal and each die member need to be defined properly, including the friction. Sometimes, it may be necessary to define contact between die members and guide pins to ensure proper movement.

Boundary Conditions. This section covers five categories:

- Draw beads
- Binder or pad pressures (or forces)
- Prescribed motions of die members
- Nodal constraints (e.g., symmetry or weld lines)
- Supplemental external pressure (used in hydroforming, superplastic forming, and other unconventional forming methods)

Numerical Controls. This section defines miscellaneous parameters that may influence the results, such as hourglass control, contact-related parameters (e.g., penetration tolerance), convergence tolerance, solver selection, and controls.

Output Controls. This section defines the specific output desired, including type of information (e.g., thickness and plastic strain), frequency of the output, and other details (such as integration points).

Auxiliary information also needs to be defined, depending on the complexity of the job. The most common ones are the curves, vectors, and coordinate system:

- Curves for information such as the constitutive relation for the material, prescribed motion curves (such as displacement or velocity), and force or pressure as a function of time
- Vectors that define the direction of motion (commonly used in line dies to define the cam trim and cam flange actions)
- Coordinate system that defines material orientation (needed for some complex material models)

Although most software define everything through a graphic user interface, it is still helpful for a novice user to understand this to make sure everything is properly defined. Sometimes, the default value defined may not be optimal. Experienced users need to have a comprehensive understanding of the job setup to make the best decision for an optimal job setup.

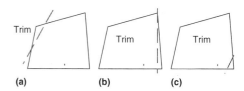

Fig. 14 Undesirable elements from trimmed mesh around a boundary. (a) A cut corner on a quad element results in a remaining element with five nodes. (b) Bad triangular elements are too stiff and often influence other elements surrounding them in the subsequent analysis. (c) Small elements slow down the calculation efficiency for dynamic explicit codes.

Boundary Conditions. A good setup must define these conditions accurately, not as a numerical tool to influence the results. It is important to define all the boundary conditions according to the physical setup. In metal-forming simulations, those conditions are usually similar and contain one or more of the following items:

- Punch travel (or die travel)
- Binder travel or force
- Draw bead
- Pad force or pressure
- Symmetrical (or other additional) constraints and blank locators or guides
- Cam or other moving element travel
- Auxiliary pressure (used in superplastic forming or hydroforming)

Not all of the items listed previously are used in a single simulation. The most common boundary conditions for a stamping simulation are the first three items in the list. It is important to model both the binder and punch (or die) movement as accurately as possible, because it may affect the binder wrap and how the sheet metal comes in contact with the die members. A good model requires:

- Accurate initial position of all die members (i.e., punch, die, binder, etc.)
- Accurate blank gaging and guides
- Gravity loading of the sheet metal
- Binder travel control (in a toggle press) or binder force (in an air-draw or transfer press)
- Punch travel (in a toggle press) or die travel (in an air-draw or transfer press)

Punch Travel Control. There are two different types of presses used for metal forming: mechanical and hydraulic. The ram velocities of these presses have very different behavior. Hydraulic presses are slow and therefore are seldom used in production. However, they are widely used in tryout for their versatile control ability. The ram velocity of a mechanical press basically moves following a sinusoidal pattern. Figure 15 shows the ram velocity and the location of the ram relative to the bottom dead center of a typical production press used in the automotive industry.

The travel can be defined using a prescribed displacement or velocity curve. The advantage of using a displacement curve is the accurate control of the location of the die members (e.g., punch). However, a simple linear displacement curve will implicitly define the punch to move at a constant high speed (as shown in "A" in Fig. 16). It may induce unwanted stress noise in the beginning and end of the operation. Hence, it is common to use a velocity curve to define the motion of the punch.

It is obviously incorrect to model the motion of the punch or the die with one-half of the sine wave. The actual velocity or displacement profile during the entire forming process is only a small portion of the sine curve that is closer to a linear line that continuously decreases to zero at the end of the forming process. However, there

are two problems with modeling the punch with such a curve. High velocity at contact often causes large disturbances that may deteriorate the solution quality. Continuously decelerating the punch in the duration of punch contact will increase the CPU time with a dynamic explicit code. It generally does not improve the solution quality that much. Figure 16 illustrates some popular curves used for prescribing the punch motion.

Some software packages have developed a built-in feature to make this easier. The user just specifies the maximum punch speed (V_{max}) and the ramp-up time (T_{rise}) needed to increase from

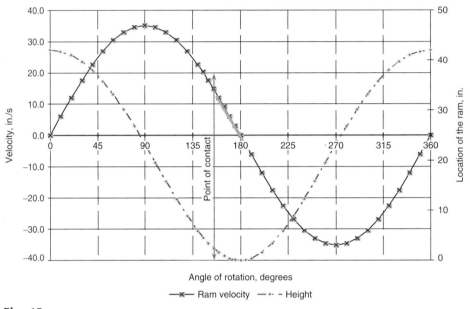

Fig. 15 Displacement and velocity curves of a mechanical press

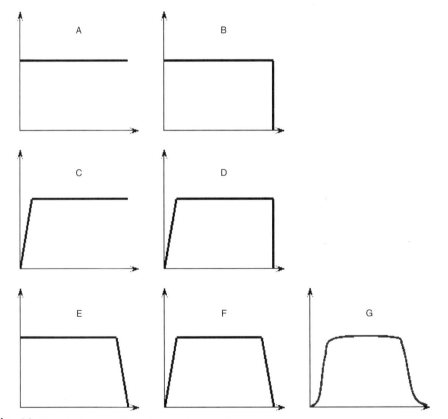

Fig. 16 Typical velocity profiles used for tamping operations

zero to V_{max}. A curve as shown in the graph on the left of Fig. 17 will be generated and used as the prescribed velocity profile for the punch. Some software even allows the user to specify the ramp-up speed (V_r) as the percentage of the maximum speed to control the transition, as shown in the graph on the right of Fig. 17. When V_r approaches, it becomes a trapezoid, as shown in "F" in Fig. 16. It is interesting to know that the total travel time (T_{total}) needed is always governed by the following equation for the profile patterns "F" and "G," regardless of the parameters chosen:

$$D_{travel} = (T_{total} - T_{rise}) \times V_{max} \qquad \text{(Eq 15)}$$

Note that neither profile shown in Fig. 17 describes the real movement of the punch (shown as the shaded area after the point of contact in Fig. 15). However, they are much faster than the real punch and do not deteriorate the results. Even the simple curve "F" (Fig. 16) gives relatively good results.

Drawbead Modeling. Draw beads are used in the majority of stamping operations. The beads are generally narrow (5 to 10 mm, or 0.2 to 0.4 in.), with two to four radii in its cross section. To model the actual bead shape for sheet metal to flow through is practically unfeasible for two reasons. First, the element size has to be less than 1 mm (0.04 in.) in the beginning. That will require extremely long CPU hours for both the binder closure and draw-forming simulations. However, it is becoming possible today (2006) with the advancement of MPP technology and the availability of low-cost high-speed CPUs. The other challenge is the reliability of the shell element used today (2006), when the size of the element becomes a fraction of its thickness. The results, especially the stress distribution, may be questionable. This approach provides an opportunity to better capture the deformation in several ways, such as the wrinkling gathered at the end of the bead.

Most popular codes developed the "line" bead in the mid-1990s. The user simply needs to provide a curve representing the center line of a bead segment instead of the true bead shape. As the sheet metal is being drawn into the die cavity, the nodes in the binder may flow across the bead line. External force will then be applied to those nodes. The force is either directly specified by the user or calculated from a more sophisticated equation. The equation could be a function of contact pressure, coefficient of friction, or other parameters.

The force vector could be in the direction against the motion of the node. It is argued that the force vector should be normal to the bead line and the magnitude should be reduced. A hybrid model may work better. However, it is not easy to verify which model really works better.

The line bead model is a big improvement over the nodal force model used in the past. First, it is much easier to set up. Second, it is much more realistic, especially for deep-drawn parts. Third, it is much more efficient than modeling

the real bead shape. It served the industry well for many years. However, as the industry shifts its focus from splitting and wrinkling prediction to springback, binder force, and sidewall curl prediction, line bead is a source of the accuracy problem.

Important Technical Issues

Units. All FEM codes use dimensionless numbers for input and calculation. However, all those numbers carry implicit dimension throughout the calculation. Hence, the units of all input numbers must be consistent to ensure meaningful results. A simple check is Newton's second law of motion ($F = ma$). Hence:

$$1 \text{ force unit} = 1 \text{ mass unit} \times 1 \text{ length unit}/(1 \text{ time unit})^2$$

All derived units must be based on the basic units. Hence:

$$1 \text{ stress or pressure unit} = 1 \text{ force unit}/(1 \text{ length unit})^2$$

$$1 \text{ density unit} = 1 \text{ mass unit}/(1 \text{ length unit})^3$$

$$1 \text{ energy or work unit} = 1 \text{ force unit} \times 1 \text{ length unit}$$

The popular units used around the world in CAE, including metal-forming simulation, are not standard centimeter-gram-second, meter-kilogram-second, Système International d'Unités (SI), or U.S. customary. It is a mixed

system that meets the needs of several considerations. Table 3 shows the systems that may be selected for use in an FEA model.

The first two systems are very popular among most types of CAE simulations. They are chosen because almost all geometry-related information such as the mesh models use millimeters. Many dynamic simulations use milliseconds, making the first system popular among analysts doing crash simulation. Many prefer the second system, because much of the test data uses MPa as the stress unit. Table 4 gives some common numbers used in a simulation that may help the user keep consistent units.

Integration Time Estimation for Dynamic Explicit Codes. Because dynamic explicit is the dominating code type used in sheet metal forming simulation, it is worthwhile and necessary to understand the CPU time needed for an analysis. The total CPU time needed for a job with the explicit integration method can be estimated using the following equation:

$$\begin{aligned} CPU &= N_{cy} \times T_{zc} \\ &= N_{cy} \times N_{el} \times \bar{t}_{zc} \end{aligned} \qquad \text{(Eq 16)}$$

where T_{zc} represents CPU time per zone cycle. It stands for the CPU time needed to process all elements per time step (or cycle). It is software and hardware dependent. Different versions of the same software may perform differently on different hardware platforms due to different compiling or system optimization.

The variable T_{zc} also depends on N_{el} (the total number of elements used for the deformable sheet metal) and $\bar{t}_{zc}$ (the average CPU time

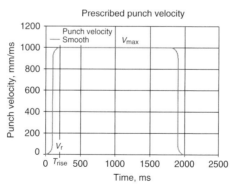
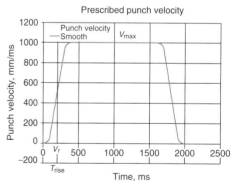

Fig. 17 Smoothed curves used in a forming simulation. V_{max}, maximum punch speed; V_r, ramp-up speed; T_{rise}, ramp-up time

Table 3 Summary of the units used for finite element modeling analysis

System	Mass (M)	Length (L)	Time (t)	Force (F)	Stress/pressure (F/L^2)	Energy (F · L)
CAE-1	kg	mm	ms	kN	GPa	kN · mm
CAE-2	tonne	mm	s	N	MPa	N · mm
CGS	g	cm	s	dyne	dyne/cm^2	erg
CGS−	g	mm	ms	N	MPa	N · mm
SI	kg	m	s	N	Pa	Joule
U.S.	lbf · s^2/in.	in.	s	lbf	psi	lbf · in.

Note: CAE, computer-aided engineering; CGS, centimeter-gram-second; SI, Système International d'Unités

needed per element). For the same element size, N_{el} is roughly proportional to the size of the blank. Conversely, N_{el} is inversely proportional to the mesh density for the same blank size. The relationship can be expressed as:

$$N_{el} \propto W \times L/l^2 \qquad \text{(Eq 17)}$$

where W is the width of the blank, L the length of the blank, and l (or h by w) the size of the element. Thus, the total number of elements could be quadrupled (assuming uniform mesh) when the element size is reduced to one-half of the original size. Therefore, a large-sized job or a dense mesh will take a considerably longer time to calculate.

The variable N_{cy} in Eq 16 is the total number of cycles (or time steps) needed to finish the job. It is approximately governed by the following equation (assuming constant time step):

$$N_{cy} = T/\Delta t \qquad \text{(Eq 18)}$$

where T is the total travel time, and Δt is the time step. The total travel time depends on travel depth D and the tool speed V, as described in Eq 19. (The equation would be exact if the speed remains constant. If users choose to gradually ramp up the die to the target speed or decelerate the die down to zero near the end of the stroke, the total time will be a little longer.) Therefore, deep draw panels such as dash panels and door rings will take considerably more CPU time than shallow panels such as roofs:

$$T \propto D/V \qquad \text{(Eq 19)}$$

The time step (Δt) used in the dynamic explicit code is a complex issue. When the time step is too large without any special treatment, the job will become unstable. Using a small time step will increase the CPU time significantly. The maximum stable time step, Δt_{cr}, in an analysis can be described by:

$$\Delta t_{cr} = \mathbf{min}(l_{ch}/C) \propto \mathbf{min}\left(l_{ch} \times \sqrt{\frac{\rho}{E}}\right) \qquad \text{(Eq 20)}$$

where l_{ch} is the characteristic length of the element, C is the wave speed, ρ is the density, and E the Young's modulus of the sheet metal. For a two-node bar element, the characteristic length is the length of the bar element. Hence, the critical time step (Δt_{cr}) is approximately 2 μs for a steel bar element of 10 mm (0.4 in.) in

length. The characteristic length for a four-node shell element is not that straightforward. For a rectangular shell element, it is the length of the shortest side. Additionally, Δt_{cr} in Eq 20 needs to be modified to consider the material bulk modulus. The actual initial time step used in the code is slightly more conservative to ensure stability.

The variable Δt in Eq 18 is governed by the smallest element of all elements in the model. Hence, a single small element will slow down the entire calculation. Similarly, if an element is severely distorted due to either stretching or wrinkling, the minimum l_{ch} (and Δt_{cr}) could be reduced significantly. This will lead to a substantial increase in CPU time.

Thus, by combining all of the previous equations, the total CPU time needed for the simulation can be estimated by:

$$CPU \propto N_{el} \times \frac{D}{V} \times \sqrt{\frac{E}{\rho}} \times \frac{1}{l_{ch}^{min}} \times \bar{t}_{zc} \qquad \text{(Eq 21)}$$

It is obvious from the previous equation that CPU time is strongly influenced by many factors. Hence, comparing CPU time between two cases without any information on the details, such as the mesh density, is totally meaningless. Finer mesh normally produces better results. However, fine mesh will require much longer CPU time. If the elements are uniformly reduced to one-half of the original size, the CPU time will increase up to 800%! Analysts must find the balance point to obtain an acceptable solution within reasonable time.

This equation is only a general guideline. The actual CPU time will be much longer because the smallest characteristic length is changing throughout the entire process. It is also much more difficult to estimate CPU time if adaptivity is used. Although the general principle still applies, the number of elements grows continuously throughout the analysis.

Metal-Forming Simulation Applications

For decades, stamping manufacturers have used their experience and knowledge to create the die face. After completion of each die design, actual die would be made, followed by the use of trial and error to troubleshoot the die set for

stamping defects such as splitting, wrinkling, and shape changes.

Forming simulation allows tool designs to be validated virtually by using incremental FEM on a computer before any tool steel is cut, which saves costly reworks that used to occur after the tool was finished. For instance, forming simulations can predict splitting (Fig. 18). In one case, for example, by adjusting the draw bead in the forming simulation of a drawn fender, it was possible to resolve a splitting problem. In addition to troubleshooting in splitting, forming simulation is also proven to be efficient in checking for other stamping defects, such as surface and shape defects. For example, the mean stress and thinning distribution from forming simulations can be used to spot potential surface defects, thus helping tool and die makers to troubleshoot potential problems on the part before the tooling is made and to make adjustments on toolmaking accordingly (Fig. 19).

In recent years, forming simulation applications have grown enormously as the benefits of troubleshooting and optimizing processes through the computer rather than through extensive shop trials have been realized. The rapid development of software technology, together with faster and lower-cost computer hardware, has enabled many metal-forming operations to be modeled cost-effectively.

Metal-Forming Simulation Systems

A typical CAE simulation begins with the geometry given and boundary conditions well defined. The analyst's job is to set up the model as defined, run the job, and compare the results against a predefined failure (or pass) criterion. Metal-forming simulation is very different from CAE simulations in any other field in several ways. First, the geometry given is only the product data and not the die face. In a large company, the die face could be designed by specialized die layout personnel using a CAD system with some special die-face design functions. Often, it is the analyst's job to create the initial die-face.

Second, the boundary conditions (binder force and bead force) are not defined either. Hence, the analyst must go through an iterative procedure to determine the optimal boundary conditions and die-face design that will make a formable part. Third, the process necessary to make a sheet metal part (especially for an automobile) may need to be defined. An experienced analyst must have the knowledge and proper tools to redefine the die operation lineup.

Generally speaking, a good metal-forming system should have all the special-purpose modules listed in Table 5 so that the analysts can do everything within a single system. The so called preprocessor for metal forming does more than the traditional ones that primarily perform mesh creation, editing, and model organization. It must have the ability to do die creation and process definition.

Table 4 Typical properties for each system (see Table 3 for base units of mass, length, force, and time)

System	Density of steel (M/L^3)	Young's modulus of steel (F/L^2)	Gravity (L/t^2)	Velocity (L/t) (35 mph)
CAE-1	7.830E−06	2.070E+02	9.806E−03	1.565E+01
CAE-2	7.830E−09	2.070E+05	9.806E+03	1.565E+04
CGS	7.830E+00	2.070E+12	9.806E+02	1.565E+03
CGS−	7.830E−03	2.070E+05	9.806E−03	1.565E+01
SI	7.830E+03	2.070E+11	9.806E+00	1.565E+01
U.S. customary	7.330E−04	3.000E+07	3.860E+02	6.160E+02

Note: CAE, computer-aided engineering; CGS, centimeter-gram-second; SI, Système International d'Unités

As analysts' jobs become more versatile, the system (including the solver) should be able to handle a variety of functions, including:

- Stamping (draw forming or crash forming, toggle draw or air draw)
- Bending or flanging
- Trimming, piercing, or hole expansion
- Sheet hydroforming (fluid forming)
- Tube multibending
- Tube hydroforming
- Springback

Some additional functions are highly desirable as the industry becomes more sophisticated and vertically integrated. Those functions include:

- Blank estimation
- Blank nesting
- Draw bead optimization
- Trim line optimization
- Smart binder and die-face creation and optimization
- Smart springback reduction
- Automatic pressure control for tube hydroforming or superplastic forming
- Automatic binder-force control
- Deformable binder

- Special forming operations such as hemming, superplastic forming, warm forming, roll forming, and rotary forming

Forming Software Overview

The following sections review the capabilities of three major systems that are popular among sheet metal forming users worldwide:

- AutoForm, developed by AutoForm Engineering (Switzerland)

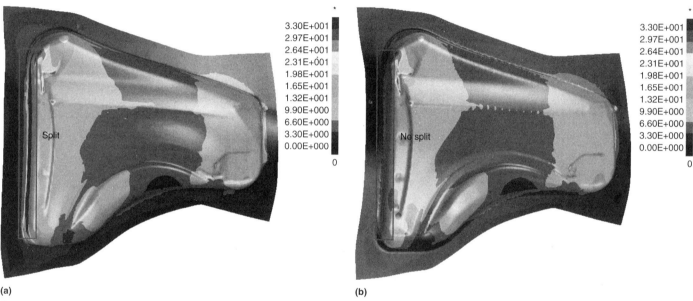

Fig. 18 Illustration showing thinning of drawn fender from NUMISHHET 2002 benchmark. (a) Split. (b) No split. Courtesy of Engineering Technology Associates, Inc.

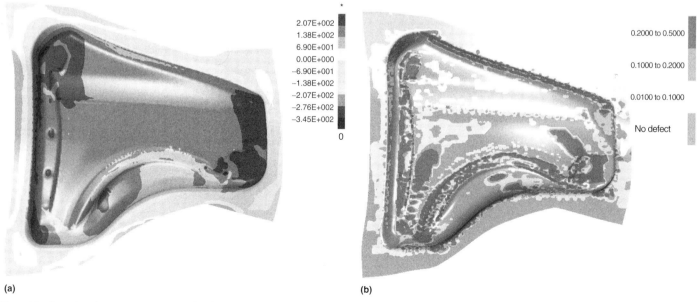

Fig. 19 Illustration showing (a) mean stress distribution and (b) stoning distribution of drawn fender of NUMISHEET 2002 benchmark. Courtesy of Engineering Technology Associates, Inc.

- PAM-STAMP 2G, developed by the ESI Group (France)
- Dynaform, developed by Engineering Technology Associates, Inc. (United States) using LS-DYNA, developed by Livermore Software Technology Corporation (United States) as the solver

Each software vendor has a particular strategy in developing their system.

PAM-STAMP 2G is one of the leading software systems designed specifically for sheet metal forming. The following sections provide an overview of this system based on information supplied by the ESI Group.

Virtual Stamping Software. The need for an integrated and scalable software for stamping, covering the entire design process from part design through die design and die evaluation, tuning, and validation of the process, is the basis for PAM-STAMP 2G. This system allows users to make decisions on-line in a continuous improvement process in a collaborative environment that connects the design engineer, the material provider, the die designer, and the tryout press shop from an early stage of the design until the production. PAM-DIEMAKER gives the design engineer the flexibility to quickly test die geometry in the initial phases, and PAM-QUIKSTAMP enables the selection of the material properties. Lastly, PAM-AUTO-STAMP focuses on part quality in the final stages of the process simulation. Each step can be iterated many times to progressively improve quality throughout simulation process without any discontinuity.

Table 5 Special-purpose modules for optimal metal-forming system

Module	Features
Preprocessor	Computer-aided design interface
	Computer-aided engineering model reader and writer
	Automeshing
	Model creation and editing
	Model morphing
	Complete process definition
	Die-face and binder creation
	Automatic positioning
	Draw bead generation
	Deck builder
Solver	Job submission and queuing
	Execution
Postprocessing	Review results (3D fringe pattern, section plots)
	Animation
	Formability assessment
	Surface-quality evaluation
	Model comparison
Special functions	Blanking and trimming with auto-remeshing
	Intelligent mesh coarsening
	Automatic morphing for springback compensation
	Results mapping
	Built-in material database supporting popular models
	Model reposition

To fulfill the new requirements, the solver was re-engineered in a new architecture with object-oriented native MPP structure, with a new data model especially designed to manage simulation data across different applications and over time. Each application has a weak link to customer applications through the use of universal files (international graphics exchange specification or visual data analysis) or strong links through application program interfaces that allow plug-ins directly to a customer's database and/or data models as special customization. All of these modules are accessible through a single graphic user interface.

The different modules allow a complete stamping simulation in one consistent environment. Integration of the modules resolves numerical transmission issues, which saves time and allows the user to focus on stamping issues instead of numerical issues. The user can simply import a CAD file and simulate the forming process step by step, from die design to an initial evaluation, then tuning and final validation.

Die-Face Design. Most commonly, a project is started with only the CAD geometry of the component. The CAD geometry is read into the program, which treats and repairs the surface data and automatically produces a clean and connected mesh. The newly imported part geometry needs to be prepared in order to serve as a reference for the die design (Fig. 20). During this part-preparation phase, the geometry will be checked for eventual problems, and eventual holes will be closed.

Flange-specific treatment allows the user either to automatically eliminate them or to hide them in order to use them as reference geometry for blank development later on. Furthermore, the user can smooth the part edge by using a rolling cylinder technology. Next to that, if the part geometry has a symmetry plane or is built as a double part, this can be defined as well.

Normally, when reading part geometry from a CAD file, this geometry is positioned in the car coordinate system (global system). Therefore,

it is necessary to define a local system that can serve as a reference for the stamping direction and also has the ability to automatically detect the optimal stamping direction based on several criteria, such as the minimum draw depth. The resulting local system can (if necessary) be further modified by manually adjusting the system.

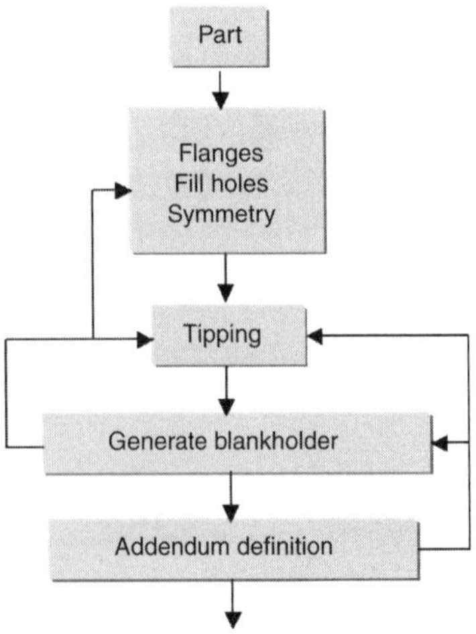

Once the part geometry is prepared, the tool design can start with an initial design of the binder surface (Fig. 21). The code has complete and intuitive functions for generating binder surface in three different methods. The user can define the binder surface using parametric profiles and sections, define the binder surface using theoretical surfaces (fitting of flat, cylindrical, or conical planes), or import an already existing surface from a CAD file. Depending on the chosen method, the binder surface can be visualized and modified in 2D and 3D simultaneously and interactively. There are also several options for verifying the binder design, such as checking the depth of draw (relative to the binder surface) or the developability of the surface.

The user can start building addendum using 2D profiles (Fig. 22). The code provides fully interactive functions that allow the user to easily position each profile. When the profiles are attached to the part geometry, they can be further

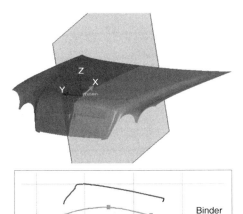

Fig. 21 Binder definition by section approach

Fig. 20 Die-face design overview

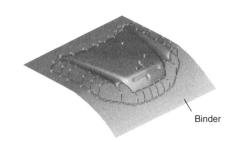

Fig. 22 Addendum design by profiles

modified by changing the orientation, resizing, extending, or shortening each profile.

The 2D geometry can be modified as well, with full control over all the parameters that construct the profile, such as angles, radii, heights and distances, and tangency. By dragging and dropping the control points of the profile, users can interactively modify the parameter values. Modifications in 2D are immediately visible in 3D, thus further enhancing the interactivity. This method is especially useful in conceptual design. The user has full control of 2D parameters by defining all parameters manually (Fig. 23). The user can also modify the design interactively. Finally, the different profiles can be saved in the database for future modification.

When all profiles are defined, the user can visualize the die entry line. This line is automatically created and can be modified manually. Once the profiles and die entry line are satisfactory, the 3D surfaces can be generated in order to have a global view of the entire tool. The 3D surface information is useful for visualization purposes and early anticipation of problems such as tangency.

Quick Evaluation. When a die design is generated (Fig. 24), the design must be checked for formability using PAM-QUIKSTAMP based on optimal hybrid formulation of explicit/implicit solvers with uncoupled membrane-bending formulation. This rapid simulation allows the user to quickly evaluate the die designs early on in the simulation process, using the extended material database and a consistent elastoplastic material model in order to simulate the blank behavior.

In addition to the binder surface and the run-off design, several process parameters can be assessed, such as the blank-cutting pattern and material, draw bead forces and positioning, and so on. All of this information can be defined and modified within the graphic user interface. Such evaluation is useful to enable decision-making early in the design process and eliminate erroneous choices (Fig. 25). However, more accurate analysis for tuning and final validation is done using the incremental solver.

Usually, the result of the forming simulation will be a plot of the deformed geometry. On top of this geometry, the user can plot iso-values such as the thickness, thickness strain, stresses, FLD, and so on. In case the forming simulation shows problems due to the die design, one can return to the PAM-DIEMAKER module and perform the appropriate changes.

It is possible to make changes in the binder region without touching the rest of the profile definition. As the binder surface is modified, all profiles will just follow the newly created binder surface. It is also possible to iterate on the profile definition itself and add or remove several profiles to the existing die design.

In cases where the die-face surface was created in a different CAD package, another method to modify the die geometry is possible by using the morphing technology (Fig. 26). This technology allows the user to manipulate the mesh in 3D by defining feature lines and feature nodes, which is especially useful if no parametric profile data are available. One very common (and useful) option of morphing is the modification of the radius of already existing fillets. Another example would be the opening of a die wall by changing the angle. It ensures the tangency and enables the export of the modified surface in order to upgrade the original CAD data. The integrated package will allow the user to progress in upgrading the stamping process design in a continuous-improvement environment without any modeling discontinuity.

Tuning and Validation. The entire process is validated using an incremental solver. PAM-AUTOSTAMP is based on a shell-element formulation with both incremental explicit and implicit solvers. Different forming stages are simulated through a sequence of data selections using the data model and object structure. The model upgrade and the data transfer from the quick solver to the incremental solver occur smoothly without any modeling process discontinuity. Both solvers share the same mesh type and requirements and the same material model and database. Consequently, the user is able to focus on solving the stamping problems without any model perturbation and artificial numerical issue related to program interfaces.

Solvers have been developed for sheet metal forming to give users details on stresses, strains, and contact phenomena. The full history of the sheet metal is simulated, taking into account blank-tool interaction (blankholders, support systems, locater pins, draw beads, trim tools, etc.), which is a pre-requisite to capture all the physics that influence the final panel quality and geometry after trimming, springback, and flanging. Figure 27 is an example of a simulated part after trimming.

After the tool design has been completely checked and iterated upon, the surface information can be used as a reference for further CAD work in order to develop a final class A surface (Fig. 28). This export operation can use either international graphics exchange specification or visual data analysis format.

Computer-Aided Die Shape Compensation. Springback prediction has been a matter of

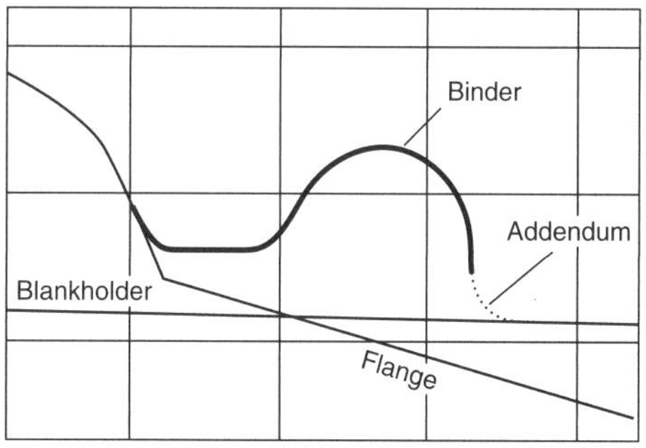

Fig. 24 Initial die-face design

Fig. 23 Details of parametric bulge profile

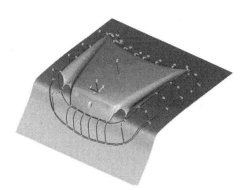
Fig. 25 Die-face design iteration

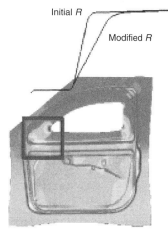
Fig. 26 Typical radius (*R*) morphing

Fig. 27 Simulated final part after trimming

interest for many years, but it has only become very useful in recent years. Prediction of springback is also only part of the problem. The difficulty lies with the proper use of the prediction. How to make corrections to minimize or even eliminate springback is the real challenge for die engineers and technicians, just as accurate prediction of springback is a critical step in the procedure. A few years ago, simulation results were not considered reliable enough to modify tooling for springback compensation; today (2006), however, with recent significant improvements in the reliability of the predictions, there has come the necessary investment in developments for simulation-based die shape compensation.

Predictive springback compensation is an important tool with the increasing use of advanced high-strength steel. In many cases, where components are to be conventionally pressed but with materials with yield strengths up to 1200 MPa (174 ksi), the magnitude of distortion due to springback is significant enough to make it impossible for corrections by restriking or overbending alone. Remachining the die face without reliable data is nearly impossible, and required tool changes may involve insertions, at a minimum, or, worst case, entirely new base castings if simply not enough cast material is available to achieve the changes in shape that may be required. This problem is significant if discovered during physical tryout.

PAM-STAMP 2G includes a module for iterative die compensation for springback. It modifies the die shape automatically, avoiding undercuts and maintaining the necessary smooth blankholder shapes without the need for any user input at all. It will also iterate, checking the results of the new forming and springback calculations until an optimized result is obtained. Typically, between three and six iterations are used to obtain the optimal die face.

The tool shape-compensation methodology existing today (2006) is based on modifications to the mesh of the tool. While this is certainly the best approach for performing iterative loops within a simulation-based environment, it still leaves the problem that, at the end of the compensation process, the user has a modified mesh of the tool, whereas he really needs a CAD model. In order to address this issue, work with surfacing specialists is being done to transform the modified tool meshes back to CADs ready for machining.

Tube Forming Simulation. Tubular hydroformed parts are beginning to account for a significant proportion of a typical vehicle structure, and with this comes an increase in the need for accurate simulation of this process. Tube forming follows a similar engineering evolution to conventional stamping, with some specific requirements.

PAM-TUBE 2G has been developed to perform a die design and forming simulation system for tube bending and tube hydroforming. It contains a similar mix of modules to address process design, tool design, fast feasibility, and accurate validation for tubular parts.

The die design module handles the design of any necessary prebending operations, including determination of the bending line, creation of bending tools, and an automatic creation of the data setup for an accurate bending simulation. The solver includes advanced tube-specific functions, such as an ovalization prediction module. The incremental solver provides the detailed validation of both bending and hydroforming processes.

LS-DYNA and DYNAFORM. LS-DYNA is based on DYNA3D, which was developed by John Hallquist at the Lawrence Livermore National Laboratory. The earlier applications of DYNA3D were primarily for the stress analysis of structures subjected to impact loadings. Hallquist later founded Livermore Software Technology Corporation (LSTC) to develop LS-DYNA for commercial applications. The vision of LSTC is to develop a single code with capabilities to solve a variety of problems involving multiphysics that may require multistages, multiprocessing, and multifunctions. Applications include crash/safety, drop test, and metal forming for a wide range of industries.

LS-DYNA is a general-purpose FEM code with a huge collection of element technology, contact models, and material models, among other FEM technology. Although it started as a dynamic explicit code, LS-DYNA has developed many efficient linear equation solvers and nonlinear implicit solution algorithms. Because of this built-in feature, solver schemes can be chosen by the user and seamlessly switched as desired. This allows complex problems such as a multistage stamping to be solved with a single code. Analysis can be done with the dynamic explicit solver primarily for forming and flanging operations. An implicit solver can be used for binder wrap and springback for better accuracy. However, LS-DYNA is only a solver. It relies on third-party software to provide the complete system solution. Among them, the most popular is DYNAFORM of Engineering Technology Associates, Inc. (ETA). The following sections provide an overview of this system, based on information provided by ETA.

Implementation and Impact of Stamping CAE. Computer technology for CAD and CAE has had a tremendous impact on the tool and die industry. Long before the implementation of CAE in the tool and die industry, CAD was adopted to create 2D and 3D die designs. However, the approach resulted in poor die design and caused many unforeseen problems during tooling tryout.

The traditional stamping manufacturing process employs CAD during the die design phrase. (Fig. 29). Upon the completion of die design, the die construction is carried out, followed by the tooling tryout to troubleshoot the new die set for major stamping defects. If the die produces parts with no stamping defects, it will be sent to the stamping plant for stamping production. On the other hand, if splitting and wrinkling are spotted during the tryout, the die set needs to be reworked. Occasionally, the die set is scrapped, and a new die design is needed. As a result, the die manufacturing time is increased as well as the cost of die making.

The advantage of stamping CAE (Fig. 30) is to effectively reduce the major stamping defects prior to the die design and construction, thus optimizing die structure design and material use. As shown in Fig. 30, the stamping CAE, including forming simulation, die-face engineering, blank-size engineering, die structural analysis, springback compensation process, and line-die simulation, are introduced to the stamping manufacturing process to achieve product manufacturing, quality, and cost-saving through optimized product and process design.

Fig. 28 Typical validated tool geometry export

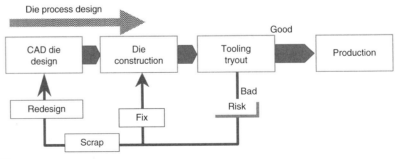

Fig. 29 Flow chart of traditional die try-out process

Die-face engineering (DFE) is a CAE-based tool developed for stamping engineers to quickly derive a suite of draw die (including binder and addendum) from the product design in the early stage of tooling design. The draw die development from DFE is then used as a baseline for tooling design and can be used for quick evaluation of tooling and formability analysis via simulation of the forming process. By integrating with forming simulation, DFE helps to reduce iteration time for tooling design in the CAD system. It can shorten the lead times for tooling design and construction by phasing out the product-intent soft tools.

Blank-size engineering (BSE) is a stamping CAE tool that allows engineers to find an efficient material utilization using blank nesting and to estimate the optimized blank size of stamped parts. In today's (2006) stamping market, an efficient material utilization plays an essential role in gaining the competitive edge for stamping manufacturers. Material utilization refers to the process of reducing scrap created by the manufacturing process, which is approximately 5% of the total part scrap.

By using BSE, stamping CAE engineers are able to optimize the material utilization according to the given constraints, such as pitch, coil width, web size, and so on. An example showing the nesting layout of blank material for a fender is illustrated in Fig. 31.

As shown in Fig. 32, BSE provides an optimized nesting layout and material utilization. Furthermore, it also allows stamping CAE engineers to estimate the cost of producing a stamped part. The BSE also allows the tool and die industry to quickly provide a stamping quotation, in addition to performing quick evaluation of tooling design and optimizing blank shape.

Die Structural Analysis (DSA). The integrity of the die structure has always been a primary concern for sheet metal suppliers in terms of part quality, weight, die life, safety, and cost-efficiency. A die structure lacking in structural integrity will produce poor-quality stamped parts due to the excessive deformation of the die during the stamping operation. This can affect the production of the entire assembly line. In addition, an ill-designed die structure can pose safety hazards to employees on the job during the maintenance operation and can increase the cost of die construction and maintenance. As a result, ensuring the integrity of the die structure is the key to successful stamping manufacturing.

The use of DSA as one of the important applications of stamping CAE has gained increasing application in the stamping manufacturing field since the early 2000s. It enables stamping CAE engineers to validate die design changes and quantify safety factors of die structures. Furthermore, DSA is used to optimize the die design for weight reduction, hence providing an opportunity for cost-savings. The DSA is also used to identify the root cause of die breakage during tryout and stamping pro-

duction. The benefits of DSA have provided significant values to stamping manufacturers in terms of die construction and stamping operations. An example showing the application of DSA to predict stresses and deflection of the upper die of an automotive body-side outer is illustrated in Fig. 33 and Fig. 34, respectively.

Springback Compensation Process. In recent years, advances in automotive manufacturing have resulted in an increasing application of lighter-gage high-strength steel and aluminum alloy sheets in automotive stampings, as opposed to the conventional mild steel grades. The application of such materials has posed

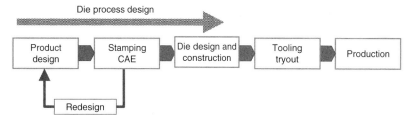

Fig. 30 Flow chart of die tryout with stamping computer-aided engineering (CAE) simulation

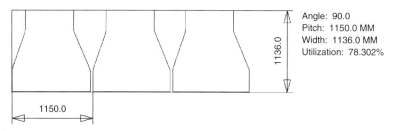

Fig. 31 Illustration showing blank-nesting layout of NUMISHEET 2002 fender using blank-size engineering

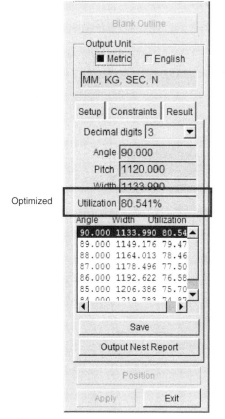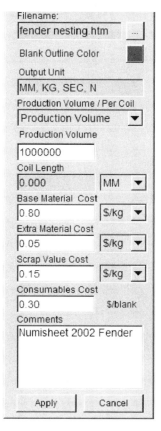

Fig. 32 Illustration showing the capabilities in blank-size engineering

challenges in sheet metal fabrication. One of the greatest challenges is springback. Springback, or shape change, is an elastic material recovery after the unloading of stamping tools and the removal of stamped parts from stamping tools. It causes undesirable final part quality and dimensional inconsistencies and creates difficulties in downstream assembly processes. In addition, it slows down new product launches while increasing changeover time and the associated cost.

Line-Die Simulation (LDS). In automotive stamping production, class A and B body panels are generally manufactured in three to four operations that consist of draw, restrike, trimming/piercing, and flanging. These operations are lined up in tandem within a press line. The quality of a stamped part relies on the success of each die operation. For example, a successfully stamped fender must be formed without major stamping defects (splitting, wrinkling, surface defects) in the draw operation. In addition, the quality of the fender also relies on the trimming and flanging operations. These secondary operations must be carried out successfully without facing other stamping defects, such as recoil, flange break line failure, springback, and so on. If the die set in each operation is not designed properly, stamping manufacturers must expand tremendous efforts to resolve the stamping defects that emerge in each operation. Such a scenario will not only affect the stamping production but also delay the subsequent vehicle assembly line. To this end, LDS saves tremendous time and monetary losses.

The application of full LDS is used to resolve stamping concerns such as formability, quality, manufacturability, and throughput requirements in each die operation, before the die sets are made. As a result, stamping manufacturers may avoid incurring extra costs during die fixing. At the present time, advanced computer technology such as MPP and faster CPUs speed up the time needed to perform a full LDS.

AutoForm. Starting from the product design data from a CAD system, AutoForm software evaluates the entire process and the details of each operation within a single system. In addition, product and process development within AutoForm includes an AutoForm-Sigma module for upfront assessment of the robustness of the product and/or process. All modules are integrated through a graphical user interface. The system has modules for various stages from concept design all the way to troubleshooting part production.

When a product design is finalized, it is released to the feasibility and tooling departments, who must then design an appropriate stamping die to fabricate the part. The traditional approach for die layout designers and die engineers (including die process, formability, and analysis functions) is to create or "build" the binder and addendum surfaces in a CAD system that can make the part. However, this method has the disadvantages of long lead time, high cost, and low throughput.

In terms of lead time, designing die faces in a conventional CAD system is very time-consuming. On average, it takes approximately 1 to 3 weeks to complete a draw-die development for typical automotive body panels and large inner panels. Even for small parts and structural components, several days are required.

The high cost of die development is due to the complexity of the process. Successful draw-die developments require the expertise of die engineers and analysts with many years of practical experience and good CAD skills. Almost always, they make several iterations of design modifications before finalizing a good stamping die design (e.g., changing radii, adjusting drawbars). Modifications for a typical nonparametric die-face design in a general-purpose CAD system are slow, cumbersome, and time-consuming. It is easy to make mistakes in the process, and it is difficult to study the design of different "what-if" concepts for the binder and addendum. Furthermore, when time runs out due to production deadlines, companies are forced to carry out tooling tryouts in the press, which is even more costly.

Low throughput is a limitation with a manual CAD-based approach, where CAE analysts

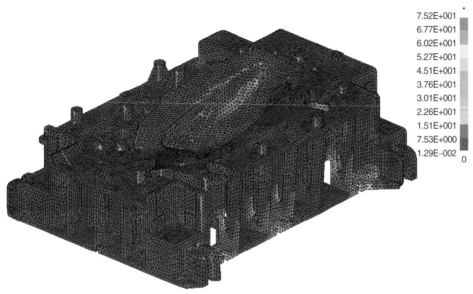

7.52E+001	*
6.77E+001	
6.02E+001	
5.27E+001	
4.51E+001	
3.76E+001	
3.01E+001	
2.26E+001	
1.51E+001	
7.53E+000	
1.29E-002	0

Fig. 33 Stress contour of upper die of body-side outer subjected to static loading. Courtesy of Engineering Technology Associates, Inc.

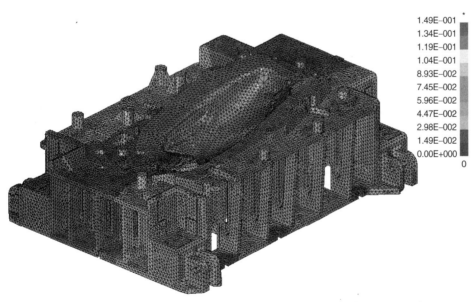

1.49E-001	*
1.34E-001	
1.19E-001	
1.04E-001	
8.93E-002	
7.45E-002	
5.96E-002	
4.47E-002	
2.98E-002	
1.49E-002	
0.00E+000	0

Fig. 34 Deflection contour of upper die of body-side outer subjected to static loading. Courtesy of Engineering Technology Associates, Inc.

must wait until the die surfaces are completed—several days or weeks—before carrying out virtual try-out simulations to check the designs for splits, wrinkles, and other stamping issues. It is far more productive and cost-effective if die-face designers or CAE analysts can immediately try out the die concepts, discard unfeasible designs early on, and concentrate on further developing of only the best design(s).

In an effort to conquer these drawbacks, several automotive companies participated in a program to develop a faster, more efficient, and less costly method for die-face design. Auto-Form-DieDesigner software is the result of a development effort led by AutoForm Engineering GmbH (developers of sheet metal forming software) in cooperation with tooling departments at BMW and Audi, and with technical feedback from DaimlerChrysler and General Motors.

The software reduces tooling development time through rapid parametric design of die faces and their immediate verification and optimization with integrated stamping simulation modules. It is specialized for generating binder and addendum surfaces and for evaluating the feasibility of draw-die developments and prototype tools. Its three most important innovations are:

- Fully parametric features
- Complete integration with virtual try-out software modules
- Automatic optimizer

The parametric approach of the software is based on analytical engineering principles. It also conforms to the die engineering practice of using surface profiles (arcs, angles, and straight sections) to design die faces and is compatible with various CAD data formats. Using the software, it takes approximately 1 h to design complete die faces, starting from only the CAD surface data of the part (such as an auto fender, Fig. 35). This includes the following steps:

- Determination of tip-angle
- Filleting of sharp edges (variable radii)
- Filling of holes
- Part modifications (morphing)
- Design of the fill surface between double-attached parts
- Overcrowning
- Binder design
- Design of outer (and, if required, inner) addendum
- Unfolding of flanges
- Generation of the tools

Because all the generated die surfaces are parametric, subsequent modifications (die engineering changes) can be done in seconds or minutes. These die surfaces and simulation results can easily be exported using international graphics exchange specification, Vereinung Deutsche Automobilindustrie Flächen Schnittstelle (German neutral file format for exchange of surface geometry), or mesh formats to other software applications for die design, die machining, and structural and crash analysis.

Part Design Modifications. Almost all part designs are subject to revisions and improvements, even after official release to the die engineering department. After the parameterized binder (Fig. 36) and addendum (Fig. 37) have been created for a specific part, when the part is replaced with another (similar) part design with a modified design, the software can automatically adjust the binder and addendum accordingly. This is a time-saving feature, because it allows the original addendum and binder geometry to be reused for future design revisions of the part. In addition, automatic filleting and the ability to vary fillet radii on the part geometry save considerable time and allow virtual tryouts to be carried out earlier in the product and tooling development cycles. The initial binder surface (light background in Fig. 36) is automatically created in a few seconds, using only the CAD surface data in (Fig. 35) as input. The user can then modify the binder geometry with surface profiles.

Parametric Linking. With an integrated system approach and full parametric linking, all die-face designs can be evaluated with one-step or incremental simulation modules. One-step simulation results are considerably more reliable when based on the complete die face—as generated by the software—rather than on the part only. An added benefit is that these refined results are made possible earlier in the design cycle. They are available within a few minutes and include the required blank outline and various stamping feasibility criteria (Fig. 38).

The incremental module is used for high accuracy and virtual tryouts of the complete stamping process. Results include predictions of wrinkles, cracks, skid and impact lines, surface quality, and so on. Incremental tryout results are typically available within 1 to 3 h, depending on the size and complexity of the part and the desired accuracy. Additional tryouts can be launched at once because of parametric linking. For example, based on the initial results, the user can make modifications to the addendum geometry, and the various tool geometries required for the next simulation are automatically updated. Similarly, if the user modifies the punch opening line, draw bead positions are automatically adjusted.

Within a few seconds, users can change "master" global profile parameters (for example, to increase the die radius). In addition, the user can also make local surface modifications to the addendum by changing local profile parameters or directly manipulating contours such as the punch opening line, bar height, and so on (Fig. 39). A drawbar can be easily created by modifying the parameters of local addendum profiles. Figure 40 shows the complete die face for the Audi TT rear fender, created in approximately 1 h using 28 customized surface profiles.

Formability Simulation. From the initial die-face concept shown in Fig. 40, a quick one-step simulation may take into account the binder and addendum. Stamping feasibility results can be calculated in minutes for various stamping feasibility criteria, typically plotted in color graphics. An example plot (in black and white only from color) is shown in Fig. 41. The circled area on the left shows a risk of wrinkles on the addendum, very near to the part boundary. The circled area on the right shows a crack.

Based on one-step results, the general die concept can be evaluated early in the design cycle, and, if necessary, modifications to the part geometry can be recommended. Furthermore, the resulting blank outline can be used to design the initial blank outline for an incremental virtual tryout.

An accurate incremental analysis of the complete stamping process, using the die faces from the initial or final die face based on the one-step results, can pinpoint the formability issues more accurately. For example, the plot in Fig. 42

Fig. 36 Automatically created initial binder surface (outer light-gray area)

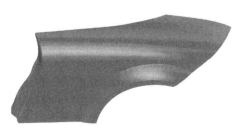

Fig. 35 Part geometry (computer-aided design surfaces) of a rear fender

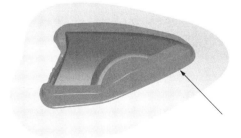

Fig. 37 View of automatically generated addendum surface (arrow), based on default profiles

(originally in color) shows the predicted thickness distribution; excessive thinning zones (originally plotted as red and yellow regions) are shown by two circled areas on the right. The simulation also shows wrinkles in the circled area on the left. These try-out results were obtained in approximately 50 min.

The die designer can then make modifications to the original die concept to smooth out the wrinkles and eliminate the cracks (for example, by changing drawbar and wall heights, radii, etc.). Furthermore, results such as wrinkling, splitting, skid mark progression, and so on can be visualized at each step of the drawing process.

Automatic Optimization of Tooling and of the Stamping Process. To further improve the die-face designs, the software includes an integrated optimizer module. The user first specifies the allowable ranges for tool geometry and stamping parameters. The optimizer then automatically determines the "best" design within these ranges through multiple one-step or incremental simulations to achieve the target function, for example, elimination of cracks and wrinkles, uniform surface stretching, or no excessive thinning. Tool geometry parameters include part, die, and drawbar radii; drawbar height; wall angles; overcrown; and so on. Stamping parameters include binder forces, draw bead strength, blank outline, and so on.

Figure 43 shows the final outcome of the optimized results by adjusting parameters (such as draw bead force) defined by the analyst. The two crack zones were eliminated, and the wrinkling zone was minimized and moved away from the part boundary. Automatic optimization can help the die designer to find the best parameters for his design. It can also help the process engineer to determine the best stamping conditions.

Stochastic Simulation for Robust Design. Manufactured part quality and cost, as well as production timelines, can be impacted by variations in the inputs to production processes. The most carefully designed and validated product and process can lead to inadequate and/or inconsistent quality of the produced part during production due to such variations. The real challenge is to neutralize or minimize the effect of production condition variations on production quality upfront during product and process design.

Robust engineering is the concept of applying appropriate safety margins in the design of product, dies, and process conditions to accommodate worst-case scenarios in terms of a limited number of specific production conditions, such as material and lubrication. This traditional approach in just applying safety margins is somewhat limited in scope and efficacy because it does not help develop an understanding of production conditions that is adequate for troubleshooting real-time production problems. In this regard, AutoForm-Sigma is a software module that enables robust design and engineering of stamped and hydroformed parts based on stochastic simulation technology. Simply stated, AutoForm-Sigma runs several simulations—approximately 100 simulations per run—over which uncontrollable noise parameters such as material properties, blank thickness, lubrication, and so on are automatically and randomly varied, and enables an assessment of applicable robustness metrics on the selected result(s). Rather than individual simulations with ideal inputs, AutoForm-Sigma can account for real product conditions and characterize virtual try-out results in terms of standard production quality metrics such as process capability index, scrap rate, and so on. Besides upfront manufacturing robustness assessment, this approach provides an opportunity for very early robust engineering (which "design" elements on product, die, and process design are most effective/influential in achieving specific quality desirables on a manufactured part) as well as robust design (what are the best ranges of these "design" variables).

Benefits of Simulation. Simulation technology enables engineers to assess product and process design from concept styling through production. In AutoForm, for example, one

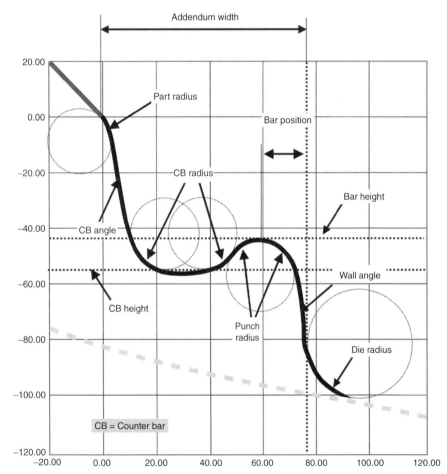

Fig. 38 Typical addendum profile (dark solid line) and its defining parameters

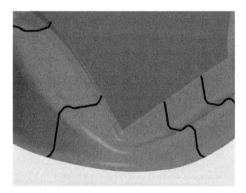

Fig. 39 Drawbar created by modifying the parameters of local addendum profiles (black lines)

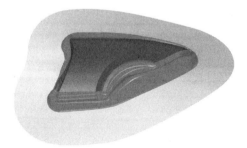

Fig. 40 Complete die face for the Audi TT rear fender

common graphic user interface provides efficient access to modules for:

- Concept styling
- Product design
- Production planning
- Process planning
- Tool design and manufacturing
- Tryout
- Production

Examples of reported savings include:

- Reduction in binder development time by 60%
- Reduction by a factor of 4 to 5 in the time from CAD part design until completed virtual tryouts
- Reduction of die-face development time from one week down to one day
- Reduction in actual die try-out time and tooling costs of nearly 50%
- Two-week reduction of tooling development time by using the generated die faces to order castings for the dies

Summary

As the focus of die development shifts away from trial and error toward CAE simulations, the burden also shifts upfront and is undertaken more by CAE analysts. For a component to be manufactured effectively, new materials or new forming processes require a much greater degree of precision and parameterization to answer the needs of forming simulation. A customizable model that keeps track of part material history and parameters such as strain rate, kinematic hardening, and so on has become a requirement in order to attain the last 10% of accuracy at the stresses level. Therefore, the need for well-trained formability engineers with in-depth understanding of CAE technology should increase.

As an essential technology in metal stamping, stamping CAE plays a central role in the die and stamping industry. The latest stamping CAE technology enables CAE analysts to do die-face engineering, blank-size engineering, forming simulation, die structural analysis, springback compensation process, and line-die simulation. By using the technology in the early stages of tooling design, manufacturers can confidently identify costly errors before die construction begins. To that end, tool and die makers can remain competitive by producing quality dies and quality parts at a lower cost.

The full implementation of stamping CAE has impacted die and stamping manufacturers in many ways, such as troubleshooting stamping defects, improving quality, reducing toolmaking lead time, optimizing die design, and so on. The application of advanced stamping CAE technology helps manufacturers reduce tryout time and improve die-development efficiency. By using stamping CAE, substantial savings can be achieved in quality, delivery, and cost of the product. It helps die and stamping manufacturers survive in the competitive stamping marketplace. With the advancement in computer hardware and software technology, the future of stamping CAE holds even greater promise as the simulation techniques mature and expand. The output of CAE for sheet forming is also being introduced to assess part performance in service, for example, the effects of work hardening, local thinning, and so on on crash worthiness.

Cross-Functional Analysis. Another outcome of using the new material is the performance of the formed part. Some material (e.g., dual-phase steel) can work harden much quicker than traditional high-strength low-alloy grades. Parts may also exhibit more localized thinning due to poor formability. Both influence the behavior of vehicle crash testing. Recent studies illustrate the impact of the manufacturing pro-

cess on the component performance. Because traditional crash simulation did not take part-material history into account, simulation has not proved to be good enough to accurately predict the performance in physical testing of critical parts and subassemblies.

A study at Volkswagen revealed that, compared to real tests, simulation results were off by 21% for maximum force and 22% for part deformation if virgin material properties were used for all metal parts in crash simulations. However, an integrated analysis that kept track of part-material history (i.e., stress and strain undergone during forming), with necessary forming simulation, followed by the crash simulation proved to be much more accurate. Thus, the use of new materials has made it a requirement for stamping simulation today (2006) to maintain the full history of a part through seamless data transfer.

Future Developments. The CAE technology for metal forming will continue to evolve to cover more applications and with more details. In recent years, metal-forming simulation has expanded from traditional failure prediction for draw (or form) operation to springback prediction after trimming, flanging, and die compensation for the draw die. Future developments may be classified into the following categories.

Complete Die Operation. The CAE technology and actual applications have been rapidly maturing for several years. Future needs will drive the technology for more advancement to achieve better accuracy, capability, and

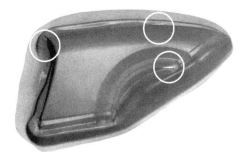

Fig. 42 Result of an accurate incremental try-out simulation of the complete stamping process. Thinning zones shown by two circled areas on the right. Wrinkling is indicated by the circled area on the left.

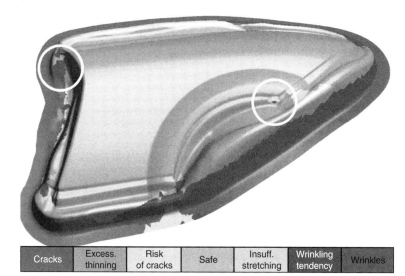

Cracks	Excess. thinning	Risk of cracks	Safe	Insuff. stretching	Wrinkling tendency	Wrinkles

Fig. 41 Result of a one-step simulation. The circled area on the left shows a risk of wrinkles on the addendum, very near to the part boundary. The circled area on the right shows a crack.

Fig. 43 Results of optimization of the initial virtual tryout based on the previous incremental analysis

usability. Once the technology is mature, simulations for the entire line-die operation described in previous sections will become common practice.

One of the most critical technology bottlenecks is the ability to model complex flanging and hemming over a tight radius. This will require a good material model with reliable parameters and advanced element formulation, among other things. Springback prediction after the final operation may become possible if the simulation can provide accurate and reliable stress distributions.

As the accuracy for springback prediction for line dies improves, die compensation for line dies besides the first draw die will become popular. The distortion induced by the flanging operation can possibly be minimized then. This may also spark interest for overall springback correction, because it may not be the optimal solution to minimize springback after each operation using die-compensation methodology.

Other Simulation Features Supporting the Line Dies. New hardware, such as the crossbar transfer press, has been created to accelerate production. This new hardware has modified the production process itself, which may create secondary problems. Some details either being omitted or simplified in current practices will be added to future CAE models for better accuracy. Examples may include the following:

- Gaging of the blank
- Actual bead geometry
- Actual trimming with solid elements
- Actual nitrogen (or air) pins instead of overall binder force
- More accurate weld-line modeling for tailor-welded blank applications
- Transferring of the sheet metal between dies
- Die flex and die wear consideration
- Scrap cutting and removal

Assembly Simulations with Formed Parts. Subassembly or vehicle assembly with "relaxed" parts (i.e., after springback) will become a popular issue. Currently, most assembly simulations rely on measurement data. This is often too late to make a big impact. If the total springback prediction is reliable, then the assembly simulation can be done much earlier in the program, without any hardware.

Another example of a new requirement for simulation can be seen from the evolution in welding technology. Whereas traditional spot welding integrated the weld and clamp and allowed for standard tolerances, the increased use of laser welding requires a more sophisticated clamping system and a very low flange tolerance. This technology has a large impact indeed on tolerance control, which translates to springback problems on the part at the forming stage. This evolution in process again illustrates the need for increased precision and tighter control during stamping simulation.

Alternative Forming Simulation and Innovative Manufacturing Simulation. Many non-traditional forming methods have been adopted in the last decade. Some can be handled fairly well with the current simulation technology. These include tube hydroforming, fluid forming (sheet hydroforming), tailored-welded blanks, superplastic forming, and warm forming. Because the business case for those methods is not as strong as for traditional stamping, the push for technology growth in these areas has not been as strong. There is room for improvement for these simulations, such as automatic pressure regulation for hydroforming. Business demands will dictate the future growth.

Other innovative simulations may include the need for surface-quality evaluation, process optimization, optimal metal flow control, process layout, automatic die-face development, and so on. Some of the calculations are very CPU-intensive; hence, the cost reduction and the growth of the MPP technology will have a strong impact on the feasibility of the aforementioned simulations. Future studies, with the help of better CAE technology, are also required to reduce the number of parts through part consolidation.

Technology Collaboration. Forming simulation in its earlier iterations has become inadequate. The focus has shifted from the initial objective for formability analysis to much broader applications, quality improvement, cost reduction, and integration. Forming simulations need to adapt to these new requirements. The most important among these are:

- Providing a streamlined, scalable environment that encompasses every step in the process, from early feasibility to final validation
- Increased precision through model parameterization to assess new materials and their properties, to accurately predict springback behavior and surface-quality problems
- Seamless simulation data transfer and management to keep track of part history at every stage in the process, in order to check the performance and the robustness of the product and its consistency as well
- Optimal exploitation of available computer resources in a customizable, extended, collaborative environment

It is not feasible for any manufacturing or CAE software company to work alone to meet future demands. More collaboration among the academia, CAE software developers, materials suppliers, and users is needed.

ACKNOWLEDGMENTS

Sheet metal forming simulation has made tremendous progress. Technology advancements have come from the academic field; researchers and users in automotive, appliance, steel, aluminum, aerospace, and other industries; developers from the software companies; and expert consultants. They should all be recognized for their contributions. Many of their names are listed in the reference section.

In particular, Fouad El Khaldi of the ESI Group, Arthur Tang and C.C. Chen of ETA, Dr. Xinahi Zhu of LSTC, and Dr. Kidambi S. Kannan of AutoForm Engineering provided valuable material and generously supplied many graphics. Thanks are due to Wei-Tsu Wu of Scientific Forming Technologies Corporation for his guidance, support, and encouragement. The author would also like to express his sincere thanks to Daniel Sa for his great effort in editing the draft manuscript, and to the ASM International staff for their support, coordination, and enhancement of the manuscript.

REFERENCES

1. F. Barlat and J. Lian, Plastic Behaviour and Stretch Ability of Sheet Metals, Part 1: A Yield Function for Orthotropic Sheets under Plane Stress Conditions, *Int. J. Plast.,* Vol 5, 1989, p 51–66
2. R. Hill, *J. Mech. Phys. Solids,* 1990, p 405–417
3. N.M. Wang and B. Budiansky, Analysis of Sheet Metal Stamping by a Finite Element Method, *J. Appl. Mech. (Trans. ASME),* Vol 45, 1978, p 73–82
4. R. Hill, *The Mathematical Theory of Plasticity,* Clarendon Press, Oxford, 1950
5. W.R. Hosford, in *Proc. Seventh North American Metalworking Conf.* (Dearborn, MI), Society of Manufacturing Engineers, 1979, p 1912–1916
6. S.P. Keeler and W.A. Backofen, Plastic Instability and Fracture in Sheets Stretched over Rigid Punches, *Trans. ASM,* Vol 56, 1963, p 25–48
7. S.P. Keeler, "Determination of Forming Limits in Automotive Stampings," Technical paper 650535, Society of Automotive Engineers, 1965
8. Z. Marciniak and K. Kuczynski, Limit Strains in the Process of Stretch-Forming Sheet Metal, *Int. J. Mech. Sci.,* Vol 9, 1967, p 609–620
9. Z. Marciniak, K. Kuczynski, and T. Pokora, Influence of the Plastic Properties of a Material on the Forming Limit Diagram for Sheet Metal in Tension, *Int. J. Mech. Sci.,* Vol 15, 1973, p 789–805
10. C.L. Chow, L.G. Yu, and M.Y. Demeri, "Prediction of Forming Limit Diagram with Damage Analysis," SAE Technical Paper 960598, 1996
11. C.L. Chow, X.J. Yang, and E. Chu, Viscoplastic Constitutive Modeling of Anisotropic Damage under Non-Proportional Loading, *J. Eng. Mater. Technol.,* Vol 3, 2001
12. W.F. Hosford, A Generalized Isotropic Yield Criterion, *J. Appl. Mech.,* Vol 39, 1972, p 607–609
13. A.F. Graf and W.F. Hosford, Calculations of Forming Limit Diagram for Changing Strain Path, *Metall. Trans. A,* Vol 24, 1993, p 2497–2501

14. T.B. Stoughton and X. Zhu, Review of Theoretical Models of the Strain-Based FLD and Their Relevance to the Stress Based FLD, *Int. J. Plast.,* Vol 20, 2004, p 1463–1486

15. J.J. Kleemola and M.T. Pelkkikangas, Effect of Predeformation and Strain Path on the Forming Limits of Steel, Copper, and Brass, *Sheet Met. Ind.,* Vol 63, 1977, p 591–599

16. R. Arrieux, C. Bedrin, and M. Boivin, "Determination of an Intrinsic Forming Limit Stress Diagram for Isotropic Metal Sheets," 12th Biennial Congress of the Int. Deep Drawing Research Group, 1982, p 61–71

17. T.B. Stoughton, A General Forming Limit Criterion for Sheet Metal Forming, *Int. J. Mech. Sci.,* Vol 42, 2000, p 1–27

SELECTED REFERENCES

Technical Papers and Presentations

- S. Aita, F. El Khaldi, L. Fontain, T. Tamada, and E. Tamur, "Numerical Simulation of a Stretch Drawn Auto Body—Part I and II: Assessment of Simulation Methodology and Modeling of Stamping Components," Technical Papers 920639 and 920640, SAE Congress, 1992

- V. Aitharaju, M. Liu, J. Dong, J. Zhang, and C.T. Wang, Integrated Forming Simulations and Die Structural Analysis for Optimal Die Design, *Proc. of NUMISHEET 2005,* 2005, p 96–100

- H. Aoh and E. Nakamachi, in *Proc. of NUMIFORM 1989,* p 357–379

- F.J. Arlinghaus, W.H. Frey, T.B. Stoughton, and B.K. Murthy, in *Computer Modeling of Sheet Metal Forming Processes,* The Metallurgical Society, 1985, p 51–64

- L. Baillet, C. Desrayaud, Y. Berthier, and M. Brunet, *Proc. of NUMISHEET 1996,* p 72–79

- F. Barlat, D. Banabic, and O. Cazacu, *Proc. of NUMISHEET 2002,* p 515–524

- O. Barlet, J.L. Batoz, Y.Q. Guo, F. Mercier, H. Naceur, and C. Knopf-Lenoir, Optimum Design of Blank Contours Using the Inverse Approach and a Mathematical Programming Technique, *Proc. of NUMISHEET Conf.* (Dearborn, MI), 1996

- R. Brun, A. Chambard, M. Lai, and P. de Luca, "Actual and Virtual Testing Techniques for a Numerical Definition of Materials," NUMIFORM Conf., 1999

- R.W. Clough, *J. Struct. Div., Proc. Second Conf. on Electronic Computation,* ASCE, 1960, p 345–378

- M. Coticchia et al., *CAD/CAM/CAE Systems: Justification, Implementation, Productivity Measurement,* 2nd ed., Marcel Dekker, 1993

- X. Deng, "Optimization of Structures Under Technological Casting Constraints," Doctoral thesis, ENS Cachan, France, 1993

- R. Dick, "Improvements of the Beverage Can Redraw Process Using LS-DYNA," Seventh International LS-DYNA Users Conf., Livermore Software Technology Corp. and Engineering Technology Associates, Inc, 2002

- E. Di Pasquale, J.-L. Duval, and F. El Khaldi, "An Integrated Approach to Sheet Metal Forming Simulation," IBEC Conf. (Detroit, MI), Society of Automotive Engineers, 1999

- E. Doege, T. El-Dsoki, and D. Seibert, in *Proc. of NUMISHEET 1993,* p 187–197

- J.L. Duncan, R. Sowerby, and E. Chu, in *Computer Modeling of Sheet Metal Forming Processes, Proc. of NUMIFORM 1989,* M. Grober and K. Gruber, Ed., p 587–600

- F. El Khaldi, "Sheet Metal Stamping: Evolve from Process Simulation to Global Virtual Try-Out Space Application," Rolling Sheet Conf. (Yokohama, Japan), 1998

- F. El Khaldi, S. Aita, L. Penazzi, T. Tamada, T. Ogawa, S. Tasaka, and E. Horie, "Industrial Validation of CAE Finite Element Simulation of Stretch Drawn Autobody Part (Front Fender Case)," International Deep Drawing Research Group. (Shenyang, China), 1992

- F. El Khaldi, R. Di Bernardi, and O. Ogura, "Sheet Metal Forming: State-of-the-Art Application Methodology and Simulation Streamlining," Technical Paper 960594, SAE Congress,1996

- F. El Khaldi et al., "Numerical Simulation of Industrial Sheet Metal Forming: Recent Trends and Advances in Application Methodology and Simulation Streamlining," Int. Symp. on Automotive Technology and Automation (Florence, Italy), 1996

- F. El Khaldi, S. Tasaka, and F. Hatt, "Sheet Metal Forming Simulation, Methodology and Material Data Base Issues," Technical Paper 940755, SAE Congress, 1994

- M. El Mouatassim, B. Thomas, J.-P. Jameux, and E. Di Pasquale, An Industrial Finite Element Code for One-Step Simulation of Sheet Metal Forming, *Simulation of Materials Processing: Theory, Methods and Applications,* S. Shen and P. Dawson, Ed., Balkema, Rotterdam, 1995

- M.J. Finn, P.C. Galbraith, L. Wu, J.O. Hallquist, L. Lum, and T.-L. Lin, "Use of a Coupled Explicit-Implicit Solver for Calculating Springback in Automotive Body Panels," Livermore Software Technology Corporation, Livermore, CA

- M. Gröber and K. Gruber, (1991) Numerical Simulation of Sheet Metal Forming of Large Car Body Components, *Int. VDI Conf.,* 1991, p 587–600

- E. Haug, D. Lefebvre, and L. Taupin, "Industrial Applications of Computer Simulation in Stamping," Calculs en Emboutissage, Publication, CETIM 1994, France, 1994

- S.S. Hecker and A.K. Ghosh, *Sci. Am.,* Vol 235, 1976, p 100–108

- H.D. Hibbitt, P.V. Marcal, and J.R. Rice, *Int. J. Solids Struct.,* 1970, p 1069–1086

- G. Hibon, G. Marron, and P. Patou, "Designing Stamped Parts in Hot-Rolled Steel Sheet," 19th International Deep Drawing Research Group Biennial Congress (Eger, Hungary), 1996

- A. Honecker and K. Mattiasson, in *Proc. of NUMIFORM 1989,* p 457–463

- W.F. Hosford and J.L. Duncan, Sheet Metal Forming: A Review, *JOM,* Vol 51 (No. 11), 1999, p 39–44

- H. Iseki and T. Murota, in *Advanced Technology of Plasticity, Proc. of the First International Conf. on Technology of Plasticity,* Japan Society for Technology of Plasticity, Tokyo, 1984, p 678–684

- N. Iwata, M. Matsui, N. Nakagawa, and S. Ikura, in *Proc. of NUMISHEET 1993,* p 303–312

- S. Kalpakian, *Manufacturing Processes for Engineering Materials,* 3rd ed., Addison-Wesley, 1997

- L. Keßler and T. Gerber, "Characterization and Behavior of Steel Sheet Materials for Linking FEM-Simulation in Successive Stamping and Crash Processes," EuroPAM Conf., 2001

- S.P. Keeler, SAE Technical Paper 650535, 1965

- S.P. Keeler and W.A. Backofen, "Plastic Instability and Fracture in Sheets Stretched over Rigid Punches," *ASM Trans. Q.,* Vol 56 (No. 1), 1964, p 25–48

- S. Kobayashi and J.H. Kim, in *Mechanics of Sheet Metal Forming,* D.P. Koistinen and N.M. Wang, Ed., Plenum Press, 1978, p 341–363

- J. Kusiak and E.G. Thompson, Optimization Techniques for Extrusion Die Shape Design, *NUMIFORM 89,* E.G. Thompson et al., Ed., Balkema, Rotterdam, 1989

- Y. Le Roch, J.L. Duval, and E. Di Pasquale, Coupled Sheet Metal Forming and Fatigue Simulation, *Proc. Int. Deep Drawing Research Group,* 1998

- Z. Li, J. Wu, and D. Zhou, "A New Concept on Stamping Die Surface Compensation," Eighth International LS-DYNA Users Conf., 2004

- R.L. Logan and W.F. Hosford, *Int. J. Mech. Sci.,* Vol 22, 1980, p 419–430

- K. Mattiasion et al., On the Use of Explicit Time Integration in Finite Element Simulation of Industrial Sheet Forming Processes, *Int. VDI Conf.* 1991, p 479–497

- H.S. Mehta and S. Kobayashi, *J. Eng. Ind. (Trans. ASME),* Vol 95, 1973, p 874–880

- N. Montmayer and C. Staub, Springback Prediction with OPTRIS, *Proc. of NUMISHEET 1999*

- Y. Nagai and A. Makinouchi, in *Proc. of NUMIFORM 1989,* p 665–676

- J.C. Nagtegaal and L.M. Taylor, in *Proc. of NUMIFORM 1989,* p 705–725

- E. Nakamachi, Y.P. Chen, N. Yokoyama, and H. Morimoto, in *Proc. of NUMISHEET 2002,* p 515–524

- K. Narasimhan, M.P. Miles, and R.H. Wagoner, in *Proc. of NUMISHEET 1993,* p 363–372

- S. Natarajan, S. Venkataswamy, and P. Bagavathiperumal, A Note on Deep

Drawing Process: Numerical Simulation and Experimental Validation, *J. Mater. Process. Technol.,* Vol 127, 2002, p 64–67

- H.D. Nine, Drawbead Forces in Sheet Metal Forming, *Mechanics of Sheet Metal Forming,* Plenum Press, 1978
- E. Onate and O.C. Zienkiewicz, in *Sheet Metal Formability, Proc. IDDRG,* Portcullis Press, Surrey, 1978, p 29–38
- C.-D. Park and S.-W. Oh, "Lead Time Reduction in Manufacturing Side Panel Outer Dies (A Project Result)," EuroPAM Conf., 2001
- R. Pearce, in *Formability of Metallic Materials,* J.R. Newby and B.A. Niemeier, Ed., ASTM, 1982, p 3–18
- M. Pietsch, S. Ohnimus, and K.-V. von Schöning, "Fast Die Design with VIKING," EuroPAM Conf., 2001
- D. Priadi et al., "Introduction of Strain Rate Effects in Constitutive Equations Suitable for Sheet Metal Stamping Applications," MECAMAT 91 (Fontainebleau, France), 1991
- N. Rebelo, J.C. Nagtegaal, and H.D. Hibbitt, in *Proc. of NUMIFORM 1989,* 1989, p 31–43
- M. Rohleder and K. Roll, "Springback on Ring Taken Out the Wall of a Deep Drawn Cup," ISMR, 1991
- G. Sachs, in *Proc. Inst. Auto. Eng.,* Vol 9, 1935, p 588–600
- K. Sato, Y. Tokita, M. Ono, and A. Yoshitake, "Dent Simulation of Automotive Outer Panel Using High Strength Steel Sheets," SAE Technical Paper 2003-01-0606, 2003
- F.I. Saunders and R.H. Wagoner, Forming of Tailor-Welded Blanks, *Metall. Mater. Trans. A,* Vol 27 (No. 9), 1996, p 2605–2616
- H. Schmidt, T.C. Vu, and L. Recke, Sheet Metal Simulation: The New Breakthrough of a New Technology, *NUMISHEET 1996* (Dearborn, MI), 1996, p 158–164
- K. Schweizerhof and J.O. Hallquist, in *Proc. of NUMIFORM 1989,* 1989, p 405–439
- V. Steininger and V. Prabhakar, Parametric Die Faces in One Hour, *Int. Sheet Met. Rev.,* 2001, p 26–39
- T.B. Stoughton, Finite Element Modeling of 1080 AK Steel Stretched over a Rectangular Punch with Bending Effects, *Computer Modeling of Sheet Metal Forming Processes,* N.M. Wang and S.C. Tang, Ed., The Metallurgical Society, 1985, p 143–159; also GM R&D Research Report PH-128
- Y.S. Suh, Y.H. Kim, and S.W. Kim, in *Proc. of NUMISHEET 1996,* p 144–150
- R. Sunkel, C. Pautsch, K. Roll, R. Toderke, F. Fuchs, and V. Steininger, in *Proc. of NUMISHEET 1996,* p 286–293
- A. Tang, W. Lee, J. He, S. Pierre, and C.-C. Chen, CAE Based Die Face Engineering Development to Contribute to the Revitalization of Tool and Die Industry, *Proc. of NUMISHEET 2005* (Detroit, MI), 2005
- A. Tang, W. Lee, J. He, J. Xu, and C.-C. Chen, Die Face Engineering Based Springback Compensation Strategy and Implementation, *Proc. of NUMISHEET 2005* (Detroit, MI), 2005
- S.C. Tang and L.B. Chappuis, A Finite Element Simulation of Sheet Metal Forming Process Using a General Non-Flat Shell Element, *Proc. of NUMIFORM 1989,* 1989, p 507–513
- S.C. Tang, L.B. Chappuis, and J. Matke, in *Proc. of NUMIFORM 1989,* p 259–258
- S.C. Tang, J. Gress, and P. Ling, in *Controlling Sheet Metal Forming Processes, Proc. IDDRG,* ASM International, 1988, p 185–193
- C. Teodosiu, D. Daniel, H.-L. Cao, and J.-L. Duval, in *Proc. of NUMISHEET 1993,* p 125–135
- O. Thorsten, S.P. Scholz, V. Schulze, and M. Taeschner, "Forming Simulation in Virtual Manufacturing," EuroPAM Conf., 2001
- M. Traversin and R. Kergen, in *Proc. of NUMISHEET 1993,* p 281–291
- H. Vegter, Y. An, H.H. Pijlman, and J. Huétink, Advanced Mechanical Testing on Aluminum Alloys and Low Carbon Steels for Sheet Forming, *Proc. of NUMISHEET 1999* (Besancon, France), J.C. Gelin and P. Picard, Ed., Vol 1, 1999, p 3–8
- R.H. Wagoner, in *Proc. of NUMISHEET 2002,* p 13–24
- C.T. Wang, in *Proc. of NUMISHEET 2002,* p 597–604
- C.T. Wang, Evolutions of Advanced Stamping CAE—Technology Adventures and Business Impact on Automotive Dies and Stamping, *Proc. of NUMISHEET 2005* (Detroit, MI), 2005, p 78–83
- N.-M. Wang, "A Mathematic Model of Sheet Metal Forming Operations," GM R&D Research Report GMR274C, April 9, 1976
- N.-M. Wang, A Rigid-Plastic Rate-Insensitive FEM for Sheet Metal Forming, *Proc. Numerical Analysis of Sheet Metal Forming,* D.P. Koistinen and N.-M. Wang, Ed., Wiley, 1984, p 117–164
- N.M. Wang, C. Du, J. Riddle, and L. Zhang, in *IBEC '94 Body Assembly and Manufacturing,* 1994, p 116–122
- N.M. Wang and S.C. Tang, *Computer Modeling of Sheet Metal Forming Processes,* The Metallurgical Society, 1985, p 1–12
- N.M. Wang, and S.C. Tang, Analysis of Bending Effects in Sheet Metal Forming Operations, *Proc. NUMIFORM,* 1986, p 71–76
- N.M. Wang and M.L. Wenner, in *Mechanics of Sheet Metal Forming,* Plenum Press, 1978, p 367–398
- M.L. Wenner, SAE Technical Paper 970431, 1997
- M.L. Wenner, Overview—Simulation of Sheet Metal Forming, *Proc. of NUMISHEET 2005,* 2005, p 3–7
- A.S. Wifi, *Int. J. Mech. Sci.,* Vol 18, 1976, p 23–31
- L. Wu and Y. Yu, in *Proc. of NUMISHEET 1996,* p 324–329
- Y. Yamada, in *Advances in Matrix Methods of Structural Analysis and Design,* R. Gallagher, T. Yamada, and J.T. Oden, Ed., The University of Alabama Press, Huntsville, 1971
- M.-C. Yang, and T.-C. Tsai, "Virtual Die Tryout of Miniature Stamping Parts," Fourth European LS-DYNA Users Conf., 2003
- H. Yao, C.-C. Chen, D.-D. Liu, K.P. Li, C. Du, and L. Zhang, "Laminated Steel Forming Modeling Techniques and Experimental Verifications," SAE Technical Paper 2003-01-0689, 2003
- O.C. Zienkiewicz, Flow Formulation for Numerical Solution of Forming Processes, *Proc. Numerical Analysis of Forming Processes,* 1984, p 1–44

Conference Proceedings and Books

- *AutoForm User's Manual and Training Manual,* AutoForm Engineering, 2004
- *eta/DYNAFORM 5.2 User's Manual,* Engineering Technology Associates, Inc., 2004
- W.F. Hosford, and R.M. Caddell, *Metal Forming: Mechanics and Metallurgy,* 2nd ed., PTR, Prentice-Hall, 1993
- *LS-DYNA Keyword User's Manual,* Livermore Software Technology Corporation, 2003
- J.T. Oden, *Finite Elements of Nonlinear Continua,* McGraw-Hill, 1973
- *Proc. of NUMIFORM 1982, the First International Conf. on Numerical Methods in Industrial Forming Processes* (Swansea, England), J.F.T. Pittman et al., Ed.
- *Proc. of NUMIFORM 1986, the Second International Conf. on Numerical Methods in Industrial Forming Processes* (Gothenburg, Sweden), K. Mattiasson et al., Ed., Balkema, Rotterdam
- *Proc. of NUMIFORM 1989, the Third International Conf. on Numerical Methods in Industrial Forming Processes* (Fort Collins, CO), E.G. Thompson et al., Ed., Balkema, Rotterdam
- *Proc. of NUMIFORM 1992, the Fourth International Conf. on Numerical Methods in Industrial Forming Processes* (Sophia-Antipolis, France), J.L. Chenot et al., Ed., Balkema, Rotterdam
- *Proc. of NUMIFORM 1995, the Fifth International Conf. on Numerical Methods in Industrial Forming Processes* (Ithaca, NY), S.F. Shen and P.R. Dawson, Ed., Balkema, Rotterdam
- *Proc. of NUMIFORM 1998, the Sixth International Conf. on Numerical Methods in Industrial Forming Processes* (Enschede, Netherlands), J. Huétink and F.P.T. Baaijens, Ed., Balkema, Rotterdam
- *Proc. of NUMIFORM 2001, the Seventh International Conf. on Numerical Methods in Industrial Forming Processes* (Toyohashi, Japan), K.-I. Mori, Ed, American Institute of Physics
- *Proc. of NUMIFORM 2004, the Eighth International Conf. on Numerical Methods in Industrial Forming Processes* (Columbus, OH), S. Ghosh, J.C. Castro, and J.K. Lee, Ed, American Institute of Physics

- *Proc. of NUMISHEET 1993, the Second International Conf. on Numerical Simulation of 3D Sheet Metal Forming Processes* (Isehara, Japan), A. Makinouchi, E. Nakamachi, E. Onate, and R.H. Wagoner, Ed., 1993
- *Proc. of NUMISHEET 1996, the Third International Conf. on Numerical Simulations of 3D Sheet Metal Forming Processes* (Dearborn, MI), J.K. Lee, G.L. Kinzel, and R.H. Wagoner, Ed., 1996
- *Proc. of NUMISHEET 1999, the Fourth International Conf. on Numerical Simulations of 3D Sheet Metal Forming Processes* (Besançon, France), J.C. Gelin and P. Picard, Ed., 1999
- *Proc. of NUMISHEET 2002, the Fifth International Conf.* (Jeju Island, Korea) D.Y. Yang, S.I. Oh, H. Huh, and Y.H. Kim, Ed., 2002
- *Proc. of NUMISHEET 2005, the Sixth International Conf.* (Dearborn, MI), L.M. Smith, F. Pourboghrat, J.-W. Yoon, and T.B. Stoughton, Ed., 2005
- *Proc. of the First International Conf. on Numerical Simulations of 3D Sheet Metal Forming Processes* (Zurich, Switzerland), VDI-Berichte 894, VDI Verlag GmbH, Dusseldorf, 1991
- A. Schuler, *Metal Forming Handbook*, Springer-Verlag, Berlin Heidelberg, 1998
- O.C. Zienkiewicz, *The Finite Element Method*, 3rd ed., McGraw-Hill, 1977

Modeling and Simulation of the Forming of Aluminum Sheet Alloys

Jeong Whan Yoon and Frédéric Barlat, Alloy Technology and Materials Research Division, Alcoa Technical Center

WITH ADVANCES in computer hardware and software, it is possible to model material processing, product manufacturing, product performance in service, and failure. Although the fine-tuning of product manufacturing and performance is empirical, modeling can be an efficient tool to guide and optimize design, to evaluate material attributes, and to predict lifetime and failure. Moreover, modeling can be used as a research tool for a more fundamental understanding of physical phenomena that can result in the development of improved or new products.

This article is concerned with the numerical simulation of the forming of aluminum alloy sheet metals. In order to design a process for a specific material, it is necessary to account for the attributes of the material in the simulations. Although the numerical methods are generic and can be applied to any material, constitutive models, that is, the mathematical descriptions of material behavior, are material-specific. Therefore, macroscopic and microscopic aspects of the plastic behavior of aluminum alloys are reviewed first. The following are then discussed to cover theoretical and implementation aspects of sheet metal forming simulation:

- Constitutive equations suitable for the description of aluminum alloy sheets
- Testing procedures and analysis methods used to measure the relevant data needed to identify the material coefficients
- Tensile and compressive instabilities in sheet forming. For tensile instability, both strain- and stress-based forming-limit curves are discussed.
- Springback analysis
- Finite Element (FE) formulation
- Stress-integration procedures for both continuum and crystal-plasticity mechanics
- Finite element design

Finally, various examples of the simulation of aluminum sheet forming are presented. These examples include earing in cup drawing, wrinkling, automotive stamping, hemming, hydroforming, and clam-shell-resistant design via FE analysis and the Taguchi (Ref 1) optimization method.

Material Modeling

Plasticity of Aluminum Alloys

Macroscopic Observations. Aspects of the plastic deformation and ductility of aluminum alloys at low and moderate strain rates and subjected to monotonic loading or to a few load cycles are briefly discussed here. The stress-strain behavior at low strain is almost always reversible and linear. The elastic range, however, is bounded by the yield limit, the stress above which permanent or inelastic deformations occur. In the plastic range, the flow stress, described by a stress-strain curve, increases with the amount of accumulated plastic strain and becomes the new yield stress if the material is unloaded.

In general, it is considered that plastic deformation occurs without any volume change and that hydrostatic pressure has no influence on yielding. Experiments conducted at high confinement pressure showed that, although very small, a pressure effect is quantifiable (Ref 2, 3). However, practically, this effect can be neglected for aluminum alloys at low confinement pressure. A feature common in aluminum alloys is the Bauschinger effect. This occurs when a material is deformed up to a given strain, unloaded, and loaded in the reverse direction—typically, tension followed by compression. The yield stress after strain reversal is lower than the flow stress before unloading from the first deformation step.

The flow stress of an alloy depends on the testing temperature. Moreover, at low absolute temperature compared to the melting point, time usually has a very small influence on the flow stress and plasticity in general. However, at higher temperatures, strain-rate effects are important. In fact, it has been observed that strain rate and temperature have virtually identical effects on plasticity. Raising the temperature under which an experiment is carried out is similar to decreasing the strain rate. Temperature has another influence on plasticity. When subjected to a constant stress smaller than the yield limit, a material can deform by creep. A similar phenomenon, called relaxation, corresponds to a decrease in the applied stress when the strain is held constant.

Microscopic Aspects. Commercial aluminum alloys used in forming operations are polycrystalline. They are composed of numerous grains, each with a given lattice orientation with respect to macroscopic axes. At low temperature compared to the melting point, metals and alloys deform by dislocation glide or slip on given crystallographic planes and directions, which produces microscopic shear deformations (Ref 4). Therefore, the distribution of grain orientations—the crystallographic texture—plays an important role in plasticity. Because of the geometrical nature of slip deformation, strain incompatibilities arise between grains and produce micro-residual stresses, which macroscopically lead to a Bauschinger effect. Slip results in a gradual lattice rotation as deformation proceeds. After slip, dislocations accumulate at microstructural obstacles and increase the slip resistance for further deformation, leading to strain hardening with its characteristic stress-strain curve.

At higher temperature, more slip systems can be available to accommodate the deformation (Ref 5), but grain-boundary sliding becomes more predominant. For instance, superplastic forming occurs mainly by grain-boundary sliding (Ref 6). In this case, the grain size and shape are important parameters. Atomic diffusion is also another mechanism that affects plastic deformation at high temperature and contributes to creep as well as the accommodation of stress concentrations that arise due to grain-boundary sliding.

Commercial aluminum alloys contain second-phase particles. These phases are present in materials by design in order to control either the microstructure, such as the grain size, or mechanical properties, such as strength (Ref 7, 8).

However, some amounts of second phases are undesired. In any case, the presence of these nonhomogeneities alters the material behavior because of their differences in elastic properties with the matrix, such as in composite materials, or because of their interactions with dislocations. In both cases, these effects produce incompatibility stresses that lead to a Bauschinger effect.

The mechanisms of failure intrinsic to materials are flow localization and fracture. Localization tends to occur in the form of shear bands, either microbands, which tend to be crystallographic, or macrobands, which are not (Ref 9). Macroscopic necking in thin sheet occurs under either three-dimensional conditions (e.g., diffuse necking) or plane-strain conditions (e.g., localized, through-thickness necking). Ductile fracture is generally the result of a mechanism of void nucleation, growth, and coalescence (Ref 10). The associated microporosity leads to volume changes, although the matrix is plastically incompressible, and hydrostatic pressure affects the material behavior. At low temperature compared to the melting point, second phases are principally the sites of damage. The stress concentration around these phases leads to void nucleation, and growth occurs by plasticity. Coalescence is the result of plastic flow microlocalization of the ligaments between voids. At higher temperature, where creep becomes dominant, cavities nucleate at grain boundaries by various mechanisms, including grain sliding and vacancy coalescence. Generally, materials subjected to creep and superplastic forming exhibit higher porosity levels than those formed at lower temperature.

Constitutive Behavior

General Approach. Constitutive laws in materials generally consist of a state equation and evolution equations. The state equation describes the relationship between the strain rate ($\dot{\varepsilon}$), stress (σ), temperature (T), and state variables (x^k), which represent the microstructural state of the material (Ref 11). The porosity of a material and a measure of the accumulated plastic deformation, such as the effective strain or the dislocation density, are a few examples of the variables x^k. The state equation can be expressed mathematically, for instance, in a scalar form for uniaxial deformation, as:

$$\dot{\varepsilon} = \dot{\varepsilon}\,(\sigma,\, T,\, x^k) \qquad \text{(Eq 1)}$$

The evolution equations describe the development of the microstructure through the change of the state variables and can take the form:

$$\dot{x}^k = \dot{x}^k(\sigma,\, T,\, x^k) \qquad \text{(Eq 2)}$$

Because slip plays a major role in plasticity, it is important to look at this mechanism in terms of both its kinetic effect on strain hardening and its geometrical effect on anisotropy. The Kocks and Mecking approach (Ref 12)

has laid the foundations for many subsequent studies by connecting the dislocation density to the flow stress. With this type of approach, it is possible to model the influence of parameters describing the microstructure, such as grain size, second phases, and solutes. Other approaches to strain hardening include dislocation dynamics and atomistic computation. However, these methods are very computationally intensive at this time and not appropriate for forming simulations.

The description of crystal plasticity has been very successful over the last few decades. This approach is based on the geometrical aspect of plastic deformation, slip and twinning in crystals, and on averaging procedures over a large number of grains. The crystallographic texture is the main input to these models, but other parameters, such as the grain shape, can also be included. It is a multiaxial approach and involves tensors instead of scalar variables. One. of the outputs of a polycrystal model is the concept of the yield surface, which generalizes the concept of uniaxial yield stress for a multiaxial stress state. Polycrystal models can be used in multiscale simulations of forming, but they are usually expensive in time, and the relevant question is to know if their benefit is worth the cost. Polycrystal modeling aspects have been treated in a large number of publications and books, such as in Ref 4, 13, and 14.

The state variables do not need to be connected to a specific microstructural feature. In this case, Eq 1 and 2 define a macroscopic model with state (or internal) variables. In fact, for forming applications, macroscopic models appear to be more appropriate. Because of the scale difference between the microstructure and a stamped part, the amount of microscopic material information necessary to store in a forming simulation would be excessively large. It is not possible to track all of the relevant microstructural features in detail. Therefore, lumping them all in a few macroscopic variables seems to be more appropriate and efficient.

Constitutive Modeling for Aluminum Alloys. At the continuum scale, for a multiaxial stress space, plastic deformation is well described with a yield surface, a flow rule, and a hardening law. The yield surface in stress space separates states producing elastic and elastoplastic deformation. It is a generalization of the tensile yielding behavior to multiaxial stress states. Plastic anisotropy is the result of the distortion of the yield surface shape due to the material microstructural state. Reference 15 discusses different phenomena attached to the yield surface shape at a macroscopic scale.

Certain properties of the continuum yield functions can be obtained from microstructural considerations. Bishop and Hill (Ref 16) showed that for a single crystal obeying the Schmid law (i.e., dislocation glide occurs when the resolved shear stress on a slip system reaches a critical value), the resulting yield surface is convex, and

the associated strain increment is normal to it. Furthermore, they extended this result to a polycrystal by averaging the behavior of a representative number of grains in an elementary volume without making any assumption about the interaction modes between grains or the uniformity of the deformation gradient. These results provide a good support for the use of the associated flow rule, which stipulates that the strain rate (increment) is normal to the yield surface.

Regardless of the shape of the yield surface, strain hardening can be isotropic or anisotropic. The former corresponds to an expansion of the yield surface without distortion. Any other form of hardening is anisotropic and leads to different properties in different directions after deformation, even if the material is initially isotropic. Whether the yield surface expands, translates, or rotates as plastic deformation proceeds, a shape must be defined to account for initial anisotropy.

If the yield surface distorts during deformation, a unique shape can still be used to describe an average material response over a certain deformation range (Ref 17). Because mechanical data are used as input, these models can still be relatively accurate when the strain is moderate. This is typically the case for sheet forming. However, for larger strains and for abrupt strain path changes, evolution is an issue (for instance, see Ref 18 to 26). Nevertheless, the description of plastic anisotropy based on the concept of yield surfaces and isotropic hardening is convenient and time-efficient for engineering applications such as forming-process simulations. Moreover, the translation of a yield surface (kinematic hardening), which captures phenomena such as the Bauschinger effect, can be easily integrated in an FE formulation.

In view of the previous section, it is obvious that it is difficult to develop constitutive models for forming applications that capture all the macroscopic and microscopic phenomena involved in plastic deformation and ductile fracture. Therefore, the discussion is mostly restricted to behavior where isotropic hardening is a good approximation. Practically, this approach is robust and reasonably accurate in many situations. In classical plasticity, the material description is fully defined using the following set of equations:

$$\Phi(\sigma_{ij}) = \bar{\sigma} \ \text{(yield function/effective stress)}$$
$$\text{(Eq 3)}$$

$$\bar{\sigma} = h(\bar{\varepsilon}) \ \text{(yield condition)} \qquad \text{(Eq 4)}$$

$$\dot{\varepsilon}_{ij} = \dot{\lambda}\frac{\partial \Phi}{\partial \sigma_{ij}} \ \text{(associated or normality flow rule)}$$
$$\text{(Eq 5)}$$

where σ_{ij} and $\dot{\varepsilon}_{ij}$ are the stress and strain-rate tensor components, and $\dot{\lambda}$ is a proportionality factor. The overdot indicates the derivative with respect to time. The variable h is the strain-hardening law (function), which expands the yield surface as plastic deformation accumulates. The symbol $\bar{\sigma}$ is the effective stress, and $\bar{\epsilon}$ is the work-equivalent effective strain defined incrementally as:

$$\bar{\sigma}\dot{\bar{\epsilon}} = \sum_{i,j} \sigma_{ij}\dot{\epsilon}_{ij} = \sigma_{ij}\dot{\epsilon}_{ij} \qquad \text{(Eq 6)}$$

The last expression is an abbreviation of the summation using the Einstein convention. Any repeated index indicates the summation of a product over 1 to 3. Note that this formulation (Eq 3 to 6) is consistent with the general framework defined by Eq 1 and 2. The problem reduces to the definitions of the hardening law and yield function, h and $\bar{\sigma}$ (or Φ), respectively.

The Swift power law (Ref 27), a very popular approximation of the hardening function h, is defined as:

$$h(\bar{\epsilon}) = K(\epsilon_0 + \bar{\epsilon})^n \qquad \text{(Eq 7)}$$

where K and n are the strength coefficient and strain-hardening exponent, respectively. The symbol ϵ_0 is another coefficient that, if equal to zero, reduces Eq 7 to the Hollomon (Ref 28) hardening law. However, one of the outcomes of the dislocation-based Kocks and Mecking approach mentioned previously is that, macroscopically, the strain-hardening behavior for aluminum alloys can be better described with the Voce law (Ref 29):

$$h(\bar{\epsilon}) = A - B \exp(-C\bar{\epsilon}) \qquad \text{(Eq 8)}$$

where A, B, and C are constant coefficients. Therefore, this equation is often preferred to represent the hardening behavior of aluminum alloys, although, as discussed in the section "Test Data Analysis" in this article, is necessary to understand the consequences of the fitting procedure on the modeled material response. It can be assumed that the effects of temperature and strain rate can be included in the formulation through h, for instance, using $h = h(\bar{\epsilon}, \dot{\bar{\epsilon}}, T)$. As an example, a practical way to include strain-rate effects is to separate this variable from strain hardening through the use of the strain-rate sensitivity parameter m:

$$h = g(\bar{\epsilon})\left(\frac{\dot{\bar{\epsilon}}}{\dot{\bar{\epsilon}}_r}\right)^m \qquad \text{(Eq 9)}$$

where g is the strain-hardening law at the reference strain rate $\dot{\bar{\epsilon}}_r$, and m is the strain-rate sensitivity parameter defined as:

$$m = \left.\frac{\partial \ln(\sigma)}{\partial \ln(\dot{\epsilon})}\right|_\epsilon \qquad \text{(Eq 10)}$$

It is worthy to note that the strain-rate sensitivity parameter m is very close to zero for most aluminum alloys at room temperature. However, m can be slightly higher at low temperature and relatively large at higher temperature (Ref 30), particularly for conditions of superplastic deformation (Ref 6).

For cubic metals, there are usually enough potentially active slip systems to accommodate

any shape change imposed to the material. Compressive and tensile yield strengths are virtually identical, and yielding is not influenced by the hydrostatic pressure. The yield surface of such materials is usually represented adequately by an even function of the principal values S_k of the stress deviator $\mathbf{s}$, such as proposed by Hershey (Ref 31) and Hosford (Ref 32), that is:

$$\phi = |S_1 - S_2|^a + |S_2 - S_3|^a + |S_3 - S_1|^a = 2\bar{\sigma}^a \qquad \text{(Eq 11)}$$

With respect to the stress tensor components σ_k, the deviatoric stresses S_k are given by:

$$S_k = \sigma_k - (\sigma_1 + \sigma_2 + \sigma_3)/3 \qquad \text{(Eq 12)}$$

The yield function in Eq 11 can be framed into the general form of Eq 3, using the simple transformation $\Phi = (\phi/2)^{1/a}$. The exponent "a" is related to the crystal structure of the lattice, that is, 6 for body-centered cubic (bcc) and 8 for face-centered cubic (fcc) materials (Ref 33, 34). This was established as a result of many polycrystal simulations. Therefore, although this model is macroscopic, it contains some information pertaining to the structure of the material.

For the description of incompressible plastic anisotropy, many yield functions have been suggested based on the isotropic-hardening assumption (Ref 35–39). Among them, Cazacu and Barlat (Ref 39) introduced a general formulation that originated from the rigorous theory of representation of tensor functions. However, with this approach, the conditions for the convexity of the yield surface are difficult to impose. As mentioned previously in this section, the convexity has a physical basis, and, in addition, this property ensures numerical stability in computer simulations. For this reason, a particular case of this general theory, which is based on linearly transformed stress components, has received more attention. Barlat et al. (Ref 37) applied this method to a general stress state in an orthotropic material, and Karafillis and Boyce (Ref 38) generalized it as the so-called isotropic plasticity equivalent theory, with a more general yield function and a linear transformation that can accommodate lower material symmetry. Cazacu et al. (Ref 40) proposed a criterion based on a linear transformation that accounts for the strength-differential effect, particularly prominent in hexagonal close-packed metals.

Because the aforementioned functions are not able to capture the anisotropic behavior of aluminum sheet to a desirable degree of accuracy, Barlat et al. (Ref 41, 42) introduced two linear transformations operating on the sum of two yield functions in the case of plane stress and general stress states, respectively. Bron and Besson (Ref 43) further extended Karafillis and Boyce's approach to two linear transformations. These recently proposed yield functions include more anisotropy coefficients and therefore give a better description of the anisotropic properties of a material. This is particularly obvious for

the description of uniaxial tension properties (Ref 42).

Extensions of Eq 11 to planar anisotropy for plane stress and general stress states are briefly summarized in the section "Anisotropic Yield Functions" in this article. Both formulations are based on two linear transformations of the stress deviator. The two linear transformations can be expressed as:

$$\tilde{\mathbf{s}}' = \mathbf{C}'\mathbf{s} = \mathbf{C}'\mathbf{T}\boldsymbol{\sigma} = \mathbf{L}'\boldsymbol{\sigma}$$
$$\tilde{\mathbf{s}}'' = \mathbf{C}''\mathbf{s} = \mathbf{C}''\mathbf{T}\boldsymbol{\sigma} = \mathbf{L}''\boldsymbol{\sigma} \qquad \text{(Eq 13)}$$

where $\mathbf{T}$ is a matrix that transforms the Cauchy stress tensor $\boldsymbol{\sigma}$ to its deviator $\mathbf{s}$:

$$\mathbf{T} = \frac{1}{3}\begin{bmatrix} 2 & -1 & -1 & 0 & 0 & 0 \\ -1 & 2 & -1 & 0 & 0 & 0 \\ -1 & -1 & 2 & 0 & 0 & 0 \\ 0 & 0 & 0 & 3 & 0 & 0 \\ 0 & 0 & 0 & 0 & 3 & 0 \\ 0 & 0 & 0 & 0 & 0 & 3 \end{bmatrix} \qquad \text{(Eq 14)}$$

In Eq 13, $\tilde{\mathbf{s}}'$ and $\tilde{\mathbf{s}}''$ are the linearly transformed stress deviators, while $\mathbf{C}'$ and $\mathbf{C}''$ (or $\mathbf{L}'$ and $\mathbf{L}''$) are the matrices containing the anisotropy coefficients. Specific forms of these matrices are given subsequently for materials exhibiting the orthotropic symmetry, that is, with three mutually orthogonal planes of symmetry at each point of the continuum, such as in sheet metals or tubes. The unit vectors describing the symmetry axes are denoted by $\mathbf{x}$, $\mathbf{y}$, and $\mathbf{z}$, which, for sheet materials, correspond to the rolling, transverse, and normal directions, respectively.

Anisotropic Yield Functions. For plane stress, these two linear transformations can be defined as:

$$\begin{bmatrix} \tilde{s}'_{xx} \\ \tilde{s}'_{yy} \\ \tilde{s}'_{xy} \end{bmatrix} = \begin{bmatrix} \alpha_1 & 0 & 0 \\ 0 & \alpha_2 & 0 \\ 0 & 0 & \alpha_7 \end{bmatrix} \begin{bmatrix} s_{xx} \\ s_{yy} \\ s_{xy} \end{bmatrix}$$

$$\begin{bmatrix} \tilde{s}''_{xx} \\ \tilde{s}''_{yy} \\ \tilde{s}''_{xy} \end{bmatrix} = \frac{1}{3}\begin{bmatrix} 4\alpha_5 - \alpha_3 & 2\alpha_6 - 2\alpha_4 & 0 \\ 2\alpha_3 - 2\alpha_5 & 4\alpha_4 - \alpha_6 & 0 \\ 0 & 0 & 3\alpha_8 \end{bmatrix}$$
$$\times \begin{bmatrix} s_{xx} \\ s_{yy} \\ s_{xy} \end{bmatrix}$$

$$\text{(Eq 15)}$$

By denoting $\tilde{S}'_j$ and $\tilde{S}''_j$ the principal values of the tensors $\mathbf{s}'$ and $\mathbf{s}''$ defined previously, the plane-stress anisotropic yield function Yld2000-2d is then defined as:

$$\phi = |\tilde{S}'_1 - \tilde{S}'_2|^a + |2\tilde{S}''_2 + \tilde{S}''_1|^a + |2\tilde{S}''_1 + \tilde{S}''_2|^a = 2\bar{\sigma}^a \qquad \text{(Eq 16)}$$

Note that this formulation is isotropic and reduces to Eq 11 if $\mathbf{C}'$ and $\mathbf{C}''$ are both equal to the identity matrix. It also reduces to the classical von Mises and Tresca yield functions if a = 2 (or 4) and a = 1 (or ∞), respectively. More details regarding Yld2000-2d are given in Ref 41 and 44.

For a full three-dimensional stress state, the linear transformations can be expressed in the most general form with the following matrices:

$$
\mathbf{C}' =
\begin{bmatrix}
0 & -c'_{12} & -c'_{13} & 0 & 0 & 0 \\
-c'_{21} & 0 & -c'_{23} & 0 & 0 & 0 \\
-c'_{31} & -c'_{32} & 0 & 0 & 0 & 0 \\
0 & 0 & 0 & c'_{44} & 0 & 0 \\
0 & 0 & 0 & 0 & c'_{55} & 0 \\
0 & 0 & 0 & 0 & 0 & c'_{66}
\end{bmatrix};
$$

$$
\mathbf{C}'' =
\begin{bmatrix}
0 & -c''_{12} & -c''_{13} & 0 & 0 & 0 \\
-c''_{21} & 0 & -c''_{23} & 0 & 0 & 0 \\
-c''_{31} & -c''_{32} & 0 & 0 & 0 & 0 \\
0 & 0 & 0 & c''_{44} & 0 & 0 \\
0 & 0 & 0 & 0 & c''_{55} & 0 \\
0 & 0 & 0 & 0 & 0 & c''_{66}
\end{bmatrix}
$$

(Eq 17)

The anisotropic yield function Yld2004-18p is defined as:

$$
\begin{aligned}
\phi = \phi(\tilde{S}'_i, \tilde{S}''_j) = & |\tilde{S}'_1 - \tilde{S}''_1|^a + |\tilde{S}'_1 - \tilde{S}''_2|^a \\
& + |\tilde{S}'_1 - \tilde{S}''_3|^a + |\tilde{S}'_2 - \tilde{S}''_1|^a + |\tilde{S}'_2 - \tilde{S}''_2|^a \\
& + |\tilde{S}'_2 - \tilde{S}''_3|^a + |\tilde{S}'_3 - \tilde{S}''_1|^a + |\tilde{S}'_3 - \tilde{S}''_2|^a \\
& + |\tilde{S}'_3 - \tilde{S}''_3|^a = 4\bar{\sigma}^a
\end{aligned}
$$

(Eq 18)

This formulation is isotropic if all the coefficients c'_{ij} and c''_{ij} reduce to unity. The reader is referred to Ref 42 for additional information about this model.

Strain-Rate Potentials. In order to represent the rate-insensitive plastic behavior of materials phenomenologically, it is typical, as explained previously, to use a yield surface, the associated flow rule, and a hardening law. Ziegler (Ref 45) and Hill (Ref 46) have shown that, based on the work-equivalence principle, Eq 6, a meaningful strain-rate potential can be associated to any convex stress potential (yield surface). Therefore, an alternate approach to describe plastic anisotropy is to provide a strain-rate potential $\Psi(\dot{\boldsymbol{\varepsilon}}) = \bar{\varepsilon}$, expressed as a function of the traceless plastic strain-rate tensor $\dot{\boldsymbol{\varepsilon}}$, the gradient of which leads to the direction of the stress deviator $\mathbf{s}$, that is:

$$
s_{ij} = \mu \frac{\partial \Psi}{\partial \dot{\varepsilon}_{ij}}
$$

(Eq 19)

In the previous equation, μ is a proportionality factor necessary to scale the stress deviator. Note the similarity between Eq 19 and Eq 5. This approach has been used for the description of the plastic behavior of fcc single crystals (Ref 47), bcc polycrystals (Ref 48–50), and cubic polycrystals (Ref 51–52). In the latter works,

the strain-rate potential $\Psi = (\psi/k)^{1/b}$ for an incompressible isotropic material was defined using:

$$
\begin{aligned}
\psi = & \left| \frac{2\dot{E}_1 - \dot{E}_2 - \dot{E}_3}{3} \right|^b + \left| \frac{2\dot{E}_2 - \dot{E}_3 - \dot{E}_1}{3} \right|^b \\
& + \left| \frac{2\dot{E}_3 - \dot{E}_1 - \dot{E}_2}{3} \right|^b = |\dot{E}_1|^b + |\dot{E}_2|^b + |\dot{E}_3|^b \\
= & \, k\dot{\bar{\varepsilon}}^b
\end{aligned}
$$

(Eq 20)

where $\dot{E}_i$ represents the principal values of the plastic strain-rate tensor $\dot{\boldsymbol{\varepsilon}}$, and $\dot{\bar{\varepsilon}}$ is the effective strain rate defined using the plastic work equivalence principle with the effective stress $\bar{\sigma}$ (Eq 6). The constant k defines the size of the potential, and the exponent "b" was shown to be $3/2$ and $4/3$ for an optimal representation of bcc and fcc materials, respectively. For orthotropic materials, the principal values $\tilde{E}_i$ of a linearly transformed plastic strain rate tensor $\dot{\tilde{\boldsymbol{\varepsilon}}}$, that is:

$$
\dot{\tilde{\boldsymbol{\varepsilon}}} = \mathbf{B}\dot{\boldsymbol{\varepsilon}}
$$

(Eq 21)

where $\mathbf{B}$ is the matrix containing six anisotropy coefficients, are substituted for $\dot{E}_1$ in Eq 20. This formulation reduces to isotropy when $\mathbf{B}$ becomes the identity matrix (i.e., $\dot{\tilde{\boldsymbol{\varepsilon}}} = \dot{\boldsymbol{\varepsilon}}$). However, it is generally accepted that six coefficients are not sufficient to describe the plastic behavior of anisotropic aluminum alloy sheets very accurately. Therefore, a formulation that accounts for two linear transformations on the traceless plastic strain-rate tensor $\dot{\boldsymbol{\varepsilon}}$ was recently introduced:

$$
\dot{\tilde{\boldsymbol{\varepsilon}}}' = \mathbf{B}'\dot{\mathbf{e}} = \mathbf{B}'\mathbf{T}\dot{\boldsymbol{\varepsilon}}
$$

(Eq 22)

$$
\dot{\tilde{\boldsymbol{\varepsilon}}}'' = \mathbf{B}''\dot{\mathbf{e}} = \mathbf{B}''\mathbf{T}\dot{\boldsymbol{\varepsilon}}
$$

(Eq 23)

where $\dot{\mathbf{e}} = \mathbf{T}\dot{\boldsymbol{\varepsilon}}$ is another expression of the traceless plastic strain-rate tensor, necessary to ensure that the strain-rate potential is a cylinder through the transformation represented by $\mathbf{T}$ defined previously. The specific strain-rate potential, called Srp2004-18p, was postulated (Ref 53):

$$
\begin{aligned}
\psi = \psi(\tilde{E}'_i, \tilde{E}''_j) = & |\tilde{E}'_1|^b + |\tilde{E}'_2|^b + |\tilde{E}'_3|^b \\
& + |\tilde{E}''_2 + \tilde{E}''_3|^b + |\tilde{E}''_3 + \tilde{E}''_1|^b + |\tilde{E}''_1 + \tilde{E}''_2|^b \\
= & (2^{2-b} + 2)\dot{\bar{\varepsilon}}^b
\end{aligned}
$$

(Eq 24)

where $\tilde{E}'_i$ and $\tilde{E}''_i$ are the principal values of the tensors $\dot{\tilde{\boldsymbol{\varepsilon}}}'$ and $\dot{\tilde{\boldsymbol{\varepsilon}}}''$. Here, the anisotropy coefficients are contained in matrices $\mathbf{B}'$ and $\mathbf{B}''$, both of a form similar to $\mathbf{C}'$ and $\mathbf{C}''$ in Eq 17.

Experiments and Constitutive Parameters

Testing Methods. Multiaxial experiments have been used to characterize a yield surface, and many issues have been addressed (Ref 54,

55). Recently, Banabic et al. (Ref 56) improved a procedure for biaxial testing of cruciform specimens machined from thin sheets, and for measuring the first quadrant of the yield locus (both stresses positive). In these tests, the onset of plastic deformation is detected from temperature measurements of specimens using an infrared thermocouple positioned at an optimized distance. This method is based on the fact that the specimen temperature drops first due to thermoelastic cooling and then rises significantly when dissipative plastic flow initiates. In spite of this and other types of improvements, multiaxial testing is tedious, difficult to conduct and interpret, and not suitable for quick characterization of anisotropy. This is more a technique for careful verifications of concepts and theories. Therefore, more practical and time-efficient methods are required to test materials and identify material coefficients in constitutive equations, in particular for sheets.

Anisotropic properties can be assessed by performing uniaxial tension tests in the $\mathbf{x}$ and $\mathbf{y}$ axes (rolling and transverse directions, respectively) and in a direction at θ degrees with respect to $\mathbf{x}$. Practically, the anisotropy is characterized by the yield stresses σ_0, σ_{45}, σ_{90}; the r-values (width-to-thickness strain ratio in uniaxial tension) r_0, r_{45}, r_{90}; and their respective averages $\bar{q} = (q_0 + 2q_{45} + q_{90})/4$ and variations $\Delta q = (q_0 - 2q_{45} + q_{90})/2$. For a better characterization of anisotropy, tensile specimens are cut from a sheet at angles of 0, 15, 30, 45, 60, 75, and 90° from the rolling direction. Tension tests are usually conducted at room temperature (RT), and standard (ASTM International) methods are used to measure the yield stresses and r-values. For some aluminum alloys, however, the stress-strain curve exhibits serrations, resulting from nonhomogeneous deformation of the tensile specimen. For instance, Fig. 1 shows the engineering tensile stress-strain curves measured in the 0, 45, and 90° directions at RT for a 5019A-O sheet sample. Because this phenomenon is due to inhomogeneous plastic deformation, the plastic strains measured with extensometers are not reliable and can lead to erroneous r-values. In order to overcome this problem, tension tests can be conducted at a somewhat higher temperature, where serrated flow is suppressed. For instance, Fig. 1 shows the engineering stress/engineering strain curves in the 0, 45, and 90° tension directions measured at RT and at a temperature of 93 °C (200 °F) for a 5019A-H48 sheet sample (Ref 57).

The balanced biaxial yield stress (σ_b) is an important parameter to measure for sheet material characterization. This stress can be obtained by conducting a hydraulic bulge test (Ref 58). Figure 2 shows a schematic diagram of this test in which a sheet blank is clamped between a die with a large circular opening and a holder. A pressure, p, is gradually applied under the blank, which bulges in a quasi-spherical shape. The radius of curvature, R, and strains at the pole of the specimen are measured independently, using mechanical or optical instruments. The stress $\sigma = pR/2t$ is simply obtained from the membrane theory using the thickness, t. This test is interesting not only because it gives information on the yield surface, but also because it allows measurements of the hardening behavior up to strains of approximately twice those achieved in uniaxial tension. This is because geometrical instabilities occur during uniaxial tension (see the section "Strain-Based Forming-Limit Curves" in this article). However, the yield point is not well defined in the bulge test because of the low curvature of the specimen in the initial stage of deformation. As for uniaxial tension, this test can be conducted at different strain rates in order to assess the strain-rate sensitivity parameter m.

Because the biaxial stress state in the bulge test is not exactly balanced, measures of the corresponding strain state may lead to substantial errors. This is because the yield locus curvature

is usually high in this stress state. Barlat et al. (Ref 41) proposed the disk-compression test, which gives a measure of the flow anisotropy for a balanced biaxial stress state, assuming that hydrostatic pressure has no influence on plastic deformation. In this test, a 12.7 mm (0.5 in.) disk is compressed through the thickness direction of the sheet. The strains measured in the **x** and **y** directions lead to a linear relationship in which the slope is denoted by r_b by analogy to the r-value in uniaxial tension:

$$r_b = d\varepsilon_{yy}^p / d\varepsilon_{xx}^p \qquad \text{(Eq 25)}$$

This parameter is a direct measure of the slope of the yield locus at the balanced biaxial stress state. Figure 3 shows deformed specimens for 6111-T4 aluminum alloy sheet samples processed with two different routes, and the corresponding strain measures performed during this test. Pöhlandt et al. (Ref 59) proposed to determine the parameter r_b using the in-plane biaxial tension test.

Other tests can be used to characterize the material behavior of sheet samples. Directional tension of wide specimens can be used to characterize plane-strain tension anisotropy (Ref 60, 61). This test does not produce a uniform state of stress within the specimen and generally leads to more experimental scatter than the uniaxial test.

Simple shear tests (Ref 62) can be carried out to characterize the anisotropic behavior of the simple shear flow stress. A relatively simple device mounted on a standard tensile machine is needed for this test. A rectangular specimen is clamped with two grips, which move in opposite directions relative to each other (Fig. 4). By simply reversing the direction of the grip displacements, forward and reverse loading sequences can be conducted in order to measure the Bauschinger effect. In-plane tension-compression tests of thin sheets have also been used for this purpose (Ref 63, 64), but special precautions need to be taken because of the

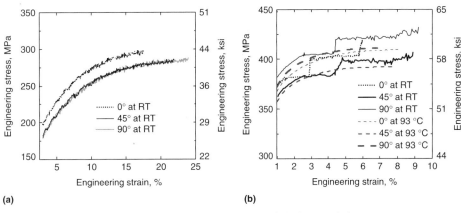

(a)

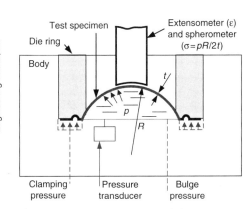

Fig. 1 Stress-strain curves. (a) Aluminum alloy 5019A-O sheet sample for three tensile directions at room temperature (RT). (b) Aluminum alloy 5019A-H48 sheet sample for three tensile directions and two temperatures. Source: Ref 57

Fig. 2 Schematic illustration of the bulge test. p, pressure; R, radius of curvature; t, thickness. Source: Ref 17

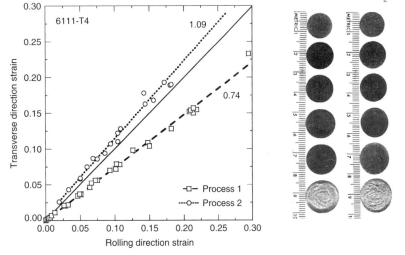

Fig. 3 Disk-compression test results for an aluminum alloy 6111-T4 sheet sample processed using two different flow paths, and selected test specimens. Source: Ref 17

Fig. 4 Simple shear test device

buckling tendency of the specimen during the compressive step.

Although the previously described tests and other experimental procedures are available to test materials in different stress states, it is not always possible to probe all of them. In this case, microstructural modeling can be used to replace the missing experimental data. For instance, the resistance to through-the-thickness shear of a sheet, σ_s, is not readily available from experimental measurements. However, crystal plasticity with a measure of the crystallographic texture of the sheet can be used to compute σ_s. A rougher approximation is to assume that this value is equal to the isotropic shear stress, which, using Eq 11 and 12, for pure shear ($\sigma_2 = -\sigma_1$, $\sigma_3 = 0$) is equal to:

$$\frac{\sigma_s}{\bar{\sigma}} = \frac{1}{(2^{a-1}+1)^{1/a}} \qquad \text{(Eq 26)}$$

Test Data Analysis. From the experimental tests, it is necessary to extract the right information that is most suitable for the identification of the constitutive parameters. For instance, the yield stresses can be used as input data to calculate the anisotropic yield function coefficients. However, as mentioned previously, the yield stress from the bulge test is not very accurate. Moreover, any stress at yield is determined in the region of the stress-strain curve where the slope is the steepest, which may involve additional inaccuracy. Finally, the yield stress is associated with a very small plastic strain and may not reflect the anisotropy of the material over a larger strain range. For these reasons, the flow stresses at equal amount of plastic work along different loading paths could be selected as input data instead of the yield stress. Figure 5 shows how the flow stresses in tension and balanced biaxial tension (bulge test) can be defined at equal amounts of plastic work ($w_p = \int \sigma_{1j}\dot{\varepsilon}_{1j}d\tau_{1j}$, where τ is the time). For many aluminum alloys, experimental observations show that after a few percent plastic strain, the flow stress anisotropy does not vary significantly, as illustrated by Fig. 6(a) and (b).

In the examples shown later in this article, the stress-strain curves employed in the forming simulations were measured using the bulge test, whenever possible. This test is important because extrapolation of uniaxial tension curves with the Voce law usually leads to an underestimation of strain hardening for strains higher than the limit of uniform elongation in tension. This is demonstrated in Fig. 6(c), which shows the Voce approximation of the bulge test data using two different strain ranges, that is, from zero up to strains of 0.2 and 0.53. These strain ranges are typical for uniaxial tension and balanced biaxial tension (bulge), respectively, for annealed aluminum alloys. This figure illustrates the risks of extrapolating flow stresses beyond the range of strains that was used for the fit. The stress-strain curve extrapolated

from the fit in the strain range 0–20% exhibits a much lower rate of strain hardening ($d\sigma/d\varepsilon$) after 25% deformation compared to the curve obtained experimentally with the bulge test.

Similar remarks hold for r-values, which can be defined as instantaneous quantities near yield or as the standard slope of the width strain/thickness strain curve over a given deformation range in tension. On one hand, the yield stresses and instantaneous r-values at yield seem to be appropriate to define the coefficients of the yield function. On the other hand, flow stresses defined at a given amount of plastic work and standard r-values can characterize the average behavior of the material over a finite deformation range. These values are more suitable and more descriptive of the average response of the material for sheet forming simulations. In this case, it would be more appropriate to talk about flow function and flow surface instead of yield function and yield surface, although, mathematically, yield or flow functions are identical concepts.

The anisotropy coefficients are calculated from flow stresses and r-values. For the yield function Yld2000-2d, the eight coefficients are computed numerically, using a Newton-Raphson numerical solver using eight input data: the flow stresses σ_0, σ_{45}, σ_{90}, σ_b and the r-values r_0, r_{45}, r_{90}, and r_b (Ref 41). For the yield function Yld2004-18p, the 18 coefficients are computed numerically with an error-minimization method using 16 experimental input data: the flow stresses σ_0, σ_{15}, σ_{30}, σ_{45}, σ_{60}, σ_{75}, σ_{90}, σ_b and the r-values r_0, r_{15}, r_{30}, r_{45}, r_{60}, r_{75}, r_{90}, and r_b. The remaining inputs, related to the through-thickness properties, are assumed to be equal to the isotropic values or are computed with a crystal-plasticity model (Ref 42). As an illustration of Yld2000-2d and Yld2004-18p, the predicted directionalities of the uniaxial flow stress and r-value for a 5019A-H48 aluminum alloy sheet sample are compared with experimental results in Fig. 7. This figure also shows that the yield function Yld91 (Ref 37), which uses only one linear transformation on the stress tensor, cannot properly capture the experimental variation of the r-value.

Tensile Instability

The success of a sheet-forming operation can be limited by several phenomena, such as plastic flow localization, fracture, and buckling. For a given forming operation, the sheet may undergo deformation up to a given strain prior to failure by one of the limiting phenomena. In this section, analyses for plastic flow localization are discussed.

Strain-Based Forming-Limit Curves. The forming-limit curve (FLC) corresponds to the maximum admissible local strains achievable just before necking. This curve is usually plotted on axes representing the major (ε_1) and the minor (ε_2) strains in the plane of a sheet. Actually, for anisotropic materials, the major strain in the

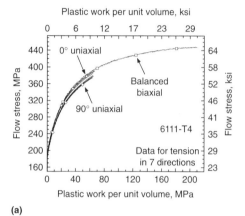

(a)

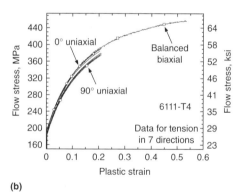

(b)

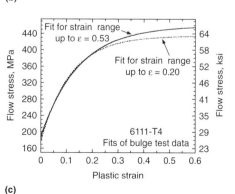

(c)

Fig. 6 Flow stress as a function of (a) plastic work or (b) plastic strain for an aluminum alloy 6111-T4 sheet sample measured in balanced biaxial tension (bulge test) and uniaxial tension for directions at every 15° from the rolling direction. (c) Fit of Voce law to bulge-test data over two different strain ranges

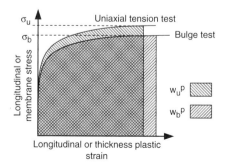

Fig. 5 Flow stresses at equivalent amount of plastic work in uniaxial tension, W_u^p, and balanced biaxial tension (bulge test), W_b^p, i.e., for $W_0^p = W_u^p$

rolling direction is different from the major strain in the transverse or in any other direction. Nevertheless, this convention is adopted in this section, although it lacks generality.

Different models have been proposed in the past to calculate the FLC. Swift (Ref 27) derived equations to predict diffuse necking. In real parts, however, the maximum admissible strains are typically limited by localized (through-thickness) necking, and Swift's equations have only limited applicability. The relationships giving the limit strain under local necking conditions were developed by Hill (Ref 65). His theory predicts that localized necking occurs in the characteristic directions of zero extension in the plane of the sheet. However, Hill's theory can only predict localized necking if one of the strains in the plane of the sheet is negative, that is, between uniaxial and plane-strain tension. This does not agree with experiments, because necking also occurs in stretching processes for which both in-plane principal strains are positive. To explain such behavior, Marciniak and Kuczynski (Ref 66) proposed an analysis in which the sheet metal is supposed to contain a region of local imperfection. Heterogeneous plastic flow develops and eventually localizes

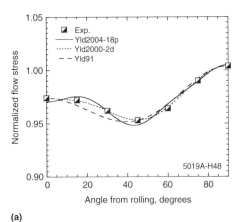

(a)

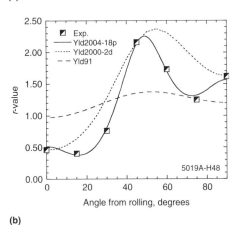

(b)

Fig. 7 Experimental results and predictions based on different yield functions of the (a) normalized tensile flow stress, and (b) r-value as a function of the angle between tensile and rolling directions for an aluminum alloy 5019A-H48 sheet sample

in the imperfection. The Marciniak and Kuczynski (MK) model can predict the FLC when minor strains are either positive or negative. It was initially developed within the flow theory of plasticity, using quadratic plastic potentials. However, for balanced biaxial stretching ($\varepsilon_2 = \varepsilon_1$) and within the classical plasticity approach (J_2 flow theory), this model largely overestimates the limit strains. Therefore, Hutchinson and Neale (Ref 67) extended this model using the deformation theory of plasticity and were able to make predictions that were in better agreement with the experiments. Many other works have been published to show the influence of various parameters on calculated forming limits (for instance, Ref 68). Stören and Rice (Ref 69) gave another description of the plastic flow localization using a bifurcation analysis.

In the MK model, an infinite sheet metal is assumed to contain an imperfection, which is represented as a linear band of infinite size whose thickness t^I is smaller than the thickness of the homogeneous region of the sheet, t. The superscript "I" is used here for all the variables in the imperfection. Quantitatively, this two-zone material is characterized by parameter D:

$$D = 1 - t^I / t \qquad \text{(Eq 27)}$$

There is no imperfection when both thicknesses are identical, that is, when parameter D reduces to 0, and no plastic flow localization can be predicted with the MK model. However, if D has a starting value larger than zero, it will increase as plastic deformation proceeds. At each step of the calculation, the ratio of the effective strain rate in the homogeneous region ($\dot{\bar{\varepsilon}}$) to the effective stain rate in the imperfection ($\dot{\bar{\varepsilon}}^I$) is evaluated. Plastic flow localization occurs when this ratio approaches 0. The basic equations of the MK model are related to equilibrium and compatibility requirements. The first condition indicates that the force perpendicular to the necking band is transmitted from the homogeneous region to the imperfection, whereas the second requirement indicates that the elongation in the direction of the necking band is identical in both regions. For materials exhibiting isotropy or planar isotropy (same properties in any direction in the plane of the sheet), the main equilibrium equation of the MK model reduces to the following form:

$$[1-D]\left[\frac{h(\bar{\varepsilon}^I)}{h(\bar{\varepsilon})}\right]\left[\frac{\bar{\sigma}_1^I}{\bar{\sigma}_1}\right] = 1 \qquad \text{(Eq 28)}$$

where $h(\bar{\varepsilon})$ is the hardening law, and $\bar{\sigma}_1 = \sigma_1/\bar{\sigma}$ is the maximum principal stress normalized by the effective stress.

The MK model also suggests the necking direction per se. In the negative ρ range ($\rho = d\varepsilon_2/d\varepsilon_1 \leq 0$) lying between uniaxial tension and plane-strain tension, the necking direction is at an angle with respect to the maximum principal strain. However, in the biaxial stretching range (both ε_1 and ε_2 are positive), necking occurs mainly in a direction perpendicular to the major

principal strain, which corresponds to the assumption of Eq 28. The product of the three quantities within the square brackets on the left side of Eq 28 must be constant. The first quantity $(1-D)$ is related to the imperfection, and it decreases when D increases. However, the two other quantities tend to balance the equation. Because of the smaller thickness, there is more plastic deformation and consequently more plastic work in the necking band than in the homogeneous region. Therefore, the ratio $h(\bar{\varepsilon}^I)/h(\bar{\varepsilon})$ becomes larger than 1. In addition, it has been shown (Ref 70) that the stress state in the imperfection moves toward a plane-strain state. Thus, the ratio $\bar{\sigma}_1^I/\bar{\sigma}_1$ also increases, because the yield stress is at a maximum stationary value for plane strain. As a result, the increasing defect size is counterbalanced by two types of material hardening. The first one is strain hardening, whereas the second one, termed yield-surface shape hardening (Ref 71), is related to the yield surface shape. In particular, it has been shown that the ratio $P = \sigma_p/\sigma_b$ of the plane-strain to balanced biaxial yield stresses is a good parameter to evaluate sheet stretchability (Ref 71). When this ratio is much larger than 1, as it is for an isotropic von Mises material ($P \approx 1.15$), the predicted limit strain for balanced biaxial stretching is very large. However, the isotropic Tresca material, for which $P = 1$, exhibits a very low forming limit for the same strain path (Ref 72). For plane-strain tension, the ratio $\bar{\sigma}_1^I/\bar{\sigma}_1$ is always equal to 1, regardless of the yield surface shape, and Eq 28 becomes:

$$(1-D)\frac{h(\bar{\varepsilon}^I)}{h(\bar{\varepsilon})} = 1 \qquad \text{(Eq 29)}$$

This explains why the plane-strain deformation state always leads to the lowest forming limit. Materials cannot take advantage of the yield-surface shape-hardening effect in this deformation state, and only strain hardening is available to counterbalance the imperfection growth.

The previous analysis can also be used to explain qualitatively the effects of strain-rate sensitivity and kinematic hardening on formability. For strain-rate-sensitive materials, the left side of Eq 28 is multiplied by an additional (strain-rate) hardening term, $(\dot{\bar{\varepsilon}}^I/\dot{\bar{\varepsilon}})^m$, where m is the strain-rate sensitivity parameter. Because the strain rate is larger in the necking band than in the homogeneous region, this parameter is larger than 1, when m is positive, and therefore tends to stabilize plastic flow. If kinematic hardening applies, the yield surface moves toward the stress or strain-rate direction without changing its shape. This has the effect of decreasing the ratio P and consequently the forming limit, as predicted by Tvergaard (Ref 73).

Needleman and Triantafyllidis (Ref 74) showed that the imperfection can be attributed to material heterogeneities. In particular, it can be viewed as a difference in surface area transmitting the load. Hence, it can be assumed that internal damage mostly is responsible for the imperfection. In this context, damage is defined as the nucleation, growth, and coalescence of microvoids in a material due to plastic deformation. Studies of damage based on microscopic observations and probability calculations have shown that the order of magnitude of D in Eq 28 for typical commercial alloys is 0.4% (Ref 75–77). In the following discussion, the value of $1-D = 0.996$ has been used to quantitatively characterize the imperfection.

Figure 8 shows an FLC computed for a 2008-T4 alloy sheet sample using two yield surfaces

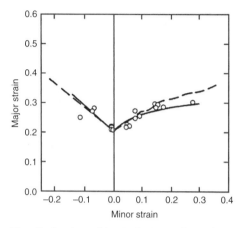

Fig. 8 Experimental forming-limit curve for an aluminum alloy 2008-T4 sheet sample (data points), and calculated curves based on the Yld89 (a special case of Yld2000-2d) yield function (solid line) and a crystal-plasticity model (dashed line). A = 447 MPa (65 ksi); B = 248 MPa (36 ksi); C = 4.3 (Voce coefficients in Eq 8). Assumed initial imperfection size, $t'/t = 0.996$. Source: Ref 78

(Ref 78). The first yield condition was described with the function Yld89 (Ref 36), a particular case of Yld2000-2d. The second yield condition was based on the Bishop and Hill (Ref 16) crystal-plasticity model. The same Voce-type work-hardening law, Eq 8, was used for both simulations. Both predictions are in good agreement with experimental data. In the stretching range, the imperfection orientation was assumed to be perpendicular to one of the orthotropic axes (rolling or transverse direction). The FLC was the lowest of the two curves calculated for the major strain either in the rolling or in the transverse direction. Experience in such computations shows that, for weakly textured materials, the forming limit is minimal for either case. However, for strongly textured aluminum alloy sheet, this is not necessarily the case. The influence of texture and microstructure on the FLC has been investigated by many authors (Ref 79–87).

Stress-Based Forming-Limit Curves. The previous discussion assumed that deformation occurs along linear strain paths, that is, $\rho = d\varepsilon_2/d\varepsilon_1$ is constant. In practice, particularly for multistep forming, this is not the case. Moreover, it was shown by Kikuma et al. (Ref 88) that nonlinear strain paths have an influence on the FLC. Kobayashi et al. (Ref 89) and Graf and Hosford (Ref 90, 91) showed that the FLC strongly depends on the strain path for steel and aluminum alloy sheets, respectively. The characterization of forming limits in strain space is therefore a practical challenge for complex forming processes due to this sensitivity.

For computational purposes, one of the most promising solutions for dealing with strain path

effects on the FLC is to use a stress-based approach, as proposed by Kleemola and Pelkkikangas (Ref 92), Arrieux et al. (Ref 93), Stoughton (Ref 94), and Stoughton and Yoon (Ref 95). These authors have shown that the FLC in stress space is path-independent and should be suitable to the analysis of any forming problem. As can be seen in Fig. 9, the FLCs described in strain space with different prestrains are mapped to a single curve in stress space. This verifies the path independence of the stress-based FLC. Recently, Kuwabara et al. (Ref 96) measured the stress state near the forming limit of tube deformed using internal pressure and end feed under proportional and nonproportional loading conditions. These authors confirmed that the forming limit as characterized by the state of stress is insensitive to the loading history.

Although not practical in a press shop, the stress-based FLC can be very effective when it is used to assess the safety margins in finite element simulations of forming processes. In order to compute the stress-based FLC, a representation of the forming-limit behavior for proportional loading in strain space, that is, the locus of principal strains, is specified as follows:

FLC (strain-based)

$$\equiv \begin{bmatrix} e_1^{FLC} \\ e_2^{FLC} \end{bmatrix} = e_1^{FLC}(\rho) \begin{bmatrix} 1 \\ \rho \end{bmatrix} \quad (Eq\ 30)$$

where e_1^{FLC} and e_2^{FLC} are the coordinates of the FLC for linear strain paths, that is, major and minor principal strains. These points can be either measured or calculated, for instance, with the MK model. The strain path $\rho = e_2^{FLC}/e_1^{FLC}$ is consistent with the definition in the previous

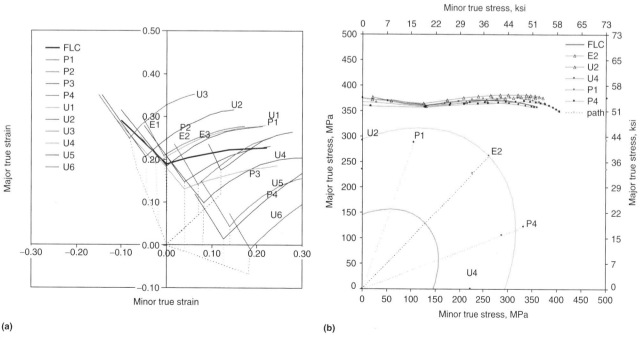

Fig. 9 Effect of strain path on forming-limit curves (FLCs). (a) Strain-based FLC (exhibiting a path effect). (b) Stress-based FLC (limited path effect). Source: Ref 95

section and can vary in the range $\rho = [-1, +1]$. Because Eq 30 is assumed to accurately characterize the strain-based forming limit under proportional loading, it can be used with a plasticity model (yield function and strain hardening) to calculate the path-independent stress-based FLC, which can be similarly represented as:

$$\text{FLC (stress-based)} = \begin{bmatrix} \sigma_1^{\text{FLC}} \\ \sigma_2^{\text{FLC}} \end{bmatrix} = \sigma_1^{\text{FLC}}(\alpha) \begin{bmatrix} 1 \\ \alpha \end{bmatrix}$$
(Eq 31)

In Eq 31, α is the ratio of the minor to the major principal stresses ($\sigma_2^{\text{FLC}}/\sigma_1^{\text{FLC}}$) of the stress-based FLC.

Although this FLC is assumed to be path-independent, the stresses calculated in the finite element (FE) analysis are path-dependent variables. Therefore, it is necessary to monitor the stress state at each step of the computation and determine if it is below or above the stress-based FLC. For this purpose, it is convenient to use a single parameter to monitor the formability margin, as follows (Ref 95):

$$\gamma_C = \frac{\bar{\sigma}(\sigma_{ij})}{h(\bar{\varepsilon}^{\text{FLC}})}$$
(Eq 32)

where:

$$\bar{\varepsilon}^{\text{FLC}} = \frac{e_1^{\text{FLC}}(1+\rho)}{\dfrac{\partial \bar{\sigma}}{\partial \sigma_{11}} + \dfrac{\partial \bar{\sigma}}{\partial \sigma_{22}}}$$
(Eq 33)

In Eq 32, γ_C is a single stress-scaling factor representing the degree of formability according to the criteria:

$$\begin{aligned} \gamma_c &< 1 \quad \text{safe} \\ \gamma_c &> 1 \quad \text{necked (failure occurs)} \end{aligned}$$
(Eq 34)

The formability margin as defined previously is based on the assumption that the forming limit is isotropic in the plane of the sheet. This assumption justifies the use of principal stresses and the representation of the forming limit as a curve in a two-dimensional diagram. In general, Graf and Hosford (Ref 90, 91) and others have shown that the forming limit is anisotropic. The method to deal with anisotropic data in stress-based FLC is explained in Ref 95.

Compressive Instability

Wrinkling occurs when a blank is subjected to compressive stresses during the forming process, as, for instance, in the flange of a cup during drawing, triggered by the elastoplastic buckling of the thin structure. Although the failure limit due to plastic flow localization can be simply defined by the FLC at each point of a continuum, the wrinkling limit cannot be defined with simple variables such as strain, stress, and thickness. Buckling is also strongly affected by the mechanical properties of the sheet material, the geometry of the blank, and contact conditions. The analysis of wrinkling initiation and growth is

therefore difficult to perform due to the complex synergistic effects of the controlling parameters. Furthermore, commonly observed in instability phenomena, small variations of the parameters can result in widely different wrinkling behaviors. In the face of these difficulties, the study of wrinkling has generally been conducted case by case. A unique wrinkling criterion, which could be used effectively for various sheet-forming processes, has not been proposed yet.

Most of the studies related to wrinkling were carried out experimentally or analytically before the numerical simulation of stamping processes (Ref 97–99). An analytical bifurcation can give a useful estimate of the elastoplastic buckling of a plate with a basic geometrical shape and subjected to simple boundary conditions. This analysis, however, cannot be employed in general sheet metal forming processes. With the rapid development of computing power, wrinkling has also been studied using the FE method. Wrinkling can be analyzed in the same way as most buckling problems, that is, assuming a nonlinear elastic material behavior and ignoring complex contact conditions. Because wrinkling occurs in the plastic region, for a more realistic approach, the computations need to be based on an elastoplastic material model and take into account the complex contact conditions inherently present in sheet-forming processes.

Two types of buckling analyses are performed with the FE method: a bifurcation analysis of a structure without imperfection (Ref 100–103) and a nonbifurcation analysis, which assumes an initial imperfection or disturbing force due to load eccentricities (Ref 104–108). Because the FE analyses of sheet metal forming processes involve strong nonlinearities in geometry, material, and contact, convergence problems are frequently observed. Nonbifurcation analyses sometimes lead to reasonable results, because all real structures have inherent imperfections, such as material nonuniformity or geometric unevenness. Thus, most wrinkling analyses have been carried out using a nonbifurcation analysis. However, the results obtained from a nonbifurcation analysis are sensitive to the amplitude of the initial imperfection, which is chosen arbitrarily. As a result, bifurcation algorithms have been implemented into the FE method in order to analyze more rigorously the wrinkling behavior of sheet metal during forming (Ref 102, 103).

A number of studies have been devoted to the plastic buckling problem as a bifurcation phenomenon. The buckling of a column or a compressed circular plate and the wrinkling of a deep-drawn cup are typical examples of bifurcation problems. Shanley (Ref 109) first showed that the buckling load of a centrally compressed short column coincides theoretically with the tangent modulus. Hill (Ref 110) later generalized Shanley's concept and established a uniqueness criterion for the mathematical solution of elastic-plastic solids. This theory is now widely accepted in the analysis of bifurcation problems. Hutchinson (Ref 111)

specialized Hill's bifurcation criterion (Ref 110) to a class of loadings characterized by a single parameter and studied the postbifurcation behavior associated with the lowest possible bifurcation load. He showed that the initial slope of the load-deflection curve governs the material behavior in only a very small neighborhood of the bifurcation point, and that the rates of change of the instantaneous moduli at bifurcation have a major effect on the postbifurcation behavior. Prebuckling and postbuckling analyses were carried out with an FE method by incorporating the arc-length control scheme (Ref 100). In the bifurcation problem (Fig. 10), the stiffness matrix of the linearized FE equation becomes singular at a bifurcation point, and the Newton-Raphson solver cannot proceed further. Riks (Ref 100) proposed the continuation method by which the postbifurcation analysis can be carried out along the secondary solution branch (path) and implemented it for the buckling of elastic shell structures. In the work of Kim et al. (Ref 103), the continuation method was introduced in order to analyze the initiation and growth of wrinkles in deep drawing processes. A brief summary of the Riks method applied to sheet metal forming simulations is summarized as follows.

For a conservative system, the change of the total potential energy, $\Delta\Pi$, due to an admissible variation $\delta\mathbf{u}$ of the displacement field $\mathbf{u}$ can be written as:

$$\Delta\Pi(\mathbf{u},\delta\mathbf{u}) = \frac{\partial\Pi}{\partial\mathbf{u}_i}\delta u_i + \frac{1}{2}\frac{\partial^2\Pi}{\partial\mathbf{u}_i\partial\mathbf{u}_j}\delta u_i\delta u_j + \cdots$$
(Eq 35)

The second variation term must be positive definite for a stable system, a condition that can be written as:

$$\frac{\partial^2\Pi}{\partial\mathbf{u}_i\partial\mathbf{u}_j}\delta u_i\delta u_j = K_{ij}\delta u_i\delta u_j > 0$$
(Eq 36)

where K_{ij} represents the component of the tangent stiffness matrix $\mathbf{K}$. Therefore, the stability limit is reached when matrix $\mathbf{K}$ ceases to be

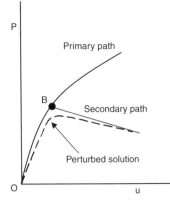

Fig. 10 Schematic illustration of solution paths in load-displacement (P-u) space for bifurcation-type problems

positive definite, that is, when:

$$\det[\mathbf{K}] = 0 \qquad (Eq\ 37)$$

In sheet metal forming process simulations, it is difficult to find the exact bifurcation point, because the control parameters in most of the problems are based on displacements. As a consequence, the determinant in Eq 37 changes sign abruptly within one incremental step in implicit analyses. The bifurcation point is, therefore, found by checking each value of the diagonal terms of a triangular form of the stiffness matrix.

The FE solution past the bifurcation point should not be the primary branch (path) but the secondary or bifurcated branch (path). In most of the bifurcation problems, the eigenvector is orthogonal to the primary branch. Thus, the increment of the nodal displacement field along the secondary branch $\Delta\mathbf{u}^s$ can be simply taken as (Ref 100–103, 112, 113):

$$\Delta\mathbf{u}^s = \chi\mathbf{v} \qquad (Eq\ 38)$$

In Eq 38, χ is a positive scalar, which should be determined, and $\mathbf{v}$ is an eigenvector at a singular point, that is, calculated from the matrix identity:

$$\mathbf{Kv} = \mathbf{0} \qquad (Eq\ 39)$$

In Eq 38, the trial increment $\Delta\mathbf{u}^s = \mathbf{v}$ is used as an initial estimate for the Newton-type solution scheme, and it is updated during the iterations. The effect of magnitude of χ on the wrinkling behavior is, therefore, eliminated. The details of this method are discussed in Ref 102 and 103.

Springback

Springback refers to the undesirable shape change due to the release of the tools after a sheet-forming operation. Previous studies indicated that the final part shape after springback depends on the amount of elastic energy stored in the part during forming (Ref 114–119). Unfortunately, the elastic energy stored is a function of the process parameters, geometry of the tools and the blank, friction conditions, and material behavior. Moreover, the prediction of springback is very sensitive to the numerical parameters used in the simulations (Ref 120, 121). Therefore, predicting and compensating springback are very complicated tasks. The analytical and FE approaches for the prediction of springback were very well summarized in Ref 122 (see also the article "Springback" in this Volume).

The elastic modulus and yield strength are the first-order material parameters that influence springback. This can be perceived intuitively by considering a stress-strain curve. The amount of recoverable stored energy, $\Delta E_r \cong \sigma\varepsilon = \sigma^2/E$, decreases when the Young's modulus increases (0 for a rigid-plastic material) and the yield strengths decreases. Aluminum alloys have similar yield strengths but lower

elastic moduli than mild steels and consequently tend to generate more springback. However, other material parameters have an effect as well. Geng and Wagoner (Ref 123) discussed the importance of plastic anisotropy and the role of the yield surface. Time-dependent springback was investigated by Wagoner et al. (Ref 124) for a 6022-T4 aluminum alloy sheet sample. This effect was incorporated in finite element (FE) model using constitutive equations involving creep deformation (Ref 125).

During springback, when the tools are removed from a formed part, the material unloads elastically and, depending on the geometry, some elements can experience reloading in the reverse direction even beyond the yield limit. This plastic deformation can, in turn, influence the amount of springback. Due to the Bauschinger effect discussed previously, that is, the yield stress for reverse loading is lower than the flow stress just before unloading, the springback can be altered. Therefore, it is often necessary to account for the Bauschinger effect in springback simulations. The isotropic hardening is no longer valid, and an effective way to model this effect is to assume that the yield surface translates in stress space. This assumption, called kinematic hardening, was introduced by Prager (Ref 126), and some modifications were proposed by Ziegler (Ref 127). In order to describe the expansion and translation of the yield surface during plastic deformation, the combination of isotropic and kinematic hardening is also commonly used (Ref 128). Other approaches based on two or multiple embedded yield surfaces (Ref 129–131) account for the Bauschinger effect as well. Because the model proposed by Chaboche built on a single yield surface, its use in FE simulations is more cost-effective compared to the more sophisticated multisurface models. In addition to the Bauschinger effect, a transient hardening behavior is also observed during reverse loading, and some of the aforementioned models can account for this phenomenon. In particular, the Bauschinger effect as well as its associated transient behavior, which result from dislocation pattern reorganization, were effectively and elegantly modeled by Teodosiu and Hu (Ref 20).

A formulation for kinematic hardening that can accommodate any yield function is well summarized in Ref 132 to 134. Assuming a homogenous yield function ϕ of degree a, translated in stress space by the backstress α, a tensorial state variable, the yield condition can be expressed as:

$$\phi(\sigma_{ij} - \alpha_{ij}) - \bar\sigma_{iso}^a = 0 \qquad (Eq\ 40)$$

where σ_{ij} are the components of the Cauchy stress tensor, and $\bar\sigma_{iso}$ is the stress measuring the size of the yield surface. The rate of plastic work, $\dot{w}$, dissipated during deformation becomes:

$$\dot{w} = \sigma_{ij}\dot\varepsilon_{ij} = (\sigma_{ij} - \alpha_{ij})\dot\varepsilon_{ij} + \alpha_{ij}\dot\varepsilon_{ij} \qquad (Eq\ 41)$$

where $\dot\varepsilon_{ij}$ are the components of the strain rate (or rate of deformation) tensor. The effective quantities are now defined considering the following modified plastic work-equivalence relationship:

$$\dot{w}_{iso} = (\sigma_{ij} - \alpha_{ij})\dot\varepsilon_{ij} = \bar\sigma_{iso}\dot{\bar\varepsilon} \qquad (Eq\ 42)$$

where $\dot{\bar\varepsilon}$ is the effective plastic strain rate. As for isotropic hardening, the normality rule is assumed to hold, and the model is complete with the definition of an evolution equation for the backstress.

Finite Element Modeling

Nonlinear FE methods are becoming very popular in sheet metal forming process simulations. A problem is nonlinear if the force-displacement relationship depends on the current state, that is, on current displacement, force, and stress-strain relationship:

$$\mathbf{P} = \mathbf{K}(\mathbf{P},\ \mathbf{u})\mathbf{u} \qquad (Eq\ 43)$$

where $\mathbf{u}$ is a displacement vector, $\mathbf{P}$ a force vector, and $\mathbf{K}$ the stiffness matrix. Linear problems form a subset of nonlinear problems. For example, in classical linear elastostatics, this relationship can be written in the form:

$$\mathbf{P} = \mathbf{Ku} \qquad (Eq\ 44)$$

where the stiffness matrix $\mathbf{K}$ is independent of both $\mathbf{u}$ and $\mathbf{P}$. If the matrix $\mathbf{K}$ depends on other state variables, such as temperature, radiation, etc., but does not depend on displacement or loads, the problem is still linear. Similarly, if the mass matrix is constant, the following dynamic problem is also linear:

$$\mathbf{P} = \mathbf{M\ddot{u}} + \mathbf{Ku} \qquad (Eq\ 45)$$

There are three sources of nonlinearities: material, geometry, and boundary condition. The material nonlinearity results from the nonlinear relationship between stresses and strains due to material plasticity. Geometric nonlinearity results from the nonlinear relationship between strains and displacements or the nonlinear relationship between stresses and forces. If the stress measure is energetically conjugate to the strain measure, both sources of nonlinearity have the same form. This type of nonlinearity is mathematically well defined but often difficult to treat numerically. Boundary conditions such as contact or friction are also sources of nonlinearities. This type of nonlinearity manifests itself in several real-life situations, for instance, in metal forming, gears, interfaces of mechanical components, pneumatic tire, and crash. A load on a structure causes nonlinearity if it changes with the displacement and deformation of the structure (such as pressure loading). Sheet metal forming processes include all of these three types of nonlinearities, in particular, elastoplastic material behavior, large rotations, and contacts between the tools and the blank.

General Kinematics

The kinematics of deformation can be described by Lagrangian, Eulerian, and Arbitrary Lagrangian-Eulerian (ALE) formulations. In the Lagrangian method, the FE mesh is attached to the material and moves through space along with the material. In the Eulerian formulation, the FE mesh is fixed in space, and the material flows through the mesh. In the ALE formulation, the grid moves independently from the material, yet in a way that spans the material at any time. The Lagrangian approach can be further classified in two categories: the total and the updated Lagrangian methods. In the total Lagrangian approach, equilibrium is expressed with the original undeformed reference frame, while, in the updated Lagrangian approach, the current configuration acts as the reference frame. In the latter, the true or Cauchy stresses and an energetically conjugate strain measure, namely, the true strain, are used in the constitutive relationships. The updated Lagrange approach is useful in:

- Analyses of shell and beam structure in which rotations are large so that the nonlinear terms in the curvature expressions may no longer be neglected
- Large strain-plasticity analyses in which the plastic deformations cannot be assumed to be infinitesimal

In general, this approach can be used to analyze structure where inelastic behavior causes large deformations. The (initial) Lagrangian coordinate frame has little physical significance in these analyses, because inelastic deformations are, by definition, permanent. Therefore, the updated Lagrangian formulation is appropriate for the simulations of sheet metal forming processes. For theses analyses, the Lagrangian frame of reference is redefined at the last completed iteration of the current increment. The variational form of the equation for the static problem in the updated Lagrangian approach is given as:

$$\int_V \frac{\partial \delta u_i}{\partial x_j} \sigma_{ij} dV - \int_\Gamma \delta u_i f_i d\Gamma = 0 \qquad \text{(Eq 46)}$$

where V is the volume considered, Γ is the surface on which the traction components f_i are imposed, and σ_{ij} is the Cauchy stress. The linearized variational form of Eq 46 needed for the Newton-Raphson numerical solver can be written as:

$$\int_V \frac{\partial \delta u_i}{\partial x_j} C_{ijkl} \frac{\partial \Delta u_k}{\partial x_l} dV + \int_V \frac{\partial \delta u_i}{\partial x_j} \sigma_{jl} \frac{\partial \Delta u_i}{\partial x_l} dV$$
$$= -\int_V \frac{\partial \delta u_i}{\partial x_j} \sigma_{ij} dV + \int_\Gamma \delta u_i f_i d\Gamma$$
$$\text{(Eq 47)}$$

The left side of Eq 47 corresponds to the material and geometric stiffness, while its right side is associated with the internal and external force vectors, respectively. In the FE method, the left side dominates the convergence rate, and the right side directly controls the accuracy of the solution.

Element Formulation

Finite element analyses of sheet metal forming processes can be broadly classified into three categories according to the element types used: membrane analysis (plane stress without bending stiffness), shell analysis (plane stress with bending stiffness), and continuum analysis (general stress state). For sheet metal forming simulations, the shell analysis is the most popular.

Wang and Budiansky (Ref 135) suggested an elastic-plastic membrane formulation based on the plane-stress assumption for forming sheet metal exhibiting normal anisotropy (also called planar isotropy). This type of anisotropy characterize a sheet with identical properties in any direction in its plane but different properties in its normal direction. These authors provided an example on the axisymmetric stretching of a sheet with a hemispherical punch. Yang and Kim (Ref 136) derived a membrane-based rigid plastic FE method incorporating material and geometric nonlinearities for materials exhibiting planar anisotropy, which was described by Hill's 1948 yield criterion. Hora et al. (Ref 137) analyzed the forming of an arbitrary-shaped autobody panel, which is a drawing-dominant process, based on membrane elements. Yoo et al. (Ref 138) suggested the bending energy augmented membrane approach in order to overcome the inherent numerical buckling occurring due to the lack of rotational stiffness in the membrane analysis. Kubli and Reissner (Ref 139) analyzed complicated panels, considering bending effects using the uncoupled solutions of membrane and bending analyses. However, the membrane analysis provides insufficient information for the treatment of bending-dominant forming processes.

Two basic approaches concerning the development of nonlinear shell FEs can be identified: classical shell elements and degenerated solid elements. The classical shell elements are directly based on the governing differential equations of an appropriate shell theory. Despite the potential economy of such elements, the development of nonlinear shell elements involves mathematical complexities. The degenerated solid element, which was initiated by Ahmad et al. (Ref 140) for the linear analysis of continuum formulation, is reduced in dimensionality by direct imposition of kinematics and constitutive constraints. The works of Ramm (Ref 141), Parish (Ref 142), Hughes and Liu (Ref 143, 144), Dvorkin and Bathe (Ref 145), and Liu et al. (Ref 146), among many others, constitute representative examples of this methodology carried out in the most general way for the nonlinear regime. The books of Bathe (Ref 147), Hughes (Ref 148), and the thesis of Stanley (Ref 149) offer a comprehensive overview of the degenerated solid approach and related methodologies. A point frequently made for the degenerated approach is that it avoids the mathematical complexities associated with the classical shell theory and hence is much easier for numerical implementation. Similarly, Belytschko et al. (Ref 150), Hallquist et al. (Ref 151), and Belytschko and Leviathan (Ref 152) developed one-point quadrature shell elements for the explicit FE methods applicable to both linear and nonlinear plate and shell analyses.

Finite element analyses using shell elements have shown a remarkable progress in the last few years with respect to both accuracy and efficiency. Recently, Cardoso and Yoon (Ref 153, 154) suggested new one-point quadrature elements considering element warping and thickness strain. In addition, Batoz et al. (Ref 155) suggested a membrane-bending FE model with membrane effects represented by constant-strain triangular elements and the bending effect represented by discrete Kirchhoff triangular elements. Alves de Sousa et al. (Ref 156) recently proposed one-point quadrature solid-shell elements that have eight nodes and consider multiple integration points through the thickness within one element layer, which can be applied for springback analysis. The application of shell elements to sheet-forming simulations can be found in Ref 154 and 157 to 161. The kinematics for a continuum-based shell is summarized as follows.

Kinematic Hypotheses. The physical spatial region of the shell element is described with coordinates (ξ, η, ζ), as shown in Fig. 11. The current position vector $\mathbf{x}$ of a generic point of the shell is defined by:

$$\mathbf{x}(\xi, \eta, \tilde{z}) = \bar{\mathbf{x}}(\xi, \eta) + \tilde{z}\check{\mathbf{x}}(\xi, \eta) \qquad \text{(Eq 48)}$$

where $\tilde{z}(\xi, \eta, \zeta) = \zeta t(\xi, \eta)/2$.

In Eq 48, the reference surface $\bar{\mathbf{x}}(\xi, \eta)$ is chosen to be the midsurface, that is $\bar{\mathbf{x}}(\xi, \eta) = \mathbf{x}(\xi, \eta, \zeta = 0)$. The symbol $\check{\mathbf{x}}$ is the unit fiber vector emanating from the midsurface, and t is the shell thickness. Equation 48 is often referred to as straight normal assumption. Equation 48 prescribes a linear $\tilde{z}$-dependence, which constrains the unit fiber vector to remain straight but not necessarily normal to the reference surface. Equation 48 can be interpolated from the node positions to any point of a shell element through the shape functions:

$$\mathbf{x}(\xi, \eta, \tilde{z}) = \sum_{p=1}^{N_{en}} N_p(\xi, \eta)\{\bar{\mathbf{x}}_p + \tilde{z}\check{\mathbf{x}}_p\} \qquad \text{(Eq 49)}$$

where:

$$\tilde{z}(\xi, \eta, \zeta) = \zeta \sum_{p=1}^{N_{en}} N_p(\xi, \eta) t_p$$

In the previous equation, the subscript "p" ranges over the number of nodes per element N_{en}. The shape functions, N_p, are typically selected from the Lagrange interpolation family. The displacement, **u**, of a generic point is the difference between its current position, **x**, and its reference position, **X**:

$$\mathbf{u}(\xi, \eta, \tilde{z}) = \mathbf{x}(\xi, \eta, \tilde{z}) - \mathbf{X}(\xi, \eta, \tilde{Z}) \qquad \text{(Eq 50)}$$

or

$$\mathbf{u}(\xi, \eta, \tilde{z}) = \bar{\mathbf{u}}(\xi, \eta) + \tilde{z}\check{\mathbf{u}}(\xi, \eta) + (\tilde{z} - \tilde{Z})\check{\mathbf{X}} \qquad \text{(Eq 51)}$$

where $\mathbf{X}(\xi, \eta, \tilde{Z}) = \bar{\mathbf{X}}(\xi, \eta) + \tilde{Z}\check{\mathbf{X}}(\xi, \eta)$.

The displacement vector, **u**, can be obtained by assuming that the fiber vector is inextensible. This assumption allows fiber vectors to rotate, but they cannot be stretched or contracted. Therefore, $\check{\mathbf{x}}$ remains a unit vector and the quantity $\tilde{z} - \tilde{Z}$ in Eq 51 vanishes. By applying the straight normal kinematics $\tilde{z} - \tilde{Z} = 0$ in Eq 51, the position of a continuum point is defined as:

$$\mathbf{u}(\xi, \eta, \tilde{z}) = \bar{\mathbf{u}}(\xi, \eta) + \tilde{z}\check{\mathbf{u}}(\xi, \eta) \qquad \text{(Eq 52)}$$

or

$$\Delta\mathbf{u}(\xi, \eta, \tilde{z}) = \Delta\bar{\mathbf{u}}(\xi, \eta) + \tilde{z}\Delta\check{\mathbf{u}}(\xi, \eta) \qquad \text{(Eq 53)}$$

with $\Delta\mathbf{u} \cdot \check{\mathbf{x}} = 0$.

Equation 52 is referred to as the incremental rigid normal assumption. In addition, the kinematics of the shell element is defined in terms of the same shape functions introduced previously in Eq 49:

$$\mathbf{u}(\xi, \eta, \tilde{z}) = \sum_{p=1}^{N_{en}} N_p(\xi, \eta)\{\bar{\mathbf{u}}_p + \tilde{z}\check{\mathbf{u}}_p\} \qquad \text{(Eq 54)}$$

where the six global incremental nodal variables at the pth node are:

$$\Delta\mathbf{a}_p = [\Delta\bar{u}_{1p} \ \Delta\bar{u}_{2p} \ \Delta\bar{u}_{3p} \ \Delta\check{u}_{1p} \ \Delta\check{u}_{2p} \ \Delta\check{u}_{3p}]^T \qquad \text{(Eq 55)}$$

Static Hypotheses. In addition to the kinematics, the static hypotheses are used to exploit the relative insignificance of the stress component acting directly on the reference surface. The first static hypothesis, zero normal stress, simply consists of eliminating this component whenever it explicitly appears in the governing equations:

$$\sigma_{33} = 0 \text{ (zero normal stress)} \qquad \text{(Eq 56)}$$

The second hypothesis is, in some sense, the time derivative of the zero normal stress because, if a quantity is identically zero, its derivative expressed in an appropriate frame should vanish as well:

$$\dot{\sigma}_{33} = 0 \text{ (zero normal stress rate)} \qquad \text{(Eq 57)}$$

For details, the reader is referred to Ref 161.

Matrix Formulation

The assembled FE matrix of the linearized variational form in Eq 47 is obtained by replacing the incremental and variational displacements ($\Delta\mathbf{u}$ and $\delta\mathbf{u}$) by their discrete element approximation and by incorporating the kinematic and static hypotheses, Eq 53 and 57. Then, the resulting equation is integrated elementwise over the entire problem domain. Using the arbitrariness of the weighting coefficients, the system of equations for the nodal displacement increments are:

$$(\mathbf{K}^{matl} + \mathbf{K}^{geom})\Delta\mathbf{a} = \mathbf{F}^{ext} - \mathbf{F}^{int} \qquad \text{(Eq 58)}$$

In Eq 58, $\mathbf{F}^{ext}$ is the assembled external force vector, $\mathbf{F}^{int}$ is the assembled internal force vector, and $\mathbf{K}^{matl}$ and $\mathbf{K}^{geom}$ are the assembled material and geometric stiffness matrices, respectively. Detailed information about Eq 58 is given in Ref 147 and 148.

Equation 58 forms a system of equations for the so-called implicit method. By ignoring the left side, that is, $\mathbf{F}^{ext} - \mathbf{F}^{int} = 0$, Eq 58 becomes an equation for the so-called explicit method. Numerical solution schemes are often referred to as being explicit or implicit. When a direct computation of the dependent variables can be made in terms of known quantities, the computation is said to be explicit. In contrast, when the dependent variables are defined by a coupled set of equations, an iterative technique is needed to obtain the solution, such as for Eq 58, and the numerical method is said to be implicit. As mentioned in the section "General Kinematics" in this article, the right side of Eq 58 controls the solution accuracy, and the left side dominates the convergence rate. Because the explicit method does not involve the left side, stiffness matrix assembling and convergence checking are not required. Therefore, it is very attractive when

convergence is a big hurdle, as it may be for complicated sheet metal stamping problems. However, the explicit solution is only stable if the time step size is smaller than a critical value:

$$\Delta t \leq \Delta t^{crit} = \frac{2}{\omega_{max}} \qquad \text{(Eq 59)}$$

which guarantees the accuracy of the solution. Here, ω_{max} is the largest natural frequency of the system (characteristics of the mesh size and material density). Usually, the explicit stable time step size is extremely small compared to the implicit solution (approximately 1000 smaller). Furthermore, the mesh size is proportional to $1/\omega_{max}$. Therefore, the stable time step size becomes smaller when the mesh size decreases. In particular, when mesh refinement is conducted during the analysis, mesh sizes become approximately one quarter smaller per one time of mesh refinement. Because the stable time step size in an explicit code is determined by the minimum element length, it causes the analysis time to be much longer. In contrast, the implicit solution is always stable; that is, the time step can be arbitrarily large for a linear problem. For nonlinear problems, the implicit time step size may need to become smaller according to convergence difficulties.

Elastic-Plastic Stress Integration

Most rate-independent plastic models are formulated in terms of rate-type constitutive equations, for which the integration method has a considerable influence on the efficiency, accuracy, and convergence of the solution. In the simulation of sheet forming processes, the constitutive equation is integrated along an assumed deformation path. Among the infinite ways to assume the deformation path, the minimum plastic work path in homogeneous deformation has been found to have several advantages. Requirements for achieving minimum plastic work paths in homogeneous deformation are well documented in Ref 162 to 165. The minimum work path, which is also the proportional logarithmic (true) strain paths, is achieved under two conditions. First, the set of three principal axes of stretching is fixed with respect to the material; second, the logarithms of the principal stretches remain in a fixed ratio. The incremental deformation theory based on the minimum plastic work path enables convenient decoupling of deformation and rotation by the polar decomposition at each process increment. The resulting incremental constitutive law is frame-indifferent (objective), because the theory uses a materially embedded coordinate system. The incremental deformation theory is useful for the FE modeling of rigid-plastic and elastoplastic constitutive formulations. In rigid plasticity, the theory was introduced for process analyses by Yang and Kim (Ref 136), Germain et al. (Ref 166), Chung and Richmond (Ref 167, 168), and Yoon et al.

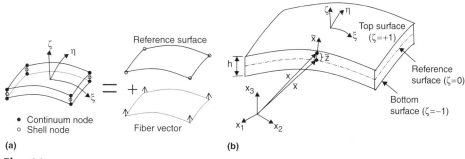

(a) **(b)**

Fig. 11 Continuum-based shell elements. (a) Construction of a typical shell element. (b) Shell geometry

(Ref 169). For elastoplastic materials, the incremental deformation theory has been successfully applied for materials exhibiting isotropy (Ref 170) and planar anisotropy (Ref 161, 171).

Continuum Models. The most popular scheme for stress integration is the predictor-corrector method (often called return mapping). This method is applied in two successive steps: the prediction step, during which a trial stress state is estimated, and the corrector step, during which a flow rule is applied by return mapping procedures in order to bring (project) the stress onto the yield surface or, in other words, to ensure the consistency condition (i.e., the stress state must be on the yield surface). During the return mapping procedure, a reasonable assumption for the deformation path must be imposed. A return mapping procedure was first introduced in the paper of Wilkins (Ref 172). The works of Ortiz and Pinsky (Ref 173) and Ortiz et al. (Ref 174) apply the closet point-projection method to perform the procedure in plane-stress conditions. Later, Ortiz and Simo (Ref 175) developed a new class of integration algorithms based on the cutting-plane approach. However, this approach has no clear physical meaning in the deformation path viewpoint. Recently, Yoon et al. (Ref 161) proposed the multistage return mapping method based on incremental deformation theory, which follows the minimum plastic work path. In this work, it was proven that, when the consistency condition and normality rule (strain increment normal to yield surface) are imposed, this new projection during the current unknown step becomes the closet point projection. The stress integration procedure is briefly summarized as follows.

The increment of the Cauchy stresses is given by applying the fourth-order elastic modulus tensor $\mathbf{C}^e$ to the incremental second-order elastic strain tensor $\Delta\boldsymbol{\varepsilon} - \Delta\boldsymbol{\varepsilon}^p$:

$$\Delta\boldsymbol{\sigma} = \mathbf{C}^e(\Delta\boldsymbol{\varepsilon} - \Delta\boldsymbol{\varepsilon}^p) \qquad \text{(Eq 60)}$$

in which $\Delta\boldsymbol{\varepsilon}$ and $\Delta\boldsymbol{\varepsilon}^p$ are the total and plastic strain increments, respectively. The relationship in Eq 60 is assumed to be expressed in a material-embedded coordinate system. Therefore, it is objective with respect to (independent of) the material rotation. In order to follow the minimum plastic work path in the incremental deformation theory, the logarithmic plastic strain needs to remain normal to the yield surface at the representative stress state, that is:

$$\Delta\boldsymbol{\varepsilon}^p = \Delta\bar{\varepsilon}^p \frac{\partial\bar{\sigma}}{\partial\boldsymbol{\sigma}} = \gamma\mathbf{m} \qquad \text{(Eq 61)}$$

where $\mathbf{m} = \partial\bar{\sigma}/\partial\boldsymbol{\sigma}$ is a symbolic notation that represents the tensor of component $m_{ij} = \partial\bar{\sigma}/\partial\sigma_{ij}$. The condition stipulating that the updated stress stays on the strain-hardening curve provides the following equation:

$$\begin{aligned} F(\lambda) &= \bar{\sigma}(\boldsymbol{\sigma}_n + \Delta\boldsymbol{\sigma}) - h(\bar{\varepsilon}_n^p + \lambda) \\ &= \bar{\sigma}[\boldsymbol{\sigma}_n + \mathbf{C}^e(\Delta\boldsymbol{\varepsilon} - \lambda\mathbf{m})] - h(\bar{\varepsilon}_n^p + \lambda) = 0 \end{aligned}$$
$$\text{(Eq 62)}$$

where the subscript "n" denote quantities at step "n" in the simulation.

The predictor-corrector scheme based on the Newton-Raphson method is generally used to solve the nonlinear system in Eq 62 for $\lambda = \Delta\bar{\varepsilon}^p$. However, while a mathematical solution to this equation does exist, it can be difficult to obtain numerically if the strain increment is not small enough. In the examples in the next section, a multistage return mapping procedure based on the control of the residual suggested by Yoon et al. (Ref 161) was employed. The proposed method is applicable to nonquadratic yield functions and general strain-hardening laws without a line search algorithm, even for a relatively large strain increment (10%). At the end of the step, when Eq 62 is solved, all kinematic variables and stresses are updated.

In order to consider the rotation of the anisotropic axes, a co-rotational coordinate system (constructed at each integration point), is defined and initially coincides with the material symmetry axes. For the examples discussed in the section "Application Examples" in this article, it is assumed that the orthogonality of the anisotropy axes is preserved during sheet forming under the isotropic hardening assumption. This assumption is generally considered as appropriate in sheet-forming process simulations. From the polar decomposition theorem, the deformation of a material element represented by the deformation gradient tensor, $\mathbf{F}$, is the combination of a pure rotation, $\mathbf{R}$, and a pure stretch, $\mathbf{U}$ ($\mathbf{F} = \mathbf{R}\mathbf{U}$). The rotation of the anisotropy axes is updated incrementally at every step by the rotation amount ($\mathbf{R}$) obtained from the polar decomposition (Fig. 12). For instance, if at the first step $^o\boldsymbol{\lambda}$ and $^o\boldsymbol{\mu}$ are the unit vectors coinciding with the rolling and transverse directions, respectively, the updated axes are given by:

$$\begin{aligned} \boldsymbol{\lambda} &= \mathbf{R}\,^o\boldsymbol{\lambda} \\ \boldsymbol{\mu} &= \mathbf{R}\,^o\boldsymbol{\mu} \end{aligned} \qquad \text{(Eq 63)}$$

Crystal-Plasticity Models. These models account for the deformation of a material by crystallographic slip and for the reorientation of the crystal lattice. The influence of crystal symmetry on elastic constants can be included, and the strain-hardening and cross-hardening effects between the slip systems can be incorporated through the use of state variables. Furthermore, a rate-dependent approach is also typically employed to relate the shear stresses and shear strains on the different slip systems. The kinematics of the model was summarized by Dao and Asaro (Ref 176). The deformation gradient, $\mathbf{F}$, is decomposed into a plastic deformation, $\mathbf{F}^p$, which is the summation of the shear strain for each slip system, and a combination of elastic deformation and rigid body motion of the crystal lattice, $\mathbf{F}^e$, as shown in Fig. 13, that is:

$$\mathbf{F} = \mathbf{F}^e\mathbf{F}^p \qquad \text{(Eq 64)}$$

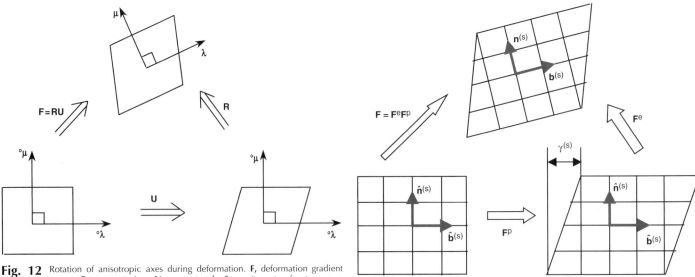

Fig. 12 Rotation of anisotropic axes during deformation. $\mathbf{F}$, deformation gradient tensor; $\mathbf{R}$, pure rotation; $\mathbf{U}$, pure stretch; $^o\boldsymbol{\mu}$, unit vector for transverse direction; $^o\boldsymbol{\lambda}$, unit vector for rolling direction

Fig. 13 Multiplicative decomposition of deformation gradient. See text for description.

Because plasticity occurs by dislocation slip, the plastic deformation rate, $\mathbf{L}^p$, is determined by the summation of the shear strain contribution over all of the slip system (Ref 176):

$$\mathbf{L}^p = \sum_{(s)} \dot{\gamma}^{(s)} \mathbf{b}^{(s)} \mathbf{n}^{(s)} \qquad \text{(Eq 65)}$$

Here, (s) denotes a slip system, $\mathbf{n}^{(s)}$ is the normal of the slip plane, and $\mathbf{b}^{(s)}$ is the vector in the slip direction. Both vectors are orthogonal, that is, $\mathbf{n}^{(s)} \times \mathbf{b}^{(s)} = \mathbf{0}$, and are assumed to rotate with the elastic spin of the lattice. In general, these vectors are not unit vectors like in a phenomenological model because they are allowed to stretch. Using Eq 65, the symmetric and skew symmetric parts of $\mathbf{L}^p$ the rate of deformation tensor $\mathbf{D}^p$ and plastic spin $\mathbf{W}^p$, respectively, can be written as:

$$\mathbf{D}^p = \tfrac{1}{2}(\mathbf{L}^p + \mathbf{L}^{pT}) = \sum_{(s)} \dot{\gamma}^{(s)} \tfrac{1}{2}(\mathbf{b}^{(s)}\mathbf{n}^{(s)} + \mathbf{n}^{(s)}\mathbf{b}^{(s)})$$

$$\text{or} \quad \mathbf{D}^p = \sum_{(s)} \dot{\gamma}^{(s)} \mathbf{P}_{(s)} \qquad \text{(Eq 66)}$$

$$\mathbf{W}^p = \tfrac{1}{2}(\mathbf{L}^p - \mathbf{L}^{pT}) = \sum_{(s)} \dot{\gamma}^{(s)} \tfrac{1}{2}(\mathbf{b}^{(s)}\mathbf{n}^{(s)} - \mathbf{n}^{(s)}\mathbf{b}^{(s)})$$

$$\text{or} \quad \mathbf{W}^p = \sum_{(s)} \dot{\gamma}^{(s)} \mathbf{W}_{(s)} \qquad \text{(Eq 67)}$$

where the tensors $\mathbf{P}_{(s)}$ and $\mathbf{W}_{(s)}$ have been introduced for notational convenience.

Usually, the Cauchy or true stress, $\boldsymbol{\sigma}$, is employed for the stress integration. However, in this section, the Kirchhoff stress, $\boldsymbol{\tau}$, is considered By ignoring the elastic volume change, the Cauchy stress is related to the Kirchhoff stress, $\boldsymbol{\tau}$, through the relationship $\boldsymbol{\tau} = J\boldsymbol{\sigma}$, where J is the determinant of $\mathbf{F}^e$. Writing the rate of deformation tensor as the sum of elastic and plastic parts, an objective stress rate, $\check{\boldsymbol{\tau}}$, can be expressed as:

$$\check{\boldsymbol{\tau}} = \mathbf{K} : \mathbf{D} - \sum_{(s)} \dot{\gamma}^{(s)}(\mathbf{K} : \mathbf{P}_{(s)} + \mathbf{W}_{(s)} \cdot \boldsymbol{\tau} - \boldsymbol{\tau} \cdot \mathbf{W}_{(s)})$$

$$= \mathbf{K} : \mathbf{D} - \sum_{(s)} \dot{\gamma}^{(s)} \mathbf{R}_{(s)} \qquad \text{(Eq 68)}$$

where $\mathbf{K}$ is the fourth-order elastic modulus tensor, and $\mathbf{D}$ is the rate of deformation tensor. Then, the resolved shear stress at the end of the time step becomes:

$$\tau^{(s)}_{t+\Delta t} = \tau^{(s)}_t + \Delta t \dot{\tau}^{(s)} = \tau^{(s)}_t + \Delta t \mathbf{R}_{(s)} : \mathbf{D} - \Delta t$$

$$\times \sum_{(\alpha)} \dot{\gamma}^{(\alpha)} \mathbf{R}_{(s)} : \mathbf{P}_{(\alpha)} \qquad \text{(Eq 69)}$$

where $\mathbf{R}_{(s)} = \mathbf{K} : \mathbf{R}_{(s)} + \mathbf{W}_{(s)} \times \boldsymbol{\tau} - \boldsymbol{\tau} \times \mathbf{W}_{(s)}$.

The resolved shear stress, $\tau^{(s)}$, also follows the rate-dependent hardening rule:

$$\tau^{(s)} = g^o \left(\frac{|\dot{\gamma}^{(s)}|}{\dot{\gamma}_o} \right)^m \text{sign}\left(\dot{\gamma}^{(s)} \right) \qquad \text{(Eq 70)}$$

where g^o is the shear yield stress on a slip system.

Finally, the resolved shear stress defined by Eq 69 and 70 must have the same value at $t + \Delta t$:

$$E^s(\dot{\gamma}^{(s)}) = \tau^{(s)}_t + \Delta t \mathbf{R}_s : \mathbf{D} - \Delta t$$

$$\times \sum_{(\alpha)} \dot{\gamma}^{(\alpha)} \mathbf{R}_s \times \mathbf{P}_\alpha - g^o \left(\frac{|\dot{\gamma}^{(s)}|}{\dot{\gamma}_o} \right)^m \text{sign}\left(\dot{\gamma}^{(s)} \right) = 0$$

$$\text{(Eq 71)}$$

Equation 71 is a nonlinear equation. It is solved with the Newton-Raphson method after linearization (Ref 177), that is:

$$E^s\left(\dot{\gamma}^{(s)} \right) + \sum_{(\alpha)} \frac{dE^s}{d\dot{\gamma}^{(\alpha)}} d\dot{\gamma}^{(\alpha)} = 0 \qquad \text{(Eq 72)}$$

Ideal Forming Design Theory

Finite element (FE) methods are used mostly for analysis of boundary-value problems. This means that the problem is well set, with known tool and blank geometries, material and interface properties, and realistic stress- and displacement-imposed boundary conditions. This type of application of the method is referred to as FE analysis. However, in practice, the final shape of the product is imposed, and the manufacturing process needs to be designed around it. Therefore, in order to improve the conventional trial-and-error-based practices for optimizing forming processes, either by experiments or FE analysis, an FE design theory, called ideal forming theory, was proposed (Ref 167, 168, 178). The FE implementation is a time-efficient one-step code, providing not only the initial blank geometry but all the intermediate shapes and the entire load history necessary to achieve it, thus providing invaluable information about the ideal process parameters. This application is referred to as FE design.

In this theory, materials are prescribed to deform following the minimum plastic work path. The final product shape is specified, and the initial blank shape is obtained from the global extremum plastic work criterion as a one-step backward solution. Although the theory is general enough to accommodate any other form of constraints, the underlying physical assumption of this extremum work condition is that the strain gradients are minimized on the overall part, thus departing as much as possible from plastic flow localization modes. In order to consider local thinning effects due to friction, a method based on a modified extremum work criterion has also been developed (Ref 178). The ideal forming theory has been successfully applied for sheet-forming processes to optimize flat blanks (Ref 179–181) and also for bulk forming in steady (Ref 182–185) and nonsteady flows (Ref 186, 187).

When a sheet (or tube) is discretized with meshes and the surface tractions are approximated by point forces, the total plastic work, W,

becomes a function of the initial and final configurations, assuming that the minimum plastic work path is imposed on each material element, that is:

$$W = W[\bar{\varepsilon}(\mathbf{x}^i, \mathbf{X}^i)] = \int\int\int \bar{\sigma}(\bar{\varepsilon})\bar{\varepsilon} dV_0 \qquad \text{(Eq 73)}$$

where $\bar{\sigma}(\bar{\varepsilon})$, $\bar{\varepsilon} = \int d\bar{\varepsilon}$, and V_0 are the effective stress, the effective strain, and the initial volume of the part, respectively. The parameters $\mathbf{x}^i$ and $\mathbf{X}^i$ are the final and initial positions of the ith node in the global Cartesian coordinate system. As mentioned previously, in this method, the part shape ($\mathbf{x}^i$) is prescribed, and the initial blank shape ($\mathbf{X}^i$) is the output of the numerical simulation. Therefore, the optimization is performed with respect to the initial configuration. In the ideal forming theory, when the minimum plastic work path is imposed for each material element, the initial blank shape is obtained by optimizing the plastic work or, more specifically, by imposing $dW/d\mathbf{X}^i = 0$. Another term is added to this equation when friction is taken into account. In this optimization, constraints pertaining to the initial shape are also imposed. For instance, the initial blank must be on a given surface, for instance, a flat sheet, a cylindrical tube, and so on. In tube hydroforming, the preform shape must be straight with a uniform cross section if the tube is extruded. Therefore, the additional constraint is that each node-set in Fig. 14 must move the same amount in the $\bar{X}_3$ direction. Because of the mathematical form of Eq 73, it is much more convenient to describe the material with a strain-rate potential (see the section "Strain-Rate Potentials" in this article), which emphasizes the importance of this type of material description, particularly applied to aluminum alloy products. The combination of plastic work optimization, material behavior, friction conditions, and initial shape constraints reduces to a system of nonlinear equations that can be solved for $\mathbf{X}^i$ (corresponding to the initial blank shape), using a Newton-Raphson solver.

Of course, this theory is based on ideal deformation conditions. More realistic simulations can be achieved with FE analysis codes, but the ideal forming theory provides a quick and excellent initial mapping of the forming process needed to manufacture a product, thus limiting the number of costly experimental and FE analysis trials.

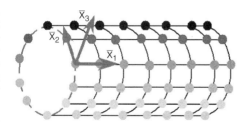

Fig. 14 Local Cartesian coordinate systems defined on an assumed cross section of the initial tube

Application Examples

Example 1: Shear Testing Simulation. This first example pertains to the simulation of the simple shear test. This example illustrates the numerical testing and possible issues, such as implementation, input data, and so on, that should be addressed before performing the simulation of a complex forming operation. Simple examples also validate the choice of the constitutive models used in the simulations. In this particular example, the simple shear test shows that the constitutive response of material subjected to a given deformation mode (e.g., shear) can be well approximated using constitutive equations identified using information measured on different stress states (e.g., uniaxial tension or balanced biaxial tension).

Simple shear tests were conducted on a commercial-purity 1050-O and automotive 6022-T4 aluminum alloy sheet samples on a device similar to that shown in Fig. 4. The testing was performed in two specimen orientations, that is, with shear in the transverse direction (TD) or at 45° from the rolling direction (RD). The

1050-O sheet sample was 3 mm (0.12 in.) thick, and the shear specimen was 40 by 40 mm (1.6 by 1.6 in.), with a shear zone width of 8 mm (0.3 in.) (Fig. 15). The 6022-T4 sheet sample was 1 mm (0.04 in.) thick, and the shear specimen was 60 by 15 mm (2.4 by 0.6 in.), with a shear zone width of 3 mm (0.12 in.) (Fig. 15). More details concerning these tests are reported in Ref 177, 188, and 189.

For 1050-O, the experimental shear stress/shear strain curves along the two shear directions showed that the material exhibits anisotropic strain hardening (Fig. 16). The strain hardening in the initial stage of plastic deformation is higher for simple shear in the TD (90°) but saturates prematurely compared to simple shear at 45° from the RD. For 6022-T4, this effect is also observed but much less pronounced. In previous work (Ref 23, 188, 189), two microstructural parameters were invoked to explain this phenomenon: dislocation cell structure and crystallographic texture. The pole figures were measured using standard x-ray diffraction techniques. The resulting (111) pole figures (Fig. 17) indicated that both as-received materi-

als exhibited a preponderant $\{100\}\langle001\rangle$ cube texture, although stronger in the 1050-O sheet sample than in the 6022-T4. As expected, the 1050-O sample developed a well-defined dislocation microstructure during deformation, principally dislocation walls on the (111) planes oriented near the planes of maximum shear stress. The 6022-T4 alloy also displayed a dislocation structure with similar features, but the walls were much fainter and the overall dislocation distribution was more uniform (Fig. 18).

Simulations. Two constitutive models, phenomenological and polycrystal, were implemented into the user material subroutines provided in FE commercial codes (Yld2004-18p yield function into MSC.Marc HYPELA2, MSC. Software Corp., and polycrystal model into ABAQUS UMAT, ABAQUS Inc.) and used for the simple shear test simulations. The meshes and boundary conditions for the shear tests are represented in Fig. 19. The left grip was fixed, while the right grip was allowed to move along the vertical direction only. The FE meshes were generated only for the shear deformation zone of the specimens (not the grip area), using 320 and 800 uniform elements for 1050-O and 6022-T4, respectively, with a single layer through the thickness direction (Fig. 19).

The input data and resulting Yld2004-18p coefficients (see the section "Anisotropic Yield Functions" in this article) for the 1050-O and 6022-T4 sheet samples are given in Tables 1 and 2, respectively. For the FE simulations performed with Yld2004-18p, isotropic hardening was assumed with the stress-strain relationships:

- 1050-O: $\bar{\sigma} = K(\bar{\varepsilon} + \bar{\varepsilon}_0)^n$, $K = 132$ MPa (19 ksi), $\varepsilon_0 = 0.0005$, and $n = 0.285$

- 6022-T4: $\bar{\sigma} = A - B\exp(-C\bar{\varepsilon})$, $A = 396$ MPa (57 ksi), $B = 234$ MPa (34 ksi), $C = 6.745$

as measured with the RD uniaxial tension and bulge tests, respectively.

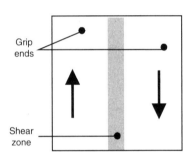

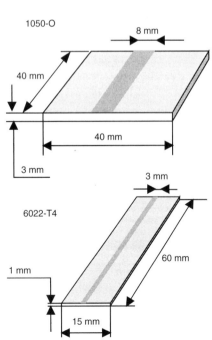

1050-O

6022-T4

Fig. 15 Shear test-specimen geometry. Source: Ref 177

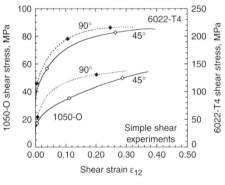

Fig. 16 Simple shear stress-strain curves measured along different directions with respect to the rolling direction for aluminum alloy 1050-O and 6022-T4 sheet samples. Source: Ref 177

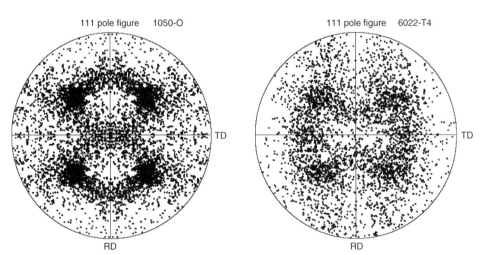

Fig. 17 111 pole figures for aluminum alloy 1050-O and 6022-T4 sheet samples. RD, rolling direction; TD, transverse direction. Source: Ref 177

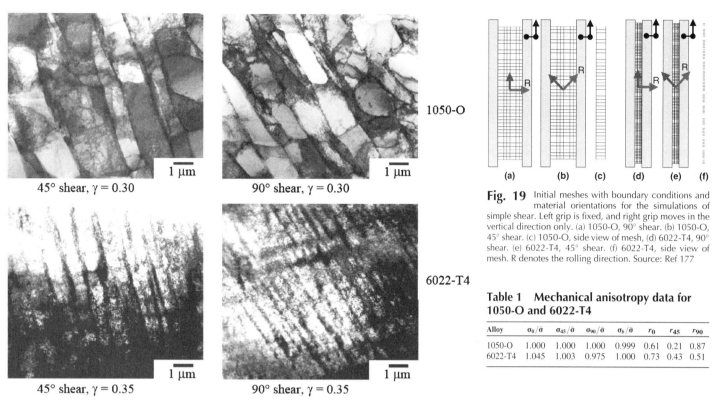

Fig. 18 Dislocation substructure developed during simple shear of aluminum alloy 1050-O and 6022-T4 sheet samples (for different amounts of shear strain: $\varepsilon_{12} = 0.5 \times \gamma$). Source: Ref 177

Fig. 19 Initial meshes with boundary conditions and material orientations for the simulations of simple shear. Left grip is fixed, and right grip moves in the vertical direction only. (a) 1050-O, 90° shear. (b) 1050-O, 45° shear. (c) 1050-O, side view of mesh. (d) 6022-T4, 90° shear. (e) 6022-T4, 45° shear. (f) 6022-T4, side view of mesh. R denotes the rolling direction. Source: Ref 177

Table 1 Mechanical anisotropy data for 1050-O and 6022-T4

Alloy	$\sigma_0/\bar{\sigma}$	$\sigma_{45}/\bar{\sigma}$	$\sigma_{90}/\bar{\sigma}$	$\sigma_b/\bar{\sigma}$	r_0	r_{45}	r_{90}
1050-O	1.000	1.000	1.000	0.999	0.61	0.21	0.87
6022-T4	1.045	1.003	0.975	1.000	0.73	0.43	0.51

Table 2 Yld2004-18p coefficients for 1050-O and 6022-T4 (exponent $a = 8$)

1050-O				6022-T4			
c'_{12}	1.019520	c''_{12}	0.830052	c'_{12}	0.755194	c''_{12}	1.120720
c'_{13}	1.094320	c''_{13}	0.850783	c'_{13}	0.799378	c''_{13}	1.056340
c'_{21}	1.270160	c''_{21}	0.537084	c'_{21}	0.773630	c''_{21}	1.146560
c'_{23}	1.127570	c''_{23}	0.753355	c'_{23}	0.865580	c''_{23}	1.132990
c'_{31}	0.794215	c''_{31}	1.119190	c'_{31}	1.047560	c''_{31}	0.763656
c'_{32}	0.829394	c''_{32}	1.009200	c'_{32}	1.088160	c''_{32}	0.954688
c'_{44}	7.713050	c''_{44}	7.717600	c'_{44}	1.016290	c''_{44}	1.009770
c'_{55}	1.004780	c''_{55}	1.009920	c'_{55}	0.993625	c''_{55}	0.994796
c'_{66}	1.282290	c''_{66}	0.302787	c'_{66}	0.624258	c''_{66}	1.208880

Figure 20 represents the effective stress contours for the shear specimens deformed up to a shear strain of $\gamma/2 = 0.25$ for the 1050-O and 6022-T4 sheet samples. It was observed that the experimental and predicted shapes of the specimen ends were very similar. Moreover, the FE simulations indicated that, for the strain range investigated, the deformation in the shear specimens was quite uniform, that is, end effects were minimal. The shear stress/shear strain curves along the two shear directions (at 45° from RD and in the TD) for the 1050-O and 6022-T4 sheet samples as predicted with Yld2004-18p are given in Fig. 21. The results were found to be in good agreement with the experimental trends (Fig. 16). For 1050-O, although the material was assumed to exhibit isotropic hardening, the simulations were able to reproduce a certain amount of anisotropic hardening observed experimentally during simple shear. For 6022-T4, although the difference in hardening between the two test orientations was noticeable and consistent with the experiments, this effect was small. The influence of the element size was evaluated by performing simulations with only one element (1 by 1 by 0.1, unit: millimeters) for both materials. It was shown that the element size had only a negligible effect on the results (Fig. 21). Therefore, the simulations using crystal plasticity were performed with only one element.

For simulations using a rate-dependent crystal-plasticity model (see the section "Crystal-Plasticity Models" in this article), a trial-and-error method was employed to find the hardening parameters that provided a good approximation of the uniaxial hardening curves (and biaxial curve for 6022-T4). The strain-rate sensitivity coefficient for 1050-O was set to 0.01, which corresponds to the experimental value. For 6022-T4, the strain-rate sensitivity coefficient was arbitrarily set to 0.001, because the experimental value for this type of material is usually close to zero or slightly negative. The textures corresponding to the (111) pole figures (Fig. 17), which include approximately 1000 grain orientations each, were used as input. The simple shear-hardening curves predicted with crystal plasticity are shown in Fig. 22 for both sheet samples. Again, the trends predicted with this model were in very good agreement with the experiments (Fig. 16). The relative shapes of the stress-strain curves predicted with crystal plasticity were similar to those predicted with the phenomenological model, but the strain-hardening difference was larger.

In summary, the simple shear stress-strain curves of 1050-O and 6022-T4 aluminum alloy sheet samples were predicted using FE analyses with two constitutive material descriptions. Although the constitutive coefficients were identified using uniaxial tension and bulge test stress-strain curves, the simple shear-hardening responses were in good agreement with the experiments. It was shown that, even for a material exhibiting an isotropic strain-hardening behavior, the simple shear behavior could display an apparent anisotropic strain-hardening behavior due to the combined effect of plastic anisotropy and realistic boundary conditions. Crystal plasticity, led to a real anisotropic-hardening effect resulting from crystallographic texture evolution. Both effects, apparent and real, could fully explain the behavior of 1050-O and 6022-T4 sheet samples in simple shear. It was also concluded that the apparent anisotropic hardening was due to texture but not to dislocation microstructure. More details can be found in Ref 177.

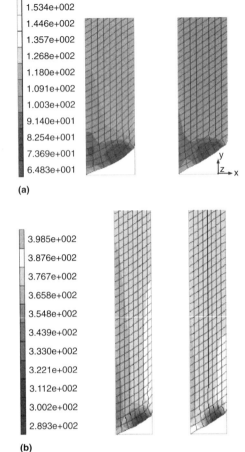

(a)

(b)

Fig. 20 Predicted effective stress superimposed on the deformed mesh for simple shear specimens of aluminum alloy sheet samples: (a) 1050-O and (b) 6022-T4. Source: Ref 177

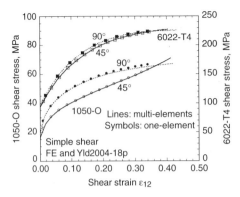

Fig. 21 Finite-element predictions of shear stress-strain curves for 1050-O and 6022-T4 sheet samples from one-element or multiple-element simulations using Yld2004-18p

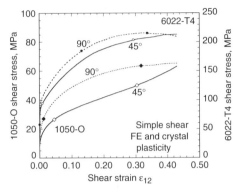

Fig. 22 Finite-element predictions of simple shear stress-strain curves for 1050-O and 6022-T4 sheet samples using a crystal-plasticity model (one-element simulations). Source: Ref 177

Table 3 Normalized yield stress input data to determine Yld2004-18p coefficients for aluminum alloy 2090-T3 sheet sample, $\bar{\sigma}/\sigma_0$

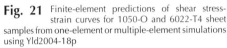

In-plane	σ_0/σ_0	σ_{15}/σ_0	σ_{30}/σ_0	σ_{45}/σ_0	σ_{60}/σ_0	σ_{75}/σ_0	σ_{90}/σ_0	σ_b/σ_0
Experimental results	1.0000	0.9605	0.9102	0.8114	0.8096	0.8815	0.9102	1.0350
Out-of-plane	$\sigma_{45}^{TD\text{-}ND}/\sigma_0$		$\sigma_{45}^{ND\text{-}RD}/\sigma_0$		$\tau^{TD\text{-}ND}/\sigma_0$		$\tau^{ND\text{-}RD}/\sigma_0$	
Crystal-plasticity results	0.92		0.89		0.48		0.47	

Table 4 r-value input data to determine Yld2004-18p coefficients for aluminum alloy 2090-T3 sheet sample

In-plane	r_0	r_{15}	r_{30}	r_{45}	r_{60}	r_{75}	r_{90}	r_b
Experimental results	0.2115	0.3269	0.6923	1.5769	1.0385	0.5384	0.6923	0.67

Example 2: Finite Element Analysis of Cup Drawing.

Cup drawing is not only a material test but also a forming operation, which accounts for the combined effects of blankholder force, friction, and sheet bending/unbending. Drawing of a circular blank with a circular punch is one of the best benchmarks to evaluate planar anisotropy. After this operation, the wall-height profile around the periphery of the resulting cup is usually not constant but exhibits peaks (also called ears) and valleys. The objective of this example is to show how FE analysis can be used to describe cup drawing for a 2090-T3 alloy sheet, which exhibits more than four ears. The Yld2004-18p yield function (see section "Material Modeling" in this article, and Ref 42) was used for the description of the plastic behavior of the aluminum alloy sheet. The FE cup-height profile was computed and compared with the results of a simple but more time-efficient analytical model as well as with experimental results.

Using the isotropic-hardening assumption, there have been previous efforts to simulate anisotropic phenomena such as the formation of ears in drawing circular blanks for aluminum alloy sheets. For this purpose, Chung and Shah (Ref 190) and Yoon et al. (Ref 169), respectively,

applied the stress and the strain-rate potentials proposed by Barlat et al. (Ref 37, 51) for a 2008-T4 aluminum alloy sheet. Andersson et al. (Ref 191) employed the criterion by Karafillis and Boyce (Ref 38) for the limiting dome height test. Yoon et al. (Ref 192) used Yld96, the yield function suggested by Barlat et al. (Ref 193) for circular deep drawing of a 2090-T3 aluminum alloy sheet. Yoon et al. (Ref 44, 194) simulated a reverse cup-drawing test using Yld2000-2d (see the section "Anisotropic Yield Functions" in this article, and Ref 41). Recently, Yoon et al. (Ref 195) successfully predicted six and eight ears in a circular cup drawing using Yld2004-18p (Ref 42).

Finite-Element-Based Earing Profile. The input data used to determine the anisotropy coefficients of the 2090-T3 aluminum alloy sheet sample are given in Tables 3 and 4. Polycrystal calculations were used to determine the out-of-plane properties given in Table 3. The out-of-plane data consisted of uniaxial tension yield stresses obtained at 45° between the symmetry axes in the planes (RD, normal direction, or ND; RD) and (TD, ND), where ND is the normal direction of the sheet, and the simple shear yield stresses in these planes. Therefore, the 18 coefficients for the yield function were optimized using 20 input data. The Yld2004-18p coefficients (see the section "Anisotropic Yield Functions" in this article) are given in Table 5. It is worth noting that, if a crystal-plasticity model were not available or if a plane-stress shell model

was employed for Yld2004-18p, the assumption $\alpha_7 = \alpha_8 = \alpha_{16} = \alpha_{17} = 1$ could be employed for the out-of-plane anisotropy coefficients. The stress-strain curve used for the simulation was given by the relationship $\bar{\sigma} = 646(0.025 + \bar{\varepsilon})^{0.227}$ (MPa). Figure 23 shows the yield surfaces of the 2090-T3 sheet sample predicted using Yld2004-18p. Figure 24 depicts the normalized yield stress and r-value anisotropies, calculated with Yld2004-18p and determined experimentally. In Fig. 24, the results of another yield function, denoted Yld96 (Ref 193), are included for comparison purposes. Yld2004-18p accounts for uniaxial properties measured every 15° in the plane of the sheet, whereas the other yield function accounts for tensile properties in only three directions. These figures show that Yld2004-18p accurately captures yield stress and r-value directionalities very well.

A sketch of the cup-drawing process that was simulated is shown in Fig. 25. The specific dimensions of the tooling and the process variables used in the simulations for the 2090-T3 sheet sample are as follows:

Punch diameter:	$D_p = 2R_p = 97.46$ mm (3.84 in.)
Punch profile radius:	$r_p = 12.70$ mm (0.50 in.)
Die opening diameter:	$D_d = 2R_d = 101.48$ mm (4.00 in.)
Die-profile radius:	$r_d = 12.70$ mm (0.50 in.)
Blank radius:	$D_b = 2R_b = 158.76$ mm (6.26 in.)
Initial sheet thickness:	1.6 mm (0.06 in.)
Coulomb coefficient of friction:	0.1
Blankholding force:	22.2 kN (2.5 tonf)

The blankholding force was selected to provide a pressure of approximately 1% of the yield stress in order to prevent wrinkling. The friction coefficient was chosen as a typical value for a well-lubricated interface between the sheet and the blankholder. Only a quarter section of the cup with the associated symmetry boundary conditions was analyzed in light of the orthotropic material symmetry. A total of 340 solid elements and 747 nodes with one layer of elements through the sheet thickness were used for the simulation. A user-material subroutine combined with MSC.Marc was employed, based on an implicit time-integration scheme. The Newton-Raphson residual norm of 0.0001 was chosen in the FE simulations for the

convergence criterion. Approximately 2 h were required to complete the FE cup-drawing analysis using an HP 730 computer (Hewlett-Packard Co.).

Figure 26 shows the deformed configurations of the completely drawn cups. The results obtained with Yld96 are included for comparison purposes. It can be seen that Yld96 predicts four ears, while Yld2004-18p predicts six. In Fig. 27, measured and predicted cup-height profiles (also called earing profiles) are compared for the 2090-T3 sheet sample. For an orthotropic material, the cup-height profile between 0 and 90° should be a mirror image of the cup-height profile between 90 and 180° with respect to the 90° axis. However, the measured earing profile slightly deviates from this condition. This deviation may have occurred because the center of the blank was not aligned properly with the centers of the die and the punch during the drawing experiment. Generally, this plot shows that the earing profile obtained from the present theory is in very good agreement with the measured 2profile. In particular, the small ears around 0° (or 360°) and 180° are well predicted.

In summary, this example indicates that an appropriate phenomenological yield function can be used in the FE simulation of the drawing process to predict an earing profile in good agreement with the experimental profile, even for a cup exhibiting more than four ears. In order to obtain this type of accuracy, the anisotropy of the tensile properties must be described very accurately. In the present case, r-values and flow stresses were determined with tensile tests conducted every 15° in the plane of the sheet and were very well captured with the yield function Yld2004-18p.

Analytical Earing Profile. Sometimes, for a quick assessment, it is advantageous to use a simple analytic equation to predict the earing profile. For this purpose, the blank of a cup is viewed as a ring (Fig. 28), the inner edge of which is drawn into the inside cavity under uniform displacement boundary control. When the ring starts to draw in, different levels of compressive strains are generated circumferentially due to planar anisotropy. The corresponding radial strains, contributing to the cup-height profile (earing profile), result from the incompressibility condition under a plane-stress state.

From Fig. 28, it is clear that the behavior of the ring in the rolling direction is controlled by the property of the material in compression in the transverse direction. Generalizing, the behavior of the ring in the direction defined by θ is controlled by the property of the material in compression in the direction defined by $\theta + 90°$. Assuming that, for a given direction, uniaxial tension and compression lead to identical properties, the r-value directionality can provide a reasonable approximation of the strains at the rim of the ring (Ref 180).

Table 5 Yield function coefficients for aluminum alloy 2090-T3 (exponent $a = 8$)

c'_{12}	-0.0699	c''_{12}	0.9812
c'_{13}	0.9364	c''_{13}	0.4767
c'_{21}	0.0791	c''_{21}	0.5753
c'_{23}	1.0031	c''_{23}	0.8668
c'_{31}	0.5247	c''_{31}	1.1450
c'_{32}	1.3632	c''_{32}	-0.0793
c'_{44}	1.0238	c''_{44}	1.0517
c'_{55}	1.0691	c''_{55}	1.1471
c'_{66}	0.9543	c''_{66}	1.4046

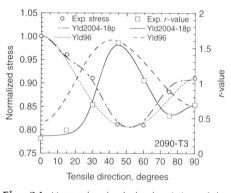

Fig. 24 Measured and calculated variations of the normalized yield stress and r-value as a function of the angle between rolling and tensile direction for aluminum alloy 2090-T3. Source: Ref 195

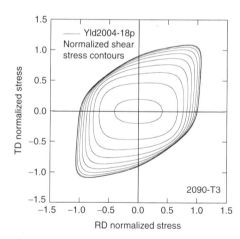

Fig. 23 Yield surface shape for an aluminum alloy 2090-T3 sheet sample. TD, transverse direction; RD, rolling direction. Source: Ref 195

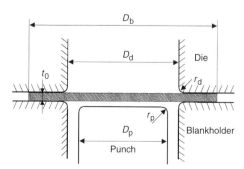

Fig. 25 Schematic illustration of the cup-drawing process. See text for description.

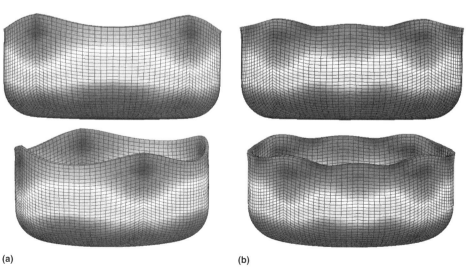

Fig. 26 Finite element predictions of the completely drawn cups for aluminum alloy 2090-T3 using (a) Yld96 and (b) Yld2004-18p yield functions

For the completely drawn cup (Fig. 29), it is assumed that no deformation of the sheet occurs at the flat punch head. Hence, the wall deformation can be mapped with the deformation of the flange, leading to an expression for the total cup height as a function of the parameter θ (Ref 194):

$$H_\theta = r_p + (R_b - R_c) + \frac{r_{\theta+90}}{(r_{\theta+90}+1)}$$
$$\times \left((R_c - R_b) + R_b \ln\left(\frac{R_b}{R_c}\right) \right) \quad \text{(Eq 74)}$$

where R_c is given by $R_c = (R_p + R_d)/2$. The parameter θ describes the angle (in degrees) between the position of the rim considered and the rolling direction. Equation 74 was derived based on a midplane geometry by neglecting the sheet thickness. Using this equation, it is possible to calculate the approximate cup height from the initial blank size, cup geometry and r-value directionality.

The analytical prediction H_θ (Fig. 27) leads to six ears, in agreement with the FE simulation results obtained with Yld2004-18p and the experimental data. It is also worth mentioning that the resulting cup-height profiles exhibit a number of peaks and valleys equal to the number of maxima and minima of the r-value directionality (Fig. 24). More precisely, peaks and valleys as defined by the angle θ in the earing profile correspond to maxima and minima, respectively, in the r-value directionality expressed as a function of $\theta + 90°$. Although additional verifications are needed, these results suggest that, for a given material, the relative earing profile and r-value directionality are "mirror images" of each other with respect to the transverse direction.

Example 3: Finite Element Simulation of Reverse Cup Drawing. Reverse cup drawing, proposed as a benchmark example at the NUMISHEET '99 conference (Ref 196) illustrates the importance of the yield function for accurate FE modeling simulations. In this process, plastic anisotropy, friction, and sheet bending/unbending are all relevant parameters. Simulation results were compared with data available from the conference proceedings (Ref 196).

The workpiece material was an aluminum alloy 6016-T4 sheet sample. The stress-strain curve and yield surfaces were described by the Swift law and the Yld2000-2d yield function (Ref 41), respectively. Because the biaxial yield stress was not available for this material, the assumption $\sigma_b = \sigma_0$ was employed, based on the properties of similar materials. The eight coefficients of Yld2000-2d were obtained with seven input data using the practical assumption $c''_{12} = c''_{21} = 0$ (see the section "Anisotropic Yield Functions" in this article). Tables 6 to 8 list the different data and coefficients used for this material. For this 6016-T4 sheet sample, an excellent agreement was observed between experimental and calculated (Yld2000-2d) variations of the r-value and the flow stress as a function of the angle between rolling and tension direction (Fig. 30). For comparison purposes, the amplitude of the flow stress anisotropy predicted with Hill's 1948 yield function, also represented in Fig. 30, is largely overestimated. This is because with this classical yield function, the anisotropy coefficients are calculated with the r-values only as input, not the flow stresses.

The specific dimensions of the tools are given in the NUMISHEET '99 proceedings (Ref 196). Only a quarter section of the cup was analyzed in light of the orthotropic material symmetry. Figure 31 shows the FE mesh used for the analysis; it comprised 1050 continuum-based shell elements (Ref 161). The process variables were as follows:

Initial sheet thickness:	1.15 mm (0.045 in.)
Coulomb coefficient of friction:	0.1
Gap between blankholder and die:	2.0 mm (0.08 in.)

The cup-drawing simulation was performed assuming isotropic hardening; that is, the yield-function coefficients were kept constant during deformation. It took 120 and 180 steps to translate the punch 60 mm (2.4 in.) in the forward and 80 mm (3.2 in.) in the reverse directions, respectively.

Experimental Validation. Figure 32 shows the deformed cups at intermediate and final stages of the forward and reverse drawing steps. Figure 33 shows that the predicted load-displacement curves agree qualitatively with the experimental curves, although the simulated results underestimate the measured loads. This can probably be attributed to the lack of transverse shear stress in the general thin-shell formulation. Figures 34(a) and (b) show the predicted and experimental thickness-strain distributions measured along the rolling and diagonal (45° from RD) directions, respectively, after the forward and reverse drawing operations. Although the simulated thickness in the flange area in the diagonal direction is slightly

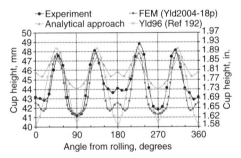

Fig. 27 Comparison of measured earing profiles for deep-drawn cups of aluminum alloy 2090-T3 with predictions from finite element simulations (using two different yield functions) and an analytical model. Source: Ref 195

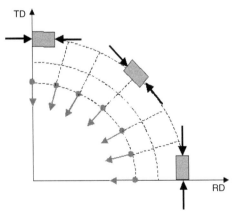

Fig. 28 Illustration of the compressive stresses developing in the elements at the rim of circular blank drawn into a circular cup. TD, transverse direction; RD, rolling direction

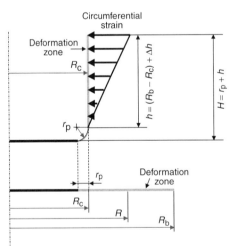

Fig. 29 Initial blank and drawn-cup geometries showing the deformation zone for an analytical solution of the earing problem

Table 6 Swift parameters for aluminum alloy 6016-T4 sheet

Strength coefficient (K)	ε_0	Strain-hardening exponent (n)
385.5	0.082	0.239

Table 7 Normalized flow stresses and r-values for aluminum alloy 6016-T4 sheet

Test	0° uniaxial	45° uniaxial	90° uniaxial	Biaxial
Flow stress	1.000	0.984	0.944	1.000
r-value	0.94	0.39	0.64	n/a

Table 8 Yield function Yld2000-2d coefficients

a	α_1	α_2	α_3	α_4	α_5	α_6	α_7	α_8
8	0.9580	1.0450	0.9485	1.0568	0.9938	0.9397	0.9200	1.1482

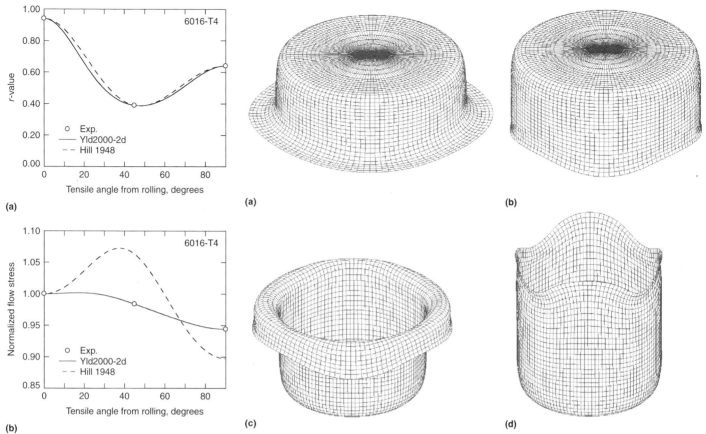

(a)   (b)

(c)   (d)

Fig. 30 Experimental and predicted (Yld2000-2d and r-value-based Hill's 1948) (a) r-value and (b) normalized flow stress as a function of tensile angle from rolling for a 6016-T4 aluminum alloy sheet sample

Fig. 32 Finite element meshes during and after (a) and (b) forward and (c) and (d) reverse drawing operations. Source: Ref 194

Fig. 31 Initial finite element mesh of the circular blank

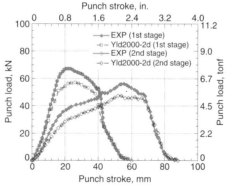

Fig. 33 Measured and FE predicted punch load versus displacement during drawing and redrawing for aluminum alloy 6016-T4 sheet. Source: Ref 194

overestimated, the agreement between predictions and measurements is generally excellent. The overestimation in the diagonal direction is presumably due to the thin-shell formulation without half-thickness consideration of the sheet material.

Nevertheless, this example shows that the material description with Yld2000-2d was able to capture experimental results very well despite the assumption of isotropic hardening, in particular, the different strain profiles in the two directions (rolling and diagonal) investigated. Moreover, separate simulations using the more classical yield function proposed by Hill (Ref 35) could not be completed because of convergence issues. It is believed that these problems were

due to the anisotropy of the sheet, which, as suggested by Fig. 30, was largely overestimated using the more classical yield function. This anisotropy was thought to be responsible for locally excessive blankholder force and element distortion, leading, in turn, to a numerical instability.

Example 4: Modeling of Wrinkling During Cup Drawing. Finite element analysis has also been applied to analyze wrinkling during cup

drawing. For this example, the dimensions of the tooling are as follows:

Punch diameter:	$D_p = 85$ mm (3.3 in.)
Punch-profile radius:	$r_p = 7$ mm (0.28 in.)
Die opening diameter:	$D_d = 88$ mm (3.5 in.)
Die-profile radius:	$r_d = 8$ mm (0.32 in.)

The FE analyses were carried out using the Yld91 (Ref 37) planar anisotropic material model (a particular case of Yld2004-18p). One-half of the circular blank was discretized into 2100 elements (30 elements in the radial direction and 70 elements in the circumferential direction) and 2201 nodes (Fig. 35). The wrinkling behavior was described using the bifurcation algorithm mentioned in the section "Compressive Instability" in this article. The process parameters and material constants used in the analysis are listed in Table 9.

Finite element analyses and experiments were carried out to determine the onset of wrinkling and the number of wrinkles for various values of the blankholding force. Figure 36 shows the measured and computed configurations of the cup after partial drawing. The experiments were carried out three times for each value of the blankholding force. When this force was set to 2 kN (0.22 tonf), 14 wrinkles resulted from the FE analysis, while 9, 11, and 12 wrinkles were

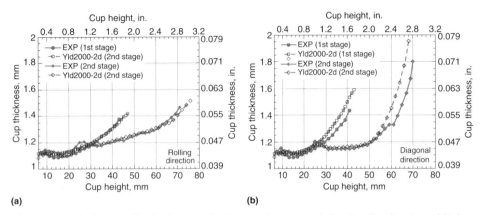

Fig. 34 Cup thickness profiles after drawing/redrawing operations measured along (a) rolling direction and (b) diagonal direction (45° from rolling direction). Source: Ref 194

Table 9 Material constants and friction coefficient for cylindrical cup drawing

Process parameters/material constants	Values
Material	Al 6111-T4
Initial blank thickness	0.9144 mm (0.036 in.)
Coulomb friction coefficient	0.1
Yield stress	185.92 MPa (26.96 ksi)
Young's modulus (E)	69 GPa (10×10^6 psi)
Poisson's ratio	0.33
Stress-strain curve	$\bar{\sigma} = 454.8 - 268.8 \exp(-6.45\bar{\varepsilon})$
r-value	$r_0 = 0.83$, $r_{45} = 0.86$, $r_{90} = 1.42$, $\bar{r} = 0.99$
Anisotropic coefficients for Barlat's yield (Ref 37)	$c_1 = 1.0406$, $c_2 = 0.9582$, $c_3 = 0.9962$, $c_4 = c_5 = c_6 = 0.9071$, $a = 8.0$

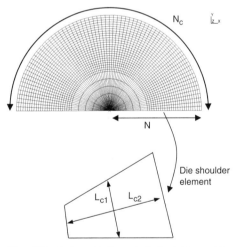

Fig. 35 Mesh used in the analysis of wrinkling and the definition of characteristic lengths

observed experimentally. When the blankholding force was equal to 5 kN (0.56 tonf), 22 wrinkles were predicted, and 17, 18, and 19 were measured. Finally, for a blankholding force of 8 kN (0.90 tonf), the predicted wrinkle number was 32, while the experimental numbers were 24, 27, and 28. The number of wrinkles obtained from the FE analyses were in reasonable agreement with the experimental results. Moreover, the larger number of wrinkles measured on the

cups drawn with a higher blankholding force was well captured by the model.

Figure 37(a) shows the critical punch stroke at which wrinkling grew abruptly, and Fig. 37(b) shows the normalized wrinkle wavelength. In Fig. 37(a), it is shown that the load corresponding to the onset of wrinkling was overestimated by the FE analyses and that the onset and growth of wrinkles was retarded as the blankholding force increased. It is thought that initial imperfections, such as eccentricity of the circular blank with the tool and the nonflatness of the initial blank, may have accelerated the initiation of wrinkling. With the application of grease as a lubricant, a lower force may have led to a larger gap between the blank and blankholder, which could have resulted in imperfections with higher initial amplitudes. Figure 37(b) shows that the normalized wrinkling wavelength decreased (more wrinkles) as the blankholding force increased. This is because, as the blankholding force increased, the lower bifurcation mode was constrained, and wrinkling took place in a higher bifurcation mode. It was also shown in Fig. 37(b) that the number of wrinkles is overestimated (wrinkling wavelength underestimated) by the FE analysis, which is also thought to be due to the effects of imperfections.

Figure 38 shows the normalized critical wrinkling stress for various blankholding forces, which was obtained by averaging the circumferential stress of the edge elements for the

punch stroke at which wrinkling grew abruptly. As can be expected, the critical wrinkling stress increased as the blankholding force increased. Considering the difficulty and sensitivity of all contributing parameters in the modeling of wrinkling, the predictions of the FE analysis are considered to be in very good agreement with the experimental results. This shows that the FE analysis, combined with a bifurcation algorithm, can be used reliably for the analysis of wrinkle initiation and growth in sheet metal forming processes.

Material and Friction Effects. Material properties such as strain-hardening rate and plastic anisotropy can have an effect on the wrinkling behavior of sheet metal. In order to investigate this effect, the initiation and growth of wrinkles in the cylindrical cup-drawing process were computed with the Swift hardening law and Hill's 1948 yield function, assuming planar isotropy (in-plane properties are the same; also called normal anisotropy). Figure 39 shows the simulated critical punch stroke at which wrinkling initiates for various values of the normal anisotropy coefficient, r, and of the strain-hardening exponent, n. In Fig. 39(a), wrinkling initiation occurred at higher punch strokes (loads) as the value of the anisotropy coefficient increased, and it disappeared when this coefficient reached a value of 2.0. Examination of the yield surface shows that the material can flow plastically with a lower pure shear stress (σ_s) in the flange when the anisotropy parameter r is higher. This means that a lower in-plane compressive stress ($\sigma = -\sigma_s$) is needed to deform the material, thus postponing wrinkling to higher loads. Figure 39(b) shows that increasing values of the strain-hardening exponent delayed the onset of wrinkling. Moreover, strain-hardening exponents of 0.25 and 0.3 inhibited the occurrence of wrinkling. This result appears to be reasonable, because a higher strain-hardening exponent leads to a larger elastoplastic modulus. All these simulation results were consistent with the experiments of Narayanasamy and Sowerby (Ref 197), who showed that sheets with higher normal anisotropy and strain-hardening rate resist wrinkling better.

Finite element simulations were also carried out with the Yld91 yield function (Ref 37) in order to assess the influence of planar anisotropy (different in-plane properties) on the wrinkling behavior. The anisotropy coefficients used in this investigation are listed in Table 9. A blankholding force of 3 kN (0.34 tonf) was imposed. Figures 40(a) and (b) show the deformed configurations analyzed with normal anistropy (Hill's 1948) and planar (Yld91) anisotropy, respectively. For normal anisotropy, wrinkling was predicted to occur at a punch stroke of 25 mm (1.0 in.), and the number of wrinkles was 20. For planar anisotropy, however, the compressive instability initiated at a punch stroke of 20 mm (0.8 in.) and produced 18 wrinkles. This effect was explained by the thickness in the flange, which is no longer uniform due to planar anisotropy. This nonuniformity could induce imperfections and modify the susceptibility to wrinkling. However, the comparison between these two simulations is not rigorous because, in addition to the two types of anisotropy (normal and planar), two yield functions were used. The difference in wrinkle number could also be influenced by the generally higher shear yield stress of Hill's 1948 yield function compared to Yld91.

Finite element analyses were performed under constant blankholder force (BHF) with various values of the Coulomb friction coefficient. The corresponding deformed configurations are shown in Fig. 41. When the friction coefficient increased, the number of wrinkles decreased, and when this coefficient reached a value of 0.2, tearing took place instead of wrinkling. Figure 42 shows the number of wrinkles, the critical punch

stroke at which wrinkling grows abruptly, and the edge draw-in for various values of the friction coefficient. It was shown that if wrinkling initiates a higher punch stroke, the number of wrinkles decreases. This is due to the fact that edge draw-in decreases because of the higher friction force, and therefore, lower compressive circumferential stress is induced.

Example 5: Finite Element Simulation of Hemming Operations. Hemming is a process used in the automotive industry to join inner and outer skins of components such as hoods, doors and deck lids. Hemming is a 180° bending operation conducted in three steps: a bending or flanging step, followed by prehemming and final hemming steps. Parameters such as the die radius, the flange radius and length, and the prehemming punch travel influence the final

product quality. Typical Al-Mg-Si 6XXX-T4 sheet processed by conventional means usually produces acceptable relieved flat hem ratings.

In this example, a specific 6XXX-T4 alloy sheet exhibited successful flat hemming when the hem line was aligned with the rolling

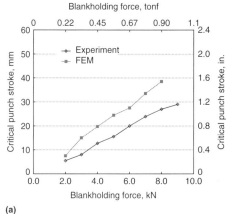

(a)

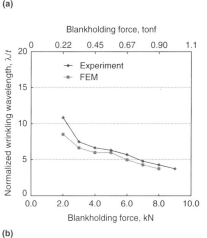

(b)

Fig. 37 Comparison of measurements and finite element (FE) predictions of (a) critical punch stroke and (b) normalized wrinkling wavelength for various levels of blankholding force. Source: Ref 103

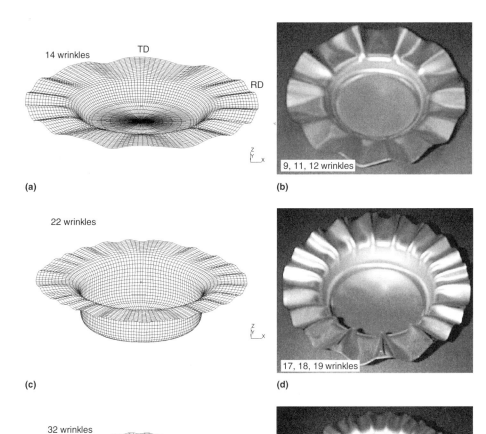

Fig. 36 Finite element predictions (FE) and experimental observations (EXP) of wrinkling in cups drawn with various levels of blankholding force (BHF). (a) FE, BHF = 2 kN (0.22 tonf), stroke = 12 mm (0.5 in.). (b) EXP, BHF = 2 kN (0.22 tonf), stroke = 10 mm (0.4 in.). (c) FE, BHF = 5 kN (0.56 tonf), stroke = 30 mm (1.2 in.). (d) EXP, BHF = 5 kN (0.56 tonf), stroke = 20 mm (0.8 in.). (e) FE, BHF = 8 kN (0.90 tonf), stroke = 40 mm (1.6 in.). (f) EXP, BHF = 8 kN (0.90 tonf), Exp, stroke = 35 mm (1.4 in.). Source: Ref 103

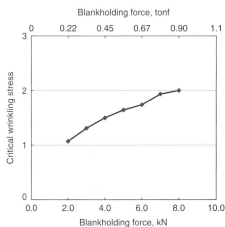

Fig. 38 Finite element predicted normalized critical wrinkling stress for various levels of blankholding force. Source: Ref 103

direction (RD-oriented hem). However, when the hem line was aligned with the transverse direction (TD-oriented hem), unacceptable hem ratings, that is, with some amount of cracking, could be observed. Because of mechanical fibering, it is counterintuitive that the RD-oriented hem leads to a better performance than the TD-oriented hem.

It is known that the stress level affects shear localization and cracking, everything else being constant. Therefore, these analyses pointed to the influence of the stress on fracture during hemming. The stress state that is prominent in hemming corresponds to plane strain, characterized by the major component σ_p. The relevant values of this stress component are associated with the RD and TD directions for the two specific cases of TD- and RD-oriented hemming, respectively. These stresses can be evaluated relative to the RD tensile yield stress, σ_y, using the yield surface concept. The plane-strain yield stress is a function of the anisotropy ratio $H = \sigma_p/\sigma_y$, which depends on the yield-surface shape and consequently on crystallographic texture. Therefore, texture analyses and crystal-plasticity yield-surface calculations were conducted for typical and experimental 6XXX-T4 sheet samples. The results showed that, in general, the yield surface was distorted (anisotropic) and that the RD plane-strain yield stress, σ_p^{RD}, was higher than the TD plane-strain yield stress, σ_p^{TD}, particularly for the experimental material. This may explain why RD-oriented hemming was acceptable (lower σ_p^{TD}) and TD-oriented hemming was more prone to fracture (σ_p^{RD} much higher) for the experimental material. Because it is usually desired to maximize σ_y for product applications, the other control parameter, H, can be modified to reduce the plane-strain yield stress. In this particular example, texture analyses and crystal-plasticity models can be used to define the appropriate texture and provide information for alloy processing. This aspect is not covered here, but the simulation of the hemming process is.

Finite element simulations of the hemming process were carried out for the 6XXX-T4 sheet sample in the RD and TD orientations using the Yld2004-18p yield function to account for plastic anisotropy. The material data for the conventionally processed alloy are shown in Ref 198. Figure 43 summarizes the results. This figure shows that the maximum plastic strain computed in the bent area is approximately the same for all specimens but one, that is, the TD-oriented hem specimen for the experimental alloy. This is in good agreement with observations, because this specimen only exhibits cracks on the hem line. Figure 44 shows a close-up view of one of the simulated hem specimens. It shows that the maximum strain is not located at 90° with respect to the plane of the sheet but in a location that agrees well with experimental observations of the cracking zone.

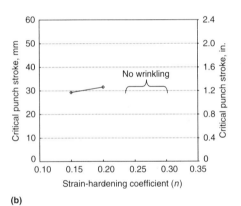

Fig. 39 Finite element (FE) predicted dependence of the critical punch stroke on (a) normal anisotropy (*r*-value) and (b) strain-hardening exponent (*n*-value in Swift law). Source: Ref 103

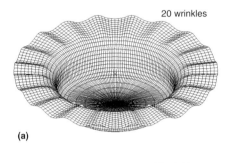

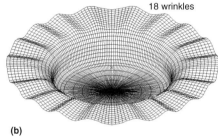

Fig. 40 FE predicted effect of planar anisotropy on wrinkling behavior. (a) Normal anisotropy (Hill's 1948 yield function, stroke = 25 mm, or 1.0 in.). (b) Planar anisotropy (Yld91 yield function, stroke = 20 mm, or 0.8 in.). Source: Ref 103

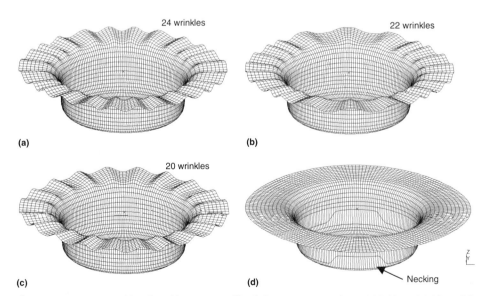

Fig. 41 FE predictions of the effect of friction on wrinkling behavior. (a) μ = 0.0. (b) μ = 0.05. (c) μ = 0.1. (d) μ = 0.2. Source: Ref 103

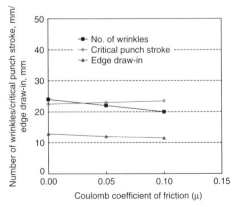

Fig. 42 Effect of Coulomb coefficient of friction on wrinkling behavior. Source: Ref 103

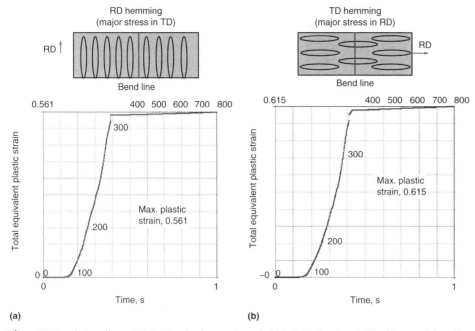

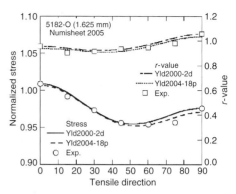

Fig. 46 Normalized flow stress and *r*-value experimentally measured and calculated using Yld2000-2d and Yld2004-18p yield functions as a function of the angle between the rolling and tensile directions for an aluminum alloy 5182-O sheet sample

Fig. 43 Simulation of hemming directionality for experimental Al-Mg-Si T4 alloy sheet. (a) Bend line parallel to rolling direction (RD). Failure resistance is lower but so is driving force (strain); better overall. (b) Bend line perpendicular to rolling direction. Failure resistance is higher, but so is driving force (strain); worse overall

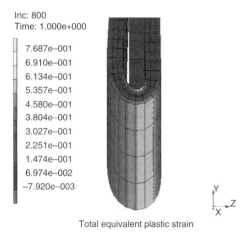

Total equivalent plastic strain

Fig. 44 Close-up view of finite element (FE) predicted plastic strain distribution after flat hemming

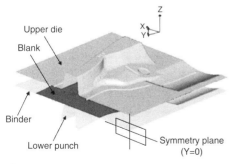

Fig. 45 Forming tool setup for an automotive underbody cross-member panel

Example 6: Finite Element Simulation of Automotive Panel Stamping. Finite element analysis can be applied to the more complex forming problems encountered in industry. The present example deals with the application of FE analysis to the stamping of an automotive component. The final shape of an underbody cross-member panel results from three major steps: forming, trimming/piercing, and springback. In this example, the final shape of the part is particularly dominated by the effect of springback. The lower punch, binder, and upper die used during forming are illustrated in Fig. 45. This tooling geometry had a *x-z* plane of symmetry at *y* = 0. A 5182-O aluminum alloy sheet, 1.6 mm (0.06 in.) thick, was used as a blank material. In order to characterize the material with Yld2000-2d and Yld2004-18p yield functions (see the section "Anisotropic Yield Functions" in this article), tensile tests conducted in different directions from rolling, with angles in 15° steps, as well as bulge and disk-compression tests were performed. Figure 46 shows that the experimental *r*-value and flow-stress directionalities are very well captured Yld2000-2d and Yld2004-18p yield functions.

The process parameters used in the forming simulation were as follows:

Total blankholder force:	400 kN (45 tonf)
Binder travel:	100 mm (4 in.)
Tool moving direction	
Upper die:	Moving in the −*z*-direction
Binder:	Moving in the −*z*-direction

In the forming simulation, the maximum downward velocity of the upper die was limited to 5 m/s (16 ft/s). The initial setup positions and physical dimensions of the draw beads were well described in the benchmark specifications (Ref 198). In the present forming simulation, line beads were used. The initial blank was modeled with coarse meshes (3124 shell elements). During the forming simulation, two-level adaptive mesh refinement was applied, and the final number of elements after forming was 27,610. Figure 47 shows that the plastic strain contours predicted with Yld2000-2d and Yld2004-18p are almost coincident. The reason is that, as suggested in Fig. 46, both yield functions predict the same plastic behavior. The calculations based on Yld2000-2d and Yld2004-18p took approximately 3 and 13 h, respectively.

Figure 48 shows the sections used for the draw-in measurements (d1, d2, etc.). In this simulation, a penalty method was used to enforce the contact conditions. The use of a very large penalty coefficient could cause partial contact of the left lower corner of the part, because of the coarseness of the mesh. This partial contact could result in very inaccurate draw-in. In order to establish the whole contact of the sheet under the binder, an appropriate penalty coefficient should be used. In this work, a finer mesh was used, which drastically reduced, the influence of the penalty parameter. Table 10 shows that the amounts of draw-in predicted by the FE simulation were in good agreement with the measurements, even though the simulation used a line-bead approach.

After forming, trimming and piercing were simulated with the aid of the IGES line data (a given international standard for surface file format). Figure 49 shows the step-by-step procedure. All the state variables were mapped without any changes during the trimming and piercing processes.

Springback. The boundary conditions for the springback analysis of the underbody cross-member panel are shown in Fig. 50. In addition to the *x-z* plane of symmetry, boundary conditions were applied at two points ("A" and "B") in

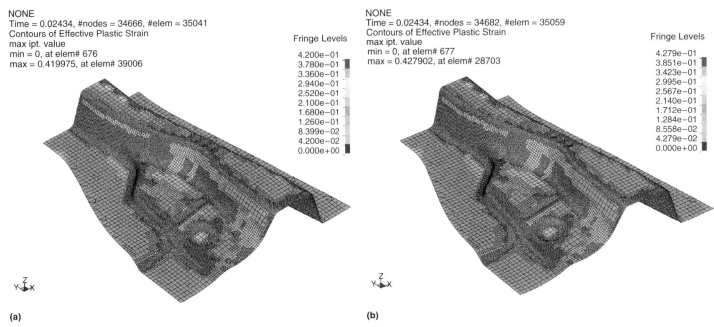

NONE
Time = 0.02434, #nodes = 34666, #elem = 35041
Contours of Effective Plastic Strain
max ipt. value
min = 0, at elem# 676
max = 0.419975, at elem# 39006

Fringe Levels
4.200e−01
3.780e−01
3.360e−01
2.940e−01
2.520e−01
2.100e−01
1.680e−01
1.260e−01
8.399e−02
4.200e−02
0.000e+00

(a)

NONE
Time = 0.02434, #nodes = 34682, #elem = 35059
Contours of Effective Plastic Strain
max ipt. value
min = 0, at elem# 677
max = 0.427902, at elem# 28703

Fringe Levels
4.279e−01
3.851e−01
3.423e−01
2.995e−01
2.567e−01
2.140e−01
1.712e−01
1.284e−01
8.558e−02
4.279e−02
0.000e+00

(b)

Fig. 47 Effective strain contours predicted using (a) Yld2000-2d and (b) Yld2004-18p. The maximum effective strain for each was 0.42

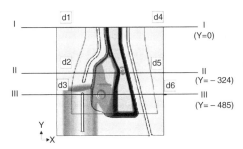

Fig. 48 Measurements of draw-in at sections I, II, and III of the automotive underbody cross-member panel

Table 10 Blank draw-in amount at various locations after forming (Fig. 48)

Measured in mm (in.)

| Method | Draw-in location | | | | | |
	d1	d2	d3	d4	d5	d6
Experiment	62.17 (2.45)	52.8 (2.08)	56.67 (2.23)	73.7 (2.90)	57.2 (2.25)	47.8 (1.88)
Simulation	60.0 (2.36)	50.7 (2.00)	56.9 (2.24)	73.5 (2.90)	57.7 (2.27)	50.5 (1.99)

order to eliminate any rigid body motion. Figures 51 and 52 show the predicted displacement magnitude and the effective stress contour before and after springback, respectively. It can be observed that the effective stress level is higher before than that after springback. This means that springback occurs in a direction that minimizes the residual stresses. The predicted amounts of draw-in and the profiles of specific sections of the panel (Fig. 53: section I, $y = 0$, and section IV, $y = -370$) after springback were compared with experimental results. Figure 54(a) shows that the prescribed profile obtained from the present simulation for section I was in excellent agreement with the experiment. In particular, the curled shape of the right wall was well predicted. Figure 54(b) shows the comparison between experiment and simulation results of section IV. The lifting of the left part and curling of the right wall were very well captured by the simulation and the predicted profile was in excellent quantitative agreement with the experiment.

Example 7: Finite Element Modeling of Tube Hydroforming Processes. Hydroforming has been an important sheet metal forming processes since it was developed before World War II for application to the German aircraft industry (Ref 199). Approximately 30 years ago, hydroformed parts were fabricated by expanding straight or prebent tubes to manufacture manifold elements and components for sanitary use. Hydroforming offers several advantages compared to conventional manufacturing via stamping and welding (Ref 200), in particular, part consolidation and tooling cost. Moreover, while saving on material and manufacturing costs, hydroforming allows the production of parts with improved performances, such as weight, strength, and stiffness, compared to conventionally processed parts in a variety of applications (Ref 201). Today (2006), many hydroformed products, including exhaust manifolds, engine cradles, exhaust pipes, space frames, and so on, are mass-produced in the automotive industry (Ref 202, 203).

Tube hydroforming is a complex process, and it is essential to understand the basic technology to take advantage of the inherent formability of the material. In this process, while the tube is pressurized, axial forces are applied to its ends in order to overcome frictional effects, which cause

premature failure by strain localization. However, if the axial forces are too large, buckling or wrinkling can occur. Consequently, a proper combination of internal pressure and axial compressive force is important to prevent necking (therefore, tearing) as well as buckling. In addition to these process parameters, it is necessary to understand the influence of the tooling geometry, lubrication, and material behavior. Designing die and preform shapes, especially in the early stage, is one of the most difficult tasks. Optimal design is essential to produce final products without defects and to effectively reduce production time.

Recently, FE analyses were applied to the design of the tube hydroforming forming process. Guan (Ref 204), Kridili et al. (Ref 205), and Koc and Altan (Ref 206) simulated axisymmetric tube bulging, while Hwang and Altan (Ref 207) simulated the crushing performance of circular tubes into folds with an overall triangular cross section. Koc et al. (Ref 208) predicted the protrusion height of T-shaped parts using a three-dimensional FE analysis. Because a prebending process often precedes hydroforming, it is important to take into account the deformation during prebending in the simulations. Using FE analyses, some researchers have investigated the cross-sectional

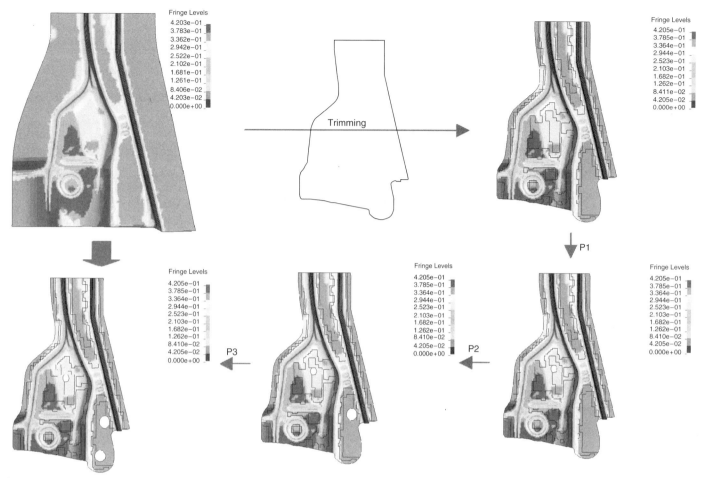

Fig. 49 Trimming and piercing procedure for the automotive underbody cross-member panel

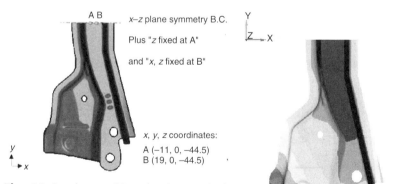

Fig. 50 Boundary conditions for the springback analysis of the automotive underbody cross-member panel

Fig. 51 Predicted displacement magnitude after springback of the automotive underbody cross-member panel

shape change and wall thinning during prebending and integrated the results in subsequent hydroforming simulations (Ref 209–211). Yang et al. (Ref 212) further included springback after the prebending operation. Berg et al. (Ref 213) developed a numerical scheme to overcome problems due to concave cross-sections introduced by local buckling during prebending and preforming processes. For the optimization

of the tube hydroforming in the design stage, several schemes based on FE analyses have been introduced (Ref 214–216). Chung and Yoon (Ref 217) applied a direct FE method (FE design) based on the ideal forming theory (see

the section "Ideal Forming Design Theory" in this article) to design the preform for tube hydroforming.

Preform Design. The use of the ideal forming theory for the optimal design of preforms is illustrated on two industrial hydroformed parts. The examples illustrate how the cross-sectional shape and the tube length can be optimized. Assuming that the final geometry is specified and, as a constraint, the initial thickness of the tube is imposed, an optimal straight, hollow preform (before bending) is determined from this design theory.

In the two examples, the isotropic strain-rate potential Srp93 (Ref 51, 52, as well as the section "Strain-Rate Potentials" in this article) was used to describe the material plasticity, with the exponent set to the recommended value of $b = 4/3$ for aluminum alloys. The stress-strain curve was approximated by the Swift law with the following parameters $\bar{\sigma} = 359.8(0.001 + \bar{\varepsilon})^{0.223}$ MPa.

For the first example, the uniform thickness of the tube preforms was assumed to be equal to 2 mm (0.08 in.). The final part shape after hydroforming is shown in Fig. 55. Only a half-section was considered in the simulation in

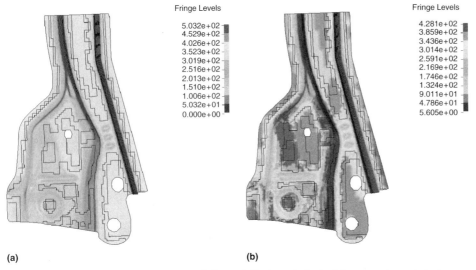

(a)   **(b)**

Fig. 52 Predicted effective stress contour (a) before and (b) after springback of the automotive underbody cross-member panel

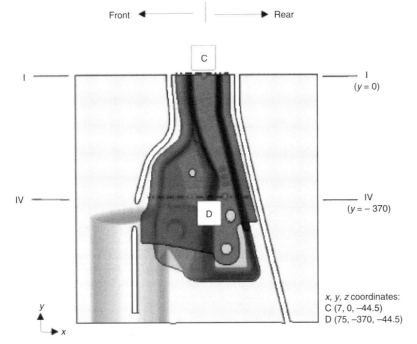

Front ◄——— | ———► Rear

x, y, z coordinates:
C (7, 0, –44.5)
D (75, –370, –44.5)

Fig. 53 Definition of sections I and IV of the automotive underbody cross-member panel

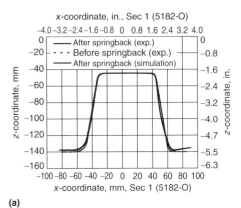

(a)

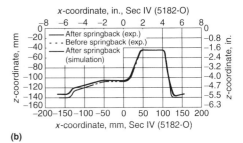

(b)

Fig. 54 Profiles of automotive underbody cross-member panels at (a) section I and (b) section IV for 5182-O sheet sample. Experimental and predicted profiles after springback are almost coincident.

view of the part symmetry. Because the solution is the result of a one-step procedure, it took only 2 min to complete the calculation of the initial preform shape. The ideal forming design code showed that the amount of optimal end feeding was 3.3 mm (0.13 in.), with an average pressure of 60.8 MPa (8.8 ksi), while the experimental feeding and pressure were 3.7 mm (0.15 in.) and 75 MPa (10.9 ksi), respectively, to successfully manufacture the final part. Figure 56 compares the thickness strains predicted with friction coefficients of 0.15 and 0.0 (no friction). This figure shows that when the frictional effect is incorporated, more thinning is observed.

In the second example, the part geometry shown in Fig. 57 was considered. Figure 58 represents the predicted optimal preform tube calculated with the ideal forming code, neglecting the influence of friction. The results showed that the optimal cross section was not circular but close to an ellipse. The predicted thickness strain contours, which are also an output of the model, are plotted on the final part geometry in Fig. 57.

These two examples illustrate that the ideal forming code can provide useful information in the early stage of design, which ultimately reduces the number of FE iterations required to finalize the design of a forming process.

Example 8: Finite Element Simulations in the Design of a Beverage Can End. In the rigid-packaging industry, products with smaller cut-edges and reduced design characteristics have been proposed to decrease overall product costs. However, for beverage can applications, clam-shell failures, as shown in Fig. 59(a), are often observed for many of the newly proposed end designs. Thus, a new geometric design that further increases the buckle pressure without clam-shell failure is an important and critical issue. In this example, finite element (FE) simulations were conducted to predict the resistance to clam-shell buckling and failure in the context of a Taguchi design approach (Ref 1). A plane stress yield function (Yld2000-2d; Ref 41), which accurately describes the behavior of aluminum alloy sheet was employed to model plastic anisotropy (see Fig. 7). The criterion for clam-shell failure was the forming limit curve (FLC) discussed in the section "Tensile Instability" in this article. Once the failure shape was properly predicted, the Taguchi method was used to extract the most important factors from the many possible design variables and to evaluate their influence on the initial buckle pressure. Finally, a new optimum design was proposed to improve the buckle pressure without clam-shell failure.

Buckle Pressure Calculation. In order to predict the buckle pressure, the center displacement

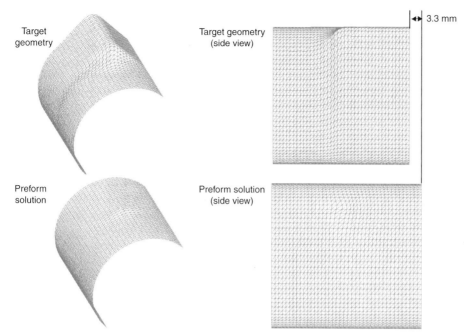

Fig. 55 Final part shape and FE predicted optimal preform shape for hydroforming of aluminum alloy 6061-T4 tube. Source: Ref 217

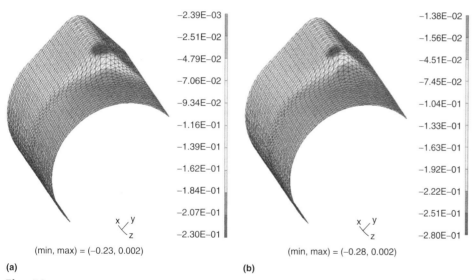

(a) (b)

Fig. 56 FE predicted thickness strain assuming (a) no friction and (b) friction coefficient μ = 0.15. Source: Ref 217

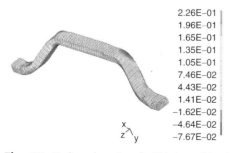

Fig. 57 Final part shape (input for FE design code) and predicted thickness strain contours for hydroforming of aluminum alloy 6061-T4 tubes. Source: Ref 217

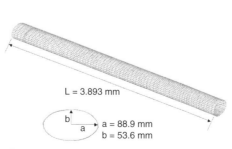

Fig. 58 Predicted optimal preform tube with near-elliptical cross section. Source: Ref 217

of the shell was obtained from FE simulations. Figure 60 shows a schematic illustration of the shell center displacement as a function of pressure at several stages. The displacement at the center increases gradually at the early stage of pressurization. After this transient period (at the end of which the clam-shell initiates), the displacement increases abruptly. The pressure associated to this rapid change is defined as the critical buckle pressure and corresponds to the intersection of the two tangent lines in Fig. 60. Figure 59 compares the observed and simulated deformed shapes of the can end, which are in excellent agreement.

Design Optimization. After confirming the applicability of the FE simulation approach, a number of numerical simulations were conducted instead of experiments to optimize the design parameters in an efficient way. Eight among the twelve geometrical parameters shown in Fig. 61 were selected as variables (denoted A to H in Table 11) to design the best geometry allowing an increase of the buckle pressure. An L_{18} orthogonal array (Ref 1, 218) was constructed to define a set of 18 simulations (Table 11), which were used to evaluate the influence of each of the eight selected design parameters on the buckle pressure (Fig. 62). By this means, Parameter *G* was found to be most critical factor influencing the buckle pressure.

Further work was conducted in investigate in more detail the influence of this parameter and other conditions. Using the FE simulations, it was found that a buckle pressure of $p_{final} = 0.64$ MPa could be obtained by judicious choice of the design geometry. This pressure was 0.07 MPa (11%) higher than the maximum pressure obtained with the original can end design. Moreover, the new design led to reduction of the maximum effective strain from 0.56 to 0.19, preventing the occurrence of clam-shell failure. Figure 63 shows the simulated and experimental deformed shapes based on the new design, which did not lead to clam-shell failure.

Conclusions

The FE method applied to sheet metal forming process analysis and design, particularly for aluminum alloys, was briefly summarized in this article. Special emphasis was paid to the constitutive modeling and its implementation in FE codes. Tensile and compressive instabilities, which limit the formability of materials, as well as springback, which compromises the overall shape stability of a part, were analyzed. A few issues in the FE method, such as kinematics, element and matrix formulation, as well as stress-integration algorithms, were discussed. The ideal forming theory was briefly introduced as a direct method for the design of sheet-forming or hydroforming processes.

In this article, specific constitutive models suitable for aluminum alloy sheets (or thin tube) were presented. These models were always able

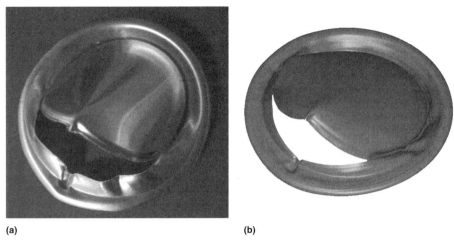

(a)   **(b)**

Fig. 59 Comparison of (a) observed and (b) predicted clam-shell failure for the original beverage-can end design

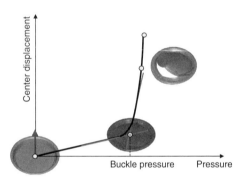

Fig. 60 Shell center displacement versus pressure curve, and the determination of the buckle pressure

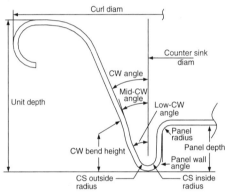

Fig. 61 Schematic illustration of design parameters for beverage-can ends. CW, chuck wall; CS, countersink

tions. In principle, modeling of sheet forming and microstructure evolution should be a concurrent process. However, in view of the size of the forming simulations and the complexity of the physical phenomena occurring during plastic deformation, it seems more efficient to use macroscopic constitutive models with one or more internal variables to account for the microstructure. Constitutive models at a finer scale are, of course, very important for the understanding of the microstructure evolution, for material design, and for providing a basis for the development of more advanced macroscopic models.

ACKNOWLEDGMENT

The authors are indebted to many of their colleagues, in particular from Alcoa Technical Center, Pittsburgh, Pennsylvania, whose contributions made this work possible. The authors also gratefully acknowledge Dr. Lee Semiatin from the US Air Force Research Laboratory, Wright-Patterson Air Force Base, Ohio, for his excellent review of the manuscript and his pertinent suggestions. Finally, the authors are obliged to ASM International personnel, in particular Mr. Steve Lampman, for the difficult task of preparing this article for publication.

REFERENCES

1. G. Taguchi, *Taguchi on Robust Technology Development,* ASME Press, 1993
2. O. Richmond and W.A. Spitzig, Pressure Dependence and Dilatancy of Plastic Flow, *IUTAM Conference, Theoretical and Applied Mechanics, Proc. 15th International Congress of Theoretical and Applied Mechanics,* North-Holland Publishers, Amsterdam, 1980, p 377–386
3. W.A. Spitzig and O. Richmond, The Effect of Pressure on the Flow Stress of Metals, *Acta Metall.,* Vol 32, 1984, p 457–463
4. U.F. Kocks, C.N. Tomé, and H.-R. Wenk, *Texture and Anisotropy,* University Press, Cambridge, 1998
5. F. Perocheau and J.H. Driver, Slip System Rheology of Al-1% Mn Crystals Deformed by Hot Plane Strain Compression, *Int. J. Plast.,* Vol 18, 2002, p 185–202
6. C.C. Bampton, J. Wadsworth, and A.K. Ghosh, Superplastic Aluminum Alloys, *Aluminum Alloys—Contemporary Research and Applications,* A.K. Vasudévan and R.D. Doherty, Ed., Academic Press, 1989
7. J.E. Hatch, Ed., *Aluminum Properties and Physical Metallurgy,* American Society for Metals, 1984
8. D.G. Altenpohl, *Aluminum: Technology, Applications, and Environment,* TMS, 1998
9. P. Perzyna, Ed., *Localization and Fracture Phenomena in Inelastic Solids,* Springer-Verlag, Wien, 1998

Table 11 Orthogonal array for L₁₈ in Taguchi method

Run #	A	B	C	D	E	F	G	H
C1	1	1	1	1	1	1	1	1
C2	1	1	2	2	2	2	2	2
C3	1	1	3	3	3	3	3	3
C4	1	2	1	1	2	2	3	3
C5	1	2	2	2	3	3	1	1
C6	1	2	3	3	1	1	2	2
C7	1	3	1	2	3	1	2	3
C8	1	3	2	3	1	2	3	1
C9	1	3	3	1	2	3	1	2
C10	2	1	1	3	2	3	2	1
C11	2	1	2	1	3	1	3	2
C12	2	1	3	2	1	2	1	3
C13	2	2	1	2	1	3	3	2
C14	2	2	2	3	2	1	1	3
C15	2	2	3	1	3	2	2	1
C16	2	3	1	3	3	2	1	2
C17	2	3	2	1	1	3	2	3
C18	2	3	3	2	2	1	3	1

Note: Numbers refer to the levels of design parameters within design range. 1, lower bound, 2, middle level, 3, upper bound

to capture experimental strain-hardening and plastic anisotropy features favorably. A variety of examples of sheet metal forming process simulations, such as cup-drawing, hemming, stamping, springback, hydroforming, and so on, were given as illustrative examples. The simulation results were generally in excellent agreement with experimental measurements.

This overall theoretical approach illustrates the importance of material and process interac-

Inside CS radius   Lower CW angle   Mid-CW angle

Outside CS radius   Panel wall angle   CW bend height

CW angle   Panel radius

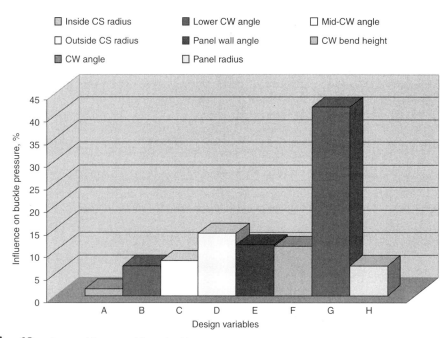

Fig. 62 Influence of design variables on buckle pressure

Fig. 63 Comparison of observed and predicted deformed shapes for the new beverage-can end design

10. P.F. Thomason, *Ductile Fracture in Metals,* Pergamon Press, 1990
11. A.S. Krausz and K. Krausz, Ed., *Unified Constitutive Laws of Plastic Deformation,* Academic Press, 1996
12. Y. Estrin, Dislocation Density-Related Constitutive Modeling, *Unified Constitutive Law of Plastic Deformation,* A.S. Krausz and K. Krausz, Ed., Academic Press, 2002, p 69–106
13. W. Gambin, *Plasticity and Texture,* Kluwer Academic Publishers, Amsterdam, 2001
14. P. Dawson, Crystal Plasticity, *Continuum Scale Simulation of Engineering Materials—Fundamentals—Microstructures—Process Applications,* D. Raabe, F. Roters, F. Barlat, and L.-Q. Chen, Ed., Wiley-VCH Verlag GmbH, Berlin, 2004, p 115–143
15. M. Życzkowski, *Combined Loadings in the Theory of Plasticity,* Polish Scientific Publisher, Warsaw, 1981
16. J.F.W. Bishop and R. Hill, A Theory of the Plastic Distortion of a Polycrystalline Aggregate under Combined Stresses, *Philos. Mag.,* Vol 42, 1951, p 414–427
17. F. Barlat, O. Cazacu, M. Zyczkowski, D. Banabic and J.W. Yoon, Yield Surface Plasticity and Anisotropy, *Continuum Scale Simulation of Engineering Materials—Fundamentals—Microstructures—Process Applications,* D. Raabe, F. Roters, F. Barlat, and L.-Q. Chen, Ed., Wiley-VCH, Verlag GmbH, Berlin, 2004, p 145–183
18. H. Sang and D.J. Lloyd, The Influence of Biaxial Prestrain on the Tensile Properties of Three Aluminum Alloys, *Metall. Trans. A,* Vol 10, 1979, p 1773–1776
19. V.N. Rao and V.V. Laukonis, Microstructural Mechanisms for the Anomalous Behavior of Aluminum Killed Steel Prestrain in Plane Strain Tension, *Mater. Sci. Eng.,* Vol 60, 1983, p 125–135
20. C. Teodosiu, and Z. Hu, Microstructure in the Continuum Modeling of Plastic Anisotropy, *Proceedings of the Risø International Symposium on Material Science: Modelling of Structure and Mechanics of Materials from Microscale to Products,* J.V. Cartensen, T. Leffers, T. Lorentzen, O.B. Pedersen, B.F. Sørensen, and G. Winther, Ed., Risø National Laboratory, Roskilde, Denmark, 1998, p 149–168
21. J.J. Gracio, A.B. Lopes, and E.F. Rauch, Analysis of Plastic Instability in Commercially Pure Aluminium, *J. Mater. Proc. Technol.,* Vol 103, 2000, p 160–164
22. F. Mollica, K.R. Rajagopal, and A.R. Srinivasa, The Inelastic Behavior of Metals Subjected to Loading Reversal, *Int. J. Plast.,* Vol 17, 2001, p 1119–1146
23. F. Barlat, J. Ferreira Duarte, J.J. Gracio, A.B. Lopes, and E.F. Rauch, Plastic Flow for Non-Monotonic Loading Conditions of an Aluminum Alloy Sheet Sample, *Int. J. Plast.,* Vol 19, 2003, p 1215–1244
24. B. Peeters, B. Bacroix, C. Teodosiu, P. Van Houtte, and E. Aernould, Work Hardening/Softening Behaviour of b.c.c. Polycrystals during Changing Strain Path, Part II: TEM Observations of Dislocation Sheets in an IF Steel during Two-Stage Strain Paths and Their Representations in Terms of Dislocation Densities, *Acta Mater.,* Vol 49, 2001, p 1621–1632
25. B. Peeters, M. Seefeldt, C. Teodosiu, S.R. Kalindindi, P. Van Houtte, and E. Aernould, Work Hardening/Softening Behaviour of b.c.c. Polycrystals during Changing Strain Path, Part II: An Integrated Model Based on Substructure and Texture Evolution, and Its Predictions of the Stress-Strain Behaviour of an IF Steel during Two-Stage Strain Paths, *Acta Mater.,* Vol 49, 2001, p 1607–1619
26. P.D. Wu, S.R. MacEwen, D.J. Lloyd, M. Jain, P. Tugcu, and K.W. Neale, On Pre-Straining and the Evolution of Material Anisotropy in Sheet Metal, *Int. J. Plast.,* Vol 21, 2005, p 723–739
27. H.W. Swift, Plastic Instability under Plane Stress, *J. Mech. Phys. Solids.,* Vol 1, 1952, p 1–18
28. J.H. Hollomon, Tensile Deformation, *Trans. ASME,* 1945, p 1–22
29. E. Voce, A Practical Strain-Hardening Function, *Metallurgica,* Vol 51, 1955, p 219–226
30. R.C. Picu, G. Vincze, F. Ozturk, J.J. Gracio, F. Barlat, and A. Maniatty, Strain Rate Sensitivity of the Commercial Aluminum Alloy AA5182-O, *Mater. Sci. Eng.,* Vol 390, 2005, p 334–343
31. A.V. Hershey, The Plasticity of an Isotropic Aggregate of Anisotropic Face Centered Cubic Crystals, *J. Appl. Mech.,* Vol 21, 1954, p 241–249
32. W.F. Hosford, A Generalized Isotropic Yield Criterion, *J. Appl. Mech. (Trans. ASME),* Vol 39, 1972, p 607–609

33. W.F. Hosford, On Yield Loci of Aniso-tropic Cubic Metals, *Proc. Seventh North American Metalworking Conf.,* SME, 1979, p 191–197

34. R.W. Logan and W.F. Hosford, Upper-Bound Anisotropic Yield Locus Calculations Assuming Pencil Glide, *Int. J. Mech. Sci.,* Vol 22, 1980, p 419–430

35. R. Hill, A Theory of the Yielding and Plastic Flow of Anisotropic Metals, *Proc. R. Soc. (London) A,* Vol 193, 1948, p 281–297

36. F. Barlat and J. Lian, Plastic Behavior and Stretchability of Sheet Metals, Part I: Yield Function for Orthotropic Sheets under Plane Stress Conditions, *Int. J. Plast.,* Vol 5, 1989, p 51–66

37. F. Barlat, D.J. Lege, and J.C. Brem, A Six-Component Yield Function for Anisotropic Metals, *Int. J. Plast.,* Vol 7, 1991, p 693–712

38. A.P. Karafillis and M.C. Boyce, A General Anisotropic Yield Criterion Using Bounds and a Transformation Weighting Tensor, *J. Mech. Phys. Solids,* Vol 41, 1993, p 1859–1886

39. O. Cazacu and F. Barlat, Generalization of Drucker's Yield Criterion to Orthotropy, *Mat. Mech. of Solids,* Vol 6, 2001, p 613–630

40. O. Cazacu, B. Plunkett, and F. Barlat, Orthotropic Yield Criterion for Hexagonal Close Packed Metals, *Int. J. Plast.,* Vol 22, 2006, p 1171–1194

41. F. Barlat, J.C. Brem, J.W. Yoon, K. Chung, R.E. Dick, D.J. Lege, F. Pourboghrat, S.H. Choi, and E. Chu, Plane Stress Yield Function for Aluminum Alloy Sheets, *Int. J. Plast.,* Vol 19, 2003, p 1297–1319

42. F. Barlat, H. Aretz, J.W. Yoon, M.E. Karabin, J.C. Brem, and R.E. Dick, Linear Transformation-Based Anisotropic Yield Functions, *Int. J. Plast.,* Vol 21, 2005, p 1009–1039

43. F. Bron and J. Besson, A Yield Function for Anisotropic Materials—Application to Aluminum Alloys, *Int. J. Plast.,* Vol 21, 2004, p 937–963

44. J.W. Yoon, F. Barlat, R.E. Dick, K. Chung, and T.J. Kang, Plane Stress Yield Function for Aluminum Alloy Sheet, Part II: FE Formulation and Its Implementation, *Int. J. Plast.,* Vol 20, 2004, p 495–522

45. H. Ziegler, *An Introduction to Thermodynamics,* North Holland Publishing Company, Amsterdam, 1977

46. R. Hill, Constitutive Dual Potentials in Classical Plasticity, *J. Mech. Phys. Solids,* Vol 35, 1987, p 22–33

47. R. Fortunier, Dual Potentials and Extremum Work Principles in Single Crystal Plasticity, *J. Mech. Phys. Solids,* Vol 37, 1989, p 779–790

48. M. Arminjon and B. Bacroix, On Plastic Potentials for Anisotropic Metals and Their Derivation from the Texture Function, *Acta Mech.,* Vol 88, 1991, p 219–243

49. P. Van Houtte, Application of Plastic Potentials to Strain-Rate Sensitive and Insensitive Materials, *Int. J. Plast.,* Vol 10, 1994, p 719–748

50. A. Van Bael and P. Van Houtte, Convex Fourth and Sixth-Order Plastic Potentials Derived from Crystallographic Texture, *Non Linear Mechanics of Anisotropic Materials, Proceedings of the Sixth European Mechanics of Materials Conference (EMMC6),* S. Cescotto, Ed., Sept 2002 (Liege, Belgium), University of Liege, 2002, p 51–58

51. F. Barlat and K. Chung, Anisotropic Potentials for Plastically Deforming Metals, *Model. Simul. Mater. Sci. Eng.,* Vol 1, 1993, p 403–416

52. F. Barlat, K. Chung, J.W. Yoon, and F. Pourboghrat, Plastic Anisotropy Modeling for Sheet Forming Design Applications, *Constitutive and Damage Modeling of Inelastic Deformation and Phase Transformation,* A.S. Khan, Ed., 1998, Neat Press, Fulton (MD), p 301–304

53. F. Barlat and K. Chung, Anisotropic Strain Rate Potential for Aluminum Alloy Plasticity, *Proc. Eight ESAFORM Conference on Material Forming,* D. Banabic, Ed., April 2005 (Cluj-Napoca, Romania), The Publishing House of the Romanian Academy, Bucharest, 2005, p 415–418

54. B. Paul, Macroscopic Criteria for Plastic Flow and Brittle Fracture, *Fracture,* Vol 1, H. Liebowitz, Ed., Academic Press, 1968, p 313–495

55. S.S. Hecker, Experimental Studies of Yield Phenomena in Biaxially Loaded Metals, *Constitutive Modeling in Viscoplasticity,* A. Stricklin and K.C. Saczalski, Ed., ASME, 1976, p 1–33

56. D. Banabic, O. Cazacu, F. Barlat, D.S. Comsa, S. Wagner, and K. Siegert, Description of Anisotropic Behaviour of AA3103-O Aluminum Alloy Using Two Recent Yield Criteria, *Non Linear Mechanics of Anisotropic Materials, Proceedings of the Sixth European Mechanics of Materials Conference (EMMC6),* S. Cescotto, Ed., Sept 2002 (Liege, Belgium), University of Liege, 2002, p 265–272

57. S.H. Choi, J.C. Brem, F. Barlat, and K.H. Oh, Macroscopic Anisotropy in AA5019A Sheets, *Acta Mater.,* Vol 48, 2000, p 1853–1863

58. R.F. Young, J.E. Bird, and J.L. Duncan, An Automated Hydraulic Bulge Tester, *J. Appl. Metalwork.,* Vol 2, 1981, p 11–18

59. K. Pöhlandt, D. Banabic, and K. Lange, Equi-Biaxial Anisotropy Coefficient Used to Describe the Plastic Behavior of Sheet Metals, *Proceedings of the Fifth ESAFORM Conference on Metal Forming,* M. Pietrzyk, Ed., April 2002 (Krakow, Poland), 2002, p 723–727

60. R.H. Wagoner and N.-M. Wang, An Experimental and Analytical Investigation of In-Plane Deformation of 2036-T4 Aluminum Sheet, *Int. J. Mech. Sci.,* Vol 21, 1979, p 255–264

61. F. Taha, A. Graf, and W.F. Hosford, Plane-Strain Tension Tests on Aluminum Alloy Sheet, *J. Eng. Mater. Technol. (Trans. ASME),* Vol 117, 1995, p 168–171

62. E.F. Rauch and J.-H. Schmitt, Dislocation Substructures in Mild Steel Deformed in Simple Shear, *Mater. Sci. Eng. A,* Vol 113, 1989, p 441–448

63. T. Kuwabara, Y. Morita, Y. Miyashita, and S. Takahashi, Elastic-Plastic Behavior of Sheet Metals Subjected to In-Plane Reverse Loading, *Dynamic Plasticity and Structural Behaviors,* S. Tanimura and A.S. Khan, Ed., Gordon and Breach Publishers, Luxembourg, 1995, p 841–844

64. R.K. Boger, R.H. Wagoner, F. Barlat, M.G. Lee, and K. Chung, Continuous, Large Strain, Tension/Compression Testing of Sheet Material, *Int. J. Plast.,* Vol 21, 2005 p 2319–2343

65. R. Hill, On Discontinuous Plastic States with Special Reference to Localized Necking in Thin Sheets, *J. Mech. Phys. Solids,* Vol 1, 1952, p 19–30

66. Z. Marciniak and K. Kuczynski, Limit Strain in the Process of Stretch Forming Sheet Metal, *Int. J. Mech. Sci.,* Vol 9, 1967, p 609–620

67. J.W. Hutchinson and K.W. Neale, Sheet Necking II: Time Independent Behavior, *Mechanics of Sheet Metal Forming,* D.P. Koistinen and N.M. Wang, Ed., Plenum Press, 1978, p 127–150

68. A. Barata da Rocha and J.M. Jalinier, Plastic Instability of Sheet Metals under Simple and Complex Strain Path, *Trans. Iron Steel Inst. Jpn.,* Vol 24, 1984, p 133–140

69. S. Stören and J.R. Rice, Localized Necking in Thin Sheet, *J. Mech. Phys. Solids.,* Vol 23, 1975, p 421–441

70. R. Sowerby and J.L. Duncan, Failure in Sheet Metal in Biaxial Tension, *Int. J. Mech. Sci.,* Vol 13, 1971, p 217–229

71. J. Lian, F. Barlat, and B. Baudelet, Plastic Behavior and Stretchability of Sheet Metals, Part II: Effect of Yield Surface Shape on Sheet Forming Limit, *Int. J. Plast.,* Vol 5, 1989, p 131–148

72. F. Barlat, Crystallographic Texture, Anisotropic Yield Surfaces and Forming Limits of Sheet Metals, *Mater. Sci. Eng.,* Vol 91, 1987, p 55–72

73. V. Tvergaard, Effect of Kinematic Hardening on Localized Necking in Biaxially Stretched Sheets, *Int. J. Mech. Sci.,* Vol 20, 1978, p 651–658

74. A. Needleman and N. Triantafyllidis, Void Growth and Local Necking in Biaxially Stretched Sheets, *J. Eng. Mech. Technol.,* Vol 100, 1978, p 164–169

75. F. Barlat, A. Barata da Rocha, and J.M. Jalinier, Influence of Damage on the Plastic Instability of Sheet Metal under Complex

Strain Path, *J. Mater. Sci.,* Vol 19, 1984, p 4133–4137

76. F. Barlat and J.M. Jalinier, Formability of Sheet Metals with Heterogeneous Damage, *J. Mater. Sci.,* Vol 20, 1985, p 3385–3399

77. F. Barlat and O. Richmond, Modeling Macroscopic Imperfections for the Prediction of Flow Localization and Fracture, *Fatigue Fract. Eng. Mater. Struct.,* Vol 26, 2003, p 311–321

78. D.J. Lege, F. Barlat, and J.C. Brem, Characterization and Modeling of the Mechanical Behavior and Formability of 2008 Autobody Sheet, *Int. J. Mech. Sci.,* Vol 31, 1989, p 549–563

79. R.J. Asaro and A. Needleman, Texture Development and Strain Hardening in Rate Dependent Polycrystals, *Acta Metall.,* Vol 33, 1985, p 923–953

80. F. Barlat, Forming Limit Diagram – Prediction Based on Some Microstructural Aspects of Materials, *Forming Limit Diagram—Concepts, Methods and Applications,* R.H. Wagoner, K.S. Chan and S.P. Keeler, Ed., TMS, 1989, p 275–301

81. Y. Zhou and K.W. Neale, Predictions of Forming Limit Diagrams Using a Rate-Sensitive Crystal Plasticity Model, *Int. J. Mech. Sci.,* Vol 37, 1995, p 1–20

82. L.S. Toth, J. Hirsch, and P. Van Houtte, On the Role of Texture Development in the Forming Limits of Sheet Metals, *Int. J. Mech. Sci.,* Vol 38, 1996, p 1117–1126

83. P.D. Wu, K.W. Neale, and E. Van der Giessen, Simulation of the Behaviour of fcc Polycrystals during Reversed Torsion, *Int. J. Plast.,* Vol 12, 1996, p 1199–1219

84. P.D. Wu, K.W. Neale, and E. Van der Giessen, On Crystal Plasticity FLD Analysis, *Proc. R. Soc. (London),* Vol 453, 1997, p 1831–1848

85. P.D. Wu, S.R. MacEwen, D.J. Lloyd, and K.W. Neale, A Mesoscopic Approach for Predicting Sheet Metal Formability, *Model. Simul. Mater. Sci. Eng.,* Vol 12, 2004, p 511–527

86. S. Hiwatashi, A. Van Bael, P. Van Houtte, and C. Teodosiu, Prediction of Forming Limit Strains under Strain-Path Changes: Applications of an Anisotropic Model Based on Texture and Dislocation Structure, *Int. J. Plast.,* Vol 14, 1998, p 647–669

87. K. Inal, N.W. Neale, and A. Aboutajeddine, Forming Limit Comparisons for fcc and bcc Sheets, *Int. J. Plast.,* Vol 21, 2005, p 1255–1266

88. T. Kikuma and K. Nakazima, Effects of Deforming Conditions and Mechanical Properties on the Stretch Forming Limits of Steel Sheets, *Trans. Iron Steel Inst. Jpn.,* Vol 11, 1971, p 827–831

89. T. Kobayashi, T. Ishigaki, and T. Abe, *Proc. Seventh Congress International Deep Drawing Research Group,* Vol 8.2 (Amsterdam), 1972

90. A. Graf and W.F. Hosford, Calculations of Forming Limit Diagrams for Changing Strain Paths, *Metall. Trans. A,* Vol 24, 1993, p 2497–2501

91. A. Graf and W.F. Hosford, Effect of Changing Strain Paths on Forming Limit Diagrams of Aluminum 2008-T4, *Metall. Trans. A,* Vol 24, 1993, p 2503–2512

92. H.J. Kleemola and M.T. Pelkkikangas, Effect of Predeformation and Strain Path on the Forming Limits of Steel, Copper, and Brass, *Sheet Met. Ind.,* Vol 63, 1977, p 591–599

93. R. Arrieux, C. Bedrin, and M. Boivin, Determination of an Intrinsic Forming Limit Stress Diagram for Isotropic Sheet Metals, *Proc. Congress International Deep Drawing Research Group,* May 1982 (S. Margherita Ligure, Italy), Working Group Volume, p 61–72

94. T.B. Stoughton, General Forming Limit Criterion for Sheet Metal Forming, *Int. J. Mech. Sci.,* Vol 42, 2000, p 1–27

95. T.B. Stoughton and J.W. Yoon, Sheet Metal Formability Analysis for Anisotropic Materials under Non-Proportional Loading, *Int. J. Mech. Sci.,* Vol 47, 2005, p 1972–2002

96. T. Kuwabara, K. Yoshida, K. Narihara, and S. Takahashi, Forming Limits of Aluminum Alloy Tubes under Axial Load and Internal Pressure, *Proceedings of Plasticity '03,* NEAT Press, 2003, p 388–390

97. T.X. Yu and W. Johnson, The Buckling of Annular Plates in Relation to the Deep Drawing Process, *Int. J. Mech. Sci.,* Vol 24, 1982, p 175–188

98. T.X. Yu, W. Johnson, and W.J. Stronge, Stamping Rectangular Plates into Doubly-Curved Dies, *Proc. Inst. Mech. Eng.,* Vol 198, 1984, p 109–125

99. W.J. Stronge, M.P.F. Sutcliffe, and T.X. Yu, Wrinkling of Elastoplastic Circular Plates during Stamping, *Exp. Mech.,* Vol 26, 1986, p 345–353

100. E. Riks, An Incremental Approach to the Solution of Snapping and Buckling Problems, *Int. J. Solids Struct.,* Vol 15, 1979, 529–551

101. X. Wang and L.H.N. Lee, Postbifurcation Behavior of Wrinkles in Square Metal Sheet under Yoshida Test, *Int. J. Plast.,* Vol 9, 1993, p 1–19

102. J.B. Kim, J.W. Yoon, and D.Y. Yang, Wrinkling Initiation and Growth in Modified Yoshida Buckling Test, *Int. J. Mech. Sci.,* Vol 42, 2000, p 1693–1714

103. J.B. Kim, J.W. Yoon, and D.Y. Yang, Investigation into the Wrinkling Behavior of Thin Sheet on the Cylindrical Cup Deep Drawing Process Using Bifurcation Theory, *Int. J. Num. Methods Eng.,* Vol 56, 2003, p 1673–1705

104. G. Powell and J. Simons, Improved Iteration Strategy for Nonlinear Structures, *Int. J. Numer. Method. Eng.,* Vol 17, 1981, p 1455–1467

105. Y. Tomita and A. Shindo, Onset and Growth of Wrinkling in Thin Square Plate Subjected to Diagonal Tension, *Int. J. Mech. Sci.,* Vol 30, 1988, p 921–931

106. S.L. Chan, A Non-Linear Numerical Method for Accurate Determination of Limit and Bifurcation Points, *Int. J. Num. Method Eng.,* Vol 36, 1993, p 2779–2790

107. A.S. Gendy and A.F. Saleeb, Generalized Mixed Finite Element Model for Pre- and Post-Quasistatic Buckling Response of Thin-Walled Frame Structures, *Int. J. Num. Method. Eng.,* Vol 37, 1994, p 297–322

108. J. Cao and M.C. Boyce, Optimization of Sheet Metal Forming Processes by Instability Analysis, *Proceedings of NUMIFORM '95,* A.A. Balkema, Rotterdam, 1995, p 675–679

109. F.R. Shanley, Inelastic Column Theory, *J. Aeronaut. Sci.,* Vol 14, 1947, p 261–268

110. R. Hill, A General Theory of Uniqueness and Stability in Elastic-Plastic Solids, *J. Mech. Phys. Solids,* Vol 6, 1958, p 236–249

111. J.W. Hutchinson, Post-Bifurcation Behavior in the Plastic Range, *J. Mech. Phys. Solids,* Vol 21, 1973, p 163–190

112. E. Riks, The Application of Newton's Method to the Problem of Elastic Stability, *J. Appl. Mech.,* Vol 39, 1972, p 1060–1066

113. H.S. Lee, D.W. Jung, J.H. Jeong, and S. Im, Finite Element Analysis of Lateral Buckling for Beam Structures, *Comput. Struct.,* Vol 53, 1994, p 1357–1371

114. F. Pourboghrat and E. Chu, Springback in Plane Strain Stretch/Draw Sheet Forming, *Int. J. Mech. Sci.,* Vol 36, 1995, p 327–341

115. F. Pourboghrat, K. Chung, and O. Richmond, A Hybrid Membrane/Shell Method for Rapid Estimation of Springback in Anisotropic Sheet Metals, *J. Appl. Mech. (Trans. ASME),* Vol 65, 1998, p 671–684

116. J.W. Yoon, F. Pourboghrat, K. Chung, and D.Y. Yang, Springback Prediction for Sheet Metal Forming Process Using a 3-D Hybrid Membrane/Shell Method, *Int. J. Mech. Sci.,* Vol 44, 2002, p 2133–2153

117. K.P. Li, W.P. Carden, and R.H. Wagoner, Simulation of Springback, *Int. J. Mech. Sci.,* Vol 44, 2002, p 103–122

118. M.C. Oliveira, J.L. Alves, B.M. Chaparro, and L.F. Menezes, Study on the Influence of the Work Hardening Models Constitutive Parameters Identification in the Springback Prediction, *Proceeding of the Sixth International Conference and Workshop on Numerical Simulation of 3-D Sheet Metal Forming Processes,* L.M. Smith, F. Pourboghrat, J.W. Yoon, and T.B. Stoughton, Ed. (Detroit, MI), 2005, p 253–258

119. D. Zeng and Z.C. Xia, An Anisotropic Hardening Model for Springback Prediction, *Proceedings of the Sixth International Conference and Workshop on Numerical Simulation of 3-D Sheet Metal Forming Processes,* L.M. Smith, F. Pourboghrat,

J.W. Yoon, and T.B. Stoughton, Ed. (Detroit, MI), 2005, p 226–241

120. S.W. Lee and D.Y. Yang, An Assessment of Numerical Parameters Influencing Springback in Explicit Finite Element Analysis of Sheet Forming Processes, *J. Mater. Process. Technol.*, Vol 80–81, 1998, p 60–67

121. K. Li and R.H. Wagoner, Simulation of Springback, *Simulation of Materials Processing*, J. Huetink and F.P.T. Baaijens, Ed., A.A. Bakema, Rotterdam, 1998, p 21–32

122. R.H. Wagoner, Sheet Springback, *Continuum Scale Simulation of Engineering Materials—Fundamentals—Microstructures—Process Applications*, D. Raabe, F. Roters, F. Barlat, L.-Q. Chen, Ed., Wiley-VCH Verlag GmbH, Berlin, 2004, p 777–794

123. L. Geng and R.H. Wagoner, Role of Plastic Anisotropy and Its Evolution on Springback, *Int. J. Mech. Sci.*, Vol 44, 2002, p 123–148

124. R.H. Wagoner, W.D. Carden, W.P. Carden, and D.K. Matlock, Springback after Drawing and Bending of Metal Sheets, *Proc. IPMM '97—Intelligent Processing and Manufacturing of Materials*, T. Chandra, S.R. Leclair, J.A. Meech, B. Verma, and M. Smith, Ed., Balachandram, University of Wollongong (Intelligent Systems Applications), 1997, p 1–10

125. J.F. Wang, R.H. Wagoner, W.P. Carden, D.K. Matlock, and F. Barlat, Creep and Anelasticity in the Springback of Aluminum Alloys, *Int. J. Plast.*, Vol 20, 2004, p 2209–2232

126. W. Prager, A New Method of Analyzing Stresses and Strains in Work-Hardening Plastic Solids, *J. Appl. Mech. (Trans. ASME)*, Vol 23, 1956, p 493–496

127. H. Ziegler, A Modification of Prager's Hardening Rule, *Q. Appl. Math.*, Vol 17, 1959, p 55–65

128. J.L. Chaboche, Time Independent Constitutive Theories for Cyclic Plasticity, *Int. J. Plast.*, Vol 2, 1986, p 149–188

129. Z. Mróz, Non-Associated Flow Laws in Plasticity, *J. Mécan.*, Vol 2, 1963, p 21–42

130. R.D. Krieg, A Practical Two Surface Plasticity Theory, *J. Appl. Mech.*, Vol 42, 1975, p 641–646

131. Y.F. Dafalias and E.P. Popov, Plastic Internal Variables Formalism of Cyclic Plasticity, *J. Appl. Mech.*, Vol 98, 1976, p 645–651

132. K. Chung, M.G. Lee, D. Kim, C. Kim, M.L. Wenner, and F. Barlat, Spring-Back Evaluation of Automotive Sheets Based on Isotropic-Kinematic Hardening Laws and Non-Quadratic Anisotropic Yield Functions, Part I: Theory and Formulation, *Int. J. Plast.*, Vol 21, 2005, p 861–882

133. M.G. Lee, D. Kim, C. Kim, M.L. Wenner, R.H. Wagoner, and K. Chung, Spring-Back Evaluation of Automotive Sheets Based on Isotropic-Kinematic Hardening Laws and Non-Quadratic Anisotropic Yield Function, Part II: Characterization of Material Properties, *Int. J. Plasticity*, Vol 21, 2005, p 883–914

134. M.G. Lee, D. Kim, C. Kim, M.L. Wenner, and K. Chung, Spring-Back Evaluation of Automotive Sheets Based on Isotropic-Kinematic Hardening Laws and Non-Quadratic Anisotropic Yield Functions, Part III: Applications, *Int. J. Plast.*, Vol 21, 2005, p 915–953

135. N.M. Wang and B. Budianski, Analysis of Sheet Metal Stamping by a Finite Element Method, *J. Appl. Mech. (Trans. ASME)*, Vol 45, 1978, p 73–82

136. D.Y. Yang and Y.J. Kim, A Rigid-Plastic Finite Element Calculation for the Analysis of General Deformation of Plastic Anisotropic Sheet Metals and Its Application, *Int. J. Mech. Sci.*, Vol 28, 1986, p 825

137. P. Hora, T.C. Vu, P.M. Wollrab, and J. Reissner, Simulation of the Forming Process for Irregularly Shaped Autobody Panels, *Advanced Technology of Plasticity (Proc. Second ICTP)*, K. Lange, Ed., Springer, Stuttgart, Germany, 1987, p 663

138. D.J. Yoo, I.S. Song, D.Y. Yang, and J.H. Lee, Rigid-Plastic Finite Element Analysis of Sheet Metal Forming Processes Using Continuous Contact Treatment and Membrane Elements Incorporating Bending Effect, *Int. J. Mech. Sci.*, Vol 36, 1994, p 513–546

139. W. Kubli and J. Reissner, Optimization of Sheet Metal Forming Processes Using the Special-Purpose Program AUTOFORM, *Proc. Second International Conference NUMISHEET '93*, A. Machinouch et al., Ed (Isehara, Japan), 1993, p 271–280

140. S. Ahmad, B.M. Irons, and O.C. Zienkiewicz, Analysis of Thick and Thin Shell Structure by Curved Finite Elements, *Int. J. Numer. Methods Eng.*, Vol 2, 1970, p 419–451

141. E. Ramm, A Plate/Shell Element for Large Deflection and Rotations, *Formulation and Computational Algorithms in Finite Element Analysis*, K.J. Bathe, J.T. Oden, and W. Wunderlish, Ed., MIT Press, 1977

142. H. Parish, Nonlinear Analysis of Shells Using Isoparametric Elements, in *ASME Conference*, T.J.R. Hughes et al., Ed., New York, 1981, p 48

143. T.J.R. Hughes and W.K. Liu, Nonlinear Finite Element Analysis of Shells. Part I: Three-Dimensional Shells, *Comput. Methods Appl. Mech. Eng.*, Vol 26, 1981, p 331–362

144. T.J.R Hughes and W.K. Liu, Nonlinear Finite Element Analysis of Shells, Part II: Two-Dimensional Shells, *Comput. Method. Appl. Mech. Eng.*, Vol 27, 1981, p 167–181

145. E.N. Dvorkin and K.J. Bathe, A Continuum Mechanics Based on Four-Node Shell Element for General Non-Linear Analysis, *Eng. Comp. (Int. J. Comput.-Aided Eng. Software)*, Vol 1, 1984, p 77–88

146. W.K. Liu, E.S. Law, D. Lam, and T. Belytschko, Resultant-Stress Degenerated Shell Element, *Comput. Methods Appl. Mech. Eng.*, Vol 55, 1986, p 259–300

147. K.J. Bathe, *Finite Element Procedures in Engineering Analysis*, Prentice-Hall, 1982

148. T.J.R. Hughes, *The Finite Element Method*, Prentice-Hall, 1987

149. G. Stanley, "Continuum-Based Shell Elements," Ph.D. dissertation, Applied Mechanics Division, Stanford University, 1985

150. T. Belytschko, J.I. Lin, and C. Tsay, Explicit Algorithms for the Nonlinear Dynamics of Shells, *Comput. Methods Appl. Mech. Eng.*, Vol 42, 1989, p 225–251

151. J.O. Hallquist, D.J. Benson, and G.L. Goudreau, Implementation of a Modified Hughes-Liu Shell into a Fully Vectorized Explicit Finite Element Code, *Finite Element Methods for Nonlinear Problems*, P. Bergan et al., Ed., Springer, Berlin, 1986, p 283

152. T. Belytschko and I. Leviatan, Physical Stabilization of the 4-Node Shell Element with One Point Quadrature, *Comput. Methods Appl. Mech. Eng.*, Vol 113, 1994, p 321–350

153. R.P.R. Cardoso and J.W. Yoon, One Point Quadrature Shell Elements for Sheet Metal Forming Analysis, *Arch. Comput. Method Eng.*, Vol 12, 2005, p 3–66

154. R.P.R. Cardoso and J.W. Yoon, A One Point Quadrature Shell Element for Through Thickness Deformation, *Comp. Methods Appl. Mech. Eng.*, Vol 194, 2005, p 1161–1199

155. J.L. Batoz, J.M. Roeland, P. Pol, and P. Duroux, A Membrane Bending Finite Element Model for Sheet Forming, *Proc. NUMIFORM '89 Conference*, A.A. Balkema, Rotterdam, 1989, p 389

156. R.J. Alves de Sousa, R.P.R. Cardoso, R.A.F. Valente, J.W. Yoon, J.J. Gracio, and R.M.N. Jorge, A New One-Point Quadrature Enhanced Assumed Strain Solid-Shell Element with Multiple Integration Points along Thickness, Part I: Geometrically Linear Applications, *Int. J. Numer. Methods Eng.*, Vol 62, 2005, p 952–977

157. M.E. Honnor and R.D. Wood, Finite Element Analysis of Axisymmetric Deep Drawing Using a Simple Two-Nodded Mindlin Shell Element, *Numerical Methods for Nonlinear Problems*, C. Taylor et al., Ed., Pineridge Press, 1987, p 440

158. A. Honecker and K. Mattiason, Finite Element Procedures for 3-D Sheet Forming Simulation, *Proc. of the NUMIFORM '89 Conference*, A.A. Balkema, Rotterdam, 1989, p 457–463

159. E. Massoni and J.L. Chenot, 3-D Finite Element Simulation of Deep-Drawing Process, *Proc. NUMIFORM '92 Conference*,

A.A. Balkema, Rotterdam, 1992, p 503–507

160. M. Kawka and A. Makinouchi, Finite Element Simulation of Sheet Metal Forming Processes by Simultaneous Use of Membrane, Shell and Solid Elements, *Proc. NUMIFORM '92 Conference*, A.A. Balkema, Rotterdam, 1992, p 491–495

161. J.W. Yoon, D.Y. Yang, and K. Chung, Elasto-Plastic Finite Element Method Based on Incremental Deformation Theory and Continuum Based Shell Elements for Planar Anisotropic Sheet Materials, *Comput. Methods Appl. Mech. Eng.*, Vol 174, 1999, p 23–56

162. R. Hill, Stability of Rigid-Plastic Solids, *J. Mech. Phys. Solids*, Vol 6, 1957, p 1–8

163. R. Hill, External Paths of Plastic Work and Deformation, *J. Mech. Phys. Solids*, Vol 34, 1986, p 511–523

164. A. Nadai, *Theory of Flow and Fracture of Solids*, Vol 2, McGraw-Hill, 1963, p 96

165. K. Chung and O. Richmond, A Deformation Theory of Plasticity Based on Minimum Work Paths, *Int. J. Plast.*, Vol 9, 1993, p 907–920

166. Y. Germain, K. Chung, and R.H. Wagoner, A Rigid-Viscoplastic Finite Element Program for Sheet Metal Forming Analysis, *Int. J. Mech. Sci.*, Vol 31, 1989, p 1–24

167. K. Chung and O. Richmond, Ideal Forming, Part I: Homogeneous Deformation with Minimum Plastic Work, *Int. J. Mech. Sci.*, Vol 34, 1992, p 575–591

168. K. Chung and O. Richmond, Ideal Forming, Part II: Sheet Forming with Optimum Deformation, *Int. J. Mech. Sci.*, Vol 34, 1992, p 617–633

169. J.W. Yoon, I.S. Song, D.Y. Yang, K. Chung, and F. Barlat, Finite Element Method for Sheet Forming Based on an Anisotropic Strain-Rate Potential and the Convected Coordinate System, *Int. J. Mech. Sci.*, Vol 37, 1995, p 733–752

170. H.J. Braudel, M. Abouaf, and J.L. Chenot, An Implicit and Incremental Formulation for the Solution of Elasto-Plastic Problems by the Finite Element Method, *Comput. Struct.*, Vol 22, 1986, p 801–814

171. J.W. Yoon, D.Y. Yang, K. Chung, and F. Barlat, A General Elasto-Plastic Finite Element Formulation Based on Incremental Deformation Theory for Planar Anisotropy and Its Application to Sheet Metal Forming, *Int. J. Plast.*, Vol 15, 1996, p 35–67

172. M.L. Wilkins, Calculation of Elasto-Plastic Flow, *Methods of Computational Physics*, B. Alder et al., Ed., Academic Press, 1964, p 3

173. M. Ortiz and P.M. Pinsky, "Global Analysis Methods for the Solution of Elastoplastic and Viscoplastic Dynamic Problems," Report UCB/SESM 81/08, Dept. Civil Eng., Univ. of California, Berkley, 1981

174. M. Ortiz, P.M. Pinsky, and R.L. Taylor, Operator Split Methods for the Numerical Solution of the Elastoplastic Dynamic Problem, *Comput. Methods Appl. Mech. Eng.*, Vol 39, 1983, p 137–157

175. M. Ortiz and J.C. Simo, An Analysis of a New Class of Integration Algorithm for Elastoplastic Relations, *Int. J. Numer. Methods Eng.*, Vol 23, 1986, p 353–366

176. M. Dao and R.J. Asaro, Localized Deformation Modes and Non-Schmid Effects in Crystalline Solids, Part I. Critical Conditions of Localization, *Mech. Mater.*, Vol 23, 1996, p 71–102

177. J.W. Yoon, F. Barlat, J.J. Gracio, and E. Rauch, Anisotropic Strain Hardening Behavior in Simple Shear for Cube Textured Aluminum Alloy Sheets, *Int. J. Plast.*, Vol 21, 2005, p 2426–2447

178. K. Chung, J.W. Yoon, and O. Richmond, Ideal Sheet Forming with Frictional Constraints, *Int. J. Plast.*, Vol 16, 2000, p 595–610

179. F. Barlat, K. Chung, and O. Richmond, Anisotropic Plastic Potentials for Polycrystals and Application to the Design of Optimum Blank Shapes in Sheet Forming, *Metall. Mater. Trans. A*, Vol 25, 1994, p 1209–1216

180. K. Chung, S.Y. Lee, F. Barlat, Y.T. Keum, and J.M. Park, Finite Element Simulation of Sheet Forming Based on a Planar Anisotropic Strain-Rate Potential, *Int. J. Plast.*, Vol 12, 1996, p 93–115

181. O. Richmond and K. Chung, Ideal Stretch Forming Processes for Minimum Weight Axisymmetric Shell Structures, *Int. J. Mech. Sci.*, Vol 42, 2000, p 2455–2468

182. O. Richmond and M.L. Devenpeck, A Die Profile for Maximum Efficiency in Strip Drawing, *Proceedings Fourth U.S. National Cong. Appl. Mech.*, 1962, p 1053–1057

183. O. Richmond and H.L. Morrison, Streamlined Wire Drawing Dies of Minimum Length, *J. Mech. Phys. Solids*, Vol 15, 1967, p 195–203

184. R. Hill, Ideal Forming Operations for Perfectly Plastic Solids, *J. Mech. Phys. Solids*, Vol 15, 1967, p 223–227

185. O. Richmond, Theory of Streamlined Dies for Drawing and Extrusion, *Mechanics of Solid State*, F.P.J Rimrott and J. Schwaighofer, Ed., Univ. of Toronto Press, 1968, p 154

186. O. Richmond and S. Alexandrov, Nonsteady Planar Ideal Plastic Flow: General and Special Analytic Solution, *J. Mech. Phys. Solids*, Vol 48, 2000, p 1735–1759

187. K. Chung, W. Lee, T.J. Kang, and J.R. Yoon, Non-Steady Plane-Strain Ideal Forming without Elastic Dead-Zone, *Fibers Polym.*, Vol 3, 2002, p 120–127

188. E.F. Rauch, J.J. Gracio, F. Barlat, A.B. Lopes, and J. Ferreira Duarte, Hardening Behaviour and Structural Evolutions upon Strain Reversal of Aluminum Alloys, *Scr. Mater.*, Vol 46, 2002, p 881–886

189. A.B. Lopes, F. Barlat, J.J. Gracio, J. Ferreira Duarte, and E.F. Rauch, Effect of Texture and Microstructure on Strain Hardening Anisotropy for Aluminum Deformed in Uniaxial Tension and Simple Shear, *Int. J. Plast.*, Vol 19, 2003, p 1–22

190. K. Chung and K. Shah, Finite Element Simulation of Sheet Metal Forming for Planar Anisotropic Metals, *Int. J. Plast.*, Vol 8, 1992, p 453–476

191. A. Andersson, C.A. Ohlsson, K. Mattiason, and B. Persson, Implementation and Evaluation of the Karafillis-Boyce Material Model for Anisotropic Metal Sheets, *Proceedings of NUMISHEET '99*, Vol 1, J.C. Gelin and P. Picart, Ed. (Besanon, France), 1999, p 115–121

192. J.W. Yoon, F. Barlat, K. Chung, F. Pourboghrat, and D.Y. Yang, Earing Predictions Based on Asymmetric Nonquadratic Yield Function, *Int. J. Plast.*, Vol 16, 2000, p 1075–1104

193. F. Barlat, Y. Maeda, K. Chung, M. Yanagawa, J.C. Brem, Y. Hayashida, D.J. Lege, K. Matsui, S.J. Murtha, S. Hattori, R.C. Becker, and S. Makosey, Yield Function Development for Aluminum Alloy Sheet, *J. Mech. Phys. Solids*, Vol 45, 1997, p 1727–1763

194. J.W. Yoon, F. Barlat, R.E. Dick, and M.E. Karabin, Prediction of Six or Eight Ears in a Drawn Cup Based on a New Anisotropic Yield Function, *Int. J. Plast.*, Vol 22, 2006, p 174–193

195. F. Barlat, H. Aretz, J.W. Yoon, M.E. Karabin, J.C. Brem, and R.E. Dick, Linear Transformation-Based Anisotropic Yield Functions, *Int. J. Plast.*, Vol 21, 2005, p 1009–1039

196. J.C. Gelin and P. Picart, *Proceedings of NUMISHEET '99*, Vol 2, J.C. Gelin and P. Picart, Ed., Sept 13–17, 1999 (Besangon, France)

197. R. Narayanasamy and R. Sowerby, Wrinkling of Sheet Metals When Drawing Through a Conical Die, *J. Mater. Process. Technol.*, Vol 41, 1994, p 275–290

198. J. Cao, M.F. Shi, T.B. Stoughton, C.T. Wang, and L. Zhang, Ed., NUMISHEET 2005 Benchmark Study, *Proceedings of the Sixth International Conference and Workshop on Numerical Simulation of 3-D Sheet Metal Forming Processes:* Part B, Aug 15–19, 2005 (Detroit, MI), 2005, p 1179–1190

199. S.H. Zhang, Developments in Hydroforming, *J. Mater. Process. Technol.*, Vol 91, 1999, p 236–244

200. M. Ahmetolgu and T. Altan, Tube Hydroforming: State-of-the-Art and Future Trends, *J. Mater. Process. Technol.*, Vol 98, 2000, p 25–33

201. Hydroforming Technology, *Adv. Mater. Process.*, Vol 151, 1997, p 50–53

202. S. Nakamura, H. Sugiura, H. Onoe, and K. Ikemoto, Hydromechanical Drawing of

Automotive Parts, *J. Mater. Process. Technol.,* Vol 46, 1994, p 491–503

203. F. Dohmann and C. Hartl, Hydroforming— A Method to Manufacture Light-Weight Parts, *J. Mater. Process. Technol.,* Vol 60, 1996, p 669–676

204. Y. Guan, "A Cross Section Analysis Finite Element Code for Tube Hydroforming," Master of Science Thesis, Michigan State University, 2000

205. G.T. Kridili, L. Bao, and P.K. Mallick, Two-Dimensional Plane Strain Modeling of Tube Hydroforming, *Proceedings of the ASME, Manufacturing in Engineering Division,* MED-Vol 11, 2000, p 629–634

206. M. Koc and T. Altan, Application of Two-Dimensional (2-D) FEA for the Tube Hydroforming Process, *Int. J. Mach. Tools Manuf.,* Vol 42, 2002, p 1285–1295

207. Y.-M. Hwang and T. Altan, FE Simulations of the Crushing of Circular Tubes into Triangular Cross-Sections, *J. Mater. Process. Technol.,* Vol 125–126, 2002, p 833–838

208. M. Koc, T. Allen, S. Jiratheranat, and T. Altan, The Use of FEA and Design of Experiments to Establish Design Guide-lines for Simple Hydroforming Parts, *Int. J. Mech. Sci.,* Vol 40, 2000, p 2249–2266

209. L. Wu and Y. Yu, Computer Simulations of Forming Automotive Structural Parts by Hydroforming Process, *Proceedings of NUMISHEET '96,* J.K. Lee, G.L. Kinzel, and R.H. Wagoner, Ed., 1996, p 324–329

210. N. Dwyer, M. Worswick, J. Gholipour, C. Xia, and G. Khodayari, Pre-Bending and Subsequent Hydroforming of Tube: Simulation and Experiment, *Proceedings of NUMISHEET 2002,* D.-Y. Yang, S.I. Oh, H. Huh, and Y.H. Kim, Ed., 2002, p 447–452

211. K. Trana, Finite Element Simulation of the Tube Hydroforming Process—Bending, Preforming and Hydroforming, *J. Mater. Process. Technol.,* Vol 127, 2002, p 410–408

212. J.-B. Yang, B.-H. Jeon, and S.-I. Oh, The Tube Bending Technology of a Hydroforming Process for an Automotive Part, *J. Mater. Process. Technol.,* Vol 111, 2001, p 175–181

213. H.J. Berg, P. Hora, and J. Reissner, Simulation of Complex Hydroforming Process Using an Explicit Code with a New Shell Formulation, *Proceedings of NUMISHEET '96,* J.K. Lee, G.L. Kinzel, and R.H. Wagoner, Ed., 1996, p 330–335

214. O. Ghouati, M. Baida, H. Lenoir, and J.C. Gelin, 3-D Numerical Simulation of Tube Hydroforming, *Proceedings of Sixth ICTP, Advanced Technology of Plasticity 1999,* M. Geiger, Ed., Springer, Berlin, 1999, p 1211–1216

215. J.-B. Yang, B.-H. Jeon, and S.-I. Oh, Design Sensitivity Analysis and Optimization of the Hydroforming Process, *J. Mater. Process. Technol.,* Vol 113, 2001, p 666–672

216. J. Kim, B.S. Kang, and S.M. Hwang, Preform Design in Hydroforming by the Three-Dimensional Backward Tracing Scheme of the FEM, *J. Mater. Process. Technol.,* Vol 130–131, 2002, p 100–106

217. K. Chung and J.W. Yoon, Direct Design Method Based on Ideal Forming Theory for Hydroforming and Flanging Processes, *Proc. NUMISHEET 2005,* L.M. Lorenzo, F. Pourboghrat, J.W. Yoon, and T.B. Stoughton, Ed., Aug 15–19, 2005 (Detroit, MI), p 40–45

218. M. Phadke, Quality Engineering Using Robust Design, Prentice Hall, 1989

Statistical Analysis of Forming Processes

Stuart Keeler, Keeler Technologies LLC

STATISTICS are extremely important tools in the operation of press shops, providing numerical process analysis capabilities that far exceed the more traditional recording of simple breakage rates. During the past decades, the prime goal of metal forming was to make stampings to part print without breakage. Today, new requirements on product consistency have greatly increased the need for process analysis capabilities.

The most common use of statistics in the press shop is statistical process control (SPC). Although used in many formats, SPC is simply the use of statistical techniques such as control charts to analyze a process or its output and thus enable appropriate actions to be taken to achieve and maintain a state of statistical control. The use of SPC instead of traditional quality-control methods, such as inspection/sorting, is beneficial in a number of ways.

Statistical process control:

- Decreases scrap, rework, and inspection costs by controlling the process
- Decreases operating costs by optimizing the frequency of tool adjustments and tool changes
- Maximizes productivity by identifying and eliminating the causes of out-of-control conditions
- Allows the establishment of a predictable and consistent level of quality
- Eliminates or reduces the need for receiving inspection by the purchaser because it produces a more reliable, trouble-free product, resulting in increased customer satisfaction

The use of statistics in analysis of the forming process goes well beyond SPC. The whole area of design of experiments (DOE) is becoming important as the interactions within the forming process are detailed and studied. Many of the more difficult forming problems result from the interaction of two or more interactive variables. Identification of these interactive variables will not be accomplished by the traditional changing of one variable at a time. The argument has been made that conducting DOE tests in the press shop is nearly impossible because of the time and difficulty for sequential grinding and welding

of the tooling, securing metals with different properties, and implementing other process changes dictated by the DOE matrix. The rapid introduction of the computerized forming simulation into the press shop since early 2000 has made DOE studies an easy task. Extremely difficult changes for the press shop are only a few simple keystrokes on the computer.

This article discusses the role of statistics in sheet metal forming operations, both in terms of SPC techniques and design of experiments. However, most of the concepts are generic and can be applied to any metalworking process.

The Forming Process

A wide range of forming processes is currently in use. Although the details of these processes differ significantly, most forming processes share certain characteristics. Each forming process, however simple, can be viewed as a system. For sheet metal forming, one common breakdown of the forming system consists of the following inputs:

- Product design
- Tooling
- Material
- Lubricant
- Press
- Operator

Each of these six major inputs can be further broken down into a multiplicity of subinputs.

The inputs of forming systems are highly interactive. Not only are the changes within a single input difficult to trace and understand, but the interaction of the inputs makes the task even more complex. Small changes made in one or more inputs of the system can cause significant changes in the output of the system (the finished part). These synergistic changes may not be predictable or even possible to anticipate.

The product design usually is not considered by press shops to be an input affecting the forming system. However, some characteristics of the product design are the root cause of the variability of both the output of the forming system and the dimensional stability of the

finished product. An example is stretching a character line with a radius-to-thickness ratio less than four. Incorporating this severe character line into the design of the tooling can produce a range of strain distributions from relatively flat to a severe gradient (Fig. 1a). In a pure stretch operation, the area under the curve is a measure of the traditional increase in length of line. Both curves have the same area under the curve with equal increase in length of line and therefore have identical formed shapes.

The solid line represents a relatively stable condition. Any change in the inputs will be evenly distributed among all elements of the stamping and will result in a small increase or decrease in the level of straining. A flat strain distribution is said to desensitize the stamping to variations in system inputs.

The severe gradient (dashed line in Fig. 1a) creates the opposite effect and is very sensitive to changes in system inputs. Because a majority of the deformation is concentrated in a small segment of the gradient, any changes in the system inputs also are concentrated in the same small segment of the gradient and therefore are magnified. The change in the gradient from maximum to minimum values (Fig. 1b) has a major impact on the stamping. The change in flow stress due to work hardening varies the amount of elastic stress, resulting in differences in springback. Even worse, the increase in localized strain creates a localized thinning. This

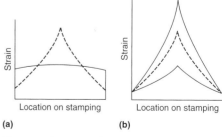

Fig. 1 Schematic (a) showing equal increase in lengths of line (average strain) created by a sharp gradient (dashed line) and a relatively flat gradient (solid line). Schematic (b) showing how a sharp strain gradient (dashed curve) concentrating deformation in a narrow band can vary radically with small changes in incoming system inputs.

local thinning acts like a hinge that allows the change in elastic stresses (a system interaction) to maximize the dimensional variation of the stamping.

Ideally, the forming system should be a continuous process. For many processes, the system appears to be continuous. The tools are inserted into the designated press and remain there for the production life cycle of the part, which often spans several years. Changes made during the production life cycle may in fact create major disruptions in the continuous nature of the process. These changes can include engineering modifications, tooling replacement, routine maintenance, process improvements, and other seemingly minor modifications to the process.

Many forming processes are conducted on a batch basis. Tooling is inserted into a press, a specific number of pieces are made, and the tooling is removed from the press. This type of process is often considered to be an interrupted or segmented form of a continuous process. In reality, however, a new forming system is created each time the tooling is inserted into the press. This often is evidenced by the extended period of trial-and-error adjustments necessary to the tooling before the production of satisfactory parts can begin (Ref 1).

The Statistical Approach

Statistical process control programs are becoming commonplace in industry. An entire science has been developed to deal with the problem of defining, analyzing, correcting, and controlling production processes. References 2 to 5 are several early references. Today (2005), many textbooks or reference books on quality control, process analysis, or process troubleshooting will detail numerous methods of statistical analysis. By providing data on the capabilities and output of a process, these statistical methods provide a rational and analytical, rather than an emotional and guessing, basis for problem solving and decision-making. Other benefits of SPC have been listed in the introduction to this article.

As a system, the forming process is amenable to system analysis and the techniques of system control. Various statistical techniques, many of which have their origins in quality-control practices, can be applied to the forming process. In this article, these statistical techniques are divided into two broad categories: historical tracking and experimental design.

Historical Tracking

Historical tracking is a process of measuring, recording, and analyzing one or more specific characteristics of a process. This record then becomes the basis on which the current state of the process can be assessed and the future state of the process predicted. Numerous stat-

istical techniques are available for historical tracking, ranging from very simple evaluation of a few pieces of data to more extensive mathematical analyses, such as control charts and statistical deformation control.

Two keys are applicable to any analysis technique. The first key is having a data acquisition system or measurement device with a measurement error approximately an order of magnitude less than the process variation being measured. Much has been written about calibration and repeatability of instruments, and again, a whole science has evolved around these procedures. Yet, passing calibration and other tests may be insufficient if the measurement error of the instrument/operator combination approaches that of the production variation of the process being measured. Assume an ultrasonic thickness gage for measuring thickness strain in a sheet metal stamping has a measurement error of ± 0.0001. Likewise, the process producing the stamping has a variation of ± 0.0001. Are the measured changes due to process variation or measurement error? In this case, no conclusions can be made relative to the source of process variation.

The second key is to analyze data at the most fundamental level. Averaging data loses information. A simple example is to average the steamy temperatures of summer with the icy temperatures of winter and conclude that, on the average, the weather in your city is comfortable.

Simple Analyses

Careful, logical acquisition of data is the heart of any statistical analysis program. The simpler the data system, the better the analysis. The following examples illustrate how averaging data loses information or how important additional information may be overlooked when press shop data are recorded. Sometimes, causes of deformation are detected by accident. More important are statistical records of when problems occur and subsequent analysis of these records. When does the problem occur? Only on rainy days? Only in the summer? Only when edge mults from the master coil of steel are being run? These are only a few additional examples of what good observation and data recording can detect.

Example 1: Breakage in One of Two Symmetrical Stampings. A complaint from a manufacturer of symmetrical stampings requested a better-quality metal because a 10% breakage was experienced during the previous run. However, the raw data from the press shop floor indicated that all breakage occurred in the left-hand stampings and none in the right-hand stampings. Because both hands were formed simultaneously in a single die, the raw data changed the focus of the investigation from the metal to the tooling and the press. Either could have caused the left-hand stamping to be more severe than the right hand. One simple test,

rotating the die 180° in the press, showed the problem was in the tooling. The high breakage rate stayed with the left-hand stamping.

Example 2: Production Run with 4% Breakage. An accurate count of each stamping that broke was carefully generated and then converted into a 4% breakage for the entire run. The breakage rate was an accurate number, but much information about the root cause of the breakage was lost because the time each broken stamping was made was not recorded. Had the breakage times been recorded and summarized, then different probable root causes could have been assigned, as shown in Table 1. The time of breakage data allowed the corrective action to be focused on one of six probable causes instead of beginning with no focus.

Example 3: Process Average Hides Process Variation. A critical part dimension was measured on stampings produced on two different press lines. All parts were reported to be within specification, and the average dimensions of parts from both presses were identical. However, if the data had been plotted on a frequency graph, a very different analysis would have emerged (Fig. 2). The graphs show all parts are within specification limits, and the average or mean values for both presses are identical. However, the dimensional distribution is very different. Press line 5 is producing more consistent parts compared to press line 2 and therefore is the preferred line for that part. This preference for press line 5, however, cannot be extrapolated to any other parts. New data are

Table 1 Data analysis based on time of stamping breakage

Time of breakage	Probable root cause
All at die transition	Insufficient die transition settings
All on one third shift	Substitute operator
All from one coil of steel	Incorrect coil of steel
All during meal breaks and shift changes	Die temperature cooled
All after die maintenance tweaked the die	Draw bead was rewelded to print
Ones and twos throughout run	Process at edge of forming limit

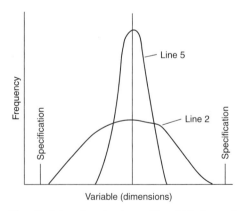

Fig. 2 Schematic showing wide distribution of dimensions for parts made on line 2, but tighter distribution for line 5

required for each part because the interactions of all the system inputs will be different.

Control Charting

The heart of many SPC systems is the control chart. The control chart is a method of monitoring process output through the measurement of a selected characteristic and the analysis of its performance over time. Because the output from one process usually is the input to the next process, the physical location of the measurement can be at either end of the transfer link. For example, the blanks from a blanking press become the input to the next operation, which usually is the forming press.

A control chart can be a very powerful statistical tool. Information from the control chart can be used to inform the operator when to adjust the process and, perhaps more important, when not to adjust the process. This allows the operator to control the process based on statistically valid

numbers instead of trial and error or, even worse, emotion.

For example, the operator is given a machine capable of producing the required parts and is then given the means to measure the characteristics of output (the finished parts) in real time. Thus, the operator knows the quality of the part coming off the machine on a real-time basis. In addition, the operator has control limits for the process. These control limits tell the operator when to adjust the process and when not to adjust the process, which permits the operator to prevent rather than detect defects. Such a system for the control of steel sheet thickness is described in the following example.

Example 4: Constructing a Control Chart for Steel Thickness Measurements. Uniformity of blank thickness is often stated to be important for high productivity in sheet metal forming operations, although published information on this subject is limited (Ref 6, 7). Once the tooling is set for a specific sheet thickness, adjustments to the tooling cannot be readily

made. A control chart can be constructed to statistically monitor variations of the incoming steel.

For each coil of steel received, five blanks are sampled at specific locations along the length of the coil, such as the head, quarter, center, quarter, and tail locations. Examples of such measurements are given in Fig. 3.

The values of $\overline{X}$ (the average thickness) and R (the range of thicknesses) are plotted on the respective graphs. After sufficient measurements are made, the $\overline{\overline{X}}$-value and the upper and lower control limits (UCL and LCL) can be calculated and plotted on the same diagram.

The control chart shown in Fig. 3 indicates that the thickness of the steel being processed is stable. A process is said to be in a state of statistical control if it is operating without special causes of variation (Ref 5). It will exhibit random variation within calculated control limits and will have predictability. Measurements taken in the future are expected to be within the same control limits. However, no prediction can

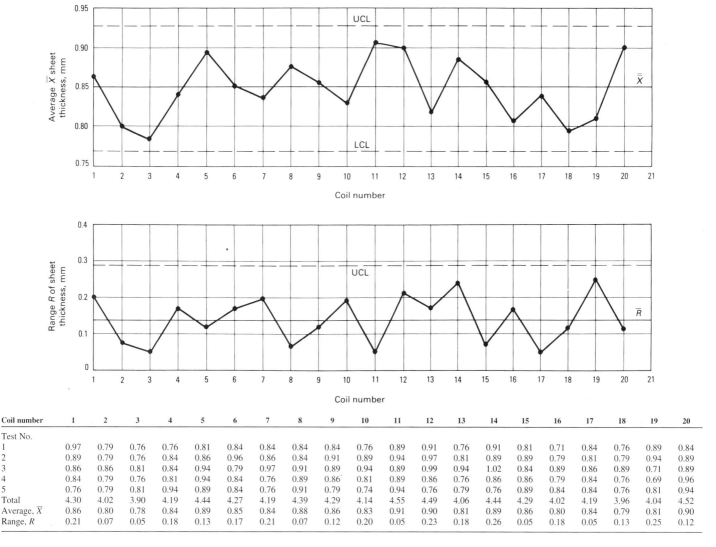

Coil number	1	2	3	4	5	6	7	8	9	10	11	12	13	14	15	16	17	18	19	20
Test No.																				
1	0.97	0.79	0.76	0.76	0.81	0.84	0.84	0.84	0.84	0.76	0.89	0.91	0.76	0.91	0.81	0.71	0.84	0.76	0.89	0.84
2	0.89	0.79	0.76	0.84	0.86	0.96	0.86	0.84	0.91	0.89	0.94	0.97	0.81	0.89	0.89	0.79	0.81	0.79	0.94	0.89
3	0.86	0.86	0.81	0.84	0.94	0.79	0.97	0.91	0.89	0.94	0.89	0.99	0.94	1.02	0.84	0.89	0.86	0.89	0.71	0.89
4	0.84	0.79	0.76	0.81	0.94	0.84	0.76	0.89	0.86	0.81	0.89	0.86	0.76	0.86	0.86	0.79	0.84	0.76	0.69	0.96
5	0.76	0.79	0.81	0.94	0.89	0.84	0.76	0.91	0.79	0.74	0.94	0.76	0.79	0.76	0.89	0.84	0.84	0.76	0.81	0.94
Total	4.30	4.02	3.90	4.19	4.44	4.27	4.19	4.39	4.29	4.14	4.55	4.49	4.06	4.44	4.29	4.02	4.19	3.96	4.04	4.52
Average, $\overline{X}$	0.86	0.80	0.78	0.84	0.89	0.85	0.84	0.88	0.86	0.83	0.91	0.90	0.81	0.89	0.86	0.80	0.84	0.79	0.81	0.90
Range, R	0.21	0.07	0.05	0.18	0.13	0.17	0.21	0.07	0.12	0.20	0.05	0.23	0.18	0.26	0.05	0.18	0.05	0.13	0.25	0.12

Fig. 3 Raw data (in millimeters; 1 in. = 25.4 mm) obtained from steel thickness measurements and the resulting $\overline{X}$-R control charts. Upper and lower control limits (UCL, LCL) can be calculated and the mean determined when sufficient data have been collected.

be made about where any individual measurement will lie within these control limits.

Unfortunately, the control limits shown in Fig. 3 are rather wide and would require frequent tooling adjustments. This means that a normal distribution of thicknesses within the control chart would be expected. However, this expected range is too large for the tooling and would necessitate tooling adjustments to accommodate changes in thickness between the extremes of the range. Therefore, it is necessary to work with the steel supplier to reduce the range of the control limits to an engineering specification window acceptable to the tooling.

Control limits are important in SPC because they define the amount of variation in a process due solely to chance causes. A process operating within these limits (Fig. 4a) is considered to be stable. A process operating outside these limits (Fig. 4b) is unstable.

The control chart represents the best that the operator of a given process can do with the process as it exists. If this is unacceptable, then the basic process must be changed. Engineering and management, not the operator, are required to implement these process changes because this may require new equipment, new tools, new processing sequences, additional stages, or even a new part design. Alternatively, the control limits of many processes can be improved by simple preventive maintenance, replacement of worn components, and bringing the tools to original performance specifications.

If the process is operating within its control limits (statistical control), then the capability of the system to meet specifications can be assessed. If the system is out of control, then changes to the process must be made to make the process stable before the capabilities can be assessed.

Control charts are most effective when monitored over an extended period of time. Then, changes in the process can be detected, often before rejects begin to occur. Two general types of changes in the process may be encountered. One is the central sample tendency, $\overline{X}$, which is monitored to detect a shift in the location of the distribution (Fig. 5). A location change would imply that the normal distribution remains the same but that the average is displaced. An example in forming may be the angle of a bend. The variation from bend to bend would remain acceptable, but the average angle of all bends would have changed. Some factors that may affect the $\overline{\overline{X}}$-value are gradual deterioration of the equipment, worker fatigue, accumulation of waste products (such as excess lubricant or flaked metallic coatings), and environment.

Another type of process change is a change in the sample range or the variation from one sample to another (Fig. 6). This range change is observed as a change in the width of the normal curve for a given area under the curve. This means increased variability. For the bending example shown in Fig. 5, the average bend angle would remain the same, but large differences would be detected among samples from the same lot or even consecutive samples. Some factors that may affect the R-chart are changes in operator skill, worker fatigue, change in the mix of components feeding an assembly line, and gradual change in quality of the incoming material.

The two key concepts that emerge from control charts are control and capability. The term *control* defines the stability of the process, which in turn means that the process is predictable. It does not necessarily mean that the process is acceptable with regard to the specification. This is in contrast to the term *capable*, which means that the process has the capability to meet specifications.

Example 5: Solving a Problem of Fender Blank Size Variation. Two fender blanks were generated with each stroke of a blanking punch (Fig. 7a). Measurements of the critical dimension y-y on each blank showed large variations in R-values, but $\overline{X}$-values were quite constant (Fig. 7b). Further analysis of individual plots of the right-side blanks and the left-side blanks had different characteristics (Fig. 8). These data showed a reduced variation in R but a larger variation in the $\overline{X}$-values. Interestingly, the $\overline{X}$-values of the right- and left-side blanks changed simultaneously but in opposite directions.

Further observation of the blanking process revealed frequent tandem adjustment of both the right- and left-side guides in response to changes in the width of the coil of steel. Even if the coil of steel did not change in width, movement of the right and left guides would cause one blank to increase in width while the other blank would decrease in width. This would cause the range to increase equal to twice the amount of the guide shift.

The first process change made was to weld the left-side guide into correct position. This meant that the left-side blank had the correct width regardless of the width of the coil. As a result, all

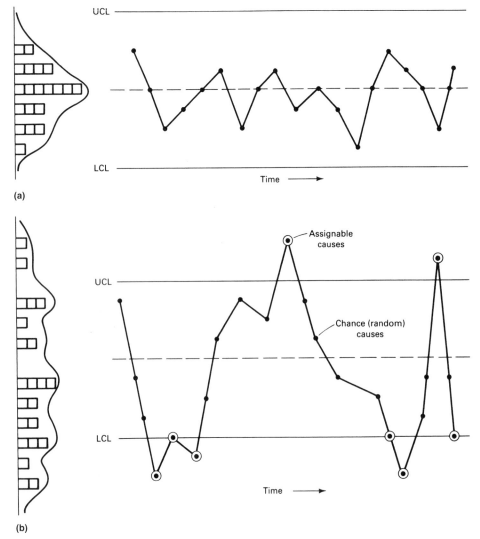

Fig. 4 Distribution curves (histograms; left) and control charts illustrating processes that are (a) in control (within control limits) and (b) out of control (outside of control limits). Variation outside of assigned control limits is always the result of some assignable cause, as shown in (b). UCL, upper control limit; LCL, lower control limit

blank width variations were associated with the right-side blank, and these variations could be related to the variations in the width of the coils of steel. These control charts could be given to the steel supplier as a performance review of the as-supplied coil width variations.

Problem Solving Using Control Charts. Control charts are only signals to operators and management of the operating conditions of the process. Personnel must respond to these signals and identify the sources of the variations. Control charts are useful in many problem-solving situations (Ref 5); they can:

- Assist in distinguishing between special and common causes of variation
- Help determine whether unacceptable variations can be improved by the personnel immediately involved in the process (such as tightening a loose tool holder) or whether they are due to the system and can be corrected only by management
- Identify trends in the process average
- Highlight increased process variability

The opportunities are limitless for using control charts as a key SPC tool. Data collection is the easiest aspect of the process. Finding the cause of a problem will often require some difficult detective work, and elimination of the problem may be a major task. For example, recording the arrival times for a scheduled shuttle bus is easy. Determining the real cause of deviation from the schedule may be difficult. Periodic high traffic density may be due to maintenance being performed on an adjacent street. To make the shuttle bus immune to delays and to improve trip time consistency would require a major study and perhaps drastic modifications to the traffic control system.

Statistical Deformation Control

Four of the six inputs of a forming system (material, lubricant, tooling, and press) can be tracked using the SPC techniques described previously. Typical measurements could include:

- Material thickness, coil/blank dimensions, and properties
- Lubricant composition, viscosity, and application thickness
- Tooling pressures, surface treatment, and dimensional accuracy
- Press speed, stroke, and ram pressures

Many of the aforementioned measurements are charted in an attempt to reduce process variability and to improve product quality. Product design changes are easily identified. On the other extreme, system interactions by the press operators and tooling maintenance personnel are very difficult to track.

The components of forming systems are complex, interactive, and synergistic. Reliable models are becoming available for predicting the output of the forming system based solely on the system inputs. Unfortunately, accurate measurements of all the necessary inputs, such as point-to-point values for both friction and workpiece temperature, are difficult, time-consuming, and even impossible to obtain. Therefore, monitoring of the final output of the system is required. Dimensional checking of the final product can be easily accomplished, but monitoring forming severity requires a different type of measurement. These include sheet metal surface strains, thickness strains, and forming severity ratings.

Numbers representing the parts per million of defects or the percentage of breakage are traditionally recorded to represent the status of the forming system. These numbers are inadequate measures of forming severity. For example, many stamping processes result in high levels of strain but not breakage. Therefore, the absence of splits gives no indication that breakage may be imminent. Some measure of performance must be sought that permits a broader range of conditions. Once breakage begins, the stamping process is out of control. General global straining ceases as the tear develops and opens. In addition, the percent breakage measurement averages the forming severity over a large number of stampings instead of determining the forming severity at a preselected location in each individual stamping. For these cases, percent breakage does not accurately define the various levels of severity.

One means of evaluating forming severity used in many sheet metal press shops is circle grid analysis and forming limit diagrams; these are described in Ref 8 to 11. The actual amount of deformation that the sheet has experienced is determined from the deformed circles (Fig. 9). The forming limit diagram shows the maximum amount of deformation a stamping can undergo before failure. Forming severity can be defined as the maximum allowable deformation minus the actual deformation. This forms the basis for monitoring and controlling stamping performance.

The process of gridding the blank before deformation and measuring the deformed circles presents logistical problems if it is used for routine analyses of a large number of stampings over an extended period of time. These problems include removing blanks from the material lifts for gridding, the time required for gridding, careful reapplication of lubrication to duplicate production levels, reinsertion of blanks into the production cycle without stopping production, proper lighting for grid reading, the time required

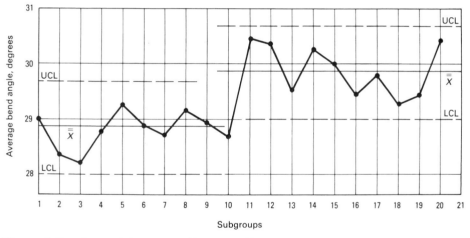

Fig. 5 Shifting of a control chart (upper and lower control limits, or UCL and LCL, and mean) due to a change in processing. Values of R, the range of thicknesses, remain the same.

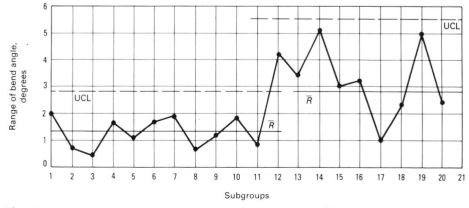

Fig. 6 Increase in the range of values within a subset (increase in R). Values of $\overline{\overline{X}}$ remain the same.

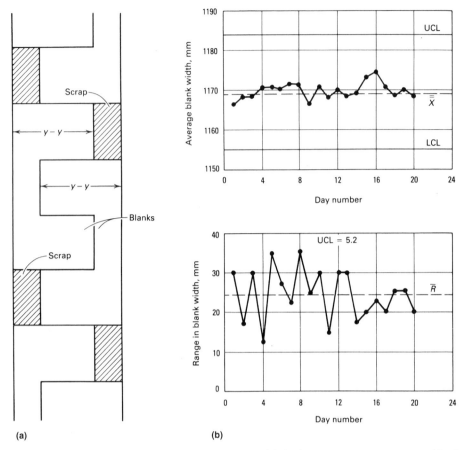

(a)
    (b)

Fig. 7 (a) Schematic showing nesting of blanks for an automobile fender. Dimension *y-y* was monitored for the statistical process control study reported in example 2. (b) $\bar{X}$-R control charts for both right-side and left-side fender blanks are taken as a single population. See also Fig. 8.

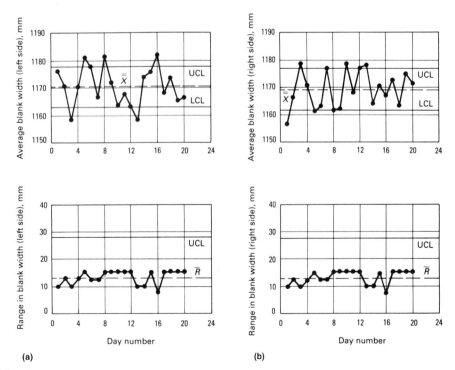

(a)
    (b)

Fig. 8 $\bar{X}$-R control charts for (a) left-side fender blanks and (b) right-side fender blanks taken as separate populations

for accurate reading, and disposal of gridded blanks after analysis. Therefore, only one or two stampings are commonly gridded to provide a single analysis (with respect to time) of the severity of the stamping. No information can be deduced about the change of the forming system with time, nor can the single stamping be characterized with respect to the dynamic changes of the system.

Dynamic changes, however, can be analyzed with SPC. A critical variable is measured at specified intervals and plotted as a function of time to generate a control chart. From this control chart, using SPC analysis techniques, the current status of the system can be identified relative to its historical performance. The dynamic variability of the system can be defined, and determinations can be made as to whether the system is in control, out of control, or changing.

Statistical deformation control (SDC) combines the best features of circle grid analysis, forming limit diagrams, and SPC (Ref 12). The deformation severity of the stamping under investigation becomes the critical variable that is tracked by SPC. To simplify the press shop procedures further, the amount of deformation experienced by the stamping is defined by the ratio of final thickness, t_f, to initial thickness, t_0. The thickness ratio is determined from ultrasonic measurements of sheet metal thickness in the most critical zone of the stamping. The ultrasonic measurements are rapid, the blanks do not have to be gridded, and the stampings are not damaged by a grid and therefore need not be scrapped after the measurement is made.

Like most analytical tools, SDC does not involve a single, invariant procedure. Several levels of complexity are possible, and each level contributes an increased degree of understanding of the forming system. These levels are:

- *Level 1:* generating a standard control chart
- *Level 2:* assigning a deformation severity value
- *Level 3:* separating material variability from process variability
- *Level 4:* determining sensitivity of the forming system to individual inputs

Again, like most analytical tools, SDC is flexible and should be applied only at the level necessary to solve the problem. Therefore, routine applications of SDC to a system that is in control will be at the level of the basic control chart.

All levels of SDC require a preliminary stamping analysis. Measurement of deformation at the critical location is a key element of SDC. Therefore, an early circle grid analysis of the stamping in question is required to identify the location, mode, and severity of strain in the most critical zone within the stamping. Repetitive ultrasonic measurement of many stampings over an extended period of time requires a template to be made for the ultrasonic probe; the simplest template is a section of an identical stamping with a hole drilled for the head of the probe.

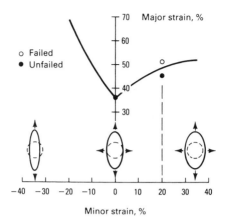

Fig. 9 Forming limit diagram showing strain states where forming is safe (below curve) and where stampings will begin to fail from localized necking during forming (above curve). The strain values are determined from deformed circle grids placed on the blank before forming.

Template holes are drilled for the most critical location and for an area of the stamping that is undeformed. The latter location determines the initial thickness of the stamping to generate the final-thickness-to-initial-thickness ratio without the inconvenience of premeasuring the blank thickness. The ability to measure critical and initial metal thicknesses on the formed stamping means that random stampings can be selected for evaluation after production.

SDC Level 1: Control Charting. Standard control-charting procedures are used. A typical application would require the removal of five stampings in sequence at the end of the press line. Measurements of the thickness in the critical zone and an undeformed zone would be made, and the thickness ratio calculated. These five ratio values would then be converted into the traditional $\overline{X}$-R values and plotted on the control chart. Standard control chart analysis procedures are used to calculate upper and lower control limits and other SPC data.

The control chart is monitored to determine the viability of the forming system and to initiate corrective actions when such actions are mandated (Ref 5). The importance of these analyses is the predictive capability, which can provide early warning of impending forming process failures. Therefore, SDC can be used to show an increase in the forming severity for a press line for which breakage is not yet a problem, as in the following example.

Example 6: Tracking a Sheet Metal Forming System by Monitoring Metal Thinning. An appliance stamping for a range top was selected for study, using sheet metal thinning as a measure of stamping severity. Measurements were conveniently made with an ultrasonic thickness gage, which permitted measured stampings to be returned to the production line without loss of the stampings. Because breakage was not a problem with this stamping, measurements were made only twice a week. The results of the measurements are shown in Fig. 10.

The stamping was shown to be out of control. However, because the amount of metal thinning did not cause breakage, the stamping was not on the list of problem stampings. After six months of good performance, the amount of metal thinning began to increase. Eventually, excessive production breakage of the stampings began to occur. Because this was a test of the SDC procedure, no attempts were made to determine the cause of the increased thinning or to correct the cause. This would have disrupted the test and would have changed the database. However, the test did show that the stampings that were not breaking could be statistically tracked over extended periods of time in order to detect changes in severity that would lead to production problems if left uncorrected. The absence of breakage in a forming process does not necessarily indicate that the process is in control.

SDC Level 2: Severity Assignment. A unique feature of SDC for formability analysis is that each control chart can be subdivided and calibrated according to forming severity for each type of stamping made in the press shop. This severity assignment is independent of whether the process is in or out of control. The severity ranges are assigned based on the thickness forming limit diagram (Ref 12). The following example shows how a process can be in control and producing good stampings and yet have an insufficient safety factor for good long-term reliability.

Example 7: Assigning a Severity Rating for Stampings with Only Sporadic Breakage. Sporadic breakage was occurring in an automotive stamping. The question was asked whether this breakage was the result of random events not typical of the normal system or whether the system was in fact critical and the breakage was a normal by-product of statistical swings in severity. To determine this, metal-thinning readings were made for several weeks (Fig. 11). The control chart showed that the system was in control.

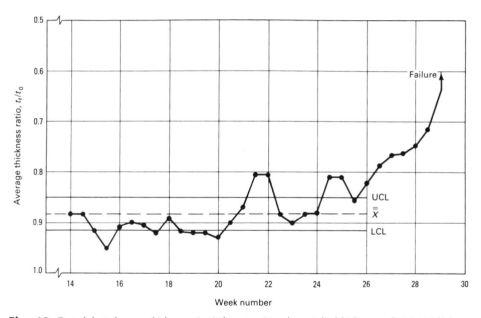

Fig. 10 Control chart of average thickness ratio t_f/t_0 for a stamping, where t_f is final thickness, and t_0 is initial thickness. Note the increase in forming severity (reduced thickness ratio) beginning at week 21 and the continuous decrease in thickness ratio beginning at week 25.

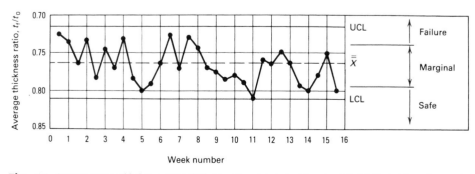

Fig. 11 Severity zones added to average thickness ratio control chart in a level 2 statistical deformation control analysis

Additional information was gained, however, when deformation severity zones were added to the control chart (Ref 12). The severity zones showed that $\overline{\overline{X}}$ was in the marginal zone and that some proportion of the control chart data points fell in the breakage zone. This meant that the sporadic breakage was normal for the system as established and could be expected to continue as long as the process was maintained as currently designed. However, the process was in control and therefore amenable to a process change that would lower the upper control limit from the failure zone into the marginal zone.

Specifications, Control Limits, and Forming Severity. An important point of understanding is the differences among engineering specifications, control limits, and forming severity. A specification may require a stamping to have a maximum metal thinning of 42% to meet in-service performance requirements. The control limits indicate whether the process is in statistical control. The state of being in statistical control means that all special causes of variation have been eliminated and that only common (random) causes remain. Special causes are intermittent sources of variation that are unpredictable or unstable. Sometimes called unassignable causes, these variations are signaled by points beyond the control limits. An example of a special cause would be inserting the wrong material into the forming process. On the other hand, common causes of variation are always present and indicate the random variation inherent in the process itself. An example of this would be excess gap in the punch guidance system. Forming severity is the proximity of a given stamping to breakage. This forming limit is independent of future in-service performance requirements or part-by-part variation over time.

Example 8: Comparing Engineering Specifications and Forming Severities to Control Charts. Engineering specifications, control limits, and forming severity for a control arm are illustrated in Fig. 12. The allowable maximum metal thinning specified by the part print (engineering specification) is rather high. In fact, the engineering specification is in the failure zone of Fig. 12. This means that the stamping would fail before attaining the maximum thinning allowed by the engineering specification. Therefore, the practical thinning limit of the process is established by forming severity rather than engineering specifications.

The control chart and its attendant control limits reflect the current operating status of the forming process. Figure 12 shows the operating status of the control arm to be in control but with a certain portion of the control chart in the failure zone. This means that some quantity of pieces will break in the forming process. Alterations to the forming process, such as modifying the lubricant or changing the die radii, could narrow the control limits shown in Fig. 12 but would not change the position of the engineering specification or the severity limits. The initial assessment of forming severity and the tracking of

the severity through the production life cycle of the stamping, especially through tooling modifications, are important aspects of SDC.

SDC Level 3: Material versus Process Variability. Statistical process control or statistical deformation control can easily identify a forming system that is out of control. Determining the cause of the variability is a more difficult problem. An important first step for most press shops is to separate variability in incoming material from process variability (for example, lubricant, tooling, press, and other in-house variables).

Statistical deformation control provides a method of separating this variability. First, a lift of blanks from the reference material is set aside. This reference material typically would be production material that has average, but consistent, properties. Ideally, this would be verified by evaluating the mechanical properties and surface characteristics of the top and bottom sheets of the lift.

Each time a control chart measurement is made with production material, an equal number of identified reference blanks are also formed, and the severity is measured. Two charts of severity versus time are maintained: one for the production material and one for the reference material (Fig. 13). Comparison of Fig. 13(a) and (b) indicates whether the system variability is due to the production material or the process.

The forming severity of the production steel in Fig. 13(a) is varying, while the forming severity of the reference steel is relatively constant. This indicates that the process parameters are producing a stamping of constant severity and that system variability is due to incoming material variability. The reverse is true in Fig. 13(b) because the variability of the production material is identical to that of the reference material. This indicates that all variation in forming severity is due to process variables.

For most forming processes, the forming severities of both the production material and the reference material will vary but will not be identical to each other. This indicates the common situation in which both the material and the process affect the output of the forming system. The variability must then be apportioned between material and process.

Assume that all the variation is caused by variation of process inputs within the stamping plant. The next step would be to further narrow the location of the variation. The aforementioned experiment is not repeated. Instead, a series of stampings are taken after the first forming operation and from the end of the line. If the variation after the first forming operation duplicates the variation at the end of the line, the cause of the variation lies within the first forming operation or earlier. If no variation is measured after the first forming operation, the search for the source of variation is focused on operations performed subsequent to the first forming operation.

SDC Level 4: Input Sensitivity. Once the forming system variability has been identified as

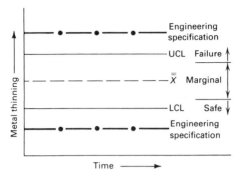

Fig. 12 Schematic control chart showing the differences among engineering specifications, control limits, and severity zones

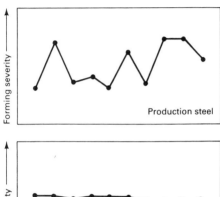

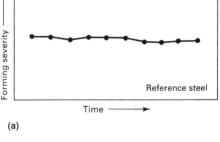

(a)

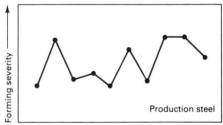

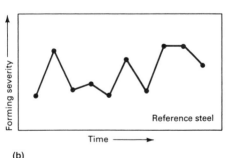

(b)

Fig. 13 Schematic control charts showing (a) the effects of materials variables and (b) process variables. The identical variations in forming severity for both production and reference steel in (b) indicate that the fluctuation in severity is caused by process variables.

being either material or process related, a more detailed analysis can be conducted to minimize this variability. A key control factor is identified and changed to determine the improvement to the system variability (or conversely, system stability). Selection of the factor should be based on experimental techniques.

There are at least two analysis possibilities. One is to process the new material according to the modified specification, for example, a new metal thickness. This new material parameter is then processed with the reference material and the severity data compared. The key is to run the reference material with the new material in order to isolate material variability from process variability.

A second technique would be to substitute the modified material (alternate parameter) for the reference material. Severity data are then collected for the regular production material and the modified material. The two severity curves can be compared to separate the process variability (Fig. 14) and to permit judgments about the relative performance of the two conditions. If sufficient data points (usually greater than 25) are collected, complete control chart information can be calculated.

The importance of SPC and SDC can be illustrated by describing two alternative methods of process control and analysis, both of which contain major flaws. One common method of process analysis is to produce a large quantity of output (in this case, sheet metal stampings)

with condition A (the first level of the parameter under investigation, such as material properties, type of lubricant, press setting, and so on). The next step is to produce a large quantity of output with condition B (the second level of the parameter). The percentage of part breakage for each is then compared. A major problem exists if breakage is zero for both levels of the parameter under study. Even if the breakage is different for the two conditions, one cannot determine if the difference is due to the change from condition A to condition B, a change in the process itself, or normal statistical variation.

A second method of process analysis is to produce one stamping with condition A and one with condition B. The stampings are compared in some manner, such as degree of tearing or visual quality (Fig. 15). The argument is made that if two sequential stampings are produced, then the process is unchanged. However, statistical information is needed to compare the relative change of the two stampings with the normal expected statistical change between any two consecutive stampings without any intentional system changes.

If the forming system is unstable, then the system must first be stabilized before any changes are made and documented. The following example illustrates how SDC can be used to identify the sources of processing difficulties.

Example 9: SDC Analysis of an Outer Side Panel of a Truck Body. The production of a

truck outer side panel was marked with sporadic breakage problems. First, high breakage rates were recorded in area A of Fig. 16 for limited but specific lots of steel brought to the press. Mechanical-property evaluation of these high-breakage lots showed certain formability parameters to be equivalent to those measured in other lots of the same steel type that did not break. Second, certain extended runs of the part experienced more incidents of breakage than other runs.

Process Analysis. This problem was selected as an early trial for SDC. The following six steps were used.

First, a circle grid analysis of the stamping was performed using a sheet of the current production material that was satisfactorily making the stamping (Ref 8–12). The most severe strain was located in area A of Fig. 16. Plotting this strain state on the forming limit diagram showed a safety margin of 8 strain percent. This means that the stamping had marginal severity but would not be expected to fracture if the current conditions of the forming system were maintained.

Second, the circle grid analysis showed no strain in area B of Fig. 16. This location was designated as the as-received sheet thickness for purposes of the ultrasonic thickness measurements. This location was ideally suited to thickness measurement because it was at the same relative edge-center-edge location in the coil as the most severe strain location.

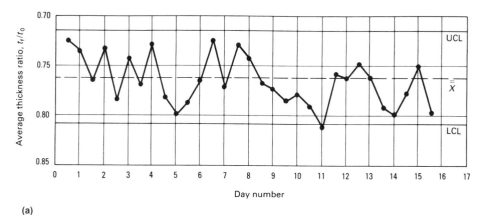

(a)

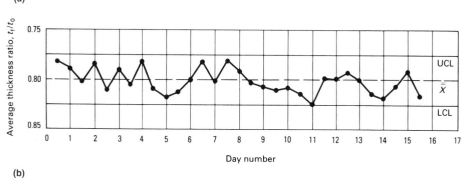

(b)

Fig. 14 Control charts showing (a) original spread of process means and (b) reduced spread of process means after a change in the process

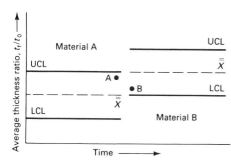

Fig. 15 Measurements from two points in time (values A and B) would suggest that material B has the lower forming severity. However, when full control charts are evaluated, material A has the lower severity.

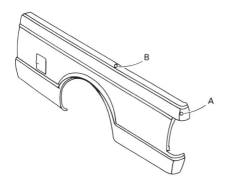

Fig. 16 Illustration of truck body side panel showing locations of critical-strain area A and zero-strain area B used for ultrasonic thickness measurements

Third, a template was prepared for the stamping by cutting a section of another stamping to encompass the two locations. A hole was drilled at each location to accommodate the head of the ultrasonic probe.

Fourth, one lot of the production steel that was successfully making the stamping was set alongside the press as a reference lot. Blank edges were marked with yellow paint to identify at the end of the line all stampings made from this reference lot.

Fifth, at the halfway point of each lot of production steel brought to the press, five of the reference blanks were formed. Stampings from these blanks, plus the next five production stampings, were removed from the end of the production line for evaluation. Thickness ratio t_f/t_0 was determined for each stamping. The $\overline{X}$ and R values were calculated for both the production steel and the reference steel.

Sixth, concurrent plots of the $\overline{X}$ and R values for the production and the reference steel lots were maintained. Sections of the plot are shown in Fig. 17. The mean, the upper control limit, and the lower control limit are not shown because the solution to the problem was achieved before sufficient data points were collected for these statistical calculations.

Solution. The left half of Fig. 17 verifies the initial circle grid evaluation that the stamping has some safety margin but that the safety margin is insufficient because the value of the thickness ratio is located in the bottom half of the marginal zone. In this case, it is possible to evaluate the severity of each reading as it is made. The quantity of data points needed to create a control chart is not required in order to determine the forming severity of each stamping.

Both the production steel and the reference steel appear to be maintaining not only a constant but also an equal forming severity. Neither the production steel nor the forming process provides indications of impending breakage.

When retry lot B (Fig. 17), previously rejected for high breakage rates, was retried, a marked increase in forming severity was noted. The formability of sheet steel is an interaction of substrate properties and the lubricity of the steel/lubricant combination. However, this lot of steel had forming properties equal to or better than the production and reference steels noted in Fig. 17. Therefore, the lubricity of the steel surface was suspected. The steel was coated on the punch side of the sheet with a corrosion-resistant paint. To test the influence of the painted surface, several blanks were turned over so that the bare side of the sheet contacted the punch. These sheets were successfully formed into stampings without breakage. Although unacceptable for production in this reversed condition, these stampings verified that the substrate had sufficient formability to produce the stamping without breakage when the interface friction was correct.

It was recommended that the painted (punch) side of the blanks be sprayed with lubricant if breakage was encountered with a specific lot of steel. If breakage persisted with the additional lubricant, the lot was to be rejected for laboratory analysis of the substrate properties.

The right-half of Fig. 17 illustrates another problem. When the tooling was reset in the press for another run, extensive breakage was encountered with many of the lots of steel brought to the press. An SDC analysis of the stamping showed that the severity of the stamping for the reference steel had increased to zero safety margin, which meant that the tooling was not performing identically to the previous run. This indicated a different setting in the press was used or modifications were made to the tooling during the period the tool was out of the press. Discussion with the tool and die staff revealed that the draw beads and binder surfaces of the tooling were reworked in an attempt to reduce low spots in the stamping. This reduced the flow of metal from the binder areas and increased the stretch component required over the punch, thus increasing the stamping severity.

The increased severity was verified when samples of the previous problem lot B were inserted in the tooling. Long splits were encountered compared to the previously encountered slight necking condition. The recommended solution was to encourage additional metal flow from the binder area through reduced binder pressure until the reworked draw beads became reduced in effectiveness through normal wear.

During the first run, production steels A, C, and D had a severity comparable to that of the reference steel. During the second run, production steel E had a lower severity than the reference steel. Therefore, using the reference steel as a common base for comparing various production runs, steel E actually had better formability than steels A, C, and D for the specific stamping under investigation.

Continued collection and monitoring of data during the entire production run was recommended. This would require a number of reference lots. However, the SDC procedure allows for this transition by the interleaf forming (alternate blanks) of both reference steels to calibrate the new reference lot relative to the previous reference lot.

A process for data-based management of the forming system was developed by a team of the Auto/Steel Partnership (Ref 13). In their analysis procedure, the line severity analysis is very similar to the SDC just described.

Experimental Design

Forming operations span an extensive production life cycle, including design, prototyping, engineering changes, tryout, and production. During this entire cycle, a pervasive question is whether the process can be modified. Process optimization is the usual motivation for this modification. This optimization may mean lower cost, faster production rates, fewer rejects, better quality, and so on.

Some process simulation techniques may be available for determining the steps needed to optimize the process and the effectiveness of the various options. However, most of these simulation techniques are mathematically based and only approximate the system under evaluation (Ref 14–16). Therefore, most optimization studies are conducted on the system itself by selectively changing the system variables and measuring the effects of the changes.

These system studies are most effectively conducted with experimental design. Experimental designs are used to increase the efficiency

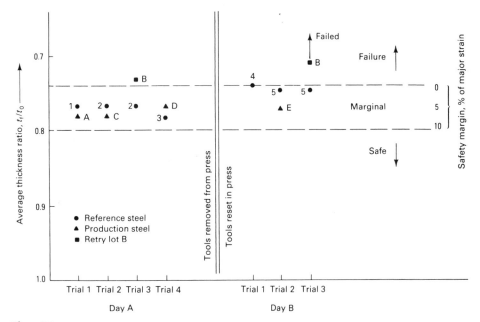

Fig. 17 Data for two days of measurements made on the truck body side panel shown in Fig. 16. See text for details.

of information acquisition. Several forms of experimental design are available, and each has advantages and disadvantages. The types of experiments discussed in this article are single-variable studies, multivariable studies, and Taguchi experiments.

Single-Variable Studies

The most common experiment conducted with forming operations is the single-variable study. Although many variables are identified as affecting the process, only one of these variables is changed at a time.

Single-variable studies are widely used because they are easy to conduct and require a minimum number of experiments. First, one variable is changed, and then another, and so on until all of the variables are tried at the second level. One such experimental design table is shown in Fig. 18. Here the (+) indicates the high level of the variable, and (−) indicates the low level of the variable. The following example outlines the procedure used in a typical single-variable experiment.

Example 10: Finding Improved Process Settings for the Production of an Automobile Inner Door. High breakage rates were observed during the die tryout of an automotive inner door. A quality group met and identified seven key factors (controlling variables) that affect this particular forming process.

Two steels and a lubricant were obtained. First, the stamping was made with steel 1 (variable +A); all other variables were kept at their existing conditions. The severity of the process was determined using the thickness ratio analysis discussed previously. The stamping was then made using steel 2 (variable −A).

Next followed the additional six changes (variables B through G) indicated in Fig. 18. The thickness ratio was determined for each change. After all eight experiments were completed, the level of each variable showing the greatest thickness ratio (least amount of metal thinning)

was selected as the new standard operating procedure.

Problems with Single-Variable Studies. In this study, one variable was changed while all other variables were kept constant. The problem is that the conclusions obtained from this experiment are valid only if all other variables are kept identical to the experimental value. This type of experiment ignores synergistic effects among process variables. In the experiment described in example 10, the effects of different materials (steels 1 and 2) can be measured only if variable C (lubricant) is maintained at its high level (with lubricant). If variable C is at its low level (no lubricant), misleading results may be obtained. In this case, one cannot deduce what variable A does for all levels of the other variables. Therefore, some type of factorial experiment (orthogonal array) is required.

Multivariable Studies

Multivariable studies are usually conducted using some form of an orthogonal array. The experimental layouts using these orthogonal arrays yield experimental results such that the effects of varying a given parameter can be separated from other effects. Figure 19 shows a complete orthogonal array, sometimes called a full factorial experiment, for seven variables. These orthogonal arrays have also been termed square games (Ref 17).

The term *orthogonal* means balanced, separable, or not mixed, and this indicates that each variable is evaluated equally for all other conditions. Therefore, for each of the two levels of any variable in the array in Fig. 19, there are 64 combinations of the other variables. For example, for level A_1 in Fig. 19, there are 64 possible combinations of variables B, C, D, E, F, and G. This also holds for level A_2. Therefore, any effect of level A_1 versus level A_2 is determined in the presence of both the high and low levels of all the other variables.

When the quantity of parameters is increased in these factorial experiments, the number of experiments required increases so rapidly that it may not be feasible to implement the experimental design. The full factorial experiment detailed previously involves 128 experiments. Increasing the number of variables by one doubles the number of experiments required to 256.

The full factorial experiment provides the effect of all the single-variable effects plus all the interactions. If some of the interactions can be ignored or deemed unimportant, then the experiment can be redesigned as a partial factorial. A partial factorial has a reduced number of required experiments. The procedure used in one full factorial analysis is outlined in the following example.

Example 11: A Full Factorial Experiment for Finding Improved Process Settings for Stamping an Automobile Inner Door. The same quality group that conducted the experiment described in example 10 realized that the synergistic effects were not being evaluated with a single-variable experiment. The group decided to design a full factorial experiment in which all interactions would be active in the study and all interactions could be calculated from the data obtained.

The full orthogonal array shown in Fig. 19 was used. The same variable assignments used in example 10 (Fig. 18) were repeated in this study. Again, ultrasonic measurements of sheet thickness were used to determine the thinnest area (maximum metal thinning) for calculation

+	−		Experimental number								
Level of variables			1	2	3	4	5	6	7	8	Process variables
A versus B			+A	−A	+A	+A	+A	+A	+A	+A	Steel
Large versus small			+B	+B	−B	+B	+B	+B	+B	+B	Blank size
Yes versus no			+C	+C	+C	−C	+C	+C	+C	+C	Lubricant
Hot versus cold			+D	+D	+D	+D	−D	+D	+D	+D	Die temperature
Normal versus offset			+E	+E	+E	+E	+E	−E	+E	+E	Blank position
High versus low			+F	+F	+F	+F	+F	+F	−F	+F	Blankholder pressure
High versus low			+G	+G	+G	+G	+G	+G	+G	−G	Ram speed

Fig. 18 Layout of an experimental design for a single-variable investigation

Fig. 19 Layout of a full factorial (multivariable) experimental design for a group of seven variables

of stamping severity. Analysis of the results showed that some of the interactions were very important and had to be considered in selecting the best operating levels of each parameter.

Taguchi Experiments

Conventional experimental design techniques were primarily developed for use in scientific research in order to determine cause-effect relationships (Ref 17). In science, there is generally only one law to explain a natural phenomenon. Therefore, the primary experimental efforts are aimed at finding the law that explains the relationships being studied.

In technological fields, however, there are numerous ways to obtain a given product design objective. In forming a sheet metal stamping, there are various steel characteristics, die designs, die steels and surface treatments, lubricants, forming sequences, and other variables to be considered. Different combinations

Factor	A	B	C	D	E	F	G	Results
No.	1	2	3	4	5	6	7	
1	1	1	1	1	1	1	1	y_1
2	1	1	1	2	2	2	2	y_2
3	1	2	2	1	1	2	2	y_3
4	1	2	2	2	2	1	1	y_4
5	2	1	2	1	2	1	2	y_5
6	2	1	2	2	1	2	1	y_6
7	2	2	1	1	2	2	1	y_7
8	2	2	1	2	1	1	2	y_8

(a)

(b)

Fig. 20 Layout of the Taguchi experimental design for seven variables

can be used to produce the same end design of stamping. A desirable goal is to find the combination that provides the most stable and reliable performance at the lowest manufacturing cost. The goal of the Taguchi experiments is to develop the most robust process possible (that is, the process that is least sensitive to the causes of variation).

In many cases, knowing the cause of a problem is insufficient for solving the problem. Removal of the cause of the problem may be too costly or even impossible. The Taguchi strategy then becomes one of finding countermeasures that do not eliminate the cause but reduce the influence of the cause on the end product. Efforts are aimed at making the final product insensitive to all process variables. Taguchi techniques are oriented more toward cost-effectiveness and marketing than are conventional experimental design techniques (Ref 17). This difference affects the nature of the parameters to be cited, the way in which the experiments are laid out, and the way in which the data are analyzed. In a sense, the Taguchi method of conducting experiments is formalized to such an extent that only minimal exchange of information is required among experimenters trained in this technique to achieve complete understanding of the experimental parameters, the experiments, the analysis of the data, and the results/conclusions.

Differences between the Taguchi approach to experimental design and the conventional approach are both philosophical and methodological (Ref 17). Philosophically, the Taguchi approach is technological rather than theoretical, inductive rather than deductive, an engineering tool rather than scientific analysis. Taguchi emphasizes productivity enhancement and cost-effectiveness with a rapid, formalized procedure that is not necessarily statistically rigorous. With regard to its methodology, the Taguchi approach emphasizes the application of orthogonal arrays, signal-to-noise ratios (in Taguchi analysis, the variables that affect a process are classified as signal, control, or noise factors), and newly developed analytical techniques (Ref 17).

One primary advantage of the special orthogonal arrays used by Taguchi is the capacity and flexibility of assigning numerous variables with a small number of experiments. An even more important advantage of the array is the reproducibility of the conclusions across many different process conditions. One criticism is that the Taguchi array is nothing more than a fractional factorial, which has long existed in conventional experimental design techniques. However, the Taguchi technique is one of philosophy and content, as explained previsously. The application is more versatile and sophisticated.

Further explanation of the Taguchi method or the experimentation/analysis procedures is beyond the scope of this article. Additional information is available in Ref 17 or from the American Supplier Institute (Dearborn, MI). The following example outlines the approach used in one Taguchi analysis.

Example 12: A Taguchi Design of Experiments for Finding Improved Process Settings for Stamping an Automobile Inner Door. The same seven variables used in examples 10 and 11 are used in this experiment. Selection of the proper experimental variables is extremely important in Taguchi experiments. The variables should be primary variables with little interaction. In studying stamping severity, for example, blankholder pressure (variable F) is not set by the tons of force shown on a load indicator, because the measured tons are highly interactive with all of the other variables of the system, such as blank thickness, metal strength, die temperature, and shims used. Instead, the blankholder pressure variable is controlled by the position of the nut (or turns of the screw) on the outer ram relative to the zero load reference position. The extra time spent identifying and selecting the important process variables is the key to the success of the Taguchi experiment.

Figure 20(a) shows the layout of the Taguchi experiment for the seven variables detailed in example 10 (Fig. 18). The array is a standard array for up to seven controlling variables or factors. The eight experiments required for a full factorial analysis are shown in Fig. 20(b). There are eight rows representing eight experiments, numbered 1 through 8. The elements of the seven columns consist of 1s and 2s. There are four 1s and four 2s in every column. In any pair of columns, there are four possible 1, 2 combinations, namely 11, 12, 21, and 22. Because each of these four combinations occurs at an equal number of times in a given pair of columns, the two columns are balanced or orthogonal. Figure 20 shows that all column combinations have an equal number of 11, 12, 21, and 22 combinations; therefore, all interactions are within the experimental design. After the eight experiments are completed, the data are analyzed according to a prescribed procedure in order to determine the level of each variable that contributes to the most robust (most insensitive process) combination of processing variables.

REFERENCES

1. *Automotive Sheet Steel Stamping Process Variation: An Analysis of Stamping Process Capability for Design, Die Tryout, and Process Control,* Auto/Steel Partnership, Southfield, MI, Jan 2000; http://www.a-sp.org
2. *Statistical Quality Control Handbook,* Western Electric Company, 1977
3. J.M. Juran, F.M. Gryna, Jr., and R.S. Bingham, Jr., *Quality Control Handbook,* McGraw-Hill, 1979
4. E.L. Grant and R.S. Leavenworth, *Statistical Quality Control,* 5th ed., McGraw-Hill, 1980

5. A.L. Strongrich, G.E. Herbert, and T.J. Jacoby, "Statistical Process Control: A Quality Improvement Tool," Paper 83099, Society of Automotive Engineers, 1983

6. J.F. Siekirk, Process Variable Effects on Sheet Metal Quality, *J. Appl. Metalwork.,* Vol 4 (No. 3), July 1986, p 262–269

7. "Final Report on Manufacturing Cost Study Conducted by Pioneering Engineering," American Iron and Steel Institute, 1986

8. S. Dinda, K. James, S. Keeler, and P. Stine, *How to Use Circle Grid Analysis for Die Tryout,* American Society for Metals, 1981

9. *Sheet Steel Formability,* American Iron and Steel Institute, 1984

10. S.P. Keeler, *Automotive Sheet Steel Stamping and Formability,* Auto/Steel Partnership, Southfield, MI, Jan 1989; http://www.a-sp.org

11. W.G. Brazier and S.P. Keeler, Relationship Between Laboratory Material Characterization and Press Shop Formability, *Proceedings of Microalloying '75,* Union Carbide, 1977, p 517–530

12. S.P. Keeler, "Statistical Deformation Control for SPQC Monitoring of Sheet Metal Forming," Paper 850278, Society of Automotive Engineers, 1985

13. *Managing the Stamping Process,* Auto/Steel Partnership, Southfield, MI, Oct 1991; http://www.a-sp.org

14. N.M. Wang and S. Tang, Ed., *Computer Modeling of Sheet Metal Forming Processes: Theory, Verification, and Application,* The Metallurgical Society, 1986

15. H. Yamasaki, T. Nishiyama, and K. Tamura, Computer Aided Evaluation Method for Sheet Metal Forming in Car Body, *Proceedings of the 14th Biennial Congress of the IDDRG* (Koln, Germany), International Deep Drawing Research Group, April 1986, p 373–382

16. K. Chung and D. Lee, Computer-Aided Analysis of Sheet Material Forming Processes, *Advanced Technology of Plasticity, Proceedings of the First International Conference of Plasticity* (Tokyo), Japan Society for Technology of Plasticity and Japan Society of Precision Engineering, 1984, p 660–665

17. Y. Wu and W.H. Moore, *Quality Engineering—Product and Process Design Optimization,* American Supplier Institute, 1985

Process and Feedback Control for Manufacturing

Mahmoud Y. Demeri, FormSys, Inc.
Jian Cao, Northwestern University
Ravi Venugopal, Sysendes, Inc.

TREMENDOUS ADVANCES in automation technology have been made since the 1990s, and the use of this technology in the area of manufacturing has yielded significant benefits. The integration of automatic process control systems in a manufacturing system ensures part quality and consistency by compensating for variations in material properties and process conditions, in addition to other benefits such as round-the-clock operation. While process control has been used extensively for chemical engineering, semiconductor manufacturing, and even processes such as welding, the application of this technology to other manufacturing processes such as sheet metal forming is relatively new. The objective of this article is to provide a basic overview of control system design and elucidate its application to various manufacturing processes.

Basic Concepts of Control Systems

A signal is defined as a representation (typically mechanical or electrical) of a physical condition, and a system is a physical entity with input and output signals. Control systems are used to drive the output signals of the system that is controlled to a desired profile, and this is done by sending mechanical or electrical signals from the controller to a device that directly affects the controlled system. The controlled system is known as the plant, while the device that directly affects it is known as the actuator. For example, the hydraulic ram of a stamping press is an actuator, with hydraulic pressure as its input and punch force as its output, while the metal that is formed is the plant, with punch force as an input and a material deformation measure such as thickness as an output. Systems may have several inputs and outputs; for instance, the blankholder force can be considered as an additional input to the plant in the previous example. The interaction between signals and systems can be represented graphically as block diagrams.

There are two main classifications of control systems, namely, open-loop control systems and feedback (or closed-loop) control systems. In open-loop control, the controller sends the actuator a predefined control signal that is not based on the output of the plant. The control signal is computed based on the desired output of the plant and models of the plant and actuator. In other words, the output has no influence on the control action taken because the output is neither measured nor fed back for comparison with the input. The obvious advantage of the open-loop control system is its easiness in implementation, faster response time, and relatively low capital investment. It is suitable for situations where variability is well under control or process design is robust, meaning that the output is not sensitive to input variations. Open-loop control requires extra demand from process design engineers. Figure 1 shows the input-output relationship of an open-loop (Fig. 1a) versus the closed-loop (Fig. 1b) control system.

When process variation is of major concern, a feedback (closed-loop) control system is highly recommended, because the control signal will be recalculated based on the current status. Feedback control is based on feedback signals from sensors that continuously monitor the output of the plant. The entire system, comprised of the controller, actuators, plant, and sensors, is known as the closed-loop system. The control signal that the controller sends to the actuator is calculated using the feedback signals from the sensors. Hence, variations in the plant or actuator will be reflected in the sensor output, and the controller will be able to compensate for these changes. In addition, the feedback sensors will also pick up the effect of external disturbances acting on the plant or actuator, and the controller can compensate for these disturbances. Open-loop controllers do not have this feedback information and thus cannot compensate for disturbances that are not modeled. Thus, it can be seen that feedback control systems are more robust than open-loop control systems and can compensate for system variations and external disturbances. However, the design of a high-performance feedback controller is not trivial, because an

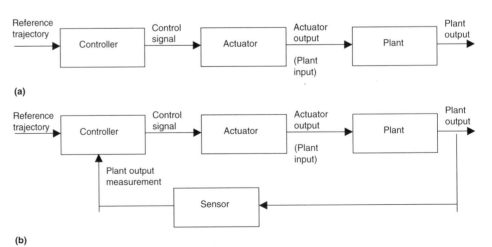

Fig. 1 Block diagrams of (a) an open-loop control system and (b) a closed-loop or feedback control system

incorrectly designed controller can cause closed-loop instability. Note that here *instability* refers to system unstable.

More sophisticated feedback control systems with the added ability to self-adjust their response by changing the gains and/or time constants in their control laws are known as adaptive control systems. Such systems usually require precise knowledge of the relationships between input parameters and the desired outputs.

Designing a Feedback Controller

The design of a feedback controller involves the following steps:

- Mathematically characterizing the input-output relationships of the plant, actuators, and sensors
- Choosing a controller structure that is appropriate for the desired control objective
- Calculating the parameters of the controller based on the control objectives and the mathematical characterizations of the components of the closed-loop system
- Ensuring closed-loop stability and robustness
- Choosing a method of controller implementation. Controllers are usually implemented electronically, either as analog circuits or digitally on microprocessors, although in rare instances, mechanical means are used.

The design methodology listed previously is a formal approach to controller design. However, it is often standard practice to use a common controller known as a proportional-integral-derivative (PID) controller and to adjust its parameters based on experimental observations. The benefits and limitations of using PID controllers are discussed later in this section.

The input-output relationships of the first step are usually represented as ordinary differential equations (ODEs) that evolve in time, although for metal forming applications, partial differential equations (PDEs) that account for both spatial and temporal variations are usually more appropriate. Most controller design techniques are based on ODE representations of systems, and thus, when the systems are characterized by PDEs, the PDEs are reduced to ODEs by using techniques such as separation of variables and by considering inputs and outputs at certain specified spatial locations of the system. The differential equations that characterize the system are obtained from the physical nature of the system; however, it is very often impossible to find the values of all the parameters in these differential equations. Thus, physical modeling can provide important information about the qualitative behavior of the system and help define the structure of the controller but is often insufficient for accurately calculating the controller parameters. In practice, system parameters are obtained using system identification, a technique by which

parameters are extracted from experimental input-output data. System identification methods can range from simple curve-fitting techniques to sophisticated methods that estimate disturbance and noise properties in addition to the system parameters. However, metal forming processes often involve complex models, and the issue of modeling metal forming processes for feedback control is still an active area of research (Ref 1, 2). Once the mathematical models have been constructed and parameterized, they are validated using computer simulations by comparing simulated outputs to experimental outputs for the same set of inputs. If the mathematical model can be reduced to a linear ODE, Laplace transform methods can be used to obtain a transfer function of the system, defined as the Laplace transform of an output signal divided by the Laplace transform of an input signal. Transfer functions are algebraic representations of linear ODEs, and there are several powerful techniques for studying system behavior based on transfer function analyses (Ref 3).

Controller Structure. The selection of a controller structure is now addressed. In the previous section, it was noted that the purpose of using an automatic control system is to drive the outputs of the plant to a desired profile or trajectory in the presence of immeasurable disturbances. This activity is referred to as command following (or tracking) with disturbance rejection. The desired output trajectory is known as the reference trajectory. As in the case of plant and actuator models, most controllers are represented as ODEs. The structure and parameters of the controller ODEs are design variables. In many cases, a linear ODE structure is chosen, and the input to the controller is the error signal, defined as the difference between the reference trajectory and the measured output signal from the plant.

Proportional Control. The simplest form of a feedback controller is a proportional or P-controller that generates a control signal proportional to the error signal. For example, consider the problem of driving a roller in a steel rolling mill at a prescribed reference rotational speed. The motor drive is the actuator, and the roller is the plant. The rotational speed is measured by an angular velocity sensor and fed back to the controller. The control signal is the current to the motor. If the speed of the roller is less than the reference speed, the error signal is positive and the control signal (current) is a factor (the P-gain) multiplied by the error. Thus, a positive current will be applied to the motor, which will speed the roller up and bring the rotational speed closer to the desired reference value. If the speed of the roller is higher than the desired reference speed, the error signal will be negative and a negative current will be applied to the motor, which will slow down the roller and bring the speed closer to the desired value. The design parameter here is the P-gain, which is usually denoted by K_p. If the P-gain is too low, the roller will take a long time to achieve the desired speed, and if it is too high, overshoots may occur,

leading to oscillations or instability in roller speed.

Derivative Control. The next extension in control structure is derivative control, or D-control, in which the control signal is proportional to the rate of change of the error signal. Derivative control adds damping to the system by ensuring that the control signal does not cause the output to change too rapidly. The factor that multiplies the derivative of the error signal to yield the control signal is known as the D-gain and is denoted by K_d. Derivative control is usually used along with proportional control in a PD control structure. While derivative control helps damp out oscillatory behavior, it is sometimes difficult to implement due to the fact that numerical differentiation of the error signal for obtaining the derivative of the error leads to amplification of noise in the error signal. Thus, a high D-gain can result in a closed-loop system that is very sensitive to noise.

Integral Control. The final element in a basic control structure is integral control, or I-control, in which the control signal is proportional to the error signal integrated over time. Integral control is usually implemented to overcome steady-state errors by incrementally adding to (or subtracting from) the control signal until the error signal is driven to zero. The factor that multiplies the integral of the error signal is known as the I-gain and is denoted by K_i.

Combination Control. For many applications, a combination of proportional, integral, and derivative (PID) control is used, and the P, I, and D gains are adjusted (tuned) either by trial and error or according to specified tuning rules. Note that tuning a PID controller does not necessarily require a mathematical model of the system, and thus, these controllers are used extensively in practice. However, the structure is constrained and is not always applicable to all systems, especially systems with multiple modes and multiple coupled inputs and outputs. The PID controllers also may not work very well for nonlinear systems.

Other controller structures are used to control more complex systems, and optimal control techniques are used to derive controller parameters based on minimizing a performance cost function involving the error signal. There are several factors that contribute to the complexity of a system, for example, a large number of inputs and outputs that are coupled, dynamics that are characterized by high-order differential equations, nonlinearities, delays, time-varying characteristics, and model uncertainty. Over the past 50 years, several advances have been made in theoretical and applied control engineering that address these issues. For example, linear-quadratic-Gaussian control is a method for designing optimal controllers for multiple-input, multiple-output linear systems of any order subjected to random immeasurable disturbances. Predictive control addresses delays in systems, while robust control techniques allow a trade-off between controller performance and system uncertainty. Adaptive control techniques modify the controller parameters during operation to adjust to system changes and effects that are not explicitly modeled. More information on these approaches is available in Ref 3 to 5. Models of the plant, actuators, and sensors are usually required to design controllers for more complex systems. For example, the autoregressive moving-average model (ARMA) has been used to identify system relationship between multi-inputs and multioutputs. The advantage of this model is its generic nature, and it has been used in various forms of time series analysis and system identification problems (Ref 6, 7). The basic format of the ARMA model can be found in Ref 8 and 9. It is also useful to note that having these mathematical models also makes tuning a PID controller much easier.

Neural Network Control. Another approach in control is to use neural network control. The concept of neural network control and potential applications are discussed in Ref 10. Briefly, a neural network mimics the structure of the brain. Through a set of carefully conducted experiments, the network is trained to identify small or large variations in the process and then act accordingly to achieve a preset goal, for example, springback control. Examples of metal forming processes that have benefited from neural network springback control include a rebar (Ref 11) and an air bending process (Ref 12). In Ref 13, a neural network has been used to determine the material properties early in the forming process of a cylindrical part. Closed-loop control of the binder force was then employed based on the theoretical wrinkle limit in the initial forming depth stage of the process and the theoretical fracture limit in the latter forming depth stage. These theoretical limits were based on an analytical calculation using the material properties computed by the neural network and a friction coefficient, which was determined throughout the forming process. A control system using artificial neural networks and a stepped binder force trajectory to minimize and control springback in a channel forming process is discussed in Ref 10, 14, and 15.

Stability and Robustness. Once the controller structure has been chosen and its parameters specified, closed-loop stability and robustness are analyzed using analytical and simulation methods. Computer-aided control system design tools are very useful for this purpose. MATLAB from The Mathworks is an example of such a tool. These tools allow the user to predict closed-loop performance based on plant, actuator, sensor, and controller parameters. Simulink, a graphical simulation tool developed by The Mathworks, allows the user to create a numerical simulation of a closed-loop system by graphically defining it in block-diagram form.

Controller Implementation. The final stage of controller development is controller implementation. This stage involves creating the physical controller with the error signal(s) as its input and the control signal(s) as its output. The controller operates in real-time and ensures that the control objective is met. The controller may

be an analog circuit characterized by the controller ODE or it may be a digital controller in which a microprocessor runs code that numerically integrates the controller ODE to calculate the control signal. In the case of digital controllers, analog sensor signals are converted to digital form using analog-to-digital converters (A/D), while the digital control output of the microprocessor is converted into an analog signal using digital-to-analog (D/A) converters.

Analog controllers have the advantage of being inexpensive and relatively simple to build. However, modifying an analog controller involves effectively rebuilding it. In contrast, digital controllers are relatively expensive and have to be carefully designed to account for issues related to converting analog signals to digital signals such as quantization and aliasing. However, they are very flexible, and changing controller parameters or structure usually involves modifying the software that runs on the microprocessor of the controller.

A new technology that has significantly reduced digital control development time is rapid control prototyping (RCP). The RCP systems consist of hardware that includes microprocessor(s); A/D, D/A, and other input/output devices; and software that allows the controller designer to define a controller in block-diagram form and have it implemented on the hardware by automatic code generation. The physical device that is controlled is interfaced to the RCP hardware by connecting the sensors to its inputs and the actuator to its outputs. The RCP software also allows real-time monitoring of sensor and actuator signals in addition to all controller variables. Controller parameters can be optimized during operation using an RCP environment, and several controller structures can literally be implemented in minutes. In contrast to traditional industrial automation systems, which allow only basic PID and switching logic controllers, RCP systems allow implementation of any control strategy. dSPACE, Inc. is a leading vendor of RCP systems that are completely integrated with The Mathworks' control system design and simulation tools, MATLAB and Simulink.

At this point, it is also important to say a few words about the generation of the reference trajectory that the control system forces the plant output to follow. This task is an important component in successfully implementing a feedback control system. For example, optimizing the punch force for a stamping process improves the consistency and quality of the formed part, as described in Ref 16. However, determining this trajectory, because a reference input to a feedback control system requires theoretical and experimental analysis of the dynamics of the process, and the actuators, sensors, and control strategy have to be chosen to ensure that this reference trajectory is tracked with sufficient accuracy. The issue of generation of system reference inputs for process control of metal forming processes is an area of active research and is discussed in Ref 16.

Application Example: Process Control for Sheet Metal Forming

Sheet metal stamping is a primary manufacturing process for high-volume production of automotive panels and components. The quality of stamped parts is important in avoiding assembly problems and in achieving product performance. Major problems with the quality of automobiles, such as wind noise, water leaks, squeaks, and fit and finish, are directly related to the quality and dimensional accuracy of stamped components. The development of production dies and the fine-tuning of the stamping process to produce high-quality parts with low scrap rate is a very expensive and time-consuming process. After production begins, disturbances to the process (material property variations, frictional interface changes, tool wear) may cause many parts to be scrapped due to tearing, wrinkling, or poor dimensional accuracy. This example illustrates a sample problem, its experimental trials and needs for a closed-loop control system, the design of the control system, and system validation.

Problem Definition. A stamping system is shown schematically in Fig. 2. F_b is the applied

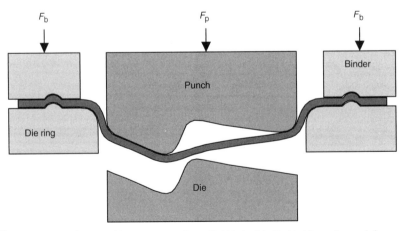

Fig. 2 Schematic of a sheet metal forming system. F_b, applied binder (blankholder) force; F_p, punch force

binder (blankholder) force, and F_p is the resulting punch force required to form the part. The binder, with or without drawbeads, is a critical part of the stamping system. While flat binders rely entirely on friction to generate binder restraining forces at the edge of the sheet metal blank, drawbeads rely on bending and unbending of the sheet metal and, to a lesser extent, on friction to regulate the amount of the blank material that flows into the die. Flat binders are more sensitive to variations in friction conditions, die misalignment, and blank thickness variations. If the restraining force is too low, too much material flows into the die, causing wrinkling in the part. If, however, the restraining force is too high, not enough material flows into the die, causing splitting in the part. The binder restraining force has also been found to have a strong effect on the final part dimension through its effect on springback. Different binder force trajectories (step increase, step decrease, and level change) were found to produce different strain histories in the blank material, leading to different formability and springback in the final part. Formability and dimensional accuracy can, therefore, be improved by the proper control of the blankholder force.

In the following example, feedback control has been applied to the process of cup drawing. The control objective here is to apply a consistent punch force, that is, to track a desired reference punch force trajectory while rejecting disturbances in the system. Disturbances in the forming system may include variations in the manufacturing parameters such as frictional behavior, blank thickness, material properties, blank placement errors, and die wear. The binder force-generating unit is used as the actuator. Successful implementation of the feedback control strategy for cup drawing entails forming cups to a specific depth without splitting. Figure 3 shows a picture of a successful and a failed cup.

Experimental Investigation of Constant Binder Force Cases. The process of deep

Fig. 3 Picture of a successful and a failed cup

drawing of conical cups involves applying a drawing force with a punch to the center of a circular blank, as shown in Fig. 4. A binder force is applied to the periphery of the blank to control the amount of material flowing into the die cavity. As the punch draws the metal, hoop compression develops in the flange area before the metal bends at the die radius and then straightens as a result of the applied tension in the cup wall. Because the cup wall transmits the punch force to the bending, straightening, compression, and friction areas, a high state of tension develops near the punch radius where the metal thins and most often fractures. For a given material, blank size, and tool geometry, the binder restraining force at the edge of the blank is determined by the level of friction at the sheet metal/tool interface. A change in the level of this force is reflected in a change in the level of the punch force required to form the cup. The punch force is the sum of the forces required to bend, straighten, and compress the metal to conform to the cup shape as well as the force needed to overcome friction between the blank and the tool. In cup drawing, the forces needed to shape the metal remain constant, but the force needed to overcome friction may vary from part to part. Differences in punch force profiles due to changes in friction conditions indicate that punch

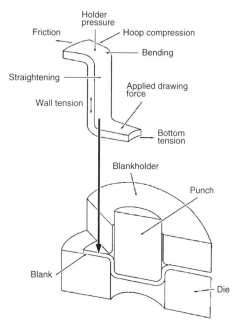

Fig. 4 Forces applied to a blank during cup forming process

force could be a good measure of the performance of the cup drawing process.

The instability that occurs in the conical cup drawing process was demonstrated in a laboratory experiment by forming 15 cups under seemingly identical conditions. The cups were made from circular blanks of DQSK-EG steel. The blanks were 0.8 mm (0.03 in.) thick and had a diameter of 200 mm (8 in.). The mechanical properties of this material are given in Table 1. The circular blanks were lightly lubricated with prelube on both sides before being drawn into conical cups. The binder force was kept constant at 575 kN (65 tonf), and the punch speed was held constant at 10 mm/s (0.4 in./s) throughout the stroke.

Experiments were performed on a double-action hydraulic press equipped with a PID digital controller. The press has a load capacity of 680 kN (76 tonf) for the punch and 700 kN (79 tonf) for the binder. The press was instrumented with sensors to measure punch and binder loads as well as punch displacement during a forming process. Additional components were added to the press to permit the construction of a closed-loop control system to track a reference punch force trajectory as the predetermined process variable. This process variable was used as feedback to modulate the binder force in real-time. Process control gains were adjusted to provide stable and acceptable tracking performance. Tooling shape and dimensions used for the conical cup tests are shown in Fig. 5.

Selection of the level of the binder force for the process instability tests was based on studying the effect of constant binder force profiles on the maximum depth of the drawn cups, as shown in Fig. 6. The maximum depth was determined when a split occurred in the wall of the drawn cup.

An examination of Fig. 6 shows that substantial instability occurs when the clamp force is at the 575 kN (65 tonf) level. The maximum depth of the drawn cups at that level ranged between 15 and 45 mm (0.6 and 1.8 in.). This indicates that the forming process, under these conditions, could not be controlled and that the success or failure of the formed cups could not be reliably predicted.

Needs for a Closed-Loop Control System. The next phase of testing to determine the instability of the cup forming process was to

Table 1 Mechanical properties of DQSK-EG steel

Direction, degrees	Yield strength		Tensile strength		Uniform elongation, %	Total elongation, %	n	R	K	
	MPa	ksi	MPa	ksi					Mpa	ksi
0	201.76	29.26	316.18	45.86	26.69	41.65	0.20	2.11	534.32	77.50
90	193.82	28.11	315.25	45.72	25.37	40.43	0.20	2.45	526.13	76.31
45	205.41	29.79	332.27	48.19	21.92	41.53	0.19	1.47	552.70	80.16
Averages	201.60	29.05	323.99	46.59	23.98	41.29	0.20	1.87	541.46	77.99

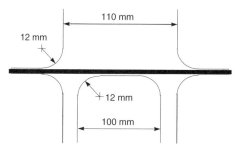

Fig. 5 Tooling shape and dimensions used for the conical cup tests

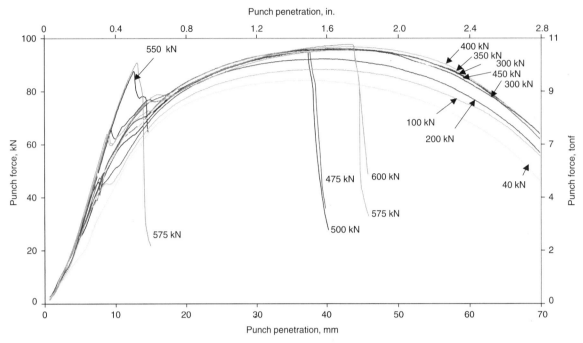

Fig. 6 Effect of constant binder force on punch force and maximum draw depth

draw 15 cups at the selected constant binder force of 575 kN (65 tonf) to find out the extent of variations in cup heights obtained without active binder force control. Similar tests have also been conducted with active binder force control to demonstrate that process control could stabilize the forming process and could produce more consistent cups. Results of the tests, as shown in Fig. 7, indicate that the forming process was not under control and that the success or failure of the formed cups could not be reliably predicted.

If the drawing process were stable, all cups would have been drawn successfully to a maximum depth of 40 mm (1.6 in.). It is evident that in-process variations caused the drawn cups to fail prematurely by tearing. For a constant binder force, tool geometry, blank size, surface finish, and sheet material, frictional conditions appear to be the most significant factor in determining the level of the binder restraining force. Because the level of the binder restraining force is reflected in the level of the punch force required to form a cup, an examination of the punch force trajectories for the 15 drawn cups should provide important information on the instability problem.

Figure 8 shows the punch force trajectories, using constant binder force of 575 kN (65 tonf), for a subset of the 15 drawn cups. This subset, which includes samples 6 through 10, was selected because it had a wide variation in the maximum draw depth obtained during the cup forming process. It is obvious from the figure that there is an inverse correlation between the magnitude of the jump in the punch force trajectory and the failure height of the drawn cups. An increase in the level of the punch force results from an increase in the level of the binder restraining

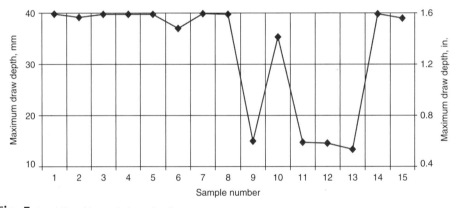

Fig. 7 Instability of the conical cup drawing process

force. The binder force increases with increase in the level of friction at the binder/sheet interface.

In order for the blank edge to move toward the punch during the cup drawing process, the force of friction between the blank and the die must be overcome. The normal force created by the binder adds significantly to the force of static friction in accordance with Coulomb's relationship:

$$\text{Friction force } (F_f)$$
$$= (\text{Friction coefficient, or } \mu)$$
$$\times (\text{Normal force, or } F_b)$$

The blank starts to move after static friction is overcome. Continuous force must be exerted to overcome sliding (dynamic) friction, and this force would be less than that needed to overcome static friction. For a constant binder force, tool geometry, blank size, surface finish, and sheet material, variations in lubrication conditions change static and dynamic frictional forces

through their effect on the coefficient of friction. It must be emphasized that lubrication conditions may not remain constant during the forming process.

Control System Design. Punch force is the sum of forces needed to bend, straighten, and compress the metal to overcome friction. Variations in the punch force observed during cup drawing are most likely due to variations in the lubrication conditions. Spikes in the punch force could be related to increases in the static frictional forces caused by changes in the friction condition in the binder area. Increasing the punch force results in a high state of tension in the cup wall near the punch radius where the metal thins out the most, thus causing instability and leading to fracture.

Because the lubrication condition changes during the cup drawing operation, a closed-loop control system must be used to control the forming process. A real-time process controller was implemented to ensure that the forming process would always give the desired successful cup regardless of friction conditions. Punch force was used as the control parameter to modulate the binder for conical cup drawing. A block diagram of the control system used for tracking the optimal punch force trajectory is shown in Fig. 9. The implementation of the process control strategy shown in the block diagram entails modeling of the forming process, designing an appropriate process controller, and determining an optimal reference punch force trajectory.

A first-order nonlinear dynamic forming process model was developed in Ref 1 to define the relationship between the binder force and the clamp force. A proportional plus integral controller with feed-forward action (PIF) was successfully implemented in the forming press (Ref 2). The controller was designed to ensure good tracking performance in the feedback control system. An optimal punch force trajectory was

selected from the punch force trajectories shown in Fig. 8. Because sample 8 provided the best results for a successfully drawn cup, its punch force trajectory was selected as the optimal reference punch force trajectory for use during real-time process control.

The effectiveness of the real-time closed-loop controller was demonstrated in the lab by forming a second set of 15 cups using active binder force control instead of the constant binder force used on the first set of 15 cups. Shortly after each experiment begins, the system shifts from a constant binder force mode to a punch-force-tracking mode following the optimal reference punch force trajectory. The process controller was operative during the entire stroke of the process. The implementation scheme of process control in the forming press is shown in Fig. 10.

The data acquisition board (DAQ) acquires data from the press digital controller and feeds

the calculated binder force command (F_{bc}) into the digital controller. The process controller constitutes the program and the DAQ blocks. The WSCI block refers to the workstation communications interface. F_p is the punch force, F_b is the binder force, X_p is the punch displacement, and F_{bc} and X_{pc} are commands to the hydraulic actuators for the binder and the punch, consecutively. The "Program" block collects data, reads, or generates the reference punch force trajectory F_{pd}, and calculates the binder force, F_{bd}, based on a PIF control algorithm.

System Validation. Results of the tests, obtained using the process controller, are shown in Fig. 11. The results indicate that the forming process was completely under control and that the success of the cups could be guaranteed. The process controller was operating during the entire forming stroke and was capable of maintaining the desired punch force trajectory regardless of the variations in the friction conditions. The punch force peaks, shown in Fig. 8, appear to have been softened by the process controller, as shown in Fig. 11.

The corresponding binder force trajectories needed to maintain the optimal punch force trajectory are shown in Fig. 12. The binder force undergoes modulation to compensate for the process variation in order to follow the reference punch force trajectory.

The conical cup deep-drawing application example showed that the control strategy increased the robustness of the conical cup drawing process by eliminating process sensitivity to variations in the lubrication condition during forming. Results showed that the implementation of active binder force control stabilized the cup forming process and produced successful cups every time.

Application Examples

Example 1: Using Neural Network for Springback Control. Springback variation in final products can be traced back to variations in incoming material, process, and lubrication. The closed-loop control system presented in the previous example of process control for sheet metal forming, using punch force as a target, can be used to control springback due to variations in binder force setup or friction conditions, as demonstrated in Ref 17. In this application example, an approach based on the use of neural networks to control springback in sheet metal forming due to process and material variations, such as thickness, coating, and material properties, is demonstrated. The approach is exemplified in a channel forming process shown in Fig. 13.

As shown in Fig. 14, a stepped binder force, starting with a lower binder force, then changing to a higher binder force at a certain punch displacement, can effectively achieve a low springback amount without overstretching the panel. When variations are present, the challenge

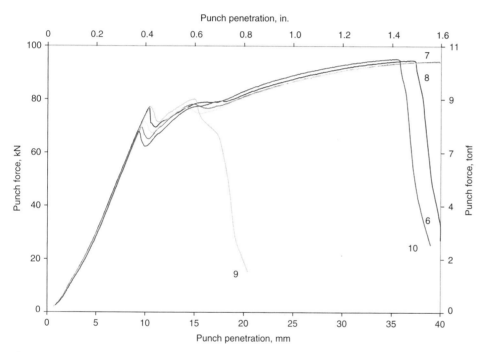

Fig. 8 Punch force trajectories for samples 6 to 10 without process control. Punch force trajectories using constant binder force of 575 kN (65 tonf)

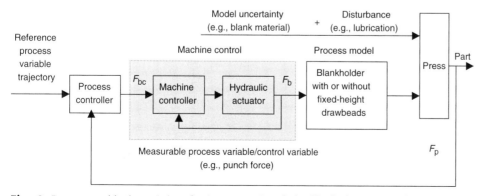

Fig. 9 Process control for the conical cup forming process. F_{bc}, calculated binder force command; F_b, binder force; F_p, punch force

is to find out when to switch to a higher binder force and how higher the binder force should be for a specific case.

Background—Neural Network. Neural networks are nonlinear analysis tools that form a highly interconnected, parallel computation structure with several simple processing elements, or neurons. Each neuron is linked to adjust downstream neurons through variable weights. These weights are calculated by an iterative method during the training process when the network is fed with a large amount of training data, input and output pairs that represent the pattern attempting to be modeled. It has been shown that a two-layer neural network, one made up of an input layer, a hidden layer, and an output layer, was sufficient to accurately model any continuous function, provided a sufficient number of hidden neurons are used (Ref 18). Therefore, this network architecture was chosen for the current example application. In a fully connected network, each neuron receives inputs from all of the elements in the preceding layer. No connections exist between neurons in the same layer. Figure 15 shows the structure of a neural network and the connect paths to and from one hidden layer neuron.

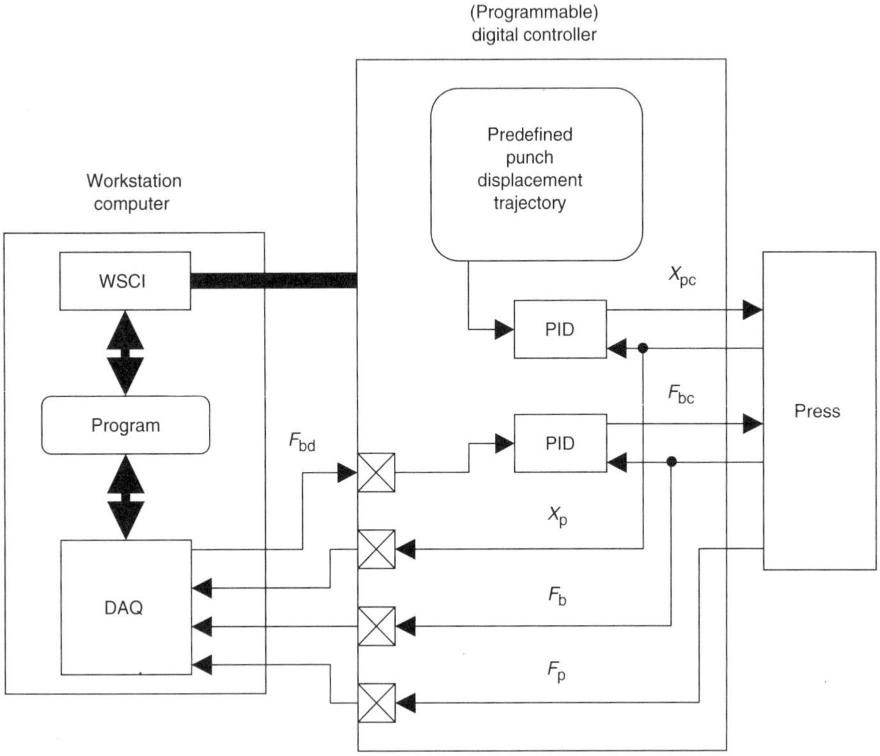

The total input to the hidden-layer neuron j, U_j, is the summation of the weight multiplied by the input value, x_i, for each connection path:

$$U_j = W_{j0} + \sum_{i=1}^{N} W_{ji} \times x_i \qquad \text{(Eq 1)}$$

where N is the number of inputs, and W_{j0} is the bias of the neuron, which is simply another weight in the network. The bias allows the neuron to have an extra degree of freedom to adjust to allow the input-output relationship to be learned accurately during training. The output from the hidden-layer neuron, Vj, is given by:

$$Vj = f(Uj) \qquad \text{(Eq 2)}$$

where f is the activation function. It has also been shown that for the hidden neurons, a sigmoidal activation function, or any other continuous, monotonically increasing, and bounded activation function, is necessary to model the continuous output function (Ref 18). However, a simple linear activation function with a slope of one is sufficient for the output neurons. With these two activation functions, a given output, y_k, based on the inputs, x_i, and the connection weights is given by:

$$y_k = \sum_{j=1}^{J} \left\{ W_{k0} + W_{kj} f(W_{j0} + \sum_{i-1}^{N} W_{ji} x_i) \right\} \qquad \text{(Eq 3)}$$

where J is the total number of hidden neurons.

The most common method to adjust the weights of the connection paths is through the back-propagation algorithm (Ref 19).

As was stated previously, neural networks must first be trained so they are able to generalize, that is, to extract the correct relationship from a finite number of corresponding input-output pairs. During training, a number of input-output pairs, Q, are given to the network, and the weights of the connection paths are adjusted. The goal of this "learning" is to get the outputs, y_k, that are calculated using the weights of the neural network as close as possible to the desired output patterns, d_k, for the training examples. A measure of how well the neural network achieves this goal is the mean square error (MSE) of the process, given by:

$$\text{MSE} = \frac{1}{QxK} \times \sum_{q=1}^{Q} \sum_{k=1}^{K} [d_k(q) - y_k(q)]^2 \qquad \text{(Eq 4)}$$

where K is the total number of outputs.

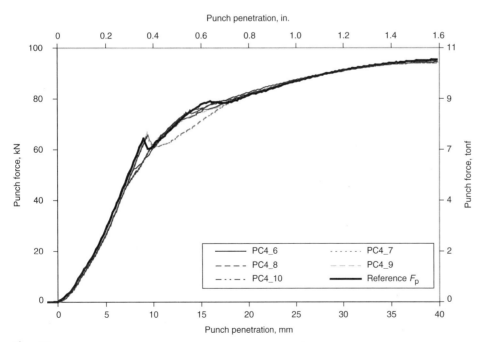

Fig. 10 Implementation of process control in the press. See text for details.

Fig. 11 Punch force plots using process control and a reference punch force (F_p) trajectory

Implementation. The control strategy is to let the drawing process start as an open-loop system until punch depth reaches 20 mm (0.8 in.). By then, a neural network will analyze the collected punch force data and provide two outputs: when to change the binder force, and how much the new binder force will be. The forming process proceeds with these two new instructions. Notice that there is no interruption of the forming process, because the calculation time needed for the neural network is negligible.

The first step to implement a neural network is to generate training data. In this example, the desired springback angle range was set to between 10 and 12°, and the training data in Tables 2 and 3 were generated via the trial-and-error approach. This springback angle range was chosen based on the capacity of the hydraulic press used in the experiment; that is, a higher binder force value would be required in order to reduce the springback angle further. A total of 20 training sets were generated, a combination of five materials with four lubricant conditions. See Table 2 for a list of all the training data, Table 3 for the obtained springback angles, and Table 4 for material constants.

Based on the training data of Table 2, the gains of a neural network were determined using the MATLAB neural network toolbox. A two-layer neural network was used with a sigmoidal activation function between the input and hidden layers and a linear activation function between the hidden and output layers. To determine the optimal number of hidden neurons to use in the neural network, a validation study was conducted. First, the training data were separated into two groups of data, a set of sixteen input-output pairs for initially training the neural network and a set of four examples for validation of the training process. These four training input-

output pairs are shown in parentheses in Tables 2 and 3. The validation study procedure is as follows. First, neural networks with varying numbers of hidden neurons are trained with the sixteen input-output pairs. Then, the inputs from the four validation examples are "fedforward;" that is the trained neural network is exposed to the input values to obtain outputs. Mean square errors are then calculated based on the difference in the output values calculated by the network and the known output data for the four validation examples. With too few hidden layer neurons, the input-output relationship would not be learned adequately, leading to a high mean square error. With too many hidden layer neurons, the neural network would be oversensitive and not adapt well to new inputs not seen during training, also leading to high mean square error. For the network, three hidden neurons were determined to be optimal to balance these two conflicting constraints.

Now that the structure for the neural network was established, that is, three inputs, three neurons at the hidden layer, and two outputs, the neural network was retrained with all twenty input and output pairs. The weights and biases were then extracted from the MATLAB neural network software and reconstructed using algebraic equations in a LabVIEW (Ref 20) process control program.

Experimental Results. With the neural network calculations now embedded in a LabVIEW process control program, experiments were conducted. Various materials with different friction conditions were formed. Curve fitting of the punch force trajectory was performed after 20 mm (0.8 in.) of forming, and three inputs from this curve fit were then used as inputs in the neural network equations to determine the higher binder force (HBF) and percentage of total punch

displacement (PD%) for controlling the springback angle. The neural network was able to determine the stepped binder force trajectory to keep springback within a reasonable range when faced with the same material and friction conditions used to train the network (Table 5). This verifies that the neural network learned the relationship of the training data well. However, in order to show that the generalized relationship between springback angle and material and process parameters is captured in the neural network, types of steel and friction condition not seen by the neural network during training were also formed. When a generic steel was formed with no lubricant, a EG250B steel with a synthetic lubrication, and EGCR3 and HDGCR3 with excessive mineral oil applied (20 mL), the neural network predicted reasonable values of HBF and PD% to produce springback only slightly outside the desired range (Table 5). If these materials and friction conditions were used in the training process as well, better results would have been obtained.

Without the neural network, the springback for the various material and friction conditions in these experiments varied from 4° to 30° for the same range of binder forces, 160 to 330 kN (18 to 37 tonf). The central processing unit was able to perform the neural network calculations fast enough to change to the HBF at the specified PD%. The time required to calculate the outputs of the neural network was 5 ms on the 233 Hz Pentium microprocessor. The network also provided good results when the tooling was disassembled and reassembled. This was done to demonstrate that the network could be used in industry where there are frequent tooling changeovers.

This example demonstrates the benefits of using neural networks to control springback in sheet metal forming. The steel channel forming process was reasonably immune to inevitable variations in process and material parameters, thus producing more consistent sheet metal products. For more details, please refer to Ref 15.

Example 2: Control Systems for Steel Rolling. Another application of feedback control in the metal forming industry is in steel rolling mills. The control objective in this application is to maintain the thickness of the rolled steel sheet. Steel rolling mills may be hot or cold, but in both cases, the steel is rolled into sheets by large rolls that apply pressure on the metal as it rolls through. The thickness of the rolled steel depends on the force applied on it by the rolls and

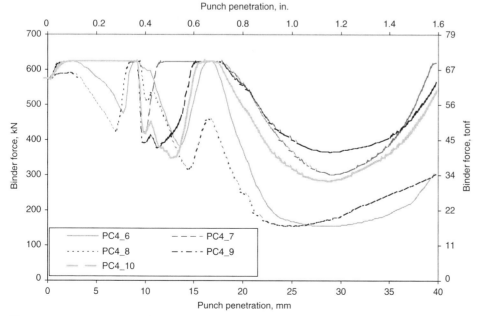

Fig. 12 Modulation of binder force to follow a predetermined punch force profile

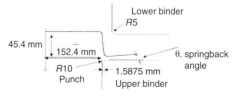

Fig. 13 Tooling geometry and springback angle in a plane-strain channel forming (only half-model is shown)

the feed rate, which is dependent on the speed of the rolls.

One of the challenges in designing a control system to minimize the thickness variation of the rolled steel is that measuring the thickness of the steel as it passes through the rolls is very difficult. Instead, piezoelectric sensors are typically used to measure the force applied by the rolls on the steel, and the resulting thickness is estimated by developing models of the effective stiffness of the system. The difference between the desired thickness and the estimated thickness is used as the error signal to the controller. The controller

typically drives two actuators, namely, a hydraulic actuator that regulates the force applied by the roll and a motor that regulates the speed of the roll. Thus, this control system can be viewed as a single-input, dual-output system. The dynamics of the system are complex, because they involve the hydraulic dynamics of the hydraulic actuator, the electrical dynamics of the motor, and the mechanical dynamics of the rolling process. In addition, in hot rolling mills, an additional control variable is the rate of flow of coolant, and, if this is the case, the thermodynamics of the process also come into play.

Several disturbances also present problems; for example, there can be variations in the properties of the steel, eccentricities in the rolls, and fluctuations in drive voltage supplies. There are also several nonlinear effects, such as backlash in the gears in the drives and flow-rate saturation limits on the hydraulic actuators.

Several control strategies ranging from simple PID controllers to intelligent adaptive controllers have been applied to the problem of controlling thickness in steel rolling mills.

Example 3: Control Systems for Resistance Spot Welding. Resistance spot welding (RSW) is a major joining operation that is used extensively in the automotive, appliance, and office furniture industries to fabricate sheet metal assemblies and components. The RSW technique involves creating coalescence between workpieces by melting the material between the electrodes through the application of a resistive electric circuit. The quality of a weld depends on the volume of melted material, which depends on the amount of generated heat.

Process variables include welding current, pressure applied on electrode, weld time, material type, and workpiece thickness. The major advantages of RSW include high speed of application, adaptability for automation, low cost, and suitability for high-volume production. A major disadvantage of RSW is the inconsistency of its welds that result from the

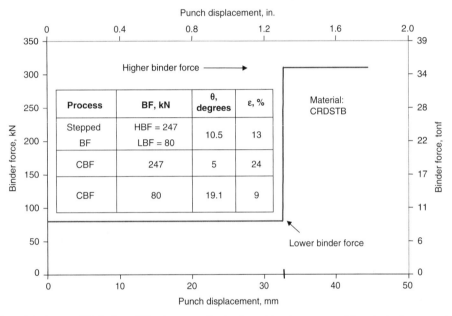

Fig. 14 A stepped binder force (BF) trajectory used in the channel forming process and its springback compared to constant binder force (CBF) cases. HBF, higher binder force; LBF, lower binder force

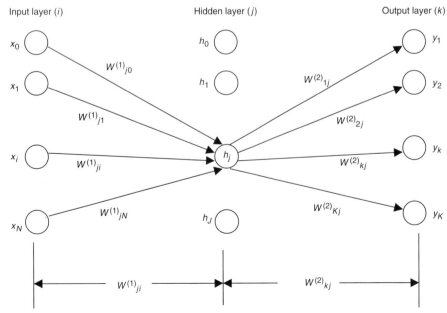

Fig. 15 Structure of a neural network

Table 2 Training data for neural network

| Material | Friction conditions | | | |
	No lubricant	Mineral oil	Prelubricant	Synthetic
High binder force, kN				
CRDSTB	231	280	264	263
EGCR3	(290)	316	312	295
EGCR3-2	163	224	167	(210)
HDGCR3	254	279.5	(268.5)	258
HSGS25S	323	(333.5)	326	326
Percentage of total punch displacement where high binder force applied				
CRDSTB	62.35	62.975	61.125	67.5
EGCR3	(61.5)	63.0	61.5	66.0
EGCR3-2	64.0	64.0	67.075	(65.75)
HDGCR3	62.25	63.75	(63.0)	62.75
HSGS25S	61.3	(60.5)	64.0	60.825

Numbers in italics are those sets used for validation.

Table 3 Resultant springback angles for the training data

| Material | Springback angles, degrees | | | |
	No lubricant	Mineral oil	Prelubricant	Synthetic
CRDSTB	10.513	10.347	10.679	10.875
EGCR3	11.234	11.098	12.001	11.965
EGCR3-2	11.001	10.789	10.453	10.791
HDGCR3	11.3	10.989	11.351	11.483
HSGS25S	10.681	11.504	12.907	11.339

Numbers in italics are those sets used for validation.

Table 4 Material properties

Steel	Thickness		Yield strength		Ultimate tensile strength		Maximum elongation, %	n	K		$\bar{R}$
	mm	in.	MPa	ksi	MPa	ksi			MPa	ksi	
CRDSTB	0.7366	0.0290	155.60	22.57	301.6	43.7	39.5	0.239	536.47	77.81	1.51
EGCR3	0.7493	0.0295	177.76	25.78	302.5	43.9	43.0	0.221	517.34	75.03	1.75
EGCR3-2	0.8509	0.0335	165.36	23.98	291.5	42.3	44.6	0.224	...	...	1.901
HDGCR3	0.8001	0.0315	194.20	28.17	354.3	51.4	37.5	0.220	623.50	90.43	1.46
HSGS25S	0.7620	0.030	187.60	27.21	306.5	44.5	40.1	0.225	537.70	77.99	1.57
EG250B	0.7542	0.0297	249.98	36.26	360.5	52.3	33.2	0.179	574.70	83.35	1.64
Generic steel	0.7014	0.0276	Unknown		Unknown		Unknown	Unknown	Unknown		Unknown

Table 5 Results of experiments with neural network control implemented
The target springback angle is between 10 and 12°.

Material	Friction	Predicted higher binder force kN	Predicted percentage of total punch displacement	Springback angle, degrees
Combinations seen by the network during training				
CRDSTB	No lubricant	247	64.0	10.523
EGCR3	No lubricant	290	63.0	11.019
EGCR3-2	No lubricant	181	69.1	10.578
HDGCR3	No lubricant	256	61.0	11.531
HSGS25S	No lubricant	309	59.8	10.809
CRDSTB	Mineral oil	291	65.1	11.179
EGCR3	Mineral oil	311	68.9	10.468
HSGS25S	Synthetic	312	61.6	10.259
Combinations not seen by the network				
HDGCR3	Excess mineral oil	304	58.9	10.163
EGCR3-2	Excess mineral oil	248	63.8	10.103
EG250B	Synthetic	329	58.9	13.670
Generic steel	No lubricant	246	65.2	12.53

complexity of the basic process as well as other sources of variation (material, thickness, coating, surface roughness, electrode wear, current, etc). Inconsistent weld quality drives up cost and therefore demonstrates the need for effective control of the welding process. Controlling the process means controlling the amount of heat generated by the electrical resistance to current and used for melting the workpiece contact interface. The heat developed can be computed from:

$$H = I^2 Rt$$

where H is the heat (joules), I is the current passing through workpieces (amperes), R is the electrical resistance in the workpieces (ohms), and t is the time of current passage (seconds).

The electrical resistance varies with location along the electrodes/workpieces setup, with the highest resistance at the workpieces interface. It has been shown that taking resistance changes during welding as a feedback parameter to control weld quality is not viable in spite of the fact that resistance is an important and measurable parameter (Ref 21). Electrode displacement caused by thermal expansion associated with nugget formation and growth during welding was found to be a reliable feedback signal for predicting and controlling weld quality. Many control systems were developed so that the actual electrode displacement tracks an optimal electrode displacement trajectory (reference trajectory). In such control systems, the welding current undergoes modulation to compensate for the process variation in order to follow the reference electrode displacement trajectory. The implementation of this tracking control strategy requires modeling of the welding process, designing an appropriate process controller, and determination of an optimal reference trajectory for each material and joint geometry. A key disadvantage for using tracking control strategies to achieve consistent weld quality is that each controller can only be used to weld a specific joint, and that limits automation of the welding process. Efforts are underway to develop intelligent control systems such as fuzzy logic and neural network to achieve effective control and automation of the resistance spot welding process.

Conclusions

In this article, the basic concepts of feedback process control have been introduced, and various examples of control system applications have been described. The examples have shown that appropriate control strategies increase the robustness of the processes by eliminating process sensitivity to system variations and external disturbances. Results have shown that the implementation of effective process control stabilized the manufacturing process and produced consistent part quality every time.

The use of effective control design, along with high-performance hardware and software for controller implementation, allows the use of feedback process control for manufacturing processes to improve part quality and consistency. The implementation of this technology requires a one-time capital cost, but the benefits from improved production efficiency are recurring. The application of feedback control in the manufacturing environment will continue to increase because of the increasing demand for better quality and more consistent products.

REFERENCES

1. C.-W. Hsu, M. Demeri, and A.G. Ulsoy, Improvement of Consistency in Stamped Parts through Process Control, *TMS Proceedings, Sheet Metal Forming Technology,* M. Demeri, Ed., TMS, 1999
2. C.-W. Hsu, A.G. Ulsoy, and M. Demeri, Modeling for Control of Sheet Metal Forming, *Proc. Japan-USA Symposium on Flexible Automation* (Kyoto, Japan), Vol 3, American Society of Mechanical Engineers, 1998, p 1143–1148
3. K. Ogata, *Modern Control Engineering,* Prentice Hall, 1997
4. G.F. Franklin, J.D. Powell, and M.L. Workman, *Digital Control of Dynamic Systems,* 2nd ed., Addison-Wesley, 1990
5. S.E. Lyshevski, *Control Systems Theory with Engineering Applications,* Birkhauser, 2001
6. N. Krishnan and J. Cao, Estimation of Optimal Blank Holder Force Trajectories in Segmented Binders Using an ARMA Model, *J. Manuf. Sci. Eng.,* Vol 125 (No. 4), Nov 2003, p 763–770
7. S.D. Fassois, K.F. Eman, and S.M. Wu, A Fast Algorithm for On-Line Machining Process Modeling and Adaptive Control, *J. Eng. Ind.* (*Trans. ASME*), Vol 111, 1989, p 133–139
8. D. Graupe, D.J. Krause, and J.B. Moore, Identification of Autoregressive Moving-Average (ARMA) Parameters of a Time Series, *IEEE Trans. Auto. Control,* 1975, p 104–106
9. S.M. Kay and S.L. Marple, Jr., Spectrum Analysis—A Modern Perspective, *Proc. IEEE,* Vol 69 (No. 11), 1981, p 1380–1414
10. B. Kinsey, J. Cao, and S. Solla, Consistent and Minimal Springback Using a Stepped Binder Force Trajectory and Neural Network Control, *J. Eng. Mater. Technol.,* Vol 122, 2000, p 113–118

11. S. Dunston, S. Ranjithan, and E. Bernold, Neural Network Model for the Automated Control of Springback in Rebars, *IEEE Expert/Intell. Syst. Applic.,* Vol 11 (No. 4), 1996, p 45–49

12. M. Inamdar, P.P. Date, K. Narasimhan, S.K. Maiti, and U.P. Singh, Development of an Artificial Neural Network to Predict Springback in Air Vee Bending, *Int. J. Adv. Manuf. Technol.,* Vol 16 (No. 5), 2000, p 376–381

13. K. Manabe, M. Yang, and S. Yoshihara, Artificial Intelligence Identification of Process Parameters and Adaptive Control System for Deep-Drawing Process, *J. Mater. Process. Technol.,* No. 1, 80–81, 1998, p 421–426

14. R. Ruffini and J. Cao, "Using Neural Network for Springback Minimization in a Channel Forming Process," Paper 98M-154, SP-1322, Developments in Sheet Metal Stamping, Society of Automotive Engineers, 1988, p 77–85

15. V. Viswanathan, B.L. Kinsey, and J. Cao, Experimental Implementation of Neural Network Springback Control for Sheet Metal Forming, *J. Eng. Mater. Technol. (Trans. ASME),* Vol 125, April 2003

16. C.-W. Hsu, A.G. Ulsoy, and M. Demeri, Optimization of the Reference Punch Force Trajectory for Process Control in Sheet Metal Forming, *Proceedings of the Japan-USA Symposium on Flexible Automation* (Ann Arbor, MI), American Society of Mechanical Engineers, July 23, 2000

17. M. Sunseri, J. Cao, A.P. Karafillis, and M. Boyce, Accommodation of Springback Error in Channel Forming Using Active Binder Force Control: Numerical Simulation and Experiments, *J. Eng. Mater. Technol. (Trans. ASME),* Vol 118, July 1996, p 426–435

18. K. Hornik, M. Stinchcombe, and H. White, Multilayer Feedforward Networks Are Universal Approximators, *Neural Networks,* Vol 2, 1989, p 359–366

19. D.E. Rumelhart, and J.L. McClelland, Learning Internal Representation by Error Propagation, *Parallel Distributed Processing,* Vol 1, MIT Press, 1986

20. LabVIEW, www.ni.com/labview/, National Instruments Corp., accessed Nov 2005

21. R.W. Messler, and M. Jou, Review of Control Systems for Resistance Spot Welding: Past and Current Practices and Emerging Trends, *Sci. Technol. Weld. Joining,* Vol 1 (No. 1), 1996, p 1–9

Rapid Prototyping for Sheet-Metal Forming

Lotta Lamminen Vihtonen, Helsinki University of Technology, Finland
Boel Wadman, IVF, Sweden
Torgeir Svinning, SINTEF, Norway
Rein Küttner, Tallinn Technical University, Estonia

RAPID PROTOTYPING (RP) techniques in the sheet-metal forming industry have been developed to meet the needs to quickly test the form and fit of new sheet-metal products on a prototype basis as well as for production runs characterized by small lot sizes. Rapid prototyping methods often place high demands on the production system as a whole. High flexibility, fast changeover, and short decision times are important. In addition, apart from adapting a manufacturing facility to low-volume production, the right forming process must be chosen. In general, considerations that are key to the design process for efficient low-volume production include:

- Reduced lead time for each product and reduced changeover time between products
- Reduced time and cost for the development and fabrication of tooling
- Flexible production units and production lines
- Reduced time between different products by using flexible tooling, for example, incremental forming or fluid forming
- Flexible and common control and materials handling system for each unit and for the whole production line

This article provides an overview of some of the technologies used for RP and low-volume production of sheet-metal parts. The discussion is summarized in two major sections, the first dealing with low-cost tooling, processing techniques, and equipment and the second with various aspects of incremental sheet forming.

Low-Cost Tooling and Flexible Sheet-Forming Processes

Low-Cost Tooling for Low-Volume Production. Low-cost tool materials and shortened tooling-manufacturing times can reduce the overall cost of tooling. Tool materials can be cast iron, aluminum, and zinc alloys with high machinability. For prototypes and mild steel, polymers and wood can also be used as tooling materials.

Rapid tool-manufacturing methods include:

- *Laminated tools:* laser or precision cut layers that are joined by bolting and/or adhesives
- *Concrete or polymeric substrates* with plated nickel surface coatings (Fig. 1)
- *Direct fabrication methods:* Laser sintering, high-speed milling, casting

Efficient systems for the design and three-dimensional (3-D) modeling of tools also are important.

Multiprocess Equipment. Most sheet-metal products are subject to processing steps in addition to the forming operation. This may include blanking/cutting before or after forming, various joining methods, processes for making holes, and so forth. For high-volume production, such operations may be performed with equipment included in the production line, for example, welding cells, flying saws, hole punching, or machining operations of various kinds. For low-volume production, on the other hand, one often has to rely on equipment that is flexible enough to be reprogrammed according to a changing product spectrum. The importance of having the same kind of control system for the combined units, that is, the production line, is obvious. Another possibility is where the production unit in itself is capable of performing various tasks.

One example of flexible manufacturing and associated equipment is laser cutting and forming (Fig. 2). A CO_2-laser can be used both to cut the sheet material and to form simple geometries. The heat induced by the laser beam bends the sheet to the calculated geometry. Metals with low thermal conductivity, for example, stainless steels, are more easily deformed. Both thin and thick gages can be formed. The method was originally used for tube bending in the shipbuilding industry. Principal recent developments in this area include models to predict the heating passes needed to produce the desired formed geometry. Laser optics to form circular and double-curved shapes have also been developed recently. Another process similar to laser forming, in which heat is used to generate internal stresses in the material and thus cause the sheet to bend, is plasma jet forming. For more information on these technologies, see the article "Thermal Forming of Sheet and Plate" in this Volume.

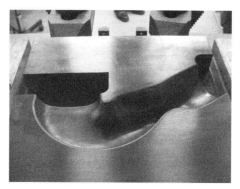

Fig. 1 Low-volume tooling produced by the rapid manufacturing method metal copy. Courtesy of Protatal AB, Sweden

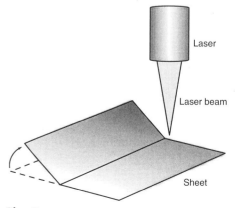

Fig. 2 Bending of sheet via laser forming

(a)

(b)

Fig. 3 Flexible multipoint forming in which the die pins replace both die and stamp. (a) Different modes of multipoint forming can be either passive or produces an active pressure toward the stamp. (b) Flexible multipoint forming of 16 mm (0.63 in.) stainless steel sheet. Courtesy of Unformtechnik Stade GmbH, Germany

Another flexible method with which several process steps can be included in the same line is roll forming. Using "start-stop" rolls or "flying tools," clinching, punching, and welding can also be performed at different stages of sequential bending operations. Special rolls may also be applied to modify the surface of the profile, for example, by "imprinting" a specified pattern. Tools for roll forming have a lower cost than complicated press tools, but a higher cost than press-brake tools, which makes the method applicable for medium-size production lots of beams and elongated panels.

Flexible Forming Media. Reducing the number of stamping dies is another way of lowering part cost. For example, advanced flexible punch-and-die methods based on multipoint methods have been developed in China. Different modes of multipoint forming are shown in Fig. 3.

In such processes, several hydraulic cylinders work against each other to shape a detail. Early try-outs of the process gave rise to poor surface finish, which was ameliorated to some extent by the use of an intermediate sheet.

In rubber-pad forming, one of the tools is replaced by a thick rubber mat that deforms during pressing. This is used, for example, for coining and embossing, in which only a small amount of metal is needed to flow into the formed region. In related hydroforming methods, a fluid replaces one of the two press tools. A method frequently used for prototypes is fluid-cell forming (Fig. 4) or flex forming. One rigid tool is used for simple geometries. The blank is formed between the tool and a rubber diaphragm that is pressurized with castor oil up to 1600 bar. The blank is not in contact with the forming oil. The process characteristics, geometric limitations, and selection criteria are:

Process characteristics:

- Large details, up to maximum press table size: 1.3×2.5 m (4.3×8.2 ft)

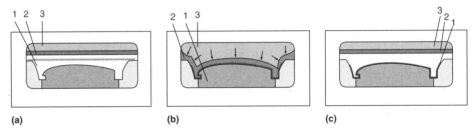

(a)

(b)

(c)

Fig. 4 Fluid-cell forming of detail with undercut. (a, b) The sheet (1) is pressed toward a rigid tool by a rubber membrane (2) filled with castor oil (3). (c) Membrane returns to original position after forming

- Surface roughness: depending on tool surface finish, $R_a = 6$ μm is normal
- Cycle time: Approximately 5 min, independent of product size. (If several tools exist, 2 to 5 smaller products can be formed simultaneously.)
- Time from computer-aided design (CAD) to product: ~2 to 4 weeks
- Tool cost: depending on size, $2000 to $5000
- Changeover time between products: 1 to 8 h, depending on tool complexity
- Cost of production line: $70K to $1M (U.S. dollars, 2006)

Geometric limitations:

- Inner (concave) radii are defined by the forming pressure and cannot reach smaller values than the sheet thickness
- The corner radii and edge radii are dependent on the product height. The lower the product height, the smaller the radii that can be formed.

Selection criteria:

- Low-volume production in which lower tool cost can balance longer cycle time
- Large products with low height and small radii
- Small radii products made from thin-gage mild steels

- Double-curved shapes and flat details that need high-pressure forming to obtain lower springback
- Negative radii—formed if the tool can be divided and withdrawn from the detail

Incremental Forming

Incremental forming of sheet metal comprises the use of rollers, pins, and so forth to form small, local regions of sheet material in an incremental fashion until the entire part is made (Fig. 5). The literature on sheet-metal forming is relatively large. However, only a limited amount of work has focused on incremental sheet forming. Most of the efforts in this area have been conducted in Japan and South Korea.

Background

Experimental investigations of incremental sheet forming have focused on the various aspects of deformation and deformation uniformity and formability. Most authors have used standard three-axis milling machines to investigate incremental forming.

Sheet-metal bulging with a spherical roller is considered in Ref 2 to 5, hammering in Ref 6 and 7, spinning in Ref 8 and 9, peen forming in Ref 10, and bulging with an elastic tool in Ref 11. Incremental forming of sheet metal with a single-point tool is studied in Ref 12 to 14. Reflective surfaces of automobile headlights were prototyped in Ref 14. A new flexible sheet-metal forming system for fabricating micro-three-dimensional structures of metallic foil materials of less than 10 nm in thickness has been developed (Ref 15). Iseki and Naganawa studied multistage forming using special tooling for straightening after incremental forming of walls (Fig. 6) and the bottom surfaces (Fig. 7) of formed parts.

Sheet formability during incremental forming has also been investigated (Ref 16–18). In Ref 16, for example, a forming tool containing a freely rotating ball was used. The results observed in the tests were quantified via grid measurement and finite-element method (FEM) analysis. A unique forming-limit curve (FLC) was obtained. It was found that the FLC is very different from that pertaining to conventional forming. The FLC appears to be a straight line with a negative slope in the positive region of minor strain. It was also observed that cracks during incremental forming occur mostly at the corners due to greater deformation in these regions.

The effects of process parameters (tool size, feed rate, plastic anisotropy) on formability during incremental forming have also been determined (Ref 17). The formability of aluminum sheet under various forming conditions is considered in Ref 18. Complex shapes (e.g., octagonal cones, stepped shapes) were produced with the proposed technique.

Different die technologies and dieless technologies for setup and small lot sizes are compared in Ref 19. The principles and features of recently developed incremental forming

processes have also been summarized (Ref 20, 21). In Ref 21, special attention is paid to spinning of cones and flow forming.

Analytical Modeling of Incremental Forming. Analyses of incremental forming have ranged from relatively simple analytical techniques to detailed FEM analyses.

An approximate plane-strain deformation analysis for the incremental bulging of sheet metal using a spherical roller was developed and applied to describe the bulge forming of non-symmetric shallow shells (Ref 2). In the model, it was assumed that the sheet metal in contact with the roller stretches uniformly. Friction at the interface between the tool and sheet, plastic anisotropy, and the Bauschinger effect were neglected. Closed-form expressions for the uniform strains and the shape of the deformed shell were developed. The tensile force was determined from the condition that the undeformed part is rigidly moved due to the stiffness of the shell. The approximate analytical results were verified by FEM analysis and validated by experiments.

Vertical-wall-surface forming of rectangular shells using multistage incremental forming has also been investigated (Ref 3). A method for calculating the approximate distribution of thickness strain and the maximum bulging height was proposed using a plane-strain deformation model with a constant strain gradient.

A relationship between the blank and its formed configuration under conditions of uniform straining was obtained in Ref 1. The following characteristics of the incremental forming process were assumed:

- The sheet is formed according to a given locus.
- The deformation of the sheet is point-by-point.
- The deformation during every step is small.

Two reasons leading to the unevenness of the wall thickness distribution were discussed in detail:

- Because of springback, the deformation at the center of the blank is smaller than that at the boundary when there is the same value of the deformation.
- For elements of the blank lying at equal distances from the center of the sheet, the portions formed later in the process undergo a certain displacement and strain hardening before the tool bulges them. Thus, these portions develop smaller plastic deformation for the same value of displacement.

The solution for both of these problems was also proposed in Ref 1, that is:

- Because the boundary of the sheet is fixed and the center is free, the stiffness of the boundary is large and the stiffness of the center is small. Therefore, incremental forming steps should be performed in the sequence from low stiffness areas to high stiffness areas. Thus, the plastic deformation of the center can be increased and the plastic deformation of the boundary can be decreased, which reduces the degree of evenness of the deformation.
- Deformation increments should be small, which makes the sheet approach the final shape of the part integrally and evenly. Thus, the strain hardening of the areas formed later in the sequence can be decreased. In practice, the sequence of forming increments should be alternated; that is, the area that is formed first in a given incremental step should be formed later during the next incremental step.

FEM Simulation of Incremental Forming. As described previously, it is difficult to predict the thickness-strain distribution in a part after the accumulation of numerous incremental deformation passes. One option is to calculate the thickness strain during the whole deformation process by using FEM analysis. However, FEM has some difficulties when applied to the incremental sheet-metal forming process. The most critical problem is the large number of calculation steps. Compared with conventional sheet-metal forming processes, incremental forming has a simple deformation mechanism, but the deformation path of its moving tool is much longer. If the entire process has to be analyzed using FEM, therefore, very long computation times may be required (Ref 22).

A simplified FEM model of incremental forming was developed assuming that all deformation occurs only by shear (Ref 22). The intermediate shape was determined from the predicted thickness strain to distribute the deformation uniformly. The proposed method was applied for the analysis of an ellipsoidal cup and a clover cup.

In their work, Kim and Park (Ref 17, 18) used the commercial FEM code PAM-STAMP (Ref 23) for the analysis of incremental forming. PAM-STAMP is a dynamic explicit program that

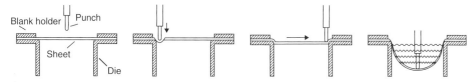

Fig. 5 Incremental sheet-metal forming process. Source: Ref 1

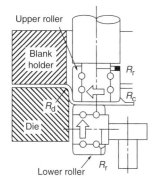

Fig. 6 Vertical wall surface forming. Source: Ref 3

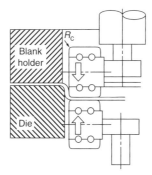

Fig. 7 Bottom surface forming. Source: Ref 3

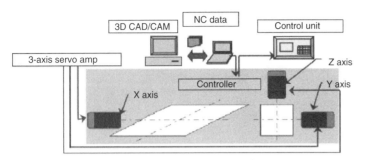

Fig. 8 The control system of a dieless numerically controlled forming machine. Courtesy of Amino Corporation

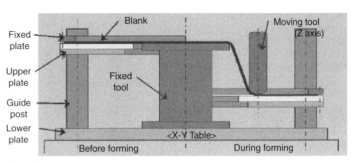

Fig. 9 Principle of dieless NC forming. Courtesy of Amino Corporation

was developed specifically for the simulation of sheet-metal forming processes. It was used to analyze the deformation that occurs in the straight-groove test. The results of this analysis provided enough information to understand observed trends (Ref 17). The model considered the tool, the blankholder, the sheet (blank), and the die, which were all modeled using 3-D shell elements. The tool was moved stepwise vertically by 0.5 mm (0.02 in.). After each vertical step, it was translated on a horizontal plane along rectangular path. Thickness distributions and strains were obtained from the analysis.

In other work, Iseki (Ref 24) used ABAQUS (Ref 25) and 3-D shell elements to simulate the incremental bulging process. A hemispherical tool was moved stepwise vertically. After each vertical step, it moved on a horizontal plane along a rectangular path. Finite-element method predictions were in good agreement with experimental measurements.

Dieless NC Forming

Dieless NC forming is a numerically controlled incremental-forming process developed in Japan that can cold form various materials into complex shapes. The method allows forming without large and expensive dies, using only a simple support tool under the formed piece. This makes the method very cost effective. The technique was developed to meet the needs of the automotive industry as an alternative manufacturing method for small production lot sizes and rapid prototyping. It has been commercialized by the Japanese company Amino Corporation.

Principle of Operation. In dieless NC forming, the geometry of the sheet-metal part is converted from CAD data through a computer-aided manufacturing (CAM) system to NC data. These data are then downloaded to a three-axis CNC machine (Fig. 8), which drives the forming tooling. The blank is clamped to a square blankholder ("work holder") such that there is no draw-in from the binders. The so-called Z-tool moves in the Z (vertical) and Y directions; the work holder is counterbalanced to the vertical movement of the tool and is translated in the vertical direction according to the descent of the Z-tool and along the X and Y directions. Some machines only allow motion along the

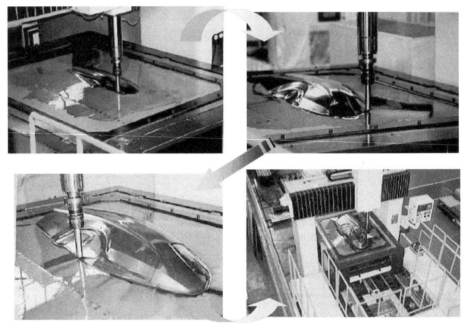

Fig. 10 Dieless NC forming process

X direction. The X and Y plane movements are syncronized with the tool motion to produce the desired form on the sheet.

In practice, a support tool is placed under the work holder and the sheet-metal blank is formed against the support tool (Fig. 9). The moving Z-tool slides over the surface of the blank and presses it into the desired form via stretching and bending. Forming starts on the top of the piece, under which the support tool is placed. The Z-tool makes a round path around the support tool and, after one round, is lowered at a defined pitch and continues forming. The tool path and the vertical pitch are defined by special conversion software based on the CAD model of the product. The tool changes its direction of motion after each round to prevent twisting of material around the Z-axis of the product. The actual forming process is illustrated in Fig. 10.

The Z-tool is made of hardened steel, and its tip is spherical. The minimum diameter of the tool is 6 mm (0.24 in.). The tool diameter affects surface quality; larger diameters result in smoother surface finishes and tool marks are smaller.

The forming force depends on the sheet thickness and material. The force has to exceed the yield strength to create plastic deformation. The forming process requires lubrication on the surface of the sheet. Lubrication decreases the friction between the tool and the sheet and absorbs the heat caused by deformation. The amount of the lubricant depends on the material and thickness of the sheet. Forming of stainless steel and thick sheets requires more lubricant than thick aluminum sheets.

Limitations. Dieless NC forming is suitable for one-piece and low-volume production. It can also be used as a prototyping method for sheet-metal products produced in moderate volumes. Maximum production capacity of a dieless NC forming machine is approximately 500 pieces/month, varying widely depending on the size and complexity of the formed product. The method is slow compared to conventional sheet-metal forming operations such as deep drawing and thus is not competitive with such techniques under full-scale production conditions. Process limitations are described in more detail as follows.

Part Geometry. Dieless NC forming is a suitable method for producing complicated three-dimensional parts. The product geometry is essentially unrestricted except for a few limitations depending on sheet thickness and the equipment used. For example, the tool size defines the minimum rounding radius, which is half of the tool diameter. The tool size depends on the sheet thickness, because small-diameter tools are not strong enough to impart the required forming forces for thick sheets. At present, the smallest tools used in the process have a diameter of 6 mm (0.24 in.); thus, the minimum rounding radius is 3 mm (0.12 in.). However, the machine manufacturer recommends that the smallest rounding radius be 5 mm (0.2 in.).

A second part limitation arises because the forming tool operates only in the Z-direction. This makes inward bent forms impossible. The wall angle is limited by the sheet thickness. The more the sheet is stretched and bent, the more it thins (Fig. 11). The material formability affects the minimum wall angle as well. Hardened sheet tends to undergo shear failure if the wall angle is too steep. Forming tests have shown that the minimum wall angle is approximately 25° for steel, 20° for pure aluminum, and 30 to 35° for heat treatable aluminum alloys.

The product geometry also defines the need and complexity of the support tool. Simple support tools can be used when the part walls do not include horizontal surfaces. In such cases, the sheet can be supported at the highest point of the product, and the walls can be formed without any extra support. Examples of such products and support tooling are shown in Fig. 12.

If there are planar surfaces that have to be accurate, the blank has to be supported from below. The Z-tool can be programmed to form a planar surface also without a support tool beneath, but this easily results in sloping. When forming proceeds to the point at which the wall bends again, the sheet bends instead of stretches and the planar surface twists if the edge is not supported. Figure 13 shows a bathtub formed with dieless NC forming and the support tool used in the forming. All of the planar surfaces and rounding radii are supported while the rest of the walls are formed freely. The more complex the part is, the more detailed the support tool has to be. In some cases, the support tool may be as detailed as the final product.

Product Size. The maximum product size is determined by the machine size. There are five different sizes of machines for research and production use. The technical characteristics of these machines are summarized in Table 1.

The maximum sheet thickness is different for each machine. The critical design factor is the force needed for forming. Because flow stress varies according to the material used, the forming force and thus the maximum sheet thickness depends on the specific material and the machine used. For example, possible material thicknesses ranges from 0.5 to 2 mm (0.02 to 0.08 in.) for stainless steel and 0.5 to 4 mm (0.02 to 0.16 in.) for aluminum (Table 1).

Dieless NC forming enables greater effective drawing ratios than conventional deep drawing. For dieless NC forming, the forming path is different, and tooling geometry does not restrict the material flow as with conventional tooling. When forming starts from the top of the piece and the center of the sheet, the material stretches and moves away from the tool as the forming proceeds. This enables greater effective drawing ratios.

Surface Quality. The Z-tool leaves marks on the formed surface during each forming iteration. The tool marks can be decreased by using tools with large diameter and by decreasing the forming pitch. However, smaller pitch leads to longer forming time, so there is a trade-off between production time and surface quality. Furthermore, if the pitch is smaller than 0.01 mm (0.0004 in.), the material is formed several times at each same spot, and it will strain harden and possibly tear. Strain hardening also increases the local strength of the sheet, which can lead to twisting.

Large, slightly curved surfaces are difficult to form. Tool marks are clear, because the horizontal movement for each pitch is large and the curve is very gentle. Tool marks can be prevented by using a sacrificial sheet on the top of the actual workpiece during the forming process. By this means, tool marks are developed on the sacrificial sheet, which in turn imparts the geometry to the actual sheet below. The geometry thus produced is not as accurate as it would be with direct tool contact, but the surface quality is better. The actual sheet can then be formed again without using a sacrificial sheet in order to sharpen the details in the product.

Hirt, Junk, and Wituski (Ref 26) have quantified how forming affects surface roughness. The forming process smoothes the surface on the forming size of the product, but the surface roughness on the other side of the sheet increases correspondingly. As the tool diameter increases, the surface roughness decreases. The surface roughness can also be decreased by changing the forming direction on each round and decreasing the forming pitch. Inasmuch as all portions of a blank are not formed equally, some regions staying completely unformed, the surface roughness is not constant over the product surface. Planar surfaces are always undeformed; thus, the surface roughness in such regions remains the same as the surface roughness of the original material. Steep wall angles require more forming and small pitch; hence, the surface roughness is decreased during the forming.

Materials. Dieless NC forming has been tested with sheets of carbon steels, alloy steels,

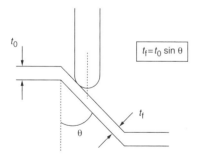

Fig. 11 Correlation between sheet thinning and bending. The final thickness (t_f) depends on initial thickness (t_0) and bending angle (θ). Source: Ref 26

$$t_f = t_0 \sin \theta$$

Fig. 13 Bathtub measuring ~1500×1000 mm (60 ×40 in.) and the corresponding support tool

Fig. 12 Simple support tool and parts made with it

Table 1 Technical data for dieless NC forming machines

	Research use			Commercial use	
	DLNC-RA	DLNC-RB	DLNC-PA	DLNC-PB	DLNC-PC
Maximum blank size, mm	400×400	600×600	1100×900	1600×1300	2100×1450
Maximum forming size, mm	300×300	500×500	1000×800	1500×1200	2000×1350
Maximum forming depth, mm	150	250	300	400	500
X-axis stroke, mm	330	550	1100	1600	2100
Y-axis stroke, mm	330	550	900	1300	1450
Z-axis stroke, mm	200	300	350	450	550
Maximum work holder size, mm	500×500	750×950	1300×110	1800×1500	2300×1650
Maximum sheet thickness, mm					
Stainless steel	0.5–1	0.5–1	0.5–2	0.5–2	0.5–2
Aluminum	0.5–3	0.5–3	0.5–4	0.5–4	0.5–4

stainless steels, aluminum alloys, titanium, and coated steels. The best results with regard to control of final geometry and springback were achieved with deep-drawing steels. Alloy steels are also easy to form, but they exhibit large springback because of their high strengths. Stainless steels require large forming forces, which lead to thinner sheets than when forming mild (carbon) steels. Stainless steels also exhibit large springback, and parts tend to twist during the forming process. In addition, stainless steel requires good lubrication during forming, and the forming speed has to remain low.

The formability of aluminum depends on the specific alloy used. Unalloyed aluminum (designation 1000 and 1100-O) is easy to form. Aluminum alloys in the 6xxx series, which contain magnesium and silicon, have poor formability and exhibit large springback. Therefore, pure aluminum is recommended for forming if it is suitable for the intended product.

Titanium heats up during forming and tends to blister. Forming coated sheets is difficult, because the coating comes off easily under the forming tool. If the sheet can be formed on the reverse side, the coating on the face side remains untouched.

It is also possible to form perforated steel using dieless NC forming. Here, the minimum wall clearance is approximately 45°. The holes stretch as the sheet stretches. However, when the clearance angle and rounding radii are kept large, the sheet does not tear during forming.

Economics of Incremental Sheet-Metal Forming. Incremental sheet forming costs have been compared with the cost of deep drawing for the simple example shown in Fig. 14. The cost comparison between these methods was based on the following relation (Ref 27, 28):

$$C = VC_{mv} + P_c R_c$$

in which C_{mv} is the cost of material per unit volume, V is the volume of material input to the process, P_c is the basic processing cost for an ideal part, and R_c is the cost coefficient for the part design that takes into account shape complexity, surface finish, and tolerances.

For the present example, the cost of material is equal for deep drawing and incremental forming and can thus be disregarded in the calculations. The basic processing cost is given by:

$$P_c = aT$$

where a is the cost of operating a specific process, including the cost of the machine and its maintenance, labor, and overhead, and T is the cycle time.

To calculate the cost of incremental forming for the example part, two different sets of processing parameters were examined, one set from experimental work done at the University of Saarland and the second set recommended by the Amino Corporation. The first set of parameters comprised a horizontal forming speed of 15 m/min (50 ft/min), vertical feed of 0.2 mm/step (0.08 in./step), and total length of the forming path of 1465 m (0.9 mi). The second set consisted of a horizontal forming speed of 30 m/min (98 ft/min), vertical feed of 0.5 mm/step (0.02 in.), and total length of the forming path of 585 m (1920 ft).

In approximate numbers, the cost of setting up and NC programming was $60 in both cases. The cost of deep-drawing dies was estimated to be $16,000. According to the chosen processing parameters, the unit price per piece can be calculated. The costs are dependent on the cost of labor, electricity, and other consumable costs that vary in each country. According to the calculations, incremental forming is profitable if the example part is made in a lot size of 1000 parts or less (Fig. 15). If the quantity is greater, deep drawing is more profitable when using the forming parameters recommended by Amino. If the research parameters are used, the break-even point is lower, that is, ~200 parts (Fig. 15).

It should be noted that cost of a production method is also highly dependent on product geometry and can vary a great deal. The previous calculation is an example and can be used as general reference, but it should not necessarily be applied to other parts without additional calculations. Nevertheless, Hirt, Ames, and Bambach (Ref 29) have estimated the economic and environmental aspects of incremental forming and found similar results.

Variants of Incremental Forming. Dieless NC forming is one variant of incremental forming. There are a few other related variants as well that are being developed in parallel with dieless NC forming.

Single-Point Forming. Dieless NC forming is sometimes called two-point forming because there are two contact points with the sheet, one between the sheet and the forming tool and the other between the sheet and the support tool. In single-point forming, the support tool is eliminated, and only the forming tool is in contact with the sheet. The principle of the method is shown in Fig. 16.

In one of the first investigations of single-point forming, Jeswiet used a three-axis CNC mill together with CAM software to control the shaping tool (Ref 12, 30). The sheet was clamped to a blankholder mounted on the mill bed. The shaping tool was a rounded rod, which was placed in the mill spindle. The blankholder remained stationary as the shaping tool pushed into the sheet, deforming it directly under the tool. During the forming, the tool was lowered in small steps after each round. Initially, Jeswiet used support under the sheet, but as the work progressed, he found that the supporting tool was not necessary. As for two-point forming, the finished part shape was created by computer-controlled tool movements.

Robot-Assisted Incremental Forming. Industrial robots can be used to control the forming tool similar to dieless NC forming (Ref 31–34). The process is comparable to, and the forming results are also very similar to, those achieved with dieless NC forming machines. The advantage of industrial robot is the possibility to turn the forming tool and thus access blind edges of parts. Also, many companies already have robots in-house, which makes it easier to take this

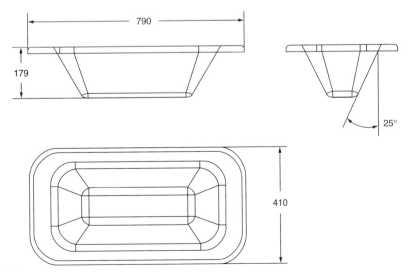

Fig. 14 Dimensions (in mm) of part used in process-economics calculations

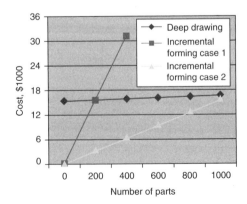

Fig. 15 Cost comparison diagram for incremental forming and deep drawing

method into use. A robot cell for incremental forming is shown in Fig. 17.

Incremental Hammering. Hammering is perhaps the oldest form of sheet-metal forming. The modern method of incremental hammering utilizes computer-controlled robots without a support tool. The sheet-metal blank is attached to a supporting frame, and the forming tool is attached to an industrial robot. The robot punches the sheet into the desired form using a circular forming path, descending a small step for each round. The tool does not touch the sheet while moving in the X-Y directions. This eliminates sheet stretching and twisting during forming, but also affects the surface quality of the product. An experimental incremental hammering setup is shown in Fig. 18.

Applications. Dieless NC forming is well suited for producing single pieces as well as small lot sizes. It can also be used in prototyping and the production of spare parts. Forming costs are approximately 5 to 10% of the costs of traditional press forming, but the production speed is also lower. Despite the slower speed, the method is more efficient when producing single parts or small quantities of parts.

When using dieless NC forming for making spare parts, considerable savings are realized because large dies need not be kept in inventory. Die storage is a problem especially in the automotive manufacturing industry, because parts are generally large and the product life cycle is relatively long. This leads to long storage times and costs. New car models are constantly developed, and the number of dies that must be stored increases.

Dieless NC forming enables the manufacture of different kinds of parts in the same machine without long lead times. The only part-specific component of the machine setup is the support tool, which is easy and quick to change. If the support tool is made of a soft material such as wood or polymer, it is also quick and affordable to manufacture.

TAI Research Center in the Helsinki University of Technology implemented a research project together with Finnish industrial partners in 2002 to assess the suitability of dieless NC forming in a real production environment. The test results showed that part geometry was important when considering the suitability of the method. The test parts in the project were real parts in production; they were not designed specifically for dieless NC forming. This fact affected the forming results. If a part is designed for the manufacturing method, the special requirements of the method can be taken into account, and the final results will be better. When applying the new method directly for an existing part, however, there are generally more difficult shapes and other features that provide challenges for the method. These problems can be avoided by designing the part specifically taking into account the capabilities and limitations of the manufacturing technique.

Unlike automobile manufacture, the manufacture of boats usually comprises low-volume production. The design of the boat (or ship) is determined by its intended use, its seagoing properties, and general characteristics regarding shape and equipment. They are typically made from sheets or plates of various thicknesses, which are formed and joined, most often by welding. The sheet segments are usually joined to an underlying frame of profiles or plate components and vary in shape from one segment to the next. The combination of flexibility in shape, adaptation to an underlying frame, and joining of adjacent segments represents a major challenge to boat builders. In order to reduce the amount of sizing after forming and distortion after welding due to residual stresses, low-cost, accurate forming methods such as dieless manufacture may be very attractive.

Conclusions

Low-volume production methods are already in use in many industries. Incremental forming is a new production method for sheet-metal components that brings several new possibilities into the sheet-metal forming field. Geometries that were previously impossible or very expensive to prototype are now possible. Prototypes can also be made from actual production materials.

One incremental-forming process, dieless NC forming, has great potential as a tool in the product development cycle. It can be used for making prototypes, but also individual parts and small lot sizes of parts. A great advantage of such techniques is that prototypes can be made from real materials with real methods and thus be tested as real products. Dieless NC forming can also be combined with other forming methods when manufacturing prototypes or short series of products.

Incremental forming, as any other forming method, has its limitations. Part geometry (e.g., wall angle restrictions, minimum rounding radii) and the number of forming steps are important considerations. On the other hand, process

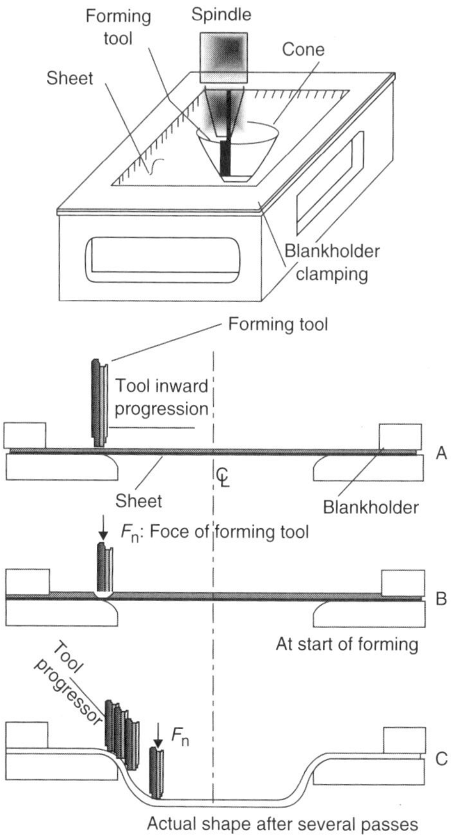

Fig. 16 Single-point forming with an industrial robot

Fig. 17 Robot cell for incremental forming

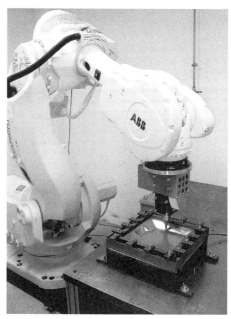

Fig. 18 Incremental hammering

design variables such as forming path can be chosen to overcome some of the limitations such as wall angle.

REFERENCES

1. K. Dai, Z.R. Wang, and Y. Fang, CNC Incremental Sheet Forming of an Axially Symmetric Specimen and the Locus of Optimization, *J. Mater. Process. Technol.*, Vol 102, 2000, p 164–167
2. M. Li et al., Multi-Point Forming: A Flexible Manufacturing Method for a 3-D Surface Sheet, *J. Mater. Process. Technol.*, Vol 87, 1999, p 277–280
3. H. Iseki and T. Naganawa, Vertical Wall Surface Forming of Rectangular Shell Using Multistage Incremental Forming with Spherical and Cylindrical Rollers, *J. Mater. Process. Technol.*, Vol 130–131, 2002, p 675–679
4. H. Iseki, K. Kato, and S. Sakamoto, Flexible and Incremental Sheet Metal Bulging Using Path-Controlled Spherical Rollers, *Trans. Jpn. Soc. Mech. Eng.*, Vol C58, 1993, p 3147–3155 (in Japanese)
5. H. Iseki, K. Kato, O. Kumon, and K. Ozaki, Flexible and Incremental Sheet Metal Bulging Using a Few Controlled Spherical Rollers, *Trans. Jpn. Soc. Mech. Eng.*, Vol C59, 1994, p 2849–2854 (in Japanese)
6. N. Nakajima, Computerizing of Traditional Sheet Metal Forming, *Trans. Jpn. Soc. Tech. Plasticity*, Vol 20, 1979, p 686–700 (in Japanese)
7. T. Hasebe and S. Shima, Study of Flexible Forming by Hammering, *J. Jpn. Soc. Tech. Plasticity*, Vol 35, 1994, p 1323–1329 (in Japanese)
8. S. Matsubara, Incremental Backward Bulge Forming of a Sheet Metal with a Hemispherical Head Tool, *J. Jpn. Soc. Tech. Plasticity*, Vol 35, 1994, p 1311–1316 (in Japanese)
9. K. Kitazawa, A. Wakabayashi, K. Murata, and J. Seino, A CNC Incremental Sheet Metal Forming Method for Producing the Shell Components Having Sharp Corners, *J. Jpn. Soc. Tech. Plasticity*, Vol 35, 1994, p 1348–1353 (in Japanese)
10. K. Kondo, S. Mtsuzaki, M. Hiraiwa, and K. Ohga, Investigation of Peen Forming, *J. Jpn. Soc. Mech. Eng.*, Vol III (No. 44), 1992, p 3663–3672 (in Japanese)
11. M. Matsubara, S. Tanaka, and T. Nakamura, Forming Process of Spherical Surface by A CNC Incremental Press Forming System, *J. Jpn. Soc. Tech. Plasticity*, Vol 35, 1994, p 1330–1335 (in Japanese)
12. J. Jeswiet, "Incremental Single Point Forming," technical paper, Society of Manufacturing Engineers, 2001
13. J. Jeswiet and E. Hagan, "Rapid Prototyping Non-Uniform Shapes from Sheet Metal Using CNC Single Point Incremental Forming," technical paper, Society of Manufacturing Engineers, 2003
14. J. Jeswiet and E. Hagan, "Rapid Prototyping of a Headlight with Sheet Metal," technical paper, Society of Manufacturing Engineers, 2002
15. Y. Saotome and T. Okamoto, An in-situ Incremental Microforming System for Three-Dimensional Shell Structures of Foil Materials, *J. Mater. Process. Technol.*, Vol 113, 2001, p 636–640
16. M.S. Shim and J.J. Park, The Formability of Aluminum Sheet in Incremental Forming, *J. Mater. Process. Technol.*, Vol 113, 2001, p 654–658
17. Y.H. Kim and J.J. Park, Effect of Process Parameters on Formability in Incremental Forming of Sheet Metal, *J. Mater. Process. Technol.*, Vol 130–131, 2002, p 42–46
18. Y.H. Kim and J.J. Park, Fundamental Studies on the Incremental Sheet Metal Forming Technique, *J. Mater. Process. Technol.*, Vol 140, 2003, p 447–453
19. T. Naganawa, Advances in Prototype and Low Volume Sheet Forming and Tooling, *J. Mater. Process. Technol.*, Vol 98, 2000, p 244–250
20. S. Shima, State of Art—Incremental Forming, Third International Conference, IPMM-2001, Intelligent Processing and Manufacturing of Materials
21. M. Strano, "Incremental Forming Processes: Current and Potential Applications," technical paper, Society of Manufacturing Engineers, 2003
22. T.J. Kim and D.Y. Yang, Improvement of Formability for the Incremental Sheet Metal Forming Process, *Int. J. Mech. Sci.*, Vol 42, 2000, p 1271–1286
23. http://www.esi-group.com/Products/Pamstamp, 2003
24. H. Iseki, An Approximate Deformation Analysis and FEM Analysis for the Incremental Bulging of Sheet Metal Using a Spherical Roller, *J. Mater. Process. Technol.*, Vol 111, 2001, p 150–154
25. http://www.abaqus.com, 2003
26. G. Hirt, S. Junk, and N. Wituski, Surface Quality, Geometric Precision and Sheet Thinning in Incremental Sheet Forming, *Materials Week 2001* (Munich, Germany), Oct 1–4, 2001, Werkstoffwoche-Partnerschaft GBr
27. K.G. Swift and J.D. Booker, op.cit.; so
28. A.J. Allen and K.G. Swift, *Proc. Inst. Mech. Eng.*, Vol 204, 1990, p 143–148
29. G. Hirt, J. Ames, and M. Bambach, "Economical and Ecological Benefits of CNC Incremental Sheet Forming," First IMEKO TC 19 Conference on Environmental Measurements (Budapest, Hungary), Oct 13–17, 2003
30. J. Jeswiet, E. Hagan, and A. Szekeres, Forming Parameters for Incremental Forming of Aluminum Alloy Sheet Metal, *Proc. Inst. Mech. Eng., Part B, J. Eng. Manuf.*, Vol 216, 2002, p 1367–1371
31. L. Lamminen and J. Tuomi, "Creating Design Rules for Incremental Sheet Forming," 18th International Conference on Production Research (ICPR-18) (Salerno, Italy), July 31 to Aug 4, 2005
32. L. Lamminen, Incremental Sheet Forming with an Industrial Robot—Forming Limits and Their Effect on Component Design, M. Geiger et al., Ed. *Proc. International Conference SheetMetal 2005 (SheMet '05)*, (Erlangen, Germany), April 5–8, 2005, Trans Tech Publications Ltd, 2005
33. L. Lamminen, T. Tuominen, and S. Kivivuori, Incremental Sheet Forming with an Industrial Robot, *Proc. Third International Conference on Advanced Materials Processing (ICAMP-3)* (Melbourne, Australia), Nov 29 to Dec 1, 2004, p 331–335
34. J. Tuomi, L. Lamminen, Incremental Sheet Forming as a Method for Sheet Metal Component Prototyping and Manufacturing, *Proc. 10th European Forum on Rapid Prototyping* (Paris, France), Sept 14–15, 2004, 7 pages

Reference Information

Glossary of Terms

A

abnormal grain growth. Rapid, nonuniform, and usually undesirable growth of one or a small fraction of grains in a polycrystalline material during annealing. The phenomenon is most frequent in fine-grained materials in which a larger-than-average grain (or grains) consumes surrounding small grains whose growth is limited by particle pinning. Also known as *secondary recrystallization.*

aging. A change in material property or properties with time. See also *quench aging* and *strain aging.*

air bend die. Angle-forming dies in which the metal is formed without striking the bottom of the die. Metal contact is made at only three points in the cross section: the nose of the male die and the two edges of a V-shaped die opening.

air cushion holddown. A pneumatic pressure device used to control pressure on the blank. See also *blank holder.*

air ejection. The ejection of small parts away from the die by a blast of air.

air-lift hammer. A type of gravity-drop hammer in which the ram is raised for each stroke by an air cylinder. Because length of stroke can be controlled, ram velocity and therefore the energy delivered to the workpiece can be varied. See also *drop hammer* and *gravity hammer.*

Almen machine. A wear test machine in which a rotating journal is held between two stationary bushings. Load is applied by a lever arrangement.

anelasticity. Material characteristic where there is no specific relationship between stress and strain but rather among time, stress, and strain. In particular, anelastic strains can occur at zero stress, whereas creep strains occur as a result of imposed stress.

angle of bite. In the rolling of metals, the location where all of the force is transmitted through the rolls; the maximum attainable angle between the roll radius at the first contact and the line of roll centers. Operating angles less than the angle of bite are termed contact angles or rolling angles.

angularity. The conformity to, or deviation from, specified angular dimensions in the cross section of a shape or bar.

anisotropy. Variations in one or more physical or mechanical properties with direction. See also *normal anisotropy, planar anisotropy,* and *plastic-strain ratio.*

annealing. Softening or strain relieving of a material by application of heat above the critical temperature for the correct time interval, then cooling slowly enough to avoid hardening.

anticlastic curvature. A configuration of a sheet or beam that is curved longitudinally in one direction and transversely in the opposite direction, as the surface of a saddle. Such shapes occur naturally in bending of thin structures because of *Poisson contraction.*

anvil. A large, heavy metal block that supports the frame structure and holds the stationary die of a forging hammer. Also, the metal block on which blacksmith forgings are made.

anvil cap. Same as *sow block.*

asperities. Protrusions rising above the general surface contours that constitute the actual contact areas between touching surfaces.

automatic press. A press with built-in electrical and pneumatic control in which the work is fed mechanically through the press in synchronism with the press action.

automatic press stop. A machine-generated signal for stopping the action of a press, usually after a complete cycle, by disengaging the clutch mechanism and engaging the brake mechanism.

Avrami plot. Plot describing the kinetics of phase transformations in terms of the dependence of fraction X of microstructure that has transformed (e.g., recrystallized, decomposed, etc.) as a function of time (t) or strain (ε). Avrami plots usually consist of a graph of log $[\ln(1/(1-X))]$ versus log t (or log ε) and are used to determine the so-called Avrami exponent n in the relation $X = 1 - \exp(-Bt^n)$, where B is an empirically derived parameter.

axial rolls. In *ring rolling,* vertically displaceable, tapered rolls mounted in a horizontally displaceable frame opposite to, but on the same centerline as, the main roll and rolling mandrel. The axial rolls control ring height during rolling.

B

backward extrusion. Same as indirect extrusion. See also *extrusion.*

bar. (1) A section hot rolled from a *billet* to a form, such as round, hexagonal, octagonal, square, or rectangular, with sharp or rounded corners or edges and a cross-sectional area of less than 105 cm^2 (16 in.^{2}). (2) A solid section that is long in relationship to its cross-sectional dimensions, having a completely symmetrical cross section and a width or greatest distance between parallel faces of 9.5 mm (^{3}/$_8$ in.) or more.

barreling. Convexity of the surfaces of cylindrical or conical bodies, often produced unintentionally during upsetting or as a natural consequence during compression testing. See also *compression test.*

Bauschinger effect. A reduction in yield strength on restraining a material in the opposite direction to the initial testing. Also used more generally to denote the changes of yield stress and strain hardening following an abrupt change of strain path.

bead. A narrow ridge in a sheet-metal workpiece or part, commonly formed for reinforcement. See also *drawbead.*

beaded flange. A flange reinforced by a low ridge, used mostly around a hole.

beading. A forming operation in which a ridge or elongated projection is raised on sheet metal.

beam. A slender, flexible body with approximately equal short dimensions and one much larger dimension. When the short dimensions are greatly different, a beam becomes a sheet or a strip.

bed. (1) Stationary platen of a press to which the lower die assembly is attached. (2) Stationary part of the shear frame that supports the material being sheared and the fixed blade.

bed deflection. The deflection of the bed as a result of the rated press load. It is usually measured in inches per foot of span between supports.

bendability. The ability of a material to be bent around a specified radius without fracture.

bend allowance. The amount of sheet metal required for making a bend around a given radius. Bend allowance (*BA*) is calculated by

$BA = D(0.01743R + 0.0078T)$, where D is the number of degrees in the bend, R is the inside radius of the bend, and T is the thickness of the metal.

bend angle. The angle through which a bending operation is performed, that is, the supplementary angle to that formed by the two bend tangent lines or planes.

bending. The deformation of a part, usually a flat sheet or strip of metal, by moving it into a circular shape. See also *bending stress.*

bending and forming. The processes of bending, flanging, folding, twisting, offsetting, or otherwise shaping a portion of a blank or a whole blank, usually without materially changing the thickness of the metal.

bending brake or press brake. A form of open-frame, single-action press that is comparatively wide between the housings, with a bed designed for holding long, narrow forming edges or dies. Used for bending and forming strip, plate, and sheet (into boxes, panels, roof decks, and so on).

bending dies. Dies used in presses for bending sheet metal or wire parts into various shapes. The work is done by the punch pushing the stock into cavities or depressions of similar shape in the die or by auxiliary attachments operated by the descending punch.

bending moment (or moment). A set of balanced torques applied to a body at some distance apart in order to induce bending without overall tension or compression of the body. It is expressed in terms of force times distance and may be computed by integrating the differential force times distance from the center of the section through the section.

bending rolls. Various types of machinery equipped with two or more rolls to form curved sheet and sections.

bending strain. The tensile and compressive strains that occur throughout the thickness of a bent beam or sheet. They are usually assumed to vary linearly with distance from the neutral axis (Z) and are equal to Z/R, where R is the radius of curvature through the neutral axis.

bending stress. A stress involving tensile and compressive forces, which are not uniformly distributed. Its maximum value depends on the amount of flexure that a given application can accommodate. Resistance to bending can be termed stiffness.

bending under tension. A forming operation in which a sheet is bent with the simultaneous application of a tensile stress perpendicular to the bend axis.

bend (longitudinal). A forming operation in which the axis is perpendicular to the rolling direction of the sheet.

bend or twist (defect). Distortion similar to warpage generally caused during forging or trimming operations. When the distortion is along the length of the part, it is termed bend; when across the width, it is termed twist. When bend or twist exceeds tolerances, it is

considered a defect. Corrective action consists of hand straightening, machine straightening, or cold restriking.

bend radius. The radius of a bend that corresponds to the curvature of a bent specimen or the bent area in a formed part. The radius is usually measured on the inside surface of the part but may refer to the center section or outer surface for some applications.

bend test. Evaluation of a sheet-metal response to a bending operation, such as around a fixed-radius tool.

bend (transverse). A forming operation in which the bend axis is parallel to the rolling direction of the sheet.

biaxial stretchability. The ability of sheet material to undergo deformation by loading in tension in two directions in the plane of the sheet.

billet. (1) A semifinished section that is hot rolled from a metal *ingot,* with a rectangular cross section usually ranging from 105 to 230 cm^2 (16 to 36 in.^{2}), the width being less than twice the thickness. Where the cross section exceeds 230 cm^2 (36 in.^{2}), the term *bloom* is properly but not universally used. Sizes smaller than 105 cm^2 (16 in.^{2}) are usually termed bars. (2) A solid semifinished round or square product that has been hot worked by forging, rolling, or extrusion. See also *bar.*

bite. Advance of material normal to the plane of deformation and relative to the dies prior to each deformation step.

blank. (1) In forming, a piece of sheet material, produced in cutting dies, that is usually subjected to further press operations. (2) A piece of stock from which a forging is made; often called a *slug* or *multiple.*

blank development. The process of determining the optimal size and shape of a blank for a specific part.

blank gridding. Imprinting a metal blank with a pattern on the sheet surface, such pattern to be used for subsequent strain measurement. See also *gridding.*

blankholder. That part of a forming die that holds the blank by pressure against a mating surface of the die to control metal flow and prevent wrinkling. The blankholder is sometimes referred to as the holddown or binder area. Pressure is applied by mechanical means, springs, air, or fluid cushions.

blankholder pressure. The pressure exerted by the blankholder against the blank. This is normally adjustable to control metal flow during the drawing.

blankholder slide. The outer slide of a double-action press operated by toggles or cams that transfers the pressure to the blankholder.

blanking. The operation of punching, cutting, or shearing a piece out of stock to a predetermined shape.

blanking die. A die used for shearing or cutting blanks, usually from flat sheets or strip. The single blanking die used for producing one blank at each stroke of the press is the simplest

of all dies, consisting essentially of punch, die block, and stripper.

blind die compression/pressing. Compression in an extrusion chamber in which the die orifice has been closed off.

blister. A local protuberance in the surface of the sheet, often elongated, resulting from an internal separation due to the expansion of entrapped gas. The gas may be entrapped during casting, pickling, annealing, or electroplating in a previously existing subsurface defect.

block. A preliminary forging operation that roughly distributes metal preparatory for *finish.*

block and finish. The forging operation in which a part to be forged is blocked and finished in one heat through the use of tooling having both a block impression and a finish impression in the same die block.

blocker dies. Dies having generous contours, large radii, draft angles of 7° or more, and liberal finish allowances. See also *finish allowance.*

blocker-type forging. A forging that approximates the general shape of the final part with relatively generous *finish allowance* and radii. Such forgings are sometimes specified to reduce die costs where only a small number of forgings are desired and the cost of machining each part to its final shape is not excessive.

block, first, second, and finish. The forging operation in which a part to be forged is passed in progressive order through three tools mounted in one forging machine; only one heat is involved for all three operations.

blocking. A forging operation often used to impart an intermediate shape to a forging, preparatory to forging of the final shape in the finishing impression of the dies. Blocking can ensure proper working of the material and can increase die life.

blocking impression. The impression that gives a forging its approximate shape.

bloom. A semifinished hot rolled product, rectangular in cross section, produced on a blooming mill. See also *billet.* For steel, the width of a bloom is not more than twice the thickness, and the cross-sectional area is usually not less than approximately 230 cm^2 (36 in.^{2}). Steel blooms are sometimes made by forging.

blooming mill. A primary rolling mill used to make blooms.

board hammer. A type of forging hammer in which the upper die and ram are attached to "boards" that are raised to the striking position by power-driven rollers and let fall by gravity. See also *drop hammer.*

bolster. Short for bolster plate. A heavy plate or block, sometimes called a die block, secured to the top of the bed of a press onto which a die is fastened. In some cases, the die is fastened directly to the bed and a bolster is not used.

boss. A relatively short, often cylindrical protrusion or projection on the surface of a forging.

bottom draft. Slope or taper in the bottom of a forge depression that tends to assist metal flow toward the sides of depressed areas.

bottoming bending. Press-brake bending process in which the upper die (punch) enters the lower die and coins or sets the material to eliminate *springback*.

bow. The tendency of material to curl downward during shearing, particularly when shearing long narrow strips.

brake forming. A forming process in which the principal mode of deformation is bending. The equipment used for this operation is commonly referred to as a press brake.

brake press. A form of open-frame, single-action press comparatively wide between the housings, with the bed designed for holding long narrow forming edges or dies. It is used for bending and forming strips and plates.

brakes. Friction brakes, set with compression springs provided on a press. Brake capacity must be sufficient to stop the motion of the slide quickly and must be capable of holding the slide and its attachments at any point in its travel.

Bravais lattices. The 14 possible three-dimensional arrays of atoms in crystals (see *space lattice*).

breakdown. (1) An initial rolling or drawing operation, or a series of such operations, for reducing an ingot or extruded shape to the desired size before the finish reduction. (2) A preliminary press-forging operation.

brick element. A common name for a solid, linear finite element with eight nodes.

Bridgman correction. Factor used to obtain the flow stress from the measured axial stress during tension testing of metals in which necking has occurred.

brittle fracture. A fracture that occurs without appreciable plastic deformation.

brittleness. A tendency to fracture without appreciable plastic deformation.

buckling. A bulge, bend, kink, or other wavy condition of the workpiece caused by compressive stresses. See also *compressive stress.*

bulge test. A test wherein the blank is clamped securely around the periphery and, by means of hydrostatic pressure, the blank is expanded. The blank is usually gridded so that the resulting strains can be measured. This test is usually performed on large blanks of 20 to 30 cm (8 to 12 in.) in diameter.

bulging. The process of increasing the diameter of a cylindrical shell (usually to a spherical shape) or of expanding the outer walls of any shell or box shape whose walls were previously straight.

bulk forming. Forming processes, such as extrusion, forging, rolling, and drawing, in which the input material is in billet, rod, or slab form and a considerable increase in surface-to-volume ratio in the formed part occurs under the action of largely compressive loading. Compare with *sheet forming.*

bull block. A machine with a power-driven revolving drum for cold drawing wire through a drawing die as the wire winds around the drum.

bulldozer. Slow-acting horizontal *mechanical press* with a large bed used for bending and straightening. The work is done between dies and can be performed hot or cold. The machine is closely allied to a forging machine.

Burgers vector. The crystallographic direction along which a *dislocation* moves and the unit displacement of dislocations; the magnitude of the Burgers vector is the smallest unit distance of slip in the direction of shear due to the movement of one dislocation.

burnishing. The smoothing of one surface through frictional contact with another surface.

burr. A thin ridge or roughness left on forgings or sheet-metal blanks by cutting operations such as slitting, shearing, trimming, blanking, or sawing.

burr side. The side or face of a blank or other stamping that comes in direct contact with the punch in a blanking operation, and the side or face of a blank or other stamping that comes in direct contact with the die because of the clearance between punch and die.

buster. A pair of shaped dies used to combine preliminary forging operations, such as edging and blocking, or to loosen scale.

C

camber. The tendency of material being sheared from sheet to bend away from the sheet in the same plane.

cam press. A mechanical press in which one or more of the slides are operated by cams; usually a double-action press in which the blankholder slide is operated by cams through which the dwell is obtained.

canned extrusion. A coextrusion process in which the billet consists of a clad material, or can, that is relatively ductile and nonreactive and the core is a reactive, brittle, powder, or other material.

canning. (1) A dished distortion in a flat or nearly flat sheet-metal surface, sometimes referred to as oil canning. (2) Distortion of a flat or nearly flat metal surface that can be deflected by finger pressure but that will return to its original position when the pressure is removed. (3) Enclosing a highly reactive metal within a relatively inert material for the purpose of hot working without undue oxidation of the reactive metal.

capacity of a press. The pressure, in tons, that the slide will safely exert at the bottom of the stroke in doing work within the range of the press.

cavitation. The formation of microscopic cavities during the cold or hot deformation of metals, generally involving a component of tensile stress. Cavities may nucleate at second-phase particles lying within grains or at grain boundaries (with or without particles) as a result of slip intersection or grain-boundary sliding. Under severe conditions, cavities may grow and coalesce to give rise to fracture.

cell. Micron-sized volume bounded by low misorientation walls comprised of dislocation tangles.

center bursting. Internal cracking due to tensile stresses along the central axis of products being extruded or drawn.

chamfer. (1) A beveled surface to eliminate an otherwise sharp corner. (2) A relieved angular cutting edge at a tooth corner.

channel forming. A typical sheet-forming operation where a top-hat section is formed from a flat sheet by pushing a long U-shaped punch into a mating die cavity. Drawbeads are not usually used for such operations.

check. (1) A crack in a die impression corner, generally due to forging strains or pressure, localized at some relatively sharp corner. Die blocks too hard for the depth of the die impression have a tendency to check or develop cracks in impression corners. (2) One of a series of small cracks resulting from thermal fatigue of hot forging dies.

chord modulus. The slope of the chord drawn between any two specific points on a stress-strain curve. See also *modulus of elasticity.*

chuckhead. See *manipulator.*

circle grid. A regular pattern of circles, often 2.5 mm (0.1 in.) in diameter, marked on a sheet-metal blank.

circle-grid analysis. The analysis of deformed circles to determine the severity with which a sheet metal blank has been deformed.

circumferential strains. The tensile and compressive strains that arise from bending and tension of a beam or sheet; they lie in the deformed sheet plane that adopts a locally circular aspect after bending.

clad. Outer layer of a coextruded or codrawn product. See also *sleeve.*

clamping pressure. Pressure applied to a limited area of the sheet surface, usually at the periphery, to control or limit metal flow during forming.

clearance. (1) In punching and shearing dies, the gap between the die and the punch. (2) In forming and drawing dies, the difference between this gap and metal thickness.

closed-die forging. The shaping of hot metal completely within the walls or cavities of two dies that come together to enclose the workpiece on all sides. The impression for the forging can be entirely in either die or divided between the top and bottom dies. Impression-die forging, often used interchangeably with the term *closed-die forging,* refers to a closed-die operation in which the dies contain a provision for controlling the flow of excess material, or *flash,* that is generated. By contrast, in flashless forging, the material is deformed in a cavity that allows little or no escape of excess material.

closed dies. Forging or forming impression dies designed to restrict the flow of metal to the cavity within the die set, as opposed to open dies, in which there is little or no restriction to lateral flow.

closed pass. A pass of metal through rolls where the bottom roll has a groove deeper than the bar being rolled and the top roll has a collar fitting into the groove, thus producing the desired shape free from *flash* or *fin*.

close-tolerance forging. A forging held to unusually close dimensional tolerances so that little or no machining is required after forging. See also *precision forging*.

cluster mill. A rolling mill in which each of two small-diameter work rolls is supported by two or more backup rolls.

clutch. The coupling mechanism used on a mechanical power press to couple the flywheel to the crankshaft, either directly or through a gear train. A full-revolution clutch is a type of clutch that, when tripped, cannot be disengaged until the crankshaft has completed a full revolution and the press slide a full stroke. A part-revolution clutch is a type of clutch that can be disengaged at any point before the crankshaft has completed a full revolution and the press slide a full stroke.

coarsening. The increase in the average size of second-phase particles, accompanied by the reduction in their number, during annealing, deformation, or high-temperature service exposure. Coarsening thus leads to a decrease in the total surface energy associated with the matrix-particle interfaces.

codrawing. The simultaneous drawing of two or more materials to form an integral product.

coefficient of friction. A measure of the ease with which one body will slide over another. It is obtained by dividing the tangential force resisting motion between the two bodies by the normal force pressing the two bodies together.

coefficient of thermal expansion. Defines the amount by which one unit length of a material changes (expands or contracts) when the temperature changes by one degree.

coextrusion. The simultaneous extrusion of two or more materials to form an integral product.

cogging. The reducing operation in working an ingot into a billet with a forging hammer or a forging press.

coil breaks. Creases or ridges that appear as parallel lines transverse to the direction of rolling and that extend across the width of the sheet. Coil breaks are caused by the deformation of local areas during coiling or uncoiling of annealed or insufficiently temper-rolled steel sheets.

coil weld. A welded joint connecting the ends of two coils to form a continuous strip.

coining. (1) A closed-die forging operation in which all surfaces of a workpiece are confined or restrained, resulting in a well-defined imprint of the die on the work. (2) A *restriking* operation used to sharpen or change an existing radius or profile. Coining can be done

while forgings are hot or cold and is usually performed on surfaces parallel to the parting line of the forging.

coining dies. Dies in which the coining or sizing operation is performed.

coin straightening. A combination coining and straightening operation performed in special cavity dies designed to impart a specific amount of working in specified areas of a forging to relieve the stresses developed during heat treatment.

cold coined forging. A forging that has been restruck cold in order to hold closer face distance tolerances, sharpen corners or outlines, reduce section thickness, flatten some particular surface, or, in non-heat-treatable alloys, increase hardness.

cold forming. See *cold working*.

cold heading. Working metal at room temperature such that the cross-sectional area of a portion or all of the stock is increased. See also *heading* and *upsetting*.

cold lap. A flaw that results when a workpiece fails to fill the die cavity during the first forging. A seam is formed as subsequent dies force metal over this gap to leave a seam on the workpiece surface. See also *cold shut*.

cold rolled sheet. A mill product produced from a hot rolled pickled coil that has been given substantial cold reduction at room temperature. After annealing, the usual end product is characterized by improved surface, greater uniformity in thickness, increased tensile strength, and improved mechanical properties as compared with hot rolled sheet.

cold shut. (1) A fissure or lap on a forging surface that has been closed without fusion during the forging operation. (2) A folding back of metal onto its own surface during flow in the die cavity; a forging defect.

cold trimming. The removal of flash or excess metal from a forging at room temperature in a trimming press.

cold working. The plastic deformation of metal under conditions of temperature and strain rate that induce *strain hardening*. Usually, but not necessarily, conducted at room temperature. Also referred to as cold forming or cold forging. Contrast with *hot working*.

combination die. See *compound die*.

compact (noun). The object produced by the compression of metal powder, generally while confined in a die.

compact (verb). The operation or process of producing a compact; sometimes called pressing.

compound die. Any die designed to perform more than one operation on a part with one stroke of the press, such as blanking and piercing, in which all functions are performed sequentially within the confines of the blank size being worked.

compression test. A method for assessing the ability of a material to withstand compressive loads.

compressive strength. The maximum compressive stress a material is capable of devel-

oping. With a brittle material that fails in compression by fracturing, the compressive strength has a definite value. In the case of ductile, malleable, or semiviscous materials (which do not fail in compression by a shattering fracture), the value obtained for compressive strength is an arbitrary value dependent on the degree of distortion that is regarded as effective failure of the material.

compressive stress. A stress that causes an elastic or plastic body to deform (shorten) in the direction of the applied load.

connections. Connecting members that convey motion and force from a rotating member to a slide or lever on a press. Also called pitmans, connecting links (or rods and straps), or eccentric straps.

constitutive equation. Equation expressing the relation between stress, strain, strain rate, and microstructural features (e.g., grain size). Constitutive equations are generally phenomenological (curve fits based on measured data) or mechanism-based (based on mechanistic model of deformation and appropriate measurements). Phenomenological constitutive equations are usually valid only within the processing regime in which they were measured, while mechanism-based relations can be extrapolated outside the regime of measurement, provided the deformation mechanism is unchanged.

contour forming. See *roll forming, stretch forming, tangent bending,* and *wiper forming*.

controlled rolling. Multistand plate or bar rolling process, typically for ferrous alloys, in which the reduction per pass, rolling speed, time between passes, and so on are carefully chosen to control recrystallization, precipitation, and phase transformation in order to develop a desired microstructure and set of properties.

controller. A control algorithm implemented in a closed-loop control system.

conventional forging. Forging process in which the work material is hot and the dies are at room temperature or slightly elevated temperature. To minimize the effects of die chilling on metal flow and microstructure, conventional forging usually involves strain rates of the order of 0.05 s^{-1} or greater. Also known as nonisothermal forging.

core. Inner material in a coextruded or codrawn product.

core rod. See *mandrel*.

coring. (1) A central cavity at the butt end of a rod extrusion; sometimes called *extrusion pipe*. (2) A condition of variable composition between the center and surface of a unit of microstructure (such as a dendrite, grain, or carbide particle); results from nonequilibrium solidification, which occurs over a range of temperature.

corrugating. The forming of sheet metal into a series of straight, parallel alternate ridges and grooves with a rolling mill equipped with matched roller dies or a *press brake* equipped with specially shaped punch and die.

corrugations. Transverse ripples caused by a variation in strip shape during hot or cold reduction.

Coulomb friction. Interface friction condition for which the interface shear stress is proportional to the pressure normal to the interface. The proportionality constant is called the Coulomb coefficient of friction (μ) and takes on values between 0 (perfect lubrication) and $1/\sqrt{3}$ (sticking friction) during metalworking. See also *friction shear factor*.

counterblow equipment. Equipment with two opposed rams that are activated simultaneously to strike repeated blows on the workpiece placed midway between them.

counterblow forging equipment. A category of forging equipment in which two opposed rams are activated simultaneously, striking repeated blows on the workpiece at a midway point. Action is vertical or horizontal.

counterblow hammer. A forging hammer in which both the ram and the anvil are driven simultaneously toward each other by air or steam pistons.

counterlock. A jog in the mating surfaces of dies to prevent lateral die shift caused by side thrust during the forging of irregularly shaped pieces.

cracked edge. A series of tears at the edge of the sheet resulting from prior processing.

crank. Forging shape generally in the form of a "U" with projections at more or less right angles to the upper terminals. Crank shapes are designated by the number of throws (for example, two-throw crank).

crank press. A mechanical press whose slides are actuated by a crankshaft.

creep. Time-dependent strain occurring under stress.

creep forming. Forming, usually at elevated temperatures, where the material is deformed over time with a preload, sometimes comprising weights placed on a part during a stress-relief cycle.

crimping. The forming of relatively small *corrugations* in order to set down and lock a seam, to create an arc in a strip of metal, or to reduce an existing arc or diameter. See also *corrugating*.

cross breaks. Visually apparent line-type discontinuities more or less transverse to the coil rolling direction resulting from bending the coil over too sharp a radius and thus kinking the metal.

crown. (1) The upper part (head) of a press frame. On hydraulic presses, the crown usually contains the cylinder; on mechanical presses, the crown contains the drive mechanism. See also *hydraulic press* and *mechanical press*. (2) A shape (crown) ground into a flat roll to ensure uniform gage control and coil tracking during rolling of cold (and hot) rolled sheet and strip.

crystal. A solid composed of atoms, ions, or molecules arranged in a pattern that is periodic in three dimensions.

crystal lattice. A regular array of points about which the atoms or ions of a crystal are centered. See also *lattice*.

crystal-plasticity modeling. Physics-based modeling techniques that treat the phenomena of deformation by way of slip and twinning in order to predict strength and the evolution of crystallographic texture during the deformation processing of polycrystalline materials. See also *deformation texture, slip, Schmid's law, Taylor factor,* and *twinning.*

crystal system. One of seven groups into which all crystals may be divided: triclinic, monoclinic, orthorhombic, hexagonal, rhombohedral, tetragonal, and cubic.

cup. (1) A sheet-metal part; the product of the first drawing operation. (2) Any cylindrical part or shell closed at one end.

cup fracture (cup-and-cone fracture). A mixed-mode fracture, often seen in tensile test specimens of a ductile material, in which the central portion undergoes plane-strain fracture and the surrounding region undergoes plane-stress fracture. One of the mating fracture surfaces looks like a miniature cup; it has a central depressed flat-face region surrounded by a shear lip. The other fracture surface looks like a miniature truncated cone.

cupping. (1) The first step in *deep drawing.* (2) Fracture of severely worked rods or wire in which one end looks like a cup and the other a cone.

cupping test. A mechanical test used to determine the ductility and stretching properties of sheet metal. It consists of measuring the maximum part depth that can be formed before fracture. The test is typically carried out by stretching the testpiece clamped at its edges into a circular die using a punch with a hemispherical end. See also *cup fracture, Erichsen test,* and *Olsen ductility test.*

curling. Forming of an edge of circular cross section along a sheet or along the end of a shell or tube, either to the inside or outside, for example, the pinholes in sheet-metal hinges and the curled edges on cans, pots, and pans.

curling die. (1) A die used when the curling is done in the die block (female part), such as curling of sheet-metal hinges and the edge plate preparatory to seaming. (2) A die that is ordinarily attached to the punch holder, such as those used in wiring of pots and pans.

curvature. The reciprocal of the radius of curvature; it expresses how curved a body is.

cushion. A mechanism mounted in or under the bed of a hydraulic press to provide a controlled resistance against the work as it is displaced by the slide. The return motion may also be used for ejecting the work.

cushion, hydraulic. A *cushion* in which a hydraulic cylinder provides the resistance. Pressure in the cylinder is developed by the main ram movement. The cushion is returned to its normal position by gas pressure acting on the hydraulic fluid in the reservoir.

cushion, hydropneumatic. A *cushion* in which a hydraulic cylinder provides the resistance.

Pressure in the cylinder is developed by the main ram movement. The cushion is returned to its normal position by gas pressure acting on the hydraulic fluid in the reservoir.

cutoff. A pair of blades positioned in dies or equipment (or a section of the die milled to produce the same effect as inserted blades) used to separate a forging from a bar after forging operations are completed. Used only when forgings are produced from relatively long bars instead of from individual, precut multiples or blanks. See also *blank* and *multiple.*

cut-off die. (1) The last die in a set of transfer dies that cuts the part loose from the scrap. Also known as a trimming die. (2) A die that cuts straight-sided blanks from a coil for later use in a draw die.

D

damage. General term used to describe the development of defects such as cavities, cracks, shear bands, and so on that may culminate in gross fracture in severe cases. The evolution of damage is strongly dependent on material, microstructure, and processing conditions (strain, strain rate, temperature, and stress state).

daylight. The maximum clear distance between the pressing surfaces of a press when the surfaces are in the usable open position. Where a *bolster* plate is supplied, it is considered the pressing surface. See also *shut height.*

DBTT. See *ductile-to-brittle transition temperature (DBTT).*

dead-metal zone. Region of metal undergoing limited or no deformation during bulk forming of a workpiece, generally developed adjacent to the workpiece-tooling interface as a result of friction, die chilling, or deformation-zone geometry.

deep drawing. Forming operation characterized by the production of a parallel-wall cup from a flat blank of sheet metal. The blank may be circular, rectangular, or a more complex shape. The blank is drawn into the die cavity by the action of a punch. Deformation is restricted to the flange areas of the blank. No deformation occurs under the bottom of the punch—the area of the blank that was originally within the die opening. As the punch forms the cup, the amount of material in the flange decreases. Also called cup drawing or radial drawing.

deflection. The amount of deviation from a straight line or plane when a force is applied to a press member. Generally used to specify the allowable bending of the bed, slide, or frame at rated capacity with a load of predetermined distribution.

deformation (adiabatic) heating. Temperature increase that occurs in a workpiece due to the conversion of strain energy, imparted during metalworking, into heat.

deformation energy method. A metal-forming analysis technique that takes into account only the energy required to deform the workpiece.

deformation limit. In *drawing,* the limit of deformation is reached when the load required to deform the flange becomes greater than the load-carrying capacity of the cup wall. The deformation limit (limiting drawing ratio, LDR) is defined as the ratio of the maximum blank diameter that can be drawn into a cup without failure, to the diameter of the punch.

deformation-mechanism map. Strain rate/ temperature map that describes forming or service regimes under which deformation is controlled by micromechanical processes such as dislocation glide, dislocation climb, and diffusional flow limited by bulk or boundary diffusion.

deformation texture. Preferred orientation of the crystals/grains comprising a polycrystalline aggregate that is developed during deformation processing as a result of slip and rotation within each crystal that comprises the aggregate.

degree of freedom. A single scalar variable that may vary at a node and for which a solution is sought. For example, if a node was allowed to move in three-dimensional space, it would have three degrees of freedom. If, in addition, the temperature was unknown and to be determined, there would be four degrees of freedom at each node.

delayed return. A controlling device for a *cushion* to prevent its return until the main ram has returned a predetermined distance.

Demarest process. A *fluid forming* process in which cylindrical and conical sheet-metal parts are formed by a modified rubber bulging punch. The punch, equipped with a hydraulic cell, is placed inside the workpiece, which in turn is placed inside the die. Hydraulic pressure expands the punch.

density. Mass per unit volume. Weight per unit volume.

design of experiments. Methodology for choosing a small number of screening experiments to establish the important material and process variables in a complex manufacturing process.

developed blank. A sheet-metal blank that yields a finished part without trimming or with the least amount of trimming.

die. (1) A tool, usually containing a cavity, that imparts shape to solid, molten, or powdered metal primarily because of the shape of the tool itself. Used in many press operations (including blanking, drawing, forging, and forming), in die casting, and in forming green powder metallurgy compacts. Die-casting and powder metallurgy dies are sometimes referred to as molds. See also *forging dies.* (2) A complete tool used in a press for any operation or series of operations, such as forming, impressing, piercing, and cutting. The upper member or members are attached to the slide (or slides) of the press, and the lower member is clamped or bolted to the bed or bolster, with the die members being so shaped as to cut or form the material placed between them when the press makes a stroke. (3) The female part of a complete die assembly as described previously.

die assembly. The parts of a die stamp or press that hold the die and locate it for the punches.

die block. A block, often made of heat treated steel, into which desired impressions are machined or sunk and from which closed-die forgings or sheet-metal stampings are produced using hammers or presses. In forging, die blocks are usually used in pairs, with part of the impression in one of the blocks and the rest of the impression in the other. In sheet-metal forming, the female die is used in conjunction with a male punch. See also *closed-die forging.*

die button. An insert in a die that matches the punch and is used for punching and piercing operations. The die button is readily removable for sharpening or replacement as an individual part of a die.

die cavity. The machined recess that gives a forging or stamping its shape.

die check. A crack in a die impression due to forging and thermal strains at relatively sharp corners. Upon forging, these cracks become filled with metal, producing sharp, ragged edges on the part. Usual die wear is the gradual enlarging of the die impression due to erosion of the die material, generally occurring in areas subject to repeated high pressures during forging.

die chill. The temperature loss experienced by a billet or preform when it contacts dies that are maintained at a lower temperature.

die clearance. Clearance between a mated punch and die; commonly expressed as clearance per side. Also called clearance or punch-to-die clearance.

die closure. A term frequently used to mean variations in the thickness of a forging.

die coating. Hard metal incorporated into the working surface of a die to protect the working surface or to separate the sheet-metal surface from direct contact with the basic die material. Hard-chromium plating is an example.

die cushion. A press accessory placed beneath or within a *bolster* plate or *die block* to provide an additional motion or pressure for stamping or forging operations; actuated by air, oil, rubber, springs, or a combination of these.

die forging. A forging that is formed to the required shape and size through working in machined impressions in specially prepared dies.

die forming. The shaping of solid or powdered metal by forcing it into or through the *die cavity.*

die height. The distance between the fixed and the moving platen when the dies are closed.

die holder. A plate or block, on which the die block is mounted, having holes or slots for fastening to the *bolster* plate or the *bed* of the press.

die impression. The portion of the die surface that shapes a forging or sheet-metal part.

die insert. A relatively small die that contains part or all of the impression of a forging or sheet-metal part and is fastened to the master *die block.*

die life. The productive life of a *die impression,* usually expressed as the number of units produced before the impression has worn beyond permitted tolerances.

die line. A line or scratch resulting from the use of a roughened tool or the drag of a foreign particle between tool and product.

die lock. A phenomenon in which the deformation is limited in a forging near the die face due to chilling of the workpiece and/or friction at the workpiece-die interface.

die lubricant. In forging or forming, a compound that is sprayed, swabbed, or otherwise applied on die surfaces or the workpiece during the forging or forming process to reduce friction. Lubricants also facilitate release of the part from the dies and provide thermal insulation. See also *lubricant.*

die match. The alignment of the upper (moving) and lower (stationary) dies in a hammer or press. An allowance for misalignment (or mismatch) is included in forging tolerances.

die pad. A movable plate or pad in a female die; usually used for part ejection by mechanical means, springs, or fluid cushions.

die proof (cast). A casting of a *die impression* made to confirm the accuracy of the impression.

die radius. The radius on the exposed edge of a deep-drawing die, over which the sheet flows in forming drawn shells.

die section. A section of a cutting, forming, or flanging die that is fastened to other sections to make up the complete working surface. Also referred to as cutting section.

die set. (1) The assembly of the upper and lower die shoes (punch and die holders), usually including the *guide pins, guide pin bushings,* and *heel blocks.* This assembly takes many forms, shapes, and sizes and is frequently purchased as a commercially available unit. (2) Two (or, for a mechanical upsetter, three) machined dies used together during the production of a *die forging.*

die shift. The condition that occurs after the dies have been set up in a forging unit in which a portion of the impression of one die is not in perfect alignment with the corresponding portion of the other die. This results in a mismatch in the forging, a condition that must be held within the specified tolerance.

die shoes. The upper and lower plates or castings that constitute a *die set* (punch and die holder). Also a plate or block on which a *die holder* is mounted, functioning primarily as a base for the complete *die assembly.* This plate or block is bolted or clamped to the *bolster* plate or the face of the press *slide.*

die sinking. The machining of the die impressions to produce forgings of required shapes and dimensions.

die space. The maximum space (volume), or any part of the maximum space, within a press for mounting a die.

die stamping. The general term for a sheet-metal part that is formed, shaped, or cut by a die in a press in one or more operations.

dimpling. (1) The stretching of a relatively small, shallow indentation into sheet metal. (2) In aircraft, the stretching of metal into a conical flange for a countersunk head rivet.

direct (forward) extrusion. See *extrusion.*

discontinuous yielding. The nonuniform plastic flow of a metal exhibiting a yield point in which plastic deformation is inhomogeneously distributed along the gage length. Under some circumstances, it may occur in metals not exhibiting a distinct yield point, either at the onset of or during plastic flow.

dishing. The formation of a shallow concave surface in which the projected area is very large compared with the depth of the impression.

dislocation. Line imperfection in an otherwise perfect crystal. Allows deformation of metals at much lower forces than would be required for perfect crystals. The two basic types are an *edge dislocation* and a *screw dislocation.*

displacement adjustment (DA) method. A *springback* compensation method characterized by a step in which the die shape is adjusted by applying error displacements computed in a springback calculation, with the objective of attaining a die shape that will produce a desired shape after springback.

double-action hydraulic press. Press equipped with two moving slide members, both located on the same side of the work. Three designs are in general use:

1. **Slide within a slide.** The outer, or blankholder, slide is guided on housings or strain rods. The inner, or punch, slide is guided in the outer slide.
2. **Slide below a slide.** The lower side is the blankholder slide. It is provided with a center hold to allow the punch to come through. Both slides are guided on housings or strain rods.
3. **Blankholder above punch slide.** Both slides are guided on housings or straight rods. Pins connect the blankholder ring (which is below the main, or punch, slide) to the blankholder slide above the main slide.

A double-action press may have a *cushion* built into the bed.

double-action mechanical press. A press having two independent parallel movements by means of two slides, one moving within the other. The inner slide or plunger is usually operated by a crankshaft; the outer or blankholder slide, which dwells during the drawing operation, is usually operated by a toggle mechanism or by cams. See also *slide.*

double-action press. A press with two mechanically or hydraulically actuated slides: an inner or forming slide and an outer blankholder slide.

double-bend die. A punch and die combination that makes two bends in one motion.

double-cone test. Simulative bulk forming test consisting of the compression of a sample shaped like a flying saucer between flat dies. The variation of strain and stress state developed across the sample is used to obtain a large quantity of data on microstructure evolution and failure in a single experiment.

double-crank press. A crank press in which the slide is driven by two connections.

double-drive eccentric. An eccentric gear drive consisting of an eccentric between two gears. It is generally made of one-piece construction and rotates on a stationary shaft.

DQSK steel. Drawing-quality, special-killed steel; a typical low-strength steel used for sheet-forming applications in the automotive industry.

draft. The amount of taper on the sides of a forging and on projections to facilitate removal from the dies; also, the corresponding taper on the sidewalls of the die impressions. In *open-die forging,* draft is the amount of relative movement of the dies toward each other through the metal in one application of power. See also *draft angle.*

draft angle. The angle of taper, usually 5 to 7°, given to the sides of a forging and the sidewalls of the die impression. See also *draft.*

drawability. A measure of the *formability* of a sheet-metal subject to a drawing process. The term is usually used to indicate the ability of a metal to be deep drawn. See also *drawing* and *deep drawing.*

drawbead. Narrow ridges along the edge of a part, and corresponding ridges in the die, to improve holding action in pressworking. The insert or riblike projection on the draw ring or holddown surfaces aids in controlling the rate of metal flow during deep-drawing operations. Drawbeads are especially useful in controlling the rate of metal flow in irregularly shaped stampings.

draw-bend. A deformation state for a sheet that is pulled over a radius such that at each location along the length the sheet undergoes bending with tension and then unbending with tension. It corresponds to sheet that is drawn into a die cavity in typical sheet-metal forming operations. Also called stretch-bend.

draw forming. A method of curving bars, tubes, or rolled or extruded sections in which the stock is bent around a rotating *form block.* Stock is bent by clamping it to the form block, then rotating the form block while the stock is pressed between the form block and a pressure die held against the periphery of the form block.

draw-in constraints. Any impediment to the draw-in of sheet into a die cavity, by, for example, drawbeads, binder friction, and so on.

drawing. A term used for a variety of forming operations, such as *deep drawing* a sheet-metal blank; *redrawing* a tubular part; and drawing rod, wire, and tube. The usual drawing process with regard to sheet-metal working in a press is a method for producing a cuplike form from a sheet-metal disk by holding it firmly between blankholding surfaces to prevent the formation of wrinkles while the punch travel produces the required shape.

drawing compound. A substance applied to prevent *pickup* and scoring during deep drawing or pressing operations by preventing metal-to-metal contact of the workpiece and die. Also known as *die lubricant.*

drawing die. A type of die designed to produce nonflat parts such as boxes, pans, and so on. Whenever practical, the die should be designed and built to finish the part in one stroke of the press, but if the part is deep in proportion to its diameter, redrawing operations are necessary.

drawing ratio. The ratio of the blank diameter to the punch diameter.

draw marks. See *scoring, galling, pickup,* and *die line.*

draw plate. A circular plate with a hole in the center contoured to fit a forming punch; used to support the *blank* during the forming cycle.

draw radius. The radius at the edge of a die or punch over which sheet metal is drawn.

draw ring. A ring-shaped die part (either the die ring itself or a separate ring) over which the inner edge of sheet metal is drawn by the punch.

draw stock. The forging operation in which the length of a metal mass (stock) is increased at the expense of its cross section; no *upset* is involved. The operation includes converting ingot to pressed bar using "V," round, or flat dies.

drop forging. The forging obtained by hammering metal in a pair of closed dies to produce the form in the finishing impression under a *drop hammer;* forging method requiring special dies for each shape.

drop hammer. A term generally applied to forging hammers in which energy for forging is provided by gravity, steam, or compressed air. See also *air-lift hammer, board hammer,* and *steam hammer.*

drop hammer forming. A process for producing shapes by the progressive deformation of sheet metal in matched dies under the repetitive blows of a gravity-drop or power-drop hammer. The process is restricted to relatively shallow parts and thin sheet from approximately 0.6 to 1.6 mm (0.024 to 0.064 in.).

drop through. The type of ejection where the part or scrap drops through an opening in the die.

dry-film lubricant. A type of lubricant applied by spraying or painting on coils or sheets prior to blanking, drawing, or stamping. The lubricant can have a wax base and be sprayed hot onto the sheet surface and solidify on cooling, or be a water-based polymer and be roll coated onto the surface (one or both sides) and be heated to cure and dry. Such lubricants have uniform thickness, low coefficients of friction, and offer protection from corrosion in transit and storage.

ductile fracture. Failure of metals as a result of cavity nucleation, growth, and coalescence. Ductile fracture may occur during metal forming at both cold and hot working temperatures.

ductile-to-brittle transition temperature (DBTT). A temperature or range of temperatures over which a material reaction to impact (high strain rate) loads changes from ductile, high-energy-absorbing to brittle, low-energy-absorbing behavior. The DBTT determinations are often done with Charpy or Izod test specimens measuring absorbed energy at various temperatures.

ductility. A measure of the amount of deformation that a material can withstand without breaking.

dummy block. In *extrusion*, a thick, unattached disk placed between the ram and the billet to prevent overheating of the ram.

dynamic friction. The friction forces between two surfaces in relative motion. See also *static friction.*

dynamic material modeling. A methodology by which macroscopic measurements of flow stress as a function of temperature and strain rate are used with continuum criteria of instability to identify regions of temperature and strain rate in which voids, cracks, shear bands, and flow localization are likely to occur.

dynamic recovery. Recovery process that occurs during cold or hot working of metals, typically resulting in the formation of low-energy dislocation substructures/subgrains within the deformed original grains. Dynamic recovery reduces the observed level of strain hardening due to dislocation multiplication during deformation.

dynamic recrystallization. The formation of strain-free recrystallized grains during hot working. It results in a decrease in flow stress and formation of equiaxed grains, as opposed to dynamic recovery in which the elongated grains remain.

dwell. Portion of a press cycle during which the movement of a member is zero or at least insignificant. Usually refers to (1) the interval when the *blankholder* in a drawing operation is holding the blank while the punch is making the draw, or (2) the interval between the completion of the forging stroke and the retraction of the ram.

E

earing. The formation of ears or scalloped edges around the top of a drawn shell, resulting from directional differences in the plastic-working properties of rolled metal with, across, and at angles to the direction of rolling.

eccentric. The offset portion of the driveshaft that governs the stroke or distance the crosshead moves on a mechanical or manual shear.

eccentric gear. A main press-drive gear with an eccentric(s) as an integral part. The unit rotates about a common shaft, with the eccentric transmitting the rotary motion of the gear into the vertical motion of the slide through a connection.

eccentric gear press. A press where the rotary motion is converted to reciprocating motion by means of an eccentric journal integral with a drive gear and rotating on a stationary shaft. Eccentric gear drives are commonly used in presses of over 300 tons capacity.

eccentric press. A *mechanical press* in which an eccentric, instead of a crankshaft, is used to move the *slide*.

eccentric shaft press. A press where the rotary motion is converted to reciprocating motion by means of an eccentric shaft. The eccentric is a crank with a pin of such size that its diameter contains or surrounds the shaft diameter.

edge dislocation. A line imperfection that corresponds to the row of mismatched atoms along the edge formed by an extra, partial plane of atoms within the body of a crystal.

edger (edging impression). The portion of a die impression that distributes metal during forging into areas where it is most needed in order to facilitate filling the cavities of subsequent impressions to be used in the forging sequence. See also *fuller (fullering impression).*

edge strain. Repetitive areas of local deformation extending inwardly from the edge of temper-rolled sheet.

edging. (1) In sheet-metal forming, reducing the flange radius by retracting the forming punch a small amount after the stroke but before release of the pressure. (2) In rolling, the working of metal in which the axis of the roll is parallel to the thickness dimension. Also called edge rolling. The result is changing a rounded edge to a square edge. (3) The forging operation of working a bar between contoured dies while turning it 90° between blows to produce a varying rectangular cross section.

effective draw. The maximum limits of forming depth that can be achieved with a multiple-action press; sometimes called maximum draw or maximum depth of draw.

effective strain. A scalar measure of strain for a general strain state that is equivalent to a tensile strain corresponding to the same material hardness. Effective strains can be similarly defined for other yield functions but are seldom used without clarification. Effective strain is a scalar conjugate to effective stress, defined in such a manner that the product of the effective stress and the effective strain increment is equal to the increment in imposed work during a deformation process. Also known as von Mises strain.

effective stress. A scalar measure of stress for a general stress state that is equivalent to a tensile stress corresponding to the same material state. Effective stresses can be similarly defined for other yield functions but are seldom used without clarification. Effective stress is a mathematical way to express a two- or three-dimensional stress state by a single number. Also known as von Mises stress.

ejection. (1) Gravity: the ejection of parts by the force of gravity. (2) Kicker: the ejection of parts by means of a mechanical device that pushes the part away from the die.

ejector. A mechanism for removing work or material from between the dies.

ejector rod. A rod used to push out a formed piece.

elastic core. The region near the neutral axis in a bent beam where the stresses are less than the yield stress, such that the material behaves elastically.

elastic deformation. Deformation that is reversed upon removal of the forces that created the deformation. Theoretically, no net work should be done during a closed path of elastic deformation.

elasticity. The property of a material that permits elastic deformation to occur.

elastic limit. The maximum stress a material can sustain without permanent strain remaining upon complete release of the stress. See also *proportional limit.*

elastic modulus. See *Young's modulus.*

elastic-plastic or elastoplastic. An adjective describing a material or constitutive equation that incorporates both elastic and plastic material properties, depending on the stress state, its magnitude, and their increments.

elastohydrodynamic lubrication. A condition of lubrication in which the friction and film thickness between two bodies in relative motion is determined by the elastic properties of the bodies in combination with the viscous properties of the lubricant. See also *thin-film lubrication.*

electric-discharge machining (EDM). Metal removal (machining) process based on the electric discharge/spark erosion resulting from current flowing between an electrode and workpiece placed in close proximity to each other in a dielectric fluid. The electrode may be a wire (as in wire EDM) or a contoured shape (so-called plunge EDM); the latter technique is used for making metalworking dies.

electromagnetic forming. A process for forming metal by the direct application of an intense, transient magnetic field. The workpiece is formed without mechanical contact by the passage of a pulse of electric current through a forming coil. Also known as magnetic pulse forming.

electron backscatter diffraction (EBSD). Materials characterization technique conducted in a scanning electron microscope (and sometimes a transmission electron microscope) used to establish the crystallographic orientation of individual (micron-sized) regions of material through analysis of

Kikuchi patterns formed by backscattered electrons. Automated EBSD systems can thus be used to determine the texture over small-to-moderate-sized total volumes of material.

elongation. A term used in mechanical testing to describe the amount of extension of a testpiece when deformed. See also *elongation, percent.*

elongation, percent. The extension of a uniform section of a specimen expressed as a percentage of the original gage length:

$$\text{Elongation}, \% = \frac{(L_x - L_o)}{L_o} \times 100$$

where L_o is the original gage length, and L_x is the final gage length.

embossing. A process for producing raised or sunken designs in sheet material by means of male and female dies, theoretically with minimal change in metal thickness. Examples are letters, ornamental pictures, and ribs for stiffening. Heavy embossing and *coining* are similar operations.

embossing die. A die used for producing embossed designs.

engineering strain. A measure of the strain proportional to extension in one direction that is obtained from: $e = (l - l_o)/l_o$, where l is the final length, and l_o is the original length in the direction of interest. Preferred term is *nominal strain.*

engineering stress. A measure of stress proportional to external load in one direction that is obtained by dividing the differential normal load carried by a cross section by the original (before deformation) differential area corresponding to that differential load. Preferred term is *nominal stress.*

equal channel angular extrusion (ECAE). Metalworking operation comprising the extrusion of a billet through two intersecting channels of identical cross section lying at a specified angle to each other, thus imparting large deformation in a single pass or multiple passes. Also sometimes referred to as equal channel angular pressing (ECAP).

Erichsen test. A *cupping test* used to assess the ductility of sheet metal. The method consists of forcing a conical or hemispherical-ended plunger into the specimen and measuring the depth of the impression at fracture.

erosion resistance. The ability of a die material to resist sliding wear and thus maintain its original dimension.

etching. Production of designs, including grids, on a metal surface by a corrosive reagent or electrolytic action.

Euler angles. Set of three angular rotations used to specify unambiguously the spatial orientation of crystallites relative to a fixed reference frame.

explicit. A type of finite element algorithm where an incremental solution for a very small step is assumed based on current conditions. No iteration is required. A solution is ensured, but the accuracy cannot be. Typical sheet-forming programs of this type are based on

momentum effects and are thus often called dynamic explicit programs.

explosive forming. The shaping of metal parts in which the forming pressure is generated by an explosive charge. See also *high-energy-rate forming.*

extruded hole. A hole formed by a punch that first cleanly cuts a hole and then is pushed farther through to form a flange with an enlargement of the original hole. This may be a two-step operation.

extrusion. The conversion of an ingot or billet into lengths of uniform cross section by forcing metal to flow plastically through a die orifice. In forward (direct) extrusion, the die and ram are at opposite ends of the extrusion stock, and the product and ram travel in the same direction. Also, there is relative motion between the extrusion stock and the die. In backward (indirect) extrusion, the die is at the ram end of the stock, and the product travels in the direction opposite that of the ram, either around the ram (as in the impact extrusion of cylinders, such as cases for dry cell batteries) or up through the center of a hollow ram. See also *hydrostatic extrusion* and *impact extrusion.*

extrusion billet. A metal slug used as *extrusion stock.*

extrusion defect. See *extrusion pipe.*

extrusion forging. (1) Forcing metal into or through a die opening by restricting flow in other directions. (2) A part made by the operation.

extrusion pipe. A central oxide-lined discontinuity that occasionally occurs in the last 10 to 20% of an extruded bar. It is caused by the oxidized outer surface of the billet flowing around the end of the billet and into the center of the bar during the final stages of extrusion. Also called *coring.*

extrusion stock. A rod, bar, or other section used to make extrusions.

eyeleting. The displacing of material about an opening in sheet or plate so that a lip protruding above the surface is formed.

F

fiber texture. Crystallographic texture in which all or a large fraction of the crystals in a polycrystalline aggregate are oriented such that a specific direction in each crystal is parallel to a specific sample direction, such as the axis of symmetry of a cylindrical object. Often found in wrought products such as wire and round extrusions that have been subjected to large axisymmetric deformation.

fillet. The concave intersection of two surfaces. In forging, the desired radius at the concave intersection of two surfaces is usually specified.

film strength. The ability of a surface film to resist rupture by the penetration of asperities during sliding or rolling of two surfaces over each other.

fin. The thin projection formed on a forging by trimming or when metal is forced under pressure into hairline cracks or die interfaces.

finish. (1) The surface appearance of a product. (2) The forging operation in which the part is forged into its final shape in the finish die. If only one finish operation is scheduled to be performed in the finish die, this operation will be identified simply as finish; first, second, or third finish designations are so termed when one or more finish operations are to be performed in the same finish die.

finish allowance. The amount of excess metal surrounding the intended final shape; sometimes called clean-up allowance, forging envelope, or machining allowance.

finisher (finishing impression). The *die impression* that imparts the final shape to a forged part.

finishing dies. The die set used in the last forging step.

finishing temperature. The temperature at which *hot working* is completed.

finish trim. Flash removal from a forging; usually performed by trimming but sometimes by band sawing or similar techniques.

finite element analysis (FEA), finite element modeling (FEM). A computer-based technique for the analysis of an arbitrary part by breaking the problem into a mesh of simplified elements and then solving the corresponding system of equations, linear or nonlinear. FEA/FEM may be used to solve a wide variety of boundary-value problems of a material continuum, including responses to loads and temperatures.

fire crack marks. A network of raised irregular lines on the sheet surface produced by hot rolling with rolls containing thermal fatigue cracks in the surface.

first block, second block, and finish. The forging operation in which the part to be forged is passed in progressive order through three tools mounted in one forging machine; only one heat is involved for all three operations.

first draw. The first drawing operation in a series, which is usually performed on a flat blank.

fishtail. (1) In *roll forging,* the excess trailing end of a forging. Before being trimmed off, it is often used as a tong hold for a subsequent forging operation. (2) In hot *rolling* or *extrusion,* the imperfectly shaped trailing end of a bar or special section that must be cut off and discarded as mill scrap.

fixture. A tool or device for holding and accurately positioning a piece or part on a machine tool or other processing machine.

flame hardening. A heat treating method for surface hardening steel of the proper specifications in which an oxyacetylene flame heats the surface to a temperature at which subsequent cooling, usually with water or air, will give the required surface hardness.

flame straightening. The correction of distortion in metal structures by localized heating with a gas flame.

flange. A projecting rim or edge of a part; usually narrow and of approximately constant width for stiffening or fastening.

flanging. A bending operation in which a narrow strip at the edge of a sheet is bent down (up) along a straight or curved line. It is used for edge strengthening, appearance, rigidity, and the removal of sheared edges. A flange is often used as a fastening surface.

flaring. The forming of an outward acute-angle *flange* on a tubular part.

flash. Metal in excess of that required to completely fill the blocking or finishing forging impression of a set of dies. Flash extends out from the body of the forging as a thin protuberance at the line where the dies meet and is subsequently removed by trimming. Because it cools faster than the body of the component during forging, flash can serve to restrict metal flow at the line where dies meet, thus ensuring complete filling of the impression. See also *closed-die forging.*

flash extension. That portion of flash remaining on a forged part after trimming; usually included in the normal forging tolerances.

flash land. Configuration in the *blocking impression* or *finisher (finishing impression)* of forging dies designed to restrict or to encourage the growth of flash at the parting line, whichever may be required in a particular case to ensure complete filling of the impression.

flash line. The line left on a forging after the flash has been trimmed off.

flash pan. The machined-out portion of a forging die that permits the flow through of excess metal.

flat-die forging. See *open-die forging.*

flattening. (1) A preliminary operation performed on forging stock to position the metal for a subsequent forging operation. (2) The removal of irregularities or distortion in sheets or plates by a method such as *roller leveling* or *stretcher leveling*. (3) For wire, rolling round wire to a flattened condition.

flattening dies. Dies used to flatten sheet-metal hems; that is, dies that can flatten a bend by closing it. These dies consist of a top and bottom die with a flat surface that can close one section (flange) to another (hem, seam).

fleck scale. A fine pattern of scale marks on the sheet surface that can be either dark or light. The dark pattern originates from scale and other impurities embedded in the strip during hot rolling that are not removed during pickling. The light pattern originates from a scale pattern imprinted on the work rolls in the finishing stands in the hot mill being printed onto the strip.

flex roll. A movable roll designed to push up against a sheet as it passes through a roller leveler. The flex roll can be adjusted to deflect the sheet any amount up to the roll diameter.

flex rolling. Passing sheets through a *flex roll* unit to minimize yield point elongation in order to reduce the tendency for *stretcher strains* to appear during forming.

floating die. (1) A die mounted in a die holder or a punch mounted in its holder such that a slight amount of motion compensates for tolerance in the die parts, the work, or the press. (2) A die mounted on heavy springs to allow vertical motion in some trimming, shearing, and forming operations.

floating plug. In tube drawing, an unsupported mandrel that locates itself at the die inside the tube, causing a reduction in wall thickness while the die is reducing the outside diameter of the tube.

flop forging. A forging in which the top and bottom die impressions are identical, permitting the forging to be turned upside down during the forging operation.

floppers. Lines or ridges that are transverse to the direction of rolling and generally confined to the section midway between the edges of the coil as rolled. They are somewhat irregular and tend toward a flat arc shape and are sometimes referred to as a shape wave.

flow curve. A curve of true stress versus true strain that shows the stress required to produce plastic deformation.

flow lines. (1) Texture showing the direction of metal flow during hot or cold working. Flow lines can often be revealed by etching the surface or a section of a metal part. (2) In mechanical metallurgy, paths followed by minute volumes of metal during deformation.

flow localization. A situation where material deformation is localized to a narrow zone. Such zones often are sites of failure. Flow localization results from poor lubrication, temperature gradients, or flow softening resulting from deformation heating, generation of softer crystallographic texture, grain coarsening, or spheroidization of second phases.

flow softening. Stress-strain behavior observed under constant strain-rate conditions characterized by decreasing flow stress with increasing strain. Flow softening may result from deformation heating as well as a number of microstructural sources, such as the generation of a softer crystallographic texture and the spheroidization of a lamellar phase.

flow stress. The uniaxial true stress required to cause plastic deformation at a particular value of strain, strain rate, and temperature.

flow through. A forging defect caused by metal flow past the base of a rib, with resulting rupture of the grain structure.

fluid-cell process. A modification of the *Guerin process* for forming sheet metal, the fluid-cell process uses higher pressure and is primarily designed for forming slightly deeper parts, using a rubber pad as either the die or punch. A flexible hydraulic fluid cell forces an auxiliary rubber pad to follow the contour of the form block and exert a nearly uniform pressure at all points on the workpiece. See also *fluid forming* and *rubber-pad forming.*

fluid forming. A modification of the Guerin process, fluid forming differs from the fluid-cell process in that the die cavity, called a pressure dome, is not completely filled with

rubber but with hydraulic fluid retained by a cup-shaped rubber diaphragm. See also *rubber-pad forming.*

fluting. A series of sharp parallel kinks or creases that can occur when sheet steel is formed cylindrically. Fluting is caused by inhomogeneous yielding of these sheets.

flying shear. A machine for cutting continuous rolled products to length that does not require a halt in rolling but rather moves along the runout table at the same speed as the product while performing the cutting, and then returns to the starting point in time to cut the next piece.

flywheel. A heavy rotating wheel, attached to a shaft, whose principal purpose is to store kinetic energy during the nonworking portion of a press cycle and to release energy during the working portion of a press cycle.

flywheel press. A mechanical power press in which the flywheel is mounted directly on the main crank or eccentric shaft without gearing.

foil. Metal in sheet form less than 0.15 mm (0.006 in.) thick.

fold. A forging defect caused by folding metal back onto its own surface during its flow in the die cavity.

follow die. A *progressive die* consisting of two or more parts in a single holder; used with a separate lower die to perform more than one operation (such as piercing and blanking) on a part in two or more stations.

foot control. A foot-operated control mechanism designed to be used with a clutch or clutch/brake control system on a press.

foot pedal (treadle). A foot-operated lever designed to operate the mechanical linkage that trips a full-revolution clutch on a press.

forgeability. Term used to describe the relative ability of material to deform without fracture. Also describes the resistance to flow from deformation. See also *formability.*

forging. The process of working metal to a desired shape by impact or pressure in hammers, forging machines (upsetters), presses, rolls, and related forming equipment. Forging hammers, counterblow equipment, and high-energy-rate forging machines apply impact to the workpiece, while most other types of forging equipment apply squeeze pressure in shaping the stock. Some metals can be forged at room temperature, but most are made more plastic for forging by heating. Specific forging processes defined in this Glossary include *closed-die forging, high-energy-rate forging, hot upset forging, isothermal forging, open-die forging, powder forging, precision forging, radial forging, ring rolling, roll forging, rotary forging,* and *rotary swaging.*

forging billet. A wrought metal slug used as *forging stock.*

forging dies. Forms for making forgings; they generally consist of a top and bottom die. The simplest will form a completed forging in a single impression; the most complex, consisting of several die inserts, may have a number of impressions for the progressive working of complicated shapes. Forging dies are usually in pairs, with part of the impression in one of the blocks and the rest of the impression in the other block.

forging envelope. See *finish allowance.*

forging machine (upsetter or header). A type of forging equipment, related to the *mechanical press,* in which the principal forming energy is applied horizontally to the workpiece, which is gripped and held by prior action of the dies.

forging plane. In forging, the plane that includes the principal die face and is perpendicular to the direction of ram travel. When the parting surfaces of the dies are flat, the forging plane coincides with the parting line. Contrast with *parting plane.*

forging quality. Term used to describe stock of sufficient quality to make it suitable for commercially satisfactory forgings.

forging rolls. Power-driven rolls used in preforming bar or billet stock that have shaped contours and notches for introduction of the work.

forging stock. A wrought rod, bar, or other section suitable for subsequent change in cross section by forging.

formability. The ease with which a metal can be shaped through plastic deformation. Evaluation of the formability of a metal involves measurement of strength, ductility, and the amount of deformation required to cause fracture. The term *workability* is used interchangeably with formability; however, formability refers to the shaping of sheet metal, while workability refers to shaping materials by *bulk forming.* See also *forgeability.*

formability parameters. Material parameters that can be used to predict the ability of sheet metal to be formed into a useful shape.

form block. Tooling, usually the male part, used for forming sheet-metal contours; generally used in *rubber-pad forming.*

form die. A die used to change the shape of a sheet-metal blank with minimal plastic flow.

forming. The plastic deformation of a workpiece to obtain a desired configuration. The shape change takes place in the solid state and does not involve material removal or addition. Metal-forming processes are typically classified as *bulk forming* and *sheet forming.* Also referred to as metalworking.

forming limit diagram (FLD) or forming limit curve (FLC). An empirical curve in which the major strains at the onset of necking in sheet metal are plotted vertically and the corresponding minor strains are plotted horizontally. The onset-of-failure line divides all possible strain combinations into two zones: the safe zone (in which failure during forming is not expected) and the failure zone (in which failure during forming is expected). Also referred to as a Keeler-Goodwin diagram.

form rolling. Hot rolling to produce bars having contoured cross sections; not to be confused with the *roll forming* of sheet metal or with *roll forging.*

forward extrusion. Same as direct extrusion. See also *extrusion.*

four-high mill. A type of rolling mill, commonly used for flat-rolled mill products, in which two large-diameter backup rolls are employed to reinforce two smaller work rolls, which are in contact with the product. Either the work rolls or the backup rolls may be driven. Compare with *two-high mill* and *cluster mill.*

fracture criterion. A mathematical relationship among stresses, strains, or a combination of stresses and strains that predicts the occurrence of ductile fracture. Should not be confused with fracture mechanics equations, which deal with more brittle types of fracture.

fracture limit line. An experimental method for predicting surface fracture in plastically deformed solids. Is related to the forming limit diagram used to predict failures in sheet forming.

fracture load. The load at which splitting occurs.

fracture-mechanism map. Strain rate/temperature map that describes regimes under which different damage and failure mechanisms are operative under either forming or service conditions.

fracture strain (ε_f). The true strain at fracture defined by the relationship:

$$\varepsilon_f = \ln\left[\frac{\text{Initial cross-sectional area}}{\text{Final cross-sectional area}}\right]$$

fracture strength. The engineering stress at fracture, defined as the load at fracture divided by the original cross-sectional area. The fracture strength is synonymous with the breaking strength.

fracture stress. The true stress at fracture, which is the load for fracture divided by the final cross-sectional area.

frame. The main structure of a *press.*

free bending. A bending operation in which the sheet metal is clamped at one end and wrapped around a radius pin. No tensile force is exerted on the ends of the sheet.

free forming. Bending over a die or shape in which the metal being bent is not considered.

friction. The resisting force tangential to the common boundary between two bodies when, under the action of an external force, one body moves or tends to move relative to the surface of the other.

friction coextrusion. Extrusion process for a solid core along with a tube made of a cladding material.

friction extrusion. A rotating round bar is pressed against a die to produce sufficient frictional heating to allow softened material to extrude through the die.

friction gouges. Line-type defects resulting from sliding sheet metal as in slipping of windings within a coil of steel. See also *scoring.*

friction hill. Shape of the normal pressure-position plot that pertains to the axisymmetric and plane-strain forging of simple and complex shapes. The pressure is approximately equal to the flow stress at the edge of the forging and increases toward the center, thus producing the characteristic hilllike shape. The exact magnitude of the increase in pressure is a function of interface friction and the diameter-to-thickness or width-to-thickness ratio of the forging.

friction shear factor. Interface friction coefficient for which the interface shear stress is taken to be proportional to the flow stress divided by $\sqrt{3}$. The proportionality constant is called the friction shear factor (or interface friction factor) and is usually denoted as *m.* The friction shear factor takes on values between 0 (perfect lubrication) and 1 (sticking friction) during metalworking. See also *Coulomb friction.*

front and back plate press. A type of hydraulic press in which the crown and bed are positioned and retained by steel plate strain members, front and back. Single rectangular openings are cut in both plates to provide entry to the work area.

Fukui cup test. A cupping test combining stretchability and drawability in which a round-nosed punch draws a circular blank into a conical-shaped die until fracture occurs at the nose. Various parameters from the test are used as the criterion of formability.

fuller (fullering impression). Portion of the die used in hammer forging primarily to reduce the cross section and lengthen a portion of the forging stock. The fullering impression is often used in conjunction with an *edger (edging impression).*

G

gage. (1) The thickness of sheet or the diameter of wire. The various standards are arbitrary and differ with regard to ferrous and nonferrous products as well as sheet and wire. (2) An aid for visual inspection that enables an inspector to determine more reliably whether the size or contour of a formed part meets dimensional requirements.

galling. A condition whereby excessive friction between high spots results in localized welding with subsequent spalling and further roughening of the rubbing surface(s) of one or both of two mating parts.

gap-frame press. A general classification of press in which the uprights or housings are made in the form of a letter "C," thus making three sides of the die space accessible.

gate or movable-barrier device. A movable barrier arranged to enclose the point of operation before a press stroke can be started.

geared press. A press whose main crank or eccentric is connected to the driving source by one or more sets of gears.

gibs. Guides or shoes that ensure the proper parallelism, squareness, and sliding fit between press components such as the slide and the frame. They are usually adjustable to compensate for wear and to establish operating clearance.

grain. An individual crystal in a polycrystalline metal or alloy.

grain boundary. The boundary between adjacent crystals/grains in a polycrystalline aggregate.

grain-boundary sliding. The sliding of grains past each other that occurs at high temperature. Grain-boundary sliding is common under creep conditions in service, thus leading to internal damage (e.g., cavities) or total failure, and during superplastic forming, in which undesirable cavitation may also occur if diffusional or deformation processes cannot accommodate the stress concentrations associated with sliding at a sufficient rate.

grain growth. The increase in the average size of grains in a crystalline aggregate during annealing (static conditions) or deformation (dynamic conditions). The driving force for grain growth is the reduction in total grain-boundary area and its associated surface energy.

grain size. A measure of the area or volume of grains in a polycrystalline material, usually expressed as an average when the individual sizes are fairly uniform.

gravity hammer. A class of forging hammer in which energy for forging is obtained by the mass and velocity of a freely falling ram and the attached upper die. Examples are the *board hammer* and *air-lift hammer*.

grease pits. Shallow, dark-colored depressions in the strip surface caused by the adhesion of grease, carbon, and so on to the work roll surface.

green. Unsintered (not sintered).

green compact. An unsintered *compact*.

green strength. (1) The ability of a *green compact* to maintain its size and shape during handling and storage prior to *sintering*. (2) The tensile or compressive strength of a green compact.

gridding. Imprinting an array of repetitive geometrical patterns on a sheet prior to forming for subsequent determination of deformation. Imprinting techniques include: (1) Electrochemical marking (also called electrochemical or electrolytic etching)—a grid-imprinting technique using electrical current, an electrolyte, and an electrical stencil to etch the grid pattern into the blank surface. A contrasting oxide usually is redeposited simultaneously into the grid. (2) Photoprint—a technique in which a photosensitive emulsion is applied to the blank surface, a negative of the grid pattern is placed in contact with the blank, and the pattern is transferred to the sheet by a standard photographic printing practice. (3) Ink stamping. (4) Lithographing.

gripper dies. The lateral or clamping dies used in a forging machine or mechanical upsetter.

groove. In *deep drawing,* the mating depression for the *drawbead.*

guard. A barrier that prevents entry of a press operator's hands or fingers into the point of operation.

Guerin process. A *rubber-pad forming* process for forming sheet metal.

guide. The parts of a drop hammer or press that guide the up-and-down motion of the ram in a true vertical direction.

guided bend test. A test in which the specimen is bent to a definite inside radius of a jig.

guide pin bushings. Bushings, pressed into a die shoe, that allow the *guide pins* to enter in order to maintain punch-to-die alignment.

guide pins. Hardened, ground pins or posts that maintain alignment between punch and die during die fabrication, setup, operation, and storage. If the press slide is out of alignment, the guide pins cannot make the necessary correction unless heel plates are engaged before the pins enter the bushings. See also *heel block.*

gutter. A depression around the periphery of a forging *die impression* outside the *flash pan* that allows space for the excess metal; surrounds the finishing impression and provides room for the excess metal used to ensure a sound forging. A shallow impression outside the parting line.

H

Hall-Petch dependence. A measure of the effect of grain size on the yield strength of a metal. Typically, the yield strength is inversely proportional to the square root of the grain size.

hammer. A machine that applies a sharp blow to the work area through the fall of a ram onto an anvil. The ram can be driven by gravity or power. See also *gravity hammer* and *power-driven hammer.*

hammer forging. Forging in which the work is deformed by repeated blows. Compare with *press forging.*

hammering. The working of metal sheet into a desired shape over a form or on a high-speed hammer and a similar anvil to produce the required dishing or thinning.

hand forge (smith forge). A forging operation in which forming is accomplished on dies that are generally flat. The piece is shaped roughly to the required contour with little or no lateral confinement; operations involving mandrels are included. The term *hand forge* refers to the operation performed, while hand forging applies to the part produced.

hand straightening. A straightening operation performed on a surface plate to bring a forging within straightness tolerance. A bottom die from a set of finish dies is often used instead of a surface plate. Hand tools used include mallets, sledges, blocks, jacks, and oil gear presses in addition to regular inspection tools.

hardness test. A test to measure the resistance to indentation of a material. Tests for sheet metal include Rockwell, Rockwell Superficial, Tukon, and Vickers.

Hartmann lines. See *Lüders lines.*

header. See *forging machine (upsetter or header).*

heading. The *upsetting* of wire, rod, or bar stock in dies to form parts that usually contain portions that are greater in cross-sectional area than the original wire, rod, or bar.

healed-over scratch. A scratch that occurred during previous processing and that was partially obliterated in subsequent rolling.

heel block. A block or plate usually mounted on or attached to a lower die that serves to prevent or minimize the deflection of punches or cams.

hemming. A bend of 180° made in two steps. First, a sharp-angle bend is made; next, the bend is closed using a flat punch and a die.

HERF. A common abbreviation for *high-energy-rate forging* or *high-energy-rate forming.*

high-angle boundary. Boundary separating adjacent grains whose misorientation is at least 15°.

high-energy-rate forging. The production of forgings at extremely high ram velocities resulting from the sudden release of a compressed gas against a free piston. Forging is usually completed in one blow. Also known as HERF processing, high-velocity forging, and high-speed forging.

high-energy-rate forming. A group of forming processes that applies a high rate of strain to the material being formed through the application of high rates of energy transfer. See also *explosive forming, high-energy-rate forging,* and *electromagnetic forming.*

holddown plate (pressure pad). A pressurized plate designed to hold the workpiece down during a press operation. In practice, this plate often serves as a *stripper* and is also called a stripper plate.

hole expansion. Forcing the metal around a hole out from the sheet surface.

hole expansion test. A formability test in which a tapered punch is forced through a punched or a drilled and reamed hole, forcing the metal in the periphery of the hole to expand in a stretching mode until fracture occurs.

hole flanging. The forming of an integral collar around the periphery of a previously formed hole in a sheet-metal part.

Hollomon hardening. See *power-law hardening.*

homogenization. Heat treatment used to reduce or eliminate nonuniform chemical composition that develops on a microscopic scale (microsegregation) during the solidification processing of ingots and castings. Homogenization is commonly used for aluminum alloys and nickel-base superalloys.

Hooke's law. Stress-strain behavior for a material in which the stress is linearly proportional to strain. See also *modulus of elasticity.*

hot brake forming. A forming process where the blank is heated up and formed in a cold tool. This is usually done very quickly, and the springback in the material is similar to a cold forming.

hot forming. See *hot working*. Similar to hot sizing; however, the forming is done at temperatures above the annealing temperature, and deformation is usually larger.

hot isostatic pressing (HIP). A process for simultaneously heating and forming a powder metallurgy compact in which metal powder, contained in a sealed flexible mold, is subjected to equal pressure from all directions at a temperature high enough for full consolidation to take place. Hot isostatic pressing is also frequently used to seal residual porosity in castings and to consolidate metal-matrix composites.

hot mill gouges. A series of short scratches caused by slippage of one surface of a coil relative to another during hot coiling.

hot rolled sheet. Steel sheet reduced to required thickness at a temperature above the point of scaling and therefore carrying hot mill oxide. The sheet may be flattened by cold rolling without appreciable reduction in thickness or by roller leveling, or both. Depending on the requirements, hot rolled sheet can be pickled to remove hot mill oxide and is so produced when specified.

hot shortness. A tendency for some alloys to separate along grain boundaries when stressed or deformed at temperatures near the melting point. Hot shortness is caused by a low-melting constituent, often present only in minute amounts, that is segregated at grain boundaries.

hot size. A process where a preformed part is placed into a hot die above the annealing temperature to set the shape and remove springback tendencies.

hot strip or pickle line scratch. Scratches that are superficially similar to slivers or skin laminations but originate from mechanical scoring of the strip in the hot mill, pickle line, or slitter.

hot trimming. The removal of *flash* or excess metal from a hot part (such as a forging) in a trimming press.

hot upset forging. A *bulk forming* process for enlarging and reshaping some of the cross-sectional area of a bar, tube, or other product form of uniform (usually round) section. It is accomplished by holding the heated forging stock between grooved dies and applying pressure to the end of the stock, in the direction of its axis, by the use of a heading tool, which spreads (upsets) the end by metal displacement. Also called hot heading or hot upsetting. See also *heading* and *upsetting*.

hot working. The plastic deformation of metal at such a temperature and strain rate that recrystallization or a high degree of recovery takes place simultaneously with the deformation, thus avoiding any *strain hardening*. Also referred to as hot forging and hot forming. Contrast with *cold working*.

hourglassing. A problem with some finite elements and some loading conditions where an element appears to be much less stiff than the actual material. The computed loads can thus be much lower than the true ones, or the computed strains can be much lower. The label refers to the appearance of the deformed finite element mesh that is sometimes observed where adjacent elements deform oppositely.

housing press. A type of hydraulic press in which the crown and bed are separated by uprights or spaces extending from front to back through which strain rods pass.

hub. A *boss* that is in the center of a forging and forms a part of the body of the forging.

hubbing. The production of die cavities by pressing a male master plug, known as a *hub,* into a block of metal.

hydraulic hammer. A gravity-drop forging hammer that uses hydraulic pressure to lift the hammer between strokes.

hydraulic-mechanical press brake. A mechanical *press brake* that uses hydraulic cylinders attached to mechanical linkages to power the ram through its working stroke.

hydraulic press. A press in which fluid pressure is used to actuate and control the ram.

hydraulic press brake. A *press brake* in which the ram is actuated directly by hydraulic cylinders.

hydraulic shear. A shear in which the crosshead is actuated by hydraulic cylinders.

hydrodynamic lubrication. A system of lubrication in which the shape and relative motion of the sliding surfaces causes the formation of a liquid film having sufficient pressure to separate the surfaces. See also *elastohydrodynamic lubrication.*

hydrostatic extrusion. A method of extruding a *billet* through a die by pressurized fluid instead of the ram used in conventional *extrusion.*

hydrostatic stress. The average value of the three normal stresses. The hydrostatic stress is a quantity that is invariant relative to the orientation of the coordinate system in which the stress state is defined.

I

IACS. See *percent IACS (%IACS).*

impact extrusion. The process (or resultant product) in which a punch strikes a slug (usually unheated) in a confining die. The metal flow may be either between punch and die or through another opening. The impact extrusion of unheated slugs is often called cold extrusion.

impact line. A blemish on a drawn sheet-metal part caused by a slight change in metal thickness. The mark is called an impact line when it results from the impact of the punch on the blank; it is called a recoil line when it results from transfer of the blank from the die to the punch during forming, or from a reaction to the blank being pulled sharply through the *draw ring*.

implicit. A type of finite element algorithm where a trial solution is proposed, the errors computed, and a better solution is proposed until a satisfactory solution is obtained. Accuracy can be ensured by such a method, but convergence to an answer cannot. Typical sheet-forming programs of this type do not consider momentum effects and thus are often called static explicit programs.

impression. A cavity machined into a forging die to produce a desired configuration in the workpiece during forging.

impression-die forging. See *closed-die forging.*

inclinable press. A small- or medium-sized press that may be inclined (tilted backward) to facilitate ejection of finished parts by gravity. Such presses are usually of the open-back, gap-frame type with right-to-left crankshaft. They are built in a maximum size of approximately 200 tons. They may be, and very often are, used in the upright or vertical position, being readily adjustable, usually by a hand mechanism, to any desired inclination up to the usual maximum of 45°.

increase in area. An indicator of sheet-metal-forming severity based on percentage increase in surface area measured after forming.

indirect (backward) extrusion. See *extrusion.*

ingot. A casting intended for subsequent rolling, forging, or extrusion.

ingot conversion. A primary metalworking process that transforms a cast ingot into a wrought mill product.

ingot metallurgy. A processing route consisting of casting an ingot that is subsequently converted into mill products via deformation processes.

integration point. Location in a finite element that is used to integrate over the volume in an approximate way to obtain the work of deformation of the element as a whole. For shell elements, the integration points through the thickness are used to obtain the sheet tension and moment by integration, with more integration points providing a better approximation to the true integral values.

interface heat-transfer coefficient (IHTC). Coefficient defined as the ratio of the heat flux across an interface to the difference in temperature of material points lying on either side of the interface. In bulk forming, the IHTC is usually a function of the die and workpiece surface conditions, lubrication, interface pressure, amount of relative sliding, and so on.

intermetallic alloy. A metallic alloy usually based on an ordered, stoichiometric compound (e.g., Fe_3Al, Ni_3Al, $TiAl$) and often possessing exceptional strength and environmental resistance at high temperatures, unlike conventional (less highly alloyed) disordered metallic materials.

ironing. An operation used to increase the length of a tube or cup through reduction of wall thickness and outside diameter, the inner diameter remaining unchanged.

isostatic pressing. A process for forming a powder metallurgy compact/metal-matrix composite or for sealing casting porosity by applying pressure equally from all directions. See also *hot isostatic pressing (HIP)*.

isothermal forging. A hot forging process in which a constant and uniform temperature is maintained in the workpiece during forging by heating the dies to the same temperature as the workpiece.

isotropy. A term indicating equal physical or mechanical properties in all directions.

J

Joffé effect. Change in mechanical properties, especially the fracture strength, resulting from testing in an environment that modifies the surface characteristics of the material.

jog. An intermittent motion imparted to a press side by momentary operation of the drive motor, after the clutch is engaged with the flywheel at rest.

joggle. (1) An offset (usually with parallel surfaces) in the surface of a sheet or part. (2) The process or act of offsetting the surface of a plate or part.

K

Keeler-Goodwin diagram. See *forming limit diagram (FLD)* or *forming limit curve (FLC)*.

kinetics. Term describing the rate at which a metallurgical process (e.g., recovery, recrystallization, grain growth, phase transformation) occurs as a function of time or, if during deformation, of strain.

kinks. Sharp bends or buckles caused by localized plastic deformation of a sheet.

klink. An internal crack caused by too rapid heating of a large workpiece.

knockout. A mechanism for releasing workpieces from a die.

knockout, hydraulic. On a hydraulic press, a mechanism actuated by a hydraulic cylinder for the purpose of ejecting work from the tools.

knockout mark. A small protrusion, such as a button or ring of flash, resulting from depression of the *knockout pin* from the forging pressure or the entrance of metal between the knockout pin and the die.

knockout, mechanical. A mechanism actuated by a hydraulic-press slide on its return stroke and arranged to eject work from the tools.

knockout pin. A power-operated plunger installed in a die to aid removal of the finished forging.

knockout, pneumatic. On a hydraulic press, a mechanism actuated by a pneumatic cylinder for the purpose of ejecting work from the tools.

knuckle-joint press. A press in which the slide is directly actuated by a single toggle (or knuckle) joint, which is closed and opened by means of a connection and a crank.

L

lancing. Cutting along a line in the workpiece without detaching a slug from the blank.

laser cutting. A cutting process that severs material with the heat obtained by directing a laser beam against a metal surface. The process can be used with or without an externally supplied shielding gas.

lateral extrusion. An operation in which the product is extruded sideways through an orifice in the container wall.

lattice. A regular geometrical arrangement of points in space. See also *point lattice* and *space lattice*.

lattice constants. See *lattice parameter*.

lattice parameter. The length of any side of a unit cell of a given crystal structure. The term is also used for the fractional coordinates x, y, z of lattice points when these are variable.

leveler breaks. Parallel surface markings perpendicular to the rolling direction caused by localized discontinuous yielding of the metal due to flexing during leveling.

leveler lines. Lines on sheet or strip running transverse to the direction of *roller leveling*. These lines may be seen upon stoning or light sanding after leveling (but before drawing) and can usually be removed by moderate stretching.

leveling. The flattening of rolled sheet, strip, or plate by reducing or eliminating distortions. See also *stretcher leveling* and *roller leveling*.

liftout. The mechanism also known as *knockout*.

limiting drawing ratio (LDR). See *deformation limit*.

linear element. A finite element that assumes that the pattern of displacements inside the element varies linearly with position inside the element.

liners. Thin strips of metal inserted between the dies and the units into which the dies are fastened.

lock. In forging, a condition in which the flash line is not entirely in one plane. Where two or more plane changes occur, it is called compound lock. Where a lock is placed in the die to compensate for die shift caused by a steep lock, it is called a counterlock.

lockbead. A ridge constructed around a die cavity to completely restrict metal flow into the die.

locked dies. Dies with mating faces that lie in more than one plane.

locking. A problem with some finite elements and some loading conditions where an element appears to be much stiffer than the actual material. The computed loads can thus be much higher than the true ones, or the computed strains can be much lower.

loop press. A type of hydraulic press in which the crown and bed are separated by straight-side members that are an integral part of an oval-shaped continuous band.

loose metal. A defect in an area of a stamping where very little contour is present. The metal in the area has not been stretched, resulting in a shape with no stiffness. The area may have waves in it or may sag so that there is a dish in an area that is intended to be flat or nearly flat. This defect differs from oil canning in that the metal cannot be snapped back into the desired shape when a load is removed or reversed on the area.

low-angle boundary. Boundary separating adjacent grains whose misorientation is less than 15°. See also *subgrain*.

lower punch. The lower part of a die that forms the bottom of the die cavity and that may or may not move in relation to the die body; usually movable in a forging die.

lubricant. A material applied to dies, molds, plungers, or workpieces that promotes the flow of metal, reduces friction and wear, and aids in the release of the finished part.

lubricant carrier. A coating, such as a phosphate or lime, used to increase the amount of lubricant available at a loaded interface.

lubricant residue. The carbonaceous residue resulting from lubricant that is burned onto the surface of a hot forged part.

Lüders lines. Elongated surface markings or depressions, often visible with the unaided eye, that form along the length of a round or sheet-metal tension specimen at an angle of approximately 55° to the loading axis. Caused by localized plastic deformation, they result from discontinuous (inhomogeneous) yielding. Also known as Lüders bands, Hartmann lines, Piobert lines, or stretcher strains.

M

major strain. Largest principal strain. For sheet-forming applications, it is often measured from the major axis of the ellipse resulting from deformation of a circular grid printed on the sheet surface.

mandrel. (1) A blunt-ended tool or rod used to retain the cavity in a hollow metal product during working. (2) A metal bar around which other metal can be cast, bent, formed, or shaped. (3) A shaft or bar for holding work to be machined.

mandrel forging. The process of rolling or forging a hollow blank over a mandrel to produce a weldless, seamless ring or tube.

manipulator. A mechanical device for handling an ingot, billet, or bar during forging. See also *chuckhead*.

Mannesmann process. A process for piercing tube billets in making seamless tubing. The billet is rotated between two heavy rolls mounted at an angle and is forced over a fixed mandrel.

Marforming process. A *rubber-pad forming* process developed to form wrinkle-free shrink flanges and deep-drawn shells. It differs from the *Guerin process* in that the sheet-metal blank is clamped between the rubber pad and the blankholder before forming begins.

master block. A forging *die block* used primarily to hold insert dies. See also *die insert.*

match. A condition in which a point in one die half is aligned properly with the corresponding point in the opposite die half, within specified tolerance.

matched edges (match lines). Two edges of the die face that are machined exactly at 90° to each other and from which all dimensions are taken in laying out the die impression and aligning the dies in the forging equipment.

matching draft. The adjustment of draft angles (usually involving an increase) on parts with asymmetrical ribs and sidewalls to make the surfaces of a forging meet at the parting line.

material heat. The pedigree of the starting stock or billet used to make a forging.

matrix phase. The continuous (interconnected) phase in an alloy with two or more phases. In cast or wrought materials, the matrix phase is often comprised of the first phase to solidify.

mechanical press. A forging press with an inertia flywheel, a crank and clutch, or other mechanical device to operate the ram.

mechanical press brake. A *press brake* using a mechanical drive consisting of a motor, flywheel, crankshaft, clutch, and eccentric to generate vertical motion.

mechanical texture. Directionality in the shape and orientation of microstructural features such as inclusions, grains, and so on.

mechanical upsetter. A three-element forging press, with two gripper dies and a forming tool, for flanging or forming relatively deep recesses.

mechanical working. The subjecting of material to pressure exerted by rolls, hammers, or presses in order to change the shape or physical properties of the material.

membrane strain. The strain at the center of a cross section of a bent beam or sheet. It is also the average strain of that cross section for an assumed linear bending strain through the thickness.

microalloyed steel. A low-to-medium-carbon steel usually containing small alloying additions of niobium, vanadium, nitrogen, and so on whose thermomechanical processing is controlled to obtain a specific microstructure and thus a suite of properties. See also *controlled rolling.*

microhardness test. An indentation test using diamond indentors at very low loads, usually in the range of 1 to 1000 g.

microstructure. The structure of polished and etched metals as revealed by a microscope.

mill. (1) A factory in which metals are hot worked, cold worked, or melted and cast into standard shapes suitable for secondary fabrication into commercial products. (2) A production line, usually of four or more *stands,* for hot or cold rolling metal into standard shapes such as bar, rod, plate, sheet, or strip. (3) A single machine for hot rolling, cold rolling, or extruding metal; examples include *blooming mill, cluster mill, four-high mill,* and *Sendzimir mill.* (4) A shop term for a milling cutter. (5) A machine or group of machines for grinding or crushing ores and other minerals.

mill digs. A defect caused by the slippage of the coil surface on the adjacent wrapped surface, sometimes caused by the use of extreme back tension and sometimes by the failure of the mandrel to grip the coil firmly due to damage of the internal wraps. Mill digs vary considerably in appearance from short, light scratches to, in extreme cases, holes in the sheet surface.

mill edge. The normal edge produced in rolling. Can be contrasted with a blanked or sheared edge that has a *burr.*

mill finish. A nonstandard (and typically nonuniform) surface finish on mill products that are delivered without being subjected to a special surface treatment (other than a corrosion-preventive treatment) after the final working or heat treating step.

mill product. Any commercial product of a *mill.*

mill scale. The heavy oxide layer that forms during the hot fabrication or heat treatment of metals.

minimum bend radius. The smallest radius about which a metal can be bent without exhibiting fracture. It is often described in terms of multiples of sheet thickness.

minor strain. Either of the two smaller principal strains. For sheet-forming applications, it is often measured from the minor axis of the ellipse resulting from deformation of a circular grid printed on the sheet surface.

mischmetal. From the German *mischmetall,* with roots *mischen* (to mix) and *metall* (metal), it is a natural mixture of rare earth metals containing approximately 50 wt% Ce, 25% La, 15% Nd, and 10% other rare earth metals, iron and silicon. It is commonly used to make rare earth additions to alloys (e.g., magnesium alloys), rather than using more expensive pure forms of the rare earth metals.

mismatch. The misalignment or error in register of a pair of forging dies; also applied to the condition of the resulting forging. The acceptable amount of this displacement is governed by blueprint or specification tolerances. Within tolerances, mismatch is a condition; in excess of tolerance, it is a serious defect. Defective forgings can be salvaged by hot reforging operations.

misorientation. Angular difference between the orientations of two grains adjacent to a grain boundary, between a twin and its parent matrix, and so on.

mixed dislocation. Any combination of a *screw dislocation* and an *edge dislocation.*

modulus of elasticity, *E.* The measure of rigidity or stiffness of a metal; the ratio of stress, below the proportional limit, to the corresponding strain. In terms of the *stress-strain diagram,* the modulus of elasticity is the slope of the stress-strain curve in the range of linear proportionality of stress to strain. Also known as *Young's modulus.* For materials that do not conform to *Hooke's law* throughout the elastic range, the slope of either the tangent to the stress-strain curve at the origin or at low stress, the secant drawn from the origin to any specified point on the stress-strain curve, or the chord connecting any two specific points on the stress-strain curve is usually taken to be the modulus of elasticity. In these cases, the modulus is referred to as the *tangent modulus, secant modulus,* or *chord modulus,* respectively.

moment of inertia. A measure of the ability of a beam or sheet to resist bending. It is a geometric property of the body and depends on a reference axis. The smallest moment of inertia is the one of interest; its reference axis passes through the centroid of the section. The area moment of inertia may be found by integrating $x^2 dA$ over the cross section, where x is the distance from the reference axis, and dA is the differential area of the beam at that distance.

Monte-Carlo modeling. Numerical modeling technique, based on statistical mechanics, that can be used to describe the migration of grain boundaries in polycrystalline aggregates during annealing or deformation processes and thus is applied to describe recrystallization, grain growth, and the accompanying evolution of texture. Also referred to as the Potts technique.

multiple. A piece of stock for forging that is cut from bar or billet lengths to provide the exact amount of material needed for a single workpiece. Also sometimes referred to as mult.

multiple-slide press. A press with individual slides, built into the main slide or connected to individual eccentrics on the main shaft, that can be adjusted to vary the length of stroke and the timing. See also *slide.*

***m*-value.** See *strain-rate sensitivity.*

N

natural draft. Taper on the sides of a forging, due to its shape or position in the die, that makes added draft unnecessary.

near-net shape forging. A forging produced with a very small finish allowance over the final part dimensions and requiring some machining prior to use.

necking. (1) The reduction of the cross-sectional area of metal in a localized area by uniaxial tension or by stretching. (2) The reduction of the diameter of a portion of the length of a cylindrical shell or tube.

necklace recrystallization. Partial static or dynamic recrystallization that nucleates heterogeneously on grain boundaries in various steels, nickel-base superalloys, and so on. A microstructure of fine (necklace-like) grains lying on the original grain boundaries is thus produced.

net shape forging. A forging produced to finished part dimensions that requires little or no further machining prior to use.

neural network. Nonlinear regression-type methodology for establishing the correlation between input and output variables in a physical system. For example, neural networks can be used to correlate processing variables to microstructural features or microstructural features to mechanical properties.

neuron. A node in a neural network system that can be considered as an internal variable and whose value is a function of the neurons in the previous layer.

neutral axis. The stress-free location in a beam or sheet subjected to bending and possibly sheet tension or compression. It may be located outside of the sheet; for pure bending, it is located at the center of the sheet. For simple cases, the neutral axis is assumed to coincide with the axis of zero extension.

node. A point of a finite element where variables such as location and force are monitored and solved for. Adjacent elements share common nodes, thus ensuring an approximation of compatibility.

no-draft (draftless) forging. A forging with extremely close tolerances and little or no draft that requires minimal machining to produce the final part. Mechanical properties can be enhanced by closer control of grain flow and by retention of surface material in the final component.

nominal strain. The unit elongation given by the change in length divided by the original length. Also called engineering strain.

nominal stress. The stress obtained when the applied load is divided by the original cross-sectional area. Also called engineering stress.

nonfill (underfill). A forging condition that occurs when the die impression is not completely filled with metal.

normal anisotropy. A condition in which a property or properties in the sheet thickness direction differ in magnitude from the same property or properties in the plane of the sheet.

notching. An unbalanced shearing or blanking operation in which cutting is done around only three sides (usually) of a punch.

n-**value.** See *strain-hardening exponent.*

O

objective function. Mathematical function describing a desired material or process characteristic whose optimization is the goal of process design. In bulk forming, typical objective functions may include forging weight (minimum usually is best), die fill (minimum underfill is best), and uniformity of strain or strain rate (maximum uniformity is best).

offal. Sheet-metal section trimmed or removed from the sheet during the production of shaped blanks or the formed part. Offal is frequently used as stock for the production of small parts.

offset. A specified distance along the strain coordinate between the initial portion of a stress-strain curve that is used to define one measure of yield stress. A line parallel to the initial elastic portion of the stress-strain curve is drawn offset by the specified strain, typically 0.2%. The intersection of this line and the stress-strain curve defines the *yield strength.* The offset yield stress is often used for materials that have no obvious *yield point.*

offset yield strength. A measure of yield stress determined by the intersection of a line parallel to the initial elastic portion of the stress-strain curve offset from the origin by the specified strain, typically 0.2%.

oil canning. Same as *canning.*

Olsen ductility test. A *cupping test* in which a piece of sheet metal, restrained except at the center, is deformed by a standard steel ball until fracture occurs. The height of the cup at the time of fracture is a measure of the ductility.

open-back inclinable (OBI) press. A gap-frame press with an opening at the back between the two side members of the frame and arranged to be inclinable to facilitate part removal by gravity. See also *inclinable press.*

open-back stationary (OBS) press. A gap-frame press with an opening at the back between the two side members of the frame and arranged to be upright or permanently inclined.

open die. Die with a flat surface that is used for preforming stock or producing hand forgings.

open-die forging. The hot mechanical forming of metals between flat or shaped dies in which metal flow is not completely/restricted. Also known as hand or smith forging. See also *hand forge (smith forge).*

open-rod press. A type of hydraulic press in which the crown and bed are positioned and retained by strain rods.

orange peel. A roughened or pebbly surface condition evident after substantial deformation that occurs with sheet metal, having an exceedingly coarse grain size. It should not be confused with a similar roughened surface resulting from fine stretcher strains.

orbital forging. See *rotary forging.*

orientation-distribution function (ODF). Mathematical function describing the normalized probability of finding grains of given crystallographic orientations/Euler angles. Because crystallographic orientations are in terms of Euler angles, the description of texture using ODFs is unambiguous, unlike pole figures. See also *texture, preferred orientation,* and *pole figure.*

Ostwald ripening. The increase in the average size of second-phase particles, accompanied by the reduction in their number, during annealing, deformation, or high-temperature service exposure. Ostwald ripening leads to a decrease in the total surface energy associated with matrix-particle interfaces. Also known as *coarsening.*

oxidation. A reaction where there is an increase in valence resulting from a loss of electrons. Such a reaction occurs when most metals or alloys are exposed to atmosphere, and the reaction rate increases as temperature increases.

oxidized surface. A tightly adhering oxide surface layer that results in modified surface color and reduced reflectivity. It is often accompanied by surface penetration of oxide that causes brittleness.

P

pack rolling. Hot, flat rolling process in which the workpiece (or a stack of workpieces) in the form of plate, sheet, or foil is encased in a sacrificial can to reduce/eliminate contamination (e.g., oxygen pickup) or poor workability due to roll chill.

pad. The general term used for that part of a die that delivers holding pressure to the metal being worked.

pad holder. Container for a rubber pad on a hydraulic press.

pancake forging. A rough forged shape, usually flat, that can be obtained quickly with minimal tooling. Usually made by upsetting a cylindrical billet to a large height reduction in flat dies. Considerable machining is usually required to attain the finish size.

parting. A shearing operation used to produce two or more parts from a stamping.

parting line. The line along the surface of a forging where the dies meet, usually at the largest cross section of the part. *Flash* is formed at the parting line.

parting plane. The plane that includes the principal die face and is perpendicular to the direction of ram travel. When parting surfaces of the dies are flat, the parting plane coincides with the parting line. Also referred to as the forging plane.

pass. (1) A single transfer of metal through a *stand* of rolls. (2) The open space between two grooved rolls through which metal is processed.

peak count. In surface measurements, the number of asperities above a given (defined) height cut-off level and within a given width cut-off. Frequency is taken at 50 μin./in.

peak density. The average number of peaks within the specified cut-off levels.

peak height. Peak-to-valley magnitude as measured by a suitable stylus instrument. Peak height is related to roughness height, depending on uniformity of surface irregularities.

peen forming. A dieless, flexible-manufacturing technique used primarily in the aerospace industry for forming sheet metals by way of the deformation imparted by the controlled-velocity impact of balls.

percent IACS (%IACS). In 1913, values of electrical conductivity were established and expressed as a percent of a standard. The standard chosen was an annealed copper wire with a density of 8.89 g/cm^3, a length of 1 m, a weight of 12 g, with a resistance of 0.1532 Ω

at 20 °C (70 °F). The 100% IACS (International Annealed Copper Standard) value was assigned with a corresponding resistivity of 0.017241 Ωmm^2/m. The percent IACS for any material can be calculated by %IACS = 0.017241 Ωmm^2/m × 100/volume resistivity.

perfectly-plastic. An idealization of plastic behavior that exhibits no strain hardening.

perforating. The punching of many holes, usually identical and arranged in a regular pattern, in a sheet, workpiece blank, or previously formed part. The holes are usually round but may be any shape. The operation is also called multiple punching. See also *piercing*.

permanent set. The deformation or strain remaining in a previously stressed body after release of the load.

persistent anticlastic curvature. *Anticlastic curvature* that remains after a bent sheet or beam is unloaded.

physical modeling. A subscale laboratory technique based on the principles of similarity used to study the effect of die design, material properties, and so forth on metal flow, defect formation, and so on during forging, extrusion, and other forming processes. The technique typically employs inexpensive die and workpiece materials. See also *visioplasticity*.

pickle patch. Localized area of tightly adhering surface oxide that results from incomplete removal of hot mill scale during pickling.

pickle stain. Discoloration of metal after oxide removal during pickling resulting from improper passivation of the sheet surface.

pickup. Small particles of oxidized metal adhering to the surface of a *mill product*.

piercing. The general term for cutting (shearing or punching) openings, such as holes and slots, in sheet material, plate, or parts. This operation is similar to *blanking*; the difference is that the slug or piece produced by piercing is scrap, while the blank produced by blanking is the useful part.

pin and bushing test. A test in which an oversized pin is driven by a hydraulic piston into a bushing. The bushing is expanded, causing an increase in pressure at the interface. The change in outside diameter of the bushing at a given pressing force is related to the coefficient of friction.

pinchers. Surface defects having the appearance of ripples or of elongated areas of variable surface texture extending at an acute angle to the sheet rolling direction and often branching. Pinchers are usually caused by poor hot rolled strip shape or improper drafting or incorrect crown on the cold reduction mill. They are sometimes referred to as feather pattern.

pinch point. Any point other than the point of operation at which it is possible for a part of a press operator's body to be caught between the moving parts of a press or auxiliary equipment, between moving and stationary parts of a press or auxiliary equipment, or between the material and a moving part or parts of the press or auxiliary equipment.

pinch trimming. The trimming of the edge of a tubular part or shell by pushing or pinching the flange or lip over the cutting edge of a stationary punch or over the cutting edge of a draw punch.

pinning. The retardation or complete cessation of grain growth during annealing or deformation by second-phase particles acting on grain boundaries.

Piobert lines. See *Lüders lines*.

pit. A small, clean depression in a sheet surface caused by the rolling-in of foreign particles such as sand, steel, and so on that subsequently fall out.

planar anisotropy. A term indicating variation in one or more physical or mechanical properties with direction in the plane of the sheet. The planar variation in plastic-strain ratio is commonly designated as Δr, given by: $\Delta r = (r_0 + r_{90} - 2r_{45})/2$. The earing tendency of a sheet is related to Δr. As Δr increases, so the tendency to form ears increases.

plane strain. Deformation in which the normal and shear components associated with one of the three coordinate directions are equal to zero. Bulk forming operations that approximate plane-strain conditions include sheet rolling and sheet drawing.

plane stress. Stress state in which the normal and shear components of stress associated with one of the three coordinate directions are equal to zero. Most sheet-forming operations are performed under conditions approximating plane stress.

planishing. Smoothing a metal surface by rolling, forging, or hammering; usually the last pass or passes of a shaping operation.

plastic anisotropy factor or plastic anisotropy ratio. See *plastic-strain ratio*.

plastic deformation. The permanent (inelastic) distortion of metals under applied stresses that strain the material beyond its *elastic limit*. The ability of metals to flow in a plastic manner without fracture is the fundamental basis for all metal-forming processes.

plastic flow. The phenomenon that takes place when metals or other substances are stretched or compressed permanently without rupture.

plastic instability. The deformation stage during which plastic flow is nonuniform and necking occurs.

plasticity. The ability of a metal to undergo permanent deformation without rupture.

plastic strain. The part of the total strain attributable to plastic, or irreversible, deformation. For simple elastic-plastic laws, it is the remainder when the elastic strains are removed from the total strains.

plastic-strain ratio (r-value). A measure of plastic anisotropy defined by the ratio of the true plastic width strain (ε_w) to the true plastic thickness strain (ε_t) in a tensile test. For large strains, the total strains are used to approximate the plastic strains. The average plastic-strain ratio, $\bar{r}$ or r_m, is determined from tensile samples taken in at least three directions from the sheet rolling direction, usually at 0, 45, and 90°. A formability parameter that relates to drawing, it is also known as the plastic anisotropy factor. A high r-value often correlates with good deep-drawing properties.

plate. See *sheet*.

platen. The sliding member, slide, or ram of a press.

plowing (ploughing). Plastic deformation of the surface of the softer component of a friction pair.

plug. (1) A rod or mandrel over which a pierced tube is forced. (2) A rod or mandrel that fills a tube as it is drawn through a die. (3) A punch or mandrel over which a cup is drawn. (4) A protruding portion of a die impression for forming a corresponding recess in the forging. (5) A false bottom in a die.

point lattice. A set of points in space located so that each point has identical surroundings. There are 14 ways of so arranging points in space, corresponding to the 14 *Bravais lattices*.

point of operation. The area of a press where material is actually positioned and work is performed during any process such as shearing, punching, forming, or assembling.

Poisson contraction. The transverse compressive strains that usually occur when a longitudinal extension is applied.

Poisson's ratio. The ratio of transverse elastic strain to longitudinal elastic strain experienced in a tensile test. Its use assumes isotropic behavior such that the transverse strains are equal for corresponding axial strain resulting from uniformly distributed axial stress below the *proportional limit* of the material in a tensile test.

pole figure. Description of crystallographic texture based on a stereographic-projection representation of the times-random probability of finding a specific crystallographic pole with a specific orientation relative to sample reference directions. For axisymmetric components, the sample reference directions are usually the axis and two radial directions; for a sheet material, the rolling, transverse, and sheet-normal directions are used. Because pole figures provide information only with regard to the orientation of one crystallographic pole, several pole figures or an orientation-distribution function (derivable from pole-figure measurements) are needed to fully describe crystallographic texture. See also *orientation-distribution function (ODF)*.

polycrystalline aggregate. The collection of grains/crystals that form a metallic material.

polygonization. A recovery-type process during the annealing of a worked material in which excess dislocations of a given sign rearrange themselves into low-energy, low-angle tilt boundaries.

porosity. Voids or pores within the workpiece; porosity is especially pertinent to powder forging.

postforming operation. Any treatment after the part has been formed, such as annealing, trimming, finishing, and so on.

powder forging. The plastic deformation of a powder metallurgy *compact* or *preform* into a fully dense finished shape by using compressive force; usually done hot and within closed dies.

powder metallurgy. A processing route consisting of the manufacture and subsequent consolidation of particulate (powder) materials to create shaped objects.

power-driven hammer. A forging hammer with a steam or air cylinder for raising the ram and augmenting its downward blow.

power-law hardening. A particularly simple form of strain hardening described by the equation $\sigma = K\varepsilon^n$, where σ is the true stress, ε is the true strain, K is called the *strength coefficient,* and the value n is a constant called the *strain-hardening exponent* or strain-hardening power. Also called *Hollomon hardening,* after J.H. Hollomon.

precision forging. A forging produced to closer tolerances than normally considered standard by the industry.

preferred orientation. Nonrandom distribution of the crystallographic orientations of the grains comprising a polycrystalline aggregate.

preform. (1) The forging operation in which stock is preformed or shaped to a predetermined size and contour prior to subsequent die-forging operations. When a preform operation is required, it will precede a forging operation and will be performed in conjunction with the forging operation and in the same heat. (2) The initially pressed powder metallurgy *compact* to be subjected to *repressing.*

prelubed sheet. A sheet or coil that has had a lubricant applied during mill processing to serve as a forming lubricant in the fabrication plant.

presence-sensing device. A device designed, constructed, and arranged to create a sensing field or area and to deactivate the clutch control of a press when an operator's hand or any other body part is within such field or area.

press. A machine tool with a stationary bed and a slide or ram that has reciprocating motion at right angles to the bed surface; the slide is guided in the frame of the machine.

press brake. An open-frame single-action press used to bend, blank, corrugate, curl, notch, perforate, pierce, or punch sheet metal or plate.

press-brake bend test. A bend test for sheet metal in which one side of the specimen is clamped and a press brake is used to bend the metal over its designed radius and with proper clearance.

press capacity. The rated force a press is designed to exert at a predetermined distance above the bottom of the stroke of the slide.

press component or mechanical power press component. The basic press (component) of the mechanical power press machine; that portion devoid of the tooling component, the safe-guarding component(s), and auxiliary feeding components. The press component is usually a multifunctional piece of equipment that is adaptable to a limitless combination of circumstances, depending on the other components incorporated with it.

press forging. The forging of metal between dies by mechanical or hydraulic pressure; usually accomplished with a single work stroke of the press for each die station.

press forming. Any sheet-metal forming operation performed with tooling by means of a mechanical or hydraulic press.

pressing. The product or process of shallow drawing sheet or plate.

press load. The amount of force exerted in a given forging or forming operation.

press slide. See *slide.*

press speed. Ram velocity (hydraulic/screw presses) or the number of strokes of the press slide in a unit of time (mechanical press). Generally expressed as strokes per minute (spm) or inches per minute (ipm).

pressure pad. A plate (flat or profiled) actuated by mechanical or fluid force to control metal movement and/or flow.

pressure plate. A plate located beneath the bolster that acts against the resistance of a group of cylinders mounted to the pressure plate to provide uniform pressure throughout the press stroke when the press is symmetrically loaded.

principal strain. The normal strain on any of three mutually perpendicular planes on which no shear strains are present.

principal strain direction. The direction perpendicular to any of the three mutually perpendicular planes on which no shear strains are present.

principal stress. One of the three normal stresses in the coordinate system in which all of the shear stresses are equal to zero.

prior particle boundary. An apparent boundary between the pre-existing powder metal particles that is still evident within the microstructure of consolidated powder metallurgy products because of the presence of carbide or other phases that form at these boundaries.

processing map. A map of strain rate versus temperature that delineates the regions that should be avoided in processing to prevent the formation of poor microstructures or voids or cracks. These maps are generally created by the dynamic material modeling method or by mapping extensive results of processing experience.

process modeling. Computer simulation of deformation, heat treating, and machining processes for the purpose of improving process yield and material properties.

profile (contour) rolling. In *ring rolling,* a process used to produce seamless rolled rings with a predesigned shape on the outside or the inside diameter, requiring less volume of material and less machining to produce finished parts.

progression. The constant dimension between adjacent stations in a progressive die.

progressive. A planned sequence to accomplish press-working operations consecutively.

progressive die. A die planned to accomplish a sequence of operations as the strip or sheet or material is advanced from station to station, manually or mechanically.

progressive forming. Sequential forming at consecutive stations with a single die or separate dies.

progressive ironing. A cold forming process in which a part is shaped by a succession of ironing operations. The successive operations may occur in one or a series of dies.

projection welding. Electrical resistance welding in which the welds are localized at embossments or other raised portions of the sheet surface.

proof. Any reproduction of a die impression in any material; often a lead or plaster cast. See also *die proof (cast).*

proof load. A predetermined load, generally some multiple of the service load, to which a specimen or structure is submitted before acceptance for use.

proof stress. (1) The stress that will cause a specified small permanent set in a material. (2) A specified stress to be applied to a member or structure to indicate its ability to withstand service loads.

proportional limit. The stress at which deviation from straight-line proportionality between stress and strain is observed. The value depends intimately on the resolution of measured stress and strain. See also *elastic limit* and *Hooke's law.*

pull-out device. A mechanism attached to an operator's hands and connected to the upper die or *slide* of a press that is intended, when properly adjusted, to withdraw the operator's hands as the dies close, if the operator's hands are inadvertently within the point of operation.

punch. (1) The male part of a die—as distinguished from the female part, which is called the die. The punch is usually the upper member of the complete die assembly and is mounted on the *slide* or in a *die set* for alignment (except in the inverted die). (2) In double-action draw dies, the punch is the inner portion of the upper die, which is mounted on the plunger (inner slide) and does the drawing. (3) The act of piercing or punching a hole. Also referred to as *punching.*

punching. The die shearing of a closed contour in which the sheared-out sheet-metal part is scrap.

punch-load/punch travel diagram. A diagram showing the relationship between the main ram (punch) load of a press and the distance the punch has traveled during metal forming.

punch nose radius. The shape of the punch end, contacting the material being formed to allow proper material flow or movement.

punch press. (1) In general, any mechanical press. (2) In particular, an endwheel gap-frame press with a fixed bed, used in piercing.

punch retainer. The detail or component of a die that holds the punch in working position and also permits easy removal.

punch section. A section of the punch used in cutting, forming, or flanging operations that is fastened to outer sections to make up the complete punch working edge.

punch slide. The main reciprocating member of a press. See *slide*.

Q

quadratic element. A finite element that assumes that the pattern of displacements inside the element varies quadratically with position inside the element.

quarter hard. A temper of nonferrous alloys and some ferrous alloys characterized by tensile strength approximately midway between that of dead-soft and half-hard tempers.

quench aging. Hardening by precipitation that results after the rapid cooling from solid solution to a temperature below which the elements of a second phase become super-saturated. Precipitation occurs after the application of higher temperatures and/or times and causes increases in yield strength, tensile strength, and hardness.

R

rabbit ear. Recess in the corner of a metal-forming die to allow for wrinkling or folding of the blank.

radial draw forming. The forming of sheet metals by the simultaneous application of tangential stretch and radial compression forces. The operation is done gradually by tangential contact with the die member. This type of forming is characterized by very close dimensional control.

radial forging. A process using two or more moving anvils or dies for producing shafts with constant or varying diameters along their length or tubes with internal or external variations in diameter. Often incorrectly referred to as *rotary forging.*

radial rolling force. The action produced by the horizontal pressing force of the rolling mandrel acting against the ring and the main roll.

radial roll (main roll, king roll). The primary driven roll of the rolling mill for rolling rings in the radial pass. The roll is supported at both ends.

radius. To remove the sharp edge or corner of forging stock by means of a radius or form tool.

ram. The moving or falling part of a drop hammer or press to which one of the dies is attached; sometimes applied to the upper flat

die of a steam hammer. Also referred to as the *slide.*

recoil line. See *impact line.*

re-coil lines. Blemishes in the form of repetitive lines of variable spacing extending transverse to the length of the coiled sheet and caused by coil breaks or leveler breaks.

recovery. Process occurring during annealing following cold or hot working of metals in which defects such as dislocations are eliminated or rearranged by way of mechanisms such as dipole annihilation, the formation of subgrains, and subgrain growth. Recovery usually leads to a reduction in stored energy, softening, reduction or elimination of residual stresses, and, in some instances, changes in physical properties. Recovery may also serve as a precursor to static recrystallization at sufficient levels of prior cold or hot work. See also *dynamic recovery.*

recrystallization. A process of nucleation and growth of new strain-free grains or crystals in a material. This process occurs upon heating above the recrystallization temperature (approximately 40% of the absolute melting temperature) during/after hot working or during annealing after cold working. Recrystallization can be dynamic (occurring during straining), static (occurring following deformation, typically during heat treatment), or metadynamic (occurring immediately after deformation due to the presence of recrystallization nuclei formed during deformation).

recrystallization texture. Crystallographic texture formed during static or dynamic recrystallization. The specific texture components that are formed are dependent on the nature of the stored work driving recrystallization and the nucleation and growth mechanisms that underlie recrystallization.

redrawing. The second and successive deep-drawing operations in which cuplike shells are deepened and reduced in cross-sectional dimensions.

reduction. (1) In cupping and deep drawing, a measure of the percentage of decrease from blank diameter to cup diameter, or of the diameter reduction in redrawing. (2) In forging, extrusion, rolling, and drawing, either the ratio of the original to the final cross-sectional area or the percentage of decrease in cross-sectional area.

reduction in area. The difference between the original cross-sectional area and the smallest area at the point of rupture in a tensile test; usually stated as a percentage of the original area.

redundant work. Energy in addition to that required for uniform flow expended during processing due to inhomogeneous deformation.

relative density. Ratio of density to pore-free density.

relief. Clearance obtained by removing material, either behind or beyond the cutting edge of a punch or die.

repressing. The application of pressure to a sintered compact; usually done to improve a physical or mechanical property or for dimensional accuracy.

rerolling quality. Rolled billets from which the surface defects have not been removed or completely removed.

reset. The realigning or adjusting of dies or tools during a production run; not to be confused with the operation setup that occurs before a production run.

residual stress. Stresses that remain in an unloaded body as a result of, for example, nonuniform plastic deformation or heating and cooling.

restriking. (1) The striking of a trimmed but slightly misaligned or otherwise faulty forging with one or more blows to improve alignment, improve surface condition, maintain close tolerances, increase hardness, or effect other improvements. (2) A sizing operation in which coining or stretching is used to correct or alter profiles and to counteract distortion. (3) A salvage operation following a primary forging operation in which the parts involved are rehit in the same forging die in which the pieces were last forged.

return die. A type of die, particularly cutting, in which the slug or blank is returned to the die surface for removal.

reverse drawing. *Redrawing* of a sheet-metal part in a direction opposite to that of the original drawing.

reverse flange. A sheet-metal flange made by shrinking, as opposed to one formed by stretching.

reverse redrawing. An operation after the first drawing operation in which the part is turned inside out by inverting and redrawing, usually in another die, to a smaller diameter.

rib. (1) A long V-shaped or radiused indentation used to strengthen large sheet-metal panels. (2) A long, usually thin protuberance used to provide flexural strength to a forging (as in a rib-web forging).

rigid. An idealized material behavior where the elastic constants are infinite such that no elastic straining occurs.

ring compression test. A workability test that uses the expansion or contraction of the hole in a thin compressed ring to measure the frictional conditions. The test can also be used to determine the flow stress.

ring rolling. The process of shaping weldless rings from pierced disks or shaping thick-wall ring-shaped blanks between rolls that control wall thickness, ring diameter, height, and contour.

rocker arm press. A press in which the force to perform work is transmitted from the driving system to the driven platen or ram by a rocker arm through a fulcrum.

rod. A solid round section 9.5 mm ($3/8$ in.) or greater in diameter whose length is great in relation to its diameter.

roll. Tooling used in the rolling process to deform material stock.

roll bending. The curving of sheets, bars, and sections by means of rolls.

rolled-in dirt. Extraneous material rolled into the surface of sheet metal.

rolled-in scale. Localized areas of heavy oxide not removed by the hot mill descaling sprays and rolled out to elongated streaks during further processing.

roller leveler breaks. Obvious transverse breaks on sheet metal usually approximately 3 to 6 mm ($\frac{1}{8}$ to $\frac{1}{4}$ in.) apart that are caused by the sheet fluting during *roller leveling*. These will not be removed by stretching.

roller leveling. *Leveling* by passing flat sheet-metal stock through a machine having a series of small-diameter staggered rolls that are adjusted to produce repeated reverse bending.

roll feed. A mechanism for feeding strip or sheet stock to a press or other machine. The stock passes between two revolving rolls mounted one above the other, which feed it under the dies a predetermined length at each stroke of the press. Two common types of drive are the oscillating lever type and the rack-and-pinion type. The single roll feed may be used to either push or pull the stock to or from the press. The double roll feed is commonly used with wider presses (left to right) or in other cases where a single roll feed is impractical.

roll flattening. The flattening of sheets that have been rolled in packs by passing them separately through a two-high cold mill with virtually no deformation. Not to be confused with *roller leveling.*

roll forging. A process of shaping stock between two driven rolls that rotate in opposite directions and have one or more matching sets of grooves in the rolls; used to produce finished parts or preforms for subsequent forging operations.

roll former. A device with three or more rolls positioned to progressively plastically form sheet or strip metal into curved or linear shapes.

roll forming. Metal forming through the use of power-driven rolls whose contour determines the shape of the product; sometimes used to denote power *spinning.*

rolling. The reduction of the cross-sectional area of metal stock, or the general shaping of metal products, through the use of rotating rolls.

rolling mandrel. In ring rolling, a vertical roll of sufficient diameter to accept various sizes of ring blanks and to exert rolling force on an axis parallel to the main roll.

rolling mills. Machines used to decrease the cross-sectional area of metal stock and to produce certain desired shapes as the metal passes between rotating rolls mounted in a framework comprising a basic unit called a *stand.* Cylindrical rolls produce flat shapes; grooved rolls produce rounds, squares, and structural shapes. See also *four-high mill, Sendzimir mill,* and *two-high mill.*

roll mark. A mark in light relief on the sheet surface produced by an indentation in the cold reduction mill work roll surface.

roll pick-up. Repetitive marks or indentations in sheets caused by extraneous material that has adhered to a work roll in the cold reduction mill or temper mill.

roll straightening. The straightening of metal stock of various shapes by passing it through a series of staggered rolls (the rolls usually being in horizontal and vertical planes) or by reeling in two-roll straightening machines.

roll threading. The production of threads by rolling the piece between two grooved die plates, one of which is in motion, or between rotating grooved circular rolls.

roping. A surface defect consisting of a series of generally parallel markings or ripples on areas of rolled formed sheet parts that have undergone substantial strain. The ripples are always parallel to rolling direction.

rotary forging. A process in which the work-piece is pressed between a flat anvil and a swiveling (rocking) die with a conical working face; the platens move toward each other during forging. Also called orbital forging. Compare with *radial forging.*

rotary shear. A sheet-metal cutting machine with two rotating-disk cutters mounted on parallel shafts driven in unison.

rotary swager. A swaging machine consisting of a power-driven ring that revolves at high speed, causing rollers to engage cam surfaces and force the dies to deliver hammerlike blows on the work at high frequency. Both straight and tapered sections can be produced.

rotary swaging. A *bulk forming* process for reducing the cross-sectional area or otherwise changing the shape of bars, tubes, or wires by repeated radial blows with one or more pairs of opposed dies.

rough blank. A blank for a forming or drawing operation, usually of irregular outline, with necessary stock allowance for process metal, which is trimmed after forming or drawing to the desired size.

roughing stand. The first stand (or several stands) of rolls through which a reheated billet or slab passes in front of the finishing stands. See also *rolling mills* and *stand.*

roughness cut-off level. Terms used in the measurement of surface roughness. (1) Width cut-off: the greatest spacing of repetitive surface irregularities used in the measurement of roughness, usually 0.75 mm (0.030 in.). (2) Height cut-off: the minimum surface irregularity in peak count determinations, usually 1.3 μm (50 μin.).

roughness height. The average height of surface irregularities with reference to a mean or nominal surface as determined by height and width cut-offs. It may be expressed as the deviation from the nominal surface, as arithmetic average, or as root mean square.

R/t **ratio.** A normalized dimensionless expression of the radius of curvature after bending (R) to the thickness of the section, t. Smaller numbers refer to more severe bending, with R/t of approximately 5 to 6 usually being considered the lower limit for application of simple shell theories of bending.

rubber forming. A sheet-metal forming process in which rubber is used as a functional die part.

rubber-pad forming. A sheet-metal forming operation for shallow parts in which a confined, pliable rubber pad attached to the press slide (ram) is forced by hydraulic pressure to become a mating die for a punch or group of punches placed on the press bed or baseplate. Developed in the aircraft industry for the limited production of a large number of diversified parts, the process is limited to the forming of relatively shallow parts, normally not exceeding 40 mm (1.5 in.) deep. Also known as the *Guerin process.* Variations of the Guerin process include the *Marforming process,* the *fluid-cell process,* and *fluid forming.*

run. The quantity produced in one setup.

*r***-value.** See *plastic-strain ratio.*

S

saddening. The process of lightly working an ingot in the initial forging operation to break up and refine the coarse, as-cast structure at the surface.

scale pattern. A transverse surface pattern on cold rolled sheet caused by intermittent removal of the scale in the scale-breaker operation prior to pickling. The result is a pattern of overpickled areas that are not eliminated in cold reduction.

Schmid factor. In a uniaxial tension test, the geometric factor that corresponds to the product of the cosine of the angle between the tension axis and the slip-plane normal and the cosine of the angle between the tension axis and the slip direction. Often denoted as *m.*

Schmid's law. Criterion that slip in metallic crystals is controlled by a critical resolved shear stress that depends on specific material, strain rate, and test temperature but is independent of the stress normal to the slip plane.

scoring. (1) The marring or scratching of any formed part by metal pickup on the punch or die. (2) The reduction in thickness of a material along a line to weaken it intentionally along that line.

scrap mark. A type of roll mark caused by rolling over ragged coiled edges or coil ends. See also *roll mark.*

scratches. Lines generally caused by sliding of the sheet surface over sharp edges of processing equipment or over other sheets.

scratch resistance. The ability of a material to resist scratching. It is a function of the material hardness, although the lubricity of the surface will also play a part.

screw dislocation. A line imperfection that corresponds to the axis of a spiral structure in a crystal and is characterized by a distortion joining normally parallel lines together to form a continuous helical ramp (with a pitch of one interplanar distance) winding about the dislocation.

screw press. A high-speed press in which the ram is activated by a large screw assembly powered by a drive mechanism.

scuffing. Localized damage caused by the occurrence of solid-phase welding between sliding surfaces. No local surface melting occurs. See also *galling* and *scoring*.

seam. A surface defect appearing as thin lines in the rolling direction of sheet metals due to voids elongated during rolling.

seaming. The process of joining sheet-metal parts by interlocking bends.

secant modulus. The slope of the secant drawn from the origin to any specified point on the stress-strain curve. See also *modulus of elasticity*.

secondary recrystallization. See *abnormal grain growth*.

secondary tensile stress. Tensile stress that develops during a bulk deformation process conducted under nominally compressive loading due to nonuniform metal flow resulting from geometry, friction, or die-chilling effects. Secondary tensile stresses are most prevalent in open-die forging operations.

secondary sheet. A shearing action that occurs between soft work metal and a cutting edge as a result of insufficient clearance.

segment die. Same as *split die*.

segregation. A nonuniform distribution of alloying elements, impurities, or microphases.

seizure. The stopping of relative motion between two bodies as the result of severe interfacial friction. Seizure may be accompanied by gross surface welding.

semifinisher. An impression in a series of forging dies that only approximates the finish dimensions of the forging. Semifinishers are often used to extend die life or the finishing impression, to ensure proper control of grain flow during forging, and to assist in obtaining desired tolerances.

seminotching. A process similar to notching except that the cutting operation is a partial one only, permitting the cut shape to remain with the blank or part. Thus, by a series of seminotching actions from coil stock, the workpiece is still retained in the strip skeleton.

Sendzimir mill. A type of *cluster mill* with small-diameter work rolls and larger-diameter backup rolls, backed up by bearings on a shaft mounted eccentrically so that it can be rotated to increase the pressure between the bearing and the backup rolls. Used to roll precision and very thin strip. Note: Sendzimir mills roll strip, not sheets.

set. The shape remaining in a stamped or press-formed part after the punch force is removed. See also *permanent set*.

severe plastic deformation. Processes of plastic deformation with accumulated natural logarithmic strains more than 4 that are usually used to change material structure and properties.

shank. The portion of a die or tool by which it is held in position in a forging unit or press.

shape distortion. A dimensional change due to warping or bending resulting mainly from thermal treatment.

shape fixability. The ability of a material to retain the shape given to it by a forming operation.

shaving. Backflow of the clad or sleeve material during hydrostatic coextrusion.

shear. (1) A machine or tool for cutting metal and other material by the closing motion of two sharp, closely adjoining edges, for example, squaring shear and circular shear. (2) An inclination between two cutting edges, such as between two straight knife blades or between the punch cutting edge and the die cutting edge, so that a reduced area will be cut each time. This lessens the necessary force but increases the required length of the working stroke. This method is referred to as angular shear. (3) The act of cutting by shearing dies or blades, as in a squaring shear. (4) The type of force that causes or tends to cause two contiguous parts of the same body to slide relative to each other in a direction parallel to their plane of contact.

shear band. Region of highly localized shear deformation developed during bulk forming (and sometimes during sheet forming) as a result of material properties (such as a high flow-softening rate and low rate sensitivity of the flow stress), metal flow geometry, friction, chilling, and so on.

shear burr. A raised edge resulting from metal flow induced by blanking, cutting, or punching.

shearing. A cutting operation in which the work metal is placed between a stationary lower blade and movable upper blade and severed by bringing the blades together. Cutting occurs by a combination of metal shearing and actual fracture of the metal.

shear strength. The maximum shear stress a material can sustain. Shear strength is calculated from the maximum load during a shear or torsion test and is based on the original dimensions of the cross section of the specimen.

shear stress. (1) A stress that exists when parallel planes in metal crystals slide across each other. (2) The stress component tangential to the plane on which the forces act.

sheet. Any material or piece of substantially uniform thickness and of considerable length and width as compared to its thickness. With regard to metal, such pieces under 6.5 mm (1/4 in.) thick are called sheets, and those 6.5 mm (1/4 in.) thick and over are called plates. Occasionally, the limiting thickness for steel to be designated as sheet steel is No. 10 Manufacturer's Standard Gage for sheet steel, which is 3.42 mm (0.1345 in.) thick.

sheet bar. Workpiece, usually with a rectangular cross section, typically used in a batch (hand mill) sheet-rolling process.

sheet forming. The plastic deformation of a piece of sheet metal by tensile loads into a three-dimensional shape, often without significant changes in sheet thickness or surface characteristics. Compare with *bulk forming*.

sheet tension. The tensile force tending to extend a sheet in the sheet plane. It is expressed in units of force and may be computed from the stress distribution through the thickness by integrating the differential force through the section.

sheet tension stress. A stresslike quantity obtained by dividing the sheet tension by the cross-sectional area on which the sheet tension force operates.

shell element. A kind of finite element that takes advantage of one dimension being much smaller than the other two by reducing the small dimension to zero while adopting an approximate relationship between location in the sheet or beam thickness and strain, usually linear. If no bending strains are allowed, that is, the strains are uniform through the sheet thickness, a shell element becomes a membrane element. Also called thin-shell element.

shell four-ball test. A lubricant test in which three balls are clamped in contact as in an equilateral triangle. The fourth ball is held in a rotating chuck and touches each of the stationary balls. Load is applied through a lever arm system that pushes the stationary balls upward against the rotating ball.

shim. A thin piece of material used between two surfaces to obtain a proper fit, adjustment, or alignment.

shrinkage. The contraction of metal during cooling after hot forging. Die impressions are made oversized according to precise shrinkage scales to allow the forgings to shrink to design dimensions and tolerances.

shrink flanging. The reduction of the length of the free edge after the flanging process.

shut height. For a press, the distance from the top of the bed to the bottom of the slide with the stroke down and adjustment up. In general, it is the maximum die height that can be accommodated for normal operation, taking the *bolster* plate into consideration.

shuttle die. A multiple-station die in which the separated workpieces are fed from station to station by bars that are positioned in the die proper. Also known as a transfer die.

side housings. Uprights or spacers extending from front to back and used to separate the crown and the bed on a hydraulic press.

side-plate press. A type of hydraulic press in which the crown and bed are separated by steel plate strain members, one on each side. The side plates are fastened to the crown and head.

sidepressing. A deformation process in which a cylinder is laid on its side and deformed in compression. It is a good test to evaluate the tendency for fracture at the center of a billet, or for evaluating the tendency to form shear bands.

side thrust. The lateral force exerted between the dies by reaction of a forged piece on the die impressions.

side-wall curl. The curving of the region of sheet drawn into the die cavity in a *channel forming* operation, when the forming forces are removed and *springback* occurs.

single-action hydraulic press. One equipped with one moving slide member. A single-action press may have a *cushion* built into the bed.

single-action press. A press having one ram or slide. Also refers to a draw die that does not have a blankholder.

single-stand mill. A rolling mill designed such that the product contacts only two rolls at a given moment. Contrast with *tandem mill*.

sinking. The operation of machining the impression of a desired forging into die blocks.

sintering. The densification and bonding of adjacent particles in a powder mass or compact by heating to a temperature below the melting point of the main constituent.

sizing. (1) Secondary forming or squeezing operations needed to square up, set down, flatten, or otherwise correct surfaces to produce specified dimensions and tolerances. See *restriking*. (2) Some burnishing, broaching, drawing, and shaving operations are also called sizing. (3) A finishing operation for correcting ovality in tubing. (4) Final pressing of a sintered powder metallurgy part.

skin lamination. A subsurface separation that can result in surface rupture during forming.

slab. Flat-shaped semifinished rolled metal stock section with a width not less than 250 mm (10 in.) and a cross-sectional area not less than 105 cm^2 (16 in.^{2}).

slabbing. The hot working of an ingot to a flat rectangular shape.

sleeve. Outer layer of a coextruded or codrawn product. See also *clad*.

slide. The main reciprocating member of a press, guided in the press frame, to which the punch or upper die is fastened; sometimes called the *ram*. The inner slide of a double-action press is called the plunger or punchholder slide; the outer slide is called the blankholder slide. The third slide of a triple-action press is called the lower slide, and the slide of a hydraulic press is often called the platen.

slide adjustment. The distance that a press slide position can be altered to change the shut height of the die space. The adjustment can be made by hand or by power mechanism.

sliding friction test—flat dies. A test in which a sheet steel sample is placed between two flat, hardened die faces and pulled through the dies under conditions that permit recording of the load applied to the dies and the force required to pull the strip.

sliding friction test—wedge dies. Similar to the flat-die test assembly except that one or both of the flat dies has a wedge configuration to confine the edges of the specimen. This permits development of unit loadings in excess of compressive strength of the specimen, and cold reduction of drawn strip is readily accomplished.

slip. Crystallographic shear process associated with dislocation glide that underlies the large plastic deformation of crystalline metals and alloys. Slip is usually observed on close-packed planes along close-packed directions, in which case it is referred to as restricted slip. In body-centered cubic materials, such as alpha iron, slip occurs along any plane containing a close-packed direction and is referred to as pencil glide.

slip-line field. Graphical technique used to estimate the deformation and stresses involved in plane-strain metal-forming processes.

slitting. Cutting or shearing along single lines to cut strips from a sheet or to cut along lines of a given length or contour in a sheet or workpiece.

sliver. A surface defect consisting of an elongated thin layer of partially attached metal.

slotting. A stamping operation in which elongated or rectangular holes are cut in a blank or part.

slug. (1) The metal removed when punching a hole in a forging; also termed punchout. (2) The forging stock for one workpiece cut to length. See also *blank*.

slugging. A sheet-metal cutting operation in which the steel punch is set to enter the sheet metal the exact amount of penetration required to complete fracturing. Due to cold welding and springback of the sheet metal, the slug remains in the sheet despite the completion of fracture.

smith forging. See *hand forge (smith forge)*.

smudge. A dark-appearing surface contamination on annealed sheet generally resulting from cold-reduction oil residues or carbon deposited from annealing gas with an unfavorable CO/CO_2 ratio. Smudge may adversely affect painting or plating but may be beneficial in the prevention of galling.

smut. A contaminant consisting of fine, dark-colored particles on the surface of pickled sheet products. This usually results from heavy oxidation of the steel surface during hot rolling.

snap through. Shock in a die due to the sudden beginning and completion of fractures in cutting dies causing the compressed punch to elastically snap into tension.

solid press. A type of hydraulic press in which the crown and bed are separated by side members that are an integral part of the crown and bed.

sow block. A block of heat treated steel placed between the anvil of the hammer and the forging die to prevent undue wear to the anvil. Sow blocks are occasionally used to hold insert dies. Also called anvil cap.

space lattice. A set of equal and adjoining parallelepipeds formed by dividing space by three sets of parallel planes, the planes in any one set being equally spaced. There are seven ways of so dividing space, corresponding to the seven *crystal system* structures. The unit parallelepiped is usually chosen as the unit cell of the system. Due to geometrical considerations, atoms can only have one of 14 possible arrangements, known as *Bravais lattices*.

spalling. The removal of small pieces of metal from the working face of the die, usually as a result of severe heat checking. Spalling is most likely in hard materials with low ductility.

spank. A press operation used to reform parts that have already had their major contour formed or drawn in the conventional manner. The spank operation is often used where it is not possible to produce the final contour, such as sharp creases or corners, in a single forming operation. It is also used at the end of a production line where large sheet-metal parts have become distorted due to previous operations such as trimming, punching, forming, and flanging. Spanking is used to bring the panels back to the desired contour. See also *restriking*.

special boundary. A boundary between two grains whose crystallographic lattices have a certain fraction ($1/N$, in which N is an integer) of coincident lattice points. Such boundaries, denoted using the notation ΣN, may have low mobility and surface energy.

specific heat. Amount of heat required to change the temperature of one unit weight of a material by one degree.

spheroidization. Process of converting a lamellar, basketweave, or acicular second phase into an equiaxed morphology via deformation, annealing, or a combination of deformation followed by annealing.

spinning. The forming of a seamless hollow metal part by forcing a rotating blank to conform to a shaped mandrel that rotates concentrically with the blank. In the typical application, a flat-rolled metal blank is forced against the mandrel by a blunt, rounded tool; however, other stock (notably, welded or seamless tubing) can be formed. A roller is sometimes used as the working end of the tool.

split die. A die made of parts that can be separated for ready removal of the workpiece. Also known as segment die.

springback. The elastically driven, usually undesirable, change of shape that occurs after forming when the external forces are removed. It is often expressed as a change of curvature, that is, the original curvature less the final curvature, which is generally greater than 1. This effect can be compensated for by overbending or by a secondary operation of *restriking*.

springback ratio. The ratio of the radius of curvature after *springback* (that is, in the unloaded state) to the radius at the end of forming (while still loaded). The ratio is generally positive.

springforward. A *springback* compensation method characterized by a step in which the forces opposite of those that induce springback are applied to induce an opposite change of shape, with the objective of attaining a die shape that will produce a desired shape after springback.

spring holddown. A general term for a spring-actuated pressure plate that could be a stripper in a cutting die, a holddown plate in a square shear, or a pad in a bending die.

stacking-fault energy (SFE). The energy associated with the planar fault formed by dissociated dislocations in crystalline materials. Low-SFE materials typically have wide stacking faults, and high-SFE materials very narrow or no stacking faults. The SFE affects a number of material properties, such as work-hardening rate and recrystallization. Materials with low SFE undergo rapid dislocation multiplication and hence show high work-hardening rates and relative ease of dynamic recrystallization because of the difficulty of dynamic recovery. Materials with high SFE energies usually exhibit low work-hardening rates because of the ease of dynamic recovery and are difficult to recrystallize.

stamping. A general term to denote all press-working. In a more specific sense, stamping is used to imprint letters, numerals, and trademarks in sheet metal, machined parts, forgings, and castings. A tool called a stamp, with the letter or number raised on its surface, is hammered or forced into the metal, leaving a depression on the surface in the form of the letter or number.

stand. A piece of rolling mill equipment containing one set of work rolls. In the usual sense, any pass of a cold or hot rolling mill. See also *rolling mills.*

static friction. The force tangential to the interface that is just sufficient to initiate relative motion between two bodies under load.

steam hammer. A type of drop hammer in which the ram is raised for each stroke by a double-action steam cylinder, and the energy delivered to the workpiece is supplied by the velocity and weight of the ram and attached upper die driven downward by steam pressure. The energy delivered during each stroke can be varied.

sticker breaks. Repetitive transverse lines, often curved, caused by localized welding of coil wraps during annealing and subsequent separation of these welded areas during an uncoiling operation. Also referred to as sticker marks.

stiffness. Resistance to elastic deformation.

stock. A general term used to refer to a supply of metal in any form or shape and also to an individual piece of metal that is formed, forged, or machined to make parts.

stop. A device for positioning stock or parts in a die.

straightening. A finishing operation for correcting misalignment in a forging or between various sections of a forging.

straight-side press. A box-shaped press having frame components of a bed, two uprights and a crown either welded together or clamped with tie rods.

strain. The unit of change in the size or shape of a body due to force, in reference to its original size or shape.

strain aging. The changes in ductility, hardness, yield point, and tensile strength that occur when a metal or alloy that has been cold worked is stored for some time. In steel, strain aging is characterized by a loss of ductility and a corresponding increase in hardness, yield point, and tensile strength.

strain hardening. An increase in hardness and strength caused by plastic deformation at temperatures below the recrystallization range. Also known as work hardening.

strain-hardening coefficient. See *strain-hardening exponent.*

strain-hardening exponent. For materials that closely obey a power-law relationship, the value n in the relationship $\sigma = K\varepsilon^n$, where σ is the true stress, ε is the true strain, K is called the *strength coefficient,* and n is a constant called the strain-hardening exponent or strain-hardening power. More generally, the slope of the logarithmic stress-strain curve, that is, $d\ln\sigma/d\ln\varepsilon$, is called the incremental n-value or instantaneous n-value. For power-law materials, n is related to the ability of a sheet material to be stretched in tension. Also called the n-value and the work-hardening exponent.

strain lines. Surface defects in the form of shallow line-type depressions appearing in sheet metals after stretching the surface a few percent of unit area or length. See also *Lüders lines.*

strain rate. The time rate of deformation (strain) during a metal-forming process.

strain-rate sensitivity. The degree to which mechanical properties are affected by changes in deformation rate. Quantified by the slope of a log-log plot of flow stress (at fixed strain and temperature) versus strain rate. Also known as the m-value.

strain-rod nuts. Elements holding the bed or the crown to strain rods on a hydraulic press.

strain rods. The tension members joining the bed and the crown of a hydraulic press. Also called tie rods or columns.

strain state. (1) an expression of the ratios of the strain components without specifying their magnitudes. (2) An expression of the strain components in toto.

strength. The ability of a material to withstand an applied force.

strength coefficient (K). The constant K in the power-law equation $\sigma = K\varepsilon^n$.

stress. The intensity of the internally distributed forces or components of forces that resist a change in the volume or shape of a material that is or has been subjected to external forces. Stress is expressed in force per unit area. Stress can be normal (tension or compression) or shear.

stress raisers. Design features (such as sharp corners) or mechanical defects (such as notches) that act to intensify the stress at these locations.

stress relaxation. Drop in stress with time when material is maintained at a constant strain. The drop in stress is a result of plastic accommodation processes.

stress state. (1) An expression of the ratios of the stress components without specifying their magnitudes. (2) An expression of the stress components in toto.

stress-strain curve. See *stress-strain diagram.*

stress-strain diagram. A plot of stress and strain from a tension, compression, torsion, or other mechanical test. Values of stress are usually plotted vertically (ordinate, or y-axis) and values of strain horizontally (abscissa, or x-axis). Also known as stress-strain curve.

stretchability. The ability of a material to undergo stretch-type deformation.

stretch-bend. See *draw-bend.*

stretcher leveling. The leveling of a piece of sheet metal (that is, removing warp and distortion) by gripping it at both ends and subjecting it to a stress higher than its yield strength.

stretcher straightening. A process for straightening rod, tubing, and shapes by the application of tension at the ends of the stock. The products are elongated a definite amount to remove warpage.

stretcher strains. Elongated markings that appear on the surface of some sheet materials when deformed just past the yield point. These markings lie approximately parallel to the direction of maximum shear stress and are the result of localized yielding. See also *Lüders lines.*

stretch flanging. The stretching of the length of the free edge after the flanging process.

stretch former. (1) A machine used to perform *stretch forming* operations. (2) A device adaptable to a conventional press for accomplishing stretch forming.

stretch forming. The shaping of a sheet or part, usually of uniform cross section, by first applying suitable tension or stretch and then wrapping it around a die of the desired shape.

stretching. The mode of deformation in which a positive strain is generated on the sheet surface by the application of a tensile stress. In stretching, the flange of the flat blank is securely clamped. Deformation is restricted to the area initially within the die. The stretching limit is the onset of metal failure.

striking surface. Those areas on the faces of a set of dies that are designed to meet when the upper die and lower die are brought together. The striking surface helps protect impressions from impact shock and aids in maintaining longer die life.

strip. A flat-rolled metal product of some maximum thickness and width arbitrarily dependent on the type of metal; narrower than *sheet.*

stripper. A plate designed to remove, or strip, sheet-metal stock from the punching members during the withdrawal cycle. Strippers are also used to guide small precision punches in close-tolerance dies, to guide scrap away from dies, and to assist in the cutting action. Strippers are made in two types: fixed and movable.

stripper punch. A punch that serves as the top or bottom of the die cavity and later moves farther into the die to eject the part or compact. See also *ejector rod* and *knockout*.

stripping. The removal of the metal strip from the punch after a cutting operation. Also a term referring to the removal of a part adhering to the punch on the upstroke after forming.

strokes per minute. The number of press-ram strokes or die closings in one minute.

stroke (up or down). The vertical movement of a ram during half of the cycle, from the full-open to the full-closed position or vice versa.

subgrain. Micron-sized volume bounded by well-defined dislocation walls. The misorientations across the walls are low angle in nature, that is, $< 15°$.

subpress die. A die that is closed by the press ram but opened by springs or other means because the upper shoe is not attached to the ram.

subsow block (die holder). A block used as an adapter in order to permit the use of forging dies that otherwise would not have sufficient height to be used in the particular unit or to permit the use of dies in a unit with different *shank* sizes.

superplastic forming. Forming using the superplasticity properties of material at elevated temperatures.

superplastic forming and diffusion bonding. The process of combining the diffusion bonding cycle into the superplastic forming.

superplasticity. The ability of certain metals to develop extremely high tensile elongations at elevated temperatures and under controlled rates of deformation. Materials that show high strain-rate sensitivity ($m \geq 0.5$) at deformation temperatures often exhibit superplasticity. The phenomenon is often developed through a mechanism of grain-boundary sliding in very fine-grained, two-phase alloys.

support plate. A plate that supports a draw ring or draw plate. It also serves as a spacer.

surface finish. The classification of a surface in terms of roughness and peak density.

surface hardness. The hardness of that portion of the material very near the surface as measured by microhardness or superficial hardness testers.

surface oxidation. Development of an oxide film or layer on the surface of metals in oxidizing environments. Oxidation at high temperatures is occasionally referred to as sealing.

surface roughness. The fine irregularities in the surface texture that result from the production process. Considered as vertical deviations from the nominal or average plane of the surface.

surface texture. Repetitive or random deviations from the nominal surface that form the pattern of the surface. Includes roughness, waviness, and flaws.

swage. (1) The operation of reducing or changing the cross-sectional area of stock by the fast impact of revolving dies. (2) The tapering of bar, rod, wire, or tubing by forging, hammering, or squeezing; reducing a section by progressively tapering lengthwise until the entire section attains the smaller dimension of the taper.

sweep device. A single or double arm (rod) attached to the upper die or slide of a press and intended to move the operator's hands to a safe position as the dies close, if the operator's hands are inadvertently within the point of operation.

Swift cup test. A simulative test in which circular blanks of various diameters are clamped in a die ring and deep drawn into a cup by a flat-bottomed cylindrical punch. The ratio of the largest blank diameter that can be drawn successfully to the cup diameter is known as the *limiting drawing ratio (LDR)* or *deformation limit.*

T

Taguchi method. A technique for designing and performing experiments to investigate processes in which the output depends on many factors (e.g., material properties, process parameters) without having to tediously and uneconomically run the process using all possible combinations of values of those variables. By systematically choosing certain combinations of variables, it is possible to separate their individual effects.

tailor-welded blank. Blank for sheet forming typically consisting of steels of different thickness, grades/strengths, and sometimes coatings that are welded together prior to forming. Tailor-welded blanks are used to make finished parts with a desirable variation in properties such as strength, corrosion resistance, and so on.

tandem die. Same as *follow die.*

tandem mill. A rolling mill consisting of two or more stands arranged so that the metal being processed travels in a straight line from stand to stand. In continuous rolling, the various stands are synchronized so that the strip can be rolled in all stands simultaneously. Contrast with *single-stand mill.*

tangent bending. The forming of one or more identical bends having parallel axes by wiping sheet metal around one or more radius dies in a single operation. The sheet, which may have side flanges, is clamped against the radius die and then made to conform to the radius die by pressure from a rocker-plate die that moves along the periphery of the radius die. See also *wiper forming (wiping).*

tangent modulus. The slope of the stress-strain curve at any specified stress or strain. See also *modulus of elasticity.*

Taylor factor. The ratio of the required stress for deformation under a specified strain state to the critical resolved shear stress for slip (or twinning) within the crystals comprising a *polycrystalline aggregate.* The determination of the Taylor factor assumes uniform and identical strain within each crystal in the aggregate and provides an upper bound on the required stresses. The Taylor factor averaged over all crystals in a polycrystalline aggregate ($=\bar{M}$) provides an estimate of the effect of texture on strength.

tearing. Failure and localized separation of a sheet metal.

temperature-compensated strain rate. Parameter used to describe the interdependence of temperature and strain rate in the description of thermally activated (diffusion-like) deformation processes. It is defined as:

$$\dot{\varepsilon}\exp(Q/RT),$$

in which $\dot{\varepsilon}$ denotes the strain rate, Q is an apparent activation energy characterizing the micromechanism of deformation, R is the gas constant, and T is the absolute temperature. Flow stress, dynamic recrystallization, and so on at various strain rates and temperatures are frequently interpreted in terms of the temperature-compensated strain rate. Also known as the *Zener-Hollomon parameter (Z).*

template (templet). A gage or pattern made in a die department, usually from sheet steel; used to check dimensions on forgings and as an aid in sinking die impressions in order to correct dimensions.

tensile ratio. The ratio of the tensile strength to yield strength. It is the inverse of the yield ratio.

tensile strength. In tensile testing, the ratio of maximum load to original cross-sectional area. Also known as ultimate strength. Compare with *yield strength.*

tension. The force or load that produces elongation.

texture. The description of the relative probability of finding the crystals comprising a polycrystalline aggregate in various orientations.

thermal conductivity. A measure of the rate at which heat is transferred through a material.

thermal-mechanical treatment. See *thermomechanical processing (TMP).*

thermocouple. A device for measuring temperature, consisting of two dissimilar metals that produce an electromotive force roughly proportional to the temperature difference between their hot and cold junction ends.

thermomechanical processing (TMP). A general term covering a variety of processes combining controlled thermal and deformation treatments to obtain synergistic effects, such as improvement in strength without loss of toughness. Same as thermal-mechanical treatment.

thick-film lubrication. A condition of lubrication in which the film thickness of the lubricant is appreciably greater than that required to cover the surface asperities when subjected to the operating load. See also *thin-film lubrication*.

thin-film lubrication. A condition of lubrication in which the film thickness of the lubricant is such that the friction and wear between the surfaces is determined by the properties of the surfaces as well as the characteristics of the lubricant. See also *elastohydrodynamic lubrication* and *thick-film lubrication*.

thin-shell element. See *shell element*.

three-point bending. The bending of a piece of metal or a structural member in which the object is placed across two supports and force is applied between and in opposition to them. See also *V-bend die*.

throw. The distance from the centerline of the crankshaft or main shaft to the centerline of the crankpin or eccentric in crank or eccentric presses. Equal to one-half of the stroke. See also *crank press* and *eccentric press*.

tilt boundary. Grain boundary for which the crystal lattices of the grains on either side of the boundary are related by a rotation about an axis that lies in the plane of the boundary.

toggle press. A *mechanical press* in which the *slide* is actuated by one or more toggle links or mechanisms.

tong hold. The portion of a forging billet, usually on one end, that is gripped by the operator's tongs. It is removed from the part at the end of the forging operation. Common to drop hammer and press-type forging.

tonnage. The rated capacity of a press to exert force in closing the die. This rating is a short distance above the bottom of the stroke in mechanical presses.

tooling marks. Indications imparted to the surface of the forged part from dies containing surface imperfections or dies on which some repair work has been done. These marks are usually slight rises or depressions in the metal.

top drive. A press having the drive mechanism, including motor, flywheel, gears, shafts, and clutch, located in, just behind, or above the crown.

torsion. A twisting deformation of a solid or tubular body about an axis in which lines that were initially parallel to the axis become helices.

torsional stress. The *shear stress* on a transverse cross section resulting from a twisting action.

total elongation. The total amount of permanent extension of a testpiece broken in a tensile test; usually expressed as a percentage over a fixed gage length. See also *elongation, percent*.

total strain. Same as *strain* but modified to emphasize that it has not been decomposed into, for example, elastic and plastic strains.

toughness. The ability of a material to resist an impact load (high strain rate) or to deform under such a load in a ductile manner, absorbing a large amount of the impact energy and deforming plastically before fracturing. Such impact toughness is frequently evaluated with Charpy or Izod notched impact specimens. Impact toughness is measured in terms of the energy absorbed during fracture. Fracture toughness is a measure of the ability of a material to withstand fracture in the presence of flaws under static or dynamic loading of various types (tensile, shear, etc.). An indicator of damage tolerance, fracture toughness is measured in terms of $MPa\sqrt{m}$ or $ksi\sqrt{in}$.

transfer. A system for moving stampings from press-to-press or station-to-station within a single large die using a grip-transfer-release motion. Activiated by the press motion.

trapped dies. Dies designed with no allowance for flash. Typical configuration consists of a ring die with top and bottom punches.

triaxiality. The ratio of the hydrostatic (mean) stress to the flow (effective) stress. Triaxiality provides a measure of the tendency for cavities to grow during deformation processing.

trimmer. The dies used to remove the flash or excess stock from a forging.

trimmer blade. The portion of the trimmers through which a forging is pushed to shear off the flash.

trimmer die. The punch-press die used for trimming flash from a forging.

trimmer punch. The upper portion of the trimmer that contacts the forging and pushes it through the trimmer blades; the lower end of the trimmer punch is generally shaped to fit the surface of the forging against which it pushes.

trimmers. The combination of trimmer punch, trimmer blades, and perhaps, trimmer shoe used to remove flash from a forging.

trimming. The mechanical shearing of flash or excess material from a forging with a trimmer in a trim press; can be done hot or cold.

trimming die. See *cut-off die*.

trimming press. A power press suitable for trimming flash from forgings.

triple-action hydraulic press. A double-action press with a third independent movable slide built into the bed. A *cushion* is not considered a third action, but a third action may be construed to also act as a cushion.

triple-action press. A mechanical or hydraulic press having three slides with three motions properly synchronized for triple-action drawing, redrawing, and forming. Usually, two slides—the blankholder slide and the plunger—are located above, and a lower slide is located within the bed of the press. See also *hydraulic press, mechanical press*, and *slide*.

triple junction/triple point. Point at which three grains meet in a polycrystalline aggregate. Also, region in which high stress concentrations may develop during hot working or elevated-temperature service, thus nucleating wedge cracking.

true strain. A measure of strain in one direction that is obtained by integrating differential material strains to obtain $\varepsilon = \ln(l/l_o)$, where l is the final length, and l_o is the original length in the direction of interest.

true stress. A measure of stress in one direction that is obtained by dividing the normal load carried by a cross section by the current area corresponding to that load.

tryout. Preparatory run to check or test equipment, lubricant, stock, tools, or methods prior to a production run. Production tryout is run with tools previously approved; new die tryout is run with new tools not previously approved.

tube stock. A semifinished tube suitable for subsequent reduction and finishing.

twinning. Also called deformation or mechanical twinning, it is a deformation mechanism, similar to dislocation slip, in which small (often plate- or lens-shaped) regions of a crystal or grain reorient crystallographically to adopt a twin relationship to the parent crystal. It is particularly common in noncubic metals (e.g., alpha-titanium and tetragonal tin) and in many body-centered cubic metals deformed at high rates and/or low temperatures. Twinning is often accompanied by an audible crackling sound, from which "crying tin" gets its name.

twist boundary. Grain boundary for which the crystal lattices of the grains on either side of the boundary are related by a rotation about an axis that lies perpendicular to the plane of the boundary.

two-hand control device. A two-hand trip that further requires concurrent pressure from both hands of the operator during a substantial part of the die-closing portion of a press stroke.

two-hand trip. A means of clutch actuation requiring the concurrent use of both hands of an operator to trip a press. Requirements for a two-hand trip are as follows: (1) The individual operator's hand controls must be protected against unintentional operation. (2) The individual operator's hand controls must be arranged, by design and construction or by separation, or both, to require the use of both hands to trip the press. (3) A control arrangement must be used that requires concurrent operation of the individual operator's hand controls.

two-high mill. A type of rolling mill in which only two rolls, the working rolls, are contained in a single housing. Compare with *four-high mill* and *cluster mill*.

TZM. A high-creep-strength titanium, zirconium, and molybdenum alloy used to make dies for the isothermal forging process.

U

U-bend die. A die, commonly used in pressbrake forming, that is machined horizontally with a square or rectangular cross-sectional opening that provides two edges over which metal is drawn into a channel shape.

ultimate strength. The maximum stress (tensile, compressive, or shear) a material can sustain without fracture; determined by dividing maximum load by the original cross-sectional area of the specimen. Also known as nominal strength or maximum strength.

ultrasonic inspection. The use of high-frequency acoustical signals for the purpose of nondestructively locating flaws within raw material or finished parts.

unbending. The straightening of a previously curved section of sheet or beam.

underdrive. A press having the drive mechanism, including motor, flywheel, gears, shafts, and clutch, located in or below the bed.

underfill. A portion of a forging that has insufficient metal to give it the true shape of the impression.

uniform elongation (e$_u$). The elongation that occurs at maximum load and immediately preceding the onset of necking in a tensile test.

upset. The localized increase in cross-sectional area of a workpiece or weldment resulting from the application of pressure during mechanical fabrication or welding.

upset forging. A forging obtained by *upset* of a suitable length of bar, billet, or bloom.

upsetter. A horizontal mechanical press used to make parts from bar stock or tubing by *upset forging,* piercing, bending, or otherwise forming in dies. Also known as a header.

upsetting. The working of metal so that the cross-sectional area of a portion or all of the stock is increased. See also *heading.*

V

vacuum forming. Sheet-forming process most commonly used for titanium in which a blank is placed into a chamber that has a heated die, and a vacuum is applied to creep form the part onto the die. The part is usually covered with an insulating material, and the bag is outside this material.

V-bend die. A die commonly used in press-brake forming, usually machined with a triangular cross-sectional opening to provide two edges as fulcrums for accomplishing *three-point bending.*

vent. A small hole in a punch or die for admitting air to avoid suction holding or for relieving pockets of trapped air that would prevent die closure or action.

vent mark. A small protrusion resulting from the entrance of metal into die vent holes.

viscosity. Bulk property of a fluid or semifluid that causes it to resist flow.

visioplasticity. A physical-modeling technique in which an inexpensive, easy-to-deform material (e.g., clay, wax, lead) is gridded and deformed in subscale tooling to establish the effects of die design, lubrication, and so forth on metal flow and defect formation by way of postdeformation examination of grid distortions. See also *physical modeling.*

von Mises strain. See *effective strain.*

von Mises stress. See *effective stress.*

W

warm working. Deformation at elevated temperatures below the recrystallization temperature. The flow stress and rate of strain hardening are reduced with increasing temperature; therefore, lower forces are required than in cold working. See also *cold working* and *hot working.*

wear plates. Replaceable elements used to face wearing surfaces on a hydraulic press.

wear resistance. Resistance of a sheet metal to surface abrasion. See also *erosion resistance.*

web. A relatively flat, thin portion of a forging that effects an interconnection between ribs and bosses; a panel or wall that is generally parallel to the forging plane. See also *rib.*

wedge compression test. A simple workability test in which a wedge-shaped specimen is compressed to a certain thickness. This gives a gradient specimen in which material has been subjected to a range of plastic strains.

Widmanstätten structure. Characteristic structure produced when preferred planes and directions in the parent phase are favored for growth of a second phase, resulting in the precipitated second phase appearing as plates, needles, or rods within a matrix.

wiper forming (wiping). Method of curving sheet-metal sections or tubing over a form block or die in which this form block is rotated relative to a wiper block or slide block.

wire. A thin, flexible, continuous length of metal, usually of circular cross section and usually produced by drawing through a die.

wire drawing. Reducing the cross section of wire by pulling it through a die.

wire drawing test. A test in which a cylindrical draw die is used to reduce the diameter of wire. The drawing force is measured and reflects lubricant effectiveness.

wire rod. Hot rolled coiled stock that is to be cold drawn into wire.

workability. See also *formability,* which is a term more often applied to sheet materials. The ease with which a material can be shaped through plastic deformation in bulk forming processes. It involves both the measurement of the resistance to deformation (the flow properties) and the extent of possible plastic deformation before fracture occurs (ductility).

work hardening. See *strain hardening.*

work-hardening exponent. See *strain-hardening exponent.*

working stroke. That portion of the press downstroke where force is being applied to shape sheet metal or retract pressure plates.

workpiece. General term for the work material in a metal-forming operation.

wrap forming. See *stretch forming.*

wrinkling. A wavy condition obtained in deep drawing of sheet metal in the area of the metal between the edge of the flange and the draw radius. Wrinkling may also occur in other forming operations when unbalanced compressive forces are set up.

wrought material. Material that is processed by plastic deformation, typically to produce a recrystallized microstructure. Cast and wrought materials are produced by ingot casting and deformation processes to produce final mill products.

Y

yield. The transition from elastic deformation, which is recoverable, and inelastic deformation, which is not.

yield point. The first stress in a particular material, usually less than the maximum attainable stress, at which an increase in strain occurs without an increase in stress. This is called the upper yield point. Also, a second, lower stress where the stress stops falling with increasing strain is known as the lower yield point. Only certain metals—those that exhibit a localized, heterogeneous type of transition from elastic to plastic deformation—produce a yield point.

yield point elongation. The extension associated with discontinuous yielding that occurs at approximately constant load following the onset of plastic flow. It is associated with the propagation of Lüder's lines or bands.

yield ratio. The ratio of the yield strength to the *tensile strength.* It is the inverse of the tensile ratio.

yield strain. The strain in a tension test corresponding to the yield stress. In principle, it may be obtained by dividing the yield stress by Young's modulus.

yield strength. The stress at which a material exhibits a specified deviation from proportionality of stress and strain. An offset of 0.2% is used for many metals. Compare with *tensile strength.*

yield stress. A stress at which a material exhibits the first measurable permanent plastic deformation.

Young's modulus. A measure of the rigidity of a metal. It is the ratio of stress, within the proportional limit, to corresponding strain. Young's modulus specifically is the modulus obtained in tension or compression.

Z

Zener-Hollomon parameter (Z). See *temperature-compensated strain rate.*

zero-extension fiber. The location through the cross section of a bent beam or sheet where the material elements neither elongate nor compress. The location may be a theoretical one outside of the sheet or beam.

Useful Formulas for Deformation Analysis and Workability Testing

Table 1 Effective stress, strain, and strain rate (isotropic material) in arbitrary coordinates

Variable or quantity	Symbol or equation
Stress tensor components	σ_{ij}
Strain increment components	$d\varepsilon_{ij}$
Strain-rate components	$\dot{\varepsilon}_{ij}$
Von Mises effective stress ($\bar{\sigma}$)	$\bar{\sigma} = \sqrt{\frac{1}{2}\left\{(\sigma_{xx}-\sigma_{yy})^2 + (\sigma_{yy}-\sigma_{zz})^2 + (\sigma_{zz}-\sigma_{xx})^2\right\} + 3\sigma_{xy}^2 + 3\sigma_{yz}^2 + 3\sigma_{zx}^2}$
	where σ_{xy}, σ_{yz}, and σ_{zx} are generalized tensor notation for shear stresses τ_{xy}, τ_{yz}, τ_{zx}, respectively
Von Mises effective strain increment ($d\bar{\varepsilon}$)	$d\bar{\varepsilon} = \sqrt{\frac{2}{9}\left\{(d\varepsilon_{xx}-d\varepsilon_{yy})^2 + (d\varepsilon_{yy}-d\varepsilon_{zz})^2 + (d\varepsilon_{zz}-d\varepsilon_{xx})^2\right\} + \frac{4}{3}d\varepsilon_{xy}^2 + \frac{4}{3}d\varepsilon_{yz}^2 + \frac{4}{3}d\varepsilon_{zx}^2}$
Von Mises effective strain rate ($\dot{\bar{\varepsilon}} = d\bar{\varepsilon}/dt$)	$\dot{\bar{\varepsilon}} = \sqrt{\frac{2}{9}\left\{(\dot{\varepsilon}_{xx}-\dot{\varepsilon}_{yy})^2 + (\dot{\varepsilon}_{yy}-\dot{\varepsilon}_{zz})^2 + (\dot{\varepsilon}_{zz}-\dot{\varepsilon}_{xx})^2\right\} + \frac{4}{3}\dot{\varepsilon}_{xy}^2 + \frac{4}{3}\dot{\varepsilon}_{yz}^2 + \frac{4}{3}\dot{\varepsilon}_{zx}^2}$

Table 2 Effective stress, strain, and strain rate (isotropic material) in principal coordinates

Variable or quantity	Symbol or equation
Principal stress components	σ_1, σ_2, σ_3
Principal strain-increment components	$d\varepsilon_1$, $d\varepsilon_2$, $d\varepsilon_3$
Principal strain-rate components	$\dot{\varepsilon}_1$, $\dot{\varepsilon}_2$, $\dot{\varepsilon}_3$
Von Mises effective stress ($\bar{\sigma}$)	$\bar{\sigma} = \sqrt{\frac{1}{2}\left\{(\sigma_1-\sigma_2)^2 + (\sigma_2-\sigma_3)^2 + (\sigma_3-\sigma_1)^2\right\}}$
Von Mises effective strain increment ($d\bar{\varepsilon}$)	$d\bar{\varepsilon} = \sqrt{\frac{2}{9}\left\{(d\varepsilon_1-d\varepsilon_2)^2 + (d\varepsilon_2-d\varepsilon_3)^2 + (d\varepsilon_3-d\varepsilon_1)^2\right\}}$
Von Mises effective strain rate ($\dot{\bar{\varepsilon}} = d\bar{\varepsilon}/dt$)	$\dot{\bar{\varepsilon}} = \sqrt{\frac{2}{9}\left\{(\dot{\varepsilon}_1-\dot{\varepsilon}_2)^2 + (\dot{\varepsilon}_2-\dot{\varepsilon}_3)^2 + (\dot{\varepsilon}_3-\dot{\varepsilon}_1)^2\right\}}$

Table 3   Formulas for compression testing of isotropic material

Variable or quantity	Symbol or relation

Uniaxial compression under uniform deformation conditions

Initial sample dimensions	Height (h_0) Diameter (d_0) Area (A_0), $A_0 = \pi d_0^2/4$
Instantaneous (final) sample dimensions	Height (h) Diameter (d) Area (A), $A = \pi d^2/4$
Crosshead speed	v
Applied load	P
Constant-volume assumption of plastic flow	$A_0 h_0 = Ah \rightarrow d_0^2 h_0 = d^2 h$
Height reduction (R), %	$R = \left[\dfrac{(h_0 - h)}{h_0}\right] \times 100$
Nominal (engineering) axial strain (e), %	$e = \left[\dfrac{(h - h_0)}{h_0}\right] \times 100$
True axial strain (ε)	$\varepsilon = \ln\left[\dfrac{h}{h_0}\right]$
True axial strain rate ($\dot{\varepsilon}$)	$\dot{\varepsilon} = -v/h$
Nominal (engineering) axial stress (S)	$S = -\dfrac{P}{A_0}$
True axial stress (σ)	$\sigma = -\dfrac{P}{A} = -\dfrac{Ph}{A_0 h_0} = \dfrac{Sh}{h_0}$
Effective stress ($\bar{\sigma}$)	$\bar{\sigma} = -\sigma$
Effective strain ($\bar{\varepsilon}$)	$\bar{\varepsilon} = -\varepsilon$

Uniaxial compression with friction correction

Friction shear factor (m_s)	$m_s \approx \sqrt{3}\,\mu$ where $\mu \equiv$ Coulomb coefficient of friction
Friction correction for flow stress	$\dfrac{\bar{\sigma}}{p_{av}} = \left(1 + \dfrac{m_s d}{(3\sqrt{3})h}\right)^{-1}$ where $\bar{\sigma}$ denotes flow stress (under homogeneous frictionless conditions) and p_{av} denotes the average pressure

Homogeneous plane-strain compression

Through-thickness true strain	ε_3
Effective strain ($\bar{\varepsilon}$)	$\bar{\varepsilon} = \dfrac{-2\varepsilon_3}{\sqrt{3}}$
True stress	σ_3
Effective stress ($\bar{\sigma}$)	$\bar{\sigma} = \dfrac{-\sigma_3 \sqrt{3}}{2}$

Table 4 Formulas for tension testing of isotropic material

Variable or quantity	Symbol or relation

Uniaxial tension under uniform deformation conditions, constant crosshead speed

Initial gage (reduced section) dimensions	Length (L_0) Diameter (D_0) Area (A_0), $A_0 = \pi d_0^2/4$
Instantaneous reduced section dimensions	Length (L) Diameter (d) Area (A), $A = \pi d^2/4$
Crosshead speed	v
Applied load	P
Constant-volume assumption of plastic flow	$A_0 h_0 = Ah \rightarrow d_0^2 h_0 = d^2 h$
Nominal (engineering) axial strain (e), %	$e = \dfrac{(L-L_0)\,100}{L_0}$
Nominal (engineering) axial strain rate ($\dot{e}$)	$\dot{e} = \dfrac{v}{L_0}$
True axial strain (ε)	$\varepsilon = \ln\left[\dfrac{L}{L_0}\right] = \ln\,(1+e)$ where e is expressed as a decimal fraction
True axial strain rate ($\dot{\varepsilon}$)	$\dot{\varepsilon} = \dfrac{v}{L}$
Nominal (engineering) axial stress (S)	$S = \dfrac{P}{A_0}$
True axial stress (σ)	$\sigma = \dfrac{P}{A} = \dfrac{PL}{A_0 L_0} = \dfrac{SL}{L_0} = S(1+e)$ where e is expressed as a decimal fraction
Effective stress ($\bar{\sigma}$)	$\bar{\sigma} = \sigma$
Effective strain ($\bar{\varepsilon}$)	$\bar{\varepsilon} = \varepsilon$
Strain-hardening exponent (n)	$n = \dfrac{\partial \ln \sigma}{\partial \ln \varepsilon}$ evaluated at a fixed strain rate and temperature
Strain-rate sensitivity exponent (m)	$m = \dfrac{\partial \ln \sigma}{\partial \ln \dot{\varepsilon}}$ evaluated at fixed strain and temperature

Postuniform deformation in uniaxial tension of round-bar samples

Initial gage (reduced section) dimensions	Length (L_0) Diameter (d_0) Area (A_0), $A_0 = \pi d_0^2/4$
Length of gage section at failure	L_f
Gage-section diameter at failure (minimum section)	d_f
Gage-section area at failure (minimum section) (A_f)	$A_f = \pi d_f^2/4$
Total elongation (e_t), %	$e_t = \dfrac{(L_f - L_0)\,100}{L_0}$
Reduction in area (RA), %	$RA = \dfrac{(A_0 - A_f)\,100}{A_0}$
True fracture strain (ε_f)	$\varepsilon_f = \ln\left[\dfrac{A_0}{A_f}\right] = 2\ln\left[\dfrac{d_0}{d_f}\right]$

Necking during tension testing of round-bar samples

Sample radius at symmetry plane of neck	a
Profile radius of neck	R
Bridgman correction for necking during tension testing of round-bar samples	$\bar{\sigma} = \dfrac{\sigma_x}{\left[1+\dfrac{2R}{a}\right]\cdot\left[\ln\left(1+\dfrac{a}{2R}\right)\right]}$ where: • $\bar{\sigma}$ denotes the flow stress • σ_x denotes the average axial stress (which is equal to applied load ÷ sample cross-sectional area at neck symmetry plane)

Table 5 Formulas for torsion testing of isotropic material (solid round-bar sample)

Variable or quantity	Symbol or relation
Reduced section dimensions	Length (L) Outer radius (R)
Radial coordinate	r
Twist (θ), in radians	θ (in radians) = twist in degrees $\times \dfrac{\pi}{180}$
Twist rate (in radians per second)	$\dot{\theta}$
Shear strain (γ)	$\gamma = \dfrac{r\theta}{L}$
Shear strain rate ($\dot{\gamma}$)	$\dot{\gamma} = \dfrac{r\dot{\theta}}{L}$
Effective strain ($\bar{\varepsilon}$)	$\bar{\varepsilon} = \gamma/\sqrt{3}$
Effective strain rate ($\dot{\bar{\varepsilon}}$)	$\dot{\bar{\varepsilon}} = \dot{\gamma}/\sqrt{3}$
Torque	M
Shear stress (τ) corresponding to shear strain/shear strain rate at $r = R$	$\tau = \dfrac{(3+n^*+m^*)\cdot M}{2\pi R^3}$ where: n^* = slope of $\ln M$-vs-$\ln \theta$ plot $\approx$ strain-hardening exponent (n) m^* = slope of $\ln M$ vs $\ln \dot{\theta}$ plot $\approx$ strain-rate sensitivity exponent (m)
Effective stress ($\bar{\sigma}$)	$\bar{\sigma} = \sqrt{3}\cdot\tau$

Table 6 Formulas related to flat (sheet) rolling

Variable or quantity	Symbol or relation
Undeformed roll radius	R
Rolling speed (roll surface velocity) (v_R), m/s	$v_R = (2\pi R)\times$(angular velocity, in revolutions per second)
Initial sheet thickness	h_0
Final sheet thickness	h_f
Draft (Δh)	$\Delta h = h_0 - h_f$
Thickness reduction, %	$\dfrac{(h_0 - h_f)100}{h_0}$
True thickness strain (ε_3)	$\varepsilon_3 = \ln\left[\dfrac{h_f}{h_0}\right]$
Rolling load (roll separating force), P_L	$P_L = \dfrac{2\bar{\sigma}}{\sqrt{3}}\left[\dfrac{1}{Q}(e^Q - 1)b\sqrt{R'\Delta h}\right]$ where $\bar{\sigma}$ denotes the flow stress under homogeneous uniaxial stress conditions $Q \equiv \mu L_p/\bar{h}$ $\mu \equiv$ Coulomb coefficient of friction $L_p \equiv$ projected contact length $\bar{h} \equiv (h_0 + h_f)/2$ $b \equiv$ sheet width $R' \equiv$ flattened roll radius
Hitchcock equation for flattened roll radius, R'	$R' = R\left\{1 + \dfrac{16(1-v^2)P_L}{b\pi E(\Delta h)}\right\}$ where: $v \equiv$ Poisson's ratio of roll material $E \equiv$ Young's modulus of roll material
Projected contact length (L_p)	$L_p = \sqrt{R'\Delta h}$
Average effective strain rate ($\dot{\bar{\varepsilon}}$)	$\dot{\bar{\varepsilon}} = \dfrac{v_R}{h_0}\sqrt{\dfrac{2(h_0 - h_f)}{R'}}$

Table 7 Formulas related to conical-die extrusion

Variable or quantity	Symbol or relation
Initial billet dimensions	Diameter (d_0) Area (A_0), $A_0 = \pi d_0^2/4$
Final billet dimensions	Diameter d_f, Area (A_f), $A_f = \pi d_f^2/4$
Die semicone angle	α
Ram speed	v
Reduction ratio	$R = A_0/A_f$
Reduction (r), %:	$r = \dfrac{(A_0 - A_f) \cdot 100}{A_0}$
Extrusion (ram) pressure (p_{av})	$\dfrac{p_{av}}{\bar{\sigma}} = a + b \cdot \ln[R]$ where $\bar{\sigma}$ denotes material flow stress, and a and b denote material-dependent constants
Average effective strain rate in deformation zone ($\dot{\bar{\varepsilon}}$)	$\dot{\bar{\varepsilon}} = \dfrac{6v d_0^2 (\tan \alpha) \ln R}{d_0^3 - d_f^3}$

Table 8 Formulas related to wire drawing

Variable or quantity	Symbol or relation
Initial wire diameter	d_0
Final wire diameter	d_f
Die semicone angle	α
Reduction ratio (R)	$R = A_0/A_f$
Reduction (r), %	$r = \dfrac{(A_0 - A_f)100}{A_0}$
Effective strain	$\ln\left[\dfrac{A_0}{A_f}\right]$
Average drawing stress (σ_{dwg})	$\dfrac{\sigma_{dwg}}{\bar{\sigma}} = \left(\dfrac{3.2}{\Delta + 0.9}\right) \cdot (\alpha + \mu)$ where: $\bar{\sigma}$ denotes material flow stress $\mu \equiv$ Coulomb coefficient of friction Δ denotes the deformation-zone geometry parameter (see below)
Deformation-zone geometry parameter (Δ)	$\Delta \equiv \dfrac{\alpha}{r}\left(1 + \sqrt{1-r}\right)^2$ where the reduction (r) is expressed as a decimal fraction, not as a percentage

Table 9 Formulas related to bending

Variable or quantity	Symbol or relation
Bending of sheet	
Bend radius	R
Sheet thickness	h
Minimum bend radius, $(2R/h)_{min}$	$(2R/h)_{min} = \dfrac{1}{\varepsilon_f} - 1$ where ε_f denotes the true fracture strain in uniaxial tension.
Bending of bars	
Bar diameter	d
Mandrel diameter	D
Strain imposed on outer fiber	$\varepsilon = \dfrac{1}{(1 + D/d)}$

Table 10 Formulas related to deep drawing of cups from sheet metal

Variable or quantity	Symbol or relation
Blank diameter	D
Cup diameter	d
Reduction (R), %	$R = \dfrac{(D-d)100}{D}$
Limiting drawing ratio (LDR)	$LDR \equiv D_{max}/d$ where D_{max} denotes the maximum blank diameter that can be drawn without cup failure

Table 11 Formulas for anisotropic sheet materials (See also Table 12)

Variable or quantity	Symbol or relation
Basic definitions	
Normal plastic anisotropy (R)	$R = d\varepsilon_w/d\varepsilon_t \approx \varepsilon_w/\varepsilon_t$ where ε_w, ε_t denote the true width and thickness strains during uniform, uniaxial tension of a sheet specimen
Average normal plastic anisotropy ($\bar{R}$)	$\bar{R} = (R_{0°} + 2R_{45°} + R_{90°})/4$
Planar plastic anisotropy (ΔR)	$\Delta R = (R_{0°} - 2R_{45°} + R_{90°})/2$
Effective stress ($\bar{\sigma}$) and effective strain increment ($d\bar{\varepsilon}$) assuming *plane-stress* conditions in principal coordinates and planar plastic isotropy ($R_{0°} = R_{90°} = R_{45°} = R$)	$\bar{\sigma} = \sqrt{\dfrac{3(1+R)}{2(2+R)}} \cdot \sqrt{(\sigma_1)^2 + (\sigma_2)^2 - \dfrac{2R}{(1+R)}(\sigma_1\sigma_2)}$ $d\bar{\varepsilon} = \sqrt{\dfrac{2(2+R)}{3(1+2R)^2}} \cdot \sqrt{(d\varepsilon_1 - Rd\varepsilon_3)^2 + (d\varepsilon_2 - Rd\varepsilon_3)^2 + R(d\varepsilon_1 - d\varepsilon_2)^2}$
Hill quadratic yield function (orthotropic texture)	$F(\sigma_{yy} - \sigma_{zz})^2 + G(\sigma_{zz} - \sigma_{xx})^2 + H(\sigma_{xx} - \sigma_{yy})^2 + 2L\sigma_{yz}^2 + 2M\sigma_{zx}^2 + 2N\sigma_{xy}^2 = 1$
Uniaxial sheet tension test assuming uniform deformation and planar isotropy ($R_{0°} = R_{90°} = R_{45°} = R$)	
Axial true stress	σ_1
Axial true strain	ε_1
Effective stress ($\bar{\sigma}$)	$\bar{\sigma} = \sqrt{\dfrac{3(1+R)}{2(2+R)}}\,\sigma_1$
Effective strain ($\bar{\varepsilon}$)	$\bar{\varepsilon} = \sqrt{\dfrac{2(2+R)}{3(1+R)}}\,\varepsilon_1$
Plane-strain compression of sheet assuming uniform deformation and planar isotropy ($R_{0°} = R_{90°} = R_{45°} = R$)	
Through-thickness true stress	σ_3
Through-thickness true strain	ε_3
Effective stress ($\bar{\sigma}$)	$\bar{\sigma} = -\sqrt{\dfrac{3(1+2R)}{2(1+R)(2+R)}}\,\sigma_3$
Effective strain ($\bar{\varepsilon}$)	$\bar{\varepsilon} = -\sqrt{\dfrac{2(2+R)(1+3R+2R^2)}{3(1+2R)^2}}\,\varepsilon_3$
Ratio of plane-strain flow stress (σ_{ps}) to uniaxial flow stress (σ_{uni}) (at equal levels of plastic work)	
Ratio of plane-strain flow stress (σ_{ps}) to uniaxial flow stress (σ_{uni})	$\dfrac{\sigma_{ps}}{\sigma_{uni}} = \dfrac{(1+R)}{\sqrt{1+2R}}$

Table 12 Barlat's anisotropic yield function Yld2000-2d for plane-stress deformation of sheet material (Ref 1, 2)

Variable or quantity	Symbol or relation(a)
Basic definitions	
Yield function (ϕ) and associated effective stress ($\bar{\sigma}$)	$\phi = \lvert \tilde{S}'_1 - \tilde{S}'_2 \rvert^a + \lvert 2\tilde{S}''_2 + \tilde{S}''_1 \rvert^a + \lvert 2\tilde{S}''_1 + \tilde{S}''_2 \rvert^a = 2\bar{\sigma}^a$
	$(\tilde{S}'_1, \tilde{S}'_2)$ and $(\tilde{S}''_1, \tilde{S}''_2)$ are the principal values of two linearly transformed stress deviators (see below).
	Recommended values for the exponent a are 6 and 8 for body-centered cubic and face-centered cubic metals, respectively.
Yield condition	$\bar{\sigma} = \left\{ \frac{\phi}{2} \right\}^{1/a} = h(\bar{\varepsilon})$
Work-equivalent effective strain ($\bar{\varepsilon}$), defined incrementally for plane stress	$d\bar{\varepsilon} = \frac{\sigma_{xx}}{\bar{\sigma}} d\varepsilon_{xx} + \frac{\sigma_{yy}}{\bar{\sigma}} d\varepsilon_{yy} + 2 \frac{\sigma_{xy}}{\bar{\sigma}} d\varepsilon_{xy}$
Hardening function [$h(\bar{\varepsilon})$]	Most prevalent choices:
	• Flow stress in uniaxial tension in the rolling direction: $d\bar{\varepsilon} = d\varepsilon_{xx}$
	• Flow stress in balanced biaxial tension ($\sigma_{yy} = \sigma_{xx}$): $d\bar{\varepsilon} = -d\varepsilon_{zz}$
First linear transformation	
Components of the transformed stress deviator	$\tilde{s}'_{xx} = \alpha_1 (2\sigma_{xx} - \sigma_{yy})/3$
	$\tilde{s}'_{yy} = \alpha_2 (2\sigma_{yy} - \sigma_{xx})/3$
	$\tilde{s}'_{xy} = \alpha_7 \sigma_{xy}$
Principal values of the transformed stress deviator	$\tilde{S}'_1 = \left\{ \tilde{s}'_{xx} + \tilde{s}'_{yy} + \sqrt{(\tilde{s}'_{xx} - \tilde{s}'_{yy})^2 + 4\tilde{s}'^2_{xy}} \right\}/2$
	$\tilde{S}'_2 = \left\{ \tilde{s}'_{xx} + \tilde{s}'_{yy} - \sqrt{(\tilde{s}'_{xx} - \tilde{s}'_{yy})^2 + 4\tilde{s}'^2_{xy}} \right\}/2$
Second linear transformation	
Components of the transformed stress deviator	$\tilde{s}''_{xx} = \left\{ (8\alpha_5 - 2\alpha_6 - 2\alpha_3 + 2\alpha_4)\sigma_{xx} + (4\alpha_6 + \alpha_3 - 4\alpha_4 - 4\alpha_5)\sigma_{yy} \right\}/9$
	$\tilde{s}''_{yy} = \left\{ (4\alpha_3 - 4\alpha_4 - 4\alpha_5 + \alpha_6)\sigma_{xx} + (8\alpha_4 + 2\alpha_5 - 2\alpha_6 - 2\alpha_3)\sigma_{yy} \right\}/9$
	$\tilde{s}''_{xy} = \alpha_8 \sigma_{xy}$
Principal values of the transformed stress deviator	$\tilde{S}''_1 = \left\{ \tilde{s}''_{xx} + \tilde{s}''_{yy} + \sqrt{(\tilde{s}''_{xx} - \tilde{s}''_{yy})^2 + 4\tilde{s}''^2_{xy}} \right\}/2$
	$\tilde{S}''_2 = \left\{ \tilde{s}''_{xx} + \tilde{s}''_{yy} - \sqrt{(\tilde{s}''_{xx} - \tilde{s}''_{yy})^2 + 4\tilde{s}''^2_{xy}} \right\}/2$
Coefficients and input data(b)(c)	
Anisotropy coefficients	α_1 to α_8
Flow stresses in uniaxial tension for directions at 0, 45, and 90° from rolling direction	$\sigma_0, \sigma_{45}, \sigma_{90}$
Bulge test flow stress (assumed to be the balanced biaxial flow stress)	σ_b
r values in uniaxial tension for directions at 0, 45, and 90° from rolling direction	r_0, r_{45}, r_{90}
Biaxial r values measured with the disk compression test	$r_b = d\varepsilon_{yy}/d\varepsilon_{xx}$

(a) The axes (x, y, z) denote the rolling direction, transverse direction, and normal direction of a sheet, respectively. (b) The eight anisotropy coefficients α_1 to α_8 can be calculated using a numerical solver (e.g., Newton-Raphson) using the eight physical parameters (flow stresses and r values) as input. (c) Note that σ_b and r_b are important experimental data to define the shape of the yield locus. However, if not available, they can be estimated using another yield function or a crystal-plasticity model.

REFERENCES

1. F. Barlat, J.C. Brem, J.W. Yoon, K. Chung, R.E. Dick, D.J. Lege, F. Pourboghrat, S.-H. Choi, and E. Chu, Plane Stress Yield Function for Aluminum Alloy Sheets—Part I: Theory, *Int. J. Plast.*, Vol 19, 2003, p 1297–1319
2. J.W. Yoon, F. Barlat, R.E. Dick, K. Chung, and T.J. Kang, Plane Stress Yield Function for Aluminum Alloy Sheets—Part II: FE Formulation and Its Implementation, *Int. J. Plast.*, Vol 20, 2004, p 495–522

Steel Hardness Conversions

FROM A PRACTICAL STANDPOINT, it is important to be able to convert the results of one type of hardness test into those of a different test. Because a hardness test does not measure a well-defined property of a material and because all the tests in common use are not based on the same type of measurements, it is not surprising that universal hardness conversion relationships have not been developed. Hardness conversions instead are empirical relationships that are defined by conversion tables limited to specific categories of materials. That is, different conversion tables are required for materials with greatly different elastic moduli or with different strain-hardening capacity.

The most reliable hardness-conversion data exist for steel that is harder than 240 HB. The indentation hardness of soft metals depends on the strain-hardening behavior of the material during the test, which in turn depends on the previous degree of strain hardening of the material before the test. The modulus of elasticity also has been shown to influence conversions at high hardness levels. At low hardness levels, conversions between hardness scales measuring depth and those measuring diameter are likewise influenced by differences in the modulus of elasticity.

Hardness conversions are covered in standards such as SAE J417, "Hardness Tests and Hardness Conversions"; ISO 4964, "Hardness Conversions—Steel"; and ASTM E 140, "Standard Hardness Conversion Tables for Metals." Conversion tables for nickel and high-nickel alloys, cartridge brass, austenitic stainless steel plate and sheet, and copper can be found in ASTM E 140. Recently, ASTM committee E-28 on indentation hardness has developed mathematical conversion formulas based on the conversion-table values fround in ASTM E 140. Over 60 conversion formulas are listed in the appendix of ASTM E 140, and these formulas can be used in place of the tables. A computer is helpful in performing the calculations quickly.

Other hardness conversion formulas for various materials have also been published, and a list of some other conversion formulas is given in Table 1. The standard procedure for reporting converted hardness numbers indicates the measured hardness and test scale in parentheses—for example, 451 HB (48 HRC). The method of conversion (table, formula, or other method) should also be defined.

When making hardness correlations, it is best to consult ASTM E 140. Tables 2 to 5, from ASTM E 140, are for conversion among Rockwell, Brinell, and Vickers hardness for heat treated carbon and alloy steels, almost all constructional alloy steels, and tool steels in the as-forged, annealed, normalized, and quenched and tempered conditions. The tables are also summarized in graphical form in Fig. 1.

Table 1 Examples of published hardness conversion equations

Steels

$HB = \dfrac{7300}{130 - HRB}$	(40–100 HRB)
$HB = \dfrac{3710}{130 - HRE}$	(30–100 HRE)
$HB = \dfrac{1,520,000 - 4500\,HRC}{(100 - HRC)^2}$	(<40 HRC)
$HB = \dfrac{25,000 - 10(57 - HRC)^2}{100 - HRC}$	(40–70 HRC)
$HRB = 134 - \dfrac{6700}{HB}$	(± 7 HRB, 95% CL)
$HRC = 119.0 - \left(\dfrac{2.43 \times 10^6}{HV}\right)^{1/2}$	(240–1040 HV)
$HRA = 112.3 - \left(\dfrac{6.85 \times 10^5}{HV}\right)^{1/2}$	(240–1040 HV)
$HR15N = 117.94 - \left(\dfrac{5.53 \times 10^5}{HV}\right)^{1/2}$	(240–1040 HV)
$HR30N = 129.52 - \left(\dfrac{1.88 \times 10^6}{HV}\right)^{1/2}$	(240–1040 HV)
$HR45N = 133.51 - \left(\dfrac{3.132 \times 10^6}{HV}\right)^{1/2}$	(240–1040 HV)
$HB = 0.951\ HV$	(steel ball, 200–400 HV)
$HB = 0.941\ HV$	(tungsten-carbide ball, 200–700 HV)

Cemented carbides

$HRC = 117.35 - \left(\dfrac{2.43 \times 10^6}{HV}\right)^{1/2}$	(900–1800 HV)
$HRA = \dfrac{211 - \left(\dfrac{2.43 - 10^6}{HV}\right)^{1/2}}{1.885}$	(900–1800 HV)

Rockwell from Knoop for steels

$HRC = 64.934 \log HK - 140.38$	(15 gf)
$HRC = 67.353 \log HK - 144.32$	(25 gf)
$HRC = 71.983 \log HK - 154.28$	(50 gf)
$HRC = 76.572 \log HK - 163.89$	(100 gf)
$HRC = 79.758 \log HK - 170.92$	(200 gf)
$HRC = 82.283 \log HK - 176.92$	(300 gf)
$HRC = 83.58 \log HK - 179.30$	(500 gf)
$HRC = 85.848 \log HK - 184.55$	(1000 gf)

White cast irons

$HB = 0.363\,(HRC)^2 - 22.515\,(HRC) + 717.8$	
$HV = 0.343\,(HRC)^2 - 18.132\,(HRC) + 595.3$	
$HV = 1.136\,(HB)^2 - 26.0$	

Austenitic stainless steel

$\dfrac{1}{HB} = 0.0001304(130 - HRB)$	(60–90 HRB, 110–192 HB)

Stable alpha-beta titanium alloys

$HRC = 0.078\ HV + 8.1$	

Table 2 Approximate Rockwell B hardness conversion numbers for nonaustenitic steels

Rockwell			Superficial Rockwell							
B, 100 kgf, 1/16 in. ball	A, 60 kgf, diamond	E, 100 kgf, 1/8 in. ball	15T, 15 kgf, 1/16 in. ball	30T, 30 kgf, 1/16 in. ball	45T, 45 kgf, 1/16 in. ball	Vickers	Knoop, 500 gf and over	Brinell, 3000 kgf, 10 mm ball	Tensile strength MPa (ksi)	Brinell, 500 kgf, 10 mm ball
100	61.5	...	93.1	83.1	72.9	240	251	240	800 (116)	201
99	60.9	...	92.8	82.5	71.9	234	246	234	787 (114)	195
98	60.2	...	92.5	81.8	70.9	228	241	228	752 (109)	189
97	59.5	...	92.1	81.1	69.9	222	236	222	724 (105)	184
96	58.9	...	91.8	80.4	68.9	216	231	216	704 (102)	179
95	58.3	...	91.5	79.8	67.9	210	226	210	690 (100)	175
94	57.6	...	91.2	79.1	66.9	205	221	205	676 (98)	171
93	57.0	...	90.8	78.4	65.9	200	216	200	648 (94)	167
92	56.4	...	90.5	77.8	64.8	195	211	195	634 (92)	163
91	55.8	...	90.2	77.1	63.8	190	206	190	620 (90)	160
90	55.2	...	89.9	76.4	62.8	185	201	185	614 (89)	157
89	54.6	...	89.5	75.8	61.8	180	196	180	607 (88)	154
88	54.0	...	89.2	75.1	60.8	176	192	176	593 (86)	151
87	53.4	...	88.9	74.4	59.8	172	188	172	579 (84)	148
86	52.8	...	88.6	73.8	58.8	169	184	169	572 (83)	145
85	52.3	...	88.2	73.1	57.8	165	180	165	565 (82)	142
84	51.7	...	87.9	72.4	56.8	162	176	162	558 (81)	140
83	51.1	...	87.6	71.8	55.8	159	173	159	552 (80)	137
82	50.6	...	87.3	71.1	54.8	156	170	156	524 (76)	135
81	50.0	...	86.9	70.4	53.8	153	167	153	503 (73)	133
80	49.5	...	86.6	69.7	52.8	150	164	150	496 (72)	130
79	48.9	...	86.3	69.1	51.8	147	161	147	482 (70)	128
78	48.4	...	86.0	68.4	50.8	144	158	144	475 (69)	126
77	47.9	...	85.6	67.7	49.8	141	155	141	469 (68)	124
76	47.3	...	85.3	67.1	48.8	139	152	139	462 (67)	122
75	46.8	...	85.0	66.4	47.8	137	150	137	455 (66)	120
74	46.3	...	84.7	65.7	46.8	135	147	135	448 (65)	118
73	45.8	...	84.3	65.1	45.8	132	145	132	441 (64)	116
72	45.3	...	84.0	64.4	44.8	130	143	130	434 (63)	114
71	44.8	100	83.7	63.7	43.8	127	141	127	427 (62)	112
70	44.3	99.5	83.4	63.1	42.8	125	139	125	421 (61)	110
69	43.8	99.0	83.0	62.4	41.8	123	137	123	414 (60)	109
68	43.3	98.0	82.7	61.7	40.8	121	135	121	407 (59)	108
67	42.8	97.5	82.4	61.0	39.8	119	133	119	400 (58)	106
66	42.3	97.0	82.1	60.4	38.7	117	131	117	393 (57)	104
65	41.8	96.0	81.8	59.7	37.7	116	129	116	386 (56)	102
64	41.4	95.5	81.4	59.0	36.7	114	127	114	...	100
63	40.9	95.0	81.1	58.4	35.7	112	125	112	...	99
62	40.4	94.5	80.8	57.7	34.7	110	124	110	...	98
61	40.0	93.5	80.5	57.0	33.7	108	122	108	...	96
60	39.5	93.0	80.1	56.4	32.7	107	120	107	...	95
59	39.0	92.5	79.8	55.7	31.7	106	118	106	...	94
58	38.6	92.0	79.5	55.0	30.7	104	117	104	...	92
57	38.1	91.0	79.2	54.4	29.7	103	115	103	...	91
56	37.7	90.5	78.8	53.7	28.7	101	114	101	...	90
55	37.2	90.0	78.5	53.0	27.7	100	112	100	...	89
54	36.8	89.5	78.2	52.4	26.7	...	111	...	...	87
53	36.3	89.0	77.9	51.7	25.7	...	110	...	...	86
52	35.9	88.0	77.5	51.0	24.7	...	109	...	...	85
51	35.5	87.5	77.2	50.3	23.7	...	108	...	...	84
50	35.0	87.0	76.9	49.7	22.7	...	107	...	...	83
49	34.6	86.5	76.6	49.0	21.7	...	106	...	...	82
48	34.1	85.5	76.2	48.3	20.7	...	105	...	...	81
47	33.7	85.0	75.9	47.7	19.7	...	104	...	...	80
46	33.3	84.5	75.6	47.0	18.7	...	103	...	...	80
45	32.9	84.0	75.3	46.3	17.7	...	102	...	...	79
44	32.4	83.5	74.9	45.7	16.7	...	101	...	...	78

Data are only approximate conversions for carbon and low-alloy steels in the annealed, normalized, and quenched-and-tempered conditions; less accurate for cold-worked condition and for austenitic steels. Source: ASTM E 140, except for values for E scale and tensile strength, which are not from standards

Table 3 Approximate Rockwell C hardness conversion numbers for nonaustenitic steels, according to ASTM E 140

C, 150 kgf, diamond	A, 60 kgf, diamond	D, 100 kgf, diamond	15 N, 15 kgf, diamond	30 N, 30 kgf, diamond	45 N, 45 kgf, diamond	Vickers	Knoop, 500 gf and over	Brinell, 3000 kgf, 10 mm ball	Tensile strength, MPa (ksi)
68	85.6	76.9	93.2	84.4	75.4	940	920	. . .	. . .
67	85.0	76.1	92.9	83.6	74.2	900	895	. . .	. . .
66	84.5	75.4	92.5	82.8	73.3	865	870	. . .	. . .
65	83.9	74.5	92.2	81.9	72.0	832	846	739(a)	. . .
64	83.4	73.8	91.8	81.1	71.0	800	822	722(a)	. . .
63	82.8	73.0	91.4	80.1	69.9	772	799	705(a)	. . .
62	82.3	72.2	91.1	79.3	68.8	746	776	688(a)	. . .
61	81.8	71.5	90.7	78.4	67.7	720	754	670(a)	. . .
60	81.2	70.7	90.2	77.5	66.6	697	732	654(a)	. . .
59	80.7	69.9	89.8	76.6	65.5	674	710	634(a)	2420 (351)
58	80.1	69.2	89.3	75.7	64.3	653	690	615	2330 (338)
57	79.6	68.5	88.9	74.8	63.2	633	670	595	2240 (325)
56	79.0	67.7	88.3	73.9	62.0	613	650	577	2158 (313)
55	78.5	66.9	87.9	73.0	60.9	595	630	560	2075 (301)
54	78.0	66.1	87.4	72.0	59.8	577	612	543	2013 (292)
53	77.4	65.4	86.9	71.2	58.6	560	594	525	1951 (283)
52	76.8	64.6	86.4	70.2	57.4	544	576	512	1882 (273)
51	76.3	63.8	85.9	69.4	56.1	528	558	496	1820 (264)
50	75.9	63.1	85.5	68.5	55.0	513	542	481	1758 (255)
49	75.2	62.1	85.0	67.6	53.8	498	526	469	1696 (246)
48	74.7	61.4	84.5	66.7	52.5	484	510	455	1634 (237)
47	74.1	60.8	83.9	65.8	51.4	471	495	443	1579 (229)
46	73.6	60.0	83.5	64.8	50.3	458	480	432	1524 (221)
45	73.1	59.2	83.0	64.0	49.0	446	466	421	1482 (215)
44	72.5	58.5	82.5	63.1	47.8	434	452	409	1434 (208)
43	72.0	57.7	82.0	62.2	46.7	423	438	400	1386 (201)
42	71.5	56.9	81.5	61.3	45.5	412	426	390	1344 (195)
41	70.9	56.2	80.9	60.4	44.3	402	414	381	1296 (188)
40	70.4	55.4	80.4	59.5	43.1	392	402	371	1254 (182)
39	69.9	54.6	79.9	58.6	41.9	382	391	362	1220 (177)
38	69.4	53.8	79.4	57.7	40.8	372	380	353	1179 (171)
37	68.9	53.1	78.8	56.8	39.6	363	370	344	1137 (166)
36	68.4	52.3	78.3	55.9	38.4	354	360	336	1110 (161)
35	67.9	51.5	77.7	55.0	37.2	345	351	327	1075 (156)
34	67.4	50.8	77.2	54.2	36.1	336	342	319	1048 (152)
33	66.8	50.0	76.6	53.3	34.9	327	334	311	1027 (149)
32	66.3	49.2	76.1	52.1	33.7	318	326	301	1006 (146)
31	65.8	48.4	75.6	51.3	32.5	310	318	294	972 (141)
30	65.3	47.7	75.0	50.4	31.3	302	311	286	951 (138)
29	64.8	47.0	74.5	49.5	30.1	294	304	279	930 (135)
28	64.3	46.1	73.9	48.6	28.9	286	297	271	903 (131)
27	63.8	45.2	73.3	47.7	27.8	279	290	264	882 (128)
26	63.3	44.6	72.8	46.8	26.7	272	284	258	861 (125)
25	62.8	43.8	72.2	45.9	25.5	266	278	253	848 (123)
24	62.4	43.1	71.6	45.0	24.3	260	272	247	820 (119)
23	62.0	42.1	71.0	44.0	23.1	254	266	243	806 (117)
22	61.5	41.6	70.5	43.2	22.0	248	261	237	792 (115)
21	61.0	40.9	69.9	42.3	20.7	243	256	231	772 (112)
20	60.5	40.1	69.4	41.5	19.6	238	251	226	758 (110)

Data are only approximate conversions for carbon and low-alloy steels in the annealed, normalized, and quenched-and-tempered conditions; less accurate for cold-worked condition and for austenitic steels. (a) Hardness values outside the recommended range for Brinell testing per ASTM E 10. Source: ASTM E 140, except for values for tensile strength, which are not from standards

Table 4 Approximate equivalent hardness numbers for Brinell hardness numbers for steel

Brinell indentation diam, mm	Brinell hardness number(a) 3000 kgf load, 10 mm ball(a)		Vickers hardness No.	Rockwell hardness No.				Rockwell superficial hardness No., diamond indenter			Knoop hardness No., 500 gf load and greater	Scleroscope hardness No.
	Standard ball	Tungsten-carbide ball		A scale, 60 kgf load, diamond indentor	B scale, 100 kgf load, 1/16 in. diam ball	C scale, 150 kgf load, diamond indenter	D scale, 100 kgf load, diamond indenter	15 N scale, 15 kgf load	30 N scale, 30 kgf load	45 N scale, 45 kgf load		
2.25	...	(745)	840	84.1	...	65.3	74.8	92.3	82.2	72.2	852	91
2.30	...	(712)	783	83.1	...	63.4	73.4	91.6	80.5	70.4	808	...
2.35	...	(682)	737	82.2	...	61.7	72.0	91.0	79.0	68.5	768	84
2.40	...	(653)	697	81.2	...	60.0	70.7	90.2	77.5	66.5	732	81
2.45	...	627	667	80.5	...	58.7	69.7	89.6	76.3	65.1	703	79
2.50	...	601	640	79.8	...	57.3	68.7	89.0	75.1	63.5	677	77
2.55	...	578	615	79.1	...	56.0	67.7	88.4	73.9	62.1	652	75
2.60	...	555	591	78.4	...	54.7	66.7	87.8	72.7	60.6	626	73
2.65	...	534	569	77.8	...	53.5	65.8	87.2	71.6	59.2	604	71
2.70	...	514	547	76.9	...	52.1	64.7	86.5	70.3	57.6	579	70
2.75	(495)	...	539	76.7	...	51.6	64.3	86.3	69.9	56.9	571	...
	...	495	528	76.3	...	51.0	63.8	85.9	69.4	56.1	558	68
2.80	(477)	...	516	75.9	...	50.3	63.2	85.6	68.7	55.2	545	...
	...	477	508	75.6	...	49.6	62.7	85.3	68.2	54.5	537	66
2.85	(461)	...	495	75.1	...	48.8	61.9	84.9	67.4	53.5	523	...
	...	461	491	74.9	...	48.5	61.7	84.7	67.2	53.2	518	65
2.90	444	...	474	74.3	...	47.2	61.0	84.1	66.0	51.7	499	...
	...	444	472	74.2	...	47.1	60.8	84.0	65.8	51.5	496	63
2.95	429	429	455	73.4	...	45.7	59.7	83.4	64.6	49.9	476	61
3.00	415	415	440	72.8	...	44.5	58.8	82.8	63.5	48.4	459	59
3.05	401	401	425	72.0	...	43.1	57.8	82.0	62.3	46.9	441	58
3.10	388	388	410	71.4	...	41.8	56.8	81.4	61.1	45.3	423	56
3.15	375	375	396	70.6	...	40.4	55.7	80.6	59.9	43.6	407	54
3.20	363	363	383	70.0	...	39.1	54.6	80.0	58.7	42.0	392	52
3.25	352	352	372	69.3	(110.0)	37.9	53.8	79.3	57.6	40.5	379	51
3.30	341	341	360	68.7	(109.0)	36.6	52.8	78.6	56.4	39.1	367	50
3.35	331	331	350	68.1	(108.5)	35.5	51.9	78.0	55.4	37.8	356	48
3.40	321	321	339	67.5	(108.0)	34.3	51.0	77.3	54.3	36.4	345	47
3.45	311	311	328	66.9	(107.5)	33.1	50.0	76.7	53.3	34.4	336	46
3.50	302	302	319	66.3	(107.0)	32.1	49.3	76.1	52.2	33.8	327	45
3.55	293	293	309	65.7	(106.0)	30.9	48.3	75.5	51.2	32.4	318	43
3.60	285	285	301	65.3	(105.5)	29.9	47.6	75.0	50.3	31.2	310	42
3.65	277	277	292	64.6	(104.5)	28.8	46.7	74.4	49.3	29.9	302	41
3.70	269	269	284	64.1	(104.0)	27.6	45.9	73.7	48.3	28.5	294	40
3.75	262	262	276	63.6	(103.0)	26.6	45.0	73.1	47.3	27.3	286	39
3.80	255	255	269	63.0	(102.0)	25.4	44.2	72.5	46.2	26.0	279	38
3.85	248	248	261	62.5	(101.0)	24.2	43.2	71.7	45.1	24.5	272	37
3.90	241	241	253	61.8	100.0	22.8	42.0	70.9	43.9	22.8	265	36
3.95	235	235	247	61.4	99.0	21.7	41.4	70.3	42.9	21.5	259	35
4.00	229	229	241	60.8	98.2	20.5	40.5	69.7	41.9	20.1	253	34
4.05	223	223	234	...	97.3	(19.0)	...	...	...	...	247	...
4.10	217	217	228	...	96.4	(17.7)	...	...	...	...	242	33
4.15	212	212	222	...	95.5	(16.4)	...	...	...	...	237	32
4.20	207	207	218	...	94.6	(15.2)	...	...	...	...	232	31
4.25	201	201	212	...	93.7	(13.8)	...	...	...	...	227	...
4.30	197	197	207	...	92.8	(12.7)	...	...	...	...	222	30
4.35	192	192	202	...	91.9	(11.5)	...	...	...	...	217	29
4.40	187	187	196	...	90.9	(10.2)	...	...	...	...	212	...
4.45	183	183	192	...	90.0	(9.0)	...	...	...	...	207	28
4.50	179	179	188	...	89.0	(8.0)	...	...	...	...	202	27
4.55	174	174	182	...	88.0	(6.7)	...	...	...	...	198	...
4.60	170	170	178	...	87.0	(5.4)	...	...	...	...	194	26
4.65	167	167	175	...	86.0	(4.4)	...	...	...	...	190	...
4.70	163	163	171	...	85.0	(3.3)	...	...	...	...	186	25
4.75	159	159	167	...	83.9	(2.0)	...	...	...	...	182	...
4.80	156	156	163	...	82.9	(0.9)	...	...	...	...	178	24
4.85	152	152	159	...	81.9	...	...	...	...	...	174	...
4.90	149	149	156	...	80.8	...	...	...	...	...	170	23
4.95	146	146	153	...	79.7	...	...	...	...	...	166	...
5.00	143	143	150	...	78.6	...	...	...	...	...	163	22
5.10	137	137	143	...	76.4	...	...	...	...	...	157	21
5.20	131	131	137	...	74.2	...	...	...	...	...	151	...
5.30	126	126	132	...	72.0	...	...	...	...	...	145	20
5.40	121	121	127	...	69.8	...	...	...	...	...	140	19
5.50	116	116	122	...	67.6	...	...	...	...	...	135	18
5.60	111	111	117	...	65.4	...	...	...	...	...	131	17

Note: Values in parentheses are beyond normal range and are given for information only. Data are for carbon and alloy steels in the annealed, normalized, and quenched-and-tempered conditions; less accurate for cold-worked condition and for austenitic steels (a) Brinell numbers are based on the diameter of impressed indentation. If the ball distorts (flattens) during test, Brinell numbers will vary in accordance with the degree of such distortion when related to hardnesses determined with a Vickers diamond pyramid. Rockwell diamond indenter, or other indenter that does not sensibly distort. At high hardnesses, therefore, the relationship between Brinell and Vickers or Rockwell scales is affected by the type of ball used. Standard steel balls tend to flatten slightly more than tungsten-carbide balls, resulting in a larger indentation and a lower Brinell number than shown by a tungsten carbide ball. Thus, on a specimen of about 539–547 HV, a standard ball will leave a 2.75 mm indentation (495 HB), and a tungsten carbide ball a 2.70 mm indentation (514 HB). Conversely, identical indentation diameters for both types of ball will correspond to different Vickers and Rockwell values. Thus, if indentation in two different specimens both are 2.75 mm diameter (495 HB), the specimen tested with a standard ball has a Vickers hardness of 539, whereas the specimen tested with a tungsten-carbide ball has a Vickers hardness of 528. Source: ASTM E 140

Table 5   Approximate equivalent hardness numbers for Vickers (diamond pyramid) hardness numbers for steel

Vickers hardness No.	Brinell hardness No., 3000 kg load, 10 mm ball		Rockwell hardness No.				Rockwell superficial (diamond pyramid) hardness No., diamond indenter			Knoop hardness No., 500 gf load and greater	Scleroscope hardness No.
	Standard ball	Tungsten-carbide ball	A scale, 60 kgf load, diamond indenter	B scale, 100 kgf load, 1/16 in. diam ball	C scale, 150 kgf load, diamond indenter	D scale, 100 kgf load, diamond indenter	15 N scale, 15 kgf load	30 N scale, 30 kgf load	45 N scale, 45 kgf load		
940	...	...	85.6	...	68.0	76.9	93.2	84.4	75.4	920	97
920	...	...	85.3	...	67.5	76.5	93.0	84.0	74.8	908	96
900	...	...	85.0	...	67.0	76.1	92.9	83.6	74.2	895	95
880	...	(767)	84.7	...	66.4	75.7	92.7	83.1	73.6	882	93
860	...	(757)	84.4	...	65.9	75.3	92.5	82.7	73.1	867	92
840	...	(745)	84.1	...	65.3	74.8	92.3	82.2	72.2	852	91
820	...	(733)	83.8	...	64.7	74.3	92.1	81.7	71.8	837	90
800	...	(722)	83.4	...	64.0	73.8	91.8	81.1	71.0	822	88
780	...	(710)	83.0	...	63.3	73.3	91.5	80.4	70.2	806	87
760	...	(698)	82.6	...	62.5	72.6	91.2	79.7	69.4	788	86
740	...	(684)	82.2	...	61.8	72.1	91.0	79.1	68.6	772	84
720	...	(670)	81.8	...	61.0	71.5	90.7	78.4	67.7	754	83
700	...	(656)	81.3	...	60.1	70.8	90.3	77.6	66.7	735	81
690	...	(647)	81.1	...	59.7	70.5	90.1	77.2	66.2	725	...
680	...	(638)	80.8	...	59.2	70.1	89.8	76.8	65.7	716	80
670	...	(630)	80.6	...	58.8	69.8	89.7	76.4	65.3	706	...
660	...	620	80.3	...	58.3	69.4	89.5	75.9	64.7	697	79
650	...	611	80.0	...	57.8	69.0	89.2	75.5	64.1	687	78
640	...	601	79.8	...	57.3	68.7	89.0	75.1	63.5	677	77
630	...	591	79.5	...	56.8	68.3	88.8	74.6	63.0	667	76
620	...	582	79.2	...	56.3	67.9	88.5	74.2	62.4	657	75
610	...	573	78.9	...	55.7	67.5	88.2	73.6	61.7	646	...
600	...	564	78.6	...	55.2	67.0	88.0	73.2	61.2	636	74
590	...	554	78.4	...	54.7	66.7	87.8	72.7	60.5	625	73
580	...	545	78.0	...	54.1	66.2	87.5	72.1	59.9	615	72
570	...	535	77.8	...	53.6	65.8	87.2	71.7	59.3	604	...
560	...	525	77.4	...	53.0	65.4	86.9	71.2	58.6	594	71
550	(505)	517	77.0	...	52.3	64.8	86.6	70.5	57.8	583	70
540	(496)	507	76.7	...	51.7	64.4	86.3	70.0	57.0	572	69
530	(488)	497	76.4	...	51.1	63.9	86.0	69.5	56.2	561	68
520	(480)	488	76.1	...	50.5	63.5	85.7	69.0	55.6	550	67
510	(473)	479	75.7	...	49.8	62.9	85.4	68.3	54.7	539	...
500	(465)	471	75.3	...	49.1	62.2	85.0	67.7	53.9	528	66
490	(456)	460	74.9	...	48.4	61.6	84.7	67.1	53.1	517	65
480	(448)	452	74.5	...	47.7	61.3	84.3	66.4	52.2	505	64
470	441	442	74.1	...	46.9	60.7	83.9	65.7	51.3	494	...
460	433	433	73.6	...	46.1	60.1	83.6	64.9	50.4	482	62
450	425	425	73.3	...	45.3	59.4	83.2	64.3	49.4	471	...
440	415	415	72.8	...	44.5	58.8	82.8	63.5	48.4	459	59
430	405	405	72.3	...	43.6	58.2	82.3	62.7	47.4	447	58
420	397	397	71.8	...	42.7	57.5	81.8	61.9	46.4	435	57
410	388	388	71.4	...	41.8	56.8	81.4	61.1	45.3	423	56
400	379	379	70.8	...	40.8	56.0	80.8	60.2	44.1	412	55
390	369	369	70.3	...	39.8	55.2	80.3	59.3	42.9	400	...
380	360	360	69.8	(110.0)	38.8	54.4	79.8	58.4	41.7	389	52
370	350	350	69.2	...	37.7	53.6	79.2	57.4	40.4	378	51
360	341	341	68.7	(109.0)	36.6	52.8	78.6	56.4	39.1	367	50
350	331	331	68.1	...	35.5	51.9	78.0	55.4	37.8	356	48
340	322	322	67.6	(108.0)	34.4	51.1	77.4	54.4	36.5	346	47
330	313	313	67.0	...	33.3	50.2	76.8	53.6	35.2	337	46
320	303	303	66.4	(107.0)	32.2	49.4	76.2	52.3	33.9	328	45
310	294	294	65.8	...	31.0	48.4	75.6	51.3	32.5	318	...
300	284	284	65.2	(105.5)	29.8	47.5	74.9	50.2	31.1	309	42
295	280	280	64.8	...	29.2	47.1	74.6	49.7	30.4	305	...
290	275	275	64.5	(104.5)	28.5	46.5	74.2	49.0	29.5	300	41
285	270	270	64.2	...	27.8	46.0	73.8	48.4	28.7	296	...
280	265	265	63.8	(103.5)	27.1	45.3	73.4	47.8	27.9	291	40
275	261	261	63.5	...	26.4	44.9	73.0	47.2	27.1	286	39
270	256	256	63.1	(102.0)	25.6	44.3	72.6	46.4	26.2	282	38
265	252	252	62.7	...	24.8	43.7	72.1	45.7	25.2	277	...
260	247	247	62.4	(101.0)	24.0	43.1	71.6	45.0	24.3	272	37
255	243	243	62.0	...	23.1	42.2	71.1	44.2	23.2	267	...
250	238	238	61.6	99.5	22.2	41.7	70.6	43.4	22.2	262	36
245	233	233	61.2	...	21.3	41.1	70.1	42.5	21.1	258	35
240	228	228	60.7	98.1	20.3	40.3	69.6	41.7	19.9	253	34
230	219	219	...	96.7	(18.0)	...	...	...	...	243	33
220	209	209	...	95.0	(15.7)	...	...	...	...	234	32
210	200	200	...	93.4	(13.4)	...	...	...	...	226	30
200	190	190	...	91.5	(11.0)	...	...	...	...	216	29
190	181	181	...	89.5	(8.5)	...	...	...	...	206	28

(continued)

Note: Values in parentheses are beyond normal range and are given for information only. Data are for carbon and alloy steels in the annealed, normalized, and quenched-and-tempered conditions; less accurate for cold-worked condition and for austenitic steels. Source: ASTM E 140

Table 5 (continued)

Vickers hardness No.	Brinell hardness No., 3000 kg load, 10 mm ball		Rockwell hardness No.				Rockwell superficial (diamond pyramid) hardness No., diamond indenter			Knoop hardness No., 500 gf load and greater	Scleroscope hardness No.
	Standard ball	Tungsten-carbide ball	A scale, 60 kgf load, diamond indenter	B scale, 100 kgf load, 1/16 in. diam ball	C scale, 150 kgf load, diamond indenter	D scale, 100 kgf load, diamond indenter	15 N scale, 15 kgf load	30 N scale, 30 kgf load	45 N scale, 45 kgf load		
180	171	171	. . .	87.1	(6.0)	. . .	. . .	. . .	. . .	196	26
170	162	162	. . .	85.0	(3.0)	. . .	. . .	. . .	. . .	185	25
160	152	152	. . .	81.7	(0.0)	. . .	. . .	. . .	. . .	175	23
150	143	143	. . .	78.7	. . .	. . .	. . .	. . .	. . .	164	22
140	133	133	. . .	75.0	. . .	. . .	. . .	. . .	. . .	154	21
130	124	124	. . .	71.2	. . .	. . .	. . .	. . .	. . .	143	20
120	114	114	. . .	66.7	. . .	. . .	. . .	. . .	. . .	133	18
110	105	105	. . .	62.3	. . .	. . .	. . .	. . .	. . .	123	. . .
100	95	95	. . .	56.2	. . .	. . .	. . .	. . .	. . .	112	. . .
95	90	90	. . .	52.0	. . .	. . .	. . .	. . .	. . .	107	. . .
90	86	86	. . .	48.0	. . .	. . .	. . .	. . .	. . .	102	. . .
85	81	81	. . .	41.0	. . .	. . .	. . .	. . .	. . .	97	. . .

Note: Values in parentheses are beyond normal range and are given for information only. Data are for carbon and alloy steels in the annealed, normalized, and quenched-and-tempered conditions; less accurate for cold-worked condition and for austenitic steels. Source: ASTM E 140

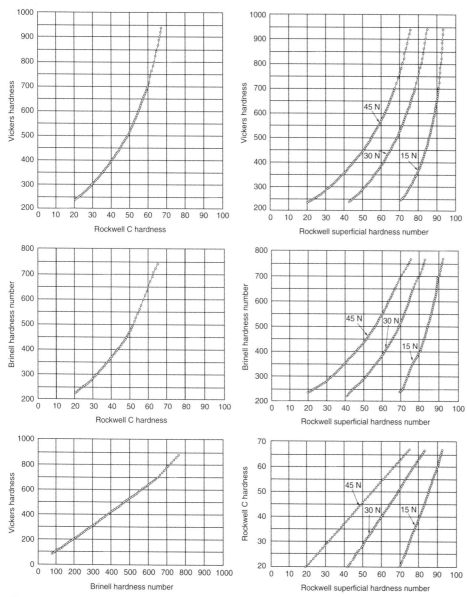

Fig. 1 Approximate equivalent hardness numbers for steel. Points represent data from the hardness conversion tables.

Nonferrous Hardness Conversions

Table 1   Approximate equivalent hardness numbers for wrought aluminum products

Brinell hardness No., 500 kgf, 10 mm ball, HBS	Vickers hardness No., 15 kgf, HV	Rock well hardness No.			Rockwell superficial hardness No.		
		B scale, 100 kgf, 1/16 in. ball, HRB	E scale, 100 kgf, 1/8 in. ball, HRE	H scale, 60 kgf, 1/8 in. ball, HRH	15T scale, 15 kgf, 1/16 in. ball, HR15T	30T scale, 30 kgf, 1/16 in. ball, HR30T	15W scale, 15 kgf, 1/8 in. ball, HR15W
160	189	91	. . .	. . .	89	77	95
155	183	90	. . .	. . .	89	76	95
150	177	89	. . .	. . .	89	75	94
145	171	87	. . .	. . .	88	74	94
140	165	86	. . .	. . .	88	73	94
135	159	84	. . .	. . .	87	71	93
130	153	81	. . .	. . .	87	70	93
125	147	79	. . .	. . .	86	68	92
120	141	76	101	. . .	86	67	92
115	135	72	100	. . .	86	65	91
110	129	69	99	. . .	85	63	91
105	123	65	98	. . .	84	61	91
100	117	60	. . .	. . .	83	59	90
95	111	56	96	. . .	82	57	90
90	105	51	94	108	81	54	89
85	98	46	91	107	80	52	89
80	92	40	88	106	78	50	88
75	86	34	84	104	76	47	87
70	80	28	80	102	74	44	86
65	74	. . .	75	100	72	. . .	85
60	68	. . .	70	97	70	. . .	83
55	62	. . .	65	94	67	. . .	82
50	56	. . .	59	91	64	. . .	80
45	50	. . .	53	87	62	. . .	79
40	44	. . .	46	83	59	. . .	77

Aluminum and aluminum alloys are tested frequently for hardness to distinguish between annealed, cold-worked, and heat treated grades. The Rockwell B scale (100 kgf load with a 1.58 mm, or 1/16 in. steel ball indenter) generally is suitable in testing grades that have been precipitation hardened to relatively high strength levels. For softer grades and commercially pure aluminum, hardness testing usually is done with the Rockwell F, E, and H scales. For hardness testing of thin gages of aluminum, the 15T and 30T scales of the Rockwell superficial tester are recommended. Source: ASTM E 140

Table 2 Approximate equivalent hardness numbers for wrought coppers (>99% Cu, alloys C10200 through C14200)

Vickers hardness No.		Knoop hardness No.		Rockwell superficial hardness No.			Rockwell hardness No.		Rockwell superficial hardness No.			Brinell hardness No.	
1 kgf, HV	100 gf, HV	1 kgf, HK	500 gf, HK	15T scale, 15 kgf, 1/16 in. (1.588 mm) ball, HR15T(a)	15T scale, 15 kgf, 1/16 in. (1.588 mm) ball, HR15T(b)	30T scale, 30 kgf, 1/16 in. (1.588 mm) ball, HR30T(b)	B scale, 100 kgf, 1/16 in. (1.588 mm) ball, HRB(c)	F scale, 60 kgf, 1/16 in. (1.588 mm) ball, HRF(c)	15T scale, 15 kgf, 1/16 in. (1.588 mm) ball, HR15T(c)	30T scale, 30 kgf, 1/16 in. (1.588 mm) ball, HR30T(c)	45T scale, 45 kgf, 1/16 in. (1.588 mm) ball, HR45T(c)	500 kgf, 10 mm diam ball, HBS(d)	20 kgf, 2 mm diam ball, HBS(e)
130	127.0	138.7	133.8	...	85.0	...	67.0	99.0	...	69.5	49.0	...	119.0
128	125.2	136.8	132.1	83.0	84.5	...	66.0	98.0	87.0	68.5	48.0	...	117.5
126	123.6	134.9	130.4	...	84.0	...	65.0	97.0	...	67.5	46.5	120.0	115.0
124	121.9	133.0	128.7	82.5	83.5	...	64.0	96.0	86.0	66.5	45.0	117.5	113.0
122	121.1	131.0	127.0	...	83.0	...	62.5	95.5	85.5	66.0	44.0	115.0	111.0
120	118.5	129.0	125.2	82.0	82.5	...	61.0	95.0	...	65.0	42.5	112.0	109.0
118	116.8	127.1	123.5	81.5	...	...	59.5	94.0	85.0	64.0	41.0	110.0	107.5
116	115.0	125.1	121.7	...	82.0	...	58.5	93.0	...	63.0	40.0	107.0	105.5
114	113.5	123.2	119.9	81.0	81.5	...	57.0	92.5	84.5	62.0	38.5	105.0	103.5
112	111.8	121.4	118.1	80.5	81.0	...	55.0	91.5	...	61.0	37.0	102.0	102.0
110	109.9	119.5	116.3	80.0	...	...	53.5	91.0	84.0	60.0	36.0	99.5	100.0
108	108.3	117.5	114.5	...	80.5	...	52.0	90.5	83.5	59.0	34.5	97.0	98.0
106	106.6	115.6	112.6	79.5	80.0	...	50.0	89.5	...	58.0	33.0	94.5	96.0
104	104.9	113.5	110.1	79.0	79.5	...	48.0	88.5	83.0	57.0	32.0	92.0	94.0
102	103.2	111.5	108.0	78.5	79.0	...	46.5	87.5	82.5	56.0	30.0	89.5	92.0
100	101.5	109.4	106.0	78.0	78.0	...	44.5	87.0	82.0	55.0	28.5	87.0	90.0
98	99.8	107.3	104.0	77.5	77.5	...	42.0	85.5	81.0	53.5	26.5	84.5	88.0
96	98.0	105.3	102.1	77.0	77.0	...	40.0	84.5	80.5	52.0	25.5	82.0	86.5
94	96.4	103.2	100.0	76.5	76.5	...	38.0	83.0	80.0	51.0	23.0	79.5	85.0
92	94.7	101.0	98.0	76.0	75.5	...	35.5	82.0	79.0	49.0	21.0	77.0	83.0
90	93.0	98.9	96.0	75.5	75.0	...	33.0	81.0	78.0	47.5	19.0	74.5	81.0
88	91.2	96.9	94.0	75.0	74.5	...	30.5	79.5	77.0	46.0	16.5	...	79.0
86	89.7	95.5	92.0	74.5	73.5	...	28.0	78.0	76.0	44.0	14.0	...	77.0
84	87.9	92.3	90.0	74.0	73.0	...	25.5	76.5	75.0	43.0	12.0	...	75.0
82	86.1	90.1	87.9	73.5	72.0	...	23.0	74.5	74.5	41.0	9.5	...	73.0
80	84.5	87.9	86.0	72.5	71.0	...	20.0	73.0	73.5	39.5	7.0	...	71.5
78	82.8	85.7	84.0	72.0	70.0	...	17.0	71.0	72.5	37.5	5.0	...	69.5
76	81.0	83.5	81.9	71.5	69.5	...	14.5	69.0	71.5	36.0	2.0	...	67.5
74	79.2	81.1	79.9	71.0	68.5	...	11.5	67.5	70.0	34.0	...	...	66.0
72	77.6	78.9	78.7	70.0	67.5	...	8.5	66.0	69.0	32.0	...	...	64.0
70	75.8	76.8	76.6	69.5	66.5	...	5.0	64.0	67.5	30.0	...	...	62.0
68	74.3	74.1	74.4	69.0	65.5	...	2.0	62.0	66.0	28.0	...	...	60.5
66	72.6	71.9	71.9	68.0	64.5	...	...	60.0	64.5	25.5	...	...	58.5
64	70.9	69.5	70.0	67.5	63.5	...	...	58.0	63.5	23.5	...	...	57.0
62	69.1	67.0	67.9	66.5	62.0	...	...	56.0	61.0	21.0	...	...	55.0
60	67.5	64.6	65.9	66.0	61.0	...	...	54.0	59.0	18.0	...	...	53.0
58	65.8	62.0	63.8	65.0	60.0	...	...	51.5	57.0	15.5	...	...	51.5
56	64.0	59.8	61.8	64.5	58.5	...	...	49.0	55.0	13.0	...	...	49.5
54	62.3	57.4	59.5	63.5	57.5	...	...	47.0	53.0	10.0	...	...	48.0
52	60.7	55.0	57.2	63.0	56.0	...	...	44.0	51.5	7.5	...	...	46.5
50	58.9	52.8	55.0	62.0	55.0	...	...	41.5	49.5	4.5	...	...	44.5
48	57.3	50.3	52.7	61.0	53.5	...	...	39.0	47.5	1.5	...	...	42.0
46	55.8	48.0	50.2	60.5	52.0	...	...	36.0	45.0	...	...	...	41.0
44	53.9	45.9	47.8	59.5	51.0	...	...	33.5	43.0	...	...	...	...
42	52.2	43.7	45.2	58.5	49.5	...	...	30.5	41.0	...	...	...	...
40	51.3	40.2	42.8	57.5	48.0	...	...	28.0	38.5	...	...	...	...

(a) For 0.010 in. (0.25 mm) strip. (b) For 0.020 in. (0.51 mm) strip. (c) For 0.040 in. (1.02 mm) strip and greater. (d) For 0.080 in. (2.03 mm) strip. (e) For 0.040 in. (1.02 mm) strip. Source: ASTM E 140

Table 3 Approximate equivalent hardness numbers for cartridge brass (70% Cu, 30% Zn)

Vickers hardness No., HV	Rockwell hardness No.		Rockwell superficial hardness No.			Brinell hardness No. 500 kgf, 10 mm ball, HBS	Vickers hardness No., HV	Rockwell hardness No.		Rockwell superficial hardness No.			Brinell hardness No. 500 kgf, 10 mm ball, HBS
	B scale, 100 kgf, 1/16 in. (1.588 mm) ball, HRF	F scale, 60 kgf, 1/16 in. (1.588 mm) ball, HRF	15T scale, 15 kgf, 1/16 in. (1.588 mm) ball, HR15T	30T scale, 30 kgf, 1/16 in. (1.588 mm) ball, HR30T	45T scale, 45 kgf, 1/16 in. (1.588 mm) ball, HR45T			B scale, 100 kgf, 1/16 in. (1.588 mm) ball, HRF	F scale, 60 kgf, 1/16 in. (1.588 mm) ball, HRF	15T scale, 15 kgf, 1/16 in. (1.588 mm) ball, HR15T	30T scale, 30 kgf, 1/16 in. (1.588 mm) ball, HR30T	45T scale, 45 kgf, 1/16 in. (1.588 mm) ball, HR45T	
196	93.5	110.0	90.0	77.5	66.0	169	116	65.0	94.5	82.0	60.0	39.0	103
194	...	109.5	...	...	65.5	167	114	64.0	94.0	81.5	59.5	38.0	101
192	93.0	...	...	77.0	65.0	166	112	63.0	93.0	81.0	58.5	37.0	99
190	92.5	109.0	...	76.5	64.5	164	110	62.0	92.6	80.5	58.0	35.5	97
188	92.0	...	89.5	...	64.0	162	108	61.0	92.0	...	57.0	34.5	95
186	91.5	108.5	...	76.0	63.5	161	106	59.5	91.2	80.0	56.0	33.0	94
184	91.0	...	...	75.5	63.0	159	104	58.0	90.5	79.5	55.0	32.0	92
182	90.5	108.0	89.0	...	62.5	157	102	57.0	89.8	79.0	54.5	30.5	90
180	90.0	107.5	...	75.0	62.0	156	100	56.0	89.0	78.5	53.5	29.5	88
178	89.0	...	...	74.5	61.5	154	98	54.0	88.0	78.0	52.5	28.0	86
176	88.5	107.0	...	...	61.0	152	96	53.0	87.2	77.5	51.5	26.5	85
174	88.0	...	88.5	74.0	60.5	150	94	51.0	86.3	77.0	50.5	24.5	83
172	87.5	106.5	...	73.5	60.0	149	92	49.5	85.4	76.5	49.0	23.0	82
170	87.0	...	...	...	59.5	147	90	47.5	84.4	75.5	48.0	21.0	80
168	86.0	106.0	88.0	73.0	59.0	146	88	46.0	83.5	75.0	47.0	19.0	79
166	85.5	...	...	72.5	58.5	144	86	44.0	82.3	74.5	45.5	17.0	77
164	85.0	105.5	...	72.0	58.0	142	84	42.0	81.2	73.5	44.0	14.5	76
162	84.0	105.0	87.5	...	57.5	141	82	40.0	80.0	73.0	43.0	12.5	74
160	83.5	...	...	71.5	56.5	139	80	37.5	78.6	72.0	41.0	10.0	72
158	83.0	104.5	...	71.0	56.0	138	78	35.0	77.4	71.5	39.5	7.5	70
156	82.0	104.0	87.0	70.5	55.5	136	76	32.5	76.0	70.5	38.0	4.5	68
154	81.5	103.5	...	70.0	54.5	135	74	30.0	74.8	70.0	36.0	1.0	66
152	80.5	103.0	...	...	54.0	133	72	27.5	73.2	69.0	34.0	...	64
150	80.0	...	86.5	69.5	53.5	131	70	24.5	71.8	68.0	32.0	...	63
148	79.0	102.5	...	69.0	53.0	129	68	21.5	70.0	67.0	30.0	...	62
146	78.0	102.0	...	68.5	52.5	128	66	18.5	68.5	66.0	28.0	...	61
144	77.5	101.5	86.0	68.0	51.5	126	64	15.5	66.8	65.0	25.5	...	59
142	77.0	101.0	...	67.5	51.0	124	62	12.5	65.0	63.5	23.0	...	57
140	76.0	100.5	85.5	67.0	50.0	122	60	10.0	62.5	62.5	...	...	55
138	75.0	100.0	...	66.5	49.0	121	58	...	61.0	61.0	18.0	...	53
136	74.5	99.5	85.0	66.0	48.0	120	56	...	58.8	60.0	15.0	...	52
134	73.5	99.0	...	65.5	47.5	118	54	...	56.5	58.5	12.0	...	50
132	73.0	98.5	84.5	65.0	46.5	116	52	...	53.5	57.0	...	...	48
130	72.0	98.0	84.0	64.5	45.5	114	50	...	50.5	55.5	...	...	47
128	71.0	97.5	...	63.5	45.0	113	49	...	49.0	54.5	...	...	46
126	70.0	97.0	83.5	63.0	44.0	112	48	...	47.0	53.5	...	...	45
124	69.0	96.5	...	62.5	43.0	110	47	...	45.0	...	...	...	44
122	68.0	96.0	83.0	62.0	42.0	108	46	...	43.0	...	...	...	43
120	67.0	95.5	...	61.0	41.0	106	45	...	40.0	...	...	...	42
118	66.0	95.0	82.5	60.5	40.0	105	...	...	...	...	...	...	...

Source: ASTM E 140

Metric Conversion Guide

This Section is intended as a guide for expressing weights and measures in the Système International d'Unités (SI). The purpose of SI units, developed and maintained by the General Conference of Weights and Measures, is to provide a basis for worldwide standardization of units and measure. For more information on metric conversions, the reader should consult the following references:

- *The International System of Units,* SP 330, 1991, National Institute of Standards and Technology. Order from Superintendent of Documents, U.S. Government Printing Office, Washington, DC 20402-9325
- *Metric Editorial Guide,* 5th ed. (revised), 1993, American National Metric Council, 4340 East West Highway, Suite 401, Bethesda, MD 20814-4411
- "Standard for Use of the International System of Units (SI): The Modern Metric System," IEEE/ASTM SI 10-1997, Institute of Electrical and Electronics Engineers, 345 East 47th Street, New York, NY 10017
- *Guide for the Use of the International System of Units (SI),* SP 811, 1995, National Institute of Standards and Technology, U.S. Government Printing Office, Washington, DC 20402

SI prefixes—names and symbols

Exponential expression	Multiplication factor	Prefix	Symbol
10^{24}	1 000 000 000 000 000 000 000 000	yotta	Y
10^{21}	1 000 000 000 000 000 000 000	zetta	Z
10^{18}	1 000 000 000 000 000 000	exa	E
10^{15}	1 000 000 000 000 000	peta	P
10^{12}	1 000 000 000 000	tera	T
10^{9}	1 000 000 000	giga	G
10^{6}	1 000 000	mega	M
10^{3}	1 000	kilo	k
10^{2}	100	hecto(a)	h
10^{1}	10	deka(a)	da
10^{0}	1	BASE UNIT	
10^{-1}	0.1	deci(a)	d
10^{-2}	0.01	centi(a)	c
10^{-3}	0.001	milli	m
10^{-6}	0.000 001	micro	μ
10^{-9}	0.000 000 001	nano	n
10^{-12}	0.000 000 000 001	pico	p
10^{-15}	0.000 000 000 000 001	femto	f
10^{-18}	0.000 000 000 000 000 001	atto	a
10^{-21}	0.000 000 000 000 000 000 001	zepto	z
10^{-24}	0.000 000 000 000 000 000 000 001	yocto	y

(a) Nonpreferred. Prefixes should be selected in steps of 10^3 so that the resultant number before the prefix is between 0.1 and 1000. These prefixes should not be used for units of linear measurement, but may be used for higher order units. For example, the linear measurement decimeter is nonpreferred, but square decimeter is acceptable.

Base, supplementary, and derived SI units

Measure	Unit	Symbol	Measure	Unit	Symbol
Base units			Force	newton	N
			Frequency	hertz	Hz
Amount of substance	mole	mol	Heat capacity	joule per kelvin	J/K
Electric current	ampere	A	Heat flux density	watt per square meter	W/m^2
Length	meter	m	Illuminance	lux	lx
Luminous intensity	candela	cd	Inductance	henry	H
Mass	kilogram	kg	Irradiance	watt per square meter	W/m^2
Thermodynamic temperature	kelvin	K	Luminance	candela per square meter	cd/m^2
Time	second	s	Luminous flux	lumen	lm
			Magnetic field strength	ampere per meter	A/m
Supplementary units			Magnetic flux	weber	Wb
			Magnetic flux density	tesla	T
Plane angle	radian	rad	Molar energy	joule per mole	J/mol
Solid angle	steradian	sr	Molar entropy	joule per mole kelvin	$J/mol \cdot K$
			Molar heat capacity	joule per mole kelvin	$J/mol \cdot K$
Derived units			Moment of force	Newton meter	$N \cdot m$
Absorbed dose	gray	Gy	Permeability	henry per meter	H/m
Acceleration	meter per second squared	m/s^2	Permittivity	farad per meter	F/m
Activity (of radionuclides)	becquerel	Bq	Power, radiant flux	watt	W
Angular acceleration	radian per second squared	rad/s^2	Pressure, stress	pascal	Pa
Angular velocity	radian per second	rad/s	Quantity of electricity, electric charge	coulomb	C
Area	square meter	m^2	Radiance	watt per square meter steradian	$W/m^2 \cdot sr$
Capacitance	farad	F			
Concentration (of amount of substance)	mole per cubic meter	mol/m^3	Radiant intensity	watt per steradian	W/sr
Current density	ampere per square meter	A/m^2	Specific heat capacity	joule per kilogram kelvin	$J/kg \cdot K$
Density, mass	kilogram per cubic meter	kg/m^3	Specific energy	joule per kilogram	J/kg
Dose equivalent, dose equivalent index	sievert	Sv	Specific entropy	joule per kiolgram kelvin	$J/kg \cdot K$
Electric charge density	coulomb per cubic meter	C/m^3	Specific volume	cubic meter per kilogram	m^3/kg
Electric conductance	siemens	S	Surface tension	newton per meter	N/m
Electric field strength	volt per meter	V/m	Thermal conductivity	watt per meter kelvin	$W/m \cdot K$
Electric flux density	coulomb per square meter	C/m^2	Velocity	meter per second	m/s
Electric potential, potential difference, electromotive force	volt	V	Viscosity, dynamic	pascal second	$Pa \cdot s$
			Viscosity, kinematic	square meter per second	m^2/s
Electric resistance	ohm	Ω	Volume	cubic meter	m^3
Energy, work, quantity of heat	joule	J	Wavenumber	1 per meter	1/m
Energy density	joule per cubic meter	J/m^3			
Entropy	joule per kelvin	J/K			

Conversion factors

To convert from	to	multiply by
Angle		
degree	rad	1.745 329 E−02
Area		
in.^{2}	mm^2	6.451 600 E+02
in.^{2}	cm^2	6.451 600 E+00
in.^{2}	m^2	6.451 600 E−04
ft^2	m^2	9.290 304 E−02
Bending moment or torque		
lbf · in.	N · m	1.129 848 E−01
lbf · ft	N · m	1.355 818 E+00
kgf · m	N · m	9.806 650 E+00
ozf · in.	N · m	7.061 552 E−03
Bending moment or torque per unit length		
lbf · in./in.	N · m/m	4.448 222 E+00
lbf · ft/in.	N · m/m	5.337 866 E+01
Current density		
A/in.^{2}	A/cm^2	1.550 003 E−01
A/in.^{2}	A/mm^2	1.550 003 E−03
A/ft^2	A/m^2	1.076 400 E+01
Electricity and magnetism		
gauss	T	1.000 000 E−04
maxwell	μWb	1.000 000 E−02
mho	S	1.000 000 E+00
Oersted	A/m	7.957 700 E+01
Ω · cm	Ω · m	1.000 000 E−02
Ω · circular-mil/ft	μΩ · m	1.662 426 E−03
Energy (impact, other)		
ft · lbf	J	1.355 818 E+00
Btu (thermochemical)	J	1.054 350 E+03
cal (thermochemical)	J	4.184 000 E+00
Cal (nutritional)	J	4.184 000 E+03
kW · h	J	3.600 000 E+06
W · h	J	3.600 000 E+03
Flow rate		
ft^3/h	L/min	4.719 475 E−01
ft^3/min	L/min	2.831 000 E+01
gal/h	L/min	6.309 020 E−02
gal/min	L/min	3.785 412 E+00
Force		
lbf	N	4.448 222 E+00
kip (1000 lbf)	N	4.448 222 E+03
tonf	kN	8.896 443 E+00
kgf	N	9.806 650 E+00
Force per unit length		
lbf/ft	N/m	1.459 390 E+01
lbf/in.	N/m	1.751 268 E+02
Fracture toughness		
ksi$\sqrt{in.}$	MPa$\sqrt{m}$	1.098 800 E+00
Heat content		
Btu/lb	kJ/kg	2.326 000 E+00
cal/g	kJ/kg	4.186 800 E+00

To convert from	to	multiply by
Heat input		
J/in.	J/m	3.937 008 E+01
kJ/in.	kJ/m	3.937 008 E+01
Impact energy per unit area		
ft · lbf/ft^2	J/m^2	1.459 002 E+01
Length		
Å	nm	1.000 000 E−01
μin.	μm	2.540 000 E−02
mil	μm	2.540 000 E+01
in.	mm	2.540 000 E+01
in.	cm	2.540 000 E+00
ft	m	3.048 000 E−01
yd	m	9.144 000 E−01
mile, international	km	1.609 344 E+00
mile, nautical	km	1.852 000 E+00
mile, U.S. statute	km	1.609 347 E+00
Mass		
oz	kg	2.834 952 E−02
lb	kg	4.535 924 E−01
ton (short, 2000 lb)	kg	9.071 847 E+02
ton (short, 2000 lb)	kg × 10^3(a)	9.071 847 E−01
ton (long, 2240 lb)	kg	1.016 047 E+03
Mass per unit area		
oz/in.^{2}	kg/m^2	4.395 000 E+01
oz/ft^2	kg/m^2	3.051 517 E−01
oz/yd^2	kg/m^2	3.390 575 E−02
lb/ft^2	kg/m^2	4.882 428 E+00
Mass per unit length		
lb/ft	kg/m	1.488 164 E+00
lb/in.	kg/m	1.785 797 E+01
Mass per unit time		
lb/h	kg/s	1.259 979 E−04
lb/min	kg/s	7.559 873 E−03
lb/s	kg/s	4.535 924 E−01
Mass per unit volume (includes density)		
g/cm^3	kg/m^3	1.000 000 E+03
lb/ft^3	g/cm^3	1.601 846 E−02
lb/ft^3	kg/m^3	1.601 846 E+01
lb/in.^{3}	g/cm^3	2.767 990 E+01
lb/in.^{3}	kg/m^3	2.767 990 E+04
Power		
Btu/s	kW	1.055 056 E+00
Btu/min	kW	1.758 426 E−02
Btu/h	W	2.928 751 E−01
erg/s	W	1.000 000 E−07
ft · lbf/s	W	1.355 818 E+00
ft · lbf/min	W	2.259 697 E−02
ft · lbf/h	W	3.766 161 E−04
hp (550 ft · lbf/s)	kW	7.456 999 E−01
hp (electric)	kW	7.460 000 E−01
Power density		
W/in.^{2}	W/m^2	1.550 003 E+03

To convert from	to	multiply by
Pressure (fluid)		
atm (standard)	Pa	1.013 250 E+05
bar	Pa	1.000 000 E+05
in. Hg (32 °F)	Pa	3.386 380 E+03
in. Hg (60 °F)	Pa	3.376 850 E+03
lbf/in.^{2} (psi)	Pa	6.894 757 E+03
torr (mm Hg, 0 °C)	Pa	1.333 220 E+02
Specific Heat		
Btu/lb · °F	J/kg · K	4.186 800 E+03
cal/g · °C	J/kg · K	4.186 800 E+03
Stress (force per unit area)		
tonf/in.^{2} (tsi)	MPa	1.378 951 E+01
kgf/mm^2	MPa	9.806 650 E+00
ksi	MPa	6.894 757 E+00
lbf/in.^{2} (psi)	MPa	6.894 757 E−03
MN/m^2	MPa	1.000 000 E+00
Temperature		
°F	°C	5/9 · (°F−32)
°R	K	5/9
K	°C	K−273.15
Temperature interval		
°F	°C	5/9
Thermal conductivity		
Btu · in./s · ft^2 · °F	W/m · K	5.192 204 E+02
Btu/ft · h · °F	W/m · K	1.730 735 E+00
Btu · in./h · ft^2 · °F	W/m · K	1.442 279 E−01
cal/cm s · °C	W/m · K	4.184 000 E+02
Thermal expansion(b)		
cm/cm · °C	m/m · K	1.000 000 E+00
in/in. · °F	m/m · K	1.800 000 E+00
Velocity		
ft/h	m/s	8.446 667 E−05
ft/min	m/s	5.080 000 E−03
ft/s	m/s	3.048 000 E−01
in./s	m/s	2.540 000 E−02
km/h	m/s	2.777 778 E−01
mph	km/h	1.609 344 E+00
Velocity of rotation		
rev/min (rpm)	rad/s	1.047 164 E−01
rev/s	rad/s	6.283 185 E+00
Viscosity		
poise	Pa · s	1.000 000 E−01
stokes	m^2/s	1.000 000 E−04
ft^2/s	m^2/s	9.290 304 E−02
in.^{2}/s	mm^2/s	6.451 600 E+02
Volume		
in.^{3}	m^3	1.638 706 E−05
ft^3	m^3	2.831 685 E−02
fluid oz	m^3	2.957 353 E−05
gal (U.S. liquid)	m^3	3.785 412 E−03
Volume per unit time		
ft^3/min	m^3/s	4.719 474 E−04
ft^3/s	m^3/s	2.831 685 E−02
in.^{3}/min	m^3/s	2.731 177 E−07

(a) kg×10^3 = 1 metric ton (tonne), or 1 megagram (Mg). (b) Preferred expression is 10^{-6}/K or 10^{-6}/°F as length units are unnecessary.

Abbreviations and Symbols

a	specimen half radius; crack length; linear distance; crystal lattice length along the a axis	CAD	computer-aided design	DMM	dynamic material modeling
		CAD/CAM	computer-aided design/computer-aided manufacturing	DMZ	dead-metal zone
A	area; heat retention factor	CAE	computer-aided engineering	DOE	design of experiment
A	ampere	CAO	computer-aided optimization	DP	design parameter
Å	angstrom	CCR	conventional controlled rolling	DRCR	dynamic recrystallization controlled rolling
$\bar{A}$	mean cross-sectional area	CDA	Copper Development Association	DRV	dynamic recovery
A_f	final cross-sectional area			DRX	dynamic recrystallization
A_1	subcritical annealing temperature	CDRX	continuous dynamic recrystallization	DSC	differential scanning colorimetry
AA	Aluminum Association			d_{ss}	steady-state subgrain size
ac	alternating current	CHR	conventional hot rolling	DTA	differential thermal analysis
ACI	Alloy Casting Institute	CIRP	College International pour l'Etude Scientifique des Techniques de Production Mecanique		
A/D	analog to digital			e	elongation; engineering, linear, strain; natural log base, 2.71828
A_i	initial cross-sectional area				
AI	artificial intelligence	cm	centimeter		
AISI	American Iron and Steel Institute	CNC	computer numerically controlled	$\dot{e}$	engineering strain rate
ALPID	Analysis of Large Plastic Incremental Deformation (bulk deformation modeling software)	COE	collaborative optimization environment	e_1	major engineering strain
				e_2	minor engineering strain
		C_p	specific heat	E	elastic modulus in axial loading (Young's modulus)
		cpm	cycles per minute		
AMS	Aerospace Material Specification (of SAE)	cps	cycles per second	E'	effective modulus for given stress or strain state, that is, $d\sigma_1/d\varepsilon_1$ in elastic region, where x_1 is the bending fiber direction. For plane strain and isotropic elasticity, $E' = \frac{E}{1-\nu^2}$.
		CPU	central processing unit		
ANSI	American National Standards Institute	CTE	coefficient of thermal expansion		
		C_v	cavity volume fraction at true strain		
API	American Petroleum Institute				
Ar$_3$	critical temperature when austenite begins to transform to ferrite upon cooling	d	grain diameter (size); density; used in mathematical expressions involving a derivative (denotes rate of change)	ECM	electrochemical machining
				EDM	electrical discharge machining
ASME	American Society of Mechanical Engineers			e_f	engineering fracture strain; elongation to fracture
		d	day		
ASTM	American Society for Testing and Materials	$d\varepsilon$	incremental strain	EHD	elastohydrodynamic
		$\bar{d}\varepsilon$	equivalent strain increment	EMC	electromagnetic casting
at.%	atomic percent	D	diffusivity; grain size; mandrel diameter	EMF	electromotive force
atm	atmospheres (pressure)			EP	extreme pressure
AWS	American Welding Society	D/A	digital-to-analog	EPA	Environmental Protection Agency
		DARPA	Defense Advanced Research Projects Agency	Eq	Equation
b	Burgers (slip) vector			ESR	electroslag remelting
b	crystal lattice length along the b axis; width or breadth	DASA	discretization, approximation, and searching	ETP	electrolytic tough pitch
				e_u	uniform elongation
B	Bridgman correction factor	DBMS	data base management system	exp	base of natural logarithms ($=2.718\ldots$)
bal	balance or remainder	DBTT	ductile-to-brittle transition temperature		
bcc	body-centered cubic				
bct	body-centered tetragonal	dc	direct current	f	local geometric inhomogeneity
BDC	bottom dead center	DC	direct chill cast	F	force
BEM	boundary element method	D_c	critical cylinder diameter (indicator of hardenability)	F_b	normalized sheet tension in draw-bend test; F_b is the tensile force/yield force
BID	blocker initial design				
B_s	bainite-start temperature	DDRX	discontinuous dynamic recrystallization process		
BVP	boundary value problems			fcc	face-centered cubic
		D_{eff}	effective diffusion coefficient	FDM	finite-difference method
c	wave speed; crystal lattice length along the c axis	DFM	design for manufacturer	FEA	finite-element analysis
		diam	diameter	FEM	finite-element method/model
C	specific heat; generalized constant			Fig.	figure
				FLD	forming limit diagram

FR	functional requirements	K	Kelvin (absolute temperature scale)	OS	operating system
FRP	fiber-reinforced plastic			OSHA	Occupational Safety and Health Administration
FSEM	finite- and slab-element modeling	k_b	Boltzmann constant		
ft	foot	kg	kilogram	oz	ounce
F_T	stress triaxiality factor	kN	kilonewton		
f_v	volume fraction; volume fraction of second-phase particles	kPa	kilopascal	p	page or pages
		ksi	1000 pounds per square inch	P	pressure; load or external force
		kW	kilowatt	Pa	pascal
g	gram			PDSR	postdynamic static recrystallization
G	free-energy density; modulus of elasticity in shear (modulus of rigidity), normalized torque softening rate under constant twisting rate	l	length		
		l_0	initial gage length	PDT	peak-ductility temperature
		L	length; liter; half sample length of four-point bend finite-element method model	PH	precipitation-hardenable
				PHD	plastohydrodynamic
				P/M	powder metallurgy
gal	gallon	lb	pound	ppb	parts per billion
GUI	graphical user interface	lbf	pound (force)	ppm	parts per million
		LDR	limiting draw ratio	psi	pounds per square inch
h	hour	ln	natural logarithm (base e)	psia	pounds per square inch (absolute)
h	height; distance, usually in thickness direction	log	logarithm to base 10	psig	pounds per square inch (gage)
H	hardness; height			PVA	plan view area
HB	Brinell hardness	m	interface; friction factor; strain-rate sensitivity exponent		
hcp	hexagonal close-packed			q	plastic constraint factor
HERF	high-energy-rate forging	M	bending moment	Q	activation energy
HIP	hot isostatic pressing	mA	milliampere		
HK	Knoop hardness	max	maximum	r	radius; radius of curvature after springback; reduction; plastic strain anisotropy ratio; thermal resistance
hp	horsepower	MC	Monte Carlo		
HR	Rockwell hardness; requires scale designation such as HRC for Rockwell C hardness	MCS	Monte Carlo step		
		MDO	multidisciplinary optimization		
		MDOL	multidisciplinary design optimization language	r'	radius of curvature in curl region of draw-bend sample, after springback
		MDRX	metadynamic recrystallization		
HSLA	high-strength, low-alloy (steel)	mg	milligram	R	extrusion ratio; radius; universal gas constant; plastic anisotropy parameter; rolling reduction ratio; tool radius in draw-bend test; radius of primary bending curvature
h_T	heat-transfer coefficient	Mg	megagram		
H_T	enthalpy at a given temperature	MIL	military		
HV	Vickers (diamond pyramid) hardness	MIL-STD	military standard		
		min	minimum; minute		
Hz	hertz	mL	milliliter	R'	radius of curvature of the region in contact with the tool, after unloading
		mm	millimeter		
I	moment of inertia	MME	material modeling environment		
I'	moment of inertia for distorted cross sections	MN	meganewton	R	Rankine
		MPa	megapascal	RA	reduction of area
IBR	integral blade and rotor	MPDO	multidiscipline process design and optimization	R_a	radii of anticlastic curvature
ID	inside diameter			r_c	critical cavity radius; critical particle size
IGES	initial graphics exchange specification				
		n	grain growth exponent; strain-hardening exponent	RCR	recrystallization controlled rolling
I/M	ingot metallurgy				
in.	inch	N	number of revolutions	r_d	tool radius in four-point bend finite-element method model
ipm	inches per minute	$\dot{N}$	twist rate		
ips	inches per second	N	Newton, unit of force	R_E	extrusion ratio
ISO	International Organization for Standardization	NASA	National Aeronautics and Space Administration	Ref	reference
				rem	remainder or balance
		NC	numerical control	RFQ	request for quotation
J	joule	NCEMT	National Center for Excellence in Metalworking Technology	R_h	radial distance
J	polar moment of inertia			RHR	root height reading
JIC	Joint Industry Conference	NDE	nondestructive evaluation	r_i	internal specimen radius
		N_{EL}	number of elements used in finite-element analysis model	r_m	average of plastic strain anisotropy of the r ratio in 0, 45 and 90 degree directions, where:
k	thermal conductivity; yield strength in pure shear; Boltzmann's constant				
		N_{IP}	number of through-thickness integration points for shell or beam element	$$r_m = \frac{r_0 + 2r_{45} + r_{90}}{4}$$	
K	strength coefficient in uniaxial plastic hardening law	No.	number		
K'	effective strength parameter, taking into account strain/ stress state and possible plastic anisotropy			r_{max}	maximum diameter of a specimen deformed to a specific height
		OBI	open-back inclinable (press)		
		OD	outside diameter		
		OEM	original equipment manufacturer	rms	root mean square
		OFHC	oxygen-free high-conductivity	rpm	revolutions per minute

s	engineering stress	VAR	vacuum arc remelted	$\propto$	varies as; is proportional to
SAE	Society of Automotive Engineers	vol%	volume percent	α	die half angle; linear coefficient of thermal expansion; flow localization parameter
SAM	successive approximation methods	w	displacement in the x, y, and z directions; width; weight or mass	α_F	draft angle of forging
SEM	scanning electron microscopy			α_P	draft angle of preform
SFE	stacking fault energy	W	watt	β	Searle's parameter ($\beta = \frac{w^2}{Rt}$)
SFEM	simplified finite-element method	W	work; width	β_t	beta transus temperature
sfm	surface feet per minute	W_f	relative portion of power consumed by friction	Γ	shear strain rate
SI	Systeme International d'Unites	$\dot{W}_i$	relative portion of power consumed by internal deformation	γ	shear strain; work-hardening coefficient; flow-softening rate; interfacial energy
SLF	slip-line field			$\dot{\gamma}$	shear strain rate
SLP	sequential linear programming			γ_f	shear strain to failure
SM	Sachs (slab) method	WIP	work-in-process	γ_i	interfacial energy of particle/matrix interface
SPC	statistical process control	w_p	width of preform		
SPD	severe plastic deformation	WRC	Welding Research Council	γ_P	interfacial energy of particle
SPF	superplastic forming	$\dot{W}_s$	relative portion of power consumed by shear or redundant work	γ_{sf}	surface shear strain at fracture
SPF/DB	superplastic forming/diffusion bonding			Δ	finite change; volume strain
SQP	sequential quadratic programming	wt%	weight percent	Δ	change in quantity, an increment, a range
SRX	static recrystallization				
SUS	Saybolt universal second (measure of viscosity)	x	parameter in Gardiner's equation ($x = \frac{R\sigma_0'}{E't}$); also longitudinal coordinate	Δ_φ	angle through which one end of gage length has been twisted relative to the other
S_v	austenite interfacial area; total effective interfacial area per unit volume	X_{tool}^i	forming tool position at i^{th} iteration	δ	deformation or elongation; crosshead speed; Kronecker delta
S_v(GB)	total effective area per unit volume from the grain-boundary contribution	y	coordinate along width direction	ε	true normal strain
S_v(IPD)	contribution from intragranular planar defect	Δy^i	the ith displacement correction	ε_b	bending strain
				ε_e	elastic strain
		z	thickness coordinate; $z = 0$ at center of thickness	ε_m	membrane strain
t	time; thickness			ε_p	plastic strain
T	absolute temperature	z_0	location of neutral axis	ε_0	zero-gage length elongation
T	processing temperature; sheet tension force	z_1^*, z_2^*	locations of elastic core on tension and compression side	$\dot{\varepsilon}$	true strain rate
				$\bar{\varepsilon}$	effective strain
$\bar{T}$	normalized sheet tension, $\frac{T}{wt\sigma_0'}$.	Z	Zener-Hollomon parameter	$\dot{\bar{\varepsilon}}$	effective strain rate; mean true strain rate
TDC	top dead center	ZDT	zero-ductility temperature		
T_{def}	deformation temperature			$\bar{\varepsilon}_f$	effective strain at fracture
t_F	thickness of forging	2-D	two-dimensional	ε_{max}	maximum tensile strain
TFP	total factor productivity	3-D	three-dimensional	ε_t	thickness strain
T_g	glass transition temperature			ε_u	true strain at maximum load; true uniform strain
T_{GC}	grain-coarsening temperature	$^\circ$	degree (angular measure)		
T_m	melting temperature	$^\circ$C	temperature, degrees Celsius (centigrade)	ε_w	width strain
TMP	thermomechanical processing			ε_θ	circumferential strain
tonf	tons of force	$^\circ$F	temperature, degrees Fahrenheit	η	coefficient of viscosity; cavity-growth-rate parameter
T_R	recrystallization temperature	$\div$	divided by		
TRIP	transformation-induced plasticity	$=$	equals	η_{APP}	apparent cavity-growth rate
T_{RXN}	recrystallization stop temperature	$\approx$	approximately equals	θ	$d\sigma/d\varepsilon$, strain-hardening rate; angle of twist in torsion
T_S	solidus temperature	$\neq$	not equal to		
tsi	tons per square inch	$>$	greater than	κ	bulk modulus or volumetric modulus of elasticity
TYS	tensile yield strength	$\gg$	much greater than		
		$\geq$	greater than or equal to	μ	shear modulus; Coulomb coefficient of friction
u	displacement	$\int$	integral of		
U	rate of dislocation generation due to strain hardening; elastic strain energy	$<$	less than	μin.	microinch
		$\ll$	much less than	μm	micron (micrometer)
UB	upper bound	$\leq$	less than or equal to	μs	microsecond
UBET	upper-bound element technique	$\pm$	maximum deviation	ν	Poisson's ratio
UBM	upper-bound method	$-$	minus; negative ion charge	π	pi (3.14159...)
UHC	ultrahigh carbon	$\times$	multiplied by; diameters (magnification)	ρ	density of a material; dislocation density
UNS	Unified Numbering System	$\cdot$	multiplied by	Σ	summation of
UTS	ultimate tensile strength	$/$	per	σ	normal stress component; true stress
		%	percent		
v	Poisson's ratio; velocity	$+$	plus; in addition to	$\sigma_1\sigma_2\sigma_3$	principal stresses
V	volume	$\sqrt{\ }$	square root of	$\bar{\sigma}$	effective or significant true stress; mean stress
V	volt	$\sim$	similar to; approximately		

σ_a	measured average stress	$\sigma_{\theta\theta}$	circumferential stress	Δ, δ	delta
σ_c	flow stress in cavity	τ	shear stress component	E, ε	epsilon
σ_d	drawing stress	τ_i	interfacial shear (or friction)	Z, ζ	zeta
σ_{ep}	stress at entrance to plane-strain		stress	H, η	eta
	cross section	τ_{max}	maximum shear stress	Θ, θ	theta
σ_1	interfacial energy per unit area of	τ_o	shear yield strength	I, ι	iota
	boundary	Φ	redundant work factor	K, κ	kappa
σ_m	hydrostatic or mean principal	φ	stress-intensification factor;	Λ, λ	lambda
	pressure		viscosity-pressure coefficient	M, μ	mu
σ_T	tensile stress	$\dot{\varphi}$	shear-strain rate	N, ν	nu
σ_{xb}	back tension	φ_r	angle between scribe line and	Ξ, ξ	xi
σ_{xf}	front tension		torsion axis at failure	O, o	Omicron
σ_u	ultimate tensile strength			Π, π	pi
σ_y	yield stress			P, ρ	rho
σ_0	flow stress; yield stress; yield			Σ, σ	sigma
	strength			T, τ	tau
σ_0'	yield stress in plane strain;			Y, υ	upsilon
	effective yield stress taking			Φ, ϕ	phi
	into account strain/stress	**Greek Alphabet**		X, χ	chi
	state and possible plastic			Ψ, ψ	psi
	anisotropy	A, α	alpha	Ω, ω	omega
		B, β	beta		
		Γ, γ	gamma		

Index